Bioconjugation

Protein Coupling Techniques for the Biomedical Sciences

Mohammed Aslam
and Alastair Dent

Bioconjugation

Protein Coupling Techniques for the Biomedical Sciences

Mohammed Aslam
and Alastair Dent

Published in the United Kingdom by
MACMILLAN REFERENCE LTD, 1998
25 Eccleston Place, London, SW1W 9NF
and Basingstoke

Companies and representatives throughout the world.

http://www.macmillan-reference.co.uk

Distributed in the UK and Europe by
Macmillan Direct,
Brunel Road, Houndmills,
Basingstoke,
Hampshire, RG21 2XS, England

ISBN 0-333-583752

A catalogue record for this book is available from the British Library.

Published in the United States and Canada by
GROVE'S DICTIONARIES INC, 1998
345 Park Avenue South, 10th Floor
New York, NY 10010-1707, USA

ISBN 1-56159-161-0

Typeset by Pure Tech Ltd, India

Printed and bound in the UK by
Bath Press Ltd, Bath

Contents

Contents

This book is dedicated to the memory of Fiona Dent and Fateh Mohammed

Acknowledgements

The preparation of this book could not have been completed without the support of Ortho-Clinical Diagnostics, and particularly of Dr. Bill Hornby and Dr. Paul Schnipelsky. Numerous invaluable suggestions and contributions have been made by many of our colleagues past and present, including Heather Shepherd, Graham Yearwood, David Lewis, Ian Lingham, John Edwards, Jennie DeVriendt, Peter Truran, Phill Dicker, Deborah Warwick-Wilson and Vanessa Lonsdale. A special mention is also due to the librarians at Nycomed Amersham plc, especially Vera Butterworth, Lynne Goodman and Sally Leeks.

The expert support of all those at Macmillan Reference Ltd is also gratefully acknowledged, in particular that of Rosemary Foster and Andrew Diggle.

Finally, the immense patience of our families and friends throughout this project must be recognised, especially that of our long-suffering partners, Mussarat Aslam and Lucy O'Reilly.

Chapter 1

Coupled Protein Reagents and Their Applications

Alastair H. Dent and Mohammed Aslam

Research and Development Department, Ortho-Clinical Diagnostics, Pollards Wood Laboratories, Nightingales Lane, Chalfont St. Giles HP8 4SP, UK

1.1 Introduction

Proteins have evolved in nature to carry out a remarkable variety of functions, many of which are complex interactive roles of exquisite subtlety and specificity. Among these functions major categories include catalysis, storage, transport, transduction of energy, defence, and structure (Franks, 1993a).

The efficiency of the evolutionary process has resulted in proteins which offer high levels of performance across this wide range, and as a consequence a bewildering number of such molecules exist *in vivo*: the human genome alone is thought to code for between 30 000 and 50 000 different proteins (Andersson *et al.*, 1980), and the number contained in the entire biosphere has been estimated at 10^{12} (Richardson, 1981). While these numbers may appear large, they represent a vanishingly small fraction of the arrangements of amino acids that could exist. Even for a modestly sized (34 kDa) protein, containing equal numbers of only twelve amino acids, there are 10^{300} possible sequence isomers (Franks, 1993).

Often the functional properties of the molecules 'chosen' by evolution are unequalled by any other approach, and it is not surprising that mankind has harnessed the more useful of these properties for a wide array of uses. In view of the origins of these molecules, it is also natural that a large proportion of this scientific interest should focus on employing the special properties of proteins to tackle problems of a biological or medical nature. These fields of study are where some of the most important and far-reaching applications are to be found.

Crude formulations of proteins have been used in medicine (with a greater or lesser degree of understanding) since the earliest days. However, the vast majority of current applications have been facilitated in the last few decades by advances in protein purification, characterisation and, increasingly, synthesis.

1.1.1 Advances in protein purification

Proteins from natural sources must inevitably be harvested from a medium containing many impurities. Some of these will be relatively small species: inorganic ions such as potassium or chloride or organic molecules such as steroid hormones, lipids and carbohydrates. Some may be from other classes of macromolecules such as polysaccharides or nucleic acids, while there will also be many proteins in addition to that targetted for purification.

These components display a wide range of solubility properties under different conditions of pH, ionic strength, temperature, dielectric constant, etc. Methods of fractional precipitation based on these variations, have existed for many years, by which partial purification of proteins can be achieved (Heide *et al.*, 1977; Ingham, 1984). In the 1940s, for example, Edward Cohn and his colleagues at Harvard University developed methods for the fractionation of human plasma proteins by manipulating pH, and by employing ethanol as a co-solvent (Cohn *et al.*, 1946). Some of the fractions obtained by his methods are reasonably pure, e.g. fraction II consists essentially of γ-globulins, while fraction V is 97–99% albumin. Variations of the Cohn method remain in use today.

Electrophoresis relies on the differential rate of movement of protein molecules under the influence of an electric field, and offers a powerful means of fractionating proteins on the basis of their charge properties (Andrews, 1986). While this has become one of the staple methods of protein analysis and characterisation, the physical constraint of heat dissipation has limited its use in large-scale preparative applications.

It is advances in chromatography that have had the greatest impact on protein purification. Variations in the interactions of protein molecules with a solid matrix in a liquid environment provide the typical means of separation. Within this broad description, a wide range of approaches can be employed which exploit differences in size, charge and hydrophobicity, or specific interactions with protein-binding ligands. (Dean *et al.*, 1985; Harris and Angall, 1989, 1990; Kenney and Fowell, 1992).

A combination of chromatographic techniques, often after a crude fractionation by a precipitation method, can yield proteins at a very high level of purity. These advances have been a major factor in widening the applicability of protein reagents.

1.1.2 Advances in protein analysis

Hand-in-hand with the evolution of protein purification techniques has come a dramatic advance in the ability to characterise the structure of these complex macromolecules.

Empirical techniques (including many of those employed for purification) can be used to characterise properties such as molecular weight (analytical ultracentrifugation, size exclusion chromatography, osmometry, sodium dodecyl sulfate-polyacrylamide gel electrophoresis (SDS-PAGE)) and charge (ion exchange chromatography, isoelectric focusing). Immunochemical techniques are also useful, using the high specificity of antibodies to demonstrate whether one protein has a structure similar to another.

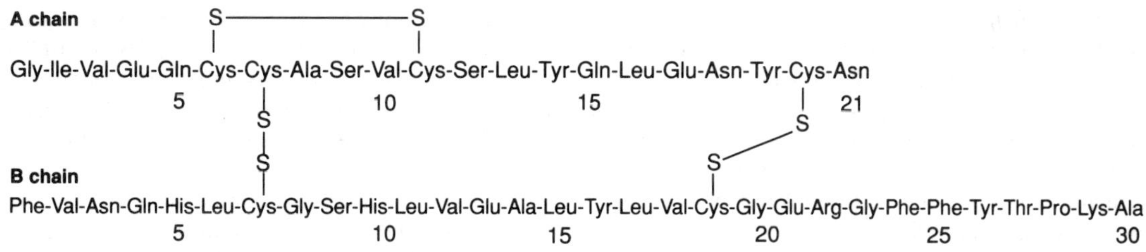

Figure 1.1 Primary structure of bovine insulin.

However, the watershed for protein characterisation was the development of techniques to determine structure at the molecular level. The pioneering studies of Fred Sanger established the amino acid sequence (primary structure) of the B-chain of the small protein insulin (Sanger and Tuppy, 1951a, b), demonstrating for the first time that proteins had defined chemical structures (Figure 1.1). Such determinations are now routine, automated laboratory operations.

Spectroscopic techniques such as circular dichroism have been developed to probe the way the amino acid chains organise themselves into helices, sheets, etc. (i.e. the secondary structure). X-ray crystallography can be employed, where pure crystalline protein is available, to determine the full three-dimensional arrangement (i.e. the tertiary structure) of the molecule, at resolutions down to the Ångstrom (10^{-10} metre) level. For a review of these procedures for investigating the solution properties of proteins, see Franks (1993b).

Finally, the molecular biology revolution has revealed the relationship between the structure of proteins and the genes which control their biosynthesis, adding a further level of understanding.

Again, the ability to characterise the structure of proteins accurately has allowed them to be used as reagents with the confidence that reproducible behaviour will be obtained, and with a chance that this behaviour can be understood to some degree at the molecular level.

1.1.3 Advances in protein synthesis

However impressive the development of techniques for purification and analysis of proteins may be, the use of material isolated from natural sources can still be a constraining factor. The quantities produced *in vivo* may be extremely small, it may prove extremely difficult to separate one component from closely-related impurities, or there may be safety aspects, such as the infectivity of the source material, which make this approach unattractive. These advances have opened up new routes for generating proteins, which do not rely on a natural source of crude material.

Once the amino acid sequence of a protein is known, it is quite possible to build up the peptide chain by chemical means, the only starting materials required being amino acids derivatised for convenient reactions. The commercialisation of instrumentation and reagents for peptide synthesis has opened up this approach for routine application, but there are a number of factors which have limited its utility to create large polypeptides.

First, each attachment of an amino acid to the growing chain is a discrete chemical reaction with a finite yield, and the result of repeating this operation many times can be dramatic. A yield of 90% of the theoretical maximum is beyond the hopes of organic chemists for many common synthetic reactions, but repeating this operation, even fifty times, gives an overall yield below 1%. To synthesise even a small protein of molecular weight 10 kDa requires more than a hundred sequential additions, and extremely careful development has been necessary to provide techniques with an acceptable output. Secondly, some chemistry is generally required after the chain has been assembled to disconnect it from the solid support on which the chemistry has been carried out, and to remove protecting groups which prevent unwanted side reactions during the synthesis. These, too, contribute to undermining the overall yield.

Synthetic peptide approaches are therefore generally too expensive and inefficient to be employed for the preparation of large proteins, although they are useful when the size of the product is more limited.

In some cases it has been possible to harness the natural processes of biosynthesis in such a way as to promote the production of one protein of interest. Notable among these has been the use of hybridoma techniques (Köhler and Milstein, 1975) to generate large quantities of homogeneous monoclonal antibodies, revolutionising the application of immunological reagents and winning the 1984 Nobel Prize for Physiology or Medicine for its inventors.

Greater potential has been realised, in the last twenty years or so, with the establishment of molecular biology techniques as everyday research tools. If the gene which codes for a protein of interest can be identified, prepared and introduced into a suitable host organism or cell, it becomes possible to generate almost limitless quantities of homogeneous protein *in vitro* (Kopetzki *et al.*, 1994).

These recombinant techniques have also, for the first time, offered a practical route to the creation of large polypeptides not found in nature—mutant proteins generated from deliberately engineered alterations to the gene stucture (El-Gewely, 1991; McPherson, 1991; Medynski, 1992), or oligomeric fusion proteins obtained by splicing the genes of the component molecules (Uhlén *et al.*, 1992; Lindbladh *et al.*, 1993; Kopetzki *et al.*, 1994). These novel reagents can be harnessed in elegant ways for biomedical applications; recently, for example, Wang *et al.* (1997) created a recombinant firefly luciferase which incorporated both a poly-histidine tag to allow convenient metal-affinity purification, and a biotin-containing domain to allow immobilisation to avidin-coated solid phases. Chimaeric antibodies which combine structural elements from two different species can be created by these means (Winter and Milstein, 1991) and have valuable properties (see Section 4.6.2.2).

The major challenges, yet to be fully overcome in this area, are the efficient purification of the recombinant material from the host culture, and attainment of the correctly constructed functional protein. In post-translational processes, such as folding and glycosylation, the subtleties achieved in nature are not always easy to mimic efficiently in the laboratory. However, these approaches are becoming more economical to apply. An additional factor is the changing regulatory climate worldwide, with ever more stringent conditions being applied to the use of material of human or mammalian origin in medical products. Therefore, it seems likely that recombinant proteins will make an increasing impact in the biomedical arena.

1.1.4 Applications of coupled protein reagents

There are countless applications in the life sciences where the protein, whether from a natural, synthetic or recombinant source, is simply used in its native form. Examples include the administration of hormones in the treatment of deficiencies and related disorders, and the use of antigenic protein material or passive antibody preparations in vaccines. However, some of the most elegant applications have resulted from the coupling of proteins to other molecules, such as a second protein, or a different class of molecule. Technologies of substantial importance in fields such as clinical diagnostics, immunology, and *in vivo* imaging rely on the use of these coupled protein reagents and it is the aim of this book to bring together the techniques which can be employed to prepare them.

Protein labelling is a commonly encountered term, but it can be confusing in its application. It could equally well be understood to mean the labelling of proteins with other species, or the labelling of other species with proteins. In terms of the preparation techniques the distinction is meaningless and there is often no value in trying to specify which component of the complex, if any, is the label. In this book, therefore, we generally refer to coupled protein reagents rather than labelled proteins, etc., except in those applications where the concept adds clarity. Also, coupled protein reagents are considered in the broadest sense, i.e. to include the coupling of a protein to any other species. Thus, the linking of proteins to solid phase materials such as latex particles is considered, as well as the preparation of conjugates the term commonly used where both partners of the complex are soluble molecular species. There have been many useful books and reviews covering various aspects of the preparation of coupled protein reagents (Lundblad and Noyes, 1984a,b; Means and Feeney, 1990; Wong, 1991; Brinkley, 1992; Mattson *et al.*, 1993; Hermanson, 1995).

The techniques for the preparation of coupled protein reagents are to a large extent independent of the use to which they will be put, but there are some characteristics which differ from one application to another. It is worth considering the uses to which such reagents are put with this in mind.

One area where the unique properties of coupled protein reagents find particularly wide application is in medical diagnosis: clinical diagnostics where the sample is generally a body fluid, histology for tissue samples, cytology for evaluation of cells, and *in vivo* imaging where analysis is carried out within the living body. These will be discussed in Sections 1.2–1.4. From the *in vivo* use of such reagents for diagnostic purposes it is a logical progression to employ them as a treatment for disease, and these therapeutic applications are also described in Section 1.4.

The majority of the remaining applications describe the use of coupled protein reagents or coupling techniques as scientific tools in the elucidation of protein structure and activity (Section 1.5), the production of antibodies with a chosen specificity (Section 1.6), and in a variety of other uses (Section 1.7).

Table 1.1 Applications of Protein Coupling

MEDICAL

Diagnostics
Therapy
Prophylaxis
Research
(see Table 1.2)

ENVIRONMENTAL

Immunoassays and biosensors for monitoring of pollutants

Immunoassays, biosensors and nucleic probe tests for monitoring of microbiological contamination of food or environment

(Reagent requirements as in equivalent categories in Table 1.4)

VETERINARY

Diagnostics especially fertility

Therapy
Prophylaxis
Research

(Reagent requirements as in equivalent categories in Tables 1.2 and 1.4)

FORENSIC

Immunoassays and biosensors for detection of explosives, etc.

Immunoassays and clinical chemistry tests for drugs of abuse (criminal and sports applications)

Nucleic acid probes for DNA fingerprinting

(Reagent requirements as in equivalent categories in Table 1.4)

MILITARY

Immunoassays and biosensors for detection of chemical and biological weapons

(Reagent requirements as in equivalent categories in Table 1.4)

INDUSTRIAL

Immobilised enzymes for industrial reactions

Immobilised proteins to form matrices for affinity chromatographic purifications

Table 1.2 Medical Applications of Protein Coupling

DIAGNOSTICS	THERAPY	PROPHYLAXIS	RESEARCH
Clinical diagnostics	Radioimmunoconjugates for targetting radioactive treatment	Protein-coupled haptens for use as vaccines	Cross-linking and derivatisation methods used in analysis of protein structure
Histochemistry	Chemoimmunoconjugates for targetting drug delivery		Derivatisation of proteins with reporter molecules to probe microenvironment
Cytochemistry	Immunoconjugates and protein microspheres for targetting gene therapy		Cross-linking and derivatisation methods for investigation of protein interactions, e.g. nearest-neighbour analyses
Imaging (see Table 1.3)	Protein drugs derivatised to alter their *in vivo* behaviour, e.g. by coupling to polyethylene glycol		Group-specific derivatisation methods for investigation of structure–activity relationships
	Immobilised enzymes for extracorporeal therapy		Immobilised enzymes for convenient separation from products of catalysed reaction
	Immobilised proteins for use as biocompatible surfaces		Immobilised proteins to provide affinity chromatography matrices for purifications.
			Immunogens for raising specific antibodies
			Research immunoassays (see Table 1.4 for reagent requirements)

Table 1.3 Diagnostic Applications of Protein Coupling

CLINICAL DIAGNOSTICS	HISTOCHEMISTRY AND CYTOCHEMISTRY	IMAGING
Clinical chemistry	Labelled antibodies or other proteins for light microscopy (gold-labelled, dye- labelled, enzyme-labelled)	Radioimmunoconjugates or other radiolabelled proteins for radioimaging
Immunoassay	Labelled antibodies or other proteins for fluorescence microscopy (fluorescent, enzyme-labelled)	Antibody-labelled gadolinium complexes for MRI
Nucleic acid probes	Labelled antibodies or other proteins for electron microscopy (uranium-labelled, gold-labelled, heavy metal-labelled via chelator groups, ferritin-labelled, enzyme- labelled)	Protein-coated magnetite particles for MRI
Blotting techniques	Labelled antibodies or other proteins for electron scattering techniques (e.g. boron-labelled for ESI)	Protein microspheres for ultrasound imaging
Chromatography	Biotinylated proteins for use in avidin–biotin complex (ABC) methods	
Biosensors (see Table 1.4)		

Table 1.4 Clinical Diagnostic Applications of Protein Coupling

CLINICAL CHEMISTRY

A few uses of immobilised proteins

NUCLEIC ACID PROBES

Enzyme-labelled oligonucleotide probes

Labelled antibodies and other binding proteins for indirect detection methods (radioactive, fluorescent, luminescent, enzyme-labelled)

BIOSENSORS

Immobilisation of proteins on biosensor surface

Labelled antibodies and other binding proteins for detection of analytes (e.g. antibody labelled with bulky refractive index probe for surface plasmon resonance-based methods)

CHROMATOGRAPHY

Immobilised proteins to provide matrices for affinity chromatographic methods

IMMUNOASSAY

BLOTTING TECHNIQUES

Blotting requires immobilisation of protein analytes

Labelled antibodies and other binding proteins for detection of blotted analytes (radioactive, fluorescent, luminescent, enzyme-labelled)

AGGLUTINATION ASSAYS

Antibody- or antigen-labelled red blood cells for haemagglutination assays

Antibody- or antigen-labelled latex particles, etc., for particle agglutination assays

TURBIDIMETRY AND NEPHELOMETRY

Antibody- or antigen-labelled particles for enhancement of signal

HETEROGENEOUS IMMUNOASSAYS

Labelled antibodies or other proteins for detection (radioactive, fluorescent, luminescent, electrochemical, dye-labelled, enzyme- labelled)

Biotinylated antibodies or other proteins for detection or immobilisation

Antibody- and enzyme-labelled haptens for detection.

Immobilised antibodies or other proteins to provide solid-phase component

HOMOGENEOUS IMMUNOASSAYS

Many different formats: typically enzyme- or enzyme-fragment labelled haptens, or reporter-labelled proteins

Table 1.5 Key Properties of Coupled Protein Reagents in Some Common Applications

APPLICATION	KEY PROPERTIES
PROTEINS IMMOBILISED TO A SOLID PHASE for affinity chromatography matrices, immobilised enzymes for industrial and research uses; solid-phase reagents for immunoassay, biosensors, etc.	1. Good retention of functional activity after immobilisation 2. Low level of leaching of protein from surface under conditions of use 3. Good *in vitro* stability 4. For biosensors, highly localised deposition techniques may be required
RADIOIMMUNOCONJUGATES for *in vitro* imaging	1. Good retention of functional activity after radiolabelling, ideally by making introduction of radioisotope the last stage before use 2. Efficient localisation at desired target 3. Moderate *in vitro* stability; high enough to allow imaging to be carried out, but with subsequent rapid excretion 4. High-energy radioisotope to allow signal to penetrate outside body for measurement 5. Radioisotope with moderate half-life (cf. 3)
RADIOIMMUNOCONJUGATES for therapy	1. Good retention of functional activity after radiolabelling, ideally by making introduction of radioisotope the last stage before use 2. Efficient localisation at desired target 3. High *in vitro* stability to allow prolonged treatment; antibody engineering or derivatisation may be used to minimise immune response 4. Low-energy radioisotope to promote absorbance by target tissue 5. Radiosotope with long half-life to allow prolonged treatment
CHEMOIMMUNOCONJUGATES for therapy	1. Good retention of functional activity of both antibody and toxin components after conjugation 2. Efficient localisation at desired target 3. High *in vitro* stability to allow prolonged treatment; antibody engineering or derivatisation may be used to minimise immune response 4. Ideally, toxin component only becomes active at target site, e.g. by controlled activation of pro-drug or photosensitizer; may require cleavable linkers

HAPTEN-LABELLED PROTEINS as vaccines or immunogens	1. Large protein component to confer immunogenicity 2. Moderate hapten density (typically around 5–10) to provide adequate epitope supply without inducing tolerogenic behaviour 3. Low *in vitro* toxicity 4. Linker group chosen to avoid bridge recognition problems 5. Hapten linked *via* appropriate position to promote antibody of desired cross-reactivity
LABELLED CONJUGATES for histochemistry	1. Small size to allow tissue penetration 2. Efficient localisation at desired target, e.g. by incorporating antibody or specific binding protein 3. Appropriate label, e.g. electronically dense (or capable of creating an electronically dense, insoluble product) for electron microscopy 4. Low background binding in the absence of specific analyte 5. Resistant to interference from other components in tissue samples
LABELLED IMMUNOCONJUGATES for immunoassay, blotting techniques and indirect measurement of nucleic acid probe assays	1. Good retention of functional activity after conjugation, both for antibody and for the signal-generating component 2. Good *in vitro* stability, ideally at working concentration 3. High specific activity of signal generation to allow good sensitivity 4. Low background binding in the absence of specific analyte 5. Resistant to interference from other sample components 6. Reasonably small size to avoid steric effects 7. Choice of linker group which avoids any potential bridge- recognition issues
ENZYME-LABELLED OLIGONUCLEOTIDES for nucleic acid probe assays	1. Good retention of oligonucleotide structure and enzyme functional activity after conjugation 2. Moderate oligonucleotide chain length to balance specificity with simplicity of preparation 3. Good *in vitro* stability, ideally at working concentration 4. High specific activity of signal generation to allow good sensitivity 5. Low background binding in the absence of specific target sequence 5. The ability to survive hybridisation conditions (e.g. 40–60°C in the presence of organic solvents or chaotropes)

The relationship of these applications is shown in Tables 1.1–1.4, along with the types of coupled protein reagents likely to be encountered in each. Table 1.5 summarises some of the attributes which define the utility of these reagents for the major applications.

1.2 Applications in Clinical Diagnosis

Clinical diagnostics, the qualitative or quantitative determination of individual components (analytes) of body fluids, can be categorised into a number of branches which differ in the technologies used, the level of sensitivity achievable, and their requirements for the use of coupled protein reagents. Of course the quantification of specific compounds for this application is no different in concept to other branches of analytical chemistry, and there is some overlap between the techniques used here and elsewhere. In particular the power of chromatography to separate individual components from complex mixtures has many diagnostic applications (Burtis and Ashwood, 1994), but there is generally little need for the use of protein coupling methods. High sensitivity can be obtained if sufficient attention is paid to developing a specific method, and particularly if the analyte has some readily detectable property such as the capability for chemiluminescence; Nakagawa and Miyazawa (1997), for example, demonstrated the detection of picomolar levels of tea catechin in plasma using reverse-phase HPLC. Similar levels of sensitivity can be achieved using detection based on strongly fluorescent tags (Kondo *et al.*, 1997) or tandem mass spectrometry (Fredline *et al.*, 1997). For many analytes, however, there is no sufficiently sensitive chromatographic procedure, and other techniques have been developed to facilitate quantification of clinical analytes.

The dominant approaches are clinical chemistry and immunoassay. More recent developments such as nucleic acid probes and biosensors have begun to have an impact in some areas, albeit much more slowly than early assessments had predicted.

There has also been some overspill of these techniques into non-clinical fields. The methods of clinical diagnostics are essentially procedures for determining minute quantities of a target substance in a complex matrix. Not surprisingly, these have attracted attention from other disciplines facing similar analytical challenges such as environmental and military monitoring. These are dealt with briefly in Section 1.7.1.

1.2.1 Clinical chemistry

In clinical chemistry, an analysis is carried out by engineering a situation where the unique properties of the analyte generate a specific, measurable signal, generally without any necessity to separate the analyte from the other components of the sample. There is a reliance on the specificity of the chemistry involved to avoid interference from other species, and these methods call on widely disparate signal generation technologies to achieve this. Examples include spectroscopic techniques for metal ions, electrochemical approaches for dissolved gases, chromogenic reactions and subsequent spectrophotometric measurement for a variety of species (Burtis and Ashwood, 1994). Proteins are frequently employed as assay reagents; enzymes in particular are often used to confer specificity, or to catalyse signal generation. However, the largely separation-free designs generally employ the protein in its native, uncoupled form and, generally, the requirements for coupled protein reagents are minimal.

The only major exception is when it is necessary to immobilise proteins onto a suitable matrix, particularly in view of the successful development of clinical chemistry systems based on solid phase formats (Libeer, 1985). A special case within this category is the preparation of immobilised dye-labelled substrates for proteolytic enzymes, whose activity is demonstrated by the release of dyed fragments into solution. Small peptides now tend to be used in this role, however (Moss and Henderson, 1994).

1.2.2 Immunoassay

There are a large number of molecules of biomedical interest that cannot conveniently be quantified by these clinical chemistry approaches, either because the analyte has no properties or reactions specific enough to form the basis of a direct test, or because it is present in such small quantities that it cannot be detected adequately.

The discipline of immunoassay has been evolved to overcome these limitations, since its description in its modern format by Berson and Yalow in the 1950s (Berson *et al.*, 1956; Yalow and Berson, 1959). The reactions between antibody and antigen are among the most specific encountered in biological systems, and their potentially quantitative nature was known by the 1920s (Heidelberger and Kendall, 1929). This led to the development of simple techniques for protein quantification based on antibody–antigen complex formation, either as a visible precipitate, or by more sensitive optical means such as nephelometry

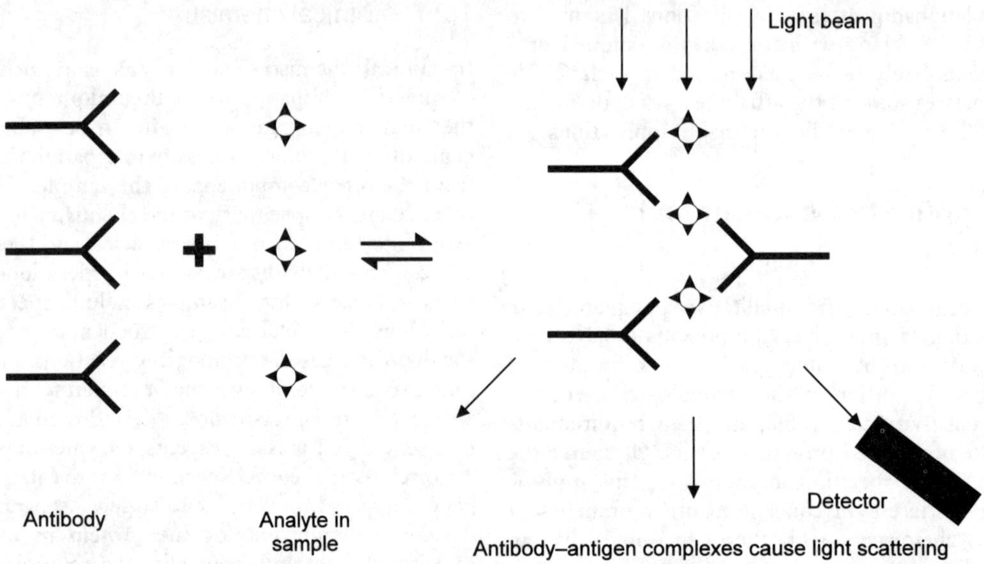

Figure 1.2 Nephelometry: an angled detector collects only light which has been scattered by antibody–antigen complexes, so the signal increases with the amount of complex formation.

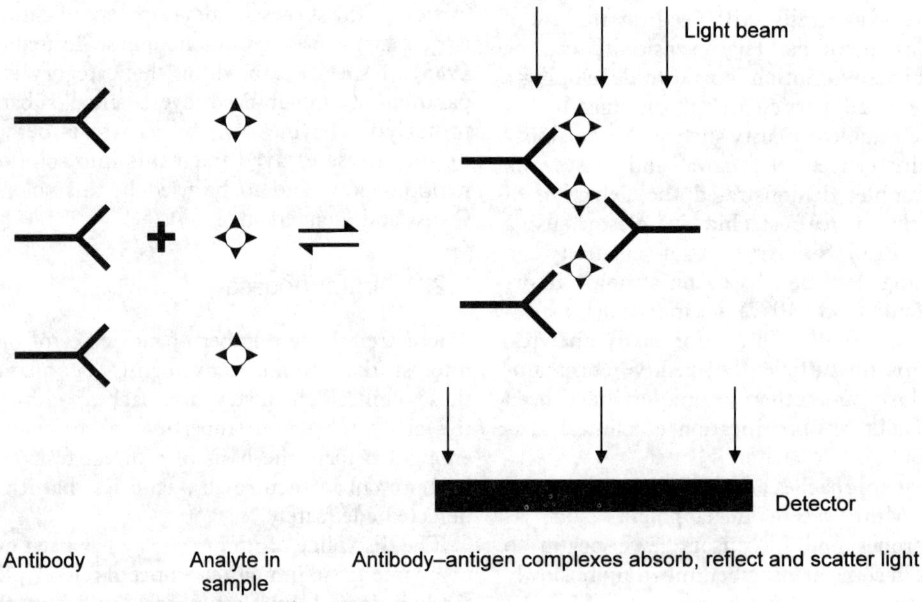

Figure 1.3 Turbidimetry: the light reaching an in-line detector is reduced by scattering, absorption and reflectance caused by antibody–antigen complexes, and hence decreases with increased complex formation.

(Figure 1.2) and turbidimetry (Figure 1.3). Many of these approaches have endured successfully over the years for low-sensitivity applications, e.g. agglutination assays, where for example antibody is bound to the surface of a large species that agglomerates in the presence of cross-linking antigen to form a precipitate. (Figure 1.4). The large species may be a red blood cell (haemagglutination) or a latex particle, and various techniques have been employed for coupling them to the antibody (Stavitsky and Arquilla, 1958; Adler and Adler, 1980). The end-point of an agglutination reaction can be determined by visual detection of a

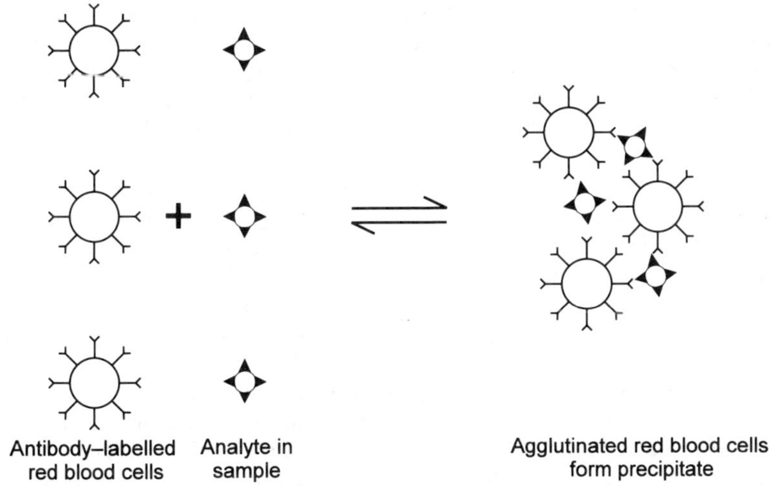

Figure 1.4 Haemagglutination: in this particular format red blood cells coated with antibody are cross-linked by specific antigen, resulting in the formation of an insoluble aggregate.

precipitate, or instrumentally *via* turbidimetry or nephelometry. The use of particle-labelled reagents to enhance signal contrast in turbidimetry is a closely related concept (Opheim *et al.*, 1984).

These agglutination assays are primitive forms of immunoassay (predating Berson and Yalow's work), where the antibody–antigen reaction is used to isolate or differentiate the analyte of interest in some way from other sample components. The agglutination assay is an example of a separation-free format; such approaches give a measurable differentiation (in this case the conferred aggregation) in the presence of unreacted species, whilst in heterogeneous formats an actual physical separation of reacted and unreacted components is employed to achieve the desired performance (see Figure 1.5). The term homogeneous is sometimes used synonymously with separation-free, but is better reserved for formats where all the reacting components remain in solution throughout the assay–not the case for agglutination assays. Some genuine homogeneous assay formats will be described later.

Once specific interactions have been exploited to achieve this differentiation or isolation, non-specific methodologies can be used to generate a detectable signal. This opens the way for the extreme sensitivity which characterises immunoassay as a technique. Because of their dependence on analyte-specific reactions, many clinical chemistry techniques are applicable only for molecules present in the sample at nanomolar levels or greater; commercial immunoassays routinely display sensitivities in the femtomolar range, and research applications have demonstrated the potential to push the limits out still further (Johannson *et al.*, 1986).

The use of these sensitive techniques is only made possible by exploiting the unparalleled specificity of the antibody–antigen reactions encountered in nature, and designing an immunoassay requires the availability of suitable antibodies to the analyte concerned. Fortunately, it is possible to obtain such antibodies for virtually any analyte, either from natural sources, (e.g. from individuals seropositive for antibodies to disease antigen), or by using well-established immunological techniques.

These tests were based initially on the use of polyclonal antisera (Bailey, 1994), relying on subtleties of the im-munisation process (and the rather unpredictable variation of host response) to generate the desired antibody characteristics. This can be a rather hit-or-miss process, and even once a suitable antiserum has been identified, the amount available is often strictly limited; the antibody profile of an immunised individual varies continually with time, so that an antiserum sample is unique to its time of removal. There is no guarantee that a replacement antiserum with identical properties will be found. Monoclonal antibodies display a much greater variety of individual behaviours, particularly with respect to the much improved cross-reactivity that they can offer. This coupled with the fact that the techniques offer a replaceable supply of homogeneous material, sparked off a revolution in the immunoassay field, and the use of these reagents quickly became commonplace. Antibodies will be discussed in much greater detail in Chapter 4.

The combination of sensitivity and specificity offered by immunoassay techniques allows them to extend the range of detection to analytes not adequately served by clinical chemistry. A wide range of immunoassay systems has been commercialised (Wild, 1994; Price and Newman, 1997), and these are now well-established diagnostic tools in such areas as the endocrinology of the thyroid and the reproductive system, the diagnosis of disease infection or immunity, therapeutic drug monitoring, and oncology.

1.2.2.1 Protein labelling in immunoassay

Immunoassay is one of the applications where the notion of protein labelling is easiest to envisage. There are many different formats by which these tests can be carried out, the major categories being competitive and immunometric designs (Figure 1.5). In almost all circumstances they rely on a label to make the antigen–antibody complex detectable.

The early heterogeneous immunoassays relied on direct labelling with the signal generating species, a radioisotope, and these labels prevailed for several decades. However, issues such as safety, cost, convenience and environmental concerns have caused a drive away from the use of radioactive molecules. The sensitivity of radioimmunoassay is also inherently limited by the detection limit of the isotope used; for example ^{125}I, one of the most commonly used radiolabels, decays at 4.8 disintegrations per minute per attomole, giving a practical detection limit of around 5–10 amol. Attempts to improve on this limit by multiple labelling can only have a relatively small effect, and can also tend to cause problems of increased radiolytic damage to the protein. Using an isotope with a more rapid disintegration rate is

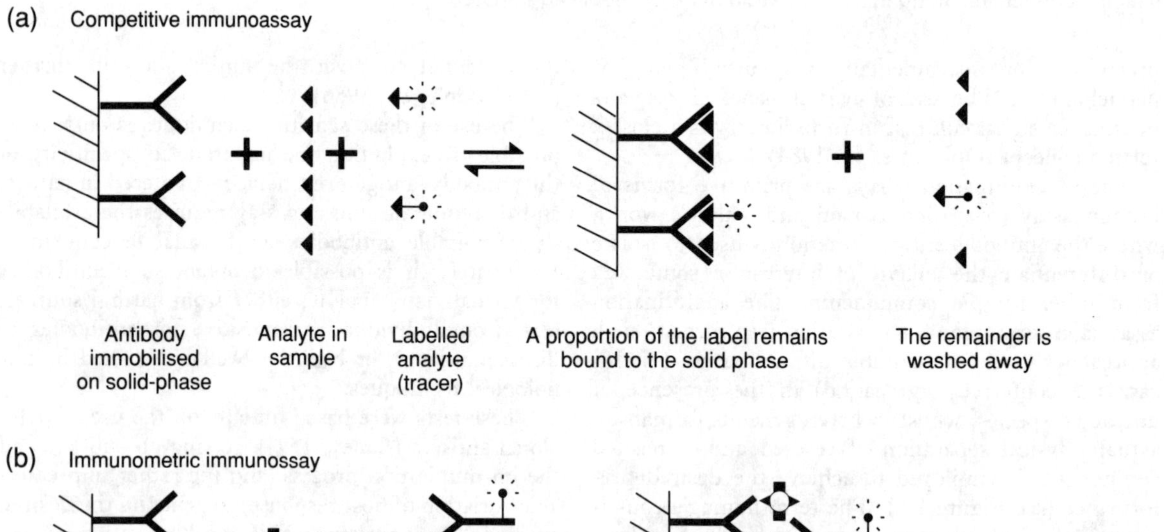

Figure 1.5 Heterogeneous immunoassay formats: (a) a competitive format, where labelled tracer competes with analyte from the sample for limited solid-phase antibody sites; the signal reduces with increasing analyte level; (b) an immunometric format, where analyte trapped by excess solid-phase antibody is detected in a sandwich with a second, labelled antibody; the signal increases with analyte level. Many variations on these basic models have been described.

an alternative strategy, but the commensurate drop in half-life can lead to stability problems. There are other theoretical factors affecting the sensitivity of radioimmunoassays, which have been discussed in depth by Ekins (1993).

Thus, although it has taken until the 1990s for the much-heralded decline of radioimmunoassay to be realised (Pettersson, 1993), replacement technologies are becoming increasingly dominant for immunoassay signal generation.

In some cases, the radiolabel has simply been replaced by direct incorporation of a different signal-generating molecule; however, it is difficult to match radioactivity in terms of direct detection limits, and this has limited the techniques which are applicable. In general only luminescent or fluorescent labels have been successfully employed in this way for high-sensitivity applications (Barnard *et al.*, 1985; Wood *et al.*, 1986; Hemmilä, 1991). Some of the most sensitive applications of direct labelling have coupled immunoassay with high-performance instrumental techniques such as capillary zone electrophoresis (Schultz and Kennedy, 1993).

A particularly widespread approach, which has the theoretical capability of exceeding the sensitivity of radioimmunoassay, is to use an enzyme label in an indirect signal generation system. The enzyme catalyses the reaction of a suitable substrate mixture to generate the signal (Blake and Gould, 1984; Ishikawa, 1987; Ishikawa and Kohno, 1989; Gosling, 1990). While there are certain disadvantages associated with the use of enzymes they offer the facility for amplification. Each molecule of the label can catalyse multiple signal generation events, contributing significantly to the overall sensitivity of the system; Johannson *et al.* (1986) demonstrated the detection of sub-attomolar levels of alkaline phosphatase, where the number of enzyme molecules in the sample was as low as 6600.

The signal itself can be colour, fluorescence, luminescence, etc. Although the latter two techniques again offer the greatest sensitivity, the enzyme-linked immunosorbent assay (ELISA) with colorimetric end-point (Engvall and Perlmann, 1971; van Weemen and Schuurs, 1971) has found widespread use, not least because the signal can be measured by simple, cheap instrumentation which is readily available in most laboratories.

The search continues for sensitive signal generation methods to match or improve on those already described. Two interesting examples from the last few years use DNA fragments as the label. In the first detection is subsequently achieved by employing the label to express a luminescence-catalysing enzyme (Christopoulos and Chiu, 1995). In the second the product of polymerase chain reaction (PCR) amplification of the label is detected by staining after gel electrophoresis (Joerger *et al.*, 1995). Electrochemiluminescence, which provides a sensitive luminescent signal without the need for an enzyme catalyst, has received considerable attention in recent years (Genesis Report, 1995). The Origen™ electrochemiluminescence technology owned by IGEN Inc. has recently been commercialised for immunoassay use by Boehringer Mannheim (Blackburn *et al.*, 1991; Hoyle *et al.*, 1996). This approach can be coupled to enzyme systems: for example hydrogen peroxide generated from a suitable oxidase-catalysed substrate reaction can be detected *via* its luminogenic reaction with electrochemically oxidised luminol (Wilson *et al.*, 1996).

Considering the possible labels described above, in combination with the possible different layouts of assay architecture, it is clear that the preparation of labelled reagents can involve all possible permutations of components. These include: protein labelled with small molecule (e.g. isoluminol-labelled antibody), small molecule labelled with protein (e.g. peroxidase-labelled thyroxine), protein labelled with protein (e.g. β-galactosidase-labelled antibody fragments), or small molecule labelled with small molecule (e.g. ^{125}I-histamine-labelled steroid). In these examples, the labelled reagent is a simple conjugate of an immunoreactive species and a signal-generating component, but indirect designs may be employed too, exploiting, for example, biotin–avidin interactions (Ternynck and Avrameas, 1990), or bispecific antibodies (Brennan *et al.*, 1985; Cook and Wood, 1994). These also rely on covalently-coupled combinations of proteins and small molecules.

Where one of the components involved is a protein, this tends to dominate the chemistry and separation strategies used, and this book attempts to cover all such situations. Thus of the four permutations described above, only the last (small molecule–small molecule) will fall outside the scope of this book. The conditions applicable to reaction between small molecules are nowhere near as homogeneous as for those involving proteins, and the chemistry and purification strategies need to be considered on an individual basis. This area is really a branch of organic synthesis and cannot be dealt with adequately by general rules.

Many of the techniques described in Chapters 5 and 6 are applicable to the preparation of labelled conjugates for immunoassay, and this is also one of the more important areas for application of the gene fusion approach to coupling described in Chapter 9 (see also Uhlén *et al.*, 1992; Lindbladh *et al.*, 1993).

The principal requirements for immunoassay conjugates are typically as follows. (1) High incorporation of signal-generating component (especially for non-enzymic systems where no amplification is achieved). (2) Retention of functional activity both of the immune and signal-generating component. (3) Reasonably small size: although this factor is less important here than in some other applications, steric hindrance effects have been shown to be important in immunoassays too (Butler *et al.*, 1986). (4) Absence of immunochemical artefacts, such as linker groups, which are recognised by antibody. These can be avoided by careful selection of coupling agents both for the conjugates used in the assay, and for the immunogens used for antibody generation (see Section 1.6). (5) Stability, such that reagents can be used over a convenient shelf-life. Storage times of many months are typical for commercial products.

1.2.2.2 Immobilised proteins in immunoassay

There are a number of immunoassay systems where no physical separation is required. Apart from the agglutination assays mentioned earlier, the first to be commercialised on a large scale was the enzyme multiplied immunoassay technique (EMIT) approach described by Rubenstein *et al.* (1972). In the best known variant of this truly homogeneous technique (Figure 1.6), a conjugate is prepared between the (small) analyte of interest and an enzyme. Binding of the analyte to antibody interferes sterically with the activity of the enzyme, and any analyte from the sample competes quantitatively with the conjugate to reduce this effect.

The homogeneous approach which is probably most widely used today is based on fluorescence polarisation (Weber, 1952), described for immunoassay purposes by Watson *et al.* in 1976 (Figure 1.7). If a small,

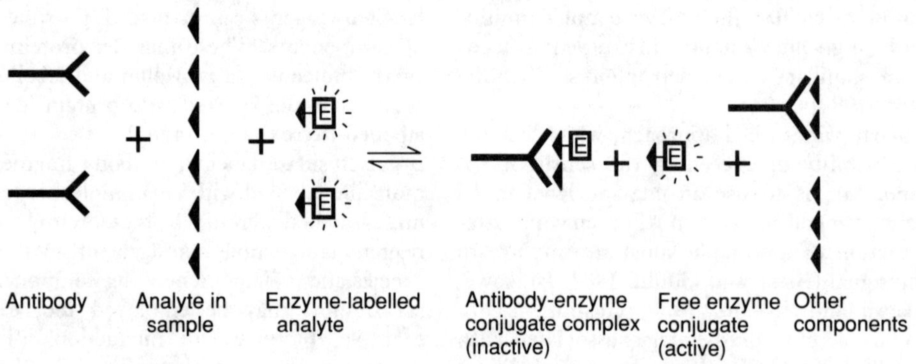

Figure 1.6 The principle of one format of the EMIT™ homogeneous assay; analyte-labelled enzyme is inactivated by the presence of analyte-specific antibody, but the degree of inactivation is reduced by free analyte from the sample. Therefore, the signal increases with the amount of analyte present.

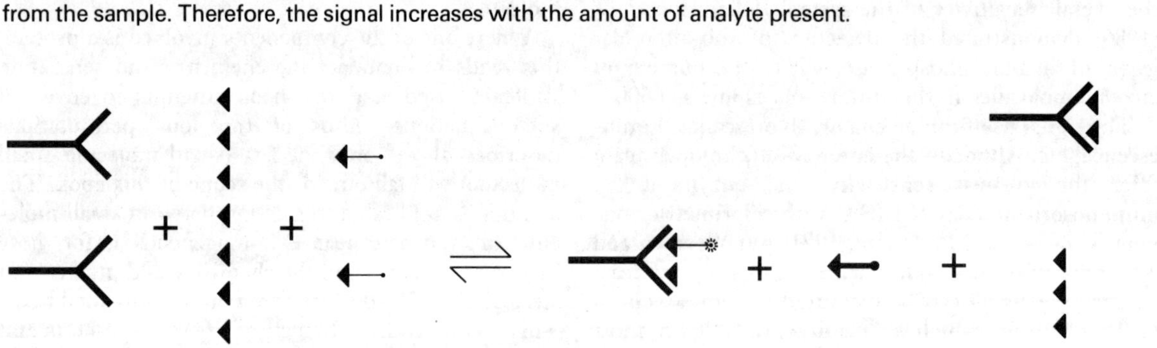

Figure 1.7 The principle of fluorescence polarisation immunoassay. Fluorescent-labelled analyte retains its plane of polarisation only when coupled to analyte-specific antibody. The presence of free analyte from the sample competes for this antibody, and retention of polarisation thus decreases with increasing analyte level.

unhindered fluorescent molecule is excited with polarised light, its rapid rotation means that the plane of polarisation of the emitted energy is different to that of the excitation signal. However, if the molecule is bound to antibody, the slower rotation of this larger species causes little change to the plane of polarisation, and analysis of the emission signal can be used to quantify the level of binding.

Both these techniques are in common use today, particularly for drug detection which requires sensitivity down to roughly the nanomolar level. A great many other separation-free formats of ingenious design have been developed with varying degrees of success; however the majority of immunoassay designs (and all those of the highest sensitivity) employ a physical separation of some sort. Therefore the second requirement in immunoassay for a coupled protein is normally a protein-coated solid phase—generally antibody or antigen attached to a solid support such as latex (magnetic or otherwise), membranes or microtitre wells. Plastics have come to dominate this field since their first description in the role, in which polypropylene or polystyrene tubes were used to immobilise antibody for a radioimmunoassay. (Catt and Tregear, 1967).

Coupling of proteins to solid phases is also a requirement in some agglutination assays, where latex particles act as a type of label.

Although the same types of chemical reaction encountered in the preparation of soluble conjugates are employed here too, there is another alternative popular due to its practical simplicity: to rely on physical adsorption of the protein to the surface. The factors controlling this type of protein binding are very different to those governing covalent reactions, and are generally less widely understood. It should also be noted that these factors may often be important, even where covalent interactions are (or are claimed as) the predominant mechanism of binding.

The techniques for protein immobilisation described in Chapter 8 find some of their most important applications in the preparation of solid-phase components for heterogeneous immunoassays.

The key aims of these immobilisation procedures are for a reasonable degree of retention of the immune activity of the immobilised species, a high level of irreversibility such that protein is not leached from the surface during the assay, and sufficient stability to allow convenient shelf-life for the coated component.

1.2.3 Nucleic acid-based techniques

The development of molecular biology techniques has opened up the potential for new high-sensitivity approaches to clinical diagnostics, sometimes offering a direct alternative to an immunoassay test, but more commonly providing complementary information. Their impact on the diagnostics field remains somewhat slower than initially predicted (Genesis Report, 1997), but they still represent a likely growth area for the future.

These techniques rely on the detection of the genetic material of viruses, etc., rather than their constituent proteins, and can also be used to detect mutations of normal human genes which are responsible for clinical conditions.

1.2.3.1 Nucleic acid probes

A labelled fragment of nucleic acid can be used as a probe to detect complementary material by specific hybridisation to the sequence of interest. The main clinical use of this approach is as a diagnostic tool for prenatal detection of hereditary genetic defects, or for disease agents. Viral genetic material in particular has received considerable attention.

Genomic DNA shows the greatest variation between different organisms, and so offers the greatest potential specificity as an analytical target, however there is frequently only a single copy of the sequence of interest per cell, and the demands on sensitivity are therefore at their greatest. Non-genomic material such as messenger RNA is present at much higher levels, and its transient nature can be useful in spotting active infection. In bacteria, ribosomal RNA is present at the highest levels of all. However, it is highly conserved between species, which can make it difficult to identify specific target sequences.

Chain length in the target sequence is another consideration, and a balance will be sought between the greater specificity of longer probes and their relative robustness to the incorporation of a label, and the relative simplicity of preparing shorter sequences and their smaller chance of including errors such as missed or incorrect bases.

These nucleic acid probes can be employed in various formats. The sample DNA or RNA is typically blotted on to a membrane, either in crude form (a dot or slot blot) or after electrophoretic separation (a Southern or northern blot). These blotting techniques are discussed in more detail in Section 1.2.4. In each of these cases the sample nucleic acid is immobilised to the solid phase, where a target sequence can then be detected by hybridisation to a suitably specific probe (Figure 1.8). Greater specificity can be achieved by the use of two probes in a sandwich format (Figure 1.9); each detects a sequence of the

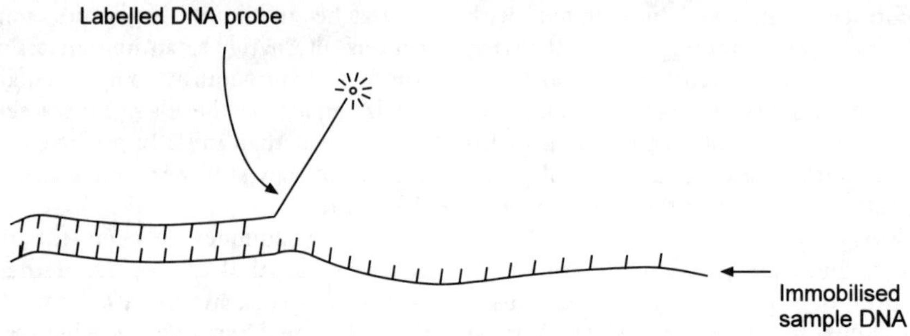

Figure 1.8 Detection of a specific nucleic acid sequence from a blot using a single complementary probe.

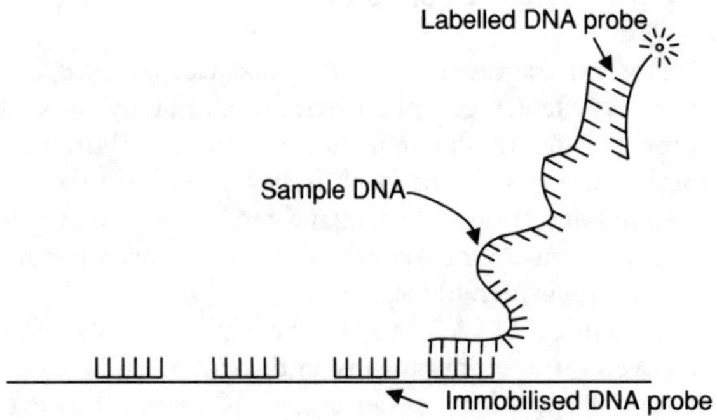

Figure 1.9 Detection of a specific nucleic acid sequence from a liquid sample using a sandwich format—one probe is immobilised, the other is labelled.

target nucleic acid, which no longer requires immobilisation.

The hybridisation of sample and probe nucleic acid is carried out under suitable conditions to ensure adequate reaction rate and specificity. Such conditions may include raised temperature (typically 40–60 °C) or the use of detergents, organic solvents or chaotropes such as urea or guanidinium isothiocyanate to disrupt hydration structures.

In these techniques it is required that the label does not interfere with the hybridisation process, and that it survives the reaction conditions. Developments in labelling have been closely analogous to those already described for immunoassay, with the same reliance of the original techniques on radiolabels and the same search for non-radioactive signal systems to equal or surpass their sensitivity.

Direct incorporation of a radiolabel into the probe is the best-known approach. The most widespread label, ^{32}P, has a number of disadvantages; its half-life is only around 14 days and even this stability may not be achieved due to radiolytic damage of the nucleotides as the label decays. Alternatives include ^{35}S and ^{125}I.

As with immunoassay, direct non-radioactive labelling strategies have been dominated by fluorescence and luminescence. Generally a chemically modified nucleotide is incorporated in the probe, giving a functional-group which can react with a suitable derivative of the fluoro- or luminophore. Great sensitivities in direct-labelling approaches have been achieved using luminescent dioxetane reagents. Bronstein *et al.* (1996), for example, reported the detection of sub-attomolar quantities of β-galactosidase in a model assay.

These radioactive and non-radioactive systems involve the labelling of nucleic acids with atoms or small molecules, and therefore fall outside the remit of this book. However, once again in close analogy to immunoassay, the use of coupled protein reagents to amplify signals has received attention. In 1984 Renz and Kurz described a colorimetric system where the probe was labelled with horseradish peroxidase coupled to a positively-charged polyethylene-imine derivative. This binds initially to the negatively charged nucleotide chain by electrostatic interaction, but then a chemical coupling agent (glutaraldehyde) is employed to give a permanent covalent link. The enzyme is stable enough to survive the hybridisation stage, and can then be used to generate an insoluble coloured product from a suitable chromophore.

In the ECL™ system developed by Amersham International, the same HRP-labelling approach is employed, but much greater sensitivity is obtained by using it to generate an enhanced chemiluminescent signal from a luminol-based system. No significant effect on hybridisation is observed despite the bulky probe, and attomolar sensitivities have been achieved. This system has been commercialised successfully as a research tool (Durrant *et al.*, 1990).

Enzymes can also be coupled directly to probes *via* amino-derivatised spacer arms. Urdea *et al.* (1988) at

Chiron, for example, have demonstrated the use of horseradish peroxidase and alkaline phosphatase-labelled probes prepared in this way.

Other approaches rely on two-stage, indirect binding of a label *via* a non-covalent protein complex; this means that the label itself does not need to be present at the hybridisation stage, thus avoiding stability and interference problems. Two of the most common systems involve: (1) derivatisation of the probe with biotin and subsequent reaction with labelled avidin or streptavidin, and (2) derivatisation of the probe with a small compound such as digoxigenin and subsequent reaction with labelled specific antibody (Figure 1.10).

The label can be any of the signal generators described already: fluorophores, luminophores and enzymes can all yield high-sensitivity nucleic acid probe assays. Nelson and Kacian (1990) have estimated the sensitivity of a luminometric system of this type using alkaline phosphatase and a dioxetane substrate at between 10^{-20} and 10^{-21} mol.

The Boehringer–Mannheim Rainbow™ system allows three different probes to be used simultaneously: one labelled with biotin, one with fluorescein and one with digoxigenin. These can be detected sequentially by suitable alkaline phosphatase conjugates, using a different coloured azo-dye signal for each stage.

Nucleic acid probes offer great potential for sensitive assays (Gravitt and Manos, 1993), and numerous diagnostics companies are now offering probe tests.

1.2.3.2 Polymerase chain reaction (PCR) and related techniques

A fairly recent development in this field has been the evolution of techniques which allow the multiplication of specific nucleic acid sequences from a sample. The polymerase chain reaction (PCR) is the best known of these, and is discussed further in Chapter 7. Its invention has had a major impact on the life sciences, and won Kary Mullis the 1993 Nobel Prize for Chemistry. This technology is undergoing widespread commercialisation, along with related techniques which have emerged partly due to the restrictive patent situation governing the use of PCR itself (Genesis Report, 1997).

It will be interesting to see how these amplification technologies affect the future of clinical diagnostics, both in the competition between the different methodologies, and in their effect on signal generation requirements: high-sensitivity detection becomes a virtual irrelevance if the level of analyte in the sample can be amplified at will.

While PCR *per se* does not require the preparation of coupled protein reagents, DNA-labelled antibodies have been employed in sensitive immunoassays based on PCR amplification of the label (Joerger *et al.*, 1995), while ELISA detection of PCR products with labelled antibodies has been commercialised by Boehringer Mannheim. The importance of this technique in the clinical diagnostics arena continues to grow (Thibodeau, 1994; McPherson, 1995) and further

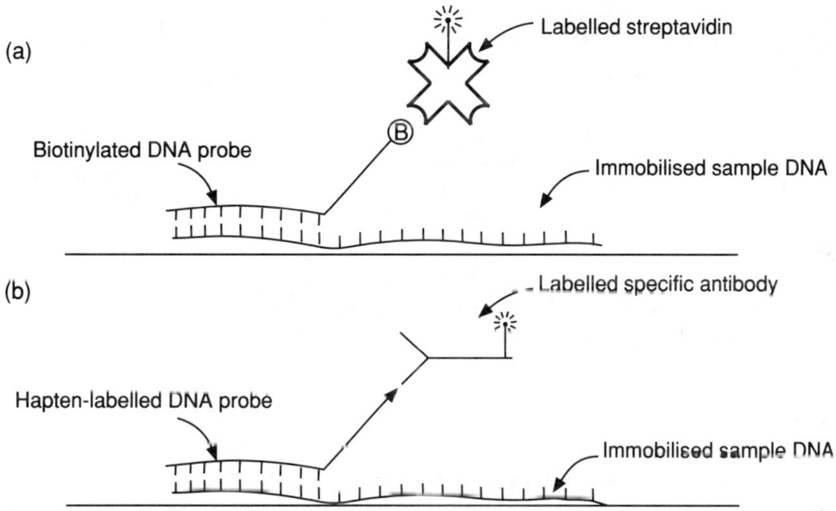

Figure 1.10 Common indirect formats for detection of specific nucleic acid sequences: (a) a biotinylated probe is detected with a labelled streptavidin reagent; (b) a hapten-labelled probe is detected with a labelled specific antibody.

applications requiring labelled proteins seem certain to emerge.

1.2.3.3 Peptide nucleic acids

Finally, an interesting new category of reagents has been developed, in which the sugar–phosphate chain which carries the base units of nucleic acids has been replaced by a backbone consisting of amino acids. Surprisingly, these peptide nucleic acids (PNAs) are capable of binding to nucleic acid targets with higher affinity and greater specificity than DNA itself (Buchardt *et al.*, 1993; Nielsen *et al.*, 1994). They are also more stable to enzymic degradation, and can be prepared using straightforward peptide synthesis methodology (Corey, 1997). Thus they show some promise as diagnostic reagents, and could allow more straightforward labelling than their nucleotide analogues.

1.2.4 Blotting techniques

In the last section it was noted that DNA samples are frequently blotted onto a membrane before detection with a probe, either by simple physical application (dot blots and related techniques), or by transfer from an electrophoresis gel. However these blotting techniques have wider applications allowing analysis of many different target categories (Hayes *et al.*, 1989; Wisdom, 1994). Often a coupled protein reagent is required for detection purposes, such as a labelled specific antibody.

These techniques are generally referred to by a system of 'geographical' terminology stemming from the name of the originator E. M. Southern, who (1975) described the technique known as Southern blotting. Single-stranded DNA fragments are immobilised following electrophoresis as described earlier, then detected by hybridisation of a suitably specific, labelled DNA-probe (Figure 1.11). Southern blotting has been widely used in molecular biology, e.g. for the identification of genetic defects associated with inherited disease states.

It is a natural extension of this approach to immobilise messenger RNA, again using a suitable DNA probe for detection. This technique (Thomas, 1980, 1983) became known as northern blotting, and is particularly useful for probing phenomena of gene translation, e.g whether cellular over-expression of a certain protein is due to proliferation of the gene, or of the messenger RNA it produces, or because excretion from the cell is impaired.

In a similar vein, western blotting (or immunoblotting) describes the case where proteins are blotted onto a membrane then detected with a labelled specific

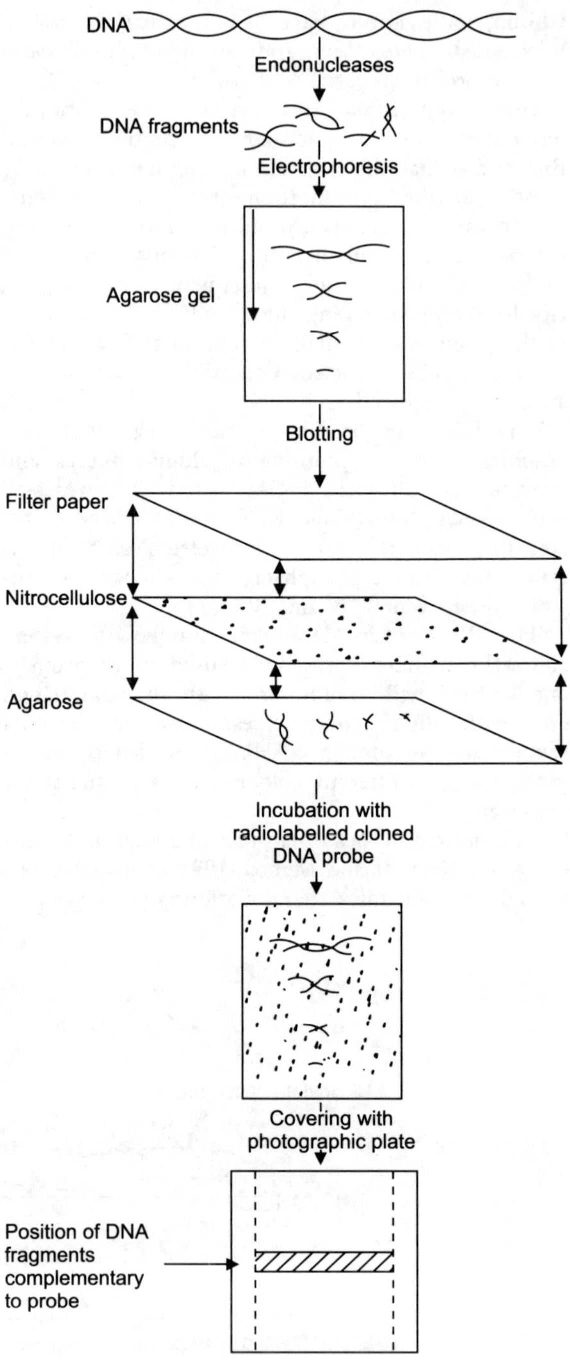

Figure 1.11 Southern blotting: following electrophoresis DNA is transferred on to a membrane and complexed by a complementary probe, in this case radiolabelled for detection with a photographic plate. (Reproduced with permission from: Hayes, P. C. *et al.* (1989). *Br. Med. J.*, **29**, 965–968.)

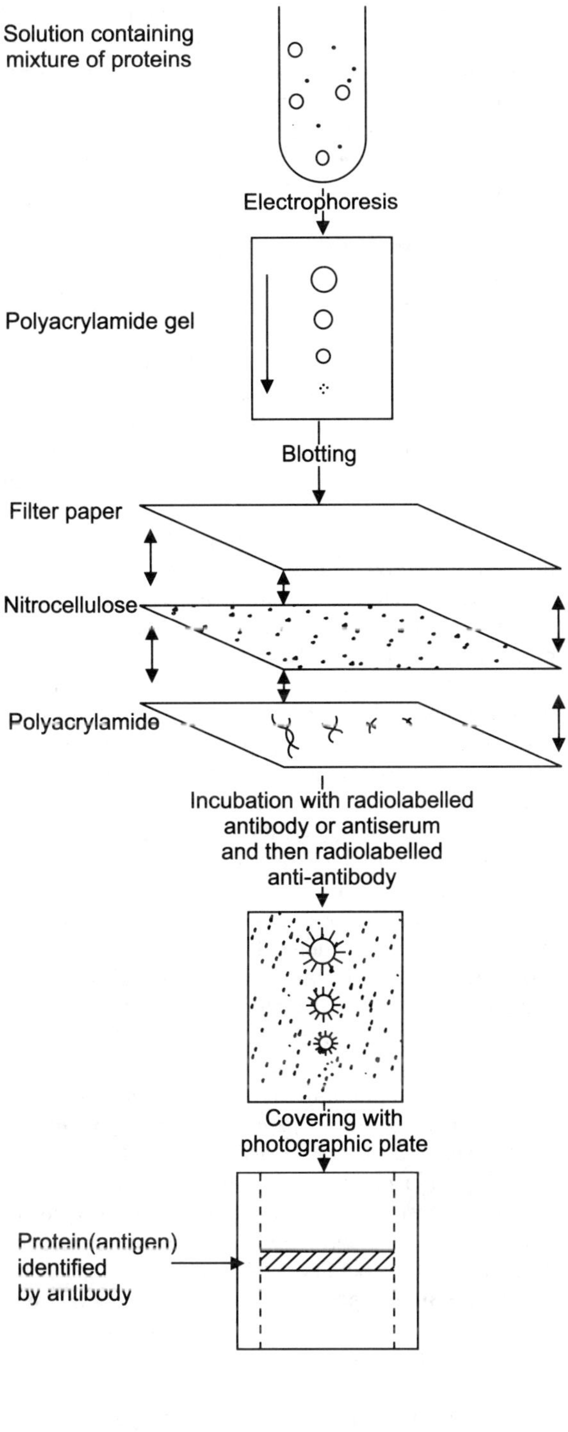

Figure 1.12 Western blotting: here it is a protein which is blotted, and detected directly or indirectly with a labelled antibody. (Reproduced with permission from: Hayes, P. C. *et al.* (1989). *Br. Med. J.*, **29**, 965–968).

antibody (Figure 1.12) (Renart *et al.*, 1979); this technique closely parallels immunoassay, and is often employed as a confirmatory test to back up assay findings (Towbin and Gordon, 1984). Ligand blotting is very similar but the detection strategy employs binding partners (ligands) other than antibodies. An example (Islam *et al.*, 1983) is the detection of immobilised thyroid stimulating hormone (TSH) receptor proteins by treating the blot with TSH as a ligand, then an anti-TSH antibody followed by a peroxidase-labelled anti-species antibody. The term eastern blotting has been applied to the transfer of proteins from an electrofocusing gel to a membrane—rather more a subset of western blotting than a separate technique.

Finally, southwestern blotting appropriately enough describes a hybrid technique in which DNA-binding proteins are immobilised then detected using a labelled DNA probe.

1.2.5 Biosensors

Another challenge to the traditional methodologies of clinical diagnostics is in the quest for cheap, fast tests, ideally with reusable components, which can be used away from laboratory locations: in the doctor's office, in the field, or even *in vivo*. Conventional diagnostic technologies have been adapted very successfully for some of these ends (for example home pregnancy and fertility monitors based on immunoassay formats with a visible coloured endpoint), but biosensors have received a particularly large amount of attention. The exact definition of a biosensor is open to debate, but the term is generally understood to describe a device which uses a transducer system to convert a biochemical event into a measurable signal of some description, a voltage or current for example (Lowe, 1985). The convenience of an electronic signal format, readily compatible with convenient data handling and display systems, largely explains the amount of interest which biosensors have attracted for 'near-patient' testing. Outside a few key applications, however, biosensors remain more a promise than a commercial reality, despite several decades of research.

The dominant application commercially is the glucose sensor. Typically, glucose oxidase is immobilised onto a suitable electrode surface, where it catalyses oxidation of glucose to generate hydrogen peroxide in a quantitative fashion, and the electrochemical system then converts the peroxide concentration into a simple digital signal. Glucose measurement for diabetes is carried out in the home on a huge scale, and current self-diagnosis methods based either on clinical chemistry or on the sensors described above require blood sampling on a daily (or more frequent) basis. A good deal of effort has been put into approaches which avoid this inconvenience, and implantable *in vivo* forms of the glucose electrode have received considerable attention. Another area where biosensors offer potential is in the monitoring of urea during blood dialysis (Thavarungkul *et al.*, 1991).

Immunosensors apply sensor-type detection to immobilised antibody–antigen reactions (North, 1985). Advanced optical techniques such as surface plasmon resonance (SPR) can detect real-time changes in refractive index at a surface (Mayo and Hallock, 1989). It is easy to see how this technology can be harnessed to detect, for example, binding of antigen to an immobilised antibody. So far the main application of SPR has been as a research tool, the BIAcore™ system developed by Pharmacia, for example, having found a niche for the measurement of affinity constants in equilibrium binding situations (Karlsson *et al.*, 1991; Karlsson and Ståhlberg, 1995).

By their nature biosensors often require immobilisation of a protein to the sensor surface, the requirements typically being similar to those of immunoassay solid phases. However, the need for highly localised deposition of proteins on the biosensor surface has led to some advanced techniques such as electropolymerisation (Scouten *et al.*, 1995), ink jet printing (Kimura *et al.*, 1988; Kobayashi and Takayama, 1997) and photolithography (Rozsnyai *et al.*, 1992). Some applications may also require the incorporation of a labelled reagent, e.g. a bulky refractive index probe to enhance sensitivity in SPR.

1.3 Applications in Histochemistry and Cytochemistry

The identification of specific components in tissue samples (histochemistry) and cells (cytochemistry) is analogous in many ways to clinical diagnostics, but it is the localisation of the components of interest within a biological structure that is required, rather than their concentration in a homogeneous medium. The components may also be much larger than the molecular analytes of clinical diagnostics, e.g. cells or muscular structures. Outside the clinical field, the identification of particular components within cells has assumed great importance in the production of recombinant proteins; screening of host cells to identify a desired product is an important step, and has extended the application of cytology methods.

These differences from clinical diagnostics dramatically alter the strategies for detection, which tend on the whole to be much more qualitative in nature and may be as simple as visual examination to identify the morphology or properties of interest. The requirement to study biological structures has meant that microscopy is the dominant technique in this area, and advances in histochemistry and cytochemistry have gone hand in hand with the development of microscope instrumentation.

The sensitivity of light microscopy is obviously limited despite an extensive repertoire of staining techniques, relatively non-specific dyes such as eosin and haematoxylin being used to enhance the contrast between features of interest and the background. A major advance was the introduction of fluorescence microscopy for histochemistry and cytochemistry following the pioneering work of Coons *et al.* (1941) at the Harvard Medical School (see also Coons and Kaplan, 1950; Coons, 1956), and this remains one of the most widely-employed approaches (Valenzuela and Deodhar, 1986). Other techniques of greater resolution such as electron microscopy, in which single protein molecules can be visualised, have also found application here (Burkholder, 1986).

Unless the component of interest is identifiable by its morphology alone, or by the fairly non-specific stains mentioned above, a signal generation reaction of some sort must be employed to indicate its presence. As it is desired to locate the physical position of the analyte in the structure, the product of the reaction is normally required to be insoluble or at least to undergo minimal diffusion.

Reactions can be used which generate a specific signal from the target molecule, in a fashion analogous to the approaches of clinical chemistry. Enzymes in particular can often be detected readily with a suitable substrate (Lojda *et al.*, 1979). Again, analogously to clinical chemistry, these approaches do not generally require the preparation of coupled protein reagents.

However there are a great many components of interest which do not give rise to a suitable reaction generating an insoluble product. Here it is attractive to employ an approach analogous to immunoassay, using the formation of an antibody–antigen complex to confer the desired specificity. Once again, this opens up the way for highly sensitive non-specific signal generation techniques to facilitate the localisation of any component for which antibodies are available.

1.3.1 Labelling for light microscopy

The simplest approach which suggests itself for visualising antibodies or other proteins used as histochemical markers is to label them with a suitable dye. Such

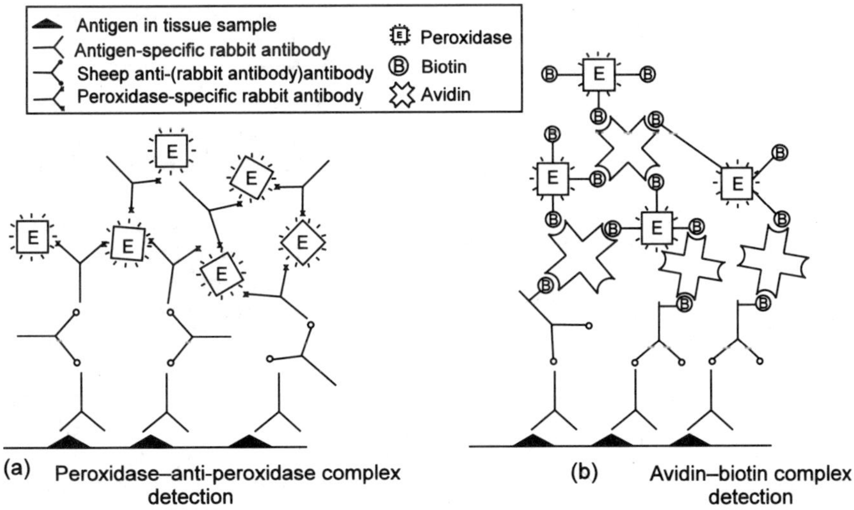

Figure 1.13 Detection systems for histo- and cytochemistry based on complex formation: (a) the peroxidase–anti-peroxidase (PAP) system; (b) the avidin–biotin complex (ABC) system.

reactions are straightforward to carry out if care is taken not to overlabel the protein, which can cause loss of biological activity or solubility problems. However, the results are often disappointing in practice. Only small reductions in light transmission are seen compared to unlabelled sample, and this approach is of limited applicability.

Highly opaque labels such as gold sols (with or without silver enhancement) are necessary for reasonable results, and these have been employed in immunogold reagents for histochemistry (Holgate *et al.*, 1983; Beesley, 1992a). These reagents are discussed in more detail in Section 1.3.3.3.

To obtain greater sensitivity, it is necessary to amplify the signal obtained from each binding event, and enzymes are employed to this end in exactly the same way as they are in immunoassay, although in this instance the reaction product should be visible and insoluble. Horseradish peroxidase is commonly used, e.g. with 3, 3′-diaminobenzidine as the substrate. This system gives an insoluble brown polymeric product which can be rendered more visible by reaction with osmium tetroxide. There are many such peroxidase substrates which can be used in this application (Avrameas, 1970).

Reagents have been developed which allow the formation of multi-layered complexes containing signal-generating species, and these have proved popular for light-based histo- and cytochemistry. Notable among these (Figure 1.13) are the peroxidase–anti-peroxidase (PAP) system (Sternberger 1967, 1986) and the avidin–biotin complex (ABC) system (Bayer *et al.*, 1976; Azar and Espinoza, 1986).

1.3.2 Labelling for fluorescence microscopy

Returning to the concept of a simple dye-labelled antibody, there is another way to increase the sensitivity of detection by reducing the light background against which the signal is viewed. The use of fluorescent labels, where light is emitted at a different wavelength from that of the illumination source, allows optical techniques to be used to reduce backgrounds to near-zero levels while giving a strong positive signal. This improves the contrast so much that a great deal of information can be gained using antibodies labelled with fluorescent molecules, at an incorporation level modest enough not to have a serious effect on binding activity. These immunofluorescence methods pioneered by Coons were responsible for a dramatic increase in the usefulness of immunocytochemistry in the 1940s and 1950s, and have remained in widespread use ever since. Fluorimetry is inherently an extremely sensitive technique, capable under suitable conditions of single molecule detection (Lu and Xie, 1997; Pirotta *et al.*, 1997).

A multitude of different fluorophores are available, but those of the fluorescein, rhodamine and cyanine groups (Figure 1.14) are among the most commonly employed. Some of the desirable properties of

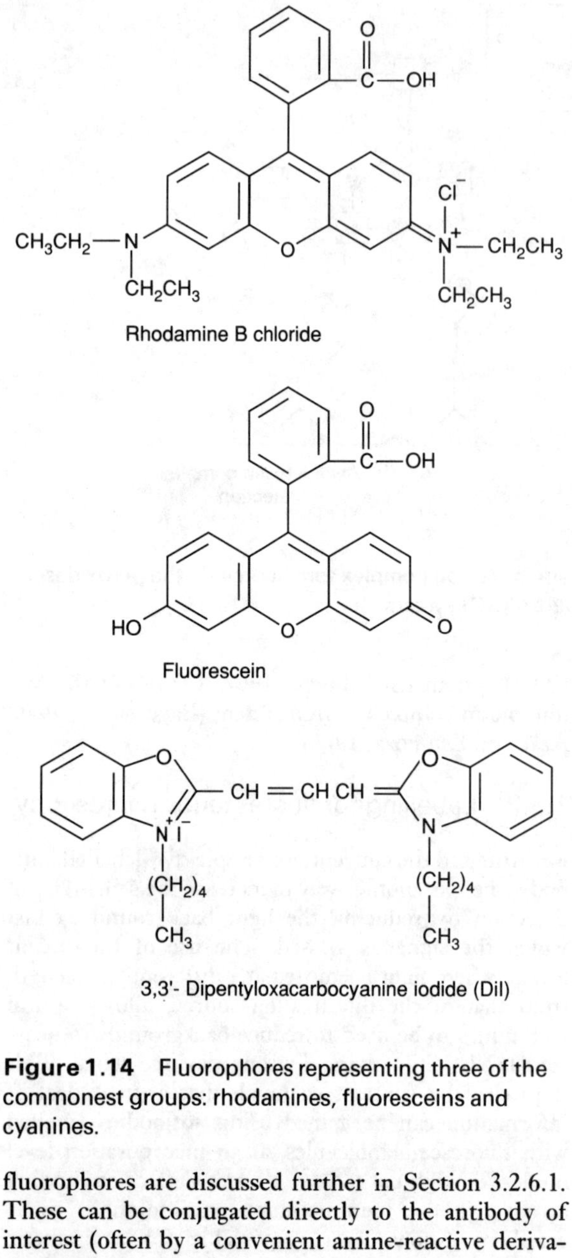

Figure 1.14 Fluorophores representing three of the commonest groups: rhodamines, fluoresceins and cyanines.

fluorophores are discussed further in Section 3.2.6.1. These can be conjugated directly to the antibody of interest (often by a convenient amine-reactive derivative such as an isothiocyanate), but as is many other applications a popular alternative is to use a second antibody approach to avoid the need for direct labelling.

Antibodies which have themselves been raised against the antibodies of another species are available commercially, with a wide variety of fluorescent labels; a reagent of this type such as a fluorescein-labelled 'donkey anti-(sheep IgG)' can be used to visualise any sheep antibody localised at the point of interest.

The spontaneously fluorescent green fluorescent protein (GFP) from the jellyfish *Aequorea victoria* (Morin and Hastings, 1971) has attracted considerable attention recently, and has been employed as a label in cell screening (Mosser *et al.*, 1997). Its primary and crystal structure are now available (Prasher *et al.*, 1992; Yang *et al.*, 1996a).

As noted earlier optical advances have made a big impact in the area of fluorescence microscopy, in particular the advent of lasers and of image analysis. Laser generation of excitation energy at high intensity and highly defined wavelengths is ideally suited to fluorimetry, offering strong signals and low backgrounds; the challenge of matching up wavelengths obtainable from lasers with suitable fluorophore excitation energies is currently an area receiving considerable attention. The digitalisation of images and subsequent processing and enhancement techniques have revolutionised this area, allowing, for example, multiple signals from different fluorophores to be built up into a single image. Commercial suppliers are constantly looking for improvements in both fluorescent reagents and optical instrumentation (Glazer and Wells, 1997).

Cells can be stained using fluorescent-labelled antibodies which bind to antigens in the surface, and flow microfluorimetry techniques can be used to count the number of labelled cells in a population. These rely on entraining the cells in a stream of liquid and allowing them (one at a time) to pass two laser systems, one of which detects the presence of the cell and its size, and the other the intensity of fluorescence being emitted. The technique can be extended very elegantly by application of an electrostatic charge to the labelled cells, allowing them to be sorted from other components (Figure 1.15). Fluorescence-activated cell sorter (FACS) systems of this type have been widely adopted in research applications, for example in isolating B-lymphocytes generating antibodies to a desired antigen from an immunisation programme.

1.3.3 Labelling for electron microscopy

The theory of the electron microscope was conceived at the beginning of the twentieth century, but was not realised until 1939. By the 1950s it was being employed for histological applications such as the elucidation of renal ultrastructure (Pease and Baker, 1950; Farquhar *et al.*, 1957). Resolution of 2.5 nm was obtainable at this time, but improved instrumentation is now capable of performance at the Ångstrom (10^{-10} m) level.

Under favourable circumstances it is possible to detect the binding of an unlabelled antibody (the

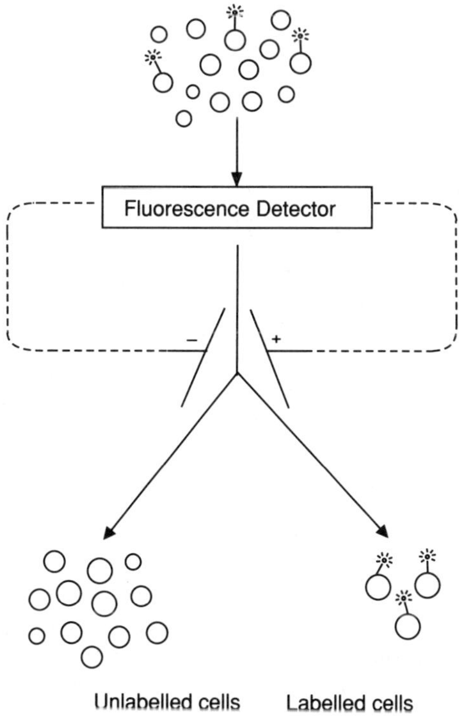

Figure 1.15 Fluorescence-activated cell sorting (FACS): detection of a fluorescent cell triggers the application of a charge which transiently redirects the eluent stream, thus sorting labelled from unlabelled cells. Multiple passages may be required to achieve an effective separation.

dimensions of which are approximately 100 Å) to its antigen directly by electron microscopy (EM); generally the binding is revealed as a reproducible thickening of the structure under observation. More normally however, the structural definition or regularity of antigen-containing features is insufficient to allow this, and to make the binding event visible by electron microscopy requires an electron dense label.

1.3.3.1 Heavy metals for EM labelling

Heavy metals are electron rich, and are attached to antibodies either *via* chelating groups, e.g. the introduction of lead atoms *via* anthranilic acid residues described by Koike *et al.* (1964). Alternatively, attachments are *via* stable organometallic molecules, e.g. such as the reaction of methylmercuric hydroxide with a thiolated antibody reported by Kendall (1965).

In the 1960s Ludwig Sternberger and his co-workers at the Johns Hopkins University described immunouranium techniques (Sternberger *et al.*, 1965; Sternberger, 1967). This metal, in the form of uranyl acetate, chelates readily to antibodies, incorporations of up to 200 atoms per moleule of IgG being achievable. The resulting electron opacity can be increased further by coupling the uranium *via* a bidentate compound such as thiocarbohydrazide to osmium tetroxide, which generates highly opaque osmium black on reduction.

1.3.3.2 Ferritin for EM labelling

Ferritin is a large (650 kDa) protein, a quarter of the mass of which is made up of an iron 'core', and Singer (1959) demonstrated its use as a convenient label for electron microscopy. Introduction of a single ferritin molecule can give substantial metal incorporation without significantly damaging the antibody's ability to bind antigen, and immunoferritin methods have found widespread application over the years (Tanaka, 1968).

Antibodies of the immunoglobulin G (IgG) class have a molecular mass of around 150 kDa, and the use of such a large label as ferritin (apart from affecting the ability of the conjugate to diffuse successfully into tissue samples) makes the consideration of steric hindrance important. For optimum performance, care must be taken to ensure that the binding sites of the antibody are not blocked by the bulkier protein, e.g. by careful control of the reaction stoichiometry.

This illustrates a general issue in protein conjugate preparation, which is particularly important where one component of the conjugate is considerably larger than the other; it applies equally, for example, to the labelling of a small steroid molecule with a substantially larger enzyme. Optimisation of stoichiometry and the choice of the right linker group between the two components can be critically important in obtaining a fully functional reagent.

1.3.3.3 Immunogold techniques for EM labelling

If chloraurie acid is reduced in low ionic strength aqueous solutions gold sols are formed, colloidal gold particles with nanometre dimensions. The particles are negatively charged, and repel each other sufficiently to give a stable suspension. They also have a propensity to adsorb macromolecules, and the resulting complexes are more stable than the sols themselves, even when the ionic strength of the solution is raised. Faulk and Taylor (1971) used a *Salmonella* antibody to stabilise a gold sol, creating a useful reagent for

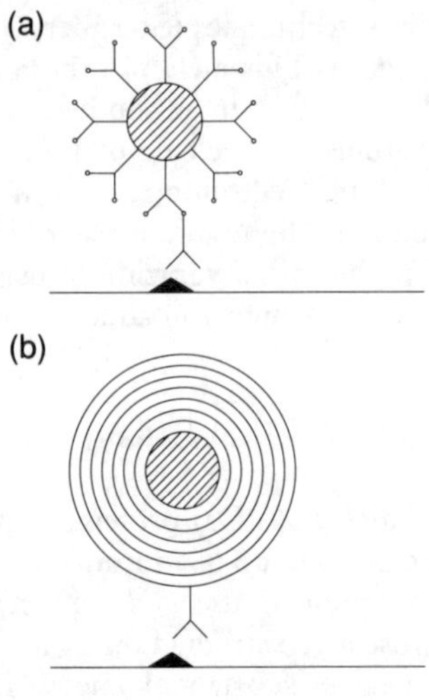

Figure 1.16 Use of an immunogold reagent for histochemical labelling: (a) the antibody-labelled gold sol binds to specific antigen, in this case *via* a second antibody; (b) a denser product can be obtained by silver enhancement, where multiple layers of silver are precipitated on to the immobilised sol.

immunocytochemistry. Labelled gold sols are capable of giving even higher degrees of contrast under electron microscopy than immunoferritin reagents. Adsorption of layers of silver to the gold sol can increase sensitivity further (Figure 1.16).

Immunogold techniques have found widespread application in electron microscopy (Beesley, 1992b), the care required to generate sols of reproducible behaviour and the limited stability of the reagents being their main disadvantages. Many users choose to get round the first by employing a second antibody format with commercially-supplied anti-species sols.

A similar approach which helps on both fronts is to use gold sols attached to Protein A, which binds with good specificity to many types of antibody molecule. Sols stabilised with this protein tend to be more robust than those bound to antibodies, and Protein A sols are commercially available. Protein G is a more recent introduction with a wider breadth of subtype recognition.

1.3.3.4 Enzyme techniques for EM labelling

The use of enzyme amplification has already been discussed for light microscopy. Engineering an electron-dense final product makes this technique applicable for electron microscopy, the best known approach being based on the peroxidase/diaminobenzidine system described above, and its analogues; subsequent reaction of the insoluble oxidation product with osmium tetroxide once again yields osmium black (Casson and Mesulam, 1982), which is electron-dense as well as opaque.

1.3.4 Recent developments in histochemistry and cytochemistry

There are numerous techniques now in existence at the boundaries of microscopy and spectroscopy which offer the potential for resolution of structures at the molecular or atomic scale. These include electron spectroscopic imaging (ESI), where X-ray photons are generated from elements in the sample by electron bombardment. The photons are captured; their energies are characteristic of the elements from which they were generated, and can be used to identify the location of particular elements in a sample (Shuman and Somlyo, 1976; Ottensmeyer and Andrew, 1980). ESI becomes less sensitive as the atomic number of the element decreases. Electron energy loss spectroscopy (EELS) is similar, except that it is the loss in the energy of the bombarding electrons that is detected (Isaacson and Johnson, 1975; Ottensmeyer and Andrew, 1980). This technique becomes more sensitive at lower atomic number and can therefore be complementary to ESI. These techniques recognise the presence of specific elements, and therefore offer the opportunity to use reagents labelled with 'unnatural' elements to detect features in a sample. Alam *et al.* (1989) described the use of immunoboron reagents in immunocytochemistry, using ESI and EELS for detection.

Finally, it is worth noting that this area has been affected by developments in molecular biology. The hybridisation techniques described in the section on nucleic acid probes are also readily applicable for histochemical or cytochemical samples (*in situ* hybridisation, ISH) so that localised information about genetic material can be obtained (Gall and Pardue, 1969; Dallman *et al.*, 1991). As might be expected it is important to keep the size of the probe down to promote accessibility to tissue components; a limit of around 500 bases has been suggested (Unger, 1992). High-sensitivity nucleic acid detection *via* PCR amplification has also been exploited for *in situ* hybridisation, for example in the detection of viruses and oncogenes (Nuovo, 1992).

1.4 *In Vivo* Applications

The applications discussed so far are concerned predominantly with clinical samples, which have been removed from the body for examination. This allows a certain flexibility in the techniques which can be used: there is generally no requirement to maintain the biological functionality of the sample, pre- treatments can be employed to remove components which may interfere with the procedure to be used, and the toxicity of the reagents used is relevant only in terms of the safety of the personnel carrying out the work in the laboratory.

There are a number of situations, however, where diagnosis must take place inside the living body—where site-specific information is required which cannot be obtained by examining body fluids, and where tissue sampling is impossible or dangerous in itself. Another important use of coupled protein reagents is in therapy, where the aim is treatment rather than diagnosis; apart from rare examples in extracorporeal therapy, these pharmaceutical reagents must of course be used *in vivo* too. For any *in vivo* reagent, the challenge is to deliver a suitable dose to the desired site while minimising harmful effects, and the barriers which must be overcome to achieve this goal have been reviewed in depth by Poznansky and Juliano (1984).

Although the desired results of imaging and therapy are different, the coupled protein reagents used are frequently similar. So too are many of the constraints which affect their design, which differ from those encountered in the *in vitro* field.

First, it is required that the reagents remain in circulation, with their functional activities substantially unaffected, long enough that they can fulfil their intended role. As soon as a foreign molecule has entered the bloodstream, it will begin to receive attention from the body's defence and excretion mechanisms. If the reagent to be introduced is of low molecular mass, the problems of *in vivo* stability can be very pronounced. Any species (including small proteins) with a molecular mass below approximately 40 kDa may be cleared *via* the kidneys within a few hours. Interestingly, small organic compounds may sometimes exhibit an unexpectedly high degree of stability, because they can be picked up by circulating serum proteins such as albumin, and therefore no longer present a small target for excretion. While larger molecules may be less prone to fast excretion, they are more likely to be immunogenic and thus attract the attention of the body's immune system; the mechanism of this process is discussed briefly in Chapter 4. Reducing the immunogenicity of *in vivo* reagents is an important factor in achieving the required stability.

Secondly, the toxicity of the reagents used is of paramount importance. It is important to ensure that any potentially harmful reagents remain in the body only for as long as is necessary for the imaging or therapeutic operation, i.e. that the rate of clearance is not too low. Achieving the right balance between these first two constraints is a fundamentally important design consideration.

Thirdly, it is not always possible to introduce these materials at the point of interest, and the challenge of enabling them to be transported to the right site, to cross delicate selective interfaces, and to localise successfully at the desired target is a large one.

In therapeutic applications it may also be the case that some dissociation of the reagent is necessary for it to be effective, and often this dissociation is required to take place only at the target site; again this is a major design challenge. Sometimes, for example, only the cleaved active component of an immunotoxin (see Section 1.4.2.4) is capable of entering a target cell, so a conjugate containing a labile disulfide bond is typically employed. In order that the bond is not cleaved too early, reagents have been developed which provide a hindered disulfide, which is more resistant to attack (Thorpe *et al.*, 1987; Blakey *et al.*, 1988; Arpicco *et al.*, 1997). Sometimes it is desired that cleavage of the conjugate takes place on endocytosis: the endosome is characterised by its low pH, and conjugate designs are therefore sometimes based on acid-labile linkers (Yang and Reisfeld, 1988; Raso, 1994).

1.4.1 Imaging

1.4.1.1 Magnetic resonance imaging (MRI) and ultrasound

There have been huge diagnostic advances in recent decades due to techniques such as ultrasound scanning, and imaging based on the nuclear magnetic resonance (NMR) signal generated by water molecules, often referred to as magnetic resonance imaging (MRI). These techniques are sometimes used in a completely non-invasive fashion, i.e. data is generated solely from signals generated by naturally-occurring components of the body, without the introduction of externally-prepared reagents.

Often, however, contrast agents are introduced to improve definition in the area of interest: for MRI, for example, contrast agents are commonly based on species which alter the nuclear relaxation times of water protons (Lauffer, 1987; Mendonca Dias and Lauterbur, 1987). Sometimes these contrast agents are coupled protein reagents. Gadolinium complexes are among the best-known MRI contrast agents, and these have been attached to antibodies with the aim of *in vivo* targetting (Shreve and Aisen, 1986; Brasch, 1992).

Alternatively, highly magnetic agents such as magnetite particles can be employed to provide contrast, and to achieve appropriate *in vivo* behaviour these have often been coated with proteins such as albumin (Renshaw *et al.*, 1986; Widder *et al.*, 1987). Again, particles of this type have been coupled to antibodies to achieve targetting (Tiefenauer *et al.*, 1993).

Contrast media for ultrasound are at a relatively early stage of development; these echocontrast agents generally rely on the formation of bubbles, or the use of 'bubble-like' structures such as air-filled albumin microspheres (Clinica, 1997).

Techniques such as MRI or ultrasound, however, are not specific or detailed enough for some of the most important diagnoses connected with life-threatening conditions; notably pinpointing the position of blood clots in the cardiovascular system, areas of damaged cardiac tissue after heart attack, irregularities in vital organs such as the brain, and tumours. Radioimaging is another powerful approach which lends itself more readily to use in these fields.

1.4.1.2 Radioimaging

Because an *in vivo* diagnostic has generally to be measured from outside the body, possible technologies are limited to those where the signal generated has the capability of passing through tissue. One of the earliest recorded properties of high-energy radiation was its ability to pass through flesh; once this phenomenon had been combined with the effect of the radiation on photographic media, the science of X-ray diagnosis was born, and it has remained an important branch of diagnostic medicine ever since. Analysis of the structure and condition of bone is relatively straightforward due to the greater opacity of the inorganic building materials compared to the surrounding tissue. Examination of some other parts of the body is made possible by the use of simple imaging agents such as the 'barium meal' which can indicate the shape of sections of the digestive tract by filling them with an opaque inorganic material. Iodinated contrast agents increase the range of X-ray diagnosis still further, allowing imaging of blood distribution; the heavy iodine atoms provide the necessary opacity. Reagents of this type have been used for half a century (Clinica, 1996).

An alternative to making a target part of the body opaque to externally generated X-rays, which is generally out of the question for sensitive organs such as the heart and brain, is to generate the signal internally and measure it from outside. High-energy emissions such as γ-radiation are the obvious choice of signal for this application; low energy radioactivity such as α-emission is rapidly absorbed by body tissue.

For imaging purposes (radioscintigraphy), it is the aim to minimise the exposure of the subject to potentially damaging radiation, and radioisotopes or nuclides are chosen which have half-lives in the order of hours, and are rapidly and completely excreted from the body. The most popular nuclides for this purpose are all γ emitters, two isotopes of iodine (^{123}I and ^{131}I) and two radiometals (indium-111 and technetium-99 m). They each have their advantages and drawbacks: iodine can be introduced directly into proteins without any appreciable change in size or behaviour, but *in vivo* deiodination processes can lead to unwanted uptake of radioactivity in the thyroid or stomach; radiometals can be incorporated by attractively simple chemistry via a suitable linking group, but are prone to ending up in the liver or kidneys. Technetium-99m has a particular advantage in that it can be conveniently generated in the laboratory as a pertechnetate solution; it is now probably the most widely-used isotope in nuclear medicine. In general radiometals are gaining in popularity, and many other metal nuclides are currently undergoing investigation.

For a radioimaging agent to provide useful information it must be localised to some extent. Sometimes this can be achieved simply by physical means, but it is often desirable to target the reagent so that it binds specifically at the site of interest. The development of radioimmunoconjugates for this purpose dates from the early 1950s (Pressman and Eisen, 1950; Pressman and Korngold, 1953). The main requirement is to combine a radionuclide with the qualities described above with an antibody, whose specificity provides the targetting capability.

An example of such a reagent is Centocor's Myoscint™, an indium-111-labelled anti-myosin monoclonal antibody which images dead heart muscle to allow assessment of damage after myocardial infarction.

Tumour imaging is another area where radioimaging has proved useful, as it provides a minimally invasive alternative to surgery for detecting the site of cancerous tissue. Radiolabelled mAbs with specificity for tumour antigens provide the reagents for this work; progress in this area has not totally matched early optimism, but this has now become an accepted branch of cancer diagnosis (von Kleist, 1993). The reasons noted by von Kleist which have prevented wider use include lack of reliable tumour markers of good specificity, high non-specific uptake of radioimmunoconjugates *in vivo* and poor conjugate stability.

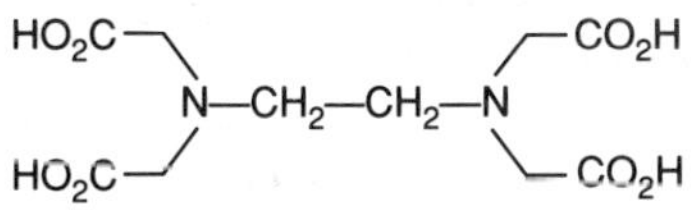

EDTA (ethylenediaminetetraacetic acid)

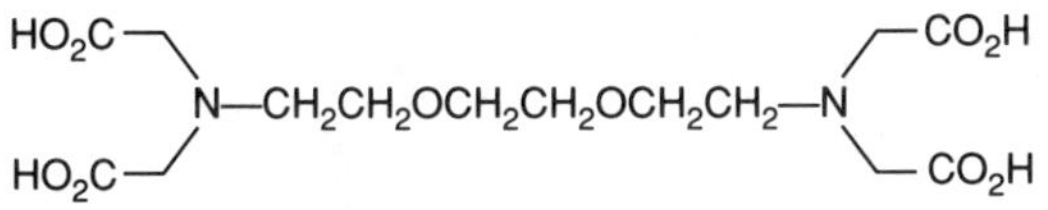

EGTA (ethylene glycol bis(β-aminoethyl ether) - *N,N,N',N'* - tetraacetic acid)

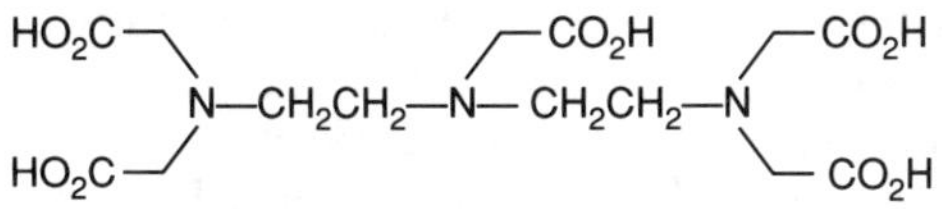

DTPA (diethylenetriaminepentaacetic acid)

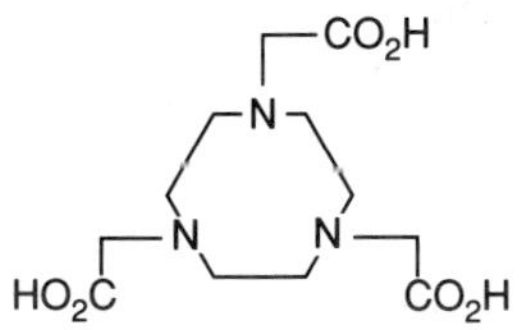

NOTA (1,4,7-triazacyclononane-*N,N',N''*-triacetic acid)

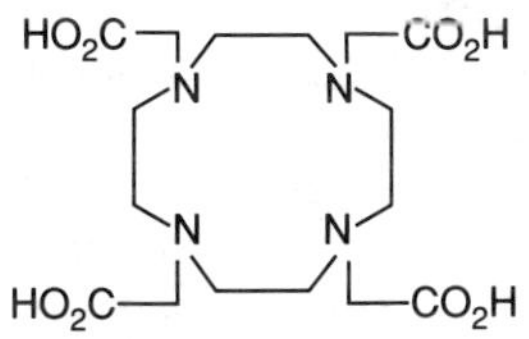

DOTA (1,4,7,10-tetraazacyclododecane-*N,N',N'',N'''*-tetraacetic acid)

Figure 1.17 Common organic chelating agents: these provide a convenient means of incorporating radioactive metal ions into radiopharmaceuticals.

A number of chemistries have been developed for the incorporation of iodine into proteins. They generally rely on oxidation of iodide (the non-volatile form in which radioactive iodine is usually supplied), and subsequent reaction of the electrophilic species formed with tyrosine or histidine groups in the protein, an extremely rapid and efficient process (for further details, see, Chapter 6, Section 6.21).

Metals on the other hand will not normally bind directly to proteins, but the class of organic compounds known as chelating agents are often employed as a convenient bridge. These are generally compounds which normally contain a number of negatively-charged groups (often carboxylic acids), or a cyclic structure containing electron-rich atoms such as nitrogen (Parker, 1990); some examples are shown in Figure 1.17. These are quite simple to attach to proteins by a number of well-established techniques, then if the dimensions of the chelator are correct, the desired metal ion can be trapped efficiently and simply through coordination with the polar groups. An advantage to this approach, pioneered by Meares and Wensel (1984), is that the bulk of the chemistry is carried out on nonradioactive materials. The radioisotope can be introduced immediately before use, simply by mixing the chelator with a solution of the radioactive ion. This not only simplifies handling during the preparation, but minimises the amount of radiolytic damage to which the antibody is subjected.

1.4.2 Pharmaceuticals

It is a short step from the *in vivo* use of imaging agents to that of pharmaceutical materials, and from diagnostics into the realms of therapy. There are a number of reasons why coupled protein reagents can provide useful therapeutic agents, and these will be discussed.

1.4.2.1 Modification of behaviour of protein pharmaceuticals

Many proteins are used as pharmaceuticals in their own right, and coupling to another species can often provide a valuable means of altering their therapeutic behaviour (Smith *et al.*, 1993; Gombotz and Pettit, 1995). The 'Ringsdorf model' is a useful generic concept for drug delivery systems, combining the active principle with a polymeric carrier chain, a spacer group to avoid steric hindrance of the drug, a targetting moiety for site-specific delivery and a hydrophilic group to confer water-solubility (Ringsdorf, 1975).

One of the earliest therapeutic applications of derivatised proteins was in the treatment of allergy: this dates back to the early part of the century (Cooke *et al.*, 1935), and is based on the administration of allergen extracts to provoke tolerance. It is thought that modified allergens are unable to bind to the immunoglobulin E (IgE) molecules which are responsible for the misplaced inflammatory response, the effects of which cause the unwelcome symptoms of allergy. At the same time they can still provoke an IgG response, thus protecting the individual from further exposure (Smith *et al.*, 1993). Many derivatised allergens have been examined in this capacity, such as pollen extract coupled to polyethylene glycols (Schäfer *et al.*, 1992), and this remains an active area of research.

In another example, tissue damage in some forms of arthritis is thought to be caused by oxygen radicals produced during the body's inflammatory response, and enzymes such as catalase, superoxide dismutase and peroxidase have been investigated as drugs to counteract this effect. The native enzymes do not bind well to the vulnerable articular structures, but increasing their positive charge by chemical derivatisation improves their retention, and their anti-arthritic capability (Schalkwijk *et al.*, 1985).

In other applications the protein may be coupled to another species to enhance its *in vivo* survival. Attachment of a polymer such as polyethylene glycol (PEG), for example, has repeatedly been shown to reduce immunogenicity; slowing down the immune response of the patient in this way is a means of increasing drug lifetimes *in vivo* (Lisi *et al.*, 1982; Harris, 1992). PEG-modified adenosine deaminase, for example, has been commercialised as Adagen™ by Enzon of New Jersey for the treatment of severe combined immunodeficiency (SCID).

As noted earlier, proteins smaller than 40 kDa are particularly prone to removal from circulation, and coupling to large polymers such as high-molecular-mass PEGs may minimise this effect. However, polymers too small to prevent excretion through their size have also been successfully employed to improve stability, by enhancing binding to serum proteins; the resulting complex is large enough to avoid excretion. The anti-cancer protein neocarzinostatin (molecular mass 12 kDa), for example, endures much longer in the circulation when conjugated to a small styrene-maleic acid/anhydride (SMA) polymer. Despite only having a molecular mass of 1.6 kDa this polymer confers the ability to be bound by albumin in the bloodstream, and the resulting drug complex (SMANCS) pioneered by Maeda *et al.* (1984) has shown sufficiently attractive properties to receive clinical attention.

Encapsulation within 'artificial cells' may provide another means of avoiding the attention of the body's clearance mechanisms such that stability is improved. Thus Siu Chong and Chang (1974) suppressed asparagine-dependent cancers by injection of microencapsulated asparaginase, while topical application of catalase in this form has been used to treat oral lesions in the congenital catalase-deficiency condition acatalasemia (Chang, 1972). There have been a number of other therapeutic applications of this approach over the years (Chang, 1995), including the preparation of encapsulated haemoglobin as a potential red blood cell substitute. The overall impact of this methodology has so far been rather limited, however.

A fascinating potential application of protein derivatisation for therapeutic purposes has been described recently by Bayley (1997). In cell membranes the protein α-haemolysin from *Staphylococcus aureus* assembles to form heptamers. These arrangements form a pore in the membrane, disrupting its normal behaviour. Derivatisation of the haemolysin molecule with a nitrobenzyl ester inhibits this pore-forming ability, but it can be reinstated by photolysis of the ester group. This reagent therefore offers the possibility of a light-triggered drug delivery system based on the controlled generation of artificial pores within target cells.

Finally a protein pharmaceutical may be targetted at the site of interest, typically by attachment to an antibody. These chemoimmunoconjugates are discussed in Section 1.4.2.4.

1.4.2.2 Modification of behaviour of non-protein pharmaceuticals

Attempts to modulate the behaviour of low-molecular-mass drugs by attaching them to large molecules date back more than half a century, e.g. the antiseptic Isodine (a polyvinyl-pyrrolidone iodine complex) is an early product of this approach which remains in use today. Again there are a variety of reasons for such modifications.

It has already been noted that small molecules are prone to excretion, so it is not surprising that attempts have been made to counter this effect by incorporating a large, stabilising carrier molecule. Also, many blood vessel surfaces are negatively charged, so to avoid electrostatic losses attention has tended to focus on neutral or slightly negatively charged carriers. However, non-proteinaceous species such as dextrans or synthetic polymers have tended to dominate approaches in this field due partially to their low immunogenicity; PEG and copolymers of HPMA (*N*-(2-hydroxypropyl)methacrylamide), for example, have been widely employed due to their biocompatibility and chemical versatility (Harris, 1992).

More usually, drugs are coupled to proteins to improve their localisation at the site of interest. This may be due to a simple physical alteration in the presentation of the drug; in the treatment of tuberculosis, for example, Hincal and Kas (1988) promoted absorption of streptomycin in the lungs by conjugating it to gelatin or human serum albumin microspheres with a diameter of around 30 μm. Mechanical filtration in the lung capillaries was postulated to explain this localisation.

This localisation of drug delivery *via* microparticulates has continued to attract interest, the attachment of antibodies offering further targetting potential (Davis, 1997).

As usual, however, it is the specificity of protein function which has led to their most important applications in therapeutic complexes, in particular using the specificity of antibodies to promote binding at a chosen site. If a drug or other agent can be localised at the point of treatment, unwanted effects on other healthy areas can be minimised and the overall dose reduced. This is particularly important where the agent is damaging in itself; the chemotherapy and radiotherapy approaches used for cancer treatment are cases in point. A substantial amount of the effort aimed at realising Paul Ehrlich's dream of the magic bullet (Ehrlich, 1906) has been in this area, in particular those forms which are least responsive to conventional therapy such as gastric, colorectal and pancreatic cancer.

Similarly in the cardiovascular arena, the use of thrombolytic agents such as t-PA (tissue plasminogen activator) in the treatment of blood clots can cause side-effects such as haemorrhaging, and this has led to research aimed at targetting their activity.

Once again monoclonal antibodies (mAbs) have been the favourite vehicle for achieving the desired localisation. In some instances the antibody itself is the therapeutic agent, for example treatment with an anti-lymphocyte mAb can clear the system of these cells, cancerous or otherwise, and anti-platelet antibodies can prevent blood clots by inhibiting aggregation.

More commonly however, the antibody's targetting ability is used in conjunction with an agent which has a more potent therapeutic effect, such as a radioisotope, a drug or a toxin.

1.4.2.3 Radioimmunoconjugates for therapy

Where radionuclides are incorporated the reagents are basically analogous to those already described for imaging, but the criteria for choosing the isotope are somewhat different. In this case it is desired that the radiation is absorbed in nearby tissue, so lower-energy Auger, α- and β-emitters such as gallium-67, yttrium-90, and astatine-211 have received attention (Schubiger *et al.*, 1996). Half-lives of the order of days are also preferred so that prolonged treatment is possible. Some nuclides such as iodine-131 or copper-67 have multiple emission properties which may enable them to be used conveniently for simultaneous imaging and therapy.

An interesting binary approach introduced by Bale and Spar (1967) is that of boron neutron capture therapy (BNCT), in which boron-10 is attached to an antibody; this isotope is not radioactive in itself, but is an α-emitter when bombarded with low energy neutrons. While these 'thermal' neutrons are not harmless, the dose can be kept reasonably low if a large number of boron atoms are incorporated in the immunoconjugate. Alpha radiation is generated only at the desired site where neutrons are introduced, minimising the damage elsewhere. Because of the large number of ^{10}B atoms required for a lethal dose, conjugates of mAbs with large ^{10}B-derivatised polymers have been described for this application (Barth *et al.*, 1994).

1.4.2.4 Chemoimmunoconjugates and related reagents for therapy

An alternative to radiation is to use a cytotoxic chemical attached to an antibody; compounds such as adriamycin or mitomycin C, for example, are beneficial in cancer treatment but have a number of unwelcome side effects which limit the maximum practical dose. If these compounds are conjugated to a suitable antibody to localise them at the site of a tumour, their action can again be concentrated at the target site (Blakey *et al.*, 1988; Koppel, 1990).

Some toxins from plants or bacteria are among the most cytotoxic compounds known, only a few molecules per cell being required for a lethal effect in some cases. Ricin and trichothecenes both interfere with protein synthesis at the ribosomal level and have been employed in therapeutic applications; so has diphtheria toxin, but the widespread immunity of the population to this compound (due to vaccination programmes) limits its usefulness. Often these toxins consist of two peptide chains, one of which confers toxicity, the other binding properties. Replacement of the latter with an antibody to the target yields an immunotoxin, and these have shown promising performance in cancer therapy.

Care must be taken to preserve the activity of the cytotoxic agent throughout the coupling process; unlike a radionuclide, the efficacy of peptide chains or organic compounds can be very sensitive to derivatisation. This fact is employed in an elegant variation of chemotherapy, where a harmless pro-drug is activated by an enzyme which is localised at the target (Figure 1.18). For example in the pioneering work of Senter *et al.* (1988, 1993), the anti-tumour agent etoposide was released from its pro-drug phosphate form by alkaline phosphatase coupled to a localising antibody. This combination of enzyme and antibody functions has been termed antibody-directed enzyme prodrug therapy, ADEPT (Bagshawe, 1994). Other enzymes have been employed similarly, for example β-lactamase (Vrudhula *et al.*, 1993). Analogously, acid-labile linker groups can be incorporated into pro-drugs, which are only rendered active by the cleavage reaction invoked by the acidic conditions (around pH 5) encountered in the intracellular environment (Raso, 1994).

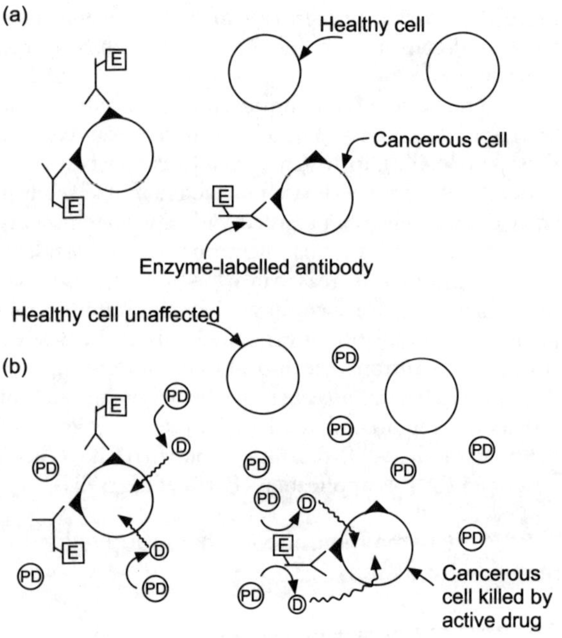

Figure 1.18 The use of targetted enzymes for pro-drug therapy: (a) cells bearing cancer antigens (shown as triangles) are selectively bound by the enzyme-labelled specific antibody; (b) inactive pro-drug (PD) is converted to its active form (D) only in the presence of enzyme, and therefore destroys cancerous cells with reasonable selectivity.

In a similar approach (Dougherty, 1987), compounds are employed which are inactive until illuminated from a light source (photodynamic therapy); once the compound has been localised at the target, a fibre optic source can be employed to activate its cytotoxic nature. Such photosensitizers have generally been employed without attachment to a targetting group such as an antibody (Levy, 1995), although Jiang *et al.* (1990) reported the use of benzoporphyrin–antibody conjugates in this application.

Two interesting approaches to improving the uptake of immunoconjugates by tumours were described by Epstein and Khawli (1991). In the first, vasoactive compounds such as interleukin-2 (IL2) are coupled to a targetting antibody; these can increase the vascular permeability of the tumour and facilitate access for therapeutic agents (see also Seymour, 1992). In the second (tumour necrosis therapy), the localising antibody is aimed at dead or degenerating cells which, although not involved in proliferation, give a much better targetting efficiency as they only tend to be found in high concentrations in malignant tumours.

1.4.2.5 Factors affecting antibody use in immunoconjugates

There are many considerations to be taken into account when designing reagents for *in vivo* use. Because antibody-based approaches occupy such a prominent place in this field, a great deal of attention has been paid to the particular requirements of reagents based on these molecules (Dyer, 1992).

As mentioned earlier, as soon as an immunoconjugate is introduced into the body, it is subject to natural processes that can lead to its removal from the circulation. This can occur either by binding to sites in organs such as the liver or lung which are receptors for parts of the antibody structure, by circulating species such as rheumatoid factors with similar binding capabilities or by neutralising antibodies arising from the patient's immune response to the foreign molecule.

Chemical and (increasingly) protein engineering approaches have been used to minimise these problems (Winter and Milstein, 1991), and are described in Section 4.6.2. Principally these are aimed at either (a) reducing the size of the antibody reagent to the minimum required for antigen recognition (Huston *et al.*, 1993) or (b) creating chimaeric antibodies in which rodent antigen-binding regions are fitted into a basically human immunoglobulin structure (see, for example, Stevenson *et al.*, 1989).

A final antibody engineering approach in this field is the use of bispecific mAbs, which have one binding site against one antigen, e.g. a target-associated component, and the second against another, e.g. a toxin. Dose of the toxin to the patient can be minimised by first administering the bifunctional antibody, allowing it to localise on the target while excess material is washed away, then introducing the toxin which will bind efficiently at the target site (Figure 1.19). Larger doses of cytotoxic activity have been localised at tumour sites by this approach than have been achieved using normal antibodies, particularly where small fragments are employed; because the circulating complex is small there is less competing uptake in the liver, etc. The bispecific mAb reagents can either be constructed chemically (Raso and Griffin, 1981; Glennie *et al.*, 1987) or using a hybrid hybridoma (Milstein and Cuello, 1983; Staerz and Bevan, 1986).

All the 'immuno' approaches described in this section rely on the use of antibodies with well-defined specificity. With some targets, such as tumours, it is often very difficult to identify antigenic structures which are unambiguously associated with them. Even some of the most commonly-employed antigens are frequently found in non-malignant tissues elsewhere in the body, while some tumours will conversely fail to display expected components. Even where the specificity of the antigen is good, it is not always easy to obtain pure material for use as an immunogen, and it may be difficult to raise good antibodies, although the ability to screen for good clones in monoclonal-raising programmes or *via* phage display is a big help.

As these conjugates are intended for *in vivo* use, the purity and reproducibility of the antibody used are particularly important, and subject to the same degree of scrutiny from regulatory organisations as other pharmaceuticals. This (and the need for extremely specific binding properties) has led to the domination of this area by monoclonal antibodies which lend themselves to reproducible manufacture. However the quality control of these materials and the requirements for comprehensive toxicology data are still very demanding. Interestingly, in some applications, reagents based on polyclonals or mixtures of mAbs

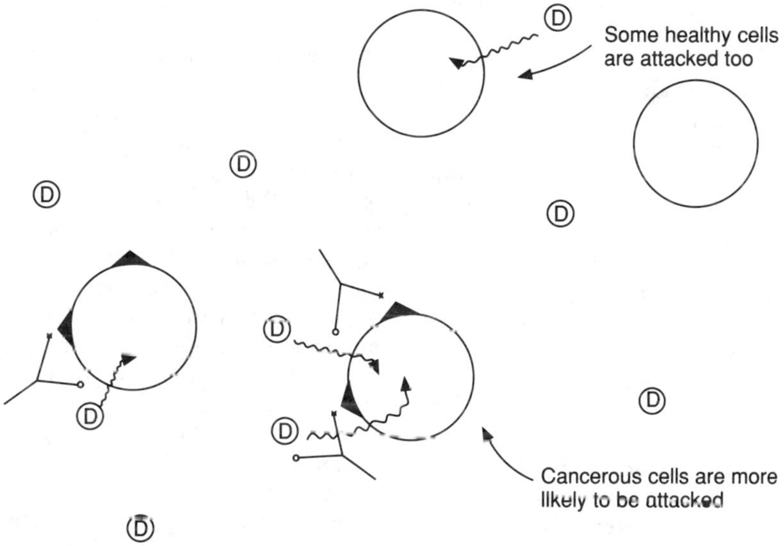

Figure 1.19 Use of a bifunctional mAb for targetted drug delivery: one arm of the antibody recognises a cancer antigen (shown as a triangle), the other is specific for a drug (D). Cross-linking of drug and target cell by the antibody provides some localisation.

have shown superior properties to single monoclonals (French *et al.*, 1991).

Finally, immunotoxins have their own problems; large cytotoxic agents such as ricin can be immunogenic themselves, and this may interfere with repeat treatments.

1.4.3 Gene therapy

One of the most exciting areas of medical research at present is gene therapy, the exploitation of molecular biology techniques to correct the genetic errors which cause disease (Friedmann, 1993). The genetic origin of a number of serious diseases has been established for many years, and some of the techniques described above have been used to pin down the genetic defects responsible for them. Every year there are new breakthroughs in identifying the genes whose malfunction is responsible for disease conditions, e.g. Duchenne muscular dystrophy, cystic fibrosis, Fragile X syndrome, Huntingdon's chorea. Dramatic though these advances have been, still greater impact may come from the identification of genetic conditions which predispose an individual to one of the big killers, such as coronary artery disease or a cancer.

Where the cells of an individual contain a faulty gene and the nature and location of the mutation are known, it is an attractive therapeutic goal to find a way to introduce the correct gene. Alternatively, nucleic acid reagents can be introduced to disable genes with an unwanted effect, for example antisense reagents which block segments of the target nucleic acids (Burger, 1993).

While techniques for introducing DNA into cells *via* suitable vector systems are now well established, they are not readily applicable to the treatment of human patients; it is a daunting task to introduce genetic material into sufficient cells in the body to have a significant effect.

Other approaches have involved targetting genetic material so that it is taken up specifically by the cells where its effect is most significant (Mulligan, 1993). Simple mechanical approaches have been employed to increase the specificity of delivery, for example by direct injection into the brain (Ram *et al.*, 1993). The lungs have been targetted by the use of adenoviruses which infect the epithelial cells of the lung lining; this route has been used for the treatment of cystic fibrosis (Rosenfeld *et al.*, 1992), which affects this organ most seriously.

Another way to introduce DNA to the desired site involves coupling the genetic material to a protein with the ability to bind to specific targets in the body, and these approaches have shown promise in animal models. Not surprisingly antibodies have been used for this purpose; Trubetskoy *et al.* (1992) targetted hepatocytes by this approach, and many more examples are bound to follow. However other protein–receptor interactions have so far received greater attention, notably those of the asialoglycoprotein receptors (ASGPr) which are highly localised in the liver (Wu *et al.*, 1991; Findeis *et al.*, 1993).

DNA is strongly negatively charged, and the ASGPr approaches have generally exploited this property in the formation of protein–DNA conjugates. As its name implies, the receptor binds to glycoproteins which do not contain the sialic or neuraminic acid group, and these can be generated from any convenient glycoprotein by the enzyme neuraminidase or by controlled acid hydrolysis. In most of the documented approaches the glycoprotein orosomucoid is treated in this way to generate asialoorosomucoid (ASOR). To confer DNA-binding ability this is linked to a positively-charged polymer, the synthetic polypeptide polylysine; the resulting conjugate forms an electrostatic complex with the DNA carrying the desired genetic information, to yield the complete reagent.

Gene therapy looks set to be a major growth area over the next few years, and is attracting major commercial interest (Evers, 1997; Langlands, 1997). An excellent summary of progress in this field was published recently (Scientific American, 1997).

1.4.4 Extracorporeal therapy

In some instances it is possible to remove fluids, cells or tissue temporarily from the body, treating them in some way to correct a failing before their return. Haemodialysis to mimic the role of the kidney in removing low-molecular-mass impurities from the blood is probably the most widely-known application of this approach. However, other blood treatments such as haemoperfusion and plasmapheresis, employing suitable techniques to remove specific components, are also important. Overall, millions of such extracorporeal perfusion operations are carried out annually. There are a couple of applications of immobilised proteins in this field, although these remain of minor significance.

The removal of urea is an important requirement of maintaining correct blood composition, and Pişkin and Chang (1980) demonstrated the use of immobilised urease in haemoperfusion columns to achieve this; the resulting ammonium ions were removed by ion exchange. Over 80% of blood urea could be converted by five minutes treatment on a perfusion column of this type. Ebert and Kim (1982) described the use of immobilised heparinase to remove the anticoagulant heparin from treated blood.

Removal of T-cells from bone marrow before transplant is another procedure which could be classed under extracorporeal therapy, and immobilised antibodies have been used for this purpose; mAb-coated magnetic particles (Hermentin *et al.*, 1990), for example, offer a convenient approach.

1.4.5 Biocompatible materials

The production of biocompatible surfaces, those which do not undergo harmful deterioration *in vivo*, is an area which has attracted enormous interest; applications range from surgical implants to contact lenses. In the main, attention has focused on hydrophilic polymers which are poor adsorbers of protein molecules (see Chapter 8), but it has also been shown that is possible to confer biocompatibility by immobilising a suitable protein onto the surface in question. The protein thrombomodulin is a natural anticoagulant and has been used for this purpose (Akashi *et al.*, 1992).

1.5 Applications in the Study of Proteins

Proteins are themselves fertile subjects for study, and the last few decades have seen enormous advances in the understanding of how these complex molecules are constructed, how they interact with their environment, and how these factors affect their functional properties.

Much of the chemistry described in later chapters is applicable to investigations in this area (Lundblad and Noyes, 1984a, b; Wong, 1991). In some cases the techniques are used to prepare reagents which serve a useful purpose in such studies, e.g. a radiolabelled antibody. In other cases, the protein of interest is itself the target for the chemistry, for example in the use of bifunctional coupling agents to demonstrate near-neighbour relationships between protein molecules in a biological setting.

There is an important distinction between this last category of applications and all the others described so far. instead of forming a component of an inherently useful product, the protein molecule forms a subject for analysis. In these cases the real output of the protein coupling reaction is in fact *information* concerning the protein, rather than a coupled protein reagent. However, the chemistry involved in achieving these two aims is extremely similar, and many coupling reagents have seen their major application in this field, hence its inclusion in this discussion.

1.5.1 Determination of structure

There are four 'layers' of protein structure which define the overall three-dimensional configuration of the molecule (see Chapter 2, Section 2.5). The primary structure of a protein describes simply the linear arrangement of amino acids that make up the peptide chain(s). The secondary structure describes how these amino acids arrange themselves in three dimensions to form helices, sheets, etc., held together by hydrogen bonding and disulfide bonds from cysteine residues (see Chapter 2, Section 2.2.8). The tertiary structure adds a further three-dimensional element in describing how the different regions of the secondary structure interact with each other to give a space-filling representation. The quaternary structure describes how in certain cases a number of subunits (each with their own tertiary structure) can associate to form an entire protein.

Primary sequence determination became much more straightforward following the introduction of Edman degradation in 1967, and its subsequent automation (see Chapter 6, Section 6.3.2.2); amino acids are removed sequentially from the N-terminus and the resulting derivatives identified to construct the full one-dimensional sequence for each chain. Similar techniques have received attention where sequencing commences instead from the C-terminus (Bailey *et al.*, 1992).

Specific covalent immobilisation methods have been employed to orient the peptide chain correctly for N or C-terminal analysis. For example, activated carboxyl matrices can be used to ensure that binding occurs *via* amines, so that the C-terminus remains free (Shenoy *et al.*, 1992); conversely, amine-bearing solid phases have been used to couple carboxylic residues if N-terminal analysis is required (Liang and Laursen, 1990).

End-group analysis is another branch of primary sequence determination which has commonly employed chemical derivatisation methods, e.g. the use of fluorescent sulfonyl chlorides to tag terminal amines (Gray, 1972).

The determination of which amino acids are juxtaposed in three dimensions, while distant in the primary sequence (e.g. where two separate helices come close to each other) is less straightforward, and cross-linking reagents have been used to elucidate these relationships. The procedure is often based on enzymic cleavage methods. If two amino acids are close enough to be linked by a bifunctional reagent (i.e. one which has two reactive sites), subsequent digestion may yield an identifiable fragment containing the bridge plus a few peptides from each of the linked portions. A knowledge of the primary structure, plus a little deduction, can identify how these spatial relationships fit together. It is simple to extend this concept to another use, in the situation where a protein molecule is cross-linked internally by a bifunctional reagent. Identification of cross-linked fragments following digestion can be used to elucidate the derivatisation sites (Yang *et al.*, 1996b). A variation on the cleavage approach is the site-specific incorporation of a chelator group into the protein molecule: this is used to complex an Fe^{2+} ion, whose redox properties promote cleavage in the immediate vicinity by active oxygen species (Rana and Meares, 1990, 1991).

The distance between individual sites within a protein molecule can also be estimated by making use of energy-transfer phenomena: typically two complementary fluorescent labels are used as reporter molecules, the emission of one providing the excitation for the other. The amount of energy transfer can be related to the distance between the labels (Leung and Meares, 1977). The use of reporter molecules as probes of protein dimensions dates to the early 1950s; Weber (1952) used the polarisation properties of fluorescent labels to make inferences about dimensional changes in the albumin molecules to which they were attached.

The cross-linking approach discussed above can be extended con-veniently to quaternary structure. The electrophoretic technique SDS-PAGE breaks proteins down into their subunits but cannot do so if they are covalently linked by a bifunctional reagent. Therefore spatial relationships between subunits can be determined by examining SDS-PAGE results before and after linking.

Another feature of three-dimensional structure which can be probed by chemical derivatisation is the differentiation of buried and exposed amino acid residues by their relative ease of reaction. In these approaches a simple monofunctional reagent is required; generally this reagent must be amenable to quantification after binding, a good example being the amine-reactive agent trinitrobenzenesulfonic acid (TNBS) which introduces a highly chromophoric trinitrophenyl group. The derivatisation may be carried out before and after denaturation to demonstrate the presence of buried residues.

1.5.2 Determination of microenvironment

In addition to being used for dimensional analysis, reporter molecules can provide information on the microenvironment of the protein, or perhaps just the derivatisation site itself. Kirtley and Koshland (1966), for example, introduced chromophoric 4-nitrophenyl groups into the enzyme glyceraldehyde-3-phosphate dehydrogenase, and observed spectral changes on the addition of the coenzyme NAD which signalled an increase in the polarity around the modified region.

Spin-labelling is an important approach in this field; the introduction of spin-label derivatives such as piperidinyl compounds and measurement of their electron spin resonance (ESR) spectrum under different conditions can yield information on protein conformation (Aplin *et al.*, 1979; Lundblad and Noyes, 1984a).

Use of a number of derivatisation reagents with different properties can also be used to make deductions about the environment of a protein. Notably, the relative efficiency of more or less hydrophobic reagents can be used to determine whether the site of interest is in a hydrophobic region (such as the cell membrane). The use of 'membrane-permeant' and 'membrane-impermeant' succinimide esters for cross-linking (Chapter 5, Section 5.3.5) is an example of this type of approach, allowing conclusions to be drawn as to whether a particular protein is in the membrane, or in the intracellular environment (Staros, 1988).

A particularly elegant application of reporter-labelling is the use of hydrogen exchange (HX) techniques (reviewed by Englander, 1993), to investigate the partially-folded forms which proteins pass through on the way to their final tertiary structure. Proteins are denatured to their fully unfolded form in a deuterium oxide (D_2O)[1] matrix, with the result that the hydrogen atoms in the NH links of the peptide chain exchange with the heavier deuterium isotope (D) to generate ND moieties. The denatured protein is then diluted into H_2O under conditions where the exchange of deuterium and hydrogen is slow; the protein starts to refold and deuterium atoms become buried in the structure. After a time the conditions are altered so that exchange becomes rapid, and all the D atoms which are accessible are replaced by H. The only deuterium atoms remaining are those which have been protected by incorporation in the newly folded regions of the protein. Following the completion of the folding process, the location of deuterium atoms can be established by instrumental techniques such as two-dimensional nuclear magnetic resonance (2D NMR), establishing a 'snapshot' of the partially-folded intermediate at the time the exchange conditions were altered.

[1] Heavy water or ^2H_2O.

1.5.3 Determination of interactions

Techniques which indicate when a complex has been formed between one protein and another are of great interest in understanding the interactions of these molecules.

In concept such experiments are a short step from those used to demonstrate quaternary structure. From cross-linking one protein subunit to another, it is simple to imagine coupling one protein to another which bears specific receptor sites or is a close neighbour (Traut and Kenny, 1977; Ji, 1979). The output of such an investigation is shown in Figure 1.20. Photoactivatable coupling reagents can be extremely useful in such experiments, so that chemical reactivity can be 'switched on' at will. A protein can be labelled with a photoactivatable group which is in itself unreactive, and allowed to bind to its receptor; activation of the group by light then allows a covalent bond to be formed to stabilise the ligand–receptor complex for subsequent analysis. A radiolabel may also be incorporated in the coupling group, and this can be a useful aid at the analytical stage.

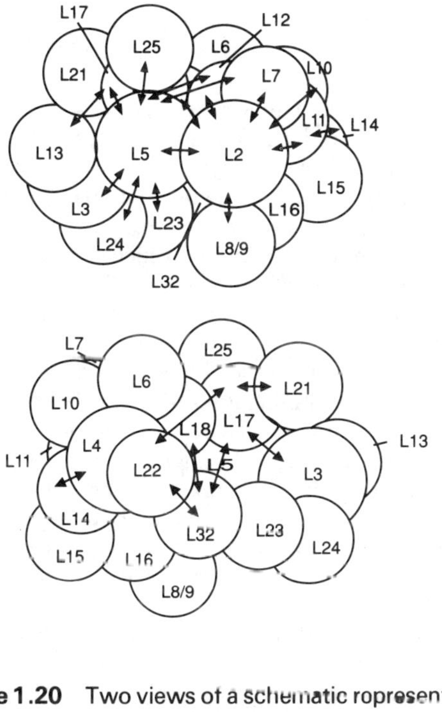

Figure 1.20 Two views of a schematic representation of the proximity relationships between 21 proteins in a subunit of the *E. coli* ribosome, determined by cross-linking and two-dimensional electrophoresis. Observed cross-links are indicated by arrows. (Reproduced with permission from: Traut, R. R. and Kenny, J. W. (1977), *Adv. Exp. Biol. Med.*, **88**, 215–233).

An interesting reporter technique for probing protein interactions in solution is based on the emission anisotropy of fluorescent labels. The degree of anisotropy is related to the relative lifetimes of the fluorescent emission and the rotation of the protein, and the latter is closely related to its molecular mass (cf. fluorescence polarisation, Section 1.2.2.2). Therefore interactions which result in a larger labelled complex can be detected by changes in anisotropy. Churchich (1993) derivatised proteins with isatoic anhydride to generate *o*-aminobenzoyl labels, which have suitable fluorescence properties for this technique. The formation of equimolar complexes between trypsin and trypsin inhibitor in solution was demonstrated in this way.

Many other techniques have been employed to indicate complex formation, some of which have no requirement for protein coupling (see Chapter 11, Section 11.1). Methods have been designed, for example, to interpret the results of size exclusion chromatography in this way (Wen *et al.*, 1996).

1.5.4 Determination of structure–activity relationships

In Chapter 2 it will be seen that the different amino acids which make up proteins may possess specific functional chemistry, allowing them to be derivatised while leaving other groups unaffected. Carrying out specific derivatisation in this way and observing the effect this has on protein behaviour can yield important information on which amino acids are functionally critical.

A large battery of group-selective reagents are available for such studies (Singer, 1967; Lundblad and Noyes, 1984a,b), and have been used very widely to probe the sites which determine protein functionality. To give a couple of simple examples, chemical conversion of the cysteine residue in the active site of papain to serine resulted in complete inactivity (Clark and Lowe, 1978), while Fohlman *et al.* (1979) demonstrated a 350-fold loss in activity of the neurotoxin taipoxin after derivatisation of a single histidine residue.

Obtaining genuinely monospecific reactions is often difficult, however, and the uncertainty of the derivatisation site has to some extent limited the applicability of the approach. The evolution of molecular biology has provided new techniques such as site-directed mutagenesis, SDM, in which manipulation of the 'parent' gene allows a single amino acid residue in the protein to be selectively replaced with another

(El-Gewely, 1991; McPherson, 1991; Medynski, 1992). Examination of the properties of the resulting mutant can provide powerful insights into structure–activity relationships, and this approach has to some extent superseded chemical derivatisation. Nevertheless SDM is a relatively time-consuming procedure, and simpler chemical methods still have their place.

Coupled to modern instrumental techniques, labelling of specific residues can provide very useful information. The power of such an approach is shown in an example from Sivaraja *et al.*, (1989), in which recombinant methods were used to prepare cytochrome *c* peroxidase containing deuterium-labelled tryptophan or methionine residues; examination of the oxidised form of these enzymes by ENDOR (electron nuclear double resonance) spectroscopy identified a Trp residue as the site of the amino acid radical used, unusually, to 'store' one oxidising equivalent.

Affinity labelling is a particularly powerful approach for determining the binding sites of proteins for a particular ligand. This is a direct extension of the approach mentioned above for identifying interactions between receptor and ligand proteins. If the analysis of the linked complex is capable of showing which residues have been coupled, it is obviously possible to make conclusions about the composition of the binding site.

Affinity labelling is more often applied to low-molecular-mass ligands (Lundblad and Noyes, 1984b). The technique employs chemically-reactive analogues of the ligand of interest; these tend to localise at the relevant binding site, where a suitable reactive group allows the formation of a covalent bond. Groups of different reactivity may allow conclusions to be drawn about the binding site chemistry, e.g. the success of a thiol-reactive affinity label may indicate the presence of a cysteine residue. However labels of this type are generally reactive *ab initio*, and some degree of non-specific labelling will occur; a proportion of the affinity label will encounter sites of suitable reactivity before reaching the specific binding site.

The technique of photoaffinity labelling, PAL (Singh *et al.*, 1962) was developed to minimise such effects; photoactivatable groups are incorporated in the affinity label, which therefore remains inert until the reactivity is switched on by light. Complex formation can therefore be allowed to progress towards equilibrium before any covalent coupling takes place, reducing the 'background' problems just described. These reagents can possess another valuable property: the species generated by photoactivation are (in the best cases) so reactive that they are capable of forming a covalent link regardless of the chemical composition of the binding site, even being capable of inserting into C–H bonds, for example.

1.6 Applications in Immunology

While the use of antibody–antigen reactions has featured heavily in virtually all the foregoing sections, coupled protein reagents are also often used in the most fundamental stage of generating the antibodies on which they rely, viz. provoking an immune response from a host animal by the use of an immunogen. This applies equally to the generation of monoclonal and polyclonal antibodies.

Where the immune response is provoked as a prophylactic measure to improve the response to subsequent infection, often in human subjects, the immunogen is termed a vaccine.

1.6.1 The generation of antibodies

The mechanisms by which antibodies (immunoglobulins) come into being are subtle and complex, and will be dealt with in more detail in Chapter 4. The following description is a brief summary of the essential stages of the process.

If a compound with a molecular mass of a few kDa or higher is introduced into a suitable vertebrate, and the host animal's lymphocytes possess the appropriate receptors to render the compound 'recognisable' or antigenic, the adaptive immune response is triggered, producing immunoglobulins which are specific to individual regions of the antigen. These regions are the size of a handful of amino acids, and are termed epitopes or antigenic determinants.

Normally the immune response to a large antigen will initiate the proliferation of a variety of different B-cells, secreting antibodies of various levels of performance against the range of epitopes present. The resulting polyclonal mixture of antibodies can be harvested from the host serum. While they might be viewed as 'old-fashioned' in the light of newer developments, polyclonal antibodies have a distinguished history as therapeutic reagents[1], and continue to find new applications (Harkonen and Gregg, 1993).

Alternatively the B-cells can be immortalised by fusion with a myeloma, cloned out by dilution and aliquoting to isolate single hybridomas from the resulting mixture (i.e. to the point of monoclonality), assayed for activity against the analyte of interest,

[1] The first ever Nobel Prize for Physiology or Medicine was awarded to Emil von Behring for immunotherapy using polyclonal antisera in 1901.

then cultured *in vivo* or *in vitro* to generate monoclonal antibodies with a reproducible level of binding to a single epitope.

Genetic engineering techniques are now beginning to make it possible to generate antibodies directly, using the gene libraries from which the organisms build up the diversity of structures they require to match all potential antigens (Winter and Milstein, 1991; Chiswell and McCafferty, 1992). However, virtually every antibody used in applications such as those discussed in earlier sections was generated by a natural or contrived immunisation process, starting with the introduction of suitable antigenic material as described at the beginning of this section.

1.6.2 Immunogens

Sometimes antibodies are obtained from sources where the immunisation process has occurred naturally, e.g. by harvesting from the serum of patients who have seroconverted after exposure to a disease. In many instances, however, antibodies will be raised deliberately in a suitable organism by introduction of an immunogen.

As mentioned earlier, suitable macromolecules such as proteins are immunogenic in their own right, and the immunogen will often simply consist of the intact protein, or possibly a subunit of it which is still large enough to trigger a response.

While the recognition of protein antigens is dependent on their three-dimensional structure, it is possible in some cases to use small, linear synthetic peptides to mimic a region of interest (Van Regenmortel, 1993). This sequence of peptides is described as a continuous epitope, as opposed to the discontinuous variety; the latter include amino acids which, although physically close to each other in the folded protein, are separated in the peptide chain. A simple fragment of the chain cannot therefore represent them.

Synthetic peptides which mimic a continuous epitope can be used to raise an immune response, and can yield antibody of greater specificity than if the whole protein is used. If the oligopeptides are larger than about 15 amino acid residues they may be immunogenic in their own right, effectively acting as a small protein (Nagy *et al.*, 1993). Below this level the immune response is unlikely to occur satisfactorily, but can still be elicited by coupling the peptide to a carrier protein.

A substance like this that is too small to elicit an immune response on its own, but which can still react with specific antibody is termed a hapten. In addition to oligopeptides, this category may include almost any small molecule. If antibodies against a hapten are required it must be introduced into a complex of suitable molecular mass, so that in effect the hapten is contributing one or more epitopes to a larger immunogenic species. A proportion of the antibodies raised will be specific for those hapten-derived epitopes.

It will be obvious from earlier sections of this chapter that the preparation of antibodies to a molecule of interest is an extremely widespread activity. For this reason, the preparation of immunogens has been a major driver in the development of methods for the conjugation of proteins to small, non-immunogenic compounds, and many of the approaches described in Chapter 6 derive from this field.

When an immunogen is introduced into a living animal, antibodies would ideally only be generated against the hapten. However this is clearly not the case: antibodies will be raised against epitopes of the macromolecule itself, which is usually a protein, and possibly against the bridge or linker group by which the hapten has been connected. The latter possibility in particular must be clearly borne in mind in subsequent screening and use. Highly misleading artefacts can be introduced into immunoassays, for example, by using a hapten conjugate, labelled *via* the same linker as the immunogen from which the binding antibody was raised. Any binding observed may be due to bridge recognition rather than the presence of the hapten, and such combinations are insensitive to the level of unlabelled hapten in the system, the parameter whose quantification is the goal of the test. To avoid such problems, a conjugate may be derivatised with different linker groups from those used in the immunogen. In other circumstances, the same phenomenon can be used to advantage to offer particular aspects of performance (Tsuji, 1981). The search for non-immunogenic linker groups remains an active research area (Boeckler *et al.*, 1996).

Similarly, part of the hapten structure must be sacrificed during the preparation of the immunogen (i.e. the atom in the molecule to which the macromolecule is attached, and the surrounding region which is sterically hindered by its proximity), and thought must be put into which position on the hapten molecule to use (Figure 1.21). To some extent this will be determined by the convenience of the chemistry, which can vary considerably from one position to another. The choice may also be affected by particular features of performance required from the antibody. For example, positions will be avoided with constituents which make a molecule different from an otherwise identical cross-reactant. Again, such factors can have a dramatic impact on subsequent behaviour of the antibody, choice of complimentary labelled hapten, etc. (Kabakoff, 1980; Yoon *et al.*, 1993). Understandably it is sometimes advantageous to distance the hapten somewhat from the carrier protein, so that steric hindrance does not interfere with recognition; a spacer length equivalent to four carbon atoms has been recommended for this purpose (Robinson *et al.*, 1975).

Epitope density is another important factor in influencing immunogen performance. Although it might be expected that a high hapten incorporation would give the best response, an optimal range is generally observed above which antibody production is less efficient. For example, Malaitsev and Azhipa (1993) identified a range of 5.8 – 14 haptens per protein molecule with dopamine-BSA immunogens. Higher incorporations may exhibit tolerogenic behaviour which prevents antibody formation, or provoke an IgM response with little or no immunological 'memory' (Klaus and Cross, 1974).

The other factors which affect the efficiency of the immune response in a given situation are not always easy to predict, although there are some largely

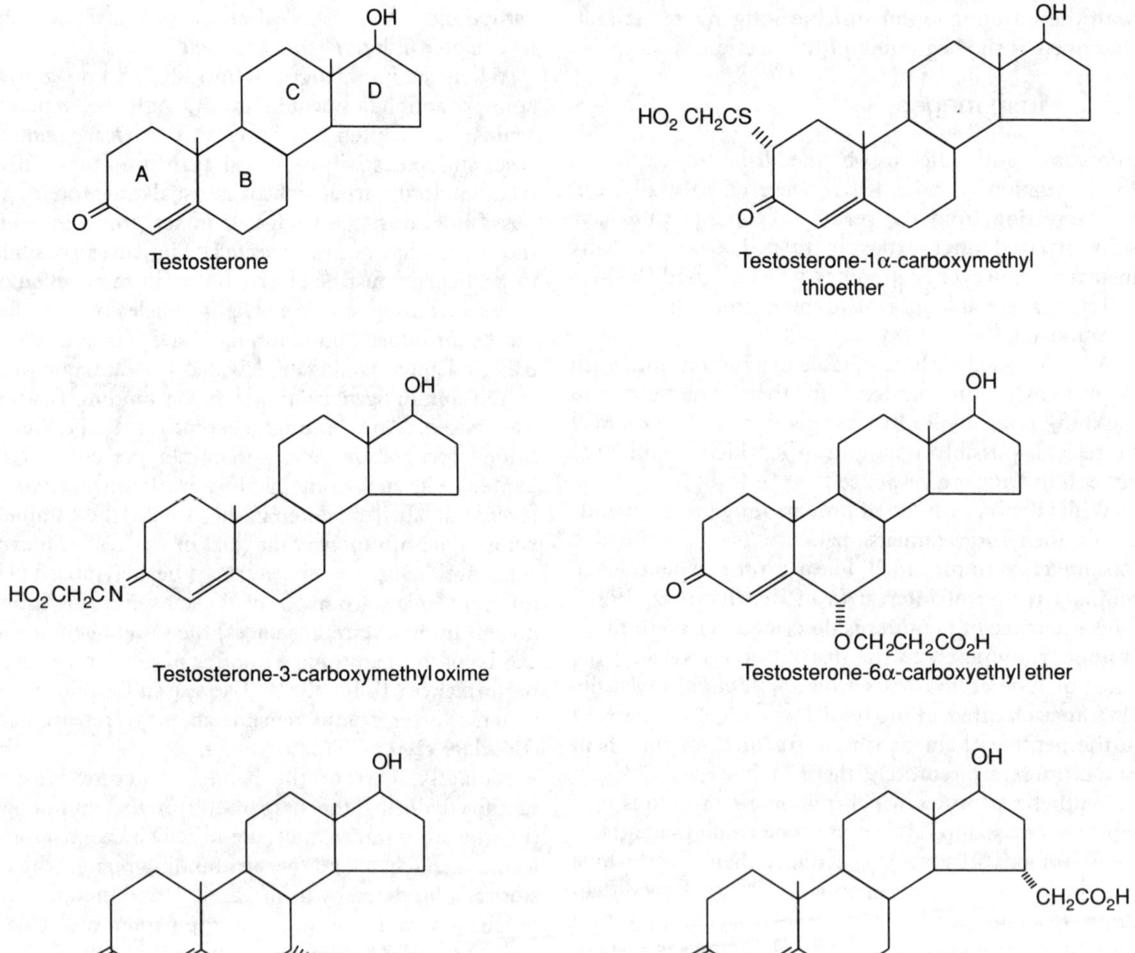

Figure 1.21 Examples of derivatives of testosterone which might be used for immunogen preparation. Linking *via* the 15-position may produce antibody with high cross-reactivity for important analogues differing only in the D-ring structure. Similarly, linking *via* the 1-or 3-position may result in poor specificity in the important A-ring region. An immunogen coupled through the 6-or 7-position in the less critical B-ring may yield an antibody with superior specificity.

empirical rules; some of these relate to the nature of the protein used. There are obvious attractions in using well-defined proteins with high solubility such as albumins. The chemistry can be carried out tidily in homogeneous solution, purification is relatively straightforward, and the resulting immunogen can be characterised quite accurately in terms of hapten incorporation.

However, these reagents are not necessarily favoured in terms of their immunogenic behaviour, and some workers favour the use of proteins such as keyhole limpet haemocyanin (KLH) which have enormous molecular masses, ill-defined structures, and less than perfect solubility—sometimes to the extent of the presence of shell fragments! Characterisation of such immunogens is much less accurate, but their immunogenic capabilities can be superior. Similar considerations can apply in terms of hapten loading, raising incorporation levels of a hydrophobic compound to the point where the protein starts to lose solubility can sometimes yield better results than a well-characterised soluble species.

1.6.3 Vaccines

The use of immunogens to generate a protective immune response in human subjects dates back to the eighteenth century, and Sir Edward Jenner's observation that milkmaids were protected from smallpox by contact with the structurally-related but harmless cowpox virus. Since then, increasingly refined techniques have been developed to provoke a protective response from pathogens while avoiding their harmful properties, from 'passaging' the organism repeatedly through a non-human host species until it mutates to a harmless form, to the use of synthetic or recombinant protein fragments which mimic parts of the pathogen structure.

These fragments are often large enough to be immunogenic in their own right, but are sometimes linked to bulky protein molecules to improve the efficiency of the process.

The use of vaccines to protect against cancer is a fascinating prospect with a chequered history (Cohen, 1993). Early approaches used crude preparations from cancerous tissue as immunogens to generate anti-tumour antibodies, which could potentially offer protection by mopping up new cancers before they could proliferate.

As the understanding of this area has increased, a number of well-defined cancer antigens have been identified. These are often carbohydrates found on the surface of cancerous cells, and vaccines have been developed employing these antigens coupled to protein carriers. Examples include a ganglioside, GM2, coupled to a KLH carrier, developed at the Memorial Sloan-Kettering Cancer Center as a melanoma vaccine, and the sialyl-Tn-KLH vaccine for a number of important cancers from Biomira Inc. in Canada (Cohen, 1993).

1.7 Other Applications of Coupled Protein Reagents

1.7.1 Non-clinical applications of immunoassay, nucleic acid probes and biosensors

Immunoassay offers one of the most sensitive analytical techniques in existence, which can be applied to virtually any target analyte to which antibodies can be obtained. It has been adopted for other uses including environmental monitoring (for example the detection of trace levels of pesticides in water), forensic investigations (such as the detection of traces of explosives on clothing) and food monitoring for microbial contamination, offering a fast alternative to traditional microbiological culture methods.

Nucleic acid probes have also been applied to the detection of pathogenic organisms in food, etc. There is also an interesting application as a forensic tool for genetic fingerprinting. Particularly variable genomic regions, first identified by Alec Jeffreys at the University of Leicester, act as a basis of identifying the origin of biological samples. Assignment of specimens to a single individual can often be made with a high degree of confidence.

The potential for carrying out rapid diagnostic determinations far from laboratory facilities has attracted interest in biosensors for a number of non-clinical applications. Particularly notable is the defence industry, where rapid detection of nerve gases, microbiological agents, etc., has an obvious importance. By the late 1980s, products such as the NAIAD nerve gas detector from Thorn EMI had found multi-million dollar markets (Owen, 1988). Other areas where biosensors have the potential to make an impact include on-site veterinary diagnosis, particularly with respect to fertility, and the forensic, environmental and food monitoring uses already mentioned for immunoassay.

1.7.2 Chromatography matrices

Chromatography, the separation of components in a mixture by exploiting differences in their interactions with a matrix, is by far the most popular approach for the purification of proteins, and of a great many other types of molecule besides.

The techniques used to purify proteins will be covered in Chapter 10. These are tools in the purification of coupled reagents protein rather than applications. However, there are a number of types of chromatography where the separation of components, proteinaceous or otherwise, is accomplished by using the special properties of a protein-labelled solid phase, and the preparation of such materials will be covered in Chapter 10. These applications fall under the general category of affinity separations, relying on specific interactions between the immobilised species and the ligand of interest in the sample. These interactions must be strong enough to retain the ligand to a useful degree, but not so strong that it cannot subsequently be removed in usable form.

Early examples included the preparation of immunoadsorbents by antigen–cellulose coupling (Campbell *et al.*, 1951) or by cross-linking antigens into an insoluble complex (Avrameas and Ternynck, 1969). The widespread adoption of this approach subsequently led to the commercialisation of many affinity matrices bearing immobilised proteins.

There are innumerable situations where a protein-labelled column is required which cannot be obtained commercially, for example affinity purification of an antibody against its specific antigen, or of a receptor protein against a hormone. In these cases there is no choice but to synthesise the specific protein-labelled solid phase. The chromatographic suppliers offer products for just this situation, for example cyanogen bromide-labelled media which will react with a wide variety of ligands under well-defined conditions.

1.7.3 Research applications of immobilised and derivatised enzymes

A number of reasons for immobilising proteins have already been touched on in previous sections: labelled latex or microtitre wells for immunoassay, enzyme immobilisation for biosensors, solid phases for chromatography. But the applications of such reagents are diverse. Where enzymes are employed for research purposes it can be convenient to use them in a form where they can be readily removed from the reaction mixture. This can be achieved by making them insoluble—either by linking them to a support, which may well be identical to those discussed in the previous section, or by cross-linking large numbers of protein molecules to each other.

Physisorption or chemisorption techniques can be used to immobilise enzymes for such uses. Chin and Lanks (1977), for example, covalently coupled lactoperoxidase to tissue culture flasks so that cellular proteins could be radioiodinated as they approached the surface. A number of immobilised enzymes such as urease are also available from major commercial suppliers for research purposes.

It should be noted that immobilisation of enzymes often leads to stabilisation, and this may represent a research topic in itself (Klibanov, 1979).

Finally, chemical derivatisation of enzymes may produce useful changes in properties. These include changes in stability, some of which are discussed in Chapter 3, or in catalytic behaviour (cf. Section 1.5.4). The creation of enzymes soluble in organic solvents by polyethylene glycol derivatisation is an interesting example (Inada *et al.*, 1986).

1.7.4 Industrial uses of immobilised enzymes

Where enzymes are used as catalysts in industrial operations, the same considerations apply, on a larger scale, and the same approaches are often employed to simplify handling. The first commercial application, by the Tanake Seiyaku Company in Japan, was the resolution of amino acids into their D and L-isomers using an aminoacylase from *Aspergillus oryzae* (Tosa *et al.*, 1967).

Weetall (1988) reviewed the commercial use of immobilised enzymes, identifying nine applications employing million-dollar quantities of enzyme. Many of these are in the food industry, including the use of fumarase to catalyse the isomerisation of fumaric acid to malic acid (an additive in fruit juices, etc.), and of amylases and glucose isomerases in the preparation of glucose and fructose from starch. The latter processes are immensely important in financial terms, accounting for around one quarter of the world's consumption of enzymes with an annual value of \$100 million (Owen, 1988).

The pharmaceutical industry is another major user of immobilised enzymes, e.g. in the preparation of the penicillin derivative 6-aminopenicillanic acid (6-APA) from native penicillins, catalysed by penicillin amidase (Lagerlöf *et al.*, 1976). The 6-APA is further derivatised chemically to yield drugs such as ampicillin and amoxicillin.

Even in the heavy chemical industry, enzymic approaches have been successfully employed, e.g. the production of acrylamide from acrylonitrile by nitrile hydratase, generated *in situ* by microorganisms such as *Rhodococcus rhodocrous* J1 (Nagasawa and Yamada, 1989).

Reviews of industrial applications of immobilised enzymes include those of Tanaka *et al.* (1993) and Katchalski-Katzir (1993).

References

Adler, F. L. and Adler, L. T. (1980). Passive hemagglutination and hemolysis for estimation of antigens and antibodies. *Methods Enzymol.*, **70**, 455–466.

Akashi, M. *et al.* (1992). Immobilization of human thrombomodulin on glass beads and its anticoagulant activity. *Bioconjugate Chem.*, **3**, 363–365.

Alam, F. *et al.* (1989). Boron containing immunoconjugates for neutron capture therapy of cancer and for immunocytochemistry. *Antibody Immunoconjugate Radiopharm.*, **2**, 145–163.

Andersson, N. L. *et al.* (1980). In *Electrophoresis '79* (Radola, B. J., ed.), p. 313, Walter de Gruyter, Berlin.

Andrews, A. T. (1986). *Electrophoresis. Theory, Techniques, and Biochemical and Clinical Applications.* Clarendon Press, Oxford.

Aplin, J. D. *et al.*, (1979). Spin-labelling of sialic acid in soluble and cell-surface glycoproteins. *Carbohydr. Res.*, **75**, 11–16.

Arpicco, S. *et al.* (1997). New coupling reagents for the preparation of disulfide cross-linked conjugates with increased stability. *Bioconjugate Chem.*, **8**, 327–337.

Avrameas, S. (1970). Immunoenzyme techniques: enzymes as markers for the localization of antigens and antibodies. *Int. Rev. Cytol.*, **27**, 349–385.

Avrameas, S. and Ternynck, T. (1969). The cross-linking of proteins with glutaraldehyde and its use for the preparation of immunoadsorbents. *Immunochemistry*, **6**, 53–66.

Azar, H. A. and Espinoza, C. G. (1986). Diagnostic immunohistochemistry and electron microscopy of 'undifferentiated' malignant tumors. In *Manual of Clinical Laboratory Immunology, 3rd edit.* (Rose, N. R. *et al.*, eds), pp. 926–930, American Society for Microbiology, Washington, DC.

Bagshawe, K. D. (1994). Antibody directed enzyme prodrug therapy. *Clin. Pharm. Concepts*, **27**, 368–376.

Bailey, G. S. (1994). The raising of a polyclonal antiserum to a protein. *Methods Mol. Biol.*, **32**, 381–388.

Bailey, J. M. *et al.* (1992). Automated carboxy-terminal sequence analysis of peptides. *Protein Sci.*, **1**, 68–80.

Bale, W. F. and Spar, I. L. (1967). Studies directed toward the use of antibodies as carriers of radioactivity for therapy. *Adv. Biol. Med. Phys.*, **5**, 285–356.

Barnard, G. J. R. *et al.* (1985). Recent advances in chemiluminescence immunoassay. In *Bioluminescence and Chemiluminescence: Instruments and Applications*, Vol. I (Van Dyke, K., ed.), pp. 151–183, CRC Press, Boca Raton, FL.

Barth, R. F. *et al.* (1994). Boronated starburst dendrimer-monoclonal antibody immunoconjugates: evaluation as a potential delivery system for neutron capture therapy. *Bioconjugate Chem.*, **5**, 58–66.

Bayer, E. A. *et al.* (1976). Affinity cytochemistry: the localization of lectin and antibody receptors via the avidin–biotin complex. *FEBS Lett.*, **68**, 240–244.

Bayley, H. (1997). Building doors into cells. *Sci. Am.*, **September 1997 issue**, 42–47.

Beesley, J. E. (1992a). Immunogold probes for light microscopy. *Methods Mol. Biol.*, **10**, 169–176.

Beesley, J. E. (1992b). Immunogold probes in electron microscopy. *Methods Mol. Biol.*, **10**, 177–185.

Berson, S. A. *et al.* (1956). Insulin-I^{131} metabolism in human subjects: demonstration of insulin binding globulin in the circulation of insulin treated subjects. *J. Clin. Invest.*, **35**, 170–190.

Blackburn, F. G. *et al.* (1991). Electroluminescence detection for development of immunoassays and DNA probes assays for clinical diagnostics. *Clin. Chem.*, **37**, 1534–1539.

Blake, C. and Gould, B. J. (1984). Use of enzymes in immunoassay techniques. A review. *Analyst*, **109**, 533–547.

Blakey, D. C. *et al.* (1988). Antibody toxin conjugates: a perspective. *Prog. Allergy*, **45**, 50–90.

Boeckler, C. *et al.* (1996). Immunogenicity of new heterobifunctional cross-linking reagents used in the conjugation of synthetic peptides to liposomes. *J. Immunol. Methods*, **191**, 1–10.

Brasch, R. C. (1992). Contrast enhancement in MRI: images of the future. *Adv. MRI Contrast*, **1**, 28–41.

Brennan, M. *et al.* (1985). Preparation of bispecific antibodies by chemical recombination of monoclonal antibody immunoglobulin G_1 fragments. *Science*, **229**, 81–83.

Brinkley, M. (1992). A brief survey of methods for preparing protein conjugates with dyes, haptens, and cross-linking reagents. *Bioconjugate Chem.*, **3**, 2–13.

Bronstein, I. *et al.* (1996). Chemiluminescence: sensitive detection technology for reporter gene assays. *Clin. Chem.*, **42**, 1542–1546.

Buchardt, O. *et al.* (1993). Peptide nucleic acids and their potential applications in biotechnology. *Trends Biotechnol.*, **11**, 384–386.

Burger, D. R. (1993). Novel antisense technology: therapeutic and diagnostic applications. *J. Clin. Immunoassay*, **16**, 224–230.

Burkholder, P. M. (1986). Electron microscopy techniques for laboratory medicine. In *Manual of Clinical Laboratory Immunology, 3rd edit.* (Rose, N. R. *et al.*, eds), pp. 931–933, American Society for Microbiology, Washington, DC.

Burtis, C. A. and Ashwood, E. R., eds (1994). *Tietz Textbook of Clinical Chemistry*, 2nd edit., W. B. Saunders Co., Philadelphia, PA.

Butler, J. E. *et al.* (1986). The immunochemistry of sandwich ELISAs. I. The binding characteristics of immunoglobulins to monoclonal and polyclonal capture antibodies adsorbed on plastic and their detection by symmetrical and asymmetrical antibody–enzyme conjugates. *Mol. Immunol.*, **23**, 971–982.

Campbell, D. H. *et al.* (1951). Immunologic adsorbents. I. Isolation of antibody by means of a cellulose–protein antigen. *Proc. Natl. Acad. Sci. USA*, **37**, 575–578.

Carson, K. A. and Mesulam, M.-M. (1982). Electron microscopic demonstration of neutral connections using horseradish peroxidase. *J. Histochem. Cytochem.*, **30**, 425–435.

Catt, K. and Tregear, G. W. (1967). Solid-phase radioimmunoassay in antibody-coated tubes. *Science*, **158**, 1570–1572.

Chang, T. M. S. (1972). Effects of local applications of microencapsulated catalase on the response of oral lesions to hydrogen peroxide in acatalasemia. *J. Dental Res.*, **51**, 319–321.

Chang, T. M. S. (1995). Artificial cells with emphasis on bioencapsulation in biotechnology. *Biotechnol. Annu. Rev.*, **1**, 267–295.

Chin, N. W. and Lanks, K. W. (1977). Covalent attachment of lactoperoxidase to polystyrene tissue culture flasks. *Anal. Biochem.*, **83**, 709–719.

Chiswell, D. J. and McCafferty, J. (1992). Phage antibodies: will new 'coliclonal' antibodies replace monoclonal antibodies? *Trends Biotechnol.*, **10**, 80–84.

Christopoulos, T. K. and Chiu, N. H. L. (1995). Expression immunoassay. Antigen quantitation using antibodies labeled with enzyme-coding DNA fragments. *Anal. Chem.*, **67**, 4290–4294.

Churchich, J. E. (1993). Fluorescence properties of *o*-aminobenzoyl-labeled proteins. *Anal. Biochem.*, **213**, 229–233.

Clark, P. I. and Lowe, G. (1978). Conversion of the active-site cysteine residue of papain into a dehydro-serine, a serine and a glycine residue. *Eur. J. Biochem.*, **84**, 293–299.

Clinica (1996). X-ray contrast agents mature into generics market. *Clinica*, **736**, 8–9.

Clinica (1997). Ultrasound is on the verge of an imaging breakthrough, thanks to a new wave of imaging agents. *Clinica*, **772**, 21–22.

Cohen, J. (1993). Cancer vaccines get a shot in the arm. *Science*, **262**, 841–843.

Cohn, E. J. *et al.* (1946). Preparation and properties of serum and plasma proteins. IV. A system for the separation into fractions of the protein and lipoprotein components of biological tissues and fluids. *J. Am. Chem. Soc.*, **68**, 459–475.

Cook, A. G. and Wood, P. J. (1994). Chemical synthesis of bispecific monoclonal antibodies: potential advantages in immunoassay systems. *J. Immunol. Methods*, **171**, 227–237.

Cooke, R. A. *et al.* (1935). Serological evidence of immunity with coexisting sensitization in a type of human allergy (hay fever). *J. Exp. Med.*, **62**, 733–750.

Coons, A. H. (1956). Histochemistry with labeled antibody. *Int. Rev. Cytol.*, **5**, 1–23.

Coons, A. H. and Kaplan, M. H. (1950). Localization of antigen in tissue cells. II. Improvements in a method for the detection of antigen by means of fluorescent antibody. *J. Exp. Med.*, **91**, 1–13.

Coons, A. H. *et al.* (1941). Immunological properties of an antibody containing a fluorescent group. *Proc. Soc. Exp. Biol. Med.*, **47**, 200–202.

Corey, D. R. (1997). Peptide nucleic acids: expanding the scope of nucleic acid recognition. *Trends Biotechnol.*, **15**, 224–229.

Dallman, M. J. *et al.* (1991). Cytokine gene expression: analysis using Northern blotting, polymerase chain reaction and *in situ* hybridization. *Immunol. Rev.*, **119**, 163–179.

Davis, S. S. (1997). Biomedical applications of nanotechnology—implications for drug targetting and gene therapy. *Trends Biotechnol.*, **15**, 217–224.

Dean, P. D. G. *et al.*, eds (1985). *Affinity Chromatography—A Practical Approach*. IRL Press, Oxford.

Dougherty, T. J. (1987). Photosensitizers: therapy and detection of malignant tumors. *Photochem. Photobiol.*, **45**, 879–889.

Durrant, I. *et al.* (1990). The application of enhanced chemiluminescence to membrane-based nucleic acid detection. *BioTechniques*, **8**, 564–570.

Dyer, M. J. S. (1992). Antibody therapy of malignancy. *Eur. J. Cancer*, **28**, 276–280.

Ebert, C. D. and Kim, S. W. (1982). Immobilized heparin: spacer arm effects on biological interactions. *Thrombosis Res.*, **26**, 43–57.

Ehrlich, P. (1906). In *Collected Studies on Immunity*, Vol. 2, p. 442, Wiley, New York.

Ekins, R. P. (1993). New perspectives in radioimmunoassay. *Nucl. Med. Commun.*, **14**, 721–735.

El-Gewely, M. R., ed. (1991). *Site-Directed Mutagenesis and Protein Engineering*, Elsevier, Amsterdam.

Englander, S. W. (1993). In pursuit of protein folding. *Science*, **262**, 848–849.

Engvall, E. and Perlmann, P. (1971). Enzyme-linked immunosorbent assay (ELISA). Quantitative assay of immunoglobulin G. *Immunochemistry*, **8**, 871–874.

Epstein, A. L. and Khawli, L. A. (1991). Tumor biology and monoclonal antibodies: overview of basic principles and clinical considerations. *Antibody Immunoconjugate Radiopharm.*, **4**, 373–384.

Erlanger, B. F. (1980). The preparation of antigenic hapten–carrier conjugates: a survey. *Methods Enzymol.*, **70**, 85–104.

Evers, P. (1997). *Gene Therapy—Current Status and Prospects.* Financial Times Healthcare, London.

Farquhar, M. G. *et al.* (1957). An electron microscope study of the glomerulus in nephrosis, glomerulonephritis and lupus erythematosus. *J. Exp. Med.*, **106**, 649–660.

Faulk, W. P. and Taylor, G. M. (1971). An immunocolloid method for the electron microscope. *Immunochemistry*, **8**, 1081–1083.

Findeis, M. A. *et al.* (1993). Targeted delivery of DNA for gene therapy via receptors. *Trends Biotechnol.*, **11**, 202–205.

Fohlman, J. *et al.* (1979). Chemical modification of taipoxin and the consequences for phospholipase activity, pathophysiology, and inhibition of high-affinity choline uptake. *Eur. J. Biochem.*, **94**, 531–540.

Franks, F. (1993a). Proteins. Description and classification. In *Protein Biotechnology. Isolation, Characterization and*

Stabilization (Franks, F., ed.), pp. 1–19, Humana Press, Totowa, NJ.

Franks, F. (1993b). Solution properties of proteins. In *Protein Biotechnology. Isolation, Characterization and Stabilization* (Franks, F., ed.), pp. 133–189, Humana Press, Totowa, NJ.

Fredline, V. F. *et al.* (1997). A reference method for the analysis of aldosterone in blood by high-performance liquid chromatography–atmospheric pressure chemical ionization–tandem mass spectrometry. *Anal. Biochem.*, **252**, 308–313.

French, R. R. *et al.* (1991). Cooperative mixtures of bispecific $F(ab')_2$ antibodies for delivering saporin to lymphoma *in vitro* and *in vivo*. *Cancer Res.*, **51**, 2353–2361.

Friedmann, T. (1993). Gene therapy—a new kind of medicine. *Trends Biotechnol.*, **11**, 156–159.

Galfrè, G. and Milstein, C. (1981). Preparation of monoclonal antibodies. Strategies and procedures. *Methods Enzymol.*, **73**, 3–46.

Gall, J. G. and Pardue, M. L. (1969). Formation and detection of RNA–DNA hybrid molecules in cytological preparations. *Proc. Natl. Acad. Sci USA*, **63**, 378–383.

Genesis Report (1995). Can Boehringer Mannheim electrify immunoassays? *Genesis Report/Dx* (Genesis Group Associates, Inc., Montclair, NJ), **September issue**, 2–5.

Genesis Report (1997). Are diagnostics not ready for the genetic revolution? *Genesis Report/Dx* (Genesis Group Associates, Inc., Montclair, NJ), **January issue**, 11–16.

Glazer, A. N. and Wells, S. (1997). Recent advances in fluorescence labeling, detection and visualization. *Bio-Radiations*, **98**, 4–8.

Glennie, M. J. *et al.* (1987). Preparation and performance of bispecific $F(ab'\gamma)_2$ antibody containing thioether-linked $Fab'\gamma$ fragments. *J. Immunol.*, **139**, 2367–2375.

Goding, J. W. (1980). Antibody production by hybridomas. *J. Immunol. Methods*, **39**, 285–308.

Gombotz, W. R. and Pettit, D. K. (1995). Biodegradable polymers for protein and peptide drug delivery. *Bioconjugate Chem.*, **6**, 332–351.

Gosling, J. P. (1990). A decade of development in immunoassay methodology. *Clin. Chem.*, **36**, 1408–1427.

Gravitt, P. E. and Manos, M. M. (1993). Nucleic acid hybridization methods to detect microorganisms. *Lab. Animal Sci.*, **43**, 5–10.

Gray, W. R. (1972). End-group analysis using dansyl chloride. *Methods Enzymol.*, **25**, 121–138.

Harkonen, W. S. and Gregg, K. A (1993). Back to the future with polyclonal antibodies. *Bio-Technology*, **11**, 1518.

Harris, E. L. V. and Angall, S., eds (1989). *Protein Purification Methods–A Practical Approach*. IRL Press, Oxford.

Harris, E. L. V. and Angall, S., eds (1990). *Protein Purification Applications—A Practical Approach*. IRL Press, Oxford.

Harris, J. M. (1992). Introduction to biotechnical and biomedical applications of poly(ethylene glycol). In *Poly (Ethylene Glycol) Chemistry: Biotechnical and Biomedical Applications* (Harris, J. M., ed.), pp. 1–14, Plenum Press, New York.

Hayes, P. C. *et al.* (1989). Blotting techniques for the study of DNA, RNA and proteins. *Br. Med. J.*, **299**, 965–968.

Heide, K. *et al.* (1977). Plasma protein fractionation. In *The Plasma Proteins. Structure, Function and Genetic Control*, 2nd Edit., Vol. III (Putnam, F. W., ed.), pp. 545–597, Academic Press, New York.

Heidelberger, M. and Kendall, F. E. (1929). A quantitative study of the precipitin reaction between type III pneumococcus polysaccharide and purified homologous antibody. *J. Exp. Med.*, **50**, 809–823.

Hemmilä, I. (1991). Applications of fluorescence in immunoassays. *Chem. Anal. (Warsaw)*, **117**, 44–52.

Hermanson, G. T. (1995). *Bioconjugate Techniques*, Academic Press, New York.

Hermentin, P. *et al.* (1990). Hinge-thiol coupling of monoclonal antibody to silanized iron oxide particles and evaluation of magnetic cell depletion. *Bioconjugate Chem.*, **1**, 411–418.

Hincal, A. A. and Kas, H. S. (1988). Targeting of immobilized drugs. In *Analytical Uses of Immobilized Biological Compounds for Detection, Medical and Industrial Uses* (Guilbault, G. G. and Mascini, M., eds), pp. 95–113, D. Reidel Publishing Co., Dordrecht, Netherlands.

Holgate, C. S. *et al.* (1983). Immunogold–silver staining. New method of immunostaining with enhanced sensitivity. *J. Histochem. Cytochem.*, **31**, 938–944.

Hoyle, N. R. *et al.* (1996). Electrochemiluminescence: leading-edge technology for automated immunoassay analyte detection. *Clin. Chem.*, **42**, 1576–1578.

Huston, J. S. *et al.* (1993). Medical applications of single-chain antibodies. *Int. Rev. Immunol.*, **10**, 195–217.

Inada, Y. *et al.* (1986). Application of polyethylene glycol-modified enzymes in biotechnological processes: organic solvent-soluble enzymes. *Trends Biotechnol.*, **4**, 190–194.

Ingham, K. C. (1984). Protein precipitation with polyethylene glycol. *Methods Enzymol.*, **104**, 351–356.

Isaacson, M. and Johnson, D. (1975). The microanalysis of light elements using transmitted energy loss electrons. *Ultramicroscopy*, **1**, 33–52.

Ishikawa, E. (1987). Development and clinical application of sensitive immunoassay for macromolecular antigens—a review. *Clin. Biochem.*, **20**, 375–385.

Ishikawa, E. and Kohno, T. (1989). Development and applications of sensitive enzyme immunoassay for antibodies: a review. *J. Clin. Lab. Anal.*, **3**, 252–265.

Islam, M. N. *et al.* (1983). Both TSH and thyroid-stimulating antibody of Graves' disease bind to an M_r 197 000 holoreceptor. *Endocrinology*, **113**, 436–438.

Ji, T. H. (1979). The application of chemical crosslinking for studies on cell membranes and the identification of surface reporters. *Biochim. Biophys. Acta*, **559**, 39–69.

Jiang, F. N. *et al.* (1990). Development of technology for linking photosensitizers to a model monoclonal antibody. *J. Immunol Methods*, **134**, 139–149.

Joerger, R. D. *et al.* (1995). Analyte detection with DNA-labeled antibodies and polymerase chain reaction. *Clin. Chem.*, **41**, 1371–1377.

Johannsson, A. *et al.* (1986). Enzyme amplification for immunoassays. Detection limit of one hundredth of an attomole. *J. Immunol. Methods*, **87**, 7–11.

Kabakoff, D. S. (1980). Chemical aspects of immunoassay. In *Enzyme-Immunoassay* (Maggio, E. T., ed.), pp. 71–104, CRC Press, Boca Raton, FL.

Kabat, E. A. (1976). Antigens. In *Structural Concepts in Immunology and Immunochemistry*, 2nd Edit., pp. 16–41, Holt, Rinehart and Winston, New York.

Karlsson, R. and Ståhlberg, R. (1995). Surface plasmon resonance detection and multispot sensing for direct monitoring of interactions involving low-molecular-weight analytes and for determination of low affinities. *Anal. Biochem.*, **228**, 274–280.

Karlsson, R. *et al.* (1991). Kinetic analysis of monoclonal antibody–antigen interactions with a new biosensor based analytical system. *J. Immunol. Methods*, **145**, 229–240.

Katchalski-Katzir, E. (1993). Immobilized enzymes—learning from past successes and failures. *Trends Biotechnol.*, **11**, 471–478.

Kendall, P. A. (1965). Labelling of thiolated antibody with mercury for electron microscopy. *Biochim. Biophys. Acta*, **97**, 174–176.

Kenney, A. and Fowell, S. (1992). *Practical Protein Chromatography*, Humana Press, Totowa, NJ.

Kimura, J. *et al.* (1988). An immobilized enzyme membrane fabrication method using an ink jet nozzle. *Biosensors*, **4**, 41–52.

Kirtley, M. E. and Koshland, D. E., Jr. (1966). The introduction of a 'Reporter' group at the active site of glyceraldehyde-3-phosphate dehydrogenase. *Biochem. Biophys. Res. Commun.*, **23**, 810–815.

Klaus, G. G. B. and Cross, A. M. (1974). The influence of epitope density on the immunological properties of hapten–protein conjugates. I. The characteristics of the immune response to hapten-coupled albumen with varying epitope density. *Cell. Immunol.*, **14**, 226–241.

Klibanov, A. M. (1979). Enzyme stabilization by immobilization. *Anal. Biochem.*, **93**, 1–25.

Kobayashi, Y. and Takayama, K. (1997). Immobilization of antigen or antibody by printing press. *Jpn. Pat.*, *97/54093*, to Adotetsuku KK.

Köhler, G. and Milstein, C. (1975). Continuous cultures of fused cells secreting antibody of predefined specificity. *Nature*, **256**, 495–497.

Koike, M. *et al.* (1964). A new immuno-electron microscope technique with use of metal chelate antibody (MCA). *J. Electron Microsc.*, **13**, 133–138.

Kondo, A. *et al.* (1997). Measurement of histidinohydroxylysinonorleucine and hydroxyproline in skin collagen by reversed-phase high-performance chromatography after 9-fluorenylmethyl chloroformate labeling. *Anal. Biochem.*, **252**, 255–259.

Kopetzki, E. *et al.* (1994). Enzymes in diagnostics: achievements and possibilities of recombinant DNA technology. *Clin. Chem.*, **40**, 688–704.

Koppel, G. A. (1990). Recent advances with monoclonal antibody drug targetting for the treatment of human cancer. *Bioconjugate Chem.*, **1**, 13–23.

Lagerlöf, E. *et al.* (1976). Production of 6-aminopenicillanic acid with immobilized *Escherichia coli* acylase. *Methods Enzymol.*, **44**, 759–768.

Langlands, B. (1997). *Gene Therapy Players*. Financial Times Healthcare, London.

Lauffer, R. B. (1987). Paramagnetic metal complexes as water proton relaxation agents for NMR imaging: theory and design. *Chem. Rev.*, **87**, 901–927.

Leung, C. S.-H. and Meares, C. F. (1977). Attachment of fluorescent metal chelates to macromolecules using 'bifunctional' chelating agents. *Biochim. Biophys. Res. Commun.*, **75**, 149–155.

Levy, J. G. (1995). Photodynamic therapy. *Trends Biotechnol.*, **13**, 14–18.

Liang, S.-P. and Laursen, R. A. (1990). Covalent immobilisation of proteins and peptides for solid-phase sequencing using prepacked capillary columns. *Anal. Biochem.*, **188**, 366–373.

Libeer, J. C. (1985). Solid phase chemistry in clinical laboratory tests: a literature review. *J. Clin. Chem. Clin. Biochem.*, **23**, 645–655.

Lindbladh, C. *et al.* (1993). Use of genetically prepared enzyme conjugates in enzyme immunoassay. *Trends Biochem. Sci.*, **18**, 279–283.

Lisi, P. J. *et al.* (1982). Enzyme therapy. I. Polyethylene glycol—glucuronidase conjugates as potential therapeutic agents in acid mucopolysaccharidosis. *J. Appl. Biochem.*, **4**, 19–33.

Lowe, C. R. (1985). An introduction to the concepts and technology of biosensors. *Biosensors*, **1**, 3–16.

Lojda, Z. *et al.*, eds (1979). *Enzyme Histochemistry. A Laboratory Manual*. Springer-Verlag, Berlin.

Lu, H. P. and Xie, X. S. (1997). Single-molecule spectral fluctuations at room temperature. *Nature*, **385**, 143–146.

Lundblad, R. L. and Noyes, C. M. (1984a). *Chemical Reagents for Protein Modification*, Vol. I, CRC Press, Boca Raton, FL.

Lundblad, R. L. and Noyes, C. M. (1984b). *Chemical Reagents for Protein Modification*, Vol. II, CRC Press, Boca Raton, FL.

Maeda, H. *et al.* (1984). Tailor-making of protein drugs by polymer conjugation for tumor targetting: a brief review on Smancs. *J. Protein Chem.*, **3**, 181–193.

Malaitsev, V. V. and Azhipa, O. Y. (1993). Influence of epitope density on immunogenic properties of hapten–protein conjugates. *Bull. Exp. Biol. Med.*, **115**, 726–728.

Mattson, G. *et al.* (1993). A practical approach to crosslinking. *Mol. Biol. Rep.*, **17**, 167–183.

Mayo, C. S. and Hallock, R. B. (1989). Immunoassay based on surface plasmon oscillations. *J. Immunol. Methods*, **120**, 105–114.

McPherson, M., ed. (1991). *Directed Mutagenesis—A Practical Approach*, IRL Press, Oxford.

McPherson, R. A. (1995). Evolution of polymerase chain reaction to a quantitative laboratory tool. *Clin. Chem.*, **41**, 1065–1067.

Means, G. E. and Feeney, R. E. (1990). Chemical modifications of proteins: history and applications. *Bioconjugate Chem.*, **1**, 2–12.

Medynski, D. (1992). Genetic approaches to protein structure and function: point mutations as modifiers of protein function. *Bio/Technology*, **10**, 1002–1006.

Meares, C. F. and Wensel, T. G. (1984). Metal chelates as probes of biological systems. *Acc. Chem. Res.*, **17**, 202–209.

Mendonca Dias, M. H. and Lauterbur, P. C. (1987). Contrast agents for nuclear magnetic resonance imaging. *Biol. Trace Element Res.*, **13**, 229–239.

Milstein, C. and Cuello, A. C. (1983). Hybrid hybridomas and their use in immunohistochemistry. *Nature*, **305**, 537–540.

Morin, J. and Hastings, J. (1971). Energy transfer in a bioluminescent system. *J. Cell Physiol.*, **77**, 313–318.

Moss, D. W. and Henderson, A. R. (1994). Enzymes. In *Tietz Textbook of Clinical Chemistry*, 2nd edit. (Burtis, C. A. and Ashwood, E. R., eds), pp. 735–896, W. B. Saunders Co., Philadelphia, PA.

Mosser, D. D. *et al.* (1997). Use of a dicistronic expression cassette encoding the green fluorescent protein for the screening and selection of cells expressing inducible gene products. *BioTechniques*, **22**, 150–161.

Mulligan, R. C. (1993). The basic science of gene therapy. *Science*, **260**, 926–932.

Nagasawa, T. and Yamada, H. (1989). Microbial transformations of nitriles. *Trends Biotechnol.*, **7**, 153–158.

Nagy, E. V. *et al.* (1993). Immunogenicity of a unique region of the human thyrotropin receptor. *J. Endocrinol. Invest.*, **16**, 485–493.

Nakagawa, K. and Miyazawa, T. (1997). Chemiluminescence–high performance liquid chromatographic determination of tea catechin, (-)-epigallocatechin 3-gallate, at picomole levels in rat and human plasma. *Anal. Biochem.*, **248**, 41–49.

Nelson, N. C. and Kacian, D. L. (1990). Chemiluminescent DNA probes: a comparison of the acridinium ester and dioxetane detection systems and their use in clinical diagnostic assays. *Clin. Chim. Acta*, **194**, 73–90.

Nielsen, P. E. *et al.* (1994). Peptide nucleic acid (PNA), A DNA mimic with a peptide backbone. *Bioconjugate Chem.*, **5**, 3–7.

North, J. R. (1985). Immunosensors: antibody-based biosensors. *Trends Biotechnol.*, **3**, 180–186.

Nuovo, G. J. (1992) *PCR* In Situ *Hybridisation: Protocols and Applications*, Raven Press, New York.

Opheim, K. E. *et al.* (1984). Particle-enhanced turbidimetric inhibition immunoassay for theophylline evaluated with the Dupont *aca*. *Clin. Chem.*, **30**, 1870–1874.

Ottensmeyer, F. P. (1982). Scattered electrons in microscopy and Microanalysis. *Science*, **215**, 461–466.

Ottensmeyer, F. P. and Andrew, J. W. (1980). High-resolution microanalysis of biological specimens by electron energy loss spectroscopy and by electron spectroscopic imaging. *J. Ultrastruct. Res.*, **72**, 336–348.

Owen, V. M. (1988). Commercial aspects of the use of immobilized compounds. In *Analytical Uses of Immobilized Biological Compounds for Detection, Medical and Industrial Uses* (Guilbault, G. G. and Mascini, M., eds), pp. 329–339, D. Reidel Publishing Co., Dordrecht, Germany.

Parker, D. (1990). Tumour targetting with radiolabelled macrocycle—antibody conjugates. *Chem. Soc. Rev.*, **19**, 271–291.

Pease, D. C. and Baker, R. F. (1950). Electron microscopy of the kidney. *Am. J. Anat.*, **87**, 349–370.

Pettersson, K. (1993). Comparison of immunoassay technologies. *Clin. Chem.*, **39**, 1359–1360.

Pirotta, M. *et al.* (1997). Fluorescence spectroscopy on the single molecule level. *Spectrosc. Eur.*, **9**, 16–21.

Pişkin, K. and Chang, T. M. S. (1980). A new combined enzyme–charcoal system formed by enzyme adsorption on charcoal followed by polymer coating. *Int. J. Artif. Organs*, **3**, 344–347.

Poznansky, M. J. and Juliano, R. L. (1984). Biological approaches to the controlled delivery of drugs: a critical review. *Pharmacol. Rev.*, **36**, 277–336.

Prasher, D. *et al.* (1992). Primary structure of the *Aequorea victoria* green-fluorescent protein. *Gene*, **111**, 229–233.

Pressman, D. and Eisen, H. N. (1950). The zone of localization of antibodies. V. An attempt to saturate antibody-binding sites in mouse kidney. *J. Immunol.*, **64**, 273–279.

Pressman, D. and Korngold, L. (1953). The *in vivo* localization of anti-Wagner-osteogenic-sarcoma antibodies. *Cancer*, **6**, 619–623.

Price, C. P. and Newman, D. J., eds (1997). *Principles and Practice of Immunoassay*, 2nd edit., Macmillan Reference Ltd, London.

Ram, Z. *et al.* (1993). *In situ* retroviral-mediated gene transfer for the treatment of brain tumors in rats. *Cancer Res.*, **53**, 83–88.

Rana, T. M. and Meares, C. F. (1990). Specific cleavage of a protein by an attached iron chelate. *J. Am. Chem. Soc.*, **112**, 2457–2458.

Rana, T. M. and Meares, C. F. (1991). Iron chelate mediated proteolysis: protein structure dependence. *J. Am. Chem. Soc.*, **113**, 1859–1861.

Raso, V. (1994). Immunotargetting intracellular compartments. *Anal. Biochem.*, **222**, 297–304.

Raso, V. and Griffin, T. (1981). Hybrid antibodies with dual specificity for the delivery of ricin to immunoglobulin-bearing target cells. *Cancer Res.*, **41**, 2073–2078.

Renart, J. *et al.* (1979). Transfer of proteins from gels to diazobenzyloxymethyl paper and detection with antisera: a method for studying antibody specificity and antigen structure. *Proc. Natl. Acad. Sci. USA*, **76**, 3116–3120.

Renshaw, P. F. *et al.* (1986). Ferromagnetic contrast agents: a new approach. *Magn. Reson. Med.*, **3**, 217–225.

Renz, M. and Kurz, C. (1984). A colorimetric method for DNA hybridization. *Nucleic Acids Res.*, **12**, 3435–3444.

Richardson, J. (1981). The anatomy and taxonomy of protein structure. *Adv. Protein Chem.*, **4**, 167–339.

Ringsdorf, H. (1975). Structures and properties of pharmacologically active polymers. *J. Polym. Sci. Polym. Symp.*, **51**, 135–153.

Robinson, J. D. *et al.* (1975). The use of rats in the screening of drug–protein conjugates for immunoreactivity. In *Radioimmunoassay in Clinical Biochemistry* (Pasternak, C. A., ed.), pp. 101–111, Heyden & Son Ltd, London.

Rosenfeld, M. A. *et al.* (1992). *In vivo* transfer of the human cystic fibrosis transmembrane conductance regulator gene to the airway epithelium. *Cell*, **68**, 143–155.

Rozsnyai, L. F. *et al.* (1992). Photolithographic immobilisation of biopolymers on solid supports. *Angew. Chem., Int. Ed. Engl.*, **31**, 759–761.

Rubenstein, K. E. *et al.* (1972). 'Homogeneous' enzyme immunoassay. A new immunochemical technique. *Biochem. Biophys. Res. Commun.*, **47**, 846–851.

Sanger, F. and Tuppy, H. (1951a). The amino acid sequence in the phenylalanine chain of insulin. 1. The identification of lower peptides from partial hydrolysates. *Biochem. J.*, **49**, 463–481.

Sanger, F. and Tuppy, H. (1951b). The amino-acid sequence in the phenylalanine chain of insulin. 2. The investigation of peptides from enzymic hydrolysates. *Biochem. J.*, **49**, 481–490.

Schäfer, T. *et al.* (1992). Two-year double-blind trial of a monomethoxy polyethylene glycol (mPEG) modified grass pollen extract at different dose levels. *Ann. Allergy*, **68**, 334–339.

Schalkwijk, J. *et al.* (1985). Cationization of catalase, peroxidase, and superoxide dismutase. Effect of improved intraarticular retention on experimental arthritis in mice. *J. Clin. Invest.*, **76**, 198–205.

Schubiger, P. A. *et al.* (1996). Vehicles, chelators, and radionuclides: choosing the 'building blocks' of an effective therapeutic radioimmunoconjugate. *Bioconjugate Chem.*, **7**, 165–179.

Schultz, N. M. and Kennedy, R. T. (1993). Rapid immunoassays using capillary electrophoresis with fluorescence detection. *Anal. Chem.*, **65**, 3161–3165.

Scientific American (1997). Making gene therapy work. *Sci. Ame.*, **June 1997 issue,** 80–103.

Scouten, W. H. *et al.* (1995). Enzyme or protein immobilization techniques for applications in biosensor design. *Trends Biotechnol.*, **13**, 178–185.

Senter, P. D. *et al.* (1988). Anti-tumor effects of antibody–alkaline phosphatase conjugates in combination with etoposide phosphate. *Proc. Natl. Acad. Sci. USA*, **85**, 4842–4846.

Senter, P. D. *et al.* (1993). Generation of cytotoxic agents by targetted enzymes. *Bioconjugate Chem.*, **4**, 3–9.

Seymour, L. W. (1992). Tumoritropic and lysosomotropic principles of macromolecular drugs. *CRC Crit. Rev. Ther. Drug Carrier Syst.*, **9**, 135–187.

Shenoy, N. R. *et al.* (1992). Carboxylic acid-modified polyethylene: a novel support for the covalent immobilization of polypeptides for C-terminal sequencing. *Protein Sci.*, **1**, 58–67.

Shreve, P. and Aisen, A. M. (1986). Monoclonal antibodies labeled with polymeric paramagnetic ion chelates. *Magn. Reson. Med.*, **3**, 336–340.

Shuman, H. and Somlyo, A. P. (1976). Electron probe X-ray analysis of single ferritin molecules. *Proc. Natl. Acad. Sci. USA*, **73**, 1193–1195.

Singer, S. J. (1959). Preparation of an electron-dense antibody conjugate. *Nature*, **183**, 1523–1524.

Singer, S. J. (1967). Covalent labeling of active sites. *Adv. Protein Chem.*, **22**, 1–54.

Singh, A. *et al.* (1962). The photolysis of diazoacetyl-chymotrypsin. *J. Biol. Chem.*, **237**, 3006–3008.

Siu Chong, E. D. and Chang, T. M. S. (1974). *In vivo* effects of intraperitoneally injected L-asparaginase solution and L-asparaginase immobilized within semipermeable nylon microcapsules with emphasis on blood L-asparaginase, 'body' L-asparaginase, and plasma L-asparaginase levels. *Enzyme*, **18**, 218–239.

Sivaraja, M. *et al.* (1989). Identification by ENDOR of Trp^{191} as the free-radical site in cytochrome *c* peroxidase compound ES. *Science*, **245**, 738–740.

Smith, R. A. G. *et al.* (1993). Chemical derivatization of therapeutic proteins. *Trends Biotechnol.*, **11**, 397–403.

Southern, E. M. (1975). Detection of specific sequences among DNA fragments separated by gel electrophoresis. *J. Mol. Biol.*, **98**, 503–517.

Staerz, U. D. and Bevan, M. J. (1986). Hybrid hybridoma producing a bispecific monoclonal antibody that can focus effector T-cell activity. *Proc. Natl. Acad. Sci. USA*, **83**, 1453–1457.

Staros, J. V. (1988). Membrane-impermeant cross-linking reagents: probes of the structure and dynamics of membrane proteins. *Acc. Chem. Res.*, **21**, 435–441.

Stavitsky, A. B. and Arquilla, E. R. (1958). Studies of proteins and antibodies by specific hemagglutination and hemolysis of protein-conjugated erythrocytes. *Int. Arch. Allergy*, **13**, 1–38.

Sternberger, L. A. (1967). Electron microscopic immunocytochemistry: a review. *J. Histochem. Cytochem.*, **15**, 139–159.

Sternberger, L. (1986). In *Immunocytochemistry*, 3rd edit., chapt. 3, John Wiley, New York.

Sternberger, L. A. *et al.* (1965). Indirect immunouranium technique for staining of embedded antigen in electron microscopy. *Exp. Mol. Pathol.*, **4**, 112–125.

Stevenson, G. T. *et al.* (1989). A chimeric antibody with dual Fc regions (bisFabFc) prepared by manipulations at the IgG hinge. *Anti-cancer Drug Des.*, **3**, 219–230.

Tanaka, A. *et al.* (eds). (1993). *Industrial Applications of Immobilized Enzymes*, Marcel Dekker, New York.

Tanaka, H. (1968). The ferritin-labelled method; its advantages and disadvantages. A methodological review. *Acta Haematol. Jpn.*, **31**, 125–137.

Ternynck, T. and Avrameas, S. (1990). Avidin–biotin systems in enzyme immunoassays. *Methods Enzymol.*, **184**, 469–481.

Thavarungkul, P. *et al.* (1991). Continuous monitoring of urea in blood during dialysis. *Biosensors Bioelectron.*, **6**, 101–107.

Thibodeau, S. N. (1994). Recent applications of PCR in clinical laboratory medicine. *Clin. Chem.*, **40**, 681–682.

Thomas, P. (1980). Hybridization of denatured RNA and small DNA fragments transferred to nitrocellulose. *Proc. Natl. Acad. Sci. USA*, **77**, 5201–5205.

Thomas, P. (1983). Hybridization of denatured RNA transferred on dotted nitrocellulose paper. *Methods Enzymol.*, **100**, 255–266.

Thorpe, P. E. *et al.* (1987). New coupling agents for the synthesis of immunotoxins containing a hindered disulfide bond with improved stability *in vivo*. *Cancer Res.*, **47**, 5924–5931.

Tiefenauer, L. X. *et al.* (1993). Antibody-magnetite nanoparticles: *in vitro* characterization of a potential tumor-specific contrast agent for magnetic resonance imaging. *Bioconjugate Chem.*, **4**, 347–352.

Tosa, T. *et al.* (1967). Studies on continuous enzyme reactions. IV. Preparation of a DEAE–Sephadex–aminoacylase column and continuous optical resolution of acyl-DL-amino acids. *Biotechnol. Bioeng.*, **9**, 603–615.

Towbin, H. and Gordon, J. (1984). Immunoblotting and dot immunobinding—current status and outlook. *J. Immunol. Methods*, **72**, 313–340.

Traut, R. R. and Kenny, J. W. (1977). Crosslinking of ribosomes by cleavable bifunctional mercaptoimidates. *Adv. Exp. Med. Biol.*, **88**, 215–233.

Trubetskoy, V. S. *et al.* (1992). Use of N-terminal modified poly(L-lysine)–antibody conjugate as a carrier for targetted gene delivery in mouse lung endothelial cells. *Bioconjugate Chem.*, **3**, 323–327.

Tsuji, A. (1981). Effects of homologous and heterologous systems on sensitivities of hapten enzyme immunoassays. In *Enzyme Immunoassay* (Ishikawa, E. *et al.*, eds), pp. 114–118, Igaku-Shoin, Tokyo.

Uhlén, M. *et al.* (1992). Fusion proteins in biotechnology. *Curr. Opin. Biotechnol.*, **3**, 363–369.

Unger, E. R. (1992). *In situ* and Northern hybridizations. Technical guidelines guiding clinical application. *Cancer*, **69**, 1532–1535.

Urdea, M. S. *et al.* (1988). A comparison of non-radioisotopic hybridization assay methods using fluorescent, chemiluminescent and enzyme labeled synthetic oligodeoxyribonucleotide probes. *Nucleic Acids Res.*, **16**, 4937–4956.

Valenzuela, R. and Deodhar, S. D. (1986). Tissue Immunofluorescence. In *Manual of Clinical Laboratory Immunology*, 3rd edit. (Rose, N. R. *et al.*, eds), pp. 923–925, American Society for Microbiology, Washington, DC.

Van Regenmortel, M. H. V. (1993). Synthetic peptides versus natural antigens in immunoassays. *Ann. Biol. Clin.*, **51**, 39–41.

van Weemen, B. K. and Schuurs, A. H. W. M. (1971). Immunoassay using antigen–enzyme conjugates. *FEBS Lett.*, **15**, 232–236.

von Kleist, S. (1993). Ten years of tumor imaging with labelled antibodies. *In Vivo*, **7**, 581–584.

Vrudhula, V. M. *et al.* (1993). Antitumor activities of a cephalosporin prodrug in combination with monoclonal antibody-β-lactamase conjugates. *Bioconjugate Chem.*, **4**, 334–340.

Wang, C.-Y. *et al.* (1997). Specific immobilization of firefly luciferase through a biotin carboxyl carrier protein domain. *Anal. Biochem.*, **246**, 133–139.

Watson, R. A. *et al.* (1976). Polarisation fluoroimmunoassay of gentamicin. *Clin. Chim. Acta*, **73**, 51–55.

Weber, G. (1952). Polarization of the fluorescence of macromolecules. 2. Fluorescent conjugates of ovalbumin and bovine serum albumin. *Biochem. J.*, **51**, 155–167.

Weetall, H. H. (1988). Enzymes immobilized on inorganic supports. In *Analytical Uses of Immobilized Biological Compounds for Detection, Medical and Industrial Uses* (Guilbault, G. G. and Mascini, M., eds), pp. 1–16, D. Reidel Publishing Co., Dordrecht.

Wen, J. *et al.* (1996). Size-exclusion chromatography with on-line light-scattering, absorbance, and refractive index detectors for studying proteins and their interactions. *Anal. Biochem.*, **240**, 155–166.

Widder, D. J. *et al.* (1987). Magnetite albumin microspheres: a new MR contrast material. *Am. J. Roentgenol.*, **148**, 399–404.

Wild, D. (1994). *The Immunoassay Handbook*, The Macmillan Press Ltd, London.

Wilson, R. *et al.* (1996). Electrochemiluminescence detection of glucose oxidase as a model for flow injection immunoassays. *Biosensors Bioelectron.*, **11**, 805–810.

Winter, G. and Milstein, C. (1991). Man-made antibodies. *Nature*, **349**, 293–299.

Wisdom, G. B. (1994). Protein blotting. *Methods Mol. Biol.*, **32**, 207–213.

Wong, S. S. (1991). *Chemistry of Protein Conjugation and Cross-Linking*, CRC Press, Boca Raton, FL.

Wood, W. G. *et al.* (1986). Luminescence assays for haptens and proteins. *Methods Enzymol.*, **133**, 354–365.

Wu, G. Y. *et al.* (1991). Receptor-mediated gene delivery *in vivo*. *J. Biol. Chem.*, **266**, 14338–14342.

Yalow, R. S. and Berson, S. A. (1959). Assay of plasma insulin in human subjects by immunological methods. *Nature*, **184**, 1648–1649.

Yang, F. *et al.* (1996a). The molecular structure of green fluorescent protein. *Nature Biotechnol.*, **14**, 1246–1251.

Yang, H. M. and Reisfeld, R. A. (1988). Doxorubicin conjugated with a monoclonal antibody directed to a human melanoma-associated proteoglycan suppresses the growth of established tumor xenografts in nude mice. *Proc. Natl. Acad. Sci. USA*, **85**, 1189–1193.

Yang, T. *et al.* (1996b). Mapping cross-linking sites in modified proteins with mass spectrometry: an application to cross-linked hemoglobins. *Anal. Biochem.*, **242**, 55–63.

Yoon, D. Y. *et al.* (1993). Influence of the conjugation site on the specificity of monoclonal antibodies to progesterone and on the performance of direct enzyme immunoassay. *Biochem. Mol. Biol. Int.*, **31**, 553–563.

Chapter 2

The Functional Chemistry of Proteins and Protein Coupling

Alastair H. Dent and Mohammed Aslam

Research and Development Department, Ortho-Clinical Diagnostics, Pollards Wood Laboratories, Nightingales Lane, Chalfont St. Giles HP8 4SP, UK

2.1 Introduction

One of the most important characteristics of proteins as a group is the subtlety of structure–activity relationships, the diversity of behaviour obtained by sometimes minor variations in molecular composition. It is a task of immense size and complexity to understand the links between the three-dimensional structure of proteins, how this structure is affected by interactions with the environment, how binding equilibria are established with other species, and how the overall combination of these factors leads to physiological or biochemical change. Only for a tiny fraction of the tens of thousands of known proteins is there anything approaching a full understanding of these factors.

Perhaps fortunately for protein chemists, these molecules are nowhere near as interesting when viewed as chemical reagents and the goal is to couple them covalently to other compounds, then purify the desired products in a usable form from the reaction mixture. Although it may seem a gross over-simplification (and a travesty considering the subtleties of these molecules), it is probably the case that greater than 90% of all protein chemistry carried out in practice can be rationalised in terms of the model shown in Figure 2.1.

There are a number of reasons for this: the chemical simplicity of the basic protein backbone and the amino acid building blocks from which it is constructed, the lack of reactivity of many of these molecules, and the severe limitations placed on the reaction conditions which can be employed without destroying the protein's functional activity.

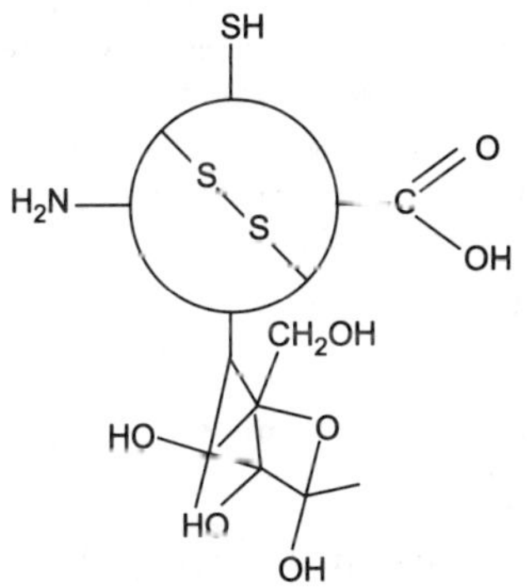

Figure 2.1 A schematic model of a protein molecule showing the functional groups most commonly employed in coupling reactions.

2.1.1 The peptide chain

What makes a protein a protein[1] is the existence of the peptide chain, a condensation polymer derived by a sequence of dehydration reactions between the amine ($-NH_2$) group of one amino acid and the carboxylic acid ($-CO_2H$) group of another, as shown in Figure 2.2. Chemically the result is a monotonous chain of amide –(CO–NH–) bonds; these make up the backbone of the protein and are unreactive under the majority of conditions encountered in protein chemistry. They are generally cleaved only by hydrolysis reactions involving strong acid or alkali, or by digestion with proteolytic reagents. The latter can display a wide range of specificity regarding the cleavage site, and are often employed to generate identifiable fragments in the analysis of peptide structure. Proteolytic enzymes are particularly useful in this respect, while selective chemical methods are a major subject in their own right (Spande *et al.*, 1970).

At each end of the peptide chain there will be an 'unused' functional group: an amine at the N terminus and a carboxylic acid at the C terminus. In the majority of cases these are available for subsequent reaction, chemically speaking at least—although there is no guarantee that their modification will not have damaging effects on protein function.

Sometimes these groups are not present in the 'normal' reactive form (Wold, 1981). In fairly rare cases the C-terminal may be derivatised in some way, for example being present as an amide. Derivatisation of the N-terminal is much commoner, particularly by acylation, and many different examples have been described. Among the best-known are N^{α}-acetylation,

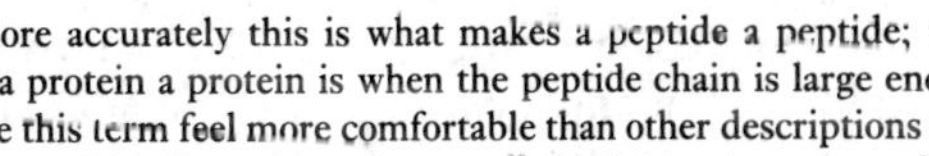

Figure 2.2 The reaction of an amine and a carboxylic acid to form an amide bond. Where the reactants are amino acids, a peptide bond results.

[1] More accurately this is what makes a peptide a peptide; what makes a protein a protein is when the peptide chain is large enough to make this term feel more comfortable than other descriptions such as 'polypeptide'. Even this rather woolly definition is not completely adequate, most workers would hesitate to describe a synthetic polymer of several thousand identical amino acids as a protein. The term generally infers that the structure is either found in nature or at least bears some reasonably close connection to something that is. Many definitions have been stated; Franks (1993b), for example, has suggested that 'protein = peptide + water', to reflect the importance of hydration in maintaining the structure.

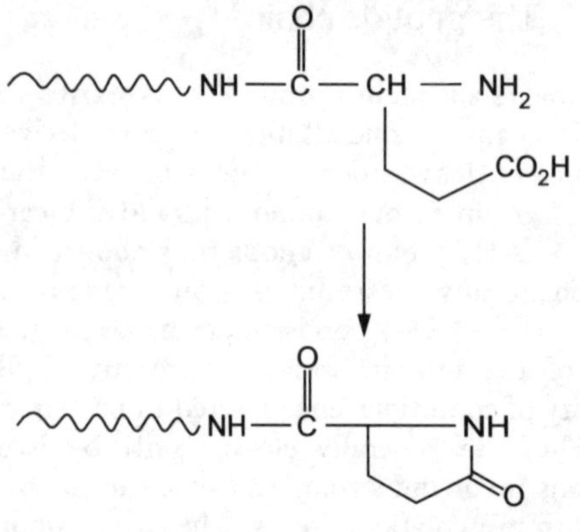

Figure 2.3 The cyclisation of an *N*-terminal glutamic acid residue to yield pyroglutamate.

by far the commonest, and the cyclisation of a terminal glutamic acid group to yield a pyroglutamate moiety (Figure 2.3), which may occur *in vivo* or as an *in vitro* artefact.

Frequently proteins are made up of more than one peptide chain: it appears that the limiting number of amino acid residues in a single chain is around 1000 (Franks, 1993a), and in molecules larger than this multiple chains are present. Therefore a number of terminal acid and amine groups may be present which offer limited potential for coupling purposes. To find other 'handles' for derivatisation reactions it is necessary to consider the individual structure of the amino acid building blocks.

There are countless amino acids in existence, and it is one of the most commonly-quoted 'remarkable facts' in biochemistry how few of them are used in nature to create proteins[1]. Only the twenty amino acids shown in Figure 2.4 are commonly encountered; relative rarities include 4-hydroxyproline, 5-hydroxylysine, 3,5-diiodotyrosine and the polymeric desmosine (Figure 2.5).

The 'normal' amino acids have a number of things roughly in common. Apart from glycine they are all optically active molecules, i.e. they can exist in two different 'mirror image' forms which rotate the plane of polarised light in different directions. Again apart from the inevitable rare exceptions in certain bacteria, they are found in only one of the two possible forms, the L-isomer.

Secondly, they are all α-amino acids, in which the amine and the carboxylic acid are attached to the same carbon atom; therefore the peptide chain described above forms a repetitive $(-\text{CHR–NH–CO–})_n$ structure regardless of which amino acids are present[2]. The R group depends on the side-chain of the amino acid, and these will now be discussed in terms of their functional chemistry. Before doing so it will be useful to consider a little of the theoretical chemistry involved.

2.2 Fundamental Chemical Concepts

Given the structure of amino acids, it is not surprising that their functional chemistry is governed to a large extent by the behaviour of acidic and basic groups. However these apparently simple terms require a little thought in their application.

A number of the most commonly encountered descriptions of chemical reactions and reagents (nucleophilicity and electrophilicity, oxidation and reduction) are closely linked, and the relationship of these will be discussed briefly.

2.2.1 Acid–base concepts: pK_a and protonation

There are numerous ways in which the concept of acids and bases can be viewed, from the classical Arrhenius definition (acids yield H^+ ions (protons) in solution, bases yield OH^- ions), to that of Brönsted and Lowry (acids are proton donors, bases are proton acceptors), to that of Lewis[3] (acids can coordinate with electron pairs, bases can provide them). The Brönsted–Lowry and Arrhenius approaches are basically equivalent under aqueous conditions, obviously the only ones of any real interest to most protein chemists. For this section the Brönsted–Lowry model will be used, but the Lewis concept will be useful later.

Taking relatively trivial examples, ammonia is a classic base in aqueous solution, accepting a proton to give ammonium and hydroxide ions:

$$NH_3 + H_2O \rightleftharpoons NH_4^+ + OH^-$$

while hydrogen chloride forms a classic aqueous acid, donating a proton:

[1] It is sometimes wrongly stated that these are the only amino acids found in nature. In fact, there are many other amino acids with important properties other than protein construction.

[2] Where proline or hydroxyproline is present, the group attached to the nitrogen is actually the 'other end' of the pyrrolidine ring rather than a hydrogen atom.

[3] 1923 was a bumper year for acid–base theories; Brönsted, Lowry and Lewis all independently put forward their models in this year.

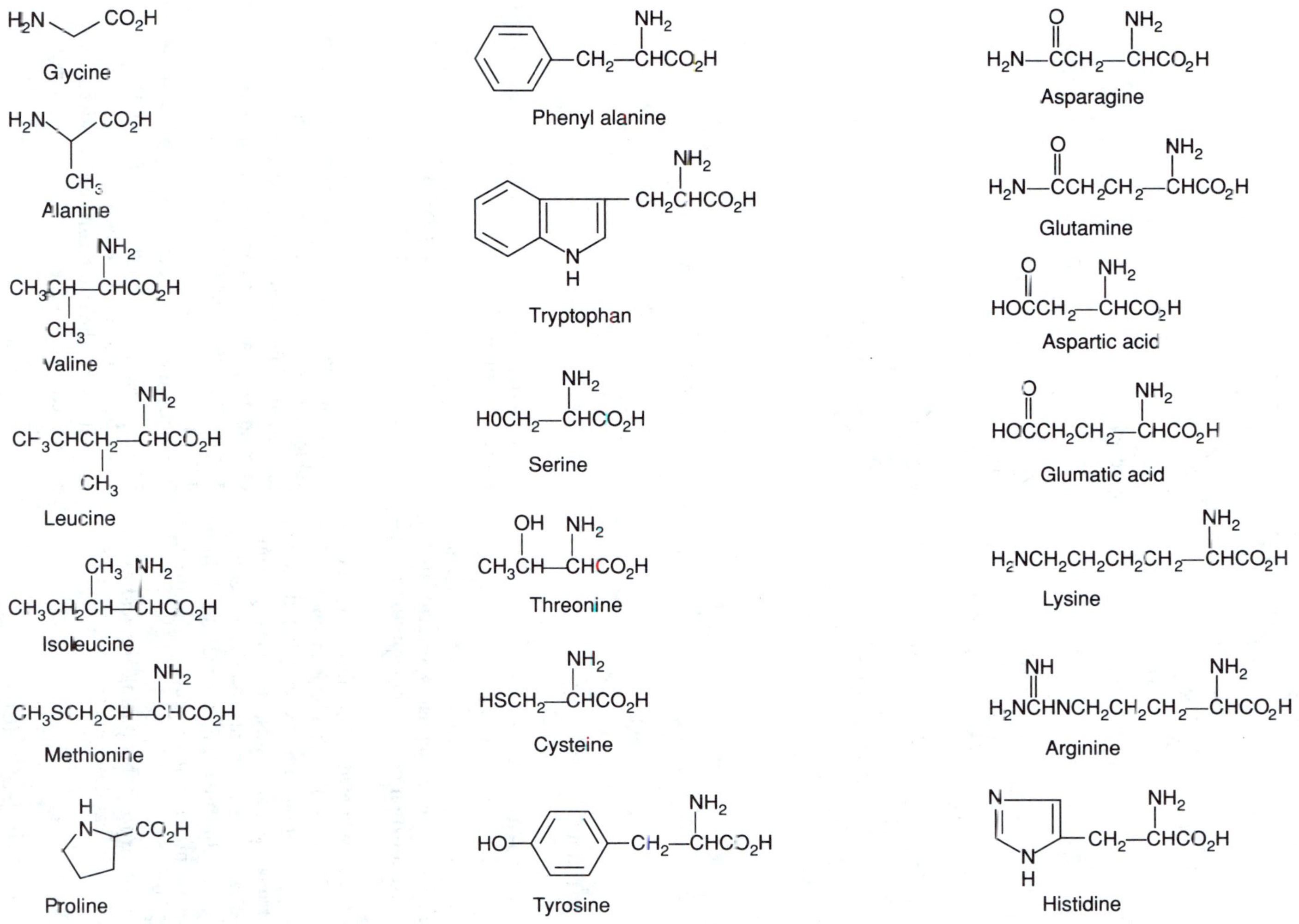

Figure 2.4 The 20 amino acids commonly encountered in proteins.

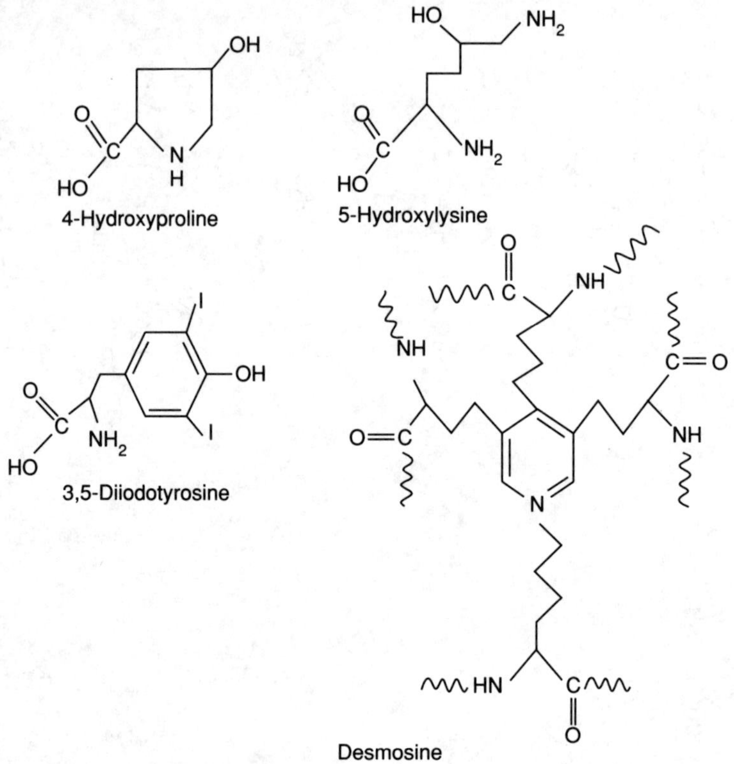

Figure 2.5 Rare amino acids.

$$HCl + H_2O \rightleftharpoons H_3O^+ + Cl^-$$

(Note straight away that the concept of acidity and basicity is relative; water is acting as an acid proton donor in the first equation but as a base proton acceptor in the second.)

These reactions are shown as equilibria, with a constant exchange of individual molecules from one 'side' of the equation to the other. This is true, but these are examples of strong bases and acids, i.e. it is very favourable for them to accept and donate protons respectively. Therefore the equilibrium will lie towards the right in both cases, and the majority of molecules will be present in the ionised form.

Another way to express this (taking the acid example) is to consider the equilibrium constant, K_a, where:

$$K_a = \frac{[H_3O^+][Cl^-]}{[H_2O][HCl]}$$

Because there is effectively a large pool of unionised water on both sides of the equation (and hence at top and bottom of the equilibrium constant expression), the reduction due to ionisation can be ignored and the expression simplifies to:

$$K_a = \frac{[H_3O^+][Cl^-]}{[HCl]}$$

The stronger the acid in water, the larger the K_a value and the more the reaction proceeds towards ionisation. These values extend across many orders of magnitude, and for convenience are often expressed as $-\log K_a$ (pK_a), a strong acid being characterised by a low pK_a.

Exactly the same logic can be applied to the ionisation of bases to give K_b and pK_b, but in practice the 'a' scales tend to be used for both. Under aqueous conditions at 25 °C, the two terms for any given compound are related by the expression:

$$pK_a + pK_b = 14.00$$

Strong acids are characterised by low pK_a values, strong bases by high; the further from 7, the stronger the acid or base. pK_a values are useful for comparing the state of ionisation of different molecules under the same circumstances, but in practice the actual amount of ionisation is often controlled by altering the availability of protons in the system, or altering the pH as it is usually described.

Returning to the expression:

$$K_a = \frac{[H_3O^+][Cl^-]}{[HCl]}$$

in the situation where $[Cl^-] = [HCl]$ (i.e. where the ionised and unionised forms of the acid are present in equal quantities) these terms cancel out, and:

$$K_a = [H_3O^+]$$

or:

$$-\log K_a = -\log[H_3O^+]$$

In aqueous systems $[H_3O^+]$ is equivalent to $[H^+]$, and therefore:

$$-\log K_a = -\log[H^+], \text{ and}$$
$$pK_a = pH$$

Therefore the pK_a of an acid can be thought of as the pH at which the protonated and unprotonated forms are present in equal quantities. This relationship is quantified in the Henderson–Hasselbach equation:

$$pH = pK_a + \log\frac{[base]}{[acid]}$$

It follows from the statement above that acids will tend to be ionised at pH values above their K_a, and unionised at values below it. Exactly the same argument applies to bases, which will tend to be ionised (protonated) at pH values below their pK_a and unionised above it. In practice, as these are equilibrium systems, there is a margin of ± 2–3 pH units either side of the pK_a where there will still be a small but appreciable fraction of the 'minority' form, so these cutoffs are gradual rather than abrupt.

These pH effects are critical in terms of the functional chemistry of acid and base groups, and also determine the charge and reactivity of a protein's constituent amino acids. Figure 2.6 for example shows the behaviour of aspartic acid across the pH spectrum.

(1) Below pH 1, the amine group is protonated but both of the acid groups are almost completely in the unionised form, so there is an overall charge of +1. (2) As the pH passes 1, the stronger of the two acid groups starts to lose a proton[1]. The NH_3^+ group aids this process by attracting electrons; it acts more strongly on the closer α-acid, so it is this group that ionises first. (3) At pH 2.09, the first acid is equally balanced between the ionised and unionised form, i.e. it is at its pK_a. (4) By pH 3 this acid is almost completely ionised, cancelling out the positive charge on the amine and giving a net charge of zero. However by this stage the second, weaker acid is beginning to ionise. (5) At pH 3.86 the second acid is at its pK_a. (6) By pH 7 the second acid is fully ionised[2], so the overall charge on the molecule is −1. (7) At pH 9 the amine group starts to deprotonate. (8) At pH 9.82 the amine is at its pK_a. (9) Above pH 12 the amine is fully unionised, so there is an overall charge of −2.

The isoelectric point (p*I*) of the amino acid is the pH at which the overall charge is zero, i.e. around position (4). The exact value is the mean of the pK_a values surrounding this isoionic region, in this case:

$$(2.09 + 3.86)/2 = 2.98$$

The same situation holds for a protein, with the pK_a of any free ionisable groups affecting the p*I*. This value has an important influence on protein behaviour, and generally falls within a range from around 4 to 10 (Righetti and Carravagio, 1976; Malamud and Drysdale, 1978). Extreme examples do however occur at both ends of the pH range (Franks, 1993c), such as pepsin (p*I* = 1) and salmine (p*I* = 12.1)

It should be understood clearly that the protein is not uncharged at its p*I*—it probably contains a great number of charged moieties, but their overall sum is zero. Because of their high content of ionisable groups, proteins can undergo very significant changes as the

[1] As this is an equilibrium system, even below this point there will be a small fraction of ionised acid.

[2] Note that at the near-neutral conditions, which are generally employed in protein chemistry, the acid is mainly in the anionic form and is often referred to as aspartate to reflect this. The same applies to glutamic acid/glutamate.

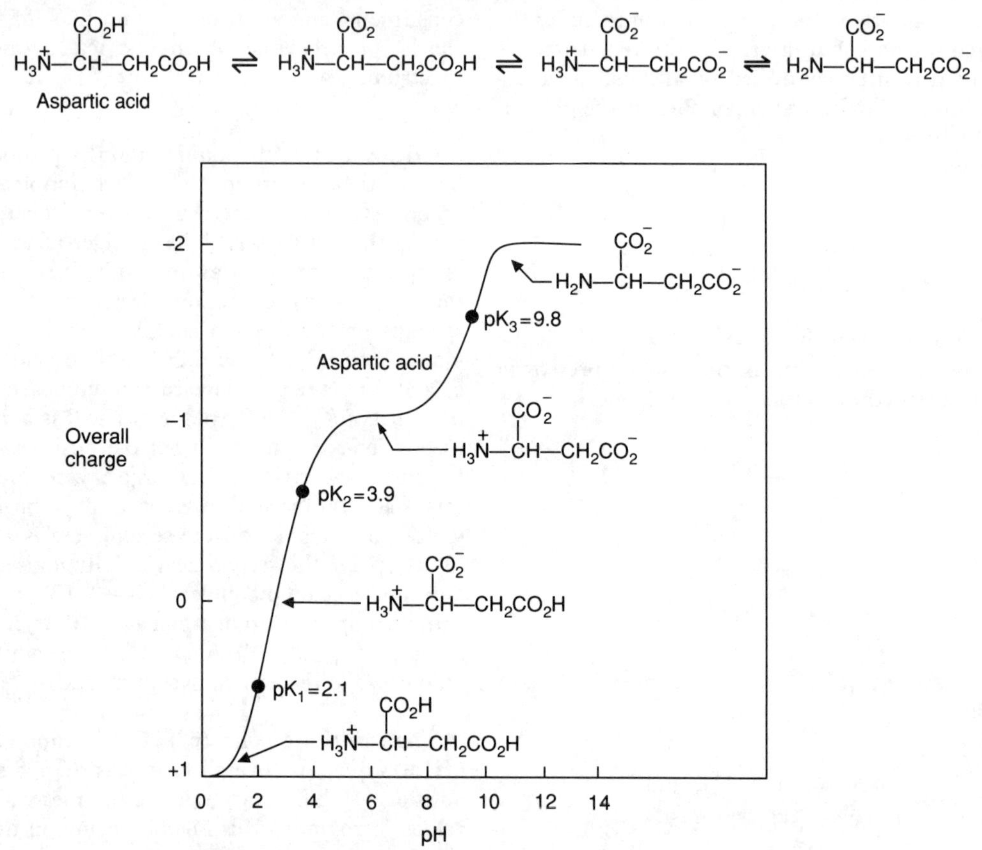

Figure 2.6 The ionisation status of aspartic acid under different pH conditions. The net charge changes from +1 to −2 as the pH rises from 1 to 12. Adapted from Conn and Stumpf (1976).

pH of their environment is altered. A pH titration can be carried out which traces the release of protons from the protein molecule as the pH is raised (cf. Figure 2.6). A typical titration curve is shown in Figure 2.7; if the amino acid sequence of the protein is known this can be used to some extent to interpret the curve. Figure 2.8 shows a stylised representation of the ionisation state of various protein functionalities over a changing pH range.

Due to these effects it is critically important to ensure that protein experiments are carried out under conditions where pH is adequately controlled. This is best achieved by the use of a buffer system, where the equilibrium of a weak acid–base conjugate pair is exploited to counter any pH shifts. Good buffering capacity is only obtained at pH values close to the pK_a, and it is worth taking a little care in choosing an appropriate buffer system. Useful systems are shown in Figure 2.9. Particularly worthy of note are the 'Good' buffers based on zwitterionic molecules (Good *et al.*, 1966), which are less prone to unwelcome interactions with the protein itself than many simple inorganic buffer matrices (Blanchard, 1984).

The effect of temperature on the buffer system should be borne in mind. The change in pH of a buffer with temperature is typically in the range 0.01–0.03 units per °C. Tris buffers, for example, display an upward shift of 0.8 pH unit in moving from 25 °C to 0 °C (Blanchard, 1984).

The pK_a of a particular group in a protein in solution will be modified by the local microenvironment, e.g. the polarity or hydrophobicity of the region. The effect of polarity changes is demonstrated in an extreme example by the effect of adding ethanol to aqueous acetic acid solutions (Lundblad and Noyes, 1984a). While the pK_a in water is an acidic 4.7, this changes to 6.9 in 80% ethanol, and 10.3 in pure ethanol. Therefore it is not always easy to predict *in*

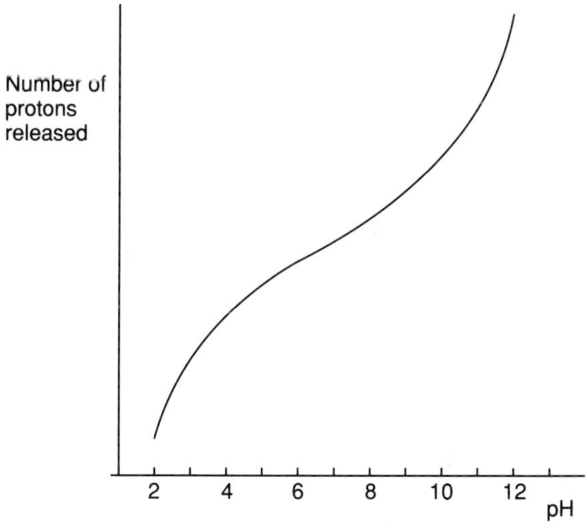

Figure 2.7 A typical titration curve for a protein, obtained by adding increasing quantities of a base to an acidic protein solution.

situ amino acid reactivities accurately. Other effects such as hydrogen bonding and the proximity of charged groups can also influence the effective pK_a.

2.2.2 Nucleophiles and electrophiles

The narrower acid–base concepts (including those of Arrhenius, Brönsted and Lowry) refer specifically to protons. They cannot therefore describe situations where movement of an H^+ ion does not take place, such as the majority of reactions in organic chemistry. As a consequence other terminologies have grown up which are more useful in the classification of reagents for these situations.

A particularly widespread occurrence in organic chemistry is the reaction of electron-rich with electron-poor species, and the terms nucleophile and electrophile respectively are commonly used to classify reagents in this way[1]. For example, the nitrogen atom in an amine group has 'unused' electrons which render it nucleophilic, while the oxygen atom in an aldehyde group attracts electrons away from the carbon to make the latter electrophilic. Amines will donate electrons to the aldehyde carbon atom, initially generating the unstable charged species shown in Figure 2.10a.

Note that while the amine is acting as a nucleophile and the carbon as an electrophile, no protons have moved and there is no concept of an acid–base reaction in the Brönsted–Lowry sense. Returning to the wider Lewis definition of acids as electron acceptors and bases as electron donors, however, the carbon is quite clearly a Lewis acid and the amine a Lewis base[2]. Electrophiles and nucleophiles can therefore be thought of as acids and bases in this sense.

The importance of the interrelationship of pH and pK_a for acids and bases, discussed in the previous section, can now be explained further. Protonation of an NH_2 group to NH_3^+ by lowering pH prevents an amine from acting as a Brönsted–Lowry base, or as a Lewis base or nucleophile—and it is in these roles that such reagents are usually employed in protein chemistry.

Thus it is generally impossible (or at least very slow) to carry out nucleophilic reactions of amines at pH values much below their pK_a. While this is an extremely important factor in choosing reaction conditions, it should not be over-emphasised. Even though at 2 pH units below the pK_a the amount of unprotonated amine is only 1% of the total, this is often quite sufficient to allow it to provide a good yield from a reaction: the equilibrium continually works to provide fresh starting material from the NH_3^+ pool as unprotonated amine is consumed in the formation of the product. Thus nucleophilic reactions of lysyl amines (pK_a around 10) can be carried out quite efficiently at or near neutrality.

The relationship of basicity and nucleophilicity is worthy of some comment—although these two properties correlate to some extent there are important differences between them. Firstly, basicity is an equilibrium property controlled by thermodynamic factors, while nucleophilicity describes the rate of reaction with an electrophilic target, and is thus kinetically controlled (see Section 3.1). In more practical terms the target of a nucleophilic attack is generally an atom forming part of a much larger structure, while for a Brönsted–Lowry base at least, the target is small (H^+, H_3O^+). Therefore a bulky anion may be a poor nucleophile due to steric effects—methoxide and *t*-butoxide (Figure 2.11) are similarly basic but the latter is poorly nucleophilic (Pine *et al.*, 1981).

Another important difference relates to the size of the group. For a series of similar species (e.g. halides) the basicity decreases as the size is increased; the same charge is spread over a wider volume and the negative character is lower at any particular point of approach of a proton. Therefore although fluoride is only a

[1] This terminology was introduced by Ingold in the 1930s.

[2] This product will rapidly rearrange to form an α-amino alcohol (Figure 2.10b); this is an internal Brönsted–Lowry *or* Lewis acid–base reaction, with the acid (nitrogen) accepting electrons from (and donating a proton to) the base (oxygen).

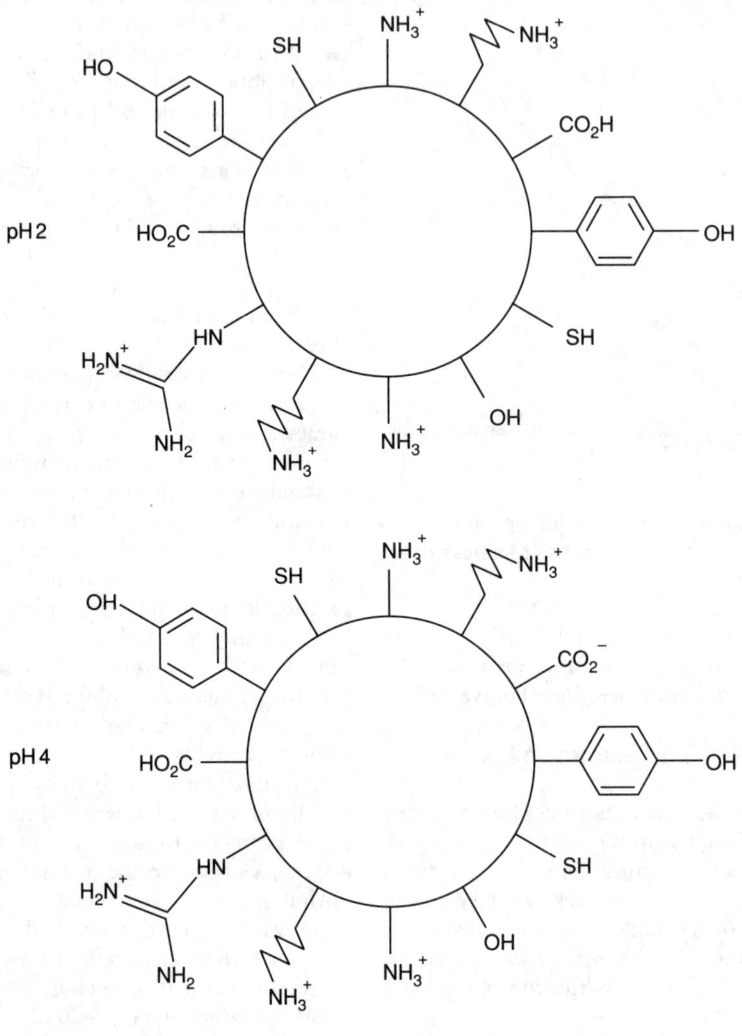

Figure 2.8 See caption page 61.

rather weak base ($pK_a = 3.17$) it is many orders of magnitude stronger than chloride ($pK_a \approx -7$), bromide ($pK_a \approx -9$) or iodide ($pK_a \approx -10$) (March, 1985a).

The wider dispersion of charge also means, however, that the ion is less strongly solvated than a smaller analogue. This factor facilitates reaction with the approaching species, and for this reason nucleophilicity *increases* with the size of a negatively charged group; iodide is therefore the most nucleophilic member of the series just discussed. However this effect is often also observed for uncharged species where solvation effects are negligible, and another explanation is invoked. This is a final acid–base model which has not been touched on yet, the concept of hard and soft acids and bases proposed by Pearson (1963). Many observed reactivities cannot be explained by Lewis basicities alone, and Pearson's approach empirically added an additional 'hardness' or 'softness' rating, hard acids preferring to react with hard bases, and soft with soft. The theory behind this empirical model is by no means defined, but is related to levels of ionic and covalent character. Small species with high charge density (hard) tend to behave in a more ionic way, while larger species with the charge more diffuse (soft) tend to bind less electrostatically and more by covalent mechanisms. Soft groups are more prone to polarisation, the distortion of

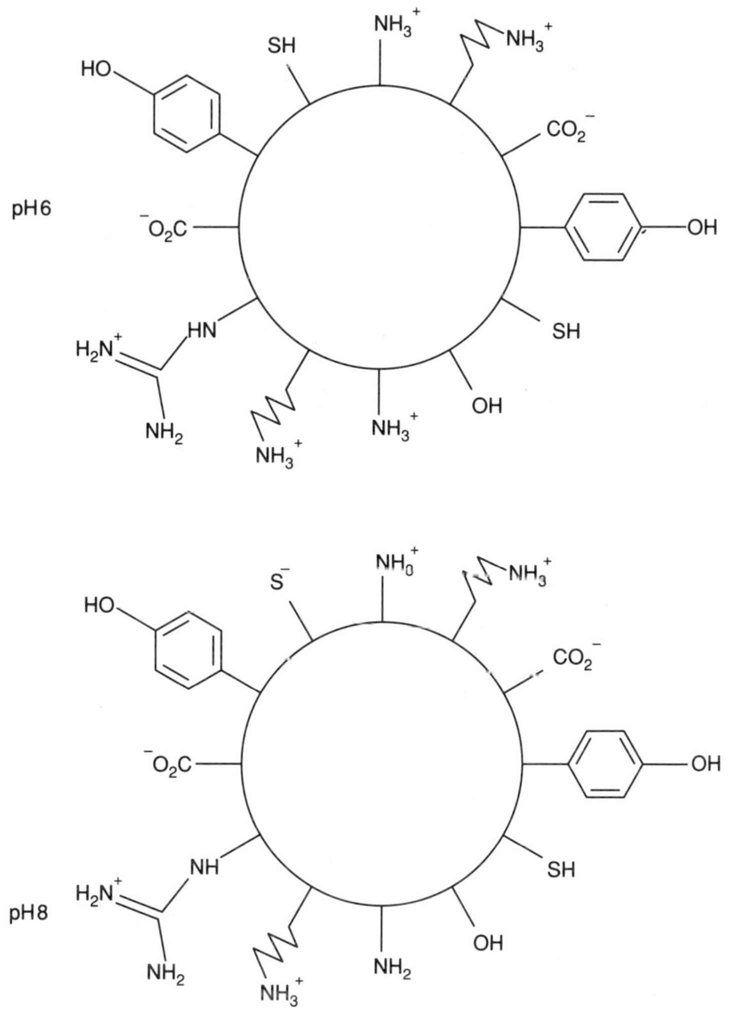

Figure 2.8 See caption page 61.

charge density caused by the approach of an attacking species.

Electrophilic targets in protein chemistry, typically including carbon or sulfur atoms, are predominantly soft in nature. Increasing the size of a nucleophilic group generally increases its softness too, and therefore its reactivity towards the soft electrophilic target in the protein.

Both this factor and the solvation effect noted above cause nucleophilicity to increase with size. Therefore, for example, reagents based on the nucleophilicity of halogen atoms predominantly employ iodine in this role. For the same reason thiolates (RS^-) are stronger nucleophiles than alkoxides (RO^-).

Based on these relationships, some simple rules of thumb can be applied to estimate the relative nucleophilicities of functional groups (Edwards and Pearson, 1962; Bunnett, 1963; March, 1985b). There are many subtle effects involved, however, and it is important not to attach too much importance to these guidelines.

For atoms of comparable size (e.g. substituting atoms from the same row of the periodic table), nucleophilicity and basicity are roughly proportional: therefore a reactive group with high pK_a can provide an indication of high nucleophilicity. The effective nucleophilicity of a reagent will always be higher at a pH above the pK_a, as the more negatively charged species will increasingly predominate.

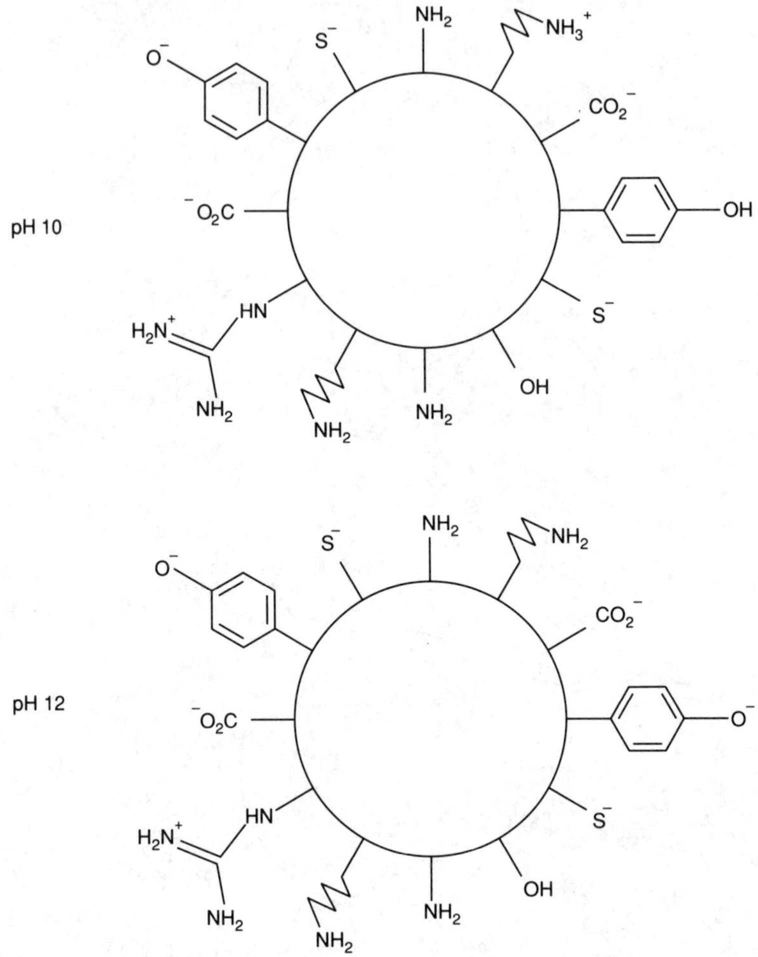

Figure 2.8 See caption page 61.

For atoms of comparable electronic status (e.g. substituting atoms from the same column of the periodic table), nucleophilicity increases as the size of the atom increases.

A rough order of nucleophilicity with some relevance for protein work can be expressed as follows (March, 1985b):

$$RS^- > CN^- > I^- > OH^- > N_3^- > Br^- > ArO^- > AcO^- > Cl^- > H_2O$$

However, this must be viewed in the light of the hard/soft model just discussed, relatively hard pairs such as amine/carboxylate and relatively soft pairs such as thiolate/maleimide will therefore tend to show higher reactivity than hard/soft combinations. Other factors may also affect this order, such as the nature of the solvent.

2.2.3 Oxidation and reduction

The definitions of oxidation and reduction have, like those of acids and bases, undergone changes over the years (Ingold, 1953). Oxidation originally described simply the addition of oxygen, e.g. atmospheric oxidation of metals, or conversion of an alcohol ($R-CH_2-OH$) to a carboxylic acid ($R-CO_2H$). This was broadened to include salts in which the role of the oxide was taken by other anions, so that the concept was no

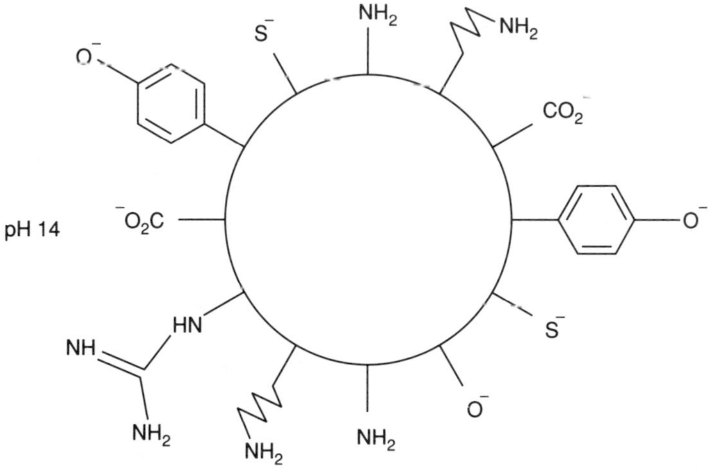

Figure 2.8 Schematic diagrams of a protein molecule, showing the change in ionisation status of functional groups under different pH conditions. Where a group is shown in two different forms at the same pH, both forms are present in appreciable quantities.

longer limited to oxygen-containing compounds. Once the electronic nature of atomic interactions was understood, the definition was widened further to include exchanges of electrons between ionic species. Looking at these reactions in reduction/oxidation (redox) terms has now become commonplace, e.g. the ferrous iron(II) ion being oxidised to ferric iron(III), or reduced to metallic iron.

It is apparent how close this is to the Lewis theory of acids and bases, and to Ingold's notion of nucleophiles and electrophiles. The ferrous ion is clearly acting as a nucleophile or Lewis base (electron donor) when it is converted to the ferric state, and as an electrophile or Lewis acid (electron acceptor) in going to the metallic form. Loosely speaking, reduction–oxidation terminology tends to be used more in situations where there is no covalent bond formed between the reacting species. There are exceptions, however—the incorporation of a thiol into a disulfide is generally described as an oxidation reaction.

All this may seem a long way removed from the practicalities of protein chemistry, but the point is that there is no fundamental concrete difference between the concepts of acids and bases, nucleophiles and electrophiles, and oxidising and reducing agents. These terms are used somewhat arbitrarily in different circumstances where one feels more comfortable than another, but they all refer fundamentally to the electronic interactions which underpin all chemical reactions.

2.2.4 Delocalisation of charge

Another chemical concept which facilitates understanding of some common reaction mechanisms, and has an impact on the strength of acidity and basicity, is that of delocalisation of charge.

The likelihood of existence of a particular molecular structure is partially governed by the relative energy content of this arrangement of atoms compared to other possible conformations. Wherever an atom is disturbed from its 'normal' elemental state by the introduction of charge, either negative or positive due to the addition or removal of electrons respectively, there is generally an increase in its local instability.

However, electrons are not discrete, localised particles; their presence should instead be thought of as diffused over a three-dimensional orbital, with the relative likelihood of the charge being found in a particular position controlled by probability functions. This diffuse nature also allows charge to be 'spread out' over different atomic positions in a molecule, and this delocalisation of the charge can have a dramatic effect on the stability of a particular molecular structure. Simply, if charge can be spread over two or more atoms instead of one, there is a large energy advantage and the structure is more likely to exist.

If a molecular structure can be drawn in more than one acceptable form simply by 'moving' electrons (not the relative positions of atoms), these are known as canonical or resonance forms, and indicate the existence of delocalisation. Different situations involving

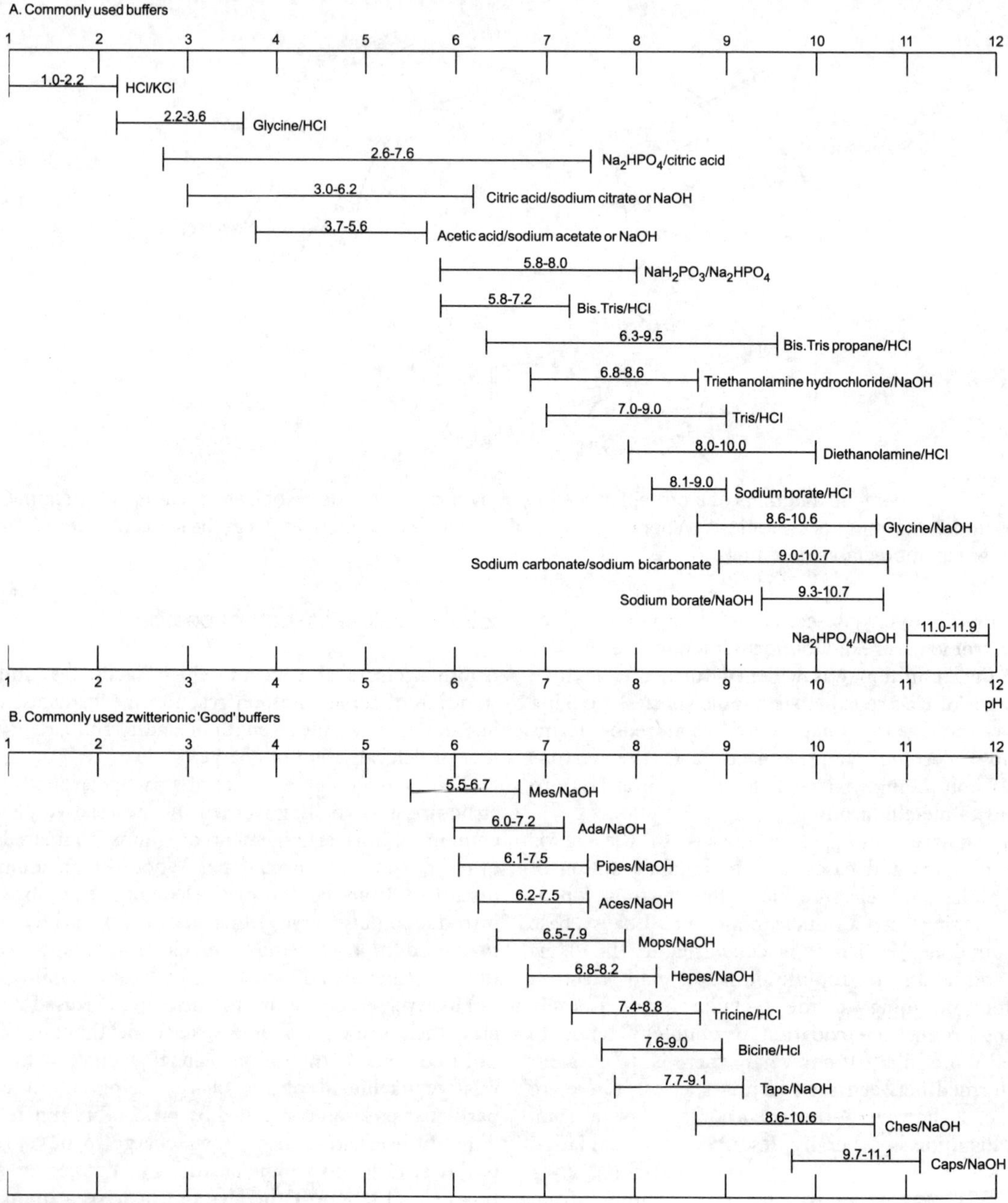

Figure 2.9 Buffer systems suitable for use over various pH ranges. A, Commonly-used buffers; B, zwitterionic 'Good' buffers. Mes, 2–(*N*-morpholino)ethanesulfonic acid; Ada, *N*-(2-acetamido)iminodiacetic acid; Pipes, piperazine-*N*, *N'*-bis-2-ethanesulfonic acid; Aces, *N*-(2-acetamido)-2-aminoethanesulfonic acid; Mops, 3-(*N*-morpholino)propanesulfonic acid; Hepes, *N*-2- hydroxyethylpiperazine-*N'*-2-ethanesulfonic acid; Tricine, *N*-tris(hydroxymethyl)methylglycine; Bicine, *N*, *N*-bis(2-hydroxyethyl)glycine; Taps, *N*-tris(hydroxymethyl)methyl-3-aminopropanesulfonic acid; Ches, 2-(cyclohexylamino)ethanesulfonic acid; Caps, 3-(cyclohexylamino)-1-propanesulfonic acid. (Reproduced with permission from: Harris, E. L. V. (1989). In *Protein Purification Methods: A Practical Approach* (Harris, E. L. V. & Angal, S., eds), pp. 6–7, IRL Press Oxford.)

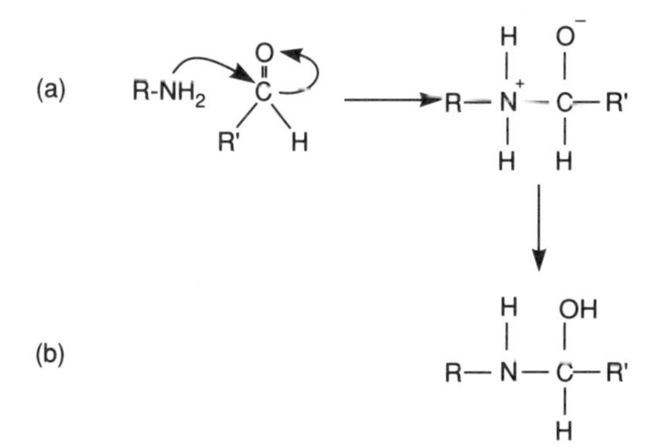

Figure 2.10 A typical reaction of a nucleophile (or Lewis base) with an electrophile (or Lewis acid). The free electrons of the amine nitrogen atom attack the electron-deficient carbon atom of the aldehyde (a). The product rearranges to an amino alcohol (b).

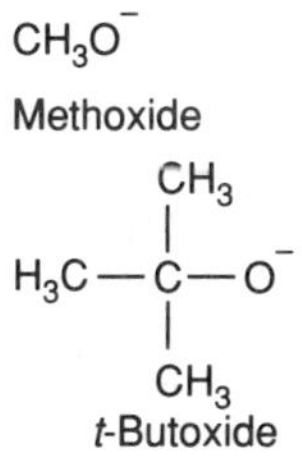

Figure 2.11 Steric effects on nucleophilicity. The bulk of the *t*-butoxide ion prevents it from behaving as a good nucleophile.

carbon–oxygen bonds are instructive (see Figure 2.12). Isolation of a negative charge on the single oxygen atom in methoxide puts a high energy burden on the structure, which is one reason why simple alcohols are unwilling to ionise in this way, and (to put it another way) are extremely weak acids.

In the case of phenol, it is possible to draw five arrangements of the electrons without disturbing the atomic links, indicating delocalisation, and the phenoxide anion is correspondingly much more stable. However, carbon atoms carry a higher energy penalty than oxygen for holding a negative charge, so the three forms containing negatively charged carbon atoms contribute only weakly to the 'average' stability. Consequently phenol is still only a weak acid, although much stronger than methanol.

Where the charge can be delocalised over a number of equivalent atoms, the stabilisation effect is greatest; thus in acetic acid, the two canonical structures for the acetate anion confer a high level of stability. Consequently this is an energetically favourable arrangement of atoms, the ionisation is likely to occur, and this is a strong acid. These delocalisation effects are reflected in the pK_a values of methanol (16), phenol (9.95) and acetic acid (4.76).

It should be noted that none of the canonical structures drawn above really exist at any given time, they simply contribute to a hybridised 'probability map' for electron location. It is also worth pointing out that the structures illustrated are only the most favourable of a multitude of possibilities. For phenol for example it is quite possible to draw structures such as those in Figure 2.13, and they are not excluded from making a contribution to delocalisation; however, they are so energetically unfavourable that they are unlikely to occur very often, and their 'existence' has a negligible effect on the stability of the structure.

One other factor determining the influence of delocalisation should be noted; the electronic orbitals can only hybridise if they are roughly in the same geometrical plane as each other. Sometimes the steric effect of a bulky substituent on a molecule can twist the structure to prevent this happening, and the arrangement will be less energetically favoured than might be expected at first glance.

2.2.5 Electronegativity and polarity

It has already been mentioned in explaining some of the key effects above that some atoms such as oxygen show a tendency to attract electrons, a property defined as electronegativity by Linus Pauling. Pauling's exact definition was the power of an atom in a molecule to attract electrons to itself; the atom in isolation does not exert this effect, but the difference in electronegativities between atoms in a molecule will help to define the electron distribution within it.

The origins of electronegativity relate to the electrostatic force exerted by the atom's nucleus on the surrounding electrons, and various approaches have been developed for calculating values for this parameter. The scale derived by Allred and Rochow in 1958 is widely accepted, and a number of Allred-Rochow electronegativities are quoted from Huheey (1975): carbon (2.50), hydrogen (2.20), oxygen (3.50), nitrogen (3.07), sulfur (2.44), chlorine (2.83), fluorine (4.10).

The higher the differential between two atoms, the more the electrons will be attracted to the atom with the higher value.

To estimate the overall, three-dimensional distribution of charge on a molecule, it is necessary to consider the symmetry of the structure. In Figure 2.14(a),

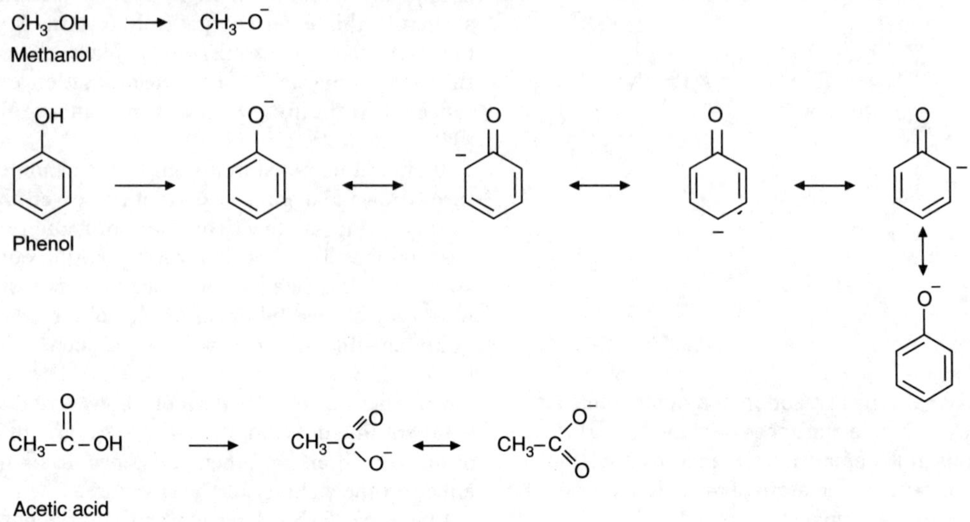

Figure 2.12 Canonical forms of methoxide, phenoxide and acetate.

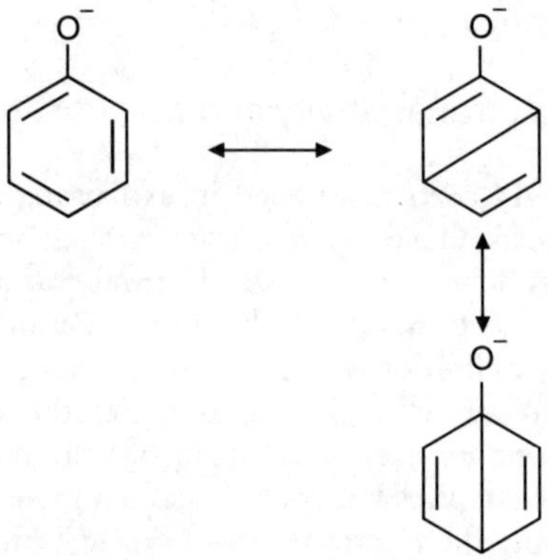

Figure 2.13 Energetically unfavoured canonical forms of phenoxide.

methane contains four hydrogen atoms arranged in a symmetrical tetrahedron around a carbon atom; the carbon is slightly more electronegative and therefore the electrons will tend to be pulled in towards the centre of the molecule. In tetrafluoromethane (Figure 2.14(b)) the hydrogen atoms are replaced by highly electronegative fluorines, and the electrons will tend to be pulled *away* from the central carbon. However in both these cases the distribution is completely symmetrical, and there is no 'end' of the molecule which holds more electrons than the other.

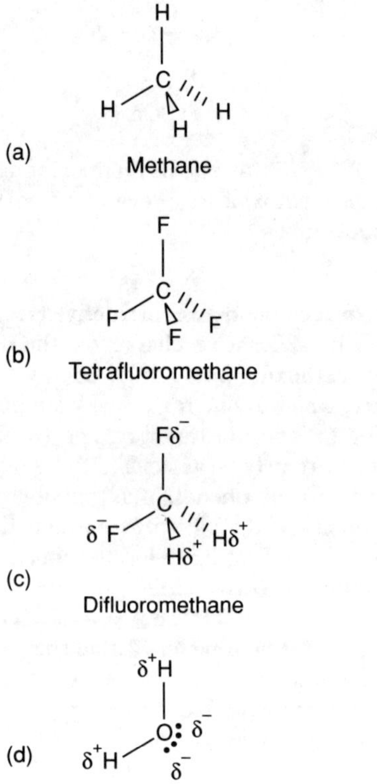

Figure 2.14 Charge distribution and polarity in methane, fluoromethanes and water.

Turning to difluoromethane (Figure 2.14(c)), there is a 'pull' of electrons from the hydrogen atoms to the carbon atom and from the carbon atom to the fluorine atoms. Therefore the fluorine end is more negative and the hydrogen end more positive (any two positions on the tetrahedron form an 'end' compared to the other two). The molecule is said to possess a dipole moment and is more polar than either of the preceding cases.

Polarity is also strongly affected by the presence of electron pairs on an atom which are not involved in bonding to other atoms, lone pairs as they are termed. The oxygen atom in water for example (Figure 2.14(d)) has two lone pairs, the other two positions in the (approximate) tetrahedron being occupied by hydrogen atoms. The electrons from the hydrogen end are attracted towards the highly electronegative oxygen atom, while the other end consists of the lone pair electrons anyway; the combination of these effects gives a very strong dipole moment, and water is an extremely polar solvent[1].

2.2.6 Leaving groups

As has been seen in Section 2.2.2, the relative nucleophilicity and electrophilicity of two potential reactive species are major factors in determining whether a coupling reaction will or will not take place. Often, however, there is another parameter which will have an influence—the requirement for a group to be expelled during the reaction.

Take, for example, the formation of a peptide bond between an amine and a carboxylic acid residue (Figure 2.2). Dehydration must occur to create the amide product *via* expulsion of a hydroxide ion, but this would require the negative charge to be carried on the rather small OH group. In practice, therefore, this reaction does not occur spontaneously. *In vivo*, enzymes have been evolved which catalyse this reaction, and these form vital links in the protein biosynthesis chain. Enzymic ap-proaches can be employed *in vitro* too, but it is also possible to influence the outcome of this reaction chemically.

The hydroxide ion in the example just discussed is a leaving group; it must leave the molecule in order for the desired reaction to take place. Increasing the ability of this group to carry a negative charge will increase the likelihood of the reaction taking place. It may be possible to do this by altering the environment. For example, the pH is lowered until the OH group is protonated to OH_2^+, which can then accept an electron to leave as neutral H_2O. This may require pH conditions too low to be practicable, however.

It is also possible to replace the OH group in the acid by another moiety that is better suited to carrying the negative charge. For example, a halogen atom can leave as a stable halide ion, while groups such as succinimide or nitrophenyl esters have electron-withdrawing components which help to disperse the charge (Figure 2.15). Thus reagents such as acyl chlorides and succinimidyl esters are capable of spontaneous reaction with amines, and are widely employed for coupling to amine moieties. Other groups such as *p*-toluenesulfonates (tosylates) and trifluoromethanesulfonates (triflates) are familiar in this role, more commonly under non-aqueous conditions (March, 1985b). These all expel very weak bases: as a rule of thumb the more basic the species to be expelled, the worse it will act as a leaving group for nucleophilic substitution.

Nucleophilic reactions employed in protein coupling can therefore be influenced by a combination of two main factors: the nucleophilicity of the attacking group, and the presence of a good leaving group to transport negative charge away from the molecule.

2.2.7 Hydrophilicity and hydrophobicity

Because water is such an important solvent in chemistry—to the virtual exclusion of all others in biochemistry—the terms hydrophilic and hydrophobic have been coined to describe molecular structures which attract and repel water respectively. These correlate by and large with polarity and non-polarity respectively, arising from the propensity of one polar molecule to associate with another (such as water) due to charge interactions.

Calculation of polarities for larger molecules is a more approximate process, but the overall effect of electronegativity and lone-pair electrons will dominate the result.

The effect of these and other factors on intra- and intermolecular bonding is immensely important in protein chemistry. Additionally, it can influence the design of reagents for protein derivatisation, e.g. the use of probes of different polarity to investigate cell membranes, which are hydrophobic in nature. Unsubstituted bis-succinimidyl esters are relatively hydrophobic and are able to enter the membrane; analogues possessing sulfonate groups, on the other hand, are too polar to do so, and can accordingly be used for selective derivatisation of surface species (Staros, 1988).

[1] If the molecule had a linear H–O–H structure the charge distribution would be symmetrical, and water would be a non-polar solvent.

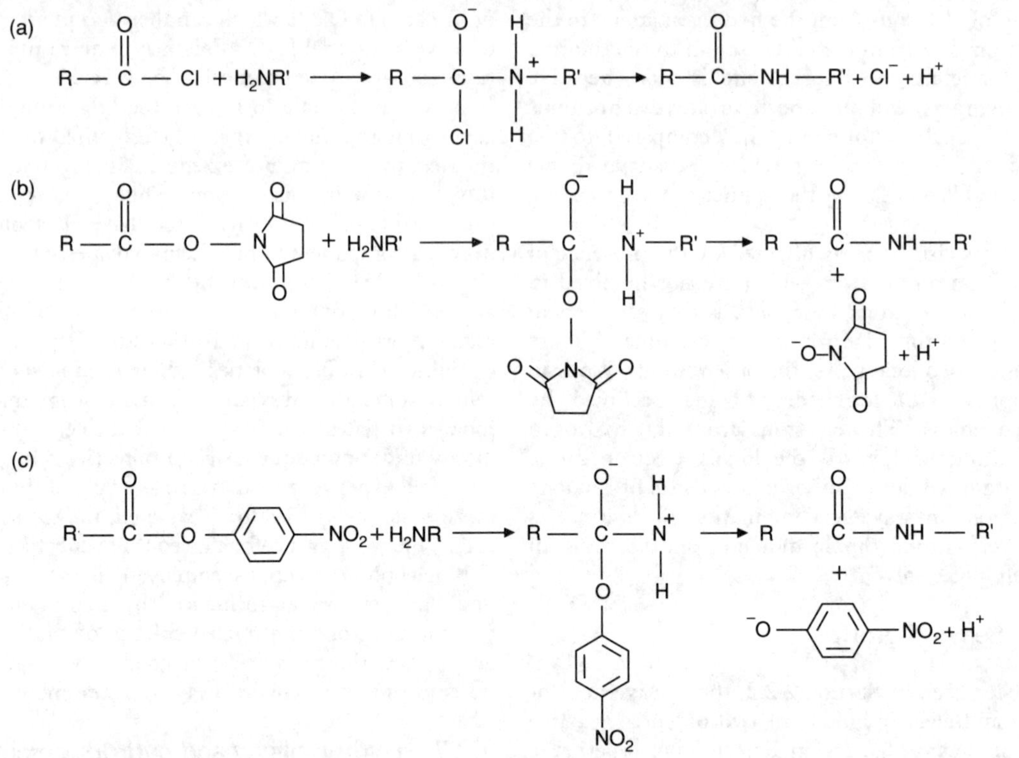

Figure 2.15 The reaction of various activated carboxyl functions with amines. All of these provide a better leaving group than the OH ion expelled from the native acid.

2.2.8 Types of non-covalent intermolecular bonding

Apart from the amide bonds which comprise the peptide chain, the only other common covalent links between amino acids in native proteins are disulfide bonds formed between the thiol groups of two cysteine residues. Much of this book will be devoted to the introduction of artificial covalent bonds, but there are many aspects of protein structure and behaviour which rely on other, non-covalent, interactions. While these are less robust in energetic terms, their influence is enormous.

The key categories of intramolecular bonding in proteins are shown in Figure 2.16.

2.2.8.1 Electrostatic forces

As noted in the discussion of p*I* above, a protein molecule may (depending of the acidity of the surrounding medium) contain a number of positively and negatively charged groups. It is axiomatic that opposite charges attract and like repel; therefore strong electrostatic or coulombic interactions can take place between oppositely charged atoms.

2.2.8.2 Hydrogen bonding

A weaker analogue of electrostatic bonding takes place between the positive and negative ends of polar molecules, weaker because shifts in electron distribution are involved rather than transfers of 'whole' electrons. Particularly in biochemical systems, hydrogen atoms tend to provide the positive regions of molecules and the term hydrogen bonding is used to describe this effect. Often the hydrogen atom concerned possesses positive character due to the presence of an electronegative neighbour such as oxygen or nitrogen, although more recently it has been demonstrated that those attached to carbon can also take part in such interactions (Wahl and Sundaralingam, 1997). The negative partner in the hydrogen bond is also often

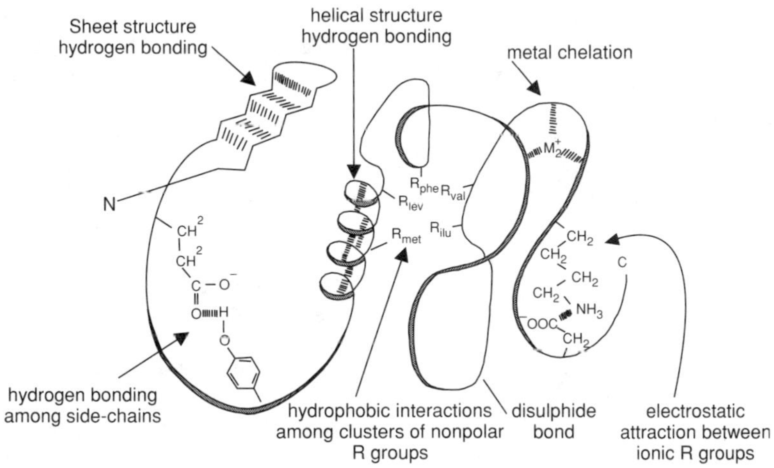

Figure 2.16 Types of bonding in proteins.

oxygen or nitrogen, atoms containing lone-pair electrons.

Hydrogen bonding, although of low energy compared to covalent links, is crucially important in protein molecules—especially in influencing the secondary structure. Arrangements of amino acids which maximise the potential for hydrogen bonding between the peptide-chain regions are often energetically favoured, and this fact is responsible for the widespread existence of regular structural motifs in proteins (Section 2.5.2).

2.2.8.3 Van der Waals forces

These are ill-defined weak forces which operate between the electron clouds surrounding atoms; their key property is their extreme distance-dependence, being proportional to $1/r^7$ where r is the separation between the two partners. They may have a role in strengthening bonds over short distances once an initial approach has been promoted by other forces.

However for proteins in an aqueous environment their effect is fairly insignificant, because the values of the Hamaker constants which control the magnitude of these interactions are similar for water and proteins (Andrade and Hlady, 1986; Norde, 1986).

2.2.8.4 Hydrophobic interactions

Polar water molecules can associate with each other in a great number of different arrangements, i.e. there is a high degree of disorder and the entropy is high. Introduction of a hydrophobic surface provokes the existence of more ordered arrangements, which allow the contact between the two components to be minimised. This increase in order decreases the entropy, and is therefore unfavourable in free energy terms, a key factor in the poor water solubility of hydrophobic molecules.

Where there are two hydrophobic moieties which must be hydrated in this unfavourable manner, an energetically-favourable increase in entropy can be achieved by the coalescence of the two components—the order around the combined product is less than the sum of the individual parts (Figure 2.17). This effect is clearly observable, for example, in the tendency of oil drops in water to coalesce.

These hydrophobic interactions are therefore widely accepted to result from the minimisation of the repulsion between water and a number of apolar components, rather than any attraction between the components themselves—although the topic has attracted considerable debate (Srinivasan and Ruckenstein, 1980; Porath, 1986). Alternative arguments based on Van der Waals attractions between the protein and the surface have been strongly refuted, for example by Franks (1993d).

Hydrophobic interactions are extremely important in protein behaviour for a number of reasons: they influence the arrangement of amino acid residues within the protein molecule and provide one of the strongest mechanisms for maintaining its three-dimensional structure. They are also primarily responsible for driving phenomena such as the physisorption of protein onto hydrophobic surfaces (Chapter 8, Section 8.2; Chapter 10, Section 10.7).

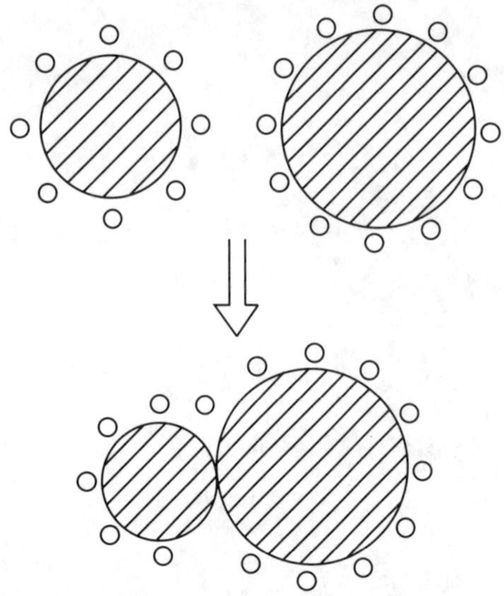

Figure 2.17 Hydrophobic interactions. The coalescence of two hydrophobic moieties (large shaded circles) in an aqueous environment results in a reduction in the number of water molecules (small open circles) forced into ordered arrangements.

2.3 Functional Chemistry of Protein-forming Amino Acids

The twenty amino acids commonly found in proteins possess a wide range of different side-chain functions, as shown in Figure 2.4. For convenience these are often referred to by abbreviations, such as the three letter codes proposed by the IUPAC-IUB Commission on Biochemical Nomenclature (1972). These are still viewed as too cumbersome for some applications, and single-letter codes have also been adopted; both sets of abbreviations are shown in Table 2.1.

There are many ways in which these amino acids can be categorised, based on parameters such as size, charge, hydrophobicity, etc. However, there is no single classification which is universally accepted.

A useful Venn diagram representation has been given by Fothergill-Gilmore (1993) and is reproduced in Figure 2.18. One of the most important distinctions in terms of protein structure is that between hydrophobic and non-hydrophobic residues, as the former are likely to be concentrated in the core of the protein where the need for hydration is minimised. Of the amino acids categorised as hydrophobic by Fothergill-Gilmore some are more so than others, and classifications vary on this point. Wong (1991a) for example included only seven residues (valine, leucine, isoleucine, methionine, phenylalanine, tryptophan and proline) in this category, while Franks (1993a) viewed just the first five of these as truly hydrophobic.

Rather than indicating any serious disagreement over an absolute definition, these differences should be taken as reflecting the real situation—that properties, such as hydrophobicity, vary over a continuum with no clear cut-off between one class and another. Under different circumstances the same amino acid may display either hydrophobic or hydrophilic aspects of behaviour.

An empirical categorisation noted by Franks (1993a) is that the amino acids can also be divided between common and rare: the former include residues such as glycine, alanine and leucine which occur in large num-

Table 2.1 Abbreviations for amino acids

Amino acid	Three-letter abbreviation	One-letter symbol	Amino acid	Three-letter abbreviation	One-letter symbol
Alanine	Ala	A	Histidine	His	H
Arginine	Arg	R	Isoleucine	Ile	I
Asparagine	Asn	N	Leucine	Leu	L
Aspartic acid	Asp	D	Lysine	Lys	K
Asparagine or aspartic acid	Asx	B	Methionine	Met	M
			Phenylalanine	Phe	F
Cysteine	Cys	C	Proline	Pro	P
Glutamine	Gln	Q	Serine	Ser	S
Glutamic acid	Glu	E	Threonine	Thr	T
Glutamine or glutamic acid	Glx	Z	Tryptophan	Trp	W
			Tyrosine	Tyr	Y
Glycine	Gly	G	Valine	Val	V

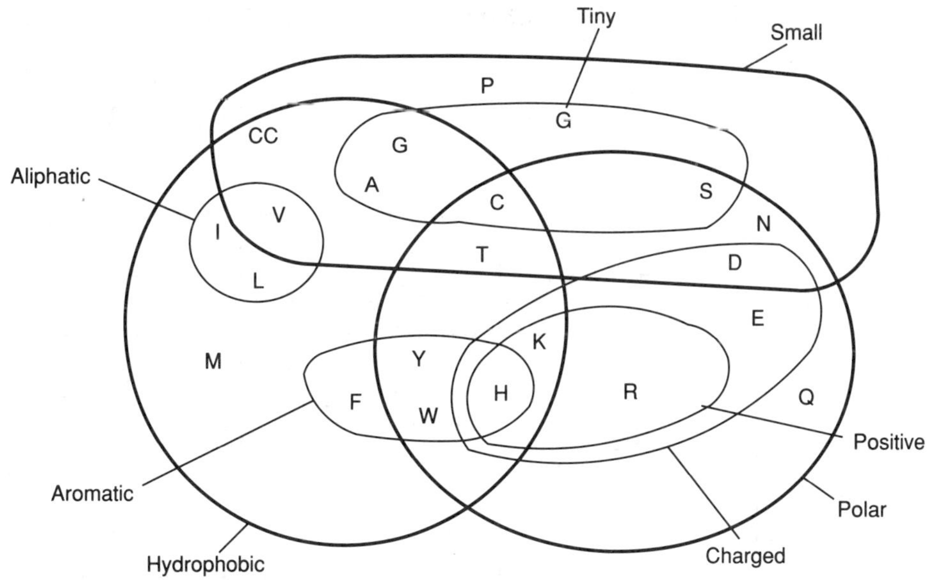

Figure 2.18 Amino acids grouped according to a selection of physico-chemical properties that are important for protein function and for protein sequencing. C, Indicates the reduced thiol form of cysteine; CC, Indicates the oxidised disulfide form, cystine. (Reproduced with permission from: Fothergill-Gilmore, L. A. (1993). In *Protein Biotechnology. Isolation, Characterisation and Stabilisation* (Franks, F., ed.), p. 257, Humana Press, Totowa, NJ.)

bers in most proteins, while the latter (e.g. histidine, tryptophan, methionine) are typically much less widely represented.

The focus of this book, however, is on the functional chemistry of proteins, and they will therefore be categorised here according to their reactive groups.

Only a brief summary of important reactions will be given here: Table 2.2 summarises the major reactivities of preparative interest, along with significant side-reactivities. More detailed information will be found in later chapters.

2.3.1 Amine groups

Amine groups in proteins are provided by the terminal amine if this has not been derivatised, and by the ϵ-amino groups of lysine.

Lysine is probably the most generally useful amino acid in terms of side-chain reactivity. The ϵ-amino group (i.e. four carbon atoms distant from that containing the acid function) is highly reactive and the $(-CH_2-)_4$ chain acts as a convenient spacer to distance the reaction site from the protein molecule.

As pointed out earlier, amine groups are electron-rich and nucleophilic (Lundblad and Noyes, 1984a). Shielded by the alkyl chain from any deactivating influences, the ϵ-amino group in lysine behaves as a typical amine; it is a strong base with a pK_a in proteins of 9–9.5 (cf. ammonia, 9.25; methylamine, 10.64). Therefore it is protonated and unreactive under non-basic conditions (see Section 2.2.1) and amine reactions are typically carried out in the pH range 7–9. More alkaline conditions would favour amine reactivity but are usually discounted to avoid any damage to protein molecules.

In the case of the terminal amine groups discussed in the previous section, the carbonyl group on the adjacent carbon atom (see Figure 2.19) tends to withdraw electrons and reduce basicity and nucleophilic reactivity[1]—pK_a values of 7.5–8 are obtained depending slightly on the amino acid concerned.

As a consequence the terminal amine will be present in the deprotonated form at somewhat lower pH than those of lysine. Some selectivity towards derivatisation of terminal amines can therefore be obtained by working under less basic conditions (see for example, Rana and Meares, 1990; Sélo *et al.*, 1996).

Alternatively, the more reactive amines can be reversibly blocked using reagents such as citraconic anhydride (Dixon and Perham, 1968), methyl acetimidate or the succinimide ester of (methylsulfonyl) ethyl

[1] The stability of peptide bonds mentioned earlier is a consequence of the strength of this effect where there is no intervening carbon atom: the NH_2 group in amides is hardly basic at all.

Table 2.2 Summary of Coupling Chemistries Applicable to Various Protein Functionalities

	(P)–NH_2	(P)–SH	(P)–CO_2H	(P)–CHO	(P)–OH	(P)–His	(P)–Tyr	(P)–Trp	(P)–Arg	(P)–Met	Non-selective
RNH_2			C	✓							
RSH		✓		(✓)							
RCO_2H	C					CH	CH				
RCOCl	✓				(✓)	H	H				
RCO–O–COR	✓				(✓)	H	H				
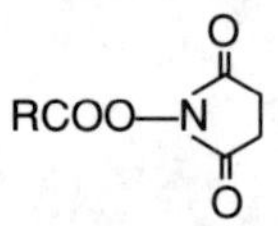	✓				(✓)	H	H				
R–S–S–R		✓									
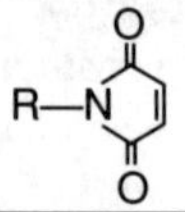	✓	✓				(✓)					
ArHal	✓	✓				✓	✓				
R–N=C=Z	✓	H			(✓)	H	H				
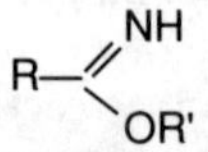	✓										
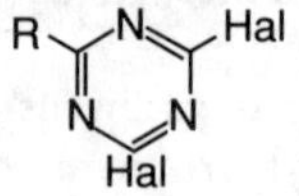	✓				✓						
$ArN{=}N^+$	✓	(✓)			(✓)	✓	✓	(✓)	(✓)		
R–CHO	✓	(✓)				(✓)	(✓)				

carbonate (Mikolajczyk *et al.*, 1996). This allows the *less* reactive amines to be targetted for derivatisation.

Derivatisation of protein amine groups will be encountered frequently in later chapters; the basic chemistries of these reactions, typically involving electrophilic attacking groups with good leaving groups, are summarised in Figure 2.20.

In protein conjugation terms there has been a historical development in the use of these reactions. Early (pre-1970s) methods employed a range of fairly non-specific chemistries such as aryl halides, isothiocyanates, etc. In the 1970s aldehyde chemistry was the dominant approach[1], and in the 1980s and 90s methods based on succinimide esters have come to dominate.

Other chemistries have been important in particular applications: for coupling with small carboxylic molecules carbodiimide methods have remained popular for many decades; coupling to glycoproteins is often

[1] Although the chemistry of the principal reagent of this category, glutaraldehyde, is very complex: see Chapter 5.

Table 2.2 (*contd.*)

	Ⓟ—NH_2	Ⓟ—SH	Ⓟ—CO_2H	Ⓟ—CHO	Ⓟ—OH	Ⓟ—His	Ⓟ—Tyr	Ⓟ—Trp	Ⓟ—Arg	Ⓟ—Met	Non-selective
HCHO	✓	(✓)				(✓)	(✓)	(✓)	(✓)		
R—N=N	▽	▽	✓		▽		(✓)				▽
$RNHNH_2$			C	✓							
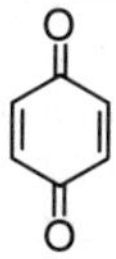	✓	✓			✓			✓			
R—CO—CH_2Hal	✓	✓				✓	✓	✓		✓	
$NaIO_4$		×			✓	×	×	×		×	
R—CO—CO—R									✓		
	✓	✓					✓				
hν		×				(×)	(×)	(×)		×	
H_2OI^+ (iodination)	H	H				✓	✓	×			
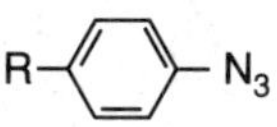	▽	▽			▽	▽	▽	▽			(▽)
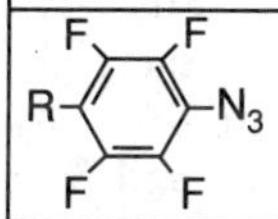	▽	▽			▽	▽	▽	▽			▽
R—SO_2Hal	✓	✓			✓	✓	✓				
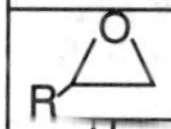	✓	✓	(✓)		✓	✓	✓				
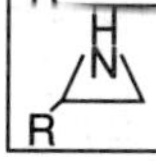	✓	✓								✓	

✓ Suitable coupling method

× Damage to residue may occur

▽ Requires photoactivation

C Requires presence of carboxyl - activating reagent such as carbodiimide, or enzymic activation

H Product is readily hydrolysed

Solid symbols denote major reactions, dotted symbols less significant pathways

Hal = halogen group of appropriate reactivity

Z = O,S

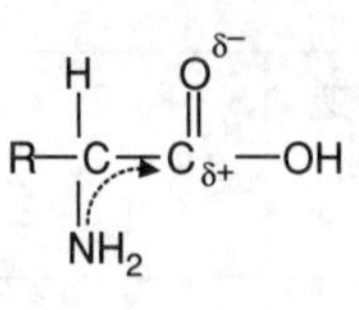

Figure 2.19 Electron withdrawal by the neighbouring carbonyl group reduces the basicity and nucleophilicity of the terminal amine.

carried out *via* reaction with aldehydes generated by oxidation of carbohydrate residues, while imidates have found a particular niche in structural studies; they are unusual in that there is no net change in protein charge during derivatisation of the amine.

Other reactions such as carbamylation by cyanate (Stark and Smyth, 1963) or guanidinylation (Habeeb, 1972) have been used in quantification and structure–activity applications.

2.3.2 Carboxylate groups

Aspartic and glutamic acid have fairly straightforward carboxyalkyl side-chains with typical pK_a values of around 4.5. The terminal carboxyl group is slightly more acidic (pK_a 3–4) but it is not generally practical to use pH to promote selectivity between these groups.

These acids will be increasingly ionised above the pK_a, but carboxylates are not in themselves electrophilic enough to react with common nucleophiles such as amines. This involves loss of the OH^- ion, a poor leaving group. The only direct reactions of carboxylic groups which have been employed to any great extent (Lundblad and Noyes, 1984b), are with esterifying agents such as diazomethane and analogues, and trialkyloxonium salts (Parsons *et al.*, 1969).

Otherwise, for reaction with amines, it is necessary to activate the acid groups. There are a number of better leaving groups which increase the reactivity compared to the acid; these have already been noted as amine-reactive species in the last section. Some of these are stable enough to exist under aqueous conditions for long enough to allow reaction to take place. These include active esters such as those containing succinimidyl groups, *O*-acylisoureas generated from carbodiimide reactions, and anhydrides.

All these approaches have been invoked in protein derivatisation. Activated acidic residues generated in this way will react with suitable nucleophiles, including the amine groups on the same protein. This can make such reactions difficult to control, as polymerisation of the activated protein will compete with the desired reaction, lowering yields and complicating purification procedures. There are a number of ways to minimise this effect. (1) By lowering the statistical likelihood of such an interaction by controlling the stoichiometry (ratio of reactants)—the more of the desired amine is present the less self-polymerisation will occur. This is most suitable for reaction with low-molecular-mass amines where it is practical to introduce huge excesses, and relatively straightforward (by size exclusion chromatography for example) to remove unreacted material after the reaction is complete. (2) By covalently blocking the amine groups on the protein before the active ester is generated, for example by reaction with a suitable low-molecular-mass species such as fluorescein isothiocyanate. Any suitable isothiocyanate could be used, as could an NHS ester or any other small amine-active compound with suitably uninteresting functionality elsewhere on the molecule. Incorporation of a readily detectable label such as fluorescein may however give a bonus in terms of subsequent detection and quantification, or to demonstrate that blocking has gone to completion. Pragmatically, it is the isothiocyanate to be found most commonly in many laboratories, and may have been employed many times for this reason alone! (3) Perhaps most elegantly, by choosing to react the activated acid selectively with a weak base such as an acid hydrazide (R–CO–NH–NH_2). Intuitively, it may seem that it should be easier to promote reaction with a strong base, but once again pH/pK_a considerations come into play. If the pH is lowered to the point where the terminal and lysine amine groups are fully protonated and unreactive towards electrophiles, a weak base will be less protonated and therefore more reactive. So if a suitably low pH is chosen, activated acids will react more or less specifically with a deliberately introduced hydrazide group and self-polymerisation can be avoided.

Carbodiimide methods provide by far the commonest approach for derivatisation of protein carboxylates, both for investigative and preparative purposes; the former are much the commoner.

2.3.3 Carboxamide groups

It is not uncommon to encounter aspartic and glutamic acid as their amide derivatives, asparagine and glutamine respectively. In sugar-containing glycoproteins, sugar chains (glycans) are normally attached to a side-chain *via* an oxygen atom (*O*-glycans) or a nitrogen atom (*N*-glycans), and an asparagine group very commonly forms the basis of the latter.

These amides are of little preparative value, but are a contributor to the heterogeneity of protein structure;

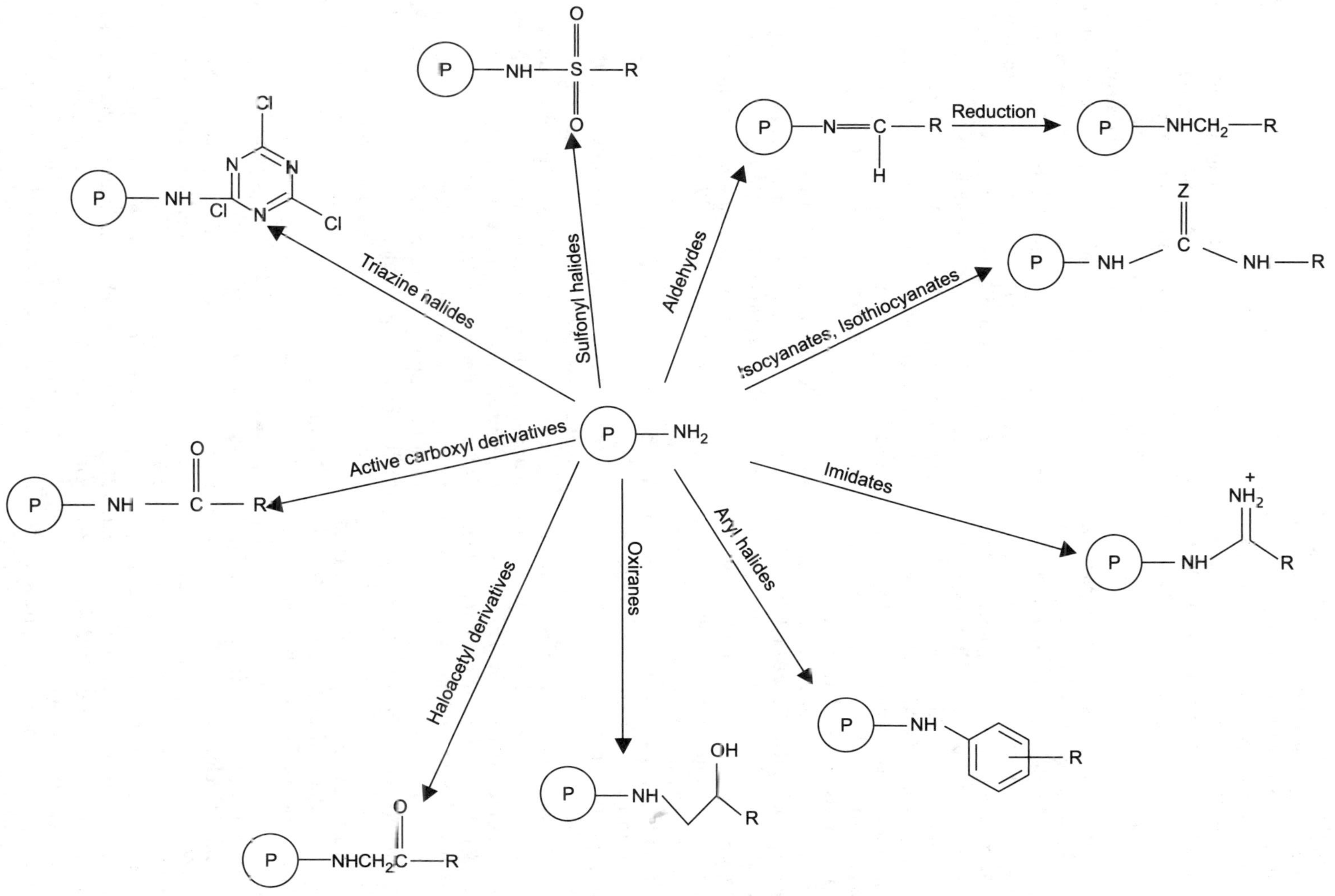

Z = O, S

Figure 2.20 Useful reactions of amine groups.

deamidation to the 'parent' acid occurs spontaneously *in vivo* and is thought to play a role in determining the lifetime of the protein molecule (Wold, 1981). Particularly for long-lived proteins, a batch isolated for *in vitro* use may contain some material which has been deamidated to various extents, the major consequence generally being a number of slight acidic shifts in the p*I*.

2.3.4 Thiol groups

Thiol or mercaptan (SH) groups have unpaired electrons which make them strong Lewis bases, giving them nucleophilic and reducing properties (Lundblad and Noyes, 1984a). This powerful reactivity means that cysteine is most commonly found in proteins in combined form, normally in the form of dimeric cystine moieties, but occasionally coupled to other groups such as sugars or porphyrins (Wold, 1981). However, cysteine is sometimes encountered in its free state, and is a valuable 'handle'. The thiolate anion, which exists predominantly above the relevant pK_a (approximately 8.5 in most proteins, but much lower in others), is much more nucleophilic than the neutral thiol.

Because of their useful reactivity, thiols are also commonly introduced artificially into proteins, so the absence of native thiol functions does not preclude approaches based on this chemistry. In fact heterobifunctional coupling strategies based on non-native thiols and complementary thiol-reactive groups are currently among the most popular routes for protein conjugation.

Thiols are soft bases, unlike amines which are hard, and therefore they do not react well with hard acids such as CO_2H groups and their derivatives. So while like amines they are strong nucleophiles, thiols tend to be employed quite differently in protein derivatisation.

The most commonly encountered reactions of thiols (Brocklehurst, 1979) are summarised in Figure 2.21. By far the most useful for coupling purposes are the reactions with haloacyl compounds, maleimides and disulfides. The specificity of reaction increases in the order just given: haloacyl compounds display a wide range of other reactivities, while disulfides (and related sulfenyl compounds, which may display higher reactivities than simple disulfides) are almost completely specific for thiols. The same is true for mercuric reagents but these are less convenient to use.

Haloacyl molecules such as iodoacetamide and iodoacetic acid have frequently been used for investigating structure–activity relationships involving cysteine, as has *N*-ethylmaleimide. These reagents do not always behave identically as thiol probes; differences have sometimes been observed in their reactivity (Hollaway *et al.*, 1964).

2.3.5 Hydroxyl groups

Serine and threonine bear straightforward aliphatic hydroxyl groups, which are passably nucleophilic and under the right conditions can react with acid derivatives to yield esters. Ideally these groups would be deprotonated to the more nucleophilic alkoxide (O^-) form, but the pK_a is so high (>13) that this approach is of little value in protein work. In any case, the reactions of these hydroxylic amino acids are subject to competition from: (1) water molecules containing OH moieties of similar reactivity, in a vast excess; (2) stronger nucleophiles such as amine groups. There are a number of other reactions of hydroxyl groups which are encountered in protein chemistry, but those based on amino acid residues are not generally favoured as preparative routes.

Occasionally, however, the microenvironment of a serine or threonine residue leads to atypical behaviour—serine residues in the active sites of enzymes, for example, have often shown unusual properties such as deprotonation at physiological pH, leading to aberrantly high reactivity (Wong, 1991a).

2.3.6 Tyrosine

Phenyl groups in themselves are only modestly reactive; phenylalanine is virtually never used as a site for protein derivatisation. In tyrosine, however, the introduction of a hydroxyl group to give a phenolic side-chain makes a dramatic difference to its reactivity; the OH group donates electrons to the phenyl ring, rendering it much more prone to electrophilic attack under gentle conditions (Lundblad and Noyes, 1984b). Due to differences in the stability of charged intermediates, it also means that the two positions on the phenyl ring adjacent (*ortho*) to the OH group are favoured for substitution.

One approach of this type which has been used for selective derivatisation of tyrosine groups is the use of diazonium salts ($Ph{-}N{\equiv}N^+$); there is also some reactivity for histidine residues. Diazonium salts can be formed from aromatic amines by reaction with nitrites in acid, and are electrophilic enough to react with tyrosine, yielding a diazo (–N=N–) bond. Therefore an aromatic group, possibly with other useful substituents attached, can be introduced into the protein at this point. The chemistry is inconvenient enough to mean that this approach will only tend to be used for specific aims, although

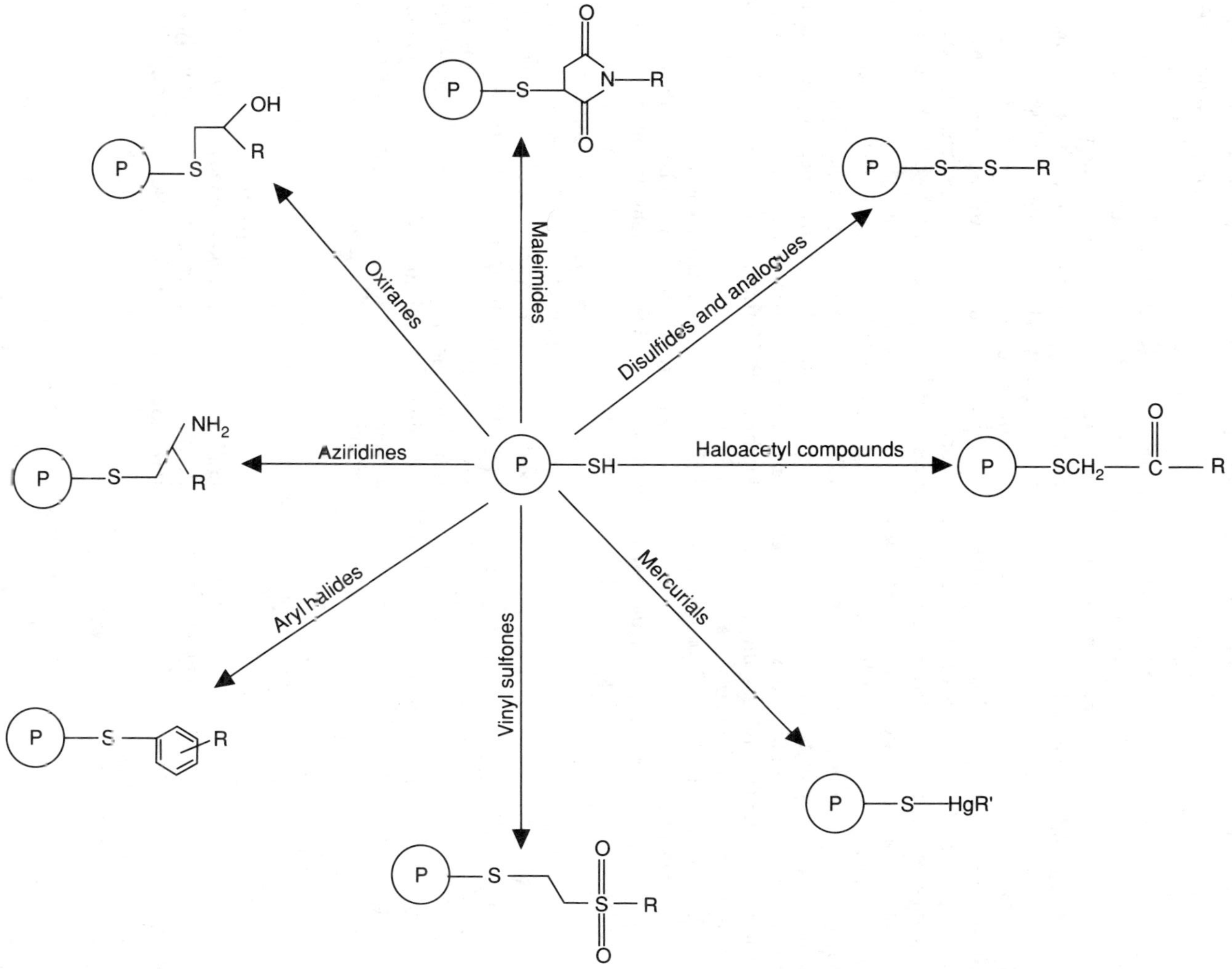

Figure 2.21 Useful reactions of thiol groups.

methods based on diazonium salts were among the most popular for protein coupling and derivatisation until the 1970s.

A much more common use of the reactivity of tyrosines towards electrophiles is in radioiodination; oxidation of radioactive sodium iodide ($Na^{125}I$) yields a complex mixture of iodine species, including H_2OI^+ which is thought to be the most significant electrophile; this reacts extremely rapidly with phenols to introduce a ^{125}I atom at one or both of the *ortho*-positions. Electrophilic nitration by tetranitromethane is another derivatisation reaction, which has been employed for structural investigations.

The hydroxy group in the phenol is a weak nucleophile which may compete in some of the reactions aimed at amines, such as those with succinimide esters; however the phenolic esters formed from this reaction are unstable, hydrolysing slowly even at neutral pH. Some workers have advocated treating protein conjugates prepared by these approaches with hydroxylamine, to break down phenolic esters which might otherwise decompose under storage and lead to a change in properties. *N*-Acetylimidazole reacts rather selectively with the hydroxyl moiety of tyrosyl groups, and has received considerable attention as an investigative reagent for this purpose (Lundblad and Noyes, 1984b).

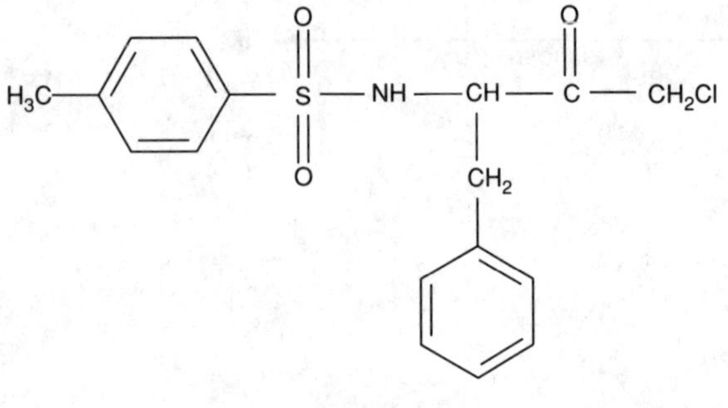

Figure 2.22 Structure of *p*-toluenesulfonyl phenylalanine chloromethyl ketone (TPCK).

2.3.7 Histidine

Histidine is most commonly encountered as a side-reactant in the derivatisation of other residues—its nitrogen atoms are mildly nucleophilic. For example it will sometimes react with haloacyl compounds, some such as *p*-toluenesulfonylphenylalanine chloromethyl ketone, TPCK, (Figure 2.22) offering some selectivity towards histidine. However this depends heavily on the microenvironment within the protein. Histidine is unusual in that the pK_a of the nitrogen atoms (6.5–7) is in the physiological range, and so is particularly susceptible to small pH shifts in this region.

The other common histidine reactions are aimed at structure–activity relationships: photooxidation in the presence of a dye such as Rose Bengal or methylene blue and reaction with diethylpyrocarbonate (Figure 2.23) (Lundblad and Noyes, 1984a; Bosshard *et al.*, 1984). Both of these approaches can also derivatise other residues, particularly methionine and tryptophan for the former, and cysteine, tyrosine and lysine for the latter. Methylation with methyl *p*-nitrobenzenesulfonate is another possibility (Nakagawa and Bender, 1970) for structure–activity work.

Other preparative approaches with histidine are based on electrophilic substitution of the imidazole ring; this can be iodinated in the same way as described for tyrosine, offering another avenue for the incorporation of ^{125}I. The reaction is much less facile than with tyrosine residues, however, and tends only to be of significance if there are none of the latter present. Similarly diazonium salts will substitute histidine residues in the same way as tyrosines residues. The ability of histidine residues to bind organometallic species such as ruthenium complexes is another useful capability, which has seen fairly wide application.

2.3.8 Tryptophan

Tryptophan is highly unreactive, its derivatisation generally requiring conditions too harsh for practical use in derivatisation. The reagents most commonly encountered for tryptophan (Lundblad and Noyes, 1984b) are

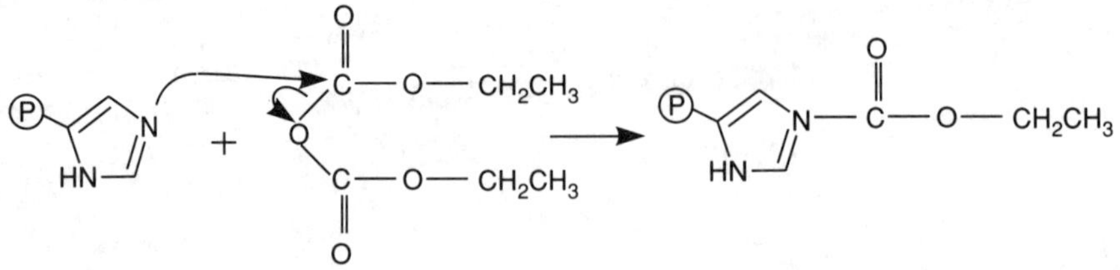

Figure 2.23 Acylation of histidine with diethylpyrocarbonate.

aimed at its oxidative destruction. Examples include *N*-bromosuccinimide and BNPS-skatole (Figure 2.24), the latter being more selective (Fontana, 1972).

Exceptions which do label the intact tryptophan molecule include benzyl bromides such as Koshland's reagent, which derivatises the indole ring as shown (Figure 2.25).

2.3.9 Methionine

The thioether (R–S–R) linkage in methionine is relatively unreactive. Iodoacetate can be used to introduce a carboxymethyl residue (Gundlach *et al.*, 1959; Stark *et al.*, 1961), and selectivity for methionine can be promoted by working below pH 3.0 (Lundblad and Noyes, 1984a). At low pH ethyleneimine may react with methionines, producing a mixture of products (Schroeder *et al.*, 1967).

Oxidation using a variety of reagents such as chloramine T or sodium periodate (Stauffer and Etson, 1969; de la Llosa *et al.*, 1980) may be useful in probing structure–activity relationships.

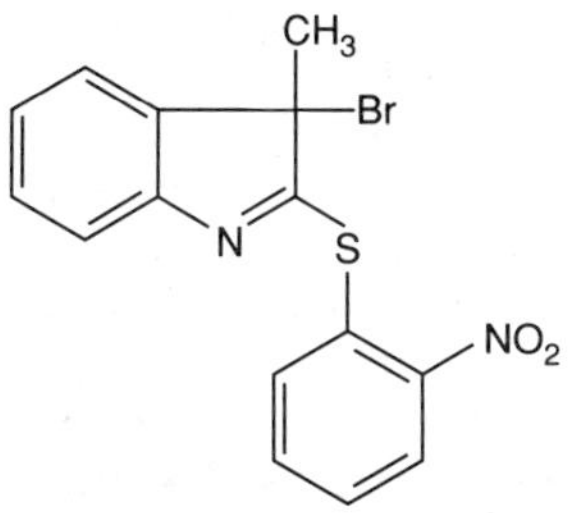

Figure 2.24 BNPS skatole.

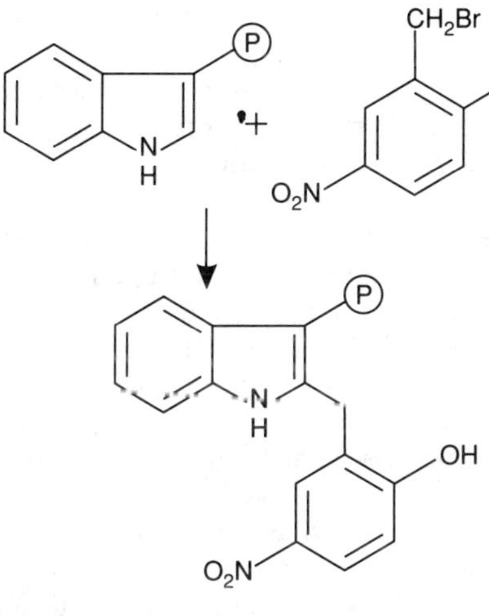

Figure 2.25 Labelling of tryptophan with Koshland's reagent.

2.3.10 Arginine

The side-chain in arginine is derived from guanidine ($HN{=}C(NH_2)_2$) which is one of the strongest organic bases known, with a pK_a too large to be measured accurately but certainly greater than 12. This is a consequence both of the absence of deactivating groups, and of delocalisation (Section 2.2.4), the ability of the molecule to spread the positive charge of its protonated form over three different nitrogen atoms (Figure 2.26). Overall the effect is to make the neutral molecule extremely prone to protonation, i.e. a very strong proton acceptor or Brönsted base.

Arginine is consequently strongly basic, and might therefore at first glance seem a useful option for derivatisation, but because the pK_a is so high, extremely alkaline conditions are necessary to deprotonate the NH_2 group and render it reactive. These are too harsh for most proteins and this approach is very rarely used for preparative purposes.

Modification of arginine for structure–activity investigations (Yankeelov, 1972; Lundblad and Noyes, 1984b) relies principally on 1,2-dicarbonyl compounds such as glyoxal—a reaction scheme is shown in Figure 2.27. There are many examples of structural investigations based on such compounds (Riordan, 1973; Rogers *et al.*, 1978; Yamasaki *et al.*, 1980).

2.3.11 Other amino acids

The side-chains of the remaining amino acids found in peptide structures (glycine, alanine, valine,

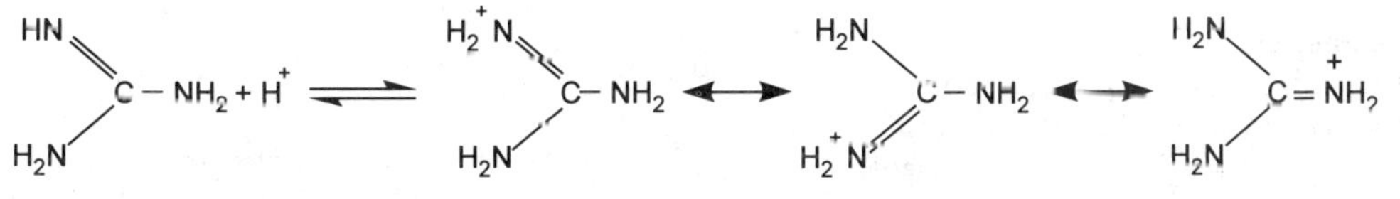

Figure 2.26 Guanidine delocalisation.

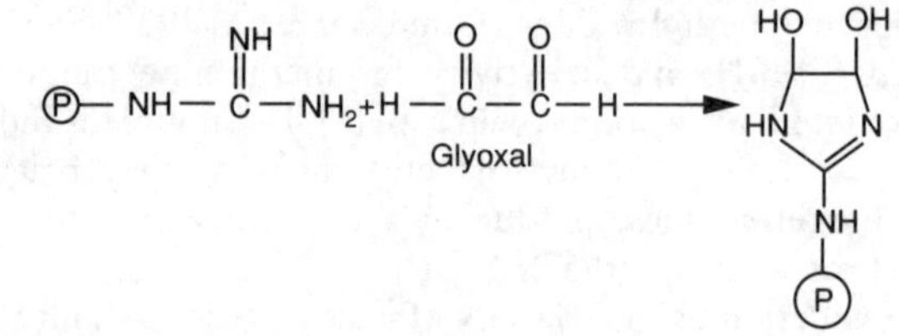

Figure 2.27 Modification of arginine residues with glyoxal.

leucine, isoleucine, proline, phenylalanine) contain only hydrocarbon groups. There are possibilities for derivatising the phenyl and benzylic groups of phenylalanine but these are rarely encountered. The reactivity of the other groups under conditions which can be tolerated by proteins is very low and to most intents and purposes they are viewed as chemically inert.

This is an unfortunate state of affairs if the aim of an experiment is to derivatise a protein, regardless of which residue comes into contact with the attacking reagent, e.g. in affinity labelling (Chapter 1, Section 1.5.4). Therefore considerable effort has been put into identifying chemistries which are truly non-selective in their reactivity.

Highly active species such as carbenes and nitrenes (Gilchrist and Rees, 1969) can be generated by the photolysis of certain precursors such as diazoesters and diazirines (for carbenes) and phenyl azides (for nitrenes). These are capable of inserting directly into C–H bonds, and have the additional advantage that this reactivity can be switched on at will by the photoactivation process. Due to re-arrangements of active species, not all of these reagents are as non-selective as was hoped, and challenges still remain in this area. The polyfluorophenyl azides seem to offer particularly attractive properties in this respect.

The side-chains of these amino acids do not possess significant dipole moments, and are predominantly hydrophobic in nature. Glycine is an exception, because the single hydrogen atom making up the side-chain is too small to shield the polar peptide region of the molecule, consequently glycine (and to a lesser extent alanine) behaves in a more hydrophilic fashion.

Proline is unusual in that its side-chain is cyclised back onto the amino group—it is in fact an imino acid rather than an amino acid. The extra rigidity that this cyclic structure imposes has very important consequences for protein structure, as will be seen in Section 2.5.2.

2.4 Other Groups Present in Proteins

The simplest proteins consist only of the amino acids described above, but many contain other components which can be pressed into service for derivatisation reactions. The vast majority of proteins, in fact, fall within this category. The groups to which the proteins are bound include carbohydrates, lipids, nucleic acids, phosphates and metals.

2.4.1 Carbohydrates

Glycoproteins have been mentioned already, and are very widespread—a substantial majority of commonly-encountered proteins contain at least a small percentage of carbohydrate. Most proteins, in fact, are believed to exist as glycoproteins *in vivo*, although the sugar groups are sometimes lost during isolation procedures (Franks, 1993a).

While their existence has been known since the 19th century, progress in understanding the function of the carbohydrate components (often termed glycans or oligosaccharides) of these proteins has been slow. Even today the overall picture is still not clearly understood, and one reason for this is that there seem to be a variety of functions which the carbohydrates perform under different circumstances. A major review of glycoproteins and other glycoconjugates (Allen and Kisailus, 1992) offers an excellent insight into this area.

Compared to amino acids there is much less variation in the identity of the sugar residues which are encountered in glycoproteins, but this is balanced by the variation in the linkages between these groups. Each carbohydrate can be coupled in many different ways, and chain branching offers the scope for a wider variety of structures. Thus while a tetrapeptide can exist as 24 sequence isomers, a tetrasaccharide can form in over 35,000 distinct arrangements.

Carbohydrate chains may reach 40 or more residues in length, and are attached to the protein either *via* the oxygen atom on the side-chain of a serine or threonine molecule (*O*-glycans) or the nitrogen atom of an asparagine[1] (*N*-glycans). The latter display a high degree of structural conformity close to the site of attachment, the core structure being shown in Figure 2.28. Reading outwards from the asparagine group, the chain

[1] Frequently this asparagine residue forms part of a sequence Asn-XXX-Thr or Asn-XXX-Ser where XXX represents any amino acid. In other words the presence of a hydroxyl-bearing residue 'next door but one' appears to be important.

I

II

III

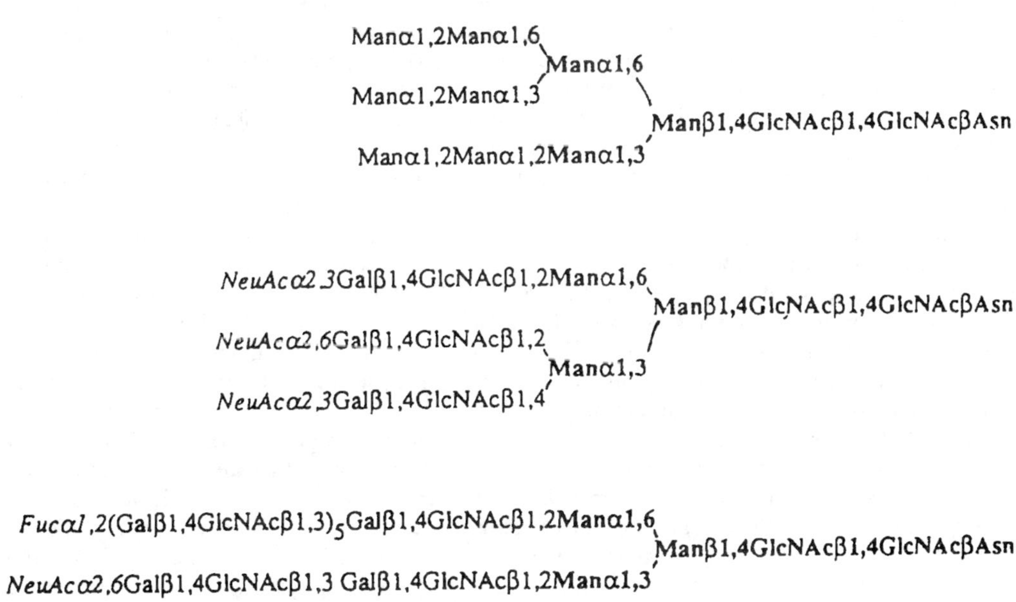

Figure 2.28 Examples of core carbohydrate structures of *N*-glycans in glycoproteins.

consists of two consecutive *N*-acetylglucosamine molecules followed by one mannose; at this point the chain splits in two to yield a biantennary structure with another mannose at the head of each new sub-chain. This $Man_3GlcNAc_2$ sequence is often extended by galactose: *N*-acetyl-glucosamine dimers with or without further branching (triantennary, etc.). Each chain may be capped by a terminal carbohydrate chosen typically from fucose, mannose, galactose, *N*-acetylglucosamine, *N*-acetylgalactosamine, or sialic (neuraminic) acid.

There is much less conformity to the structure of *O*-glycans, although the first sugar in the chain is usually *N*-acetylgalactosamine, and the Gal:GlcNAc motif remains widespread as a means for chain extension. Paulson (1989) has described five basic classes of *O*-glycan with the core structures GalNAc, GalNAc(GlcNAc)GlcNAc, GalNAc.Gal, GalNAc.GlcNAc, and GalNAc(GlcNAc)Gal.

Sometimes the presence of the oligosaccharide moiety of a glycoprotein is necessary for it to function properly, and sometimes it does not appear to be. For example among recombinant glycoproteins for therapeutic use, tissue plasminogen activator (TPA) and erythropoietin (EPO) require the presence of the glycans for *in vivo* activity, while others such as interleukin-2 (IL2) and gamma-interferon (IFN) function perfectly well without (Vapnek, 1990). Therefore the contributions of the glycans to a glycoprotein are not necessarily critical to its function, at least under the conditions of interest. This is obviously an important factor in the recombinant production of glycoproteins (Jenkins *et al.*, 1996), and will no doubt continue to attract significant attention.

Putative roles for the oligosaccharide components of glycoproteins (Olden *et al.*, 1982; Varki, 1993) fall roughly into two camps: (1) to confer physicochemical characteristics to the protein exploiting the hydrophilic, charge or steric properties of the carbohydrate, or (2) to carry specific information which controls the response of receptors, etc. to the presence of the protein. Failure of a non-glycosylated glycoprotein to function *in vivo* may be due to effects in either or both of these categories.

Considering the 'non-specific' category first, the ability of glycosylation to confer certain properties has been known for many years (West, 1986). These include improved solubility, resistance to heat denaturation, and protection from proteolytic enzymes. Perhaps most important though is the maintenance of a stable microenvironment around the protein; the bulky hydrophilic carbohydrate chains generate extensive hydration layers which limit the free movement of molecules in that region, minimising the impact of diffusion and convection on disturbing the solvent matrix 'seen' by the protein. Effectively the protein is cocooned against damaging temporary swings in the properties of this matrix, such as pH and ionic strength, and the stability to such changes may be enhanced.

Similar effects may well be important in the cell surface where glycoproteins are abundant, establishing a microenvironment around the membrane which may be quite different to the bulk medium; such carbohydrates may also mediate binding properties in a non-specific fashion due to their hydrophilic nature, and may also serve as a non-immunogenic protective coating for antigenic species (see Chapter 4).

The steric and charge properties of glycans are also thought to be important in promoting the correct folding of glycoproteins, a finding sparked off by the observation that the *en bloc* transfer of the Glc_3Man_9 $GlcNAc_2$ complex (from which *N*-glycans are derived) to the protein takes place in the endoplasmic reticulum (ER) of the cell, before folding is complete. The introduction of inhibitors of N-glycosylation such as tunicamycin prevents some proteins from being successfully expressed from the ER, although others seem to be unaffected (Kornfeld and Kornfeld, 1985).

The microheterogeneity of many glycoproteins is strongly indicative of non-specific roles—the frequent existence of a multitude of isoforms with slight variations in the oligosaccharide composition (Oshiro and Eylar, 1968; Vaughan *et al.*, 1982) does not suggest a highly-defined receptor function. Similarly, Stanley (1984) selected mutant cells on the basis of resistance to carbohydrate-binding lectins, forcing extensive glycosylation changes, and there was generally no great impact on their behaviour or proliferation.

Turning however to the second category of glycan function, there are a number of fairly well-documented examples of specific signal roles for oligosaccharides. Notably, Sly and Fisher (1982) described the role of mannose-6-phosphate in enzymes destined for the lysosome; this sugar is recognised by receptors in the Golgi apparatus of the cell, or on the cell membrane, which subsequently route the enzymes to their lysosomal target. Similarly, variations in the level of *N*-acetylgalactosamine sulfate in reproductive hormones such as luteinising hormone (LH) and follicle stimulating hormone (FSH) have been suggested to assist in their localisation in the pituitary gland (Smith and Baenziger, 1988).

Variations in glycan structure have been shown to affect the *in vivo* survival of serum glycoproteins (Ashwell and Harford, 1982); removal of terminal sialic acid groups to expose galactose, for example, promoting rapid removal by the liver. A suitable galactose receptor has been identified on hepatocytes which may be implicated. The cytolytic function of natural killer (NK) cells in the immune system also appears to be controlled to some extent by carbohydrate interactions (McCoy and Chambers, 1991), as are other phenomena as fundamental as the adhesion of sperm to egg (Wassarman, 1987). It seems likely that further important signalling roles of carbohydrates will become clear as our knowledge widens.

The presence of oligosaccharide chains opens up the protein to the techniques of carbohydrate chemistry. These approaches may be very useful as alternatives to the amino acid based chemistry outlined above, for example where reaction with the amine groups in a protein causes a serious loss of activity and there are few other useful 'handles' on the peptide chain.

Some reactions will take place directly with carbohydrate residues, for example derivatisation by epoxides, benzoquinone or cyanogen bromide. These have all been employed primarily in the preparation of solid phases for chromatography. Acidic groups originating from glycans such as sialic acid offer another site for derivatisation, using chemistries similar to those in Section 2.3.2 (Aplin *et al.*, 1979). More options are opened up, however, by exposing new reactive groups in the glycan by suitable treatment.

Undoubtedly the most commonly encountered reaction of this type is periodate oxidation. Periodate is a powerful oxidising agent, and undergoes a reasonably specific reaction with sugar molecules containing hydroxyl groups on adjacent carbon atoms (vicinal diol), cleaving the carbohydrate ring and generating two aldehyde groups in each case. Aldehydes undergo a dehydration reaction with amines to yield an imine or Schiff base (HC=N–), so this is a means of coupling an amine-containing molecule to a sugar chain; for greater stability this bond is generally reduced to a substituted amine bond ($-H_2C-NH-$). It should be noted that periodate oxidation is quite a harsh technique, and under the wrong conditions can result in partial deglycosylation of the protein. Milder oxidising agents have been examined, including enzymic examples such as galactose oxidase.

An alternative to aminolysis of the oxidised material is to react the aldehyde with hydrazides, the resulting hydrazones being quite stable. Working at a low pH allows this type of reaction to be carried out selectively in the presence of amines (*cf.* Section 2.3.2).

In terms of purification, the most important property of glycoproteins is their ability to bind to lectins, and this forms the basis of a major branch of affinity chromatography.

Finally methods for the removal of carbohydrate residues have proved useful: procedures employing trifluoromethanesulfonic acid, TFMS, have proved popular in comparison with those based on methanolysis or hydrofluoric acid (Edge *et al.*, 1981; Sojar and Bahl, 1987).

2.4.2 Lipids

Lipids are commonly encountered in living systems: compounds which are hydrophobic in nature, and consequently more soluble in non-polar solvents than in aqueous systems. The most familiar are probably the esters of 'fatty' acids, whose long hydrophobic side-chains confer this property, and in particular the triglyceride esters formed with the tribasic alcohol glycerol. These are of great dietary importance and provide one of the main means of energy storage within the body. The free fatty acids themselves are also lipids, and are among the most metabolically active. Others are based on the steroid cholesterol, which is a lipid in itself but can also be esterified with fatty acids

An important class of compounds are the phospholipids, where the structure includes a polar phosphate group. Often these are also based on glycerol (e.g. cardiolipin), while another group employ the alcohol sphingosine (e.g. sphingomyelins). Other hydrophilic groups are also often incorporated: sugar molecules in glycolipids such as the cerebrosides and gangliosides associated with brain function, or bases such as choline which are charged at physiological pH, and many lipid molecules contain more than one of these types (e.g. phosphatidylcholines or lecithins).

The combination of the hydrophobic lipid with these more polar groups leads to detergent or amphipathic properties, i.e. the molecule will tend to align itself across lipid–aqueous interfaces. In a wholly aqueous environment, the molecules will tend to form spherical micelles where the majority of the hydrophobic groups point inwards. In this way they are shielded most efficiently from the polar medium which surrounds them. It is a fairly simple extension of this idea to imagine two aqueous environments separated by a double layer of phospholipids, again with the hydrophobic 'ends' pointing inwards away from the solvent, and this lipid bilayer underlies the construction of cell membranes in biological systems. Transport of molecules across the cell membrane is to a large degree controlled by the subtleties of hydrophobic and hydrophilic interaction offered by the bilayer, and other species such as proteins embedded in it.

In such applications it is obviously required that the lipid is insoluble, however this is not the case where the molecules are being transported in the bloodstream. Once again the introduction of more polar groups such as phosphates, sugars, or bases will reduce the overall hydrophobicity of the lipid, but it can still be a problem for living systems to keep such species in solution. The cardiovascular problems associated with the deposition of lipids in human blood vessels, for example, are implicated in many of the most life-threatening medical conditions.

One approach which living systems employ to keep lipids in solution is to couple them to a hydrophilic protein to give a lipoprotein. Breakthroughs in the separation of lipoproteins by electrophoresis and ultracentrifugation in the 1940s ushered in an era of intensive investigation of the human plasma lipoproteins, which constitute around 8% of total plasma protein. The subsequent discovery that the concentrations of some of these molecules correlated with the incidence of coronary artery disease, one of the biggest killers in 'developed' societies, assured them the high level of attention they have received to this day.

The nature of the protein–lipid link in these molecules has been extensively investigated. Segrest *et al.* (1974) proposed the existence of amphipathic helices in the apoproteins, in which the polar and non-polar amino acid residues are concentrated on opposite sides of the helix. These regions therefore exhibit detergent-like properties, and provide a convenient interface between the lipid molecules and the surrounding aqueous medium. The non-polar regions are thought to penetrate into the lipid surface, and the nature of this interaction is reflected in the relative ease of displacement of lipids of differing hydrophobicity. Cholesterol can be displaced quite readily when presented with an alternative binding site, but the more hydrophobic cholesteryl esters cannot be shifted without a specific transfer factor.

In terms of coupling chemistry, standard methods have largely been employed to derivatise the apolipoprotein component: even where this only constitutes a small fraction of the molecule, it tends to be found on the surface and is therefore accessible for reaction. Care must obviously be taken, however, not to disturb the amphipathic nature of the protein by neutralising too many charged groups.

The hydrophobic nature of the lipid groups has been exploited in labelling strategies, however (Via and Smith, 1986)—fluorescent tagging can be achieved for example by using lipophilic reagents such as fluorescein coupled to a long-chain fatty acid or to a cholesterol derivative (see Figure 2.29). The former type can bind to the lipids already present on the lipoprotein, while the latter can act as lipid analogues to be incorporated during lipoprotein synthesis. A single incubation can replace nearly half of the cholesteryl ester content of low-density lipoprotein (LDL) with labelled analogues.

Some lipophilic fluorescent molecules based on pyrene, notably PMCA (3′-pyrenylmethyl-23,24-*dinor*-5-cholen-22-oate-3β-ol), show emission wavelength maxima related to concentration; this allows their

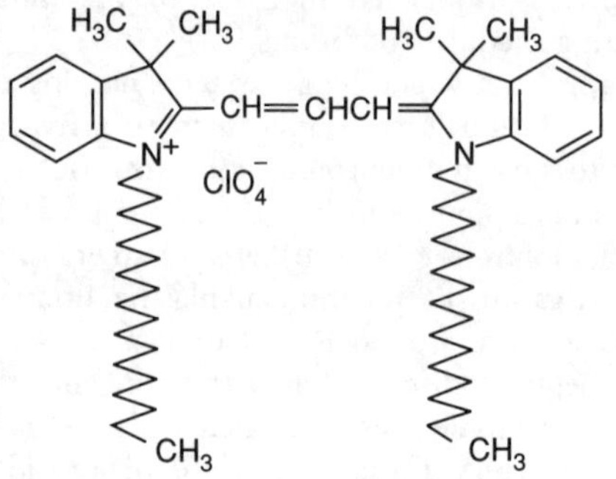

Figure 2.29 1, 1′-Dioctadecyl-3,3,3′,3′-tetramethylindocarbocyanine perchlorate.

rate of transfer from a lipoprotein to be conveniently monitored, for example. 1,1′-Dioctadecyl-3,3,3′,3′-tetramethylindo-carbocyanine perchlorate) is an extremely hydrophobic fluorophore which has been used extensively to label lipoproteins both *in vivo* and *in vitro*. This reagent is shown in Figure 2.29.

While the wide size range of the plasma lipoproteins has made them a natural target for separation by size exclusion chromatography, their hydrophobicity has also been exploited in the application of reverse phase HPLC.

2.4.3 Metal ions

Many proteins, including some of those with the most elaborate functions, contain other groups besides those covered above. Frequently metal ions are incorporated in the structure, particularly in enzymes where they are often central to the catalytic function.

Sometimes the metal in these metalloproteins is bound directly to the peptide chain, for example in the proteolytic enzyme carboxypeptidase A where a zinc atom is held in a pocket between two histidines and a glutamate. This arrangement obviously leaves enough access to the metal for it to take part in the proteolysis reaction, but the precise choice of such binding arrangements may be much more subtle in their effect; spectroscopic studies on metals in this position suggest that they may be present in a strained or entatic structure which might facilitate the desired reaction. Many metals have been identified in metalloproteins including iron, copper, manganese, magnesium, zinc, vanadium, molybdenum and selenium. Typically the metal content is very low, as only one or two ions are required per molecule to fulfil a specific function. Exceptions include transport or storage proteins such as ferritin, with an iron content of 23%. This property of ferritin has been exploited in its use as a label for electron microscopy and related techniques.

Metalloproteins can offer the possibility of radiolabelling without loss of activity using a suitable isotope of the metal, and this approach can occasionally be useful. In rare cases a different metal can be incorporated without destroying activity, e.g. cobalt can be used to replace zinc in carboxypeptidase-A, and horseradish peroxidase containing manganese instead of iron retains substantial enzyme activity; this may widen labelling possibilities.

2.4.4 Other constituents of proteins

Other non-metallic moieties which may also be incorporated covalently into proteins include phosphate and halogens (Wold, 1981). Phosphoproteins may incorporate an acyl phosphate attached to carboxylate residues, for example in ATPases; a phosphoamidate attached to lysine or histidine, for example in histones, or (most commonly) a *O*-monoester on serine or threonine in a wide range of important proteins such as phosphatases and casein.

Haloproteins may incorporate chlorine, bromine or iodine. As might be expected chemically, these atoms are found predominantly on tyrosine and (to a lesser extent) histidine residues. Examples include thyroglobulin, which is involved in the biosynthesis of iodine-containing thyroid hormones (Wold, 1981).

In other cases evolution has favoured the incorporation of a discrete organic molecule or prosthetic group to house the metal ion. Probably the best known are the porphyrin molecules which contain the iron atom in such proteins as haemoglobin, cytochrome-*c* and peroxidases. This particular combination of iron with protoporphyrin IX (Figure 2.30) is sufficiently common to warrant the common names haem (for the group) and haemoproteins. These groups not only provide a suitable location for the metal ion but also have the facility for taking part in redox reactions themselves; the system of linked pyrrole groups offers the possibility of delocalising charges over a large 'molecular area'. The ensuing stability gains may facilitate reactions, and subtleties of behaviour, which could not happen with an isolated metal atom.

The flavoproteins are another group, critical in many metabolic processes, which contain a prosthetic group based on the heterocyclic compound 6,7-dimethyl-*iso*-alloxazine (Figure 2.31).

In terms of derivatisation chemistry it is understandably more usual to avoid these prosthetic groups than to employ them as a site for linkage. They are

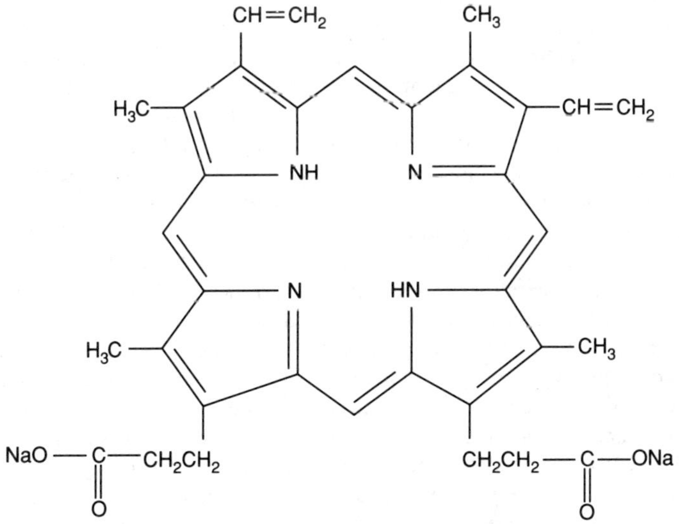

Figure 2.30 Protoporphyrin IX.

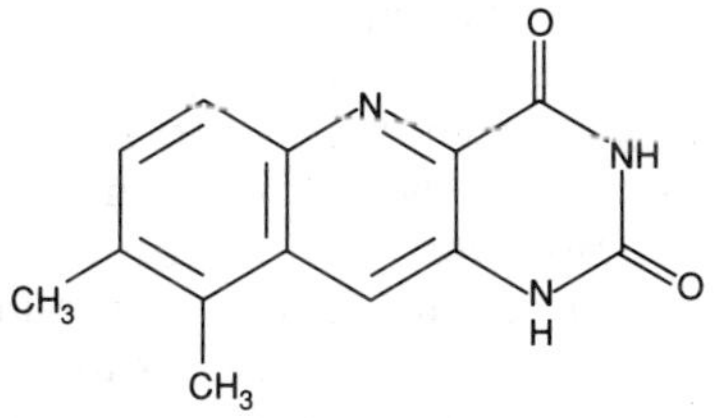

Figure 2.31 6,7-Dimethylisoalloxazine.

generally after all the focus of the protein's functionality, and therefore the region most sensitive to inactivation.

Derivatisation of such groups is generally encountered in structural studies where retention of protein functionality is not necessary. Specific experiments may determine the effect of such substitutions on the binding or activity of the prosthetic group, e.g. the effect of derivatisation of various positions of the haem group on its ability to bind to the apoprotein (protein without prosthetic group).

One other essential component of protein molecules has not been discussed in this section: this is water, whose role will be considered as part of the overall discussion of protein structure in Section 2.5.5.

2.5 Protein Structure

Proteins are complex molecules, and this is reflected in the multiple layers of structure which they possess. These are fundamental to the understanding of protein behaviour, and will be discussed briefly here.

2.5.1 Primary structure

2.5.1.1 The amino acid sequence

In Section 2.1 the peptide chain was introduced, a sequence of amide links between amino acids drawn from the pool of 20 shown in Figure 2.4, and connected by carbon atoms bearing a variable side-chain:

$$H_2N.CHR^a.CO\ldots.NH.CHR^b.CO.NH.CHR^c.$$
$$CO\ldots.NH.CHR^d.CO_2H$$

This sequence, extending between the N and C termini, represents the primary structure of a protein—or at least of one chain of a protein. It is only necessary to define all the R groups to assign a reliable identity to this primary structure, and this amino acid sequence is a fundamental signature of a protein molecule. This can be expressed using either of the recognised abbreviation systems shown in Table 2.1:

...Cys – Tyr – Met – Lys – Ala – Ala – Gly–
Leu – Gly – Ile...

or

...CYMKAAGLGI...

Sequences such as this are known for a wide range of proteins, until recently almost always determined by a number of complementary techniques such as the analysis of products from Edman degradation, or from proteolytic digests.

Such techniques are still widely used, but are being gradually overtaken by approaches rooted in molecular biology. The amino acid sequence is itself derived from the gene which controls production of the particular protein concerned, and consists of an analogous DNA sequence of purine and pyrimidine bases. Each set of three bases constitutes a codon, which specifies a single amino acid. The DNA sequence in fact codes for a sequence of amino acids rather longer than that observed in the native protein. All protein genes commence with the universal initiation codon AUG which codes for methionine, the first amino acid in a signal sequence which is involved in the transport of the protein from the ribosome where it is produced, but is normally no longer present in the final product. Other sequences may also be lost from the amino acid chain during the processing of the new protein, for example when cleavage of a single chain takes place to produce a two-chain product.

The genetic information required to produce a given amino acid sequence is all present in this DNA code, so identifying the gene for a protein of interest is an alternative method for determining its complete primary structure. This approach of course has its own pitfalls, but with improvements in instrumentation has begun to compete favourably with protein sequencing methods. The codon 'language' is so universal that the correct sequence will be obtained even if the DNA is transferred to an organism which is completely unrelated to that from which it originated. Thus it is possible to generate 'human' amino acid sequences in primitive organisms such as yeast or the bacterium *Escherichia coli*, using recombinant DNA in which the correct base sequence has been artificially incorporated.

An amino acid sequence is not a protein, however; two main activities need to take place to convert the nascent peptide chain(s) into a functional protein.

2.5.1.2 Protein folding

Firstly only the very shortest oligopeptides are capable of fulfilling their biological function in a simple linear conformation. Any sizeable polypeptide must be organised into the correct three-dimensional structure before it can start to be considered as a functional protein. The processes which control the folding of a protein molecule are far from understood, and this remains a highly active area of research (Creighton, 1997; Gross and Plaxco, 1997; Marchler-Bauer and Bryant, 1997; Netzer and Hartl, 1997). A recent breakthrough has seen the complete *de novo* design of a desired three-dimensional protein structure by computational means (Dahiyat and Mayo, 1997), yielding an artificial 3.5 kDa molecule with 'zinc finger' morphology. The spontaneous formation of disulfide bridges between suitably located cysteine residues is a closely-related activity (Creighton *et al.*, 1995).

To some extent protein molecules *per se* represent autonomous self-assembly systems; it will be seen in the next section how certain regular three-dimensional structures are energetically favoured simply by the geometry of the peptide chain. However other agents are involved in the folding process, including protein molecules such as the chaperonins (Ellis, 1987; Frydman *et al.*, 1994). These may prevent unwanted folding events from taking place during the production of the proteins in the ribosome.

Folding processes take place so quickly (on a micro- to millisecond timescale) that ingenious techniques are required to investigate them (Englander, 1993; Ballew *et al.*, 1996). However, it appears that folding is already occurring cotranslationally as the incomplete protein molecule is expressed, with secondary structure appearing within microseconds. Both folding and disulfide bond formation may also continue as post-translational events, however (Creighton *et al.*, 1995; Netzer and Hartl, 1997).

2.5.1.3 Post-translational modification

The processes involved in protein folding appear to be highly conserved, so that it is often feasible to produce a correctly folded molecule in a foreign organism (but see Netzer and Hartl, 1997). However, this does not apply to the incorporation of the non-amino acid components described in Section 2.4. Depending on the organism used, there is no guarantee that post-translational processes such as glycosylation or attachment of prosthetic groups will take place correctly. This places limitations on the approaches for recombinant protein production: it is well known, for example, that prokaryotic organsims such as *E. coli* do not glycosylate their protein products, while the glycosylation patterns in lower eukaryotes may only partly match those of vertebrates. There may also be variation in other post-translational phenomena such as deamidation.

2.5.2 Secondary structure

The secondary structure of a protein describes the organisation of the peptide chain into three-dimensional arrangements, which are more favourable in energetic terms than a structureless random coil. These are driven by the torsional angles which minimise unfavourable interactions between neighbouring groups, and by the additional stabilisation obtained from the formation of hydrogen bonds. Although these bonds are at least an order of magnitude smaller than covalent links (typically 10 kJ mol^{-1} compared to >200 kJ mol^{-1}), they are critical in maintaining protein structure.

2.5.2.1 The α-helix

Ignoring the influence of side-chains, the lowest-energy arrangement of a peptide chain is an α-helix (Figure 2.32). Regular arrangements of this type are characterised by the symmetry index, η_m, where η is the number of residues which form a perfect repeat after m turns. This α-helix is a right-handed 18_5 example, i.e. the structure repeats after five turns, and requires 18 amino acid residues for this to be achieved. It can be rapidly deduced that there are 3.6 residues per turn of the helix. In dimensional terms, this corresponds to a rise per turn of 5.4 Å (Franks, 1993b).

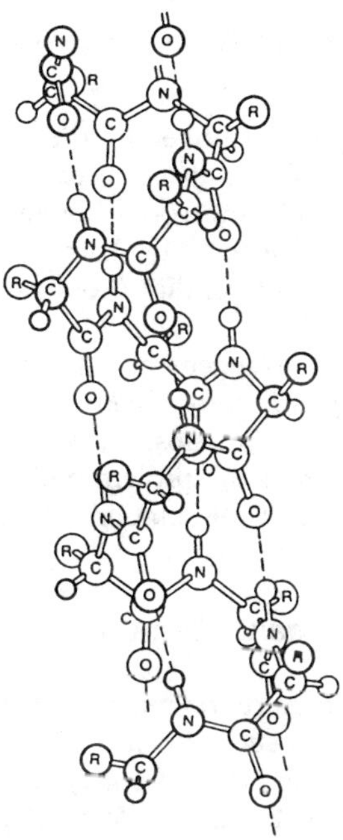

α-Helix (right-handed) of a protein chain.

Figure 2.32 The α-helix structure encountered in proteins. (Reproduced with permission from Low and Edsall, 1956. In *Currents in Biochemical Research* (Green, D.E., ed.), Interscience Publishers, Inc., New York).

It can be seen from Figure 2.32 that a hydrogen bond is formed between the NH group of one residue and the CO group of its fourth-closest neighbour. These bonds, parallel to the axis of the helix, are important stabilising factors.

The rigidity of proline residues excludes them from taking part in the α-helix arrangement; other residues also tend to be associated with breaks in the α-helix structure, notably those bearing charged or polar groups such as arginine, lysine, aspartic acid, serine and threonine and (for less obvious reasons) glycine and isoleucine (Harper, 1979).

2.5.2.2 The β-sheet

The β-sheet structures tend to be found where there are constraints preventing α-helix formation. They are right-handed 2_1 arrangements; as the name implies they consist of flat sheets of peptide chains running parallel to each other, in either a parallel or antiparallel form (Figure 2.33).

2.5.2.3 Other secondary structures

The majority of regular domains in proteins are α-helices and β-sheets; between these there are unstructured loops or turns which may constitute over 30% of the protein content.

Other regular structures do exist in specialised circumstances. For example there are multiple proline residues (notably in collagen) where the normal α-helix and β-sheet arrangements cannot be accommodated and other patterns are observed. There are two polyproline structures of roughly similar energy: a right-handed 10_3 helix (polyproline I) and a left-handed 3_1 arrangement (polyproline II). These are also described as *cis* and *trans* arrangements on account of the relative positions of nitrogen atoms from adjacent residues: solvent changes may cause *cis–trans* isomerism, where the structure flips from one form to the other.

2.5.3 Supersecondary and tertiary structure

Another level of organisation is represented by the arrangement of the α and β-domains with respect to

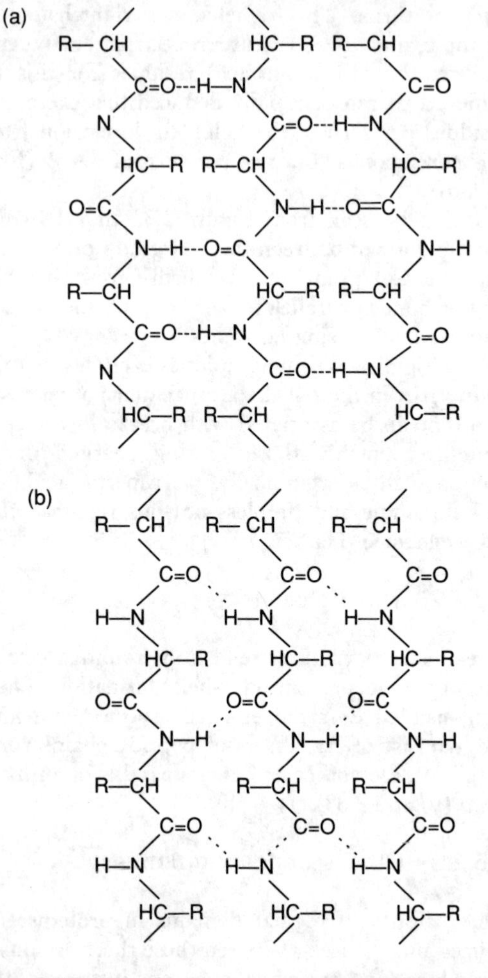

Figure 2.33 The β-sheet structures commonly found in proteins. (a) Antiparallel, (b) parallel. (Reproduced with permission from: Franks, F. (1993b). In *Protein Biotechnology. Isolation, Characterisation and Stabilisation* (Franks, F., ed.), Humana Press, Totowa, NJ.)

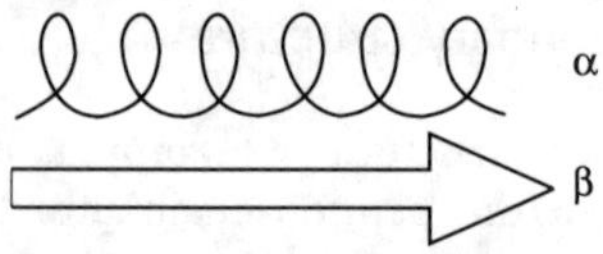

Figure 2.34 Cartoon representations of α and β domains.

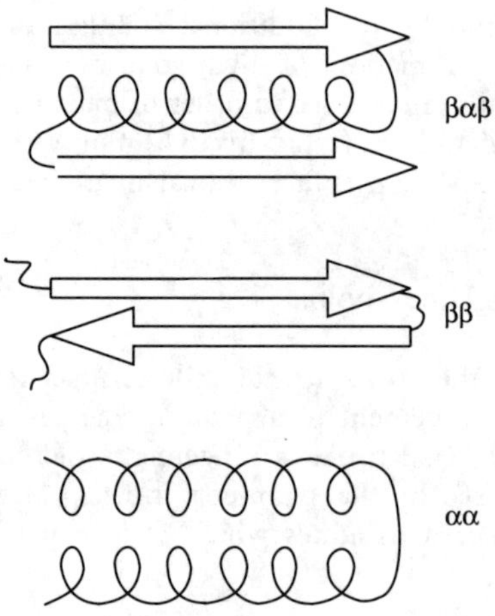

Figure 2.35 α and β domain combinations.

each other. These can be drawn in 'cartoon' fashion as shown in Figure 2.34, and various combinations of these domains are shown in Fig. 2.35. The interactions which maintain this organisation are even weaker than those of the secondary structure (typically only 5 kJ mol^{-1}).

Proteins may contain only α or β-domains (plus loop regions of course), or a mixture of both. Simple combinations of these domains are described as the supersecondary structure, e.g. $\beta\beta$-sheets give rise to the antiparallel arrangement shown in Figure 2.33, while $\beta\alpha\beta$ sequences allow parallel β-sheet structures to form. Larger assemblies of these domains occur frequently enough to have been given names of their own, such as the $\beta\alpha\beta\alpha\beta$ Rossman fold, and the $\beta\beta\beta\beta$ Greek key, or the TIM barrel which consists of a hollow cylinder of parallel β-domains separated by α-helices.

The overall arrangement of these regions of supersecondary structure, along with unstructured loops, prosthetic groups, etc., represents the tertiary structure of the protein molecule. Some cartoon or 'ribbon' representations of protein tertiary structure.

2.5.4 Quaternary structure

Quaternary structure refers to the arrangement of individual protein subunits into a larger complex; the number of subunits involved may range from two to several thousand. Formation of these complexes generally occurs spontaneously by a self-assembly process, and is believed to be driven predominantly by hydrophobic interactions.

Investigation of quaternary structure has provided a major impetus for the development of cross-linking reagents: bifunctional compounds capable of coupling residues in neighbouring subunits have provided valuable insights into the topology of protein complexes.

2.5.5 Hydration

The presence of water is critical to the maintenance of protein structure. The water content of proteins may approach 40% (Finney, 1979). Water molecules tend to be closely associated with a number of typical locations (Franks, 1993d). These include active sites of enzymes, coordination sites on metal ions, and locations between polar side-chain groups and the peptide chain amides, or between nearby charged side-chains. Franks has proposed three categories of hydration: the solvation of specific ions such as protons which are associated with the structure, of hydrophilic groups in the protein (particularly the peptide chain amides), and of hydrophobic side-chain moieties. As noted earlier, it is the minimisation of this latter, energetically unfavourable hydration that is believed to drive hydrophobic interactions.

2.5.6 Overall protein structure

All the features described in the preceding sections, including the presence of other groups such as sugars and metal ions, combine to produce the overall protein structure. As well as the many factors already discussed, a number of broad statements can be made to complete the general 'picture' of how protein molecules behave.

(1) Proteins typically possess dimensions in the 10–1000 Å (1–100 nm) range (Figure 2.36); some examples, however, exceed these limits significantly, such as the elastic muscle protein titin (connectin) with μm dimensions (Tskhovrebova *et al.*, 1997). Larger proteins tend to employ multi-unit or multi-domain structures: for single chains around 1000 residues appears to be the upper limit (Franks, 1993a), while single-domain proteins rarely exceed 200 residues. Bussemaker *et al.* (1997) have recently presented calculations which suggest that this latter limit may be driven by the sensitivity of larger domains to point mutations.

(2) Proteins display an average packing density of around 0.75 (Andrade and Hlady, 1986), comparable to a close-packed arrangement of spheres. So their internal structure is typically highly compact.[1] It can, however, expand somewhat under the influence of pH changes and consequent alterations in charge distribution.

(3) The packing density is also essentially identical whether the protein molecule is in solution or in the crystalline state.

(4) To minimise the need for hydrophobic hydration, apolar residues tend to be associated with the interior of protein molecules; conversely formally charged groups tend to be localised near the surface. These hydrophobic residues are also critical in maintaining structure, in globular proteins there appears to be a requirement for them to constitute around 50% of the total amino acid content. They are also more strongly conserved between species than the outer regions (Franks, 1993b), probably reflecting their structural significance. This distribution rule is a broad approximation however (Finney, 1979), and has been over-emphasised in the past; there is in practice a fairly wide spread of residues throughout the structure, with examples of all types present to some extent near the surface.

(5) The conformations of protein molecules are only marginally stable, the energies of the bonds maintaining three-dimensional structure being comparable to the fluctuations due to thermal buffeting (Andrade and Hlady, 1986). Indeed the native states of proteins display some characteristics of 'chaotic' systems in this respect (Bussemaker *et al.*, 1997). The conformation in solution should be thought of, therefore, as in a permanent state of flux; changes to individual molecules are constantly occurring which will transiently reveal non-optimal conformational features.

In summary, protein molecules in solution can typically be thought of as dynamic structures, tightly-packed but flexible, built around a predominantly hydrophobic core but displaying a variety of surface regions possessing different properties of charge, hydrophobicity, chemical reactivity, etc.

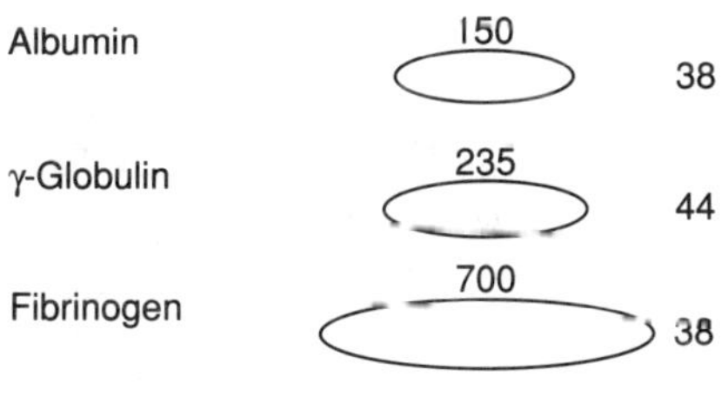

Figure 2.36 Approximate dimensions (Å) of some human serum proteins (from Oncley *et al.*, 1947).

[1] 'an extremely well-fitting three dimensional jigsaw puzzle' (Finney, 1979).

2.6 Approaches to Protein Coupling

2.6.1 Categories of protein coupling reactions

While there are a fair number of different chemistries used in protein coupling, and a plethora of different specific reagents for exploiting them, they can be categorised usefully into three classes.

2.6.1.1 Direct coupling

In a few cases, atoms naturally present in one molecule can be made to react directly with atoms naturally present in another without the overall introduction of any additional atoms—in fact normally with the loss of a few. It may be necessary to activate one or more of the groups by chemical derivatisation, but if extra atoms are introduced they are lost again when the intermolecular bond is formed.

An example of this direct coupling approach is the formation of a peptide bond between the amine group of one protein and the carboxyl group of another (Figure 2.2). As noted earlier, the acid group will not react directly in this way because the hydroxide ion is a poor leaving group; however, introduction of a succinimide ester provides a more favourable situation and the reaction can proceed. The resulting peptide bond contains only atoms that were present in the original protein molecules. In fact, two hydrogen atoms and one oxygen atom have been eliminated, and although the 'extra' succinimide group had to be introduced, it did not appear in the final product.

Apart from limitations on the available chemistries for this type of coupling, it has other disadvantages. Not only may this approach require fairly specific alignment of the reactants in order for reaction to take place (observed in practice as a low reaction rate and/or a poor yield), but the two molecules will be 'cramped' in the final product. The behaviour of one or both of them may be adversely affected by this close proximity. It may well be better to give the components of the product a little 'breathing space' by introducing some extra atoms between them—a linker group. More chemical options are opened up by this approach too.

2.6.1.2 Homobifunctional coupling

A linker molecule must by definition be (at least) bifunctional, each end reacting ideally with a separate molecule to form the product. Some of its atoms will probably be lost as the bonds are formed, but some atoms not present in the starting materials are included in the final product.

In the simplest cases a group of the same reactivity is employed at both ends of the linker, and such reagents are termed homobifunctional. It is a short step from the direct peptide bond formation described above, to imagine a linker molecule consisting of *two* succinimide esters joined by an unreactive bridge. This can clearly be used to join the amine group of one protein to that of another (Figure 2.37), and is a perfect example of a homobifunctional reagent.

Because the same reactive group is being derivatised in both reactants in a homobifunctional approach, it follows that the formation of dimers of each of the starting materials is unavoidable. Reaction times, concentrations, and rates and orders of addition can be optimised to reduce this effect to a minimum but it can never be avoided completely.

It is also very likely that each molecule will contain more than one of the reactive group, and this opens up further opportunities for heterogeneity in the product. Once such a dimer has formed it may still be capable of further conjugation and the end result is not difficult to work out—the product will consist of a range of oligomers, varying both in the number of protein

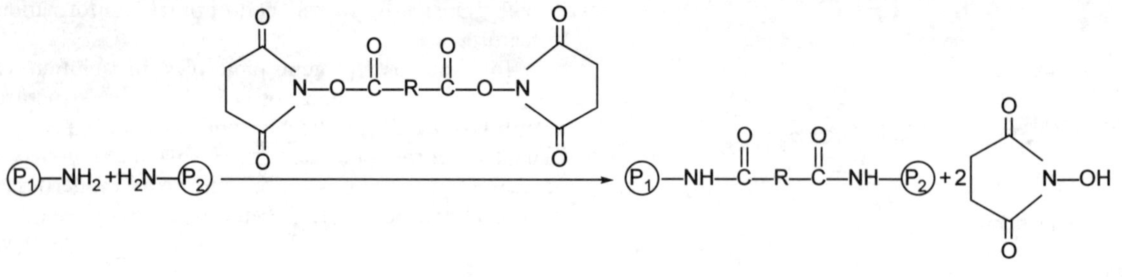

Figure 2.37 An example of a homobifunctional reagent. A bis-succinimide ester can be used to link the amine groups of two protein molecules.

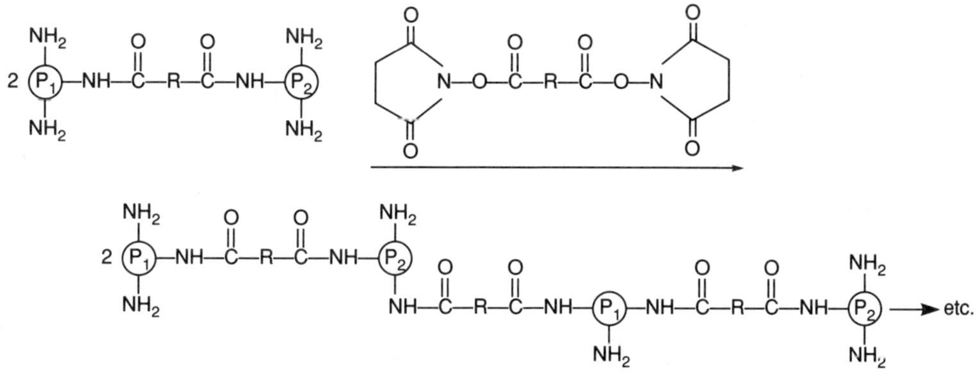

Figure 2.38 The potential for the formation of oligomers from the homobifunctional coupling reaction shown in Figure 2.37.

'units' they contain and the relative content of each starting material (Figure 2.38).

Again careful optimisation can mimimise the scale of this effect, but it is a common result of homobifunctional coupling methods. It is an equally valid concern in direct coupling approaches where there are multiple reacting groups on one or more of the reactants.

2.6.1.3 Heterobifunctional coupling

In some applications 'messy' conjugates of the type just described may be perfectly adequate, but it is often desired to synthesise a product of more defined stoichiometry. To do this it is necessary to control the number of reactive groups on each of the protein starting materials to the desired level.

It is simplest to imagine an 'ideal' A–B conjugate of 1:1 stoichiometry. If both A and B contain a single similar reactive group each, e.g. an amine, then there is no chance that a homobifunctional reagent can create anything larger than a dimer. However, as noted above, the reaction mixture will undoubtedly contain A–A and/or B–B dimer as well as the desired product.

If A and B each contain a single complimentary reactive group, for example an amine on A and a succinimide ester on B, then (apart from unreacted starting materials) the only possible product is the desired A–B complex. This is a potential outcome of the direct coupling approach discussed earlier, but it is very unusual for a situation to be encountered where each reactant bears a single complimentary reactive group. To approach this ideal in practice, it is necessary to engineer this situation by the controlled introduction of 'new' reactive functionalities.

If, for example, a single thiol group is introduced onto each molecule of A and a single maleimide group onto each molecule of B, the specific reaction between these two moieties can only yield an A–B complex[1]. So we need a reagent which can 'convert' one functionality already present into another, and this aim requires the use of reagents which possess groups of different reactivity at each end. For example, a molecule with a succinimide ester at one end and an SH group at the other will 'convert' an amine into a thiol ($A\text{-}NH_2 \rightarrow A\text{-}SH$), and such a reagent is termed heterobifunctional.

Importantly, we have now obtained a measure of control over the amount of the new group that we introduce. Assuming in the thiolation example that there are a number of NH_2 groups of similar reactivities on A, it is not possible to convert a single amine in every molecule—chemical reactions are controlled by probability arguments and there will always be a statistical distribution of products. However it is possible to approximate to this goal by altering the stoichiometry of the conversion reaction. The amount of the heterobifunctional reagent can be optimised to give a predominance of A-SH, although the product will undoubtedly contain a proportion of $A\text{-}(SH)_2$, $A\text{-}(SH)_3$, etc. as well as some unreacted A (see Sectin 2.6.2).

It is equally possible to exert this control on the introduction of maleimide into B, and the predominant product of the reaction between the two 'activated' starting materials will be the A–B complex.

In many cases this is about as close as it is possible to get to a rigorously defined conjugation product. The potential heterogeneity of conjugates is conveniently

[1] Assuming that (1) there are no free thiols present in either of the starting materials, and (2) that conditions leading to the formation of disulfide bonds between two thiols are avoided.

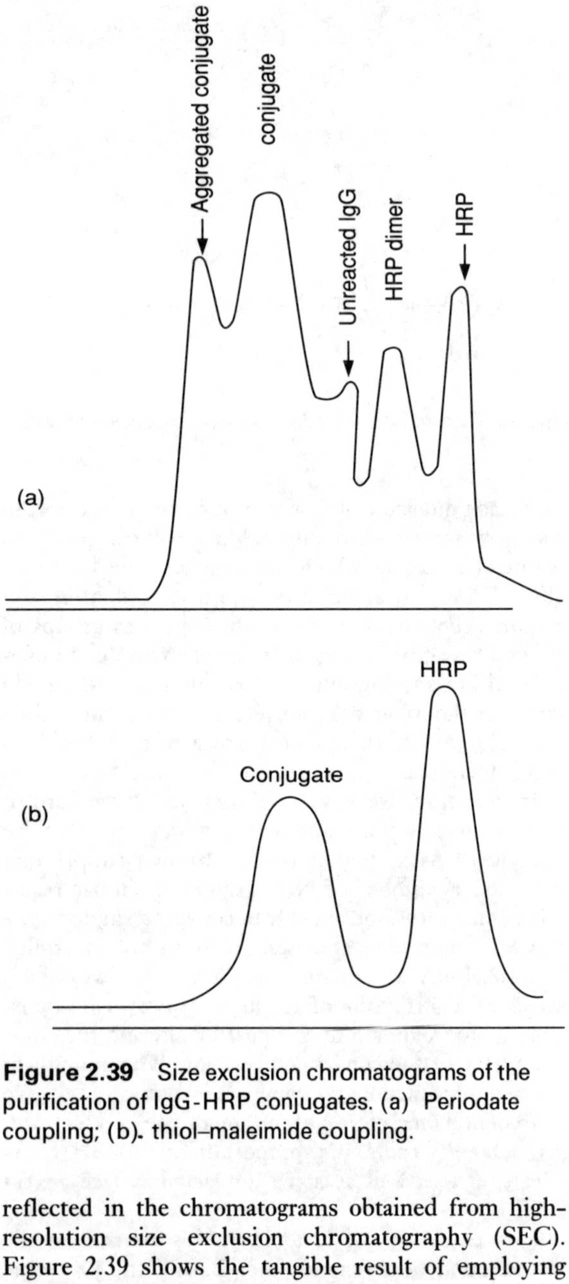

Figure 2.39 Size exclusion chromatograms of the purification of IgG-HRP conjugates. (a) Periodate coupling; (b). thiol–maleimide coupling.

reflected in the chromatograms obtained from high-resolution size exclusion chromatography (SEC). Figure 2.39 shows the tangible result of employing an 'uncontrolled' direct-coupling reaction and an optimised heterobifunctional approach to the same task, the preparation of a conjugate between immunoglobulin G (IgG: molecular mass 150 kDa) and horseradish peroxidase (HRP: molecular mass 44 kDa) for use in an immunoassay.

Figure 2.39a shows the SEC chromatogram of the product of periodate oxidation, a direct-coupling technique between carbohydrates and amines. The chromatogram shows the presence of unreacted HRP (molecular mass 44 kDa), HRP dimer (molecular mass 88 kDa) and possibly trimer (molecular mass 132 kDa), HRP-IgG (molecular mass 194 kDa) and then a range of larger conjugates (HRP_xIgG_y)—the heaviest of these are too large to be resolved by the SEC column and are excluded as a single 'aggregate' peak.

By contrast Figure 2.39b shows the product of a thiol-maleimide reaction of the same reagents, where thiols are introduced into the HRP and maleimides into the IgG in a controlled fashion. The chromatogram shows only unreacted HRP-SH, HRP-IgG and probably a small amount of larger conjugates. There are no conjugates large enough to exclude from the column.

Note the presence of unreacted HRP (or HRP-SH) in both these traces; this results from a deliberate choice to include an excess of this reagent. There are a number of reasons for this decision. First, including an excess of one component is the best way to drive a reaction to its conclusion in the shortest possible time—the more activated HRP molecules are present, the more chance each activated IgG molecule has of colliding and reacting with one in a given time.

Second, driving the reaction in this direction (as opposed to using an excess of activated IgG) ensures that the product contains no unreacted IgG, which can have a harmful effect on the performance of the immunoassay; the presence of HRP is less damaging.

Third, unreacted HRP is simple to remove by SEC due to its low molecular mass, while unreacted IgG is uncomfortably close to the IgG-HRP conjugate; complete separation of components of 150 kDa and 200 kDa molecular mass is not generally achievable by preparative SEC, and other methods must be employed.

Finally, HRP is inexpensive and readily obtainable compared to the antibody, and it is more important to ensure that all the IgG is gainfully employed than to worry about 'wasting' HRP.

An excess of one component is nearly always employed in protein coupling reactions for one or more of these reasons, but the choice of which component will depend on the consideration of individual factors such as eventual use of the product, purification issues, cost and availability.

2.6.1.4 Genetic engineering approaches to coupling

There is a fourth approach to the generation of coupled protein reagents—genetic or protein engineering. If molecular biology techniques are applied to identify the DNA sequences which code for the proteins of interest, and to introduce the linked sequences into suitable host cells, a 'ready-made' conjugate or fusion protein can be expressed. There will be no doubt as to the stoichiometry of the conjugate, as it will be created to a defined pattern, usually 1 : 1. This approach has seen increasing application over the last few years (Uhlén *et al.*, 1992), in particular in diagnostics (Lindbladh *et al.*, 1993; Carrier *et al.*, 1995), and in the preparation of pharmaceuticals such as cytotoxic agents (Kiyokawa *et al.*, 1991).

However, the techniques of this approach are so radically different from the chemical strategies just discussed that they merit separate treatment; they are covered in Chapter 9.

2.6.2 Stoichiometry of protein coupling

2.6.2.1 The choice of stoichiometry

In the last section the whole discussion assumed, for the sake of simplicity, that a 1 : 1 A–B complex was the desired outcome of a conjugation. This is by no means always the case.

Where, as in the example above, an enzyme label is to be introduced onto an antibody, it is generally desired to obtain the maximum enzyme activity per conjugate molecule. This may drive the strategy towards incorporating as many enzymes as possible. However a level of incorporation will probably be reached where the performance of the antibody is compromised by the steric effect of too many enzyme molecules, and there will be an optimum stoichiometry balancing the two effects.

A number of factors will affect the position of this optimum. There may be more or less reactive groups in the antibody binding sites, for example, and these groups may be more or less reactive than those in the rest of the molecule. In a 'lucky' case, there may be no reactive groups in the binding sites, or those that are there may be less reactive than those in the rest of the molecule; in an 'unlucky' situation the reverse may hold. Often however there will be little or no differential distribution or reactivity of such groups, and the resulting conjugate is bound to contain a proportion of material whose antibody binding sites are disabled to some extent. Attempting to increase the enzyme incorporation may therefore result in reduced overall conjugate performance due to impaired binding.

This problem can be avoided by derivatising the antibody at a site known to be removed from the antigen-binding regions, e.g. by using the carbohydrate groups contained in other regions of the molecule. However it is still possible that the activity of the antibody will be hindered, simply by the bulk of the enzymes or by more subtle mechanisms—a change in the charge or polarity of a conjugate compared to the native antibody may prevent it from approaching the antigen closely, for example. These factors too may affect the optimum conjugate stoichiometry.

Finally, there may be an effect of 'diminishing returns' in incorporating more enzyme molecules. The fact that twice as many enzymes are present on a conjugate does not necessarily mean that there will be twice as much signal obtained from it—there may be quenching effects where some of the extra fluorescence or luminescence produced is absorbed, for example, or the problem of overcoming diffusion limitations to supply the enzymes with enough substrate in their microenvironment may be heightened[1].

The factors described for antibody–enzyme conjugates are generally mirrored in other applications, but their impact may vary dramatically from case to case. One of the most important factors is the relative size of the two partners: for example it may be possible to label a very large protein with many smaller groups without hindering its activity, and with sufficient space between the labels to avoid quenching. High incorporations of small fluorophores can still display quenching problems, however (Midoux *et al.*, 1987). The effect of these factors is difficult to quantify in a given situation, and, although experience can yield guidelines, a degree of trial and error may well be necessary to obtain the best result.

Where there is a desired 'target' stoichiometry, its production in the conjugation reaction can be promoted by manipulation of reaction ratios. Again the most controlled approach is to use heterobifunctional reagents, e.g. to obtain a 3 : 1 conjugate (HRP_3IgG) in the example above, the derivatisation of the IgG activation will be optimised to give a predominance of material containing three maleimide groups, while the HRP activation will be aimed at one thiol per molecule.

Finally, if a 1 : 1 conjugate does represent the 'Holy Grail' in a particular application, and it is critical to avoid the presence of any larger components in the product, it may be advantageous to use sub-stoichiometric quantities (e.g. 0.25 molar equivalents) of the heterobifunctional activating agents. Under the right

[1] Both of these effects may be exacerbated if enzyme dimers or polymers are used as the labelling reagent, to increase the incorporation from the same number of reactive sites on the antibody. Some enzyme activity may well be lost in preparing the dimer or polymer as well. In practice, although exceptions exist, these approaches rarely give the increase in performance that might be hoped for.

conditions this will result in a mixture of (for example) HRP-SH and unreacted HRP, but virtually no HRP-$(SH)_2$, etc. If IgG is derivatised by maleimide in the same fashion, HRP-IgG is the largest product that can be formed.

The obvious disadvantages of this strategy are two-fold: the waste of material which will not end up in the conjugate, and the problem of separating the product from large quantities of both unreacted starting materials, or (generally more difficult) separating activated from unactivated proteins before the reaction is carried out.

2.6.2.2 Statistical analysis of product stoichiometry

The empirically-determined stoichiometries of protein conjugates or derivatives encountered in the literature are, by necessity, mean values. Although some degree of separation between product fractions (and more often between product and unreacted starting materials) may be obtained, the sub-populations are seldom resolved sufficiently to allow individual analyses to be carried out.

Simple statistical techniques can be applied to obtain some insight into the distribution of stoichiometries within a population of coupled protein products. In the literature such calculations have been used primarily in the investigation of energy transfer between protein-bound ligands (Green, 1964; Weber and Daniel, 1966; Gennis and Cantor, 1972), but the statistics apply equally to any protein derivatisation reactions.

An assumption must be made that each potential derivatisation site on the protein is equally and independently reactive. It is likely that neither of these conditions will be met in many cases—the reactivity of sites may vary with the protein microenvironment, and the derivatisation of one site may hinder another due to steric constraints. Nevertheless, this approach can provide a broad guide to likely product distributions.

With a mean incorporation, μ (Weber and Daniel, 1966), the probability of finding a protein molecule containing n ligands out of a possible total of N is given by the binomial equation:

$$P(n, \mu) = N!/[n!(N-n)!] \cdot (\mu/N)^n \cdot [1 - (\mu/N)]^{N-n}$$

Thus in a situation where a protein has four equally-reactive amine groups available for derivatisation by a thiolation reagent ($N = 4$) and the mean thiol incorporation is 1.0 ($\mu = 1$), the actual fraction of protein molecules containing one thiol group ($n = 1$) is given by:

$$\begin{aligned} P(1,1) &= 4!/1! \cdot 3! \times (2/4)^1 \times [1 - (2/4)]^3 \\ &= 4 \times 0.25 \times 0.422 \\ &= 42.2\% \end{aligned}$$

The full distribution for all possible values of n in this example is shown in Figure 2.40.

This approach requires a knowledge of the total number of sites available for reaction. Where this is unknown, other authors have employed the Poisson distribution for the labelled complexes—an approximation of the binomial distribution where the probability of an event occuring (in this case a thiol being incorporated) is low (Green, 1964; Gennis and Cantor, 1972). Situations where the probability of derivatisation at a given site is too high may not fit this model well. However, this approach can again offer a rough guide to likely distribution patterns, particularly for large proteins with many potential derivatisation sites. This distribution is described by the equation:

$$P(n, \mu) = \frac{\mu^n \cdot e^{-\mu}}{n!}$$

Where the mean incorporation is 1, the Poisson distribution estimates that 37% of the protein molecules are unlabelled, 37% contain a single thiol molecule,

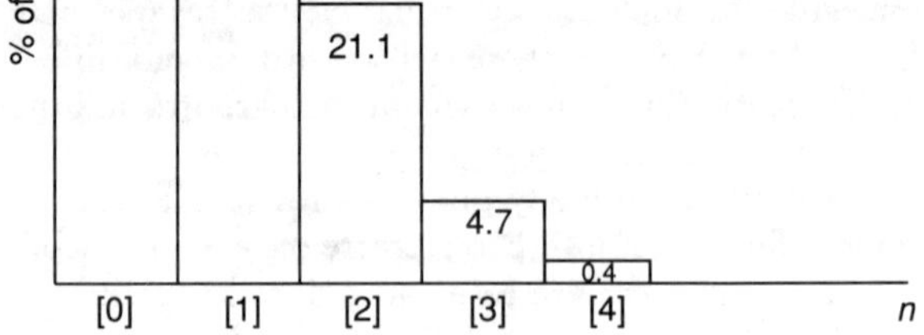

Figure 2.40 Binomial distribution of thiol incorporation (n) in a protein with four derivatisation sites ($N = 4$), where the mean incorporation is 1 ($\mu = 1$).

18% contain two, 6% contain three, 1.5% contain four, etc. (Figure 2.41).

Åkerblom *et al.* (1993) analysed a series of monomeric immunotoxins by electrophoresis to determine the distribution of toxin incorporations. Some examples of these distributions are shown in Figure 2.42, and can be compared with Figures 2.40 and 2.41.

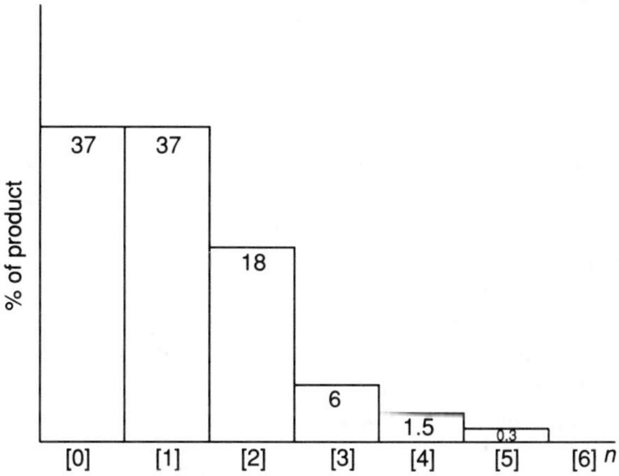

Figure 2.41 Poisson distribution of thiol incorporation in a derivatised protein, where the mean incorporation is 1 ($\mu = 1$).

When interpreting the relevance of mean incorporations it is tempting to assume a more-or-less homogeneous population. The shape of Figures 2.40 and 2.41 should be borne in mind on these occasions! Another point to consider is that the techniques used to measure incorporations, like any other measurement method, have their own limitations of sensitivity and precision. Some understanding of these limitations is essential in the interpretation of whether apparent differences are indeed significant.

2.6.3 The limitations of protein-coupling chemistry

While the techniques used in the derivatisation of proteins are derived from broader applications in synthetic organic chemistry, they form a very limited subset indeed. The need to minimise damage to the remainder of the molecule during the reaction rules out conditions which are ubiquitous in organic synthesis—raised temperatures and/or pressures, organic solvent systems, extremes of pH, etc.

As a broad generalisation, only reactions which can take place in a predominantly aqueous system, in a pH range of 3–11 and at temperatures below 40–50°C are of interest to protein chemists if the product is to retain its functionality. These constraints clearly arise

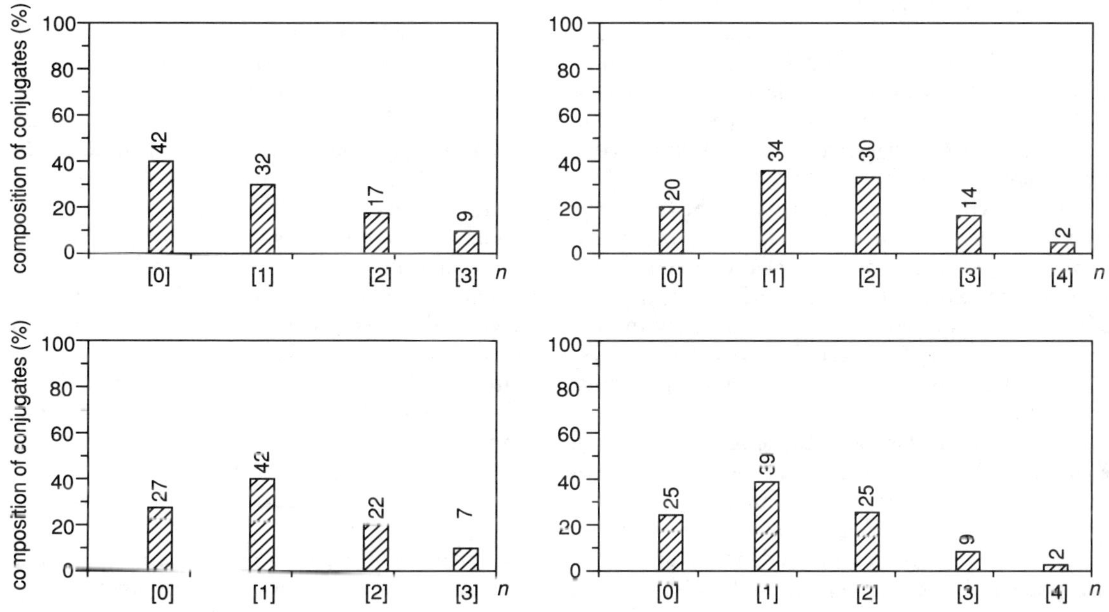

Figure 2.42 The distribution of incorporations in a number of immunotoxin preparations, analysed by SDS-PAGE; *n* is the number of toxin groups incorporated per antibody molecule. (Reproduced with permission from: Åkerblom *et al.* (1993). *Bioconjugate Chem.*, **4**, 455–466.)

because evolution has favoured structures which are best suited to their natural surroundings, most of which fall within the criteria just described.

This reasoning also gives the best clue to the source of exceptions to these generalisations. Where organisms have been faced with environments chronically outside these ranges, evolution has promoted traits which improve efficiency under the ambient conditions. Unless the organism can protect its internal structures from the environment, this will often require the mutation of proteins to give acceptable stability and performance, and proteins of this type may well survive (or indeed function best in) conditions outside the ranges above.

Because of their rapid replication capabilities microorganisms such as bacteria adapt most efficiently to such challenges, and wherever extreme environments have existed for long enough, there are 'extremophilic' bacteria attempting to exploit them. The proteins used by these organisms can have interesting properties: the polymerase chain reaction which has had such an impact on molecular biology was made possible by the use of heat-resistant polymerases from thermophilic bacteria found in hot springs. Other applications are more homely, for example the use of heat- and alkali-resistant enzymes in detergents.

Even within an organism existing under 'normal' conditions, more extreme microenvironments may have evolved, and tolerant proteins along with them. The mammalian digestive system is a classic example—strongly acidic conditions favour breakdown of food by hydrolytic reactions, while acid-tolerant proteolytic enzymes such as pepsin have been evolved to work as a cooperative digestive mechanism in the same environment.

So for specific proteins there may be a greater tolerance on temperature, pH, etc. than the broad ranges quoted, but the vast bulk of protein chemistry has to operate under very gentle conditions.

Finally, many proteins are also poisoned by specific reagents, e.g. urease by silver, haemoproteins by cyanide or carbon monoxide. These considerations must also be taken into account.

A related point is the possibility of interference with the coupling chemistry rather than the protein molecule itself. For example, sodium azide is commonly encountered in protein chemistry as a bacteriostat, but it is not always realised that the azide ion is a powerful nucleophile and can compete strongly with amines, etc. for electrophilic derivatisation reagents. Simple buffer components are another source of potential pitfalls, particularly those containing amines (e.g. Tris) or carboxylates (e.g. acetate). Many a promising coupling reaction has come to grief because such factors have been neglected!

2.6.4 Other factors affecting the choice of coupling reactions

Within the limitations described in the previous section, there remain a very large number of possible approaches for coupling one molecule to another. A range of factors affect the choice of method:

2.6.4.1 Choice of molecules to be coupled

Ignoring for a moment the chemistry required for coupling, the choice of molecules to be linked obviously stems directly from the individual aim of the experimenter. In some cases this may leave no room for choice. If the aim of the experiment is to create a conjugate of β-lactamase and a specific monoclonal antibody then no further consideration is necessary on this point, beyond the usual precautions of using pure, well-characterised and well-quantified starting materials.

However there are cases where there is still some choice, for example if the aim is to produce a preparation of an enzyme immobilised onto latex beads. Generally the only requirement of the experimenter is that the beads should bind protein, and be insoluble and of suitable dimensions. Within this limitation a wide choice of beads of different composition is likely to be available, and it is worth putting some thought at this stage into which might be best in terms of the exact immobilisation properties, rather than starting with a preconception or using what is most readily to hand. Chapter 8 discusses many of the factors which might influence such a choice.

Similarly if the experimenter simply requires to label a specific protein with an enzyme that can be conveniently quantified, and is flexible in the choice of enzyme, then the comparisons made between major candidates in Chapter 3 may be helpful.

In general though it is not the aim here to influence the experimenter in what to couple, only in how to couple it.

2.6.4.2 Choice of coupling chemistry

The intention of this book is to present a wide range of methods, which may be useful in any situation where a protein molecule is to be coupled—regardless of what it is to be coupled to.

Ideally the choice among these methods is made with as much knowledge as possible of the functional chemistry of the two components, although it is generally reasonable to make assumptions about the available groups on the protein(s) involved.

Section 2.6.1 has already discussed the differences between direct and linker methods, and this should be part of the choice. It has been seen how much more control can be gained by using heterobifunctional approaches. If a particular direct coupling chemistry is chosen from those presented in this book, then the optimisation of this method for the individual task is all that remains to be achieved.

With bifunctional approaches, however, one further choice can be made: even though the reactive groups at each end of the molecule may have been decided already, they could conceivably be joined by any suitably inert linker group, and this can have an important influence on the properties of the product.

2.6.4.3 Length, flexibility and polarity of linker groups

As noted above, a linker group can provide a useful spacer between one conjugation partner and another (Figure 2.43a). If such a spacer is required to overcome steric effects it should be long enough to achieve this purpose; while it is obviously impossible to generalise too much a distance of around 10 Å is a good starting choice. This is roughly equivalent to a seven or eight-carbon bridge.

However, if the linker groups are very flexible (e.g. if there are no double bonds or cyclic moieties in the structure), they may have little real impact in providing a separation. With protein–protein conjugates, for example, the two molecules may simply drift together, and coalesce *via* hydrophobic interactions (Figure 2.43b). This effect may be heightened if the linker itself is hydrophobic[1], as there is a high likelihood that this too will bind to hydrophobic regions in the protein molecule(s).

The presence of a flexible, hydrophobic linker could in fact be looked at as providing the optimal conditions for hydrophobic aggregation of the conjugation partners. The two molecules are tethered closely together so that thermal collisions between them are likely to occur very frequently; the flexibility of the tether,

[1] There are of course occasions where a hydrophobic reagent is deliberately chosen, for example as a probe for lipoproteins.

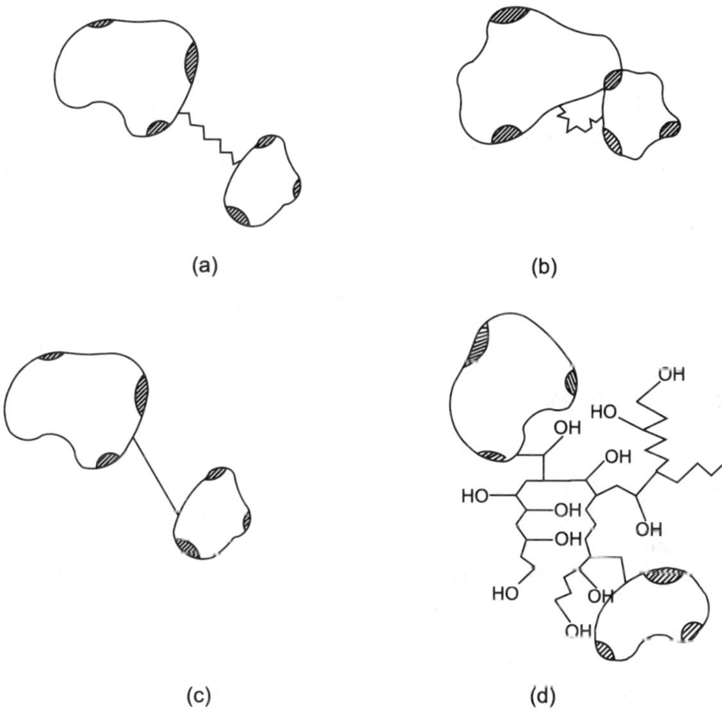

Figure 2.43 Behaviour of linker groups. A linker group may be long enough to separate two protein molecules (a), but if it is flexible, interactions between hydrophobic regions may still occur (b). This may be avoided by the use of a linker which is rigid (c) or bulky and/or polar (d).

however, allows many different protein–protein interactions to be sampled during these collisions, with a resulting high probability that a hydrophobic interaction will eventually be provoked.

Examination of many of the coupling reagents described in this book will reveal that they are in fact both flexible and hydrophobic in nature, and their ability to create a real (as opposed to theoretical) separation between partners is perhaps rather questionable. In these terms ideal linkers ought to possess sufficient rigidity to prevent the drift together of conjugation partners (Figure 2.43c), or be polar and/or bulky enough to discourage such interactions (Figure 2.43d). If it is genuinely desired to create a permanent separation between two components, it may be worth considering the chemistry carefully, including somewhat more unusual options such as coupling *via* a bulky hydrophilic polymer, a dextran or polyethyleneglycol (PEG) for example.

2.6.4.4 Stability of linker groups

Once a coupled protein reagent has been prepared, it is generally desired that the link should remain stable in solution for a substantial period of preparation, and this may affect the choice of coupling methods too.

The long-term stability of conjugate solutions is an area which has received surprisingly little detailed attention in the literature, but it is certainly true that products from some common approaches (e.g. ureas from isothiocyanate coupling) are rather labile compared to others (e.g. amide bonds from carbodiimide methods or succinimidyl esters). Wherever possible the discussion of individual chemistries in this book has included a consideration of the likely stability of the link.

2.6.4.5 Cleavability of linker groups

Conversely, in certain circumstances it is desirable to use a linker group which can be cleaved at will, under moderate conditions. Examples include structural studies where protein sub-units have been linked, and it is desired to uncouple them for analytical reasons, or pharmaceutical applications where disintegration of a conjugate is desired *in vivo*. Linker cleavability can have a significant effect on the *in vivo* behaviour of protein conjugates (Bridger *et al.*, 1996; Patel *et al.*, 1996).

The cleavability concept can be applied very elegantly. Ito *et al.* (1994), for example, described a device for releasing insulin in response to the presence of glucose, in which glucose oxidase was coupled to insulin *via* a disulfide link. Glucose is oxidised by the enzyme, this reaction generates reducing species which are capable of cleaving the S–S bond, and the insulin is accordingly released.

A number of cleavable groups and the conditions required for cleavage are given in Table 2.3; the latter are taken from Wong (1991b) unless otherwise stated. Many of these require somewhat harsh conditions which can harm proteins, basic hydrolysis or hydroxylaminolysis for example. Disulfides are by far the most commonly encountered cleavable groups, but the acid-labile group derived from citraconic anhydride is useful in some circumstances, e.g. where it is desired to cleave a bond in acidic intracellular endosomes.

More sophisticated examples of this approach exploit the incorporation of groups cleavable by specific enzymes such as chymotrypsin (Imai *et al.*, 1990) or β-lactamase (Senter *et al.*, 1995). *In vivo* cleavage of reagents is difficult to achieve in a controlled fashion, and this remains a challenge (Arano *et al.*, 1996). Finally, the photolabile properties of nitrobenzyl esters (Patchornik *et al.*, 1970) offer some interesting possibilities.

2.6.4.6 Iodinatability of linker groups

One final feature of linkers is important in certain structural applications: the ability for it to be readily iodinated for radiolabelling purposes. Typically this is achieved by incorporation of a phenolic group—many commercial reagents offer this facility.

2.6.4.7 Optimisation of coupling conditions

Although general rules are given wherever possible throughout this book, to achieve the best performance it will still often be necessary to optimise a particular chemistry for the application in hand.

There are certain 'tricks' that can be applied in particular circumstances to improve product properties, e.g. protecting the active regions of a protein during conjugation (Ramjeesingh *et al.*, 1990), or opening up new sites for derivatisation by provoking conformational changes (Degani and Heller, 1988; Weston, 1995).

In general however, optimisation will require a study of key process variables. In terms of influencing product properties reagent stoichiometry is generally the best place to start. This is followed by pH, reaction time and temperature in that general order, plus choice and concentration of organic cosolvent if one is required. A well-designed study of this type will

Table 2.3 Some Useful Cleavage Reactions

Cleavable group	Products		Cleavage reagent
R—S—S—R' (Disulfide)	R—SH	HS—R'	Reducing agent
R—S—S—R' (Disulfide)	R—S—SO_3^-	^-O_3S—S—R'	Sulfite
R—S—Hg—R' (Mercurial)	R—SH	R"S—Hg—R'	R"SH
R—C(OH)—C(OH)—R' (glycol)	R—C(=O)—H	H—C(=O)—R'	Periodate (eg 15 mM $NaIO_4$, pH 7.5, 4 h
R—N=N—R' (Diazo compound)	R—NH_2	H_2N—R'	Dithionite (eg 100 mM $Na_2S_2O_4$, pH 8.0, 25min)
R—S(=O)(=O^-)—R' (Sulfone)	R—OH	SO_3^-—R'	Base (eg pH 11-12, 2 h, 37 °C)
R—C(=O)—O—R' (Ester)	R—C(=O)OH	HO—R'	Acid or base hydrolysis (eg 100 mM NaOH, Hydroxylamine (eg 1M NH_2OH, pH7.5-8.5, 25-37 °c, 3-6h)
R—C(=O)—S—R' (Thioester)	R—C(=O)—NHR"	HS—R'	R"—NH_2
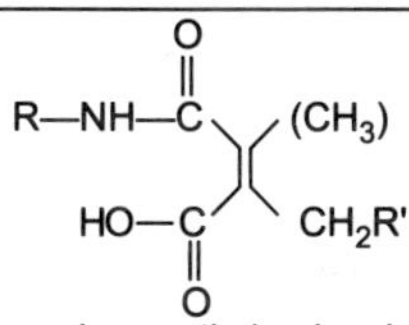 (Citraconyl or methylmaleoyl amide)	R—NH_2	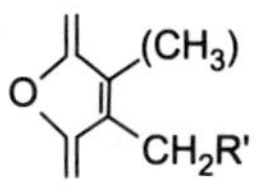	Mild acid (pH 4.5)
R—O—CHR"—O—R' (Acetal)	R—OH	HO—R'	Mild acid (pH 5.5)
R—O—CR"$_2$—O—R' (Ketal)	R—OH	HO—R'	Mild acid (pH 5.5)
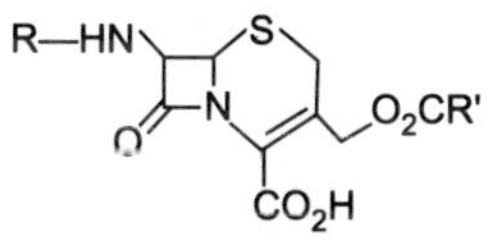 (β-Lactam derivative)	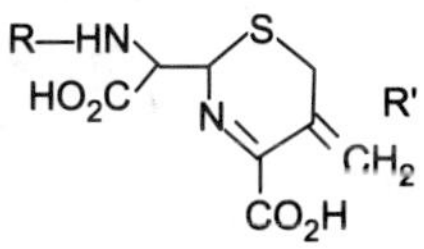	HO_2C—R'	β-Lactamase (Senter *et al.*, 1995)
	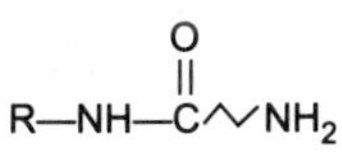	R"–NH–C(=O)–R'	Chymotrypsin, tyrosinamide (R"–NH_2) (Imai *et al.*, 1990)
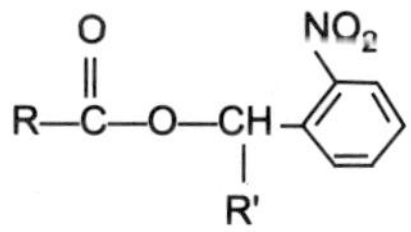	R—C(=O)—OH	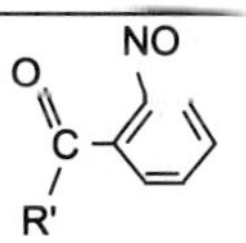	Light (320 nm) (Patchornik *et al.*, 1970)

often provide acceptable performance without too much effort.

A final point will be discussed further in Chapter 13. If the coupling process is to be carried out frequently with the goal of providing a reproducible product, the robustness of the method should also be considered at this stage. Optimisation for product properties is not always the same as optimisation for manufacturability, and trade-offs may need to be made.

References

Åkerblom, E. *et al.* (1993). Preparation and characterization of conjugates of monoclonal antibodies and staphylococcal enterotoxin A using a new hydrophilic cross-linker. *Bioconjugate Chem.*, **4**, 455–466.

Allen, H. J. and Kisailus, E. C. (1992). Editors of *Glycoconjugates. Composition, Structure and Function*, Marcel Dekker, Inc., New York.

Andrade, J. D. and Hlady, V. (1986). Protein adsorption and materials biocompatability: a tutorial review and suggested hypotheses. *Adv. Polym. Sci.*, **79**, 1–63.

Aplin, J. D. *et al.* (1979). Spin-labelling of sialic acid in soluble and cell-surface glycoproteins. *Carbohydr. Res.*, **75**, 11–16.

Arano, Y. *et al.* (1996). Assessment of radiochemical design of antibodies using an ester bond as the metabolizable linkage: evaluation of maleimidoethyl 3-(tri-*n*-butylstannyl)hippurate as a radioiodination reagent of antibodies for diagnostic and therapeutic applications. *Bioconjugate Chem.*, **7**, 628–637.

Ashwell, G. and Harford, J. (1982). Carbohydrate-specific receptors of the liver. *Annu. Rev. Biochem.*, **51**, 531–554.

Ballew, R. M. *et al.* (1996). Direct observation of fast protein folding: the initial collapse of apomyoglobin. *Proc. Natl. Acad. Sci. USA*, **93**, 5759–5764.

Blanchard, J. S. (1984). Buffers for enzymes. *Methods Enzymol.*, **104**, 404–414.

Bosshard, H. R. *et al.* (1984). The cytochrome *c* peroxidase–cytochrome *c* electron transfer complex. The role of histidine residues. *J. Biol. Chem.* **259**, 5683–5690.

Bridger, G. J. *et al.* (1996). A comparison of cleavable and noncleavable hydrazinopyridine linkers for the ^{99m}Tc labeling of Fab′ monoclonal antibody fragments. *Bioconjugate Chem.*, **7**, 255–264.

Brocklehurst, K. (1979). Specific covalent modification of thiols: applications in the study of enzymes and other biomolecules. *Int. J. Biochem.*, **10**, 259–274.

Bunnett, J. F. (1963). Nucleophilic reactivity. *Annu. Rev. Phys. Chem.*, **14**, 271–290.

Bussemaker, H.J. *et al.* (1997). Thermodynamic stability of folded proteins against mutations. *Phys. Rev. Lett.*, **79**, 3530–3533.

Carrier, A. *et al.* (1995). Recombinant antibody–alkaline phosphatase conjugates for diagnosis of human IgGs: application to anti-HBsAg detection. *J. Immunol. Methods*, **181**, 177–186.

Conn, E. E. and Stumpf, P. K. (1976). Amino acids and proteins. In *Outlines of Biochemistry*, 4th edit., pp. 73–110. John Wiley & Sons Inc., New York.

Creighton, T. E. (1997). How important is the molten globule for correct protein folding? *Trends Biochem. Sci.*, **22**, 6–10.

Creighton, T. E. *et al.* (1995). Mechanisms and catalysts of disulphide bond formation in proteins. *Trends Biotechnol.*, **13**, 18–23.

Dahiyat, B. I. and Mayo, S. L. (1997). *De novo* protein design: fully automated sequence selection. *Science*, **278**, 82–87.

Degani, Y. and Heller, A. (1988). Direct electrical communication between chemically modified enzymes and metal electrodes. 2. Methods for bonding electron-transfer relays to glucose oxidase and D-amino acid oxidase. *J. Am. Chem. Soc.*, **110**, 2615–2620.

De la Llosa, P. *et al.* (1980). Oxidation of methionine residues in lutropin. *Can. J. Biochem.*, **58**, 745–748.

Dixon, H. B. F. and Perham, R. N. (1968). Reversible blocking of amino groups with citraconic anhydride. *Biochem. J.*, **109**, 312–314.

Edge, A. S. B. *et al.* (1981). Deglycosylation of glycoproteins by trifluoromethanesulfonic acid. *Anal. Biochem.*, **118**, 131–137.

Edwards, J. O. and Pearson, R. G. (1962). The factors determining nucleophilic reactivities. *J. Am. Chem. Soc.*, **84**, 16–24.

Ellis, J. (1987). Proteins as molecular chaperones. *Nature*, **328**, 378–379.

Englander, S. W. (1993). In pursuit of protein folding. *Science*, **262**, 848–849.

Finney, J. L. (1979). The organization and function of water in protein crystals. In *Water – A Comprehensive Treatise* (Franks, F., ed.), vol. 6, pp. 47–122, Plenum Press, New York.

Fontana, A. (1972). Modification of tryptophan with BNPS-skatole (2-(2-nitrophenylsulfenyl)-3-methyl-3-bromoindolenine. *Methods Enzymol.*, **25**, 419–423.

Fothergill-Gilmore, L. A. (1993). Protein fragmentation. In *Protein Biotechnology. Isolation, Characterization and Stabilization* (Franks, F., ed.), p. 257, Humana Press, Totowa, NJ.

Franks, F. (1993a). Proteins. Description and classification. In *Protein Biotechnology. Isolation, Characterization and Stabilization* (Franks, F., ed.), pp. 1–19, Humana Press, Totowa, NJ.

Franks, F. (1993b). Internal structure and organization. Relationship to function. In *Protein Biotechnology. Isolation, Characterization and Stabilization* (Franks, F., ed.), pp. 91–132, Humana Press, Totowa, NJ.

Franks, F. (1993c). Solution properties of proteins. In *Protein Biotechnology. Isolation, Characterization and Stabilization* (Franks, F., ed.), pp. 133–189, Humana Press, Totowa, NJ.

Franks, F. (1993d). Protein hydration. In *Protein Biotechnology. Isolation, Characterization and Stabilization* (Franks, F., ed.), pp. 437–465, Humana Press, Totowa, NJ.

Frydman, J. *et al.* (1994). Folding of nascent polypeptide chains in a high molecular mass assembly with molecular chaperones. *Nature*, **370**, 111–117.

Gennis, R. B. and Cantor, C. R. (1972). Use of nonspecific dye labeling for singlet energy-transfer measurements in complex systems. A simple model. *Biochemistry*, **11**, 2509–2517.

Gilchrist, T. L. and Rees, C. W. (1969). *Carbenes, Nitrenes and Arynes*, Thomas Nelson and Sons Ltd, London.

Good, N. E. *et al.* (1966). Hydrogen ion buffers for biological research. *Biochemistry*, **5**, 467–477.

Green, N. M. (1964). Avidin. 5. Quenching of fluorescence by dinitrophenyl groups. *Biochem. J.*, **90**, 564–568.

Gross, M. and Plaxco, K. W. (1997). Reading, writing and redesigning. *Nature*, **388**, 419–420.

Gundlach, H. G. *et al.* (1959). The reaction of iodoacetate with methionine. *J. Biol. Chem.*, **234**, 1761–1764.

Habeeb, A. F. S. A (1972). Guanidination of proteins. *Methods Enzymol.*, **25**, 558–566.

Harper, H. A. *et al.* (1979). Proteins. In *Review of Physiological Chemistry*, 17th edn, pp. 37–47, Lange Medical Publications, Los Altos, CA.

Harris, E. L. V. (1989). Initial planning. Basic preparations. In *Protein Purification Methods A Practical Approach* (Harris, E. L. V. and Angal, S., eds), pp. 6–7, IRL Press, Oxford.

Hollaway, M. R. *et al.* (1964). The chemical reactivity of the thiol group in the active centre of ficin. *Biochim. Biophys. Acta*, **92**, 111–124.

Huheey, J. E. (1975). *Inorganic Chemistry: Principles of Structure and Reactivity (SI Units Edition)*, pp. 160–162, Harper & Row, London.

Imai, N. *et al.* (1990). Photoaffinity heterobifunctional cross-linking reagents based on *N*-(azidobenzoyl)tyrosines. *Bioconjugate Chem.*, **1**, 138–143.

Ingold, C. K. (1953). *Structure and Mechanism in Organic Chemistry*, pp. 198–199, G. Bell and Sons, London.

Ito, Y. *et al.* (1994). Design and synthesis of a protein device that releases insulin in response to glucose concentration. *Bioconjugate Chem.*, **5**, 84–87.

IUPAC-IUB Commission on Biochemical Nomenclature (1972). Symbols for amino acids and peptides. Recommendations (1971). *J. Biol. Chem.*, **247**, 977–983.

Jenkins, N. *et al.* (1996). Getting the glycosylation right: implications for the biotechnology industry. *Nature Biotechnol.*, **14**, 975–981.

Kiyokawa, T. *et al.* (1991). Protein engineering of diphtheria-toxin-related interleukin-2 fusion toxins to increase cytotoxic potency for high-affinity IL-2-receptor-bearing target cells. *Protein Eng.*, **4**, 463–468.

Kornfeld, R. and Kornfeld, S. (1985). Assembly of asparagine-linked oligosaccharides. *Annu. Rev. Biochem.*, **54**, 631–664.

Kurumbail, R. G. *et al.* (1996). Structural basis for selective inhibition of cyclooxygenase-2 by anti-inflammatory agents. *Nature*, **384**, 644–648.

Lindbladh, C. *et al.* (1993). Use of genetically prepared enzyme conjugates in enzyme immunoassay. *Trends Biochem. Sci.*, **18**, 279–283.

Low, B. W. and Edsall, J. T. (1956). Aspects of protein structure. In *Currents in Biochemical Research*, pp. 378–433. Interscience Publishers, Inc., New York.

Lundblad, R. L. and Noyes, C. M. (1984a). *Chemical Reagents for Protein Modification*, vol. 1, CRC Press, Boca Raton, FL.

Lundblad, R. L. and Noyes, C. M. (1984b). *Chemical Reagents for Protein Modification*, vol. 2, CRC Press, Boca Raton, FL.

Malamud, D. and Drysdale, J. W. (1978). Isoelectric points of proteins: a table. *Anal. Biochem.*, **86**, 620–647.

March, J. (1985a). Acids and bases. In *Advanced Organic Chemistry. Reactions, Mechanisms and Structure*, 3rd edn., pp. 218–236, John Wiley & Sons, New York.

March, J. (1985b). Aliphatic nucleophilic substitution. In *Advanced Organic Chemistry. Reactions, Mechanisms and Structure*, 3rd edn, pp. 255–446, John Wiley & Sons, New York.

Marchler-Bauer, A. and Bryant, S. H. (1997). A measure of success in fold recognition. *Trends Biochem. Sci.*, **22**, 236–240.

McCoy, J. P., Jr. and Chambers, W. H. (1991). Carbohydrates in the functions of natural killer cells. *Glycobiology*, **1**, 321–328.

Midoux, P. *et al.* (1987). Quantitation of the binding, uptake and degradation of fluoresceinylated neoglycoproteins by flow cytometry. *Cytometry*, **8**, 327–334.

Mikolajczyk, S. D. *et al.* (1996). Dextran modification of a Fab′–β-lactamase conjugate modulated by variable pretreatment of Fab′ with amine-blocking reagents. *Bioconjugate Chem.*, **7**, 150–158.

Nakagawa, Y. and Bender, M. L. (1970). Methylation of histidine in α-chymotrypsin by methyl *p*-nitrobenzenesulfonate. A new approach to enzyme modification. *Biochemistry*, **9**, 259–267.

Netzer, W. J. and Hartl, F. U. (1997). Recombination of protein domains facilitated by co-translational folding in eukaryotes. *Nature*, **388**, 343–349.

Olden, K. *et al.* (1982). Function of the carbohydrate moieties of glycoproteins. *J. Cell. Biochem.*, **18**, 313–335.

Oncley, J. L. *et al.* (1947). Physical-chemical characteristics of certain of the proteins of normal human plasma. *J. Phys. Colloid Chem.*, **51**, 184–198.

Oshiro, Y. and Eylar, E. H. (1968). Physical and chemical studies on glycoproteins. III. The microheterogeneity of fetuin, a fetal calf glycoprotein. *Arch. Biochem. Biophys.*, **127**, 476–489.

Parsons, S. M. *et al.* (1969). The nature of amino acid side chains which are critical for the activity of lysozyme. *Biochemistry*, **8**, 700–712.

Patchornik, A. *et al.* (1970). Photosensitive protecting groups. *J. Am. Chem. Soc.*, **92**, 6333–6335.

Patel, V. F. *et al.* (1996). Novel acid labile COL1 trityl-linked difluoronucleoside immunoconjugates: synthesis, characterization, and biological activity. *Bioconjugate Chem.*, **7**, 497–510.

Paulson, J. C. (1989). Glycoproteins: what are the sugar chains for? *Trends Biochem. Sci.*, **14**, 272–276.

Pearson, R. G. (1963). Hard and soft acids and bases. *J. Am. Chem. Soc.*, **85**, 3533–3539.

Pine, S. H. *et al.* (1981). Nucleophilic substitutions at saturated carbon. *Organic Chemistry*, 4th edn, pp. 380–382, McGraw-Hill, London.

Porath, J. (1986). Salt-promoted adsorption: recent developments. *J. Chromatogr.*, **376**, 331–341.

Ramjeesingh, M. *et al.* (1990). Antigen protection of monoclonal antibodies undergoing labelling. *J. Immunol. Methods*, **133**, 159–167.

Rana, T. M. and Meares, C. F. (1990). N-Terminal modification of immunoglobulin peptide chains tagged with isothiocyanato chelates. *Bioconjugate Chem.*, **1**, 357–362.

Riek, R. *et al.* (1996). NMR structure of the mouse prion protein domain PrP(121–231). *Nature*, **382**, 180–182.

Righetti, P. G. and Caravaggio, T. (1976). Isoelectric points and molecular weights of proteins. A table. *J. Chromatogr.*, **127**, 1–28.

Riordan, J. F. (1973). Functional arginyl residues in carboxypeptidase A. Modification with butanedione. *Biochemistry*, **12**, 3915–3922.

Rogers, T. B. *et al.* (1978). Chemical modification of arginines in transferrins. *Biochemistry*, **17**, 1105–1109.

Russo, A. A. *et al.* (1996). Crystal structure of the $p27^{Kip1}$ cyclin-dependent-kinase inhibitor bound to the cyclin A-Cdk2 complex. *Nature*, **382**, 325–331.

Scheffzek, K. *et al.* (1996). Crystal structure of the GTPase-activating domain of human p120GAP and implications for the interaction with Ras. *Nature*, **384**, 591–596.

Schroeder, W. A. *et al.* (1967). Modification of methionyl residues during aminoethylation. *Biochim. Biophys. Acta*, **147**, 590–592.

Segrest, J. P. *et al.* (1974). A molecular theory of lipid–protein interactions in the plasma lipoproteins. *FEBS Lett.*, **38**, 247–253.

Sélo, I. *et al.* (1996). Preferential labeling of α-amino N-terminal groups in peptides by biotin: application to the detection of specific anti-peptide antibodies by enzyme immunoassays. *J. Immunol. Methods*, **199**, 127–138.

Senter, P. D. *et al.* (1995). Poly(ethylene glycol)–doxorubicin conjugates containing β-lactamase-sensitive linkers. *Bioconjugate Chem.*, **6**, 389–394.

Sly, W. S. and Fisher, H. D. (1982). The phosphomannosyl recognition system for intracellular and intercellular transport of lysosomal enzymes. *J. Cell Biochem.*, **18**, 67–85.

Smith, P. L. and Baenziger, J. U. (1988). A pituitary N-acetylgalactosamine transferase that specifically recognizes glycoprotein hormones. *Science*, **242**, 930–933.

Sojar, H. T. and Bahl, O. P. (1987). A chemical method for the deglycosylation of proteins. *Arch. Biochem. Biophys.*, **259**, 52–57.

Spande, T. F. *et al.* (1970). Selective cleavage and modification of peptides and proteins. *Adv. Protein Chem.*, **24**, 97–260.

Srinivasan, R. and Ruckenstein, E. (1980). Role of physical forces in hydrophobic interaction chromatography. *Sep. Purif. Methods*, **9**, 267–370.

Stanley, P. (1984). Glycosylation mutants of animal cells. *Annu. Rev. Genet.*, **18**, 525–552.

Stark, G. R. and Smyth, D. G. (1963). The use of cyanate for the determination of NH_2-terminal residues in proteins. *J. Biol. Chem.*, **238**, 214–226.

Stark, G. R. *et al.* (1961). Relationships between the conformation of ribonuclease and its reactivity toward iodoacetate. *J. Biol. Chem.*, **236**, 436–442.

Staros, J. V. (1988). Membrane-impermeant cross-linking reagents: probes of the structure and dynamics of membrane proteins. *Acc. Chem. Res.*, **21**, 435–441.

Stauffer, C. E. and Etson, D. (1969). The effect of subtilisin activity on oxidizing a methionine residue. *J. Biol. Chem.*, **244**, 5333–5338.

Tskhovrebova, L. *et al.* (1997). Elasticity and unfolding of single molecules of the giant muscle protein titin. *Nature*, **387**, 308–312.

Uhlén, M. *et al.* (1992). Fusion proteins in biotechnology. *Curr. Opin. Biotechnol.*, **3**, 363–369.

Vapnek, D. (1990). A biotechnology view of glycobiology. *Glycobiology*, **1**, 3–4.

Varki, A. (1993). Biological roles of oligosaccharides: all of the theories are correct. *Glycobiology*, **3**, 97–130.

Vaughan, L. *et al.* (1982). α_1–Antitrypsin microheterogeneity. Isolation and physical significance of isoforms. *Biochim. Biophys. Acta*, **701**, 339–345.

Via, D. P. and Smith, L. C. (1986). Fluorescent labeling of lipoproteins. *Methods Enzymol.*, **129**, 848–857.

Wahl, M. C. and Sundaralingam, M. (1997). C–H···O hydrogen bonding in biology. *Trends Biochem. Sci.*, **22**, 97–102.

Wassarman, P. M. (1987). Early events in mammalian fertilization. *Annu. Rev. Cell. Biol.*, **3**, 109–142.

Weber, G. and Daniel, E. (1966). Cooperative effects in binding by bovine serum albumin. II. The binding of 1-anilino-8- naphthalenesulfonate. Polarization of the ligand fluorescence and quenching of the protein fluorescence. *Biochemistry*, **5**, 1900–1907.

West, C. M. (1986). Current ideas on the significance of protein glycosylation. *Mol. Cell. Biochem.*, **72**, 3–20.

Weston, S. A. *et al.* (1995). Effect of β-mercaptoethanol on the detection of biotinylated proteins. *Anal. Biochem.*, **225**, 28–33.

Wold, F. (1981). *In vivo* chemical modification of proteins (post-translational modification). *Annu. Rev. Biochem.*, **50**, 783–814.

Wong, S. S. (1991a). Reactive groups of proteins and their modifying agents. In *Chemistry of Protein Conjugation and Cross-Linking*, pp. 7–48, CRC Press, Boca Raton, FL.

Wong, S. S. (1991b). Choice and design of cross-linking agents. In *Chemistry of Protein Conjugation and Cross-Linking*, pp. 49–73, CRC Press, Boca Raton, FL.

Yamaguchi, H. and Hendrickson, W. A. (1996). Structural basis for activation of human lymphocyte kinase Lck upon tyrosine phosphorylation. *Nature*, **384**, 484–490.

Yamasaki, R. B. *et al.* (1980). Modification of available arginine residues in proteins by *p*-hydroxyphenylglyoxal. *Anal. Biochem.*, **109**, 32–40.

Yankeelov, J. A., Jr (1972). Modification of arginine by diketones. *Methods Enzymol.*, **25**, 566–579.

Chapter 3

Enzymes

Alastair H. Dent and Mohammed Aslam

Research and Development Department, Ortho-Clinical Diagnostics, Pollards Wood Laboratories, Nightingales Lane, Chalfont St. Giles HP8 4SP, UK

The preceding chapter has described generic properties of proteins which affect the chemistry used to couple them to each other, or to other molecules, and the means of separating the products of these reactions from other components. It is clearly impossible to give more specific information on the thousands of different proteins which might be encountered in the fields addressed by this book; however, there are two classes of protein whose properties are so useful that they have cropped up again and again in the chapter describing applications. These are enzymes and antibodies, and in the next two chapters these will be considered in greater detail.

Enzymes are catalysts. The classical definition of a catalyst can be worded in a number of different ways, but the essence is as follows:

Catalysts accelerate the rate of reactions. Their chemical state is identical at the beginning and the end of the reaction.

Thus while the catalyst has to take part in at least one stage of the reaction (otherwise it can have no effect on the rate!), it is always regenerated at a subsequent stage and is therefore available to be used again. So a characteristic of catalysts is that they can be used in quantities much less than the stoichiometric equivalent of the reagents which *are* chemically altered at the end of the reaction.

Catalysts are, of course, not magically resistant to reactions other than the one they are catalysing, and they may appear to be 'used up'. The metals which catalyse hydrogenation of alkenes, for example, may gradually lose their activity during the reaction – but this is due to a competing reaction with a 'poison' such as a sulfide. Purely in terms of the reaction being catalysed, the metal ends up in exactly the same chemical state as it began.

3.1 The Theoretical Basis of Catalysis

To understand catalysis it is necessary to review very briefly the factors which determine whether a chemical reaction 'goes' or not, and if so how quickly.

3.1.1 Thermodynamics

Thermodynamics considers whether one arrangement of a set of atoms is more or less likely than another to exist, based on the energy tied up in each arrangement. To form a covalent bond between two atoms requires that energy is put into the system, and the same amount of energy is released when the bond is broken. Each bond in an arrangement of atoms (i.e. one or more molecules) contributes a discrete amount of energy, and these amounts can be summed to give the overall heat content or enthalpy (H). When the heat contents of two arrangements of the same atoms are compared, the difference (ΔH) is the amount of energy released or consumed in going from one to the other. In many situations this parameter is the most important factor in determining the most stable (i.e. low energy) arrangement, and consequently the likely course of a reaction. The majority of favoured reactions result in a net decrease in the enthalpy of the system; these generate kinetic energy (heat) and are described as exothermic. However due to a second effect there are many reactions which are still thermodynamically favoured while consuming kinetic energy—endothermic reactions.

This second effect is entropy, which is a measure of the degree of disorder in an arrangement of atoms. The most disordered system, i.e. that with the highest entropy term, is the most favoured. Because the importance of entropy is related to the ability of the atoms to move, this term contains a temperature component (T) which expresses the kinetic energy available in the system. Where a reaction results in no change to the number of molecules, to their phase, or the degree to which they are associated with solvent molecules, the entropy change (ΔS) will be small. However, where this is not the case, it can be an important parameter. Situations where the disorder is significantly increased, such as the dissolution of a solid, may be thermodynamically favoured even though there is no decrease in enthalpy. An interesting case is where a protein is adsorbed from solution onto a solid phase; although the disorder of the macromolecule is reduced by its immobilisation, this is generally more than outweighed by the increase in possible arrangements of the water molecules. Therefore there is a net increase in entropy, and this is often a driving force in immobilisation reactions.

Correcting the enthalpy change of a potential reaction for entropy effects gives a parameter called the Gibbs free energy change (ΔG), the relationship being described as follows:

$$\Delta G = \Delta H - (T\Delta S)$$

It is the change in free energy that determines whether or not a reaction is likely to proceed in a given direction, rearrangements of atoms resulting in a negative ΔG being favoured. Notice that it is only a likelihood or probability that is predicted in this way; in real systems both forward and reverse reactions occur, and the free energy change determines the relative frequency of these.

For a generalised reaction:

$$x\mathrm{A} + \mathrm{y}B \Rightarrow z\mathrm{C}$$

it is a condition of the equilibrium law that the ratio of products to reactants remains constant. Therefore, an equilibrium constant, K_{eq}, can be defined where:

$$K_{eq} = \frac{k_1}{k_2} = \frac{[\mathrm{C}]^z}{[\mathrm{A}]^x[\mathrm{B}]^y}$$

It can be shown[1] that this constant is related to the free energy change of the reaction by the equation:

$$\Delta G^0 = -2.303RT\ \log K_{eq}$$

where ΔG^0 is the standard free energy change, the free energy change at an arbitrary set of conditions where all reactants and products are present in molar concentration. R is the gas constant and T the absolute temperature.

If the equilibrium constant is greater than one, the forward reaction is favoured thermodynamically; if it is less than one the reverse reaction predominates. Where K_{eq} of a reaction is very large, i.e. where there is a substantial reduction in free energy, the number of reverse reactions may be vanishingly small, and to all intents and purposes the reaction goes 'from left to right.' Where ΔG is small (and K_{eq} is close to one) the numbers of each will be comparable. Examples of such situations are shown in Figure 3.1.

Importantly, these thermodynamic considerations only determine the *relative* frequency of forward and reverse reactions in a given scheme. They say nothing about the actual number of such reactions, and therefore nothing about the rate at which the conversion of starting materials to products occurs. Many thermodynamically favoured 'reactions' are so slow that in practice they do not occur at all.

3.1.2 Kinetics

The reason for this is kinetic in nature, and is pictured in Figure 3.2. The potential product state (C + D) is more stable and has a lower free energy content, and therefore the reaction is thermodynamically favourable. But between the two (A . . . B) is a less stable intermediate, and energy has to be put into the system to convert the starting materials to this form. Intermediates of this type generally have extremely transitory existences, as the reaction of (A . . . B) to C + D is clearly very favourable in thermodynamic terms; so the limiting stage in the reaction is the process of pushing enough molecules of A and B up to the top of the 'energy barrier'.

The energy required to achieve this is termed the activation energy, and the size of this parameter is fundamental in determining the rate of a reaction. If the energy barrier is too high, i.e. the molecules of the starting reagents have insufficient kinetic energy to

[1] In fact this is an approximation, ignoring the contributions of non-ideal behaviour (see, for example, Hamill *et al.*, 1966).

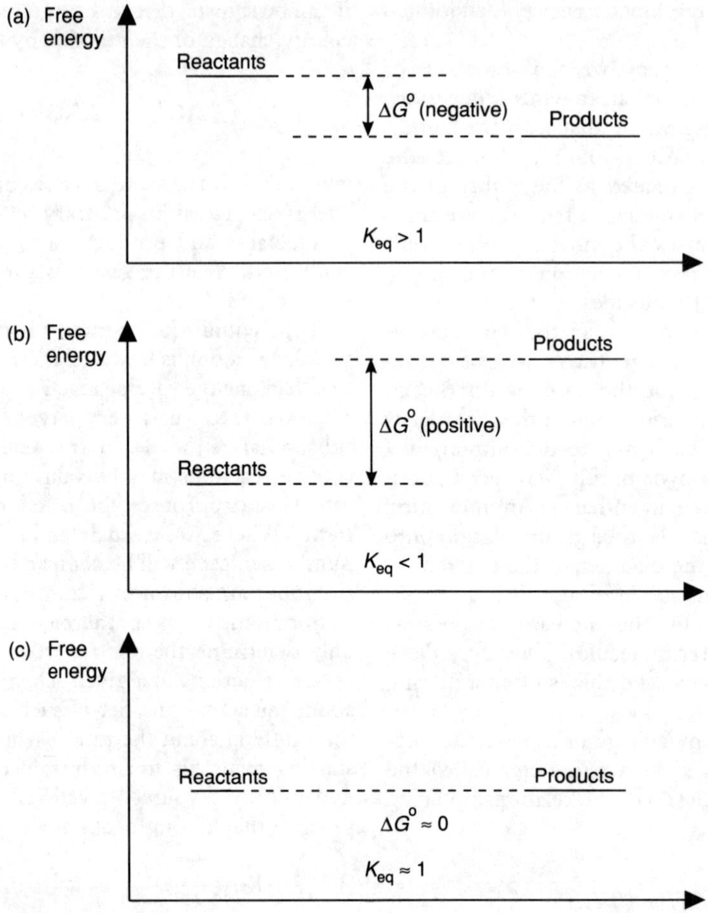

Figure 3.1 Comparison of the free energy states of reactants and products for energetically favourable (a) and unfavourable (b) reactions, and for a borderline case (c).

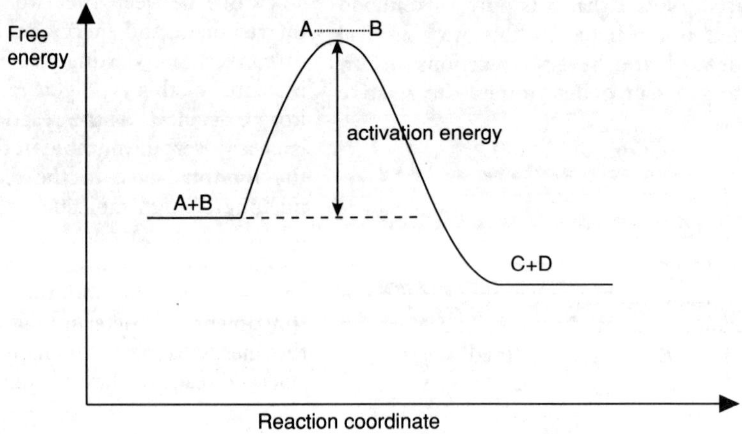

Figure 3.2 Activation energy: a barrier to the progress of a reaction.

promote formation of the intermediate when they collide, no reaction will occur. One way to overcome this situation is obviously to supply more kinetic energy, which is why the application of heat is an almost universal means of increasing the rate of chemical reactions[1].

The dependence of reaction rate on temperature is described for all simple reactions by the Arrhenius equation, first derived empirically by Svante Arrhenius in 1889 and subsequently explained theoretically, where:

$$k = Ae^{-E_a/RT}$$

k is the rate constant at a given temperature T, R is the gas constant and A and E_a are constant for a given reaction. E_a is found to correspond to the activation energy. In practical terms if $\log k$ is plotted against $1/T$ for a range of temperatures, a straight line should be obtained where the slope is $-E_a/2.303R$.

One corollary of this relationship is that the increase in reaction rate observed from a given increase in temperature varies with the activation energy of the reaction concerned; this fact should be borne in mind when applying 'rules of thumb' sometimes encountered in the life sciences, for example in the justification of 'accelerated stability studies' which extrapolate the rate of decomposition reactions from behaviour at a raised temperature. Assumptions of this type, e.g. 'the rate of reaction doubles for each 10°C increase in temperature' may be very valuable in a class of similar reactions where there is some empirical and/or theoretical justification for the activation energies being similar, but they should not be applied indiscriminately.

[1] In thermodynamic terms increasing the temperature may or may not make the reaction more favourable, depending on whether the entropy term is positive or negative; reactions where entropy is reduced are *less* favoured thermodynamically at higher temperatures.

Returning to Figure 3.2 however, another means of reducing the energy barrier may be to introduce another component so that a different intermediate or (more likely) series of intermediates is formed (Figure 3.3) *en route* to the same products. If the overall activation energy is lower than in the original case, the rate of the reaction will be increased. The 'extra' participants, which are in the same state at the beginning and the end of the reaction, are catalysts, and (as might be suspected from the exponential form of the Arrhenius equation) their effect on reaction rates can be dramatic.

The development of efficient catalysts in the chemical industry has literally changed the world around us. Many of the polymers which we encounter every day can only be prepared economically by the use of catalysts, for example the Ziegler-Natta and metallocene reagents which facilitate the production of polyethylene and polypropylene[2]. However the achievements of these man-made catalysts are dwarfed by their counterparts in nature, the enzymes.

[2] Before the introduction of such catalysts these polymers could only be produced under the influence of extremely high temperatures and pressures (typically 2000 atmospheres and 300 °C). It is an indication of the importance of these new materials that, in the original (1930s) ICI process for polyethylene production, the destruction of the manufacturing plant by explosion was apparently accepted as a fairly regular feature!

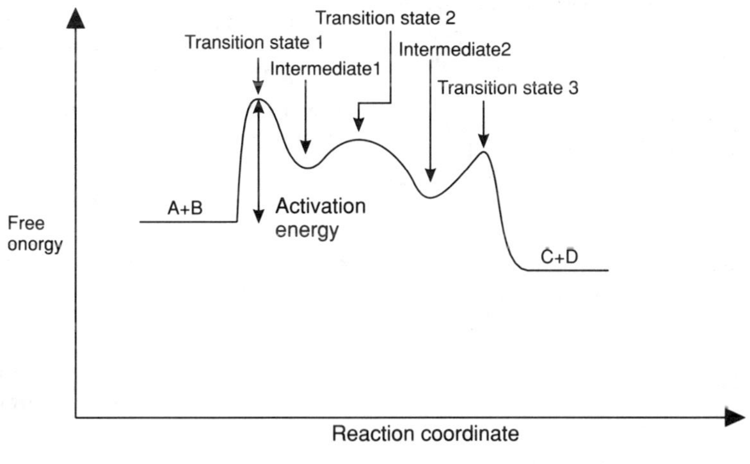

Figure 3.3 Lowering the energy barrier by using an alternative reaction pathway. The activation energy is smaller than that in Figure 3.2, and thus the rate of reaction is faster while the same products are obtained.

3.2 Catalysts in Nature: The Enzymes

3.2.1 The definition of an enzyme

It is relatively straightforward to think of enzymes[1] as biological catalysts. However, defining in molecular terms what makes a particular catalyst an enzyme is not totally straightforward, and has in fact become more difficult over the last decade or so.

Until the early 1980s all known species of this type were proteins, and an enzyme could be defined as a protein catalyst (see for example Harper *et al.*, 1979). However, in 1982 Tom Cech at the University of Colorado and Sydney Altman at Yale University made the discovery that certain RNA molecules could act in a catalytic fashion, and many examples of this behaviour have now been identified (Bartel and Szostak, 1993). Clearly the action of a nucleic acid in this way is very closely related to protein catalysis, and these ribozymes are sometimes classed as enzymes too,[2] in which case the definition needs to be wider. Perhaps the term biological macromolecular catalyst is adequate.

Even amongst proteins, it has become harder to differentiate between those which are enzymes and those which are not—while antibodies would not commonly be thought of as enzymes, some examples (abzymes) are capable of catalysing reactions (Tramontano *et al.*, 1986; Jacobs *et al.*, 1987). These molecules can have useful applications, Kricka and Ji (1993), for example, showed that antibodies raised against the aminophthalate intermediate of luminol oxidation were capable of catalysing the luminogenic luminol-peroxide reaction. On the whole, however, their catalytio ability is rather limited (Holliger and Hoogenboom, 1995); and a further twist to this topic has been introduced by the observation that simple proteins such as albumins can sometimes rival carefully designed antibodies in this respect (Hollfelder *et al.*, 1996).

Another challenge to a clear definition comes from catalytically active synthetic oligopeptides, which were first identified in the work of Sheehan and his colleagues at M. I. T. (see, for example, Sheehan *et al.*, 1966). More recent work along these lines has yielded larger and more complex examples such as the 'oxaldies' of Johnsson *et al.* (1993), and the 73-residue 'chymohelizyme' of Hahn *et al.* (1990).

[1] The word enzyme literally means 'in yeast', and enzyme activity was originally believed to be a property of intact cells.

[2] In fact they are beginning to compete with enzymes in developments for commercial application (Genesis Report, 1997).

While enzymes are not surprisingly most commonly encountered in soluble form in aqueous environments, they can be made to perform adequately in other formats—as cross-linked crystals (Margolin, 1996), and/or in organic solvents (Bartling *et al.*, 1974; Kazandjian *et al.*, 1986; Klibanov, 1997).

3.2.2 The role of enzymes *in vivo* and *in vitro*

3.2.2.1 Enzymes *in vivo*

The practical constraints on protein chemistry in the laboratory have already been discussed, and if anything the limitations on biosynthesis within living organisms are even more stringent. The fact that an extraordinary range of biochemical molecules is manufactured *in vivo*, in aqueous systems, under strict limitations of ionic strength and pH, and using only reagents derived from food and the atmosphere, is a daunting one—particularly perhaps to a synthetic chemist. In a chemical laboratory without these limitations, using the entire arsenal of synthetic methods and possible starting materials, the preparation of a single biochemical may take weeks of effort; living systems carry out millions of such operations every day.

Similarly, nature has developed an elaborate interlinked system of reactions to exploit the energy contained in foodstuff molecules. Proteins, fats and carbohydrates can all be broken down in a variety of ways to free this energy, often storing it temporarily in convenient 'energy-rich' compounds such as adenosine triphosphate (ATP). The stored energy can be released as necessary to the system's advantage, to maintain a constant body temperature, to carry out physical work, etc. Metabolism of a single glucose molecule under aerobic conditions can generate up to 38 ATPs, storing nearly half of the carbohydrate's available energy in a usable form. Once again, to recreate the chemistry employed *in vivo* in the laboratory would be a lengthy process, and one that in practical terms would use far more energy than it generated.

The prodigious efficiency of natural systems as synthesisers of complex molecules, and as exploiters of nutritional energy, is made possible only by the engineered existence of enzymes, and their importance in living systems cannot be overestimated.

The origin of life is a fascinating topic for discussion, although unfortunately somewhat beyond the remit of this book! It is interesting however to speculate that the development of biologically interesting reactions must have required the concurrent evolution of catalysts to make them happen in what was presumably an unfavourable chemical environment. So the existence of crude precursors of enzymes to make unlikely reactions slightly more favourable could be looked on as one of the earliest milestones in the development of life. As organisms have evolved to a bewildering variety of forms and specialisations, so too a multitude of different enzyme structures have developed to offer efficient catalysis in a host of applications.

For a given reaction, the evolution of an enzyme to offer greater and greater efficiency of catalysis is likely to have rendered it more and more specific in its action. The more it has been honed to perfection for this application, the less of an effect it is likely to have in another. Also, it is often to the advantage of the organism to develop enzymes which only catalyse a very limited range of reactions. Therefore the overall evolutionary pressure has been towards specificity, and this is indeed a key characteristic of these catalysts.

The degree of specificity may vary. For example, there is no particular need for 'crude' digestive enzymes such as pepsin and trypsin to be particularly specific and they will cleave the peptide bonds of food proteins in a relatively non-selective way; moving up the sensitivity scale, carboxypeptidases and aminopeptidases will only cleave the bonds adjacent to the C terminus and N terminus of peptide chains respectively, but regardless of the specific amino acid to be removed; chymotrypsin preferentially cleaves bonds where the carbonyl group derives from phenylalanine, tryptophan or tyrosine. Many enzymes are specific to a single molecule, and generally only one optical isomer of the structure will be favoured.

3.2.2.2 Enzymes *in vitro*

The reason that enzymes have found such widespread application as reagents in the life sciences is partly due to this specificity, which allows them to be used to catalyse one particular reaction among a complex mixture of components. The other factor of course is amplification; since they are catalysts they can act over and over again to facilitate a signal generating reaction. A luminescent, radioactive or chromogenic label can generate at best a single 'signal event' per molecule, and may often give much less than this in practice, for example, through losses of light signal by quenching phenomena, or through the statistical limitations on the incidence of radioactive decay of an isotope with an acceptably long half-life.

In contrast, an enzyme label with a suitable excess of substrate can catalyse a virtually unlimited number of signal events, and turnover numbers of hundreds and thousands of reactions per second can be obtained. This obviously offers enormous potential in terms of sensitivity, and the combination of this property with the specificity described above explains their widespread exploitation in the applications discussed in Chapter 1.

These enzymes all originate from natural sources and many are still prepared by extraction from plant or animal material. Where the amino acid sequence has been determined recombinant techniques for production are sometimes applicable, and, excitingly, the prospect of modifying the enzyme's properties by introducing mutations into the structure. This offers not only the capability of gaining insights into the mechanism of catalytic action (for example, the work described by Edwards *et al.* in 1987 on cytochrome *c* peroxidase), but also the potential of creating new enzymes with superior properties.

3.2.2.3 Substrate *versus* enzyme labelling

There are two possible approaches where it is desired to employ an enzyme-based system for signal generation: both will be encountered in practice.

The first technique is to use an enzyme substrate as the label; the subsequent addition of excess enzyme initiates the turnover of this substrate, generating a signal which can be used to quantify the amount of label. The major disadvantage of this approach is that the amplification power of the enzyme is wasted, because the substrate is the limiting reagent and, as noted above, each label can only give rise to a single signal generating event. Also, because substrates are commonly fairly small organic molecules, it can be difficult to attach them to the target of interest without significant losses of their activity.

A further problem arises where the signal generated is of a transient nature. With a colorimetric or fluorimetric system, the product generated by enzyme-catalysed substrate turnover is generally long-lived, so that the entire yield of the signal reaction is 'integrated' by conversion to a stable, measurable form. Therefore (within reasonable constraints) measurement can be carried out more-or-less at leisure once the signal generation reaction is completed. In effect this means that the instrumentation for colorimetric or fluorimetric systems can be relatively simple—rapid scanning of a 96-well microtitre plate with a purpose-built spectrophotometer, for example.

In contrast, the light signal generated from a luminometric system must be detected as it is formed; there is no integrating effect which allows the yield to be accumulated. In a substrate-labelling system the addition of the enzyme can provoke a rapid release of the light signal. For example in an HRP-based immunoassay employing an antibody labelled with a luminol analogue, the addition of the peroxidase/peroxide mixture generates a short-lived intense 'spike' of light, typically within milliseconds. This necessitates the peroxidase/peroxide mixture being added with the reaction vessel already in place under the light-detection system, and leads to complexity in the design of instrumentation, especially if it is desired to read a large number of samples in a short time. This either requires slow repetition of the signal generation and measurement operation for each sample, rather complicated engineering to carry out these operations on multiple samples simultaneously, or a compromise between these two approaches.

The second approach, more attractive in many respects, is to use the enzyme as the label, adding substrate at the signal generation stage. With a large substrate excess, the amplification properties of the enzyme can then be exploited by repeated cycling. Also, as there is no need to conjugate the substrate to another molecule, activity loss on derivatisation is not a problem; with a little care it is not normally too difficult to find a means of coupling the enzyme without large activity losses.

This approach is also much more amenable to the transient signal of luminometric systems: once the enzyme has reached an equilibrium state in the signal reagent (i.e. it is turning over substrate at its limiting rate) the luminescence generated during a fixed time window should remain constant—at least until the substrate excess nears exhaustion or the enzyme starts to be inactivated. Therefore, although the sensitivity gain by cycling the enzyme repeatedly is counteracted by a detection method that only 'samples' the light signal, this format offers a substantial advantage in allowing the photometric step to be carried out at any time while the enzyme is operating at a steady state.

As might be expected from this brief comparison, enzyme-labelling approaches are by far the more common. Luminometric formats have provided something of an exception, with substrate-labelling approaches predominating for many years—the reason for this apparent paradox being a lack of good luminogenic substrates for the common labelling enzymes. Thus with HRP, for example, simple turnover of excess luminol by a limiting quantity of enzyme generates a poor signal level; greater sensitivity could be obtained by releasing all of the signal at once in a substrate-labelling format. This need has been reduced by the advent of improved substrate systems such as enhanced luminescence for HRP, and dioxetanes for all the common enzymes.

3.2.2.4 Coupled-enzyme approaches

The detection power achievable from enzyme-based systems can be magnified still further by using a format combining more than one enzyme. Potentially, multi-enzyme systems have two major attractions. On the one hand, combining two amplification steps can yield prodigious sensitivity, and some exquisite detection limits have been achieved by this means.

Secondly, it may be possible to harness a good signal-generating system for a greater variety of applications; peroxidase-based detection, for example, can be coupled to any other enzyme which catalyses the production of hydrogen peroxide from the substrate
of interest. Probably the best-known manifestation of this approach is the coupled-enzyme detection of glucose: glucose oxidase generates hydrogen peroxide from this substrate in a dose-dependent fashion, and this is quantified by 'channelling' it into a suitable HRP/substrate system.

There are many different ways in which coupled-enzyme approaches have been employed, some of these are dealt with briefly in Section 3.6.

3.2.3 Enzyme kinetics

Returning to the generalised reaction discussed above, and introducing an enzyme catalyst, E:

$$x\mathrm{A} + y\mathrm{B} + \mathrm{E} \Rightarrow x\mathrm{A}\ldots\mathrm{E}\ldots y\mathrm{B} \Rightarrow z\mathrm{C} + \mathrm{E}$$

the overall equilibrium constant is now given by:

$$K_{\mathrm{eq}} = \frac{[\mathrm{C}]^z[\mathrm{E}]}{[\mathrm{A}]^x[\mathrm{B}]^y[\mathrm{E}]}$$

but the enzyme concentrations cancel out and the term reduces to that seen earlier, i.e. the involvement of the enzyme has no effect on the equilibrium constant, and, q.e.d., the free energy change and overall direction of reaction. Therefore the effect of an enzyme has to be considered in terms of rates, not equilibrium constants.

Consider an even more simplified general reaction

$$\mathrm{E} + \mathrm{S} \Rightarrow \mathrm{E}\ldots\mathrm{S} \Rightarrow \mathrm{E} + \mathrm{P}$$

where S is the substrate and P the product or products. Plotting the initial rate of reaction (at a fixed enzyme concentration) against [S], as the latter is increased, often yields a graph of the hyperbolic type shown in Figure 3.4. As the substrate concentration rises to a larger and larger excess, the initial rate becomes independent of this parameter; under these conditions where the plot is horizontal, the velocity is at its maximum level, $V_{\max}$ and each enzyme molecule is saturated with substrate. Where exactly half of the catalyst molecules are saturated, it follows that the velocity will be $V_{\max}/2$, and the substrate concentration which yields this initial rate is termed the Michaelis constant, K_{m}.

For many enzymes, the effect of varying substrate concentration follows this graphical representation, and is described by the Michaelis-Menten equation:

$$v = \frac{V_{\max}[\mathrm{S}]}{K_{\mathrm{m}} + [\mathrm{S}]}$$

Often it is impractical to obtain $V_{\max}$ directly, and various arrangements of the Michaelis-Menten equation have been derived which do not require knowledge of this parameter. Particularly useful among these is the expression:

$$\frac{1}{v} = \frac{K_{\mathrm{m}}}{V_{\max}} \frac{1}{[\mathrm{S}]} + \frac{1}{V_{\max}}$$

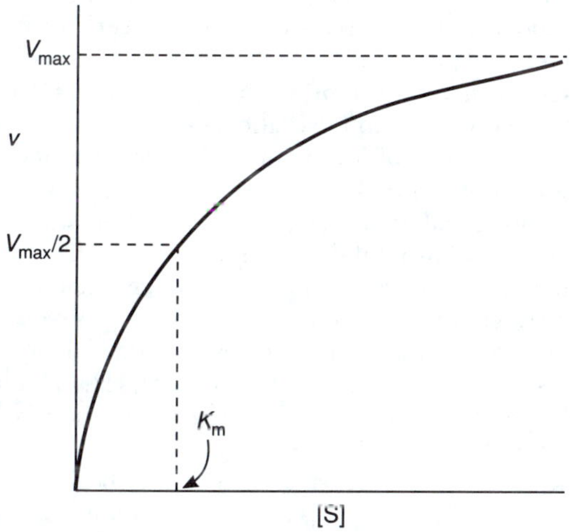

Figure 3.4 A plot of the reaction velocity, *V*, as a function of the substrate concentration, [S], for an enzyme that obeys Michaelis-Menten kinetics (V_{max} is the maximal velocity and K_m is the Michaelis constant).

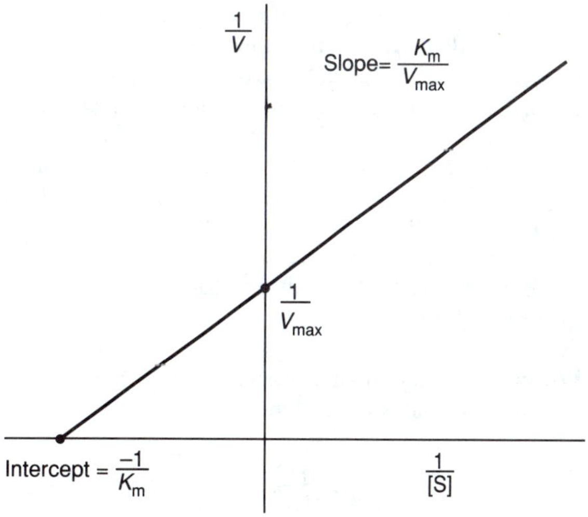

Figure 3.5 A typical Lineweaver-Burk plot.

This is a linear equation of the familiar $y = mx + c$ form, which is amenable to simple graphical treatment. Carrying out initial rate measurements at a range of substrate concentrations, a double reciprocal plot of $1/v$ against $1/[\mathrm{S}]$ is constructed (Figure 3.5), termed a Lineweaver-Burk plot. The value of $V_{\max}$ is the reciprocal of the y intercept, and the value of K_{m} can either be obtained as the negative reciprocal of the x intercept, or from the slope $K_{\mathrm{m}}/V_{\max}$.

A knowledge of these parameters for a given enzyme–substrate system can be useful in a number of ways.

3.2.3.1 Choice of conditions for enzyme quantification

Under conditions where the substrate is in a large excess, $K_{\mathrm{m}}+$ [S] in the Michaelis-Menten equation approximates to [S]. Therefore:

$$v \approx \frac{V_{\max}[\mathrm{S}]}{[\mathrm{S}]} = V_{\max}$$

and the initial rate (as seen in Figure 3.4) is the maximum possible for that enzyme concentration. Because the substrate is in excess, each increase in the enzyme concentration leads directly to an increase in this maximum rate, and therefore the initial rate v is proportional to [E].

At substrate concentrations of 1, 10 and 100 K_{m} respectively, the velocity obtained will be 0.5, 0.9 and 0.99 $V_{\max}$. Further increases in [S] can have virtually no effect on the rate of reaction[1], so a substrate loading of 100 or 1000 K_m is an economical choice for such work. This rule also holds true where enzymes are being used in preparative or other non-quantitative applications.

The precautions required to obtain a linear dose-response from increasing enzyme concentration have been clearly and simply discussed by Murachi (1981).

3.2.3.2 Choice of conditions for substrate quantification

Where [S] is very small compared to K_m, the K_m+ [S] term approximates to K_m, and:

$$v \approx \frac{V_{max}}{K_m}[S]$$

Both V_{max} and K_m are fixed for a given amount of enzyme, and therefore the initial rate is proportional to the substrate concentration. Typically values of [S] $< 0.01K_m$ are employed where it is desired to quantify substrate using a fixed amount of enzyme.

3.2.3.3 Investigation of inhibitor behaviour

Various mechanisms of enzyme inhibition will be discussed in Section 3.2.5, and it will be seen that the effect of inhibitors on Lineweaver-Burk plots can provide useful information as to which type of mechanism is involved.

Some enzymes show a different dependence of initial rate on [S], for example (see Figure 3.6) those displaying sigmoidal saturation kinetics. These generally indicate that multiple binding sites are present, and that binding is cooperative, i.e. the binding at one site is affected by that at another. Predictably the kinetics of such enzymes are more complex to describe, although different approaches can be used to determine substrate concentrations analogous to K_m.

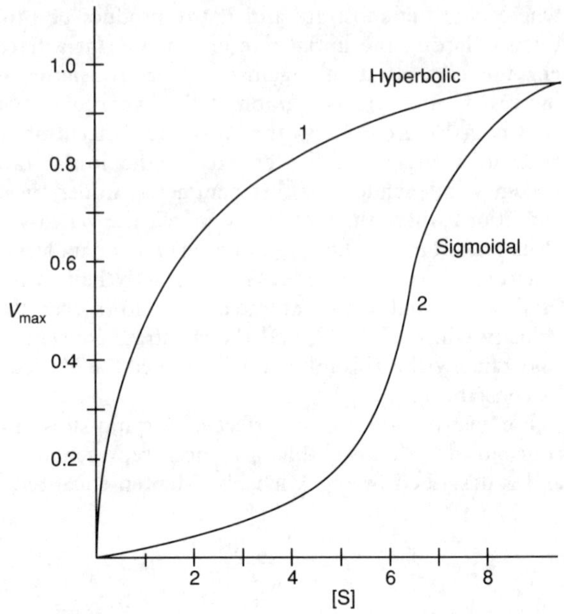

Figure 3.6 The effect of substrate concentration on the velocity of a reaction catalysed by: (1) a Michaelis-Menten type of enzyme and (2) an allosteric enzyme with sigmoidal kinetics.

3.2.4 Enzyme nomenclature: the enzyme commission (EC) system

In the 1950s the number of known enzymes had begun to escalate significantly, and it became obvious that their nomenclature was becoming a problem. Some enzymes had been given a variety of names by different authors, whereas in other cases the same name was in use for different species. In 1955 the International Union of Biochemistry (IUB) set up a commission to address this area, and since then they have periodically issued a master list of all sufficiently characterised enzymes, numbered by a rational system.

This consists of four figures, the first three indicating the class, sub-class, and sub-sub-class to which it belongs (based on the nature of the reaction catalysed), and the fourth its unique number within that group. Thus cytochrome *c* peroxidase, which catalyses the oxidation of ferrocytochrome *c* by hydrogen peroxide, is placed in class 1 which contains all the oxidoreductases which catalyse redox reactions, subclass 11 which includes all those where hydrogen peroxide is the hydrogen acceptor (oxidising agent), and sub-subclass 1 as there is no further categorisation. It is the fifth listed enzyme within this group, and its full identification is EC 1.11.1.5.

There is also a systematic name for each enzyme under the EC system, in this case ferrocytochrome *c*: hydrogen peroxide oxidoreductase. Many of these systematic names are rather cumbersome, and there has been considerable resistance to their everyday use; in

[1] If measurements are to be made before an appreciable amount of substrate has been consumed. If the reaction time is sufficiently long that [S]/K_m is substantially reduced linearity will be lost; the same also applies if there is sufficient time for enzyme inhibition to become significant, for example as a result of product inhibition.

response to this, the IUB classification also allows for the use of recommended names. These are often widely-accepted trivial names (e.g. cytochrome *c* peroxidase is an acceptable recommended name), but the system prevents ambiguity in their use; this seems to be a compromise acceptable to most parties. One final point to be noted is that neither the EC number nor the systematic or recommended name signify a unique molecular structure or source; a number of similar enzymes from different sources with common substrate specificity may well be classified under a single number.

Briefly, the main classes of enzymes under this system are as follows:

(1) Oxidoreductases. As mentioned above, this includes all enzymes catalysing redox reactions. The substrate to be oxidised is defined as the hydrogen donor, and the oxidising agent as the acceptor, nomenclature follows the format donor: acceptor oxidoreductase as seen above.
(2) Transferases. These catalyse the transfer of a group from one compound (donor) to another (acceptor), such as methyltransferases.
(3) Hydrolases. These catalyse hydrolysis reactions, for example of amide bonds or carbohydrate chains, and include glycosidases, ribonucleases, and carboxypeptidases.
(4) Lyases. These catalyse elimination and addition reactions, i.e. those generating or reacting with unsaturated systems (i.e. those containing C≡C bonds or aromatic rings). For example, aspartase catalyses the removal of ammonia from aspartate to leave the unsaturated fumarate.
(5) Isomerases. These catalyse geometric or structural changes within a single molecule, for example alanine racemase which promotes the formation of the D-enantiomer of this amino acid from the L-form.
(6) Ligases. These catalyse the coupling together of two molecules accompanied by hydrolysis of a pyrophosphate bond. There are a group of ligases, for example, which catalyse the attachment of amino acids to the appropriate transfer-RNA, each of these reactions being fuelled by the hydrolysis of an ATP molecule.

3.2.5 Structural features of enzymes, and their influence on derivatisation sites

In common understanding enzymes are proteins, and the chemistry used to derivatise them is of course subject to the general constraints discussed earlier. However there are a number of additional considerations related to their catalytic behaviour, the nature of which will now be dealt with briefly.

In the classic 'lock and key' analogy proposed by Emil Fischer, also referred to as the rigid template model, the key characteristic of specificity is explained by the enzyme possessing a fixed 'active' site. The physical and chemical characteristics of this site are such that only the molecule of interest (substrate) can 'fit' in a way that facilitates its subsequent reaction. Molecules whose shape does not allow them to fit the 'lock' are excluded and hence their reaction is not catalysed.

Since the 1960s, this has been viewed increasingly as an over-simplistic approach, and the induced fit model proposed by Koshland has received widespread support; under this hypothesis, the approach of the substrate molecule causes an alteration in the three-dimensional structure of the enzyme, and this conformational change helps to promote its catalytic nature. Such structural changes have been observed by spectroscopic and other means for a number of enzymes, and the validity of this model is now beyond doubt.

Due to the folding of the protein, the catalytic site where the substrate is bound may obviously include amino acids widely separated in the primary structure; however under the Fischer model it is at least simple to picture them as being close to each other in the three-dimensional layout. One corollary of the Koshland approach is to complicate this simple picture, as amino acids which are physically distant in the 'resting' structure may be brought together on the approach of the substrate, so (even if one has the luxury of knowing a detailed three-dimensional structure) the effect of labelling the enzyme *via* a particular amino acid is not always straightforward to predict.

Nor is interference with the catalytic site the only way in which enzyme behaviour can be modified by labelling, as shown by experiments with inhibitors. As might be expected from either of the models above, the catalytic activity of an enzyme can often be blocked by substrate analogues, inhibitors which are sufficiently close in structure to the real substrate to compete for the binding site and swamp out the desired effect. This phenomenon can be particularly dramatic if the inhibitor binds irreversibly. In some cases, however, the enzyme activity can be inhibited by a structurally unrelated molecule, and it can be shown that the catalytic site is not involved in this interaction, i.e. the inhibitor is acting on a different part of the enzyme structure.

Such allosteric interactions are often employed in nature as control mechanisms, for example with the

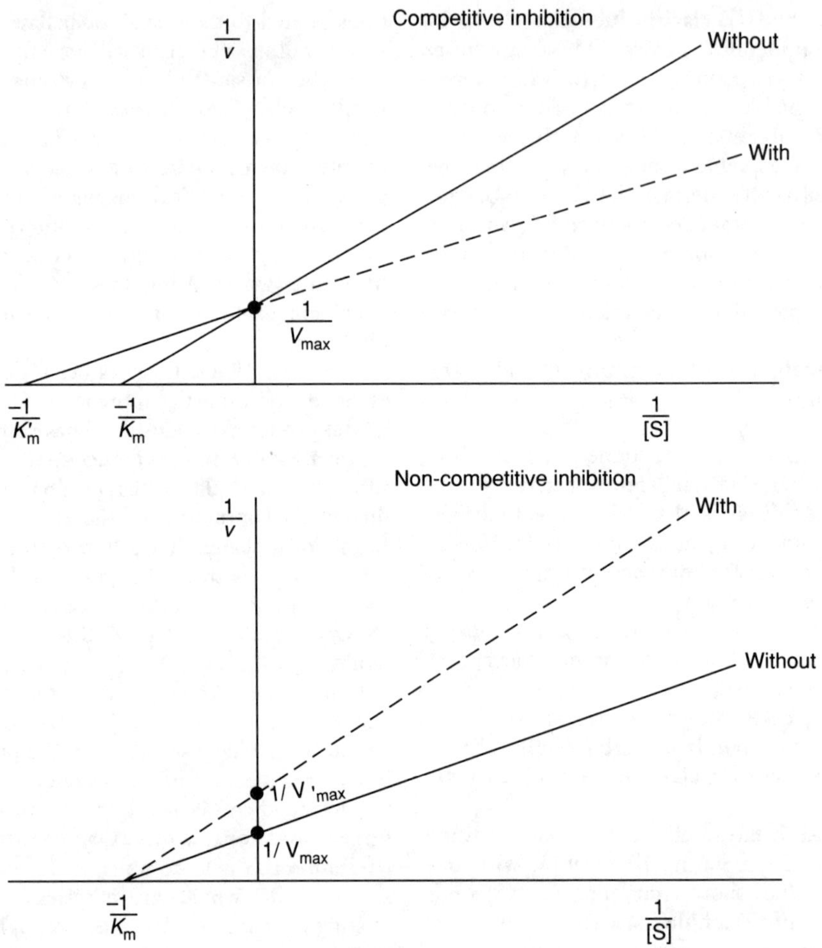

Figure 3.7 The effect of competitive and non-competitive inhibitors on Lineweaver-Burk plots.

enzyme aspartate transcarbamoylase which is involved in the biosynthesis of pyrimidine. Phosphate released from the reaction catalysed by this enzyme leads to an increase in cytidine triphosphate (CTP) which acts as an allosteric effector, binding to a receptor on the enzyme to reduce the activity. This offers a negative feedback mechanism to regulate the extent of the catalysed reaction[1]. Labelling at an allosteric site may have unpredictable effects on enzyme behaviour.

[1] In other circumstances positive feedback mechanisms are employed, often where it is the aim to generate a large amount of an enzyme very rapidly. Many of the digestive enzymes behave this way—they are stored as inactive proenzymes, and are autocatalytic. Inactive pepsinogen for example is converted slowly to pepsin under the influence of proteolytic enzymes or acid conditions; this conversion however is also catalysed by pepsin itself, so the rate of formation increases astronomically until the supply of pepsinogen begins to dry up or other factors intervene.

While the catalytic activity may be unaffected, unexpected changes can occur such as loss of thermal stability in the presence of the effector.

The effect of competitive and non-competitive inhibitors on Lineweaver-Burk plots is shown in Figure 3.7: competitive examples such as substrate analogues increase the apparent value of K_m while having no effect on V_{max}; non-competitive inhibitors lower V_{max} but have no effect on K_m.

Prosthetic groups have been mentioned already, organic molecules forming a permanent and essential part of the enzyme structure, such as haem in haemoglobin. As part of the enzyme they are in the same chemical state at the beginning and the end of the substrate reaction. They are often central to the catalytic site, and predictably their chemical modification often destroys enzyme activity.

'Coenzymes' are a rather poorly named group whose nature is sometimes confused with prosthetic groups, indeed the terms are often used more synonymously than they should be. They are related in that they are generally organic molecules and are essential for activity, but they are not a permanent component of the enzyme structure; in fact they often end up chemically different after the substrate reaction, and so cannot possibly be viewed as part of the enzyme. Generally they are better described as cosubstrates, although the term is less commonly encountered.

Whatever the terminology, these species must bind to the enzyme in much the same way as a substrate. Whether the portion of the protein where this occurs is described as a coenzyme binding region or as part of the catalytic site, chemical modification can obviously have harmful effects here too.

In summary then, there may be many sites on an enzyme where attempts to attach another molecule may cause unwanted changes in behaviour. Other specific chemical interactions can also be vitally important even though they do not affect the peptide structure. For example, many enzymes require the presence of metal ions, and are sensitive to chelating agents such as ethylenediamine tetraacetic acid (EDTA) which remove these species from the system.

3.2.6 Functional features of enzymes affecting their use as labels

As observed in Chapter 1, there are many uses of enzymes; these molecules are increasingly important in catalysing industrial processes, including such important applications as polymer synthesis (Dordick, 1992); other examples are also commercially important as components of soap powders, etc. There are many preparative applications in the protein labelling field too, and some are mentioned in this book. For example, the use of proteases to catalyse protein conjugation (Chapter 5, Section 5.2.4), or of lactoperoxidase to catalyse the oxidation of radioactive iodide (Chapter 6, Section 6.21.3.6).

However, the most commonly-employed aspect of enzyme behaviour is the ability of these molecules to generate a large signal numerically related to their concentration, and therefore to the concentration of any other species to which the enzyme is quantitatively linked by one means or another. The use of enzymes as labels is thus particularly widespread, especially in the diagnostic field. There are a number of factors which affect the suitability of enzymes for labelling purposes, and these are discussed below.

3.2.6.1 Choice of signal type

Probably the most important choice in designing an enzyme-labelled detection system is the type of signal which is to be quantified.

Colorimetric approaches are based simply on the formation of a product which can be detected by spectrophotometry; to be useful the product must absorb either more strongly or at a different wavelength to the substrate. Colorimetric methods are very convenient, but are limited by the sensitivity and range of the spectrophotometric detection methods by which the signal is quantified. The absorbance range for reliable quantitation of UV/visible signals varies with the instrument concerned, but limits of 0.1–1.0 are not unusual, i.e. the working range covers only a single order of magnitude. The sensitivity of colorimetric approaches is generally also rather poor, although detection of enzymes down to sub-femtomolar levels has been demonstrated in specialised formats. In immunoassay terms, the limit of analyte detection using colorimetric systems is generally in the nanomolar to picomolar range.

Colorimetric products are quantified by the amount of light (from an external source) which they absorb. Greater sensitivity is achievable where light of a specific wavelength is actually generated from the sample, e.g. in fluorimetric detection systems. Fluorogenic substrates produce a product which fluoresces, i.e. it absorbs light at one wavelength and emits light at another. For a fluorogenic substrate to be useful, the difference in λ_{max} for excitation and emission (the Stokes shift) must be as wide as possible, otherwise there is a danger that the light used for excitation will cause background problems (Figure 3.8). Fluorogenic substrates with better properties are constantly being sought. Many organic compounds provide adequate performance, but the fluorescence of chelated rare earth metals such as europium displays unusually large Stokes shifts, and has formed the basis of commercial immunoassay systems.

Another issue with fluorimetric systems is the background fluorescence observed from sample components such as serum proteins. This can be minimised by a suitable choice of emission wavelength, but differences in the emission lifetime have also been exploited. Time-resolved fluorimetry employs gated detection techniques to eliminate the short-lived background fluorescence from the measured signal—again europium chelates display desirable properties with regard to fluorescence life times.

It is a reasonable rule of thumb that moving from a colorimetric to a fluorimetric substrate can yield a

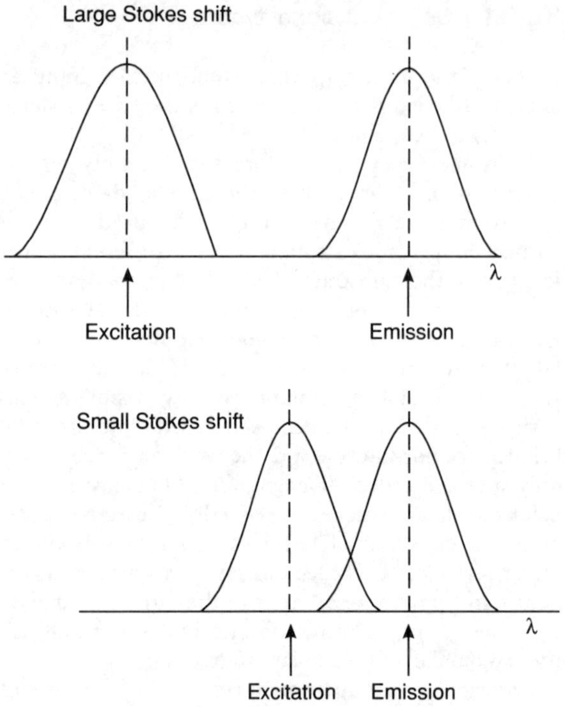

Figure 3.8 Large and small Stokes shifts for fluorophores. With a small shift, the light used for excitation may cause background problems in the detection of the emitted signal.

1000-fold increase in sensitivity, and this obviously offers an attractive alternative. Many commercial immunoassay systems are based on fluorimetry, for example, and this is also a popular approach for histochemistry and cytochemistry.

Luminometric methods offer still greater potential, in that light is generated from the sample without the necessity for an excitation signal. It is, instead, produced from a suitable chemical reaction. This can allow great range and sensitivity to be achieved; the signal can be detected against a dark background using photomultiplier or photon counting techniques which are inherently more sensitive than absorbance measurement, and work over a much greater operating range.

Moving away from light-based systems, there has been considerable recent interest in electrochemical methods, where an electroactive product is generated from an electroinactive substrate. This product can be detected *via* its oxidation at a suitable electrode—quantification methods based on either current (amperometry) or electrode potential (voltammetry) can be employed. Typically for example a phenolic compound is released from a substrate under the catalytic influence of the enzyme, and the oxidation of this phenol at an electrode produces a detectable current. Choice of an appropriate voltage is required to provide the necessary electrochemical selectivity: too high a voltage may result in oxidation of the substrate too.

3.2.6.2 Enzyme availability

Many enzymes are only encountered at extremely low concentrations in living systems, and it is virtually impossible to isolate large quantities of such materials reliably and economically. Although recombinant techniques are offering a way out of this limitation, it is still an important consideration (not least in terms of cost), and some of the most commonly encountered enzymes owe their popularity to this factor. Peroxidases for example are found throughout the plant and animal kingdoms, but in the root of the horseradish, *Armoracia rusticana*, the enzyme is found at extravagantly high concentrations which make supply and subsequent purification relatively straightforward. It is this peroxidase that has predominated in labelling applications. Because microorganisms can be cultured cheaply and rapidly, enzymes from these sources are also favoured in terms of availability, e.g. β-D-galactosidase from *E. coli*.

3.2.6.3 Specific activity

The efficiency of an enzyme as a catalyst is obviously a key factor; it should be noted however that this parameter can only be quantified for a given substrate and reaction conditions. Units of activity under these constraints are generally defined rather arbitrarily and are of little use in comparing one enzyme to another. Similarly, particular care should be taken where more than one activity system is in use for a single enzyme; there may be little correlation between activities with different substrates.

The turnover number is equally dependent on the choice of substrate, but offers a slightly greater insight into comparative behaviour. Ignoring the actual amount of signal generated it quantifies the number of substrate reactions catalysed in a given time, and typical values obtained are between 10^4 and 10^6 per minute.

3.2.6.4 pH optimum

Enzyme activity can show a strong dependence on pH, with activity often tailing off rapidly away from the optimum. Problems can occur where this optimum is incompatible with other pH-dependent parameters in the system, e.g. antigen–antibody binding (generally best at around physiological pH), or the degree of ionisation of substrate molecules or intermediates. Where the pH dependence of the enzyme is not too narrow, it may be possible to choose conditions which provide a compromise between these differing needs.

3.2.6.5 Substrate suitability

There are many features of the substrate which will affect the overall success of the detection method, and part of the choice of enzyme will be dictated by the availablity of substrates with suitable properties.

In order to control the signal generation reaction it is generally required that there are no other substrates competing for catalysis. In many applications the format is arranged so that as few other species as possible are present at the signal generation stage; this is not always possible however, and any competing substrates may significantly alter the observed signal. Therefore the choice of an enzyme for which there are likely to be no endogenous substrates in the sample is advantageous.

Sensitivity is one of the most important characteristics governing the choice of enzyme/substrate pairing, typically measured by the change in signal for a given concentration of enzyme. Often this signal (measured under standard conditions) is converted into nominal units of enzyme activity, and this allows the relative strength and/or purity of enzyme formulations to be expressed, e.g. as 'units/mg'. Often there are multiple systems in existence for expressing these activities, and care must be taken when comparing them, not just because the actual numbers will be different, but because not all isoenzymes behave the same with all substrates. Two batches of material with similar specific activities under one system may be quite different to each other under another.

In histochemical applications a 'hard copy' of the stained sample is generally obtained photographically, and the contrast obtained from the oxidation product is an important factor; this may not correlate completely with sensitivity in a solution format, as not all colours photograph with equal clarity against a tissue background.

Stability is important in two main respects. While many of these reagents are quite stable if stored as solids, the situation may change dramatically once a solution has been prepared. Particularly for oxidising enzymes such as HRP, where the substrates are chosen partly for their susceptibility to oxidation, is not surprising that some non-enzymic oxidation may occur under the influence of dissolved oxygen—in some cases exacerbated by the presence of trace metals. Therefore it is not uncommon for some colorimetric substrate solutions to have short shelf-lives and deteriorating performance throughout this period. Deoxygenation, for example by bubbling an inert gas through the solution, may be beneficial.

As would be expected from the catalytic nature of their reactions, contamination by minute levels of enzyme can rapidly ruin substrate solutions. Clearly the cleanliness of vessels and equipment is of paramount importance, single-use containers being the best option in many cases. An obvious point that is sometimes overlooked in this field is the need for care in weighing out and handling enzymes, particularly light, fluffy powders such as HRP; atmospheric contamination of a working laboratory can have disastrous consequences. Isolated weighing cabinets or a remote weighing facility are generally a good investment, coupled of course with careful handling techniques.

The second aspect of stability is that of the oxidation product. Various mechanisms (some promoted by the influence of light) can cause bleaching of the coloured solution. It is generally important to control accurately the 'reading time' between substrate addition and spectrophotometric measurement, because the level of oxidised product will continue to rise until either the substrate is exhausted or (more usually) the enzyme becomes inactivated. For convenience a 'stopping solution' is generally added after a fixed time which prevents further reaction (by a pH change or addition of an inhibitor), and this allows some leeway in when the spectrophotometry is carried out. However bleaching effects can mean that the reading drops slowly after the 'stopper' is added, so timing of the photometry is still important.

Solubility of the product is another factor, and the desired property is dependent largely on the field of application. For solution-phase reactions such as those encountered in clinical diagnostics a highly soluble product is obviously required. Even if widespread precipitation does not occur, the formation of invisible microparticulates can cause gross interferences with spectrophotometric measurements. However if the application is histochemistry, where localisation of the signal is desired, a completely insoluble oxidation product is ideal.

Safety is a salient concern, particularly in the field of peroxidase colorimetry. Many good HRP chromogens have structures based on aromatic amines, and it is an unfortunate fact that some of these molecules display carcinogenic traits as well. The use of some formerly common peroxidase substrates has declined drastically as a result of these considerations.

3.2.6.6 Enzyme stability

Enzymes for labelling should show reasonable stability to loss of activity, in both native and conjugated form. Instability *in vitro* is frequently quoted as a disadvantage of enzyme labelling, but can be overestimated; even quite dilute solutions of some enzyme conjugates can be kept refrigerated in a usable form for several years, so long as bacterial contamination is avoided.

In vivo stability of proteins is an altogether more complex subject, and enzymes are as prone to degradation as other examples. Surprisingly simple rules connecting *in vivo* protein stability with the identity of the N-terminal residue have been demonstrated in some cellular environments (Gonda *et al.*, 1989); the degradation rate of β-galactosidase in yeast cells, for example, could be modulated by modifications in this region.

3.2.6.7 Enzyme size

In some applications such as histochemistry, the ability of an enzyme to penetrate tissue samples to the site of interest can be critical, and smaller enzymes have been favoured over larger ones. HRP (molecular mass 44 kDa) is commonly encountered in histochemistry, whereas β-D-galactosidase (molecular mass 465 kDa) is too large for some situations. For *in vivo* reagents, the importance of size in determining their survival in the body has been discussed in Chapter 1; again smaller enzymes may have a greater half-life in such applications as enzymic activation of pro-drugs.

3.2.6.8 Chemical reactivity, and retention of activity after derivatisation

Some of the factors which may affect successful labelling with an enzyme have already been discussed. Most enzymes offer at least a number of amine groups which can be used as 'handles' without destroying activity, while some have features of particular utility such as the carbohydrate groups in glycoprotein enzymes like HRP, or the free thiol groups in β-D-galactosidase.

Obviously too much loss of enzyme activity on derivatisation cannot be tolerated. Guesdon *et al.* (1979) demonstrated (Figure 3.9) that glucose oxidase and horseradish peroxidase were unaffected by almost exhaustive reaction of amines with a succinimide ester; β-D-galactosidase lost almost half of its activity under these circumstances, while alkaline phosphatase lost two-thirds.

Looking at it another way, succinimide ester treatment of peroxidase using a 100-fold excess over amines caused no loss in activity, while even a twofold excess reduced the phosphatase activity by more than half.

The effect of derivatisation on an enzyme can be hard to predict; for example Ullman and Maggio (1980) made interesting observations on the derivatisation of malate dehydrogenase. Labelling the enzyme's amine groups with two molecules of a morphine derivative had virtually no effect, while a similar level of derivatisation with a thyroxine analogue destroyed 90% of the enzyme's activity. It has been surmised (Maggio, 1980) that highly hydrophobic ligands such as thyroxine have a more disruptive effect on protein conformation, as the surface region is generally hydrophilic in nature.

Derivatisation can sometimes cause an increase in activity too: when Ullman and Maggio (1980) attached three equivalents of a morphine derivative to the enzyme *via* cysteine residues, a 20% rise in activity was noted. Suzuki *et al.* (1970) observed that the activity of Taka-amylase was almost doubled by the derivatisation of one amine group with the thiolating agent SAMSA, but was virtually unaffected by a similar level of succinylation (Figure 3.10).

3.2.6.9 Absence of sample interference

Avoiding situations where substrates from the sample may cause interference has already been mentioned. Obviously if the analysis being performed is quantified by a signal proportional to the amount of enzyme present, any endogenous enzyme from the sample can also give a spurious result.

Again, the importance of this factor depends on the format of the application; in heterogeneous immunoassays where most non-specifically bound molecules are washed away before the signal is generated it may be unimportant, but in a separation-free format it may be critical. This was one reason for the replacement of lysozyme, which was used as the label in the first EMIT™ immunoassays—it catalyses the formation of fragments from cell wall materials, and these can be detected by spectrophotometry. Sometimes an inactivation step can be carried out before the enzyme reaction, e.g. treating a histology sample with azide to inactivate endogenous peroxidases before investigation with an HRP-labelled probe.

The most thorough method of avoiding such interferences is to choose an enzyme whose substrate

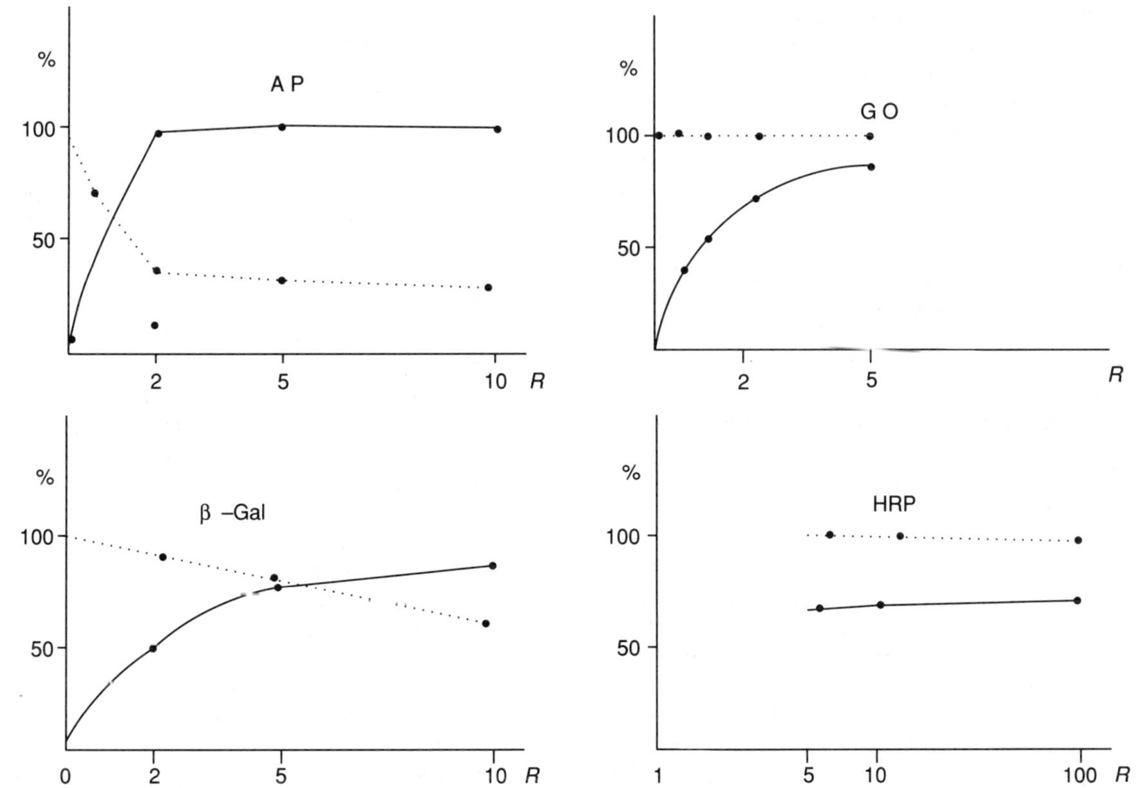

Figure 3.9 The percentage of amino groups substituted (●—●) and of remaining catalytic activity (●· · ·●) were determined after biotin labelling of alkaline phosphatase (AP), glucose oxidase (GO), β-galactosidase (β-Gal) and horseradish peroxidase (HRP). Abscissas indicate the molar ratio (R) of biotin-NHS to amino groups used in the labelling of the enzymes. (Reproduced with permission from: Guesdon *et al.* (1979). *J. Histochem. Cytochem.* **27**, 1131–1139.)

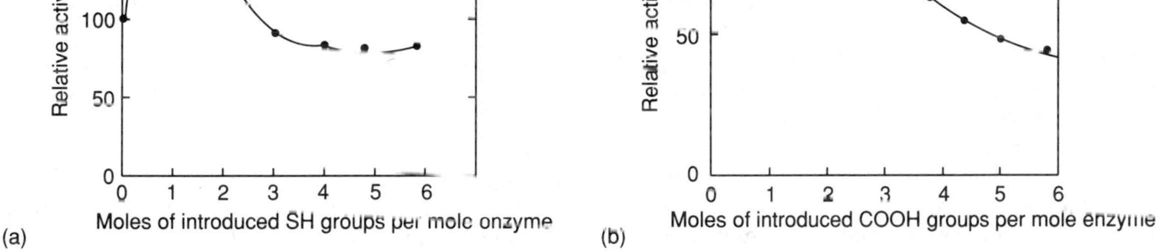

Figure 3.10 The activities of mercaptosuccinyl (a) and succinyl (b) Taka-amylase A. (Reproduced with permission from: Suzuki *et al.* (1970). *Bull. Chem. Soc. Japan*, **43**, 3849–3852.)

activity is not encountered in the samples of interest (e.g. β-D-galactosidase for human serum samples).

There are many other possible interferences with enzyme activity that may be encountered, and overcoming these is one of the key challenges in obtaining accurate test results regardless of other components in the sample. Some human samples for example contain antibodies to peroxidases and these can interfere with the signal generated, while some samples may contain inhibitors such as azide (commonly employed as a bacteriostat but also a peroxidase inhibitor) or EDTA (e.g. in plasma samples; as mentioned above this is a chelator and will inactivate many enzymes requiring the presence of metal ions). Some enzymes are also susceptible to inactivation on immobilisation (see Chapter 8).

3.2.7 Commonly-encountered labelling enzymes

A wide variety of enzymes have found labelling applications. These include oxidoreductases such as glucose-6-phosphate dehydrogenase (EC 1.1.1.49; Ullman and Maggio, 1980), xanthine oxidase (EC 1.1.3.22; Luong *et al.*, 1993) and laccase (EC 1.10.3.2; Skorobogat'ko *et al.*, 1994), transferases such as hexokinase (EC 2.7.1.1; Maeda *et al.*, 1993a) and pyruvate kinase (EC 2.7.1.40; Fricke *et al.*, 1982), and hydrolases such as ribonuclease (EC 3.1.27.5; Ullman and Maggio, 1980), β-lactamase (EC 3.5.2.6; Yolken *et al.*, 1984) and inorganic pyrophosphatase (EC 3.6.1.1; Peuravuori and Korpela, 1993).

A small number of enzymes, however, have gained such widespread use as labels that they deserve more detailed attention. In a review by Gosling (1990) of the immunoassay field for example, more than 90% of the 1989 applications employed one of the 'top three' enzymes, horseradish peroxidase (HRP), alkaline phosphatase (AP), and β-D-galactosidase (β-Gal), and they still dominate many fields of use. Direct comparisons between these enzymes have been made for specific needs (Porstmann *et al.*, 1985), but of course it is really a particular combination with the chosen substrate and conditions of use that is being compared, rather than the enzyme *per se*. As might be expected from their popularity, all these enzymes meet several of the key criteria with regard to the properties outlined in Section 3.2.6. They are all readily available, and can be used with a range of good substrates to generate signals of different types with high specific activity. The differences observed relate primarily to pH optimum, enzyme size, availability of reactive groups, robustness to derivatisation and potential for interference from sample components.

These 'top three' enzymes will now be considered individually in Sections 3.3, 3.4 and 3.5 respectively.

3.3 Horseradish Peroxidase

Horseradish peroxidase (HRP) is categorised under the Enzyme Commission (EC) system as a donor: hydrogen-peroxide oxidoreductase (EC 1.11.1.7).

Although the term horseradish peroxidase is frequently used without further qualification, this is not really a complete or unambiguous description. Even the term horseradish is not without problems: the plant currently termed *Armoracia rusticana* has been described variously as *Armoracia lapathifolia*, *Nasturtium armoracia* or *Cochlearia armoracia*. However these all refer to the same plant, originating from south-eastern Europe but introduced to many other parts of the world as a consequence of its culinary properties.

The early history of HRP is dealt with in an interesting review by Paul (1986), from which the following brief summary has been made. Planche (1810) observed a blue colour when fresh horseradish root was soaked in 'tincture of guaiacum', an anti-tubercular preparation derived from Caribbean hardwoods. The tincture displayed a similar phenomenon with milk, and the reaction was also used as a test for the presence of pus.

Linossier (1898) made some critical observations on the peroxidase activity of leucocyte preparations, in particular that the preparation was active only in the presence of oxygen or peroxides. Linossier probably used the term peroxidase for the first time, and recognised the separate existence of oxidase and peroxidase activities. He also described a number of peroxidase substrates which displayed similar properties to guaiacum extract, including *o*- and *p*-toluidine, *p*-phenylenediamine, and pyrogallol.

A few years later the high peroxidase content of horseradish root was established (Bach and Chodat, 1903; Chodat and Bach, 1903; Bach, 1904), and the next few decades saw a gradual improvement in techniques for the extraction of peroxidase at high purity and in the understanding of its catalytic properties. A watershed was reached in the 1940s: Theorell (1942) was able to produce HRP of crystallisable purity using a combination of precipitation steps and electrophoresis, while Chance (1943) published an extensive investigation of reaction kinetics.

The evolution of protein chromatography resulted in improved purifications of HRP and other peroxidases, exploiting techniques such as ion exchange (Paul, 1958), hydroxyapatite (Beaudreau and Yasunobu, 1966) or affinity methods using lectins (Darbyshire, 1973) and substrate analogues (Reimann and Schonbaum, 1978).

HRP is now one of the most widely used labelling enzymes; annual world production was estimated (Krell, 1991) at 30 GU, or around 100 kg. Based on Boehringer Mannheim's figures from the same article about 6000 tonnes of horseradishes are required to support this. By far the largest quantities of the Boehringer material were at this time employed in clinical chemistry tests for glucose (68%) and cholesterol (14%), detecting hydrogen peroxide released by the specific substrate oxidase.

The root of the horseradish plant contains an abnormally high content of peroxidase enzymes, making it a convenient source of this material. The peroxidase content is by no means homogeneous; more than thirty different forms (isoenzymes or isozymes) have been identified, differing widely in electrophoretic properties.

This heterogeneity is common among higher plant peroxidases, and appears to be due to a number of factors. First, peroxidases are versatile enzymes capable of a number of related functions within the plant (Gaspar *et al.*, 1991). Those postulated include catabolism of the growth regulator (auxin) indole acetic acid (IAA), both during normal growth and as a wound response; a role in the biosynthesis of ethylene; a number of redox roles including hydrogen peroxide production in cell membranes, and key functions in the various stages of cell wall construction which provide the plant with its physical strength. Therefore a plant is likely to need a number of different peroxidases with appropriate functional specificities, and this requirement is reflected in a multiplicity of peroxidase gene loci. Yamazaki and Nakajima (1986) observed differences in the redox behaviour of the active site of HRP isoenzymes, consistent with discrete substrate specificities.

Less easy to rationalise is the widespread existence of duplicated gene copies with minor allelic variations, leading to families of closely-related isoenzymes of more or less identical behaviour; similarly there are many seemingly accidental variations in post-translational processes such as glycosylation. These add to the multiplicity of peroxidase structures within the plant apparently without any noticeable increase in functional capability.

Plant and animal peroxidases are viewed as separate superfamilies (Welinder, 1991), with little structural similarity despite a distant common ancestry. Plant peroxidases have been categorised by Welinder (1992) as Class I (intracellular, prokaryotic), Class II (extracellular, fungal) and Class III (extracellular, higher plants); these differ in structural and functional aspects. HRP is an example of a Class III peroxidase.

The HRP isoenzymes have been loosely classified as acidic (A), neutral (B, C) and basic (D, E), with p*I* values in the ranges 3–5, 5–10 and >10 respectively (Paul, 1958; Shannon *et al.*, 1966). There has been little attempt to correlate these accurately with their *in vivo* functions. These groups have been further classified as B1, B2, B3, etc. (probably reflecting allelic variations), and there are multiple components (due mostly to post-translational differences) even within these groupings.

Table 3.1 summarises the amino acid composition of acidic, neutral and basic isoenzymes of HRP. As might be expected, the trend moving from low to high p*I* is for higher levels of arginine and lysine, and lower levels of free aspartate and glutamate. (Over half of the 'acidic' groups in the C isoenzyme are actually present as neutral amides, either as linkage sites for carbohydrate chains, or as simple asparagine or glutamine entities).

The largely monomorphic component of HRP with p*I* around 7–9, described as isoenzyme C, accounts for about 50% of the activity in the horseradish root. It is this material that has received the greatest attention, is most widely available from commercial sources, and is usually intended when HRP is referred to without further qualification. The classification of the dominant isoenzyme described in the studies below is strictly C1; allelic variants (C2, C3) of the controlling gene exist

Table 3.1. Amino acid content of horseradish peroxidase isoenzymes C, A2 and E5

Amino acid	Number A2	C	E5	Amino acid	Number A2	C	E5
Asp/Asn	48	48	43	Met	2	4	4
Thr	24	25	22	Ile	14	13	14
Ser	37	25	23	Leu	34	35	33
Glu/Gln	21	20	15	Tyr	2	5	5
Pro	12	17	18	Phe	19	20	21
Gly	24	17	19	His	3	3	3
Ala	29	23	25	Lys	5	6	7
Cys	8	8	8	Arg	11	21	26
Val	14	17	19	Trp	1	1	1
Free acids (Asp/Glu, excluding pyroglutamate)					[a]	28	26

[a] Not known

The C and E5 figures have been taken directly from published amino acid sequences, those for A2 rounded to whole numbers from quantitative analysis. (Taken or adapted from Welinder *et al.* (1976); Stoffer *et al.* (1991); Morita *et al.* (1991)).

(Fujiyama *et al.*, 1988) although their products have not been observed *in vivo*. The pH optimum of HRP C is around 6 (Boehringer Mannheim, 1990), but is broad. This allows compromise conditions to be used—for example with luminol (Section 3.3.7), where the requirement for the substrate to exist in the anionic form leads to the use of a matrix at around pH 8–9.

3.3.1 Structural features of horseradish peroxidase

3.3.1.1 Amino acids

The complete peptide sequence of horseradish peroxidase isoenzyme C (HRP-C) was established by Welinder (1976; see Table 3.1); it consists of a single chain of 308 amino acid residues with a calculated molecular mass of 33.89 Da. In addition the molecule contains eight carbohydrate chains, attached *via* asparagine residues, a haemin (iron protoporphyrin IX) prosthetic group, and two calcium ions (at least one of which is essential for maintaining the structure of the active site). The overall molecular mass has been estimated at between 40 and 45 kDa; recently the figure of 44 kDa seems to have gained fairly wide acceptance.

The N-terminal amino acid is glutamate, but this is present in a cyclised pyroglutamate or pyrrolidone form so there is no free amine group from this source. There are six lysine residues in the sequence, however in practice only around one to two of these are generally available for reaction. This is probably due to a combination of two factors; the first is common to most proteins, namely that some of these are likely to be 'buried' deep in the protein where they are too sterically hindered for other reagents to gain access. This does not appear to be a particularly powerful factor with HRP, however, as Ugarova *et al.* (1978) were able to access all six lysine residues at 40 °C (see below).

The second factor is more idiosyncratic: the component of the horseradish responsible for its characteristic pungent flavour, sinigrin, is prone to releasing allyl isothiocyanate (Stoll and Seebeck, 1948). Unless precautions are taken to the contrary, these isothiocyanate is likely to react with the enzyme's amine groups. Any resulting thioureas will be quite stable, so these lysines are effectively lost as useful 'handles' for derivatisation. Perhaps most importantly, changes in extraction conditions are likely to lead to batch-to-batch variation in the number of reactive amines that survive the process. While this may not necessarily be the cause, such differences in amine availability have indeed been observed in practice with some suppliers. It hardly needs to be said that such differences make it difficult to establish reaction conditions which can be used for more than one batch of HRP, and this should be borne in mind when optimising production procedures, etc.

Rennke and Venkatachalam (1979) showed that derivatisation of HRP with succinic anhydride to yield a polycarboxylated molecule (pK_a < 4) had no detectable effect on the pH optimum or enzyme activity in a colorimetric assay. Guesdon *et al.* (1979) observed similar retention of activity during extensive biotinylation (see Figure 3.9).

Studies by Ugarova *et al.* (1978, 1979a, b) investigated the effect of derivatising the amine groups of HRP-C in more detail. Anhydrides of mono- and dicarboxylic acids were capable of acylating four lysine residues at 0 °C and pH 8.0, while the more bulky trinitrobenzenesulfonic acid (TNBS) modified three. Increasing temperature facilitated further derivatisation: TNBS modified four groups at 22 °C and pH 9.5, and all six lysine residues at 40 °C and pH 8.5.

Enzyme activity was remarkably stable to lysine modification, no loss being detected under any of the conditions described above. The introduction of extra negative charges by succinylation caused a broadening of the pH dependence around the same optimum (5.5–6.0), and additionally showed a greater resistance to thermal inactivation than native HRP (Ugarova *et al.*, 1979b). This effect could be quite dramatic—the succinylated derivative showed a 90-fold increase in stability at 56 °C over the native enzyme, which was attributed to a reduction in mobility around the haem group. Similarly, Ryan *et al.* (1994) observed substantial stabilisation at 75 °C when HRP was derivatised with *bis*-succinimide esters; these were able to derivatise five lysine residues at room temperature, and no cross-linking was observed under the conditions used.

On the basis of findings such as these, derivatisation of the amine groups offers an attractive route for HRP conjugate preparation.

Despite the high proportion of asparagine and glutamine residues noted above, HRP-C still contains 28 free carboxyl groups which offer another route to derivatisation. Rennke and Venkatachalam (1979) also prepared a polyaminated HRP (pK_a 8.4–9.2) by reaction with hexanediamine and a carbodiimide, and again noted no change in pH optimum or enzyme activity with a colorimetric substrate. Interestingly polymerisation of the HRP appears to have been avoided in this preparation, although it might have been expected from the conditions used.

The thiol groups of all eight cysteine residues are used intramolecularly in four disulfide linkages, therefore these do not provide a naturally occurring handle for derivatisation.

3.3.1.2 Carbohydrates

HRP-C contains around 20% of carbohydrate by weight, made up of around fifty sugar residues in eight groups. Each is attached at an asparagine residue forming part of a sequence Asn-XXX-Ser or Asn-XXX-Thr (XXX = any residue) (see Section 2.4.1). Only one such sequence in the structure is unglycosylated, presumably due to inaccessibility to the transferase responsible. Lew and Shannon (1973) demonstrated, with radiolabelled sugars, that glycosylation took place after the production of the peptide portion.

The carbohydrate of HRP consists principally of mannose, *N*-acetylglucosamine, xylose and fucose. The major N-linked oligosaccharide of HRP has the structure:

Manα3(Manα6)(Xylβ2)Manβ4GlcNAcβ4(Fucα3) GlcNAc

(McManus *et al.*, 1988; Harthill *et al.*, 1989), while minor components contain an extra mannose or lack a mannose, fucose or xylose residue (Tams and Welinder, 1995). The major oligosaccharide is shared by many other plant enzymes, probably helping to explain some of the immunological cross-reactivity observed with anti-peroxidase antibodies (Conroy *et al.*, 1982; McManus *et al.*, 1988).

The carbohydrate residues of HRP obviously provide a convenient site for derivatisation, and this has been exploited particularly in the periodate oxidation approach to protein coupling. If over-oxidation by periodate is avoided, these residues can be derivatised without any significant loss in activity (Schalkwijk *et al.*, 1985).

Deglycosylation work supports the relative insensitivity of HRP to modification of the carbohydrate moieties. Tams and Welinder (1991) derived conditions for deglycosylation of HRP-C; there was no effect on specific activity or thermal stability, but a decay in activity during the enzyme reaction was observed. Egorov *et al.* (1993) examined a recombinant HRP from *E. coli*, lacking sugars due to its prokaryotic origin, and observed a threefold lower stability at 56 °C than the native enzyme. Like Tams and Welinder (1991), these authors also reported a faster decay of the enzyme during catalysis, and postulated that the sugar chains might protect the protein from substrate radicals.

Later studies examined deglycosylated HRP purified: (1) on a substrate affinity matrix to remove inactive enzyme and (2) by ion exchange to remove a large proportion of more negatively charged species (Tams and Welinder, 1995). This material showed unchanged amino acid content, p*I*, electronic absorption spectrum, and enzyme kinetics. The major difference from native HRP-C was a much lower solubility in a salt solution.

3.3.1.3 Haem and the active site

The presence of the haem group seems to have an important structural function as well as being the catalytic centre; there are a number of changes to the accessibility of amino acid residues when the haem-free apoenzyme is formed, and these are too widely dispersed to be explained simply by the removal of the haem from a fixed structure. The iron in porphyrin groups such as haem is coordinated to the four pyrrole nitrogen atoms, leaving two axial coordination positions free; when the porphyrin is linked to a protein, these two positions are referred to as proximal and distal, facing respectively towards and away from the bulk of the protein in the binding site. In HRP-C the proximal position is taken up by the histidine at position 170 in the amino acid sequence, but the distal position appears to be vacant, unlike some similar proteins, e.g. metmyoglobin, where it is occupied by a water molecule. Close to the distal site, however, is another region of the peptide chain consisting of the Arg-Leu-His-Phe-His-Asp sequence (residues 38–43), which is involved in the reaction of the enzyme with its substrate.

The HRP-C gene has been sequenced, and recombinant material has been prepared by a number of groups (see for example Smith *et al.*, 1990). It is no easy matter to create functional recombinant enzyme in good yields; folding the protein to give the correct conformation is a delicate operation, although there have been advances in understanding in this area (Pappa and Cass, 1993; Henriksen *et al.*, 1995). This does not appear to offer an attractive commercial route at present.

Site-directed mutagenesis has allowed investigation of mutations in the active site (Phe41 → Val, Arg38 → Lys) and in the access channel (Phe142 → Ala); all of these changes affected the behaviour of the recombinant enzyme to some extent (Smith *et al.*, 1992a, b; Veitch *et al.*, 1995; Rodriguez-Lopez *et al.*, 1996).

Above around pH 11 there is a change in the conformation of HRP, evidenced by differences in electronic spin state and kinetics. This change is thought to be associated with one of the distal histidines coordinating to the vacant sixth site.

The haem group has a strong absorbance in the visible region with a maximum at 403 nm. The apoprotein (like other proteins) absorbs most strongly below 300 nm, and the relationship of absorbance at 403 and 280 nm is commonly used as an indication of peroxidase purity. The parameter A_{403}/A_{280} is termed the Reinheitzahl (Rz) and can approach a value of around 3.55 in material of the highest purity—good quality HRP should certainly have an $Rz > 3.0$. It should be noted that the position of the wavelength maximum and the magnitude of the absorbance can both be affected by a number of factors, e.g. pH changes and chemical reactions. Therefore care needs to be taken in interpretation of Rz data other than in the very simplest cases.

3.3.1.4 Calcium

Analysis for calcium has indicated that at least two Ca^{2+} ions are present per HRP-C molecule (Haschke and Friedhoff, 1978). These have been shown to be located either side of the haem group, one each in the distal and proximal domains (Henriksen, 1997). It appears that calcium needs to be present for the protein to fold into the correct conformation to accept the haem group, and at least one Ca^{2+} ion is involved in maintaining the structure of the active site (Ogawa *et al.*, 1979). Calcium-free HRP shows substantially reduced thermal stability compared to the native enzyme, and there are suggestions that the electronic state of the iron is also different. Chelating agents such as EDTA can remove this calcium, destroying the enzyme activity; unless a denaturing agent such as urea or guanidine is present however, this effect is quite slow at moderate temperatures.

For many plant peroxidases it has been demonstrated that calcium ions induce binding to cell structures (Penel, 1986, 1991); it is therefore possible that its presence in HRP-C has multiple functions.

3.3.1.5 Three-dimensional structure of HRP

The heterogeneity of HRP-C, particularly the existence of multiple glycosylation variants, has hindered crystallography studies. Braithwaite (1976) crystallised native HRP-C, but the unit cell obtained was too complex to allow structural analysis. Until fairly recently, the only detailed structural information available was from the Class I cytochrome *c* peroxidase from yeast (Poulos *et al.*, 1980; Finzel *et al.*, 1984). A low-resolution structure of the HRP-E5 isozyme followed (Morita *et al.*, 1991), as well as the structures of some Class II (fungal) peroxidases (Poulos, 1993; Poulos *et al.*, 1993; Doyle and Smith, 1996).

Recently, however, recombinant HRP-C has been obtained from *E. coli* and used to generate a crystal structure at 2.1 Å resolution (Henriksen *et al.*, 1995; Henriksen, 1997).

This shows HRP-C to possess the basic structure typical of plant peroxidases, with the haem group sandwiched between a distal and a proximal domain. The former contains five α-helices, the latter eight α-helices and two short β-strands. A key aspect of the structure is a channel connecting the distal haem region with the surface of the molecule, and providing an entry route for substrates. Hydrophobic phenylalanine residues feature strongly around this channel, and the Phe-179 residue in particular has been found to hinder access to the active site. These observations help to explain the preference of HRP-C for small, aromatic substrates.

3.3.2 The mechanism of HRP catalysis

Simply, HRP-C and other peroxidases catalyse the oxidation of a wide range of organic and inorganic compounds by peroxides (Dunford, 1986; Ortiz de Montellano, 1992). The first stage of the catalysed reaction is a two-electron oxidation of the haem group by the peroxide. This stage is fairly specific; it is generally stated that only hydrogen peroxide (H_2O_2) and simple alkyl hydroperoxides (ROOH) can act as substrates. In terms of direct conversion of HRP to compound I this is close to the truth, although other peroxides such as peracids can also fulfil this role (Baek and Van Wart, 1992). However the oxidation can in fact be effected by apparently unrelated agents such as bromate and periodate (George, 1953). These oxidations may involve the generation of hydroperoxides *in situ*, or other more complex routes which are mentioned below.

In the formation of compound I, two different oxidation reactions occur on the haem group. In one the ferric Fe^{III} ion is oxidised to a ferryl ($Fe^{IV}{=}O$) group; the oxidation is facilitated by hydrogen transfer to the His42 residue. In the second an electron is abstracted from the porphyrin molecule to generate a π-radical cation; it is likely that this reaction is facilitated by the distal amino acid residues Asp43 and Arg38. The peroxidase converted to this form is known as compound I, the greenish colour of which is indicative of the spectral shifts which accompany the change.

The conversion of HRP to compound I is now in fact believed to involve more than one stage, and the identity of the intermediate 'compound 0' is a source of interest. Baek and Van Wart (1992) have suggested an Fe^{V}-oxo complex or a π-dication radical as possible structures.

Compound I is a powerful oxidising agent, and will react with a large range of oxidisable substrates. Some of these such as the iodide and bisulfite ions are able to reduce compound I directly back to the native form, i.e. a two-electron reduction. In most cases however, two successive single-electron reductions occur *via* a second intermediate form of the enzyme, the reddish-coloured compound II. The classic triangular reaction scheme incorporating the key species is shown in Figure 3.11. There have been some conflicting results on the structure of compound II but current opinion favours the continued presence of the ferryl moiety, consistent with the reduction from compound I taking place on the porphyrin site.

Compound II may also be formed directly from HRP *via* a single-electron oxidation. It is thought that the formation of compound I from HRP and hexachloroiridate(IV) actually occurs *via* intermediate formation of compound II (Hayashi and Yamazaki, 1979).

The substrates for compounds I and II are highly varied, but many have phenolic structures. One of the *in vivo* functions of peroxidases is the lignification of cell walls; lignin, which confers structural robustness to the cells, is formed by the peroxidase-catalysed polymerisation of phenols such as coniferyl alcohol (Figure 3.12) in the presence of hydrogen peroxide. Many other phenolic compounds behave in a similar fashion, for example *p*-cresol and L-tyrosine, but other substrates have unrelated structures, such as ferrocyanide and *p*-aminobenzoic acid (PABA). Some substrates are particularly useful where the peroxidase is employed as a label, generating oxidation products (notably coloured or luminescent) which are readily measurable. Some of these will be described in detail later on.

The oxidation of some compounds by molecular oxygen (in the absence of hydrogen peroxide) is also catalysed by these enzymes, behaving in this instance as oxidases. These include important naturally occurring reducing agents such as reduced nicotinamide adenine dinucleotide (NADH). The initial presence of hydrogen peroxide converts this to $NAD^{\cdot}$ radicals *via* the normal peroxidase-catalysed route, but these can then act as oxidase substrates, converting molecular oxygen to superoxide ($O_2^{\cdot-}$) radical anions which can combine under acid conditions to produce hydrogen peroxide:

$$O_2^{\cdot-} + O_2^{\cdot-} + 2H^+ = H_2O_2 + O_2$$

These also convert the peroxidase to an enzymatically-inactive form, compound III, which can be viewed

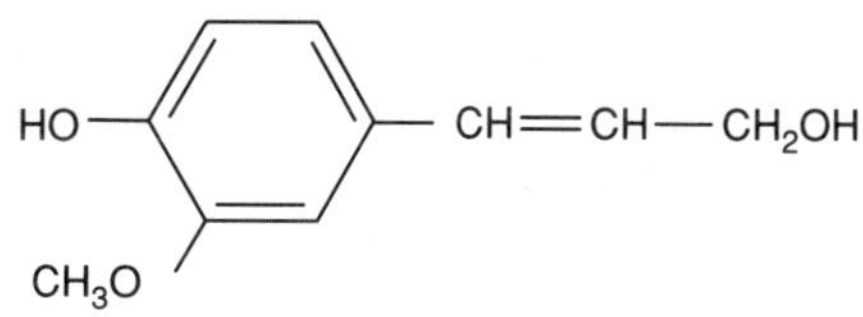

Figure 3.12 Coniferyl alcohol.

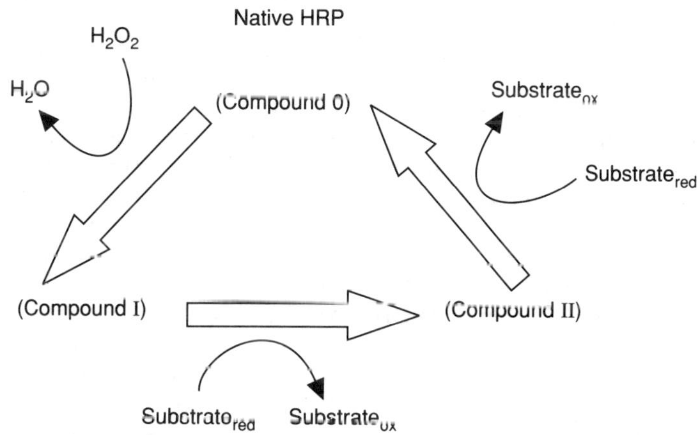

Figure 3.11 A simplified view of HRP catalysis. The native enzyme is first oxidised to Compound I, which catalyses two successive substrate oxidation reactions. In so doing, the enzyme is converted first to Compound II, then back to the native Compound 0.

either as a native HRP: superoxide complex (containing $Fe^{3+} \cdot O_2^-$), or a reduced HRP: oxygen complex (containing $Fe^{2+} \cdot O_2$). This form of the enzyme can also be formed by reaction of compound II with an excess of hydrogen peroxide.

Compound III decays slowly to regenerate the native enzyme; however certain compounds (again including phenols) can increase the rate of this breakdown and, therefore, the overall rate of superoxide and hydrogen peroxide production.

3.3.3 Stability of HRP

Native HRP is reasonably stable up to around 40 °C: above 60 °C loss of activity is observed within minutes (Boehringer Mannheim, 1990). As discussed earlier, acylation of the lysine residues of HRP results in increased thermostability compared to the native enzyme, a serendipitous advantage of conjugation techniques involving this chemistry.

The pH stability of HRP is very good, in a three-hour incubation at 25 °C virtually no loss of activity is observed over the range 4 to 11 (Boehringer Mannheim, 1990).

Although it seems intuitively unlikely that enzymes would function effectively in non-aqueous media, the retention of a number of water molecules may provide a sufficiently 'normal' microenvironment to permit this to occur, provided the enzyme itself is not denatured. HRP is rather robust towards organic solvents (Bartling *et al.*, 1974; Slepnev *et al.*, 1995), and has been shown to retain a considerable amount of its functional activity in non-aqueous environments (Kazandjian *et al.*, 1986): in diethyl ether the level of enzyme activity was 25% of that in water.

A wide variety of reagents have been claimed as stabilisers of HRP or HRP conjugates, both in the liquid and solid formulations. Stabilisers in aqueous solution include polyvinyl alcohol (Boyd and Yamazaki, 1994). For lyophilised formulations, trehalose and lactalbumin-sucrose mixtures have been described (Nielsen, 1995).

Although HRP is extremely stable in its native form, it is rather more prone to deterioration during signal generating reactions than its major competitors—due to some extent to the powerful redox species involved in the mechanism.

3.3.4 Activators and inhibitors of HRP

The best-known activators of HRP are the 'enhancers' which radically increase the turnover rate of substrate oxidation. This effect can be observed in many signal systems, but has seen the greatest application in luminol-based luminometric formats. These compounds will be dealt with briefly in Section 3.3.7.

Guilbault *et al.* (1968b) listed the following substances which inhibited HRP at milli- to micromolar levels, using the fluorogenic substrate homovanillic acid: Mn^{2+}, S^{2-}, Co^{2+}, $Cr_2O_7^{2-}$, hydroxylamine, Pb^{2+}, Fe^{2+}, Fe^{3+}, Cu^{2+}, CN^-, SO_3^{2-}, cystine, Cd^{2+}, Ni^{2+}, *p*-aminobenzoic acid. Other known HRP inhibitors include azide, sulfide, cyanide, fluoride and hydroxide ions (Boehringer Mannheim, 1990). Inactivation by some nucleophilic species involves reaction with the porphyrin ring, and this has been used in elucidation of the mechanism of enzyme action (Ator *et al.*, 1987; Ortiz de Montellano *et al.*, 1988; Ortiz de Montellano, 1992).

Ascorbate is a powerful reducing agent, capable of converting substrate radicals rapidly back to their original state. Therefore it interferes strongly with HRP-catalysed oxidations, and has in fact been used to maintain substrate concentration in kinetic investigations (Goodwin *et al.*, 1995).

3.3.5 Colorimetric substrates for HRP

As mentioned above, many compounds (chromogens) have been identified which generate coloured products on peroxidase-catalysed oxidation, and are therefore useful for quantification of enzyme labels. A number of factors have affected their relative popularity.

Some of the most commonly encountered colorimetric substrates of HRP are shown in Figure 3.13, and their properties summarised below. They all require the provision of hydrogen peroxide in the assay formulation.

As noted in the Introduction, colorimetric substrates such as guaiacol (the active component of guaiacum) and pyrogallol had been identified before the end of the nineteenth century. The first two have remained in circulation as accepted substrates for quantifying peroxidase activity, guaiacol and purpurogallin[1] units still being commonly encountered as measures of HRP potency. The structure of the coloured product of guaiacol oxidation has recently been elucidated (Doerge *et al.*, 1997).

Many other substrates derive from tests for the presence of blood, which posseses considerable peroxidase activity. Benzidine has been used in this application since the start of the century (Adler and Adler, 1904; Goodpasture, 1919). It was therefore an obvious

[1] Purpurogallin (Figure 3.14) is the product of pyrogallol oxidation.

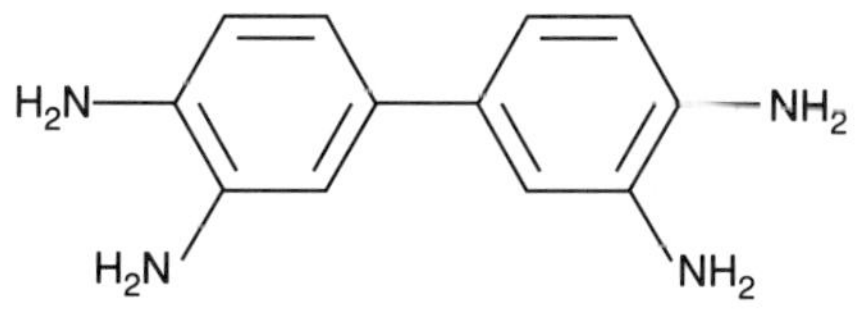

3,3'-Diaminobenzidine (DAB)

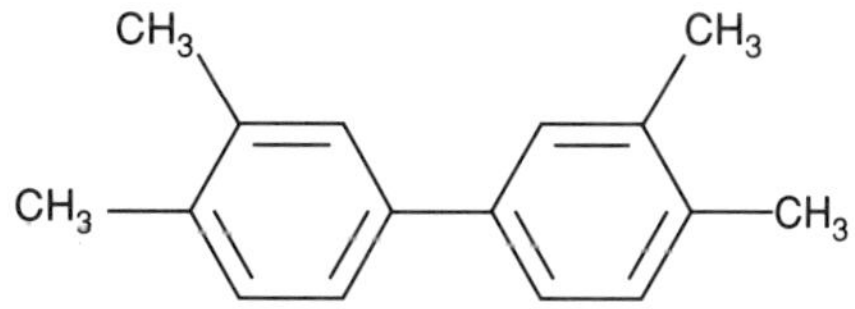

3,3',5,5'-Tetramethylbenzidine (TMB)

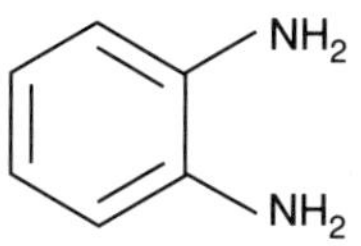

o-Phenylenediamine (OPD)

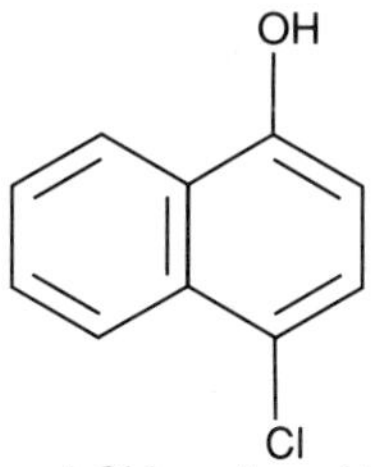

4-Chloro-1-naphthol (4CN)

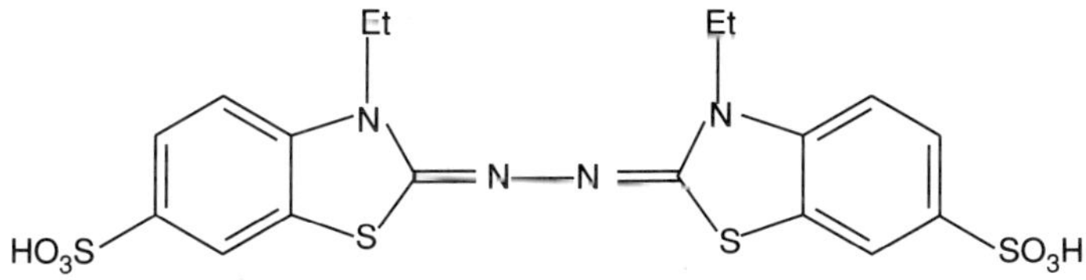

2,2'-Azino-di-[3-ethylbenzthiazoline sulfonate] (ABTS)

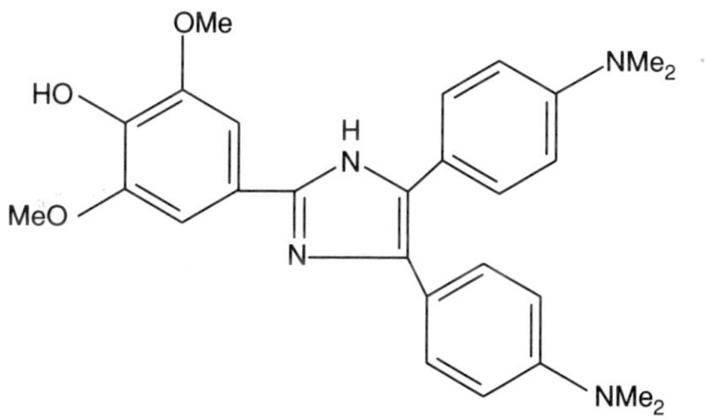

2-(3,5-Dimethoxy-4-hydroxyphenyl)-4,5-bis(4-[dimethylamino]phenyl)imidazole

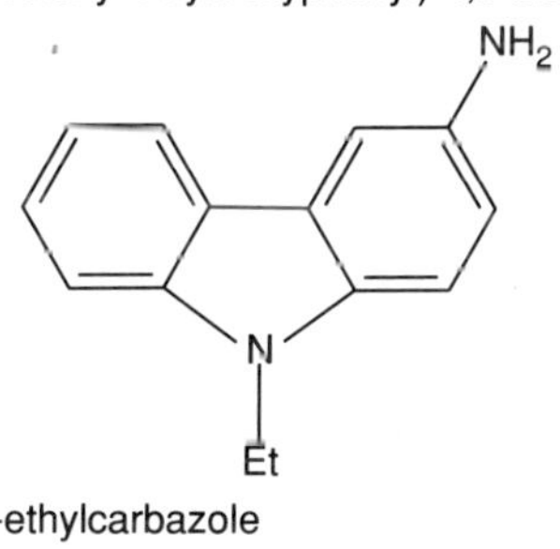

3-Amino-9-ethylcarbazole

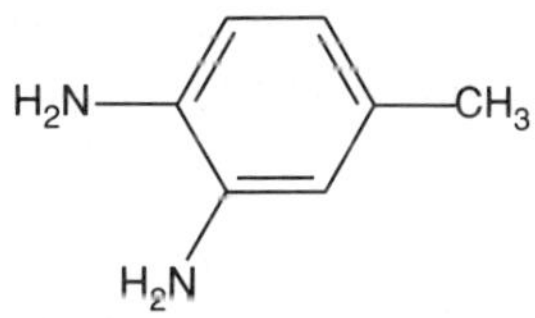

3,4-Diaminotoluene

Figure 3.13 Common colorimetric substrates of HRP.

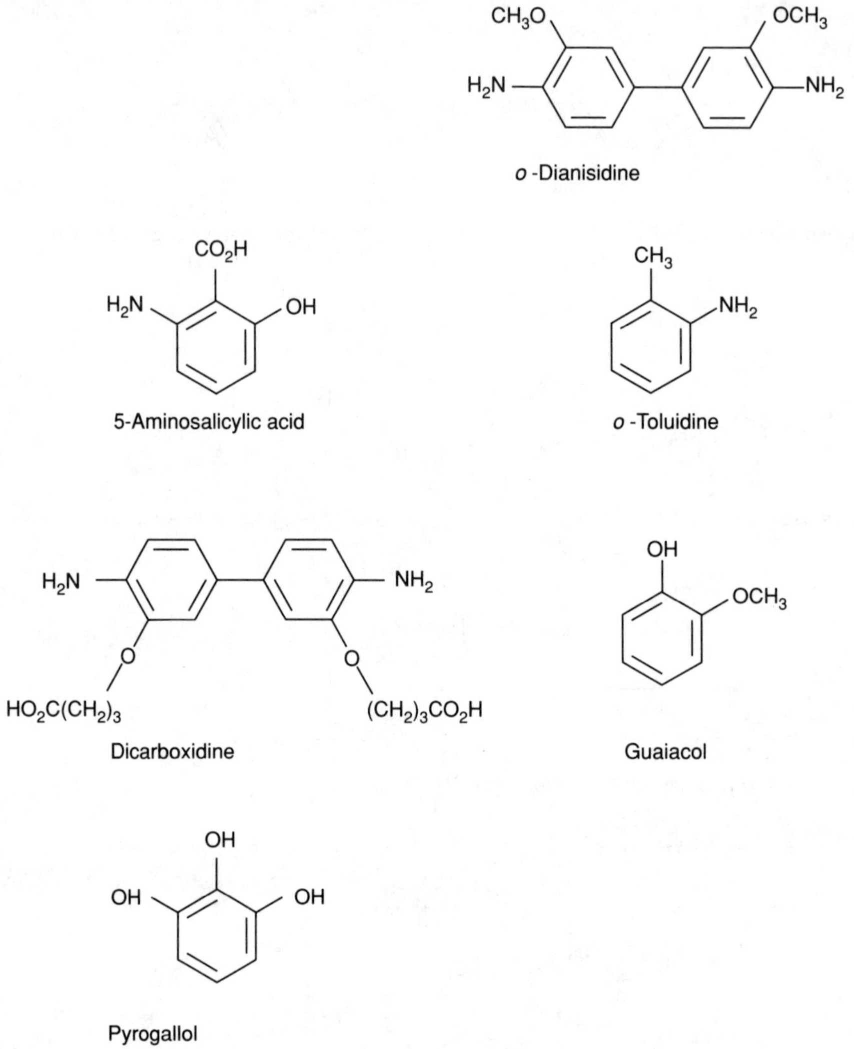

Figure 3.13 Contd.

choice for histochemical detection of HRP, providing a blue product; a fairly detailed study of this reaction was carried out by Straus (1964). However benzidine is a powerful carcinogen and mutagen (Holland *et al.*, 1974), and should no longer be encountered in practical use.

It has long been recognised that the presence of cyanide compounds such as sodium nitroprusside stabilise the oxidation products of benzidine and its analogues (Goodpasture, 1919). However this too is a highly toxic compound: Albers *et al.* (1984) used potassium ferricyanide as a safer alternative.

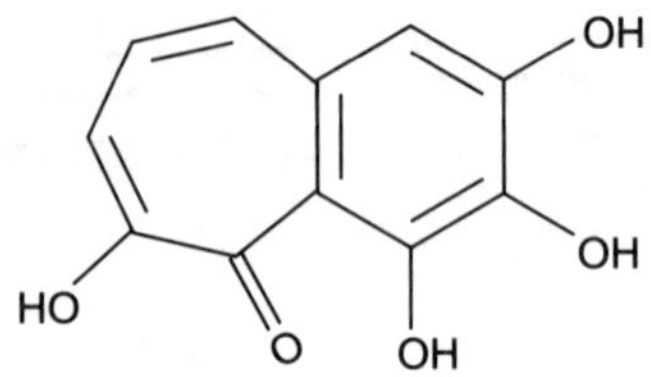

Figure 3.14 Purpurogallin, the product of pyrogallol oxidation.

Many derivatised benzidines have been examined in this role as well, including 3,3-dimethylbenzidine or *o*-tolidine; this was a common early substrate which is somewhat safer than benzidine itself (Holland *et al.*, 1974). Similarly 3, 3′-dimethoxybenzidine or *o*-dianisidine was originally described as an HRP substrate for use in test strips for glucose (Keston, 1956); the products of its oxidation are were studied in detail by Claiborne and Fridovich (1979a,b).

3, 3′-Diaminobenzidine (DAB) was first proposed as a histochemical substrate by Graham and Karnovsky (1966), yielding an insoluble brown product; conditions for yielding a soluble product (involving the addition of gelatin) were described by Herzog and Fahimi (1973). The absorption maximum of the product in the homogeneous format is at 465 nm, and the optimal pH is 4.3. Sensitivity is good; detection of around 1 ng (25 fmol) of HRP is achievable. One disadvantage in histochemical applications is the poor contrast obtained from the brown product in photographic reproduction. Osmication with osmium tetroxide yields a darker product (Lojda, 1979a), while Young (1989) described the use of a cocktail of DAB with 4CN (see below) to give a high-contrast black product with good sensitivity.

These benzidine reagents all suffer from some mutagenicity problems, and should no longer be in common use. The major exception in this category is 3, 3′, 5, 5′-tetramethylbenzidine (TMB) which was described for blood detection by Holland *et al.* (1974), for immunoperoxidase staining by Mesulam (1978), and for immunoassay use by Bos *et al.* (1981). TMB is considerably less hazardous than other benzidine substrates, repeated studies having failed to demonstrate mutagenic or carcinogenic activity (Holland *et al.*, 1974; Garner *et al.*, 1975). This is attributed to the inability of TMB to undergo *o*-hydroxylation reactions, a recognised cause of aromatic amine toxicity (Miller *et al.*, 1961).

Nanomolar levels of HRP can be detected, and the formation of the oxidised blue product follows linear kinetics for up to three minutes; some other substrates only demonstrate linearity for seconds. The absorption maximum of the blue-green product is 655 nm, but a stronger absorption is obtained at 450 nm in the presence of sulfuric acid. The pH optimum of the HRP reaction is around 5–6. On the whole, TMB offers excellent chromogenic properties both for solution and histochemical applications (Bos *et al.*, 1981; Morrell *et al.*, 1981; Carson and Mesulam, 1982), in combination with its lack of toxicity problems this represents one of the best HRP substrates.

One of the disadvantages of TMB is its low solubility in aqueous solution, crystallisation sometimes occurring at concentrations suitable for assay. Weavers and Powell (1989), of Boots-Celltech, patented the use of a TMB/β-cyclodextrin complex to get round this problem. Pauly and Schwarz (1990) of Behringwerke have patented the use of penicillins to stabilise TMB solutions in the presence of hydrogen peroxide.

Phenylenediamines are another category of HRP substrate which have found widespread use, which again have been dogged by toxicity problems. *o*-Phenylenediamine (OPD) yields an orange product which is highly water-soluble; absorbance is measured at 475 nm, or 490 nm after stopping with 2 M sulfuric acid. OPD has enjoyed considerable popularity as an HRP substrate (Wolters *et al.*, 1976; Schall *et al.*, 1978); it was described as by far the best HRP chromogen before TMB (Bos *et al.*, 1981), but has also been demonstrated to possess mutagenic properties (Voogd *et al.*, 1980).

1-Naphthol derivatives such as 4-chloro-1-naphthol (4CN) can take part in the NaDi reaction, combining with a suitable derivative of *p*-phenylenediamine to generate a deeply-coloured precipitate of indophenol blue. While this reaction was in histochemical use in the last century, backgrounds were high and sensitivity low; however Conyers and Kidwell (1991) have modified the NaDi approach for membrane staining of peroxidase. Quantities of 5 pg (120 amol) of HRP could be localised by using combinations such as *N*,*N*-dimethyl-*p*-phenylenediamine (DMPDA) with 4CN at pH 6, or 3-methyl-2-benzothiazolinone hydrazone (MBTH) with 4CN at pH 4.

4CN itself yields an insoluble blue product; while this gives good contrast in photographic reproduction, the sensitivity is not good.

2, 2′-Azino-di-[3-ethylbenzthiazoline sulfonate] (ABTS™) (Boehringer Mannheim) yields a soluble blue-green product with an absorbance maximum at 414 nm. ABTS has demonstrated excellent chromogenic properties (Groome, 1980) but been shown to be mutagenic (Voogd *et al.*, 1980).

Triarylimidazole leuco-dyes such as 2-(3,5-dimethoxy-4-hydroxyphenyl)-4,5-bis-(4-{dimethyl-amino}phenyl) imidazole have also been employed as HRP substrates, for example as described by Bruschi (1978) and subsequently employed in Eastman Kodak systems. McClune (1989) described the use of phenols and anilines to enhance the rate of these reactions.

Other colorimetric substrates which have been employed for HRP include 3-amino-9-ethylcarbazole,

3,4-diaminotoluene, Hanker-Yates reagent (insoluble products), *o*-dianisidine, 5-aminosalicylic acid, *N,N*-dimethyl-*p*-phenylenediamine (Yalçin *et al.*, 1993) and dicarboxidine (soluble products). Commercial products of undisclosed structure are also available for both liquid- and solid-phase use, for example the Indophane™ series from Vio-Medics.

3.3.6 Fluorimetric substrates for HRP

HRP is unusual among the commonly-encountered labelling enzymes in that it is not generally employed in a fluorimetric format; there are no fluorogenic substrates for peroxidases that are in very widespread use.

However, Guilbault *et al.* (1967, 1968b) used homovanillic acid for the fluorimetric determination of HRP: the phenolic substrate is dimerised to a highly fluorescent product (Figure 3.15). Excitation is carried out at 315 nm and emission is recorded at 425 nm. Approximately 2×10^{-13} moles of HRP could be detected by this technique.

Ishikawa (1987) at Miyazaki used the analogous 3-(4-hydroxyphenyl)propionic acid (Figure 3.16) as an HRP fluorogen. The latter substrate was capable of detecting 0.03 amol (18 000 molecules) of peroxidase in a small-volume format.

3.3.7 Luminometric substrates for HRP

By far the most common luminescent substrates for HRP are based on cyclic diacylhydrazides such as luminol (5-amino-2,3-dihydro-1,4-phthalazinedione (Figure 3.17). Harvey (1940) noted that peroxidases produced light in the presence of peroxide and luminol, but Dure and Cormier (1964) made the first detailed report of its peroxidase-catalysed oxidation. Luminol oxidation can be initiated by the presence of oxygen and a base in aprotic solvents, but in aqueous systems a catalyst is required (Schroeder and Yeager, 1978), a number of inorganic reagents are applicable in this role as well as enzymes such as HRP.

Various mechanisms have been described for the simple (unenhanced) HRP–luminol reaction (Misra and Squatrito, 1982; Thorpe and Kricka, 1986; Easton *et al.*, 1996), but the fundamentals are captured in a simplified scheme in Figure 3.18. Turnover of HRP Compound I or II extracts an electron from the anionic form of luminol (LH^-), whose presence is promoted by using the sodium salt. This generates the anion radical (LH^-), two of which can disproportionate to yield one molecule each of LH^- and the diazaquinone (LH). The latter reacts with peroxide ion to form an endoperoxide; this in turn expels nitrogen to yield 3-aminophthalate in an excited state, and it is the decay of this species to the ground state which produces light. The pH optimum of the HRP/luminol system is around 9.5, and the light emission maximum is at 425 nm.

Unfortunately with horseradish peroxidase and luminol the levels of luminescence obtained in an enzyme-labelled immunoassay at steady state are far too low to be of practical use, precluding the application of this format. This led to the adoption of systems

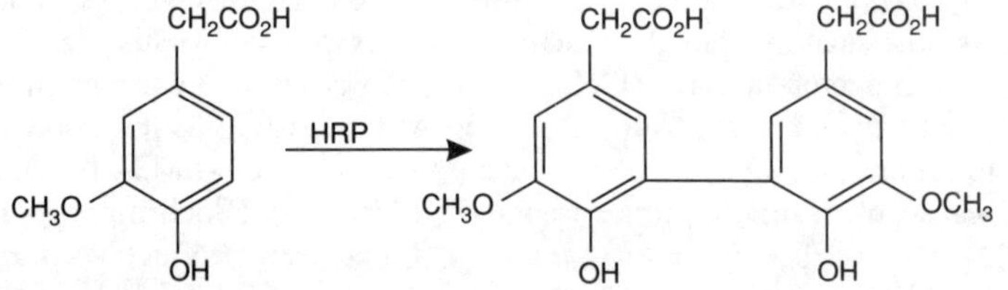

Figure 3.15 Homovanillic acid and its fluorescent product.

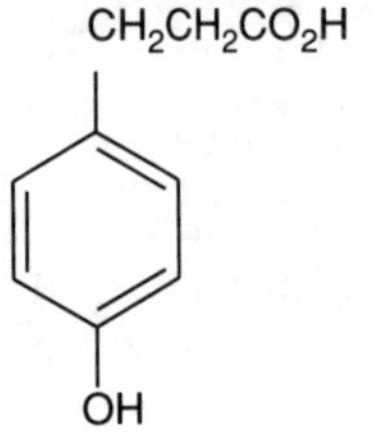

Figure 3.16 3-(4-Hydroxyphenyl)propionic acid.

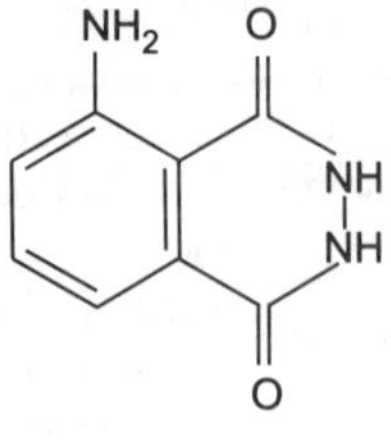

Figure 3.17 Luminol.

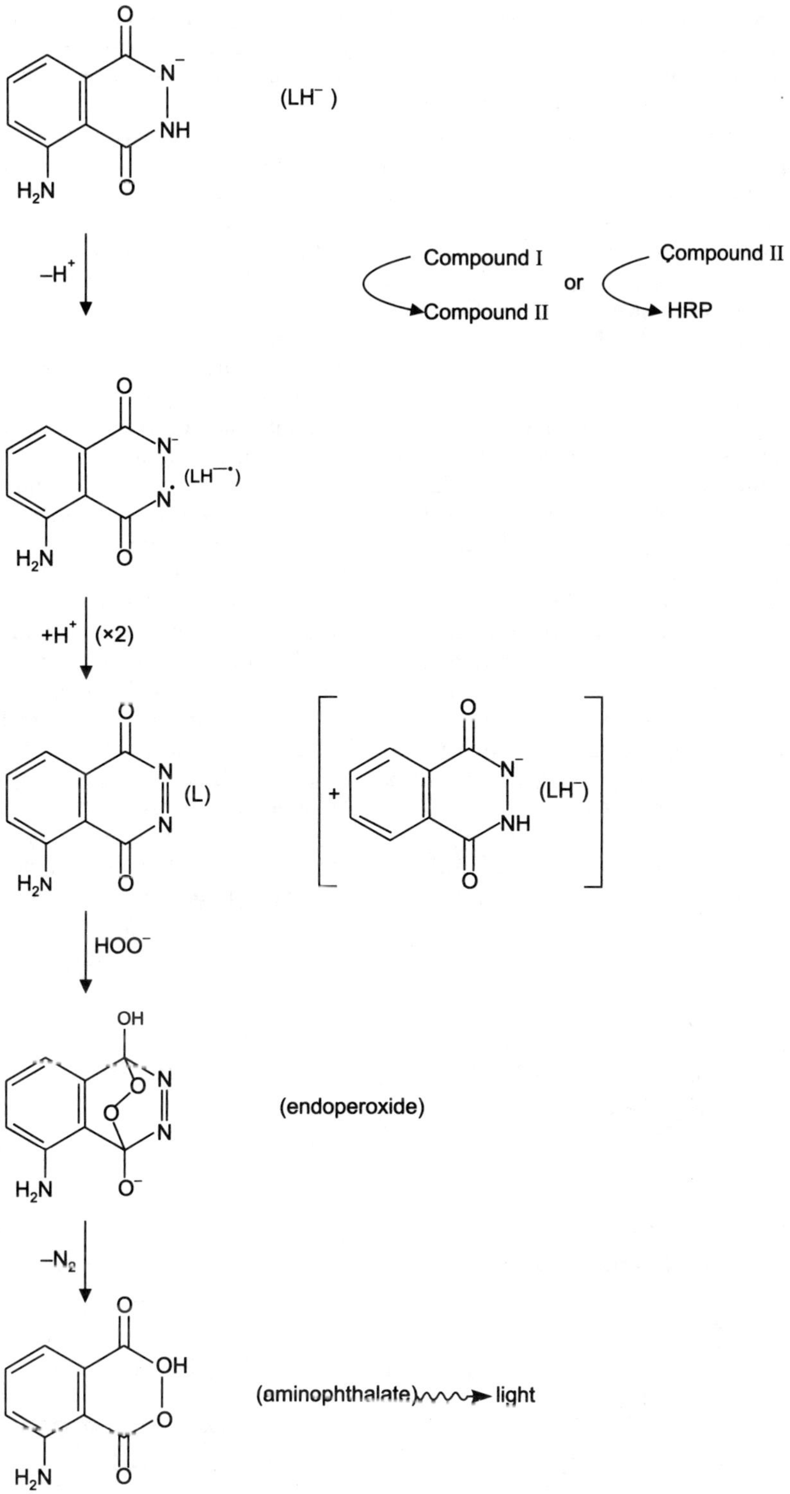

Figure 3.18 The mechanism of the HRP-luminol reaction.

based on substrate-labelling (see Section 3.2.2.3), where the entire light yield could be generated over an extremely short (millisecond range) timescale. While luminol itself gives the most powerful luminescence in its unconjugated form, it tends to lose a substantial proportion of its activity when attached *via* the amine group to another molecule. Isoluminol, although a less powerful luminescer in itself, shows a greater robustness to derivatisation (Schroeder and Yeager, 1978). For this reason it has been more widely employed as a luminescent label; similarly, *N*-(6-aminobutyl)-*N*-ethylisoluminol (ABEI) has found widespread application.

In the early 1980s, at the Wolfson Research Laboratories in Birmingham, UK, the phenomenon of enhanced luminescence was discovered (Whitehead *et al.*, 1984; Thorpe and Kricka, 1986). The addition of certain 'enhancers' such as substituted phenols and benzothiazole derivatives caused dramatic increases, sometimes by factors of thousands, in the level of light output from HRP/luminol systems. This phenomenon is largely specific for neutral/basic isoenzymes, does not affect the wavelength of emitted light, and shifts the pH optimum to 8.6. The mechanism of enhanced luminescence has received considerable attention, and remains a subject of discussion (Lundin and Hallander, 1986; Vlasenko *et al.*, 1989; Egorov *et al.*, 1993; Easton *et al.*, 1996).

The signal enhancement obtained increases the output from a steady-state peroxidase-labelled system to practical levels, allowing the advantages of this format to be exploited for the first time in a luminol/peroxidase system. This technique has been employed successfully in commercial systems for both immunoassay (the Ortho-Clinical Diagnostics Amerlite and Vitros ECi ranges) and for nucleic acid labelling applications (the Amersham Life Sciences ECL system).

Luminol analogues offering greater levels of light emission have been described recently, often employing an enhancer of the type described. Ii *et al.* (1993) reported the use of a phenylpyridyl analogue of luminol (Figure 3.19) in the presence of a novel enhancer (Figure 3.20); twice the sensitivity of a comparable luminol system (also using the new enhancer) was claimed in an immunoassay. Ternaux and Chamoin (1994) described a dimethylamino analogue of luminol which gave a threefold increase in sensitivity over luminol itself, again in the presence of enhancers. Lumigen, Inc. described the use of hydroxy-analogues of luminol in a 1994 patent (Akhavan-Tafti, 1994); one of these was compared to luminol by Kricka *et al.* (1996) and showed poorer performance. Finally, Pierce have started marketing enhanced luminol substrate reagents, SuperSignal™ and SuperSignal Ultra™, these showed higher sensitivity than the ECL system in a recent comparison (Mattson and Bellehumeur, 1996).

Luminol-based systems have really dominated luminometric HRP detection, but a few other examples have been described. Some interesting

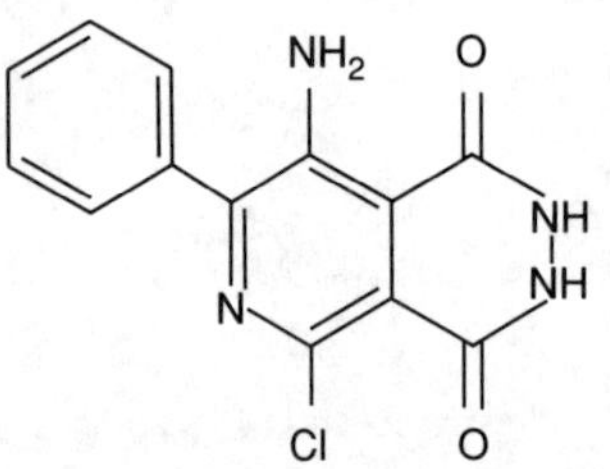

Figure 3.19 8-Amino-5-chloro-7-phenyl-pyrido[3,4*d*]pyridazine-1,4-(2H,3H)-dione.

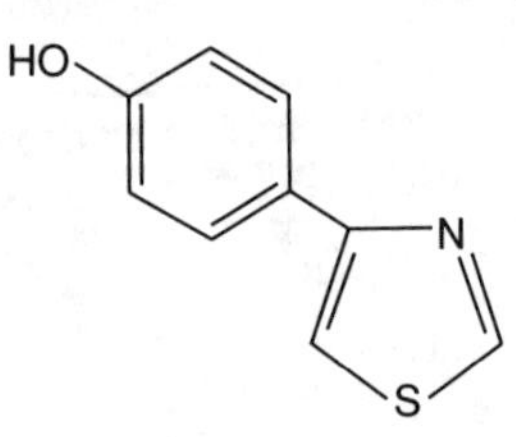

Figure 3.20 4-(4-Hydroxyphenyl)thiazole.

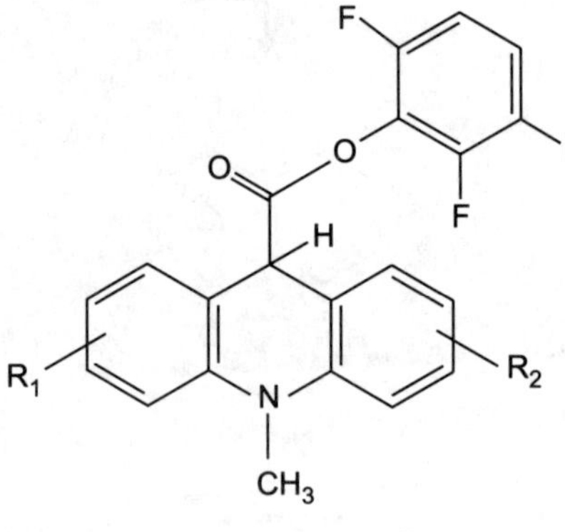

	R_1	R_2
Lumigen PS-1	3-OCH_3	H
Lumigen PS-2	1-OCH_3	6-OCH_3
Lumigen PS-3	H	H

Figure 3.21 Lumigen™ PS HRP substrates.

light-emitting reactions of HRP have been described (Bechara *et al.*, 1979; Baader *et al.*, 1985; Dunford, 1986), in which isobutyraldehyde undergoes autoxidation to yield a peracid; this is capable of oxidising HRP to compound I. The enol form of the aldehyde provides a substrate for compounds I and II, yielding triplet acetone *via* a dioxetane intermediate. Light is produced by the decay of the triplet form to the ground state.

Akhavan-Tafti *et al.* (1994, 1995) described a promising new series of HRP substrates under the tradename Lumigen™ PS (Figure 3.21). In the presence of peroxide, HRP promotes the formation of an acridinium ester, which breaks down to an acridone product with the emission of light (Figure 3.22). This breakdown can either take place continually (PS-1) or be triggered by a pH shift (PS-2) to provide flash kinetics. Detection of less than 10^{-19} mol of HRP was claimed from the PS-1 system.

3.3.8 Electrochemical substrates for HRP

The redox nature of HRP catalysis provides another route for quantification of the enzyme. Typically an oxidised HRP substrate is reduced at a suitable electrode to provide a measurable current: any suitable substrate can be employed, though most approaches of this type have employed polyphenols, polyamines, or iodide (Krishnan *et al.*, 1995).

HRP itself can be incorporated in an electrode to mediate the reduction of hydrogen peroxide (Ho *et al.*, 1995), and this has formed the basis of coupled systems (see Section 3.6.2).

3.3.9 Other labelling applications of HRP

An interesting technique for amplification of HRP signal was described by DuPont (Bobrow *et al.*, 1989; Bobrow and Litt, 1993). Described as catalysed reporter deposition (CARD), it involves immobilising HRP in a standard immunometric format then using biotinylated phenols as substrates. On oxidation these are postulated to react with amino acids in the immobilised protein layer, effectively giving a surface concentration of biotin related to the original HRP level. Incubation with a second, streptavidin-labelled enzyme (HRP, β-galactosidase or alkaline phosphatase) gives an amplified enzyme concentration which can be detected by standard signal generation.

3.3.10 Peroxidases from other sources

Although horseradish peroxidase dominates the field in labelling applications, enzymes of this type are

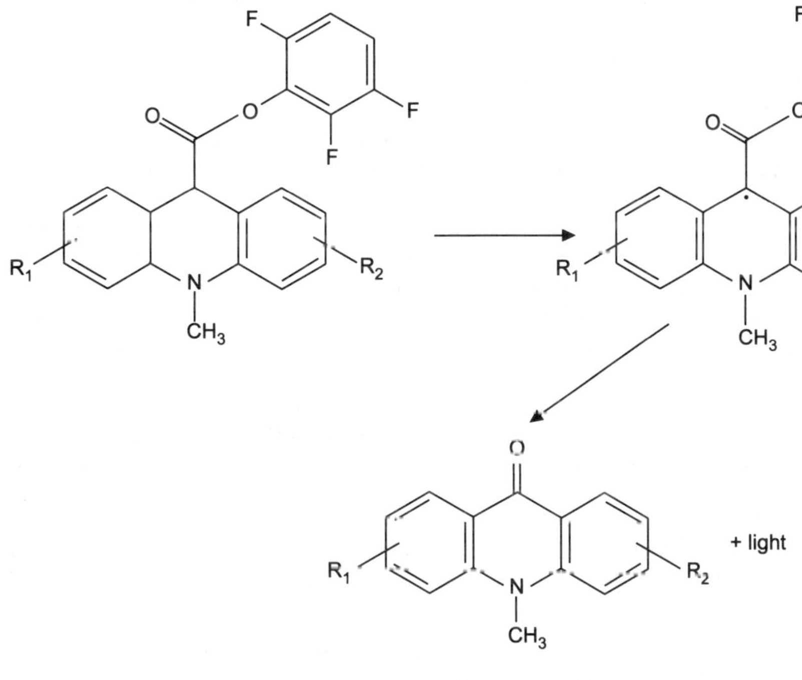

Figure 3.22 Chemiluminescent reaction of Lumigen™ PS with HRP.

prodigiously distributed in nature and it is not surprising that other sources have received some examination. Among the higher plants most attention has focused on peroxidases from turnip and Japanese radish; due to the existence of multiple isoenzymes it is difficult to generalise in terms of properties, but Aibara *et al.* (1981) compared enzymes from these sources with HRP isoenzymes and the differences between species were no greater than those between the horseradish isoenzymes. The amino acid composition of HRP isoenzyme E4, for example, is much more closely related to that of Japanese radish isoenzyme *c* than to HRP E1, E2, E5 or C, while HRP E6 is as similar to turnip peroxidase isoenzyme P7 as it is to HRP C. Overall there does not seem yet to be any particular reason to favour their use over HRP; horseradish is a widespread and easily-grown plant with a high peroxidase yield, and there are no properties of the related enzymes which would make their lower extraction yield worthwhile. Soybean peroxidase has become commercially available, however, and may receive more attention as a result of possible stability advantages.

There are still attractions, however, in identifying peroxidases from more diverse sources which might offer improved properties over those commonly in use—particularly if they originate from a microbiological organism which would allow the enzyme to be produced by cell culture in fermenters.

In the last decade a lot of attention has been focused on peroxidases from fungal sources, notably from the organisms described as *Arthromyces ramosus, Coprinus cinereus,* and *Coprinus macrorhizus*. Taxonomy of such fungi is notoriously difficult and the situation here is by no means clear: '*C. macrorhizus*' is now classified as a form of *C. cinereus*, and doubts have been expressed as to whether even *A. ramosus* is a genuinely different organism.

Kjalke *et al.* (1992) showed that the peroxidases from these organisms were identical[1] in terms of amino acid composition, molecular mass (38 kDa), p*I* (3.8), immunochemical activity and kinetics and pH optima of guaiacol oxidation; the only difference observed was in glycosylation levels. *Arthromyces ramosus* peroxidase (ARP) was reported by Akimoto *et al.* (1990) to show interesting differences from HRP-C, more than a hundredfold greater light output being obtained from luminol oxidation under comparable conditions. There has been sufficient interest in this fungal peroxidase and the virtually identical *Coprinus* enzymes[2] to lead to their commercial availability from a number of suppliers.

Krell (1991), of Boehringer Mannheim, described a programme of screening microorganisms for desirable thermostability properties; 5000 strains were screened, of which 40% showed peroxidase activity and four showed sufficient heat stability to merit further work.

Cytochrome *c* peroxidase (CCP) from yeast mitochondria has already been mentioned above. The availability of crystalline CCP of sufficiently high quality for X-ray crystallography has provided good structural information for this enzyme (Poulos *et al.*, 1980; Finzel *et al.*, 1984). More is known about the three-dimensional structure of CCP than any other peroxidase, and it has been used as a model for understanding the structure of related enzymes. It has even been possible to convert entire crystals of CCP to the compound I form (Edwards *et al.*, 1987) and carry out crystallography on these, helping to pin down the oxidation sites.

It should be noted that there are substantial differences in the behaviour of CCP and HRP: while one electron is lost from the iron when both these peroxidases are oxidised to compound I, the second electron comes from the haem group in HRP but from the peptide chain in CCP. There also appear to be differences in the compound II structures.

While cytochrome *c* peroxidase is readily obtainable from yeast it is unusual among peroxidases in its high substrate specificity, only catalysing to any great extent the conversion of ferrocytochrome *c* to the 'ferri-' form. This does not offer any great potential as a diagnostic label.

Cytochrome *c* is itself a haemoprotein, and if subjected to enzymic cleavage can yield an oligopeptide chain of eleven amino acids, linked *via* two cysteine thiols to the haem group (Figure 3.23). This fragment of the protein has been shown to display peroxidase-like activity and is described as a microperoxidase. These pseudoenzymes can catalyse many of the same substrate oxidations as HRP, including chromogenic reductants like TMB, and luminogens like luminol. They have been claimed to show advantages in terms of broad pH specificity, stability and rapid kinetics, but there are potential disadvantages too, such as a dramatically increased level of binding to serum proteins (Bundgaard and Møller, 1981).

Many other peroxidases may be encountered; some are haemoproteins while others rely on different

[1] Conversely, the variability of peroxidases can be employed to investigate the taxonomy of plants, Krzakowa (1991) used electrophoretic analysis of peroxidases in determining species relationships in bryophytes.

[2] Including some non-biomedical uses, such as dealing with the problems caused by a stray red sock in white laundry (Coghlan, 1997)!

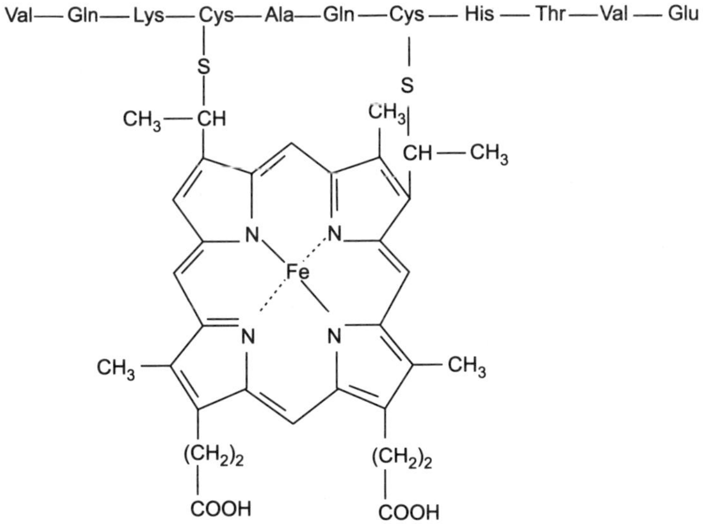

Figure 3.23 A microperoxidase, obtained by enzymic cleavage of cytochrome *c* peroxidase.

groups to provide the active site. These include lactoperoxidase from milk, which is used as an oxidising agent for $Na^{125}I$ in radioiodination reactions; myeloperoxidase which is used by leucocytes in the disabling of foreign organisms by phagocytosis; thyroid peroxidase which takes part in the biosynthesis of the hormones T3 (3,3′,5′-triiodo-L-thyronine) and T4 (L-thyroxine); the selenium-containing glutathione peroxidase, and haloperoxidases which catalyse halide oxidation such as the vanadium-containing bromoperoxidases from various brown algae. The latter are beginning to attract interest for luminescent signal generation (Friedman *et al.*, 1997).

3.4 Alkaline Phosphatase

Alkaline phosphatase (AP) is categorised under the EC system as an orthophosphoric-monoester phosphohydrolase (alkaline optimum), EC 3.1.3.1.

Enzymes which hydrolyse phosphate esters have been known since around the beginning of the century. It was soon discovered that various examples of these enzymes operated at different pH optima, and the phosphatases were classified loosely on this basis. Some of the enzymes from mammalian sources such as milk and intestinal mucosa showed optimal activity at pH 9–10 and were classed as alkaline phosphatases.

Alkaline phosphatases are ubiquitous in the animal kingdom, being present in both vertebrates and invertebrates (Fernley, 1971; Zueva *et al.*, 1993). While they do not appear to have a significant role in higher plants, they are found in fungi and bacteria, although their expression in the latter is often repressed by the presence of phosphate ions. In prokaryotic organisms AP is located in the periplasmic space of the cell, while in mammalian and other eukaryotic species it is generally linked to the cytoplasmic membrane *via* a glycosyl phosphatidylinositol moiety (Millán, 1992). Unlike some enzymes, there is little difference in activity between free mammalian enzyme and that still bound to the membrane (Chappelet-Tordo *et al.*, 1974).

An individual organism may possess a number of different AP isoenzymes which carry out different functions, and which are often localised to some extent at different sites. These functions are multifarious; alkaline phosphatase activity in mammals for example is associated with bone formation (ossification), with tissue at an early stage of development, and with the hepatobiliary system. Nutrient transport is probably the predominant function, however; it has been observed that AP is virtually absent from erythrocytes, muscles, and other tissue which are not involved in this activity (Zueva *et al.*, 1993).

The relative levels of these isoenzymes may vary with states such as disease and pregnancy, and the diversity of human AP isoenzymes has received substantial attention (Boyer, 1961; Moss and King, 1962; McKenna *et al.*, 1979; Stinson and Seargeant, 1981). While human AP is not in use as a labelling reagent, the attention it has received has thrown light on the complexity of alkaline phosphatase activities *in vivo*, a factor worth bearing in mind in the use of AP from other sources.

Interestingly the placental form of human AP bears a binding site for the Fc region of the immunoglobulin G molecule (Makiya and Stigbrand, 1992); it is probably involved in the transport of functional antibodies to the fetus.

By far the commonest source of AP as a labelling reagent is bovine intestinal mucosa. More structural research however has been carried out on alkaline phosphatase from *Escherichia coli*, the expression of which is promoted by phosphate starvation, and which has been known since the late 1950s (Garen and Levinthal, 1960). This enzyme compares closely with the intestinal material in terms of catalytic properties, pH–rate relation, and intermediate formation (Schwartz and Lipmann, 1961; Schwartz *et al.*, 1963); in view of the more definitive state of knowledge on this material, it will be discussed first.

Much of the information below is taken from the review by Coleman (1992), to which the reader is referred for more detailed mechanistic information.

3.4.1 Structural features of alkaline phosphatase from *E. coli*

AP from *E. coli* exists as three closely-related dimeric isozymes. In isozyme 1 there is an N-terminal arginine residue on both subunits, which is lacking from the subunits of isozyme 3. Isozyme 2 is a heterodimer containing one subunit of each type (Bradshaw *et al.*, 1981).

E. coli AP has a molecular mass of around 94.2 kDa based on the amino acid sequence (Bradshaw *et al.*, 1981) and the metal content. Garen and Levinthal (1960) reported a p*I* of 4.5.

3.4.1.1 Amino acids

AP (isozyme 3) from *E. coli* consists of two identical subunits each consisting of 449 amino acids (Bradshaw *et al.*, 1981). The sequence is depicted in Figure 3.24 and the content summarised in Table 3.2. A nucleotide sequence has also been published (Chang *et al.*, 1986). The four cysteine residues are present as intrachain disulfides.

As shown in Figure 3.9, Guesdon *et al.* (1979) demonstrated that the activity of AP from *E. coli* was seriously reduced by derivatisation of the amine groups; exhaustive reaction resulted in a loss of 60–70%. Therefore, incorporations should be kept as low as practicable if this approach is taken. Coupled with the lack of free thiol groups, this is a limitation on the use of AP in protein coupling applications. The use of phosphate buffers may protect the active site during coupling reactions.

3.4.1.2 Carbohydrates

As would be expected from its origin the *E. coli* enzyme, unlike its mammalian counterparts, contains no carbohydrate.

3.4.1.3 Metal ions

This enzyme is a metalloprotein, containing one magnesium and two zinc ions in each subunit. The three sites are termed A and B for the zinc ions and C for the magnesium; the normal form of the enzyme is $Zn_AZn_BMg_C$. Removal of the C ion does not have a dramatic effect on activity—the Zn_AZn_B enzyme is still quite active. Replacement of one zinc with magnesium in the $Zn_AMg_BMg_C$ form reduces hydrolase activity but has little effect on transphosphorylation (see below).

Even replacement of these metals with similar ions can maintain the functionality of the enzyme; replacement of zinc by cobalt for example retains up to nearly one third of the activity and the pH profile remains unchanged.

Cadmium substitution reduces the catalytic rate dramatically, but all the stages of the mechanism appear to mimic the native enzyme, therefore it has been employed in structural and mechanistic studies. This substitution has been useful for ^{113}Cd NMR determinations, e.g. definitively demonstrating the existence of three discrete sites (Gettins and Coleman, 1983).

3.4.1.4 Three-dimensional structure of AP from *E. coli*

The crystal structures of *E. coli* AP and two of its intermediates have been determined to 2.0 or 2.5 Å resolution (Kim and Wyckoff, 1989, 1991). Each subunit contains a central ten-stranded β-sheet surrounded by fifteen α-helices; a three-stranded sheet and one further helix form a 'lid' to the molecule.

The three metals form a triangle with sides of around 7 Å, 5 Å and 4 or 5 Å depending on the subunit, near the C terminus end of the central β-sheet. Zn_A is octahedrally coordinated to two oxygen atoms from Asp327, two nitrogen atoms from His331 and His412 and probably two water molecules; Zn_B is

10 20
Thr–Pro–Glu–Met–Pro–Val–Leu–Glu–Asn–Arg–Ala–Ala–Gln–Gly–Asn–Ile–Thr–Ala–Pro–Gly–

30 40
Gly–Ala–Arg–Arg–Leu–Thr–Gly–Asp–Gln–Thr–Ala–Ala–Leu–Arg–Asn–Ser–Leu–Ser–Asp–Lys–

50 60
Pro–Ala–Lys–Asn–Ile–Ile–Leu–Leu–Ile–Gly–Asp–Gly–Met–Gly–Asp–Ser–Glu–Ile–Thr–Ala–

70 80
Ala–Arg–Asn–Tyr–Ala–Glu–Gly–Ala–Gly–Gly–Phe–Phe–Lys–Gly–Ile–Asp–Ala–Leu–Pro–Leu–

90 100
Thr–Gly–Gln–Tyr–Thr–His–Tyr–Ala–Leu–Asn–Lys–Lys–Thr–Gly–Lys–Pro–Asp–Tyr–Val–Thr–

110 120
Asp–Ser–Ala–Ala–Ser–Ala–Thr–Ala–Trp–Ser–Thr–Gly–Val–Lys–Thr–Tyr–Asn–Gly–Ala–Leu–

130 140
Gly–Val–Asp–Ile–His–Glu–Lys–Asp–His–Pro–Thr–Ile–Leu–Glu–Met–Ala–Lys–Ala–Ala–Gly–

150 160
Leu–Ala–Thr–Gly–Asn–Val–Ser–Thr–Ala–Glu–Leu–Gln–Asp–Ala–Thr–Pro–Ala–Ala–Leu–Val–

170 180
Ala–His–Val–Thr–Ser–Arg–Lys–Cys–Tyr–Gly–Pro–Ser–Ala–Thr–Ser–Gln–Lys–Cys–Pro–Gly–

190 200
Asn–Ala–Leu–Glu–Lys–Gly–Gly–Lys–Gly–Ser–Ile–Thr–Glu–Gln–Leu–Leu–Asn–Ala–Arg–Ala–

210 220
Asp–Val–Thr–Leu–Gly–Gly–Gly–Ala–Lys–Thr–Phe–Ala–Glu–Thr–Ala–Thr–Ala–Gly–Glu–Trp–

230 240
Gln–Gly–Lys–Thr–Leu–Arg–Glu–Gln–Ala–Gln–Ala–Arg–Gly–Tyr–Gln–Leu–Val–Ser–Asp–Ala–

250 260
Ala–Ser–Leu–Asn–Ser–Val–Thr–Glu–Ala–Asn–Gln–Gln–Lys–Pro–Leu–Leu–Gly–Leu–Phe–Ala–

270 280
Asp–Gly–Asn–Met–Pro–Val–Arg–Trp–Leu–Gly–Pro–Lys–Ala–Thr–Tyr–His–Gly–Asn–Ile–Asp–

290 300
Lys–Pro–Ala–Val–Thr–Cys–Thr–Pro–Asn–Pro–Gln–Arg–Asn–Asp–Ser–Val–Pro–Thr–Leu–Ala–

310 320
Gln–Met–Thr–Asp–Lys–Ala–Ile–Glu–Leu–Leu–Ser–Lys–Asn–Glu–Lys–Gly–Phe–Phe–Leu–Gln–

330 340
Val–Glu–Gly–Ala–Ser–Ile–Asp–Lys–Gln–Asp–His–Ala–Ala–Asn–Pro–Cys–Gly–Gln–Ile–Gly–

350 360
Glu–Thr–Val–Asp–Leu–Asp–Glu–Ala–Val–Gln–Arg–Ala–Leu–Glu–Phe–Ala–Lys–Lys–Glu–Gly–

370 380
Asn–Thr–Leu–Val–Ile–Val–Thr–Ala–Asp–His–Ala–His–Ala–Ser–Gln–Ile–Val–Ala–Pro–Asp–

390 400
Thr–Lys–Ala–Pro–Gly–Leu–Thr–Gln–Ala–Leu–Asn–Thr–Lys–Asp–Gly–Ala–Val–Met–Val–Met–

410 420
Ser–Tyr–Gly–Asn–Ser–Glu–Glu–Asp–Ser–Gln–Glu–His–Thr–Gly–Ser–Gln–Leu–Arg–Ile–Ala–

430 440
Ala–Tyr–Gly–Pro–His–Ala–Ala–Asn–Val–Val–Gly–Leu–Thr–Asp–Gln–Thr–Asp–Leu–Phe–Tyr–

449
Thr–Met–Lys–Ala–Ala–Leu–Gly–Leu–Lys

Figure 3.24 The amino acid sequence of *E. coli* alkaline phosphatase (isozyme 3). (Reproduced with permission from: Bradshaw *et al.* (1981). *Proc. Natl. Acad. Sci. USA*, **78**, 3473–3477.)

tetrahedrally coordinated to one oxygen atom each from Asp51 and Asp369, one nitrogen atom from His370 and probably the hydroxyl oxygen atom of Ser102; Mg_C is octahedrally coordinated to oxygen atoms from Asp51 (which thus forms a bridge to the Zn_B atom), Glu322, Thr155 and three water molecules.

Not surprisingly, alteration of the residues which are responsible for metal binding can have a serious effect on the enzyme's properties (Xu and Kantrowitz, 1993; Tibbitts *et al.*, 1994; Xu *et al.*, 1994).

3.4.2 Structural features of bovine intestinal alkaline phosphatase

In contrast with the relatively simple isoenzyme status of the *E. coli* enzyme, the situation with bovine intestinal AP (bIAP) is less well-defined. The age of the source animal(s) is of particular importance, due to apparent changes in enzyme production on tissue maturation: the existence of multiple forms of calf IAP has been known for many years (Engström, 1961; Khattab and Pfleiderer, 1976), and Besman and

Table 3.2. Amino acid content of the *E. coli* alkaline phosphatase dimer

Amino acid	Number	(%)	Amino acid	Number	(%)
Asp	52	5.8	Val	46	5.1
Asn	44	4.9	Met	16	1.8
Thr	80	8.9	Ile	32	3.6
Ser	44	4.9	Leu	78	8.7
Glu	46	5.1	Tyr	22	2.4
Gln	46	5.1	Phe	16	1.8
Pro	42	4.7	His	20	2.2
Gly	90	10.0	Lys	56	6.2
Ala	128	14.3	Arg	26	2.9
Cys	8	0.9	Trp	6	0.7
Free acids	98	10.9			

Source: Bradshaw *et al.* (1981).

Coleman (1985) observed that only one of the major forms was still apparent in adult-derived enzyme. Commercial 'calf' or 'bovine' enzyme preparations undoubtedly contain varying quantities of these isoenzymes.

Khattab and Pfleiderer (1976) recorded p*I* values of 4.4 and 5.9 for calf IAP components, Lazdunski *et al.* (1965) having noted a p*I* of around 5.7 for unresolved calf IAP.

Determinations of molecular mass and carbohydrate content of bovine AP preparations are summarised in Table 3.3. These findings are reasonably consistent; in summary: (1) adult bIAP has a molecular mass of about 125 kDa and contains around 10% carbohydrate; (2) calf IAP contains two isozymes, one (C1) of similar size and carbohydrate content to bIAP, the other (C2) a more highly-glycosylated form with a molecular mass of about 140 kDa. Commercial calf IAP will display properties dependent on the isozyme balance of the preparation.

Table 3.3. Molecular mass and carbohydrate content determinations of bovine intestinal alkaline phosphatase preparations.

Source/ isoenzyme	Mol. mass (kDa)	Carbohydrate (%)	Ref.
Calf IAP	140	12	Fosset *et al.* (1974)
Calf IAP C-1	123	10	Besman and Coleman (1985)
Calf IAP C-2	141	22	Besman and Coleman (1985)
Bovine IAP	127	8	Hua *et al.* (1986)
Bovine IAP	123	12	Besman and Coleman (1985)

3.4.2.1 Amino acids

Amino acid compositions of bIAP (Besman and Coleman, 1985; Hua *et al.*, 1986) and sequences of the N-terminal fragment (Culp *et al.*, 1985; Hua *et al.*, 1986) were followed by a full description of the bIAP gene and amino acid sequence (Weissig *et al.*, 1993). A summary of the composition is given in Table 3.4.

The comments on the susceptibility of AP to derivatisation, noted for the *E. coli* enzyme, should be borne in mind for material of bovine origin too.

3.4.2.2 Carbohydrates

As noted above bIAP is quite heavily glycosylated, containing around 10–20% carbohydrate depending on the isozyme. Besman and Coleman (1985) observed the presence of two *N*-glycan and one or more *O*-glycan residues per subunit: galactose, mannose and *N*-acetylglucosamine were predominant, with smaller quantities of *N*-acetylgalactosamine, sialic acid and fucose.

3.4.2.3 Metal ions

The metal ion dependence of mammalian APs such as bIAP is less well characterised than that of the *E. coli* enzyme. However the presence of three zinc atoms per

Table 3.4. Amino acid content of the bovine intestinal alkaline phosphatase dimer

Amino acid	Number	(%)	Amino acid	Number	(%)
Asp	66	6.4	Val	82	8.0
Asn	46	4.5	Met	28	2.7
Thr	74	7.2	Ile	28	2.7
Ser	52	5.1	Leu	84	8.2
Glu	40	3.9	Tyr	40	3.9
Gln	52	5.1	Phe	28	2.7
Pro	62	6.0	His	30	2.9
Gly	86	8.4	Lys	42	4.1
Ala	130	12.6	Arg	40	3.9
Cys	10	1.0	Trp	8	0.8
Free acids	106	10.3			

Source: Weissig *et al.* (1993).

subunit has been demonstrated (Besman and Coleman, 1985), and magnesium causes an increase in activity even when a full complement of zinc is present. This effect is more dramatic in mammalian APs (Coleman, 1992). The amino acids involved in coordination of active-site metals have been identified (Weissig *et al.*, 1993), and predominantly consist of serine, histidine and carboxylic residues.

3.4.3 Differences between AP from bovine intestine and from *E. coli*

A number of important structural differences between *E. coli* AP and bIAP have been noted in the preceding sections. The most important of these is the glycosylation of bIAP, but the amino acid content is significantly different too.

Comparisons have also been made between the kinetic properties of the different APs. Chappelet-Tordo *et al.* (1974) observed that a significantly higher maximal activity could be obtained from calf intestinal AP (500 μmol min^{-1} mg^{-1} *versus* 35 μmol min^{-1} mg^{-1} for *E. coli* AP). The effect of pH was also different: for *E. coli* enzyme the activity reached a plateau at pH 8.5, while the activity of the calf intestinal enzyme continued to increase up to pH 10.5.

3.4.4 The mechanism of AP catalysis

AP catalyses (Coleman, 1992) the hydrolysis of organic oxyphosphate monoesters (R-O-PO_3^{2-}), *O*-phosphorothioates (R-O-PSO_2^{2-}), *S*-phosphorothioates (R-S-PO_3^{2-}), and phosphoramidates (R-NH-PO_3^{2-}), and inorganic thiophos-phates (H-S-PO_3^{2-}), and phosphates (H-O-PO_3^{2-}); the discussion below relates primarily to oxyphosphate monoesters, which will be represented for simplicity by R-O-P. Inorganic phosphate is denoted by P_i. While the enzyme is highly non-specific with regard to the non-phosphoryl region of the substrate, phosphodi- and triesters are not hydrolysed by AP, nor are mixed esters such as that shown in Figure 3.25 (Fernley, 1971).

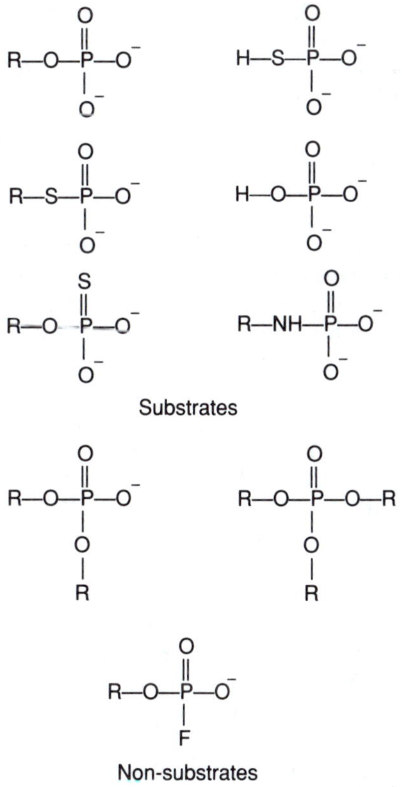

Figure 3.25 Alkaline phosphatase substrates and non-substrate.

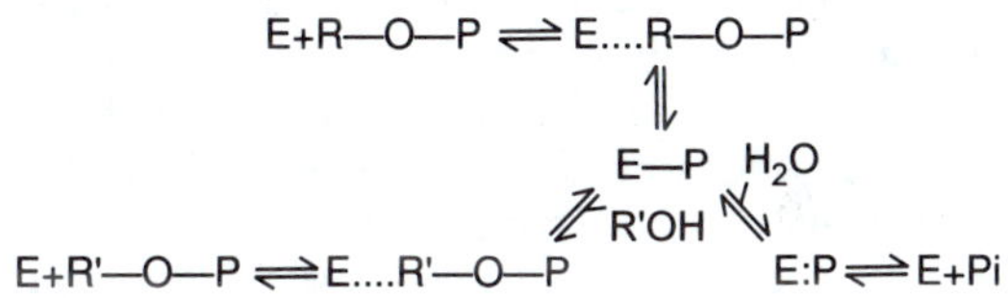

Figure 3.26 Hydrolase and phosphotransferase activity of alkaline phosphatase. The E–P complex can either react with water to generate inorganic phosphate (Pi), or with an alcohol molecule to generate a new phosphate ester (R′–O–P).

The stages of the hydrolysis reaction are shown in Figure 3.26. The enzyme E first forms a non-covalently bound complex E... R–O–P. In this complex, the ester oxygen atom complexes to Zn_A, a second oxygen atom to Zn_B, and the remaining two are hydrogen-bonded to Arg166. This brings the phosphorus atom close to Ser102.

In the second stage, the Ser102 hydroxyl oxygen atom binds to the phosphorus, displacing the R–O$^-$ group. The phosphate is now linked covalently through Ser102 to give the E–P complex; the serine oxygen is also coordinated to Zn_B, and two of the other oxygens remain hydrogen-bonded to Arg166. A water molecule takes up the vacant position on Zn_A, losing a proton to balance the R-O group and yield an OH^- moiety.

In the hydrolase mode of the enzyme this hydroxide ion then attacks the phosphorus, displacing the serine oxygen atom to yield a non-covalent E... P_i complex. The phosphate group is coordinated to both zinc ions and remains hydrogen bonded to arginine. Dissociation of this complex regenerates the original enzyme E plus inorganic phosphate P_i.

As mentioned above P_i and thiophosphate can also act as AP substrates (demonstrable for P_i only by isotopic exchange with ^{18}O, as the substrate and product are chemically identical). In these cases it appears that the formation of E–P is rate-limiting. However for oxyphosphate esters, the rate-determining step of this process appears to be dependent on pH, and involves either the hydrolysis/dephosphorylation of E–P to $E \ldots P_i$, or the subsequent dissociation of P_i. This is consistent with the observation that the nature of R is not critical to an acceptable rate of hydrolysis, so that an enormous range of potential substrates exist (Garen and Levinthal, 1960; Chappelet-Tordo *et al.*, 1974; McComb, 1979). Large variations in charge of the dephosphorylated product may have little effect on reaction rate. Chappelet-Tordo *et al.* (1974) observed comparable rates of hydrolysis at pH 8.0 for *p*-nitrophenyl phosphate (pK_a 7.15) and 2,4-dinitrophenyl phosphate (pK_a 4.1).

This lack of substrate specificity has been largely responsible for the great popularity of AP as a reagent. Virtually any molecule whose properties are modified by the removal of a phosphate group can be employed as a substrate, and this change in behaviour can range from triggering the generation of luminescence in a clinical assay to releasing the active form of a cytotoxic drug *in vivo*.

^{31}P NMR can be conveniently used to monitor the equilibria between the different phosphate-containing intermediates, and indicates strong pH dependency. As the pH rises from 5 to 7, the amount of $E \ldots P_i$ present compared to E–P increases; above pH 7 there is virtually no E–P detectable. The rate of dissociation of $E \ldots P_i$ to generate P_i, however, is relatively independent of pH. The overall effect is that dephosphorylation of E–P to $E \ldots P_i$ is rate-limiting at lower pH, while dissociation of $E \ldots P_i$ becomes limiting as the pH increases and dephosphorylation becomes more rapid.

Hydrolysis implies that the phosphate group has been transferred from the ester to a water molecule. However the enzyme is quite capable of transferring the group to almost any alcohol, generating a new ester. Therefore the enzyme is also capable of acting as a phosphotransferase. In this mode the E–P intermediate undergoes alcoholysis to yield the products *via* the non-covalent E . . . R′–O–P complex. Using α-amino alcohols above their hydroxyl pK_a as receptors can enhance the catalytic rate, making the use of buffers containing such compounds (e.g. tris-(hydroxymethyl)aminomethane ('Tris'), 2-amino-2-methyl-1-propanol (2A2M1P) and ethylaminoethanol (EAE); McComb and Bowers, 1972; Chappelet-Tordo *et al.*, 1974) popular for AP determinations.

There has been a great deal of debate over whether both subunits of the AP molecule are phosphorylated simultaneously, and many theories invoke negative cooperativity, the presence of one substrate discouraging the incorporation of a second, and a 'flip-flop' mechanism with the phosphate being transferred intramolecularly from one subunit to the other (Lazdunski *et al.*, 1971). However some doubt has been cast on this hypothesis, as both contamination with inorganic phosphate and the presence of substoichiometric amounts of metal ions may have clouded the findings (Coleman, 1992). Ammonium sulfate, widely used as a precipitant in AP preparation, is an excellent agent for metal removal!

Although there is still some evidence for negative cooperativity in unequal binding constants for the phosphates in each subunit, there is now considerable support for the view that both subunits may be present in the phosphorylated form simultaneously. Debate continues however (Zueva *et al.*, 1993).

AP from *E. coli* is generally very forgiving to mutations where a single amino acid is replaced; the central Ser102 group can be substituted by Cys and retain a large amount of activity, presumably forming a phosphorothioate intermediate (Ghosh *et al.*, 1986). Some activity remains even on incorporation of non-nucleophilic groups such as Ala and Leu (Butler- Ransohoff *et al.*, 1992); water may directly attack the E . . . R–O–P intermediate without involving a covalent intermediate. Mutants demonstrating increased catalytic activity have been patented by Mandecki *et al.* (1991).

3.4.5 Stability of AP

Fosset *et al.* (1974) reported that calf intestinal AP retained full activity for at least two months at pH 8.0 and 4 °C. The storage buffer was 0.1 M Tris containing 0.1 M NaCl, and the enzyme concentration was 1 mg/ml. The greatest stability was observed in the pH range 7.5 to 9.5; acidic conditions caused rapid denaturation although some protection was afforded by the presence of inorganic phosphate. Activity drops off rapidly below pH 4 and above pH 9 (180 minutes, 25 °C); Boehringer Mannheim, 1990). Mammalian intestinal APs are readily inactivated by heat (McComb, 1979, p. 405), half-lives of around 15 minutes at 55 °C being typical. However, AP is better than many enzymes in this respect, and has therefore been widely employed in applications such as DNA probes, where hybridisation temperatures must be tolerated.

Phosphatases are sometimes more stable in their native environment. For example, Garen and Levinthal (1960) reported full retention of activity of crude *E. coli* cell extracts after 30 minutes at 85 °C, while the purified enzyme was unstable under the same conditions. The presence of suitable quantities of Mg^{2+} ion was beneficial to stability.

Alkaline phosphatase tends to display good stability during substrate reactions, allowing long incubation periods to be employed.

3.4.6 Activators and inhibitors of AP

As noted earlier, α-amino alcohols can enhance the rate of AP hydrolysis (McComb and Bowers, 1972; Chappelet-Tordo *et al.*, 1974). The latter authors also noted that increasing ionic strength could yield rate increases at pH 8.0 for both calf intestinal and *E. coli* AP; this effect could not be observed at pH 10, however.

Probably the most important inhibitor of AP is the product P_i, which competes with an affinity comparable to that of good substrates (Fernley, 1971). For this reason Fernley recommended that initial rate measurements should not be performed beyond the point where 10% of the product has been hydrolysed. A number of other inorganic anions are also inhibitory towards AP, including pyrophosphate, borate, carbonate and arsenate (Zueva *et al.*, 1993). The choice of buffer for AP work can obviously therefore be critical.

Other inhibitors of AP include—not surprisingly in view of its essential metal content—chelating agents such as EDTA, α,α'-dipyridyl and *o*-phenanthroline, and competing metal ions such as Be^{2+} and Cu^{2+}. Low levels of mild chelating agents such as amino acids and cyanides can actually have an activating effect, presumably by mopping up traces of Cu^{2+}, etc.

L-Phenylalanine is a strong non-competitive inhibitor (Fernley, 1971) which may act by preventing the breakdown of the phosphorylated enzyme; L-leucine and L-homoarginine behave in the same fashion but less strongly (Zueva *et al.*, 1993).

Certain compounds closely related to AP substrates are also inhibitors, including estradiol-3,17-diphosphate, polyestradiol phosphate (Fernö *et al.*, 1958), and bis-(*p*-nitrophenyl phosphate). The latter compound has been identified as an impurity of the common substrate *p*-nitrophenyl phosphate (McComb, 1979, p. 307).

3.4.7 Colorimetric substrates for AP

The earliest AP determinations were typically based on quantifying the amount of P_i released from a suitable substrate, such as β-glycerophosphate (Folley and Kay, 1936). One particularly enduring methodology of this type is based on the reaction of phosphate with molybdate (MoO_4^{2-}) ions to yield phosphomolybdate, which can be reduced to the readily measurable molybdenum blue product (Bodansky, 1933; Kay and Graham, 1933; cf. the Folin–Ciocalteu reagent for protein determination, Chapter 11, Section 11.13.1.3). Although the signal obtainable from this approach is rather weak, it has been discovered that phosphomolybdate causes characteristic shifts in absorption properties of many dyes such as Malachite Green and Methyl Green 00, and colorimetric methods have been described based on this phenomenon (Itaya and Ui, 1966; Van Belle, 1970).

In histochemical applications, the Gomori method (Gomori, 1939) has also retained long-standing popularity: phosphate is precipitated by a calcium salt, the calcium exchanged for cobalt (Co^{2+}) and a black precipitate of cobalt sulfide formed with ammonium sulfide.

An alternative approach was described by King and Armstrong (1934), which was the predecessor of some of the commonest methods in use today. Instead of measuring the phosphate released, these authors used a phenyl phosphate as substrate, quantifying the product with the phenol-specific Folin–Ciocalteu reagent. This offered greater sensitivity than could be achieved by simple phosphate assays, although it was necessary to remove protein by precipitation before the phenol analysis was carried out. Kind and King (1954) described the use of 4-aminoantipyrine to quantify phenol in the same way, but without the requirement for protein precipitation.

Another method for phenol determination that has been employed quite commonly is the reaction with a diazonium salt to form an azo-dye. These compounds are strongly chromophoric, particularly for multicyclic examples such as those based on naphthols. They are also often rather insoluble, and have therefore predominantly been encountered in solid-phase applications such as histochemistry (Menten *et al.*, 1944; Burstone, 1960) and staining of electrophoresis gels (Boyer, 1961; Smith *et al.*, 1968). The use of more soluble examples allows this approach to be used in solution applications. For example, this has remained, for many years, a common method for selective determination of acid phosphatases in serum (Babson, 1984). It is also still recommended as an AP determination method, using Fast Red TR salt (Boehringer Mannheim, 1990).

From the detection of phenol using a secondary reaction, it is a short step to 'self-indicating' phenyl phosphates which generate a characteristic change in

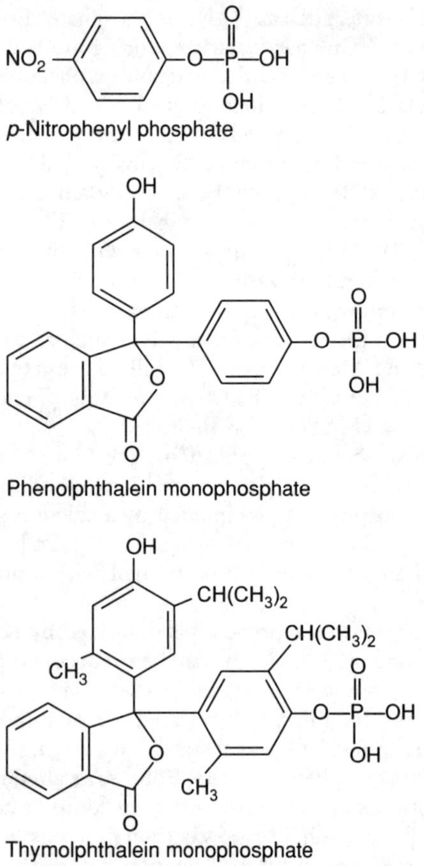

Figure 3.27 Some common colorimetric substrates for alkaline phosphatase.

absorbance properties on removal of the phosphate moiety; some common colorimetric substrates of AP are shown in Figure 3.27. *p*-Nitrophenyl phosphate (PNP-P; Ohmori, 1937; Bessey *et al.*, 1946) is one of the best known of these substrates, giving a product absorbing strongly at 400–410 nm (typically 402 nm). Although there is little difference in the low-pH spectra of *p*-nitrophenol and *p*-nitrophenylphosphate, under basic conditions the phenolate form of the hydrolysed product will predominate. Not only does this have a molar extinction coefficient roughly double that of the phenylphosphate, but the λ_{max} is shifted upwards by 80 nm, to a point where there is virtually no interference from unhydrolysed PNP-P.

This substrate has remained popular for over 50 years, providing a standard AP substrate for many different applications (Bowers and McComb, 1975; Boehringer Mannheim, 1990). It has the added advantage (particularly when compared to many HRP substrates) that it has not demonstrated mutagenic properties (Voogd *et al.*, 1980). PNP-P is capable of sufficiently high sensitivity to make immunoassay applications feasible: under conditions of high surface area and long incubation, Shalev *et al.* (1980) estimated that 24 million molecules (40 amol) of labelled immunoglobulin could be detected. Ishikawa (1987) recorded the detection of 10 fmol in ten minutes under more standard conditions.

Other self-indicating AP substrates have been introduced to avoid interference in serum phosphatase assays based on PNP-P: bilirubin absorbs similarly to the phenolate product and can therefore cause problems. These include the phosphates of phenolphthalein (Bray and King, 1943; Comfort and Campbell, 1966), which are hydrolysed more slowly than PNP-P (Bessey *et al.*, 1946), but which provide a pink product (λ_{max} approximately 550 nm) under alkaline conditions. This can be measured without bilirubin interference. The diphosphate can cause linearity problems in the assay, therefore the monophosphate has been recommended (Babson *et al.*, 1966; Wilkinson and Vodden, 1966). The monophosphates of thymolphthalein (Coleman, 1966; Roy, 1970) and thymol blue (Dalal *et al.*, 1975; Elser, 1976) have also been described in this role, yielding products with λ_{max} of approximately 600 nm. There is no particular reason to employ these substrates where interference from absorbing species is not a problem.

Indolyl phosphates can be cleaved by AP to generate the free indoxyl, which can then either dimerise to form indigo (Seligman *et al.*, 1954; Tsou and Su,

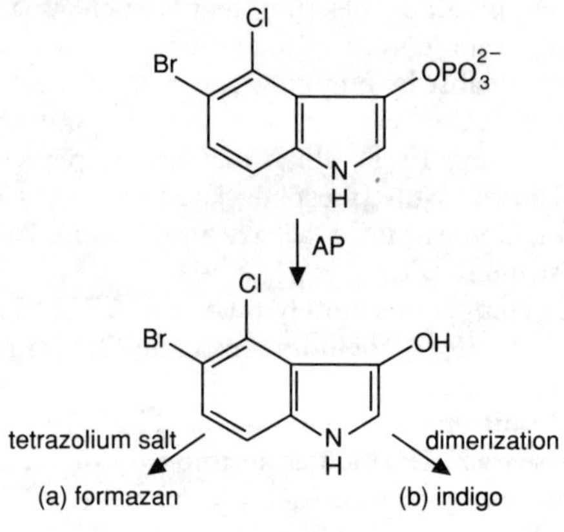

Figure 3.28 Cleavage of 5-bromo-4-chloro-3-indolyl phosphate (BCIP) followed by dimerisation of the product to form an insoluble indigo analogue (b), or further reaction with a tetrazolium salt to yield a formazan (a).

1965), or react with tetrazolium salts to yield a formazan (Figure 3.28). The commonest example of this category is 5-bromo-4-chloro-3-indolyl phosphate (BCIP), shown in Figure 3.28, which was pioneered by Horwitz *et al.* (1966) and Wolf *et al.* (1968). This remains a useful AP substrate (Boehringer Mannheim, 1990), particularly in applications where an insoluble product is desired.

3.4.8 Fluorimetric substrates for AP

Unlike peroxidase there is no shortage of fluorogenic substrates for AP, generally sharing the common design of a fluorophore, the activity of which is either significantly reduced by conjugation to a labile phosphate group, or whose emission wavelength is altered by such a derivatisation.

Simple naphthyl phosphates have been described in this role (Moss, 1960; Greenberg, 1962): the emission wavelength of the free naphthol is strongly shifted compared to the phosphate substrate so that the AP-catalysed hydrolysis can be quantified. A series of derivatives of 3-hydroxy-2-naphthanilide (naphthol AS) has also been described, some of which have superior fluorescence properties (Burstone, 1960; Johnson, 1969; Vaughan *et al.*, 1971).

4-Methylumbelliferyl phosphate (Figure 3.29) was described as a fluorogenic substrate for AP by Fernley and Walker (1965) and is still widely employed. The phosphate is virtually non-fluorescent, but the free umbelliferone has useful properties (excitation 340 nm, emission 440 nm). Great sensitivities have been obtained with this substrate: under forcing conditions of surface area and time, Shalev *et al.* (1980) claimed detection of 24 000 molecules (0.04 amol) of an immunoglobulin. Ishikawa (1987) was able to detect 1 amol of the enzyme using this substrate. Analogous umbelliferyl phosphates have also been described as AP substrates (Guilbault *et al.*, 1968a; Koller and Wolfbeis, 1985).

Other fluorescent AP substrates include phosphates of fluorescein (Rotman *et al.*, 1963; Hill *et al.*, 1968) and 3-hydroxyflavone (Land and Jackim, 1966).

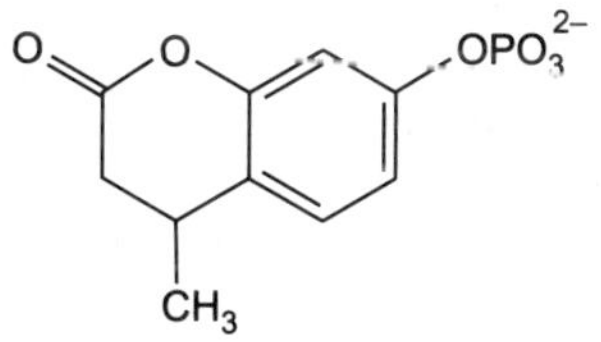

Figure 3.29 4-Methylumbelliferyl phosphate (MUP).

3.4.9 Luminometric substrates for AP

The most significant luminogenic substrates for AP belong to the class of 1,2-dioxetanes. Oxalate esters have long been known to generate luminescence with hydrogen peroxide (Chandross, 1963; Mohan and Rauhut, 1984), and this has been used *per se* as an immunoassay signal generation technique (Arakawa *et al.*, 1985), and in 'lightsticks' such as the Cyalume™ from American Cyanamid Co. The transient intermediate for these reactions is the peroxyoxalate moiety (Figure 3.30), an example of a 1,2-dioxetane.

Two groups in the USA (Schaap *et al.*, 1987; Bronstein *et al.*, 1995) have developed stable dioxetane compounds, which can be destabilised by the action of a suitable enzyme. This allows the dioxetane chemistry to be exploited directly *via* common signal enzymes. An AP substrate of this kind was described by Schaap *et al.* (1987). This was the phosphate-substituted adamantyl xanthenyl dioxetane shown in Figure 3.31a; the phosphate group can be removed by AP at pH 10.3 and the resulting phenoxide is unstable, generating strong luminescence at a wavelength of around 480 nm as the dioxetane cleaves to generate 2-adamantanone and the xanthenone derivative. Schaap (1988) also demonstrated a similar phosphatase-catalysed reaction with the phenyl phosphate dioxetane in Figure 3.31b, allowing detection of less than 1 amol of AP. This compound has been commercialised in the Lumi-Phos™ 530 reagent from Lumigen Inc., employed in Boehringer Mannheim's Genius™ system for nucleic acid labelling.

The same compound has been described by Bronstein *et al.* (1989) as adamantyl-1,2-dioxetane phosphate (AMPPD), claiming detection of 0.001 amol (10^{-21} mol) of AP, and has been marketed by Tropix for a variety of applications including immunoassay, nucleic acid probes and cytochemistry.

An analogue with a chloro-substituted adamantyl group, CSPD™ (disodium 3-chloro-3-methoxyspiro{1,2-dioxetane-3,2′-(5′-chloro)tricyclo[3.3.1.1^{3,7}]-decan}-4-yl)phenyl phosphate) (Figure 3.31c), has also been described by the Tropix team (Bronstein *et al.*, 1992). The use of luminescent dioxetanes in energy-transfer formats with fluorescers such as AttoPhos™ (JBL Scientific) has also been described in the patent literature (Bronstein *et al.*, 1995).

Surfactant additives such as polyalkylammonium salts can be used to improve performance in these systems (Schaap, 1992; Bronstein *et al.*, 1992, 1996).

There has been a substantial amount of patent litigation over the rights to these dioxetane compounds

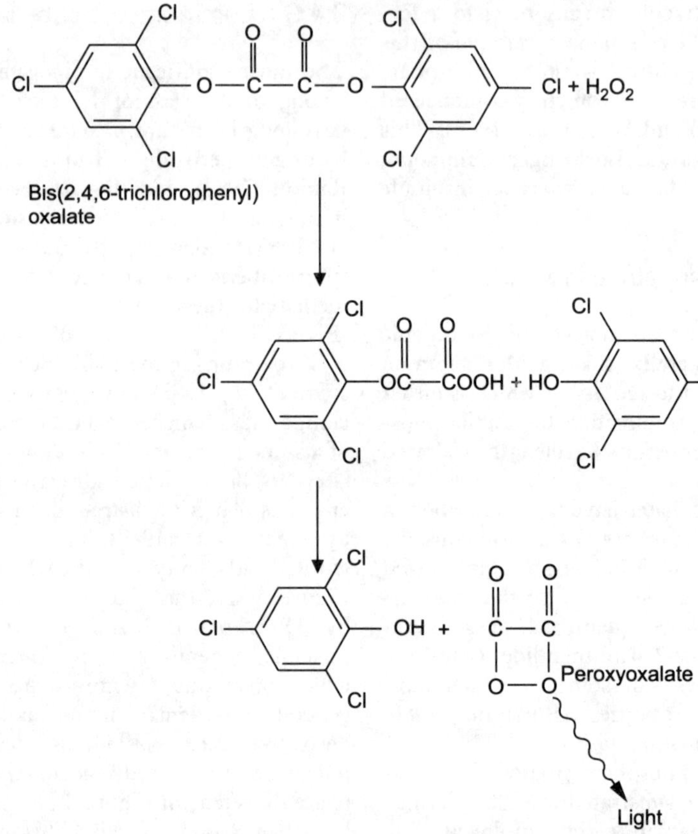

Figure 3.30 Formation of peroxyoxalate moiety in the chemiluminescent reaction of an oxalate ester.

in the last decade, a final legal settlement apparently having been reached in 1995 (Genesis Report, 1995). The introduction of these compounds is among the most important recent developments in the use of enzyme-based detection.

Lucigenin (*N,N'*-dimethyl-9,9'-bisacridinium dinitrate) (Figure 3.32) yields luminescence under the influence of hydrogen peroxide or organic reducing agents. AP assays have been described exploiting this reaction: in one example (Maeda *et al.*, 1991) ascorbic acid-2-*O*-phosphate formed the AP substrate and the dephosphorylated acid provided the reducing agent (Figure 3.33).

A number of enzyme-coupled approaches for luminometric AP detection are discussed in Section 3.6.

3.4.10 Electrochemical substrates for AP

Phenols are conveniently detected by their ability to be oxidised at a suitable electrode, and this oxidation can be quantified *via* either current or voltage measurements. Therefore the generation of an electroactive phenol from an electroinactive substrate forms the basis of a number of amperometric or voltammetric methods for enzyme quantification (cf. Section 3.5.8).

With AP the commonest electrochemical substrate is *p*-aminophenyl phosphate, PAPP (Kulys *et al.*, 1980): the aminophenol product of AP-catalysed hydrolysis is readily oxidised to the *p*-quinoneimine (Figure 3.34). This and similar substrates have been exploited as the basis for a number of immunoassay designs, (Jenkins *et al.*, 1988; Thompson *et al.*, 1991; Niwa *et al.*, 1993).

3.4.11 Radiometric substrates for AP

An AP-based immunoassay system, of interest at the time for its sensitivity, combined enzyme amplification with radiometric detection. Ultrasensitive enzymatic radioimmunoassay (USERIA), described by Harris *et al.* (1979), employed an AP conjugate to release

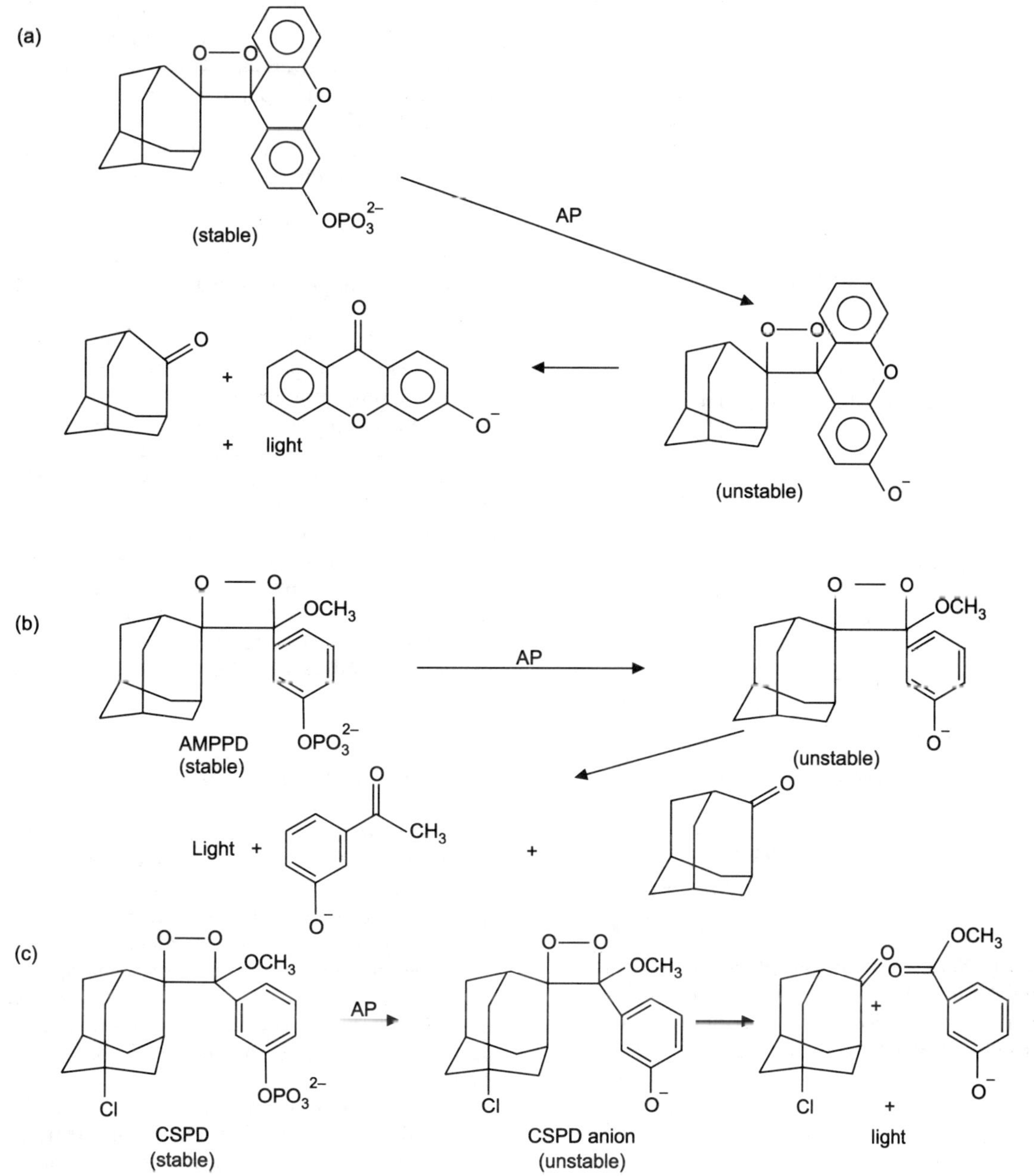

Figure 3.31 Chemiluminescence from 1,2-dioxetanes: (a) a xanthenyl dioxetane; (b) AMPPD; (c) CSPD.

^{3}H-adenosine from tritiated AMP. The adenosine was quantified radiometrically after ion exchange separation from the mixture, and allowed detection of less than 0.01 amol of AP. Applied with a long incubation time to the quantification of cholera toxin, as little as 10^{-21} mol (600 molecules) could be detected.

3.4.12 Other labelling applications of AP

An interesting everyday application of AP assays has remained in use since its discovery in the 1930s. Milk is rich in phosphatase activity, but this is lost during heat treatment; therefore the absence of AP can be

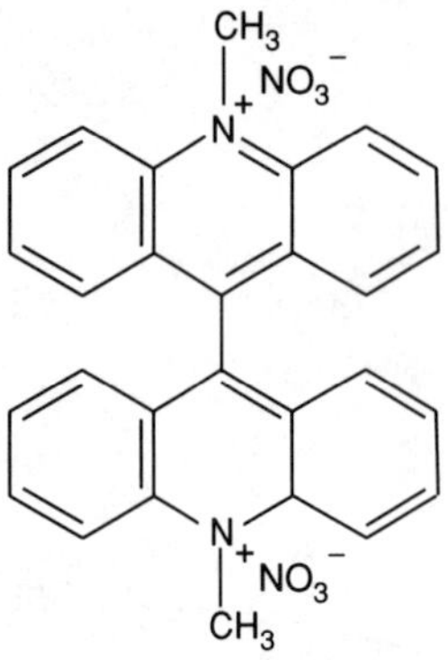

Figure 3.32 Structure of lucigenin.

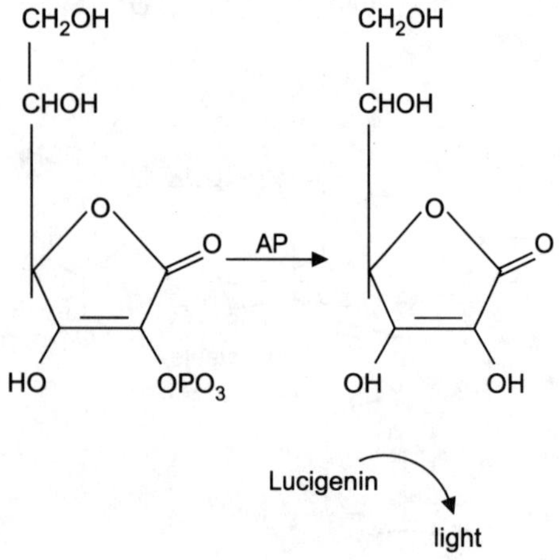

Figure 3.33 Alkaline phosphatase catalyses the dephosphorylation of ascorbic acid 2-*O*-phosphate; the product acts as a reducing agent for lucigenin, resulting in light generation.

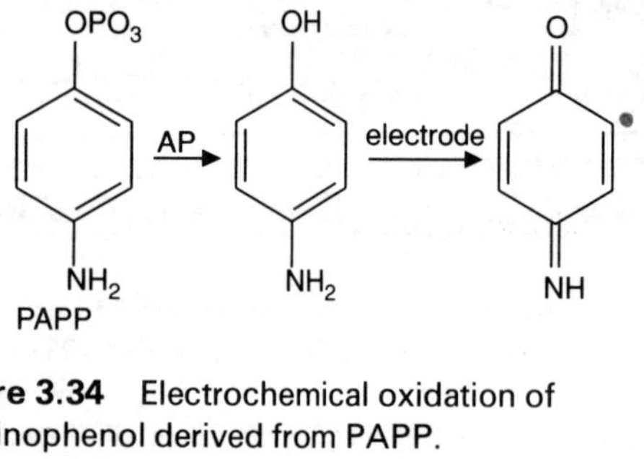

Figure 3.34 Electrochemical oxidation of *p*-aminophenol derived from PAPP.

used as a 'label' for the pasteurization process: Kay and Graham (1933) exploited this fact to design a test for successful pasteurization. With modifications, this remained the most popular test in recent times (McComb, 1979).

At the other extreme of technical complexity, AP has also been employed in therapeutic applications in the field of prodrugs.

Since the original description of the use of this enzyme to activate the relatively harmless cytotoxin derivative etoposide phosphate (Senter *et al.*, 1988), the approach has been applied to other drugs such as mitomycin C, phenol mustard and doxorubicin. In all these cases the activation is effected by AP, conjugated to a 'targetting' antibody to direct it to the desired site of action. The enzyme catalyses the hydrolysis of a phosphate group to generate the active toxin.

One disadvantage of alkaline phosphatase is that it is widespread in the body, so a substantial proportion of the prodrug may be activated before it reaches the targeted site. Therefore prodrug research may focus more in future on enzymes not occurring naturally in human tissue. Many enzymes have now been employed in prodrug applications, including carboxypeptidase A and G, cytosine deaminase and penicillin amidase.

3.5 β-D-Galactosidase

β-D-Galactosidase (β-Gal or GZ) is categorised under the EC system as a β-D-galactoside galactohydrolase (EC 3.2.1.23). It is also known as lactase.

β-Gal is a bacterial protein responsible *in vivo* for the hydrolysis of lactose to yield glucose and galactose. Some of the groundbreaking experiments on the genetic regulation of protein synthesis focussed on this enzyme, which is encoded by the first structural gene (*lacZ*) of the *lac* operon of *E. coli* (Jacob and Monod, 1961).

The material most commonly employed as a label is derived from *E. coli* and is a tetramer of identical subunits. It has a p*I* of 4.61 (Boehringer Mannheim, 1990) and a molecular mass of about 465 kDa (Fowler and Zabin, 1978; Kalnins *et al.*, 1983; Jacobson *et al.*, 1994), considerably larger than most labelling proteins. This size factor can rule it out for some applications, it is often too large to penetrate tissue samples in histochemistry for example. It does however have an advantage in that (unlike peroxidase and phosphatase) there is no human enzyme of comparable activity to compete with it in clinical samples, etc. Human β-galactosidases are ubiquitous in lysosomes, and are found in high concentrations in organs such as the kidney, liver and uterus (Lojda *et al.*, 1979b). However, these enzymes display acidic pH optima (3–5) and therefore cause little interference under the neutral conditions generally employed for the *E. coli* enzyme. The pH optimum of β-Gal is around 8 (Boehringer Mannheim, 1990).

Being another hydrolase, there are many similarities between the substrates of this enzyme and those of alkaline phosphatase, with galactosides replacing phosphates as the chromo-, fluoro- or luminogenic species. Like many other enzymes it is also capable of catalysing the reverse of its 'usual' reaction, and can lead to the formation of intersaccharide bonds (Huber *et al.*, 1994).

In a review (Gosling, 1990) of enzymes employed in new immunoassays, β-Gal had grown to the third most popular, employed in something under 10% of applications. The increase in its popularity seems to have tailed off somewhat in the last few years, and it seems set to lag behind HRP and AP in many applications.

3.5.1 Structural features of β-D-galactosidase from *E. coli*

The first reported preparations of β-Gal date from the 1950s (Kuby and Lardy, 1953; Cohn, 1957), and some early studies of enzyme from the ML309 strain investigated specificity, kinetics, and subunit interactions (Wallenfels *et al.*, 1960; Weber *et al.*, 1964). Craven *et al.* (1965) published a definitive description of highly-purified material from the K12 *E. coli* strain, and noted that it appeared to contain non-identical peptide chains (Steers *et al.*, 1965). The gene responsible for β-Gal production in *E. coli* is prone to many mutations, and the early literature is complicated by strain-to-strain differences. The data of Craven *et al.* (1965) appears to be representative of 'normal' β-galactosidase, and describes the presence of a large number of free thiol moieties. Their molecular mass estimate of 540 kDa has been revised downwards to 465 kDa in the light of later work, however.

3.5.1.1 Amino acids

Fowler and Zabin (1978) published the primary sequence of the β-Gal monomer, but agreed with later revisions by Kalnins *et al.* (1983) based on the nucleotide sequence: this revised 1023-residue sequence is shown in Figure 3.35 and the composition is summarised in Table 3.5. Compared to most proteins, β-galactosidase contains a relatively high proportion of tryptophan and arginine, and is poor in lysine and tyrosine.

Craven *et al.* (1965) observed that mercaptoethanol exerted a powerful protective effect on enzyme activity, consistent with the presence of essential free thiols. Their results in fact suggested that all the cysteine residues of the molecule were present in the sulfhydryl form, and that around half could be derivatised without loss of activity. Later, Jörnvall *et al.* (1978) carried out investigations on the reactivity of the β-Gal cysteine groups. They observed that only a mean level of 3.2 of the 16 residues in each subunit could be derivatised with iodoacetate, and that 90% of activity was retained following this treatment. The predominant derivatisation was at Cys498 and Cys1019. Hashinaka *et al.* (1994) were able to conjugate up to 4.4 maleimide groups per monomer using *o*-phenylenedimaleimide (see Chapter 5, Section 5.4.1), resulting in a 28% loss of enzyme activity.

When coupling *via* amines was carried out using MHS (see Chapter 5, Section 5.7.1), an incorporation of 0.4 per subunit was accompanied by the loss of 8% of the galactosidase activity. Abuknesha and Exley (1978) observed a fairly sharp loss of activity when more than a hundredfold excess of hapten was used to derivatise β-Gal in a carbodiimide reaction, the incorporation at this loading being around 0.05 per subunit. At excesses higher than around 650-fold, about a quarter of the enzyme activity remained. Using a biotin succinimide ester, Guesdon *et al.* (1979) demonstrated fairly severe inactivation of β-Gal on derivatisation of amine groups (see Figure 3.9). Around 40% of the activity was lost following near-exhaustive reaction. Overall, it appears that derivatisation of either amines or thiols should be kept to a moderate level for reasonable retention of β-Gal activity.

3.5.1.2 Carbohydrates

Being of prokaryotic origin, β-galactosidase from *E. coli* is not glycosylated.

3.5.1.3 Metal ions

β-Gal is activated by the monovalent cations Na^+ and K^+, and divalent cations (Mg^{2+} or Mn^{2+}) are also important for activity; the roles of these metals are not well understood, however (Huber *et al.*, 1994). In the native enzyme Mg^{2+} appears to be bound to a key glutamate residue (Glu461); Selwood and Sinnott (1990) have proposed a catalytic role for this ion, but Huber *et al.* (1994) consider its role to be limited to conformational effects.

1	Thr	Met	Ile	Thr	Asp	Ser	Leu	Ala	Val	Val	Leu	Gln	Arg	Arg	Asp	Trp	Glu	Asn	Pro	Gly
21	Val	Thr	Gln	Leu	Asn	Arg	Leu	Ala	Ala	His	Pro	Pro	Phe	Ala	Ser	Trp	Arg	Asn	Ser	Glu
41	Glu	Ala	Arg	Thr	Asp	Arg	Pro	Ser	Gln	Gln	Leu	Arg	Ser	Leu	Asn	Gly	Glu	Trp	Arg	Phe
61	Ala	Trp	Phe	Pro	Ala	Pro	Glu	Ala	Val	Pro	Glu	Ser	Trp	Leu	Glu	Cys	Asp	Leu	Pro	Glu
81	Ala	Asp	Thr	Val	Val	Val	Pro	Ser	Asn	Trp	Gln	Met	His	Gly	Tyr	Asp	Ala	Pro	Ile	Tyr
101	Thr	Asn	Val	Thr	Tyr	Pro	Ile	Thr	Val	Asn	Pro	Pro	Phe	Val	Pro	Thr	Glu	Asn	Pro	Thr
121	Gly	Cys	Tyr	Ser	Leu	Thr	Phe	Asn	Val	Asp	Glu	Ser	Trp	Leu	Gln	Glu	Gly	Gln	Thr	Arg
141	Ile	Ile	Phe	Asp	Gly	Val	Asn	Ser	Ala	Phe	His	Leu	Trp	Cys	Asn	Gly	Arg	Trp	Val	Gly
161	Tyr	Gly	Gln	Asp	Ser	Arg	Leu	Pro	Ser	Glu	Phe	Asp	Leu	Ser	Ala	Phe	Leu	Arg	Ala	Gly
181	Glu	Asn	Arg	Leu	Ala	Val	Met	Val	Leu	Arg	Trp	Ser	Asp	Gly	Ser	Tyr	Leu	Glu	Asp	Gln
201	Asp	Met	Grp	Arg	Met	Ser	Gly	Ile	Phe	Arg	Asp	Val	Ser	Leu	Leu	His	Lys	Pro	Thr	Thr
221	Gln	Ile	Ser	Asp	Phe	His	Val	Ala	Thr	Arg	Phe	Asn	Asp	Asp	Phe	Ser	Arg	Ala	Val	Leu
241	Glu	Ala	Glu	Val	Gln	Met	Cys	Gly	Glu	Leu	Arg	Asp	Tyr	Leu	Arg	Val	Thr	Val	Ser	Leu
261	Trp	Gln	Gly	Glu	Thr	Gln	Val	Ala	Ser	Gly	Thr	Ala	Pro	Phe	Gly	Gly	Glu	Ile	Ile	Asp
281	Glu	Arg	Gly	Gly	Tyr	Ala	Asp	Arg	Val	Thr	Leu	Arg	Leu	Asn	Val	Glu	Asn	Pro	Lys	Leu
301	Trp	Ser	Ala	Glu	Ile	Pro	Asn	Leu	Tyr	Arg	Ala	Val	Val	Glu	Leu	His	Thr	Ala	Asp	Gly
321	Thr	Leu	Ile	Glu	Ala	Glu	Ala	Cys	Asp	Val	Gly	Phe	Arg	Glu	Val	Arg	Ile	Glu	Asn	Gly
341	Leu	Leu	Leu	Leu	Asn	Gly	Lys	Pro	Leu	Leu	Ile	Arg	Gly	Val	Asn	Arg	His	Glu	His	His
361	Pro	Leu	His	Gly	Gln	Val	Met	Asp	Glu	Gln	Thr	Met	Val	Gln	Asp	Ile	Leu	Leu	Met	Lys
381	Gln	Asn	Asn	Phe	Asn	Ala	Val	Arg	Cys	Ser	His	Tyr	Pro	Asn	His	Pro	Leu	Trp	Tyr	Thr
401	Leu	Cys	Asp	Arg	Tyr	Gly	Leu	Tyr	Val	Val	Asp	Glu	Ala	Asn	Ile	Glu	Thr	His	Gly	Met
421	Val	Pro	Met	Asn	Arg	Leu	Thr	Asp	Asp	Pro	Arg	Trp	Leu	Pro	Ala	Met	Ser	Glu	Arg	Val
441	Thr	Arg	Met	Val	Gln	Arg	Asp	Arg	Asn	His	Pro	Ser	Val	Ile	Ile	Trp	Ser	Leu	Gly	Asn
461	Glu	Ser	Gly	His	Gly	Ala	Asn	His	Asp	Ala	Leu	Tyr	Arg	Trp	Ile	Lys	Ser	Val	Asp	Pro
481	Ser	Arg	Pro	Val	Gln	Tyr	Glu	Gly	Gly	Gly	Ala	Asp	Thr	Thr	Ala	Thr	Asp	Ile	Ile	Cys
501	Pro	Met	Tyr	Ala	Arg	Val	Asp	Glu	Asp	Gln	Pro	Phe	Pro	Ala	Val	Pro	Lys	Trp	Ser	Ile
521	Lys	Lys	Trp	Leu	Ser	Leu	Pro	Gly	Glu	Thr	Arg	Pro	Leu	Ile	Leu	Cys	Glu	Tyr	Ala	His
541	Ala	Met	Gly	Asn	Ser	Leu	Gly	Gly	Phe	Ala	Lys	Tyr	Trp	Gln	Ala	Phe	Arg	Gln	Tyr	Pro
561	Arg	Leu	Gln	Gly	Gly	Phe	Val	Trp	Asp	Trp	Val	Asp	Gln	Ser	Leu	Ile	Lys	Tyr	Asp	Glu
581	Asn	Gly	Asn	Pro	Trp	Ser	Ala	Tyr	Gly	Gly	Asp	Phe	Gly	Asp	Thr	Pro	Asn	Asp	Arg	Gln
601	Phe	Cys	Met	Asn	Gly	Leu	Val	Phe	Ala	Asp	Arg	Thr	Pro	His	Pro	Ala	Leu	Thr	Glu	Ala
621	Lys	His	Gln	Gln	Gln	Phe	Phe	Gln	Phe	Arg	Leu	Ser	Gly	Gln	Thr	Ile	Glu	Vel	Thr	Ser
641	Glu	Tyr	Leu	Phe	Arg	His	Ser	Asp	Asn	Glu	Leu	Leu	His	Trp	Met	Val	Ala	Leu	Asp	Gly
661	Lys	Pro	Leu	Ala	Ser	Gly	Glu	Val	Pro	Leu	Asp	Val	Ala	Pro	Gln	Gly	Lys	Gln	Leu	Ile
681	Glu	Leu	Pro	Glu	Leu	Pro	Gln	Pro	Glu	Ser	Ala	Gly	Gln	Leu	Trp	Leu	Thr	Val	Arg	Val
701	Val	Gln	Pro	Asn	Ale	Thr	Ala	Trp	Ser	Glu	Ala	Gly	His	Ile	Ser	Ala	Trp	Gln	Gln	Trp
721	Arg	Leu	Ala	Glu	Asn	Leu	Ser	Val	Thr	Leu	Pro	Ala	Ala	Ser	His	Ala	Ile	Pro	His	Leu
741	Thr	Thr	Ser	Glu	Met	Asp	Phe	Cys	Ile	Glu	Leu	Gly	Asn	Lys	Arg	Trp	Gln	Phe	Asn	Arg
761	Gln	Ser	Gly	Phe	Leu	Ser	Gln	Met	Trp	Ile	Gly	Asp	Lys	Lys	Gln	Leu	Leu	Thr	Pro	Leu
781	Arg	Asp	Gln	Phe	Thr	Arg	Ala	Pro	Leu	Asp	Asn	Asp	Ile	Gly	Val	Ser	Glu	Ala	Thr	Arg
801	Ile	Asp	Pro	Asn	Ala	Trp	Val	Glu	Arg	Trp	Lys	Ala	Ala	Gly	His	Tyr	Gln	Ala	Glu	Ala
821	Ala	Leu	Leu	Gln	Cys	Thr	Ala	Asp	Thr	Leu	Ala	Asp	Ala	Val	Leu	Ile	Thr	Thr	Ala	His
841	Ala	Trp	Gln	His	Gln	Gly	Lys	Thr	Leu	Phe	Ile	Ser	Arg	Lys	Thr	Tyr	Arg	Ile	Asp	Gly
861	Ser	Gly	Gln	Met	Ala	Ile	Thr	Val	Asp	Val	Glu	Val	Ala	Ser	Asp	Thr	Pro	His	Pro	Ala
881	Arg	Ile	Gly	Leu	Asn	Cys	Gln	Leu	Ala	Gln	Val	Ala	Glu	Arg	Val	Asn	Trp	Leu	Gly	Leu
901	Gly	Pro	Gln	Glu	Asn	Tyr	Pro	Asp	Arg	Leu	Thr	Ala	Ala	Cys	Phe	Asp	Arg	Trp	Asp	Leu
921	Pro	Leu	Ser	Asp	Met	Tyr	Thr	Pro	Tyr	Val	Phe	Pro	Ser	Glu	Asn	Gly	Leu	Arg	Cys	Gly
941	Thr	Arg	Glu	Leu	Asn	Tyr	Gly	Pro	His	Gln	Trp	Arg	Gly	Asp	Phe	Gln	Phe	Asn	Ile	Ser
961	Arg	Tyr	Ser	Gln	Gln	Gln	Leu	Met	Glu	Thr	Ser	His	Arg	His	Leu	Leu	His	Ala	Glu	Glu
981	Gly	Thr	Trp	Leu	Asn	Ile	Asp	Gly	Phe	His	Met	Gly	Ile	Gly	Gly	Asp	Asp	Ser	Trp	Ser
1001	Pro	Ser	Val	Ser	Ala	Glu	Phe	Gln	Leu	Ser	Ala	Gly	Arg	Tyr	His	Tyr	Gln	Leu	Val	Trp
1021	Cys	Gln	Lys	***	***	***														

Figure 3.35 Amino acid sequence of β-galactosidase.

Table 3.5. Amino acid content of the β-galactosidase tetramer

Amino acid	Number	(%)	Amino acid	Number	(%)
Asp	256	6.3	Val	256	6.3
Asn	188	4.6	Met	92	2.2
Thr	224	5.5	Ile	156	3.8
Ser	240	5.9	Leu	384	9.4
Glu	248	6.1	Tyr	124	3.0
Gln	232	5.7	Phe	152	3.7
Pro	248	6.1	His	136	3.3
Gly	284	6.9	Lys	80	2.0
Ala	308	7.5	Arg	264	6.5
Cys	64	1.6	Trp	156	3.8
Free acids	508	12.4			

Source: Kalnins *et al.* (1983).

3.5.1.4 Three-dimensional structure of β-Gal

The three-dimensional structure of β-Gal was published by Jacobson *et al.* (1994), at 3.5 Å resolution. The structure features a 51-residue N-terminal 'complementation' sequence and five compact domains, all of largely β-structure; the active site is contained within the third of these, a distorted TIM-barrel motif. The enzymically important residues (Glu461, Met502, Tyr503 and Glu537) are clustered around the substrate-binding site, a deep pit at the end of the barrel.

The structure described by Jacobson *et al.* (1994) also contains 148 water molecules and two Mg^{2+} ions.

The large size of the β-Gal molecule has been the subject of much discussion, considering its relatively simple function. Examination of the primary sequence by Fowler and Zabin (1978) suggested a relatively arbitrary evolution of the *lacZ* gene: the duplication of an ancestral gene to form a fused dimer, itself fused to the gene for a completely unrelated *E. coli* protein, dihydrofolate reductase. However, Jacobson *et al.* (1994) noted that all regions of the molecule contributed either to the active site or to the interfaces between subunits.

The interactions between monomer units are quite complex, including links between the complementation regions of neighbouring units, the formation of a four-helix bundle from components of adjacent third domains, and the interaction of a second-domain loop with a neighbouring third domain.

The involvement of the complementation sequence is essential for enzyme activity. Mutant galactosidase (α-acceptor) lacking residues in this region is dimeric and inactive, but can be converted to active enzyme by addition of a suitable peptide (α-donor) containing at least residues 3–41 (Langley *et al.*, 1975).

3.5.2 The mechanism of β-Gal catalysis

The mechanism of β-Gal catalysis has been reviewed by Huber *et al.* (1994), and appears to be rather complex. Hydrolysis takes place of the bond between two saccharide residues (in natural substrates such as lactose) or between a saccharide and a non-sugar (aglycone) portion in many synthetic substrates. In each case the hydrolysis is accompanied by the covalent attachment of the galactose unit to the enzyme, and the release of the other sugar or aglycone. The galactose is subsequently displaced to regenerate the enzyme.

Many observed differences between substrates are difficult to rationalise in terms of their chemical properties, and Huber *et al.* (1994) have proposed that non-covalent interactions are important in determining their behaviour. Indeed it is likely that changes to the conformation of the enzyme itself take place, dependent on substrate structure. Two glutamate residues in the active site (Glu461 and Glu537) appear to take part in the stabilisation of a transition state, in which one end of the saccharide ring is held in a planar orientation.

A neighbouring tyrosine residue, Tyr503, is thought to play an acid catalysis role, providing a proton to the departing sugar or aglycone as the galactose moiety is coupled to the carboxyl group of Glu537. There is good evidence that the galactose residue passes through a positively charged transition state stabilised by Glu461 (Huber *et al.*, 1994; Richard, 1995; Richard *et al.*, 1996) which is rapidly trapped by Glu537, the carboxylate acting as a nucleophile.

Subsequent base catalysis by the deprotonated tyrosine then facilitates the release of the galactose and the regeneration of the enzyme.

3.5.3 The stability of β-Gal

Abuknesha and Exley (1978) showed that β-Gal lost activity under acidic conditions, particularly on prolonged incubation: significant losses were observed at pH 6.5 after 30 hours, for example. At pH 5 almost half the activity was lost after six hours. The pH stability of this enzyme is in fact rather narrow (Boehringer Mannheim, 1990), only in the range 6–7 was activity fully maintained after three hours at 25 °C.

Thermal stability is also rather poorer than HRP or AP (Boehringer Mannheim, 1990), the activity dropping rapidly after even ten minutes at 50 °C. Many other microorganisms besides *E. coli* produce β-galactosidases (Erickson and Steers, 1970); those from thermophilic species have received some attention due to their high temperature optima (65–80 °C: Buonocore *et al.*, 1980; Jagota *et al.*, 1981).

3.5.4 Activators and inhibitors of β-Gal

As mentioned above both divalent and monovalent cations are activators of β-Gal, and chelating agents might therefore be expected to have a harmful effect. Because of the essential role of a tyrosine residue, substitution reactions such as iodination cause inactivation (Ring and Huber, 1993); similarly carboxyl-reactive groups such as epoxides cause loss of activity *via* a critical glutamate residue (Herrchen and Legler, 1984). Many different carbohydrate inhibitors of β-Gal have been identified, including amino- and triazeno-sugars (Huber and Gaunt, 1982; Mega *et al.*, 1990), and have been employed in the elucidation of the mechanism of enzyme activity (Huber *et al.*, 1994).

3.5.5 Colorimetric substrates for β-Gal

The substrates employed for β-Gal are strongly reminiscent of those for AP, with galactosides replacing phosphates. Aryl galactosides are generally better substrates than alkyl galactosides or disaccharides (Wallenfels *et al.*, 1960). In histochemical applications where a solid product is required, naphthyl galactosides have been employed, the resulting naphthols being converted to azo-dyes by a diazonium salt (Cohen *et al.*, 1952; Gossrau, 1973a,b; Lojda *et al.*, 1973). Alternatively, direct oxidative coupling of the naphthol can be employed to yield a coloured precipitate (Boehringer Mannheim, 1990).

Probably the most common histochemical substrate for this enzyme, however, is the analogue of BCIP, 5-bromo-4-chloro-3-indolyl-β-D-galactopyranoside, X-Gal (Figure 3.36), which provides a blue precipitate

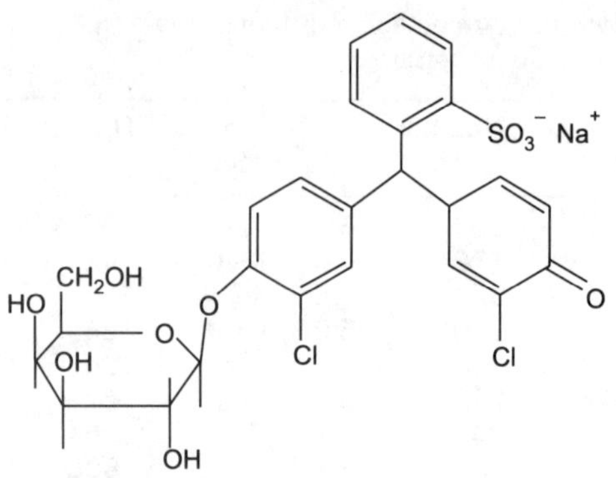

Figure 3.37 Chlorophenol red-β-D-galactopyranoside.

of indigo (Pearson *et al.*, 1963; Yarborough *et al.*, 1967; Lojda *et al.*, 1973; Sanes *et al.*, 1986; Boehringer Mannheim, 1990)

The β-Gal equivalent of the alkaline phosphatase substrate PNP-P is *p*-nitrophenyl-β-D-galactoside (PNPG), yielding the same product, *p*-nitrophenol (Craven *et al.*, 1965; Boehringer Mannheim, 1990). The absorbance is therefore measured at the same wavelength, 405–410 nm. The *o*-equivalent ONPG is also employed in the same role, and is capable of detecting 100 amol of β-Gal with a 100-minute incubation (Ishikawa, 1987). The absorbance characteristics are similar to the *p*-isomer.

Chlorophenol red-β-D-galactopyranoside (Figure 3.37) has also found some application as a colorimetric substrate for β-Gal, yielding a soluble product absorbing at around 576 nm (Kuhr *et al.*, 1987; Boehringer Mannheim, 1990).

3.5.6 Fluorimetric substrates for β-Gal

Once again the most commonly encountered fluorogen for β-Gal is analogous to an AP substrate: 4-methylumbelliferyl-β-D-galactoside (MUG; Boehringer Mannheim, 1990): the product is identical to that

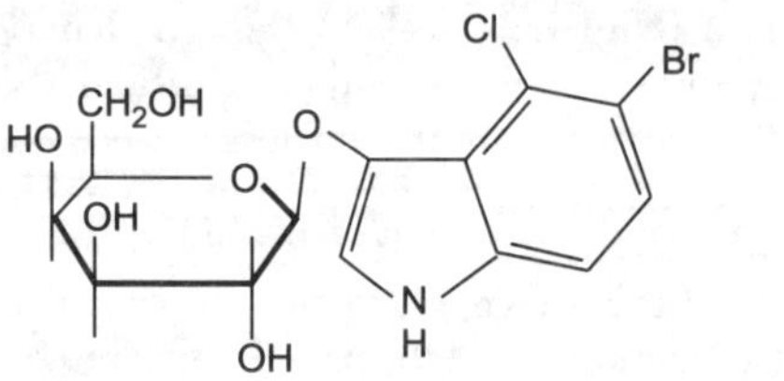

Figure 3.36 5-Bromo-4-chloro-3-indolyl-β-D-galactoside (X-Gal).

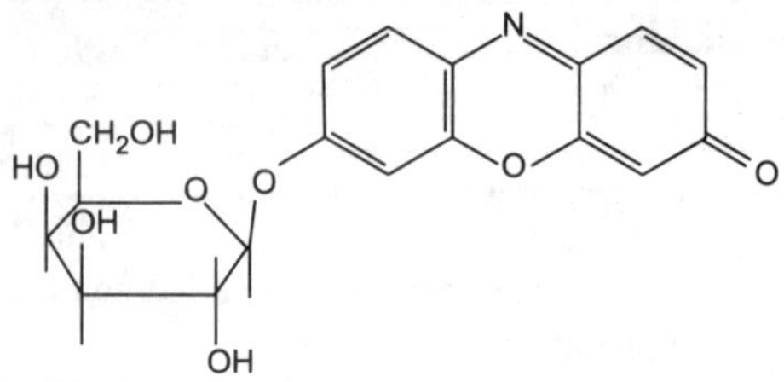

Figure 3.38 Resorufin-β-D-galactopyranoside.

obtained by AP-catalysed hydrolysis of MUP, and so consequently are the fluorescence conditions (excitation 340 nm, emission 440 nm). Using a very small sample size, Ishikawa (1987) has shown detection of 0.0003 amol (180 molecules) of the enzyme with this substrate.

Resorufin-β-D-galactopyranoside (Figure 3.38) has also found some use as a fluorescent substrate for this enzyme (Hofmann and Sernetz, 1984; Klein *et al.*, 1985); so too have galactosides of fluorescein and other fluorans (Rotman, 1961; Rotman *et al.*, 1963).

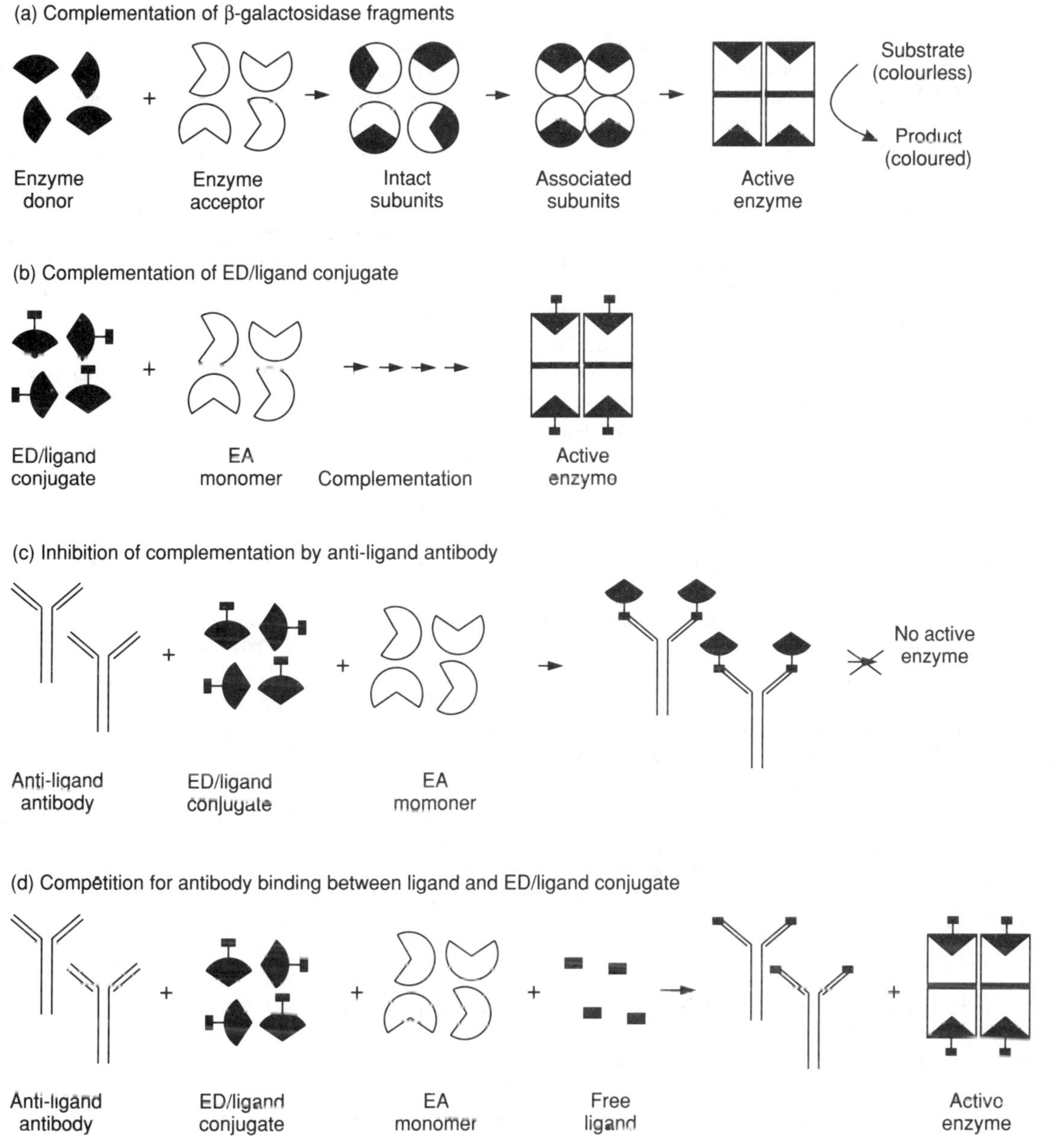

Figure 3.39 The principle of CEDIA assays. Enzyme fragments can reassemble to form active enzyme (a), even if one fragment is labelled with the analyte to be detected (b). The presence of specific antibody, however, interferes with enzyme reassembly (c). The presence of free analyte competes for this antibody, and can be quantified by the amount of enzyme formed (d)—the more free analyte, the more enzyme activity.

3.5.7 Luminometric substrates for β-Gal

Tropix Inc. market two substrates for β-Gal based on 1,2-dioxetane chemistry, Galacton™ (3-(4-methoxyspiro{1,2-dioxetane-3,2′-(5′-chloro)tricyclo[3.3.1.1^{3,7}]decan}-4-yl)phenyl β-D-galactopyranoside) and Galacton Plus™ (3-chloro-5-(4-methoxyspiro{1,2-dioxetane-3,2′-(5′-chloro)tricyclo[3.3.1.1^{3,7}]decan}-4-yl)phenyl β-D-galactopyranoside) (Bronstein *et al.*, 1996). These are directly analogous to the AP substrates AMPPD and CSPD (see Section 3.4.9).

Indirect, enzyme-coupled approaches for β-Gal detection are described in Section 3.6.

3.5.8 Electrochemical substrates for β-Gal

It has already been mentioned that the generation of a phenol from an electroinactive substrate can form the basis of detection systems for enzymes such as AP (Section 3.4.10). Exactly the same approach can be applied to β-Gal, using substrates such as the PAPP analogue *p*-aminophenyl-β-D-galactopyranoside, PAPG (Másson *et al.*, 1995), or naphthyl galactosides (Brown *et al.*, 1995).

3.5.9 Other labelling applications of β-Gal

The Microgenics Corp. of California developed a successful homogeneous immunoassay format based around a combination of recombinant donor and acceptor fragments of β-galactosidase, the latter constituting the vast majority of the native structure (Henderson *et al.*, 1986; Khanna *et al.*, 1989). This approach has been used in a number of applications (Kaspar and Khanna, 1991; Kopetzki *et al.*, 1994; Brown *et al.*, 1995), termed cloned enzyme-donor immunoassays (CEDIA™ Microgenics Corp.; Figure 3.39).

The sodium dependency of β-Gal has also been exploited in the design of serum sodium assays (Quiles *et al.*, 1993) using ONPG as the enzyme substrate.

3.6 Coupled-enzyme Systems

As discussed in Section 3.2.2.4, systems employing more than one enzyme provide the potential for both higher sensitivity and a wider range of applications. The range of substrates for AP and β-Gal is wide, and these enzymes are particularly suited to detection *via* coupled approaches, the hydrolysis product obtained from a suitable substrate being 'channelled' into a second enzyme reaction (Figure 3.40).

3.6.1 Enzyme-coupled detection of HRP

The products of HRP oxidation are less amenable to such formats, so this enzyme is more commonly encountered in the final stage of an enzyme-channelling format. The best-known examples include the detection of analytes such as glucose and cholesterol, by channelling the H_2O_2 produced from their oxidase-catalysed oxidation into an HRP-substrate system, typically a luminol/HRP mixture for luminometric detection. This approach can be applied to a very wide range of analytes, as there are a large number of compounds for which there is a specific oxidase.

Many other coupled-enzyme systems have been described, employing luminol-peroxidase elements: Maeda *et al.* (1989, 1993a), for example, described an assay for NAD(P)H, using the design shown in Figure 3.41. In the presence of the electron mediator 1-methoxy-5-methylphenazinium methyl sulfate, NAD(P)H produces hydrogen peroxide from molecular oxygen, and the peroxide is channelled into a microperoxidase-isoluminol system.

Other uses of HRP in the final stage of enzyme-coupled AP and β-Gal detection will be mentioned in the following sections.

3.6.2 Enzyme-coupled detection of AP

An interesting multi-enzyme system for AP was first described by Self of the Hammersmith Hospital, London and commercialised by the IQ(Bio) group of Cambridge (Self, 1985; Johannson *et al.*, 1986); AP forms the primary label, and dephosphorylates NADP to NAD. The NAD then activates a redox cycle employing diaphorase and alcohol dehydrogenase to convert *p*-iodonitrotetrazolium violet (INT-Violet) to an intensely coloured formazan, detectable at 492 nm. Thus there is a double amplification effect, each label invoking a large number of signal-generating events, and detection of AP has been claimed at 0.01 amol (6000 molecules), a prodigious sensitivity for a colorimetric system. An assay for TSH (thyroid stimulating hormone) was sensitive down to 0.0013 μIU/ml (0.46 amol).

Cook and Self (1993) have described an extension of this system where the chromogen INT-Violet has been replaced by the fluorogen resazurin (Figure 3.42). Diaphorase catalyses the reduction of this compound to the fluorescent resorufin (cf. Figure 3.38), which has maxima for excitation at 570 nm and emission at 590 nm. This has allowed the detection of 0.6 zmol (6×10^{-22} mol, around 350 molecules) of AP. Applied

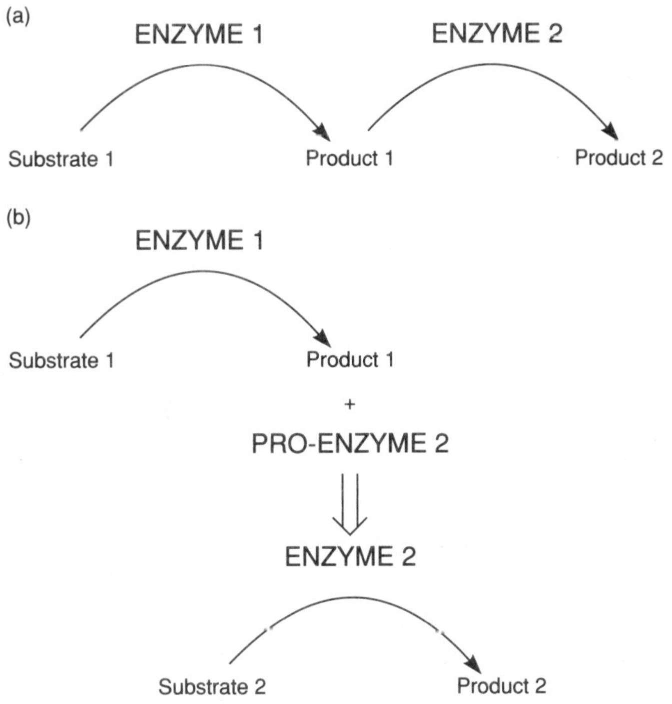

Figure 3.40 Coupled-enzyme detection systems. One substrate undergoes an enzyme-catalysed reaction to produce a product, which itself provides a substrate (a) or a cofactor (b) for a second enzyme reaction.

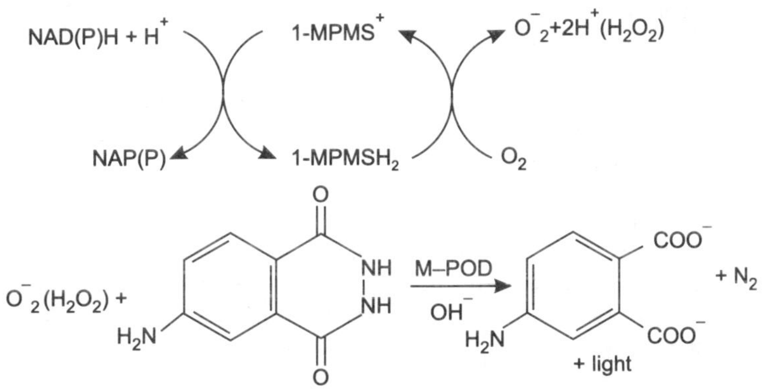

Figure 3.41 Assay for NAD(P)H using a coupled-enzyme system. 1-MPMS, 1-methoxy-5-methylphenazinium methyl sulfate; M-POD, microperoxidase.

to an immunoassay for proinsulin, 3.4 amol of the analyte were detected in a ten-minute format.

A luminometric detection system for AP was described by Arakawa *et al.* (1991), and was based on the BCIP colorimetric reaction described in Section 3.4.7. The indoxyl dimerisation generates hydrogen peroxide as a by-product, and this peroxide can be quantified by a peroxidase/luminol system. The authors used isoluminol/microperoxidase and were able to detect 0.7 amol of AP. Using the same approach, Ikegami *et al.* (1994) obtained higher sensitivity from 3-indolylphosphate disodium salt, detecting half this quantity of AP. Ho *et al.* (1995) reported the use of an HRP electrode for amperometric

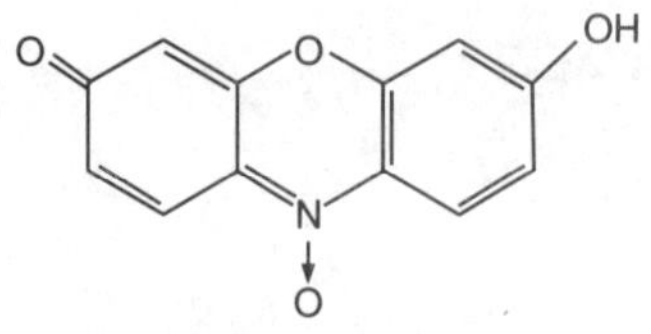

Figure 3.42 Structure of resazurin.

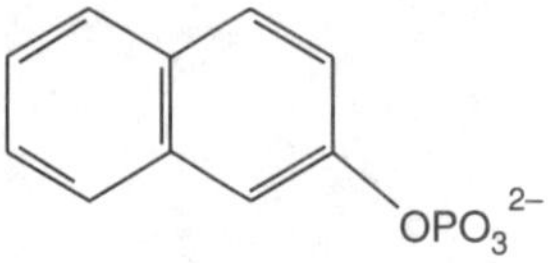

Figure 3.43 A protected enhancer; the phosphate is removed by AP, yielding an enhancer for HRP.

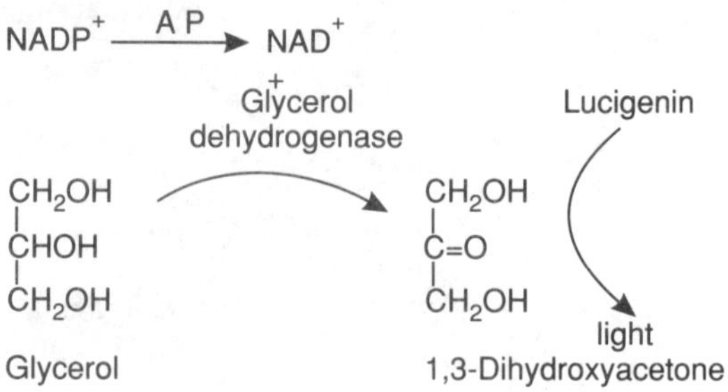

Figure 3.44 AP catalyses the production of NAD from NADP; NAD is a coenzyme for glycerol dehydrogenase, converting glycerol to 1,3-dihydroxyacetone. This is capable of reducing lucigenin in a light-generating reaction.

detection of hydrogen peroxide, generated by AP from BCIP in the same way.

In another HRP-coupled application, 0.1 amol of AP could be detected by an interesting variation on the enhanced luminescence theme (see Section 3.3.7). Protected enhancers such as those shown in Figure 3.43 are released by AP, and promote the desired signal generation in the HRP system (Akhavan-Tafti *et al.*, 1994).

Rabin and his colleagues at London Biotechnology Ltd. have developed AP assays based on the principle of prosthetogenesis, in which an AP substrate is dephosphorylated to provide the prosthetic group for another enzyme. They used the dephosphorylation of FADP to yield FAD, the prosthetic group for D-amino acid oxidase, which yields hydrogen peroxide with a suitable amino acid substrate such as proline (Harbron *et al.*, 1992, 1993). The hydrogen peroxide can then be harnessed to generate a colorimetric signal from HRP and a suitable substrate, allowing 0.1 amol of AP to be detected by a 30-minute assay. In a more recent report, Fisher *et al.* (1995) detected 0.4 amol of AP in a five-minute assay using an HRP/luminol/*p*-hydroxycinnamic acid enhanced luminescence format.

In another format (Maeda *et al.*, 1993b) $NADP^+$ was dephosphorylated by AP; the resultant NAD^+ formed the cosubstrate for the oxidation of glycerol by glycerol dehydrogenase (Figure 3.44), and the product of this reaction (1,3-dihydroxyacetone) acted to reduce lucigenin (see Section 3.4.9).

Another multi-enzyme assay system based on AP was described by Schulte *et al.* (1989). A carboxylesterase inhibitor, blocked from inhibiting by phosphorylation, was unblocked by AP and was then free to inhibit the esterase-catalysed deesterification of an indophenyl derivative: 80 fmol of AP could be detected in this way.

Finally, Miska and Geiger (1987) described the use of D-luciferin-*O*-phosphate as a substrate for AP, the luciferin released becoming available as a luminogenic substrate for the enzyme luciferase. A sensitivity 60-fold higher than that obtained with *p*-nitrophenyl phosphate was claimed.

3.6.3 Enzyme-coupled detection of β-Gal

β-Gal activity can be measured by indirect luminometric techniques. For example, the hydrogen peroxide generated by the dimerisation of indoxyl substrates such as X-Gal can be detected using a suitable enzyme system such as as microperoxidase with an isoluminol substrate. Arakawa *et al.* (1991) detected 0.1 amol of β-Gal by this method.

Hydrolysis of galactoside-protected enhancers has been described (Akhavan-Tafti *et al.*, 1994), allowing detection of 0.56 amol of β-Gal in an enhanced chemiluminescence format.

A more complicated enzyme-channelling procedure based on β-Gal (Tanaka and Ishikawa, 1986) yielded prodigious sensitivity. The D-galactose released by β-Gal from ONPG was employed as a substrate for galactose dehydrogenase, at the same time converting NAD^+ to NADH. The NADH was used to convert FMN to $FMNH_2$ in the presence of NAD(P)H:FMN oxidoreductase, and the $FMNH_2$ was employed with myristic aldehyde to generate light in the presence of luciferase. Detection of 0.2 zeptomoles (120 molecules) of β-Gal was reported from this assay.

References

Abuknesha, R. and Exley, D. (1978). Design and development of oestradiol-17β enzyme immunoassay. In *Enzyme Labelled Immunoassay of Hormones and Drugs. Proceedings of the International Symposium on Enzyme Labelled Immunoassay of Hormones and Drugs, Ulm, West Germany, July 10 and 11, 1978* (Pal, S. B., ed.), pp. 139–152, Walter de Gruyter & Co., Berlin.

Adler, O. and Adler, R. (1904). Über das Verhalten Gewisser Organischer Verbindungen gegenüber Blut mit Besonderer Berücksichtigung des Nachweises von Blut. *Hoppe-Seyler's Z. Physiol. Chem.*, **41**, 59–67.

Aibara, S. *et al.* (1981). Isolation and properties of basic isoenzymes of horseradish peroxidase. *J. Biochem.*, **90**, 489–496.

Akhavan-Tafti, M. H. (1994). Enzyme-catalyzed chemiluminescence from hydroxyaryl cyclic diacylhydrazide compounds. *World Patent WO 94/10337*, to Lumigen, Inc.

Akhavan-Tafti, H. *et al.* (1994). Lumigen™ PS: new chemiluminescent substrates for the detection of horseradish peroxidase. Product Literature, Lumigen, Inc. Southfield, MI.

Akhavan-Tafti, H. *et al.* (1995). Lumigen™ PS: chemiluminescent detection of horseradish peroxidase by enzymatic generation of acridinium esters. *Clin. Chem.*, **41**, 1368–1369.

Akimoto, K. *et al.* (1990). Luminol chemiluminescence reaction catalyzed by a microbial peroxidase. *Anal. Biochem.*, **189**, 182–185.

Albers, F. J. *et al.* (1984). Potassium ferricyanide: a stabilizer for horseradish peroxidase. *J. Histochem. Cytochem.*, **32**, 1005–1008.

Arakawa, H. *et al.* (1985). Chemiluminescence enzyme immunoassay for thyroxin with use of glucose oxidase and a bis(2,4,6-trichlorophenyl)oxalate–fluorescent dye system. *Clin. Chem.*, **31**, 430–434.

Arakawa, H. *et al.* (1991). Chemiluminescent assay of various enzymes using indoxyl derivatives as substrate and its applications to enzyme immunoassay and DNA probe assay. *Anal. Biochem.*, **199**, 238–242.

Ator, M. A. *et al.* (1987). Structure and catalytic mechanism of horseradish peroxidase. Regiospecific meso alkylation of the prosthetic heme group by alkylhydrazines. *J. Biol. Chem.*, **262**, 14954–14960.

Baader, W. J. *et al.* (1985). Peroxidase-catalyzed formation of triplet acetone and chemiluminescence from isobutyraldehyde and molecular oxygen. *J. Biol. Chem.*, **260**, 10217–10225.

Babson, A. L. (1984). α-Naphthyl phosphate: the preferred substrate for acid phosphatase. *Clin. Chem.*, **30**, 1418–1419.

Babson, A. L. *et al.* (1966). Phenolphthalein monophosphate as a substrate for serum alkaline phosphatase. *Clin. Chem.*, **12**, 482–490.

Bach, A. (1904). Über die Wirkungweise der Peroxydase bei der Reaction zwischen Hydroperoxyd und Jodwasserstoffsäure. *Ber. Dtsch. Chem. Ges.*, **37**, 3785–3800.

Bach, A. and Chodat, R. (1903). Untersuchungen über die Rolle der Peroxyde in der Chemie der lebende Zelle. IV. Über Peroxydase. *Ber. Dtsch. Chem. Ges.*, **36**, 600–605.

Baek, H. K. and Van Wart, H. E. (1992). Elementary steps in the reaction of horseradish peroxidase with several peroxides: kinetics and thermodynamics of formation of compound 0 and compound I. *J. Am. Chem. Soc.*, **114**, 718–725.

Bartel, D. P. and Szostak, J. W. (1993). Isolation of new ribozymes from a large pool of random sequences. *Science*, **261**, 1411–1418.

Bartling, G. J. *et al.* (1974). Preparation and properties of horseradish peroxidase cross-linked in nonaqueous media. *Int. J. Peptide Protein Res.*, **6**, 287–294.

Beaudreau, C. and Yasunobo, K. T. (1966). Heme proteins. VI. Crystalline pineapple peroxidase B. *Biochemistry*, **5**, 1405–1412.

Bechara, E. J. H. *et al.* (1979). Peroxidase catalyzed generation of triplet acetone. *Photochem. Photobiol.*, **30**, 101–110.

Besman, M. and Coleman, J. E. (1985). Isozymes of bovine intestinal alkaline phosphatase. *J. Biol. Chem.*, **260**, 11190–11193.

Bessey, O. A. *et al.* (1946). A method for the rapid determination of alkaline phosphatase with five cubic millimeters of serum. *J. Biol. Chem.*, **164**, 321–329.

Bobrow, M. N. and Litt, G. J. (1993). Method for the detection or quantitation of an analyte using an analyte dependent enzyme activation system. *US Patent 5,196,306*, to E.I. Du Pont de Nemours and Co.

Bobrow, M. N. *et al.* (1989). Catalyzed reporter deposition, a novel method of signal amplification. Application to immunoassays. *J. Immunol. Methods*, **125**, 279–285.

Bodansky, A. (1933). Phosphatase studies. II. Determination of serum phosphatase. Factors influencing the accuracy of the determination. *J. Biol. Chem.*, **101**, 93–104.

Boehringer Mannheim (1990). *Immunologicals for the Diagnostic Industry 1990/91*, Boehringer Mannheim Biochemica, Mannheim, Germany.

Bos, E. S. *et al.* (1981). 3,3′,5,5′-Tetramethylbenzidine as an Ames test negative chromogen for horse-radish peroxidase in enzyme-immunoassay. *J. Immunoassay*, **2**, 187–204.

Bowers, G. N., Jr and McComb, R. B. (1975). Measurement of total alkaline phosphatase activity in human serum. *Clin. Chem.*, **21**, 1988–1995.

Boyd, S. and Yamazaki, H. (1994). Use of polyvinyl alcohol as a stabilizer of peroxidase–antibody conjugate for enzyme immunoassay. *Biotechnol. Techn.*, **8**, 123–128.

Boyer, S. H. (1961). Alkaline phosphatase in human sera and placentae. *Science*, **134**, 1002–1004.

Bradshaw, R. A. *et al.* (1981). Amino acid sequence of *Escherichia coli* alkaline phosphatase. *Proc. Natl. Acad. Sci. USA*, **78**, 3473–3477.

Braithwaite, A. (1976). Unit cell dimensions of crystalline horseradish peroxidase. *J. Mol. Biol.*, **106**, 229–230.

Bray, J. and King, E. J. (1943). The phosphatase reaction as an aid to the identification of micro-organisms using phenolphthalein phosphate as substrate. *J. Pathol. Bacteriol.*, **55**, 315–320.

Bronstein, I. *et al.* (1988). Novel chemiluminescent enzyme substrates and their application in immunoassays. *J. Biolumin. Chemilumin.*, **2**, 186.

Bronstein, I. *et al.* (1989). 1,2-Dioxetanes: novel chemiluminescent enzyme substrates. Applications to immunoassays. *J. Biolumin. Chemilumin.*, **4**, 99–111.

Bronstein, I. *et al.* (1995). Chemiluminescent energy transfer assays. *World Patent WO 95/17672*, to Tropix, Inc.

Bronstein, I. *et al.* (1996). Chemiluminescence: sensitive detection technology for reporter gene assays. *Clin. Chem.*, **42**, 1542–1546.

Bronstein, I. Y. *et al.* (1992). Chemiluminescent 3-(substituted adamant-2′-ylidene) 1,2-dioxetanes. *US Patent 5,112,960*, to Tropix, Inc.

Brown, M. E. *et al.* (1995). Electrochemical enzymatic complementation immunoassay. *US Patent 5,427,912*, to Boehringer Mannheim.

Bruschi, B. J. (1978). Compositions for the detection of hydrogen peroxide. *US Patent 4,089,747*, to Eastman Kodak Co.

Bundgaard, M. and Møller, M. (1981). Horseradish peroxidase and microperoxidase. Their purity and binding to serum proteins. *J. Histochem. Cytochem.*, **29**, 331–336.

Buonocore, V. *et al.* (1980). A constitutive β-galactosidase from the extreme thermoacidophile archaebacterium *Caldariella acidophila*: properties of the enzyme in the free state and in immobilized whole cells. *J. Appl. Biochem.*, **2**, 390–397.

Burstone, M. S. (1960). Postcoupling, noncoupling, and fluorescence techniques for the determination of alkaline phosphatase. *J. Natl. Cancer Inst.*, **24**, 1199–1207.

Butler-Ransohoff, J. E. *et al.* (1992). Active-site mutagenesis of *E. coli* alkaline phosphatase: replacement of serine-102 with nonnucleophilic amino acids. *J. Org. Chem.*, **57**, 142–145.

Carson, K. A. and Mesulam, M.-M. (1982). Electron microscopic demonstration of neural connections using horseradish peroxidase. A comparison of the tetramethylbenzidine procedure with seven other histochemical methods. *J. Histochem. Cytochem.*, **30**, 425–435.

Chance, B. (1943). The kinetics of the enzyme–substrate compound of peroxidase. *J. Biol. Chem.*, **151**, 553–577.

Chandross, E. A. (1963). A new chemiluminescent system. *Tetrahedron Lett.*, **12**, 761–765.

Chang, C. N. *et al.* (1986). Nucleotide sequence of the alkaline phosphatase gene of *Escherichia coli*. *Gene*, **44**, 121–125.

Chappelet-Tordo, D. *et al.* (1974). Intestinal alkaline phosphatase. Catalytic properties and half of the sites reactivity. *Biochemistry*, **13**, 1788–1795.

Chodat, R. and Bach, A. (1903). Untersuchungen über die Rolle der Peroxyde in der Chemie der lebende Zelle. V. Zerlegung der sogenannten Oxydasen in Oxygenasen und Peroxydasen. *Ber. Dtsch. Chem. Ges.*, **36**, 606–608.

Claiborne, A. and Fridovich, I. (1979a). Chemical and enzymatic intermediates in the peroxidation of *o*-dianisidine by horseradish peroxidase. 1. Spectral properties of the products of dianisidine oxidation. *Biochemistry*, **18**, 2324–2329.

Claiborne, A. and Fridovich, I. (1979b). Chemical and enzymatic intermediates in the peroxidation of *o*-dianisidine by horseradish peroxidase. 2. Evidence for a substrate radical-enzyme complex and its reaction with nucleophiles. *Biochemistry*, **18**, 2329–2335.

Coghlan, A. (1997). No more washday whites that come out pink. *New Sci.*, 28 June 1997, 18.

Cohen, R. B. *et al.* (1952). The colorimetric estimation and histochemical demonstration of β-D-galactosidase. *J. Biol. Chem.*, **195**, 239–249.

Cohn, M. (1957). Contributions of studies on the β-galactosidase of *Escherichia coli* to our understanding of enzyme synthesis. *Bacteriol. Rev.* **21**, 140–168.

Coleman, C. M. (1966). The synthesis of thymolphthalein monophosphate, a new substrate for alkaline phosphatase. *Clin. Chim. Acta*, **13**, 401–403.

Coleman, J. E. (1992). Structure and mechanism of alkaline phosphatase. *Annu. Rev. Biophys. Biomol. Struct.*, **21**, 441–483.

Comfort, D. and Campbell, D. J. (1966). One-step automated alkaline phosphatase analysis. *Clin. Chim. Acta*, **14**, 263–269.

Conroy, J. M. *et al.* (1982). Homology of plant peroxidases. An immunochemical approach. *Plant Physiol.*, **69**, 28–31.

Conyers, S. M. and Kidwell, D. A. (1991). Chromogenic substrates for horseradish peroxidase. *Anal. Biochem.*, **192**, 207–211.

Cook, D. B. and Self, C. H. (1993). Detection of one thousandth of an attomole (1 zeptomole) of alkaline phosphatase: application in an immunoassay of proinsulin. *Clin. Chem.*, **39**, 965–971.

Craven, G. E. *et al.* (1965). Purification, composition, and molecular weight of the β-galactosidase of *Escherichia coli* K12. *J. Biol. Chem.*, **240**, 2468–2477.

Culp, J. S. *et al.* (1985). The active-site and amino-terminal amino acid sequence of bovine intestinal alkaline phosphatase. *Biochim. Biophys. Acta*, **831**, 330–334.

Dalal, F. R. *et al.* (1975). Thymol blue monophosphate: a new substrate for kinetic alkaline phosphatase determinations. *Clin. Chem.*, **21**, 969.

Darbyshire, B. (1973). The glycoprotein nature of indoleacetic acid oxidase/peroxidase fractions and their development in pea roots. *Physiol. Plant.*, **29**, 293–297.

Doerge, D. R. *et al.* (1997). Identification of the colored guaiacol oxidation product produced by peroxidases. *Anal. Biochem.*, **250**, 10–17.

Dordick, J. S. (1992). Enzymatic and chemoenzymatic approaches to polymer synthesis. *Trends Biotechnol.*, **10**, 287–293.

Doyle, W. A. and Smith, A. T. (1996). Expression of lignin peroxidase H8 in *Escherichia coli*: folding and activation of the recombinant enzyme with Ca^{2+} and haem. *Biochem. J.*, **315**, 15–19.

Dunford, H. B. (1986). Catalytic mechanisms of plant peroxidases: with emphasis on reactions of compounds I and II. In *Molecular and Physiological Aspects of Plant Peroxidases* (Greppin, H. *et al.*, eds), pp. 15–24, University of Geneva, Geneva.

Dure, L. S. and Cormier, M. J. (1964). Studies on the bioluminescence of *Balanoglossus biminiensis* extracts. III. A kinetic comparison of luminescent and nonluminescent peroxidation reactions and a proposed mechanism for peroxidase action. *J. Biol. Chem.*, **239**, 2351–2359.

Easton, P. M. *et al.* (1996). Quantitative model of the enhancement of peroxidase-induced luminol luminescence. *J. Am. Chem. Soc.*, **118**, 6619–6624.

Edwards, S. L. *et al.* (1987). Crystal structure of cytochrome *c* peroxidase compound I. *Biochemistry*, **26**, 1503–1511.

Egorov, A. M. *et al.* (1993). Fundamental and applied aspects of reaction of enhanced luminescence. In *Bioluminescence and Chemiluminescence. Status Report. Proceedings of the VIIth International Symposium on Bioluminescence and Chemiluminescence, Banff, March 1993* (Szalay, A. A. *et al.*, eds), pp. 286–290, John Wiley & Sons, New York.

Elser, R. C. (1976). Serum alkaline phosphatase measured by using thymol blue monophosphate and a bichromatic analyzer. *Clin. Chem.*, **22**, 1737–1740.

Engström, L. (1961). Studies on calf-intestinal alkaline phosphatase. I. Chromatographic purification, microheterogeneity and some other properties of the purified enzyme. *Biochim. Biophys. Acta*, **52**, 36–48.

Erickson, R. P. and Steers, E., Jr. (1970). Comparative study of isoenzyme formation of bacterial β-galactosidase. *J. Bacteriol.*, **102**, 79–84

Fernley, H. N. (1971). Mammalian alkaline phosphatases. In *The Enzymes*, 3rd edit., vol. 4, pp. 417–447, Academic Press, New York.

Fernley, H. N. and Walker, P. G. (1965). Kinetic behaviour of calf-intestinal alkaline phosphatase with 4-methylumbelliferyl phosphate. *Biochem. J.*, **97**, 95–103.

Fernö, G. *et al.* (1958). High molecular weight enzyme inhibitors. III. Polyestradiol phosphate (P.E.P.), a long-acting estrogen. *Acta Chem. Scand.*, **12**, 1675–1689.

Finzel, B. C. *et al.* (1984). Crystal structure of yeast cytochrome *c* peroxidase refined at 1.7-Å resolution. *J. Biol. Chem.*, **259**, 13027–13036.

Fisher, M. *et al.* (1995). An amplified chemiluminescent assay for the detection of alkaline phosphatase. *Anal. Biochem.*, **227**, 73–79.

Folley, S. J. and Kay, H. D. (1936). The phosphatases. *Ergeb. Enzymforsch.*, **5**, 159–212.

Fosset, M. *et al.* (1974). Intestinal alkaline phosphatase. Physical properties and quaternary structure. *Biochemistry*, **13**, 1783–1788.

Fowler, A. V. and Zabin, I. (1978). Amino acid sequence of β-galactosidase. XI. Peptide ordering procedures and the complete sequence. *J. Biol. Chem.*, **253**, 5521–5525.

Fricke, H. *et al.* (1982). Enzyme enhanced luminescence immunoassay for the determination of transferrin concentrations in serum. *J. Clin. Chem. Clin. Biochem.*, **20**, 91–94.

Friedman, A. E. *et al.* (1997). Use of vanadium bromoperoxidase as a signal-generating enzyme for chemiluminescent systems: test kits and analytical methods. *World Patent 97/09447*, to Johnson & Johnson Clinical Diagnostics Inc.

Fujiyama, K. *et al.* (1988). Structure of the horseradish peroxidase isozyme C genes. *Eur. J. Biochem.*, **173**, 681–687.

Garen, A. and Levinthal, C. (1960). A fine-structure genetic and chemical study of the enzyme alkaline phosphatase of *E. coli*. I. Purification and characterization of alkaline phosphatase. *Biochim. Biophys. Acta*, **38**, 470–483.

Garner, R. C. *et al.* (1975). Testing of some benzidine analogues for microsomal activation to bacterial mutagens. *Cancer Lett.*, **1**, 39–42.

Gaspar, T. *et al.* (1991). Peroxidases in plant-growth, differentiation and development processes. In *Biochemical, Molecular, and Physiological Aspects of Plant Peroxidases* (Lobarzewski, J. *et al.*, eds), pp. 249–280, University M. Curie-Sklodowska, Lublin, Poland/University of Geneva, Switzerland.

Genesis Report (1995). Chemiluminescence: the dust has settled. *Genesis Report/Dx*, May 1995, 13–20, Genesis Group Associates, Inc., Montclair, NJ.

Genesis Report (1997). Will ribozymes produce a better diagnostic? *Genesis Report/Dx*, March/April 1997, 15–19. Genesis Group Associates, Inc., Montclair, NJ).

George, P. (1953). Intermediate compound formation with peroxidase and strong oxidizing agents. *J. Biol. Chem.*, **201**, 413–426.

Gettins, P. and Coleman, J. E. (1983). ^{113}Cd nuclear magnetic resonance of Cd(II) alkaline phosphatases. *J. Biol. Chem.*, **258**, 396–407.

Ghosh, S. S. *et al.* (1986). Modification of the active site of alkaline phosphatase by site-directed mutagenesis. *Science*, **231**, 145–148.

Gomori, G. (1939). Microtechnical demonstration of phosphatase in tissue sections. *Proc. Soc. Exp. Biol. Med.*, **42**, 23–26.

Gonda, D. K. *et al.* (1989). Universality and structure of the N-end rule. *J. Biol. Chem.*, **264**, 16700–16712.

Goodpasture, E. W. (1919). A peroxidase reaction with sodium nitroprusside and benzidine in blood smears and tissues. *J. Lab. Clin. Med.*, **4**, 442–444.

Goodwin, D. C. *et al.* (1995). Determination of rate constants for rapid peroxidase reactions. *Anal. Biochem.*, **231**, 333–338.

Gosling, J. P. (1990). A decade of development in immunoassay methodology. *Clin. Chem.*, **36**, 1408–1427.

Gossrau, R. (1973a). Über den Histochemischen und Mikrochemischen Nachweis der β-Galactosidase mit 1-Naphthyl-β-Galactopyranosid. *Histochemie*, **35**, 199–218.

Gossrau, R. (1973b). Splitting of naphthol AS-BI β-galactopyranoside by acid β-galactosidase. *Histochemie*, **37**, 89–91.

Graham, R. C., Jr and Karnovsky, M. J. (1966). The early stages of absorption of injected horseradish peroxidase in the proximal tubules of mouse kidney. ultrastructural cytochemistry by a new technique. *J. Histochem. Cytochem.*, **14**, 291–302.

Greenberg, L. J. (1962). Fluorometric measurement of alkaline phosphatase and aminopeptidase activities in the order of 10^{-14} mol. *Biochem. Biophys. Res. Commun.*, **9**, 430–435.

Groome, N. P. (1980). Superiority of ABTS over Trinder reagent as chromogen in highly sensitive peroxidase assays for enzyme linked immunoadsorbent assay. *J. Clin. Chem. Clin. Biochem.*, **18**, 345–349.

Guesdon, J.-L. *et al.* (1979). The use of avidin–biotin interaction in immunoenzymatic techniques. *J. Histochem. Cytochem.*, **27**, 1131–1139.

Guilbault, G. G. *et al.* (1967). A new substrate for fluorometric determination of oxidative enzymes. *Anal. Chem.*, **39**, 271.

Guilbault, G. G. *et al.* (1968a). Umbelliferone phosphate as a substrate for acid and alkaline phosphatase. *Anal. Lett.*, **1**, 333–345.

Guilbault, G. G. *et al.* (1968b). Homovanillic acid as a fluorometric substrate for oxidative enzymes. Analytical applications of the peroxidase, glucose oxidase, and xanthine oxidase systems. *Anal. Chem.*, **40**, 190–196.

Hahn, K. W. *et al.* (1990). Design and synthesis of a peptide having chymotrypsin-like esterase activity. *Science*, **248**, 1544–1547.

Hamill, W.H. *et al.* (1966). Equilibria in solutions of nonelectrolytes. In *Principles of Physical Chemistry*, 2nd edit., pp. 196–213, Prentice-Hall, Inc., Englewood Cliffs, NJ.

Harbron, S. *et al.* (1992). Amplified assay of alkaline phosphatase using flavin-adenine dinucleotide phosphate as substrate. *Anal. Biochem.*, **206**, 119–124.

Harbron, S. *et al.* (1993). A model for prosthetogenic enzyme amplification assays. *Biochim. Biophys. Acta*, **1161**, 311–316.

Harper, H. A. *et al.* (1979). General properties of enzymes. In *Review of Physiological Chemistry*, 17th edit., pp. 48–59, Lange Medical Publications, Los Altos, CA.

Harris, C. C. *et al.* (1979). Ultrasensitive enzymatic radioimmunoassay: application to detection of cholera toxin and rotavirus. *Proc. Natl. Acad. Sci. USA*, **76**, 5336–5339.

Harthill, J. E. *et al.* (1989). The structure of the N-linked oligosaccharides of horseradish peroxidase. *Proc. X Int. Symp. on Glycoconjugates, September 10–15, 1989, Jerusalem* (Sharon, N. *et al.*, eds), p.147, Kenes Ltd., Tel Aviv, Israel.

Harvey, E. N. (1940). *Living Light*, p.148, Princeton University Press, Princeton, NJ.

Haschke, R. H. and Friedhoff, J. M. (1978). Calcium-related properties of horseradish peroxidase. *Biochem. Biophys. Res. Commun.*, **80**, 1039–1042.

Hashinaka, K. *et al.* (1994). Conjugation of recombinant reverse transcriptase of HIV-1 to β-D-galactosidase from *Escherichia coli* for ultrasensitive enzyme immunoassay (immune complex transfer enzyme immunoassay) of anti-HIV-1 IgG. *J. Immunol. Methods*, **172**, 179–187.

Hayashi, Y. and Yamazaki, I. (1979). The oxidation-reduction potentials of compound I and compound II/ferric couples of horseradish peroxidases A_2 and C. *J. Biol. Chem.*, **254**, 9101–9106.

Henderson, D. R. *et al.* (1986). CEDIA™, a new homogeneous immunoassay system. *Clin. Chem.*, **32**, 1637–1641.

Henriksen, A. (1997). Structure–mechanism relationships in plant peroxidases. *Ph.D. Thesis*, University of Copenhagen.

Henriksen, A. *et al.* (1995). Crystallization and preliminary X-ray studies of recombinant horseradish peroxidase. *Acta Crystallogr.*, **D51**, 121–123.

Herrchen, M. and Legler, G. (1984). Identification of an essential carboxylate group at the active site of *lacZ* β-galactosidase from *Escherichia coli. Eur. J. Biochem.*, **138**, 527–531.

Herzog, V. and Fahimi, H. D. (1973). A new sensitive colorimetric assay for peroxidase using 3,3′-diaminobenzidine as hydrogen donor. *Anal. Biochem.*, **55**, 554–562.

Hill, H. D. *et al.* (1968). An automated fluorometric assay for alkaline phosphatase using 3-*O*-methylfluorescein phosphate. *Anal. Biochem.*, **24**, 9–17.

Ho, W. O. *et al.* (1995). Amperometric detection of alkaline phosphatase activity at a horseradish peroxidase enzyme electrode based on activated carbon: potential application to electrochemical immunoassay. *Biosensors Bioelectron.*, **10**, 683–691.

Hofmann, J. and Sernetz, M. (1984). Immobilized enzyme kinetics analyzed by flow-through microfluorimetry. Resorufin-β-D-galactopyranoside as a new fluorogenic substrate for β-galactosidase. *Anal. Chim. Acta*, **163**, 67–72.

Holland, V. R. *et al.* (1974). A safer substitute for benzidine in the detection of blood. *Tetrahedron*, **30**, 3299–3302.

Hollfelder, F. *et al.* (1996). Off-the-shelf proteins that rival tailor-made antibodies as catalysts. *Nature*, **383**, 60–63.

Holliger, P. and Hoogenboom, H. R. (1995). Artificial antibodies and enzymes: mimicking nature and beyond. *Trends Biotechnol.*, **13**, 7–9.

Horwitz, J. P. *et al.* (1966). Substrates for cytochemical demonstration of enzyme activity. II. Some dihalo-3-indolyl phosphates and sulfates. *J. Med. Chem.*, **9**, 447.

Hua, J.-C. *et al.* (1986). Partial sequencing of human adult, human fetal, and bovine intestinal alkaline phosphatases: comparison with the human placental and liver isozymes. *Proc. Natl. Acad. Sci. USA*, **83**, 2368–2372.

Huber, R. E. and Gaunt, M. T. (1982). The inhibition of β-galactosidase (*Escherichia coli*) by amino sugars and alcohols. *Can. J. Biochem.*, **60**, 608–612.

Huber, R. E. *et al.* (1994). The active site and mechanism of the β-galactosidase from *Escherichia coli. Int. J. Biochem.*, **26**, 309–318.

Ii, M. *et al.* (1993). Improved enzyme immunoassay for human basic fibroblast growth factor using a new enhanced chemiluminescence system. *Biochem. Biophys. Res. Commun.*, **193**, 540–545.

Ikegami, T. *et al.* (1994). Chemiluminescent enzyme immunoassay using an alkaline phosphatase label and an indolyl phosphate substrate. *Anal. Sci.*, **10**, 831–834.

Ishikawa, E. (1987). Development and clinical application of sensitive enzyme immunoassay for macromolecular antigens—a review. *Clin. Biochem.*, **20**, 375–385.

Itaya, K. and Ui, M. (1966). A new micromethod for the colorimetric determination of inorganic phosphate. *Clin. Chim. Acta*, **14**, 361–366.

Jacob, F. and Monod, J. (1961). Genetic regulatory mechanisms in the synthesis of proteins. *J. Mol. Biol.*, **3**, 318–356.

Jacobs, J. *et al.* (1987). Catalytic antibodies. *J. Am. Chem. Soc.*, **109**, 2174–2176.

Jacobson, R. H. *et al.* (1994). Three-dimensional structure of β-galactosidase from *E. coli. Nature*, **369**, 761–766.

Jagota, S. K. *et al.* (1981). Beta-galactosidase of *Streptococcus cremoris* H. *J. Food Sci.*, **46**, 161–168.

Jenkins, S. H. *et al.* (1988). Extending the detection limit of solid-phase electrochemical enzyme immunoassay to the attomole level. *Anal. Biochem.*, **168**, 292–299.

Johannson, A. *et al.* (1986). Enzyme amplification for immunoassays. Detection limit of one hundredth of an attomole. *J. Immunol. Methods*, **87**, 7–11.

Johnson, R. B., Jr. (1969). A new fluorometric method for the estimation or detection of total and fractionated alkaline phosphatase. *Clin. Chem.*, **15**, 108–123

Johnsson, K. *et al.* (1993). Synthesis, structure and activity of artificial, rationally designed catalytic polypeptides. *Nature*, **365**, 530–532.

Jörnvall, H. *et al.* (1978). Probe of β-galactosidase structure with iodoacetate. Differential reactivity of thiol groups in wild-type and mutant forms of β-galactosidase. *Biochemistry*, **17**, 5160–5164.

Kalnins, A. *et al.* (1983). Sequence of the *lacZ* gene of *Escherichia coli. EMBO J.*, **2**, 593–597.

Kaspar, P. and Khanna, P. L. (1991). Convenient therapeutic drug monitoring. *Eur. Clin. Lab.*, **10**, 24.

Kay, H. D. and Graham, W. R., Jr. (1933). Phosphorus compounds of milk. VI. The effect of heat on milk phosphatase. A simple method for distinguishing raw from pasteurised milk, raw from pasteurised cream, and butter made from raw cream from that made from pasteurised cream. *J. Dairy Res.*, **5**, 63–74.

Kazandjian, R. Z. *et al.* (1986). Enzymatic analyses in organic solvents. *Biotechnol. Bioeng.*, **28**, 417–421.

Keston, A. S. (1956). Specific colorimetric enzymatic analytical reagents for glucose. *Abstracts 129th ACS Meeting, Dallas, TX* (eds), pp. 31C–32C, American Chemical Society, Washington, DC.

Khanna, P. L. *et al.* (1989). A new homogeneous enzyme immunoassay using recombinant enzyme fragments. *Clin. Chim. Acta*, **185**, 231–240.

Khattab, M. and Pfleiderer, G. (1976). Alkaline phosphatase of human and calf small intestine. Purification and immunochemical characterization. *Hoppe-Seyler's Z. Physiol. Chem.*, **357**, 377–391.

Kim, E. E. and Wyckoff, H. W. (1989). Structure of alkaline phosphatases. *Clin. Chim. Acta*, **186**, 175–188.

Kim, E. E. and Wyckoff, H. W. (1991). Reaction mechanism of alkaline phosphatase based on crystal structures. Two-metal ion catalysis. *J. Mol. Biol.*, **218**, 449–464.

Kind, P. R. N. and King, E. J. (1954). Estimation of plasma phosphatase by determination of hydrolysed phenol with amino-antipyrine. *J. Clin. Pathol.*, **7**, 322–326.

King, E. J. and Armstrong, A. R. (1934). A convenient method for determining serum and bile phosphatase activity. *Can. Med. Assoc. J.*, **31**, 376–381.

Kjalke, M. *et al.* (1992). Comparison of structure and activities of peroxidases from *Coprinus cinereus, Coprinus macrorhizus* and *Arthromyces ramosus. Biochim. Biophys. Acta*, **1120**, 248–256.

Klein, C. *et al.* (1985). Glycoside von Resorufin-Derivaten, Verfahren zu deren Herstellung sowie deren Verwendung zur Bestimmung der Aktivität von Glycosidasen. *European Patent 156,347*, to Boehringer-Mannheim.

Klibanov, A. M. (1997). Why are enzymes less active in organic solvents than in water? *Trends Biotechnol.*, **15**, 97–101.

Koller, E. and Wolfbeis, O. S. (1985). Syntheses and spectral properties of longwave absorbing and fluorescing substrates for the direct and continuous kinetic assay of carboxyesterases, phosphatases, and sulfatases. *Monatsh. Chem.*, **116**, 65–75.

Kopetzki, E. *et al.* (1994). Enzymes in diagnostics: achievements and possibilities of recombinant DNA technology. *Clin. Chem.*, **40**, 688–704.

Krell, H.-W. (1991). Peroxidase: an important enzyme for diagnostic test kits. In *Biochemical, Molecular, and Physiological Aspects of Plant Peroxidases* (Lobarzewski, J. *et al.*, eds), pp. 469–478, University M. Curie-Sklodowska, Lublin, Poland/University of Geneva, Switzerland.

Kricka, L. J. and Ji, X. (1993). Catalytic antibody for the chemiluminescent luminol-peroxide reaction. In *Bioluminescence and Chemiluminescence. Status Report. Proceedings of the VIIth International Symposium on Bioluminescence and Chemiluminescence, Banff, March 1993* (Szalay, A. A. *et al.*, eds), pp. 286–290, John Wiley & Sons, New York.

Kricka, L. J. *et al.* (1996). Comparison of 5-hydroxy-2,3-dihydrophthalazine-1,4-dione and luminol as co-substrates for detection of horseradish peroxidase in enhanced chemiluminescent reactions. *J. Immunoassay*, **17**, 67–83.

Krishnan, R. *et al.* (1995). Fast amperometric immunoassay utilizing highly dispersed electrode material. *Anal. Lett.*, **28**, 2459–2474.

Krzakowa, M. (1991). Peroxidases in genetic and taxonomic investigations. In *Biochemical, Molecular, and Physiological Aspects of Plant Peroxidases* (Lobarzewski, J. *et al.*, eds), pp. 15–20, University M. Curie-Sklodowska, Lublin, Poland/University of Geneva, Switzerland.

Kuby, S. A. and Lardy, H. A. (1953). Purification and kinetics of β-D-galactosidase from *Escherichia coli*, strain K-12. *J. Am. Chem. Soc.*, **75**, 890–896.

Kuhr, M. *et al.* (1987) Phenolsulphonphthaleinyl-β-D-galactosidases and diagnostic agents containing them. *US Patent 4,668,622*, to Boehringer Mannheim

Kulys, J. *et al.* (1980). Electrochemical oxidation of catechol and *p*-aminophenol esters in the presence of hydrolases. *Bioelectrochem. Bioenerg.*, **7**, 11–24.

Land, D. B. and Jackim, E. (1966). A new fluorescence-yielding substrate for alkaline and acid phosphatase. *Anal. Biochem.*, **16**, 481–486.

Langley, K. E. *et al.* (1975). Molecular basis of β-galactosidase α-complementation. *Proc. Natl. Acad. Sci. USA*, **72**, 1254–1257.

Lazdunski, M. *et al.* (1965). Étude des Effets Électrostatiques sur le Mécanisme d'Action Catalytique de la Phosphatase Alcaline Intestinale de Veau. *Can. J. Chem.*, **43**, 2222–2235.

Lazdunski, M. *et al.* (1971). Flip-flop mechanisms in enzymology. A model: the alkaline phosphatase of *Escherichia coli. Eur. J. Biochem.*, **20**, 124–139.

Lew, J. Y. and Shannon, L. M. (1973). Incorporation of carbohydrate residues into peroxidase isoenzymes in horseradish roots. *Plant Physiol.*, **52**, 462–465.

Linossier, G. (1898). Contribution à l'Étude des Ferments Oxydants. Sur la Peroxydase de Pus. *C. R. Soc. Biol.*, **50**, 373–375.

Lojda, Z. *et al.* (1973). Synthetic substrates in the histochemical demonstration of intestinal disaccharidases. *Histochemie*, **34**, 361–369.

Lojda, Z. *et al.* (1979a). Peroxidase. *Enzyme Histochemistry. A Laboratory Manual*, pp. 248–256, Springer-Verlag, Berlin.

Lojda, Z. *et al.* (1979b). Acid β-galactosidase. *Enzyme Histochemistry. A Laboratory Manual*, pp. 145–154, Springer-Verlag, Berlin.

Lundin, A. and Hallander, L. O. B. (1986). Mechanisms of horseradish peroxidase catalyzed luminol reaction in the presence and absence of various enhancers. *J. Biolumin. Chemilumin.*, **1**, 186–192.

Luong, J. H. T. *et al.* (1993). Electrochemical preparation of 1,1′-dimethylferricinium from a water-soluble 1,1′-dimethylferrocene-2-hydroxypropyl-β-cyclodextrin complex and its applications in enzyme assay. *Anal. Biochem.*, **212**, 269–276.

Maeda, M. *et al.* (1989). Chemiluminescent assay of various enzyme activities and its application to enzyme immunoassays. *J. Biolumin. Chemilumin.*, **4**, 140–148.

Maeda, M. *et al.* (1991). Chemiluminescent assay of alkaline phosphatase using ascorbic acid 2-phosphate as substrate and its application to chemiluminescent enzyme immunoassays. In *Bioluminescence and Chemiluminescence. Current Status. Proceedings of the VIth International Symposium on Bioluminescence and Chemiluminescence, Cambridge, September 1990* (Stanley, P. E. and Kricka, L. J., eds), pp. 119–122, John Wiley & Sons, Chichester.

Maeda, M. *et al.* (1993a). Flow injection determination of glucose, bile acid and ATP using immobilized enzyme reactor and chemiluminescent assay of NAD(P)H. *J. Biolumin. Chemilumin.*, 8, 241–246.

Maeda, M. *et al.* (1993b). New chemiluminescence assay of alkaline phosphatase using lucigenin and its application to enzyme immunoassay. In *Bioluminescence and Chemiluminescence. Status Report. Proceedings of the VIIth International Symposium on Bioluminescence and Chemiluminescence, Banff, March 1993* (Szalay, A.A. *et al.*, eds), pp. 356–360, John Wiley & Sons, New York.

Maggio, E. T. (1980). Enzymes as immunochemical labels. In *Enzyme-Immunoassay* (Maggio, E. T., ed.), pp. 53–70, CRC Press, Boca Raton, FL.

Makiya, R. and Stigbrand, T. (1992). Placental alkaline phosphatase has a binding site for the human immunoglobulin-G Fc portion. *Eur. J. Biochem.*, **205**, 341–345.

Mandecki, W. *et al.* (1991). Alkaline phosphatase enzymes having improved specific activity for use in indicator reagents. *European Patent 441,252*, to Abbott Labs.

Margolin, A. L. (1996). Novel crystalline catalysts. *Trends Biotechnol.*, **14**, 223–230.

Másson, M. *et al.* (1995). Immunosensing with amperometric detection, using galactosidase as label and *p*-aminophenyl-β-D-galactopyranoside as substrate. *Anal. Chim. Acta*, **304**, 353–359.

Mattson, D. L. and Bellehumeur, T. G. (1996). Comparison of three chemiluminescent horseradish peroxidase substrates for immunoblotting. *Anal. Biochem.*, **240**, 306–308.

McClune, G. J. (1989). Use of phenols and anilines to increase the rate of peroxidase catalyzed oxidation of leuco dyes. *US Patent 4,828,983*, to Eastman Kodak Co.

McComb, R. B. (1979). *Alkaline Phosphatase*. Plenum Press, New York.

McComb, R. B. and Bowers, G. N., Jr. (1972). Study of optimum buffer conditions for measuring alkaline phosphatase activity in human serum. *Clin. Chem.*, **18**, 97–104.

McKenna, M. J. *et al.* (1979). Comparison of human alkaline phosphatase isoenzymes. Structural evidence for three protein classes. *Biochem. J.*, **181**, 67–73.

McManus, M. T. *et al.* (1988). Identification of a monoclonal antibody to abscission tissue that recognises xylose/fucose-containing N-linked oligosaccharides from higher plants. *Planta*, **175**, 506–512.

Mega, T. *et al.* (1990). Active-site-directed inactivation of *Aspergillus oryzae* β-galactosidase with β-D-galactopyranosyl methyl-*p*-nitrophenyltriazene. *J. Biochem.*, **107**, 641–644.

Menten, M. L. *et al.* (1944). A coupling histochemical azo dye test for alkaline phosphatase in the kidney. *J. Biol. Chem.*, **153**, 471–477.

Mesulam, M.-M. (1978). Tetramethyl benzidine for horseradish peroxidase neurohistochemistry: a non-carcinogenic blue reaction-product with superior sensitivity for visualizing neural afferents and efferents. *J. Histochem. Cytochem.*, **26**, 106–117.

Millán, J. L. (1992). Alkaline phosphatase as a reporter of cancerous transformation. *Clin. Chim. Acta*, **209**, 123–129.

Miller, E. C. *et al.* (1961). *N*-Hydroxy-2-acetylaminofluorene: a metabolite of 2-acetylaminofluorene with increased carcinogenic activity in the rat. *Cancer Res.*, **21**, 815–824.

Miska, W. and Geiger, R. (1987). Synthesis and characterization of luciferin derivatives for use in bioluminescence enhanced enzyme immunoassays. *J. Clin. Chem. Clin. Biochem.*, **25**, 23–30.

Misra, H. P. and Squatrito, P. M. (1982). The role of superoxide anion in peroxidase-catalyzed chemiluminescence of luminol. *Arch. Biochem. Biophys.*, **215**, 59–65.

Mohan, A. G. and Rauhut, M. M. (1984). Aqueous chemical light. A new horizon. *Naval Res. Rev.* 16–21.

Morita, Y. *et al.* (1991). Primary and crystal structures of horseradish peroxidase isozyme E5. In *Biochemical, Molecular, and Physiological Aspects of Plant Peroxidases* (Lobarzewski, J. *et al.*, eds), pp. 81–88, University

M. Curie-Sklodowska, Lublin, Poland/University of Geneva, Switzerland.

Morrell, J. I. *et al.* (1981). Comparison of horseradish peroxidase visualization methods: quantitative results and further technical specifics. *J. Histochem. Cytochem.*, **29**, 903–916.

Moss, D. W. (1960). An improved spectrofluorimeter for biochemical analysis. *Clin. Chim. Acta*, **5**, 283–288.

Moss, D. W. and King, E. J. (1962). Some properties of alkaline phosphatase iso-enzymes. *Biochem. J.*, **85**, 19P–20P.

Murachi, T. (1981). Enzyme reactions. In *Enzyme Immunoassay* (Ishikawa, E. *et al.*, eds), pp. 5–13, Igaku Shoin, Tokyo.

Nielsen, K. (1995). Stability of freeze dried horseradish peroxidase conjugated monoclonal antibodies used in diagnostic serology. *J. Immunoassay*, **16**, 183–197.

Niwa, O. *et al.* (1993). Small-volume voltammetric detection of 4-aminophenol with interdigitated array electrodes and its application to electrochemical enzyme immunoassay. *Anal. Chem.*, **65**, 1559–1563.

Ogawa, S. *et al.* (1979). Calcium binding by horseradish peroxidase C and the heme environmental structure. *Biochem. Biophys. Res. Commun.*, **90**, 674–678.

Ohmori, Y. (1937). Über die Phosphomonoesterase. *Enzymologia*, **4**, 217–231.

Ortiz de Montellano, P. R. (1992). Catalytic sites of hemoprotein peroxidases. *Annu. Rev. Pharmacol. Toxicol.*, **32**, 89–107.

Ortiz de Montellano, P. R. *et al.* (1988). Mechanism-based inactivation of horseradish peroxidase by sodium azide. Formation of meso-azido protoporphyrin IX. *Biochemistry*, **27**, 5470–5476.

Pappa, H. S. and Cass, A. E. (1993). A step towards understanding the folding mechanism of horseradish peroxidase. Tryptophan fluorescence and circular dichroism equilibrium studies. *Eur. J. Biochem.*, **212**, 227–235.

Paul, K. G. (1958). Die Isolierung von Meerrettichperoxydase. *Acta Chem. Scand.*, **12**, 1312–1318.

Paul, K. G. (1986). Peroxidases: historical background. In *Molecular and Physiological Aspects of Plant Peroxidases* (Greppin, H. *et al.*, eds), pp. 1–14, University of Geneva, Geneva.

Pauly, H. E. and Schwarz, H. (1990). Agent for the determination of peroxidase activity, with stabilizer, a process for its preparation and its use. *US Patent 4,891,314*, to Behringwerke Aktiengesellschaft.

Pearson, B. *et al.* (1963). A comparative study of a series of new indolyl compounds to localize β-galactosidase in tissues. *Lab. Invest.*, **12**, 1249–1259.

Penel, C. (1986). The role of calcium in the control of peroxidase activity. In *Molecular and Physiological Aspects of Plant Peroxidases* (Greppin, H. *et al.*, eds), pp. 155–164. University of Geneva, Geneva.

Penel, C. (1991). Ca^{2+}-mediated binding of peroxidases to cell structures. In *Biochemical, Molecular, and Physiological Aspects of Plant Peroxidases* (Lobarzewski, J. *et al.*, eds), pp. 225–236, University M. Curie-Sklodowska, Lublin, Poland/University of Geneva, Switzerland.

Peuravuori, H. and Korpela, T. (1993). Pyrophosphatase-based enzyme-linked immunosorbent assay of total IgE in serum. *Clin. Chem.*, **39**, 846–851.

Planche, L. A. (1810). Note sur la Sophistication de la Résine de Jalap et sur les Moyens de la Reconnaître. *Bull. Pharm.*, **2**, 578–580.

Poulos, T. L. (1993). Peroxidases. *Curr. Opin. Biotechnol.*, **4**, 484–489.

Poulos, T. L. *et al.* (1980). The crystal structure of cytochrome *c* peroxidase. *J. Biol. Chem.*, **255**, 575–580.

Poulos, T. L. *et al.* (1993). Crystallographic refinement of lignin peroxidase at 2 Å. *J. Biol. Chem.*, **268**, 4429–4440.

Porstmann, B. *et al.* (1985). Which of the commonly used marker enzymes gives the best results in colorimetric and fluorimetric enzyme immunoassays: horseradish peroxidase, alkaline phosphatase or β-galactosidase? *J. Immunol. Methods*, **79**, 27–37.

Quiles, R. *et al.* (1993). Automated enzymatic determination of sodium in serum. *Clin. Chem.*, **39**, 500–503.

Reimann, L. and Schonbaum, G. R. (1978). Purification of plant peroxidases by affinity chromatography. *Methods Enzymol.*, **52**, 514–521.

Rennke, H. G. and Venkatachalam, M. A. (1979). Chemical modification of horseradish peroxidase. Preparation and characterization of tracer enzymes with different isoelectric points. *J. Histochem. Cytochem.*, **27**, 1352–1353.

Richard, J. P. (1995). On the mechanism of galactosyl transfer catalyzed by β-galactosidase from *E. coli*. *Abstr. 209th Am. Chem. Soc. Natl. Meeting*, p. ORGN 139, American Chemical Society, Washington, DC.

Richard, J. P. *et al.* (1996). Structure–reactivity relationships for β-galactosidase (*Escherichia coli*, *lac Z*). 4. Mechanism for reaction of nucleophiles with the galactosyl–enzyme intermediates of E461G and E461Q β-galactosidases. *Biochemistry*, **35**, 12387–12401.

Ring, M. and Huber, R. E. (1993). The properties of β-galactosidases (*Escherichia coli*) with halogenated tyrosines. *Biochem. Cell Biol.*, **71**, 127–132.

Rodriguez-Lopez, J. N. *et al.* (1996). Role of arginine 38 in horseradish peroxidase. A critical residue for substrate binding and catalysis. *J. Biol. Chem.*, **271**, 4023–4030.

Rotman, B. (1961). Measurement of activity of single molecules of β-D-galactosidase. *Proc. Natl. Acad. Sci. USA*, **47**, 1981–1991.

Rotman, B. *et al.* (1963). Fluorogenic substrates for β-D-galactosidases and phosphatases derived from fluorescein (3,6-dihydroxyfluoran) and its monomethyl ether. *Proc. Natl. Acad. Sci. USA*, **50**, 1–6.

Roy, A. V. (1970). Rapid method for determining alkaline phosphatase activity in serum with thymolphthalein monophosphate. *Clin. Chem.*, **16**, 431–436.

Ryan, O. *et al.* (1994). Thermostabilized chemical derivatives of horseradish peroxidase. *Enzyme Microb. Technol.*, **16**, 501–505.

Sanes, J. R. *et al.* (1986). Use of a recombinant retrovirus to study post-implantation cell lineage in mouse embryos. *EMBO J.*, **5**, 3133–3142.

Schaap, A. P. (1988). Chemical and enzymatic triggering of 1,2-dioxetanes. *Photochem. Photobiol.*, **47**, (Suppl.), 50S.

Schaap, A. P. (1992). Method and composition providing enhanced chemiluminescence from 1,2-dioxetanes. *European Patent 518,387*, to The Board of Governors of Wayne State University.

Schaap, A. P. *et al.* (1987). Chemical and enzymatic triggering of 1,2-dioxetanes. 3: Alkaline phosphatase-catalyzed chemiluminescence from an aryl phosophate-substituted dioxetane. *Tetrahedron Lett.*, **28**, 1159–1162.

Schalkwijk, J. *et al.* (1985). Cationization of catalase, peroxidase, and superoxide dismutase. Effect of improved intraarticular retention on experimental arthritis in mice. *J. Clin. Invest.*, **76**, 198–205.

Schall, R. F., Jr. *et al.* (1978). A sensitive manual enzyme immunoassay for thyroxine. *Clin. Chem.*, **24**, 1801–1804.

Schroeder, H. R. and Yeager, F. M. (1978). Chemiluminescence yields and detection limits of some isoluminol derivatives in various oxidation systems. *Anal. Chem.*, **50**. 1114–1120.

Schulte, T. *et al.* (1989). The development of a novel immunoassay amplification system and its use in viral detection. *Ann. Biol. Clin.*, **47**, 535–540.

Schwartz, J. H. and Lipmann, F. (1961). Phosphate incorporation into alkaline phosphatase of *E. coli. Proc. Natl. Acad. Sci. USA*, **47**, 1996–2005.

Schwartz, J. H. *et al.* (1963). The amino acid sequence of a tetrapeptide containing the reactive serine in *E. coli* alkaline phosphatase. *Proc. Natl. Acad. Sci. USA*, **49**, 722–729.

Self, C. H. (1985). Enzyme amplification—a general method applied to provide an immunoassisted assay for placental alkaline phosphatase. *J. Immunol. Methods*, **76**, 389–393.

Seligman, A. M. *et al.* (1954). The histochemical demonstration of alkaline phosphatase with indoxyl phosphate. *J. Histochem. Cytochem.*, **2**, 441–442.

Selwood, T. and Sinnot, M. L. (1990). The solvent–isotope effect study of proton transfer during catalysis by *Escherichia coli (lacZ)* β-galactosidase. *Biochem. J.*, **268**, 317–323.

Senter, P. D. *et al.* (1988). Anti-tumor effects of antibody–alkaline phosphatase conjugates in combination with etoposide phosphate. *Proc. Natl. Acad. Sci. USA*, **85**, 4842–4846.

Shalev, A. *et al.* (1980). Detection of attograms of antigen by a high-sensitivity enzyme-linked immunoabsorbent assay (HS-ELISA) using a fluorogenic substrate. *J. Immunol. Methods*, **38**, 125–139.

Shannon, L. M. *et al.* (1966). Peroxidase isozymes from horseradish roots. I. Isolation and physical properties. *J. Biol. Chem.*, **241**, 2166–2172.

Sheehan, J. C. *et al.* (1966). Synthetic peptide models of enzyme active sites. III. Stereoselective esterase models. *J. Am. Chem. Soc.*, **88**, 3455–3456.

Skorobogat'ko, O. V. *et al.* (1994). Laccase as a new enzymatic label for enzyme immunoassay. *Anal. Lett.*, **27**, 2997–3012.

Slepnev, V. I. *et al.* (1995). Fatty acid acylated peroxidase as a model for the study of interactions of hydrophobically-modified proteins with mammalian cells. *Bioconjugate Chem.*, **6**, 608–615.

Smith, A. T. *et al.* (1990). Expression of a synthetic gene for horseradish peroxidase C in *Escherichia coli* and folding and activation of the recombinant enzyme with Ca^{2+} and heme. *J. Biol. Chem.*, **265**, 13335–13343.

Smith, A. T. *et al.* (1992a). Probing the mechanism of horseradish peroxidase by site-directed mutagenesis. *Biochem. Soc. Trans.*, **20**, 340–345.

Smith, A. T. *et al.* (1992b). Characterisation of a haem active-site mutant of horseradish peroxidase, Phe41→Val, with altered reactivity towards hydrogen peroxide and reducing substrates. *Eur. J. Biochem.*, **207**, 507–519.

Smith, I. *et al.* (1968). Separation of human tissue alkaline phosphatases by electrophoresis on acrylamide disc gels. *Clin. Chim. Acta*, **19**, 499–505.

Steers, E., Jr *et al.* (1965). Evidence for nonidentical chains in the β-galactosidase of *Escherichia coli* K12. *J. Biol. Chem.*, **240**, 2478–2484.

Stinson, R. A. and Seargeant, L. E. (1981). Comparative studies of pure alkaline phosphatases from five human tissues. *Clin. Chim. Acta*, **110**, 261–272.

Stoffer, B. *et al.* (1991). Amino acid sequence of acidic horseradish peroxidase HRP A. In *Biochemical, Molecular, and Physiological Aspects of Plant Peroxidases* (Lobarzewski, J. *et al.*, eds), pp. 43–47, University M. Curie-Sklodowska, Lublin, Poland/University of Geneva, Switzerland.

Stoll, A. and Seebeck, E. (1948). Die Isolierung von Sinigrin als genuine, krystallisierte Muttersubstanz des Meerrettichöls. *Helv. Chim. Acta*, **31**, 1432–1434.

Straus, W. (1964). Factors affecting the cytochemical reaction of peroxidase with benzidine and the stability of the blue reaction product. *J. Histochem. Cytochem.*, **12**, 462–469.

Suzuki, S. *et al.* (1970). The chemical modification of biopolymers. III. The introduction of SH groups into Taka-amylase A. *Bull. Chem. Soc. Jpn.*, **43**, 3849–3852.

Tams, J. W. and Welinder, K. G. (1991). Deglycosylation of horseradish peroxidase. In *Biochemical, Molecular, and Physiological Aspects of Plant Peroxidases* (Lobarzewski, J. *et al.*, eds), pp. 111–113, University M. Curie-Sklodowska, Lublin, Poland/University of Geneva, Switzerland.

Tams, J. W. and Welinder, K. G. (1995). Mild chemical deglycosylation of horseradish peroxidase yields a fully active, homogeneous enzyme. *Anal. Biochem.*, **228**, 48–55.

Tanaka, K. and Ishikawa, E. (1986). A highly sensitive bioluminescent assay of β-D-galactosidase from *Escherichia coli* using 2-nitrophenyl-β-D-galactopyranoside as a substrate. *Anal. Lett.*, **19**, 433–444.

Ternaux, J.-P. and Chamoin, M.-C. (1994). Enhanced chemiluminescent assays for acetylcholine. *J. Biolumin. Chemilumin.*, **9**, 65–72.

Theorell, H. (1942). The preparation and some properties of crystalline horse-radish peroxidase. *Ark. Kemi Mineral. Geol.*, **16A**, 1–11.

Thompson, R. Q. *et al.* (1991). Comparison of methods for following alkaline phosphatase catalysis: spectrophotometric versus amperometric detection. *Anal. Biochem.*, **192**, 90–95.

Thorpe, G. H. G. and Kricka, L. J. (1986). Enhanced chemiluminescent reactions catalyzed by horseradish peroxidase. *Methods Enzymol.*, **133**, 331–353.

Tibbitts, T. T. *et al.* (1994). Kinetics and crystal structure of a mutant *Escherichia coli* alkaline phosphatase (Asp-369→Asn): a mechanism involving one zinc per active site. *Protein Sci.*, **3**, 2005–2114.

Tramontano, A. *et al.* (1986). Catalytic antibodies. *Science*, **234**, 1566–1570.

Tsou, K. C. and Su, H. C. F. (1965). A new colorimetric method for the determination of alkaline phosphatase with indoxyl phosphate. *Anal. Biochem.*, **11**, 54–64.

Ugarova, N. N. *et al.* (1978). Chemical modification of the ϵ-NH_2 groups of lysine residues in horseradish peroxidase. Accessibility of these groups to various modifying agents. *Biokhimiya (English Edn)*, **43**, 629–632.

Ugarova, N. N. *et al.* (1979a). Chemical modification of the ϵ-NH_2 groups of lysine in horseradish peroxidase. Its influence on the catalytic structure of the enzyme. *Biokhimiya (English Edn)*, **43**, 983–989.

Ugarova, N. N. *et al.* (1979b). Chemical modification of the ϵ-NH_2 groups of lysine residues in horseradish peroxidase and its effect on the catalytic properties and thermostability of the enzyme. *Biochim. Biophys. Acta*, **570**, 31–42.

Ullman, E. F. and Maggio, E. T. (1980). Principles of homogeneous enzyme-immunoassay. In *Enzyme-Immunoassay* (Maggio, E. T., ed.), pp. 105–134, CRC Press, Inc., Boca Raton, Fl.

Van Belle, H. (1970). New and sensitive reaction for automatic determination of inorganic phosphate and its application to serum. *Anal. Biochem.*, **33**, 132–142.

Vaughan, A. *et al.* (1971). Fluorometric methods for analysis of acid and alkaline phosphatase. *Anal. Chem.*, **43**, 721–724.

Veitch, N. C. *et al.* (1995). Solution characterisation by NMR spectroscopy of two horseradish peroxidase isoenzyme C mutants with alanine replacing either Phe142 or Phe143. *Eur. J. Biochem.*, **233**, 650–658.

Vlasenko, S. B. *et al.* (1989). Investigation of the catalytic mechanism of enhanced chemiluminescence. Analytical applications of this reaction. *J. Biolumin. Chemilumin.*, **4**, 164–176.

Voogd, C. E. *et al.* (1980). On the mutagenic action of some enzyme immunoassay substrates. *J. Immunol. Methods*, **36**, 55–61.

Wallenfels, K. *et al.* (1960). Untersuchungen über Milchzuckerspaltende Enzyme. VII. Die Spezifität der β-Galaktosidase von E. coli ML 309. *Biochem. Z.*, **333**, 209–225.

Weavers, E. A. and Powell, M. J. (1989). Improvements relating to assay reagents. *UK Patent 2,208,712*, to Boots-Celltech Diagnostics Ltd.

Weber, K. *et al.* (1964). Über die Art der Bindung zwischen den Untereinheiten im Molekül der β-Galaktosidase aus *E. coli. Biochem. Z.*, **339**, 498–500.

Weissig, H. *et al.* (1993). Cloning and expression of the bovine intestinal alkaline phosphatase gene: biochemical characterization of the recombinant enzyme. *Biochem. J.*, **290**, 503–508.

Welinder, K. G. (1985). Plant peroxidases: their primary, secondary, and tertiary structures, and relation to cytochrome *c* peroxidase. *Eur. J. Biochem.*, **151**, 497–504.

Welinder, K. G. (1991). The plant peroxidase superfamily. In *Biochemical, Molecular, and Physiological Aspects of Plant Peroxidases* (Lobarzewski, J. *et al.*, eds), pp. 3–13, University M. Curie-Sklodowska, Lublin, Poland/University of Geneva, Switzerland.

Welinder, K.G. (1992). Superfamily of plant, fungal and bacterial peroxidases. *Curr. Opin. Struct. Biol.*, **2**, 388–393.

Whitehead, T. P. *et al.* (1984). Enhanced luminescent or luminometric assay. *European Patent 116,454*, to the Government of the UK.

Wilkinson, J. H. and Vodden, A. V. (1966). Phenolphthalein monophosphate as a substrate for serum alkaline phosphatase. An appraisal. *Clin. Chem.*, **12**, 701–708.

Wolf, P. L. *et al.* (1968). A new histochemical stain for neutrophilic leukocyte alkaline phosphatase. *Enzymologia*, **35**, 154–156.

Wolters, G. *et al.* (1976). Solid-phase enzyme-immunoassay for detection of hepatitis B surface antigen. *J. Clin. Pathol.*, **29**, 873–879.

Xu, X. and Kantrowitz, E. R. (1993). Binding of magnesium in a mutant *Escherichia coli* alkaline phosphatase changes the rate-determining step in the reaction mechanism. *Biochemistry*, **32**, 10683–10691.

Xu, X. *et al.* (1994). Probing the role of histidine-372 in zinc binding and the catalytic mechanism of *Escherichia coli* alkaline phosphatase by site-specific mutagenesis. *Biochemistry*, **33**, 2279–2284.

Yalçin, A. S. *et al.* (1993). Simple colorimetric method for the determination of peroxide. *Clin. Chem.*, **39**, 2534–2535.

Yamazaki, I. and Nakajima, H. (1986). Physico-chemical comparison between horseradish peroxidases A and C. In *Molecular and Physiological Aspects of Plant Peroxidases* (Greppin, H. *et al.*, eds), pp. 71–84, University of Geneva, Geneva.

Yarborough, D. J. *et al.* (1967). Histochemistry of macrophage hydrolases. *J. Reticuloendothel. Soc.*, **4**, 390–408.

Yolken, R. H. *et al.* (1984). The use of beta-lactamase in enzyme immunoassays for detection of microbial antigens. *J. Immunol. Methods*, **73**, 109–123.

Young, P. R. (1989). Enhancement of immunoblot staining using a mixed chromogenic substrate. *J. Immunol. Methods*, **121**, 295–296.

Zueva, N. N. *et al.* (1993) Properties, preparation, and practical use of alkaline phosphatase. *Biochemistry (Russia)*, **58**, 702–712.

Chapter 4

Antibodies

Alastair H. Dent and Mohammed Aslam

Research and Development Department, Ortho-Clinical Diagnostics, Pollards Wood Laboratories, Nightingales Lane, Chalfont St. Giles HP8 4SP, UK.

4.1 Introduction: The Immune Response

Along with enzymes, the group of proteins which crop up in the greatest number of coupled applications are the antibodies or immunoglobulins. This popularity has arisen from their ability to bind with extraordinary specificity to other molecular structures, ranging from small organic compounds, through proteins, to larger components of living organisms.

This property of immunoglobulins has been evolved by animal species as a defence mechanism against invasion by harmful microorganisms, a constant threat to the existence and well-being of living creatures. Immunoglobulins are found in vertebrates and only in vertebrates; they are produced in some form by the lowliest members of this class, and represent an enhancement of the more primitive defence mechanisms relied upon by lower animal forms. In order to understand the function of antibodies, it is worth digressing briefly to examine the more primitive defence mechanisms, which still have a crucial function even in the highest mammals.

4.1.1 The innate immune response

These lower, innate, mechanisms include physical prevention of entry by the skin (not just as a barrier but because secretions such as sweat create an acidic environment), by mucus (containing the bactericidal enzyme lysozyme which breaks cell walls) which protects exposed epithelial cells, and by the washing action of tears, saliva, urine, etc. Even the presence of other harmless microorganisms may form part of the body's defences; these may compete for nutrients, or again generate an acidic environment.

If a microorganism manages to overcome these hurdles and enter the body, it is faced by a range of mechanisms by which it can be disabled. In plasma, interferons react to viral infection by rendering neighbouring cells incapable of supporting viral replication, and promoting the destruction of the infected cell by natural killer cells.

C-reactive protein (CRP), fibrinogen and caeruloplasmin are examples of acute phase proteins, which increase dramatically after infection or injury, and are involved in these processes; CRP for example binds to microorganism membranes and the resulting complex activates the complement sequence (see below). This ultimately results in the destruction of the organism, for example through phagocytosis by macrophages or polymorphs.

In order for phagocytosis to take place, the microorganism must bind to the surface of one of these phagocytes. Once this event has occurred the destruction process may begin, ingestion taking place to form a phagosome containing the foreign species. The next stage is fusion with lysosomal granules which bring a battery of harmful agents into action, including superoxide (generated in the 'oxygen' or 'respiratory burst' from glucose and NADP) and other oxygenous radicals, hydrogen peroxide, hypochlorite from the action of myeloperoxidase and H_2O_2 on chloride, lysozyme, and membrane-damaging proteins such as phagocytin.

Some microorganisms will provoke phagocytosis by binding directly to the cell surface, but with the huge range of organisms with which the phagocyte may be faced this is not a sufficiently reliable approach. Consequently the complement pathways offer a variety of routes for dealing with diverse organisms in the bloodstream. One of the end results may be to increase the efficiency of phagocytosis, while other complement components such as C8 and C9 are microbicidal in themselves.

A key stage of these pathways is the fragmentation, catalysed by a convertase, of C3 to the daughter fragments C3a and C3b. C3a promotes two activities of mast cells in the skin: (1) the release of a chemotactic agent which attracts polymorphs to the site, and (2) the release of histamine which, amongst other effects, amplifies this process by increasing vascular permeability.

C3b meanwhile remains on the surface of the microorganism, where it is recognised and bound by receptor sites on the polymorph. Not only does this allow binding to phagocytes of a wider range of microorganisms, but the specific C3b/receptor interaction is stronger than the non-specific attractions between the phagocyte and a native microorganism.

Production of the C3 convertases referred to above can be initiated by the influence of cell-wall components of the microorganisms themselves, but this route, referred to as the alternative pathway, is relatively non-specific. Higher organisms have developed mechanisms for invoking the sequence of complement fixation and phagocytosis in a more specific way, that can deal more efficiently with a wider range of invading microorganisms or harmful substances. These involve the second route to a C3 convertase, the so-called classical pathway, and it is this mechanism that involves antibodies.

4.1.2 The adaptive immune response

In order to bind successfully to *any* harmful organisms or substances that may invade the system, it is not possible to rely on the non-specific binding capabilities of the complement components. A wide range of binding molecules with prodigious variations in their 'target' structure needs to be available; not only that but these molecules should ideally be available immediately to deal with the invader, rather than having to be 'designed' after the event.

Having bound successfully they should also have the ability to disable the invader, either by an independent mechanism or by invoking those already provided by the innate response—the complement fixation process and subsequent phagocytosis. It is also a desirable function to be able to deal with particular efficiency with repeated challenge by a particular organism, thus focusing the energies of the response on the most commonly encountered targets. These properties are fulfilled by the adaptive or acquired immune response, fundamental to which is the production of antibodies by certain categories of lymphocyte cells.

4.1.2.1 Antibodies

The term antibody was coined by Emil von Behring in 1890[1] to describe the antitoxin proteins, discovered by himself and Shibasaburo Kitasato, which appeared in the bloodstream after infections. Behring's subsequent development of a diphtheria antiserum resulted in him receiving the first Nobel prize for Medicine in 1901.

One of the biggest unanswered questions in immunology in the first half of the twentieth century was how an animal could generate antibodies of sufficient diversity to deal with all the different target structures that threatened it. This requirement for a staggeringly wide spectrum of specificity led to the template theory (originally postulated by Felix Haurowitz in the 1930s), calling for an 'unformed' antibody molecule to mould itself around a challenging antigen to yield the appropriate binding characteristics. These held sway until around the early 1960s.

In the 1950s an alternative model was suggested: that the fully developed organism is equipped with the capability to produce specific antibodies covering all the different binding characteristics available, and that challenge by an antigen selectively promotes production of the relevant type. Later theories proposed that each lymphocyte bore a single type of antibody on its surface, and that the binding of the appropriate antigen was the trigger for proliferation. These were borne out in subsequent investigation and they now form the basis of current understanding.

The genetic origin of antibody diversity is discussed in Section 4.4. As a result of this diversity, the physicochemical properties of immunoglobulins are uniquely variable: unlike the majority of serum proteins, the electrophoretic mobility of immunoglobulins in any individual is spread over an extensive electrophoretic range (from the cathode virtually to the anode for IgG), and reflects great variation in primary sequence.

[1] The history of the development of immunology over the past century has been concisely reviewed by Nossal (1993).

Not all regions of the molecule are prone to this variation, however; the more fixed regions possess effectors for defence mechanisms which are invoked once the specific invader has been bound. There is a region capable of binding the complement component C1q which initiates the classical pathway to another C3 convertase, and the C3 fragmentation processes described above subsequently lead to phagocytosis (and the other defensive processes carried out by complement components); also there is a site which can bind directly to phagocytes and promote their activity. These binding capabilities provide two routes by which antibodies can facilitate the phagocytosis of unwelcome invaders, strengthening the innate system described above.

Antibodies offer a number of other enhancements to the innate system. Some microorganisms are resistant to the phagocytic process because their surface structure is not recognised by the alternative complement pathway, for example because it is coated by disguising features such as a capsule of carbohydrate molecules. Antibodies can recognise some of these features, and their binding will initiate the classical pathway. Other organisms are equipped to interfere in some way with the phagocytic process itself. Streptococci for example contain an inhibitory protein termed the M-protein, while the capsules of the anthrax bacillus release a powerfully debilitating toxin. Antibodies can neutralise these components and allow phagocytosis to proceed[2].

The cells which are responsible for antibody production develop from the astonishingly versatile haematopoietic stem cells, which make up a small fraction of the cells in bone marrow (around 1 in 2000 in mice). These possess the ability to generate the entire repertoire of blood and immune cells. The stem cells which are destined to an immune role differentiate *via* two separate routes during foetal development. These yield T-lymphocytes (or T-cells), whose development is controlled by the thymus, a small organ in the chest, and B-lymphocytes or B-cells, whose early foetal development in humans occurs in the liver but which subsequently migrate to the bone marrow[3].

A freshly-created B-lymphocyte carries copies of its own specific antibodies on its surface. It will not begin to create new antibody molecules until it is required to do so; a system which required large quantities of millions of different antibodies to be permanently

[2] Antibodies are also prone to interferences too: *Staphylococcus aureus*, for example, produces Protein A to bind to immunoglobulin receptors and prevent their activation of phagocyte activity. These proteins are now in widespread use for the purification of antibodies.

[3] In birds these cells originate from a lymphoid organ termed the Bursa of Fabricius, hence the 'B' designation.

circulating in the bloodstream was not surprisingly passed over as an evolutionary option. Many B-cell types will go through the entire life of the organism without being required, staying permanently in the 'resting' state. Therefore although the specific antibody for a 'new' invader (and the cell latently capable of creating it) already exist when it enters the system, a mechanism is required to 'turn on' the B-cell to create working quantities of the antibody.

4.1.2.2 Mechanisms for B-cell activation and function

As proposed by Jerne it is the appearance of the antigen itself which is central to this event, in a gross simplification being 'spotted' by the resting B-lymphocytes which recognise it best; this recognition instructs the B-cells to start generating antibodies to deal with the intruder. As might be expected the real chain of events is rather more involved.

As will be discussed below, the immune response cannot be initiated by the presentation of a single antibody-specific site (determinant) to the appropriate specific lymphocyte. Most antigens are termed thymus-dependent, as the immune response to them operates *via* a more complex mechanism which involves T-cells.

For such antigens, the first key stage of the response is their capture by a dendritic macrophage, which unlike other macrophages is not primarily destructive. Instead it presents the antigen determinants on its surface in combination with other proteins (the Class II proteins of the major histocompatibility complex, MHC). This combination of antigen and MHC Class II proteins is recognised by determinant-specific T-lymphocytes of the class termed helper or CD4 T-cells; the latter definition refers to the surface receptor protein termed CD4 which these cells bear. These T-cells are only capable of recognising an antigen fragment if it is presented in a groove in the MHC protein structure.

Binding of the CD4 T-cell to the antigen/MHC complex provokes the release of interleukin-1 (IL-1) from the macrophage; the interleukin-activated T-cells then have the capability of releasing factors which will bind to receptors on the B-cell. Helper T-cells are also changed by IL-1 to lymphoblasts which have a receptor for interleukin-2 (IL-2), produced by another subset of T-cells. This leads to proliferation of the helper cells, amplifying the response to antigen.

Once a determinant-specific B-cell has been activated by the thymus-dependent or independent mechanisms outlined above, it generates receptors for a B-cell growth factor, produced by activated T-helpers; this promotes rapid proliferation. A second T-helper-derived factor (B-cell maturation factor) then switches on antibody production, and the resulting mature, immunoglobulin-secreting B-lymphocyte is known as a plasma cell or plasmablast. These cells are prodigiously productive; a single cell is capable of generating ten million immunoglobulin molecules per hour.

Obviously there comes a time when antibody secretion is no longer required, and the system responds to falling antigen concentration by a number of mechanisms which curtail production of new lymphocytes, and induce apoptosis in many of those which already exist. These involve a number of different T-cell types of the class known as T-suppressors; some of these are antigen-specific, but they recruit the assistance of other, non-specific T-suppressors in their function.

Not all of the B-cells produced by the clonal proliferation process are switched on for antibody production; some remain dormant as memory cells and form the basis of the secondary response, the intensified mechanism which is provoked by a repeat challenge by the same antigen. They remain in the system, often 'returning' to the bone marrow, and offer a responsive pool in the event of re-infection; not only are they present in greater numbers than the 'virgin' B-lymphocytes, but are also at a more advanced stage of development. The combination of these factors provides a rapid, intense response which is often enough to disable an infection before it can take hold in the system.

4.1.2.3 The requirement for multiple determinants for B-cell activation

The lymphocyte activation mechanism outlined above appears to be triggered by the need for a number of receptors on the B-cell to be bound simultaneously. Some antigens have repeating antibody-specific sites or determinants, and a number of these can bind simultaneously to the immunoglobulins on the cell surface – thereby fulfilling the triggering requirement (Figure 4.1(a)). Other antigens, while only having a single determinant for the B-cell's immunoglobulin, also possess components which can bind to other surface receptors; again the triggering requirement is satisfied (Figure 4.1(b)). These two categories of antigen which can directly activate B-cells are termed thymus-independent.

However the bulk of antigens do not possess these qualities, and as noted above, are unable to activate B-cells without the assistance of T-lymphocytes which

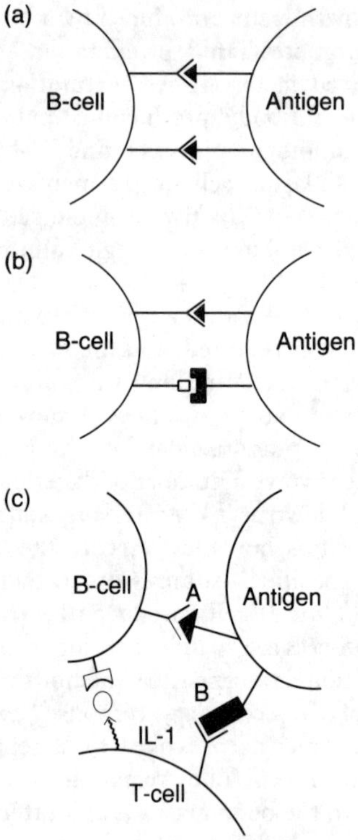

Figure 4.1 At least two receptors must be bound for activation of a B-cell; this can occur in isolation if the antigen possesses repeating determinants (a) or a single determinant plus another binding group (b). In most cases, however, activation is mediated by a T-cell with different specificity (c).

also bear specific antigen-binding groups. These are the thymus-dependent antigens.

A single determinant on such an antigen cannot simultaneously bind to the specific B-cell immunoglobulin, and to the similar receptor on an appropriate T-cell; therefore a cooperative system operates *via* other determinants on the antigen (Figure 4.1(c)). Note that although for simplicity the antigen is shown here in its intact form, it has already in fact been 'processed' by the mechanism described earlier so that the determinants are presented as complexes with MHC proteins.

In the example shown the antigen possesses two different determinants D_1 and D_2. The D_1-determinant may be bound by a D_1-specific B-cell, leaving the D_2 determinant vacant. Therefore a D_2 specific T-cell can bind to the remaining determinant, where it can release its receptor-binding factors as discussed above. This completes the triggering mechanism for B-cell activation.

4.1.2.4 Antigens which do not provoke an immune response

If a thymus-dependent antigen contains only a single, non-repeating determinant, this T-cell cooperation cannot take place, and the B-cell will not be activated. Such antigens are described as haptens, and are non-immunogenic; in order to raise antibodies against them in the laboratory, it is necessary to couple them to another molecule which can provide the necessary 'carrier' determinants for T-cell cooperation, hence the need for immunogens (see Chapter 1, Section 1.6.2).

There are other categories of antigen which do not provoke an immune response, notably the 'self' antigens which are properly found within the organism itself. It is obviously of critical importance that these do *not* elicit an immune response, and the elucidation of these mechanisms has occupied immunologists until the present day. In the 1950s Sir Peter Medawar at University College, London showed that introduction of foreign antigens into developing embryos resulted in a lack of immune response to those antigens in later life, demonstrating that tolerance was 'learnt' during early development. Sir Frank Burnet of the Hall Institute in Melbourne had postulated this theory earlier but failed to demonstrate it in practice; Medawar and Burnet shared a Nobel prize for this breakthrough (Nossal, 1993).

Later Joshua Lederberg proposed the theory now described as clonal deletion, in which antigens binding to immature lymphocytes provoked their destruction, so that the cells which recognised self components would be switched off during the development of the organism (Marrack and Kappler, 1993). The suicide (apoptosis) of unwanted cells is effected by the activation of DNA-cleaving enzymes.

It has taken many years to prove that this deletion mechanism does indeed operate, but the absence of such lymphocytes has now been demonstrated in many cases. It has also become clear that this process operates on lymphocytes produced in adulthood, as the cell forms it will be switched off if it binds an antigen present at the time. Only cells which develop without encountering their specific antigen will survive to maturity. This means that new lymphocytes recognising an antigen from an infectious agent will be destroyed if they are created during an infection, a price the organism pays for avoiding self-directed cells. The immune response is mounted by B- and T-cells which had matured before the infectious agent was present.

Some antigens exist solely within cells, so that they are unable to be presented to developing lymphocytes, yet they still do not provoke an immune response. Other mechanisms than clonal deletion must therefore be involved, and this has been the focus of much research. It appears that certain presentations of antigens, rather than provoking the immune response, cause the lymphocyte to enter an inactive state termed anergy, and considerable progress has been made in determining the interactions which cause such differentiations of behaviour.

These breakthroughs are beginning to shed some light on phenomena known for years in laboratory immunisation. Procedures such as repeated challenge with low doses of antigen, or use of a wholly soluble immunogen can cause tolerisation rather than antibody production. Hence it is possible that an immunisation programme can yield precisely the reverse of the desired effect if the wrong conditions are employed. Therefore care is taken to present the immunogen in a strongly antigenic form, using empirically derived adjuvant reagents to provoke a potent immune response. Where it is the aim to produce antibodies with a particularly low cross-reactivity against antigens of a closely similar structure, deliberate tolerisation with the cross-reactant may be attempted by using 'non-antigenic' presentations before the immunisation proper. It is likely that these differences in immune response with varying antigen presentation are closely related to the differences in lymphocyte behaviour just described.

There are certain molecular structures which are not readily bound by the combinations of amino acids available to the immunoglobulin variable regions, notably highly polar species such as carbohydrates. As a consequence (although there are many exceptions) the immune response to such groups is often poor, for reasons separate from the structure-independent causes just outlined.

4.1.2.5 Cell-mediated immunity

There is a second branch of the adaptive immune response which does not involve antibodies; it is rather outside the scope of the section but will be described briefly for completeness.

Some organisms thrive in the lysosomes of the phagocytic cells which are designed to destroy them, such as the protozoan parasite *Leishmania*. These are beyond the reach of circulating antibody molecules, however their presence is indicated by the presentation of their peptide fragments by the MHC Class II molecules possessed by phagocytes. When these are recognised by a passing CD4 T-cell (again, so long as the dual signal criteria are met), cytokines are released which prompt a more vigorous attack by the phagocyte on its invader. Interestingly the nature of these cytokines can vary from individual to individual, which can lead to different responses to the same organism. There are therapeutic opportunities in this field: some interleukins for example promote the production of γ-interferon which stimulates intensified phagocytosis, and may be beneficial in vaccines for diseases such as leprosy.

Besides the CD4 helper class, there are other types of T-cells which can be activated in different ways. One class known as killer T-cells possess CD8 receptors, and recognise antigen fragments of intracellular origin presented by class I MHC proteins; these are present in almost all cell types, unlike the class II proteins mentioned already which are basically limited to phagocytes. These T-cells are invoked to deal with invading organisms such as viruses, which have the capability of thriving within the cytoplasm of cells, thus escaping the attentions of antibodies (which cannot penetrate the hydrophobic cell wall) or phagocytes. Therefore the response of the T-cells under these circumstances is to promote the death of the cell; this action can control a viral infection if cell death occurs before the virus has a chance to complete its replication.

The cytotoxic strategy is less beneficial in the presence of a benign or weakly harmful virus, where continued attacks on cells can cause more damage than the infectious agent itself. The human hepatitis B virus is an example; chronic carriers suffer liver damage due to the attentions of CD8 T-cells although the virus is fairly harmless in itself. There are some indications that the organism can respond to this situation by switching off the response where it does more harm than good.

CD8 cells also recognise cellular material from 'foreign' individuals by variations in the major histocompatibility complex, the name of the latter deriving from this effect, and their defensive response is the cause of medical phenomena such as rejection of tissue grafts or transplanted organs.

The mechanisms discussed briefly in this section come under the category of cell-mediated immunity, and form the second, antibody-independent branch of the adaptive response.

4.1.3 Polyclonal antibodies

Returning to the very start of the antibody-generation process, it is useful to consider the factors which determine the actual composition of the population of activated B-cells, and therefore the population of antibodies which will be produced. With such an enormous library of available antibody specificities, it is an overwhelming certainty that there will be more than one type of B-cell displaying immunoglobulins which are capable of binding each determinant of the antigen. In fact there will be a continuum of antibodies ranging from those with an extremely good 'fit' for the determinant, through those which bind with decreasing efficiency to those which do not bind at all.

The binding of antibody to antigen is an equilibrium process, and this presumably allows for some level of 'shuffling' in the B-cells to which antigen is bound: if the antigen binds initially to a B-cell displaying antibodies of poor fit, the complex may well dissociate before the activation process can be triggered. The better the fit of an interaction, the longer the antigen is likely to remain in contact with the B-cell, and the more likely it is that activation will be initiated. So the overall effect will be that the antigen is likely to end up bound to the B-cells with the best fit for its determinants.

Like all equilibrium processes, whether a particular event occurs or not is a matter of probability rather than immutable certainty. Therefore there will always be some B-cells which end up being activated despite having relatively low specificity. This situation may well represent the optimum for the defence of the organism. Clearly it is beneficial that antibodies with the best fit are activated, but the additional enlistment of imperfect binders may provide better protection against slight variations in antigen structure.

It is interesting to consider the consequences of challenging the B-cell system with different amounts of antigen. An initial 'dose' of antigen will be distributed as outlined above, with the equilibrium shuffling process favouring the ultimate activation of the 'best-fit' cells. However there will only be a limited number of these cells, and once they have been committed to activation the remaining antigen will tend to associate with the B-cells of slightly worse fit. The less favourable equilibrium situation will mean that dissociation will occur more frequently, but activation will take place eventually. This process continues with cells of less and less perfect fit being enlisted as the better ones are removed from the pool.

If the initial dose of antigen is larger, this process will carry on to enlist more B-cells, but of poorer and poorer fit. The consequence, borne out in the practice of laboratory immunisation, is that antibodies with the best fit are obtained by using a low dose of immunising antigen; a higher dose will tend to give a higher yield of antibody, but with a poorer average fit.

Another experimentally-observed immunological phenomenon is the maturation of the response to immunisation, where the affinity or avidity of the antibodies generated rises over periods of weeks or months. To some extent this effect is provoked by continued challenge of the organism by antigen, by repeated injection, and/or by the 'depot' effect of adjuvant reagents, which act as a slow-release antigen store. Once the B-cells generating antibody of the highest affinity have been activated by the processes just described, proliferation will occur and there will obviously be more of these cells available in the bloodstream than there were at the initial immunisation. So a subsequent dose of antigen will tend to be taken up by these high affinity cells rather than being forced to invoke those of poorer fit, and the overall effect is for the response to become more and more dominated by high-affinity antibodies.

Other more complex effects are sometimes observed, e.g. reductions in affinity following the initial maturation gains, and other factors must be involved here. These may include tolerisation phenomena, or effects from the increasing serum concentration of antibody interfering with continued immunisation.

Because a number of different B-cells are activated by immunisation, each undergoing clonal proliferation to generate a population of mature antibody-secreting plasma cells, the resulting mixture of antibodies is termed polyclonal.

4.1.4 Monoclonal antibodies

A polyclonal mixture of antibodies is made up of sub-populations of identical structure, each originating from a set of identical clones of a single activated B-cell, i.e. each of these sub-populations is monoclonal in origin. Obviously the behaviour of the polyclonal mixture is an 'average' of that of the individual sub-populations, so while a particular immunoglobulin molecule may exhibit very particular properties these may be insignificant in terms of the overall performance.

Cancers can occur (myelomas) in which one particular set of B-cell clones proliferate in an uncontrolled fashion, generating large quantities of an antibody of monoclonal origin in the bloodstream. These antibodies were very important in early structural research, being the only source of reasonably large quantities of a single molecular structure. However until the 1970s there was no way to direct this phenomenon towards a given antibody specificity. Köhler and Milstein at Cambridge revolutionised the immunology world with a 1975 paper describing a technique for achieving just this.

The stages of the process are as follows:

(1) A polyclonal mixture of B-cells of interest is generated by a standard immunisation programme, and extracted (along with all the other B-cells of unwanted activity) from the animal.

(2) The mixture is fused with B-cells originating from a myeloma, this cancerous origin giving them the property of reproducing themselves indefinitely. Particular myelomas are chosen generating B-cells which do not secrete antibody themselves, and also which suffer from deficiency of a particular enzyme used in purine production. The fusion process can be carried out by a number of different methods, but the use of polyethylene glycol (PEG) is widespread. This process is somewhat hit-or-miss, but a number of fused (hybridoma) cells should be produced, combining properties of the myeloma (notably 'immortality') and one of the B-cells from the immunised animal (ideally secretion of an antibody of interest). At this stage the fusion mixture will also contain unfused myeloma cells, unfused cells from the immunised animal, and fusions producing unwanted antibodies.

(3) The fusion mixture is diluted and cloned out into microtitre wells. The dilution used is so large that there is likely to be no more than one cell per well. The HAT medium in the wells contains hypoxanthine, aminopterin and thymidine which block the alternative purine production pathway of the myeloma cells; therefore these die off. Fused cells will survive because the B-cell from the immunised animal is able to create purine by the usual pathway, unaffected by this medium. The unfused B-cells are short-lived anyway and also die off; therefore the only cells which thrive are those which are the result of a successful fusion.

(4) The hybridomas secrete their specific monoclonal antibodies into the supernatant of the wells, where their presence can be detected by a suitable test such as an immunoassay. This allows fusions which are producing antibodies with the desired properties to be identified; this identification can only be as sensitive as the test itself, so it is critical that sufficient care is put into its design.

(5) Hybridomas of interest are cultured to produce a stable cell line. Generally at an early stage in this process the culture is cloned out again, to confirm that it is genuinely monoclonal. Cell line samples can be stored at low temperatures (generally in liquid nitrogen) and revived as necessary.

(6) If it is required to produce large amounts of antibody from a cell-line seedstock, this can be achieved either *in vitro* by scaling up the culture of the cells in appropriate media, or *in vivo* by inoculating them into the peritoneal cavity of mice, where under the right conditions they will form a tumour secreting ascitic fluid which contains the antibodies.

Many of the benefits which monoclonal antibodies can offer have already been alluded to in the first chapter. Notable among these are: (1) the ability to offer specificity for a given antigen determinant not achievable with an 'averaged' polyclonal population; (2) an infinite supply of this antibody, unlike polyclonals where the vagaries of the individual immune response mean it may be impossible to reproduce the results of a particular immunisation, or even of a particular bleed from that immunisation; (3) the ability to generate specific antibodies from an immunogen of low purity; cell-lines generating antibodies against unwanted components can be selected out at the cloning stage.

The frequent reference to monoclonal antibodies in Chapter 1 is an indication, if any further were needed, of the impact that the discovery of Köhler and Milstein has had in the life sciences.

In some circumstances there may be disadvantages connected with the use of these reagents, e.g. in the detection of an antigen with a variable structure; while a polyclonal antiserum may well contain a sufficient spread of antigen-binding sites to accommodate the variations, a monoclonal antibody may be too precise in its recognition capabilities and fail to pick up some of the variants. This is a particular issue in areas where antigen structure is sometimes poorly defined, such as the detection of viruses or tumour markers.

Finally it has been noted already that a single antibody molecule can coincidentally possess the capability of binding antigens of unrelated structure – Rosenstein *et al.* (1972), for example, described a mouse myeloma antibody, which bound both a dinitrophenyl lysine compound and the menadione molecule (see Figure

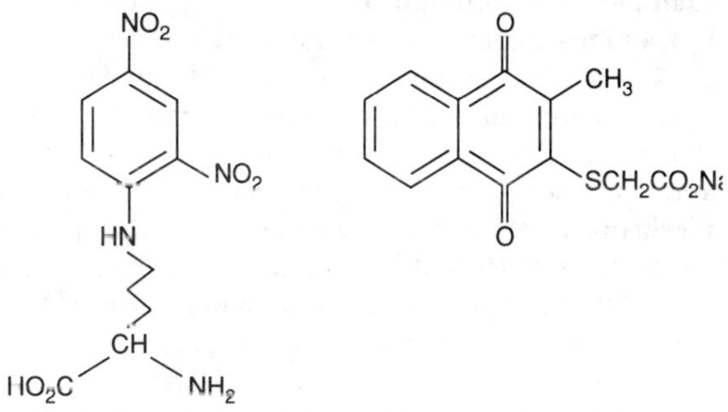

(a) 2,4-Dinitrophenyl lysine (b) Menadione

Figure 4.2 Two dissimilar haptens bound by the same antibody.

4.2). The binding sites for these antigens appeared to be different, in that under certain conditions each could be selectively disabled while leaving the other intact.

In a polyclonal population the number of antibodies with a coincidental binding capability of this type will be very small, and this ability will be statistically 'lost' in the overall performance. However if a monoclonal antibody with this behaviour is used as a reagent, the unexpected cross-reactivity may be significant.

An interesting variation on monoclonal antibody technology is the elaboration of hybrid hybridomas, where a second fusion is carried out between a stable hybridoma and B-cells specific for a second antigen. This can generate bispecific antibodies capable of simultaneously recognising (and binding) two different antigens[1].

4.2 Antigen–Antibody Binding

The binding of antibody to antigen is entirely non-covalent and neither molecule is altered as a result of binding. Therefore the reaction is reversible, and an equilibrium will exist between the bound and unbound states. As with other such interactions, hydrogen bonding, polar or hydrophobic bonding, ionic or coulombic interaction and van der Waals forces all play a role (see Chapter 2, Section 2.2.8).

Despite the weakness of these short-range individual bonds, the large numbers involved can result in strong binding. This is particularly so when there is a close conformational fit between the antigenic determinant and the antigen binding site.

The formation of a complex between an antibody and a protein antigen can lead to visible effects such as precipitation, the precipitin reaction which has been known since the early decades of the century. For example if an optimum concentration of soluble protein antigen is mixed with its homologous antiserum at 37 °C, the solution quickly becomes cloudy and after approximately an hour a precipitate forms at the bottom of the bottom of the tube. At lower or higher antigen concentrations no precipitate is formed. The mechanism behind this phenomenon is that protein antigens are often multivalent, i.e. they possess a relatively large number of antibody binding sites. When the antibody is in too high an excess only a fraction of the binding sites on the antibody are occupied, there is little chance of cross-linking, and the small complexes remain in solution. Similarly, when antigen is in too high an excess most will remain unlinked, and again no precipitation occurs. However, when the two reactants are present in optimal concentrations, known as the zone of equivalence, extensive cross-linking can occur. As the complex gradually becomes larger it starts to precipitate out of solution.

Another example of the consequences of antigen–antibody binding is the phenomenon of agglutination, whereby large species such as bacteria are cross-linked *via* surface antigens in the presence of specific antibody. *In vivo* the formation of immune complexes may, as discussed in Section 4.1, activate defensive mechanisms such as the complement system.

[1] Similar reagents can be created by the chemical coupling of antibody fragments of appropriate specificity by some of the methods described in Chapter 5.

4.2.1 Affinity, avidity and valency

The 'goodness of fit' of a particular antibody for an antigenic determinant is termed its affinity. As mentioned above this parameter is reflected in the equilibrium between bound and unbound antigen and monovalent antibody, and affinities tend to be expressed in terms of equilibrium constants, in this application described as affinity constants.

Taking a straightforward situation such as the interaction of a single monoclonal antibody and its specific antigen, the equilibrium between the bound complex and the unbound components is described very simply by:

$$[\mathrm{AbAg}] = k[\mathrm{Ab}][\mathrm{Ag}],$$

or

$$k = \frac{[\mathrm{AbAg}]}{[\mathrm{Ab}][\mathrm{Ag}]}$$

k being the affinity constant. Methods of measuring k will be discussed in the following sections.

The affinity constant is expressed in terms of inverse concentration, and units of $\mathrm{l\,mol^{-1}}$ are generally employed. The equivalent term $\mathrm{M^{-1}}$ is often encountered too, and corresponding antibody 'molarities' are often quoted based slightly loosely on these terms; thus an affinity of $10^9\ \mathrm{M^{-1}}$ will be referred to as 'nanomolar' and any affinity higher than this as 'subnanomolar'. An antibody of extremely good fit may display an affinity of the order of $10^{11}\ \mathrm{l\,mol^{-1}}$; lower values correlate with lower degrees of fit. Affinities of around $10^9\ \mathrm{l\,mol^{-1}}$ are fairly normal, while anything much below $10^6\ \mathrm{l\,mol^{-1}}$ is an antibody of poor affinity.

The strongest biological interactions are of the order of $10^{12}\,\mathrm{l\,mol^{-1}}$, for example the binding of biotin by avidin. These are bordering on the irreversible, i.e. the probability of dissociation is so small as to be negligible.

A second measure of binding effectiveness is often used for polyclonal antisera, where the presence of many antibodies of different individual affinity complicates the situation. Thus the term avidity is used to describe the overall equilibrium between a number of antibody and antigen molecules, i.e. where:

$$x\mathrm{Ab} + y\mathrm{Ag} \rightarrow \mathrm{Ab}_x\mathrm{Ag}_y$$

and the more complex equilibrium expression is given by:

$$k = \frac{[\mathrm{Ab}_x\mathrm{Ag}_y]}{[\mathrm{Ab}]^x[\mathrm{Ag}]^y}$$

where k in this case can be thought of loosely as an avidity constant.

An interesting situation arises where more than one antibody : determinant interaction may occur simultaneously on an antigen (see Figure 4.3). All naturally-occurring antibody molecules have two or more binding sites, i.e. they are bi- or multivalent. The largest immunoglobulin molecules may contain as many as ten.

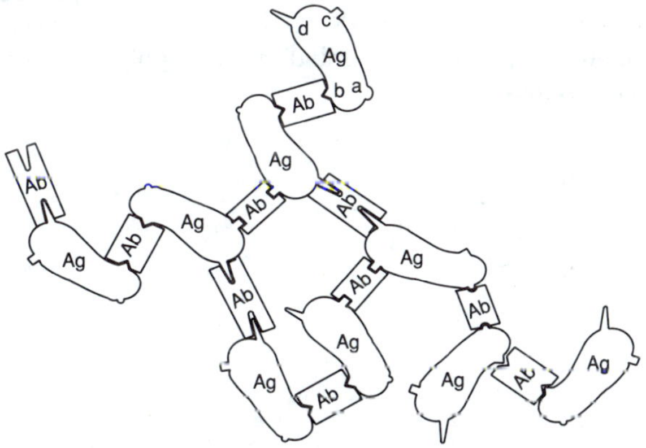

Figure 4.3 Because antibodies are at least divalent, they are capable of cross-linking antigen molecules. Where the antigen possesses multiple determinants (repeating or—as shown here—of separate specificity), aggregation can occur. Dissociation of an antigen molecule may therefore require more than one antibody–antigen bond to be broken.

Therefore an antibody molecule is able to cross-link two antigen molecules *via* the relevant antigenic determinants, in fact this aggregation process is critical to subsequent events such as complement binding. A second antibody may also cross-link the antigen through another pair of determinants, and it is found that the combined binding effect of two such interactions is at least tenfold greater than that from a single antibody.

The explanation of this 'bonus effect' again lies in equilibrium arguments. Antibody : antigen interactions are not so strong as to be irreversible, and therefore once such a complex has formed there is a chance that dissociation will occur (the lower the affinity, the greater the likelihood). However if two such interactions are operating, they must both dissociate simultaneously for the overall Ag:Ab_n:Ag complex to disintegrate; the chances of this are obviously much smaller, hence the cooperative effect.

The pitfalls of applying simple affinity calculations to complex situations such as these have been discussed by Mattes (1997).

4.2.2 The calculation of affinity, avidity and valency

All methods of quantifying the binding reaction rely on analysing the amount of the antigen–antibody complex formed at equilibrium in a given situation, either directly or indirectly, or measuring the rate of formation of the complex.

A range of methods have been developed for determining the equilibrium quantities of bound and unbound antigen (or antibody) in a mixture, and will be discussed briefly in the next section. However these data are generated, they can be analysed by various standard approaches, the most well-known of which is Scatchard analysis (Scatchard, 1949).

4.2.2.1 Scatchard plots

The affinity equation:

$$k = \frac{[\mathrm{AbAg}]}{[\mathrm{Ab}][\mathrm{Ag}]}$$

can be re-expressed as follows:

$$k = \frac{b}{A_{\mathrm{F}}c}$$

where, at equilibrium:

b = concentration of bound antigen, A_F = concentration of unbound antibody sites, c = concentration of free antigen.

However, A_F can also be expressed as follows:

$$A_F = nA_T - b$$

where:
n = the number of binding sites per antibody molecule (i.e. the valency), A_T = total concentration of antibody.

Thus:

$$k = \frac{b}{(nA_T - b)c}$$

Dividing by A_T:

$$k = \frac{b/A_T}{cnA_T/A_T - cb/A_T}$$

b/A_T is the ratio of moles of antigen bound per mole of antibody, which we will call r. Thus:

$$k = \frac{r}{cn - cr}$$

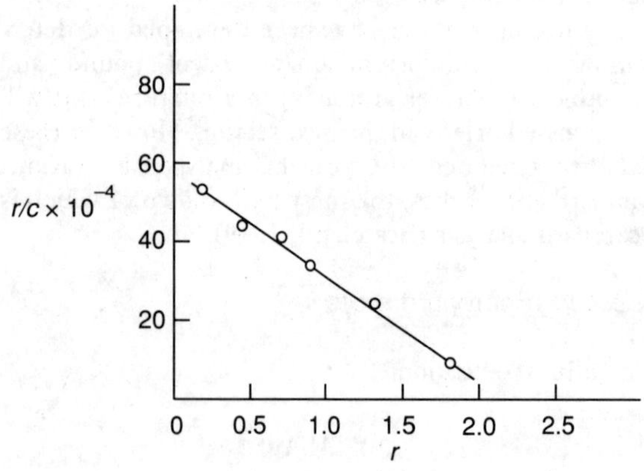

Figure 4.4 A plot of r/c *versus* r for a pure immunoglobulin. The intercept on the ordinate is 2.0, and indicates the valency of the immunoglobulin. Reading of the abscissa where $r = 1$ will yield the affinity of the antibody, in this case $\sim 3 \times 10^5$ l mol^{-1}. (Reproduced with permission from: Tizard, I.R. (1984). *Immunology: An Introduction*, p. 110, Saunders College Publishing, New York.)

which can be rearranged to:

$$r/c = -kr + kn$$

In this system, both r and c are variables, while k and n are constants. This equation is thus a form of the linear equation $y = mx + c$, where m is the slope and c is the intercept on the abscissa. Therefore if r/c is plotted against r:

the slope = $-k$, the intercept on the abscissa = kn, and the intercept on the ordinate = n.

So if the bound Ag:Ab ratio is established at a range of different antigen concentrations, it is possible to determine both the affinity constant and the valency of the antibody concerned. This form of data expression is known as a Scatchard plot, an example of which is shown in Figure 4.4.

The plot shown displays 'ideal' Scatchard behaviour, as may be obtained with a completely homogeneous immunoglobulin and antigen. When an antigen reacts with a polyclonal antiserum, in contrast, a less ideal plot will result (Figure 4.5). This is of course due to the variety of different affinities displayed by the antibody population. In cases of this type the convention is to read the slope at the point where $r = 1$, thereby measuring the average affinity of the antibody population.

The ways in which the data from such analyses are expressed have caused a great deal of discussion over the years, particularly the pitfalls of misinterpreting Scatchard plots, etc. (Klotz, 1989). Such dangers should be borne in mind when applying these approaches.

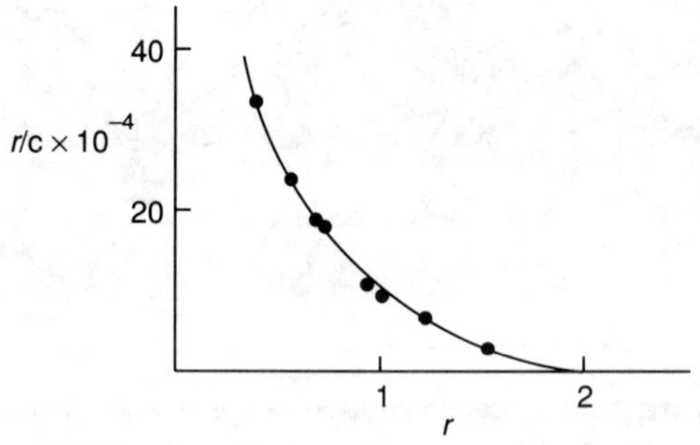

Figure 4.5 A plot of r/c *versus* r for a polyclonal antibody. The variety of individual affinities produce a curved plot. (Reproduced with permission from: Eisen, H. N. (1974). *Immunology*, p. 363, Harper and Row, Hagerstown, MD.)

4.2.3 The detection and quantification of antigen–antibody complexes

As noted earlier the formation of antibody–antigen complexes can sometimes be accompanied by readily detectable effects such as precipitation; however these are rarely quantitative enough to provide useful information on affinity constants. Consequently, a vast number of different approaches have been applied to the generation of equilibrium data for these determinations (Klotz, 1989; Neri *et al.*, 1996). These include partition methods where one component is immobilised in some way, and techniques based on the perturbation of properties of one or other of the components by the binding event.

For small antigens, antibody affinity can be conveniently measured by the technique of equilibrium dialysis. A mixture of antibody and (ideally univalent) antigen is confined in solution within a semi- permeable dialysis membrane. The antigen, which is tagged with a suitable label such as a radioisotope, can diffuse freely through the semi-permeable barrier while the antibody remains inside the sealed membrane. Samples are taken over a period of time from both sides of the membrane, and when equilibrium is reached the amount of free antigen present will be equivalent on each side. However, any antigen which is bound to the antibody will also be retained inside the membrane; thus the difference in concentration of labelled antigen between the inside and outside of the membrane represents the bound fraction. Typically this experiment will be repeated on a number of occasions with different quantities of antigen, such that a range of equilibrium distributions is obtained.

Many techniques have been developed in which the antibody or antigen[1] is immobilised by other means, for example by adsorption to a solid phase format such as a plastic tube, microtitre well, or particulate latex. This may provide a more convenient vehicle for the determination of the bound antigen content: the supernatant can simply be decanted or aspirated away to leave (after washing) only bound antigen. Such approaches are also applicable for large antigens which are not amenable to dialysis methods. They are subject to a number of well-known pitfalls, however, and these must be borne in mind: (1) With formats such as wells or tubes the time required to achieve equilibrium may be very slow, as the interaction of antigen with an adsorbed antibody (or *vice versa*) may be limited by transport effects such as the establishment of diffusion layers. (2) Due to steric effects and conformational changes on adsorption, the number of binding sites available may be very much lower than that expected from the amount of antibody immobilised, and can only be accurately determined by saturation binding. (3) The behaviour of immobilised proteins does not always reflect their solution properties; while antibodies are fairly robust to such effects, this may be a serious issue for conformationally labile antigens.

These formats are of course widely used in techniques such as immunoassay (Chapter 1, Section 1.2.2) which can be applied directly to affinity determinations (Larvor *et al.*, 1994), but are more usually applied to the quantification of antigen or antibody levels in a clinical sample. For these applications some of the constraints above may be less problematic; the failure to establish an equilibrium, for example, may be acceptable provided the assay is carried out under suitably controlled conditions.

Other approaches to the determination of antibody affinities include quantitative immunoelectrophoresis (Birkmeyer *et al.*, 1981), calorimetry (Livingstone, 1996), size-exclusion chromatography (Sanny and Price, 1997) and biosensors. The last method may be based on surface plasmon resonance (SPR) such as the BIAcore instrument developed by Pharmacia (Karlsson *et al.*, 1991) or integrated optical (IO) grating couplers (Polzius *et al.*, 1997). These biosensor approaches offer some potential, particularly in terms of ease of use; concerns have been raised over the effect of varying solute concentrations, however (Hall *et al.*, 1996).

An interesting approach to the characterisation of antibody affinities is an analogue of chromatographic affinity elution (Section 10.9.2), carried out in microtitre wells (Pullen *et al.*, 1986; Macdonald *et al.*, 1988). Monoclonal or polyclonal antibody was coupled to a series of antigen-coated wells, each of which was then incubated with a different, increasing concentration of thiocyanate, a powerful chaotrope. The resistance of the antibody to desorption, determined *via* a suitable enzyme-labelled anti-IgG conjugate, gave an indication of its affinity or avidity.

Ferreira and Katzin (1995) used this approach to characterise the heterogeneity of a polyclonal population: the results were plotted as a histogram to demonstrate the spread of affinities (Figure 4.6).

[1] Because equilibrium dialysis is applicable only to small antigens it is the antibody which must be immobilised by containment within the semi-permeable membrane. However in many other techniques there is no reason why the antigen should not be immobilised rather than the antibody, the same affinity information can be obtained by appropriate analysis.

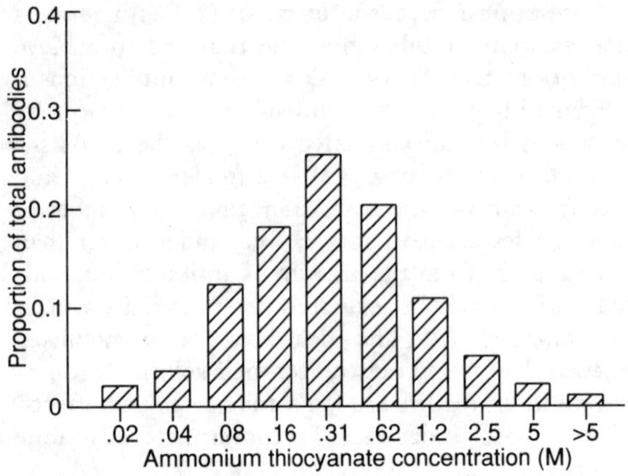

Figure 4.6 Graphical distribution of antibody affinities. (Reproduced with permission from: Ferreira and Katzin (1995). *J. Immunol. Methods*, 187, 297–305.)

4.2.4 Kinetics of antigen–antibody binding

Antigen–antibody binding events are extremely rapid, and sophisticated kinetic methods such as stopped-flow and temperature-jump relaxation are used for their study—although the biosensor techniques noted above are beginning to have an impact here too (Adamczyk *et al.*, 1997; Zeder-Lutz *et al.*, 1997). These methods show that the majority of antigen–antibody systems studied have association rate constants of around $10^6 M^{-1} s^{-1}$, while the dissociation rate constants can vary considerably. Thus the stability of an antibody–antigen complex—and hence its affinity constant—is largely determined by its dissociation rate.

4.2.5 The quantification of antibodies

Many different techniques have been applied to the quantification of antibody samples in various matrices, ranging from polyclonal mixtures in complex biological fluids such as serum, to solutions of pure monoclonal antibody in a simple buffer.

The simplest approach for pure solutions of suitably high concentration is to use UV spectrophotometry at 280 nm (see Chapter 11, Section 11.13.2.2). The absorbance value obtained can either be converted into a concentration by comparison with a standard, or, more commonly, by applying a conversion factor based on typical antibody properties. For example, IgG solutions typically display an A_{280} value of 1.4–1.5 per mg ml^{-1} of concentration, and nominal concentrations are often based on this relationship. As with other UV determinations care should be taken to avoid artefacts such as interference from light-scattering due to microparticulates in the sample, or losses on sample filtration.

Alternatively, the antibody can be treated as a target for analytical chromatography using many of the methods discussed in Chapter 10, typically by comparing peak area with that obtained from a standard of known concentration. In simple matrices it may be possible to apply relatively non-specific techniques such as ion exchange or size exclusion chromatography; in complex media it is usually necessary to exploit the specificity of affinity methods with ligands such as Protein A or G, or specific antigen. All these chromatographic approaches can yield useful estimates of antibody concentration, but it should be remembered that the value obtained will reflect the mechanism and effectiveness of the separation technique employed. Taking the quantification of a mouse monoclonal IgG in ascites, for example, an affinity chromatography estimate based on Protein A may well be higher than one obtained using immobilised specific antigen, because it will also pick up non-specific 'host Ig'. On the other hand, a *lower* value may be observed if the monoclonal is of a sub-class poorly bound by Protein A.

Immunochemical techniques can provide information on the quantity of specific antibody present, often comparable with that obtained by affinity chromatography with a specific ligand: these include immunoassays, discussed earlier, and radial immunodiffusion (RID) and reversed rocket immunoelectrophoresis, described in Chapter 11, Sections 11.11.1 and 11.11.2 respectively.

In all such techniques, it is possible with sufficient attention to relate the result obtained to an actual or nominal Ig concentration. However, a series of similar antibody-containing samples are often compared semi-quantitatively in terms of their titre, a measure of how much the sample can be diluted before a defined response is lost.

Screening a range of antisera for specific Ig content is a typical application of this approach. Using an appropriate test system such as an immunoassay, for example, a series of dilutions of the serum under investigation are made up and assayed for activity in the test system. The definition of titre is totally subjective and dependent on the performance of the test system: under some commonly encountered conventions it may be defined as the reciprocal of the highest dilution giving a positive reaction, or the dilution giving 50% of the maximum response of the assay system (see Figure 4.7).

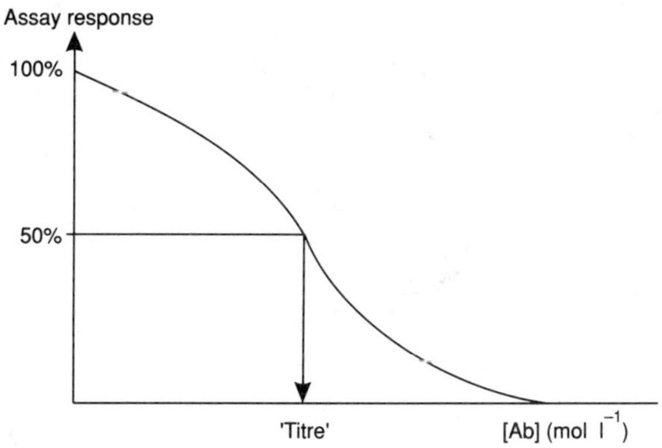

Figure 4.7 The dose–response curve of a competitive immunoassay at different concentrations of a candidate antibody; a nominal 'titre' for the specific antibody is defined as the concentration giving 50% of the maximum assay response. For other assay designs (e.g. immunometric), note that the response may increase with concentration.

Returning to techniques that do provide an estimate of the actual antibody concentration, it is worth noting the variance in values which may be obtained from different methodologies. While comparisons of values obtained by a single technique are limited only by the (commonly ignored) inherent precision of the method, attempts to correlate between, say, RID and Protein A chromatography should be interpreted with great caution. Once again the design and effectiveness of the quantification systems may lead to apparently discrepant results from a single sample.

4.3 The Structure of Antibodies

4.3.1 The basic structure of antibody molecules

All antibodies are made up of a number of peptide chains, linked through disulfide bonds, and the basic model for their structure was proposed by Rodney Porter of the University of Oxford (Porter, 1962). This describes a molecule consisting of four peptide chains of two separate types, and this model has been fully borne out as more structural information has become available. There are two predominant types of chain, heavy and light with a molecular mass of 50–80 kDa and 23 kDa respectively. The basic immunoglobulin 'unit' consists of two heavy chains surrounded by two light (Figure 4.8). Some types of antibody consist of only a single four-chain unit, while others exist as dimers or pentamers; these oligomeric examples contain joining regions as well as heavy and light chains.

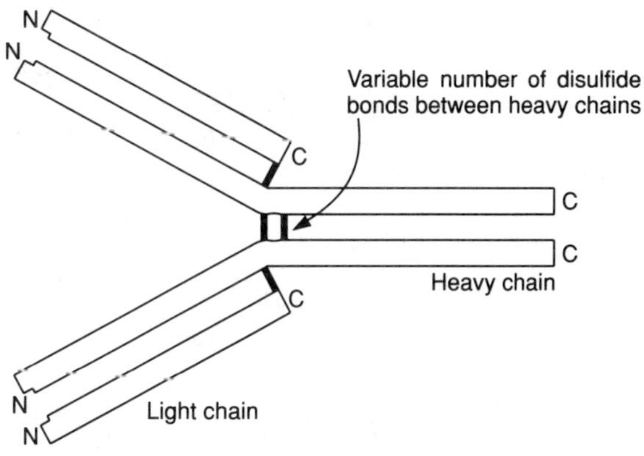

Figure 4.8 The basic immunoglobulin structure. C and N indicate the C- and N-termini of the chains.

4.3.1.1 Classes and subclasses

The heavy chains occur in a number of different forms, which define the class of an immunoglobulin molecule, and possess specific antigenic, immunological and chemical characteristics. The heavy chains characteristic of the major immunoglobulin classes are designated by appropriate Greek letters, γ chains for the IgG class, α for IgA, μ for IgM, δ for IgD and ϵ for IgE.

Some classes of immunoglobulin are further divided into subclasses, depending on structural variations in the heavy chains. For example, in the human IgG molecule the heavy chain exists in four different forms (γ_1, γ_2, γ_3, γ_4) giving rise to four subclasses (IgG_1, IgG_2, IgG_3 and IgG_4). These different subclasses show small variations heavy-chain structure. The differences were originally identified by antigenic analysis; they can be differentiated by specific antisera. Subsequent chemical analysis has shown that the subclasses have substantial differences in amino acid structure. The number and position of the interchain disulfide linkages may vary between different subclasses, as shown in Figure 4.9.

The structure of immunoglobulin classes and subclasses, then, is defined principally by the identity of the heavy chains, see Table 1.1. The Ig classes will be discussed in more detail in Section 4.3.2.

The antibody light chains exist in only two types, κ or λ, common to all Ig classes. Both contain two

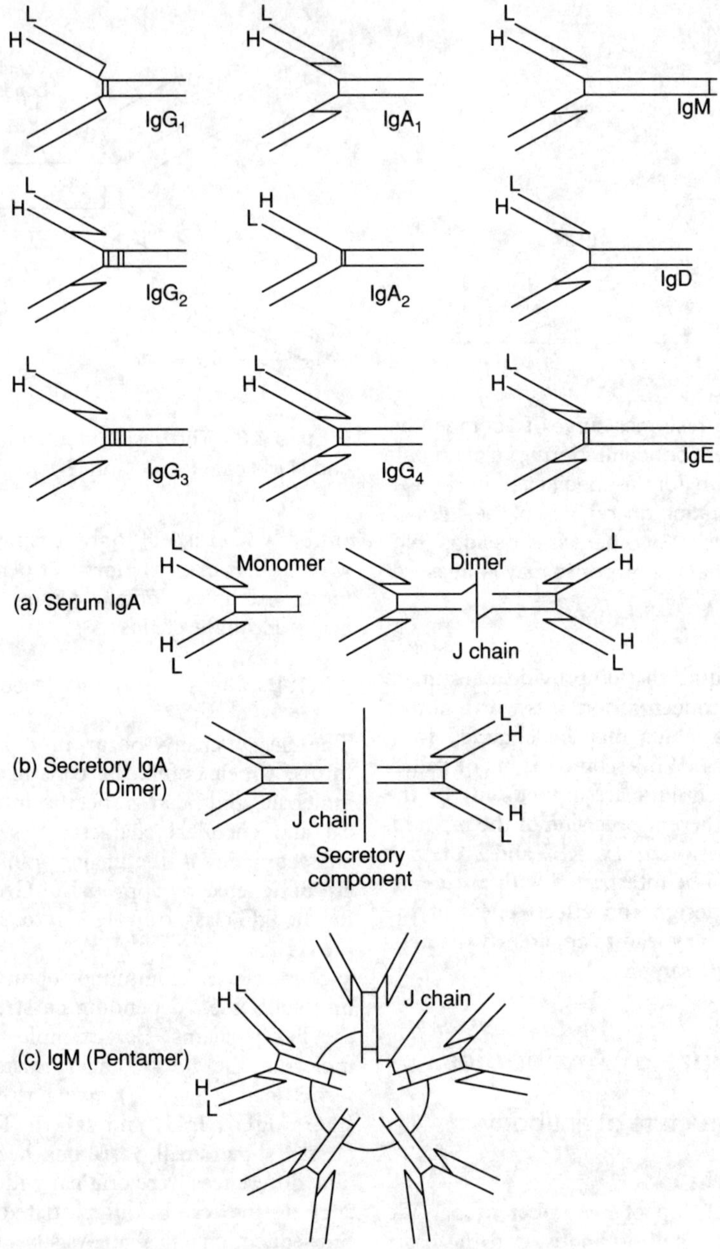

Figure 4.9 Distribution of interchain disulfide bonds in various human immunoglobulin classes and subclasses.

inter-chain disulfide bonds at almost exactly the same position, with the amino acid cysteine at the C-terminus. Their relative incidence varies from species to species and from one Ig class to another; in humans for example IgG_4 contains around 80–90% κ, IgG_2 approximately equal quantities of each, and IgD around 80% λ. Overall κ predominates (60–70%) in total human immunoglobulin, while other species range from one extreme to the other, from mouse (>90% κ) to horse (>90% λ) in fact. A single antibody unit will only contain one or the other of these types.

Table 4.1 Heavy chains differ from one immunoglobulin to another and determine the class and subclass

Basic formula	Chain composition
IgG_1	$(L\ H\gamma 1)_2$
IgG_4	$(L\ H\gamma 4)_2$
IgM	$[(L\ H\mu)_2]_5\ J$
IgA	$(L\ H\alpha)_2$ or $[(L\ H\alpha)_2]_2\ J$
IgE	$(L\ H\varepsilon)_2$
IgD	$(L\ H\delta)_2$

4.3.1.2 Constant, variable and hypervariable regions

The first complete amino acid sequence of an immunoglobulin was determined by Edelman and coworkers in 1969 on an IgG_1 myeloma protein. The position of all the disulfide bonds and the arrangement of the light and heavy chains were reported. The complete sequences for the light and heavy chains are shown in Figures 4.10 and 4.11.

On studying a number of different IgG molecules, it was found that the amino acids in the C-terminal half of the light chains of different myelomas were virtually identical, while the sequences in the N-terminal half were different for each myeloma light chain studied. Thus the two regions of the light chain were referred to as the constant (C_L) and variable (V_L) regions. Similarly analysis of heavy chains showed that they contained a variable N-terminal region (V_H) and a larger and relatively invariant C_H portion.

The existence of these constant and variable segments has been shown to be a basic feature of all immunoglobulins, although more detailed understanding has resulted in the definition of other regions too. Thus each immunoglobulin chain whether heavy or light has a variable (V) region and a constant (C) region; these are connected by joining (J) regions in the light chains, and joining and 'diversity' (D) regions in the heavy chains. The variable region always starts at the N terminus and the constant region ends at the C terminus. The constant region contains the apparatus for binding to receptors such as those of phagocytes or complement components, while the V, J and D regions all contribute to the diversity of binding sites which can be obtained from the basic structure.

Further to this broad identification of constant and variable regions in the heavy and light chains, comparison of large numbers of amino acid sequences of variable regions led to the important observation that the variation is largely restricted to certain key areas. Thus, within the V/J/D segment of each chain are three regions which show the greatest variation in amino acid structure between individual antibodies;

1 10
Asp–Ile–Gln–Met–Thr–Gln–Ser–Pro–Ser–Thr–Leu–Ser–Ala–Ser–Val–Gly–Asp–Arg–Val–

20 30
Thr–Ile–Thr–Cys–Arg–Ala–Ser–Gln–Ser–Ile–Asn–Thr–Trp–Leu–Ala–Trp–Tyr–Gln–Gln–

40 50
Lys–Pro–Gly–Lys–Ala–Pro–Lys–Leu–Leu–Met–Tyr–Lys–Ala–Ser–Ser–Leu–Glu–Ser–Gly–

60 70
Val–Pro–Ser–Arg–Phe–Ile–Gly Ser Gly–Ser–Gly–Thr–Glu–Phe–Thr–Leu–Thr–Ile–Ser–

80 90
Ser–Leu–Gln–Pro–Asp–Asp–Phe–Ala–Thr–Tyr–Tyr–Cys–Gln–Gln–Tyr–Asn–Ser–Asp–Ser–

100 110
Lys–Met–Phe–Gly–Gln–Gly–Thr–Lys–Val–Glu–Val–Lys–Gly–Thr–Val–Ala–Ala–Pro–Ser–

120 130
Val–Phe–Ile–Phe–Pro–Pro–Ser–Asp–Glu–Gln–Leu–Lys–Ser–Gly–Thr–Ala–Ser–Val–Val–

140 150
Cys–Leu–Leu–Asn–Asn–Phe–Tyr–Pro Arg–Glu–Ala–Lys–Val–Gln–Trp–Lys–Val–Asp–Asn–

160 170
Ala–Leu–Gln–Ser–Gly–Asn–Ser–Gln–Glu–Ser–Val–Thr–Glu–Gln–Asp–Ser–Lys–Asp–Ser–

180 190
Thr–Tyr–Ser–Leu–Ser–Ser–Thr–Leu–Thr–Leu–Ser–Lys–Ala–Asp–Tyr–Glu–Lys–His–Lys–

200
Val–Tyr–Ala–Cys–Glu–Val–Thr–His–Gln–Gly–Leu–Ser–Ser–Pro–Val–Thr–Lys–Ser–Phe–

210 211
Asn–Arg–Gly–Glu–Cys

Figure 4.10 Amino acid sequence of the human IgG_1 Eu light chain. Lysine residues are in bold. (Reproduced with permission from: Edelman, G. M. *et al.* (1960), *Proc. Natl. Acad. Sci. USA*, **63**, 78–85.

1 10 20
Pca–Val–Gln–Leu–Val–Gln–Ser–Gly–Ala–Glu–Val–Lys–Lys–Pro–Gly–Ser–Ser–Val–Lys–Val–

30 40
Ser–Cys–Lys–Ala–Ser–Gly–Gly–Thr–Phe–Ser–Arg–Ser–Ala–Ile–Ile–Trp–Val–Arg–Gln–Ala–

50 60
Pro–Gly–Gln–Gly–Leu–Glu–Trp–Met–Gly–Gly–Ile–Val–Pro–Met–Phe–Gly–Pro–Pro–Asn–Tyr–

70 80
Ala–Gln–Lys–Phe–Gln–Gly–Arg–Val–Thr–Ile–Thr–Ala–Asp–Glu–Ser–Thr–Asn–Thr–Ala–Tyr–

90 100
Met–Glu–Leu–Ser–Ser–Leu–Arg–Ser–Glu–Asp–Thr–Ala–Phe–Tyr–Phe–Cys–Ala–Gly–Gly–Tyr–

110 120
Gly–Ile–Tyr–Ser–Pro–Glu–Glu–Tyr–Asn–Gly–Gly–Leu–Val–Thr–Val–Ser–Ser–Ala–Ser–Thr–

130 140
Lys–Gly–Pro–Ser–Val–Phe–Pro–Leu–Ala–Pro–Ser–Ser–Lys–Ser–Thr–Ser–Gly–Gly–Thr–Ala–

150 160
Ala–Leu–Gly–Cys–Leu–Val–Lys–Asp–Tyr–Phe–Pro–Glu–Pro–Val–Thr–Val–Ser–Trp–Asn–Ser–

170 180
Gly–Ala–Leu–Thr–Ser–Gly–Val–His–Thr–Phe–Pro–Ala–Val–Leu–Gln–Ser–Ser–Gly–Leu–Tyr–

190 200
Ser–Leu–Ser–Ser–Val–Val–Thr–Val–Pro–Ser–Ser–Ser–Leu–Gly–Thr–Gln–Thr–Tyr–Ile–Cys–

210 220
Asn–Val–Asn–His–Lys–Pro–Ser–Asn–Thr–Lys–Val–Asp–Lys–Arg–Val–Glu–Pro–Lys–Ser–Cys–

230 240
Asp–Lys–Thr–His–Thr–Cys–Pro–Pro–Cys–Pro–Ala–Pro–Glu–Leu–Leu–Gly–Gly–Pro–Ser–Val–

250 260
Phe–Leu–Phe–Pro–Pro–Lys–Pro–Lys–Asp–Thr–Leu–Met–Ile–Ser–Arg–Thr–Pro–Glu–Val–Thr–

270 280
Cys–Val–Val–Val–Asp–Val–Ser–His–Glu–Asp–Pro–Gln–Val–Lys–Phe–Asn–Trp–Tyr–Val–Asp–

290 300
Gly–Val–Gln–Val–His–Asn–Ala–Lys–Thr–Lys–Pro–Arg–Glu–Gln–Gln–Tyr–Asx–Ser–Thr–Tyr–

310 320
Arg–Val–Val–Ser–Val–Leu–Thr–Val–Leu–His–Gln–Asn–Trp–Leu–Asp–Gly–Lys–Glu–Tyr–Lys–

330 340
Cys–Lys–Val–Ser–Asn–Lys–Ala–Leu–Pro–Ala–Pro–Ile–Glu–Lys–Thr–Ile–Ser–Lys–Ala–Lys–

350 360
Gly–Gln–Pro–Arg–Glu–Pro–Gln–Val–Tyr–Thr–Leu–Pro–Pro–Ser–Arg–Glu–Glu–Met–Thr–Lys–

370 380
Asn–Gln–Val–Ser–Leu–Thr–Cys–Leu–Val–Lys–Gly–Phe–Tyr–Pro–Ser–Asp–Ile–Ala–Val–Glu–

390 400
Trp–Glu–Ser–Asn–Asp–Gly–Glu–Pro–Glu–Asn–Tyr–Lys–Thr–Thr–Pro–Pro–Val–Leu–Asp–Ser–

410 420
Asp–Gly–Ser–Phe–Phe–Leu–Tyr–Ser–Lys–Leu–Thr–Val–Asp–Lys–Ser–Arg–Trp–Gln–Glu–Gly–

430 440
Asn–Val–Phe–Ser–Cys–Ser–Val–Met–His–Glu–Ala–Leu–His–Asn–His–Tyr–Thr–Gln–Lys–Ser–

446
Leu–Ser–Leu–Ser–Pro–Gly

Figure 4.11 Amino acid sequence of the human IgG_1 Eu heavy chain. (Reproduced with permission from: Edelman, G. M. *et al.*, *Proc. Natl. Acad. Sci. USA*, **63**, 78–85.)

these are called the hypervariable or complementarity determining regions (CDRs) and are responsible for the actual antigen binding function. The more invariant sequences joining them are called framework regions.

The variable segments of both κ and λ light chains contain three hypervariable regions, while those of the heavy chains contain four. The framework regions of all chains under closer examination are not totally constant, but show some degree of variability – they can in fact can be classified into subgroups. In human IgG, for example, V_κ possesses three subgroups, V_λ five and V_H three. The number of subgroups varies enormously between species and because each is quite distinct, it is thought that V-region subgroups are coded for by separate V-region genes.

Wu and Kabat (1970) developed a method for plotting the variability of each amino acid position in the variable regions, defined as follows[1]:

[1] The denominator value is obtained by taking the number of times the most common amino acid occurs in the particular position in a range of different immunoglobulins, and dividing by the number of examples studied.

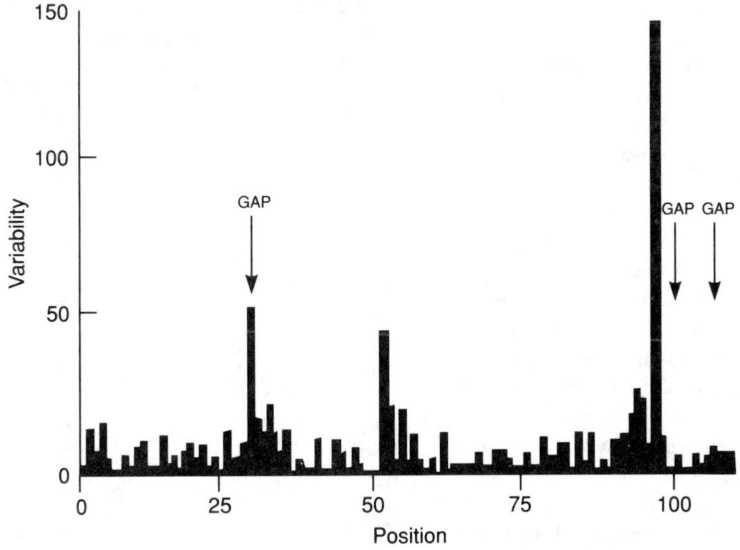

Figure 4.12 The variability at each different amino acid position for the variable region of human immunoglobulin light chains. (The variability is calculated as the number of different amino acids at a given position divided by the frequency of the most common amino acid at that position. An absolutely invariant residue has a value of 1.) Note the three hypervariable regions. GAP indicates positions at which deletrons have been made from some sequences to attain maximum homology. (Reproduced with permission from: Wu, T. T. and Kabat, E. A. (1970). *J. Exp. Med.* **132**, 237.)

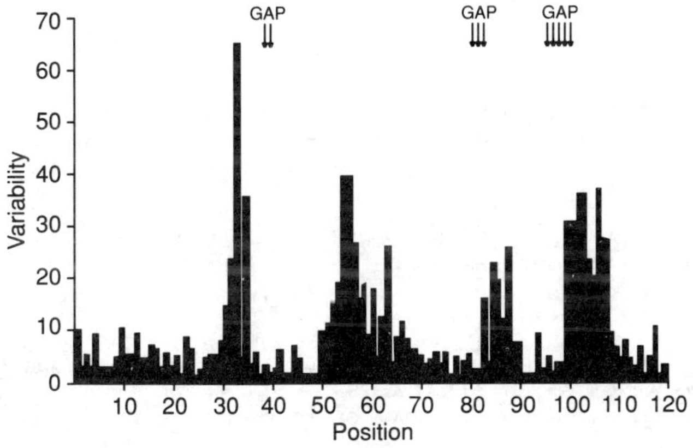

Figure 4.13 The variability at each different amino acid position for the variable region of human immunoglobulin heavy chains. See Figure 4.12 for further details. (Reproduced with permission from: Wu, T. T and Kabat, E. A. (1970). *J. Exp. Med.* **132**, 237.)

$$\frac{\text{Number of different amino acids observed at a given position}}{\text{Frequency of the most common amino acid at that position}}$$

An absolutely invariant residue has a variability value of 1. A plot of variability for human immunoglobulin light chains is given in Figure 4.12. An analogous plot for the variability of the human heavy chain V regions is shown in Figure 4.13.

4.3.1.3 Homology regions and domains

Turning to the constant regions of the heavy and light chains, detailed examination of the sequences revealed that each chain contains a set of repeated linear segments. Thus the 110 residue segment of the C_L region is closely mirrored in the three distinct segments (also of about 110 residues) making up the C_H region. These repeating sequences are known as the homology regions, and are thought to have evolved by duplication of the original single gene coding for the basic 110 amino acid sequence. Edelman's comparison of the amino acid sequence of C_{H1}, C_{H2} and C_{H3} of the myeloma IgG molecule referred to above is shown in Figure 4.14 and summarised in Table 4.2. In a stretch of 100 residues, any two regions are typically identical in 29 to 34 positions.

Thus the structure of the C_H and C_L regions may be broken down further into these repeating 110–120 residue sections of high homology, which along with the similarly-sized V_H and V_L regions, are termed domains. Light chains comprise only two domains, which correspond to the entire V_L and C_L regions. Heavy chains contain four or five domains depending on the class, e.g. four for IgG and five for IgM. These are designated V_H and C_{H1}, C_{H2}, C_{H3}, etc. The domains for the human Ig classes are shown in Table 4.3.

Table 4.2 The amino acid content of the three C_H domains of a human IgG_1 (En)

	C_{H1} residues 119–220	C_{H2} residues 234–341	C_{H3} residues 342–446
Ser	19	7	13
Thr	10	7	7
Lys	7	12	6
Gly	8	5	5
Pro	9	9	9
Val	12	15	8
Phe	3	3	4
Leu	8	8	8
Ala	5	4	2
Glu	2	5	9
Asp	2	6	5
Met	–	1	2
Ile	1	3	1
Arg	1	3	3
Gln	2	5	5
Asn	4	4	5
Cys	3	2	2
Tyr	3	4	5
His	2	3	3
Trp	1	2	2

Worked out from a sequence by Edelman *et al.* (1969).

120 130

C_{H1} (Residues 119–220) Ser Thr Lys Gly Pro Ser Val Phe Pro Leu Ala Pro Ser Ser Lys Ser – – Thr Ser Gly Gly Thr

C_{H2} (Residues 234–341) Leu Leu Gly Gly Pro Ser Val Phe Leu Phe Pro Pro Lys Pro Lys Asp Thr Leu Met Ile Ser Arg Thr

C_{H3} (Residues 342–446) Gln Pro Arg Glu Pro Gln Val Tyr Thr Leu Pro Pro Ser Arg Glu Glu – – Met Thr Lys Asn Gln

140 150 160

Ala Ala Leu Gly Cys Leu Val Lys Asp Tyr Phe Pro Glu Pro Val Thr Val – – Ser Trp Asn Ser – Gly Ala Leu Thr Ser

Pro Glu Val Thr Cys Val Val Val Asp Val Ser His Glu Asp Pro Gln Val Lys Phe Asn Trp Tyr Val Asp Gly – Val Gln Val

Val Ser Leu Thr Cys Leu Val Lys Gly Phe Tyr Pro Ser Asp Ile Ala Val – – Glu Trp Glu Ser Asn Asp – Gly Glu Pro

170 180 190

Gly – Val His Thr Phe Pro Ala Val Leu Gln Ser – Ser Gly Leu Tyr Ser Leu Ser Ser Val Val Thr Val Pro Ser Ser Ser

His Asn Ala Lys Thr Lys Pro Arg Gln Gln Gln Tyr – Asp Ser Thr Tyr Arg Val Val Ser Val Leu Thr Val Leu His Gln Asn

Glu Asn Tyr Lys Thr Thr Pro Pro Val Leu Asp Ser – Asp Gly Ser Phe Phe Leu Tyr Ser Lys Leu Thr Val Asp Lys Ser Arg

200 210

Leu Gly Thr Gln – Thr Tyr Ile Cys Asn Val Asn His Lys Pro Ser Asn Thr Lys Val – Asp Lys Arg Val – – Glu Pro

Trp Leu Asp Gly Lys Glu Tyr Lys Cys Lys Val Ser Asn Lys Ala Leu Pro Ala Pro Ile – Glu Lys Thr Ile Ser Lys Ala Lys

Trp Gln Glu Gly Asn Val Phe Ser Cys Ser Val Met His Glu Ala Leu His Asn His Tyr Thr Gln Lys Ser Leu Ser Leu Ser Pro

220

Lys Ser Cys

Gly

Gly

Figure 4.14 Amino acid sequences of the C_{H1}, C_{H2} and C_{H3} domains of a human IgG_1 myeloma protein (Eu), aligned so as to maximise homologies. (Reproduced with permission from: Edelman, G. M. *et al.*, *Proc. Natl. Acad. Sci. USA*, **63**, 78–85.

Table 4.3 The domain composition of human immunoglobulin classes

Antibody	Chain domains
IgA, IgD, IgG	$2 \times V_L$ $2 \times V_H$ $2 \times C_L$ $2 \times C_{H1}$ $2 \times C_{H2}$ $2 \times C_{H3}$
IgM	$10 \times V_L$ $10 \times V_H$ $10 \times C_L$ $10 \times C_{H1}$ $10 \times C_{H2}$ $10 \times C_{H3}$ $10 \times C_{H4}$
IgE	$2 \times V_L$ $2 \times V_H$ $2 \times C_L$ $2 \times C_{H1}$ $2 \times C_{H2}$ $2 \times C_{H3}$ $2 \times C_{H4}$

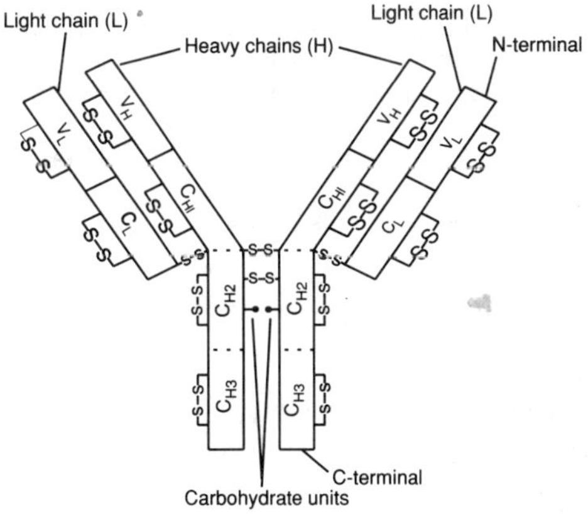

Figure 4.15 The domain structure of human IgG, a typical immunoglobulin.

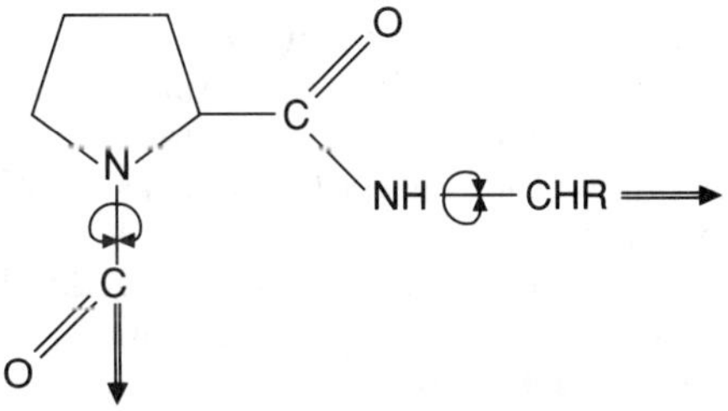

Figure 4.16 Proline introduces a right-angle into polypeptide chains due to the rotational configuration of its substituents. This can cause breaks in the regular secondary structures such as α-helices.

The domains are paired in the intact immunoglobulin molecule to form functional units. Thus the V_H and V_L units interact to form the antigen binding region, while the paired C_{H1} and C_L units act to stabilise the critical structure of the binding site. The remaining heavy chain C regions pair with each other to construct the characteristic Y-shaped molecule, and are the site of the various *in vivo* effector functions. For example, the paired C_{H2} regions of IgG contain the activation site for the complement cascade, and the paired C_{H3} regions contain a site responsible for the mediation of IgG adherence to phagocyte receptors. Figure 4.15 shows the domain structure of IgG, a typical immunoglobulin.

Within the Y-shaped Ig unit the antigen-binding 'arms' swing readily around a central hinge region, which is made up of about 15 amino acids situated between C_{H1} and C_{H2}. The hinge residues are very variable and the sequences are specific to each immunoglobulin class and subclass. However, a dominant feature is the presence of large numbers of hydrophilic residues, notably proline. The unique configuration of proline results in a right angled bend in the polypeptide chain (see Figure 4.16), and therefore this residue acts as a natural break to typical compact secondary structures such as the α-helix; regions possessing many proline residues are unusually flexible in nature. Since amino acids are also free to rotate around peptide bonds, the high concentration of prolines in the hinge region results in a 'universal joint', which allows great rotational freedom to the arms, presumably a key factor in efficient antigen binding. Sequences of the hinge region for the four subclasses of human IgG are shown in Figure 4.17.

The flexible and hydrophilic nature of the hinge area also makes it particularly accessible to proteolytic cleavage by enzymes such as pepsin and papain (see Section 4.5.1). For reasons which will be explained in this section, this has led to a commonly encountered nomenclature for the major portions of the Ig unit, Fab (fragment–antigen binding) for a binding arm, and Fc (fragment–crystalline) for the central 'stem'.

γ_1 Lys Val Asp Lys Lys (Arg) – – – – – Val Glu Pro Lys Ser Cys Asp Lys Thr His Thr Cys Pro Pro Cys Pro

γ_2 Lys Val Asp Lys Thr – – – – – Val Glu Arg Lys Cys Cys – – – Val Glu Cys Pro Pro Cys Pro

γ_3 Pro Glu Glu Asx Thr Cys Thr His Arg Cys Pro Glu Pro Lys Ser Cys Asp Thr Pro Pro Pro Cys Pro Arg Cys Pro

γ_4 Lys Val Asp Lys Arg – – – – – Val Glu Ser Lys Ser – – Tyr Gly Pro Pro Cys Pro Pro Cys Pro

Figure 4.17 Sequences of the hinge region for the four subclasses of human IgG.

4.3.1.4 Subtypes: isotypic and allotypic variation

While there is only one basic gene type for each of the heavy and light chain constant regions, subtypes do exist which create minor differences in the resulting antibodies. For example, in the human λ chain constant region, there are two positions where amino acid variation is seen, the Oz and Kern markers, as they are termed. Normal individuals display both types of each of these markers, so these constitute isotypic variation, no different in concept to the range of IgG subclasses encountered.

In addition to the primary antigenic determinants found on immunoglobulins of all individuals of a species, variable markers are encountered which only occur in certain individuals. These are controlled by inherited allelic differences in the genes and are referred to as allotypic. Like other such genetic features, Mendelian inheritance can lead to an individual who is homozygous (having genes from both parents coding for the same marker type, which will therefore occur throughout the entire antibody population) or heterozygous (having a different marker-coding gene from each parent, and an equal population of the two antibody types). Again the differences in these markers are small, generally one or two amino acid groups at specific sites are replaced by others. Over twenty different allotypic variations of the human gamma heavy chain, described as Gm groups, have been identified, while there are three Km allotypes of the human λ light chain.

In effect these minor isotypic and allotypic variations produce small 'variable regions' within the otherwise fixed C_H and C_L structures, generally with little impact on the behaviour of the molecule or its chemistry.

4.3.1.5 Paratopes, epitopes and idiotopes

Each pair of heavy and light chains operates as a single antigen-binding entity, and the antigen is bound by six separate peptide loops, three from each chain. The binding region created by these loops is termed the paratope, and the region of the antigen to which it binds is called the epitope; a single antigen may contain many different epitopes spread over its structure, and these may be identical (repeating epitopes) or dissimilar. In the latter case a range of monoclonal antibodies, each randomly recognising one of these sites on the antigen, can be used to define how many different epitopes exist; Mabs for the same epitope will interfere with each others' binding and, conversely, a lack of interference indicates specificity for a second binding site. This technique is described as epitope mapping.

The binding regions of antibodies can themselves be antigenic. In fact as Niels Jerne first pointed out, the very diversity of the antigen-binding capabilities provided by the B-cell library makes it almost inevitable that the paratope region of one antibody will match, perfectly or imperfectly, the binding site of another. Epitopes which form part of the binding site of an antibody in this way are termed idiotopes, the antibodies bearing them idiotypic, and the antibodies which recognise them anti-idiotypic. Quite apart from the usefulness of anti-idiotypes as reagents to recognise specific antibody binding regions, the interactions of idiotypes and anti-idiotypes are undoubtedly of some criticality in the immune response. The production of antibody in response to an antigen, for example, can be dramatically enhanced by the presence of anti-idiotypes, while higher doses can lead to tolerisation effects.

4.3.1.6 The presence of carbohydrate in immunoglobulin molecules

Carbohydrate, although a general feature of all immunoglobulins, is not essential for the antibody function and has not been shown to be directly involved in any of the other biological activities. The total carbohydrate content of human immunoglobulin varies from 3% for IgG_1 to 13% for IgE. Carbohydrate is characteristically present on the constant regions of H chains, but may however be adventitiously attached to occasional variable regions of L and H chains owing to the presence of the obligatory sequence Asn-XXX-Thr/Ser (see Section 4.3.4).

4.3.1.7 Three-dimensional structure of immunoglobulins

In 1970 Wu and Kabat proposed that the hypervariable amino acid sequences of each chain were in some way responsible for the extraordinary diversity in antigen recognition by antibodies. The segments are quite distant from each other in terms of the primary sequence, but the authors proposed that in the native immunoglobulin, these might be placed in close proximity to each other by the folding of the polypeptide chains. This arrangement, it was proposed, would form the antigen binding site, and the highly variable primary structure in this region would provide the molecular basis for functional diversity.

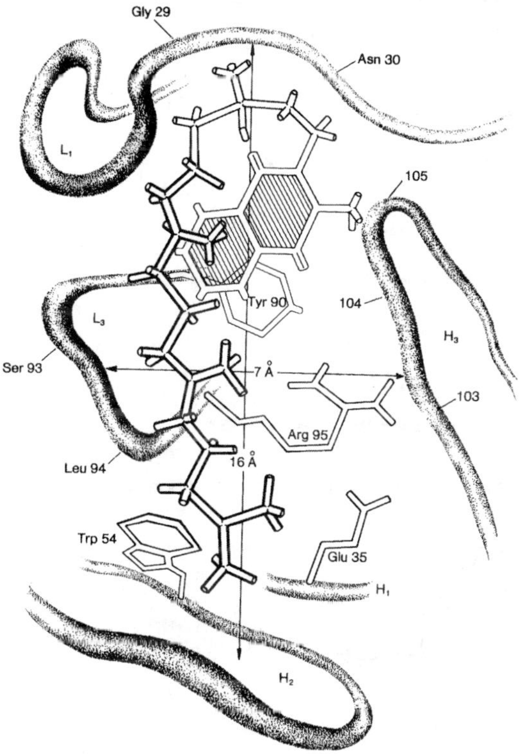

Figure 4.18 Antigen-combining site of an antibody molecule. (Reproduced with permission from: Amzel, L. M. *et al.* (1974). *Proc. Natl. Acad. Sci. USA*, **71**, 1427–1430.)

X-ray crystallography has been an essential tool in the detailed evaluation of the 3D structure of immunoglobulins. For the principles of X-ray diffraction see the reviews by Dickerson (1964), Eisenberg (1970) and Matthews (1975). Such studies on crystal structures over the past twenty years have confirmed that the three hypervariable segments are indeed located on the surface of each N-terminal domain of both the light and heavy chains. The sequences are arranged in adjacent loops which are joined to give a unified surface on each domain.

Wu and Kabat's hypothesis was confirmed further as data from the high resolution structures of Fab–antigen complexes became available (Amzel *et al.*, 1974; Padlan *et al.*, 1976). An example is shown in Figure 4.18. The determination of the Fc region came later (Deisenhofer, 1981), finally making possible the structural definition of the complete immunoglobulin molecule.

The majority of studies on 3D structure have been based on fragments because the flexibility of the hinge region tends to preclude crystal formation of whole IgG. However there are exceptions, and the 3D structure of intact molecules has been elucidated for antibodies with a hinge deletion (Sarma *et al.*, 1971; Sarma and Laudin, 1982; Guddat *et al.*, 1993); the structures of some such examples have been determined to 6 Å and 3.2 Å respectively. Work by Harris *et al.* (1992) has demonstrated the 3.5 Å resolution of a complete anti-canine lymphoma monoclonal. Despite the existence of a normal hinge region, this showed the presence of well ordered Fab and Fc regions.

As with many areas of immunology research, early work concentrated on the use of myeloma immunoglobulins, since a very homogeneous source material is needed for crystal growth. With the advance of hybridoma technology (Köhler and Milstein, 1975), a new source of suitably pure material opened up. The crystal structures of a number of Fab fragments from monoclonals have since been determined, either in their unbound state or complexed with antigen (see for example Davies *et al.*, 1975, 1979; Amzel and Poljak, 1979; Davies and Metzger, 1983; Alzari *et al.*, 1988; Colman, 1988; Davies and Padlan, 1990; Wilson, 1990).

More recently, the smaller Fv fragments obtained by recombinant or enzymatic means (see Section 4.6.2.1) have received some attention too (Bhat *et al.*, 1990; Zhao-Chang *et al.*, 1992).

The conserved three-dimensional structures of the variable and constant domains of the light and heavy chains are composed of an approximately cylindrical array of β-pleated sheets. Each domain contains two layers of β-pleated sheet enclosing a core of hydrophobic amino acid side-chains. One sheet is made up of four, and the other of three antiparallel strands. Figure 4.19 shows a schematic representation of the variable and constant domains in a single light chain of an IgG molecule. The domains are connected by a short length of irregularly folded chain. Following the chain along from the N-terminus down to the C-terminus, it can be seen that there is a remarkable similarity in the folding patterns of the domains. Also note in Figure 4.19 the three loops in close proximity in the N-terminal region of the variable domain, which in line with Wu and Kabat's predictions, form the basis of the antigen binding site.

The conservation of the three-dimensional structure, despite the differences in primary structure, is attributable to the stabilising effect of the intra-chain disulfide bonds, and to the large numbers of hydrophobic amino acid residues in the interior of the folds. These hold the anti-parallel strands in a tightly packed mass. The fact that glycine residues are numerous in the regions where the chain folds back on itself has suggested that their small side-chains may provide the flexibility needed to maintain the folded structure,

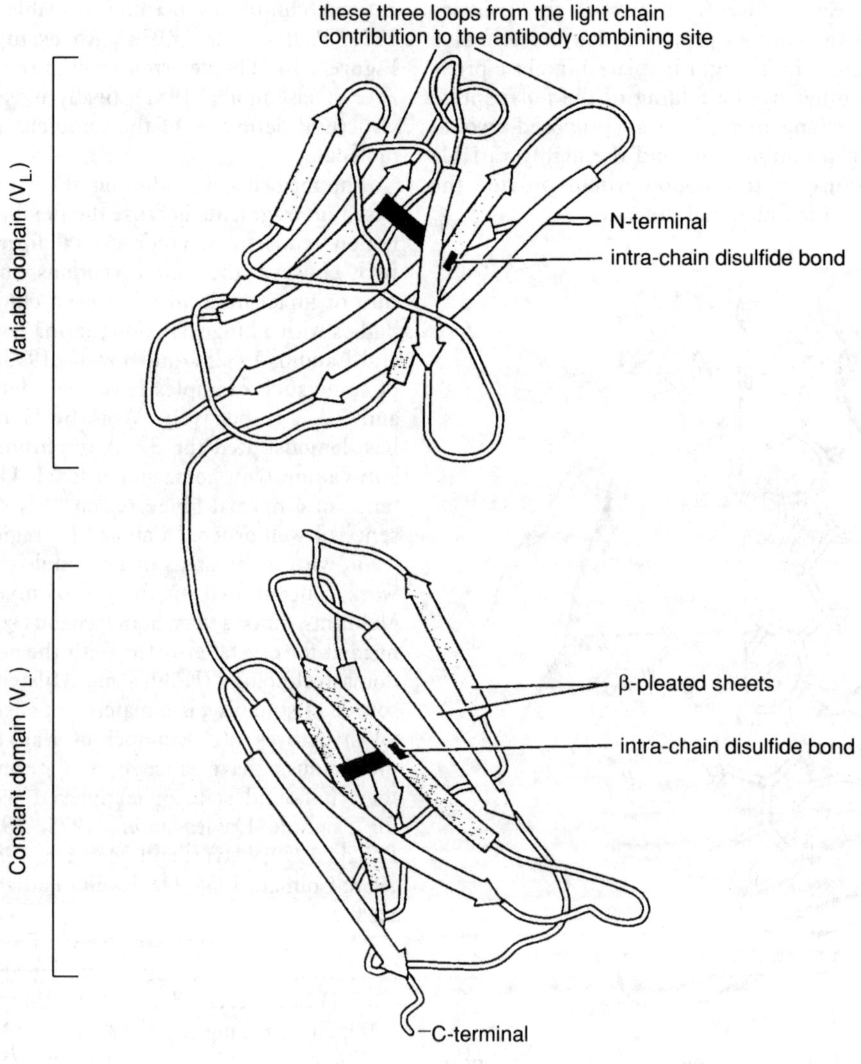

Figure 4.19 Tertiary structure of variable and constant domains in a single light chain of an IgG molecule. The arrows indicate the folding of the chain into similar domains of β-pleated sheets. (Reproduced with permission from: Open University (1984). *Biochemistry and Molecular Biology*, p. 18, Open University Press, Milton Keynes.)

regardless of the variety of the residues in the hypervariable regions.

Using the example of IgG discussed above, the four domains of the heavy chain (V_H, C_{H1}, C_{H2} and C_{H3}) are also folded in the same way as the light chain domains. Each domain is stabilised by an intra-chain disulfide bond, and each is connected to the next domain by a short sequence of irregularly folded chain. The light and heavy chains spiral round each other, and hydrophobic bonding is again critical in holding this structure together. This is particularly important at the end of the Fab arms, where a rigid conformation must be maintained.

Unlike the well conserved overall structure of the V and C domains, the functions of the three and four-stranded β-pleated sheets have diverged. Edmundson *et al.* (1975) considered the V and C domains as rotational allomers, defined as molecules related by rotation with similar 3D structure, but with different amino acid sequences.

The four-stranded layers of the C domains face each other and are tightly paired by hydrophobic bonding of amino acid side-chains, the exception being the C_{H2} unit where the carbohydrate moiety is located across the four-stranded layer and prevents close C_{H2}–C_{H2} pairing (Deisenhofer *et al.*, 1976; Deisenhofer, 1981).

In the V domains, the three-stranded layers face each other, while the side-chains project into the binding cavity situated between the paired V domains at the end of the Fab arms. The amino acid sequences in the areas lining the binding region can be extremely variable, while the residues in the hydrophobic interior sites are highly conserved. The short lengths of chain connecting the domains are typically the sites of additions or deletions.

This highly conserved domain structure appears to be virtually universal. However, an exception has been reported (Wang *et al.*, 1979; Furey *et al.*, 1979). Crystallographic analysis of this example revealed a loop at one end jutting out at a virtual right angle to the pleated sheet structure. Consequently, the binding cavity and the interactions between the domains were unlike those of other immunoglobulins.

4.3.2 Immunoglobulin classes

The different antibody classes will now be discussed individually. As reagents for the applications discussed in Chapter 1 the IgG class vastly predominate, with IgM constituting almost all the remainder of uses. The other classes of immunoglobulin are only encountered in very specialised applications.

Table 4.4 The different immunoglobulin classes and subclasses found in the serum of different animals

Animal	Immunoglobulin	
	Class	Subclass
Human	IgA	IgA_1 IgA_2
	IgD	None
	IgE	IgE_1 IgE_2
	IgG	IgG_1 IgG_2 IgG_3 IgG_4
	IgM	None
Mouse	IgA	None
	IgD	None
	IgE	None
	IgG	IgG_1 IgG_{2a} IgG_{2b} IgG_3
	IgM	None
Rat	IgA	None
	IgD	None
	IgE	None
	IgG	IgG_1 IgG_{2a} IgG_{2b}
	IgM	None
Cow	IgA	None
	IgD	None
	IgE	None
	IgG	IgG_1 IgG_{2a} IgG_{2b}
	IgM	None
Rabbit	IgA	None
	IgD	None
	IgE	None
	IgG	IgG_1 IgG_2
	IgM	None
Sheep	IgA	None
	IgD	None
	IgE	None
	IgG	IgG_1 IgG_{2a} IgG_{2b}
	IgM	None

The existing classes and subclasses for a range of commonly-encountered species are given in Table 4.4.

4.3.2.1 Immunoglobulin G (IgG)

The IgG molecule consists of the 'standard' four-chain unit with two gamma heavy chains and two κ or λ light chains joined by disulfide bridges. Each gamma heavy chain has a molecular mass of around 50–60 kDa, and each light chain contributes 23 kDa; therefore the overall molecular mass of an IgG molecule is around 150–170 kDa.

The basic structure is shown in Figure 4.20. As with other Ig units, the molecule is Y-shaped, with the antigen binding sites situated towards the end of the 'arms,' and the hinge of the 'Y' is flexible. One

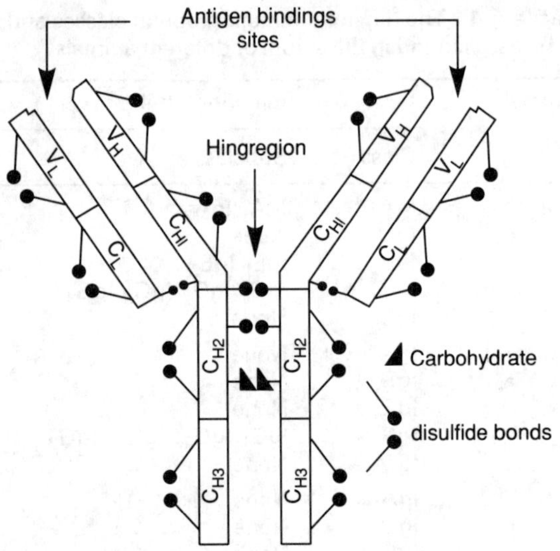

Figure 4.20 The structure of IgG. The number of disulfide bonds between the heavy chains depends on the subclass (see Figure 4.9).

disulfide bond links each light chain to the neighbouring heavy chain, while the heavy chains may be linked to each other by between one and five of these bonds depending on the subclass. Additionally there are a number of intrachain disulfide bonds, between each of which there is a globular domain formed by folding of the peptide loop. These domains occur in four pairs as shown, three of these interacting by hydrophobic bonding: the V_H/V_L, C_{HI}/C_L and C_{H3}/C_{H3} pairs; the remaining C_{H2} domains do not interact with each other in the same way.

The IgG molecule also contains around 2–4% of carbohydrate, predominantly on the C_{H2} domain of the Fc region.

IgG is the major immunoglobulin of the secondary immune response. Apart from the usual complement and phagocyte-binding sites, which are found on the C_{H2} and C_{H3} domains respectively, it also has receptors which allow it to cross the placenta. IgG is therefore probably critical in neonatal protection. It is by far the most abundant immunoglobulin in normal human serum, being present at levels of 8–16 mg/ml and making up around 15% of the total protein content.

As can be seen from Table 4.4, this class of immunoglobulins is the most prone to the existence of multiple subclasses. These occur in varying quantities: the human IgG subclasses, for example, are dominated by IgG_1 (65% in normal serum), followed by IgG_2 (23%), IgG_3 (8%) and IgG_4 (4%). There are some differences in the behaviour of these subclasses, and IgG_3 is probably the most atypical: apart from the molecular mass difference noted above it is much more prone to spontaneous aggregation and is not bound by Protein A from *Staphylococcus aureus.*

4.3.2.2 Immunoglobulin M (IgM)

IgM consists of a pentamer of five basic subunits, each of molecular mass ca 180 kDa, yielding a total molecular mass of ca 850–1000 kDa. Each subunit consists of two κ or two λ light chains and two μ heavy chains. IgM, unlike all other immunoglobulin classes except IgE, has a fourth C_H domain. The monomers are joined together through disulfide bridges in this C_{H4} domain. The extra C_{H4} domains overlap each other and together with the C_{H3} form a central disc, with five pairs of Fab arms radiating out in a star formation (Figure 4.21). Therefore the theoretical valency of IgM is ten, although in practice a valency of five is commonly observed due to steric effects. This high valency makes the IgM an efficient cytolytic and agglutinating agent.

The molecule is rich in carbohydrate, around 12% by weight.

IgM is predominantly produced as a major immunoglobulin class in a primary immune response. Its production during a secondary response is commonly masked by the production of large amounts of IgG. Due to the nature of size, IgM molecules are essentially restricted to the blood vascular system and therefore provide little protection in tissue fluids or body secretions.

IgM constitutes the majority of the immunoglobulin carried on the surface of resting B-cells (although IgG and IgD are also found), and is the first antibody to be generated in quantity on proliferation of these cells. In the secondary response this role is taken over by IgG: many of the 'memory' cells undergo a class switch (see Section 4.4.3) where the μ gene is 'switched off' and one of the γ genes 'switched on'. All the other content of the original genetic combination is retained, however, so the specificity of the antibody is maintained.

On the surface of the B-cell the IgM is anchored by an extra peptide region which is expressed in the material secreted by the proliferated plasma cells. IgM is found at levels of 0.5–2 mg/ml in normal human serum. In most commonly encountered species there is only a single subclass of IgM.

4.3.2.3 Immunoglobulin A (IgA)

IgA constitutes 20% of all serum immunoglobulins and like IgG has three C_H domains. 80% of IgA is in

Figure 4.21 The structure of IgM. (Reproduced with permission from: Turner, M. W. (1981). In *Structure and Function of Antibodies* (Glyn, L. E. and Steward, M. W., eds), J. Wiley & Sons, New York.)

the form of a monomer of molecular mass 160 kDa. The remaining 20% exists as polymers; these can consist of two, three, or five monomers held together by disulfide bridges and a single J chain linked to the Fc regions. Figure 4.22 shows the structure of the tetrameric form of IgA.

An additional molecule that is often covalently linked to the dimer is the secretory piece found predominantly in external secretions such as saliva, colostrum, milk and genitourinary fluids. Secretory IgA consists of four components: a dimer of Ig molecules, a 70 kDa secretory component and a 15 kDa J chain. A diagram of the association of secretory components and J chain with the IgA molecule is shown in Figure 4.23. The secretory component provides the dimer with some protection against proteolysis as well as helping its selective transport to mucous membranes in the body.

Carbohydrate constitutes 7–11% by weight of the IgA molecule.

IgA occurs in human serum in substantial quantities (1.4–4 mg/ml) with a half-life of six days, but its function is still somewhat unclear. It is found predominantly in the sero-mucous secretions of the body, e.g. saliva, tears, sweat, nasal mucus, and lung secretions, where it forms a 'front-line' defence against microorganisms by inhibiting their binding to mucosal cells. It can activate phagocytic activity in polymorphs and can also activate the alternative complement pathway.

In some mammalian species, multiple subclasses of IgA are encountered. In man for example, differences in heavy-chain structure give rise to the subclasses IgA_1 and IgA_2, where IgA_2 exist as two variants. One variant is held together non-covalently as it lacks disulfide bonds between the heavy and the light chains. The other variant contains interchain disulfide bonds, as does IgA_1.

4.3.2.4 Immunoglobulin E (IgE)

The IgE molecule is another monomeric four-chain unit (Figure 4.24). It consists of pairs of κ or λ light chains and two ϵ heavy chains. Like IgM each heavy

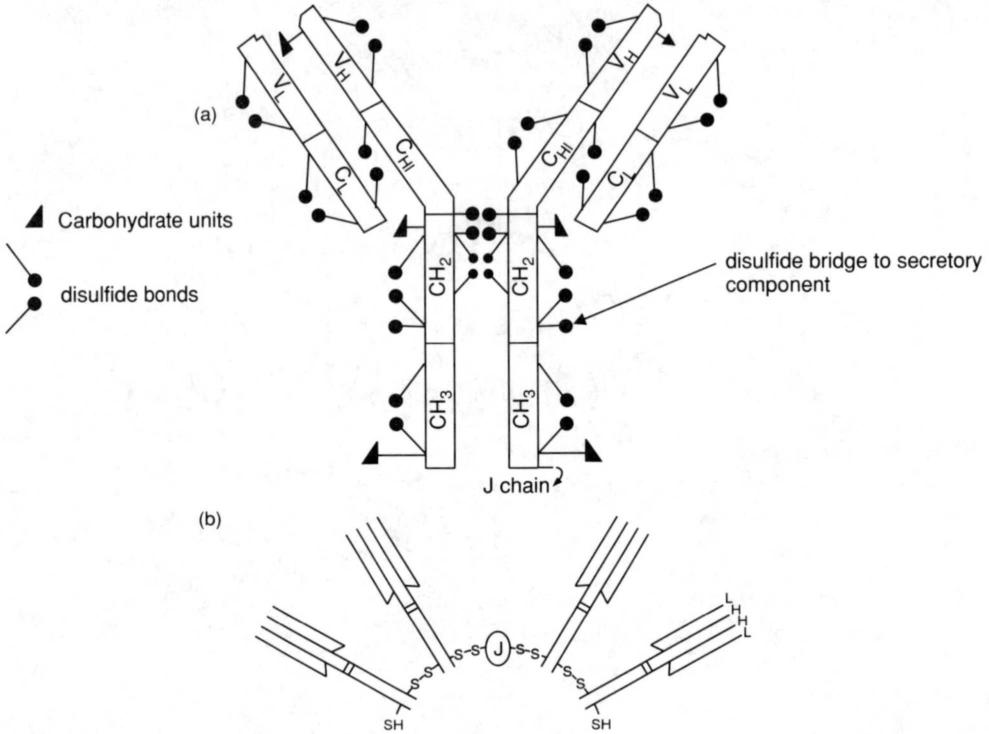

Figure 4.22 (a) The structure of IgA. The number of disulfide bonds between the heavy chains depends on the subclass (see Figure 4.9); (b) an IgA tetramer containing a J-chain.

chain contains four C_H regions, and as a result the molecular mass is rather higher than typical for both the heavy chain (72 kDa) and for the whole molecule (180–190 kDa).

The IgE molecule has a carbohydrate content of around 12–15%.

It is located primarily on basophils and mast cells, and its role is concerned with the triggering of the acute inflammatory response which activates vasoactive and chemotactic factors (see Section 4.1.1). These promote phagocytosis and IgG response when invaders have penetrated the mucosal defences of the body (IgA, lysozyme, etc.)

It is perhaps best known among humans for provoking this response in circumstances which are more of a hardship than a benefit to the host, e.g. the binding of pollen allergens which provokes asthma and hay fever, but it is also effective in recognising more harmful invaders such as flatworms.

IgE has the lowest concentration in serum, only 5 ng/ml, yet it has a profound effector function. The Fc portion binds strongly to a receptor on mast cells and in this manner carries out the molecule's effector function. IgE has a short half-life and is readily broken down by mild heat treatment.

Like IgA, IgE occurs in two subclasses in human serum, although only a single subclass has been identified for many other species.

4.3.2.5 Immunoglobulin D (IgD)

IgD was discovered as a myeloma protein by Rowe and Fahey (1965). The IgD molecule is a single four-chain unit with an extended hinge region: it consists of two d heavy chains and two κ or λ light chains, giving it an approximate molecular mass of 180–185 kDa (Figure 4.25). It is unique amongst the immunoglobulin classes in having only two C_H domains, and a single disulfide bridge between the heavy chains.

The carbohydrate content is around 9–14%.

The concentration of IgD in plasma is low (0.03 mg/ml). It is prone to rapid proteolysis in the bloodstream, and it has a circulating half-life of three days. Little is known about its precise functional importance compared to other classes. Apart from its low concentration, studies on this class are further

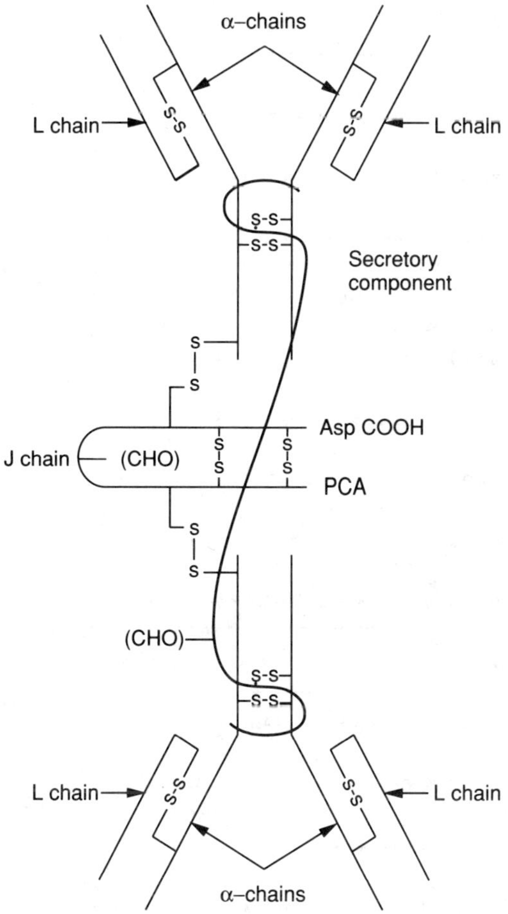

Figure 4.23 Diagram showing the possible arrangement of secretory IgA, combining the two IgA monomers with the secretory component and the J chain. (Reproduced with permission from: Turner, M. W. (1981). In *Structure and Function of Antibodies* (Glynn, L. E. and Steward, M. W., eds), p. 27, John Wiley & Sons, New York.)

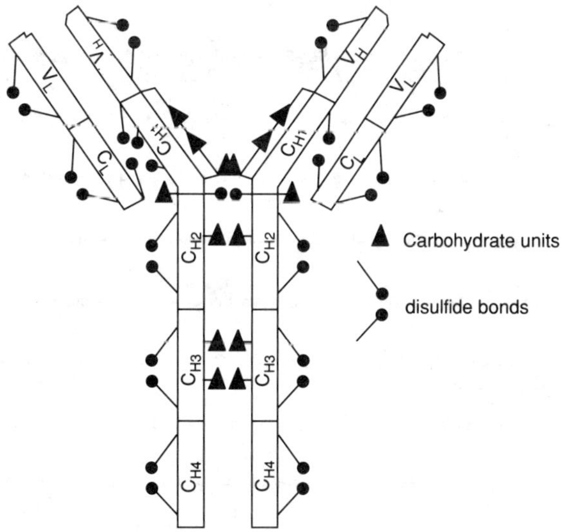

Figure 4.24 Polypeptide chain structure of IgE.

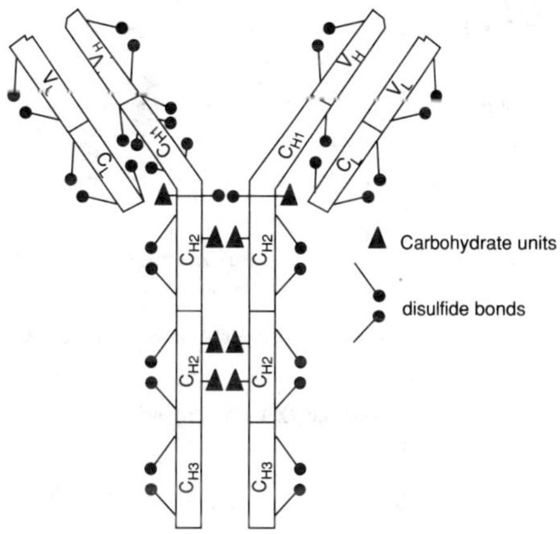

Figure 4.25 Polypeptide chain structure of IgD.

hampered by the fact that the molecule is extremely labile to enzymatic degradation. Since proteolytic enzymes are released during blood clotting, IgD is often absent from serum samples. IgD is also susceptible to heat degradation, and contains an exposed hinge region.

Its main function appears to be as a receptor on lymphocytes—it is found to be present in association with IgM on the surface membranes of B lymphocytes. Its role in this capacity is unclear but it does not appear to be used as an 'active' antibody.

Only a single IgD subclass is generally encountered.

4.3.3 The amino acid composition of immunoglobulins

As noted earlier, the first complete amino acid sequence of an immunoglobulin molecule was determined by Edelman *et al.* (1969), for an IgG. Since then the primary sequences of large numbers of immunoglobulins have been established, initially using pathological myeloma proteins and more recently the monoclonal antibodies generated *in vitro* by hybridoma techniques. For example, the complete amino acid sequence of the μ heavy chain of a human IgM immunoglobulin, including the location of all disulfide

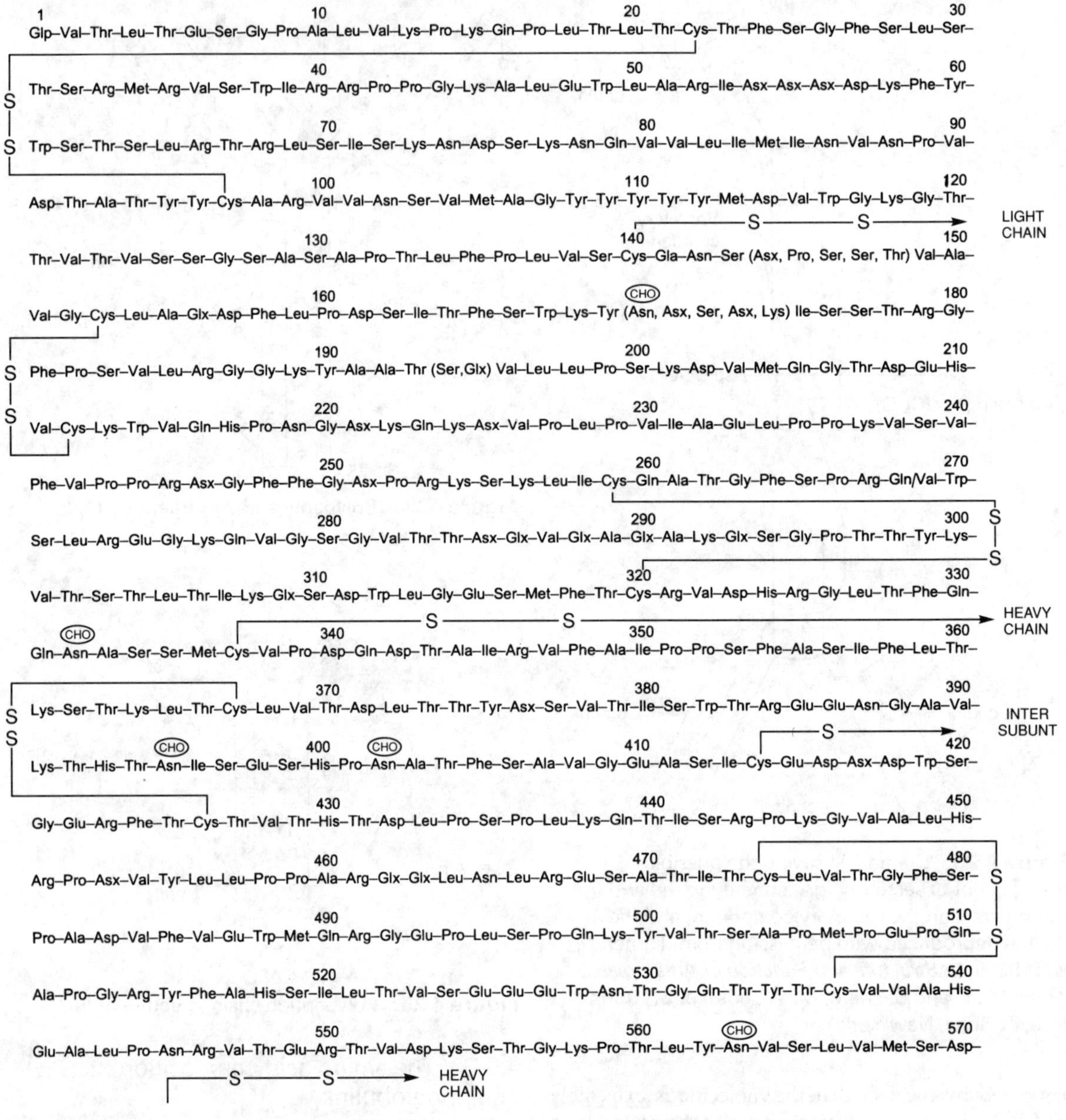

Figure 4.26 Amino acid sequence of the μ heavy chain of IgM OU. (Reproduced with permission from: Putnam, F. W. *et al.* (1973). *Science*, **182**, 287–291.)

bridges and oligosaccharides, was determined by Putnam *et al.* in 1973 (see Figure 4.26).

Given the basic model of construction of immunoglobulin molecules, it is clear that the overall amino acid composition will be determined by a number of factors. There will be considerable homology between the constant and framework regions of antibody molecules within a common subclass, but a degree of variation will be introduced by the composition of the antigen-specific hypervariable regions, and by other minor structural differences such as the isotypic and allotypic effects described in Section 4.3.1.4.

Table 4.5 Comparison of the constant regions of the heavy chains of the five human classes; positions with four or five homologous residues show these boxed

δ	H T Q P L G V Y L L T P A V Q D – L W L R D K A T F T C F V V G S D L K D A – – – – H L T W E V A G K V P T G
μ	Q D T A I R V F A I P P S F A S – I F L I K S T K L T C L V T – D L T Y D S – – V T I S W T R Q D G E A V K
γ	L L G G P S V F L F P P K P K D T L M I S R T P E V T C V V V – – D V S H E D P Q V K F N W Y V – D G V Q V H
α	S C C H P R L S L H R P A L Q D – L L L G S E A N L T C T L T – – G L R D A S G – V T F T W P S T S G K – – –
ε	N P R G V S A Y L S R P S P F D – L F I R K S P T I T C L V V – – D L A P S K G T V N L T W S R A S G K P V –
δ	G V E E G L L E R H S N G S Q S Q H S R L T L P R S L W N A G T S V T C T L N H P S L P P Q R L M A L R E P A
μ	T H T B – I S E S H P B A T F S A V G E A S I C E B B W N S G E R F T C T V T H T D L P S P L K Q T I S R P K
γ	N A K T K P R E Q Q Y B S T Y R V V S V L T V L H Q N W L D G K E Y K C K V S N K A L P A P I E K T I S K A K
α	S A V Q G P P E R D L C G C Y S V S S V L P G C A E P W N H G K T F T C T A A Y P E S K T P L T A T L S K – S
ε	N H S T R K E E K Q R N G T L T V T S T L P V G T R D W I E G E T Y G C R V T H P H L P R A L M R S T T K T S
δ	A Q A P V K L S L N L L A S S D P P E A A S – – – W L L C E V S G F S P P N I L L M W L E D Q R – E V N T S G
μ	G V A L H R P D V Y L L P P A R E Q L N L R E S A T I T C L V T G F S P A D V F V Q W Q M Q R G Q P L S P F K
γ	G Q P R E – P Q V Y T L P P S R E E M T K N Q – V S L T C L V K G F Y P S D I A V E W E S N D G E – – – P E N
α	G – N T F R P Q V H L L P P P S Z Z L A L B Z L V T L T C L A R G F S P K D V L V R W L – Q G S Q E L P R E K
ε	G – P R A A P E V Y A F A T P E W P G S R D K – R T L A C L I Q N F M P E D I S V Q W L H N E V Q – L P D A R
δ	F – – A P A R P P P Q P G S T T F W A W S V L R V P A P P S P Q P A T Y T C V V S H E D S R T L L N A S R S L
μ	Y V T S A P M P E P Q – A P G R Y F A H S I L T V S E E E W N T G E T Y T C V V A H E A L P N R V T – E R T V
γ	Y K T T P P V L D – S D G S – – F F L Y S K L T V D K S R W Q Q G N V F S C S V M H E A L H N H Y T – Q K S L
α	Y L T W A S R Q E P S Q G T T T F A V T S I L R V A A E D W K K G D T F S C M V G H Z A L P L A F T – Q K T I
ε	H S T T Q P R K – – T K G S G – F F V F S R L E V T R A E W Q E K D E F I C R A V H E A A S P S Q T V Q R A V
δ	E V S Y V T D H G P M
μ	D K S T G K P T L Y B V S L V M S B T A G T C Y
γ	S L S P G
α	D R L A Q K P T H V N V S V V M A E V D G T C Y
ε	S V N P G K

Single-letter notation is used here for clarity: see Table 2.1. (Redrawn with permission from Lin, L.-C. and Putnam, F. W. (1981), *Proc. Natl. Acad. Sci. USA*, **78**, 504–508.)

In fact there is also a high degree of homology between the constant regions of different subclasses within a class (typically around 95%), and even between different classes (around 30%). Thus there are many similarities between the amino acid compositions of the IgG classes, e.g. Table 4.5 shows the constant region compositions of human IgG chains. There are seventeen positions at which all five classes have identical amino acids, while approximately 80% of residues show homology between two, three or four classes.

Immunoglobulin classes from other mammalian species have been studied in some detail. Figure 4.27 details the sequence of the κ constant region of four different species: rat, mouse, human and rabbit. Once again there is tremendous homology between species, pointing to their common evolutionary origin. Predictably the closest homology is between mouse and rat

Species	110–149
Rat	R A N A A P T V S T F P P S T E Q L A T G G A S V V C L M N X F Y P R D I S V K
Mouse	R A D A A P T V S I F P P S S E Q L T S G G A S V V C F L N N F Y P K D I N V K
Human	R T V A A P S V F I F P P S D E Q L K S G T A S V V C L L N N F Y P R E A K V Q
Rabbit	G D P V A P T V L I F P P A A D Q V A T G T V T I V C V A N K Y F P R D V T V T

Species	150–189
Rat	X K I D G T E R R N G V L N S V T B Q D S K D S T Y S H S S T L S L T X A D Y Q
Mouse	X K I D G S E R Q N G V L N S D I Z X D S K D S T Y S M S S T I T L T K D E Y E
Human	X K V D N A I Q S G N S Q E S V T Z O D S K D S T Y S L S S T L T L S K A D Y E
Rabbit	X E V D G T T Q T T G I E N S K T P Q D S A D C T Y N L S S T L T L T S T Q Y N

Species	190–214
Rat	S H N L Y T C Q V V H K I S S S P V V K N F N R N E C
Mouse	R H N S Y T C E A T H K T S T S P T V K S F N R N E C
Human	X H X L Y A C E V I H Q G L S S P V T K S F N R G F C
Rabbit	S H X E Y T C X V T H Q G T T S P V V Q S F N R G D C

Figure 4.27 Comparison of the κ light chain constant regions of four different species; positions with three or four homologous residues show these boxed. (Drawn with permission from: Kindt, T. J. and Capra, J. D. (1984). In *The Antibody Enigma*, Plenum Press, New York, pp. 61–93.)

Table 4.6 Typical numbers of the more useful amino acid residues likely to be encountered in whole molecules of human IgG and IgM (data adapted from a number of sources)

	IgG	IgM
Lys	90	350
Glu+Asp	120	620
Cys	30–40	180
Tyr	55	240
His	25	130

Note that most cysteine molecules are present in the disulfide (cystine) form.

which are from the same phylogenetic order. In general, immunoglobulins from the same class show approximately 60% homology between species. The heavy chain of IgM is an exception to this, where human, dog and mouse chains show greater than 90% homology.

While it is impossible to provide accurate information on specific immunoglobulin molecules, Table 4.6 gives some indicative figures for the number of 'useful' amino acids likely to be encountered in whole molecules of human IgG and IgM.

4.3.4 The carbohydrate composition of immunoglobulins

All immunoglobulin molecules are glycoproteins and contain a proportion of carbohydrate, whose content varies from one class to another. The carbohydrates are present in the form of simple or complex side-chains that are covalently linked to amino acids present in the polypeptide chains.

4.3.4.1 The site of carbohydrate attachment

In general the attachment of carbohydrate is to the constant portion of the heavy chain by means of an *N*-glycosidic bond, typically between an *N*-acetylglucosamine residue of the sugar side chain and an asparagine residue of the polypeptide chain. Figure 4.28 shows a common linkage involving C(1) of *N*-acetylglucosamine and the amide group of asparagine, forming 2–acetamido-*N*-(2-aspart-4-oyl)-2-deoxy-*β*-D-glucopyranosylamine. This product is more commonly known as asparaginyl-*N*-acetylglucosamine or GlcNAc.Asn. This is typical of glycoproteins, where the linkage of oligosaccharide to Asn often occurs when the sequence Asn-XXX-Ser/Thr is present (see Section 2.4.1).

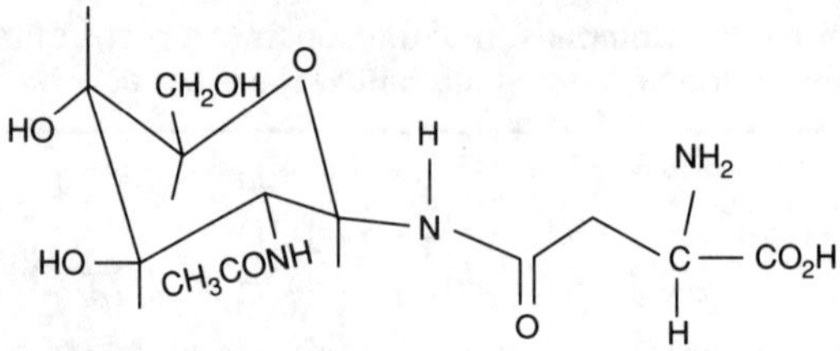

Figure 4.28 Structure of 2-acetamido-*N*-(2-aspart-4-oyl)-2-deoxy-*β*-D-glucopyranosylamine.

The *N*-acetylglucosamine residue offers a point of further attachment of other monosaccharides, to form oligosaccharide moieties termed glycans. These are typically branched chains where sialic acid, galactose and *N*-acetylglucosamine occupy the end residues. Generally, there are around five oligosaccharide residues in IgM and IgE, whereas IgG contains only one and IgA two or three. This is in agreement with the overall carbohydrate content by weight of immunoglobulins (see Table 4.7).

Sugar linkage can also occur through C(1) by *O*-glycosidic bonds to hydroxylated side-chains such as serine and threonine.

As mentioned earlier, the carbohydrate moieties of immunoglobulin molecules are typically associated with the non-antigen binding regions, the Fc fragment of IgG and its equivalent in other classes.

However, it has been known for some time that sugars are sometimes found in atypical positions in the Ig molecule: there is an approximately 30% chance of Fab fragments derived from human myeloma IgG

Table 4.7 Molecular mass and carbohydrate content of major human immunoglobulin classes

Subclass	IgG_1	IgA_2	IgG_3	IgG_4	IgM	IgD	IgA_1	IgA_2	IgA_{sec}	IgE
Molecular mass (×1000)	146	146	170	146	970	185	160	160	385	184
Carbohydrate (%)	2–3	2–3	2–3	2–3	12	13	7–11	7–11	7–11	9–14

being glycosylated (Abel *et al.*, 1968). More recently Rademacher and Dwek (1984) have found that pooled human IgG consistently contained greater than the expected number of carbohydrate chains per molecule. Rabbit IgG has also been shown to display an approximately 30% incidence of Fab glycosylation. These examples of sugars in atypical positions may be associated with the fortuitous presence of the Asn-XXX-Ser/Thr sequence mentioned above.

Non-enzymatic (chemical) glycosylation of proteins including immunoglobulins can also occur *in vivo*, as seen in hyperglycaemia patients (Brownlee *et al.*, 1984; Kaneshige, 1987). The covalent linkage of glucose can change the structural and functional properties of immunoglobulin, and this can lead to various complications in diabetic patients. Oligosaccharide residues can also be introduced synthetically into Ig molecules.

4.3.4.2 Microheterogeneity of immunoglobulin glycosylation

The carbohydrate composition of immunoglobulins is nowhere near as predictable as the amino acid sequence. Considerable heterogeneity has been found in the composition of sugar residues among immunoglobulins from the same species and class (Kornfeld *et al.*, 1971), and in fact the exact content of sugar units commonly varies even within a single preparation of a given monoclonal immunoglobulin. Analysis of myeloma and synthetic monoclonal antibodies has shown that each clone generates a heterogeneous distribution of oligosaccharides (Mizuochi *et al.*, 1982).

This microheterogeneity of immunoglobulin molecules is now a well-accepted phenomenon, and is clearly evident from analytical techniques such as isoelectric focussing (Chapter 11, Section 11.8): the small variations in molecular charge deriving from differences in sugar content[1] are revealed by the existence of multiple bands over a p*I* range of a unit or so.

From the literature it appears that there is no correlation between variation in sugar heterogeneity and IgG subclass.

4.3.4.3 Carbohydrate composition of immunoglobulins

Despite this microheterogeneity, considerable work has been carried out to elucidate the carbohydrate composition of immunoglobulins. Oligosaccharide structures of examples of all five classes of human immunoglobulins have been determined: IgG (Kornfeld *et al.*, 1971; Clamp and Putnam, 1964); IgD (Clamp and Johnson, 1972); IgE (Kochwa *et al.*, 1971; Baenziger and Kornfeld, 1974a,b); IgA (Baenziger and Kornfeld, 1974c,d; Dawson and Clamp, 1968); and IgM (Hickman *et al.*, 1972; Clamp and Johnson, 1971).

IgG, IgM, IgE and IgA contain one or more complex oligosaccharides which are always linked to Asn residues in the C_H region. The sugar moiety consists of L-Fuc, D-Gal, D-Man, D-GlcNAc and *N*-acetylneuraminic acid in different proportions depending on the site. The structure of the oligosaccharide portion of IgG, IgM, IgE and IgA_1 together with residues and the domains to which the oligosaccharide is attached are shown in Figure 4.29.

In IgM and IgE there is a second type of oligosaccharide moiety that consists of D-Man and D-GlcNAc. A considerable homology exists in the location of the carbohydrate in these two classes where the complex oligosaccharides occur toward the N-terminus, while the second type with D-Man and D-GlcNAc occurs closer to the C-terminus. In IgM the position 402 is homologous with IgE position 384. Similarly position 453 of IgA_1 is homologous with position 563 in IgM, both being 14 residues from their respective C-termini.

The IgA_1 also contains an additional third type of oligosaccharide where D-GalNAc is linked to serine. There are five such sites all of which are located in the hinge region.

IgA_1 and IgD are unusual in that *N*-acetylgalactosamine and *N*-acetylglucosamine-containing oligosaccharide units appear in both of these immunoglobulins.

It also appears that only IgM and IgE contain the high mannose saccharides: IgM contains two high mannose oligosaccharide units, compared to IgE where only one is present per heavy chain. The structure of the high mannose unit in IgM differs in several respects from the one observed in IgE. With IgM, microheterogeneity is seen that results in structures based on a core which consists of the sequence GlcNAc⟶GlcNAc⟶Asn. On the other hand the high mannose oligosaccharides of IgE display no detectable microheterogeneity and contain the following core sequence: Man(β1,4) ⟶ GlcNAc(α1,3) ⟶ Man(β1,4) ⟶ GlaNAc(β) ⟶ Asn; the presence of an α-acetylglucosamine making this a unique sequence.

As shown in Figure 4.29 the complex oligosaccharide units of IgG, IgA_1 and IgE are very closely related. Differences are the presence or absence of fucose in the core, the presence or absence of a non-reducing

[1] And other variants such as deamidated forms, where some amide-containing residues such as asparagine have undergone hydrolysis.

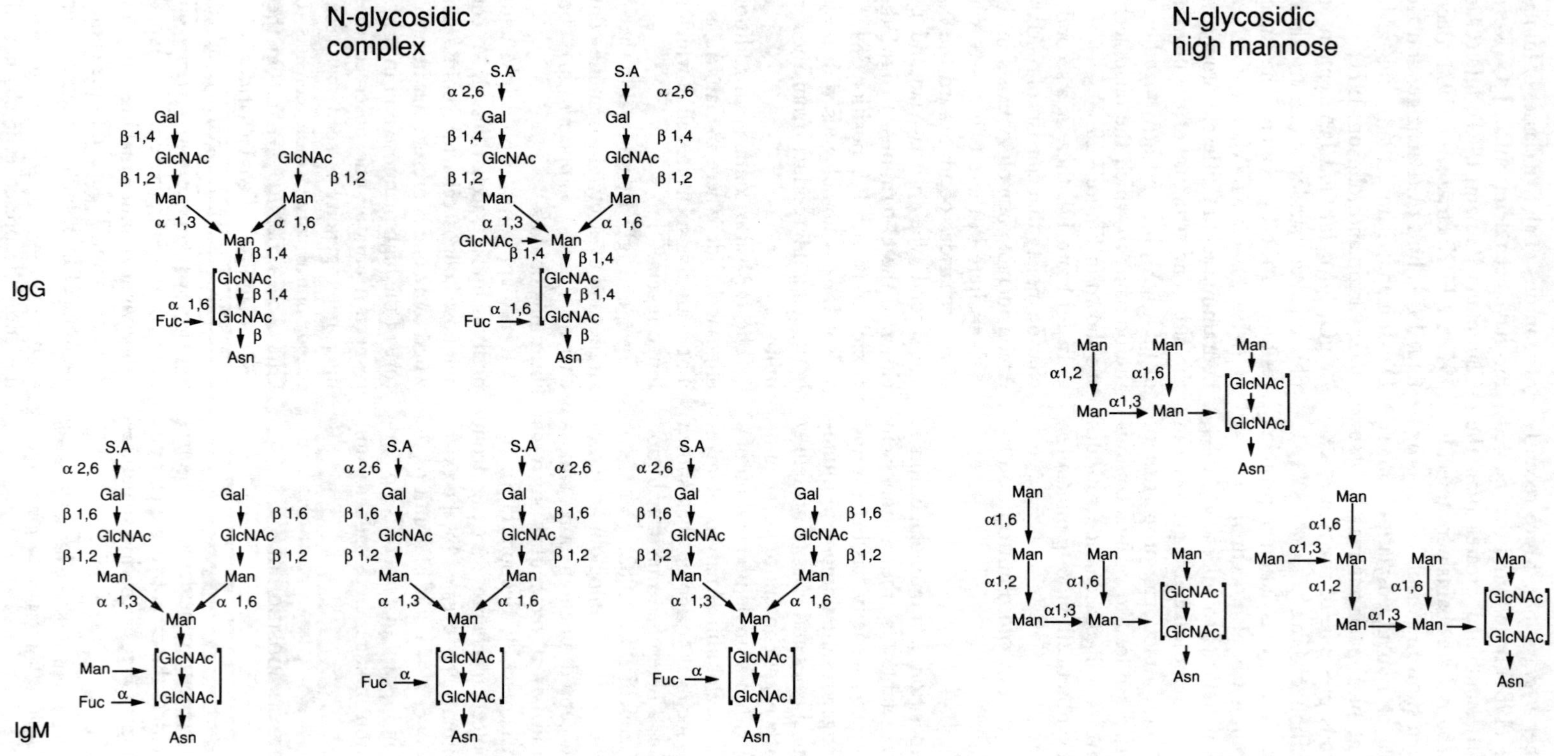

N-glycosidic complex
N-glycosidic high mannose
IgG
IgM
S.A
α 2,6
Gal
β 1,4
β 1,6
GlcNAc
β 1,2
Man
α 1,3
α 1,6
β 1,4
α 1,6
Fuc
β
Asn
α
α1,2
α1,3
α1,6

Figure 4.29

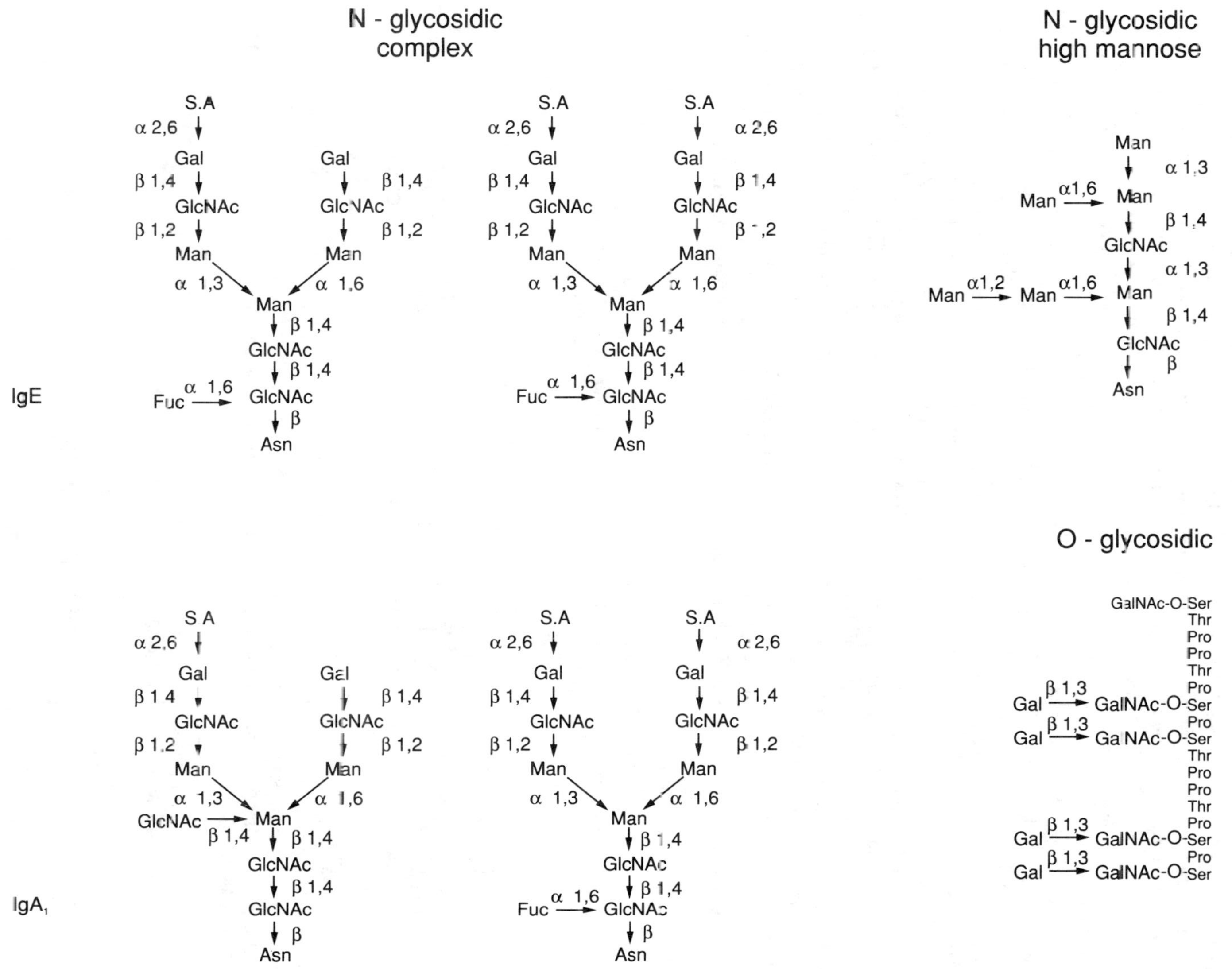

Figure 4.29 Summary of oligosaccharide structures present on human IgG, IgM, IgA and IgE myeloma proteins. (Reproduced with permission from: Baeziger, J. and Kornfeld, S. (1974d). *J. Biol. Chem.*, **249**, 7270–7281.)

terminus containing *N*-acetylglucosamine that is linked to the core mannose, and a number of sialic acid and galactose residues.

The complex oligosaccharides of IgM by contrast appear to differ primarily in the presence of peripheral *N*-acetylglucosamine residues which are substituted at position 6 of the sugar ring rather than position 4. In addition, the complex oligosaccharide appears to arise from the Fd portion of the μ chain that contains fourth residue mannose arising from the core.

Kornfeld *et al.* (1971) isolated IgG glycopeptides from myelomas and from normal serum. The distribution of each glycopeptide was calculated and the structure of the oligosaccharide portion of the molecule was established. The carbohydrate composition of the glycopeptide was found to consist of sialic acid, fucose, galactose, *N*-acetylglucosamine and mannose. It was shown that the variability in the number of sialic acids, fucose and galactose residues was responsible for the microheterogeneity of the oligosaccharides. An additional oligosaccharide chain was also seen in two of the seven myeloma proteins. After a detailed study on the structure and sequences of the glycopeptides Kornfeld *et al.* (1971) proposed the structure of the glycopeptides shown in Figure 4.30.

Kabat *et al.* (1982) carried out structural studies on the carbohydrate moiety of human IgG by sequential exoglycosidase digestion of oligosaccharides. It was concluded that all the carbohydrate chains of normal human IgG have Manα1 $\longrightarrow$ 6(Manα1 $\longrightarrow$ 3)Manβ1 $\longrightarrow$ 4GlcNcβ1 $\longrightarrow$ 4(+/−Fucα1 $\longrightarrow$ 6)GlcNAc as their core portion.

A common evolutionary feature of IgG structure is the presence of two sugar chains attached to Asn297 of the C_{H2} domains. These sugar chains consist of about 20 oligosaccharide moieties which are based on a mannosyl-chitobiose core as shown in Figure 4.31. As this diagram shows four types of mannosyl-chitobiose cores are found (+/−*N*-acetylglucosamine and +/−fucose). Outer-chain differences include the presence or absence of galactose and sialic acid.

In rabbit Fc the neutral and monosialylated carbohydrate is in the proportion of 88 and 12% (Rademacher and Dwek, 1984). The level of neutral to acid carbohydrates is conserved amongst rabbit, human and bovine Fc; on the other hand the exact distribution of

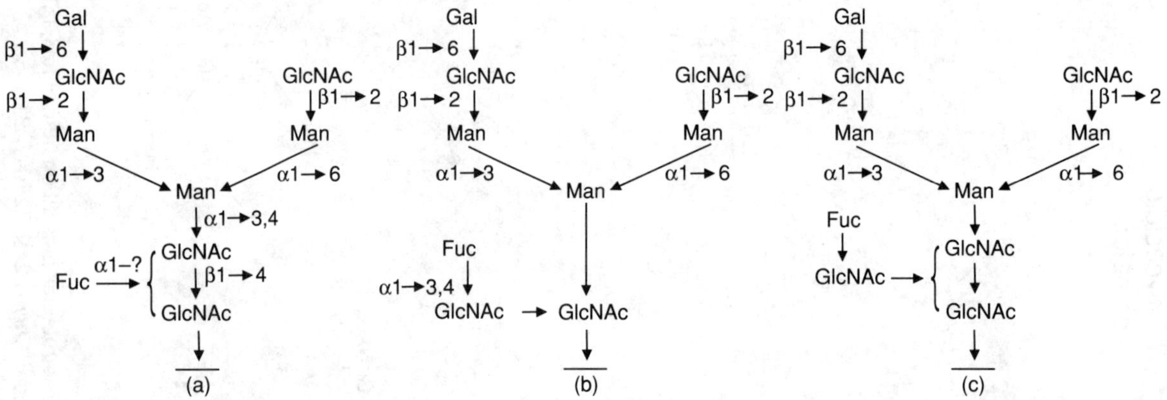

Figure 4.30 Proposed structure of the IgG glycans. (a) Structure of M'M and Kel glycan; (b) structure of Lang and Hum A glycan; (c) structure of Red glycan. (Reproduced with permission from: Kornfeld, R. J. *et al.* (1971). *J. Biol. Chem.*, **246**, 3259–3268.)

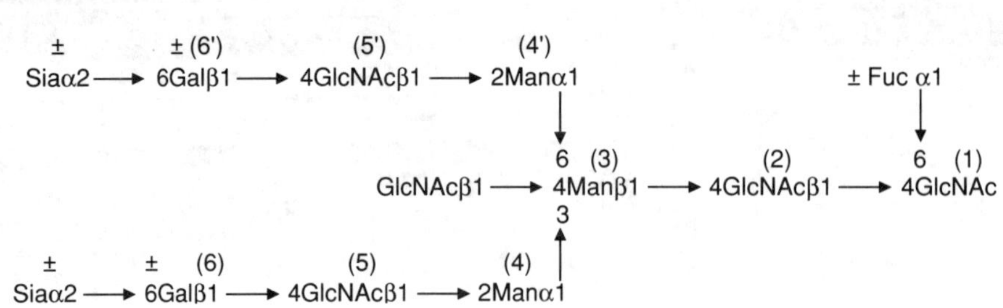

Figure 4.31 Core structure of IgG glycans based on mannosyl-chitobiose.

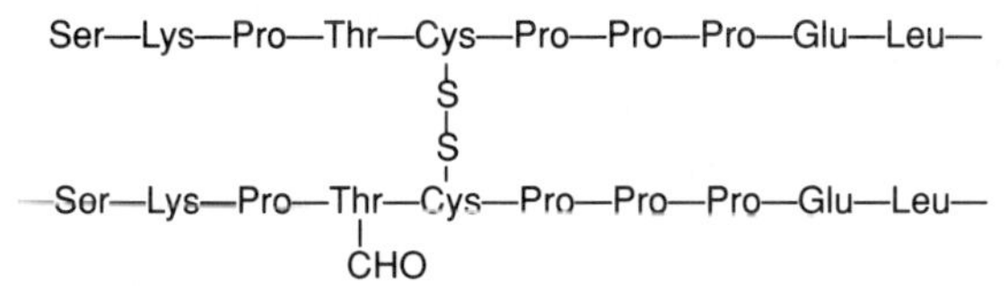

Figure 4.32 Hinge region of rabbit γ-chain showing attachment of carbohydrate.

neutral carbohydrates is species-dependent. In rabbit Fc the variable residues in the neutral oligosaccharides are fucose (32% incidence), *N*-acetylglucosamine (38% incidence) and galactose (two residues, 19% incidence; one residue, 40% incidence). Mouse Fc N-linked carbohydrate lacks the *N*-acetylglucosamine completely. Human Fc contains about 38% of the *N*-acetylglucosamine structure and approximately 80% fucose.

Rabbit γ-chains have been shown to contain a carbohydrate chain that is covalently linked to Thr226 in the hinge region as well as the N-linked sugars (Fanger and Smyth, 1972). The oligosaccharide is linked to only one γ-chain in an intact IgG molecule as shown in Figure 4.32. The average composition of the carbohydrate is one galactosamine residue (presumably N-acetylated) linked to Thr226, one galactose and either one or two sialic acid residues.

N-linked oligosaccharides derived from rabbit Fab regions are found to possess more acidic residues than their Fc counterparts, containing 25% monosialylated, 9% disialylated and 66% neutral species.

4.4 Antibody Genetics

4.4.1 The genetic basis of antibody diversity

The key property of the adaptive immune system is its ability to generate specific antibodies to a vast range of complementary antigens of differing structure. Understandably in view of this diversity of binding behaviour, it was originally assumed that the antibody molecule was inherently non-specific, and was 'moulded' in some way around the presenting antigen (the template theory). However it later became apparent that the ability to recognise specific antigens was innately present, and that the mechanism for creating each specific antibody structure was present in the organism, regardless of whether a foreign antigen of complementary structure had ever been encountered.

In practice an organism can produce antibodies of approximately 10^8 different specificities. Consequently if the antibody-synthesising cells are pre-committed for each given antibody structure, the organism needs to contain all the genetic information for these 10^8 different antibodies, even though only a fraction of these would ever be expressed. If each variation in specificity was due to a totally independent genetic origin, an extraordinary number of individual genes would be required to code for all the possible structures—in fact around 500 times more than the total DNA in a single cell. Clearly another genetic mechanism must be at work.

An alternative method would be the rearrangement of a much more limited pool of mini-genes to give a large array of combinations, and this is in fact the approach encountered *in vivo*. This mechanism is known as somatic recombination (Tonegawa, 1983).

The embryonic precursors of the antibody-producing cell contain less than 1000 mini-genes, each coding for a different part of the immunoglobulin molecule. Rearrangement of this gene pool results in a population of lymphocytes of extraordinary diversity, each having a specific combination of the available genes. Thus only a handful of lymphocytes will be genetically identical in each individual, but the potential exists to generate vast numbers of specific antibodies. Each one of the antibody 'designs' generated by the B-cell population may offer binding sites for a number of structurally unrelated molecules, extending the capabilities of the response; finally, it also appears that the genes concerned are particularly prone to somatic mutation, increasing the diversity of available antibodies still further. Thus the ability to generate an enormous range of different specific antibody molecules can be explained, without invoking a mechanism which places a crippling genetic load on the organism.

While all individuals of an organism carry similar sizes of gene and B-cell libraries, their exact specificities are not identical, and the immune response to a particular invading organism is variable.

4.4.2 The structure of antibody-coding genes

Immunoglobulins are coded for by three families of genes, each located on a different chromosome, one family coding for the immunoglobulin heavy chain and one each for the κ and λ light chains. Each of these three families of germ line genes contain separate genes for the variable and constant regions of immunoglobulin chains and are shown diagrammatically in Figure 4.33.

It can been seen that the variable (V) and constant (C) genes (or gene segments) are separated by so-called

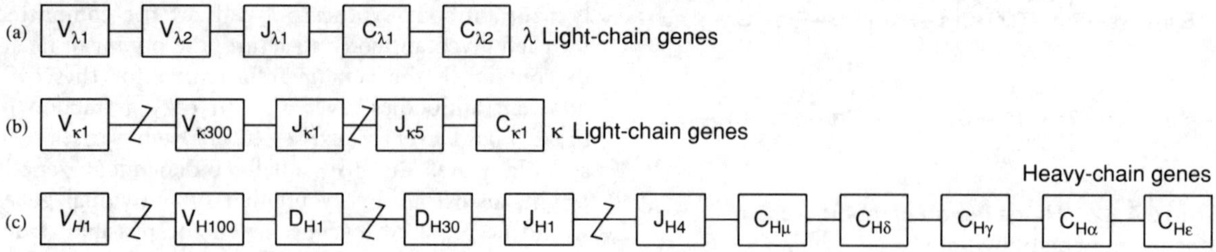

Figure 4.33 The three immunoglobulin gene families, coding for (a) λ-light chains, (b) κ-light chains and (c) heavy chains.

J (for joining) genes in the two types of light chains, and J and D (diversity) genes in the heavy-chain coding sequence. J and D genes code for parts of the hypervariable loops within the V region.

The vast majority of the gene combinations only cause variations in specific regions of the antibody, not surprisingly those responsible for antigen binding. However, the different classes and subclasses of Ig discussed in Section 4.3 are of course another result of this genetic diversity. Thus while the genes for the V, D and J regions are responsible for the diversity of antibody specificity, it is the C genes which have the greatest impact on the overall molecular structure, and the C_H genes which encode for the constant region of the heavy chain in particular.

The C_H pool in mice, for example, consists of eight genes, termed alpha, gamma_1, gamma_{2a}, gamma_{2b}, gamma_3, delta, epsilon and mu, and the presence of these genes in the overall combination give rise to the molecules classed as IgA, IgG_1, IgG_{2a}, IgG_{2b}, IgG_3, IgD, IgE and IgM respectively. In other words this gene population correlates with the classes and subclasses of immunoglobulin discussed above. Figure 4.33 shows five separate constant-region genes coding for the heavy chain.

As discussed earlier, the different C regions of the immunoglobulin family possess great structural similarity; Edelman's comparison of the amino acid sequence of C_{H1}, C_{H2} and C_{H3} of a human IgG molecule, for example, is shown in Figure 4.14. Antibody structures, therefore, are thought to have evolved by replication of an original single gene coding for a basic amino acid sequence of around 110 residues. It is postulated that the first duplication of this gene occurred early in the evolutionary sequence, resulting in a prototype 'light chain', i.e. a linked constant and variable region. Subsequent duplication(s) of the constant region probably allowed the heavy chains of various classes to be built up from the same common origin.

During replication of the germ-line cells and their eventual differentiation into B-lymphocytes, the wide ranges of alternative V, D and J possibilities are each reduced to a single functional option as shown in Figure 4.34. A gene segment from each group is selected to remain active in the DNA of a particular lymphocyte, defining its individual specificity.

Using the known diversity of the gene pool, the number of different combinations of alternative genes can be determined. Any one of approximately 300 V_κ genes from the κ light chain can combine with one of four J genes ($J_{\kappa 3}$ being inactive) giving a total of 1200 different light chain variable regions. The heavy chain has even more diversity as a result of the D genes, since one of approximately 100 V_H genes combines

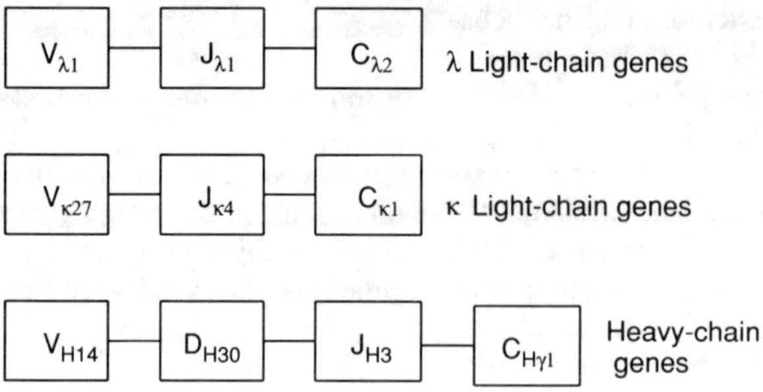

Figure 4.34 The immunoglobulin genes remaining active in a single typical B-cell—in this case producing IgG_1.

with one of at least 30 D_H genes and one of four J_H genes, to give 12,000 different possible combinations. Since the antibody combining site is formed from the variable region of one light and one heavy chain, the 1200 light chain possibilities can combine randomly with the 12,000 heavy chain options to give about 10^7 different products. In addition the variations in the precise length of V, D and J coding sequences, and the points of attachment between them, further increase the diversity, as do somatic mutations during the development of the lymphocytes.

So the key to the diversity of antibody specificity lies in the 'sharing out' of alternative germ-line genes coding for the variable regions, so that lymphocytes of a particular antibody specificity have a unique combination of genes. As a result approximately 10^8 different genetic combinations exist within the total pool of lymphocytes, allowing for the synthesis of this many different antibody specificities.

4.1.3 Class-switching

As discussed, the class of an antibody will depend on which of the five heavy-chain constant-region genes in the germ-line DNA is transcribed in the specific lymphocyte. It is observed however that class-switches take place, where an individual lymphocyte product retains its binding characteristics but shifts from one class to another. This is an important phenomenon *in vivo*, but can also be exploited *in vitro* to manipulate antibody behaviour.

This class-switching process is mediated by a DNA rearrangement process known as S-S recombination (Honjo *et al.*, 1981; Cory and Adams, 1980). These rearrangements do not alter the variable region or affect the antigen binding domain. Instead, one heavy-chain constant region is replaced with another. The rearrangement occurs downstream of the J region: aligned here in germ-line DNA (Figure 4.35) is a tandem array of heavy-chain constant region genes. The human sequence, for example, consists of μ-chain (IgM), δ-chain (IgD), γ_3-chain (IgG_3), γ_1-chain (IgG_1), γ_{2b}-chain (IgG_{2b}), γ_{2a}-chain (IgG_{2a}), ϵ-chain (IgE) and α-chain (IgA).

Class-switching characteristically moves 'downstream' through this sequence. IgM antibody is typically produced first because the μ C_H gene is the closest to the J_H segments. Analysis of myeloma DNAs producing different immunoglobulin classes revealed that specific C_H genes may subsequently be deleted, and their place taken by segments further down the sequence (Honjo and Kataoka, 1978). So for example a cell may switch from IgM to IgD production (Neuberger and Rajewski, 1981), from IgM to IgG (Thammana and Scharff, 1983) or from IgG_{2b} to IgG_{2a} (Francus *et al.*, 1978). Class-switching is a common phenomenon of immune systems.

In general, spontaneous *in vitro* class-switching in hybridoma and myeloma cell lines is a low frequency event, varying from 10^{-7} and 10^{-5} per cell per generation. Usually, therefore, at least 10^6 cells are required for initial screening of cell lines for spontaneous class-switches. To enhance the frequency of switching events a number of mutagens have been applied to myeloma cells (Preud'homme *et al.*, 1975; Yelton and Scharff, 1982). Recently Faguet and Agee

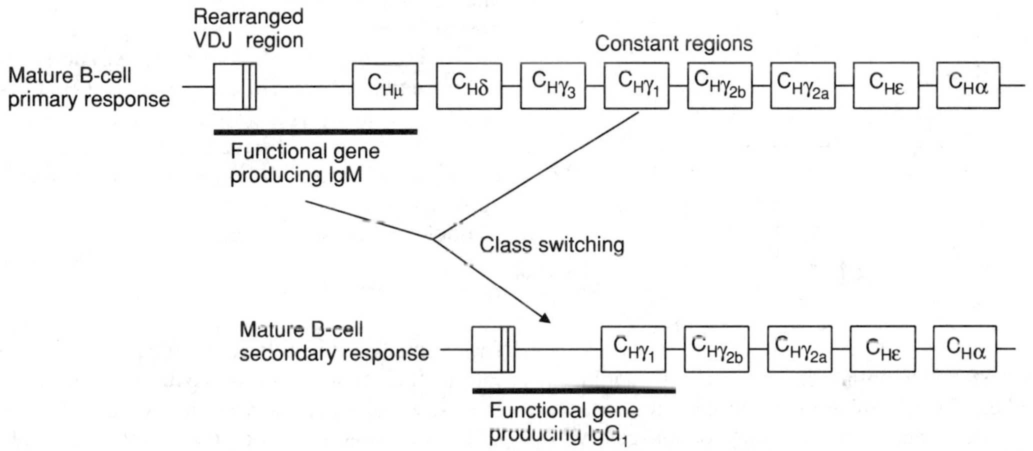

Figure 4.35 Class switching: the primary response produces IgM, but subsequent activations of a 'downstream' gene can yield a different class or subclass—in this case IgG_1.

(1993) described a simple and rapid method for enriching downstream switch variant sub-clones, based on lysing the 'unswitched' fraction of a target hybridoma population.

In order to exploit class-switching phenomena, it is necessary to detect and isolate the cells producing the desired products. Detection is relatively straightforward: any technique which can differentiate antibody classes can be employed to analyse the cells or supernatant fluid, such as fluorescent staining (Liesegang *et al.*, 1978; Radbruch *et al.*, 1982), agar overlay methods incorporating specific antisera (Yelton and Scharff, 1982; Coffino *et al.*, 1971), or immunoassays (Spira *et al.*, 1984; Muller and Rajewsky, 1983).

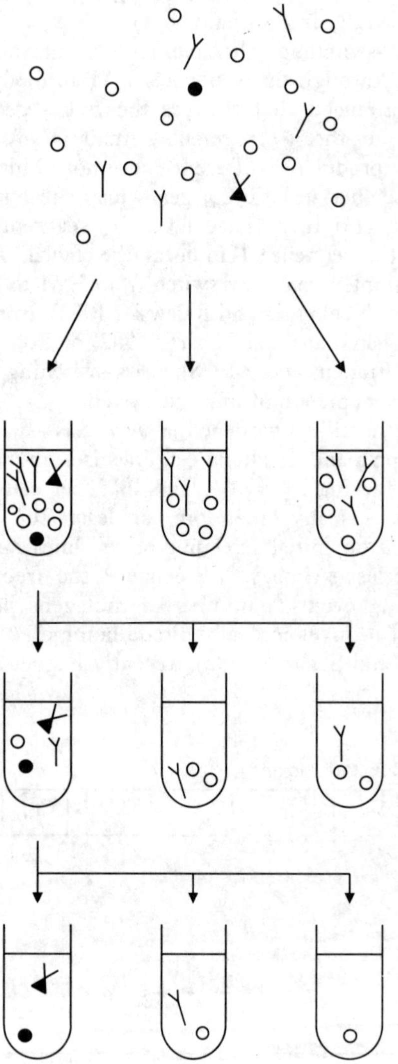

Figure 4.36 Sequential sub-lining or sub-cloning: repeated dilution of a mixture of non-secreting, secreting non-specific and secreting specific cells will eventually yield single clones—for example one of the desired specificity (black circle), yielding only the required antibody (marked by black triangle).

The isolation of class-switched cells is achieved by one of two major approaches. The first is the traditional method of sequential sub-lining or sub-cloning, which uses limiting dilution techniques to select and enrich the required sub-clones (Cavalli-Sfrorza and Lederberg, 1956). The general principle of the method is outlined in Figure 4.36.

Less labour-intensive is the use of the fluorescence-activated cell sorter (FACS), which can simultaneously recognise and isolate hybridoma cells bearing the required immunoglobulin on their surface (Radbruch *et al.*, 1980; Dangel *et al.*, 1982). Fluorescence-activated cell sorting is among the most powerful tools for the enrichment of rare cells. The principle is shown in Figure 1.15.

The characterisation of class-switch antibodies has employed standard techniques such as SDS-PAGE (Francus *et al.*, 1978; Radbruch *et al.*, 1980) and amino acid sequencing (Sablitzky *et al.*, 1982).

4.5 The Generation of Antibody Fragments

It has been seen that antibodies are multi-functional molecules, combining antigen-binding regions with areas bearing effectors for processes such as phagocytosis. Knowledge of the structure of immunoglobulins, and of structure—activity relationships, has been greatly increased by controlled degradation of these molecules—most notably with proteolytic enzymes.

Subsequently some of the fragments generated from antibody proteolysis have been shown to possess useful properties *per se*, and have become valuable immunochemical reagents. More recently molecular biology approaches have been harnessed to create recombinant analogues of these fragments.

4.5.1 Proteolysis of antibodies

In landmark studies, Porter (1958, 1959) showed that rabbit IgG treated with papain yielded three fragments, which could be separated on a carboxymethylcellulose column. Two fragments were of equal molecular mass (ca 50 kDa); these retained the ability to bind antigen, and were therefore designated Fab, fragment–antigen binding. The third had a molecular

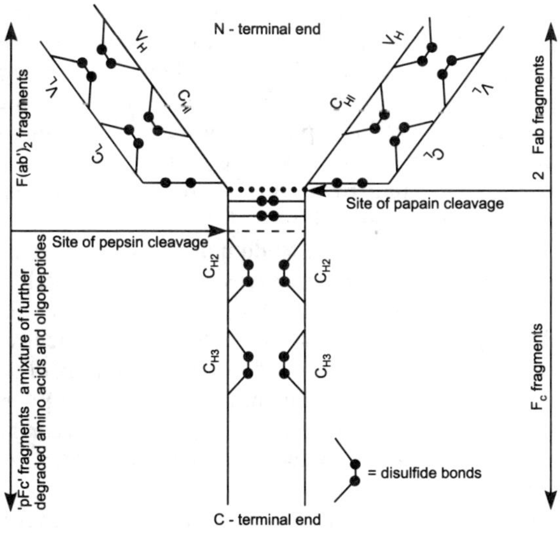

Figure 4.37 Diagram showing pepsin and papain cleavage sites and the fragments obtained.

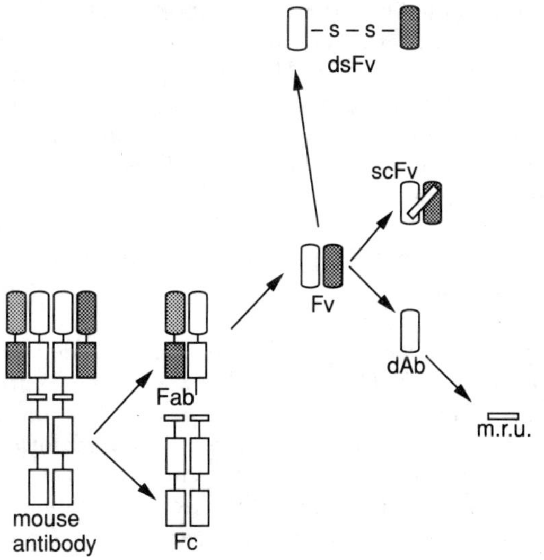

Figure 4.38 Varieties of recombinant antibody reagents. (Adapted from Winter, G. and Milstein, C. (1991). *Nature*, **349**, 293–299.)

mass of ca 80 kDa; it did not bind antigen but contained most of the class-specific features and could be crystallised. This was termed Fc, fragment–crystalline.

Pepsin, another proteolytic enzyme, was shown by Nisonoff *et al.* (1960) to produce a single definable fragment and numerous small pieces from IgG. The major fragment retained antigen-binding capacity, but unlike the papain products could still precipitate antigen. It was determined that the enzyme degraded the heavy chain starting at the C-terminal end, and proceeding as far as the interchain disulfide bridge region. This yielded a product analogous to a Fab dimer, termed $F(ab')_2$. In pepsin digestion the class-specific determinants are lost through digestion of the Fc region, the resulting mixture of amino acids and oligopeptides sometimes being termed pFc′. A schematic representation of the proteolysis of IgG by papain and pepsin, showing the cleavage sites, is given in Figure 4.37.

Other fragments of the IgG molecule have been named: the N-terminal half of the heavy chain is designated Fd, and can be produced by reduction and alkylation of the $F(ab')_2$ fragment. It is of significance in that it contains part of the antigen binding site. So too does the Fv fragment, which consists of just the variable region of a chain, and is generally prepared by recombinant means.

Because most fragmentation work has been carried out on IgG antibodies, the terms noted above are generally applied without further qualification to IgG fragments. Analogous fragments obtained from other classes are generally denoted by a suitable modifier, such as Fabμ from IgM, or Fcα from IgA.

Since the pioneering work by Porter, Nisonoff and colleagues, extensive studies have been carried out on the fragmentation of immunoglobulins using various proteolytic enzymes and reducing agents. Because of their difference in properties from the parent immunoglobulins, principally size and binding characteristics, the antigen-binding fragments produced in these studies have found useful applications in many areas as immunochemical tools, for example as diagnostic reagents and therapeutics (Mage, 1980; Parham, 1983, 1986; Lamoyi, 1986).

The yield, nature, precise point of cleavage and properties of the fragments generated depend on the enzyme used, the conditions of digestion, and the class, subclass and species origin of the immunoglobulin. For example, human IgG_1 and IgG_3 are sensitive to papain digestion in the absence of reducing agent, IgG_4 becomes sensitive only in the presence of 0.01M cysteine, and IgG_2 remains resistant even under reducing conditions.[1] Digestion conditions, therefore, may have to be extensively optimised for a given application (Potter, 1967, Gorini *et al.*, 1969; Parham, 1983, 1986).

A survey of the later literature on proteolysis shows that it is dominated by studies on mouse monoclonal

[1] Incidentally providing the basis for a subclass determination method.

immunoglobulins, primarily because of the availability of large quantities of homogeneous antibody from hybridoma production. Most of the work has concentrated on the effects of papain and pepsin, their broad substrate specificity making them suitable for a wide range of immunoglobulins of different classes and species origin.

4.5.1.1 Fragmentation of immunoglobulins using papain

Papain cleavage of IgG is the standard method of preparing Fab fragments (Stanworth and Turner, 1978). Papain is a thiol protease and thus the SH group at the active site must be in reduced form for enzyme activity. The reducing agents such as cysteine, mercaptoethanol and dithiothreitol, which are employed for activation, may also aid the digestion by their effects on the substrate. Thiols are readily oxidised in air, and it is essential to exclude oxygen during the reaction to ensure reproducible results.

Since papain is inactivated by heavy metals, it is often supplied complexed with mercury (mercuripapain); this protects the enzyme prior to use. Preparation of the enzyme for proteolysis (with or without the presence of mercury) can be accomplished by a brief incubation in neutral buffer (e.g. 0.1 M Tris-HCl, pH 8.0) containing 2 or 3 mM EDTA and a reducing agent such as 1 mM dithiothreitol or 50 mM cysteine.

The digestion reaction can be terminated by irreversible alkylation of the active thiol with a 2:1 molar excess of iodoacetamide. Alternatively, reversible inactivation is achieved by lowering the pH to acid conditions.

Digestion should result in the complete disappearance of the intact IgG molecule and the appearance of the 50 kDa Fab fragment. Samples are taken at intervals and can be analysed by a variety of methods such as SDS-PAGE or gel filtration.

The degree of digestion of mouse and rat IgG is proportional to the concentration of reducing agent used. It is essential to monitor the completeness of the reaction as individual IgG subclasses vary in their susceptibility to digestion. Mouse IgG_{2a} and IgG_{2b} are very susceptible to cleavage, while IgG_1 will require a minimum of 100 mM dithiothreitol. By contrast, all rat IgG subclasses are readily cleaved by papain in the presence of reducing agent: Rousseaux *et al.* (1980) found that Fab fragments could be produced from all rat subclasses by treatment with 1% w/w papain at pH 7.0 with 10 mM cysteine for 2–4 hours at 37 °C.

Separation of the Fab and Fc fragments from the intact molecule can be accomplished by a number of methods. The Fc regions of IgG subclasses 2a, 2b and 3 (mouse) and 2c (rat) have a strong affinity for staphylococcal Protein A (see Chapter 10, Section 10.9.4). Goding *et al.* (1978) described the recovery of Fab from a Protein A-Sepharose column: the desired Fab product passes through the column while Fc and undigested IgG are bound. The column can be regenerated by elution of the bound material with 0.1 M glycine-HCl, pH 2.5.

For IgG subclasses which do not bind to Protein-A, ion-exchange chromatography such as DEAE-cellulose chromatography may be used (Parham *et al.*, 1982). The digest is transferred to low molarity neutral Tris-HCl by gel filtration, dialysis or dilution. The Fab fragments are generally more basic than the Fc or intact molecule and will emerge in the 'drop-through', or early in the salt gradient. The other species emerge at higher salt concentrations.

Occasionally Fab and Fc fragments have a similar charge, and this indicates the use of an alternative method, especially for subclasses such as mouse IgG_1 where Protein A is inappropriate. Analogous matrices such as Protein G or A/G may be useful in such circumstances. Size exclusion chromatography offers another alternative, although high resolution HPLC matrices are required to effect a good separation. The main limitation is the capacity of these columns. Protocols have been developed for obtaining milligram quantities of purified Fab, Fc and intact molecules with relatively little dilution (see for example Parham (1986) and references cited therein).

Papain can also be used for the preparation of Fc and Fab like fragments from other immunoglobulin classes, which in general tend to be more susceptible. Using a limited incubation time, Fab and Fc type fragments can be obtained from myeloma IgD (Rowe and Fahey, 1965). Analogous fragments can also be produced by papain digestion of the 7S subunit prepared after reductive cleavage of human IgM (Onoue *et al.*, 1967). By contrast, using the standard conditions of papain digestion with two myeloma IgA proteins led to destruction of the Fc region (Bernier *et al.*, 1965). Papain cleavage has also been studied with IgE myeloma proteins, but under the usual conditions significant amounts of undigested immunoglobulin can remain (Bennich and Johansson, 1967).

4.5.1.2 Fragmentation of immunoglobulins using pepsin

The other widely used non-specific protease is pepsin, used for the preparation of $F(ab')_2$ fragments. It is only active at acid pH and is irreversibly denatured at neutral or alkaline pH. Treatment of human or rabbit immunoglobulin with pepsin at pH 4.5 at 37 °C usually results in satisfactory generation of the $F(ab')_2$ fragment (Stanworth and Turner, 1978).

However, the standard procedures are frequently not successful when applied to mouse and rat IgG. For IgG_2 antibodies the classical method works well, provided a fairly high pepsin/Ab ratio is used. By contrast, this protocol has little effect on many mouse IgG_1 proteins.

Lamoyi and Nisonoff (1983) optimised peptic digestion of mouse monoclonals of each subclass. $F(ab')_2$ fragments were obtained in good yield from IgG_1, IgG_{2a} and IgG_3. The relative rates of digestion were $IgG_3 > IgG_{2a} > IgG_1$. Optimal digestion times varied, and incubation times resulting in no undigested IgG led to somewhat lower yields, presumably due to over-digestion of the $F(ab')_2$ into smaller fragments. It was not possible to find suitable conditions for the IgG_{2b} subclass. Similar results were observed by Parham (1983).

The production of $F(ab')_2$ fragments in rat IgG has similarly been studied by Rousseaux *et al.*, (1980). As with the above two studies, the results serve to underline the need to optimise the method used for the specific proteins under consideration, and not to rely on the standard methods developed with polyclonal material.

The effect of peptic digestion has also been studied for other immunoglobulin classes (see for example Seligman and Mihaesco, 1967; Wilson and Williams, 1969; Bennich and Johansson, 1971).

4.5.1.3 Fragmentation of immunoglobulins using other proteolytic enzymes

Trypsin proteolysis of immunoglobulins results in similar fragments to papain cleavage although the method has not been used so extensively. Schrohenloher demonstrated that digestion of an IgG solution at 37 °C for three days led to two major fragments which were designated Fab(t) and Fc(t) to distinguish them from the papain fragments (Schrohenloher, 1963).

Trypsin can be used for the production of Fabμ fragments from IgM. Digestion at 37 °C produced four well resolved peaks when separated on Sephadex G-200; the first peak contained heterogeneous material, the second was similar to the peptic $F(ab')_2\mu$ fragment, the third was comparable to the papain Fabμ product, and the fourth contained peptides (Miller and Metzger, 1966).

If however, digestion is carried out at 65 °C, then Fabμ fragments are generated and a 340 kDa polymer containing five Fcμ fragments joined by the inter-subunit disulfide bonds. Because of its size, this fragment is easily separated by gel filtration, after which it will crystallise spontaneously. Reduction and separation in a non-dissociating solvent produced homogeneous 67 kDa fragments (Plaut and Tomasi, 1970).

To a much lesser extent, other enzymes have been used for fragment preparation. Some of these will be mentioned briefly. Cederblad *et al.* (1969) first looked at the effect of chymotrypsin on pooled IgG, obtaining a mixture of fragments. A cleavage pattern with characteristics of both papain and pepsin was observed, producing the following species: an $F(ab')_2$-like fragment, Fab-like fragments and an Fc′-like fragment. With IgM, a different pattern was obtained, Fabμ; a Fabμ dimer with part of the Fcμ region and a heterogeneous mixture of IgM with one or more of the $F(ab')_2\mu$ fragments hydrolysed (Chen *et al.*, 1969).

A method for digestion of rabbit IgG by the plasma enzyme plasmin was developed by Connell and Porter (1971). This resulted in the formation of a large Facb fragment (the $F(ab')_2$ region together with the N-terminal half of the Fc region) and a smaller fragment comparable to the pFc′ fragment produced by peptic cleavage of rabbit IgG.

An enzyme produced by enteric streptococci, IgA protease, has been found to have proteolytic activity specific to the human IgA_1 subclass, but not to IgA_2 (Plaut *et al.*, 1974). The protease was inactive against all other protein substrates examined, including the other classes of human immunoglobulin. The IgA_1 was cleaved in the hinge region, producing Faba and Fca fragments.

A comparative study of the effect on mouse IgG of a range of narrow-specificity proteolytic enzymes was carried out recently by Yamaguchi *et al.* (1995). A panel of mouse monoclonal switch variant antibodies (see Section 4.4.3) of IgG_1, IgG_{2a} and IgG_{2b} subclasses were examined. Cleavage sites for clostripain, lysyl endopeptidase, metalloendopeptidase and V8 protease were mapped on the hinge region of each of the IgG subclasses. No effect was seen with IgG_1, except with metalloendopeptidase when the IgG had been reduced and alkylated with iodoacetamide. IgG_{2a} was cleaved by clostripain, but only in the presence of 5 mM dithiothreitol. Lysyl endopeptidase however, cleaved both IgG_{2a} and IgG_{2b} without perturbing the interchain disulfide bridges, resulting in $F(ab')_2$ and Fc fragments.

A summary of digestion methods for the different immunoglobulin classes of a number of species in given in Table 4.8.

Table 4.8 Summary of antibody digestion methods

Enzyme	Species	Class	Subclass	Fragments produced	Conditions	Reference
Papain	Rabbit	IgG		Fab & Fc	(P) pH 7.0, 10 mM cysteine, 2 mM EDTA, E:Ab (1: 100), 16 h, 37 °C	Porter (1958, 1959)
Pepsin	Rabbit?	IgG		$F(ab')_2$	10 mM acetate, pH 4.8, E 28 mg/ml + reducing agent	Nisonoff *et al.* (1960)
Papain	Mouse/rat	IgG	IgG_1	Fab	100 mM acetate pH 5.5, 3 mM EDTA, E:IgG (1:20), 37 °C	Stanworth and Turner (1978)
Papain	Rat	IgG	All	Fab	1% enzyme, pH 7.0, 10 mM cysteine, 2–4 h, 37 °C	Rousseaux *et al.* (1980)
Papain	Human	IgD		Fab, Fc (type)		Rowe and Fahey (1965)
Papain	Human	IgM		Fab, Fc (type)	100 mM (P), pH 7.0, E:IgG (1: 100), 10 mM cysteine, 2 mM EDTA, 37 °C	Onoue *et al.* (1967)
Papain	Human	IgA		Fab, Fc destroyed	Standard conditions	Bernier (1965)
Papain	Human?	IgE			Standard conditions	Bennich and Johansson (1967)
Pepsin	Human, rabbit		IgG	Invariably $F(ab')_2$	Acetate pH 4.5, 37 °C	Stanworth and Turner (1978)
Pepsin	Human, rabbit, goat, mouse					Nisonoff *et al.* (1960)
Pepsin	Mouse	IgG	$IgG_{1,\ 2a,\ 3}$, not for IgG_{2b}	$F(ab')_2$	100 mM acetate, pH 4.2, E:IgG (1:33), 37 °C	Lamoyi and Nisonoff (1983)
Pepsin	Mouse?	Other than IgG		$F(ab')_2$		Seligman and Mihaesco (1967); Wilson and Williams (1969); Bennich and Johansson (1967)
Trypsin	Human	IgG		Fab (t) & Fc (t)	(P) pH 7.6, E:IgG (1: 100), 3 days 37 °C	Schrohenloher (1963)
Trypsin	Human	IgM		Fabμ, $F(ab')_{2\mu}$	pH 8.0, 37 °C	Miller and Metzger (1966)
Trypsin	Human	IgM		Fab & Fc (5) polymer	50 mM Tris-HCl pH 8.1, 30 min, E:IgG (1: 25), 65 °C	Plaut and Tomassi (1970)
Chymotrypsin	Human	IgG		$F(ab')_2$, Fab & Fc like fragments	10 mM Tris-HCl, pH 8.5, 37 °C, E:IgG (1:40)	Cederblad *et al.* (1969)
Chymotrypsin	Human	IgM		Fabμ, $F(ab')_2$, Fcμ and $F(ab')_{2\mu}$ hydrolysed	100 mM (P) pH 8.0, 1–2% enzyme by wt.	Chen *et al.* (1969)
IgA protease	Human	IgA	IgA_1 only not IgA_2	Fabα, Fcα	50 mM Tris-HCl pH 8.1, E:IgG (1:1), 16 h, 37 °C	Plaut *et al.* (1974)
Lysyl endopeptidase	Mouse Mab switch variant	IgG	$IgG_{2a,\ 2b}$ not IgG_1	$F(ab')_2$, Fc	50 mM Tris-HCl pH 8.5, E:IgG (1: 50–1000), 37 °C, 1–6 h	Yamaguchi *et al.* (1995)
Papain	Mouse	IgG	$IgG_{1,\ 2a,\ 2b}$	$F(ab')_2$, Fc	75 mM (P) pH 7.0, 2 mM EDTA, 10 mM cysteine, E:IgG (1: 500–100), 37 °C, 2–8 h	Yamaguchi *et al.* (1995)
Clostripain	Mouse	IgG	Not IgG_{2a}	$F(ab')_2$, Fc	100 mM Tris-HCl pH 7.5, 5 mM EDTA, E:IgG (1:100), 37 °C, 2–18 h	Yamaguchi *et al.* (1995)
Metalloendopeptidase	Mouse	IgG	IgG-CM, not IgG_1	$F(ab')_2$, Fc	100 mM glycine pH 8.5, E:IgG (1:100), 37 °C, 4 h	Yamaguchi *et al.* (1995)
V8 protease	Mouse	IgG	IgG_{2b}-CM, not $IgG_{1,\ 2a}$	Fab, Fc	100 mM (P) pH 7.8, 2 mM EDTA, E:IgG (1: 30–100), 1–24 h, 37 °C	Yamaguchi *et al.* (1995)
Papain	Rabbit	IgG		$F(ab')_2$, Fc	100 mM Tris-HCl pH 8.5, 2 mM EDTA, cysteine, E:IgG (1:20), 37 °C	Cebra *et al.* (1961)
Pepsin	Mouse	IgG	$IgG_{1,\ 3,\ 2a}$, not IgG_{2b}	$F(ab')_2$	100 mM acetate pH 4.2–4.5, E:IgG (1:33), 37 °C	Lamoyi (1986)
Papain	Human myeloma	IgE		Fab, Fc	(P) pH 7.0, 10 mM cysteine, 2 mM EDTA, 1% papain, 37 °C, 40 min–3 h	Ishizaka *et al.* (1970)
Papain	Rabbit	IgG		Fab, Fc	10 mM (P) pH 7.3, 0.5 mM cysteine, E:IgG (1:100), 1 h, 37 °C	French *et al.* (1986)
Papain	Mouse	IgG	IgG_1	$F(ab')_2$	Cysteine activation, acetate pH 5.5, 3 mM EDTA, 37 °C, E:IgG (1:10), 1 h	Fredrickson *et al.* (1989)

Enzyme	Species	Class	Subclass	Fragments produced	Conditions	Reference
Pepsin	Mouse	IgG	IgG_{2b}	Fab, Fc	Acetate pH 4.8, E:IgG (1:30), 37 °C, 4 h	Demignot *et al.* (1989)
Pepsin	Mouse	IgG	IgG_{2a}, IgG_3, IgG_1	$F(ab')_2$	500–800 mM ammonium sulfate, 20 mM citrate pH 4.5	Ultee and Rea (1993)
Bromelain	Mouse	IgG	IgG_1	$F(ab')_2$	50 mM Tris-HCl pH 7.0, 2 mM EDTA, 1 mM cysteine, E:IgG (1:50), 37 °C, 8 h	Mariani *et al.* (1991)
Ficin	Mouse	IgG	IgG_1	$F(ab')_2$	50 mM Tris-HCl pH 7.0, 2 mM EDTA, 1 mM cysteine, E:IgG (1:30), 37 °C	Mariani *et al.* (1991)
Papain	Mouse	IgG	IgG_1	$F(ab')_2$, Fab	20 mM acetate pH 4.0, 225 U/mg IgG, 37 °C, 16 h	Ying *et al.* (1995)
Papain	Mouse	IgG	IgG_1	$F(ab')_2$	100 mM acetate pH 5.5, 50 mM cysteine, 3 mM EDTA, 37 °C, 8 h	Ying *et al.* (1995)
Bromelain	Mouse	IgG	IgG_1	$F(ab')_2$, Fab	50 mM Tris pH 7.0, 2 mM EDTA, 0.2 U/mg IgG, cysteine activation, 4 h, 37 °C	Ying *et al.* (1995)
Ficin	Mouse	IgG	IgG_1	$F(ab')_2$, Fab	50 mM Tris pH 7.0, 2 mM EDTA, 0.06 U/mg IgG, cysteine activation, 4 h, 37 °C	Ying *et al.* (1995)
Trypsin	Mouse	IgM		Small active fragments	50 mM Tris-HCl pH 8.0, E:IgG (1:100), 5 h, 10 mM mercaptoethanol, 37 °C	Matthew and Reichardt (1982)
Pepsin	Human	IgG	IgG_1	$F(ab')_2$, Fab	Acetate pH 4.0, E:IgG (1:33), 4 h, 37 °C	Vihko and Kurkela (1988)
Papain	Human	IgG	IgG_1	$F(ab')_2$	Acetate pH 5.5, 50 mM cysteine, 3 mM EDTA, E:IgG (1:10), 37 °C	Vihko and Kurkela (1988)
Pepsin	Mouse	IgG	IgG_1	$F(ab')_2$	Acetate pH 4.5, 4–24 h, 37 °C	Colcher *et al.* (1989)
Papain	Mouse	IgG	IgG_1	$F(ab')_2$	Acetate pH 5.5, 3 mM EDTA, 50 mM cysteine, 5% E, 1.5 h, 37 °C	Colcher *et al.* (1989)
Bromelain	Mouse	IgG	IgG_1	$F(ab')_2$	50 mM Tris-HCl pH 7.0, 2 mM EDTA, 10 mM cysteine, 2% E, 37 °C, 24 h	Colcher *et al.* (1989)
Papain	Mouse	IgG	$IgG_{1,2}$	Fc	(P) pH 8.0, ascitic fluid, cysteine, EDTA, 4 h, 37 °C	Harris *et al.* (1975)
Armillaria melleaprotease	Human	IgG	$IgG_{1,3,4}$, not IgG_2	Fab, Fc	10 mM Tris-HCl pH 7.0, E:IgG (1:100), rapid, 37 °C	Hunneyball and Stanworth (1975)
Trypsin	Human	IgG		Fab	50 mM Tris-HCl pH 8.1, 10 mM cysteine, 37 °C, 16–24 h	Waller *et al.* (1969)
Chymotrypsin	Human	IgG		Fab	80 mM Tris-HCl pH 7.8, 10 mM cysteine, 37 °C, 18–24 h	Waller *et al.* (1969)
Subtilisin	Human	IgG		Fab	25 mM Tris-HCl pH 7.5, 10 mM cysteine, 37 °C, 16–20 h	Waller *et al.* (1969)
Elastase	Human	IgG		Fab	20 mM Tris-HCl pH 8.8, 10 mM cysteine, 37 °C, 18–20 h	Waller *et al.* (1969)
Lysyl endopeptidase	Human	IgG		$F(ab')_2$-like fragment	100 mM ammonium carbonate, E:IgG (1:100), 37 °C, 8 h	Fisch *et al.* (1992)
Pepsin	Mouse	IgG	$IgG_{2b>2a>1}$	$F(ab')_2$	Acetate pH 4.0, ascites fluid, 37 °C, 8 h	Parham (1983)
Pepsin	Rat	IgM		$F(ab')_2$, L chain	100 mM citrate pH 4.5, E:IgG (1:200), 37 °C, 2 h	Inouye and Morimoto (1994)

4.6 Alternative Methods for Generating Antibodies and Antibody Fragments

As noted earlier, the introduction of hybridoma technology in the 1970s represented a quantum leap in immunology. For the first time reasonable quantities of essentially monotypic immunoglobulins could be obtained without relying on the products of naturally occurring myelomas, and, most importantly, suitable combinations of immunisation and selection could provide antibodies of exquisitely defined specificity.

However, it is now nearly 25 years since this revolution, and further advances have taken place to widen the range of available antibodies beyond those obtainable by hybridoma technology alone. These will be reviewed briefly here.

4.6.1 Variations on hybridoma technology

4.6.1.1 Somatic mutations

Mutations in protein synthesis occur naturally during cell culture, and these changes have been harnessed to obtain modified products from hybridomas. The early 1980s saw this approach used to generate mutants with modified specificity (Winter and Milstein, 1991), allowing some widening of the properties obtainable. 'Class switching' to obtain antibodies of the *same* specificity but with a different heavy chain is another example of somatic mutation, and has already been discussed in Section 4.4.3.

4.6.1.2 Hybrid hybridomas

Hybridomas are obtained by fusing an antibody-producing lymphocyte cell to an 'immortal' myeloma cell. An extension of this principle is to carry out a *further* fusion operation on such a hybridoma, so that it acquires the properties of a second antibody-producing cell of different specificity. A fraction of the immunoglobulin product of such cells (hybrid hybridomas or quadromas) may be bispecific, combining both specificities in a single molecule. This technique was introduced by Milstein and Cuello (1983), who demonstrated it by the production of a Mab combining specificity for horseradish peroxidase and somatostatin.

This provides an alternative to chemical methods for the production of bispecific antibodies by conjugation of fragments from separate monoclonals.

4.6.2 Recombinant techniques for antibody production

The molecular biology revolution has opened up the possibility of much greater engineering of protein synthesis, and the genes responsible for antibody production have provided key targets for manipulation of this type (Winter and Milstein, 1991; Owens and Young, 1994). The use of amplification techniques such as the polymerase chain reaction (PCR) has facilitated such studies, allowing reasonable quantities of genetic material to be obtained from a sparse original source.

Early approaches involved obtaining genes from existing hybridomas, cloning them into suitable vectors such as plasmids, and expressing whole antibody chains in mammalian cells of lymphoid or other origin (Neuberger, 1983). However, genetic engineering has been employed to generate a wider and more interesting range of products from recombinant routes (Figure 4.38).

4.6.2.1 Recombinant antibody fragments

The gene sections responsible for antibody fragment production can be cloned into microbial cells (Boss *et al.*, 1984; Cabilly *et al.*, 1984) and these approaches have been refined to produce fully active products (Skerra and Plückthun, 1988). As shown in Figure 4.38 a wide range of different species can be obtained in this way. In particular attention has been focussed on molecules of increasingly small size which retain their antigen-binding capability: from Fab, through Fv which contains only the light and heavy chain variable regions, to domain antibodies (dAbs) which consist of only a single variable region, and finally to minimal recognition units (MRUs) which possess only the peptide regions essential for binding (Winter and Milstein, 1991; Skerra, 1993).

The light and heavy chain units of Fv are only linked by non-covalent forces, and a useful variant is the single-chain Fv (scFv) fragment, obtained by engineering a flexible, hydrophilic peptide region between the two units (Huston *et al.*, 1988, 1993). Alternatively, a disulfide bond can be engineered between the V_H and V_L units, yielding disulfide-stabilised Fv (dsFv) fragments (Reiter *et al.*, 1996).

Another variation on the recombinant production of native fragments is to use genetic manipulation to introduce new functional regions into the molecule. For example flag sequences (*cf.* Chapter 10, Section 10.9.41) can facilitate affinity purification of the products (Schmidt and Skerra, 1993), while terminal cysteine residues can facilitate covalent attachment to solid phases (Kipriyanov and Little, 1997).

The introduction of antibody-producing genes into plants is another interesting route for producing these molecules. Such transgenic plants can be engineered to produce single types of heavy or light chain, for example; then standard plant hybridisation procedures can be employed to combine these capabilities to construct more complex molecules.[1] Ma and Hein (1995) described a transgenic *Nicotiana* capable of expressing

[1] A fascinating combination of traditional and 'high-tech' genetic engineering techniques!

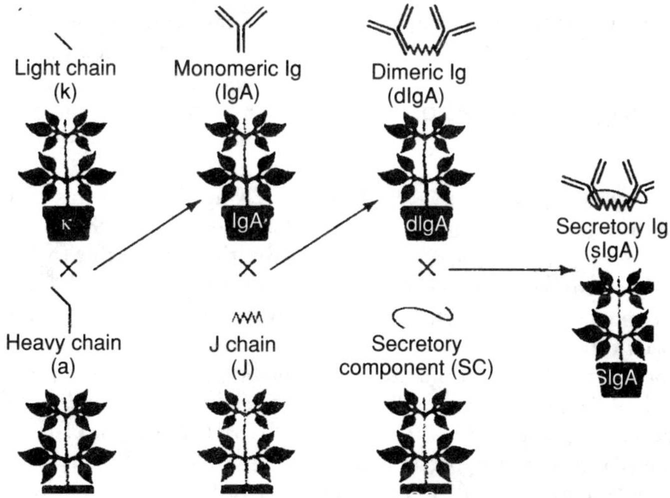

Figure 4.39 Assembly of secretory IgA in transgenic plants. The genes encoding the four polypeptides of sIgA are accumulated in a single plant by successive cross-fertilisation of single transgenic plants and their filial recombinants. The molecular structure of the antibody molecules in each generation is depicted above each plant. (Reproduced with permission from: Ma, J. K.-C. and Hein, M. B. (1995). *Trends Biotechnol.*, **13**, 522–527.)

intact secretory IgA, which was obtained by hybridisation of individual plants encoding for heavy chain, light chain, J-chain, and secretory fragment (Figure 4.39).

4.6.2.2 Chimaeric antibodies

A major target for genetic manipulation has been to resolve the paradox that most monoclonal antibodies derive from mice, while most therapeutic uses are in humans. Unwanted immunological responses often result from the use of mouse antibodies *in* (human) *vivo*, the generation of human anti-mouse antibodies (HAMA) presenting a serious obstacle to continued treatment with such reagents. Much effort has therefore gone into 'humanising' mouse antibody molecules by introducing as much human character as possible; these are termed chimaeric antibodies in reference to their dual-species origin.

Thus genetic manipulation has been used to produce antibodies or fragments which contain binding domains of rodent origin grafted onto human frameworks (Better *et al.*, 1988; Winter and Milstein, 1991; Owens and Young, 1994). In the extreme approach only the hypervariable regions of the molecule do not derive from human immunoglobulin—reshaped or CDR-grafted antibodies; with these reagents the anti-mouse response is minimised (see Figure 4.38).

The use of transgenic plants and animals also lends itself to the production of engineered antibodies of this type. Recently for example large sections of the human gene loci for immunoglobulin production have been transferred into the germ line of mice using yeast artificial chromosomes (YACs). The intrinsic Ig-producing genes of these transgenic mice can be selectively deactivated, so they offer another vehicle for the convenient production of antibodies compatible with human use (Neuberger and Brüggemann, 1997).

4.6.2.3 Fusion proteins

Chapter 9 describes in detail how genetic engineering can be used as an alternative to chemical coupling of protein molecules, the entire conjugate instead being expressed as a fusion protein from a combined gene. Needless to say the production of fusion proteins containing an antibody fragment has been reported on numerous occasions (Uhlén *et al.*, 1992), and offers another means by which the available range of antibody reagents can be extended by recombinant techniques.

In fact the scFv fragments referred to above are fusion proteins, and the concept can be extended further to the production of bispecific analogues, so-called diabodies, which combine fragments of more than one affinity in a single fusion construct (Zhu *et al.*, 1996; Coloma and Morrison, 1997). Other examples from this field include immunotoxins combining an antibody fragment with a cytotoxic species such as *Pseudomonas* exotoxin A (Chen *et al.*, 1997).

4.6.3 Surface display of antibodies on cells and microorganisms

Another application of gene fusion involves linking peptide sequences of interest to the surface (coat) proteins of cells or microorganisms. These are then amenable to screening procedures, allowing selection of those displaying the sequence of interest.

This general approach can be applied to eukaryotic cells, where the anchor is typically a phosphatidylinositol-glycan (P-IG) moiety (Whitehorn *et al.*, 1995), to yeast cells *via* agglutinins (Schreuder *et al.*, 1993), to both Gram-negative and Gram-positive bacteria *via* a range of surface proteins (Georgiou *et al.*, 1997; Ståhl and Uhlén, 1997), and to viruses (Grabherr *et al.*, 1997). In the antibody field it is the display of polypeptides on the surface of bacteriophages, the viruses which infect bacteria, which have attracted most attention.

4.6.3.1 Phage display of antibody fragments

Most commonly, polypeptide sequences representing antibody fragments are displayed on filamentous bacteriophages such as M13 or fd, in particular the cpIII proteins which are expressed at the tip of the phage and mediate binding to the *E. coli* host (Smith, 1985). McCafferty *et al.* (1990) first described the display of scFv fragments at the surface of an fd phage; the labelled phages exhibited similar specificity to the antibody fragments themselves and were termed phage antibodies. Barbas *et al.* (1991) displayed Fab fragments on M13 phages.

This approach offers a number of features which are worthy of interest (Chiswell and McCafferty, 1992; Winter *et al.*, 1994; Nissim *et al.*, 1995). The phage antibodies[1] themselves can be multiplied readily by culture in *E. coli*; those of desired specificity can be selected by affinity chromatography or panning, using a solid phase labelled with the antigen of interest. Therefore a workable quantity of phage antibody can be obtained at high purity simply by culturing the product of an affinity selection procedure.

PCR has been combined with this technique to build up large libraries of phages displaying scFvs made up of random combinations of V_H and V_L fragments, obtained by amplification of the genetic material from lymphocytes. In effect this approach mimics the operation of the immune system in creating antibodies with varying specificities. Once these libraries have been established, it is facile to isolate any phage antibodies of the desired specificity by the procedures just described.

Using lymphocytes from an immunised mouse, Clackson *et al.* (1991) isolated phage antibodies with affinities comparable to those obtained from an *in vivo* immune response (10^{-8} M^{-1}). Thus the necessity for hybridoma production can be bypassed; this is particularly advantageous for the derivation of human antibodies (see, for example, Siegel *et al.*, 1997) where hybridoma techniques are difficult to apply satisfactorily. A comparison of the 'traditional' and phage display routes to these reagents is shown in Figure 4.40.

Most excitingly, with a large enough library of scFv-bearing phages it should be possible to identify examples of desired specificity without the enrichment afforded by the immunisation procedure. Using such a 'naive' library of 3×10^7 clones bearing random combinations of human V_H and V_L domains, Marks *et al.* (1991) identified phage antibodies with specificity for over 14 different targets, including self antigens which cannot be obtained by immunisation. Challenges remain in this area to reproduce the high affinities obtained by the natural 'maturation' of the immune response, but progress has been made in mimicking this effect *in vitro* by shuffling the chains of the phages originally identified (Marks *et al.*, 1992). No doubt progress will continue to be made in this respect (Hoogenboom, 1997); Vaughan *et al.* (1996) reported antibodies of affinities greater than 10^9 M^{-1} from a naive library of over 10^{10} scFv fragments.

Applications of phage display in this field are becoming commoner all the time (see for example McGuinness *et al.*, 1996; Hansen *et al.*, 1997; Lindner *et al.*, 1997; Pereira *et al.*, 1997), and this technology looks set to have a major impact on the identification of antibodies with a particular, desired specificity. With large libraries in place, considerable time could be saved in this process compared to 'traditional' hybridoma methodologies, as well as potentially avoiding animal usage completely. There may be a greater chance of obtaining antibodies of a desired specificity as well.

4.6.3.2 Other surface display applications

Phage display can be used 'in reverse' for the selection of antibodies of a desired specificity. Lijnen *et al.* (1997),

[1] Chiswell and McCafferty's suggestion of the term 'coliclonal' to describe such antibodies does not seem to have been widely adopted.

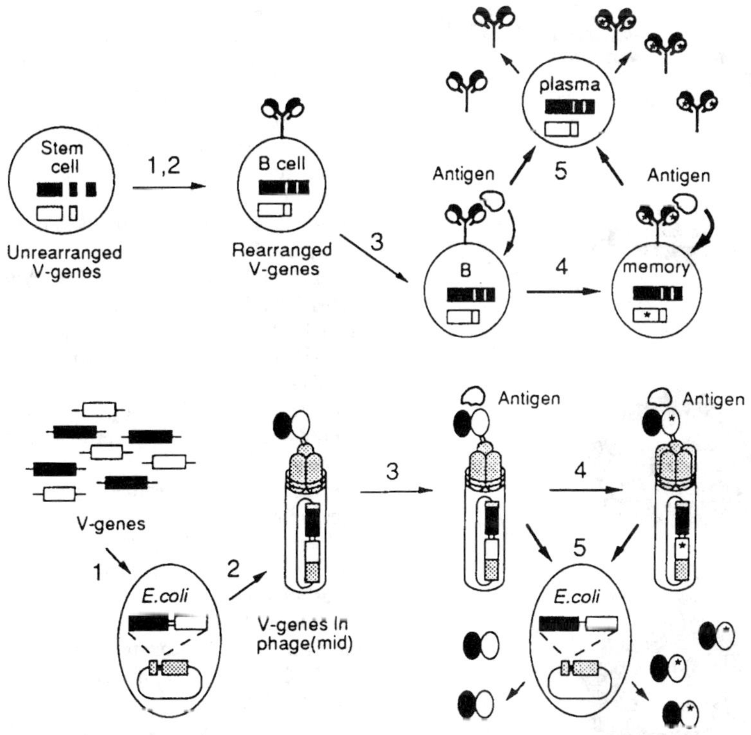

Figure 4.40 Generation of antibodies by the immune system and by phage technology. Steps: 1, rearrangement of assembly; 2, surface display of antibody (fragment); 3, antigen-driven or affinity selection; 4, affinity maturation; 5, production of soluble antibody (fragment). (Reproduced with permission from: Nissim *et al.* (1995). *Tumor Targeting*, **1**, 13–16.)

for example, used a panel of phage-displayed antigens to screen antibodies coated to microtitre wells, an anti-phage-HRP conjugate being used to detect 'hits'.

Finally, it has been noted in the introduction that bacteriophages represent only one type of vehicle for the general surface display approach, and antibody fragments have also been displayed on the exterior of bacteria such as *E. coli* and *Staphylococcus* (Georgiou *et al.*, 1997; Ståhl and Uhlén, 1997).

4.6.4 Molecular imprinting

A completely different approach to the generation of antibody-like properties is based on molecular imprinting (Wulff, 1995; Ansell *et al.*, 1996; Mosbach and Ramström, 1996). In this technique an antigen is used as a 'template' around which a suitable monomer is polymerised[1]. Once the reversible links to the template antigen have been removed, an extremely stable polymeric material with antibody-like binding characteristics is obtained (Figure 4.41). Binding constants in the nanomolar range have been reported (Ansell *et al.*, 1996), although a range of affinities are typically observed within the product. Current research is aimed at reducing this 'polyclonal' behaviour.

Typical monomers are methacrylates such as ethylene glycol dimethacrylate (EDMA) or trimethylol propane trimethacrylate. These are mixed with the imprinting antigen in a porogenic solvent such as chloroform or acetonitrile, which promotes porosity in the polymer structure. Following polymerisation, extensive washing and extraction procedures remove the template to yield the product (Ansell *et al.*, 1996).

Due to the use of organic solvents to carry out the polymerisation reaction this approach is so far limited to haptens; work continues to enable the use of protein antigens (Ansell *et al.*, 1996).

[1] A fascinating echo of the template theory of antibody generation mentioned earlier.

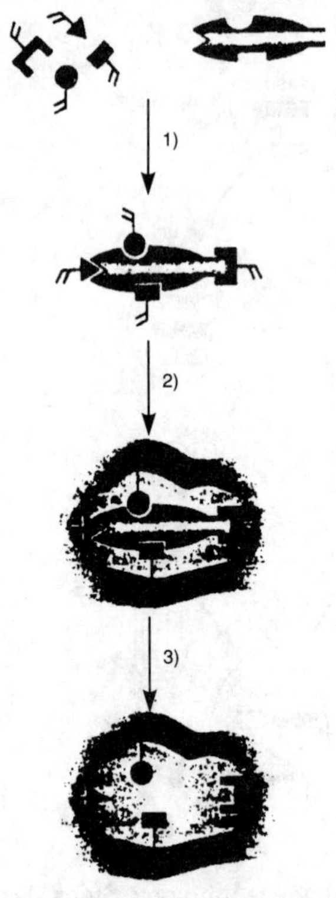

Figure 4.41 Schematic representation of the preparation of artificial antibodies by molecular imprinting. (1) Functional monomers interact reversibly with an imprint antigen to give a complex in solution; (2) cross-linker and initiator are added and polymerization takes place, retaining in the rigid polymer the interactions between the imprint antigen and the monomers; (3) the imprint antigen is removed, leaving in the polymer binding sites complementary to the imprint antigen in the positioning of functional groups and shape. (Reproduced with permission from: Ansell *et al.* (1996). *Clin. Chem.*, **42**, 1506–1512.

The high stability of these molecularly imprinted polymers (MIPs) is their greatest attraction, and applications under stringent conditions can be envisaged.

References

Abel, C. A. *et al.* (1968). The carbohydrate content of fragments and polypeptide chains of human γG-myeloma protein of different heavy-chain subclasses. *Biochemistry*, **7**, 1271–1278.

Adamczyk, M. *et al.*, (1997). Immunoassay reagents for thyroid testing. 2. Binding properties and energetic parameters of a T_4 monoclonal antibody and its Fab fragment with a library of thyroxine analog biosensors using surface plasmon resonance. *Bioconjugate Chem.*, **8**, 133–145.

Alzari, P. M. *et al.* (1988). three-dimensional structure of antibodies. *Annu. Rev. Immunol.*, **6**, 555–580.

Amzel, L. M. and Poljak, R. J. (1979). Three-dimensional structures of immunoglobulin. *Annu. Rev. Biochem.*, **48**, 961–997.

Amzel, L. M. *et al.* (1974). The Three-dimensional structure of a combining region-ligand complex of immunoglobulin NEW at 3.5 Å resolution. *Proc. Natl. Acad. Sci. USA*, **71**, 1427–1430.

Ansell, R. J. *et al.* (1996). Towards artificial antibodies prepared by molecular imprinting. *Clin. Chem.*, **42**, 1506–1512.

Baeziger, J. and Kornfeld, S. (1974a). Structure of the carbohydrate units of IgE immunoglobulin (Part I). *J. Biol. Chem.*, **249**, 1889–1896.

Baeziger, J. and Kornfeld, S. (1974b). Structure of the carbohydrate units of IgE immunoglobulin (Part II). *J. Biol. Chem.*, **249**, 1897–1903.

Baeziger, J. and Kornfeld, S. (1974c). Structure of the carbohydrate units of IgA_1 (Part I). *J. Biol. Chem.*, **249**, 7260–7269.

Baeziger, J. and Kornfeld, S. (1974d). Structure of the carbohydrate units of IgA_1 (Part II). *J. Biol. Chem.*, **249**, 7270–7281.

Barbas, C. F., III *et al.* (1991). Assembly of combinatorial antibody libraries on phage surfaces: the gene III site. *Proc. Natl. Acad. Sci. USA*, **88**, 7978–7982.

Bennich, H. and Johansson, S. G. O. (1967). Studies on a new class of immunoglobulins (II). In *Nobel Symposium in Gamma Globulins: Structure and Control of Biosynthesis* (Killander, J., ed.), p. 199, Almqvist and Wiskell, Stockholm.

Bennich, H. and Johansson, S. G. O. (1971). Structure and function of human imunoglobulin E. *Adv. Immunol.*, **13**, 1–55.

Bernier, G. M. *et al.* (1965). Structural studies of the immunoglobulins (II). Antigenic and chemical properties of gamma-A myeloma globulins. *Biochemistry*, **4**, 2072–2079.

Better, M. *et al.* (1988). *Escherichia coli* secretion of an active chimeric antibody fragment. *Science*, **240**, 1041–1043.

Bhat, T. N. *et al.* (1990). Small rearrangements in structures of Fv and Fab D1.3 on antigen binding. *Nature*, **347**, 483–485.

Birkmeyer, R. C. *et al.* (1981). Determination of relative antigen–antibody affinities by quantitative immunoelectrophoresis. *J. Immunol. Methods*, **44**, 271–284.

Boss, M. A. *et al.* (1984). Assembly of functional antibodies from heavy and light chains synthesised in *E. coli*. *Nucleic Acids Res.*, **12**, 3791–3806.

Brownlee, M. *et al.* (1984). Nonenzymatic glycosylation and pathogenesis of diabetic complications. *Ann. Intern. Med.*, **101**, 527–537.

Cabilly, S. *et al.* (1984). Generation of antibody activity from immunoglobulin polypeptide chains produced in *Escherichia coli. Proc. Natl. Acad. Sci. USA*, **81**, 3273–3277.

Cavalli-Sfrorza, L. L. and Lederberg, J. (1956). Isolation of pre-adaptive mutants in bacteria by sib selection. *Genetics*, **41**, 367–382.

Cebra, J. J. *et al.* (1961). A two-stage cleavage of rabbit γ-globulin by a water-insoluble papain preparation followed by cysteine. *J. Biol. Chem.*, **236**, 1720–1725.

Cederblad, G. *et al.* (1969). Immunological investigation of human γ-globulin degraded by chymotrypsin. *Clin. Chim. Acta*, **8**, 78–85.

Chen, J. P. *et al.* (1969). Studies on the chymotrypsin C and papain fragments of human immunoglobulin M. *Biochemistry*, **8**, 2246–2251.

Chen, S.-Y. *et al.* (1997). Potent antitumor activity of a new class of tumor-specific killer cells. *Nature*, **385**, 78–80.

Chiswell, D. J. and McCafferty, J. (1992). Phage antibodies: will new 'coliclonal' antibodies replace monoclonal antibodies? *Trends Biotechnol.*, **10**, 80–84.

Clackson, T. *et al.* (1991). Making antibodies using phage libraries. *Nature*, **352**, 624–628.

Clamp, J. R. and Putnam, F. W. (1964). The carbohydrate prosthetic group of human γ-globulin. *J. Biol. Chem.*, **239**, 3233–3240.

Clamp, J. R. and Johnson, I. (1971). The oligosaccharide units of a human type L immunoglobulin M (macroglobulin). *Biochem. J.*, **123**, 739–745.

Clamp, J. R. and Johnson, I. (1972). In *Glycoproteins, Part A* (Gottschalk, A., ed.), 2nd edit., pp. 630–631, Elsevier, Amsterdam.

Coffino, P. *et al.* (1971). Cloning of mouse myeloma cells and detection of rare variants. *J. Cell. Physiol.*, **79**, 429–435.

Colcher, D. *et al.* (1989). Comparison of methods for the generation of immunoreactive fragments of a monoclonal antibody (B72.3) reactive with human carcinomas. *J. Immunol. Methods*, **120**, 71–83.

Colman, P. M. (1988). Structure of antibody–antigen complexes: implications for immune recognition. *Adv. Immunol.*, **43**, 99–132.

Coloma, M. J. and Morrison, S. L. (1997). Design and production of novel tetravalent bispecific antibodies. *Nature Biotechnol.*, **15**, 159–163.

Connell, G. E. and Porter, R. R. (1971). A new enzymic fragment (Facb) of rabbit immunoglobulin G. *Biochem. J.*, **124**, 53–59.

Cory, S. and Adams, J. M. (1980). Deletions are associated with somatic rearrangement of immunoglobulin heavy chain genes. *Cell*, **19**, 37–43.

Dangel, A. A. *et al.* (1982). Rapid isolation of cloned isotype switch variants using fluorescence-activated cell sorting. *Cytometry*, **2**, 395–399.

Davies, D. R. and Metzger, H. (1983). Structural basis of antibody function. *Annu. Rev. Immunol.*, **1**, 87–117.

Davies, D. R. and Padlan, E. A. (1990). Antibody–antigen complexes. *Annu. Rev. Biochem.*, **59**, 439–473.

Davies, D. R. *et al.* (1975). Three-dimensional structure of immunoglobulins. *Annu. Rev. Biochem.*, **44**, 639–667.

Davies, D. R. *et al.* (1979). Immunoglobulin structures at high resolution. *Annu. Rev. Biochem.*, **48**, 961–997.

Dawson, G. and Clamp, J. R. (1968). Investigation on the oligosaccharide units of an myeloma globulin. *Biochem. J.*, **107**, 341–352.

Deisenhofer, J. (1981). Crystallographic refinement and atomic models of a human Fc fragment and its complex with fragment B of protein A from *Staphylococcus aureus* at 2.9 Å and 2.8 Å resolution. *Biochemistry*, **20**, 2361–2370.

Deisenhofer, J. *et al.* (1976). Crystallographic structural studies of a human Fc fragment: a complete model based on a fourier map at 0.35 nm resolution. *Hoppe-Seyler's Z. Physiol. Chem.*, **357**, 1421–1426.

Demignot, S. *et al.* (1989). Mouse IgG_{2b} monoclonal antibody fragmentation: preparation and purification of Fab, Fab/c fragments. *J. Immunol. Methods*, **121**, 209–217.

Dickerson, R. E. (1964). In *The Proteins* (Neurath, H., ed.), Vol. 2, pp. 603–778, Academic Press, New York.

Edelman, G. M. *et al.* (1969). The covalent structure of an entire γG immunoglobulin molecule. *Proc. Natl. Acad. Sci. USA*, **63**, 78–85.

Edmundson, A. B. *et al.* (1975). Rotational allomerism and divergent evolution of domains in immunoglobulin light chains. *Biochemistry*, **14**, 3953–3960.

Eisen, H. N. (1974). Antibody–antigen reactions. In *Immunology*, pp 360–404, Harper & Row, Hagerstown, MD.

Eisenberg, D. (1970). In *The Enzymes* (Boyer, P. D., ed.), Vol. 1, pp. 1–89, Academic Press, New York.

Faguet, G. B. and Agee, J. F. (1993). A simple technique for the rapid enrichment of class and subclass hybridoma switch variants. *J. Immunol. Methods*, **165**, 217–224.

Fanger, F. W. and Smyth, D. G. (1972). The oligosaccharide units of rabbit immunoglobulin G. *Biochem. J.*, **127**, 757–774.

Ferreira, M. U. and Katzin, A. M. (1995). The assessment of antibody affinity distribution by thiocyanate elution: a simple dose-response approach. *J. Immunol. Methods*, **187**, 297–305.

Fisch, I. *et al.* (1992). Site-specific modification of a fragment of a chimeric monoclonal antibody using reverse proteolysis. *Bioconjugate Chem.*, **3**, 147–153.

Francus, T. *et al.* (1978). IgG_{2a}-producing variants of an IgG_{2b}-producing mouse myeloma cell line. *J. Exp. Med.*, **147**, 1535–1540.

Fredrickson, R. A. *et al.* (1989). Improved procedure for preparation of $F(ab')_2$ fragments of mouse IgGs by papain digestion. *J. Immunol. Methods*, **120**, 51–56.

French, M. F. *et al.* (1986). Isolation and partial characterisation of a fragment corresponding to the dimeric form of the C_H2 domain of rabbit IgG. *Mol. Immunol.*, **23**, 489–502.

Furey, W. *et al.* (1979). Phase extension and refinement of Bence-Jones protein Rhe (0.19 nm). *Acta Crystallogr. Sect. A*, **35**, 810–815.

Georgiou, G. *et al.* (1997). Display of heterologous proteins on the surface of microorganisms: from the screening of combinatorial libraries to live recombinant vaccines. *Nature Biotechnol.*, **15**, 29–34.

Goding, J. W. (1978). Use of staphylococcal Protein A as an immunological reagent. *J. Immunol. Methods*, **20**, 241–250.

Golub, E. S. (1987). *Immunology: A Synthesis*, Sinauer Associates, Inc., Sunderland, MA.

Gorini, G. *et al.* (1969). Heterogeneity of mouse myeloma gamma globulins as revealed by enzymatic proteolysis. *J. Immunol.*, **103**, 1132–1142.

Grabherr, R. *et al.* (1997). Expression of foreign proteins on the surface of *Autographa californica* nuclear polyhedrosis virus. *BioTechniques*, **22**, 730–735.

Guddat, L. W. *et al.* (1993). Three-dimensional structure of a human immunoglobulin with a hinge deletion. *Proc. Natl. Acad. Sci. USA*, **90**, 4271–4275.

Hall, D. R. *et al.* (1996). Demonstration of an upper limit to the range of association rate constants amenable to study by biosensor technology based on surface plasmon resonance. *Anal. Biochem.*, **235**, 175–184.

Hansen, M. H. *et al.* (1997). Detection of PNA/DNA hybrid molecules by antibody Fab fragments isolated from a phage display library. *J. Immunol. Methods*, **203**, 199–207.

Harris, J. *et al.* (1992). Three-dimensional structure of an intact monoclonal antibody for canine lymphoma. *Nature*, **360**, 369–372.

Harris, T. N. *et al.* (1975). A preparation of Fc fragment of normal mouse IgG_1 for production of rabbit anti-mouse IgG_1 serum. *J. Immunol. Methods*, **8**, 203–212.

Hickman, S. *et al.* (1972). The structure of the glycopeptides of a human γM-immunoglobulin. *J. Biol. Chem.*, **247**, 2156–2163.

Honjo, T. and Kataoka, T. (1978). Organisation of immunoglobulin heavy chain genes and allelic deletion model. *Proc. Natl. Acad. Sci. USA*, **75**, 2140–2144.

Honjo, T. *et al.* (1981). Rearrangement of immunoglobulin genes during differentiation and evolution. *Immunol. Rev.*, **59**, 33–68.

Hoogenboom, H. R. (1997). Designing and optimizing library selection strategies for generating high-affinity antibodies. *Trends Biotechnol.*, **15**, 62–70.

Hunneyball, I. and Stanworth, D. R. (1975). Fragmentation of human IgG by a new protease isolated from the basidiomycete *Armillaria mellea*. *Immunology*, **29**, 921–931.

Huston, J. *et al.* (1988). Protein engineering of antibody binding sites: recovery of specific activity in an anti-digoxin single-chain Fv analogue produced in *Escherichia coli*. *Proc. Natl. Acad. Sci. USA*, **85**, 5879–5883.

Huston, J. S. *et al.* (1993). Medical applications of single-chain antibodies. *Int. Rev. Immunol.*, **10**, 195–217.

Inouye, K. and Morimoto, K. (1994). Preparation of $F(ab')_{2\mu}$ fragments from rat IgM monoclonal antibodies and their application to the enzyme immunoassay of mouse interleukin-6. *J. Immunol. Methods*, **171**, 239–244.

Ishizaka, K. *et al.* (1970). Biologic function of the Fc fragments of E myeloma protein. *Immunochemistry*, **7**, 687–702.

Kabat, E. A. *et al.* (1982). Structural and numerical variations of the carbohydrate moiety of immunoglobulin G. *J. Immunol.*, **129**, 2016–2020.

Kaneshige, A. (1987). Non-enzymatic glycosylation of serum IgG and its effect on antibody activity in patients with diabetes mellitus. *Diabetes*, **36**, 822–828.

Karlsson, R. *et al.* (1991). Kinetic analysis of monoclonal antibody–antigen interactions with a new biosensor based analytical system. *J. Immunol. Methods*, **145**, 229–240.

Kindt, T. J. and Capra, J. D. (1984). The biochemist's approach to the problem. In *The Antibody Enigma*, pp. 61–93, Plenum Press, New York.

Kipriyanov, S. M. and Little, M. (1997). Affinity purification of tagged recombinant proteins using immobilized single chain Fv fragments. *Anal. Biochem.*, **244**, 189–191.

Klotz, I. M. (1989). Ligand–protein binding affinities. In *Protein Function: A Practical Approach* (Creighton, T. E., ed.), pp. 25–54, IRL Press, Oxford.

Kochwa, S. *et al.* (1971). Structural studies of immunoglobulin E: 1. Physicochemical studies of the IgE molecule. *Ann. NY Acad. Sci.*, **190**, 49–70.

Köhler, G. and Milstein, C. (1975). Continuous cultures of fused cells secreting antibody of predefined specificity. *Nature*, **256**, 495–497.

Kornfeld, R. J. *et al.* (1971). The structure of the glycopeptide of human γG myeloma proteins. *J. Biol. Chem.*, **246**, 3259–3268.

Lamoyi, E. (1986). Preparation of $F(ab')_2$ fragments from mouse IgG of various subclasses. *Methods Enzymol.*, **121**, 652–663.

Lamoyi, E. and Nisonoff, A. (1983). Preparation of F(ab′)2 fragments from mouse IgG of various subclasses. *J. Immunol. Methods*, **56**, 235–243.

Larvor, M.-P. *et al.* (1994). Measurement of the dissociation rate constant of antigen/antibody complexes in solution by enzyme-linked immunosorbent assay. *J. Immunol. Methods*, **170**, 167–175.

Liesegang, B. *et al.* (1978). Isolation of myeloma variants with predefined variant surface immunoglobulin by cell sorting. *Proc. Natl. Acad. Sci. USA*, **75**, 3901–3905.

Lijnen, H. R. *et al.* (1997). Screening panels of monoclonal antibodies using phage-displayed antigen. *Anal. Biochem.*, **248**, 211–215.

Lin, L.-C. and Putnam, F. W. (1981). Primary structure of the FC region of human immunoglobulin D: implications for evolutionary origin and function. *Proc. Natl. Acad. Sci. USA*, **78**, 504–508.

Lindner, P. *et al.* (1997). Specific detection of His-tagged proteins with recombinant anti-His tag scFv-phosphatase or scFv-phage fusions. *BioTechniques*, **22**, 140–149.

Livingstone, J. R. (1996). Antibody characterization by isothermal titration calorimetry. *Nature*, **384**, 491–492.

Ma, J. K.-C. and Hein, M. B. (1995). Immunotherapeutic potential of antibodies produced in plants. *Trends Biotechnol.*, **13**, 522–527.

Macdonald, R. A. *et al.* (1988). The measurement of relative antibody affinity by ELISA using thiocyanate elution. *J. Immunol. Methods*, **106**, 191–194.

Mage, M. G. (1980). Preparation of Fab fragments from IgGs of different animal species. *Methods Enzymol.*, **70**, 142–150.

Mariani, M. *et al.* (1991). A new enzymatic method to obtain high-yield $F(ab')_2$ suitable for clinical use from mouse IgG_1. *Mol. Immunol.*, **28**, 69–77.

Marks, J. D. *et al.* (1991). By-passing immunisation. Human antibodies from V-gene libraries displayed on phage. *J. Mol. Biol.*, **222**, 581–597.

Marks, J. D. *et al.* (1992). By-passing immunisation: building high affinity human antibodies by chain shuffling. *Bio/Technology*, **10**, 779–783.

Marrack, P. and Kappler, J. W. (1993). How the immune system recognizes the body. *Sci. Am.*, September issue, 49–55.

Mattes, M. J. (1997). Binding parameters of antibodies reacting with multivalent antigens: functional affinity or pseudo-affinity. *J. Immunol. Methods*, **202**, 97–101.

Matthew, W. D. and Reichardt, L. F. (1982). Development and application of an efficient procedure for converting mouse IgM into small, active fragments. *J. Immunol. Methods*, **50**, 239–253.

Matthews, B. W. (1975). In *The Proteins* (Neurath, H. and Hill, R. L., eds.), Vol. 3, Academic Press, New York.

McCafferty, J *et al.* (1990). Phage antibodies: filamentous phage displaying antibody variable domains. *Nature*, **348**, 552–554.

McGuinness, B. T. *et al.* (1996). Phage display repertoires for selection of large numbers of bispecific antibody fragments. *Nature Biotechnol.*, **14**, 1149–1154.

Miller, F. and Metzger, H. (1966). Characterisation of a human macroglobulin (III). The products of tryptic digestion, *J. Biol. Chem.*, **241**, 1732–1740.

Milstein, C. and Cuello, A. C. (1983). Hybrid hybridomas and their use in immunohistochemistry. *Nature*, **305**, 537–540.

Mizuochi, T. *et al.* (1982). Structural and numerical variations of the carbohydrate moiety of immunoglobulin G. *J. Immunol.*, **129**, 2016–2020.

Mosbach, K. and Ramström, O. (1996). The emerging technique of molecular imprinting and its future impact on biotechnology. *Bio/Technology*, **14**, 163–170.

Muller, C. and Rajewsky, K. (1983). Isolation of immunoglobulin class-switch variants from hybridoma lines secreting anti-idiotype antibodies by sequential sublining. *J. Immunol.*, **131**, 877–881.

Neri, D. *et al.* (1996). Biophysical methods for the determination of antibody–antigen affinities. *Trends Biotechnol.*, **14**, 465–470.

Neuberger, M. S. (1983). Expression and regulation of immunoglobulin heavy chain gene transfected into lymphoid cells. *EMBO J.*, **2**, 1373–1378.

Neuberger, M. and Bruggemann, M. (1997). Mice perform a human repertoire. *Nature*, **386**, 25–26.

Neuberger, M. S. and Rajewsky, K. (1981). Switch from hapten-specific IgM to IgD in a hybrid mouse cell line. *Proc. Natl. Acad. Sci. USA*, **78**, 1138–1142.

Nisonoff, A. *et al.* (1960). Separation of univalent fragments from bivalent antibody molecule by reduction of disulfide bonds. *Arch. Biochem. Biophys.*, **89**, 230–244.

Nissim, A. *et al.* (1995). Mimicking the immune system using phage antibody technology. *Tumor Targeting*, **1**, 13–16.

Nossal, G. J. V. (1993). Life, death and the immune system. *Sci. Am.*, September issue, 21–30.

Onoue, K. *et al.* (1967). Papain fragmentation of the subunits of human macroglobulin. *J. Immunol.*, **98**, 303–313.

Open University (1984). Molecular immunology. In *Biochemistry and Molecular Biology*, Unit 13, pp. 4–70, Open University Press, Milton Keynes.

Owens, R. J. and Young, R. J. (1994). The genetic engineering of monoclonal antibodies. *J. Immunol. Methods*, **168**, 149–165.

Padlan, E. A. *et al.* (1976). Structural basis for the specificity of phosphorylcholine-binding immunoglobulins. *Immunochemistry*, **13**, 945–949.

Parham, P. (1983). On the fragmentation of monoclonal IgG_1, IgG_{2a} and IgG_{2b} from balb/c mice. *J. Immunol.*, **131**, 2895–2902.

Parham, P. (1986). Preparation and purification of active fragments from mouse monoclonal antibodies. In *Handbook of Experimental Immunology* (Weir, D.M., ed.), 4th edit., pp. 14.1–14.23, Blackwell Scientific Publications, Oxford.

Parham, P. *et al.* (1982). Monoclonal antibodies: purification, fragmentation and application to structural and functional studies of class 1 MHC antigens. *J. Immunol. Methods*, **33**, 133–140.

Pereira, S. *et al.* (1997). A model system for detection and isolation of a tumor cell surface antigen using antibody phage display. *J. Immunol. Methods*, **203**, 11–24.

Plaut, A. G. and Tomasi, T. B. J. (1970). Immunoglobulin M: pentameric Fc mu fragments released by trypsin at higher temperatures. *Proc. Natl. Acad. Sci. USA*, **65**, 318–322.

Plaut, A. G. *et al.* (1974). Differential susceptibility of human IgA to streptococcal IgA protease. *J. Clin. Invest.*, **54**, 1295–1300.

Poljak, R. J. *et al.* (1973). Three-dimensional structure of the Fab′ fragment of a human immunoglobulin at 2.8-Å resolution. *Proc. Natl. Acad. Sci. USA*, **70**, 3305–3310.

Polzius, R. *et al.* (1997). Real-time observation of affinity reactions using grating couplers: determination of the detection limit and calculation of kinetic rate constants. *Anal. Biochem.*, **248**, 269–276.

Porter, R. R. (1958). Separation and isolation of fractions of rabbit γ-globulin containing the antibody and the antigenic combining sites. *Nature*, **182**, 670–671.

Porter, R. R. (1959). The hydrolysis of rabbit gamma globulin and antibodies with crystalline papain. *Biochem. J.*, **73**, 119–127.

Porter, R. R. (1962). In *Basic Problems in Neoplastic Disease* (Gellhorn, A. and Hirschberg, E., eds.), pp. 176–194, Columbia University Press.

Potter, M. (1967). The plasma cell tumours and myeloma proteins of mice. *Methods Cancer Res.*, **105**, 257–261.

Preud'homme J.-L. *et al.* (1975). Variants of a mouse myeloma cell line that synthesise immunoglobulin. *Proc. Natl. Acad. Sci. USA*, **72**, 1427–1430.

Pullen, G. R. *et al.* (1986). Antibody avidity determination by ELISA using thiocyanate elution. *J. Immunol. Methods*, **86**, 83–87.

Putnam, F. W. *et al.* (1973). Complete amino acid sequence of the mu heavy chain of a human IgM immunoglobulin. *Science*, **182**, 287–291.

Radbruch, A. *et al.* (1980). Isolation of variants of mouse myeloma X63 that express changed immunoglobulin class. *Proc. Natl. Acad. Sci. USA*, **77**, 29009–29013.

Radbruch, A. *et al.* (1982). Isolation of immunoglobulin class-switch and variable region variants from mouse myeloma and hybridoma cell lines. In *Stability and Switching in Cellular Differentiation* (Clayton, R. and Truman, D., eds.), pp. 93–106, Plenum Press, New York.

Rademacher, T. W. and Dwek, R. A. (1984). Structural, functional and conformational analysis of immunoglobulin G-derived asparagine-linked oligosaccharides. *Prog. Immunol.*, **5**, 95–112.

Reiter, Y. *et al.* (1996). Engineering antibody Fv fragments for cancer detection and therapy: disulfide-stabilized Fv fragments. *Nature Biotechnol.*, **14**, 1239–1245.

Rosenstein, R. W. *et al.* (1972). Contact regions for dinitrophenyl and menadione haptens in an immunoglobulin binding more than one antigen. *Proc. Natl. Acad. Sci. USA*, **69**, 877–881.

Rousseaux, J. *et al.*, (1980). The differential enzyme sensitivity of rat immunoglobulin G subclasses to papain and pepsin. *Mol. Immunol.*, **17**, 469–476.

Rowe, D. S. and Fahey, J. L. (1965). A new class of human immunoglobulins (I). A unique myeloma protein. *J. Exp. Med.*, **121**, 171–176.

Sablitzky, F. *et al.* (1982). Spontaneous immunoglobulin class-switching in myeloma and hybridoma cell lines differs from physiological class-switching. *Immunol. Rev.*, **67**, 59–80.

Sanny, C. G. and Price, J. A. (1997). Analysis of antibody–antigen interactions using size-exclusion high- performance (pressure) liquid chromatography. *Anal. Biochem.*, **246**, 7–14.

Sarma, R. *et al.* (1971). The three-dimensional structure at 6.0 Å resolution of a human γG1 immunoglobulin molecule. *J. Biol. Chem.*, **246**, 3753–3759.

Sarma, R. and Laudin, A. G. (1982). The three-dimensional structure of a human IgG_1 immunoglobulin at 4 Å resolution: a computer fit of various structural domains on the electron density map. *J. Appl. Crystallogr*, **15**, 476–481.

Scatchard, G. (1949). The attractions of proteins for small molecules and ions. *Ann. NY Acad. Sci.*, **51**, 660–672.

Schmidt, T. G. M. and Skerra, A. (1993). The random peptide library-assisted engineering of a C-terminal affinity peptide, useful for the detection and purification of a functional Ig Fv fragment. *Protein Eng.*, **6**, 109–122.

Schreuder, M. P. *et al.* (1993). Targeting of a heterologous protein to the cell wall of *Saccharomyces cerevisiae*. *Yeast*, **9**, 399–409.

Schrohenloher, R. E. (1963). The degradation of human γ-globulin by trypsin. *Arch. Biochem. Biophys.*, **101**, 456–465.

Seligman, M. and Mihaesco, C. (1967). Studies on the structure of human IgM globulins. In *Nobel Symposium in Gamma Globulins: Structure and Control of Biosynthesis* (Killander, J., ed.), p. 169, Almqvist and Wiskell, Stockholm.

Siegel, D. L. *et al.* (1997). Isolation of cell surface-specific human monoclonal antibodies using phase display and magnetically-activated cell sorting: applications in immunohematology. *J. Immunol. Methods*, **206**, 73–85.

Skerra, A. (1993). Bacterial expression of immunoglobulin fragments. *Curr. Opin. Immunol.*, **5**, 256–262.

Skerra, A. and Plückthun, A. (1988). Assembly of a functional immunoglobulin Fv fragment in *Escherichia coli*. *Science*, **240**, 1038–1041.

Smith, G. P. (1985). Filamentous fusion phage: novel expression vectors that display cloned antigens on the virion surface. *Science*, **228**, 1315–1317.

Spira, G. *et al.* (1984). The identification of monoclonal class-switch variants by sib selection and ELISA assay. *J. Immunol. Methods.*, **74**, 307–312.

Ståhl, S. and Uhlén, M. (1997). Bacterial surface display: trends and progress. *Trends Biotechnol.*, 185–192.

Stanworth, D. R. and Turner, M. W. (1978). Immunochemistry. In *Handbook of Experimental Immunology* (Weir, D. M., ed.), 3rd edit., Vol. 1, chapt. 6, Blackwell Scientific Publications, Oxford.

Thammana, P. and Scharff, M. D. (1983). Immunoglobulin heavy chain class switch from IgM to IgG in a hybridoma. *Eur. J. Immunol.*, **13**, 614–619.

Tizard, I. R. (1984). The combination of antigen with antibody. In *Immunology: An Introduction*, pp. 105–131, Saunders College Publishing, New York.

Tonegawa, S. (1983). Somatic generation of antibody diversity. *Nature*, **302**, 575–582.

Turner, M. W. (1981). Structure and function of immunoglobulins. In *Structure and Function of Antibodies* (Glynn, L. E. and Steward, M. W., eds), pp. 1–57, J. Wiley & Sons, New York.

Uhlén, M. *et al.* (1992). Fusion proteins in biotechnology. *Curr. Opin. Biotechnol.*, **3**, 363–369.

Ultee, M. E. and Rea, D. W. (1993). A novel method for controlling the pepsin digestion of antibodies. *J. Immunol. Methods*, **157**, 165–173.

Vaughan, T. J. *et al.* (1996). Human antibodies with sub-nanomolar affinities isolated from a large non-immunized phage display library. *Nature Biotechnol.*, **14**, 309–314.

Vihko, P. and Kurkela, R. (1988). Preparation of $F(ab')_2$ fragments from monoclonal mouse IgG_1 suitable for use in radioimaging. *J. Immunol. Methods*, **110**, 229–236.

Waller, M. *et al.* (1969). Immunochemical and serological studies of enzymatically fragmented human IgG globulins: II. Hydrolysis with subtilisin, elastase, trypsin and chymotrypsin. *Immunochemistry*, **6**, 207–214.

Wang, B.-C. *et al.* (1979). Crystal structure of Bence-Jones protein Rhe (0.3 nm) and its unique domain–domain association. *J. Mol. Biol.*, **129**, 657–661.

Whitehorn, E. A. *et al.* (1995). A generic method for expression and use of 'tagged' soluble versions of cell surface receptors. *Bio/Technology*, **13**, 1215–1219.

Wilson, I. D. and Williams, R. C., Jr. (1969). Two distinct groups of immunoglobulin A (IgA) revealed by peptic digestion. *J. Clin. Invest.*, **48**, 2409–2415.

Wilson, I. A. *et al.* (1990). Crystal structures of an antibody to a peptide and its complex with peptide antigen at 2.8 Å. *Science*, **248**, 712–719.

Winter, G. and Milstein, C. (1991). Man-made antibodies. *Nature*, **349**, 293–299.

Winter, G. *et al.* (1994). Making antibodies by phage display technology. *Annu. Rev. Immunol.*, **12**, 433–455.

Wu, T. T. and Kabat, E. A. (1970). An analysis of the sequences of the variable regions of Bence-Jones proteins and myeloma light chains and their implications for antibody complementarity. *J. Exp. Med.*, **132**, 211–250.

Wulff, G. (1995). Molecular imprinting in cross-linked materials with the aid of molecular templates — a way towards artificial antibodies. *Angew. Chem., Int. Ed. Engl.*, **34**, 1812–1832.

Yamaguchi, Y. *et al.* (1995). Proteolytic fragmentation with high specificity of mouse immunoglobulin G. *J. Immunol. Methods*, **181**, 259–267.

Yelton, D. and Scharff, M. D. (1982). Mutant monoclonal antibodies with alterations in biological functions. *J. Exp. Med.*, **156**, 1131–1137.

Ying, Z. *et al.* (1995). Comparison of four methods to generate immunoreactive fragments of a murine monoclonal antibody OC859 against human ovarian epithelial cancer antigen. *Chin. Med. Sci. J.*, **10**, 78–81.

Zeder-Lutz, G. *et al.* (1997). Thermodynamic analysis of antigen–antibody binding using biosensor measurements at different temperatures. *Anal. Biochem.*, **246**, 123–132.

Zhao-Chang, F. *et al.* (1992). Three-dimensional structure of an Fv from a human IgM immunoglobulin. *J. Mol. Biol.*, **228**, 188–207.

Zhu, Z. *et al.* (1996). High level secretion of a humanized bispecific diabody from *Escherichia coli*. *Bio/Technology*, **14**, 192–196.

Chapter 5

The Preparation of Protein–Protein Conjugates

Alastair H. Dent and Mohammed Aslam

Research and Development Department, Ortho-Clinical Diagnostics, Pollards Wood Laboratories, Nightingales Lane, Chalfont St. Giles HP8 4SP, UK

5.1 Introduction

Often one of the main features of a protein's behaviour is its ability to form complexes with other such molecules, the complexes formed between antibodies and complementary protein antigens, for example, or the binding of a macromolecular hormone to its receptor. Even such 'artificial' reagents as enzyme–antibody conjugates are found *in vivo* on occasion (Wilding *et al.*, 1964; Nagamine and Ohkuma, 1975). The binding of such complexes depends most frequently on the non-covalent interactions outlined in Chapter 2, Section 2.2.8. These are strong enough to fulfil the requirements for complex formation in living systems, and are a fundamental *in situ* feature of many laboratory applications such as immunoassay.

However, while there are exceptions such as the biotin–avidin complex, or Sternberger's stable peroxidase–anti-peroxidase reagent, they are rarely robust enough to be employed where they must survive intensive physical manipulations or an unnatural chemical environment.

5.1.1 Covalent coupling of proteins

Covalent coupling of proteins also occurs *in vivo*, the cross-linking of fibrin to produce blood clots, for example, or of collagen to produce connective tissue. Deleterious cross-linking of proteins also appears to be one of the manifestations of the ageing process (Graham and Gallop, 1994). Such complexes are generally much less prone to dissociation than the non-covalent type, especially under the mild conditions tolerated by the proteins themselves.

The preparation of protein–protein complexes *in vitro* has tended therefore to rely on such covalent techniques, where electrons are transferred between atoms to generate 'chemical' bonds. While the reactions forming such bonds are always theoretically reversible (see Chapter 3, Section 3.1), the probability of the reverse reaction occuring in practice is often so low that the products can be thought of as permanent.

Some of the reactions encountered are readily reversible given a change in conditions. For example, the formation of a Schiff base linking an amine to an aldehyde is a very commonly used reaction and will occur readily in a concentrated mixture of reagents containing such functionalities. The reverse reaction (hydrolysis of the Schiff base) may occur frequently under these conditions, but will be 'invisible'; there will always be another suitable reaction partner nearby, and the net result is a predominance of product formation.

However a simple operation such as dilution can alter the situation radically. The rate of hydrolysis is largely unaffected by such a change, as there is no less chance of a suitably energetic encounter between ions from the solvent and the C=N bond. But the chance of the forward reaction reoccurring is reduced due to the greater distance between protein molecules, and the overall result may be that the conjugate slowly dissociates on storage in dilute solution.

Changes in pH can also have a drastic effect on the stability of covalent bonds; returning to the Schiff base example, lowering the pH increases the number of H^+ ions available for electrophilic attack on the nitrogen atom and results in rapid hydrolysis.

Schiff bases are often reduced to substituted amines to avoid these stability issues. Some other bonds generated in conjugation reactions are robust enough to be viewed as permanent for practical purposes, i.e. they are unlikely to dissociate significantly under the conditions they will encounter. Due to the stability limitations of the protein itself these may not place a particularly heavy onus on an artificially introduced bond; e.g. most amide bonds can be viewed as permanent to the extent that any conditions causing their dissociation would be likely to destroy the protein structure anyway.

5.1.2 The formation and purification of protein–protein conjugates

The conjugation of one protein to another is a key stage of the majority of the applications described in Chapter 1, and tens of thousands of such reactions have been described in the scientific literature. Yet the chemistry of this category of reactions is probably the simplest to define, limited as it is by the restrictions described in Chapter 2, Section 2.6.3.

In a large number of applications it reduces to the various ways and means of forming links between the amine groups available on each protein; of the other likely functionalities, carboxyl groups necessitate less convenient chemistry, and thiols and carbohydrates may not be present at all. This is not to say that these, and other more idiosyncratic routes, do not have their applications.

All in all, however, a fairly limited repertoire of chemical reactions underpins the vast majority of protein-coupling techniques.

In contrast, the purification of protein–protein conjugates is probably the most demanding among the categories described in this book. The most facile purifications of protein reagents generally exploit differences in size between the protein and other components of the reaction mixture. Thus, a passable separation of a few mg of labelled protein from low-molecular-mass impurities can be achieved in ten minutes from start to finish, using a small 'desalting' size exclusion column. However the reaction mixture from a protein–protein conjugation may contain a number of unwanted components of around the same molecular mass as the product, and other techniques must be brought to bear. Often a sequential treatment by more than one approach may be required for the best results. Purification of protein–protein complexes will be covered in Chapter 10.

This chapter lists many hundreds of reagents which can be used to form links between one protein molecule and another, and it may appear a daunting choice as to which method to use in a particular situation. However, a few simple guidelines may help to narrow this decision considerably.

As discussed in Chapter 2, heterobifunctional approaches offer the greatest degree of control in obtaining a conjugate of defined stoichiometry. Therefore for high-performance preparative applications, a multi-step route based on two complementary heterobifunctional reagents is probably a good place to start.

The degree of control is greatest if the two reactive moieties are otherwise unrepresented in the proteins to be coupled, and this narrows the choice down even more. The most straightforward way to achieve this situation in practice is to introduce a thiol-reactive moiety into one protein molecule (typically *via* the amine functions, using reagents of the type described in Section 5.7), and to couple this to thiol groups in the second protein. These thiol groups may be native to the protein, they may be generated from it by reduction of cystine disulfides, or they may be artificially introduced using one of the reagents described in Section 5.12. This simple logic has led to the predominance of methods based on combinations such as thiol–maleimide and thiol–disulfide; over the last 15 years these have come to dominate the field in high-perfomance applications such as diagnostics and therapeutics.

In some *in vivo* applications the intention is to keep the conjugate intact for as long as possible, and the chemical nature of the conjugate bond will be chosen to provide the greatest possible stability, the thioether linkages obtained from maleimides and haloalkyl methods, for example. However, for many pharmaceutical uses the intention is that the conjugate is cleaved at a moderate rate *in vivo*, so that an active component is freed when the conjugate reaches its target. Sometimes this may be triggered by pH changes, in which case acid-sensitive groups such as acetals may be incorporated in the linker. More often, however, disulfide links provide an acceptable level of gradual cleavage, and this has partly fuelled the popularity of disulfide-based reagents such as SPDP in this field. Recent coupling agents of this type have been designed to yield hindered disulfide bonds in the conjugate, allowing the control of *in vivo* stability to be tuned still further.

As an alternative complementary system, the biotin–(strept)avidin system (Section 5.11) offers a considerable number of potential advantages. This is particularly true if it is possible to use a common (strept)avidin-labelled component in a number of applications, even more so if such a reagent can be bought inexpensively. For example if a series of alkaline phosphatase (AP)-labelled conjugates are to be prepared, a convenient approach might be to biotinylate the proteins of interest and couple them to commercially obtained streptavidin–AP. This minimises the amount of chemistry which must be carried out; it is only necessary to optimise the biotinylation step in each case, typically using a long-chain biotin active ester of the type described in Section 5.11.4. The large size of the avidin or streptavidin molecule may argue against this approach in some applications.

With the ready availability of amines on virtually all protein molecules, other amine-based coupling routes might be considered for preparative applications, e.g. using the homobifunctional reagents described in Section 5.3. However the lack of control with homobifunctional methods (Chapter 2, Section 2.6.1.2) limits their potential; these rarely offer optimal performance, and, having dominated the field until around 1980, are now really best viewed as second-line approaches.

Where both proteins to be coupled possess thiol residues (or where these can be generated readily by reduction, for example), the prospects for homobifunctional coupling are improved. Because the number of thiol groups is likely to be very limited, reagents such as those described in Section 5.4 are capable of giving a high degree of control, the use of bis-maleimides to prepare superbly defined conjugates between β-galactosidase and antibody fragments, for example. Such routes have the advantage that the orientation of the linkage can be much better defined than when amines are used. In cases where both proteins bear free thiols, simple oxidative formation of disulfide bridges (Section 5.2.1) may even be an option.

Where one of the conjugation partners is a glycoprotein, it is possible to use the selectivity of the periodate oxidation method (Section 5.2.3) to yield conjugates where the orientation of the linkage is at least partly known. This can be very powerful, for example, in labelling IgG molecules *via* the carbohydrate moieties which are concentrated away from the binding sites. Periodate oxidation methods have sometimes acquired a poor reputation for reliability, but the more recent methodologies use more carefully controlled oxidation conditions, and are less likely to cause problems. Coupling specifically *via* carbohydrate groups can only really be achieved by these oxidative methods, and the utility of this approach seems too great to be ignored.

Moving to structural applications, where the aim of the conjugation is to provide information rather than a stable reagent, the requirements are very different. The overriding characteristics of a successful coupling agent are its length and its ability to form inter-protein links with a high degree of efficiency. Here amine-reactive homobifunctionals come into their own, as they have a reasonable chance of being able to form a bond, those relying on less common moieties being far less likely to succeed. Even the availability of amines is too low to give high efficiencies, leading to a search for reactive groups which will react in a totally non-selective fashion; ideally any part of the protein then forms a possible linking site, even unpromising alkyl side-chains such as those of leucine or valine. The photoactivatable reagents (Section 5.10) are the most successful manifestation of this approach so far. Other secondary requirements of structural investigations have influenced the design of coupling agents, the ability to permeate cell membranes, to be conveniently radiolabelled, and to be cleaved under controlled conditions to regenerate the original proteins.

Of course for some structural investigations it is the aim to demonstrate the proximity of groups of a particular reactivity, and then heterobifunctional agents, or homobifunctional agents targetted at less numerous residues may be employed.

There are many classes of reagent described in this chapter which have not been mentioned in this summary, because they do not provide an obvious first-choice route to a common application. However they may all be useful under specialised circumstances.

For example, it is sometimes the case that the most reactive groups perform a critical function in a protein, or are close enough to an active site to interfere with it if derivatised. The most convenient means to bypass the problem altogether may be to exploit an alternative chemistry[1].

[1] An alternative answer to this problem has recently entered the realms of reality: genetic engineering techniques can be used to replace awkward residues by another less-reactive function. Benhar *et al.* (1994), for example, have described replacing two lysine residues in a recombinant immunotoxin with arginine, dramatically reducing the degree of inactivation observed on derivatisation with amine-specific reagents.

Also (despite the wish of the authors to present the science of conjugation chemistry as one with a consistently solid theoretical footing!), it has to be admitted that trial- and-error approaches will sometimes generate the best conjugate. If a battery of different conjugation methods is applied to the preparation of an enzyme-labelled antibody for immunoassay, for example, a relatively 'primitive' technique may sometimes yield the best performance in the assay, despite the elegance on paper of other techniques, or the fact that they display a more attractive chromatography profile. It is often advisable to examine a number of different chemistries for this reason if for no other. Making the covalent linkage through a different functional group on one of the proteins is obviously a good bet for altering the conjugate behaviour in some way, although substantial variations in behaviour can sometimes be observed by using different chemistries or linker groups on the same functionality.

Finally, to use a well-worn cliché, the applications of such reagents are bounded only by the imagination and ingenuity of the investigators, and it is hoped that this summary will help to facilitate new uses.

5.2 Direct Coupling Approaches

Direct coupling relies on forming a bond between two proteins using only the available functional groups which will either react together spontaneously, or can be persuaded to do so by an enhancement of their reactivity.

5.2.1 Thiol–thiol linkage by direct coupling

Obviously proteins are designed to exist *in vivo* in a reasonably stable form, so examples of the spontaneous type are few and far between. The only groups likely to be encountered in proteins which are prone to useful reaction without activation are free thiols from cysteine residues. Even here, in fact, activation is necessary—but because this requires only the presence of atmospheric oxygen it is frequently described as a spontaneous reaction of the thiol function. The activation stage involves the formation of a thiyl radical *via* electron transfer between oxygen and the anionic thiolate (S^-) form of the thiol. Basic conditions are favoured as they increase the proportion of thiolate present. Thiol–thiol reactivity is of course normally employed by the protein to create the disulfide bonds that help define its secondary and tertiary structure, but many proteins contain free thiols, protected by a microenvironment which prevents them from reacting with other such groups that they may encounter[1].

5.2.1.1 Thiol–thiol coupling by atmospheric oxidation

If proteins containing free thiols undergo denaturation to disrupt this microenvironment, or if reactive thiols are deliberately introduced, oxidation by dissolved oxygen can rapidly lead to dimerisation reactions. In fact in any cases where free thiols are present in an accessible position, the problem is often to prevent reaction rather than to provoke it. The dimerisation reaction is strongly promoted by some metal ions, so it is a wise precaution to include a chelating agent such as ethylenediaminetetraacetic acid (EDTA) to remove these from circulation; this approach for thiol stabilisation has been in use for many years (Habeeb, 1972). These chelating reagents are often used in a similar way during the preparation of recombinant proteins, to achieve the correct folding of the newly expressed protein. It may be a useful approach to employ a slowly diminishing concentration of EDTA to promote the gentle initiation of disulfide formation.

Where it is the aim to produce intermolecular disulfide bonds between two reactive thiols, oxidation by dissolved oxygen may still be a sufficiently efficient reaction (White, 1972a). An application of this approach which received early attention was the combination of antibody fragments containing free thiols, Nisonoff and his colleagues at the University of Illinois being responsible for much of the pioneering work. They described the combination of the 3.5S (Fab′) fragments of IgG molecules by stirring at pH 8 under a stream of oxygen (Nisonoff and Rivers, 1961; Nisonoff and Mandy, 1962). This simple reaction, resulting in hybrid molecules of dual specificity, has latterly been used to produce reagents for pharmaceutical targetting. Raso and Griffin (1981), for example, allowed Fab′ fragments of anti-IgG and anti-ricin antibodies to combine to produce a proportion of a bispecific reagent. Recently Wang *et al.* (1997) studied the optimisation of conditions for the dimerisation of cysteine-containing antibody fragments by this route. They recommended conditions of 4 °C and pH 8.5–9.0. However, more efficient reaction can be promoted by the use of either catalysts or stronger oxidants.

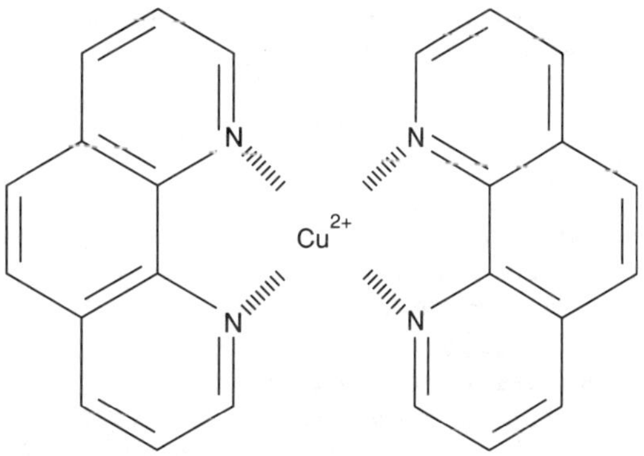

Figure 5.1 Copper complex of 1,10-phenanthroline (Cu–P).

Copper ions, even in some chelated forms, appear to be particularly effective catalysts for atmospheric oxidation of thiols. Simple reagents such as copper(II) chloride have been used (Mahbouba and Smith, 1977), but the copper(II) complex with 1,10-phenanthroline (Cu–P) (Figure 5.1) has received particular attention. Cu–P complexes were discovered to be responsible for the inactivation of rabbit muscle aldolase (Kobashi and Horecker, 1967), oxidation of eight of the enzyme's free thiol groups to form cystine linkages being sufficient for complete inactivation. Kobashi (1968) investigated this reaction further, finding that the Cu–P interaction was quite specific: the presence of either competing metals (e.g. Zn^{2+}, Ni^{2+}) or competing chelators (e.g. EDTA, histidine) prevented the oxidation. Tris gave partial protection, but Tris–Cu^{2+} complexes appear to catalyse oxidation of thiols under some circumstances (Takahashi *et al.*, 1981). Kobashi proposed a role for the Cu^{+} ion in the mechanism of oxidation; the Cu^{+}-specific reagent 2,2′-biquinoline prevented the reaction from occurring.

Cu–P complexes have sometimes been employed as catalytic cross-linking agents, generally in structural studies rather than in the preparation of conjugates. Steck (1972), for example, used Cu–P in studying the spatial arrangement of proteins within human erythrocyte membranes. The disulfide bond formed is readily cleaved by reducing agents such as mercaptoethanol or dithiothreitol, an attractive feature for such investigations. EDTA was again observed to inhibit the catalytic effect of Cu–P. The occurrence of the proteins vimentin and desmin within single filaments from hamster muscle cells was demonstrated by Cu–P-

[1] The protective effect of protein microenvironment can be very striking: one of the best known examples is haemoglobin, where the iron atom is maintained in the Fe^{2+} form despite the inherent instability of the Fe^{2+}: protoporphyrin IX (haem) complex. Free haem is instantly oxidised to the Fe^{3+} (haematin) form in the same environment, and it is only the hydrophobic environment in which it is located in the haemoglobin molecule that prevents this occurrence.

catalysed formation of intermolecular disulfide bonds (Quinlan and Franke, 1982).

5.2.1.2 Thiol–thiol coupling by other oxidising agents

Rather than catalysing oxidation by atmospheric oxygen, it is quite possible to employ alternative oxidising agents to promote disulfide formation. Reported examples include potassium periodate (Steck, 1972), Chloramine-T (Steck, 1972), potassium ferricyanide (Mahbouba and Smith, 1977), sodium tetrathionate (Haest *et al.*, 1977), and diamide (Figure 5.2, Haest *et al.*, 1977; Sato *et al.*, 1995).

The use of such reagents appears to have received relatively little attention, presumably because they are more likely to cause oxidative damage to proteins than the milder techniques based on dissolved oxygen.

5.2.1.3 Thiol–thiol coupling *via* disulfide exchange

Formation of disulfide-coupled protein conjugates by the oxidations described in Sections 5.2.1.1 and 5.2.1.2 is a predominantly random process, with links being formed between any sufficiently reactive thiol moieties. To avoid this problem nature has had to evolve highly specific and complex circumstances under which nascent proteins are assembled, to ensure that only the correct disulfide bonds are formed. One of the greatest challenges of protein synthesis by artificial means is to achieve this process satisfactorily.

Similarly, unless both partners are identical and contain a single reactive thiol, preparative protein conjugation by random disulfide formation is inherently inefficient, and a mixed product is inevitable. Different approaches must be employed to direct disulfide formation, and these are often based around the disulfide exchange reaction (Figure 5.3). The sulfur atoms in a disulfide bond tend to display electrophilic reactivity; therefore they can undergo nucleophilic attack by suitable species such as thiols. The net result of such a reaction is to replace one S–R component of the disulfide with that originating from the free thiol. Such reactions are often readily reversible, and the kinetics of a mixed disulfide exchange can be complex (Brocklehurst, 1979). However, disulfide exchange is one of the most specific reactions known for thiols, and has proved an extremely useful tool. The pH optimum is typically in the range 4–7.5 (Carlsson *et al.*, 1978), although the variability of individual protein thiols may provide exceptions.

In protein chemistry, the exchange reaction can be encountered in many forms. The starting disulfide may for example be a naturally-occuring cystine function, reaction with a suitable thiol leading to cleavage of the S–S bond (Figure 5.4(a)). An example is the formation of Fab′ fragments from $F(ab')_2$ by treatment with mercaptoethylamine (Nisonoff and Mandy, 1962).

However, the controlled formation of cystine links can also be accomplished *via* disulfide exchange (Figure 5.4(b)), and in this case the structure of the starting disulfide can be chosen to give optimal reactivity. As it is the electrophilic nature of the S–S bond which drives the reaction, electron-withdrawing substituents have generally been incorporated to heighten this characteristic. Two categories of activated disulfides are particularly noteworthy, those based on thionitrobenzoates, and those based on pyridylsulfides.

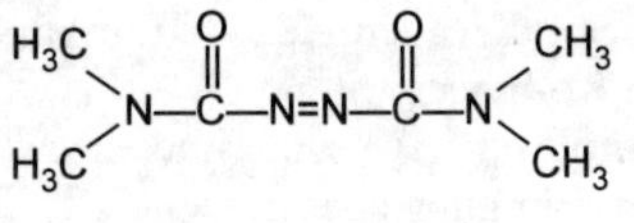

Figure 5.2 Diamide.

$$R^1SH + R^2{-}S{-}S{-}R^2 \underset{}{\overset{k_1}{\rightleftharpoons}} R^1S{-}S{-}R^2 + R^2SH + R^1SH$$

$$\updownarrow k_2$$

$$R^1S{-}S{-}R^1 + R^2SH$$

Figure 5.3 Disulfide exchange reaction.

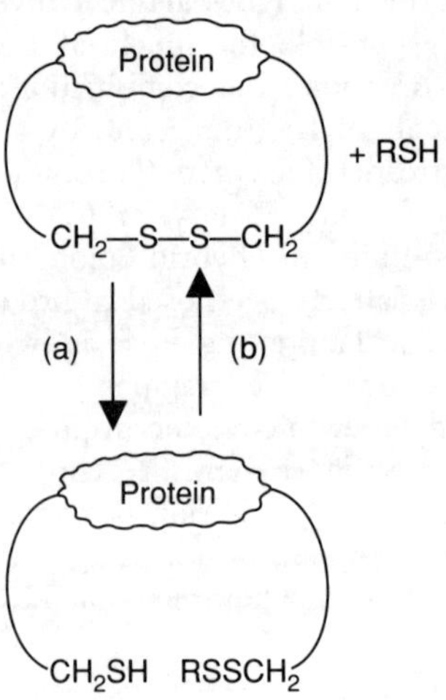

Figure 5.4 Thiols can react with protein disulfide bonds from cystine residues (a), resulting in cleavage; however, activation of a cysteine thiol by disulfide can also be used to promote cystine formation (b).

Disulfide exchange using thionitrobenzoates

There are of course many examples of disulfides bearing aromatic end-groups which could be employed in this application; the main attributes which have been sought from such reagents are solubility in aqueous systems, and the generation of a characteristic chromophore on exchange. There are still many classes of reagent which meet this description, e.g. dithio-bis-benzenesulfonates (Smith *et al.*, 1964). However, one compound has enjoyed a particularly predominant role since its introduction in the late 1950s: 5,5′-dithio-bis(2-nitrobenzoic acid) or Ellman's reagent (Ellman, 1959; Habeeb, 1972). This is commonly encountered as a reagent for the quantification of thiols; on undergoing the exchange reaction it releases the chromophoric 2-nitro-5-mercaptobenzoate anion (Figure 5.5), which absorbs strongly at 412 nm. If a thiol-containing protein is derivatised with this reagent, the resulting ('Ellmanised') disulfide is both activated towards reaction with another thiol and inhibited from self-coupling, a convenient combination of effects. Brennan *et al.* (1985) used this approach in the controlled synthesis of a bispecific $F(ab')_2$ reagent from monoclonal mouse Fab′ fragments: the thiol groups of one Fab′ fragment were Ellmanised to yield an activated disulfide, which was then reacted with the second (underivatised) Fab′ (Figure 5.6).

Because the conjugation can be carried out under non-oxidising conditions (in the presence of EDTA) there is no coupling of underivatised Fab′, while the Ellmanised Fab′ is inherently protected from dimerisation as noted above. Therefore the only product favoured is the desired asymmetrical disulfide.

Another interesting stratagem involving this chemistry was exploited in Brennan's work. Because the Fab′ fragment from mouse IgG contains three free thiols, intra-chain formation of disulfides following reduction was suspected as the cause of low yields. To overcome this, the reduction was carried out in the presence of the reagent sodium arsenite which binds specifically to vicinal dithiols, blocking intra-chain reaction (Figure 5.7). The arsenite-blocked Fab′ was then reacted with Ellman's reagent to yield a stable derivative, from which the free thiols could be regenerated (where necessary) by mercaptoethylamine treatment, immediately before use.

The Ellmanisation approach can be employed wherever two thiols are to be linked: Ghetie *et al.* (1990), for example, derivatised deglycosylated, reduced ricin A chain with Ellman's reagent in order to promote controlled coupling with thiolated recombinant CD4. Wilson *et al.* (1991) linked HIV peptides to Fab′ fragments in the same way.

Disulfide exchange using pyridylsulfides

The 2- and 4-pyridyldisulfides have some interesting properties which have led to their widespread use. Firstly, the mercaptopyridines released by the exchange reaction can exist as a number of tautomeric

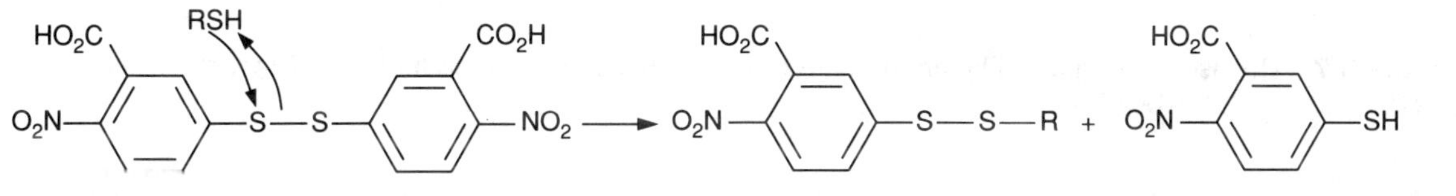

Figure 5.5 Disulfide exchange using Ellman's reagent.

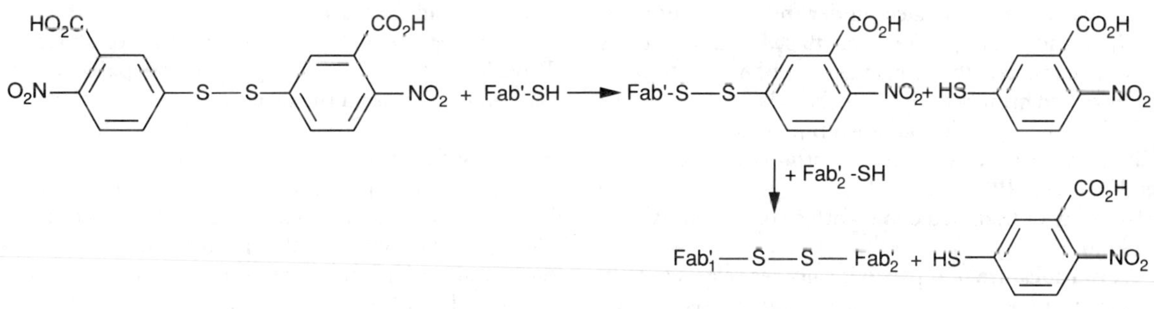

Figure 5.6 Synthesis of a bispecific Fab′ reagent using Ellman's reagent.

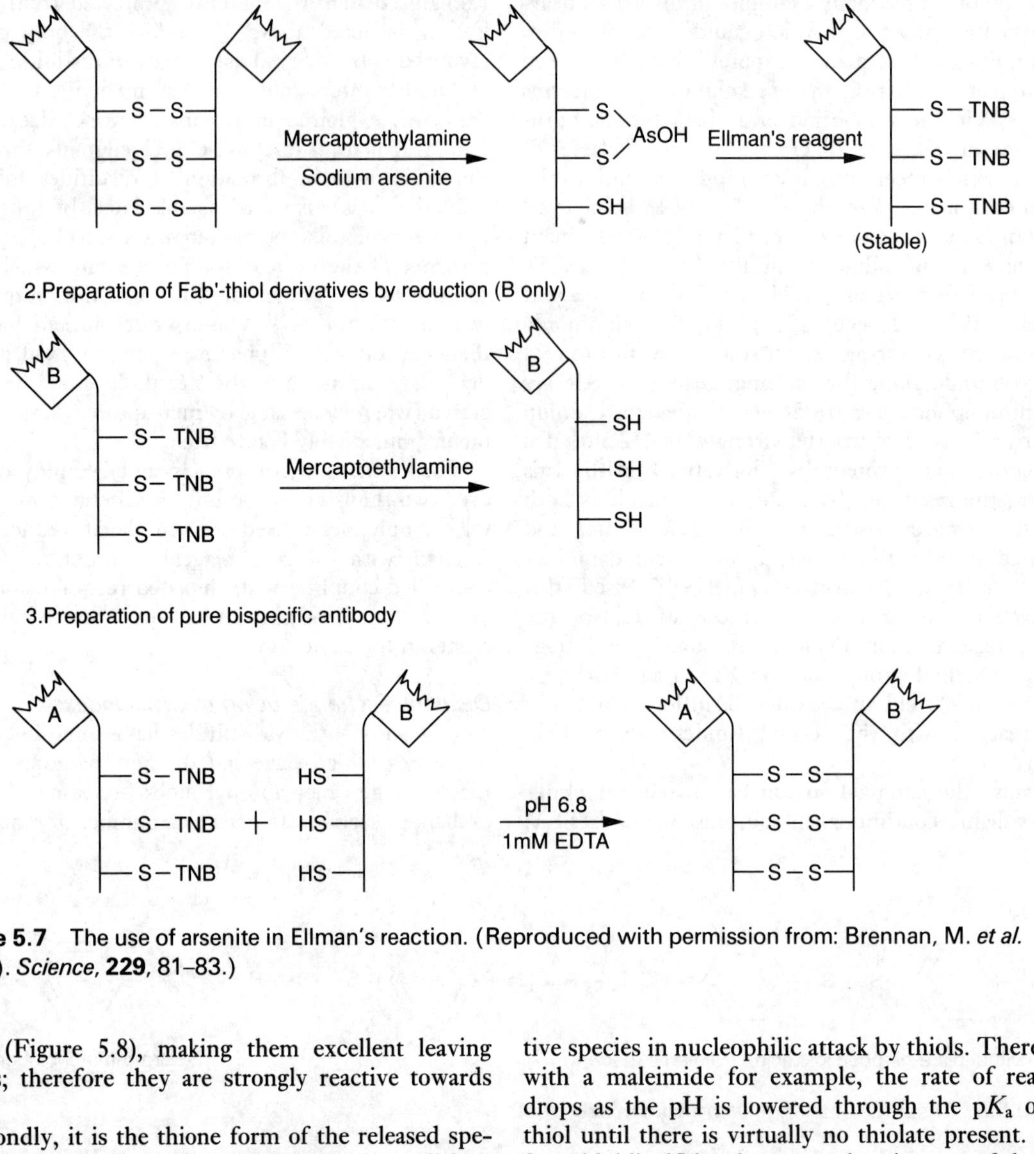

Figure 5.7 The use of arsenite in Ellman's reaction. (Reproduced with permission from: Brennan, M. *et al.* (1985). *Science*, **229**, 81–83.)

forms (Figure 5.8), making them excellent leaving groups; therefore they are strongly reactive towards thiols.

Secondly, it is the thione form of the released species which predominates under normal conditions (Albert and Barlin, 1959), and this displays a strong UV absorbance; the exchange reaction can thus be followed colorimetrically. This has also led to the use of 2,2′- and 4,4′-dipyridyldisulfide as alternatives to Ellman's reagent for thiol determination (Grassetti and Murray, 1967), the 4,4′-isomer giving the greater sensitivity of thiol detection, with a convenient λ_{max} at 324 nm.

A third feature of dipyridyldisulfides relates to their reactivity under acid conditions. As discussed earlier, it is the thiolate anion which is the predominant reactive species in nucleophilic attack by thiols. Therefore, with a maleimide for example, the rate of reaction drops as the pH is lowered through the pK_a of the thiol until there is virtually no thiolate present. With 2-pyridyldisulfides, however, the nitrogen of the pyridine ring can be protonated with a pK_a of about 2.5 (Brocklehurst, 1979). This protonation pulls further electron density from the disulfide bond, increasing its electrophilicity, and therefore rendering the reagent more reactive towards thiols as the pH decreases through this second pK_a. This feature allows pyridyldisulfide reagents to be used under acidic conditions where reaction with simple disulfides would not take place, and this in turn offers some selectivity towards low-pK_a thiols such as those found in the active sites of papain and ficin (Brocklehurst, 1979).

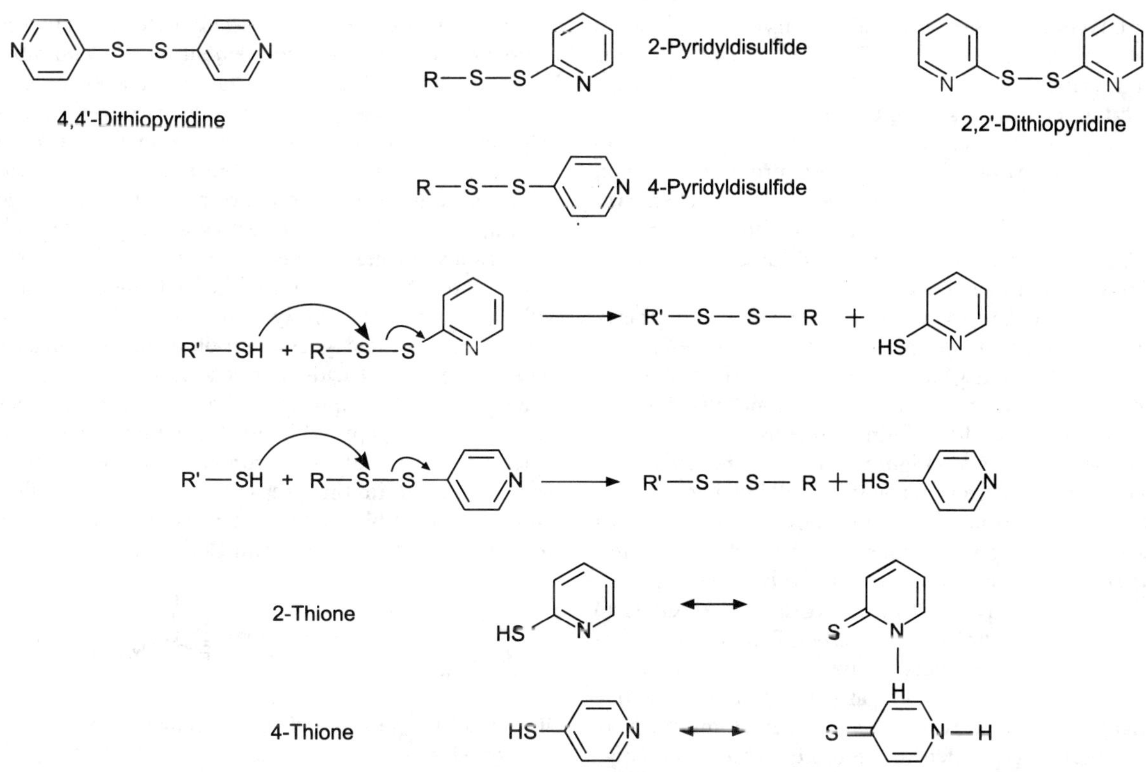

Figure 5.8 Pyridyldisulfides and their reaction in disulfide exchange. Tautomeric forms of 2- and 4-pyridylsulfides are also shown.

Like Ellman's reagent, dipyridyldisulfides can be used to promote controlled formation of disulfide-linked protein conjugates. King and Kochoumian (1979), for example, thiolated HRP in the presence of 4,4′-dipyridyldisulfide; the activated enzyme was then coupled to a thiolated antibody. Stevenson *et al.* (1989) used 2,2′-dipyridyldisulfide to mediate the controlled dimerisation of antibody fragments *via* hinge thiols.

Disulfide analogues

A number of other categories of reagent exist which are analogous to disulfides in this application. These are sulfenyl compounds which share the common formula R-$S^{\delta+}$-$X^{\delta-}$. Many such compounds are known in organic chemistry applications (Brois *et al.*, 1970); examples of X include SOR (sulfoxides), SO_2R (alkanethiosulfates), SO_3 (thiosulfates), CN (thiocyanates), SCO_2R (sulfenyl thiocarbonates) and SSO_3 (sulfenyl thiosulfates).

Some of these have been applied in the protein field: sulfitolysis of disulfides yields an equilibrium mixture of thiol and thiosulfate, for example (Figure 5.9; Swan,

$$\mathrm{RSSR} + \mathrm{SO_3^{2-}} + \mathrm{H^+} \rightleftharpoons \mathrm{RSH} + \mathrm{RSSO_3^-}$$

Figure 5.9 Conversion of disulfide to thiosulfate by sulfitolysis.

1959), but if the thiol is continuously returned to the disulfide form it is possible to obtain complete conversion to the thiosulfate. This can then be used analogously to a mixed disulfide; see for example, Masuho *et al.* (1979).

Reactions of this type are more typically met in analytical applications, and there are a number of others which may be encountered in this role: cyanolysis to yield thiocyanates (Swan, 1959; Degani and Degani, 1979) and reaction with tetrathionate to yield sulfenyl thiosulfates (Pihl and Lange, 1962).

5.2.2 Amine–carboxyl linkage by direct coupling

Despite the inherent inability of these groups to react together spontaneously, the coupling of amines to carboxylic acids is carried out millions of times a day in the human body, in the construction of peptide chains in the ribosomes. Like virtually all *in vivo* reactions, it is catalysed in this setting by enzymes.

In vitro, chemical routes tend to be employed to render the carboxyl group more prone to reaction with the amine. It is worth examining the chemistry of this group briefly to understand how this can be achieved. The concepts employed have already been introduced in Chapter 2.

The amine group is rich in electrons, and is therefore a powerful nucleophile. It is likely to react with electron-deficient groups such as the carbon of a carbonyl C=O group, where the electronegative oxygen atom has pulled the bonding electrons towards itself, e.g. in an aldehyde (Figure 5.10(a)). The carboxylic acid carries a second oxygen to pull further electron density away from the carbon; however a conflicting delocalisation or mesomeric effect also operates, whereby the lone pair electrons of this oxygen interact with the carbonyl electrons, and the net result is to increase the electron density around the carbon (Figure 5.10(b)). Even worse, above its pK_a (around pH 4) the carboxyl group will be ionised and carry a negative charge delocalised over the two oxygen atoms, pushing still more electron density onto the carbon atom (Figure 5.10(c)). So the carboxylate anion is a very poor target for nucleophilic attack[1].

The first feature of any strategy to activate the carboxyl group is the replacement of the ionisable hydrogen atom which is the source of acidity. Any group which is less likely to lose an electron than H will improve the situation. The second feature is to reduce the mesomeric effect—but not too far. In acid chlorides (Figure 5.11), the electrons of the chlorine atom are not particularly effective at interacting with the carbonyl orbitals. This group is therefore extremely reactive towards nucleophiles, so much so that it will react instantaneously with less electron-rich targets such as the oxygen atom of water. This makes *in situ* generation of acid chloride groups on proteins in aqueous solution impossible. There are many other approaches to peptide bond formation which are equally inappropriate for protein work. Many have been employed in the preparation of oligopeptides, where it is possible to work in non-aqueous solvents (for a review, see Klausner and Bodansky, 1972).

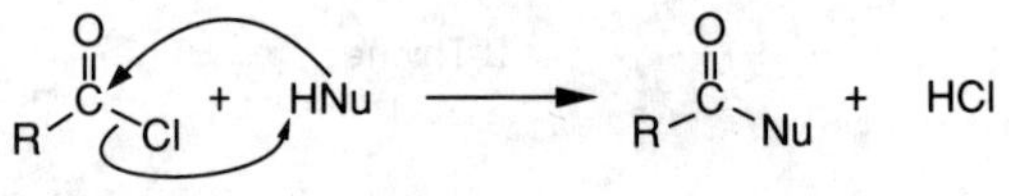

Figure 5.11 Reaction of acid chlorides with nucleophiles.

[1] Carrying out the reaction below the pK_a of the acid might appear to offer an improved chance of promoting reaction; unfortunately, at this pH the amine group will be well below its pK_a, and thus thoroughly protonated. The positive charge on the NH_3^+ group removes all the nucleophilic character of the amine.

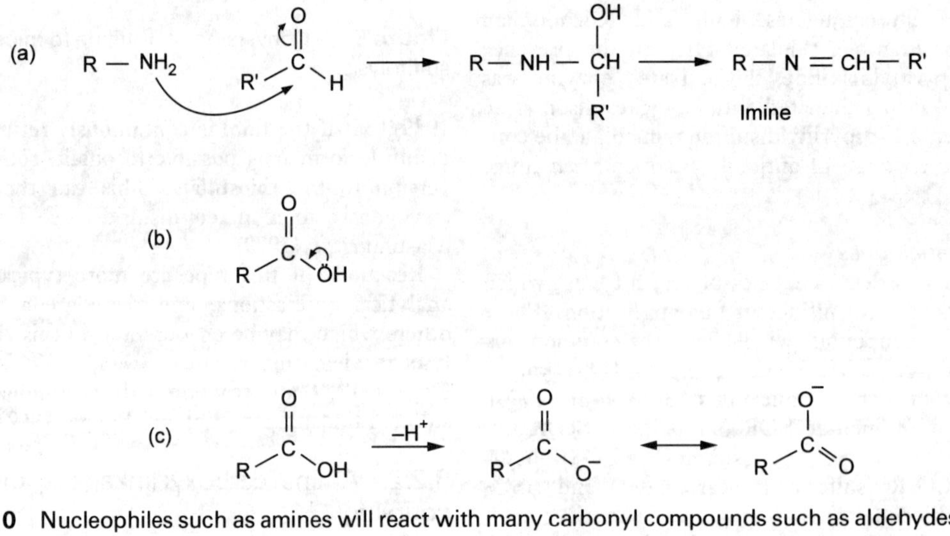

Figure 5.10 Nucleophiles such as amines will react with many carbonyl compounds such as aldehydes (a), the hydroxyamine product dehydrating to an imine; with carboxylic acids, however, the mesomeric effect from the hydroxyl group (b) and, particularly, the likely predominance of the anionic carboxylate form (c) prevent reaction with amines occurring spontaneously.

For protein chemistry, a balance is required between reactivity with amines and survival of the activated carboxyl group in water, and the reagents discussed below fulfil this requirement. However, hydrolysis will always compete to some extent with the formation of a peptide bond.

Acid anhydrides

These groups (Figure 5.12) are somewhat less reactive than acid chlorides: the central oxygen atom will still exert a mesomeric effect, but the *two* carbonyl groups compete for this to reduce its impact. Acid anhydrides will still react with water (some strongly, depending on other effects within the molecule) but sufficiently slowly to allow them to be used in protein derivatisation. They are formed from acids by dehydration, and quantitative *in situ* generation of such groups in an aqueous environment is clearly impractical.

They can, however, be generated indirectly *via* chloroformates (Figure 5.13); these can be regarded as acid chlorides which have been deactivated by the mesomeric effect of the neighbouring oxygen atom, an effect heightened by the electron-donating effect of the adjacent alkyl group. They are slightly stable in aqueous systems, while still being potent enough to react with more powerful nucleophiles such as the O^- atom of the carboxylate group. The result is to generate an acid anhydride *in situ*, which can take part in direct coupling reaction with the amines of another protein. A number of side reactions are possible, and this rather inefficient approach has been applied only infrequently to protein–protein coupling (Avrameas and Ternynck, 1967; Patramani *et al.*, 1969).

Some other reagents occasionally encountered in protein coupling can operate *via* similar mixed anhydride intermediates (Figure 5.14), e.g. 1-ethoxycarbonyl-2-ethoxy-1,2-dihydroquinoline (EEDQ; Belleau and Malek, 1968), carbonyldiimidazole (CDI) and diethylpyrocarbonate (Wolf *et al.*, 1970). Sometimes the stability of the protein to organic solvents is sufficient to allow such reactions to be carried out under non-aqueous conditions, obviously an advantage in terms of avoiding competitive hydrolysis. Bartling *et al.* (1974) prepared an insoluble cross-linked peroxi-

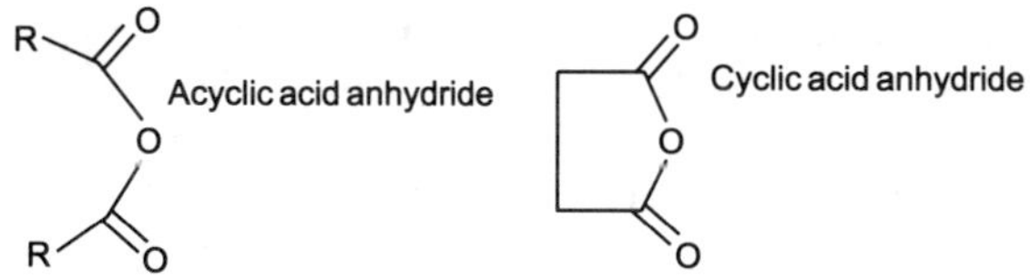

Figure 5.12 Acid anhydrides.

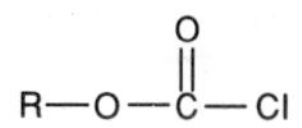

Figure 5.13 A chloroformate.

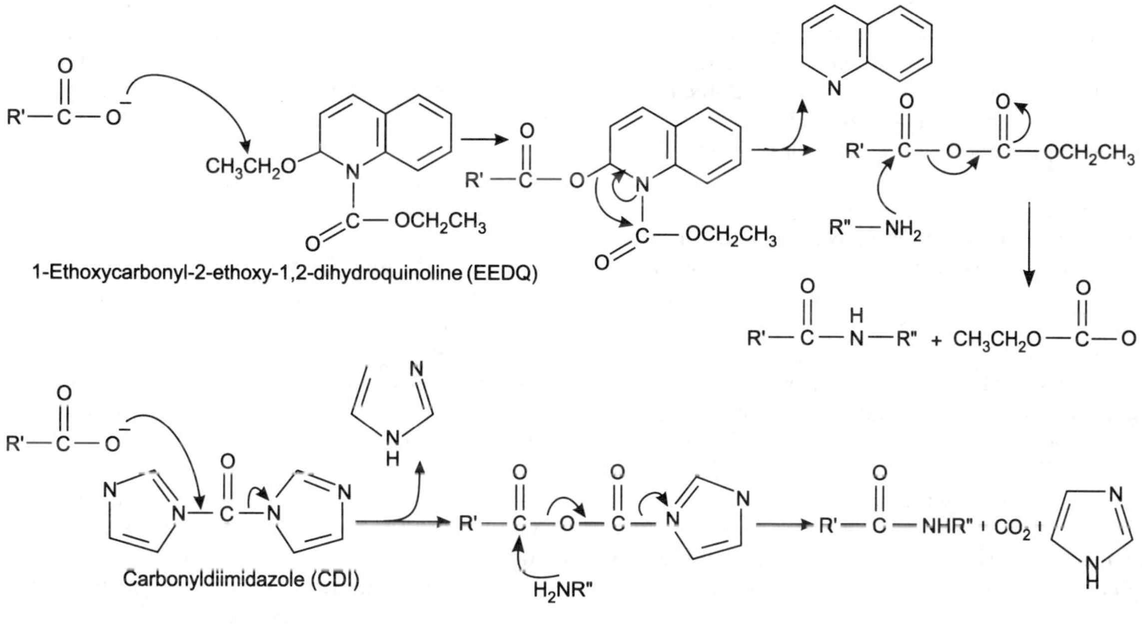

Figure 5.14 Cross-linking using EEDQ and CDI, which feature anhydride intermediates.

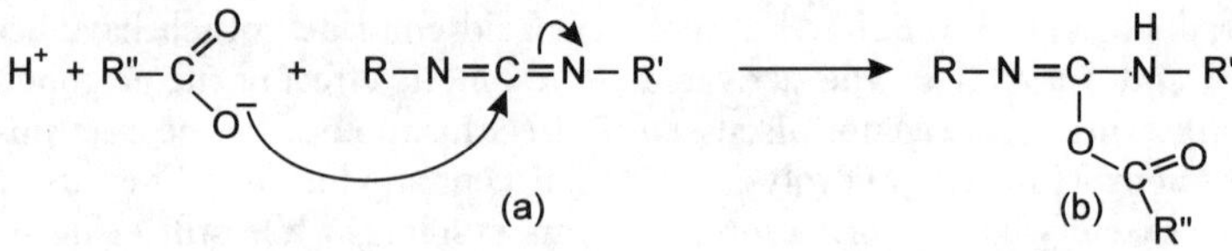

Figure 5.15 A carbodiimide (a) and its reaction with carboxylate to form a reactive *O*-acylisourea (b).

dase reagent by refluxing the dry enzyme in dichloromethane (boiling point 39 °C) for 18 hours with CDI. The fact that some enzyme activity was retained is a tribute to the robustness of HRP, but many proteins are unlikely to survive this degree of insult.

Carbodiimides

Carbodiimides (Figure 5.15(a)) contain a carbon atom made electron-deficient by the adjacent nitrogen atoms. This will therefore react with the nucleophilic O^- atom of a carboxyl group to generate an O-acylisourea (Figure 5.15(b)), which is reactive towards nucleophiles such as amines.

Carbodiimide chemistry is most efficient if the activation of the carboxyl function can be carried out without the risk of cross-linking to amines in the same molecule, and especially if this reaction can be executed under non-aqueous conditions. Neither of these conditions are easy to meet in the field of protein–protein conjugation, and carbodiimide chemistry is therefore most widely used for coupling low-molecular-mass carboxylic acids to proteins. It will be dealt with in considerably more detail in Chapter 6, Section 6.1.1. The choice of buffer can be important: many common examples such as phosphate, acetate and Tris interfere with the reaction. Mattson *et al.* (1993) have observed that 4-morpholineethanesulfonic acid (MES) is a good choice for such reactions. In the commonest one-step applications the choice of pH must reflect a compromise between the acidic optimum of the carbodiimide activation, and the availability of unprotonated amine groups; the range 6–7 is commonly encountered.

Some early applications of carbodiimide chemistry for protein cross-linking used 'insoluble' examples such as DCC[1] (Maclaren, 1958). However, the water-soluble analogues have not surprisingly seen wider use in this application; CMC was used by Sheehan and Hlavka (1957) to cross-link gelatin, and by Marfey *et al.* (1965) to cross-link synthetic polypeptides.

EDAC has probably seen the greatest use. Philpott *et al.* (1980), for example, used EDAC to couple avidin to glucose oxidase, claiming a more efficient conjugation than that obtained with glutaraldehyde or dimethyl suberimidate. Equal masses of both proteins were dissolved at 10 mg/ml in water, then an equal weight of EDAC added. After ten minutes the solution was adjusted to pH 5–6, then to 8.5 after a further ten minutes, these adjustments were followed by a 30-minute incubation at room temperature with dialysis. Good retention of enzyme activity and biotin-binding capacity was obtained. The use of EDAC approaches has maintained some popularity up to the present day, e.g. for preparing cytochrome *c*–plastocyanin conjugates (Zhou *et al.*, 1992) or coupling polylysine-based copolymers to various proteins (Lamture and Wensel, 1995).

Levy and Liu (1989) reported a variation on this theme, using a polar organic solvent as the reaction medium. EDAC was used to couple dispersions of IgG to alkaline phosphatase in dimethylsulfoxide; the resulting conjugate was not characterised, however.

A number of other reagents (including other examples of the groups above) have been employed infrequently for direct coupling of proteins *via* amine and carboxyl functions, but are of little everyday interest. The reader is referred to the exhaustive review by Wong (1991) for details of these. Enzymic approaches to the formation of direct amine-carboxyl linkages are discussed in Section 5.2.4.

5.2.3 Amine–aldehyde linkage by direct coupling

The polysaccharide chains found on glycoproteins offer a route to direct coupling which includes a 'built-in' linker group; the glycan chain can act as a spacer between two proteins without the necessity of introducing any extra atoms.

Of the carbohydrate molecules generally encountered in glycoproteins, sialic acid contains carboxyl groups which might form a site for derivatisation; glucosamine possesses an amine group, but is generally encountered as the *N*-acetyl derivative, an unreactive amide.

All such molecules possess hydroxyl functions, but any group capable of reacting with these is likely to do so far more often with the huge excess of water mole-

[1] See Section 6.1.1 for the identity of these compounds.

cules with similar reactivity. However there is a specific reaction of molecules containing two hydroxyl groups[1] on neighbouring carbon atoms, vicinal diols as they are termed, which has spawned a useful and widely applied protein coupling method.

5.2.3.1 Periodate oxidation of carbohydrate residues: the Malaprade reaction

This is the periodate oxidation reaction first described in 1928 by Malaprade, who observed that mannitol was oxidised by periodic acid to yield a mixture of formaldehyde and formic acid (Figure 5.16). Fleury and Lange (1932) reported the selectivity of this oxidation for vicinal diols, a feature of the reaction which was quickly exploited in the determination of sugar structures (Jackson and Hudson, 1937). This technique remained one of the most important tools in carbohydrate analysis until the evolution of instrumental techniques such as NMR. The application of this approach had extended to the analysis of glycoproteins by the 1960s (Rothfus and Smith, 1963), and was later adapted for protein coupling by Paul Nakane and his colleagues at the University of Colorado Medical Centre (Kawaoi and Nakane, 1972; Nakane and Kawaoi, 1974).

So long as the two vicinal hydroxyl groups are able to orientate themselves correctly, a complex can be formed between them and the periodate ion. The formation of this complex facilitates the cleavage of the carbon carbon bond, and generates an aldehyde group at the site of each hydroxyl group (Figure 5.17).

Molecules containing multiple aldehyde and hydroxyl groups are prone to a number of intramolecular reactions, and it is highly likely that the oxidised sugars present after periodate treatment are rapidly converted into more stable forms. Rowen *et al.* (1951) observed that periodate-oxidised cellulose showed no infrared absorbance characteristic of aldehyde groups, and proposed structures such as that shown in Figure 5.18, resulting from: (1) intramolecular formation of hemiacetal functions by attack of a hydroxyl group on an aldehyde, and (2) attack on the second aldehyde group by water to yield a hydrate.

Hurd *et al.* (1953) proposed a similar set of reactions for the oxidation product of methyl α-D-glucopyrano side (Figure 5.19): one aldehyde of the oxidised product (Figure 5.19(a)) forms a hemiacetal with the free hydroxyl group (Figure 5.19(b)); this generates another hydroxyl moiety which reacts with the second aldehyde group (Figure 5.19(c)), and finally one of the acetal bonds is hydrated to yield a product (Figure 5.19(d)) analogous to Figure 5.18(b). Barry and Mitchell (1953) invoked similar mechanisms in proposing a seven-membered ring structure from an oxidised cellulose.

[1] The specificity of the periodate oxidation reaction is actually not quite this simple; one of the vicinal hydroxyl groups can be replaced by an amine, aldehyde or ketone group, while α-amino aldehydes, vicinal diketones and some activated methylene (CH_2) groups will also undergo oxidation (Bobbitt, 1956). The oxidation of such groups has recently been exploited in protein–protein conjugation (see Section 5.3.3.2).

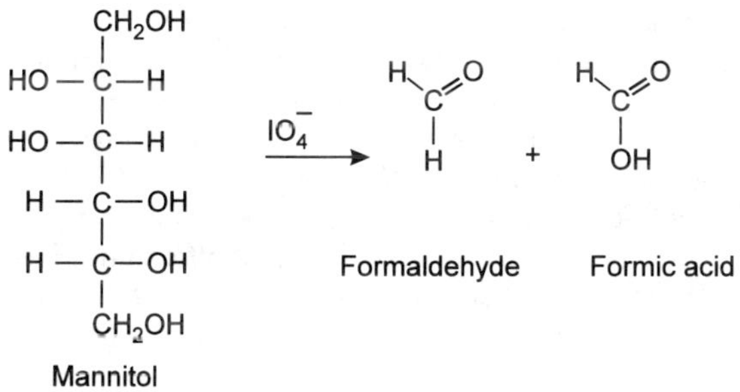

Figure 5.16 Oxidation of mannitol with periodate.

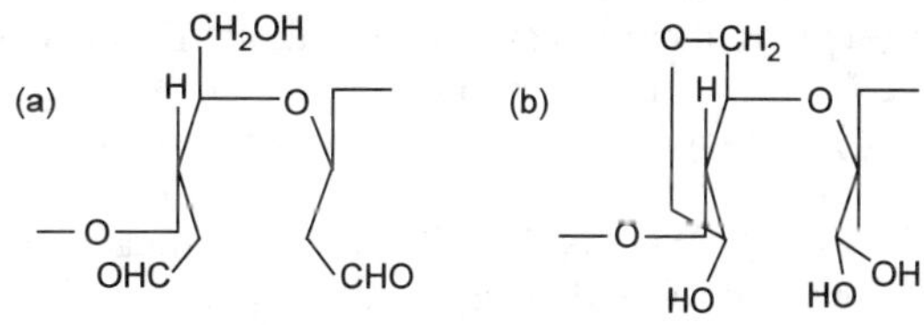

Figure 5.18 Rearrangement of periodate-oxidised cellulose dialdehyde (a) proposed by Rowen *et al.* (1951): one aldehyde forms a hemiacetal with the free hydroxyl group and the second is hydrated. The product (b) contains no free aldehyde groups.

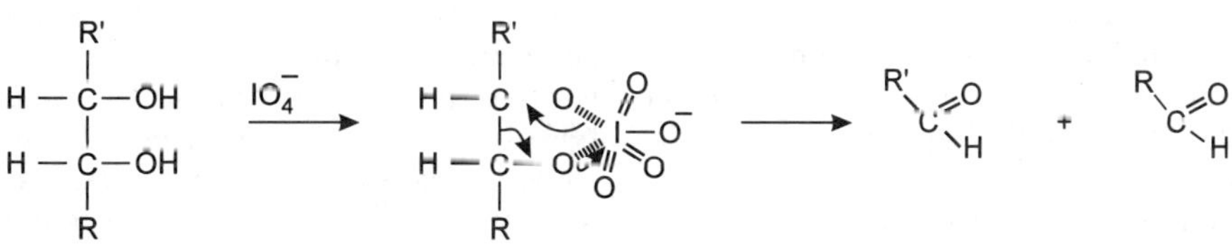

Figure 5.17 Oxidation of vicinal hydroxyl groups.

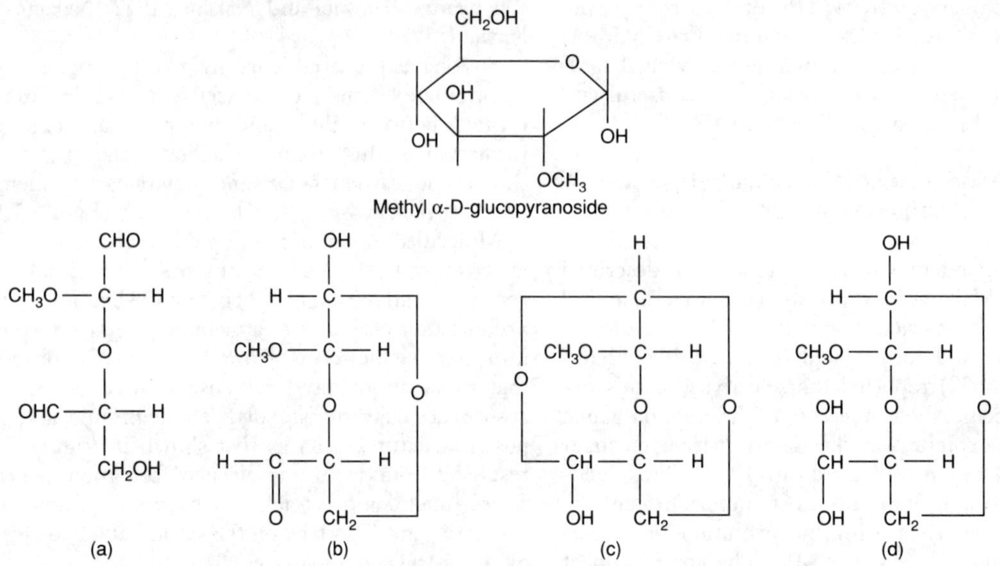

Figure 5.19 Oxidation of methyl α-D-glucopyranoside; see text for details.

Determining the exact structure of periodate-oxidised carbohydrate functions in glycoproteins does not appear to have received a great deal of attention; in terms of the subsequent chemistry, these rearrangements are probably not particularly important, as both hemiacetals and aldehyde hydrates react rather similarly to the parent aldehyde.

As mentioned in the introduction of this chapter, aldehydes (and the rearrangement products just discussed) contain a highly electron-deficient carbon atom, and are therefore an excellent target for nucleophilic attack. If this nucleophile is an amine (Figure 5.10(a)) an α-hydroxyamine intermediate is produced, which dehydrates spontaneously to form an imine or Schiff base ($R_2>C{=}N{-}R$). Therefore a conjugate can be formed between the carbohydrate portion of one protein and the amine functions of another.

The optimum pH of the periodate oxidation reaction is 3–5 (Dyer, 1956). For protein chemistry applications the reaction can be difficult to control under these conditions, although an advantage is that the protonation of amines will prevent self-coupling. A greater degree of reproducibility may be gained, at the expense of reaction rate, by working under neutral or alkaline conditions. However the amine groups of the protein must be blocked if self-reaction is to be avoided.

Occasionally, oligomerisation of the oxidised protein *via* its own amine groups has been used to bring about deliberate cross-linking. Kozulic and Barbaric (1988), for example, produced a stabilised oligomeric form of yeast invertase by periodate oxidation using a long incubation, albeit at a pH where the number of deprotonated amines must have been very small (24 hours, 4 °C, pH 4.6).

Periodate decomposes under the influence of light, generally to produce more reactive species (Dyer, 1956). To obtain reproducible results it is therefore beneficial to keep periodate solutions and reaction mixtures in the dark.

Periodate activity can be assayed by titration with arsenite and iodine (Dyer, 1956). A spectrophotometric method (Dixon and Lipkin, 1954) may be useful under certain circumstances, based on the absorbance of periodate at 223 nm. Morehead *et al.* (1991) used aldehyde-reactive dyes to monitor the extent of glycoprotein oxidation by periodates; Wolfe and Hage (1994) have described the automation of this technique in a flow injection analysis system.

Reduction of Schiff bases

Schiff bases are prone to hydrolysis, particularly under acid conditions, and it is customary to follow Nakane and Kawaoi's example in employing a reducing agent to convert them to stable dialkylamine moieties (R–NH–R). The reducing agent employed by these authors was sodium borohydride ($NaBH_4$), which had been described for imine reduction in other applications such as conjugation of nucleotide sugars to amines (Khym, 1963), and was already in use for reductive amination of proteins (Means and Feeney, 1968).

This formation of a Schiff base followed by reduction to a substituted amine is really a special case of the reductive alkylation approach. This has most usually been applied to the methylation of protein amine groups by formaldehyde (Figure 5.20), a convenient means of introducing a radiolabel *via* ^{14}C or ^{3}H-labelled formaldehyde, amongst other applications.

The use of borohydride (Means and Feeney, 1968) and cyanoborohydride (Friedman *et al.*, 1974; Dottavio-Martin and Ravel, 1978; Jentoft and Dearborn, 1979) originates from this methodology. Sodium cyanoborohydride ($NaCNBH_3$) shows greater selectivity than $NaBH_4$, which can also reduce aldehydes, disulfides, and even the amides of the peptide chain (Paz *et al.*, 1970). It also has a considerably wider stable pH range—borohydride activity begins to be lost below around pH 9[1], while cyanoborohydride is stable down to around pH 3–4 (Borch *et al.*, 1971) and shows broadly constant activity between pH 5 and 9 (Geoghegan *et al.*, 1981). Release of toxic cyanide can, however, be a concern with the use of this reagent.

[1] Solutions of sodium borohydride in water display a rather greater stability than might be expected, as such solutions tend to be quite strongly alkaline. A 17 μM solution, for example, displayed a pH of 10.2 (Lin and Lee, 1994).

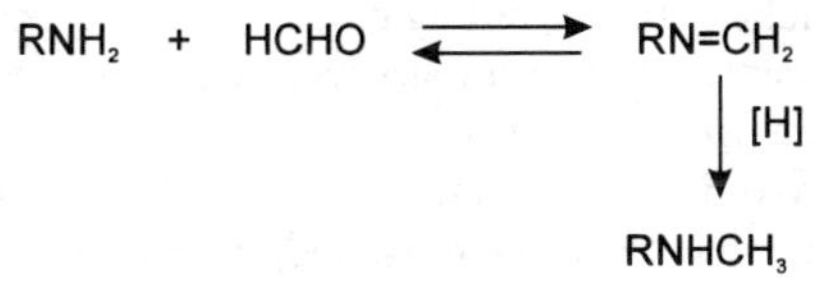

Figure 5.20 Reductive alkylation with formaldehyde.

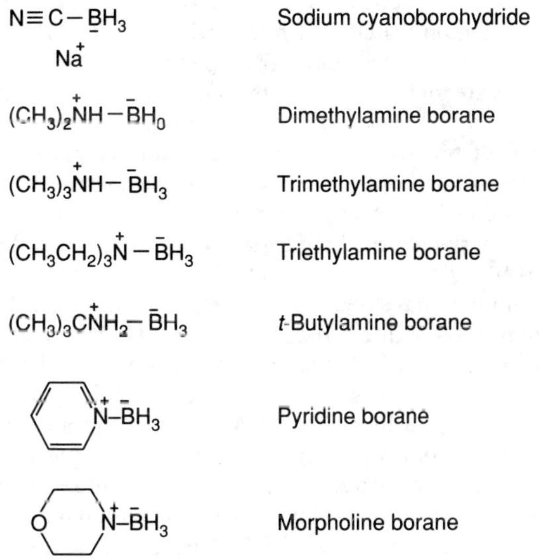

Figure 5.21 Amine boranes.

Amine boranes such as pyridine borane (Figure 5.21) have also been employed in reductive amination reactions (Geoghegan *et al.*, 1981; Cabacungan *et al.*, 1982), showing good stability under acid conditions while avoiding the toxicity of cyanoborohydride. They do not appear, however, to have found their way into popular use as reducing agents for periodate-oxidised protein conjugates.

The potency of these reducing agents can be determined colorimetrically using trinitrobenzoic acid (Means and Feeney, 1968). Lin and Lee (1994) have described a colorimetric technique based on neocuproine, which claims a more stable colour than the trinitrobenzoic acid method.

A number of alternatives to borohydride and its analogues have been examined as reducing agents following periodate oxidation and Schiff base formation: Henning and Nielsen (1987) obtained HRP–IgG conjugates of greater stability than those reduced with $NaBH_4$ by employing ethanolamine or ascorbic acid as the reducing agent.

Overall, the robustness of the linkages in reduced periodate conjugates is rather good. For example, when the long-term stability of borohydride-reduced periodate HRP–IgG conjugates has been examined, loss of both enzyme activity and immunoreactivity has been reported (Presentini and Terrana, 1995), but no cleavage of the interprotein link.

The Amadori rearrangement

There is another mechanism which may be of relevance to the stabilisation of Schiff base linkages between proteins. The Amadori rearrangement (Hodge, 1955) can occur where a Schiff base is adjacent to a carbon bearing a hydroxyl group—as will frequently be the case in the setting of periodate-oxidised protein conjugates—and results in the formation of a ketoamine (Figure 5.22). This product is considerably more stable to acid hydrolysis than the Schiff base.

Amadori rearrangement products are found in naturally-occurring glycoproteins, such as mucoproteins (Hodge, 1955) and the minor haemoglobin component HbA_{1c} (Bunn *et al.*, 1975). They are also formed when sugars react with protein amine groups *in vivo*. These non-enzymatic glycation reactions[2] are

[2] These are often referred to as Maillard reactions (after the author of the original report of their occurence in 1912), and are also responsible for a number of ageing phenomena in the food industry, browning and flavour spoilage, for example.

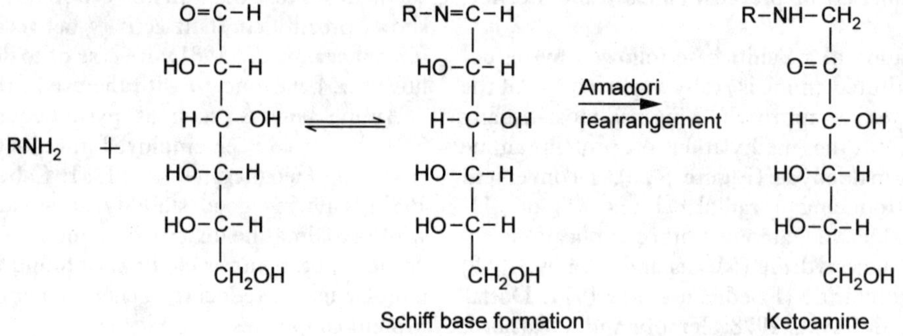

Figure 5.22 Amadori rearrangement.

generally very slow, but can result in the cross-linking of proteins. The Amadori products (and the products of subsequent rearrangements) are termed advanced glycosylation end-products (AGEs), and have received increasing levels of attention as a result of their implication in cell ageing, in some of the harmful consequences of diabetes (Makita *et al.*, 1992; Wolff *et al.*, 1991) and other diseases (Harrington and Colaco, 1994).

The same rearrangements might be expected where aldehydes of suitable structure react with protein amines *in vitro*, and this has indeed been observed. Acharya and Manning (1980a,b) showed that Amadori products were formed during the reaction of glyceraldehyde with a haemoglobin derivative, the ketoamine product being detected by reaction with phenylhydrazine.

Imines generated from periodate-oxidised glycoproteins will often have hydroxyl groups in the α-position, and are therefore highly likely to undergo Amadori rearrangements to produce stable ketoamine links. Therefore some stable product may be formed from periodate conjugations without the addition of any reducing agent: support for this mechanism was observed by Kozulic and Barbaric (1988) during the cross-linking of periodate-oxidised invertase.

Changes to protein structure during periodate oxidation

Damage to the carbohydrate moieties is obviously unavoidable during periodate oxidation reactions, but the destruction of amino acids can also occur, even at relatively low periodate levels: Clamp and Hough (1965) showed substantial reaction of native and substituted amino acids with 15 mM periodate over periods of a few hours. Cysteine, methionine, tryptophan, tyrosine and histidine were all extensively oxidised even when C and N-substituted to mimic a peptide environment.

Other workers have observed a much greater degree of stability in intact glycoproteins, particularly at low periodate concentrations, e.g. Krotoski and Weimer (1966) observed that treatment of orosomucoid with 120 mM periodate for 6–12 hours was required to destroy tyrosine, cysteine and methionine, while other amino acids remained intact. Oxidation of cysteine, methionine and tryptophan residues has been reported in a number of proteins (O'Shannessy and Wilchek, 1990), but again at rather higher periodate concentrations than are required for generating aldehydes from glycoproteins.

Hydroxylic N-terminal amino acids such as serine and threonine are prone to periodate oxidation, and this has indeed been used as a route to conjugation (Section 5.2.3.3). On the whole however, gross changes to peptide structure are unlikely under moderate conditions of glycoprotein activation. Christensen and Ruud (1975) have demonstrated that tetramethyl urea (TMU) can offer dramatic protection from peptide damage during periodate oxidation of some enzymes. At periodate levels of 23 mM, 40% TMU allowed retention of 97% of trypsin activity after 24 hours at pH 4.5, in contrast to complete destruction in the control.

Interestingly changes in protein properties observed after periodate oxidation may sometimes bring about fortuitous benefits: Haug *et al.* (1989) of Boehringer Mannheim have demonstrated a patented reduction in immunoassay interferences when the HRP is oxidised before use. An interesting observation in a pharmaceutical application was also made by Thorpe *et al.* (1985a), in an investigation of the effect of an oxidation/reduction cycle on the carbohydrate portion of ricin. In order to avoid reaction of the oxidised sugars with amine groups, they carried out periodate oxidation in the presence of $NaCNBH_3$, and (perhaps contrary to intuition) it proved feasible to carry out a 'one-pot' oxidation/reduction cycle. The clearance of the toxin from the bloodstream was dramatically reduced, so that although the modification rendered the toxin up to 90% less active *in vitro*, the *in vivo* toxicity was increased up to fourfold. Also, Houba *et al.* (1996) observed greatly improved *in vivo* stability of an antibody–glucuronidase conjugate for prodrug therapy when the enzyme was deglycosylated with periodate.

5.2.3.2 Periodate coupling techniques

The preparation of HRP–IgG conjugates has been a key application of periodate oxidation, *via* the HRP sugar residues, and the development of this approach will be discussed in some detail. Many of the findings are equally applicable to glycoproteins other than HRP, although their sensitivity to oxidation may vary.

'Nakane-type' methods for HRP–IgG conjugates
Nakane and Kawaoi's (1974) method was applied to the preparation of HRP–IgG conjugates, and features the following stages. (1) Dissolution of HRP in 0.3 M sodium bicarbonate, pH 8.1. (2) Blocking of amine groups by reaction with 1% ethanolic FDNB for one hour. (3) Oxidation by 0.04–0.08 M aqueous sodium periodate for 30 minutes. (4) Quenching of unreacted periodate with 0.16 M aqueous ethylene glycol for one hour. (5) Dialysis against 0.01 M sodium carbonate buffer (pH 9.5) at 4 °C to remove low-molecular-mass components. (6) Reaction with IgG (HRP:IgG ratio 4:1) for two to three hours. (7) Reduction of imine bonds with sodium borohydride for three hours or overnight. (8) Dialysis against phosphate-buffered saline solution. (9) Purification of HRP–IgG component by size exclusion on Pharmacia Sephadex™ G-100 or G-200.

By 1978, Nakane's group were no longer recommending blocking of amines or quenching with ethylene glycol; instead Wilson and Nakane (1978) advocated the use of acidic conditions (pH 4–5) which represent the optimum oxidation pH (Dyer, 1956). Their protocol seemingly relied on a suitable pH being obtained by dissolution of the HRP and periodate in water for the actual oxidation stage, before dialysing into 1 mM sodium acetate buffer (pH 4.4).

Immediately before addition to IgG (at a lower HRP:IgG ratio of about 2:1) the pH is raised above 9 to deprotonate the amines and promote the Schiff base reaction. By this modification, 5% self-coupling of HRP was observed compared to 35% by the 1974 method. Wilson and Nakane also made a number of other observations. (1) That varying the oxidation time from 20 to 60 minutes made little difference to the resulting conjugate. (2) That increasing the excess of periodate over HRP increased the incorporation of enzyme in the conjugate, and the percentage of aggregate formed. Increasing the ratio above 200:1 showed little effect. (3) That the pH optimum for the HRP–IgG reaction was around 9.5, and that at this pH reaction was largely complete after around four hours. Around 60–80% of the enzyme could be incorporated into the product. (4) That the number of aldehydes generated per enzyme molecule (determined by the incorporation of ^{3}H-labelled glycine) was 18, although steric hindrance would prevent many of these reacting with IgG.

Many authors have examined these procedures, or close modifications, and the reported results have been highly variable. The extent of this variation can be observed in the observed losses of both enzyme activity and immunoreactivity in the conjugates prepared.

Reported losses of HRP activity have been insignificant (Nakane and Kawaoi, 1974), 7–14% (Imagawa *et al.*, 1982), 30–40% (Boorsma and Streefkerk, 1979), 45% (Boorsma and Streefkerk, 1976), 55% (Nygren, 1982), 63% (Porstmann *et al.*, 1985) and 70% (Boorsma *et al.*, 1976). Losses of immunoreactivity have ranged from insignificant (Nakane and Kawaoi, 1974) to 80% (Johnson *et al.*, 1980) or 94% (Boorsma and Streefkerk, 1976), based on changes in effective titre.

It should be borne in mind that, particularly ten or twenty years ago, a high degree of variability in commercial HRP performance was often observed between batches. In some cases, this factor may well have contributed at least as much variation to conjugate performance as the coupling method, and may explain conflicting observations by different authors.

Another factor in this variation is undoubtedly the high level of aggregation observed from these conjugations; large quantities of high-molecular-mass material have frequently been reported. This feature of Nakane-type methods is another disadvantage which led to further developments of the periodate technology.

'Tijssen-type' methods for HRP–IgG conjugates
Tijssen and Kurstak (1984) reported modifications to the Nakane approach for HRP–IgG coupling. They turned away from Wilson and Nakane's acidic conditions on the grounds that it promoted over-oxidation, and looked instead to high pH and low periodate concentration to control oxidation. They observed significant loss of enzyme activity at concentrations greater than around 2 mM (see Figure 5.23). In their preferred protocol, therefore, they employed periodate at 4–8 mM (the lowest level that would yield sufficient oxidation) for two hours at pH 8.1.

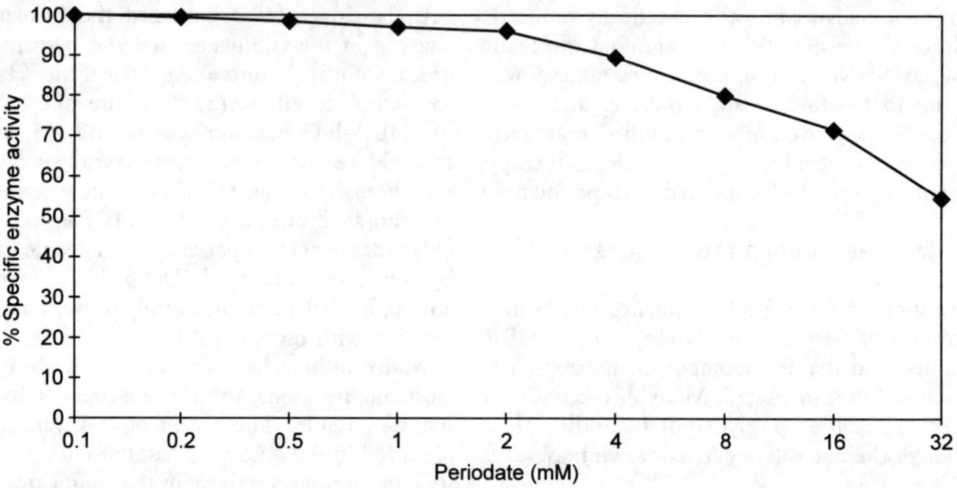

Figure 5.23 Effects of periodate oxidation on activity of horseradish peroxidase (carried out at pH 8.1, 2 h, 20 °C in the dark). Data from Tijssen and Kurstak (1984), Table 3.

Tijssen and Kurstak also recommended carrying out the oxidation in the dark, and highlighted the importance of water quality and HRP homogeneity, mentioned earlier.

These authors obtained conjugates exhibiting substantial incorporation of HRP and retention of enzyme function. At 4 mM periodate 94% of the enzyme was incorporated at 90% of its specific activity. Jeanson *et al.* (1988a) compared Tijssen and Kurstak and Nakane and Kawaoi's protocols for the preparation of mouse monoclonal IgG–HRP conjugates; the former gave better retention of both antibody and enzyme activity.

Overview on periodate oxidation conjugations for HRP–IgG conjugates

Variations of the methods of Nakane and Kawaoi, Wilson and Nakane and Tijssen and Kurstak are still widely encountered; a summary of the key quantitative differences between them is shown in Table 5.1. Most recently, Tsang *et al.* (1995) have carried out optimisation studies on this reaction, their preferred route featured a short reaction time, counterbalanced by low pH, high periodate concentration, and a raised temperature (37 °C). With the exception of the HRP:IgG ratio, the conditions noted in the table all refer to the oxidation stage; there is little variation in the conditions of the conjugation reaction, which typically takes place at pH 9.5 and 2–8 °C for a couple of hours or more.

Compared to the polyclonal IgG used in most of this work, the more idiosyncratic behaviour of monoclonal antibodies is likely to necessitate reoptimisation of reaction conditions, for example with respect to HRP:IgG ratio. Madersbacher *et al.* (1992), for example, reported optimal immunoassay performance from a number of monoclonals by returning to the 'old' high ratio approaches of Nakane and Kawaoi.

Table 5.1 Variation in key process variables for the periodate oxidation of HRP in IgG–HRP conjugations

Method	IO_4^-:HRP XS	$[IO_4^-]$ (mM)	Time (min)	Temp. (°C)	pH	HRP:IgG ratio
Nakane and Kawaoi (1974) (high)	640	38	30	rt	8.1	4
Nakane and Kawaoi (1974) (low)	320	14	30	rt	8.1	4
Wilson and Nakane (1978)	200	17	20	rt	acid	2
Porstmann *et al.* (1984) (P1)	684	30	30	rt	8.1	10
Porstmann *et al.* (1984) (P2)	70	10	20	rt	PBS	10
Tijssen and Kurstak (1985) (high)	64	8	120	rt	8.1	1.33
Tijssen and Kurstak (1985) (low)	32	4	120	rt	8.1	1.33
Tsang *et al.* (1995)	40	140	5	37	5.0	6

On the whole, the milder Tijssen-type methods appear to offer good all-round retention of activity, and are recommended as a 'first shot' for IgG–HRP conjugations. Periodate conjugation of HRP to IgG under controlled conditions is a fairly straightforward procedure yielding stable products, and it will undoubtedly continue to find application. Much of its reputation for variability probably stems from poor control of materials and processes.

However, it cannot match heterobifunctional approaches in terms of the control of incorporation that can be obtained, and for high-performance applications it is best viewed as a 'reserve' method for cases which are not amenable to more modern techniques.

Periodate oxidation of immunoglobulins

Immunoglobulins contain between 4 and 18% carbohydrate. Their antigen-binding ability is generally unaffected by alterations to the sugar moieties, which are almost exclusively found in the Fc regions away from the binding sites. Therefore these molecules are an attractive target for conjugation methods based on periodate oxidation.

The conditions for oxidation of immunoglobulins are not dissimilar to those employed for HRP. Again it is likely that individual optimisation may be necessary, for monoclonals in particular. Although some of these examples do not derive from protein–protein conjugations this is unimportant in terms of the antibody oxidation, which is subject to much the same constraints regardless of the subsequent conjugation partner.

Murayama *et al.* (1978) oxidised IgG with 20 mM periodate (30 minutes, 22 °C, pH 4.0) before coupling to HRP, obtaining successful immunoassay conjugates. O'Shannessy and Quarles (1985, 1987) described general conditions for immunoglobulin oxidation, employing 10 mM periodate at pH 5.5 for 20–30 minutes at 0 °C. Roberts *et al.* (1990) oxidised rabbit polyclonal IgG for eight hours at pH 5.5 at a periodate concentration of 5 mM, retaining 90% of immunological activity.

In examples involving oxidation of human and mouse IgM monoclonals, Chua *et al.* (1984) used 9 mM periodate for 30–240 minutes at pH 5.3. Chua's methods demonstrated the introduction of 5–35 aldehyde groups per subunit, with an increase in incorporation apparent on longer incubation even after two hours. No reduction in antibody affinity was noted for one of the murine fragments after 30 minutes treatment.

Rodwell *et al.* (1986) used 10–30 mM periodate for one hour at pH 6.0 on mouse IgM and rat and mouse IgG_{2b} antibodies; *in vitro* assessment of the IgM showed no loss of affinity after treatment in this way. In contrast, derivatisation of carboxyl or amine groups using carbodiimide approaches caused substantial reductions in antigen-binding capability.

Periodate oxidation of alkaline phosphatase

Alkaline phosphatase from *E. coli*, in common with other proteins of bacterial origin such as β-D-galactosidase, does not contain carbohydrate groups. AP from mammalian sources, however, does: bovine intestine AP contains between 8 and 22% according to different authors, and depending on the isozyme(s) under consideration (Hua *et al.*, 1986; Besman and Coleman, 1985). Periodate oxidation therefore offers a feasible route to the preparation of AP conjugates, although it has received considerably less attention than HRP in this respect, and the conjugates reported have generally offered no advantage over other methods.

Beyzavi *et al.* (1987) adapted both Nakane and Kawaoi's, and Wilson and Nakane's HRP protocol, using the same mass of enzyme in the oxidation stage (giving lower periodate:enzyme ratios), and an AP:IgG ratio of approximately 5:1. The resulting conjugates had extremely low activity in an immunoassay.

Jeanson *et al* (1988a,b) used periodate coupling with four different mouse monoclonal IgGs; they observed little loss of enzyme activity, but the overall performance of the conjugates varied considerably from one Mab to another. AP:IgG ratios of 1:1 and 4:1 were examined. In three cases, the 4:1 ratio was considerably better, but for one mAb there was little difference. Better conjugates were obtained in each case by using the heterobifunctional reagent SPDP.

Overall, the periodate method does not seem to be particularly successful in the preparation of AP conjugates.

General conditions for periodate oxidation of glycoproteins

Clearly the ability of different glycoproteins to survive the periodate oxidation process will vary, but general conditions are again close to those discussed in the applications above.

Ingham and Brew (1981) described the following guidelines for periodate oxidation of glycoproteins before fluorescent labelling: the protein is dissolved at 1–5 mg/ml in 0.1 M sodium acetate/0.1 M sodium chloride, pH 5.0–5.6. Periodate is added to a final concentration of 1–20 mM, and the mixture allowed to incubate for 10–20 minutes at 0–10 °C. Excess periodate is destroyed by the addition of a twofold molar excess of glucose.

O'Shannessy and Wilchek (1990) have suggested 1 mM periodate for 20 minutes at pH 5.5 and 0–4 °C as 'safe' starting conditions, with the options of increasing $[IO_4^-]$, time and temperature.

Some selectivity for particular carbohydrate moieties can also be sought by modifying conditions: Van Lenten and Ashwell (1971) described conditions for the selective oxidation of terminal sialic acid residues in glycoproteins. Using 12 mM periodate for 10 minutes at pH 5.6 and 0 °C these authors avoided damage to other saccharide residues. Reuter *et al.* (1989) carried out detailed studies on *N*-acetyl and *N*-glycoloylsialic acids in glycoproteins. These authors showed significant differences in the behaviour of different sialic acid types: for example, the glycolyl acid could be completely oxidised to the C-7 form with a threefold excess of periodate (one hour, 4 °C, in the dark, pH 5.5), while the acetyl analogue required a tenfold excess.

A generic procedure for periodate coupling of proteins is decribed in Protocol 5.3.

5.2.3.3 Periodate oxidation of amino acid residues

N-Terminal serines and threonines, and hydroxylysine, contain the α-amino alcohol moiety which can be oxidised very readily with periodate. Geoghegan and Stroh (1992) oxidised murine interleukin-1α, which bears an N-terminal serine, with a threefold excess of periodate (30 minutes, 0 °C, pH 7.0) and observed quantitative and specific oxidation. The periodate concentration was around 18 mM. In other experiments with peptides, some oxidation of methionine residues was noted by these authors.

In an application aimed at prodrug therapy, Werlen *et al.* (1994) oxidised a carboxypeptidase G2 mutant containing N-terminal threonine (ten minutes, room temperature, pH 8.3) using a tenfold excess of 20 mM periodate, and again observed specific oxidation. The reaction was carried out in the presence of an excess of methionine to minimise the potential oxidation of such residues, and the activity of the mutant enzyme was not altered by the oxidation process. The resulting aldehydic enzyme was coupled (60 hours, room temperature, pH 4.6) to an insulin hydrazide prepared by reverse proteolysis.

5.2.3.4 Galactose oxidase oxidation of carbohydrate residues

The only other approach for carbohydrate oxidation which has found significant application in protein chemistry employs the enzyme galactose oxidase, which reacts with terminal galactose and *N*-acetylgalactosamine residues to yield an aldehyde functionality. The enzyme also catalyses the oxidation of other saccharides including glucose, xylose, and arabinose (Avigad *et al.*, 1962)

Galactose is often the penultimate rather than the terminal sugar residue in a glycan chain, particularly in mammalian glycoproteins; galactose oxidase oxidation is often carried out in conjunction with neuraminidase treatment, which removes the sialic acid groups commonly occupying the terminal position.

Heitzmann and Richards (1974) compared periodate and galactose oxidase for the oxidation of saccharides in erythrocyte membrane proteins, prior to coupling with ferritin as an electron microscopy label. Both were successful although the periodate route gave a denser stain. Brumeanu *et al.* (1995) used this enzyme to activate desialylated IgG for coupling to peptides.

Galactose oxidase treatment is certainly milder than the use of periodate: oxidation of IgMs fragments (four hours at pH 6 and 37 °C, after neuraminidase treatment for four hours at pH 5 and 37 °C) had no effect on the affinity of the antibodies (Chua *et al.*, 1984). For general purposes, however, galactose oxidase does not generally offer any significant advantages over the periodate route, and, unless immobilised, is considerably more difficult to separate from the product.

Therefore, although references are occasionally encountered, it is not surprising that the approach has failed to achieve widespread popularity in protein conjugation. It may have particular value where the target protein is unusually sensitive to chemical oxidation.

5.2.4 Enzymic direct coupling methods

5.2.4.1 Transpeptidation

The enzyme transglutaminase, which is widely distributed in mammalian tissue, has long been known (Wajda *et al.*, 1971) to be capable of replacing the amide NH_2 group of glutamine residues with other NH_2-containing species (Figure 5.24(a)). Obviously the propensity of proteins to act as a substrate for this enzyme varies with their glutamine content, and is most efficient with glutamine-rich examples such as casein, β-lactoglobulin and the chromogranins.

Generally this route to site-specific introduction of amines has employed low-molecular-mass compounds, but the amine groups of the protein itself are also susceptible to incorporation (Figure 5.24(b)). *In vivo*, for example, cross-linking of fibrin to form blood clots is catalysed by transglutaminase. *In vitro*,

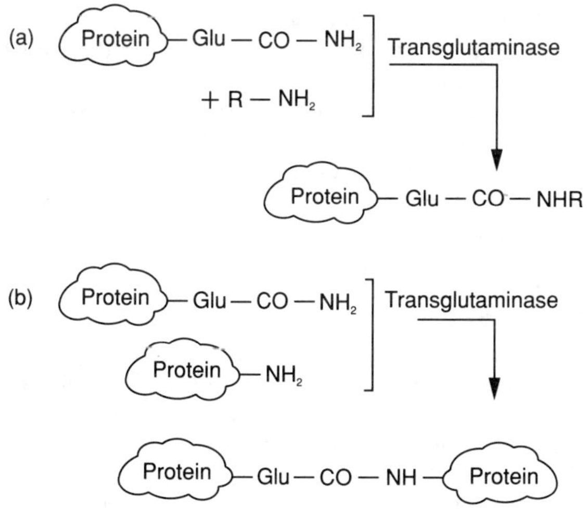

Figure 5.24 Enzymic direct coupling using transglutaminase.

Cooke and Holbrook (1974) observed cross-linking of β-casein during a transpeptidation reaction; this could be prevented by prior acetylation of the protein.

As discussed by Uy and Wold (1977) it is also possible to imagine glutamate-based homo- or hetero-bifunctional reagents whose attachment to protein lysine moieties is catalysed by the presence of the enzyme. The realisation of this possibility does not appear to have been reported, however.

5.2.4.2 Reverse proteolysis

Another closely linked approach, albeit of limited potential for direct protein–protein conjugation, exploits the capacity of proteolytic enzymes to work in reverse. This capability has been known for many decades (Bergmann and Fruton, 1938).

In the presence of a suitable excess of an amine, amide or hydrazide, and sometimes at a pH well removed from the optimum for proteolysis, dehydration reactions with terminal carboxyl functions can be catalysed by enzymes such as carboxypeptidase Y (Schwarz *et al.*, 1990) or lysyl endopeptidase (Rose *et al.*, 1989).

This results in specific linkage of such moieties to the C-terminus (Figure 5.25), and where the nucleophile is the NH_2 group of another amino acid, this provides a practical means of peptide synthesis (Oka and Morihara, 1977; Kasche, 1989; Bongers *et al.*, 1992). However the direct coupling of proteins *via* this route is generally impractical (Rose *et al.*, 1991), and the relevance of this technique to protein–protein

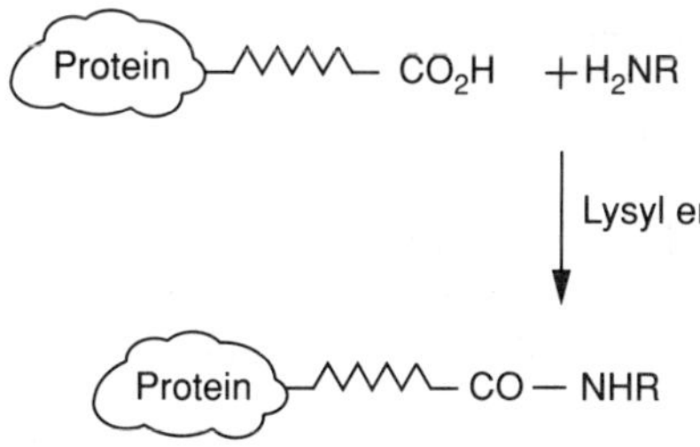

Figure 5.25 Direct coupling by lysyl endopeptidase.

coupling is seen predominantly in heterobifunctional applications, where it is used as a means to couple various activating groups to carboxyl moieties (see Section 5.9.5).

5.2.5 Other direct coupling methods

The mechanisms described so far in Section 5.2 have been based on well-defined interactions of known functional groups. It is possible however to generate highly reactive species from a wide range of amino acids, which can then react with each other or with other residues to form crosslinks. These approaches involve high-energy species and it is often impossible to define the exact nature of the reactants or products.

5.2.5.1 Peroxydisulfate coupling

The peroxydisulfate anion (Figure 5.26) is one of the strongest oxidising agents in aqueous solution, decomposing by a multi-step process to generate a number of radicals such as $OH^{\cdot}$ and $SO_4^{\cdot-}$. These can react with a variety of residues such as histidine, proline, cystine, methionine, tyrosine and tryptophan (Needles, 1977). The oxidation of histidine, tyrosine or tryptophan can result in quinonoid intermediates which are capable of reacting with nucleophilic residues such as lysine or arginine (cf. benzoquinone, Section 5.3.6). This offers a rather aggressive and non-specific route to protein coupling which has found some application in the modification of textile properties. It should be noted that degradation of proteins can also occur on peroxydisulfate treatment, as reactive species cleave the links of peptide chains.

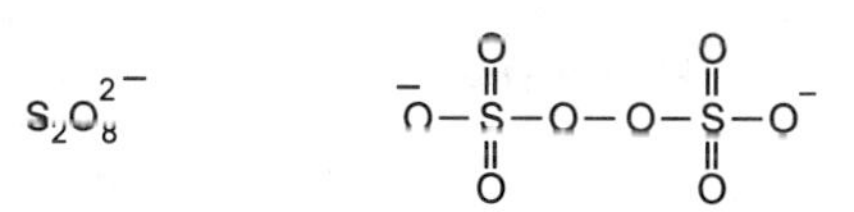

Figure 5.26 Peroxydisulfate anion.

5.2.5.2 γ-Irradiation

Somewhat similar is the effect of γ-irradiation and other forms of ionising radiation on proteins (Yamamoto, 1977). The high energy of the radiation can displace electrons from the protein's amino acid residues and from the solvent if one is present, again generating reactive radicals. Thiyl (S˙) and peptidyl (NH–C˙R–CO) radicals predominate in the dry state and result in crosslinking between molecules; in the aqueous state the OH˙ radicals discussed above are also involved and quinonoid intermediates probably take part in coupling reactions. Again protein degradation is a competing process as peptide chain links are destroyed.

Irradiation and peroxydisulfate treatment of proteins is of little interest for controlled preparative or analytical applications, and the reader is referred to the reviews by Needles and Yamamoto for a more detailed treatment of the reactions involved.

5.3 Homobifunctional Coupling of Amines

Homobifunctional approaches rely on linking similar functionalities in the proteins to be coupled, and can be categorised quite simply in terms of the target group. Amine–amine linkers predominate, while thiol–thiol reagents have also found widespread application. The bis-diazonium salt reagents which predominantly link tyrosines and histidines held an important place in the early development of protein conjugation, but are nowadays outmoded for virtually all uses.

A number of different groups capable of reacting with amines have been described in Chapter 2, Section 2.3.1. Any molecule containing two such groups has the potential to serve as an amine-targetted homobifunctional reagent, and categories of these are based on aldehydes, succinimide (NHS) esters, imidates, aryl halides, isocyanates and isothiocyanates, among others.

Often the reactivity of the two amine-reactive groups is identical, but occasionally differential behaviour can be obtained by manipulation of the reaction conditions, e.g. benzoquinone will only react with one amine group at low pH. This effect can sometimes be exploited to give a greater degree of control than that obtainable from the use of straightforward homobifunctionals.

5.3.1 Bis-aldehydes

The reaction of an aldehyde with an amine to form a Schiff base is shown in Figure 5.27, and has already been encountered as a direct coupling method ensuing from periodate oxidation (Section 5.2.3). Dialdehydes may employ this reaction in homobifunctional mode, and glutaraldehyde (Figure 5.28) is the most commonly encountered example of such a reagent. This compound is sometimes referred to as glutardialdehyde, and almost never by its systematic name of pentanedial (pentane-1,5-dial).

5.3.1.1 The chemistry of glutaraldehyde coupling

The chemistry of glutaraldehyde is far more complicated than the neat mechanism of Figure 5.28 would suggest. Quiocho and Richards (1966) observed that the effects of glutaraldehyde on carboxypeptidase-A could not be reversed by semicarbazide, a reagent capable of cleaving imine bonds under many circumstances. Also, lysine groups were 'lost' from the amino acid composition of the glutaraldehyde-derivatised protein even after the acid hydrolysis necessary for the analysis—conditions which would be expected to hydrolyse the imine rapidly back to the original lysine.

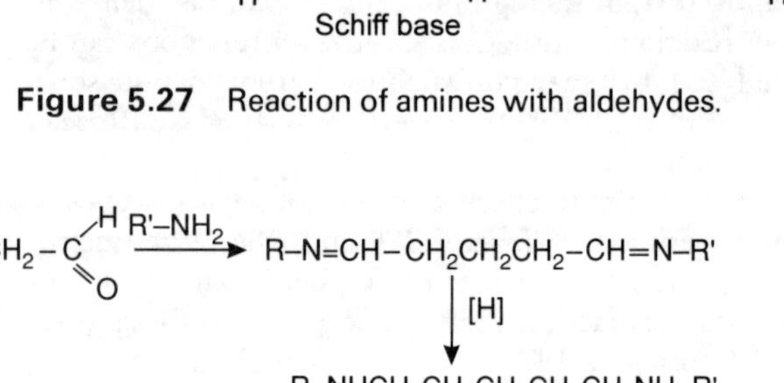

Figure 5.27 Reaction of amines with aldehydes.

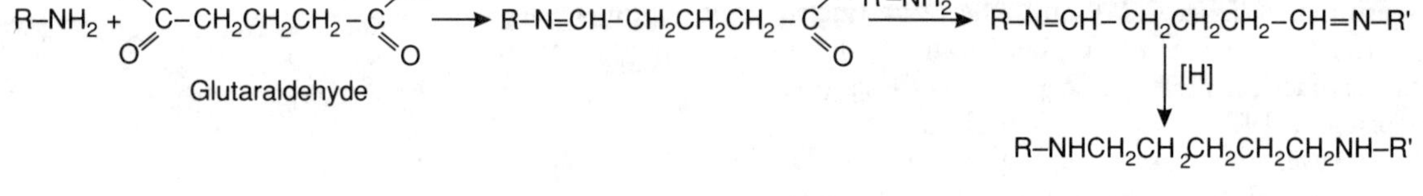

Figure 5.28 Use of glutaraldehyde as a homobifunctional reagent.

Reduction with borohydride before hydrolysis, which would be expected to generate more stable substituted amine moieties, showed no difference from the unreduced material. The authors concluded that simple Schiff base formation could not explain the observed behaviour.

Richards and Knowles (1968) carried out ^{1}H NMR on commercial glutaraldehyde solutions, and observed strong evidence for the presence of α,β-unsaturated aldehydes, generated by aldol condensations between molecules. These authors proposed both linear and cyclic structures based on a combination of intra- and intermolecular reactions (see Figure 5.29). The presence of these structures offers two separate opportunities for the formation of conjugates, more stable than those expected from simple Schiff base conjugation of the 'textbook' glutaraldehyde structure.

First, the 'new' double bond is conjugated to the C=N bond of any Schiff base formed from the aldehyde, delocalising the negative charge (see Chapter 2, Section 2.2.4) and reducing its reactivity. Such a species will therefore be more stable than a simple aliphatic Schiff base. Secondly, the carbon of an aldehyde group is sufficiently electron-withdrawing to render the C=C bonds susceptible to nucleophilic attack, and it was suggested that this can also provide the site for attack by protein amines (Figure 5.30). The product of such a reaction, a 1,4-addition or Michael reaction, is a substituted amine—again, a species whose stability is more consistent with the observed behaviour of glutaraldehyde-linked proteins.

Hardy *et al.* (1969) proposed an alternative explanation for the NMR data, concluding that the solution contained mainly the mono- and dihydrate of glutaraldehyde, and a hemiacetal formed from the monohydrate (Figure 5.31). One of the NMR peaks attributed by Richards and Knowles to the aldol condensation product was shown by these authors to be due to an impurity, which could be removed by further purification. This purer material exhibited a stable composition over several weeks at room temperature. In weakly alkaline solution (pH 8), glutaraldehyde was observed by Hardy *et al.* (1969) to generate a polymer, whose properties did correspond to unsaturated aldol condensation products. Glutaraldehyde could not be regenerated from this polymer.

Hopwood *et al.* (1970) observed that the kinetics of glutaraldehyde–protein reactions were inconsistent with classical Schiff base formation, showing both an unusual pH-dependence and order with respect to the aldehyde.

Korn *et al.* (1972), again on the basis of NMR data, proposed a mixture of free glutaraldehyde, the cyclic hemiacetal proposed by Hardy *et al.* (1969), polymers

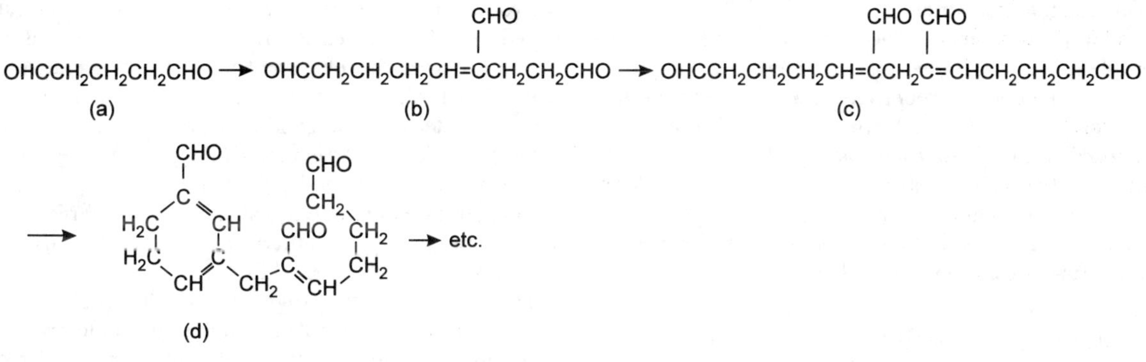

Figure 5.29 Aldol condensation of glutaraldehyde.

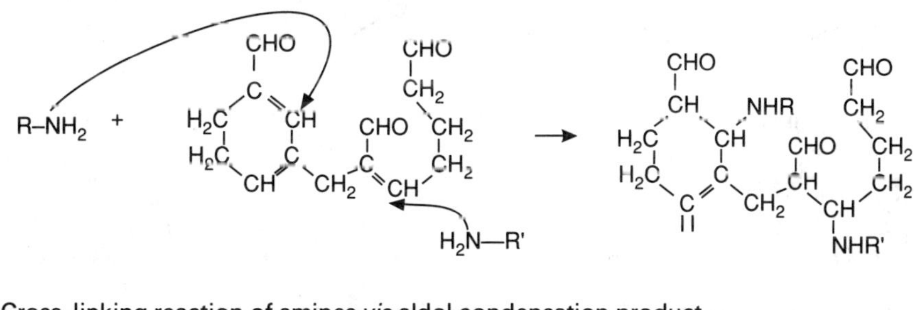

Figure 5.30 Cross-linking reaction of amines *via* aldol condensation product.

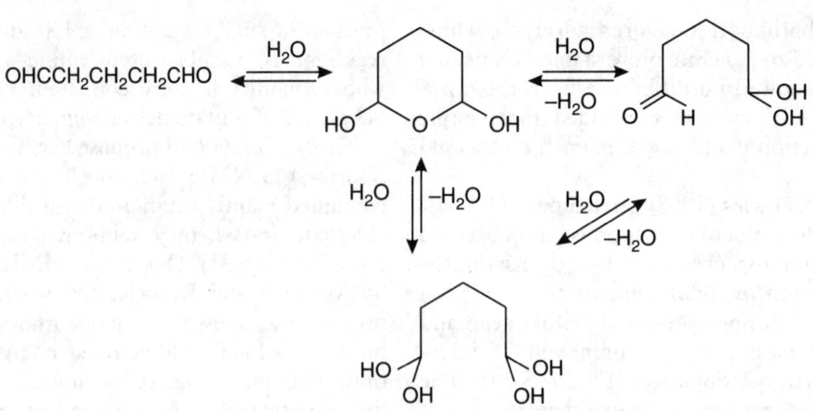

Figure 5.31 Formation of mono-and dihydrate and hemiacetal of glutaraldehyde.

of this species, and methanol as an impurity. They noted that the composition of the solution varied dramatically as the temperature changed, but that all the NMR data could be explained on the basis above. These authors also observed that around four moles of glutaraldehyde were consumed per mole of lysine during coupling reactions, and did not comment on the surprising stability of the products formed[1].

Gillett and Gull (1972) noted that all commercially-supplied glutaraldehyde solutions contained impurities absorbing at 235 nm, consistent with the present of unsaturated polymers.

Whipple and Ruta (1974) applied ^{13}C NMR to this problem, and concluded that commercial 25% glutaraldehyde solution at room temperature consisted predominantly of *cis* and *trans*-isomers of the cyclic hemiacetal (35–36% each) of Hardy *et al.* (1969), plus smaller quantities of the monohydrate (16%), dihydrate (9%), and native glutaraldehyde (4%) (Figure 5.32). As noted by Korn *et al.* (1972), the composition varied considerably with temperature, the level of native glutaraldehyde in particular increasing at the expense of hemiacetals at raised temperatures.

Rasmussen and Albrechtsen (1974) noted that the concentration of the 235 nm-absorbing impurity increased exponentially with both pH and temperature.

Monsan *et al.* (1975) confirmed Whipple and Ruta's general picture of the composition of commercial aqueous glutaraldehyde, noting that the pH of the solutions was around 3.1. At higher pH these authors observed the aldol condensation products described earlier by Richards and Knowles (1968). As glutaraldehyde coupling reactions are generally carried out at raised pH, this seemed to offer an explanation consistent both with the observed composition of (acidic) commercial solutions, and the behaviour of conjugates formed under alkaline conditions where the α,β-unsaturated aldehyde structure could provide the reactive species. Using a model aldehyde of this type, these authors also demonstrated that the product of reaction with lysine was the α,β-unsaturated imine rather than the amine from Michael addition.

Hardy *et al.* (1976a) suggested another mechanism for the formation of stable conjugates from glutaraldehyde: in model work with small amines they isolated polymeric species with behaviour consistent for the

[1] Aldehyde hydrates and hemiacetals exhibit very similar reactivity to the native aldehyde, cf. products of periodate oxidation of carbohydrates.

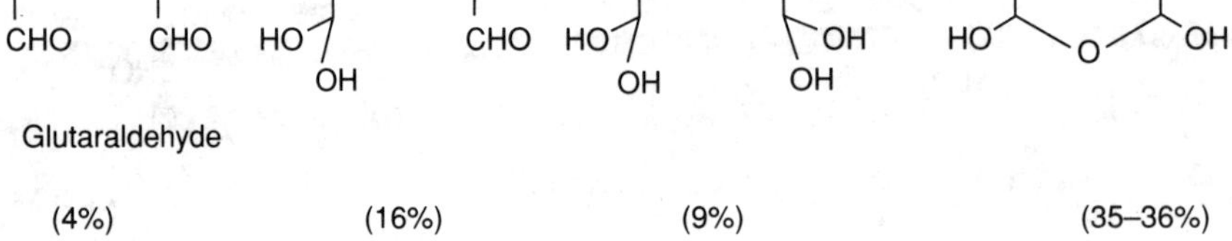

Figure 5.32 Composition of commercial 25% glutaraldehyde as determined by Whipple and Ruta (1974).

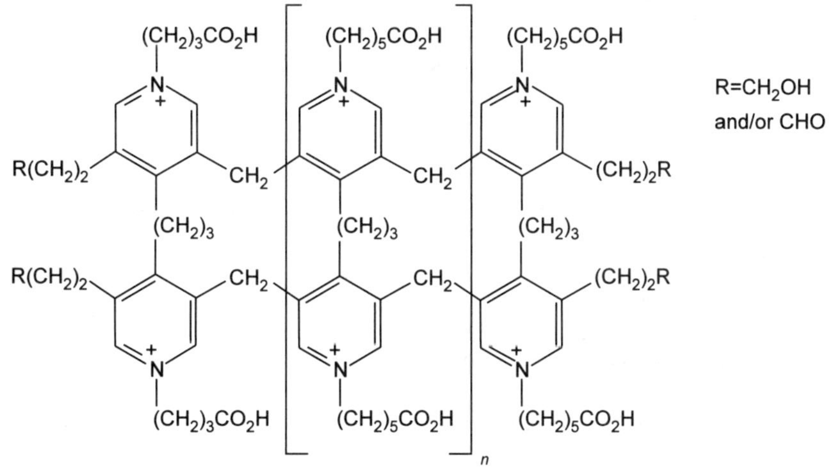

Figure 5.33 Polymeric pyridinium salts identified by Hardy *et al.* (1976a) from the reaction of glutaraldehyde with small amines.

presence of pyridinium salts (Figure 5.33), formed by the reaction of three molecules of glutaraldehyde with the amine. This hypothesis was followed up (Hardy *et al.*, 1976b) by the isolation of a pyridinium derivative from the reaction of glutaraldehyde and ovalbumin, a species consistent with cyclisation of a simple Schiff base. This molecule showed UV absorbance at 250–265 nm, a characteristic of glutaraldehyde-coupled protein products.

Ford and Pesce (1981) analysed the evidence to date, and concluded that models involving α, β-unsaturated polymers were unable to explain the observed data, e.g. the levels of polymer present were generally very low; also, glutaraldehyde showed high reactivity compared to monoaldehydes which might be expected to behave similarly in a condensation model. The pyridinium salt model of Hardy *et al.* (1976b) appeared most convincing as the predominant mechanism, although the overall complexity of the situation was again stressed.

Kirkeby *et al.* (1987) partially supported this view with regard to the low levels of polymerised material; they estimated from UV measurements that practical grade 25% glutaraldehyde solution might contain up to 1 mol% of unsaturated components such as the aldol condensation product; NMR showed less than 5% of dimeric and polymeric species, and the presence of native dialdehyde, hydrate, dihydrate and cyclic hemiacetal was again confirmed. Purer (electron microscopy grade) material showed the same basic composition, but there was no evidence for either unsaturated or dimeric or polymeric species.

While the two grades of material showed only slight differences in composition, there were dramatic changes in the UV spectra of the products formed on reaction with amino acids. With alanine for example, the practical grade material gave a range of products depending on the glutaraldehyde:amino acid ratio, at 10:1 the maximum absorbance was at 230 nm, at 100:1 λ_{max} fell at around 265 nm (cf. Hardy *et al.*, 1976a,b), while at higher ratios absorbances above 300 nm prevailed. The purer material, by contrast, gave a single predominant absorbance at around 230 nm at all glutaraldehyde:amino acid ratios. The authors proposed a possible mechanism to explain this observation based on aldol-type condensation of two molecules of Schiff base, followed by loss of a substituted amine (Figure 5.34). This again results in the formation of a stable α, β-unsaturated Schiff base linkage.

Okuda *et al.* (1991) carried out further studies which again failed to support the unsaturated polymer model, and these authors backed the possibility of pyridinium salt links.

Ukeda *et al.* (1993) carried out an examination of the reaction of bovine serum albumin (BSA) with a glutaraldehyde-derivatised solid phase. This revealed complex kinetics and at least two different binding mechanisms; the findings were consistent with a fast initial binding to an ionic species (again, possibly Hardy's pyridinium salt) and a slower covalent reaction. Although blocking the solid phase with ethanolamine caused the disappearance of the carbonyl absorbance in the IR spectrum, the amount of BSA bound was reduced by only 40%; reduction of the

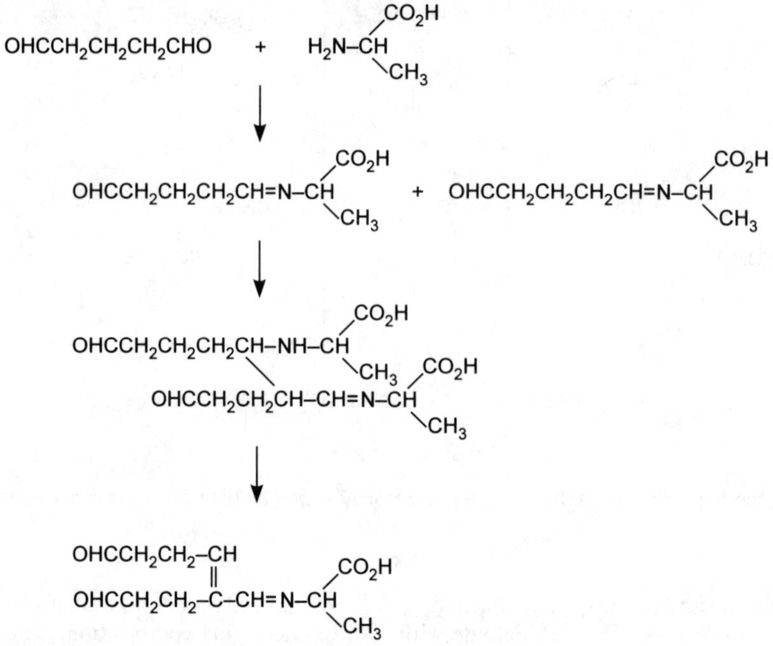

Figure 5.34 Proposed reaction of glutaraldehyde with alanine (Kirkeby *et al.*, 1987).

solid phase with borohydride reduced the BSA binding by around 70%.

Overview of glutaraldehyde composition and reactivity

The chemistry of glutaraldehyde and its reactions with protein amine groups, as correctly predicted by Quiocho and Richards (1966), is complex, and heavily dependent on a number of factors such as reagent purity and pH.

The following list attempts to summarise the current state of understanding in this area. (1) Typically, commercial 25% aqueous glutaraldehyde solutions have a pH of around 3, and consist primarily of a mixture of native dialdehyde, monohydrate, dihydrate, and isomeric forms of a cyclic hemiacetal. The quantities of these components vary with temperature (see Figure 5.32). (2) They may also contain small quantities of other components such as dimeric and polymeric hemiacetals, and α,β-unsaturated aldehydes from aldol condensation. These may have a dramatic effect on the products of reaction with proteins. (3) Once these solutions are introduced into an alkaline environment, the concentration of aldol condensation products increases. (4) The following reactions may occur with protein amine groups. (a) The native dialdehyde (or its hydrate or hemiacetal) may form an unstable imine bond *via* an amino alcohol intermediate. (b) The remaining aldehyde group may react similarly with a second protein to yield an unstable bis-imine link. (c) The amino alcohol from (a) may cyclise and dehydrate to form an unreactive pyridinium structure, this will not result in a protein–protein link. (c) The imine from (a) may undergo condensation with other molecules of glutaraldehyde (native or partially reacted), then undergo multiple cyclisations to form stable, linked pyridinium structures (Hardy *et al.*, 1976a,b). (d) This imine may condense with another such molecule to form a stable α,β-unsaturated imine (Kirkeby *et al.*, (1987)). (e) The α,β-unsaturated aldehyde of the aldol condensation product may form a stable α,β-unsaturated imine. (f) The α,β-unsaturated aldehyde of the aldol condensation product may undergo Michael addition to form a stable substituted amine.

Which of these reactions predominates will depend on many factors. The bulk of the evidence suggests that the aldol condensation products play only a minimal role, despite being quoted as the major factor in recent reviews (Jayakrishnan and Jameela, 1996). However at alkaline pH (and depending on how long the glutaraldehyde is stored in alkaline buffer before use!) the aldol condensation routes may become increasingly important. The pyridinium salt route appears to play a key role, but it seems likely that reactions *via* the classical Schiff base generally play a significant role too. Finally, the observations of Kirkeby *et al.* (1987) using glutaraldehyde solutions of different purity suggest that there are subtleties yet to be understood in this area.

Although it reacts principally with the amines of lysine groups and terminal residues, glutaraldehyde can also react efficiently with the thiol group of cysteine (Okuda *et al.*, 1991), and to some extent with the imidazole ring of histidine and with tyrosine residues (Habeeb and Hiramoto, 1968).

5.3.1.2 The development of glutaraldehyde coupling methods

Glutaraldehyde has been used to couple proteins in an applied fashion for many decades, its early history being linked to the leather tanning industry. Some investigations on its reactivity were carried out in this connection in the 1950s (e.g. Fein *et al.*, 1959).

Similarly, glutaraldehyde became popular in the 1960s as a fixative for electron microscopy and cytochemistry (Sabatini *et al.*, 1963), insolubilising proteins without serious effects on enzyme activity.

In the 1960s, it received attention in connection with the physical stabilisation of protein crystals used in X-ray diffraction studies. Quiocho and Richards (1964) observed that small molecules such as glutaraldehyde could penetrate between the individual molecules of the crystal structure, and the cross-links formed by the subsequent reaction led to a dramatic improvement in physical stability, in some cases without any serious effect on the diffraction pattern.

Crystals of carboxypeptidase A treated in this way lost about half their enzyme activity (already 300-fold lower than the aqueous form), but showed no change in activity over a three-month period of use in a flow-through column (Quiocho and Richards, 1966).

Habeeb and Hiramoto (1968) first described the production of links between dissolved proteins with glutaraldehyde, using this approach to yield an ovalbumin–bovine serum albumin conjugate. The subsequent development of glutaraldehyde conjugation techniques was championed in the histochemical and cytochemical field, in particular by Therese Ternynck and Stratis Avrameas in France.

Both one-step and two-step glutaraldehyde coupling methods have been described; in the former both proteins are reacted with the glutaraldehyde simultaneously, while in the latter one protein is derivatised, then purified to remove excess coupling agent before the second protein is added.

The fact that the two-step procedure results in conjugation of course indicates that reactive groups from the glutaraldehyde remain after the initial incubation. This procedure has probably been most popular where the first protein used is horseradish peroxidase, the low amine content of this enzyme facilitating controlled reaction.

However two-step procedures have been applied to many other proteins with high amine contents, and, if the production of a single 'macroconjugate' is to be avoided, the success of these must rely on stopping the first stage well before the possibilities of cross-linking reactions have been exhausted.

5.3.1.3 Two-step glutaraldehyde coupling methods

The method of Avrameas and Ternynck (1971) describes a two-step procedure for coupling horseradish peroxidase with sheep IgG or Fab fragments. It consists of the following stages. (1) HRP is dissolved at $50\,mg\,ml^{-1}$ in 0.1 M phosphate buffer (pH 6.8), containing 1.25% glutaraldehyde, and the mixture allowed to react for 18 hours at room temperature. (2) The mixture is chromatographed in 0.15 M sodium chloride solution on a column of Sephadex™ G-25, to remove unreacted glutaraldehyde and the HRP concentration adjusted to around 10 mg/ml. (3) At an HRP:IgG ratio of 8:1, IgG is added at 5 mg/ml in 0.15 M sodium chloride solution, followed by 0.1 ml of 1 M carbonate–bicarbonate buffer (pH 9.5). The mixture is allowed to react for 24 hours at 4 °C. (4) An amount (0.1 ml) of 0.2 M aqueous lysine solution is added to quench unreacted sites on the activated HRP, allowing two hours reaction at 4 °C. (5) The resulting conjugate is purified by dialysis and ammonium sulfate precipitation or size exclusion.

Mannik and Downey (1973) used Avrameas and Ternynck's conditions to conjugate rabbit and goat Fab fragments to HRP, but investigated the effect of using higher ratios of enzyme to antibody. The highest yield of monosubstituted conjugate was obtained at the standard 8:1 ratio. Higher incorporations could be obtained as the ratio was increased, but only at the expense of aggregation. Antibody activity was significantly reduced by conjugation, particularly where more than one HRP molecule was contained in the complex: 50% was retained in Fab–HRP_1 conjugate, 35% in Fab–HRP_2 and 24% in Fab–HRP_3.

By 1978, Avrameas *et al.* (1978) had introduced only minor changes to their original protocol, a reduction in the glutaraldehyde concentration to 1%, addition of

0.2 ml of 0.5 M carbonate–bicarbonate buffer (cf. 0.1 ml of 1 M), addition of a greater quantity of lysine at a controlled pH (0.1 ml of 1 M lysine pH 7, cf. 0.1 ml of 0.2 M).

Molin *et al.* (1978) investigated the effect of varying reaction conditions on Ternynck and Avrameas' methodology. Their experiments differ in using a much lower HRP concentration (4 mg/ml); their key conclusions were as follows. (1) The activation of HRP by glutaraldehyde did not continue beyond around four hours, around 30% of the HRP was activated at this time, and no increase was observed up to 18 hours. (2) The enzyme activity of the HRP was adversely affected by the glutaraldehyde treatment; 20% was lost after two hours, and 50% after 18 hours. (3) The efficiency of the glutaraldehyde activation was roughly constant over the pH range 3–8 but increased quite sharply above this value. This could be due to increased deprotonation of the lysine amine groups (pK_a around 10.5), but was felt by the authors to be related to increased polymerisation of glutaraldehyde at high pH leading to more effective interprotein linkage. (4) There was no difference in the efficiency of activation in increasing the glutaraldehyde level from 0.25% (100:1 excess over HRP) to 2.5% (1000:1 excess). On the basis of these findings, Molin *et al.* (1978) recommended HRP activation conditions of two hours at room temperature and pH 9.5, with a glutaraldehyde excess of 100:1.

Porstmann *et al.* (1984) compared 'Avrameas' and 'Molin'-type conditions in the preparation of sheep IgG–HRP conjugates. The Avrameas conditions employed HRP at 4.35 mg/ml in 0.1 M phosphate buffer (pH 6.8) and glutaraldehyde at 0.16 M, representing an excess over HRP of about 1500:1. The reaction time was 18 hours at 20 °C; the activated enzyme was coupled to the IgG, apparently at an HRP:IgG ratio of around 17:1.

The Molin conditions employed HRP at 3.26 mg/ml in 0.1 M borate buffer (pH 9.5) and glutaraldehyde at 0.025 M, representing an excess over HRP of about 300:1. The reaction time was two hours at 20 °C and the same HRP:IgG ratio was employed.

The yield and Rz of these two conjugates were virtually identical, while the Molin conjugate displayed a significantly lower enzyme activity and a higher percentage of uncoupled IgG. However, in an immunoassay the two conjugates behaved very similarly.

Lannér and Bergquist (1978) described an unusual variation on the protocol of Avrameas and Ternynck (1971) where the conjugation was carried out during size exclusion chromatography. This modification reduced the degree of aggregation, but the product contained large amounts both of unreacted HRP and IgG.

Overall, despite the control which it offers compared to one-step methods, the two-step glutaraldehyde approach still tends to yield HRP conjugates with a high degree of aggregation. Losses of both HRP and IgG activity can be severe: Madore and Baumgarten (1979) saw enzyme activity reduced by up to 90%, and Mannik and Downey (1973) observed up to 75% loss of immunoreactivity.

Two-step coupling for alkaline phosphatase was originally described by Engvall and Perlmann (1971) in the paper which first described the enzyme-linked immunosorbent assay (ELISA); it gave higher yields than the one-step protocol.

AP (calf intestine), 18.75 mg/ml in 0.1 M sodium phosphate buffer (pH 6.8), was activated with glutaraldehyde (to 0.2%) for one hour, then IgG (1.25 mg/ml in the same buffer) was added and allowed to incubate, presumably for one or two hours. This resulted in a 13% yield of IgG in the conjugate, compared to 0.7–10% for one-step protocols.

Engvall (1978) activated AP (Sigma Type VII; 5 mg/ml in phosphate buffered saline) with glutaraldehyde (to 0.2%) overnight at room temperature before conjugation to Protein A. Dialysis was carried out both before and after the activation, in the first instance to remove any traces of ammonium sulfate, and in the second to remove unreacted glutaraldehyde. Reis *et al.* (1988) used identical conditions to activate AP (Sigma Type VII-S) for conjugation to streptococcal Fc receptors.

Jeanson *et al.* (1988a,b) used conditions recommended by Boehringer Mannheim GmbH in the preparation of AP conjugates with mouse monoclonal IgG. AP at 2.24 mg/ml in 0.05 M phosphate buffer (pH 7.2) was reacted with glutaraldehyde at 0.16% for 50 minutes at 24 °C, no separation of unreacted glutaraldehyde was carried out, and the mixture was allowed to react with IgG (at an AP:IgG ratio of around 2.5:1) for 75 minutes at 24 °C. This approach allowed slightly better retention of enzyme activity than other approaches (periodate, disulfide coupling), but the latter offered the best performance overall.

A method for glutaraldehyde activation of various proteins was described by Zegers *et al.* (1990), dialysing a 10 mg/ml solution of the protein against 0.2% glutaraldehyde in 0.01 M phosphate buffer (pH 7.2) for 16 hours at 4 °C, then dialysing against three changes of buffer before adding the reagent to be conjugated.

5.3.1.4 One-step glutaraldehyde coupling methods

One-step preparation of HRP–IgG conjugates dates from the work of Avrameas (1968, 1969), before their advocation of two-step methods. Clyne *et al.* (1973) used conditions similar to those of Avrameas with goat IgG, following up the dialysis with fractionation of the conjugate on Sephadex™ G-200. This revealed a high degree of polymerisation in the product, which gave good histochemical performance.

Boorsma and Kalsbeek (1975) followed exactly the same methodology with rabbit IgG and obtained comparable results. This conjugate compared poorly with the product of a two-step protocol, however, and examples of such approaches become increasingly infrequent after this time.

Avrameas (1969) used the following conditions for the preparation of IgG–AP conjugates. (1) Human IgG (5 mg) and 10 mg of AP (*E. coli*) were dissolved in 1.3 ml of 0.1 M phosphate (pH 6.8), and 0.10 ml of 1% aqueous glutaraldehyde added. (2) After two hours at room temperature, the reaction mixture was dialysed against 2 × 5 litres of phosphate-buffered saline at 4 °C overnight. Precipitated material was removed by centrifugation.

Minor variations were introduced to obtain optimal performance, depending on the source of the proteins. For sheep IgG and AP from *E. coli* a volume of 2 ml of buffer was used at the first stage, and a larger amount of glutaraldehyde solution added (0.15 ml). For human IgG and AP from chicken intestine, 12 mg of IgG and 50 mg of AP were dissolved in 2 ml of the phosphate buffer, and 0.4 ml of the glutaraldehyde solution added.

Engvall *et al.* (1971) coupled 0.5 mg of rabbit IgG and 1.5 mg of AP (calf intestine) in phosphate-buffered saline solution (around 200 ml), using 10 ml of 4.2% glutaraldehyde to give a final concentration of 0.2%, after two hours at room temperature the mixture was diluted with buffer, dialysed and fractionated on Sepharose 6B using 0.05 M Tris buffer (pH 8.0).

Ford *et al.* (1978) coupled 12 mg of AP (calf intestine) with 4 mg of goat IgG in phosphate-buffered saline solution using 0.5 2.5 μmol of glutaraldehyde. Two hours reaction was allowed before dilution and dialysis. The loss of enzyme activity varied from 18 to 76% after this period, worsening with increased glutaraldehyde concentration, but larger amounts of enzyme were incorporated at the higher concentration. Therefore the overall efficiency of incorporating enzyme activity into the conjugate remained roughly constant at around 0.4%. Antibody function was also damaged more by greater glutaraldehyde concentrations, the loss ranging from 73 to 90%. A greater degree of high-molecular-mass material was observed in the high-glutaraldehyde conjugates.

Avrameas *et al.* (1978) recommended some modifications to the 1969 protocol, a volume of 2 ml, the use of 0.05 ml of 1% glutaraldehyde solution, a reaction time of three hours, and the use of 0.1 ml of 1 M lysine solution (pH 7) in a two-hour quenching stage.

Jeanson (1988a,b) used the Engvall and Perlmann (1971) conditions to prepare AP conjugates from four different mouse monoclonal antibodies. Performance varied widely from one mAb to another, the resulting conjugate gave the best combination of enzyme and antibody titre from several different methods in one case, and the worst in another.

Cameron and Erlanger (1976) used a modification of Avrameas's peroxidase method to couple β-Gal to sheep IgG. β-Gal (20 mg) and 10 mg of IgG were coupled with 125 μl of 5% glutaraldehyde using the usual two hour reaction at room temperature, followed by dialysis and centrifugation. The product was used in an immunoassay for lymphocyte surface receptors.

Weltman *et al.* (1976) reported conjugation of 0.2 mg of β-Gal with 0.5 mg of rabbit IgG in phosphate-buffered saline solution (pH 7.2; no volume given) using 1 mmol of glutaradehyde. After six hours reaction the mixture was dialysed and the conjugate used in an IgE assay.

Avrameas *et al.* (1978) recommended reacting 10 mg of β-Gal with 5 mg of IgG (or 2.5 mg of Fab) in 2 ml of 0.1 M phosphate buffer (pH 6.8), using 0.1 ml of 1% glutaraldehyde. After three hours at room temperature, 0.1 ml of 1 M lysine (pH 7) is added and allowed to react for two hours before dialysis of the conjugate.

Deelder and De Water (1981) coupled 2 mg sheep IgG with 5 mg β-Gal in 2 μl of 0.1 M phosphate buffer (pH 6.8), using 125 ml of 1% glutaraldehyde with a three hour reaction time. Purification was again by dialysis. High-molecular-mass material predominated, and this was reflected by high backgrounds in an immunoassay. Heterobifunctional approaches gave smaller conjugates and improved assay performance.

Many other protein conjugations have been performed by the one-step glutaraldehyde method. Despite the high degree of aggregation and heterogeneity generally observed, this approach has remained in use, presumably due to the simplicity of the methodology, to the present day. For example, Peuravuori and Korpela (1993) used this technique to prepare conjugates of human IgE with pyrophosphatase for immunoassay.

5.3.1.5 Overview of glutaraldehyde coupling techniques

Glutaraldehyde coupling offered a major advance over the techniques available beforehand (Ishikawa *et al.*, 1983), and was arguably the most widely used approach in the 1970s. Although it is still frequently encountered, its importance has declined as other methods have become available. For peroxidase conjugates, periodate oxidation has generally offered better performance, and in all high-performance applications heterobifunctional reagents have been more widely favoured in the 1980s and 1990s.

One-step approaches are frequently characterised by a substantial degree of polymerisation to give a heterogeneous, high-molecular-mass conjugate, by a low incorporation of the activated component, and by losses of functional activity. These losses are predominantly related to the polymerised nature of the conjugates which makes steric hindrance a serious concern. Enzymes may retain their activity, as their substrates are generally small enough to gain access despite the polymerisation—better than 60% retention is achievable in many cases (Engvall *et al.*, 1971; Ford *et al.*, 1978). Protein-binding activity is often seriously compromised, however, and antibody function may be reduced by as much as 99% (Engvall *et al.*, 1971).

Two-step procedures rely on incomplete reaction or paucity of amine groups to limit polymerisation of the activated protein. HRP is particularly suited to this approach with its low amine content, and can yield conjugates of greater homogeneity and lower molecular mass (Avrameas and Ternynck, 1971) than the one-step procedure will allow. Although incorporation of enzyme may still be low (3–5%: Avrameas and Ternynck, 1971), the smaller size of the conjugate allows more than half of the antibody activity to be retained (Boorsma and Kalsbeek, 1975; Boorsma and Streefkerk, 1976; Mannik and Downey, 1973).

Although other enzymes do not have the same low amine content as HRP, the use of a two-step protocol generally results in better retention of protein function, e.g. in the preparation of AP–Protein A conjugates (Engvall, 1978), where both the enzyme activity and the antibody-binding properties of the Protein A were well-maintained. This effect is understandable: if the enzyme is allowed to polymerise first it may tend to form a 'core' to the conjugate while remaining accessible to small substrate molecules. The second protein may then form a 'mantle' around the core, where its protein-binding functions will be relatively unimpaired by steric effects (Figure 5.35).

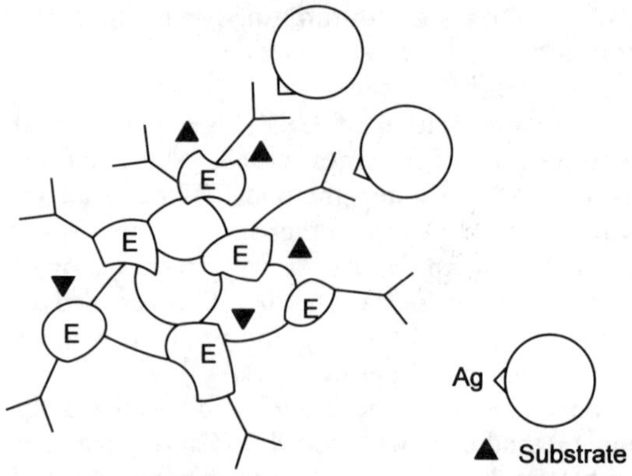

Figure 5.35 In a two-step coupling procedure, the initial stage may form a 'core' of enzyme molecules (E). The outer 'mantle' of antibodies is accessible to bulky antigens, while small substrate molecules can still penetrate the structure to reach the enzyme.

Even if glutaraldehyde behaved as the simple monomeric molecule pictured in catalogues, it would be not be very likely to offer an efficient route to protein–protein conjugation, simply by virtue of its homobifunctional nature. As noted in the introduction to this chapter, this category of reagents is really best suited to structural investigations, etc. But because of the uncertainty over the exact nature and length of the protein–protein bond in glutaraldehyde conjugates, its use in investigations of this type is of limited value, particularly with respect to distance determinations (Peters and Richards, 1977).

For preparative applications, this coupling method is hampered by its inherent lack of control, and the serious issues around variability and stability of glutaraldehyde as a raw material. The only fully agreed conclusions on the reaction mechanism are that it is complex and highly dependent on material quality. It does not lend itself to the reproducible production of conjugates.

Glutaraldehyde can also display irritant and allergenic properties.

In the light of these conclusions glutaraldehyde seems best suited to its original applications where a 'protein glue' is really all that is required—fixing, tanning, etc. In these uses the control of the reaction and the nature of the bond formed are relatively immaterial, and the vigorous reactivity of this reagent meets the major need. Its only advantage for more demanding applications is its practical simplicity.

As noted earlier, protein coupling chemistry is still a field with a reasonable degree of unpredictability, and very occasionally a method such as glutaraldehyde coupling has been known to give, serendipitously, the best results, therefore it should not be ruled out altogether: a generic two-step procedure is described in Protocol 5.2. However, the days of glutaraldehyde coupling as a first-choice method for reproducible, high-performance applications should really be over.

5.3.1.6 Other dialdehydes and related reagents

A number of other dialdehydes have been used relatively infrequently for protein coupling; most are fairly close aliphatic analogues of glutaraldehyde such as adipaldehyde (Figure 5.36(a)) which are likely to display the same mechanistic complexities. Some of the smaller examples such as glyoxal (Figure 5.36(b)) are undoubtedly simpler in terms of their chemistry (not being prone to aldol condensations, for example), but probably give too short a bridge for optimal performance in many applications. Other reagents use aromatic aldehydes, such as dimeric derivatives of pyridoxal (Figure 5.36(c)). Diketones such as hexane-2,5-dione (Figure 5.36(d)) can react similarly and have also been employed as coupling agents. The reader is referred to Wong (1991), Chapter 4, for details of these rarely-encountered reagents and references to their use.

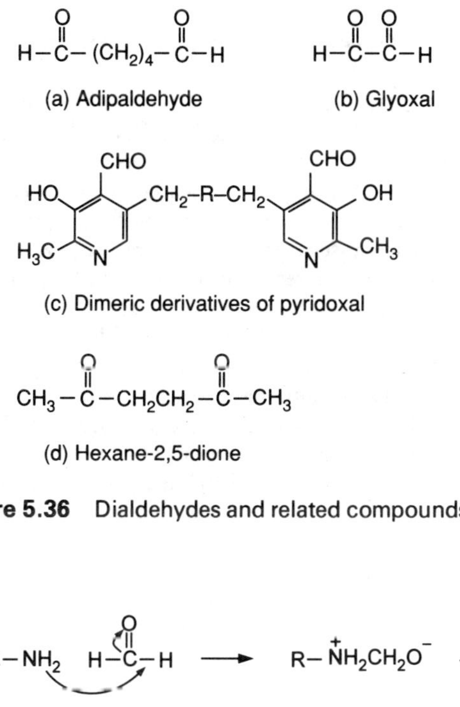

Figure 5.36 Dialdehydes and related compounds.

Cai and Keana (1991a) described some interesting glutaraldehyde derivatives incorporating photoactivatable moieties, which were proposed as potential cross-linking agents or fixatives for electron microscopy: these might offer better fixative properties than glutaraldehyde itself, particularly towards unreactive groups such as lipids. These reagents are dealt with in Sections 5.10.1 and 5.10.4.

PEG–bis-aldehydes are supplied by Shearwater Polymers.

5.3.1.7 Formaldehyde

Formaldehyde is an interesting coupling agent, the stabilising effect of its cross-linking properties having been exploited since at least the last century for tissue preservation. It exists in concentrated solution (formalin) as oligomers of the form $H(OCH_2)_nOH$ which dissociate to release the monomer on dilution.

The aldehyde group of formaldehyde (Figure 5.37(a)) will yield a Schiff base on reaction with an amine (cf. periodate oxidation, Section 5.2.3). As mentioned in that section, this reaction followed by reduction is a commonly used approach to the alkylation of amines (Means and Feeney, 1968). However, another route can be followed from this point which results in a protein-coupling reaction. Below its pK_a the base will be protonated, and the positive charge of this immonium salt pulls electrons away from the adjacent carbon atom. This can be attacked by a nucleophile such as an amine group from a second protein (Figure 5.37(b)); the net result is a methylene ($-CH_2-$) bridge, one of the simplest linkers imaginable, between the two proteins. Apart from its amine-derivatising properties, formaldehyde has some reactivity for cysteine, tyrosine, histidine, tryptophan and arginine (Ji, 1983).

Cross-linking of erythrocyte membrane proteins using formaldehyde was carried out in a structural investigation by Steck (1972). More extensive reaction was observed when formaldehyde concentration

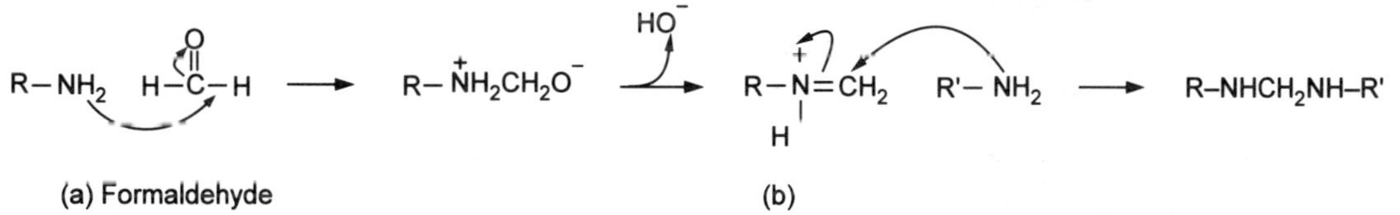

Figure 5.37 Reaction of amines with formaldehyde.

(5–15 mM), pH (6–8), temperature, reaction time or ionic strength was increased.

5.3.2 Bis-imidates

Imidates or imidoesters (Figure 5.38) have a pK_a in the region of 5–7 (Hunter and Ludwig, 1962; Hand and Jencks, 1962). The positively charged nitrogen in the protonated form has a strong electron-withdrawing effect, resulting in a highly electrophilic carbon atom susceptible to attack by nucleophiles such as amines (Hunter and Ludwig, 1972). The reaction pH must be low enough to maintain at least a fraction of the protonated imidate, but quite alkaline conditions can be employed in practice. The 'textbook' product of an amine–imidate reaction is an amidine group (Figure 5.39), a highly basic functionality (pK_a 11–13) unlikely to be encountered in deprotonated form[1].

A key characteristic of forming an amidine group from an amine is that (below the pK_a of the amine) there is no change to the charge carried by the protein. This can be contrasted, for example, with the formation of an amide group, which is so weakly basic that it is uncharged under all but the most strongly acidic conditions. A protein whose amine groups have been converted to amides will carry less positive charges under neutral conditions and may display a lower p*I* (see Chapter 2, Section 2.2.1) than the native molecule.

[1] This group is not quite as basic as the guanidine group found in arginine (Section 2.3.2), where the delocalisation which can be achieved over the three nitrogen atoms results in greater stability of the protonated form.

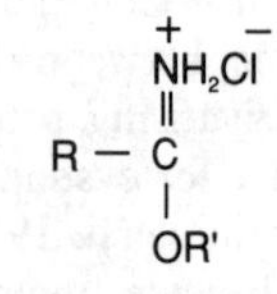

Figure 5.38 Imidate or imidoester shown here as the chloride salt.

Derivatisation of a protein *via* amidine groups, therefore, results in less charge modification than many other approaches; consequently there may be less change to tertiary or quaternary structure, and for this reason imidates have been popular in structural investigations. Wofsy and Singer (1963) showed that amidination of 85–90% of the available residues in BSA and rabbit γ-globulin caused negligible changes in electrophoretic properties, sedimentation behaviour, optical rotation or ultraviolet absorbance, and even exhaustive reaction had only small effects.

There are other advantages of imido ester reagents, they are generally used as hydrochlorides, whose charge renders them highly water-soluble, and among protein functionalities they react with great specificity with amines (Hunter and Ludwig, 1962). The amidine bonds which they form are stable to the acidic conditions under which peptide hydrolysis is carried out, allowing them to remain intact during subsequent analytical investigations.

However the chemistry of imidate reactions is much more complex than the simple formulae above would suggest, and features such as rapid hydrolysis, and strong pH and temperature dependences are clearly disadvantageous.

5.3.2.1 The chemistry of imidate coupling

Given the ready availability of bis-imidates, their specificity and their water-solubility, it is perhaps surprising that they have not been used more in the preparation of protein conjugates such as enzyme-labelled antibodies. References to such applications are virtually absent from the literature. One reason which may have deterred workers is the large number of reactions which may compete with the ideal mechanism shown in Figure 5.40. These have been discussed by Browne and Kent (1975) and Peters and Richards (1977), and have painted a more complex picture than earlier reports. Some of the major competing reactions are listed below. (1) The diaminoester intermediate (Figure 5.39(b)) may lose ammonia to

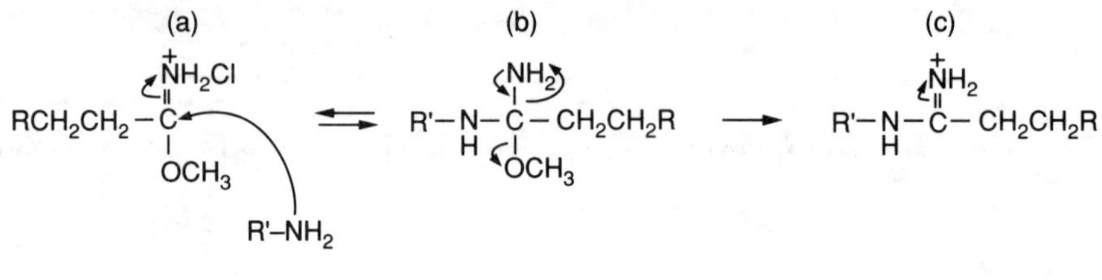

Figure 5.39 Reaction of an amine with an imidate (a) to form an amidine (c) *via* a diamino ester intermediate (b).

generate a different linkage to the protein, still containing the imido ester functionality (Figure 5.40(a)) and therefore still reactive. (2) This imido ester can react with a second protein amine to form another link (Figure 5.40(b)). (3) Ammonia from (1) can react with the starting imido ester to form an unsubstituted amidine, no protein–protein link results (Figure 5.40(c)). (4) The starting imido ester can react with water to generate an alcohol and an amide, this predominates at high pH (Figure 5.40(d)). (5) The starting imido ester can react with water to generate ammonia and an ester, predominant at low pH (Figure 5.40(e)).

Other possibilities are described in Peters and Richards' review.

The reaction pH has a complex effect on the course of the aminolysis reaction. It is necessary to strike a balance between a pH low enough to maintain a supply of protonated imidate, and high enough to provide sufficient unprotonated amine for successful competition with imidate hydrolysis. This balance will depend heavily on the exact structure of the reagents, and the reaction intermediates may also display their own pH-dependent behaviour. Overall, then, pH dependence of imidate reactions is complicated and difficult to predict.

In general, the greatest efficiency in protein work appears to be obtained by working at around pH 10, where the hydrolysis of the imidate is slow (half-life is about 30 minutes) and the formation of the aminolysis product rapid (90% complete in three minutes: Peters and Richards, 1977; Browne and Kent, 1975). At least a 100-fold excess of imidate is generally employed at a concentration of 0.1–10 mM, and incremental addition can be employed to reduce the effects of hydrolysis. Hunter and Ludwig (1962) observed a strong temperature-dependence of the reaction between 0 and 40 °C,

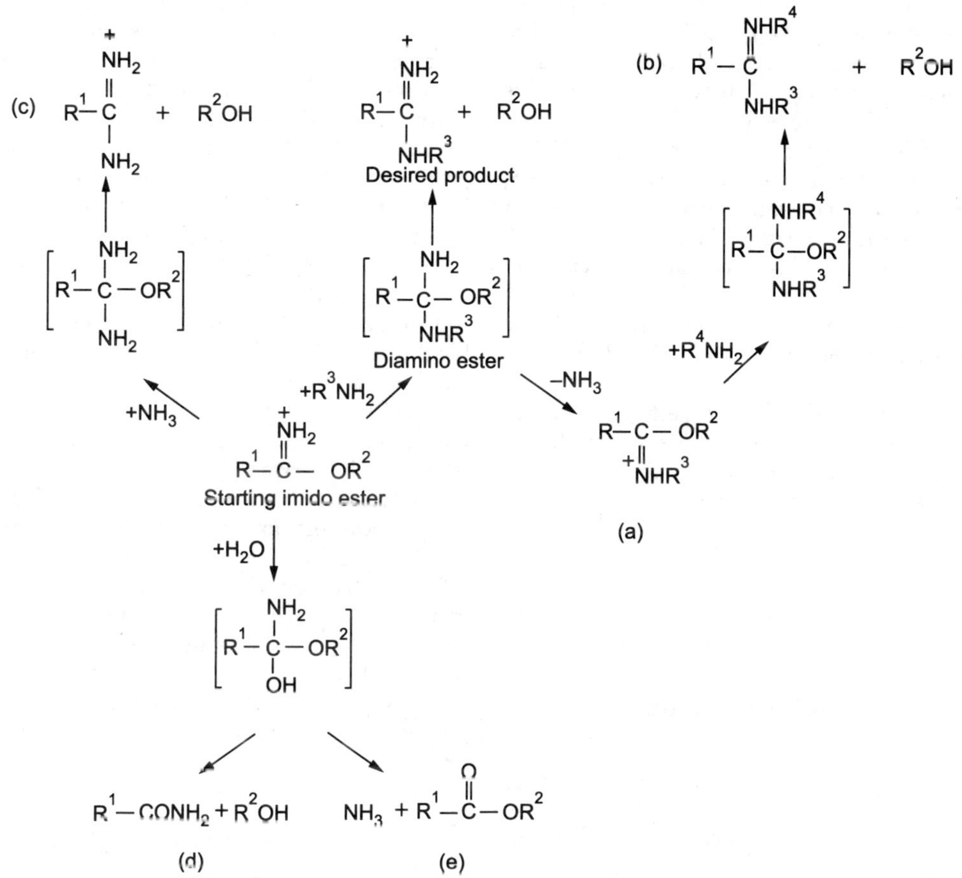

Figure 5.40 Competing reactions of imidates. See text for details. R^3NH_2 and R^4NH_2 represent protein molecules.

though Swaney (1986) saw little effect between 4 and 30 °C in the cross-linking of liporoteins.

If the protein of interest cannot withstand pH 10 even for the few minutes required for reaction, working at lower pH is prone to increasing complexity of mechanism and, despite the greater level of protonated imidate, decreasing efficiency. Reaction below around pH 8 is seldom encountered. This requirement for high pH is another factor militating against more widespread use of these reagents in the preparation of protein conjugates, and has led to increased popularity of bis-succinimide esters (Section 5.3.5) in studies of protein structure and function.

5.3.2.2 Bis-imidates in studies of protein structure

Interest in imidates for protein modification dates from the early 1960s, and the detailed reports by Hunter and Ludwig (1962), Hand and Jencks (1962), and Wofsy and Singer (1963). Coupling two imidate groups together to form a homobifunctional reagent was an almost inevitable consequence, and was reported by Singer's group at Yale University (Dutton *et al.*, 1966). These authors synthesised diethyl malonimidate (DEM; Figure 5.41) and used it (pH 8.5, 0 °C, 24 hours) to demonstrate some cross-linking of BSA or ferritin with human γ-globulin. Again good retention of protein activity was noted even with 80–90% of amines derivatised. BSA and human γ-globulin displayed undiminished antigenicity, while a specific anti-hapten antibody retained full binding capabilities.

Davies and Stark (1970) used dimethyl suberimidate (DMS; Figure 5.42) in their pioneering studies of quaternary structure; limited cross-linking (usually at pH 8.5) followed by denaturation and molecular mass analysis by SDS-PAGE revealed the existence of four similar subunits of glyçeraldehyde-3-phosphate dehydrogenase, for example. Carpenter and Harrington (1972) applied the same reagent to the elucidation of subunit relationships in leucine aminopeptidase from bovine lens, and imidates were by this stage firmly established as quaternary structure probes. 0.2 M triethanolamine buffer (pH 8.5) was the standard buffer used in these investigations.

Dutton *et al.* (1966) had predicted that bifunctional imidates of different lengths might be useful in mapping distances between lysine residues in proteins. Advanced studies of this type followed, using series of diimidates of increasing length to probe changes in sub-unit proximity. Hajdu *et al.* (1979) demonstrated that the addition of AMP to glycogen phosphorylase b altered the degree of cross-linking observed with diimidates whose length varied from 3.7 to 14.5 Å.

It can sometimes offer further structural information if the linkage formed between protein subunits or neighbouring protein molecules can be cleaved at a later stage (see Section 2.6.4.5). A number of bis-imidates have been synthesised which offer this possibility (see for example Schramm and Dülffer, 1977); the most commonly used is dimethyl-3,3′-dithio-bis(-propionimidate) (DTBP; Figure 5.43), introduced by Wang and Richards (1974, 1975) and commercially available. Cross-linking of human erythrocytes was carried out using DTBP in sodium phosphate buffer (pH 8.0), for 30–60 minutes at room temperature. SDS-PAGE before and after reduction of the disulfide bond offered detailed information on membrane protein topography. The disulfide linkage was cleaved by suitable thiol-containing reagents (e.g. 10% 2-mercaptoethanol). Other bis-imidates contain vicinal diols which can be cleaved by periodate oxidation.

D'Albis and Gratzer (1976) used DTBP and a range of non-cleavable analogues to cross-link myosin molecules (50 mM ethanolamine and 500 mM sodium chloride (pH 8.0); 90 minutes in the cold). Dimethylmalonimidate gave no cross-linking, but longer-chain analogues such as the adipimidate bridged myosin heads successfully.

Staros *et al.* (1981) introduced a bis-imidate containing extra charged moieties, to discourage the

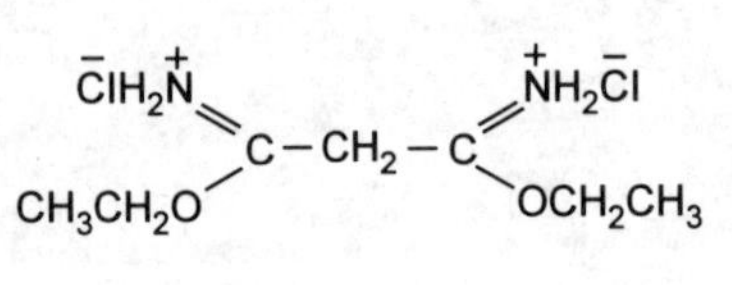

Figure 5.41 Diethyl malonimidate (DEM), dichloride.

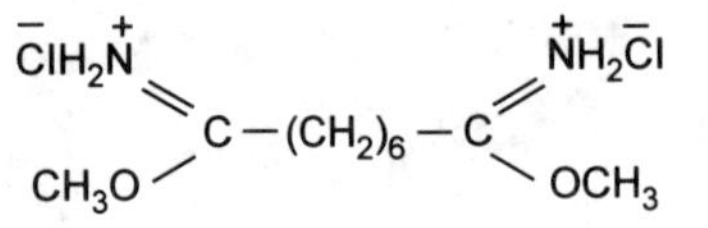

Figure 5.42 Dimethyl suberimidate (DMS), dichloride.

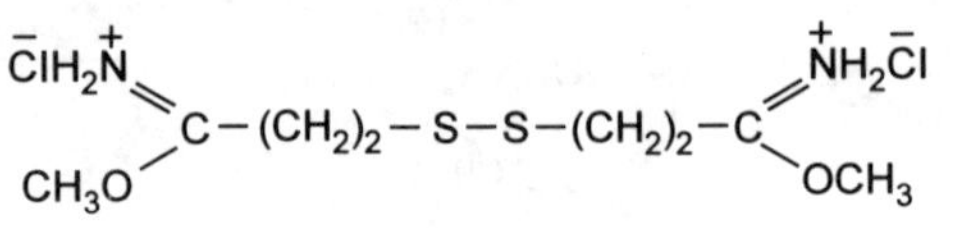

Figure 5.43 Dimethyl-3,3′-dithio-bis-(propionimidate) (DTBP), dichloride.

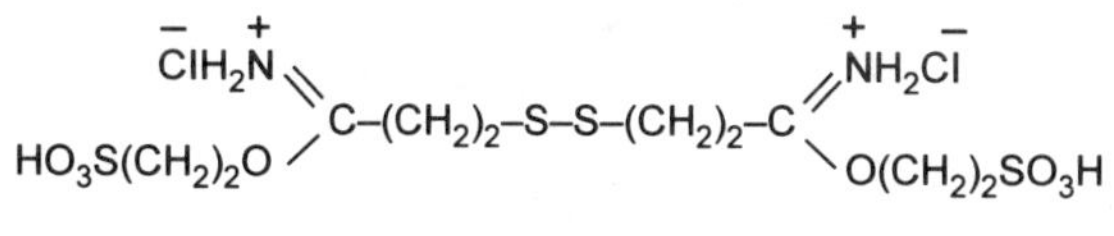

Figure 5.44 Diisethionyl-3,3′-dithio-bis-(propionimidate) (DIIT), dichloride.

molecule from crossing cell membranes. This compound, diisethionyl-3,3′-dithio-bis(propionimidate) (DIDIT; Figure 5.44), is also water-soluble due to the presence of the sulfonyl groups, and cleavable by reduction at the dithiol linkage. (See Section 5.3.5 for a fuller discussion of membrane-impermeant coupling reagents; bis-succinimides have increasingly come to dominate this field.)

A range of homobifunctional imidate reagents are now commercially available, generally as their dihydrochloride salts. Of these DMS probably remains the most commonly encountered, and was listed by Means and Feeney (1990) in a table of nine 'widely used and important' cross-linking agents. Other modifications to the chemistry of the linker group have also been tried, the incorporation of multiple hydroxyl residues to increase hydrophilicity, for example, and the introduction of chromophores. Almost thirty different bis-imidates are listed in a review (Wong, 1991; Chapter 4).

5.3.2.3 Synthesis of bis-imidates

Imidates can be prepared by the Pinner reaction from nitriles and alcohols (Figure 5.45), while other mechanisms are discussed by McElvain and Schroeder (1949) and Hunter and Ludwig (1962). Such syntheses are often quite straightforward, making the preparation of 'customised' reagents a feasible option; how-

Figure 5.46 Isocyanates and isothiocyanates, generically referred to in the text as N = C = X reagents.

ever, enough of these reagents are now commercially available to make this an infrequent necessity.

5.3.3 Bis-isocyanates and isothiocyanates

Isocyanate and isothiocyanate groups (Figure 5.46) contain a carbon which is activated towards nucleophiles by the electron-withdrawing effect of the adjacent nitrogen, sulfur and/or oxygen atoms. It can therefore act as a site for attack by an amine group (Figure 5.47), the product being a substitued urea or thiourea.

As with many other electrophilic reagents, isocyanates and isothiocyanates are also prone to attack by water, and hydrolysis is a significant side-reaction. Wold (1972) noted a half-life of less than two minutes for aliphatic isocyanates at pH 7.6.

Sulfur is less electronegative than oxygen (see Section 2.2.5), and isothiocyanates are therefore less reactive towards amines than isocyanates (Satchell and Satchell, 1975). The N=C=X reagents may also react with cysteine, histidine and tyrosine groups but the products undergo rapid hydrolysis, so they can be considered as primarily amine-specific. Isocyanates can also react with hydroxyl groups (including some serine residues) however (see Section 5.9.3).

The pH optimum for reaction with amines is around 9 to 9.5, at lower pH there will be little unprotonated amine present, while hydrolysis becomes more significant under more basic conditions. Many

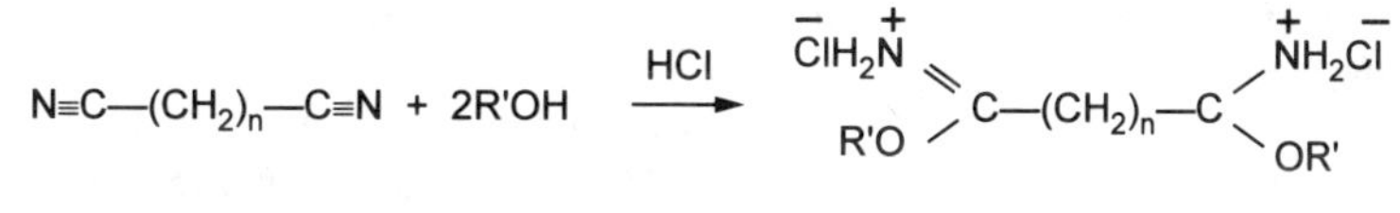

Figure 5.45 Synthesis of bis-imidates by the Pinner reaction.

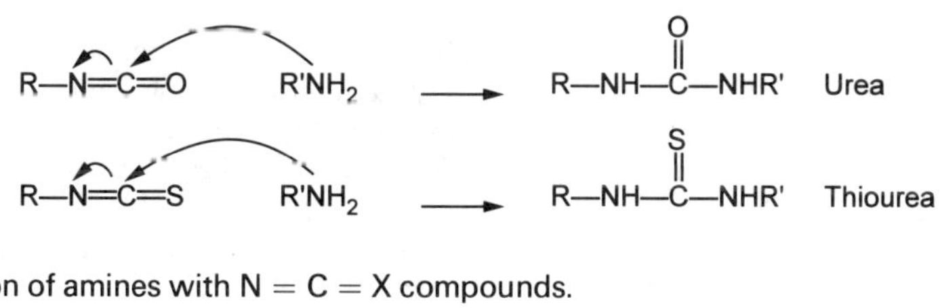

Figure 5.47 Reaction of amines with N = C = X compounds.

reactions are carried out closer to neutrality, however, for the sake of protein stability.

Isocyanate chemistry has been used in protein derivatisation since the earliest days of labelling for histochemistry, the pioneering studies of Coons *et al.* (1941) employing fluorescein isocyanate in this capacity. Their use in bifunctional reagents for protein coupling dates from the work of Singer (1959) at Yale University, and their heyday was in the 'pre-glutaraldehyde' period of the 1960s. Singer employed a two-stage protocol, using *m*-xylene diisocyanate (XC; Figure 5.48) to activate ferritin (45 minutes at pH 9.5 and 0 °C) before conjugation to rabbit γ-globulin (two days at pH 9.5 and 6 °C). Borek and Silverstein (1961) showed that this method yielded conjugates free from unreacted ferritin or antibody.

Schick and Singer (1961), however, encountered problems of over-reaction with XC: incorporations observed on activating BSA indicated that polymerisation of the diisocyanate was occurring. A mechanism was suggested involving hydrolysis of an isocyanate group to the carbamic acid, and thence to an amine which could react with remaining isocyanate moieties. These authors looked for more controllable reagents; also, as it had been shown that antibody was deactivated by direct reaction with XC, there was a particular interest in those which included N=C=X functions of differential reactivity. This feature facilitated two-stage reactions.

The result of these investigations, and perhaps the most widely-employed diisocyanate reagent, was toluene-2,4-diisocyanate (TDIC; Figure 5.49), in which the 4-group reacts relatively readily and will predominate in a first stage carried out under moderate conditions (0 °C or pH 7.5). The sterically-hindered 2-group adjacent to the methyl moiety is relatively unreactive under these conditions, but increasing this reactivity by raising the temperature to 38 °C (or the pH to 9.5) permits coupling to a second protein to take place (Schick and Singer, 1961; Sternberger, 1967).

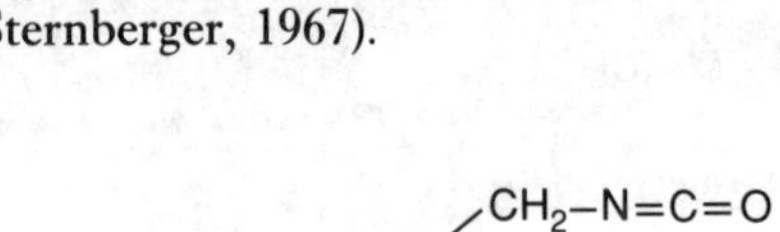

Figure 5.48 *m*-Xylene diisocyanate (XC).

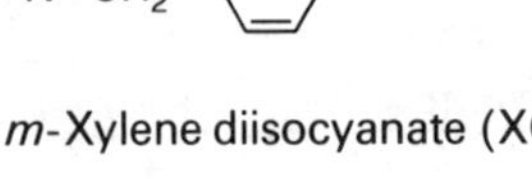

Figure 5.49 Toluene-2,4-diisocyanate (TDIC).

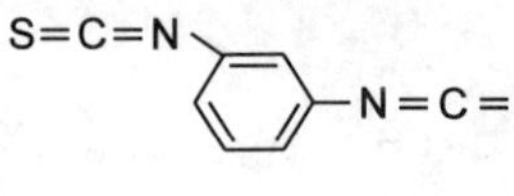

Figure 5.50 *m*-Phenylene diisothiocyanate.

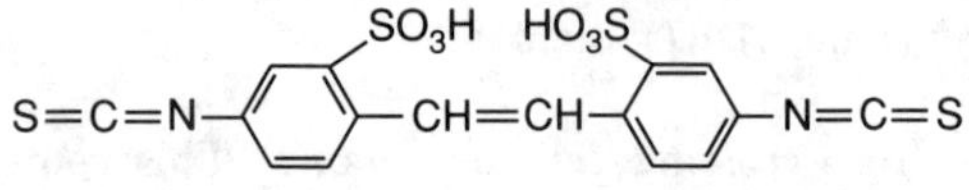

Figure 5.51 4,4′-Diisothiocyanate-2,2′-diisosulfonic acid stilbene (DIDS).

Bifunctional N=C=X reagents generally offer low yields and many have poor water-solubility; isocyanates in particular are also inherently unstable, and there are some indications that the links formed from isothiocyanate–amine reactions are not robust. In most applications they offer no particular advantage over more modern reagents and their use for protein coupling has become fairly rare since the early 1970s.

Homobifunctional N=C=X compounds have been employed in structural determinations too; Brandenburg *et al.* (1971), for example, used *m*-phenylene diisothiocyanate (Figure 5.50) to couple the A and B chains of insulin through different positions, and noted varying degrees of structural perturbation as a result. Again these compounds offer little advantage over more modern reagents for these studies. They do however offer one of the most simple ways to confer amine-reactivity to aromatic molecules, and reagents such as 4,4′-diisothiocyanato-2,2′-disulfonic acid stilbene (DIDS; Figure 5.51) have been useful in the investigation of specific binding sites (Rao *et al.*, 1979).

In the protein field N=C=X chemistry is most commonly encountered in its original monofunctional application of labelling with small molecules.

5.3.4 Bis-aryl halides

The benzene ring in itself is too electron-rich to react with nucleophiles such as amines, and must be activated by the presence of electron-withdrawing substituents such as nitro or carboxyl groups. The hydrogen atoms of the ring are also poor Lewis bases (see Chapter 2, Section 2.2.1) and are difficult to displace; halogen atoms are much more inclined to leave in an anionic state and are therefore a better target for nucleophilic substitution. So an aromatic molecule containing both halide and electron-withdrawing groups offers suitable chemistry for reaction with amines, and certain reagents of this type have been

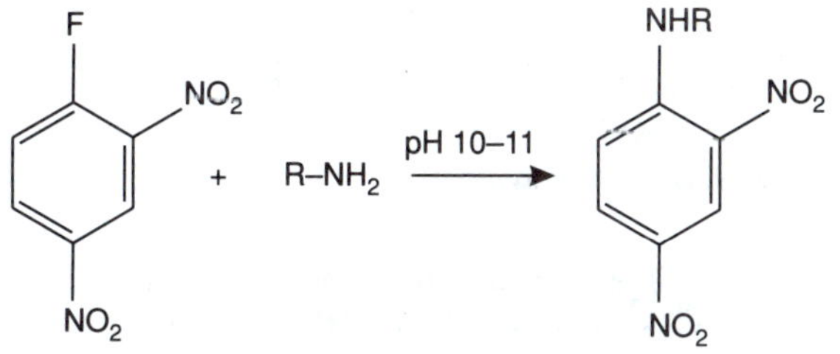

Figure 5.52 Reaction of amines with FDNB.

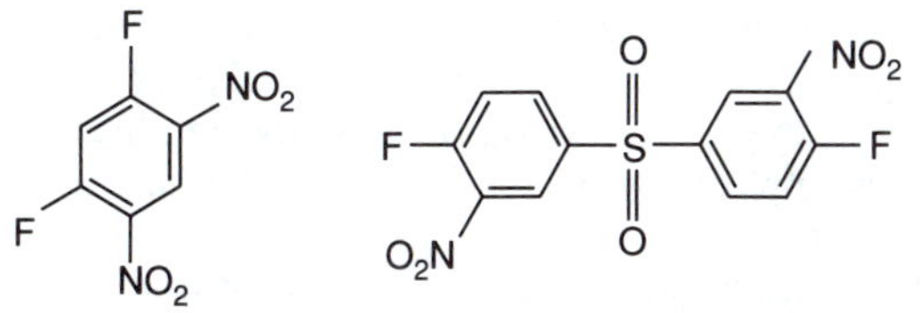

Figure 5.53 1,5-Difluoro-2,4-dinitrobenzene (DFDNB) and 4,4′-difluoro-3,3′-dinitrophenylsulfone (FNPS).

employed in important applications in protein derivatisation: fluoro-2,4-dinitrobenzene, FDNB, for example was used by Fred Sanger in his historic studies on the structure of insulin (Sanger, 1945; Sanger and Tuppy, 1951). The mechanism is shown in Figure 5.52: fluorine is the most reactive halogen (chlorine and bromine being increasingly less effective) and is encountered most often in this role.

Bifunctional application of the approach can rely either on the presence of two halides on the same ring (e.g. 1,5-difluoro-2,4-dinitrobenzene, DFDNB) or two linked rings (e.g. 4,4′-difluoro-3,3′-dinitrophenyl sulfone, FNPS; Figure 5.53).

As might be expected, such reagents require alkaline conditions for efficient reaction with amines; pH 10–11 is optimal, although a more neutral environment is often employed to preserve protein activity. Again, competing hydrolysis is an important consideration, fluorine atoms are rapidly lost under these conditions. Reaction with cysteine, histidine and tyrosine groups may also occur (Jackson and Young, 1986).

Differences in reactivity of the halogen atoms can sometimes be observed, in DFDNB, for example, the fluorine atoms act to enhance each other's reactivity due to their own electron-withdrawing capacity. As a result the protein-substituted monofluoro-derivative is less reactive than DFDNB itself (Wold, 1972). Attempts to exploit such effects in two-stage reactions to minimise self-coupling are limited by rapid hydrolysis problems, however.

Aryl halides are generally insoluble in water, and must be introduced into protein reactions in a suitable organic co-solvent such as ethanol. The successful incorporation of nitroaryl groups into a conjugate can often be detected by ultraviolet spectroscopy, such moieties being strongly chromophoric at wavelengths not often encountered in native proteins.

DFDNB and FNPS were employed in some of the pioneering applications of homobifunctional reagents, and indeed, of chemical derivatisation for the elucidation of protein structure and reactivity. In common with a number of early protein coupling agents[1], these compounds were first investigated in an industrial application: in this case, in attempts to modify the behaviour of textiles by chemical derivatisation.

In 1950, Helmut Zahn of the University of Heidelberg used DFDNB to cross-link the proteins of wool and silk fibres, observing an increased resistance to treatment by acid, alkali or phenol (Zahn, 1950). These studies, and the dramatic advances in the understanding of protein structure and analysis, led Zahn and his co-workers to investigate the chemistry of such phenomena. Zahn (1955) demonstrated that DFDNB and FNPS formed links between lysine, hydroxylysine, cysteine and tyrosine residues in keratin, fibroin and collagen, and these aryl halides were transformed from agents for the empirical modification of natural products to sensitive probes of protein chemistry and reactivity. Within a few years, these compounds were being employed in quite advanced structural studies, demonstrating, for example, that the two chains of insulin could be cross-linked (Zahn and Meienhofer, 1958).

Use of aryl halides in preparative protein coupling followed; in 1963, for example, FNPS was being used in the synthesis of antibody–ferritin conjugates (24 hours, 2–4 °C, pH 10.5 giving a 53% yield; Ram *et al.*, 1963), and Marfey *et al.* (1965) used DFDNB to couple synthetic polypeptides. Despite very low yields in some cases (less than 1% in an IgG-HRP conjugation; Modesto and Pesce, 1971), these reagents held some importance as coupling agents before the advent of glutaraldehyde. Like N=C=X reagents, however, their application in this role has been rare since the 1970s.

Where they have found continued application is in structural determinations. The small size of DFDNB (around 4 Å) coupled with its facility for spectroscopic detection have favoured its use in nearest neighbour

[1] Glutaraldehyde and formaldehyde are other examples, their applications in tanning and preservation of animal products predating their use as protein reagents by many decades.

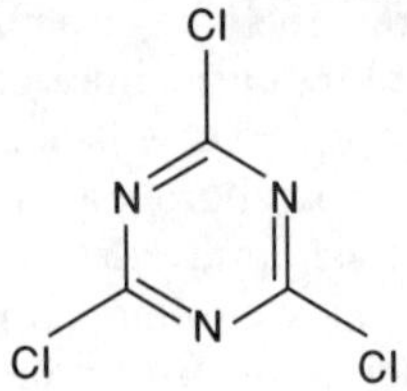

Figure 5.54 Cyanuric chloride or 2,4,6-trichloro-*s*-triazine.

analysis. A moderately recent review described it as the reagent of choice for this technique for lipoproteins (Swaney, 1986), an application where the hydrophobic nature of this molecule may also be beneficial compared to more polar compounds. Conditions for the cross-linking reaction were quoted as follows: DFDNB in ethanol is added to a solution of the lipoprotein (1–2 mg/ml in 25 mM sodium bicarbonate, pH 8.8) to give a final concentration of 20 μg/ml. After six hours at room temperature the reaction is stopped by the addition of the detergent SDS to give a concentration of 1%.

The monofunctional reagent fluoro-2,4-dinitrobenzene (FDNB) has found indirect applications in protein coupling: as a blocker to prevent self-coupling, for example during periodate coupling (Nakane and Kawaoi, 1974), and as an indicator of whether monoclonal antibodies are likely to lose their immunoreactivity on conjugation *via* amines (Burkot *et al.*, 1985). Jeanson *et al.* (1988a,b) did not observe the correlation observed by Burkot, however.

Cyanuric chloride or 2,4,6-trichloro-*s*-triazine (Figure 5.54) is a related compound with chlorines activated towards nucleophilic displacement, in this case by the electron-withdrawing effect of the neighbouring nitrogen atoms and the chlorines themselves. Analogously to DFDNB, the loss of the first chlorine reduces the reactivity of the second, and the loss of the second reduces that of the third. As well as amines, cyanuric chloride will cross-link cysteine and tyrosine residues and carbohydrate groups.

This compound and its analogues react similarly to the aryl halides, suffer from the same drawbacks in terms of competing hydrolysis and poor yields, and are now similarly outmoded.

5.3.5 Bis-succinimide esters, etc.

A number of techniques for activating carboxyl groups to nucleophilic attack were discussed in Section 5.2.2. Under the direct coupling category being discussed in that section, the carboxyl groups of the protein itself were the target for activation, and the only usable chemistries were those where: (1) the activating agent survived aqueous conditions long enough to achieve conversion of the carboxyl function, and (2) the activated acid resisted hydrolysis long enough to react with protein amines. Carbodiimides are the only class of compounds which have been successful enough in meeting these two criteria to earn any significant use in this application, and these are still rather poorly suited to the job.

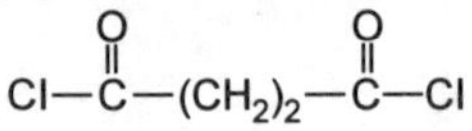

Figure 5.55 Succinyl chloride.

For homobifunctional linked coupling, however, the first constraint is removed; it is straightforward to create organic compounds containing two activated carboxyl groups using the full arsenal of synthesis techniques—anhydrous conditions, raised temperatures, etc. Chemistries become available which are impractical under the constraint of protein compatibility, and, so long as the second criterion is still met, the compounds obtained can be used as amine-directed homobifunctional coupling agents.

Many active carboxyl derivatives are still ruled out on these grounds; compounds containing two acid chloride groups, for example, are readily obtainable (e.g. succinyl chloride, Figure 5.55) and make excellent homobifunctional agents for amines under anhydrous conditions. However their rapid reaction with water renders them more or less useless in aqueous systems.

5.3.5.1 Active ester chemistry

For an electrophilic reagent to function successfully as a coupling agent under aqueous conditions, it must display a high degree of selectivity toward amines; in other words the rate of aminolysis must outweigh that of hydrolysis. The most successful approach of this type has been the development of 'active esters'. As mentioned in Section 5.2.2, the first requirement for activating a carboxyl group to nucleophilic attack is to replace the proton with a group which is not ionised under protein-compatible conditions, so that the negatively-charged carboxylate is not formed. Many carboxyl derivatives fulfil this criterion, amides for example. These however are still unreactive to nucleophiles, due to the mesomeric effect of the nitrogen atom adjacent to the carbonyl group.

The oxygen atom of esters displays less of a mesomeric effect, and simple esters such as methyl carboxylates are more susceptible to nucleophilic attack than the acid itself. This reaction was described by Curtius at the beginning of the century, but drastic conditions are required to promote aminolysis by these compounds.

Interest in efficient methods for generating amide links burgeoned in the 1950s, as the increased understanding of protein structure led to the new techniques of peptide synthesis. Although coupling reactions of small peptides could be carried out under anhydrous conditions, it was observed that routinely-used reagents such as DCC resulted in some loss of optical activity (racemisation) of the product (Anderson and Callahan, 1958) *via* the formation of oxazolone structures. It was perceived that increasing the speed of peptide formation would reduce the chance of oxazolone formation, and therefore minimise racemisation. Approaches in which the acid function was activated by esterification with a suitable alcohol were prime candidates. In some cases the active ester was isolated and added directly to the target peptide, in others the alcohol was added to the DCC reaction, so that an active ester was formed and consumed *in situ*.

A number of groups examined such approaches in the 1950s, using a variety of activating functions. Some of the early approaches were based on thioesters (R–S–C(=O)–R) rather than esters: Farrington *et al.* (1955, 1957), for example, employed the *p*-nitrothiophenyl function. However, the removal of thiol byproducts was troublesome, and when it became apparent that the presence of the sulfur was unnecessary, attention switched to simple oxygen esters with electron-withdrawing substituents.

Schwyzer *et al.* (1955) recommended cyanomethyl esters; Bodanszky (1960) pioneered the use of the *p*-nitrophenyl function, and *p*-nitrophenyl esters were the first such reagents to gain widespread acceptance. Pentachlorophenyl esters (Kovacs *et al.*, 1967) also enjoyed some popularity, but by now attention was turning to heterocyclic compounds containing electron-withdrawing functionalities, such as derivatives of phthalimide, succinimide, and benzotriazole (Nefkens and Tesser, 1961; Anderson *et al.*, 1963; König and Geiger, 1970). The thiazolidine-2-thione amides of carboxylic acids (Figure 5.56) appeared to behave analogously to active esters, and also received a modest amount of attention in this application (Yajima *et al.*, 1980).

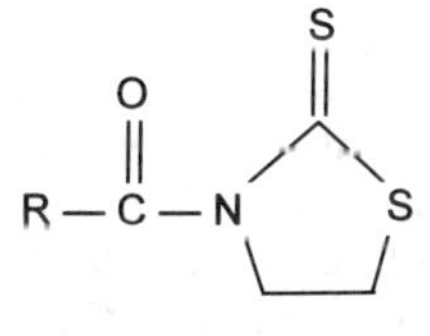

Figure 5.56 Thiazolidine-2-thione derivative of a carboxylic acid.

The more successful of the active ester approaches outlined above have found their way into protein-coupling chemistry. A number of homobifunctional reagents were introduced in the 1960s and 1970s based on *p*-nitrophenyl esters, e.g. bis-(*p*-nitrophenyl) adipate (Figure 5.57). This compound was used by Brandenburg (1972) to link the A1 glycine and B29 lysine groups in insulin. The crystal structure of insulin revealed these two residues to be separated by 8–10 Å, and Brandenburg used the 8 Å span of the adipyl group to compare solution conformations with the solid state.

Busse and Carpenter (1974) used a homobifunctional nitrophenyl ester derivative of a bis-methionine compound to carry out the same linkage. This generated a single chain whose dithiol linkages could be cleaved and reoxidised to generate the native conformation, thus acting as an analogue to the proinsulin molecule from which insulin is derived *in vivo* (Busse *et al.*, 1974). Native insulin could be regenerated from the bis-methionyl derivative by treatment with cyanogen bromide.

Plotz (1977) carried out some interesting work which combined the antigenicity of the nitrophenyl group with its role in an active ester. Antibodies specific for *p*-nitrophenyl groups were allowed to complex with bis(*p*-nitrophenyl) esters, the active functionality subsequently reacting with nearby amine groups to create a covalent bond between the hapten and the active site.

Tesser *et al.* (1975) described a number of homobifunctional reagents of this type, including a bis-(*p*-nitrophenyl) ester and some analogous compounds activated with *N*,*N*-dimethylaminopyridine (Figure 5.58). The latter reagents suffered from rapid hydrolysis.

5.3.6.2 The chemistry of bis-succinimide esters

Active ester coupling agents based on nitrophenyl esters are rarely encountered nowadays. There are many alternatives to the nitrophenyl group in this

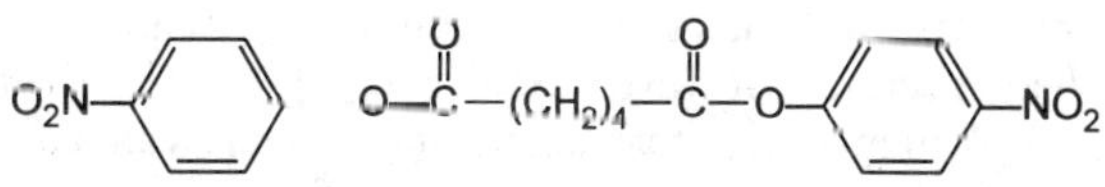

Figure 5.57 Bis(*p*-nitrophenyl) adipate.

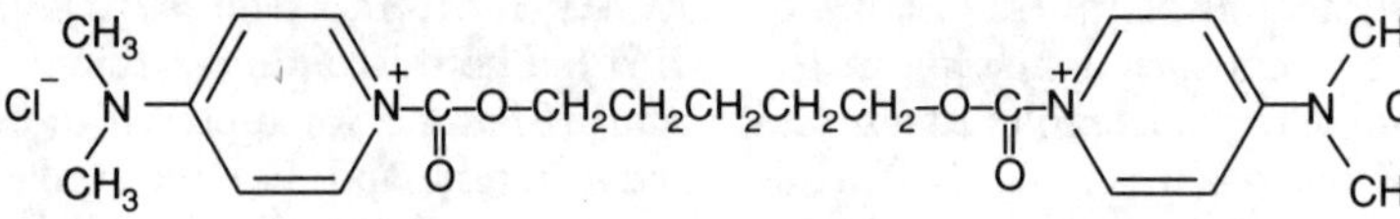

1,5-Bis(4-diethylaminopyridinocarbonyloxy)pentane dichloride

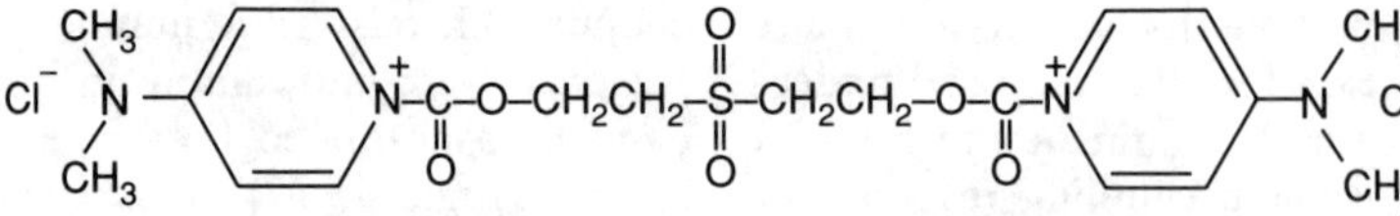

Bis-[2-(4-dimethylaminopyridinocarbonyloxyl)ethyl]sulfone dichloride

Figure 5.58 Two homobifunctional compounds activated with *N, N*-dimethylaminopyridines, described by Tesser *et al.* (1995).

role, and a variety of examples will still be encountered in isolated cases; Yang *et al.* (1996), for example, used dibromosalicylate esters in a homobifunctional reagent. However, by far the most successful reagents of this type to enter the protein chemistry field are the succinimide esters, whose selectivity for amines over hydroxyl groups allows them to be used quite efficiently under aqueous conditions. The *N*-succinimidyl group contains two powerfully electron-withdrawing carbonyl functions, and therefore forms strongly activated esters. To synthesise active esters it is still necessary to activate the carboxylic acid itself. Succinimide esters are usually prepared using a carbodiimide such as DCC to activate the acid (see Section 5.2.2) under anhydrous conditions, in the presence of the appropriate alcohol, *N*-hydroxysuccinimide (NHS). As a result of this origin, the term NHS ester will often be encountered referring to these reagents.

With regard to protein derivatisation, these compounds can also react weakly with thiols as well as amines; they also react more strongly with the imidazolyl group of histidine, but the product is very readily hydrolysed (Cuatrecasas and Parikh, 1972). The net effect of histidine can in fact be catalytic, increasing the overall hydrolysis rate of the ester; it has been suggested that higher excesses of succinimide esters should be used where histidines are particularly accessible (Mattson *et al.*, 1993). Cuatrecasas and Parikh (1972) also observed little or no reaction of a succinimide-derivatised solid phase with methionine, tyrosine, tryptophan, serine, arginine, glutamine or phenylalanine. However, Miller *et al.* (1992, 1994) have demonstrated that succinimide esters can react with hydroxyl-containing residues, if these are activated by a nearby histidine group.

Obviously if a dicarboxylic acid is derivatised in this way, a bis-succinimide ester is obtained. As dry chemicals these compounds are stable for months or years in the active form, and thus offer a highly convenient approach to amine–amine conjugation. As they are purely homobifunctional in nature, they generally offer no great utility in the preparation of protein conjugates. Their primary application is in cross-linking experiments to determine structure and activity. Bis-imidates have dominated this field for many years, but succinimide esters have gained inroads by virtue of their greater stability in aqueous systems, and their more predictable chemistry. They do however yield a derivative with one less positive charge (under all but the most basic conditions), a change not seen with imidate reagents.

The absorbance of the free *N*-hydroxysuccinimide group at 260 nm can offer a useful means of tracking hydrolysis and/or aminolysis of these compounds (Mattson *et al.*, 1993); molar extinction coefficients at a range of pH values have been reported (Carlsson *et al.*, 1978; Abdella *et al.*, 1979).

Succinimide esters react rapidly with protein amine groups in the pH range 6–9, with an optimum between 7 and 8. Raising the pH increases the amount of unprotonated amine available for reaction, but dramatically reduces the necessary selectivity for amines, at pH 7.5 the desired amine reaction can go to completion in 10–20 minutes (despite the huge excess of water molecules present) while the competing hydrolysis reaction has a half-life of 4–5 hours. At pH 8.6, in contrast, a hydrolysis half-life of less than ten minutes has been observed (Ji, 1983; Lomant and Fairbanks, 1976). An excess of around tenfold over the available amine groups is generally sufficient to give complete reaction where the protein concentration is 50–500

μM, and temperatures between 4 and 25 °C give satisfactory results.

Typical conditions for a water-soluble succinimide ester cross-linking reaction are given by Swaney (1986): BS^3 (10 mM in 50 mM sodium phosphate buffer, pH 7.4) is added to a solution of the protein in the same buffer to give a final concentration of 1–2 mM. After six hours reaction at room temperature, the reaction is stopped by the addition of 1/6 volume of 50 mM ethanolamine in the same buffer.

5.3.5.3 Examples of bis-succinimide esters for protein coupling

The commonly encountered examples of this class reflect their major field of application in structural studies. Apart from the length of the linker group, there are two main features of interest. (1) Whether the reagent is capable of being cleaved by a suitably mild chemistry, so that any cross-linked species can be separated during subsequent analysis. This consideration is determined by the presence of a bond within the linker which possesses the latent capability to be cleaved at will. (2) Whether the reagent is likely to pass through the cell membrane or not. Where the topography of cell membrane proteins is being investigated by cross-linking, it is useful to know that the coupling agent remains outside the cell rather than being transported across the membrane into the cytoplasm. James Staros and his colleagues at the Vanderbilt University at Nashville, Tennessee have been central in this field; the use of membrane-impermeable compounds in this way has been reviewed by Staros and Anjaneyulu (1989).

This capability is defined partly by the dimensions of the reagent, and partly by its hydrophobicity or hydrophilicity. Unless the reagent is particularly bulky, a highly polar group such as a sulfonate must be incorporated to confer membrane-impermeability, so that this capability frequently correlates with water solubility.

There is another potential advantage to the use of these sulfonated reagents. The compounds discussed earlier are not directly soluble in aqueous solutions, and require the use of a co-solvent such as DMF or DMSO. The introduction of sulfonate groups, which are negatively charged under all feasible conditions, increases the polarity of the molecules sufficiently (in most cases) for them to become water-soluble. This can offer advantages simply in terms of convenience, or where the protein concerned is particularly susceptible to organic solvents. There is a downside to this approach as well, however: water-soluble compounds (particularly those prone to hydrolysis!) are generally more difficult to purify efficiently than their hydrophobic analogues, and it is not always straightforward to obtain high-purity materials of this type.

Common bis-succinimides are categorised by spacer length and other key properties in Figure 5.59.

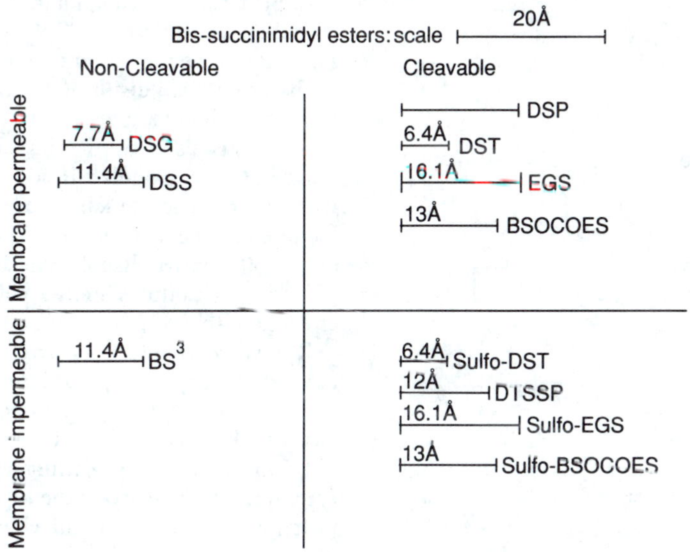

Figure 5.59 A number of readily available bis-succinimide esters, showing the length of spacer group which they incorporate, and categorised in terms of membrane permeability and cleavability. See text for details of compounds.

Non-cleavable membrane-permeant bis-succinimide esters

This category contains some of the simplest examples, derived from straight-chain aliphatic bis-carboxylic acids. The earliest report of a reagent of this type (Lindsay, 1971) was the bis-succinimide ester of succinic acid (Figure 5.60), which was used to couple the A and B chains of insulin.

The suberoyl analogue was synthesised by Pilch and Czech (1979), disuccinimidyl suberate (DSS; Figure 5.61a), and was used by these authors to probe the glucose transport system of the rat adipocyte. This reagent, reported almost simultaneously by Hill *et al.* (1979), was singled out as the most widely used of this class in a 1990 review (Means and Feeney, 1990).

Hill *et al.* (1979) also prepared DSS homologues containing two, four and eight-carbon spacers, yielding spans between 6 and 14 Å: the last of these, disuccinimidyl sebacate is commercially available, as is the five-carbon disuccinimidyl glutarate (DSG; Figure 5.61(b)).

Cleavable membrane-permeant bis-succinimide esters

Bragg and Hou (1975) used dithio-bis(succinimidyl propionate) (DSP or Lomant's reagent; Figure 5.62), to investigate the arrangement of subunits in Ca^{2+} and Mg^{2+}-activated ATPase from *E. coli*. Dithiothreitol was used to cleave the dithiol bond.

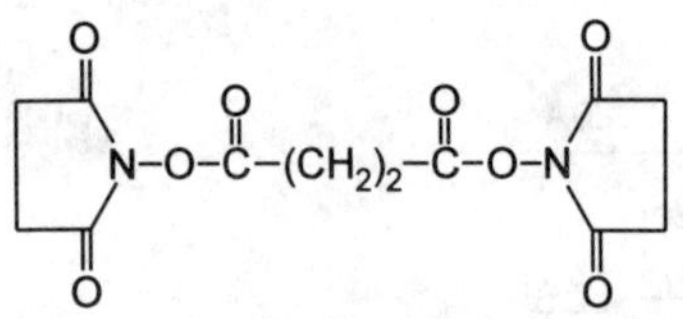

Figure 5.60 Succinate bis(*N*-hydroxysuccinimide ester)

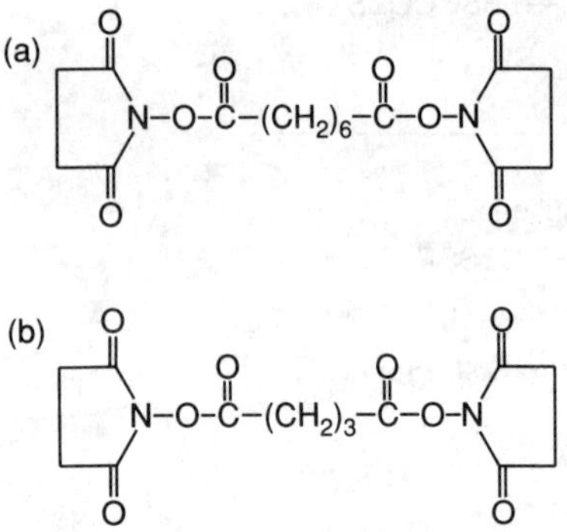

Figure 5.61 (a) Disuccinimidyl suberate (DSS); (b) disuccinimidyl glutarate (DSG).

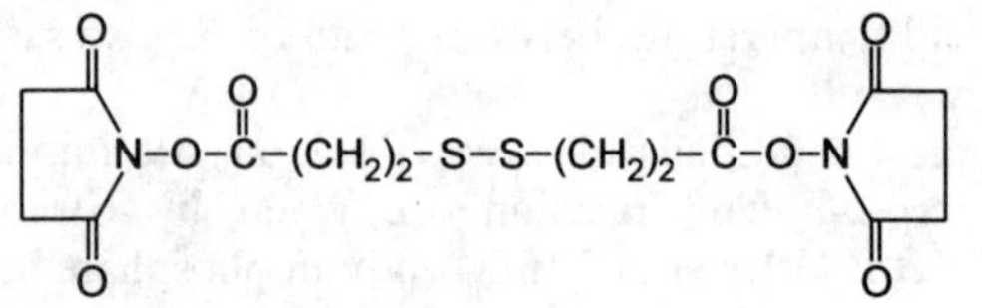

Figure 5.62 Dithio-bis(succinimidyl propionate) (DSP; Lomant's reagent).

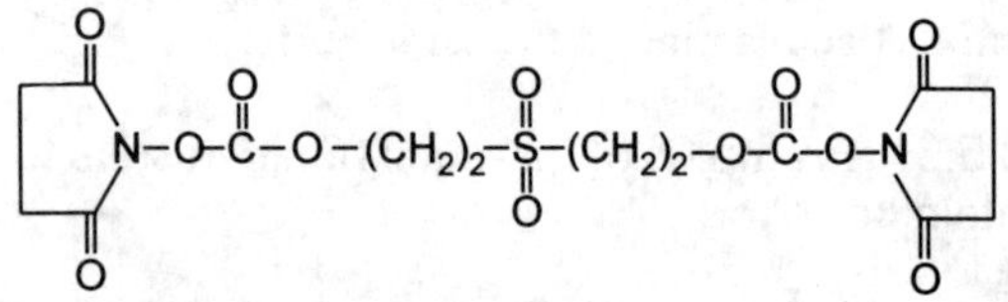

Figure 5.63 Bis [2-(succinimidooxycarbonyloxy) ethyl] sulfone (BSOCOES).

This compound, with a 12 Å span, was already commercially obtainable at the time, and it has maintained its popularity for such applications to the present day. Reaction with amine groups is rapid, especially in comparison with competing hydrolysis, Lomant and Fairbanks (1976) used a ^{35}S-labelled analogue of DSP to demonstrate complete reaction with aminated glass within two or three minutes, while the hydrolytic half-life was four to five hours. Quantitative cleavage of the S–S bonds was observed on treatment of DSP-treated haemoglobin with 10 mM dithioerythritol for 30 minutes at 37 °C. Lomant and Fairbanks also noted that the reagent could form covalent bonds to the protein *via* disulfide exchange with thiols in the protein, giving complex kinetics.

DSP was used by d'Albis and Gratzer (1976) to cross-link the globular heads of myosin molecules, with an efficiency much higher than the bis-imidates included in the same study. The reaction was carried out in 50 mM triethanolamine buffer, 500 mM NaCl, pH 8.0 (90 minutes in the cold), and cleavage was effected with 5% 2-mercaptoethanol.

Tesser *et al.* (1975) prepared bis[2-(succinimidooxycarbonyloxy)ethyl] sulfone (BSOCOES; Figure 5.63) which is partially water-soluble, reactive at 0 °C and can be cleaved by base (typically two hours at pH 11.6 and 37 °C). Zarling *et al.* (1980) used this compound to investigate the topography of membrane proteins in mouse cells infected with murine leukaemia virus.

Smith *et al.* (1978) synthesised disuccinimidyl tartrate (DST; Figure 5.64), a reagent with a 6 Å span which is stable to reducing thiols, but can be cleaved by periodate oxidation. This and a longer-chain (18 Å) analogue were used to examine near-neighbour relationships of polypeptides within the bovine ubiquinone–cytochrome *c* reductase complex. To effect the cleavage, electrophoresis gels were soaked with 15 mM sodium periodate and 0.1% SDS in 0.02 M sodium phosphate buffer (pH 7.0) for two hours. Again, both DST and sulfo-DST are commercially available.

Hill *et al.* (1979) described a number of DSS analogues containing vicinal diols (cleavable by periodate) or an ethylenic bond (cleavable by a periodate-permanganate mixture). In studies with myosin, the periodate cleavage was successful, but the periodate–permanganate mixture caused degradation of the protein; the latter approach has not been pursued.

Abdella *et al.* (1979) prepared ethylene glycol bis-(succinimidylsuccinate) (EGS; Figure 5.65), which is cleavable by hydroxylamine treatment (1 M NH_2OH at pH 8.5, six hours at 37 °C). These authors observed 60% retention of lactate dehydrogenase activity following hydroxylamine treatment, and no effect on trypsin function. They also used this approach to immobilise trypsin reversibly to to an amine-substituted solid phase, hydroxylamine treatment resulting in substantial release of the enzyme.

Koch *et al.* (1990) described a series of bis-succinimides containing selenide links, including disuccinimidyl selenodipropionate (DSE; Figure 5.66). These reagents can readily be oxidised to selenoxides under very mild conditions (e.g. Iodo-Beads™ or diisopropylamine). The selenoxides can rearrange to cleave the linkage (Figure 5.67): in the example shown, elimination occurs to yield a selenenic acid and an alkene. The facility of this cleavage reaction is clouded slightly by the need to include an amine, which reacts with the selenenic acid to prevent recombination with the alkene. These reagents are commercially available, for example from ProChem, Inc.

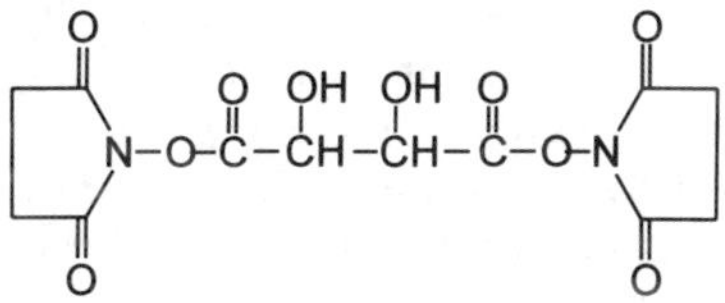

Figure 5.64 Disuccinimidyl tartrate (DST).

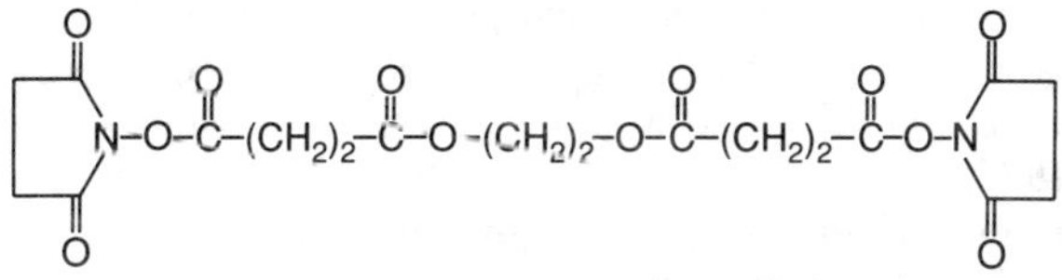

Figure 5.65 Ethylene glycol-bis(succinimidyl-succinate) (EGS).

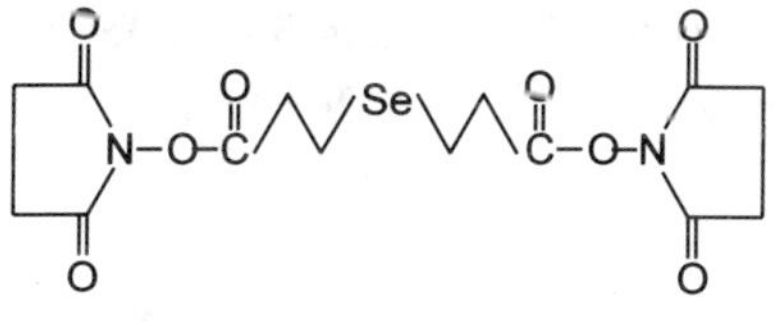

Figure 5.66 Disuccinimidyl selenodipropionate (DSE).

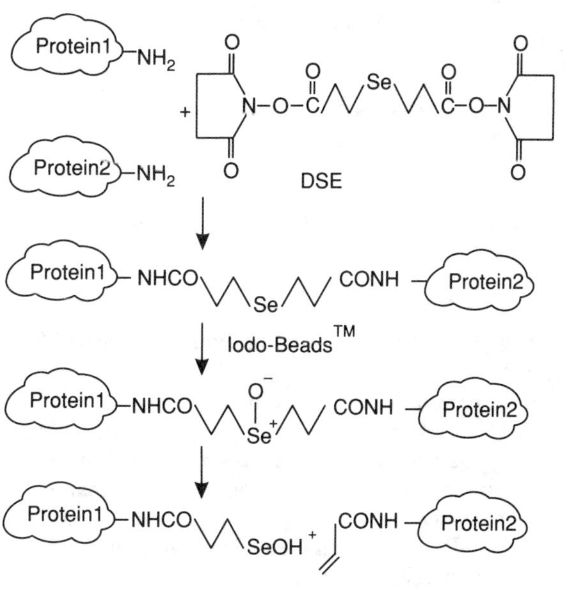

Figure 5.67 Oxidation of the bond formed using DSE yields a selenoxide, which cleaves to form a seleninic acid and an alkene.

Non-cleavable membrane-impermeant bis-succinimide esters

The impermeant bis-succinimide esters are generally sulfonated analogues of the permeant compounds discussed above. Staros (1982) introduced bis(sulfosuccinimidyl) suberate, (BS^3; Figure 5.68), an analogue of DSS.

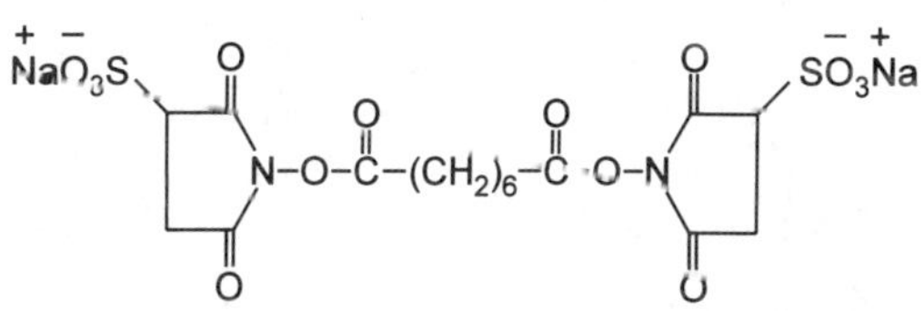

Figure 5.68 Bis(sulfosuccinimidyl) suberate (BS^3).

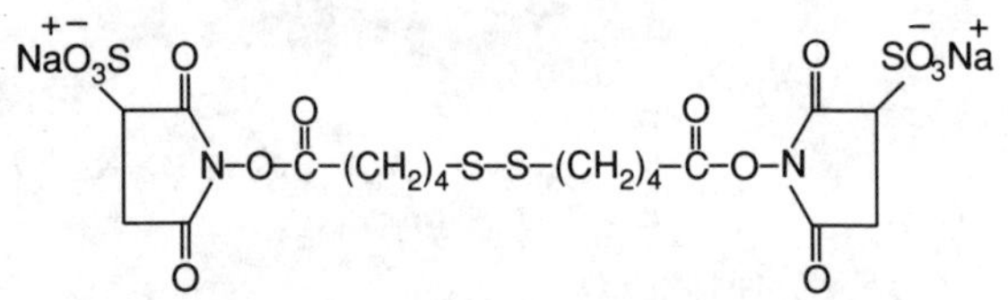

Figure 5.69 Dithio-bis(sulfosuccinimidyl) suberate (DTSSP).

Cleavable membrane-impermeant bis-succinimide esters

Staros (1982) also described the DSP analogue dithio-bis(sulfosuccinimidyl) suberate (DTSSP; Figure 5.69). This and BS[3] were successfully used in structural investigations on the anion exchange channel in human erythrocytes (Staros and Kakkad, 1983) and were rapidly commercialised. The sulfo-analogues of many bis-succinimides are now commercially available, including sulfo-DST, sulfo-BSOCOES, sulfo-EGS and sulfo-DSE.

Other bis-succinimide esters

Where a spin label is to be used as a reporter group to determine the rotational behaviour of the protein to which it is attached, it is important that the link is inflexible so that the label experiences (as closely as possible) the same movement as the macromolecule. Bifunctional spin-label reagents which can be anchored to the protein at both ends offer a means of generating an inflexible label. Similarly, where spin studies are used to determine the orientation of amphiphilic reagents in contact with cell membranes, reactive end-groups can fix the probe at the site of interest so that the complex remains intact during subsequent examination. Gaffney *et al.* (1983) introduced a series of bis-succinimide esters with this aim in mind, containing between 15 and 30 atoms. An example is shown in Figure 5.70.

Other spin-labelled bis-succinimides were developed later, again largely within Staros' group at Vanderbilt University. Often, these incorporated sulfonate moieties to confer membrane-impermeance,

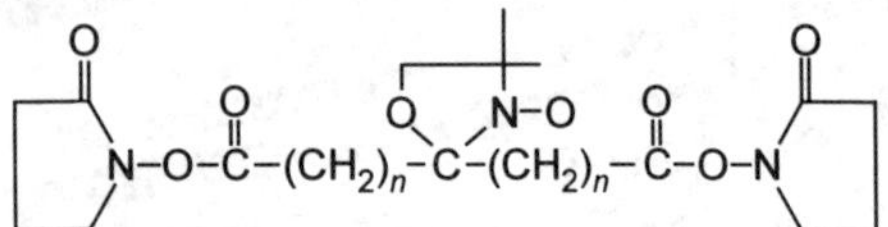

Figure 5.70 Structure of bis-succinimide spin labels described by Gaffney *et al.* (1983).

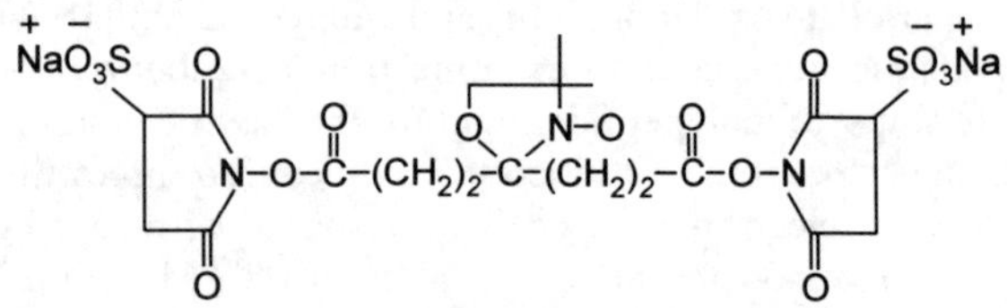

Figure 5.71 Bis(sulfo-*N*-succinimidyl)-doxyl-2-spiro-4′-pimelate (BSSDP).

for example bis(sulfo-*N*-succinimidyl)-doxyl-2-spiro-4′-pimelate (BSSDP; Figure 5.71). The development of membrane-impermeant homobifunctional agents, both with and without spin labels, has been summarised by Staros in a 1988 review.

Polyethylene glycol (PEG) has some very useful properties (see Chapter 8, Section 8.5.2); PEG-bis-succinimide esters are available from Pro Chem, Inc. and from Shearwater Polymers.

Other infrequently-encountered bis-succinimidyl compounds have been synthesised and used for protein coupling; the reader is referred to a review (Wong, 1991, pp. 81–85 and 98–99) for details of these.

5.3.5.4 Bis-acylazides

Bis-acylazides are closely analogous to active esters; the azide (N_3) moiety is a good leaving group. Lutter *et al.* (1974) developed a series of hydrophilic homobifunctional reagents based on this moiety, containing a vicinal diol function which allows them to be cleaved by periodate oxidation. These authors used bis-acylazides of varying length to investigate the distribution of proteins in the *E. coli* ribosome. Tartryl diazide (TDA; Figure 5.72) is an example of such a reagent.

Similar compounds incorporating a cleavable azo group and polyproline spacers to confer hydrophilicity were also described by the same group (Fasold *et al.*, 1977).

Acylazides can lose nitrogen and undergo an intramolecular rearrangement to generate an isocyanate (the Curtius rearrangement); to avoid unwanted reactions such as this, the azide is generated *in situ* by treatment of a stable acyl hydrazide with sodium nitrite and hydrochloric acid. This inconvenience obviously limits the attractiveness of the approach and it is very rarely encountered.

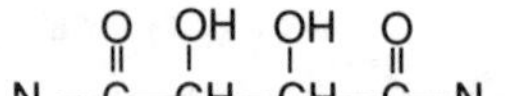

Figure 5.72 Tartryl diazide (TDA).

5.3.5.5 *In situ* activation of bis-carboxylic acids

Although the use of ready-activated carboxyl derivatives such as bis-succinimides is extremely convenient, it is also possible to activate bis-carboxylic acids *in situ* using a carbodiimide. The homobifunctional reagent here is really the bis-*O*-acylisourea intermediate.

Ito *et al.* (1994) coupled the methyl ester of insulin to glucose oxidase (GOD) *via* 5,5′-dithiobis(2-nitrobenzoic acid), both stages of the two-step reaction employing EDAC as the activating reagent.

5.3.6 *p*-Benzoquinone

In general the C=C double bond characteristic of alkenes and their derivatives is unreactive to nucleophiles, as would be expected from the high electron density at such sites. However the presence of electron-withdrawing moieties in neighbouring positions can cause sufficient charge movement to render the C=C bond electrophilic, and such activated double bonds offer potential for reaction with amines.

1,4- or *p*-benzoquinone (BQ) (Figure 5.73(a)) contains two double bonds activated by adjacent carbonyl groups, and shows substantial bifunctional reactivity with amines, as well as other nucleophiles such as thiols and the hydroxyl groups of carbohydrates. The reactivity towards the latter is particularly high—benzoquinone is a convenient reagent for coupling proteins to polysaccharide solid phases (Brandt *et al.*, 1975). This can obviously be a significant property in reactions with glycoproteins.

Benzoquinone is also a quite powerful oxidising agent, capable for example of thiol oxidation; the product of two-electron reduction is a hydroquinone structure (Figure 5.73(b)), while a transient intermediate semiquinone radical (Figure 5.73(c)) may be formed by single-electron reduction or by fission of the hydroquinone. BQ is not an aromatic molecule, but tautomerism is possible to yield species with a positive charge on the ring (Figure 5.73(d)). Any substituent groups on the molecule which are capable of donating electrons will stabilise these resonance structures, lowering the oxidising potential of the quinone.

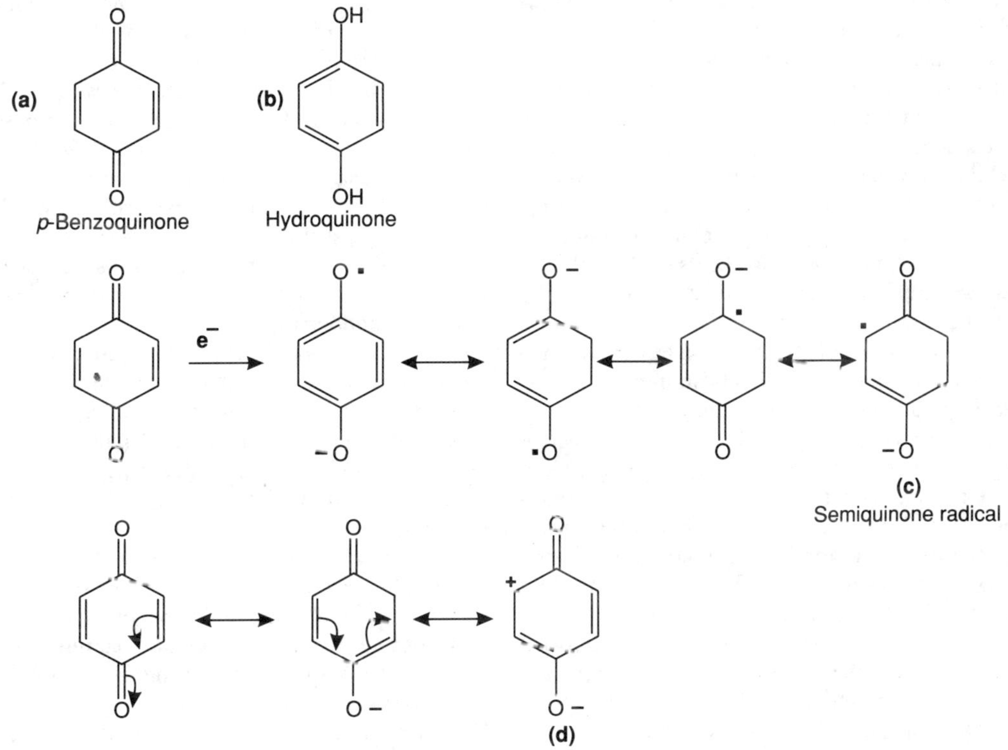

Figure 5.73 (a) *p*-Benzoquinone, (b) its reduced form hydroquinone, and some of its structural forms. See text for details.

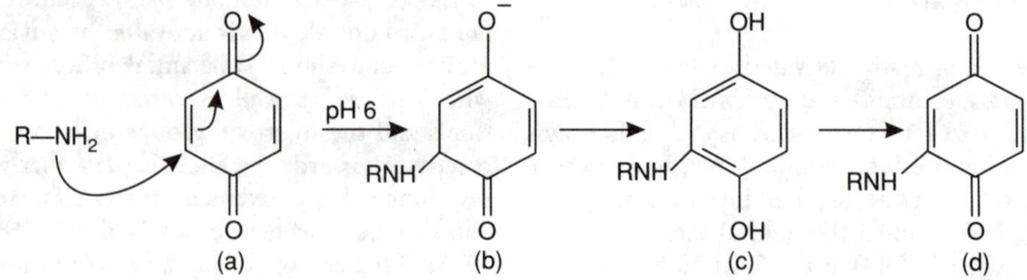

Figure 5.74 Reaction of benzoquinone with amines.

The initial reaction of the quinone molecule with the amine group (Figure 5.74(a)) is generally agreed to be a 1,4 addition. The product (Figure 5.74(b)) exists predominantly in the aromatic hydroquinone form (Figure 5.74(c)), which does not contain an activated double bond and is unreactive to nucleophiles. However the derivatised hydroquinone has sufficient reducing power to be reoxidised by benzoquinone (or dissolved oxygen), thus regenerating a quinone structure (Mason, 1955; Figure 5.74(d)). Although the electron-donating effect of the amine substituent deactivates the double bonds somewhat, this species is still capable of reacting with a second protein; the reoxidation therefore offers the capability of bifunctionality. Further reoxidation to a quinone may then occur, but steric hindrance and the cumulative electron-donating effects of the amine groups presumably prevent any further reaction with proteins.

The ability of *p*-benzoquinone to react with proteins was recognised from the early years of the century, many such reactions generating coloured products. Like other bifunctional coupling agents such as glutaraldehyde, much of the early investigative work on BQ was carried out in connection with its use as a tanning agent for animal skins. There is also an extensive early literature relating to its many effects on living systems, which facilitated development of the recognition that quinones can function as inhibitors of enzymic activity. BQ has frequently been employed as a model quinone for investigations in this area. Webb (1966) has reviewed both the early development of benzoquinone biochemistry, and the work carried out up to that time on its inhibitory functions. While protein–protein links were undoubtedly formed in many of the early applications of BQ, this capability was not exploited preparatively until the 1970s, and was often not evident under the reaction conditions used. Morrison *et al.* (1969), for example, described the reaction of BQ with cytochrome *c* at pH 7.4 for up to 72 hours and observed little formation of protein–protein conjugates.

Its first reported use as a preparative coupling agent was in the formation of protein-carbohydrate linkages by Brandt *et al.* (1975). Insoluble agarose and dextran carriers were activated by derivatisation with benzoquinone at pH 8.0, then incubated with protein solutions at the same pH. High protein incorporations were observed. The efficiency of the activation reaction increased dramatically over the pH range 6 to 8, decreasing again above this value. The reaction with protein was much slower, and the pH dependence varied dramatically from one protein to another, it was surmised that conformational changes in protein structure complicated the chemically-defined pH response which would otherwise predominate.

Its preparative use for protein conjugation was first described by Ternynck and Avrameas (1976) of the Institut Pasteur in Paris in the preparation of enzyme-labelled antibodies. They employed a two-step approach based on the reduction in reactivity of the molecule following monoderivatisation, noted above. By carrying out the first step at a relatively low pH, of value 6 where the amine groups are substantially protonated, the reaction is prevented from progressing beyond monoderivatisation (Figure 5.75(a)). Following purification of a protein activated in this way, the second protein can be added and the pH raised to 9. This increases the concentration of deprotonated amine and promotes the second (conjugate-forming) step (Figure 5.75(b)). This differential reactivity gives a degree of heterobifunctionality to benzoquinone as a coupling agent.

The semiquinone intermediate may have more than a negligible role in the mechanism (Webb, 1966; Ternynck and Avrameas, 1977), dimerisation or reaction with other species is possible, complicating side-reactions. A solution containing 100 mmol/l each of *p*-benzoquinone and its hydroquinone analogue has been calculated to contain 4 μmol/l of the semiquinone (Webb, 1966). Because of side reactions such as those involving the semiquinone it is advisable to make up BQ solutions shortly before use. Deoxygena-

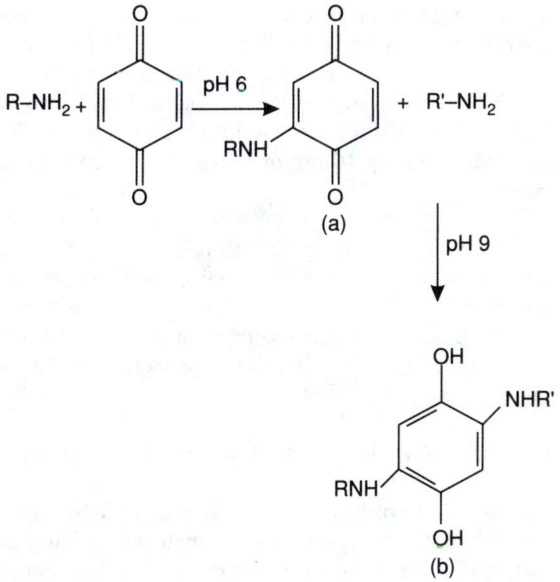

Figure 5.75 Benzoquinone as a quasi-heterobifunctional agent.

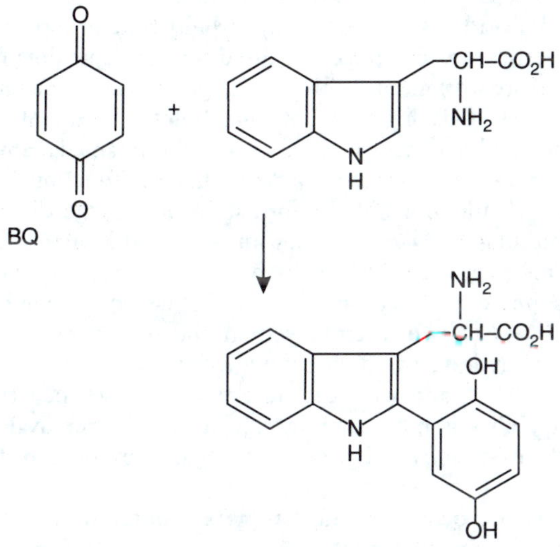

Figure 5.76 Reaction of benzoquinone with indolyl group of tryptophan.

tion and protection from light may also be beneficial in preventing unwanted reactions at this stage.

Commercially-supplied BQ can contain impurities which reduce the yield from protein-coupling reactions. Webb has described vacuum sublimation as a useful method of purifying the reagent for use.

Morrison *et al.* (1969) noted that amine-substituted benzoquinone derivatives, unlike BQ itself, are capable of reduction by ascorbic acid to produce hydroquinones which react in a quantifiable way with oxygen. They used this phenomenon to monitor the progress of benzoquinone reactions.

Apart from the facile reaction with amine and hydroxyl groups already described, BQ can react with heterocycles such as the indolyl group of tryptophan and with thiols (Mason, 1955). In the former case indolyl derivatives are formed which form intensely blue or green coloured salts (Figure 5.76); free thiols react analogously to amines, i.e. by nucleophilic substitution at the activated double bonds. However it should be borne in mind that oxidation to form intermolecular disulfide links is a possibility.

5.3.6.1 Benzoquinone coupling methods

Ternynck and Avrameas' original benzoquinone method was described in detail for the preparation of sheep or rabbit IgG–HRP conjugates for histo- and cytochemistry. It featured the following stages. (1) Addition of 0.2 ml of BQ (30 mg/ml in ethanol) and 0.1 ml of 1 M phosphate buffer, pH 6, to 0.7 ml of IgG (4.3 mg/ml in 0.15 M sodium chloride solution). This represents a BQ:IgG ratio of around 2800:1. The mixture was incubated in the dark for one hour at room temperature. (2) Purification of the activated IgG on a Sephadex™ G25 Fine column with 0.15 M sodium chloride solution as eluent, the protein elutes immediately before the yellow BQ and low-molecular-mass byproducts. (3) HRP solution (0.12 ml of 10 mg/ml in presumably 0.15 M sodium chloride solution) was added, and 1 M carbonate–bicarbonate buffer, pH 9, to give a final molarity of 0.1 M. This represents an HRP:IgG ratio of around 1.6:1. The mixture was incubated for 15 hours at room temperature. (4) The reaction was quenched by the addition of 1 M lysine solution, pH 7.5, to give a final concentration of 0.1 M; at least four hours incubation was allowed. (5) The conjugate was purified by dialysis against phosphate-buffered saline.

Where the Fab fragment of the IgG was used the masses of reagent were kept the same, indicating a BQ:Fab ratio of about 700:1 and an HRP:Fab ratio of 0.4:1. The latter may be erroneous: later, the authors described the general employment of a more likely HRP:Fab ratio of 4:1.

During the establishment of this method, it was observed: (a) that a BQ:IgG ratio of at least 200:1 was necessary, (b) that activation at pH 5 gave very poor results, and (c) that the use of buffer rather than ethanol to dissolve the BQ was less effective under all pH conditions examined.

A second paper by Ternynck and Avrameas (1977) reversed the technique so that the HRP was activated by BQ before reaction with IgG or Fab. Conditions of an optimised procedure were as follows. (1) HRP solution (0.4 ml of 10 mg/ml in 0.15 M sodium chloride solution), 0.05 ml of 1 M potassium phosphate, pH 6, and 0.1 ml of BQ solution (30 mg/ml in ethanol) were incubated in the dark at room temperature (<22 °C) for one hour. This represents a BQ:HRP ratio of around 280:1. (2) The activated HRP was purified on Sephadex G25 Fine as above, around 1 ml of product was collected. Fab (1 mg of 4–5 mg/ml in 0.15 M sodium chloride solution) was added. From procedures elsewhere in the paper it is inferred that 4 mg of IgG could be substituted, in either case giving an HRP:IgG or Fab ratio of 4:1. (3) One tenth volume of 1 M sodium bicarbonate solution was added and the mixture incubated for 48 hours at 4 °C. (4) One volume of 1 M lysine solution in phosphate-buffered saline solution was added and the mixture incubated for four hours before dialysis against phosphate-buffered saline solution.

BQ activation of HRP (in this case at pH 8) resulted in 92% retention of activity, while this dropped to around 80% after conjugation using a 4:1 HRP:IgG ratio; around 60% of antibody activity was retained. Of the antibody and HRP, 60% and 15% respectively, could be incorporated into the conjugate.

A paper by the same group (Avrameas *et al.*, 1978) essentially repeated the two methods detailed above, except that the protein for activation was dissolved directly in 0.1 M phosphate, pH 6. Purification of the conjugate by size exclusion on Sephadex G200 or by ammonium sulfate precipitation was also described.

Nygren (1982) modified the activated enzyme approach slightly, increasing HRP and BQ concentrations to 50 mg/ml and 60 mg/ml respectively, employing a BQ:HRP ratio of about 220:1 and eluting the Sephadex column with 0.05 M carbonate–bicarbonate buffer, pH 9.5. After this stage the HRP concentration was adjusted to 20 mg/ml, and Fab at 20 mg/ml in 0.05 M borate buffer, pH 9.5, added. Many of Ternynck and Avrameas' findings were essentially confirmed, the 4:1 molar ratio giving the best yield of low-molecular-mass conjugate, 83% of enzyme activity remaining after the activation stage, and good immunochemical performance.

The papers from the Institut Pasteur group all include ratios optimised for AP and β-Gal as well as for HRP. The protocols for both the activated-enzyme and activated-antibody approaches remain as described by Avrameas *et al.* (1978), but the enzyme quantities are as follows. For 1 mg of BQ-activated Fab, 8 mg of AP (*E. coli*) or 40 mg of β-Gal are employed. For 4 mg of BQ-activated AP, 0.5 mg of Fab or 2 mg of IgG are employed. For 4 mg of BQ-activated β-Gal, 0.07 mg of Fab or 0.3 mg of IgG are employed.

No loss of enzyme activity was observed after the activation of AP or β-Gal at pH 6 (Ternynck and Avrameas, 1977), while the final conjugates retained around 60–80%. It appeared that AP retained greater activity in Fab conjugates compared to those prepared from whole IgG, this effect was not seen for HRP and not evaluated for β-Gal.

5.3.6.2 Overview on benzoquinone coupling

Benzoquinone has some advantages as a linking agent: enzyme activity is generally retained well during the conjugation process, and reasonable yields can be obtained without the significant production of high-molecular-mass material. BQ itself has desirable properties compared to some other coupling agents, it is inexpensive and has a high stability.

Its readiness to react with carbohydrate groups can be either an advantage or a disadvantage depending on the circumstances. With heavily glycosylated proteins such as HRP this no doubt contributes substantially to the yield and nature of the conjugate obtained. However, as this reagent is not truly heterobifunctional, it is difficult to avoid the formation of enzyme dimers and oligomers, sometimes in significant quantities. This may complicate subsequent purification stages. It should also be borne in mind that benzoquinone has been successfully employed in the derivatisation of carbohydrate-based chromatography media (Brandt *et al.*, 1975), and will react in the same way when this capability is not required, e.g. during the removal of BQ from activated proteins by Sephadex or Sepharose™ size exclusion.

While benzoquinone has never found widespread application as a protein conjugation reagent, it can sometimes offer a useful back-up when more established methods prove unsuccessful. A generic protocol is described in Protocol 5.4.

5.3.7 Equilibrium transfer alkylating reagents

An interesting series of compounds were described by Mitra and Lawton (1979) in the role of equilibrium transfer alkylating cross-linkers (ETAC), which are capable of forming links between nucleophilic groups. Amine–amine bonds are among the most likely out-

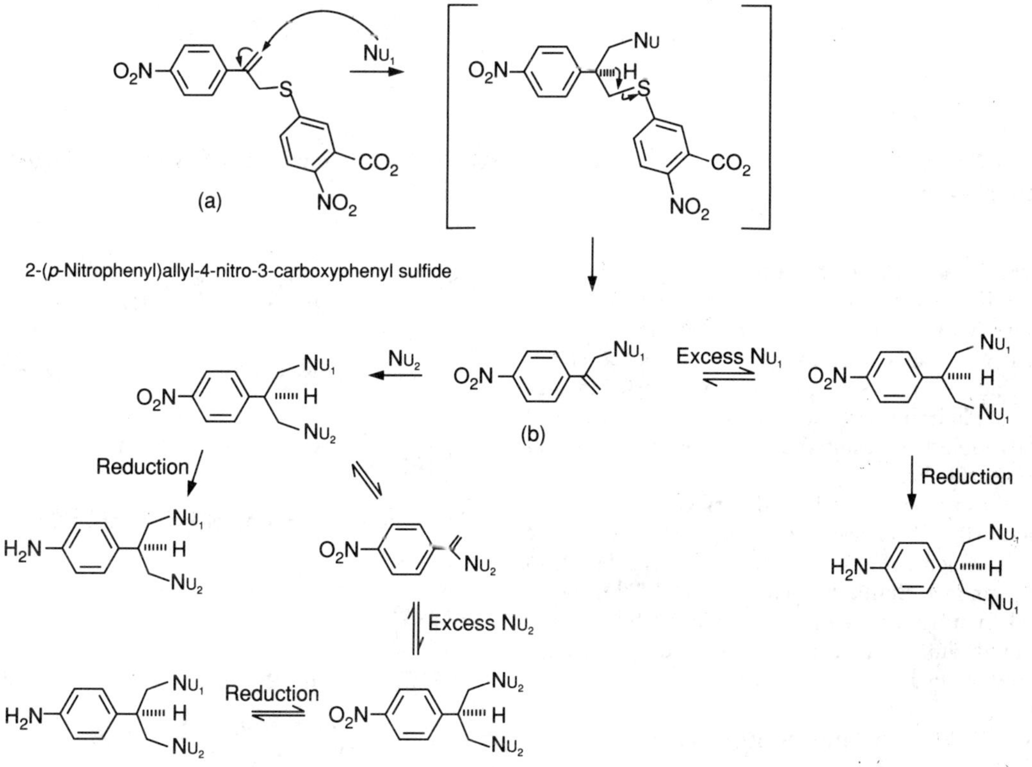

Figure 5.77 Chemistry of ETAC agents.

comes of their use, hence their inclusion in this section. One of the first examples and its mode of action is shown in Figure 5.77. The key components of these reagents are: (1) an electrophilic double bond (in this case activated by the nitrophenyl group) and (2) a good electron acceptor at the β-position relative to the double bond. This electron acceptor, which will act as a leaving group, is provided in this case by the benzenethiol; the sulfur is electron-deficient due to the electron-withdrawing effect of the nitro and carboxyl functions.

The activated double bond of this compound will undergo the usual Michael addition reaction with a suitable nucleophile such as an amine (Figure 5.77(a)), but this generates a second active bond with the loss of the benzenethiol (Figure 5.77(b)). This in turn can undergo a similar reaction with a second nucleophile, another amine, perhaps, or a thiol, to generate a stable conjugate (Figure 5.77(c)). However, if the first nucleophile was relatively weak it may be ejected gradually as an electron-accepting leaving group, again regenerating an active double bond (Figure 5.77(d)). This process can be repeated until the most stable set of linkages has been obtained. This equilibrium configuration can be 'fixed' by dithionite reduction of the nitrophenol, removing the activating influence on the double bond (Figure 5.77(e)).

This rather neat approach, therefore, drives the reaction towards forming the most stable cross-linked product(s), a useful capability both in terms of obtaining a robust conjugate and identifying which residues generate the most stable bonds. Mitra and Lawton (1979), for example, found that the equilibrium mixture for cross-linked pancreatic ribonuclease contained bonds between two separate pairs of lysine residues, Lys7/Lys37 and Lys31/Lys41.

Liberatore *et al.* (1990) developed more efficient ETAC agents, which incorporate: (1) substituted acetophenones as more powerful activating groups, which can be conveniently reduced to alcohols to quench reaction; (2) a latent double bond which is generated *in situ* by base catalysis, and (3) bulky substituted phenylsulfones as leaving groups. The latter two features are aimed at mimimising unwanted side-reactions of the reactive double bond. One of these compounds is shown in Figure 5.78.

Figure 5.78 An ETAC reagent described by Liberatore *et al.* (1990); the active double bond is generated *in situ* by base catalysis.

These authors made the interesting proposal that cysteine links which had been disrupted might be reformed by ETAC reagents, on the basis that the original S–S bonds were likely to be favoured structures which would be 'homed in' on by the transfer reaction. Their investigations with partially reduced antibody molecules resulted in a mixture of intra- and intermolecular bonds, however.

Del Rosario *et al.* (1990) described examples of trifunctional ETAC reagents incorporating an amine function, which could be used to anchor probes such as fluorophores. Similarly, Wilbur *et al.* (1994) incorporated an arylstannane moiety into an acetophenone-bis-phenylsulfone ETAC reagent, providing a site for facile radioiodination.

5.3.8 Other homobifunctional amine reagents

There are a number of powerfully electrophilic groups which are capable of reacting with amines and other nucleophiles, e.g. epoxides, vinylsulfones, and aziridines. Sulfonyl chlorides too react with amines, although they are generally difficult to dissolve in aqueous systems and hydrolyse rapidly once this has been achieved.

Homobifunctional examples of these compounds are clearly capable of forming protein–protein links, and such behaviour has been known for many years (Goodlad, 1957; Wold, 1972); examples of such reagents are shown in Figure 5.79. There is little to recommend these reagents for preparative use: by the early 1970s, most had been reduced to footnote status in reviews of protein-coupling reagents. (Likhite and Sehon, 1971; Wold, 1972). Occasional reports of such reagents persist however, for example the use of bis-acrylamide for antibody coupling (Houen *et al.*, 1997).

It should also not be forgotten that, although maleimides and haloalkyl compounds are dealt with in Section 5.4 as 'thiol-reactive' moieties, they are quite capable too of reaction with amines. This is particularly true of highly reactive alkyl halides such as the nitrogen and sulfur mustards (Olcott and Fraenkel-Conrat, 1947; Ross, 1953). The selectivity of these

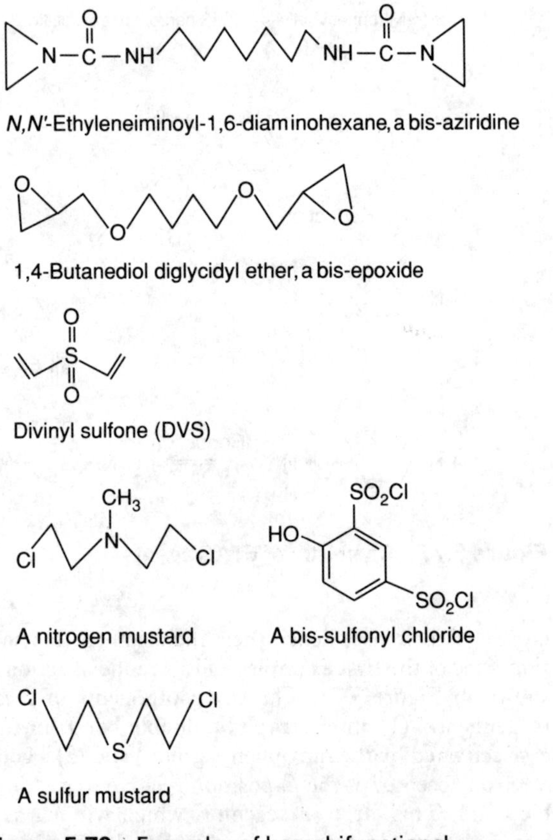

Figure 5.79 Examples of homobifunctional epoxides, sulfones, aziridines, mustard compounds and sulfonyl chlorides.

compounds is extremely low, some of the mustards are even capable of reacting directly with carboxyl functions, for example.

'Semi-homobifunctional' reagents have occasionally been described, with two different groups of similar specificity: Kraehenbuhl *et al.* (1974) described, for example, the preparation of the succinimide ester of *p*-formylbenzoic acid (Figure 5.80), the two functionalities offering differential reactivity towards amines to allow a two-stage reaction. Thus 8-microperoxidase (8-MP, haem-octapeptide) could be derivatised by

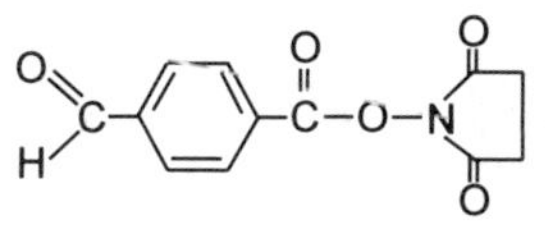

Figure 5.80 *N*-Succinimidyl *p*-formylbenzoate.

the succinimide ester (pyridine, 12 hours at room temperature: 80% yield), leaving the aldehyde free for subsequent reaction with the amine groups of antibody fragments (borate buffer, pH 9.5, five hours at 37 °C).

5.4 Homobifunctional Coupling of Thiols

Thiol groups are strongly nucleophilic, and where present they will react with many of the reagents already described for the formation of amine–amine bonds. However, there are a number of approaches which are specific (to a greater or lesser degree) for thiol moieties, and these will be dealt with in this section.

Thiol–thiol techniques do not have quite the same widespread applicability as amine–amine approaches; while there are virtually no proteins lacking a reasonable population of amine groups, there are very many which contain no free thiols. As discussed earlier, this is not surprising in view of their tendency to undergo dimerisation under oxidising conditions: for free thiols to exhibit long-term stability in an oxidising environment, there must be features of the molecular structure which reduce their reactivity. These may involve electronic effects which alter the inherent behaviour of the thiol function, or factors which prevents the access of potentially reactive groups.

The pK_a of a thiol is critical in determining its reactivity. Above the corresponding pH its nucleophilicity increases dramatically as the thiolate (S^-) form replaces the protonated SH species.

Thiols in isolated organic compounds can exhibit pK_a values over an extremely wide range, depending on other substituents: where electron-donating substituents are present the pK_a may be as high as 10 or 11 (e.g. mercaptoacetate, 10.2), while electron-withdrawing substituents may give a value below 2 (e.g. 4-mercaptopyridine, 1.43 (Albert and Barlin, 1959)). The thiol in an isolated cysteine molecule has a pK_a of around 8.6, and the pK_as of many protein thiols are around this value or slightly lower (Torchinsky, 1981a). However, the effects of microenvironment can sometimes result in anomalously low pK_a values: examples include thiols in papain with pK_a 3.38 (Roberts *et al.*, 1986), in bovine serum albumin with pK_a < 5 (Lewis *et al.*, 1980) and in phosphoenolpyruvate carboxykinase with pK_a around 7 (Lewis *et al.*, 1989). Proximity to acidic groups can sometimes increase the pK_a of a thiol compared to typical examples; the lack of reactivity of one of the three thiols in lactose repressor protein from *E. coli*, for example, is attributed to this effect (Brown and Matthews, 1979).

The complexity of microenvironmental charge effects is well demonstrated in the work of Roberts *et al.* (1986) on papain. The anomalously acidic thiol group is from a cysteine residue (Cys25) in the active site. There is a histidine residue (His159) nearby; if this is protonated the thiol shows the pK_a of 3.38 already mentioned. However if this histidine is deprotonated, the pK_a of the thiol rises dramatically to 7.59. Similarly the pK_a of the histidine nitrogen is affected by the charge state of the thiol, 4.35 with the SH form, 8.56 where the thiolate is present. Such cooperative effects can obviously complicate any assessment of reactivity.

In the majority of examples, these microenvironmental charge variations tend to result in thiols of greater rather than lower reactivity—the low pK_as resulting in greater-than-typical proportions of thiolate at neutral pH—and do not help to explain the lack of reactivity of some protein thiol groups. Steric effects are much more important in creating the stable environment in which these free thiols can exist. Such effects are obviously critically dependent on the three-dimensional structure of the protein, and can be modified by conformational alterations. Bednar (1990) carried out an interesting study on the reactivity of a thiol residue in the enzyme chalcone isomerase, from the soybean *Glycine max*. Above pH 8, this thiol exhibits a rate constant for reaction with *N*-ethylmaleimide which is around 100,000-fold lower than would be expected from a simple thiol of the same pK_a. Under more acid conditions however, there appears to be a change to the protein conformation which disables the protective effect; the rate constant increases 1000-fold to a more normal level.

Hydrophobic interactions of thiols with apolar residues within the protein have also been proposed as an explanation of low reactivity. Torchinsky (1981b) has written an excellent summary of this and other aspects of differential thiol reactivity in proteins.

The rarity of free thiol groups in proteins can be advantageous in limiting or directing the degree of derivatisation during a conjugation reaction. The

Fab′ fragment of an immunoglobulin molecule, for example, may contain a single thiol group located away from the antigen-recognition site; reaction with a maleimide compound leads to derivatisation at this single site, with a degree of control and specificity not generally obtainable from amine coupling.

The use of oxidative disulfide formation in protein coupling has already been covered in Section 5.2, as it does not involve the incorporation of any 'foreign' atoms. In homobifunctional and heterobifunctional reagents, the thiol reactions which are predominantly exploited are the Michael addition to maleimides, the substitution of halogens in acyl and alkyl halides, and the disulfide exchange reaction. The rather specific interaction of thiols with mercury derivatives has also seen some limited application.

5.4.1 Bis-maleimides

The maleimide moiety (Figure 5.81(a)) contains a double bond activated to nucleophilic attack by the two neighbouring carbonyl functions, and is therefore capable of undergoing Michael addition reactions (cf. benzoquinone, Section 5.3.6)[1]. The product with thiols is a thioether (R–S–R) attached to a saturated succinimide ring (Figure 5.81(b)). This is generally a very stable linkage, although cleavage can occur under certain conditions (Presentini and Terrana, 1995).

[1] Both maleimides and benzoquinone contain α,β-unsaturated carbonyl functions. The regioselectivity of nucleophilic attack on such species is a common topic in organic chemistry (1,2-addition at the carbonyl carbon *versus* 1,4 Michael addition); however in both these cases Michael addition is strongly favoured because of the double activating effect of the two carbonyl groups, and because the carbonyl groups themselves are relatively unreactive ketones.

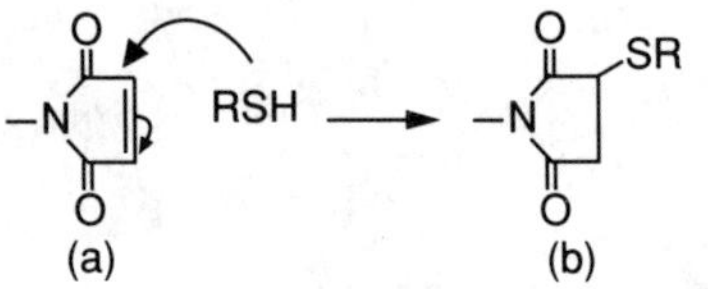

Figure 5.81 The maleimide moiety and its reaction with thiols.

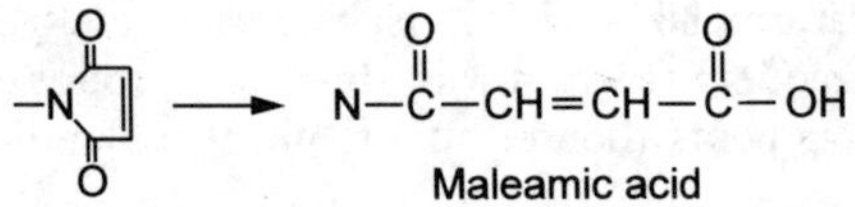

Figure 5.82 Hydrolysis of maleimide group to maleamic acid.

As noted above, the rate of reaction of thiols increases at high pH due to the increased concentration of thiolate anion. However, there are two drawbacks to carrying out the thiol–maleimide reaction under alkaline conditions. The first is the hydrolysis of the maleimide group to maleamic acid (Figure 5.82), which becomes a significant side-reaction at high pH; the second is the possibility of competing reactions with other functional groups in the protein. Maleimides will react with both lysine and histidine residues, and both these reactions become more significant at higher pH (Brewer and Riehm, 1967).

At neutral or slightly acidic pH, the rate of maleimide hydrolysis is low. Palumbo (1993), for example, followed the hydrolysis of a maleimide-containing reagent (30 minutes, 30 °C) and observed losses of 0, 2, 14 and 56% at pH 5, 6, 7 and 8 respectively (Figure 5.83). Similarly, Yoshitake *et al.* (1979a) demonstrated that maleimide groups showed no loss on storage for up to 45 hours at 4 °C and pH 5. Higher pH and/or

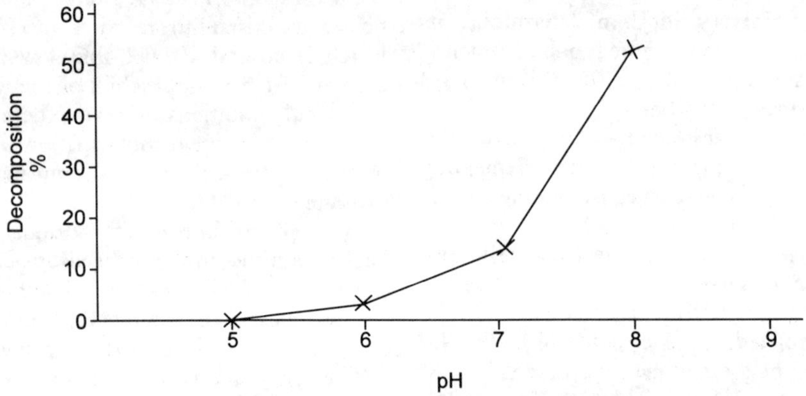

Figure 5.83 Hydrolysis of maleimide group from an *N*-phenylmaleimide compound (data from Palumbo, 1993).

temperature led to significant loss of maleimides, at pH 6 and 30 °C, for example, half of these groups were lost within two hours. At pH 6.5 and 4 °C the loss of maleimide groups was small enough to be unimportant (10% loss after 16 hours) and these conditions were adopted as a good compromise of maleimide and protein stability.

Under less basic conditions such as these, the majority of amine groups are also predominantly present in the unreactive NH_3^+ form, and the reaction of maleimide with thiols is 1000-fold faster than with amines (Ji, 1983). Zahn and Lumper (1968), for example, observed quantitative reaction of a bis-maleimide with the cysteine residues of BSA at pH 7.2, whilst only 7% of the much more abundant lysines were derivatised. Therefore, the optimal pH for thiol-maleimide reactions is generally in the range 6–7, despite the scarcity of thiolate anions (Riordan and Vallee, 1972a; Manecke and Middeke, 1980).

Using low-molecular-mass models, Bednar (1990) observed that the reactivity of a protonated thiol with *N*-ethylmaleimide was negligible compared to that of the thiolate anion, the rate constant being lower by a factor of 5×10^{10}. Therefore even at acidic pH it is probably the small fraction of thiolate which is responsible for the maleimide reaction. These findings are consistent with the conclusions of other authors (Brocklehurst, 1979) and appear to supersede the observations of Gorin *et al.* (1966) on the kinetics of the reaction of *N*-ethylmaleimide with cysteine; these authors attributed a small pH-independent term in the rate equation to the contribution from the SH form.

Homobifunctional applications of this reaction rely, predictably enough, on molecules containing two maleimide moieties connected by a suitable linker function. Reagents of this type were in use in the 1950s for protein cross-linking: *N,N′-o*-phenylenedimaleimide (Figure 5.84) was first synthesised by Moore and Ward in 1956, and used by them to dimerise bovine serum albumin and to cross-link wool proteins. This reagent has since been employed quite widely in more demanding applications.

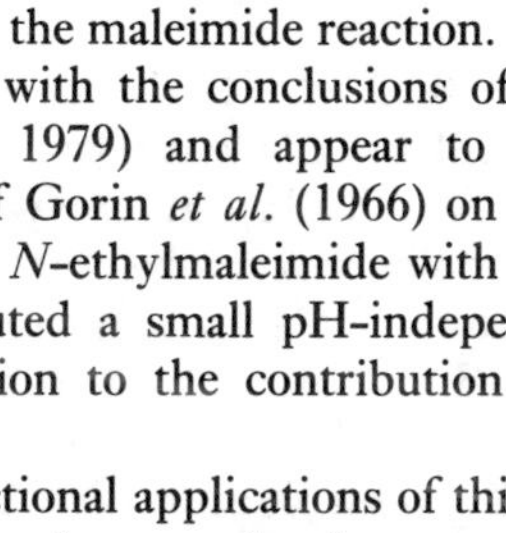

Figure 5.84 *N, N-o*-Phenylenedimaleimide.

Since the bis-maleimide approach requires the presence of a free thiol on each protein to be conjugated, there are certain applications to which it is particularly suited. An example is the preparation of conjugates between β-galactosidase and the Fab′ fragments of immunoglobulins. Both the enzyme and the Fab′ contain thiol groups, the latter being particularly attractive as a conjugation site as it is remote from the antigen-binding sites. This application has attracted a particularly large amount of interest in the use of bis-maleimide coupling. Eiji Ishikawa's group at the Medical College of Miyazaki in Japan have been instrumental in the development of these methods, and of maleimide-based conjugation methods in general.

5.4.1.1 *N,N′-o*-Phenylenedimaleimide coupling methods

The first paper by this group (Kato *et al.*, 1975a) using *N,N′-o*-phenylenedimaleimide describes the conjugation of thiolated insulin to β-Gal. This approach was then applied to the preparation of Fab′ and IgG–β-Gal conjugates; these have been used in some very sensitive immunoassay formats, and will be dealt with in some detail.

The early development of the methodology for coupling IgG, Fab′ fragments or 'half-molecules[1]' is described in a series of papers from the Miyazaki group (Kato *et al.*, 1975b,c, 1976a,b; Ishikawa and Kato, 1978). Hamaguchi *et al.* (1979) and Yoshitake *et al.* (1979a) reported improvements to this approach, aimed principally at answering doubts over the stability of the thiolated and maleimide-derivatised intermediates, and improving incorporations. Their conditions were maintained in later applications (Imagawa *et al.*, 1982b), and protocols based on this work were included in a review by Ishikawa *et al.* (1983).

These offer methods for the preparation of Fab′–β-Gal conjugates by two separate dimaleimide approaches. The first derives from those described above, where the antibody fragment is derivatised with the maleimide reagent before conjugation with the enzyme; in the second it is the enzyme which is derivatised first. This can be more convenient where the quantities of antibody available are small.

The conjugates made by these Ishikawa protocols showed excellent properties, no formation of Fab′ dimers or polymers; > 99% of the β-Gal incorporated in the conjugate with no loss of enzyme activity; 1–4.5

[1] The disulfide bond which links the two halves of rabbit IgG is more easily reduced than the other S–S bonds in the molecule, and can therefore be selectively cleaved (Nisonoff and Dixon, 1964).

Fab′ groups per β-Gal molecule depending on molar ratio, corresponding to incorporation of about 30–45% of the available Fab′. In the whole IgG methods, incorporation was 1–2 IgG molecules per β-Gal and IgG recovery was 23–29%.

Examination of the conjugates by size exclusion after one year at pH 6–7 and 4 °C showed no breakdown of the thioether link. Similar conjugates showed unchanged performance in an immunoassay after four years storage under the same conditions. It should be noted that other authors have seen less satisfactory stability of the thioether linkage: Presentini and Terrana (1995) observed that half the thioether bonds in an IgG–HRP conjugate had been cleaved after 15 days at 37 °C and pH 7.8. This might be expected from the higher pH, but the loss was still significant at pH 6.5; the addition of substantial quantities of glycerol was required to achieve full retention. The possibility of thioether cleavage of this type should be therefore perhaps be borne in mind for thiol–maleimide conjugates.

Some problems in applying the method to other enzymes were noted by Ishikawa *et al.*: α-D-glucosidase lost 30% of its activity on incubation with dimaleimide, while alkaline phosphatase lost 20–40% on thiolation and was inactivated by the presence of EDTA. The observation was also made that azide rapidly destroys maleimide functions, a valuable point to be borne in mind considering the widespread use of this reagent as a bacteriostat in stored reagents.

Later developments to the activated enzyme approach concentrated on improving the efficiency of the antibody fragment preparation and purification either by micro-scale approaches (Inoue *et al.*, 1985) or by eliminating unnecessary size exclusion stages (Hashida *et al.*, 1985). Some of the most sensitive immunoassays ever reported employed these dimaleimide techniques: Ruan *et al.* (1987) described the detection of 1 zmol (600 molecules) of ferritin using rabbit Fab′–β-Gal conjugates prepared by Ishikawa's activated-enzyme protocol. This reagent has remained in routine use for such applications right up to the present day (Hashida *et al.*, 1993), in this example 14 maleimide groups per molecule were introduced into β-galactosidase by treatment with the dimaleimide.

Of course there are other applications where the creation of thiol–thiol linked conjugates is convenient, many of them related to antibodies. The creation of bispecific antibodies by the combination of two different fragments is one such application with therapeutic potential. Glennie *et al.* (1987), used *N,N′-o*-phenylenedimaleimide to produce reagents of this type from Fab′ fragments. These authors used 50 mM sodium acetate buffer, pH 5.3 (0.5 mM EDTA) for the activation (30 minutes at 4 °C) and coupling (18 hours at 4 °C) reactions; the dimaleimide was dissolved at 12 mM in chilled DMF. DeSilva and Wilson (1995) used the same reagent to form an bifunctional antibody in a 'solid phase synthesis' approach.

5.4.1.2 Other bis-maleimides

Weston *et al.* (1980) introduced *N,N′*-(oxydimethylene)dimaleimide (Figure 5.85), which is more soluble in acetone than phenylene dimaleimides. These authors used this reagent to couple peroxidase and other enzymes to IgG after thiolation of both proteins with heterobifunctional reagents. This reagent appears to offer little clear advantage over the phenylene dimaleimides. The latter show quite adequate solubility in DMF, and are usually employed in this medium.

Straight-chain aliphatic bis-maleimides have found some application: 1,6-bis(*N*-maleimido)hexane (BMH; Figure 5.86) is commercially available; Stalteri and Mather (1995) used BMH to cross-link Fab′ fragments. Longer homologues such as 1,8-bis-(*N*-maleimido)octane and 1,12-bis(*N*-maleimido)dodecane have also been described (Heilmann and Holzner, 1981).

Albarella and Buckler (1989) described bis-maleimido derivatives of polyalkylene glycols, which, in contrast to all those described so far, are water-soluble and result in a hydrophilic spacer group. An example, 1,17-bis(*N*-maleimido)-3,6,9,12,15-pentaoxaheptadecane, is shown in Figure 5.87. Larger and more hydrophilic still are polyethylene glycol (PEG)-bis-maleimides, which are commercially available from Pro Chem, Inc. and from Shearwater Polymers.

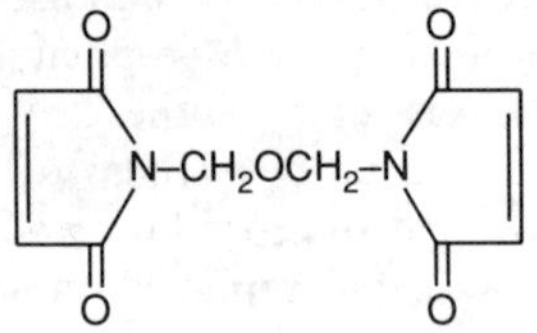

Figure 5.85 *N, N′*-(Oxydimethylene)dimaleimide.

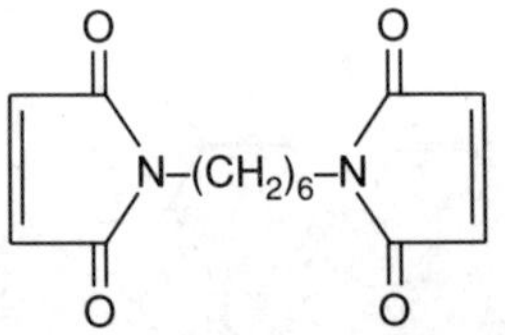

Figure 5.86 1,6-Bis(*N*-maleimido)hexane (BMH).

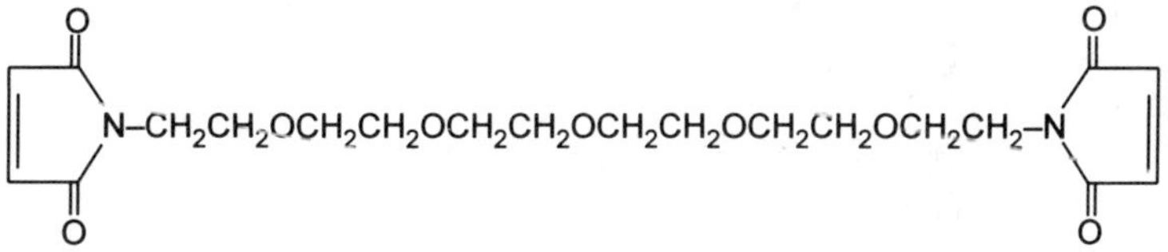

Figure 5.87 1,17-Bis(*N*-maleimido)-3,6,9,12,15-pentaoxaheptadecane.

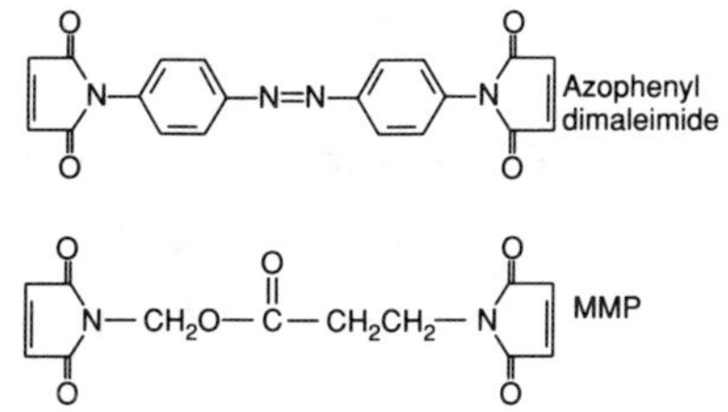

Figure 5.88 Cleavable bis-maleimide reagents, azophenyldimaleimide and maleimidomethyl 3-maleimido-propionate (MMP).

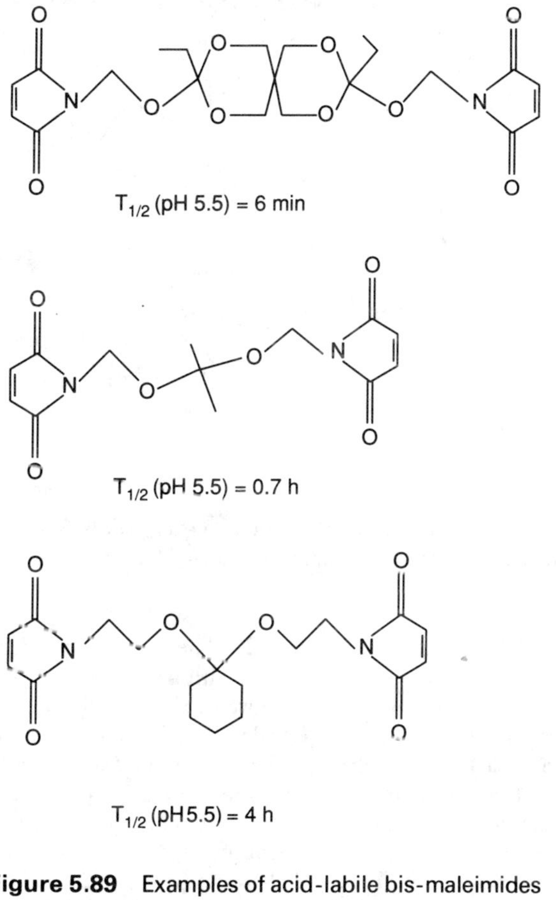

Figure 5.89 Examples of acid-labile bis-maleimides containing ortho ester or ketal moieties.

Cleavable bis-maleimide reagents have also been in use since the 1960s, such as azophenyldimaleimide (Fasold *et al.*, 1963), the azo bond of which can be cleaved by dithionite treatment. Another cleavable bis-maleimide is maleimidomethyl 3-maleimidopropionate, MMP (Sato and Nakao, 1981) whose ester linkage is susceptible to alkaline hydrolysis. These authors used a five minute incubation with 0.1 M sodium hydroxide at room temperature. These cleavable reagents are shown in Figure 5.88.

Acid-cleavable bis-maleimides have also been described, for the production of immunotoxins which are cleaved in the microenvironment of the endosomes by which they are internalised by cells (Srinivasachar and Neville, 1989; Neville *et al.*, 1989). These compounds, some of which are shown in Figure 5.89, incorporate ortho-ester or ketal moieties to confer acid-lability. These functionalities show a tenfold increase in hydrolysis for each single-unit drop in pH, and at roughly endosomal pH (5.5) display half-lives of minutes for the ortho-esters, and hours for the ketals.

Recently there has been some effort put into the development of trifunctional and tetrafunctional maleimide reagents; in particular, the aim has been to create oligofunctional antibody reagents for pharmaceutical and imaging uses which display improved biodistribution properties. The compounds shown in Figure 5.90a were described by Ahlem *et al.* (1991), Anderson *et al.* (1993) and Schott *et al.* (1993). Similarly, Ali and Quadri (1996) described bis-maleimides incorporating a polycarboxylate chelating agent, which could be used to create a bispecific antibody fragment dimer bearing a radiometal label. An example is shown in Figure 5.90(b).

5.4.2 Bis-alkyl halides

Despite their very different structure, α-carbonyl alkyl halides have rather similar reactivity to maleimides, the CH_2 group being activated to nucleophilic attack both by the carbonyl group and the halogen atom. The halogen is lost during the reaction (Figure 5.91). Many different groups in this category are reactive towards thiols, most typically haloacetic acids, haloacetamides

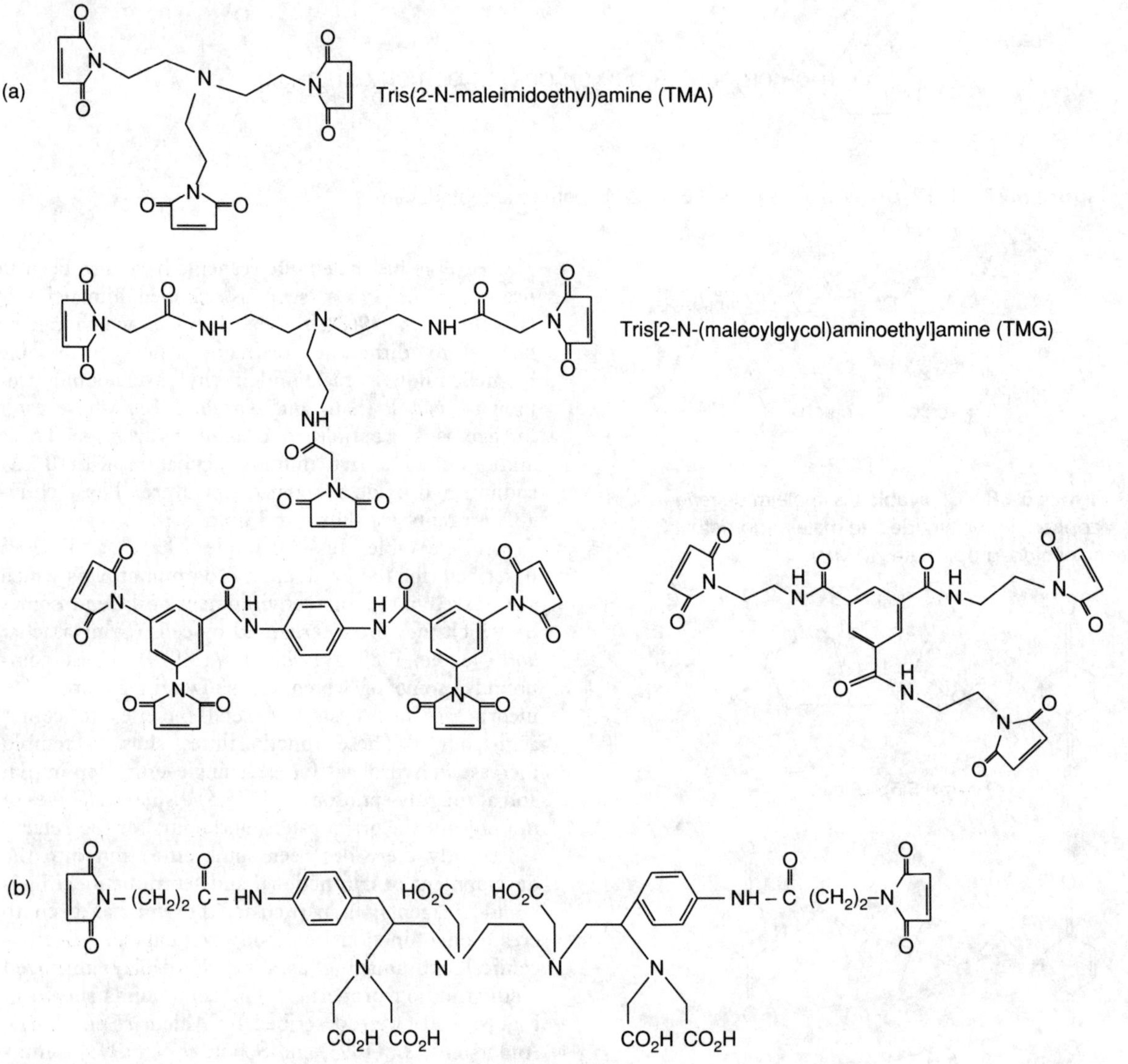

Figure 5.90 (a) Tri- and tetrafunctional maleimide reagents (Ahlem *et al.*, 1991; Anderson *et al.*, 1993; Schott *et al.*, 1993). (b) 2,9-Bis{*p*-[(β-maleimidopropionyl)amino]benzyl}triethylaminetetraamine-*N,N,N′,N″,N‴,N‴*-hexaacetic acid.

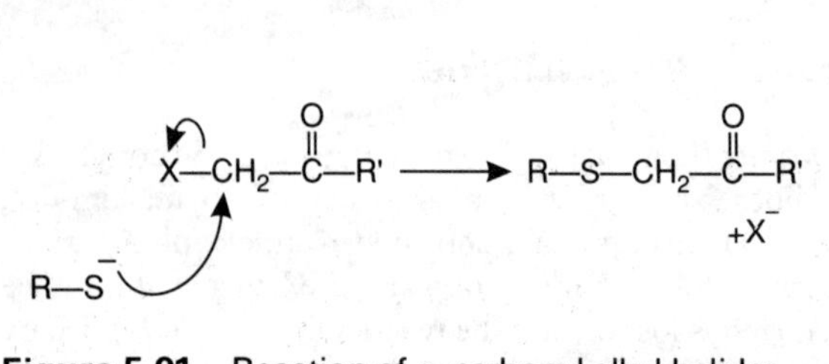

Figure 5.91 Reaction of α-carbonyl alkyl halides with a nucleophile (thiolate).

and haloketones (Figure 5.92). They are not particularly selective—they will react with free amine groups, histidine, methionine and more rarely tyrosine or even glutamic acid residues (Gurd, 1972). Selectivity can be modulated somewhat by pH: the rate of reaction with methionine is more or less independent of pH and is therefore favoured under acid conditions where amines and thiols are predominantly present in their less reactive NH_3^+ and SH forms (e.g. pH 2.8: Gundlach *et al.*, 1959). Of course methionine derivatisation

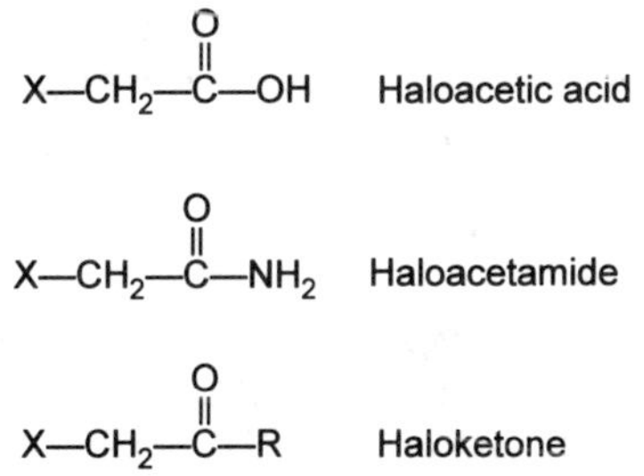

Figure 5.92 Halocarbonyl compounds.

can still be observed under neutral conditions if steric factors are favourable (Lawson and Schramm, 1962). Reaction with amines becomes predominant in more alkaline media; Mattson *et al.* (1993) have reported, for iodoacetyl compounds, that the specificity for thiols is highest at pH 8.3.

Among the halogens, the reactivity of these compounds decreases in the order iodine > bromine > chlorine. Cysteine reacts with iodoacetate twice as fast as with bromoacetate, and 20–100 times as fast as with chloroacetate (Lundblad and Noyes, 1984). These halo compounds can display photolability, and it is generally advisable to handle them in the dark. As the reaction releases a strong acid, a suitably robust buffer formulation should be employed, alternatively a pH-stat can be used, in which case the amount of base added to keep the pH constant is a useful means by which to follow the course of the reaction.

Once again it is the negatively-charged thiolate species which is most reactive with activated alkyl halides, so that the reaction rate increases as pH is increased towards and above the pK_a of the thiol. However, it has already been noted that competition from amine groups also becomes significant under alkaline conditions; so does hydrolysis to yield halide plus a glycollic acid (Figure 5.93). The pH optimum for reaction with protein thiol groups is generally neutral to mildly basic, and can depend on the nature of the reagent used. Chaiken and Smith (1969a,b) for example saw an optimum of around 7 for the reaction of chloroacetic acid with papain, compared to above 9 for chloroacetamide. Haloacetamides are generally more reactive than haloacetic acids.

These authors also noted interesting effects of the protein microenvironment on thiol reactivity: the active-site papain thiol reacts with chloroacetamide over ten times more quickly than free cysteine.

A number of other compounds containing halogens can react similarly (Figure 5.94), benzyl halides (Horton and Koshland, 1972), and sulfur and nitrogen mustards (see Section 5.3.8), for example, and these have been employed in protein derivatisation and linking. The broad reactivity towards nucleophiles of such groups generally precludes any selectivity for thiols; Hiremath and Day (1964), in fact, used bifunctional benzyl halides at pH 9.1 to cross-link lysine residues in lysozyme. The most useful specificity of benzyl halides is perhaps that towards tryptophan in acidic media, which can be used as a basis for quantification of these residues (Horton and Koshland, 1972). On the whole the 'non-carbonyl' bis-alkyl halides are now rarely encountered as coupling agents, although occasional exceptions exist such as the fluorescent dibromobimane compounds (Kim and Raines, 1995).

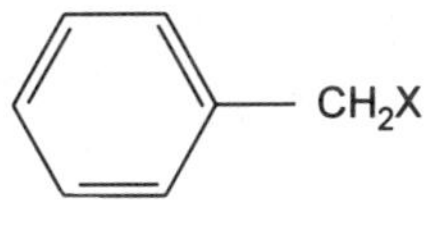

Figure 5.94 Benzyl halides.

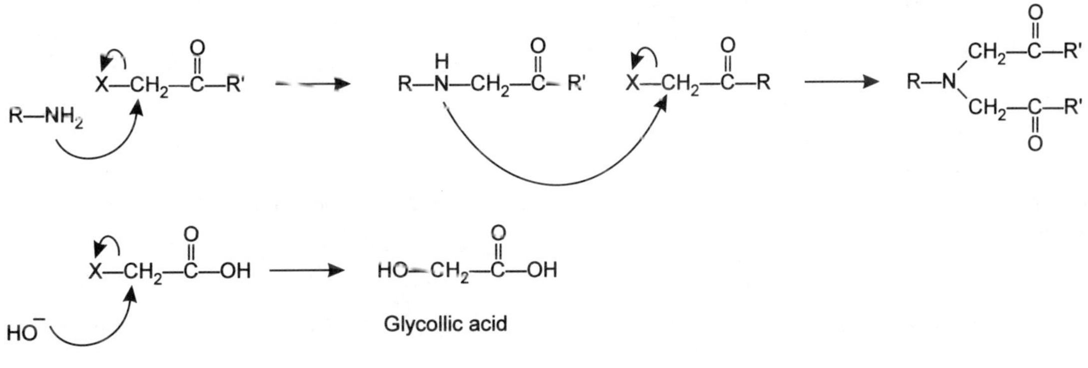

Figure 5.93 Reactions of halocarbonyl compounds under alkaline conditions: aminolysis and hydrolysis compete increasingly with the thiol reaction.

While monofunctional compounds containing labile halogens have been very widely used in thiol derivatisation, there are much fewer examples of homobifunctional reagents. However, this chemistry was employed in a very early application in the textile industry. Harris *et al.* (1942) used methylene bromide (CH_2Br_2) to form cross-links between the thiols of reduced wool proteins.

Fasold *et al.* introduced the reagent 2,2′-dicarboxy-4,4′-diiodoacetamidoazobenzene (Figure 5.95) in 1964; the azo group can be cleaved by dithionite. This compound reacted quantitatively with reduced glutathione at pH 8. Similar compounds incorporating polyproline moieties to confer hydrophilicity were introduced later by the same group (Fasold *et al.*, 1977).

Ozawa (1967) synthesised a number of types of *N,N′*-polymethylene-bis(iodoacetamide) (Figure 5.96), and used them to cross-link the subunits of rabbit muscle aldolase, in 100 mM phosphate buffer, pH 8.0. Reaction times were of the order of days. The $(CH_2)_2$ homologue with a span of 7.5 Å was capable of linking two subunits, whereas the $(CH_2)_0$ compound (diiodoacetamide) was not. The $(CH_2)_{11}$ homologue suffered from solubility problems, even when organic cosolvents were employed.

Husain and Lowe (1968) used 1,3-dibromoacetone to form an intramolecular cross-link between a cysteine and a neighbouring histidine residue in the active sites of ficin and stem bromelain. This resulted in irreversible inhibition of enzyme activity within a few minutes at pH 5.6.

Segal and Hurwitz (1976) synthesised a bivalent affinity label which combined a 2,4-dinitrophenyl (DNP) hapten function with a reactive bromoacetyl group (Figure 5.97(a)). Anti-DNP antibodies were covalently cross-linked by this reagent (IgG at 10 mg/ml in 20 mM sodium borate, 150 mM sodium chloride, pH 8.7, bifunctional reagent in DMSO; 24 hours at 30 °C). The efficiency of the cross-linking reaction was low.

Wilchek and Givol (1977) also worked with DNP affinity reagents. Previously it had been shown that two different bromoacyl-DNP compounds reacted at specific sites of an antibody molecule, one with a tyrosine residue on the light chain, and one with a lysine residue on the heavy chain. These authors designed a bifunctional molecule whose span was suitable for linking the two reactive sites (Figure 5.97(b)), and used it successfully to cross-link the heavy and light chains.

Ludueña *et al.* (1981) introduced the reagent *N,N′*-bis(iodoacetyl)-2,2′-dithio-bis(ethylamine) (BIDBE; Figure 5.98) to cross-link thiol groups in aldolase and tubulin. This reagent allowed groups separated by 18 Å to be coupled; with aldolase, reaction at pH 6.4 gave selective reaction of sulfhydryl groups. Dimethyl sulfoxide was used to dissolve the coupling reagent.

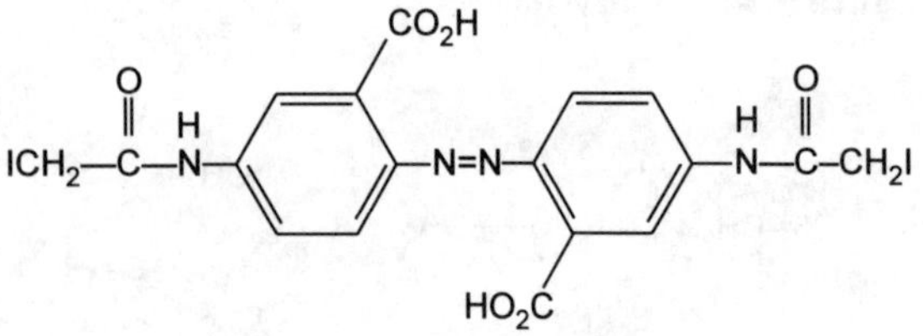

Figure 5.95 2,2′-Dicarboxy-4,4′-diiodoacetamidobenzene.

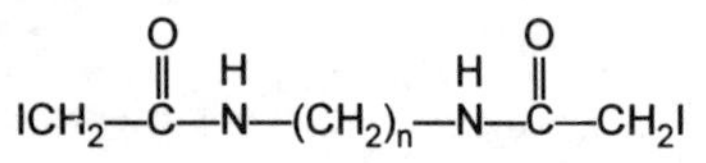

Figure 5.96 *N, N′*-Polymethylene-bis-(iodoacetamide).

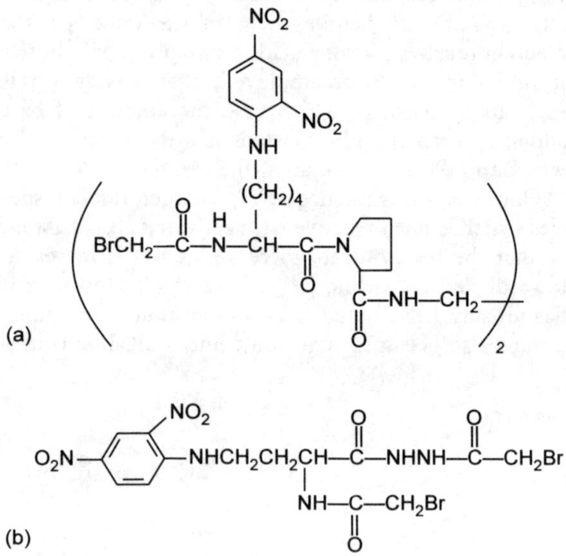

Figure 5.97 Bivalent affinity labels containing DNP and bromoacetyl groups.

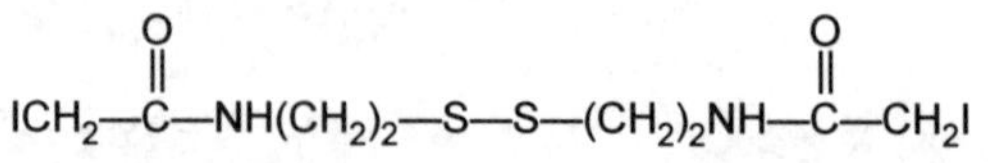

Figure 5.98 *N, N′*-Bis(iodoacetyl)-2, 2′-dithio-bis-(ethylamine) (BIDBE).

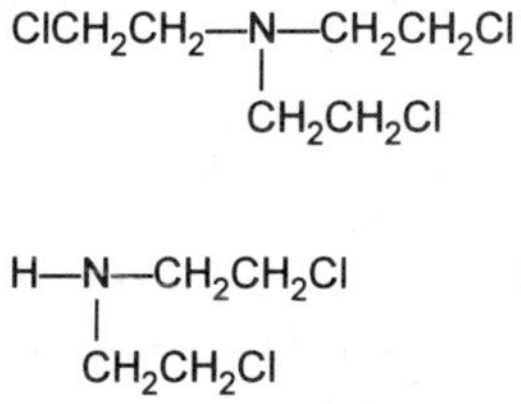

Tri-(2-chloroethyl)amine (TCEA)

Bis-(2-chloroethyl)amine(BCEA)

Figure 5.99 Bifunctional and trifunctional nitrogen mustard compounds.

A trifunctional nitrogen mustard, tri(2-chloroethyl)amine (TCEA), was used by Hiratsuka (1988) to cross-link three domains of the S-1 subfragment of myosin, indicating that the 26 kDa and 50 kDa domains were close to the reactive SH_1 thiol group on the remaining (20 kDa) domain. The bifunctional analogue, BCEA, was also employed in this investigation. These compounds are shown in Figure 5.99.

Although they are of no interest as preparative reagents for protein coupling, the chloroalkylnitrosoureas have received a considerable amount of attention as a result of their anti-tumour activity. A bifunctional example is shown in Figure 5.100; while other such compounds may possess only a single alkylating function, they appear to decompose *in vivo* to produce a number of reactive bifunctional species such as chloroalkyl isocyanates and chloroalkyl carbonium ions. The former can cross-link proteins or form protein–DNA complexes, while the latter are also capable of producing DNA–DNA links (Montgomery *et al.*, 1966; Ewig and Kohn, 1978; Ali-Osman *et al.*, 1985). This is a powerful combination of effects, not only interfering with DNA to prevent cell replication, but also apparently capable of disabling the DNA repair enzymes which counteract this nucleic acid damage (Calabresi and Chabner, 1990).

5.4.3 Bis-disulfides, etc.

The usefulness of the disulfide exchange reaction has been mentioned already. In terms of cross-linking and conjugation, there have been applications in lending control to direct thiol–thiol coupling (Section 5.2.1.3) but the heterobifunctional reagent SPDP (Section 5.7.3) has seen its most widespread exploitation. Homobifunctional uses are few and far between, perhaps surprisingly so considering the high selectivity of the reaction.

At least one such compound is commercially available: 1,4-di-[3′-(2′-pyridyldithio)propionamido]-butane (DPDPB; Figure 5.101). A bis-pyridyl disulfide PEG derivative is also commercially available from Shearwater Polymers.

A spin-labelled reagent of this type was described by Gaffney *et al.* (1983) and is shown in Figure 5.102; like Ellman's reagent it releases the chromophoric 2-nitro-5-mercaptobenzoate group on reaction with a thiol.

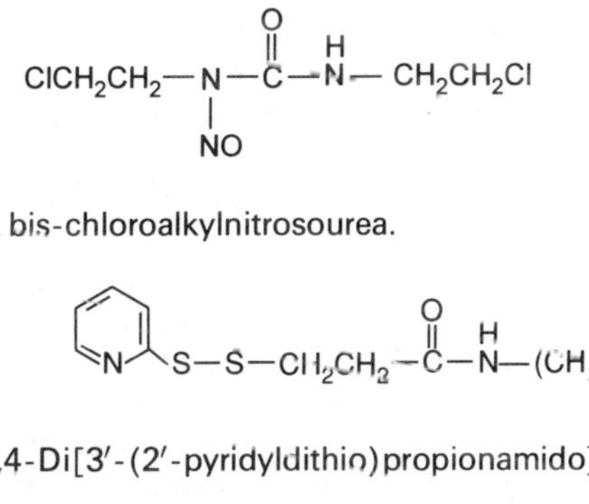

1,3-Bis(2-chloroethyl)-1-nitrosourea

Figure 5.100 A bis-chloroalkylnitrosourea.

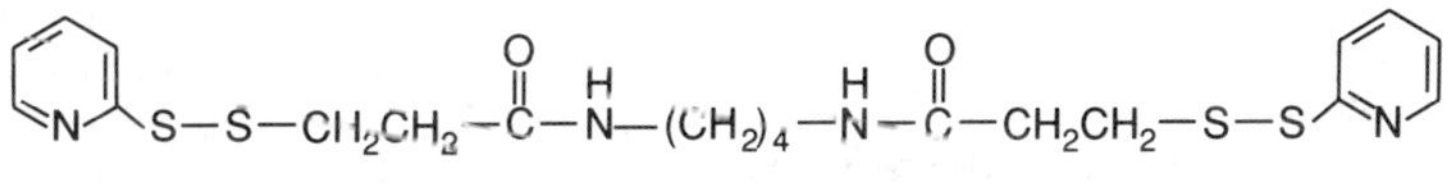

Figure 5.101 1,4-Di[3′-(2′-pyridyldithio)propionamido]butane (DPDPB).

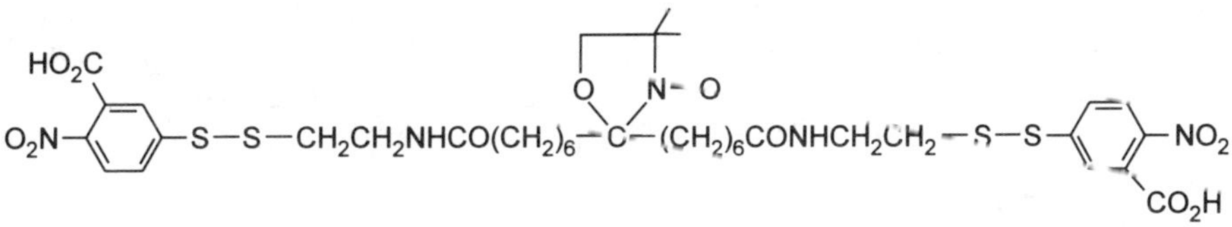

Figure 5.102 3-Oxy-2,2′-bis{[({2-[(3- carboxy-4-nitrophenyl)dithio]ethyl}amino)carbonyl]hexanyl}-4,4′-dimethyloxazolidine.

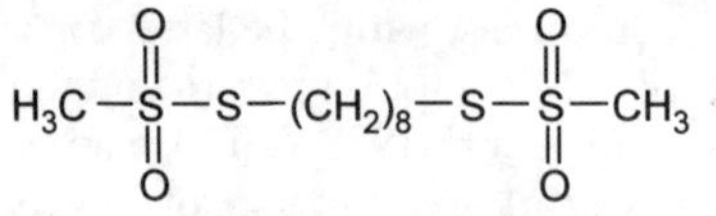

Figure 5.103 8,8′-Octamethylene-bis-(methanethiosulfonate).

It is the electron-deficient nature of the disulfide group which renders it susceptible to attack by nucleophiles such as thiols, and it has been noted in Section 5.2.1.3 that analogous sulfenyl compounds containing 'positive' sulfur atoms can be used similarly. Electron-withdrawing groups obviously increase the electrophilic character of the sulfur atom: for example, while disulfides generally require an alkaline pH to promote the exchange reaction, disulfide monosulfoxides (R–S(=O)–S–R′) will react at neutral pH (Steinman and Richards, 1970). Within this category, a series of bifunctional methanethiosulfonates have been described by Bloxham and Sharma (1979) for protein cross-linking. One of these, 8,8′-octamethylene-bis-(methanethiosulfonate) is shown in Figure 5.103 and was used by Bloxham and Cooper (1982) to probe subunit interactions in rabbit muscle glyceraldehyde-3-phosphate dehydrogenase.

5.4.4 Bis-mercurial reagents

The Hg^{2+} ion has a particularly strong affinity for thiol groups, and its capability for interaction with proteins has been recognised since the 1940s. In the same group that pioneered blood fractionation techniques at Harvard, albumin from human plasma was crystallised by the addition of mercuric chloride (Hughes, 1947). Edelhoch *et al.* (1953) carried out an extensive study of the kinetics and stoichiometry of this reaction. The rate decreased as the pH was raised above the isoelectric point of the protein (4.75), reflecting the increased electrostatic work required to bring the two protein monomers together. However, increased reactivity has been observed for the majority of proteins under acidic conditions (Riordan and Vallee, 1972a). The same authors noted that high concentrations of anions such as chloride and phosphate can retard the reaction of mercuric compounds with thiols.

Organomercurials display similar behaviour, and compounds such as 3,6-bis(mercurimethyl)dioxane salts (Figure 5.104) have also been used to dimerise human serum albumin (Edsall *et al.*, 1954). The rate of dimerisation was much larger than for simple mercuric salts, a finding explained by the greater electrostatic work necessitated by the closer approach (around 10 Å) required in the latter case. Nisonoff and Rivers (1961) used one of these reagents to polymerise thiolated antigens, yielding an insoluble complex.

Although mercurial reagents have seen little use in recent decades, Vas and Csanády (1987) used both mercuric chloride and an acyclic organomercurial, 1,4-bis(bromomercuri)butane (Figure 5.105), in investigations of the inhibition of 3-phosphoglycerate kinase. Either reagent could simultaneously block two reactive thiols in this enzyme; the small span of the $HgCl_2$ in particular (around 3 Å) was good evidence that the two reactive SH groups were very close to each other in the protein structure.

The unusually small span of Hg^{2+} as a homobifunctional agent may be useful in similar applications where the juxtaposition of thiols is to be demonstrated.

5.4.5 Other homobifunctional thiol reagents

Where free thiols are present they are often the strongest nucleophiles available. Many electrophilic homobifunctional agents have a broad range of reactivity covering thiols, amines and hydroxyl groups, and thiol–thiol links may contribute to the conjugates that they form. Such links can be encouraged by the choice of pH; around neutrality most amines and hydroxyls will be protonated and relatively unreactive. However, it is generally impossible to conclude from the reported applications exactly which types of bond are formed, and these reagents are best viewed as having generic reactivity for nucleophiles. On the basis that amine–amine links are probably the most frequent result of their use in protein coupling, they are dealt with in Section 5.3. Examples of this category (Figure 5.105) include cyanuric chloride and related triazines, bis-aziridines, bis-epoxides, and divinylsulfone, DVS. Aryl halides too have substantial reactivity for thiols, and bifunctional examples have sometimes been listed as predominantly thiol-reactive (Mattson *et al.*, 1993).

With DVS, Zahn and Lumper (1968) characterised the product of BSA cross-linking and observed some specificity for the thiol groups, particularly at less

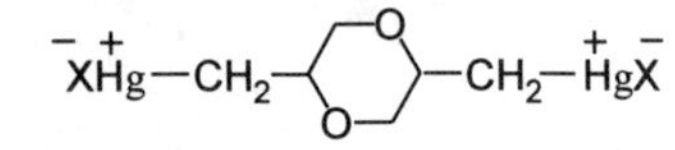

Figure 5.104 3,6-Bis(mercurimethyl)dioxane salt.

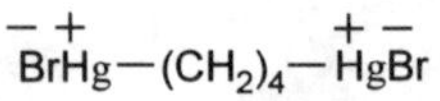

Figure 5.105 1,4-Bis(bromomercuri)butane.

basic pH. Such reagents still find some use, Houen and Jensen (1995), for example, used DVS for protein conjugation reactions and sought thiol specificity by using a relatively low pH of 8.

An early homobifunctional reagent described by Zahn and Waschka (1956), bis-diazohexane, also produced thiol–thiol links as well as coupling acid and tyrosine groups. The ETAC reagents described in Section 5.3.7 are also capable of generating thiol–thiol linkages.

5.5 Homobifunctional Coupling of Other Groups

5.5.1 Homobifunctional coupling of carboxyl groups

The carboxylate anion is nucleophilic, but at neutral to alkaline pH its reactivity for electrophiles is generally insignificant compared to that of amines and thiols. Under modestly acidic conditions, however, the predominance of the protonated NH_3^+ and SH forms of these groups is reduced so dramatically that the acid can compete. At pH conditions above the pK_a of the carboxylate (generally around 3 to 5), reaction can occur with such reagents as nitrogen and sulfur mustards (Figure 5.106). These reactions have seen little practical application.

β-Carbonyldiazo compounds have also been employed as coupling agents under such conditions. Although they can be photoactivated to generate carbenes (see Section 5.10), these compounds can also undergo a dark reaction which is thought to target nucleophiles. Husain *et al.* (1971) described the synthesis and use of 1,1-bis(diazoacetyl)-2-phenylethane to probe the active site of pepsin. It had already been shown that the monodiazo analogue reacted with a catalytically active aspartate residue, and this bifunctional reagent was synthesised to couple it to similar groups in the vicinity. It was more effective in deactivating the enzyme than the monofunctional diazoacetate, and may have formed intramolecular links between neighbouring carboxyl moieties.

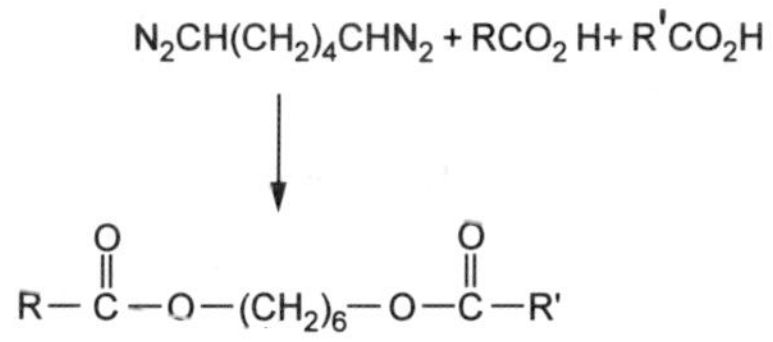

Figure 5.107 Straight chain bis-diazo compounds forming carboxyl–carboxyl links.

Diazomethane is a well-known agent for the esterification of carboxyl groups. Consequently, straight-chain aliphatic bis-diazo compounds are capable of forming (among other products) direct carboxyl–carboxyl links (Figure 5.107), and such products were observed in the cross-linking of silk fibroin (Zahn and Waschka, 1956).

In the leather industry a number of tanning agents are employed to modify the properties of the native hide, which is primarily composed of the protein collagen. Crosslinking the collagen units promotes the desired physical properties, hence the close relationship of tanning agents with some of the coupling agents of protein chemistry. Formaldehyde and glutaraldehyde are probably the best known tanning chemicals, but a number of commonly employed 'mineral' tanning agents are based on metal salts, notably chromium(III) sulfate complexes[1].

The 33% basic chromium sulfate solution typically employed in tanning contains a variety of chromium complexes such as those shown in Figure 5.108(a). Protein carboxylate groups can act as ligands for these complexes, which can therefore form cross-links between collagen molecules (Harlan and Feairheller, 1977). Apart from the formation of simple coordination complexes containing two protein carboxylates (Figure 5.108(b)), cross-linking can be achieved by dehydration between two single-carboxylate complexes (olation, Figure 5.108(c)). As the number of olate bridges increases, further dehydrations can

[1] A third class of tanning agents are the tannins extracted from plants and their synthetic analogues, the syntans. These are high-molecular-mass polyols which do not appear to form chemical cross-links.

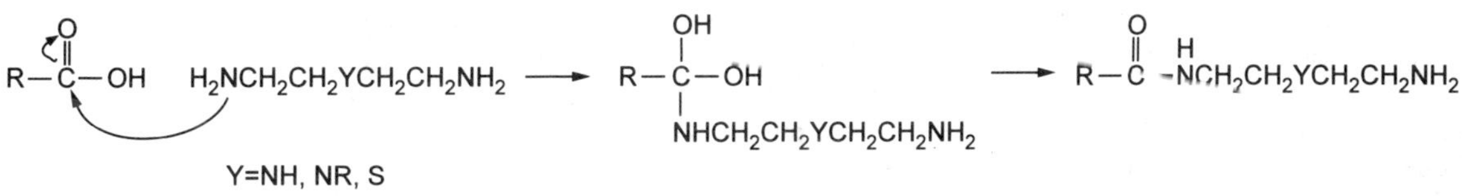

Figure 5.106 Reaction of carboxylate with mustards.

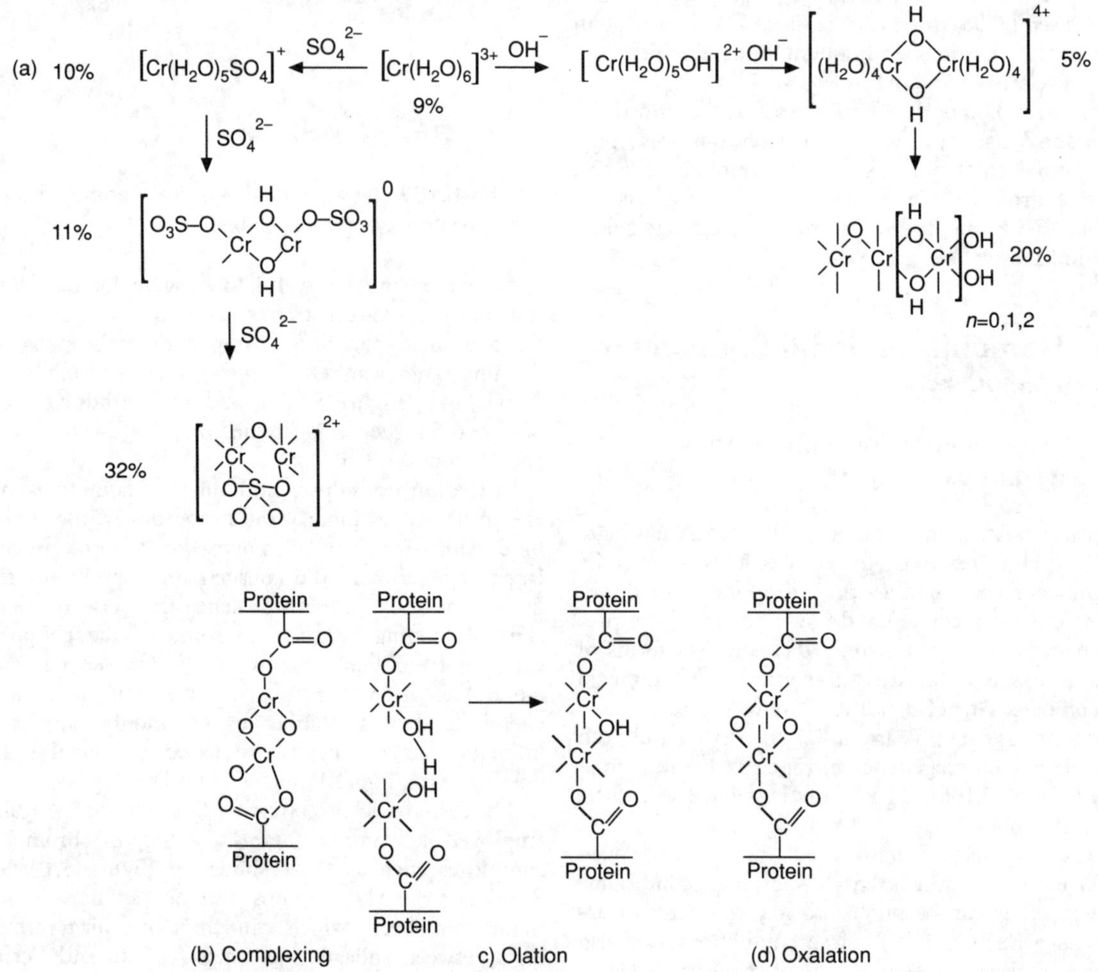

Figure 5.108 (a) Chromium complexes in 33% basic chromium sulfate solution and (b) three mechanisms for the formation of carboxyl–carboxyl links in their presence.

occur (oxalation, Figure 5.108(d)) to form more stable 'oxalate' links. These structures offer a simplified version of what is undoubtedly a complex situation; there is evidence for example that sulfate groups also play a part in the final tanned structure.

Chromium salts have been used for other protein applications, such as cross-linking of gelatin (Robinson, 1964). Although this is a rather 'old-fashioned' method, it is still employed in current applications such as enzyme immobilisation (Elçin and Akbulut, 1992).

The approaches above are based on reaction with the native carboxyl function. More usually in protein chemistry, of course, the electrophilic properties of the carboxyl group are enhanced by the formation of suitable derivatives which, as discussed in Section 5.2.2, promote reactivity with nucleophiles. If protein carboxyl functions are activated in this way, any suitable bis-nucleophile will function as a homobifunctional coupling agent: diamines and bis-hydrazides are examples which spring readily to mind. Unless the amine groups of the protein are first blocked it is impossible to avoid cross-linking under such circumstances, and it is therefore a rarely-encountered method for protein coupling.

5.5.2 Homobifunctional coupling of arginine residues

As noted in Chapter 2, the guanidine group of arginine undergoes reasonably specific reaction with 1,2-diones

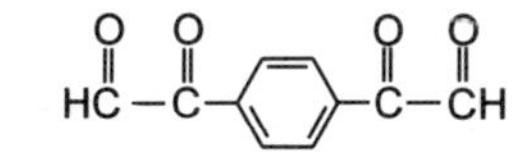

Figure 5.109 *p*-Phenylenediglyoxal.

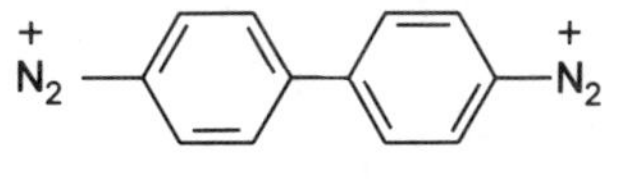

Figure 5.111 Bis-diazotised benzidine.

such as glyoxal. A homobifunctional reagent, *p*-phenylenediglyoxal (Figure 5.109) has been described for coupling nucleic acids (Wagner and Garrett, 1978; Hancock and Wagner, 1982). As far as the authors are aware, however, this reagent has not been used to couple proteins *via* arginine residues.

5.5.3 Homobifunctional coupling of tyrosine and histidine residues

These amino acids contain phenol and imidazolyl groups respectively, which are electron-rich aromatic systems prone to substitution by suitable electrophiles. The hydroxyl group of tyrosine donates electrons mesomerically to the phenyl ring, both activating it to electrophilic attack and directing the substitution to the *ortho* positions.

5.5.3.1 Bis-diazonium reagents

Aromatic diazonium salts have seen the most widespread application in the electrophile role, the positively charged $N{\equiv}N^+$ group reacting readily with activated aromatic groups; mono- and bis-azo derivatives are obtainable from both histidine and tyrosine (Figure 5.110) and, to a large extent, can be distinguished spectroscopically (Riordan and Vallee, 1972b).

Diazonium salts are not particularly stable, and are generally prepared from the corresponding amine by treatment with acidic sodium nitrite (Figure 5.110). While reaction with tyrosine and histidine predominates, diazonium salts are capable of diazotising other suitably nucleophilic residues such as terminal amines[1], lysine, cysteine, arginine, the imine group of proline and even tryptophan residues (Higgins and Harrington, 1959); aliphatic hydroxylic residues may also be reactive, possibly undergoing oxidation (Riordan and Vallee, 1972b). As might be expected the reactivity of various residues is strongly affected by the relation of their pK_a to the ambient pH; the optimum for the diazotisation itself is generally around pH 9.

Homobifunctional reagents based on diazonium salts have been in use since the 1940s: Pressman *et al.*[2] (1942) for example reported the attachment of ovalbumin to rabbit erythrocytes using bis-diazotised benzidine (Figure 5.111).

Howard and Wild (1957) described a historically interesting two-stage application of this chemistry. One protein was activated with tetrazotised *p*-phenylenediamine under acid conditions, where only one diazo function is reactive. The second protein was derivatised using a heterobifunctional reagent, pyrrole-α-carboxylic acid azide, the pyrrole moiety being capable of reacting with the remaining diazonium group. It was therefore possible to derivatise one molecule with the tetrazo compound without significant cross-linking, then to couple this relatively cleanly to a second (pyrrole-derivatised) protein (Figure 5.112). The defined chemistry and logic of this approach were quite advanced compared to most other protein-coupling techniques of the time.

Borek described the use of bis-diazotised or tetrazotised *o*-dianisidine (Figure 5.113) for the preparation of ferritin-antibody conjugates in 1961, and this compound remained in reasonably frequent use for protein conjugation in the 'pre-glutaraldehyde' era. A form stabilised by zinc chloride is referred to as Diazo Blue B or Fast Blue B (Avrameas, 1968).

Wang and Richards (1974) reported cross-linking of aldolase using the thiol-cleavable reagent 4,4′-bis(phe-

[1] Terminal glycines are diazotised, while the reaction of other NH_2-terminal residues tends to result in deamination.

[2] Including Linus Pauling in one of the lesser-known aspects of a spectacularly wide-ranging career.

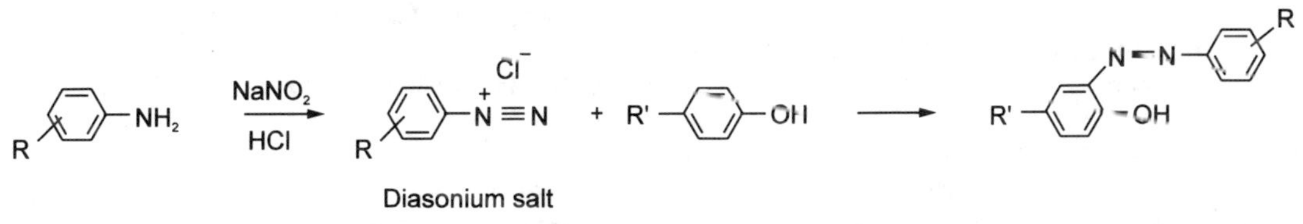

Figure 5.110 Formation and reaction of diazonium salts.

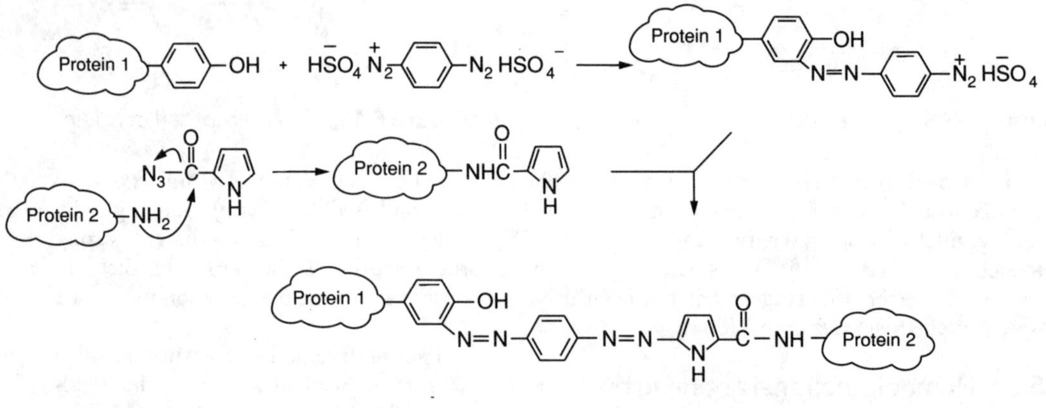

Figure 5.112 Tetrazotised-*p*-phenylenediamine and pyrrole-α-carboxylic acid azide.

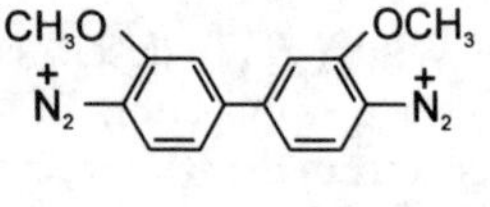

Figure 5.113 Bis-diazotised dianisidine.

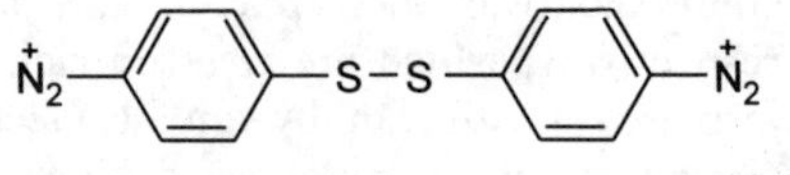

Figure 5.114 4,4′-Bis(phenyldiazonium) disulfide.

nyldiazonium) disulfide fluoroborate (Figure 5.114); this protein was chosen for examination due to its high content of free thiols, which could potentially cause problems from the disulfide exchange reaction. In practice cross-linking and cleavage could be carried out successfully, but the results obtained were less satisfactory than those with the bis-imidate, DTBP (see Section 5.3.2).

Like most of the reagents in use for protein coupling at this time, however, these diazonium approaches exhibited serious drawbacks in terms of yield and retention of activity. Avrameas (1968) noted considerable loss of enzyme activity in many cases, as well as over-reaction to form insoluble products (Avrameas *et al.*, 1978). Multimeric aromatic amines have been diazotised and the resulting reagents used for the formation of large protein complexes (Anderer and Schlumberger, 1969), sometimes the intention has been to yield an insoluble product (Riesel and Katchalski, 1964).

The limitations of this methodology, in particular the need to generate the diazotised reagents *in situ*, have rendered diazonium approaches largely obsolete for protein coupling applications.

5.5.3.2 Other methods for coupling *via* tyrosine and histidine residues

It should be noted that the hydroxyl group of tyrosine residues (pK_a around 10) is nucleophilic, particularly at raised pH, and is therefore a potential competitor for most of the electrophilic reagents described for amines and thiols. To a lesser extent, the same applies to the weakly basic nitrogen atoms of histidine, and the lower pK_a (6–7) means that a large fraction of these groups will be deprotonated at commonly encountered pH values. It is not surprising, therefore, that by-products derived from these residues have often been observed in most of the common protein derivatisation reactions. With tyrosine, some degree of specificity can be sought from broadly reactive reagents such as divinyl sulfone by working at a higher pH (see for example Houen and Jensen, 1995). On the whole, though, the prevalence of reaction with tyrosine is often limited by two factors, the tendency for these residues to be found in the interior of the protein molecule, and the instability of the products formed. In particular the tyrosyl esters formed from active carboxyl functions are readily hydrolysed. This may occur slowly at moderate pH, and hydroxylamine treatment is sometimes employed to ensure complete removal of such moieties.

Finally, tetranitromethane is a nitrating agent for groups such as tyrosine, and has been shown to yield cross-linked products by a poorly understood mechanism (Boesel and Carpenter, 1970).

5.5.4 Homobifunctional coupling of hydroxyl groups

The hydroxyl groups of serine and threonine, and of carbohydrates in glycoproteins, are available for homobifunctional reactions. Many of the electrophilic reagents described in Sections 5.3 and 5.4 also have the documented capability of reacting with hydroxyl moieties, bis-epoxides, cyanuric chloride and benzoquinone for example. Under favourable circumstances where the hydroxyl function is particularly reactive, this range can be extended to include other categories such as succinimide esters

Under normal conditions there will be powerful competition from amines and any free thiols that are present. Therefore it is not generally practical to carry out specific coupling through hydroxyl functions using these reagents (Miller *et al.*, 1992, 1994).

5.5.5 Homobifunctional coupling of aldohydes

While the hydroxyl functions of saccharides do not offer much scope for homobifunctional coupling *per se*, periodate oxidation to yield aldehydes opens up some opportunities. Any diamine, for example, could be used as a homobifunctional reagent for coupling oxidised glycoproteins; in an analogous approach Ey (1993) oxidised polysaccharides in the presence of 1,6-diaminohexane, and similar reactions of glycoproteins would undoubtedly yield some carbohydrate–carbohydrate links.

However any native amines in the glycoprotein will of course be reactive towards these aldehydes too, and, unless these have been protected first, uncontrolled polymerisation may follow to yield high-molecular-mass aggregates.

To avoid this problem rather elegantly, hydrazides (R–CO–NH–NH_2) can be employed in homobifunctional reagents. These compounds show greater nucleophilicity than amines, and have significantly lower pK_a values of around 2 to 3 (Inman and Dintzis, 1969; O'Shannessy and Wilchek, 1990). At pH 4 to 5, where the other nucleophiles in the protein will be heavily protonated and unreactive, the hydrazide will exist predominantly in the deprotonated state; carrying out coupling under these conditions, therefore, favours selective reaction with the hydrazide. The hydrazone product (Figure 5.115(a)) is also much more stable than a Schiff base, and reduction is not generally necessary (O'Shannessy and Wilchek, 1990).

Kozulic *et al.* (1984, 1987) used bis-hydrazides in this way to couple whole molecules or subunits of acid phosphatase and other proteins *via* their carbohydrate moieties. The reagents they employed were adipic acid dihydrazide (AADH) and malonic acid dihydrazide (MADH). Cross-linked acid phosphatase prepared by this route showed dramatically improved stability at alkaline pH.

The same group (Heimgartner *et al.*, 1990) introduced a cleavable analogue of these compounds, dithiodipropionic acid dihydrazide (DTDPADH), using it to carry out intramolecular cross-linking of immunoglobulin heavy chains *via* the glycan groups. By this approach the authors were able to demonstrate the proximity of the carbohydrate groups to each other. The hydrazide-coupling reactions were carried out at pH 4.5.

These reagents are shown in Figure 5.115(b).

Macromolecular polyhydrazides have been described, based for example on dextran and polybenzyl glutamate; these reagents may be soluble or insoluble depending on the macromolecule concerned and the degree of derivatisation. PEG-bis-hydrazides are commercially available from Shearwater Polymers. The soluble examples may be viewed as homofunctional coupling agents, while the insoluble polyhydrazides offer a route to protein-labelled solid phases. These reagents have been reviewed by Wilchek and Bayer (1987).

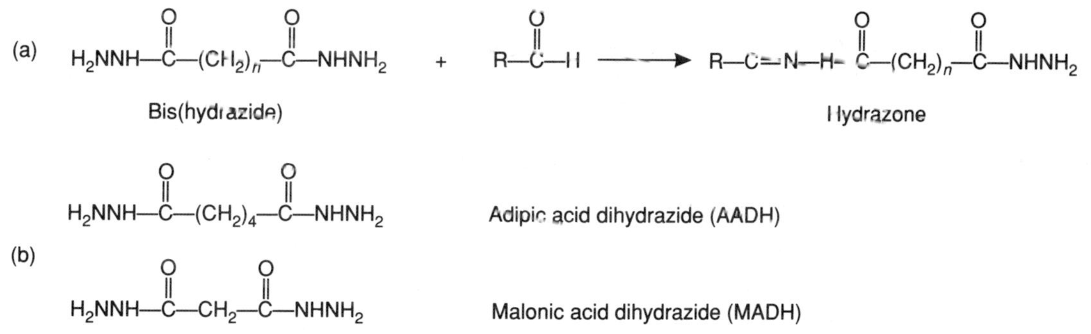

Figure 5.115 (a) Reaction of a bis(hydrazide) with an aldehyde and (b) examples of bis(hydrazides).

5.5.6 Homobifunctional photoactivatable reagents

Photoaffinity labelling was invented by Westheimer and co-workers (Singh *et al.*, 1962), and exploited the fact that certain groups such as diazo esters generate extremely reactive carbene intermediates on irradiation with UV light. These were capable of relatively non-selective reaction with protein residues. Fleet *et al.* (1969) introduced the use of aryl azides in this technique; on irradiation these compounds generate nitrenes, which also react with a wide range of molecules.

In protein coupling terms, this chemistry has been applied predominantly in heterobifunctional formats, one end of the reagent being photoactivatable and the other a functional group which allows incorporation into the protein. The chemistry of these compounds will be dealt with more fully in Section 5.10, but there are a few homobifunctional reagents which incorporate them.

Mikkelsen and Wallach (1976) described three compounds of this type, including 4,4′-dithio-bis(phenylazide) (DTPA; Figure 5.116) for coupling within cell membranes; the lipophilic nature of the compound promotes localisation within the apolar environment of the membrane (Dentler *et al.*, 1980) while the disulfide linkage confers cleavability. This reagent has proved useful enough to be commercialised, as have the analogous 4,4′-diazidodiphenyl ether (DADE), and 4,4-diazodiphenylethane (BioAffinity Systems).

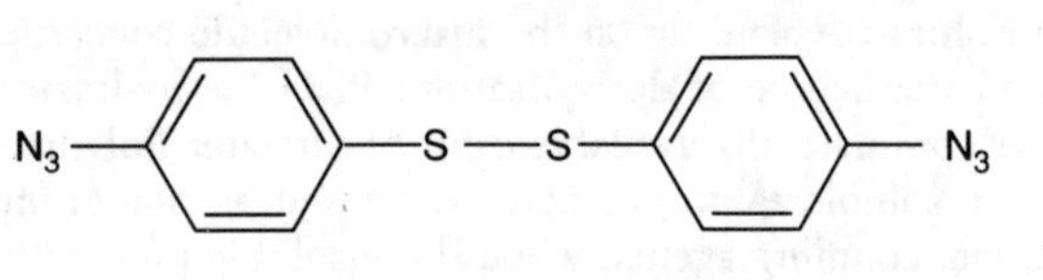

Figure 5.116 4, 4′-Dithio-bis(phenylazide) (DTPA).

Huang and Richards (1976) described the use of bis[*N*-(2-nitro-4-azidophenyl)]cystamine-*S,S*-dioxide (DNCO; Figure 5.117) as a photoactivatable thiol reagent, but in the absence of free thiols it could act as a photoactivatable homobifunctional.

Bis-[β-(4-azidosalicylamido)ethyl] disulfide (BASED; Figure 5.118) is another cleavable bis-phenyl azide which is also readily radioiodinatable. This has been developed and commercialised by Pierce Chemical Co.

Heterobifunctionality can be introduced if two different photoactivatable groups are incorporated—see for example Figure 6.138.

5.5.7 Homobifunctional biotin reagents

Reagents based on the extremely useful biotin–avidin interaction will be dealt with in some detail in Section 5.11, as they are virtually always heterobifunctional in nature. Homobifunctional biotin reagents, i.e. bis-biotin compounds, are of rather limited interest but are included here for completeness.

Green *et al.* (1971) synthesised a series of bis-biotin reagents incorporating linkers varying in length from 11 to 31 Å, and used then to probe the binding sites in avidin. It was deduced that a linker of at least 15 Å was required to allow intermolecular coupling, and that a 28 Å linker allowed intramolecular bridging to take place.

A polyethylene glycol (PEG)-bis-biotin reagent, commercially available from ProChem Inc., seems likely to have some interesting applications.

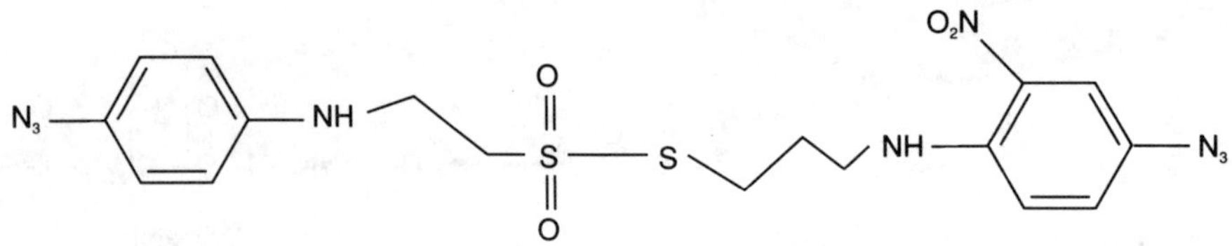

Figure 5.117 Bis[*N*-(2-nitro-4-azidophenyl)]cystamine-*S, S*-dioxide (DNCO).

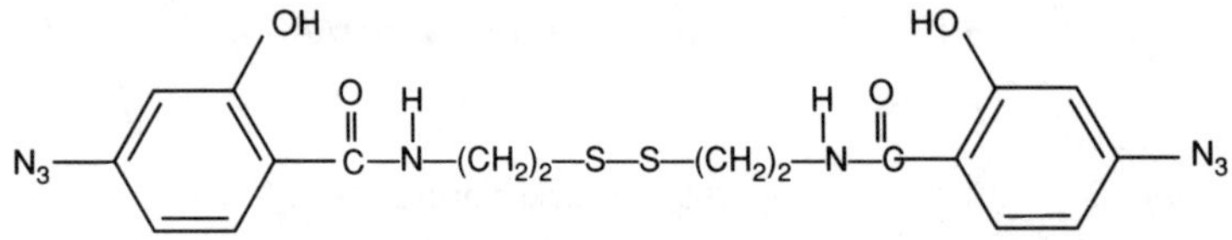

Figure 5.118 Bis[β-(4-azidosalicylamido)ethyl] disulfide (BASED).

5.6 Introduction to Heterobifunctional Coupling Approaches

The direct coupling approaches discussed in Section 5.2 are severely limited in terms of their applicability to protein conjugation, partly by the lack of suitably reactive groups, and partly by the difficulty of controlling the stoichiometry of the reaction. Apart from periodate oxidation and the occasional 'assisted' disulfide approach they are very rarely ecountered in preparative uses.

The homobifunctional techniques of Sections 5.3–5.5 target the most abundant reactive groups (predominantly amines) so that availability is not usually a constraint. Virtually all mixtures of protein molecules will be coupled to some extent by a bis-succinimide ester. However it is still not easy to control the extent of reaction, because both proteins to be coupled must by definition contain groups of similar reactivity. It is generally necessary to choose conditions which strike a balance between incomplete reaction of starting materials, self-coupling of starting materials, and over-reaction to give aggregated product.

Many of the reagents discussed in the last section are not strictly homobifunctional. Conditions are employed which introduce some degree of differential reactivity to the functional groups, pH variation to activate successively the two double bonds of the *p*-benzoquinone molecule, for example, and this is exploited to allow a greater degree of control on the reaction stoichiometry. True heterobifunctional approaches extend the same concept further, allowing a different reactive functionality to be introduced in controlled amounts into each protein. With some optimisation it is possible to obtain conjugates of a consistent stoichiometry within a chosen range.

Where coupling reagents are used in studies of protein topography it is often sufficient to show some reaction between neighbouring molecules, the nature of the link being less important. Once again the amine-directed homobifunctionals are dominant due to the abundance of these groups Non-photoactivatable heterobifunctional reagents offer opportunities for determining more specific information about the spatial relationships of different groups, or may happen to give the best coupling in a particular situation. On the whole, however, they are most often employed in preparative conjugations.

The photoactivatable heterobifunctional reagents, on the other hand, are ideally suited to receptor binding studies. A photoaffinity probe is prepared from the ligand of interest, using an amine or thiol-reactive group to introduce the photoactive group. The labelled ligand is then allowed to bind to its receptor, then (all stages so far having been in darkness) the sample is irradiated, the photoactive group becomes intensely reactive and fixes the ligand at the site of interest (Ji, 1976). The great advantage of these reagents, of course, is that they can be made reactive just at the site of interest, a feature unobtainable from other simple reagents. This represents one of the most important applications of heterobifunctional protein-coupling agents.

5.6.1 Types of coupling reaction using heterobifunctional reagents

Heterobifunctional reagents are occasionally used in single-step coupling reactions, mainly in non-photoactivated cross-linking studies; Hingorani *et al.* (1988), for example, used the thiol–amine heterobifunctional maleimidobenzoyl-*N*-hydroxysuccinimide ester (MBS) to demonstrate cross-linking of bovine eye proteins. However it is much more common that two or three stages are involved, and indeed these are necessary to exploit the control of conjugation that these reagents offer. A typical example is shown in Figure 5.119, one or both of the proteins to be conjugated have new groups introduced, which then take part in the final coupling reaction.

The chemistry involved in the use of these reagents, both in their coupling to proteins and to each other, will virtually all have been encountered earlier in the book. It is only the way in which these chemistries are employed that is different.

5.6.2 Classification of heterobifunctional reagents for protein–protein coupling

It is simplest to categorise these compounds in terms of the reactivity of their two functional groups. Most of the compounds encountered in Sections 5.10–5.12 actually incorporate amine or thiol-reactive groups, but are more conveniently classified there than in Sections 5.8 or 5.9.

5.7 Heterobifunctional Coupling of Amines to Thiols

As might be expected, the functional groups incorporated in these reagents are largely culled from those already encountered in the corresponding

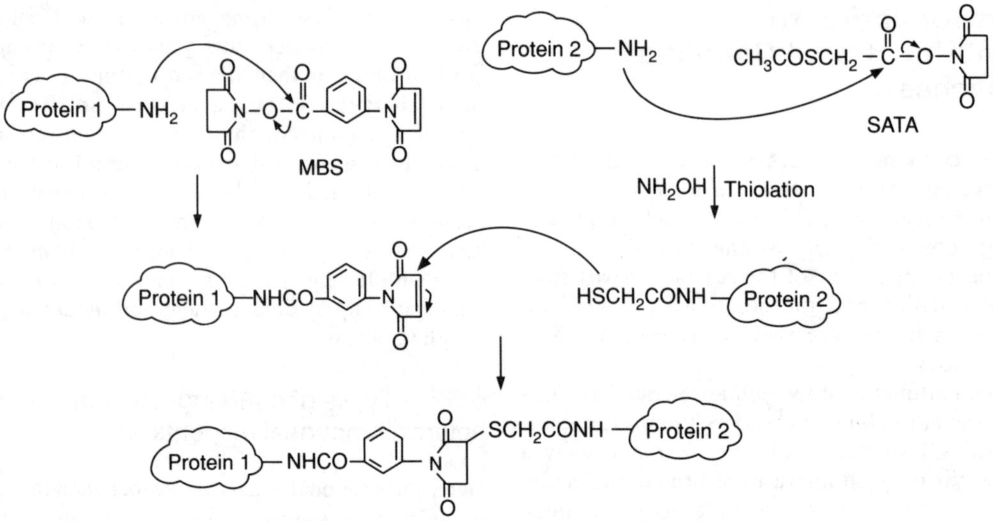

Figure 5.119 A typical application of heterobifunctional coupling. A maleimide group is introduced into Protein 1 using the heterobifunctional reagent MBS (see Section 5.7.1.1). A protected thiol group is introduced into Protein 2 using the heterobifunctional reagent SATA (see Section 5.12.1.8), and the thiol group is de-protected with hydroxylamine. The two newly introduced groups react to form a conjugate, linked *via* a thioether moiety.

homobifunctionals: active esters and imidates at the 'amine end', and maleimides, alkyl halides and disulfides at the 'thiol end'. Disulfides are much more commonly encountered in heterobifunctional reagents, however.

The usual issues of specificity must be borne in mind when using these reagents: active esters may react with histidine (although the products are readily hydrolysed) and with some serine residues; maleimides can react with lysine or histidine, particularly at raised pH, and alkyl halides may derivatise lysine, tyrosine, histidine or methionine residues. Imidates and disulfides are more or less specific for their intended targets.

For convenience these reagents will be discussed in subclasses based on the thiol-reactive component.

5.7.1 Maleimide-based reagents for coupling of amines to thiols

Reagents in this category first appeared in the mid-1970s. Rich *et al.* (1975) used *in situ* activation of maleimido acids to couple them to peptide amines, and, almost in parallel, Keller and Rudinger (1975) described similar syntheses in which succinimidyl esters of maleimido acids were isolated. As suggested by the authors, compounds of this type were taken up by other groups as heterobifunctional coupling agents for protein modification, and an important new class of reagent was born.

While a variety of such compounds have been adopted for routine use, the reaction conditions used are, not surprisingly, very similar. The activation reaction of the succinimide ester with a protein amine is rapid at pH 7, while competing hydrolysis is slow; this pH is often also suitable for subsequent reaction of the maleimide with thiols. These conditions tend to dominate the applications listed below.

The major variable to be optimised for the activation is the heterobifunctional:protein ratio, and the level chosen will often be a compromise between maximising maleimide incorporation, and minimising any adverse effects on protein functionality through loss of amine groups. Generalisations can be made for the more commonly encountered proteins, but the sensitivity of some species to changes in this ratio is great. For individual monoclonal antibodies, for example, very careful optimisation may be needed in some cases to obtain acceptable performance, presumably where reactive amines are an essential part of the antigen-binding site. On the other hand, the absence of critical amine groups can allow large ratios to be employed with impunity. Bieniarz *et al.* (1992b) used a 100:1 ratio of SMCC to β-lactamase, for example, and higher ratios have been used in specific applications in the authors' laboratories.

The reagents listed below represent some of the most commonly-encountered examples of this class. They are categorised by water-solubility[1], and by the nature of the group to which the maleimide function is attached, aromatic or aliphatic. The latter is an important factor in determining stability.

5.7.1.1 Water-insoluble succinimide esters of aromatic maleimides

These represent the earliest of these reagents to gain widespread attention. Kitagawa and Aikawa (1976) introduced one of the most enduring examples, *m*-maleimidobenzoyl succinimide ester (MBS; Figure 5.120), which they employed to couple insulin to β-D-galactosidase. The insulin (in 50 mM phosphate buffer, pH 7.0) was activated by reaction with a 20% excess of MBS (in tetrahydrofuran) for 30 minutes at room temperature; the reaction of the maleimide with thiols from β-Gal (two hours at room temperature) took place in the same phosphate buffer.

The utility of a heterobifunctional reagent of this type was rapidly recognised, and it was soon in widespread use, particularly in immunoassay development. O'Sullivan *et al.* (1978) described the preparation of IgG–β-Gal conjugates using this reagent. MBS solution (50 μl, 14.6 mg/ml in dioxane) was added to 7.0 ml of donkey IgG solution (0.5 mg/ml in 0.1 M phosphate buffer, 0.09 M NaCl, pH 7.0), and the mixture allowed to react for one hour at 30 °C. This represents an MBS:IgG ratio of 100:1, and resulted in the incorporation of three maleimide groups per molecule.

The activated IgG was purified on Sephadex™ G-25 using 0.1 M phosphate buffer (pH 7.0) as eluent, then allowed to incubate with an equal weight of β-Gal for one hour at 30 °C (both proteins at 0.25 mg/ml in 10 mM phosphate buffer (pH 7.0) containing 10 mM $MgCl_2$). The reaction was stopped by the addition of 2-mercaptoethanol (to 10 mM), and the resulting conjugate purified by ion exchange on DEAE-agarose.

The resulting conjugate showed no loss of enzyme activity, and 55% of the β-Gal was incorporated in the conjugate. Enzyme and antibody activity (85%) was retained after storage for four months at 4 °C. In immunoassay applications, β-Gal conjugates prepared by this approach gave better performance than those obtained using *N,N′-o*-phenylenedimaleimide (O'Sullivan *et al.*, 1979a), and were comparable to HRP conjugates prepared by the periodate route (O'Sullivan *et al.*, 1979b)

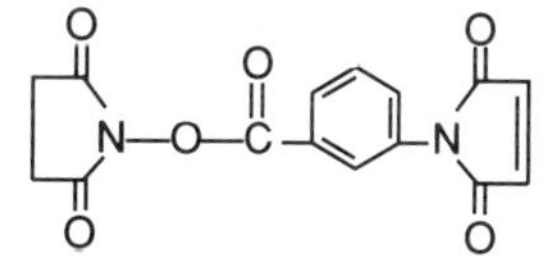

Figure 5.120 *m*-Maleimidobenzoyl succinimide ester (MBS).

Kitagawa *et al.* (1981) carried out a comparitive study of a number of heterobifunctionals of this type, including MBS. They showed that MBS had the most stable maleimide group and the most reactive succinimide function of the three (*o*, *m* and *p*) isomers. Incorporation of an electron-donating methoxy group did not significantly improve either of these attributes.

Yoshitake *et al.* (1982a,b) activated HRP (in 100 mM phosphate buffer, pH 7.0) with a 50 to 100-fold excess of MBS (dissolved in DMF), allowing the reaction to proceed for 30–60 minutes at room temperature. Around 0.6–0.7 maleimide molecules per enzyme molecule were incorporated under these conditions. Although activation at pH 7.0 was optimal, the use of prolonged times (over one hour) resulted in a decrease in performance; so did raising the temperature from 25 to 30 °C. These observations were attributed to the poor stability of the aromatic maleimide group under neutral to basic conditions, which had already been noted by Kitagawa *et al.* (1981). Chromatography was carried out at pH 5 for the same reason. The superior stability of reagents containing aliphatic maleimides such as SMCC (q.v.) led the Miyazaki group to neglect MBS in most subsequent applications. Detailed conditions for HRP activation by this route are given by Ishikawa *et al.* (1983) but the superiority of SMCC is again stressed. Gadow *et al.* (1984) came to similar conclusions in a comparative study on the preparation of pyruvate kinase–IgG conjugates: SMCC was again superior to MBS. Hashida *et al.* (1984) also recorded that it showed no advantage over aliphatic analogues.

Despite these limitations, and the necessity of employing an organic co-solvent, MBS has retained a considerable degree of popularity; Means and Feeney (1990) still included it in a list of 'widely used and most important' cross-linking reagents in a review.

It has continued to be employed not only in structural applications where yield and completeness of reaction are of secondary importance (Hingorani *et al.*, 1988), but also in preparative methods: Myers *et al.*

[1] Because these reagents are probably used more commonly in preparative applications than in structural studies, this definition is considered more useful than membrane permeability (cf. bis-succinimides, Section 5.3.5.3). For these reagents however, water solubility and membrane impermeability correlate fairly well.

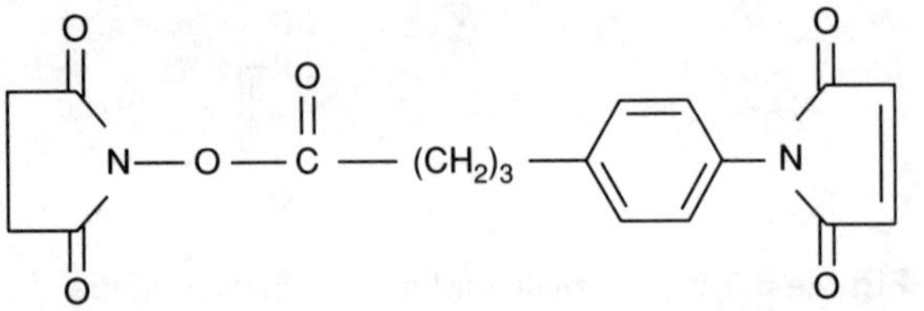

Figure 5.121 Succinimidyl 4-(*p*-maleimidophenyl)-butyrate (SMPB).

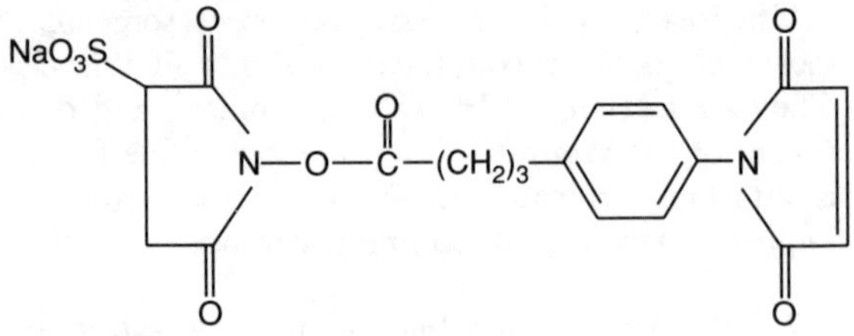

Figure 5.123 Sulfo-SMPB, sodium salt.

(1989) prepared immunotoxins using a variety of reagents of this type, and observed better yields from MBS than from the aliphatic GMBS (q.v.). Water-soluble aromatic reagents such as sulfo-MBS and sulfo-SMPB (q.v.) gave better yields still, however.

Another reagent containing a maleimide function attached to an aromatic group is succinimidyl 4-(*p*-maleimidophenyl)butyrate (SMPB; Figure 5.121). This differs from MBS by a three-carbon spacer group. In common with other aromatic reagents of this type, neither Hashida *et al.* (1984) nor Gadow *et al.* (1984) observed any advantage in using this compound in the preparation of enzyme–antibody conjugates.

Gitman *et al.* (1985) used a 5:1 excess of SMPB in phosphate-buffered saline (one hour at 20 °C) to introduce maleimide groups into insulin, bovine serum albumin and IgG; the activated proteins were subsequently coupled to Sendai virus glycoproteins which had been reduced to provide free thiols.

Teale and Kierney (1986) used SMPB to activate alkaline phosphatase for reaction with goat Fab′, while Bagshawe (1991) used this reagent to derivatise carboxypeptidase G2.

5.7.1.2 Water-soluble succinimide esters of aromatic maleimides

The sulfonated analogue of MBS is sulfo-MBS (Figure 5.122), commercialised by the Pierce Chemical Co. and included by Hashida *et al.* (1984) in a comparitive study on introducing maleimide moieties into HRP. No advantages were noted over aliphatic analogues. The literature on the use of this reagent for protein conjugation is fairly sparse.

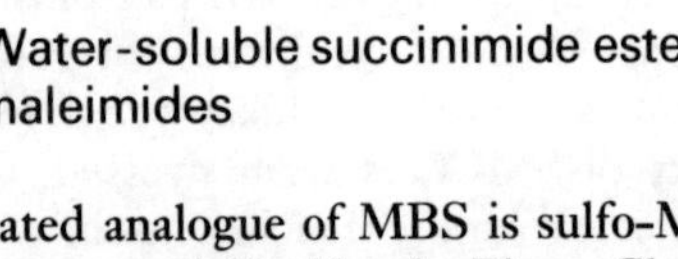

Figure 5.122 Sulfo-MBS, sodium salt.

Myers *et al.* (1989) observed an approximate doubling in yield of immunotoxin when sulfo-MBS replaced MBS as the ricin-activating reagent. Conditions were identical except that the reagent could be dissolved in water rather than DMF.

Another sulfonated reagent commercialised by Pierce is the water-soluble analogue of SMPB, sulfo-SMPB (Figure 5.123). Again it was included by Hashida *et al.* in their 1984 study without any advantageous behaviour being noted. Bangs *et al.* (1986) used sulfo-SMPB as a membrane-impermeant coupling agent for trypanosome variant surface glycoproteins from *Trypanosoma brucei*.

Myers *et al.* (1989) obtained the highest yield of immunotoxin when sulfo-SMPB was used in a comparative study with MBS, sulfo-MBS and GMBS.

5.7.1.3 Water-insoluble succinimide esters of aliphatic maleimides

Following the observation that *N*-ethylmaleimide was more stable in neutral solution than compounds in which the maleimide was attached to an aromatic group, Ishikawa and his colleagues at Miyazaki synthesised succinimide-containing reagents of this type: succinimidyl 4-(*N*-maleimidomethyl) cyclohexane-1-carboxylate (SMCC; Figure 5.124(a)) and succinimidyl 4-(*N*-maleimidomethyl) benzoate (Figure 5.124(b)). Over two hours at 30 °C and pH 7, these showed hydrolysis rates of 4 and 15% respectively (Ishikawa *et al.*, 1978). At 4 °C, SMCC was reported to be stable for 64 hours (Yoshitake *et al.*, 1979b; Ishikawa *et al.*, 1981).

SMCC was used by Yoshitake *et al.* (1979b) to produce conjugates of glucose oxidase and fragments of rabbit IgG containing free thiols. The enzyme activation used a 680:1 excess of SMCC in dioxan, and was carried out at 30 °C at pH 7.0. A variety of times and enzyme:antibody ratios were used, and the results were generally quite good. Enzyme and antibody activity was fairly well retained, incorporations were reasonable, and there was little self-coupling. A good deal

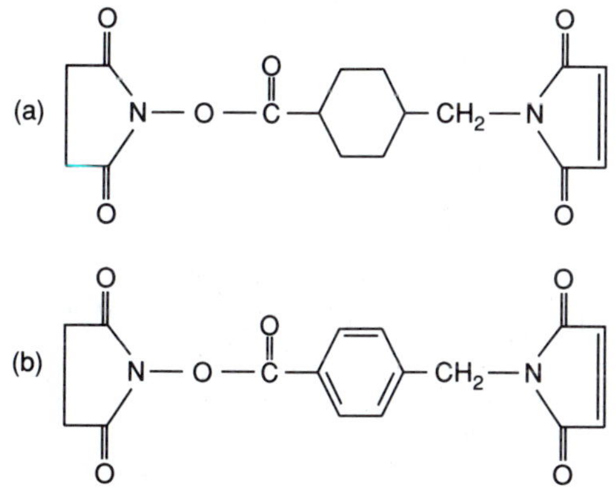

Figure 5.124 (a) Succinimidyl 4-(*N*-maleimidomethyl)cyclohexane-1-carboxylate (SMCC); (b) succinimidyl 4-(*N*-maleimidomethyl)-benzoate.

of the subsequent development of this approach, again, principally in Ishikawa's group, was aimed at horseradish peroxidase–Fab′ conjugates for immunoassay. Two papers by Yoshitake *et al.* (1982a,b) describe the first uses of SMCC in this application, in both the HRP is activated by incorporation of maleimide groups, then allowed to react with the thiol group of the rabbit Fab′.

The protocol in the 1982a paper is as follows. (1) HRP solution (1.0 ml of 6 mg/ml in 0.1 M sodium phosphate buffer, pH 7.0) was stirred with 40 ml of SMCC solution (60 mg/ml in DMF) for 60 minutes at room temperature. This represents a 50-fold excess of the heterobifunctional reagent. (2) After centrifugation to remove precipitates, the product was chromatographed on a 1 × 45 cm column of Sephadex™ G-25, using the same phosphate buffer as eluent. The product fractions were concentrated using a Collodion bag. Between 1.0 and 1.2 maleimide molecules per enzyme molecule were incorporated. (3) The activated HRP solution (100 μl of 15 mg/ml in the same buffer) and 150 μl of rabbit Fab′ (11.3 mg/ml in 0.1 M sodium phosphate buffer, pH 6.0, 5 mM EDTA) were incubated at 4 °C for 20 hours. The HRP:Fab′ ratio was 1:1 with both proteins being present at around 15 μmol/l. (4) The resulting conjugate was purified on a 1.5 × 45 cm column of Sephacryl™ S-200 using 0.1 M sodium phosphate buffer (pH 6.5) as eluent.

The conjugate obtained was largely homogeneous as indicated by the chromatography trace, and contained virtually no aggregated material. The Fab′:HRP ratio was around 0.94:1, the recovery of HRP was 66 to 87%, the recovery of Fab′ 75 to 82%, and the enzyme activity was unaltered by the process. Antibody activity was also well retained as indicated by immunoassay performance.

These reports represented a major step forward in the preparation of HRP conjugates, offering a considerably more homogeneous and less aggregated product, with better incorporations and retention of enzyme activity than the major existing methods (glutaraldehyde and periodate). The incorporations were also better than those seen using the earlier, more readily hydrolysed MBS. The conjugate displayed approximately 70% recovery of both HRP and Fab′, complete retention of HRP activity, and predominantly a 1:1 ratio of HRP to Fab′. No loss of Fab′ was observed after the conjugate was stored for six months at 4 °C in a 0.1 M sodium phosphate buffer (pH 6.5) including 0.1% BSA and 0.002% thiomersal.

Activation at pH 6.5 and 7.0 was compared: the higher pH gave a faster reaction with saturation (1.0–1.2 maleimide molecules per HRP) being achieved within 60 minutes; at the lower pH the incorporation was still rising after two hours.

In an assessment of a number of different *N*-succinimidyl maleimides, Hashida *et al.* (1984) reported reasonably good performance for SMCC, although maleimide incorporations were lower than with straight-chain aliphatic analogues. Subsequent reports from the Miyazaki group thus tend to use MHS (q.v.) in the preparation of HRP–Fab′ conjugates.

Ravinder (1991) described somewhat modified conditions in the SMCC activation of HRP for coupling to antibody: a 15:1 excess of SMCC was incubated with HRP at 13 mg/ml in 0.1 M potassium phosphate buffer, pH 7.4 + 0.1 M NaCl for 60 minutes at 21–25 °C, followed by purification using Sephadex G-50. Subsequent reaction of a fourfold excess of the activated HRP with thiolated IgG (two hours at pH 7.4) yielded a conjugate containing three enzyme molecules per IgG. Sequential quenching of excess maleimide and thiol groups was carried out using cysteine and iodoacetamide respectively.

Ishikawa *et al.* (1978) observed that calf intestine AP suffered fairly serious inactivation on derivatisation with SMCC, or on thiolation with SAMSA, 20 to 40% of activity was lost. This was considerably worse than the behaviour observed with other enzymes such as β-Gal.

Mahan *et al.* (1987) derivatised AP with SMCC as follows: nine 12.5 μl aliquots of SMCC solution (8.9 mM, 2.98 mg/ml in dioxane) were added at five minute intervals to 2.5 ml of AP solution (around 8 mg/ml in 0.1 M sodium phosphate buffer, pH 7.0).

After a total reaction time of 100 minutes the activated enzyme was purified on Sephadex G-25 using 0.1 M sodium phosphate buffer (pH 7.0), 1 mM EDTA as eluent. Subsequent reaction with human Fab′ (20 hours at 4 °C in the same buffer) was followed by quenching of maleimides with 0.5 mM 2-mercapto-ethylamine. The retained activity of the enzyme was not assessed quantitatively.

As β-galactosidase contains free thiol groups, it is not usual to encounter maleimide derivatisation of this enzyme; more often it is the conjugation partner that is derivatised in this way to facilitate reaction with β-Gal.

Ishikawa *et al.* (1983) described conditions for the SMCC activation of polyclonal IgG for subsequent reaction with β-Gal. They employed a 15-fold excess of the heterobifunctional agent (0.9 mg/ml in DMF), while the IgG was present at 2.8 mg/ml in 0.1 M sodium phosphate buffer (pH 7.0). The mixture was incubated for 30 minutes at 30 °C before purification on Sephadex G-25 with a 0.1 M phosphate buffer (pH 6.5), and resulted in the incorporation of around 3.2 maleimide molecules per IgG molecule.

Donkey IgG was activated by SMCC at excesses of between 1.2 and 10:1 (Gadow *et al.*, 1984). The activation was carried out in 20 mM potassium phosphate buffer (pH 7), 0.15 M NaCl for 45 minutes at room temperature. The activated antibody was then coupled to pyruvate kinase. The degree of derivatisation of the antibody, and presumably of the resulting enzyme–antibody conjugate, was reflected in changes in the enzyme kinetics.

Freytag *et al.* (1984) derivatised rabbit $F(ab')_2$ with a 20-fold excess of SMCC in phosphate-buffered saline (pH 7.0). The SMCC was dissolved at 60 mM (20.1 mg/ml) in DMF, and the reaction was allowed to continue for one hour at 20–23 °C. The subsequent conjugation to β-Gal was carried out in 50 mM Tris (pH 7.5, 16–20 hours at 4 °C).

Lambert *et al.* (1985) reported that, with a mouse monoclonal IgG_{2a}, the maleimide incorporation could be modulated by the initial SMCC concentration. The relationship was roughly linear. Under their conditions (IgG at 1 mg/ml in 100 mM phosphate buffer (pH 7.0), 0.5 mM EDTA, SMCC at 10 mM (3.3 mg/ml) in dioxane; 30 minutes at 30 °C), one maleimide could be incorporated per 40 μM SMCC concentration. This corresponds to one maleimide per sixfold excess of SMCC.

A generic procedure for thiol–maleimide coupling incorporating SMCC is described in Protocol 5.1.

Moving to less commonly-encountered members of this class, the simplest aliphatic succinimidyl maleimide which has seen widespread use is succinimidyl 4-maleimidobutyrate or N-γ-maleimidobutyryloxysuccinimide (GMBS; Figure 5.125).

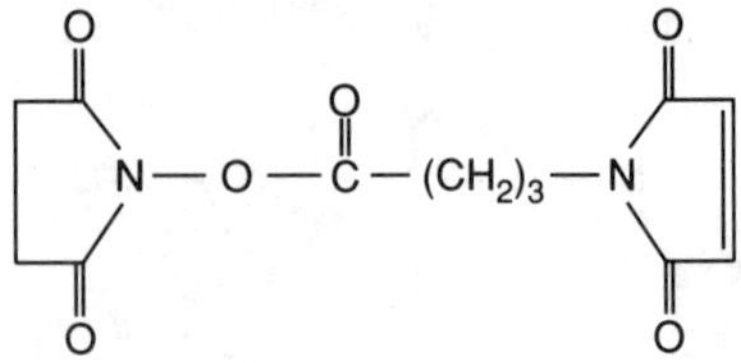

Figure 5.125 Succinimidyl 4-maleimidobutyrate or N-γ-maleimidobutyryloxysuccinimide (GMBS).

Hashida *et al.* (1984) rated this compound as one of the most useful succinimidyl maleimides: it could be obtained in high purity, showed reasonable resistance to hydrolysis of the maleimide (79% remaining after six hours at 30 °C and pH 7.0), and gave the highest incorporation into HRP and one of the best HRP recoveries in a Fab′ conjugate.

Finally, one of the succinimidyl maleimides described in the original paper by Keller and Rudinger (1975) was 6-maleimidohexanoyl-N-succinimide (MHS; Figure 5.126), two carbon atoms longer than its homologue GMBS. This compound is commercially available; it sometimes appears under the name N-(ϵ-maleimidocaproyloxy)succinimide (EMCS).

Kitagawa *et al.* (1981) investigated the reactivity and stability of this reagent in comparison to a number of aliphatic and aromatic reagents of this type. In common with other aliphatic compounds its reactivity towards amines in model amino acids was lower than the best aromatics such as MBS; its stability above pH 7 was considerably superior, however (MHS: 8% loss of maleimides after 30 minutes at 30 °C and pH 8; MBS: 44% loss under the same conditions).

Hashida *et al.* (1984) rated this compound with GMBS as one of the most useful of this class, with similar properties to those discussed for this compound. Among the Miyazaki group and other Japanese workers this appears to have remained the reagent of choice for many applications such as activation of (strept)avidin (Hashida *et al.*, 1991; Tsuruta *et al.*,

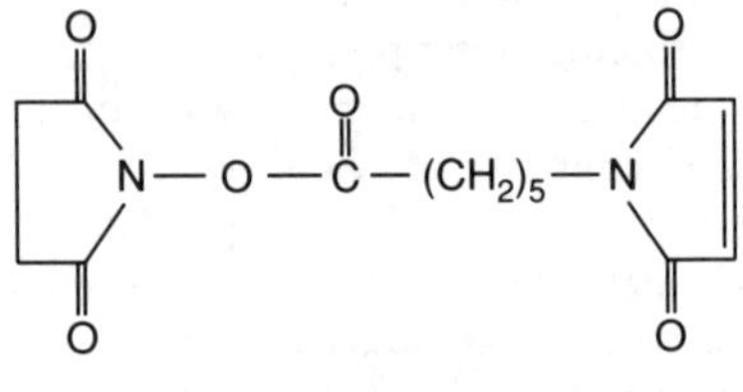

Figure 5.126 6-Maleimidohexanoyl-N-succinimide (MHS).

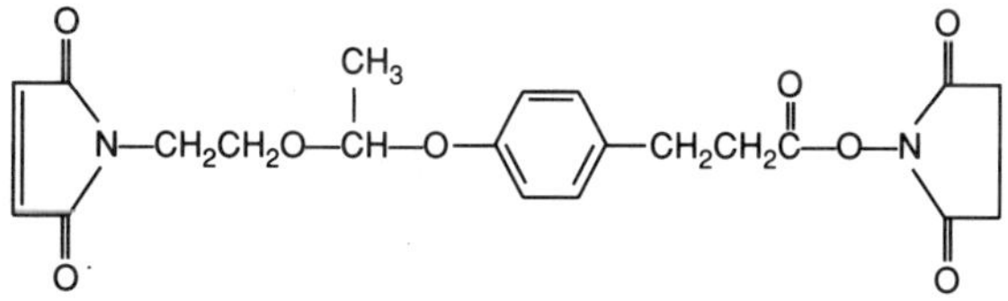

Figure 5.127 Example of cleavable succinimidyl maleimide incorporating an acetal function.

1994), or of HRP or albumin for coupling to rabbit Fab′ (Kohno *et al.*, 1989) or recombinant HIV proteins (Hashida *et al.*, 1993; Hashinaka *et al.*, 1994; Hashida *et al.*, 1995).

Ghosh *et al.* (1990) used a 50-fold excess of MHS to derivatise calf intestine alkaline phosphatase (30 minutes, 23 °C, pH 8.5). The MHS was dissolved in acetonitrile, the final concentration of cosolvent in the reaction mixture being 10%. A molar incorporation of 6.2 maleimide groups was obtained.

With avidin (Hashida *et al.*, 1991) 1.5 maleimide groups per molecule were incorporated by treatment with a fivefold excess of MHS for 30 minutes at 30 °C. The MHS was prepared at 5 mM in DMF and the reaction buffer was 100 mM sodium phosphate buffer (pH 7.0).

MHS probably deserves more attention than it has received so far in these applications.

In a rare example of a cleavable succinimidyl maleimide, Srinivasachar and Neville (1989) described a succinimidyl maleimide incorporating an acetal function, which could be cleaved slowly under acidic conditions (cf. the analogous bis-maleimides, Section 5.4.1). The acetal moiety of this compound (Figure 5.127) has a half-life of around six days at pH 5.5.

5.7.1.4 Water-soluble succinimide esters of aliphatic maleimides

The water-soluble analogue of SMCC (Figure 5.128) was developed by the Pierce Chemical Co. and its use was first described by Hashida *et al.* (1984). In common with other sulfonated agents examined, its purity was rather lower than that obtained for SMCC, but it showed excellent stability to maleimide hydrolysis; there was no detectable loss after six hours at pH 7 and 30 °C. Performance in the preparation of an HRP–Fab′ conjugate was comparable to that obtained with the un-sulfonated analogue.

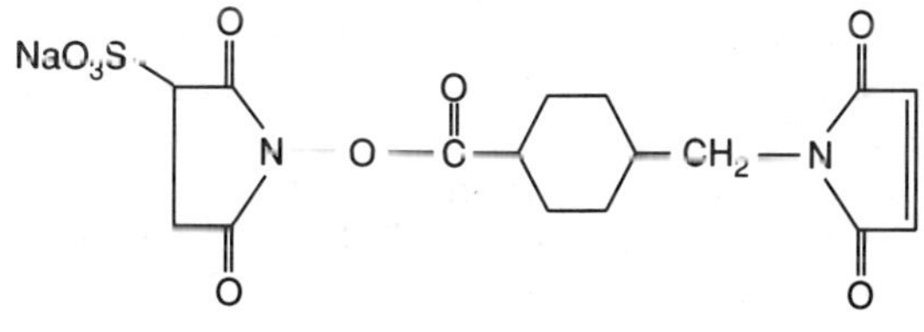

Figure 5.128 Sulfo-SMCC, sodium salt.

The conditions of use of sulfo-SMCC predictably reflect those of other sulfosuccinimidyl maleimides, with no need for an organic cosolvent: Meyer *et al.* (1992), for example, derivatised β-lactamase with a twofold excess of this reagent (one hour, room temperature) in a borate–chloride buffer (pH 8.2), before gel filtration of the product and subsequent reaction with Fab′ fragments.

The sulfonated analogues of GMBS and MHS are also available, although references to their use are relatively scarce. Pierce also market a number of commonly used proteins which have already been activated with sulfo-SMCC.

5.7.1.5 Long-chain succinimide esters of maleimides

Christopher Bieniarz and his colleagues at Abbott Laboratories have developed a series of reagents in which maleimide is coupled to succinimide *via* a methylcyclohexyl group and one, two or three 6-aminocaproyl residues (Figure 5.129(a)). This results in heterobifunctional agents giving a 16, 23 or 30-atom linker group respectively (Bieniarz *et al.*, 1991, 1996), and the length and polar nature of the polyamide chains are claimed to be advantageous in avoiding steric problems in immobilised and coupled proteins. These reagents, in particular the 30-atom compound, succinimidyl 4-[(*N*-maleimidomethyl)tricaproamido]-

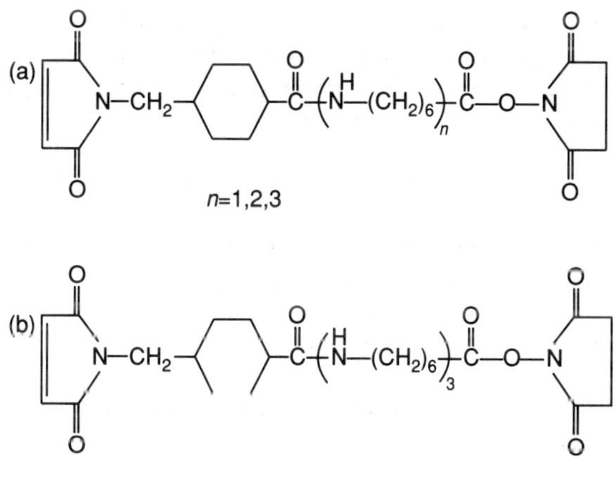

Figure 5.129 (a) Long-chain *N*-succinimidyl maleimides; (b) SMTCC.

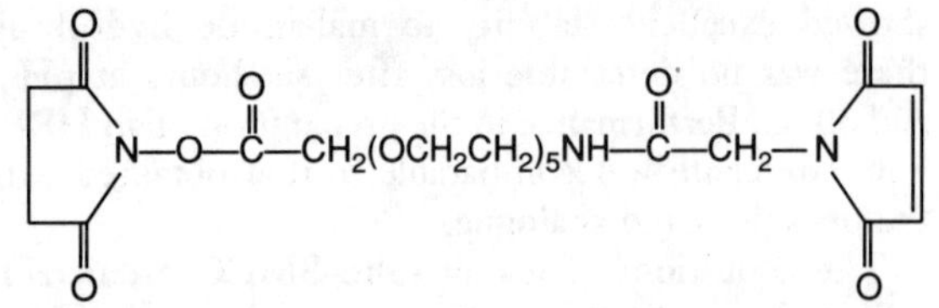

Figure 5.130 Succinimidyl maleimide compound containing hydrophilic ethylene oxide units.

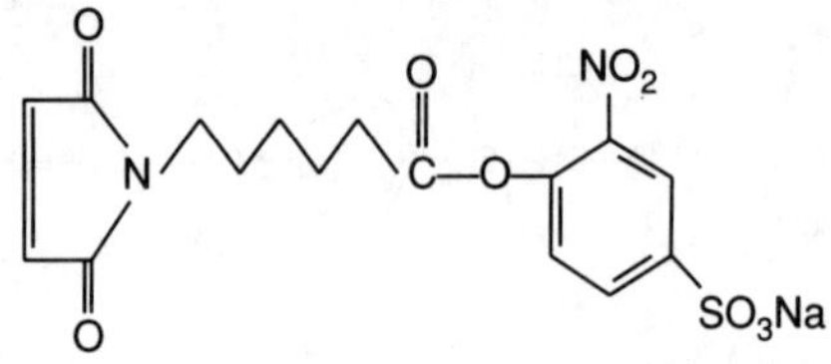

Figure 5.131 A water-soluble maleimidyl nitrophenyl ester, which releases a chromophore on reaction.

cyclohexane-1-carboxylate (SMTCC; Figure 5.129(b)), have been used by the Abbott group in the preparation of alkaline phosphatase–IgG conjugates (Bieniarz *et al.*, 1992a, 1996; Husain and Bieniarz, 1994). Typically AP was activated by a 30-fold excess of SMTCC (30 minutes, ambient temperature, pH 7.0).

Åkerblom *et al.* (1993) described a succinimidyl maleimide containing a hydrophilic spacer of repeated ethylene oxide units (Figure 5.130). To avoid maleimide hydrolysis problems, derivatisation of mAbs was recommended to be carried out for 15 minutes at pH 7.5, with a purification time of less than two hours. A wide range of time and pH conditions, using heterobifunctional excesses of between 7:1 and 20:1, gave maleimide incorporations of 2 to 9 for a number of different mAbs. Coupling to thiolated toxins was generally carried out at pH 5.4 or 6.0 for 17 to 23 hours.

Even larger, and highly hydrophilic, is a maleimide-PEG-succinimide commercially available from Pro Chem, Inc.

Smaller polyethylene oxide linkers have been included in a maleimide-acid reagent described by Frisch *et al.* (1996), the aim being to minimise immunogenic effects *in vivo*—this could also provide the basis for a succinimide ester reagent. Johnson *et al.* (1997) described the preparation of water-soluble maleimide–succinimide ester reagents containing propylene and butylene oxide groups.

5.7.1.6 Succinimide esters of bis-maleimides

In a recent patent, Markert-Hahn *et al.* (1994) described reagents containing two maleimide groups and one succinimide ester, allowing trimeric conjugates to be prepared for immunoassay use.

5.7.1.7 Other amine-reactive maleimide reagents

Although succinimides have dominated this field, other 'older' activating groups such as phenyl esters are still sometimes encountered. One such reagent (Bjorn *et al.*, 1986; Aldwin and Nitecki, 1987; Figure 5.131) features both water solubility, and the convenient release of a chromophore on reaction.

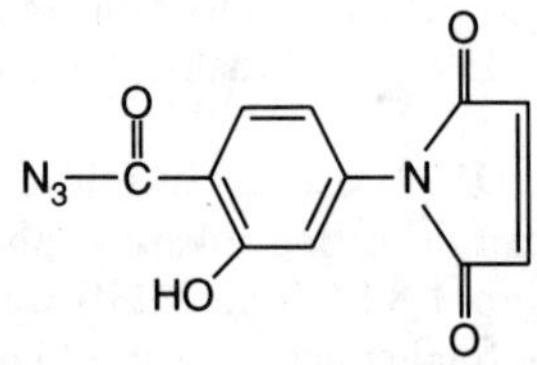

Figure 5.132 2-Hydroxy-4-(*N*-maleimido)-benzoylazide.

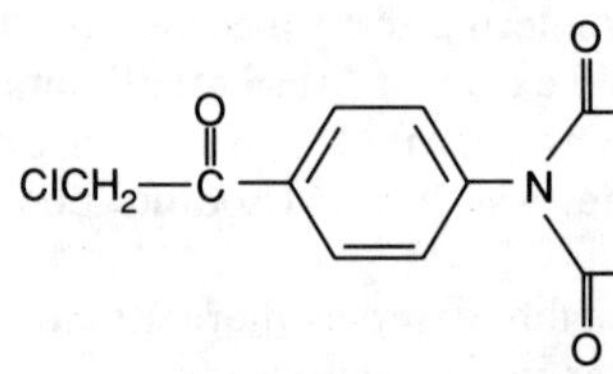

Figure 5.133 4-Chloroacetylphenylmaleimide.

A maleimide-acylazide compound (Figure 5.132) was described by Trommer *et al.* (1977); the azide moiety is a good leaving group (cf. Section 5.3.5) and this reagent reacts analogously to active esters. These authors also described an unusual use of a haloacetyl function with the compound shown in Figure 5.133 (Trommer and Hendrick, 1973), which was used to cross-link Cys165 and Lys179 of lactate dehydrogenase. Although both ends of the molecule would generally be regarded as thiol-reactive, the reactivity of the maleimide group is so much higher that the cysteine thiol reacts almost exclusively with this function. This leaves the chloroacetyl group to react with other targets such as amines.

Blättler *et al.* (1985) described a novel heterobifunctional reagent containing a maleimide linked through a chain containing both aliphatic and aromatic portions to the citraconic anhydride (2-methylmaleic

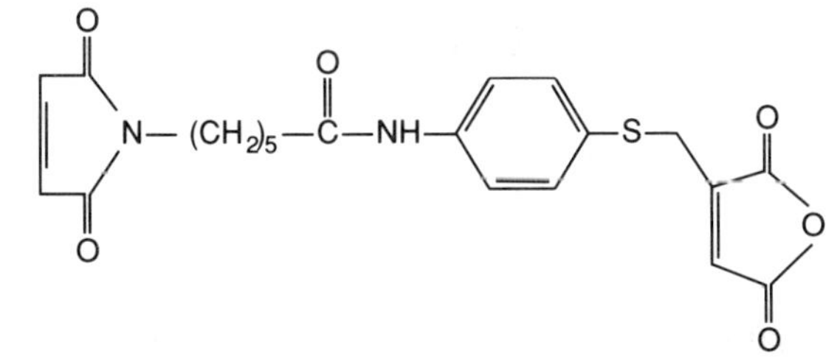

Figure 5.134 A heterobifunctional reagent which introduces a maleimide moiety *via* the citraconic anhydride group, which yields an acid-labile link.

anhydride) moiety (Figure 5.134). As expected this anhydride reacts with amine groups, but the resulting amide is unusually susceptible to acid hydrolysis[1] (Dixon and Perham, 1968); the bond can be cleaved by treatment at pH 3 to 4[2].

The cleavage of Blättler's reagent, used to couple a monoclonal antibody to the ribosome-inactivating protein gelonin, was incomplete (around 75%) even after 24 hours incubation under these conditions. However, unusually for a cleavable cross-linker, the amine-containing protein is regenerated in the native form with no extra atoms. Thus, although the ability of gelonin to shut down protein synthesis was drastically reduced in the Mab conjugate, this activity was completely restored on cleavage. This reagent may have some potential in cases where it is important to regenerate the protein in its fully native form. An analogous reagent incorporating an iodoacetyl function was described by McIntyre *et al.* (1994) and is covered in Section 5.7.2.

p-Maleimidophenyl isocyanate (Palumbo, 1993; Annunziato *et al.*, 1993) is undoubtedly capable of forming amine–thiol links between proteins. However it was introduced as a hydroxyl–thiol reagent, and is dealt with in Section 5.12.

[1] The lability of amide-linked maleoyl groups at low pH is considered to be due to intramolecular acid catalysis by the protonated carboxyl group. The degree of lability is influenced by substituents on the maleyl function, electron-donating substituents lower the acidity, thus raising the pK_a, increasing the proportion of protonated acid groups at a moderate pH, and promoting amidolysis. Thus amides formed from maleic anhydride are inconveniently difficult to cleave, while those from 2,3-dimethylmaleic anhydride are rather too labile. Citraconic anhydride offers a convenient level of cleavability.

[2] An interesting use of citraconic anhydride was discussed by Atassi (1977); acylation of the amine groups of lysozyme with this reagent induced sufficient perturbation of structure to allow access to proteolytic reagents without rupturing the protein's disulfide bonds, and the amine could be regenerated by mild acid treatment. Therefore tryptic fragments could be prepared which reflected the structure of the intact protein.

Other amine-reactive maleimides are noted by Wong (1991, Chapter 5), including examples bearing acyl chloride groups.

5.7.2 Haloalkyl-based reagents for coupling of amines to thiols

5.7.2.1 Active esters of haloacetyl-containing acids

Lorand *et al.* (1962) described the synthesis of *p*-nitrophenyl iodoacetate and its hydrolysis properties, which were unremarkable. A similar bromoacetate was described by Lawson and Schramm (1962); see Figure 6.98. Rector *et al.* (1978) prepared the succinimidyl ester of iodoacetic acid (Figure 5.135) and used it to derivatise IgG (10 mM sodium phosphate, 2% NaCl, 1 mM EDTA, pH 7.5; 15 minutes at room temperature). Using a molar excess between 2:1 and 40:1, these authors observed incorporations of 1.2 to 17.4 iodoacetate groups per molecule. The activated IgG was subsequently reacted with thiolated ovalbumin (overnight at room temperature in the buffer described for iodoacetylation) to yield a conjugate.

Broker *et al.* (1978) used the succinimidyl ester of bromoacetic acid to activate ferritin for reaction with thiolated avidin. The reaction was carried out (one hour, room temperature) between ferritin (20 mg/ml in 0.3 M potassium borate buffer, pH 9.3) and the succinimide ester (10 mg/ml in DMSO). Active ester:-ferritin ratios of around 100–120:1 resulted in incorporations of 10–20 bromoacetyl groups per molecule.

McIntyre *et al.* (1994) used the succinimide ester of iodoacetic acid to derivatise gelonin for reaction with thiolated interleukin-2, the aim being to target this toxin to T cells; in common with other succinimide-based reagents, however, this resulted in a serious loss of gelonin toxicity.

Weltman *et al.* (1983) described the preparation and use of *N*-succinimidyl (4-iodoacetyl)aminobenzoate (SIAB; Figure 5.136), a non-cleavable cross-linker which was commercialised by Pierce at around the same time as its literature description. The sulfonated analogue is also available.

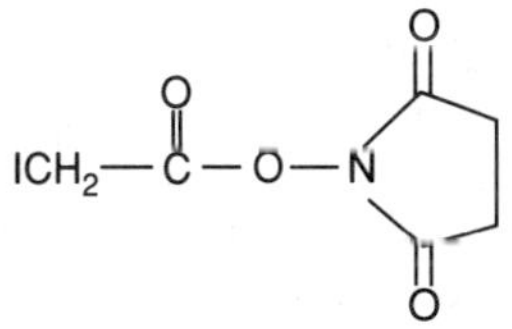

Figure 5.135 Succinimidyl ester of iodoacetic acid.

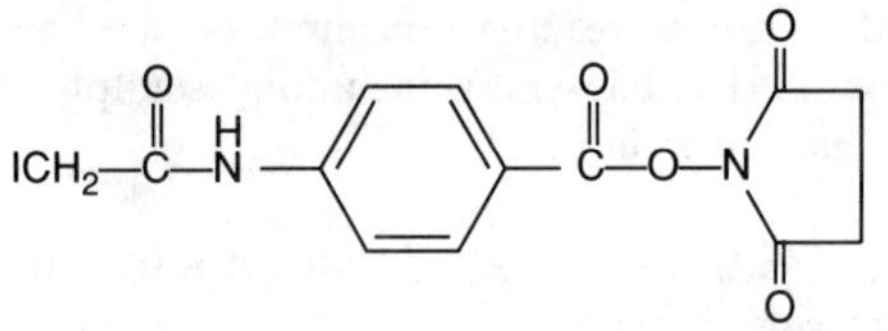

Figure 5.136 *N*-Succinimidyl (4-iodoacetyl)aminobenzoate (SIAB).

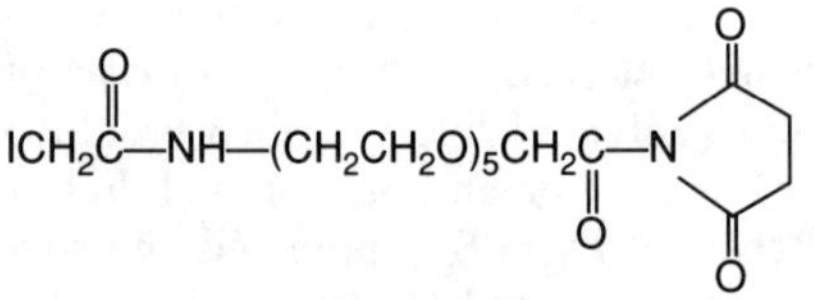

Figure 5.137 A succinimidyl iodoacetamide reagent incorporating a hydrophilic linker (Åkerblom *et al.*, 1993).

Worrell *et al.* (1986) described a novel synthesis of SIAB and used it to prepare immunotoxins. Compared to disulfide-linked conjugates, the thioether-linked SIAB reagent showed a lower rate of long-term removal *in vivo*.

Åkerblom *et al.* (1993) described the preparation and use of a hydrophilic reagent containing repeating ethylene oxide units (Figure 5.137). This was used to derivatise Mabs at pH 8.0, excesses of between 10:1 and 28:1 giving an incorporation of five to 17 iodoacetyl groups. Coupling to thiolated toxins was carried out at pH 7.5. While the activation of the mAb proceeded at a similar rate to *N*-succinimidyl iodoacetic acid, the coupling to thiolated toxin took place considerably faster, the longer span and more polar nature of this linker presumably facilitating access. Frisch *et al.* (1996) described a bromoacetyl–acid reagent containing three ethylene oxide units which could form the basis of a succinimide analogue.

5.7.2.2 Other amine-reactive haloalkyl reagents

There are many other infrequently-encountered reagents in this category, see Wong (1991) for further examples. They rarely offer any great advantage over better-known reagents. For example, Singh *et al.* (1980) prepared the succinimide ester of *N*-bromoacetylglycine and used it for the activation of glucose 6-phosphate dehydrogenase (three hours, 0 °C, pH 8.0). The modified enzyme, which in this case was coupled to gentamicin, lost 70% of its activity.

McIntyre *et al.* (1994) described the synthesis of an acid-labile compound of this type, designed for the

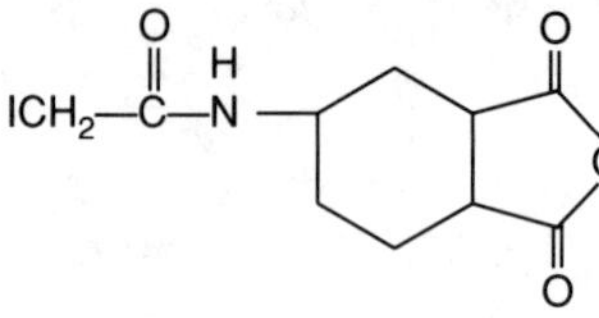

Figure 5.138 4-(Iodoacetamido)-1-cyclohexane-1,2-dicarboxylic anhydride.

preparation of targetted toxin reagents which would cleave to release the free toxin on entering an acidic intracellular environment. In this study the compound, 4-(iodoacetamido)-1-cyclohaxane-1,2-dicarboxylic anhydride (Figure 5.138), inactivated the toxin, presumably due to the alteration of charge caused by neutralising an amine group and introducing an extra carboxyl function. This compound is analogous to the maleimide-citraconic anhydride reagent (Blättler *et al.*, 1985) described above.

5.7.3 Disulfide-based reagents for coupling of amines to thiols

Disulfides offer the most specific reactivity for thiols, and it is perhaps surprising that so few coupling reagents have exploited this route. However, one heterobifunctional reagent of this type has been particularly successful, and is one of the most commonly-encountered conjugation reagents of the 1980s and 1990s.

5.7.3.1 SPDP

N-Succinimidyl 3-(2-pyridyldithio)propionate (SPDP; Figure 5.139) was introduced by Jan Carlsson and his colleagues at the University of Uppsala in Sweden in 1978 (Carlsson *et al.*, 1978), following on from earlier applications of the same chemistry in preparing thiol-reactive solid phases for affinity chromatography (Brocklehurst *et al.*, 1973; Carlsson and Svenson, 1974). The thiol-reactive functionality is provided by the 2-pyridyldisulfide group; the pH dependence of this reaction is not strong, and good results can be obtained over the range 4–7.5 (Carlsson *et al.*, 1978). These authors reported general conditions for protein derivatisation as follows:

SPDP (0.2–6 μmol) in 99.5% ethanol is added dropwise to 0.05–2 μmol of the protein in 2 ml of 0.1 M sodium phosphate buffer (pH 7.5), 0.1 M NaCl, and the mixture incubated for 30 minutes at 23 °C before gel filtration on Sephadex™ G-25.

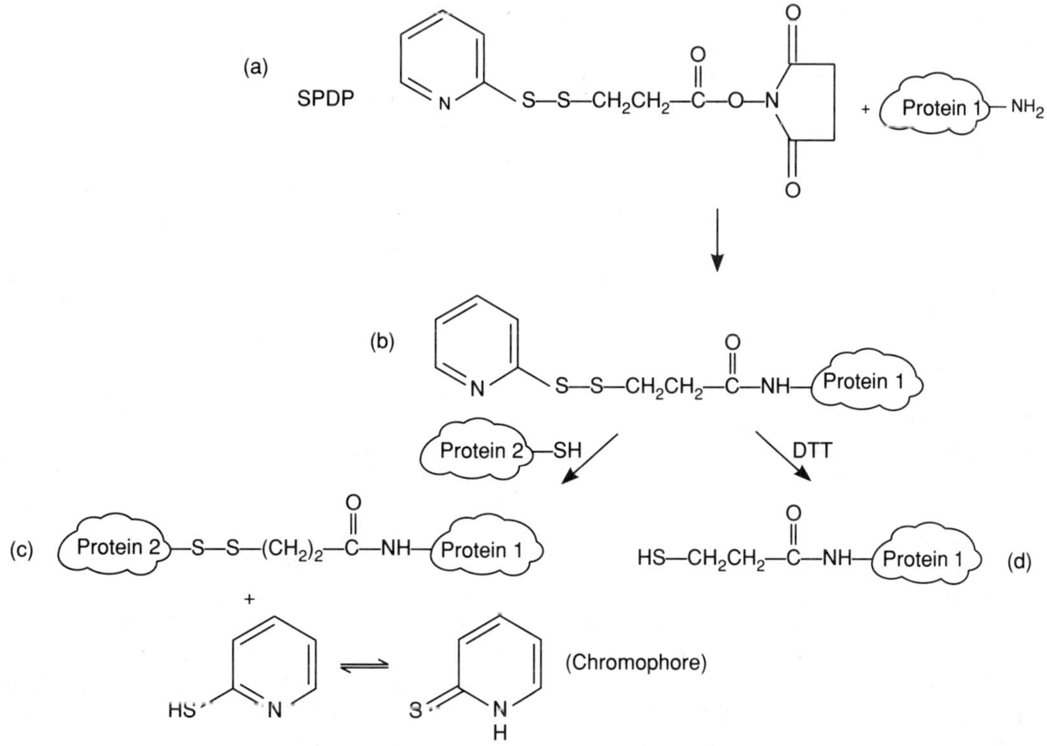

Figure 5.139 *N*-Succinimidyl-3-(2-pyridyldithio)propionate (SPDP): this reagent (a) can be used to introduce a pyridyl disulfide moiety into proteins (b). This can then react with a protein thiol to create a protein–protein link (c), or provide a thiol on reduction (d).

Reduction to thiol could be achieved by treatment with 25 mM dithiothreitol in sodium acetate buffer (pH 4.5) containing 0.1 M NaCl; these conditions allow reduction of the pyridyldisulfide without cleavage of native disulfide bonds.

Particular examples were described, covering the preparation of ribonuclease–mercaptalbumin, α-amylase–urease and HRP–IgG conjugates. In the peroxidase conjugate, the protocol above yielded a pyridyldisulfide incorporation of 2.3 per molecule for the HRP (using an SPDP:HRP ratio of 25:1), and 1.7 per molecule for the IgG (using an SPDP:IgG ratio of 2:1). The derivatised IgG was reduced with dithiothreitol for 30 minutes as outlined above and purified by size exclusion.

The two activated proteins were then coupled: peroxidase-2-pyridyldisulfide (2.6 mg, 64 nmol) in 1.2 ml of 0.1 M sodium phosphate buffer (pH 7.5), 0.1 M NaCl was incubated with the thiolated IgG (11 mg, 68 nmol) in 0.9 ml of the same buffer without sodium chloride, for 20 hours at 23 °C. The resulting conjugate appeared largely homogeneous from the size exclusion chromatogram. About 90% of the enzyme activity was retained after the SPDP activation stage, and about 40% after the disulfide exchange. Of this, slightly less than half was contained in the conjugate.

α-Amylase was less affected by both stages of the reaction (92% activity retained after activation, 83% after exchange), while ribonuclease was seriously damaged during activation (54% retained) but virtually unaffected by exchange (50% retained).

5.7.3.2 Conditions for SPDP activation of proteins

Many other reports of protein activation using SPDP followed the Carlsson paper; the use of a phosphate buffer with pH around 7.5 and an ethanol solution of the coupling agent appear to be consistent features, but approaches differ over the ratio of SPDP to protein employed.

Nilsson *et al.* (1981) carried out some optimisation on the procedure. SPDP:HRP ratios of 5, 10 and 20:1

were employed at a pH of 7.4, and it was noted that saturation had been achieved (about 1.6 pyridyl disulfide residues per HRP molecule) at the 10:1 excess. This is consistent with the low amine content of this enzyme.

Other authors reported similar procedures using SPDP:HRP ratios which varied from 12:1 (Pain and Surolia, 1981) to 50:1 (Imagawa *et al.*, 1982a; Ishikawa *et al.*, 1983; Gadkari *et al.*, 1985). Overall a 50:1 ratio appears to have gained a reasonable degree of concensus, and has offered satisfactory performance on many occasions.

Jeanson *et al.* (1988a,b) used SPDP:AP ratios of up to 60:1 in the preparation of Mab conjugates, they reported that immunoassay performance continued to improve at least up to this level.

Subsequent to Carlsson's original report using SPDP:rabbit IgG ratios of between 2 and 3.5:1, other authors have examined a wide variety of ratios. Presumably because quantifying the incorporation of pyridyldisulfide groups is so simple, an unusually abundant amount of information is available, albeit using IgG from a range of sources, and this is plotted in Figure 5.140. The data used originate from the following references: Battelli *et al.*, 1990; Thorpe *et al.*, 1981; Staerz *et al.*, 1985; Ebert and Spryn, 1990; Lambert *et al.*, 1985; Nilsson *et al.*, 1981; Cumber and Wawrzynczak, 1992; Dewerchin *et al.*, 1989; Dosio *et al.*, 1989; Jou *et al.*, 1985. Typically the incorporations are around a third of the input ratio.

Much lower incorporations have occasionally been reported (Deelder and de Water, 1981; Gadkari *et al.*, 1985: denoted as outliers in Figure 5.140): one is tempted to suspect the quality of the raw material in such preparations.

Incorporations above around 2–3 pyridyldisulfide groups per IgG molecule have been reported to cause inactivation and/or precipitation on some occasions (Cumber and Wawrzynczak, 1992; Cook and Wood, 1994). In comparative studies, SPDP preparations of IgG conjugates have shown better performance than most other methods (Jeanson *et al.*, 1988a,b).

Thorpe *et al.* (1985b) noted that N_3^- ions reacted with pyridyldisulfide groups, and recommended omitting azide salts from reaction buffers.

5.7.3.3 SPDP analogues

A number of SPDP analogues have been investigated, particularly for immunotoxins where the *in vivo* survival and efficacy of the reagent can depend heavily on the nature of the antibody–toxin link.

Worrell *et al.* (1986), for example, compared ricin A–IgG conjugates incorporating bridges with varying resistance to reduction. An SPDP analogue containing a hindering methyl group, *N*-succinimidyl-3-(2-pyridyldithio)butyrate (SPDB; Figure 5.141(a)) displayed approximately tenfold slower reduction (of the free acid) by dithiothreitol, and this increased stability was reflected to some extent *in vivo*. Attempts to synthesise the succinimide ester of the dimethyl analogue, which showed much slower reduction kinetics, met with failure.

Along the same lines, Thorpe *et al.* (1987b, 1988) synthesised another hindered SPDP analogue, 4-succinimidyloxycarbonyl-α-methyl-α-(2-pyridyldi-

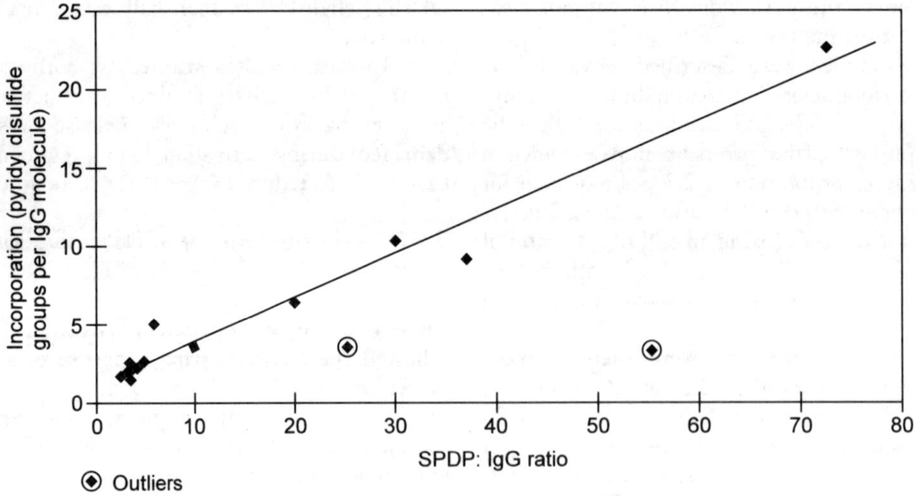

Figure 5.140 Incorporation ratios of pyridyl disulfides from SPDP activation of IgG.

(a)

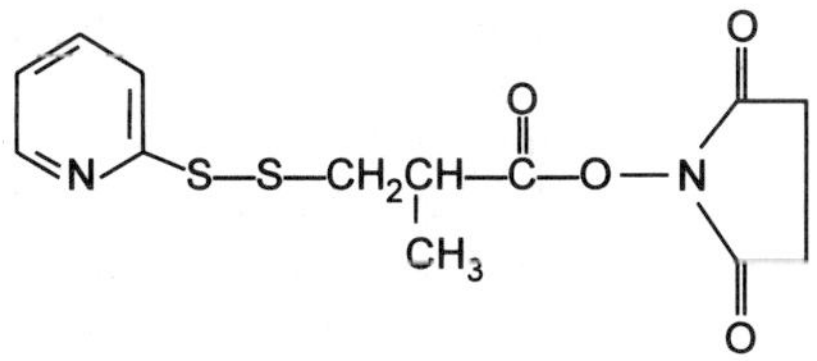

N-Succinimidyl 3-(2-pyridyldithio)butyrate (SPDB)

(b)

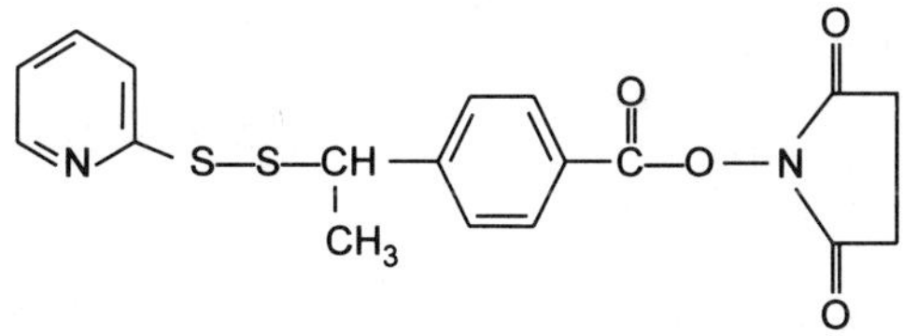

4-Succinimidyloxycarbonyl-α-methyl-α-(2-pyridyldithio)toluene (SMPT)

(c)

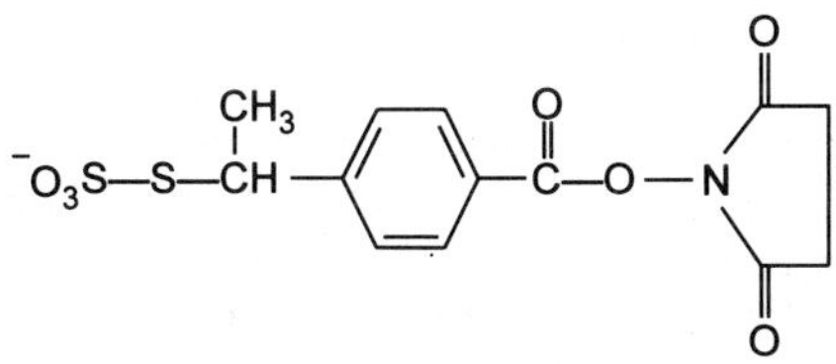

4-Succinimidyloxycarbonyl-α–methyltoluene–α–thiosulfate (SMBT)

Figure 5.141 Structures of SPDB, SMPT and SMBT, hindered analogues of SPDP.

thio)toluene (SMPT; Figure 5.141(b)), and its thiosulfate analogue SMBT (Figure 5.141(c)). These were used in the preparation of ricin A immunotoxins and displayed considerably higher *in vivo* stability than unhindered conjugates such as those prepared using 2-IT (q.v., Section 5.12.1). Both SMPT and a sulfonated long-chain analogue are commercially available.

Dosio *et al.* (1989) described the synthesis and use of two SPDP analogues containing phenyldisulfide as opposed to pyridyldisulfide moieties (Figure 5.142). These gave a higher degree of incorporation into mouse monoclonal IgG than SPDP itself (14 to 15 groups per molecule at a 20:1 excess, *versus* 6.3 for SPDP under the same conditions) and displayed better stability of the protected thiol on storage.

Åkerblom *et al.* (1993) described a hydrophilic reagent of this type (Figure 5.143) incorporating repeating ethylene oxide units. This was used to derivatise staphylococcal enterotoxin A: a 4.5-fold excess gave a disulfide incorporation of 2.9. Larger analogues of this compound were described by Haselgrübler *et al.*

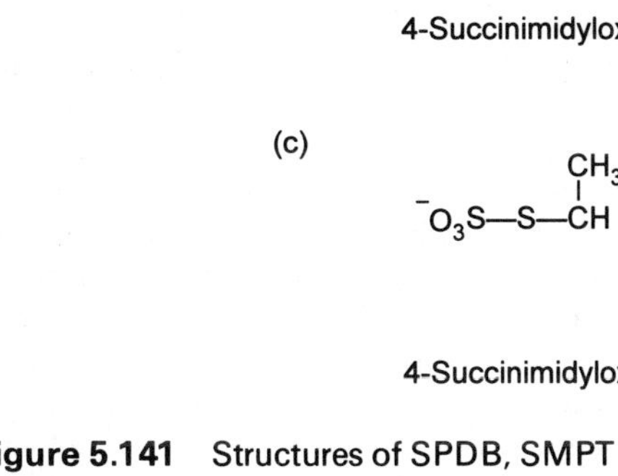

N-Succinimidyl-3-phenyl-dithiopropionate (SPheDP)

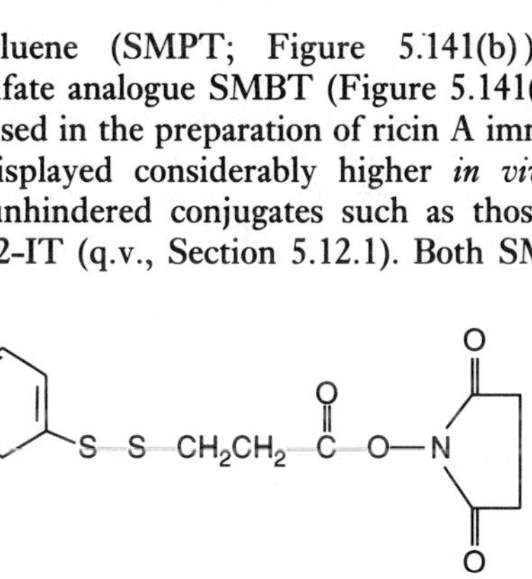

N-Succinimidyl-2-phenyl-dithioacetate (SPheDA)

Figure 5.142 SPDP analogues containing phenyldisulfide moieties.

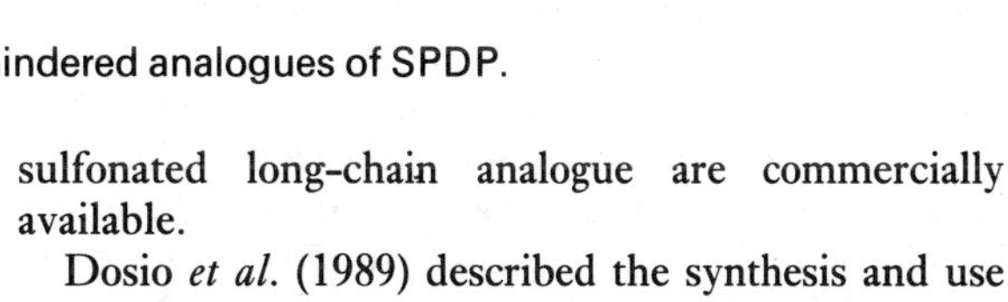

Figure 5.143 Hydrophilic analogue of SPDP containing ethylene oxide units.

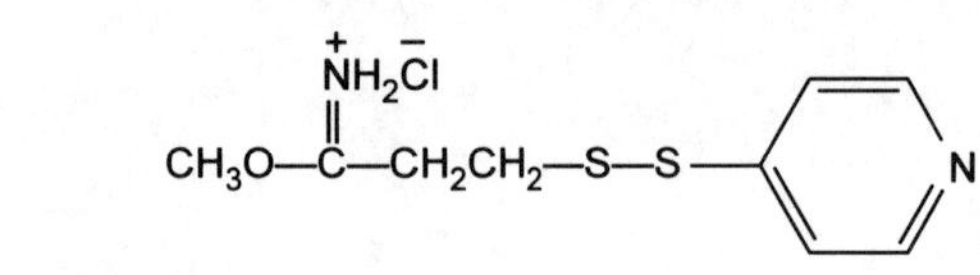

Figure 5.144 Methyl 3-(4-pyridyldithio)-propionimidate hydrochloride.

(1995), based on derivatisation of a diamino-polyethylene glycol with molecular mass 800 Da. A smaller disulfide–acid reagent was described by Frisch *et al.* (1996), forming a possible basis for a succinimide analogue.

Long-chain analogues of SPDP incorporating an extra caproamide spacer are available commercially in both sulfonated and unsulfonated forms, for example from Pierce as sulfo-LC-SPDP and LC-SPDP respectively.

Another disulfide which received some attention for protein coupling is methyl 3-(4-pyridyldithio)-propionimidate hydrochloride (Figure 5.144) reported by King *et al.* in 1978. Incorporations of 0.8, 2.2 and 3 pyridyldisulfide groups per molecule were obtained when bovine plasma albumin was derivatised by a 50-fold excess of this reagent for two hours at room temperature, at a pH of 7.2, 8.1 and 9.0 respectively.

A number of cleavable amine-reactive homobifunctionals include disulfide linkages and are therefore capable of undergoing exchange reactions with free protein thiols. Thus in the cross-linking of haemoglobin by DSP (Section 5.3.5), Lomant and Fairbanks (1976) reported the presence of some amine–thiol linked product in addition to the predominant amine–amine species.

The thiols can also be liberated by a suitable reducing agent such as dithiothreitol: Broker *et al.* (1978) used the dithiodiacetic acid analogue to thiolate avidin in this way (Figure 5.145). Note that some protein will obviously be cross-linked by the use of such reagents, but this has no effect on the product after reduction.

5.7.4 Other reagents for coupling of amines to thiols

N-Succinimidyl-*p*-(2-nitrovinyl)benzoate (Figure 5.146) features an electrophilic double bond, activated by the adjacent nitro group (Fujii *et al.*, 1984); the reactivity of this moiety is comparable to that of a maleimide. These authors demonstrated that the nitrovinyl group could be conjugated to thiols within 30 minutes at pH 3, where the succinimidyl group should be resistant to hydrolysis and aminolysis. Subsequent reaction with the amine group of alanine under slightly basic conditions was slow, however. The analogous 2,4-dinitrophenyl-*p*-(2-nitrovinyl)-benzoate (Figure 5.147) was described later by the same group (Fujii *et al.*, 1985a,b), and was claimed to yield more rapid aminolysis with bovine serum albumin, four hours *versus* eight hours for the succinimide at pH 7.5. Overall the reaction times for aminolysis of succinimide esters obtained with these compounds seem anomalously long. An analogous vinylpyridire reagent is shown in Figure 6.6.9.

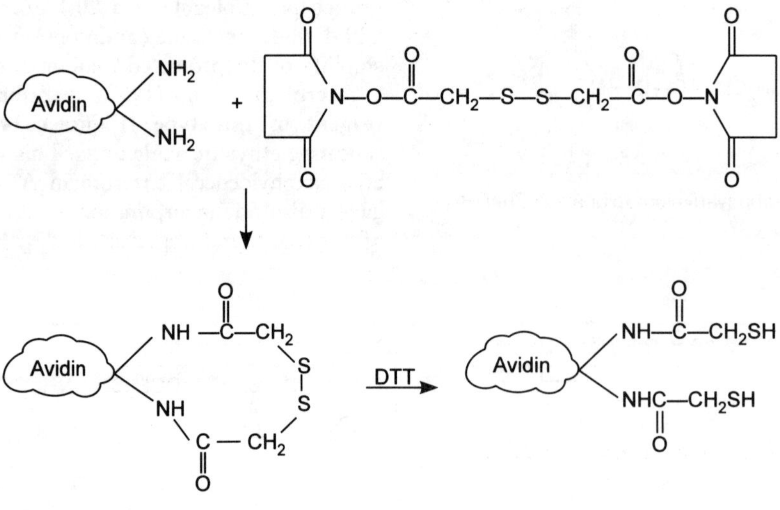

Figure 5.145 Thiolation of protein by reaction with DSP and reduction with dithiothreitol.

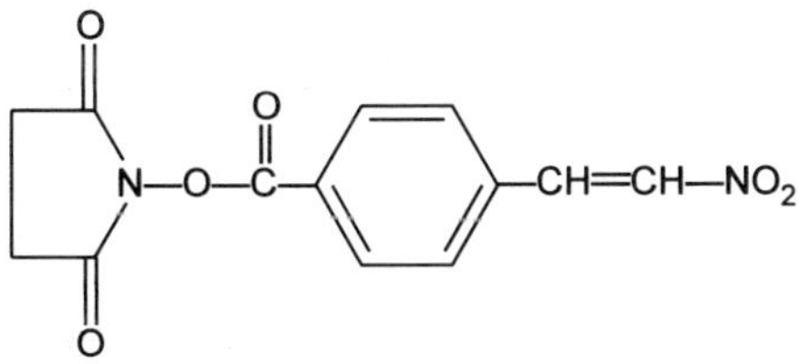

Figure 5.146 *N*-Succinimidyl *p*-(2-nitrovinyl) benzoate.

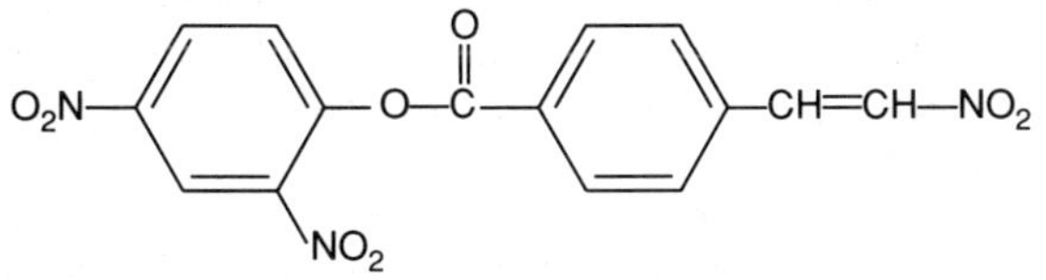

Figure 5.147 2,4-Dinitrophenyl *p*-(2-nitrovinyl)benzoate.

Non-specific electrophilic reagents such as cyanuric chloride and the nitrogen and sulfur mustards may result in amine–thiol linkages, albeit in a largely uncontrolled fashion. The ETAC reagents (Section 5.3.7) may also yield amine–thiol products in some cases, if they are energetically favoured.

5.8 Heterobifunctional Coupling of Amines to Other Groups

Following amine–thiol reagents, the most important amine-reactive heterobifunctionals in protein chemistry are those employed in photoactivation, biotinylation and thiolation reactions. For convenience these are dealt with separately in Sections 5.10, 5.11, and 5.12 respectively. A number of other categories exist however, and are dealt with here.

5.8.1 Reagents for coupling of amines to carboxyl groups

In terms of reactive groups that are encountered in native proteins, carboxyl functions offer the most likely alternative to thiols as a target for heterobifunctional coupling with amines. Usually linkages between these two functions are formed, without the incorporation of extra atoms, by dehydration to an amide (Section 5.2.2). The carbodiimide reagents commonly used to facilitate amidation can be thought of loosely as 'unincorporated' heterobifunctional amine-carboxyl reagents.

A hypothetical group of reagents which would meet the criteria for this class would contain any compound

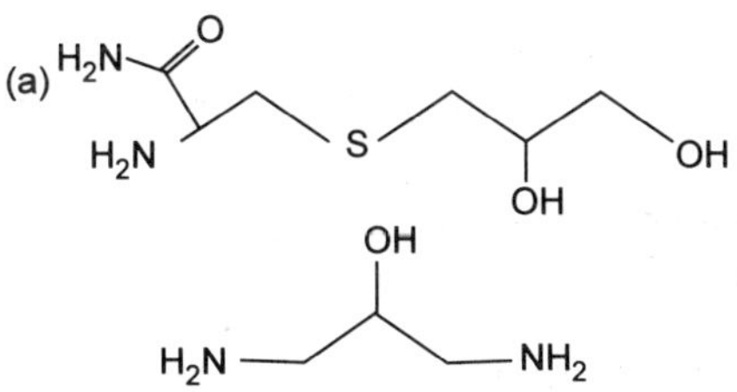

Figure 5.148 Reagents containing vicinal diols and amino alcohols, which can be used in the coupling of amines to carboxyl groups.

$$R{-}CH{=}N_2 \longleftrightarrow R{-}CH_2{-}N_2$$

Figure 5.149 Diazo reagents: unprotonated and protonated forms.

bearing an amine at one end (allowing reaction with a suitably activated carboxyl function) and an amine-reactive moiety at the other. Cyclisation or cross-linking reactions would render such reagents hopelessly unstable, but an interesting approach is based on this route. Reagents coupling amines to the precursor of an amine-reactive function have been described (Offord *et al.*, 1987; Rose *et al.*, 1989): these contain vicinal diols or α-amino alcohols (Figure 5.148(a)) which can readily be oxidised to aldehydes. Therefore the amine end of the linker can be coupled to a carboxyl residue (these authors used reverse proteolysis), followed by generation of the amine-reactive aldehyde by mild periodate oxidation.

There are also *bona fide* heterobifunctionals for achieving an amine–carboxyl linkage, although they are infrequently encountered; these are diazo esters (Figure 5.149). They rely on the fact that the carboxylate anions encountered increasingly as the pH rises above the pK_a of 3–4.5 are weakly nucleophilic; at neutral to alkaline pH this reactivity is swamped by the stronger thiols and amines, but at pH 5, for example, they may be the strongest nucleophiles present. Because of the weakness of this property, strong electrophiles are required to exploit it for protein coupling.

Examples of heterobifunctional reagents which could operate *via* this route are *p*-nitrophenyl diazoacetate (Shafer *et al.*, 1966) and diazopyruvate (Harrison *et al.*, 1989). Although the diazo-functions have primarily been employed for their photoreactive properties (Section 5.10), they will also react spontaneously with carboxyl functions (Lundblad and Stein, 1969). At the mildly acidic pH required to exploit the nucleophilicity of carboxylate ions, the diazoacetate nitrogen

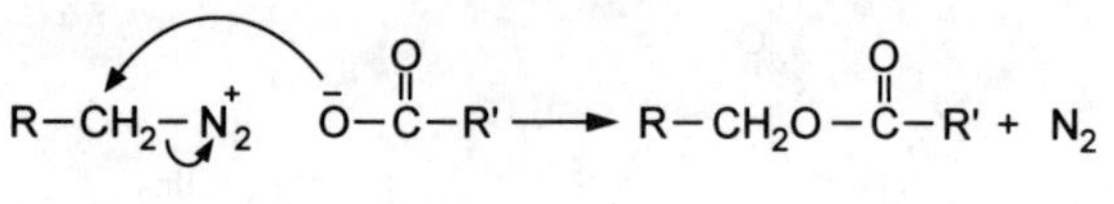

Figure 5.150 Reaction of carboxylate with diazo reagent.

is protonated; acting in concert with the carbonyl group, this generates a powerfully electron-deficient CH_2 group which is thought to couple to carboxylate with the loss of nitrogen gas (Figure 5.150).

The high reactivity required for this reaction partially explains the lack of popularity of such approaches, as diazo-esters will also react strongly with water and many inorganic ions. Specificity within the protein may also be imperfect due to interference from thiols, especially those with a low pK_a.

5.8.2 Reagents for coupling of amines to tyrosine and histidine residues

The reagent *N*-(aminobenzoyloxy)succinimide (Figure 5.151) was described by Fujiwara *et al.* in 1990; after reaction of the succinimide with an amine-containing species, the aminobenzoyl function was diazotised to confer reactivity with tyrosine or histidine residues. However this work was aimed at the preparation of protein–hapten conjugates where the derivatised hapten could be diazotised without harm, such an approach is less attractive for protein–protein work where one protein must survive the diazotisation conditions intact, and has not been reported.

5.8.3 Reagents for coupling of amines to carbohydrates

Coupling of amines to carbohydrates may well occur as a side-reaction in the use of powerful homo-

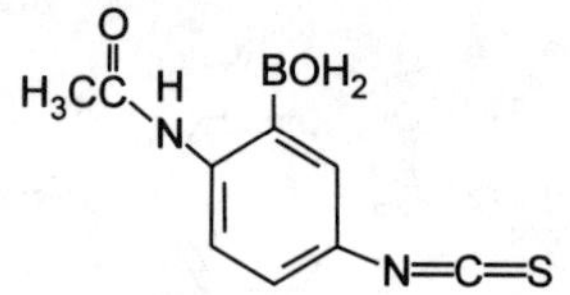

Figure 5.152 A heterobifunctional reagent described by Cai and Keana (1991b) using a boronate function for coupling to vicinal diols.

bifunctional amine reagents such as benzoquinone, and direct coupling of amines to carbohydrates is of course widely employed in the periodate oxidation approach.

Cai and Keana (1991b) described an interesting heterobifunctional reagent combining an amine-reactive isothiocyanate group with a phenylboronate capable of reacting with vicinal diols (Figure 5.152).

5.9 Heterobifunctional Coupling of Other Groups

With amine-reactive heterobifunctionals dealt with in the last two sections, the remaining reagents nearly all involve thiol or carboxyl derivatisation. This category involves some interesting approaches, but none have yet found widespread application.

5.9.1 Reagents for coupling of thiols to aldehydes

Webb and Kaneko (1990) reported the use of compounds combining an amine function with a nitropyridyldisulfide group; one of these, 1-(aminooxy)-4-[(3-nitro-2-pyridyl)dithio]butane hydrochloride, is shown in Figure 5.153. This reacts with the aldehyde groups

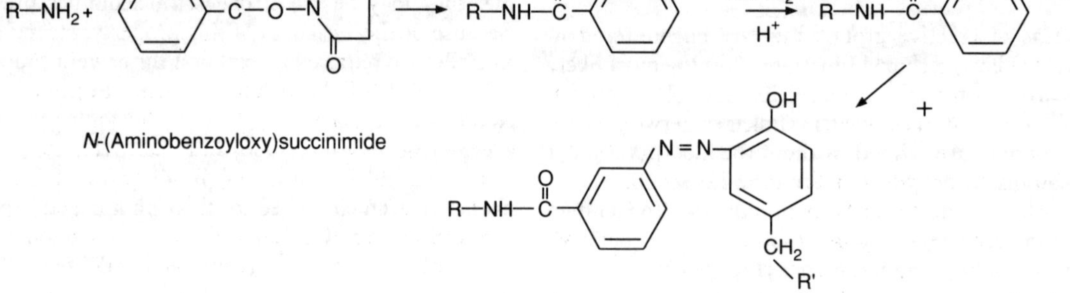

Figure 5.151 Use of *N*-(aminobenzoyloxy)succinimide to couple amines to tyrosines (shown) or histidines.

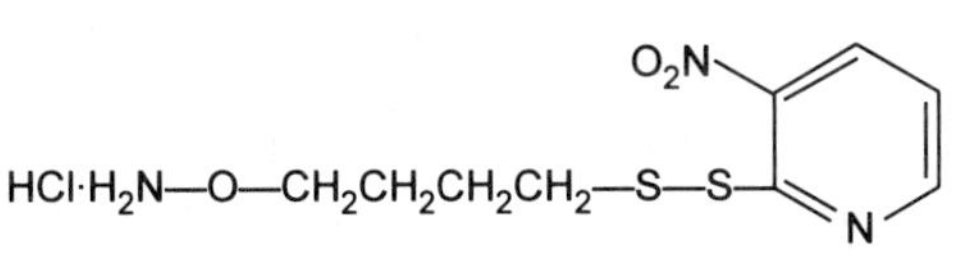

Figure 5.153 Aminooxy-4-[(3-nitro-2-pyridyl)-dithio]butane hydrochloride

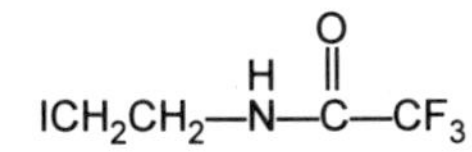

Figure 5.154 *N*-(2-Iodoethyl)trifluoroacetamide

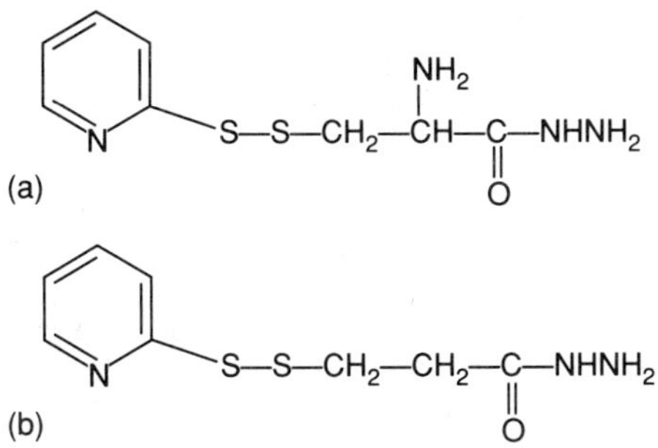

Figure 5.155 (a) *S*-(2-Thiopyridyl)-1-cysteine hydrazide (TPCH); (b) 3-(2-pyridyldithio)propionyl hydrazide (PDPH).

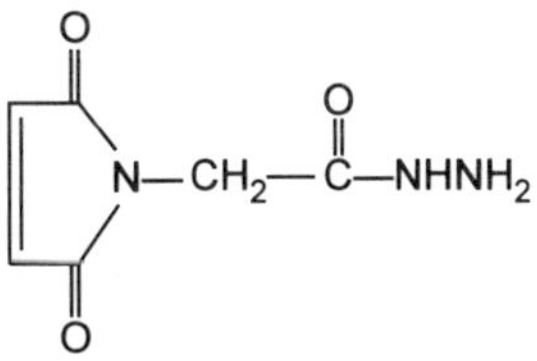

Figure 5.156 Maleimidoacetic acid hydrazide.

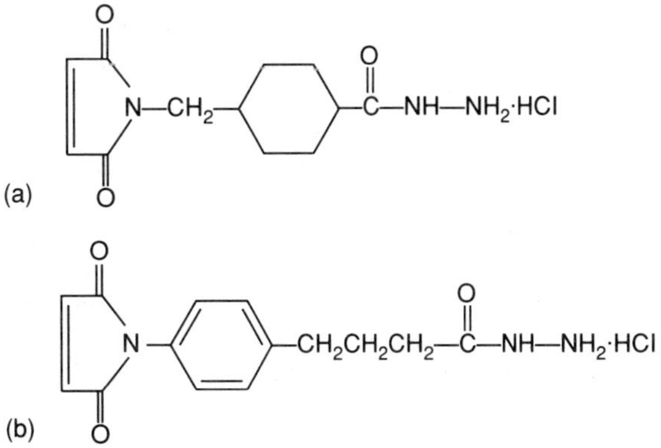

Figure 5.157 (a) 4-(*N*-Maleimidomethyl) cyclohexane-1-carboxylhydrazide (M_2C_2H), hydrochloride; (b) 4-(4-*N*-maleimidophenyl)butyric acid hydrazide, hydrochloride (MPBH).

of periodate-oxidised glycoproteins to activate them towards thiols. A maleimide-containing analogue was described by Mikolajczyk *et al.* (1994).

Pierce's Aminoethyl-8™ reagent, *N*-(2-iodoethyl)-trifluoroacetamide (Figure 5.154), replaces thiol functions with amines at pH 8.1 to 8.6: the trifluoromethyl group is hydrolysed under these conditions.

Zara *et al.* (1991) described two heterobifunctional reagents combining a pyridyldisulfide moiety with a hydrazide group, and therefore capable of coupling thiols to periodate-oxidised carbohydrates. The more reactive of these, *S*-(2-thiopyridyl)-L-cysteine hydrazide (TPCH) is shown in Figure 5.155(a). Immunoconjugates were prepared by the periodate oxidation of immunoglobulin, derivatisation with TPCH, and subsequent reaction with a suitable toxin containing free thiols, reduced ricin A-chain and thiolated barley toxin were used in this example, the latter giving a higher incorporation presumed to be due to the greater number of thiols present. Neither the activity of the toxin nor that of the antibody were impaired by this conjugation procedure.

Pierce Chemical Co. and BioAffinity Systems sell the similar 3-(2-pyridyldithio)propionyl hydrazide (PDPH; Figure 5.155(b)), while a long-chain analogue was described by Kondejewski *et al.* (1994).

Heindel *et al.* (1991) described a reagent combining a hydrazide with a maleimide function, maleimidoacetic acid hydrazide (Figure 5.156), this was used to couple antibodies to a derivatised carbohydrate, but could equally well be employed in protein–protein conugations of the type described by Zara *et al.*

Two similar compounds are commercially available from Pierce Chemical Co. and from BioAffinity Systems Inc.: 4-(*N*-maleimidomethyl)cyclohexane-1-carboxylhydrazide hydrochloride (M_2C_2H) and 4-(4-*N*- maleimidophenyl)butyric acid hydrazide hydrochloride (MPBH). These are shown in Figure 5.157. BioAffinity Systems also sell the analogous maleimido-hydrazides of caproic and benzoic acid, the latter as the *m*-isomer analogous to MBS. ProChem Inc. market a maleimide-PEG-hydrazide.

5.9.2 Reagents for coupling of thiols to carboxyl groups

The amine and hydrazide-containing reagents described in the previous section also have the potential

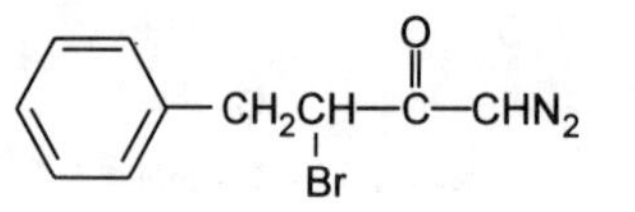

Figure 5.158 Diazoacetyl-1-bromo-2-phenylethane.

for reaction with carboxyl functions, if a suitable activating agent such as a carbodiimide is incorporated.

Husain *et al.* (1971) described a reagent combining a diazoketone function with a bromoalkyl moiety (Figure 5.158), this was designed to couple an active aspartate in pepsin to neighbouring nucleophiles such as thiols. It was less effective in deactivating the enzyme than mono- or bifunctional diazoketones.

Similarly, Henkin (1977) described a compound combining pyridyldisulfide and diazoacetate moieties. This was used in a photoactivated mode, the resulting carbene losing any specificity for carboxyl groups; in fact the major product indicated thiol–hydroxyl coupling. In a dark reaction this reagent could potentially form thiol–carboxyl links, however.

5.9.3 Reagents for coupling of thiols to hydroxyl groups

Some amine-reactive functionalities (e.g. isocyanates) are also reactive towards other nucleophiles such as hydroxyls, and a heterobifunctional reagent linking a maleimide to an isocyanate was described (Annunziato *et al.*, 1993; Palumbo, 1993) for linking thiol-containing proteins to hydroxyl-containing haptens. The reagent, *p*-maleimidophenyl isocyanate (PMPI), is shown in Figure 5.159 and is commercially available. A serine-containing oligopeptide was among the haptens derivatised by these authors, and where-suitably specialised requirements dictated, PMPI could potentially be used to form thiol–hydroxyl links between proteins. Its primary use, however, is likely to be for coupling proteins to haptens or polysaccharides.

Thiol-reactive moieties have been introduced into tosylated cellulose, by transesterification with thiosulfate (Izard and Morgan, 1949; Schwenker *et al.*, 1962) to form a Bunte salt (R–S–S(=O)$_2$Na). Such compounds can be viewed as disulfides with an enhanced reactivity towards thiols (cf. methanethiosulfonates, Section 5.4.3). There is no record of this approach having been applied to preparative protein chemistry; it might offer a useful route to directed disulfide formation.

A possible drawback is the possibility of disulfide formation by the reaction of two Bunte salts; this can be induced deliberately by mild oxidation (Izard and Morgan, 1949) but can also occur under ambient conditions. This was attributed tentatively to the liberation of small quantities of free thiol by acid catalysis or radical cleavage (Schwenker *et al.*, 1962). Conditions avoiding such effects would clearly be beneficial where the aim is to activate a protein for reaction with a second, thiol-containing species.

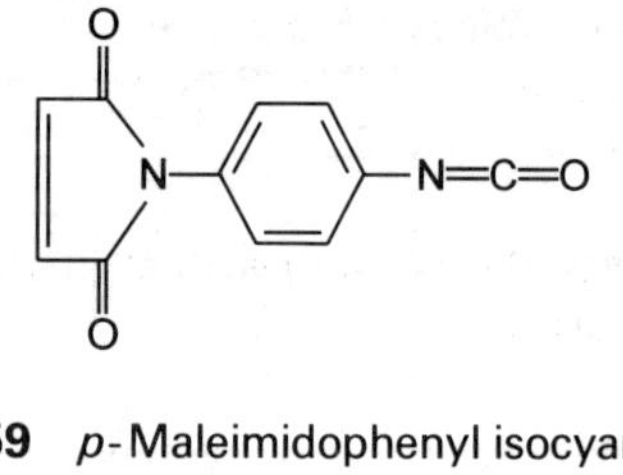

Figure 5.159 *p*-Maleimidophenyl isocyanate (PMPI).

5.9.4 Reagents for coupling of thiols to tyrosine and histidine residues

Fujiwara and Saita (1986) described the preparation of *N*-(4-diazophenyl)maleimide (DPM; Figure 5.160(a)), a reagent for incorporating maleimide groups at the site of histidine or tyrosine residues. A stable aminophenyl precursor is diazotised by acidic sodium nitrite before use to yield the bifunctional reagent, which was used to couple the oligopeptide hormone TRH (thyrotropin-releasing hormone) to thiolated albumin.

In response to the reported instability of maleimide groups attached directly to aromatic molecules (Ishikawa *et al.*, 1978; cf. Section 5.7.1) Fujiwara and Saita subsequently (1987) described the synthesis of an analogue where the maleimide is attached to an aliphatic moiety, *N*-[β-(4-diazophenyl)ethyl] maleimide

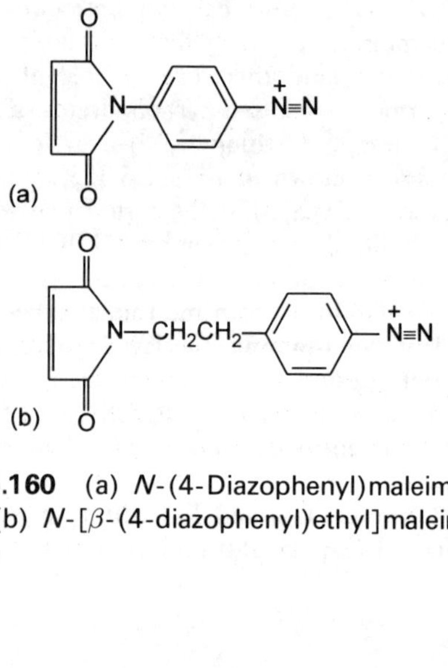

Figure 5.160 (a) *N*-(4-Diazophenyl)maleimide (DPM); (b) *N*-[β-(4-diazophenyl)ethyl]maleimide (DPEM).

(DPEM; Figure 5.160(b)). This reagent has since been used, for example, to couple β-galactosidase to antibodies (Fujiwara *et al.*, 1988).

The conjugates obtained in this way showed good retention of protein functionality: no loss of β-galactosidase activity was observed, while only 8–10% of the antibody activity was lost.

5.9.5 Reagents for coupling of carboxyl groups to aldehydes

The concept of coupling amine-containing aldehyde precursors to protein carboxyl functions by reverse proteolysis was discussed in Section 5.8.1, and the same approach is applicable for other reagents too. The introduction of hydrazide moieties is convenient for conferring reactivity towards aldehydes, typically those obtained from the oxidation of glycoproteins. Because hydrazides will also undergo reverse proteolysis coupling, the homobifunctional carbohydrazide (Figure 5.161) can rather neatly be used to 'convert' a carboxyl function to a hydrazide.

Thus, Rose *et al.* (1989, 1991) derivatised insulin (from which the AlaB30 residue had been removed to leave a C-terminal lysine) with carbohydrazide using porcine trypsin, or lysyl endopeptidase from *Achromobacter lyticus*. At pH 5.5 the *Achromobacter* protease gave extremely rapid incorporation of the hydrazide: around an 80% yield was obtained within 15 minutes. The trypsin gave similar performance over the 5 to 6.5 pH range although a small amount of cleavage at the ArgB22 residue was observed; use of the lysyl endopeptidase removes this danger. The resulting insulin-hydrazide was used to couple small aldehyde molecules.

Rose's group (Werlen *et al.*, 1994) extended this approach to the preparation of conjugates between enzymes and antibody fragments: the Mab was first incubated with lysyl endopeptidase (20 hours, 37 °C, pH 8.4) to generate a $F(ab')_2$-like fragment with terminal lysines, then activated by reverse proteolysis with carbohydrazide (three hours, room temperature, pH 5.5) using the same enzyme. This allowed the Mab fragment to be coupled to an oxidised glycoprotein.

Of course reverse proteolysis is not the only way of coupling hydrazides to carboxyl functions. Gershoni *et al.* (1985), for example, coupled adipic dihydrazide to AP and HRP using carbodiimide activation.

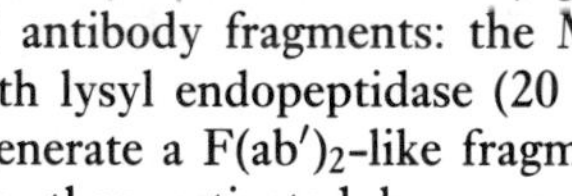

Figure 5.161 Carbohydrazide.

5.9.6 Reagents for coupling of carbohydrates to histidine residues

An interesting approach was used by Jin *et al.* (1995) to couple histidine-tagged recombinant proteins to HRP: the enzyme was derivatised with α-*N*,*N*-bis(carboxymethyl)lysine *via* periodate oxidation of the carbohydrate residues, and a nickel chelate of the resulting poly(bis-carboxylate) derivative used to bind to histidines.

5.10 Photoactivatable Heterobifunctional Reagents

For the vast majority of reagents described in this chapter, selective reactivity of the functional groups is a key parameter; the ability to couple the derivatising compound to more-or-less specific residues in the target protein can confer many advantages. For preparative work it may be possible to avoid reaction with residues which are essential for protein function, while sequential conjugation procedures based on defined reactivities offer controlled product stoichiometry. The least selective chemistries encountered in protein conjugations, such as those based on nitrogen and sulfur mustards, have largely been dropped on account their uncontrollable nature.

For structural studies, too, it is often a key aim of the experiment to define the site of any derivatisation, and defined chemistry can only make this task easier.

In affinity labelling, however, the selectivity of the derivatising reagent for the binding site is achieved through the specificity of the ligand–receptor interaction (Figure 5.162(a)). The binding of the affinity reagent to the site of interest would ideally be followed by a rapid 'freezing' reaction, which anchored it permanently to the receptor site for subsequent analysis and handling. This freezing reaction would require two main attributes: (1) a broad reactivity so that binding took place regardless of which amino acid residues were nearby, and (2) the ability to be initiated when (and only when) the ligand–receptor complex had formed, so that freezing did not occur until the ratio of specific to non-specific interactions was high enough to generate a meaningful result, i.e. until the situation depicted in Figure 5.162(c) had been achieved. Rapidity of reaction could also be advantageous in some instances, for example to capture a 'snapshot' of a transient (non-equilibrium) situation. The interactions within cell membranes are considered to be dynamic in nature, for example (Ji, 1979) and

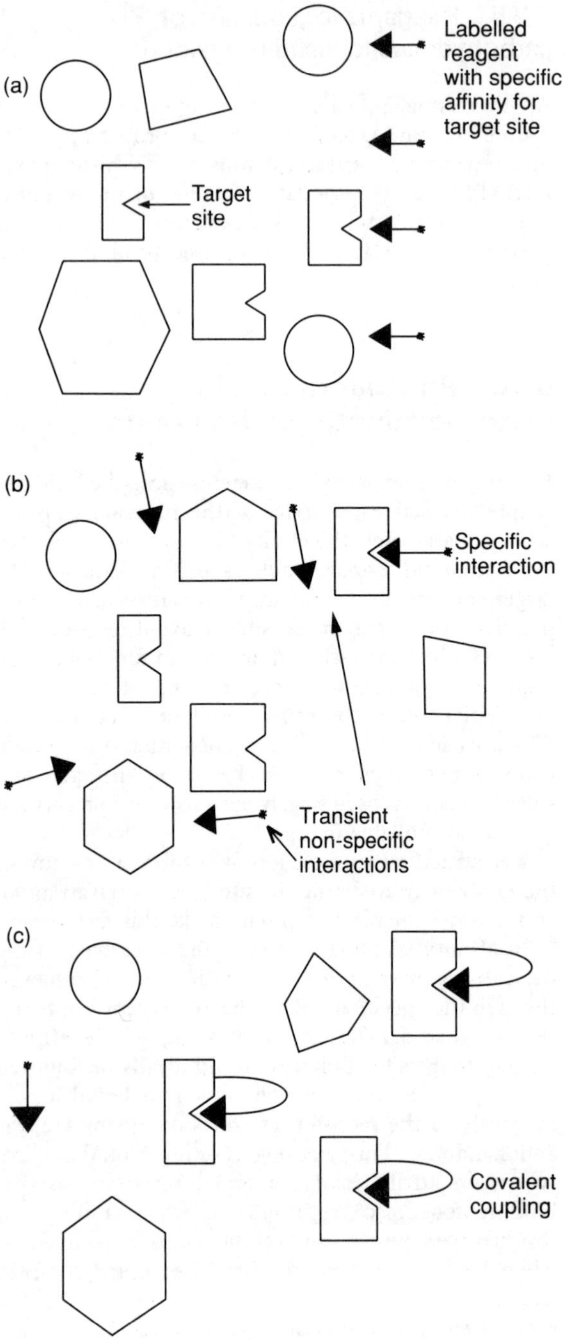

changes can only be captured by a suitably rapid means.

In 1962 Westheimer's group at Harvard University described experiments aimed at achieving this ideal (Singh *et al.*, 1962; Shafer *et al.*, 1966), which introduced the technique of photoaffinity labelling (PAL). The affinity reagent in Singh's original paper was *p*-nitrophenyl diazoacetate, which reacted stoichiometrically at the active site serine of the enzyme α-chymotrypsin (Figure 5.163). After ten hours of incubation, the diazoacetylated enzyme was isolated and photolysed.

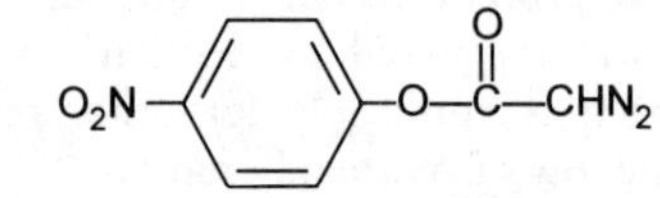

Figure 5.163 *p*-Nitrophenyl diazoacetate.

Around 75% of the ester was released on photolysis with sun lamps, and this loss was accompanied by a commensurate return of enzyme activity. However, the other 25% remained associated with the enzyme, and could not be removed by de-esterification with hydroxylamine. It was therefore concluded that reactive carbenes generated by photolysis had resulted in new, stable bonds between the protein and the ligand.

Using a similar approach Vaughan and Westheimer (1969) isolated a small yield of product from the insertion of the carbene intermediate into an alanine residue in trypsin, the first concrete evidence for labelling of residues lacking a readily reactive functionality.

Summarising the early work of Westheimer's group on trypsin and chymotrypsin, the PAL approach was able to show that the active site serine was close in space to histidine and alanine residues, and to a disulfide bridge (Hexter and Westheimer, 1971). These findings were consistent with previous crystallographic evidence for the proximity of the active site to a region of the proteins' structure termed the 'histidine loop' (Figure 5.164).

The technique of photoaffinity labelling is thus a powerful tool for the elucidation of ligand–receptor interactions in protein systems (Knowles, 1972; Das and Fox, 1979; Bayley and Staros, 1984), and has been

Figure 5.162 Scheme for ideal affinity labelling. The labelling reagent bears a group with high (non-covalent) affinity for the target site. On addition to a sample of specific receptors mixed with other proteins (a), some specific interactions will occur but there will also be transient, non-specific interactions with other proteins (b). After a time, however, diffusion and the high affinity of the reagent will ensure that all the specific receptors are labelled (c), at which point a 'freezing' reaction is executed to provide a permanent covalent bond between the label and its receptor.

Chymotrypsin

-Cys-Gly-Gly-Ser-Leu-Ile-Asn-Glu-Asn-Trp-Val-Val-Thr-Ala-Ala-His-Cys-

Trypsin

-Cys-Gly-Gly-Ser-Leu-Ile-Asn-Ser-Gln-Trp-Val-Val-Ser-Ala-Ala-His-Cys-

Figure 5.164 The histidine loop regions of chymotrypsin and trypsin, which contain histidine, alanine and cystine groups. Hexter and Westheimer (1971) used PAL to show that this was close in space to the active site of each enzyme.

very widely employed. The very special requirement of PAL reagents, to incorporate a functional group of rapid, non-selective reactivity which can be turned on at will, has led to the growth of a new class of photoactivatable heterobifunctional reagents for the preparation of affinity reagents.

These incorporate a photoactivatable group, in practice limited almost exclusively to a diazo moiety of the type just described or an aryl azide, and a conventional reactive function for introducing it (in the dark!) into the affinity reagent[1]. A plethora of photoactivatable heterobifunctionals has been developed, incorporating all the usual reactive groups for derivatisation *via* amines, thiols, etc., and these will be dealt with in this section.

Note that there is little call for homobifunctional reagents of this type—sequential reaction of two different functionalities is the key advantage of photoactivation. However such reagents do exist (see Section 5.5.6).

5.10.1 Photoactivatable heterobifunctional reagents based on β-carbonyldiazo groups

As mentioned above, the earliest PAL reagents incorporated the diazoacetate function, an example of the β-carbonyldiazo family; the adjacent electron-withdrawing group is necessary to render the diazo group stable. On irradiation this group loses nitrogen to yield a carbene, a highly electrophilic species which can undergo a variety of reactions with protein constituents. These include insertion into carbon–carbon and carbon–hydrogen bonds (Shafer *et al.*, 1966; Knowles, 1972). The most potent carbenes react more-or-less equally well with primary, secondary and tertiary C–H bonds, while less reactive examples show increasing selectivity towards the latter: the carbene generated from diazoacetate, for example, reacts three times faster with tertiary than primary bonds.

Carbenes also react rapidly with nucleophiles, which (for example in the case of solvent water molecules) can be a disadvantage. Using a diazoacetate to photolabel creatine kinase, Henkin (1977) observed that 10% of the carbene reacted with water and 30% with threonine and serine residues in the enzyme. Carbenes can also undergo a Wolff rearrangement to generate a ketene C=C=O function, itself highly reactive towards a variety of groups including water and protein nucleophiles. Figure 5.165 shows a number of the possible reactions of carbenes.

Vaughan and Westheimer (1969) used another diazoester, ethyl *p*-nitrophenyl diazomalonate (Figure 5.166), to label trypsin, and, as mentioned above, showed that insertion into an alanine residue had taken place. Hexter and Westheimer (1971) identified another product from this reaction, which indicated cleavage of a disulfide bridge by the carbene.

Introduction of an α-trifluoromethyl function into diazoesters has a number of beneficial effects (Chowdhry *et al.*, 1976): not only does this reduce the level of Wolff rearrangement that occurs, but the acid stability of the reagents is dramatically improved. The amine-active reagent *p*-nitrophenyl-2-diazo-3,3,3-trifluoropropionate (PNP-DTP or NDTFP, Figure 5.167) has found some application and remains commercially available.

Cai and Keana (1991a) described the synthesis of a glutaraldehyde analogue containing the same 2-diazo-3,3,3-trifluoromethyl moiety (Figure 5.168): this compound behaved similarly to glutaraldehyde itself in its fixative properties, preventing haemolysis of erythrocytes.

[1] In Westheimer's pioneering studies the heterobifunctional photoaffinity reagent was used directly for labelling the site of interest in the enzyme, an approach made possible by the high reactivity of the active-site serine. More commonly, however, site-specific binding is achieved by labelling a ligand, which is allowed to bind to its receptor before photoactivation. These heterobifunctional reagents, then, are normally used to label a ligand, a protein or small molecule, rather than the binding site itself.

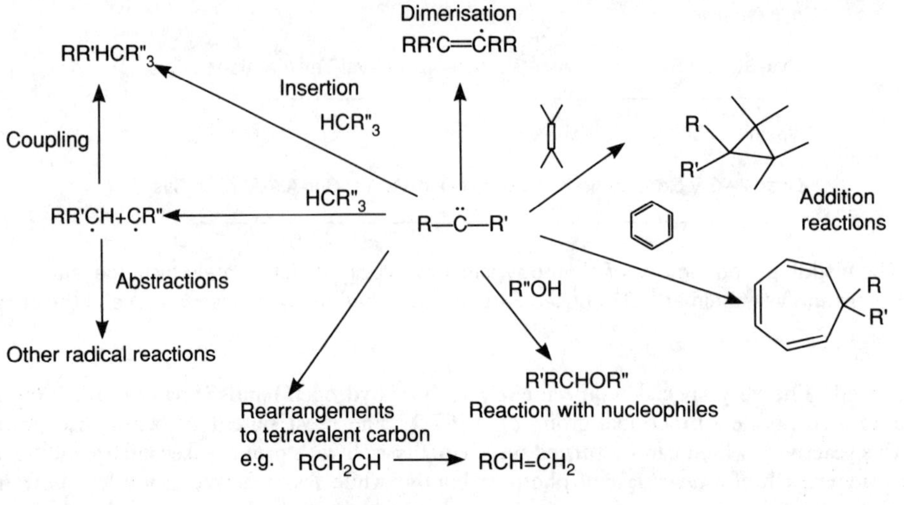

Figure 5.165 Typical reactions of carbene.

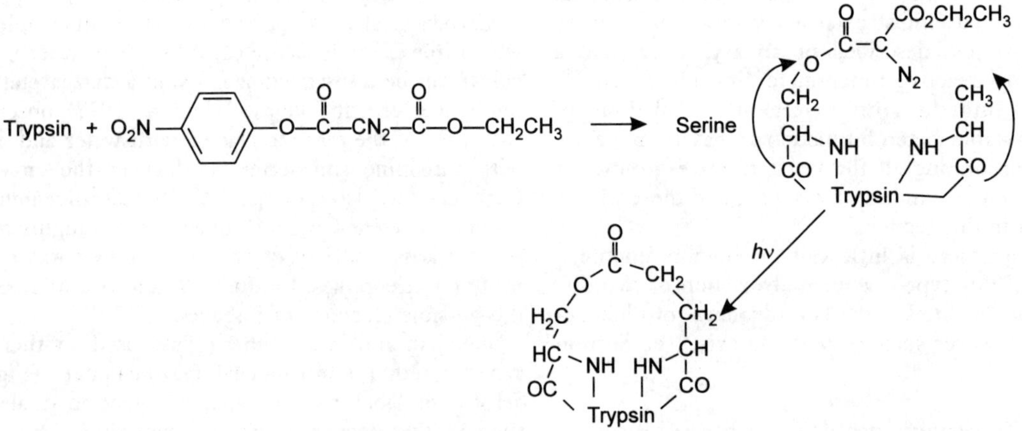

Figure 5.166 Labelling of trypsin using ethyl *p*-nitrophenyl diazomalonate. An alanine from the histidine loop region (Figure 5.164) is coupled to the active site serine.

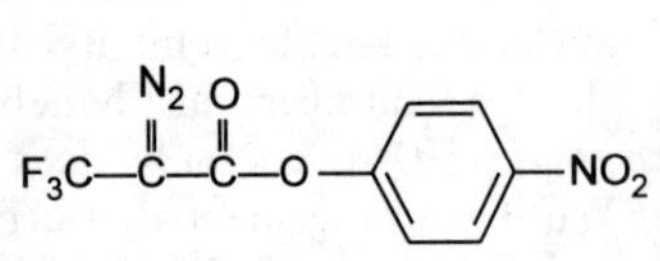

Figure 5.167 *p*-Nitrophenyl-2-diazo-3,3,3-trifluoropropionate (PNP-DTP).

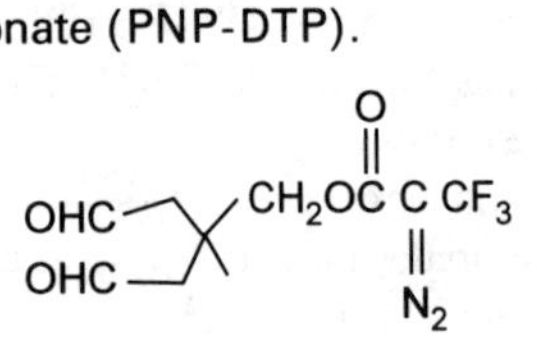

Figure 5.168 Glutaraldehyde analogue containing diazotrifluoromethyl moiety.

As mentioned earlier in the chapter (Sections 5.5.1, 5.9.2), β-carbonyldiazo compounds can also undergo non-photoactivated reactions under certain circumstances, principally with carboxyl residues.

5.10.2 Photoactivatable heterobifunctional reagents based on diazirines

Another closely-related group capable of carbene generation is the diazirine function (Figure 5.169). While this moiety has been incorporated more frequently into small molecules for reaction with proteins, diazirine-containing heterobifunctional reagents have been described.

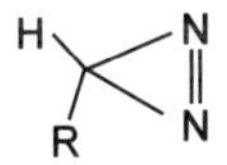

Figure 5.169 Diazirine group.

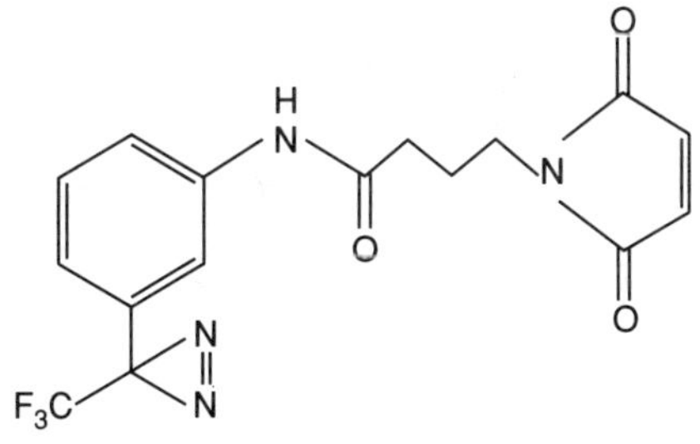

Figure 5.170 *N*-(4-Maleimidobutyryl)-*N*-{*m*-[3-(trifluoromethyl)diazirin-3-yl]phenyl}amide (MAD).

Collioud *et al.* (1993), for example, reported the preparation of *N*-(4-maleimidobutyryl)-*N*-{*m*-[3-(trifluoromethyl)diazirin-3-yl]phenyl}amide (MAD; Figure 5.170), a thiol-reactive photolabelling reagent. This reagent, used in this case for solid-phase derivatisation, could be activated at low-energy wavelengths (350 nm) avoiding damage to proteins.

Platz *et al.* (1991) noted that the products formed from the reaction of trifluoromethylaryl diazirines of this type with amines could be labile.

5.10.3 Photoactivatable heterobifunctional reagents based on phenyl azides

Fleet *et al.* (1969) introduced aryl azides as photoactivatable reagents. In a manner analogous to the diazo compounds described above, these generate nitrenes, which are also highly reactive. Other categories of azide were considered, but suffered from disadvantages: acyl azides require irradiation of such high energy that the protein itself could be damaged, are too reactive for convenience, and undergo rearrangements to the much less reactive isocyanate on photolysis. In addition they have a significant dark reactivity towards amines (Trommer *et al.*, 1977; see Section 5.7.1). Alkyl azides again require high-energy irradiation, and are unstable.

As described by Fleet, aryl azides display attractive characteristics: they are highly stable, they can be activated at high wavelengths which are less harmful to protein structure, particularly if electron-withdrawing substituents such as nitro groups are incorporated on the ring, and they do not undergo rearrangement to isocyanate. Incorporation of electron-withdrawing groups is doubly advantageous in that an increase in reactivity is also obtained. These authors successfully used a nitrophenyl azide-lysine hapten to affinity-label the antigen-binding sites of an antibody, raised against the nitrophenyl azide structure.

The photochemistry of aryl azides has been extensively reviewed by Schuster and Platz (1992), and is a complex field where, even in simple matrices, many issues remain unresolved regarding reactive intermediates, etc. The situation in a proteinaceous reaction mixture is even less defined, but certain key conclusions have emerged: molecular orbital considerations indicate that the singlet phenyl nitrene generated by irradiation can easily undergo ring expansion to the ketenimine shown in Figure 5.171, and this in fact takes place so quickly under ambient conditions (10–100 ps) that the singlet nitrene cannot be detected by trapping experiments[1].

Ketenimines are not ideal reactive intermediates, as they react moderately slowly, and only with nucleophiles rather than by insertion into C–H bonds (De Graff *et al.*, 1974). Amines for example yield substituted azepines (Figure 5.171). These ketenimines will also react with the aryl azides themselves (Li *et al.*, 1988), leading to the formation of polymeric tars. They are also somewhat unstable under acid conditions, which may be problematic if it is desired to carry out Edman degradation to analyse the structure of photolabelling products (Kim *et al.*, 1995).

Thus while simple phenyl azides have dominated the photolabelling field, their ability for truly non-selective reactivity is not optimal. The possible reactions of singlet nitrenes outlined by Knowles (1972), and still widely referred to, are for the most part unlikely to occur with unsubstituted phenyl azides.

Nitrophenyl azides are employed in many such reagents, principally to allow photoactivation at longer, less-damaging wavelengths. These may also give better performance in terms of reactivity: thus while an 80% yield of azepine was obtained from the reaction of phenyl azide with diethylamine, introduction of *m*- and *p* nitro groups reduced this to 10% and $<3\%$ respectively (Li *et al.*, 1988). The yield of the aniline, produced from the singlet nitrene *via* a triplet intermediate, rose accordingly for the nitro compounds. Different reaction routes may also become available when the electron-withdrawing nitro group

[1] Different reactive intermediates are formed under other circumstances: at 10–77 K the triplet nitrene is generated, while in the gas phase the ring-contracted cyanocyclopentadienyl radical is obtained. These are also shown in Figure 5.171.

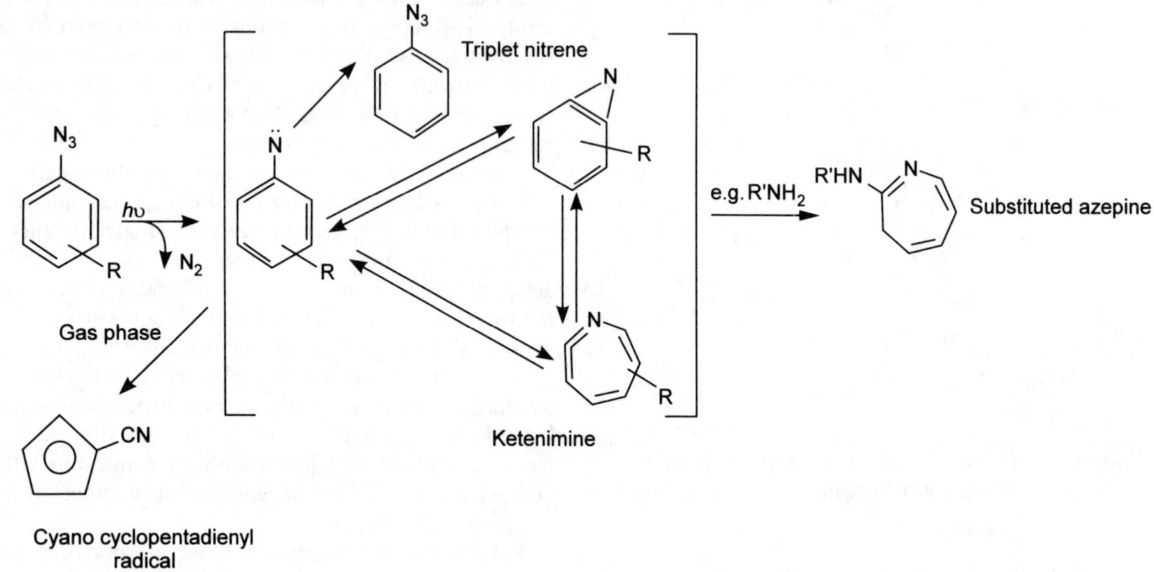

Figure 5.171 Generation and reaction of a nitrene.

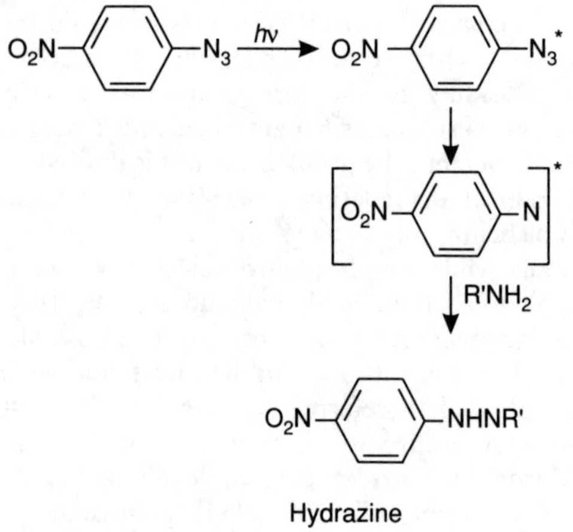

Figure 5.172 Hydrazine formation from *p*-nitrophenylazide compound.

is introduced (Schuster and Platz, 1992); the *p*-nitro compound, for example, can yield a hydrazine (Li *et al.*, 1988), as shown in Figure 5.172.

The reaction of aryl azides with thiols is a potential problem for approaches using these reagents (Staros *et al.*, 1978); this is particularly rapid with dithiols such as DTT (Bayley *et al.*, 1978), but also occurs more slowly with simple SH moieties.

The search for further improvements has led to some recent interest in fluorinated analogues, which will be discussed later.

5.10.3.1 Photolysis conditions for phenyl azides

As mentioned earlier, substituents to the aryl ring cause an alteration in the optimum energy for photoactivation, electron-withdrawing substituents allowing longer wavelengths to be used. Therefore the wavelength of light used must take the structure of the reagent into account; loosely speaking however, unsubstituted phenyl azides require irradiation in the region of 260–290 nm, while the presence of a nitro group introduces new absorptions at around 320–350 nm.

These higher wavelengths are inherently less likely to cause damaging effects to the photosensitive amino acid residues of the protein. Galardy *et al.* (1974) showed that irradiation (five hours, 4 °C, pH 7.4) at $\geq$280 nm destroyed 45% of cysteine, 30% of methionine, 3% of tyrosine and 3% of tryptophan in a mixture of free amino acids, less damage was caused to an intact protein (BSA) but 40% of cysteine was still lost. Irradiation at $\geq$320 nm for two hours at 4 °C and pH 7.4 caused no loss of amino acids either in free form or in BSA.

Because the longer-wave absorptions have lower extinction coefficients, however, photolysis requires a longer irradiation time. There is a balance between whether more damage will be caused by short exposure to low wavelengths, or by prolonged irradiation at a lower energy (Ji, 1983). Thermal damage due to heating of the solution can also be a factor in prolonged irradiation. In general Ji favoured flash photolysis.

Prolonged irradiation can be supplied from a simple UV light source such as the Mineralight™ type. Using a Mineralight USV-11 source, Ji recommended 3–5 minutes irradiation at a distance of 15 cm from the sample for reaction mixtures containing proteins or membranes. UV or direct viewing light for 5–10 minutes is recommended by the Pierce Chemical Co.

Flash photolysis using a more powerful source allows the reaction to proceed much more quickly, in the order of milliseconds. Ji quoted 1–10 flashes as a typical requirement in the presence of membranes, and suggested the use of inexpensive photographic flash units for such a purpose; Pierce recommend 5–10 flashes.

In terms of handling the unreacted azides, Pierce recommend the use of a darkroom with a red safety light. For unsubstituted phenyl azides where the wavelength is further from those of visible light, subdued lighting may be sufficient.

5.10.3.2 Amine reactive phenyl azides

Not surprisingly the most common heterobifunctionals incorporating the aryl azide function are designed to react with amines, and a bewildering array of these reagents have been tried out: Wong (1991, pp. 168–176) lists over 40 such compounds based on active esters, imidates, aryl halides and isothiocyanates. Only the most commonly-encountered or those with particularly useful attributes will be dealt with here.

They will be categorised by membrane-permeability and cleavability, reflecting their major application in structural studies. Another useful feature of the aryl azide ring is its capability for radioiodination, if a hydroxyl group is introduced to activate it towards electrophilic attack. The iodination can be carried out as a dark reaction with both the active ester and azide functionalities remaining essentially intact; this is particularly convenient for affinity labelling, facilitating the localisation of the probe after binding to the receptor has taken place.

Radiolabelling *after* binding can be carried out under more straightforward conditions, but obviously provides less unambiguous data due to potential iodination of histidines and tyrosines in the native protein. 'Cold' (inactive) iodination of the protein(s) before reaction will minimise this effect (although some exchange of active and inactive iodine atoms may occur), but leads to different problems, notably whether the introduction of large, fairly hydrophobic iodine atoms has affected the binding characteristics of the system under investigation.

Non-cleavable, membrane-permeable succinimide esters of phenyl azides

Succinimidyl-4-azidobenzoate (HSAB; Figure 5.173) is one of the simplest of photoactivatable active esters, and was first prepared by Galardy *et al.* (1974). It has an 8 Å span and is commercially available. Dark reaction conditions for HSAB are typical for succinimide esters: Ballmer-Hofer *et al.* (1982) for example derivatised human IgG and BSA with a 30-fold excess of HSAB (30 minutes, 23 °C, pH 7.4).

Succinimidyl 4-azidosalicylate (NHS-ASA; Figure 5.174) was described as a radioiodinatable photoactivatable reagent by Ji and Ji (1982). The presence of the hydroxyl group lowers the energy required for photoactivation so that the absorbance maximum is at 305 nm.

Radioiodination of this compound under non-aqueous conditions is desirable to avoid hydrolysis of the succinimide group. Early studies in acetone (Ji and Ji, 1982) gave a very low yield, but subsequent work (Ji *et al.*, 1985) identified a largely organic solvent system (acetonitrile–0.5 M phosphate (pH 7.0)–DMF, 24:5:1) which gave good results. A 63% incorporation of ^{125}I was obtained.

As mentioned above, the incorporation of a nitro group into the aryl ring allows irradiation to be carried out at a higher, less-damaging wavelength. *N*-5-Azido-2-nitrobenzoyloxysuccinimide (ANB-NOS; Figure 5.175) was introduced by Lewis *et al.* (1977) on this basis, and can be activated at 320–350 nm. It has a linker span of 7.7 Å.

Lewis used a tenfold excess of ANB-NOS in dioxane to derivatise phospholipase A_2 (four hours, room

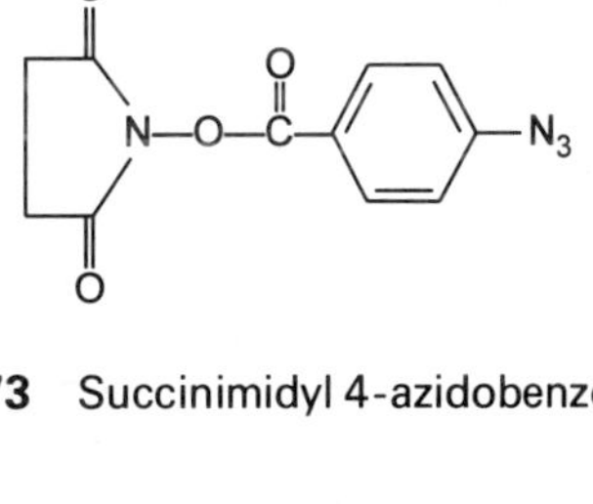

Figure 5.173 Succinimidyl 4-azidobenzoate (HSAB).

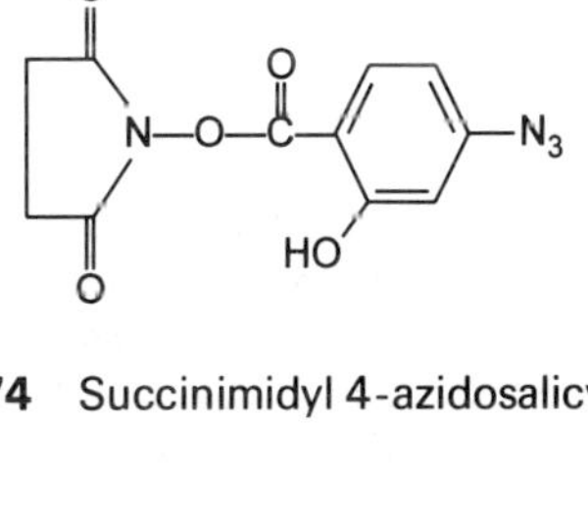

Figure 5.174 Succinimidyl 4-azidosalicylate (NHS-ASA).

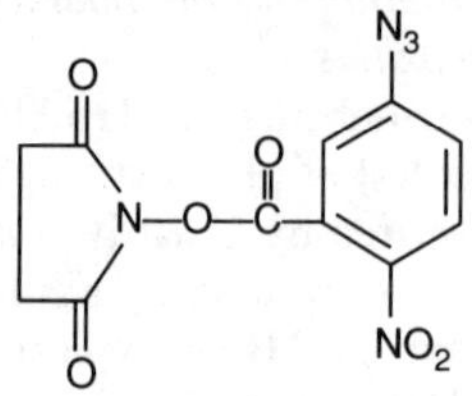

Figure 5.175 *N*-5-Azido-2-nitrobenzoyloxy-succinimide (ANB-NOS).

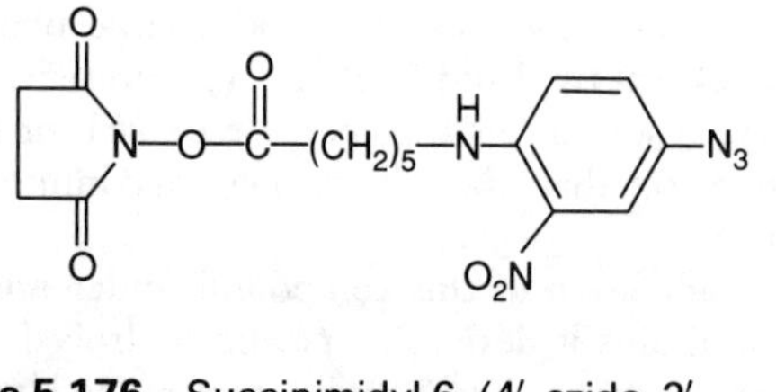

Figure 5.176 Succinimidyl 6-(4′-azido-2′-nitrophenylamino)hexanoate (SANPAH).

temperature, pH 9.0) with no detectable loss of enzyme activity; irradiation of the derivatised protein resulted in an 80–90% yield of cross-linked oligomers. Around half the enzyme activity was lost during the cross-linking procedure.

Moving on to longer-chain reagents, succinimidyl 6-(4′-azido-2′-nitrophenylamino)hexanoate (SANPAH; Figure 5.176) is a simple nitrophenyl azide, an 18 Å homologue of ANB-NOS.

Ballmer-Hofer *et al.* (1982) used a 30-fold excess of SANPAH to derivatise human IgG and BSA (30 minutes, 23 °C, pH 7.4): around 12 molecules were incorporated into IgG and 20 into BSA under these conditions. These reagents were used in the generation of anti-SANPAH antibodies, which could be used to isolate the products of photoaffinity labelling experiments. Erythrocyte ghosts were derivatised (30 minutes, 37 °C, pH 7.4), bound to lectins by photoactivation, and precipitated using these antibodies.

Wood and O'Dorisio (1985) used SANPAH to couple radiolabelled vasoactive intestinal peptide to receptors on lymphoblasts (20 minutes, on ice, pH 7.4), identifying a 47 kDa receptor protein by subsequent analysis.

Cleavable, membrane-permeable succinimide esters of phenyl azides

Cleavability is another desirable feature for some applications of these reagents. Typically the incorporation of a disulfide link has been employed to confer this property, and a number of such reagents were described by Vanin and Ji (1981).

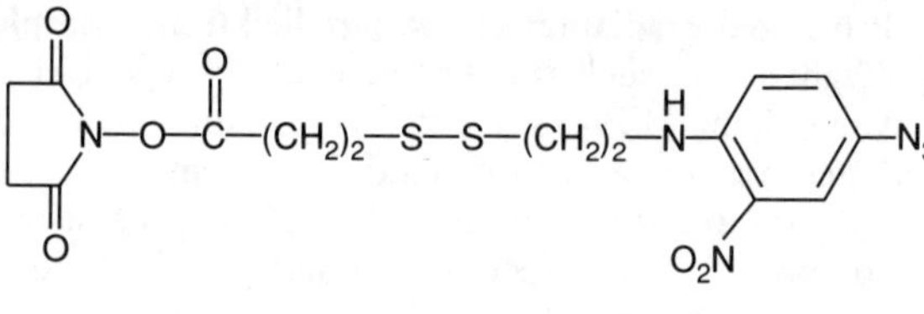

Figure 5.177 3-[(2-Nitro-4-azidophenyl)-2-aminoethyldithio]-*N*-succinimidylpropionate (NAP-AEDSP).

Succinimidyl (4-azidophenyl)-1-(3′-dithiopropionate) (SADP) is an example of this type, with a span of 13.9 Å. SADP (Figure 5.183(a)) was synthesised by Zarling *et al.* (1982) and used to derivatise Sendai virions (30 minutes, room temperature, PBS). The virions were quite sensitive to derivatisation, even at a fivefold loading of SADP there was a 30% reduction in bioactivity, but they were still able to infect cells.

Among the nitrophenyl azides, Schwartz *et al.* (1982) described 3-[(2-nitro-4-azidophenyl)-2-aminoethyldithio]-*N*-succinimidyl propionate (NAP-AEDSP; Figure 5.177).

Koch *et al.* (1990) described succinimidyl aryl azides containing a selenol function cleavable by mild oxidation (cf. similar bis-succinimidyl esters, Section 5.3.5).

Non-cleavable, membrane-impermeable succinimide esters of phenyl azides

The sulfonated analogues of HSAB and NHS-ASA are available, as well as a sulfonated long-chain caproamido analogue (18 Å), also first described by Ji and Ji (1982), and sulfosuccinimidyl 4-(*p*-azidophenyl)-butyrate (sulfo-SAPB; Figure 5.178), a 12.8 Å analogue.

The 7-azido-4-methylcoumarin-3-acetate (AMCA) group contains the photoactivatable phenyl azide function but is also a fluorophore, allowing its position to be identified without additional labelling. Two reagents combining succinimide esters with the AMCA group are available, one non-cleavable and

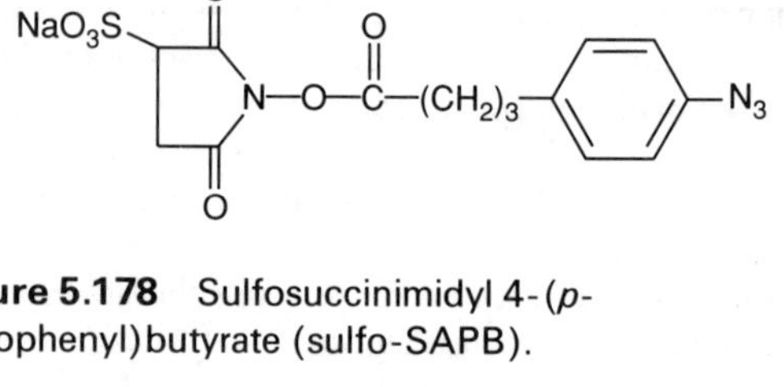

Figure 5.178 Sulfosuccinimidyl 4-(*p*-azidophenyl)butyrate (sulfo-SAPB).

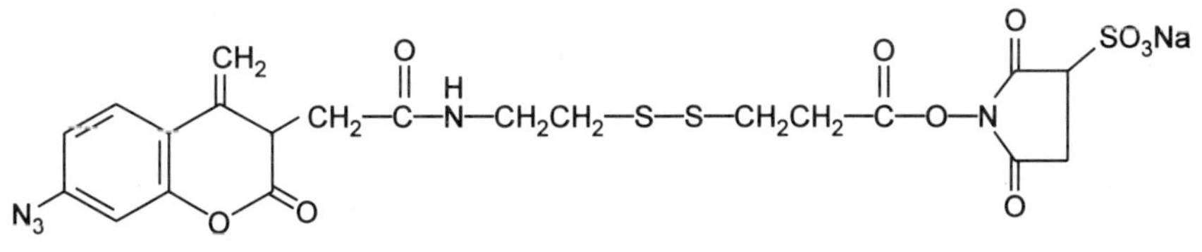

Figure 5.179 Sulfosuccinimidyl 2-(7-azido-4-methylcoumarin-3-acetamido)-1,3′-dithiopropionate (SAED), sodium salt.

one cleavable. The former is sulfosuccinimidyl 2-(7-azido-4-methylcoumarin-3-acetamido)-1,3′-dithiopropionate (SAED; Figure 5.179).

In the nitrophenyl azide category, the sulfosuccinimidyl analogue of SANPAH has been commercialised by Pierce but references to its use are scarce.

Cleavable, membrane-impermeable succinimide esters of phenyl azides

Sulfosuccinimidyl 2-(*p*-azidosalicylamido)ethyl-1-(3′-dithiopropionate) (SASD; Figure 5.180) is both cleavable and radioiodinatable, with an 18.9 Å span. Despite its sulfonate group SASD is only sparingly soluble in aqueous systems, and an organic cosolvent is generally employed. This reagent offers the interesting capability of transferring the radiolabel from an affinity ligand to its receptor (Figure 5.181): the ligand is labelled with ^{125}I-SASD, coupled to the receptor following photoactivation, then cleaved to free the ligand while leaving the radiolabel attached.

Sorenson *et al.* (1986), for example, used this capability to identify the receptor protein for mouse interleukin-3. The radiolabelled SASD was dissolved in acetone[1] and conjugated to the interleukin (30 minutes, 23 °C, pH 9.0). Photoactivation of the azide function and reduction with mercaptoethanol left the radiolabel attached to a 67 kDa receptor protein.

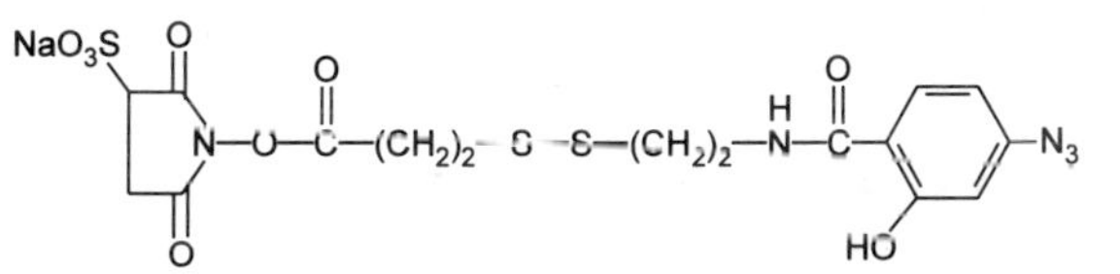

Figure 5.180 Sulfosuccinimidyl 2-(*p*-azidosalicylamide)ethyl-1-(3′-dithiopropionate) (SASD), sodium salt.

[1] Shephard *et al.* (1988) later found SASD to be insoluble in acetone: this may be connected with variable purity of the reagent. Similar purity-dependent batch-to-batch variations in the solubility of sulfosuccinimidyl reagents have been noted in the authors' laboratory.

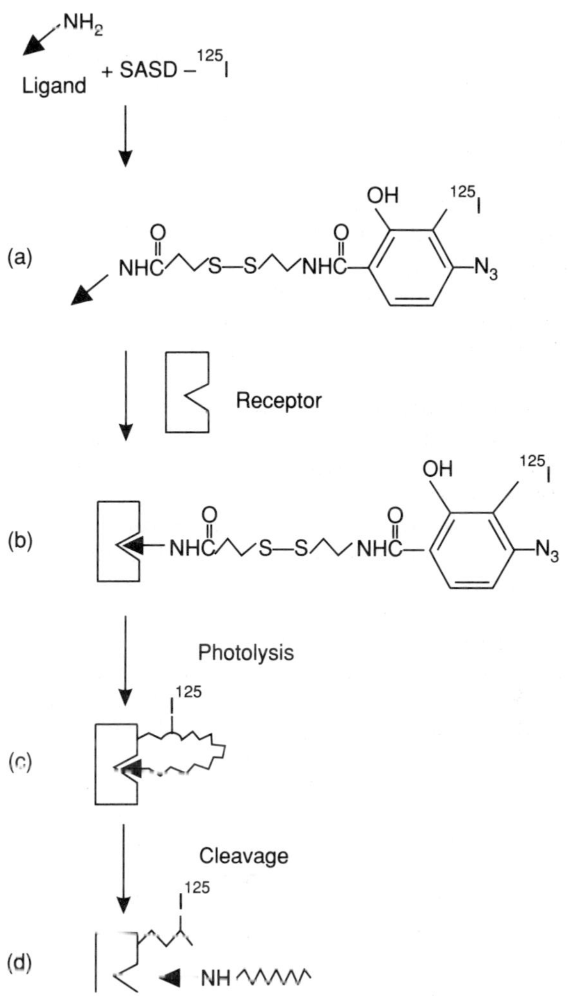

Figure 5.181 Use of SASD to transfer radiolabel from a ligand to its receptor: the ligand is labelled with SASD *via* an amine group (a), and allowed to bind to its receptor (b). Photoactivation creates a covalent link through the phenyl moiety (c), and subsequent cleavage of the disulfide link allows the ligand to desorb while leaving the ^{125}I-labelled phenyl residue attached to the receptor (d).

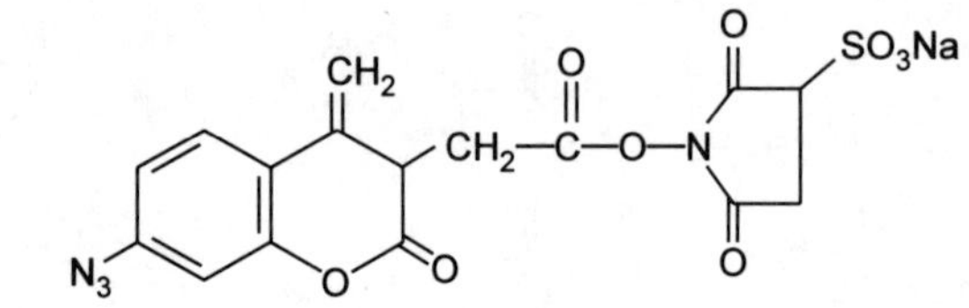

Figure 5.182 Sulfosuccinimidyl 7-azido-4-methylcoumarin-3-acetate (sulfo-SAMCA), sodium salt.

Similarly, Shephard *et al.* (1988) identified receptors for phytohaemagglutinin in human mononuclear cell membranes. Again a radiolabel was transferred to the receptors by SASD derivatisation (30 minutes, 4 °C, PBS), irradiation and reduction.

The sulfonated analogue of SADP is available; so is a cleavable AMCA reagent, sulfosuccinimidyl 7-azido-4-methylcoumarin-3-acetate (sulfo-SAMCA; Figure 5.182). In the nitrophenyl azide category, sulfosuccinimidyl 2-(*m*-azido-*o*-nitrobenzamido)ethyl-1-(3′-dithiopropionate) (SAND), is also available.

Other amine-reactive phenyl azides

While succinimide esters have come to predominate among amine-active groups, imidates do have the advantage that their incorporation does not alter the charge of the protein under normal pH conditions. In view of the importance of correct binding behaviour for affinity probes, it may sometimes be important to take this factor into account. Many reagents combining aryl azide and imidate functions are listed by Wong (1991, pp. 168–176) and in earlier reviews (Ji, 1979).

The simplest of these is methyl 4-azidobenzimidate (MABI; Figure 5.184), which was introduced by Ji (1976) for use in probing lectin–receptor interactions in human erythrocyte ghosts. ^{125}I-Concanavalin A was derivatised by MABI at pH 9.5 (two hours at 25 °C) before incubation with the ghosts and subsequent photoactivation.

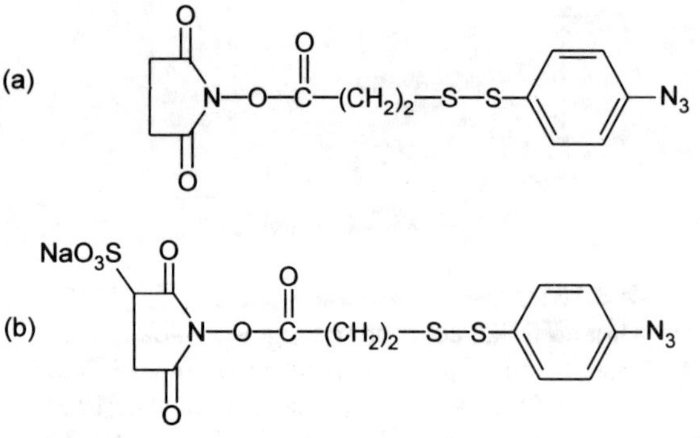

Figure 5.183 (a) Succinimidyl (4-azidophenyl)-1-(3′-dithiopropionate (SADP) and (b) the sodium salt of its sulfonated analogue, sulfo-SADD.

Another imidate is methyl 4-(6-formyl-3-azidophenoxy)butyrimidate (FAPOB; Figure 5.185); the formyl group can be used to incorporate radioactivity *via* reductive alkylation with sodium borotritiide (NaB^3H_4), or can form the basis of further linkage to amine residues (Maassen *et al.*, 1981).

Vanin and Ji (1981) described the synthesis of methyl 3-[(4-azidophenyl)dithio]propionimidate (MADP; Figure 5.186) which is a cleavable example from this category.

Other amine-reactive agents of this type include isothiocyanates (Sigrist *et al.*, 1982) and aryl fluorides (Bisson *et al.*, 1978; Burleigh *et al.*, 1978). These are highly hydrophobic reagents, and can be used to probe apolar environments. Structures of such compounds are shown in Figure 5.187.

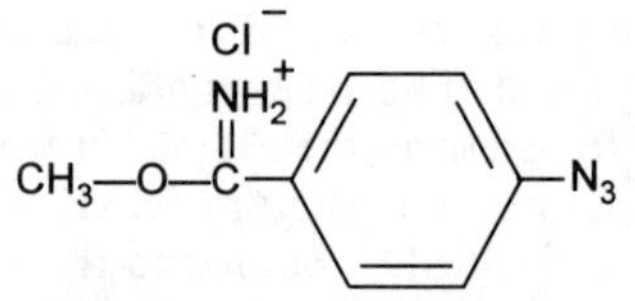

Figure 5.184 Methyl 4-azidobenzimidate (MABI).

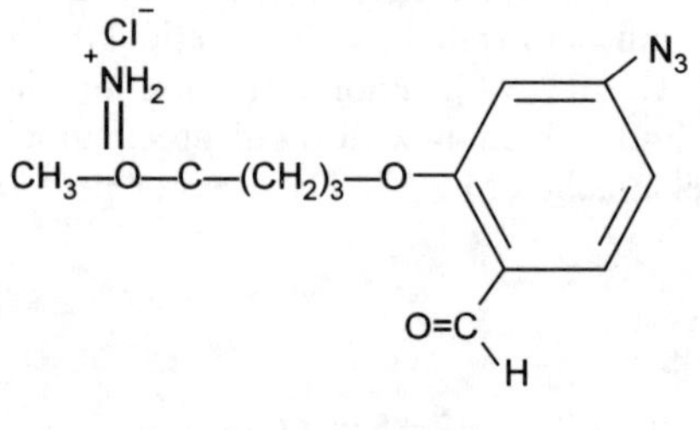

Figure 5.185 Methyl 4-(6-Formyl-3-azidophenoxy)-butyrimidate (FAPOB).

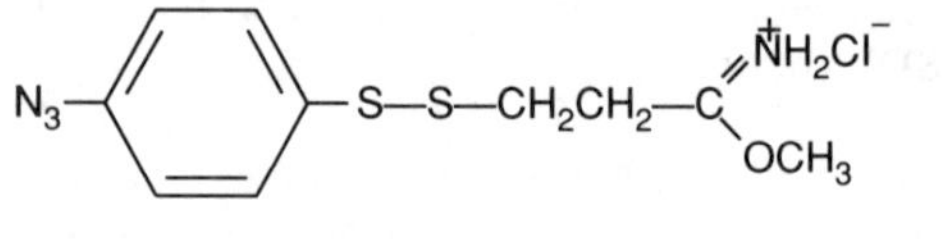

Figure 5.186 Methyl 3-[(4-azidophenyl)dithio] propionimidate (MADP).

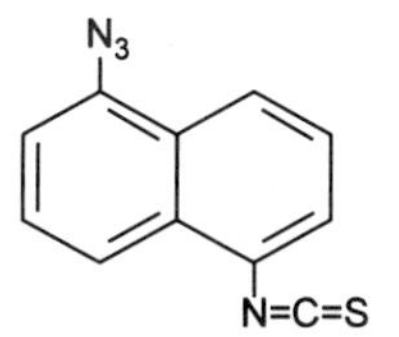

5-Isothiocyanato-1-naphthalene azide ($N_3NapNCS$)

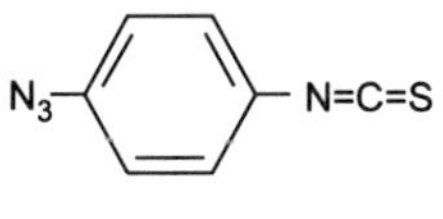

p-Azidophenylisothiocyanate (N_3PhNCS)

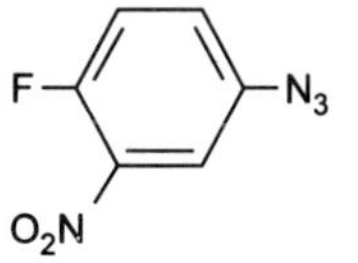

4-Fluoro-3-nitrophenylazide

Figure 5.187 Hydrophobic, amine-reactive phenyl azides.

5.10.3.3 Thiol-reactive phenyl azides

Phenyl azide reagents based on a number of thiol-reactive groups are available. Those which generate an S–S bond in the product (disulfides or analogous sulfenyl reagents) have a particular advantage if cleavability is require, and examples of these are common.

N-[4-(*p*-Azidosalicylamido)butyl]-3′-(2′-pyridyldithio)propionamide (APDP; Figure 5.188) combines the thiol reactivity of the pyridyldisulfide moiety with the photoactivatable, iodinatable azidosalicyl function. It is commercially available from Pierce. A shorter analogue was described by Ebright *et al.* (1996): *S*-[2-(4-azidosalicylamido)ethylthio]-2-thiopyridine (AET).

Another pyridyldisulfide–phenyl azide reagent was described by Chong and Hodges (1981): *N*-(4-azidobenzoylglycyl)-*S*-(2-thiopyridyl)cysteine (AGTC; Figure 5.189). It contains a glycine moiety for inexpensive incorporation of a radiolabel, and the intact cysteine carboxyl function to confer water-solubility.

Huang and Richards (1976) described a bis-nitrophenyl azide containing an S–S(=O)$_2$ linkage, effectively a disulfide bond activated towards reaction with thiols. This compound, bis[*N*-(2-nitro-4-azidophenyl)] cystamine-*S*,*S*-dioxide (DNCO; Figure 5.190), is highly lipophilic and is therefore suitable for probing apolar areas such as the interior of cell membranes. It clearly also has the capability for homobifunctional reactions (see Section 5.4.3). Moreland *et al.* (1982) described a thiophthalimide reagent, *N*-(4-azidophenylthio)phthalimide (APTP) for incorporating phenyl

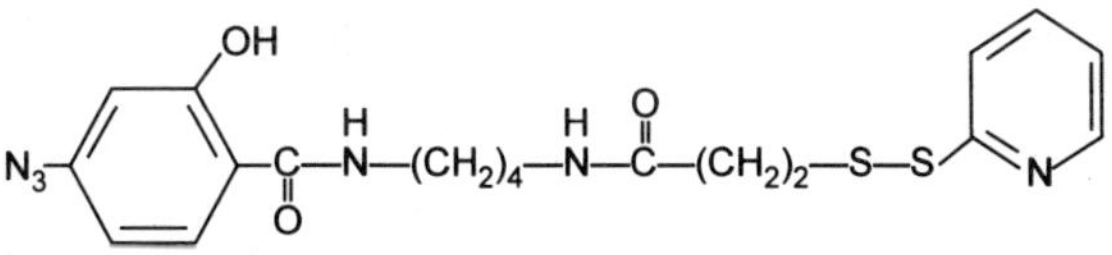

Figure 5.188 *N*-[4-(*p*-Azidosalicylamido)butyl]-3′-(2′-pyridyldithio)propionamide (APDP).

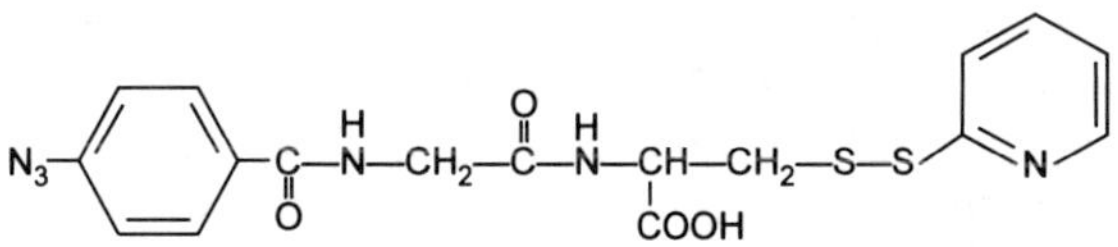

Figure 5.189 *N*-(4-Azidobenzoylglycyl)-*S*-(2-thiopyridyl)cysteine (AGTC).

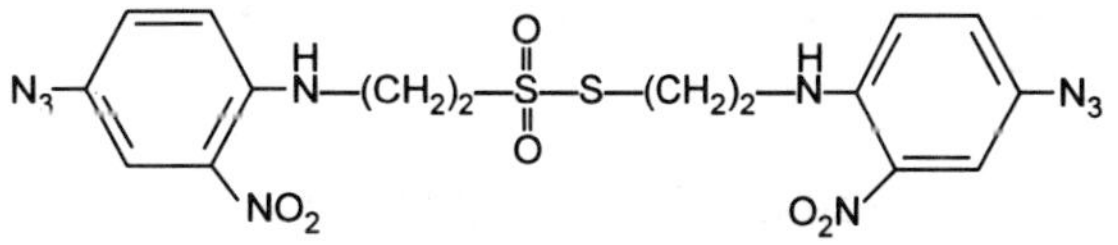

Figure 5.190 Bis[*N*-(2-nitro-4-azidophenyl)] cystamine-*S*,*S*-dioxide (DNCO).

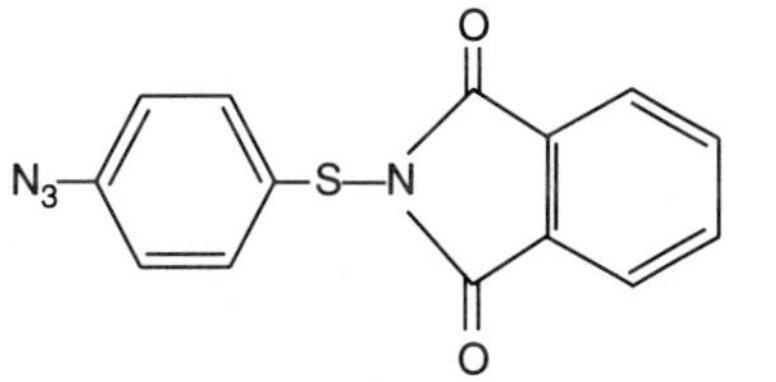

N-(4-Azidophenylthio) phthalimide (APTP)

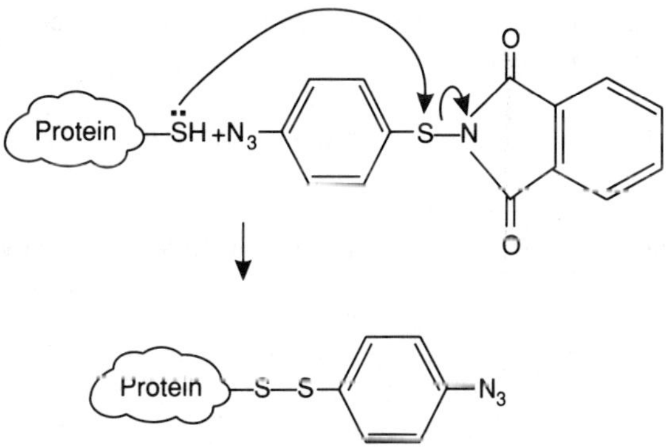

Figure 5.191 Reaction of APTP with free thiol groups in proteins.

azide; it too reacts with thiols to form a disulfide linkage (Figure 5.191).

Moving away from disulfide-forming reagents, Trommer *et al.* (1977) describes the simple 4-azidophenylmaleimide (Figure 5.192), although it was not favoured due to the high-energy light required for activation. The preparation of this compound was described by Trommer and Hendrick (1973). *p*-Azidophenacyl bromide (Figure 5.193) was described by Hixson and Hixson (1975); it was used, for example, to label the free thiol of cytochrome *c* (Erecińska, 1977). An iodoacetyl aryl azide is commercially available: 1-(*p*-azidosalicylamido)-4-(iodoacetamido)butane (ASIB; Figure 5.194).

An interesting photoactivatable reagent based on a selenol ester was described by Trommer *et al.* (1977), 2-carboxyethyl selenol-4-azido-2-nitrobenzoate (Figure 5.195). Selenol esters react with thiols to yield thioesters and selenols (Figure 5.196), the latter dimerising readily to give a favourable equilibrium.

5.10.3.4 Carboxyl-reactive phenyl azides

4-(*p*-Azidosalicylamido)butylamine (ASBA; Figure 5.197) has a 16.3 Å span and can be coupled to carboxyl functions *via* a suitable dehydrating agent such as EDAC. Similar reagents which link amines to nitrophenyl azide groups have been described (Darfler and Marinetti, 1977; Das and Fox, 1979), for example *N*-(4-azido-2-nitrophenyl)diamine (ANPEDA; Figure 5.198).

5.10.3.5 Aldehyde-reactive phenyl azides

The aryl azide–hydrazide reagent 4-azidobenzoyl hydrazide (ABH; Figure 5.199) can be used to incorporate the photoactivatable group into periodate-oxidised carbohydrate moieties on glycoproteins.

5.10.3.6 Arginine-reactive phenyl azides

The glyoxal group has a high degree of selectivity for arginine, and APG, *p*-azidophenylglyoxal monohydrate (Figure 5.200), has been used to introduce aryl azides into proteins by this route.

This reagent was described by Vanin *et al.* (1981) and Ngo *et al.* (1981), the latter authors demonstrated that a number of enzymes containing active-site arginines were inactivated by treatment with APG in the dark (1–2 hours, pH 7.5), and that subsequent irradiation resulted in the formation of oligomers.

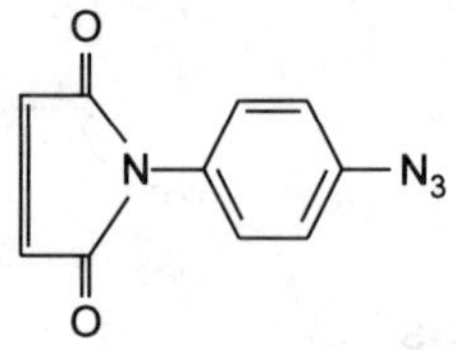

Figure 5.192 4-Azidophenylmaleimide (APM).

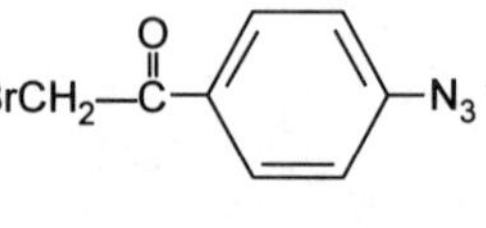

Figure 5.193 *p*-Azidophenacyl bromide (APB).

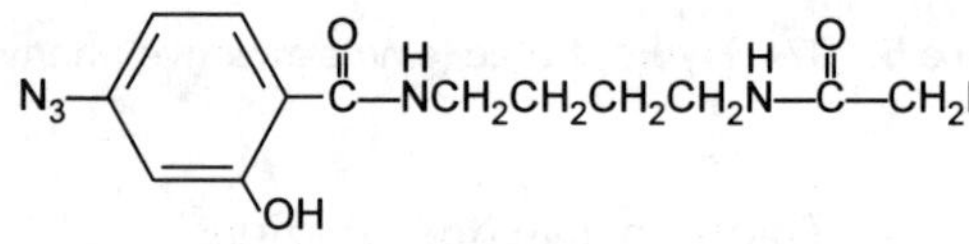

Figure 5.194 1-(*p*-Azidosalicylamido)-4-(iodoacetamido)butane (ASIB).

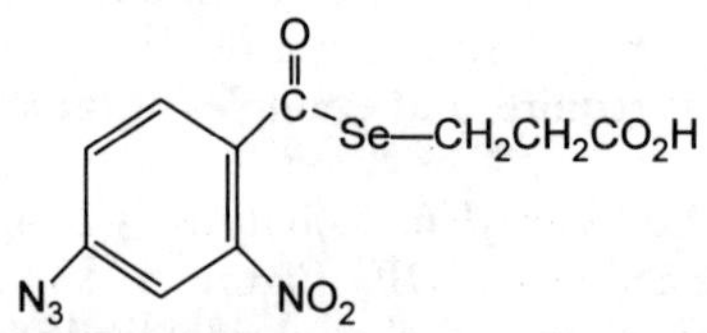

Figure 5.195 2-Carboxyethyl selenol-4-azido-2-nitrobenzoate.

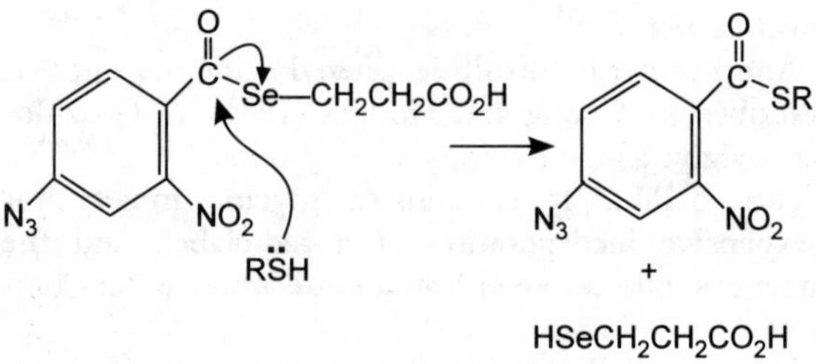

Figure 5.196 Reaction of selenol esters with thiols.

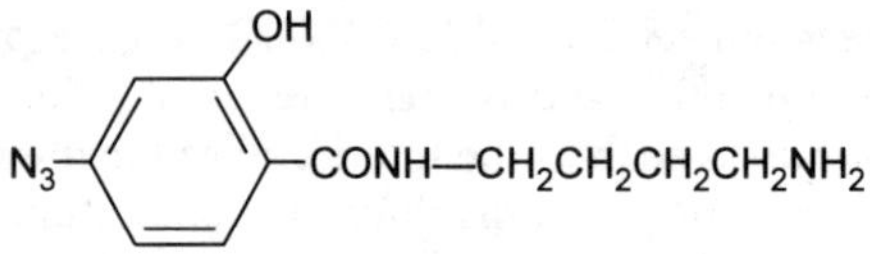

Figure 5.197 4-(*p*-Azidosalicylamido)butylamine (ASBA).

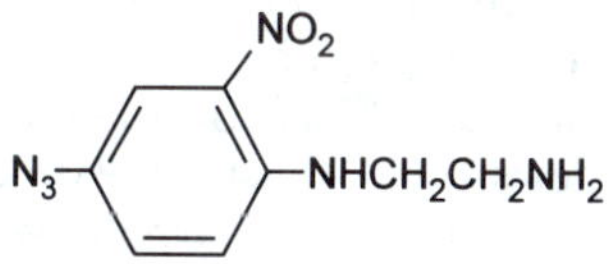

Figure 5.198 *N*-(4-Azido-2-nitrophenyl)ethylenediamine (ANPEDA).

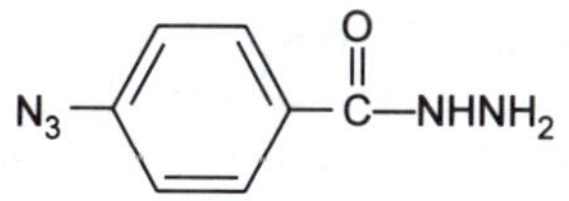

Figure 5.199 4-Azidobenzoyl hydrazide (ABH).

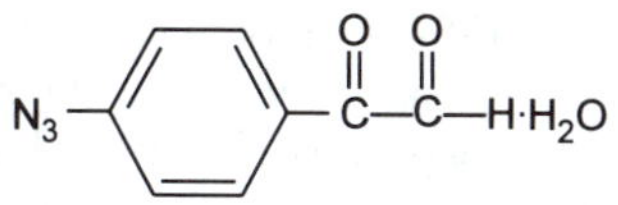

Figure 5.200 *p*-Azidophenylglyoxal monohydrate.

5.10.4 Photoactivatable heterobifunctional reagents based on polyfluorophenyl azides

While most of the reagents discussed above probably rely on ketenimine intermediates, a good deal of attention in recent years has focussed on polyfluorophenyl azides (PFPAs). Work in organic systems suggested that these reagents were particularly effective in facilitating nitrene reactions: for example Leyva *et al.* (1986) from the Ohio State University reported the formation of nitrene-derived products from the reaction of pentafluoro- and 2,6-difluorophenyl azide with toluene at low temperatures.

The chemistry of pentafluorophenyl azide has been thoroughly reviewed by Schuster and Platz (1992), although much of the work described is in systems very different to those encountered in protein chemistry. In summary, both singlet and triplet nitrenes can be involved in photoactivated reactions, and additives to the reaction mixture can cause dramatic changes in the products obtained. In contrast to the unsubstituted phenyl azide, ring-expansion to yield ketenimines is not generally favoured: molecular orbital effects make this a much less facile reaction than the unsubstituted case, and thermodynamic factors may also be involved. Overall, it can be stated that polyfluorophenyl azides offer a much greater chance of obtaining genuine nitrene reactions than their unsubstituted analogues.

More recent work from Platz's Ohio group has elucidated the situation further (Schnapp *et al.*, 1993; Schnapp and Platz, 1993). It has been shown that 2,6-difluoro substitution of the phenyl azide is necessary to give the desired retardation of ring-opening, and consequent promotion of nitrene reactions. In hydrophobic environments the singlet nitrene is favoured, leading to the classical reactions with C–C and C–H bonds, as well as with nucleophiles, and it is suggested that these reagents will work well in PAL applications in hydrophobic regions. A hydrophilic environment, on the other hand, promotes the generation of triplet nitrenes, whose free-radical-like reactivity is predicted to lead to cleavage of amino acids rather than labelling.

Workers at the University of Oregon have described many PFPA reagents for labelling purposes, including 4-azidotetrafluorobenzoyl chloride (Keana and Cai, 1989), succinimidyl 4-azido-2-iodotrifluorobenzoate (Cai and Keana, 1989; Cai *et al.*, 1992), 4-azidotetrafluorobenzaldehyde, succinimidyl 4-azidotetrafluorobenzoate, 4-azidotetrafluorophenylbenzylamine, 4-azidotetrafluorobenzoylimidazole, 4-azidotetrafluorophenacyl bromide (Keana and Cai, 1990), a range of 4-azidotetrafluorophenylglutaraldehyde reagents (Cai and Keana, 1991a), succinimidyl 3-(4-azidotetrafluorobenzamido)-5-iodobenzoate (Cai *et al.*, 1992), 4-azidotetrafluorophenyl-maleimide compounds (Aggeler *et al.*, 1992) and longer-chain succinimido-PFPAs such as succinimidyl 5-(4-azidotetrafluorobenzamido)pentanoate (Yan *et al.*, 1994). The iodine-containing compounds were designed with the introduction of a radiolabel in mind.

The tetrafluorophenyl azide maleimides were used to study subunit topography in *E. coli* ATP synthase mutants (Aggeler and Capaldi, 1992; Aggeler *et al.*, 1992).

At the University of Kentucky, Crocker *et al.* (1990) described succinimidyl *N*-(4-azidotetrafluorobenzoyl)tyrosinate and succinimidyl 2-(4-azidotetrafluorophenyl)thiazole-4-carboxylate as photolabelling reagents convenient for the introduction of ^{125}I and ^{35}S radiolabels respectively. These reagents were used to investigate the binding of calmodulin to membrane proteins.

Some of the PFPA-based heterobifunctionals described above are shown in Figure 5.201.

Although PFPA labelling reagents for protein cross-linking have not yet been widely employed outside these pioneering groups, their superior photochemical properties suggest that it is only a matter of time before they gain a broader acceptance, and commercialisation of some of the more useful compounds seems very likely.

5.10.5 Other photoactivatable heterobifunctional reagents

On irradiation at 340–400 nm, the coumarin molecule undergoes photodimerisation *via* the 3,4-double bonds. Heterobifunctional maleimide-derivatised coumarins have been in use for many years for reaction with thiols (Sutoh and Matsuzaki, 1980), while Baranowska-Kortylewicz and Kassis (1993) described the synthesis and use of reagents combining a coumarin ring with a haloalkyl group; an example is shown in Figure 5.202.

Benzophenones are also photoactivated at conveniently high wavelengths (>320 nm), undergoing insertion reactions into C–H bonds which are relatively unaffected by interference from nucleophiles (Dormán and Prestwich, 1994); they therefore seem to offer considerable potential in photolabelling applications. Heterobifunctional reagents based on benzophenones have been described, with 'handles' such as maleimides (Tao *et al.*, 1984; Leszyk *et al.*, 1987) or succinimide esters (Olszewski *et al.*, 1995; Hypolite *et al.*, 1997). One such reagent is shown in Figure 5.203. Photoheterobifunctional reagents also exist—see Section 6.16.5.

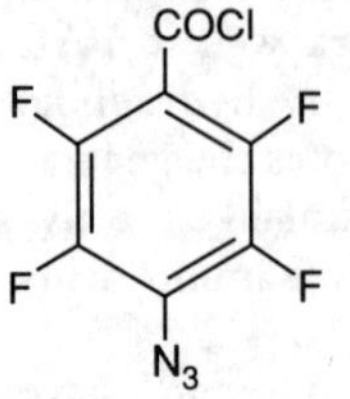

4-Azidotetrafluorobenzoyl chloride

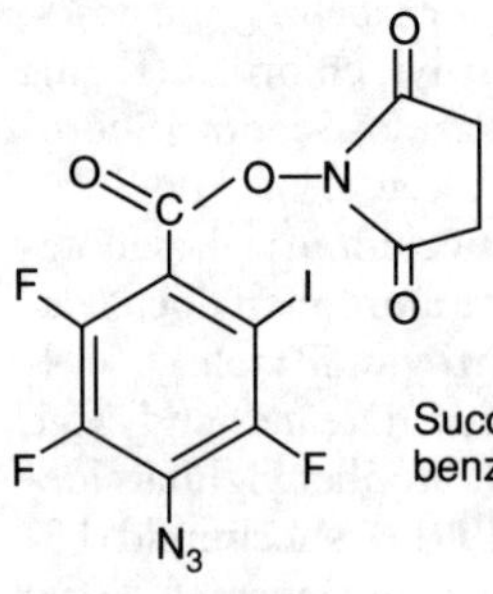

Succinimidyl 4-azido-2-iodotrifluorobenzoate

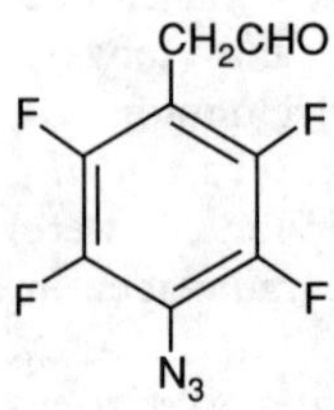

4-Azidotetrafluorobenzaldehyde

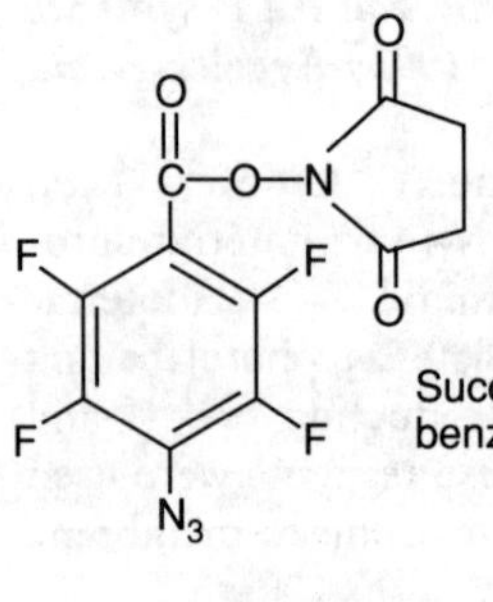

Succinimidyl 4-azidotetrafluorobenzoate

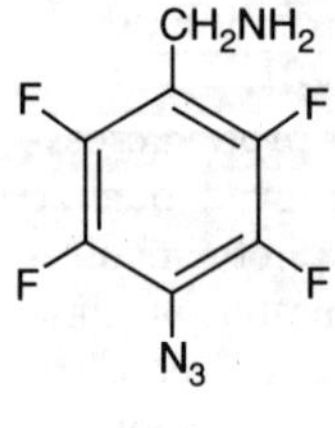

4-Azidotetrafluorophenylbenzylamine

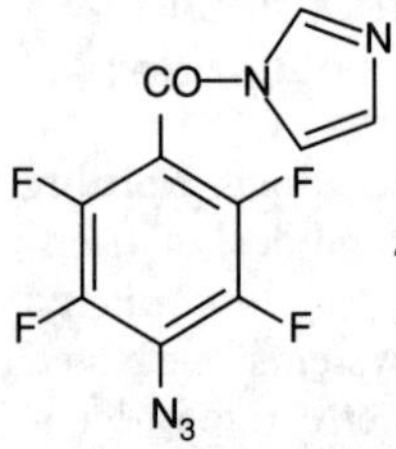

4-Azidotetrafluorobenzoylimidazole

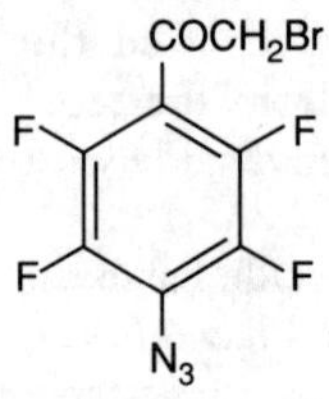

4-Azidotetrafluorophenacyl bromide

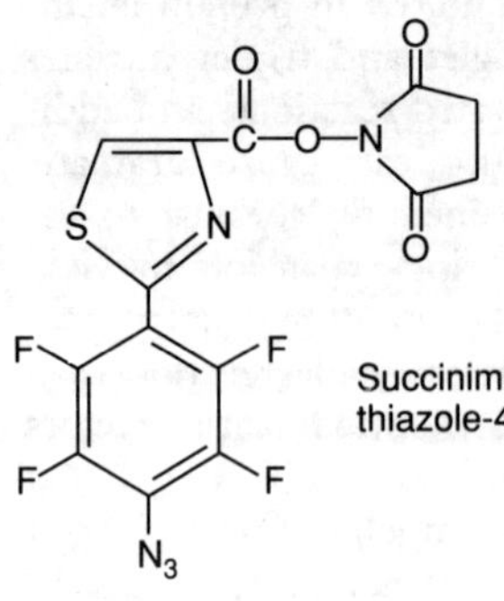

Succinimidyl 2-(4-azidotetrafluorophenyl)-thiazole-4-carboxylate

Figure 5.201 Some PFPA-based heterobifunctional reagents.

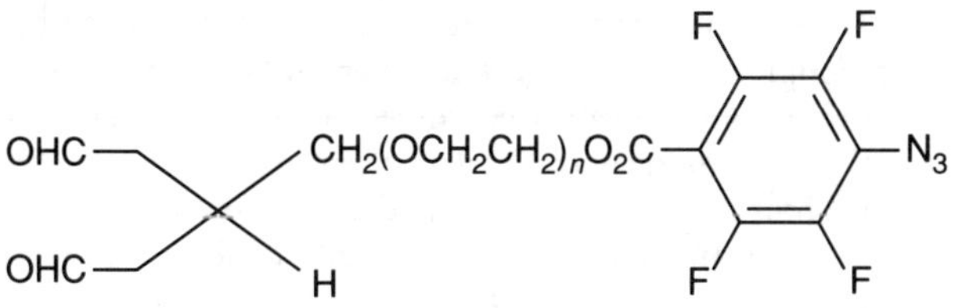

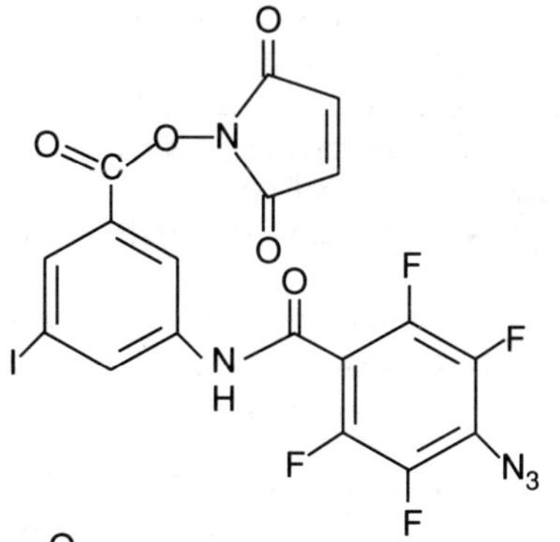

Succinimidyl 3-(4-azidotetrefluorobenzamido)-5-iodobenzoate

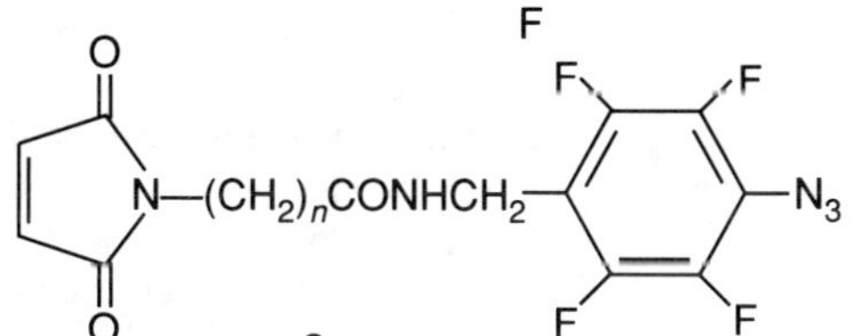

n = 2
n = 5

4-Azidotetrafluorophenyl-maleimide compounds

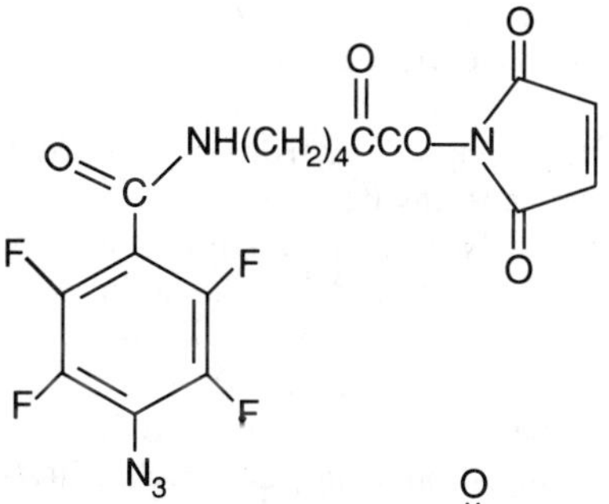

Succinimidyl 5-(4-azidotetrafluorobenzamido)pentanoate

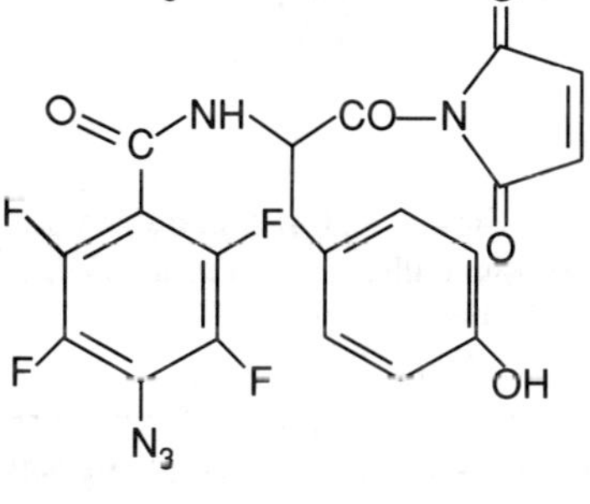

Succinimidyl *N*-(4-azidotetrafluorobenzoyl)tyrosinate

Figure 5.201 (contd.)

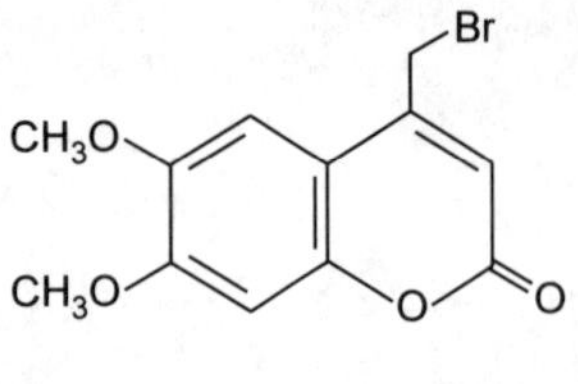

Figure 5.202 A haloalkyl coumarin derivative.

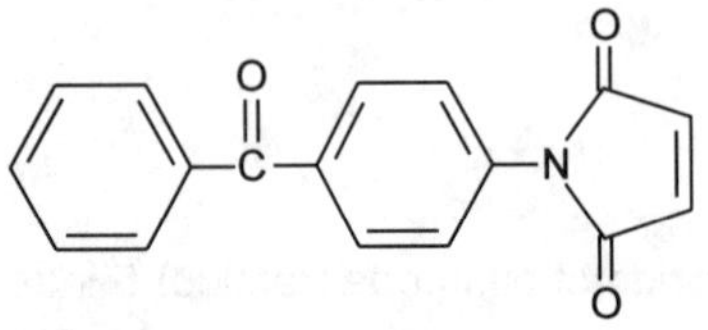

Figure 5.203 4-Maleimidobenzophenone.

5.11 Heterobifunctional Reagents Based on the Biotin–Avidin Interaction

Although they do not involve the formation of any covalent bonds, the reactions between the small vitamin biotin and the proteins avidin (from chicken egg white) and streptavidin (from *Streptomyces avidinii*) are so strong that the resulting complexes can often be viewed as irreversibly bound; the affinity constants ($K_a = 10^{15}\,M^{-1}$) for these interactions are among the highest observed in biological systems (Green, 1975).

The structure of biotin and the binding proteins and the nature of their interaction are described in more detail below; at the simplest level, however, each avidin or streptavidin molecule possesses four identical subunits containing a biotin-binding site. The four binding sites are cooperative in nature: the affinity constant of monovalent avidin is around $10^7\,M^{-1}$, providing reversible biotin binding which can be preparatively useful (Kohanski and Lane, 1990). Although each molecule theoretically binds four biotins, it has frequently been observed that lower valencies are obtained from commercial material: Shao *et al.* (1990) quoted a typical range of 2.8–3.2.

In terms of a practical tool for protein chemistry, the key features of the biotin–avidin system have been summarised by Wilchek and Bayer (1990a) as follows. (1) High affinity and stability of avidin–biotin complexes. (2) Straightforward methods for labelling other molecules with biotin, with physical characteristics and biological activity generally retained after biotinylation. (3) Multiply-biotinylated molecules plus tetrameric binding capacity of avidin can lead to amplification of binding events. (4) Amenable to double-labelling and kinetic studies. (5) Versatility through use of a variety of avidin-labelled probes to analyse one biotinylated molecule, or *vice versa*. (6) Wide commercial availability of biotinylation reagents, and biotin- and avidin-labelled compounds.

The availability of 'off-the-shelf' avidin or streptavidin-labelled reagents (e.g. streptavidin–HRP, streptavidin–AP) means that the only chemistry which needs to be carried out to create a specific labelled protein may be a simple biotinylation. Not only are efficient biotinylation reagents readily available for this purpose, but the purification of the product is often straightforward due to the small size of the biotin derivative. A simple size exclusion step is frequently all that is required.

Coupled with the potential for amplification noted above, it is not surprising that the biotin–avidin system has had a dramatic impact on protein chemistry techniques.

In 1975, a major review by Michael Green of the National Institute of Medical Research in London delineated the key attributes of the biotin–avidin interaction, and helped to provide the impetus for a plethora of biomedical applications: cytochemistry (Heitzmann and Richards, 1974; Bayer *et al.*, 1976), gene mapping (Manning *et al.*, 1975; Broker *et al.*, 1978), immunoassay (Guesdon *et al.*, 1979), immunotherapy (Philpott *et al.*, 1980), etc. Virtually every area where protein chemistry is applied has seen an application of this technology—the 1990 review mentioned earlier lists 60 prior articles in the *Methods in Enzymology* series alone.

The authors of this review, Edward Bayer and Meir Wilchek of the Weizmann Institute in Rehovot, Israel, have become particularly associated with promoting the utility of the technique, and have written many other articles on this topic (Bayer and Wilchek, 1978, 1980, 1990, 1997; Wilchek and Bayer, 1984, 1990a,b). They and their colleagues have also introduced a number of the biotinylation reagents which will be discussed below.

5.11.1 Biotin

The structure of biotin (Figure 5.204) was determined by du Vigneaud and his colleagues at the Cornell University Medical College in New York in the 1940s (Hofmann *et al.*, 1942; du Vigneaud *et al.*, 1942). Synthetic routes to this compound were subsequently derived by workers at Merck and Co. in New Jersey (Harris *et al.*, 1943, 1944).

Biotin is a bicyclic molecule consisting of fused imidazolinone and thiophan rings, the latter bearing a pentanoate (valerate) side-chain; this carboxyl group clearly offers a convenient site for the formation of reactive derivatives. The predominant chemical attribute of the rest of the molecule is its lack of reactivity; there are, however, three asymmetric carbon atoms (C2, C3, C4) which confer chirality, the stereochemistry of the naturally occuring D-biotin is that depicted in Figure 5.204.

The *in vivo* function of biotin (vitamin H) was investigated in the 1930s and 1940s (György, 1954), following the observation that rats obtaining all their protein from egg-white developed a form of dermatitis which could be counteracted by a component of many foodstuffs. Like many other vitamins it was observed that this compound performed a coenzyme function *in vivo*, and biotin groups were subsequently found (Wakil *et al.*, 1958) to be covalently incorporated in some carboxylases and transcarboxylases.

In the ubiquitous pyruvate carboxylase, for example, biotin acts as a cofactor by binding carbon dioxide and transferring it to pyruvate to form oxaloacetate. Transcarboxylase, found in propionic acid bacteria, catalyses the transfer of the carboxyl group of methylmalonyl CoA to pyruvate, again producing oxaloacetate. This enzyme contains a number of subunits and a total of six biotin groups (Green *et al.*, 1972).

Avidin affinity columns have been employed in the purification of transcarboxylase subunits (Berger and Wood, 1975), while Rylatt *et al.* (1977) resolved biotin-containing tryptic peptides of pyruvate carboxylase by virtue of their avidin-binding capability. Avidin-based affinity chromatography remains a common approach for purification of biotin-containing enzymes (Purcell and Wallace, 1996).

Although all parts of the biotin molecule exhibit some binding to avidin, it is the imidazolidone ring which is the most important feature in this respect (Green, 1975; Bayer and Wilchek, 1978). The thiophan and valerate moieties can both be altered substantially while retaining the avidin-binding capability of the molecule, again favouring the valerate arm as a site for derivatisation. Some modification of the imidazolinone ring is also possible: replacement of the C=O function with a C=NH moiety, for example, yields iminobiotin (Figure 5.205) which retains a strong, though reduced, affinity for the binding proteins.

This lower affinity is especially pronounced at low pH, and can be used to advantage where a reversible interaction is desired, the prodigious affinity of the native biotin molecule in this case being a positive disadvantage. In affinity purification, for example, the harsh conditions required for dissociation of the biotin–avidin complex (6 M guanidine-HCl at pH 1.5; Cuatrecasas and Wilchek, 1968) were a disincentive to early development (Bayer and Wilchek, 1978). Iminobiotin can be dissociated from avidin at around pH 4, and has been employed, for example, in the affinity purification of streptavidin and other proteins (Hofmann *et al.*, 1980; Orr, 1981; Zeheb *et al.*, 1983).

The small size of the biotin molecule is undoubtedly an advantage for its use as a protein label, the functional activity of the labelled molecule generally being well retained. Nevertheless, it should be borne in mind that biotinylation—like any other derivatisation procedure—can cause functional changes in some cases. Storm *et al.* (1996), for example, observed changes in binding of a number of proteins to cell surfaces on biotinylation; the effect was accompanied by a large increase in hydrophobicity of the target proteins, suggesting that this was an important factor.

4-Hydroxyazobenzene-2′-carboxylic acid (HABA; Figure 5.206) has a reasonably high affinity for avidin and streptavidin, although it does show some binding to albumins, and forms the basis of a colorimetric assay for biotin binding or capacity (Green, 1965, 1970). The 348 nm absorbance of the free dye almost disappears on binding, while a strong new band appears at 500 nm. Livnah *et al.* (1993b) and Weber *et al.*

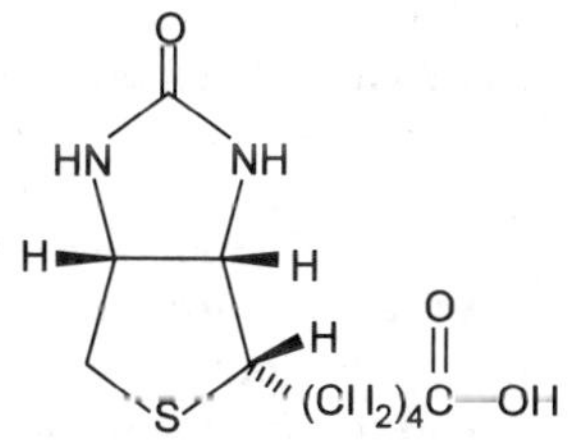

Figure 5.204 Structure of D-biotin.

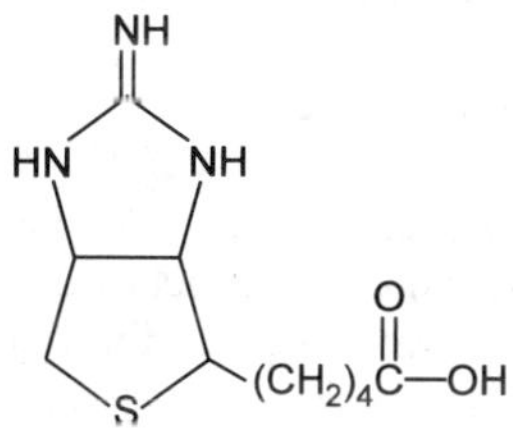

Figure 5.205 Iminobiotin.

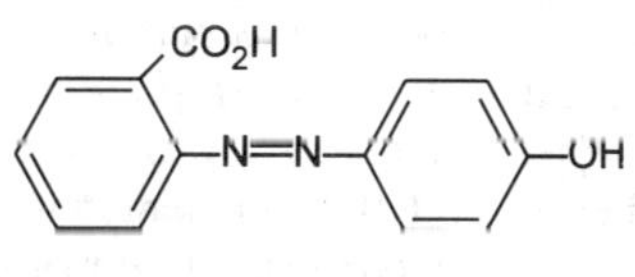

Figure 5.206 4-Hydroxyazobenzene-2′-carboxylic acid (HABA).

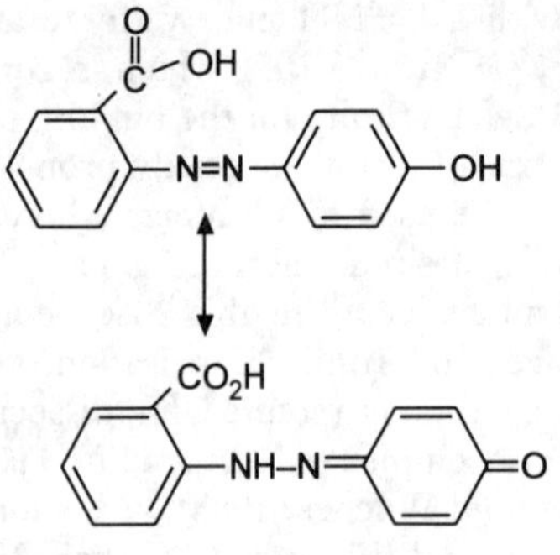

Figure 5.207 Hydrazone tautomer of HABA.

(1992a) have carried out crystallographic studies on avidin– and streptavidin–HABA complexes, observing some similarities with the biotin–avidin interaction: a benzoate oxygen atom in HABA occupies an equivalent position to the anionic ureido oxygen of biotin (see below) and is identically hydrogen-bonded to adjacent residues. Also, as with biotin, avidin shows a stronger interaction than streptavidin, in this case due to an interaction with the Phe72 and Ser73 residues.

It appears that HABA exists in these complexes as the hydrazone tautomer (Figure 5.207), which allows more efficient stabilisation of the charge of the carboxylate.

Another route to the detection of biotin–avidin interactions exploits the red shift which accompanies binding (Green, 1970), and can be detected by monitoring A^{233}.

Recently, Rao *et al.* (1997) described a method for the determining biotinylation ratios based on biotin enhancement of the fluorescence of fluoresceinated streptavidin. Hydrolysis of the biotinylated proteins was employed to avoid steric problems.

5.11.2 Avidin

As noted in Wilchek and Bayer's historical review of biotin–avidin technology (Wilchek and Bayer, 1990a), 'toxic substances' were observed in egg white by Bateman (1916). The work of Eakin and his colleagues established that a protein was responsible, and that its toxicity was related to its ability to bind biotin with a high affinity (Eakin *et al.*, 1941). They named the protein avidin, but it was not isolated until the early 1950s (Fraenkel-Conrat *et al.*, 1952a,b). In 1963 Green confirmed quantitatively the high affinity of avidin for biotin, and this phenomenon was quickly applied to the affinity purification of avidin on biotin–cellulose (McCormick, 1965; Cuatrecasas and Wilchek, 1968).

Over subsequent decades the level of knowledge of this protein gradually increased: improved purification and crystallisation procedures and the amino acid composition were published by Green and Toms (1970), and the amino acid sequence was established by De Lange and Huang (1971). The following brief summary is derived partially from Green's 1975 review and the references therein.

Avidin and proteins with similar activity are found in the eggs and oviducts of many species of birds, reptiles and amphibians, and are also produced by several species of *Streptomyces* (see streptavidin, below). In vertebrates they constitute only a small fraction of the total protein content, and are thought to act primarily as antibacterial agents by blocking biotin uptake; this is to some extent supported by the rather unexpected *Streptomyces* connection, such organisms being adept at creating antibiotic systems. Avidin production in the chicken is induced by progesterone, or by tissue inflammation (Gope *et al.*, 1987).

Avidin is a remarkably stable protein; biotin-binding ability is retained between pH values of 2 and 13, while the tertiary structure is not disrupted at temperatures up to 85 °C. The avidin-biotin complex is even more stable, remaining undisrupted up to a startling 132 °C, and is also resistant to attack by many proteolytic enzymes. Tryptophan residues are instrumental to biotin-binding, and as a result (see Chapter 2, Section 2.3.8) avidin is somewhat sensitive to oxidising conditions including photooxidation.

Chicken-egg avidin is a glycoprotein with a molecular mass of 67 kDa and a p*I* of 10; it contains four essentially identical subunits, each bearing a single oligosaccharide chain. The subunit amino acid sequence (De Lange and Huang, 1971) is shown in Figure 5.208 and the content summarised in Table 5.2.

The two cysteine residues (Cys4 and Cys83) in avidin form a disulfide link, hence there are no free thiols present in the avidin subunit. The basic character of the protein is attributable to a relatively high lysine and arginine content and a substantial degree of amidation of the acid residues. The oligosaccharide chain is attached *via* Asn17 and contains mannose and *N*-acetylglucosamine.

The tryptophan content is abnormally high, and as mentioned earlier, Trp residues have been shown to be essential for biotin-binding activity. Oxidation of two tryptophan residues by *N*-bromosuccinimide or periodate (Green, 1963) was sufficient to destroy this capability, while in the biotin–avidin complex these groups were resistant to such treatment.

The high tryptophan content gives avidin a strong UV absorbance at 282 nm; $\epsilon_{282} = 24,000$ for each

10 Carb. 20 30
H_2N–Ala–Arg–Lys–Cys–Ser–Leu–Thr–Gly–Lys–Trp–Thr–Asn–Asp–Leu–Gly–Ser–Asn–Met–Thr–Ile–Gly–Ala–Val–Asn–Ser–Arg–Gly–Glu–Phe–Thr–Gly–Thr–

Thr/Ile 40 50 60
Tyr–(Thr/Ile)–Thr–Ala–Val–Thr–Ala–Thr–Ser–Asn–Glu–Ile–Lys–Glu–Ser–Pro–Leu–His–Gly–Thr–Glu–Asn–Thr–Ile–Asn–Lys–Arg–Thr–Gln–Pro–Thr–Phe–

70 80 90
Gly–Phe–Thr–Val–Asn–Trp–Lys–Phe–Ser–Glu–Ser–Thr–Thr–Val–Phe–Thr–Gly–Gln–Cys–Phe–Ile–Asp–Arg–Asn–Gly–Lys–Glu–Val–Leu–Lys–Thr–Met

100 110 120 128
Trp–Leu–Leu–Arg–Ser–Ser–Val–Asn–Asp–Ile–Gly–Asp–Asp–Trp–Lys–Ala–Thr–Arg–Val–Gly–Ile–Asn–Ile–Phe–Thr–Arg–Leu–Arg–Thr–Gln–Lys–Glu–CO_2H

Figure 5.208 The complete amino acid sequence of egg white avidin. (Reproduced with permission from: De Lange, R.J. and Huang, T.-S. (1971). *J. Biol. Chem.* **246**, 698–709.)

Table 5.2 Amino acid content of chicken-egg avidin subunit (AV) (De Lange and Huang, 1971) and *Streptomyces avidinii* streptavidin subunit (SA) (Argaraña *et al.*, 1986)

Amino acid	AV	SA	Amino acid	AV	SA
Asp	5	8	Val	7	10
Asn	10	10	Met	2	0
Thr	20–21	19	Ile	7–8	4
Ser	9	14	Leu	7	8
Glu	7	5	Tyr	1	6
Gln	3	6	Phe	7	2
Pro	2	4	His	1	2
Gly	11	18	Lys	9	8
Ala	5	25	Arg	8	4
Cys	2	–	Trp	4	6

subunit in 0.05 M sodium phosphate (pH 7), this absorbance is unaffected by biotin binding.

The histidine, tyrosine and cysteine residues in intact avidin are unreactive and probably inaccessible. The carbohydrate portion of the molecule is unnecessary for biotin binding (Hiller *et al.*, 1990).

Using Raman spectroscopy Honzatko and Williams (1982) found that avidin consisted of 55% β-structure and 10% α-helix, more-or-less in line with theoretical calculations from the amino acid sequence (Argaraña *et al.*, 1986).

X-ray crystallography has been carried out on chicken-egg avidin by Pinn *et al.* (1982), Gatti *et al.* (1984), and Livnah and Sussman (1990). Under different crystallisation conditions it appeared either that all four subunits formed asymmetric units, or that there were two pairs of asymmetric dimers. These differences were tentatively attributed to carbohydrate heterogeneity.

Higher-resolution studies on this protein and its biotin complexes were reported independently by Livnah *et al.* (1993a) and Pugliese *et al.* (1993, 1994), the former on deglycosylated and the latter on intact avidin. These revealed the avidin monomer to have an eight-stranded antiparallel β-barrel structure with extended loops between the strands. The biotin-binding site is buried in a deep pocket at the centre of the barrel, and incorporates a highly hydrophobic region (Trp70, Phe72, Phe79, Trp97); a further tryptophan residue (Trp110) from a neighbouring subunit partially closes the pocket. The polar regions of the biotin molecule are hydrogen-bonded to adjacent residues: the carboxyl group forms five bonds, the ureido-oxygen three (suggesting an oxyanion) and the ureido-nitrogen atoms one each.

The avidin gene has been isolated, sequenced and used to express recombinant avidin in *E. coli* (Gope *et al.*, 1987; Chandra and Gray, 1990).

Pierce market a deglycosylated avidin (NeutrAvidin™), with a reduced p*I* of 6.5, the lysine content being conserved. The molecular mass is 60 kDa, and the specific activity is close to the theoretical maximum. Belovo Chemicals of Belgium sell a largely deglycosylated avidin as Lite Avidin.

5.11.3 Streptavidin

In the early 1960s Chaiet and his colleagues at Merck Sharp & Dohme isolated an avidin-like protein with antibiotic properties from strains of *Streptomyces* (Chaiet *et al.*, 1963; Chaiet and Wolf, 1964), naming it streptavidin. Its intact molecular mass is around 66–75 kDa (Bayer *et al.*, 1990); like avidin it consists of four identical subunits each binding one biotin molecule.

Green (1975) published an amino acid sequence for streptavidin. Later work by Argaraña *et al.* (1986), however, suggests that this may have been based on partially degraded material. These authors observed varying amounts of such degradation in commercial preparations of streptavidin, and surmised that this protein is particularly susceptible to proteolysis in the terminal regions. The primary structure from Argaraña *et al.*, determined from the cloned DNA sequence, is shown in Figure 5.209 and summarised in Table 5.2.

									1					
									Asp	Pro	Ser	Lys	Asp	Ser
			10										20	
Lys	Ala	Gln	Val	Ser	Ala	Ala	Glu	Ala	Gly	Ile	Thr	Gly	Thr	Trp
								30						
Tyr	Asn	Gln	Leu	Gly	Ser	Thr	Phe	Ile	Val	Thr	Ala	Gly	Ala	Asp
			40										50	
Gly	Ala	Leu	Thr	Gly	Thr	Tyr	Glu	Ser	Ala	Val	Gly	Asn	Ala	Glu
								60						
Ser	Arg	Tyr	Val	Leu	Thr	Gly	Arg	Tyr	Asp	Ser	Ala	Pro	Ala	Thr
			70										80	
Asp	Gly	Ser	Gly	Thr	Ala	Leu	Gly	Trp	Thr	Val	Ala	Trp	Lys	Asn
								90						
Asn	Tyr	Arg	Asn	Ala	His	Ser	Ala	Thr	Thr	Trp	Ser	Gly	Gln	Tyr
			100										110	
Val	Gly	Gly	Ala	Glu	Ala	Arg	Ile	Asn	Thr	Gln	Trp	Leu	Leu	Thr
								120						
Ser	Gly	Thr	Thr	Glu	Ala	Asn	Ala	Trp	Lys	Ser	Thr	Leu	Val	Gly
			130										140	
His	Asp	Thr	Phe	Thr	Lys	Val	Lys	Pro	Ser	Ala	Ala	Ser	Ile	Asp
								150						
Ala	Ala	Lys	Lys	Ala	Gly	Val	Asn	Asn	Gly	Asn	Pro	Leu	Asp	Ala
Val	Gln	Gln												

Figure 5.209 Amino acid sequence of streptavidin derived from the DNA sequence. (Reproduced with permission from: Argaraña, C.E. *et al.* (1986). *Nucleic Acids Res.* **14**, 1871–1882.)

While sharing certain features of the composition of avidin, notably the characteristically high tryptophan content, there are many differences. Streptavidin contains an extra 31 residues (159 *versus* 128), is completely devoid of sulfur-containing residues, and contains no carbohydrate. It has both a lower content of lysine and arginine, and more unamidated acid residues than avidin, resulting in a much lower p*I*.

Argaraña *et al.* noted a high degree of similarity in the predicted secondary structures of avidin and streptavidin, both of which contain a high percentage of β features.

A processed streptavidin has been prepared by Apcel Ltd. (later part of Celltech Ltd.) in the UK, and has been investigated by Pähler *et al.* (1987). Terminal sequences have been removed to yield a core molecule, representing residues 13 or 14 to 138 or 139 of the full streptavidin structure, and almost identical in size to the protein portion of chicken-egg avidin. This has fully maintained biotin-binding capacity, and was crystallised to form the basis of some preliminary X-ray crystallography. Bayer *et al.* (1990) have discussed the lability of the terminal regions of streptavidin to proteolysis; they also noted the excellent stability of the core portion, and the fact that its biotin affinity was even higher than the native molecule. Green (1990) observed that its solubility in water was also higher than streptavidin itself, and that it resembled avidin in many features of its secondary structure.

Higher-resolution crystallography studies revealed an eight-stranded antiparallel β-barrel structure for the subunit, very similar to that determined later for avidin. The four subunits were assembled in an hourglass formation. Again the biotin-binding site was deeply buried, with a hydrophobic region formed in this case from three tryptophans from one subunit and one from another. The hydrogen-bonding of the ureido group was similar to avidin, but the valerate formed only two H-bonds as opposed to five (Hendrickson *et al.*, 1989; Livnah *et al.*, 1993a).

Weber *et al.* (1987, 1989) carried out complimentary work, largely on the 120-residue core fragment. The overall structure seen by Hendrickson was confirmed, and a change in the quaternary structure on biotin-binding was also noted: as well as the hydrogen bonding changes, two of the interstrand loops became more ordered so that a glutamate–arginine salt bridge could be formed. Overall these effects resulted in a flattening and tightening of the barrel structures.

Weber *et al.* (1992b) have identified oligopeptides which exhibit binding to streptavidin and examined the crystal structure of their complexes.

As well as the 'processed' streptavidin described earlier, other derivatised streptavidins have been described with useful properties. Reznik *et al.* (1996) noted that the introduction of covalent cross-links between sub-units led to increased stability, while Morag *et al.* (1996) nitrated the tyrosine residues to lower the affinity for biotin—providing a reversible binder for affinity applications (*cf.* iminobiotin above).

5.11.4 Amine-reactive biotin reagents

Apart from the biotin-containing enzymes mentioned above, biotin must be introduced artificially into proteins of interest. Therefore a series of heterobifunctional reagents have been developed which contain the biotin moiety attached to a suitable reactive group (Wilchek and Bayer, 1990b; Bayer and Wilchek, 1990). Not surprisingly the most commonly encountered of these are targetted at amine groups in the protein, and these are dealt with below.

The simplest biotinylation reagents contain activated derivatives of the native carboxyl group. In 1971 Becker *et al.* synthesised the *p*-nitrophenyl ester of biotin (Figure 5.210) and used it as an affinity label to probe biotin transport in yeast. In this application it was more effective than the succinimide ester (q.v.) but the acidic conditions employed (pH 4) are atypical

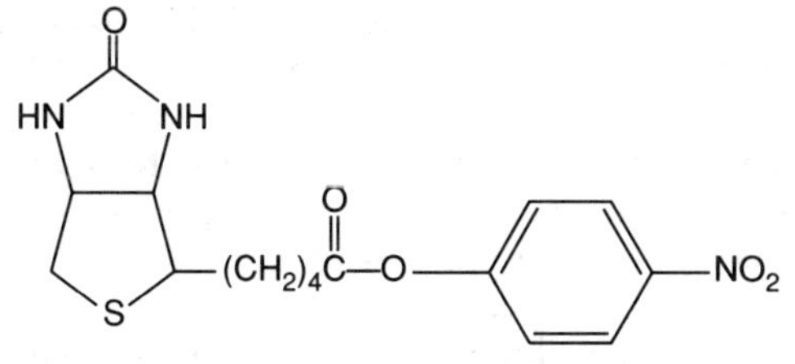

Figure 5.210 Biotin *p*-nitrophenyl ester.

of those used for active esters in preparative protein chemistry.

In general protein chemistry succinimides have come to dominate the active ester field, and this is true for biotinylation reagents also; the nitrophenyl ester is seldom encountered. Bayer and Wilchek (1992a) observed that essentially identical levels of biotinylation were obtained using *p*-nitrophenyl and succinimide esters.

5.11.4.1 Succinimidyl biotin

Succinimidyl biotin (biotin-NHS, BNHS; Figure 5.211), also synthesised by Becker *et al.* (1971) and used in the yeast study mentioned above, is now used extremely widely and is available from a number of commercial sources.

It is fairly insoluble in aqueous buffers and a suitable co-solvent such as DMF or DMSO must be employed. Solutions of at least 0.1 M (34.1 mg/ml) can be obtained in DMF; some authors recommend heating briefly (hot tap water, around 60 °C) to facilitate dissolution (Bayer and Wilchek, 1992a). Avoiding a cosolvent is not impossible, however: Jasiewicz *et al.* (1976) obtained a 1 mg/ml solution in phosphate-buffered saline by heating at 40 °C for two minutes.

Bayer *et al.* (1976) derivatised concanavalin A with various ratios of BNHS, in 0.1 M sodium bicarbonate/

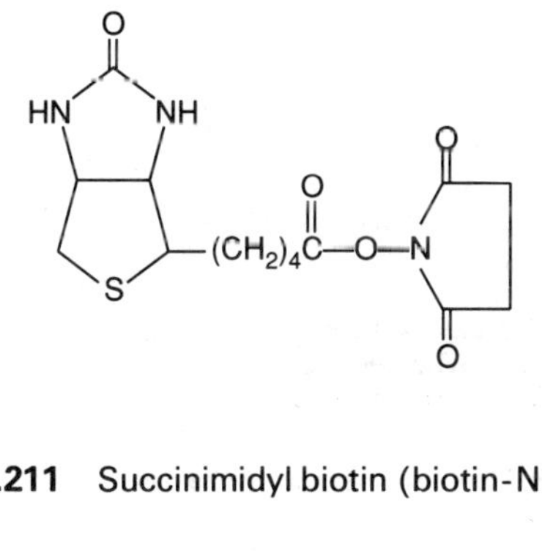

Figure 5.211 Succinimidyl biotin (biotin-NHS, BNHS).

1 M NaCl (four hours at room temperature). The resulting conjugates were assessed for their biotin content and for their ability to be retained by an avidin column.

It was noted that complete binding of the biotinylated Con A to the avidin was not obtained until 15 biotin groups had been introduced per dimer molecule, despite the high affinity of the biotin–avidin interaction. This may have reflected the presence of inaccessible biotin groups in the conjugate.

Avrameas's group at the Institut Pasteur in Paris used BNHS in the preparation of reagents for immunoassay and immunohistochemistry (Guesdon *et al.*, 1979). Using a 0.1 M solution of BNHS in DMF, these authors derivatised a range of proteins dissolved at 10 mg/ml in 0.1 M sodium bicarbonate solution (one hour at room temperature). The proteins varied in their capacity for biotinylation, even when the excess of BNHS used was calculated on the basis of the number of available amine groups. Thus alkaline phosphatase from *E. coli* was exhaustively biotinylated by a fivefold excess over available amines, while only 80% incorporation could be achieved with HRP even at a 100:1 ratio.

The effect of biotinylation on enzyme activity was also quite varied (see Figure 3.9): while HRP and glucose oxidase were more-or-less unaffected by this derivatisation, β-galactosidase showed a gradual loss with increasing biotin loadings and alkaline phosphatase was inactivated quite drastically.

When a sheep IgG was biotinylated, there was no increase in avidin-binding capability between loadings of 1:1 and 10:1 over available amines (8.5:1 and 85:1 in molecular terms), these resulted in 42% and 100% derivatisation respectively. The antigen-binding capacity was unaffected by a 0.2:1 loading over amine content, but was reduced quite markedly by loadings of 3:1 or 8:1.

Flanders *et al.* (1982) biotinylated an acetimidate derivative of the peptide hormone glucagon using a 25-fold excess of biotin-NHS (four hours, 25 °C, DMF–HCl–potassium phosphate, apparent pH 8.2); over 90% of the product obtained was singly biotinylated at the terminal amino group.

Ngo *et al.* (1982) used a DMSO solution of BNHS to derivatise glucose-6-phosphate dehydrogenase at a 100:1 excess (30–120 minutes, room temperature, pH 7.2) obtaining incorporations between 16 and 24 biotin molecules per enzyme molecule.

Kendall *et al.* (1983) derivatised goat IgG with biotin-NHS for use in an immunoassay, and obtained the best performance with BNHS:IgG ratios between 1:1 and 4:1. A higher ratio (8:1) resulted in precipitation, and both higher and lower ratios caused reductions in sensitivity.

To give low incorporation and retain immunoreactivity, Frobert and Grassi (1992) recommended a general BNHS:protein ratio according to the following formula: 5 equivalents of BNHS for each 10 kDa in 1 equivalent of protein (e.g. a 15:1 molar ratio for a 30 kDa antigen). These authors employed unusually basic conditions for an active ester reaction (borate buffer, pH 9, 30 minutes at room temperature), claiming a greater efficiency than using phosphate at pH 7. For the biotinylation of acetylcholinesterase, Frobert and Grassi used a BNHS:protein ratio of around 310:1.

Bayer and Wilchek (1992a) recommended the use of 25 ml of BNHS solution (2 mg/ml in DMF) for each ml of a 2 mg/ml protein solution in 0.1 M sodium bicarbonate, pH 8.5. with a reaction time of one hour. For an IgG molecule this represents a BNHS:protein ratio of around 11:1.

Pereira *et al.* (1993) observed inactivation of two mouse monoclonal antibodies on biotinylation with BNHS (ratio not given), the activity was retained when the most reactive amine groups were reversibly protected with dimethylmaleic anhydride (Endo *et al.* 1987).

5.11.4.2 Sulfosuccinimidyl biotin

For many years the sulfosuccinimidyl analogue of BNHS (Figure 5.212) has been commercially available, originally from Pierce; this reagent obviates the need for an organic cosolvent and has achieved some popularity. Even in the presence of cosolvents, the retention of activity in the target protein may be greater than with BNHS—see Figure 5.216, for example.

LaRochelle and Froehner (1986) biotinylated a Mab with BNHSS at an 85-fold excess (four hours, room temperature, pH 8.0), obtaining a successful immunoassay conjugate. Lisanti *et al.* (1988, 1989) used BNHSS to biotinylate cell surface proteins (30 minutes, 4 °C, PBS).

Yem *et al.* (1989) used this reagent to biotinylate recombinant human interleukin-1β. Treatment with a tenfold excess of the active ester (two hours, 25 °C, pH 7.5) resulted in a conjugate which was resolved by cation exchange chromatography and individual components analysed by peptide mapping. This showed that the derivatisation occurred at the N-terminal alanine and at three specific lysine residues. No reaction with hydroxylic residues was noted (cf. Miller *et al.*, 1992, 1994, see below).

5.11.4.3 Succinimidyl 6-(biotinamido) hexanoate and analogues

The use of active esters of the native biotin carboxyl group results in a relatively short link between the biotin moiety and the derivatised protein, and long-chain analogues have been synthesised to increase this separation. Such reagents were first described by Bayer *et al.* (1975), who introduced a series of homologous *p*-nitrophenyl esters to probe the effect of their structure on the ability to inactivate biotin transport in yeast. This study showed up interesting differences in reactivity: those compounds incorporating the native valeryl side-chain, or the shorter butyryl chain of norbiotin, appeared to form bonds with cysteine or histidine. The longer caproyl group of homobiotin, on the other hand, promoted reaction with lysine residues (Figure 5.213). Longer analogues showed the ability to react with both types of residue.

However, the most commonly employed reagent of this type is a succinimide ester which incorporates an additional 6-carbon chain in the form of a 6-aminohexanoyl (or ϵ-aminocaproyl) group, succinimidyl

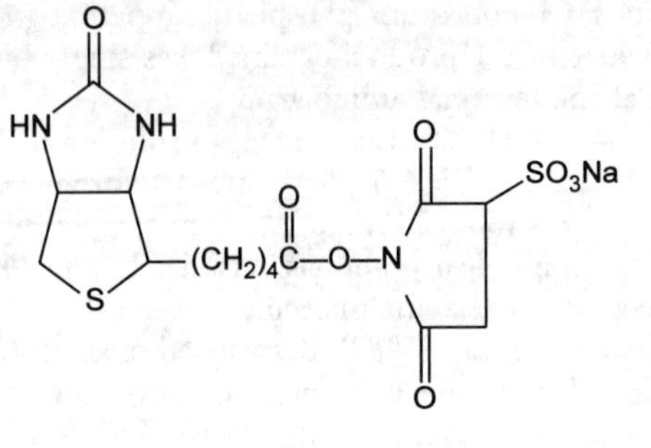

Figure 5.212 Sulfo-NHS-biotin, sodium salt.

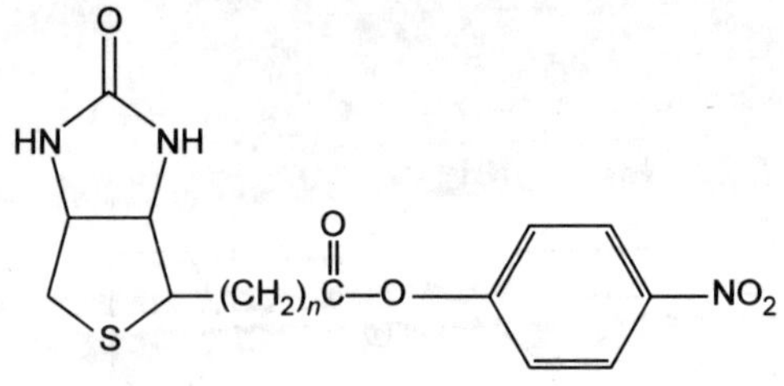

Figure 5.213 *p*-Nitrophenyl esters of biotin homologues.

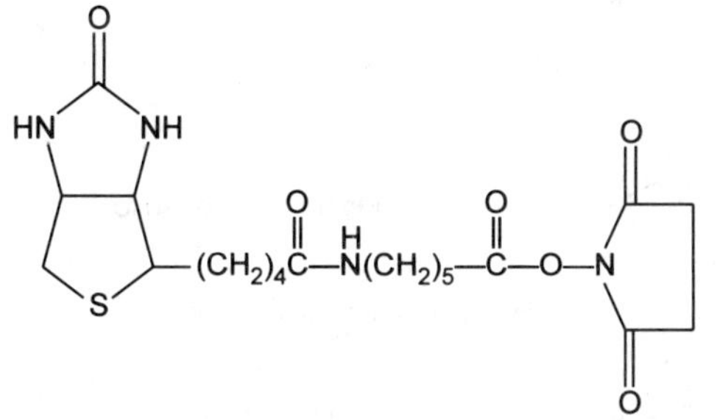

Figure 5.214 NHS-LC-biotin.

6-(biotinamido)hexanoate (Figure 5.214). This reagent is sold under a number of pseudonyms, for example NHS-LC-Biotin II™ (Pierce Chemical Co.) or Biotin-X-NHS™ (Calbiochem Biochemicals); it increases the span of NHS-biotin by around 9 Å.

This compound was first described by Costello *et al.* (1979), and used by these authors in a 100-fold excess to biotinylate various proteins (one hour, room temperature, pH 7.4). Albumin, with 59 amino groups, showed a biotin incorporation of 37, while ribonuclease (eleven amine groups) gave an incorporation of about seven.

Hochhaus *et al.* (1988) used a twofold excess of biotin-X-NHS to derivatise human β-endorphin (30 minutes, room temperature, pH 8.3), obtaining a range of monosubstituted products which were separated by HPLC. Dimethylsulfoxide was used as solvent for the active ester.

Miller *et al.* (1992, 1994) observed that this compound and its sulfonated analogue reacted not only with amine groups, but with active serine moieties in peptide hormones, contrary to the expected specificity of succinimide esters. It is likely that this reactivity is common to other succinimide esters rather than being specific to these reagents.

Other long-chain biotin succinimide esters have been described in a recent patent, for example describing the condensation of a long-chain aminobiotin with the bis-succinimide ester DSS (Huber *et al.*, 1994) to produce the compound shown in Figure 5.215. A biotin succinimide ester with a 13-atom spacer is also commercially available as biotin-XX-NHS from Calbiochem.

A biotin-polyethylene glycol (PEG)-succinimide is commercially available from ProChem Inc.

5.11.4.4 Sulfosuccinimidyl 6-(biotinamido) hexanoate

The sulfonated analogue of succinimidyl 6-(biotinamido)hexanoate is also commercially available (e.g. Pierce NHS-LC-Biotin™, Calbiochem biotin-X-NHS (water soluble)).

Gretch *et al.* (1987) studied the performance of a number of Mabs biotinylated (one hour, room temperature) with various excesses of this reagent (used as a 1.63 mg/ml DMSO solution). The performance of the biotinylated Mabs improved with increasing biotin loading, then deteriorated as overlabelling became a problem. For six of the seven Mabs studied, performance declined above an excess of 15:1, and in one case a 53-fold excess resulted in the loss of 50% of binding capacity.

Yem *et al.* (1989) used this reagent to biotinylate recombinant human IL-1β, noting slight differences in product composition from that obtained with sulfosuccinimidyl biotin (see above). Similarly, De Jong *et al.* (1995) observed better retention of interleukin-2 activity with this reagent than with shorter biotinylating agents (see Figure 5.216), particularly when the use of an organic cosolvent was avoided.

Grumbach and Veh (1991) investigated the hydrolysis rates of this compound, the half-life was

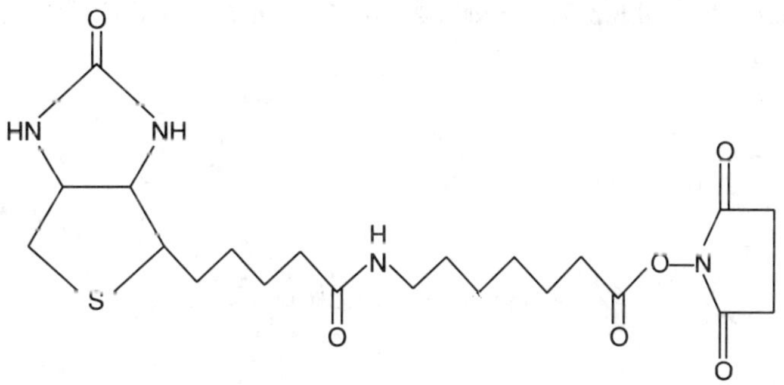

Figure 5.215 Long-chain succinimide ester.

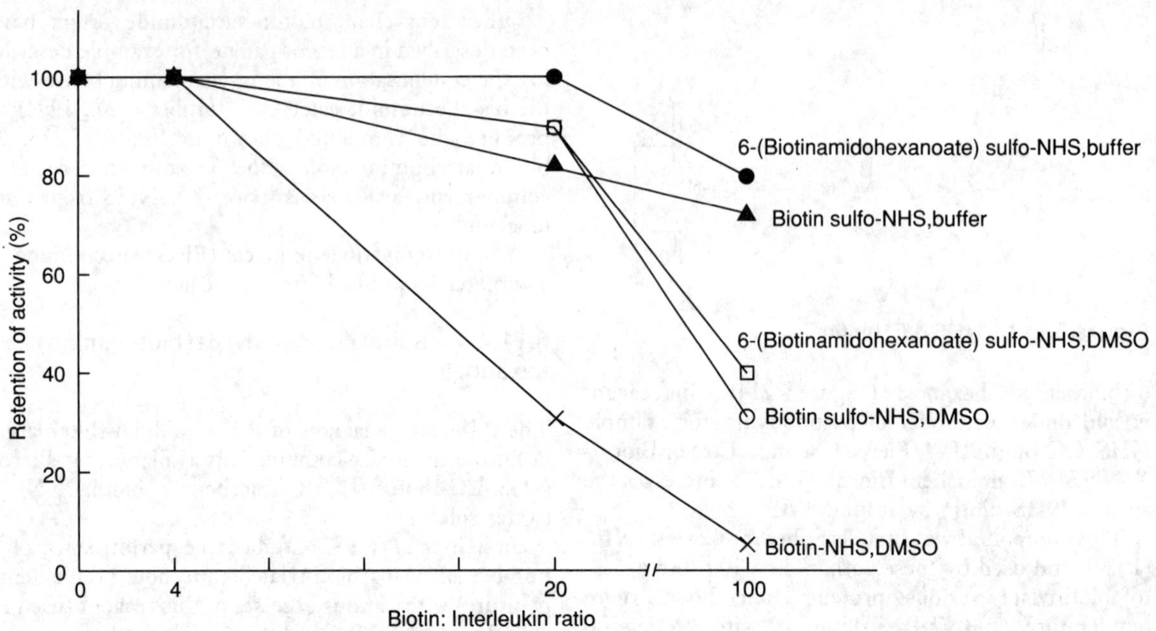

Figure 5.216 Retention of interleukin-2 activity after treatment with various biotinylation reagents in buffer (0.1 M carbonate–bicarbonate, pH 8.5) or DMSO. Data from De Jong *et al.*, 1990 (Table 2).

15 minutes at pH 8, and over two hours at pH 6.5. Although reaction rates with amines were also raised at the higher pH, this did not compensate for the increased hydrolysis, and neutral conditions were recommended for such reactions.

5.11.4.5 Other biotin succinimide esters

As with other types of coupling agent, the facility of cleaving a bond at will is sometimes useful, and biotin succinimide esters incorporating this property have been described.

Sulfosuccinimidyl 2-(biotinamido)ethyl-1,3-dithiopropionate (NHS-SS-biotin; Figure 5.217) is a commercially-available reagent yielding a cleavable, 24.3 Å linkage. This reagent is useful, for example, in recovering bound materials from avidin affinity columns (Shimkus *et al.*, 1985).

Gretch *et al.* (1987) used NHS-SS-biotin in DMSO to derivatise a number of Mabs (one hour, room temperature) at different loadings: a 2.2-fold excess gave a biotin incorporation of about two, while other loadings and incorporations [bracketed] were as follows: 13.2 [8]; 22 [14]; 53 [25].

Only 75% of the Mab–$biotin_2$ conjugate bound to streptavidin, while 100% binding was obtained with the Mab–$biotin_8$ and Mab–$biotin_{14}$ complexes. Based on Gretch's results with this reagent and with biotin-

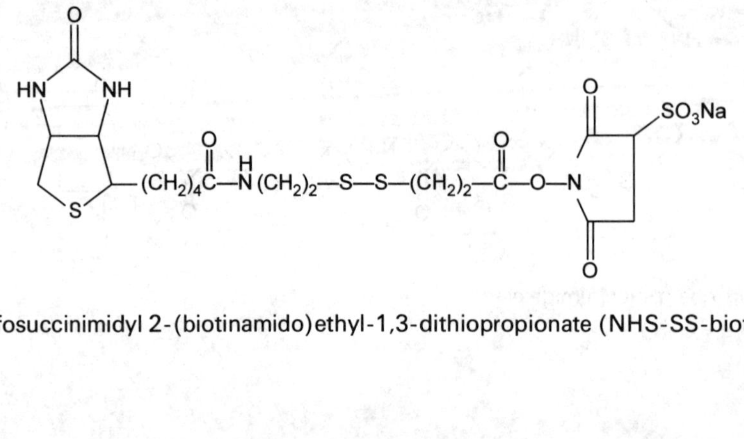

Figure 5.217 Sulfosuccinimidyl 2-(biotinamido)ethyl-1,3-dithiopropionate (NHS-SS-biotin).

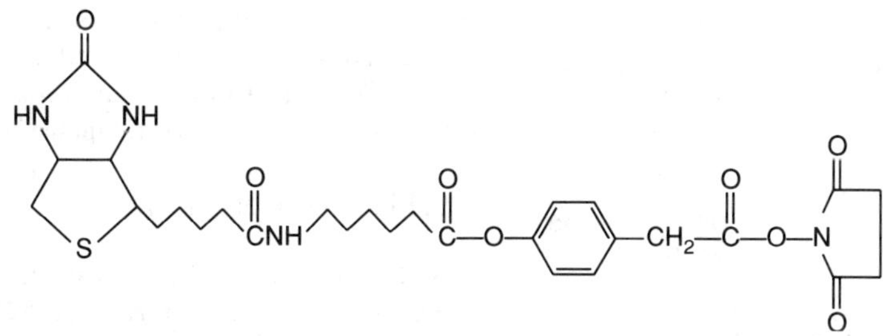

Figure 5.218 Succinimidyl 3-[4-(*N*-biotinyl-6-aminocaproyloxy)phenyl]propionate (BPE).

LC-NHS (above), it appears that optimal biotinylation of Mabs can be obtained by aiming for an incorporation of around 6–8.

Ghebrehiwet *et al.* (1988) observed that derivatisation (four hours, room temperature, pH 7.4) of human C1q with NHS-SS-biotin in DMF resulted in significant impairment of haemolytic activity only at initial ratios of 50:1 (8% impairment) and 100:1 (15% impairment). At initial loadings from 5 to 100:1 the biotin incorporation achieved was always 40–60% of the amount added.

Another cleavable biotin ester, succinimidyl 3-[4-(*N*-biotinyl-6-aminocaproyloxy)phenyl]propionate, BPE (Mouton *et al.*, 1982) is shown in Figure 5.218. This incorporates a phenyl ester function which can be cleaved with hydroxylamine.

Another useful property for a derivatisation reagent is the capability for convenient detection and quantification. In a recent patent, Ghazarossian *et al.* (1997) described biotin succinimide esters, incorporating fluorescent moieties in the linker group, for this purpose.

5.11.4.6 Succinimide esters of iminobiotin

The succinimide ester of iminobiotin (IBNHS, Figure 5.219) is also available commercially, having been synthesised originally by Orr (1981) who used it for the derivatisation of insulin, fetuin and alkaline phosphatase.

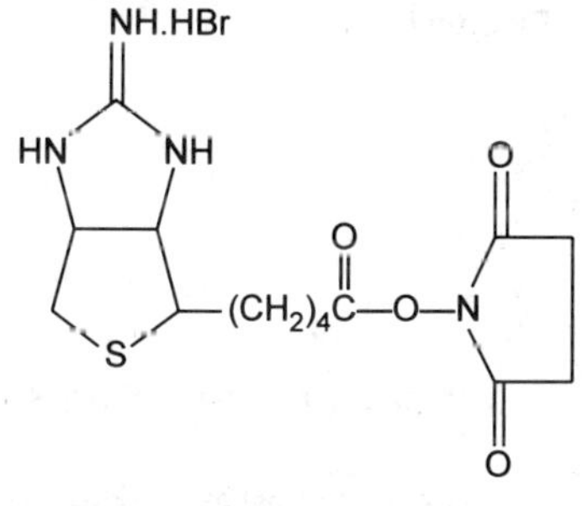

Figure 5.219 NHS-iminobiotin (IBNHS), hydrobromide.

Hofmann *et al.* (1982) synthesised the succinimide esters of both iminobiotin and desthiobiotin, with and without a caproyl spacer group, and coupled them to insulin. The iminobiotinyl insulin derivatives, however, did not bind satisfactorily to avidin even at pH 9.0. Zeheb *et al.* (1983), however, observed useful binding behaviour with iminobiotin-labelled BSA, which precipitated avidin strongly at pH 10 but much less efficiently as conditions became less basic: at pH 6 no precipitation was observed. The biotinylation was carried out using a 40-fold excess of IBNHS as a DMF solution (two hours, 4 °C, pH 8.0).

5.11.4.7 Biotin–aldehyde reagents

Musso *et al.* (1992) described the synthesis of a biotin–acetal reagent, *N*-[6-(ethylenedioxy)hexyl]biotinamide (Figure 5.220), which generates an aldehyde group on treatment with acid (80% aqueous acetic acid, three hours, 37 °C).

5.11.5 Carboxyl-reactive biotin reagents

Biocytin (Figure 5.221(a)) is a condensation product of the ε-amino group of L-lysine with the pentanoate carboxyl function of biotin; it can be prepared by the reaction of L-lysine hydrochloride and BNHS in the presence of copper(II) carbonate (Wilchek and Bayer, 1987). Biocytin moieties are present in the naturally occuring biotin-containing enzymes. Protection of the lysine carboxyl function can be achieved by esterification (Wilchek and Bayer, 1987) or by amidation (Hofmann and Kiso, 1976), yielding a reagent which can be used for derivatising active carboxyl groups. Hofmann and Kiso attached biocytinamide (Figure 5.221(b)) to the carboxyl groups of the peptide hormone ACTH by this approach.

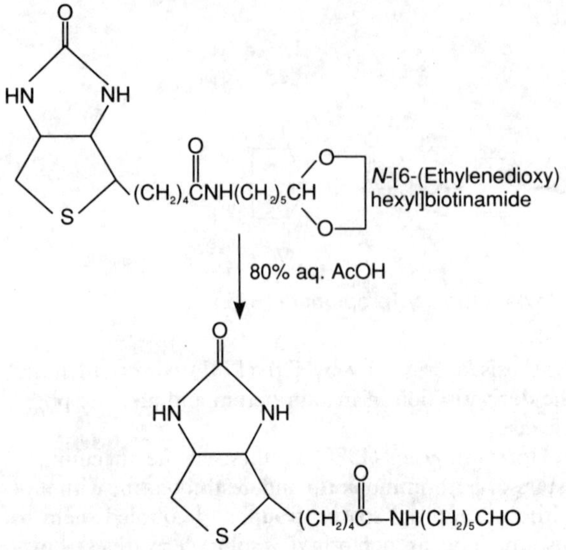

Figure 5.220 *N*-[6-(Ethylenedioxy)hexyl] biotinamide, which generates an aldehyde on acid treatment.

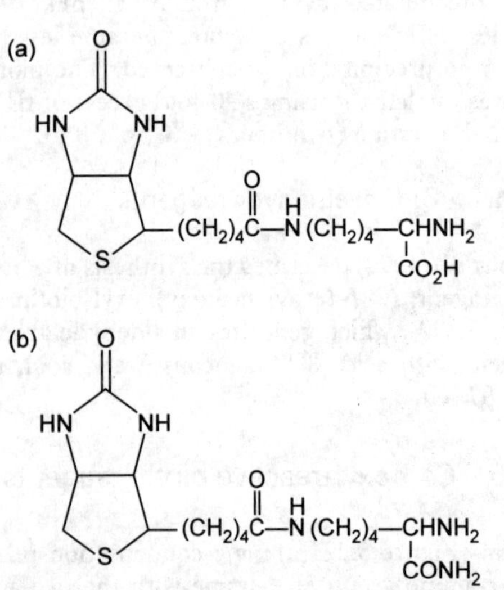

Figure 5.221 (a) Biocytin; (b) biocytin amide.

An interesting alternative means of attaching biocytinamide specifically to the C-terminus of proteins is reverse proteolysis using carboxypeptidase Y (Schwarz *et al.*, 1990).

Basak *et al.* (1994) described the preparation of a series of aminobiotins, bearing aminoethyl, amino-*n*-butyl, amino-*n*-hexyl and aminocyclohexyl arms attached *via* the valerate moiety. These reagents were used to derivatise oligopeptides, but could also be attached to suitably-activated protein carboxyl moieties. A number of companies market other aminobiotins, such as 5-biotinamidopentylamine.

5.11.6 Aldehyde-reactive biotin reagents

Biotin hydrazide (BHZ; Figure 5.222(a)) was first described by Hofmann *et al.* (1942) and has frequently been employed for coupling to periodate-oxidised carbohydrates (Heitzmann and Richards, 1974; Wynne *et al.*, 1976; Bayer and Wilchek, 1980; O'Shannessy *et al.*, 1984). It is commercially available, for example from BioAffinity Systems. The longer, lysine-containing analogue biocytin hydrazide (Figure 5.222(b)) was described by Wilchek and Bayer (1987), while biotin aminocaproyl hydrazide (BACH; Figure 5.222(c)) was synthesised by O'Shannessy *et al.* (1987) and is commercially available, with a span of 24.7 Å. Calbiochem also sell a longer-chain analogue, biotin-XX-hydrazide.

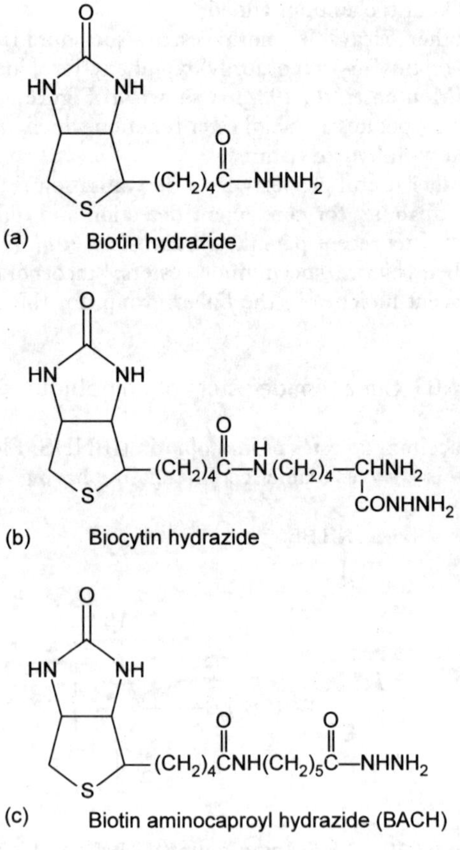

Figure 5.222 Biotin hydrazide derivatives.

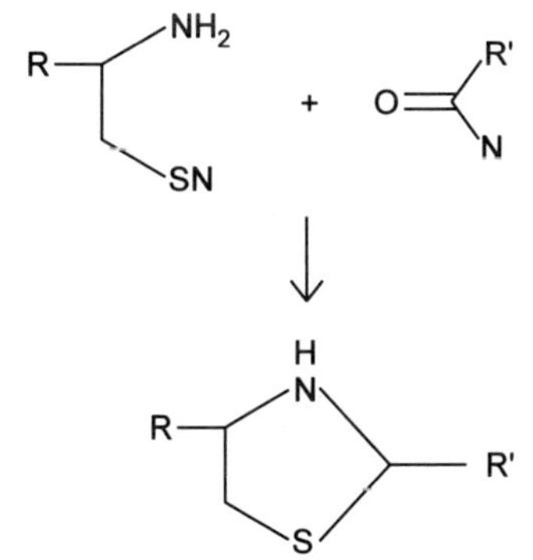

Figure 5.223 The reaction of a 1,2-aminothiol with an aldehyde to generate a stable thiazolidine linkage.

Gretch *et al.* (1987) used BHZ to derivatise a number of Mabs: in all cases the performance was poorer than that obtained with succinimide esters, and similar conclusions were reached by Diamandis and Christopoulos (1991). Lisanti *et al.* (1989) used this reagent successfully, to label periodate-oxidised cell-surface glycoproteins (30 minutes, 4 °C, pH 5.5).

O'Shannessy *et al.* (1987) compared BHZ with BACH for derivatisation of oxidised glycoproteins on electroblots, observing approximately twofold greater sensitivity with the long-chain reagent. Labelling was carried out for one hour at room temperature and pH 5.5.

Orr (1981) used iminobiotin hydrazide for reductive alkylation of periodate-oxidised erythrocytes (three hours, 30 °C, pH 6.5); the Schiff base product was reduced with sodium borohydride. A biotin-PEG-hydrazide is commercially available from ProChem Inc.

These hydrazides could also be used as carboxyl-reactive reagents if a suitable activating agent such as a carbodiimide were employed.

An interesting biotin derivative (Figure 6.85) was described by Zhang and Tam (1996) which incorporates a 1,2-aminothiol function; this reacts with aldehydes to form a stable thiazolidine product (Figure 5.223).

5.11.7 Thiol-reactive biotin reagents

Wilchek and Givol (1977) described the preparation of the thiol-reactive bromoacetyl (Figure 5.224) and mercuribenzoate derivatives of biotin hydrazide. The former had been employed earlier as an unsuccessful inactivation agent for the yeast biotin transport system (Becker *et al.*, 1971).

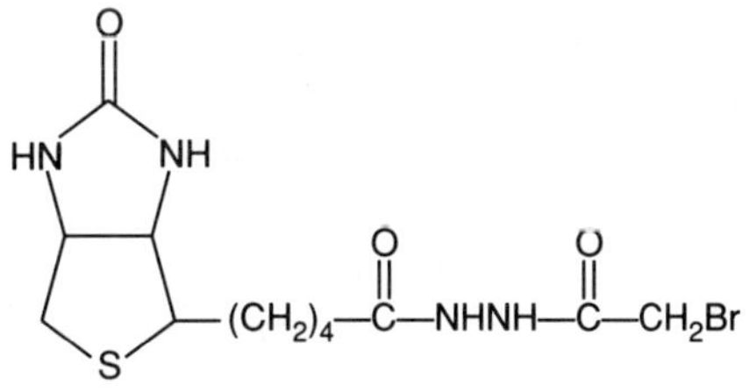

Figure 5.224 Bromoacetylbiotin hydrazide.

Yamamoto *et al.* (1984) described the derivatisation of a myosin fragment with a fourfold excess of biotin iodoacetate (2.5 hours, 0 °C, pH 8.5).

Biotinylation reagents based on all three main thiol-reactive chemistries are commercially available, and are typified by 1-biotinamido-4-[4′-(maleimidomethyl) cyclohexanecarboxamido]butane (biotin-BMCC), *N*-(6-biotinamidohexyl)-3′-(2′-pyridyldithio)propionamide (biotin-HPDP) and *N*-biotinamido-*N*′-iodoacetylhexane-1,6-diamine (iodoacetyl-LC-biotin). The structures of these compounds are depicted in Figure 5.225, and another compound of this type is shown in Figure 6.68. The use of these and similar thiol-reactive biotin derivatives has been described in many patents and reports: see, for example, Sigler (1988); Wilchek and Bayer (1990b); Marui *et al.* (1995). New reagents of this type are still being developed, for example a long-chain biotin–maleimide incorporating three ethylene oxide groups, described by Wilbur *et al.* (1996). A biotin–PEG–maleimide is commercially available from ProChem Inc., while a biotin–PEG–pyridyl disulfide reagent was described by Kaiser *et al.* (1997).

5.11.8 Tyrosine- and histidine-reactive biotin reagents

In 1980 Bayer and Wilchek described the synthesis of biotinyl-*p*-diazoanilide (Figure 5.226(a)) for derivatisation of phenol and imidazole-containing molecules. Similarly in 1986 Wilchek *et al.* described the synthesis of *p*-diazobenzoyl biocytin (Figure 5.226(b)), which was successfully employed to biotinylate model proteins. Like other diazonium reagents they must be prepared immediately before use from the corresponding aniline, which is commercially available for the biocytin derivative.

5.11.9 Photoactivatable biotin reagents

A number of companies sell a nitrophenyl azide derivative of biotin termed photobiotin (Figure 5.227(a))

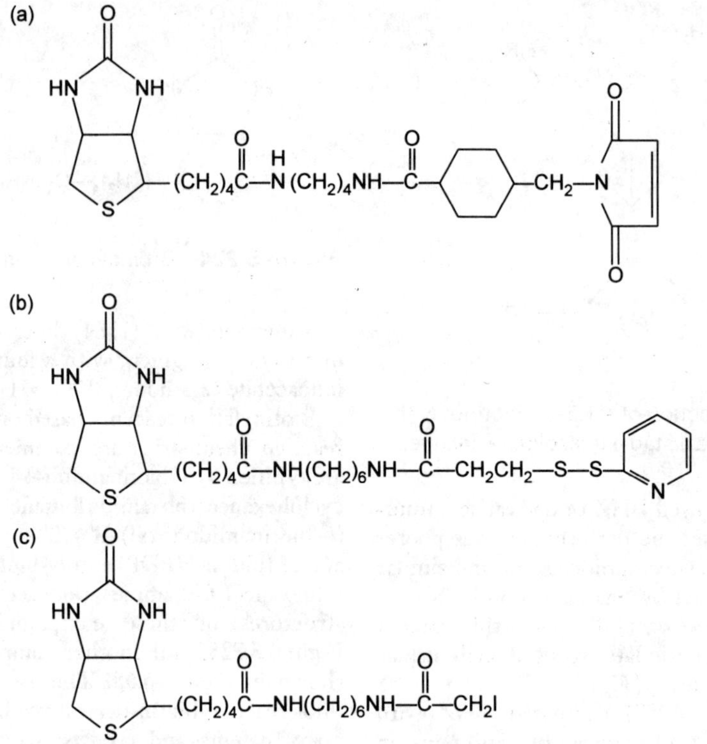

Figure 5.225 (a) Biotin-BMCC; (b) biotin-HPDP; (c) iodoacetyl-LC-biotin.

principally for labelling nucleic acids (McInnes *et al.*, 1990). However, this reagent has also been used in other applications where conventional labeling is difficult, such as biotinylation of carbohydrate antigens (Hylkema *et al.*, 1994). Protein-labelling applications

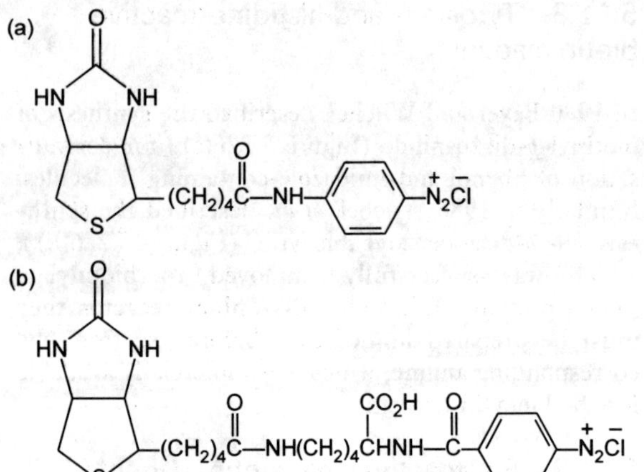

Figure 5.226 (a) Biotinyl-*p*-diazoanilide; (b) *p*-diazobenzoyl biocytin.

of photobiotin are mentioned in the review by McInnes.

Recently Pierce have also introduced a cleavable trifunctional reagent combining a biotin group, a phenyl azide, and a succinimide ester: 2-[6-biotinamido-2-(*p*-azidobenzamido)-hexanamido]ethyl-1, 3′-dithiopropionate (sulfo-SBED; Figure 5.227(b)). A suggested application scheme describes the transfer of biotin from an antibody to its antigen (cf. SASD, Figure 5.181). See also the angiotensin II derivative shown in Figure 6.137.

5.11.10 Avidin reagents

As avidin is a protein of fairly typical structure, the techniques for coupling it to other proteins are those described throughout this chapter. Obviously one of the great beauties of the biotin–avidin system is the ability to avoid the preparation of avidin-labelled reagents, as a wide range of commonly-employed reagents of this type are commercially available. As a result it is largely superfluous to describe routes to the preparation of common avidin reagents.

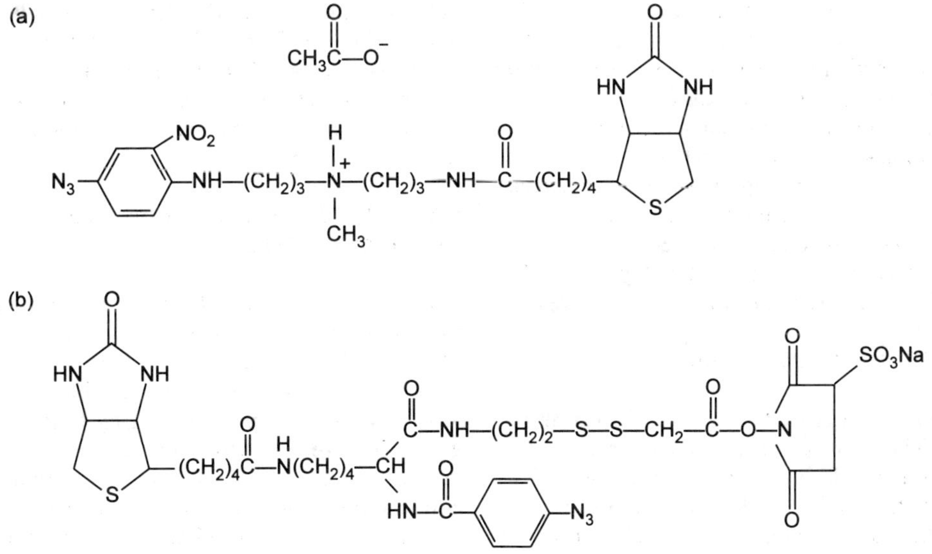

Sulfo-SBED

Figure 5.227 (a) Photobiotin, acetate salt; (b) 2-[6-biotinamido-2-(*p*-azidobenzamido)-hexanamido]ethyl-1,3′-dithiopropionate (sulfo-SBED), sodium salt.

Bayer and Wilchek (1992b) have described protocols for avidin conjugations, but, as the authors themselves state, it is much more convenient to obtain these commercially. A protocol for introducing maleimide groups into avidin with MHS has already been described in the text (Section 5.7.1; Hashida *et al.*, 1991), as have two rather idiosyncratic methods for thiolation of avidin with a bis-succinimidyl disulfide (Section 5.3.5; Broker *et al.*, 1978), and for carbodiimide coupling to glucose oxidase (Section 5.2.2; Philpott *et al.*, 1980).

5.12 Reagents for Protein Thiolation

Although some proteins contain free thiols which can be used for coupling reactions, it is a common requirement to introduce such groups artificially. The compounds which are employed to accomplish these thiolation reactions are essentially heterobifunctional in nature.

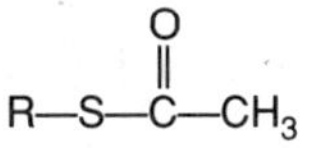

Figure 5.228 The *S*-acetyl thiol group.

Because thiols are prone to atmospheric oxidation they are often introduced into proteins in an unreactive protected form, the reactive species being generated by a suitable deprotection step immediately before use. Protected thiol derivatives include the disulfides encountered already, in particular those with good leaving groups such as pyridyldisulfides, and thioesters such as *S*-acetyl compounds (Figure 5.228).

5.12.1 Reagents for thiolation *via* amines

Predictably the most commonly-encountered protein thiolation reagents are those which exhibit reactivity towards amines; other routes to thiol incorporation are encountered extremely rarely.

This class of compounds offers an interesting example of the effect of charge changes brought about by protein derivatisation. Common thiolation reagents can replace the amine group, positively charged under most conditions, with a roughly equivalent amidine residue (e.g. 2-IT: $\Delta = 0$), with a neutral amide (e.g. SATA: $\Delta = -1$), or with a negatively charged carboxylate (e.g. SAMSA: $\Delta = -2$).

In the immunotoxin field particularly, it has been found that the use of reagents which change the charge of the toxin often results in deactivation. 2-IT and its analogues are especially popular in this application.

Thiols can be generated from protein disulfides, either those contained in the native structure, or introduced *de novo* (Section 5.12.1.1). Often however, other reagents are used to introduce thiols directly, or in protected forms; the most common of these are *S*-acetyl derivatives (thioesters), from which the thiol can be released by a suitable nucleophile such as hydroxylamine. The amine-reactive function may be drawn from the usual choices of succinimidyl esters and imidates; however some common reagents employ acid anhydrides or sulfur-containing heterocyclic structures.

5.12.1.1 Amine-reactive disulfides

All the compounds discussed in Section 5.7.3 can also be viewed as thiolation reagents, as the SH group can be freed from the disulfide by simple reduction. This applies to SPDP, methyl 3-(4-pyridyldithio)propionimidate hydrochloride, and the disulfide-containing amine-reactive homobifunctionals such as DSP. For the latter reagents, thiol release is the result of the cleavage reaction anyway; exactly the same chemistry, therefore, can be viewed either as a thiolation technique, or as a means for sequential cross-linking and cleavage.

Reagents for the reduction of protein disulfides are described in Section 5.12.5.

5.12.1.2 Thioglycolides

Thioglycolic acid (Figure 5.229(a)) was commonly employed as a reducing agent for the cleavage of disulfide links in proteins in the 1940s and 1950s, but was sometimes found to increase the sulfur content (White, 1960). This was attributed to the formation of thioester bonds which resulted in linear or cyclic polymers (Figure 5.229(b)) termed polythioglycolides (PTG); the thioester linkages were susceptible to aminolysis by protein amines, effectively resulting in thiolation (Figure 5.229(c)). This reaction had been described by Schöberl for preparative use in 1948; however its main significance has been as an unwanted complication to disulfide reduction, and it is of historical interest only.

5.12.1.3 *N*-Acetyl-DL-homocysteine thiolactone (AHTL)

Benesch and Benesch (1956) introduced the thiolating reagent *N*-acetyl-DL-homocysteine thiolactone (AHT or AHTL) (Figure 5.230(a)); as in thioglycolides the carbonyl group is susceptible to nucleophilic attack by amines, opening the ring to generate a thiol group (Figure 5.230(b)). The presence of the acetyl group prevents intermolecular reactions between amine and thiolactone groups, as demonstrated earlier by du Vigneaud *et al.* (1938) with the benzoyl derivative.

The structure of this compound has some advantages in this application: the labile thiol is not generated until the derivatisation reaction takes place so that AHT is quite stable, while the sulfur activates the carbonyl to nucleophilic attack.

Alkaline conditions (pH 9.6) were employed by Benesch and Benesch to provide a suitably high con-

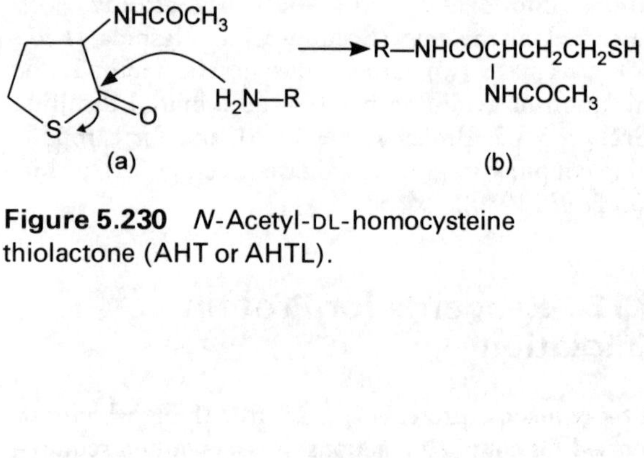

Figure 5.230 *N*-Acetyl-DL-homocysteine thiolactone (AHT or AHTL).

Figure 5.229 Polythioglycolides in protein thiolation.

centration of unprotonated amine groups; hydrolysis competed significantly at room temperature, but this could be suppressed by performing the reaction at 0 °C. Some early work was carried out using the *N*-benzoyl derivative, but the acetyl compound was favoured due to its excellent aqueous solubility. Reaction times of several hours were required for significant thiolation (White, 1972b).

The reaction of AHTL with amines is made considerably faster by the addition of silver ions which form an AHTL:Ag_2^+ complex (Benesch and Benesch, 1958, 1959), and this allows the reaction pH to be dropped to neutrality. At pH 7.5 and room temperature, for example, a lysine residue essential for ribonuclease activity could be thiolated within 30 minutes in the presence of silver nitrate (Shall and Barnard, 1969; White, 1972b). Exhaustive measures were taken to preclude the presence of halide and phosphate to prevent these complexing with the Ag^+ ions. The presence of silver ions lends another advantage to this approach, by protecting the free thiol groups as they are formed.

Abadi and Wilcox (1960) investigated the AHTL thiolation of α-chymotrypsinogen using the slower silver-free conditions. Their aim was to determine conditions under which a single amine group only was derivatised, but all attempts yielded a mixed product. Their report also includes some detailed data on rates of room temperature hydrolysis of AHTL under different pH conditions. Dramatic changes were seen between pH 7 and 8.

Acyl imidazoles (Figure 5.231) are powerful acylating agents (Stadtman and White, 1953). Klotz and Elfbaum (1964) investigated using these and related species as intermediates in the AHTL thiolation reaction. Imidazole is a nucleophile which reacts with the thiolactone, freeing the thiol function and generating an acyl imidazole. The latter reacts rapidly with other nucleophiles such as protein amines, incidentally regenerating the imidazole, which is therefore acting as a classical catalyst (see Chapter 3) although being used in stoichiometric quantities. Other nucleophiles proved successful in the imidazole role: *N*-methylmorpholine, tetramethylethylenediamine, triethylenediamine, and 2-pyridine-aldoxime methiodide. The latter generates an active ester intermediate (Figure 5.232) analogous to the popular succinimide reagents.

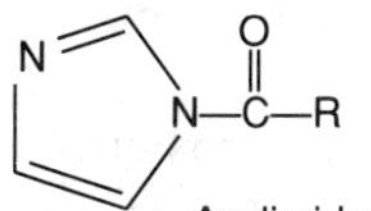

Acylimidazole

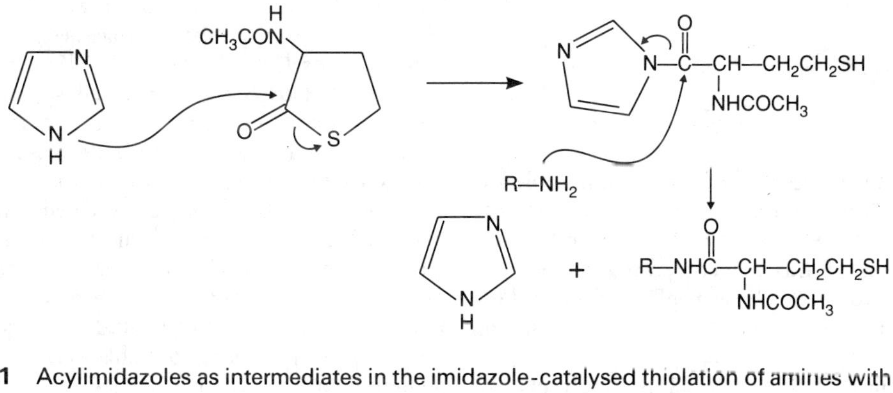

Figure 5.231 Acylimidazoles as intermediates in the imidazole-catalysed thiolation of amines with AHTL.

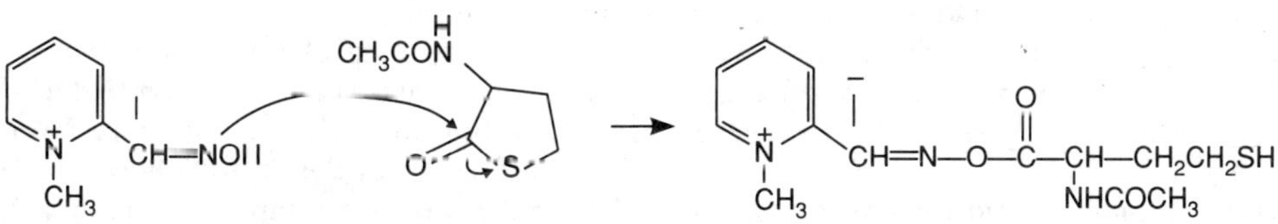

Figure 5.232 The use of 2-pyridinealdoxime methiodide to generate an active ester intermediate.

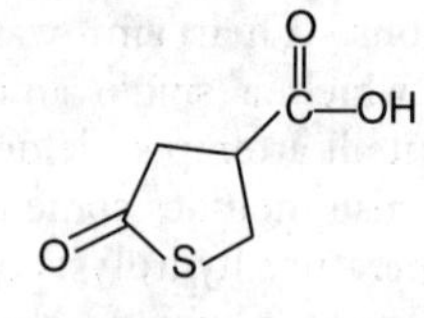

Figure 5.233 Thioparaconic acid.

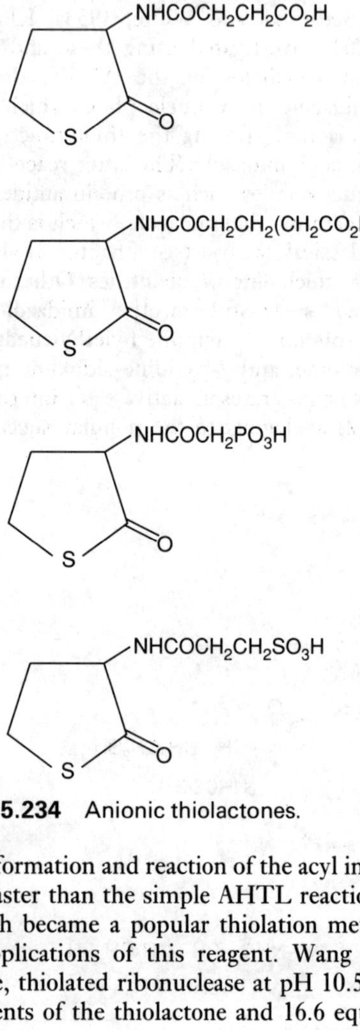

Figure 5.234 Anionic thiolactones.

The formation and reaction of the acyl imidazole are much faster than the simple AHTL reaction, and this approach became a popular thiolation method in the later applications of this reagent. Wang (1979), for example, thiolated ribonuclease at pH 10.5 using 14.6 equivalents of the thiolactone and 16.6 equivalents of imidazole.

Thioparaconic acid (Figure 5.233) is closely analogous to the preceding compound and was examined by Klotz and Elfbaum (1964) with the nucleophilic catalysts just discussed. It does not seem to offer any particular advantages over AHTL and was never widely adopted.

More convenient modern thiolation reagents, not requiring catalysts or extended reaction times, have rendered AHTL much less popular in recent years.

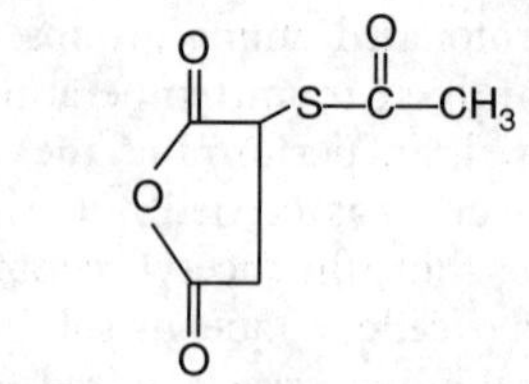

Figure 5.235 *S*-Acetyl-mercaptosuccinic anhydride (SAMSA).

Leanza *et al.* (1992) described anionic analogues of AHTL containing carboxylic, sulfonic or phosphonic acid groups (Figure 5.234); these could be employed to counterbalance the introduction of excess positive charge into proteins by derivatisation with polyarginyl compounds, for example.

5.12.1.4 *S*-Acetyl-mercaptosuccinic anhydride (SAMSA)

One of the most enduring thiolation reagents is *S*-acetyl-mercaptosuccinic anhydride (SAMSA; Figure 5.235), introduced by Irving Klotz and his colleagues at the Northwestern University in Illinois (Klotz and Heiney, 1959, 1962). These authors added solid SAMSA portionwise to stirred, deoxygenated, aqueous solutions of proteins under nitrogen over periods of 15–60 minutes, keeping pH constant with sodium hydroxide. The reaction rate was strongly dependent on pH below 7, but constant between 7 and 9. Temperature changes between 0 and 30 °C had little effect, behaviour similar to that seen with succinic anhydride itself (Habeeb *et al.* 1958). It should be noted that Habeeb *et al.* observed significant effects on the physical properties of proteins after succinylation, increased viscosity and a decreased sedimentation constant suggesting an expansion or unfolding of the molecule. This is not unexpected given the dramatic effect of exhaustive succinylation on overall charge; at moderate pH each neutral or positive NH_2 or NH_3^+ moiety is replaced by a negative COO^- function. This consideration should be borne in mind when using this route to thiolation, and it may be advisable to avoid high succinate incorporations for sensitive proteins.

The ring-opening of the anhydride can yield two isomeric succinates (Figure 5.236) of which Klotz and Heiney estimated (a) to be the more favoured; it is likely that mixtures are formed, however. Hydrolysis of the anhydride is a competing reaction; the authors also observed some deacetylation of the thiol, probably due to local pH ‘hot-spots’, close to the droplets of

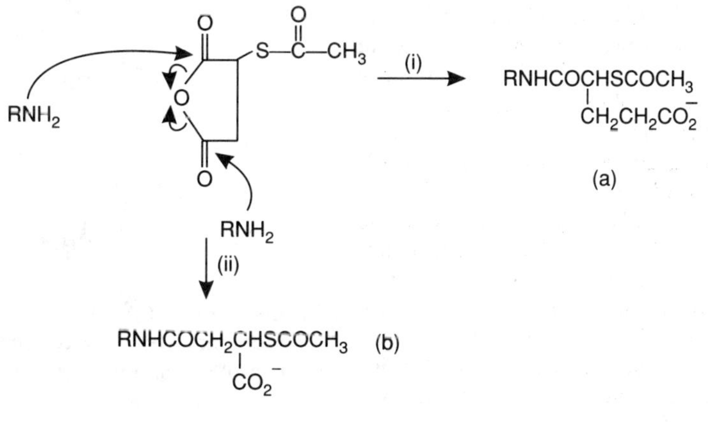

Figure 5.236 Ring-opening of SAMSA by two routes yielding two isomeric succinates.

sodium hydroxide added (in separate experiments the thioester was shown to be stable for several days at pH 9; hydrolysis was only observed above pH 10).

The obvious alternative of carrying out the thiolation reaction in a buffer was tried successfully, but high molarities were required due to the generation of carboxyl groups during the reaction, even a 125 mM phosphate buffer, pH 6.8, became significantly more acidic under the conditions used by these authors.

Hydrolysis of the thioester to free the thiol was readily achievable under basic conditions (60 minutes in 0.01 M NaOH, pH 11.5), but milder reagents were examined to avoid possible protein damage. Nitrogenous nucleophiles such as imidazole were known to catalyse hydrolyses of esters (Bruice and Schmir, 1957; Bender and Turnquest, 1957), a reaction thought to be relevant to the role of histidine residues in esterase enzymes. Accordingly, Klotz and Heiney assessed a number of such compounds for thioester hydrolysis, of which hydroxylamine (NH_2OH) gave outstandingly good performance. At pH 7.2 to 8, deacetylation was achieved at a rate comparable to the use of sodium hydroxide at the same concentration. With 0.1 M hydroxylamine the hydrolysis was complete within minutes, and this method for deacetylation of thioesters has remained in use up to the present day. Interestingly, the authors noted that longer incubation with hydroxylamine resulted in the loss of thiol groups at a rate of 10% per hour.

Hydroxylamine can also catalyse the hydrolysis of *O*-acetyl compounds such as phenolic esters under mildly acidic conditions, while more stable *N*-acetyl derivatives (amides), which may be present in the carbohydrate moieties of glycoproteins, are reported to be unaffected (Smyth, 1967).

Klotz and Stryker (1959) demonstrated that SAMSA could also react with hydroxyl groups, successfully derivatising polyols such as dextran and polyvinyl alcohol. Reaction of this type with OH groups in proteins does not, however, seem to occur to a significant extent (Klotz and Heiney, 1962; Suzuki *et al.*, 1970). By analogy with succinic anhydride (Habeeb *et al.*, 1958), reaction may occur with any thiols already present, but is of little practical impact.

A SAMSA analogue containing an extra methyl group was described later by the Northwestern group (Klotz *et al.*, 1965), but does not appear to have been adopted for wider use. SAMSA was rapidly taken up by other groups for protein thiolation: Nisonoff and Rivers (1961), for example, used SAMSA in the preparation of insoluble antigen complexes.

Suzuki *et al.* (1970) thiolated Taka-amylase A with SAMSA: the protein was present in 150 mM phosphate buffer, pH 8.0, while the SAMSA (in a 10 to 500-fold excess over the enzyme) was dissolved in dioxane. After 30 minutes the thiol was deprotected by treatment with 330 mM hydroxylamine solution, pH 7.5, for 10 minutes. Again, temperature had little effect on the thiolation reaction between 2 and 25 °C.

Rector *et al.* (1978) used SAMSA to thiolate ovalbumin: 50 mg of the protein in 1.25 ml of 10 mM sodium phosphate buffer, 2% NaCl, 1 mM EDTA (pH 7.5), was stirred at room temperature under nitrogen for one hour, then 967 μg of SAMSA added, equivalent to a fivefold molar excess. The mixture was allowed to incubate for a further hour before gel filtration on Sephadex™ G-25, with the pH being maintained at 7.5–8.0 with 1 M NaOH. Deacetylation was achieved by carrying out the subsequent conjugation with iodoacetylated IgG (in the phosphate/NaCl/EDTA buffer just described) in the presence of 10 mM hydroxylamine. This gave an incorporation of 0.6–0.8 thiol groups per ovalbumin molecule.

Ishikawa *et al.* (1978) described a general protocol for thiolation of peptides and enzymes using SAMSA: up to 0.5 mg of the peptide or enzyme in 1 ml of 0.1 M sodium phosphate buffer (pH 6–7), was stirred for 30 minutes with 2–5 mg of SAMSA. Deacetylation was achieved by incubation for 20 minutes at 30 °C with 0.5 ml of 0.5 M hydroxylamine (pH 7.0).

The Miyazaki group also (Ishikawa *et al.*, 1983) described detailed protocols for thiolation of HRP and IgG. For HRP, the conditions were as follows. (1) Activation: 10 μl (10 μmol) of SAMSA solution (174 mg/ml in DMF) was added to 250 μl (50 nmol) of HRP solution (8 mg/ml in 100 mM sodium phosphate buffer, pH 7.5), and the mixture stirred at 30 °C for 30 minutes. This represents a SAMSA:HRP excess of 200:1. (2) Deacetylation: 50 μl of 100 mM Tris-HCl buffer, 5 μl of 100 mM EDTA and 50 μl of 1 M hydroxylamine (all pH 7.0) were added, and the mixture incubated at 30 °C for five minutes. (3) Purification: the mixture was fractionated on a 1 × 45 cm Sephadex™ G-25 column using 100 mM sodium phosphate buffer, 5 mM EDTA, pH 6.0 as eluent. An incorporation of 2.2 thiol groups per peroxidase molecule was obtained.

The protocol for IgG was as follows. (1) Activation: 10 μl (3.4 μmol) of SAMSA solution (60 mg/ml in DMF) was added to 500 μl (33 nmol) of IgG solution (10 mg/ml in 100 mM sodium phosphate buffer, pH 6.5) and the mixture stirred at room temperature for 30 minutes. This represents a SAMSA:HRP excess of approximately 100:1. (2) Deacetylation: 100 μl of 100 mM Tris-HCl buffer, 20 μl of 100 mM EDTA and 100 μl of 1 M hydroxylamine (all pH 7.0) were added, and the mixture incubated at 30 °C for four minutes. (3) Purification: as for HRP above. An incorporation of 4.3–7.3 thiol groups per IgG molecule was obtained, while at a 50:1 SAMSA excess, 2.3 thiols were introduced.

These protocols or very close modifications continue to be used in later publications from the Miyazaki group up to the present day, both for HRP, IgG and Fab′, and for other proteins such as bovine serum albumin and thyroglobulin (Kohno *et al.*, 1989).

5.12.1.5 Methyl 3-thiopropionimidate

In 1971 Perham and Thomas described the synthesis of methyl 3-thiopropionimidate (Figure 5.237a), a thiolation reagent which maintains the charge of the

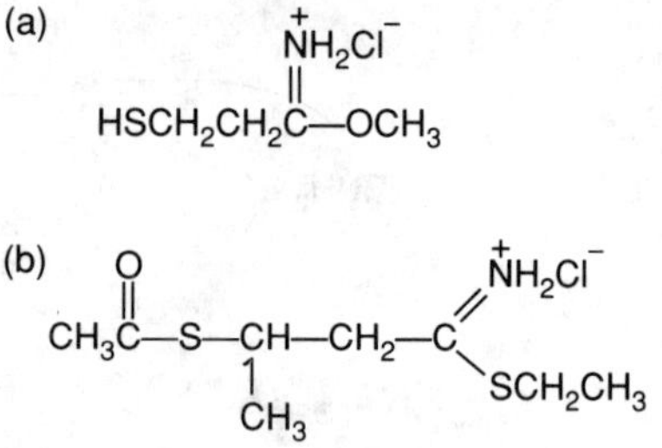

Figure 5.237 (a) Methyl 3-thiopropionimidate; hydrochloride (b) Ethyl *S*-acetyl-3-mercaptobutyrothioimidate, hydrochloride.

protein by incorporating a positively-charged amidine group at the site of the derivatised amine. Perham and Thomas used this reagent to thiolate tobacco mosaic virus; the thiol was used as a binding site for mercury to facilitate X-ray diffraction experiments. Traut *et al.* (1973) described a modified synthesis, and employed the reagent to elucidate subunit topography in the *E. coli* ribosome. Recently, protected-thiol reagents based on the analogous thioimidate group have been described (Arpicco *et al.*, 1997); an example is shown in Figure 5.237(b). These reagents offer a sterically hindered linker for *in vivo* stability, as well as the charge-preserving property of amidination.

5.12.1.6 2-Iminothiolane (2-IT)

Traut *et al.* (1973) also synthesised what was believed to be a longer homologue of the preceding compound, methyl 4-thio (or 4-mercapto) butyrimidate (MMBI; Figure 5.238(a)) and described its use for activating *E. coli* ribosome subunits towards oxidative cross-linking. Under the conditions employed (50 mM triethanolamine buffer, pH 8.0, containing 1 mM magnesium acetate, 50 mM KCl and 3% 2-mercaptoethanol) it

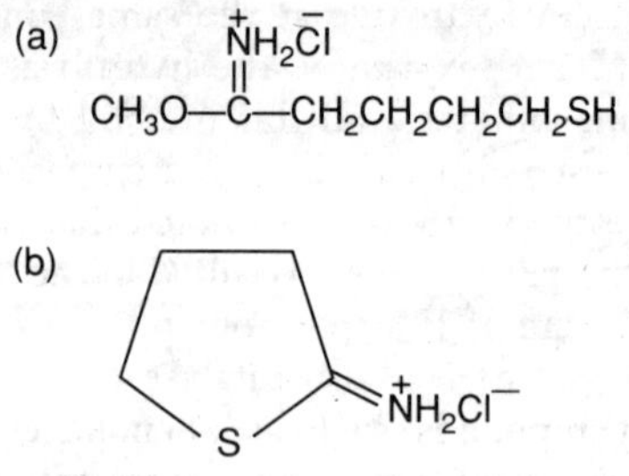

Figure 5.238 (a) Methyl 4-thiobutyrimidate (MMBI); (b) 2-iminothiolane (2-IT); both shown as hydrochloride.

showed better performance than the shorter propionimidate. Oxidative coupling was achieved by dialysis into buffer not containing mercaptoethanol, and the addition of hydrogen peroxide to a concentration of 40 mM.

Cleavage of the linked subunits was achieved by reverting to the original mercaptoethanol buffer (30 minutes at 37 °C). These authors alkylated any free thiol groups in the derivatised or cleaved proteins before electrophoresis, to prevent any changes in behaviour due to charge modification from thiolate moieties.

Over the next few years, Traut's group at the University of California carried out further studies on the use of 'methyl 4-mercaptobutyrimidate' in determining ribosome topography. These are reviewed by Traut and Kenny (1977).

Simultaneously with Traut and Kenny's review, Schramm and Dülffer (1977) reported the use of a cyclic imidate, 2-iminothiolane (2-IT; Figure 5.238(b)) in similar applications. Carrying out characterisation work on MMBI, it was subsequently recognised by Traut's group that this was in fact the true nature of the compound which they had been using (Jue *et al.*, 1978), cyclisation of the butyrimidate having occurred during synthesis or storage (Figure 5.239). King *et al.* (1978) also observed that 2-IT was formed from MMBI on storage and that it too could serve as a thiolation reagent.

The resulting compound was considerably more stable at acid to neutral pH than a linear imidate. It was also less susceptible to side-reactions during protein derivatisation, while still retaining the key advantage of maintaining a positive charge at the site of modification. These properties have made it a popular thiolation reagent right up to the present day.

It has been known for many years that 2-IT can also react with hydroxyl groups at high pH, and with free thiols (Alagon and King, 1980), but these side-reactions are often of little importance. Reaction with glycoprotein carbohydrates has, however, been invoked as a possible explanation for increased p*I* after 2-IT treatment (Bush and Winkler, 1989)

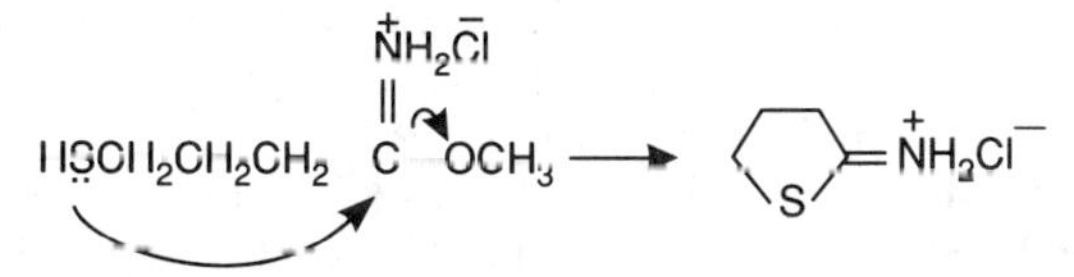

Figure 5.239 Cyclisation of methyl 4-thiobutyrimidate to produce 2-IT.

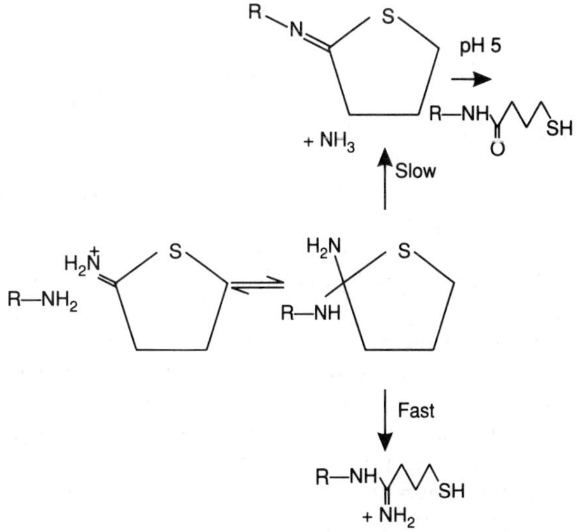

Figure 5.240 Alternative product (classical product) formation with 2-IT upon protein thiolation.

Recent studies by Singh *et al.* (1996) have tarnished the clean reputation of 2-IT somewhat further. They observed that an alternative product could be obtained from reaction with amines, an unreactive iminothiolane derivative formed from the expulsion of ammonia. This product was gradually formed from the initial thiol derivative; it could be ring-opened to provide a thiol by weak acid treatment, but the product is linked to the protein *via* an amide rather than an amidine, so a decrease in p*I* will be observed. Singh *et al.* recommended that 2-IT activated proteins should be kept at low temperature and used rapidly to prevent this side-reaction. These reactions are summarised in Figure 5.240.

Gnemmi *et al.* (1978) used 2-IT to derivatise donkey IgG: the buffer used was 50 mM *N*-ethylmorpholine pH 7.5, and the thiolating agent was dissolved in 100 mM sodium carbonate solution. The antibody was derivatised while complexed to an immunoadsorbent gel, to avoid reaction of residues in the binding sites.

Blättler *et al.* (1985) thiolated a mouse monoclonal antibody with 2-IT: the antibody (2 mg/ml in 60 mM triethanolamine HCl buffer, 7 mM potassium phosphate, 100 mM NaCl, 1 mM EDTA, pH 8.0) was degassed then incubated with 1 mM 2-IT for 90 minutes at 0 °C under nitrogen. The reaction was stopped by gel filtration on a Sephadex™ G-25F column using a 5 mM Bistris-acetate buffer, 50 mM NaCl, 1 mM EDTA (pH 5.8), as eluent. This resulted in the incorporation of 2.0 thiol groups per molecule.

The same group used similar conditions to thiolate the toxin gelonin (Lambert *et al.*, 1985; McIntyre *et al.*, 1994) without significant loss of activity. This appears to be a good example of the benefit of 2-IT in retaining the charge of the native protein, derivatisation with succinimide esters or acid anhydrides which neutralised or reversed the charge of the reacting amines caused drastic reductions in toxicity.

King *et al.* (1978) thiolated a number of proteins with a thousandfold excess of 2-IT at room temperature: bovine plasma albumin yielded a thiol incorporation of 4.5 per molecule at pH 8.0, and 6.5 at pH 9.0, both after one hour; ribonuclease yielded a single thiol after one hour at pH 8 or 9; sheep antibody gave 3.9 thiols per molecule after two hours at pH 8, and 4.8 after two hours at pH 8.5. The iminiothiolane was observed to be less reactive than open-chain imidates, which was convenient for obtaining a controlled reaction.

King and Kochoumian (1979) used 2-IT excesses of 10,000 and 1000:1 respectively to thiolate goat antibody (4.0 thiols/molecule after two hours at pH 8) and HRP (1.2 thiols/molecule after two hours at pH 9). The HRP was thiolated in the presence of 4,4′-dipyridyldisulfide (4,4′-dithiodipyridine) to yield protected thiol groups (see Section 5.2.1). The activated proteins were coupled (pH 6.6, room temperature, 17 hours) to yield a low-molecular-mass conjugate which retained 90% of enzyme activity and performed well in an immunoassay.

Thorpe *et al.* (1987a) used a 12:1 excess of 2-IT over mouse monoclonal IgG using 0.05 M borate buffer (pH 9.0), as the reaction medium. Reaction was allowed to continue for one hour at room temperature.

Marsh (1988) used a fivefold excess of 2-IT to thiolate diphtheria toxin, using a 160 mM borate buffer (pH 8.0), as the reaction medium. Thiolation was allowed to continue for 90 minutes at room temperature.

Ebert and Spryn (1990) used a 50-fold excess of 2-IT over barley toxin and gelonin (100 mM phosphate buffer, pH 8.0, 1 mM EDTA; 10–30 minutes at 30 °C), giving incorporations of one to two thiol groups per molecule.

Ghosh *et al.* (1990) thiolated HRP with a 100-fold excess of 2-IT (one hour, 23 °C, pH 9.0), achieving an incorporation of 1.9.

Ravinder (1991) used the following conditions for the thiolation of IgG: a 15-fold molar excess of 2-IT was incubated with a 20 mg/ml IgG solution in 100 mM phosphate buffer, 150 mM NaCl, 2 mM EDTA (pH 7.4), for one hour at 22 °C.

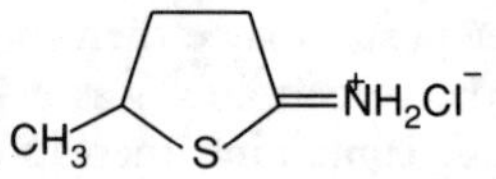

5-Methyl-2-iminothiolane (M2IT),hydrochloride

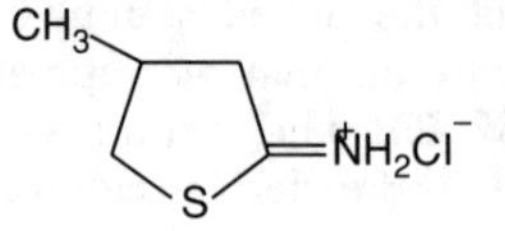

4-Methyl-2-iminothiolane,hydrochloride

Figure 5.241 Substituted 2-iminothiolanes.

5.12.1.7 Substituted iminothiolanes

In 1990 Goff and Carroll described the synthesis of a series of 4- and 5-substituted 2-iminothiolanes, two of which are shown in Figure 5.241. These were used to yield model disulfides which were considerably more stable to reduction than those formed with 2-IT, offering the potential for immunoconjugates with improved *in vivo* stability.

5-Methyl-2-iminothiolane (M2IT) was used by Bernhard *et al.* (1994) to prepare immunotoxins from ribosome-inactivating proteins. Either Mab or RIP (3 mg/ml in 25 mM triethanolamine-HCl, 150 mM NaCl, pH 8.0) was incubated with M2IT to a concentration of 0.5 mM, and 5,5′-dithio-bis(2-nitrobenzoic acid) to 1 mM, for three hours at 25 °C. The disulfide protects the protein thiols by exchanging with them as they are formed; it is removed by reduction with 5 mM dithiothreitol (90 minutes at 25 °C)

5.12.1.8 *N*-Succinimidyl *S*-acetylthioacetate (SATA)

Given the ascendancy of active esters for amine derivatisation since the late 1970s, it is not surprising that thiolating agents based on this chemistry have found some popularity. SPDP – in which the thiol must be released by reduction of a disulfide – has already been mentioned; the other common reagent of this type is *N*-succinimidyl-*S*-acetylthioacetic acid (NHS-SATA or SATA; Figure 5.242). This reagent was first described by Duncan *et al.* in 1983.

Dosio *et al.* (1989) used SATA to thiolate Mabs, and observed that the *S*-acetyl group was considerably more stable than the pyridyldisulfide moiety of SPDP. No deacetylation was observed after 30 days at 4 °C, while over half of the pyridyldisulfide groups had been

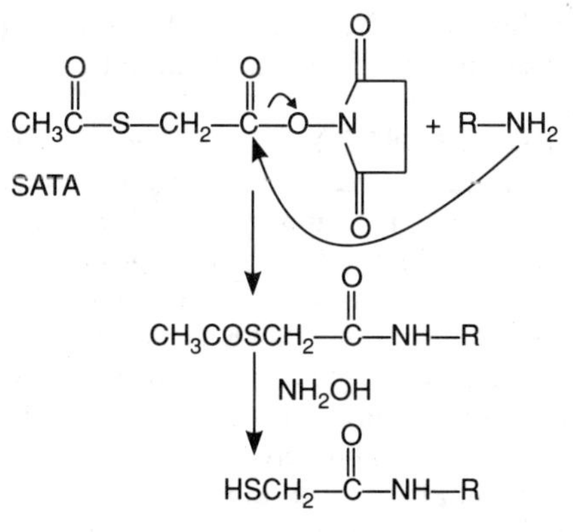

Figure 5.242 Thiolation with SATA.

cleaved, a process accompanied by polymerisation of the antibody molecules.

As with SAMSA, the acetyl function can be removed by treatment with hydroxylamine. Two advantages are claimed over the anhydride, however: first there is less disruption to the electrostatic charge of the protein as the introduction of a free carboxyl function is avoided, and secondly only one isomeric form of the thiolated product is possible.

Duncan *et al.* (1983) dissolved SATA at up to 150 mM (34.7 mg/ml) in DMF, and added this reagent to solutions of protein in 50 mM sodium–potassium phosphate buffer, 1 mM EDTA (pH 7.5). The amount of DMF added was kept below 1% (v/v). The excess of SATA required was obviously related to the amine content of the protein: for HRP no further incorporation could be obtained by increasing the SATA:protein ratio above 15:1, while IgG and thyroglobulin showed increasing thiolation at least up to a 20–fold excess.

With a number of different polyclonal and monoclonal IgG samples there was no variation in this incorporation behaviour, and it was also noted that the presence of at least six thioacetate functions per IgG molecule could be tolerated without loss of antibody functionality. The IgG concentration used had a significant effect on incorporation: with all other conditions the same, altering [IgG] from 12 mM (1.8 μg/ml) to 60 μM (9 mg/ml) resulted in an increase in thiol incorporation from 1.84 to 4.0 per molecule.

Deacetylation using essentially the same conditions as Klotz and Heiney (1962) exposed 90% of the thiol groups within one hour at room temperature, and there was no further reaction on storage for 24 hours at 4 °C. Reaction of thiolated HRP with maleimide-derivatised IgG was carried out at various ratios, and resulted in very well-defined conjugates of 1.5 to 5.4 enzyme molecules per HRP.

SATA is commercially available from a number of sources, and has been widely adopted for protein thiolation. Recently, for example, it has been used in prodrug studies by Bagshawe *et al.* (1991) to activate monoclonal antibodies for conjugation to carboxypeptidase G2, and by the Miyazaki group (Hashida *et al.*, 1991, 1993; Hashinaka *et al.*, 1994) to thiolate 2,4-dinitrophenyl bovine serum albumin and recombinant HIV proteins.

In the last example Hashinaka *et al.* used an approximately eightfold excess of SATA over recombinant HIV reverse transcriptase, and achieved an incorporation of 1.5 thiol groups per molecule. Reaction conditions were typical, 30 minutes at 30 °C, with the protein dissolved in a phosphate buffer (pH 7.0), and the thiolating agent dissolved in DMF.

A generic procedure for thiol–maleimide coupling incorporating SATA is described in Protocol 5.1.

5.12.1.9 Other active-ester-based thiolating reagents

The propionyl homologue of SATA was described by Fujii *et al.* (1985a), as was an analogous compound in which the thiol was protected by an *S*-(*p*-methoxybenzoyl) function rather than the *S*-acetyl moiety (Figure 5.243(a)). Advantages were claimed for this protecting group in avoiding the hydroxylaminolysis step generally employed for deacetylation, the methoxybenzoyl group is instead removed by treatment with trifluoromethanesulfonic acid–thioanisole–trifluoroacetic acid (ice-bath, 60 minutes). This might

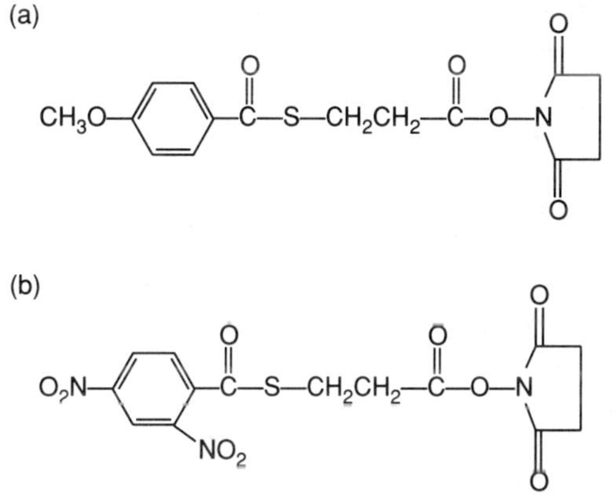

Figure 5.243 (a) Succinimidyl *S*-(*p*-methoxybenzoyl)thiopropionate (analogous to SATP); (b) 2,4-dinitrophenyl ester of *S*-acetylthiopropionate.

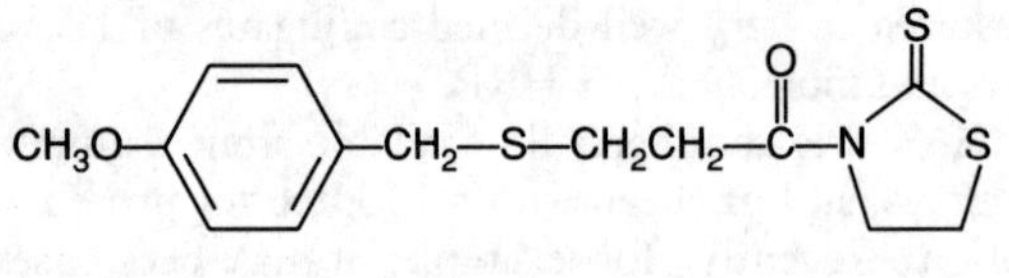

Figure 5.244 3-(3-*p*-Methoxybenzylthiopropionyl)thiazolidine-2-thione (MTPTT).

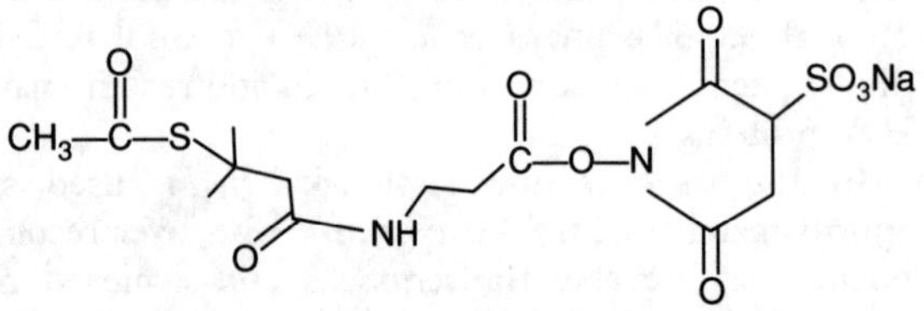

Figure 5.245 Sulfosuccinimidyl *N*-[3-(acetylthio)-3-methylbutyryl]-β-alanine (sNHS-ATMBA), sodium salt.

provide an alternative approach for proteins which are sensitive to hydroxylamine treatment; however the latter is a convenient and mild reaction which is not noted for such problems. The hydroxylaminolysis time (four hours) reported by Fujii *et al.* with low-molecular-mass models is long compared to those routinely employed in protein chemistry with SATA.

These authors also described a 2,4-dinitrophenyl ester of the *S*-acetylthiopropionate (Figure 5.243(b)), presumably with the aim of achieving more rapid aminolysis than with the succinimidyl ester (cf. Section 5.7.4).

Fujii *et al.* (1985a,b) also described two thiolation reagents using the thiazolidine-2-thione moiety as the activating function (Yajima *et al.*, 1980), again with *S*-acetyl and *S*-*p*-methoxybenzyl protecting groups (Figure 5.244). They do not appear to offer any great advantage over more readily-available reagents.

Greenfield *et al.* (1990) described a longer-chain sulfosuccinimidyl analogue of SATA, which features an *S*-acetyl function hindered by two α-methyl groups. This compound, sulfosuccinimidyl *N*-[3-(acetylthio)-3-methylbutyryl]-β-alanine (sNHS-ATMBA), is shown in Figure 5.245. The hindered immunotoxins produced using this reagent showed an improved *in vivo* stability compared to unhindered analogues. A very similar linker developed with pharmacokinetic behaviour in mind is succinimidyl *N*-[3-(acetylthio)-3-methylbutyryl]glycinate (Govindan *et al.*, 1996).

5.12.1.10 Cyanuric chloride/NaSH

Sodium hydrosulfide or thiolate, NaSH, is capable of replacing some labile species such as halogen atoms, the SH^- ion acting as a powerful nucleophile. Cyanuric chloride (2,4,6-trichloro-1,3,5-triazine) has already been mentioned as a homobifunctional reagent

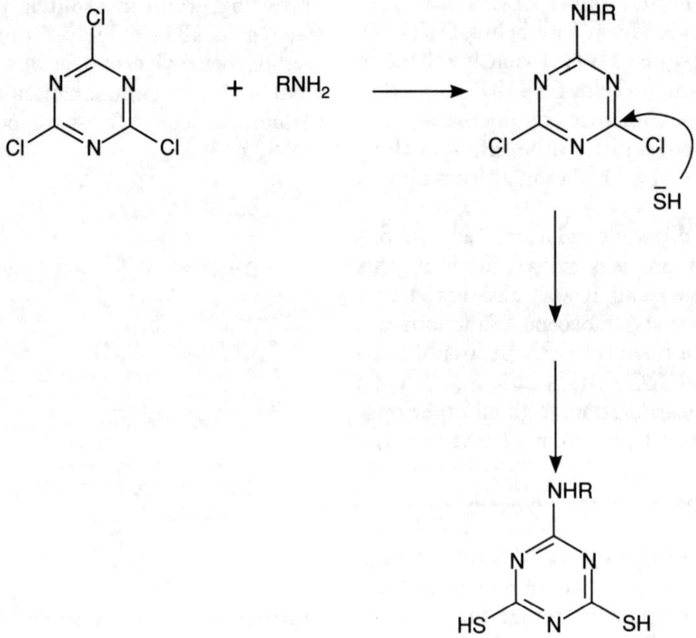

Figure 5.246 Reaction of cyanuric chloride with NaSH.

for nucleophiles. If only a single residue is allowed to react, however, the resulting dichlorotriazine can be thiolated with NaSH as indicated in Figure 5.246 (Levchenko *et al.*, 1975; Torchinsky, 1981c).

5.12.1.11 Formation of dithiocarbamates with carbon disulfide

Dithiocarbamates of secondary amines (R_2>N–C(=S)SH) have some properties analogous to thiols, for example the ability to bind specifically to mercury. Szurdoki *et al.* (1995) used carbon disulfide to generate these groups from secondary amines.

Dithiocarbamate-derivatised proteins have been observed to undergo cross-linking (Valentine *et al.*, 1992), both by disulfide formation, and *via* isothiocyanates generated by decomposition of the dithiocarbamate group.

5.12.2 Reagents for thiolation *via* carboxyls

Gaertner *et al.* (1994) used reverse proteolysis (cf. Section 5.9.5) to couple homocysteine thiolactone to the C-terminal lysine residues of polypeptide fragments, generating a free thiol by subsequent treatment with hydroxylamine (Figure 5.247). This was coupled to haloacetylated polypeptides.

Other chemical methods for peptide bond formation can probably be applied in this way too.

5.12.3 Reagents for thiolation *via* aldehydes

Bieniarz *et al.* (1991) described methods for thiolation of antibodies and other glycoproteins *via* aldehydes generated by periodate oxidation of the carbohydrate moieties. In one example, cystamine was used as the thiolating reagent, the resulting Schiff base being reduced with sodium cyanoborohydride. In another, hydrazidolipoamide was used to derivatise the oxidised glycoprotein; thiol groups were freed enzymatically from the lipoamide using lipoamide dehydrogenase.

Further details of this approach were described by Husain and Bieniarz (1994). A number of polyclonal and monoclonal antibodies were oxidised with periodate (one hour, 5 °C, pH 8.0), incubated with a large excess of cystamine dihydrochloride and sodium cyanoborohydride (overnight, room temperature, pH 7.0), then reduced with 2 mM DTT (15 minutes, room temperature, pH 7.0). Between three and ten thiol groups were incorporated per IgG molecule, and the immunoreactivity of one such Mab was virtually unaffected. The thiolated IgG was subsequently coupled (overnight, 2–8 °C, pH 7.0) to maleimide-derivatised AP.

An analogous indirect approach was described by Pollner *et al.* (1996): periodate-oxidised protein was reacted with a diamine, and the resulting amine groups derivatised with 2-IT.

5.12.4 Reagents for thiolation *via* hydroxyls

5.12.4.1 Tosylation and transesterification with thioacetate

Hydroxyl groups can be thiolated *via* a tosyl (*p*-toluenesulfonyl) intermediate. Reactions of this type (followed by oxidation to yield disulfide bridges) have been in use to modify the properties of cellulose-based textiles since the 1940s (Izard and Morgan, 1949; Schwenker *et al.*, 1962; Schwenker and Lifland, 1963). The first stage involves reaction of a hydroxyl group with tosyl chloride to yield the

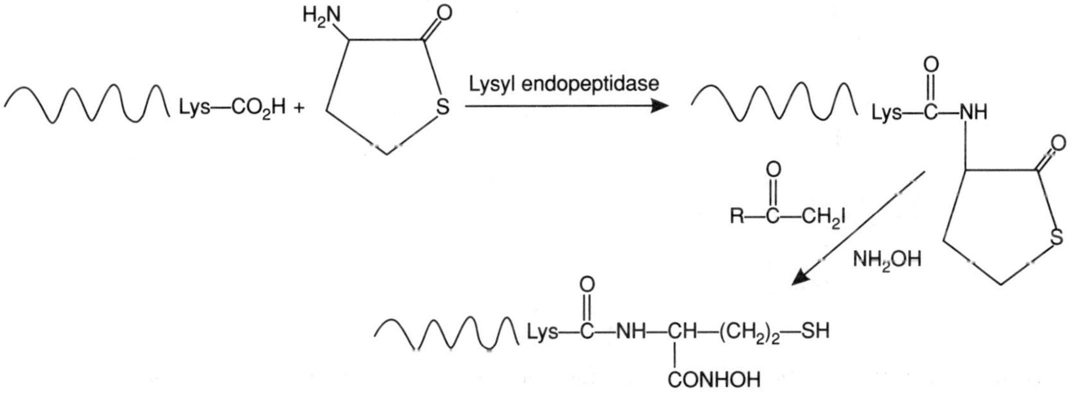

Figure 5.247 Reverse proteolysis using homocysteine thiolactone.

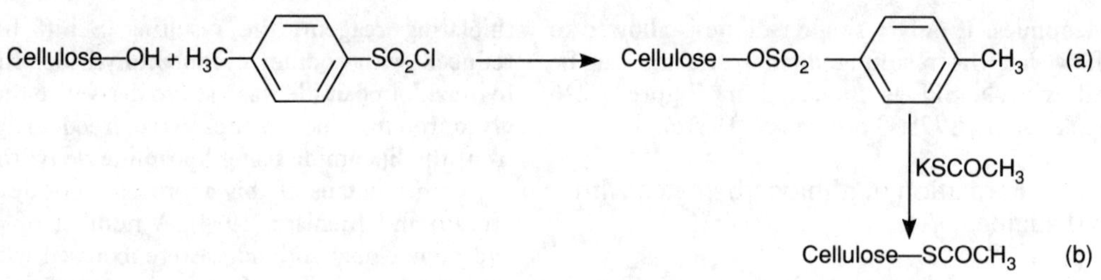

Figure 5.248 Tosylation and transesterification with thioacetate.

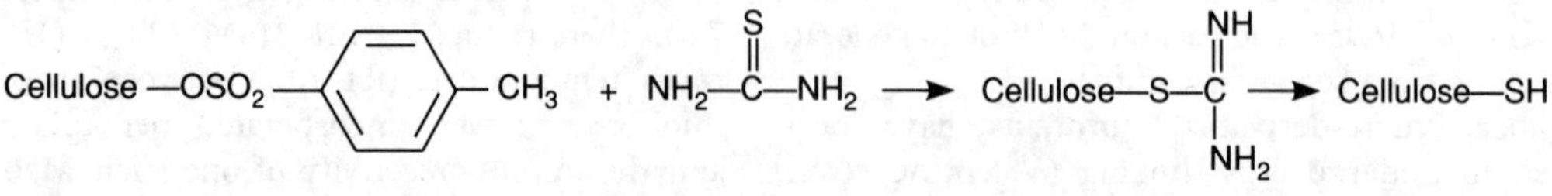

Figure 5.249 Reaction of tosylated cellulose with thiourea, and hydrolysis of the isourea product to yield a thiol.

tosylate (Figure 5.248(a)), which can subsequently be displaced by potassium thioacetate to yield an acetylated thiol (Figure 5.248(b)). Reaction of the tosylate with thiourea followed by alkaline hydrolysis (Figure 5.249) is another possibility (Izard and Morgan, 1949).

Photaki and her colleagues investigated the application of such reactions to amino acids and oligopeptides, principally with the aim of interconverting serine and cysteine residues (Zervas and Photaki, 1960; Photaki and Bardakos, 1965). It was shown that tosylation of serine followed by treatment with thioacetate or thiobenzoate resulted in the formation of the *S*-acylcysteine derivative. The free thiol is readily liberated from such compounds (cf. deprotection of SATA and SAMSA derivatives above). Under the conditions employed there was little competition from the β-elimination reaction which yields dehydroalanine (Figure 5.250).

Interest in the derivatisation of serine residues in protein research grew with the recognition of the importance of their role in the active sites of many enzymes, such as α-chymotrypsin. Koshland and his colleagues (Strumeyer *et al.*, 1963; Weiner *et al.*, 1966) demonstrated that tosylation could be carried out on this protein in aqueous solution, subsequently using

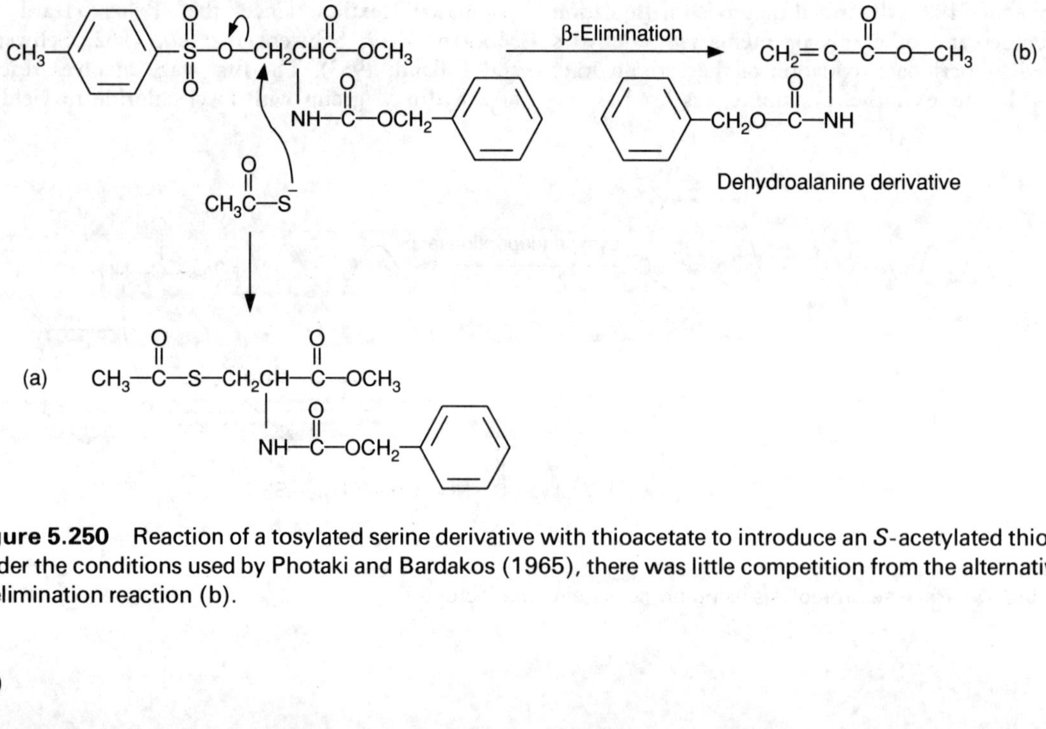

Figure 5.250 Reaction of a tosylated serine derivative with thioacetate to introduce an *S*-acetylated thiol (a). Under the conditions used by Photaki and Bardakos (1965), there was little competition from the alternative β-elimination reaction (b).

the elimination just mentioned to yield 'anhydro-chymotrypsin'.

Tosylation of proteins followed by thioacetate displacement therefore offers an interesting route to thiolation, and Ebert *et al.* (1977) demonstrated this approach with insoluble collagen and silk fibroin, and soluble BSA. Collagen was treated with tosyl chloride in a 2:1 mixture of a phosphate buffer (pH 8–9) and dioxane, then transesterification with 0.5 M thioacetate solution was carried out at pH 5.5. The resulting thioacetate was deacetylated with 0.5 M sodium methoxide. This sequence resulted in the conversion of 40–50% of the serine groups into cysteine. With fibroin only 13% of the serines were converted, a result attributed to the inaccessibility of the densely-packed β-pleated structure.

With BSA, tosylation was carried out as described for collagen, transesterification was performed at pH 9, and deacetylation was achieved by incubation at pH 12 for one hour. The thiol content rose by around 20 fold and intermolecular aggregates were formed. While some serine had been lost from the BSA, there were also significant losses of lysine and tyrosine.

5.12.4.2 PMSF and transesterification with thioacetate

Phenylmethanesulfonyl fluoride, PMSF (Figure 5.251) is another reagent for activating nucleophiles such as hydroxyl groups. Gold and Fahrney (1964) used this reagent to inactivate α-chymotrypsin *via* the active-site serine, for example.

This reaction can also provide a route to thiolation: Polgar and Bender (1966, 1967) converted the serine group of subtilisin to cysteine, using PMSF to activate the hydroxyl moiety. This reagent was reported to give a much more rapid reaction than tosyl chloride. Transesterification with 0.7 M thioacetate (pH 5.2–5.3, 72 hours at 25 °C) yielded the *S*-acetylthiol, which was in this case conveniently deprotected by the enzymic activity of the subtilisin itself.

5.12.4.4 Thiophosphorylation *via* hydroxyl groups

Another interesting route to the incorporation of reactive sulfur into serine-containing proteins was described by Facemyer and Cremo (1992). A number of such serine residues can be phosphorylated by ATP under the influence of an appropriate kinase enzyme. Many of these enzymes will accept the sulfur-containing analogue adenosine-5′-*O*-(3-thiotriphosphate), ATPγS, as a substrate, resulting in the incorporation of a phosphorothioate function (Figure 5.252). The sulfur of this group is nucleophilic enough to react with reagents such as haloacetates, so the ATPγS acts as a heterobifunctional reagent.

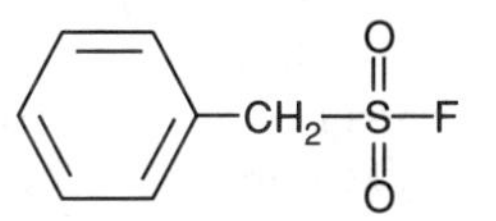

Phenylmethanesulfonyl fluoride (PMSF)

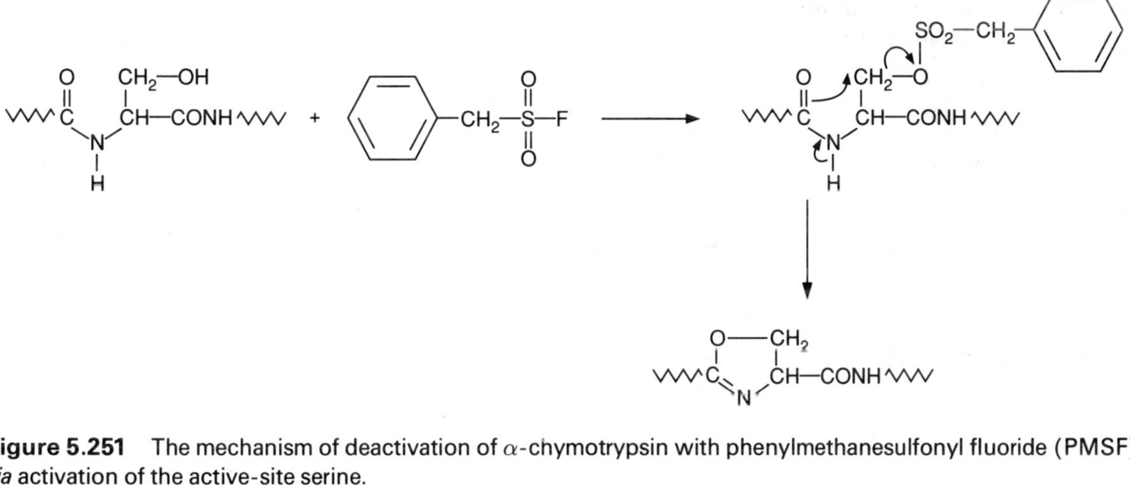

Figure 5.251 The mechanism of deactivation of α-chymotrypsin with phenylmethanesulfonyl fluoride (PMSF) *via* activation of the active-site serine.

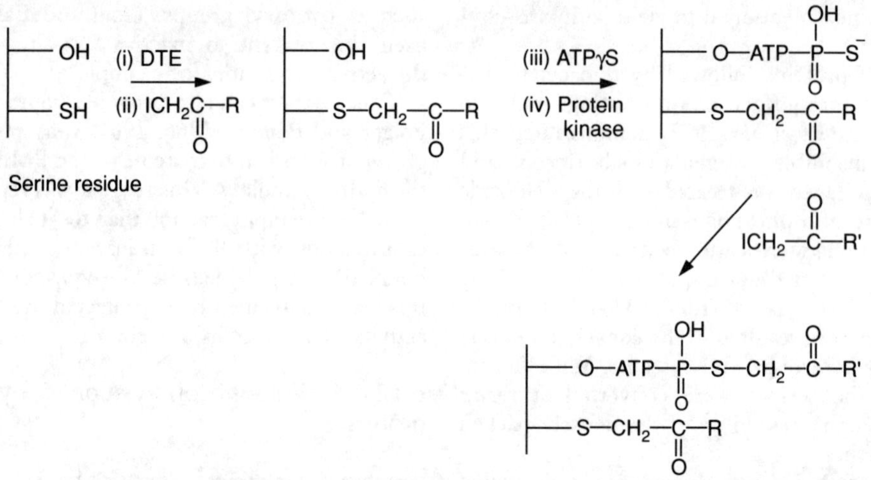

Figure 5.252 Outline of the method for specifically labelling protein serines *via* enzymic thiophosphorylation.

Facemyer and Cremo thiophosphorylated the Ser19 residue of a myosin subunit RLC20, and labelled it with a haloacetate-containing fluorophore. There is no reason why the same approach could not be used with a suitably derivatised protein to give a highly specific conjugate linkage. The complexity of the approach, however, would clearly limit its application in the protein coupling field to very specific needs.

5.12.5 Reagents for the reduction of disulfides

The reduction reactions described here can be used to generate thiols from disulfide groups, either those native to the protein, or those which have been artificially introduced *via* one of the reagents just mentioned. The only difference will be in terms of stability—the latter are likely to be somewhat more readily reduced.

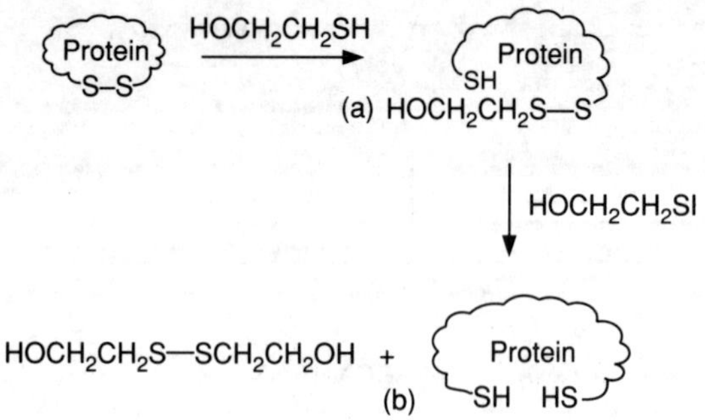

Figure 5.253 Reduction of protein disulfides with 2-mercaptoethanol.

Any suitably reactive water-soluble thiol can be used to reduce protein disulfides: after all this is simply the reverse application of the reaction of protein thiols with disulfides such as SPDP. However, these are likely to yield a mixture of thiol and mixed disulfide products (Figure 5.253(a)).

Completion can still be attained by simple thiols if their disulfide formation is more favourable than that of the cystine link, in which case the result shown for 2-mercaptoethanol will eventually predominate (Figure 5.253(b)). However the stability of this disulfide is very close to that of cystine (Konigsberg, 1972), and therefore a very large excess is required to avoid a mixed product. The use of this reagent is nevertheless very common (White, 1972a; Habeeb, 1972). Interestingly, it has been shown that the presence of EDTA in

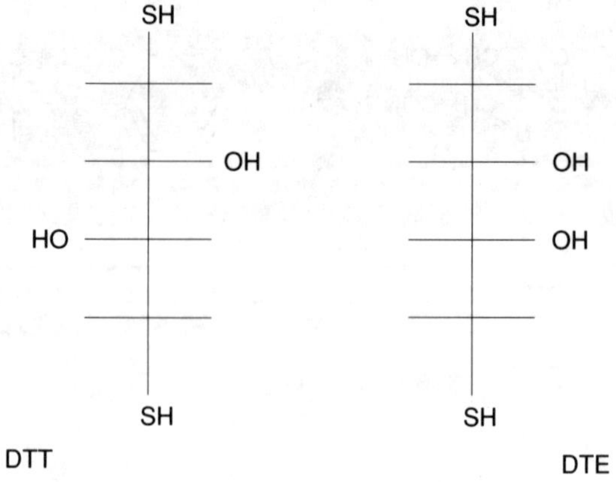

Figure 5.254 DTT and DTE (Cleland's reagents).

the reduction medium does not interfere with thiol yields in such cases (Yoshitake *et al.*, 1979a,b).

The reduction can be effected more efficiently by a dithiol such as dithiothreitol, DTT or its close analogue dithioerythritol, DTE (Figure 5.254). These are sometimes referred to as Cleland's reagents (Cleland, 1964). The great advantage of these compounds is the overriding tendency of their mixed disulfide products to self-cleave, yielding more thiol and a highly stable cyclic disulfide byproduct (Figure 5.255). Therefore the reaction can be driven readily to completion.

They are usually employed at an enormous excess, often of the order of 10^6:1, and the exchange reaction is generally complete within 20–40 minutes at room temperature.

If the aim is to reduce an artificially introduced disulfide, the main concern is usually to avoid disruption of the native cystine bonds in the protein. As it is the thiolate species which drives the exchange reaction (Torchinsky, 1981a) a low pH can be used to limit the degree of reaction, and Carlsson *et al.* (1978) recommended using an acetate buffer, pH 4.5, for the reduction of SPDP-substituted proteins.

Suitable conditions are typified by the use of 25 mM DTT or DTE in 0.1 M sodium acetate (pH 4.5), 0.1 M NaCl, with 20–30 minutes incubation at room temperature (Deelder and de Water, 1981; Jeanson *et al.*, 1988b). Where reduction of protein disulfides is not an issue, a higher pH can be used to drive the reaction more vigorously, Lomant and Fairbanks (1976) for example used 10 mM DTE at pH 8.5 and 37 °C to

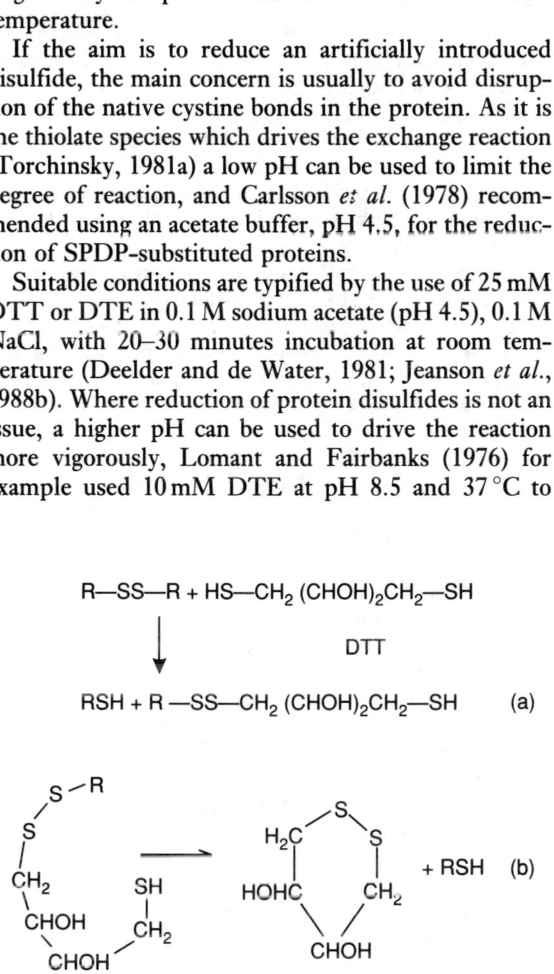

Figure 5.255 Reaction of DTT with a disulfide yields the expected mixed disulfide (a); however, this is prone to self-cleavage and cyclisation to a stable product (b). Conversion of all the initial disulfide to thiols is the end result.

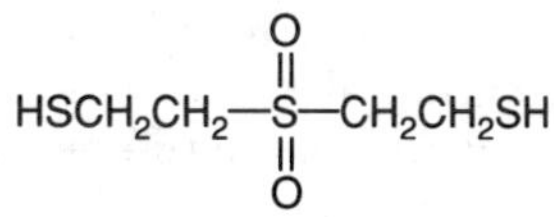

Figure 5.256 Bis(2-mercaptoethyl) sulfone (BMS).

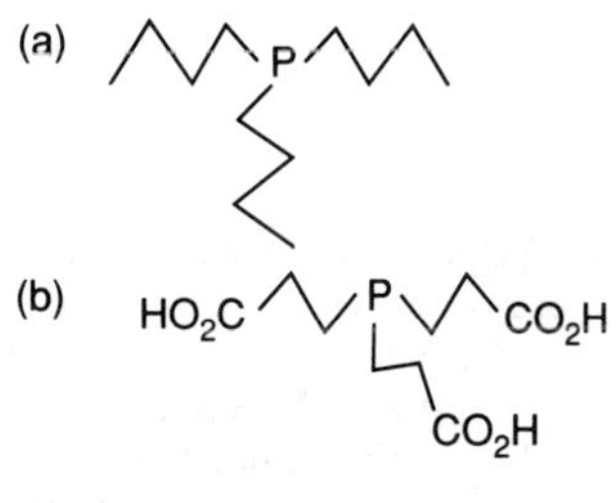

Figure 5.257 (a) Tris(*n*-butyl)phosphine; (b) tris-(2-carboxyethyl)phosphine (TCEP).

cleave disulfides in DSP-linked haemoglobin before electrophoresis.

Other cyclisable bis-thiol reagents such as bis-(2-mercaptoethyl)sulfone (BMS; Figure 5.256) have been reported to display faster hydrolysis of some disulfides than DTT (Lamoureux and Whitesides, 1993). BMS is commercially available from Calbiochem.

Rüegg and Rudinger (1974, 1977) described the use of tri-*n*-butylphosphine (Figure 5.257(a)) for disulfide reduction, which proceeds according to the following equation:

$$Bu_3P + R-S-S-R \Rightarrow Bu_3P=O + 2R-SH$$

This compound has some definite advantages: only a slight excess (5 to 20%) over the stoichiometric quantity is required, reaction times are similar to DTT and DTE, and reaction appears to be highly specific for cystine moieties. The toxicity of tributylphosphine and its unpleasant smell have probably prevented its more widespread use in this capacity. The analogous tris(2-carboxyethyl)phosphine (TCEP; Figure 5.257(b)) displays similarly powerful reducing properties but is also odourless (Han and Han, 1994). It is commercially available as the hydrochloride, for example from Pierce or Calbiochem.

Other chemical reducing agents have also been described in this role, such as sodium borohydride (Habeeb, 1972), Singh and Kats (1995) have noted that selenol acts as a catalyst for this reaction, as well as for DTT reduction. A recent patent has also described the use of lipoic acid and its analogues in an alkaline cyanide matrix, for the quantitative removal of disulfide-bound ligands from a solid phase (Hoyle, 1994). Finally, the small protein thioredoxin can (in its reduced, bis-thiol form) be used for disulfide reduction (Buchanan *et al.*, 1994). It is commercially available in recombinant form, e.g. from Calbiochem.

References

Abadi, D. M. and Wilcox, P. E. (1960). Chemical derivatives of α-chymotrypsinogen. III. Reaction with *N*-acetyl-DL-homocysteine thiolactone. *J. Biol. Chem.*, **235**, 396–404.

Abdella, P. M. *et al.* (1979). A new cleavable reagent for cross-linking and reversible immobilisation of proteins. *Biochem. Biophys. Res. Commun.*, **87**, 734–742.

Acharya, A. S. and Manning, J. M. (1980a). Reactivity of the amino groups of carbonmonoxyhemoglobin S with glyceraldehyde. *J. Biol. Chem.*, **255**, 1406–1412.

Acharya, A. S. and Manning, J. M. (1980b). Amadori rearrangement of glyceraldehyde–hemoglobin Schiff base adducts. A new procedure for the determination of ketoamine adducts in proteins. *J. Biol. Chem.*, **255**, 7218–7224.

Aggeler, R. and Capaldi, R. A. (1992). Cross-linking of the γ subunit of the *Escherichia coli* ATPase (ECF_1) *via* cysteines introduced by site-directed mutagenesis. *J. Biol. Chem.*, **267**, 21355–21359.

Aggeler, R. *et al.* (1992). Introduction of reactive cysteine residues in the ϵ subunit of *Escherichia coli* F_1 ATPase, modification of these sites with tetrafluorophenyl azide-maleimides, and examination of changes in the binding of the ϵ subunit when different nucleotides are in catalytic sites. *Biochemistry*, **31**, 2956–2961.

Ahlem, C. N. *et al.* (1991). Tris-maleimido compounds as intermediates in trifunctional antibody synthesis. *European Patent 446,071*, to Hybritech Inc.

Åkerblom, E. *et al.* (1993). Preparation and characterization of conjugates of monoclonal antibodies and staphylococcal enterotoxin A using a new hydrophilic cross-linker. *Bioconjugate Chem.*, **4**, 455–466.

Alagon, A. C. and King, T. P. (1980). Activation of polysaccharides with 2-iminothiolane and its uses. *Biochemistry*, **19**, 4341–4345.

Albarella, J. P. and Buckler, R. T. (1989). Enzyme-labeled antibody reagent with polyalkyleneglycol linking group. *US Patent 4,810,638*, to Miles Inc.

Albert, A. and Barlin, G. B. (1959). Ionization constants of heterocyclic substances. Part III. Mercapto-derivatives of pyridine, quinoline and isoquinoline. *J. Chem. Soc.*, 2384–2396.

Aldwin, L. and Nitecki, D. E. (1987). A water-soluble, monitorable peptide and protein crosslinking agent. *Anal. Biochem.*, **164**, 494–501.

Ali, M. S. and Quadri, S. M. (1996). Maleimido derivatives of diethylenetriaminepentaacetic acid and triethylenetetraaminehexaacetic acid: their synthesis and potential for specific conjugation with biomolecules. *Bioconjugate Chem.*, **7**, 576–583.

Ali-Osman, F. *et al.* (1985). Chemical structure of carbamoylating groups and their relationship to bone marrow toxicity and antiglioma activity of bifunctionally alkylating and carbamoylating nitrosoureas. *Cancer Res.*, **45**, 4185–4191.

Anderer, F. A. and Schlumberger, H. D. (1969). Antigenic properties of proteins cross-linked by multidiazonium compounds. *Immunochemistry*, **6**, 1–10.

Anderson G. W. and Callahan, F. M. (1958). Racemization by the dicyclohexylcarbodiimide method of peptide synthesis. *J. Am. Chem. Soc.*, **80**, 2902–2903.

Anderson, G. W. *et al.* (1964). The use of esters of *N*-hydroxysuccinimide in peptide synthesis. *J. Am. Chem. Soc.*, **86**, 1839–1842.

Anderson, L. D. *et al.* (1993). Method for the synthesis of trifunctional maleimide–antibody complex. *US Patent 5,262,524*, to Hybritech Inc.

Annunziato, M. E. *et al.* (1993). *p*-Maleimidophenyl isocyanate: a novel heterobifunctional linker for hydroxyl to thiol coupling. *Bioconjugate Chem.*, **4**, 212–218.

Argaraña, C. E. *et al.* (1986). Molecular cloning and nucleotide sequence of the streptavidin gene. *Nucleic Acids Res.*, **14**, 1871–1882.

Arpicco, S. *et al.* (1997). New coupling reagents for the preparation of disulfide cross-linked conjugates with increased stability. *Bioconjugate Chem.*, **8**, 327–337.

Atassi, M. Z. (1977). Chemical strategy for studying the antigenic structures of disulfide-containing proteins: hen egg-white lysozyme as a model. *Adv. Exp. Med. Biol.*, **86A**, 89–137.

Avigad, G. *et al.* (1962). The D-galactose oxidase of *Polyporus circinatus*. *J. Biol. Chem.*, **237**, 2736–2743.

Avrameas, S. (1968). Détection d'anticorps et d'antigènes a l'aide d'enzymes. *Bull. Soc. Chim. Biol.*, **50**, 1169–1178.

Avrameas, S. (1969). Coupling of enzymes to proteins with glutaraldehyde: use of the conjugates for the detection of antigens and antibodies. *Immunochemistry*, **6**, 43–52.

Avrameas, S. and Ternynck, T. (1967). Biologically active water-insoluble protein polymers. 1. Their use for isolation of antigens and antibodies. *J. Biol. Chem.*, **242**, 1651–1659.

Avrameas, S. and Ternynck, T. (1971). Peroxidase labelled antibody and Fab conjugates with enhanced intracellular penetration. *Immunochemistry*, **8**, 1175–1179.

Avrameas, S. *et al.* (1978). Coupling of enzymes to antibodies and antigens. *Scand. J. Immunol.*, **8** (Suppl. 7), 7–23.

Bagshawe, K. D. *et al.* (1991). Antibody directed enzyme prodrug therapy (ADEPT): clinical report. *Disease Markers*, **9**, 233–238.

Ballmer-Hofer, K. *et al.* (1982). Isolation of *in situ* cross-linked ligand–receptor complexes using an anticrosslinker specific antibody. *Anal. Biochem.*, **126**, 246–250.

Bangs, J. D. *et al.* (1986). Posttranslational modification and intracellular transport of a trypanosome variant surface glycoprotein. *J. Cell Biol.*, **103**, 255–263.

Baranowska-Kortylewicz, J. and Kassis, A. I. (1993). Labeling of immunoglobulins with bifunctional, sulfhydryl-selective, and photoreactive coumarins. *Bioconjugate Chem.*, **4**, 300–304.

Barry, V. C. and Mitchell, P. W. D. (1953). Properties of periodate-oxidised polysaccharides. Part II. The structure of some nitrogen-containing polymers. *J. Chem. Soc.*, 3631–3635.

Bartling, G. J. *et al.* (1974). Preparation and properties of horseradish peroxidase cross-linked in nonaqueous media. *Int. J. Peptide Protein Res.*, **6**, 287–294.

Basak, A. *et al.* (1994). Biotinylation of an enkephalin-containing heptapeptide *via* various spacer arms. Synthesis, comparitive binding studies towards avidin, and application as substrates in enzymatic reactions. *Bioconjugate Chem.*, **5**, 301–305.

Bateman, W. G. (1916). The digestibility and utilization of egg proteins. *J. Biol. Chem.*, **26**, 263–291.

Battelli, M. G. *et al.* (1990). Toxicity of, and histological lesions caused by, ribosome-inactivating proteins, their IgG conjugates, and their homopolymers. *APMIS*, **98**, 585–593.

Bayer, E. A. and Wilchek, M. (1977). The biotin transport system. *Methods Enzymol.*, **46**, 613–617.

Bayer, E. A. and Wilchek, M. (1978). The avidin–biotin complex as a tool in molecular biology. *Trends Biochem. Sci.*, **3**, N257–N259.

Bayer, E. A. and Wilchek, M. (1980). The use of the avidin–biotin complex as a tool in molecular biology. *Methods Biochem. Anal.*, **26**, 1–45.

Bayer, E. A. and Wilchek, M. (1990). Protein biotinylation. *Methods Enzymol.*, **184**, 138–160.

Bayer, E. A. and Wilchek, M. (1992a). Avidin–biotin technology. Preparation of biotinylated probes. *Methods Mol. Biol.*, **10**, 137–142.

Bayer, E. A. and Wilchek, M. (1992b). Avidin–biotin technology. Preparation of avidin conjugates. *Methods Mol. Biol.*, **10**, 143–148.

Bayer, E. A. *et al.* (1975). The use of a homologous series of affinity labeling reagents in the study of the biotin transport system in yeast cells. *FEBS Lett.*, **60**, 309–312.

Bayer, E. A. *et al.* (1976). Affinity cytochemistry: the localization of lectin and antibody receptors on erythrocytes *via* the avidin–biotin complex. *FEBS Lett.*, **68**, 240–244.

Bayer, E. A. *et al.* (1990). Isolation and properties of streptavidin. *Methods Enzymol.*, **184**, 80–89.

Bayley, H. and Staros, J. V. (1984). Photoaffinity labeling and related techniques. In *Azides and Nitrenes. Reactivity and Utility* (Scriven, E. F. V., ed.), pp. 433–490, Academic Press, New York.

Bayley, H. *et al.* (1978). Propane-1,3-dithiol: a selective reagent for the efficient reduction of alkyl and aryl azides to amines. *Tetrahedron Lett.*, **39**, 3633–3634.

Becker, J. M. *et al.* (1971). Irreversible inhibition of biotin transport in yeast by biotinyl-*p*-nitrophenyl ester. *Proc. Natl. Acad. Sci. USA*, **68**, 2604–2607.

Bednar, R. A. (1990). Reactivity and pH dependence of thiol conjugation to *N*-ethylmaleimide: detection of a conformational change in chalcone isomerase. *Biochemistry*, **29**, 3684–3690.

Belleau, B. and and Malek, G. (1968). A new convenient reagent for peptide synthesis. *J. Am. Chem. Soc.*, **90**, 1651–1652.

Bender, M. L. and Turnquest, B. W. (1957). The imidazole-catalyzed hydrolysis of *p*-nitrophenyl acetate. *J. Am. Chem. Soc.*, **79**, 1652–1655.

Benesch, R. and Benesch, R. E. (1956). Formation of peptide bonds by aminolysis of homocysteine thiolactones. *J. Am. Chem. Soc.*, **78**, 1597–1599.

Benesch, R. and Benesch, R. E. (1958). Thiolation of proteins. *Proc. Natl. Acad. Sci. USA*, **44**, 848–853.

Benesch, R. and Benesch, R. E. (1959). The introduction of new sulfhydryl groups and disulfide bonds into proteins. In *Sulfur in Proteins* (Benesch, R. *et al.*, eds), pp. 15–24, Academic Press, New York.

Benhar, I. *et al.* (1994). Mutations of two lysine residues in the CDR loops of a recombinant immunotoxin that reduces its sensitivity to chemical derivatization. *Bioconjugate Chem.*, **5**, 321–326.

Berger, M. and Wood, H. G. (1975). Purification of the subunits of transcarboxylase by affinity chromatography on avidin–Sepharose. *J. Biol. Chem.*, **250**, 927–933.

Bergmann, M. and Fruton, J. S. (1938). Some synthetic and hydrolytic experiments with chymotrypsin. *J. Biol. Chem.*, **124**, 321–329.

Bernhard, S. L. *et al.* (1994). Cysteine analogs of recombinant barley ribosome inactivating protein form antibody conjugates with enhanced stability and potency *in vitro*. *Bioconjugate Chem.*, **5**, 126–132.

Besman, M. and Coleman, J. E. (1985). Isozymes of bovine intestinal alkaline phosphatase. *J. Biol. Chem.*, **260**, 11190–11193.

Beyzavi, K. *et al.* (1987). Comparison of horseradish peroxidase and alkaline phosphatase-labelled antibodies in enzyme immunoassays. *Ann. Clin. Biochem.*, **24**, 145–152.

Bieniarz, C. *et al.* (1991). Covalent attachment of antibodies and antigens to solid phases using extended length heterobifunctional coupling agents. *US Patent 5,063,109*, to Abbott Laboratories.

Bieniarz, C. *et al.* (1992a). Site-specific conjugation of immunoglobulins and detectable labels. *World Patent WO 92/10507*, to Abbott Laboratories.

Bieniarz, C. *et al.* (1992b). Chromogenic redox assay for β-lactamases yielding water-insoluble products. *Anal. Biochem.*, **207**, 329–334.

Bieniarz, C. *et al.* (1996). Extended length heterobifunctional coupling agents for protein conjugations. *Bioconjugate Chem.*, **7**, 88–95.

Bisson, R. *et al.* (1978). Interaction of cytochrome *c* with cytochrome *c* oxidase. *J. Biol. Chem.*, **253**, 1874–1880.

Bjorn, M. J. *et al.* (1986). Antibody–*Pseudomonas* exotoxin A conjugates cytotoxic to human breast cancer cells *in vitro*. *Cancer Res.*, **46**, 3262–3267.

Blättler, W. A. *et al.* (1985). New heterobifunctional protein cross-linking reagent that forms an acid-labile link. *Biochemistry*, **24**, 1517–1524.

Bloxham, D. P. and Cooper, G. K. (1982). Formation of a polymethylene bis(disulfide) intersubunit cross-link between cysteine-281 residues in rabbit muscle glyceraldehyde-3-phosphate dehydrogenase using octamethylene bis (methane [^{35}S] thiosulfonate). *Biochemistry*, **21**, 1807–1812.

Bloxham, D. P. and Sharma, R. P. (1979). The development of *S,S'*-polymethylenebis(methanethiosulfonates) as reversible cross-linking reagents for thiol groups and their use to form stable catalytically active cross-linked dimers within glyceraldehyde-3-phosphate dehydrogenase. *Biochem. J.*, **181**, 355–366.

Bobbitt, J. M. (1956). Periodate oxidation of carbohydrates. *Adv. Carbohydr. Chem.*, **11**, 1–41.

Bodanszky, M. (1960). Stepwise synthesis of peptides by the nitrophenyl ester method. *Ann. N. Y. Acad. Sci.*, **88**, 655–664.

Boesel, R. W. and Carpenter, F. H. (1970). Crosslinking during the nitration of bovine insulin with tetranitromethane. *Biochem. Biophys. Res. Commun.*, **38**, 678–682.

Bongers, J. *et al.* (1992). Semisynthesis of human growth hormone-releasing factor by trypsin catalyzed coupling of leucine amide to a C-terminal precursor. *Int. J. Peptide Protein Res.*, **40**, 268–273.

Boorsma, D. M. and Kalsbeek, G. L. (1975). A comparitive study of horseradish peroxidase conjugates prepared with a one-step and a two-step method. *J. Histochem. Cytochem.*, **23**, 200–207.

Boorsma, D. M. and Streefkerk, J. G. (1976). Peroxidase conjugate chromatography. Isolation of conjugates prepared with glutaraldehyde or periodate using polyacrylamide-agarose gel. *J. Histochem. Cytochem.*, **24**, 481–486.

Boorsma, D. M. and Streefkerk, J. G. (1979). Periodate or glutaraldehyde for preparing peroxidase conjugates? *J. Immunol. Methods*, **30**, 245–255.

Boorsma, D. M. *et al.* (1976). Peroxidase and fluorescein isothiocyanate as antibody markers. A quantitative comparison of two peroxidase conjugates prepared with glutaraldehyde or periodate and a fluorescein conjugate. *J. Histochem. Cytochem.*, **24**, 1017–1025.

Borch, R. F. *et al.* (1971). The cyanohydridoborate anion as a selective reducing agent. *J. Am. Chem. Soc.*, **93**, 2897–2904.

Borek, F. (1961). A new two-stage method for cross-linking proteins. *Nature*, **191**, 1293–1294.

Borek, F. and Silverstein, A. M. (1961). Characterization and purification of ferritin–antibody globulin conjugates. *J. Immunol.*, **87**, 555–561.

Bragg, P. D. and Hou, C. (1975). Subunit composition, function and spatial arrangement in the Ca^{2+} and Mg^{2+}-activated adenosine triphosphatases of *Escherichia coli* and *Salmonella typhimurium*. *Arch. Biochem. Biophys.*, **167**, 311–321.

Brandenburg, D. (1972). Peptides 87: preparation of N^{aA1}, N^{eB29}-adipoyl-insulin, an intramolecularly cross-linked derivative of beef insulin. *Hoppe-Seyler's Z. Physiol. Chem.*, **353**, 869–873.

Brandenburg, D. *et al.* (1971). Structure function studies with derivatives and analogs of insulin and its chains. *Excerpta Med. Int. Congr. Ser.*, **231**, 363–376.

Brandt, J. *et al.* (1975). Covalent attachment of proteins to polysaccharide carriers by means of benzoquinone. *Biochim. Biophys. Acta*, **386**, 196–202.

Brennan, M. *et al.* (1985). Preparation of bispecific antibodies by chemical recombination of monoclonal immunoglobulin G_1 fragments. *Science*, **229**, 81–83.

Brewer, C. F. and Riehm, J. P. (1967). Evidence for possible non-specific reactions between *N*-ethylmaleimide and proteins. *Anal. Biochem.*, **18**, 248–255.

Brocklehurst, K. (1979). Specific covalent modification of thiols: applications in the study of enzymes and other biomolecules. *Int. J. Biochem.*, **10**, 259–274.

Brocklehurst, K. *et al.* (1973). Covalent chromatography. Preparation of fully active papain from dried papaya latex. *Biochem. J.*, **133**, 573–584.

Brois, S. J. *et al.* (1970). A new pathway to unsymmetrical disulfides. The thiol-induced fragmentation of sulfenyl thiocarbonates. *J. Am. Chem. Soc.*, **92**, 7629–7631.

Broker, T. R. *et al.* (1978). Electron microscopic visualisation of tRNA genes with ferritin–avidin: biotin labels. *Nucleic Acids Res.*, **5**, 363–384.

Brown, R. D. and Matthews, K. S. (1979). Chemical modification of lactose repressor protein using *N*-substituted maleimides. *J. Biol. Chem.*, **254**, 5128–5134.

Browne, D. T. and Kent, S. B. H. (1975). Formation of non-amidine products in the reaction of primary amines with imido esters, *Biochem. Biophys. Res. Commun.*, **67**, 126–132.

Bruice, T. C. and Schmir, G. L. (1957). Imidazole catalysis. I. The catalysis of the hydrolysis of phenyl acetates by imidazole. *J. Am. Chem. Soc.*, **79**, 1663–1667.

Brumeanu, T.-D. *et al.* (1995). Enzymatically mediated, glycosidic conjugation of immunoglobulins with viral epitopes. *J. Immunol. Methods*, **183**, 185–197.

Buchanan, B. B. *et al.* (1994). Thioredoxin: a multifunctional regulatory protein with a bright future in technology and medicine. *Arch. Biochem. Biophys.*, **314**, 257–260.

Bunn, H. F. *et al.* (1975). Further identification of the nature and linkage of the carbohydrate in hemoglobin A_{1c}. *Biochem. Biophys. Res. Commun.*, **67**, 103–109.

Burkot, T. R. *et al.* (1985). Use of fluorodinitrobenzene to identify monoclonal antibodies which are suitable for conjugation to periodate-oxidized horseradish peroxidase. *J. Immunol. Methods*, **84**, 25–31.

Burleigh, B. D. *et al.* (1978). Photocoupling of the subunits of ovine lutropin using a specific aryl azide derivative of the β subunit. *J. Biol. Chem.*, **253**, 7179–7185.

Bush, D. A. and Winkler, M. A. (1989). Isoelectric focusing of cross-linked monoclonal antibody heterodimers, homodimers and derivatized monoclonal antibodies. *J. Chromatogr.*, **489**, 303–311.

Busse, W.-D. and Carpenter, F. H. (1974). Carbonylbis(L-methionine *p*-nitrophenyl ester). A new reagent for the reversible intramolecular cross-linking of insulin. *J. Am. Chem. Soc.*, **96**, 5947–5949.

Busse, W.-D. *et al.* (1974). Carbonylbis(L-methionyl)insulin. a proinsulin analog which is convertible to insulin. *J. Am. Chem. Soc.*, **96**, 5949–5950.

Cabacungan, J. C. *et al.* (1982). Amine boranes as alternative reducing agents for reductive alkylation of proteins. *Anal. Biochem.*, **124**, 272 278.

Cai, S. X. and Keana, J. F. W. (1989). 4-Azido-2-iodo-3,5,6-trifluorophenylcarbonyl derivatives. A new class of functionalized and iodinated perfluorophenyl azide photolabels. *Tetrahedron Lett.*, **30**, 5409–5412.

Cai, S. X. and Keana, J. F. W. (1991a). Diazo- and azido-functionalized glutaraldehydes as cross-linking reagents and potential fixatives for electron microscopy. *Bioconjugate Chem.*, **2**, 38–43.

Cai, S. X. and Keana, J. F. W. (1991b). *o*-Acetamidophenylboronate esters stabilized toward hydrolysis by an intramolecular O–B interaction: potential linkers for selective bioconjugation *via* vicinal diol moieties of carbohydrates. *Bioconjugate Chem.*, **2**, 317–322.

Cai, S. X. *et al.* (1992). Toward the development of radiolabeled fluorophenyl azide-based photolabeling reagents: synthesis and photolysis of iodinated 4-azidoperfluorobenzoates and 4-azido-3,5,6-trifluorobenzoates. *J. Org. Chem.*, **57**, 1299–1304.

Calabresi, P. and Chabner, B. A. (1990). Antineoplastic agents. In *The Pharmacological Basis of Therapeutics* (Goodman, L. S. *et al.*, eds), 8th edn, chapt. 52, Pergamon Press, New York.

Cameron, D. J. and Erlanger, B. F. (1976). An enzyme-linked procedure for the detection and estimation of surface receptors on cells. *J. Immunol.*, **116**, 1313–1318.

Carlsson, J. and Svenson, A. (1974). Preparation of bovine mercaptalbumin by means of covalent chromatography. *FEBS Lett.*, **42**, 183–186.

Carlsson, J. *et al.* (1978). Protein thiolation and reversible protein–protein conjugation. *N*-Succinimidyl 3-(2-pyridyldithio) propionate. A new heterobifunctional reagent. *Biochem. J.*, **173**, 723–737.

Carpenter, F. H. and Harrington, K. T. (1972). Intermolecular cross-linking of monomeric proteins and cross-linking of oligomeric proteins as a probe of quaternary structure. Application to leucine aminopeptidase (bovine lens), *J. Biol. Chem.*, **247**, 5580–5586.

Chaiet, L. and Wolf, F. J. (1964). The properties of streptavidin, a biotin-binding protein produced by *Streptomycetes*. *Arch. Biochem. Biophys.*, **106**, 1–5.

Chaiet, L. *et al.* (1963). Antibiotic MSD-235. II. Separation and purification of synergistic components. *Antimicrob. Agents Chemother.*, **3**, 28–32.

Chaiken, I. M. and Smith, E. L. (1969a). Reaction of chloroacetamide with the sulfhydryl group of papain. *J. Biol. Chem.*, **244**, 5087–5094.

Chaiken, I. M. and Smith, E. L. (1969b). Reaction of the sulfhydryl group of papain with chloroacetic acid. *J. Biol. Chem.*, **244**, 5095–5099.

Chandra, G. and Gray, J. G. (1990). Cloning and expression of avidin in *Escherichia coli*. *Methods Enzymol.*, **184**, 70–79.

Chong, P. C. S. and Hodges, R. S. (1981). A new heterobifunctional cross-linking reagent for the study of biological interactions between proteins. Design, synthesis and characterization. *J. Biol. Chem.*, **256**, 5064–5070.

Chowdhry, V. *et al.* (1976). 2-Diazo-3,3,3-trifluoropropionyl chloride: reagent for photoaffinity labeling. *Proc. Natl. Acad. Sci. USA*, **73**, 1406–1408.

Christensen, T. B. and Ruud, E. (1975). Protection of polypeptides by tetramethyl urea against denaturation by periodate. *J. Carbohydr. Nucleosides Nucleotides*, **2**, 133–139.

Chua, M.-M. *et al.* (1984). Attachment of immunoglobulin to liposomal membrane *via* protein carbohydrate. *Biochim. Biophys. Acta*, **800**, 291–300.

Clamp, J. R. and Hough, L. (1965). The periodate oxidation of amino acids with reference to studies on glycoproteins. *Biochem. J.*, **94**, 17–24.

Cleland, W. W. (1964). Dithiothreitol, a new protective reagent for SH groups. *Biochemistry*, **3**, 480–482.

Clyne, D. H. *et al.* (1973). Antibody enzyme conjugates. The preparation of intermolecular conjugates of horseradish peroxidase and antibody and their use in immunohistology of renal cortex. *J. Histochem. Cytochem.*, **21**, 233–240.

Collioud, A. *et al.* (1993). Oriented and covalent immobilization of target molecules to solid supports: synthesis and application of a light-activatable and thiol-reactive cross-linking reagent. *Bioconjugate Chem.*, **4**, 528–536.

Cook, A. G. and Wood, P. J. (1994). Chemical synthesis of bispecific monoclonal antibodies: potential advantages in immunoassay systems. *J. Immunol. Methods*, **171**, 227–237.

Cooke, R. D. and Holbrook, J. J. (1974). Calcium and the assays of human plasma clotting factor XIII. *Biochem. J.*, **141**, 71–78.

Coons, A. H. *et al.* (1941). Immunological properties of an antibody containing a fluorescent group. *Proc. Soc. Exp. Biol. Med.*, **47**, 200–202.

Costello, S. M. *et al.* (1979). Enhancement of immune cellular agglutination by use of an avidin–biotin system. *Clin. Chem.*, **25**, 1572–1580.

Crocker, P. J. *et al.* (1990). Heterobifunctional cross-linking agents incorporating perfluorinated aryl azides. *Bioconjugate Chem.*, **1**, 419–424.

Cuatrecasas, P. and Parikh, I. (1972). Adsorbents for affinity chromatography. Use of *N*-hydroxysuccinimide esters of agarose. *Biochemistry*, **11**, 2291–2299.

Cuatrecasas, P. and Wilchek, M. (1968). Single-step purification of avidin from egg-white by affinity chromatography on biocytin–Sepharose columns. *Biochem. Biophys. Res. Commun.*, **33**, 235–239.

Cumber, A. J. and Wawrzynczak, E. J. (1992). Preparation of cytotoxic antibody–toxin conjugates. *Methods Mol. Biol.*, **10**, 283–293.

D'Albis, A. and Gratzer, W. (1976). Chemical cross-linking of myosin. Disposition of the globular heads. *J. Biol. Chem.*, **251**, 2825–2830.

Darfler, F. J. and Marinetti, G. V. (1977). Synthesis of a photoaffinity probe for the beta-adrenergic receptor. *Biochem. Biophys. Res. Commun.*, **79**, 1–7.

Das, M. and Fox, C. F. (1979). Chemical cross-linking in biology. *Annu. Rev. Biophys. Bioeng.*, 8, 165–193.

Davies, G. E. and Stark, G. R. (1970). Use of dimethyl suberimidate, a cross-linking reagent, in studying the sub-unit structure of oligomeric proteins. *Proc. Natl. Acad. Sci. USA*, **66**, 651–656.

Deelder, A. M. and De Water, R. (1981). A comparitive study on the preparation of immunoglobulin–galactosidase conjugates. *J. Histochem. Cytochem.*, **29**, 1273–1280.

Degani, Y. and Degani, C. (1979). Subunit-selective chemical modifications of creatine kinase. Evidence for asymmetrical association of the subunits. *Biochemistry*, **18**, 5917–5923.

De Graff, B. A. *et al.* (1974). Phenyl nitrene. A flash photolytic investigation of the reaction with secondary amines. *J. Am. Chem. Soc.*, **96**, 7491–7496.

De Jong, M. O. *et al.* (1995). Biotinylation of interleukin-2 (IL-2) for flow cytometric analysis of IL-2 receptor expression. Comparison of different methods. *J. Immunol. Methods*, **184**, 101–112.

De Lange, R. J. and Huang, T.-S. (1971). Egg white avidin. III. Sequence of the 78-residue middle cyanogen bromide peptide. Complete amino acid sequence of the protein subunit. *J. Biol. Chem.*, **246**, 698–709.

Del Rosario, R. B. *et al.* (1990). Sulfhydryl site-specific cross-linking and labeling of monoclonal antibodies by a fluorescent equilibrium transfer alkylation cross-link reagent. *Bioconjugate Chem.*, **1**, 51–59.

Dentler, W. L. *et al.* (1980). Microtubule-membrane interactions in cilia. II. Photochemical cross-linking of bridge structures and the identification of a membrane-associated dynein-like ATPase. *J. Cell Biol.*, **84**, 381–403.

DeSilva, B. S. and Wilson, G. S. (1995). Solid phase synthesis of bifunctional antibodies. *J. Immunol. Methods*, **188**, 9–19.

Dewerchin, M. *et al.* (1989). Biochemical properties of conjugates of urokinase-type plasminogen activator with a monoclonal antibody specific for cross-linked fibrin. *Eur. J. Biochem.*, **185**, 141–149.

Diamandis, E. P. and Christopoulos, T. K. (1991). The biotin–(strept)avidin system: principles and applications in biotechnology. *Clin. Chem.*, **37**, 625–636.

Dixon, H. B. F. and Perham, R. N. (1968). Reversible blocking of amino groups with citraconic anhydride. *Biochem. J.*, **109**, 312–314.

Dixon, J. S. and Lipkin, D. (1954). Spectrophotometric determination of vicinal glycols. *Anal. Chem.*, **26**, 1092–1093.

Dormán, G. and Prestwich, G. D. (1994). Benzophenone photophores in biochemistry. *Biochemistry*, 33, 5661–5673.

Dosio, F. *et al.* (1989). Antibody–toxin conjugate preparation: effects of the linker on activity and stability. *Antibody, Immunoconj. Radiopharm.*, **2**, 101–115.

Dottavio-Martin, D. and Ravel, J. M. (1978). Radiolabeling of proteins by reductive alkylation with $[^{14}C]$formaldehyde and sodium cyanoborohydride. *Anal. Biochem.*, **87**, 562–565.

Duncan, R. J. S. *et al.* (1983). A new reagent which may be used to introduce sulfhydryl groups into proteins, and its use in the preparation of conjugates for immunoassay. *Anal. Biochem.*, **132**, 68–73.

Dutton, A. *et al.* (1966). Bifunctional imidoesters as cross-linking reagents. *Biochem. Biophys. Res. Commun.*, **23**, 730–739.

Du Vigneaud, V. *et al.* (1938). Opening of the ring of the thiolactone of homocysteine. *J. Biol. Chem.*, **126**, 217–231.

Du Vigneaud, V. *et al.* (1942). The structure of biotin: a study of desthiobiotin. *J. Biol. Chem.*, **146**, 475–485.

Dyer, J. R. (1956). Use of periodate oxidations in biochemical analysis. *Methods Biochem. Anal.*, **3**, 111–152.

Eakin, R. E. *et al.* (1941). The concentration and assay of avidin, the injury-producing protein in raw egg white. *J. Biol. Chem.*, **140**, 535–543.

Ebert, C. *et al.* (1977). On the introduction of disulfide crosslinks into fibrous proteins and bovine serum albumin. *Adv. Exp. Med. Biol.*, **86A**, 235–245.

Ebert, R. F. and Spryn, L. A. (1990). Immunotoxin construction with a ribosome-inactivating protein from barley. *Bioconjugate Chem.*, **1**, 331–336.

Ebright, Y. W. *et al.* (1996). *S*-[2-(4-Azidosalicylamido)ethylthio]-2-thiopyridine: radioiodinatable, cleavable, photoactivatable cross-linking agent. *Bioconjugate Chem.*, **7**, 380–384.

Edelhoch, H. *et al.* (1953). Dimerization of serum mercaptalbumin in presence of mercurials. I. Kinetic and equilibrium studies with mercuric salts. *J. Am. Chem. Soc.*, **75**, 5058–5072.

Edsall, J. T. *et al.* (1954). Dimerization of serum mercaptalbumin in presence of mercurials. II. Studies with a bifunctional organic mercurial. *J. Am. Chem. Soc.*, **76**, 3131–3138.

Elçin, Y. M. and Akbulut, U. (1992). Polyester film strips coated with photographic gelatin containing immobilized glucose oxidase hardened by chromium(III) sulphate. *Biomaterials*, **13**, 156–161.

Ellman, G. L. (1959). Tissue sulfhydryl groups. *Arch. Biochem. Biophys.*, **82**, 70–77.

Endo, N. *et al.* (1987). A novel covalent modification of antibodies at their amino groups with retention of antigen-binding activity. *J. Immunol. Methods*, **104**, 253–258.

Engvall, E. (1978). Preparation of enzyme-labelled staphylococcal protein A and its use for detection of antibodies. *Scand. J. Immunol.*, 8 (Suppl. 7), 25–31.

Engvall, E. and Perlmann, P. (1971). Enzyme-linked immunosorbent assay (ELISA). Quantitative assay of immunoglobulin G. *Immunochemistry*, **8**, 871–874.

Engvall, E. *et al.* (1971). Enzyme-linked immunosorbent assay. II. Quantitative assay of protein antigen, immunoglobulin G, by means of enzyme-labelled antigen and antibody-coated tubes. *Biochim. Biophys. Acta*, **251**, 427–434.

Erecińska, M. (1977). A new photoaffinity-labeled derivative of mitochondrial cytochrome *c*. *Biochem. Biophys. Res. Commun.*, **76**, 495–501.

Ewig, R. A. G. and Kohn, K. W. (1978). DNA–protein cross-linking and DNA interstrand cross-linking by haloethylnitrosoureas in L1210 cells. *Cancer Res.*, **38**, 3197–3203.

Ey, P. L. (1993). Use of stable 6-aminohexyl derivatives for labelling polysaccharides with haptens and for preparing polysaccharide immunoadsorbents. *J. Immunol. Methods*, **160**, 135–137.

Facemyer, K. C. and Cremo, C. R. (1992). A new method to specifically label thiophosphorylatable proteins with extrinsic probes. Labeling of serine-19 of the regulatory light chain of smooth muscle myosin. *Bioconjugate Chem.*, **3**, 408–413.

Farrington, J. A. *et al.* (1955). The preparation of *p*-nitrophenyl thiolesters and their application to peptide synthesis. *Chem. Ind.*, 601–602.

Farrington, J. A. *et al.* (1957). Peptides. Part VII. The preparation and use of *p*-nitrophenyl thioesters. *J. Chem. Soc.*, 1407–1413.

Fasold, H. *et al.* (1963). Azophenyl-Dimaleimide als spaltbare Peptidbrücken-bildende Reagentien zwischen Cysteinresten. *Biochem. Z.*, **337**, 425–430.

Fasold, H. *et al.* (1964). Synthese und Reaktionen eines Wasserlöslichen, spaltbaren Reagens zur Verknüpfung frier SH-Gruppen in Proteinen. *Biochem. Z.*, **339**, 487–490

Fasold, H. *et al.* (1977). Comparison of hydrophobic and strongly hydrophilic cleavable crosslinking reagents in intermolecular bond formation in aggregates of proteins or protein–RNA. *Adv. Exp. Med. Biol.*, **86A**, 207–214.

Fein, M. L. *et al.* (1959). Tanning with glutaraldehyde. I. Rate studies. *J. Am. Leather Chem. Assoc.*, **54**, 488–502.

Flanders, K. C. *et al.* (1982). Semisynthetic derivatives of glucagon: (Des-His^1) N^ϵ-acetimidoglucagon and N^α-biotinyl-N^ϵ-acetimidoglucagon. *Biochemistry*, **21**, 4244–4251.

Fleet, G. W. J. *et al.* (1969). Affinity labelling of antibodies with aryl nitrene as reactive group. *Nature*, **224**, 511–512.

Fleury, P. and Lange, J. (1932). Sur l'oxydation des acides alcools et des sucres par l'acide periodique. *C. R. Acad. Sci.*, **195**, 1395–1397.

Ford, D. J. and Pesce, A. J. (1981). Reaction of glutaraldehyde with proteins. In *Enzyme Immunoassay* (Ishikawa, E. *et al.*, eds), pp. 54–66, Igaku-Shoin, Tokyo.

Ford, D. J. *et al.* (1978). Characterization of glutaraldehyde coupled alkaline phosphatase–antibody and lactoperoxidase–antibody conjugates. *Immunochemistry*, **15**, 237–243.

Fraenkel-Conrat, H. *et al.* (1952a). Avidin. I. Isolation and characterization of the protein and the nucleic acid. *Arch. Biochem. Biophys.*, **92**, 80–96.

Fraenkel-Conrat, H. *et al.* (1952b). Avidin. II. Composition and mode of action of avidin A. *Arch. Biochem. Biophys.*, **92**, 97–107.

Freytag, J. W. *et al.* (1984). Affinity-column-mediated immunoenzymometric assays: influence of affinity-column ligand and valency of antibody–enzyme conjugates. *Clin. Chem.*, **30**, 1494–1498.

Friedman, M. *et al.* (1974). Reductive alkylation of proteins with aromatic aldehydes and sodium cyanoborohydride. *Int. J. Peptide Protein Res.*, **6**, 183–185.

Frisch, B. *et al.* (1996). Synthesis of short polyoxyethylene-based heterobifunctional cross-linking reagents. Application to the coupling of peptides to liposomes. *Bioconjugate Chem.*, **7**, 180–186.

Frobert, Y. and Grassi, J. (1992). Screening of monoclonal antibodies using antigens labeled with acetylcholinesterase. *Methods Mol. Biol.*, **10**, 65–78.

Fujii, N. *et al.* (1984). Studies on peptides. CXXII. *N*-Succinimidyl-*p*-(2–nitrovinyl)-benzoate and its *m*-isomer, as heterobifunctional conjugating reagents for immunoassay. *Chem. Pharm. Bull.*, **32**, 5036–5039.

Fujii, N. *et al.* (1985a). Studies on peptides. CXXV. 3-(3-*p*-Methoxybenzylthiopropionyl)-thiazolidine-2-thione and its analogues as reagents for the introduction of the mercapto group into peptides and proteins. *Chem. Pharm. Bull.*, **33**, 362–367.

Fujii, N. *et al.* (1985b). Studies on peptides. CXXVIII. Application of new heterobifunctional crosslinking reagents for the preparation of neurokinin (A and B)–BSA (bovine serum albumin) conjugates. *Int. J. Peptide Protein Res.*, **26**, 121–129.

Fujiwara, K. and Saita, T. (1986). Novel enzyme immunoassay for thyrotropin-releasing hormone using *N*-(4-diazophenyl)maleimide as a coupling agent. *FEBS Lett.*, **202**, 197–201.

Fujiwara, K. and Saita, T. (1987). The use of *N*-[β-(4-diazophenyl)ethyl]maleimide as a heterobifunctional agent in developing enzyme immunoassay for neurotensin. *Anal. Biochem.*, **161**, 157–163.

Fujiwara, K. *et al.* (1988). The use of *N*-[β-(4-diazophenyl)-ethyl]maleimide as a coupling agent in the preparation of enzyme–antibody conjugates. *J. Immunol. Methods*, **110**, 47–53.

Fujiwara, K. *et al.* (1990). The use of *N*-(aminobenzoyloxy)-succinimide as a two-level heterobifunctional agent for the preparation of hapten–protein conjugates. *J. Immunol. Methods*, **134**, 227–235.

Gadkari, D. A. *et al.* (1985). Enzyme–antibody conjugation by a heterobifunctional reagent and its application in enzyme-linked immunosorbent assay (ELISA) for the detection of hepatitis B surface antigen. *J. Virol. Methods*, **10**, 215–224.

Gadow, A. *et al.* (1984). Synthesis and evaluation of luminescent tracers and hapten–protein conjugates for use in luminescence immunoassays with immobilised antibodies and antigens. A critical study of macro solid phases for use in immunoassay systems, Part II. *J. Clin. Chem. Clin. Biochem.*, **22**, 337–347.

Gaertner, H. F. *et al.* (1994). Site-specific religation of G-CSF fragments through a thioether bond. *Bioconjugate Chem.*, **5**, 333–338.

Gaffney, B. J. *et al.* (1983). Synthesis and membrane interactions of spin-label bifunctional reagents. *Biochemistry*, **22**, 881–892.

Galardy, R. E. *et al.* (1974). Photoaffinity labeling of peptide hormone binding sites. *J. Biol. Chem.*, **249**, 3510–3518.

Gatti, G. *et al.* (1984). Crystallization of hen egg-white avidin in a tetragonal form. *J. Mol. Biol.*, **178**, 787–789.

Geoghegan, K. F. and Stroh, J. G. (1992). Site-directed conjugation of nonpeptide groups to peptides and proteins *via* periodate oxidation of a 2-amino alcohol. Application to modification at N-terminal serine. *Bioconjugate Chem.*, 3, 138–146.

Geoghegan, K. F. *et al.* (1981). Alternative reducing agents for reductive methylation of amino groups in proteins. *Int. J. Peptide Protein Res.*, 17, 345–352.

Gershoni, J. M. *et al.* (1985). Blot analyses of glycoconjugates: enzyme-hydrazide: a novel reagent for the detection of aldehydes. *Anal. Biochem.*, **146**, 59–63.

Ghazarossian, V. *et al.* (1997). Chromophoric reagents for incorporation of biotin or other haptens into macromolecules. *US Patent* 5,614,368, to Molecular Devices Corporation.

Ghebrehiwet, B. *et al.* (1988). Reversible biotinylation of C1q with a cleavable biotinyl derivative. Application in C1q receptor (C1qR) purification. *J. Immunol. Methods*, **110**, 251–260.

Ghetie, V. *et al.* (1990). Preparation and characterization of conjugates of recombinant CD4 and deglycosylated ricin A chain using different cross-linkers. *Bioconjugate Chem.*, 1, 24–31.

Ghosh, S. S. *et al.* (1990). Use of maleimide–thiol coupling chemistry for efficient syntheses of oligonucleotide–enzyme conjugate hybridization probes. *Bioconjugate Chem.*, 1, 71–76.

Gillett, R. and Gull, K. (1972). Glutaraldehyde: its purity and stability. *Histochemie*, **30**, 162–167.

Gitman, A. G. *et al.* (1985). Use of virus-attached antibodies or insulin molecules to mediate fusion between Sendai virus envelopes and neuraminidase-treated cells. *Biochemistry*, **24**, 2762–2768.

Glennie, M. J. *et al.* (1987). Preparation and performance of bispecific $F(ab')_2$ antibody containing thioether-linked Fab′ fragments. *J. Immunol.*, **139**, 2367–2375.

Gnemmi, E. *et al.* (1978). A sensitive immunoenzymometric assay (IEMA) to quantitate hormones and drugs. In *Enzyme Labelled Immunoassay of Hormones and Drugs. Proceedings of the International Symposium on Enzyme Labelled Immunoassay of Hormones and Drugs, Ulm, West Germany, July 10 and 11, 1978* (Pal, S.B., ed.), pp. 29–41, Walter de Gruyter & Co. Berlin.

Goff, D. A. and Carroll, S. F. (1990). Substituted 2-iminothiolanes: reagents for the preparation of disulfide cross-linked conjugates with increased stability. *Bioconjugate Chem.*, 1, 381–386.

Gold, A. M. and Fahrney, D. (1964). Sulfonyl fluorides as inhibitors of esterases. II. Formation and reactions of phenylmethanesulfonyl α-chymotrypsin. *Biochemistry*, 3, 783–791.

Goodlad, G. A. J. (1957). Cross linking of collagen by S- and N-mustards. *Biochim. Biophys. Acta*, 25, 202–203.

Gope, M. L. *et al.* (1987). Molecular cloning of the chicken avidin cDNA. *Nucleic Acids Res.*, 15, 3595–3606.

Gorin, G. *et al.* (1966). Kinetics of the reaction of *N*-ethylmaleimide with cysteine and some congeners. *Arch. Biochem. Biophys.*, **115**, 593–597.

Govindan, S. V. *et al.* (1996). Thiolations, ^{99m}Tc labelings, and animal *in vivo* biodistributions of divalent monoclonal antibody fragments. *Bioconjugate Chem.*, 7, 290–297.

Graham, L. and Gallop, P. M. (1994). Covalent protein crosslinks: general detection, quantitation, and characterization *via* modification with diphenylborinic acid. *Anal. Biochem.*, **217**, 298–305.

Grassetti, D. R. and Murray, J. F., Jr. (1967). Determination of sulfhydryl groups with 2,2′- or 4,4′-dithiodipyridine. *Arch. Biochem. Biophys.*, **119**, 41–49.

Green, N. M. (1963). Avidin. 3. The nature of the biotin-binding site. *Biochem. J.*, **89**, 599–609.

Green, N. M. (1965). A spectrophotometric assay for avidin and biotin based on binding of dyes by avidin. *Biochem. J.*, **94**, 23c–24c.

Green, N. M. (1970). Spectrophotometric determination of avidin and biotin. *Methods Enzymol.*, **18A**, 418–424.

Green, N. M. (1975). Avidin. *Adv. Protein Chem.*, **29**, 85–133.

Green, N. M. (1990). Avidin and streptavidin. *Methods Enzymol.*, **184**, 51–67.

Green, N. M. and Toms, E. J. (1970). Purification and crystallization of avidin. *Biochem. J.*, **118**, 67–70.

Green, N. M. *et al.* (1971). The use of bifunctional biotinyl compounds to determine the arrangement of subunits in avidin. *Biochem. J.*, **125**, 781–791.

Green, N. M. *et al.* (1972). Transcarboxylase. XI. Electron microscopy and subunit structure. *J. Biol. Chem.*, **247**, 6284–6298.

Greenfield, L. *et al.* (1990). Thiol-containing cross-linking agent with enhanced steric hindrance. *Bioconjugate Chem.*, 1, 400–410.

Gretch, D. R. *et al.* (1987). The use of biotinylated monoclonal antibodies and streptavidin affinity chromatography to isolate herpesvirus hydrophobic proteins or glycoproteins. *Anal. Biochem.*, **163**, 270–277.

Grumbach, I. M. and Veh, R. W. (1991). Sulpho-*N*-hydroxysuccinimide activated long chain biotin. *J. Immunol. Methods*, **140**, 205–210.

Guesdon, J.-L. *et al.* (1979). The use of avidin–biotin interaction in immunoenzymatic techniques. *J. Histochem. Cytochem.*, **27**, 1131–1139.

Gundlach, H. G. *et al.* (1959). The reaction of iodoacetate with methionine. *J. Biol. Chem.*, **234**, 1761–1764.

Gurd, F. R. N. (1972). Carboxymethylation. *Methods Enzymol.*, **25**, 424–438.

György, P. (1954). Biotin. II. Chemistry. In *The Vitamins: Chemistry, Physiology, Pathology. Volume I* (Sebrell, W. H., Jr. and Harris, R. S., eds), pp. 527–571, Academic Press, New York.

Habeeb, A. F. S. A. (1972). Reaction of protein sulfhydryl groups with Ellman's reagent. *Methods Enzymol.*, **25**, 457–464.

Habeeb, A. F. S. A. and Hiramoto, R. (1968). Reaction of proteins with glutaraldehyde. *Arch. Biochem. Biophys.*, **126**, 16–26.

Habeeb, A. F. S. A. *et al.* (1958). Molecular structural effects produced in proteins by reaction with succinic anhydride. *Biochim. Biophys. Acta*, **29**, 587–593.

Haest, C. W. M. *et al.* (1977). Intra-and intermolecular cross-linking of membrane proteins in intact erythrocytes and ghosts by SH-oxidizing agents. *Biochim. Biophys. Acta*, **469**, 226–230.

Hajdu, J. *et al.* (1979). Structural changes in glycogen phosphorylase as revealed by cross-linking with bifunctional diimidates: phosphorylase b. *Biochemistry*, **18**, 4037–4041.

Hamaguchi, Y. *et al.* (1979). Improved procedure for the conjugation of rabbit IgG and Fab′ antibodies with β-D-galactosidase from *Escherichia coli* using *N,N′-o*-phenylenedimaleimide. *J. Biochem.*, **85**, 1289–1300.

Han, J. C. and Han, G. Y. (1994). A procedure for quantitative determination of tris(2-carboxyethyl)phosphine, an odorless reducing agent more stable and effective than dithiothreitol. *Anal. Biochem.*, **220**, 5–10.

Hancock, J. and Wagner, R. (1982). A structural model of 5S RNA from *E. coli* based on intramolecular crosslinking evidence. *Nucleic Acids Res.*, **10**, 1257–1269.

Hand, E. S. and Jencks, W. P. (1962). Mechanism of the reaction of imido esters with amines. *J. Am. Chem. Soc.*, **84**, 3505–3514.

Hardy, P. M. *et al.* (1969). The nature of glutaraldehyde in aqueous solution. *J. Chem. Soc., Chem. Commun.*, 565–566.

Hardy, P. M. *et al.* (1976a). The nature of the cross-linking of proteins by glutaraldehyde. Part I. interaction of glutaraldehyde with the amino-groups of 6-aminohexanoic acid and of α-*N*-acetyl-lysine. *J. Chem. Soc., Perkin Trans. 1*, 958–962.

Hardy, P. M. *et al.* (1976b). Formation of quaternary pyridinium compounds by the action of glutaraldehyde on proteins. *J. Chem. Soc., Chem. Commun.*, 157–158.

Harlan, J. W. and Feairheller, S. H. (1977). Chemistry of the crosslinking of collagen during tanning. *Adv. Exp. Med. Biol.*, **86A**, 425–440.

Harrington, C. R. and Colaco, C. A. L. S. (1994). Alzheimer's disease: a glycation connection. *Nature*, **370**, 247–248.

Harris, M. *et al.* (1942). Elasticity of wool as related to its chemical structure. *J. Res. Natl. Bur. Stand.*, **29**, 73–86.

Harris, S. A. *et al.* (1943). Synthetic biotin. *Science*, **97**, 447–448.

Harris, S. A. *et al.* (1944). Biotin. II. Synthesis of biotin. *J. Am. Chem. Soc.*, **66**, 1756–1757.

Harrison, J. K. *et al.* (1989). Development of a novel photoreactive calmodulin derivative: cross-linking of purified adenylate cyclase from bovine brain. *Biochemistry*, **28**, 6023–6027.

Haselgrübler, T. *et al.* (1995). Synthesis and applications of a new poly(ethylene glycol) derivative for the crosslinking of amines with thiols. *Bioconjugate Chem.*, **6**, 242–248.

Hashida, S. *et al.* (1984). More useful maleimide compounds for the conjugation of Fab′ to horseradish peroxidase through thiol groups in the hinge. *J. Appl. Biochem.*, **6**, 56–63.

Hashida, S. *et al.* (1985). A simple method for the conjugation of affinity-purified Fab′ to horseradish peroxidase and β-D-galactosidase from *Escherichia coli*. *J. Immunoassay*, **6**, 111–123.

Hashida, S. *et al.* (1991). Detection of one attomole of [Arg⁸]-vasopressin by novel noncompetitive enzyme immunoassay (hetero-two-site complex transfer enzyme immunoassay). *J. Biochem.*, **110**, 486–492.

Hashida, S. *et al.* (1993). Detection of antibody IgG to HIV-1 in urine by sensitive enzyme immunoassay (immune complex transfer enzyme immunoassay) using recombinant proteins as antigens for diagnosis of HIV-1 infection. *J. Clin. Lab. Anal.*, **7**, 353–364.

Hashida, S. *et al.* (1995). Anti-HIV-1 IgG in urine detected by sensitive enzyme immunoassay (immune complex transfer enzyme immunoassay) using recombinant proteins as antigens for diagnosis of HIV-1 infection. *Clin. Chem. Enzymol. Commun.*, **6**, 277–290.

Hashinaka, K. *et al.* (1994). Conjugation of recombinant reverse transcriptase of HIV-1 to β-D-galactosidase from *Escherichia coli* for ultrasensitive enzyme immunossay (immune complex transfer enzyme immunoassay) of anti-HIV-1 IgG. *J. Immunol. Methods*, **172**, 179–187.

Haug, H. *et al.* (1989). Conjugate for enzyme immuno determinations. *US Patent 4,843,001*, to Boehringer Mannheim.

Heilmann, H.-D. and Holzner, M. (1981). The spatial organisation of the active sites of the bifunctional oligomeric enzyme tryptophan synthase: cross-linking by a novel method. *Biochem. Biophys. Res. Commun.*, **99**, 1146–1152.

Heimgartner, U. *et al.* (1990). Reversible and irreversible cross-linking of immunoglobulin heavy chains through their carbohydrate residues. *Biochem. J.*, **267**, 585–591.

Heindel, N. D. *et al.* (1991). A novel heterobifunctional linker for formyl to thiol coupling. *Bioconjugate Chem.*, **2**, 427–430.

Heitzmann, H. and Richards, F. M. (1974). Use of the avidin–biotin complex for specific staining of biological membranes in electron microscopy. *Proc. Natl. Acad. Sci. USA*, **71**, 3537–3541.

Hendrickson, W. A. *et al.* (1989). Crystal structure of core streptavidin determined from multiwavelength anomalous diffraction of synchrotron radiation. *Proc. Natl. Acad. Sci. USA*, **86**, 2190–2194.

Henkin, J. (1977). Photolabeling reagent for thiol enzymes. studies on rabbit muscle creatine kinase. *J. Biol. Chem.*, **252**, 4293–4297.

Henning, D. and Nielsen, K. (1987). Peroxidase-labelled monoclonal antibodies for use in enzyme immunoassay. *J. Immunoassay*, **8**, 297–307.

Hexter, C. S. and Westheimer, F. H. (1971). *S*-Carboxymethylcysteine from the photolysis of diazoacyl trypsin and chymotrypsin. *J. Biol. Chem.*, **246**, 3934–3938.

Higgins, H. G. and Harrington, K. J. (1959). Reaction of amino acids and proteins with diazonium compounds. II. Spectra of protein derivatives. *Arch. Biochem. Biophys.*, **85**, 409–425.

Hill, M. *et al.* (1979). Disuccinimidyl esters as bifunctional cross-linking reagents for proteins. *FEBS Lett.*, **102**, 282–286.

Hiller, Y. *et al.* (1990). Nonglycosylated avidin. *Methods Enzymol.*, **184**, 68–70.

Hingorani, V. N. *et al.* (1988). Chemical cross-linking of bovine retinal transducin and cGMP phosphodiesterase. *J. Biol. Chem.*, **263**, 6916–6926.

Hiratsuka, T. (1988). Cross-linking of three heavy-chain domains of myosin adenosinetriphosphatase with a trifunctional alkylating reagent. *Biochemistry*, **27**, 4110–4114.

Hiremath, C. B. and Day, R. A. (1964). Introduction of covalent cross-linkages into lysozyme by reaction with α,α'-dibromo-*p*-xylenesulfonic acid. *J. Am. Chem. Soc.*, **86**, 5027–5028.

Hixson, S. H. and Hixson, S. S. (1975). *p*-Azidophenacyl bromide, a versatile photolabile bifunctional reagent: reaction with glyceraldehyde 3-phosphate dehydrogenase. *Biochemistry*, **14**, 4251–4255.

Hochhaus, G. *et al.* (1988). Biotinylated human β-endorphins as probes for the opioid receptor. *J. Biol. Chem.*, **263**, 92–97.

Hodge, J. E. (1955). The Amadori rearrangement. *Adv. Carbohydr. Chem.*, **10**, 169–205.

Hofmann, K. and Kiso, Y. (1976). An approach to the targeted attachment of peptides and proteins to solid supports. *Proc. Natl. Acad. Sci. USA*, **73**, 3516–3518.

Hofmann, K. *et al.* (1942). Adipic acid as an oxidation product of the diamino-carboxylic acid derived from biotin. *J. Biol. Chem.*, **144**, 513–518.

Hofmann, K. *et al.* (1980). Iminobiotin affinity columns and their application to retrieval of streptavidin. *Proc. Natl. Acad. Sci. USA*, **77**, 4666–4668.

Hofmann, K. *et al.* (1982). Avidin binding of carboxyl-substituted biotin and analogues. *Biochemistry*, **21**, 978–984.

Honzatko, R. B. and Williams, R. W. (1982). Raman spectroscopy of avidin: secondary structure, disulfide conformation, and the environment of tyrosine. *Biochemistry*, **21**, 6201–6205.

Hopwood, D. *et al.* (1970). The reactions between glutaraldehyde and various proteins. An investigation of their kinetics. *Histochem. J.*, **2**, 137–150.

Horton, H. R. and Koshland, D.E., Jr. (1972). Modification of proteins with active benzyl halides. *Methods Enzymol.*, **25**, 468–482.

Houba, P. H. J. *et al.* (1996). Improved characteristics of a human β-glucuronidase–antibody conjugate after deglycosylation for use in antibody-directed prodrug therapy. *Bioconjugate Chem.*, **7**, 606–611.

Houen, G. and Jensen, O. M. (1995). Conjugation to preactivated proteins using divinylsulfone and iodoacetic acid. *J. Immunol. Methods*, **181**, 187–200.

Houen, G, *et al.* (1997). Conjugation to preadsorbed preactivated proteins and efficient generation of anti-peptide antibodies. *J. Immunol. Methods*, **206**, 125–134.

Howard, A. N. and Wild, F. (1957). A two-stage method of crosslinking proteins suitable for use in serological techniques. *Br. J. Exp. Pathol.*, **38**, 640–643.

Hoyle, N. R. (1994). Process for liberating an analyte from its binding protein. *US Patent 5,332,678*, to Boehringer Mannheim.

Hua, J.-C. *et al.* (1986). Partial sequencing of human adult, human fetal, and bovine intestinal alkaline phosphatases: comparison with the human placental and liver isozymes. *Proc. Natl. Acad. Sci. USA*, **83**, 2368–2372.

Huang, C.-K. and Richards, F. M. (1976). An unsymmetrical lipid soluble cleavable crosslinking reagent for use in membrane systems. *Fed. Proc.*, **35**, 1378.

Huber, E. *et al.* (1994). New biotinylation reagent. *World Patent WO 94/17072*, to Boehringer Mannheim.

Hughes, W. L. (1947). An albumin fraction isolated from human plasma as a crystalline mercuric salt. *J. Am. Chem. Soc.*, **69**, 1836–1837.

Hunter, M. J. and Ludwig, M. L. (1962) The reaction of imidoesters with proteins and related small molecules. *J. Am. Chem. Soc.*, **84**, 3491–3504.

Hunter, M. J. and Ludwig, M. L. (1972). Amidination. *Methods Enzymol.*, **25**, 585–596.

Hurd, C. D. *et al.* (1953). Cyclic modifications of the dialdehyde from periodate oxidation of methyl α-D-glucopyranoside. *J. Org. Chem.*, **18**, 186–191.

Husain, M. and Bieniarz, C. (1994). Fc site-specific labeling of immunoglobulins with calf intestinal alkaline phosphatase. *Bioconjugate Chem.*, **5**, 482–490.

Husain, S. S. and Lowe, G. (1968). Evidence for histidine in the active sites of ficin and stem-bromelain. *Biochem. J.*, **110**, 53–57.

Husain, S. S. *et al.* (1971). Bifunctional inhibitors of pepsin. *Proc. Natl. Acad. Sci. USA*, **68**, 2765–2768.

Hylkema, M. N. *et al.* (1994). A new ELISA for the detection of anti-heparan sulfate reactivity, using photobiotinylated antigen. *J. Immunol. Methods*, **176**, 33–43.

Hypolite, C. L. *et al.* (1997). Formation of microscale gradients of protein using heterobifunctional photolinkers. *Bioconjugate Chem.*, **8**, 658–663.

Imagawa, M. *et al.* (1982a). Characteristics and evaluation of antibody–horseradish peroxidase conjugates prepared by using a maleimide compound, glutaraldehyde, and periodate. *J. Appl. Biochem.*, **4**, 41–57.

Imagawa, M. *et al.* (1982b). Development of a highly sensitive sandwich enzyme immunoassay for human ferritin using affinity-purified anti-ferritin labelled with β-D-galactosidase from *Escherichia coli*. *Clin. Chim. Acta*, **121**, 277–289.

Ingham, K. C. and Brew, S. A. (1981). Fluorescent labeling of the carbohydrate moieties of human chorionic gonadotropin and α_1–acid glycoprotein. *Biochim. Biophys. Acta*, **670**, 181–189.

Inman, J. K. and Dintzis, H. M. (1969). The derivatization of cross-linked polyacrylamide beads. Controlled introduction of functional groups for the preparation of special-purpose, biochemical adsorbents. *Biochemistry*, **8**, 4074–4082.

Inoue, S. *et al.* (1985). A micro-scale method for the conjugation of affinity-purified Fab′ to β-D-galactosidase from *Escherichia coli*. *J. Biochem.*, **98**, 1387–1394.

Ishikawa, E. and Kato, K. (1978). Ultrasensitive enzyme immunoassay. *Scand. J. Immunol.*, 8 (Suppl. 7), 43–55.

Ishikawa, E. *et al.* (1978). Enzyme-labelling with maleimides and its application to the immunoassay of peptide hormones, in *Enzyme Labelled Immunoassay of Hormones and*

Drugs. Proceedings of the International Symposium on Enzyme Labelled Immunoassay of Hormones and Drugs, Ulm, West Germany, July 10 and 11, 1978 (Pal, S.B., ed.), pp. 43–57, Walter de Gruyter & Co., Berlin.

Ishikawa, E. *et al.* (1981). A more stable maleimide, N-(4-carboxycyclohexylmethyl)maleimide, for enzyme labeling. in *Enzyme Immunoassay* (Ishikawa, E. *et al.*, eds), pp. 91–105, Igaku-Shoin, Tokyo.

Ishikawa, E. *et al.* (1983). Enzyme-labeling of antibodies and their fragments for enzyme immunoassay and immunohistochemical staining. *J. Immunoassay*, **4**, 209–325.

Ito, Y. *et al.* (1994). Design and synthesis of a protein device that releases insulin in response to glucose concentration. *Bioconjugate Chem.*, **5**, 84–87.

Izard, E. F. and Morgan, P. W. (1949). Synthesis of disulfide cross links in polyvinyl alcohol and cellulose derivatives. *Ind. Eng. Chem.*, **41**, 617–621.

Jackson, E. L. and Hudson, C. S. (1937). Studies on the cleavage of the carbon chain of glycosides by oxidation. a new method for determining ring structures and alpha and beta configurations of glycosides. *J. Am. Chem. Soc.*, **59**, 994–1003.

Jackson, G.E.D. and Young, N. M. (1986). Determination of chemical properties of individual histidine and tyrosine residues of concanavalin A by competitive labeling with 1-fluoro-2,4-dinitrobenzene. *Biochemistry*, **25**, 1657–1662.

Jasiewicz, M. L. *et al.* (1976). Selective retrieval of biotin-labeled cells using immobilized avidin. *Exp. Cell Res.*, **100**, 213–217.

Jayakrishnan, A. and Jameela, S. R. (1996). Glutaraldehyde as a fixative in bioprostheses and drug delivery matrices. *Biomaterials*, **17**, 471–484.

Jeanson, A. *et al.* (1988a). Comparison of conjugation procedures for the preparation of monoclonal antibody–enzyme conjugates. *J. Immunol. Methods*, **111**, 261–270.

Jeanson, A. *et al.* (1988b). Preparation of reproducible alkaline phosphatase–antibody conjugates for enzyme immunoassay using a heterobifunctional linking agent. *Anal. Biochem.*, **172**, 392–396.

Jentoft, N. and Dearborn, D. G. (1979). Labeling of proteins by reductive methylation using sodium cyanoborohydride. *J. Biol. Chem.*, **254**, 4359–4365.

Ji, I. *et al.* (1985). Radioiodination of a photoactivatable heterobifunctional reagent. *Anal. Biochem.*, **151**, 348–349.

Ji, T. H. (1976). Crosslinking of lectins and receptors in membranes with hetero-bifunctional crosslinking reagents. In *Membranes and Neoplasia: New Approaches and Strategies* (Marchesi, V. T., ed.), pp. 171–178, Alan R. Liss, Inc., New York.

Ji, T. H. (1979). The application of chemical crosslinking for studies on cell membranes and the identification of surface reporters. *Biochim. Biophys. Acta*, **559**, 39–69.

Ji, T. H. (1983). Bifunctional reagents. *Methods Enzymol.*, **91**, 580–609.

Ji, T. H. and Ji, I. (1982). Macromolecular photoaffinity labeling with radioactive photoactivable heterobifunctional reagents. *Anal. Biochem.*, **121**, 286–289.

Jin, L. *et al.* (1995). Use of α-*N*, *N*-bis[carboxymethyl]lysine modified peroxidase in immunoassays. *Anal. Biochem.*, **229**, 54–60.

Johnson, G. M. *et al.* (1997). Heterobifunctional crosslinkers containing 4,9-dioxa-1, 12-dodecanediamine spacers. *Bioconjugate Chem.*, **8**, 447–452.

Johnson, R. B. *et al.* (1980). Comparison of glucose oxidase and peroxidase as labels for antibody in enzyme-linked immunosorbent assay. *J. Immunoassay*, **1**, 27–37.

Jou, Y.-H. *et al.* (1985). Monoclonal antibodies and a heterobifunctional reagent: a novel approach to the vectorial labeling of selected membrane proteins. *Immunol. Commun.*, **11**, 357–375.

Jue, R. *et al.* (1978). Addition of sulfhydryl groups to *Escherichia coli* ribosomes by protein modification with 2-iminothiolane (methyl 4-mercaptobutyrimidate). *Biochemistry*, **17**, 5399–5406.

Kaiser, K. *et al.* (1997). Basic studies on heterobifunctional biotin–PEG conjugates with a 3-(4-pyridyldithio) propionyl marker on the second terminus. *Bioconjugate Chem.*, **8**, 545–551.

Kasche, V. (1989). Proteases in peptide synthesis. In *Proteolytic Enzymes. A Practical Approach.* (Beynon, R. J. and Bond, J. S., eds), pp. 125–143, IRL Press, Oxford, UK.

Kato, K. *et al.* (1975a). Enzyme-linked immunoassay. I. Novel method for synthesis of the insulin–β-D-galactosidase conjugate and its applicability for insulin assay. *J. Biochem.*, **78**, 235–237.

Kato, K. *et al.* (1975b). Enzyme-linked immunoassay. II. A simple method for synthesis of the rabbit antibody–β-D-galactosidase complex and its general applicability. *J. Biochem.*, **78**, 423–425.

Kato, K. *et al.* (1975c). Coupling Fab′ fragment of rabbit anti-human IgG antibody to β-D-galactosidase and a highly sensitive immunoassay of human IgG. *FEBS Lett.*, **56**, 370–372.

Kato, K. *et al.* (1976a). Enzyme-linked immunoassay. Conjugation of rabbit anti-(human immunoglobulin G) antibody with β-D-galactosidase from *Escherichia coli* and its use for human immunoglobulin G assay. *Eur. J. Biochem.*, **62**, 285–292.

Kato, K. *et al.* (1976b). Enzyme-linked immunoassay: conjugation of the Fab′ fragment of rabbit IgG with β-D-galactosidase from *E. coli* and its use for immunoassay. *J. Immunol.*, **116**, 1554–1560.

Kawaoi, A. and Nakane, P. K. (1972). An improved method of conjugation of peroxidase with proteins. *Fed. Proc.*, **32**, 840.

Keana, J. F. W. and Cai, S. X. (1989). Functionalized perfluorophenyl azides: new reagents for photoaffinity labeling. *J. Fluorine Chem.*, **43**, 151–154.

Keana, J. F. W. and Cai, S. X. (1990). New reagents for photoaffinity labeling: synthesis and photolysis of functionalized perfluorophenyl azides. *J. Org. Chem.*, **55**, 3640–3647.

Keller, O. and Rudinger, J. (1975). Preparation and some properties of maleimido acids and maleimido derivatives of peptides. *Helv. Chim. Acta*, **58**, 531–541.

Kendall, C. *et al.* (1983). Utilization of the biotin/avidin system to amplify the sensitivity of the enzyme-linked immunosorbent assay (ELISA). *J. Immunol. Methods*, **56**, 329–339.

Khym, J. X. (1963). The reaction of methylamine with periodate-oxidized adenosine 5′-phosphate. *Biochemistry*, **2**, 344–350.

Kim, J.-S. and Raines, R. T. (1995). Dibromobimane as a fluorescent crosslinking reagent. *Anal. Biochem.*, **225**, 174–176.

King, T. P. and Kochoumian, L. (1979). A comparison of different enzyme–antibody conjugates for enzyme-linked immunosorbent assay. *J. Immunol. Methods*, **28**, 201–210.

King, T. P. *et al.* (1978). Preparation of protein conjugates *via* intermolecular disulfide bond formation. *Biochemistry*, **17**, 1499–1506.

Kirkeby, S. *et al.* (1987). Glutaraldehyde: 'pure and impure'. A spectroscopic investigation of two commercial glutaraldehyde solutions and their reaction products with amino acids. *Anal. Lett.*, **20**, 303–315.

Kitagawa, T. and Aikawa, T. (1976). Enzyme coupled immunoassay of insulin using a novel coupling reagent. *J. Biochem.*, **79**, 233–236.

Kitagawa, T. *et al.* (1981). Preparation and characterization of hetero-bifunctional cross-linking reagents for protein modifications. *Chem. Pharm. Bull.*, **29**, 1130–1135.

Klausner, Y. S. and Bodansky, M. (1972). Coupling reagents in peptide synthesis. *Synthesis*, 453–463.

Klotz, I. M. and Elfbaum, S. G. (1964). Activated acyl compounds as intermediates in the thiolation of proteins. *Biochim. Biophys. Acta*, **86**, 100–105.

Klotz, I. M. and Heiney. R. E. (1959). A new method for the introduction of thiol groups into proteins. *J. Am. Chem. Soc.*, **81**, 3802–3803.

Klotz, I. M. and Heiney, R. E. (1962). Introduction of sulfhydryl groups into proteins using acetylmercaptosuccinic anhydride. *Arch. Biochem. Biophys.*, **96**, 605–612.

Klotz, I. M. and Stryker, V. E. (1959). Introduction of sulfhydryl groups into macromolecules. *Biochem. Biophys. Res. Commun.*, **1**, 119–123.

Klotz, I. M. *et al.* (1965). Sulfur-containing derivatives of proteins. *Biochim. Biophys. Acta*, **100**, 104–110.

Knowles, J. R. (1972). Photogenerated reagents for biological receptor-site labeling. *Acc. Chem. Res.*, **5**, 155–160.

Kobashi, K. (1968). Catalytic oxidation of sulfhydryl groups by *o*-phenanthroline copper complex. *Biochim. Biophys. Acta*, **158**, 239–245.

Kobashi, K. and Horecker, B. L. (1967). Reversible inactivation of rabbit muscle aldolase by *o*-phenanthroline. *Arch. Biochem. Biophys.*, **121**, 178–186.

Koch, T. *et al.* (1990). The oxidative cleavability of protein cross-linking reagents containing organoselenium bridges. *Bioconjugate Chem.*, **1**, 296–304.

Kohanski, R. A. and Lane, M. D. (1990). Monovalent avidin affinity columns. *Methods Enzymol.*, **184**, 194–200.

Kohno, T. *et al.* (1989). Measurement of anti-thyroglobulin IgG in serum by novel and sensitive immune complex transfer enzyme immunoassay. *Clin. Biochem.*, **22**, 277–284.

Kondejewski, L. H. *et al.* (1994). Synthesis and characterization of carbohydrate-linked murine monoclonal antibody K20–human serum albumin conjugates. *Bioconjugate Chem.*, **5**, 602–611.

König, W. and Geiger, R. (1970). Eine neue Methode zur Synthese von Peptiden: Aktivierung der Carboxylgruppe mit Dicylohexylcarbodiimid unter Zusatz von 1-Hydroxybenzotriazolen. *Chem.´Ber.*, **103**, 788–798.

Konigsberg, W. (1972). Reduction of disulfide bonds in proteins with dithiothreitol. *Methods Enzymol.*, **25**, 185–188.

Korn, A. H. *et al.* (1972). Glutaraldehyde: nature of the reagent. *J. Mol. Biol.*, **65**, 525–529.

Kovacs, J. *et al.* (1967). On the optical purity of peptide active esters prepared by N,N'-dicyclohexylcarbodiimide and 'complexes' of N,N'-dicyclohexylcarbodiimide–pentachlorophenol and N,N'-dicyclohexylcarbodiimide–pentafluorophenol. *J. Am. Chem. Soc.*, **89**, 183–184.

Kozulic, B. and Barbaric, S. (1988). Stabilisation of glycoproteins by periodate oxidation. *UK Patent 2,225,324*.

Kozulić, B. *et al.* (1984). Study of the carbohydrate part of yeast acid phosphatase. *Biochem. Biophys. Res. Commun.*, **122**, 1083–1090.

Kozulić, B. *et al.* (1987). Preparation of the stabilized glycoenzymes by cross-linking their carbohydrate chains. *Appl. Biochem. Biotechnol.*, **15**, 265–278.

Kraehenbuhl, J. P. *et al.* (1974). Preparation and characterization of an immunoelectron microscope tracer consisting of a heme-octapeptide coupled to Fab. *J. Exp. Med.*, **139**, 208–223.

Krotoski, W. A. and Weimer, H. E. (1966). Peptide-associated and antigenic changes accompanying periodic acid oxidation of human plasma orosomucoid. *Arch. Biochem. Biophys.*, **115**, 337–344.

Lambert, J. M. *et al.* (1985). Purified immunotoxins that are reactive with human lymphoid cells. Monoclonal antibodies conjugated to the ribosome-inactivating proteins gelonin and the pokeweed antiviral proteins. *J. Biol. Chem.*, **260**, 12035–12041.

Lamoureux, G. V. and Whitesides, G. M. (1993). Synthesis of dithiols as reducing agents for disulfides in neutral aqueous solution and comparison of reduction potentials. *J. Org. Chem.*, **58**, 633–641.

Lamture, J. B. and Wensel, T. G. (1995). Intensely luminescent immunoreactive conjugates of proteins and dipicolinate-based polymeric Tb^{III} chelates. *Bioconjugate Chem.*, **6**, 88–92.

Lannér, M. and Bergquist, R. (1978). A chromatographic labelling procedure for the preparation of conjugates with small enzyme–immunoglobulin complexes. In *Immunofluorescence and Related Techniques* (Knapp, W. *et al.*, eds), pp. 237–244; Elsevier/North Holland Medical Press, Amsterdam.

LaRochelle, W. J. and Froehner, S. C. (1986). Determination of the tissue distributions and relative concentrations of the postsynaptic 43-kDa protein and the acetylcholine receptor in *Torpedo*. *J. Biol. Chem.*, **261**, 5270–5274.

Lawson, W. B. and Schramm, H. J. (1962). Modification of a methionine residue near the active site of chymotrypsin. *J. Am. Chem. Soc.*, **84**, 2017–2018.

Leanza, W. J. *et al.* (1992). Acidic derivatives of homocysteine thiolactone: utility as anionic linkers. *Bioconjugate Chem.*, **3**, 514–518.

Leszyk, J. *et al.* (1987). Cross-linking of rabbit skeletal muscle troponin with the photoactive reagent 4-maleimidobenzophenone: identification of residues in troponin I that are close to cysteine-98 of troponin C. *Biochemistry*, **26**, 7042–7047.

Levchenko, L. A. *et al.* (1975). The structural investigation of biological systems by the method of electron-density labels. *Bioorg. Khim*, **1**, 1778–1782.

Levy, J. G. and Liu, D. (1989). Anhydrous enhanced coupling of proteins. *US Patent 4,843,147*, to the University of British Columbia.

Lewis, C. T. *et al.* (1989). Cysteine 288: an essential hyperreactive thiol of cytosolic phosphoenolpyruvate carboxykinase (GTP). *J. Biol. Chem.*, **264**, 27–33.

Lewis, R. V. *et al.* (1977). Photoactivated heterobifunctional cross-linking reagents which demonstrate the aggregation state of phospholipase A_2. *Biochemistry*, **16**, 5650–5654.

Lewis, S. D. *et al.* (1980). Determination of interactive thiol ionizations in bovine serum albumin, glutathione, and other thiols by potentiometric difference titration. *Biochemistry*, **19**, 6129–6137.

Leyva, E. *et al.* (1986). High yields of formal CH insertion products in the reaction of polyfluorinated aromatic nitrenes. *J. Am. Chem. Soc.*, **108**, 8307–8309.

Li, Y.-Z. *et al.* (1988). 1,2-Didehydroazepines from the photolysis of substituted aryl azides: analysis of their chemical and physical properties by time-resolved spectroscopic methods. *J. Am. Chem. Soc.*, **110**, 8092–8098.

Liberatore, F. A. *et al.* (1990). Site-directed chemical modification and cross-linking of a monoclonal antibody using equilibrium transfer alkylating cross-link reagents. *Bioconjugate Chem.*, **1**, 36–50.

Likhite, V. and Sehon, A. (1971). Protein–protein conjugation. *Methods Immunol. Immunochem.*, **1**, 150–167.

Lin, D. P. and Lee, Y. C. (1994). Determination of boron-based reducing agents. *Anal. Biochem.*, **219**, 157–159.

Lindsay, D. G. (1971). Intramolecular cross-linked insulin. *FEBS Lett.*, **21**, 105–108.

Lisanti, M. P. *et al.* (1988). Polarized apical distribution of glycosyl-phosphatidylinositol-anchored proteins in a renal epithelial cell-line. *Proc. Natl. Acad. Sci. USA*, **85**, 9557–9561.

Lisanti, M. P. *et al.* (1989). Steady-state distribution and biogenesis of endogenous Madin–Darby canine kidney glycoproteins: evidence for intracellular sorting and polarized cell surface delivery. *J. Cell Biol.*, **109**, 2117–2127.

Livnah, O. and Sussman, J. L. (1990). Crystal forms of avidin. *Methods Enzymol.*, **184**, 90–93.

Livnah, O. *et al.* (1993a). Three-dimensional structures of avidin and the avidin–biotin complex. *Proc. Natl. Acad. Sci. USA*, **90**, 5076–5080.

Livnah, O. *et al.* (1993b). The structure of the complex between avidin and the dye, 2-(4′-hydroxyazobenzene)-benzoic acid (HABA). *FEBS Lett.*, **328**, 165–168.

Lomant, A. J. and Fairbanks, G. (1976). Chemical probes of extended biological structures: synthesis and properties of cleavable protein cross-linking reagent [^{35}S]dithiobis(succinimidyl propionate). *J. Mol. Biol.*, **104**, 243–261.

Lorand, L. *et al.* (1962). Thrombin-catalyzed hydrolysis of *p*-nitrophenyl esters. *Arch. Biochem. Biophys.*, **96**, 147–151.

Ludueña, R. F. *et al.* (1981). *N,N*-Bis(α-iodoacetyl)-2,2′-dithiobis(ethylamine), a reversible crosslinking reagent for protein sulfhydryl groups. *Anal. Biochem.*, **117**, 76–80.

Lundblad, R. L. and Noyes, C. M. (1984). The modification of cysteine. In *Chemical Reagents for Protein Modification*, Volume I, CRC Press, Boca Raton, FL, pp. 55–93.

Lundblad, R. L. and Stein, W. H. (1969). On the reaction of diazoacetyl compounds with pepsin. *J. Biol. Chem.*, **244**, 154–160.

Lutter, L. C. *et al.* (1974). Use of a new series of cleavable protein cross-linkers on the *Escherichia coli* ribosome. *FEBS Lett.*, **48**, 288–292.

Maassen, J. A. *et al.* (1981). Structural analysis of ribosomal protein L7/L12 by the heterobifunctional cross-linker 4-(6-formyl-3-azidophenoxy)butyrimidate. *Biochemistry*, **20**, 1020–1025.

Maclaren, J. A. (1958). Attempts to introduce amide cross-links into wool. *Textile Res. J.*, **28**, 946–949.

Madersbacher, S. *et al.* (1992). Increased ELISA sensitivity using a modified method for conjugating horseradish peroxidase to monoclonal antibodies. *J. Immunol. Methods*, **152**, 9–13.

Madore, H. P. and Baumgarten, A. (1979). Enzyme-linked protein A: an enzyme-linked immunosorbent assay reagent for detection of human immunoglobulin G and virus-specific antibody. *J. Clin. Microbiol.*, **10**, 529–532.

Mahan, D. E. *et al.* (1987). Phase change enzyme immunoassay. *Anal. Biochem.*, **162**, 163–170.

Mahbouba, M. and Smith, H. J. (1977). Thiolation and disulphide cross-linking of insulin to form macromolecules of potential therapeutic value. *Adv. Exp. Med. Biol.*, **86A**, 247–260.

Makita, Z. *et al.* (1992). Immunochemical detection of advanced glycosylation end products *in vivo*. *J. Biol. Chem.*, **267**, 5133–5138.

Malaprade, L. (1928). Oxydation de quelques polyalcools par l'acide periodique. Applications. *C. R. Acd. Sci.*, **186**, 382–384.

Manecke, G. and Middeke, H.-J. (1980). Reaktive Träger mit Maleimidengruppen zur Enzymimmobilisierung. *Angew. Makromol. Chem.*, **91**, 179–201.

Mannik, M. and Downey, W. (1973). Studies on the conjugation of horseradish peroxidase to Fab fragments. *J. Immunol. Methods*, **3**, 233–242.

Manning, J. E. *et al.* (1975). A new method of *in situ* hybridization. *Chromosoma*, **53**, 107–117.

Marfey, P. S. *et al.* (1965). Crosslinked synthetic polypeptides. I. Preparation and characterisation of crosslinked poly $Glu^{51}Lys^{33}Tyr^{16}$ and of poly $Glu^{52}Lys^{33}Tyr^{15}$ No. 3. *Biopolymers*, **3**, 27–41.

Markert-Hahn, C. *et al.* (1994). Immunologisch Aktive Konjugate und ein Verfahren zu ihrer Herstellung. *European Patent 618,231*, to Boehringer Mannheim.

Marsh, J. W. (1988). Antibody-mediated routing of diphtheria toxin on murine cells results in a highly efficacious immunotoxin. *J. Biol. Chem.*, **263**, 15993–15999.

Marui, Y. *et al.* (1995). Enzyme immunoassay for antigen and solid phase used therefore. *US Patent 5,378,608*, to Nippon Shoji Kabushiki Kaisha.

Mason, H. S. (1955). Comparative biochemistry of the phenolase complex. *Adv. Enzymol.*, **16**, 105–184.

Masuho, Y. *et al.* (1979). Preparation of a hybrid of fragment Fab′ of antibody and fragment A of diphtheria toxin and its cytotoxicity. *Biochem. Biophys. Res. Commun.*, **90**, 320–326.

Mattson, G. *et al.* (1993). A practical approach to crosslinking. *Mol. Biol. Rep.*, **17**, 167–183.

McCormick, D. B. (1965). Specific purification of avidin by column chromatography on biotin-cellulose. *Anal. Biochem.*, **13**, 194–198.

McElvain, S. M. and Schroeder, J. P. (1949). Orthoesters and related compounds from malono- and succinonitriles. *J. Am. Chem. Soc.*, **71**, 40–46

McInnes, J. L. *et al.* (1990). Preparation and uses of photobiotin. *Methods Enzymol.*, **184**, 588–601.

McIntyre, G. D. *et al.* (1994). Preparation and characterization of interleukin–2-gelonin conjugates made using different cross-linking reagents. *Bioconjugate Chem.*, **5**, 88–97.

Means, G. E. and Feeney, R. E. (1968). Reductive alkylation of amino groups in proteins. *Biochemistry*, **7**, 2192–2201.

Means, G. E. and Feeney, R. E. (1990). Chemical modification of proteins: history and applications. *Bioconjugate Chem.*, **1**, 2–12.

Meyer, D. L. *et al.* (1992). Preparation and characterization of a β-lactamase–Fab′ conjugate for the site-specific activation of oncolytic agents. *Bioconjugate Chem.*, **3**, 42–48.

Mikkelsen, R. B. and Wallach, D. F. H. (1976). Photoactivated cross-linking of proteins within the erythrocyte membrane core. *J. Biol. Chem.*, **251**, 7413–7416.

Mikolajczyk, S. D. *et al.* (1994). High yield, site-specific coupling of N-terminally modified β-lactamase to a proteolytically derived single-sulfhydryl murine Fab′. *Bioconjugate Chem.*, **5**, 636–646.

Miller, B. T. *et al.* (1992). The occurrence of *O*-acylation during biotinylation of gonadotropin-releasing hormone and analogs. Evidence for a reactive serine. *J. Biol. Chem.*, **267**, 5060–5069.

Miller, B. T. *et al.* (1994). Identification and characterization of *O*-biotinylated hydroxy amino acid residues in peptides. *Anal. Biochem.*, **219**, 240–248.

Mitra, S. and Lawton, R. G. (1979). Reagents for the crosslinking of proteins by equilibrium transfer alkylation. *J. Am. Chem. Soc.*, **101**, 3097–3110.

Modesto, R. R. and Pesce, A. J. (1971). The reaction of 4,4′-difluoro-3,3′-dinitro-diphenyl sulfone with γ-globulin and horseradish peroxidase, *Biochim. Biophys. Acta*, **229**, 384–395.

Molin, S.-O. *et al.* (1978). A kinetic study of the reaction between glutaraldehyde and horseradish peroxidase. *J. Histochem. Cytochem.*, **26**, 1053–1056.

Monsan, P. *et al.* (1975). Étude du mécanisme d'établissement des liaisons glutaraldéhyde–protéines. *Biochimie*, **57**, 1281–1292.

Montgomery, J. A. *et al.* (1966). The modes of decomposition of 1,3-bis(2-chloroethyl)-1-nitrosourea and related compounds. *J. Med. Chem.*, **10**, 668–674.

Moore, J. E. and Ward, W. H. (1956). Cross-linking of bovine plasma albumin and wool keratin. *J. Am. Chem. Soc.*, **78**, 2414–2418.

Morag, E. *et al.* (1996). Immobilized nitro-avidin and nitrostreptavidin as reusable affinity matrices for application in avidin–biotin technology. *Anal. Biochem.*, **243**, 257–263.

Morehead, H. W. *et al.* (1991). Optimization of oxidation of glycoproteins: an assay for predicting coupling to hydrazide chromatographic supports. *J. Chromatogr.*, **587**, 171–176.

Moreland, R. B. *et al.* (1982). Synthesis and characterization of *N*-(4-azidophenylthio)phthalimide: a cleavable, photoactivatable crosslinking reagent that reacts with sulfhydryl groups. *Anal. Biochem.*, **121**, 321–326.

Morrison, M. *et al.* (1969). The reaction of benzoquinone with amines and proteins. *Arch. Biochem. Biophys.*, **134**, 515–523.

Mouton, C. A. *et al.* (1982). A reagent for covalently attaching biotin to proteins *via* a cleavable connector arm. *Arch. Biochem. Biophys.*, **218**, 101–108.

Murayama, A. *et al.* (1978). Modification of immunoglobulin G using specific reactivity of sugar moiety. *Immunochemistry*, **15**, 523–528.

Musso, G. F. *et al.* (1992). Synthesis of *N*-[6-(ethylenedioxy)hexyl]biotinamide: a biotinyl aldehyde precursor for labeling hydrazine-modified biomolecules. *Bioconjugate Chem.*, **3**, 88–90.

Myers, D. E. *et al.* (1989). The effects of aromatic and aliphatic maleimide crosslinkers on anti-CD5 ricin immunotoxins. *J. Immunol. Methods*, **121**, 129–142.

Nagamine, M. and Ohkuma, S. (1975). Serum alkaline phosphatase isoenzymes linked to immunoglobulin G. *Clin. Chim. Acta*, **65**, 39–46.

Nakane, P. K. and Kawaoi, A. (1974). Peroxidase-labeled antibody: a new method of conjugation, *J. Histochem. Cytochem.*, **22**, 1084–1091.

Needles, H. L. (1977). Peroxydisulfate anion-induced cross-linking of proteins. *Adv. Exp. Med. Biol.*, **86A**, 549–556.

Nefkens, G. H. L. and Tesser, G. I. (1961). A novel activated ester in peptide syntheses. *J. Am. Chem. Soc.*, **83**, 1263.

Neville, D.M., Jr *et al.* (1989). Enhancement of immunotoxin efficacy by acid-cleavable cross-linking agents utilizing diphtheria toxin and toxin mutants. *J. Biol. Chem.*, **264**, 14653–14661.

Ngo, T. T. *et al.* (1981). *p*-Azidophenylglyoxal. A heterobifunctional photoactivatable cross-linking reagent selective for arginyl residues. *J. Biol. Chem.*, **256**, 11313–11318.

Ngo, T. T. *et al.* (1982). Biotinyl-glucose-6-phosphate dehydrogenase. Preparation, kinetics and modulation by avidin. *Appl. Biochem. Biotechnol.*, **7**, 443–454.

Nilsson, P. *et al.* (1981). A technique for preparing defined conjugates of horseradish peroxidase and immunoglobulin. *J. Immunol. Methods*, **41**, 81–93.

Nisonoff, A. and Mandy, W. J. (1962). Quantitative estimation of the hybridization of rabbit antibodies. *Nature*, **194**, 355–359.

Nisonoff, A. and Rivers, M. M. (1961). Recombination of a mixture of univalent antibody fragments of different specificity. *Arch. Biochem. Biophys.*, **93**, 460–462.

Nygren, H. (1982). Conjugation of horseradish peroxidase to Fab fragments with different homobifunctional and heterobifunctional cross-linking reagents. *J. Histochem. Cytochem.*, **30**, 407–412.

Offord, R. E. *et al.* (1987). Press-stud protein conjugates. In *Peptides 1986* (Theodoropoulos, D., ed.), pp. 279–281, Walter de Gruyter & Co., Berlin.

Oka, T. and Morihara, K. (1977). Trypsin as a catalyst for peptide synthesis. *J. Biochem.*, **82**, 1055–1062.

Okuda, K. *et al.* (1991). Reaction of glutaraldehyde with amino and thiol groups. *J. Ferment. Bioeng.*, **71**, 100–105.

Olcott, H. S. and Fraenkel-Conrat, H. (1947). Specific group reagents for proteins. *Chem. Rev.*, **41**, 151–197.

Olszewski, J. D. *et al.* (1995). Tethered benzophenone reagents for the synthesis of photoactivatable ligands. *Bioconjugate Chem.*, **6**, 395–400.

Orr, G. A. (1981). The use of the 2-iminobiotin–avidin interaction for the selective retrieval of labeled plasma membrane components. *J. Biol. Chem.*, **256**, 761–766.

O'Shannessy, D. J. and Quarles, R. H. (1985). Specific conjugation reactions of the oligosaccharide moieties of immunoglobulins. *J. Appl. Biochem.*, **7**, 347–355.

O'Shannessy, D. J. and Quarles, R. H. (1987). Labeling of the oligosaccharide moieties of immunoglobulins. *J. Immunol. Methods*, **99**, 153–161.

O'Shannessy, D. J. and Wilchek, M. (1990). Immobilization of glycoconjugates by their oligosaccharides: use of hydrazido-derivatized matrices. *Anal. Biochem.*, **191**, 1–8.

O'Shannessy, D. J. *et al.* (1984). A novel procedure for labeling immunoglobulins by conjugation to oligosaccharide moieties. *Immunol. Lett.*, **8**, 273–277.

O'Shannessy, D. J. *et al.* (1987). Quantitation of glycoproteins on electroblots using the biotin–streptavidin complex. *Anal. Biochem.*, **163**, 204–209.

O'Sullivan, M. J. *et al.* (1978). A simple method for the preparation of enzyme–antibody conjugates. *FEBS Lett.*, **95**, 311–313.

O'Sullivan, M. J. *et al.* (1979a). Comparison of two methods of preparing enzyme–antibody conjugates: application of these conjugates for enzyme immunoassay. *Anal. Biochem.*, **100**, 100–108.

O'Sullivan, M. J. *et al.* (1979b). A comparison of the ability of β-galactosidase and horseradish peroxidase enzyme–antibody conjugates to detect specific antibodies. *J. Immunol. Methods*, **31**, 247–250.

Ozawa, H. (1967). Bridging reagent for protein. II. The reaction of N,N'-polymethylene-bis(iodoacetamide) with cysteine and rabbit muscle aldolase. *J. Biochem.*, **62**, 531–536.

Pähler, A. *et al.* (1987). Characterization and crystallisation of core streptavidin. *J. Biol. Chem.*, **262**, 13933–13937.

Pain, D. and Surolia, A. (1981). Preparation of protein A: peroxidase monoconjugate using a heterobifunctional reagent, and its use in enzyme immunoassays. *J. Immunol. Methods*, **40**, 219–230.

Palumbo, P. S. (1993). Novel compounds and conjugates. *World Patent WO 93/22677*, to PB Diagnostic Systems, Inc.

Patramani, I. *et al.* (1969). Glutamic-aspartic transaminase–antitransaminase interaction: a method for antienzyme purification. *Eur. J. Biochem.*, **11**, 28–36.

Paz, M. A. *et al.* (1970). Amino alcohols as products of a reductive side reaction of denatured collagen with sodium borohydride. *Biochemistry*, **9**, 2123–2127.

Pereira, A. P. *et al.* (1993). Monoclonal antibody-based immunoenzymometric assays of retinol-binding protein. *Clin. Chem.*, **39**, 472–476.

Perham, R. N. and Thomas, J. O. (1971). Reaction of tobacco mosaic virus with a thiol-containing imidoester and a possible application to X-ray diffraction analysis. *J. Mol. Biol.*, **62**, 415–418.

Peters, K. and Richards, F. M. (1977). Chemical cross-linking: reagents and problems in studies of membrane structure. *Annu. Rev. Biochem.*, **46**, 523–551.

Peuravuori, H. and Korpela, T. (1993). Pyrophosphatase-based enzyme-linked immunosorbent assay of total IgE in serum. *Clin. Chem.*, **39**, 846–851.

Philpott, G. W. *et al.* (1980). Selective binding and cytotoxicity of rat basophilic leukemia cells (RBL-1) with immunoglobulin E: biotin and avidin–glucose oxidase conjugates. *J. Immunol.*, **125**, 1201–1209.

Photaki, I. and Bardakos, V. (1965). Transformation of L-serine peptides to L-cysteine peptides. *J. Am. Chem. Soc.*, **87**, 3489–3492.

Pihl, A. and Lange, R. (1962). The interaction of oxidized glutathione, cystamine monosulfoxide, and tetrathionate with the -SH groups of rabbit muscle D-glyceraldehyde 3-phosphate dehydrogenase. *J. Biol. Chem.*, **237**, 1356–1362.

Pilch, P. F. and Czech, M. P. (1979). Interaction of cross-linking agents with the insulin effector system of isolated fat cells. *J. Biol. Chem.*, **254**, 3375–3381.

Pinn, E. *et al.* (1982). Crystallization and preliminary X-ray investigation of avidin. *Eur. J. Biochem.*, **123**, 545–546.

Platz, M. *et al.* (1991). Photolysis of 3-aryl-3-(trifluoromethyl)diazirines: a caveat regarding their use in photoaffinity probes. *Bioconjugate Chem.*, **2**, 337–341.

Plotz, P. H. (1977). Bivalent affinity labeling haptens in the formation of model immune complexes. *Methods Enzymol.*, **46**, 505–508.

Polgar, L. and Bender, M. L. (1966). A new enzyme containing a synthetically formed active site. Thiol-subtilisin. *J. Am. Chem. Soc.*, **88**, 3153–3154.

Polgar, L. and Bender, M. L. (1967). The reactivity of thiolsubtilisin, an enzyme containing a synthetic functional group. *Biochemistry*, **6**, 610–620.

Pollner, R. *et al.* (1996). Process for preparing conjugates consisting of a specific binding partner and a carbohydrate-containing protein. *US Patent 5,534,414*, to Behringwerke Aktiengesellschaft.

Porstmann, B. *et al.* (1984). Direct two-site binding immunoassay for human α-1-fetoprotein: part 1: comparison of different peroxidase–IgG conjugates and various reaction conditions, *Z. Med. Labor.-Diagn.*, **25**, 5–13.

Porstmann, B. *et al.* (1985). Which of the commonly used marker enzymes gives the best results in colorimetric and fluorometric enzyme immunoassays: horseradish peroxidase, alkaline phosphatase or β-galactosidase? *J. Immunol. Methods*, **79**, 27–37.

Presentini, R. and Terrana, B. (1995). Influence of the antibody–peroxidase coupling methods on the conjugates stability and on the methodologies for the preservation of the activity in time. *J. Immunoassay*, **16**, 309–324.

Pressman, D. *et al.* (1942). The agglutination of intact azoerythrocytes by antisera homologous to the attached groups. *J. Immunol.*, **44**, 101–105.

Pugliese, L. *et al.* (1993). Three-dimensional structure of the tetragonal crystal form of egg-white avidin in its functional complex with biotin at 2.7 Å resolution. *J. Mol. Biol.*, **231**, 698–710.

Pugliese, L. *et al.* (1994). Crystal structure of apo-avidin from hen egg-white. *J. Mol. Biol.*, **235**, 42–46.

Purcell, A. W. and Wallace, J. C. (1996). The accessibility of the prosthetic group biotin during monomeric avidin affinity chromatography. *Anal. Biochem.*, **238**, 213–216.

Quinlan, R. A. and Franke, W. W. (1982). Heteropolymer filaments of vimentin and desmin in vascular smooth muscle tissue and cultured baby hamster kidney cells demonstrated by chemical crosslinking. *Proc. Natl. Acad. Sci. USA*, **79**, 3452–3456.

Quiocho, F. A. and Richards, F. M. (1964). Intermolecular cross linking of a protein in the crystalline state: carboxypeptidase A. *Proc. Natl. Acad. Sci. USA*, **52**, 833–839.

Quiocho, F. A. and Richards, F. M. (1966). The enzymic behavior of carboxypeptidase A in the solid state. *Biochemistry*, **5**, 4062–4076.

Ram, J. S. *et al.* (1963). Preparation of antibody–ferritin conjugates for immuno-electron microscopy, *J. Cell Biol.* **17**, 673–675.

Rao, A. *et al.* (1979). Location of the stilbenedisulfonate binding site of the human erythrocyte anion-exchange system by resonance energy transfer. *Biochemistry*, **18**, 4505–4516.

Rao, S. V. *et al.* (1997). Determination of the extent of protein biotinylation by fluorescence binding assay. *Bioconjugate Chem.*, **8**, 94–98.

Rasmussen, K.-E. and Albrechtsen, J. (1974). Glutaraldehyde. The influence of pH, temperature, and buffering on the polymerization rate. *Histochemistry*, **38**, 19–26.

Raso, V. and Griffin, T. (1981). Hybrid antibodies with dual specificity for the delivery of ricin to immunoglobulin-bearing target cells. *Cancer Res.*, **41**, 2073–2078.

Ravinder, K. G. (1991). Improved antibody–enzyme direct conjugates and method of making same. *World Patent WO 91/13356*, to Coulter Corporation.

Rector, E. S. *et al.* (1978). A method for the preparation of protein–protein conjugates of predetermined composition. *J. Immunol. Methods*, **24**, 321–336.

Reis, K. J. *et al.* (1988). Enzyme-labeled type III bacterial Fc receptors. A versatile tracer for immunoassay. *J. Immunol. Methods*, **107**, 273–280.

Reuter, G. *et al.* (1989). A detailed study of the periodate oxidation of sialic acids in glycoproteins. *Glycoconjugate J.*, **6**, 35–44.

Reznik, G. O. *et al.* (1996). Streptavidins with intersubunit crosslinks have enhanced stability. *Nature Biotechnol.*, **14**, 1007–1011.

Rich, D. H. *et al.* (1975). Alkylating derivatives of amino acids and peptides. Synthesis of *N*-maleoylamino acids, [1-(*N*-maleoylglycyl)cysteinyl]oxytocin, and [1-(*N*-maleoyl-11-aminoundecanoyl)cysteinyl]oxytocin. Effects on vasopressin-stimulated water loss from isolated toad bladder. *J. Med. Chem.*, **18**, 1004–1010.

Richards, F. M. and Knowles, J. R. (1968). Glutaraldehyde as a protein cross-linking reagent. *J. Mol. Biol.*, **37**, 231–233.

Riesel, E. and Katchalski, E. (1964). Preparation and properties of water-insoluble derivatives of urease. *J. Biol. Chem.*, **239**, 1521–1524.

Riordan, J. F. and Vallee, B. L. (1972a). Reactions with *N*-ethylmaleimide and *p*-mercuribenzoate. *Methods Enzymol.*, **25**, 449–456.

Riordan, J. F. and Vallee, B. L. (1972b). Diazonium salts as specific reagents and probes of protein conformation. *Methods Enzymol.*, **25**, 521–531.

Roberts, D. D. *et al.* (1986). Reactivity of small thiolate anions and cysteine-25 in papain toward methyl methanethiosulfonate. *Biochemistry*, **25**, 5595–5601.

Roberts, J. C. *et al.* (1990). Using Starburst dendrimers as linker molecules to radiolabel antibodies. *Bioconjugate Chem.*, **1**, 305–308.

Robinson, I. D. (1964). Rate of crosslinking of gelatin in aqueous solution. *J. Appl. Polym. Sci.*, **8**, 1903–1918.

Rodwell, J. D. *et al.* (1986). Site-specific covalent modification of monoclonal antibodies: *in vitro* and *in vivo* evaluations. *Proc. Natl. Acad. Sci. USA*, **83**, 2632–2636.

Rose, K. *et al.* (1989). Attachment of linker groups to carboxyl termini using enzyme-assisted reverse proteolysis. In *Peptides 1988* (Jung, G. and Bayer, E., eds), pp. 274–276, Walter de Gruyter & Co., Berlin.

Rose, K. *et al.* (1991). Preparation of well-defined protein conjugates using enzyme-assisted reverse proteolysis. *Bioconjugate Chem.*, **2**, 154–159.

Ross, W. C. J. (1953). The chemistry of cytotoxic alkylating agents. *Adv. Cancer Res.*, **1**, 397–449.

Rothfus, J. A. and Smith, E. L. (1963). Glycopeptides. IV. The periodate oxidation of glycopeptides from human γ-globulin. *J. Biol. Chem.*, **238**, 1402–1410.

Rowen, J. W. *et al.* (1951). Spectrophotometric evidence for the absence of free aldehyde groups in periodate-oxidized cellulose. *J. Am. Chem. Soc.*, **73**, 4484–4487.

Ruan, K.-H. *et al.* (1987). A small-scale sandwich enzyme immunoassay for macromolecular antigens using β-D-galactosidase from *Escherichia coli* and horseradish peroxidase as labels. *Anal. Lett.*, **20**, 587–601.

Rüegg, U. T. and Rudinger, J. (1974). Reaction of cysteine thiol groups with 1,3-propane sultone: *S*-3-sulphopropyl as a modifying group for protein chemistry. *Int. J. Peptide Protein Res.*, **6**, 447–456.

Rüegg, U. T. and Rudinger, J. (1977). Reductive cleavage of cystine disulfides with tributylphosphine. *Methods Enzymol.*, **47**, 111–116.

Rylatt, D. B. *et al.* (1977). Pyruvate carboxylase: isolation of the biotin-containing tryptic peptide and the determination of its primary sequence. *Arch. Biochem. Biophys.*, **183**, 113–122.

Sabatini, D. D. *et al.* (1963). Cytochemistry and electron microscopy. The preservation of cellular ultrastructure and enzymatic activity by aldehyde fixation. *J. Cell Biol.*, **17**, 19–58.

Sanger, F. (1945). The free amino groups of insulin. *Biochem. J.*, **39**, 507–515.

Sanger, F. and Tuppy, H. (1951). The amino acid sequence in the phenylalanine chain of insulin. I. The identification of lower peptides from partial hydrolysates. *Biochem. J.*, **49**, 463–490.

Satchell, D. P. N. and Satchell, R. S. (1975). Acylation by ketens and isocyanates. A mechanistic comparison. *Chem. Soc. Rev.*, **4**, 231–250.

Sato, N. *et al.* (1995). Thiol-mediated redox regulation of apoptosis. Possible roles of cellular thiols other than glutathione in T cell apoptosis. *J. Immunol.*, **154**, 3194–3203.

Sato, S. and Nakao, M. (1981). Cross-linking of intact erythrocyte membrane with a newly synthesized cleavable bifunctional reagent. *J. Biochem.*, **90**, 1177–1185.

Schick, A. F. and Singer, S. J. (1961). On the formation of covalent linkages between two protein molecules. *J. Biol. Chem.*, **236**, 2477–2485.

Schnapp, K. A. and Platz, M. S. (1993). A laser flash photolysis study of di-, tri- and tetrafluorinated phenylnitrenes; implications for photoaffinity labeling. *Bioconjugate Chem.*, **4**, 178–183.

Schnapp, K. A. *et al.* (1993). Exploratory photochemistry of fluorinated aryl azides. Implications for the design of photoaffinity labeling reagents. *Bioconjugate Chem.*, **4**, 172–177.

Schöberl, A. (1948). Eine neue Methode zur Einführung von Schwefel in Eiweistoffe. *Angew. Chem.*, **60**, 7–9.

Schott, M. E. *et al.* (1993). Preparation, characterization and *in vivo* biodistribution properties of synthetically cross-linked multivalent antitumor antibody fragments. *Bioconjugate Chem.*, **4**, 153–165.

Schramm, H. J. and Dülffer, T. (1977). Synthesis and application of cleavable and hydrophilic crosslinking reagents. *Adv. Exp. Med. Biol.*, **86A**, 197–206.

Schuster, G. B. and Platz, M. S. (1992). Photochemistry of phenyl azide. *Adv. Photochem.*, **17**, 69–143.

Schwartz, M. A. *et al.* (1982). A new radioactive cross-linking reagent for studying the interactions of proteins. *J. Biol. Chem.*, **257**, 2343–2349.

Schwarz, A. *et al.* (1990). Enzymatic C-terminal biotinylation of proteins. *Methods Enzymol.*, **184**, 160–162.

Schwenker, R. F., Jr and Lifland, L. (1963). 6,6′-Disulfide cross-links in cotton cellulose. Part II: Effects of different cross-linking conditions. *Textile Res. J.*, **33**, 107–117.

Schwenker, R.F., Jr *et al.* (1962). 6,6′-Disulfide cross-links in cotton cellulose; effects on fabric wrinkle recovery. *Textile Res. J.*, **32**, 797–804.

Schwyzer, R. *et al.* (1955). Über aktivierte Ester. III. Umsetzungen aktivierter Ester von Aminosäure- und Peptid-Derivaten mit Aminen und Aminosäureestern. *Helv. Chim. Acta*, **38**, 83–91.

Segal, D. M. and Hurwitz, E. (1976). Dimers and trimers of immunoglobulin G covalently cross-linked with a bivalent affinity label. *Biochemistry*, **15**, 5253–5258.

Shafer, J. *et al.* (1966). Products from the photolysis of diazoacetyl chymotrypsin. *J. Biol. Chem.*, **241**, 421–427.

Shall, S. and Barnard, E. A. (1969). Heavy atom-labelled derivatives of bovine pancreatic ribonuclease. I. Specific reactions with *N*-acetylhomocysteine thiolactone and silver ion. *J. Mol. Biol.*, **41**, 237–251.

Shao, M.-C. *et al.* (1990). Complex neoglycoproteins. *Methods Enzymol.*, **184**, 653–659.

Sheehan, J. C. and Hlavka, J. J. (1957). The cross-linking of gelatin using a water-soluble carbodiimide. *J. Am. Chem. Soc.*, **79**, 4528–4529.

Shephard, E. G. *et al.* (1988). The use of sulfosuccinimidyl-2-*p*-azidosalicylamido)-1,3′-dithiopropionate as a cross-linking reagent to identify cell surface receptors. *Anal. Biochem.*, **168**, 306–313.

Shimkus, M. *et al.* (1985). A chemically cleavable biotinylated nucleotide: usefulness in the recovery of protein–DNA complexes from avidin affinity columns. *Proc. Natl. Acad. Sci. USA*, **82**, 2593–2597.

Sigler, G. F. (1988). Biotinylating agents. *European Patent 279,365*, to Hoechst Celanese Corporation.

Sigrist, H. *et al.* (1982). 5-Isothiocyanato-1-naphthalene azide and *p*-azidophenylisothiocyanate. *Eur. J. Biochem.*, **125**, 197–201.

Singer, S. J. (1959). Preparation of an electron-dense antibody conjugate. *Nature*, **183**, 1523–1524.

Singh, A. *et al.* (1962). The photolysis of diazoacetyl-chymotrypsin. *J. Biol. Chem.*, **237**, 3006–3008.

Singh, P. *et al.* (1980). A method for controlled coupling of amino compounds to enzymes. A homogeneous enzyme immunoassay for gentamicin. *Anal. Biochem.*, **104**, 51–58.

Singh, R. and Kats, L. (1995) Catalysis of reduction of disulfide by selenol. *Anal. Biochem.*, **232**, 86–91.

Singh, R. *et al.* (1996). Formation of *N*-substituted 2 iminothiolanes when amino groups in proteins and peptides are modified by 2-iminothiolane. *Anal. Biochem.*, **236**, 114–125.

Smith, H. A. *et al.* (1964). Mercaptan–disulfide interchange reactions. III. Reaction of cysteine with 4,4′-dithiobis-(benzenesulfonic acid). *J. Org. Chem.*, **29**, 1484–1488.

Smith, R. J. *et al.* (1978). Cross-linking of ubiquinone cytochrome *c* reductase (complex III) with periodate-cleavable bifunctional reagents. *Biochemistry*, **17**, 3719–3723.

Smyth, D. G. (1967). Acetylation of amino and tyrosine hydroxyl groups. Preparation of inhibitors of oxytocin with no intrinsic activity on the isolated uterus. *J. Biol. Chem.*, **242**, 1592–1598.

Sorenson, P. *et al.* (1986). Identification of the interleukin-3 receptor using an iodinatable, cleavable, photoreactive cross-linking agent. *J. Biol. Chem.*, **261**, 9094–9097.

Srinivasachar, K. and Neville, D. M., Jr (1989). New protein cross-linking reagents that are cleaved by mild acid. *Biochemistry*, **28**, 2501–2509.

Stadtman, E. R. and White, F. H., Jr (1953). The enzymatic synthesis of *N*-acetylimidazole, *J. Am. Chem. Soc.*, **75**, 2022.

Staerz, U. D. *et al.* (1985). Hybrid antibodies can target sites for attack by T cells. *Nature*, **314**, 628–631.

Stalteri, M. A. and Mather, S. J. (1995). A cross-linked monoclonal antibody fragment for improved tumor targeting. *Bioconjugate Chem.*, **6**, 179–186.

Staros, J. V. (1982). *N*-Hydroxysulfosuccinimide active esters: bis(*N*-hydroxysulfosuccinimide) esters of two dicarboxylic acids are hydrophilic, membrane-impermeant protein cross-linkers. *Biochemistry*, **21**, 3950–3955.

Staros, J. V. (1988). Membrane-impermeant cross-linking reagents: probes of the structure and dynamics of membrane proteins. *Acc. Chem. Res.*, **21**, 435–441.

Staros, J. V. and Anjaneyulu, P. S. R. (1989). Membrane-impermeant cross-linking reagents. *Methods Enzymol.*, **172**, 609–628.

Staros, J. V. and Kakkad, B. P. (1983). Cross-linking and chymotryptic digestion of the extracytoplasmic domain of the anion exchange channel in intact human erythrocytes. *J. Membrane Biol.*, **74**, 247–254.

Staros, J. V. *et al.* (1978). Reduction of aryl azides by thiols: implications for the use of photoaffinity reagents. *Biochem. Biophys. Res. Commun.*, **80**, 568–572.

Staros, J. V. *et al.* (1981). A membrane-impermeant, cleavable cross-linker: dimers of human erythrocyte band 3 subunits cross-linked at the extracytoplasmic membrane face. *J. Biol. Chem.*, **256**, 5890–5893.

Steck, T. L. (1972). Cross-linking the major proteins of the isolated erythrocyte membrane. *J. Mol. Biol.*, **66**, 295–305.

Steinman, H. M. and Richards, F. M. (1970). Participation of cysteinyl residues in the structure and function of muscle aldolase. Characterization of mixed disulfide derivatives. *Biochemistry*, **22**, 4360–4372.

Sternberger, L. A. (1967). Electron microscopic immunocytochemistry: a review. *J. Histochem. Cytochem.*, **15**, 139–159.

Stevenson, G. T. *et al.* (1989). A chimeric antibody with dual Fc regions (bisFabFc) prepared by manipulations at the IgG hinge. *Anti-Cancer Drug Des.*, **3**, 219–230.

Storm, D. *et al.* (1996). Biotinylation of proteins via amino groups can induce binding to U937 cells, HL-60 cells, monocytes and granulocytes. *J. Immunol. Methods*, **199**, 87–99.

Strumeyer, D. H. *et al.* (1963). Role of serine in chymotrypsin action. Conversion of the active serine to dehydroalanine. *Proc. Natl. Acad. Sci. USA*, **50**, 931–935.

Sutoh, K. and Matsuzaki, F. (1980). Millisecond photocross-linking of protein components in vertebrate striated muscle thin filaments. *Biochemistry*, **19**, 3878–3882.

Suzuki, S. *et al.* (1970). The chemical modification of biopolymers. III. The introduction of SH groups into Taka-amylase A. *Bull. Chem. Soc. Japan*, **43**, 3849–3852.

Swaney, J. B. (1986). Use of cross-linking reagents to study lipoprotein structure. *Methods Enzymol.*, **128**, 613–626.

Swan, J. M. (1958). Chemical modification of thiol and disulfide groups in proteins and peptides. In *Sulfur in Proteins* (Benesch, R. *et al.*, eds), pp. 3–14, Academic Press, New York.

Szurdoki, F. *et al.* (1995). Development of rapid mercury assays. Synthesis of sulfur- and mercury-containing conjugates. *Bioconjugate Chem.*, **6**, 145–149.

Takahashi, M. *et al.* (1981). Tris-induced cross-linking of thyalkoid peptides; thiol oxidation catalyzed by tris–Cu^{2+} complexes as a possible mechanism. *J. Biochem.*, **90**, 87–94.

Tao, T. *et al.* (1984). Studies on the proximity relationships between thin filament proteins using benzophenone-4-maleimide as a site-specific photoreactive crosslinker. *Biophys. J.*, **45**, 261a.

Teale, J. M. and Kearney, J. F. (1986). Clonotypic analysis of the fetal B cell repertoire: evidence for an early and predominant expression of idiotypes associated with the V_H 36–60 family. *J. Mol. Cell. Immunol.*, **2**, 283–292.

Ternynck, T. and Avrameas, S. (1976). A new method using *p*-benzoquinone for coupling antigens and antibodies to marker substances. *Ann. Immunol. (Inst. Pasteur)*, **127C**, 197–208.

Ternynck, T. and Avrameas, S. (1977). Conjugation of *p*-benzoquinone treated enzymes with antibodies and Fab fragments. *Immunochemistry*, **14**, 767–774.

Tesser, G. I. *et al.* (1975). Some new bifunctional reagents for the conjugation of a protein with an amino compound in water. *Hoppe-Seyler's Z. Physiol. Chem.*, **356**, 1625–1629.

Thorpe, P. E. *et al.* (1981). Cytotoxicity acquired by conjugation of an anti-$Thy_{1.1}$ monoclonal antibody and the ribosome-inactivating protein, gelonin. *Eur. J. Biochem.*, **116**, 447–454.

Thorpe, P. E. *et al.* (1985a). Modification of the carbohydrate in ricin with metaperiodate–cyanoborohydride mixtures. Effects on toxicity and *in vivo* distribution. *Eur. J. Biochem.*, **147**, 197–206.

Thorpe, P. E. *et al.* (1985b). An immunotoxin composed of monoclonal anti-Thy 1.1 antibody and a ribosome-inactivating protein from *Saponaria officinalis*: potent antitumor effects *in vitro* and *in vivo*. *J. Natl. Cancer Inst.*, **75**, 151–159.

Thorpe, P. E. *et al.* (1987a). Comparison of two anti-Thy 1.1–abrin A-chain immunotoxins prepared with different cross-linking agents: antitumor effects, *in vivo* fate, and tumor cell mutants. *J. Natl. Cancer Inst.*, **79**, 1101–1111.

Thorpe, P. E. *et al.* (1987b). New coupling agents for the synthesis of immunotoxins containing a hindered disulfide bond with improved stability *in vivo*. *Cancer Res.*, **47**, 5924–5931.

Thorpe, P. E. *et al.* (1988). Improved antitumor effects of immunotoxins prepared with deglycosylated ricin A-chain and hindered disulfide linkages. *Cancer Res.*, **48**, 6396–6403.

Tijssen, P. and Kurstak, E. (1984). Highly efficient and simple methods for the preparation of peroxidase and active peroxidase–antibody conjugates for enzyme immunoassays. *Anal. Biochem.*, **136**, 451–457.

Torchinsky, Yu. M. (1981a). *Sulfur in Proteins*, pp. 3–71, Pergamon Press, New York.

Torchinsky, Yu. M. (1981b). *Sulfur in Proteins*, pp. 133–142, Pergamon Press, New York.

Torchinsky, Yu. M. (1981c). *Sulfur in Proteins*, pp. 232–235, Pergamon Press, New York.

Traut, R. R. *et al.* (1973). Methyl 4-mercaptobutyrimidate as a cleavable cross-linking reagent and its application to the *Escherichia coli* 30S ribosome. *Biochemistry*, **12**, 3266–3273.

Traut, R. R. and Kenny, J. W. (1977). Crosslinking of ribosomes by cleavable bifunctional mercaptoimidates. *Adv. Exp. Med. Biol.*, **86A**, 215–233.

Trommer, W. and Hendrick, M. (1973). The formation of maleimides by a new mild cyclization procedure. *Synthesis*, 484–485.

Trommer, W. E. *et al.* (1977). Synthesis and application of new bifunctional reagents. *Adv. Exp. Med. Biol.*, **86A**, 187–195.

Tsang, V. C. W. *et al.* (1995). Optimization of the covalent conjugating procedure ($NaIO_4$) of horseradish peroxidase to antibodies for use in enzyme-linked immunosorbent assay. *J. Immunoassay*, **16**, 395–418.

Tsuruta, H. *et al.* (1994). Detection of the products of polymerase chain reaction by an ELISA system based on an ion sensitive field effect transistor. *J. Immunol. Methods*, **176**, 45–52.

Ukeda, H. *et al.* (1993). Dynamic analysis of binding process of bovine serum albumin on glutaraldehyde-activated support. *Anal. Sci.*, **9**, 617–623.

Uy, R. and Wold, F. (1977). Introduction of artificial crosslinks into proteins. *Adv. Exp. Med. Biol.*, **86A**, 169–186.

Valentine, W. M. *et al.* (1992). Covalent cross-linking of proteins by carbon disulfide. *Chem. Res. Toxicol.*, **5**, 254–262.

Vanin, E. F. and Ji, T. H. (1981). Synthesis and application of cleavable photoactivable heterobifunctional reagents. *Biochemistry*, **20**, 6754–6760.

Vanin, E. F. *et al.* (1981). *p*-Azidophenylglyoxal: a heterobifunctional photosensitive reagent. *FEBS Lett.*, **124**, 89–92.

Van Lenten, L. and Ashwell, G. (1971). Studies on the chemical and enzymatic modification of glycoproteins. A general method for the tritiation of sialic acid-containing glycoproteins. *J. Biol. Chem.*, **246**, 1889–1894.

Vas, M. and Csanády, G. (1987). The two fast-reacting thiols of 3-phosphoglycerate kinase are structurally juxtaposed. Chemical modification with bifunctional reagents. *Eur. J. Biochem.*, **163**, 365–368.

Vaughan, R. J. and Westheimer, F. H. (1969). A method for marking the hydrophobic binding sites of enzymes. An insertion into the methyl group of an alanine residue of trypsin. *J. Am. Chem. Soc.*, **91**, 217–218.

Wagner, R. and Garrett, R. A. (1978). A new RNA–RNA crosslinking reagent and its application to ribosomal 5S RNA. *Nucleic Acids Res.*, **5**, 4065–4075.

Wajda, I. J. *et al.* (1971). Chromogranins as substrate for transglutaminase. *Biochem. Pharmacol.*, **20**, 3197–3206.

Wakil, S. J. *et al.* (1958). Evidence for the participation of biotin in the enzymic synthesis of fatty acids. *Biochim. Biophys. Acta*, **29**, 225–226.

Wang, D. (1979). Preparation of the bifunctional enzyme ribonuclease–deoxyribonuclease by cross-linkage. *Biochemistry*, **18**, 4449–4452.

Wang, D. *et al.* (1997). Optimization of conditions for formation and analysis of anti-CD19 FVS191 single-chain Fv homodimer $(scFv')_2$. *Bioconjugate Chem.*, **8**, 64–70.

Wang, K. and Richards, F. M. (1974). The behavior of cleavable crosslinking reagents based on the disulfide group. *Israel J. Chem.*, **12**, 375–389.

Wang, K. and Richards, F. M. (1975). Reaction of dimethyl-3,3′-dithiobispropionimidate with intact human erythrocytes. Cross-linking of membrane proteins and hemoglobin. *J. Biol. Chem.*, **250**, 6622–6626.

Webb, J. L. (1966). *Enzyme and Metabolic Inhibitors*, vol. 3, pp. 420–594, Academic Press, New York.

Webb, R. R., II and Kaneko, T. (1990). Synthesis of 1-(aminooxy)-4-[(3-nitro-2-pyridyl)dithio]butane and 1-(aminooxy)-4-[(3-nitro-2-pyridyl)dithio]but-2-ene, novel heterobifunctional cross-linking reagents. *Bioconjugate Chem.*, **1**, 96–99.

Weber, P. C. *et al.* (1987). Crystallographic data for *Streptomyces avidinii* streptavidin. *J. Biol. Chem.*, **262**, 12728–12729.

Weber, P. C. *et al.* (1989). Structural origins of high-affinity biotin binding to streptavidin. *Science*, **243**, 85–88.

Weber, P. C. *et al.* (1992a). Crystallographic and thermodynamic comparison of natural and synthetic ligands bound to streptavidin. *J. Am. Chem. Soc.*, **114**, 3197–3200.

Weber, P. C. *et al.* (1992b). Crystal structure and ligand-binding studies of a screened peptide complexed with streptavidin. *Biochemistry*, **31**, 9350–9354.

Weiner, H. *et al.* (1966). The formation of anhydrochymotrypsin by removing the elements of water from the serine at the active site. *J. Am. Chem. Soc.*, **88**, 3851–3859.

Weltman, J. K. *et al.* (1976). A galactosidase immunosorbent test for human immunoglobulin E. *J. Allergy Clin. Immunol.*, **58**, 426–431.

Weltman, J. K. *et al.* (1983). N-Succinimidyl (4-iodoacetyl) aminobenzoate: a new heterobifunctional crosslinker. *Biotechniques*, **1**, 148–152.

Werlen, R. C. *et al.* (1994). Site-specific conjugation of an enzyme and an antibody fragment. *Bioconjugate Chem.*, **5**, 411–417.

Weston, P. D. *et al.* (1980). Conjugation of enzymes to immunoglobulins using dimaleimides. *Biochim. Biophys. Acta*, **612**, 40–49.

Whipple, E. B. and Ruta, M. (1974). Structure of aqueous glutaraldehyde. *J. Org. Chem.*, **39**, 1666–1668.

White, F. H., Jr (1960). Regeneration of enzymatic activity by air-oxidation of reduced ribonuclease with observations on thiolation during reduction with thioglycolate. *J. Biol. Chem.*, **235**, 383–389.

White, F. H., Jr (1972a). Reduction and reoxidation at disulfide bonds. *Methods Enzymol.*, **25**, 387–392.

White, F. H., Jr (1972b). Thiolation. *Methods Enzymol.*, **25**, 541–546.

Wilbur, D. S. *et al.* (1994). Monoclonal antibody Fab′ fragment cross-linking using equilibrium transfer alkylation reagents. A strategy for site-specific conjugation of diagnostic and therapeutic agents with $F(ab')_2$ fragments. *Bioconjugate Chem.*, **5**, 220–235.

Wilbur, D. S. *et al.* (1996). Antibody fragments in tumor pretargeting. Evaluation of biotinylated Fab′ colocalization with recombinant streptavidin and avidin. *Bioconjugate Chem.*, **7**, 689–702.

Wilchek, M. and Bayer, E. A. (1984). The avidin–biotin complex in immunology. *Immunol. Today*, **5**, 39–43.

Wilchek, M. and Bayer, E. A. (1987). Labeling glycoconjugates with hydrazide reagents. *Methods Enzymol.*, **138**, 429–442.

Wilchek, M. and Bayer, E. A. (1990a). Introduction to avidin–biotin technology. *Methods Enzymol.*, **184**, 5–13.

Wilchek, M. and Bayer, E. A. (1990b). Biotin-containing reagents. *Methods Enzymol.*, **184**, 123–138.

Wilchek, M. and Givol, D. (1977). Affinity cross-linking of heavy and light chains. *Methods Enzymol.*, **46**, 501–504.

Wilchek, M. *et al.* (1986). *p*-Diazobenzoyl biocytin: a new biotinylating reagent for the labeling of tyrosines and histidines in proteins. *Biochem. Biophys. Res. Commun.*, **138**, 872–879.

Wilding, P. *et al.* (1964). Globulin-bound amylase: a cause of persistently elevated levels in serum. *Ann. Intern. Med.*, **60**, 1053–1059.

Wilson, K. M. *et al.* (1991). Rapid whole blood assay for HIV-1 seropositivity using an Fab–peptide conjugate. *J. Immunol. Methods*, **138**, 111–119.

Wilson, M. B. and Nakane, P. K. (1978). Recent developments in the periodate method of conjugating horseradish peroxidase (HRPO) to antibodies. In *Immunofluorescence and Related Staining Techniques* (Knapp, W. *et al.*, eds), pp. 215–224, Elsevier/North Holland Medical Press, Amsterdam.

Wofsy, L. and Singer, S. J. (1963). Effects of the amidination reaction on antibody activity and on the physical properties of some proteins. *Biochemistry*, **2**, 104–116.

Wold, F. (1972). Bifunctional reagents. *Methods Enzymol.*, **25**, 623–651.

Wolf, B. *et al.* (1970). A mechanism of the irreversible inactivation of bovine pancreatic ribonuclease by diethylpyrocarbonate. A general reaction of diethylpyrocarbonate with proteins. *Eur. J. Biochem.*, **13**, 519–525.

Wolfe, C. A. C. and Hage, D. S. (1994). Automated determination of antibody oxidation using flow injection analysis. *Anal. Biochem.*, **219**, 26–31.

Wolff, S. P. *et al.* (1991). Protein glycation and oxidative stress in diabetes mellitus and ageing. *Free Rad. Biol. Med.*, **10**, 339–352.

Wong, S. S. (1991). *Chemistry of Protein Conjugation and Cross-Linking*, CRC Press, Boca Raton, FL.

Wood, C. L. and O'Dorisio, M. S. (1985). Covalent cross-linking of vasoactive intestinal polypeptide to its receptors on intact human lymphoblasts. *J. Biol. Chem.*, **260**, 1243–1247.

Worrell, N. R. *et al.* (1986). Effect of linkage variation on pharmacokinetics of ricin A chain–antibody conjugates in normal rats. *Anti-Cancer Drug Des.*, **1**, 179–188.

Wynne, D. *et al.* (1976). A chemical approach for the localization of membrane sites involved in lymphocyte activation. *Biochem. Biophys. Res. Commun.*, **68**, 730–739.

Yajima, H. *et al.* (1980). Studies on peptides. XCIV. Thiazoline-2-thione as a carboxyl-activating reagent for peptide synthesis. *Chem. Pharm. Bull.*, **28**, 3140–3142.

Yamamoto, K. *et al.* (1984). Spatial relationship between SH_1 and the actin binding site on myosin subfragment-1 surface. *FEBS Lett.*, **176**, 75–78.

Yamamoto, O. (1977). Ionizing radiation-induced crosslinking in proteins. *Adv. Exp. Med. Biol.*, **86A**, 509–547.

Yan, M. *et al.* (1994). *N*-Hydroxysuccinimide ester functionalized perfluorophenyl azides as novel photoactive heterobifunctional cross-linking reagents. The covalent immobilization of biomolecules to polymer surfaces. *Bioconjugate Chem.*, **5**, 151–157.

Yang, T. *et al.* (1996). Mapping cross-linking sites in modified proteins with mass spectrometry: an application to cross-linked hemoglobins. *Anal. Biochem.*, **242**, 55–63.

Yem, A. W. *et al.* (1989). Biotinylation of reactive amino groups in native recombinant human interleukin-1β. *J. Biol. Chem.*, **264**, 17691–17697.

Yoshitake, S. *et al.* (1979a). Efficient conjugation of rabbit Fab′ with β-D-galactosidase from *Escherichia coli*. *Scand. J. Immunol.*, **10**, 81–86.

Yoshitake, S. *et al.* (1979b). Conjugation of glucose oxidase from *Aspergillus niger* and rabbit antibodies using *N*-hydroxysuccinimide ester of *N*-(carboxycyclohexylmethyl)-maleimide. *Eur. J. Biochem.*, **101**, 395–399.

Yoshitake, S. *et al.* (1982a). Efficient preparation of rabbit Fab′: peroxidase conjugates using maleimide compounds and its use for enzyme immunoassay. *Anal. Lett.*, **15**, 147–160.

Yoshitake, S. *et al.* (1982b). Mild and efficient conjugation of rabbit Fab′ and horseradish peroxidase using a maleimide compound and its use for enzyme immunoassay. *J. Biochem.*, **92**, 1413–1424.

Zahn, H. (1950). Versuche zur Stabilisierung von Wolle und Seide mit 1,3-Difluor-4,6-dinitrobenzol. *Melliand Textilber.*, **31**, 762–763.

Zahn, H. (1955). Brückenreaktionen an Aminosäuren und Faserproteinen. *Angew. Chem.*, **67**, 561–572.

Zahn, H. and Lumper, L. (1968). Über die Spezifität bifunktioneller SH-reagenzieren und die Synthese eines definierten Dimeren des Rinderserumalbumins. *Hoppe-Seyler's Z. Physiol. Chem.*, **349**, 485–488.

Zahn, H. and Meienhofer, J. (1958). Reaktionen von 1,5-Difluor-2,4-dinitrobenzol mit Insulin 2. Mitt. Versuche mit Insulin. *Makromol. Chem.*, **26**, 153–156.

Zahn, H. and Waschka, O. (1956). Reaktionen von Bisdiazohexan mit Aminosäuren, Mercaptanen, Wollkeratin und Seidenfibroin. *Makromol. Chem.*, **18**, 201–216.

Zara, J. J. *et al.* (1991). A carbohydrate-directed heterobifunctional cross-linking reagent for the synthesis of immunoconjugates. *Anal. Biochem.*, **194**, 156–162.

Zarling, D. A. *et al.* (1980). Mapping of lymphocyte surface polypeptide antigens by chemical cross-linking with BSOCOES. *J. Immunol.*, **124**, 913–920.

Zarling, D. A. *et al.* (1982). Association of Sendai virion envelope and a mouse surface membrane polypeptide on newly infected cells: lack of association with H-2K/D or alteration of viral immunogenicity. *J. Immunol.*, **128**, 251–257.

Zegers, N. *et al.* (1990). An improved conjugation method for controlled covalent coupling of synthetic peptides to proteins using glutaraldehyde in a dialysis method. *J. Immunol. Methods*, **130**, 195–200.

Zeheb, R. *et al.* (1983). An analytical method for the selective retrieval of iminobiotin-derivatized plasma membrane proteins. *Anal. Biochem.*, **129**, 156–161.

Zervas, L. and Photaki, I. (1960). Über Cystein- und Cystinpeptide. *Chimia*, **14**, 375–377.

Zhang, L. and Tam, J. P. (1996). Thiazolidine formation as a general and site-specific conjugation method for synthetic peptides and proteins. *Anal. Biochem.*, **233**, 87–93.

Zhou, J. S. *et al.* (1992). Metalloprotein complexes for the study of electron-transfer reactions. Characterization of diprotein complexes obtained by covalent cross-linking of cytochrome *c* and plastocyanin with a carbodiimide. *Bioconjugate Chem.*, **3**, 382–390.

Chapter 6

The Preparation of Protein–Small Molecule Conjugates

Alastair H. Dent and Mohammed Aslam

Research and Development Department, Ortho-Clinical Diagnostics, Pollards Wood Laboratories, Nightingales Lane, Chalfont St. Giles HP8 4SP, UK

The previous chapter described the techniques which can be employed to form conjugates of two or more protein molecules. Many of the same chemistries can be employed where the aim is to couple the protein to a small target, a drug molecule, for example, or a fluorophore.

The reactive groups available from the protein are obviously unchanged, as are the limited conditions which these rather delicate molecules will stand. Small organic compounds are in general much more robust, and can be exposed to wider variations of temperature (and to some extent pH) without harm. Most importantly they are normally soluble in inert, non-aqueous solvents, which allow moisture-sensitive groups to be handled without fear of hydrolysis. The greatest flexibility in chemical manipulation is therefore obtained before the involvement of the protein.

Sometimes the small molecule of interest will already contain a suitable functionality ('handle') for reaction with proteins, an amine or a carboxylic acid for example. If this group can be used for derivatisation without harmful effects, a straightforward one-step conjugation to the protein target may be all that is required.

This may or may not be the case. A simple illustration concerns the luminescent compounds luminol and isoluminol (Figure 6.1): both contain an amine handle,

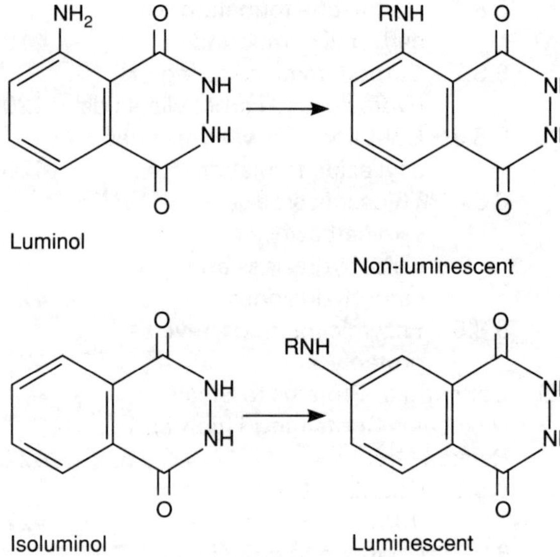

Figure 6.1 Derivatisation of the amine group of luminol results in significant loss of luminescence, whereas isoluminol retains substantial activity. The latter is therefore more popular for direct labelling.

but *N*-alkylation of luminol causes a large loss of luminescence due to steric hindrance. Alkylation of the isoluminol amine, on the other hand, can actually increase the quantum yield dramatically (Rongen *et al.*, 1994), and is therefore much more attractive as a means of introducing a luminescent label into a protein.

So even if the structure of the small molecule contains a likely group for derivatisation it may prove impractical to use it. In many cases, moreover, there is no ready-made handle, and it is necessary to introduce one.

Coupling of small molecules to proteins, therefore, will often consist of two stages. In the first, and often the most time-consuming by far, the entire arsenal of techniques for organic synthesis available to yield a derivative of the small molecule containing a suitable handle; in the second, this derivative is coupled to the target using the rather more limited range of protein-compatible chemistries.

The number of possible approaches to the first stage is enormous; there is an unlimited supply of target molecules of interest, and an almost unlimited number of reactions for manipulating their structure to obtain a protein-reactive derivative. This book cannot attempt to do justice to this topic. Individual problems will require a specific knowledge of the chemistry of the particular molecule concerned.

The bulk of this chapter will be devoted to the second stage, i.e. to the methods of coupling a small molecule, suitably derivatised to provide a reactive functionality if necessary, to a target protein. It is necessary to set some sort of rule for where the first stage ends and the second begins, as this is not always clear: for example, does the use of a heterobifunctional reagent such as MBS or SMCC for the introduction of a thiol into a small molecule constitute part of the organic synthesis, or part of the coupling procedure? For the purposes of this chapter, any procedure in which a derivative of the small molecule is prepared and isolated will be considered part of the synthesis stage, and therefore generally excluded from treatment.

An important difference from protein–protein coupling relates to solubility. While proteins are almost universally more soluble in aqueous than in organic solvents, this is by no means true for small molecules. In fact the reverse is true in very many cases. Therefore there can be a mismatch between the optimum solvent polarity for the protein and the organic molecule to which it is to be coupled, and the choice of conditions may require some care. Often an organic cosolvent is employed to provide an appropriate milieu, and in extreme cases it may be possible to dispense with the aqueous component altogether: Jalalian *et al.* (1995), for example, described the preparation of immunoconjugates in a medium of diethyleneglycol monoethyl ether.

The possible reasons for coupling small molecules to a protein are manifold, but fall loosely into two categories. On the one hand are reagents such as enzyme-labelled steroids for immunoassay, radiolabelled antibodies for histochemistry, oligopeptide-labelled immunogens for the generation of antibodies, and radioimmunoconjugates for therapeutic purposes. In these applications it is the identity of the small molecule that is paramount; the nature of the chemistry used to attach these molecules to the protein, while often important, is somewhat secondary.

At the other extreme are applications where the aim is to derivatise a particular functionality of the protein in a specific fashion, where the reactivity of the small molecule is the important parameter and its precise structure may be fairly unimportant. The elucidation of protein structure–activity relationships, for example, has owed much to observing the effect of specific derivatisation reactions. The active site of the enzyme papain contains an essential cysteine residue, and the enzyme can thus be inactivated by a wide range of thiol-reactive species: chloroacetamide and chloroacetic acid (Chaiken and Smith, 1969a,b), 2-bromoacetamido-4-nitrophenol (Lewis and Shafer, 1974), methyl-2-pyridyldisulfide (Kimura *et al.*, 1982), *O*-methylisourea (Banks and Shafer, 1972). Similarly, chymotrypsin is stoichiometrically inactivated by the reaction of *p*-nitrophenyl acetate (Balls and Wood, 1956), but this inactivation can equally well be achieved by other active esters such as the photoactivatable diazoacetate analogue (Singh *et al.*, 1962).

More mundanely, it may be desired to block particular reactive sites of a protein to prevent unwanted reaction during subsequent stages; for example Nakane and Kawaoi (1974) blocked the amine functions of HRP with 2,4-dinitrofluorobenzene before periodate oxidation, to prevent self-coupling through the newly generated aldehydes. Again the exact nature of the blocking reagent is relatively unimportant.

Whatever the intent of the experiment, the chemistries underlying protein derivatisation with small molecules are predominantly the same. Applications of all types will be covered in this chapter.

Categorisation of methods for coupling proteins to small molecules

The availability of useful functionalities on protein molecules is at least reasonably consistent. The presence of lysine, aspartate, glutamate, tyrosine and histidine residues can normally be relied upon, plus free cysteine residues or saccharides in many cases.

By their nature small molecules are likely to contain very few reactive moieties, often only one. Their exact structure is also very likely to be known, so the problem faced by a worker is generally of the form, 'how can I use the functionality which is present (or available by fairly simple derivatisation) to couple this compound to a protein?' Therefore, unlike the last chapter, it is most useful to structure this section in terms of the available reactive groups in the small molecule.Containing Carboxyls or Carboxyl-derived Groups

In any case, homo- and heterobifunctional approaches are much less commonly encountered here; direct coupling methods are much more popular for coupling small molecules to proteins than for protein–protein conjugation, and dominate the field heavily. There are two main reasons for this.

First, the relative ease of handling small molecules often makes it possible to carry out direct activation reactions with a much greater efficiency than can be obtained with proteins. Carbodiimide activation of a protein carboxyl function, for example, must be carried out in fairly dilute aqueous media under constant competition from hydrolysis. With a small molecule on the other hand, this can generally be carried out at high concentrations in a water-free organic solvent, resulting in a much smoother reaction and an often quantitative yield.

Secondly, activation with a bifunctional reagent (and with heterobifunctional linkers in particular) often necessitates a purification step, to remove unreacted coupling agent from the activated species. With proteins this can be rapidly and simply achieved by 'desalting' on a size exclusion matrix, but with low-molecular-mass compounds this separation is much more troublesome. While it is possible to develop suitable methods to achieve such separations, it is normally simpler to avoid the problem altogether by finding another coupling route.

Linked methods do offer some advantages however, in particular by introducing a spacer between the protein and the small molecule without the need for time-consuming organic synthesis.

6.1 Coupling of Proteins to Small Molecules Containing Carboxyls or Carboxyl-derived Groups

Given the ready availability of amines in proteins, it is not surprising that many derivatisation routes rely on amine-reactive functionalities. Among these, conjugation *via* suitably activated carboxylic acid groups is probably the commonest methodology. These are all direct approaches; there are very few homo- or heterobifunctional reagents targetted at carboxyl groups.

6.1.1 *In situ* carbodiimide activation

The central carbon atom of carbodiimides (Figure 6.2) is rendered electrophilic by the electron-withdrawing effect of the adjacent nitrogens, and will therefore react with nucleophiles (Figure 6.3) such as water, amines, thiols, phenols, and, most usefully, carboxylic acids (Khorana, 1953; Kurzer and Douraghi-Zadeh, 1967). While most of these products are stable, the *O*-acylisourea obtained from carboxylates contains a carbonyl group which is highly activated towards other nucleophiles[1].

R—N=C=N—R'

Figure 6.2 Carbodiimides.

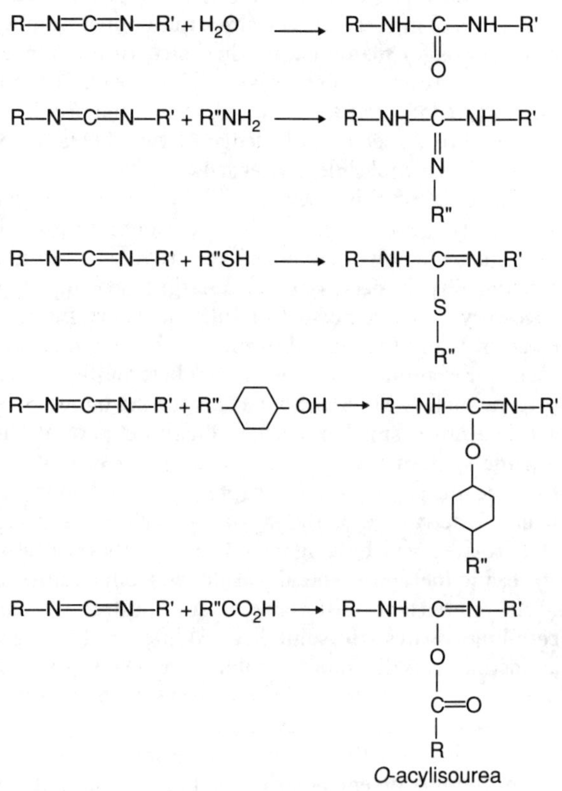

Figure 6.3 Reactions of carbodiimides with water, amines, thiols, alcohols and carboxylic acids.

[1] The carbon atom derived from the carbodiimide is still electrophilic, pulling electrons away from the adjacent oxygen and reducing its mesomeric deactivation of the carbonyl carbon.

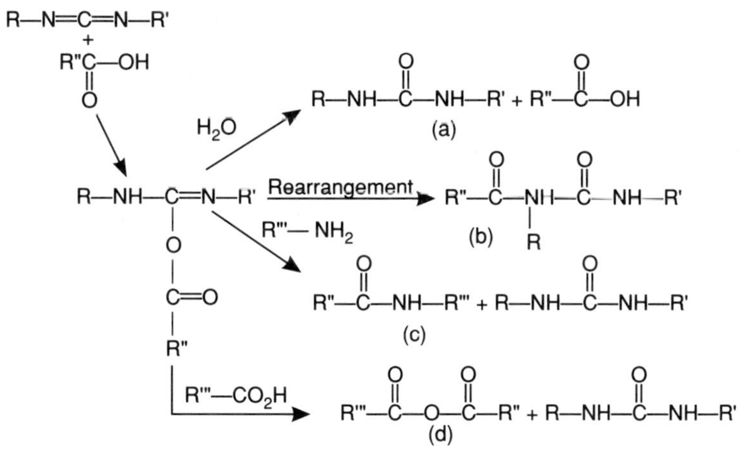

Figure 6.4 Reactions of *O*-acylisoureas.

There is good evidence that the carbodiimide is first protonated (Khorana, 1953; Hoare and Koshland, 1967), the resulting carbocation providing the site for nucleophilic attack. This model is supported by the observation of Nakajima and Ikada (1995) that formation of the *O*-acylisourea displays a pH optimum at around pH 4. Low-pH suppression was attributed to unavailability of ionised carboxylate, and high-pH suppression to the relative scarcity of H^+.

Once the acylurea has been formed, the nucleophile which reacts with it may be (Figure 6.4): (1) water, in which case the urea is formed and the acid regenerated; (2) the NH group of the intermediate itself, producing a stable *N*-acylurea; (3) another nucleophilic reagent such as an amine (Figure 6.4(c)), providing a useful amide link to the original acid; (4) another molecule of the acid (Figure 6.4(d)), yielding an acid anhydride; this too can react with nucleophiles such as amines. The product is therefore the same amide, but the reaction stoichiometry is obviously affected. The balance between the two routes to amidation will depend on the reagent ratios and reaction times employed, higher acid:carbodiimide ratios and reaction times obviously favouring the anhydride mechanism.

The first major application of these reagents was for peptide synthesis, pioneered by Sheehan and his colleagues at the Massachussetts Institute of Technology in the 1950s. Later the use of carbodiimides became popular for the quantification of protein carboxyl groups (Hoare and Koshland, 1967; Carraway and Koshland, 1972), and for studies of the role of such groups in enzyme function (Riehm and Scheraga, 1966; Carraway *et al.*, 1969; Carraway and Koshland, 1972). Despite occasional applications for protein– protein conjugation (see Chapter 5, Section 5.2.2), the most widespread use in the bioconjugation field has been for coupling proteins to small molecules.

A number of different carbodiimides are in use, their utility depending on a combination of reactivity, solubility and stability. The introduction of electron-donating *N*-substituents such as alkyl or alicyclic groups lowers reactivity, but improves stability to a convenient level. The simple non-polar carbodiimide dicyclohexylcarbodiimide, DCC (Figure 6.5(a)), first synthesised in the 1930s, is extremely popular for use in organic media, giving high efficiencies and yielding a conveniently insoluble urea byproduct[1].

Because carboxylates are such powerful nucleophiles they can be activated quite readily even in aqueous media, and water-soluble carbodiimides have been developed for this purpose. Far and away the most popular of these reagents are 1-ethyl-3-(3-dimethylaminopropyl)carbodiimide hydrochloride (EDAC or EDC; Sheehan *et al.*, 1961; Figure 6.5(b)) and 1-cyclohexyl-3-[2-(4-methylmorpholin-4-yl)ethyl]carbodiimide *p*-toluenesulfonate (CMC; Sheehan and Hlavka, 1956; Figure 6.5(c)).

Carbodiimide chemistry has been applied widely to the coupling of proteins to carboxyl-bearing small molecules in both one- and two-step formats, and the simplicity of such techniques has rendered them extremely popular in applications such as the prepara-

[1] It should be noted that dicyclohexylurea (DCU) is not completely insoluble in all organic media, using DMF, for example, filtration cannot be relied upon to remove all the DCU.

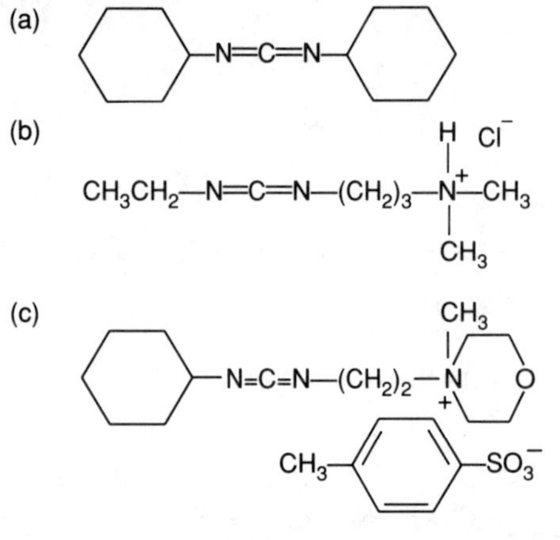

Figure 6.5 Structures of (a) DCC, (b) EDAC hydrochloride and (c) CMC tosylate.

tion of immunogens (Bauminger and Wilchek, 1980) and hapten conjugates for immunoassay (O'Sullivan *et al.*, 1979; Blake and Gould, 1984).

Note that the same chemistry can be used to couple amine-bearing small molecules to the carboxylate residues of proteins; these methods will be covered in Section 6.7.1. Sometimes, with oligopeptides, for example, more than one of these routes may be available simultaneously. Unless one functionality is blocked this can lead to inefficient coupling, as oligomerisation of the small molecules will begin to occur as activated carboxylates are generated. Reaction of carbodiimides with amines and other residues may reduce efficiency still further.

A convenient means to prevent crosslinking of this type (Atassi and Habeeb, 1972; Briand *et al.*, 1985) is to block the carboxyl residues of the peptide with citraconic anhydride. This blocker can be removed by acid hydrolysis (e.g. dialysis against 5% acetic acid; Briand *et al.*, 1985) after the coupling reaction, which is carried out under unusually basic conditions (pH 8) to avoid deprotection at this stage.

6.1.1.1 One-step carbodiimide methods

The one-step approach closely parallels protein–protein coupling methods, in that both protein and small molecule are present during the activation of carboxylate, hence the need for an aqueous medium and for a water-soluble carbodiimide such as EDAC or CMC. The one-step methodology is somewhat inelegant, for a number of reasons. (1) The presence of water will result in hydrolysis of the carbodiimide itself, and of the *O*-acylisourea intermediate. (2) The carbodiimide will activate carboxylic residues in the protein as well as in the target molecule, leading to protein cross-linking. Davis and Preston (1981) compared one and two-step EDAC protocols for coupling hippuric acid to IgG: one-step protocols resulted in two-thirds or more of the protein being cross-linked, while no more than 10% cross-linking was observed in the equivalent two-step experiment. This dramatic difference resulted from an activation step of only two minutes. (3) Other groups in the protein such as amines and thiols may react with the carbodiimide, yielding unreactive products and possibly altering the protein's properties. (4) The optimum pH for the activation of carboxyl groups is 3.5–4.5, while the requirement for unionised amine groups means that the coupling reaction is most efficient above around pH 8; therefore the pH of a single-stage reaction can only offer a compromise. (5) Many buffer components are reactive towards carbodiimides. The choice of buffer for aqueous carbodiimide reactions has received surprisingly little attention: while it may be obvious that matrices such as acetate ought to be avoided, orthophosphates are also susceptible to carbodiimide activation (Brown, 1963) and their use may hinder efficiency. In spin-labelling experiments Aplin *et al.* (1979) noted that EDAC activated the phosphate groups of DNA, as well as glycoprotein carboxyl groups from amino acid and sialic acid residues. Gilles *et al.* (1990) observed, too, that EDAC was inactivated to some extent by the presence of phosphate or ATP. Amine-containing buffers should also be avoided as these too can react with the carbodiimide; amines have been shown to cause losses of carbodiimide (Gilles *et al.*, 1990).

One-step methods using EDAC

EDAC is by far the most commonly encountered water-soluble carbodiimide. Its use in the preparation of protein–small molecule conjugates dates from the early 1960s, notably the work of Goodfriend and his colleagues on the preparation of peptide immunogens.

The difficulty in choosing an optimal pH for one-step carbodiimide reactions is reflected in the wide range of conditions reported in the literature. Nakajima and Ikada (1995) observed an optimum of pH 5 for a one-step EDAC coupling of ethylenediamine to carboxylated resins, and the 5–6 range is probably the most commonly encountered. However a pH as high as 8.5 has been successfully employed (Ferrua *et al.* (1986), see below). Swaisgood and Natake (1973), activating acetic acid with EDAC at pH 4.75 and 7, made some observations which may explain the moderate pH sensitivity of this method. Although the rate of formation of *O*-acylisourea was 25–30 times lower at pH 7, this was offset by a six to tenfold reduction in the rate of isourea hydrolysis and by the absence of rearrangement to the *N*-acylurea at the higher pH.

(a) Methods without pH control

Until around 1970 it was common to carry out the reaction in water, without pH control. Early methods also often also employed enormous EDAC excesses and long reaction times.

Thus Goodfriend *et al.* (1964) used EDAC to prepare immunogens of the synthetic oligopeptides angiotensin and bradykinin: 10 mg of the peptide and 20 mg of rabbit serum albumin (RSA) were dissolved in 0.5 ml of water, then 0.25 ml of water containing 100–200 mg of EDAC added. The unadjusted pH was 6–8, and the reaction was allowed to proceed for 5–30 minutes before (gradual) stopping by dialysis. With angiotensin the peptide:EDAC:RSA ratio employed was 25:(1670 to 3330):1.

Similar conditions were used by the same group (McGuire *et al.*, 1965) to couple α-melanocyte stimulating hormone (α-MSH) to RSA using an α-MSH:EDAC:RSA ratio of 30:6700:1. Similarly Goodfriend *et al.* (1966) obtained a molar incorporation of six angiotensin molecules in an RSA immunogen, using a 100:10000:1 ratio and reaction times of 1–12 hours.

Later uses of the uncontrolled-pH approach demonstrated that such high excesses of EDAC were not necessary for a reasonable amount of reaction: Cessi and Fiume (1969) coupled β-amanitin to RSA (24 hours, room temperature, water) using a hapten:EDAC:RSA ratio of 25:40:1 and obtained a molar incorporation of 2.3. Bauminger *et al.* (1969) coupled genesteic acid to an alanine/lysine polymer using an acid:EDAC:polymer ratio of 233:218:1 and obtained a molar incorporation of 22.

Faulstich and Fiume (1985) described one-step EDAC coupling of fungal toxins to BSA; in a typical example β-amanitin, EDAC and BSA were reacted at a 38:90:1 ratio (24 hours, room temperature, water). Amanitin incorporation of 1.6 was obtained in a similar preparation. These compounds also contain a phenol moiety which can be esterified by EDAC and acid groups from the protein; phenolic esters (cleavable by hydroxylaminolysis) represented 13–20% of the product obtained[1]. As frequently observed with one-step reactions of this type there was considerable cross-linking of protein, only 30% of the albumin remaining monomeric.

(b) Methods with pH controlled at 5–6

Despite recent uncontrolled examples it has been common for many years to introduce some element of pH control: Oliver *et al.* (1968), for example controlled pH at 5.5 for the reaction of 3-*O*-succinyldigitoxigenin with human serum albumin (room temperature, 18 hours). Using a steroid:EDAC:HSA ratio of about 64:300:1, an incorporation of at least five digoxigenins per molecule was obtained. Similar conjugation of succinyl cyclic AMP to HSA (Steiner *et al.*, 1969) gave a molar incorporation ratio of 5–6.

Spector and Parker (1970) coupled 3-*O*-carboxymethylmorphine to BSA (overnight, room temperature, pH 5.5) using an acid:EDAC:BSA ratio of around 175:365:1, and obtained a morphine incorporation of three to four.

Jaffe *et al.* (1971) coupled prostaglandins to human serum albumin (24 hours, 20 °C, pH 5.5) using an acid:EDAC:protein ratio of approximately 100:180:1. Incorporations varied between 1.5 and 3.3 prostaglandins per albumin.

Knoll and Wisser (1973) coupled succinyl-dimethoxy-phenylethylamine to bovine γ-globulin (17 hours, room temperature, pH 5.5) at an acid:EDAC:protein ratio of about 500:5000:1, and obtained a molar incorporation of 21. Doubling the EDAC loading resulted in an incorporation of 29.

Davis and Preston (1981) carried out one-step couplings of hippuric acid to IgG at pH 5 (overnight, room temperature), using an acid:EDAC:IgG ratio of 100:1000:1 and obtained incorporations of around four. As noted earlier, high levels of protein cross-linking were observed, 66% at pH 5, and 80% in the unbuffered experiment.

Lewis and Elder (1985) coupled cortisol-3-carboxymethyloxime to thyroglobulin by this route: the steroid was solubilised in a 1:1 mixture of DMSO and dioxane, and the product diluted with water before addition of aqueous solutions of the protein and of EDAC. The steroid:protein ratio was 2000:1. The reaction was allowed to proceed overnight (4 °C, pH 5.6) before dialysis, and yielded a successful solid phase conjugate for an immunoassay.

Flurkey *et al.* (1985) described the coupling of an acid derivative of haloperidol to BSA or whelk haemocyanin to provide a successful immunogen: the acid was dissolved in dimethylacetamide and added dropwise to an aqueous EDAC solution. After adjustment of the pH to 5.0, a solution of the protein in

[1] Phenolic esters cleavable by hydroxyalminolysis may also have been derived from the reaction of the activated carboxyl group with tyrosines in the protein.

50 mM phosphate buffer, pH 6.0, was added dropwise and the mixture stirred at 4 °C overnight before dialysis.

Degani and Heller (1987, 1988) coupled ferrocene-carboxylic acid to glucose oxidase using EDAC. In one preparation an acid:EDAC:oxidase ratio of 920:1390:1 was employed, and the coupling reaction (15 hours, ice–water, pH 7.2–7.3) yielded a molar incorporation ratio of 12.0. The presence of the ferrocene groups allowed direct electron transfer between the enzyme, which retained something over half of its original activity, and a metal electrode. Using identical conditions, Suzawa *et al.* (1994) coupled 3–8 ferrocenes to the enzyme.

Boersma *et al.* (1988) used EDAC in a mannitol buffer (pH 5.0) for coupling synthetic peptides to carrier proteins (overnight, 4 °C), but observed high levels of neodeterminant formation rather than peptide incorporation. Use of an *N*-methylimidazole buffer, pH 6.0 (30 minutes, room temperature) reduced this problem, probably by promoting the rapid formation of an imidazolide intermediate (Deen *et al.*, 1990; Zegers *et al.*, 1990).

(c) Methods with pH controlled at a higher value

Raising the pH reduces the efficiency of the carbodiimide activation but increases the availability of unprotonated amine groups in the protein.

Levine and Van Vunakis (1970) coupled prostaglandins to polylysine at pH 6.8 (overnight, room temperature): one mole of prostaglandin E_1 was incorporated per 10–20 lysine residues.

Jaffe *et al.* (1970) coupled gastrin oligopeptides to proteins at pH 6.5–7.0 (room temperature, several hours) using a peptide:EDAC:protein ratio of 30:90:1, and obtained molar incorporations of 8–12.

Preston *et al.* (1981) coupled a carboxyl derivative of α-amanitin (ABGG) to BSA using an ABGG:EDAC:BSA ratio of 11:4285:1. The coupling reaction (24 hours, room temperature, pH 7.2) yielded a molar incorporation of 2.9.

Kulkarni *et al.* (1981) coupled methotrexate to IgG (two hours, 4 °C, pH 7.1) using a methotrexate:EDAC:IgG ratio of around 730:100:1. An incorporation of around 14 was obtained, rather lower than that achieved using the succinimide ester.

Ferrua *et al.* (1986) dissolved thyroxine and BSA in a 10:1 mixture of 0.01 M sodium hydroxide–DMF, the rather harsh solvent conditions being necessary due to the insolubility of thyroid hormones of this type, then solid EDAC was added and the pH of the mixture adjusted to 8.5. Overnight reaction at room temperature followed by dialysis yielded a conjugate which incorporated 5.6 thyroxine molecules per BSA.

One-step methods using CMC

CMC does not appear to offer any great advantage over EDAC, and is less frequently encountered. It has found its place in immunochemistry, however, where it may be necessary to avoid the use of the same reagent to prepare both immunogen and tracer due to bridge recognition[1]. With direct reactions such as these, where the coupling agent is eliminated during the reaction, the nature of the main product should be identical regardless of the carbodiimide used. However some rearrangement of the *O*-acylisourea intermediate to a stable *N*-acylurea is probably inevitable, and these moieties can lead to the bridge recognition problems discussed.

CMC is a useful partner to EDAC in the preparation of heterologous pairs of immunogen and tracer, where there is no danger of spurious results due to recognition of *N*-acylureas. There appears to be little difference in the conditions applied for CMC use.

Goodfriend *et al.* (1964, 1966) used CMC as well as EDAC to prepare peptide immunogens; the methodologies were identical to those described earlier. Isturiz *et al.* (1978) followed the same conditions to couple formylated oligopeptides to IgG and IgM in PBS.

Jaffe *et al.* (1970) coupled gastrin peptides to polylysine, polyglutamate[2] and bovine pancreatic ribonuclease carriers, obtaining incorporations of 10–17, 20–27 and 4 respectively. The coupling reaction (several hours, room temperature, pH 6.5–7.0) employed a peptide:CMC:carrier ratio of 30:90:1 in each case.

Tateishi *et al.* (1976) coupled testosterone-3-*O*-carboxymethyloxime to glucoamylase using a one-step CMC protocol (three hours, room temperature, pH 5.5). Full enzyme activity was retained, compared to the loss of around 97% using mixed anhydride coupling, and the conjugate was successful in an immunoassay.

Faulstich and Fiume (1985) used a one-step CMC protocol to couple fungal toxins to BSA (24 hours, room temperature, in water) at a toxin:CMC:BSA

[1] Failure to consider this point can result in the generation of antibodies which recognise the coupling group rather than the hapten of interest. This result can be compounded dangerously by using a tracer containing the same linker, which will therefore be bound by the antibody in a manner independent of the presence or absence of the hapten.

[2] The peptides contained both amine and carboxyl moieties. The polyglutamate conjugate was presumably coupled predominantly *via* peptide amine rather than carboxyl groups, and therefore should properly belong in Section 6.7.1.

ratio of 38:65:1. In a similar preparation an incorporation of 0.6 was obtained. The conjugates of β-amanatoxin obtained with CMC were considerably less hepatotoxic than the equivalent EDAC conjugates.

6.1.1.2 Two-step carbodiimide methods

Where the small molecule does not contain any nucleophilic groups which will cross-link to the activated carboxylate, it is possible to allow the activation of the carboxyl to go to completion before introducing the protein target.

The carbodiimide reaction can take place within minutes, and some of the one-step protocols described above actually go some distance towards this approach, simply by making the addition of the protein the last step (e.g. Flurkey *et al.*, 1985).

However protocols have arisen where a very definite two-stage format is employed. The rapidity of formation of *O*-acylisourea means that it is generally disadvantageous to employ activation times greater than a few minutes. Depending on the matrix, rearrangement to the *N*-acylurea, formation of anhydrides and hydrolysis can all reduce efficiency. This is particularly important if aqueous systems are employed at the activation stage: Nakajima and Ikada (1995) observed a rapid decline in reactivity towards an amine when carboxylated beads were activated for more than five minutes in an acetate matrix: the amide yield was reduced to around two-thirds of its maximum after 30 minutes, and about one-third after an hour.

Two-step protocols using EDAC

Dray *et al.* (1975) coupled a carboxyl-containing progesterone analogue to β-galactosidase: the steroid was activated with EDAC (30 minutes, room temperature, pH 4.7) before coupling to the enzyme (twelve hours, 4 °C, pH 5.5). Incorporations were not calculated but 26% of the conjugate was bound by anti-progesterone antiserum, forming the basis of a sensitive immunoassay. Around 90% of enzyme activity was retained.

Mattox *et al.* (1979) activated cortisone 21-glucosiduronic acid with an equimolar quantity of EDAC in DMF (30 minutes, 25 °C) before coupling to BSA (2.5 to 4 hours, 25 °C, pH 5.5). Soluble conjugates containing up to two steroid molecule per molecule of conjugate were obtained, but losses of protein, presumably due to cross-linking and precipitation, were up to 90%. Similar observations were made by Lau *et al.* (1981) in the preparation of aflatoxin–BSA immunogens, although details of the method were not recorded.

Davis and Preston (1981) demonstrated a dramatic reduction in protein cross-linking by using a two-step protocol for coupling hippuric acid to IgG. Due to the reactivity of carbodiimides with phosphate the molarity of the phosphate buffers employed was used to help control the reaction: ionic strength was low at the activation stage (12.5 mM *vs.* 1–100 mM carbodiimide), and high during coupling (200 mM) to encourage reaction with any remaining carbodiimide. Activation was carried out at pH 5 (two minutes, room temperature) and conjugation at pH 8.0 (overnight, room temperature).

EDAC:hippuric acid ratios from 1:1 to 100:1 were employed, with a constant 100-fold excess of hippuric acid over IgG, and resulted in the incorporations illustrated in Figure 6.6. There was no benefit in using a ratio greater than 10:1, higher ratios in fact leading to somewhat lower incorporation and to some (13 to 20%) protein cross-linking.

Properties of hippuric acid immunoglobulin G conjugates prepared by different coupling methods

Method	Concentration (mM) EDAC	HA	IgG	Product HA/IgG ratio	Protein crosslinking (%)
Water	10	1	0.01	ND	80
pH 5	10	1	0.01	ND	66
pH 5	6.5	0.65	0.0065	4.2	ND
pH 5/pH 8	1	1	0.01	1.6	≤10
pH 5/pH 8	5	0.5	0.01	2.6	≤10
pH 5/pH 8	10	0.5	0.01	3.0	≤10
pH 5/pH 8	10	1	0.01	6.2	<10
pH 5/pH 8	10	2	0.01	14.6	≤10
pH 5/pH 8	20	1	0.01	6.3	≤10
pH 5/pH 8	50	1	0.01	5.8	13
pH 5/pH 8	100	1	0.01	4.9	20

Figure 6.6 Properties of hippuric acid–IgG conjugates made by EDAC coupling under various conditions (Davis and Preston, 1981).

Increasing the hippuric acid:IgG ratio from 50:1 to 200:1, on the other hand, resulted in a steady increase in incorporation from 3.0 to 14.6, even with a lower carbodiimide excess at the higher ratios.

Canfi *et al.* (1989) activated testosterone-3-carboxymethyloxime (in DMF) with EDAC (in a 2:1 mixture of DMF and 0.05 M phosphate buffer, pH 5.6) for 30 minutes at 10 °C, before addition of human serum albumin (in 0.05 M phosphate buffer, pH 7.5). Two hours reaction at 10 °C was allowed before purification by dialysis.

Khosravi and Morton (1991) activated cortisol-21-hemisuccinate with a 1.2-fold excess of EDAC in a 3:1 DMSO–water mixture (30 minutes, room temperature), coupling the product to streptavidin at ratios up to ten steroid molecules per amino group (90 minutes, room temperature, pH 8.0). Incorporations up to 16.8 were obtained. The effect of the incorporation level on a number of key parameters is shown in Figure 6.7. The biotin-binding activity of streptavidin was unaffected up to an incorporation of around three, but was lowered by around 20% for all higher levels. Ability to bind an anti-cortisol Mab increased over the whole incorporation range, roughly linearly from three to nine. Protein recovery began to decline significantly at incorporations above around six, presumably due to overlabelling and precipitation.

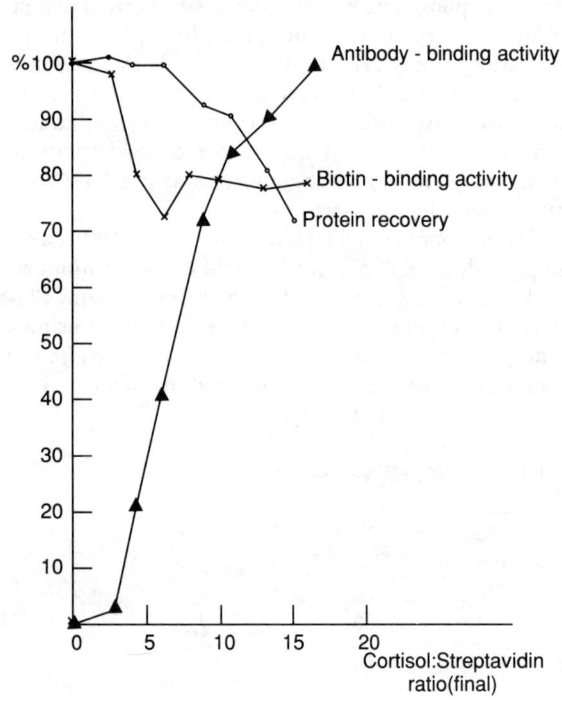

Figure 6.7 Properties of streptavidin–cortisol conjugates of various incorporations, as measured by absorbance at 242 nm: protein recovery determined by radiometry; biotin-binding activity determined fluorimetrically by the ability to bridge biotinylated proteins; antibody-binding activity determined fluorimetrically by the ability to bridge biotinylated protein to an anti-cortisol coated well.

Two-step protocols using CMC

Rosenfeld *et al.* (1975) activated a carboxylated steroid with CMC (35 minutes, room temperature, pyridine–water (1:1)) before coupling to BSA (150 minutes, room temperature, pyridine–water (1:2)). A molar incorporation ratio of 13 was obtained.

Carboxylated estradiol derivatives have been coupled to β-D-galactosidase by this means. Exley and Abuknesha (1977) activated estradiol-6-carboxymethyloxime with CMC in a 10 mM acetate buffer, pH 4.7 (30 minutes, 10–15 °C) before coupling to the enzyme in 100 mM acetate buffer, pH 5.8 (overnight, 4 °C). At least 90% of the enzyme activity was retained and, despite low incorporation, the conjugate was used in a successful immunoassay. A similar preparation by the same authors (Exley and Abuknesha, 1978) employed a 200 mM phosphate buffer (pH 6.7, plus 100 mM NaCl) rather than the acetate buffers just described. The conditions, while sub-optimal for carbodiimide chemistry both in terms of pH and competition from phosphate ions, resulted in virtually complete retention of galactosidase activity.

In the same application, these authors explored the effect of steroid incorporation on enzyme activity in some detail (Abuknesha and Exley, 1978). Steroid:enzyme ratios between 20:1 and 1380:1 were employed: above around 100:1 (corresponding to an incorporation of five steroid molecules per enzyme molecule) the activity of the enzyme began to decline. At an initial ratio of 500:1 almost half of the enzyme activity was lost.

Mattox *et al.* (1979) used a two-step CMC protocol to couple cortisone 21-glucosiduronic acid to BSA, under the same conditions and with the same results as with EDAC (see above).

Eruk *et al.* (1984) activated thyroxine for three minutes with a 2.08:1 excess of CMC in 0.01 M sodium hydroxide solution, before coupling to alkaline phosphatase (five hours, room temperature, pH 9.1).

Flurkey *et al.* (1985) used a two-step CMC protocol to couple *N*-carboxymethyldopamine to keyhole limpet haemocyanin (KLH). The hapten was activated in DMF by a 2:1 excess of CMC (60 minutes, room temperature). The CMC was added in two portions, the second halfway through the reaction period.

A portion of the activated acid was coupled to KLH in a carbonate–phosphate buffer, pH 9.0, containing 10 μM sodium bisulfite (one hour at room temperature) before purification by dialysis. Despite failing as an immunogen, the preparation was chemically successful: it resulted in the incorporation of about 20 dopamine molecules per KLH subunit. The carboxymethyldopamine:CMC:KLH ratio in this experiment was approximately 750:1500:1.

Two-step protocols using DCC

An activation stage in a non-aqueous medium opens up the possibility of using non-water-soluble carbodiimides, in particular dicyclohexylcarbodiimide (DCC, Figure 6.5(a)). This approach was employed by Ranadive and Sehon (1967a), describing the first stage as the formation of the acid anhydride.

Fuchs and Fuchs (1969) used similar conditions to couple indoleacetic acid (IAA) and gibberellic acid to carrier proteins, e.g. an IAA–BSA conjugate was prepared using an IAA–DCC–BSA ratio of 345:162:1. The activation of the acid was carried out for 30 minutes at room temperature in dry dioxane, before conjugation to BSA (overnight, 4 °C, pH 8.5).

Buckley and Searle (1984) activated DTPA with a slightly less than equimolar quantity of DCC (overnight, DMSO) before coupling at a tenfold excess to IgG. Although the incorporation of the conjugate was not calculated directly, the radioactivity introduced by chelation of ^{111}In was around 40% of that obtained by *in situ* generation of an active ester by DCC/NHS (see Section 6.1.2). The low efficiency was probably exacerbated by consumption of the acid in anhydride formation.

6.1.2 *In situ* active ester formation *via* carbodiimides

While carbodiimide-activated carboxylates will react directly with amines, it has been found that more rapid and efficient reaction can generally be obtained by employing an intermediate stage, the formation of an active ester. Carbodiimide activation of a carboxyl group in the presence of *N*-hydroxysuccinimide (NHS), for example, conveniently generates the succinimide ester.

This product can be isolated to provide a stable reagent (Section 6.1.4), but the crude active ester solution is commonly used without further processing. Two-step procedures allow the formation of the active ester without competition from protein amines, but the rather less elegant one-step approach is often encountered too.

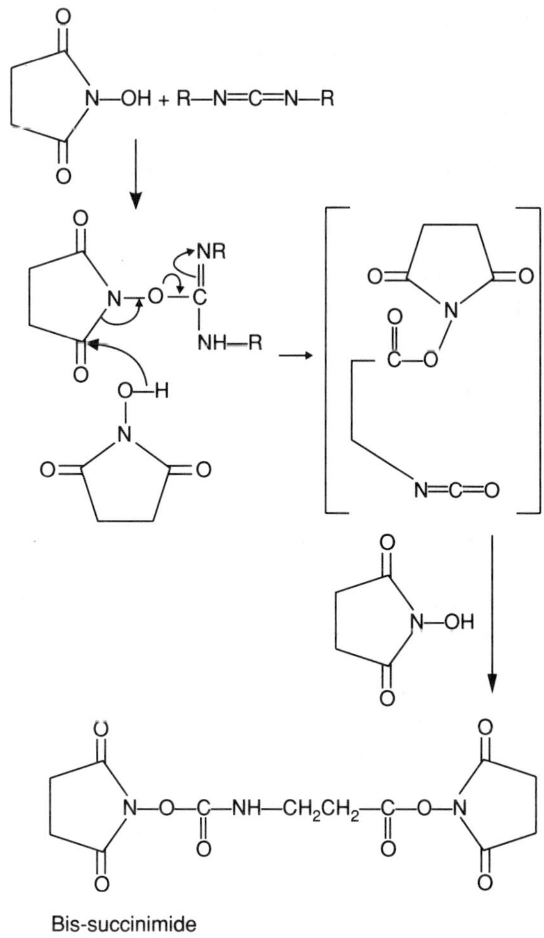

Figure 6.8 Reaction of DCC with NHS to yield bis-succinimide.

Gross and Bilk (1968) reported an interesting reaction of DCC with three molecules of NHS, resulting in the bis-succinimide ester shown in Figure 6.8. Problems from potential protein cross-linking *via* this species can be minimised by avoiding unnecessarily high quantities of NHS, and by limiting reaction times.

6.1.2.1 One-step carbodiimide/active ester methods

As with straight carbodiimide methods, one-step approaches of this type fail to exploit some key potential advantages, such as avoiding the need for aqueous systems at the activation stage. Again, there is a conflict between the pH optima required for carbodiimide activation of the carboxyl, reaction of the resulting *O*-acylisourea with the alcohol, availability of protein amines, and hydrolysis of the succinimide ester.

Their simplicity has assured them of some use, however. As an aqueous system is necessitated by the approach, the carbodiimide and succinimide used must of course be water soluble.

One-step methods using active ester formation via EDAC

In 1986 Staros *et al.* introduced the use of *N*-hydroxysulfosuccinimide (NHSS) to enhance the efficiency of EDAC reactions. Coupling 3.1 mM glycine to 0.7 mg/ml keyhole limpet haemocyanin (35 mM) in water with 100 mM EDAC, these authors observed a 15-fold increase in incorporation when 5 mM NHSS was included in the reaction mixture. The overall glycine:EDAC:NHSS:haemocyanin ratio was around 90:2900:140:1.

Sehgal and Vijay (1994) carried out some interesting optimisation of one-step EDAC/NHSS coupling of *N*-carboxypentyl-1-deoxynojirimycin to amine-containing solid phases. In summary there was a quite sharp pH optimum at pH 6.5; an EDAC:NHSS ratio of at least 5:1 was required, and coupling continued to increase for twelve hours. In a typical optimised protocol, the acid was incubated (24 hours, room temperature, pH 6.5) with EDAC and NHSS (1:670:45) and the solid phase.

Hendrickson *et al.* (1994) used this protocol to form immunogens from the peptide leucine-enkephalin and IgG molecules: a peptide:EDAC:NHS:IgG ratio of 90:3000:150:1 was employed

6.1.2.2 Two-step carbodiimide/active ester methods

Two-step protocols allow the activation to be carried out at greater efficiency in a non-aqueous medium, thus enabling non-water-soluble carbodiimides such as DCC to be employed. Aqueous systems are still sometimes encountered in the activation stage, an unnecessary source of low yield unless the acid is insoluble in organic solvents.

Two-step methods using active ester formation via EDAC

Schmidt *et al.* (1990) activated a carboxyl-containing derivative of the antibiotic ivermectin with EDAC/NHS in a minimal volume of DMF (overnight at room temperature) before coupling to carrier proteins (BSA, KLH, conalbumin). The ivermectin:EDAC:NHS:protein ratio was between 20:40:40:1 and 30:60:60:1 and the products were fractionated before characterisation: incorporations varied within the range of 5–21 haptens per protein molecule.

Mueller *et al.* (1990) activated carboxyl-containing doxorubicin derivatives with EDAC in dimethylacetamide (18 hours, in the dark, 4 °C) before coupling to the Mab (three hours, pH 7.8–8.0). Molar incorporation ratios of 3–6 were obtained from an initial acid: EDAC:NHS:Mab ratio of 30:30:30:1.

Smith *et al.* (1993) used a two-step EDAC/NHSS procedure to couple a carboxy-containing macrocyclic ligand to Mabs; the ligand was activated with EDAC/NHSS (one hour, room temperature, pH 7.0) then coupled to the Mab at a 64:1 excess (one hour, room temperature, pH 7.0).

Lewis *et al.* (1994) activated DOTA (1,4,7,10-tetraazacyclododecane-1,4,7,10-tetrayltetraacetic acid) with equimolar EDAC and NHSS in water (30 minutes, 4 °C), before coupling at 10–100:1 to cytochrome *c* (18 to 24 hours, 4 °C, pH 9.0). The molar incorporations observed are shown in Table 6.1.

Table 6.1 Molar incorporations of DOTA obtained from EDAC/NHSS coupling to cytochrome *c*, at different input ratios

DOTA : protein	Incorporation
10:1	2.64
20:1	4.27
50:1	6.78
100:1	8.79

An interesting variation on the reaction stoichiometry was investigated, where DOTA:NHSS:EDAC ratios between 2:2:1 and 10:10:1 were employed. The aim was to limit carbodiimide to avoid multiple activation of DOTA carboxyl groups, but to maintain high NHSS levels to avoid rearrangement of the activated *O*-acyl urea. While this made little difference to the degree of incorporation obtained, the conjugates manufactured with low carbodiimide levels showed improved kinetic stability, possibly due to the absence of destabilising by-products.

Two-step methods using succinimide ester formation via DCC

Bauminger *et al.* (1973) activated prostaglandin E_2 with DCC/NHS before coupling to BSA. The activation was carried out in DMF (30 minutes, room temperature) before coupling to BSA in 0.1 M sodium bicarbonate (two hours, 4 °C). A molar incorporation ratio of around 15 was obtained: the initial acid: DCC:NHS:BSA ratio was around 79:81:193:1. Similar conditions were employed by Van Vunakis *et al.* (1975) to couple carboxylated phenylpiperidine drugs to haemocyanin.

Mattox *et al.* (1979) coupled cortisone 21-glucosiduronic acid to BSA using DCC/NHS: the steroid was activated with equimolar quantities of both reagents in DMF (one hour, 25 °C) before coupling to BSA (four hours, 25 °C, pH 6.5–7.0). Steroid excesses of 0.7, 2.0 and 6.0-fold over the number of protein amine groups gave incorporations of around 13, 33 and 43 respectively.

Kulkarni *et al.* (1981) prepared the succinimide ester of methotrexate by activation with equimolar quantities of DCC and NHS (DMF, one hour at room temperature then 18 hours at 4 °C, in the dark). After removal of the urea by filtration, the resulting solution (stable for at least 30 days at 4 °C) was used to derivatise IgG (four hours, 4 °C, in the dark, pH 7.1) at an excess of up to 170-fold. At this level an incorporation of around 18 was obtained. This was around threefold higher than that obtained with a one-step EDAC reaction using the same quantity of methotrexate.

The incorporation obtained increased with pH: at an 80-fold excess of the ester, coupling at pH 6.1 gave an incorporation of 5, while carrying out the reaction at pH 8.6 resulted in about 25 methotrexate molecules per IgG.

Schroeder (1982) coupled a carboxylated isoluminol derivative (CM-AHEI) to IgG by a two-step DCC/NHS protocol. CM-AHEI and NHS were dissolved in dry DMF under nitrogen, and the resulting solution cooled to −10 °C before addition of DCC. The activation was allowed to continue overnight at 0 °C, then the mixture was added to IgG in a phosphate–NaCl buffer (pH 8.0). Conjugation was for 2.5 hours at 4 °C followed by 30 minutes at 20 °C, and the resulting conjugates, purified by size exclusion, incorporated about seven isoluminol groups per IgG molecule. The initial CM-AHEI:DCC:NHS:IgG ratio for these conjugates was 60:69:60:1.

Schroeder noted that higher incorporation could be obtained, not unexpectedly, by a higher loading of the isoluminol derivative, or by a more alkaline pH; this however resulted in problems with loss of immunoreactivity of the IgG, loss of chemiluminescence, and increased non-specific binding in an immunoassay.

Buckley and Searle (1984) activated DTPA with DCC/NHS (overnight, DMSO) at a DTPA:DCC: NHS ratio of 1:0.9:1 before coupling to IgG (two hours, room temperature, pH 8.0). Increasing incorporation was apparent as the initial DTPA:IgG ratio was raised up to 15:1, and levelled off above this loading.

Wood *et al.* (1986) described general conditions for labelling proteins with similar carboxylated isoluminol derivatives. Activation was carried out using DCC and NHS in dry DMF (overnight, 4 °C, in the dark) before addition to the protein in 0.05 M phosphate buffer (pH 8.0). A carboxylated isoluminol:DCC:NHS:protein ratio of 20:30:20:1 was recommended, overnight conjugation at 4 °C in the dark being predicted to give incorporations of around 2–5 labels per protein molecule.

Carboxylated thyroxine (C-T4) was activated with DCC/NHS in DMF for three hours at room temperature (Sturgess *et al.*, 1986), and conjugated to rabbit IgG in PBS (room temperature, 30 minutes) before purification by size exclusion. An incorporation of two moles of thyroxine per IgG molecule was reported.

Gendloff *et al.* (1986) used equimolar quantities of DCC and NHS to couple carboxyl-containing mycotoxins to polylysine. The activation was carried out in dry tetrahydrofuran (30–60 minutes at room temperature) and the conjugation in a DMF–sodium bicarbonate mixture (30 minutes).

Sauer *et al.* (1986) activated progesterone 11-glucuronide with DCC/NHS for one hour at 20–22 °C in dry dioxane, before coupling to alkaline phosphatase in phosphate buffer (two hours, 20–22 °C, pH 7.0). The overall acid:DCC:NHS:protein ratio was 2:2.2:2.2:1; the incorporation of the resulting conjugate was not calculated but this product was used in a successful immnuoassay.

Yoon *et al.* (1993) coupled carboxyl-containing progesterone derivatives to HRP by a DCC/NHS protocol. The steroid was activated in DMF (room temperature, two hours) before coupling to HRP in a carbonate buffer, pH 8.8 (room temperature, one hour), yielding successful conjugates for immunoassay. The steroid:DCC:NHS:HRP ratio was 93:97:171:1.

Adamczyk *et al.* (1994) described typical conditions for DCC/NHS activation of carboxylic haptens, the acids were activated with a 1.1-fold excess of both DCC and NHS in DMF (24 hours), filtered, then coupled to BSA at a 38-fold excess (pH 7.8, overnight). Analysis of the resulting conjugates by mass spectometry showed molar incorporations between 5 and 27. Interestingly, however, some conjugates contained a small sub-population containing fully derivatised BSA with a molar incorporation of 60. The latter phenomenon was not fully discussed, but might reflect the presence of unfolded protein in the reaction mixture.

A generic two-step DCC/NHS procedure for coupling carboxylic molecules to proteins is described in Protocol 6.1.

Two-step methods using succinimide ester formation via diisopropylcarbodiimide

The only other commonly-encountered carbodiimide for these applications is diisopropylcarbodiimide. Paxton *et al.* (1985) used this reagent with NHS to activate DTPA (1.5 hours, 15 °C, acetonitrile) before coupling to monoclonal antibodies (one hour, pH 7.0). The carbodiimide:NHS:DTPA ratio was 0.5:0.5:1, limiting the degree of activation of the pentacarboxylic acid and discouraging cross-linking of the protein.

Varadarajan and Hawthorne (1991) coupled carboxyl-containing carboranylpeptides to monoclonal antibodies by carbodiimide activation in the presence of NHS or NHSS. Water-soluble carbodiimides caused precipitation problems due to ionic interactions, while DCC solutions in DMF separated from the aqueous phase necessary for solubility of the carboranes. Diisopropylcarbodiimide was found to avoid these problems, solutions in DMF remaining miscible with aqueous buffers.

With one peptide, a peptide:carbodiimide:NHS: Mab ratio of 32:33:62:1 was employed; the activations were carried out (one hour, room temperature) with the peptide dissolved in 0.1 M phosphate buffer, pH 4.35, NHS in water and the carbodiimide in DMF (solvent ratio 80:18:8). Coupling in 50 mM phosphate, pH 7.4 or 100 mM carbonate–bicarbonate, pH 9.3 (one hour, room temperature) gave incorporations of 8.5 and 9.4 respectively.

Modifications to the experimental conditions were examined with a second peptide: broadly speaking, coupling in the phosphate or carbonate–bicarbonate buffers just described appeared to be superior to 0.1 M HEPES, pH 8.2 or 0.1 M sodium bicarbonate, pH 8.5.

Paxton *et al.* (1992) coupled carboranylpeptides to monoclonal antibodies using 1,3-diisopropylcarbodiimide/NHSS. The peptide activation was carried out in a phosphate buffer (pH 4.35), the diimide was added in DMF, and the reaction allowed to proceed for 55 minutes at room temperature. The activated peptide was then coupled to the Mab in a bicarbonate buffer (pH 8.5, 30 minutes at room temperature), the overall peptide:diimide:NHSS:IgG ratios used were 15:150:220:1 and 10:100:140:1 in two separate preparations. Successful incorporation was indicated by immmunoassay results, but large amounts of aggregate were formed. This was attributed to disruption of the Mab structure by the ambiphilic carboranylpeptide.

6.1.3 *In situ* active ester formation via other reagents

6.1.3.1 Disuccinimidyl carbonate

N,N′-Disuccinimidyl carbonate (DSC) can be used to produce the succinimide esters of carboxylic acids (Figure 6.9). Ogura *et al.* (1979) described the preparation of 80 to 100% yields of succinimides from the reaction of amino acids with equimolar quantities of DSC and pyridine in acetonitrile.

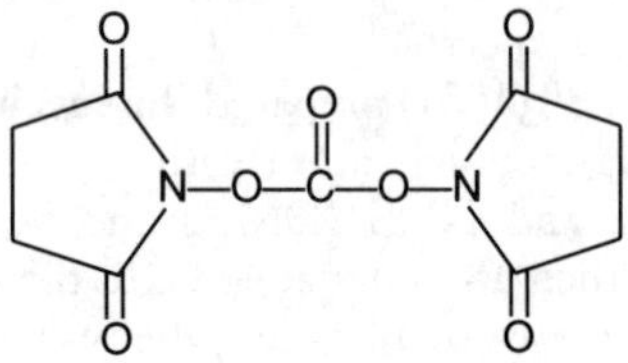

Figure 6.9 *N,N′*-Disuccinimidyl carbonate (DSC).

6.1.3.2 Phosphonium and carbonium compounds

Rather analogous to *N*-hydroxysuccinimide is 1-hydroxybenzotriazole (HOBt; Figure 6.10(a), and the formation of benzotriazole esters has been widely employed in peptide synthesis since its introduction (König and Geiger, 1970)[1].

Derivatives of this compound based on phosphonium and uronium salts were subsequently developed for use in this field, which provided a convenient means of esterification, notably benzotriazol-1-yl-oxy-tris (dimethylamino)phosphonium hexafluorophosphate (BOP; Figure 6.10(b); Castro *et al.*, 1975; Fournier *et al.*, 1988) and *O*-benzotriazolyl-*N,N,N′,N′*-tetramethyluronium hexafluorophosphate (HBTU; Figure 6.10(c), Dourtoglu *et al.*, 1978; Knorr *et al.*, 1989b).

These compounds contain a positively charged carbon or phosphorus atom which behaves analogously to the electron-deficient carbon in carbodiimides, and release HOBt when this group reacts with carboxylates. They therefore represent a 'one-molecule' parallel to carbodiimide–HOBt activation.

With BOP, for example (Figure 6.11), the carboxyl group is converted to an acyloxyphosphonium salt, which can then react with HOBt[2] to produce a

[1] The introduction of unnatural groups into synthetic peptides has been achieved using this chemistry. Sluka *et al.* (1987), attached tribenzyl-EDTA to a 52-residue peptide by DCC/HOBt activation.

[2] Excess HOBt can be introduced to supplement that released by the activation reaction.

benzotriazolyl ester, or with further carboxylate to yield a symmetrical anhydride (Hudson, 1988). Either of these products will react efficiently with amine-containing species to produce the desired peptide link.

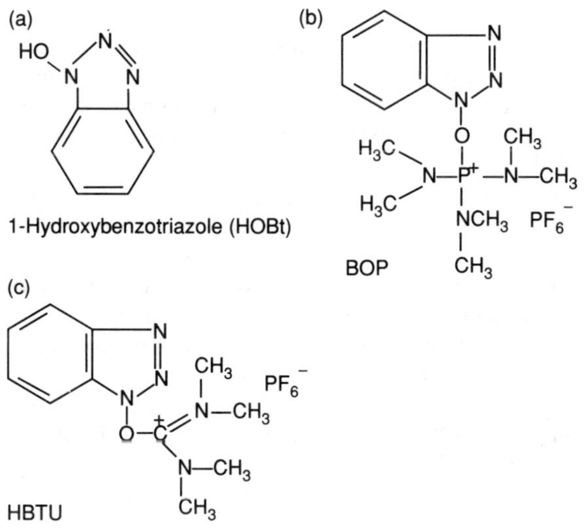

Figure 6.10 1-Hydroxybenzotriazole and phosphonium and uronium derivatives.

Benzotriazolyl esters are too reactive to be used conveniently in aqueous systems, and there are also toxicity issues with BOP and its analogues due to the possible release of hexamethylphosphoric triamide (HMPA), a potent carcinogen. However, succinimide equivalents have been developed which can act as alternatives to carbodiimide–succinimide activation, notably *N,N,N′,N′*-tetramethyl(succinimido)uronium tetrafluoroborate, (TSTU: Figure 6.12; Bannwarth *et al.*, 1988; Knorr *et al.*, 1989a, 1991). Bannwarth *et al.* (1988) formed the succinimide esters of ruthenium-bathophenanthroline carboxylic acids by using a 10% excess of TSTU in DMF (two hours, room temperature). Bannwarth and Knorr (1991) produced biotin-NHS by the reaction of biotin, TSTU and diisopropylethylamine (ratio 1:1.25:3) in a DMF–dioxane–water (2:2:1) mixture. Complete reaction was achieved in ten minutes.

These more modern reagents for carboxyl activation have already gained widespread acceptance in the

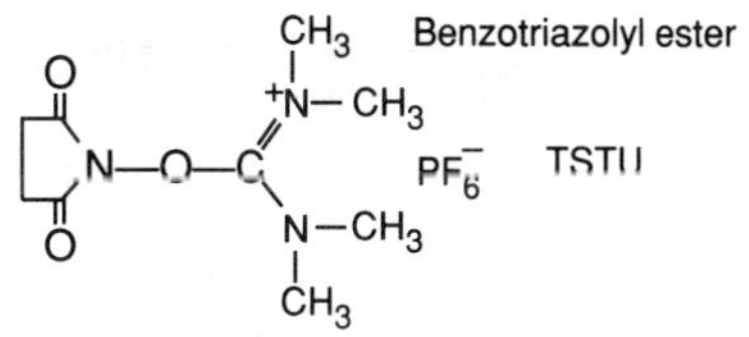

Figure 6.12 Succinimide equivalent TSTU to HBTu.

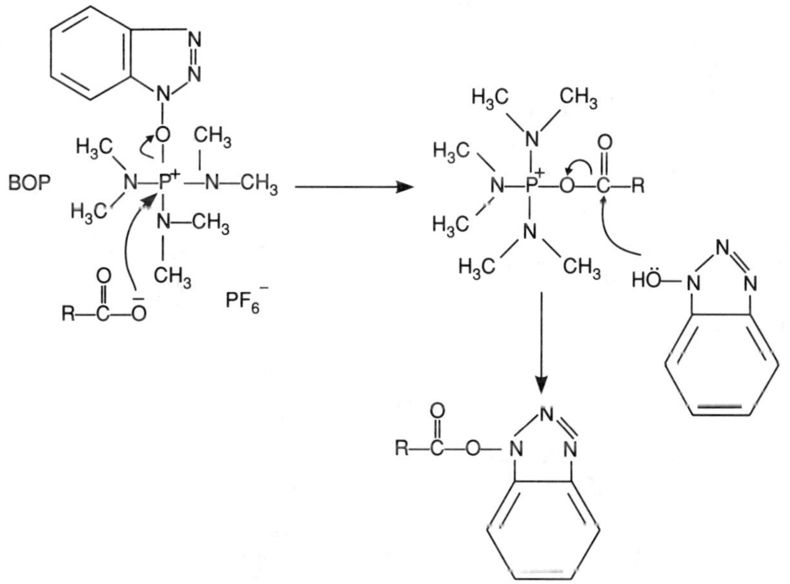

Figure 6.11 Reaction of carboxylate with BOP to produce an acyloxyphosphonium salt; this will react with HOBt released from the reaction to yield an amine-reactive benzotriazolyl ester.

peptide synthesis field, and are widely viewed as possessing superior properties to carbodiimides. There seems no reason why they should not achieve similar popularity for the non-aqueous activation of carboxyl-bearing compounds for coupling to proteins, as an alternative to DCC.

6.1.3.3 Oxazolium compounds

The reactions of *N*-alkylisoxazolium salts were investigated by Woodward and Olofson (1961), and these studies resulted in the introduction (Woodward *et al.*, 1961) of *N*-ethyl-5-phenylisoxazolium-3′-sulfonate (Woodward's reagent K; Figure 6.13) for activation of carboxyl functions.

Under alkaline conditions the isoxazolium ring opens to yield an electrophilic ketoketenimine (Figure 6.13(a)), rather analogous to a carbodiimide. Following attack by the carboxylate anion, a rearrangement occurs to produce an enol ester (Figure 6.13(b)), which is sufficiently activated for reaction with amines. Other groups may react with Reagent K, such as thiols and amines, but the products of these reactions are reversible by borohydride treatment (Puri and Colman, 1996).

Subramanian and Wolf (1990) used Woodward's Reagent K to couple triethylenetetraminehexaacetic acid (TTHA) to human serum albumin (overnight, 22 °C, pH around 7.5). The TTHA:Reagent K:HSA ratio was 10:30:1. A molar incorporation ratio of 3.9 was obtained.

Any isoxazolium salt bearing a hydrogen atom on the 3-carbon (adjacent to the nitrogen) should be capable of creating a ketoketenimine. Kemp and Woodward (1965) studied *N*-ethylbenzisoxazolium salts, which yield more reactive ketenimines than the compunds described above. *N*-Ethylbenzisoxazolium tetrafluoroborate was consequently proposed as a reagent for activation of carboxylates towards nucleophiles.

Goodfriend *et al.* (1966) used an 8.5-fold excess of this reagent to activate the oligopeptide angiotensin for coupling to polylysine (two days, room temperature, pH 8.5). An incorporation of 13 peptide groups per polylysine molecule (17 kDa) was obtained. Bodlaender *et al.* (1968) used these compounds to inactivate trypsin at pH 4.75.

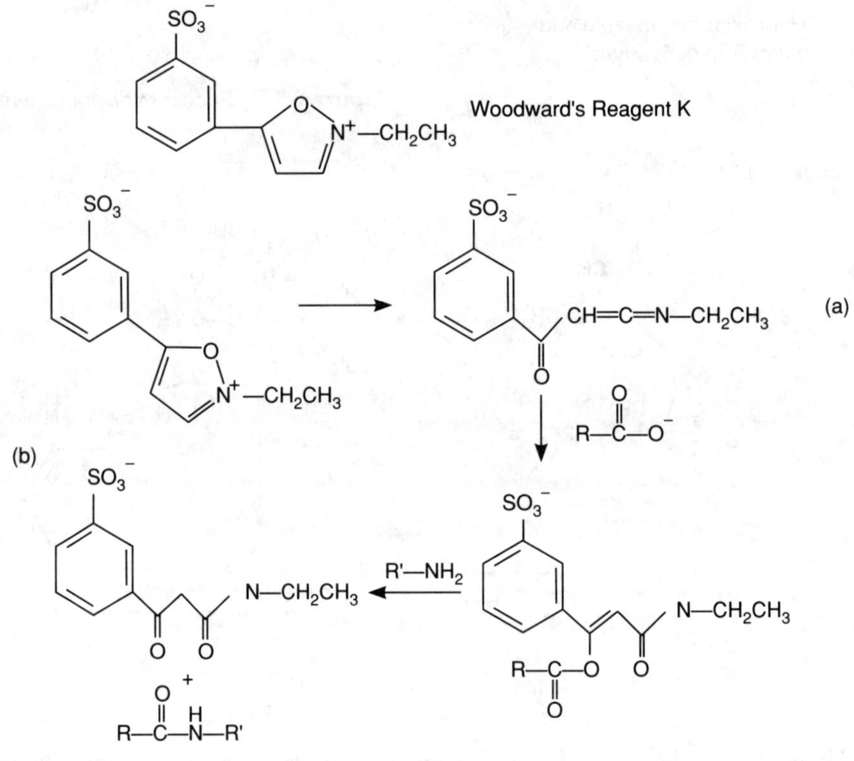

Figure 6.13 Woodward's reagent K and its reactions with acids and amines under alkaline conditions.

6.1.4 Active esters

Sections 6.1.2 and 6.1.3 described methods in which a succinimide ester is generated from a carboxyl group *in situ*, facilitating reaction with nucleophiles. With small molecules that can be handled in non-aqueous systems, it is quite possible to isolate a solid, stable succinimide ester from such a reaction. This may be done relatively crudely, for example by lyophilising the entire product of DCC/NHS activation (Wood *et al.*, 1986), or by isolating pure succinimide ester. In either case the solid ester can be used in a very convenient one-stage reaction to couple to protein amine groups.

6.1.4.1 Succinimide esters

The convenience of succinimide esters, whose chemistry is discussed in Chapter 5, Section 5.3.5, has made their use one of the most popular tools in protein chemistry. A wide array of these esters are sold commercially, including many of the bifunctional reagents discussed in Chapter 5; their stability is good provided they are protected from moisture.

There is very little difference in the conditions employed for derivatisation with monofunctional succinimide esters: 30 to 60 minutes reaction at around neutral pH and room temperature is typical. If a pure succinimide ester is available this approach offers one of the cleanest and most reproducible routes to the incorporation of small molecules into proteins.

There are few serious pitfalls with this methodology, but one was reported by Goon *et al.* (1994). A common approach to the introduction of a carboxyl function into an amine-bearing small molecule is to derivatise it with succinic anhydride, yielding the succinamyl or hemisuccinate group (Figure 6.14(a)). Again, a common means of activating such species towards reaction with proteins is to produce the succinimide ester, by the DCC/NHS route for example (Figure 6.14(b)). Goon, however, observed poor yields in one such activation, and subsequently determined that the carbodiimide was in fact facilitating cyclisation of the succinamyl group to produce an unreactive succinimide (Figure 6.14(c)). By using the homologous glutaramyl group, containing an extra CH_2 unit, Goon *et al.* observed good yields, cyclisation to a six-membered ring being less favourable.

The reverse of this cyclisation reaction, ring-opening of stable succinimides by treatment with a nucleophile, has been employed as a means for the introduction of carboxyl residues, for example into sugars (Andersson and Oscarson, 1992).

Succinimidyl esters of a number of common fluorophores are commercially available and some of these are shown in Figure 6.15. Apart from variations in absorbance and emission properties, some of these exhibit interesting characteristics. When bound to proteins, long-chain esters such as SFX may suffer less from quenching than their shorter analogues; fluorescein diacetates such as CFSE are non-fluorescent until

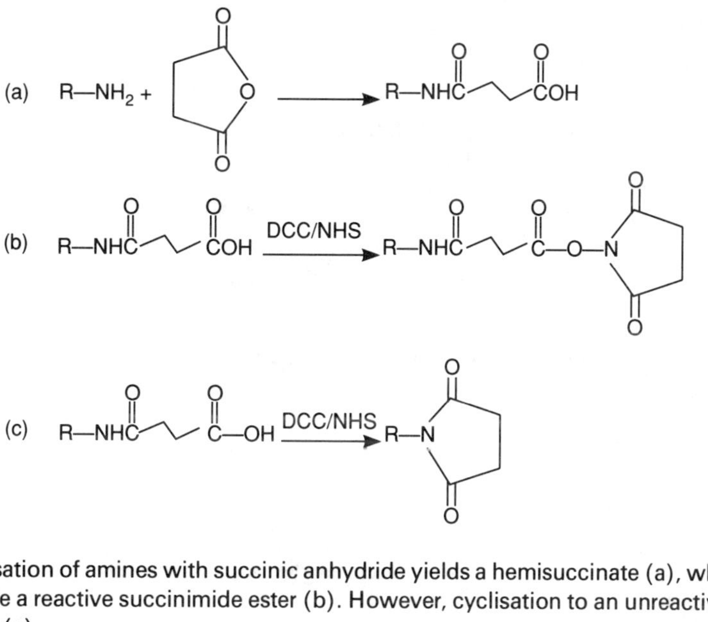

Figure 6.14 Derivatisation of amines with succinic anhydride yields a hemisuccinate (a), which can be activated by DCC/NHS to provide a reactive succinimide ester (b). However, cyclisation to an unreactive alkyl succinimide is a competing reaction (c).

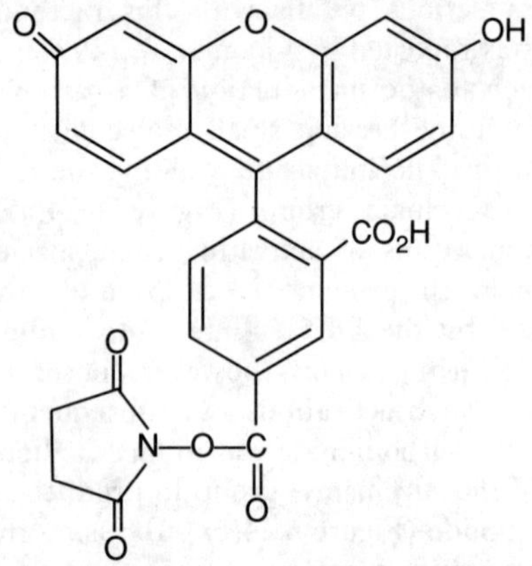

5-Carboxyfluorescein succinimidyl ester

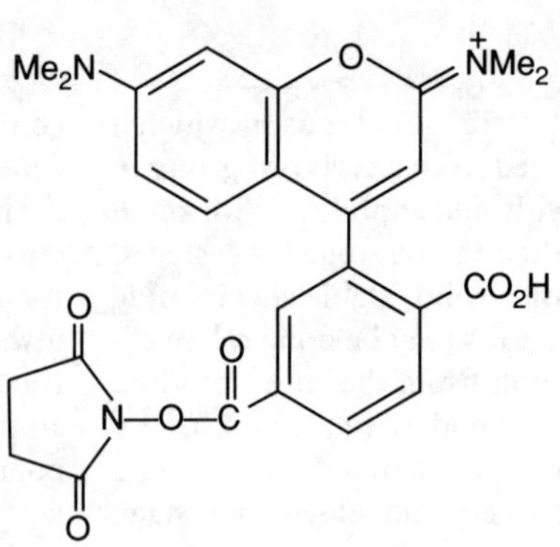

6-Carboxytetramethylrhodamine succinimidyl ester

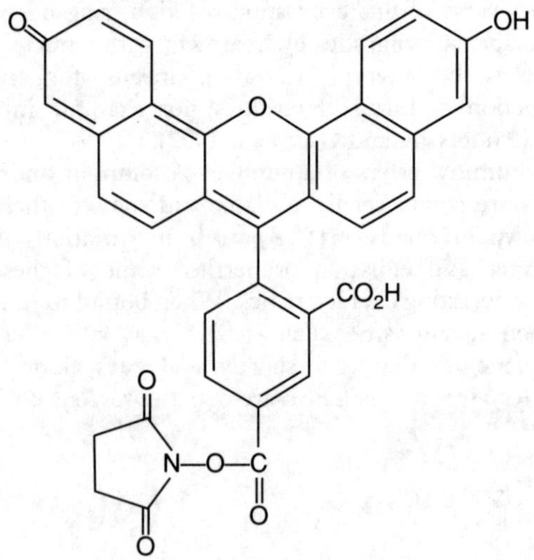

5-Carboxynaphthofluorescein succinimidyl ester

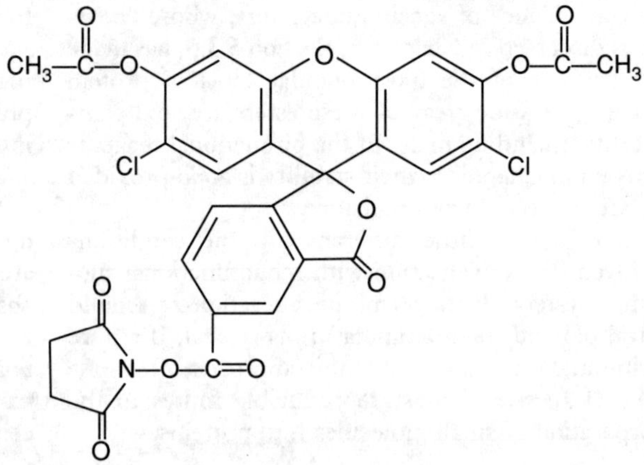

5-Carboxy-2',7'-dichlorofluorescein diacetate succinimidyl ester

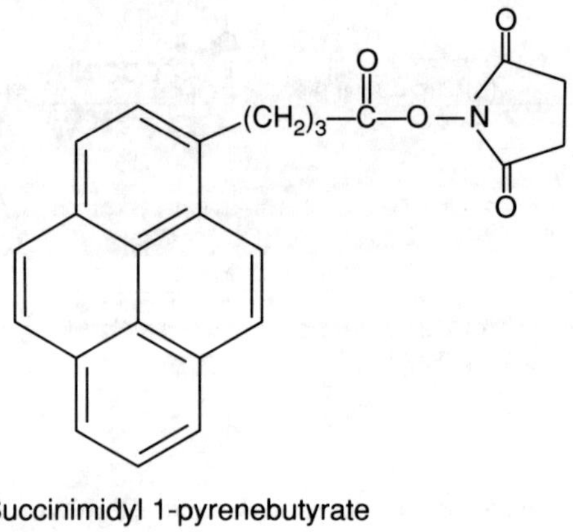

Succinimidyl 1-pyrenebutyrate

Figure 6.15 Succinimidyl esters of common fluorophores.

the acetate groups are removed, for instance by intracellular esterases, and can therefore show selective intracellular fluorescence; similarly, dihydrofluoresceins are non-fluorescent unless oxidised, and can be used to detect the oxygen burst of phagocytosis; NBD derivatives are essentially non-fluorescent until bound to hydrophobic moieties; pyrenebutyrate has a long fluorescent lifetime making it useful for fluorescence polarisation experiments; the DABCYL group is non-fluorescent but is a broad-range energy acceptor useful for quenching studies.

CFSE has been shown to display both faster derivatisation and a more stable product than the commonly encountered FITC, when used for fluorescein labelling of proteins (Banks and Paquette, 1995).

Hosoda *et al.* (1979a) and Nambara and Hosoda (1981) prepared succinimide esters of a number of steroids (using a 40% excess of EDAC and NHS in dioxane), isolating them before coupling to BSA. The conjugation reaction was carried out (twelve hours, 4 °C) in a mixture of 50 mM phosphate buffer (pH 7.4), and various organic co-solvents (pyridine, dioxane, DMF). The use of pyridine gave a slightly higher incorporation than dioxane or DMF. Incorporation also varied from one steroid ester to another, with some evidence for better incorporation where there was a longer spacer between the steroid and the ester group. Using various steroid:BSA ratios they obtained the incorporations shown in Table 6.2.

Workers from the same group (Hosoda *et al.*, 1979b; Niwa *et al.*, 1994) prepared steroid-β-galactosidase conjugates by similar chemistry (using dioxane as co-solvent) but these were not well characterised.

Ullman *et al.* (1979) conjugated a succinimide ester derivative of thyroxine to malate dehydrogenase at pH 9.1, and observed over 90% inhibition of enzyme activity with an incorporation of two. This inhibition could be reversed by anti-thyroxine antibody, engendering one variation of Syva's homogeneous EMIT™ immunoassays.

Table 6.2 Molar incorporations of steroids obtained from EDAC/NHS coupling to BSA, at different input ratios

Steroid : BSA	Incorporation
60:1	22–33
20:1	12–18
10:1	5.9–9.0
5:1	3.0–4.6
3:1	2.8

Eshhar *et al.* (1981) coupled the succinimide ester of a progesterone metabolite to BSA (30 minutes, 0.13 M sodium bicarbonate solution). The ester was used as a DMF solution: an initial excess of around 70:1 resulted in the incorporation of eleven steroids per albumin molecule.

Gadow *et al.* (1984) described the synthesis of the succinimide ester of a carboxylated isoluminol derivative, which was used at ratios between 30 and 120:1, for luminescent labelling of IgG (24 hours, 4 °C, pH 7). The 60:1 conjugate gave the best immunoassay performance. Strasburger and Kohen (1990) coupled the same compound to streptavidin at a 75-fold excess (overnight, 4 °C, pH 8.3), obtaining an incorporation of 11–13.

Weeks *et al.* (1983, 1986) derivatised a mouse Mab and polyclonal guinea-pig IgG with the succinimide derivative of a luminescent acridinium ester, 4-(2-succinimidyloxycarbonylethyl)phenyl-10-methylacridinium-9-carboxylate fluorosulfonate (Figure 6.16). At pH 8.0 in a phosphate–NaCl buffer, reaction was 95% complete within ten minutes. A range of initial ratios were employed with the two antibodies, resulting in the incorporations shown in Figure 6.17. There was no apparent loss of chemiluminescence in the range shown, but ratios above about 1: 1 resulted in protein

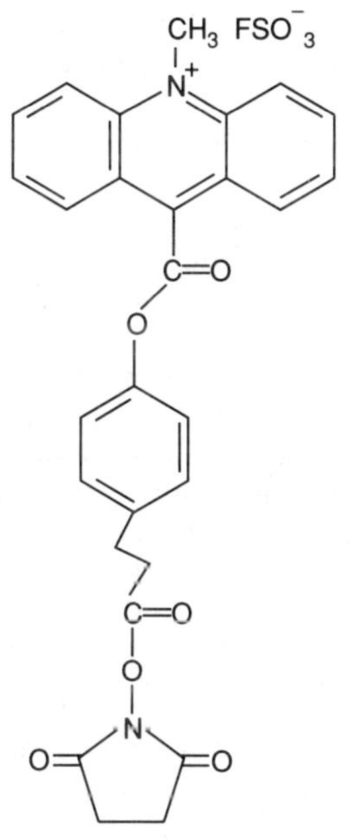

Figure 6.16 4-(2-Succinimidyloxycarbonylethyl)phenyl-10-methylacridinium-9-carboxylate fluorosulfonate.

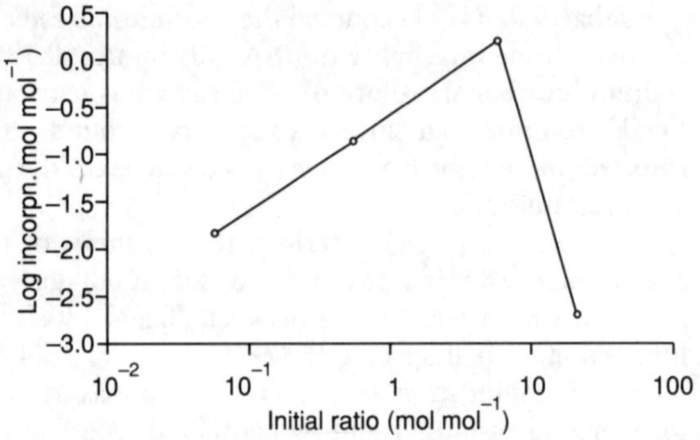

Figure 6.17 Incorporation of acridinium succinimide esters into IgG from a range of initial ratios. (Reproduced with permission from: Weeks (1983). *Methods Enzymol.* **33**, 366–387.)

precipitation. A 21:1 ratio resulted in the loss of 90% of the guinea-pig IgG.

Eskola *et al.* (1985) coupled a succinimide ester of cortisol to ovalbumin at a 1.3:1 ratio: the ester was dissolved in dioxane and coupled to the protein at pH 7.3 for 20 hours at 4 °C, resulting in an incorporation of about 1.

Hummelen *et al.* (1986) recommended general conditions for labelling proteins with succinimide esters of 1,2-dioxetanes: the protein was allowed to react for one hour at 25 °C in 100 mM borate buffer (pH 8.5), with a solution of the active ester in dioxane. It was recommended that the dioxane content of the solution was kept at 5–10% for antibodies, to avoid loss of immunoreactivity.

Succinimidyl 2,2,5,5-tetramethyl-3-pyrroline-1-oxyl-1-carboxylate (Figure 6.18) is a useful reagent for spin-labelling experiments, and is commercially available.

Dillman *et al.* (1988) prepared an acid-labile succinimide ester from daunorubicin by reacting a free amine with *cis*-aconitic anhydride and attaching a succinimide group to the resulting carboxylate *via* DCC/NHS chemistry. This active ester was coupled to an aminated Mab (18 hours, room temperature, PBS) giving an incorporation of 25 with minimal loss of immunoreactivity.

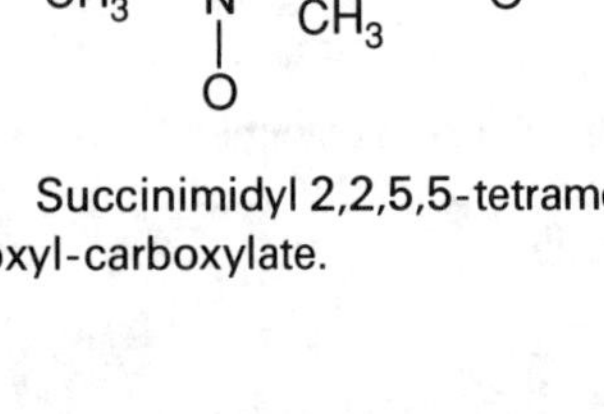

Figure 6.18 Succinimidyl 2,2,5,5-tetramethyl-3-pyrroline-1-oxyl-carboxylate.

Huber *et al.* (1991) employed a sequential addition approach for the preparation of steroid–HRP conjugates. For example, testosterone-3-dimethylcarboxymethyloxime succinimide ester was dissolved in DMSO and added to a peroxidase solution in six aliquots over approximately 2.5 hours. The reaction mixture (in 0.1 M phosphate buffer (pH 8.0)–DMSO, approximately 1:1) was stirred overnight at 4 °C.

McAdam *et al.* (1992) coupled succinimide derivatives of the pesticide fenitrothion to proteins. Ratios of 25:1 for chicken IgG and ovalbumin and 40:1 for HRP were employed, and the reaction was carried out overnight at 4 °C in a phosphate buffer (pH 9.1). The fenitrothion derivative was dissolved in DMF.

Using succinimide esters of sulfoindocyanine dyes, Mujumdar *et al.* (1993) studied the effect of pH on rates of hydrolysis and reaction with IgG: hydrolysis became much more rapid above pH 7.5 (Figure 6.19), but the correspondingly faster aminolysis resulted in virtually complete reaction within minutes (Figure 6.19). The authors therefore recommended pH 9.4 for this reaction.

A succinimide ester which has been used particularly widely for radiolabelling of proteins is the Bolton–Hunter reagent, *N*-succinimidyl 3-(4-hydroxy-3-[^{125}I]iodophenyl)propionate (Figure 6.20) (Bolton and Hunter, 1973; Rudinger and Ruegg,

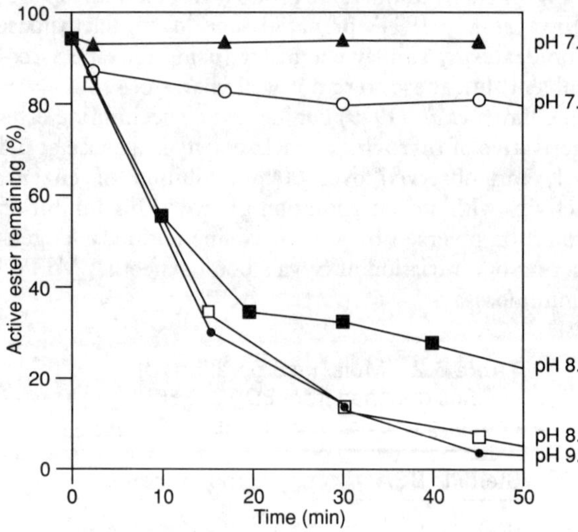

Figure 6.19 Stability of an active ester in aqueous solution under a range of pH conditions. (Reproduced with permisson from: Mujumdar *et al.* (1993). *Bioconjugate Chem.* **4**, 105–111.)

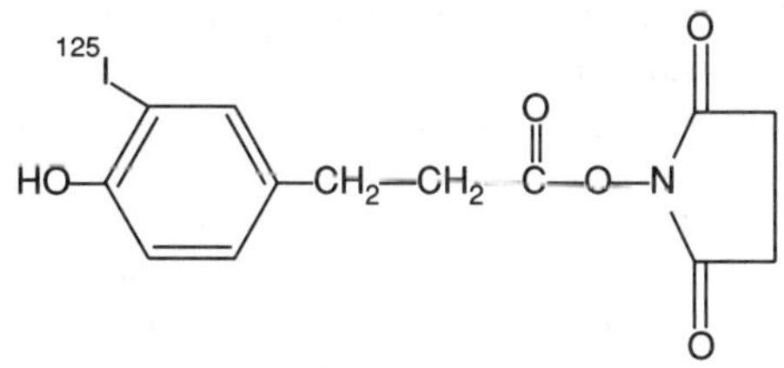

Figure 6.20 Bolton–Hunter reagent.

1973). Conditions of use are typical for succinimide esters: O'Kennedy and Keating (1993), for example, coupled Bolton–Hunter reagent in DMF to IgG in 0.1 M borate buffer (pH 8.5), for two hours on ice before purification: incorporations of up to 23 labels per IgG molecule were obtained.

Analogues of the Bolton–Hunter reagent have been described, including some derived from destannylation of organotin precursors (Vaidyanathan *et al.*, 1997). See also Section 6.21.4.

6.1.4.2 Other active esters

As discussed in the previous chapter, a wide variety of active esters have been examined over the years: despite the current predominance of succinimides, some other types have found their way into protein derivatisation. *p*-Nitrophenyl esters are probably the commonest category: biotin nitrophenyl esters, for example, were used (Becker *et al.*, 1971; Bayer *et al.*, 1975) as affinity labels for the biotin transport system of yeast cells. Like succinimide esters, these compounds can occasionally react with active hydroxyl groups as well as amines, e.g. the well-studied reaction of *p*-nitrophenyl esters with the active-site serine of chymotrypsin (Balls and Wood, 1956; Oosterbaan *et al.*, 1962).

Phenolic esters continue to be employed as alternatives to those based on succinimides in this application—for example those derived from pentafluorophenol (Schubiger *et al.*, 1996), 4-(methylthio) phenol (Bedel-Cloutour *et al.*, 1996) and *o*-nitrophenol (Safavy *et al.*, 1997).

6.1.5 *In situ* mixed anhydride formation

Chloroformates (chlorocarbonates) react with carboxyl groups to yield mixed anhydrides, which are sufficiently activated for reaction with nucleophiles such as amines (Figure 6.21); this reaction is sometimes described as the Schotten–Baumann reaction. Chloroformates react readily with water, so the activation is most efficiently carried out in an inert organic medium: an organic base is normally used to neutralise the hydrochloric acid formed during the reaction. The anhydride products are hydrolysed sufficiently slowly that they can be used to derivatise proteins in an aqueous environment.

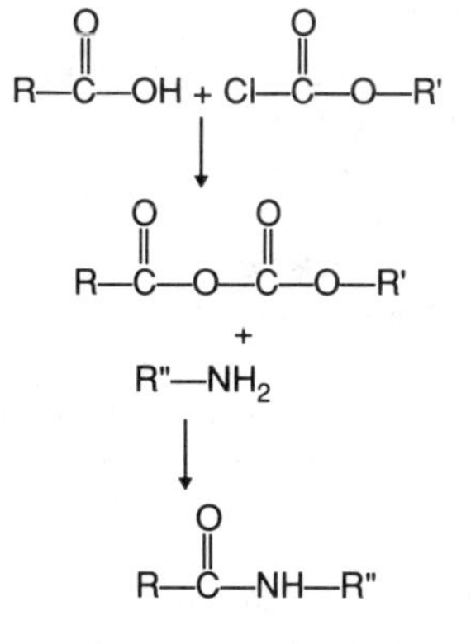

Figure 6.21 Reaction of chloroformates to form mixed anhydrides followed by reaction with amines to form amides (Schotten–Baumann reaction).

Apart from the obvious danger of crosslinking, the presence of amine groups in the compound to be activated can also lead to the formation of carbamate esters (Adams and Baron, 1965).

Emery and Gold (1950) investigated the factors affecting mixed anhydride reactivity under various conditions, particularly with respect to controlling which of the two different acyl groups is dominant in reaction with amines. As might be expected, both steric hindrance and electron donation from neighbouring groups reduce the attractiveness of a carbonyl carbon to nucleophilic attack.

In terms of activation of carboxyl compounds for protein derivatisation, it is clearly desired to minimise the amount of acylation from the chloroformate carbonyl. In keeping with the factors noted above, reagents containing branched alkyl groups such as *sec*-butyl and isobutyl chloroformate (IBCF) proved to give the best results[1] (Vaughan and Osato, 1951), although up to 8% of the secondary product could still be observed with the latter reagent (Bodanszky and Tolle, 1977).

Vaughan and Osato (1952) optimised the conditions for carboxyl activation with *sec*-butyl chloroformate. The reaction was quite temperature-sensitive, being

[1] Although alkyl chains offering even greater bulk and electron donation (e.g. *t*-butyl) are attractive on selectivity grounds, the steric hindrance is so great that the formation of the mixed anhydride becomes inefficient (Bodanszky and Tolle, 1977). IBCF appears to offer the best compromise and has remained the most popular reagent of this type for over 40 years.

inefficient below −20 °C or above 15 °C. At −5 °C the anhydride formation was complete within 30 minutes; overnight standing was recommended for the aminolysis step. Moisture in the activation solvents caused a 10–15% lowering of yield.

Under polar conditions the selectivity of acylation is lessened (Emery and Gold, 1950; Vaughan and Osato, 1951) and the formation of secondary products is quite likely. Gendloff *et al.* (1977) warned of the problems of unexpected isobutyl epitopes which could be introduced by using IBCF activation in the preparation of immunogens.

At low temperatures only lysine and tyrosine residues in the protein are likely to react with mixed anhydrides, and tyrosines may not always do so (Vaughan and Osato, 1952). Phenolic esters formed from acylation of tyrosine are readily cleaved, by hydroxylamine for example (Riordan and Vallee, 1972b).

6.1.5.1 Activation with isobutyl chloroformate

Erlanger *et al.* (1957, 1959, 1967) coupled carboxysteroids to BSA using IBCF: the steroid and tri-*n*-butylamine were dissolved in dioxane and the solution cooled to 4–11 °C before addition of the chloroformate. After 20–35 minutes at this temperature, the resulting mixed anhydride solution was added to BSA in a basic aqueous dioxane mixture: the pH was maintained between 7 and 9.5 over a total reaction period of 4–4.5 hours. Steroid:BSA ratios of between 40 and 60:1 were employed, and the ratios of tributylamine, IBCF and steroid were generally 1:1:1 or 2:1:1. Incorporations of around 20–35 steroid groups per BSA molecule were obtained. The methods described by Erlanger have seen very little change up to the present day.

Oliver *et al.* (1968) used virtually identical conditions to couple a carboxylated digitoxigenin derivative to BSA, obtaining an incorporation of at least 13.

Van Weemen and Schuurs (1972) used these conditions to produce oestrogen–BSA conjugates, an initial oestrogen excess of 60:1 resulted in incorporations of 25–32.

Wainer *et al.* (1972) coupled morphine-3-hemisuccinate to BSA by the mixed anhydride method. The hemisuccinate was activated by equimolar amounts of isobutyl chloroformate and tributylamine in dioxane (30 minutes, below 20 °C), then coupled at a 100:1 excess to BSA in a 1:1 water–dioxane mixture at pH 8.5 (four hours, 5 °C). This resulted in the incorporation of around 6.5 morphine molecules per BSA.

Rowley *et al.* (1975) activated a carboxymethylmorphine derivative with equimolar IBCF (one hour, −5 °C, DMF) before coupling to malate dehydrogenase (pH 8.8–9.0, 30 minutes at 0 °C, 30 minutes at room temperature). A 65-fold excess of mixed anhydride gave a morphine incorporation of 22, of which around four could be removed by hydroxylaminolysis. While the incorporation of morphine (particularly at lower levels) gave only moderate inactivation of the enzyme, complexing to anti-morphine antibody resulted in much greater losses (Rubenstein *et al.*, 1972). This formed the basis of one form of the Syva EMIT™ homogeneous immunoassay.

Similarly, Rubenstein and Ullman (1974) used a roughly 110-fold excess of carboxymethylmorphine to obtain an incorporation of 47 in a BSA immunogen; numerous other mixed anhydride conjugations of opiates to enzymes are described in this patent. Unusually these preparations by the Syva group did not employ a base to neutralise acid formation.

Quinn *et al.* (1979) activated a carboxyl-containing antiviral guanine derivative with IBCF/triethylamine at room temperature for one hour, then coupled the mixed anhydride to rabbit serum albumin, RSA (overnight, room temperature, pH 8.15) and obtained a molar incorporation ratio of 6.3. The acid:RSA ratio used was 200:1.

Krejcarek and Tucker (1977) coupled the penta-(triethylammonium) salt of DTPA to HSA using IBCF: the pentavalent acid was activated with a fivefold excess of IBCF (ice-bath, 30 minutes) before coupling to the protein (overnight, 4 °C, pH 7.0–8.0). Khaw *et al.* (1980) and Halpern *et al.* (1983) used the same approach to label Mabs and Fab fragments with DTPA.

Mattox *et al.* (1979) used IBCF to couple cortisone 21-glucosiduronic acid to BSA. The acid was activated with equivalent quantities of IBCF and tributylamine in DMF (10 minutes, ice-bath) before coupling to BSA (four hours, ice-bath, pH around 8). Acid excesses of 0.7, 2.0, 6.0 and 12.0-fold over the available amine content gave incorporations of 3, 17, 19 and 30 respectively, and the resulting conjugates were stable on storage.

Lau *et al.* (1981) used identical conditions to produce an aflatoxin–BSA immunogen with an incorporation of 10–12. The aflatoxin acid:IBCF:TEA:BSA ratio was 60:100:100:1.

Gendloff *et al.* (1986) used IBCF to couple carboxyl-containing mycotoxins to polylysine: the activation was carried out in the presence of triethylamine in dry tetrahydrofuran (20 minutes at −5 °C) before conjugation in a 2:1 water–pyridine mixture (30 minutes at 4 °C then overnight at room temperature). These authors observed problems with the incorporation of isobutyl groups into the polylysine, due to the side-reaction discussed earlier.

Niwa *et al.* (1994) coupled a cholic acid derivative to BSA using IBCF/TBA (acid:IBCF:TBA:BSA ratio 50:115:115:1) and obtained an incorporation of 13. The activation was carried out in DMF (30 minutes, ice-cooling) and the coupling in a basic DMF–water mixture (three hours, ice-cooling).

Ferrua *et al.* (1994) used only ten minutes activation time to activate a thymidine analogue with IBCF/TBA (dioxane, 0 °C) before coupling to BSA and HRP (two hours, 4 °C or room temperature, pH 8–9): the conjugates were not characterised, however.

A generic procedure for coupling carboxylic molecules to proteins using IBCF is described in Protocol 6.2.

6.1.5.2 Activation with other chloroformates

Since the earliest experiments on the mixed anhydride method, *sec*-butyl chloroformate demonstrated similar properties to IBCF, and has often been used in an identical fashion. Following the IBCF conditions described by Erlanger *et al.* (1967), for example, Dawson *et al.* (1978) obtained an incorporation of 20 estradiol moieties in a BSA conjugate, from an initial 60-fold excess.

Jaffe *et al.* (1971) used equimolar quantities of ethyl chloroformate (ECF) and triethylamine in dioxane to activate prostaglandins (15 minutes) for coupling to KLH (one hour, 0.1 M sodium bicarbonate).

Shimada *et al.* (1984) used ECF and triethylamine to couple digoxin analogues to BSA; the steroid:ECF: triethylamine:BSA ratio was around 35:525:525:1, and incorporations of 11–15 were obtained.

Faulstich and Fiume (1985) coupled fungal toxins to BSA using ECF and *N*-methylmorpholine in DMF (15 minutes, −15 °C): the toxin:ECF:morpholine:BSA ratio was 20:22:22:1, and incorporations of 1 and 8 were obtained for toxins bearing short and long carboxyl-bearing chains respectively. There was no evidence for the formation of phenolic esters.

6.1.5.3 Activation with other reagents

Mixed anhydrides can be formed, rather less efficiently than with chloroformates, by heating an acid with a symmetrical acid anhydride (Figure 6.22). Thus Burstein and Knapp (1977) heated methotrexate with a large excess of acetic anhydride (100 °C, 30 minutes) before coupling to γ-globulin (18 hours, room temperature, pH 8.5). Very high molar incorporations (up to 120) were reported, possibly due to polymerisation

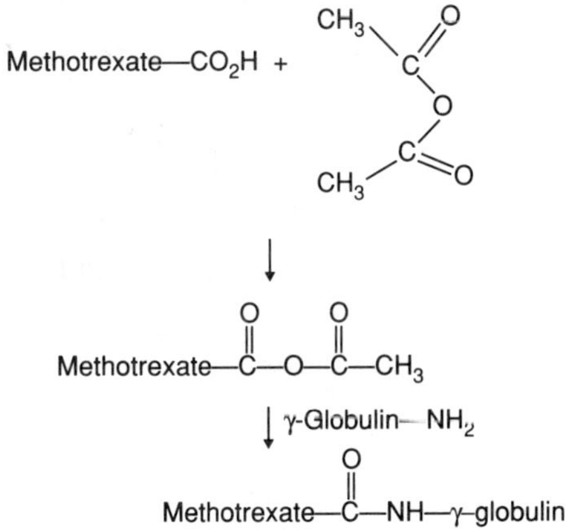

Figure 6.22 Formation of mixed anhydride of methotrexate with acetic anhydride followed by reaction with γ-globulin.

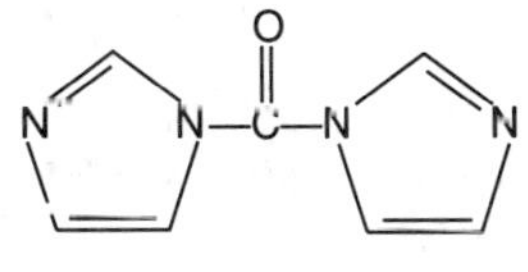

Figure 6.23 1,1′-Carbonyldiimidazole (CDI).

of the methotrexate. Kulkarni *et al.* (1981), however, could only obtain methotrexate incorporations of 3 by the same approach.

Another route to mixed anhydride formation employs 1,1′-carbonyldiimidazole (CDI; Figure 6.23), which yields an imidazolyl anhydride. Like many other reagents in this section, CDI has its roots in peptide synthesis (Paul and Anderson, 1962). It was adapted for coupling prostaglandins to BSA by Axen (1974); in an unoptimised procedure, prostaglandin E_2 was activated with CDI in DMF (20 minutes) before coupling to BSA (five hours, DMF/water).

Mattox *et al.* (1979) used an equimolar quantity of CDI to activate cortisone 21-glucosiduronic acid in DMF (20 minutes, 25 °C) before coupling the resulting imidazolyl anhydride to BSA (four hours, 25 °C). The presence of 85mM sodium dodecyl sulfate was required to keep the reactants in solution. Steroid excesses of 0.7, 2.0, 6.0, and 12.0-fold over available amine groups resulted in incorporations of around 7, 16, 37 and 49 respectively.

Unlike other coupling routes (including IBCF and DCC/NHS), however, some of the conjugate was unstable: after 18 hours at 4 °C, approximately 25% of the steroid had dissociated. The authors suspected this lability to be attributable to reaction of the

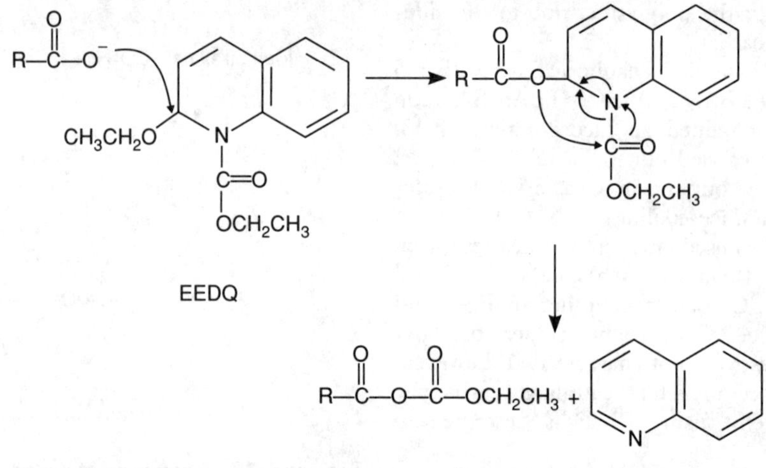

Figure 6.24 Reaction of acid with EEDQ to form mixed anhydride.

anhydride with tyrosine groups to form unstable phenolic esters.

Wring *et al.* (1994) coupled a carboxy derivative of the antiviral agent lamivudine to BSA using CDI; the hapten was activated with CDI in DMF (four hours, room temperature) before coupling to the protein (15 hours, room temperature, pH 8.0). The hapten:CDI: protein ratio was 50:100:1, and an incorporation of 7.6 was observed.

Another reagent for carboxyl activation *via* a mixed anhydride is *N*-(ethoxycarbonyl)-2-ethoxy-1,2-dihydroquinoline (EEDQ, Figure 6.24, Belleau *et al.*, 1969). The carboxylate first acylates the heterocyclic ring, then a rearrangement occurs which results in the formation of a mixed anhydride and quinoline. Pratt *et al.* (1978) used EEDQ to activate a mixture of carboxyl derivatives of luminol and testosterone (five minutes, DMSO) for coupling to ovalbumin (24 hours, 6 °C, pH 7.8).

Phosphinyl chlorides have also been described for forming mixed anhydrides, in peptide synthesis at least (Poulos and Ramage, 1989). They may have utility for protein-labelling applications.

6.1.6 Acid anhydrides

Acid anhydrides are also excellent reagents for the derivatisation of protein amine groups, and the generation of mixed anhydrides *in situ*, using a chloroformate under non-aqueous conditions, was discussed in the preceding section.

Due to their reactivity with water the derivatisation of proteins by anhydrides of all types is rather inefficient in terms of incorporation of the acid into the product. Where the starting material is scarce, other routes such as active ester coupling are probably more appropriate.

Under forcing conditions anhydrides are capable of acylating other residues such as histidines, tyrosines, tryptophans, thiols and hydroxylic groups (Olcott and Fraenkel-Conrat, 1947). Activated serine groups may be capable of acylation under mild conditions (Moorman and Abeles, 1982).

6.1.6.1 Acyclic acid anhydrides

Simple acyclic anhydrides such as acetic anhydride are convenient for processes such as radiolabelling, blocking of amine groups or investigating the effect of their modification on protein behaviour, and have a long history of use in such roles (Landsteiner, 1945a; Fraenkel-Conrat *et al.*, 1949; Grossberg and Pressman, 1963; Riordan and Vallee, 1972a; Buechler *et al.*, 1989; Marini, 1989; Seligman, 1993).

In an interesting application of acyclic anhydrides, Kaplan and his colleagues described the technique of competitive labelling (Kaplan *et al.*, 1971; Kaplan, 1972; Hasnain *et al.*. 1977), used to determine the reactivity of individual amine groups within a protein. The protein was incubated with a limiting amount of radiolabelled acetic anhydride (5–10 minutes, 10 °C) over a wide range of pH, and the location of the radioactivity in the protein determined by pepsin digestion and electrophoresis. The variation of individual reactivities with pH was used to estimate the pK_as of the amine groups. An adaptation of this approach was described by Bosshard *et al.* (1978): they compared the reactivities of amine groups in aminoacyl-tRNA synthetase in the native enzyme, and in a complex with tRNA. Their observation of differential reactivity of three amine groups was used to interpret the nature of the tRNA complex.

Despite their utility in these applications, acyclic anhydrides are less favoured for preparative coupling due to the poor yield (theoretically 50%) obtainable from the starting acid; this is of course compounded by the general inefficiency of anhydride reactions in an aqueous environment.

Finally, it may be noted that the common activation of carboxylic compounds by a carbodiimide (Section 6.1.1) may involve some formation of anhydrides, especially where the carbodiimide activation is carried out as a discrete stage under non-aqueous conditions. In some cases the anhydride may in fact represent the primary amine-reactive species, as described by Ranadive and Sehon (1967a).

6.1.6.2 Cyclic acid anhydrides

Cyclic anhydrides of compounds containing more than one acid group do not suffer from the disadvantage of yields limited to 50%, and have found widespread use in preparative applications.

The intramolecular or cyclic anhydrides of bis-carboxylic acids initially found application in protein chemistry because of their capacity to reverse the positive charge of a protein amine group, by replacing it with a free carboxylate (Habeeb *et al.*, 1958; Klapper and Klotz, 1972; Shiao *et al.*, 1972). Extensive changes in protein behaviour were observed on derivatisation with succinic anhydride, consistent with unfolding or expansion of the structure. Maurer and Lebovitz (1956) observed that succinylated gelatin reacted only very slightly with antisera raised against the native molecule.

Such alterations in protein behaviour were not promising for the use of intramolecular anhydrides in most applications as derivatisation reagents. However, their convenience has made them quite popular despite this limitation, e.g. the thiolation reagent SAMSA (Chapter 5, Section 5.12.1.4).

An application where cyclic anhydrides have been particularly useful is the introduction of metal ions *via* chelating groups, for example in the preparation of radioimmunoconjugates for therapeutic or imaging purposes *in vivo*, or of rare-earth labelled reagents for time-resolved fluorescence. Polycarboxylic acids are often excellent chelators of metal ions (Gansow, 1991), and it is often simple to generate intramolecular anhydrides of such compounds. Eckelman *et al.* (1975) for example described the preparation of bis-anhydrides (EDTAA, DTPAA) of ethylenediaminetetra-

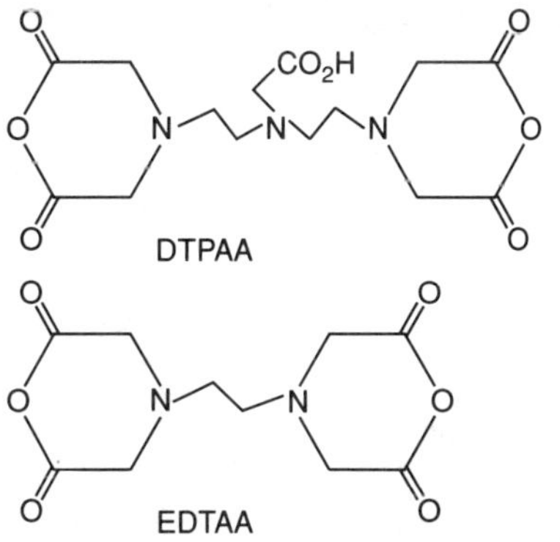

Figure 6.25 Bis-anhydride derivatives of EDTA and DTPA.

acetic acid (EDTA) and diethylenetriaminepenta-acetic acid (DTPA), shown in Figure 6.25. Many such reagents are commercially available.

Hnatowich *et al.* (1982) investigated the reaction of DTPAA with human serum albumin. Solid anhydride was added to the HSA solution (0.1 M HEPES, pH 7.0) at various ratios, and the resulting reaction (presumably limited by anhydride hydrolysis) was rapid: there was no difference in coupling efficiency between overnight reaction and that stopped after five minutes. Incorporations of up to five DTPA groups per HSA molecule could be obtained using a 50-fold excess of anhydride. Coupling in other buffers resulted in lower efficiency than that obtained with HEPES (65%): borate pH 8.0 (42%) and acetate pH 6.0 (37%) possibly due to sub-optimal pH, and phosphate pH 7.0 (45%) perhaps due to phosphate ions being responsible for either catalysis of anhydride hydrolysis, or for interference with the method for determining incorporation.

The same group (Layne *et al.*, 1982) used an even shorter reaction time of one minute for derivatisation of fibrinogen: incorporations of 0.13 and 0.7 were obtained from initial anhydride loadings of 1:1 and 5:1 respectively. With IgG, Hnatowich *et al.* (1983) observed an incorporation of 0.7 from a 1:1 loading, using a one minute reaction time and a bicarbonate buffer, pH 7.0.

Paik *et al.* (1983) carried out further optimisation of IgG derivatisation with DTPAA (one hour, 24 °C), at a lower IgG concentration (300 μg/ml) than those employed by Hnatowich and colleagues. These authors found that bicarbonate buffer (pH 8.2) and borate buffer (pH 8.6) gave the highest efficiencies, although these were still low: a hundredfold excess gave an incorporation of 1.7.

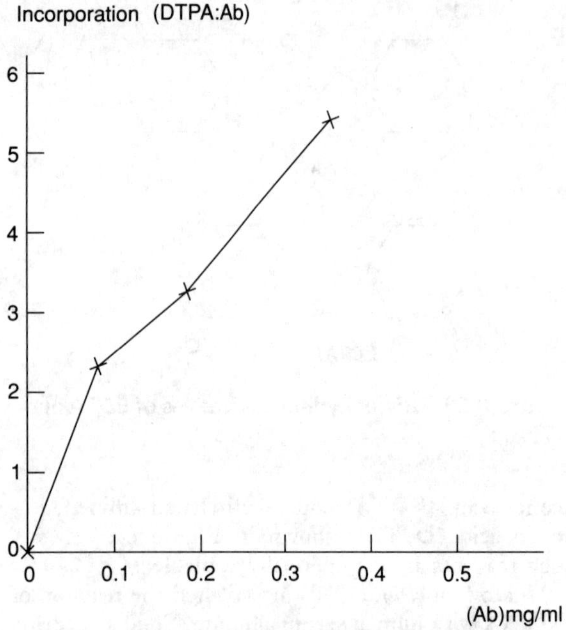

Figure 6.26 The effect of antibody concentration on the incorporation of DTPA obtained from reaction with the anhydride (250:1 excess). Incorporations were estimated radiometrically.

Sakahara *et al.* (1985) coupled DTPAA (added solid) to a Mab (one hour, room temperature, 0.1 M sodium bicarbonate): the antibody concentration again had a significant effect as shown in Figure 6.26. At a Mab concentration of 5 mg/ml, a 33-fold excess of DTPAA resulted in the incorporation of around 12–13 groups per molecule. The binding capacity of the antibody was strongly affected by incorporations above 2.4; at 5.5 DTPA groups per IgG molecule, the capacity was reduced to 1% of the underivatised value.

Hnatowich *et al.* (1987) coupled DTPAA to avidin and streptavidin (0.05 M sodium bicarbonate, pH 8.4), achieving a DTPA incorporation of eleven without affecting biotin-binding behaviour. The resulting conjugate was used as a ^{111}In chelate for *in vivo* imaging.

Canfi *et al.* (1989) used DTPAA to form a useful terbium chelate with 4-aminosalicylic acid (Figure 6.27(a)). The unreacted anhydride group was then coupled to a variety of protein–polylysine conjugates (Figure 6.27(b)). The reaction with protein was carried out overnight at room temperature in a phosphate buffer, with the DTPA derivative dissolved in DMSO.

Subramanian and Wolf (1990) coupled ETDA and DTPA anhydrides to HSA (240 hours, 22 °C, HEPES, pH 7.0) and obtained high incorporations in keeping

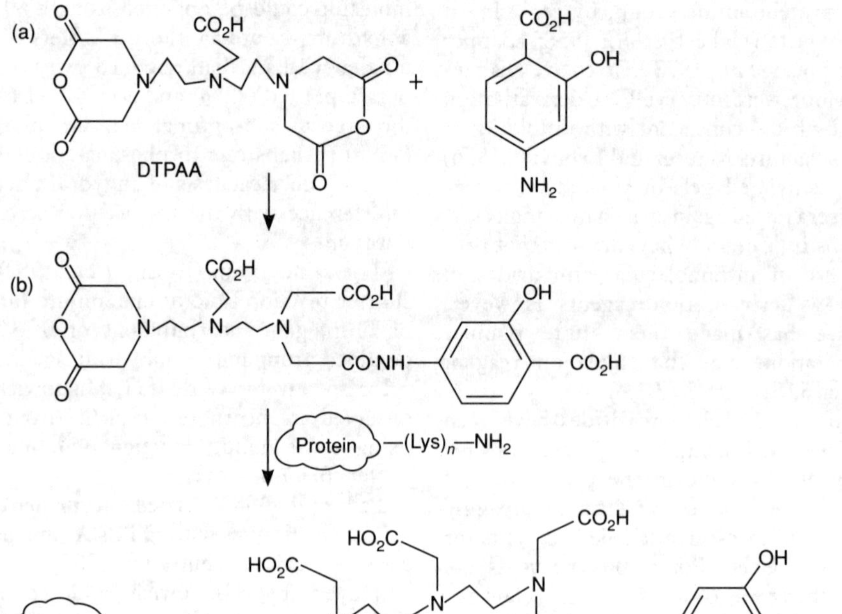

Figure 6.27 Formation of protein–polylysine conjugates with DTPAA and 4-aminosalicylic acid.

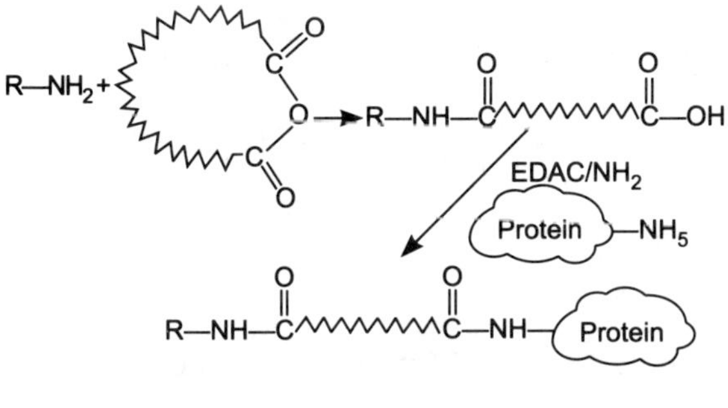

Figure 6.28 Use of cyclic anhydrides as homofunctional species.

with the concentrated (20 mg/ml) protein solution used: 8.5 (EDTA) and 7.8 (DTPA) from a tenfold excess of anhydride.

Bickel *et al.* (1994) coupled DTPAA to native or cationised IgG to facilitate ^{111}In-labelling for brain imaging. A 30:1 molar excess of DTPAA (in DMSO) was added to the IgG (in 0.1 M aqueous sodium bicarbonate) and allowed to react for 60 minutes at room temperature before purification by ultrafiltration.

Many other such compounds have been used in specific applications, for example the use of isatoic anhydride as an inactivator of serine protease (Moorman and Abeles, 1982).

Cyclic anhydrides can also be used in combination with carbodiimide chemistry, effectively giving a two-stage homobifunctional coupling of amine-containing species (Figure 6.28). This approach has been used, for example, to couple daunomycin to protein carriers *via* the acid-labile *cis*-aconityl group (Shen and Ryser, 1981; Diener *et al.*, 1986).

An interesting class of reagents are the cyclic anhydrides of *N*-carboxy-α-amino acids. These Leuchs' anhydrides have been used in peptide synthesis under aqueous conditions (Hirschmann *et al.*, 1967); they incorporate a conveniently-protected amine group which can be regenerated in the product by acid treatment, to yield a dipeptide for example (Figure 6.29).

They can also be used to couple peptide moieties to proteins (Sela and Arnon, 1972). Stahmann and Becker (1952), for example, coupled *N*-carboxyglycine anhydride to albumin and chymotrypsin (several hours, 4 °C then warming to room temperature, pH 7.4): a 4.8-fold molar excess of anhydride resulted in the addition of 171 glycines per albumin molecule, while a 1.8-fold excess incorporated 65 glycines per chymotrypsin.

Finally, maleic anhydride and 2-methylmaleic or citraconic anhydride have attracted attention due to their readily cleavable nature (Butler and Hartley, 1972; Atassi and Habeeb, 1972).

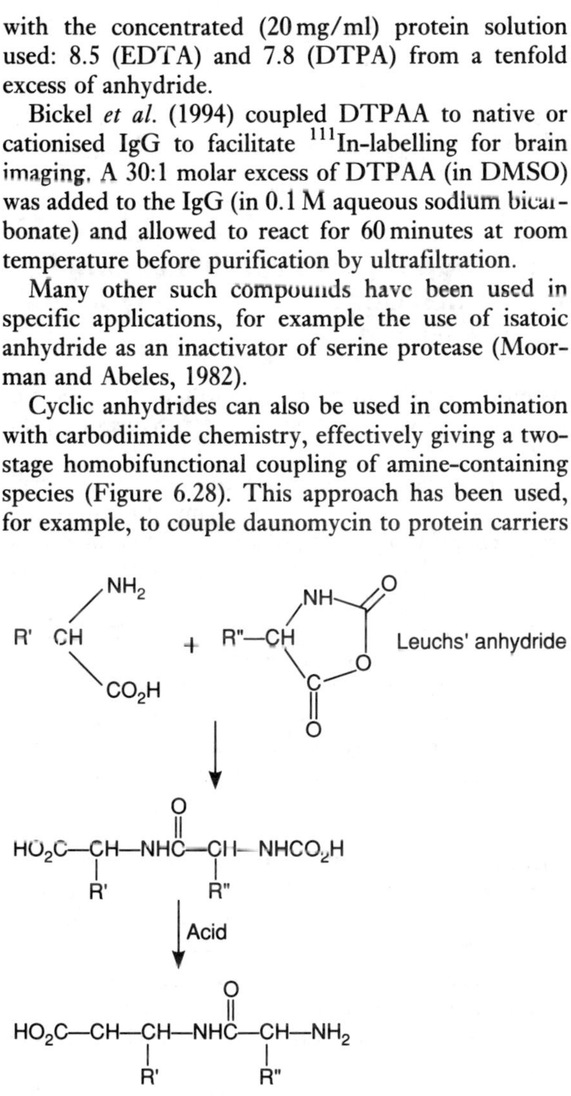

Figure 6.29 Coupling using Leuchs' anhydrides.

6.1.7 Acid chlorides

Acid chlorides react vigorously with water and cannot be generated *in situ* in aqueous systems. However, addition of preformed acid chlorides to proteins is feasible, if inefficient in terms of usage of the acid derivative. Reaction with amines, the Schotten–Baumann reaction, of course predominates.

Some acid chlorides are commercially available, such as the isonicotinoyl chloride used by Degani and Heller (1988) to introduce pyridyl residues into glucose oxidase; this was used as a means of coupling ruthenium complexes. However, these compounds can also be readily obtained from carboxylic acids in good yield, by treatment with reagents such as thionyl chloride. This approach has been used to prepare conjugates of such carboxyl-containing molecules as aspirin (Weiner *et al.*, 1963; Wicher *et al.*, 1968; Hoffman and Campbell, 1969), and analogues of the insecticides DDT and malathion (Haas and Guardia, 1968).

Alternatively, acid chlorides can be generated from hydroxylic compounds by reaction with the bis-chloride of a dicarboxylic acid, as described by Bailey and Butler (1967) for cholesterol.

However prepared, the acid chlorides are typically reacted with protein amines at a mildly basic pH. The generation of HCl during the reaction is often countered by the addition of base so that this pH can be maintained. If a 'pH-stat' approach is employed the course of the reaction can be followed by the uptake of base (Wicher *et al.*, 1968).

An interesting application of the acid chloride approach was described by Slepnev *et al.* (1995). HRP was acylated with stearoyl chloride in a reversed-micelle system, using the detergent Aerosol OT in a borate buffer–octane medium. This allowed controlled, low-level modification of the enzyme to be achieved.

Where the compound to be coupled to the protein contains a hydroxyl function, an acyl chloride function can be generated directly by reaction with phosgene, although this is a highly unpleasant reagent. This route, which provides a chloroformate (cf. Section 6.1.5) was used by Teale *et al.* (1974) to couple tetrahydrocannabinol to BSA for use as an immunogen.

6.1.8 Other activated carboxyl reagents

Acyl azides ($R-CO-N_3$) display adequate reactivity for reaction with protein amines, and have been used for derivatisation reactions. They can be generated by diazotisation of a hydrazide (Dixon *et al.*, 1975; Tse *et al.*, 1976), or by reaction of an acid chloride with azide ions (Butler *et al.*, 1940).

As for acid chlorides, the reaction with proteins is typically carried out under mildly basic conditions, extra base being added during the reaction to maintain pH as acid is released.

N-Acylimidazoles are reactive enough to be used for acylation, although they are not as powerful in this respect as anhydrides, for example. This allows some selectivity in their use: *N*-acetylimidazole (Figure 6.30) has been described for the selective acylation of the hydroxyl functions of tyrosine (Riordan *et al.*, 1965; Riordan and Vallee, 1972b). The resulting tyrosyl esters are fairly labile, and can be readily hydrolysed with hydroxylamine; this has been used for the reversible inactivation of enzymes with active-site tyrosine residues (Pugh and Horecker, 1967). However this reagent is capable in many cases of acylating lysine residues too (Perlmann, 1966; Cuatrecasas *et al.*, 1968).

6.1.9 Coupling of amides by reverse proteolysis

Reverse proteolysis is more commonly encountered for the introduction of amine-containing molecules,

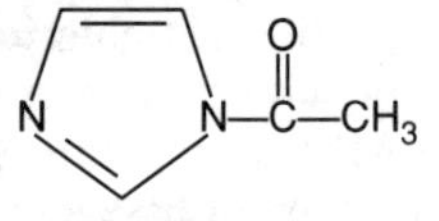

Figure 6.30 *N*-Acetylimidazole.

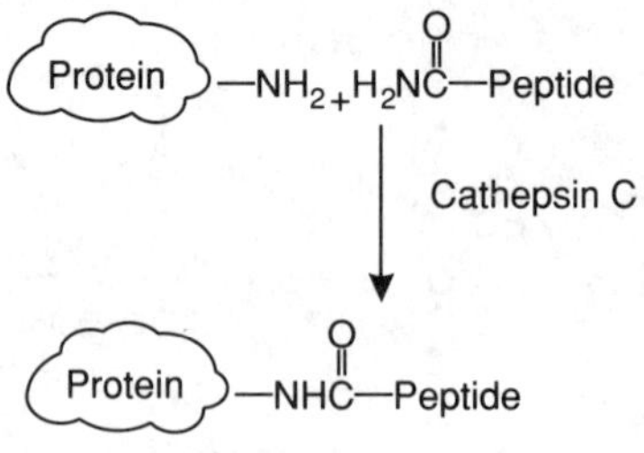

Figure 6.31 Use of Cathepsin C to form oligopeptides.

and is discussed more fully in Sections 5.2.4.2 and 6.7.10.2.

Gittel and Schmidtchen (1995), however, described the use of Cathepsin C to attach dipeptide amides to oligopeptides (Figure 6.31) such as myosine light-chain kinase inhibitor (one hour, 24 °C, pH 7.9). This approach will almost certainly find application for larger polypeptides too.

6.2 Coupling of Proteins to Small Molecules Containing Aldehyde or Ketone Groups

Chapter 5 cites numerous instances of the reaction between aldehydes and amines (or hydrazides) to form Schiff base linkages, which are often subsequently reduced to improve their stability. Small molecules containing similar carbonyl functions can also be coupled to proteins by this reductive alkylation approach, which is by far the commonest method based on aldehyde reactivity.

Bowman and Stroud (1950) described a method for reductive dimethylation of amino acids, by reaction with formaldehyde and catalytic hydrogenation of the resulting Schiff base. Fischer *et al.* (1958) reported the use of sodium borohydride for stabilising Schiff base intermediates from enzyme reactions. Khym (1963) coupled periodate-oxidised nucleosides to small amines, reducing the products with borohydride. Erlanger and Beiser (1964) adapted this approach to couple oxidised nucleosides to BSA (see below). Means and Feeney (1968) pioneered the widespread application of the reductive alkylation approach to protein chemistry. Formaldehyde, acetaldehyde and acetone were coupled to proteins in borate buffer, pH 9.0 (0 °C, 30 minutes) in the presence of sodium borohydride.

In a model reaction with butylamine, Means and Feeney showed that the pH optimum for reductive alkylation was at least 10, but recommended a slightly lower pH for the sake of protein stability. With formaldehyde, the reaction resulted almost exclusively in dimethylation, very little monomethyl product was observed at any pH.

This conjugation procedure has seen two main areas of application. The first (Section 6.2.1) is as a convenient means for radiolabelling, *via* reductive methylation with formaldehyde and a suitable reducing agent. Secondly reductive alkylation has been employed as a general procedure for coupling aldehyde-containing molecules to proteins, either native aldehydes (Section 6.2.2) or those derived from other functionalities such as sugars (Section 6.2.3).

The occasional examples of Schiff base formation from amines without subsequent reduction are dealt with in Section 6.2.4, and the analogous reactions of hydrazides in Section 6.2.5. A couple of specific reactions of dicarbonyl compounds are mentioned in Section 6.2.6.

6.2.1 Radiolabelling *via* reductive alkylation with formaldehyde, etc.

The main variations in this approach are related to the nature of the reducing agent employed.

6.2.1.1 Reduction with borohydride

Reaction of proteins with formaldehyde–borohydride allows radiolabels to be introduced without significant alteration of properties. *N*-Methylation of lysines is the most important reaction, although formaldehyde is prodigiously reactive and can react with cysteine, arginine, tyrosine, histidine, glutamine, asparagine, and even with the amide bonds of the peptide chain itself (Galembeck *et al.*, 1977). With lysine an *N*-hydroxymethyl derivative is formed initially (Figure 6.32), but a small amount of dehydrated imine is present in equilibrium. In the presence of a reducing agent, the formation of the methylamine from the imine fraction drives the equilibrium towards further imine formation and subsequently to complete methylation.

Borohydride has been shown to cause alterations to protein structure, including cleavage of disulfides and of peptide bonds, and reduction of amino acids to α amino alcohols (Crestfield *et al.*, 1963; Paz *et al.*, 1970). Paz *et al.* observed reduction of seryl, threonyl, lysyl, aspartyl and asparaginyl residues in collagen and trypsin, as a result of borohydride treatment (80 minutes, room temperature, pH 8–9). Therefore, although

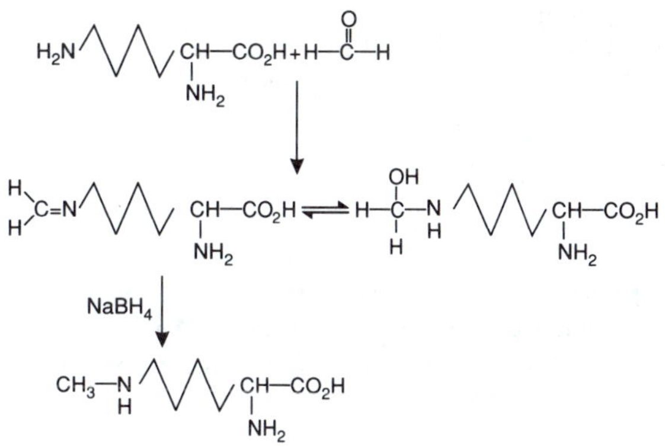

Figure 6.32 Methylation of lysine with formaldehyde.

borohydride frequently has no observable effect on protein behaviour (see for example Autor and Fridovich, 1970), reduction times should be kept to a practical minimum.

Both [^{14}C]formaldehyde and sodium borotritiide ([^{3}H]borohydride) have been used to provide the radiolabel. The methylation of amine-containing residues has little effect on the protein structure, either sterically or in charge terms—the pK_a of a dimethylamine group differs by only 0.5 from the native amine.

Rice and Means (1971) described general conditions for rapid reductive methylation of proteins with [^{14}C]formaldehyde: the protein was incubated with [^{14}C]formaldehyde at a four- or fivefold excess over the amine content (30 seconds, ice-bath, pH 9.0) before reduction by an equivalent excess of sodium borohydride[1]. A further, slightly larger excess of borohydride was added a minute later to ensure complete reduction. Similar conditions were used by Gualerzi *et al.* (1973) to label ribosome proteins.

Moore and Crichton (1973) obtained around 10–15% dimethylation of lysines in ribosome proteins using a double cycle of aminolysis and reduction at pH 8.5 and 0 °C. The proteins were allowed to react sequentially with formaldehyde (30 seconds), [^{3}H]borohydride (one minute), formaldehyde (30 seconds) and [^{3}H]borohydride (ten minutes).

In competitive labelling studies, Kraal and Hartley (1978) used a substoichiometric quantity of ^{14}C-formaldehyde (0.4 equivalents of the amine content) to investigate lysine reactivities within the bacterial elongation factor EF-Tu. A roughly 4.5-fold excess

[1] Equivalent to a 0.25 molar quantity due to the tetravalent reactivity of the borohydride.

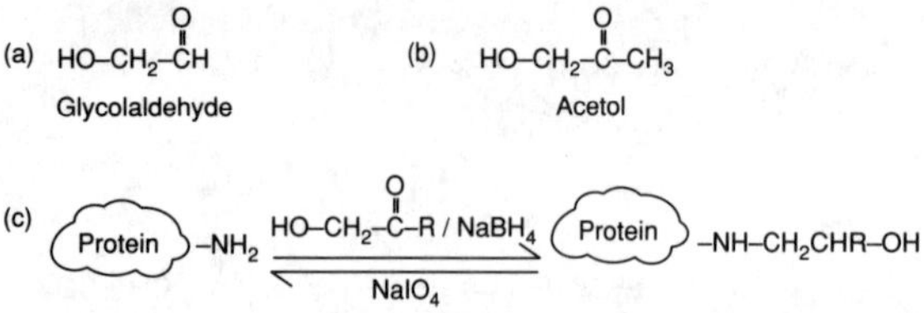

Figure 6.33 Reductive alkylation of protein with glycolaldehyde (a) or acetol (b) results in a derivative which can be cleaved by periodate to regenerate the original structure (c).

of borohydride was used as the reductant. Changes in lysine reactivity were observed when the protein was bound to nucleotides and nucleic acids.

In the absence of a reducing agent, a number of slower, irreversible reactions take place: Galembeck *et al.* (1977) showed that, with lysozyme and ribonuclease, these resulted in the incorporation of one or two formaldehyde residues. Little reaction took place below pH 7; at pH 9 reaction was still incomplete after 50 minutes. The rapid reaction times generally described for reductive methylation probably preclude any significant contribution from these side-reactions.

An interesting reversible variant of the reductive alkylation approach was described by Geoghegan *et al.* (1979). β-Hydroxy-ethylamines are readily oxidised by periodate to generate the free amine, and therefore the reductive monoalkylation of amines with compounds such as glycolaldehyde and acetol yields a periodate-cleavable product (Figure 6.33).

6.2.1.2 Reduction with cyanoborohydride

Sodium borohydride has a number of disadvantages in this application. The first is its instability in neutral or acidic systems, hence the need for a basic buffer which is detrimental to formaldehyde stability; even under these conditions the borohydride is often also added incrementally to minimise the effects of loss. Secondly, borohydride reduces aldehydes and ketones to alcohols, and simultaneous Schiff base formation and borohydride reduction are therefore inefficient. Thirdly, as noted above, it can cause modifications of protein structure such as reduction of amino acids and cleavage of disulfide and peptide bonds.

Borch *et al.* (1971) described the properties of sodium cyanoborohydride, $NaBH_3CN$, which is stable down to pH 3. This compound was shown to reduce aldehydes and ketones only under quite strongly acidic conditions, generally around pH 3 or 4. Imines, on the other hand, could be reduced at pH 6–7 *via* the protonated imminium ion. Reductive alkylation of amines in a one-pot reaction was therefore feasible, and was duly demonstrated with a range of model amines and suitably-reactive carbonyl compounds (aldehydes or unhindered ketones). Although pH 6 to 8 was quoted as optimal, successful reactions were carried out from pH 4 to pH 10, the major requirement being for sufficient H^+ ions to be available to protonate the imine.

Friedman *et al.* and Gray independently used cyanoborohydride in 1974 for general reductive alkylation (see below), and in 1978 Dottavio-Martin and Ravel adopted this reagent for reductive radiomethylation with [^{14}C]formaldehyde. These authors compared a simple cyanoborohydride reduction at pH 7 with incremental addition of borohydride at pH 9, and observed 6 to 45-fold greater incorporation with the former. There was quite a strong temperature dependence, with approximately equivalent incorporation being obtained over 24 hours at 0 °C, 4–6 hours at 25 °C or one hour at 37 °C.

MacKeen *et al.* (1979) and Jentoft and Dearborn (1979) reported similar improvements in incorporation over $NaBH_4$ reductions. MacKeen *et al.* used equimolar formaldehyde and $NaBH_3CN$ at a 39 to 87-fold molar excess over protein (60 minutes, room temperature, pH 7.5). Three to 20-fold improvements in incorporation were seen over borohydride reduction, although smaller excesses of formaldehyde were employed.

Jentoft and Dearborn studied reaction conditions in some detail. The reaction was shown to be more-or-less insensitive to cyanoborohydride concentration[1], but high levels (above around 100 mM) showed some inhibition which could be lessened but not eliminated by recrystallisation. This was attributed to an impurity or to a small degree of reduction of the carbonyl compound by cyanoborohydride. The pH optimum was around 8, but was only significant for short reaction times: incorporation at pH 6–9 was roughly equivalent after two hours. An excess of twelve formaldehydes per lysine gave 95–100% of theoretical dimethylation after 24 hours.

These authors noted that there was little or no crosslinking of protein despite the propensity of formaldehyde for forming such products, the rapid reduction of the Schiff base probably preventing further reaction. Many enzymes were inactivated by this treatment: alkaline phosphatase lost 36% of its

[1] As the formation of the Schiff base is the slower, rate-limiting step.

activity (10 mM formaldehyde, 12 mM cyanoborohydride, pH 7.5, 90 minutes), and β-galactosidase 16% (4 mM formaldehyde, 25 mM cyanoborohydride, pH 7.0, 95 minutes), while glucose-6-phosphate dehydrogenase was completely inactivated (10 mM formaldehyde, 12 mM cyanoborohydride, pH 7.5, 14 minutes).

Jentoft and Dearborn (1980) reported further refinements to the cyanoborohydride approach, centred around the effect of cyanide ions. Cyanide may be present in the cyanoborohydride as an impurity, and is unavoidably generated from the reduction reaction anyway; it can compete for reaction with formaldehyde to yield the cyanohydrin (Figure 6.34) and therefore reduce the efficiency of the reaction with protein amines. Jentoft and Dearborn observed improvements in efficiency either by recrystallisation of the cyanoborohydride, or by incorporation of a complexing agent such as nickel (II). The formaldehyde excess needed to obtain maximum incorporation could be reduced from five- to tenfold to three- to fourfold by taking these precautions against cyanide interference.

Another application of the reductive methylation approach was reported by Jentoft *et al.* (1979), [^{13}C]-formaldehyde was used to incorporate enriched levels of this isotope at ribonuclease amine groups, where it could be detected by NMR. The presence of similar chemical functions in different microenvironments was demonstrated by the coalescence of their signals in denatured protein.

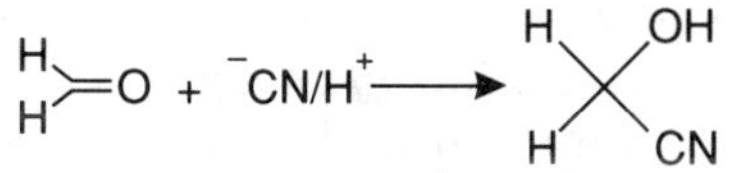

Figure 6.34 Formation of cyanohydrin from formaldehyde and cyanide

6.2.1.3 Reduction with other reagents

The use of cyanide-containing compounds is not always welcome; apart from the interferences with the formaldehyde reaction noted above, there are general safety considerations[1], while some metalloproteins are also adversely affected by cyanide ions.

Geoghegan *et al.* (1981) and Cabacungan *et al.* (1982) assessed amine boranes as alternatives to cyanoborohydride. Pyridine borane is a slightly stronger reducing agent, while alkylamine boranes such as tri- and dimethylamine borane are somewhat weaker. They exhibit similar pH specificity and reactivity to proteins to cyanoborohydride.

In a typical experiment, treatment of turkey ovomucoid with 15 mM dimethylamine borane and 50 mM formaldehyde (two hours, room temperature, pH 7) gave almost quantitative dimethylation of lysine residues.

[1] The release of hydrogen cyanide gas on acidification is a small but serious risk, and the cyanoborohydride itself is toxic.

6.2.1.4 Reductive non-alkylation

Closely related to the reductive alkylation approaches outlined above is the reduction of oxidised glycoproteins with radioactive borohydride, etc. This leads to the incorporation of a radiolabel without the incorporation of an alkylating agent such as formaldehyde.

Oxidation of the glycoprotein is typically carried out using periodate (Van Lenten and Ashwell, 1971; Gahmberg and Andersson, 1977) or galactose oxidase (Morell *et al.*, 1966; Gahmberg and Hakomori, 1973, 1976; Pfannschmidt and Schauer, 1980). While the latter approach is milder, it often requires the removal of terminal sialic acid residues to reveal a galactose substrate.

6.2.2 General coupling of aldehydes and ketones to proteins *via* reductive alkylation

Frey *et al.* (1971) coupled 5-nitrosalicylaldehyde to an active-site lysine in acetoacetate decarboxylase (five minutes, ice bath, pH 6.0), observing spectrophotometric evidence for imine formation. This was 'trapped' by treatment with borohydride for 5–10 minutes.

Friedman *et al.* (1974) alkylated wool and soluble proteins with nitrobenzaldehydes in the presence of sodium cyanoborohydride. A 1:1 mixture of methanol and lithium acetate buffer (pH 5.2), was used for the two-hour coupling reaction, which resulted in the modification of between 10 and 40% of the available lysine groups, depending on the protein concerned.

As might be expected from the ability of borohydride to reduce aldehydes, a higher level of derivatisation of wool proteins was obtained by sequential coupling and reduction (30% of lysines derivatised) *versus* simultaneous reaction (5 to 10%).

Joniau *et al.* (1976) coupled a formyl derivative of the anti-tuberculosis drug rifampicin to albumin, at a 20:1 excess (one hour, room temperature, pH 9.2, dioxane); the product was reduced with borohydride.

Andree and Zantema (1978) coupled a ketone-containing spin-label to glutamate dehydrogenase: the

Schiff base was formed using a large excess of the ketone (30 minutes, ice-bath, pH 8.0), followed by borohydride reduction (several hours, room temperature, pH < 8.5). The results suggested that the spin-label was linked stoichiometrically *via* the Lys126 residue of the enzyme.

Rodgers *et al.* (1978) coupled an aromatic aldehyde derivative of tetrahydrocannabinol to BSA in the presence of sodium cyanoborohydride (four days at 4 °C and two days at room temperature, pH 7.0). Using a 55–fold molar excess of aldehyde and a 77–fold molar excess of cyanoborohydride, these authors observed an incorporation of four to five haptens per albumin molecule.

Gaur *et al.* (1981) coupled aflatoxin B2a to BSA or HRP in a methanol–phosphate buffer mixture (30 minutes, 37 °C, pH 7.2), followed by borohydride reduction (30 minutes, 4 °C). Incorporations of 12 and 2.5 were obtained for BSA and HRP respectively.

Li *et al.* (1991) required antibodies to the 4-nitrophenol moiety, and prepared immunogens and solid-phase-coating conjugates with BSA, KLH and ovalbumin using 3-hydroxy-6-nitrobenzaldehyde. The Schiff base was formed in a 30 minute reaction at 37 °C, in 0.2 M phosphate–saline buffer (pH 7.5) with the benzaldehyde dissolved in methanol. The product was reduced with sodium borohydride: antibody generation from these immunogens was poor, however.

6.2.3 Reaction of aldehydes and ketones generated from other groups

Where a small molecule contains a suitable saccharide structure, this can be oxidised to generate reactive aldehydes. Thus Erlanger and Beiser (1964) coupled oxidised nucleosides or nucleotides to BSA to form immunogens: the nucleoside was oxidised with 0.1 M periodate (20 minutes, room temperature, water, stopped with ethylene glycol), coupled to BSA (45 minutes, pH 9–9.5) and the product reduced with sodium borohydride (18 hours). The pH was then lowered with formic acid to destroy excess borohydride, followed by neutralisation with ammonia. Molar incorporations of 20–30 were observed.

There are a class of pharmacologically active steroids, the cardiac glycosides, which contain an oligosaccharide moiety. Examples include digoxin and digitoxin. Oxidation of the sugars followed by reductive alkylation offers a convenient route to the formation of protein conjugates, and this was exploited by Butler and colleagues to produce digoxin immunogens (Butler and Chen, 1967; Smith *et al.*, 1970). Reaction conditions were very similar to those of Erlanger and Beiser above. In the later paper an incorporation of 6.7 digoxin molecules per albumin was reported.

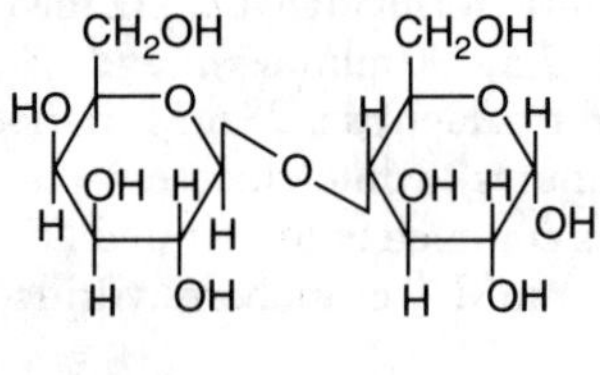

Figure 6.35 Lactose.

Similarly, Lauer and Erlanger (1974) oxidised adenosine (100 mM periodate, 30 minutes in the dark, 4 °C) and coupled it to β-galactosidase (90 minutes, 4 °C, pH 9), the product was reduced with sodium borohydride.

Hurwitz *et al.* (1975) oxidised the saccharide ring of adriamycin and daunomycin with a slight molar excess of 100 mM periodate (one hour, room temperature, in the dark, pH 7.2) then coupled the product to IgG (one hour, room temperature, pH 9.5). Reduction with borohydride (two hours, 37 °C, pH 9.5) followed. The incorporation obtained was 2–5.

While carbohydrate-containing molecules can be oxidised to yield aldehydes, they may also provide a small equilibrium quantity of aldehyde in their native state (e.g. lactose, Figure 6.35). This has also been used as a basis for coupling saccharides to proteins, albeit rather slowly: Gray (1974) coupled lactose to BSA by this route and stabilised the resulting Schiff base *in situ* with sodium cyanoborohydride[1]. Reaction for two weeks resulted in a lactose incorporation of around four.

Thong *et al.* (1985) described another route to the generation of aldehydes. In addition to an oligosaccharide moiety, the cardiac glycosides mentioned earlier contain a lactone function; Thong converted the lactone of digoxin to its ozonide by ozone treatment (Figure 6.36(a)), then reduced this with dimethyl sulfide to an aldehyde-bearing ester (Figure 6.36(b)). While hydrolysis to the hydroxyketone (Figure 6.36(c)) was shown to ensue, rapid introduction of a protein (methylated BSA) allowed the aldehyde to be trapped in a more stable ester conjugate (Figure 6.36(d)), which was subsequently reduced with cyanoborohydride. Using an initial 250-fold excess of digoxin a conjugate was obtained containing 40 to 50 steroid residues per albumin molecule. The saccharide moieties were reported to remain intact throughout this procedure.

[1] Cyanoborohydride reduces Schiff bases at pH 6–8, but reduction of aldehydes is negligible under these conditions.

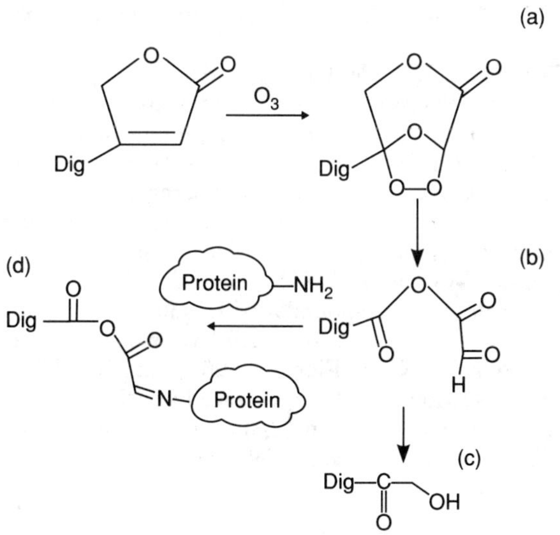

Figure 6.36 Conversion of digoxin-lactone to ozonide, aldehyde-ester and subsequent hydrolysis and aminolysis reactions.

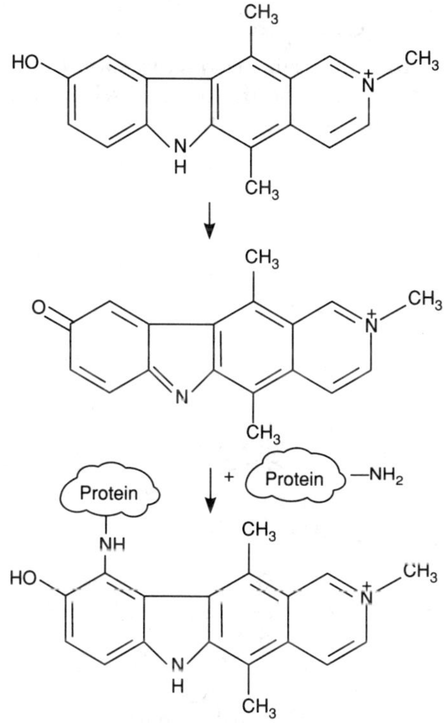

Figure 6.37 Enzyme-catalysed oxidation of the ellipticine derivative (a) yields a quinonoid structure (b), which can react with protein amine groups by Michael addition.

An interesting, enzyme-catalysed conjugation has been described (Auclair and Paoletti, 1981; Auclair *et al.*, 1983) using the antitumour drug 2-*N*-methyl-9-hydroxyellipticine: this can be catalytically oxidised by HRP and other peroxidases to yield a quinonoid molecule (Figure 6.37), which can then couple to protein amine groups in a manner analogous to benzoquinone (cf. Section 6.7.3, and Chapter 5, Section 5.3.6).

6.2.4 General coupling of aldehydes and ketones to proteins *via* Schiff base formation without reduction

Although there is a clear rationale for omitting reduction in cases where hydrazides are employed (see Section 6.2.5), this approach is occasionally taken too in amine-based preparations. Acharya and Manning (1980a), for example, derivatised haemoglobin S with glyceraldehyde, to prevent the sickling phenonenon responsible for sickle-cell anaemia. The Schiff base formation was carried out using an approximately 20-fold excess of [^{14}C]glyceraldehyde in phosphate-buffered saline (pH 7.4), at 37 °C. The course of the reaction was followed by monitoring incorporation of the radiolabel, and indicated essentially complete reaction after around 18 hours.

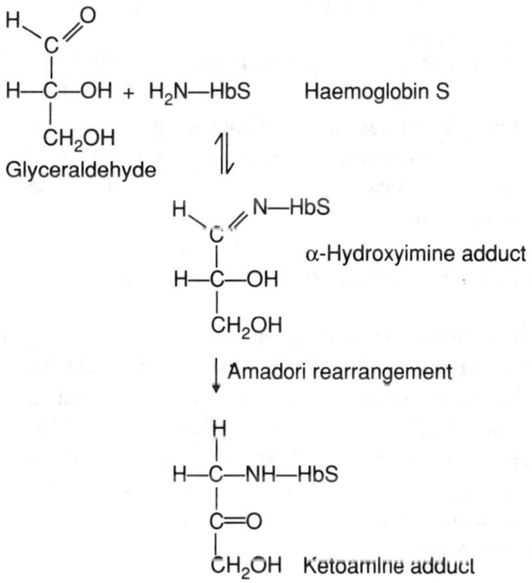

Figure 6.38 Formation of α-hydroxyimine by reaction of haemoglobin and glyceraldehyde and subsequent Amadori rearrangement to ketoamine.

About twelve glyceraldehyde molecules per haemoglobin molecule could be coupled by this route. The same authors (Acharya and Manning, 1980b) noted that the α-hydroxyimine formed could undergo an Amadori rearrangement to a ketoamine (Figure 6.38).

Zaugg *et al.* (1977) investigated the reaction (30 minutes, 37 °C, pH 7.2) of a series of aldehydes and ketones with haemoglobin, again in search of antisickling effects. Aromatic aldehydes were more reactive than their aliphatic analogues, while ketones were unreactive[1].

6.2.5 Coupling of aldehydes to protein hydrazides

Although (for obvious reasons) the normal nucleophilic target for aldehydes or ketones is an amine function, it has been mentioned earlier that hydrazides are capable of fulfilling the same role. The hydrazones obtained by reaction with carbonyl compounds are quite stable, and although reduction to hydrazide is quite simple, the need for reduction is largely obviated. Obviously the protein-hydrazide must be generated *a priori*, for example by carbodiimide-mediated attachment of hydrazine to carboxylic residues. This approach was used by Clark *et al.* (1977) to couple the ketone moiety of haloperidol to BSA; presumably due to the low reactivity of the ketone, a seven-day reaction time was employed.

Rose and his colleagues (Rose *et al.*, 1991; Fisch *et al.*, 1992; Vilaseca *et al.*, 1993) produced protein hydrazides *via* reverse hydrolysis, and coupled these (17–20 hours, room temperature, pH 4.6) to aldehyde-containing species, such as derivatives of the chelator ferrioxamine.

6.2.6 Reactions of dicarbonyl compounds

The commonest application of a dicarbonyl compound in protein coupling is the use of glutaraldehyde as a homobifunctional coupling agent for amines (see Section 6.2.6, and Chapter 5, Section 5.3.1). However, there are a couple of special cases of dicarbonyl reactivity, which, although their use is generally analytical rather than preparative, deserve brief mention here.

[1] Without subsequent reduction to drive the equilibrium, the reversibility of the Schiff base formation will prevent the appearance of product from less reactive carbonyl compounds. Under different circumstances ketones can form Schiff base products with amines (e.g. Geoghegan *et al.*, 1979).

6.2.6.1 Aromatic α,β-dicarbonyl compounds

o-Phthaldialdehyde (OPA, Figure 6.39(a)) generates a usefully fluorescent product from amine-containing compounds, including obviously proteins and peptides (Roth, 1971; Weidekamm *et al.*, 1973); the presence of a thiol is also required for the fluorogenic reaction to occur. The mechanism involves the creation of an isoindole fluorophore, as shown in Figure 6.40 (Alvarez-Coque *et al.*, 1989; Haugland, 1992).

Analogous compounds with superior properties have been developed, such as naphthalene-2,3-dialdehyde (NDA, Figure 6.39(b)) and the ATTO-TAG™ (Molecular Probes) reagents such as 3-(*p*-carboxybenzoyl) quinoline-2-aldehyde (CBQ; Figure 6.39(c)). See Haugland (1992) for further details and references.

6.2.6.2 Aliphatic α,β-dicarbonyl compounds

Vicinal dicarbonyl compounds display an unusual specificity for arginine residues, and have been employed as group-specific analytical reagents. The commonest examples are phenylglyoxal (Figure 6.41(a); Takahashi, 1968) and butane-2,3-dione (Figure 6.41(b)), although the reactive form of the latter has been shown to be the trimeric self-condensation product shown in Figure 6.41(c) (Yankeelov *et al.*,

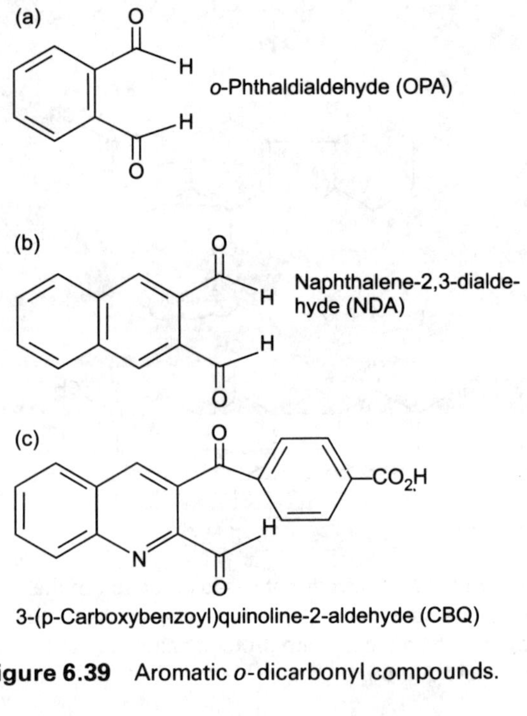

Figure 6.39 Aromatic *o*-dicarbonyl compounds.

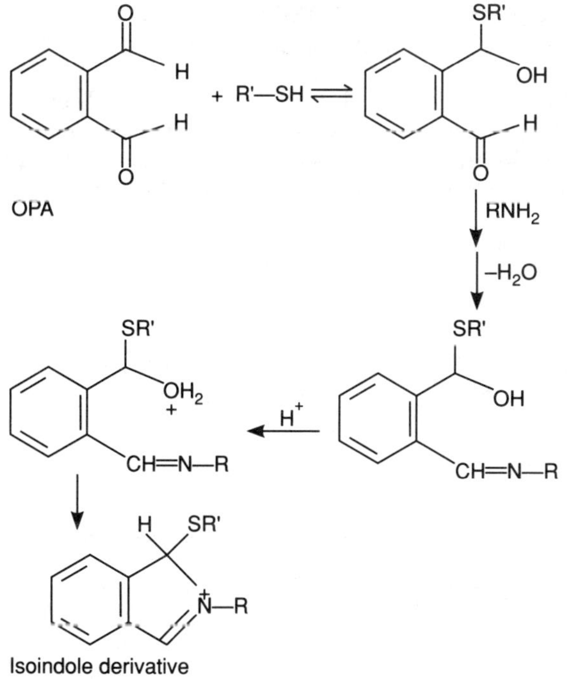

Figure 6.40 Mechanism of reaction of *o*-phthaldialdehyde with amines in the presence of thiol to create an isoindole fluorophore.

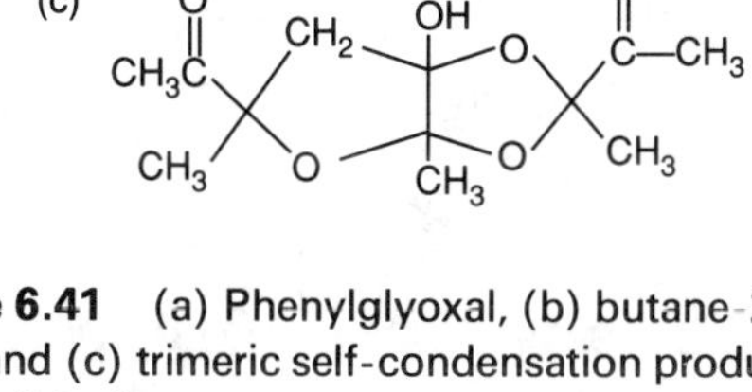

Figure 6.41 (a) Phenylglyoxal, (b) butane-2,3-dione and (c) trimeric self-condensation product of butane-2,3-dione.

1968). The commonest application of such reagents has been the demonstration of essential arginine residues (Kasher *et al.*, 1986; Konishi and Fujioka, 1987).

6.3 Coupling of Proteins to Small Molecules Containing Isocyanate or Isothiocyanate Groups

Isocyanates and isothiocyanates react primarily with protein amine functions to yield substituted ureas and thioureas respectively (Figure 6.42; see Chapter 5, Section 5.3.3 for a fuller discussion of their chemistry). They are generally poorly soluble in aqueous systems and require the use of an organic cosolvent. Some instability of protein–fluorescein conjugates linked *via* thioureas has been noted (Haugland, 1992). The and Feltkamp (1970b) observed some loss of performance on storage at room temperature in daylight for seven days; in the dark no such effect was noted even at 37 °C.

While these compounds can react with other groups such as hydroxyls, thiols and carboxyls, the effects are probably negligible under the conditions of protein labelling; amines are the only relevant groups which react more strongly than water (Satchell and Satchell, 1975).

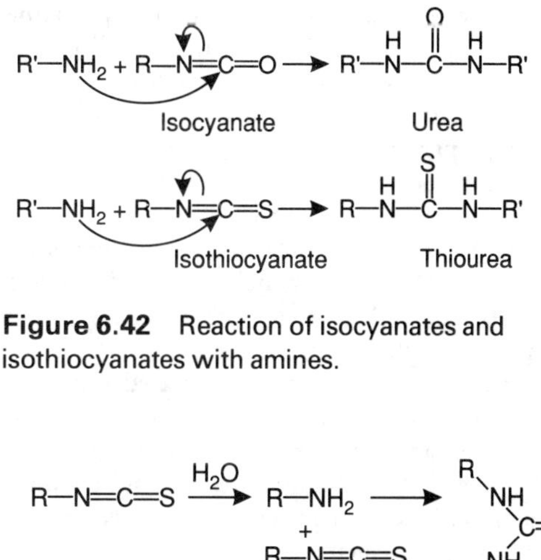

Figure 6.42 Reaction of isocyanates and isothiocyanates with amines.

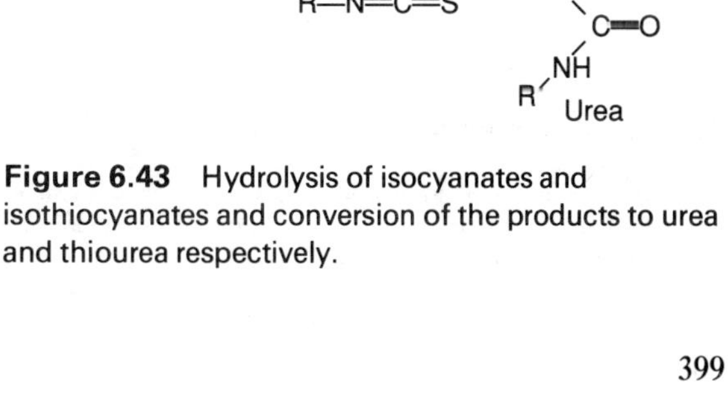

Figure 6.43 Hydrolysis of isocyanates and isothiocyanates and conversion of the products to urea and thiourea respectively.

Hydrolysis can yield anilines, which can then react with a second molecule of the N=C=X compound to give a urea or thiourea; thioureas can oxidise to ureas (Figure 6.43)[1].

Isothiocyanate groups can be quantitatively determined by reaction with a vicinal dithiol such as benzene-1,2-dithiol (Zhang *et al.*, 1996).

6.3.1 Coupling of proteins to small molecules containing isocyanates

Isocyanates are less stable in aqueous systems than isothiocyanates, and are thus encountered rather infrequently. However they hold an important place historically, having been employed in some of the earliest protein derivatisation experiments. Creech and Jones (1940, 1941) and Coons *et al.* (1941) derivatised albumin and rabbit antibodies with fluorescent hydrocarbon isocyanates such as β-anthryl isocyanate (Figure 6.44), demonstrating that their immunological properties remained essentially unchanged. Reactivity was principally determined by the solubility of the particular isocyanate and the fraction of organic co-solvent (dioxane) employed: between four and 30 hydrocarbon groups could be incorporated per albumin molecule, for example. Landsteiner (1945b) noted that antigens derivatised with isocyanates could still be recognised by antisera to the native protein. Early immunogen preparations also exploited this chemistry, for example the coupling of estrone to albumin *via* its 17-isocyanate (Sehon *et al.*, 1958).

McDevitt *et al.* (1963) observed that fluorescein isocyanate yielded approximately twofold greater emission efficiency than the isothiocyanate, when used to label horse diphtheria antitoxin.

If required, isocyanates can be generated from acyl azides by the Curtius rearrangement, and a few useful examples of the latter are commercially available. On the whole however, isothiocyanates are much more convenient to handle, and provide an important class of protein-labelling reagents.

The reactivity of isocyanates for hydroxylic compounds has been exploited in a heterobifunctional reagent (see Section 6.14).

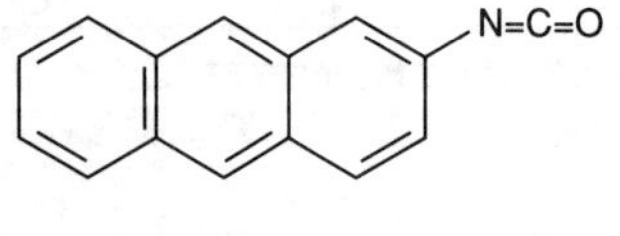

Figure 6.44 β-Anthryl isocyanate.

[1] Such byproducts of isothiocyanate chemistry are well-known artefacts of amino acid sequencing by the Edman degradation technique, described briefly in Section 6.3.2.2.

6.3.2 Coupling of proteins to small molecules containing isothiocyanates

Fluorescein 5- and 6-isothiocyanate, FITC Isomer I and II (references below) and tetramethylrhodamine 5- and 6-isothiocyanate, TRITC Isomer G and R (Felton and McMillion, 1961; Smith *et al.*, 1962) represent

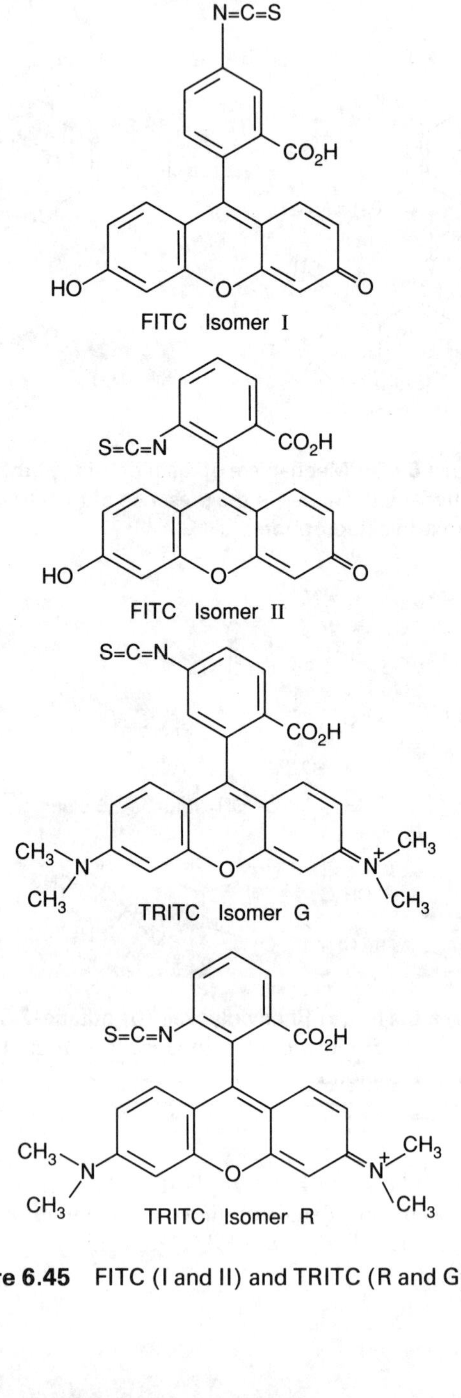

Figure 6.45 FITC (I and II) and TRITC (R and G).

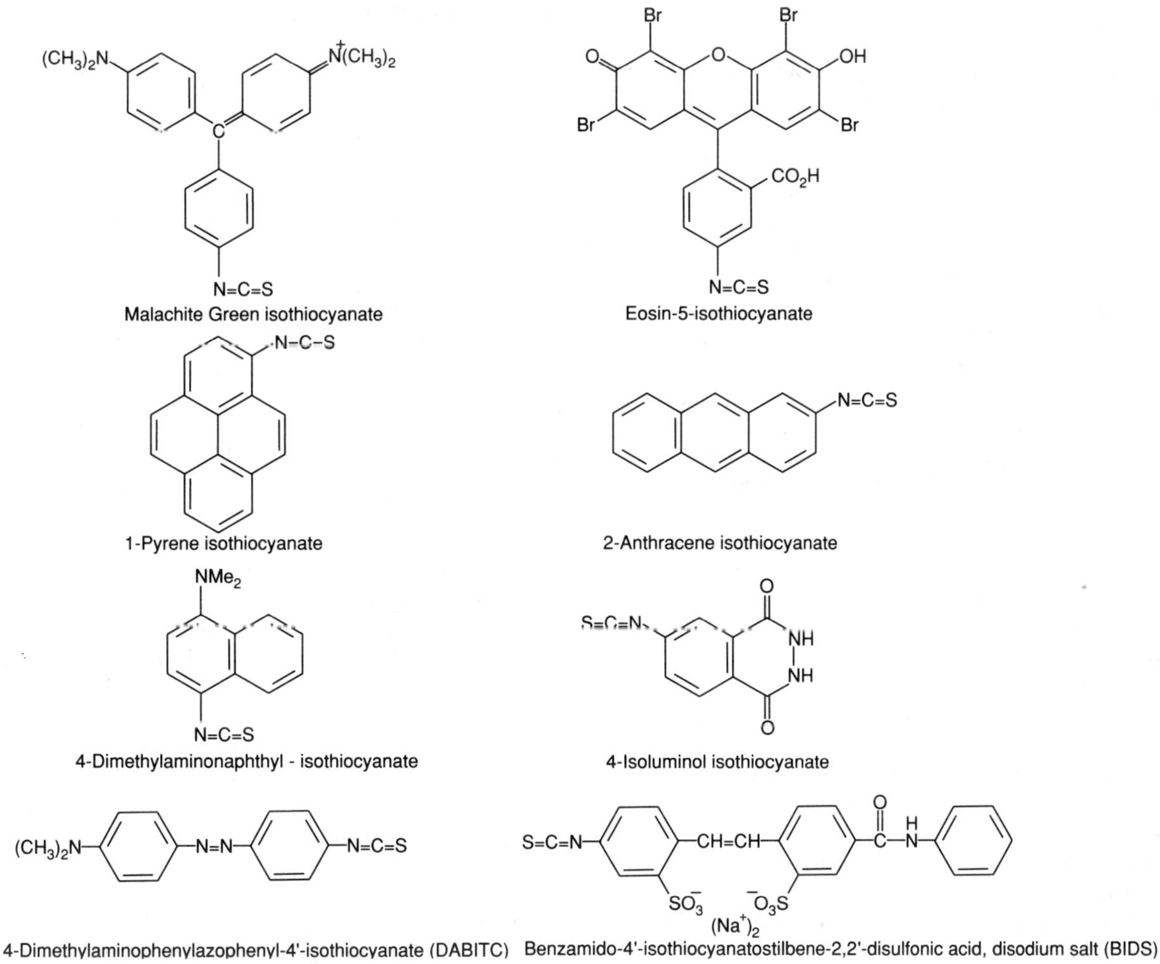

Figure 6.46 Fluorescent isothiocyanates.

some of the most commonly employed reagents for fluorescent labelling. They are shown in Figure 6.45. Many other fluorescent isothiocyanates are commercially available (see Figure 6.46).

The choice of isomer for fluorescent labelling is not always straightforward: for example, Steinbach and von Mayersbach (1976) observed that a stronger emission was obtained with FITC Isomer I when conjugated to human and rabbit γ-globulin, but that Isomer II gave superior results with BSA.

6.3.2.1 Direct labelling of protein amines with isothiocyanates

In terms of the conditions used for coupling to proteins, the FITC isomers are representative of their class, and are dealt with in some detail here.

It should be noted that quite significant batch-to-batch variations of organic isothiocyanates have been observed since early reports, notably between amorphous and crystalline forms (Goldstein *et al.*, 1961; McDevitt *et al.*, 1963; Amante *et al.*, 1972) and in terms of purity (Felton and McMillion, 1961; McKinney *et al.*, 1964). This makes quantitative comparison between early references somewhat imprecise.

Goldstein *et al.* (1961) carried out some optimisation on the reaction of FITC with globulins for histochemistry. Contemporary procedures recommended an FITC–IgG loading of 50 mg/g (approximately 20:1), but Goldstein observed high non-specific staining from such conjugates. Instead a ratio of 2.4–3.2:1 was recommended, yielding incorporations of 4–7 (24 hours, 6 °C, pH 9.0–9.5). Reduction of the reaction time to two hours resulted in approximately one-third less incorporation.

McDevitt *et al.* (1963) obtained broadly similar results, observing the best signal-to-noise ratio in staining experiments with conjugates bearing a single fluorescein molecule. Their reactions with protein were carried out at pH 7, probably resulting in the lower incorporations observed (cf. Goldstein *et al.*).

McKinney *et al.* (1964) observed extensive variations in FITC Isomer I purity from one source to another, as determined by infrared spectroscopy. These authors also showed that protein concentration, time, pH and temperature had significant effects on reaction rate. In 0.05 M phosphate buffer (pH 9.5), with a protein concentration of 25 mg/ml, reaction was essentially complete within 30 minutes at 25 °C. At a 2.4-fold initial loading, incorporations of around four could be obtained with rabbit γ- globulin.

Wood *et al.* (1965) recommended a tenfold FITC loading, using subsequent fractionation to obtain optimal globulin conjugate performance. It was noted that allowing the pH to drop below 9 caused a loss in efficiency.

Spendlove (1966) recommended a loading of 20 mg/g (8:1) for staining of mammalian cells with FITC–globulin conjugates.

The and Feltkamp (1970a,b) optimised FITC labelling of IgG for histochemistry. The method of choice used a 3.85-fold excess of FITC (60 minutes, room temperature, pH 9.5) and resulted in a molar incorporation of around 2.5. Boorsma *et al.* (1976) followed this procedure and obtained incorporations between 2.4 and 3.2.

Bergquist and Schilling (1970) described FITC-labelling of IgG (80 minutes, room temperature, pH 9.5) using a loading of 33 mg/g (13:1). This gave an incorporation of five.

Reisberg *et al.* (1970) described conditions for FITC labelling of IgG, which was adsorbed onto antigen-labelled bentonite to prevent reaction of binding-site amines. Conjugation was carried out in a carbonate–bicarbonate buffer (pH 9.5) at 0 to 2 °C; the FITC was dissolved in acetone, diluted into 20 volumes of the carbonate–bicarbonate buffer and added to the protein solution at a rate of 1 μl/sec. Thirty to sixty minutes reaction was allowed. Molar FITC excesses up to 25:1 gave incorporations up to about six.

Instability of the FITC in solution was noted by these authors, particularly in the first hour after dissolution: after 60 minutes at least twice as much FITC was required to obtain the same incorporation.

Aalberse (1973) obtained a fluorescein incorporation of 2–5 by derivatisation of sheep IgG with a 68-fold excess of FITC (two hours, room temperature, pH 9.6).

Brandtzaeg (1973a,b) derivatised IgG with FITC (one hour, 22–24 °C, pH 9.5), the isothiocyanate being employed as a solid suspension. Two major populations of conjugate were obtained, with fluorescein incorporations of less than 2.5, and 3.8 to 5.3. Some batch-to-batch difference in the FITC itself was observed, as well as changes on long-term FITC storage.

Rapid procedures for derivatisation of proteins before electrophoresis have been described, using pH or temperature conditions rather more severe than those encountered in preparative experiments: 30 minutes, room temperature, pH 11.1 (Strottmann *et al.*, 1983), or ten minutes, 50 °C, pH 9.5 (Muramoto *et al.*, 1977).

Haaijman *et al.* (1986) used four- to twentyfold excesses of FITC to produce immunogens from various proteins (pH 9.0: two hours, room temperature, followed by eight hours at 4 °C). Incorporations between one and four were obtained.

Bailey and Rasums described FITC-labelling of antibodies in a 1989 patent, using a FITC:IgG ratio of 130:1 (35 minutes, 37 °C, pH 9.5). DMF was used to dissolve the FITC. These conditions gave an incorporation of 28.2 while retaining 90.7% of immunoreactivity. Higher FITC:IgG ratios or reaction times resulted in fairly small increases in incorporation while seriously affecting the immunological properties of the antibody.

Banks and Paquette (1995) reported that a fluorescein succinimidyl ester gave both faster reaction and a more stable product than FITC in the derivatisation of myoglobin; this lack of stability has also been noted elsewhere (Haugland, 1992).

Patel *et al.* (1982) prepared a luminol–isothiocyanate reagent, and coupled it at a twofold excess to rabbit IgG (twelve hours, pH 9.5), obtaining a molar incorporation of 0.2. The luminescent efficiency of the reagent was unaltered.

Isothiocyanates have also been employed for introducing polycarboxylic chelating agents into proteins for radiopharmaceutical purposes (Gansow, 1991). Thus Meares *et al.* at the University of California (1984) described the preparation of (*S*)-4-{2, 3-bis [bis(carboxymethyl)amino]propyl}phenylisothiocyanate (CITC; Figure 6.47), and its use for labelling monoclonal antibodies. A CITC:IgG ratio of around 9:1 was employed under typical reaction conditions (two hours, 37 °C, pH 9.0–9.5). It was noted that incorporation of up to three or four chelators of this type did not seriously affect immunoreactivity.

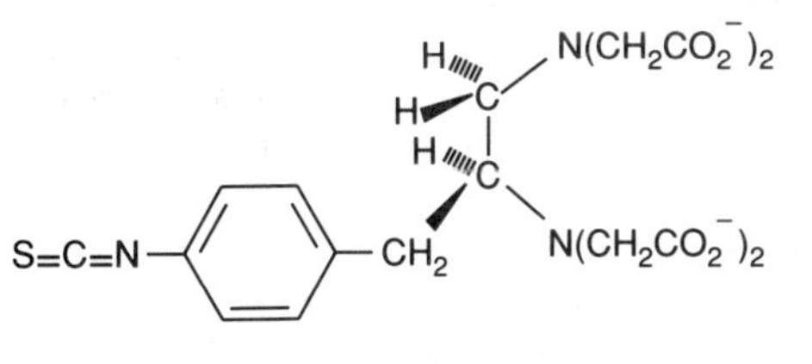

Figure 6.47 (*S*)-4-{2,3-Bis[bis-(carboxymethyl)amino]propyl}phenylisothiocyanate (CITC).

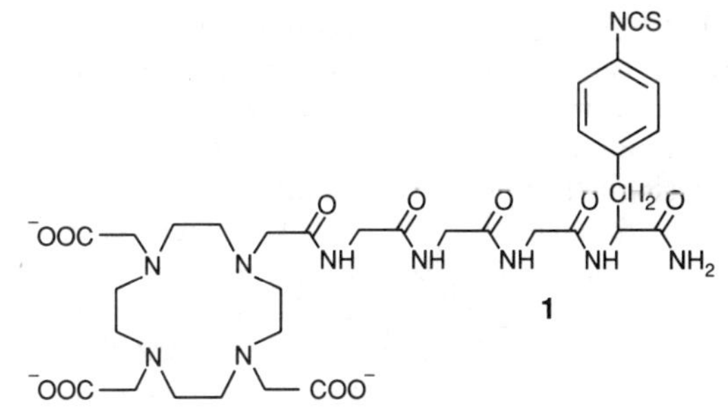

Figure 6.48 DOTA-peptide polycarboxylic isothiocyanate.

Brechbiel *et al.* (1986) described the preparation of isothiocyanatobenzyl derivatives of EDTA and DTPA, again using them to label monoclonal antibodies (two hours, pH 8.5). An initial ratio of 3:1 resulted in an incorporation of 1.4. Westerberg *et al.* (1988) described the derivatisation of a Mab (three hours, 37 °C, pH 8.5) with similar compounds, and obtained an immunochemically-optimal incorporation of 3 from a fivefold loading.

Other polycarboxylic isothiocyanates have since been introduced for this purpose, for example the DOTA-peptide (Figure 6.48) described by the California group (Li and Meares, 1993; Li *et al.*, 1994). Reaction conditions of 1–2.5 hours at 37 °C and pH 8.9–9.5 were described for labelling Mabs with this reagent. Gansow and Kumar (1990) similarly described a DOTA–phenylisothiocyanate with ^{212}Bi radioimmunotherapy in mind.

Linder *et al.* (1991) synthesised an isothiocyanate derivative of a BATO (boronic acid adduct of a technetium dioxime) and used it to label a Mab (two hours, 37 °C, pH 9.5). This offered a route to a ^{99m}Tc-radioimmunoconjugate.

Eskola *et al.* (1985) coupled a europium-containing chelate to sheep IgG, using a 400-fold molar excess (20 hours, 4 °C, pH 9.0) and obtained a europium incorporation of around 17.7 per IgG molecule. Kankare *et al.* (1992) produced terbium chelates of polyclonal antisera (overnight, pH 9.5), obtaining an incorporation of four from a 60-fold loading. The resulting conjugates were used in an electroluminescence immunoassay.

Smith *et al.* (1978) prepared isothiocyanate derivatives of oligosaccharides and coupled them to BSA (18 hours, room temperature, pH 9.5)

Andersson and Oscarson (1993) synthesised an isothiocyanate derivative of lactose (Figure 6.49) and coupled it to BSA (48 hours, room temperature, pH 8.5) to yield a neoglycoprotein. The isothiocyanate was generated *in situ* from an amine derivative by treatment with thiophosgene—a commonly employed synthetic strategy for the preparation of isothiocyanates in general (Schubiger *et al.*, 1996).

Isothiocyanate derivatives remain a fairly popular route for the introduction of small molecules into proteins, both for affinity labelling (Jacobson *et al.*, 1995) and for preparative work such as the incorporation of metal complexes for energy transfer studies (Youn *et al.*, 1995; Li and Selvin, 1997).

6.3.2.2 Other approaches to protein labelling with isothiocyanates

Isothiocyanates can be converted by hydrazine treatment to thiosemicarbazides, which react similarly to hydrazides (see Section 6.8.4).

Another important use of isothiocyanates in protein chemistry ought to be mentioned briefly. Edman degradation (Edman, 1950, 1953, 1956; Edman and Begg, 1967), a procedure for the determination of N-terminal amino acid residues, is based on the derivatisation of amine groups with phenyl isothiocyanate

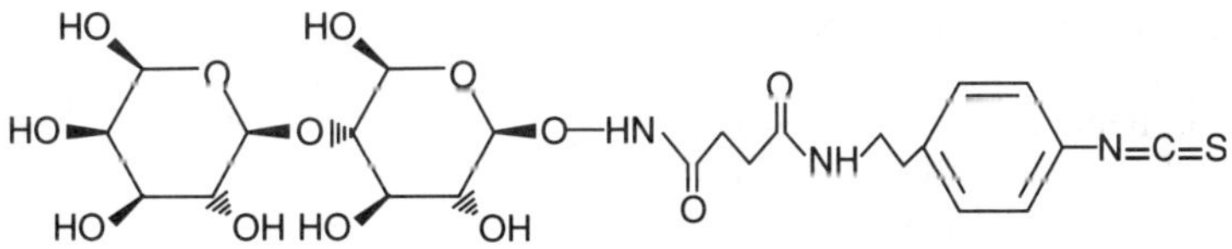

Figure 6.49 An isothiocyanate derivative of lactose.

(PITC) at around pH 9 (Figure 6.50(a)). Anhydrous strong acids such as trifluoroacetic acid are then employed to cleave the terminal residue in the form of an unstable anilinothiazolinone (Figure 6.50(b)), exposing a 'new' terminal amine.

The thiazolinone is converted by aqueous acid to a stable phenylthiohydantoin (Figure 6.50(c)), for which chromatographic identification techniques are readily available. At each repetition of the derivatisation–cleavage–identification cycle (within certain constraints) the identity of another residue in the amino acid sequence is established. PITC analogues such as 4-dimethylamino-azobenzene-4′-isothiocyanate (DABITC) have also been used (Chang and Creaser, 1976; Chang, 1977), this particular reagent giving brightly coloured phenylthiohydantoin derivatives which are convenient for analysis.

Kubota and Tsugita (1980) described a modification of this technique for the determination of protein subunit sizes. The protein's amine groups were saturated with an isothiocyanate (e.g. PITC), subjected to one cycle of Edman degradation to cleave the terminal amine, then the newly exposed α-NH_2 derivatised with a second readily-detectable reagent (e.g. [^{14}C]PITC). The incorporation of the second reagent gives an estimate of molecular weight and/or the number of subunits present.

Jay (1984) and Jue and Doolittle (1985) used very similar approaches to end-label proteins with Bolton–Hunter reagent and FITC respectively. Selective proteolysis and fragment analysis then gave information on the location of specific amino acids in relation to the N-terminus.

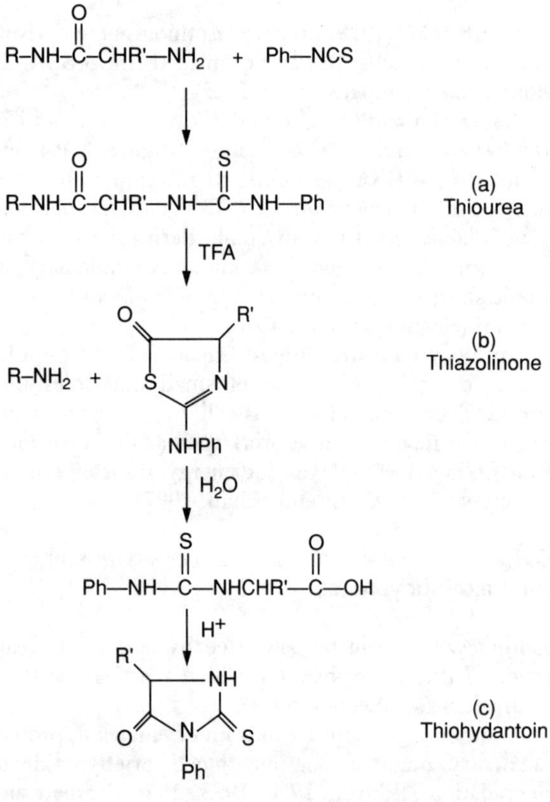

Figure 6.50 Edman degradation. Phenyl isothiocyanate reacts with the terminal amine of a peptide to generate the usual thiourea (a). Under the influence of anhydrous trifluoroacetic acid (TFA), however, this cleaves off an anilinothiazolinone (b), reducing the peptide chain by one residue and exposing a new terminal amine. The thiazolinone is converted to a more stable thiohydantoin by aqueous acid treatment (c), and the latter is characterised to identify the terminal residue *via* the unique side-chain R′.

6.4 Coupling of Proteins to Small Molecules Containing Sulfonyl Halide Groups

6.4.1 Sulfonyl chlorides

Sulfonyl chlorides are the sulfonic acid analogues of carboxylic acid chlorides, reacting with amines to form highly stable sulfonamides (Figure 6.51). They can also react with tyrosine, histidine and cysteine residues, and with hydroxyl moieties from serine or from saccharides.

An important reagent in this class is 5-dimethylaminonaphthalene-1-sulfonyl chloride, dansyl chloride (Figure 6.52), which was developed by Weber in 1952. This compound is non-fluorescent, but the sulfonamides formed from its reaction with amines fluoresce strongly at around 578 nm in water. It has been used for example in the investigation of active-site residues (Hartley and Massey, 1956; Renthal *et al.*, 1987), determination of the N-terminal residue in proteins (Gray, 1972), and as a fluorescent label in many

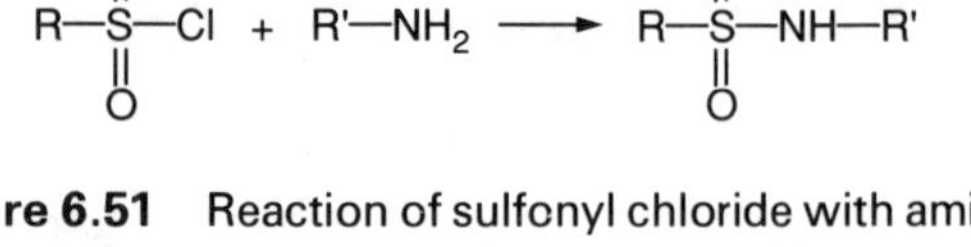

Figure 6.51 Reaction of sulfonyl chloride with amine to form a sulfonamide.

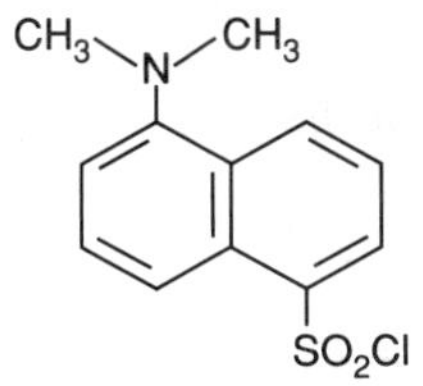

Figure 6.52 Dansyl chloride.

chromatographic and electrophoretic detection procedures (Talbot and Yphantis, 1971; Stephens, 1975; Tijssen and Kurstak, 1979).

Weber coupled dansyl chloride to ovalbumin and BSA (five to twelve hours, 0–3 °C, pH 7.5 to 8), the sulfonyl chloride was used as an acetone solution. Incorporations of 1.7–2.4 (ovalbumin) and 1–3 (BSA) were obtained. Such long reaction times are probably unnecessary however: Cuatrecasas *et al.* (1967) derivatised *Staphylococcus aureus* nuclease (5 mg/ml) with an acetone solution of dansyl chloride (10–20 minutes, 25 °C, pH 8.2): a 12% excess of the chloride resulted in incorporations between 0.8 and 1.2, without loss of nuclease activity.

Lissamine™ Rhodamine B sulfonyl chloride[1] (Figure 6.53), emitting at 576 nm, was used by Brandtzaeg (1973a) to derivatise IgG (one hour, 22–24 °C, pH 9.5). At low excesses the increase in incorporation was linear, up to about two fluorophores per IgG molecule at a fivefold loading. Above this level the amount of labelling obtained was relatively smaller, presumably due to differential amine reactivities in the protein. Brandtzaeg noted that storage of this reagent at −20 °C for nine weeks resulted in a halving of the conjugation efficiency. Texas Red™ from Molecular Probes Inc. (Figure 6.54) is a fluorescent sulfonyl chloride whose sulfonamide conjugates emit at around 615 nm (Titus *et al.*, 1982).

The europium chelator 4,7-bis(chlorosulfophenyl)-1,10-phenanthroline-2,9-dicarboxylic acid, BCPDA (Figure 6.55) has been used to label thyroglobulin *en route* to the production of streptavidin reagents containing extremely high Eu^{3+} loadings[2] for time-resolved fluorescence immunoassays (Diamandis *et al.*, 1989; Diamandis, 1991). The coupling reaction (one hour, room temperature, pH 9.1) was carried out using an ethanol solution of the sulfonyl chloride, added dropwise in five portions.

Ci *et al.* (1995) prepared sulfonyl chloride derivatives of β-diketone europium chelators, and coupled them to BSA (one hour, room temperature, carbonate buffer). Using an initial loading of 100:1, incorporations of between 18 and 63 were observed.

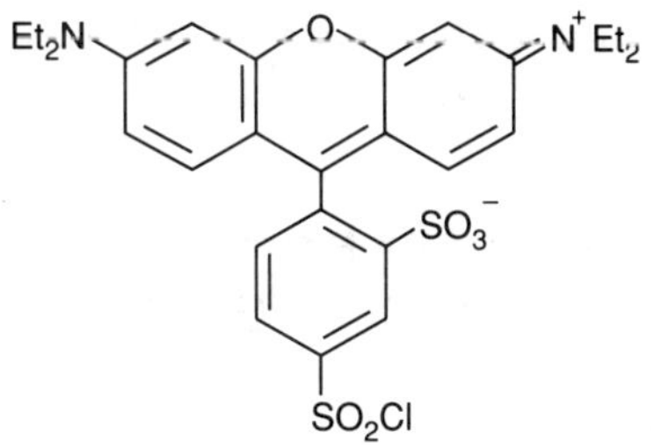

Figure 6.53 Lissamine™ Rhodamine B sulfonyl chloride.

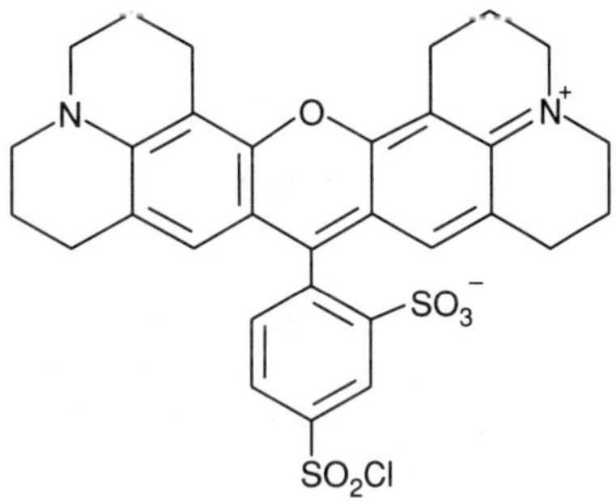

Figure 6.54 Texas Red™Sulfonyl chloride.

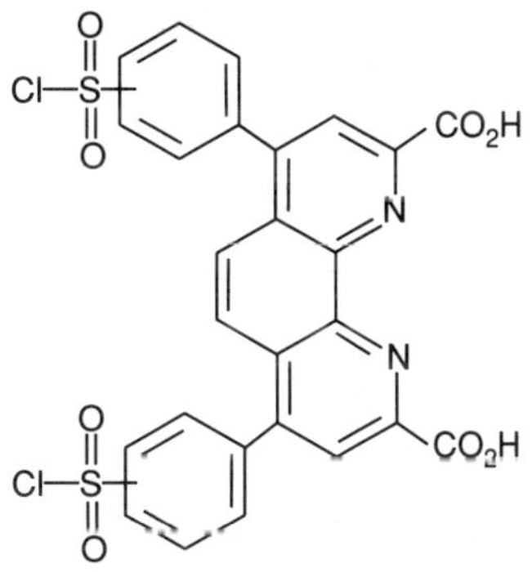

Figure 6.55 4,7-Bis(chlorosulfophenyl)-1,10-phenanthroline-2,9-dicarboxylic acid (BCPDA).

[1] Lissamine is a trademark of ICI.

[2] Each thyroglobulin molecule could be labelled with 160 phenanthroline moieties, each capable of bearing up to six Eu^{3+} ions. After conjugation to streptavidin, this conjugate was used to detect antibodies labelled with up to ten biotin groups per molecule, giving a theoretical 9600-fold amplification of each antibody-binding event.

6.4.2 Sulfonyl fluorides

Sulfonyl fluorides are less reactive than sulfonyl chlorides towards amine groups, and can be used for selective labelling of active-site serine residue (see also Section 5.12.4.2): Polgar and Bender (1966), for example, derivatised the active-site serine of subtilisin with phenylmethanesulfonyl fluoride. The reaction took place almost instantaneously at pH 7.0. Radiolabelled reagents containing sulfonyl fluoride groups have also been used to demonstrate derivatisation of both tyrosine and lysine residues (Esch and Allison, 1978; Zoller and Taylor, 1979).

6.5 Coupling of Proteins to Small Molecules Containing Aryl Halide Groups

6.5.1 Phenyl halides

Phenyl halides react readily with amine groups, eliminating the halogen as a halide leaving group to yield an *N*-phenyl derivative.

Due to their convenient chromophoric properties, nitrophenyl compounds have received attention since the earliest days of protein derivatisation, and nitrophenyl halides offer a convenient route to their introduction. As early as 1923, Aberhalden and Stix used 1-chloro-2,4-dinitrobenzene in an attempted characterisation of silk fibroin hydrolysates. In his historic work on insulin structure Sanger (1945) adapted this approach, employing the more reactive 1-fluoro-2,4-dinitrobenzene (FDNB; Figure 6.56(a)) to identify amine-containing residues.

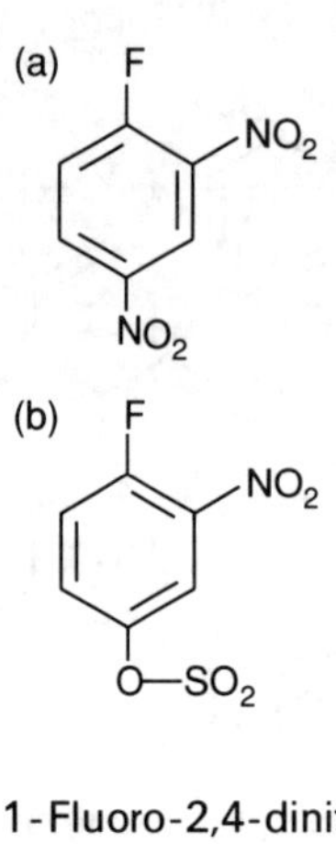

Figure 6.56 (a) 1-Fluoro-2,4-dinitrobenzene (FDNB) and (b) analogue (SNFB).

Sanger derivatised insulin for two hours in sodium bicarbonate solution, using an ethanolic FDNB solution. By analysis of the product he was able to show that the protein contained six amino groups, two ϵ-NH_2 groups from lysines, and four α-NH_2 groups, two each from glycine and phenylalanine. The multi-chain structure of insulin was inferred from these data. The method was widely adopted for demonstrating the identity of N-terminal residues, α-dinitrophenylated amino acids being identified by chromatography after hydrolysis. Hirs (1956), for example, demonstrated that lysine was the only α-dinitrophenylated amino acid in FDNB-treated oxidised ribonuclease A, and therefore formed the N-terminal residue.

FDNB also reacts with the hydroxyl group of tyrosine, the imidazole ring of histidine, and the thiol group of cysteine (Sanger, 1945; Whitaker and Jandorf, 1956). With model peptides, Hirs *et al.* (1965) showed that the OH group of tyrosine reacted with FDNB five to seven times faster than terminal or lysyl amines, or histidines. In ribonuclease A, however, FDNB treatment (15 °C, pH 8.0) resulted in derivatisation of two lysines within 60 minutes, reaction with histidines and tyrosines only began to become significant after three to four hours.

François *et al.* (1972) coupled FDNB (as a dioxane solution) to HRP at a 22-fold excess (20 hours, room temperature, pH 8). A molar incorporation of two was obtained.

Nakane and Kawaoi (1974) used an ethanol solution of FDNB to block the amine groups of HRP before coupling to antibodies *via* carbohydrate moieties. The reaction was carried out for one hour at room temperature in 0.3M sodium bicarbonate (pH 8.1).

Burkot *et al.* (1985) used FDNB to assess the susceptibility of monoclonal antibodies to derivatisation *via* their amine groups.

The reaction was carried out in 0.01 M phosphate-buffered-saline, (pH 7.4) for one hour at room temperature. Five out of eight Mabs were inactivated by this procedure, and also by subsequent coupling to periodate-oxidised HRP. Jeanson *et al.* (1988) applied the same approach but were not able to detect a clear correlation in performance.

Mullink *et al.* (1987) used aryl halides such as FDNB to haptenise monoclonal antibodies, rendering them immunoreactive towards the appropriate anti-hapten antibody: the resulting reagents were used in immunohistochemical staining procedures. The aryl halides were dissolved in 0.1 M sodium bicarbonate solution and coupled to proteins. The conditions used for the protein coupling reaction (45–60 minutes, 37 °C, in the dark, pH 8.0–8.5) represented the best compromise between under- and overlabelling for histochemical purposes.

Carty and Hirs (1968) described the FDNB analogue 4-sulfonyloxy-2-nitrofluorobenzene (SNFB; Figure 6.56(b)), claiming advantages over FDNB in terms of improved aqueous solubility and more convenient spectral properties. The lower reactivity also resulted in a greater selectivity for amine groups, and offered the possibility of obtaining selective reactivity at a higher alkalinity. Carty and Hirs derivatised ribonuclease A with SNFB (ten days, 3.6 °C, pH 8.1) and observed reaction of the terminal amine group and the Lys41 residue.

Affinity reagents incorporating nitrophenyl halides have been described, allowing an active-site lysine to be identified in aspartate aminotransferase for example (Riva *et al.*, 1980).

Aryl halides continue to be encountered in the derivatisation of small molecules for attachment to proteins, for example the fluoro- and nitrofluorophenyl porphyrin compounds described by Bedel-Cloutour *et al.* (1996).

It should be noted that aryl halides can undergo photochemical activation (Bayley, 1983a). This rather limited field is discussed briefly in Section 6.16.3.3.

6.5.2 Triazine halides

Triazine halides display powerful reactivity towards amines, for example in the coupling reagent 1,3,5-trichlorotriazine (cyanuric chloride; see Section 5.3.4). This approach has been used as the basis for small-molecule labelling, most notably in the fluorescent label dichlorotriazine fluorescein (DTAF; Figure 6.57) (Wadsworth and Salmon, 1986). Banks and Paquette (1995) demonstrated that the product of DTAF labelling of myoglobin was more stable than that obtained with FITC, but that the rate of reaction was slower than that of a succinimide ester derivative.

The preparation of chlorotriazinyl derivatives of small molecules for coupling to proteins is occasionally encountered, for example in the preparation of antibody conjugates of phenylethynylpyridines (Karsilayan *et al.*, 1997).

6.6 Coupling of Proteins to Small Molecules Containing Other Amine-reactive Groups

6.6.1 Imidates

Monofunctional imidates had been in use since at least the early 1960s as amine-blocking reagents, for example to demonstrate that peptide-albumin conjugates prepared by carbodiimide methods involved protein amine groups (Goodfriend *et al.*, 1964).

Benisek and Richards (1968) introduced heavy metal atoms into lysozyme for X-ray crystallography, using methyl picolinimidate (24 hours, room temperature, pH 8.0); the resulting picolinamidine derivative (Figure 6.58) is a chelator for transition metals.

Montgomery *et al.* (1975) synthesised an imidate derivative of diphenylhydantoin (Figure 6.59), 3-N-(methoxycarbimidomethyl)diphenylhydantoin, and coupled it to BSA to produce an immunogen. Rowley *et al.* (1975) coupled a morphine-acetimidate compound to malate dehydrogenase (three hours, 0 °C, pH 8.5) and obtained incorporations of up to around eight.

Imidate derivatives of sugars have been used to allow their incorporation into neoglycoproteins (Marshall, 1978).

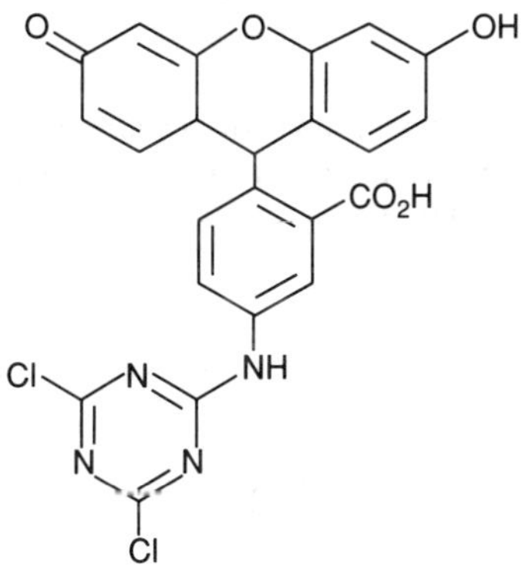

Figure 6.57 DTAF.

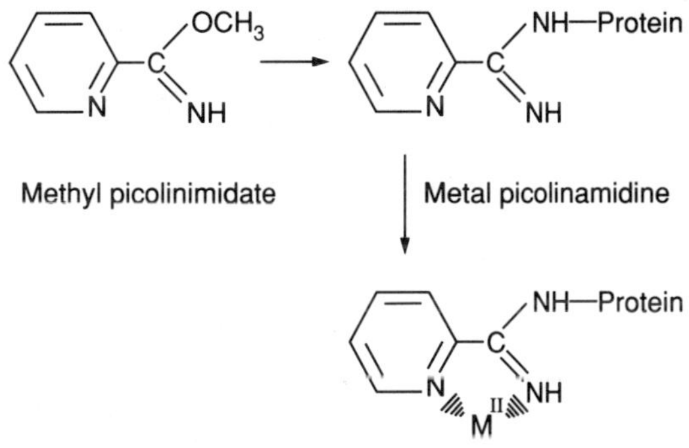

Figure 6.58 Introduction of heavy metal atoms into protein using methyl picolinimidate.

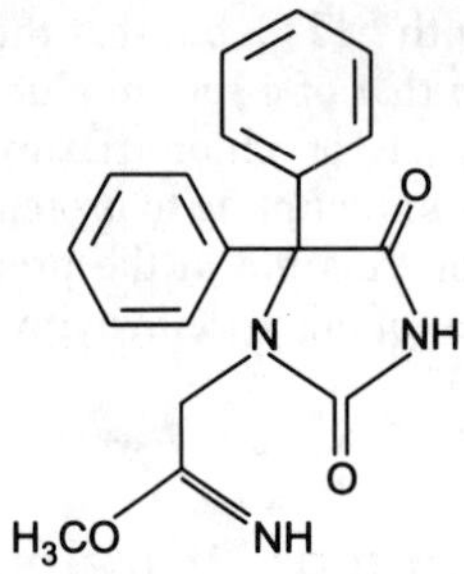

Figure 6.59 3-*N*-(Methoxycarbimidomethyl)-5,5-diphenylhydantoin.

Paik *et al.* (1980) used an imidate derivative of a chelator to attach it to lysine residues in HSA; a diazonium salt derivative of the chelator was linked to a benzimidate to provide this reagent. Scheinberg *et al.* (1982), however, obtained poor results when an imidate derivative of EDTA was coupled to monoclonal antibodies.

The use of bifunctional imidates to couple amine-containing compounds to proteins is covered in Section 6.7.6.

6.6.2 Ninhydrin, fluorescamine and analogues

Ninhydrin (1,2,3-triketohydrindene hydrate; Figure 6.60(a)) was first synthesised in 1910 by Ruhemann, who shortly afterwards discovered its colour reaction with amine-containing compounds such as amino acids and peptides (West, 1965). The reaction involves the formation of highly-coloured dimers such as that shown in Figure 6.60(b).

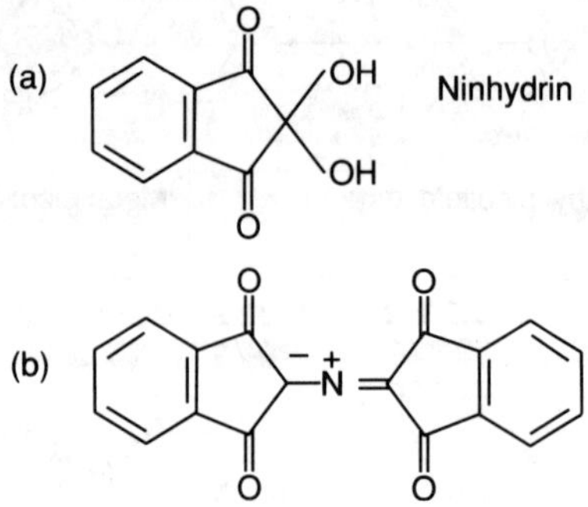

Figure 6.60 (a) Ninhydrin and (b) its highly coloured dimer.

It later became apparent that fluorescent products could be obtained under certain conditions, offering the possibility of far more sensitive analytical applications. In particular, assays for phenylalanine were developed which depended on the presence of a peptide for high levels of fluorescence (Lowe *et al.*, 1958; McCaman and Robins, 1962).

Studies by Samejima *et al.* (1971a,b) established that phenylacetaldehyde, the product of the deamination and oxidative decarboxylation of phenylalanine by ninhydrin, was one of the reactive species, and that many different amines could provide the third component. These authors therefore proposed a fluorometric procedure for peptides, amino acids, amines and amino sugars, based on treatment with ninhydrin and phenylacetaldehyde.

Weigele *et al.* (1972a) established the structure of the primary fluorescent product of this reaction (Figure 6.61), and subsequently (1972b) proposed a single reagent, fluorescamine (Figure 6.62), which resulted in the same product with amines. Fluorescamine (Fluram™, Hoffmann-LaRoche) is an excellent reagent for amine determination (Udenfriend *et al.*, 1972): at mildly basic pH (typically pH 9) reaction with amines occurs in seconds, while competing hydrolysis of the reagent is complete within minutes. As neither fluorescamine nor its hydrolysis products are fluorescent,

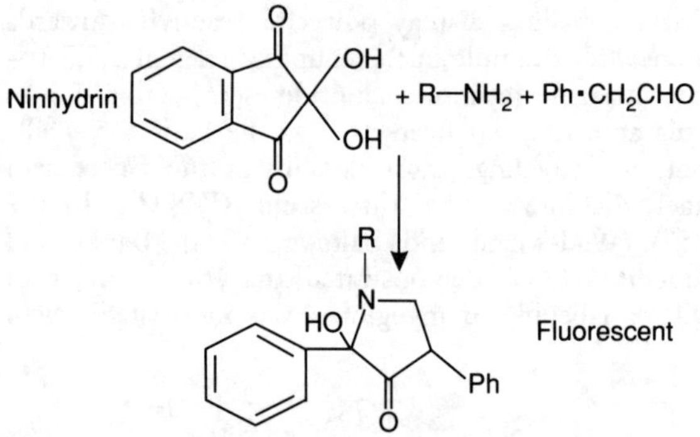

Figure 6.61 The fluorescent product formed from the reaction of ninhydrin with an amine in the presence of phenylacetaldehyde.

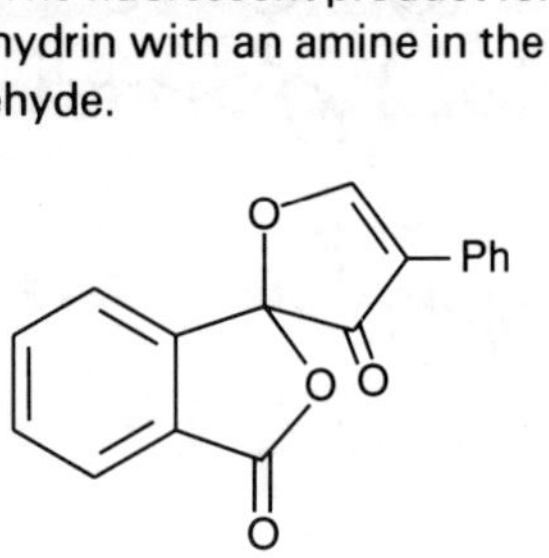

Figure 6.62 Fluorescamine.

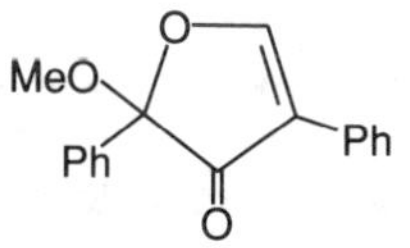

Figure 6.63 MDPF.

the amine derivative can be quantified rapidly and simply. Peptides and proteins fluoresce more strongly at pH 7 while free amino acids show maximum emission at pH 9.

Because of its rapid reaction with water fluorescamine is generally employed in an aprotic solvent such as DMSO or acetone.

Katsh *et al.* (1974) derivatised guinea pig aspermatogenic antigen with a 3.2-fold molar excess of fluorescamine (used as an acetone solution, 0.3 mg/ml). At room temperature and pH 8.3, reaction was complete within one minute; the pH was adjusted to 7.1–7.4 for measurement of fluorescence.

Sung *et al.* (1978) labelled antibodies (20 minutes, room temperature, pH 8.0) with a 140-fold excess of a fluorescamine solution, 10 mg/ml in acetonitrile. The fluorescence of cytochemical preparations stained with these reagents was stable for over 9 months.

There have been many reports of the use of fluorescamine for staining proteins on electrophoresis gels (Ragland *et al.*, 1974; Eng and Parkes, 1974; Jackowski and Liew, 1980), detection being achieved at picomolar levels (Vandekerckhove *et al.*, 1985).

A fluorescamine analogue lacking the lactone ring, 2-methoxy-2,4-diphenyl-3-(2*H*)-furanone (MDPF; Figure 6.63), was described by Weigele *et al.* (1973). The fluorescence obtained from MDPF-labelled protein fragments was reported to be 2.5-fold higher than with fluorescamine (Barger *et al.*, 1976). Labelling conditions are similar: Chen-Kiang *et al.* (1979) derivatised aldolase fragments with a methanol solution of MDPF at pH 9.5 and room temperature for 20 minutes.

6.6.3 Other amine-reactive groups

Under nucleophilic attack, sulfonate groups can function as leaving groups in much the same way as halogen atoms (March, 1985), so aryl sulfonates can display reactivity for amines. 2,4,6-Trinitrobenzenesulfonic acid (TNBS; Figure 6.64) reacts in this way, it was developed for the quantification of amines (Okuyama and Satake, 1960; Satake *et al.*, 1960) and is commonly employed for this purpose (Fields, 1972). It does not react with histidine, tyrosine, arginine or hydroxylic residues, and the reaction with thiols is much slower than with amines and yields a labile product (Haynes *et al.*, 1967).

Mullink *et al.* (1987) successfully used 4-ethoxymethylene-2-phenyloxazolone to haptenise antibodies for immunohistochemical use. The oxazolone was used as a sodium bicarbonate solution, and coupled to the antibodies for 90 minutes at 37 °C and pH 8.0–8.5, in the dark.

Carboxymethyl dithioesters react readily with lysine residues (Figure 6.65) and have been used (Souppe *et al.*, 1988) to derivatise HRP (carboxymethyl dithiotridecanoate, 18 hours, room temperature, pH 8.5) and papain (carboxymethyl dithiobenzoate, two hours, 40 °C, pH 8.5).

Pyrylium salts react with protein amines to generate the corresponding pyridinium salt (O'Leary and Samberg, 1971; Katritzky *et al.*, 1984). Salmain *et al.* (1994) used this approach to introduce a metal-carbonyl label into proteins. The pyrylium salt shown in Figure 6.66 was synthesised and coupled to BSA (24 hours, room temperature, pH 9.0). Coupling

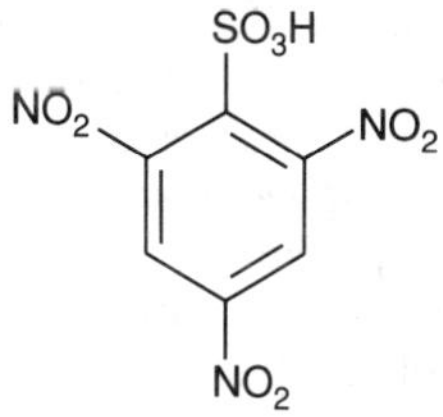

Figure 6.64 TNBS.

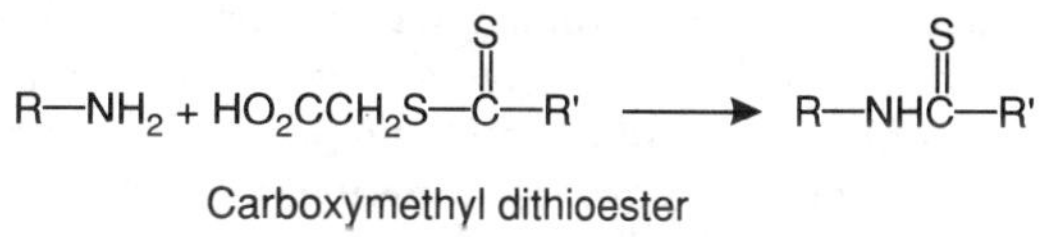

Figure 6.65 Use of carboxymethyl dithioesters for labelling amines.

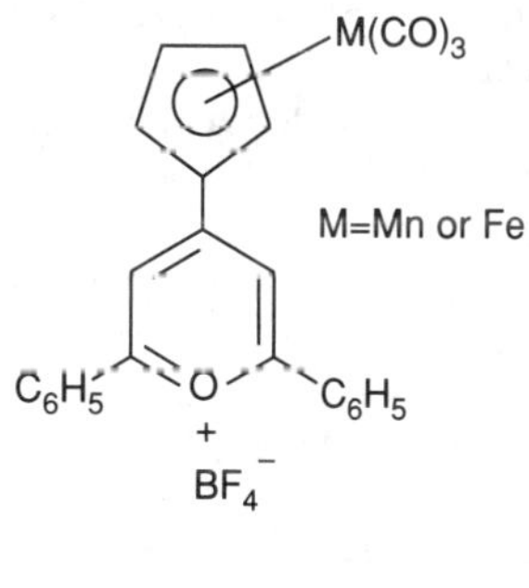

Figure 6.66 Pyrylium salt.

efficiency was very high: incorporations >5 were obtained from an initial sixfold excess.

The primarily thiol-reactive haloalkyl compounds also display substantial reactivity towards amines (see Section 6.11).

6.7 Coupling of Proteins to Small Molecules Containing Amine Groups

The methods for coupling small amine-containing molecules to proteins are of course closely related to those targetted at protein amine groups, and because the latter are so common the formation of amine–amine links *via* suitable bifunctional reagents is commonly exploited. Virtually all the techniques described in this section have already been met in another application.

However there is more variation in the characteristics of the amines which may be encountered in small molecules. The aliphatic amines of proteins have high pK_a values: 7–8 for those at the N-terminus, where the electron-withdrawing effect of the neighbouring carbonyl group lowers the basicity, and around 10 for the more isolated ε-amine groups of lysine.

Other amine groups not encountered in native protein structures (especially aromatic amines) may have pK_a values as low as 3–5. Such groups are therefore unprotonated and reactive as nucleophiles at neutral or even mildly acidic pH (cf. hydrazides, Section 6.8). Although they are weaker bases, they may therefore react more vigorously under such conditions than their aliphatic analogues. Abraham and Low (1980), for example, noted that aromatic amine fluorophores such as fluoresceinamine and 9-aminoacridine coupled readily to aldehyde groups in periodate-oxidised erythrocytes at pH 7.0, while aliphatic reagents such as dansylcadaverine were unreactive. Similarly, Nakajima and Ikada (1995) observed substantially greater incorporation of benzylamine (pK_a 4.4) than ethylenediamine (pK_{a_1} 7.15) into carbodiimide-activated carboxylic gels.

Aromatic amines undergo the useful diazotisation reaction, generating an electrophilic species coupling readily with lysine, tyrosine and histidine residues. This chemistry has seen very widespread application for the conjugation of such molecules.

6.7.1 Carbodiimide methods

The methods described in Section 6.1.1 for coupling carboxyl-containing small molecules to protein amine groups are equally applicable to the reverse reaction, particularly for carboxyl-rich proteins such as alkaline phosphatase.

Because it is the protein molecule which is activated in these approaches, the carbodiimides used are limited to the water-soluble variety. Once again, both one- and two-step protocols have been employed. Again because it is the protein that is being activated, the potential cross-reactivity of carbodiimides for other functionalities should not be overlooked; neither should interferences from buffer components (see Section 6.1.1).

Carraway and Koshland (1968) demonstrated that tyrosine residues could be activated by EDAC, and that the reaction was reversible by hydroxylamine. Carraway and Triplett (1970) demonstrated that more than 95% of the thiol content of BSA was derivatised by treatment with a 670-fold excess of EDAC (30 minutes, room temperature, pH 5.0). Furthermore the thiols could not be regenerated by hydroxylaminolysis or reduction with mercaptoethanol. Perfetti *et al.* (1976) observed reaction of cysteine, tyrosine and carboxylic residues when papain was treated with EDAC or EDAC/ethyl glycinate (one hour, 25 °C, pH 4.75).

Phosphate groups too can be activated by carbodiimides, and in phosphoproteins these positions may too be derivatised by this approach (Wang and Giese, 1995).

The potential reactivity of amines (and hence the likelihood of protein cross-linking) tends to be ignored due to their high degree of protonation under the acidic conditions employed. Timkovich (1977), however, observed cross-linking yields of up to 20% when various proteins were treated with EDAC alone: this was attributed tentatively to unfavourable routes such as the reaction of the activated carboxyl groups with the mildly nucleophilic hydroxyl groups of serine, threonine or tyrosine. Inclusion of a strong nucleophile (semicarbazide) in the reaction mixture reduced this cross-linking but did not eliminate it in all cases.

6.7.1.1 One-step carbodiimide methods

Lin (1970) used derivatisation with a 140-fold excess of EDAC/glycinamide to probe carboxyl behaviour in lysozyme and α-lactalbumin. At pH 4.75 (25 °C, 400 minutes) twenty residues in lactalbumin were amidated in the native form and a further 0.6 after denaturation with guanidine; eight of the eleven acid groups in lysozyme were derivatised and a further two after denaturation. Repeating the work at pH 7.0 led to substantially lower rates of glycinamide incorporation.

Using a more reactive aromatic amine, sulfanilic acid, as the nucleophile, Kramer and Rupley (1973) were able to derivatise all the carboxyl groups of lysozyme (six hours, pH 4.9). EDAC gave superior incorporation (9.1) to either CMC (8.0) or BDC (6.0) in this study.

Aplin *et al.* (1979) coupled the spin label TEMPO-amine to glycoproteins at an amine:EDAC:protein ratio of 100:100:1 (three hours, room temperature, pH 4.5–6.0). Although the aim was selective labelling of sialic acid residues in the saccharide moieties, coupling to carboxylic amino acids was also observed.

Yamada *et al.* (1981) coupled a number of small amines to lysozyme with EDAC, using sub-stoichiometric carbodiimide loadings to encourage selectivity for the reactive carboxyl groups in the enzyme's active site. In one example lysozyme was activated with a 3.7-fold excess of EDAC before derivatisation with ethanolamine, 14, 140 and 1400-fold loadings of the amine in the coupling reaction (two to three hours, room temperature, pH 5) resulted in increasing yields of an Asp 101 labelled derivative. Most amines examined displayed similar selectivity, but D-glucosamine and histamine gave a more complex mixture of products, tentatively attributed to non-covalent binding of these amines to the substrate-binding site, leading in turn to hindrance of the carbodiimide reaction. Atassi *et al.* (1974) showed that two acid residues (Asp119 and the C-terminal leucine) of lysozyme were selectively derivatised by treatment with EDAC/methyl glycinate or histidinate (two hours, room temperature, pH 4.75).

Swaisgood and Natake (1973) carried out similar studies with EDAC/methyl glycinate on glutamate dehydrogenase. By modulating pH, denaturant and temperature conditions they were able to obtain incorporations between 3 (four hours, 0 °C, pH 7.0) and 65 (four hours, room temperature, 6 M guanidine, pH 4.7).

Hurwitz *et al.* (1975) coupled adriamycin and daunomycin to IgG using EDAC (four hours, room temperature, phosphate, pH 7.2): a 100-fold excess of the drug and a 360-fold excess of EDAC were employed. The low incorporation obtained (four drug molecules per IgG) probably reflects the high pH employed, and possibly the phosphate medium.

Bickel *et al.* (1994) cationised IgG by derivatisation with 1,6-diaminohexane (three hours, room temperature, pH 6.0), using a diamine:EDAC:IgG ratio of 120 000:420:1. The p*I* of various antibodies was altered from 5.5–7.5 to 9 or greater by this procedure, lower ratios or higher pH resulted in less efficient reaction. Immunoreactivity was retained in the cationised IgG.

Gadow *et al.* (1984) coupled gentamicin to BSA (overnight, 4 °C, pH 4.5) using a gentamicin:CMC:BSA ratio of 80:60:1, obtaining an incorporation of 23–28 gentamicin molecules per BSA.

Hoare and Koshland (1966) used *N*-benzyl-*N'*-3-(dimethylamino) propylcarbodiimide, BDC, to facilitate modification of carboxyl groups in lysozyme, trypsin and chymotrypsin with glycine methyl ester hydrochloride. At pH 4.75 and 25 °C, reaction continued for over an hour: incorporations after 5–6 hours were uniformly higher than at 60 minutes, and approached quantitative levels in some cases. Similar work by Lin and Koshland (1969) demonstrated that all but one of the carboxyl groups of lysozyme could be derivatised by BDC or EDAC plus a suitable nucleophile.

6.7.1.2 Two-step carbodiimide methods

Gadow *et al.* (1984) activated transferrin (five minutes, room temperature, pH 6.0) with EDAC before coupling to the amine groups of thyroxine (overnight, 4 °C, pH 6.0–6.5). The thyroxine:EDAC:protein ratio was 200:50:1, the high thyroxine loading being employed to counteract its poor solubility under non-basic conditions. An incorporation of 24 thyroxine groups per transferrin molecule was obtained. Using CMC in an analogous fashion, an incorporation of 16 was obtained.

Rodwell *et al.* (1986) activated the carboxyl functions of IgG by treatment with EDAC (two hours, room temperature, pH 5.9) before coupling to an amine-containing chelator, 1,6-diaminohexyldi-*o*-hydroxyphenylacetic acid (two hours, room temperature, pH 5.9).

6.7.1.3 Methods using carbodiimides with NHS or NHSS

In situ conversion of the protein carboxyl moieties to succinimide esters can improve the efficiency of reactions of this type, and both NHS and its sulfonated analogue NHSS have been used to achieve this (cf.Section 6.1.2). Both one- and two-step protocols can again be employed.

In a one-step example, Battaglini *et al.* (1994) coupled daunomycin and dopamine to glucose oxidase using EDAC/NHSS at an amine:EDAC:NHSS:protein ratio of 25:10 000:500:1 (daunomycin: four hours, 4 °C, pH 7.0) or 100:2000:200:1 (dopamine: twelve hours, 4 °C, pH 7.0). Incorporations of 2.5 and 4, respectively, were obtained.

In a two-step procedure, Matthiasson *et al.* (1978) coupled gentamicin to catalase (pH 6.5, overnight) following an initial EDAC/NHS activation step (pH 6.5, 15 minutes). The gentamicin:EDAC:NHS:catalase ratio was around 100:300:600:1.

6.7.1.4 Anhydride derivatisation followed by carbodiimide coupling to protein amines

In a variation of the carbodiimide approach for amine-containing molecules, Yang and Reisfeld (1988) coupled doxorubicin to a Mab by a three-stage reaction. First, the drug was derivatised with *cis*-aconitic anhydride to incorporate an acid-sensitive linker. Then, without intermediate purification, the resulting carboxy compound was activated by EDAC (30 minutes, 25 °C, pH 7.0), then coupled to Mab (three hours, 25 °C, pH 7.0). An incorporation of 10 was obtained.

6.7.2 Periodate oxidation methods

Glycoproteins can be rendered amine-reactive by periodate oxidation (Chapter 5, Section 5.2.3), and, while best-known for protein-protein conjugation, this approach has been used for coupling glycoproteins to small amine-containing compounds as well (O'Shannessy and Quarles, 1987). Conditions for periodate oxidation of proteins are discussed in more detail in Section 5.2.3.

Willan *et al.* (1977) oxidised IgG with periodate (3 mM, pH 5.5, eight hours at room temperature) then coupled it to a 25-fold excess of the spin-label TEMPO-amine (six hours, room temperature pH 6). Reduction by a 40-fold excess of sodium cyanoborohydride followed. Nezlin and Sykulev (1982) carried out the same reaction at pH 8 (15 hours, 4 °C), again followed by sodium cyanoborohydride reduction.

Abraham and Low (1980) oxidised the sialic acid or galactose residues of human erythrocytes with periodate or galactose oxidase respectively, then coupled the resulting aldehydes (two hours, 37 °C, pH 7.0–7.4) to amine-containing fluorophores such as fluoresceinamine and 3-aminofluoranthene.

Ingham and Brew (1981) oxidised α_1-acid glycoprotein and human chorionic gonadotrophin for labelling with amine-containing dyes, fluoresceinamine and dansylethylenediamine. The coupling reaction (15–30 minutes, room temperature) was carried out at pH 7.1 in the presence of sodium cyanoborohydride, and was stopped with sodium borohydride. Three dansyl groups were incorporated into the gonadotrophin, and eight into the acid glycoprotein. Wisely these authors performed control experiments to confirm that the proposed coupling mechanism was in operation, showing that there was little coupling without periodate oxidation. As a contrast, albumin, which contains virtually no carbohydrate, showed an incorporation of around two dansyl groups, but this level of labelling was also seen without oxidation of the protein and could be reversed by size exclusion chromatography after denaturation. This labelling was therefore shown to be attributable to non-covalent binding.

Rodwell *et al.* (1986) coupled 1,6-diaminohexylethylenediaminedi-*o*-hydroxyphenylacetic acid to periodate-oxidised IgG (one hour, room temperature, pH 6.0) using a 270-fold molar excess of amine over IgG. Sodium cyanoborohydride was used to reduce the conjugate (four hours, room temperature, pH 6.0). Five chelate groups per IgG molecule were introduced. Similarly, two amine-containing DTPA derivatives were coupled to IgG by these authors, using a 1000-fold excess over antibody. Radioimmunoconjugates prepared from these oligosaccharide-modified reagents showed better retention of affinity than those where the antibody was derivatised *via* its amine or carboxyl functions.

Schrenk and Kürzinger (1986) described the coupling of thyroxine to periodate-oxidised HRP, although detailed conditions were not given.

Oseroff *et al.* (1986) coupled chlorin-e_6 to monoclonal antibodies *via* two reductive alkylations. A dextran was oxidised with periodate, coupled to a mixture of the chlorin and ethylenediamine, and reduced with borohydride. A 125-fold molar excess of the aminated dextran was then coupled to periodate-oxidised IgG, and the resulting Schiff base was reduced with a 140-fold excess of cyanoborohydride.

Dillman *et al.* (1988) aminated a mouse monoclonal IgG, by oxidation of the carbohydrate groups followed by incubation with a 200 to 400-fold excess of ethylenediamine (one hour, room temperature, pH 6.5). The Schiff base product was reduced with cyanoborohydride.

This approach for coupling small amines to glycoproteins and their glycans remains popular, recent examples include the incorporation of nickel-chelating groups into HRP (Jin *et al.*, 1995), and the fluorescent labelling of glycans with 2-aminobenzamide (Bigge *et al.*, 1995).

An interesting variation on this theme was described recently by Zhang and Tam (1996), using 1,2-aminothiols to generate a thiazolidine; see Section 6.9.3.

6.7.3 Benzoquinone methods

Again better known for protein–protein conjugation, *p*-benzoquinone has been used for small molecules too: Guesdon *et al.* (1986) coupled histamine and histidine to BSA or ovalbumin using this route. The proteins were activated (one hour, room temperature, in the dark) in 0.1 M phosphate buffer (pH 4.5) by a large excess (ca. 5000-fold) of benzoquinone, dissolved in ethanol. The ethanol content of the reaction mixture was 15%. After purification, the activated protein was coupled to histidine or histamine at pH 8.5 (20 hours, room temperature, in the dark). Using histamine, incorporation into BSA varied from 0.6 (using an initial 58:1 ratio) to 17 (using a 3694:1 ratio).

Optimisation of the method indicated that the best incorporations were obtained when the protein activation was carried out for one hour at 4 °C and pH 4.5, and the hapten coupling for four hours at 37 °C and pH 9.0. Temperature effects were quite pronounced at both stages: the activation reaction was much reduced at higher temperatures.

6.7.4 Glutaraldehyde methods

Given the popularity of glutaraldehyde as a reagent for protein–protein conjugation, particularly before the availability of useful heterobifunctional agents, it is not surprising that this approach has also been widely used for coupling small amine-containing molecules to proteins as well. The complexity of glutaraldehyde chemistry has been discussed in depth in Chapter 5, Section 5.3.1, and will not be repeated here. A few uses of glutaraldehyde coupling only will be described here, to exemplify conditions employed in such applications.

Hurwitz *et al.* (1975) coupled adriamycin and daunomycin to IgG with a one-step glutaraldehyde procedure. A roughly 50-fold excess of both drug and coupling agent was incubated with the protein (15 minutes, room temperature, pH 7.2). An incorporation of 7–10 was obtained, but there was a high degree of aggregation.

Miyai *et al.* (1981) coupled thyroxine (T4) to alkaline phosphatase (AP) using a one-step glutaraldehyde procedure (2.5 hours, 25 °C, pH 6.8). The T4:glutaraldehyde:AP ratio was around 45:1000:1; no indication of incorporation was given but the conjugate was used successfully in an immunoassay.

Eruk *et al.* (1984) also coupled T4 to AP in a one-step glutaraldehyde reaction. The enzyme (in a bicarbonate–barbitone buffer mixture, pH *ca.* 9.4) was allowed to react with a 100–fold excess of T4 (in 0.01 M NaOH) and a 67-fold excess of glutaraldehyde (in water) for 2.5 hours at room temperature. Potassium borohydride was used to reduce the resulting conjugate. Apparent T4 incorporations of 0.1 to 0.6 were obtained depending on the purification procedures employed.

Briand *et al.* (1985) coupled peptides to carrier proteins in a one-step procedure. A 20 to 40-fold excess of the peptide was allowed to react (one hour, 4 °C, pH 7.35) with varying excesses of glutaraldehyde. Interestingly, BSA conjugates prepared using 1% glutaraldehyde (3300-fold excess over protein) appeared to display excellent stability, no loss of peptide after 200 days at 4 °C. At half this glutaraldehyde excess, much worse stabilities were observed (50% loss at 200 days for one peptide, 70% loss at 40 days for another).

Using a two-step protocol Ferrua *et al.* (1986) activated HRP with 1% glutaraldehyde, then coupled the enzyme to T4. The coupling reaction (five hours at room temperature then overnight at 4 °C, pH 9.5) resulted in an incorporation of 1.2.

Zegers *et al.* (1990) coupled synthetic peptides to proteins using a two-step protocol: the protein carrier (keyhole limpet haemocyanin) was activated by dialysis against 0.2% glutaraldehyde (16 hours, 4 °C, pH 7.2) then coupled to a 100-fold excess of the synthetic peptide (16 hours, 4 °C, pH 7.2). The resulting conjugate was used as an efficient immunogen for anti-peptide antibodies.

6.7.5 Bis-active ester methods

Homobifunctional active esters can obviously couple amine-containing small molecules to proteins: indeed, one of the earliest descriptions of such compounds used the conjugation of tryptophan to albumin to demonstrate their performance (Tesser *et al.*, 1975). Equimolar amounts of tryptophan and a bis-(4-dimethylaminopyridyl) ester were coupled to a quantity of albumin containing the stoichiometric number of lysine groups (15 minutes, pH 10.2, dropping to 9.2 over the reaction period). Approximately one third of the available lysine residues were derivatised by tryptophan, resulting in an incorporation of 17.

Paek *et al.* (1993) used disuccinimidyl suberate (DSS) to couple an amine-containing progesterone derivative to HRP, using a steroid:DSS:HRP ratio of 20:40:1. The steroid and the DSS were dissolved in DMSO and allowed to incubate for 30 minutes at room temperature, then the reaction mixure added dropwise to the enzyme, dissolved in phosphate-buffered saline. Two hours coupling at room temperature was allowed, resulting in conjugates bearing one or two progesterone molecules per enzyme: intensive characterisation was carried out to identify two different monosubstituted conjugates, presumably representing alternative derivatisation positions, as well as a homogeneous disubstituted species.

Chen, C.-J. *et al.* (1994) used a 50-fold excess of DSS to activate a carboranyl oligophosphate CB 10 (two hours, room temperature, pH 7.0, DMF), using ethyl acetate extraction to remove unreacted DSS. The resulting solution was coupled to a Mab (two hours, room temperature, PBS) at a five to 25-fold excess: the resulting incorporations ranged from around 2 to 11.

Linker groups with very different properties can also be introduced by using suitable bis-succinimides, e.g. polyethylene glycol bis-esters can be used to produce a long and highly hydrophilic link (see also Chapter 8, Section 8.5.2). This approach was used by Sato *et al.* (1995), for example, to couple deferoxamine to superoxide dismutase for therapeutic purposes; PEG confers useful *in vivo* properties to the conjugate.

6.7.6 Bis-imidate methods

While bis-imidates have rarely been employed for preparative protein–protein conjugation (Chapter 5, Section 5.3.2), they have found some use in small-molecule applications.

O'Sullivan *et al.* (1978) activated triiodothyronine (T3) with dimethyladipimidate (DMA) in dry methanol containing 5% (w/v) *N*-ethylmorpholine (ten minutes, room temperature) before coupling to β-galactosidase at pH 9.9 (90 minutes, room temperature). Equimolar T3 and DMA were employed, but the excess over β-Gal was varied from 100 to 5000:1. There was a steady increase in the amount of conjugate which could be bound by anti-T3 antiserum right up to the highest loading, although around 25% of the enzyme activity was lost. At the 1000:1000:1 ratio, eight T3 groups were incorporated per enzyme molecule. Al-Bassam *et al.* (1978) used essentially the same conditions to couple a nortriptyline derivative to β-galactosidase.

Blake *et al.* (1982) prepared thyroxine–AP conjugates by the same route using an activation time of 30 minutes and a T4:DMA:AP ratio of 1667:3333:1. Blake *et al.* also pepared a triiodothyronine–β-Gal conjugate using a T3:DMA:β-Gal ratio of 35 000:57 000:1. The conjugates were not characterised, but were used successfully in a simultaneous immunoassay for the two thyroid hormones.

Flurkey *et al.* (1985) coupled serotonin to BSA using dimethyl pimelimidate, with a serotonin:DMP:BSA ratio of 80: 73:1 (40 minutes, room temperature, under nitrogen, pH 10). The resulting conjugate displayed a serotonin incorporation of around 50.

Varenne *et al.* (1995) coupled amine-containing carbamazepine derivatives to BSA using a two-step reaction; a 90-fold excess of the hapten was first activated with equimolar quantities of dimethyl adipimidate and triethylamine (15 minutes, room temperature, methanol) before reaction with BSA (2.5 hours, room temperature, pH 8.5, methanol). An incorporation of 33 was observed.

6.7.7 Other homobifunctional methods for amines

There are many more homobifunctional reagents for amines, mostly dating from the 'pre-glutaraldehyde' era, which have also found their way into small molecule labelling.

Spragg *et al.* (1966), for example, used toluene-2,4-diisocyanate to couple bradykinin to polylysine, while Eruk *et al.* (1984) used cyanuric chloride to couple thyroxine (T4) to alkaline phosphatase.

Even if a homobifunctional approach is favoured, methods of this type have been rendered largely obsolete by such reagents as bis-succinimide esters.

6.7.8 Methods using amine-reactive heterobifunctional reagents

Chapter 5 introduced a multitude of heterobifunctional reagents incorporating an amine-reactive moiety, and these offer a further option for conjugating small amine-bearing compounds to proteins. Such approaches are much less common than in the field of protein–protein coupling.

However, this route may have some attractions where it is desired to couple the small molecule to free thiol groups on the target protein. Not surprisingly, the reagents discussed in Section 5.7 also feature heavily here.

Kitagawa *et al.* (1976), for example, coupled the antibiotic viomycin to β-galactosidase using the heterobifunctional reagent MBS in a two-step procedure (Figure 6.67). Viomycin sulfate was incubated with MBS for 30 minutes at 30 °C and pH 7.0 before purification on Sephadex LH-20. The maleimide-containing viomycin derivative was then coupled at a 25-fold excess to β-galactosidase (overnight, 4 °C, pH 7.0).

Similarly, Kitagawa *et al.* (1978) activated ampicillin with an equimolar amount of MBS (50 minutes, 30 °C, pH 7.0), removing unreacted MBS by organic solvent

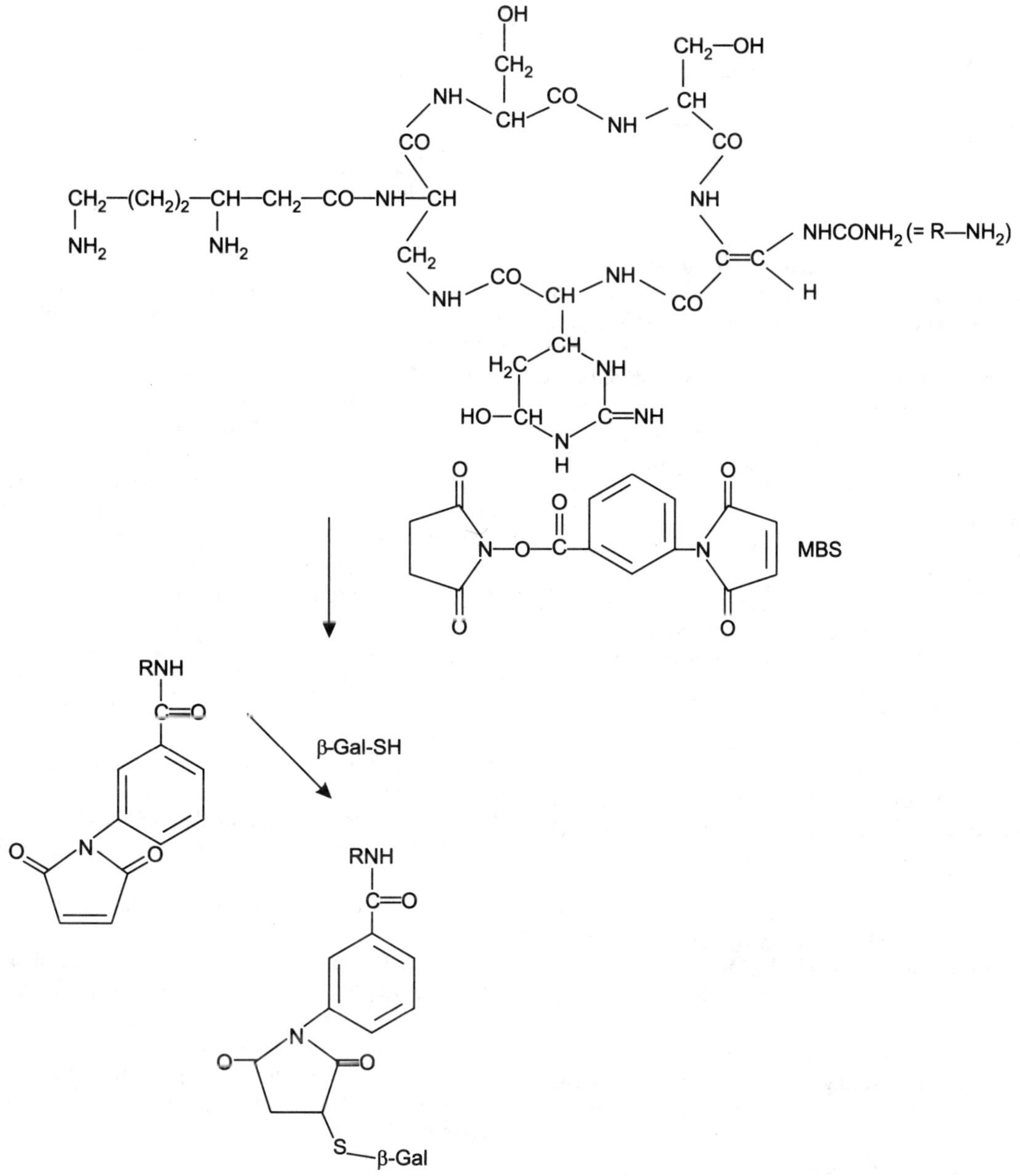

Figure 6.67 Cross-linking of viomycin to β-galactosidase using MBS.

extraction before coupling to the thiol groups of reduced BSA (two hours, room temperature, pH 7.0). A roughly 33-fold excess of the maleimide-containing ampicillin derivative was employed, and resulted in an incorporation of around 15.

These authors also coupled ampicillin to β-galactosidase using a different heterobifunctional reagent, *N*-(3-maleimidopropionyl-glycyloxy)succinimide (MPGS; Figure 6.68), to avoid bridge recognition problems with antibodies raised using the MBS–BSA immunogen. The activation reaction was similar to that just described, while coupling to the enzyme (overnight, room temperature, pH 6.0) was carried out using an elevenfold excess of the maleimidyl derivative. No decrease in enzyme activity was observed.

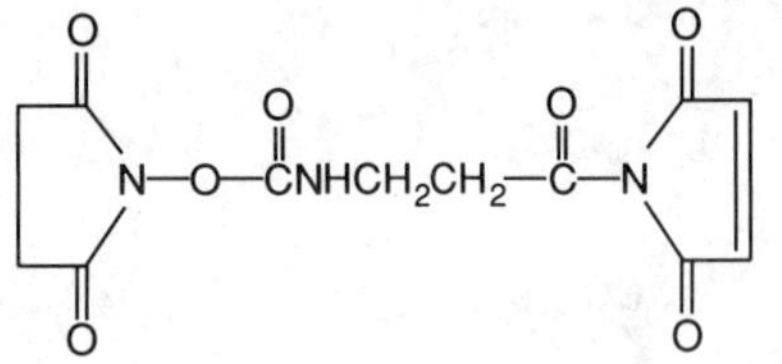

Figure 6.68 MPGS.

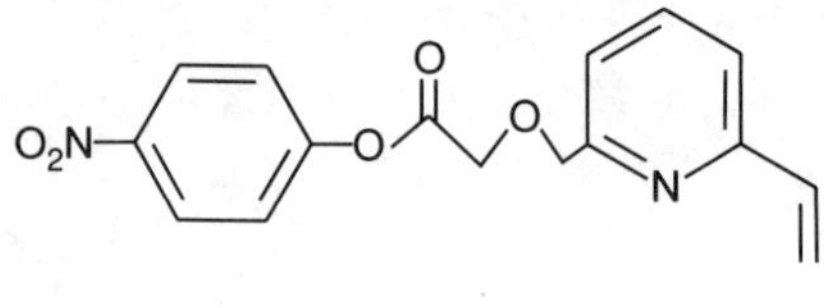

Figure 6.69 Heterobifunctional reagent based on vinylpyridine and *p*-nitrophenyl ester moieties.

Fujiwara *et al.* (1981) used GMBS to activate daunomycin (30 minutes, room temperature, pH 6.0) for coupling to thiolated BSA (30 minutes, room temperature, pH around 7.0). The GMBS was dissolved in tetrahydrofuran, and unreacted material was removed from the activated daunomycin by solvent extraction. Daunomycin was also coupled to β-Gal using MBS as described above (activation and coupling both 30 minutes, room temperature, pH 6.0). 95% of the enzyme activity was retained.

Kitagawa *et al.* (1982) used similar approaches to produce blasticidin S conjugates as immunogens (MBS, reduced BSA) and labelled immunoassay reagents (GMBS, β-Gal). Conditions were similar to those described above, but the coupling of GMBS-derivatised blasticidin to β-Gal was carried out at around pH 7.1 for two hours at 30 °C.

Chen, C.-J. *et al.* (1994) used a 50-fold excess of MBS to activate a carboranyl oligophosphate CB10 (two hours, room temperature, pH 7.0, DMF) for coupling to Fab′ fragments (tenfold excess of carborane, two hours, room temperature, pH 6.3–7.0). An extraction with ethyl acetate was used to remove excess MBS from the activated carborane. The best incorporation of CB10 obtained was around 30%.

Yamamoto *et al.* (1981) used SMCC to couple thyronines to β-galactosidase: the thyronine was activated with SMCC (90 minutes, 30 °C, pH 7.0) using DMF as co-solvent. After stopping the reaction with glycine, the maleimide-containing thyroid hormone was then conjugated to the free thiols of the enzyme (30 minutes, 30 °C, pH 7.0).

Fert and Baret (1990) coupled thyroxine to thiolated xanthine oxidase using SMCC. The hormone was activated (90 minutes, 35 °C, pH 7.3) with an equimolar amount of SMCC; the reaction was stopped by the addition of lysine, and no attempt was made to separate the product components before coupling (overnight, 4 °C, pH 7.3) at a 2.5, 5 or 10-fold excess with thiolated enzyme. Incorporations of 1.2, 1.6 and 2 thyroxines per enzyme molecule respectively were obtained by this procedure. The native enzyme contained sufficient free thiols to allow introduction of 1.1 thyroxines from a fivefold excess.

Although many heterobifunctional reagents are commercially available which are suitable for coupling amine-containing ligands to protein thiols, new reagents are occasionally synthesised with this aim in mind: for example, Morphy *et al.* (1988) prepared the reagent shown in Figure 6.69 which combines a vinylpyridine group with a *p*-nitrophenyl ester.

While strictly outside the remit imposed in the introduction to this chapter, another common use of these reagents is to prepare small compounds containing a thiol-reactive group for subsequent conjugation. *m*-Maleimidobenzoyl chloride was used (Monji *et al.*, 1978, 1980; Monji and Castro, 1980) to incorporate the maleimide moiety into the hormones before coupling to β-galactosidase. Similarly, Kriwaczek *et al.* (1978b) prepared maleimidyl peptides using the *p*-nitrophenyl ester of 6-maleimidocaproic acid. Aikawa et *al.* (1979) used MBS to derivatise angiotensin I, and Cox *et al.* (1990) used *N*-succinimidyl 3-maleimidopropionate to derivatise chelating agents. However as the maleimide derivatives were isolated and characterised before coupling to the protein, the use of these reagents is dealt with in Section 6.10.

The use of heterobifunctional reagents for coupling small amine-containing molecules to proteins has become quite popular, and the variety of readily available reagents for this purpose is convenient. Rose *et al.* (1995), for example, used combinations of different proteins and heterobifunctional reagents (SMCC, GMBS, SMPB) to provide optimal immunogens and coating conjugates for an immunoassay for the cephalosporin ceftiofur.

6.7.9 Diazo-coupling methods for aromatic amines

Aromatic amines can be treated with acidic nitrite solutions to yield electrophilic diazonium salts (Figure 6.70). Protein–protein coupling *via* bis-diazo reagents (Chapter 5, Section 5.5.3) was an important method of the pre-glutaraldehyde era.

Derivatisation of proteins with small diazonium salts dates back to the start of the century (Pauly, 1904) and the involvement of tyrosine and histidine

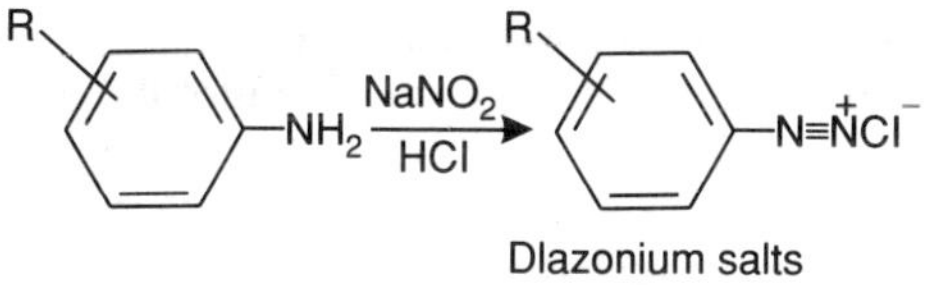

Figure 6.70 Diazonium salts.

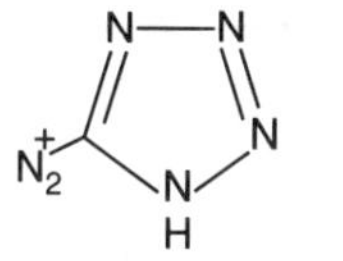

Figure 6.71 Diazonium 1*H*-tetrazole.

was established within a short time (Pauly, 1915). Some of the earliest studies in immunochemistry described the retention of immunoreactivity in diazotised antisera (Landsteiner and Lampl, 1918; Reiner, 1930). The chemistry of diazo coupling has been reviewed by Zollinger (1961).

Howard and Wild (1957) carried out further investigations into the reactivity of diazonium salts with proteins and came to a number of conclusions: reaction took place with a stoichiometry of one (cysteine, tryptophan), two (tyrosine, histidine, lysine, terminal glycine) or three (arginine). N-Terminal residues other than glycine were deaminated.

Horinishi *et al.* (1964) introduced diazonium-1*H*-tetrazole (Figure 6.71) as a reagent for the determination of reactive residues in proteins, and the use of this reagent was developed (despite its explosive nature) to allow the presence of tyrosyl and histidyl residues of different reactivities to be visualised by the spectral properties of the mono and bis-azo derivatives. Reaction with other residues yielded colourless products which did not interfere with the analysis. Reinterpretation of spectral data led to some revision of the procedures (Sokolovsky and Vallee, 1966; Takenaka *et al.*, 1969). The standard derivatisation procedure of Takenaka *et al.* involved reaction with the protein for one hour at room temperature and pH 8.6–9.2. Analysis of insulin demonstrated that the reactivity of the amino acid residues was not as simple as first thought, one tyrosine and one histidine residue giving only the monoazo derivative.

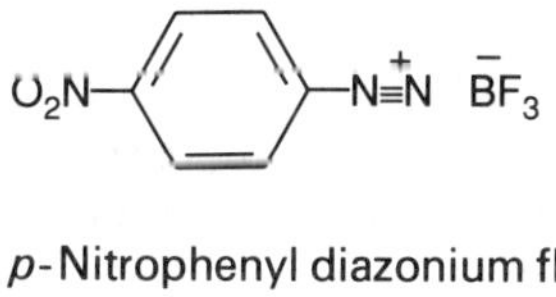

Figure 6.72 *p*-Nitrophenyl diazonium fluoroborate.

Diazonium salts were also used in some of the earliest applications of the highly successful affinity labelling methodology. For example, Metzger *et al.* (1963) used *p*-nitrophenyl diazonium fluoroborate as an affinity label for anti-2,4-dinitrophenyl antibodies[1] (Figure 6.72) and observed the hoped-for attributes: (1) reaction took place much more rapidly than if the binding sites were blocked; (2) the loss of binding capacity correlated closely with the appearance of azotyrosine, the only diazotisation product observed; (3) the rate of reaction was first-order with respect to the reversible antibody–antigen complex. The diazotisation reaction was carried out at pH 5.0 and 2–4 °C: after 200 minutes an incorporation of around 0.75 was obtained (cf. 0.15 for antibody complexed to a 2,4-dinitrophenyl compound). Similar results were obtained with diazonium-based affinity analogues of other haptens, such as *p*-azobenzenearsonate (Wofsy *et al.*, 1962) and *p*-azophenyltrimethylammonium (Fenton and Singer, 1965). The consistent appearance of tyrosine residues in the binding sites for these differently-charged haptens was a particularly striking finding, which was maintained even across immunoglobulins from different species (Fenton and Singer, 1971).

Carrying out the labelling reaction at a higher pH obviously increases the likelihood of lysine reactivity: Kagan and Vallee (1969) coupled a *p*-diazobenzenearsonate salt to carboxypeptidase A (40 minutes, 0 °C, pH 8.8) and observed the derivatisation of both tyrosyl and lysyl residues.

Gadow *et al.* (1984) coupled diazoluminol and-isoluminol to IgG. The diazotisation was carried out by dropwise addition (over 15–20 minutes) of a 50% excess of cold potassium nitrite solution to 1 mmol of luminogen in 1 M HCl.

The coupling reaction was carried out under various conditions: pH 4 or 9, reaction times of 6, 24 or 72 hours, and luminogen:IgG ratios between 8:1 and 200:1, all at 4 °C. Diazoisoluminol is much more reactive than diazoluminol, as the latter is stabilised by intramolecular charge effects; no isoluminol conjugates prepared in this study retained their antigen-binding ability, presumably due to over-reaction. Diazoluminol gave no usable conjugates at pH 4; at pH 9 the optimum coupling performance was obtained with 24 hours reaction and a luminogen:IgG ratio of 40:1.

The luminescent properties of luminol are badly affected by diazo coupling however; only 1.3% of the

[1] The more obvious affinity reagent 2,4 dinitrophenyldiazonium fluoborate was too reactive for convenient use. This is due to the extra electron-withdrawing power of the second nitro group, which enhances the electrophilic character of the diazonium salt.

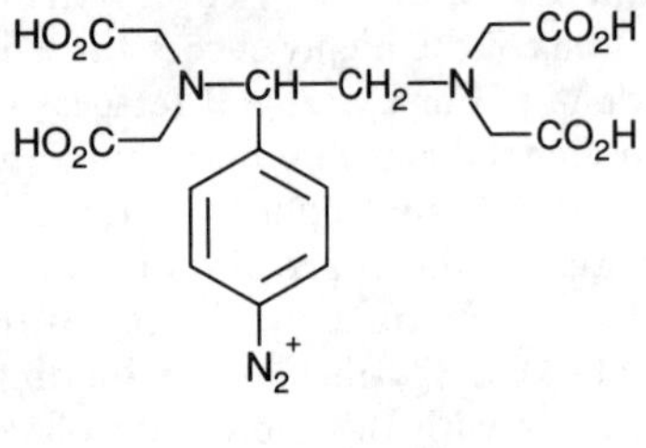

Figure 6.73 1-(*p*-Benzenediazonium)ethylenediaminetetraacetic acid.

original quantum yield is retained after diazotisation, dropping to 0.7% on overnight storage or coupling to IgG (30:1 excess, 18 hours, 4 °C, pH 8.5; Simpson *et al.*, 1979).

Diazo methods have also been used for the introduction of chelating groups for labelling with radiometals: Sundberg *et al.* (1974) described the introduction of ^{111}In into fibrinogen and albumin *via* 1-(*p*-benzenediazonium)ethylenediamine tetraacetic acid (Figure 6.73). The resulting conjugates were fairly stable *in vivo*. The same reagent was also used by this group (Leung and Meares, 1977) to introduce terbium and europium ions into albumin for fluorescent energy-transfer studies: in these studies the coupling reaction (several hours, 4 °C, pH 8) used a tenfold excess of diazonium salt and resulted mainly in coupling to lysine and histidine residues. In similar work, Leung *et al.* (1978) observed that only a little over 10% of the albumin derivatisation occurred at tyrosine or histidine residues at pH 8; raising the pH to 10 improved this fraction to around 70%

Milby and Zare (1984) described time-resolved fluorescence immunoassays in which terbium had been introduced *via* this reagent. The azophenyl group had a useful energy-transfer effect, absorbing He–Cd laser light at 325 nm and emitting at a longer wavelength suitable for terbium excitation.

Stavrianopoulos (1990) diazotised 4-aminothiophenyl-*trans*-diaminocyclohexanetetraacetic acid with equimolar sodium nitrite in approximately 0.1 M hydrochloric acid (30 minutes, 0 °C). The resulting diazo compound was coupled to BSA (one hour, 4 °C, pH 6.7) and gave an incorporation of 1.7.

McAdam *et al.* (1992) coupled an anilino analogue of the pesticide fenitrothion to various proteins, by diazotisation with a 30% excess of sodium nitrite in 0.33 M hydrochloric acid (10 minutes, 0 °C) followed by coupling at pH 9.1 and 0 °C. Incorporations of 18, 112 and 0.5–2.5 fenitrothion groups per molecule were obtained for ovalbumin, keyhole limpet haemocyanin and HRP respectively.

Homobifunctional bis-diazo reagents have occasionally been used for protein–small molecule coupling (see Section 6.15).

6.7.10 Enzymic methods

The concepts of transpeptidation and reverse proteolysis for protein–protein coupling were introduced in Chapter 5, Section 5.2.4, and have also been exploited for the conjugation of low-molecular-mass amines. Both lead to the incorporation of the small molecule at reasonably specific sites in the protein.

6.7.10.1 Transpeptidation

Mycek *et al.* (1959) observed that the enzyme they called transglutaminase (Factor XIIIa) could catalyse the incorporation of cadaverine, putrescine, methylamine, histamine, ammonia or hydroxylamine at the glutamine residue of insulin (one hour, 38 °C, pH 8.0). The formation of a hydroxamate from hydroxylamine could be monitored spectrophotometrically.

Wajda *et al.* (1971) carried out similar introductions of a range of amines into high-glutamine proteins (one hour, 37 °C, pH 8.3). As might be expected from an enzymic approach, variations in the amine structure led to differential reactivity: β-lactoglobulin, for example, showed incorporations ranging from 1.33 (5-hydroxytryptamine) to 5.95 (spermine). Cooke and Holbrook (1974) introduced ethyl glycinate into casein using the same enzyme (15 minutes, 37 °C, pH 7.5).

6.7.10.2 Reverse proteolysis

If proteolytic enzymes are used under certain conditions (typically in the presence of a large excess of amine, and often at a pH not optimal for proteolysis), they can be used instead for conjugation. Many of these enzymes show predominantly exopeptidase activity, and therefore offer a high degree of selectivity towards terminal modifications.

Moving on from studies on peptide synthesis using reverse proteolysis (Wong *et al.*, 1979; Breddam *et al.*, 1980), it proved possible to attach amino acids to the C-terminus of proteins such as insulin. Using both the proteolytic and reverse-proteolytic functions of enzymes such as carboxypeptidase Y (CPD-Y), it was even possible to exchange the terminal residue of porcine insulin (alanine) to the human equivalent (threonine) (Breddam *et al.*, 1981). Immobilised trypsin has also been used to achieve the same conversion (Könnecke *et al.*, 1990), which has obvious commercial possibilities.

These reverse proteolysis methods have achieved some popularity in the last few years due to their selectivity for the C terminus. Berne *et al.* (1992) described conditions for CPD-Y-catalysed labelling of proteins with isobutyl glycinate or valin amide (four minutes, 37 °C, pH 9.5–11.0).

Extending the scope beyond the introduction of simple protected amino acids, Buckler *et al.* (1993) used CPD-Y to couple the fluorescent probe 3-(2-naphthyl)-L-alaninamide to ribonuclease (1.8 hours, room temperature, pH 5.8). The reaction only took place in the presence of 5 M urea between pH 5.5 and 6.5. The product consisted of enzyme in which one or two residues at the C-terminus had been replaced by the fluorescent peptide.

6.8 Coupling of Proteins to Small Molecules Containing Hydrazides and Related Groups

Hydrazines ($RNHNH_2$), acyl hydrazides ($RC(=O)NHNH_2$) and sulfonyl hydrazides (RSO_2NHNH_2) show basically similar reactivity to amines, in fact they are somewhat more nucleophilic. Their much lower pK_a means that higher concentrations of the unprotonated form are present at a lower pH; therefore nucleophilic coupling can be carried out under less basic conditions.

6.8.1 Hydrazone formation with carbonyl compounds

Most commonly, hydrazines and hydrazides have been used in dehydration reactions with aldehydes generated by oxidation of glycoprotein carbohydrates (Wilchek and Bayer, 1987; O'Shannessy and Quarles, 1987).

Working at a low pH (ca. 5) can be particularly useful for avoiding unwanted protein crosslinking, which would normally occur between these aldehydes and the native amine groups of the protein. Under these conditions the amines will be protonated and unreactive, making the desired hydrazide reaction the only likely outcome. Also hydrazone products are generally quite stable, and may not require reduction like Schiff bases derived from amines (O'Shannessy and Quarles, 1987). Acyl hydrazones (and the related compounds described in Section 6.8) are prone to slow hydrolysis under acidic conditions, however. This fact has been exploited in the incorporation of acid-labile acyl hydrazone linker groups in

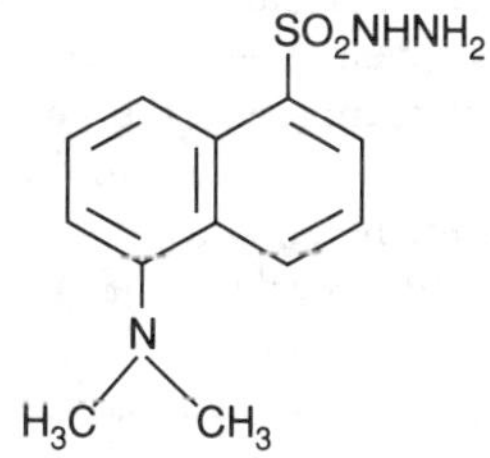

Figure 6.74 Dansylhydrazide.

immunoconjugates (Greenfield *et al.*, 1990; Kaneko *et al.*, 1991).

Itaya *et al.* (1975) used[^{35}S]methionine sulfone hydrazide to label oxidised erythrocyte glycoproteins (2–3 hours, 37 °C, pH 6.0) The reaction rate was enhanced significantly by the presence of Mn^{2+} ions, possibly by protecting nucleophilic amine groups from protonation under the acidic conditions.

Weber and Hof (1975) labelled oxidised glycoproteins (25 °C, pH 5.9) with dansylhydrazide (Figure 6.74), a widely used sulfonyl hydrazide fluorophore. With ethanol as cosolvent two hours reaction was required to achieve a plateau of fluorescent incorporation, while the use of methoxyethanol reduced this time to around 20 minutes.

Wynne *et al.* (1976) created mitogenic sites on oxidised lymphocyte membranes by reaction with 2,4-dinitrophenylhydrazine (30 minutes, room temperature, pH 7.2). Similar conditions were used by Ravid *et al.* (1978) to derivatise lymphocytes with dinitrophenyl-2,4-diaminobutyric acid hydrazide, and by Wilchek *et al.* (1980) for general fluorescent labelling with fluorescein and rhodamine acyl hydrazides. Rotman *et al.* (1980) used rather milder conditions (15 minutes, 5 °C, pH 7.2) for labelling oxidised platelet membrane glycoproteins with 3,5-diiodotyrosine hydrazide.

Eckhardt *et al.* (1976) used dansylhydrazide to label oxidised glycoproteins on electrophoresis gels (two hours, 60 °C, acidic aqueous DMSO). Shaper and Stryer (1977) incorporated galactose moieties enzymatically into rhodopsin, oxidised these specifically with galactose oxidase, and labelled them with a 20-fold excess of dansylhydrazide (6 8 hours, pH 5.6). Incorporation of 0.35 was observed, and Mn^{2+} ions had almost no effect on this result (cf. Itaya *et al.*, above).

Ingham and Brew (1981) coupled dansylhydrazide to the periodate-oxidised saccharides of human chorionic gonadotrophin (hCG) and α_1-acid glycoprotein, but the resulting conjugates were unstable on storage in neutral phosphate buffer, even after borohydride reduction to the hydrazine. At 37 °C, the dansylhydrazine-labelled hCG lost around 80% of its fluorescence within five hours. Even at −20 °C the lability of such conjugates remained apparent; the authors tentatively attributed this unusual instability to the electron-withdrawing effect of the neighbouring sulfonyl group.

O'Shannessy and Quarles (1987) quoted general conditions (1–24 hours, room temperature, sodium acetate buffer, pH 5.5) for hydrazone formation from oxidised glycoproteins and hydrazides.

Geoghegan and Stroh (1992) reported that α-amino alcohol moieties in proteins, encountered in N-terminal serine or threonine residues or in hydroxylysine, can be readily oxidised by periodate to yield aldehydes. Using *N*-seryl peptides and proteins as models, they demonstrated that the oxidised serine could be coupled to hydrazides such as biotin-X-hydrazide, Lucifer Yellow CH and Cascade Blue hydrazide. Oxidations were carried out using periodate excesses of between 1.5 and 3.1, for 4–30 minutes at pH 7.0; the hydrazone reaction was performed for 1–16 hours, 22 °C, pH 4.5.

The effect of pH on the oxidation products of one of the peptides (also containing tyrosyl, methionyl, histidyl and tryptophan residues) was reported; below around pH 5 oxidation of the methionine predominated, but at pH 7–8 over 90% of the product displayed only serine oxidation.

The biotin hydrazone of this peptide was almost completely stable for at least eleven hours in the pH range 6–10, but was labile under more acidic conditions. Around 10% of the conjugate had dissociated after eleven hours at pH 4, while almost complete cleavage occurred at pH 2.

Hydrazide derivatives of small molecules continue to be prepared for coupling to proteins, such as the benzoylecgonine derivatives conjugated to periodate-oxidised HRP by Gowda *et al.* (1996).

6.8.2 Carbodiimide coupling of hydrazides to carboxylic acids

Hydrazines and hydrazides will react with activated acid groups in the same way as amines. Again the low pK_a values of these compounds are convenient, as the conflict between the acidic optimum of the carbodiimide activation and the alkaline optimum for unprotonated amine availability is removed; this makes one-step reaction somewhat more attractive for hydrazides.

Semicarbazide ($NH_2CONHNH_2$), for example could be used in structural studies to derivatise the carboxylic residues of chymotrypsin at pH 3.7, conditions under which an amine would be completely unreactive (Fersht and Sperling, 1973). Carbodiimide coupling of dansyl hydrazide has also been employed to derivatise carboxylic residues in structural studies (Renthal *et al.*, 1987).

Preparative examples are encountered too: for example Tse *et al.* (1976) coupled isonicotinic acid hydrazide to HSA using a one-step EDAC reaction, while Roffman *et al.* (1980) used a similar approach to link adipic acid dihydrazide to ferritin.

6.8.3 Coupling of acyl hydrazides *via* acyl azide formation

Acyl hydrazides can be undergo the equivalent of the diazotisation reaction of aromatic amines (Section 6.7.9): treatment with acidic nitrite yields acyl azides, which form amide links to protein amines (Section 6.1.8).

This approach was in use as early as the 1930s: Clutton *et al.* (1938) synthesised a hydrazide derivative of diiodothyronine, diazotised it to the acyl azide, and coupled the unisolated product to various proteins (30 minutes, on ice, pH 5).

Stavrianopoulos (1990) treated 5-hydroxy-*trans*-diaminocyclohexanetetraacetic acid-4-β-thiopropionic

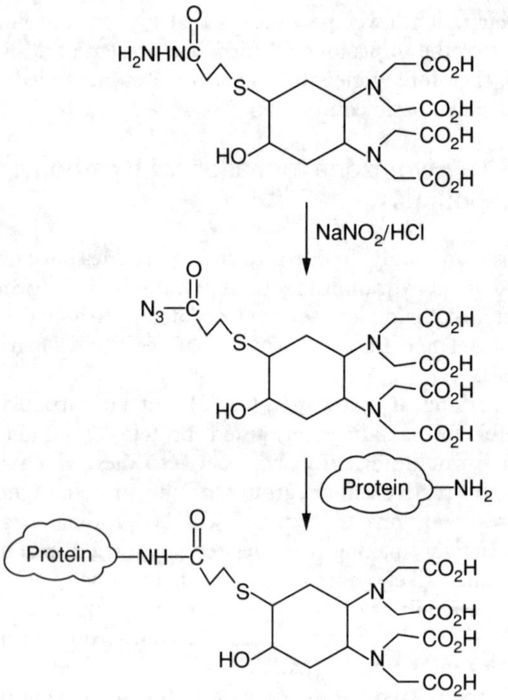

Figure 6.75 Diazotisation of a hydrazide yields an acyl azide, which is reactive towards amines.

acid hydrazide (Figure 6.75) with one equivalent of acidic sodium nitrite at 0 °C and coupled the unisolated azide to IgG (overnight, 0 °C, pH 8.5). Rotman *et al.* (1980) labelled platelet membrane proteins with iodinated phenols by this approach.

6.8.4 Thiosemicarbazides, semicarbazides, carbohydrazides and carbohydrazones

Thiosemicarbazides (RNHC(=S)NHNH$_2$) display similar reactivity to hydrazides, and can be prepared from isothiocyanates by treatment with hydrazine (Figure 6.76). This widens the range of available hydrazide reagents considerably.

Bauman *et al.* (1981) prepared the thiosemicarbazides of FITC and TRITC (hydrazine–DMSO–pyridine, four hours, room temperature) for labelling of RNA. Duijndam *et al.* (1988) coupled these reagents, which were stable at −30 °C for several weeks, to periodate-oxidised IgG (four hours, room temperature, phosphate-buffered saline) to yield conjugates with 30–40% of the fluorochrome content obtained by direct isothiocyanate labelling.

Semicarbazides (RNHC(=O)NHNH$_2$) can be generated by hydrazinolysis of carbamate esters, and a fluorescent labelling reagent of this type (proflavine monosemicarbazide, Figure 6.77) has been described (Reines and Cantor, 1974).

Carbohydrazides (RNHNHC(=O)NHNH$_2$) offer another category of similar reactivity. Lucifer Yellow CH (Stewart, 1978, 1981) is a useful fluorescent reagent which can be viewed essentially as a carbohydrazide (Figure 6.78). It is intensely fluorescent, highly soluble, and has a low toxicity; these characteristics have rendered it suitable for *in vivo* experiments, for example in examining the behaviour of nerve cells.

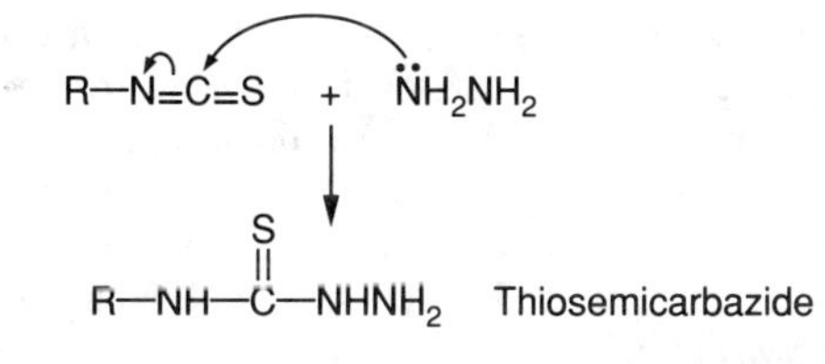

Figure 6.76 Preparation of thiosemicarbazides.

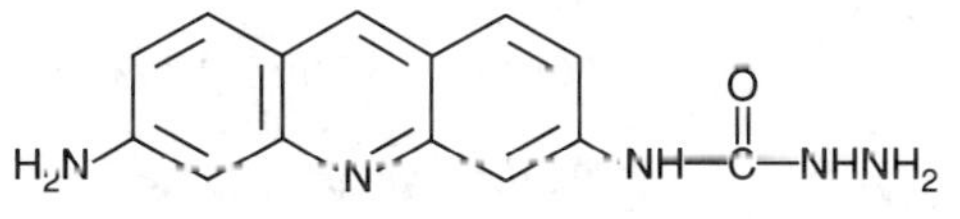

Figure 6.77 Proflavine monosemicarbazide.

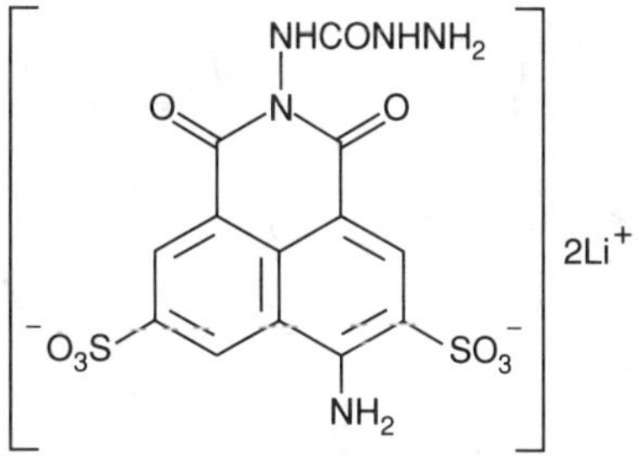

Figure 6.78 Lucifer Yellow CH.

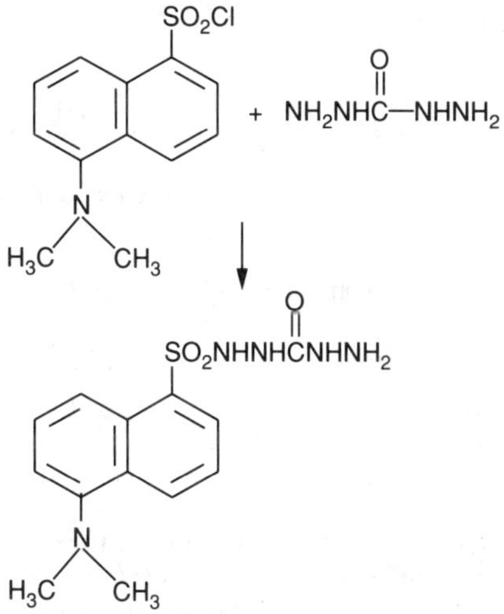

Figure 6.79 Preparation of dansylcarbohydrazide.

Lee and Fortes (1985) prepared dansylcarbohydrazide (Figure 6.79), a sulfonyl carbohydrazide, from dansyl chloride and carbohydrazide, using it to label erythrocyte ghosts. Dicarbohydrazido-triazinylamino fluorescein was also described by these authors. They described general conditions for use of these hydrazide reagents (two hours, 37 °C, pH 6.0) to label erythrocyte ghosts.

Carbohydrazones (RC=NNHC(=O)NHNH$_2$) are generated by the reaction of carbohydrazide with aldehydes, Reines and Cantor (1974) described the preparation of anthracene-9-carbohydrazone (Figure 6.80) for fluorescent labelling.

6.8.5 Enzymic methods: reverse proteolysis

Hydrazides act as substrates for reverse proteolysis using the lysyl endopeptidase from *Achromobacter lyticus*.

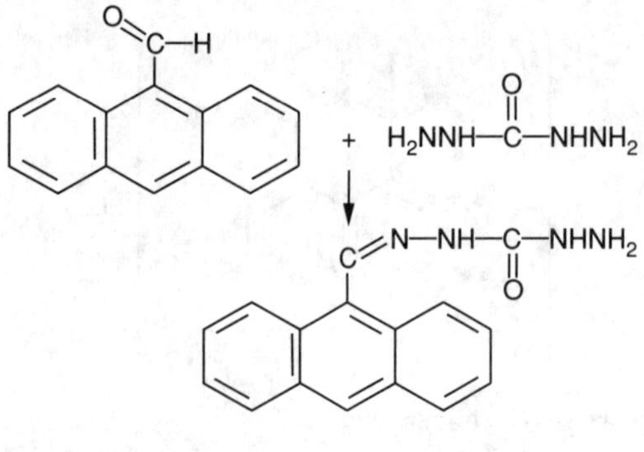

Figure 6.80 Preparation of anthracene-9-carbohydrazone.

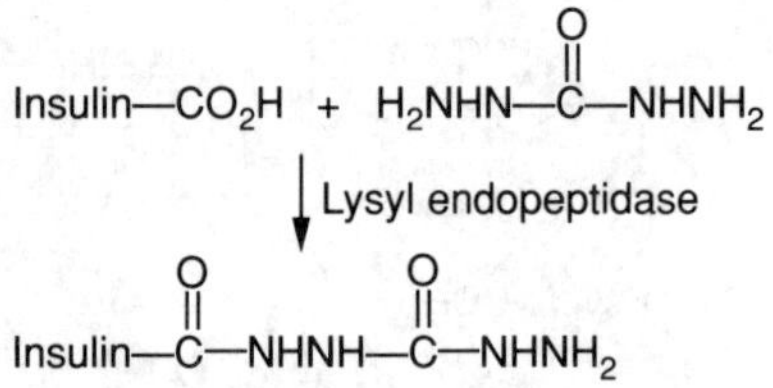

Figure 6.81 The C-terminal lysine of des-Ala^{B30} insulin can be coupled to carbohydrazide by reverse proteolysis.

Rose *et al.* (1991) and Vilaseca *et al.* (1993) used this enzyme to couple hydrazides to proteins: Rose *et al.* for example coupled carbohydrazide to the C-terminus of des-Ala^{B30} insulin (Figure 6.81).

Fisch *et al.* (1992) used this method to produce the C-terminal carbohydrazide of antibody fragments (23 °C, pH 5.5), reaction was substantially complete in around four hours. Gaertner *et al.* (1992) described similar work with recombinant analogues of granulocyte colony stimulating factor; Boc-hydrazide and -carbohydrazide were coupled (15 hours, room temperature, pH 5.5) to the C-terminal lysine of a fragment, generated by the use of the same enzyme in its proteolytic capacity (eight hours, 37 °C, PBS).

6.9 Coupling of Proteins to Small Molecules Containing Thiols and Related Groups

Although direct, oxidative thiol–thiol coupling is feasible (Chapter 5, Section 5.2.1) it is not an attractive route for conjugation of low-molecular-mass thiols.

More commonly, the protein is derivatised with a heterobifunctional reagent (Chapter 5, Section 5.7) to introduce thiol-reactive groups such as maleimides, disulfides or haloalkyls.

In this section the reactions are categorised with respect to the nature of the thiol-reactive species.

6.9.1 Coupling to native protein thiols

This is probably the simplest approach for coupling where a native thiol is available, but is not very commonly encountered. Even where this approach is applied it is common to increase the reactivity of the thiol-containing molecule by converting it to a disulfide, properly dealt with in Section 6.12.

For example, Rowley *et al.* (1975) prepared a mercaptoethyl derivative of morphine, converted it to the disulfide using 5,5′-dithio-bis(2-nitrobenzoic acid) at pH 9.5, and coupled it to the cysteine groups of malate dehydrogenase (20 hours, pH 9.5). A 20-fold excess of the morphine compound resulted in an incorporation of 14.

It is far commoner for thiols to be used for coupling to artificially introduced thiol-reactive groups such as maleimides and haloacetyl functions.

6.9.2 Coupling to artificially introduced thiol-reactive groups

Bieniarz *et al.* (1994) described the preparation of a mercaptoacetyl derivative of fluorescein, which was coupled to maleimide-derivatised IgG (overnight, 5 °C, pH 7.0). Use of a 25-fold excess of the thiol allowed the incorporation of two fluorescein molecules per IgG molecule.

An interesting self-catalysing conjugation was also described by these authors using a phosphorothioate derivative of the mercaptoacetyl fluorescein. Maleimide-derivatised alkaline phosphatase catalysed the hydrolysis of the phosphoryl group (with a half-life of around 15 minutes) to release the mercapto-compound (Figure 6.82), making it available for reaction with the maleimide. Phosphorothioate excesses of 27, 55 and 82:1 (48 hours, 5 °C, pH 7.0) resulted in fluorescein incorporations of seven to eight. The enzyme lost a considerable amount of activity during this process, however: around 80% of the phosphatase activity was retained after derivatisation with maleimide, but dropped to 30% after the enzymatically catalysed fluorescein incorporation.

Markert-Hahn *et al.* (1994) prepared a bis-mercaptoacetyl derivative of bisphosphonate, and, following deacetylation with 70 mM hydroxylamine (one hour, 25 °C, pH 7.5), coupled it to maleimide-derivatised peroxidase (90 minutes, 25 °C, pH 7.5).

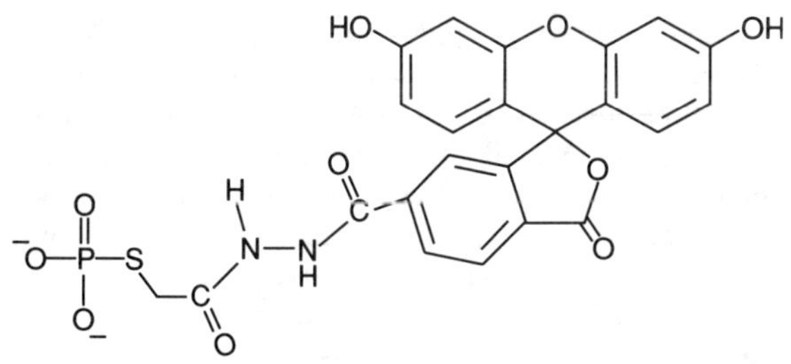

Figure 6.82 Phosphorothioate derivative of mercaptoacetyl fluorescein.

Hanin *et al.* (1994) used sulfo-SMCC to activate BSA or thyroglobulin, then coupled cysteine-containing peptides to these carrier proteins. Green *et al.* (1982) used MBS in a similar approach to couple viral peptides to KLH, and Raymond *et al.* (1995) coupled a cysteine-containing bradykinin derivative to maleimide-activated keyhole limpet haemocyanin (two hours, room temperature, pH 7.2).

Rowley *et al.* (1980) coupled a thiolated morphine derivative to bromoacetylated glucose-6-phosphate dehydrogenase. The thiol was added in small portions at 0 °C to a deaereated solution of the derivatised enzyme; 30 minutes reaction at 0 °C was followed by 75 minutes at room temperature and 14 days at 4 °C under argon.

Bernatowicz and Matsueda (1986) synthesised *N*-succinimidyl bromoacetate and used it to activate various proteins (two additions, 2 × 30 minutes, 4 °C to room temperature, pH 7.0). Around three-quarters of the available amines were bromoacetylated in one case. The resulting solution was used for coupling to cysteinyl-peptides (two hours, room temperature, pH 7.0. High incorporations of peptide were obtained: with BSA a 100-fold excess of one peptide gave an incorporation of 53. These authors demonstrated that the bromoacetate reacted specifically with thiols: there was no reaction at pH 7.0 with a peptide containing lysine, tyrosine, and histidine.

Houen and Jensen (1995) coupled cysteine, glutathione and a 12-residue cysteine-containing oligopeptide to iodoacetylated BSA (overnight, pH 8) and observed incorporations of between 12.8 and 35.4.

6.9.3 Coupling of 1,2-aminothiols to aldehydes

Being strongly nucleophilic, thiols react with aldehyde groups in a manner analogous to amines. The initial hydroxylic products of both reactions are generally unstable towards dissociation in aqueous media

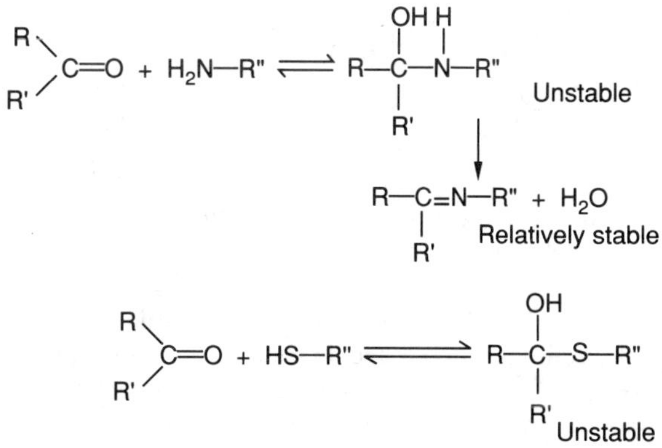

Figure 6.83 Reaction of thiols with aldehydes in comparison to amines.

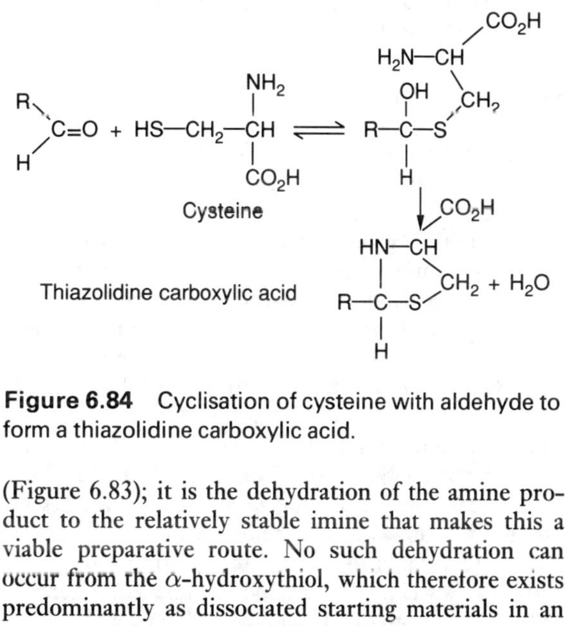

Figure 6.84 Cyclisation of cysteine with aldehyde to form a thiazolidine carboxylic acid.

(Figure 6.83); it is the dehydration of the amine product to the relatively stable imine that makes this a viable preparative route. No such dehydration can occur from the α-hydroxythiol, which therefore exists predominantly as dissociated starting materials in an aqueous environment (Schubert, 1936).

With 1,2-aminothiols such as cysteine, however, the presence of the α-amino group leads to rapid intramolecular cyclisation of the hydroxythiol (Schubert, 1936; Ratner and Clarke, 1937) to yield a stable thiazolidine (Figure 6.84). Additionally, since amines of this type are only weakly basic, they are partially unprotonated at quite acidic pH (cf. hydrazides) and the reaction can therefore be carried out quite selectively under these conditions.

Mitchell *et al.* (1984) used this approach to radiolabel oxidised cell-surface glycoproteins with [^{35}S]cysteine methyl ester (30 minutes, 37 °C, PBS) and obtained favourable results.

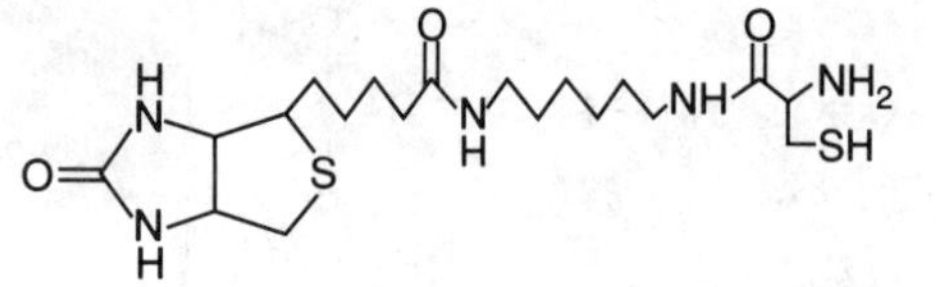

Figure 6.85 A biotinylation reagent incorporating a 1,2-aminothiol function

Zhang and Tam (1996) also applied this methodology, and described reagents such as the biotin derivative shown in Figure 6.85, which are designed for site-specific incorporation by this route. See also Figure 5.223.

6.10 Coupling of Proteins to Small Molecules Containing Maleimides and Related Groups

6.10.1 Maleimides

Maleimide-containing small molecules, as well as the haloalkyl and disulfide compounds described in Sections 6.11 and 6.12, can be coupled conveniently to protein thiol groups, which may be either native or artificially introduced. Maleimide chemistry is described in detail in Chapter 5, Section 5.4.1. In summary the reaction with thiols can take place within minutes at pH 6–7 (Gregory, 1955), competing hydrolysis becoming significant only at higher pH. The reaction is not wholly thiol-specific, derivatisation of histidines and terminal amines having been noted (Smyth *et al.*, 1964; Bernatowicz and Matsueda, 1986).

A number of maleimide derivatives of fluorescent compounds are available, being particularly well-suited to this application as many are non-fluorescent until the succinimide product is formed. Examples include *N*-(1-pyrene)maleimide (Haugland, 1970; Weltman *et al.*, 1973), *N*-(4-anilinonaphthyl)maleimide (ANM; Kanaoka *et al.*, 1973), *N*-(7-dimethylamino-4-methylcoumarinyl)maleimide (DACM), eosin- 5-maleimide, fluorescein-5-maleimide, etc. Some of these compounds and their fluorescence properties are shown in Figure 6.86.

N-(1-Pyrene)maleimide[1] is particularly useful as a proximity probe, as it forms spectrally-recognisable dimers if the labels are close enough together; eosin maleimide is used primarily for studies of protein rotation (Haugland, 1992).

[1] As pointed out by Haugland (1992), this compound is often wrongly referred to as the 3-isomer.

Yamamoto *et al.* (1977, 1978) used DACM as a fluorescent tracer for protein thiols. Actin was derivatised (one hour, 0 °C, pH 6.85) with a 40% excess of DACM and yielded an incorporation of 0.97. Interestingly, these authors noted a gradual ring-opening of the fluorescent succinimide thioether product to yield a stable succinamate derivative, also fluorescent (Figure 6.87). This ring-opening was detected across the pH range 6.5–9.0, the rate increasing with alkalinity. At pH 7–8 ring-opening was essentially complete after 24 hours. Ring-opening of protein–maleimide conjugates has been described by other authors, attributed both to hydrolysis (Ishii and Lehrer, 1986) and intramolecular aminolysis (Wu *et al.*, 1976).

β-Galactosidase, containing thiol groups which are inessential for enzyme activity, is an obvious target for

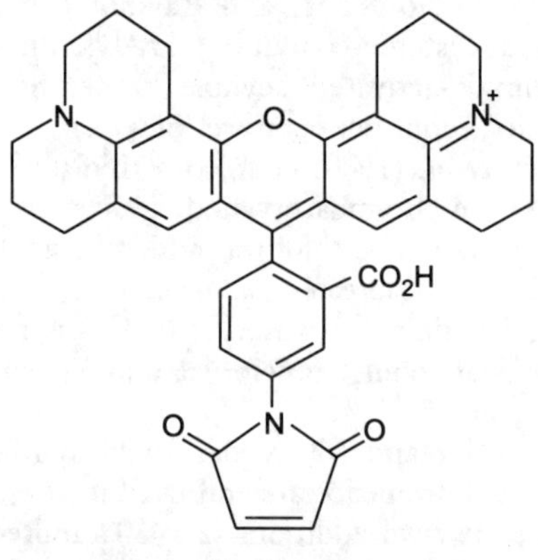

Rhodamine-5-maleimide

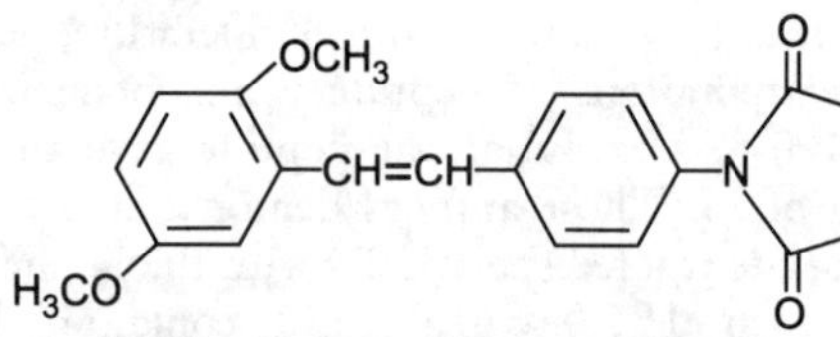

2,5-Dimethyloxystilbene-4-maleimide

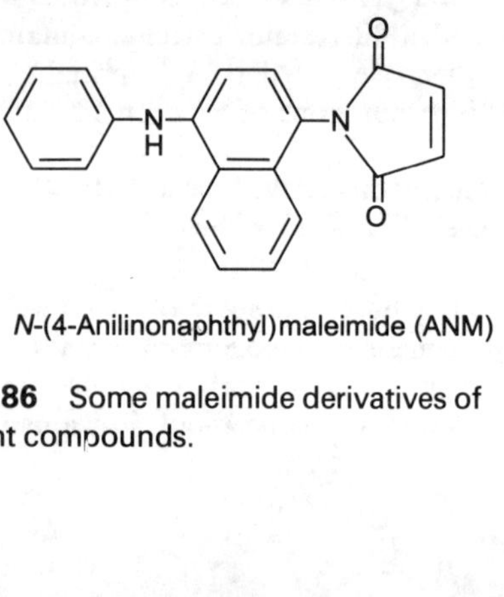

N-(4-Anilinonaphthyl)maleimide (ANM)

Figure 6.86 Some maleimide derivatives of fluorescent compounds.

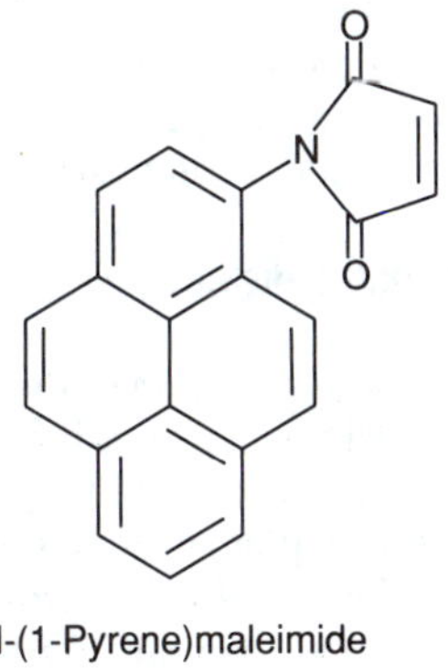

N-(1-Pyrene)maleimide

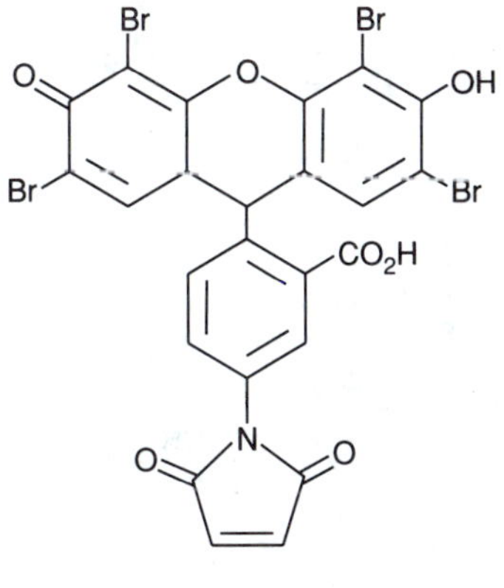

Eosin-5-maleimide

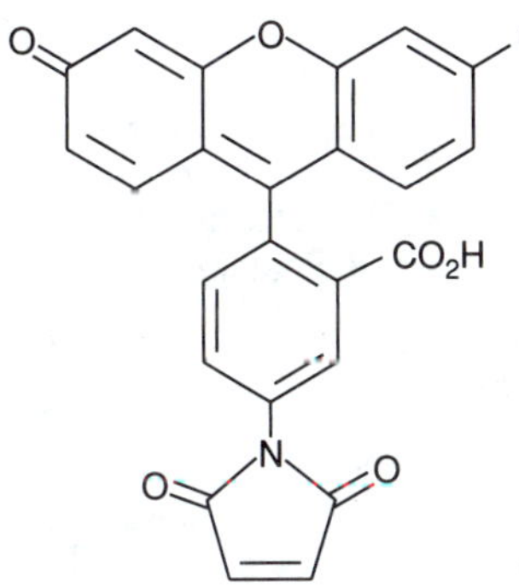

Fluorescein-5-maleimide

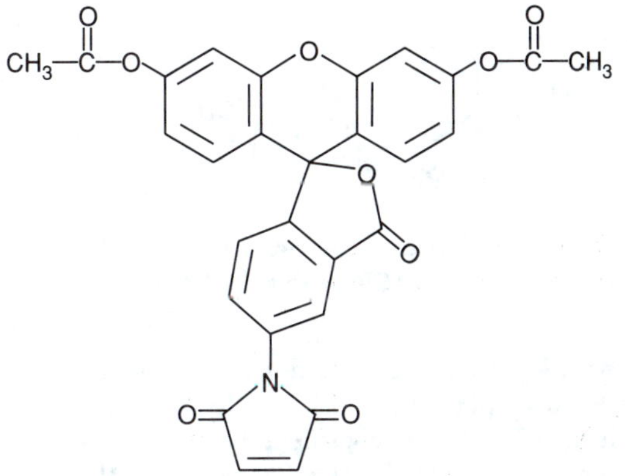

Fluorescein-5-maleimide diacetate

Figure 6.86 (contd.)

derivatisation with maleimide-containing compounds. Monji *et al.* (1978) prepared a maleimidyl derivative of thyroxine and coupled it (two hours, room temperature, pH 7.0) at an elevenfold excess to the enzyme. The thyroxine incorporation was not calculated, but enzyme activity was fully retained and the conjugate was used in a successful immunoassay. The same conditions were used for the preparation of β-galactosidase conjugates with maleimidyl derivatives of cortisol (Monji and Castro, 1980) and digoxin (Monji *et al.*, 1980); in the latter case, the absence of free thiols in the product suggested a quantitative incorporation of ten steroids per enzyme molecule.

Aikawa *et al.* (1979) derivatised β-galactosidase with a five, ten or 15-fold excess of a maleimidyl angiotensin (two hours, room temperature, pH 7.0). This resulted in labelling of 20, 28 or 36% of the enzyme respectively.

Kriwaczek *et al.* (1978b) prepared a maleimide derivative of the oligopeptide angiotensin, and coupled it at a tenfold excess to thiolated tobacco mosaic virus (one hour, 20 °C, pH 3); the low pH was necessitated by the poor solubility of the peptide. The thiolated virus contained around 600 SH groups, and the crude conjugate displayed an incorporation of 1140 angiotensin molecules. Electrodialysis removed non-covalently bound material to reduce the incorporation to 635, but the virus was seriously damaged in the process.

Brown and Matthews (1979) coupled *N*-ethylmaleimide, two hydrophilic maleimidyl spin reagents and a hydrophobic maleimidyl fluorophore to lactose repressor protein, which contains three thiol groups. With the first three maleimide compounds, increasing the maleimide:protein ratio up to around 30:1 resulted in labelling of two cysteine residues, predominantly Cys

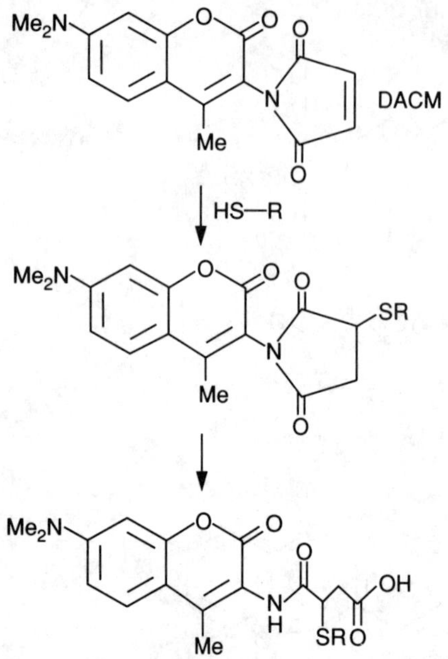

Figure 6.87 Ring opening of the succinimide product of the reaction of DACM with a thiol.

107 and Cys 140 (coupling conditions: four hours, room temperature, pH 7.0). With the hydrophobic pyrene-maleimide there was considerably more reaction with the remaining cysteine, Cys281, although the other residues still predominated.

Cox *et al.* (1990) described the preparation of maleimidyl derivatives of polyazamacrocycles, aimed at the introduction of radiometals into antibodies. One such compound is shown in Figure 6.88. Trubetskoy and Torchilin (1995) described DTPA-maleimide derivatives for a similar application.

Hermentin *et al.* (1990) prepared maleimidyl derivatives of rhodosaminylanthracyclinones *via* their saccharide moieties, and coupled these to the thiols of reduced Mabs (one hour, in the dark, pH 6.0). An

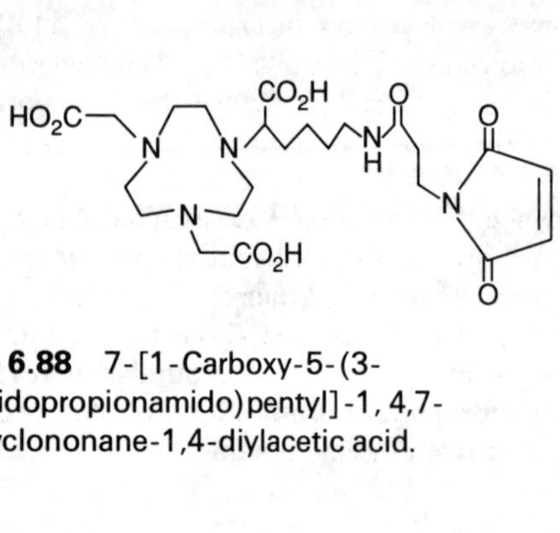

Figure 6.88 7-[1-Carboxy-5-(3-maleimidopropionamido)pentyl]-1, 4,7-triazacyclononane-1,4-diylacetic acid.

initial loading of 50:1 resulted in incorporations of 4.8–6.8.

Gleizes *et al.* (1994) coupled a maleimidyl derivative of digoxigenin to recombinant basic fibroblast growth factor (five hours, 5 °C, pH ca 7) to produce a predominantly monovalent conjugate.

6.10.2 Isomaleimides

Isomaleimides (Figure 6.89(a)) display similar thiol reactivity to maleimides, the initial product rearranging to give the same succinimide. Isomaleimides have occasionally been exploited in this way for labelling; Stewart and Diakiw (1978), for example, described the use of *N*-pyrroloisomaleimide as a means of introducing mercury atoms for electron microscopy (Figure 6.89(b)).

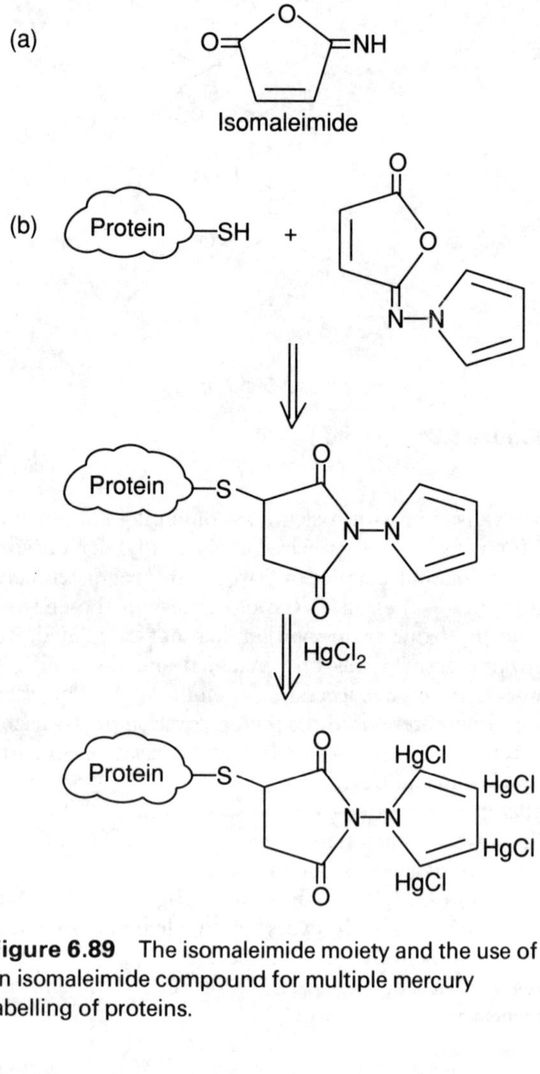

Figure 6.89 The isomaleimide moiety and the use of an isomaleimide compound for multiple mercury labelling of proteins.

6.11 Coupling of Proteins to Small Molecules Containing Haloalkyl Groups

The chemistry of haloalkyl compounds is dealt with in Chapter 5, Section 5.4.2 and is only summarised briefly here. With a suitable nucleophile alkylation takes place, with the elimination of the hydrogen halide. In protein derivatisation, haloalkyl compounds generally react preferentially with thiol groups (particularly at neutral to acidic pH), but can also couple to lysyl, histidyl, tyrosyl and methionyl residues. The reactivity of such compounds follows the order iodo > bromo >> chloro, and is enhanced by electron-withdrawing components in the molecule. Thus reactivity also follows the series:

$RCHXCOOR > RCHXCONH_2 > RCHXCOOH > RCH_2CH_2X > RCHXCH_2OH$

(Dahl and McKinley-McKee, 1981). These reagents are photolabile, particularly in solution, and should be protected from exposure to light. They are also hydrolysed by base to yield the alcohol.

6.11.1 Specific coupling of alkyl halides to protein thiols

The most obvious targets for the derivatisation of proteins with alkyl halides are thiols, either from free native cysteine residues, from reduced cystine groups, or from thiolation with heterobifunctional reagents. The use of such heterobifunctional reagents without isolation of the thiolated protein is covered in Section 6.11.3.

6.11.1.1 Alkyl iodides

As might be expected from the relative reactivities outlined above, iodoalkyl, and in particular iodoacetyl, compounds probably constitute the most commonly used category of such reagents.

The early history of protein derivatisation with iodoacetyl compounds was concerned with thiol quantification, iodoacetate and iodoacetamide offering convenient combinations of reactivity and solubility. Originally, protein disulfides were often oxidised with performic acid (see, for example, Singer, 1949), but side-reactions of methionine and tryptophan were inconvenient. Reduction of the disulfides was an alternative, but the resulting thiols were unstable: *S*-carboxymethylation overcame this problem (Anfinsen and Haber, 1961; Crestfield *et al.*, 1963). As well as recommending 2-mercaptoethanol as the reducing agent, Anfinsen and Haber described conditions (tenfold excess of iodoacetate, 12–15 minutes, room temperature, pH 8.5) which favoured alkylation of cysteine over other residues such as methionine and lysine.

The same techniques were later employed to demonstrate differential reactivity of protein thiols: Jörnvall *et al.* (1978) showed that a mutant form of β-galactosidase displayed modified thiol accessibility, as well as altered enzyme activity. Excesses of 10–10 000:1 of iodoacetate were employed in the derivatisation reaction (90 minutes, pH 8.1), giving incorporations of up to 3.2 in the native enzyme and 6.0 in the mutant.

Iodoacetyl chemistry has also been used extensively in preparative protein labelling, and there are many commercial reagents of this type. A number of commonly used fluorescent probes fall in this class, and are shown in Figure 6.90; among the most commonly encountered are 5-iodoacetamidofluorescein (5-IAF; Hartig *et al.*, 1977; Aguirre *et al.*, 1986), and *N*-iodoacetylaminoethyl)-5-naphthylamine-1-sulfonic acid (5-IAEDANS; Wu and Stryer, 1972; Hudson and Weber, 1973). There is an enormous literature for the use of such compounds (Haugland, 1992) and a few examples only can be given here to demonstrate typical coupling conditions.

Kwok and Churchich (1980) derivatised pyridoxal kinase with 5-IAEDANS (10 hours, 4 °C, pH 6.8), a loading of 30:1 giving an incorporation of around one. Pyridoxine-5-P-oxidase was also derivatised with a 30-fold excess of 5-IAF (2 hours, 4 °C, pH 6.8), again giving an incorporation of about 1.

Gorman (1987) compared 5-IAF and 5-IAEDANS for labelling proteins prior to electrophoresis: the proteins were reduced with DTT beforehand to provide thiols. Labelling (two hours, in the dark, 22 °C) was carried out in 0.1 M ammonium bicarbonate. Greater sensitivity of detection was obtained with 5-IAEDANS.

Sinha *et al.* (1991, 1994) used 5-IAF labelling (15 minutes in 50% aqueous ethanol) to demonstrate the topography of thiol distribution in spermatazoa, relating the findings to infertility problems.

Andree and Zantema (1978) attached the spin label, *N*-(1-oxyl-2,2,6,6-tetramethyl-4-piperidinyl)iodoacetamide, to glutamate dehydrogenase (14 hours, room temperature, pH 8) giving an incorporation of 2.5 labels per enzyme molecule. These included one non-cysteinyl residue, the derivatisation of which resulted in inactivation of the enzyme.

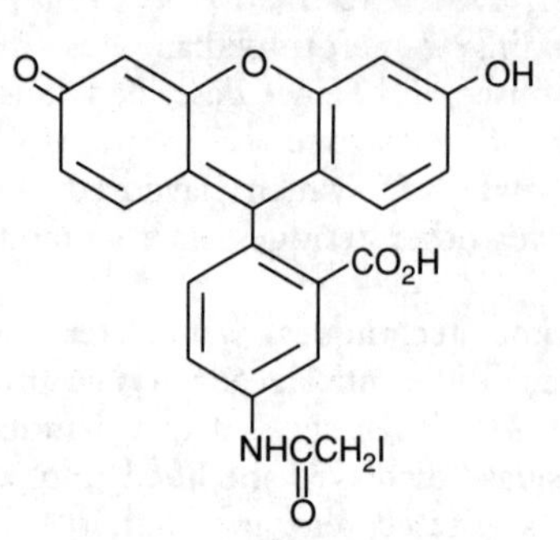

5-Iodoacetamidofluorescein
(5-IAF)

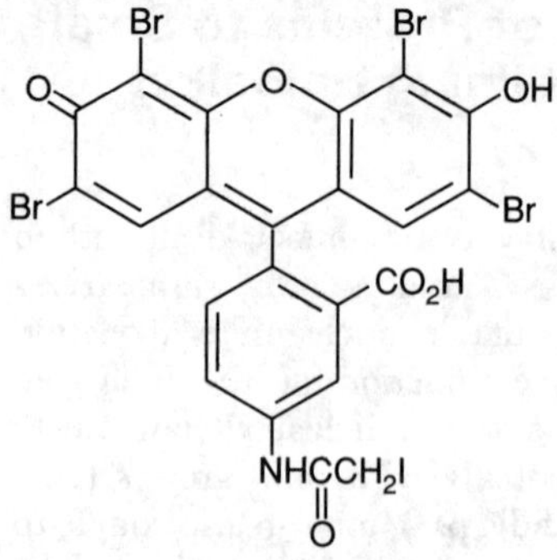

Eosin-5-iodoacetamide

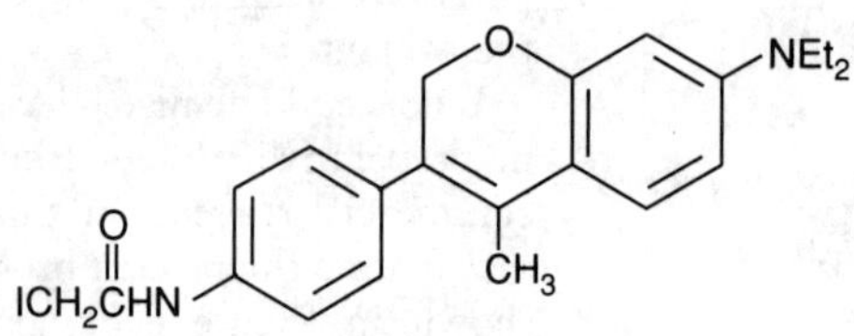

7-Diethylamino-3-[(4'-iodoacetylamino)phenyl]-4-methylcoumarin (DCIA)

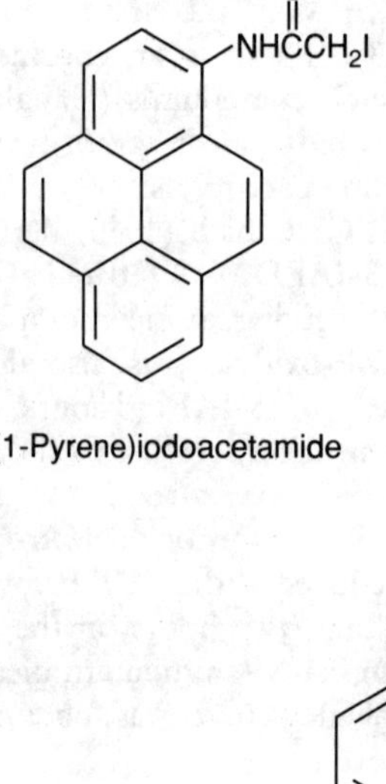

N-(1-Pyrene)iodoacetamide

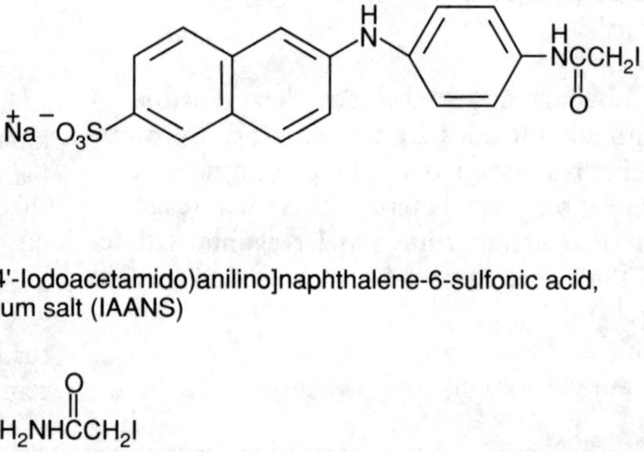

2-[(4'-Iodoacetamido)anilino]naphthalene-6-sulfonic acid, sodium salt (IAANS)

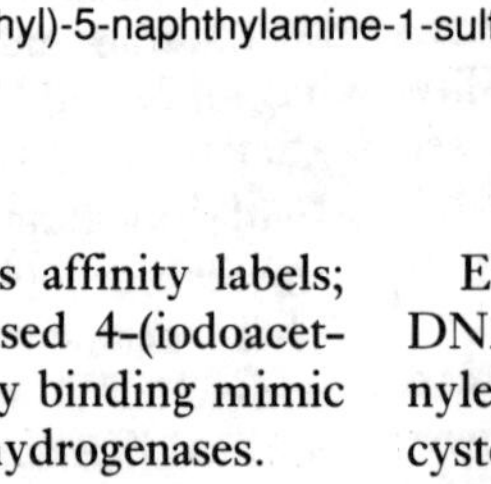

N-(Iodoacetylaminoethyl)-5-naphthylamine-1-sulfonic acid (5-IAEDANS)

Figure 6.90

Iodoacetamides have been used as affinity labels; Baker *et al.* (1961), for example, used 4-(iodoacetamido)-salicylic acid as an irreversibly binding mimic of salicylate in the inactivation of dehydrogenases.

Ebright *et al.* (1993) introduced radiometals into a DNA-binding protein, Cro, with *N*-iodoacetyl-*p*-phenylenediamine-EDTA. The coupling, to an accessible cysteine residue introduced *via* site-directed mutagenesis, was carried out using a 75-fold excess of the haloacetamide (three hours, 23 °C, pH 8.0, then 15 hours, 4 °C, pH 8.0, all in the dark). An incorporation of 0.60–0.84 radiometals per protein molecule was achieved by this route, compared to incorporations of less than 0.01 for the wild-type Cro.

6.11.1.2 Alkyl bromides

The difference in reactivity between alkyl bromides and alkyl iodides is not enormous, iodides showing rates approximately 1.4–2.4 times greater than the corresponding bromides (Dahl and McKinley-McKee, 1981). Therefore there are many examples of protein derivatisation reagents based on bromoalkyl chemistry, those with an α-carbonyl group again being the most commonly encountered. Bromoacetyl-based fluorescent labels have been described, for example those shown in Figure 6.91(a), recently reported by Pouchnik *et al.* (1996). These are good energy-transfer receptors for natural tryptophan fluorescence, and their long fluorescent lifetimes make them good candidates for rotational investigations.

A highly activated reagent of this type is 3-bromo-1,1,1-trifluoroacetone (Figure 6.91(b)); Huestis and Raftery (1972) labelled the β-chain thiols of haemoglobin by using a 2.5-fold excess (over available thiols) of the reagent under mild conditions (30 minutes, room temperature, pH 7.15). This reaction was later shown to be specific for the β93 cysteine (Huestis and Raftery, 1978), but Knowles (1975) presented evidence that the thiol residue had in fact undergone oxidation.

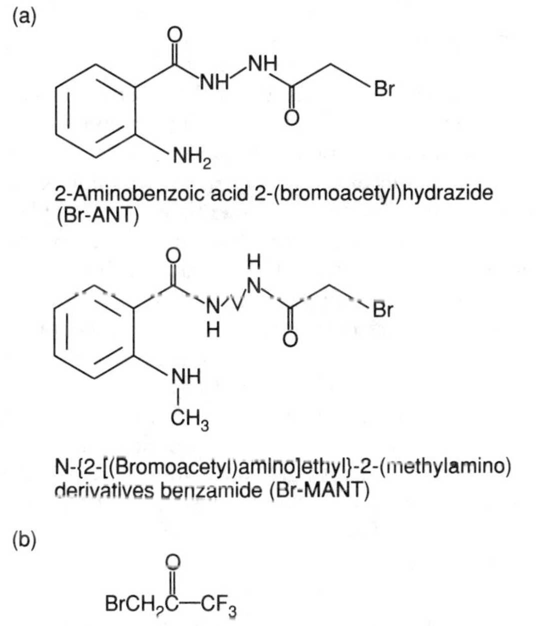

Figure 6.91 Bromoacetyl derivatives: (a) benzamide derivates described by Pouchnik *et al.* (1996) for energy transfer work; (b) 3-bromo-1,1,1-trifluoroacetone.

Cuatrecasas *et al.* (1969) used a range of bromoacetyl substrate analogues to affinity-label staphylococcal nuclease; it was shown that the derivatives labelled lysine, tyrosine and possibly methionine residues in the enzyme.

Kriwaczek *et al.* (1978a) derivatised thiolated tobacco mosaic virus with a bromoacetyl derivative of the oligopeptide melanotropin (four hours, 20 °C, pH 8.7–9.0). Five hundred peptide residues were introduced into the virus, which contained five to six hundred thiol functions.

Hoebeke *et al.* (1978) synthesised a bromohydrin ($RCH(OH)CH_2Br$) derivative of the β-adrenergic antagonist alprenolol, and coupled it to reduced BSA (48 hours, 4 °C, in the dark, pH 8.2). The resulting conjugate was not characterised but was used successfully to raise specific antibodies.

Terouanne *et al.* (1980) coupled a bromoacetyl derivative of progesterone to thiolated Δ5,3-ketosteroid isomerase: sixteen hours reaction at room temperature and pH 8.2 gave an incorporation of two steroids per enzyme molecule.

Luedtke *et al.* (1988) conjugated bromoacetyl derivatives of spiroperidol analogues to BSA and keyhole limpet haemocyanin (sixteen hours, 4 °C, pH 10).

Yeh *et al.* (1979) coupled an equimolar quantity of 1-(*p*-bromoacetamidophenyl)-EDTA to HSA (six hours, 34 °C, pH 7.6) and observed complete reaction of the free thiol group. The conjugate was used to chelate ^{111}In. The bromoacetamidobenzyl analogue was described by DeRiemer *et al.* (1981), and used to prepare a complex (BLEDTA II) with the antibiotic bleomycin A_2.

Meares *et al.* (1984) coupled a tenfold excess of *N*-4-{2,3-bis[bis(carboxymethyl)amino]propyl}phenyl bromoacetamide (BABE) to monoclonal IgG (two hours, 37 °C, pH 9), giving an incorporation of 3.5 chelate groups per IgG molecule. This conjugate was used for the preparation of ^{111}In-labelled radioimmunoconjugates.

Rana and Meares (1990, 1991) have used BABE to introduce Fe^{2+} ions into BSA and carbonic anhydrase, where they mediate proteolysis by active oxygen species generated from peroxide and ascorbate. The

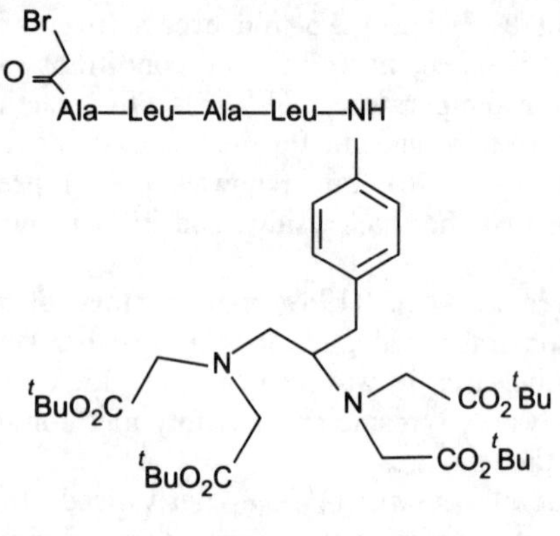

Figure 6.92 {(*p*-[(Bromoacetyl)-Ala-Leu-Ala-Leu-amido]benzyl}-EDTA, tetra-*t*-butyl ester).

albumin derivatisation (five hours, 37 °C, pH 8.0) was carried out using a twofold excess of BABE. By this technique the albumin and anhydrase were cleaved into three and two fragments respectively.

Studer and Meares (1992) described the preparation of a reagent (Figure 6.92) combining a number of interesting features for use in radiopharmaceutical immunoconjugates: (1) a bromoacetyl group for reactivity with thiols, e.g. from antibody Fab fragments; (2) the tetrapeptide sequence Ala-Leu-Ala-Leu which is known to be cleaved by liver peptidases, thus allowing efficient excretion of radiolabel; (3) an EDTA group protected as the tetra-*t*-butyl ester, which is thus safe from accidental metal chelation. When required the acid can be released by treatment with trifluoroacetic acid.

Among examples not possessing an α-carbonyl function, a useful class of fluorescent labels are the bromobimanes (Kosower *et al.*, 1979); in contrast to the parent bimanes, these are non-fluorescent so (in common with many maleimides, for example) fluorescence is only generated once coupling has been achieved. An example of a bromobimane is shown in Figure 6.93. Proteins such as haemoglobin were readily labelled under mild conditions (one hour, 23 °C, pH 7.4).

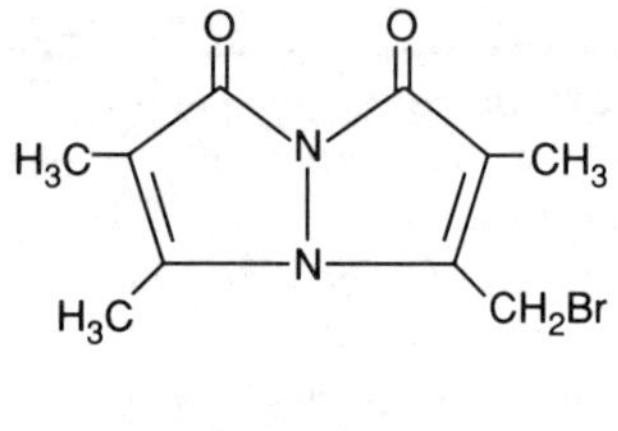

Figure 6.93 A monobromobimane.

Nitrobenzyl bromides were coupled to ovalbumin, BSA and keyhole limpet haemocyanin (two hours, room temperature, pH 7.5) by Li *et al.* (1991) but the resulting immunogens were unsuccessful in raising useful antibodies.

Katzenellenbogen *et al.* (1981) prepared a number of bromo-derivatives of hexestrol, an oestrogen receptor ligand. Reaction with the receptor (room temperature, pH 7.4) appeared to be complete within around one hour for the bromoacetyl derivatives, within two hours for a bromohydrin, and within around eight hours for a methyl bromide.

6.11.1.3 Alkyl chlorides

With their poor electrophilic character, these reagents have found limited use in the investigation of protein thiol properties, except in cases where the microenvironment results in cysteine groups of unusually high nucleophilicity (Hollaway *et al.*, 1964; Gerwin, 1967). Gerwin's work showed the complexity of microenvironmental effects on reactivity, the rate of alkylation of a single cysteine group in streptococcal proteinase showing dramatic differences in pH dependence with chloroacetate and chloroacetamide (Figure 6.94). No such difference was observed with simple thiols, hence the variation was attributed to electrostatic changes in the vicinity of the reactive cysteine.

The low reactivity of chloroalkyl compounds has often been demonstrated: Bindal *et al.* (1987) demonstrated that the binding to estrogen receptor of 11β-chloromethyl estradiol, potentially an alkylating affinity reagent, was in fact fully reversible. Similarly Blakeslee and Kennedy (1974) and De Weger *et al.* (1982) described only non-covalent attachment of the anti-cancer drug chlorambucil (Figure 6.95) to antibodies, after incubation with a 750-fold excess of the drug (ten minutes, 37 °C, pH 10.5).

Nevertheless some examples of this category are powerful enough to react with proteins, such as the highly unpleasant nitrogen and sulfur mustards (Figure 6.96; Alexander, 1954; Burnop *et al.*, 1957; Burke *et al.*, 1966). These react with many nucleophilic groups including amines; thiol products may often be a minor component (Alexander, 1954).

Holowka (1981) described the preparation of *N*-chloroacetyl-[^{125}I]iodotyramine (Figure 6.97) for radiolabelling of protein thiols. Alkylation of reduced and unreduced ribonuclease (nine hours, room temperature, pH 8.0) gave equivalent incorporations to iodoacetamide.

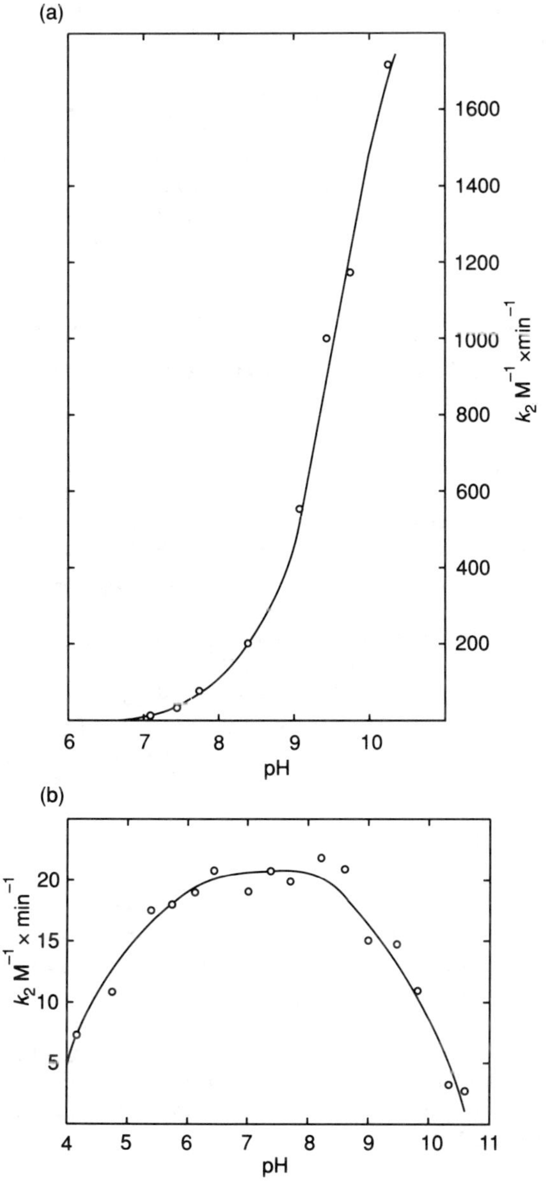

Figure 6.94 The pH dependence of second-order rate constants for the alkylation of the sulfhydryl of streptococcal proteinase by: (a) chloroacetamide and (b) chloroacetic acid. (Reproduced with permission from: Gerwin (1967). *J. Biol. Chem.* **242**, 451–456.)

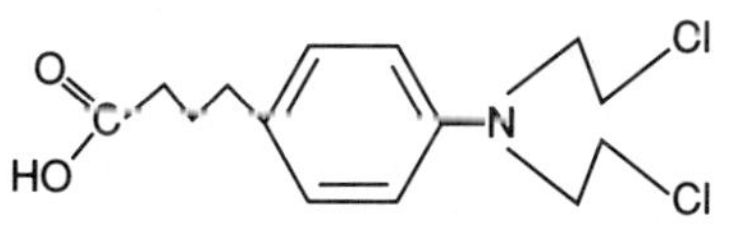

Figure 6.95 Chlorambucil.

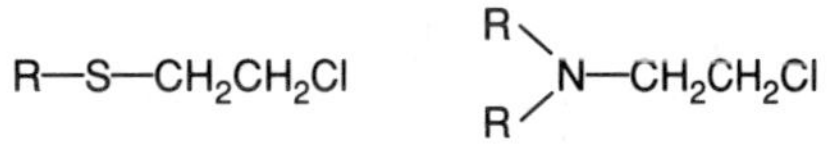

Figure 6.96 Mustards.

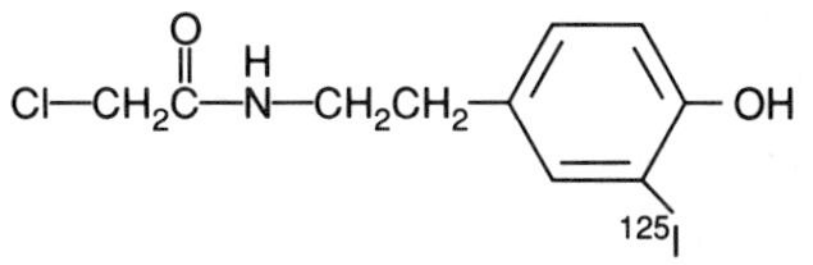

Figure 6.97 *N*-Chloroacetyl[^{125}I]iodotyramine.

6.11.2 Specific coupling of alkyl halides to other residues

As well as thiols, haloacetamides are capable of reacting with a number of other protein residues, and it is sometimes possible to confer some selectivity by an appropriate choice of conditions.

Amines are quite strongly reactive towards haloalkyl compounds, and this route can be particularly useful where there is no competition from thiols: Lamture and Wensel (1995), for example, describe the reaction of 4-(iodoacetamido)-2,6-diethylpyridine dicarboxylate with polylysine (overnight, 55 °C, pH 9) to produce a chelating conjugate. At a 50% excess of the iodoacetamide, >80% of the lysine groups were derivatised.

The reactivity of alkyl halides with histidine residues was exploited in a selective fashion by Liu (1967): the thiol group of streptococcal proteinase was reversibly protected by treatment with tetrathionate, to yield a negatively charged *S*-sulfenylsulfonate. This allowed a 500-fold excess of a positively-charged alkylating agent, α-bromoacetylarginine methyl ester, to react selectively with a nearby histidine residue (18 hours, 25 °C, pH 6.7).

Substrate-analogue chloromethyl ketones have been shown to react with active-site histidine residues in chymotrypsin and trypsin (Shaw, 1972); quite sharp pH optima for these reactions were observed at around pH 7.0–7.5.

Without an appropriate α-substituent the specificity for histidine is lost, and reaction with methionine may be observed, especially at acidic pH. Lawson and Schramm (1965) used a heterobifunctional active ester-alkyl halide to label both a serine and a methionine moiety in the active site area of chymotrypsin (Figure 6.98). The derivatisation of methionine took 3–4 hours at pH 5.

Finally 2-hydroxy-5-nitrobenzyl bromide (Figure 6.99) is specific for tryptophan, having been employed

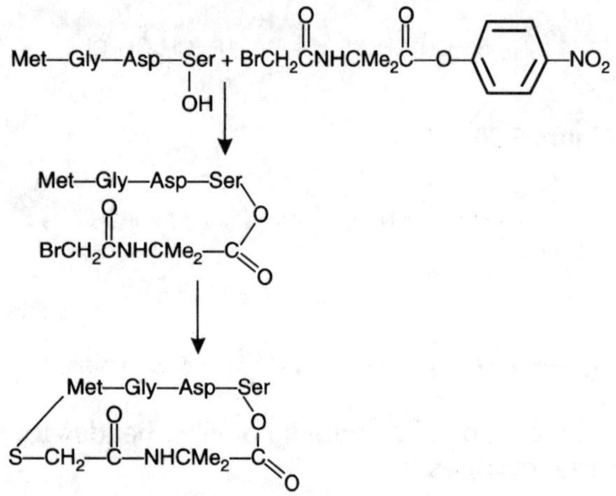

Figure 6.98 Intramolecular cross-linking of methionine and serine with a bromoacetyl-containing heterobifunctional reagent.

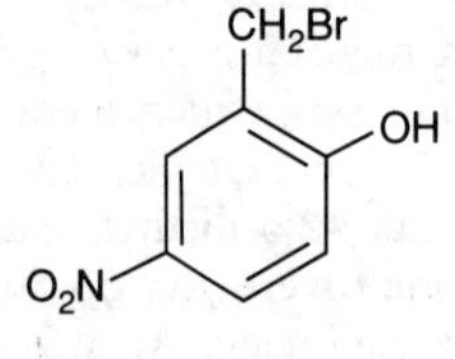

Figure 6.99 2-Hydroxy-5-nitrobenzyl bromide.

as the basis of a colorimetric test for these residues (Barman and Koshland, 1967).

6.11.3 Coupling of alkyl halides to amines *via* heterobifunctional ragents

Reagents more commonly known for the thiolation of proteins (Section 5.12) can also be employed to couple alkyl halides to protein amines. The approach is

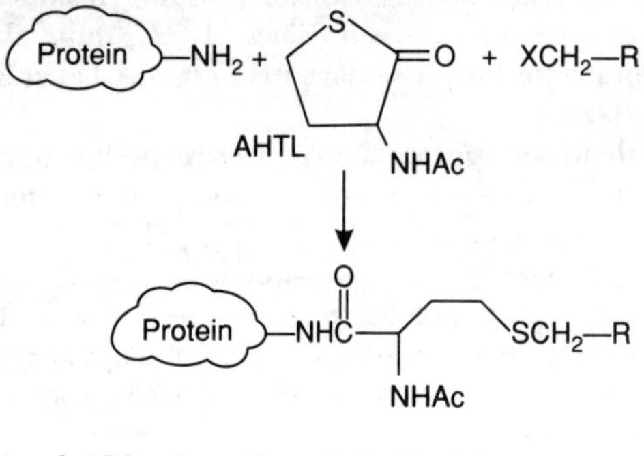

Figure 6.100 Coupling of alkyl halides to proteins using thiolation reagent.

depicted in Figure 6.100 using *N*-acetylhomocysteine thiolactone (AHTL) as an example.

Feng *et al.* (1990, 1992) used both AHTL and *S*-acetylmercapto-succinic anhydride (SAMSA) in this role; they coupled chloroacetanilide herbicides to BSA, HSA or IgG at an alkyl chloride:heterobifunctional:protein ratio of 25:25:1 (pH 11[1]; 15 minutes at 0 °C then two hours at 50 °C). Incorporations of 12 (BSA), 11–14 (HSA) and 12–21 (IgG) were observed.

6.12 Coupling of Proteins to Small Molecules Containing Disulfides and Related Groups

Besides offering a convenient means of generating thiols from a stable precursor, disulfides offer an extremely selective route for coupling to protein thiols—more so than either maleimides or alkyl halides. Consequently, the reaction of low-molecular-mass disulfides with proteins has been widely employed as a means of thiol quantitation, using such compounds as Ellman's reagent, 5,5′-dithio-bis(2-nitrobenzoic acid) and 2,2′-dithio-bis(5-nitropyridine). These reagents have also been employed to control the conjugation of one thiol-containing protein to another, and have therefore been dealt with in Section 5.2.1.3.

6.12.1 Disulfides as an *in situ* source of thiols

A methyldithio derivative of gentamicin was reduced (equimolar dierythritol in 50% aqueous THF, 61 hours at 4 °C) to generate thiols for reaction with bromoacetylated glucose-6-phosphate dehydrogenase (Rowley *et al.*, 1980; Singh *et al.*, 1980). A 50-fold excess of the thiol was added in small portions over five hours at 0 °C, and coupling took place over five days at 4 °C and pH 7.3.

6.12.2 Coupling of disulfides and related groups to protein thiols

Disulfides which release a chromophoric group on reaction with thiols have proved very useful for quantitation of these moieties in proteins. Similar compounds have also been used to investigate biochemical properties, Grassetti *et al.* (1969) used 6,6′-dithiodinicotinic acid to modify the metabolism

[1] The conditions are sufficiently basic to cause hydrolysis of the *S*-acetyl function derived from SAMSA, which would otherwise interfere with the alkylation reaction.

of Ehrlich ascites cells. The reagent was also used for thiol quantitation (pH 7.5, room temperature); the necessary reaction time varied from five minutes (BSA) to three hours (aldolase).

Disulfide derivatives of the antitumour antibiotic adriamycin were described by workers from Bristol-Myers Squibb (Greenfield *et al.*, 1990; Kaneko *et al.*, 1991), and coupled to a thiolated monoclonal antibody (overnight, 4 °C, pH 7.4) in equivalent quantities to the available thiol groups. Incorporations from 2–8 were observed under these conditions, but could be raised to 20 by increasing the thiol availability on the antibody. The linkers used by this group incorporated acylhydrazones and related groups, designed to confer lability under the acidic conditions (pH 5.0–5.5) of cell lysosomes.

Ebright *et al.* (1992) described the preparation of a pyridyldisulfide analogue of EDTA (Figure 6.101), which was used for the selective derivatisation of artificially introduced cysteine residues in DNA-binding proteins. Incorporation of Fe^{3+} ions provided a reagent for investigation of the ability of EDTA–iron complexes to facilitate the cleavage of DNA bound nearby.

Ermácora *et al.* (1992) reported similar studies with (EDTA-2-aminoethyl)-2-pyridyldisulfide, and observed an incorporation of 0.9 in a staphylococcal nuclease mutant. A twofold excess of the disulfide was used in the coupling reaction (one hour, room temperature, pH 7.2). Other pyridyl disulfide derivatives of polycarboxylate chelators have been described by Heyduk and Heyduk (1997).

As observed in Chapter 2, Section 2.3.4, sulfenyl compounds bearing electron-withdrawing substituents adjacent to a sulfur atom are more reactive than the analogous disulfides. Pihl and Lange (1962), for example, showed that cystamine monosulfoxide reacted with enzyme thiol groups that were stable to cystamine itself (Figure 6.102). Other examples have also been employed for derivatisation of protein thiols, such as alkylthiosulfonates (Bruice *et al.*, 1976) and sulfenyl halides (Fontana and Scoffone, 1972); preparative applications are rare, however.

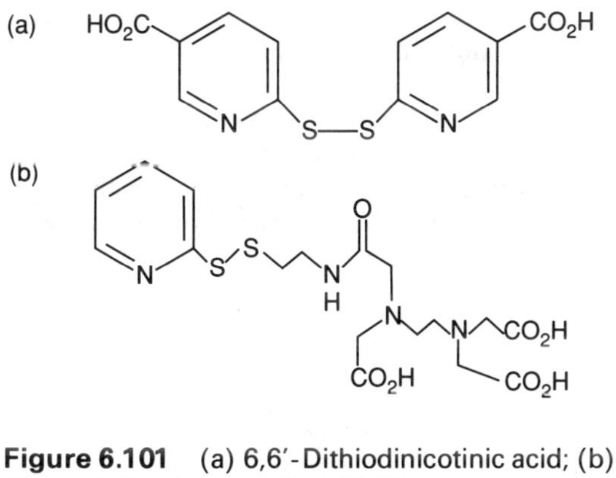

Figure 6.101 (a) 6,6′-Dithiodinicotinic acid; (b) *S*-(2-pyridylthio)cysteaminyl-EDTA.

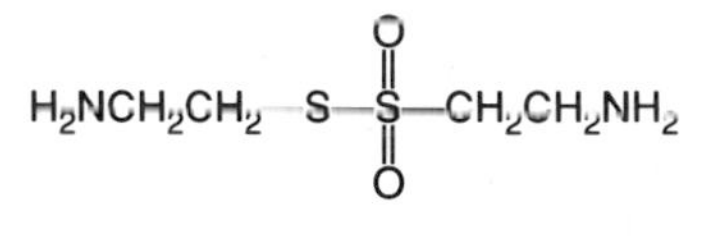

Figure 6.102 Cystamine monosulfoxide.

6.13 Coupling of Proteins to Small Molecules Containing Other Thiol-reactive Groups

6.13.1 Other groups containing active double bonds

There are many other compounds which possess double bonds analogous to that in the maleimide moiety, and which react in a similar fashion.

The double bond of vinyl ($RCH{=}CH_2$) compounds may be activated enough to take part in Michael addition reactions. The most common application of this approach in protein chemistry is probably in the pyridylethylation of cysteine residues for amino acid analysis (Friedman *et al.*, 1970; Mak and Jones, 1978), using 4-vinylpyridine (Figure 6.103). Vinylpyridine-based heterobifunctional reagents have been described.

The sulfonyl group of vinyl sulfones is strongly activating, allowing reaction with any suitably nucleophilic groups. This moiety has been incorporated into

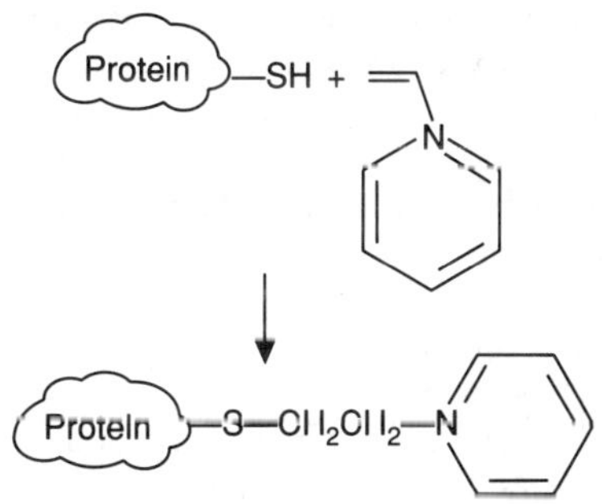

Figure 6.103 Protein pyridylethylation using 4-vinylpyridine.

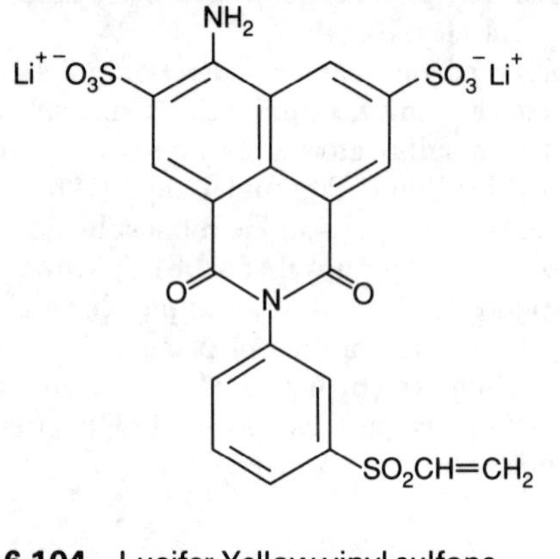

Figure 6.104 Lucifer Yellow vinyl sulfone.

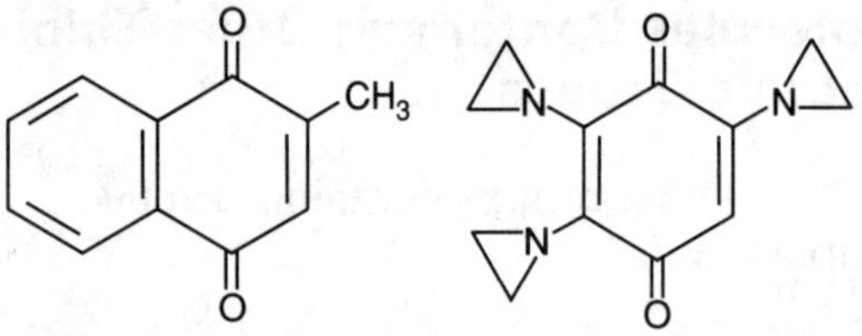

Menadione Triaziquone

Figure 6.105 Reagents containing quinone moieties reactive towards thiols.

reagents such as the fluorescent dye Lucifer Yellow vinyl sulfone (VS; Figure 6.104) (Stewart, 1981). Bailey *et al.* (1983, 1984) coupled Lucifer Yellow VS to albumin (two to four hours, room temperature, 0.2 M sodium bicarbonate) to yield a fluorescent reagent for immunoassay.

Divinylsulfone (DVS; $CH_2{=}CH{-}SO_2{-}CH{=}CH_2$) has been employed as a homobifunctional cross-linking agent (Houen and Jensen, 1995). Peptides were coupled to DVS-derivatised BSA or carboxymethyl-BSA at a pH determined by the target moieties in the peptide: 8 for thiols, 9 for amines and 10 for tyrosine hydroxyls. Incorporations of 7.4–23 were obtained for native BSA and 20.3–58.9 for the carboxymethylated protein. These higher incorporations were attributed to the greater degree of denaturation of the latter providing increased access for DVS activation, which was virtually quantitative.

The double bonds of quinones are reactive in the same way (Friedmann *et al.*, 1948), and this has been used as the basis for coupling compounds containing such structures to thiol-bearing proteins. Examples (Figure 6.105) include menadione (Johnston and Eisen, 1974) and triaziquone (Linford and Froese, 1978).

6.13.2 Mercurial reagents

Thiols combine readily with certain metal atoms such as gold, silver and mercury (cf. Section 5.4.4). This fact has been exploited in electrochemical methods for thiol determination based on silver (Benesch *et al.*, 1955) or mercury (Kolthoff *et al.*, 1954).

The ready availability of organomercurial reagents opened up other routes to thiol determination; Boyer (1954) described a spectrophotometric assay based on *p*-mercuribenzoate, for example, which could identify three different classes of thiol reactivity in ovalbumin. Reaction at pH 4.6 was much faster and more complete than at pH 7.0. Duke *et al.* (1971) used this reagent for kinetic studies on the derivatisation of thiols in methemerythrin.

Fluorescent labels based on mercury have been described, such as *S*-mercuric-*N*-dansylcysteine (Figure 6.106) (Leavis and Lehrer, 1974). The approach has also been used for affinity labels such as the NAD^+ analogue shown in Figure 6.107 (Woenckhaus and Duchmann, 1975).

The fact that mercury is a heavy metal gives these approaches an advantage for techniques based on X-rays or electron beams; the mercury atom can act as both linker and label. Thus McMurray and Trentham (1969) described the use of mercurinitrophenols to derivatise protein thiol sites for X-ray crystallography. An incorporation of two was obtained by treatment of glyceraldehyde 3-phosphate dehydrogenase with a large excess of 2-chloromercuri-4-nitrophenol (Figure 6.108) at pH 5.8. Marshall and Cohen (1980) used the same reagent to study the essential thiol group of ornithine transcarbamylases.

Mercury atoms are also convenient labels for electron microscopy probes; Azadova *et al.* (1965), for example, labelled antibodies with *p*-aminophenylmercuryl acetate to study the topography of Sendai virus proteins.

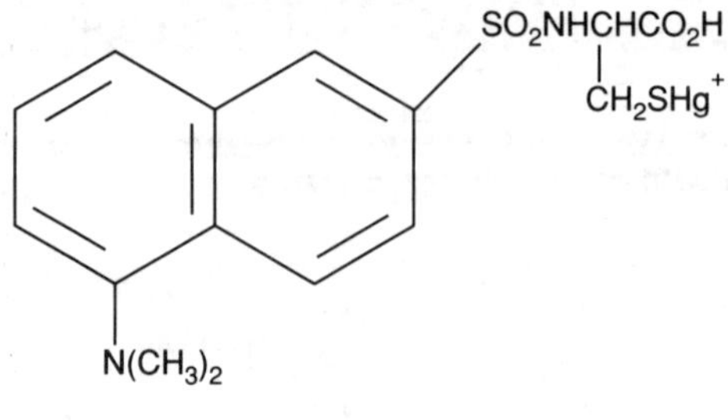

Figure 6.106 *S*-Mercuric-*N*-dansylcysteine.

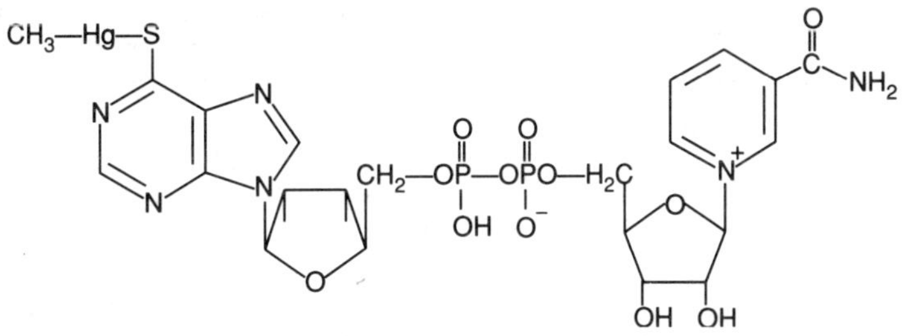

Figure 6.107 Affinity label of NAD$^+$ analogue containing a mercuric group.

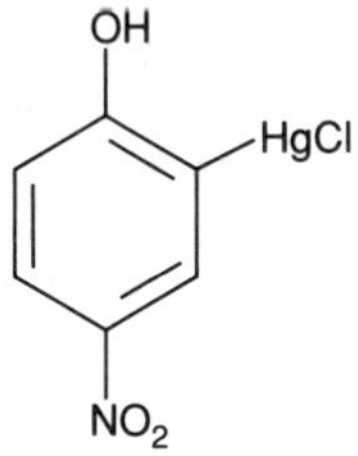

Figure 6.108 2-Chloromercuri-4-nitrophenol.

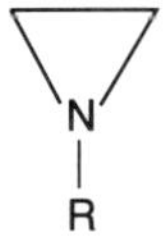

Figure 6.109 The aziridine function

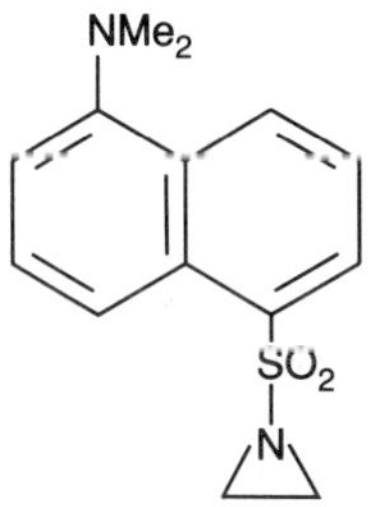

Figure 6.110 *N*-Dansylaziridine.

6.13.3 Aziridines

Aziridines (Figure 6.109) are highly-reactive electrophiles capable of reaction with thiols and amines. *N*-Dansylaziridine (Figure 6.110) is a thiol-specific fluorescent label based on this chemistry (Scouten *et al.*, 1974). Reaction with methionine is also possible with this reagent (Johnson and Schwartz, 1978).

Aziridines have found a role in affinity labelling experiments, particularly to provide reactive mimics of ligands containing related groups such as dimethylamines or pyrroles. A number of ligands to the estrogen receptor (ER) fall in this category, and the aziridines of tamoxifen (Robertson *et al.*, 1981; Katzenellenbogen *et al.*, 1983; Salituro *et al.*, 1986) and desmethylnafoxidine (Simpson *et al.*, 1987) have been used to radiolabel the active site. The suitability of aziridines as structural mimics for these compounds is obvious from Figure 6.111; the aziridine analogue of tamoxifen in fact bound to ER with a higher affinity than the parent dimethylamine (Robertson *et al.*, 1981).

Good results have also been obtained where there is less structural similarity, but the aziridine group is distanced from the body of the molecule, e.g. with ketononestrol (Elliston *et al.*, 1987; Zablocki *et al.*, 1987).

Harlow *et al.* (1989) showed that aziridines of both agonist and antagonist ligands coupled to a reactive cysteine residue (Cys530) in the ER molecule. Typically these labelling reactions were carried out for 0.5 to 3 hours at pH 7.4 and 4 °C, using DMF to dissolve the aziridines.

6.13.4 Epoxides

Epoxides have broad reactivity towards nucleophiles, and have been used for example to label active-site carboxylic acid residues (Schray *et al.*, 1973; Eshdat *et al.*, 1973). However they have also been used for labelling cysteines (Klinman, 1975; Bruice *et al.*, 1978).

6.13.5 Halobenzoxadiazoles

The halobenzoxadiazoles (more specifically 7-halobenzo-2-oxa-1,3-diazoles; Figure 6.112), are closely analogous to simple aryl halides in their reactions, but yield fluorescent products with amines, obviously a useful characteristic. Ghosh and Whitehouse (1968) reported 7-chloro-4-nitrobenzo-2-oxa-1,3-diazole (NBD-Cl; Figure 6.113), noting that it reacted with

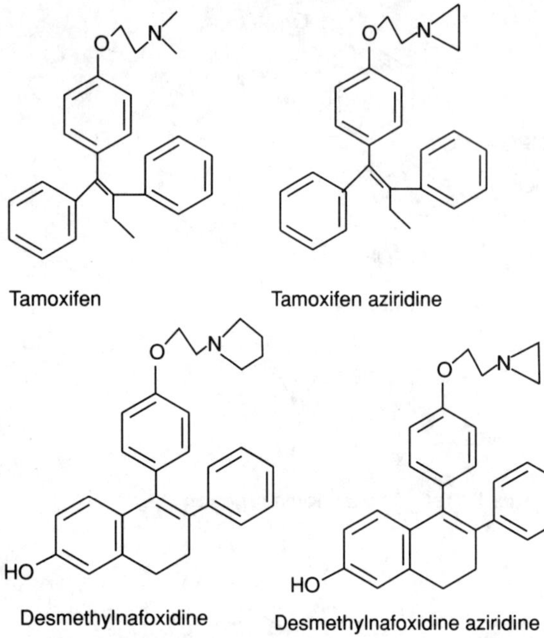

Figure 6.111 Some examples of aziridines used for radiolabelling active sites to mimic ligands.

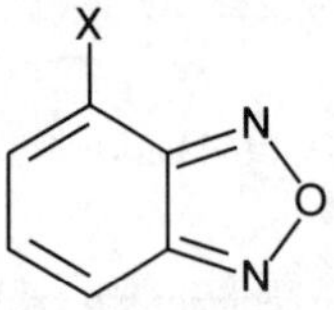

Figure 6.112 Halobenzoxadiazoles.

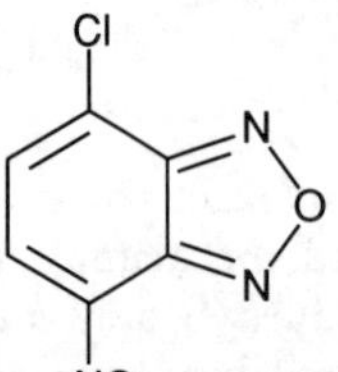

Figure 6.113 NBD-Cl.

both the thiol and amine groups of cysteine methyl ester.

Compounds of this type have been used for both amine and thiol derivatisation, their selectivity being modulated both by the structure of the benzoxadiazole itself, and by the reaction conditions. Although it will react to some extent with the primary amine groups of most amino acids, NBD-Cl shows rather higher reactivity with secondary amines, a particularly useful phenomenon since most reagents for amino acid derivatisation react poorly with such groups. This reagent, therefore, has been used as the basis of methods for determination of proline and hydroxyproline in protein hydrolysates (Roth, 1978; Ahnoff *et al.*, 1981). Later it was demonstrated (Imai and Watanabe, 1981; Watanabe and Imai, 1982) that the fluoro analogue, NBD-F, is more reactive, and provides a useful reagent for this application. NBD-F treatment (one minute, 60 °C, pH 8.0) could derivatise amino acids of all categories in protein hydrolysates, including the secondary amines of proline and hydroxyproline.

Other protein residues have also been shown to react with NBD-Cl, such as tyrosine (Ferguson *et al.*, 1974). With intact proteins, however, the predominant application of benzoxadiazoles has been in the derivatisation of thiols.

Working with NBD-Cl at pH 7.0–7.5, Birkett *et al.* (1970, 1971) were able to obtain specificity for thiols. With phosphorylase *b*, they observed differential reactivity of the thiol groups present, and a probable conformational change of the enzyme at around 11 °C.

Shipton *et al.* (1976) established that NBD-Cl existed in monocationic form below pH 8.0, and observed variations in thiol reactivity and kinetics for three different enzymes. This allowed useful conclusions to be made about the microenvironment of these active-site thiols. Evidence was also presented for the existence of Meisenheimer complexes as intermediates in the reaction (Figure 6.114). Baines *et al.* (1977) noted that sites within the benzoxadiazole structure other than the chloro-substituted 4-position were also susceptible to nucleophilic attack, adding a further degree of complexity to the reactivity of this compound.

Lad *et al.* (1977) derivatised a single cysteine residue on rabbit muscle phosphofructokinase with a 2.5-fold excess of NBD-Cl (three hours, 25 °C, pH 7.0).

Nitta *et al.* (1979) reported anomalies in the reaction of NBD-Cl with thiols, concluding that at least two

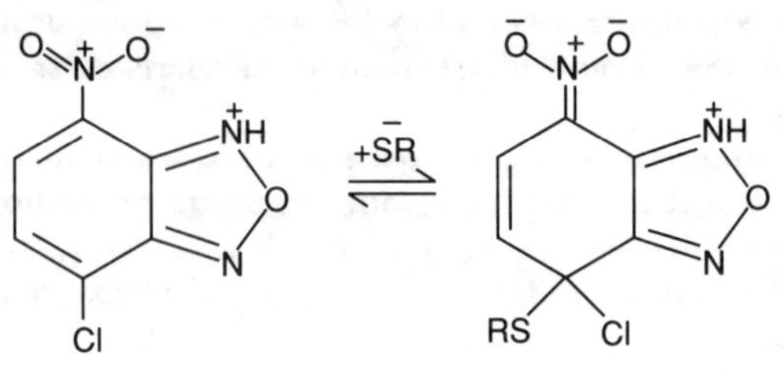

Figure 6.114 Formation of Meisenheimer complex between NBD-Cl and thiol.

other products must be formed in addition to the 'NBD-S' conjugate. One of these was postulated to be a Meisenheimer complex, the other was able to react with proteins but was not identified.

Due to the complexity of thiol derivatisation with NBD-Cl, and other disadvantages such as poor aqueous solubility, cytotoxicity and mutagenicity, analogous compounds have been sought with improved properties. Rapid reaction has also been sought to minimise the risk of thiol oxidation.

These reagents include 4-chloro-4-sulfonylbenzo-2-oxa-1,3-diazole (SBD-Cl; Andrews *et al.*, 1982; Nanami *et al.*, 1993); 7-fluoro-4-sulfonylbenzo-2-oxa-1,3-diazole (SBD-F; Imai *et al.*, 1983; Sueyoshi *et al.*, 1985); 4-aminosulfonyl-7-fluoro-benzo-2-oxa-1,3-diazole (ABD-F; Toyo'oka and Imai, 1984, 1985; Kirley, 1989a,b), and 4-(dimethylamino-sulfonyl)-7-fluoro-benzo-2-oxa-1,3-diazole (DBD-F; Toyo'oka *et al.*, 1989). With the last compound thiols could be determined quantitatively by ten minutes reaction at pH 8.0 and 50 °C.

A useful feature of these reagents is their ability to be used in the presence of reducing agents such as tributylphosphine, meaning that the total thiol and disulfide content can be determined in a one-pot reaction (Kirley, 1989a).

While the thiobenzoxadiazole products of the reactions described above are highly stable in an acidic environment, Treuheit and Kirley (1993) reported that the thiol could be partially regenerated under basic and/or reducing conditions.

6.14 Coupling of Proteins to Small Molecules Containing Hydroxyl Groups

6.14.1 Heterobifunctional hydroxyl reagents

A heterobifunctional reagent has been developed (Palumbo, 1993; Annunziato *et al.*, 1993) for coupling hydroxylic compounds to protein thiols. The isocyanate group of *p*-maleimidophenyl isocyanate (PMPI) was shown to react with hydroxyl groups in serine-containing oligopeptides, vitamin B_{12} and folic acid derivatives, steroids and thyroid hormones. Typical activations employed equimolar PMPI in DMSO (overnight, room temperature) although higher PMPI loadings and/or shorter times were used in some cases. In these studies the maleimidyl derivatives were isolated and characterised before use (and therefore belong properly in Section 6.10) but it seems likely that PMPI-derivatised compounds will generally be used without this step. Coupling of the maleimidyl groups to thiolated proteins was carried out at pH 7.3 (one hour at room temperature, then overnight at 4 °C).

Generally however, the hydroxyl function is converted synthetically to a more reactive moiety before protein coupling is carried out.

6.14.2 Activation of hydroxyl groups for derivatisation

The oxidation of small hydroxylic compounds to generate carbonyl groups has often been exploited, and such approaches are covered in Section 6.2.3. Alternatively conversion of the hydroxylic group (under non-aqueous conditions) to introduce a good leaving group such as a tosylate or tresylate may offer a useful route.

Hydroxyl groups can also be used as a site for reaction with activated carboxyl groups, for example allowing a free carboxyl residue to be introduced by reaction with succinic anyhdride under forcing conditions. Similarly heterobifunctional acid chlorides have been used to introduce maleimide moieties (Hermentin *et al.*, 1990).

Another alternative for hydroxyl activation is to generate a chloroformate by reaction with phosgene (Erlanger *et al.*, 1957; Teale *et al.*, 1974), this can then be used directly for reaction with protein amines.

6.15 Coupling of Proteins to Small Molecules Containing Imidazole, Phenolic and Related Groups

The diazo coupling technique, described in Section 6.7.9 for linking amine-bearing molecules to proteins, can also be utilised 'in reverse'. Thus, small molecules containing imidazolyl, phenolic or other diazotisable moieties can be coupled to proteins using a suitable homo- or heterobifunctional diazonium reagent (cf. Chapter 5, Section 5.5.3).

6.15.1 Homobifunctional diazo coupling

Bassiri and Utiger (1972) coupled a histidine-containing peptide to BSA using bis-diazotised benzidine, BDB (two hours, 5 °C, pH 9.0), while Tamura and Bauer (1982) used the same approach with a histidine-containing example.

Briand *et al.* (1985) used similar conditions to attach a 30-fold excess of histidine or tyrosine-containing peptides to carrier proteins. Under these conditions lysine groups will react; those in the peptide were sometimes blocked with citraconic anhydride, but in both cases a good deal of the protein derivatisation probably occurred at lysyl residues.

With two different peptides, Briand *et al.* obtained incorporations of 13 and 19 in BSA conjugates, but over 75% of the peptide content was lost on storage at 4 °C for 40 days.

As well as ring substitution reactions, diazonium salts are capable of derivatising some hydroxylic functions; an example (Johnson and Smith, 1972) is the hydroxypyrazole in Figure 6.115, which was coupled to protein as shown using BDB (30 minutes, room temperature, pH 7.3).

6.15.2 Heterobifunctional diazo coupling

The most obvious route for heterobifunctional coupling of compounds containing histidine or tyrosine residues is *via* a reagent coupling the diazonium function to an amine-reactive group, but examples of this approach are not common. Koch *et al.* (1973) used diazonium salts of phenylacetic acid for this purpose, although the carboxylic derivatives were isolated before activation with carbodiimide (Figure 6.116). Levy *et al.* (1976) used a similar approach to couple the alkaloid reserpine to BSA, using a mixed anhydride procedure for the final coupling to protein.

Berkowitz *et al.* (1975) described the use of ethyl-*N*-carbamylcyanomethyl acetimidate (DAZ reagent; Figure 6.117) to derivatise BSA; this was then

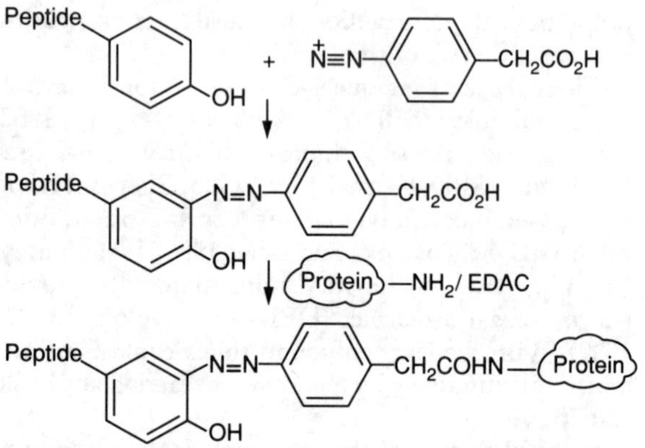

Figure 6.116 Use of phenylacetic acid diazonium salt as a heterobifunctional reagent to cross-link tyrosine residues with proteins.

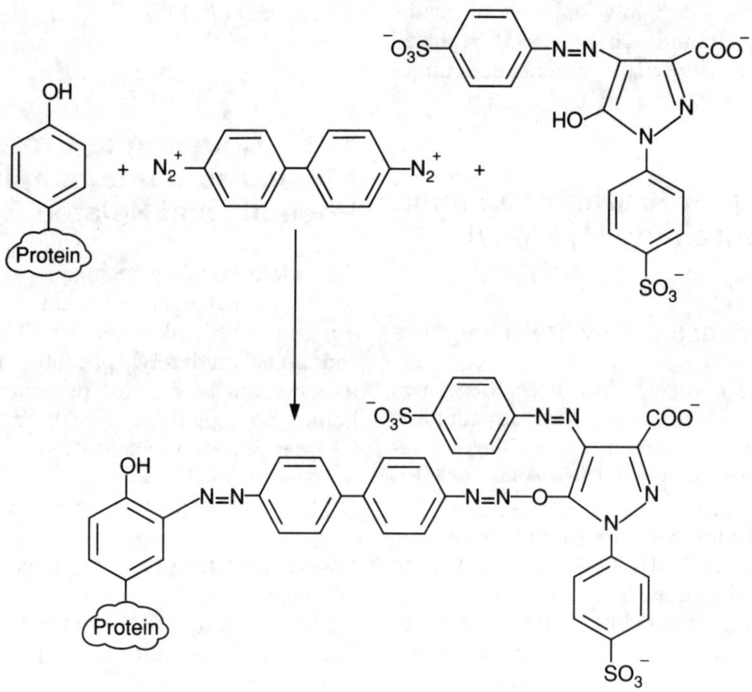

Figure 6.115 BDB coupling of a hydroxypyrazole to protein.

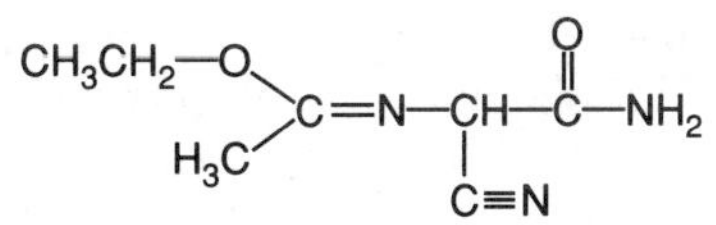

Figure 6.117 DAZ reagent.

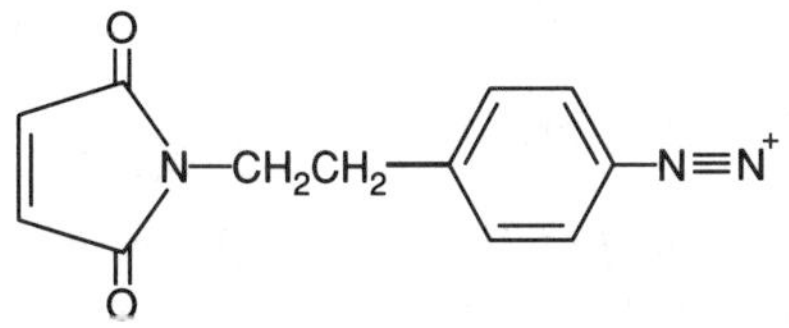

Figure 6.118 *N*-[β-(4-diazophenyl)ethyl]maleimide (DPEM).

diazotised, and coupled to the phenolic groups of the narcotic antagonist naloxone.

Fujiwara and Saita (1987) described the use of *N*-[β-(4-diazophenyl)ethyl]maleimide (DPEM; Figure 6.118) for coupling BSA and β-galactosidase to the tyrosine residues of the oligopeptide neurotensin. Diazotisation of the aniline precursor was carried out by a twofold excess of acidic sodium nitrite in aqueous DMF (15 minutes, 0 °C).

The activation of the peptide was carried out in a borate–phosphate buffer, pH 9.0 (15 minutes, room temperature), excess DPEM removed by ethyl acetate extraction, and the resulting peptide solution coupled directly to thiolated proteins (15–20 minutes, room temperature, pH 6–7). The resulting conjugates were successful as neurotensin immunogens (BSA) or in an immunoassay (β-Gal).

6.15.3 Conversion of diazo derivatives to amines

Although not strictly within the remit of this chapter, the reduction of diazo derivatives to the corresponding amine with dithionite (Gorecki *et al.*, 1971) is a potentially useful reaction, and is shown in Figure 6.119.

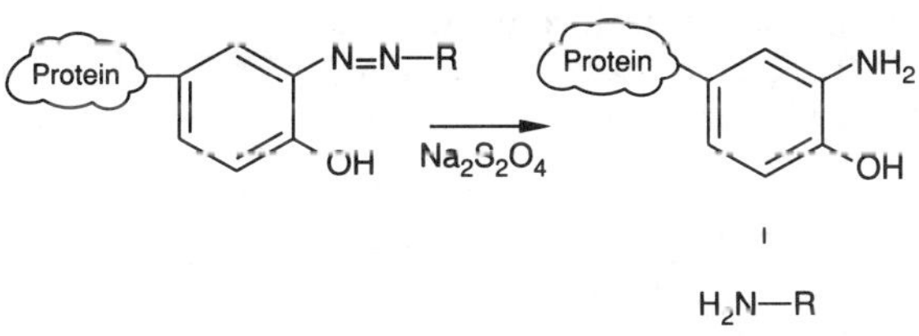

Figure 6.119 Reduction of diazo derivative to corresponding amine with dithionite.

6.16 Coupling of Proteins to Small Molecules Containing Photoactivatable Groups

Photoactivatable reagents have become very popular in studying receptor–ligand interactions, and Chapter 5, Section 5.10 described the formation of protein–protein links by this methodology; however, the photoaffinity labelling approach was originally demonstrated using low-molecular-mass ligands (Singh *et al.*, 1962), and there have been numerous applications where photoactivatable analogues of small compounds are coupled to proteins.

The use of photogenerated reagents is an extensive subject in its own right (Bayley, 1983a,b), and only a brief consideration of the different photoactivatable moieties will be attempted here. The major categories are: (1) carbene-generating reagents, principally diazocarbonyl compounds and diazirines, (2) nitrene-generating reagents, notably aryl azides, and (3) free-radical-generating reagents such as α,β-unsaturated ketones. The first two classes are by far the most widely encountered.

The chemistry of most of these reagents and their photoactivation conditions are discussed in Section 5.10, and will not be repeated here.

It is worth noting that the choice of reagent for a particular application, typically affinity labelling, is influenced by more than one factor. Although it may be straightforward to rank the relative photochemical advantages of the different categories just discussed, the other properties of photoactivatable derivatives may override this; problems with solubility or with retention of functional activity may lead to the selection of a derivative with sub-optimal photochemical properties. Therefore it is likely that all these categories will remain useful for certain applications.

6.16.1 Carbene-generating photoactivatable reagents

6.16.1.1 Diazocarbonyl compounds

The pioneering photoaffinity labelling experiments used diazoesters such as *p*-nitrophenyl diazoacetate (Singh *et al.*, 1962) or diazomalonate (Vaughan and Westheimer, 1969) to label chymotrypsin and trypsin.

Katzenellenbogen and his colleagues at the University of Illinois carried out extensive investigations of estrogen binding proteins using PAL. Their early work (Katzenellenbogen *et al.*, 1973a,b) described

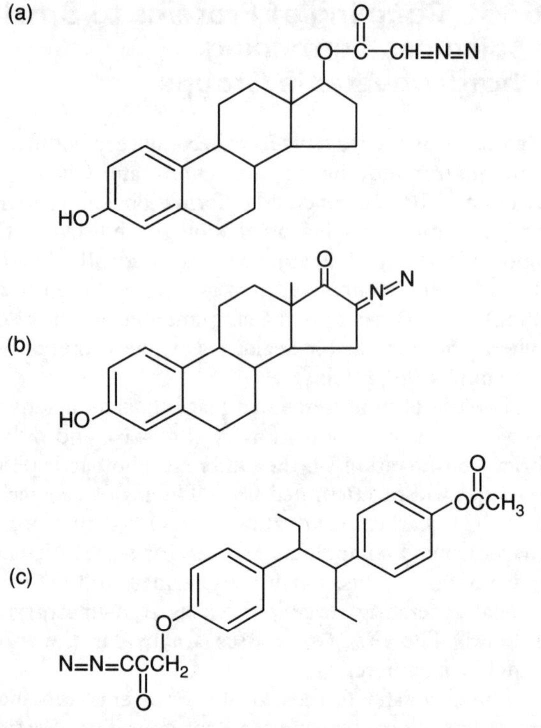

Figure 6.120 Derivatives of steroids and steroid analogues used for PAL studies.

steroid derivatives such as those shown in Figure 6.120(a) and 6.120(b), while later studies focussed on analogues of the non-steroidal estrogen hexestrol such as Figure 6.120(c) (Katzenellenbogen *et al.*, 1977a,b). In the last of these papers photocoupling was carried out at 254 nm (ten minutes, 0–3 °C, pH 7.4), although reaction did not appear to be complete within this time.

Casanova *et al.* (1984) used *N*-(2-diazo-3,3,3-trifluoropropionyl)-3,3′,5-triiodothyronine (Figure 6.121) to probe thyroid hormone receptors. Photocoupling was at 254 nm (two minutes, 4 °C, isotonic saline).

Klueppelberg *et al.* (1990) used an N-terminal *N*-2-diazo-3,3,3-trifluoropropionyl oligopeptide for photocoupling to a pancreatic receptor, although the yield

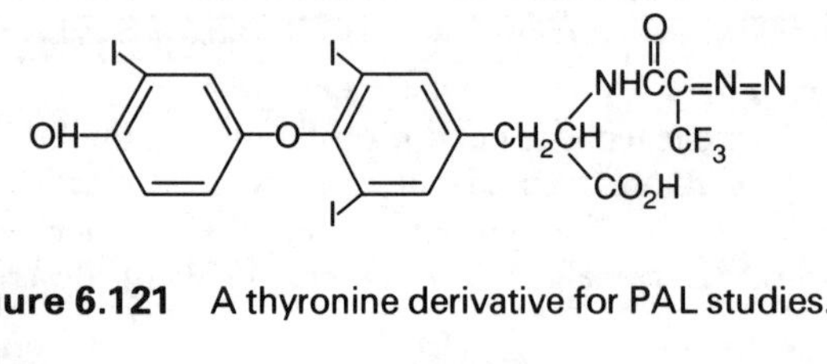

Figure 6.121 A thyronine derivative for PAL studies.

was low compared to all other photoactive groups studied (aryl azides, nitrophenols, benzophenones).

Many other diazoacetyl-based photoaffinity reagents have been described, including, for example derivatives of choline (Frank and Schwyzer, 1970; Waser *et al.*, 1970) and rhodopsin (Sen *et al.*, 1982, 1984; Ok *et al.*, 1988).

6.16.1.2 Diazirines

Diazirines (Figure 6.122) are closely analogous to the diazo-compounds just discussed, generating carbenes on irradiation at the higher wavelength range of around 350 nm. Much of the basis of diazirine chemistry was established in the 1960s (Church *et al.*, 1965; Church and Weiss, 1970), and Smith and Knowles (1973, 1975) first proposed these compounds as photoactivatable reagents.

These simple diazirines suffered from conversion to linear diazo compounds on photolysis, which, although still photoactive, are much more long-lived at the low-energy wavelengths suited to diazirines. In fact most compounds of this category, including simple examples such as dialkyldiazirines (Modarelli *et al.*, 1992) undergo rapid intramolecular reactions on irradiation. Although carbene products can be isolated in low yield, they are therefore of little use for the intermolecular reactions of photoaffinity labelling. Exceptions are adamantane derivatives which generate a trappable adamantylidene carbene (Bayley and Knowles, 1978, 1980; Morgan *et al.*, 1991) and compounds containing perfluorinated groups; the C–F bonds of the latter are strong enough to withstand intramolecular rearrangement, particularly if stabilising groups such as phenyl, methoxy, or chlorine are in adjacent positions (Modarelli *et al.*, 1992).

Brunner *et al.* (1980) prepared 3-phenyl-3-trifluoromethyl-diazirine compounds, which produced good yields of direct carbene product. Lecithin analogues incorporating this group were more efficient than aryl azides in coupling to lipids, but less so in labelling the oligopeptide gramicidin A (Brunner and Richards, 1980). Shih and Bayley (1985) synthesised a phenylalanine derivative incorporating this group.

Van Ceunebroeck *et al.* (1986) prepared a radiolabelled hydrophobic reagent of this type, 3-trifluor-

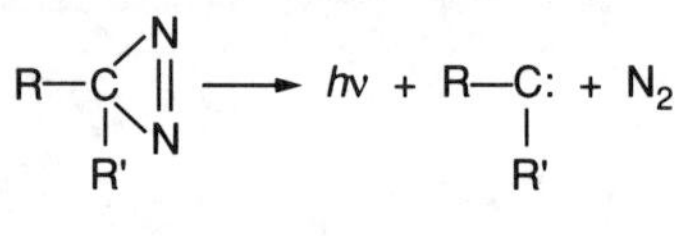

Figure 6.122 Diazirines generate carbenes on pholotysis.

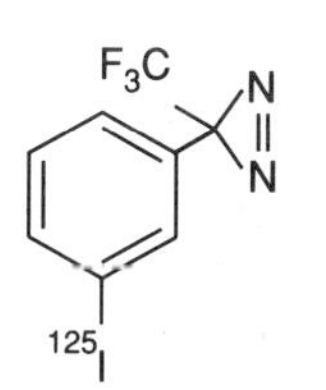

Figure 6.123 [^{125}I] TID, a radiolabelled lipophilic PAL reagent.

omethyl-3-(*m*-[^{125}I]iodophenyl)diazirine ([^{125}I]TID; Figure 6.123), and used its lipophilic properties to probe binding sites in bovine α-lactalbumin. Alterations in ionic strength and protein concentration changed the accessibility of hydrophobic regions within the protein. Photoactivation was for ten minutes at 315–435 nm.

Other groups have used this reagent in similar applications, for example in studies on the acetylcholine receptor (White and Cohen, 1988; White *et al.*, 1991). Irradiation was for 10–30 minutes using a 365 nm lamp, and specific binding-site labelling was observed as indicated by competition from agonists.

Nakayama and Khorana (1990) synthesised 3-trifluoromethyl-3-aryl diazirine derivatives of 11-*cis*-retinal, the chromophore of the photoreceptor protein rhodopsin.

Delfino *et al.* (1993) prepared 3-trifluoromethyl-3-phenyl derivatives of long-chain phospholipids, and used them to photolabel (ca 360 nm) the hydrophobic core of membranes. These reagents contained a potentially cleavable thioacetal group.

Platz *et al.* (1991) noted that N–H insertion reactions of 3-trifluoromethyl-3-aryl diazirines could be reversed by an elimination and hydrolysis sequence (Figure 6.124), and warned of the possible instability of photolabelled protein adducts obtained using this chemistry.

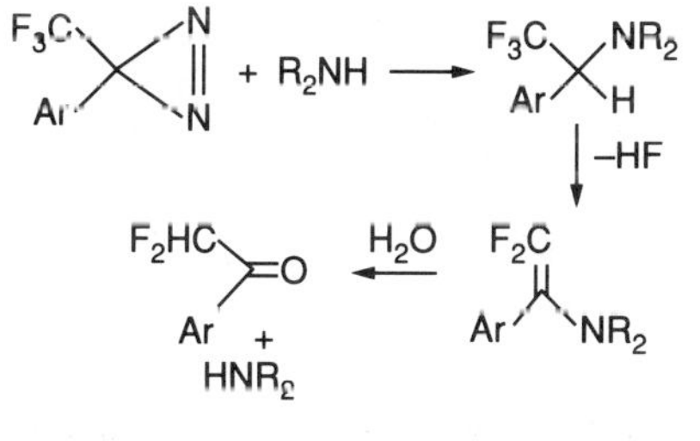

Figure 6.124 Reversal of N-H insertion by elimination and hydrolysis.

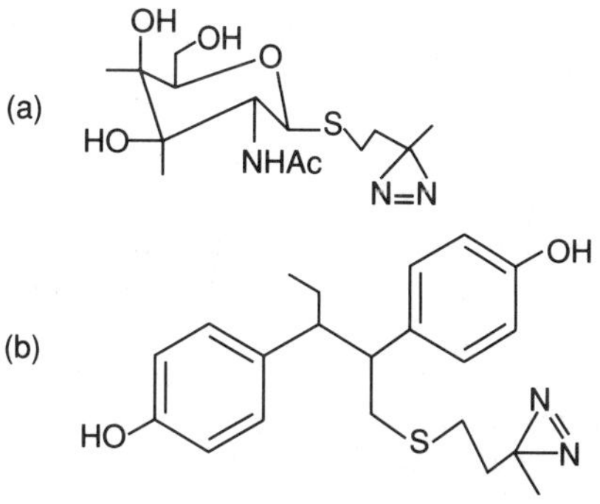

Figure 6.125 Diazirine derivatives of glycoside and hexestrol for PAL studies.

White *et al.* (1991) also used another hydrophobic diazirine probe, tritiated adamantyl diazirine, which labelled the receptor in a non-specific fashion. Morgan *et al.* (1991) demonstrated the reaction of the adamantylidene carbene with pyridine, although other reactions could be attributed to the existence of a transient precursor.

Kuhn *et al.* (1992) described photolabelling of β-hexosaminidose A using diazirine-containing glycoside analogues such as that shown in Figure 6.125(a). Successful results were obtained, although it seems unlikely that the reaction involves carbenes (see above). Bergmann *et al.* (1994) prepared similar derivatives of the estrogen receptor ligand hexestrol (Figure 6.125(b)), and again these compounds displayed selective and efficient photolabelling.

6.16.1.3 Diazocyclohexadienones

Diazocyclohexadienones lose nitrogen on irradiation at reasonably high wavelengths (ca. 350 nm) to generate carbenes (Figure 6.126). This group has occasionally been incorporated in photoaffinity reagents (Goeldner *et al.*, 1989; Hawkinson *et al.*, 1991). Despite their interesting photochemical properties, a comparative study by the same group (Kapfer *et al.*, 1994) displayed poor selectivity compared to PFPA analogues (see below).

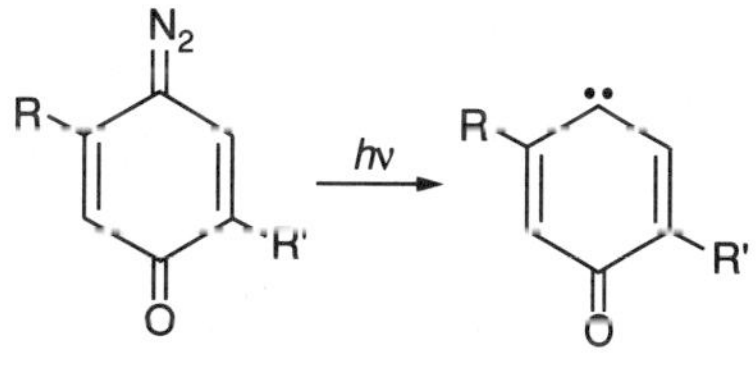

Figure 6.126 Diazocyclohexadienones generate carbenes on photolysis.

6.16.2 Nitrene-generating photoactivatable reagents

Nitrene-generating reagents were examined as alternatives to diazo compounds in the 1960s: Fleet *et al.* (1969) used 4-fluoro-3-nitrophenyl azide to photolabel antibodies. Due to the unsuitable properties of other nitrene sources, this field has been dominated by aryl azides. These may be simple phenyl derivatives, or nitrophenyl azides which can be activated at a longer, less damaging wavelength. More recently, considerable attention has been focussed on polyfluorophenyl azides (PFPAs), which provide better yields of product derived from nitrenes rather than less-reactive rearrangement products such as ketenimines.

6.16.2.1 Phenyl and nitrophenyl azides

Katzenellenbogen *et al.* (1973a,b) synthesised aryl azide analogues of steroidal and non-steroidal estrogen receptor (ER) ligands; examples are shown in Figure 6.127. The hexestrol derivative was later used for photoaffinity labelling of ER (Katzenellenbogen *et al.*, 1977a,b), irradiation being carried out at 254 nm. The affinity of hexestrol for the receptor was reduced much less than an analogous diazo compound, albeit coupled through a different position. Tritium-labelling of this compound reduced its binding however, and the labelling obtained was relatively non-specific.

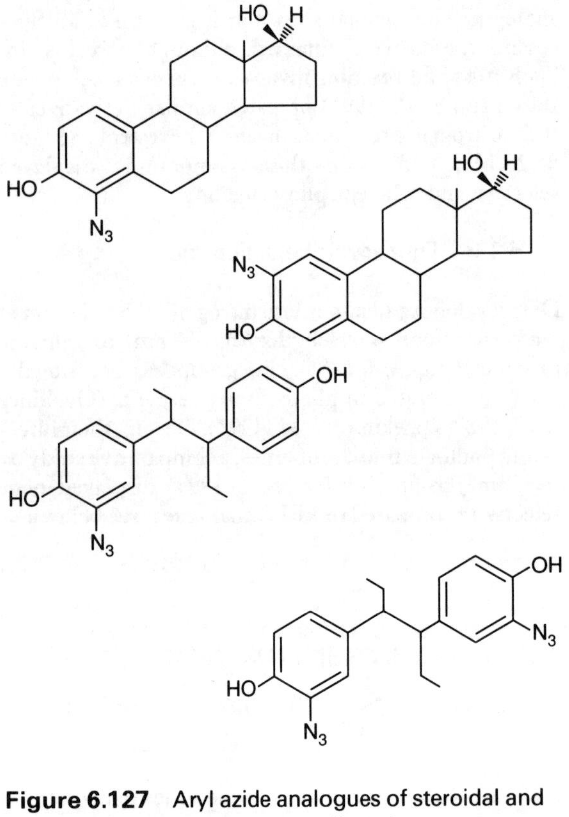

Figure 6.127 Aryl azide analogues of steroidal and non-steroidal estrogen receptor ligands.

Katzenellenbogen's group has continued to develop photoaffinity reagents for ER and the progesterone receptor (Pinney and Katzenellenbogen, 1991; Pinney *et al.*, 1991), and described a phenyl azide derivative of another non-steroidal ligand based on a benzo[*b*]thiophene structure (Figure 6.128). This displayed the best efficiency and selectivity for ER photoaffinity labelling described to date. Kym *et al.* (1995) described a phenyl azide derivative of progestin ketal (Figure 6.129), a similarly efficient photoaffinity reagent for progesterone receptor.

Galardy *et al.* (1974) synthesised succinimide ester derivatives of azidobenzoic and azidonitrobenzoic acids, and coupled them to pentagastrin. These photoaffinity reagents displayed some selectivity for binding sites in BSA.

Hydrophobic examples of this category such as 1-azidonaphthalene and 1-azido-4-iodobenzene have been used to label the lipid core of cell membranes (Klip and Gitler, 1974).

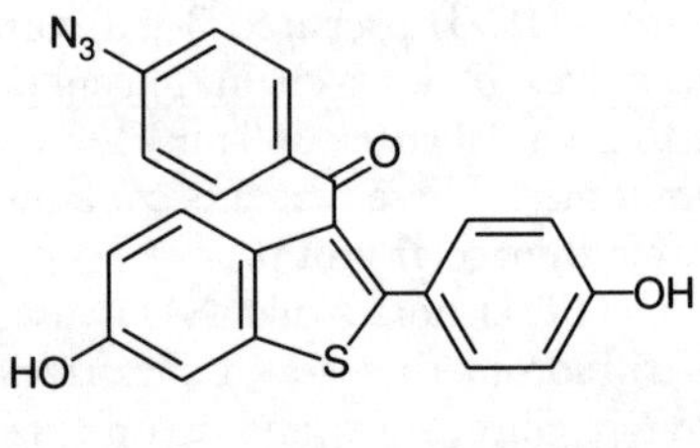

Figure 6.128 Phenyl azide based on the benzo[*b*]thiophene structure: 3-(4-azidobenzoyl)-6-hydroxy-2-(4-hydroxyphenyl)benzo[*b*]thiophene.

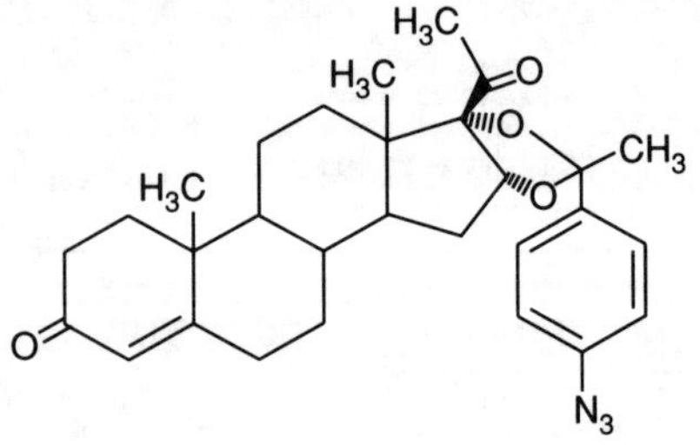

Figure 6.129 Phenyl azide derivative of progestin ketal: 16α,17α-[(*R*)-1′-(4-azidophenyl)-ethylidenedioxy]pregn-4-ene-3,20-dione.

Dockter (1979) described an azidonaphthalene disulfonate which possessed useful properties: although non-fluorescent *per se*, the photolabelled products displayed fluorescence.

Brunner and Richards (1980) used simple phenyl azide derivatives of lecithin to photolabel the peptide gramicidin A, and observed attack predominantly on the tryptophan residues.

Baenziger and Fiete (1982) used commercial succinimide esters of aryl azides (SANPAH, SASD and HSAB; see Section 5.10) to prepare photoactivatable glycopeptide probes. These reagents showed good selectivity for lectin binding sites.

Similarly, Shanahan *et al.* (1985) used NHS-ASA to prepare a labelled glucose analogue as a probe for the human erythrocyte monosaccharide transporter protein. A mixture of specific and non-specific labelling was observed, the former partially blockable by glucose.

Nitrophenyl azide groups attached to vitamin D_3 have been used to investigate its binding properties (Kutner *et al.*, 1986; Link *et al.*, 1987).

Klueppelberg *et al.* (1990) described comparitive experiments with cholecystokinin photoaffinity reagents of different types; an azidophenyl derivative gave considerably better incorporation than nitrophenyl, diazo ester or benzophenone reagents. Around 40 minutes ultraviolet irradiation was required for the reaction to reach its maximum yield.

Non-fluorinated phenyl azides are still often employed in the preparation of photoactivatable reagents. Although the superiority of fluorinated analogues continues to receive confirmation in recent studies (Kapfer *et al.*, 1995), examples where simple phenyl azides are superior are also reported (Golinski *et al.*, 1995a).

6.16.2.2 Polyfluorophenyl azides

The recent developments in polyfluorophenyl azide chemistry have been discussed in Section 5.10.4. Predictably, this approach has also been applied for photocoupling of small molecules to proteins, notably in the preparation of photoaffinity reagents analogous to those described in the remainder of Section 6.16. In brief, the longer-lived nitrenes obtained from these reagents are more capable of achieving the desired photolabelling, rather than decomposing into less reactive species.

A tetrafluorophenyl azide analogue of the reagent shown in Figure 6.128 was described simultaneously by Katzenellenbogen's group (Pinney and Katzenellenbogen, 1991; Pinney *et al.*, 1991). Somewhat

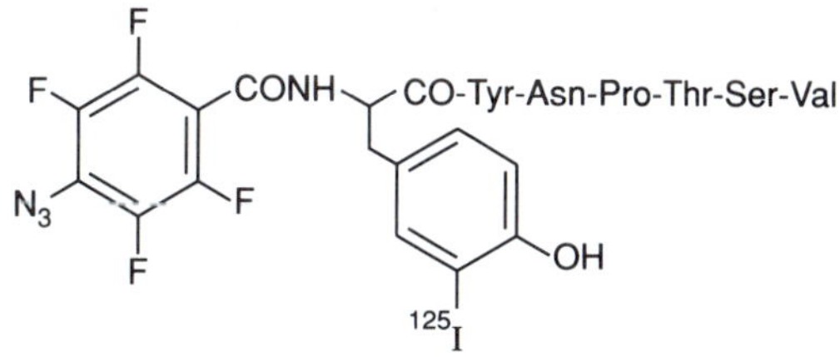

Figure 6.130 A tetrafluorophenylazide derivative of a radioiodinated oligopeptide.

contrary to expectations, in comparison to the simple phenyl analogue the fluoro reagent displayed poorer affinity for the ER binding site, and generally comparable labelling efficiency.

Drake *et al.* (1992) described a photoaffinity reagent for *O*-mannosyltransferase, an oligopeptide labelled with a tetrafluorophenyl azide and a [^{125}I]iodotyrosyl residue (Figure 6.130). Results were mixed, although specific labelling of a protein was observed there were problems with reduction of the azide moiety by thiols, whose presence was required for the stability of the enzyme. Dithiothreitol caused less problems than 2-mercaptoethanol. Also, the observed reactivity of the photoactivated species was consistent with a more long-lived intermediate than the expected nitrene.

Soundarajan *et al.* (1993) described the preparation of tetrafluorophenyl-azide-based photoaffinity reagents for α-chymotrypsin, such as those shown in Figure 6.131. Binding to the enzyme was similar to that observed for non-fluorinated control compounds.

This chemistry has attracted a considerable amount of attention: Kapfer *et al.* (1994, 1995), for example, described the successful use of tetrafluorophenyl-azide affinity reagents for the $GABA_A$ receptor, while a study by Pandurangi *et al.* (1995) demonstrated very high incorporations of model PFPAs into IgG (82%) and HSA (78%).

6.16.2.3 Other azides

It should be noted that groups other than phenyl (or phenyl-containing polycyclic moieties such as naphthalenes) can provide photoactivatable azides. Some of these are closely analogous heterocyclic compounds: Kerlavage and Taylor (1980), for example, described the use of 8-azido-AMP, in which it is the purine group of adenosine which bears the azide function.

There are many categories of azides other than those attached to an aromatic ring, but they are generally of little use in this application (Bayley, 1983b).

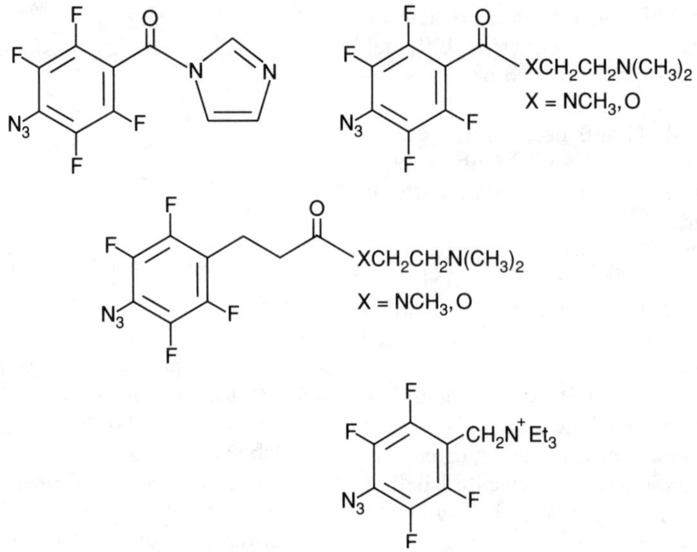

Figure 6.131 Tetrafluorophenyl-azide-based photoaffinity reagents for chymotrypsin.

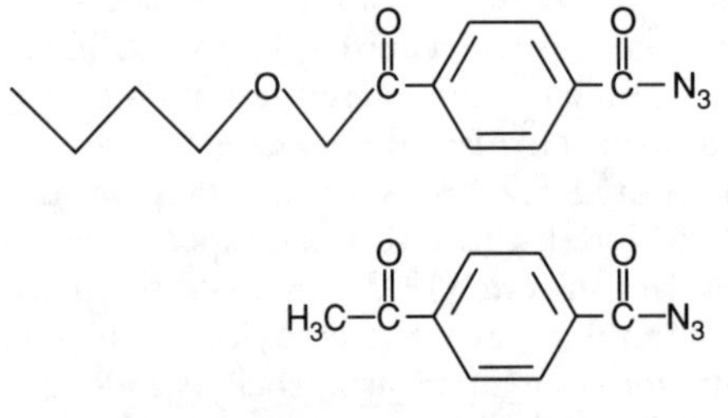

Figure 6.132 Aromatic acyl azides.

Alkyl azides are poor absorbers of ultraviolet light, and their nitrenes rearrange to readily hydrolysed imines. Acyl azides generate potent nitrenes, but have significant dark reactivity, they have been employed for the acylation of amine groups (Section 6.1.8).

Aromatic acyl (aroyl) azides have been proposed as photolabelling reagents: Melvin and Schuster (1990) described a series of hydrophobic aroyl azides for labelling lipophilic species. Two of these are shown in Figure 6.132.

6.16.3 Photoactivatable reagents generating free radicals or excited states

A wide range of compounds can be photoactivated to produce a more reactive species of some form (Bayley, 1983a). The activation may be brought about by homolytic fission to produce free radicals—photochemical activation has long been associated with free radical mechanisms in all fields of chemistry—or simply by exciting the target compound to a more energetic state. The two categories are closely related, as triplet states can behave in a fashion similar to free radicals.

Groups listed by Bayley (1983a) under this category include α,β-unsaturated ketones, aryl halides, nitroaryl compounds, purines, pyrimidines, and psoralens.

6.16.3.1 α,β-Unsaturated ketones

These reagents are generally activated initially to a triplet state (Bayley, 1983a), which abstracts a hydrogen atom from a suitable donor to produce two radicals capable of coupling to each other (Figure 6.133).

Galardy *et al.* (1973, 1974) introduced this approach, favouring bis-aromatic ketones such as

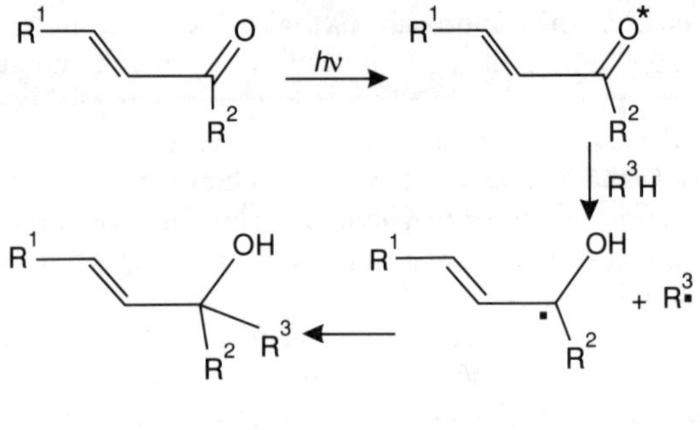

Figure 6.133 α,β-Unsaturated ketones.

benzophenone on account of the longer wavelengths which could be used for photoactivation (>320 nm *versus.* >280 nm for acetophenone). Coupling to the α-carbon of a model target, *N*-acetyl methyl glycinate, was observed even under aqueous conditions. Acetylbenzoylgastrin and benzoylbenzoylgastrin were both coupled to BSA and lysozyme.

Klueppelberg *et al.* (1990) included a benzoylbenzoyl derivative of cholecystokinin in their comparitive study of receptor labelling, along with representatives of the other main photoactivatable categories. While the overall level of incorporation was unexceptional, the selectivity for the receptor was the highest observed.

A phenothiazine derivative possessing a benzophenone group displayed the best functional performance in a comparative study with phenyl and fluorophenyl azides (Golinski *et al.*, 1995a); this compound retained a greater ability to inhibit the activation of phosphodiesterase by calmodulin than other derivatives.

6.16.3.2 Nitrophenyl compounds

Nitrophenyl compounds are thought to react *via* a triplet excited state, and have seen some use in photoaffinity reagents; they are not efficient enough to have achieved widespread popularity, however.

The nitrophenyl compounds included in the comparitive study by Klueppelberg *et al.* (1990) displayed indifferent behaviour, being bettered both in terms of overall labelling yield and selectivity by other reagents. Katzenellenbogen *et al.* (1974) observed no reaction of a 315 nm-photoactivated nitrophenyl derivative of hexestrol with estrogen receptor proteins.

One of the more useful applications of this approach is for photocleavable links. Barltrop *et al.* (1966) described the photolytic cleavage of 2-nitrobenzyloxycarbonyl (NBOC) moieties; Patchornik *et al.* (1970) used these as protecting functions for both amine and carboxyl groups, which could be cleaved by irradiation at >320 nm. The mechanisms of the cleavage reactions are shown in Figure 6.134. Rich and Gurwara (1975) exploited this chemistry for photocleavage (350 nm) of synthetic peptides from the solid support (Figure 6.135). Other applications of this methodology in the protein chemistry field are readily imaginable.

Finally with regard to nitrophenyl compounds, it should be borne in mind that the mechanism discussed in this section may have some influence in reactions of the many aryl azide reagents which bear a nitro group.

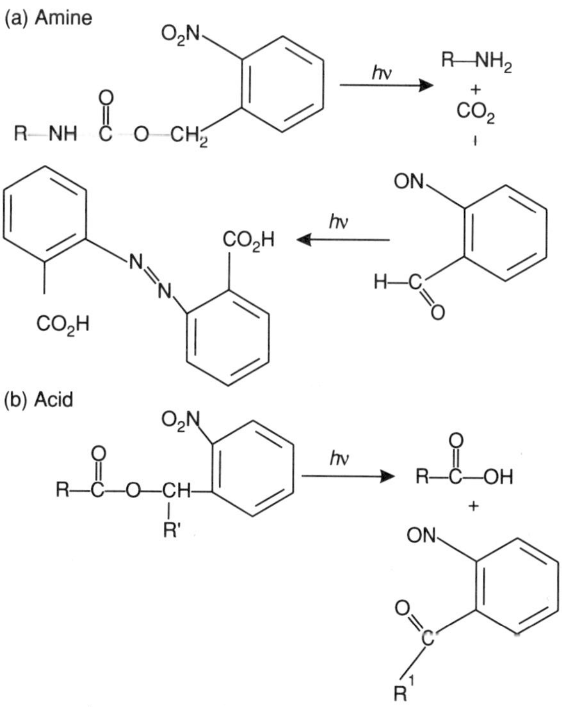

Figure 6.134 Mechanisms of cleavage of NBOC moieties by irradiation, from (a) protected amines and (b) protected acids.

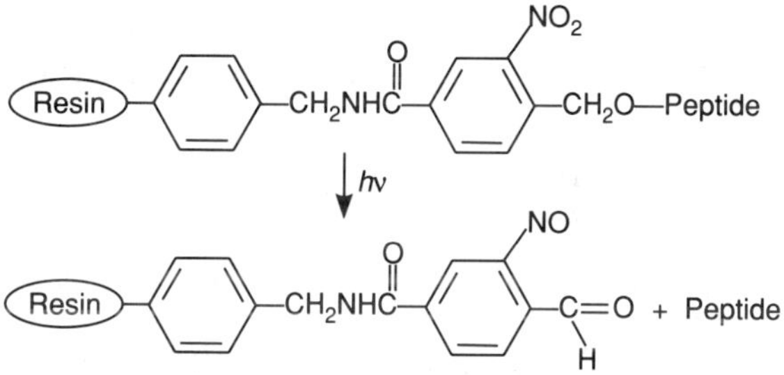

Figure 6.135 Photocleavage of synthetic peptides from a solid support *via* photocleavable NBOC links.

6.16.3.3 Other photoactivatable reagents

Iodophenyl derivatives of hexestrol were examined for photolabelling estrogen receptor proteins (Katzenellenbogen *et al.*, 1974; Katzenellenbogen and Hsiung, 1975). Despite the observation of photolytic behaviour in organic solutions, no coupling to ER proteins was observed on irradiation at 254 or 315 nm.

Katzenellenbogen *et al.* (1974) did observe quite extensive photocoupling of triarylethylene compounds

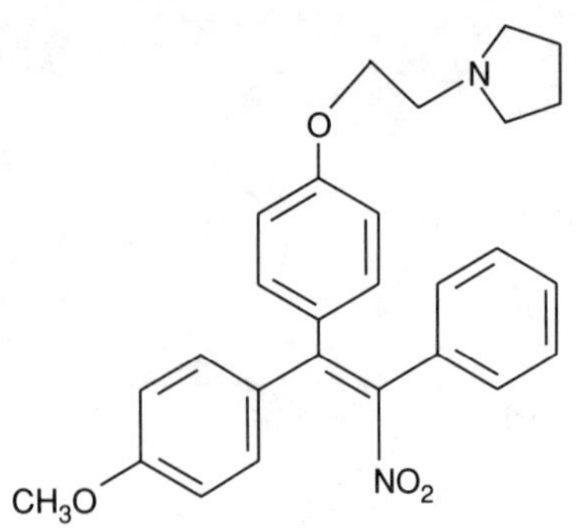

Figure 6.136 Triarylethylene photocoupling reagent.

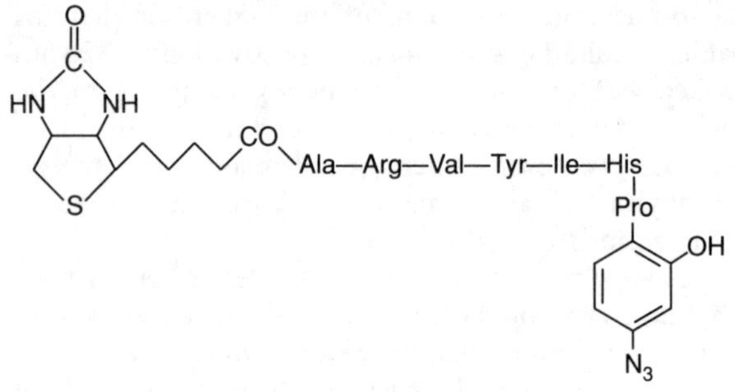

Figure 6.137 Biotinylated, radioiodinatable photoaffinity analogue of angiotensin II.

to ER: an example of these reagents is shown in Figure 6.136.

Katzenellenbogen *et al.* (1975) also observed that estradiol itself could undergo photocoupling to the receptor, rapidly at 254 nm and more slowly at >315 nm. Coupling was also noted when estradiol or diethylstilbestrol were irradiated in the presence of BSA or ovalbumin.

6.16.4 Biotinylated photoactivatable reagents

Seyer *et al.* (1989) described the synthesis and characterisation of a biotinylated, radioiodinatable photoaffinity analogue of angiotensin II (Figure 6.137). Similarly, DeLaLuz *et al.* (1995) described a biotinylated phenothiazine derivative incorporating a phenyl azide function.

6.16.5 Photobifunctional reagents

Finally, compounds have been described which combine two moieties, capable of photoactivation at different wavelengths: Golinski *et al.* (1995b) synthesised phenothiazine derivatives such as that shown in Figure 6.138, possessing both a phenyl azide and a benzophenone moiety.

6.17 Coupling of Proteins to Small Molecules Containing *N*-Nitrosoamide Groups

An interesting class of reagents, albeit with a fairly specialised field of application to date, are the *N*-nitrosoamides (Figure 6.139(a)). Cleavage of the amide bond by a suitable nucleophile results in the formation of an unstable *N*-hydroxydiazo compound (Figure 6.139(b)), which decomposes to yield a carbocation (White *et al.*, 1968, 1975). The reactivity of this carbocation is on a par with the photoaffinity reagents described in the last section, and it is capable of labelling nucleophilic protein moieties—not just side-chain groups such as cysteine thiols, but both the nitrogen and oxygen atoms of the peptide-chain amides (Donadio *et al.*, 1985; White *et al.*, 1990).

These compounds are elegant affinity probes for proteases such as chymotrypsin, the enzymatic cleavage of the amide bond releasing the carbocation as a

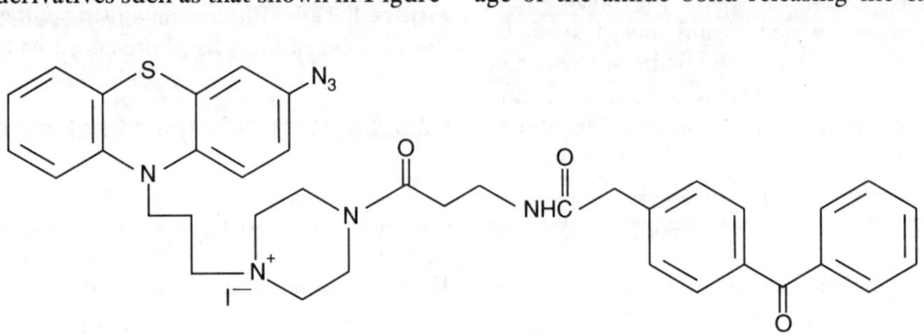

Figure 6.138 Photobifunctional phenothiazine derivative containing a phenyl azide and a benzophenone moiety.

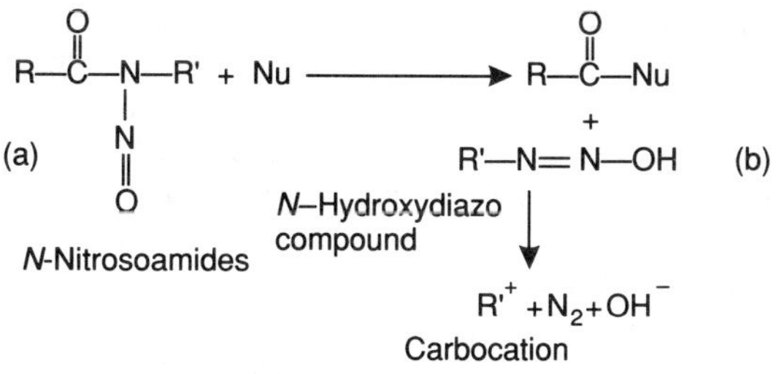

Figure 6.139 *N*-Nitrosoamide and its cleavage of the amide bond with nucleophile leading to carbocation formation.

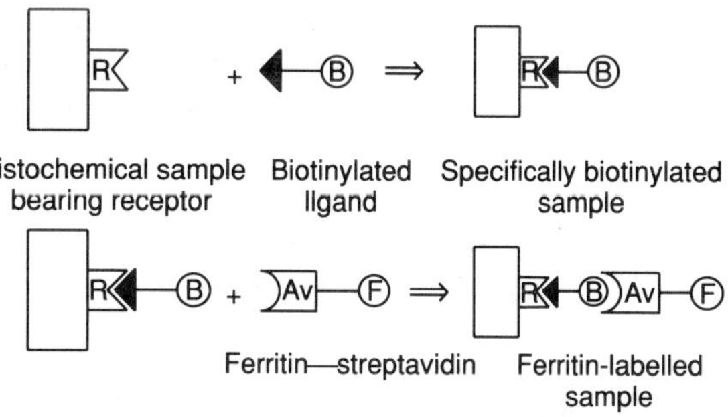

Figure 6.140 Indirect ferritin labelling of a receptor *via* the biotin–streptavidin reaction.

powerful labelling species at the active site. Li and White (1994) described compounds of this type incorporating fluorophores in the reactive fragment, and their use for inactivation of α-chymotrypsin.

6.18 Coupling of Proteins to Small Molecules by Non-covalent Interactions

Of course, many small molecules naturally bind to proteins with higher or lower affinities. It is possible to exploit these tendencies by derivatisation, for example the hydrophobic tetraphenylporphyrin derivative prepared by Tsuchida *et al.* (1997) which displays binding to human serum albumin. For non-covalent interactions to be employed for the preparation of stable conjugates, however, the affinity must of course be very high. As elsewhere, the biotin–(strept) avidin interaction is by far the most common approach of this type, but other examples are occasionally encountered.

6.18.1 Coupling using the biotin–(strept)avidin interaction

The use of the biotin–avidin – or – streptavidin interaction as a means for protein-protein conjugation has been discussed at length in Section 5.11: one protein is biotinylated using one of the reagents described in that section, and can then be conveniently coupled to streptavidin or a streptavidin-labelled protein. Biotin–avidin interactions can also be used to produce protein–small molecule conjugates, generally by coupling a biotinyl derivative of the small compound to (strept)avidin or a (strept)avidin-linked protein.

Shao *et al.* (1990), for example, produced model glycoproteins by attaching biotinylated saccharides to avidin and streptavidin. As usual with the biotin–avidin systems the coupling reaction was facile and essentially irreversible; no loss of glycan was detected after dialysis for two weeks.

In other examples, protein–small molecule conjugates have been formed as part of a more complex reaction sequence: Atlas *et al.* (1978), for example, effectively produced a propranolol–ferritin conjugate for histochemistry by this route, but the actual sequence of reactions was: (1) incubation of the histochemical sample with biotinylpropranolol to label adrenergic receptors, then (2) incubation of the labelled sample with streptavidin–ferritin (Figure 6.140).

Biotinylated adrenergic antagonists were also used in receptor studies by Meier and Ruoho (1990). These authors noted that better binding was observed when a spacer group was incorporated between the biotin and the ligand; similar results were observed by Redeuilh *et al.* (1990) when biotinylestradiol reagents were used to produce affinity resins for estrogen receptor purification. The incorporation of long spacer groups is a recurring feature of reagents of this type (Huber *et al.*, 1993; Mares *et al.*, 1995); these observations are consistent with the rather deep cleft of the biotin-binding site.

The synthesis of biotinylated oligopeptides has been described by many authors (see for example Geahlen *et al.*, 1992; Leftheris and Mapelli, 1993; Basak *et al.*, 1994), and such compounds have also been used for probing receptor interactions of ligands such as gonadotropin-releasing hormone (Childs, 1990) and angiotensin II (Bonnafous *et al.*, 1988).

6.18.2 Coupling of lipophilic molecules to lipoproteins

Lipoproteins contain phospholipid groups which bind strongly to suitable lipophilic ligands, offering a potential route for binding of small molecules to these proteins. Jasanada *et al.* (1996) prepared *bis*-stearyl derivatives of the chelating agent DTPA, and were able to use these for ^{111}In-labelling of lipoproteins.

6.19 Coupling of Proteins to Metals and Metal Complexes

Many proteins, of course, contain metal ions in their native structure, where they are often vital to functional activity. These metal–protein bonds are beyond the scope of this book; see Bezrukov *et al.* (1982) for a review of such interactions.

However, the deliberate incorporation of non-native metals into proteins has had a number of important applications. Radiometal-labelled antibodies have seen a number of exciting *in vivo* applications, both diagnostic and therapeutic. Electron microscopy for histochemistry and cytochemistry has relied on the presence of electron-opaque metal atoms to confer high resolution. Immunoassays based on time-resolved fluorimetry have been developed, and this has again necessitated the preparation of metal-containing conjugates, in this case rare earths such as europium and terbium.

Direct labelling of protein thiols by radiometals such as ^{183}Re and ^{99m}Tc is a useful approach (Schubiger *et al.*, 1996; Capala *et al.*, 1997).

Most commonly, however, these metals have been introduced *via* a suitable chelating group which binds the ion strongly (Schubiger *et al.*, 1996). Many different techniques for conjugating the protein to the chelator, either before or after incorporation of the metal, have been described in the earlier sections of this chapter; compounds containing a protein-reactive moiety and a chelating group can be viewed as heterobifunctional reagents for creating protein–metal bonds. Occasionally convenient naturally occurring metal complexes have been employed too, for example the iron-containing protein ferritin.

The immunouranium technique (Sternberger, 1967) for electron microscopic histochemistry relied on the ability of antibody molecules to bind up to 300 uranium ions per molecule (even after protection of the antigen-binding sites) when treated with uranyl acetate. This technique has been in use for many decades (McClintock and Friedman, 1945), and has more recently been combined with one or more stages employing lead salts, which enhance the contrast obtained (Daddow, 1983).

Another useful interaction is that between proteins and gold sols, which has formed the basis of many labelling techniques; the ability of gold to catalyse the reduction of silver solutions allows further contrast to be obtained *via* the deposition of silver metal on the gold sol. Binding of proteins to gold and other metals has been investigated by studying the behaviour of surface-displayed polypeptides selected from an *E. coli* library (Brown, 1997); high-binding examples were isolated by this approach, particularly where repeat sequences were incorporated to amplify the avidity.

Histidine residues are often particularly important in terms of protein–metal interactions and certain metal complexes will combine covalently with these groups. Aquopentammineruthenium(II), $a_5Ru^{II}H_2O$ has been employed extensively for histidine derivatisation (although lysines are sometimes labelled too), and the complexes formed are stable on oxidation to Ru^{III}. The a_5Ru^{III}His complexes of many proteins have been isolated and their properties examined, including ribonuclease A (Matthews *et al.*, 1978), lysozyme (Reccia *et al.*, 1982), glucose oxidase (Degani and Heller, 1988) and cytochrome *c* peroxidase (Fox *et al.*, 1994). The electron transfer properties of ruthenium-labelled proteins has received considerable attention (Winkler and Gray, 1992).

6.20 Coupling of Proteins to Small Molecules Using the Mannich Reaction

The Mannich reaction is a useful (if fairly slow) technique which can be applied to the conjugation of a wide variety of small molecules, including many which are rather difficult to couple by other techniques. The essential ingredients are an amine or amide, an aldehyde (normally formaldehyde), and a compound containing an 'active' hydrogen atom, loosely speaking one with some acidic character. Some of the groups which can provide this active hydrogen (March, 1985; Agababyan *et al.*, 1982) are shown in Figure 6.141, and any compounds of this type can be coupled to the amine groups of an amino acid or protein in the presence of formaldehyde. As shown in Figure 6.142, the resulting conjugate contains a methylene linker group; the net result is (substituted) aminomethylation of the group containing the active hydrogen.

Ranadive and Sehon (1967b) used the Mannich reaction to couple serotonin analogues to BSA, these compounds contain an active indole NH hydrogen (Figure 6.141), and the proposed product with 5-hydroxytryptamine is shown in Figure 6.143. The

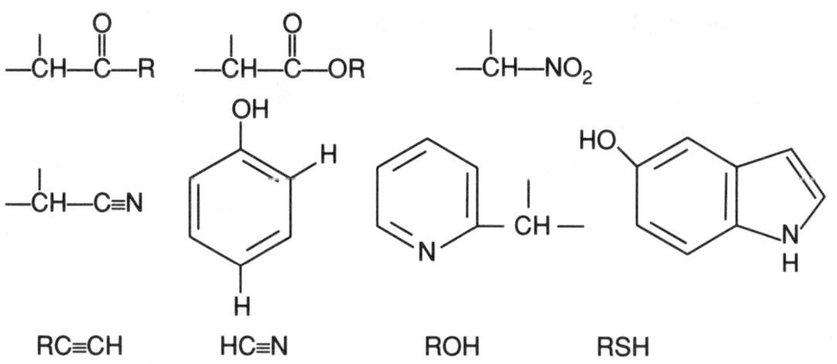

Figure 6.141 Moieties providing active hydrogen for the Mannich reaction.

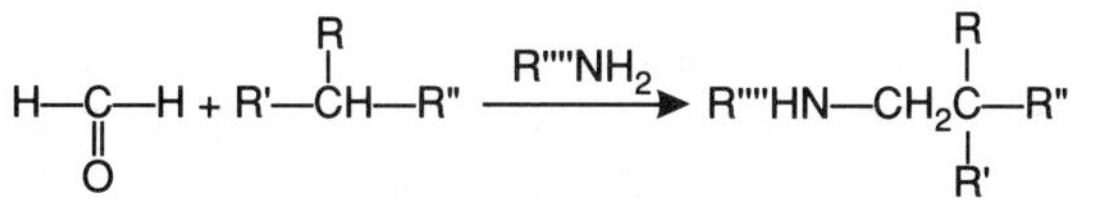

Figure 6.142 The Mannich reaction.

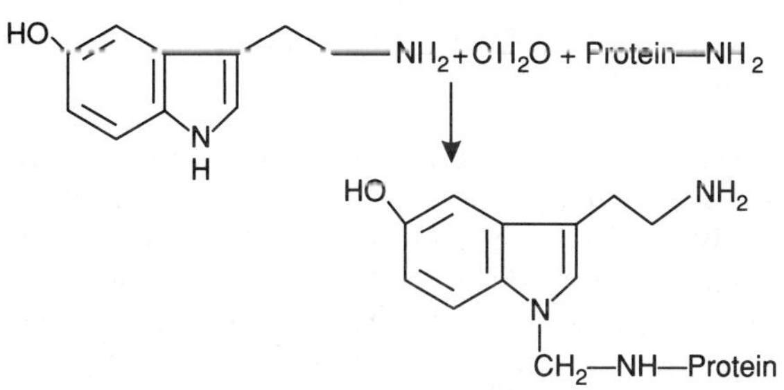

Figure 6.143 Preparation of protein–5HT conjugate by the Mannich reaction.

reactions were carried out in sodium acetate solution for up to 72 hours, the pH sometimes being adjusted to 6 to maintain solubility.

Grota and Brown (1974) carried out similar work; while observing good incorporations from serotonin and *N*-acetylserotonin, melatonin gave very little reaction (Figure 6.144). It therefore seemed to be the case that these compounds coupled primarily *via* the phenolic ring hydrogens. Grota and Brown used a hapten:formaldehyde:BSA ratio of around 100:2230:1; reaction was carried out at room temperature in sodium acetate solution, giving a pH of 6–6.8, and appeared to be nearing completion after an hour. Extraordinarily high incorporations of up to 200 were reported, perhaps indicating some oligomerisation of the haptens *via* the amine or amide moieties.

Grota and Brown (1976) extended this approach to catecholamine analogues. In one example *p*-tyramine was coupled to BSA at a hapten:formaldehyde:BSA ratio of about 163:6700:1 (50 hours, room temperature,

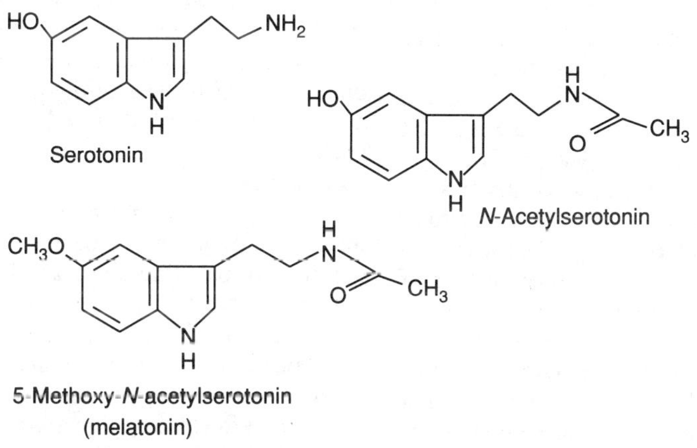

Figure 6.144 Of the compounds shown, melatonin failed to give a Mannich product—evidence in this case for reaction *via* the phenolic hydrogen.

pH 6.0–6.5) and an incorporation of about 100 was observed.

Levy *et al.* (1976) coupled the alkaloid reserpine to BSA using a hapten:formaldehyde:BSA ratio of around 67:3500:1 (overnight, room temperature), and obtained an incorporation of 5.

Miwa *et al.* (1977) also used this approach for producing catecholamine–BSA conjugates. The amine functions were protected with maleic anhydride to prevent cross-linking, then the coupling reaction with BSA carried out under nitrogen (three days, 18–23 °C, pH 6.0). The hapten:formaldehyde:BSA ratio was about 33:1700:1, and incorporations of 15–30 were observed. Changing pH between 5.5 and 7.6 had no significant effect on the reaction.

Faraj and Ali (1980) used the Mannich reaction to produce a tetracycline–BSA immunogen. The BSA was incubated (two hours, room temperature) with a 49-fold excess of tetracycline hydrochloride and a 2330-fold excess of formaldehyde, in 0.67 M sodium acetate buffer at pH 6–7. This resulted in the incorporation of 30 to 40 equivalents of tetracycline, and the immunogen was used successfully to raise antibodies.

Flurkey *et al.* (1985) coupled serotonin to BSA or human γ-globulin by this route. With BSA, incubation (30 minutes in an unspecified acetate buffer) with 5-hydroxytryptamine–creatinine sulfate and formaldehyde resulted in the incorporation of 10 to 30 serotonin groups per molecule.

A generic procedure for Mannich coupling is described in Protocol 6.3.

6.21 Radioiodination

A great many of the techniques discussed in this chapter can be applied in such a way as to introduce a radiolabel. Some of the more popular of these are listed below, and the reader is referred to the relevant sections for further details. (1) Acylation of amine groups with suitably activated acid derivatives such as anhydrides or succinimide esters. The Bolton–Hunter reagent (Bolton and Hunter, 1973) has been adopted particularly widely as a convenient means of radiolabelling *via* amines. Other amine-reactive groups are occasionally employed, such as imidates (Wood *et al.*, 1975). (2) Incorporation of radiometals bound to a chelating group, which may be coupled to protein amines, thiols, etc., *via* a wide range of reactive moieties such as cyclic anhydrides, isothiocyanates or bromoacetamides. (Schubiger *et al.*, 1996). Direct metal labelling of thiols is also occasionally encountered. (3) Reduction or reductive alkylation of oxidised glycoproteins using radioactive formaldehyde and/or borohydride.

The incorporation of suitable 'tags' into recombinant proteins allows other routes to be adopted, for example radiophosphorylation of peptide sequences which are substrates for kinase enzymes (Li *et al.*, 1989; Neri *et al.*, 1996).

The most widespread approach to radiolabelling of proteins however is based on the use of radioactive iodine; this will derivatise suitably reactive residues within the protein such as tyrosine and (to a lesser extent) histidine.

This approach also underlies the Bolton–Hunter reagent, and many other examples in Chapters 5 and 6 where an iodinatable group is deliberately introduced *via* an artificial reagent. Iodination chemistry has only been dealt with tangentially, however, and is discussed more fully here for completeness.

6.21.1 Halogenation of aromatic compounds

Halogenation of aromatic compounds can occur by free radical routes, but the mechanism involved is usually electrophilic substitution (cf. reaction with diazonium salts). Benzene itself will only react with elemental halogens in the presence of a Lewis acid, which polarises the halogen to provide a more positively charged 'attacking end'.

Electron-donating substituents such as hydroxyl or amine groups increase the electron density of the aromatic ring, which is therefore activated towards electrophilic attack— hence the reactivity of tyrosine, the main site of protein iodination (Hughes, 1950). Due to stabilisation of electron-deficient intermediates by the electron-donating group, attack on such molecules is directed towards the *o*- and *p*-positions. The phenolate ion is a much more activated species than the phenol itself, and therefore iodination is generally favoured by alkaline conditions.

For heterocyclic aromatic species, activating and directing influences are more complicated: the nitrogen atom of pyridine, for example, pulls electron density away from the ring and electrophilic substitution is extremely difficult. In pyrrole, on the other hand, an aromatic system can only be established by donation of charge by the nitrogen, and the molecule is strongly susceptible to electrophilic attack. The imidazole ring of histidine and its analogues is an intermediate case; the net result is that electrophilic substitution takes place, but higher pH and oxidant levels are typically

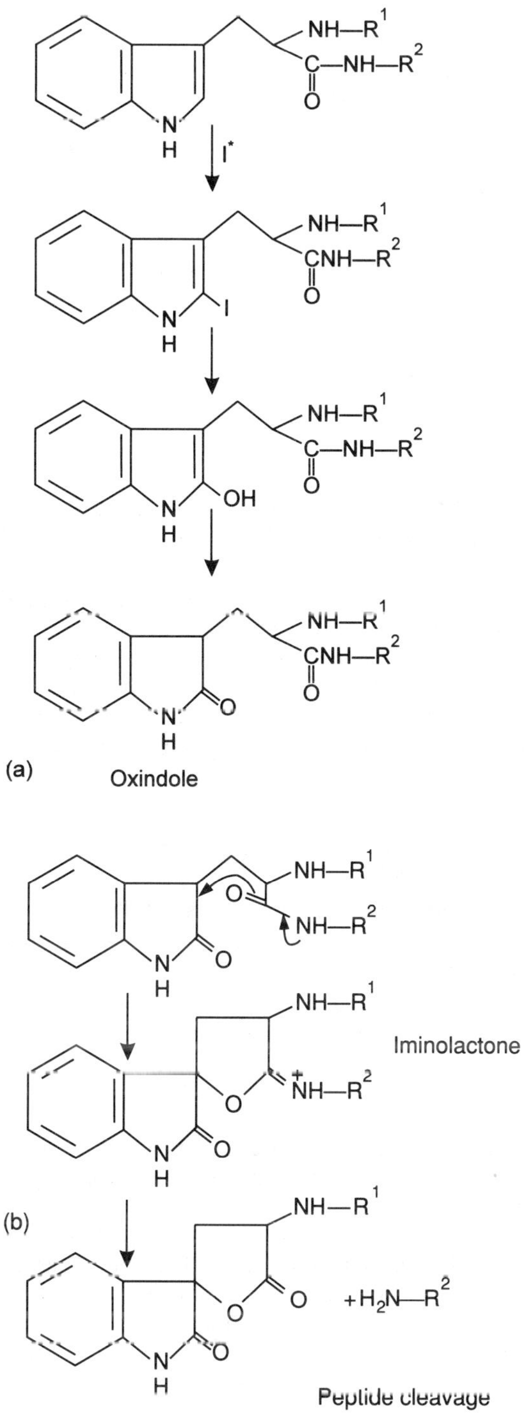

Figure 6.145 Iodination of tryptophan can result in the formation of an oxindole (a) or clevage *via* an iminolactone (b).

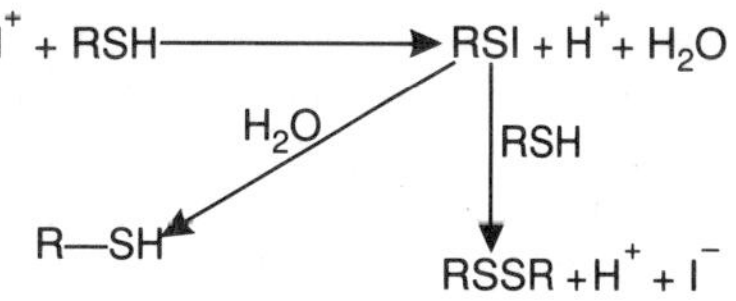

Figure 6.146 Iodination of cysteine generates a sulfenyl iodide, which may react with another thiol to yield a disulfide, or hydrolyse to regenerate thiol.

required (Brunings, 1947; Holohan *et al.*, 1973, 1974). Iodine may also reversibly derivatise amine groups, including the secondary amine of histidine (Hughes, 1950), and this may occur as a precursor to C-iodination of the histidine ring; the formation of this unstable intermediate has been invoked to explain labile product in iodinated polypeptides (Lamoureux *et al.*, 1967). Where the position of iodination of histidine has been investigated, it appears that reaction takes place at the 4-position (Holohan *et al.*, 1974).

Tryptophan possesses an indole ring with a pyrrole-like nitrogen. It is less well recognised that this amino-acid also undergoes ready iodination (Roche and Michel, 1948; Hartdegen and Rupley, 1967), but the accompanying hydrolysis and oxidation reactions result in formation of an oxindole (Figure 6.145(a)) possibly (at low pH) followed by cleavage of the peptide chain (Fig 6.145(b)) (Blake, 1967; Alexander, 1973, 1974). Therefore tryptophan (or primary protein structure) may be lost during iodination reactions but no iodine is stably incorporated into the product.

Reaction with cysteine residues is also possible (Hughes, 1950): initially a sulfenyl iodide is formed (Figure 6.146) but this is unstable. In the presence of an accessible thiol the result may be dimerisation *via* a disulfide bond, otherwise hydrolysis may regenerate the thiol. In either case iodine is consumed but is not incorporated stably in the product.

6.21.2 The radioisotopes of iodine

The most abundant naturally-occurring isotope of iodine is the non-radioactive ^{127}I. However there are a number of other possible isotopes, three of which have properties which make them of interest for radiolabelling.

^{131}I was the commonest isotope in use for many years; it has a halflife of 8.04 days and provides high-energy emission, both β (principal emission 606 keV) and γ (principal emission 364 keV). This isotope is long-lived enough to be used conveniently for *in vitro* purposes, and powerful enough to be detected *in vivo*. However it has a number of disadvantages: the highly penetrating emissions require extensive shielding, and are not counted with a particularly high efficiency. Also, ^{131}I is not generally available in 'carrier-free' form[1], so it may be difficult to achieve high specific activity iodinations.

Fortunately there are two other isotopes which offer suitable properties for these uses, and both of which are available carrier-free. ^{123}I is a less energetic emitter of γ radiation (principal emission 159 keV), with a halflife of 13.2 hours. This combination of moderate energy and high turnover makes this an ideal choice for radiopharmaceuticals (Stöcklin, 1977; Seevers and Counsell, 1982).

For *in vitro* use, ^{125}I has a more convenient halflife of 60.14 days; its single emission energy (35 keV) is readily counted and is low enough to allow relatively safe handling.

Therefore ^{131}I has gradually come to be replaced in many *in vivo* applications by ^{123}I, and in many *in vitro* uses by ^{125}I.

6.21.3 Radioiodination of proteins

The iodination of protein residues is quite a complex area, and mechanistic descriptions often appear contradictory. Most studies have focussed on tyrosine, and a number of constant facts emerge from these reports. The first is that it is the phenolate ion which predominates in the iodination reaction (Li, 1942; Mayberry *et al.*, 1965; Baldas *et al.*, 1981); this is entirely consistent with expectations, as the electron-donating power of the phenoxide is much greater than that of the unionised phenol.

The second constant observation is that molecular iodine (I_2) is involved in the iodination reaction, but the exact nature of the attacking species has been the subject of considerable debate. Mechanisms based on I_2 itself have been supported by kinetic studies (Mayberry *et al.*, 1965; Baldas *et al.*, 1981), and the mechanism depicted in Figure 6.147 has been invoked. However, as observed with other phenols (Grovenstein *et al.*, 1973), studies of kinetic isotope effects have revealed that the extraction of a proton by base is generally the rate-determining step (Baldas *et al.*, 1981); therefore the exact nature of the iodinating species would not be expected to affect the kinetics so long as the incorporation of iodine is relatively rapid.

[1] i.e. without significant contamination from non-radioactive iodine isotopes.

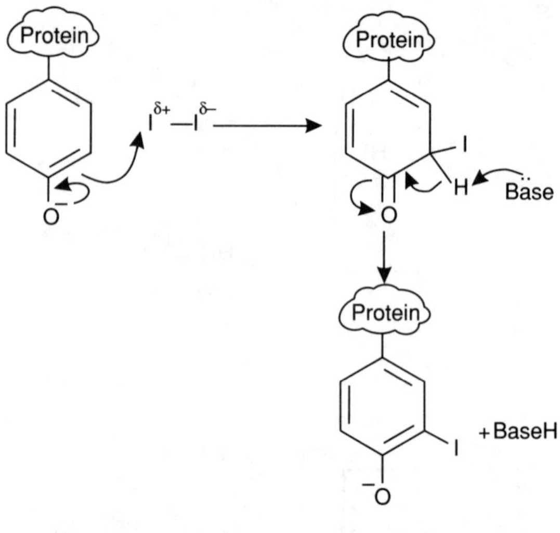

Figure 6.147 Mechanism of iodination of protein tyrosines based on molecular iodine.

In general the consensus is that, at least in mildly alkaline solution which is commonly the milieu for reaction, the predominant attacking species is the H_2OI^+ ion (Li, 1942; Hughes, 1950; Roholt and Pressman, 1972; Kabat, 1976). This ion, the protonated form of hypoiodous acid, is generated from the equilibrium:

$$I_2 + H_2O \rightleftharpoons H_2OI^+ + I^-$$

At a higher pH the acid is ionised ($pK_a = 11$), and the resulting hypoiodite disproportionates to iodate and iodide:

$$HOI \rightleftharpoons H^+ + OI^-$$
$$3OI^- \Rightarrow IO_3^- + 2I^-$$

This reaction places an upper pH limit on iodination: above around pH 10 the loss of the active species is too great.

One final equilibrium is of relevance; because molecular iodine is only sparingly soluble in water it is often used in the presence of iodide ions, which form soluble triiodide (I_3^-) complexes:

$$I_2 + I^- \rightleftharpoons I_3^-$$

The presence of an excess of iodide also acts as a buffer to provide a constant, small proportion of reactive free iodine.

Once a protein has been radioiodinated, it is generally necessary to separate it from the other reactants. This is normally achieved quite straightforwardly by size-based methods such as size exclusion chromatography (Chapter 10, Section 10.5), as iodide and the oxidant species are of low molecular mass. An exception is where lactoperoxidase is used to catalyse oxidation; here other methods such as ion exchange may need to be employed, unless the enzyme can be employed in immobilised form (Morrison, 1980).

Iodide can sometimes bind quite strongly to proteins; an interesting possibility for its removal in difficult cases is adsorption onto hydrogen-activated palladium (Tóth, 1968; Jász and Tóth, 1968; Szendröi *et al.*, 1973).

6.21.3.1 Early radioiodination methods

For practical reasons radioactive iodine is usually supplied as the iodide; as noted above the solubility of molecular iodine is poor, and its volatility is a serious safety hazard. For iodination it is necessary to convert this initially into I_2. In the earliest methods the triiodide equilibrium just noted was exploited (Pressman and Keighley, 1948; Hughes and Straessle, 1950; Masouredis *et al.*, 1951); labelled iodide was allowed to equilibrate with unlabelled iodine, providing a small quantity of labelled iodine.

This approach, however, was extremely inefficient in terms of radioactive iodine incorporation, and it soon became commonplace to incorporate an oxidising agent to convert iodide to iodine rather than relying on equilibrium conversion. Early methods of this type exploited oxidants such as hydrogen peroxide (McFarlane, 1956), acidic iodate (Francis *et al.*, 1951; McFarlane, 1956; Utiger *et al.*, 1962), acidic nitrite (Eisen and Keston, 1949; Johnson *et al.*, 1960), ferric sulfate (Stadie *et al.*, 1952) and ammonium persulfate (Gilmore *et al.*, 1954).

Many of these are powerful oxidising agents, capable of considerable damage to protein molecules. Sometimes this was avoided by carrying out a separate oxidation step, the molecular iodine product being removed by distillation for use in iodination. However this is an inconvenient and hazardous procedure; additionally it is wasteful in terms of labelled iodine usage, as half of the oxidised material will be converted back to iodide during the formation of the active species. More modern methods have concentrated on the use of oxidising agents which are mild enough to be tolerated by the protein; their continued presence means that the iodide product is reconverted to iodine as it is formed, so 100% incorporation is theoretically possible. Most of the methods about to be discussed are variants on this theme, but one other approach achieved widespread popularity, the use of iodine monochloride.

6.21.3.2 Iodine monochloride

The iodine monochloride (ICl) approach was pioneered by McFarlane (1958, 1963). This compound bears a slight positive charge on the iodine atom, and ICl may sometimes represent the active electrophilic species (Seevers and Counsell, 1982); however the 'usual' H_2OI^+ form is probably involved in many circumstances.

As with other iodination reagents based on this species, derivatisation occurs principally on tyrosine residues and less so on histidines. Reaction with cysteine is another possibility (Parker and Allison, 1969); the product is rather labile and iodine may be lost more readily on storage (Krohn *et al.*, 1977).

Typically a stable stock solution of cold ICl (principally present as the ICl_2^- ion) in HCl–NaCl is employed. When radioiodide is added this exchanges rapidly such that all the radioactive iodine ends up as ICl, and this is converted to the reactive form by raising the pH to around 8. For trace iodination experiments where the protein is present in a large excess over ICl, it is therefore possible to achieve 100% uptake of radioactive iodine, although the product will of course incorporate unlabelled iodine too. Procedures have been described for accurately quantifying the amount of radioiodinating species obtained from ICl, such that this information can be used for calculation of uptake (Roholt and Pressman, 1972).

Helmkamp *et al.* (1967a) published a fairly detailed analysis of the ICl iodination procedure, in particular attempting to maximise the incorporation. Decomposition of the active iodinating species was avoided by adding the ICl to a buffered solution of protein and radioiodide, and hydrogen peroxide (present as an impurity in many ^{131}I preparations) was removed with a trace of catalase. These authors (Helmkamp *et al.*, 1967b) also described an oxidative technique for preparing labelled ICl according to the equation:

$$2I^- + IO_3^- + 6HCl \rightleftharpoons 3ICl_2^- + 3H_2O$$

By this method (resurrected by Doran and Spar, 1980) the radioactive content of the ICl can reach two-thirds, considerably higher than is achievable by the original exchange methods. This was reflected in incorporations reaching 110 mCi/mg, compared with levels of <10 mCi/mg typical from ICl iodination using the exchange method (Kolb and Grodsky, 1971).

For pharmaceutical applications ^{123}I is typically the isotope of choice, and the availability of 'carrier-free' ^{123}ICl from the reaction of $^{123}I_2$ with chlorine (Stöcklin, 1977) allows the possibility of very high incorporations.

As it is the only common method avoiding the presence of 'extra' oxidising reagents, the ICl method has been demonstrated to yield comparatively less damaged product (Krohn *et al.*, 1977; Doran and Spar, 1980). Principally for this reason the technique has retained some support into fairly recent times (Contreras *et al.*, 1983).

6.21.3.3 Chloramine-T

Chloramine-T (CT), the sodium salt of *N*-chloro-*p*-toluenesulfonamide (Figure 6.148(a)), has been used as an oxidant for radioiodination since the early 1960s (Hunter and Greenwood, 1962; Greenwood *et al.*, 1963). The iodination of human growth hormone described in these reports was carried out in the presence of the oxidising agent, which was therefore still available for reoxidation of iodide byproducts; complete incorporation of the radioactive iodine therefore became a realistic possibility, and the specific activities obtainable by this route (590 mCi/mg or greater) reflect this.

Chloramine-T undergoes a sequence of reactions in aqueous systems (Figure 6.148(b)) *via* the *N*-dichloro form, to produce hypochlorite, ClO^- (Jennings, 1974); this is probably the main oxidising species[1].

CT is a relatively unstable compound, particularly if exposed to light; activity can be lost quite quickly from both solid and solution forms unless the material is stored in darkness (Jennings, 1974). If protected, however, the rate of loss from either form is around 0.02% per month (Jennings, 1974). Thermal instability is also observed at temperatures around 40 °C.

The CT iodination methods in the literature generally fall into one of two categories. At one extreme the original methods described by Hunter, Greenwood *et al.* employ large concentrations of oxidant to promote high incorporations in the product, the aim being to produce labelled proteins of the highest possible specific activity. An optimum target of one iodine atom per protein molecule is often employed as a rule of thumb in such applications. Due to the rapidity

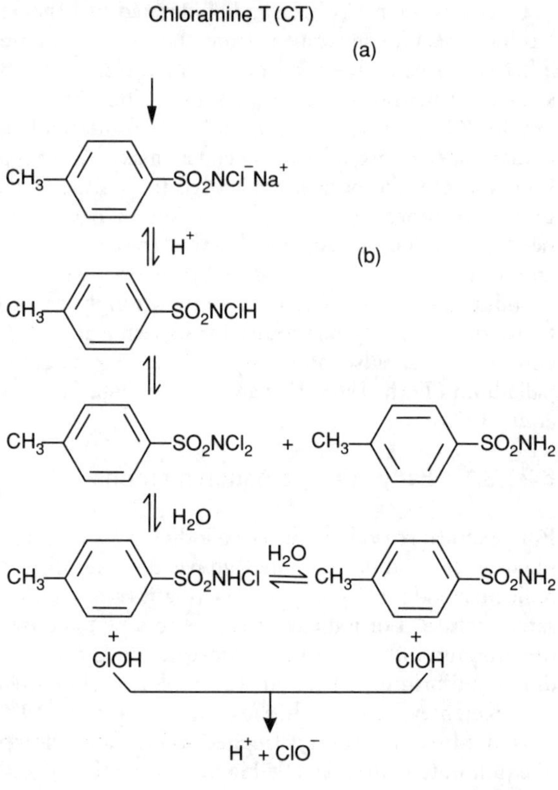

Figure 6.148 Chloramine-T and its sequence of reactions in aqueous systems.

of iodination under these conditions, reaction times may be as low as a few seconds.

At the other, much lower CT concentrations are employed with the aim of effecting 'trace' iodination, where only a very small proportion of the protein molecules are labelled, perhaps as low as 1 in 10 000. In methods of this type (McConahey and Dixon, 1966, 1980; Sonoda and Schlamowitz, 1970) the reaction times employed are much longer, typically ten minutes or more.

High specific activity labelling using CT

The methods for achieving high specific activity product have really altered very little since their original description more than 30 years ago. Their popularity remains high for applications such as radioimmunoassay.

With low reaction times it may well be possible to avoid significant damage to the protein molecule: Hunter and Greenwood (1962) for example observed that 95% of the labelled growth hormone was still

[1] Hypochlorite has been employed *per se* as an oxidant for radioiodination (Redshaw and Lynch, 1974).

recognised by specific antiserum so long as incorporations were kept below around 300 mCi/mg. Nevertheless reports of inactivation by such procedures are common, and not surprisingly the severity of the effect varies widely from protein to protein: in an example of a sensitive case (Kolb and Grodsky, 1971), fructose-1,6-diphosphatase lost 95% of its functional activity during a 60-second CT iodination, although the specific activity achieved was only 32 mCi/mg. Although loss of functional activity may be easy to detect, some changes are more insidious: Comens *et al.* (1982), for example, observed that a proportion of epidermal growth factor labelled by a CT procedure became capable of spontaneous (and specific) covalent coupling to receptors.

The reasons for protein modification during CT iodinations can vary; the oxidising nature of CT can promote various side-reactions such as iodination or dimerisation of cysteine thiols (Hayes and Goldstein, 1975), formation of the *S*-oxide from methionine (Stagg *et al.*, 1970), and, below pH 6.5, destruction

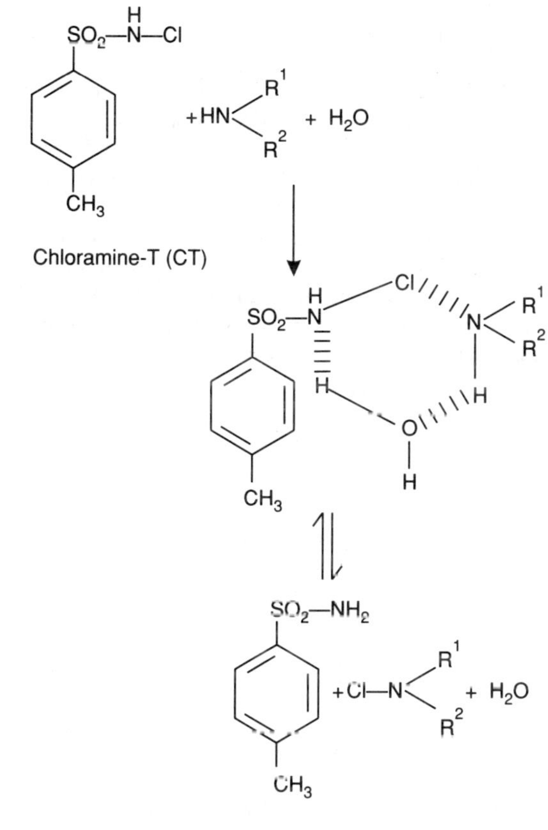

Figure 6.149 Transfer of chlorine from CT to a secondary amine.

of tryptophans (Alexander, 1973, 1974). Metabisulfite is a powerful reducing agent, and has also been shown to cause changes to protein structure (Kagan and Glick, 1974; Binion and Rodkey, 1983).

The chlorine of CT may also participate in reactions with amino acid residues. Ring-chlorinated products have been observed in CT-mediated radioiodination of small molecules such as chloroquine (Bakker and Kaspersen, 1978), uracil (Bakker and Kaspersen, 1981), and fluorescein (Baldwin and Lin, 1981), and it is possible that such products are obtained from tyrosine and histidine also.

Also, CT has been shown to be capable of transferring chlorine to the nitrogen atom of secondary amines (Dannan *et al.*, 1982) and amino acids (Hussain *et al.*, 1993) *via* a cyclic transition state incorporating water (Figure 6.149). It is quite likely, therefore, that *N*-chlorination of protein nitrogens may reduce the efficiency of the iodination reaction (Chen, P. *et al.*, 1994), although subsequent hydrolysis probably means there is no permanent modification of protein structure. These authors described increased radiochemical yields from CT iodination of model tyrosine-containing peptides, by using dimethylamine as a chlorine-transfer agent.

A procedure for minimisation of CT-related damage has been described, based on monitoring of the reaction potential (Hunter, 1970), but does not seem to have been widely employed.

Trace labelling using CT

Where high specific activity is not required, there are more possibilities for optimising the conditions of iodination to avoid damage. Trace labelling methods of this type were first proposed by McConahey and Dixon (1966) and a simple protocol for such work has been published (McConahey and Dixon, 1980).

Sonoda and Schlamowitz (1970) carried out a major study to optimise the conditions of CT oxidation and trace iodination; the following observations were made. (1) Iodide was oxidised within a few seconds to iodine, this reaction being largely independent of pH over the range 6–8, and temperature from 2 to 24 °C. (2) The maximum yield of iodine was observed when a half-stoichiometric amount of CT was employed. At higher CT:I^- ratios, increasing conversion of the iodine to higher oxidation states such as IO_3^- and IO_4^- was observed. This conversion to non-electrophilic forms was more pronounced at lower pH: a fivefold excess of CT was required for complete iodine oxidation (ten minutes, 2 °C) at pH 8, compared to a twofold excess at pH 6. (3) At 24 °C, the proportion of reactive product appeared to drop rapidly with time: at pH 7 with a fivefold CT excess, no H_2OI^+ or I_2 was detectable after five minutes. This probably reflected over-oxidation to IO_3^- and IO_4^-. The loss was much less pronounced at 2 °C, as predicted by the Nernst equation.

These observations were made using unlabelled iodide, and are incompatible with the conditions of 'carrier-free' radioiodination of proteins, where CT excesses must be much higher if a kinetically-suitable concentration is to be maintained. These excesses are typically in the order of $10^4 - 10^5$:1, so over-oxidation would seem to be extremely likely.

Under such conditions (60 000 fold excess, 2 °C), it was observed that the maximum amount of H_2OI^+ formed at the oxidation stage decreased as the pH was raised from 6 to 8. However at pH 6 the yield rose to a maximum at around ten minutes then dropped sharply; the yield at pH 7 was more or less constant up to 80 minutes oxidation. Working at pH 7 and 2 °C therefore appeared optimal in terms of iodine production.

A further series of observations were made concerning the iodination step itself, using IgG as the target protein. (1) The incorporation of active iodine into IgG was optimal at pH 7, conveniently the same as for iodine production, and increased up to about 40 minutes. (2) The incorporation decreased with increasing ionic strength ($\tau/2 = 0.05–0.5$ M) of the reaction buffer. (3) The protein consumed a quantity of CT (of the order of one molar equivalent, but variable between different proteins and different lots of a given protein) which was therefore unavailable for iodide oxidation. An excess of CT over protein was therefore required to effect iodination. Pre-oxidation of the protein with CT removed this effect. (4) At IgG concentrations up to 1 mg/ml, iodine incorporation increased linearly; above this level incorporation began to taper off. Therefore working below this value was best in terms of specific activity, while working above it gave the highest overall incorporation.

A procedure for IgG iodination was proposed based on these observations, using a tenfold excess of CT over protein in a 50 mM buffer (pH 7.0, 30 minutes, 2–3 °C). This yielded material with a specific activity of 0.4 to 4 μCi/mg, which retained almost complete recognition of antigen. Specific activity was deliberately kept low, less than one in 4000 IgG molecules being labelled. As a consequence the presence of tracer molecules containing more than one iodine atom could be discounted, i.e. the [^{125}I] IgG, although only present in small quantities, was highly homogeneous in nature.

6.21.3.4 Iodo-beads

Iodo-beads™ (Pierce) essentially represent an immobilised presentation of chloramine-T, consisting of hydrophilised polystyrene beads labelled with *N*-chlorobenzenesulfonamide (Figure 6.150). They offer certain advantages: (1) facile control of stoichiometry to allow different levels of labelling, and (2) removal of the need for metabisulfite reduction, the reaction being stopped instead by physical separation of solution from the beads[1].

Iodination using Iodo-beads demonstrated a fairly broad pH optimum over the range 5.5–8.5, the peak falling at around 6.5 (Markwell, 1982). Specific activities in excess of 150 mCi/mg were obtainable.

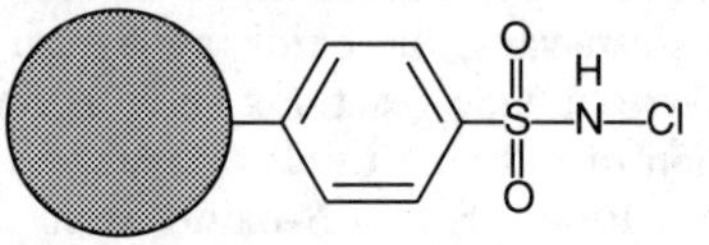

Figure 6.150 Iodo-beads™.

6.21.3.5 Iodogen

1,3,4,6-Tetrachloro-3α,6α-diphenyl glycoluril (Figure 6.151) was first described as a mild oxidising agent for radioiodination by Fraker and Speck (1978), and Markwell and Fox (1978); it was commercialised by Pierce as Iodogen™. Another *N*-chloro compound, this reagent is highly insoluble in aqueous systems and is typically used as a film 'plated' onto the reaction vessel surface from an organic solution. The reaction mixture can therefore be separated from the oxidant by decantation if required.

Early reports suggested that labile proteins such as pituitary hormones could be labelled safely by this

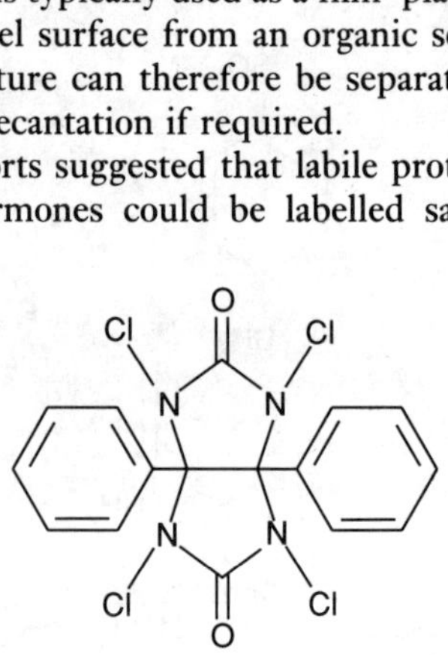

Figure 6.151 Iodogen™.

[1] Although it might be expected that this would leave the active iodinating species in the reaction mixture, it appeared instead that these remained closely associated with the beads. It was proposed that this was probably due to formation of the *N*-iodosulfonamide.

route (Ferguson and Jeffcoate, 1979; Wood *et al.*, 1980), although with an extremely sensitive case such as fibrinogen damage was still greater than with the non-oxidative ICl method (Knight *et al.*, 1981). A comparative study with CT and lactoperoxidase methods demonstrated less loss of functional activity for most proteins when Iodogen was employed (Salacinski *et al.*, 1981). Only a small excess of Iodogen over iodide was required for full incorporation into a typical protein, and the reaction was complete within around five minutes.

High specific activity products are achievable from Iodogen-mediated iodinations: Zalutsky *et al.* (1985), for example, obtained incorporations of one ^{125}I atom per F(ab′)$_2$ molecule.

6.21.3.6 Lactoperoxidase

Peroxidases are capable of mediating iodide oxidation, and have been known to catalyse the iodination of proteins for over half a century (Keston, 1944). Lactoperoxidase was demonstrated to be particularly effective for iodide oxidation (Morrison, 1968, 1970), and, when examined in the search for alternatives to chloramine-T, proved effective in catalysing the radioiodination of proteins (Marchalonis, 1969; Morrison *et al.*, 1971). The products of these preparations possessed fairly low specific activity, but Thorell and Johansson (1971) demonstrated that activities of 350 mCi/mg could be reached—high enough to satisfy virtually all needs.

Although the enzyme requires the presence of hydrogen peroxide, the level of this reagent can be kept fairly low; accordingly this is a mild technique allowing the functional activity of sensitive proteins such as pituitary hormones to be maintained (Miyachi *et al.*, 1971).

Although the pH optimum of iodide oxidation by lactoperoxidase is 5.0, the activity is relatively well maintained over the range 4–8.5 (Morrison *et al.*, 1971). The reaction is extremely rapid, allowing reaction times of a minute or less to be employed: iodination yields greater than 80% have been observed from two second incubations (Thorell and Johansson, 1971).

An immobilised presentation of the enzyme was an obvious development for preparative purposes, and Sepharose™-4B-lactoperoxidase was described by David (1972). As with many immobilised enzymes the pH optimum appeared to be shifted, in this case conveniently to around 6.5–7. However, self-iodination proved to be a significant source of inefficiency: even where the lactoperoxidase represented only 0.1% of the total protein, 18% of the ^{125}I was incorporated into the enzyme, in competition with the desired target IgG. Where the relative enzyme concentration was higher, losses were as high as 64%.

A procedure for high-activity labelling of proteins using soluble lactoperoxidase is given by Morrison (1980), as well as instructions for separating the product from enzyme under various circumstances.

6.21.3.7 Electrolytic methods

An interesting alternative to chemical or enzymic oxidation methods is the electrolytic approach first described by Rosa *et al.* (1964), employing an electrolytic cell to oxidise the iodide. This approach has been demonstrated to allow very high specific acitivities to be achieved (up to 25 iodine atoms per molecule of HSA; Rosa *et al.*, 1967), and has consistently given product with better retention of protein properties than other methods (Rosa *et al.*, 1964; Katz and Bonorris, 1968; Donabedian *et al.*, 1972). It can be carried out at pH 7 with quantitative incorporation of iodine from dilute solutions.

Despite these advantages, the electrolytic approach does not appear to have achieved widespread popularity, presumably due to a general lack of familiarity with electrochemical procedures and the modest requirement for specialised equipment.

The procedure has also been adapted for small molecules containing phenolic groups such as estradiol (Moore and Wolf, 1978).

6.21.4 Heterobifunctional reagents for radioiodination of proteins

In the introduction to this section it was noted that one approach to the introduction of radioactive iodine was to label the protein with an iodinated (or readily iodinatable) reagent. Such compounds are therefore heterobifunctional, typically combining a phenolic or similar group with a moiety reactive towards protein amines or thiols. The chemistry of coupling such reagents to proteins has already been dealt with adequately in relevant sections of this chapter, but a brief summary of some of the more important reagents of this type is presented in Figure 6.152.

These include diazonium salts for coupling to tyrosine or histidine residues, imidates which retain the native charge of the protein when coupled to a lysine residue, maleimides for coupling to thiols, and hydrazides for coupling to oxidised glycoproteins. In keeping with current trends in protein labelling, however, most are succinimidyl or sulfosuccinimidyl esters.

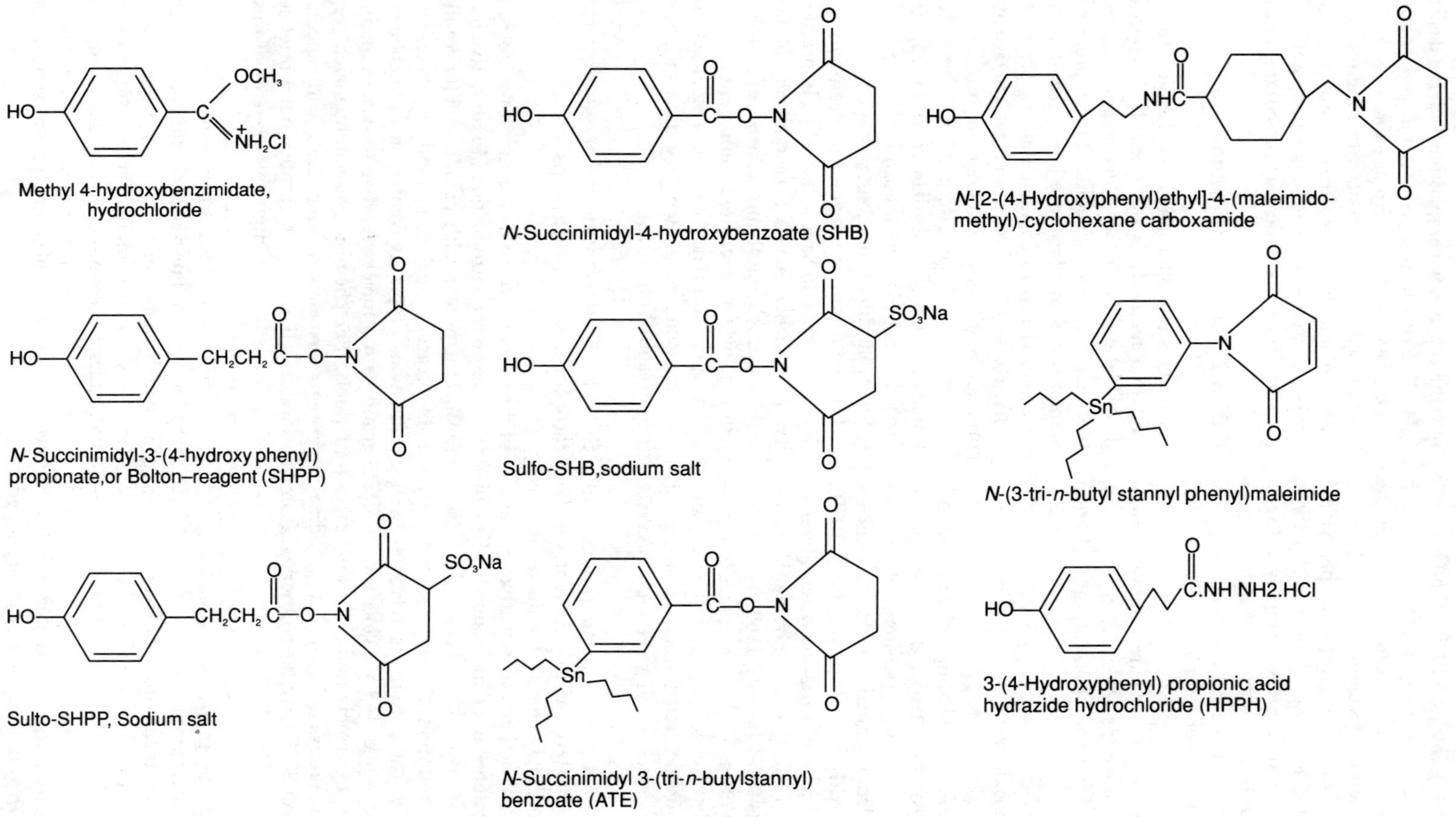

Figure 6.152 Some heterobifunctional compounds employed in radioiodination. These couple typical reactive groups such as imidates, succinimide esters, maleimides and hydrazides with iodinatable moieties such as phenols or phenyl alkyltins.

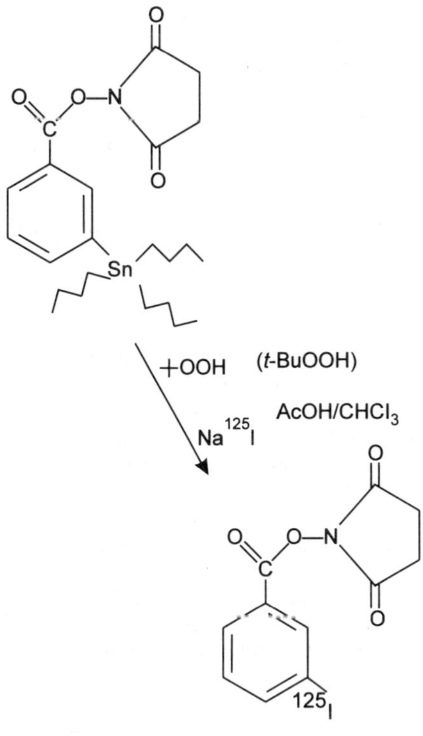

Figure 6.153 Iodination by destannylation of alkyltin groups in the presence of hydroperoxide.

Generally these reagents are predictable combinations of phenols with reactive functionalities, but one category is worthy of further comment: the organotin derivatives. The usual phenolic groups are iodinated in the *o*-position, making the products analogous to naturally occurring iodotyrosines such as the thyroid hormone thyroxine and its analogues. There are a number of enzymes present *in vivo* (Smallridge *et al.*, 1981; Koehrle *et al.*, 1986) which are capable of deiodinating such compounds, and are therefore an unwelcome cause of loss of activity in radiopharmaceuticals.

Alternative iodinatable groups have therefore been sought, and the most successful of these are based on destannylation of alkyltin groups in the presence of hydroperoxide (Figure 6.153) (Zalutsky and Narula, 1987; Zalutsky *et al.*, 1989). These allow radioiodinated aromatic groups to be introduced which are more resistant to *in vivo* deiodination than *o*-iodophenols.

References

Aalberse, R. C. (1973). Quantitative fluoroimmunoassay. *Clin. Chim. Acta.*, **48**, 109–111.

Aberhalden, E. and Stix, W. (1923). Weiterer Beitrag zur Frage der Konstitution der Proteine. Untersuchungen über den Aufbau des Seidenfibroins. *Hoppe-Seyler's Z. Physiol. Chem.*, **129**, 143–156.

Abraham, G. and Low, P. S. (1980). Covalent labelling of specific membrane carbohydrate residues with fluorescent probes. *Biochim. Biophys. Acta*, **597**, 285–291.

Abuknesha, R. and Exley, D. (1978). Design and development of oestradiol-17β enzyme-immunoassay. In *Enzyme Labelled Immunoassay of Hormones and Drugs. Proceedings of the International Symposium on Enzyme Labelled Immunoassay of Hormones and Drugs, Ulm, West Germany, July 10 and 11, 1978* (Pal, S. B., ed.), pp. 139–152, Walter de Gruyter & Co., Berlin.

Acharya, A. S and Manning, J. M. (1980a). Reactivity of the amino groups of carbonmonoxyhemoglobin S with glyceraldehyde. *J. Biol. Chem.*, **255**, 1406–1412.

Acharya, A. S. and Manning, J. M. (1980b). Amadori rearrangement of glyceraldehyde–hemoglobin Schiff base adducts. *J. Biol. Chem.*, **255**, 7218–7224.

Adamczyk, M. *et al.* (1994). Characterization of protein–hapten conjugates. 1. Matrix-assisted laser desorption ionization mass spectrometry of immuno BSA–hapten conjugates and comparison with other characterization methods. *Bioconjugate Chem.*, **5**, 631–635.

Adams, P. and Baron, F. A. (1965). Esters of carbamic acids. *Chem. Rev.*, **65**, 567–602.

Agababyan, A. G. *et al.* (1982). Amino acids as the amine component in the Mannich reaction. *Russ. Chem. Rev.*, **51**, 387–396.

Aguirre, R. *et al.* (1986). Interaction of fluorescently labeled myosin subfragment 1 with nucleotides and actin. *Biochemistry*, **25**, 6827–6835.

Ahnoff, M. *et al.* (1981). Derivatization with 4-chloro-7-nitrobenzofuran for liquid chromatographic determination of hydroxyproline in collagen hydrolysate. *Anal. Chem.*, **53**, 485–489.

Aikawa, T. *et al.* (1979). Enzyme immunoassay of angiotensin I. *Endocrinology*, **105**, 1–6.

Al-Bassam, M. N. *et al.* (1978). Nortryptiline enzyme-immunoassay. In *Enzyme Labelled Immunoassay of Hormones and Drugs. Proceedings of the International Symposium on Enzyme Labelled Immunoassay of Hormones and Drugs, Ulm, West Germany, July 10 and 11, 1978* (Pal, S. B., ed.), pp. 375–386, Walter de Gruyter & Co., Berlin.

Alexander, N. M. (1973). Oxidation and oxidative cleavage of tryptophanyl peptide bonds during iodination. *Biochem. Biophys. Res. Commun.*, **54**, 614–621.

Alexander, N. M. (1974). Oxidative cleavage of tryptophanyl peptide bonds during chemical- and peroxidase-catalyzed iodinations. *J. Biol. Chem.*, **249**, 1946–1952.

Alexander, P. (1954). The reaction of carcinogens with macromolecules. *Adv. Cancer Res.*, **2**, 1–72.

Amante, L. *et al.* (1972). The conjugation of immunoglobulins with tetramethylrhodamine isothiocyanate. A comparison between the amorphous and the crystalline fluorochrome. *J. Immunol. Methods*, **1**, 289–301.

Andersson, M. and Oscarson, S. (1992). Synthesis of *O*-glycopyranosyl-*N*-hydroxysuccinimides of glucose and lactose and their opening by nucleophiles into prespacer glycosides. *Glycoconjugate J.*, **9**, 122–125.

Andersson, M. and Oscarson, S. (1993). Synthesis of glycoconjugates by covalent coupling of *O*-glycopyranosyl-*N*-hydroxysuccinimide derivatives of lactose to proteins and lipids and polymerisation of their parent acryloyl derivatives into acrylamide polymers. *Bioconjugate Chem.*, **4**, 246–249.

Andree, P. J. and Zantema, A. (1978). Electron spin resonance and nuclear relaxation studies on spin-labeled glutamate dehydrogenase. *Biochemistry*, **17**, 778–782.

Andrews, J. L. *et al.* (1982). Ammonium 4-chloro-7-sulfobenzofurazan: a new fluorigenic thiol-specific reagent. *Arch. Biochem. Biophys.*, **214**, 386–396.

Anfinsen, C. B. and Haber, E. (1961). Studies on the reduction and re-formation of protein disulfide bonds. *J. Biol. Chem.*, **236**, 1361–1363.

Annunziato, M. E. *et al.* (1993). *p*-Maleimidophenyl isocyanate: a novel heterobifunctional linker for hydroxyl to thiol coupling. *Bioconjugate Chem.*, **4**, 212–218.

Aplin, J. D. *et al.* (1979). Spin-labelling of sialic acid in soluble and cell-surface glycoproteins. *Carbohydr. Res.*, **75**, 11–16.

Atassi, M. Z. and Habeeb, A. F. S. A. (1972). Reaction of proteins with citraconic anhydride. *Methods Enzymol.*, **25**, 446–453.

Atassi, M. Z. *et al.* (1974). Enzymic and immunochemical properties of lysozyme: VIII. Modification of two carboxyl groups and their role in the antigenic reactivity. *Immunochemistry*, **11**, 495–500.

Atlas, D. *et al.* (1978). Ultrastructural probing of β-adrenoreceptors on cell surfaces. *FEBS Lett.*, **95**, 173–176.

Auclair, C. and Paoletti, C. (1981). Bioactivation of the antitumor drugs 9-hydroxyellipticine and derivatives by a peroxidase–hydrogen peroxide system. *J. Med. Chem.*, **24**, 289–295.

Auclair, C. *et al.* (1983). Peroxidase-catalysed covalent binding of the antitumor drug 2*N*-methyl-9-hydroxy-ellipticine to proteins. *Biochem. Pharmacol.*, 32, 3883–3886.

Autor, A. P. and Fridovich, I. (1970). The interaction of acetoacetate decarboxylase with carbonyl compounds, hydrogen cyanide, and an organic mercurial. *J. Biol. Chem.*, **245**, 5214–5222.

Axen, U. (1974). *N,N′*-Carbonyldiimidazole as coupling reagent for the preparation of bovine serum albumin conjugates. *Prostaglandins*, **5**, 45–47.

Azadova, N. B. *et al.* (1965). Action of proflavine on the development of Sendai virus in HEp 2 cells. *Br. J. Exp. Pathol.*, **47**, 121–124.

Baenziger, J. U. and Fiete, D. (1982). Photoactivatable glycopeptide reagents for site-specific labeling of lectins. *J. Biol. Chem.*, **257**, 4421–4425.

Bailey, A. and Rasums, A. (1989). Dye labelled antibodies as reagents for use in immunoassay systems. *US Patent 4,863,875* to GIA Research Company.

Bailey, J. M. and Butler, J. (1967). Synthetic cholesterol-ester antigens in experimental atherosclerosis. In *The Reticuloendothelial System and Atherosclerosis* (Di Luzio, N. R. and Paoletti, R., eds.), pp. 433–441, Plenum Press, New York.

Bailey, M. P. *et al.* (1983). Use of Lucifer Yellow VS as a label in fluorescent immunoassays illustrated by the determination of albumin in serum. *Ann. Clin. Biochem.*, **20**, 213–216.

Bailey, M. P. *et al.* (1984). Homogeneous fluoroimmunoassay using Lucifer Yellow VS: determination of albumin in plasma. *Ann. Clin. Biochem.*, **21**, 59–63.

Baines, B. S. *et al.* (1977). The highly electrophilic character of 4-chloro-7-nitrobenzofurazan and possible consequences for its application as a protein-labelling reagent. *Biochem. J.*, **163**, 189–192.

Baker, B. R. *et al.* (1961). Potential anticancer agents. LXVI. Non-classical antimetabolites. III. 4-(Iodoacetamido)-salicylic acid, an *exo*-alkylating irreversible inhibitor of glutamic dehydrogenase. *J. Am. Chem. Soc.*, **83**, 3713–3714.

Bakker, C. N. M. and Kaspersen, F. M. (1978). Labeling with $^{1}31$ I of chloroquine-analogues for the detection of ocular melanoma. *J. Labelled Compd. Radiopharm.*, **15**, 681–691.

Baldas, J. *et al.* (1981). The tritium kinetic isotope effect for the iodination of L-(5-^{3}H)-3-iodotyrosine. *Aust. J. Chem.*, **34**, 1147–1151.

Baldwin, R. M. and Lin, T. H. (1981). Chlorination as a side-reaction in radioiodination with Chloramine-T. *J. Radioanal. Chem.*, **65**, 163–165.

Balls, A. K. and Wood, H. N. (1956). Acetyl chymotrypsin and its reaction with ethanol. *J. Biol. Chem.*, **219**, 245–256.

Banks, P. R. and Paquette, D. M. (1995). Comparison of three common amine reactive fluorescent probes used for conjuagtion to biomolecules by capillary zone electrophoresis. *Bioconjugate Chem.*, **6**, 447–458.

Banks, T. E. and Shafer, J. A. (1972). Inactivation of papain by *S*-methylation of its cysteinyl residue with *O*-methylisourea. *Biochemistry*, **11**, 110–114.

Bannwarth, W. and Knorr, R. (1991). Formation of carboxamides with *N,N,N′,N′*-tetramethyl (succinimido) uronium tetrafluoroborate in aqueous/organic solvent systems. *Tetrahedron Lett.*, **32**, 1157–1160.

Bannwarth, W. *et al.* (1988). Bathophenanthroline-ruthenium (II) complexes as non-radioactive labels for oligonucleotides which can be measured by time-resolved fluorescence techniques. *Helv. Chim. Acta*, **71**, 2085–2099.

Barger, B. O. *et al.* (1976). Estimation of molecular weight by polyacrylamide gel electrophoresis using heat stable fluorophors. *Anal. Biochem.*, **70**, 327–335.

Barltrop, J. A. *et al.* (1966). Photosensitive protective groups. *J. Chem. Soc., Chem. Commun.*, 822–823.

Barman, T. E. and Koshland, D. E., Jr (1967). A colorimetric procedure for the quantitative determination of tryptophan residues in proteins. *J. Biol. Chem.*, **242**, 5771–5776.

Basak, A. *et al.* (1994). Biotinylation of an enkephalin-containing heptapeptide *via* various spacer arms. Synthesis, comparitive binding studies toward avidin, and application as substrates in enzymatic reactions. *Bioconjugate Chem.*, **5**, 301–305.

Bassiri, R. M. and Utiger, R. D. (1972). The preparation and specificity of antibody to thyrotropin releasing hormone. *Endocrinology*, **90**, 722–727.

Battaglini, F. *et al.* (1994). Targeting glucose oxidase at aspartate and glutamate residues with organic two-electron redox mediators. *Bioconjugate Chem.*, **5**, 430–435.

Bauman, J. G. J. *et al.* (1981). Cytochemical hybridization with fluorochrome-labeled RNA. *J. Histochem. Cytochem.*, **29**, 227–237.

Bauminger, S. and Wilchek, M. (1980). The use of carbodiimides in the preparation of immunizing conjugates. *Methods Enzymol.*, **70**, 151–159.

Bauminger, S. *et al.* (1969). Antibodies to a phyto-oestrogen: antigenicity of genistein coupled to a synthetic polypeptide. *J. Endocrinol*, **44**, 567–578.

Bauminger, S. *et al.* (1973). Radioimmunological assay of prostagladin synthetase activity. *Prostaglandins*, **4**, 313–324.

Bayer, E. A. *et al.* (1975). The use of a homologous series of affinity labelling reagents in the study of the biotin transport system in yeast cells. *FEBS Lett.*, **60**, 309–312.

Bayley, H. (1983a). Photogenerated reactive intermediates and their properties. In *Photogenerated Reagents in Biochemistry and Molecular Biology*, pp. 8–24, Elsevier, Amsterdam.

Bayley, H. (1983b). Reagents for photoaffinity labeling. In *Photogenerated Reagents in Biochemistry and Molecular Biology*, pp. 25–65, Elsevier, Amsterdam.

Bayley, H. and Knowles, J. R. (1978). Photogenerated reagents for membrane labeling. 2. Phenylcarbene and adamantylidene formed within the lipid bilayer. *Biochemistry*, **17**, 2420–2423.

Bayley, H. and Knowles, J. R. (1980). Photogenerated reagents for membranes: selective labeling of intrinsic nembrane proteins in the human erythrocyte membrane. *Biochemistry*, **19**, 3883–3892.

Becker, J. M. *et al.* (1971). Irreversible inhibition of biotin transport in yeast by biotinyl-*p*-nitrophenyl ester. *Proc. Natl. Acad. Sci. USA*, **68**, 2604–2607.

Bedel-Cloutour, C. H. *et al.* (1996). Syntheses of functionalized indium porphyrins for monoclonal antibody labeling. *Bioconjugate Chem.*, **7**, 617–627.

Belleau, B. *et al.* (1969). The mechanism of irreversible adrenergic blockade by *N*-carbethoxydihydroquinolines: model studies with typical serine hydrolases. *Biochem. Pharmacol.*, **18**, 1039–1044.

Benesch, R. E. *et al.* (1955). The sulfhydryl groups of crystalline proteins. I. Some albumins, enzymes and hemoglobins. *J. Biol. Chem.*, **216**, 663–676.

Benisek, W. F. and Richards, F. M. (1968). Attachment of metal-chelating functional groups to hen egg white lysozyme. *J. Biol. Chem.*, **243**, 4267–4271.

Bergmann, K. E. *et al.* (1994). Hexrestrol diazirine photoaffinity labeling reagent for the estrogen receptor. *Bioconjugate Chem.*, **5**, 141–150.

Bergquist, N. R. and Schilling, W. G. E. E. (1970). Preparation of anti-human immunoglobulin for indirect fluorescent tracing of auto-antibodies. In *Standardization in Immunofluorescence* (Holborrow, E. J., ed.), pp. 171–176, Blackwell Scientific Publications, Oxford.

Berkowitz, B. A. *et al.* (1975). Disposition of naloxone: use of a new radioimmunoassay. *J. Pharmacol. Exp. Ther.* **195**, 499–504.

Bernatowicz, M. S. and Matsueda, G. R. (1986). Preparation of peptide–protein immunogens using *N*-succinimidyl bromoacetate as a heterobifunctional crosslinking reagent. *Anal. Biochem.*, **155**, 95–102.

Berne, P.-F. *et al.* (1992). Carboxypeptidase Y-catalyzed transpeptidation of esterified oligo- and polypeptides and its use for the specific carboxyl-terminal labeling of proteins. *J. Am. Chem. Soc.*, **114**, 2603–2610.

Bezrukov, M. G. *et al.* (1982). The interaction of metals with proteins. *Russ. Chem. Rev.*, **51**, 696–711.

Bickel, U. *et al.* (1994). Development and *in vitro* characterization of a cationized monoclonal antibody against βA4 protein: a potential probe for Alzheimer's disease. *Bioconjugate Chem.*, **5**, 119–125.

Bieniarz, C. *et al.* (1994). Thiolate and phosphothiorate functionalized fluoresceins and their use as fluorescent labels. *Bioconjugate Chem.*, **5**, 31–39.

Bigge, J. C. *et al.* (1995). Nonselective and efficient fluorescent labeling of glycans using 2–amino benzamide and anthranilic acid. *Anal. Biochem.*, **230**, 229–238.

Bindal, R. D. *et al.* (1987). 11β-Chloromethyl-[^{3}H]estradiol-17β: a very high affinity, reversible ligand for the estrogen receptor. *J. Steroid Biochem.*, **28**, 361–370.

Binion, S. B. and Rodkey, L. S. (1983). Destruction of antibody idiotopes with ultra-low concentrations of reducing agents. *Mol. Immunol.*, **20**, 475–482.

Birkett, D. J. *et al.* (1970). The reactivity of SH groups with a fluorogenic reagent. *FEBS Lett.*, **6**, 346–348.

Birkett, D. J. *et al.* (1971). Probes for the conformational transitions of phosphorylase *b*. *Eur. J. Biochem.*, **20**, 494–508.

Blake, C. and Gould, B. J. (1984). Use of enzymes in immunoassay techniques. A review. *Analyst*, **109**, 533–547.

Blake, C. *et al.* (1982). Simultaneous enzyme immunoassay of two thyroid hormones. *Clin. Chem.*, **28**, 1469–1473.

Blake, C. C. F. (1967). A crystallographic study of the oxidation of lysozyme by iodine. *Proc. Roy. Soc. Ser. B*, **167**, 435–438.

Blakeslee, D. and Kennedy, J. C. (1974). Factors affecting the noncovalent binding of chlorambucil to rabbit immunoglobulin G. *Cancer Res.*, **34**, 882–885.

Bodanszky, M. and Tolle, J. C. (1977). Side reactions in peptide synthesis. V. A reexamination of the mixed anhydride method. *Int. J. Peptide Protein Res.*, **10**, 380–384.

Bodlaender, P. *et al.* (1968). Isoxazolium salts for carboxyl-group modification in proteins. *Fed. Proc.*, 27, 291.

Boersma, W. J. A. *et al.* (1988). Antibodies to short synthetic peptides for specific recognition of partly denatured protein. *Anal. Chim. Acta*, **213**, 187–197.

Bolton, A. E. and Hunter, W. M. (1973). The labelling of proteins to high specific radioactivities by conjugation to a ^{125}I-containing acylating reagent. *Biochem. J.*, **133**, 529–539.

Bonnafous, J-C. *et al.* (1988). New probes for angiotensin II receptors. *Biochem. J.*, **251**, 873–880.

Boorsma, D. M. *et al.* (1976). Peroxidase and fluorescein isothiocyanate as antibody markers. A quantitative comparison of two peroxidase conjugates prepared with glutaraldehyde or periodate and a fluorescein conjugate. *J. Histochem. Cytochem.*, **24**, 1017–1025.

Borch, R. F. *et al.* (1971). The cyanohydridoborate anion as a selective reducing agent. *J. Am. Chem. Soc.*, **93**, 2897–2904.

Bosshard, H. R. *et al.* (1978). The aminoacyl-tRNA synthetase-tRNA complex: detection by differential labelling of lysine residues involved in complex formation. *J. Mol. Biol.*, **119**, 377–389.

Bowman, R. E. and Stroud, H. H. (1950). N-substituted amino acids. Part I. A new method of preparation of dimethylamino acids. *J. Chem. Soc.* 1342–1345.

Boyer, P. D. (1954). Spectrophotometric study of the reaction of protein sulfhydryl groups with organic mercurials. *J. Am. Chem. Soc.*, **76**, 4331–4337.

Brandtzaeg, P. (1973a). Conjugates of immunoglobulin G with different fluorochromes. I. Characterization by anionic-exchange chromatography. *Scand. J. Immunol.*, **2**, 273–290.

Brandtzaeg, P. (1973b). Conjugates of immunoglobulin G with different fluorochromes. II. Specific and nonspecific binding properties. *Scand. J. Immunol.*, **2**, 333–348

Brechbiel, M. W. *et al.* (1986). Synthesis of 1-(*p*-isothiocyanatobenzyl) derivatives of DTPA and EDTA. Antibody labeling and tumor-imaging studies. *Inorg. Chem.*, **25**, 2772–2781.

Breddam, K. *et al.* (1980). Influence of the substrate structure on carboxypeptidase Y catalyzed peptide bond formation. *Carlsberg Res. Commun.*, **45**, 361–367.

Breddam, K. *et al.* (1981). Carboxypeptidase Y catalyzed C-terminal modification in the B-chain of porcine insulin. *Carlsberg Res. Commun.*, **46**, 361–372.

Briand, J. P. *et al.* (1985). Synthetic peptides as antigens: pitfalls of conjugation methods. *J. Immunol. Methods*, **78**, 59–69.

Brown, D. M. (1963). Phosphorylation. *Adv. Org. Chem.*, **3**, 75–157.

Brown, R. D. and Matthews, K. S. (1979). Chemical modification of lactose repressor protein using *N*-substituted maleimides. *J. Biol. Chem.*, **254**, 5128–5134.

Brown, S. (1997). Metal-recognition by repeating polypeptides. *Nature Biotechnol.*, **15**, 269–272.

Bruice, P. Y. *et al.* (1978). Inactivation of glyceraldehyde-3-phosphate dehydrogenase and yeast alcohol dehydrogenase by arene oxides. *Biochemistry*, **17**, 1662–1669.

Bruice, T. W. *et al.* (1976). Reversible derivatization of cysteinyl sulfhydryls to generate new substrates for trypsin and chymotrypsin. *Fed. Proc.*, **35**, 1475.

Brunings, K. J. (1947). Preparation and properties of the iodohistidines. *J. Am. Chem. Soc.*, **69**, 205–208.

Brunner, J. and Richards, F. M. (1980). Analysis of membranes photolabeled with lipid analogues. *J. Biol. Chem.*, **255**, 3319–3329.

Brunner, J. *et al.* (1980). 3-Trifluoromethyl-3-phenyldiazirine. *J. Biol. Chem.*, **255**, 3313–3318.

Buckler, D. R. *et al.* (1993). C-terminal labeling of ribonuclease A with an extrinsic fluorescent robe by carboxypeptidase Y-catalyzed transpeptidation in the presence of urea. *Anal. Biochem.*, **209**, 20–31.

Buckley, R. G. and Searle, F. (1984). An efficient method for labelling antibodies with ^{111}In. *FEBS Lett.*, **166**, 202–204.

Buechler, J. A. *et al.* (1989). Differential labeling of the catalytic subunit of cAMP-dependent protein kinase with acetic anhydride: substrate-introduced conformational changes. *Biochemistry*, **28**, 3018–3024.

Burke, J. F. *et al.* (1966). The effect of antibody to l-phenylalanine mustard conjugate on malignant cells selectively marked through "early inflammatory like" vascular permeability. *Cancer Res.*, **26**, 1893–1904.

Burkot, T. R. *et al.* (1985). Use of fluorodinitrobenzene to identify monoclonal antibodies which are suitable for conjugation to periodate oxidized horseradish peroxidase. *J. Immunol. Methods*, **84**, 25–31.

Burnop, V. C. E. *et al.* (1957). Studies with ^{15}N-labelled nitrogen mustards. The combination of di-2-(chloroethyl)-methylamine with proteins. *Biochem. J.*, **66**, 504–515.

Burstein, S. and Knapp, R. (1977). Chemotherapy of murine ovarian carcinoma by methotrexate–antibody conjugates. *J. Med. Chem.*, **20**, 950–952.

Butler, G. C. *et al.* (1940). Studies in synthetic immunochemistry. 5. Observations on antisera against aspirin–protein complexes. *Biochem. J.*, **34**, 838–845.

Butler, P. J. G. and Hartley, B. S. (1972). Maleylation of amino groups. *Methods Enzymol.*, **25**, 191–199.

Butler, V. P., Jr and Chen, J. P. (1967). Digoxin-specific antibodies. *Proc. Natl. Acad. Sci. USA*, **57**, 71–78.

Cabacungan, J. C. *et al.* (1982). Amine boranes as alternative reducing agents for reductive alkylation of proteins. *Anal. Biochem.*, **124**, 272–278.

Canfi, A. *et al.* (1989). Multiple labelling of immunoglobulin G, albumin and testosterone with a fluorescent terbium chelate for fluorescence immunoassays. *Analyst*, **114**, 1407–1411.

Capala, J. *et al.* (1997). Radiolabeling of epidermal growth factor with ^{99m}Tc and *in vivo* localization following intracerebral injection into normal and glioma-bearing rats. *Bioconjugate Chem.*, **8**, 289–295.

Carraway, K. L. and Koshland, D. E., Jr (1968). Reaction of tyrosine residues in proteins with carbodiimide reagents. *Biochim. Biophys. Acta*, **160**, 272–274.

Carraway, K. L. and Koshland, D. E., Jr (1972). Carbodiimide modification of proteins. *Methods Enzymol.*, **25**, 616–623.

Carraway, K. L. and Triplett, R. B. (1970). Reaction of carbodiimides with protein sulfhydryl groups. *Biochim. Biophys. Acta*, **200**, 564–566.

Carraway, K. L. *et al.* (1969). Carboxyl group modification in chymotrypsin and chymotrypsinogen. *J. Mol. Biol.*, **42**, 133–137.

Carty, R. P. and Hirs, C. H. W. (1968). Modification of bovine pancreatic ribonuclease A with 4-sulfonyl-2-nitro-fluorobenzene. *J. Biol. Chem.*, **243**, 5244–5253.

Casanova, J. *et al.* (1984). Photoaffinity labeling of thyroid hormone nuclear receptors. *J. Biol. Chem.*, **259**, 12084–12091.

Castro, B. *et al.* (1975). Réactifs de couplage peptidique IV (1) – L'hexafluorophosphate de benzotriazolyl *N*-oxytris-diméthylamino phosphonium (B.O.P.) *Tetrahedron Lett.*, **14**, pp 1219–1222.

Cessi, C. and Fiume, L. (1969). Increased toxicity of β-amanitin when bound to a protein. *Toxicon*, **6**, 309–310.

Chaiken, I. M. and Smith, E. L. (1969a). Reaction of chloro-acetamide with the sulfhydryl group of papain. *J. Biol. Chem.*, **244**, 5087–5094.

Chaiken, I. M. and Smith, E. L. (1969b). Reaction of the sulfhydryl group of papain with chloroacetic acid. *J. Biol. Chem.*, **244**, 5095–5099.

Chang, J. Y. (1977). High-sensitivity sequence analysis of peptides and proteins by 4-*N'N-dimethylaminoazobenzene 4'-isothiocyanate. Biochem. J.*, **163**, *517–520.*

Chang, J. Y. and Creaser, E. H. (1976). A novel manual method for protein-sequence analysis. *Biochem. J.*, **157**, 77–85.

Chen, C.-J. *et al.* (1994). Synthesis and characterization of oligomeric *nido*-carboranyl phosphate diester conjugates to antibody and antibody fragments for potential use in boron neutron capture therapy of solid tumors. *Bioconjugate Chem.*, **5**, 557–564.

Chen, P. *et al.* (1994). An improved method of radio-iodination with chloramine-T. *Anal. Biochem.*, **219**, 159–161.

Chen-Kiang, S. *et al.* (1979). Gel electrophoresis of fluorescent labeled cyanogen bromide cleavage products at the submicrogram level. *Anal. Biochem.*, **95**, 122–126.

Childs, G. V. *et al.* (1990). Localization of gonadotropin-releasing hormone receptors. *Methods Enzymol.*, **184**, 395–404.

Church, R. F. R. and Weiss, M. J. (1970). Diazirines. II. Synthesis and properties of small functionalized diazirine molecules. Some observations on the reaction of a diaziridine with the iodine–iodide ion system. *J. Org. Chem.*, **35**, 2465–2471.

Church, R. F. R. *et al.* (1965). Diazirines. I. Some observations on the scope of the ammonia-hydroxylamine-*O*-sulfonic acid diaziridine synthesis. The preparation of certain steroid diaziridines and diazirines. *J. Am. Chem. Soc.*, **87**, 2665–2671.

Ci, Y.-X. *et al.* (1995). Fluorescence labelling with eropium chelate of β-diketones and application in time-resolved fluoroimmunoassays (TR-FIA). *J. Immunol. Methods*, **179**, 233–241.

Clark, B. R. *et al.* (1977). Radioimmunoassay of haloperidol in human serum. *Life Sci.*, **20**, 319–326.

Clutton, R. F. *et al.* (1938). Studies in synthetic immunochemistry. III. Preparation and antigenic properties of thyroxyl derivatives of proteins, and physiological effects of their antisera. *Biochem. J.*, **32**, 1119–1132.

Comens, P. G. *et al.* (1982). Direct linkage of ^{125}I-EGF to cell surface receptors. A useful artefact of chloramine-T treatment. *J. Biol. Chem.*, **257**, 42–45.

Contreras, M. A. *et al.* (1983). Iodine monochloride (ICl) iodination techniques. *Methods Enzymol.*, **92**, 277–292.

Cooke, R. D. and Holbrook, J. J. (1974). Calcium and the assays of human plasma clotting factor XIII. *Biochem. J.*, **141**, 71–78.

Coons, A. H. *et al.* (1941). Immunological properties of an antibody containing a fluorescent group. *Proc. Soc. Exp. Biol. Med.*, **47**, 200–202.

Cox, J. P. L. *et al.* (1990). Synthesis of *C*- and *N*-functionalised derivatives of 1,4,7-triazacyclononane-1,4,7- triyltriacetic acid (NOTA), 1,4,7,10-tetra-azacyclododecane-1, 4,7,10–tetrayltetraacetic acid (DOTA) and diethylenetriaminepenta-acetic acid (DTPA): bifunctional complexing agents for the derivatisation of antibodies. *J. Chem. Soc., Perkin Trans. 1*, 2567–2576.

Creech, H. J. and Jones, R. N. (1940). The conjugation of horse serum albumin with 1,2-benzanthryl isocyanates. *J. Am. Chem. Soc.*, **62**, 1970–1975.

Creech, H. J. and Jones, R. N. (1941). The conjugation of horse serum albumin with isocyanates of certain polynuclear aromatic hydrocarbons. *J. Am. Chem. Soc.*, **63**, 1661–1669.

Crestfield, A. M. *et al.* (1963). The preparation and enzymatic hydrolysis of reduced and *S*-carboxymethylated proteins. *J. Biol. Chem.*, **238**, 622–627.

Cuatrecasas, P. *et al.* (1967). Fluorescence studies of the interaction of nucleotides with the active site of the nuclease of *Staphylococcus aureus. Proc. Natl. Acad. Sci. USA*, **58**, 2043–2050.

Cuatrecasas, P. *et al.* (1968). The effect of a competitive inhibitor on the acetylation of tyrosyl and lysyl residues of staphylococcal nuclease. *Biochim. Biophys. Acta*, **159**, 417–419.

Cuatrecasas, P. *et al.* (1969). Affinity labeling of the active site of staphylococcal nuclease. *J. Biol. Chem.*, **244**, 4316–4329.

Daddow, L. Y. M. (1983). A double lead stain method for enhancing contrast of ultrathin sections in electron microscopy: a modified multiple staining technique. *J. Microsc.*, **129**, 147–153.

Dahl, K. H. and McKinley-McKee, J. S. (1981). The reactivity of affinity labels: a kinetic study of the reaction of alkyl halides with thiolate anions: a model reaction for protein alkylation. *Bioorg. Chem.*, **10**, 329–341.

Dannan, H. *et al.* (1992). Kinetics and mechanism of chlorine exchange between Chloramine-T and secondary amines. *J. Pharm. Sci.*, **81**, 652–656.

David, G. S. (1972). Solid state lactoperoxidase: a highly stable enzyme for simple, gentle iodination of proteins. *Biochem. Biophys. Res. Commun.*, **48**, 464–471.

Davis, M.-T. B. and Preston, J. F. (1981). A simple modified carbodiimide method for conjugation of small-molecular-weight compounds to immunoglobulin G with minimal protein crosslinking. *Anal. Biochem.*, **116**, 402–407.

Dawson, E. C. *et al.* (1978). A simple and efficient method for raising steroid antibodies in rabbits. *Steroids*, **31**, 357–366.

Deen, C. *et al.* (1990). A novel carbodiimide coupling method for synthetic peptides. Enhanced anti-peptide antibody responses. *J. Immunol. Methods*, **129**, 119–125.

Degani, Y. and Heller, A. (1987). Direct electrical communication between chemically modified enzymes and metal electrodes. 1. Electron transfer from glucose oxidase to metal electrodes *via* electron relays, bound covalently to the enzyme. *J. Phys. Chem.*, **91**, 1285–1289.

Degani, Y. and Heller, A. (1988). Direct electrical communication between chemically modified enzymes and metal electrodes. 2. Methods for bonding electron-transfer relays to glucose oxidase and D-amino acid oxidase. *J. Am. Chem. Soc.*, **110**, 2615–2620.

DeLaLuz, P. J. *et al.* (1995). Synthesis and use of a biotinylated 3-azidophenothiazine to photolabel both amino- and carboxyl-terminal sites in calmodulin. *Bioconjugate Chem.*, **6**, 558–566.

Delfino, J. M. *et al.* (1993). Design, synthesis, and properties of a photoactivatable membrane-spanning phospholipidic probe. *J. Am. Chem. Soc.*, **115**, 3458–3474.

DeRiemer, L. H. *et al.* (1981). BLEDTA II: synthesis of a new tumor-visualizing derivative of CoIII-bleomycin. *J. Labelled Compd. Radiopharm.*, **18**, 1517–1533.

De Weger, R. A. *et al.* (1982). Eradication of murine lymphoma and melanoma cells by chlorambucil antibody complexes. *Immunol. Rev.*, **62**, 29–45.

Diamandis, E. P. (1991). Multiple labeling and time-resolvable fluorophores. *Clin. Chem.*, **37**, 1486–1491.

Diamandis, E. P. *et al.* (1989). Multiple fluorescence labeling with europium chelators. Application to time-resolved fluoroimmunoassays. *Anal. Chem.*, **61**, 48–53.

Diener, E. *et al.* (1986). Specific immunosuppression by immunotoxins containing daunomycin. *Science*, **231**, 148–150.

Dillman, R. O. *et al.* (1988). Superiority of an acid-labile daunorubicin–monoclonal antibody immunoconjugate compared to a free drug. *Cancer Res.*, **48**, 6097–6102.

Dixon, W. R. *et al.* (1975). Radioimmunoassay of chlordiazepoxide in plasma. *J. Pharm. Sci.*, **64**, 937–939.

Dockter, M. E. (1979). Fluorescent photochemical surface labeling of intact human erythrocytes. *J. Biol. Chem.*, **254**, 2161–2164.

Donabedian, R. K. *et al.* (1972). Micro-electrolytic iodination of polypeptide hormones for radio-immunoassay. *Clin. Chim. Acta*, **36**, 517–520.

Donadio, S. *et al.* (1985). Alkylation of amide linkages and cleavage of the C chain in the enzyme-activated-substrate inhibition of α-chymotrypsin with *N*-nitrosamides. *Biochemistry*, **24**, 2447–2458.

Doran, D. M. and Spar, I. L. (1980). Oxidative iodine monochloride iodination technique. *J. Immunol. Methods*, **39**, 155–163.

Dottavio-Martin, D. and Ravel, J. M. (1978). Radiolabeling of proteins by reductive alkylation with [^{14}C]formaldehyde and sodium cyanoborohydride. *Anal. Biochem.*, **87**, 562–565.

Dourtoglu, V. *et al.* (1978). L'hexafluorophosphate de *O*-benzotriazolyl-*N,N*-tétraméthyluronium: un réactif de couplage peptidique nouveau et efficace. *Tetrahedron Lett.*, **15**, 1269–1272.

Drake, R. R. *et al.* (1992). Application of an *N*-(4-azido-2,3,5,6-tetrafluorobenzoyl)tyrosine-substituted peptide as a heterobifunctional cross-linking agent in a study of protein *O*-glycosylation in yeast. *Bioconjugate Chem.*, **3**, 69–73.12

Dray, F. *et al.* (1975). Enzyme immunoassay of progesterone at the picogram level using β-galactosidase as label. *Biochim. Biophys. Acta*, **403**, 131–138.

Duijndam, W. A. L. *et al.* (1988). A simple method for labelling the carbohydrate moieties of antibodies with fluorochromes. *J. Immunol. Methods*, **109**, 289–290.

Duke, J. *et al.* (1971). The reaction of *p*-mercuribenzoate with methemerythrin. *Biochim. Biophys. Acta*, **229**, 155–166.

Ebright, Y. W. *et al.* (1992). Incorporation of an EDTA–metal complex at a rationally selected site within a protein: application to EDTA–iron DNA affinity cleaving with catabolite gene activator protein (CAP) and Cro. *Biochemistry*, **31**, 10664–10670.

Ebright, Y. W. *et al.* (1993). *N*-(Iodoacetyl)-*p*-phenylenediamine–EDTA: a reagent for high-efficiency incorporation of an EDTA–metal complex at a rationally selected site within a protein. *Bioconjugate Chem.*, **4**, 219–225.

Eckelman, W. C. *et al.* (1975). New compounds: fatty acid and long chain hydrocarbon derivatives containing a strong chelating agent. *J. Pharm. Sci.*, **64**, 704–706.

Eckhardt, A. E. *et al.* (1976). A sensitive fluorescent method for the detection of glycoproteins in polyacrylamide gels. *Anal. Biochem.*, **73**, 192–197.

Edman, P. (1950). Method for determination of the amino acid sequence in peptides. *Acta Chem. Scand.*, **4**, 283–293.

Edman, P. (1953). Note on the stepwise degradation of peptides *via* phenylthiohydantoins. *Acta Chem. Scand.*, **7**, 700–701.

Edman, P. (1956). On the mechanism of the phenyl isothiocyanate degradation of peptides. *Acta Chem. Scand.*, **10**, 761–768.

Edman, P. and Begg, G. (1967). A protein sequenator. *Eur. J. Biochem.*, **1**, 80–91.

Eisen, H. N. and Keston, A. S. (1949). The immunologic reactivity of bovine serum albumin labelled with trace amounts of radioactive iodine (I^{131}). *J. Immunol.*, **63**, 71–80.

Elliston, J. F. *et al.* (1987). Ketononestrol aziridine, an agonistic estrogen receptor affinity label: study of its bioactivity and estrogen receptor covalent labeling. *Endocrinology*, **121**, 667–676.

Emery, A. R. and Gold, V. (1950). Quantitative studies of the reactivities of mixed carboxylic anhydrides. Parts I–III. *J. Chem. Soc.*, 1443–1460.

Eng, P. R. and Parkes, C. O. (1974). SDS electrophoresis of fluorescamine-labeled proteins. *Anal. Biochem.*, **59**, 323–325.

Erlanger, B. F. and Beiser, S. M. (1964). Antibodies specific for ribonucleosides and ribonucleotides and their reaction with DNA. *Proc. Natl. Acad. Sci. USA*, **52**, 68–74.

Erlanger, B. F. *et al.* (1957). Steroid–protein conjugates. I. Preparation and characterization of conjugates of bovine serum albumin with testosterone and with cortisone. *J. Biol. Chem.*, **228**, 713–727.

Erlanger, B. F. *et al.* (1959). Steroid–protein conjugates. *J. Biol. Chem.*, **234**, 1090–1094.

Erlanger, B. F. *et al.* (1967). The preparation of steroid–protein conjugates to elicit antihormonal antibodies. *Methods Immunol. Immunochem.*, **1**, 144–150.

Ermácora, M. R. *et al.* (1992). Conformation-dependent cleavage of staphylococcal nuclease with a disulfide-linked iron chelate. *Proc. Natl. Acad. Sci. USA*, **89**, 6383–6387.

Eruk, A. A. *et al.* (1984). The preparation and characterization of alkaline phosphatase–thyroxine conjugates and their use in the enzyme immunoassay of thyroxine. *Ann. Clin. Biochem.*, **21**, 434–443.

Esch, F. S. and Allison, W. S. (1978). Identification of a tyrosine residue at a nucleotide binding site in the β subunit of the mitochondrial ATPase with *p*-fluorosulfonyl[^{14}C]benzoyl-5′-adenosine. *J. Biol. Chem.*, **253**, 6100–6106.

Eshdat, Y. *et al.* (1973). Identification of aspartic acid 52 as the point of attachment of an affinity label in hen egg white lysozyme. *J. Biol. Chem.*, **248**, 5892–5898.

Eshhar, Z. *et al.* (1981). Use of monoclonal antibodies to pregnanediol-3α-glucuronide for the development of a solid phase chemiluminescence immunoassay. *Steroids*, **38**, 89–109.

Eskola, J. U. *et al.* (1985). Direct solid-phase time-resolved immunofluorometric assay of cortisol in serum. *Clin. Chem.*, **31**, 1731–1734.

Exley, D. and Abuknesha, R. (1977). The preparation and purification of a β-D-galactosidase-oestradiol-17β conjugate for enzyme immunoassay. *FEBS LETT.*, **79**, 301–304.

Exley, D. and Abuknesha, R. (1978). A highly sensitive and specific enzyme-immunoassay method for oestradiol-17β. *FEBS Lett.*, **91**, 162–165.

Faraj, B. A. and Ali, F. M. (1980). Development and application of a radioimmunoassay for tetracycline. *J. Pharmacol. Exp. Ther.*, **217**, 10–14.

Faulstich, H. and Fiume, L. (1985). Protein conjugates of fungal toxins. *Methods Enzymol.*, **112**, 225–237.

Felton, L. C. and McMillion, C. R. (1961). Chromatographically pure fluorescein and tetramethylrhodamine isothiocyanates. *Anal. Biochem.*, **2**, 178–187.

Feng, P. C. C. *et al.* (1990). Development of an enzyme-linked immunosorbent assay for alachlor and its application to the analysis of environmental water samples. *J. Agric. Food Chem.*, **38**, 159–163.

Feng, P. C. C. *et al.* (1992). A general method for developing immunoassays to chloroacetanilide herbicides. *J. Agric. Food Chem.*, **40**, 211–214.

Fenton, J. W., II, and Singer, S. J. (1965). Affinity labeling of antibodies to the *p*-azophenyltrimethylammonium hapten and a structural relationship among antibody active sites of different specificities. *Biochem. Biophys. Res. Commun.*, **20**, 315–320.

Fenton, J. W., II, and Singer, S. J. (1971). Affinity labeling of antibodies to the *p*-azophenyltrimethylammonium hapten, and a comparison of affinity-labeled antibodies of two different specificities. *Biochemistry*, **10**, 1429–1437.

Ferguson, K. M. and Jeffcoate, S. L. (1979). The effect of purity of LH tracer on the radioimmunoassay of LH in human serum. *Acta Endocrinol.*, 225 **(Suppl)**, 130.

Ferguson, S. J. *et al.* (1974). An unusual and reversible chemical modification of soluble beef heart mitochondrial ATPase. *FEBS Lett.*, **38**, 234–236.

Ferrua, B. *et al.* (1986). A novel enzyme immunoassay for total thyroxine using immobilized antibodies and hydrophobic chromatography purified thyroxine–peroxidase conjugate. *J. Immunol. Methods*, **87**, 137–143.

Ferrua, B. *et al.* (1994). Measurement of the anti-HIV agent 2′,3′-didehydro-2′,3′-dideoxythymidine (D4T) by competitive ELISA. *J. Immunol. Methods*, **176**, 103–110.

Fersht, A. R. and Sperling, J. (1973). The charge relay system in chymotrypsin and chymotrypsinogen. *J. Mol. Biol.*, **74**, 137–149.

Fert, V. and Baret, A. (1990). Preparation and characterization of xanthine oxidase–antibody and –hapten conjugates for use in sensitive chemiluminescent immunoassays. *J. Immunol. Methods*, **131**, 237–247.

Fields, R. (1972). The rapid determination of amino groups with TNBS. *Methods Enzymol.*, **25**, 464–468.

Fisch, I. *et al.* (1992). Site-specific modification of a fragment of a chimeric monoclonal antibody using reverse proteolysis. *Bioconjugate Chem.*, **3**, 147–153.

Fischer, E. H. *et al.* (1958). The reaction of sodium borohydride with muscle phosphorylase. *J. Am. Chem. Soc.*, **80**, 2906–2907.

Fleet, G. W. J. *et al.* (1969). Affinity labelling of antibodies with aryl nitrene as reactive group. *Nature*, **224**, 511–512.

Flurkey, K. *et al.* (1985). Preparation and characterization of antisera and monoclonal antibodies to serotonergic and dopaminergic ligands. *J. Neuroimmunol.*, **8**, 115–127.

Fontana, A. and Scoffone, E. (1972). Sulfenyl halides as modifying reagents for polypeptides and proteins. *Methods Enzymol.*, **25**, 482–494.

Fournier, A. *et al.* (1988). Applications of BOP reagent in solid phase synthesis. *Int. J. Peptide Protein Res.*, **31**, 86–97.

Fox, T. *et al.* (1994). Derivatization of yeast cytochrome *c* peroxidase with pentammineruthenium III. *Bioconjugate Chem.*, **5**, 14–20.

Fraenkel-Conrat, H. *et al.* (1949). Essential groups for the interaction of ovomucoid (egg white trypsin inhibitor) and trypsin, and for tryptic activity. *J. Biol. Chem.*, **177**, 385–403.

Fraker, P. J. and Speck, J. C., Jr. (1978). Protein and cell membrane iodinations with a sparingly soluble chloroamide, 1,3,4,6-tetrachloro-3α,6α-diphenylglycouril. *Biochem. Biophys. Res. Commun.*, **80**, 849–857.

Francis, G. E. *et al.* (1951). Labelling of proteins with iodine-131, sulphur-35 and phosphorus-32. *Nature*, **167**, 748–751.

François, D. F. *et al.* (1972). Immunoperoxidase localization of antihapten antibodies in rats responding unequally to antigenic stimulation. *J. Histochem. Cytochem.*, **20**, 257–535.

Frank, J. and Schwyzer, R. (1970). Diazoacetyl choline bromide. *Experientia*, **26**, 1207–1209.

Frey, P. A. *et al.* (1971). A reporter group at the active site of acetoacetate decarboxylase. I. Ionization constant of the nitrophenol. *J. Am. Chem. Soc.*, **93**, 7266–7269.

Friedman, M. *et al.* (1970). The chromatographic determination of cystine and cysteine residues in proteins as *S*-*β*-(4-pyridylethyl) cysteine. *J. Biol. Chem.*, **245**, 3868–3871.

Friedman, M. *et al.* (1974). Reductive alkylation of proteins with aromatic aldehydes and sodium cyanoborohydride. *Int. J. Peptide Protein Res.*, **6**, 183–185.

Friedmann, E. *et al.* (1948). Sulphydryl addition compounds of some quinones and related substances in their action on the growth of normal cells. *Br. J. Pharmacol.*, **3**, 335–340.

Fuchs, S. and Fuchs, Y. (1969). Immunological assay for plant hormones using spcific antibodies to indoleacetic acid and gibberellic acid. *Biochim. Biophys. Acta*, **192**, 528–530.

Fujiwara, K. *et al.* (1981). Novel preparation method of immunogen for hydrophobic hapten, enzyme immunoassay for daunomycin and adriamycin. *J. Immunol. Methods*, **45**, 195–203.

Fujiwara, K. and Saita, T. (1987). The use of *N*-[*β*-(4-diazophenyl)ethyl]maleimide as a heterobifunctional agent in developing enzyme immunoassay for neurotensin. *Anal. Biochem.*, **161**, 157–163.

Gadow, A. *et al.* (1984). Synthesis and evaluation of luminescent tracers and hapten–protein conjugates for use in luminescence immunoassays with immobilised antibodies and antigens. *J. Clin. Chem. Clin. Biochem.*, **22**, 337–347.

Gaertner, H. F. *et al.* (1992). Construction of protein analogues by site-specific condensation of unprotected fragments. *Bioconjugate Chem.*, **3**, 262–268.

Gahmberg, C. G. and Andersson, L. C. (1977). Selective radioactive labeling of cell surface sialoglycoproteins by periodate-tritiated borohydride. *J. Biol. Chem.*, **252**, 5888–5894.

Gahmberg, C. G. and Hakomori, S. (1973). External labeling of cell surface galactose and galactosamine in glycolipid and glycoprotein of human erythrocytes. *J. Biol Chem.*, **248**, 4311–4317.

Gahmberg, C. G. and Hakomori, S. (1976). Organization of glycoprotein and glycolipid in the plasma membrane of normal and transformed cells as revealed by galactose oxidase. *Biomembranes*, **8**, 131–165.

Galardy, R. E. *et al.* (1973). Benzophenone triplet: a new photochemical probe of biological ligand–receptor interactions. *Nature New Biol.*, **242**, 127–128.

Galardy, R. E. *et al.* (1974). Photoaffinity labeling of peptide hormone binding sites. *J. Biol. Chem.*, **249**, 3510–3518.

Galembeck, F. *et al.* (1977). Reaction of proteins with formaldehyde in the presence and absence of sodium borohydride. *J. Agric. Food Chem.*, **25**, 238–245.

Gansow, O. A. (1991). Newer approaches to the radiolabeling of monoclonal antibodies by use of metal chelates. *Nucl. Med. Biol.*, **18**, 369–381.

Gansow, O. A. and Kumar, K. (1990). Process for synthesizing macrocyclic chelates. *US Patent 4,923,985*, to the US Department of Health & Human Services.

Gaur, P. K. *et al.* (1981). Production and characterization of aflatoxin B_{2a} antiserum. *Appl. Environ. Microbiol.*, **41**, 478–482.

Geahlen, R. L. *et al.* (1992). A general method for preparation of peptides biotinylated at the carboxy terminus. *Anal. Biochem.*, **202**, 68–70.

Gendloff, E. H. *et al.* (1986). Hapten–protein conjugates prepared by the mixed anhydride method. Cross-reactive antibodies in heterologous antisera. *J. Immunol. Methods*, **92**, 15–20.

Geoghegan, K. F. and Stroh, J. G. (1992). Site-directed conjugation of nonpeptide groups to peptides and proteins *via* periodate oxidation of a 2-amino alcohol. Application to modification at N-terminal serine. *Bioconjugate Chem.*, **3**, 138–146.

Geoghegan, K. F. *et al.* (1979). Reversible reductive alkylation of amino groups in proteins. *Biochemistry*, **18**, 5392–5399.

Geoghegan, K. F. *et al.* (1981). Alternative reducing agents for reductive methylation of amino groups in proteins. *Int. J. Peptide Protein Res.*, **17**, 345–352.

Gerwin, B. (1967). Properties of the single sulfhydryl group of streptococcal proteinase. *J. Biol. Chem.*, **242**, 451–456.

Ghosh, P. B. and Whitehouse, M. W. (1968). 7-Chloro-4-nitrobenzo-2–oxa-1,3-diazole: a new fluorogenic reagent for amino acids and other amines. *Biochem. J.*, **108**, 155–156.

Gilles, M. A. *et al.* (1990). Stability of water-soluble carbodiimides in aqueous solution. *Anal. Biochem.*, **184**, 244–248.

Gilmore, R. C., Jr. *et al.* (1954). Labeling bovine and human albumin with I^{131}. *Nucleonics*, **12**, 65–68.

Gittel, C. and Schmidtchen, F. P. (1995). Directed N-terminal elongation of unprotected peptides catalyzed by cathepsin C in water. *Bioconjugate Chem.*, **6**, 70–76.

Gleizes, P-E. *et al.* (1994). Labeling of basic fibroblast growth factor with digoxigenin: a nonradioactive probe for biochemical and cytological applications. *Anal. Biochem.*, **219**, 360–367.

Goeldner, M. P. *et al.* (1989). Diazocyclohexadienones as photoaffinity ligands: syntheses of trioxabicyclooctane probes for the convulsant binding site of the $GABA_A$ receptor. *Tetrahedron Lett.*, **30**, 823–826.

Goldstein, G. *et al.* (1961). Studies on fluorescent antibody staining. I. Non-specific fluorescence with fluorescein-coupled sheep anti-rabbit globulins. *J. Exp. Med.*, **114**, 89–110.

Golinski, M. *et al.* (1995a). Synthesis, binding affinity and cross-linking of monodentate photoactive phenothiazines to calmodulin. *Bioconjugate Chem.*, **6**, 549–557.

Golinski, M. *et al.* (1995b). Synthesis and binding affinity of bidentate phenothiazines with two different photoactive groups. *Bioconjugate Chem.*, **6**, 567–572.

Goodfriend, T. L. *et al.* (1964). Antibodies to bradykinin and angiotensin: a use of carbodiimides in immunology. *Science*, **144**, 1344–1346.

Goodfriend, T. *et al.* (1966). Immunochemical studies of angiotensin. *Immunochemistry*, **3**, 223–231.

Goon, D. J. W. *et al.* (1994). Glutaramyl-β-alanyl spacer group for haptenic coupling to proteins. Preparation of immunogens for antibody production against polychlorinated biphenyls. *Bioconjugate Chem.*, **5**, 418–422.

Gorecki, M. *et al.* (1971). The conversion of 3-monoazotyrosine to 3-aminotyrosine in peptides and proteins. *Biochim. Biophys. Acta*, **229**, 590–595.

Gorman, J. J. (1987). Fluorescent labeling of cysteinyl residues to facilitate electrophoretic isolation of proteins suitable for amino-terminal sequence analysis. *Anal. Biochem.*, **160**, 376–387.

Gowda, D. C. *et al.* (1996). Benzoylecgonine hydrazides: synthesis, coupling to horseradish peroxidase, and characterization of the conjugates. *Bioconjugate Chem.*, **7**, 265–270.

Grassetti, D. R. *et al.* (1969). The interaction of 6,6′-dithiodinicotinic acid with thiols and with Ehrlich ascites tumor cells. *Biochem. Pharmacol.*, **18**, 603–611.

Gray, G. R. (1974). The direct coupling of oligosaccharides to proteins and derivatized gels. *Arch. Biochem. Biophys.*, **163**, 426–428.

Gray, W. R. (1972). End-group analysis using dansyl chloride. *Methods Enzymol.*, **25**, 121–138.

Green, N. *et al.* (1982). Immunogenic structure of the influenza virus hemagglutinin. *Cell*, **28**, 477–487.

Greenfield, R. S. *et al.* (1990). Evaluation *in vitro* of adriamycin immunoconjugates synthesized using an acid-sensitive hydrazone linker. *Cancer Res.*, **50**, 6600–6607.

Greenwood, F. C. *et al.* (1963). The preparation of ^{131}I-labelled human growth hormone of high specific radioactivity. *Biochem. J.*, **89**, 114–123.

Gregory, J. D. (1955). The stability of N-ethylmaleimide and its reaction with sulfhydryl groups. *J. Am. Chem. Soc.*, **77**, 3922–3923.

Gross, H. and Bilk, L. (1968). Zur Reaktion von *N*-Hydroxysuccinimid mit Dicyclohexylcarbodiimid. *Tetrahedron*, **24**, 6935–6939.

Grossberg, A. L. and Pressman, D. (1963). Effect of acetylation on the active site of several antihapten antibodies: further evidence for the presence of tyrosine in each site. *Biochemistry*, **2**, 90–96.

Grota, L. J. and Brown, G. M. (1974). Antibodies to indolealkylamines: serotonin and melatonin. *Can. J. Biochem.*, **52**, 196–202.

Grota, L. J. and Brown, G. M. (1976). Antibodies to catecholamines. *Endocrinology*, **98**, 615–622.

Grovenstein, E., Jr. *et al.* (1973). Aromatic halogenation. IV. Kinetics and mechanism of iodination of phenol and 2,6-dibromophenol. *J. Am. Chem. Soc.*, **95**, 4261–4270.

Gualerzi, C. *et al.* (1973). Lack of stable initiation factor 3 (IF-3) binding to dimers of the 30 S ribosomal subunits. *FEBS Lett.*, **35**, 313–316.

Guesdon, J.-L. *et al.* (1986). Monoclonal anti-histamine antibody. Preparation, characterization and application to enzyme immunoassay of histamine. *J. Immunol. Methods*, **87**, 69–78.

Haaijman, J. J. *et al.* (1986). Fluorescein and tetramethyl rhodamine as haptens in enzyme immunohistochemistry. *Histochemistry*, **84**, 363–370.

Haas, G. J. and Guardia, E. J. (1968). Production of antibodies against insecticide–protein conjugates. *Proc. Soc. Exp. Biol. Med.*, **129**, 546–551.

Habeeb, A. F. S. A. *et al.* (1958). Molecular structural effects produced in proteins by reaction with succinic anhydride. *Biochim. Biophys. Acta* , **29**, 587–593.

Halpern, S. E. *et al.* (1983). Stability, characterization, and kinetics of ^{111}In-labeled monoclonal antitumor antibodies in normal animals and nude mouse–human tumor models. *Cancer Res.*, **43**, 5347–5355.

Hanin, V. *et al.* (1994). Characterization of a monoclonal antibody produced in an attempt to mimic the active site of HIV aspartyl protease using haptens based on inhibitor models. *J. Immunol. Methods*, **173**, 139–147.

Harlow, K. W. *et al.* (1989). Identification of cysteine 530 as the covalent attachment site of an affinity-labeling estrogen (ketonestrol aziridine) and antiestrogen (tamoxifen aziridine) in the human estrogen receptor. *J. Biol. Chem.*, **264**, 17476–17485.

Hartdegen, F. J. and Rupley, J. A. (1967). The oxidation by iodine of tryptophan 108 in lysozyme. *J. Am. Chem. Soc.*, **87**, 1743–1745.

Hartig, P. R. *et al.* (1977). 5-Iodoacetamidofluorescein-labeled chloroplast coupling factor. 1: Conformational dynamics and labeling-site characterization. *Biochemistry*, **16**, 4275–4282.

Hartley, B. S. and Massey, V. (1956). The active centre of chymotrypsin, I. Labelling with a fluorescent dye. *Biochim. Biophys. Acta*, **21**, 58–70.

Hasnain, S. *et al.* (1977). Topography of *Escherichia coli* ribosomal protein L12 *in situ*. *Eur. J. Biochem.*, **80**, 35–41.

Haugland, R. P. (1970). PhD thesis, Stanford University, Palo Alto, CA.

Haugland, R. P. (1992). *Handbook of Fluorescent Probes and Research Chemicals* (Larison, K. D., ed.), Molecular Probes Inc., Eugene, OR.

Hawkinson, J. E. *et al.* (1991). Photoaffinity ligands for the [^{3}H]TBOB binding site of the $GABA_A$ receptor. *J. Receptor Res.*, **11**, 391–405.

Hayes, C. E. and Goldstein, I. J. (1975). Radioiodination of sulfhydryl-sensitive proteins. *Anal. Biochem.*, **67**, 580–584.

Haynes, R. *et al.* (1967). Modification of amino groups in inhibitors of proteolytic enzymes. *Biochemistry*, **6**, 541–547.

Helmkamp, R. W. *et al.* (1967a). ^{131}I-labeling of proteins by the iodine monochloride method. *Int. J. Appl. Radiat. Isotopes*, **18**, 737–746.

Helmkamp, R. W. *et al.* (1967b). ^{131}I-labeling of proteins at high activity level with ^{131}ICl produced by oxidation of total iodine in Na^{131}I preparations. *Int. J. Appl. Radiat. Isotopes*, **18**, 747–754.

Hendrickson, T. L. *et al.* (1994). Enhanced immunogenicity of leucine enkephalin following coupling to anti-immunoglobulin and anti-CD3 antibodies. *J. Immunol. Methods*, **172**, 165–172.

Hermentin, P. *et al.* (1990). Attachment of rhodosaminylanthracyclinone-type anthracyclines to the hinge region of monoclonal antibodies. *Bioconjugate Chem.*, **1**, 100–107.

Heyduk, E. and Heyduk, T. (1997). Thiol-reactive, luminescent europium chelates: luminescence probes for resonance energy transfer distance measurements in biomolecules. *Anal. Biochem.*, **248**, 216–227.

Hirs, C. H. W. (1956). The oxidation of ribonuclease with performic acid. *J. Biol. Chem.*, **219**, 611–621.

Hirs, C. H. W. *et al.* (1965). Dinitrophenylation and inactivation of bovine pancreatic ribonuclease A. *Arch. Biochem. Biophys.*, **111**, 209–222.

Hirschmann, R. *et al.* (1967). The controlled synthesis of peptides in aqueous medium. III. Use of Leuchs' anhydrides in the synthesis of dipeptides. Mechanism and control of side reactions. *J. Org. Chem.*, **32**, 3415–3425.

Hnatowich, D. J. *et al.* (1982). The preparation and labeling of DTPA-coupled albumin. *Int. J. Appl. Radiat. Isotopes*, **33**, 327–332.

Hnatowich, D. J. *et al.* (1983). Radioactive labeling of antibody: a simple and efficient method. *Science*, **220**, 613–615.

Hnatowich, D. J. *et al.* (1987). Investigations of avidin and biotin for imaging applications. *J. Nucl. Med.*, *28*, 1294–1302.

Hoare, D. G. and Koshland, D. E. (1966). Procedure for the selective modification of carboxyl groups in proteins. *J. Am. Chem. Soc.*, **88**, 2057–2058.

Hoare, D. G. and Koshland, D. E., Jr (1967). A method for the quantitative modification and estimation of carboxylic acid groups in proteins. *J. Biol. Chem.*, **242**, 2447–2453.

Hoebeke, J. *et al.* (1978). The production and characterization of antibodies gainst β-adrenergic antagonists. *Biochem. Pharmacol.*, **27**, 1527–1532.

Hoffman, D. R. and Campbell, D. H. (1969). Model systems for the study of drug hypersensitivity. I. The specificity of the rabbit anti-aspiryl system. *J. Immunol.*, **103**, 655–661.

Hollaway, M. R. *et al.* (1964). The chemical reactivity of the thiol group in the active centre of ficin. *Biochim. Biophys. Acta*, **92**, 111–124.

Holohan, K. N. *et al.* (1973). Enzymic iodination of the histidyl residue of secretin: a radioimmunoassay of the hormone. *Biochim. Biophys. Acta*, **322**, 178–180.

Holohan, K. N. *et al.* (1974). The site of substitution in the imidazole nucleus after the lactoperoxidase-catalysed iodination of histidine residues in polypeptides. *Biochem. Soc. Trans.*, **2**, 739–740.

Holowka, D. (1981). *N*-Chloroacetyl-[^{125}I]iodotyramine: an alkylating agent with high specific activity. *Anal. Biochem.*, **117**, 390–397.

Horinishi, H. *et al.* (1964). States of amino acid residues in proteins. III. Histidine residues in insulin, lysozyme, albumin and proteinases as determined with a new reagent of diazo-1*H*-tetrazole. *Biochim. Biophys. Acta*, **86**, 477–489.

Hosoda, H. *et al.* (1979a). The preparation of steroid *N*-hydroxysuccinimide esters and their reactivities with bovine serum albumin. *Chem. Pharm. Bull.*, **27**, 742–746.

Hosoda, H. *et al.* (1979b). Enzyme labeling of steroids by the activated ester method. *Chem. Pharm. Bull.*, **27**, 2147–2150.

Houen, G. and Jensen, O. M. (1995). Conjugation to preactivated proteins using divinylsulfone and iodoacetic acid. *J. Immunol. Methods*, **181**, 187–200.

Howard, A. N. and Wild, F. (1957). The reactions of diazonium compounds with amino acids and proteins. *Biochem. J.*, **65**, 651–659.

Huber, E. *et al.* (1991). Hapten–protein conjugates and methods of use in immunoassays. *US Patent 4,990,443*, to Boehringer Mannheim GmbH.

Huber, E. *et al.* (1993), Hapten–biotin conjugates and their use. *US Patent 5,219,764*, to Boehringer Mannheim.

Hudson, D. (1988). Methodological implications of simultaneous solid-phase peptide synthesis. 1. Comparison of different coupling procedures. *J. Org. Chem.*, **53**, 617–624.

Hudson, E. N. and Weber, G. (1973). Synthesis and characterization of two fluorescent sulfhydryl reagents. *Biochemistry*, **12**, 4154–4161.

Huestis, W. H. and Raftery, M. A. (1972). A study of cooperative interactions in hemoglobin using fluorine magnetic resonance. *Biochemistry*, **11**, 1648–1654.

Huestis, W. H. and Raftery, M. A. (1978). Bromotrifluoroacetone alkylates hemoglobin at cysteine β93. *Biochem. Biophys. Res. Commun.*, **81**, 892–899.

Hughes, W. L. (1950). The chemistry of iodination. *Ann. NY Acad. Sci.*, **70**, 3–18.

Hughes, W. L., Jr and Straessle, R. (1950). Preparation and properties of serum and plasma proteins. XXIV. Iodination of human serum albumin. *J. Am. Chem. Soc.*, **72**, 452–457.

Hummelen, J. C. *et al.* (1986). Stable 1,2–dioxetanes as labels for thermochemiluminescent immunoassay. *Methods Enzymol.*, **133**, 531–557.

Hunter, W. M. and Greenwood, F. C. (1962). Preparation of iodine-131 labelled human growth hormone of high specific activity. *Nature*, **194**, 495–496.

Hunter, R. (1970). Standardization of the chloramine-T method of protein iodination. *Proc. Soc. Expt. Biol. Med.*, **133**, 989–992.

Hurwitz, E. *et al.* (1975). The covalent binding of daunomycin and adriamycin to antibodies, with retention of both drug and antibody activities. *Cancer Res.*, **35**, 1175–1181.

Hussain, A.A. *et al.* (1993). Chloramine-T in radiolabeling techniques. I. Kinetics and mechanism of the reaction between Chloramine-T and amino acids. *Anal. Biochem.*, **214**, 495–499.

Imai, K. and Watanabe, Y. (1981). Fluorometric determination of secondary amino acids by 7-fluoro-4-nitrobenzo-2-oxa-1,3-diazole. *Anal. Chim. Acta*, **130**, 377–383.

Imai, K. *et al.* (1983). A novel fluorogenic reagent for thiols: ammonium 7-fluorobenzo-2-oxa-1,3-diazole-4-sulfonate. *Anal. Biochem.*, **128**, 471–473.

Ingham, K. C. and Brew, S. A. (1981). Fluorescent labeling of the carbohydrate moieties of human chorionic gonadotropin and α_1-acid glycoprotein. *Biochim. Biophys. Acta*, **670**, 181–189.

Ishii, Y. and Lehrer, S. S. (1986). Effects of the state of the succinimido-ring on the fluorescence and structural properties of pyrene maleimide-labeled $\alpha\alpha$-tropomyosin. *Biophys. J.*, **50**, 75–80.

Isturiz, M. A. *et al.* (1978). Chemotactic antibody. *Science*, **200**, 554–556.

Itaya, K. *et al.* (1975). Cell surface labeling of erythrocyte glycoproteins by galactose oxidase and Mn^{2+}-catalyzed coupling reaction with methionine sulfone hydrazide. *Biochem. Biophys. Res. Commun.*, **64**, 1028–1035.

Jackowski, G. and Liew, C. C. (1980). Fluorescamine staining of nonhistone chromatin proteins as revealed by two-dimensional polyacrylamide gel electrophoresis. *Anal. Biochem.*, **102**, 321–325.

Jacobson, K. A. *et al.* (1995). "Cleavable trifunctional" approach to receptor affinity labeling: chemical regeneration of binding to A_1-adenosine receptors. *Bioconjugate Chem.*, **6**, 255–263.

Jaffe, B. M. *et al.* (1970). The effect of carriers on the production of antibodies to the gastrin tetrapeptide. *Immunochemistry*, **7**, 715–725.

Jaffe, B. M. *et al.* (1971). Radioimmunoassay for prostaglandins. *Science*, **171**, 494–496.

Jalalian, M. *et al.* (1995). Process for preparing immunoconjugates. *US Patent 5,399,672*, to Merck.

Jasanada, F. *et al.* (1996). Indium-111 labeling of low density lipoproteins with DTPA–bis (stearylamide): evaluation as a potential radiopharmaceutical for tumor localization. *Bioconjugate Chem.*, **7**, 72–81.

Jász, Á. and Tóth, G. (1968). Entfernung Anorganischer Jodverunreinigungen aus ^{131}J-Markiertem Serum-Albumin durch Adsorption. *Isotopenpraxis*, **4**, 404–405.

Jay, D. G. (1984). A general procedure for the end labeling of proteins and positioning of amino acids in the sequence. *J. Biol. Chem.*, **259**, 15572–15578.

Jeanson, A. *et al.* (1988). Comparison of conjugation procedures for the preparation of monoclonal antibody–enzyme conjugates. *J. Immunol. Methods*, **111**, 261–270.

Jennings, V. J. (1974). Analytical applications of chloramine-T. *CRC Crit. Rev. Anal. Chem.*, **3**, 407–419.

Jentoft, J. E. *et al.* (1979). ^{13}C NMR studies of ribonuclease A methylated with [^{13}C]formaldehyde. *J. Biol. Chem.*, **254**, 4366–4370.

Jentoft, N. and Dearborn, D. G. (1979). Labeling of proteins by reductive methylation using sodium cyanoborohydride. *J. Biol. Chem.*, **254**, 4359–4365.

Jentoft, N. and Dearborn, D. G. (1980). Protein labeling by reductive methylation with sodium cyanoborohydride: effect of cyanide and metal ions on the reaction. *Anal. Biochem.*, **106**, 186–190.

Jin, L. *et al.* (1995). Use of α-*N*,*N*-bis(carboxymethyl)lysine-modified peroxidase in immunoassays. *Anal. Biochem.*, **229**, 54–60.

Johnson, A. *et al.* (1960). The effect of iodination on antibody activity. *J. Immunol.*, **84**, 213–220.

Johnson, H. M. and Smith, B. G. (1972). Haptenic relationships of *p*-azobenzenesulfonate and some structurally-related food dyes. *Immunochemistry*, **9**, 253–261.

Johnson, J. D. and Schwartz, A. (1978). Dansylaziridine-labeled troponin C. *J. Biol. Chem.*, **253**, 5243–5246.

Johnston, M. F. M. and Eisen, H. N. (1974). Preparation and characterization of antibodies to menadione. *Biochemistry*, **13**, 5547–5552.

Joniau, M. *et al.* (1976). Immunochemical studies on rabbit antibodies which bind rifampicin. *Immunochemistry*, **13**, 715–720.

Jörnvall, H. *et al.* (1978). Probe of β-galactosidase structure with iodoacetate. Differential reactivity of thiol groups in wild-type and mutant forms of β-galactosidase. *Biochemistry*, **17**, 5160–5164.

Jue, R. A. and Doolittle, R. F. (1985). Determination of the relative positions of amino acids by partial specific cleavages of end-labeled proteins. *Biochemistry*, **24**, 162–170.

Kabat, E. A. (1976). Antigens. In *Structural Concepts in Immunology and Immunochemistry*, 2nd edn., pp. 16–41, Holt, Rinehart and Winston, New York.

Kagan, A. and Glick, S. M. (1974). Oxytocin. In *Methods of Hormone Radioimmunoassay* (Jaffe, B. M. and Behrman, H. R., eds), pp. 173–185, Academic Press, New York.

Kagan, H. M. and Vallee, B. L. (1969). Environmental sensitivity of azo chromophores in arsanilazocarboxypeptidase. *Biochemistry*, **8**, 4223–4230.

Kanaoka, Y. *et al.* (1973). Fluorescent thiol reagents. VI. *N*-(1-Anilinonaphthyl-4)maleimide; a fluorescent hydrophobic probe directed to thiol groups in protein. *Biochim. Biophys. Acta*, **317**, 563–568.

Kaneko, T. *et al.* (1991). New hydrazone derivatives of adriamycin and their immunoconjugates: a correlation between acid stability and cytotoxicity. *Bioconjugate Chem.*, **2**, 133–141.

Kankare, J. *et al.* (1992). Immunoassay by time-resolved electrogenerated luminescence. *Anal. Chim. Acta*, **266**, 205–212.

Kapfer, I. *et al.* (1994). Photoactivatable 2-(4′-azidotetrafluorophenyl)-5-*tert*-butyl-1,3-dithiane-bis-sulfone and related compounds as candidate irreversible probes for the GABA-gated chloride channels. *J. Med. Chem.*, **37**, 133–140.

Kapfer, I. *et al.* (1995). Comparative photoaffinity labeling study between azidophenyl, difluoroazidophenyl, and tetrafluoroazidophenyl derivatives for the GABA-gated chloride channels. *Bioconjugate Chem.*, **6**, 109–114.

Kaplan, H. (1972). Determination of the ionization constants and reactivities of the amino-termini of α-chymotrypsin. *J. Mol. Biol.*, **72**, 153–162.

Kaplan, H. *et al.* (1971). Competitive labelling, a method for determining the reactivity of individual groups in proteins. *Biochem. J.*, **124**, 289–299.

Karşilayan, H. *et al.* (1997). Influence of coupling method on the luminescence properties, coupling efficiency, and binding affinity of antibodies labeled with europium (III) chelates. *Bioconjugate Chem.*, **8**, 71–75.

Kasher, J. S. *et al.* (1986). Characterization of an essential arginine residue in the plasma membrane H^+-ATPase of *Neurospora crassa*. *J. Biol. Chem.*, **261**, 10808–10813.

Katritzky, A. R. *et al.* (1984). Pyrylium-mediated transformations of natural products. Part 5. Reactions of gelatin and chymotrypsin with 4-(4-methoxy-3-sulphophenyl)-2,6-bis-(4-sulphophenyl)pyrilium perchlorate. *J. Chem. Soc., Perkin Trans. 2*, 875–878.

Katsh, S. *et al.* (1974). A simple, rapid fluorometric assay for antigens. *J. Immunol. Methods*, **5**, 179–187.

Katz, J. and Bonorris, G. (1968). Electrolytic iodination of proteins with ^{125}I and ^{131}I. *J. Lab. Clin. Med.*, **72**, 966–970.

Katzenellenbogen, J. A. and Hsiung, H. M. (1975). Iodohexestrols. I. Synthesis and photoreactivity of iodinated hexestrol derivatives. *Biochemistry*, **14**, 1736–1741.

Katzenellenbogen, J. A. *et al.* (1973a). Reagents for photoaffinity labeling of estrogen binding proteins. Synthesis of some azide and diazo derivatives of estradiol, estrone, and hexestrol. *J. Org. Chem.*, **38**, 3525–3533.

Katzenellenbogen, J. A. *et al.* (1973b). Photoaffinity labels for estrogen binding proteins of rat uterus. *Biochemistry*, **12**, 4085–4091.

Katzenellenbogen, J. A. *et al.* (1974). Photoreactivity of some light-sensitive estrogen derivatives. Use of an exchange assay to determine their photointeraction with the rat uterine estrogen binding protein. *Biochemistry*, **13**, 2986–2994.

Katzenellenbogen, J. A. *et al.* (1975). Ultraviolet photosensitivity of the estrogen binding protein from rat uterus. wavelength and ligand dependence. Photocovalent attachment of estrogens to protein. *Biochemistry*, **14**, 2310–2316.

Katzenellenbogen, J. A. *et al.* (1977a). Estrogen photoaffinity labels. 1. Chemical and radiochemical synthesis of hexestrol diazoketone and azide derivatives; photochemical studies in solution. *Biochemistry*, **16**, 1964–1970.

Katzenellenbogen, J. A. *et al.* (1977b). Estrogen photoaffiity labels. 2. Reversible binding and covalent attachment of photosensitive hexestrol derivatives to the uterine estrogen receptor. *Biochemistry*, **16**, 1970–1976.

Katzenellenbogen, J. A. *et al.* (1981). Chemically reactive estrogens: synthesis and estrogen receptor interactions of hexestrol ether derivatives and 4-substituted deoxyhexestrol derivatives bearing alkylating functions. *J. Med. Chem.*, **24**, 435–450.

Katzenellenbogen, J. A. *et al.* (1983). Efficient and highly selective covalent labeling of the estrogen receptor with [^{3}H]tamoxifen aziridine. *J. Biol. Chem.*, **258**, 3487–3495.

Kemp, D. S. and Woodward, R. B. (1965). The *N*-ethylbenzisoxazolium cation: I. Preparation and reactions with nucleophilic species. *Tetrahedron*, **21**, 3019–3035.

Kerlavage, A. R. and Taylor, S. S. (1980). Covalent modification of an adenosine 3′:5′-monophosphate binding site of the regulatory subunit of cAMP-dependent protein kinase II with 8-azidoadenosine 3′:5′-monophosphate. *J. Biol. Chem.*, **255**, 8483–8488.

Keston, A. S. (1944). The Schardinger enzyme in biological iodinations. *J. Biol. Chem.*, **153**, 335–336.

Khaw, B. A. *et al.* (1980). Myocardial infarct imaging of antibodies to canine cardiac myosin with indium-111–diethylenetriamine pentaacetic acid. *Science*, **209**, 295–297.

Khorana, H. G. (1953). The chemistry of carbodiimides. *Chem. Rev.*, **53**, 145–166.

Khosravi, M. J. and Morton, R. C. (1991). Novel application of streptavidin–hapten derivatives as protein–tracer conjugate in competitive-type immunoassays involving biotinylated detection probes. *Clin. Chem.*, **37**, 58–63.

Khym, J. X. (1963). The reaction of methylamine with periodate-oxidized adenosine 5′-phosphate. *Biochemistry*, **2**, 344–350.

Kimura, T. *et al.* (1982). New reagents for the introduction of the thiomethyl group at sulfhydryl residues of proteins with concomitant spectrophotometric titration of the sulfhydryls: methyl 3-nitro-2-pyridyl disulfide and methyl 2-pyridyl disulfide. *Anal. Biochem.*, **122**, 274–282.

Kirley, T. L. (1989a). Reduction and fluorescent labeling of cyst(e)ine-containing proteins for subsequent structural analyses. *Anal. Biochem.*, **180**, 231–236.

Kirley, T. L. (1989b). Determination of three disulfide bonds and one free sulfhydryl in the β subunit of (Na,K)-ATPase. *J. Biol. Chem.*, **264**, 7185–7192.

Kitagawa, T. *et al.* (1976). Enzyme-coupled immunoassay of viomycin. *J. Antibiotics*, **29**, 1343–1345.

Kitagawa, T. *et al.* (1978). A new method for preparation of an antiserum to penicillin and its application for novel enzyme immunoassay of penicillin. *J. Biochem.*, **84**, 491–494.

Kitagawa, T. *et al.* (1982). Enzyme immunoassay of blasticidin S with high sensitivity: a new and convenient method for preparation of immunogenic (hapten–protein) conjugates. *J. Biochem.*, **92**, 585–590.

Klapper, M. H. and Klotz, I. M. (1972). Acylation with dicarboxylic acid anhydrides. *Methods Enzymol.*, **25**, 531–536.

Klinman, J. P. (1975). The interaction of an epoxide with yeast alcohol dehydrogenase: evidence for binding and the modification of two active site cysteines by styrene oxide. *Biochemistry*, **14**, 2568–2574.

Klip, A. and Gitler, C. (1974). Photoactive covalent labeling of membrane components from within the lipid core. *Biochem. Biophys. Res. Commun.*, **60**, 1155–1162.

Klueppelberg, U. G. *et al.* (1990). The efficiency of covalent labeling of the pancreatic cholecystokinin receptor using a battery of crosslinkable and photolabile probes. *Receptor*, **1**, 1–11.

Knight, L. C. *et al.* (1981). Comparison of the iodogen and iodine monochloride methods for radioiodination of fibrinogen. *J. Labelled Comp. Radiopharm.*, **18**, 50–51.

Knoll, E. and Wisser, H. (1973). Gewinnung und Charakterisierung von Antikörpern gegen 3,4-Dimethoxyphenyläthylamin. *Clin. Chim. Acta*, **48**, 183–192.

Knorr, R. *et al.* (1989a). New coupling reagents in peptide chemistry. *Tetrahedron Lett.*, **30**, 1927–1930.

Knorr, R. *et al.* (1989b). New coupling reagents in peptide chemistry. In *Peptides 1988* (Jung, G. and Bayer, E., eds), pp. 37–39, Walter de Gruyter & Co., Berlin.

Knorr, R. *et al.* (1991). 1,1,3,3-Tetramethyluronium compounds as coupling reagents in peptide and protein chemistry. In *Peptides 1990* (Giralt, E. and Andreu, D., eds), pp. 62–64, ESCOM Science Publishers, Amsterdam.

Knowles, F. C. (1975). The site of modification of hemoglobin A by bromotrifluoroacetone. *Biochem. Biophys. Res. Commun.*, **65**, 1155–1162.

Koch, Y. *et al.* (1973). Production and characterization of an antiserum to synthetic gonadotropin-releasing hormone. *Biochem. Biophys. Res. Commun.*, **55**, 616–622.

Koehrle, J. *et al.* (1986). Rat liver iodothyronine monodeiodinase. *J. Biol. Chem.*, **261**, 11613–11622.

Kolb, H. J. and Grodsky, G. M. (1971). High specific activity iodination and displacement radioimmunoassay of fructose-1,6-diphosphatase. *Proc. Soc. Exp. Biol. Med.*, **137**, 464–468.

Kolthoff, I. M. *et al.* (1954). Amperometric mercurimetric titration of sulfhydryl groups in biologically important substances. *Anal. Chem.*, **26**, 366–372.

König, W. and Geiger, R. (1970). Eine neue Methode zur Synthese von Peptiden: Aktivierung der Carboxylgruppe mit Dicyclohexylcarbodiimid und 3-Hydroxy-4-oxo-3,4-dihydro-1,2,3-benzotriazin. *Chem. Ber.*, **103**, 2034–2040.

Konishi, K. and Fujioka, M. (1987). Chemical modification of a functional arginine residue of rat liver glycine methyltransferase. *Biochemistry*, **26**, 8496–8502.

Könnecke, A. *et al.* (1990). Direct conversion of porcine insulin into human insulin catalyzed by immobilized trypsin. *Tetrahedron Lett.*, **31**, 989–990.

Kosower, N. S. *et al.* (1979). Bimane fluorescent labels: labeling of normal human red cells under physiological conditions. *Proc. Natl. Acad. Sci. USA*, **76**, 3382–3386.

Kraal, B. and Hartley, B. S. (1978). Reactivity of amino groups in various complexes of the peptide chain elongation factor EF-Tu from *Escherichia coli*. A new method of competitive labelling using reductive methylation. *J. Mol. Biol.*, **124**, 551–564.

Kramer, K. J. and Rupley, J. A. (1973). Sulfanilation of lysozyme by carbodiimide reaction. Reactivities of individual carboxyl groups. *Arch. Biochem. Biophys.*, **156**, 414–425.

Krejcarek, G. E. and Tucker, K. L. (1977). Covalent attachment of chelating groups to macromolecules. *Biochem. Biophys. Res. Commun.*, **77**, 581–585.

Kriwaczek, V. M. *et al.* (1978a). Tobacco mosaic virus as a carrier for small molecules. I. The preparation and characterization of a TMV/α-melanotropin conjugate. *Helv. Chim. Acta*, **61**, 1232–1240.

Kriwaczek, V. M. *et al.* (1978b). Tobacco mosaic virus as a carrier for small molecules. II. Cooperative affinity labeling of membrane vesicles with a TMV angiotensin conjugate. *Helv. Chim. Acta*, **61**, 1241–1245.

Krohn, K. A. *et al.* (1977). Differences in the sites of iodination of proteins following four methods of radioiodination. *Biochim. Biophys. Acta*, **490**, 497–505.

Kubota, I. and Tsugita, A. (1980). A new semi-empirical method for the determination of the subunit molecular weight of a protein. *Eur. J. Biochem.*, **106**, 263–268.

Kuhn, C-S. *et al.* (1992). Efficient photoaffinity labeling of human β-hexosaminidase A. Synthesis and application of 3-azi-1-[(2-acetanido-2-deoxy-1-β-D-glucopyranosyl)-thio] and galactopyranosyl)-thio]butane. *Bioconjugate Chem.*, **3**, 230–233.

Kulkarni, P. N. *et al.* (1981). Covalent binding of methotrexate to immunoglobulins and the effect of antibody-linked drug on tumor growth *in vivo*. *Cancer Res.*, **41**, 2700–2706.

Kurzer, F. and Douraghi-Zadeh, K. (1967). Advances in the chemistry of carbodiimides. *Chem. Rev.*, **67**, 107–152.

Kutner, A. *et al.* (1986). Photoactivatable analogs for labeling 25-hydroxyvitamin D_3 serum binding protein and for 1,25-dihydroxyvitamin D_3 intestinal receptor protein. *Bioinorg. Chem.*, **14**, 134–147.

Kwok, F. and Churchich, J. E. (1980). Interaction between pyridoxal kinase and pyridoxine-5-P oxidase, two enzymes involved in the metabolism of vitamin B_6. *J. Biol. Chem.*, **255**, 882–887.

Kym, P. R. *et al.* (1995). Evaluation of a highly efficient aryl azide photoaffinity labeling reagent for the progesterone receptor. *Bioconjugate Chem.*, **6**, 115–122.

Lad, P. M. *et al.* (1977). Properties of rabbit muscle phosphofructokinase modified with 7-chloro-4-nitrobenz-2-oxa-1,3-diazole. *Biochemistry*, **16**, 4802–4806.

Lamoureux, G. *et al.* (1967). Experimental allergic encephalomyelitis: properties of an iodinated encephalitogenic polypeptide. *Immunochemistry*, **4**, 273–281.

Lamture, J. B. and Wensel, T. G. (1995). Intensely luminescent immunoreactive conjugates of proteins and dipicolinate-based polymeric Tb^{III} chelates. *Bioconjugate Chem.*, **6**, 88–92.

Landsteiner, K. (1945a). *The Specificity of Serological Reactions*, p. 51., Harvard University Press, Cambridge, MA, USA.

Landsteiner, K. (1945b). *The Specificity of Serological Reactions*, p. 52, Harvard University Press, Cambridge, MA, USA.

Landsteiner, K. and Lampl, H. (1918). Über die Abhängigkeit der Serologischen Spezifität von der chemischen Struktur. (Darstellung von Antigenen mit bekannter Chemischer Konstitution der spezifischen Gruppen). XII. Mitteilung über Antigene. *Biochem. Z.*, **86**, 343–394.

Langone, J. J. (1980). Radioiodination by use of the Bolton–Hunter and related reagents. *Methods Enzymol.*, **70**, 221–247.

Langone, J. J. (1981). Radioiodination by use of the Bolton–Hunter and related reagents. *Methods Enzymol.*, **73**, 112–127.

Lau, H. P. *et al.* (1981). Preparation and characterization of aflatoxin B_{2a}-hemiglutarate and its use for the production of antibody against aflatoxin B_1. *J. Food Safety*, **3**, 1–13.

Lauer, R. C. and Erlanger, B. F. (1974). An enzyme-immunoassay of antibody specific for adenosine using β-galactosidase. *Immunochemistry*, **11**, 533–536.

Lawson, W. B. and Schramm, H.-J. (1965). Modification of a methionine residue near the active site of chymotrypsin by *p*-nitrophenyl bromoacetyl-α-aminoisobutyrate. *Biochemistry*, **4**, 377–386.

Layne, W. W. *et al.* (1982). Evaluation of the viability of in-labelled DTPA coupled to fibrinogen. *J. Nucl. Med.*, **23**, 627–630.

Leavis, P. C. and Lehrer, S. S. (1974). A sulfhydryl-specific fluorescent label, *S*-mercuric-*N*-dansylcysteine. Titrations of glutathione and muscle proteins. *Biochemistry*, **13**, 3042–3048.

Lee, J. A. and Fortes, P. A. G. (1985). Labeling of the glycoprotein subunit of (Na, K)ATPase with fluorescent probes. *Biochemistry*, **24**, 322–330.

Leftheris, K. and Mapelli, C. (1993). Facile methods for the C-terminal biotinylation of peptides. In *Peptides 1992 (Proceedings of the 22nd European Peptide Symposium)* (Schneider, C. H. and Eberles, A. N., eds), pp. 349–350, ESCOM, Leiden.

Leung, C. S.-H. and Meares, C. F. (1977). Attachment of fluorescent metal chelates to macromolecules using "bifunctional" chelating agents. *Biochem. Biophys. Res. Commun.*, **75**, 149–155.

Leung, C. S-H. *et al.* (1978). The attachment of metal-chelating groups to proteins: tagging of albumin by diazonium coupling and use of the products as radiopharmaceuticals. *Int. J. Appl. Radiat. Isotopes*, **29**, 687–692.

Levine, L. and Van Vunakis, H. (1970). Antigenic activity of prostaglandins. *Biochem. Biophys. Res. Commun.*, **41**, 1171–1177.

Levy, A. *et al.* (1976). Radioimmunoassay for reserpine. *Life Sci.*, **19**, 1421–1430.

Lewis, J. G. and Elder, P. A. (1985). An enzyme-linked immunosorbent assay (ELISA) for cortisol. *J. Steroid Biochem.*, **22**, 673–676.

Lewis, M. R. *et al.* (1994). A facile, water-soluble method for modification of proteins with DOTA. Use of elevated temperature and optimized pH to achieve high specific activity and high chelate stability in radiolabeled immunoconjugates. *Bioconjugate Chem.*, **5**, 565–576.

Lewis, S. D. and Shafer, J. A. (1974). Ionization of a nitrophenol-containing reporter group at the active site of papain. *Biochemistry*, **13**, 690–698.

Li, B. L. *et al.* (1989). Creation of phosphorylation sites in proteins: construction of a phosphorylatable human interferon α. *Proc. Natl. Acad. Sci. USA*, **86**, 558–562.

Li, C. H. (1942). Kinetics and mechanism of 2,6-diiodotyrosine formation. *J. Am. Chem. Soc.*, **64**, 1147–1152.

Li, M. and Meares, C. F. (1993). Synthesis, metal chelate stability studies, and enzyme digestion of a peptide-linked DOTA derivative and its corresponding radiolabeled immunoconjugates. *Bioconjugate Chem.*, **4**, 275–283.

Li, M. and Selvin, P. R. (1997). Amine-reactive forms of a luminescent diethylenetriaminepentaacetic acid chelate of terbium and europium: attachment to DNA and energy transfer measurements. *Bioconjugate Chem.*, **8**, 127–132.

Li, M. and White, E. H. (1994). Fluorogenic *N*-nitrosoamides: active-site labeling reagents for chymotrypsin-like proteases. *Bioconjugate Chem.*, **5**, 454–458.

Li, M. *et al.* (1994). Labeling monoclonal antibodies with ^{90}yttrium and ^{111}indium-DOTA chelates: a simple and efficient method. *Bioconjugate Chem.*, **5**, 101–104.

Li, Q. X. *et al.* (1991). Development of enzyme-linked immunosorbent assays for 4-nitrophenol and substituted 4-nitrophenols. *J. Agric. Food Chem.*, **39**, 1685–1692.

Lin, T.-Y. (1970). Lactose synthetase. Modification of carboxyl groups in α-lactalbumin. *Biochemistry*, **9**, 984–995.

Lin, T.-Y. and Koshland, D. E., Jr (1969). Carboxyl group modification and the activity of lysozyme. *J. Biol. Chem.*, **244**, 505–508.

Linder, K. E. *et al.* (1991). Technetium labeling of monoclonal antibodies with functionalized BATOs. 1. $TcCl(DMG)_3PITC$. *Bioconjugate Chem.*, **2**, 160–170.

Linford, J. H. and Froese, G. (1978). Comparisons of the chemical and biologic properties of triaziquone and triaziquone–protein conjugates. *J. Natl. Cancer Inst.*, **60**, 307–316.

Link, R. P. *et al.* (1987). Photoaffinity labeling of serum vitamin D binding protein by 3-deoxy-3-azido-25-hydroxyvitamin D_3. *Biochemistry*, **26**, 3957–3964.

Liu, T.-Y. (1967). Demonstration of the presence of a histidine residue at the active site of streptococcal proteinase. *J. Biol. Chem.*, **242**, 4029–4032.

Lowe, I. P. *et al.* (1958). The fluorometric measurement of glutamic decarboxylase and its distribution in brain. *J. Neurochem.*, **3**, 8–18.

Luedtke, R. R. *et al.* (1988). Monoclonal antibodies with high affinity for spiroperidol. *J. Neurochem.*, **50**, 1253–1262.

MacKeen, L. A. *et al.* (1979). Reductive methylation of IF-3 and EFTu with [^{14}C]formaldehyde and sodium cyanoborohydride. *FEBS Lett.*, **101**, 387–390.

Mak, A. S. and Jones, B. L. (1978). Application of *S*-pyridylethylation of cysteine to the sequence analysis of proteins. *Anal. Biochem.*, **84**, 432–440.

March, J. (1985). *Advanced Organic Chemistry. Reactions, Mechanisms and Structure*, 3rd edn., pp. 800–802, Wiley, New York.

Marchalonis, J. J. (1969). An enzymic method for the trace iodination of immunoglobulins and other proteins. *Biochem. J.*, **113**, 299–305.

Mares, A. *et al.* (1995). Synthesis of a novel biotin-estradiol conjugate and its use for the development of a direct, broad range enzyme immunoassay for plasma estradiol. *J. Immunol. Methods*, **183**, 211–219.

Marini, S. *et al.* (1989). A simple method for increasing hapten immunogenicity by a specific structural modification of the carrier. *J. Immunol. Methods*, **120**, 57–63.

Markert-Hahn, C. *et al.* (1994). Immunologisch aktive Konjugate und ein Verfahren zu ihrer Herstellung. *European Patent 618,231* to Boehringer-Mannheim.

Markwell, M. A. K. (1982). A new solid-state reagent to iodinate proteins. I. Conditions for the efficient labeling of antiserum. *Anal. Biochem.*, **125**, 427–432.

Markwell, M. A. K. and Fox, C. F. (1978). Surface-specific iodination of membrane proteins of viruses and eukaryotic cells using 1,3,4,6–tetrachloro-3α, 6α-diphenylglycouril. *Biochemistry*, **17**, 4807–4817.

Marshall, J. J. (1978). Manipulation of the properties of enzymes by covalent attachment of carbohydrate. *Trends Biochem. Sci.*, 3, 79–83.

Marshall, M. and Cohen, P. P. (1980). The essential sulfhydryl group of ornithine transcarbamylases. *J. Biol. Chem.*, **255**, 7296–7300.

Masouredis, S. P. *et al.* (1951). Specificity of radioiodinated (^{131}I) immune globulin as determined by quantitative precipitin reaction. *J. Immunol.*, **66**, 297–302.

Matthews, C. R. *et al.* (1978). Synthesis of pentaammine-ruthenium–histidine complexes in ribonuclease A. *J. Am. Chem. Soc.*, **100**, 2260–2262.

Matthiasson, B. *et al.* (1978). Non-equilibrium enzyme immunoassay of gentamicin. *Clin. Chem.*, **24**, 1770–1773.

Mattox, V. R. *et al.* (1979). A comparison of procedures for attaching steroidal glucosiduronic acids to bovine serum albumin. *J. Steroid Biochem.*, **10**, 167–172.

Maurer, P. H. and Lebovitz, H. (1956). Studies on the antigenicity of modified fluid gelatin. *J. Immunol.*, **76**, 335–341.

Mayberry, W. E. *et al.* (1965). Kinetics of iodination. IV. A comparison of the kinetics of iodination of L-tyrosine and some derivatives. *Biochemistry*, **4**, 2606–2611.

McAdam, D. P. *et al.* (1992). Mono- and polyclonal antibodies to the organophosphate fenitrothion. 1. Approaches to hapten–protein conjugation. *J. Agric. Food Chem.*, **40**, 1466–1470.

McCaman, M. W. and Robins, E. (1962). Fluorimetric method for the determination of phenylalanine in serum. *J. Lab. Clin. Med.*, **59**, 885–890.

McClintock, L. A. and Friedman, M. M. (1945). Utilization of antibody for the localization of metals and dyes in the tissues. *Am. J. Roentgenol. Radium Therapy*, **54**, 704–706.

McConahey, P. J. and Dixon, F. J. (1966). A method of trace iodination of proteins for immunologic studies. *Int. Arch. Allergy*, **29**, 185–189.

McConahey, P. J. and Dixon, F. J. (1980). Radioiodination of proteins by the use of the Chloramine-T method. *Methods Enzymol.*, **70**, 210–213.

McDevitt, H. O. *et al.* (1963). Purification and analysis of fluorescein-labeled antisera by column chromatography. *J. Immunol.*, **90**, 634–642.

McFarlane, A. S. (1956). Labelling of plasma proteins with radioactive iodine. *Biochem. J.*, **62**, 135–143.

McFarlane, A. S. (1958). Efficient trace-labelling of proteins with iodine. *Nature*, 182, p. 53.

McFarlane, A. S. (1963). *In vivo* behavior of ^{131}I-fibrinogen. *J. Clin. Invest.*, **42**, 346–361.

McGuire, J. *et al.* (1965). The experimental generation of antibodies to α-melanocyte stimulating hormone and adrenocorticotropic hormone. *J. Clin. Invest.*, **44**, 1672–1678.

McKinney, R. M. *et al.* (1964). Factors affecting the rate of reaction of fluorescein isothiocyanate with serum proteins. *J. Immunol.*, **93**, 232–242.

McMurray, C. H. and Trentham, D. R. (1969). A new class of chromophoric organomercurials and their reactions with D-glyceraldehyde 3-phosphate dehydrogenase. *Biochem. J.*, **115**, 913–921.

Means, G. E. and Feeney, R. E. (1968). Reductive alkylation of amino groups in proteins. *Biochemistry*, **7**, 2192–2201.

Meares, C. F. *et al.* (1984). Conjugation of antibodies with bifunctional chelating agents: isothiocyanate and bromoacetamide reagents, methods of analysis, and subsequent addition of metal ions. *Anal. Biochem.*, **142**, 68–78.

Meier, K. E. and Ruoho, A. E. (1990). Probing β-adrenergic receptors. *Methods Enzymol.*, **184**, 660–668.

Melvin, T. and Schuster, G. B. (1990). The photochemistry of acetyl-substituted aroyl azides: the design of photolabeling agents for inert sites in hydrophobic regions. *Photochem. Photobiol.*, **51**, 155–160.

Metzger, H. *et al.* (1963). Affinity labeling of the active sites of antibodies to the 2,4-dinitrophenyl hapten. *Biochemistry*, **2**, 979–988.

Milby, K. H. and Zare, R. N. (1984). Antibodies, lasers and chromatography. *Int. Clin. Prod. Rev.*, March/April, 10–20.

Mitchell, R. N. *et al.* (1984). The use of radioactive cysteine methyl ester for labeling glycosylated molecules oxidized by periodate or neuraminidase plus galactose oxidase. *Arch. Biochem. Biophys.*, **229**, 544–554.

Miwa, A. *et al.* (1977). Preparation of specific antibodies to catecholamines and L-3,4-dihydroxyphenylalanine. I. Preparation of the conjugates. *Chem. Pharm. Bull.*, **25**, 1904–1910.

Miyachi, Y. *et al.* (1971). Enzymatic radioiodination of gonadotropins. *J. Clin. Endocrinol. Metab.*, **34**, 23–28.

Miyai, K. *et al.* (1981). Heterologous enzyme immunoassay for measurement of thyroxine in dried blood sample on filter paper. In *Enzyme Immunoassay* (Ishikawa, E. *et al.*, eds), pp. 186-192, Igaku-Shoin, Tokyo.

Modarelli, D. A. *et al.* (1992). Carbene formation, hydrogen migration, and fluorescence in the excited states of dialkyldiazirines. *J. Am. Chem. Soc.*, **114**, 7034–7041.

Monji, N. and Castro, A. (1980). Synthesis of maleimide derivative of cortisol for enzyme coupling in cortisol enzyme immunoassay. *Experientia*, **36**, 347–349.

Monji, N. *et al.* (1978). Maleimide derivative of hapten for coupling to enzyme: a new method in enzyme immunoassay. *Biochem. Biophys. Res. Commun.*, **85**, 671–677.

Monji, N. *et al.* (1980). Quantification of digoxin by enzyme immunoassay: synthesis of a maleimide derivative of digoxigenin succinate for enzyme coupling. *Experientia*, **36**, 1141–1143.

Montgomery, M. R. *et al.* (1975). Determination of diphenylhydantoin in human serum by spin immunoassay. *Clin. Chem.*, **21**, 221–226.

Moore, D. H. and Wolf, W. (1978). Electrochemical radioiodination of estradiol. *J. Labelled Comp. Radiopharm.*, **15**, 443–450.

Moore, G. and Crichton, R. R. (1973). Reductive methylation: a method for preparing functionally active radioactive ribosomes. *FEBS Lett.*, **37**, 74–78.

Moorman, A. R. and Abeles, R. H. (1982). A new class of serine protease inactivators based on isatoic anhydride. *J. Am. Chem. Soc.*, **104**, 6785–6786.

Morell, A. G. *et al.* (1966). Physical and chemical studies on ceruloplasmulin. IV. Preparation of radioactive sialic acid-free ceruloplasmulin labeled with tritium on terminal D-galactose residues. *J. Biol. Chem.*, **241**, 3745–3749.

Morgan, S. *et al.* (1991). Laser flash photolysis study of adamantylidene. *J. Am. Chem. Soc.*, **113**, 2782–2783.

Morphy, J. R. *et al.* (1988). Antibody labelling with functionalised cyclam macrocycles. *J. Chem. Soc., Chem. Commun.*, 156–158.

Morrison, M. (1968). Iodination reactions catalyzed by lactoperoxidase. *Gunma Symp. Endocrinol.*, **5**, 239–246.

Morrison, M. (1970). Iodination of tyrosine: isolation of lactoperoxidase (bovine). *Methods Enzymol.*, **17**, 653–657.

Morrison, M. (1980). Lactoperoxidase-catalyzed iodination as a tool for investigation of proteins. *Methods Enzymol.*, **70**, 214–220.

Morrison, M. *et al.* (1971). Use of lactoperoxidase catalyzed iodination in immunochemical studies. *Immunochemistry*, **8**, 289–297.

Mueller, B. M. *et al.* (1990). Antibody conjugates with morpholinodoxorubicin and acid-cleavable linkers. *Bioconjugate Chem.*, **1**, 325–330.

Mujumdar, R. B. *et al.* (1993). Cyanine dye labeling reagents: sulfo-indocyanine succinimidyl esters. *Bioconjugate Chem.*, **4**, 105–111.

Mullink, H. *et al.* (1987). A rapid and simple hapten conjugation method for monoclonal antibodies to be used in immunoenzyme single and double staining procedures. *J. Immunol. Methods*, **99**, 199–204.

Muramoto, K. *et al.* (1977). Application of fluorescein-isothiocyanate labeling of protein to SDS-electrophoresis and subsequent sequence analysis. *Agric. Biol. Chem.*, **41**, 2059–2062.

Mycek, M. J. *et al.* (1959). Amine incorporation into insulin as catalyzed by transglutaminase. *Arch. Biochem. Biophys.*, **84**, 528–540.

Nakajima, N. and Ikada, Y. (1995). Mechanism of amide formation by carbodiimide for bioconjugation in aqueous media. *Bioconjugate Chem.*, **6**, 123–130

Nakane, P. K. and Kawaoi, A. (1974). Peroxidase-labeled antibody: a new method of conjugation. *J. Histochem. Cytochem.*, **22**, 1084–1091.

Nakayama, T. A. and Khorana, H. G. (1990). Synthesis of a new photoactivatable analogue of 11-*cis*-retinal. *J. Org. Chem.*, **55**, 4953–4956.

Nambara, T. and Hosoda, H. (1981). Preparation of testosterone–enzyme conjugates and their reactivities with anti-testosterone antibody. In *Enzyme Immunoassay* (Ishikawa, E. *et al.*, eds), pp. 106–113, Igaku-Shoin, Tokyo.

Nanami, M. *et al.* (1993). Preparation of fluorescence labeled insulins, sulfobenzoxadiazolyl-insulins, for fluorescence immunoassay. *Biol. Pharm. Bull.*, **16**, 99–102.

Neri, D. *et al.* (1996). Radioactive labeling of recombinant antibody fragments by phosphorylation using human casein kinase II and [$\gamma -^{32} P$]-ATP. *Nature Biotechnol.*, **14**, 485–490.

Nezlin, R. S. and Sykulev, Y. K. (1982). Structural studies of immunoglobulins spin-labeled at the carbohydrate moiety. *Mol. Immunol.*, **19**, 347–356.

Nitta, K. *et al.* (1979). Anomalous reaction of 4-chloro-7-nitrobenzofurazan with thiol compounds. *Biochem. J.*, **177**, 385–392.

Niwa, T. *et al.* (1994). Preparation of antiserum for use in the immunoassay of unconjugated and conjugated 4β-hydroxycholic acids. *Anal. Sci.*, **10**, 349–351.

Ogura, H. *et al.* (1979). A novel active ester synthesis reagent (*N,N′*-disuccinimidyl carbonate). *Tetrahedron Lett.*, **49**, 4745–4746.

Ok, H. *et al.* (1988). Synthesis of optically active 3-diazoacetylretinals with triisopropylphenylsulfonylhydrazone. *Tetrahedron Lett.*, **29**, 2275–2278.

O'Kennedy, R. and Keating, P. (1993). The optimisation of a novel iodide microassay and its application in an immunoassay for human antibody levels in serum. *J. Immunol. Methods*, **163**, 225–231.

Okuyama, T. and Satake, K. (1960). On the preparation and properties of 2,4,6-trinitrophenyl-amino acids and -peptides. *J. Biochem.*, **47**, 454–466.

Olcott, H. S. and Fraenkel-Conrat, H. (1947). Specific group reagents for proteins. *Chem. Rev.*, **41**, 151–197.

O'Leary, M. H. and Samberg, G. A. (1971). Chemical modification of proteins by pyrylium salts. *J. Am. Chem. Soc.*, **93**, 3530–3532.

Oliver, G. C., Jr. *et al.* (1968). The measurement of digitoxin in human serum by radioimmunoassay. *J. Clin. Invest.*, **47**, 1035–1042.

Oosterbaan, R. A. *et al.* (1962). An acetyl-peptide from acetylchymotrypsin. *Biochim. Biophys. Acta*, **63**, 204–206.

Oseroff, A. R. *et al.* (1986). Antibody-targeted photolysis: selective photodestruction of human T-cell leukemia cells using monoclonal antibody–chlorin e_6 conjugates. *Proc. Natl. Acad. Sci. USA*, **83**, 8744–8748.

O'Shannessy, D. J. and Quarles, R. H. (1987). Labeling of the oligosaccharide moieties of immunoglobulins. *J. Immunol. Methods*, **99**, 153–161.

O'Sullivan, M. J. *et al.* (1978). An enzyme-immunoassay for triiodothyronine. In *Enzyme Labelled Immunoassay of Hormones and Drugs. Proceedings of the International Symposium on Enzyme Labelled Immunoassay of Hormones and Drugs, Ulm, West Germany, July 10 and 11, 1978* (Pal, S. B., ed.), pp. 301–310, Walter de Gruyter, Berlin.

O'Sullivan, M. J. *et al.* (1979). Enzyme immunoassay: a review. *Ann. Clin. Biochem.*, **16**, 221–239.

Paek, S.-H. *et al.* (1993). Defined analyte-enzyme conjugates as signal generators in immunoassays. *Anal. Biochem.*, **210**, 145–154.

Paik, C. H. *et al.* (1980). Synthesis, plasma clearance, and in vitro stability of protein containing a conjugated indium-111 chelate. *J. Radioanal. Chem.*, **57**, 553–564.

Paik, C. H. *et al.* (1983). Factors influencing DTPA conjugation with antibodies by cyclic DTPA anhydride. *J. Nucl. Med.*, **24**, 1158–1163.

Palumbo, P. S. (1993). Novel compounds and conjugates. *World Patent WO 93/22677* to PB Diagnostic Systems, Inc.

Pandurangi, R. S. *et al.* (1995). High efficiency photolabeling of human serum albumin and human γ-globulin with [^{14}C]methyl 4-azido-2,3,5,6-tetrafluorobenzoate. *Bioconjugate Chem.*, **6**, 630–634.

Parker, D. J. and Allison, W. S. (1969). The mechanism of inactivation of glyceraldehyde 3-phosphate dehydrogenase by tetrathionate, *o*-iodosobenzoate, and iodine monochloride. *J. Biol. Chem.*, **244**, 180–189.

Patchornik, A. *et al.* (1970). Photosensitive protecting groups. *J. Am. Chem. Soc.*, **92**, 6333–6335.

Patel, A. *et al.* (1982). A new chemiluminescent label for use in immunoassay. *Biochem. Soc. Trans.*, **10**, 224–225.

Paul, R. and Anderson, G. W. (1962). *N,N′*-Carbonyldiimidazole in peptide synthesis. III. A synthesis of isoleucine-5 angiotensin II amide-1. *J. Org. Chem.*, **27**, 2094–2099.

Pauly, H. (1904). Über die Konstitution des Histidins. *Hoppe-Seyler's Z. Physiol.*, **42**, 508–518.

Pauly, H. (1915). Zur Kenntnis der Diazoreaktion des Eiweisses. *Hoppe-Seyler's Z. Physiol.*, **94**, 294–290.

Paxton, R. J. *et al.* (1985). High-specific-activity ^{111}In-labeled anticarcinoembryonic antigen monoclonal antibody: improved method for the synthesis of diethylenetriaminepentaacetic acid conjugates. *Cancer Res.*, **45**, 5694–5699.

Paxton, R. J. *et al.* (1992). Carboranyl peptide–antibody conjugates for neutron-capture therapy: preparation, characterization and *in vivo* evaluation. *Bioconjugate Chem.*, **3**, 241–247.

Paz, M. A. *et al.* (1970). Amino alcohols as products of a reductive side reaction of denatured collagen with sodium borohydride. *Biochemistry*, **9**, 2123–2127.

Perfetti, R. B. *et al.* (1976). The chemical modification of papain with 1-ethyl-3-(3-dimethylaminopropyl)carbodiimide. *Biochemistry*, **15**, 1735–1743.

Perlmann, G. E. (1966). Acetylation of pepsin and pepsinogen. *J. Biol. Chem.*, **241**, 153–157.

Pfannschmidt, G. and Schauer, R. (1980). The role of membrane sialyl and galactosyl residues in regulation of the life-time of rabbit erythrocytes. *Hoppe-Seyler's Z. Physiol. Chem.*, **361**, 1683–1695.

Pihl, A. and Lange, R. (1962). The interaction of oxidized glutathione, cystamine monosulfoxide, and tetrathionate with the -SH groups of rabbit muscle D glyceraldehyde 3-phosphate dehydrogenase. *J. Biol. Chem.*, **237**, 1356–1362.

Pinney, K. G. and Katzenellenbogen, J. A. (1991). Synthesis of a tetrafluoro-substituted aryl azide and its protio analogue as photoaffinity labeling reagents for the estrogen receptor. *J. Org. Chem.*, **56**, 3125–3133.

Pinney, K. G. *et al.* (1991). Efficient and selective photoaffinity labeling of the estrogen receptor using two nonsteroidal ligands that embody aryl azide or tetrafluoroaryl azide photoreactive functions. *Biochemistry*, **30**, 2421–2431.

Platz, M. *et al.* (1991). Photolysis of 3-aryl-3-(trifluoromethyl)diazirines: a caveat regarding their use in photoaffinity probes. *Bioconjugate Chem.*, **2**, 337–341.

Polgar, L. and Bender, M. L. (1966). A new enzyme containing a synthetically formed active site. Thiol-subtilisin. *J. Am. Chem. Soc.*, **88**, 3153–3154.

Pouchnik, D. J. *et al.* (1996). Synthesis and spectral characterization of sulfhydryl-reactive fluorescent probes. *Anal. Biochem.*, **235**, 26–35.

Poulos, C. and Ramage, R. (1989). The use of phosphinyl chlorides for carboxyl activation and Nα-amino protection in peptide synthesis. In *Peptides 1988* (Jung, G. and Bayer, E., eds), pp. 34–36, Walter de Gruyter & Co., Berlin.

Pratt, J. J. *et al.* (1978). Chemiluminescence-linked immunoassay. *J. Immunol. Methods*, **21**, 179–184.

Pressman, D. and Keighley, G. (1948). The zone of activity of antibodies as determined by the use of radioactive tracers. *J. Immunol.*, **59**, 141–146.

Preston, J. F. *et al.* (1981). Preparation and characterization of an α-amanitin azo derivative with a free carboxyl group and its bovine serum albumin conjugate. *Arch. Biochem. Biophys.*, **209**, 63–71.

Pugh, E. L. and Horecker, B. L. (1967). The effect of acetylation on the activity of rabbit muscle fructose diphosphate aldolase. *Biochem. Biophys. Res. Commun.*, **26**, 360–365.

Puri, R. N. and Colman, R. W. (1996). A novel method for chemical modification of functional groups other than a carboxyl group in proteins by *N*-ethyl-5-phenylisooxazolium-3'-sulfonate (Woodward's reagent-K): inhibition of ADP-induced platelet responses involves covalent modification of aggregin, an ADP receptor. *Anal. Biochem.*, **240**, 251–261

Quinn, R. P. *et al.* (1979). A sensitive radioimmunoassay for the antiviral agent BW248U [9-(2-hydroxyethoxymethyl)-guanine]. *Anal. Biochem.*, **98**, 319–328.

Ragland, W. L. *et al.* (1974). Fluorometric scanning of fluorescamine-labeled proteins in polyacrylamide gels. *Anal. Biochem.*, **59**, 24–33.

Rana, T. M. and Meares, C. F. (1990). Specific cleavage of a protein by an attached iron chelate. *J. Am. Chem. Soc.*, **112**, 2457–2458.

Rana, T. M. and Meares, C. F. (1991). Iron chelate mediated proteolysis: protein structure dependence. *J. Am. Chem. Soc.*, **113**, 1859–1861.

Ranadive, N. S. and Sehon, A. H. (1967a). Antigenicity of 5-hydroxyindole-3-acetic acid, a derivative of serotonin. 1. Preparation of protein conjugates of 5-hydroxyindole-3-acetic acid. *Can. J. Biochem.*, **45**, 1681–1688.

Ranadive, N. S. and Sehon, A. H. (1967b). Antibodies to serotonin. *Can. J. Biochem.*, **45**, 1701–1710.

Ratner, S. and Clarke, H. T. (1937). The action of formaldehyde upon cysteine. *J. Am. Chem. Soc.*, **59**, 200–206.

Ravid, A. *et al.* (1978). Grafting of triggering sites onto lymphocytes. Requirement of multivalency in the stimulation of dinitrophenyl-modified thymocytes by anti-dinitrophenyl antibody. *Eur. J. Immunol.*, **8**, 289–294.

Raymond, P. *et al.* (1995). Quantification of des-Arg^9-bradykinin using a chemiluminescence enzyme immunoassay:

application to its kinetic profile during plasma activation. *J. Immunol. Methods*, **180**, 247–257.

Reccia, J. *et al.* (1982). Interresidue distance measurements in proteins. fluorescent energy transfer between tryptophans and a $Ru^{III}(NH_3)_5$–histidine complex in α-lytic protease and lysozyme. *Biochim. Biophys. Acta*, **702**, 105–111.

Redeuilh, G. *et al.* (1990). Biotinylestradiol for purification of estrogen receptor. *Methods Enzymol.*, **184**, 292–300.

Redshaw, M. R. and Lynch, S. S. (1974). An improved method for the preparation of iodinated antigens for radioimmunoassay. *J. Endocrinol.*, **60**, 527–528.

Reiner, L. (1930). On the chemical alteration of purified antibody-proteins. *Science*, **72**, 483–484.

Reines, S. A. and Cantor, C. R. (1974). New fluorescent hydrazide reagents for the oxidized 3′-terminus of RNA. *Nucleic Acids Res.*, **1**, 767–786.

Reisberg, M. A. *et al.* (1970). A method for preparing specific fluorescein-conjugated antibody reagents using bentonite immunoadsorbents. *J. Immunol.*, **105**, 1151–1161.

Renthal, R. *et al.* (1987). Fluorescent labeling of bacteriorhodopsin: implications for helix connections. *Biochim. Biophys. Acta*, **897**, 384–394.

Rice, R. H. and Means, G. E. (1971). Radioactive labeling of proteins *in vitro*. *J. Biol. Chem.*, **246**, 831–832.

Rich, D. H. and Gurwara, S. K. (1975). Preparation of a new *o*-nitrobenzyl resin for solid-phase synthesis of *tert*-butyloxycarbonyl-protected peptide acids. *J. Am. Chem. Soc.*, **97**, 1575–1579.

Riehm, J. P. and Scheraga, H. A. (1966). Structural studies of ribonuclease. XXI. The reaction between ribonuclease and a water-soluble carbodiimide. *Biochemistry*, **5**, 99–115.

Riordan, J. F. and Vallee, B. L. (1972a). Acetylation. *Methods Enzymol.*, **25**, 494–499.

Riordan, J. F. and Vallee, B. L. (1972b). *O*-Acetyltyrosine. *Methods Enzymol.*, **25**, 500–506.

Riordan, J. F. *et al.* (1965). *N*-Acetylimidazole: a reagent for the determination of 'free' tyrosyl residues of proteins. *Biochemistry*, **4**, 1758–1765.

Riva, F. *et al.* (1980). Different reactiviy of mitochondrial and cytoplasmic aspartate aminotransferases toward an affinity labeling reagent analog of the coenzyme. *J. Biol. Chem.*, **255**, 9230–9235.

Robertson, D. W. *et al.* (1981). Tamoxifen aziridines: effective inactivation of the estrogen receptor. *Endocrinology*, **109**, 1298–1300.

Roche, J. and Michel, R. (1948). L'action de l'iode sur le tryptophane dans les protéines et a l'état libre. *Biochim. Biophys. Acta*, **2**, 97–103.

Rodgers, R. *et al.* (1978). Homogeneous enzyme immunoassay for cannabinoids in urine. *Clin. Chem.*, **24**, 95–100.

Rodwell, J. D. *et al.* (1986). Site-specific covalent modification of monoclonal antibodies: *in vitro* and *in vivo* evaluations. *Proc. Natl. Acad. Sci. USA*, **83**, 2632–2636.

Roffman, E. *et al.* (1980). Ferritin hydrazide, a novel covalent electron dense reagent for the ultrastructural localization of glycoconjugates. *Biochem. Biophys. Res. Commun.*, **97**, 1192–1198.

Roholt, O. A. and Pressman, D. (1972). Iodination: isolation of peptides from the active site. *Methods Enzymol.*, **25**, 438–449.

Rongen, H. A. H. *et al.* (1994). Chemiluminescence and immunoassays. *J. Pharm. Biomed. Anal.*, **12**, 433–462.

Rosa, U. *et al.* (1964). Labelling of human fibrinogen with ^{131}I by electrolytic iodination. *Biochim. Biophys. Acta*, **86**, 519–526.

Rosa, U. *et al.* (1967). Chemical and biological effects of iodination on human albumin. *Biochim. Biophys. Acta*, **133**, 486–498.

Rose, B. G. *et al.* (1995). Competitive indirect ELISA for ceftiofur sodium and the effect of different immunizing and coating antigen concentrations. *Bioconjugate Chem.*, **6**, 529–535.

Rose, K. *et al.* (1991). Preparation of well-defined protein conjugates using enzyme-assisted reverse proteolysis. *Bioconjugate Chem.*, **2**, 154–159.

Rosenfeld, R. S. *et al.* (1975). Preparation and antigenic properties of androsterone-7–BSA conjugate. *Steroids*, **25**, 153–162.

Roth, M. (1971). Fluorescence reaction for amino acids. *Anal. Chem.*, **43**, 880–882.

Roth, M. (1978). Fluorimetric determination of free hydroxyproline and proline in blood plasma. *Clin. Chim. Acta*, **83**, 273–277.

Rotman, A. *et al.* (1980). Specific labeling of platelet membrane glycoproteins. *FEBS Lett.*, **120**, 85–88.

Rowley, G. L. *et al.* (1975). Mechanism by which antibodies inhibit hapten–malate dehydrogenase conjugates. *J. Biol. Chem.*, **250**, 3759–3766.

Rowley, G. L. *et al.* (1980). Method for conjugating to polyamino compounds employing haloacyl groups and compositions prepared thereby. *US Patent 4,220,722* to Syva Co.

Rubenstein, K. E. and Ullman, E. F. (1974). Enzyme amplification assay. *US Patent 3,817,837* to Syva Co.

Rubenstein, K. E. *et al.* (1972). "Homogeneous" enzyme immunoassay. A new immunochemical technique. *Biochem. Biophys. Res. Commun.*, **47**, 846–851.

Rudinger, J. and Ruegg, U. (1973). Preparation of *N*-succinimidyl 3-(4-hydroxyphenyl)propionate. *Biochem. J.*, **133**, 538–539.

Safavy, A. *et al.* (1997). Conjugation of unprotected trisuccin, *N*-[tris[2-[(*N*- hydroxyamino) carbonyl]ethyl]-methyl]succinamic acid, to monoclonal antibody CC49 by an improved active ester protocol. *Bioconjugate Chem.*, **8**, 766–771.

Sakahara, H. *et al.* (1985). Effect of DTPA conjugation on the antigen binding activity and biodistribution of monoclonal antibodies against α-fetoprotein. *J. Nucl. Med.*, **26**, 750–755.

Salacinski, P. R. P. *et al.* (1981). Iodination of proteins, glycoproteins and peptides using a solid-phase oxidizing reagent, 1,3,4,6–tetrachloro-3α, 6α-diphenyl glycoluril. *Anal. Biochem.*, **117**, 136–146.

Salmain, M. *et al.* (1994). [η^5-Cyclopentadienyl]metal tricarbonyl pyrylium salts: novel reagents for the specific conjugation of proteins with transition organometallic labels. *Bioconjugate Chem.*, **5**, 655–659.

Salituro, F. G. *et al.* (1986). [^{125}I]Iododesethyl tamoxifen aziridine: synthesis and covalent labeling of the estrogen receptor with an iodine-labeled affinity label. *Steroids*, **48**, 287–313.

Samejima, K. *et al.* (1971a). Condensation of ninhydrin with aldehydes and primary amines to yield highly fluorescent ternary products. I. Studies on the mechanism of the reaction and some characteristics of the condensation product. *Anal. Biochem.*, **42**, 222–236.

Samejima, K. *et al.* (1971b). Condensation of ninhydrin with aldehydes and primary amines to yield highly fluorescent ternary products. I. Application to the detection and assay of peptides, amino acids, amines and amino sugars. *Anal. Biochem.*, **42**, 237–247.

Sanger, F. (1945). The free amino groups of insulin. *Biochem. J.*, **39**, 507–515.

Sanger, F. (1949). Fractionation of oxidized insulin. *Biochem. J.*, **44**, 126–128.

Satake, K. *et al.* (1960). The spectrophotometric determination of amine, amino acid and peptide with 2, 4, 6-trinitrobenzene 1-sulfonic acid. *J. Biochem.*, **47**, 654–660.

Satchell, D. P. N. and Satchell, R. S. (1975). Acylation by ketens and isocyanates. A mechanistic comparison. *Chem. Soc. Rev.*, **4**, 231–250.

Sato, H. *et al.* (1995). Synthesis and characterization of superoxide dismutase–deferoxamine conjugate *via* polyoxyethylene: a new molecular device for removal of a variety of reactive oxygen species. *Bioconjugate Chem.*, **6**, 249–254.

Sauer, M. J. *et al.* (1986). Use of progesterone 11-glucuronide-alkaline phosphatase conjugate in a sensitive microtitre-plate enzymeimmunoassay of progesterone in milk and its application to pregnancy testing in dairy cattle. *J. Reprod. Fert.*, **76**, 375–391.

Scheinberg, D. A. *et al.* (1982). Tumor imaging with radioactive metal chelates conjugated to monoclonal antibodies. *Science*, **215**, 1511–1513.

Schmidt, D. J. *et al.* (1990). Monoclonal antibodies for immunoassay of avermectins. *J. Agric. Food chem.*, **38**, 1763 1770.

Schray, K. J. *et al.* (1973). Inactivation of muscle triose phosphate isomerase by D- and L-glycidol phosphate. *J. Biol. Chem.*, **248**, 2214–2218.

Schrenk, W. J. and Kürzinger, K. (1986). Purpose-made enzyme conjugates for quantitative immunoassays. *Med. Lab. Sci.*, **43**, 269–276

Schroeder, H. R. (1982). Luminescent immunoassays and binding assays monitored by chemilumigenic labels. In *Luminescent Assays: Perspectives in Endocrinology and Clinical Chemistry* (Serio, M. and Pazzagli, M., eds), pp. 129–146, Raven Press, New York.

Schubert, M. P. (1936) Compounds of thiol acids with aldehydes. *J. Biol. Chem.*, **114**, 341–350.

Schubiger, P. A. *et al.* (1996). Vehicles, chelators, and radio nuclides: choosing the 'building blocks' of an effective therapeutic radioimmunoconjugate. *Bioconjugate Chem.*, **7**, 165–179.

Scouten, W. H. *et al.* (1974). *N*-Dansylaziridine: a new fluorescent modification for cysteine thiols. *Biochim. Biophys. Acta*, **336**, 421–426.

Seevers, R. H. and Counsell, R. E. (1982). Radioiodination techniques for small organic molecules. *Chem. Rev.*, **82**, 575–590.

Sehgal, D. and Vijay, I. K. (1994). A method for the high efficiency of water-soluble carbodiimide-mediated amidation. *Anal. Biochem.*, **218**, 87–91.

Sehon, A. H. *et al.* (1958). Preparation of an estrone–protein conjugate. In *Abstracts of Papers 133rd Meeting American Chemical Society, San Francisco*, p. 19C, American Chemical Society, Washington, DC.

Sela, M. and Arnon, R. (1972). Reaction with *N*-carboxy-α-amino acid anhydrides. *Methods Enzymol.*, **25**, 553–558.

Seligman, S. J. (1993). Radiolabeling of synthetic peptides by acetylation of the N-terminal amino group. *Anal. Biochem.*, **211**, 324–325.

Sen, R. *et al.* (1982). Synthesis and binding studies of a photoaffinity label for bovine rhodopsin. *J. Am. Chem. Soc.*, **104**, 3214–3216.

Sen, R. *et al.* (1984). Photoaffinity labeling of bovine rhodopsin. *Tetrahedron*, **40**, 493–500.

Seyer, R. *et al.* (1989). Synthesis of a biotinylated, iodinatable, and photoactivatable probe for angiotensin receptors. *Int. J. Peptide Protein Res.*, **34**, 235–245.

Shanahan, M. F. *et al.* (1985). Photoaffinity labeling of the human erythrocyte monosaccharide transporter with an aryl azide derivative of D-glucose. *J. Biol. Chem.*, **260**, 10897–10900.

Shao, M.-C. *et al.* (1990). Complex neoglycoproteins. *Methods Enzymol.*, **184**, 653–659.

Shaper, J. H. and Stryer, L. (1977). Accessibility of the carbohydrate moiety of membrane-bound rhodopsin to enzymatic and chemical modification. *J. Supramol. Struct.*, **6**, 291–299.

Shaw, E. (1972). Site-specific reagents for chymotrypsin, trypsin, and other serine proteases. *Methods Enzymol.*, **25**, 655–660.

Sheehan, J. C. and Hlavka, J. J. (1956). The use of water-soluble and basic carbodiimides in peptide synthesis. *J. Org. Chem.*, **21**, 439–441.

Sheehan, J. C. *et al.* (1961). A convenient synthesis of water-soluble carbodiimides. *J. Org. Chem.*, **26**, 2525–2528.

Shen, W.-C. and Ryser, H. J.-P. (1981). *cis*-Aconityl spacer between daunomycin and macromolecular carriers: a model of pH-sensitive linkage releasing drug from a lysosomotropic conjugate. *Biochem. Biophys. Res. Commun.*, **102**, 1048–1054.

Shiao, D. D. F. *et al.* (1972). Modification of protein properties by change in charge. Succinylated chymotrypsinogen. *Eur. J. Biochem.*, **29**, 377–385.

Shih, L. B. and Bayley, H. (1985). A carbene-yielding amino acid for incorporation into peptide photoaffinity reagents. *Anal. Biochem.*, **144**, 132–141

Shimada, K. *et al.* (1984). Preparation of specific antisera to digoxin using hapten-[C-17]-and [C-12]-bovine serum albumin conjugates. *Chem. Pharm. Bull.*, **32**, 2301–2306.

Shipton, M. *et al.* (1976). 4-Chloro-7-nitrobenzo-2-oxa-1,3-diazole as a reactivity probe for the investigation of the thiol proteinases. *Biochem. J.*, **159**, 235–244.

Simpson, D. M. *et al.* (1987). Desmethylnafoxidine aziridine: an electrophilic affinity label for the estrogen receptor with high efficiency and selectivity. *J. Steroid Biochem.*, **28**, 233–245.

Simpson, J. S. A. *et al.* (1979). A stable chemiluminescent-labelled antibody for immunological assays. *Nature*, **279**, 646–647.

Singh, A. *et al.* (1962). The photolysis of diazoacetylchymotrypsin. *J. Biol. Chem.*, **237**, 3006–3008.

Singh, P. *et al.* (1980). A method for controlled coupling of amino compounds to enzymes: a homogeneous enzyme immunoassay for gentamicin. *Anal. Biochem.*, **104**, 51–58.

Sinha, S. *et al.* (1991). Over-expression of superoxide dismutase and lack of surface-thiols in spermatozoa: inherent defects in oligospermia. *Biochem. Biophys. Res. Commun.*, **174**, 510–517.

Sinha, S. *et al.* (1994). The topography of thiol distribution on sperm surface: a sensitive marker for human fertility diagnosis. *Clin. Chim. Acta*, **225**, 83–84.

Slepnev, V. I. *et al.* (1995). Fatty acid acylated peroxidase as a model for the study of interactions of hydrophobically-modified proteins with mammalian cells. *Bioconjugate Chem.*, **6**, 608–615.

Sluka, J. P. *et al.* (1987). Synthesis of a sequence-specific DNA-cleaving peptide. *Science*, **238**, 1129–1132.

Smallridge, R. C. *et al.* (1981). 3′,5′-Diiodothyronine to 3′-monoiodothyronine conversion in the fed and fasted rat: enzyme characteristics and evidence for two distinct 5′-deiodinases. *Endocrinology*, **108**, 2336–2345.

Smith, A. *et al.* (1993). Preclinical evaluation of ^{67}Cu-labeled intact and fragmented anti-colon carcinoma monoclonal antibody mAb35. *Cancer Res.*, **53**, 5727–5733.

Smith, D. F. *et al.* (1978). Carbohydrate antigens: coupling of oligosaccharide phenethylamine-isothiocyanate derivatives to bovine serum albumin. *Methods Enzymol.*, **50**, 169–171.

Smith, M. L. *et al.* (1962). Modification of fluorescent-antibody procedures employing crystalline tetramethylrhodamine isothiocyanate. *J. Bacteriol.*, **83**, 1358–1359.

Smith, R. A. G. and Knowles, J. R. (1973). Aryldiazirines. Potential reagents for photolabeling of biological receptor sites. *J. Am. Chem. Soc.*, **95**, 5072–5073.

Smith, R. A. G. and Knowles, J. R. (1975). The preparation and photolysis of 3-aryl-3*H*-diazirines. *J. Chem. Soc., Perkin Trans. 2*, 686–694.

Smith, T. W. *et al.* (1970). Characterization of antibodies of high affinity and specificity for the digitalis glycoside digoxin. *Biochemistry*, **9**, 331–337.

Smyth, D. G. *et al.* (1964). Reactions of *N*-ethylmaleimide with peptides and amino acids. *Biochem. J.*, **91**, 589–595.

Sokolovsky, M. and Vallee, B. L. (1966). The reaction of diazonium-1*H*-tetrazole with proteins. Determination of tyrosine and histidine content. *Biochemistry*, **5**, 3574–3581.

Sonoda, S. and Schlamowitz, M. (1970). Studies of ^{125}I trace labeling of immunoglobulin G by Chloramine-T. *Immunochemistry*, **7**, 885–898.

Soundarajan, N. *et al.* (1993). Synthesis and binding of new polyfluorinated aryl azides to α-chymotrypsin. New reagents for photoaffinity labeling. *Bioconjugate Chem.*, **4**, 256–261.

Souppe, J. *et al.* (1988). Application of the reaction of dithioesters with ε-amino groups in lysine to the chemical modification of proteins. *Biochim. Biophys. Acta*, **957**, 254–257.

Spector, S. and Parker, C. W. (1970). Morphine: radioimmunoassay. *Science*, **168**, 1347–1348.

Spendlove, R. S. (1966). Optimal labeling of antibody with fluorescein isothiocyanate. *Proc. Soc. Exp. Biol. Med.*, **122**, 580–588.

Spragg, J. *et al.* (1966). Production of antibody against bradykinin: demonstration of specificity by complement fixation and radio-immunoassay. *J. Immunol.*, **96**, 865–871.

Stadie, W. C. *et al.* (1952). Studies of insulin binding with isotopically labeled insulin. *J. Biol. Chem.*, **199**, 729–739.

Stagg, B. H. *et al.* (1970). Iodination and the biological activity of gastrin. *Nature*, **228**, 58–59.

Stahmann, M. A. and Becker, R. R. (1952). A new method for adding amino acids and peptides to proteins. *J. Am. Chem. Soc.*, **74**, 2695–2696.

Staros, J. V. *et al.* (1986). Enhancement by *N*-hydroxysulfosuccinimide of water-soluble carbodiimide-mediated coupling reactions. *Anal. Biochem.*, **156**, 220–222.

Stavrianopoulos, J. G. (1990). Detectable molecules, method of preparation and use. *US Patent 4,943,523* to Enzo Biochem, Inc.

Steinbach, G. and von Mayersbach, H. (1976). Characterization of fluorescein isothiocyanate. II. Absorption and fluorescence after conjugation to human- and rabbit-γ-globulin and bovine serum albumin. *Acta Histochem.*, **55**, 110–123.

Steiner, A. L. *et al.* (1969). Radioimmunoassay for the measurement of adenosine 3′,5′-cyclic phosphate. *Proc. Natl. Acad. Sci. USA*, **64**, 367–373.

Stephens, R. E. (1975). High-resolution preparative SDS-polyacrylamide gel electrophoresis: fluorescent visualization and electrophoretic elution-concentration of protein bands. *Anal. Biochem.*, **65**, 369–379.

Sternberger, L. A. (1967). Electron microscopic immunocytochemistry: a review. *J. Histochem. Cytochem.*, **15**, 139–159.

Stewart, M. and Diakiw, V. (1978). Electron microscopic location of protein thiol residues. *Nature*, **274**, 184–186.

Stewart, W. W. (1978). Functional connections between cells as revealed by dye-coupling with a highly fluorescent naphthalimide tracer. *Cell*, **14**, 741–759.

Stewart, W. W. (1981). Lucifer dyes: highly fluorescent dyes for biological tracing. *Nature*, **292**, 17–21.

Stöcklin, G. (1977). Bromine-77 and iodine-123 radiopharmaceuticals. *Int. J. Appl. Radiat. Isotopes*, **28**, 131–147.

Strasburger, C. J. and Kohen, F. (1990). Two-site and competitive chemiluminescent immunoassays. *Methods Enzymol.*, **184**, 481–496.

Strottmann, J. M. *et al.* (1983). Advantages of preelectrophoretic conjugation of polypeptides with fluorescent dyes. *Anal. Biochem.*, **132**, 334–337.

Studer, M. and Meares, C. F. (1992). A convenient and flexible approach for introducing linkers on bifunctional chelating agents. *Bioconjugate Chem.*, **3**, 420–423.

Sturgess, M. L. *et al.* (1986). Chemiluminescent labeled-antibody assay for thyroxin in serum, with magnetic separation of the solid-phase. *Clin. Chem.*, **32**, 532–535.

Subramanian, K. M. and Wolf, W. (1990). A new radiochemical method to determine the stability constants of metal chelates attached to a protein. *J. Nucl. Med.*, **31**, 480–488.

Sueyoshi, T. *et al.* (1985). Application of a fluorogenic reagent, ammonium 7-fluorobenzo-2-oxa-1,3-diazole-4-sulfonate for detection of cystine-containing peptides. *J. Biochem.*, **97**, 1811–1813.

Sundberg, M. W. *et al.* (1974). Selective binding of metal ions to macromolecules using bifunctional analogs of EDTA. *J. Med. Chem.*, **17**, 1304–1307.

Sung, M. T. *et al.* (1978). *In situ* atigen localization with fluorescamine-conjugated antibodies: a new method. *Anal. Biochem.*, **84**, 225–230.

Suzawa, T. *et al.* (1994). Multilabeling of ferrocenes to a glucose oxidase–digoxin conjugate for the development of a homogeneous electroenzymatic immunoassay. *Anal. Chem.*, **66**, 3889–3894.

Swaisgood, H. and Natake, M. (1973). Effect of carboxyl group modification on some of the enzymatic properties of L-glutamate dehydrogenase. *J. Biochem.*, **74**, 77–86.

Szendröi, Z. *et al.* (1973). Isotope scanning of the prostate. *Lancet*, **2nd June 1973 issue**, 1252–1253.

Takahashi, K. (1968). The reaction of phenylglyoxal with arginine residues in proteins. *J. Biol. Chem.*, **243**, 6171–6179.

Takenaka, A. *et al.* (1969). States of amino acid residues in proteins. XVIII. A revised way of using diazonium-1*H*-tetrazole for reactivity examination of histidine and tyrosine residues. *Biochim. Biophys. Acta*, **194**, 293–300.

Talbot, D. N. and Yphantis, D. A. (1971). Fluorescent monitoring of SDS gel electrophoresis. *Anal. Biochem.*, **44**, 246–253.

Tamura, T. and Bauer, H. (1982). Monoclonal antibody against the carboxy terminal peptide of pp60^{src} of Rous sarcoma virus reacts with native pp60^{src}. *EMBO J.*, **12**, 1479–1485.

Tateishi, K. *et al.* (1976). Enzyme immunoassay of testosterone using the testosterone–glucoamylase complex. *J. Biochem.*, **80**, 191–194.

Teale, J. D. *et al.* (1974). Production of antibodies to tetrahydrocannabinol as the basis for its radioimmunoassay. *Nature*, **249**, 154–155.

Terouanne, B. *et al.* (1980). Affinity chromatography: a tool for enzyme immunoassay. *J. Immunol. Methods*, **35**, 277–284.

Tesser, G. I. *et al.* (1975). Some new bifunctional reagents for the conjugation of a protein with an amino compound in water. *Hoppe-Seyler's Z. Physiol. Chem.*, **356**, 1625–1629.

The, T. H. and Feltkamp, T. E. W. (1970a). Conjugation of fluorescein isothiocyanate to antibodies. I. Experiments on the conditions of conjugation. *Immunology*, **18**, 865–874.

The, T. H. and Feltkamp, T. E. W. (1970b). Conjugation of fluorescein isothiocyanate to antibodies. II. A reproducible method. *Immunology*, **18**, 875–881.

Thong, B. *et al.* (1985). Lack of specificity of current anti-digoxin antibodies, and preparation of a new, specific polyclonal antibody that recognizes the carbohydrate moiety of digoxin. *Clin. Chem.*, **31**, 1625–1631.

Thorell, J. I. and Johansson, B. G. (1971). Enzymatic iodination of polypeptides with ^{125}I to high specific activity. *Biochim. Biophys. Acta*, **251**, 363–369.

Tijssen, P. and Kurstak, E. (1979). Simple and sensitive method for the purification and peptide mapping of proteins solubilized from densonucleosis virus with sodium dodecyl sulfate. *Anal. Biochem.*, **99**, 97–104.

Timkovich, R. (1977). Polymerization side reactions during protein modifications with carbodiimide. *Biochem. Biophys. Res. Commun.*, **74**, 1463–1468.

Titus, J. A. *et al.* (1982). Texas red, a hydrophilic, red-emitting fluorophore for use with fluorescein in dual parameter flow microfluorometric and fluorescence microscopic studies. *J. Immunol. Methods*, **50**, 193–204.

Tóth, G. (1968). Untersuchung der Jodadsorption an Palladium-oberflächen. *Isotopenpraxis*, **4**, 60–63.

Toyo'oka, T. and Imai, K. (1984). New fluorogenic reagent having halogenobenzofurazan structure for thiols: 4-(aminosulfonyl)-7-fluoro-2,1,3-benzoxadiazole. *Anal. Chem.*, **56**, 2461–2464.

Toyo'oka, T. and Imai, K. (1985). Isolation and characterization of cysteine-containing regions of proteins using 4-(aminosulfonyl)-7-fluoro-2,1,3-benzoxadiazole and high performance liquid chromatography. *Anal. Chem.*, **57**, 1931–1937.

Toyo'oka, T. *et al.* (1989). Fluorogenic reagent for thiols: 4-(*N,N*-dimethylaminosulphonyl)-7-fluoro-2,1,3-benzoxadiazole. *Analyst*, **114**, 413–419.

Treuheit, M. J. and Kirley, T. L. (1993). Reversibility of cysteine labeling by 4-(aminosulfonyl)-7-fluoro-2,1,3-benzoxadiazole. *Anal. Biochem.*, **212**, 138–142.

Trubetskoy, V. S. and Torchilin, V. P. (1995). Fast and specific labeling of antibody fragments with multiple atoms of heavy metal radioisotopes. *Anal. Biochem.*, **229**, 345–347.

Tse, K. S. *et al.* (1976). Antibodies specific to isoniazid and isonicotinic acid. *J. Immunol.*, **116**, 965–969.

Tsuchida, E. *et al.* (1997). Properties of and oxygen binding by albumin–tetraphenylporphyrinatoiron (II) derivative complexes. *Bioconjugate Chem.*, **8**, 534–538.

Udenfriend, S. *et al.* (1972). Fluorescamine: a reagent for assay of amino acids, peptides, proteins, and primary amines in the picomole range. *Science*, **178**, 871–872.

Ullman, E. F. *et al.* (1979). Mechanism of inhibition of malate dehydrogenase by thyroxine derivatives and reactivation by antibodies. *Biochim. Biophys. Acta*, **567**, 66–74.

Utiger, R. D. *et al.* (1962). Studies on human growth hormone. I. a radioimmunoassay for human growth hormone. *J. Clin. Invest.*, **41**, 254–261.

Vaidyanathan, G. *et al.* (1997). Method for radioiodination of proteins using *N*-succinimidyl 3-hydroxy-4-iodobenzoate. *Bioconjugate Chem.*, **8**, 724–729.

Van Ceunebroeck, J. C. *et al.* (1986). Study of a hydrophobic site on bovine α-lactalbumin by labeling with [^{125}I]-TID. *Biochem. Biophys. Res. Commun.*, **138**, 604–610.

Vandekerckhove, J. *et al.* (1985). Protein-blotting on Polybrene-coated glass-fiber sheets. *Eur. J. Biochem.*, **152**, 9–19.

Van Lenten, L. and Ashwell, G. (1971). Studies on the chemical and enzymatic modification of glycoproteins. A general method for the tritiation of sialic acid-containing glycoproteins. *J. Biol. Chem.*, **246**, 1889–1894.

Van Vunakis, H. *et al.* (1975). Radioimmunoassay for anileridine, meperidine and other N-substituted phenylpiperidine carboxylic acid esters. *Res. Commun. Chem. Pathol. Pharmacol.*, **12**, 379–387.

Van Weemen, B. K. and Schuurs, A. H. W. M. (1972). Immunoassay using hapten–enzyme conjugates. *FEBS Lett.*, **24**, 77–81.

Varadarajan, A. and Hawthorne, M. F. (1991). Novel carboranyl amino acids and peptides: reagents for antibody modification and subsequent neutron-capture studies. *Bioconjugate Chem.*, **2**, 242–253.

Varenne, A. *et al.* (1995). Production of specific antibodies and development of a non-isotopic immunoassay for carbamazepine by the carbonyl metallo-immunoassay (CMIA) method. *J. Immunol. Methods*, **186**, 195–204.

Vaughan, J. R., Jr. and Osato, R. L. (1951). Preparation of peptides using mixed carboxylic anhydrides. *J. Am. Chem. Soc.*, **73**, 5553–5555.

Vaughan, J. R., Jr. and Osato, R. L. (1952). The preparation of peptides using mixed carbonic–carboxylic anhydrides. *J. Am. Chem. Soc.*, **74**, 676–678.

Vaughan, R. J. and Westheimer, F. H. (1969). A method for marking the hydrophobic binding sites of enzymes. An insertion into the methyl group of an alanine residue of trypsin. *J. Am. Chem. Soc.*, **91**, 217–218.

Vilaseca, L. A. *et al.* (1993). Protein conjugates of defined structure: synthesis and use of a new carrier molecule. *Bioconjugate Chem.*, **4**, 515–520.

Wadsworth, P. and Salmon, E. D. (1986). Preparation and characterization of fluorescent analogs of tubulin. *Methods Enzymol.*, **134**, 519–528.

Wainer, B. H. *et al.* (1972). Morphine-3-succinyl-bovine serum albumin: an immunogenic hapten–protein conjugate. *Science*, **176**, 1143–1145.

Wajda, I. J. *et al.* (1971). Chromogranins as substrate for transglutaminase. *Biochem. Pharmacol.*, **20**, 3197–3206.

Wang, P. and Giese, R. W. (1995). Phosphate-specific fluorescence labeling of pepsin by BO-IMI. *Anal. Biochem.*, **230**, 329–332.

Waser, P. G. *et al.* (1970). "Affinity labelling" des cholinergischen Rezeptors der motorischen Endplatte mit Diazoacetylcholin. *Experientia*, **26**, 1342–1343.

Watanabe, Y. and Imai, K. (1982). Pre-column labelling for high-performance liquid chromatography of amino acids with 7-fluoro-4-nitrobenzo-2-oxa-1,3-diazole and its application to protein hydrolysates. *J. Chromatogr.*, **239**, 723–732.

Weber, G. (1952). Polarization of the fluorescence of macromolecules. 2. Fluorescent conjugates of ovalbumin and bovine serum albumin. *Biochem. J.*, **51**, 155–167.

Weber, P. and Hof, L. (1975). Introduction of a fluorescent label into the carbohydrate moiety of glycoconjugates. *Biochem. Biophys. Res. Commun.*, **65**, 1298–1302.

Weeks, I. *et al.* (1983). Acridinium esters as high-specific-activity labels in immunoassay. *Clin. Chem.*, **29**, 1474–1479.

Weeks, I. *et al.* (1986). Immunoassays using acridinium esters. *Methods Enzymol.*, **133**, 366–403.

Weidekamm, E. *et al.* (1973). A new sensitive, rapid fluorescence technique for the determination of proteins in gel electrophoresis and in solution. *Anal. Biochem.*, **54**, 102–114.

Weigele, M. *et al.* (1972a). The fluorogenic ninhydrin reaction. Structure of the fluorescent principle. *J. Am. Chem. Soc.*, **94**, 4052–4054.

Weigele, M. *et al.* (1972b). A novel reagent for the fluorometric assay of primary amines. *J. Am. Chem. Soc.*, **94**, 5927–5928.

Weigele, M. *et al.* (1973). Fluorescent labeling of proteins. A new methodology. *Biochem. Biophys. Res. Commun.*, **54**, 899–906.

Weiner, L. M. *et al.* (1963). The detection of humoral antibodies directed against salicylates in hypertensive states. *J. Immunol.*, **90**, 788–792.

Weltman, J. K. *et al.* (1973). *N*-(3-Pyrene)maleimide: a long lifetime fluorescent sulfhydryl reagent. *J. Biol. Chem.*, **248**, 3173–3177.

West, R. (1965). Siegfried Ruhemann and the discovery of ninhydrin. *J. Chem. Educ.*, **42**, 386–387.

Westerberg, D. A. *et al.* (1988). Synthesis of novel bifunctional chelators and their use in preparing monoclonal antibody conjugates for tumor targeting. *J. Med. Chem.*, **32**, 236–243.

Whitaker, J. R. and Jandorf, B. J. (1956). Specific reactions of dinitrofluorobenzene with active groups of chymotrypsin. *J. Biol. Chem.*, **223**, 751–764.

White, B. H. and Cohen, J. B. (1988). Photolabeling of membrane-bound *Torpedo* nicotinic acetylcholine receptor with the hydrophobic probe 3-trifluoromethyl-3-(*m*-[^{125}I]iodophenyl) diazirine. *Biochemistry*, **27**, 8741–8751.

White, B. H. *et al.* (1991). The hydrophobic photoreagent 3-(trifluoromethyl)-3-*m*-([^{125}I]iodophenyl)diazirine is a novel noncompetitive antagonist of the nicotinic acetylcholine receptor. *J. Biol. Chem.*, **266**, 21595–21607.

White, E. H. *et al.* (1968). Interception of carbonium ions in the deamination of 1-norbornylamine. *J. Am. Chem. Soc.*, **90**, 4734–4736.

White, E. H. *et al.* (1975). Active site directed inhibition of enzymes utilizing deaminatively produced carbonium ions. Application to chymotrypsin. *J. Am. Chem. Soc.*, **97**, 2290–2291.

White, E. H. *et al.* (1990). Multiple alkylation and mapping of the active site of α-chymotrypsin by carbonium ions generated with active-site-directed enzyme-activated nitrosoamide substrates. *J. Am. Chem. Soc.*, **112**, 1956–1961.

Wicher, K. *et al.* (1968). Immunologic studies on aspirin. I. Antibodies to aspiryl–protein conjugates. *J. Immunol.*, **101**, 342–348.

Wilchek, M. and Bayer, E. A. (1987). Labeling glycoconjugates with hydrazide reagents. *Methods Enzymol.*, **138**, 429–442.

Wilchek, M. *et al.* (1980). Fluorescent reagents for the labeling of glycoconjugates in solution and on cell surfaces. *Biochem. Biophys. Res. Commun.*, **92**, 1215–1222.

Wilcox, P. E. (1972). Esterification. *Methods Enzymol.*, **25**, 596–615.

Willan, K. J. *et al.* (1977). Specific spin labelling of the Fc region of immunoglobulins. *FEBS Lett.*, **80**, 133–136.

Winkler, J. R. and Gray, H. B. (1992). Electron transfer in ruthenium-modified proteins. *Chem. Rev.*, **92**, 369–379.

Woenckhaus, C. and Duchmann, H. (1975). Reactions of essential sulfhydryl residues of dehydrogenases with nicotinamide-(*S*-methylmercury-thioinosine) dinucleotide. *Z. Naturforsch.*, **30c**, 562–564.

Wofsy, L. *et al.* (1962). Affinity labeling – a general method for labeling the active sites of antibody and enzyme molecules. *Biochemistry*, **1**, 1031–1039.

Wong, C.-H. *et al.* (1979). Enzymic synthesis of opioid peptides. *Biochim. Biophys. Acta*, **576**, 247–249.

Wood, B. T. *et al.* (1965). Fluorescent antibody staining. III. Preparation of fluorescein-isothiocyanate-labeled antibodies. *J. Immunol.*, **95**, 225–229.

Wood, F. T. *et al.* (1975). The radioactive labeling of proteins with an iodinated amidination reagent. *Anal. Biochem.*, **69**, 339–349.

Wood, W. G. *et al.* (1980). Comparison of ^{125}I-labelling of polypeptides using Chloramine-T and 1,3,4,6-tetrachloro-3α,6α-diphenyl glycoluril (Iodogen) with subsequent separation of iodide and labelled protein on cation-exchange dextran. *Fresenius Z. Anal. Chem.*, 301, 119.

Wood, W. G. *et al.* (1986). Luminescence assays for haptens and proteins. *Methods Enzymol.*, **133**, 354–365.

Woodward, R. B. and Olofson, R. A. (1961). The reaction of isoxazolium salts with bases. *J. Am. Chem. Soc.*, **83**, 1007–1009.

Woodward, R. B. *et al.* (1961). A new synthesis of peptides. *J. Am. Chem. Soc.*, 83, 1010–1012.

Wring, S. A. *et al.* (1994). The production and evaluation of a radioligand and antiserum for the radioimmunoassay of sub-nanogram per millilitre concentrations of lamivudine. *J. Pharm. Biomed. Anal.*, **12**, 1573–1583.

Wu, C.-W. and Stryer, L. (1972). Proximity relationships in rhodopsin. *Proc. Natl. Acad. Sci. USA*, **69**, 1104–1108.

Wu, C.-W. *et al.* (1976). *N*-(1-Pyrene)maleimide: a fluorescent cross-linking reagent. *Biochemistry*, **15**, 2863–2868.

Wynne, D. *et al.* (1976). A chemical approach for the localization of membrane sites involved in lymphocyte activation. *Biochem. Biophys. Res. Commun.*, **68**, 730–739.

Yamada, H. *et al.* (1981). Selective modification of aspartic acid-101 in lysozyme by carbodiimide reaction. *Biochemistry*, **20**, 4836–4842.

Yamamoto, K. *et al.* (1977). Fluorescent thiol reagents. XII. *Anal. Biochem.*, **79**, 83–94.

Yamamoto, K. *et al.* (1978). Fluorescent tracer method for protein SH groups. II. *Anal. Biochem.*, **84**, 313–318.

Yamamoto, R. *et al.* (1981). Enzyme immunoassay for thyroxine and triiodothyronine in human serum, with use of a covalent chromatographic separation method. *Clin. Chem.*, **27**, 1721–1723.

Yang, H. M. and Reisfeld, R. A. (1988). Doxorubicin conjugated with a monoclonal antibody directed to a human melanoma-associated proteoglycan suppresses the growth of established tumor xenografts in nude mice. *Proc. Natl. Acad. Sci. USA*, **85**, 1189–1193.

Yankeelov, J. A., Jr. *et al.* (1968). A simple trimerization of 2,3-butanedione yielding a selective reagent for the modification of arginine in proteins. *J. Am. Chem. Soc.*, **90**, 1664–1666.

Yeh, S. M. *et al.* (1979). Decomposition rates of radiopharmaceutical indium chelates in serum. *J. Radioanal. Chem.*, **53**, 327–336.

Yoon, D. Y. *et al.* (1993). Influence of the conjugation site on the specificity of monoclonal antibodies to progesterone and on the performance of direct enzyme immunoassay. *Biochem. Mol. Biol. Int.*, **31**, 553–563.

Youn, H. J. *et al.* (1995). Fluorescence energy transfer immunoassay based on a long-lifetime luminescent metal–ligand complex. *Anal. Biochem.*, **232**, 24–30.

Zablocki, J. A. *et al.* (1987). Estrogenic affinity labels: synthesis, irreversible receptor binding, and bioactivity of aziridine-substituted hexestrol derivatives. *J. Med. Chem.*, **30**, 829–838.

Zalutsky, M. R. and Narula, A. S. (1987). A method for the radiohalogenation of proteins resulting in decreased thyroid uptake of radioiodine. *Appl. Radiat. Isotopes.*, **38**, 1051–1055.

Zalutsky, M. R. *et al.* (1985). Radioiodinated B6.2 monoclonal antibody: further characterization of a potential radiopharmaceutical for the identification of breast tumors. *Int. J. Nucl. Med. Biol.*, **12**, 227–233.

Zalutsky, M. R. *et al.* (1989). Enhanced tumor localization and *in vivo* stability of a monoclonal antibody radioiodinated using *N*-succinimidyl 3-(tri-*n*-butylstannyl)benzoate. *Cancer Res.*, **49**, 5543–5549.

Zaugg, R. H. *et al.* (1977). Schiff base adducts of hemoglobin. Modifications that inhibit erythrocyte sickling. *J. Biol. Chem.*, **252**, 8542–8548.

Zegers, N. *et al.* (1990). An improved conjugation method for controlled covalent coupling of synthetic peptides to proteins using glutaraldehyde in a dialysis method. *J. Immunol. Methods*, **130**, 195–200.

Zhang, L. and Tam, J. P. (1996). Thiazolidine formation as a general and site-specific conjugation method for synthetic peptides and proteins. *Anal. Biochem.*, **233**, 87–93.

Zhang, Y. *et al.* (1996). Quantitative determination of isothiocyanates, dithiocarbamates, carbon disulfide, and related thiocarbonyl compounds by cyclocondensation with 1, 2-benzenedithiol. *Anal. Biochem.*, **239**, 160–167.

Zoller, M. J. and Taylor, S. S. (1979). Affinity labeling of the nucleotide binding site of the catalytic subunit of cAMP-dependent protein kinase using *p*-fluorosulfonyl[^{14}C]-benzoyl 5′-adenosine. *J. Biol. Chem.*, **254**, 8363–8369.

Chapter 7

The Preparation of Protein–Nucleic Acid Conjugates

Adrian Simmonds and Martin Cunningham

Nycomed Amersham plc, White Lion Road, Amersham HP7 9LL, UK

7.1 Introduction

7.1.1 The structure of nucleic acids

The nucleic acids, i.e. deoxyribonucleic acid (DNA) and ribonucleic acid (RNA), are polymers each composed of an extended sequence of four nucleotide monomers. Figure 7.1(a) shows a short stretch of DNA. Each nucleotide is comprised of a pentose sugar phosphate linked to an organic base. The sugar is deoxyribose or ribose for DNA or RNA respectively. The organic base is a purine, viz. adenine (A) or guanine (G), or a pyrimidine, viz. cytosine (C), thymine (T) or uracil (U) (Figure 7.1(b)). Thymine is found only in DNA and uracil is found only in RNA. Nucleotide subunits comprise the sugar linked to the base at the 1′ carbon, and to the phosphate at the 5′ carbon. In the polymerised form the sugar and phosphate form a 'backbone' such that each nucleotide is linked to the 3′-position of the adjacent sugar. Thus the orientation of the chain is described according to the position of 3′-and 5′-hydroxyl groups. The bases project from the sugar–phosphate backbone and are linked to the sugar through the N^9-position of purines and the N^1-position of pyrimidines.

DNA itself is normally found in a double-stranded form in which the two strands wind around each other in a double helix as illustrated in Figure 7.1(c). The

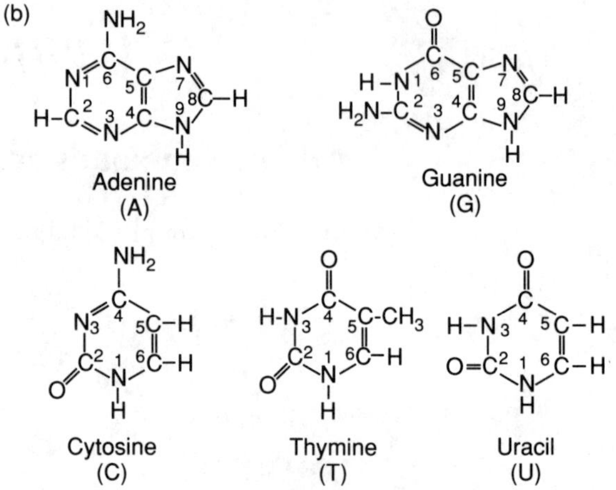

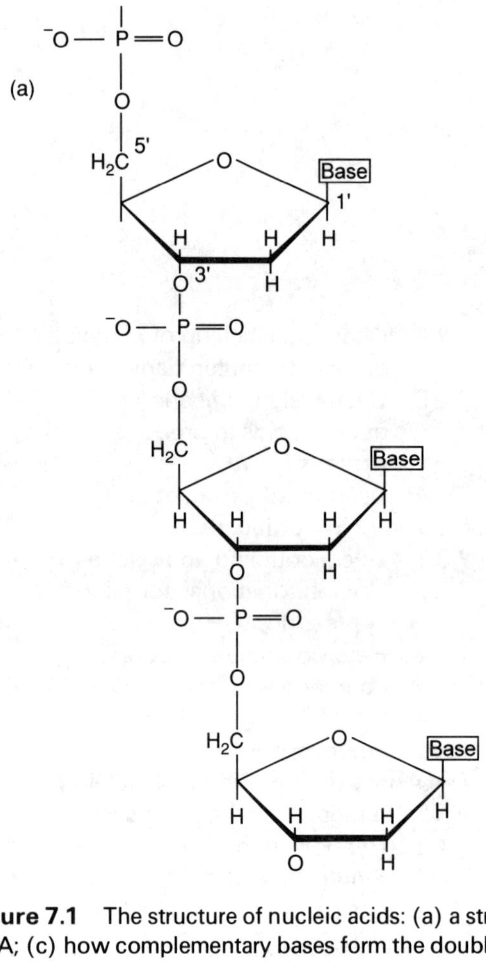

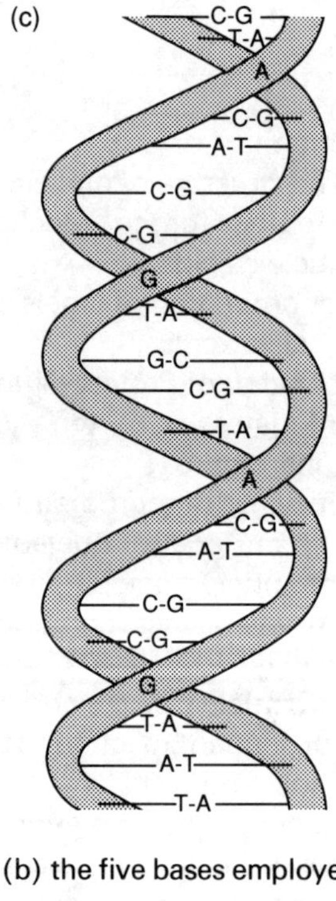

Figure 7.1 The structure of nucleic acids: (a) a stretch of a DNA chain; (b) the five bases employed in DNA and RNA; (c) how complementary bases form the double helix of DNA.

strands are held together primarily due to hydrogen bonding between purine and pyrimidine bases of complementary strands, C bonding to G and A bonding to T. The bases also stack on top of each other and are held together by Van der Waals and hydrophobic interactions (see Chapter 2).

In the human genome there are approximately 3×10^9 base-pairs, i.e. in each nucleated cell of the human body there is a copy of a particular sequence of DNA made up of 3×10^9 base-pairs. Whilst much of the DNA is common to every human being, there are differences which denote gender, hair colour, etc. and the sequence of an individual person's DNA will be unique for that person. The more distantly related individuals are, the greater the divergence in the sequence of their DNA. This divergence is even greater between species so that stretches of DNA (genes) coding for proteins of the same function will have significant sequence variation even between relatively closely related species.

It is possible to distinguish between genes (or other DNA of interest) on the basis of the sequence by matching a known piece of DNA to the unknown. This is done experimentally by hybridising a complementary piece of single-stranded DNA, called the probe, to the unknown target, effectively recreating *in vitro* the double stranded structure of the nucleic acid *via* hydrogen-bonding between complementary bases (see Figures 1.8 and 1.9). Statistics suggest that it is necessary to use a DNA probe which is 17 bases or more in length in order to detect a single sequence in the human genome. Specific nucleic acid sequences are detected by the addition of a label to the probe. Methods which aim to link labels to DNA probes must be compatible with the recognition process and the formation of a double helix which takes place during hybridisation.

While RNA largely occurs naturally as a single-stranded molecule, it too can hybridise with a complementary sequence of either RNA or DNA to form a double-stranded hybrid. In this case U base pairs with A on the complementary strand. It is therefore possible to use RNA as either probe or target in an *in vitro* hybridisation reaction.

The specificity of the hybridisation reaction is governed by the sequence of the nucleic acid. Discrimination between two short 'oligonucleotide' probes (composed typically of 15 to 20 bases) differing only in one base can be achieved. In theory the specificity obtained with nucleic acid hybridisation is greater than that achieved with antibody–antigen interactions where similar epitopes can sometimes be detected in otherwise dissimilar proteins.

7.1.2 Methods for the detection of nucleic acids

The traditional method of detecting nucleic acid probes is through the use of radioactivity. Radioactive phosphorus isotopes are particularly useful in this regard since they can be incorporated into nucleic acids (see below) with no perturbation in the properties of the nucleic acid. They also have a major advantage in that they can be detected at high sensitivity and are in practice the standard against which any alternative methods are compared. Sensitivity is generally a more important parameter in nucleic acid analysis than in immunoassay because levels of nucleic acid are generally much lower than levels of protein; from a single copy gene many RNA transcripts can be produced, each of which may be translated into protein a number of times. Additionally, because of the lower specificity of the antibody–antigen interaction, non-specific interactions are often the limiting factor in any immunoassay before detection sensitivity becomes limiting. Radioactivity has therefore played a major role in the application of hybridisation techniques to nucleic acid analysis.

However, other detection methods are becoming increasingly important in this area, firstly because of the development of more sensitive non-radioactive methods and secondly because molecular biology has been revolutionised by the introduction of nucleic acid amplification techniques, particularly the polymerase chain reaction (PCR) (McPherson *et al.*, 1991). This is a technique which amplifies regions of DNA by DNA polymerase mediated copying (see also Section 9.2.8). The technique uses two oligonucleotide primers which are complementary to opposite strands at the ends of the DNA sequence to be amplified; one oligonucleotide being complementary to the 5′ terminal sequence of the first strand and the other oligonucleotide being complementary to the 5′ terminal sequence of the second strand. The DNA polymerase recognises the primer/template hybrid and adds nucleotides to produce a new strand which is complementary in sequence to the old strand. This part of the process is called the extension reaction and it is followed by heating to denature the double-stranded DNA to individual single strands. Temperature cycling allows hybridisation of primers. extension and denaturation of product and the process is repeated typically up to 25 or 30 times. Depending on the efficiency at each stage, an almost exponential increase in DNA copy number can be achieved for many cycles. This technique has allowed, for example, the amplification of target nucleic acid sequences prior to detection, so that the need for high sensitivity has been reduced.

A final factor which may push people towards the use of non-radioactive methods for nucleic acid detection is regulatory pressure. The potential hazards associated with radioactivity generally are causing authorities to restrict their disposal in conventional waste sites and, although the phosphorus isotopes have fairly short half-lives (14 days for ^{32}P and 25 days for ^{33}P), they too can be included under these restrictions.

Of the various non-radioactive detection methods available, enzymatic methods tend to be among the more sensitive because of the amplification afforded by their catalytic turnover. This, allied with chemiluminescent or fluorescent substrates, provides a very respectable level of sensitivity. Although in most instances not equivalent to that of the phosphorus isotopes, this is sufficient to allow the use of enzymatic methods in a variety of common molecular biology techniques. The main type of protein–nucleic acid complex which will be discussed in this chapter is the enzyme–nucleic acid complex or probe which is used to allow detection of specific nucleic acid sequences. The experimental formats in which these probes can be used to detect nucleic acids are similar to those by which proteins have been traditionally detected with antibodies. Quantification of specific proteins can be achieved by ELISAs which are usually carried out in tubes or microtitre plates. Additional information can be obtained by the use of alternative formats such as Western blotting. The additional information relates to the size of the protein, or immunohisto- and immunocytochemistry, where information can be obtained about localisation of the protein within a tissue or cell.

Similarly, with nucleic acid probes, different experimental formats give different types of information. Blotting has been the most widely used technique in research laboratories for a number of years. In this technique the nucleic acid target, which may be a small fraction of a nucleic acid sample, is immobilised on a membrane and a probe labelled with a marker is used to pick out the specific target nucleic acid by means of hybridisation to its complementary sequence. Southern blotting is the name given to this technique where DNA fragments are first separated according to their size by electrophoresis through a gel matrix and then transferred to a membrane. In Northern blotting RNA fragments are separated and transferred in the same way. Other techniques involving hybridisation have been slower to catch on although they are now becoming increasingly popular, particularly in the clinical area. These consist of solution hybridisation, usually in a tube or microtitre plate, and *in situ* hybridisation which involves localisation of nucleic acid within a tissue or within a chromosome preparation.

7.1.3 Incorporation of labels into nucleic acids

DNA polymerases are enzymes that are involved in the replication of DNA. During this process the two strands of the DNA duplex are separated and two new strands are synthesised using nucleotides complementary to each of the strands. The final result is two identical duplexes. One strategy for labelling DNA involves the attachment of a label to a nucleotide and subsequent incorporation into DNA by a polymerase. Any such modifications to a nucleotide must therefore allow its recognition by the polymerase otherwise it will not be incorporated into the growing DNA strand. The introduction of a ^{32}P or ^{33}P-labelled nucleotide in effect involves no chemical modification of the nucleotide itself and these compounds therefore represent enzymologically ideal labels. No such option is available for non-radioactive labelling. The types of label used for this purpose tend to be small molecules such as haptens, often with a spacer arm between the label and the nucleotide. This has the effect of removing the label away from the active site of the polymerase to avoid interference in nucleotide recognition. Large molecules such as enzymes have not been used in this way. Similar considerations apply to the labelling of RNA using RNA polymerase.

The description of the labelling of nucleic acids with proteins, typically enzymes as reporters, will be divided into indirect and direct labelling methods (see Figure 7.2). Indirect methods (Section 7.2) are those where the protein is linked to an antibody or other binding protein, and the nucleic acid is labelled with a hapten or ligand to the binding protein. Binding of the protein to the nucleic acid therefore relies on the interaction between hapten and antibody or binding protein and partner. (In future references for the sake of convenience we shall include biotin, a frequently used low-molecular-mass label, under the term hapten, even though in almost all cases its binding protein is streptavidin or avidin. Biotin–avidin techniques are discussed in detail in Section 5.11).

Direct methods (Section 7.3), in contrast, involve covalent attachment of the protein to the nucleic acid probe itself.

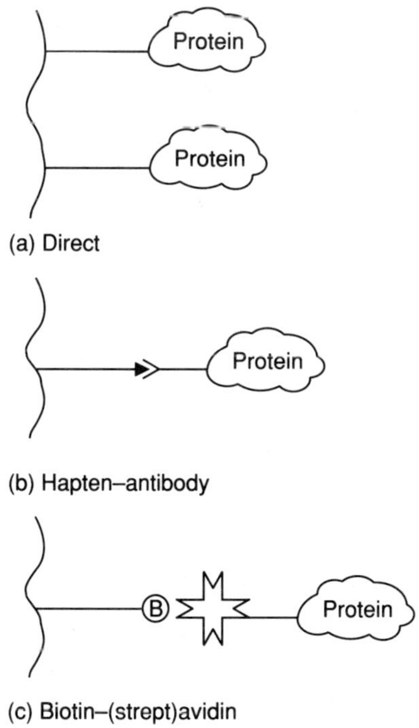

Figure 7.2 Approaches to protein labelling of nucleic acids: the protein may be attached *via* a direct covalent linkage (a), or indirectly *via* a non-covalent binding pair such as a hapten and its specific antibody (b), or biotin and avidin or streptavidin (c).

7.2 The Preparation of Protein–Nucleic Acid Conjugates *via* Hapten–Antibody Interactions, etc.

The haptens most widely used in the labelling of nucleic acid probes are biotin, fluorescein and digoxigenin (Keller and Manak, 1989). This choice has come about to some extent due to their commercial availability and in the case of biotin its high binding coefficient and its proven use in immunoassay. However, it is also important that the hybridisation properties of nucleic acid probes labelled with a hapten are not significantly affected. The parameters which are important in looking at the efficiency of hybridisation are the reassociation rate of complementary nucleic acid strands and the melting temperature, T_m of a nucleic acid. T_m is defined as the temperature at which half of a double-stranded nucleic acid becomes single-stranded. Whilst reassociation rates and T_m are reduced by the addition of hapten groups to nucleic acids, in practice the fidelity of hybridisation does not seem to be significantly altered (Langer *et al.*, 1981; Gebeyehu *et al.*, 1987).

Once hybridisation with a nucleic acid probe has taken place and appropriate washes have been carried out to reduce background binding of surplus probe, it is necessary to apply the antibody or binding protein conjugated to the enzyme. The preparation of these conjugates is described in Chapter 5.

7.2.1 Enzymatic incorporation of hapten-labelled nucleotides into nucleic acids

Nucleotide–hapten conjugates probably provide the easiest means of labelling nucleic acids with haptens. They are incorporated into nucleic acids using polymerase enzymes in much the same way as are radiolabelled nucleotides. For experienced molecular biologists this therefore provides an easy entry into non-radioactive labelling methods.

7.2.1.1 Factors affecting hapten–nucleotide incorporation by enzymes

The first important property of any nucleotide–hapten conjugate is its polymerase acceptability. For probe labelling methods such as nick translation or random prime reactions, the polymerase most widely used is the *E. coli* DNA polymerase I (for random prime labelling the Klenow fragment is used; this lacks the 5′ to 3′ exonuclease activity of the intact enzyme, which progressively excises nucleotides from the DNA chain starting at the 5′-end). More recently, thermostable polymerases such as *Taq* DNA polymerase have been used to generate hapten-labelled probes during PCR reactions. For labelling of RNA the polymerases most widely used are the bacteriophage T7, T3 and SP6 RNA polymerases. For oligonucleotide labelling the enzyme most widely used is terminal deoxynucleotidyl transferase (TdT). In order for a hapten–nucleotide conjugate to be accepted by a polymerase, the position on the nucleotide at which the hapten is chemically linked is of utmost importance. Success has been achieved by labelling through the base, both for purines and pyrimidines. However the position of labelling is also important, because of the orientation of the bases in double-stranded nucleic acids in the centre of the double helix, and the desirability of maintaining the hydrogen bonding between bases on complementary strands.

The hapten whose linkage to nucleoside triphosphates has been the most widely researched is biotin. Of the various positions that can be labelled, the pyrimidine nucleotides dUTP (an analogue of TTP) modified at the C-5 position, dCTP modified at the N^4-position, and the purine nucleotide dATP modified at the N^6-position have been the most widely studied for incorporation into nucleic acids (Langer *et al.*, 1981; Gebeyehu *et al.*, 1987). The length and chemical composition of the linker between the biotin and the base have also been found to be important, not only in the ability of the conjugate to be accepted by the polymerase but also in the final sensitivity of detection, which is presumably related to the ease of binding of streptavidin or avidin (Brigati *et al.*, 1983; Gebeyehu *et al.*, 1987). Both of these proteins have binding sites buried within the structure and so a certain minimum length of linker is necessary for optimal binding. It appears that for U, C or A derivatives, a six to nine carbon atom linker is optimal. Derivatives with longer linkers are not incorporated so readily whilst shorter derivatives may suffer detection problems.

Information relating to other haptens is more limited. Nevertheless, what is known about these other systems tends to back up what is known about the biotin system. Despite the more exposed hapten-binding site of an antibody, a six carbon linker between allyluracil and hapten has been found to give better sensitivities than no linker, but a further increase in linker length provides no additional benefit (Muhlegger *et al.*, 1989; Vincent *et al.*, 1982). One difference between biotin and most of the other haptens used so far is that, while in the biotin system it is sometimes possible to substitute completely the unmodified nucleotide with the nucleotide–hapten conjugate (Langer *et al.*, 1981), when using other haptens it is usually necessary to add a proportion of the unmodified nucleotide along with the conjugate in order to obtain acceptable quantities of probe from the polymerase labelling reaction. This has also been found with less readily incorporated analogues of biotin nucleotides (Gillam and Tener, 1986). In practice though, this makes little difference to the detectability of the probe, since only a proportion of the modified bases can accept an antibody or binding protein due to the relatively large 'footprint' of the protein. Even for biotin the inclusion of more than 35 biotins per kilobase of DNA does not increase the sensitivity of subsequent detection (Gebeyehu *et al.*, 1987). The inclusion of manganese ions in place of the magnesium normally required in polymerase reactions can be used to decrease fidelity of base incorporation and aid the inclusion of modified nucleotides (Gillam and Tener, 1986)

A range of haptens has now been conjugated to nucleotides, predominantly U-derived bases, although some of these 'haptens' have been added merely for their fluorescence and not for their antibody binding properties, e.g. tetramethylrhodamine, aminomethylcoumarinacetic acid (Dirks *et al.*, 1993). It has been suggested that some haptens intercalate into DNA (Davis *et al.*, 1991; Vincent *et al.*, 1982) and that a rigid linker may help to stop this, but as yet there appears to be no direct evidence that this is the case.

7.2.1.2 Methods for hapten–nucleotide incorporation by enzymes

There are a number of polymerase-mediated reactions in common use for the incorporation of nucleotide–hapten conjugates into nucleic acid probes. These are listed in Table 7.1, together with some of the properties of the probes produced. In all of these reactions the proportion of modified to native nucleotide has to be optimised to give a result which meets requirements both in terms of the quantity of probe produced (too little of the native nucleotide will result in a poor yield of probe) and in the extent of modification of the probe (too low a level of modification will result in poor sensitivity whereas too great a level may result in added background, depending on the properties of the hapten). Commercial kits are now available which seek to provide ready-optimised systems for the most popular labelling methods. In particular, there are kits for DNA labelling by nick translation, random prime labelling and oligonucleotide tailing, and RNA labelling by *in vitro* transcription.

The nick translation reaction (Figure 7.3(a)) involves the reaction of two enzymes, pancreatic deoxyribonuclease I (DNase I) and *E. coli* DNA polymerase I (Rigby *et al.*, 1977). The double-stranded DNA template is first 'nicked' by the action of DNase I to produce single stranded breaks, the complementary strands remaining hybridised. This then allows the 5′ to 3′ exonuclease activity of DNA polymerase I to remove nucleotides from the 5′-ends of the fragments and the polymerase activity of the same enzyme to replace nucleotides at the 3′-end of the adjacent fragment. The reaction proceeds in the presence of all four dNTPs and any modified dNTPs will also be incorporated. The size of the probe produced by this reaction will depend on the relative amounts of DNase and polymerase activity present, but is typically in the order of 200 to 500 bases. The flexibility in being able to select the size range of the probes produced makes this an ideal method for labelling probes for *in situ* hybridisation, where the size of the probe is critical for tissue penetration and may be dependent on choice of fixation/permeabilisation protocols.

Table 7.1 Nucleic acid labelling methods

	Random primer	Nick translation	Phage polymerase	Terminal transferase
Enzyme	Klenow polymerase	DNA polymerase I/ DNase I	SP6/T7/T3 RNA polymerase	Terminal deoxynucleotidyl transferase
Template	ssDNA and denatured dsDNA	dsDNA	dsDNA	ssDNA preferred, oligonucleotide
Form of template	Linear	Linear or circular	Linear	Linear
Amount of template	50–500 ng	0.1–1 μg	1–2 μg	~ 10 pmol ends
Reaction time	1–2 hours	1–3 hours	1–2 hours	1 hour
Nature of probe	DNA	DNA	RNA	DNA/oligonucleotide
Insert specific	No	No	Yes	Not applicable
Position of label	Uniform	Uniform	Uniform	3′-end
Competing probe strand	Yes	Yes	No	No
Probe length	~150–500 bases	~200–500 bases	Defined	Defined
Amount of probe generated	150–250 ng plus template	No net synthesis	250–500 ng	No net synthesis

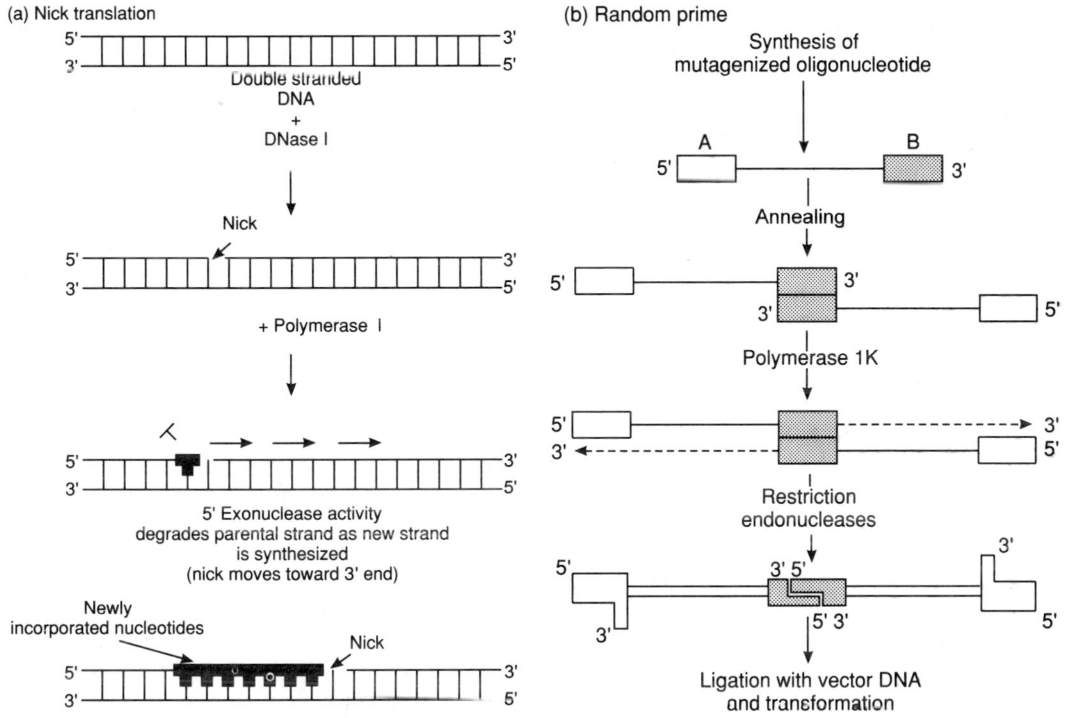

Figure 7.3 Principle of (a) nick translation and (b) random prime methods for incorporation of new nucleotides into nucleic acids.

The random prime reaction (Figure 7.3(b)) involves the use of a single-stranded template (e.g. heat denatured double-stranded DNA) to which are hybridised random hexamers or nonamers (Feinberg and Vogelstein, 1984). The Klenow fragment of DNA polymerase I which lacks the 5′ to 3′ exonuclease activity is then used to extend from the annealed primers along these double-stranded templates. The high specific activities conventionally attainable with this method have resulted in its use in a variety of techniques, but particularly in Southern and Northern blotting. Low amounts of template (25 to 50 ng) can be used and there is a net increase in probe concentration during labelling.

Terminal deoxynucleotidyl transferase (TdT) is used to label oligonucleotides at the 3′ end in a reaction known as 'tailing' (Bollum, 1974). Here, the template (which is generally single-stranded) typically receives an extension of 10 to 100 native nucleotides, depending on the conditions and concentration of the reactants. Hapten-modified nucleotides can be incorporated in a similar manner although the length of the tail is somewhat shorter. Igloi and Schiefermayr (1993) found only one or two biotin or fluorescein-dUTP residues and only a single biotin or fluorescein-UTP residue to be incorporated. Typically, tailed oligonucleotides are used purely for hybridisation and subsequent detection of target nucleic acid. In most instances the tail is found not to interfere with the specificity of hybridisation. However, by using biotin or fluorescein-UTP derivatives, Igloi and Schiefermayr have been able to use oligonucleotides labelled by TdT as substrates for polymerases, extension of the oligonucleotide taking place from the 3′ (hapten-labelled) end. For the assured addition of a single nucleotide it is possible to use a [2′,3′]dideoxy nucleotide derivative in place of the usual deoxy derivative (Schmitz *et al.*, 1991), but of course these products cannot be extended further by polymerases as there is no available 3′-hydroxyl group.

RNA probes (Melton *et al.*, 1984) are generated by the use of RNA polymerase enzymes and plasmid (or other appropriate) constructs containing a promoter element specific for a particular RNA polymerase. The plasmid acts as a template and is supplied in a double-stranded form; the polymerase produces single-stranded RNA transcribed from only one of the strands. By using a plasmid carrying two promoters either side of a cloned template sequence, an antisense probe is made that will hybridise to mRNA. A control (sense) probe can also be produced that should show no hybridisation signal, thus confirming the specificity of the hybridisation process. This additional control is of particular use for *in situ* hybridisation as a false positive signal is frequently encountered in this application. RNA probes also have the practical advantage over most DNA probes that they are single stranded, so that they are not removed from the hybridisation due to strand reannealing. Additionally, RNA–DNA and RNA–RNA hybrids are somewhat more stable than DNA–DNA hybrids in the presence of formamide[1], although this can sometimes lead to difficulties in determining the optimum conditions for hybridisation and washing.

There are three commonly used RNA polymerases which have been isolated from the bacteriophages T3, T7 and SP6. Each has a distinct and highly specific promoter sequence, so that the transcription reaction is specific for each enzyme. The labelling reaction is simple to perform and reliable, provided that precautions are taken to reduce exposure of the reaction mixture to exogenous RNase to a minimum. The use of sterile pipette tips, reaction tubes and solutions and the wearing of protective gloves are recommended. The labelling reaction was originally developed for the introduction of radioactive labels and has been adapted for the production of non-radioactive probes. The modified reactions are efficient processes, leading to high levels of incorporation of hapten-labelled nucleotide and relatively large amounts of synthesised probe.

The highest yielding of methods for synthesising non-radioactive probes is the polymerase chain reaction (PCR) which can either be used to incorporate modified nucleotides or to incorporate labelled primers (Innis *et al.*, 1990). This has the advantage of being usable with low levels of template.

7.2.2 Enzymatic incorporation of readily-derivatised nucleotides into nucleic acids, and subsequent reaction with hapten derivatives

Chemically, nucleic acids are relatively stable and robust materials. Although there are a number of ways of modifying them chemically, it is necessary to use specialist conditions to obtain selectivity of modification (Hall, 1971). The first method described here avoids this problem. It relies on the incorporation of a modified nucleotide into the nucleic acid, and the modification allows chemical labelling to take place under mild conditions. The modified nucleotide used can be N^6-6-aminohexyl)(d)ATP, 8-(6-aminohexyl)(d)ATP, N^6-6-aminohexyl)(d)CTP or 5-(3-aminoallyl)(d)UTP. After polymerase incorporation under conditions similar to those described in the previous section, these provide a nucleic acid probe with reactive primary amines. Unlike the amines present on native nucleic acids which are deactivated by the purine or pyrimidine ring systems, these amines are reactive under mild conditions with many groups such as isothiocyanates, *N*-hydroxysuccinimidyl esters and sulfonyl halides (see Section 2.3.1). The most common haptens are available in at least one of these three forms, and the *N*-hydroxysuccinimidyl esters are available for biotin, fluorescein and digoxigenin. (Dinitrophenyl moieties can also be added using

[1] A reagent commonly employed for the modification of stringency.

fluorodinitrobenzene (FDNB) but this is a very hazardous material to handle). Examples are given for the use of these modified nucleotides with DNA polymerase I by nick translation (Vincent *et al.*, 1982; Gillam and Tener, 1986), with terminal transferase by tailing (Vincent *et al.*, 1982), and with T7 RNA polymerase by *in vitro* transcription (Folsom *et al.*, 1989).

Methods of labelling which involve polymerase enzymes are normally carried out on a relatively small scale (e.g. 50 to 500 ng template for random prime or nick translation) and can become expensive if large amounts of probe are required. Various chemical methods have therefore been investigated, directed particularly towards large scale labelling. In order to incorporate chemically a reactive primary amine into DNA one can use (1) sulfonation of cytosine residues at the N^4-position with bisulfite in the presence of a diamine, e.g. ethylenediamine (Viscidi *et al.*, 1986) or diaminohexane (Gillam and Tener, 1986), or (2) bromination of guanine (C-8 position) and cytosine (C-5 position) followed by displacement with a diamine (Gillam and Tener, 1986; Keller and Manak, 1989). The resulting material can then be modified with amine-active hapten derivatives as indicated above, though to avoid the extra step it is possible to carry out the bromine displacement with a hapten labelled with a suitable primary amine (Keller *et al.*, 1988). These procedures can readily be carried out on a 20 μg scale but some of the components are toxic or corrosive, particularly the diamines. The multiple steps involved have also limited the popularity of these methods.

7.2.3 Chemical incorporation of hapten-labelled nucleotides into synthetic nucleic acids

By far the simplest way of labelling oligonucleotides with haptens is during the synthesis of the oligonucleotide itself. Oligonucleotides are made by automated techniques in which phosphoramidite precursors are joined together in the desired sequence. They are synthesised on a solid support resin such that after synthesis they can be removed from the resin, purified by HPLC if desired and deprotected to remove chemical protecting groups present from the automated synthesis. Phosphoramidite precursors are now commercially available which incorporate some of the more common haptens such as biotin and fluorescein. At present there is no digoxigenin phosphoramidite available. The hapten linked phosphoramidites are of two types, those which contain a base and those which do not. Those which contain a base, such as the biotinyl and dinitrophenyl derivatives synthesised by Roget *et al.* (1989), are labelled on the base and provide a final oligonucleotide product which is similar to that which might be obtained by polymerase incorporation of labelled nucleotides. The potential problem with this strategy is that the position of labelling in the oligonucleotide is dictated by the sequence and the availability of a derivative with the correct base. However, multiple labelling is possible and the label can be incorporated at internal sites on the oligonucleotide. The alternative strategy of using non-nucleosidic phosphoramidites (Misiura *et al.*, 1990; Nelson *et al.*, 1989) has the advantage of not interfering with the specificity of hybridisation. These derivatives are usually incorporated at the terminus of an oligonucleotide to ensure minimal perturbation of hybrid formation. A fairly recent publication describes the synthesis of a phosphoramidite which may provide benefits in improved internucleosidic distances for internal labelling (Nelson *et al.*, 1992). Commonly, the label is incorporated at the 5′-terminus of the oligonucleotide (Misiura *et al.*, 1990), although it has been shown to be possible also to place the label directly on the 3′-terminus (Nelson *et al.*, 1992).

In instances where the hapten of choice is not available as a phosphoramidite, the usual strategy is to incorporate a primary amine into the oligonucleotide, again with a modified phosphoramidite, allowing subsequent derivatisation with the appropriate amine-active hapten derivative. The amine is most often added at the 5′ terminus to avoid interference of the subsequent label with hybridisation (Agrawal *et al.*, 1986) although methods for 3′-amino label incorporation with non-nucleosidic phosphoramidites (Nelson *et al.*, 1989) or internal amine incorporation with nucleosidic derivatives (Cook *et al.*, 1988) have also been used.

Many of the other purely chemical methods for labelling nucleic acids (described in the next section) have in the past been applied to oligonucleotide probes. However, the sequence specificity of these methods has made them generally unreliable for this application. Attempts have been made to circumvent this problem by incorporating a tail of the readily modified base prior to labelling. However, the newer methods, which provide better characterised probes more simply, have made these other approaches unnecessary.

7.2.4 Chemical labelling of nucleic acids with hapten derivatives

7.2.4.1 Photochemical labelling of nucleic acids

One of the few chemical methods to be commercialised is photolabelling. This procedure was first reported by Forster *et al.* (1985) who synthesised an aryl azide derivative of biotin (photobiotin) containing a tertiary amine grouping to aid localisation on the negatively charged nucleic acid. On irradiation with visible light (maximum absorbance seen at 473 nm), this forms reactive species (see Section 5.10) which attack the bases of nucleic acids, preferentially adenine, then to a lesser extent cytosine and thymine. A DNP equivalent has also been prepared and, although the structure omitted the use of the tertiary amine grouping, similar results were obtained in a side-by-side comparison (Keller *et al.*, 1989).

Another photochemical method for biotinylation was published by Dattagupta *et al.* (1989). This uses a biotinylated derivative of angelicin, which is related to the psoralens but produces monoadducts and therefore cannot cross-link nucleic acids. The psoralens intercalate into DNA but the angelicin derivative was shown to label both double-stranded and alkali-denatured DNA. Labelling was carried out with a long UV wavelength transilluminator (light output over the 300 nm to 400 nm range with a maximum at 365 nm) for one hour. In general, care must be taken with photoactivatable methods to calibrate lamps and optimise irradiation times; results can otherwise be less reproducible than with enzyme-mediated methods.

7.2.4.2 Electrostatic labelling and cross-linking of nucleic acids

Other biotin adducts which have been used in the chemical labelling of DNA include biotinylated cytochrome *c* (Manning *et al.*, 1975), biotinylated histones (Renz, 1983) and biotinylated polyethyleneimine (Al-Hakim and Hull, 1986). These methods rely on the electrostatic interaction between the negatively charged nucleic acid and the positively charged adduct to form a complex, which is then stabilised by cross-linking using formaldehyde or glutaraldehyde.

7.2.4.3 Intercalation of platinum complexes

More recently, the DNA binding properties of platinum-containing compounds have been tested for labelling purposes. The anti-tumour drug Cisplatin is the *cis*-isomer of diamminedichloroplatinum (DDP; Figure 7.4). This is thought to act by forming bidentate complexes with DNA so preventing replication (Hambley, 1991). A monofunctional biotinylated platinum derivative has been investigated in the detection of DNA (van Belkum *et al.*, 1994). The labelling method preferentially modifies guanine residues. The potential advantage of such a method is its simplicity.

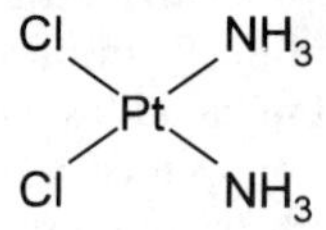

Figure 7.4 *cis*-Diamminedichloroplatinum (DDP)

7.2.5 Chemical modification of nucleic acids to produce haptens in site

A number of chemical procedures have been used in which a chemical modification of the nucleic acid itself provides the hapten. However, these chemical modifications tend to take longer than enzymatic incorporation of hapten–nucleotide conjugates. One such system, which was commercialised by Orgenics, came from original work by Sverdlov *et al.* (1974) in which the cytosine residues were sulfonated using a mixture of sodium bisulfite and methoxyamine. Efficient labelling of low (1 pg/ml) or high (1 mg/ml) concentrations of DNA is claimed to be possible, though it is probably best to use at least μg levels to avoid too much loss during the subsequent purification by precipitation. Labelling is carried out overnight, resulting in sulfonation of 10 to 15% of the cytosine residues (Nur and Herzberg, 1992). Detection is carried out using a monoclonal antibody directed against the sulfonated DNA.

Both *cis*-and *trans*-diaminodichloroplatinum (DDP) adducts of DNA have been found to be highly immunogenic. The antibodies produced react only with the adduct, and not with either the DNA or DDP components individually. Antibodies against the *cis*-DDP modified DNA were most active against denatured DNA and reacted poorly with double-stranded DNA (Kiseleva *et al.*, 1991a). The *trans*-DDP system was therefore used in detection on blots (Kiseleva *et al.*, 1991b, 1992). Probe labelling is carried out with *trans*-DDP in the presence of perchlorate for up to 48 hours. The degree of modification is very important since *trans*-DDP forms cross-links between separate nucleotides in the same chain, resulting in loops which prevent proper hybridisation. Optimal hybridisation was found with a degree of base modification estimated at 3 to 4%.

A method whose use has probably been limited by its potentially carcinogenic components is the modification of DNA by *N*-acetoxy-2-acetylaminofluorene (AAAF). The labelling reaction with AAAF takes one hour at 37 °C and a monoclonal antibody to AAF is then used for detection. Concentrated sulfuric acid has been used to destroy unreacted AAAF (Larzul *et al.*, 1987) but toxicity is more of a concern with this method than with other contenders.

7.3 The Preparation of Covalent Protein–Nucleic Acid Conjugates

Despite the similarity in use of hapten-labelled probes and radioactive probes, the additional steps for the application of enzyme conjugate and subsequent washing can be inconvenient and an additional source of background. The ideal then is a system in which the signal-generating system (i.e. the enzyme) is attached to the nucleic acid probe throughout the hybridisation.

The necessity for the enzyme to retain its activity through the hybridisation process has perhaps limited interest in this approach, but it has been applied successfully on a number of occasions. The methods employed for these conjugations are mainly drawn from those described in Chapters 5 and 6; while they are predominantly encountered in the preparation of enzyme–nucleic acid conjugates, they are equally applicable to other proteins. The highly anionic nature of nucleic acids has also been exploited in electrostatic coupling approaches (cf. biotinylation *via* positively-charged carriers, Section 7.2.4.2)

As in the earlier chapters, the techniques used for preparing protein–nucleic acid conjugates can be split into direct methods, in which the conjugate contains only atoms originating from the component species, and methods which incorporate a linker with two functional groups of similar (homobifunctional) or different (heterobifunctional) reactivity. The methods based on electrostatic interactions are sufficiently different in mechanism to be regarded as a separate category here.

7.3.1 Direct coupling approaches

Nucleic acids are inherently unreactive, and electrostatic approaches are the only common preparative methods which do not require the incorporation of reactive moieties into the nucleic acid. However, fairly non-specific high-energy methods can be applied to promote direct coupling: photochemically induced cross-linking can occur, either in visible light with some form of sensitiser such as methylene blue (Lalwani *et al.*, 1990) or in UV light in the absence of sensitisers (Hockensmith *et al.*, 1991). These methods are generally encountered in structural investigations rather than preparative methods.

7.3.2 Homobifunctional coupling approaches

A number of methods of directly labelling premodified oligonucleotides have been successfully used which are based on homobifunctional coupling reagents. These are analogous to the protein–protein approaches discussed in Sections 5.3–5.5.

7.3.2.1 Homobifunctional coupling of amines

The majority of these have used oligonucleotides which have been modified with an amine-containing group at the 5′-end (cf. Section 7.2.2), opening up the possibility of using the multitude of amine-reactive homobifunctionals described in Section 5.3. The synthesis of an RNA oligonucleotide bearing a 3′-terminal alkylamine has also been described recently (Igloi, 1996).

Probably the most widely used of these methods for coupling *via* terminal amines is that developed by Jablonski *et al.* (1986) and commercialised as the SNAP™ (Synthetic Nucleic Acid Probe) technology (Syngene Inc.). In this method the homobifunctional cross-linker, disuccinimidyl suberate (DSS), is used (cf. Section 5.3.5). This is reactive with primary amines and provides a 12-carbon linker between oligonucleotide and enzyme. The oligonucleotide is first modified with DSS under conditions which favour a 1:1 reaction (to avoid cross-linking the oligonucleotide). The product is then purified on a gel filtration column and lyophilised. A twofold molar excess of alkaline phosphatase is then added and the products of this reaction again purified by gel filtration chromatography. This step removes unreacted oligonucleotide from oligonucleotide–enzyme conjugate and free enzyme. The free enzyme can then be separated from the conjugate by anion exchange chromatography. Overall yield with respect to oligonucleotide was reported to be 30 to 50%. Characterisation of the conjugate by spectrophotometry suggested a 1:1 oligonucleotide to enzyme ratio.

Another homobifunctional cross-linker which has been used is *p*-phenylene diisothiocyanate (cf. Section 5.3.3), which also reacts with amines (Urdea *et al.*, 1988). Again, the strategy was to derivatise the oligonucleotide (containing a 5′-alkylamine group) with the cross-linker and purify. In this case it was possible to carry out a simple *n*-butanol extraction to remove the non-polar cross-linker, followed by lyophilisation and the addition of enzyme. Both of the commonly used enzymes, horseradish peroxidase and calf intestinal alkaline phosphatase, can be conjugated with these amine-reactive homobifunctional cross-linkers. Many other amine-reactive homobifunctionals are described in Section 5.3, all of which could potentially be used for forming links between proteins and aminated nucleic acids.

7.3.2.2 Homobifunctional coupling of thiols

It is also possible to incorporate thiol residues into synthetic oligonucleotides, Connolly (1985), for example, described the introduction of a 5′-thiol *via* a non-nucleosidic phosphoramidite containing a six carbon atom linker arm. Another strategy developed by Murakami *et al.* (1989) was to introduce thiol-containing groups into the oligonucleotide by chemical derivatisation, an alternative to the incorporation of the thiol at the synthesis stage. The oligonucleotide was first phosphorylated, then derivatised with cystamine to introduce the thiol. A similar approach was used by Ghosh *et al.* (1990).

The presence of a thiol opens up the possibility of reaction with a thiol-reactive derivative of the enzyme, generally prepared using one of the amine-thiol heterobifunctionals described in Section 5.7; these methods will be discussed in the next section. However, an alternative approach based on a homobifunctional coupling agent was described by Ghosh *et al.* (1990), who used *N*,*N*′-*o*-phenylenedimaleimide (cf. Section 5.4.1) to couple oligonucleotides bearing a 5′-terminal thiol group to thiol-bearing enzymes.

7.3.3 Heterobifunctional coupling approaches

The majority of these approaches exploit oligonucleotides bearing amine or thiol groups (Figure 7.5), allowing the use of many of the coupling techniques described in Sections 5.6–5.12.

7.3.3.1 Heterobifunctional approaches based on thiolated oligonucleotides

Once a thiolated nucleotide has been prepared, it is generally coupled to the amine groups of proteins derivatised using one of the reagents described in Section 5.7, which introduces a thiol-reactive moiety such as a maleimide, a haloalkyl function or a disulfide.

A method of this type has been used in the formation of alkaline phosphatase labelled probes, for example (Alves *et al.*, 1988). In this instance the alkaline phosphatase was first derivatised with *m*-maleimidobenzoyl-*N*-hydroxysuccinimide ester (MBS). Analogous reagents described in Section 5.7.1 have also been used in this role, such as 6-maleimidohexanoyl-*N*-hydroxysuccinimide ester, MHS (Ghosh *et al.*, 1990).

A similar approach has been used for labelling with horseradish peroxidase (Fowler *et al.*, 1990), which was successful for small probes where the Renz and Kurz approach (see below) was inefficient. A 5′-thiol group is introduced into the oligonucleotide using the

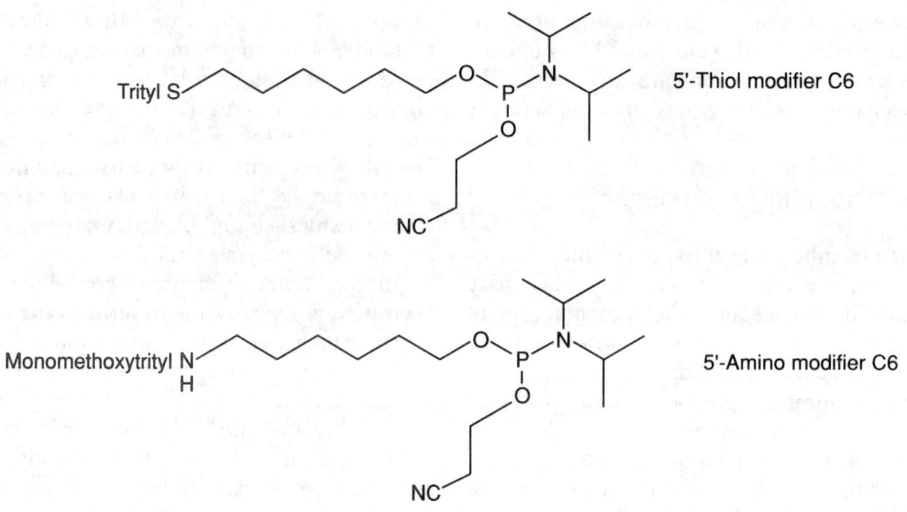

Figure 7.5 Examples of protected phosphoramidites used to introduce free thiols or amines into oligonucleotides.

method of Connolly (1985). The horseradish peroxidase is available commercially in a modified form which is reactive with the thiol-modified oligonucleotides. Incubation of enzyme and oligonucleotide is carried out at room temperature for one hour, or at 4 °C overnight. Purification of the probe is unnecessary if previously purified oligonucleotide has been used since greater than 85% of the probe is labelled in this situation. Decreases in detection sensitivity can be seen later if unpurified thiol-modified oligonucleotide is used in the labelling, since 'failure' sequences which are not full length (and do not contain the thiol modification) will compete with the labelled probe for the target.

Murakami *et al.* (1989) used the pyridyldisulfide-based heterobifunctional SPDP (Section 5.7.3) in the linking role, producing a conjugate linked by a disulfide bond. This method however involves a number of steps and unless the disulfide bond is required, e.g. as a means of cleaving enzyme from oligonucleotide at a later stage, is probably less useful than some of the simpler methods described.

7.3.3.2 Heterobifunctional approaches based on aminated oligonucleotides

The heterobifunctional reagents described in Section 5.7 are equally applicable for linking amine-containing oligonucleotides to SH-containing proteins. Examples include coupling to thiolated antibodies with sulfo-SMCC (Joerger *et al.*, 1995), and to thiolated ricin A chain with SPDP (Wu *et al.*, 1995).

Alternatively a combination of heterobifunctional reagents can be employed: Hendrickson *et al.* (1995) activated an aminated oligonucleotide with the thiolating agent SATA (Section 5.12.1.8), and coupled this to IgG derivatised with the sulfo-SMCC to introduce maleimide groups. The chemical incorporation of cystamine into aminated oligonucleotides (Section 7.3.2) is closely analogous to the activation step here.

7.3.3.3 Photoactivatable heterobifunctional approaches

Photoactivatable heterobifunctional cross-linkers have also been used. SANPAH (see Section 5.10.3.2) is amine-reactive at one end, and capable of reacting with many groups (including amines) in a light-dependent fashion at the other. The enzyme is first derivatised with SANPAH and purified, all in the dark. Oligonucleotide is then added and covalently linked to the SANPAH-enzyme using the light output from a sun-lamp (Keller and Manak, 1989).

7.3.4 Electrostatic coupling approaches

The electrostatic approach was used by Renz and Kurz (1984), although the resulting conjugates were further stabilised by glutaraldehyde cross-linking. The enzymes they used were horseradish peroxidase and calf intestinal alkaline phosphatase; more is now known about the peroxidase system (Pollard-Knight *et al.*, 1990) and this is therefore discussed here in more detail. In order to cross-link the enzymes effectively to the nucleic acid Renz and Kurz found it necessary to form a conjugate of the enzyme with a positively charged macromolecule. This appears to be necessary although the predominant isoenzyme of HRP is basic: it has a very weak positive charge at neutral pH and its electrostatic interaction with negatively charged nucleic acid is limited. By first linking HRP to a macromolecule with a strong positive charge the electrostatic interaction is increased considerably. The macromolecule that was found to be most effective was a low-molecular-mass (about 1400 Da) polyethyleneimine (PEI) which contained equal amounts of primary, secondary and tertiary amine groups. This was linked to horseradish peroxidase in a mass ratio of 150:1 (HRP:PEI) by means of benzoquinone, a small cross-linking molecule which reacts both with carbohydrate and primary amines (see Section 5.3.6). Sizing chromatography has shown that the conjugate is a mixture of HRP monomers, dimers and trimers. The peroxidase–polyethyleneimine conjugate is very stable and can be stored at 4 °C for several months.

In order to label a DNA probe with the HRP–PEI conjugate the DNA is first denatured by boiling for 5–10 minutes to render it single-stranded. (This is carried out at this stage rather than after labelling to avoid damage to the enzyme.) The HRP–PEI conjugate is then added under low ionic strength conditions which favour electrostatic binding to the DNA. Incubation at 37 °C for ten minutes in the presence of glutaraldehyde stabilises the interaction between the DNA and the HRP–PEI conjugate by covalent bonding. It was found that about 25% of the peroxidase activity was bound to the DNA, corresponding to a protein to DNA mass ratio of about 30 (Renz and Kurz, 1984). The trimeric HRP conjugate appeared to be largely responsible for labelling the DNA (Pollard-Knight *et al.*, 1990). Whilst it is possible to purify the probe away from unreacted peroxidase conjugate at this point, it does not appear to be necessary. Labelled probes may be stored in 50% glycerol at −20 °C for up to six months with no loss of activity.

Probes from 50 bases up to 3.5 kilobases have been labelled by the Renz and Kurz method: a 20mer was not successfully labelled (Pollard-Knight *et al.*, 1990).

7.4 The Application of Protein–Nucleic Acid Conjugates as Diagnostic Probes

The commonest application of protein–nucleic acid conjugates is in the preparation of nucleic acid probes for diagnostic purposes. The use of nucleic acid probes invariably involves two key stages: hybridisation to the target sequence, and, following a wash step, the detection of the hybridised complex.

7.4.1 Hybridisation

The aim of hybridisation with any labelled probe is to ensure specific binding to complementary sequences in the target nucleic acid, with the minimum of non-specific binding to other target sequences or to experimental matrices.

There are three major experimental formats in which hybridisation is used. These are blotting, solution and *in situ* hybridisation (Table 7.2). The basic factors which affect hybridisation are common to all applications but each format has specific requirements.

Blotting covers a range of techniques but all are based around filters or 'membranes' which provide a solid support or matrix (see also Section 1.2.4). These tend to be of three types: nylon which may have a net neutral charge or may be derivatised so as to have a net positive charge to ensure good electrostatic binding of nucleic acids; nitrocellulose which tends to be used less than nylon because of its poor retention of nucleic acids (Gingeras *et al.*, 1987) and its poorer handling properties (it becomes very brittle when dry); and polyvinyldifluoride (PVDF) which is not used very widely for nucleic acids because it is hydrophobic and requires wetting with methanol or ethanol prior to wetting with aqueous buffers. For all types of membrane there are procedures developed for 'fixing' the nucleic acid to the membrane to prevent loss of the nucleic acid during subsequent manipulations, though as indicated earlier some membranes are more efficient at retention than others. 'Blots' can take the form of dot blots in which nucleic acid is simply dotted onto the membrane, typically in 1 or 2 μl aqueous volumes. Slot blots are similar to dot blots except that a larger volume may be used and may be drawn through the slots of a template overlaying the membrane with a vacuum manifold. Southern blots and Northern blots are formed when DNA or RNA respectively is size-separated by agarose gel electrophoresis and then transferred to a membrane. The transfer may be carried out by capillary action, using paper towels to draw liquid through the gel and at the same time drawing up the nucleic acid onto a membrane, or by vacuum blotting or electroblotting. Colony or plaque lifts are specialised blotting applications carried out during screening of bacterial clones for specific sequences, the filter being placed directly onto agar plates containing colonies of lysed cells in order to pick up their nucleic acid. In all these applications the nucleic acids are retained on the membrane in a denatured (*i.e.* single-stranded) form to allow subsequent hybridisa-

Table 7.2 Applications for protein-mediated nucleic acid detection

Format	Application	Detection method
Blotting	High sensitivity Southern, *e.g.* single copy mammalian gene	Luminol or dioxetane chemiluminescence
	Intermediate sensitivity Southern, *e.g.* single copy *Drosophila* gene	Luminol or dioxetane chemiluminescence
	Low sensitivity, *e.g. E. coli* gene	Chemiluminescent or chromogenic detection (*e.g.* NBT/BCIP)
	Northern hybridisation	Luminol or dioxetane chemiluminescence
	Plaque and colony hybridisation	Chemiluminescent or chromogenic detection
In situ	Chromosome mapping/painting	Fluorescence *via* a direct or indirect label
	Message detection	
		Chromogenic detection *e.g. via* streptavidin or antibody–AP conjugate and NBT/BCIP
Solution	Probe assays (PCR products)	Chemiluminescent or chromogenic detection

tion to a nucleic acid probe. Detailed protocols for carrying out blotting are available in a number of texts including Sambrook *et al.* (1989).

Solution hybridisation assays (see Section 1.2.3) often take the form of a sandwich format. After hybridisation of the probe in solution, the hybrid is captured onto a solid support to allow separation from the non-hybridised probe, and eventual detection. Membranes can be used in this application, but commonly capture is carried out either in a tube or microplate format to allow high throughput and the possibility of automation.

In situ hybridisation (see Section 1.3.4) is carried out on isolated chromosomes, nuclei, cells or tissue sections. In this situation the accessibility of the nucleic acid to labelled probes is a crucial factor: experimental conditions must be optimised to render the preparation permeable to probe and yet retain its structural morphology. Microscopy is used to visualise the result.

The main experimental variables that are used to control hybridisation processes are probe concentration, temperature and salt or buffer concentration. Together temperature and salt concentration control the stringency[1] of hybridisation. High stringency conditions employ high temperatures and low salt concentrations, with the aim of melting off probe that has hybridised incorrectly.

When using radiolabelled probes under conditions where target concentration is high and there is no problem with sensitivity (for example plaque or colony lifts) the concentration of the probe can be kept as low as possible (1 to 2 ng/ml). This favours specific hybridisation and also tends to avoid general matrix binding. For applications requiring higher sensitivity (such as single copy gene detection on Southern blots) a higher probe concentration may be needed. In the case of hapten- or enzyme-labelled probes where the endpoint is non-radioactive, the probe concentration is usually higher and depends very much on the sensitivity of the endpoint as well as the method of labelling. With the more sensitive endpoints such as chemiluminescence, long probes tend to be used at concentrations in the region of 5 to 50 ng/ml (although 2 ng/ml can be acceptable for some high target applications), whereas the less sensitive endpoints require up to several hundred ng/ml.

Oligonucleotide probes tend to be used at concentrations of 5 to 20 ng/ml but, because of their smaller size, this represents a significantly higher molar concentration than used for long probes. In the case of *in situ* hybridisation, probe concentrations tend to be much higher than those used for blotting because of the problems of accessibility of the target. Solution hybridisation is more similar to blotting with respect to probe concentrations.

The temperature at which hybridisation is carried out is extremely important as it is one of the main determinants of stringency. The optimal temperature for long (native DNA) probes appears to be 25 °C lower than the melting temperature of the probe (T_m, temperature at which half of the DNA is single stranded). For average sequences a temperature of 65 ±2 °C is normally used. It is however possible to modify the T_m of a probe by incorporating a chaotropic agent into the hybridisation medium. Formamide is conventionally used in this role, 50% formamide allowing a drop in the hybridisation temperature to 42 °C.

Hapten-labelled probes tend to have a slightly lower T_m compared to their native counterparts: For biotin, a drop of 5 °C has been observed on total replacement of thymine residues in one strand by biotin-11-dUTP, although smaller changes were seen with lower degrees of substitution (Langer *et al.*, 1981). Since most hapten-labelled nucleotides are used in combination with the equivalent native nucleotide so that relatively low degrees of substitution might be expected, the T_m of the labelled probe is unlikely to be altered significantly. However, some of the chemical labelling methods such as *trans*-DDP labelling may be more likely to affect DNA structure and therefore T_m (Sundquist *et al.*, 1986). Hybridisation buffers optimised for hapten-labelled probes tend to be similar to those optimised for ^{32}P labelled probes, since the hapten is normally inert. The main cause of background is generally the antibody–enzyme or binding protein–enzyme conjugate and it is crucial to achieve efficient blocking of the membrane with suitable blocking agents.

Hybridisation temperatures and buffers used with enzyme-labelled probes must be compatible with the stability of the enzyme. For overnight hybridisations, temperatures in the order of 42 °C are usually used in combination with a T_m modifying agent (Renz and Kurz, 1984). Formamide which would normally be used to reduce T_m with radioactive probes is detrimental to HRP and alkaline phosphatase. Urea has therefore been used in its place with HRP-labelled

[1] Stringency is a difficult term to define succinctly, but is essentially the degree to which hybridisation, in particular non-specific hybridisation, is discouraged by the operating conditions. Under low stringency conditions (low temperature, high salt) some non-complementary hybridisation may be observed; however if stringency is increased too much (by raising temperature and/or lowering salt concentration), even perfectly complementary binding may be inhibited.

probes (Pollard-Knight *et al.*, 1990) although even urea affects alkaline phosphatase activity (Simmonds and Cunningham, unpublished observation). For short hybridisations or washes, it is possible to expose HRP or alkaline phosphatase labelled probes to more adverse conditions: 65 °C for HRP (Fowler *et al.*, 1990) and 50 °C in 1% sodium dodecyl sulfate (SDS) for alkaline phosphatase (Jablonski *et al.*, 1986).

The direct labelling of long probes with horseradish peroxidase has been shown to have little effect on T_m (Pollard-Knight *et al.*, 1990). Similarly, oligonucleotide probes labelled at the 5′-end with HRP showed no significant change in T_m compared to ^{32}P labelled probes (Fowler *et al.*, 1990) although it was reported that alkaline phosphatase-labelled probes had a significantly lower T_m than the control (Jablonski *et al.*, 1986). The latter result may have been affected by the lability of alkaline phosphatase at the higher temperatures used.

Optimum hybridisation temperatures for oligonucleotide probes are usually about 5 to 10 °C lower than their T_m. A rough guide to T_m can be gained from the equation:

$$T_m = (4 \times \text{number of G + C bases}) + (2 \times \text{number of A + T bases})$$

However, for shorter probes the actual sequence can be important in determining T_m and it is often advisable to determine optimum hybridisation temperatures by experimentation.

The specificity of probe binding is usually controlled by a series of stringency washes. Rather than using temperature to determine stringency, which may be difficult to control in practice, salt concentration is usually decreased until an appropriate specificity is attained.

7.4.2 Signal generation

7.4.2.1 Enzymes

By far the most widely used enzymes for labelling nucleic acids are horseradish peroxidase and alkaline phosphatase, which are discussed in detail in Chapter 3. Both of these enzymes can be used with a variety of endpoints, though the main applications in molecular biology are colorimetric and chemiluminescent. Others becoming more widely used include fluorescence and amperometry. The choice of endpoint, and to some extent the choice of enzyme, depends on the sensitivity required and the format of the experiment.

Horseradish peroxidase is a glycoprotein. Isoenzyme C which is the predominant form has a molecular mass of 44 kDa, an isoelectric point of about 9 and relatively little surface charge, typically with only one or two amino groups available for coupling at the surface (see Section 3.3.1). Possibly because of these physical characteristics it tends to show relatively low background binding to blots and other experimental matrices.

Alkaline phosphatase from calf intestine, however, has a molecular mass of 140 kDa, an isoelectric point which is somewhat acidic and a relatively high surface charge, with about 20 amino groups available at the surface for modification (see Section 3.4.2). Background binding of alkaline phosphatase can be a problem, although it can be minimised by careful choice of wash buffers which should include detergents. Possibly a greater problem with alkaline phosphatase is the presence of this enzyme in bacteria. In practice this means that extreme care must be taken to use buffers without even the smallest level of contamination. Phosphatases are also present in cells and tissues so that during the process of *in situ* hybridisation, particularly in certain tissues such as placenta, care must be taken to remove as much endogenous enzyme as possible.

Whilst there is also peroxidase activity in tissue sections and in bacteria, this can either be readily destroyed or is sufficiently substrate specific not to be detected. The smaller size of horseradish peroxidase makes it more useful in permeating tissue sections for *in situ* hybridisation but its sensitivity with precipitating colorimetric substrates tends to be lower than alkaline phosphatase systems. Provided with suitable conditions, alkaline phosphatase can if required continue turning over for days. Horseradish peroxidase tends to lose activity over shorter time periods, although it can often produce rapid results.

7.4.2.2 Substrates

The substrate used depends firstly on whether localisation of the signal is important, for example for blotting or *in situ* hybridisation. If it is, a precipitating product of the reaction is necessary for colorimetric or fluorescence detection, while for chemiluminescence detection light emission is in any case normally restricted to the immediate environment of the enzyme. The colorimetric substrates most widely used with alkaline phosphatase for localisation purposes are nitro blue tetrazolium (NBT) and bromochloroindolyl phosphate (BCIP). Diaminobenzidine or 4-chloronaphthol with peroxidase are sometimes used for less sensitive work, or where background phosphatase activity is a problem. The diaminobenzidine precipitate can be amplified by the use of a silver precipitation technique which deposits specifically on the product of the colorimetric reaction. For solution assays where localisation is not required, substrates which give a soluble product are used: 3,3′,5,5′-tetramethylbenzidine (TMB) or 2,2′-azino-bis(3-ethylbenzthiazoline-6-sulfonic acid (ABTS) for peroxidase; *p*-nitrophenyl phosphate (pNPP) for alkaline phosphatase.

There are two main classes of enzyme-mediated chemiluminescent substrate for enzyme-labelled probes; see Chapter 3 for more detailed information. The first is the enhanced chemiluminescent system for horseradish peroxidase which uses luminol as the chemiluminescent agent (Thorpe and Kricka, 1986). The second is the enzyme-triggered dioxetane family of substrates for alkaline phosphatase (Schaap *et al.*, 1989; Bronstein *et al.*, 1993), of which the most widely used include AMPPD and CDP Star™ (Tropix Inc.) (Voyta *et al.*, 1994). The main application of these chemiluminescent substrates in molecular biology has been in blotting, but they are now becoming more widely used in solution assays.

Fluorescent enzyme substrates have not been widely used in molecular biology, partly because the instrumentation is not generally available but also because they have not been sufficiently sensitive. A relatively new substrate has appeared called Attophos™ (JBL) which does appear to be more sensitive in solution assays (Cano *et al.*, 1992), and the ECF substrate (Amersham Pharmacia Biotech UK Ltd) is being used in blotting applications in conjunction with imaging instruments such as Fluorimager™ or Storm™ (Molecular Dynamics).

7.4.2.3 Signal amplification techniques

Non-radioactive detection methods tend to be less sensitive than their radioactive counterparts. This has led to attempts to amplify the signal from non-radioactive methods in different ways. The simplest way of increasing sensitivity is to increase the number of enzymes present, either through use of multiple valency as encountered in the biotin–streptavidin system, by use of enzyme–antienzyme complexes (such as peroxidase anti-peroxidase, or PAP) or by the use of polymerised forms of enzymes. In a hapten-based system, the simplest way to increase the number of enzymes present is to add multiple layers of species-specific or haptenated antibody, or binding protein. However, care must be taken to avoid an increase in noise due to non-specific binding of the enzyme.

Another way to increase sensitivity is by use of a cascade system whereby the product of the primary reaction is a cofactor for a second enzyme (see Section 3.6). If coupled to a suitable signal generating system this strategy can be very effective. The two most successful examples of this both use alkaline phosphatase as the primary enzyme. In the first method the substrate is NADP, the NAD produced by dephosphorylation acting as a cofactor for a second enzyme, alcohol dehydrogenase, which oxidises ethanol and reduces the NAD to NADH. The NADH is in turn recycled back to NAD by the enzyme diaphorase with the concomitant reduction of a signal generating substrate such as INT to a coloured formazan product (Johansson *et al.*, 1986).

In the second example of this type using alkaline phosphatase, the primary substrate is FADP. The dephosphorylated form of this, FAD, converts the enzyme apo-D-amino acid oxidase to the active form or holoenzyme. In the presence of a suitable substrate such as D-proline, hydrogen peroxide is formed which can be detected by a suitable horseradish peroxidase-based detection system (Harbron *et al.*, 1992). In general, these methods are only successful in solution assays because they lack the ability to localise the signal, necessary in experimental formats such as blotting or *in situ* hybridisation.

To amplify a localised signal, a method based on the enzyme horseradish peroxidase has been used to increase the number of haptens surrounding the target. It was originally developed for use on blots (Bobrow *et al.*, 1989; Bobrow *et al.*, 1991). It does this by using biotinylated tyramine as the substrate for HRP. On oxidation by the peroxidase, the substrate forms a radical which reacts with other proteins on the blot. Due to turnover by the enzyme, many biotins may be deposited in this way around the target. By further addition of streptavidin–enzyme conjugates, the number of enzymes per target molecule is therefore increased. So far, the method has been developed for nitrocellulose matrices, and, more recently, for use on tissue sections.

Other methods use DNA itself as a non-specific structure to which labels may be attached. Two methods have been developed in this area. The first involves the plasmid M13 which is often used for cloning purposes. This particular phage is single-stranded during a part of its life cycle and if prepared in pure form is thus available for hybridisation with other probes. The amplification system involves the cloning of the sequence complementary to the target into this plasmid and using the whole plasmid as the probe. The probe is then detected by means of multiple enzyme-labelled oligonucleotides directed against different regions of the M13 (Chiswell, 1987). M13 is sufficiently long to provide sites for up to a dozen different oligonucleotides in this way with an equivalent theoretical increase in sensitivity. A second approach is by means of branched oligonucleotides (Horn and Urdea, 1989). A solution hybridisation system incorporating these structures has been developed for a number of infectious diseases, including hepatitis B and C, cytomegalovirus and HIV1. These assays include two means of signal amplification. The first is through hybridisation of probes directed against multiple target sequences within the target organism. The second uses the branched oligonucleotide principle: an overhang sequence from the primary probes is hybridised to the stem of a second oligonucleotide which is branched. The branches have a comb-like structure, with the 'teeth' of the comb containing identical sequences. To each of these is hybridised another linear oligonucleotide which is labelled directly with an enzyme, normally alkaline phosphatase. This strategy has a detection sensitivity of about 3000 molecules of target nucleic acid (Urdea *et al.*, 1989), although this figure may be reduced further.

7.5 Other Applications of Protein–Nucleic Acid Conjugates

DNA–enzyme complexes used for *in vitro* analyses are but one category of many DNA–protein complexes found in nature or generated through experimental manipulation.

Within eukaryotic cells, DNA is found associated with a basic family of proteins called histones in structures called nucleosomes. This complex of DNA and protein is called chromatin and great changes in the structure are seen during the active transcription of genes. Transcription itself is regulated by more transient interactions between the DNA and other proteins called transcription factors. This area is of particular interest at the moment with attempts to find ways of manipulating gene expression for clinical applications. Antigene (anti-DNA) and antisense (anti-RNA) technologies involving the delivery of oligonucleotides to cells, to knock out transcription or translation respectively, are receiving much attention. One of the major problems with this approach *in vivo* is ensuring the selective delivery of oligonucleotides to target cells. One approach which has had some recent success is the use of DNA–protein conjugates where the protein has an affinity for receptors on a particular cell type (Findeis *et al.*, 1993).

As well as transcription factors, other regulatory factors and enzymes play important roles in the synthesis, modification and repair of DNA. Research into many of these processes has involved the development of procedures to form covalent cross-links, particularly to characterise weak or transient interactions. Chemical cross-linkers have been used such as formaldehyde (Schouten, 1985) or platinum-containing derivatives (Chu and Orgel, 1992). Some degree of selectivity has been sought with bisulfite catalysed transamination (Shapiro and Gazit, 1977) and the use of bifunctional agents where one group can be induced to react preferentially with DNA and the other with protein (Fink *et al.*, 1980).

Photochemically induced cross-linking between DNA and protein (see Section 7.3.1) has also been exploited in this field (Lalwani *et al.*, 1990; Hockensmith *et al.*, 1991). The recent use of short (5 ns) bursts of laser light has enabled transient DNA–protein interactions to be studied, and with biotinylated DNA, the resulting complexes can be isolated (Ho *et al.*, 1994).

Nucleic acid–protein complexes therefore have an important role to play in many areas of biology and medicine. The area that we have concentrated on is the *in vitro* use of DNA–enzyme complexes which are used for analytical purposes. However, there is much overlap in the techniques used for different applications and future benefit may arise from comparisons between the two.

References

Agrawal, S. *et al.* (1986). Efficient methods for attaching non-radioactive labels to the 5′ ends of synthetic oligodeoxyribonucleotides. *Nucleic Acids Res.*, **14**, 6227–6245.

Al-Hakim, A. H. and Hull, R. (1986). Studies toward the development of chemically synthesized non-radioactive biotinylated nucleic acid hybridisation probes. *Nucleic Acids Res.*, **14**, 9965–9976.

Alves, A. M. *et al.* (1988). Hybridisation detection of single nucleotide changes with enzyme labelled oligonucleotides. *Nucleic Acids Res.*, **16**, 8722.

Bobrow, M. N. *et al.* (1989). Catalyzed reporter deposition, a novel method of signal amplification. Application to immunoassays. *J. Immunol. Methods*, **125**, 279–285.

Bobrow, M. N. *et al.* (1991). Catalyzed reporter deposition, a novel method of signal amplification. II. Application to membrane immunoassays. *J. Immunol. Methods*, **137**, 103–112.

Bollum, F. J. (1974). Terminal deoxynucleotidyl transferase. In *The Enzymes* (Boyer, P.D., Ed.), vol. 10, p. 145, - Academic Press, London.

Brigati, D. J. *et al.* (1983). Detection of viral genomes in cultured cells and paraffin-embedded tissue sections using biotin-labeled hybridisation probes. *Virology*, **126**, 32–50.

Bronstein, I. *et al.* (1993). Detection of DNA in Southern blots with chemiluminescence. *Methods Enzymol.*, **217**, pp. 398–414.

Cano, R. J. *et al.* (1992). Detection of salmonellas by DNA hybridisation with a fluorescent alkaline phosphatase substrate. *J. Appl. Bacteriol.*, **72**, 393–399.

Chiswell, D. J. (1987). Detecting polynucleotide sequence: using probe having sequences complementary to target and secondary probe. *US Patent 4,716,106*, to Amersham International plc.

Chu, B. C. F. and Orgel, L. E. (1992). Crosslinking transcription factors to their recognition sequences with Pt^{II} complexes. *Nucleic Acids Res.*, **20**, 2497–2502.

Connolly, B. (1985). Chemical synthesis of oligonucleotides containing a free sulphydryl group and subsequent attachment of thiol specific probes. *Nucleic Acids Res.*, **13**, 4485–4502.

Cook, A. F. *et al.* (1988). Synthesis of a series of biotinylated oligonucleotides. *Nucleic Acids Res.*, **16**, 4077–4095.

Dattagupta, N. *et al.* (1989). Rapid identification of microorganisms by nucleic acid hybridisation after labeling the test sample. *Anal. Biochem.*, **177**, 85–89.

Davis, L. M. *et al.* (1991). Rapid DNA sequencing based upon single molecule detection. *Gen. Anal. Tech. Appl.*, **8**, 1–7.

Dirks, R. W. *et al.* (1993). Methodologies for specific intron and exon RNA localisation in cultured cells by haptenized and fluorochromized probes. *J. Cell Sci.*, **104**, 1187–1197.

Feinberg, A. P. and Vogelstein, B. (1984). A technique for radiolabeling DNA restriction endonuclease fragments to high specific activity (addendum). *Anal. Biochem.*, **137**, 266–267.

Findeis, M. A. *et al.* (1993). Targeted delivery of DNA for gene therapy *via* receptors. *Trends Biotechnol.*, **11**, 202–205.

Fink, G. *et al.* (1980). Reagents suitable for the crosslinking of nucleic acids to proteins. *Anal. Biochem.*, **108**, 394–401.

Folsom, V. *et al.* (1989). Detection of DNA targets with biotinylated and fluoresceinated RNA probes. *Anal. Biochem.*, **182**, 309–314.

Forster, A. C. *et al.* (1985). Non-radioactive hybridisation probes prepared by the chemical labelling of DNA and RNA with a novel reagent, photobiotin. *Nucleic Acids Res.*, **13**, 745–761.

Fowler, S. J. *et al.* (1990). Labeling of oligonucleotides with horseradish peroxidase and detection using enhanced chemiluminescence. *Technique*, **2**, 261–267.

Gebeyehu, G. *et al.* (1987). Novel biotinylated nucleotide: analogs for labeling and colorimetric detection of DNA. *Nucleic Acids Res.*, **15**, 4513–4534.

Ghosh, S. S. *et al.* (1990). Use of maleimide–thiol coupling chemistry for efficient syntheses of oligonucleotide–enzyme conjugate hybridization probes. *Bioconjugate Chem.*, **1**, 71–76.

Gillam, I. C. and Tener, G. M. (1986). N^4-(6-Aminohexyl) cytidine and -deoxycytidine nucleotides can be used to label DNA. *Anal. Biochem.*, **157**, 199–207.

Gingeras, T. R. *et al.* (1987). Hybridisation properties of immobilised nucleic acids. *Nucleic Acids Res.*, **15**, 5373–5390.

Hall, R. H. (1971). Chemical reactions that can be applied selectively to nucleic acids. In *The Modified Nucleosides in Nucleic Acids* (Hall, R.H., ed.), pp. 347–400, Academic Press, New York.

Hambley, T. W. (1991). New approaches to platinum anticancer drugs. *Chem. Aust.* **58**, 154–156.

Harbron, S. *et al.* (1992). Amplified assay of alkaline phosphatase using flavin-adenine dinucleotide phosphate as substrate. *Anal. Biochem.*, **206**, 119–124.

Hendrickson, E. R. *et al.* (1995). High sensitivity multianalyte immunoassay using covalent DNA-labeled antibodies and polymerase chain reaction. *Nucleic Acids Res.*, **23**, 522–529.

Ho, D. T. *et al.* (1994). Detection and isolation of DNA-binding proteins using single-pulse ultraviolet laser crosslinking. *Anal. Biochem.*, **218**, 248–254.

Hockensmith, J. W. *et al.* (1991). Laser cross-linking of protein–nucleic acid complexes. *Methods Enzymol.*, **208**, 211–236.

Horn, T. and Urdea, M. S. (1989). Forks and combs and DNA: the synthesis of branched oligodeoxyribonucleotides. *Nucleic Acids Res.*, **17**, 6959–6967.

Igloi, G. L. (1996). Nonradioactive labeling of RNA. *Anal. Biochem.*, **233** 124–129.

Igloi, G. L. and Schiefermayr, E. (1993). Enzymatic addition of fluorescein- or biotin-riboUTP to oligonucleotides results in primers suitable for DNA sequencing and PCR. *Biotechniques*, **15**, 486–496.

Innis, M. A. et al. (1990). *PCR Protocols: A Guide to Methods and Applications*, Academic Press, San Diego, CA.

Jablonski, E. *et al.* (1986). Preparation of oligodeoxynucleotide–alkaline phosphatase conjugates and their use as hybridisation probes. *Nucleic Acids Res.*, **14**, 6115–6128.

Joerger, R. D. *et al.* (1995). Analyte detection with DNA-labeled antibodies and polymerase chain reaction. *Clin. Chem.*, **41**, 1371–1377.

Johansson, A. *et al.* (1986). Enzyme amplification for immunoassays. Detection limit of one hundredth of an attomole. *J. Immunol. Methods*, **87**, 7–11.

Keller, G. H. and Manak, M. M. (1989). *DNA Probes*, Macmillan, London.

Keller, G. H. *et al.* (1988). A chemical method for introducing haptens onto DNA probes. *Anal. Biochem.*, **170**, 441–450.

Keller, G. H. *et al.* (1989). Labeling of DNA probes with a photoactivatable hapten. *Anal. Biochem.*, **177**, 392–395.

Kiseleva, V. I. *et al.* (1991a). Detection of hybridisation probes using antibodies of adducts of DNA with *cis*- and *trans*-diamminedichloroplatinum(II). I. Immunochemical characteristics of DNA modified by *cis*- and *trans*-diamminedichloroplatinum(II). *Mol. Biol.*, **25**, 503–507.

Kiseleva, V. I. *et al.* (1991b). Detection of hybridisation probes using antibodies of adducts of DNA with *cis*- and *trans*-diamminedichloroplatinum(II). II. Use of antibodies to DNA modified with *trans*-diamminedichloroplatinum(II) for identifying specific DNA sequences. *Mol. Biol.*, **25**, 508–514.

Kiseleva, V. I. *et al.* (1992). *trans*-Diamminedichloroplatinum(II): modified probes for detection of picogram quantities of DNA. *Anal. Biochem.*, **206**, 43–49.

Lalwani, R. *et al.* (1990). Visible light induced DNA–protein cross-linking in DNA–histone complex and sarcoma-180 chromatin in the presence of methylene blue. *J. Photochem. Photobiol.*, **7**, 57–73.

Langer, P. A. *et al.* (1981). Enzymatic synthesis of biotin-labeled polynucleotides: novel nucleic acid affinity probes. *Proc. Natl. Acad. Sci. USA*, **78**, 6633–6637.

Larzul, D. *et al.* (1987). Non-radioactive hepatitis B virus DNA probe for detection of HBV-DNA in serum. *J. Hepatol.*, **5**, 199–204.

Manning, J. E. *et al.* (1975). A new method of *in situ* hybridisation. *Chromosoma*, **53**, 107–117.

McPherson, M. J. *et al.* (eds). (1991). *PCR: A Practical Approach*, Oxford University Press, Oxford.

Melton, D. A. *et al.* (1984). Efficient *in vitro* synthesis of biologically active RNA and RNA hybridisation probes from plasmids containing a bacteriophage SP6 promoter. *Nucleic Acids Res.*, **12**, 7035–7056.

Misiura, K. *et al.* (1990). Biotinyl and phosphotyrosinyl phosphoramidite derivatives useful in the incorporation of multiple reporter groups on synthetic oligonucleotides. *Nucleic Acids Res.*, **18**, 4345–4354.

Muhlegger, K. *et al.* (1989). Synthesis and use of new digoxigenin-labeled nucleotides in non-radioactive labeling and detection of nucleic acids. *Nucleosides Nucleotides*, **8**, 1161–1163.

Murakami, A. *et al.* (1989). Highly sensitive detection of DNA using enzyme-linked DNA probe. 1. Colorimetric and fluorometric detection. *Nucleic Acids Res.*, **17**, 5587–5595.

Nelson, P. S. *et al.* (1989). A new and versatile reagent for incorporating multiple primary aliphatic amines into synthetic oligonucleotides. *Nucleic Acids Res.*, **17**, 7179–7186.

Nelson, P. S. *et al.* (1992). Oligonucleotide labeling methods. 3. Direct labeling of oligonucleotides employing a novel, non-nucleosidic, 2-aminobutyl-1,3–propanediol backbone. *Nucleic Acids Res.*, **20**, 6253–6259.

Nur, I. and Herzberg, M. (1992). The sulfone system. In *Nonradioactive Labeling and Detection of Biomolecules* (Kessler, C., ed.), Springer-Verlag, Berlin.

Pollard-Knight, D. *et al.* (1990). Nonradioactive nucleic acid detection by enhanced chemiluminescence using probes directly labeled with horseradish peroxidase. *Anal. Biochem.*, **185**, 84–89.

Renz, M. (1983). Polynucleotide–histone H1 complexes as probes for blot hybridisation. *EMBO J.*, **2**, 817–822.

Renz, M and Kurz, C. (1984). A colorimetric method for DNA hybridisation. *Nucleic Acids Res.*, **12**, 3435–3444.

Rigby, P. W. J. *et al.* (1977). Labeling deoxyribonucleic acid to high specific activity *in vitro* by nick translation with DNA polymerase I. *J. Mol. Biol.*, **113**, 237–251.

Roget, A. *et al.* (1989). Synthesis and use of labelled nucleoside phosphoramidite building blocks bearing a reporter group: biotinyl, dinitrophenyl, pyrenyl and dansyl. *Nucleic Acids Res.*, **17**, 7643–7651.

Sambrook, J. *et al.* (1989). *Molecular Cloning, A Laboratory Manual*, 2nd edn, Cold Spring Harbor Laboratory Press, Cold Spring Harbor, NY.

Schaap, A. P. *et al.* (1989). Chemiluminescent substrates for alkaline phosphatase: application to ultrasensitive enzyme-linked immunoassays and DNA probes. *Clin. Chem.*, **35**, 1863–1864.

Schmitz, G. G. *et al.* (1991). Nonradioactive labeling of oligonucleotides *in vitro* with the hapten digoxigenin by tailing with terminal transferase. *Anal. Biochem.*, **192**, 222–231.

Schouten, J. P. (1985). Hybridisation selection of covalent nucleic acid–protein complexes. *J. Biol. Chem.*, **260**, 9929–9935.

Shapiro, R. and Gazit, A. (1977). Crosslinking of nucleic acids and proteins by bisulfite. *Advn. Exp. Med. Biol.*, **86A**, 633.

Sundquist, W. I. *et al.* (1986). Binding of *cis*- and *trans*-diamminedichloroplatinum(II) to deoxyribonucleic acid exposes nucleosides as measured immunochemically with anti-nucleoside antibodies. *Biochemistry*, **25**, 1520–1524.

Sverdlov, E. D. *et al.* (1974). Modification of cytidine residues with a bisulfite–*O*-methylhydroxylamine mixture. *Biochim. Biophys. Acta*, **340**, 153–165.

Thorpe, G. H. G. and Kricka, L. J. (1986). Enhanced chemiluminescent reactions catalyzed by horseradish peroxidase. *Methods Enzymol.*, **133**, 331–353.

Urdea, M. S. *et al.* (1988). A comparison of non-radioisotope hybridisation assay methods using fluorescent, chemiluminescent and enzyme labeled synthetic oligodeoxyribonucleotide probes. *Nucleic Acids Res.*, **16**, 4937–4956.

Urdea, M. S. *et al.* (1989). Application of a rapid non-radioisotopic nucleic acid analysis system to the detection of sexually transmitted disease-causing organisms and their associated antimicrobial resistances. *Clin. Chem.*, **35**, 1571–1575.

van Belkum, A. *et al.* (1994). Non-isotopic labeling of DNA by newly developed hapten-containing platinum compounds. *Biotechniques*, **16**, 148–153.

Vincent, C. *et al.* (1982). Synthesis of 8-(2,4-dinitrophenyl-2,6-aminohexyl)amino-adenosine-5′-triphosphate: biological properties and potential uses. *Nucleic Acids Res.*, **10**, 6787–6796.

Viscidi, R. P. *et al.* (1986). Novel chemical method for the preparation of nucleic acids for nonisotopic hybridisation. *J. Clin. Microbiol.*, **23**, 311–317.

Voyta, J. C. *et al.* (1994). Chemiluminescent detection of DNA and protein with CDP™ and CDP-Star™ 1,2-dioxetane substrates. *J. Biolumin. Chemilumin.*, **9**, 317.

Wu, M. *et al.* (1995). Cell-specific delivery of bacteriophage-encapsidated ricin A chain. *Bioconjugate Chem.*, **6**, 587–595.

Chapter 8

Other Categories of Protein Coupling

Alastair H. Dent and Mohammed Aslam

Research and Development Department, Ortho-Clinical Diagnostics, Pollards Wood Laboratories, Nightingales Lane, Chalfont St. Giles HP8 4SP, UK

8.1 Introduction

The previous three chapters have outlined the vast array of methods which have been employed to couple proteins to the commonest conjugation targets: other proteins, small organic molecules, or nucleic acids. There remain a number of coupling targets which have not been dealt with under these categories, and which will form the subject of this chapter.

8.1.1 Coupling of proteins to solid phases

The first category, and by far the most commonly encountered, is the coupling of proteins to solid-phase materials. Such immobilised protein reagents are largely incompatible with *in vivo* applications[1], and are principally encountered in such areas as diagnostics (coated latex, beads, tubes, microtitre wells or biosensor surfaces) and chromatography (protein-derivatised media for affinity purification). Many of the chemistries encountered in earlier chapters have also been used for protein immobilisation. However, the surfaces of solid materials may have electrostatic, hydrophobic or hydrophilic properties which make non-covalent binding of proteins a realistic approach. These methods are characterised by their practical simplicity, but lack the potential specificity of covalent techniques.

The simplicity of solid-phase labelling techniques, whether covalent or non-covalent, is enhanced by the convenience of the subsequent 'purification' step; this is often as straightforward as a series of washes of the solid surface to remove unbound protein. This facility often extends to the end-use of the coated component too: the recognition that antibodies could be directly coated onto plastic tubes (Catt and Tregear, 1967) was a major milestone in immunoassay development.

The coupling of proteins to the surface of a solid is an interfacial phenomenon, requiring the transport of the dissolved protein molecules through any interfacial layers to the surface itself, where adsorption may take place—either physisorption or chemisorption depending on whether the bonds formed are non-covalent or covalent in nature. These will be dealt with in Sections 8.2 and 8.3 respectively; however it should be borne in mind that physisorption processes are virtually always involved in a coating procedure, even if the intended mechanism of immobilisation is covalent in nature.

Some methods for immobilisation of proteins other than those involving simple surface coating are described in Section 8.4.

8.1.2 Coupling of proteins to cells, etc.

The second category of targets for protein coupling consists of macromolecular biological structures such as cells and viruses. At the molecular level, the targets for such conjugations are almost always proteinaceous, and the methods employed are therefore closely related to those described in Chapter 5. There are numerous applications within that chapter where cell membrane proteins are attached to each other or to extraneous proteins, the intention generally being to glean information on their structural relationships or binding activities. In Chapter 8, however, methods are described for the preparation of intact protein–cell complexes for use as reagents in themselves.

Probably the most important application of such reagents has been the use of antibody or antigen-labelled red blood cells (erythrocytes) in diagnostic techniques such as haemagglutination or complement-dependent haemolysis. These constitute some of the earliest immunodiagnostic methodologies, dating back over fifty years (Pressman *et al.*, 1942). Numerous methods for the preparation of protein–erythrocyte complexes have been described (Jou *et al.*, 1983).

The attachment of proteins to synthetic liposomes, whose membrane behaviour is analogous to that of cells, will also be covered under this second category. Liposomes can provide a convenient means for transporting molecules within an aqueous environment, and have found *in vivo* applications in drug delivery, etc.

[1] Exceptions include adjuvants for use in vaccination, which may be protein coated (see, for example, Frey *et al.*, 1997), and nanoparticles of sub-μm dimensions which have applications in drug delivery and gene therapy (Davis, 1997).

The methods for attachment of proteins to cells and their artificial analogues are covered in Sections 8.6 and 8.7.

8.1.3 Coupling of proteins to soluble polymers, carbohydrates, etc.

The third and final category to be covered in Chapter 8 is a catch-all for macromolecular species which are neither proteins nor nucleic acids: these include natural polysaccharides such as dextrans, and synthetic polymers exemplified by poly(ethylene glycol) (PEG). Complexes of proteins with these soluble macromolecules have typically been employed in pharmaceutical applications, the use of such carriers influencing key properties such as the rate of clearance from the bloodstream. Clearly, increasing the *in vivo* half-life of a protein can be of enormous significance in exploiting any therapeutic potential.

Polymers bearing large numbers of convenient reactive groups have been developed to allow high labelling efficiencies in conjugation reactions, for example the Starburst™ dendrimers from Dendritech (Roberts *et al.*, 1990; Singh *et al.*, 1996; Capala *et al.*, 1997) which are amine-rich spherical polymers of methyl acrylate and ethylenediamine.

The methods for coupling proteins to these various macromolecules are very similar to those described in Chapter 6, the size of the target molecule being the only significant difference. They are described in Section 8.5.

8.2 Physisorption of Proteins to Solid Surfaces

The spontaneous adsorption of proteins to surfaces is a subject of great interest in the biomedical field: as an immobilisation technique, as a key component of many fundamental *in vivo* processes, and as a phenomenon to be avoided in devices which need to come into prolonged contact with body fluids. Only the first of these categories is relevant to this book[1]. However the quest for understanding in the other two fields has led to great advances in the general state of knowledge in this area, particularly in the 1980s and 1990s. Much of this information is captured in major reviews by Norde (1986), Andrade and Hlady (1986), and Butler (1992), the latter particularly for immunochemical applications. A useful brief introduction to this area was provided by Wahlgren and Arnebrant (1991).

The adsorption of even simple solutes onto a solid surface can incorporate a complex chain of events, and numerous models have been developed to describe such interactions. With polymers such as proteins the situation is complicated further, both by possible heterogeneity of the protein and (even for a totally homogeneous protein population) the large number of alternative configurations which are possible at the solid–liquid interface. The energy required to induce conformational changes in proteins in solution is considerably lower than the thermal energy of their environment (Cooper, 1980; Andrade and Hlady, 1986), and conformational heterogeneity must therefore be expected in the protein population. Large, hydrophobic moieties such as aromatic residues have been shown to flip in and out of the protein surface in solution (Jardetzky, 1981), undoubtedly affecting their instantaneous binding properties. As will be seen later, conformational changes after binding may be even more significant.

As a consequence of these factors, the achievement of a state of equilibrium at the surface can be exceedingly slow (Adamson, 1990): the adsorption of organic polymers has been demonstrated to continue increasing for several weeks (Peterson and Kwei, 1961). Admittedly only denatured proteins are likely to display the random-coil flexibility of such polymers, but long equilibration times are still observed even for more rigid examples such as native globular proteins.

Because the three-dimensional structure of protein molecules is rather fragile, surface conformation changes may well result in denaturation; consequently it is common to lose a large proportion of a protein's functionality on immobilisation. They also often mean that the final surface conformation is very different in adsorption characteristics from the native state.

The adsorption of protein molecules to surfaces is often largely irreversible (Adamson, 1990): except with highly hydrophilic surfaces little protein can be removed by equilibrating the surface with pure solvent[2]. Hysteresis of this type has been observed

[1] In fact chromatography should properly be included in this list. Most branches of this methodology rely on the reversible adsorption of proteins onto a solid phase, the goal being to induce differential adsorption behaviour among the solutes by a careful choice of conditions. Protein chromatography is dealt with in Chapter 10.

[2] This irreversibility makes non-covalent adsorption of proteins a viable means of immobilisation, although desorption still remains a problem under certain conditions. The simplicity of physisorption approaches has made them very popular, despite the large loss of functionality which generally results.

since the earliest detailed studies of protein immobilisation (Bull, 1956), with albumin on powdered glass[1]; Andrade and Hlady (1986) have observed that the majority of such work has been carried out under conditions of at least partial irreversibility.

This observation can be explained on the basis of the conformational changes just described, as the properties of the molecule may have been altered substantially between adsorption and (potential) desorption. On the other hand, this irreversibility can be rationalised without invoking conformational changes (Norde, 1986), simply as a consequence of multiple binding sites (Z) between the protein and the surface. Although each individual site in the protein may spend a considerable proportion of time dissociated from the surface, raising the Z value makes it increasingly unlikely that all the sites will be dissociated at a single timepoint. Therefore a protein bound by a large number of sites will exhibit a strong tendency to remain immobilised, despite the inherent reversibility of each individual binding interaction.

As a consequence of this model it might be expected that, all other things being equal, larger molecules would prove harder to dissociate than small ones due to the larger Z value achievable. Such effects are indeed widespread: the avidity of individual proteins often correlates well with molecular mass (van Oss and Singer, 1966; Canterero *et al.*, 1980; Butler, 1992), although exceptions are not uncommon either (Pesce *et al.*, 1977[2]). With mixed solutions of proteins or polydisperse synthetic polymers, high-molecular-mass species eventually dominate the adsorption process (Andrade and Hlady, 1986; see the Vroman effect (Section 8.2.1.5)). Because of the multi-site nature of protein binding, it is also more correct to refer to avidity rather than affinity to describe the strength of these combined interactions (cf. antibodies: Chapter 4).

In summary, the factors outlined above make it a formidable challenge to attempt any realistic, detailed physicochemical treatment of protein adsorption at a solid–liquid interface. Nevertheless these approaches can often yield useful descriptions of adsorption behaviour, Adamson (1990) reflected that the success of the most simplistic models for describing polymer adsorption was 'almost embarrassing', given the complexity of some other treatments.

[1] By this time Henry Bull had been involved in studies of protein immobilisation for at least two decades (Neurath and Bull, 1938), and the University of Iowa has remained a centre of expertise for such studies to the present day.

[2] But see Walsh and Gosling (1986) for a possible rationalisation of these results.

8.2.1 Models of protein physisorption behaviour

The models below are presented predominantly in terms of adsorption isotherms; the amount of solute bound in a given time is generally a function of temperature and the solute concentration in the liquid phase, and an isotherm describes the relationship between adsorption and solute concentration at a fixed time and temperature.

Although it may (or may not) allow some conclusions to be made about the processes involved, the empirical demonstration of a particular variety of isotherm does not require any prior knowledge of the mechanism of adsorption. However before embarking on the various isotherm models that have been described, it is worth describing in very broad terms the 'typical' mechanism that drives immobilisation.

8.2.1.1 The general mechanism of protein physisorption

Non-covalent (passive) protein immobilisation is a relatively facile process: as Norde (1986) has observed, 'most proteins interact with most surfaces'.

All such processes can be described by five stages: (1) transport of molecules to the surface; (2) adsorption of molecules to the surface; (3) rearrangement of adsorbed molecules; (4) desorption of adsorbed molecules; (5) transport of molecules away from the surface.

In terms of controlling the overall adsorption rate, either (1) or (2) may be the predominant factor, and both may be involved at different stages of the same adsorption process. The effect of (3) is more subtle: rearrangement sometimes affects the overall adsorption rate, but may also alter the conformation of bound molecules. Thus its effects will often be seen in changes in the properties of the immobilised protein, rather than in the adsorption rate. It has already been discussed that protein immobilisation is generally carried out under conditions where a large part of the binding is irreversible, therefore (4) and (5) are rather less relevant than the other three stages in this application.

(1) Transport of molecules to the surface

There are four major mechanisms which govern transport of molecules in solution (Andrade and Hlady, 1986): diffusion, thermal convection, flow or convective transport, and coupled transport, such as convection–diffusion. Diffusion is always important, as there is a boundary layer at each surface which can only be crossed by this mechanism. This layer may be up to 1 mm thick in unstirred systems, but can be reduced to 10 μm by vigorous stirring (MacRitchie, 1978). Thermal, flow and coupled effects may also be relevant in many systems.

Diffusion rates are governed (MacRitchie, 1978) by a relationship of the form:

$$\frac{\mathrm{d}n}{\mathrm{d}t} = C_0(D/\pi t)^{1/2}$$

where n = the number of molecules, t = time, C_0 = the initial concentration, D = the diffusion coefficient.

Therefore diffusion rates are proportional to the initial concentration, to the square root of the diffusion coefficient, and to the inverse square root of time.

Realistic analysis of transport behaviour, particularly in the early stages of an adsorption process, requires consideration of the other possible mechanisms such as convection–diffusion. Such situations have been analysed, but the reader is referred to Andrade and Hlady (1986) and references therein for more detailed discussion. For the preparative applications relevant to this book, diffusion only will be considered.

Proteins are bulky molecules with low diffusion constants ($2 - 6 \times 10^{-7}$ cm s^{-1}; Andrade and Hlady, 1986), and it is therefore quite likely that adsorption processes will be limited by this factor, especially at low concentrations. Such behaviour has been noted by many authors (Norde, 1986, and references therein), typified by the dependence of adsorption rate on $t^{1/2}$.

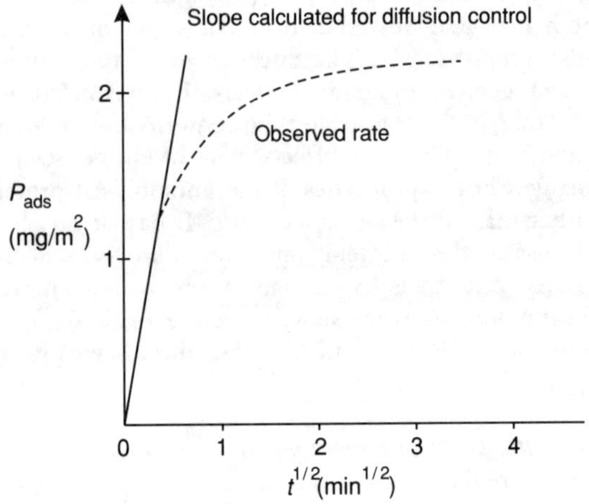

Figure 8.1 Adsorption of ^{125}I-HPA as a function of $t^{1/2}$ on positively charged polystyrene surfaces. (Adapted with permission from: van Dulm and Norde (1983). *J. Colloid Interface Sci.*, **91**, 249–255.)

As a general guideline, rates greater than this indicate the involvement of other transport processes such as convection, while lower rates suggest that the adsorption process itself is rate-limiting (Andrade and Hlady, 1986).

Adsorption processes which start off diffusion-controlled may later come to be governed by the adsorption rate, as the surface coverage increases and subsequent adsorption becomes more difficult (van Dulm and Norde, 1983). An example of such a case is shown in Figure 8.1, where the adsorption of albumin to positively charged polystyrene at pH 4 is plotted *versus* $t^{1/2}$. The process appears to be diffusion-limited for around the first 30 seconds, before adsorption becomes the governing factor.

In summary, diffusion may well be significant in controlling the rates of adsorption processes, particularly in the early stages. If this factor is limiting, it can be overcome or minimised by increasing the protein concentration in solution, or by reducing the thickness of the boundary layer by mixing.

It should be noted that most of the models to be discussed in the later sections assume that the rate of reaction is *not* limited by transport of molecules to the surface.

(2) Adsorption of molecules to the surface

A number of different forces can drive protein–solid phase interactions (providing, incidentally, the basis for some of the most useful forms of protein purification; see Chapter 10). Potentially the strongest of these are electrostatic or coulombic forces driven by charged groups, and the hydrophobic interactions resulting from the existence of groups which interact poorly with the aqueous solvent.

The former, however, are generally insignificant in buffer systems with a reasonably high ionic strength, say at molarities above around 50–100 mM. Even under favourable ionic strength conditions, electrostatic forces may not provide a strong binding mechanism. Bagchi and Birnbaum (1981), for example, coupled IgG to anionic latex particles at low ionic strength, under pH conditions where the overall charge of the protein was negative, positive or neutral. The highest levels of adsorption were obtained at the p*I* of the protein, where the overall charge was zero, and similar observations have been made by other groups. Although there are other properties of protein molecules in the charged state which may contribute to these results, notably increased repulsion of a lateral (Shastri and Roe, 1970) or intramolecular (Norde, 1986) nature, they demonstrate the weakness of coulombic forces as a driving force for binding.

Electrostatic binding mechanisms should not be discounted altogether; understandably they are most likely to be relevant under conditions where the possibilities of hydrophobic interactions are limited. Thus with haematite (a strongly hydrophilic surface) and ribonuclease (a protein with a very cohesive structure, where exposure of hydrophobic regions from the core is unlikely), binding is controlled principally by charge effects (Norde, 1986). Charge effects on binding may not be simple to predict: van Dulm *et al.* (1981), for example, observed that in some circumstances albumin above its pK_a (i.e. with a negative charge) could bind to negative surfaces, *via* cationic counter ions acting as an electrostatic bridge.

In general, however, electrostatic effects are probably more significant as inhibitors of protein binding, presenting a barrier to the hydrophobic adsorption process. Thus, under different conditions, van Dulm and Norde (1983) observed that binding of negatively charged albumin to an anionic surface was retarded compared to other charge combinations. Of course this view of electrostatic effects, too, is consistent with maximal binding at the p*I* of the protein, where charge effects are weakest (Andrade and Hlady, 1986).

In most circumstances, then, hydrophobic interactions drive the immobilisation of proteins. More specifically, if the solid surface and the protein both possess hydrophobic regions, then before immobilisation these must be solvated by water molecules from the solvent. Because contacts between water molecules are much more favourable than those with hydrophobic groups, the number of the latter will tend to be minimised by maintaining the smallest possible interface with the solvent. This is achieved (Figure 8.2) by 'rejecting' hydrophobic species from the aqueous environment, with the net effect of promoting an interaction between the protein and the surface (cf. Figure 2.17). Thus it is the great strength of water–water interactions that drives this mechanism, rather than any powerful affinity of the hydrophobic groups for each other.

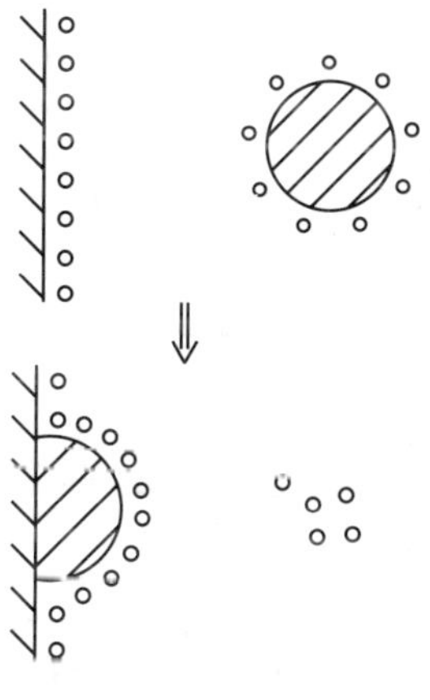

Figure 8.2 The coalescence of the hydrophobic regions of a protein (large circle) with a hydrophobic surface reduces the number of water molecules (small circles) which must adopt unfavourable solvation positions, and results in an overall increase in entropy.

Looking at this process thermodynamically the overall change in enthalpy during the immobilisation is small, although favourable ΔH_{ads} values have been observed in some cases (Chiu *et al.*, 1976; Norde, 1986). These are attributed to attractions between the adsorbed molecules.

More generally, however, there is a large increase in the entropy of the solvent molecules which are freed from the solvation structure, and it is this factor which is generally held to be the major contributor to the favourable free energy change driving binding, the reduction in entropy of the immobilised protein molecule being outweighed by the solvent effects. This entropic mechanism is consistent with the observed endothermic behaviour of many immobilisation processes (Norde, 1986).

Although proteins will bind to hydrophilic materials, the strength of such interactions tends to be rather low: on a simplistic basis the more a surface resembles water in energetic terms, the less of a driving force there is for adsorption (Andrade and Hlady, 1986). For very hydrophilic surfaces such as wettable quartz, even strongly hydrophobic proteins such as albumin bind in a truly reversible fashion (Morrissey, 1977) and such situations offer little practical potential for immobilisation. Indeed, the development of hydrophilic polymers such as hydrogels has been driven by the opposite desire, to produce biocompatible surfaces which exhibit insignificant protein binding properties, for applications such as contact lenses or *in vivo* medical devices.

Passive immobilisation of proteins, therefore, predominantly employs surfaces which encourage hydrophobic interactions, and powerful binding of an irreversible nature can be observed on highly apolar surfaces such as polystyrene. However, proteins possess a combination of polar and hydrophobic regions, and it seems intuistically unlikely that a wholly hydrophobic surface should offer optimal binding properties.

Baszkin and Lyman (1980) bound proteins to polymer films possessing a range of polar and non-polar properties, and observed maximum adsorption (and minimum desorption) for surfaces where the polar

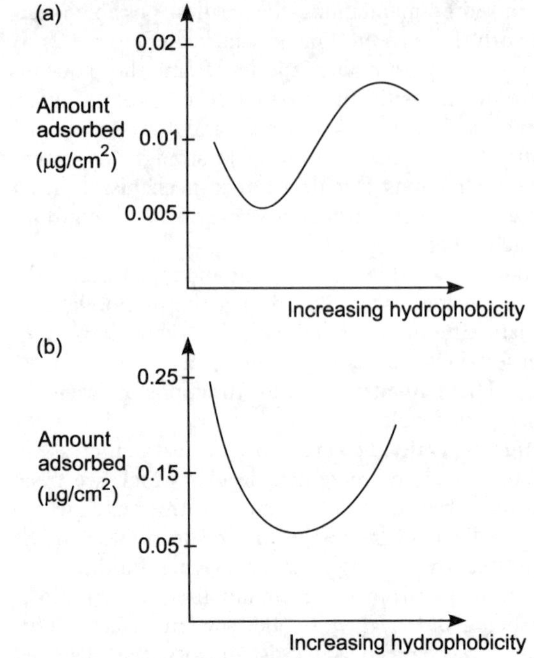

Figure 8.3 The approximate pattern of adsorption of proteins to HEMA–EMA polymers of increasing hydrophobicity; adapted from Horbett (1981). Fibrinogen, IgG and haemoglobin showed maximum binding at an intermediate hydrophobicity (a), while the greatest amount of albumin bound to the most hydrophilic polymer (b).

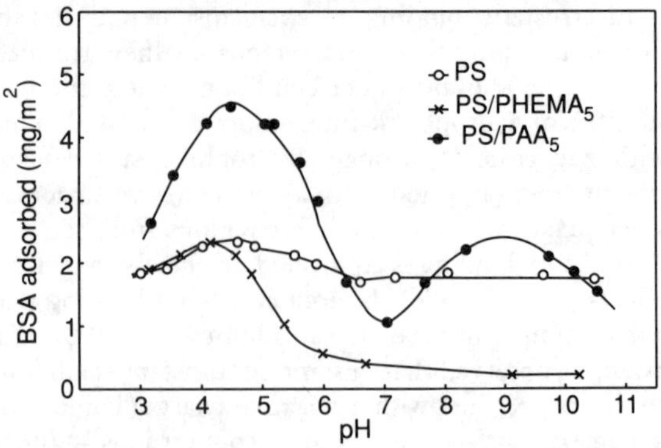

Figure 8.4 pH dependence of BSA adsorption onto various polymer latices at ionic strength 0.1, 25 °C. (Reproduced with permission from: Shirahama and Suzawa (1985). *Colloid Polym. Sci.*, **263**, 141–146.)

and non-polar components were roughly equal. Thus it would appear that the optimum surface for passive protein immobilisation may be one which allows hydrophobic binding, but which also possesses an appreciable hydrophilic content.

Similar studies using hydroxyethylmethacrylate–ethylmethacrylate (HEMA–EMA) copolymers yielded broadly consistent results (Horbett and Weathersby, 1981; Horbett, 1981): fibrinogen, IgG and haemoglobin all showed maxima at compositions of intermediate hydrophobicity (Figure 8.3(a)), although albumin bound more strongly at both the polar and hydrophobic extremes (Figure 8.3(b)). Lee and Ruckenstein (1988) observed the highest binding of BSA to a surface of intermediate hydrophobicity.

Shirahama and Suzawa (1985) cast further light on this topic by comparing the binding of BSA to various polystyrene-containing polymers. A polystyrene–HEMA polymer (PS/$HEMA_5$), considerably more hydrophilic than polystyrene itself (PS) but with similar charge density, showed binding which was similar to or lower than the native material (Figure 8.4). When the polystyrene was copolymerised with acrylic acid (PS/PAA_5), however, the binding was strongly enhanced, particularly at the p*I* of the protein. This effect was attributed to hydrogen bonding with the acrylic acid polymer.

Finally, it has frequently been observed that γ-irradiation or plasma discharge treatment of polystyrene and other polymers leads to increased protein uptake; similarly extremes of hydrophobicity or hydrophilicity introduced by such means generally lower protein binding (Ekdahl *et al.*, 1993). Under atmospheric conditions these processes promote the incorporation of oxygen-containing polar groups into the polymer surface (Hansen *et al.*, 1965; Gerenser *et al.*, 1985; Egitto and Matienzo, 1993; Morra *et al*; 1993). Again these findings tend to support the hypothesis that a balance of hydrophobic and hydrophilic properties is required for maximal protein binding[1].

(3) Rearrangement of adsorbed molecules

Rearrangement of bound molecules could be expected to occur to a lower or higher degree. There could be simple movement of molecules on the surface, and this might be accompanied by the formation of new structures as a result of lateral interactions. Alternatively, or in parallel with these movements, there could be changes to the conformation of the protein molecules—alterations to their secondary, tertiary or quaternary structure.

[1] The recent description of surfaces whose hydrophobicity can be modulated by light (Wang *et al.*, 1997) must surely have some fascinating applications in the biomedical field!

Simple surface diffusion of molecules is a well-characterised phenomenon, and the evidence for movement taking place among immobilised proteins is discussed more fully in Section 8.2.1.5. Various models of physisorption behaviour have incorporated the rearrangement of randomly bound molecules to more ordered structures, such as two-dimensionsal crystals or aggregated clusters; again these will be discussed in Section 8.2.1.5.

It is worth spending a little more time here to discuss conformational changes, both the driving forces which are likely to underlie them, and the evidence for their existence. Models incorporating conformation changes are discussed in Section 8.2.1.6.

The likely causes of these changes are principally related to the hydrophobic interactions which dominate the binding mechanism. As might be expected for molecules evolved to exist in aqueous media, proteins generally contain the majority of their most hydrophobic residues 'buried' in the interior of the structure. The outer regions normally contain a higher content of the more polar amino acid residues, and, frequently, highly hydrophilic carbohydrate chains too.

As discussed, however, binding of protein molecules *via* hydrophilic regions is weak, and, while the number of accessible hydrophobic patches may be small, it is these which are likely to provide the first bonds to the surface (Figure 8.5(a)). If the protein molecule is not desorbed quickly from this tenuous hold, the constant thermal movement of the unbound parts of the molecule may bring other hydrophobic regions in contact with the surface, where they can strengthen the foothold (Figure 8.5(b)). Also, however, the thermal buffeting of the molecule is constantly provoking transient conformational changes, which may occasionally reveal a hydrophobic region near enough to the surface for binding to occur.

In solution these equilibrium changes are fully reversible, with the molecule spending the bulk of its time in its most stable conformation. However the generation of links to an immobile solid phase distorts the normal equilibrium situation: conformations incorporating a bindable hydrophobic moiety will tend to be 'frozen' by surface capture (Figure 8.5(c)), and the normal shift back to the more stable arrangement will be hindered. Although the individual hydrophobic interactions with the surface are likely to be reversible too, the free energy gain from binding may be greater than that of a return to the native conformation; this will be reflected by a predominance of equilibrium events resulting in binding.

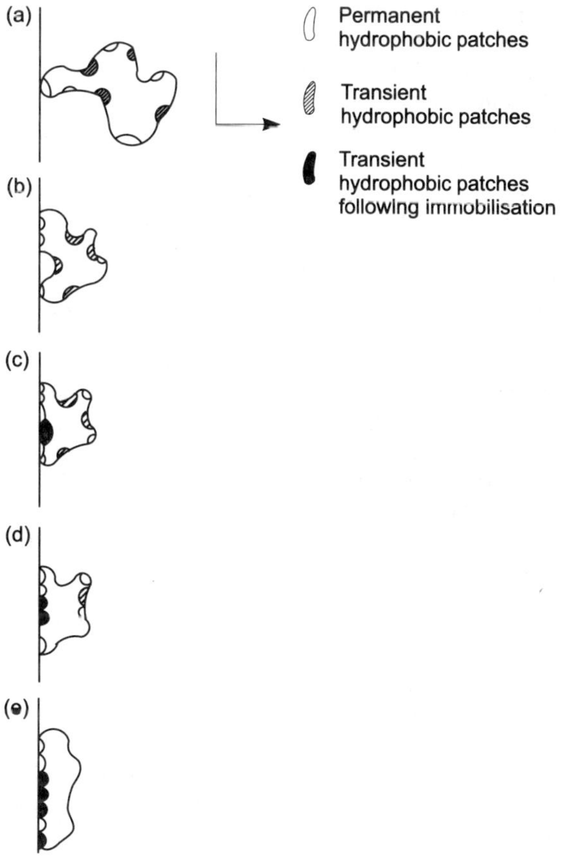

Figure 8.5 Rearrangement of a protein molecule on a hydrophobic surface; see text for details.

Once a number of such bonds have been formed, their arrangement may begin to impact the sensitive balance of forces which govern the stability of the native solution conformation. This breakdown may result in the formation of new conformational arrangements, which would have been too unstable to exist in the solution phase. If these reveal more hydrophobic regions in bindable positions (Figure 8.5(d)), the sequence may be repeated over and over again: new surface links favouring new conformations, which in turn reveal new bindable regions (Figure 8.5(e)). In an extreme case one can imagine the protein molecule being virtually 'turned inside-out', with the hydrophobic regions originating in the core eventually being concentrated at the surface–water interface.

This hypothetical sequence of events would be expected to result in drastic changes to protein behaviour, and complete loss of functional activity might well be the consequence of an extreme case. It is worth considering now the evidence for such events occurring to a significant degree in practice.

As discussed in the introduction to Section 8.2, the substantial irreversibility of protein immobilisation has been reported since the earliest quantitative studies. It is tempting to attribute this observation to conformational changes, but as discussed this can be explained adequately in terms of the avidity effect of multiple binding sites, without necessarily invoking any change in the protein structure.

Loss of functional activity on binding, too, is widely reported, but is not in itself proof of conformational changes. Even with complete retention of the protein structure, a considerable degree of loss might be expected statistically due to steric hindrance of functional sites as a result of random-orientation binding (Lu *et al.*, 1996). The expected degree of loss will be dependent on the relative areas of the molecule and the functional site, and on the accessibility of this site at a given binding orientation. While it is difficult to quantify such expectations, one might reasonably expect a substantial percentage of activity to be retained if this were the only cause of loss. Observed data, however, tends to suggest that other mechanisms are implicated in losses of activity. Butler *et al.* (1992) measured the retained binding activity of antibodies immobilised to polystyrene, and recorded losses of greater than 90% in all cases. Polyclonal antibody preparations were rather more stable than monoclonals: for the latter category less than 3% of activity could be retained in the best cases, and in a number of examples there were no detectable binding properties after physisorption. Data such as these are strongly indicative of surface denaturation, purely on statistical grounds.

Direct evidence for conformational changes on the surface can be sought using suitable instrumental techniques, and has undoubtedly confirmed the occurrence of surface denaturation in some cases – and confirmed its absence in others.

Compelling evidence comes from circular dichroism (CD) of desorbed material (Soderquist and Walton, 1980; Chan and Brash, 1981) and Fourier transform infrared (FT-IR) spectroscopy of *in situ* immobilised proteins (Lenk *et al.*, 1989). Both have revealed changes in the secondary structure, some still taking place ten days after coating (Soderquist and Walton, 1980).

Other physicochemical techniques have also supported the existence of surface changes, such as total internal reflectance (TIRF) spectroscopy (Axelrod *et al.*, 1984). The fluorescence maximum of tryptophan residues has been shown to shift substantially on binding of albumin to quartz, consistent with a more hydrophobic environment for these residues (Andrade *et al.*, 1984; Andrade and Hlady, 1986). Benko *et al.* (1975) observed conformational changes in polystyrene-bound haemoglobin by the use of NMR methods, while as early as 1959, Chattoraj and Bull observed that the electrophoretic behaviour of immobilised albumins suggested changes from the native conformation. Kochwa *et al.* (1967) noted that charged groups were exposed on adsorption of γ-globulin to polystyrene particles, sufficient to cause a drift in the pH of the coating solution.

Finally, a number of immunochemical techniques have revealed changes in the antigenicity of bound proteins compared to their form in solution (Kennel, 1982; Djavadi-Ohaniance *et al.*, 1984) or when immobilised indirectly (Dierks *et al.*, 1986). While the loss of epitopes could be explained by steric effects, the appearance of neoepitopes which did not exist in the native molecule (Friguet *et al.*, 1984; Stevens *et al.*, 1986) can really only be rationalised on the basis of conformational changes. The pitfalls that these effects can introduce into analytical methods using immobilised proteins (e.g. immunoassays, biosensors) are readily imaginable.

There are many occasions, however, where retention of conformation has been observed on binding: using TIRF, Brash and Lyman (1969) reported that the globular form of a number of plasma proteins was maintained on binding to hydrophobic polymers; using IR spectroscopy Morrisey and Stromberg (1974) saw no change in the conformation of fibrinogen, albumin or prothrombin adsorbed to silica; Sarver and Kreuger (1993) observed little change in FT-IR studies of a number of proteins bound to polyethylene sheets. Some studies showed differences between proteins: using CD, McMillin and Walton (1974) observed differences between the solution and adsorbed state for factor XII but not for fibrinogen. Overall, the evidence is overwhelming that surface conformation changes do commonly take place, but their existence and magnitude appear to be affected by a number of factors: (a) The rigidity of the protein molecule. The hypothetical description of changes on adsorption, above, assumes a reasonable degree of flexibility in the molecule, and it has been observed that rigid molecules such as albumin do in fact display less modification than more flexible examples such as IgG (Morrissey, 1977; Butler, 1992). Sarver and Krueger (1993) also observed that the helix and turn regions of proteins were more prone to distortion on binding than the β-sheet areas. (b) The hydrophobicity of the surface. As might be expected from the model discussed, more severe conformational changes can be expected on surfaces demonstrating a higher degree of hydrophobicity. Elwing *et al.* (1988) demonstrated this effect immunochemically, using complement factor C3 bound to a surface with a polarity gradient: greater changes were observed at the hydrophobic end of the surface. (c) The surface concentration. The greatest changes seem to occur at low surface concentrations (Kochwa *et al.*, 1967; Nyilas *et al.*, 1974; Burghardt and Axelrod, 1981; Norde, 1986), where there is little chance of stabilising effects from neighbouring molecules. As an alternative to increasing the concentration of the protein of interest, it has been demonstrated that addition of a 'filler' protein can result in improved activity: Dierks *et al.* (1986), for example, obtained better retention of IgG immuno reactivity when microtitre well coating was carried out in the presence of albumin, while Jitsukawa *et al.* (1989) observed dramatic improvements in the binding of interferon under similar conditions.

Interestingly, when a hydrophilic surface is employed, this concentration effect may be reversed: Koutsoukos *et al.* (1983), for example, showed that albumin bound to haematite at pH 5 showed greater IR evidence for conformational changes at higher surface concentrations. Presumably on a hydrophilic surface where there is little tendency for denaturation (see (b) above), lateral intermolecular effects provide a greater driver towards structural change.

Finally, it is maybe worth pointing out that conformation changes on immobilisation can occasionally be beneficial rather than destructive: enzyme activity is sometimes retained better in immobilised preparations than in the native form (Manecke *et al.*, 1970; Dimarzio and Bishop, 1974; Klibanov, 1979; Schellenberger and Ulbrich, 1989). The last authors suggested that immobilisation might block a likely nucleus for unfolding, thereby providing a stabilising influence.

(4) Desorption of adsorbed molecules and (5) Transport away from the surface

As already discussed, many protein immobilisation processes are irreversible to a large extent, at least in the absence of competing species. Therefore the desorption process is often of somewhat secondary interest, but a number of the models discussed below take desorption into account, with or without changes in the binding properties of the desorbed material. On the whole it seems likely that desorbed material, even if it initially shows strongly decreased binding behaviour (Soderquist and Walton, 1980), will gradually recover its native conformation and properties (Andrade and Hlady, 1986).

Where competing proteins are present, physisorbed material is certainly not irreversibly bound: the phenomena observed with mixed protein solutions will be discussed further in Section 8.2.1.5. The factors controlling such effects may be complex, in some circumstances depending on shear forces in a moving matrix, and in others being independent of these (Chan and Brash, 1981; Norde, 1986). Again some of the models to be discussed below incorporate the effects of competing protein molecules.

8.2.1.2 The Langmuir adsorption isotherm

The simplest isotherms reflect the situation where only a single layer of molecules can be deposited on the surface; these are described by the Langmuir adsorption equation and its variants.

Before progressing any further it should be pointed out that the Langmuir equation was developed in the rather different circumstance of adsorption of gas molecules onto a solid phase. This model incorporates a number of assumptions. (1) that the binding is reversible (and, as a consequence, that there is no change to the structure on adsorption); (2) that there are a number of fixed, identical binding sites on the solid phase; (3) that only one molecule can be adsorbed per site; (4) that there are no lateral interactions between bound molecules; (5) that the adsorbing molecules are all identical; (6) that the concentration of adsorbing molecules in the surrounding matrix is low.

Several of these assumptions are likely to be invalid for protein immobilisation processes, in particular (1), (2), and (4). Therefore, while the Langmuir isotherm may offer a convenient empirical description of adsorption behaviour, such an observation should not be over-interpreted: as pointed out by Norde (1986), for example, calculations of ΔG_{ads} based on such models are without physical meaning.

In one form the Langmuir equation for formation of a protein monolayer can be expressed as:

$$\frac{P_{\text{ads}} = S_{\text{ads}} b[\mathrm{P}]}{1 + b[\mathrm{P}]}$$

where P_{ads} is the amount of protein adsorbed per unit area[1], S_{ads} is the total number of adsorption sites per unit area, [P] is the concentration of protein in solution[2], and b is a constant.

[1] Or another convenient quantity proportional to surface area, such as the mass of adsorbent beads.

[2] Properly activities should be used rather than concentrations, but the approximation is good for dilute solutions, as is often the case for protein adsorption.

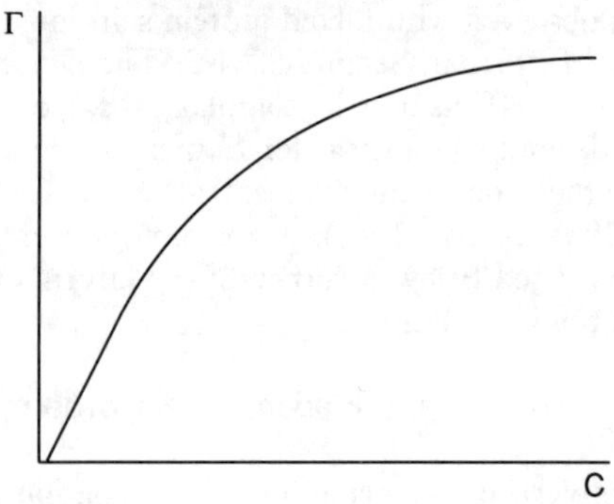

Figure 8.6 Isotherm obeying Langmuir equation.

The shape of an isotherm obeying the Langmuir equation is shown in Figure 8.6; it is readily understood in terms of a saturation (monolayer) level of protein adsorption, which is approached more rapidly from a more concentrated protein solution. The key features are that the isotherm tends towards linearity over a region approaching the origin (an effectively constant fraction of the solute protein being bound in this concentration range), and that a plateau is observed at high concentrations. As usual there is a linear form of the equation which is convenient for graphical treatment of data:

$$\frac{[P]}{P_{ads}} = \frac{1}{S_{ads}b} + \frac{[P]}{S_{ads}}$$

Plotting $[P]/P_{ads}$ *versus* [P] allows S_{ads} to be determined from the slope, and then b from the intercept.

The term θ is often encountered to describe the fraction of the available sites occupied (P_{ads}/S_{ads}). The equation above can therefore also be stated as:

$$\theta = \frac{b[P]}{1 + b[P]}$$

It can be deduced from this equation that a high degree of surface coverage at a given protein concentration is obtained where b is large. The value of b depends on the relative ease of adsorption of solute and solvent molecules on the surface in question, and this parameter, as discussed earlier, is principally influenced by hydrophobic interactions. With simple solutes such as a homologous series of aliphatic carboxylic acids, it was demonstrated over a hundred years ago that the degree of adsorption could be quite neatly predicted on this basis (Adamson, 1990). With a non-polar surface (carbon) and a polar solvent (aqueous) the least polar acid, butyric acid, was adsorbed most strongly. With a polar surface (silica) and a non-polar solvent (toluene) the situation was reversed, and formic acid was bound most readily. This general relationship is termed Traube's rule, after the scientist who originally described it in 1891, and remains broadly applicable to protein immobilisation applications.

Another useful plot which can be constructed to describe adsorption behaviour is one of the percentage of protein bound against the protein concentration, i.e. $100P_{ads}/[P]$ *versus* [P] or $\log[P]$. Rearranging the Langmuir equation as shown earlier yields:

$$\frac{100\,P_{ads}}{[P]} = \frac{S_{ads}100b}{1 + b[P]}$$

Where the solid phase and the coating conditions are otherwise constant, $100P_{ads}/[P]$ tends towards a constant value at low protein concentrations, i.e. where $1 + b[P] \approx 1$. This plot is therefore useful in demonstrating the extent of the region of independence, in which this approximation does hold[1] (Canterero *et al.*, 1980; Butler, 1992). Examples of such plots for a number of proteins on polystyrene are shown in Figure 8.7 (Butler, 1992).

A number of other useful parameters emerge from such plots. Proteins with a high avidity for the surface display a higher constant percentage, and, with the exception of ovalbumin, this correlates well with molecular mass (cf. the multiple binding site model discussed in the introduction to Section 8.2). Also (with the same exception), the region of independence ends at a lower concentration for high-avidity binders.

This can be rationalised on the basis of a finite area where the binding kinetics fit the constant-fraction model: most simply it is convenient to relate this to the area required for monolayer coverage (Butler *et al.*, 1992[2]), and the concept that subsequent binding is less efficient. An alternative explanation might be suggested from the Langmuir derivation above, despite

[1] It should be pointed out that, although the constancy of $100\,P_{ads}/[P]$ at low concentrations can be derived from the Langmuir equation, Butler and his colleagues have never suggested that demonstration of such a relationship is indicative of Langmuir kinetics. It is presented simply as a useful empirical description of adsorption behaviour.

[2] Butler's group have taken various positions on the relationship between the end of the region of independence and monolayer coverage (see for example Butler, 1992), but with this paper returned to their original perception of a correspondence between the two events.

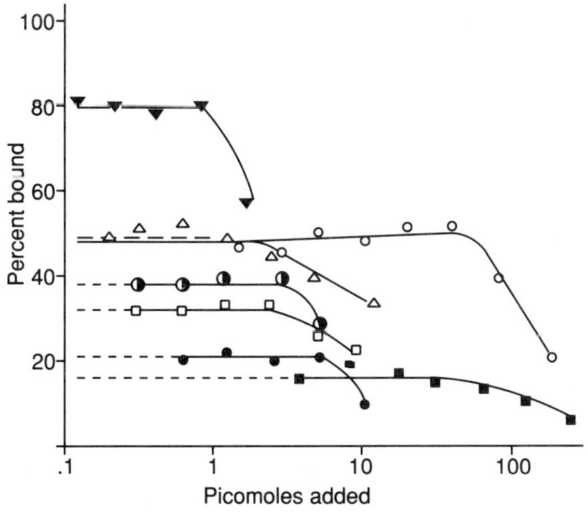

Figure 8.7 Adsorption on polystyrene of several different proteins. ▼, Bovine IgM; △, Bovine SIgA; ○, ovalbumin; ◑, bovine IgG_1; □, bovine IgG_{2a}; ●, BSA; ■, bovine α-lactalbumin. (Reproduced with permission from: Butler (1992). In *Structure of Antigens* (Van Regenmortel, M.H.V., ed.), CRC Press, Boca Raton, FL.)

the well-recognised limitations of this model for protein immobilisation: simply, that the approximation $1 + b[\text{P}] \approx 1$ breaks down earlier for high-avidity adsorbers where the value of b is large.

As Andrade and Hlady (1986) have pointed out, the assumptions underlying Langmuir adsorption are common to other examples of reversible (equilibrium) binding; in particular, they are directly analogous to those governing reversible ligand–receptor interactions in solution. Under these conditions, the solid phase can simply be viewed in the same way as a binding protein, and the usual techniques for analysis of binding equilibria can be applied. Andrade and Hlady have described the use of Scatchard plots to analyse such situations; useful information on surface interactions, heterogeneity, etc. can be obtained by this approach, in the rare examples of protein immobilisation which are truly reversible. These authors have also graphically demonstrated the failings of this approach where the fundamental criterion of reversibility is not met. Scatchard plots generated under such conditions are not readily analysed, and Andrade and Hlady noted that this approach was inadequate in such situations.

Wilkins Stevens and Kelso (1995) described a model for estimating binding capacities using the data from sequential immunoassay experiments, which was based on Langmuir-type binding principles. Again, the model was stated to be inapplicable for examples displaying different adsorption behaviour. The approach can also be used for tracking the subsequent desorption of protein (Wilkins Stevens *et al.*, 1995).

8.2.1.3 Models incorporating surface heterogeneity

The Langmuir isotherm assumes homogeneity of both binding sites and solute molecules. Even if the second condition is met by using a homogeneous protein solution for coating, most surfaces are heterogeneous to some extent, and in such situations Langmuir behaviour is unlikely to be observed. The Freundlich isotherm may describe such situations (Adamson, 1990). The equations which describe Freundlich behaviour can be expressed as:

$$P_{\text{ads}} = S_{\text{ads}} a[\text{P}]^{1/n}$$

or:

$$\theta = a[\text{P}]^{1/n}$$

where a and n are constants, with $n > 1$.

These isotherms differ from the Langmuir model in that they remain convex to the concentration axis right down to the origin, and that they do not reach a saturation value (Figure 8.8). A plot of $\log P_{\text{ads}}$ *versus* $\log[\text{P}]$ can allow aS_{ads} and $1/n$ to be estimated from the intercept and slope, corresponding roughly to adsorbent capacity and binding strength, respectively (Adamson, 1990).

The Freundlich equation is wholly empirical, and observation of Freundlich behaviour can only really be used as mild supporting evidence for a heterogeneous

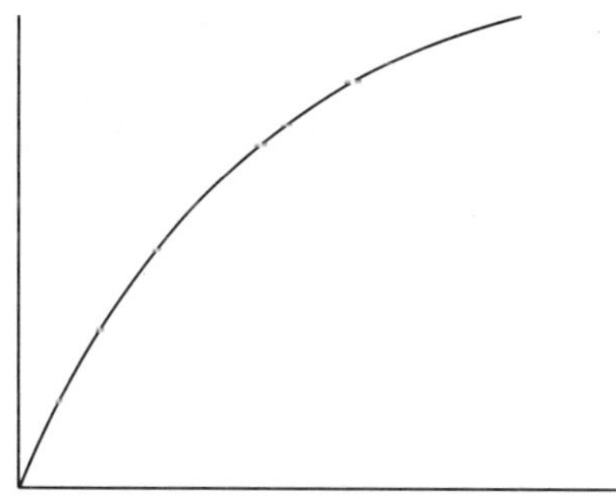

Figure 8.8 Freundlich isotherm.

surface. Horbett *et al.* (1977) for example observed Freundlich-type isotherms for haemoglobin, albumin and fibrinogen on polyethylene, while IgG displayed Langmuir behaviour on the same surface.

Bornzin and Miller (1982) interpreted the binding of proteins to a number of surfaces, on the basis of a combination of regions displaying reversible and irreversible binding properties. Their model was described kinetically, with separate rate terms for the diffusion-controlled irreversible binding, and the equilibrium-modified reversible component. Sevastianov *et al.* (1983) developed a similar model, with centres of irreversible and reversible adsorption; the latter, however, were described as centres of irreversible *de*sorption, as the desorbed protein was considered to be unable to readsorb.

8.2.1.4 Models incorporating multilayer formation

It was apparent from fairly early studies (Oreskes and Singer, 1961; Singer *et al.*, 1963) that many examples of protein adsorption exhibited biphasic properties. While the behaviour was basically of a Langmuir type, two distinct regions could be observed: above a certain concentration, binding with a lower avidity was observed. There was also evidence, in this example at least, that the material bound during the secondary phase was more readily removed by washing. Two possible mechanisms were proposed. (1) the initial deposition of a protein–solid phase monolayer was followed by the addition of a second, more loosely bound protein–protein layer (Figure 8.9(a)); (2) the initial deposition of a loosely dispersed protein–solid phase monolayer was followed by reorganisation to a closer-packed but less tightly bound arrangement (Figure 8.9(b)).

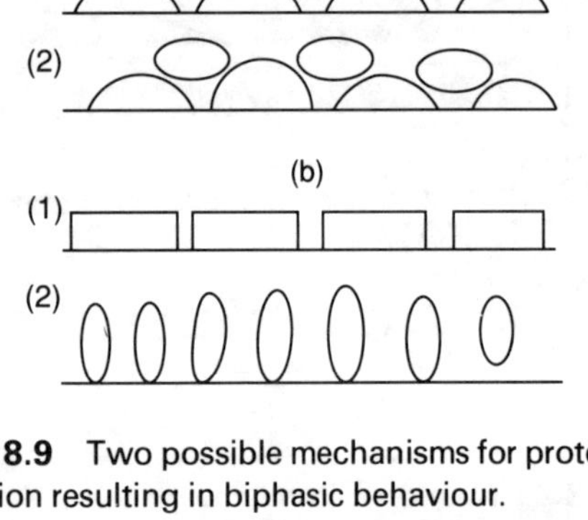

Figure 8.9 Two possible mechanisms for protein adsorption resulting in biphasic behaviour.

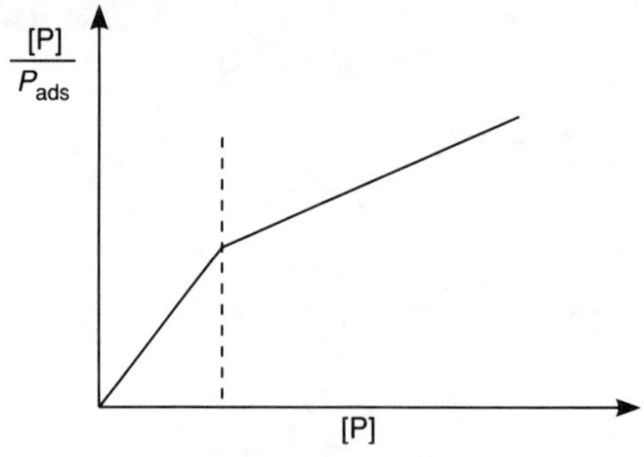

Figure 8.10 Biphasic adsorption.

This biphasic behaviour is extremely widespread, and various attempts have been made to include the possibility of multilayers or conformation changes in adsorption models. In the standard $[P]/P_{ads}$ *versus* [P] plot, the biphasic behaviour appears as line segments of different slope (Figure 8.10). The surface concentrations S_{ads} and relative binding affinities b of these two regions can be derived independently from the two regions of the graph, regardless of which model is adopted.

The existence of protein bilayers is not unreasonable given the ability of hydrophobic interactions to extend to distances as great as 140 Å, in excess of the diameter of small proteins (Kaelble and Moacanin, 1977; Young *et al.*, 1988a) (see Figure 8.11). A certain amount of data supports this concept.

Fluorescence quenching data from Walton and Maenpa (1979) on the binding of albumin to polyamino acid substrates supported the idea of two layers, one tightly bound and the second more diffuse. Interestingly, Arnebrant *et al.* (1985) observed clear bilayer behaviour for proteins on a hydrophilic chromium surface, but monolayer adsorption when the surface was rendered hydrophobic.

Young *et al.* (1988a,b) observed biphasic behaviour for a number of proteins including albumin, transferrin, fibrinogen, macroglobulin and IgG Mabs, on a number of different surfaces, and interpreted it on the basis of two layers with capacity $S_{ads}(1)$ and $S_{ads}(2)$ respectively. For many combinations of protein and surface, the ratio $S_{ads}(2)/S_{ads}(1)$ was roughly constant at 0.59 ± 0.20, suggesting a fairly consistent geometrical mechanism for this effect. It seems less likely that reorganisation to a closer-packed structure would occur at such a consistent ratio for the range of proteins studied, which cover a variety of shapes. Although the polar polymer PEUU showed a high tendency towards bilayer formation (cf. Arnebrant,

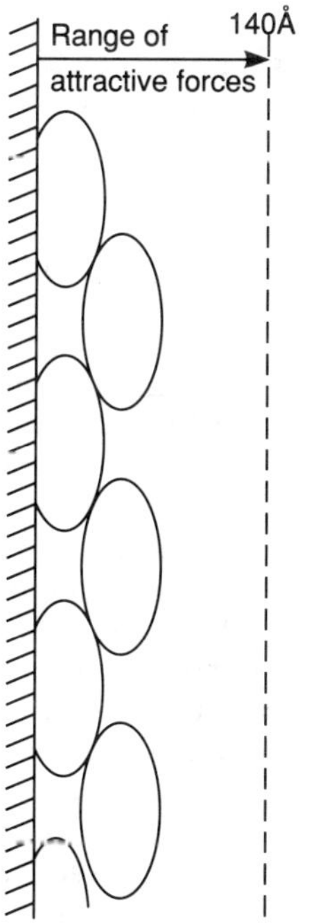

Figure 8.11 As proteins typically have widths of a few tens of Ångstroms, it is conceivable that hydrophobic interactions may still drive binding even for a second layer.

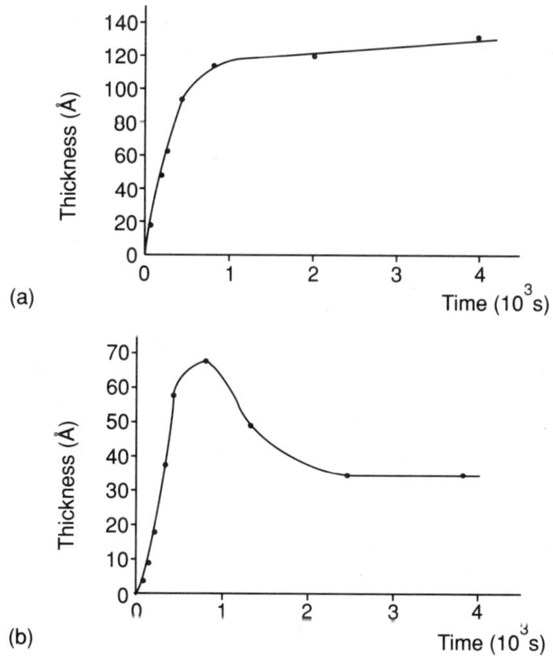

Figure 8.12 Behaviour of thickness as a function of time during the adsorption of fibrinogen (10 μg/ml) in 10mM Tris-HCl buffer pH 7.0 onto a chromium oxide hydrophilic (a) and hydrophobic (b) surface. (Reproduced with permission from: Cuypers *et al.* (1977). *Ann. NY Acad. Sci.*, **283**, 77–87.)

above), so too did the rather hydrophobic silicone rubber.

Measurement of the thickness of adsorbed layers might give some insight into these phenomena. This can be achieved by a number of techniques such as ellipsometry, light-scattering or viscometry. Ellipsometry yields the optical thickness of the layer, which represents a mean value; the other two techniques give the hydrodynamic thickness, which is defined principally by the outermost parts of the layer.

At low coverages these thicknesses are generally consistent with a monolayer of 'sideways-on' molecules: Uzgiris and Fromageot (1976), for example, measured the hydrodynamic thickness of albumin on polystyrene at around 30–40 Å. Under high-coverage conditions Morrissey and Han (1978) measured γ-globulin layers on polystyrene using light-scattering, and recorded a hydrodynamic thickness of 216 Å, consistent with an 'end-on' monolayer rather than a 'sideways-on' bilayer. Similar studies with ellipsometry (Morrissey, 1977) demonstrated an increase of 180 Å in the optical radius of γ-globulin on silica when the surface coverage was increased from 170 to 340 ng cm^{-2} by using a more concentrated solution.

Therefore the available evidence on layer thicknesses tends to support the concept of a differently organised monolayer under high-coverage conditions, rather than multi-layer formation. Actual reorganisation on the surface cannot be inferred from these experiments: the results may just reflect that the protein is deposited in a different initial conformation when the solution concentration is higher.

More direct evidence comes from timecourse examinations of thickness (Cuypers *et al.*, 1977). As shown in Figure 8.12(a), fibrinogen on a hydrophilic chromium oxide surface forms a fairly stable diffuse layer, of width about 130 Å. Rather different behaviour is observed on a hydrophobic chromium surface (Figure 8.12(b)): the fibrinogen initially builds up a 70 Å layer, but after ten minutes the layer begins to compress, reaching a plateau thickness of about 30 Å after 40 minutes. The refractive index increases in parallel with this reduction in thickness, indicating that material is being reorganised into a dense layer rather than being lost.

In summary, bilayer or multilayer models have been difficult to prove unequivocally, and more attention has been paid to the reorganisation of bound molecules on the surface.

8.2.1.5 Models incorporating surface reorganisation

Obviously, reorganisation of bound molecules on the surface cannot occur if they are completely immobilised. In view of the large number of separate binding interactions which have been mentioned earlier, and the consequent irreversibility on dilution in many examples, it may seem improbable that surface reorganisations can take place.

However, it has been noted that each binding interaction may in itself be reversible, and it is relatively straightforward to imagine a 'creeping' movement (Figure 8.13), where peripheral binding sites at one edge of the protein dissociate while new bonds are formed at the opposite boundary. Thus a protein molecule can undergo two-dimensional diffusion while continually remaining bound to the surface. Surface diffusion has been demonstrated for albumin on glass and various polymers (Michaeli *et al.*, 1980; Burghardt and Axelrod, 1981; Tilton *et al.*, 1990). Coagulation effects have also been noted (MacRitchie and Owens, 1969; MacRitchie, 1972). The formation of two-dimensional protein crystals on surface films of lipids and other compounds has also been demonstrated (Darst *et al.*, 1991), even in the absence of specific binding ligands in the film (Furuno and Sasabe, 1993).

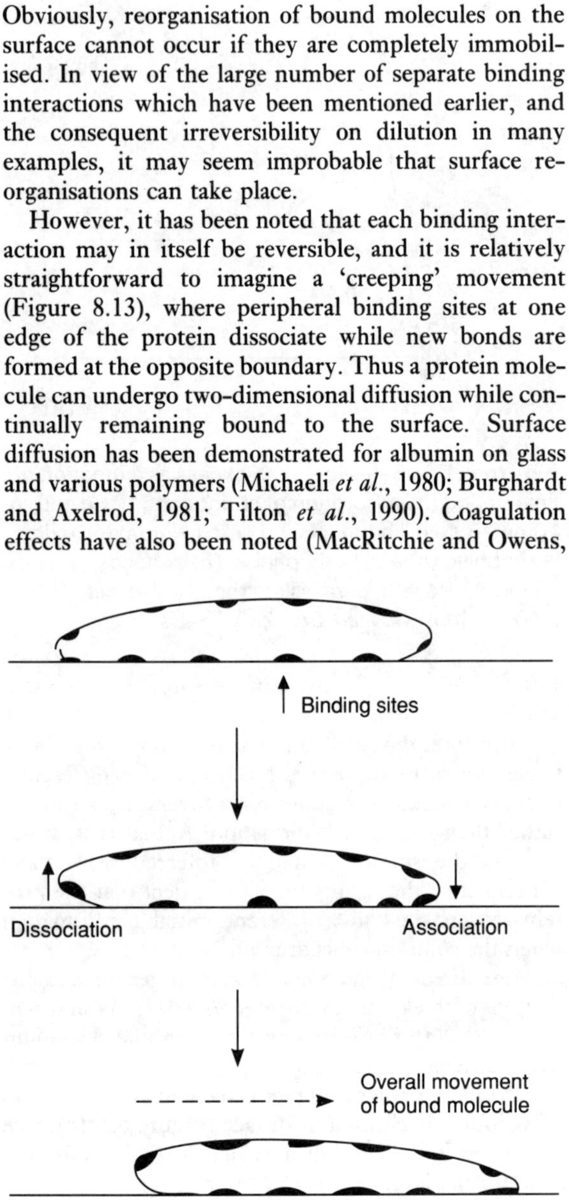

Figure 8.13 Creep of a protein molecule on a surface, due to a fortuitous combination of association and dissociation at opposite ends of the molecule.

The mobility of the films themselves may aid the rearrangement of the protein molecules under these conditions, however.

Further strong evidence for surface movement of bound proteins comes from the examination of mixed solutions, notably the Vroman effect which describes the adsorption of plasma proteins (Vroman and Adams, 1969; Andrade and Hlady, 1986). A sequence of events is observed whereby the surface is initially covered by high-concentration proteins of low avidity. As time progresses, however, these are gradually displaced by stronger-binding proteins which are present at lower concentrations, and the sequence continues until the proteins with the highest binding affinities become bound, regardless of their solution concentration. For example fibrinogen is one of the earliest proteins to be adsorbed to a surface, but in a protein mixture it is soon displaced by competition from more avid components; the most powerfully bound species are thought to include high-molecular-mass kininogen (HMWK). The exact breakdown of binding is dependent on the polymer under examination (Lee *et al.*, 1974; Norde, 1986); the relative order of binding can be reversed in some cases.

Clearly displacement events such as these can only take place if the early binders can be dislodged from the surface. Given the multiple binding site model this can only occur by a gradual sequence of events, the competing protein first gaining a small 'foothold', then covering new positions as these are vacated by the equilibrium dissociation of peripheral sites in the original protein. The low-avidity proteins are gradually 'unpeeled' from the surface.

As discussed earlier, large molecules possess high avidity due primarily to the high number of binding sites: it is unlikely that the potential binding regions of larger molecules generally display greater individual affinity for the surface than their counterparts on smaller molecules. Thus the primacy of larger molecules is driven, predominantly, by the greater probability of maintaining enough binding sites simultaneously to avoid desorption.

Pursuing this argument a little further, similar competition between identical molecules will also take place: when binding occurs from a homogeneous solution, a number of molecules initially gain a small foothold and then gradually spread out as new sites become available (this model is supported by calculations of the

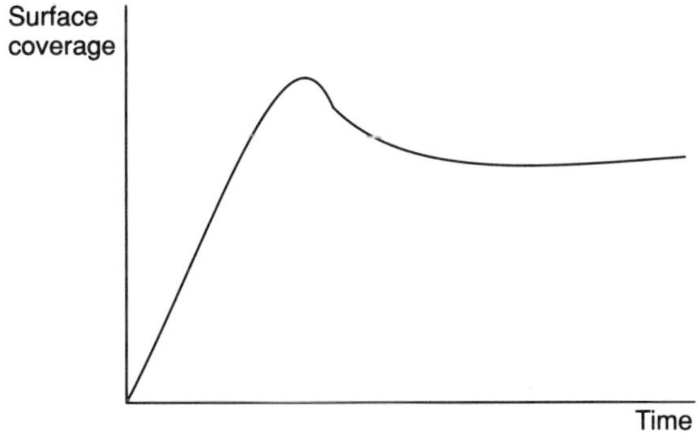

Figure 8.14 The overshoot phenomenon. See text for details.

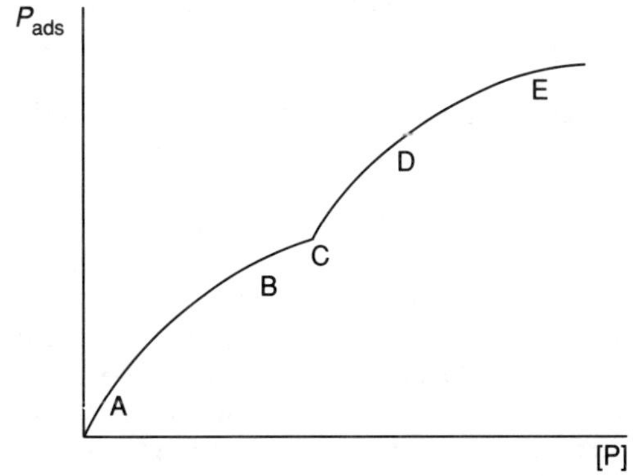

Figure 8.15 Bimodal adsorption behaviour for albumin and γ-globulin on polystyrene.

areas required for initial adsorption (MacRitchie, 1978; Andrade and Hlady, 1986), which are typically (1) much smaller than, and (2) roughly independent of, the cross-sectional area of the protein).

Although the affinity constants for binding individual regions of the protein to the surface are identical, some molecules may lose their footholds simply due to a sequence of 'failures' in competition for sites. Exchange of radiolabelled albumin with a concentrated solution of unlabelled albumin has confirmed this effect (Brash *et al.*, 1974): after one hour 10% of the labelled material had been exchanged, and after nine days the turnover level had reached 85%. This expulsion of some bound molecules offers one possible explanation for 'overshoot' phenomena (e.g. van Dulm and Norde, 1983), where the surface coverage passes through a maximum then declines after a longer adsorption time (Figure 8.14); the overall coverage may be reduced as individual molecules displace 'competitors' to widen their own surface contact areas.

Fair and Jamieson (1980) observed bimodal adsorption behaviour for albumin and γ-globulin on polystyrene. Light-scattering experiments revealed no change in the thickness of the bound layer between the first and second phases of the isotherm, eliminating both the formation of a second layer or a shift from 'sideways-on' to 'end-on' alignment as explanations. They therefore proposed a 'crystallisation' model (Figure 8.15). In the first phase the surface layer is composed of random isolated molecules, with a high content of unoccupied space (A). As the concentration is raised, the capacity for this type of binding begins to become exhausted and the isotherm begins to level off (B); at a certain stage (C), however,the coverage becomes sufficient to allow intermolecular interactions, and nucleation to a more close-packed pseudo-crystalline structure begins to occur. The isotherm therefore goes through an intermediate 'glassy' phase (D), with the coverage gradually increasing as the ordered structure becomes more widespread. Eventually the surface becomes saturated even with close-packing, and the isotherm levels off again (E). One interesting property of this model is that, unlike multilayer approaches, it predicts *lower* desorbability of protein bound at higher concentrations, which has been observed under some circumstances (Norde, 1986).

There have been a certain number of observations which suggest that clustering of adsorbed molecules may take place on hydrophobic surfaces. Ratner *et al.* (1981) observed using ESCA that haemoglobin adsorbs homogeneously onto platinum, but in 'islands' on polytetrafluoroethylene. Nygren (1988), too, observed from electron microscopy that the early binding of ferritin onto hydrophobic quartz took place in discrete clusters.

Børmer (1989) reported that immobilised mouse IgG was able to fix complement factors C′3 and C1q, a property consistent with clustered antibody.

Butler *et al.* (1992) noted that immobilised polyclonal antibodies behaved as if they were cross-linking multivalent antigens, even though the proportion of antibody retaining its binding activity was small (5–10%). This was rationalised on the basis that the surviving functional antibody was present in active clusters, surrounded by areas of denatured material. Some confirmation for this hypothesis was observed in scanning electron micrographs of the coated surface, which showed large, heterogeneously dispersed areas of binding activity.

Similarly, Davies *et al.* (1995) observed by scanning tunnelling microscopy that only a small fraction of an antibody-coated polystyrene surface retained its immunological activity; also, various partially denaturing pretreatments yielded aggregates whose immunological activity was retained.

Cluster formation could take place in a number of ways, particularly in cases involving protein–protein interactions as described by Butler. Either the adsorbed protein could spontaneously form clusters on the surface, or this aggregation could be prompted by complexation with other proteins (e.g. the formation of antibody–antigen complexes). Alternatively, aggregation could take place in the coating solution itself (consistent with the common observation of a binding maximum at the p*I* of the protein, where aggregate formation is favoured), followed by preferential adsorption due to the greater molecular mass of the aggregated species. The state of knowledge in this area is insufficient to judge between these possibilities, but in any case the concept of isolated, active clusters adds an interesting extra dimension to the understanding of protein adsorption.

8.2.1.6 Models incorporating surface conformation changes

Butler's report of isolated, active clusters also noted the deactivation of the other bound material, which can be understood therefore to have undergone conformational changes of the type discussed in Section 8.2.1.1. A number of theoretical approaches have been developed which attempt to add the capacity for surface conformation changes to basic Langmuir-type models.

Beissinger and Leonard (1982) proposed a fairly simple model (Figure 8.16) whereby the bound protein could undergo an irreversible transition to a second form with a different dissociation constant. This was based on classical Langmuir–Hinshelwood kinetics, the conversion of the bound protein to a form with different desorption properties being viewed as analogous to a chemical reaction catalysed by a solid phase.

Data from the (separate) adsorption of albumin and γ-globulin to quartz could be adequately fitted by the model, and suggested that the desorption rate of the modified conformation was around 100-fold lower than that of the initial state. With γ-globulin, this change was accompanied by a reduction in the surface area per molecule, a finding reminiscent of the model proposed by Fair and Jamieson which has just been discussed.

Soderquist and Walton (1980) described a slightly more complex model in which the reduction in dissociation constant could continue in a time-dependent fashion, eventually to a point where the binding was irreversible (Figure 8.17). Desorbed material was considered more or less unable to readsorb.

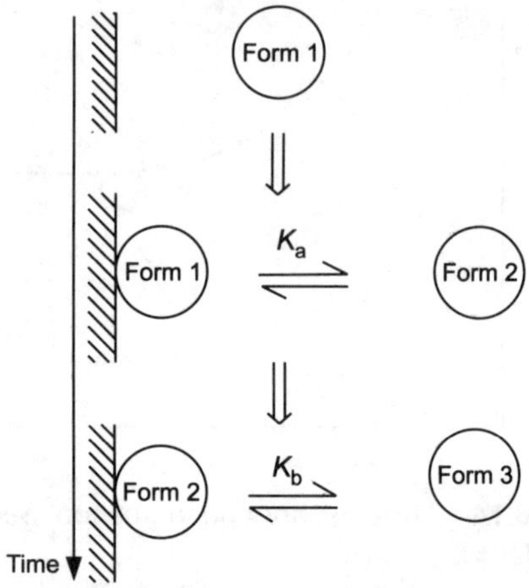

Figure 8.16 A model for surface conformation changes described by Beissinger and Leonard (1982).

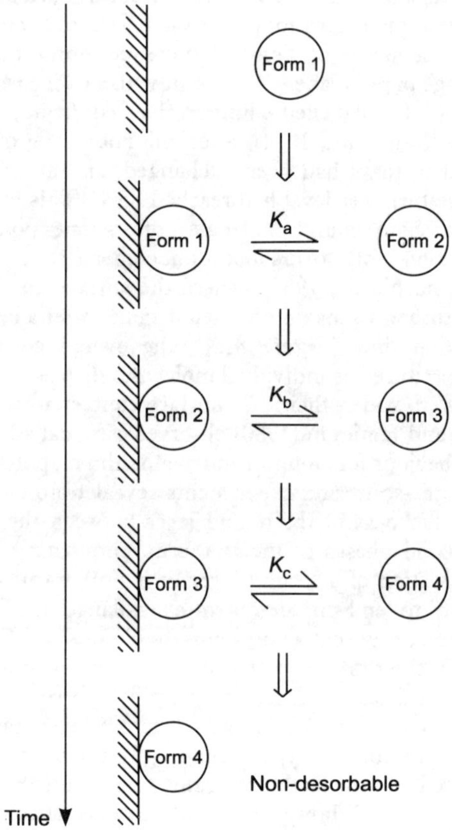

Figure 8.17 A model for surface conformation changes described by Soderquist and Walton (1980).

An equation was derived combining the binding rate with the decreasing rate of desorption: it tends towards the Langmuir equation at time zero, and towards fully irreversible binding at infinity. Soderquist and Walton's description of the processes involved allows for many observed binding phenomena: ordering of bound molecules at higher surface concentrations leading to closer packing (cf. Fair and Jamieson), optimisation of surface interactions with some loss of 'unlucky' molecules (cf. the overshoot phenomenon), and reduced desorbability of those retained.

Although this approach has been criticised for its failure to describe much experimental data in detail (Norde, 1986), Andrade and Hlady (1986) have suggested that the realistic nature of its general properties makes development of this theme worthwhile. They proposed a general model based on this approach, but incorporating rapid renaturation of desorbed molecules. Another model based on the gradual reduction of the dissociation constant with time was presented by Lundström and Elwing (1990), this time focussing more strongly on the exchange of bound molecules.

8.2.1.7 Overview of models of protein physisorption

Many different events are possible during protein physisorption, resulting in heterogeneity in the state of the bound protein. As noted by Horbett and Brash (1987), these variations can be caused by a variety of factors: occupancy effects, adsorbate–adsorbate interactions, conformational heterogeneity of the protein in solution, conformational changes during or after binding, differences in orientation of the adsorbed molecules, or in their number of links to the surface. Examination of coated surfaces with techniques such as electron microscopy has revealed further complexities, for example the reticulated surface films observed with some proteins (Eberhart *et al.*, 1987).

It is not surprising, therefore, that it has proved difficult to find satisfactory models of physisorption behaviour which deal with even a few of these complexities. Even for the relatively simple situation of adsorption of a single protein from solution, the various models which attempt to take into account changes in surface conformation, desorbability, resorbability, etc., have all been criticised for artificiality in certain aspects. Although they have successfully explained many observed phenomena, this has often been achieved to a large extent using simple, Langmuir-based approaches.

As the quantitative aspects of these models are still open to some question, they are probably best viewed as qualitative explanations of physisorption behaviour. As Andrade and Hlady have observed, a useful general model incorporates initial reversible binding, time-dependent conformational or organisational changes which result in decreased desorbability, and renaturation of the desorbed protein in solution.

These models are mainly limited to explaining the amount of protein on the surface under a given set of conditions. The description of changes in protein functionality during physisorption has barely been touched, and is much more complex still. Retention of functional activity will be influenced by steric effects during binding and surface reorganisation (including aggregation), and by conformational changes which are themselves controlled by many individual factors.

The quantitative description of protein physisorption behaviour seems likely to provide scope for research for many years to come.

8.2.2 Key process variables of protein physisorption

Many of the variables which control protein physisorption have already been touched on in the preceding sections, but it is useful to draw them together briefly here, along with some which have not been dealt with so far.

8.2.2.1 The nature of the protein

Where the aim is to use physisorption of protein as a physical technique, irreversible binding is obviously the ideal. The main requirement is that the protein possesses enough hydrophobic character to promote interactions with the surface, which is true in nearly all cases. The main exceptions are highly glycosylated proteins, which are so polar (and strongly solvated) that the chance of hydrophobic interactions occurring are very low. Similarly, variations in the degree of protein hydration have been related to the strength of hydrophobic binding interactions (van Oss and Singer, 1966); these can be estimated from the solubility of the protein in dehydrating media (Keller and Block, 1960).

A number of techniques have been employed for increasing the binding of proteins, which are probably related to increasing the hydrophobic character: Ishikawa *et al.* (1980) observed that IgG displayed better binding to polystyrene after treatment at pH 2.5, where a conformation change of the Fc region has been shown to occur (Hong and Nisonoff, 1965). Similarly, Vandenbranden *et al.* (1981, 1982) observed increased binding of a fraction of IgG to model lipid membranes after treatment at pH 2–3, and postulated that the molecule could exist in two isomeric forms, one possibly possessing more exposed hydrophobic groups in the hinge region.

Conradie *et al.* (1983) confirmed these observations, showing that 2.5 represented the optimum pH for the promotion of IgG immobilisation. They also noted enhanced levels of binding following partial denaturation by 6 M urea, or heat-treatment at 70°C. The treated proteins were investigated using 1-anilino-8-naphthalene sulfonate (ANS), their increased hydrophobicity being demonstrated by changes in the fluorescence properties of the bound ligand.

Less rigid proteins are more likely to be able to unfold to some extent on the surface, which will promote the maximisation of contact area to give stronger binding. Morrissey and Stromberg (1974), for example, showed that the binding to silica of albumin, a rigid protein, was increased by the reduction of its disulfide links, presumably *via* the increase in flexibility. The conformational changes which accompany such increases in binding may have serious effects on functionality, however, see Section 8.2.1.1). So more compact, non-deformable proteins may show a higher retention of activity even if their binding levels are lower.

Another characteristic of the protein which may be relevant is its p*I*, as maximum binding is often observed at this pH. Away from these conditions, intra- and intermolecular repulsions are likely to interfere with surface coverage, and there may also be repulsion from the surface if it bears similarly charged groups too.

The molecular mass of the protein may also be relevant: in competitive situations smaller proteins will display a greater tendency to be displaced. Even in the absence of competition, larger proteins will tend to show less desorption due to the larger number of binding sites likely to be operating.

The purity of the protein may also have a significant effect on its binding properties, particularly the presence of bound ligands or prosthetic groups. For example the binding of albumin is affected by its content of bilirubin or fatty acids, while the properties of haemoglobin depend on the coordination state of the haem group (Andrade and Hlady, 1986).

Derivatisation of the protein can obviously alter its binding properties, and the dangers of such effects in experiments using labelled proteins will be dealt with more fully in Chapter 11. Common labelling procedures such as radioiodination have been shown to interfere with binding, particularly at high ^{125}I incorporations (Walsh and Gosling, 1986).

Detailed knowledge of the three-dimensional structure of many proteins is now available from X-ray crystallography, and this can be manipulated to allow predictions of surface behaviour (Andrade and Hlady, 1986). Site-directed mutagenesis can be employed to investigate the effect on surface properties of changing individual amino acid residues (Horbett, 1988). Conclusions from work of this kind can demonstrate large changes in behaviour from small structural modifications, e.g. the dramatic differences in surface activity at the air–water interface, observed among haemoglobin variants differing sometimes by only a single residue (Horbett and Brash, 1987).

8.2.2.2 The nature of the surface

As discussed earlier, it would appear that the best surfaces for passive adsorption offer a combination of hydrophilic and hydrophobic properties. Within this broad rule of thumb, one might expect the exact surface composition to have a significant effect on protein binding, particularly the distribution of hydrophilic and hydrophobic regions at the Ångstrom level. However, to date there has been very little work in this field.

While electrostatic effects are less important they should not be ignored; in particular charged groups will require the incorporation of counter ions into the adsorbed phase (Norde, 1986), and this is a highly unfavourable process energetically. Therefore surfaces with a low electrokinetic charge (the charge of the surface itself, plus its associated solvent molecules) are favoured for binding.

The macroscopic heterogeneity of a surface may complicate attempts to model adsorption behaviour, but is generally unlikely to have a dramatic effect on preparative physisorption, unless surface differences are so large as to cause batch-to-batch variation in coating behaviour.

More likely to be a problem for protein immobilisation is surface contamination: binding based on hydrophobic interactions is unavoidably sensitive to agents such as detergents, which act to reverse the surface properties. Contamination with hydrophobic species such as hydrocarbons may also interfere with binding, as indicated by improved performance of polystyrene plates after degreasing with methanol (van Oss and Singer, 1966). The cleanliness of the surface is therefore an important factor in obtaining optimal performance, particularly in achieving within and between-batch consistency.

8.2.2.3 Protein concentration

The concentration of the protein solution may affect the coating process in a number of ways. First, the kinetics of the overall process will be strongly dependent on whether the transport or adsorption stage is rate-limiting, and the likelihood of the former can be minimised by working at a higher protein concentration. Obviously the concentration at which diffusion control is overcome will depend totally on the system concerned, but as a rule of thumb protein levels of 10–100 μg/ml are generally sufficient. Herrmann and Collins (1976), for example, observed that a 100 μg/ml solution of IgG showed no increase in binding to polystyrene between five and 18 hours after coating was started. A 1 μg/ml solution, however, showed a 27% increase over this period, suggesting that transport to the surface remained significant over this timescale.

As a consequence, too, mixing is unlikely to have an impact on the adsorption rate at high protein concentrations[1], whereas it may be a significant factor under more dilute conditions.

Second, as seen in the various isotherm models discussed earlier, the protein concentration may affect the kinetics of the adsorption stage. Often bimodal behaviour is observed, a different relationship between solution concentration and surface coverage becoming apparent at higher concentrations.

Tied in with this is the effect that concentration may have on the conformation of the bound protein. Compare rapid adsorption from a concentrated solution and slower adsorption from a more dilute one: in the latter case there will be more time for individual molecules to maximise their interaction with the surface without competition. One can imagine, therefore, that binding from the lower concentration may result in a lower level of surface coverage, but stronger binding of each individual molecule. Initially at least, binding from a more concentrated solution may result in a higher coverage of less firmly bound protein; however, surface competition may subsequently result in some molecules being expelled as others increase their contact area (the overshoot phenomenon, see above). With time, therefore, the surface may come more and more to resemble that obtained from the more dilute coating solution.

Other models of concentration-dependent conformation changes have been discussed earlier too, for example the formation of quasi-crystalline structures at high concentrations (Fair and Jamieson, 1980).

Finally, for proteins which are unstable in solution it may be advantageous to work at a high concentration, both because the stability of the coating solution is likely to be greatest under these circumstances, and because this may facilitate short coating times.

8.2.2.4 Coating time

Again, the effect of coating time can be looked at both in terms of surface coverage and conformational effects. With respect to the former, time dependence is intimately tied up with concentration; as discussed earlier diffusion control may be significant under low-concentration conditions, and coating rate will be proportional to $t^{1/2}$. At higher concentrations the diffusion element is minimised and the process is limited only by the adsorption step itself. For binding to hydrophobic surfaces this is a rather rapid process. At high protein concentrations, therefore, the bulk of the adsorption often takes place in the first few minutes, and good surface coverages can be obtained in run times under an hour (see, for example, Pesce *et al.*, 1977). However, small changes in coverage can continue to be observed even after many hours of coating: Young *et al.* (1988b), for example, observed many examples of adsorption where material was continuing to be deposited after ten hours.

The effect of time on surface conformation (and therefore on protein functionality) is harder to predict. Although the changes to the surface concentration over long coating periods may be relatively small, conformational changes of the type described in the last section may occur to a lesser or greater degree for a substantial period. Careful consideration therefore needs to be given to coating time as a process variable.

Needless to say the possibilities for long-term effects of all kinds become even greater if mixed protein solutions are coated, the Vroman effect in general promoting a gradual move towards predominance of large molecules.

Whether a single protein or a mixture is employed, further complexity may be introduced by any changes which take place in the coating solution with time; events such as aggregation may obviously cause a gradual alteration of the binding properties, and result in 'run drift' effects.

[1] Note that experiments on the effects of mixing need to be interpreted with care. In situations where only part of the solid surface is in contact with the coating solution (e.g. partially-filled microtitre wells), mixing may result in increased adsorption simply because the turbulence increases the contact area. Also, very vigorous mixing can cause temperature rises in the solution, again with possible effects on binding.

8.2.2.5 Coating solution pH

As discussed earlier, the optimum binding performance is often obtained at the p*I* of the coating protein, and various explanations have been proposed for suboptimal performance under conditions where the protein bears an overall charge.

It has also been mentioned above that pH-dependent conformation changes can be exploited to promote binding, for example the exposure of new hydrophobic regions in immunoglobulins by acid treatment. The arrangement of chargeable groups within the protein molecule has a significant effect on pH-sensitivity (Norde, 1986), *via* the variation in attractive and repulsive effects seen as the pH is altered: thus some proteins (e.g. albumin) show considerably greater expansion than others (e.g. ribonuclease) as the pH moves away from the p*I*.

In the absence of phenomena of this type the effect of pH is often quite small (Butler, 1992). It may be more significant for heavily glycosylated proteins (Elkins *et al.*, 1990), or where the ionic strength of the coating solution is too low to suppress electrostatic effects. The pH-dependence of coating may vary with protein concentration: Butler *et al.* (1992) observed that rabbit IgG at about 700 ng/ml showed 60% binding regardless of pH changes from 4.5 to 9.6. In the same system but with an IgG concentration of 1 ng/ml, the binding rose from 10 to 85% as the pH was increased.

A degree of buffer capacity is advised for coating solutions, as pH changes can be observed in unbuffered matrices due to exposure of charged groups in the adsorbed protein (Kochwa *et al.*, 1967).

Permanent, pH-dependent changes to the surface rather than the protein are rarely considered, but interesting effects have been observed. Kuen *et al.* (1993), for example, observed that microtitre wells from a variety of sources showed higher non-specific binding of serum IgM if they had been treated at high pH (9.6), even though they had subsequently been blocked with BSA.

8.2.2.6 Coating solution ionic strength

As would be expected from a process driven by hydrophobic interactions, a moderately high ionic strength facilitates binding. This effect can be significant even if the subsequent binding is covalent (Smalla *et al.*, 1988), presumably providing a valuable initial impetus to localise the protein molecules near the reactive surface.

The medium effect of transfer of ions from the solution to the adsorbed protein environment (Section 8.2.2.2) is minimised where they bear multiple charges (Norde, 1986), therefore the presence of high-valency ions may promote adsorption.

Some ions may have specific affinities for proteins, and may therefore affect binding behaviour: Zn^{2+}, for example, stabilises the oligomerisation of insulin, and causes increased binding to chromium surfaces (Arnebrant and Nylander, 1988).

8.2.2.7 Other features of coating solution composition

It has been noted above that contamination of the surface by detergents can interfere with binding, and this is equally true where the detergent is contained in the coating solution (Kenny and Dunsmoor, 1983; Gardas and Lewartowska, 1988). This situation may crop up with membrane proteins which require the presence of detergents for solubilisation; covalent coupling techniques may avoid these problems (Newman *et al.*, 1981), although acceptable coating of such materials can be obtained in the presence of some proteins (Palfree and Elliott, 1982).

The presence of chaotropic agents is also predicted to interfere with hydrophobic binding (Norde, 1986), and van Oss and Singer (1966) observed interference from thiocyanate in the binding of IgG to polystyrene latex. However if the structure is disrupted enough to cause denaturation, the reverse effect may be observed due to the exposure of hydrophobic regions (see Section 8.2.2.1). So chaotropic agents such as thiocyanate have also been observed to promote increased binding (Muñoz *et al.*, 1986).

8.2.2.8 Coating temperature

Hydrophobic interactions are promoted by raised temperature, and it is therefore not unexpected that increased adsorption is often observed as the temperature is raised (Dillman and Miller, 1973; Norde, 1986). However the opposite effect may also be observed (Wahlgren and Arnebrant, 1991).

Norde (1986) has noted that there is an interaction between pH and temperature, in that temperature effects are sometimes more pronounced at a pH away from the isoelectric point.

In a manner analogous to pH-dependent conformational shifts, temperature-dependent changes in protein structure can have a dramatic effect on coating behaviour. β-Lactoglobulin, for example, shows a drastic increase in binding to chromium at the denaturation temperature, probably related to surface aggregation (Nylander, 1987, described in Wahlgren and Arnebrant, 1991).

8.2.2.9 Post-coating processes

After the coating stage itself, a number of processes may be required before the coated component is in a suitable state for use (or, more often long-term storage and use). They have received very little attention in the literature, and only very broad guidelines can be offered here.

Aspiration

The means of removing the 'spent' coating solution (and solutions for other purposes which are about to be described) will depend totally on the geometry of the solid phase. The simplest approaches such as decanting of liquid from components coated by immersion, or centrifugation and decanting of particulate suspensions, are unlikely to have a significant effect on the protein molecules. However, those which involve air flow (such as filtration of particulates or vacuum aspiration of microtitre wells) have the potential to cause harmful drying effects.

Washing

Often a number of washing stages are incorporated after the coating, the physical method again being defined principally by the solid phase geometry. Not only do these remove residual traces of the coating solution, but, remembering that protein physisorption is only partially irreversible, they may be employed to remove material which could otherwise be released during the use of the component. Unsurprisingly, the composition of the wash buffer and the mechanics of washing can have a critical effect: Lehtonen and Viljanen (1980) reported that changes in the level of BSA adsorbed to polystyrene were observable after six washes, and interestingly that the difference between coating at 0.1 and 10 mg/ml was almost obliterated after this treatment (Table 8.1).

Table 8.1 BSA detachment during repeated washes of polystyrene cuvettes.

Number of washes	BSA concentration on the surface (ng cm^{-2})[a]	
	10 mg ml^{-1} [b]	100 μg ml^{-1} [b]
2	570 ± 160	130 ± 23
3	120 ± 20	58 ± 8.5
4	85 ± 22	48 ± 9.4
5	55 ± 4.0	46 ± 7.9
6	46 ± 5.4	32 ± 7.8

(Reproduced with permission from: Lehtonen and Viljanen (1980). *J. Immunol. Methods*, **34**, 61–70.)
[a] Measured in triplicate. Mean ± S.D. are given.
[b] Concentration of BSA in the coupling solution.

It is common to include detergents to desorb weakly-bound protein, say 0.1% of Tween™ 20. The incorporation of a detergent has also been shown to lead to improved functional performance of bound proteins (Wilkins Stevens *et al.*, 1995), again emphasising the complexity of this topic.

Overcoating

During or after washing, an overcoating stage may be incorporated, where another coating protein is introduced. This may reduce subsequent non-specific binding, or have a beneficial effect on the stability of the component. Comparitive studies are few and far between, but Vogt *et al.* (1987) demonstrated significant differences between proteins in this role: casein and dried milk powder gave the best performance. Graves (1988) showed large differences in non-specific binding of IgG, to surfaces coated with proteins and polypeptides with different charge properties, anionic materials giving the lowest levels of adsorbance.

Drying

Next, it is generally necessary to dry the coated components for storage, and this can be carried out by combinations of air flow, warming, humidity reduction, etc. It is fairly straightforward to dry residual quantities of wash buffers (bearing in mind, of course, that any non-volatile components of these buffers will be dried down onto the surface), but some care must be taken to ensure that damage to the protein is not caused by over-drying. This is best achieved by gentle drying procedures and careful examination of other parameters such as overcoating.

Storage

Finally the dried components may need to be stored for prolonged periods under variable conditions of humidity (which can be controlled to some extent by careful packing) or temperature.

These post-coating processes of aspiration, washing, overcoating, drying and storage all have their own key process variables, which may have a critical effect on the functional performance of the immobilised protein. Detailed investigations of these effects are virtually absent from the literature; this seems a fertile area of investigation for optimising the performance of protein-coated solid phases. The importance of this area is also largely independent of the means of coating employed, these processes can be equally critical where covalent (chemisorption) methods are used.

8.2.3 Surface concentrations achievable by protein physisorption

It is convenient to have a rough rule of thumb to predict how much protein can be expected to adsorb to a given area of surface. Such estimates are most easily used for surfaces which are fairly smooth, such as moulded polystyrene. For other solid phases such as porous particles or membranes the actual area may be difficult to estimate from geometrical dimensions, and special techniques may be required to estimate this parameter. Nyholm and Ramlau (1988) have estimated that the actual surface area of a membrane may be up to 1000-fold greater than the planar dimensions. Surface areas of particles may be of the order of a few hundred m^2 g^{-1}.

Calculations of the surface concentration resulting from monolayer coverage can be made, if the cross-sectional area of the molecule and its molecular mass are known. Estimates of the molecular area vary fairly widely depending on conditions and the measurement techniques employed: the reported dimensions of IgG, for example, yield 'sideways-on' areas of 6000 Å^2 (electron microscopy in the dry state; Edelman and Gall, 1969), 9300 Å^2 (hydrodynamic measurements; Walsh and Gosling, 1986); 10300 Å^2 (hydrodynamic measurements; Oncley *et al.*, 1947), 14400 A^2 (hydrodynamic measurements; Valentine and Green, 1967). The threefold variation apparent in this range obviously leads to some uncertainty, but the 'sideways-on' monolayer coverage can be calculated as between 230 and 560 ng cm^{-2} using the extreme values[1]. Similar figures are obtained for other moderately sized proteins (Young *et al.*, 1988a): 200 ng cm^{-2} for human albumin, 270–670 ng cm^{-2} for human fibrinogen. The higher values presented here are based on theoretical close-packing of dehydrated proteins; for proteins immobilised from an aqueous solution, the larger hydrodynamic dimensions might be expected to apply, at least in the absence of lateral interactions such as surface aggregation. Therefore coverages at the lower end of these ratios probably reflect the most realistic estimates of a 'sideways-on' monolayer.

Higher incorporations are clearly possible if a smaller fraction of the protein molecule is in contact with the surface. MacRitchie (1978) has shown that the area required to provide a foothold for a protein molecule is around 100–200 Å^2, more or less independently of the molecular mass—lysozyme (15 kDa) required 100 Å^2, myosin (600 kDa) 145 Å^2. For moderately-sized proteins this area is considerably smaller than the 'end-on' dimensions of the molecule (typically 1000 Å^2 or higher) and it can be safely assumed that 'foothold availability' will not limit the maximum coverage achievable[2]. It can be assumed, therefore, that a close-packed 'end-on' arrangement represents the likely upper limit of monolayer coverage: for IgG this yields surface concentration in the range of 1000–2400 ng cm^{-2}, depending on the dimensions assumed: again the lower end of the range is probably more realistic.

Generalising for proteins of moderate size and asymmetry, then a surface concentration in the region of 100–1000 ng cm^{-2} is consistent with monolayer coverage, depending on the orientation of the molecules. Values significantly lower than this range reflect incomplete reaction (for example due to diffusion limitation) or inefficiencies in the surface packing, while higher values could be explained on the basis of multilayers, 'end-on' packing of highly asymmetric molecules, reduction of effective hydrodynamic radius due to surface aggregation, etc. A final possibility is inadequate washing of the coated surface before measurement[3].

In practice, observed coverages for very many proteins on a wide range of surfaces fall within a range reassuringly consistent with these calculations: 400 ng cm^{-2} for albumin on glass (Bull, 1956); 80–150 ng cm^{-2} for albumin, γ-globulin and fibrinogen on a variety of hydrophobic polymers (Brash and Lyman, 1969); 390–510 ng cm^{-2} for fibrinogen on surfaces as varied as platinum, quartz and polyethylene (Morrissey, 1977); 60–580 ng cm^{-2} for viral antigens and immunoglobulins on polystyrene (Salonen and Vaheri, 1979; Nieto *et al.*, 1986); 180 ng cm^{-2} for a haptenated albumin on polystyrene (Nieto *et al.*, 1986); 40–700 ng

[1] In molar terms this corresponds to a range of about 1–4 pmol cm^{-2}.

[2] If the protein molecules were deformable enough to allow binding of a close-packed monolayer based on this area, prodigiously high coverages could potentially be achieved, 70 μg cm^{-2} for a large molecule such as myosin, for example.

[3] As noted above, the washing stages of coating experiments have received very poor attention in many reported studies. With the partially reversible nature of protein physisorption, it is by no means straightforward to define a 'correct' washing protocol, but for comparison the methods should at least be consistent. The descriptions of such stages in most reports are very rarely detailed enough to allow this conclusion to be made. The results reported by Lehtonen and Viljanen (1980) for binding levels of BSA on polystyrene after multiple washes (see Table 8.1) are cautionary in this respect.

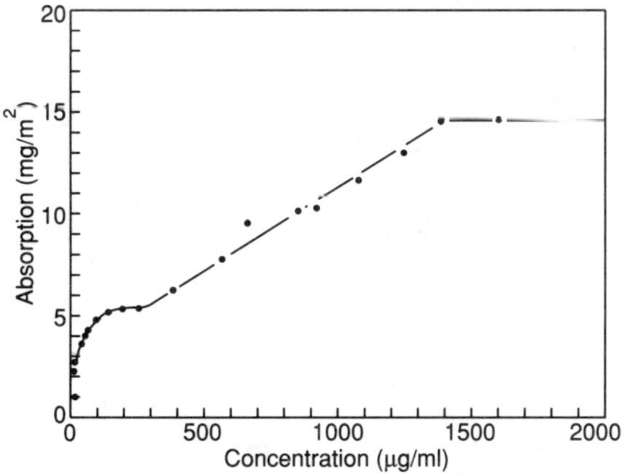

Figure 8.18 Adsorption isotherm for γ- globulin at pH 7.4. (Reproduced with permission from: Fair and Jamieson (1980). *J. Colloid Interface Sci.*, 77, 525–534.)

cm^{-2} for monoclonal antibodies on polystyrene latex (Elgersma *et al.*, 1991, 1992).

In summary, then, the observed adsorption of a wide range of proteins on a wide range of surfaces is broadly consistent with monolayer coverage. The uncertainties of molecular dimensions and orientations make it impossible to preclude the possibility of more than one layer being formed (Section 8.2.1.4); from these results, however, they certainly do not seem to suggest the formation of deep multilayers.

It is also fair to say that true 'end-on' packing is very rarely observed. An exception is probably seen in the work of Fair and Jamieson (1980), where coverages of γ-globulin on polystyrene latex reached an apparent plateau at around 1400 ng cm^{-2}. The bimodal isotherm observed by these authors (Figure 8.18) suggests to some extent why this behaviour has not been reported more frequently: working at protein concentrations up to around 250 μg ml^{-1} would suggest that saturation had been reached at around 550 ng cm^{-2}, the final plateau only being observed where solutions of greater than 1 mg ml^{-1} were employed.

8.2.4 Variations on protein physisorption

8.2.4.1 Multi-stage physisorption

As observed in the last discussion, there is little evidence that non-specific protein–protein interactions typically lead to the formation of multiple layers on the coated surface. However, there are of course a multitude of specific ligand–binder interactions which proteins may take part in, and these have sometimes been used to immobilise proteins indirectly *via* a multi-stage process. A number of possible benefits can be obtained from such an approach.

First, the use of a binder layer may make it possible to bind proteins which would otherwise be difficult or impossible to immobilise (Figure 8.19(a)). For example, Robinson *et al.* (1990) observed no direct binding of detergent-solubilised HIV-1 glycoproteins to polystyrene microtitre wells, but were able to immobilise them successfully *via* a layer of the lectin concanavalin A. Second, the selectivity of the binder layer may make it possible to coat only the desired component of a mixture, so that a prior purification step is rendered unnecessary (Figure 8.19(b)). In the example just described, Robinson *et al.* were able to coat the viral glycoproteins direct from a serum-free culture medium.

Third, if the selectivity of the binder is reasonably broad, a single coated solid phase may be usable for many different individual applications, for example Protein A-coated gels or glass beads as universal carriers for IgG (Gersten and Marchalonis, 1978; Phillips *et al.*, 1985).

Finally, it has been seen how both steric effects and conformational changes on the surface can lead to drastic losses of protein functionality after immobilisation (Figure 8.19(c)). Butler *et al.* (1992) observed virtually complete inactivation of some monoclonal antibodies, while Berkowitz and Webert (1981) observed 90% loss of HRP activity. Changes in antigenicity are another symptom of surface conformational changes which can be problematic.

The conformational element of this inactivation can potentially be minimised by coating the surface with a suitable binding protein, which then serves to immobilise the protein of interest (Figure 8.19(d)). The interaction with the binding protein is likely to be non-denaturing, and consequently there will be little loss of functionality or antigenicity. Ngai *et al.* (1993), for example, showed that immobilising native and denatured cytochrome c *via* Protein A and a specific antibody preserved their antigenic differences.

Butler and his colleagues as the University of Iowa have carried out much more detailed quantitative investigations in this area (Suter and Butler, 1986; Suter *et al.*, 1989; Butler *et al.*, 1992). Using a solid phase coated sequentially with a biotinylated carrier protein and streptavidin (the Protein Avidin Biotin Capture system, PABC) they were able to immobilise monoclonal antibodies with up to 400-fold greater retention of immunoreactivity; in some cases greater than 25% of the initial activity could be preserved.

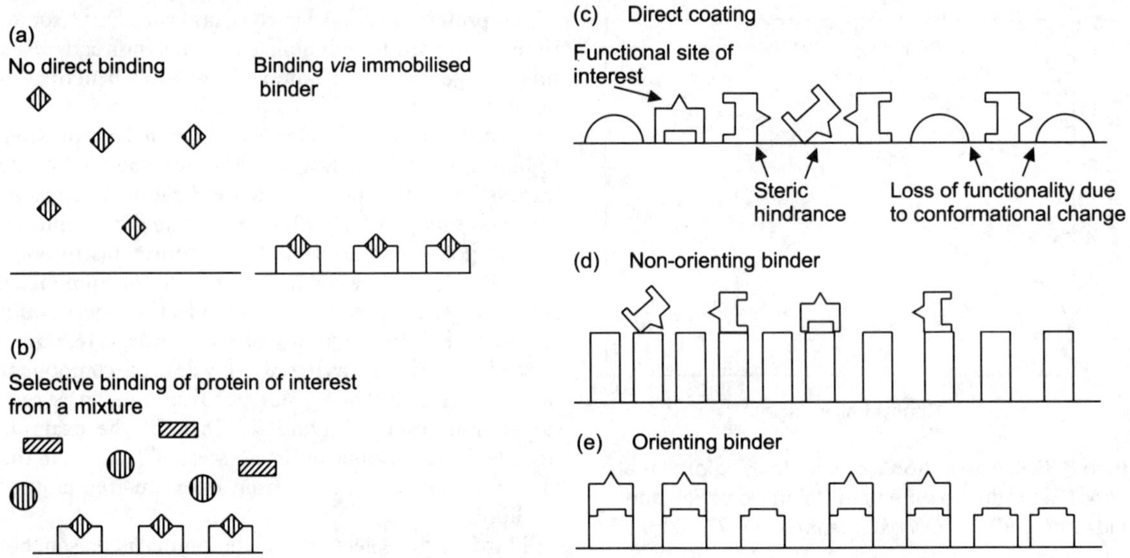

Figure 8.19 Mechanisms by which multiple adsorption strategies can lead to improved surface properties. A binding layer can allow immobilisation of a protein which will not bind *per se* (a), and if the interaction is specific this can even allow selective binding from a mixture (b). Direct binding (c) may result in losses of functionality due to conformational changes or simple steric hindrance. A non-orienting binder layer may prevent the former (d), while an orienting layer may prevent both effects (e).

Although these improvements are impressive, there is still likely to be some inefficiency due to steric effects: the random biotinylation of the antibody molecules will result in some immobilisation *via* the immunoreactive regions.

Even better in principle, then, is the possibility of choosing a binder which facilitates the desired orientation of the protein of interest, too (Figure 8.19(e)); in this way the remaining steric limitations can be minimised as well. Butler *et al.* (1992), for example, used a polyclonal anti-globulin antibody as a binder for monoclonals; the specificity of the polyclonal is predominantly towards the constant (non-binding) regions of the monoclonal, which is therefore most likely to be bound in the correct orientation for retention of immunoreactivity. Using this approach, over 70% of the activity of bound monoclonal could be retained in some cases.

The comparison of these different approaches should not be made strictly on the basis of percentage retained activity, however. The binder layer is of course prone to the same steric and conformational losses shown in Figure 8.19(c), with the result that the number of available binding sites for the protein of interest may be very small. Therefore, although the percentage of immobilised material that remains reactive is high, the actual surface concentration may be very low. When this factor was taken into account with Butler's results, the surface concentration of active monoclonal was actually lower than with the PABC system, and even, in some cases, than with direct coating.

The denaturation of binder protein was also presumably behind the sequential biotinylated carrier/streptavidin architecture of the PABC system, as it was reported that much worse performance was observed when streptavidin was coated directly (Suter and Butler, 1986). Streptavidin posseses four binding sites which make steric hindrance unlikely, so the loss of functionality was most likely due to surface conformational changes. Denaturation of a biotinylated protein layer, on the other hand, is likely to be harmless: the biotin moieties will remain intact, and denaturation of the carrier protein will if anything improve its adherence to the surface.

A problem with this multi-stage approach is that the binding of the protein of interest to the binder layer is likely to be genuinely reversible, with the consequence that a reasonably high solution concentration may be required to achieve sufficient equilibrium immobilisation. This represents a further inefficiency of such systems; unless the affinity of the binder–ligand system is so high as to be effectively irreversible (as is the case with the biotin–avidin system), the usage of the protein of interest may be quite high. Again this is apparent in the results of Butler *et al.* (1992): in terms of monoclonal usage the antiglobulin approach generally fared badly even against direct coating. The PABC system offered the best economy in most cases.

Another problem of this reversibility is the potential for unwanted desorption of the protein of interest during use; for this reason it is common in immunoassay, for example, to carry out the second stage of the binding operation *in situ*, where the solution concentration will maintain an equilibrium level of protein on the solid phase. Thus microtitre wells may be supplied which are coated with a polyclonal anti-globulin, and the immunoassay matrix will contain the specific antibody against the analyte of interest; the immobilisation of the specific antibody therefore takes place during the assay incubation.

Biotin–avidin based approaches such as Butler's PABC system again offer the potential for lower desorbability of the protein of interest due to their effective irreversibility. Of course, desorption of the entire multi-stage complex is still a possibility, but the choice of a strongly-binding biotinylated carrier protein for streptavidin immobilisation can minimise this risk. Chemical cross-linking after binding (cf. Section 8.2.4.2) is another possibility here, as exemplified by the use of bifunctional bis-imidate reagents to stabilise Protein A–IgG matrices (Schneider *et al.*, 1982; Sisson and Castor, 1990).

In summary then, multi-stage approaches can sometimes offer benefits in achieving immobilisation of 'difficult' proteins, in selective binding from mixtures, in allowing the use of a common solid phase for multiple applications, in improved retention of protein activity, and in the economy of protein usage. Their chief drawback (apart from manufacturing complexity) is the reversibility of the binding, both as a contributor to the inefficiency of usage, and as a potential cause of unwanted desorption at a later stage. Biotin–avidin systems offer a particularly good combination of properties for protein immobilisation due to their effective irreversibility.

8.2.4.2 Physisorption followed by chemical cross-linking

An alternative approach to the desorption problem is to introduce chemical cross-links between the bound protein molecules after immobilisation. Methodologically this is quite a simple modification to the process, only requiring an incubation of the coated solid phase with a suitable bifunctional reagent.

Reisberg *et al.* (1970), for example, adsorbed proteins onto bentonite to produce immunoadsorbents, then stabilised the protein layer by treatment with a carbodiimide.

Goldman *et al.* (1968) adsorbed papain passively onto collodion membranes, but observed that the enzyme was leached out after a few minutes incubation in water. When cross-linking was carried out using a bis-diazonium compound, however, there was no apparent leaching after four months storage under water. Also, provided the concentration of the cross-linking agent was kept low enough, there was no significant loss of enzyme activity; higher levels led to irreversible inactivation.

Similar 'fixing' of membrane-bound proteins using glutaraldehyde is common in blotting procedures: see for example Gershoni and Palade (1982).

In view of the methodological simplicity of this approach it is perhaps surprising that it is not encountered more frequently as a means of countering desorption.

8.2.4.3 Heterofunctional reagents for physisorption

An interesting approach to passive adsorption was reported in a patent by Pegg and Saunders (1994): reagents such as that shown in Figure 8.20 were described, which possess a phenylpentyl group for binding to hydrophobic polymers (the alkyl moiety being claimed to anchor the reagent by intercalating the polymer groups), and a number of amine-reactive succinimide ester groups for covalent binding of proteins, coupled *via* polar linker regions.

In the example quoted, a considerable increase in the detectable amount of a rabbit polyclonal IgG bound to polystyrene was observed, compared to direct coating. This may be attributed either to an

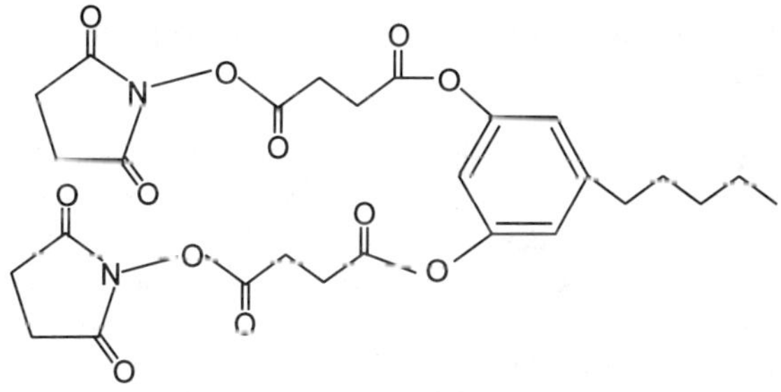

Figure 8.20 A reagent combining amine-reactive succinimidyl esters with a phenylpentyl group, which can bind to hydrophobic surfaces.

increase in surface binding capacity due to the multiple linker chains, or to a reduction in steric hindrance and/or denaturation as a consequence of the protein molecules being held away from the surface by the polar linker groups.

Somewhat similarly, Saito and Nagai (1983) coated polystyrene and polypropylene tubes with octadecylamine, which bound hydrophobically to the surface and provided a layer of amines for subsequent covalent reactions. Kennel (1982) used amine-containing lipids the same way.

Li *et al.* (1996) prepared Pluronic™ (BASF Corp.) surfactants incorporating reactive groups such as active esters and pyridyl disulfides and used these to couple proteins to polystyrene. Pande *et al.* (1996) prepared reagents combining hydrophobic tails with a biotin group, and used them to couple enzymes to a hydrophobic surface.

Chu and Orgel (1997) identified polyanionic peptides which bind strongly to hydroxyapatite, and used derivatives of these to mediate the binding of proteins: dihydrofolate reductase and avidin, for example, were coupled *via* methotrexate- and biotin-derivatised peptides, respectively.

8.2.4.4 Physisorption *via* specific binding domains

An interesting approach to passive protein adsorption is to exploit the binding properties of certain natural proteins, for example the cellulose-binding domains (CBDs) of bacterial cellulases. These bind strongly to microcrystalline cellulose matrices, and can be incorporated in recombinant fusion proteins to allow direct adsorption to materials of this type (Ong *et al.*, 1989a, b). Adapting this approach to a broader application, Ramirez *et al.* (1993) constructed a CBD–Protein A fusion, offering an alternative to existing means of Protein A immobilization.

8.2.4.5 Mixed physisorption/chemisorption approaches

A number of mixed approaches have been described where a protein is covalently coupled to a passively-adsorbed layer. Papadea *et al.* (1985), for example, prepared BSA-coated microtitre wells by physisorption, then activated these with glutaraldehyde for coupling of antibodies.

Synthetic polypeptides can be used the same way (Kennel, 1982; Hobbs, 1989), the incorporation of a large number of lysine groups offering a convenient 'handle' for covalent reactions. This is really a borderline case between physisorption of one protein followed by covalent coupling of another, and polymer-coating of a surface to introduce reactive groups. Where organic polymers are used for this purpose, the subsequent immobilisation of a protein falls truly in the realm of chemisorption (see also Section 8.3.1.1, for example).

Finally proteins can be bound by non-covalent interactions to a covalently-bound hapten, for example Varga *et al.* (1990) bound a specific monoclonal antibody to irradiated polystyrene *via* chemically-bound dinitrophenyl groups.

8.2.5 Overview of protein physisorption

For all the complexity of protein physisorption at the molecular level, its pros and cons as a preparative technique are very simple to list.

By far the biggest advantage of this approach is its simplicity: in essence the coating process requires little more than bringing the solid phase into contact with a protein solution for a short time.

There are only two major disadvantages to passive adsorption – denaturation and reversibility. Because the major drivers for the immobilisation process are hydrophobic interactions, surface conformational changes are widespread. Consequently it has been seen that a loss in activity of the bound protein is often observed, even to the extent of complete inactivation. This can make passive adsorption a very inefficient process in terms of protein usage, which may be an important consideration for scarce or costly materials. As discussed in Section 8.2.4 these losses of activity can be reduced to some extent by modified approaches.

Although passive adsorption generally displays a high degree of irreversibility, it is sometimes forgotten that small but significant percentages of the bound protein are often desorbable, and that much larger effects are likely to be observed when the coated component is used (as is often the case) in a proteinaceous medium. These can have serious effects, particularly in terms of reproducibility if the amount of material desorbed varies from one occasion to another. Again, multi-stage systems employing high-affinity ligand–binder pairs have the potential to minimise these desorbability problems; so can chemical cross-linking after coating.

In practice, the simplicity of passive adsorption methods has allowed them to retain their popularity, despite the obvious drawbacks just mentioned. There are many areas where the current level of understanding is not good, notably in the influence of solid phase ultrastructure, and in the post-coating operations discussed in Section 8.2.2.9. And while much more effort has been put into describing the behaviour of proteins during the actual coating stage, the models which have been developed are generally still rather artificial in some aspects. Therefore there still seems to be considerable scope for improving understanding in this field, which may help to overcome the limitations of this otherwise attractive methodology.

8.3 Chemisorption of Proteins to Solid Phases

As discussed above, the major drawbacks of passive adsorption are reversibility leading to desorption, and loss of activity due to conformational changes. An obvious answer to the first of these problems is to employ a covalent chemical bond to attach the protein to the surface of interest, and this is the basis of covalent binding or chemisorption techniques.

Indirectly, these methods can also (if used correctly) provide a solution to the second problem too. Because it is no longer necessary to promote hydrophobic interactions as a driver for protein adsorption, the surface can now be made hydrophilic; consequently there will be very little interaction with the protein molecule, and denaturation is unlikely, the only expected losses of activity are from steric effects, and from the chemical derivatisation of the protein. If covalent binding is intended to be the only mechanism for immobilisation, in fact, it is essential that the surface is rendered hydophilic: otherwise passive adsorption will still undoubtedly take place too, with the accompanying issues of desorbability, etc. It is not unusual for 'activated' solid phases, marketed on the basis of providing a convenient matrix for covalent binding, to possess substantial hydrophobic character. Although some covalent binding may be seen with such materials, it is easy to be misled into presuming that this is the only mechanism; it should be borne in mind that the majority of immobilisation may still be non-covalent unless the surface is hydrophilic in nature.

A final advantage which may be obtained from covalent approaches is the possibility of orienting the bound protein by using chemistry specific for a non-functional site in the molecule (Lu *et al.*, 1995, 1996). The binding of immunoglobulins *via* carbohydrate groups in the Fc region is a good example.

As with the conjugation procedures for proteins described in earlier chapters, these will be approached in terms of the groups which may be available for coupling (Section 8.3.1) and the chemistries available for creating the covalent links between the protein and the solid phase (Section 8.3.2). Finally in Section 8.3.3, examples of some of the more commonly encountered solid phase formats will be described, which are commercially available in a suitable form for covalent linking of proteins.

One specific application of covalent binding of proteins to solid phase materials is the preparation of immobilised protein matrices for affinity chromatography. Because this is a relatively specialised use, the methods and materials employed in this field will be described separately in Chapter 10, Section 10.9.3. There is however a considerable degree of overlap with Section 8.3.2.

8.3.1 Reactive groups available in the solid phase for protein chemisorption

The reactive groups in the protein which are available for chemisorption purposes are obviously no different to those described in earlier chapters: amines, carboxyl groups, thiols, and carbohydrate moieties offer the most widely used 'handles' for covalent immobilisation. A very wide range of solid phases may be encountered however, including inorganic ceramics such as glass, silica and alumina, insoluble carbohydrates such as dextrans, and organic polymers such as polystyrene, nylon and polyacrylamide.

It is useful, therefore, to review briefly the types of reactive groups which are likely to be encountered in solid phases, either in their native form, or after derivatisation of some type.

8.3.1.1 Amines

As a quick examination of earlier chapters will confirm, a large percentage of the chemistry of protein conjugation has been developed around amine groups, and it is not surprising that solid phases bearing these groups have been found useful. Conveniently, many useful solid-phase formats are commercially available with amine groups already present: microtitre wells, glass beads, magnetic particles, etc. These will be covered in Section 8.3.3.

Organic polymers can also be synthesised which contain amines, for example the glycosyl aminopolyacrylamide copolymers described by Hořejší *et al.* (1982); these can be coated onto solid components such as polystyrene wells to provide an amine-containing surface (Hořejší and Matoušek, 1985). Polypeptide copolymers containing lysine or amino-phenylalanine can also be prepared, and be used *per se* for protein immobilisation (Riesel and Katchalski, 1964) or for surface coating (Wood and Gadow, 1983; Hobbs, 1989).

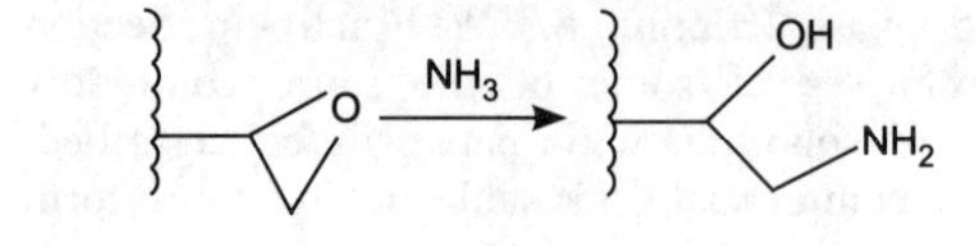

Figure 8.21 Conversion of poly(glycidyl methacrylate) to yield amine groups.

Also, polymers can be prepared containing reactive groups which can readily be converted to amines: Hosaka *et al.* (1983), for example, produced microspheres of poly(glycidyl methacrylate) which yield amine groups on simple ammonia treatment (Figure 8.21).

Often it is more convenient to use components made from common polymers, in which case chemical derivatisation of some form is necessary to introduce amine moieties.

Polystyrene can be nitrated with fuming nitric acid, and the resulting NO_2 groups reduced to amines with dithionite (Figure 8.22(a)) (Gyenes and Sehon, 1960; Filippusson and Hornby, 1970; Chin and Lanks, 1977; Reimer *et al.*, 1978).

Alternatively it can be chlorosulfonated followed by sulfonamide formation with a diamine or hydrazine (Figure 8.22(b)) (Chu and Tarcha, 1987).

Another approach with polystyrene is to introduce carboxyl residues (Section 8.3.1.3) then derivatise these with a diamine *via* carbodiimide activation (Zammatteo *et al.*, 1996). Finally, amine groups can be introduced by the irradiation or plasma treatment of polystyrene (and other polymers) in an ammonia atmosphere (Hollahan *et al.*, 1969; Jimbo and Saito, 1988; Hayat *et al.*, 1992; Matson *et al.*, 1994).

Aromatic amines such as those generated by these methods are fairly unreactive *per se*, and it is a common practice to derivatise them further with a diazotisation reaction (Section 8.3.2.1) which allows reaction with lysines, tyrosines and histidines (Gyenes and Sehon, 1960; Riesel and Katchalski, 1964; Hornby and Filippusson, 1970; Wood and Gadow, 1983).

Chloromethylated polystyrene–divinylbenzene copolymers are useful supports for the immobilisation of amino acids during peptide synthesis—the Merrifield resins (Merrifield, 1964; Horiki *et al.*, 1978)—but can also be derivatised to yield carboxyl or hydroxyl groups (Bayer *et al.*, 1971).

Polyamides such as nylon can simply be hydrolysed to generate amine groups (Figure 8.23(a)) (Sundaram and Hornby, 1970; Horvath and Solomon, 1972; Bunting and Laidler, 1974, Sundaram, 1978; Zucchini

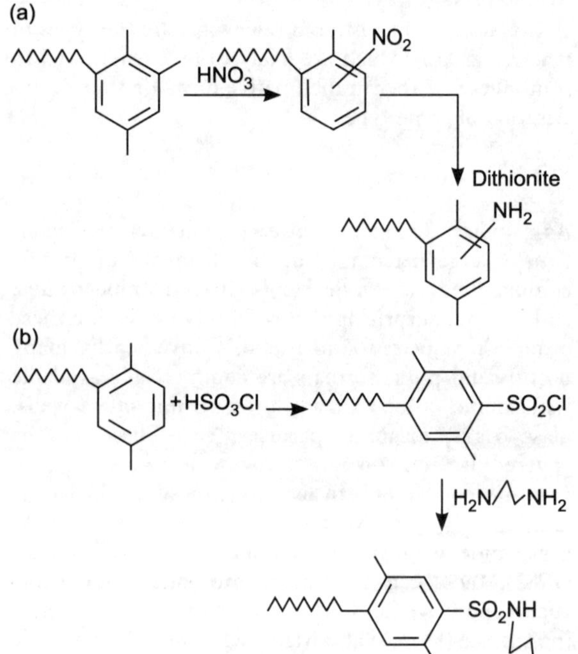

Figure 8.22 Conversion of polystyrene to amine groups.

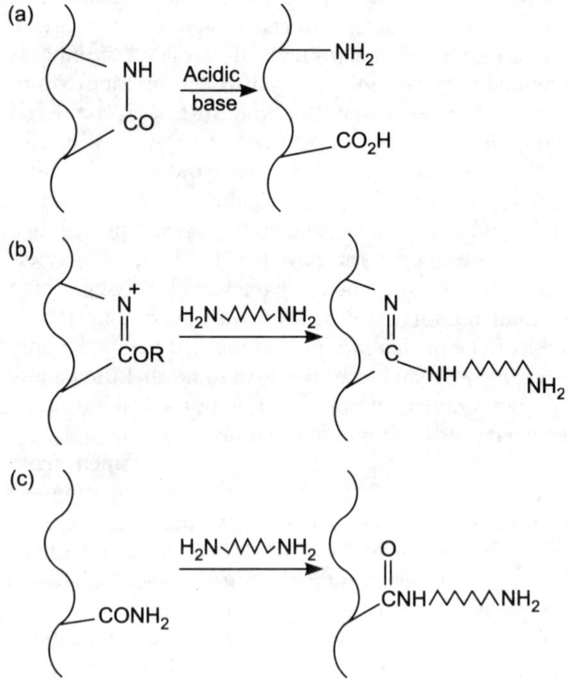

Figure 8.23 Generation of amine groups from polyamide.

et al., 1989), or the carboxylic groups exposed by the same process derivatised by carbodiimide coupling to a diamine (Hornby and Filippusson, 1970).

Alternatively, nylon alkylated to produce imidate groups (Section 8.3.1.7) can be reacted with diamines such as lysine or hexamethylene diamine (Figure 8.23(b)) (Hornby *et al.*, 1974; Campbell *et al.*, 1975; Zucchini *et al.*, 1989). It is also possible under forcing conditions to substitute amide groups directly with amines, so use of a diamine again provides an amine-bearing surface (Figure 8.24(c)), e.g. polyacrylamide and ethylenediamine (Inman and Dintzis, 1969), nylon and propanediamine (Inman and Hornby, 1974).

Hydroxylic matrices such as nitrocellulose can be derivatised with powerful bifunctional electrophiles such as divinylsulfone or a bis-epoxide, followed by treatment with a diamine to leave a free amine group at the surface (Lauritzen *et al.*, 1990). Arylamines can also be introduced into matrices of this type, for example *p*-nitrobenzoyl chloride will react with cellulose under suitably forcing conditions, reduction being required as mentioned earlier to generate the amine form. This approach has been in use for many decades (Campbell *et al.*, 1951). Acyl chlorides will also react readily with aliphatic amines, and it is therefore possible to 'convert' matrices of this type to an arylamine form, where diazotisation is the preferred chemistry for protein immobilisation (Inman and Dintzis, 1969; Weetall, 1970).

Matrices containing carboxylic groups, e.g. those based on methacrylic acid, can be coupled to a diamine *via* carbodiimide activation (Rembaum *et al.*, 1976).

Commonly encountered matrices, such as glass and silica, which possess silanol functions, can be derivatised with amine-containing silanes (Weetall, 1970, 1988; see Section 8.3.1.6).

Various methods for the determination of amine groups on solid phases have been described (Lee and Loudon, 1979).

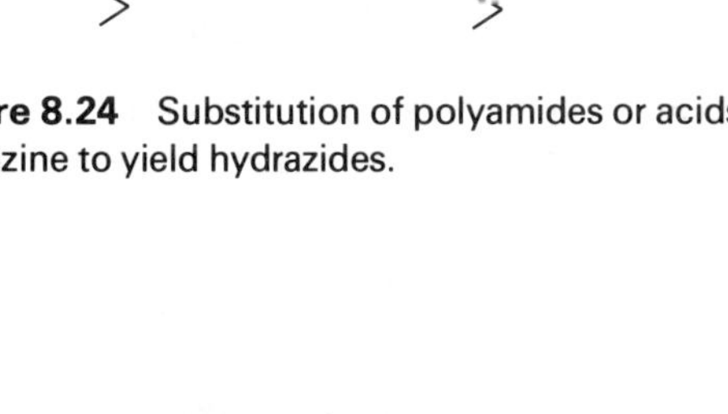

Figure 8.24 Substitution of polyamides or acids with hydrazine to yield hydrazides.

8.3.1.2 Hydrazides

Hydrazides resemble amines in their reactivity, but have a couple of useful advantages: first, their very much lower pK_a values (typically around 3) mean that they are almost always present in their reactive unprotonated form; second, they form stable hydrazones with aldehydes, unlike the rather labile Schiff bases formed with amines (Inman and Dintzis, 1969; O'Shannessy and Wilchek, 1990).

Again a number of commercial hydrazide-derivatised solid phases are available; they can also be prepared by a number of synthetic routes.

Analogously to the aminolysis reaction described in the last section, polyamides can be substituted with hydrazine to yield hydrazides (Figure 8.24(a)) (Inman and Dintzis, 1969; Khrapko *et al.*, 1991). Hydrazine can also be coupled to carboxylic groups *via* a suitable activating agent such as a carbodiimide, offering an indirect route for creating hydrazide-derivatised polystyrene (Figure 8.24(b)) (Thomas *et al.*, 1990); matrices bearing active esters can be derivatised the same way, for example succinimide-activated agarose (O'Shannessy and Hoffman, 1987; Hoffman and O'Shannessy, 1988).

Hydrazides can be diazotised in the same way as aromatic amines, generating acyl azides which are highly reactive towards amines (Inman and Dintzis, 1969).

8.3.1.3 Carboxyls

There are a number of commercially available components bearing carboxyl groups, which will be covered in Section 8.3.3. Carboxylic acids can also be incorporated in polymer formation, for example the use of methacrylic acid in the formation of hydrophilic copolymer microspheres for protein binding (Montagne *et al.*, 1992). Acids can also be grafted onto inert polymers (e.g. the attachment of crotonic acid to PVC and polystyrene) (Larsson *et al.*, 1987); similarly, carboxylic latex solutions can be dried down onto such materials (Foster and Casey, 1988; Blanchard *et al.*, 1990).

Polystyrene and other polymers such as polyethylene terephthalate (PET) can be carboxylated by permanganate oxidation (Marchand-Brynaert *et al.*, 1995; Zammatteo *et al.*, 1996).

Polyamides such as nylon or polyacrylamide release carboxyls on hydrolysis (Inman and Dintzis, 1969; Sundaram, 1978); in order to increase the carboxyl content the hydrolysed polymer can be treated with succinic anhydride, which carboxylates the corres-

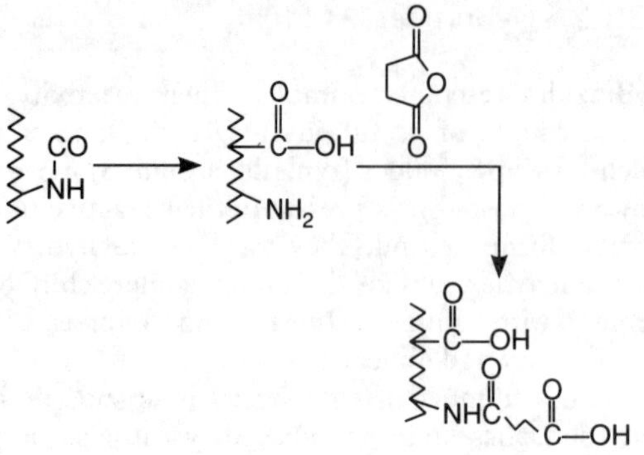

Figure 8.25 Hydrolysis and succinylation of polyamides.

ponding amine groups (Figure 8.25). Acetic anhydride can be used similarly, simply to block the amine functions (Hendry and Herrmann, 1980).

Carboxyl groups are also frequently introduced as a secondary stage following the generation of amine or hydrazide groups (Sections 8.3.1.1, 8.3.1.2); again succinic anhydride is a convenient reagent for such conversions (Inman and Dintzis, 1969; Weetall, 1988; Quash *et al.*, 1989). Succinic anhydride also reacts with hydroxyl groups (Section 8.3.1.5), providing a means of carboxylating materials of this type too (Cuatrecasas and Parikh, 1972; Niveleau *et al.*, 1993).

An alternative means of introducing carboxyl groups is to generate an amine-reactive moiety in the solid phase (e.g. cyanogen bromide activation of hydroxylic material), and then to react this with an amino acid (Ghosh and Musso, 1987).

8.3.1.4 Thiols

Thiols are rarely encountered in naturally occurring or polymeric solid phases, and must generally therefore be introduced in a derivatisation reaction.

Most commonly this is carried out after amine groups have been introduced (Section 8.3.1.1); a number of reagents for thiolation of amines exist (see Chapter 5, Section 5.12 for a full discussion), and of these 2-iminothiolane appears to have had the most widespread recent use for solid-phase thiolation (Hermentin *et al.*, 1990; Bieniarz *et al.*, 1991).

Alternative approaches include the reaction of epoxide-containing polymers with hydrogen sulfide (Hosaka *et al.*, 1983), or for silanol-containing matrices such as glass, derivatisation with a mercapto-alcohol under forcing conditions (Figure 8.26) (Cormier and Gagnon, 1979), or with a thiol-containing silane

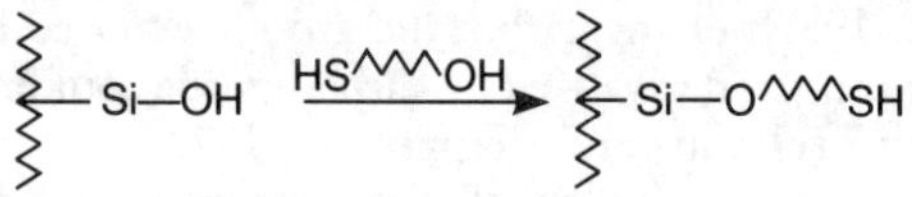

Figure 8.26 Introduction of thiols into glass.

(Weetall, 1976). Thiols can also be obtained from the reduction of disulfides (Hermentin *et al.*, 1990).

8.3.1.5 Hydroxys

In contrast to thiols, hydroxyl groups are among the most commonly-encountered functional groups in naturally occurring solid phases, notably the carbohydrate class which were dominant for many years as the basis of chromatography matrices (see Chapter 10). Apart from these polysaccharide materials (cellulose, agarose, dextrans, etc.), hydroxyl groups are found in a number of commercially available products such as magnetic beads. Irradiation of polystyrene generates nucleophilic groups including hydroxyls (Varga *et al.*, 1990), and commercial 'activated' polystyrene microtitre wells sometimes contain these functionalities, which can be used as a basis for derivatisation (Niveleau *et al.*, 1993, 1995).

Hydroxylic groups can occur in polymers, for example where a hydroxylic monomer such as hydroxyethyl methacrylate is incorporated (Montagne *et al.*, 1992), or those whose synthesis is based on epoxide reactions (Hosaka *et al.*, 1983). Epoxide-containing silanes can also be used to coat glass, silica, etc., yielding a hydroxylic surface on hydrolysis (Regnier and Noel, 1976; Roederer and Baastians, 1983), and long-spacer hydroxyl groups can be synthesised from these by sulfuric acid-catalysed reaction with other alcohols (Southern and Maskos, 1988; Southern *et al.*, 1992). An example of one of these groups is shown in Figure 8.27.

Finally, the organosilanes which are well-known for derivatisation of silanol groups, and are about to be discussed, can also be used with hydroxylic matrices,

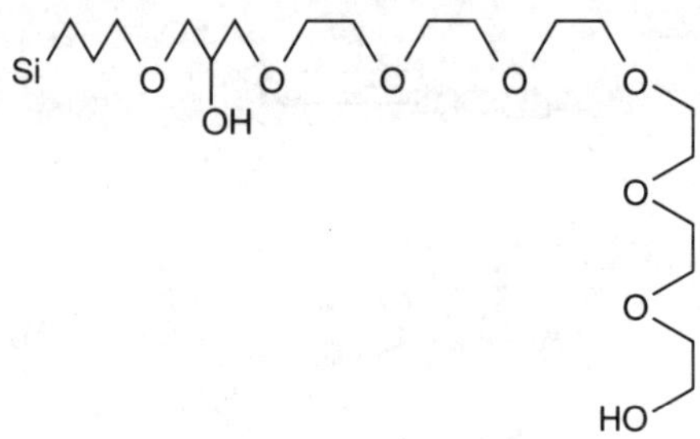

Figure 8.27 An example of a long-spacer hydroxysilane compound for coating on glass surfaces.

although the stability of C—O—Si bonds to hydrolysis is rather poor. Pope *et al.* (1993) used this approach to introduce reactive groups into cellulose hollow-fibre dialysers, and reasonable short-term stability was observed, at least.

8.3.1.6 Silanols, etc.

Silanol (Si—OH) groups are widely encountered in silica in its various forms, which include amorphous silica particles, quartz, and the major component of most common glasses. The most useful reactions of these groups are those with organosilane compounds, which form stable siloxane (Si—O—Si) bonds; a typical example, γ-aminopropyltriethoxy silane, is shown in Figure 8.28. These reactions can be carried out under either aqueous or organic conditions.

Howard Weetall of Corning was instrumental in the development of these techniques, which are equally applicable to other ceramics and metal oxides such as alumina, titanium dioxide and iron oxide (Weetall, 1970, 1976, 1988; Hermentin *et al.*, 1990).

Investigations have shown that residual silanol groups may well remain after the silanization process has been carried out (Čukman *et al.*, 1987).

Using an aminosilane such as that shown in Figure 8.28, it is clear how covalent coupling to such surfaces is facilitated, with or without further intermediate

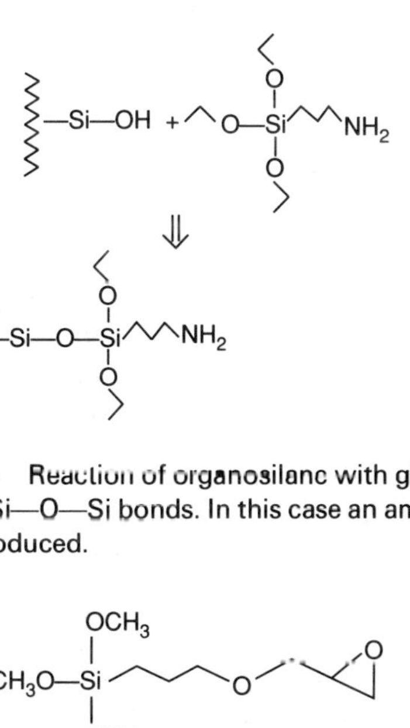

Figure 8.28 Reaction of organosilane with glass to form stable Si—O—Si bonds. In this case an amine group is introduced.

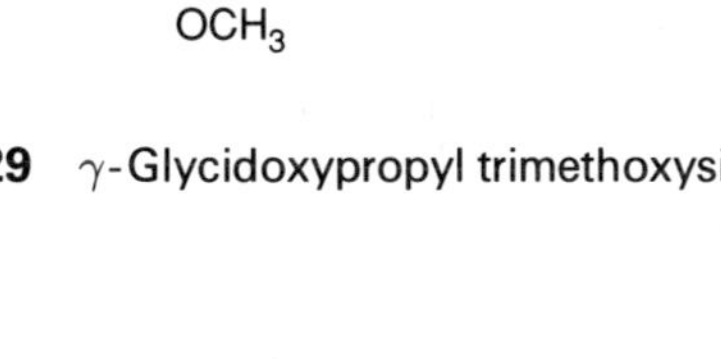

Figure 8.29 γ-Glycidoxypropyl trimethoxysilane.

derivatisation (Weetall, 1988; Frey *et al*; 1997). Epoxide-containing silanes such as γ-glycidoxypropyl-trimethoxysilane (Figure 8.29) are sometimes employed in the activation of silica for affinity chromatography; their hydrolysis, on the other hand, yields polar glyceryl functions, the hydrophilic properties of which have been commonly employed to provide a surface resistant to non-specific protein binding.

Silanol functions will also react with other reagents such as cyanogen bromide and diazonium salts (Weetall, 1976). Under suitably forcing conditions, they will also react with alcohols: Cormier and Gagnon (1979) derivatised glass beads with a 2-mercaptoethanol (five hours reflux at 165–175 °C).

8.3.1.7 Amine-reactive groups

Not surprisingly amine-reactive solid phases are a popular and convenient route to protein immobilisation. Generally the methods fall into two categories: direct derivatisation of the surface with a reagent containing the amine-reactive group, or amination of the surface followed by reaction with a suitable homobifunctional. The latter approach is extremely popular; in fact solid phases bearing amines are nearly always converted to an amine-reactive form before use for chemisorption.

Aldehydes occur in polymers containing monomers such as acrolein (Margel *et al.*, 1981; Montagne *et al.*, 1992). Like hydroxyl groups they may be introduced by irradiation of polystyrene, etc. (Niveleau *et al.*, 1993, 1995), or they may be generated by oxidation of hydroxyl functions (Hosaka *et al.*, 1983). A number of commercial aldehydic matrices are available, such as activated agarose gels.

Aldehyde groups can readily be attached to amine-derivatised solid phases *via* glutaraldehyde, whose use in this role has enjoyed prolonged popularity (see, for example, Robinson *et al.*, 1971; Weetall, 1976; Wood and Gadow, 1983; Thomas, 1988; Onyezili, 1988; Lauritzen *et al.*, 1990; Hayat *et al.*, 1992; Pundir *et al.*, 1993). In the field of protein–protein coupling, glutaraldehyde was one of the most popular reagents for many years. While it is undoubtedly a powerful cross-linker, it has many drawbacks: the composition and properties of 'glutaraldehyde' change dramatically on storage due to condensation reactions (see Ordronneau *et al.*, 1991 for an example of these effects in a chemisorption application), the presence of impurities can affect reactivity, and the mechanisms of the conjugation reactions themselves are complex and poorly understood[1]. Its use for protein conjugation has

[1] See Section 5.3.1.1 for a more detailed discussion.

declined significantly over the last fifteen years, partly due to these problems and partly due to a general shift away from homobifunctional methods. For protein chemisorption however, glutaraldehye appears to have maintained its popularity up to the present day, despite the availability of more predictable alternatives such as bis-succinimides.

Interestingly, glutaraldehyde has also been used to activate polymers such as polystyrene and polypropylene directly, without the presence of amine groups (Boenisch, 1976; Howell *et al.*, 1981; Ordronneau *et al.*, 1991). The mechanism is open to debate. It is likely that this in fact represents a hydrophobic binding of glutaraldehyde to the surface (cf. Section 8.2.4.3), particularly larger products of glutaraldehyde condensation reactions, perhaps, which will be less prone to desorption.

Isocyanates can be introduced directly into plastic surfaces by direct grafting of an organic solution of a bis-isocyanate compound (Saito and Nagai, 1983). Alternatively, amine derivatisation can be followed by direct conversion to an isothiocyanate using thiophosgene (Figure 8.30(a)) (Weetall, 1976, 1988), although this is a highly toxic and unpleasant compound. More conveniently, a bis-isothiocyanate can be coupled to the amine residues, leaving one N=C=S group free for further reaction (Figure 8.30(b)) (Aebersold *et al.*, 1988). Isothiocyanate-bearing supports have often been used to immobilise protein samples for sequence analysis.

Imidates, probably the most amine-specific of the groups discussed in this section, can be generated from amides by *O*-alkylation with an agent such as dimethyl sulfate or triethyloxonium tetrafluoroborate (Figure 8.31). This offers a neat route to the activation of nylon, for example (Benson and Cairns, 1951; Morris *et al.*, 1975; Zucchini *et al.*, 1989); one advantage of this method is that no cleavage of the polymer structure is required. Bis-imidates such as dimethyl suberimidate can also be used with an aminated solid phase.

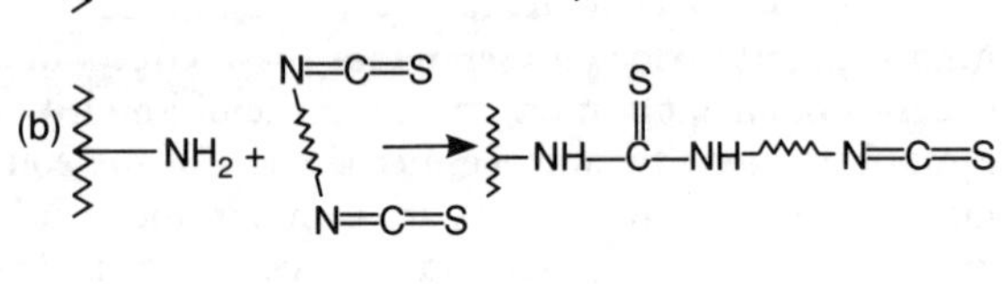

Figure 8.30 Introduction of isothiocyanates to plastic surfaces.

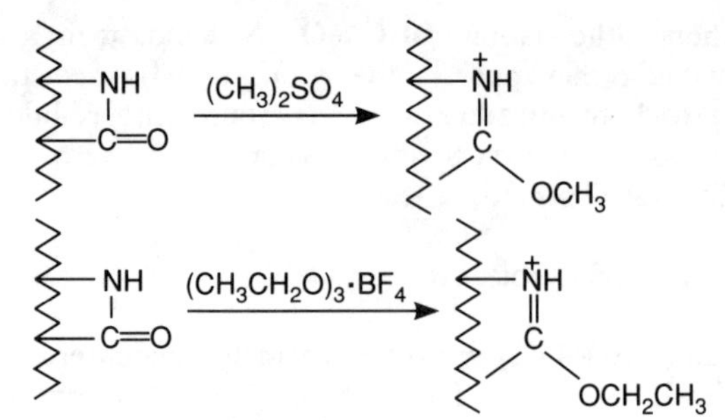

Figure 8.31 Generation of imidates from polyamides.

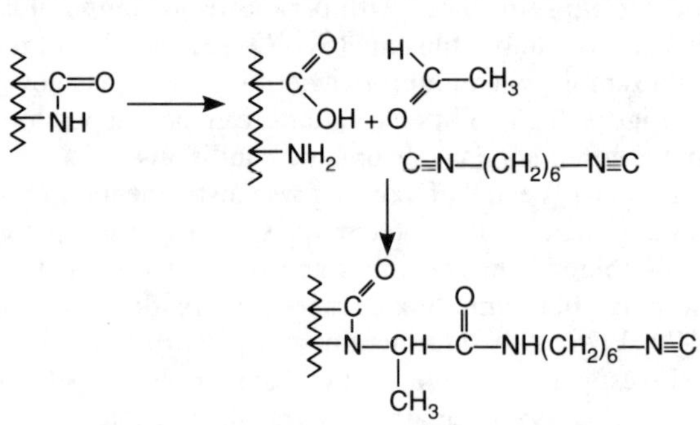

Figure 8.32 Generation of isonitrile moiety from hydrolysed polyamides by Ugi reaction.

Bis-isonitriles can be coupled to amines and carboxylates in the presence of an aldehyde (Ugi, 1962; Axén *et al.*, 1971) to yield a reactive isonitrile-derivatised surface (Figure 8.32). Zucchini *et al.* (1989) used this approach to derivatise hydrolysed nylon.

Epoxides, already discussed as a route to many other reactive groups, are also amine-reactive in their own right. They are contained in some commercially available matrices, and can also be incorporated in polymer preparations (Hosaka *et al.*, 1983), or introduced *via* reagents such as glycidoxysilanes (Roederer and Baastians, 1983; Pope *et al.*, 1993). Bis-epoxides such as butane-1,4-diol diglycidyl ether (Figure 8.33) can also be employed with aminated solid phases.

Amine-reactive carboxyl derivatives can easily be prepared from carboxylated solid phases: examples include acyl chlorides (Weetall, 1976), anhydrides (Sundaram, 1974, 1978), and succinimide esters (Cuatrecasas and Parikh, 1972; Weetall, 1976). Similarly reactive functionalities can be introduced by other means too: acyl azides can be generated from hydrazides (Inman and Dintzis, 1969; Erlanger *et al.*, 1970; Blum and Coulet, 1984).

Polymers are commercially available containing azlactone moieties (Figure 8.34), while maleic anhyd-

Figure 8.33 Butane-1,4-diol diglycidyl ether.

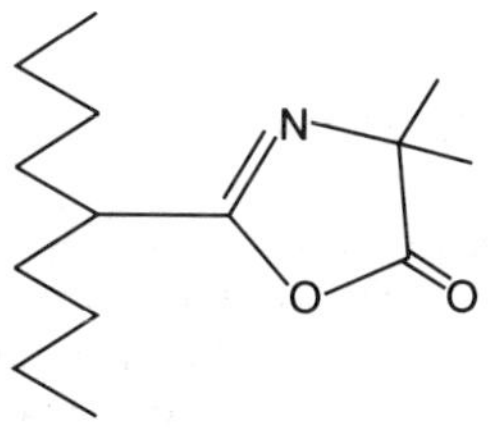

Figure 8.34 The azlactone moiety.

ride can be incorporated directly into polymer preparations (Beddows *et al.*, 1986). Again, bis-succinimides can be used to convert aminated solid phases to an amine-reactive surface (Hobbs, 1989), although, as noted at the start of this section, this approach is much rarer than might be expected.

Although the intermediates of carbodiimide activation of carboxyls are fairly labile, it is possible to isolate them briefly. Shenoy *et al.* (1992) carried out a detailed study on different activating agents for a carboxylated polyethylene film: a number of carbodiimides and analogues were compared, and DCC in DMF gave the best performance. The optimal activation time was around 1–2 hours.

Cyanate esters can be generated from the reaction of cyanogen bromide with hydroxylic matrices, see Chapter 10, Section 10.9 for a fuller discussion. Although probably best known for the preparation of chromatographic affinity media, there are many other applications such as the preparation of 'activated paper' (Ceska and Lundkvist, 1972; Lehtonen and Viljanen, 1980).

Many other amine-reactive functionalities have been employed for protein immobilisation: commercially available materials contain groups such as aryl halides and imidazolyl carbamates (generated using carbonyldiimidazole, CDI, see Chapter 5), while matrices have been described bearing groups such as triazines (Horvath and Solomon, 1972; Weetall, 1976), alkyl halides (Pope *et al.*, 1993), benzoquinone (Brandt *et al.*, 1975; Kučera, 1986), fluoromethylpyridinium (FMP) tosylates (Ngo, 1986), tresylates (Nilsson and Mosbach, 1984; Leckband and Langer, 1991) and other sulfonyl derivatives (Sundaram, 1978).

8.3.1.8 Thiol-reactive groups

There are four primary categories of thiol-reactive groups, two highly selective (disulfides, organomercurials) and two reasonably so under controlled conditions (maleimides, acyl halides); all of these have been employed in the activation of solid phases towards thiol-containing proteins. It should be noted that thiols themselves (Section 8.3.1.4) represent a fifth class of thiol-reactive moiety.

The typical means of introducing such groups is with a suitable amine-thiol heterobifunctional following the introduction of amines or hydrazides into the solid phase. This has been carried out with a number of reagents of this type containing maleimides (Plant *et al.*, 1988; Hermentin *et al.*, 1990; Bieniarz *et al.*, 1991; Niveleau *et al.*, 1995), acyl halides (Peterman *et al.*, 1988; Hermentin *et al.*, 1990; Spitznagel *et al.*, 1993) and disulfides (Hermentin *et al.*, 1990; Lu *et al.*, 1995).

Approaches for hydroxylic media include direct reaction of haloacyl halides (Figure 8.35) (Jagendorf

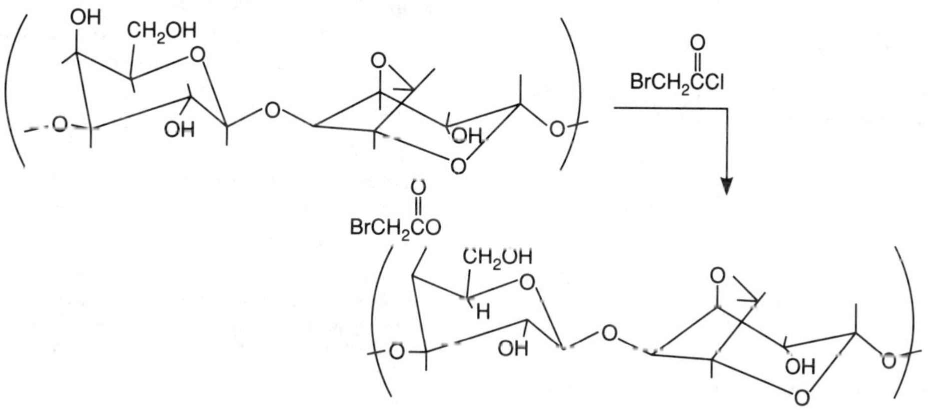

Figure 8.35 Bromoacetylation of cellulose.

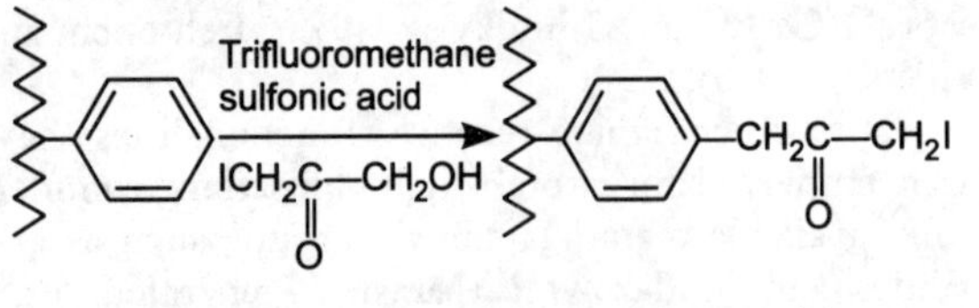

Figure 8.36 Introduction of an iodoacetamide group into polystyrene *via* Friedel–Crafts alkylation.

et al., 1963; Leckband and Langer, 1991), and a multi-stage synthesis *via* active ester formation and an amino-disulfide (Caldwell *et al.*, 1996).

A patent by Okrongly (1990) described the introduction of haloacyl and other groups into polystyrene *via* Friedel–Crafts alkylation with *N*-hydroxymethyl-acetamides (Figure 8.36).

Media are also commercially available containing thiol-reactive groups such as iodoacetates and mercuribenzoates. Finally it should be noted that some 'amine-reactive' moieties such as epoxides, alkyl halides and triazines are in fact broadly reactive to nucleophiles, including thiols.

8.3.1.9 Photoactivatable groups

The introduction of photoactivatable groups into the surface provides a highly non-selective route for immobilisation of proteins. Groups employed include benzophenones (Rozsnyai *et al.*, 1992) and perfluorophenyl azides (Yan *et al.*, 1993). Stein and Gerisch (1996), for example, coupled a hydrophobic amino-perfluorophenyl azide to a carboxymethyl-dextran-coated biosensor (Figure 8.37), to allow immobilisation of a lipoprotein.

An interesting advantage of these approaches is that, by masking during the photoactivation process, the target protein can be immobilised in a chosen pattern on the surface[1].

8.3.1.10 Biotin

The incorporation of biotin into a polymer surface obviously provides a convenient basis for pseudo-covalent binding of proteins. Hengsakul and Cass (1996) described the use of photobiotin for this purpose—irradiation of a dried film of this reagent deposited on polystyrene or nitrocellulose resulted in coupling of the biotin. Again, protein patterning is achievable by this route.

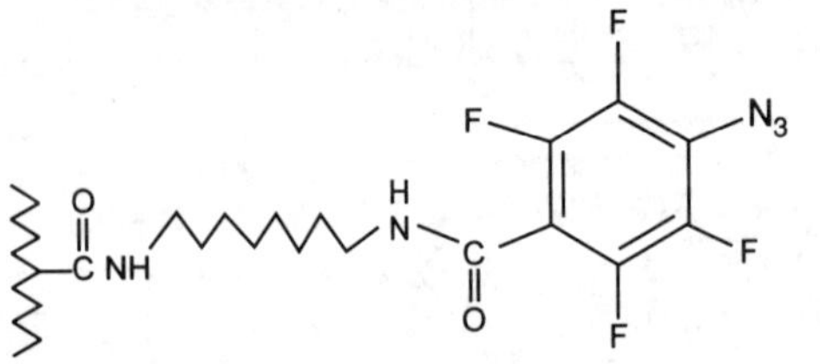

Figure 8.37 Introduction of a photoactivatable hydrophobic amino-perfluorophenyl azide onto a carboxymethyldextran-coated surface.

8.3.2 Chemistries for protein chemisorption

The 'purification' of an activated solid phase is unusually facile, requiring only the decanting or aspiration of unreacted reagent solution followed by simple washing. Therefore it is a simple matter to isolate a convenient protein-reactive solid phase, for example those described in Sections 8.3.1.7 and 8.3.1.8. Indeed, these protein-reactive solid phases are often sufficiently stable that they can be stored for long periods, and many are commercially available in convenient formats. With solid phases of these categories, the coupling to protein becomes a simple one-step incubation.

With the surfaces bearing groups which are not directly reactive towards proteins (amines, carboxyl and hydroxyl groups etc.) the chemistry requires slightly more thought. However, with very few exceptions the methods employed are identical to those discussed in earlier chapters, particularly Chapter 5. The choice of conditions for these methods has been discussed in detail there. This section, therefore, will only provide a brief summary of their application for covalent immobilisation.

Sometimes there may be charge effects which interfere with surface interactions, and the relationship between these and the p*I* of the protein may affect the choice of pH for the coupling reaction (Frost *et al.*, 1981). Reaction times may also need to be longer to overcome any diffusion limitations, but the use of reasonably concentrated reactants can minimise this factor. On the whole there is little need for changes from the conditions used for solution-phase reactions.

8.3.2.1 Chemisorption to surface amines

As noted earlier, aminated solid phases are often derivatised further before use, most commonly by

[1] Another interesting means of achieving this goal is to incorporate a blocked reactive group in the surface, which can be deprotected by light; the best known example is the use of caged biotin which incorporates a photolabile nitrobenzyl group (Sundberg *et al.*, 1995; Pirrung and Huang, 1996). The use of accurate physical deposition techniques such as ink jet printing is also having an impact on this area, following early demonstrations of feasibility (Kimura *et al.*, 1988)

glutaraldehyde. Because it is such a straightforward operation, the activated surface is nearly always isolated before use, and surfaces of this type are therefore dealt with in the relevant section for the final reactive group. Glutaraldehyde-activated solid phases, therefore, are dealt with in Section 8.3.2.5.

Occasionally, however, a one-step reaction is carried out using a mixture of protein and a homobifunctional reagent: Chin and Lanks (1977) for example added a mixture of lactoperoxidase and a bis-imidate simultaneously to aminated polystyrene flasks. Unless the reactive groups formed at the surface are exceptionally labile, there seems little advantage to this approach, as it is almost certain to lead to cross-linking of protein molecules in solution. Some of these oligomers may then become immobilised, bringing unwanted[1] heterogeneity to the coated surface.

Another way of using the surface amines directly is *via* carbodiimide activation of the protein carboxyl functions; the chemistry involved is familiar from coupling low-molecular-mass compounds to proteins, and is discussed more fully in Chapter 6, Sections 6.1.1 and 6.7.1. Typically such reactions are carried out at around pH 5–6; this represents a compromise between the optimum pH for carboxyl activation (3.5–4.5) and the availability of unprotonated amines. Clearly the lower the pK_a of the amine groups, the less of a compromise there has to be: Fernandez-Lafuente *et al.* (1993) described solid phases bearing amines with pK_a values as low as 6.8, which obviously facilitate reactions of this type.

Akashi *et al.* (1992) described the immobilisation of thrombomodulin to aminated glass by a one-step EDAC approach. The reaction time was 48 hours at 0 °C, typical of these one-step methods; reasonably long incubations are probably advisable, as it has been shown that reaction with carbodiimide-activated carboxyl groups can continue at least beyond four hours (Shenoy *et al.*, 1992). Reasonable results have been obtained, however, from one hour incubations (Niveleau *et al.*, 1995).

Two-step approaches might offer some potential for efficiency improvement, although protein cross-linking is a possibility in either case. Søndergård-Andersen *et al.* (1990) carried out a two-step coupling of oligopeptides to aminated polystyrene wells: in the first the peptides were activated with EDAC/NHS in water, before coupling to the wells at pH 8.6. It was noted that this gave a much higher yield than simple EDAC coupling, and that the optimum time for the activation stage was 30 minutes.

[1] Not always: see Rotman and Delwel (1983) in Section 8.3.2.3.

Surface amines can be coupled directly to carbohydrate residues, either *via* periodate oxidation, or using a powerful bifunctional agent such as divinylsulfone (Hatakeyama *et al.*, 1996).

Aromatic amines can be converted to diazonium salts (see Section 5.5.3 for more details), and this route has been employed quite frequently for immobilisation of proteins. This occurs principally *via* tyrosines and histidines, although other residues can react. Typically the diazotisation is performed using acidic sodium nitrite (0 °C, 10–15 minutes), and the reaction with protein is carried out immediately afterwards at at the optimal pH, which is around 9. This approach has been used, for example, with arylamine-derivatised glass (Weetall, 1971) and polyacrylamide (Inman and Dintzis, 1969).

8.3.2.2 Chemisorption to surface hydrazides

Hydrazides can be coupled to protein carboxyls in the same way as amines, for example by carbodiimide activation; in fact the low pK_a of hydrazides means there is less conflict between the pH requirements, and greater efficiencies can potentially be observed.

Most typically, however, hydrazide matrices are used to couple proteins *via* oxidised carbohydrate moieties. Periodate oxidation of proteins is discussed in detail in Chapter 5, Section 5.2.3; 'safe' conditions have been described by O'Shannessy and Wilchek (1990) which may provide a starting point for investigation, viz. 1 mM periodate for 20 minutes at pH 5.5 and 0–4 °C. The subsequent reaction with the surface hydrazides may typically also be carried out at an acidic pH: O'Shannessy and Hoffman (1987) observed a maximum at around pH 3.5 for a reaction of this type. This route is particularly attractive for orientation of antibodies *via* Fc carbohydrates (Lu *et al.*, 1996), and this approach has been used with solid phases such as beads (Blanchard *et al.*, 1990) and microtitre plates (Brillhart and Ngo, 1991).

8.3.2.3 Chemisorption to surface carboxyls

The commonest means of exploiting surface carboxyl groups is to activate them towards protein amines. Where the activated derivative is stable enough, the reactive solid phase can be isolated, and the use of these reagents is therefore dealt with in Section 8.3.2.8.

One of the commonest techniques, however, is the use of carbodiimides and their analogues, and the activated solid phase is frequently not isolated in these methods. Simple one step reactions using

EDAC at around pH 5–6 are commonly encountered, essentially identical in practice to those described for amine solid phases above. Again, long reaction times at low temperatures are typically employed, although one hour at 37 °C has also been advocated as sufficient (Foster and Casey, 1988). For examples of these one-step approaches, see also Rotmans and Delwel (1983), Larsson *et al.* (1987), Blanchard *et al.* (1990) and Niveleau *et al.* (1993). As noted earlier, cross-linking of protein is inevitable under such circumstances, and has indeed been advocated as a method of increasing the uptake from a mixed antigen solution (Rotmans and Delwel, 1983): low-molecular-mass antigens which bound poorly by passive means were successfully immobilised by this approach.

Two-step reactions are more elegant in design, and offer the possibility of carrying out the activation under conditions which are not compatible with the presence of the protein. Thus the solid phase can be activated in an organic solvent and washed to provide an amine-reactive surface (see Sections 8.3.1.7 and 8.3.2.8).

Cross-linking can be completely avoided by such methods. A compromise approach is to carry out the activation stage in an aqueous buffer, so that the second stage can be initiated simply by addition of the protein solution (see Niveleau *et al.* (1995) for an oligopeptide example: ten minutes activation of carboxylated plates with EDAC was followed by a one-hour incubation with peptide).

8.3.2.4 Chemisorption to surface thiols

Surface thiols are generally only used to yield disulfide linkages to proteins also containing free thiols, either in the native structure, freed by reduction of cystine links, or artificially introduced. Moderate conditions are suitable for forming the disulfide link (see also Chapter 5, Section 5.2.1). An oxidising agent may or may not be incorporated, atmospheric oxygen may be sufficient in itself. Cormier and Gagnon (1979) used sodium selenate to catalyse the reaction of thiolated α-amylase with thiol-containing glass beads (18 hours, 5 °C, pH 7.5). Hermentin *et al.* (1990) coupled antibody Fab fragments to thiolated iron oxide particles without a catalyst (one hour, room temperature, pH 7.2)

8.3.2.5 Chemisorption to surface amine-reactive groups

As discussed in Section 8.3.1.7, there are many amine-reactive groups which can be incorporated into solid phases. The conditions for coupling these to protein amines will be comparable in one respect: it is the deprotonated form of the amine which is reactive towards these electrophiles, and therefore the reactions are likely to be carried out at a mildly basic pH where there is at least a small fraction of deprotonated lysines. The terminal amine(s) will be more fully deprotonated due to the lower pK_a.

For glutaraldehyde-activated solid phases, typical protein immobilisation conditions employ a pH of around 6.5–7.5, rather lower than that used in many protein conjugation techniques. In one study where the effect of pH was examined (Onyezili, 1988), an optimum of immobilized urease activity was observed at 6.5, but this was attributed to instability of the enzyme itself. In the absence of such effects, it seems likely that working under more basic conditions would give better incorporations. The deprotonated fraction of lysines at pH 6.5 is extremely low. With a more typical protein, BSA, a pH optimum of 8 was observed by Ukeda *et al.* (1993), and values as high as pH 10 have sometimes been employed (Matsumoto *et al.*, 1986).

Even at low pH Onyezili (1988) observed that the reaction went to completion in less than three hours at 37 °C, and Hobbs (1989) successfully used a 30 minute incubation at pH 7.5 and 25 °C. Many other workers have employed longer reaction times (overnight at 4 °C is typical).

Sometimes a reducing agent such as sodium borohydride or cyanoborohydride is employed to stabilise the Schiff bases purportedly formed from the coupling reaction (Wood and Gadow, 1983; Thavarungkul *et al.*, 1991). However (see Section 5.3.1) imine behaviour is not generally observed in the products of glutaraldehyde coupling, so this precaution may well be unnecessary. Blocking of unreacted aldehyde sites with a small amine such as lysine or aminoethanol is, however, advisable. The conditions for coupling of amines to other active solid phases are fairly unremarkable; more detail is contained in Chapter 5, Section 5.3. With isothiocyanates fairly basic conditions (pH 8.5–9) are often used (Weetall, 1970, 1976; Aebersold *et al.*, 1988).

Active esters are typically employed at around neutrality to minimise hydrolysis (Leckband and Langer, 1991); even with these reagents the greater availability of NH_2 moieties at higher pH can lead to higher coupling levels despite the increase in hydrolysis (Frost *et al.*, 1981). On the other hand, surprisingly good incorporations have been observed when proteins are immobilised to a freshly-prepared succinimide ester surface at pH 5, even with reaction times of only a few minutes (Johne *et al.*, 1993).

Otherwise incubation at around pH 7.5–8.5 is likely to be a good starting point for most of these reagents, whether they be aryl halides (Blanchard *et al.*, 1990), cyanate esters (Lehtonen and Viljanen, 1980), epoxides (Leckband and Langer, 1991), quinones (Brandt *et al.*, 1975; Kučera, 1986), triazines (Horvath and Solomon, 1972; Weetall, 1976), etc.

8.3.2.6 Chemisorption to surface thiol-reactive groups

Conditions for coupling thiolated proteins to maleimides, disulfides and haloacyl compounds are, again, rather unremarkable. Incubation for a number of hours at a roughly neutral pH is generally effective for all these types of reagent (Peterman *et al.*, 1988; Hermentin *et al.*, 1990; Bieniarz *et al.*, 1991; Spitznagel *et al.*, 1993; Lu *et al.*, 1995).

With reagents which are broadly reactive with nucleophiles, some specificity for thiols can sometimes be obtained by working at a moderately low pH: Leckband and Langer (1991) for example carried out coupling to surface epoxides at pH 7 to encourage reaction with protein thiols, much more basic conditions often being used where amine coupling is the aim (e.g. pH 9.6; Pope *et al.*, 1993).

Where recombinant proteins are being used, it is possible to introduce reactive thiols *via* extra cysteine residues; these can then be used to provide specific, oriented immobilisation to a thiol-reactive support. Ljungquist *et al.* (1989) used this approach to couple recombinant Protein A to a disulfide-bearing matrix.

8.3.2.7 Chemisorption to other surfaces *via* photoactivation

If a protein is activated with a photoactivatable reagent such as a phenyl azide, it can be covalently linked to surfaces which are generally unreactive, such as alkylated glass (Tseng *et al.*, 1993).

8.3.2.8 Graft copolymerisation

If the protein is derivatised with a group which can take part in a polymerisation reaction, it can be incorporated directly into the polymer as it is formed. D'Angiuro *et al.* (1978, 1980), for example, prepared vinyl derivatives of enzymes which were copolymerised with a vinyl monomer onto cellulose and other solid matrices.

This offers an interesting alternative to the more traditional covalent approaches.

8.3.3 Commercially available solid phases for protein chemisorption

Many convenient particulate solid phases bearing reactive functionalities are commercially available, and these are listed in Table 8.2. Note that these are commonly encountered as reactive supports for affinity chromatography—see Secton 10.9. Note also that similar matrices may be marketed in different ways, e.g. amine-derivatised matrices for carbonyl coupling or for the Mannich reaction.

8.3.4 Variations on protein chemisorption

8.3.4.1 Multi-stage chemisorption

The use of multi-stage approaches has already been discussed where all the binding steps are non-covalent in nature (Section 8.2.4.1), and where a mixture of covalent and non-covalent binding is employed (Section 8.2.4.5). Of course it is equally feasible for multi-stage coating to be used where all the binding is covalent, and such systems should be mentioned for completeness. The obvious main difference with this approach is that unwanted desorption is more or less ruled out, while it remains a distinct possibility for multi-stage processes based on passive or mixed mechanisms.

An example of a wholly-covalent multi-stage approach was described by Habeeb (1981). An agarose gel was activated with cyanogen bromide, allowing gelatin to be bound covalently to the solid phase; then as a second stage, the thiol groups of reduced BSA or lysozyme were coupled covalently to this matrix using SPDP.

8.3.5 Overview of protein chemisorption

Recalling the limitations of physisorption techniques, there are really three questions which can be asked to determine what advantages might be gained from a chemisorption approach. (1) Can more protein be bound? (2) Can protein be bound with less loss of activity? (3) Can protein be bound more irreversibly? The answer to all these questions, at least sometimes, is 'yes'; but this simple conclusion requires a little discussion.

8.3.5.1 Can more protein be bound?

It has already been discussed that the amount of protein which can be bound by physisorption to a hydrophobic surface is broadly consistent with a monolayer.

Table 8.2 Commercially available particulate matrices bearing reactive groups

Name	Reactive functionality	Matrix	Primary target	Supplier
AminoLink	—CHO	Agarose	Amine	Pierce
UltraLink	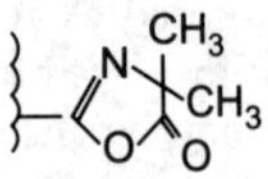	Bis-acrylamide azlactone copolymer	Amine	Pierce
Reacti-Gel CDI	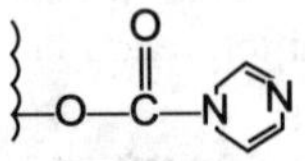	Trisacryl GF-2000, Agarose, TSK HW-65F	Amine	Pierce
CarboLink	$—C(=O)NHNH_2$	Agarose	Carboxyl, aldehyde	Pierce
UltraLink Hydrazide	$—C(=O)NHNH_2$	Bis-acrylamide azlactone copolymer	Carboxyl, aldehyde	Pierce
SulfoLink	$—NH\,C(=O)\,CH_2I$	Agarose	Sulfhydryl	Pierce
UltraLink Iodoacetyl	$—NH\,C(=O)\,CH_2I$	Bis-acrylamide azlactone copolymer	Sulfhydryl	Pierce
Pharma Link	$—CH_2\,NH_2$	Agarose	Mannich substrates (in presence of formaldehyde)	Pierce
Immobilsed TNB-Thiol	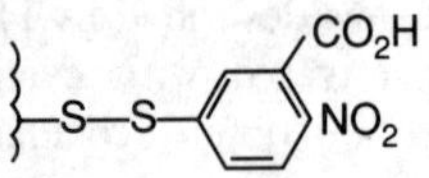	Agarose	Sulfhydryl	Pierce
UltraLink Carboxy	$—CO_2H$	Bis-acrylamide azlactone copolymer	Amine	Pierce
Amino Spherilose	$—CH_2NH_2$	Cellulose	Carboxyl, aldehyde	Isco Inc.
Carboxy Spherilose	$—CO_2H$	Cellulose	Amine	Isco Inc.
Formyl Spherilose	—CHO	Cellulose	Amine	Isco Inc.
FMP-Activated Spherilose	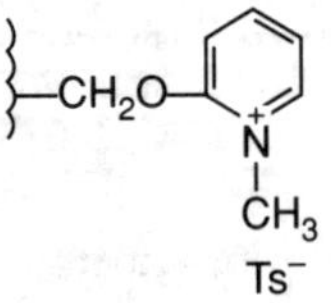	Porous cellulose	Sulfhydryl amine	Isco Inc.
Protein-Pak Epoxy Activated	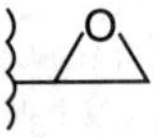	Silica	Amine, hydroxyl, sulfhydryl	Waters

Affi-Gel Hz	$-C(=O)NHNH_2$	Agarose	Carboxyl, aldehyde	BioRad
Affi-Gel 102	$-NH_2$	Agarose	Carboxyl, aldehyde	BioRad
Affi-Prep 10	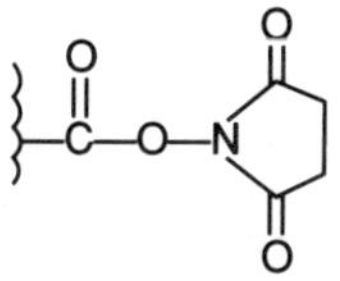	Acrylic	Amine	BioRad
Ultraffinity-FP	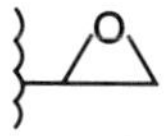	Silica	Amine, hydroxyl, sulfhydryl	Beckman Inc.
Toyo Pearl AF-Tresyl-650 M	$-O-S(=O)_2-CH_2CF_3$	Toyo Pearl HW65	Amine, sulfhydryl	Toso Haas
Toyo Pearl AF-Epoxy-650 M	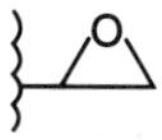	Toyo Pearl HW65	Amine, hydroxyl, sulfhydryl	Toso Haas
Toyo Pearl AF Formyl-650M	$-CHO$	Toyo Pearl HW65	Amine	Toso Haas
Toyo Pearl AF-Amino-650M	$-NH_2$	Toyo Pearl HW65	Aldehyde, carboxyl	Toso Haas
Toyo Pearl AF-Carboxy-650 M	$-CO_2H$	Toyo Pearl HW65	Amine	Toso Haas
Poros AL20	$-CHO$	Polysterene/Divinyl benzene	Amine	Boehringer Mannheim
Poros EP20, 50	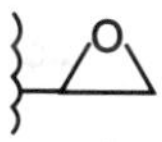	Polystyrene/ Divinyl benzene	Amine, hydroxyl, sulfhydryl	Boehringer Mannheim
Poros NH20	$-NH_2$	Polystyrene/ Divinyl benzene	Carboxyl, aldehyde	Boehringer Mannheim
Bakerbond Glutoraldehyde-P	$-CHO$	Silica	Amine	J. T.Baker
Glu-Activated Trisacryl	$-CHO$	Trisacryl GF-2000	Amine	Biosepra
TSK Tresyl-5PW	$-O-S(=O)_2-CH_2CF_3$	TSK-PW	Amine, sulfhydryl	Toso Haas
Affi-Gel 10, 15	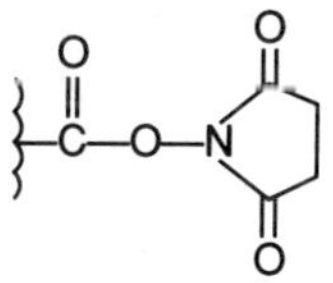	Agarose	Amine	BioRad

Table 8.2 (contd.)

Name	Reactive functionality	Matrix	Primary target	Supplier
Affi-Prep Hz	–C(=O)NHNH$_2$	Polymer	Carboxyl, aldehyde	BioRad
CM-BioGel A	–CO$_2$H	Agarose	Amine	BioRad
HiTrap NHS Activated Sepharose	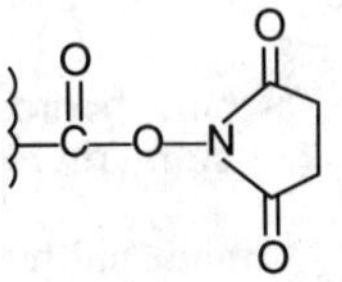	Agarose	Amine	APB
Activated CH-Sepharose 4B	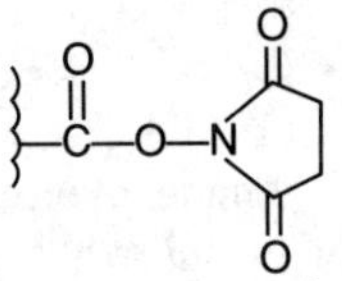	Agarose	Amine	APB
CNBr Activated Sepharose 4B, 4 Fast Flow, 6MB	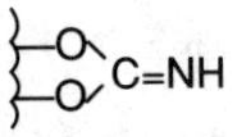	Agarose	Amine, sulfhydryl	APB
ECH Sepharose 4B	–CO$_2$H	Agarose	Amine	APB
EAH Sepharose 4B	–NH$_2$	Agarose	Carboxyl, aldehyde	APB
Epoxy-Activated Sepharose 6B	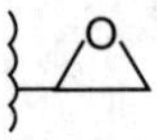	Agarose	Carboxyl, hydroxyl, sulfhydryl	APB
Agarose, Adipic Acid Hydrazide	–NHNH$_2$	Agarose	Carboxyl, aldehyde	APB
Thiopropyl Agarose 6B	–SH	Agarose	Sulfhydryl	APB
Activated Thiol Sepharose 4B	–S–S–R	Agarose	Sulfhydryl	APB

Note: APB = Amersham Pharmacia Biotech UK Ltd.

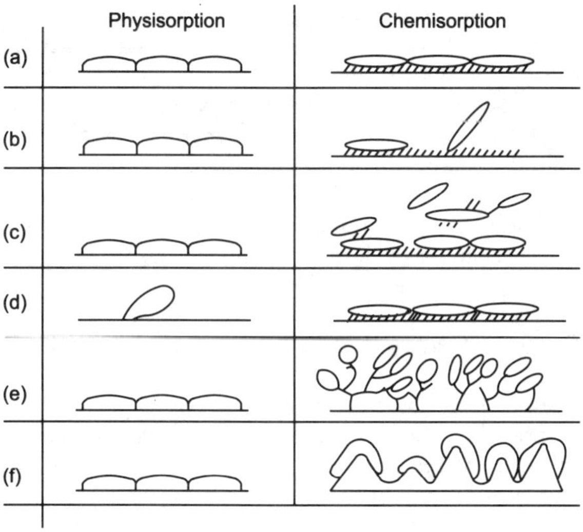

Figure 8.38 Physisorption vs chemisorption: for a hydrophobic or other high-binding surface (a), it is unlikely that chemisorption of a monolayer will result in any increase in binding; in fact, the level of binding may even be reduced if the coupling efficiency is low (b). Chemisorption may show an increase if there is a possibility of interprotein links (c), if the activated surface has low adsorption properties *per se* (d), if large multivalent linkers are used (e) or if the surface area is actually increased by the activation procedure.

Therefore, even if the entire activated surface possessed a dense coverage of reactive moieties, there would be no reason to expect that a greater level of protein adsorption should be obtained by covalent means.

Where the concentration of reactive groups has been calculated for an activated surface, values up to around a couple of nmol cm^{-2} have been observed (Larsson *et al.*, 1987; Søndergård-Andersen *et al.*, 1990). Referring back to the quantities of a typical protein molecule such as IgG contained in a monolayer (Section 8.2.3), this suggests that hundreds or thousands of reactive groups are available per potential binding site. So even allowing for some inefficiency in the coupling chemistry, it would seem likely that covalent binding under such circumstances would yield very similar levels of coverage to that seen from physisorption (Figure 8.38(a)). Values in the right sort of range have been seen from chemisorption experiments, for example with antibody fragments on semiconductors (50 ng cm^{-2}; Jimbo and Saito, 1988) or on silica (125 ng cm^{-2}; Lu *et al.*, 1995), or antibodies on glass beads or cellulose (around 800 ng cm^{-2}; Pope *et al.*, 1993).

Low levels of surface activation or inefficient chemical coupling might even be expected to show a reduction from the level achievable from passive adsorption, assuming that physisorption has itself been discouraged from taking place (Figure 8.38(b)). Such behaviour has been observed—for example when a hapten–BSA conjugate was bound to an activated nylon support (Wood and Gadow, 1983), glutaraldehyde decreased the binding capacity of an aminated surface. This was attributed to cross-linking of surface amines by the coupling agent.

In a large number of reported cases, however, increased binding is observed *versus* physisorption (see for example Hendry and Herrmann, 1980; Saito and Nagai, 1983; Larsson *et al.*, 1987; Peterman *et al.*, 1988; Pope *et al.*, 1993). There are several different potential causes of such behaviour, depending on the circumstances. (1) If the design allows inter-protein coupling to occur as well as surface–protein reactions, then multiple layer formation and binding of aggregates may give higher levels (Figure 8.38(c)). This has been postulated as the cause of high binding from a one-step carbodiimide coupling, for example (Timkovich, 1977; Rotmans and Scheven, 1984; Larsson, *et al.*, 1987). (2) If the activated surface also has low physisorption properties, e.g. for hydrophilic materials, then covalent coupling would be expected to give an increase in adsorption (Figure 8.38(d)). This may explain some of the reports above which used matrices such as nylon, cellulose or PVC (Hendry and Herrmann, 1980; Larsson *et al.*, 1987; Pope *et al.*, 1993). (3) If the activation of the surface involves the introduction of large, multivalent groups it may be possible for an increased level of binding to take place *via* this bridge (Figure 8.38(e)). This may explain raised levels observed where polylysine or polylysine–polyphenylalanine linker layers were employed (Hendry and Herrmann, 1980; Wood and Gadow, 1983). (4) If the activation has actually resulted in an increase in surface area as well, then higher adsorption is a natural consequence (Figure 8.38(f)). This could explain the increase seen when polystyrene and polypropylene tubes are grafted with an organic bis-isocyanate solution, which is known to erode the polymer surface (Saito and Nagai, 1983), or where chlorosulfonation has been employed (Peterman *et al.*, 1988).

Therefore there are many reasons why chemisorption can give higher levels of binding than physisorption under particular circumstances, but there is no reason to expect this as a general consequence of the approach. If the surface is hydrophobic, its area is unaffected by the activation, the activating groups are short and/or monovalent, and there is no chance of protein cross-linking, the expected level of adsorption is no greater than that observed from passive coating.

8.3.5.2 Can protein be bound with less loss of activity?

Again there is no inherent reason why covalent coupling *per se* should result in greater retention of activity in the bound protein. However, if covalent binding is combined with the presence of a hydrophilic surface, the prospects for improvement are good: provided the chemical derivatisation of the protein does not cause inactivation, there should be little conformational change of the type driven by hydrophobic interactions. Secondly, if the chemisorption approach allows the protein molecules to be bound specifically in a favourable orientation, the chances are better still that a greater proportion of functional activity will be retained.

There is certainly a great deal of evidence that such improvements are observed in practice: increases of functional activity compared to passive adsorption have been reported for many different proteins and different conditions. The typical 'improvement factor' in the presence of orienting effects has been estimated to be in the range of 2–8 (Lu *et al.*, 1996), which appears to be broadly consistent with the improvement that might be expected from steric considerations. It is sometimes difficult to deconvolute the results of increased uptake and increased activity. However, many reports describe improvements within this general range (Rotmans and Delwel, 1983; Larsson *et al.*, 1987; Peterman *et al.*, 1988; Brillhart and Ngo, 1991; Niveleau *et al.*, 1993).

Where oriented and random chemisorption have been compared, factors towards the bottom end of this range have been reported, for example antibodies bound *via* Fc carbohydrate moieties showed around threefold greater retention of activity than those immobilised *via* randomly-distributed amines (O'Shannessy and Hoffman, 1988).

Thus chemisorption certainly offers the potential to improve the retention of activity over physisorption methods. Orienting methods can in themselves offer moderate improvements over random ones, simply on steric grounds, and this applies whether the random method is covalent or non-covalent. But the greatest improvements in retention are likely to be obtained where this steric improvement is combined with a reduction in conformational changes, due to hydrophilisation of the surface.

Sometimes, of course, chemical derivatisation of a protein can have a drastic effect on its activity, and this can lead to poor results from chemisorption methods. Leckband and Langer (1991) for example reported up to 100% loss of activity when heparinase was immobilised *via* amine moieties, these losses, however, could be reduced to manageable levels by lowering the incorporation. And it should be recalled from the Section 8.2 that similarly devastating losses can be observed from physisorption techniques.

8.3.5.3 Can protein be bound more irreversibly?

Here at least there is an inherent reason why chemisorption should be better than physisorption, simply in terms of the relative stability of a suitably chosen chemical linkage to the surface. If the stability of this bond is as good as those which hold the protein itself together, by no means an impossible target, then the binding is as good as irreversible in practical terms.

The quantification of this improvement is strongly dependent on how the measurement is carried out, so it is very difficult to make meaningful comparisons between reports; different studies might investigate the desorbability of bound protein using simple buffer washes, detergent solutions, acid treatments, or prolonged soaking in protein solutions where surface exchange reactions are likely.

However, within experiments it is often clear that covalent binding gives a more desorption-resistant coating than physisorption: for antibodies on polystyrene, 0.01 M hydrochloric acid removed 28% of the physisorbed protein *versus* 3–14% for a chemisorption method; 6 M urea and 10% SDS removed 59% of the physisorbed material and 15–20% of that from chemisorption (Blanchard *et al.*, 1990).

With albumin on polystyrene, Larsson *et al.* observed 53% loss of physisorbed material on strong detergent treatment *versus* 17% for a chemisorption method. With milder treatments it is sometimes observed that there is no loss of covalently coupled protein, as seen by Peterman *et al.* (1988) with IgG on bromoacetyl polystyrene. Under the same conditions (0.05% Tween 20 in phosphate buffered saline, 12 hours) around 25% of physisorbed material was desorbed from polystyrene.

In general, it can certainly be stated that greater irreversibility can be obtained from covalent methods. However it is easy to assume that the binding from such methods may be completely irreversible, and this is very often not the case. Unless the surface is completely resistant to physisorption this is still likely to be a factor, even on a highly hydrophilic matrix such as glutaraldehyde-activated agarose, non-covalent binding effects have been observed (Ukeda *et al.*, 1993). Typical results from chemisorption studies indicate that a certain proportion of the immobilised protein is often in fact still bound by a non-covalent mechanism.

This issue has attracted great interest in the field of affinity chromatography, where it is clearly a great advantage if the only binding taking place to the matrix is that of the desired, specific variety.

8.3.5.4 Summary

It can be concluded from the preceding sections that chemisorption approaches can produce a wide range of differing effects, dependent on the materials involved and the design of the process. It is a mistake to assume that a complete change in performance from physisorbed material is likely to be observed, simply by switching to a chemisorption method.

If conditions are chosen carefully enough, however, it is not impossible to approach 'ideal' chemisorption behaviour. It should not be expected that a greater surface coverage of protein should be obtained than from a typical physisorption process. However, great potential is offered by a hydrophilic surface which is highly resistant to physisorption, while possessing functional groups of suitable reactivity to allow immobilisation in a favourable orientation. Not only is the protein likely to be bound in an irreversible fashion with a high degree of retained functional activity; a surface of this kind is also likely to display less subsequent non-specific binding of proteins.

Chemisorption approaches which do not meet all these criteria are, of course, only likely to yield a proportional improvement over passive methods.

8.4 Other Methods of Protein Immobilisation

There are a number of other approaches to the immobilisation of proteins, which do not rely on adsorption of the molecules to a surface. Either the solid phase is created by reactions between the protein molecules themselves, or the protein molecules are physically trapped within a polymer lattice, sphere or microcapsule.

All of these approaches have been applied particularly within the field of enzyme immobilisation (Manzoni, 1988; Pişkin, 1988a; Katchalski-Katzir, 1993) and, more recently, for controlled release of protein pharmaceuticals (Gombotz and Pettit, 1995). They are really at the limits of the scope of this book, and will only be mentioned here briefly for completeness.

8.4.1 Formation of insoluble protein aggregates

One side-effect of protein aggregation is generally a reduction in solubility. It is therefore possible to create immobilised protein matrices simply by allowing cross-linking reactions to continue until this stage is reached.

Any of the chemistries described in Chapter 5 can potentially be used for this approach; however the aim here is produce a large number of intermolecular bonds, and 'simple' homobifunctional reagents such as glutaraldehyde are generally better suited to this application than multi-stage heterobifunctional routes.

There is bound to be some loss of activity by this route, and although simple this approach is rather inefficient in terms of protein usage compared to adsorption methods. Some optimisation of conditions may be needed to minimise the loss of activity while maintaining sufficient cross-linking to cause insolubilisation.

Avrameas and Ternynck (1967, 1969) produced water-insoluble aggregates of antibodies and albumin by adding chloroformates or glutaraldehyde to stirred solutions of the proteins; Patramani *et al.* (1969) used a similar approach to form insoluble transaminase, or 'copolymers' of the enzyme with albumin or globulins.

Gelatin has some useful properties as a carrier protein, as it can be deposited in reasonably stable layers on solid components. Hardening of gelatin is promoted by chemical cross-linking, and it is possible to include other proteins in the setting matrix. For example, Akbulut and colleagues have described the copolymerisation of glucose oxidase and gelatin, which could then be coated conveniently onto solid components (Elçin and Akbulut, 1992; Sungur *et al.*, 1992).

8.4.2 Formation of microspheres from protein emulsions

Another approach where the protein plays a bulk role is to form an emulsion by stirring an aqueous protein solution in an oil matrix, yielding microspheres. For added stability the emulsion droplets can then be stabilised by cross-linking in an organic solvent, for example with a bifunctional aldehyde reagent.

Scheffel *et al.* (1972) described the basic procedure, producing HSA microspheres for investigating the function of the reticuloendothelial system. However it was a simple step to include other useful components in the emulsion, and this general approach has been used in many ways (Poznansky and Juliano, 1984). Yoshioka *et al.* (1981) included mitomycin in the formulation, providing a means of delivering this drug to the liver, spleen and lungs *via* the RES, while Hincal and Kas (1988) produced streptamycin-loaded albumin and gelatin microspheres for tuberculosis treatment.

In another application Widder *et al.* (1981) incorporated iron oxide and Protein A to produce magnetic immunoglobulin-binding microspheres.

8.4.3 Immobilisation of proteins by microencapsulation

Chang and colleagues (Chang, 1964; Chang *et al.*, 1966; Siu Chong and Chang, 1974) pioneered the production of polymeric microcapsules by carrying out polymerisation reactions in stirred solutions. For example, nylon microcapsules of various dimensions could be created by mixing diamines and bis-acyl chlorides. The dimensions of these microcapsules could be varied by altering the experimental conditions; typical examples described by Siu Chong and Chang had a mean diameter of 80 μm, while the membrane was 200 Å thick, with pores of 32 Å diameter.

If the polymerisation was carried out in a protein solution, a proportion of the protein molecules could be trapped within the microcapsule as it formed. The semi-permeable nature of the membrane allowed the transport of small molecules such as enzyme substrates, while retaining the protein content. Asparaginase immobilised within nylon microcapsules was used, for example, for the treatment of enzyme deficiencies.

Again, this approach has found many applications in the drug delivery field (Poznansky and Juliano, 1984).

More than one protein can be encapsulated simultaneously, allowing some fascinating applications to be realised: Gu and Chang (1990), for example, immobilised a combination of enzymes and dextran-NAD within a nylon–polyethyleneimine microcapsule, forming an 'artificial cell' which was capable of transforming ammonia and urea into essential amino acids. This area was reviewed recently by Chang (1995).

8.4.4 Immobilisation of proteins within polymer lattices

Similar in principle to the formation of microcapsules is the creation of a cross-linked polymer matrix in which protein molecules are trapped. A major area of interest for this approach is in providing controlled-release formulations of protein pharmaceuticals (Gombotz and Pettit, 1995): biodegradable polymers release the therapeutic agent as they are broken down *in vivo*. The polymers may undergo bulk degradation, such as the poly(lactic-co-glycolic acid), PLGA polymers which dominate the field, or they may be designed to erode slowly from the surface. Some PLGA formulations of small peptide hormones have entered the marketplace, and further applications seem set to follow.

The most convenient methods for trapping the protein molecules are those where the polymerisation can be initiated without the addition of chemicals, for example photo-cross-linking of polyvinyl alcohol containing a benzoate sensitiser (Imai *et al.*, 1986) or a proportion of styrylpyridinium groups (Ichimura and Watanabe, 1978, 1980; Ichimura, 1984). The porosity of the resulting matrices could be improved by incorporating a water-soluble reagent such as glycerol during the polymerisation, which could subsequently be washed out (Ichimura, 1987).

Buonocore *et al.* (1980) trapped entire cells of a thermoacidophilic bacterium in a polyacrylamide lattice, where it could be employed as an immobilised source of a thermostable β-galactosidase.

Yokoyama *et al.* (1988) trapped enzymes within a polymeric bilayer system, creating an interesting membrane format.

Radiation can also be used to promote cross-linking of polymeric matrices (Pişkin, 1988b).

8.5 The Preparation of Protein Conjugates With Soluble Polymers and Polysaccharides

8.5.1 Polysaccharides

Many proteins are, of course, already conjugated to polysaccharide moieties in their native state: glycoproteins are extremely common in nature. The role of the carbohydrate groups has been the subject of much debate (see Chapter 2, Section 2.4.1), but their presence certainly seems to enhance the stability of the protein in many cases (Arnold 1969; Biondi *et al.*,1980; Olden *et al.*, 1982).

The preparation of synthetic analogues of glycoproteins, neoglycoproteins, has attracted considerable attention, both as a model of glycoprotein behaviour (Stowell and Lee, 1980), and as a possible means of stabilisation in itself. This area has been reviewed by Lee and Lee (1992).

Artificial glycosylation of proteins has indeed been shown to demonstrate a stabilising effect, for example with respect to thermal degradation (Mitz and Summaria, 1961; Blomhoff and Christensen, 1982, 1983). Other attractive properties of neoglycoproteins relate to *in vivo* use; in particular there is often a general increase in the circulatory half-life (Marshall *et al.*, 1977; Marshall, 1978), caused partly by a reduction in the immunological response which the artificially-introduced protein molecules provoke.

Soluble dextrans are convenient polysaccharides for protein glycosylation. Being hydroxylic in nature, they and other polysaccharides are more or less inert to direct reaction with proteins, and some form of activation must be carried out to facilitate conjugation.

The available options are very similar to those already discussed for the activation of hydroxylic solid phases.

8.5.1.1 Introduction of amine-reactive groups into polysaccharides

Cyanogen bromide activation is a popular approach (see Chapter 10 for further details of this methodology). It is readily employed for dextrans (Vegarud and Christensen, 1975; Marshall and Rabinowitz, 1976; Blomhoff and Christensen, 1983) and is equally applicable to other carbohydrate-containing molecules such as bacterial lipopolysaccharides (Lange *et al.*, 1983).

The carboxymethylation of dextrans has been carried out since the 1950s (Berger and Novak, 1958), and activation of the resulting carboxyl moieties offers an alternative approach to protein conjugation. Heindel *et al.* (1994) described another variation on this theme: they carried out a cyclodehydration reaction on the carboxymethyl dextran by heating in anhydrous solvents, and obtained a lactone derivative, albeit with reduced solubility, which could be used for direct reaction with amines. Introduction of active ester moieties can also be carried out using chloroformates such as 4-nitrophenyl chloroformate (Figure 8.39), which react directly with hydroxylic species in the presence of a catalyst such as 4-(dimethylamino)pyridine, DMAP, to yield carbonates (Herman *et al.*, 1993).

Periodate oxidation of soluble carbohydrates to produce aldehydes is another possibility (Bernstein *et al.*, 1979), for example to create polyaldehyde dextran (PAD) which can be used for direct reaction with protein amines (Hurwitz *et al.*, 1985). Alternatively periodate-oxidised carbohydrate can be used as the basis for other routes; for example to couple oxidised glycoproteins *via* a diamine linker (Shih *et al.*, 1990), or to introduce a biotin moiety *via* a hydrazide (Luk *et al.*, 1995).

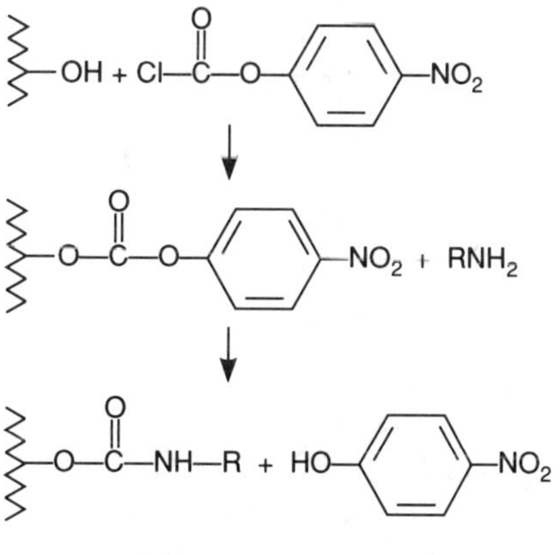

Figure 8.39 Introduction of a reactive carbonate into a polysaccharide using 4- nitrophenyl chloroformate.

8.5.1.2 Introduction of thiol-reactive groups into polysaccharides

Further derivatisation of aldehyde-containing polysaccharide such as PAD with a heterobifunctional maleimido-hydrazide compound, yields a thiol-reactive derivative, which can be used for example for direct coupling to reduced immunoglobulins (Figure 8.40)

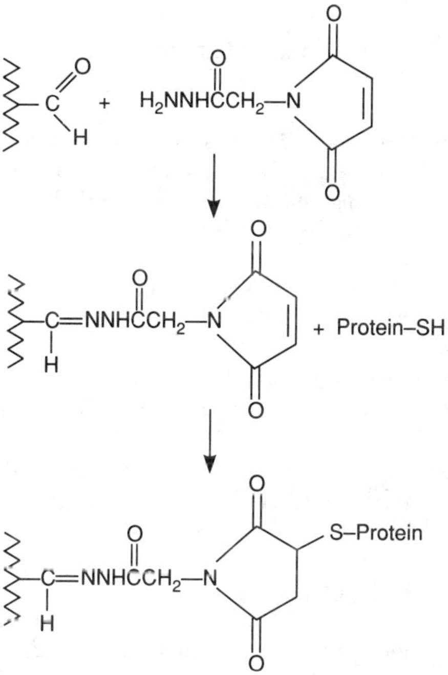

Figure 8.40 Introduction of thiol-reactive maleimide groups into an oxidised polysaaccharide aldehyde moiety.

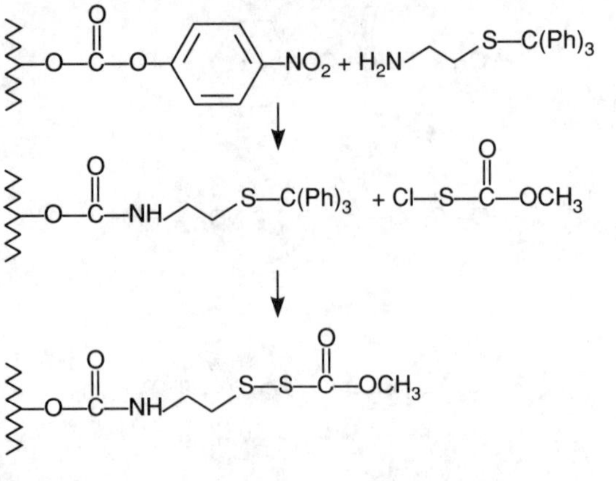

Figure 8.41 Introduction of thiol-reactive moiety by exploiting *S*-tritylated aminothiol chemistry.

(Heindel *et al.*, 1991). The preparation of disulfide-activated dextrans has also been described (Herman *et al.*, 1993): an active carbonate derivative was reacted first with an *S*-tritylated aminothiol, which was converted to a reactive disulfide using (methoxycarbonyl)-sulfenyl chloride (Figure 8.41).

Gedda *et al.* (1996) recently described the use of an allyl-derivatised dextran for coupling to thiol-containing moieties.

8.5.1.3 Introduction of amines into polysaccharides

Carboxymethylated dextrans can be converted into an amine-bearing form by carbodiimide-mediated coupling with a diamine such as hexamethylenediamine (Rosemeyer and Seela, 1984).

8.5.1.4 Coupling of activated polysaccharides to proteins

There is really very little remarkable in the chemistry used to couple these activated polysaccharides to proteins. The conditions required are closely comparable to those described for activated solid phases (Section 8.3.2), and, for further details, the reader is referred there and to relevant sections of the earlier chapters.

8.5.2 Poly(ethylene glycol)

The beneficial properties discussed at the start of Section 8.5.1 are not just limited to polysaccharides; other hydrophilic macromolecules of synthetic origin are capable of showing similar behaviour. Poly(ethylene glycol), PEG, has dominated this field (Harris, 1992).

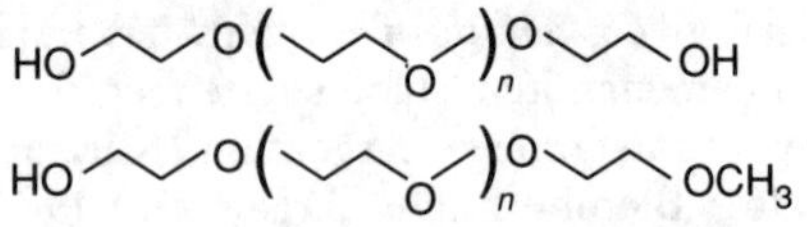

Figure 8.42 PEG and monomethyl PEG.

PEGs (Figure 8.42(a)) are soluble in both water and organic media[1], and for the applications discussed here molecular masses of between a few hundred and 20,000 Da are generally employed. PEG is also sometimes referred to by a variety of other names: poly(-ethylene oxide) (PEO), poly(oxyethylene) (POE), and polyoxirane; PEO is sometimes (but not always) reserved for materials with molecular mass in excess of 20,000 Da. Like polysaccharides, PEGs have been shown to reduce the immunogenicity of proteins to which they are conjugated (Abuchowski *et al.*, 1977; Sehon, 1992), and in general they are unusually effective at excluding other proteins from their surroundings (Harris, 1992). A great deal of interest has therefore also centred around the use of PEG-coated surfaces in applications where protein adsorption is to be discouraged, notably for improving the biocompatibility of *in vivo* devices (Gombotz *et al.*, 1992). The search for alternatives to PEG in this role is an active research area (Deng *et al.*, 1996).

The most commonly used forms of PEG are the monomethyl ethers (MPEG), in which one terminal is a hydroxyl moiety and the other a methoxy function (Figure 8.42(b)).

Many different routes have been used to activate PEGs for reaction with proteins (Zalipsky and Lee, 1992; Zalipsky, 1995), and these will be summarised briefly below; Zalipsky's 1995 review is particularly comprehensive in describing the available PEG derivatives. Virtually all categories are commercially available.

8.5.2.1 Introduction of amine groups into PEG

PEGs bearing amine groups have been prepared by a number of routes, including a number where the amine is a straight replacement for the terminal hydroxyl group. This can be achieved by substitution with a halide or tosylate function, which can then be displaced by ammonia to yield the amine (Bückmann *et al.*, 1981; Leonard and Dellacherie, 1988; Figure 8.43(a)); alternatively the halide can be replaced with

[1] Leading also to biotechnology applications, such as the preparation of active enzymes with good solubility in organic solvents (Inada *et al.*, 1986)

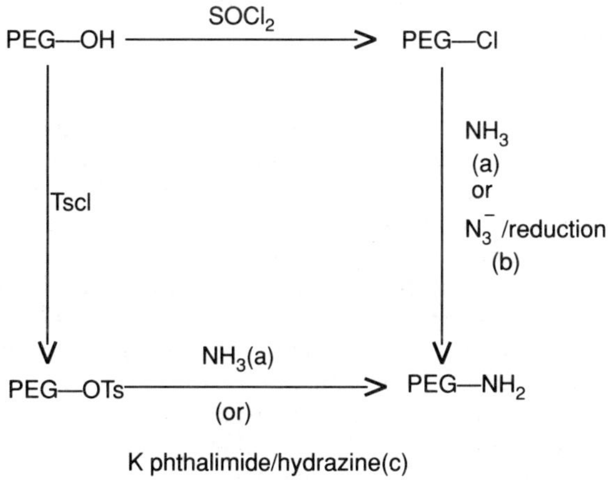

Figure 8.43 Introduction of amine groups into PEG.

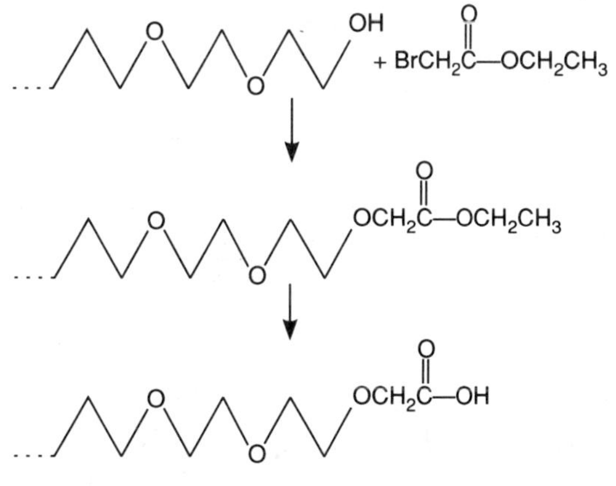

Figure 8.44 Introduction of carboxyl group into PEGs using ethyl bromoacetate.

azide followed by reduction (Zalipsky *et al.*, 1994; Figure 8.43(b)), or the tosylate can be replaced with potassium phthalimide followed by hydrazinolysis (Pillai *et al.*, 1980; Figure 8.43c).

Other methods introduce a spacer group, for example reaction of the terminal hydroxyl with bromoethyl phthalimide followed by hydrazinolysis (Geckeler, 1979), carbonyldiimidazole-mediated direct coupling with diamines (Ranucci and Ferruti, 1990) or DMAP-catalysed esterification with an amino acid (Zalipsky *et al.*, 1984).

Aromatic amines can also be introduced, typically by the reaction of nitrophenyl derivatives such as *p*-nitrobenzyl bromide or *p*-nitrobenzoyl chloride, followed by reduction of the nitro groups (Pollak and Whitesides, 1976; Rozzell, 1987).

8.5.2.2 Introduction of hydrazide groups into PEG

Hydrazido PEGs can be prepared by hydrazinolysis of PEG esters (Andresz *et al.*, 1978; Zalipsky *et al.*, 1992a) or reaction of activated hydroxyl groups (*e.g.* tresylates) with bis-hydrazides (Persson and Olde, 1988).

8.5.2.3 Introduction of carboxyl groups into PEG

Carboxy PEGS are produced by the reaction of PEG alkoxides with haloesters or haloacids such as ethyl bromoacetate (Figure 8.44) (Bückmann *et al.*, 1981), or by succinylation of PEG with succinic anhydride (Ferruti *et al.*, 1981; Zalipsky *et al.*, 1983). Succinylation of an amino-PEG is an alternative, yielding an amide link rather than an ester (Bückmann *et al.*, 1981). A lysine-PEG_2 reagent has been described (Monfardini *et al.*, 1995), the carboxyl function remaining available for reaction.

8.5.2.4 Introduction of thiol groups into PEG

Thiol PEGs can be prepared by a number of routes directly from PEG: tosylation followed by reaction with potassium ethyl xanthate (Zalipsky *et al.*, 1987), with thioacetate (Woghiren *et al.*, 1993) or with hydrosulfide, thiourea or trithiocarbonate (Harris *et al.*, 1992). Disulfide formation is a common problem which can be cured by reduction (Zalipsky, 1995). Further derivatisation of amino-PEGs can also be used to introduce thiols: direct thiolation using reagents of the type described in Chapter 5, Section 5.12 does not appear to have been described, but Herron *et al.* (1994) described the use of disulfide-containing heterobifunctionals followed by reduction.

8.5.2.5 Introduction of amine-reactive groups into PEG

Amine-reactive PEGs form, predictably, one of the most widely used categories.

Aldehyde PEGs can be prepared, for example by direct oxidation of PEG (Chamow *et al.*, 1994), by reaction of PEG with a protected bromoacetaldehyde (Paley and Harris, 1987), or by reaction of a mercapto-PEG with a protected chloro-propionaldehyde (Harris *et al.*, 1992). The reactivity of some aldehyde PEGs

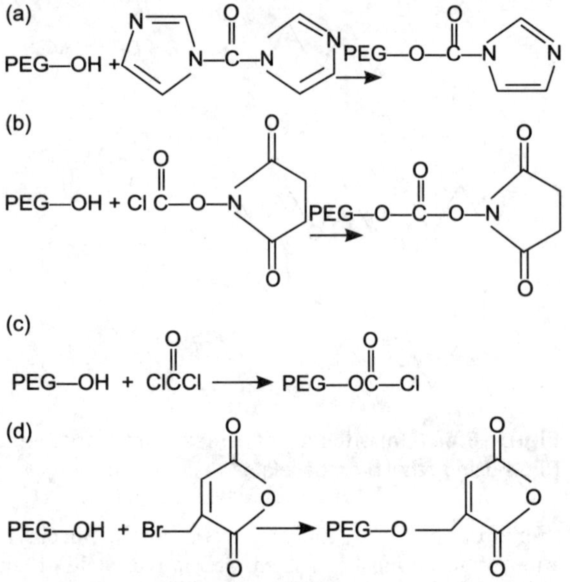

Figure 8.45 Some routes for the formation of activated carboxyl functions.

can be problematic, as condensation products are readily formed (Zalipsky, 1995).

Activated carboxyl functions can be formed from the carboxy-PEGs described above, typically by carbodiimide chemistry (Zalipsky *et al.*, 1983).

Similar groups can also be introduced directly into PEG using reagents such as bromoalkyl derivatives of maleic anhydride (Garman and Kalindjian, 1987), carbonyldiimidazole (Beauchamp *et al.*, 1983; Brygier *et al.*, 1993), active esters of chloroformic acid (Veronese *et al.*, 1985, 1992; Zalipsky *et al.*, 1992b), or phosgene (Galin *et al.*, 1965) (Figure 8.45). The succinimidyl carbonate shown in Figure 8.45(b) has achieved particular popularity (Chiu *et al.*, 1993; Miron and Wilchek, 1993). Acyl azides have also been described (Rubinstein *et al.*, 1978), and a thiazolidine-2-thione-derivatised PEG was developed recently by Greenwald *et al.* (1996).

Imidate PEG derivatives (Ueno and Fujino, 1987) offer good selectivity for amines. Other groups have also been employed to confer reactivity with amines and other nucleophiles, including triazines (Abuchowski *et al.*, 1977; Shafer and Harris, 1986; Jackson *et al.*, 1987), tresylates (Nilsson and Mosbach, 1984), aryl halides (Ladd and Snow, 1993) and epoxides (Mathis *et al.*, 1985; Bergstrom *et al.*, 1992). The non-selectivity of the triazine derivatives in particular can be problematic (Zalipsky, 1995), but they have remained popular.

8.5.2.6 Introduction of thiol-reactive groups into PEG

Thiol-reactive PEGs have received relatively little attention, although the broad-reactivity electrophiles just discussed will also react with thiols (Wieder *et al.*, 1979). So too will disulfides prepared directly from thiols (Woghiren *et al.*, 1993) or by reaction of heterobifunctional reagents with amino-PEGs (Herron *et al.*, 1994). Maleimide groups can also be introduced by the latter approach (Goodson and Katre, 1990), or by cyclisation of maleamyl derivatives (Albarella and Buckler, 1989) and it is simple to imagine other derivatisations of amino-PEGs using reagents described in Chapter 5. Vinyl sulfone PEG derivatives are now commercially available too, for example from Shearwater Polymers; the preparation of a reagent of this type is described by Morpurgo *et al.* (1996).

8.5.2.7 Other reactive PEG derivatives

Arginine-reactive PEGs bearing bis-carbonyl moieties have been described (Pande *et al.*, 1980).

The PEG derivatives discussed so far are generally monofunctional derivatives of MPEG or homobifunctional PEG derivatives. Recently, however, there has been some interest in the preparation of well-defined heterobifunctional PEGs, for example α-carboxyl-ω-hydroxyl-PEG (Zalipsky and Barany, 1990), α-amino-ω-hydroxyl-PEG (Harris *et al.*, 1992; Cammas *et al.*, 1995), α-formyl-ω-hydroxyl-PEG (Nagasaki *et al.*, 1995) and α-amino-ω-carboxyl-PEG, available from Shearwater Polymers.

Copolymers of PEG with other compounds offer further possibilities, for example those with diamines such as lysine: examples with (activated) carboxyl, amino, hydroxyl and aldehyde moieties have been described (Nathan *et al.*, 1993). Branched copolymers with maleic anhydride have shown useful protein-stabilising effects, as well as possessing intrinsic reactivity for amines (Kodera *et al.*, 1994).

8.5.2.8 Coupling of activated PEGs to proteins

As with polysaccharides, there is little remarkable about the chemistries employed to couple these active PEG derivatives to proteins. Site-specific derivatisation methods have received some attention, however, using protein thiols (Benhar *et al.*, 1994) or selective activation of the N-terminal residue (Gaertner and Offord, 1996).

8.5.3 Other soluble polymers

The Pluronic™ surfactants (BASF Corp.) are poly (ethylene oxide) – poly (propylene oxide) – poly (ethylene oxide) block copolymers, which possess useful properties such as enhancing the therapeutic effect of drugs and binding to hydrophobic surfaces. Li *et al.* (1996) described the preparation of reactive derivatives of Pluronics incorporating active esters or pyridyl disulfides, and used these to bind β-galactosidase to polystyrene.

A relatively small amount of work has been carried out on soluble polyacrylamides. Heimgartner *et al.* (1989, 1990) described polyhydrazide derivatives of a polyacrylamide, which could be used for specific coupling of oxidised glycoproteins. Mage *et al.* (1990) described the use of partially hydrolysed polyacrylamides as soluble carriers for proteins, the conjugations being carried out using carbodiimide chemistry on the exposed carboxyl groups.

Many polymers show interesting temperature-dependent solubility properties, becoming dehydrated and considerably less soluble at the lower critical solution temperature (LCST). PEGs, for example, typically have LCST values in the region of 100 °C (Harris, 1992), and there is considerable interest in developing polymers which exhibit this property at a temperature more compatible with protein stability. The usefulness of being able to precipitate soluble polymer–protein complexes simply by raising the temperature a few degrees is clearly apparent, for example in affinity purification methods or in immunoassay.

Some polyacrylamides fulfil this requirement, particularly *N*-isopropylacrylamide oligomers and polymers, which may have LCST values as low as 13 °C (Heskins *et al.*, 1968; Monji *et al.*, 1994). These have been coupled to proteins *via* succinimide esters, either generated through copolymerisation (Chen *et al.*, 1990; Monji *et al.*, 1994), or DCC/NHS activation of carboxyl groups (Takei *et al.*, 1993a,b, 1994). Maleimide-containing analogues have also been described (Chilkoti *et al.*, 1994), allowing oriented coupling to cysteine residues.

Outside of the protein-coupling field, interesting applications of these thermoresponsive polymers continue to emerge, such as catalytic reactions that exhibit their own temperature control *via* polymer-bound catalyst or substrate (Bergbreiter and Caraway, 1996). Changes in pH cause these polymers to alter their shape too, and an intriguing 'self-oscillating gel' has been produced, which spontaneously undergoes cyclical volume changes when coupled to the self-oscillating Belousov-Zhabotinsky reaction (Yoshida *et al.*, 1996).

There are countless other soluble polymers which can be employed for conjugation to polymers: to give a couple of examples only polyvinyl alcohol (PVA) has attracted a fair amount of interest (see, for example, Hixson, 1973; Jiang *et al.*, 1990), while electrically conductive polythiophene and polyaniline polymers have been coupled to proteins as the basis of a novel immunoassay system (Englebienne and Weiland, 1996).

8.6 Coupling of Proteins to Cells

Some of the oldest immunochemical procedures are based on the aggregation of red blood cells (erythrocytes) by specific antibodies. Application of this haemagglutination approach to the detection of a wide range of antigens and antibodies is possible, but requires the coupling of molecules with the correct specific binding properties to the erythrocyte surface (Figure 1.4).

A number of variations of this general format exist, for example detecting antigens by inhibition of agglutination, or promoting lysis of immune complex-labelled cells by the addition of complement.

The role of erythrocytes in some of these assays can be taken by other agglutinable materials such as polystyrene latex particles (Marumoto *et al.*, 1978) or other polymeric microspheres (Hosaka *et al.*, 1983); protein labelling of these materials is carried out using the techniques described in Sections 8.2 and 8.3 of this chapter. However, the use of erythrocyte-based formats remains popular.

Erythrocytes are also of interest for pharmaceutical applications, as they have a long circulatory half-life. They therefore have potential as a stable carrier for therapeutic proteins (Nicolau *et al.*, 1993).

While some bacterial polysaccharide antigens coat directly to red blood cells (Middlebrook and Dubos, 1948), proteins cannot be adsorbed in this way—as might be suspected from the *in vivo* existence of these cells in a protein-rich medium. Various methods have therefore been developed for the attachment of protein molecules to erythrocytes (Jou *et al.*, 1983), and will be reviewed briefly here.

8.6.1 Tannic acid

Tannic acid has an extensive history of application to protein modification (in the leather industry for example), and has been known since the 1920s to cause

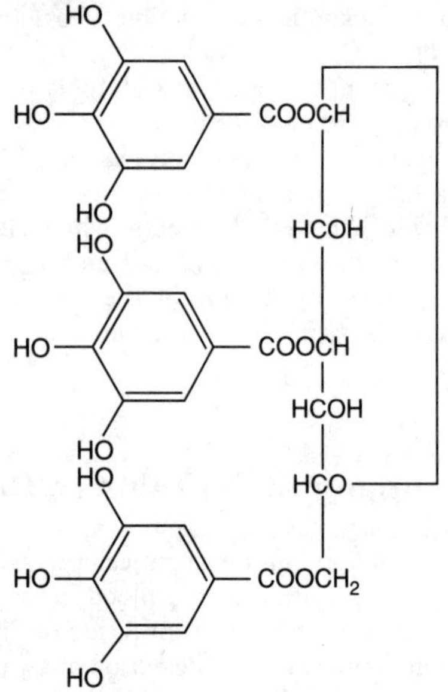

Figure 8.46 Corilagin, a component of tannic acid.

the aggregation of red blood cells. Tannic acid may contain a number of different components (tannins), such as corilagin (Figure 8.46).

Carefully controlled treatment with this reagent was therefore examined by Boyden (1951) as a possible means of rendering the cells more receptive to protein adsorption. The experiments were successful, and it was found that tanned erythrocytes could be coated with protein within 15 minutes at room temperature (Chen and Meyer, 1954).

Tanned erythrocytes are fragile entities: they could not, for example, be used in the lysis-based assay formats mentioned above as they could be lysed by complement alone (Stavitsky and Arquilla, 1958). Their robustness can, however be increased by the use of a cross-linking reagent such as formaldehyde or glutaraldehyde (Adler and Adler, 1980), and such preparations can even be frozen or lyophilised successfully. See Adler and Adler (1980) for a typical preparation protocol for tanned erythrocytes.

8.6.2 Bis-diazonium salts

Another chemical means of attaching proteins to erythrocytes dates from roughly the same time; the use of bis-diazo compounds to achieve the necessary links. This was first described by Pressman *et al.* (1942), and used by Stavitsky and Arquilla (1958) to avoid the complement-sensitivity of tanned cells; bis-diazotised benzidine (BDB) was the reagent commonly employed in these early studies, and, despite its carcinogenic nature, has remained in use at least into the 1980s (Adler and Adler, 1980). The order of addition of reagents can be important in these preparations (Gold *et al.*, 1966). Also, the amount of the coupling agent in such methods needs to be kept reasonably low to avoid spontaneous lysis of the cells, and again the robustness of BDB-cells can be increased by cross-linking procedures (Adler and Adler, 1980).

The chemistry of protein conjugation *via* bis-diazonium salts has been dealt with in Chapter 5, and will not be discussed further here. Again, see Adler and Adler (1980) for a typical preparation protocol.

8.6.3 Carbodiimides

Simple one-step procedures using water-soluble carbodiimides such as EDAC were introduced in the 1960s (Johnson *et al.*, 1966; Golub *et al.*, 1968) and remained in use virtually without change at least into the 1980s (Humphries and McConnell, 1974; Levin *et al.*, 1977; Borsos and Langone, 1981). Carbodiimide is added to a mixture of cells and protein and incubated typically for 30–60 minutes; see Borsos and Langone (1981) for a protocol.

The pH used for these methods is often in the region of 7.2–7.4, rather higher than the optimum for carbodiimide reactions. Some later methods for coupling succinylated proteins to erythrocytes use more typical conditions (pH 6) for this reaction (Jou *et al.*, 1983), so there appears to be no overriding reason why the more basic conditions remained in use for so long.

8.6.4 Chromium chloride

Chromium(III) salts have been used as tanning agents for many decades, and form inter-protein cross-links by forming complexes between carboxyl functions (see Chapter 5). Chromium salts have been employed since the 1950s for coupling proteins to erythrocytes (Kritzman, 1958; Gold and Fudenberg, 1967). Again these methods have stayed in popular use for many years (Jerne *et al.*, 1974; Molinaro *et al.*, 1981; Galfrè and Milstein, 1981), see any of these references for typical protocols. One point to note is that phosphate should not be included in buffers for chromium chloride coupling, as it interferes with the complexation reaction.

Jerne *et al.* (1974) noted that this technique was simpler and more efficient in terms of protein use than the BDB or carbodiimide procedures.

8.6.5 Amine-reactive homobifunctional reagents

Not surprisingly this category of coupling agents has also been exploited for linking proteins to or within the membranes of cells such as erythrocytes. Examples include bifunctional aryl halides (Ling, 1961), bis-isocyanates (Gyenes and Sehon, 1964), bis-imidates (Wang and Richards, 1975), bis-succinimides (Anjaneyulu *et al.*, 1989), benzoquinone (Ternynck and Avrameas, 1976), and glutaraldehyde (Steck, 1972; Lemieux *et al.*, 1974).

It has already been mentioned that reagents such as glutaraldehyde were already in use for stabilising erythrocytes by cross-linking of surface proteins, and it is difficult to separate this role from the binding to other proteins in solution which may also occur. Some methods have described combinations of these homobifunctional reagents to carry out the combined role of stabilisation and conjugation (Mahan and Copeland, 1978).

8.6.6 Periodate oxidation

Keren *et al.* (1986) used mild oxidation with periodate to generate aldehyde groups from cell surface glycoproteins; these were then coupled to a hydrazide-derivatised enzyme. Similarly, Heitzmann and Richards (1974) coupled biotin hydrazide to periodate-oxidised erythrocytes, then used this to bind a ferritin–avidin conjugate.

8.6.7 Heterobifunctional reagents

Jou *et al.* (1983) have pointed out the lack of control achievable from homobifunctional coupling methods, and described the use of a heterobifunctional approach using SPDP (see also Chapter 5). Disulfides in the cell surface proteins are reduced with dithiothreitol, and allowed to react with disulfides introduced into the protein using SPDP (Figure 8.47).

Obviously many other heterobifunctional routes can be imagined using reagents described in any of the relevant categories of Chapter 5, including those based on the biotin–avidin reaction (Heitzmann and Richards, 1974; Bayer *et al.*, 1976).

8.6.8 Electroinsertion

An interesting alternative to chemical methods for forming protein–erythrocyte conjugates is electroinsertion, in which the protein is immobilised directly into the cell membrane by the application of a pulsed electric field. The mechanism of the insertion process, first described by Blobel in 1980, is still the subject of some debate (Singer, 1990; Nicolau *et al.*, 1993).

However the methodology is very simple, and avoids any chemical modification of the protein

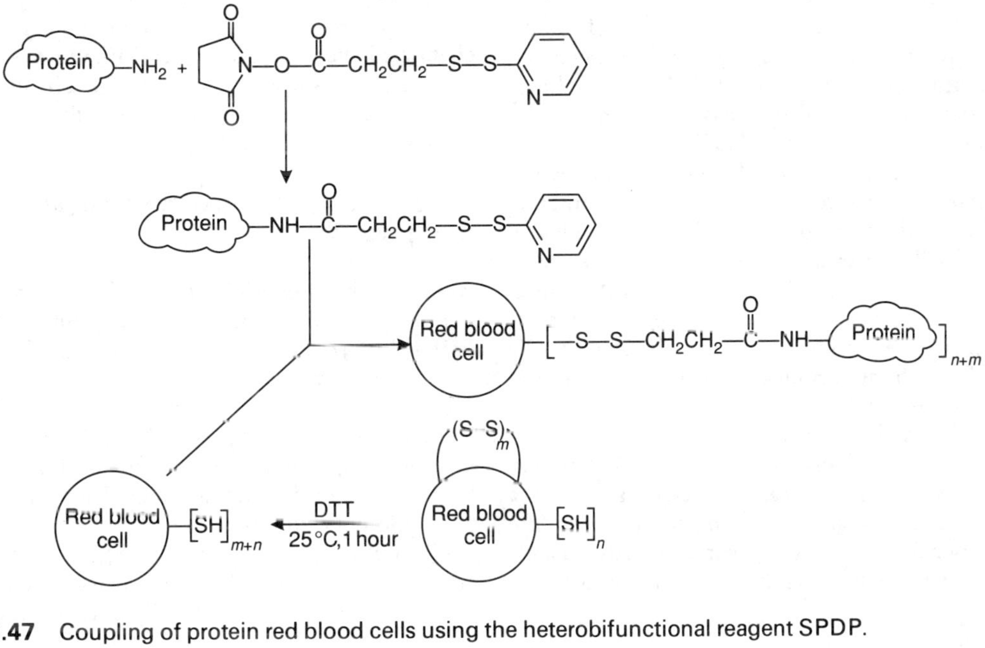

Figure 8.47 Coupling of protein red blood cells using the heterobifunctional reagent SPDP.

molecule. It therefore seems to have considerable potential as a method for the formation of erythrocyte-bound protein complexes. See Nicolau *et al.* (1993) for a typical electroinsertion procedure.

8.7 Coupling of Proteins to Liposomes

Liposomes, which form an example of a self-assembling system (Kossovsky *et al.*, 1993), are created when certain lipid mixtures are dispersed in aqueous media. They were described in detail by Bangham *et al.* in 1974, and can consist of single or multiple, concentric lipid bilayers containing a portion of aqueous solution. Small liposomes (of the order of 200 Å) are likely to have only one bilayer (small unilamellar vesicles, SUV); larger examples may reach diameters of several microns and may possess one (large unilamellar vesicles, LUV) or multiple bilayers (multilamellar vesicles, MLV).

Liposomes have numerous important applications, and there have been many volumes dealing entirely with their use (see for example Gregoriadis, 1984; Ostro, 1988, Lopez-Berestein and Fidler, 1989).

Some of these require that the liposomes should be associated with proteins; they may merely require a quantity of protein to be contained within the liposome, such as their use as adjuvants to improve the effectiveness of vaccines (Gregoriadis, 1990). For other uses, it is necessary to coat proteins to the liposome surface, these include the use of immunoliposomes for targetted drug delivery (Huang *et al.*, 1980; Martin *et al.*, 1981), for cell isolation (Golubev and Sidorova, 1989), and for lysis-based immunoassays (Ishimori and Rokugawa, 1993).

As might be expected from the hydrophobic nature of the liposome surface, some hydrophobic interactions with proteins are observed: Vandenbranden *et al.* (1981, 1982), for example, reported the binding of hydrophobic conformations of acid-treated IgG directly to liposomes. However it is often desirable for proteins to be bound more strongly than can be achieved by such interactions, and two approaches can be taken to this problem.

Both rely on forming a covalent link between a component of the liposome itself, and the protein label. Either the protein can be derivatised with a lipophilic group which will be incorporated in the liposome membrane, or lipids bearing reactive 'handles' can be incorporated, followed by reaction with the protein.

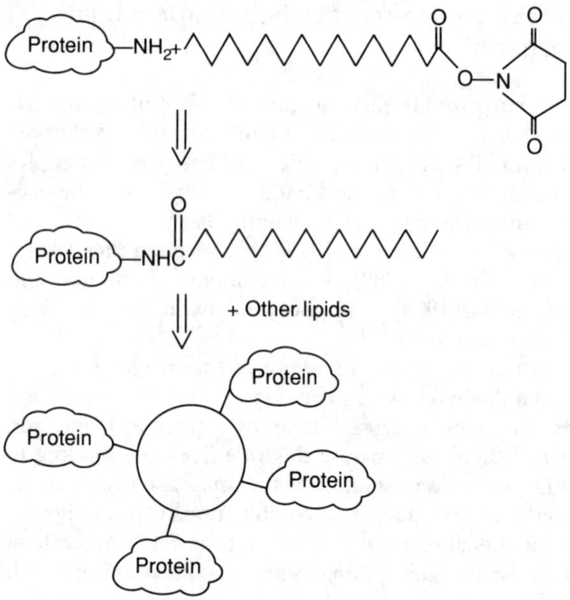

Figure 8.48 Formation of a protein-labelled liposome from protein-substituted lipids.

8.7.1 Protein derivatisation with lipid groups

In the first approach any suitable chemistry can be employed to introduce a hydrophobic group into the protein of interest.

Typically, the amine groups of the protein are allowed to react with the succinimide ester of a long-chain acid such as palmitic acid (Figure 8.48) (Huang *et al.*, 1980; Wang and Huang, 1987). The lipophilised protein can then be incorporated directly into the liposome preparation.

8.7.2 Incorporation of reactive groups into the liposome

Alternatively, chemically reactive lipids can be incorporated into the liposome for subsequent reaction with protein molecules. Many examples of this approach have been described.

Amine-containing lipids such as phosphatidylethanolamine and its analogues (Figure 8.49) are readily incorporated, and are then available for the use of homobifunctional coupling agents such as glutaraldehyde (Golubev and Sidorova, 1989).

Chua *et al.* (1984) described the use of hydrazide-containing lipids, which allowed oriented immobilisation of IgG *via* oxidised carbohydrate moieties.

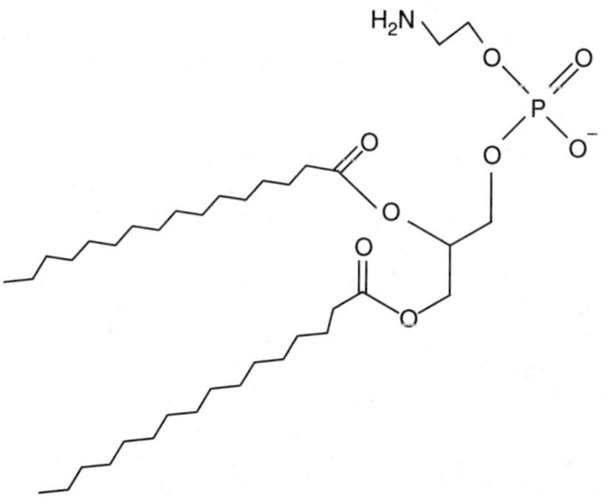

Figure 8.49 Phosphatidylethanolamine.

There have been a large number of reports of thiol-reactive groups incorporated into liposome membranes, including disulfides (Martin *et al.*, 1981; Umeda *et al.*, 1986; Boeckler *et al.*, 1996; Frisch *et al.*, 1996), maleimides (Martin and Papahadjopoulos, 1982; Matthay *et al.*, 1986; Boeckler *et al.*, 1996; Ansell *et al.*, 1996; Frisch *et al.*, 1996) and haloacyl groups (Ishimori and Rokugawa, 1993; Boeckler *et al.*, 1996; Frisch *et al.*, 1996).

Sänger *et al.* (1992) described a photoactivatable lipid for liposome formation. Finally, PEG-derivatised lipids have been utilised to modify the properties of the liposome, in particular to avoid aggregation problems (Harasym *et al.*, 1995) or to improve the *in vivo* half-life (Sakai *et al.*, 1977). Some of these PEG-lipids also possess reactive groups such as hydrazides for linking to proteins, etc. (Zalipsky, 1993).

References

Abuchowski, A. *et al.* (1977). Alteration of immunological properties of bovine serum albumin by covalent attachment of polyethylene glycol. *J. Biol. Chem.*, **252**, 3578–3581.

Adamson, A. W. (1990). The solid–liquid interface: adsorption from solution. in *Physical Chemistry of Surfaces*, 5th edn, 421–459, John Wiley and Sons, Chichester.

Adler, F. L. and Adler, L. T. (1980). Passive hemagglutination and hemolysis for estimation of antigens and antibodies. *Methods Enzymol.*, **70**, 455–466.

Aebersold, R. H. *et al.* (1988). Covalent immobilization of proteins for high-sensitivity sequence analysis: electroblotting onto chemically activated glass from sodium dodecyl sulfate–polyacrylamide gels. *Biochemistry*, **27**, 6860–6867.

Akashi, M. *et al.* (1992). Immobilization of human thrombomodulin on glass beads and its anticoagulant activity. *Bioconjugate Chem.*, **3**, 363–365.

Albarella, J. P. and Buckler, R. T. (1989). Enzyme-labeled antibody reagent with polyalkyleneglycol linking group. *US Patent* 4,810,638, to Miles Inc.

Andrade, J. D. and Hlady, V. (1986). Protein adsorption and mbterials biocompatibility: a tutorial review and suggested hypotheses. *Adv. Polym. Sci.*, **79**, 1–63.

Andrade, J. D. *et al.* (1984). Effects of plasma adsorption on protein conformation and activity. *Pure Appl. Chem.*, **56**, 1345–1350.

Andresz, H. *et al.* (1978). Synthese verzweigte Polysaccharide. 5. Koplung von Oligosacchariden und Amilose an verschiedene Trager durch Hydrazonbindung. *Makromol. Chem.*, **179**, 301–312.

Anjaneyulu, P. S. R. *et al.* (1989). Bis (Súlfo-*N*-succinimidyl) doxyl-2-spiro-5'-azelate: synthesis, characterization, and reaction with the anion-exchange channel in intact human erythrocytes. *Biochemistry*, **28**, 6583–6590.

Ansell, S. M. *et al.* (1996). 3-(2-Pyridyldithio) propionic acid hydrazide as a cross-linker in the formation of antibody–liposome conjugates. *Bioconjugate Chem.*, **7**, 490–496.

Arnold, W. N. (1969). Heat inactivation kinetics of yeast β-fructofuranoside. A polydisperse system. *Biochim. Biophys. Acta*, **178**, 347–353.

Arnebrant, T. and Nylander, T. (1988). Adsorption of insulin on metal surfaces in relation to surface behavior. *J. Colloid Interface Sci.*, **122**, 557–566.

Arnebrant, T. *et al.* (1985). Bilayer formation at adsorption of proteins from aqueous solutions at metal surfaces. *Prog. Colloid Polym. Sci.*, **70**, 62–66.

Avrameas, S. and Ternynck, T. (1967). Biologically active water-insoluble protein polymers. *J. Biol. Chem.*, **242**, 1651–1659.

Avrameas, S. and Ternynck, T. (1969). The cross-linking of proteins with glutaraldehyde and its use for the preparation of immunoadsorbents. *Immunochemistry*, **6**, 53–66.

Axelrod, D. *et al.* (1984). Total internal reflection fluorescence. *Annu. Rev. Biophys. Bioeng.*, **13**, 247–268.

Bagchi, P. and Birnbaum, S. M. (1981). Effect of pH on the adsorption of immunoglobulin G on anionic poly (vinyltoluene) model latex particles. *J. Colloid Interface Sci.*, **83**, 460–478.

Bangham, A. D. *et al.* (1974). Preparation and use of liposomes as models of biological membranes. *Methods Membrane Biol.*, **1**, 1–68.

Baszkin, A. and Lyman, D. J. (1980). The interaction of plasma proteins with polymers. I. Relationship between polymer surface energy and protein adsorption/desorption. *J. Biomed. Mater. Res.*, **14**, 393–403.

Bayer, E. *et al.* (1971) Synthese des *C*-Terminalen Hexapeptidamids von Secretin an einem Neuen Festen Träger. *Hoppe-Seyler's Z. Physiol. Chem.*, **352**, 759–760.

Bayer, E. A. *et al.* (1976). Affinity cytochemistry: the localization of lectin and antibody receptors on erythrocytes via the avidin–biotin complex. *FEBS Lett.*, **68**, 240–244.

Beauchamp, C. O. *et al.* (1983). A new procedure for the synthesis of polyethylene glycol–protein adducts: effects on function, receptor recognition, and clearance of superoxide dismutase, lactoferrin and α_2-macroglobulin. *Anal. Biochem.*, **131**, 25–33.

Beddows, C. G. *et al.* (1986). Immobilization of BSA, enzymes and cells of *Bacillus stearothermophilus* onto cellulose, polygalacturonic acid and starch based graft copolymers containing maleic anhydride. *Biotechnol. Bioeng.*, **28**, 51–57.

Beissinger, R. L. and Leonard, E. F. (1981). Sorption kinetics of binary protein solutions: general approach to multicomponent systems. *J. Colloid Interface Sci.*, **52**, 444–451.

Benhar, I. *et al.* (1994). Pseudomonas exotoxin A mutants. Replacement of surface exposed residues in domain III with cysteine residues that can be modified with polyethylene glycol in a site specific manner. *J. Biol. Chem.*, **269**, 13398–13404.

Benson, R. E. and Cairns, T. L. (1951). *O*-Methylcaprolactim. *Org. Synth.*, **31**, 72–74.

Bergbreiter, D. E. and Caraway, J. W. (1996). Thermoresponsive polymer-bound substrates. *J. Am. Chem. Soc.*, **118**, 6092–6093.

Berger, C. and Novak, L. J. (1958). Ferrous carboxymethyl dextran. *US Patent 2,862,920*, to the commonwealth engineering company.

Bergstrom, K. *et al.* (1992). Reduction of fibrinogen adsorption on PEG-coated polystyrene. *J. Biomed. Mater. Res.*, **26**, 779–790.

Berkowitz, D. B. and Webert, D. W. (1981). The inactivation of horseradish peroxidase by a polystyrene surface. *J. Immunol. Methods*, **47**, 121–124.

Bernstein, A. *et al.* (1978). Higher antitumor efficacy of daunomycin when linked to dextran: *in vivo* and *in vitro* studies. *J. Natl. Cancer Inst.*, **60**, 379–384.

Bieniarz, C. *et al.* (1991). Covalent attachment of antibodies and antigens to solid phases using extended length heterobifunctional coupling agents. *US Patent 5,063,109*, to Abbott Laboratories.

Biondi, L. *et al.* (1980). Modification of the properties of bovine pancreatic ribonuclease a by covalent attachment of D-gluconyl-glycine residues. *Int. J. Peptide Protein Res.*, **15**, 253–260.

Blanchard, G. C. *et al.* (1990). Regeneration of immunosorbent surfaces used in clinical, industrial and environmental biosensors. Role of covalent and non-covalent interactions. *J. Immunol. Methods*, **130**, 263–275.

Blobel, G. (1980). Intracellular protein topogenesis. *Proc. Natl Acad. Sci. USA*, **77**, 1496–1500.

Blomhoff, H. K. and Christensen, T. B. (1982). Comparison of the thermal stability of acid phosphatase conjugated with Ficoll and with modified Ficolls. *Acta Chem. Scand.*, **36**, 130–132.

Blomhoff, H. K. and Christensen, T. B. (1983). Effect of dextran and dextran modifications on the thermal and proteolytic stability of conjugated bovine testis β-galactosidase and human serum albumin. *Biochim. Biophys. Acta*, **743**, 401–407.

Blum, L. J. and Coulet, P. R. (1984). Bioluminescent determination of reduced nicotinamide adenine dinucleotide with immobilized bacterial luciferase and flavin mononucleotide oxidoreductase on collagen film. *Anal. Chim. Acta*, **161**, 355–358.

Boeckler, C. *et al.* (1996). Immunogenicity of new heterobifunctional cross-linking reagents used in the conjugation of synthetic peptides to liposomes. *J. Immunol. Methods*, **191**, 1–10.

Boenisch, T. (1976). Improved enzymeimmunoassay for trace proteins. In *Protides of the Biological Fluids. Proceedings of the 24th Colloquium, Brugge, 1976* (Peeters, H., ed.), pp. 743–749, Pergamon Press, Oxford.

Bornzin, G. A. and Miller, I. F. (1982). The kinetics of protein adsorption on synthetic and modified natural surfaces. *J. Colloid Interface Sci.*, **86**, 539–558.

Borsos, T. and Langone, J. J. (1981). Detection of antigens and haptens by inhibition of passive immune hemolysis. *Methods Enzymol.*, **74**, 161–165.

Boyden, S. V. (1951). The adsorption of proteins on erythrocytes treated with tannic acid and subsequent hemagglutination by antiprotein sera. *J. Exp. Med.*, **93**, 107–120.

Brandt, J. *et al.* (1975). Covalent attachment of proteins to polysaccharide carriers by means of benzoquinone. *Biochim. Biophys. Acta*, **386**, 196–202.

Brash, J. L. and Lyman, D. J. (1969). Adsorption of plasma proteins in solution to uncharged, hydrophobic polymer surfaces. *J. Biomed. Mater. Res.*, **3**, 175–189.

Brash, J. L. *et al.* (1974). Exchange of albumin adsorbed on polymer surfaces. *Trans. Am. Soc. Artif. Intern. Organs*, **20**, 69–76.

Brillhart, K. L. and Ngo, T. T. (1991). Use of microwell plates carrying hydrazide groups to enhance antibody immobilization in enzyme immunoassays. *J. Immunol. Methods*, **144**, 19–25.

Brygier, J. *et al.* (1993). Covalent attachment of poly (ethylene glycol) to peptides and proteins: reevaluation of the synthesis, properties and usefulness of carbonylimidazol-1-yl-methoxypolyethylene glycol. *Appl. Biochem. Biotechnol.*, **42**, 127–135.

Bückmann, A. *et al.* (1981). Functionalization of poly (ethylene glycol) and monomethoxy-poly (ethylene glycol). *Makromol. Chem.*, **182**, 1379–1384.

Bull, H. B. (1956). Adsorption of bovine serum albumin on glass. *Biochim. Biophys. Acta*, **19**, 464–471.

Bunting, P. S. and Laidler, K. J. (1974). Flow kinetics of L-asparaginase attached to nylon tubing. *Biotechnol. Bioeng.*, **16**, 119–134.

Burghardt, T. P. and Axelrod, D. (1981). Total internal reflection fluorescence photobleaching recovery study of serum albumin adsorption dynamics. *Biophys. J.*, **33**, 455–468.

Buonocore, V. *et al.* (1980). A constitutive β-galactosidase from the extreme thermoacidophile Archaebacterium *Caldariella acidophila*: properties of the enzyme in the free state and in immobilized whole cells. *J. Appl. Biochem.*, **2**, 390–397.

Butler, J. E. (1992). The behaviour of antigens and antibodies immobilized on a solid phase. In *Structure of Antigens*

(Van Regenmortel, M. H. V., ed.), vol. 1, pp. 209–259, CRC Press, Boca Raton, FL.

Butler, J. E. *et al.* (1992). The physical and functional behavior of capture antibodies adsorbed on polystyrene. *J. Immunol. Methods*, **150**, 77–90.

Caldwell, K. M. E. *et al.* (1996). Coating of hydrophobic surfaces to render them protein-resistant while permitting covalent attachment of specific ligands. *US Patent 5;516,703*, to University of Utah.

Cammas, S. *et al.* (1995). Heterobifunctional poly(ethylene oxide): synthesis of α-methoxy-ω-amino and α-hydroxy-ω-amino PEOs with the same molecular weights. *Bioconjugate Chem.*, **6**, 226–230.

Campbell, D. H. *et al.* (1951). Immunologic adsorbents. I. Isolation of antibody by means of a cellulose–protein antigen. *Proc. Natl. Acad. Sci. USA*, **37**, 575–578.

Campbell, J. *et al.* (1975). The preparation of several new nylon tube–glucose oxidase derivatives and their incorporation into "eeagentless" automated analysis of glucose. *Biochim. Biophys. Acta*, **384**, 307–316.

Canterero, L. A. *et al.* (1980). The adsorptive characteristics of proteins for polystyrene and their significance in solid-phase immunoassays. *Anal. Biochem.*, **105**, 375–382.

Capala, J. *et al.* (1996). Boronated epidermal growth factor as a potential targeting agent for boron neutron capture therapy of brain tumors. *Bioconjugate Chem.*, **7**, 7–15.

Catt, K. and Tregear, G. W. (1967). Solid-phase radioimmunoassay in antibody-coated tubes. *Science*, **158**, 1570–1572.

Ceska, M. and Lundkvist, U. (1972). A new and simple radioimmunoassay method for the determination of IgE. *Immunochemistry*, **9**, 1021–1030.

Chamow, S. M. *et al.* (1994). Modification of CD4 immunoadhesin with monomethoxypoly (ethylene glycol) aldehyde *via* reductive alkylation. *Bioconjugate Chem.*, **5**, 133–140.

Chan, B. M. C. and Brash, J. L. (1981). Conformational change in fibrinogen desorbed from glass surface. *J. Colloid Interface Sci.*, **84**, 263–265.

Chang, T. M. S. (1964) Semipermeable microcapsules. *Science*, **146**, 524–525.

Chang, T. M. S. (1966). Semipermeable aqueous microcapsules. I. Preparation and properties. *Can. J. Physiol. Pharmacol.*, **44**, 115–128.

Chang, T. M. S. (1995). Artificial cells with emphasis on bioencapsulation in biotechnology. *Biotechnol. Annu. Rev.*, **1**, 267–295.

Chattoraj, D. K. and Bull, H. B. (1959). Electrophoresis of adsorbed protein. *J. Am. Chem. Soc.*, **81**, 5128–5133.

Chen, J. P. *et al.* (1990). Polymer–protein conjugates I. effect of protein conjugation on the cloud point of poly (*N*-isopropylacrylamide). *Biomaterials*, **11**, 625–630.

Chen, T. H. and Meyer, K. F. (1954). A hemagglutination test with the protein fraction of *Pasturella pestis*. *J. Immunol.*, **72**, 282–298.

Chilkoti, A. *et al.* (1994). Site-specific conjugation of a temperature-sensitive polymer to a genetically-engineered protein. *Bioconjugate Chem.*, **5**, 504–507.

Chin, N. W. and Lanks, K. W. (1977). Covalent attachment of lactoperoxidase to polystyrene tissue culture flasks. *Anal. Biochem.*, **83**, 709–719.

Chiu, H-C. *et al.* (1993). Enzymatic activity of chymotrypsin and its poly(ethylene glycol) conjugates toward low and high molecular weight substrates. *Bioconjugate Chem.*, **4**, 290–295.

Chiu, T.-H. *et al.* (1976). Thermodynamics of native protein/foreign surface interactions. IV. Calorimetric and microelectrophoretic study of human fibrinogen sorption onto glass and LTI-carbon. *Trans. Am. Soc. Artif. Intern. Organs*, **22**, 498–513.

Chu, B. C. F. and Orgel, L. E. (1997). Presentation of ligands on hydroxylapatite. *Bioconjugate Chem.*, **8**, 103–105.

Chu, V. P. and Tarcha, P. J. (1987). Protein-reactive, molded polystyrene surfaces having applications to immunoassay formats. *J. Appl. Polym. Sci.*, **34**, 1917–1924.

Chua, M.-M. *et al.* (1984). Attachment of immunoglobulin to liposomal membrane *via* protein carbohydrate. *Biochim. Biophys. Acta*, **800**, 291–300.

Conradie, J. D. *et al.* (1983). ELISA solid phase: partial denaturation of coating antibody yields a more efficient solid phase. *J. Immunol. Methods*, **59**, 289–299.

Cooper, A. (1980). Conformational fluctuation and change in biological macromolecules. *Sci. Prog.*, **66**, 473–497.

Cormier, R. A. and Gagnon, C. F. (1979). Enzyme immobilization with a disulfide bridge. *US Patent 4,176,006*, to Redpath Sugars Limited.

Cuatrecasas, P. and Parikh, I. (1972). Adsorbents for affinity chromatography. Use of *N*-hydroxysuccinimide esters of agarose. *Biochemistry*, **11**, 2291–2299.

Čukman, D. *et al.* (1987). Characterization of active sites at chemically modified glass surfaces. *J. Colloid Interface Sci.*, **115**, 357–361.

Cuypers, P. A. *et al.* (1977). Ellipsometric study of protein film on chromium. *Ann. NY Acad. Sci.*, **283**, 77–87.

D'Angiuro, L. *et al.* (1978). Enzyme grafting on cellulose by Fe^{2+}–H_2O_2 redox system. *Cellulose Chem. Technol.*, **12**, 685–692.

D'Angiuro, L. *et al.* (1980). Preparation and characterization of enzymes immobilized by graft copolymerization to different polysaccharides. *Biotechnol. Bioeng.*, **22**, 2251–2272.

Darst, S. A. *et al.* (1991). Two-dimensional crystals of streptavidin on biotinylated lipid bilayers and their interactions with biotinylated macromolecules. *Biophys. J.*, **59**, 387–396.

Davies, J. *et al.* (1995). Scanning tunnelling microscopy and dynamic contact angle studies of partial denaturation on immunoassay solid phase immunoassay. *J. Immunol. Methods*, **186**, 111–123.

Davis, S. S. (1997). Biomedical applications of nanotechnology—implications for drug targeting and gene therapy. *Trends Biotechnol.*, **15**, 217–224.

Deng, L. *et al.* (1996). Self-assembled monolayers of alkanethiolates presenting tri(propylene sulfoxide) groups resist the adsorption of protein. *J. Am. Chem. Soc.*, **118**, 5136–5137.

Dierks, S. E. *et al.* (1986). Altered recognition of surface-adsorbed compared to antigen-bound antibodies in the ELISA. *Mol. Immunol.*, **23**, 403–411.

Dillman, W. J., Jr and Miller, I. F. (1973). On the adsorption of serum proteins on polymer membrane surfaces. *J. Colloid Interface Sci.*, **44**, 221–241.

Dimarzio, E. A. and Bishop, M. (1974). The nature of the helix–random coil transition of biological macromolecules attached to a surface. *Biopolymers*, **13**, 2331–2348.

Ding, Z. *et al.* (1996). Synthesis and purification of thermally sensitive oligomer–enzyme conjugates of poly (*N*-isopropylacrylamide)–trypsin. *Bioconjugate Chem.*, **7**, 121–125.

Eberhart, R. C. *et al.* (1987). Organization of albumin on polymer surfaces. In *Proteins at Interfaces. Physicochemical and Biochemical Studies.* (Brash, J. L. and Horbett, T. A., eds.), 378–400, American Chemical Society, Washington, DC.

Egitto, F. D. and Matienzo, L. J. (1993). Plasma modification of polymer surfaces. In *Society Vacuum Coaters, 36th Annual Technical Conference Proceedings*, (Society of Vacuum Coaters Staff, eds), Society of Vacuum Coaters, Albuquerque, NM, pp. 10–22.

Ekdahl, K. N. *et al.* (1993). Complement activation on radio frequency plasma modified polystyrene surfaces. *J. Colloid Interface Sci.*, **158**, 121–128.

Elcin, Y. M. and Akbulut, U. (1992). Polyester film strips coated with a photographic gelatin containing immobilized glucose oxidase hardened by chromium(III) sulfate. *Biomaterials*, **13**, 156–161.

Elgersma, A. V. *et al.* (1991). The adsorption of different types of monoclonal immunoglobulin on positively and negatively charged polystyrene lattices. *Colloids Surf.*, **54**, 89–101.

Elgersma, A. V. *et al.* (1992). Adsorption competition between albumin and monoclonal immuno-gamma-globulins on polystyrene latices. *J. Colloid Interface Sci.*, **152**, 410–428.

Elwing, H. *et al.* (1988). Conformational changes of a model protein (complement factor 3) adsorbed on hydrophilic and hydrophobic solid surfaces. *J. Colloid Interface Sci.*, **125**, 139–145.

Englebienne, P. and Weiland, M. (1996). Water-soluble conductive polymer homogeneous immunoassay (SOPHIA). A novel immunoassay capable of automation. *J. Immunol. Methods*, **191**, 159–170.

Erlanger, B. F. *et al.* (1970). Insoluble bacterial luciferases: A new approach to some problems in bioluminescence. *Biochem. Biophys. Res. Commun.*, **40**, 70–76.

Fair, B. D. and Jamieson, A. M. (1980). Studies of protein adsorption on polystyrene latex surfaces. *J. Colloid Interface Sci.*, **77**, 525–534.

Fernandez-Lafuente, R. *et al.* (1993). Preparation of activated supports containing low p*K* amino groups. A new tool for protein immobilization via the carboxyl coupling method. *Enzyme Microb. Technol.*, **15**, 546–550.

Ferruti, P. *et al.* (1981). Succinic half-esters of poly(ethylene glycol)s and their benzotriazole and imidazole derivatives as oligomeric drug-binding matrices. *Makromol. Chem.*, **182**, 2183–2192.

Filippusson, H. and Hornby, W. E. (1970). The preparation and properties of yeast β-fructofuranosidase chemically attached to polystyrene. *Biochem.J.*, **120**, 215–219.

Foster, T. L. and Casey, R. C. (1988). Covalent solid phase for enzyme immunoassay. *Am. Clin. Prod. Rev.* **February 1988 issue**, 32–35.

Frey, A. *et al.* (1997). Peptomer aluminium oxide nanoparticle conjugates as systemic and mucosal vaccine candidates: synthesis and characterization of a conjugate derived from the C4 domain of HIV-1_{MN} Gp120. *Bioconjugate Chem.*, **8**, 424–433.

Friguet, B. *et al.* (1984). Some monoclonal antibodies raised with a native protein bind preferentially to the denatured antigen. *Mol. Immunol.*, **21**, 673–677.

Frisch, B. *et al.* (1996). Synthesis of short polyoxyethylene-based heterobifunctional cross-linking reagents. Application to the coupling of peptides to liposomes.

Frost, R. G. *et al.* (1981). Covalent immobilization of proteins to *N*-hydroxysuccinimide ester derivatives of agarose. Effect of protein charge on immobilization. *Biochim. Biophys. Acta*, **670**, 163–169.

Furuno, T. and Sasabe, H. (1993). Two-dimensional crystallization of streptavidin by nonspecific binding to a surface film: Study with a scanning electron microscope. *Biophys. J.*, **65**, 1714–1717.

Gaertner, H. F. and Offord, R. E. (1996). Site-specific attachment of functionalized poly (ethylene glycol) to the amino terminus of proteins. *Bioconjugate Chem.*, **7**, 38–44.

Galfrè, G. and Milstein, C. (1981). Preparation of monoclonal antibodies: Strategies and procedures. *Methods Enzymol.*, **73**, 3–46.

Galin, J.-C. *et al.* (1965). Préparation des chaînes macromoléculaires dotées d'extrémités fonctionalles réactives. *C. R. Acad. Sci. Paris*, **260**, 5558–5561.

Gardas, A. and Lewartowska, A. (1988). Coating of proteins to polystyrene ELISA plates in the presence of detergents. *J. Immunol. Methods*, **106**, 251–255.

Gedda, L. *et al.* (1996). Development and *in vitro* studies of epidermal growth factor–dextran conjugates for boron neutron capture therapy. *Bioconjugate Chem.*, **7**, 584–591.

Geckeler, K. (1979). Functionalization of soluble polymers. 1. Replacement of the hydroxyl groups of poly (oxyethylene) by amino groups. *Polym. Bull.*, **1**, 427–431.

Gerenser, L. J. *et al.* (1985). E.s.c.a. studies of corona-discharge-treated polyethylene surfaces by use of gas-phase derivatization. *Polymer*, **26**, 1162–1166.

Gershoni, J. M. and Palade, G. E. (1982). Electrophoretic transfer of proteins from sodium dodecyl sulfate–polyacrylamide gels to a positively charged membrane filter. *Anal. Biochem.*, **124**, 396–405.

Gersten, D. M. and Marchalonis, J. J. (1978). A rapid, novel method for the solid-phase derivatization of IgG antibodies for immune-affinity chromatography. *J. Immunol. Methods*, **24**, 305–309.

Ghosh, S. S. and Musso, G. F. (1987). Covalent attachment of oligonucleotides to solid supports. *Nucleic Acids Res.*, **15**, 5353–5372.

Gold, E. R. and Fudenberg, H. H. (1967). Chromic chloride: A coupling agent for passive hemagglutination reactions. *J. Immunol.*, **99**, 859–866.

Gold, P. *et al.* (1966). The effect of the sequence of erythrocyte sensitization on the bis-diazotized benzidine hemagglutination reaction. *J. Allergy*, **37**, 311–318.

Goldman, R. *et al.* (1968). Papain–collodion membranes. I. Preparation and properties. *Biochemistry*, **7**, 486–500.

Golub, E. S. *et al.* (1968). A modification of the hemolytic plaque assay for use with protein antigens. *J. Immunol.*, **100**, 133–137.

Golubev, A. and Sidorova, E. (1989). The use of targeted liposomes to isolate cells bearing immunoglobulin receptors. *J. Immunol. Methods*, **125**, 29–34.

Gombotz, W. R. and Pettit, D. K. (1995). Biodegradable polymers for protein and peptide drug delivery. *Bioconjugate Chem.*, **6**, 332–351.

Gombotz, W. R. *et al.* (1992). Protein adsorption to and elution from polyether surfaces. In *Poly(Ethylene Glycol) Chemistry: Biotechnical and Biomedical Applications* (Harris, J. M., ed.), pp. 247–261, Plenum Press, New York.

Goodson, R. J. and Katre, N. V. (1990). Site directed PEGylation of recombinant interleukin-2 at its glycosylation site. *Bio/Technology*, **8**, 343–346.

Graves, H. C. B. (1988). The effect of surface charge on non-specific binding of rabbit immunoglobulin G in solid-phase immunoassays. *J. Immunol. Methods*, **111**, 157–166.

Greenwald, R. B. *et al.* (1996). PEG thiazolidone-2-thione, a novel reagent for facile protein modification: conjugation of bovine hemoglobin. *Bioconjugate Chem.*, **7**, 638–641.

Gregoriadis, G., Ed. (1984). *Liposome Technology*, vols 1–3, CRC Press, Boca Raton, FL.

Gregoriadis, G. (1990). Immunological adjuvants: a role for liposomes. *Immunol. Today*, **11**, 89–97.

Gu, K. F. and Chang, T. M. S. (1990). Conversion of ammonia or urea into essential amino acids, L-leucine, L-valine and L-isoleucine using artificial cells containing an immobilized multienzyme system and dextran-NAD^+. L-Lactic dehydrogenase for coenzyme recycling. *Appl. Biochem. Biotechnol.*, **26**, 115–124.

Gyenes, L. and Sehon, A. H. (1960). Preparation and evaluation of polystyrene–antigen conjugates for the isolation of antibodies. *Can. J. Biochem. Physiol.*, **38**, 1235–1248.

Gyenes, L. and Sehon, A. H. (1964). The use of tolylene-2,4-diisocyanate as a coupling reagent in the passive hemagglutination reaction. *Immunochemistry*, **1**, 43–48.

Habeeb, A. F. S. A. (1981). Controlled coupling of mildly reduced proteins to sepharose gelatin by heterobifunctional reagent. *Biochem. Biophys. Res. Commun.*, **100**, 1154–1166.

Hansen, R. H. *et al.* (1965). Effect of atomic oxygen on polymers. *J. Polym. Sci., Part A*, **3**, 2205–2214.

Harasym, T. O. *et al.* (1995). Poly(ethylene glycol)-modified phospholipids prevent aggregation during covalent conjugation of proteins to liposomes. *Bioconjugate Chem.*, **6**, 187–194.

Harris, J. M. (1992). Introduction to biotechnical and biomedical applications of poly(ethylene glycol). In *Poly (Ethylene Glycol) Chemistry: Biotechnical and Biomedical Applications* (Harris, J. M., ed.), pp. 1–14, Plenum Press, New York.

Harris, J. M. *et al.* (1992). Synthesis of new poly(ethylene glycol) derivatives. In *Poly(Ethylene Glycol) Chemistry: Biotechnical and Biomedical Applications* (Harris, J. M., ed.), pp. 371–381, Plenum Press, New York.

Hatakeyama, T. *et al.* (1996). An assay for lectin activity using microtiter plate with chemically immobilized carbohydrates. *Anal. Biochem.*, **237**, 188–192.

Hayat, U. *et al.* (1992). ESCA investigation of low-temperature ammonia plasma-treated polyethylene substrate for immobilization of protein. *Biomaterials*, **13**, 801–806.

Heimgartner, U. *et al.* (1989). Polyacrylic polyhydrazides as reagents for detection of glycoproteins. *Anal. Biochem.*, **181**, 182–189.

Heimgartner, U. *et al.* (1990). Polyacrylic polyhydrazides as novel reagents for detection of antibodies in immunoblotting Assays. *J. Immunol. Methods*, **132**, 239–245.

Heindel, N. D. *et al.* (1991). A novel heterobifunctional linker for formyl to thiol coupling. *Bioconjugate Chem.*, **2**, 427–430.

Heindel, N. D. *et al.* (1994). Carboxymethyldextran lactone: a preactivated polymer for amine conjugations. *Bioconjugate Chem.*, **5**, 98–100.

Heitzmann, H. and Richards, F. M. (1974). Use of the avidin–biotin complex for specific staining of biological membranes in electron microscopy. *Proc. Natl. Acad. Sci. USA*, **71**, 3537–3541.

Hendry, R. M. and Herrmann, J. E. (1980). Immobilization of antibodies on nylon for use in enzyme-linked immunoassay. *J. Immunol. Methods*, **35**, 285–296.

Hengsakul, M. and Cass, A. E. G. (1996). Protein patterning with a photoactivatable derivative of biotin. *Bioconjugate Chem.*, **7**, 249–254.

Herman, S. *et al.* (1993). Synthesis of dextran derivatives with thiol-specific reactive groups for the preparation of dextran–protein conjugates. *Bioconjugate Chem.*, **4**, 402–405.

Hermentin, P. *et al.* (1990). Hinge-thiol coupling of monoclonal antibody to silanized iron oxide particles and evaluation of magnetic cell depletion. *Bioconjugate Chem.*, **1**, 411–418.

Herrmann, J. E. and Collins, M. F. (1976). Quantitation of immunoglobulin adsorption to plastics. *J. Immunol. Methods*, **10**, 363–366.

Herron, J. N. *et al.* (1994). Antibodies as targeting moieties: affinity measurements, conjugation chemistry, and applications in immunoliposomes. *J. Controlled Release*, **28**, 155–166.

Heskins, M. *et al.* (1968). Solution properties of poly(*N*-isopropylacrylamide). *J. Macromol. Sci. Chem.*, **A2**, 1441–1455.

Hincal, A. A. and Kas, H. S. (1988). Targeting of immobilized drugs. In *Analytical Uses of Immobilized Biological Compounds for Detection, Medical and Industrial Uses*

(Guilbault, G. G. and Mascini, M., eds), pp. 95–113, D. Reidel Publishing Co., Dordrecht, Netherlands.

Hixson, H. F. (1973). Water-soluble enzyme–polymer grafts: thermal stabilization of glucose oxidase. *Biotechnol. Bioeng.*, **15**, 1011–1016.

Hobbs, R. N. (1989). Solid-phase immunoassay of serum antibodies to peptides. Covalent antigen binding to adsorbed phenylalanine–lysine copolymers. *J. Immunol. Methods*, **117**, 257–266.

Hoffman, W. L. and O'Shannessy, D. J. (1988). Site-specific immobilization of antibodies by their oligosaccharide moieties to new hydrazide derivatized solid supports. *J. Immunol. Methods*, **112**, 113–120.

Hollahan, J. R. *et al.* (1969). Attachment of amino groups to polymer surfaces by radiofrequency plasmas. *J. Appl. Polym. Sci.*, **13**, 807–816.

Hong, R. and Nisonoff, A. (1965). Relative labilities of the two types of interchain disulfide bond of rabbit γ-globulin. *J. Biol. Chem.*, **240**, 3883–3891.

Horbett, T. A. (1981). Adsorption of proteins from plasma to a series of hydrophilic–hydrophobic copolymers. II. Compositional analysis with the prelabeled protein technique. *J. Biomed. Mater. Res.*, **15**, 673–695.

Horbett, T. A. (1988). Molecular origins of the surface activity of proteins. *Protein Eng.*, **2**, 172–174.

Horbett, T. A. and Brash, J. L. (1987). Proteins at interfaces: current issues and future prospects. In *Proteins at Interfaces. Physicochemical and Biochemical Studies.* (Brash, J. L. and Horbett, T. A., eds), pp. 1–33, American Chemical Society, Washington, DC.

Horbett, T. A. and Weathersby, P. K. (1981). Adsorption of proteins from plasma to a series of hydrophilic–hydrophobic copolymers. I. Analysis with the *in situ* radioiodination technique. *J. Biomed. Mater. Res.*, **15**, 403–423.

Horbett, T. A. *et al.* (1977). The preferential adsorption of hemoglobin to polyethylene. *J. Bioeng.*, **1**, 61–78.

Hořejši, V. and Matoušek, V. (1985). Equilibrium in the protein–immobilized-ligand–soluble-ligand system: Estimation of dissociation constants of protein–soluble-ligand complexes from binding-inhibition data. *Mol. Immunol.*, **22**, 125–133.

Horiki, K. *et al.* (1978). Synthesis of the Merrifield resin esters of N-protected amino acids with the aid of hydrogen bonding. *Chem. Lett.*, 165–168.

Hornby, W. E. and Filippusson, H. (1970). The preparation of trypsin chemically attached to nylon tubes. *Biochim. Biophys. Acta*, **220**, 343–345.

Horvath, C. and Solomon, B. A. (1972). Open tubular heterogeneous enzyme reactors: preparation and kinetic behaviour. *Biotechnol. Bioeng.*, **14**, 885–914.

Hosaka, S. *et al.* (1983). Preparation of microspheres of poly(glycidyl methacrylate) and its derivatives as carriers for immobilized proteins. *Immunol. Commun.*, **12**, 509–517.

Howell, E. E. *et al.* (1981). Coated tube enzyme immunoassay: factors affecting sensitivity and effects of reversible protein binding to polystyrene. *J. Immunoassay*, **2**, 205–225.

Huang, A. *et al.* (1980). Monoclonal antibody covalently coupled with fatty acid. A reagent for *in vitro* liposome targeting. *J. Biol. Chem.*, **255**, 8015–8018.

Humphries, G. K. and McConnell, H. M. (1974). Immune lysis of liposomes and erythrocyte ghosts loaded with spin label. *Proc. Natl. Acad. Sci. USA*, **71**, 1691–1694.

Hurwitz, E. *et al.* (1984). The covalent linking of two nucleotide analogues to antibodies. *J. Med. Chem.*, **28**, 137–140.

Ichimura, K. (1984). A convenient photochemical method to immobilize enzymes. *J. Polym. Sci.*, **22**, 2817–2828.

Ichimura, K. (1987). Effect of poly(ethylene glycol) on the photochemical immobilization of an enzyme in photocrosslinkable poly(vinyl alcohol). *Makromol. Chem.*, **188**, 763–768.

Ichimura, K. and Watanabe, S. (1978). Immobilization of enzymes with use of photosensitive polymers bearing stilbazolium groups. *Chem. Lett.*, 1289–1292.

Ichimura, K. and Watanabe, S. (1980). Immobilization of enzymes with use of photosensitive polymers having the stilbazolium group. *J. Polym. Sci.*, **18**, 891–902.

Imai, K. *et al.* (1986). Immobilization of enzyme into poly (vinyl alcohol) membrane. *Biotechnol. Bioeng.*, **28**, 1721–1726.

Inada, Y. *et al.* (1986). Application of polyethylene glycol-modified enzymes in biotechnological processes: organic solvent-soluble enzymes. *Trends. Biotechnol.*, **4**, 190–194.

Inman, D. J. and Hornby, W. E. (1974). Preparation of some immobilized linked enzyme systems and their use in the automated determination of disaccharides. *Biochem. J.*, **137**, 25–32.

Inman, J. K. and Dintzis, H. M. (1969). The derivatization of cross-linked polyacrylamide beads. Controlled introduction of functional groups for the preparation of special-purpose, biochemical adsorbents. *Biochemistry*, **8**, 4074–4082.

Ishikawa, E. *et al.* (1980). An improved preparation of antibody-coated polystyrene beads for sandwich enzyme immunoassay. *J. Immunoassay*, **1**, 385–398.

Ishimori, Y. and Rokugawa, K. (1993). Stable liposomes for assays of human sera. *Clin. Chem.*, **39**, 1439–1443.

Jackson, C.-J. C. *et al.* (1987). Synthesis, isolation, and characterization of conjugates of ovalbumin with monomethoxypolyethylene glycol using cyanuric chloride as the coupling agent. *Anal. Biochem.*, **165**, 114–127.

Jagendorf, A. T. *et al.* (1963). Use of antibody bound to modified cellulose as an immunospecific adsorbent of antigens. *Biochim. Biophys. Acta*, **78**, 516–528.

Jardetzky, O. (1981). NMR studies of macromolecular dynamics. *Acc. Chem. Res.*, **14**, 291–298.

Jerne, N. K. *et al.* (1974). Plaque forming cells: methodology and theory. *Transplant. Rev.*, **18**, 130–191.

Jiang, F. N. *et al.* (1990). Development of technology for linking photosensitizers to a model monoclonal antibody. *J. Immunol. Methods*, **134**, 139–149.

Jimbo, Y. and Saito, M. (1988). Orientation-controlled immobilization of protein molecules on thin organic films deposited by the plasma technique. *J. Mol. Electron.*, **4**, 111–118.

Jitsukawa, T. *et al.* (1989). Increased coating efficiency of antigens and preservation of original antigenic structure after coating in ELISA. *J. Immunol. Methods*, **116**, 251–257.

Johne, B. *et al.* (1993). Epitope mapping and binding kinetics of monoclonal antibodies studied by real time biospecific interaction analysis using surface plasmon resonance. *J. Immunol. Methods*, **160**, 191–198.

Johnson, H. M. *et al.* (1966). The use of a water-soluble carbodiimide as a coupling reagent in the passive hemagglutination test. *J. Immunol.*, **97**, 791–796.

Jou, Y.-H. *et al.* (1983). Methods for the attachment of haptens and proteins to erythrocytes. *Methods Enzymol.*, **92**, 257–276.

Kaelble, D. H. and Moacanin, J. (1977). A surface energy analysis of bioadhesion. *Polymer*, **18**, 475–482.

Katchalski-Katzir, E. (1993). Immobilized enzymes – learning from past successes and failures. *Trends Biotechnol.*, **11**, 471–478.

Keller, S. and Block, R. J. (1960). Fractionation of proteins by adsorption and ion exchange. A. The fractionation and purification of proteins by adsorption methods. In *Analytical Methods of Protein Chemistry* (Alexander, P. and Block, R. J., eds), vol. 1, pp. 67–87, Pergamon Press, New York.

Kennel, S. J. (1982). Binding of monoclonal antibody to protein antigen in fluid phase or bound to solid supports. *J. Immunol. Methods*, **55**, 1–12.

Kenny, G. E. and Dunsmoor, C. L. (1983). Principles, problems and strategies in the use of antigenic mixtures for the enzyme-linked immunosorbent assay. *J. Clin. Microbiol.*, **17**, 655–665.

Keren, Z. *et al.* (1986). Identification of cell surface glycoproteins by periodate–alkaline phosphatase hydrazide. *Anal. Biochem.*, **155**, 182–187.

Khrapko, K. R. *et al.* (1991). A method for DNA sequencing by hybridization with oligonucleotide matrix. *DNA Sequence*, **1**, 375–388.

Kimura, J. *et al.* (1988). An immobilized enzyme membrane fabrication method using an ink jet nozzle. *Biosensors*, **4**, 41–52.

Klibanov, A. M. (1979). Enzyme stabilization by immobilization. *Anal. Biochem.*, **93**, 1–25.

Kochwa, S. *et al.* (1967). Adsorption of proteins by polystyrenep particles. I. Molecular unfolding and acquired immunogenicity of IgG. *J. Immunol.*, **99**, 981–986.

Kodera, Y. *et al.* (1994). Stabilization of L-asparaginase modified with comb-shaped poly(ethylene glycol) derivatives, *in vivo* and *in vitro*. *Bioconjugate Chem.*, **5**, 283–286.

Kossovsky, N. *et al.* (1993). Self-assembling nanostructures. *Bio/Technology*, **11**, 1534–1536.

Koutsoukos, P. G. *et al.* (1983). Protein adsorption on Hematite (α-Fe_2O_3) surfaces. *J. Colloid Interface Sci.*, **95**, 385–397.

Kritzman, J. (1958). Studies of rheumatoid serum employing a modified Coombs' slide test. *J. Lab. Clin. Med.*, **52**, 328–334.

Kučera, J. (1986). The polymeric *p*- and *o*-quinones as the reactive supports for the enzymes immobilization. *Biotechnol. Bioeng.*, **28**, 110–111.

Kuen, L. S. *et al.* (1993). Background noise in ELISA procedures. Influence of the pH of the coating buffer and correlations with serum IgM concentration. *J. Immunol. Methods*, **163**, 277–278.

Ladd, D. L. and Snow, R. A. (1993). Reagents for the preparation of chromophorically labeled polyethylene glycol–protein conjugates. *Anal. Biochem.*, **210**, 258–261.

Lange, M. *et al.* (1983). Covalent coupling of antigens to chemically activated lipopolysaccharide: a tool for *in vivo* and *in vitro* specific B cell stimulation. *J. Immunol. Methods*, **63**, 123–131.

Larsson, P. H. *et al.* (1987). Covalent binding of proteins to grafted plastic surfaces suitable for immunoassays. I. Binding capacity and characteristics of grafted polymers. *J. Immunol. Methods*, **98**, 129–135.

Lauritzen, E. *et al.* (1990). Dot immunobinding and immunoblotting of picogram and nanogram quantities of small peptides on activated nitrocellulose. *J. Immunol. Methods*, **131**, 257–267.

Leckband, D. and Langer, R. (1991). An approach for the stable immobilization of proteins. *Biotechnol. Bioeng.*, **37**, 227–237.

Lee, C. C. Y. and Loudon, G. M. (1979). Quantitative determination of amino groups on derivatized controlled pore glass: a comparison of methods. *Anal. Biochem.*, **94**, 60–64.

Lee, R. G. *et al.* (1974). Competitive adsorption of plasma proteins onto polymer surfaces. *Thromb. Res.*, **4**, 485–490.

Lee, S.-H. and Ruckenstein, E. (1988). Adsorption of proteins onto polymeric surfaces of different hydrophilicities – a case study with bovine serum albumin. *J. Colloid Interface Sci.*, **125**, 365–379.

Lee, Y. C. and Lee, R. T. (1992). Synthetic glycoconjugates. In *Glycoconjugates. Composition, Structure and Function* (Allen, H. J. and Kisailus, E. C., eds), pp. 121–165, Marcel Dekker, Inc., New York.

Lehtonen, O.-P. and Viljanen, M. K. (1980). Antigen attachment in ELISA. *J. Immunol. Methods*, **34**, 61–70.

Lemieux, S. *et al.* (1974). Local hemolysis plaque assay using a new method of coupling antigens on sheep erythrocytes by glutaraldehyde. *Immunochemistry*, **11**, 261–269.

Lenk, T. J. *et al.* (1989). IR spectral changes of bovine serum albumin upon surface adsorption. *J. Biomed. Mater. Res.*, **23**, 549–569.

Leonard, M. and Dellacherie, E. (1988). Synthesis and characterization of a polyoxyethylene derivative for the affinity labeling of human hemoglobin. *Makromol. Chem.*, **189**, 1809–1817.

Levin, D. *et al.* (1977). An improved plaque Aassay for mouse myeloma (MOPC 315) cells for use in studies of humoral and cell-mediated immunity. *J. Immunol. Methods*, **17**, 101–116.

Li, J.-T. *et al.* (1996). Chemical modification of surface active poly(ethylene oxide) – poly (propylene oxide) triblock copolymers. *Bioconjugate Chem.*, **7**, 592–599.

Ling, N. R. (1961). The coupling of protein antigens to erythrocytes with difluorodinitrobenzene. *Immunology*, **4**, 49–54.

Ljungquist, C. *et al.* (1989). Thiol-directed immobilization of recombinant IgG-binding receptors. *Eur. J. Biochem.*, **186**, 557–561.

Lopez-Berestein, G. and Fidler, I. J. (eds). (1989). *Liposomes in the Therapy of Infectious Diseases and Cancer*, Alan R. Liss Inc., New York.

Lu, B. *et al.* (1995). Oriented immobilization of Fab′ fragments on silica surfaces. *Anal. Chem.*, **67**, 83–87.

Lu, B. *et al.* (1996). Oriented immobilization of antibodies and its applications in immunoassays and immunosensors. *Analyst*, **121**, 29R–32R.

Luk, J. M. *et al.* (1995). Biotinylated lipopolysaccharide binds to endotoxin receptor in endothelial and monocytic cells. *Anal. Biochem.*, **232**, 217–224.

Lundström, I. and Elwing, H. (1990). Simple kinetic models for protein exchange reactions on solid surfaces. *J. Colloid Interface Sci.*, **136**, 68–84.

MacRitchie, F. (1972). The adsorption of proteins at the solid/liquid interface. *J. Colloid Interface Sci.*, **38**, 484–488.

MacRitchie, F. (1978). Proteins at interfaces. *Adv. Protein Chem.*, **32**, 283–326.

MacRitchie, F. and Owens, N. F. (1969). Interfacial coagulation of proteins. *J. Colloid Interface Sci.*, **29**, 66–71.

Mage, M. *et al.* (1990). Avidin and streptavidin modified water-soluble polymers and soluble, multivalent macromolecular conjugates thereof. *International Patent PCT WO 90/13815*, to the US Department of Commerce.

Mahan, D. E. and Copeland, R. L., Jr (1978). A method using toluene-2,4-diisocyanate and glutaraldehyde to stabilize and conjugate antigens to erythrocytes for use in passive hemagglutination tests. *J. Immunol. Methods*, **19**, 217–225.

Manecke, G. *et al.* (1970). Enzymes covalently bound to various polymers and the effects on the properties of these enzymes. *J. Polym. Sci., Part C*, **30**, 607–613.

Manzoni, A. (1988). Application of immobilized enzymes in diagnostics. In *Analytical Uses of Immobilized Biological Compounds for Detection, Medical and Industrial Uses* (Guilbault, G. G. and Mascini, M., eds), pp. 71–82, D. Reidel Publishing Co., Dordrecht.

Marchand-Brynaert, J. *et al.* (1995). Surface functionalisation of poly(ethylene terephthalate) film and membrane by controlled wet chemistry: chemical characterization of carboxylated surfaces. *J. Colloid Interface Sci.*, **173**, 236–244.

Margel, S. *et al.* (1981). A novel synthesis of polyacrolein microspheres and their application for cell labeling and cell separation. *Immunol. Commun.*, **10**, 567–575.

Marshall, J. J. (1978). Manipulation of the properties of enzymes by covalent attachment of carbohydrate. *Trends Biochem. Sci.*, 3, 79–83.

Marshall, J. J. and Rabinowitz, M. L. (1976). Preparation and characterization of a dextran–trypsin conjugate. *J. Biol. Chem.*, **251**, 1081–1087.

Marshall, J. J. *et al.* (1977). Attachment of carbohydrate to enzymes increases their circulatory lifetimes. *FEBS Lett.*, **83**, 249–252.

Martin, F. J. and Papahadjopoulos, D. (1982). Irreversible coupling of immunoglobulin fragments to preformed vesicles. An improved method for liposome targeting. *J. Biol. Chem.*, **257**, 286–288.

Martin, F. J. *et al.* (1981). Immunospecific targeting of liposomes to cells: A novel and efficient method for covalent attachment of Fab′ fragments via disulfide bonds. *Biochemistry*, **20**, 4229–4238.

Marumoto, K. *et al.* (1978). Synthesis and properties of polymeric latex particles and their conjugates with human immunoglobulin G. *Polymer*, **19**, 867–871.

Mathis, R. *et al.* (1985). Polyethylene oxide: A ligand for mild hydrophobic interaction chromatography? *J. Chromatogr.*, **347**, 291–296.

Matson, R. S. *et al.* (1994). Biopolymer synthesis on polypropylene supports. I. Oligonucleotides. *Anal. Biochem.*, **217**, 306–310.

Matsumoto, K. *et al.* (1986). Amperometric flow injection determination of fructose with an immobilized fructose 5-dehydrogenase reactor. *Anal. Chem.*, **58**, 2732–2734.

Matthay, K. K. *et al.* (1986). Antibody-directed kiposomes: comparison of various ligands for association, endocytosis, and drug delivery. *Cancer Res.*, **46**, 4904–4910.

McMillin, C. R. and Walton, A. G. (1974). A circular dichroism technique for the study of adsorbed protein structure. *J. Colloid Interface Sci.*, **48**, 345–349.

Merrifield, R. B. (1964). Solid phase peptide synthesis. II. The synthesis of bradykinin. *J. Am. Chem. Soc.*, **86**, 304–305.

Michaeli, I. *et al.* (1980). Diffusion of adsorbed protein within the plane of adsorption. *J. Colloid Interface Sci.*, **77**, 586–587.

Middlebrook, G. and Dubos, R. J. (1948). Specific serum agglutination of erythrocytes sensitized with extracts of tubercle bacilli. *J. Exp. Med.*, **88**, 521–528.

Miron, T. and Wilchek, M. (1993). A simplified method for the preparation of succinimidyl carbonate polyethylene glycol for coupling to proteins. *Bioconjugate Chem.*, **4**, 568–569.

Mitra, S. P. and Chattoraj, D. K. (1978). Some thermodynamic aspects of expanded and condensed films of BSA adsorbed at the alumina–water interface. *Indian J. Biochem. Biophys.*, **15**, 147–152.

Mitz, M. A. and Summaria, L. J. (1961). Synthesis of biologically active cellulose derivatives of enzymes. *Nature*, **189**, 576–577.

Molinaro, G. A. *et al.* (1981). Antigen quantitation by a reverse hemolytic assay. *Methods Enzymol.*, **73**, 319–326.

Monfardini, C. *et al.* (1995). A branched monomethoxypoly(ethylene glycol) for protein modification. *Bioconjugate Chem.*, **6**, 62–69.

Monji, N. *et al.* (1994). Activated, *N*-substituted acrylamide polymers for antibody coupling: application to a novel membrane-based immunoassay. *J. Biomater. Sci., Polym. Ed.*, **5**, 407–420.

Montagne, P. *et al.* (1992). Use of microsphere–antibody conjugates in microparticle-enhanced nephelometric immunoassay. *Bioconjugate Chem.*, **3**, 504–509.

Morpurgo, M. *et al.* (1996). Preparation and characterization of poly (ethylene glycol) vinyl sulfone. *Bioconjugate Chem.*, **7**, 363–368.

Morra, M. *et al.* (1993). Surface analysis of medical plastics. *Spectrosc. Eur.*, **5**, 10–16.

Morris, D. L. *et al.* (1975). A chemistry for the immobilization of enzymes on nylon. *Biochem. J.*, **147**, 593–603.

Morrissey, B. W. (1977). The adsorption and conformation of plasma oroteins: A physical approach. *Ann. NY Acad. Sci.*, **283**, 50–64.

Morrissey, B. W. and Han, C. C. (1978). The conformation of γ-globulin adsorbed on polystyrene latices determined by quasielastic light scattering. *J. Colloid Interface Sci.*, **65**, 423–431.

Morrissey, B. W. and Stromberg, R. R. (1974). The conformation of adsorbed blood proteins by infrared bound fraction measurements. *J. Colloid Interface Sci.*, **46**, 152–164.

Muñoz, C. *et al.* (1986). New experimental criteria for optimization of solid-phase antigen concentration and stability in ELISA. *J. Immunol. Methods*, **94**, 137–144.

Nagasaki, Y. *et al.* (1995). Formyl-ended heterobifunctional poly(ethylene oxide): synthesis of poly(ethylene oxide) with a formyl group at one end and a hydroxyl group at the other end. *Bioconjugate Chem.*, **6**, 231–233.

Nathan, A. *et al.* (1993). Copolymers of lysine and polyethylene glycol: a new family of functionalized drug carriers. *Bioconjugate Chem.*, **4**, 54–62.

Neurath, H. and Bull, H. B. (1938). The surface activity of proteins. *Chem. Rev.*, **23**, 391–435.

Newman, P. J. *et al.* (1981). Detection and characterization of monoclonal antibodies to platelet membrane proteins. *J. Cell Biol.*, **90**, 249–253.

Ngai, P. K. M. *et al.* (1993). Protein A antibody-capture ELISA (PACE): an ELISA format to avoid denaturation of surface-adsorbed antigens. *J. Immunol. Methods*, **158**, 267–276.

Ngo, T. T. (1986). Facile activation of sepharose hydroxyl groups by 2-fluoro-1 methylpyridinium toluene-4-sulfonate: preparation of affinity and covalent chromatographic matrices. *Bio/Technology*, **4**, 134–137.

Nicolau, C. *et al.* (1993). Electroinsertion of proteins in the plasma membrane of red blood cells. *Anal. Biochem.*, **214**, 1–10.

Nieto, A. *et al.* (1986). Adsorption–desorption of antigen to polystyrene plates used in ELISA. *Ann. Inst. Pasteur Immunol.*, **137C**, 161–172.

Nilsson, K. and Mosbach, K. (1984). Immobilization of ligands with organic sulfonyl chlorides. *Methods Enzymol.*, **104**, 56–69.

Niveleau, A. *et al.* (1993). Covalent linking of haptens, proteins and nucleic acids to a modified polystyrene support. *J. Immunol. Methods*, **159**, 177–187.

Niveleau, A. *et al.* (1995). Grafting peptides onto polystyrene microplates for ELISA. *J. Immunol. Methods*, **182**, 227–234.

Norde, W. (1986). Adsorption of proteins from solution at the solid – liquid interface. *Adv. Colloid Interface Sci.*, **25**, 267–340.

Nygren, H. (1988). Experimental demonstration of lateral cohesion in a layer of adsorbed protein and in layers of gold–antibody complexes bound to surface immobilised antigen. *J. Immunol. Methods*, **114**, 107–114.

Nyholm, L. and Ramlau, J. (1988). Nitrocellulose membranes as solid phase in immunoblotting. In *Handbook of Immunoblotting of Proteins* (Bjerrum, O. J. and Heegaard, N. H. H., eds), vol. 1, p. 101, CRC Press, Boca Raton, FL.

Nyilas, E. *et al.* (1974). Thermodynamics of native protein/foreign surface interactions. I. Calorimetry of the human α-globulin/glass system. *Trans. Am. Soc. Artif. Intern. Organs*, **20**, 480–490.

Nylander, T. (1987). PhD Thesis, University of Lund, Sweden.

Okrongly, D. (1990). Covalent attachment of macromolecules on substrate surfaces. *European Patent 390,500*, to Applied Immunesciences, Inc.

Oncley, J. L. *et al.* (1947). Physical-chemical characteristics of certain of the proteins of normal human plasma. *J. Phys. Colloid Chem.*, **51**, 184–198.

Ong, E. *et al.* (1989a). Enzyme immobilization using the cellulose-binding domain of a *Cellulomonas fimi* exoglucanase. *Bio/Technology*, **7**, 604–607.

Ong, E. *et al.* (1989b). The cellulose-binding domains of cellulases: tools for biotechnology. *Trends Biotechnol.*, **7**, 239–243.

Onyezili, F. N. (1988). The enzyme coupling process in urease immobilization on *O*-alkylated nylon tubes. *J. Biochem. Biophys. Methods*, **16**, 255–262.

Ordronneau, P. *et al.* (1991). An efficient enzyme immunoassay for glutamate using glutaraldehyde coupling of the hapten to microtiter plates. *J. Immunol. Methods*, **142**, 169–176.

Oreskes, I. and Singer, J. M. (1961). The mechanism of particulate carrier reactions. I. Adsorption of human γ-globulin to polystyrene latex particles. *J. Immunol.*, **86**, 338–344.

O'Shannessy, D. J. and Hoffman, W. L. (1987). Site-directed immobilization of glycoproteins on hydrazide-containing solid supports. *Biotechnol. Appl. Biochem.*, **9**, 488–496.

O'Shannessy, D. J. and Wilchek, M. (1990). Immobilization of glycoconjugates by their oligosaccharides: use of hydrazido-derivatized matrices. *Anal. Biochem.*, **191**, 1–8.

Ostro, M. (ed.). (1988). *Liposomes: From Biophysics to Therapeutics*. Marcel Dekker, New York.

Paley, M. S. and Harris, J. M. (1987). Synthesis of the aldehyde of oligomeric polyoxyethylene. *J. Polym. Sci., Polym. Chem.*, **25**, 2447–2454.

Palfree, R. G. E. and Elliott, B. E. (1982). An enzyme-linked immunosorbent assay (ELISA) for detergent solubilized Ia glycoproteins using nitrocellulose membrane discs. *J. Immunol. Methods*, **52**, 395–408.

Pande, C. S. *et al.* (1980). Camphorquinone-10-sulfonic acid and derivatives: convenient reagents for reversible modification of arginine residues. *Proc. Natl. Acad. Sci. USA*, **77**, 895–899.

Pande, R. *et al.* (1996). A biotinylated undecylthiophene copolymer bioconjugate for surface immobilization: creating an alkaline phosphatase chemiluminescence-based biosensor. *Bioconjugate Chem.*, **7**, 159–164.

Papadea, C. *et al.* (1985). Monoclonal antibody-based solid-phase immunoenzymometric assays for quantifying human immunoglobulin G and its subclasses in serum. *Clin. Chem.*, **31**, 1940–1945.

Patramani, I. *et al.* (1969). Glutamic–aspartic transaminase–antitransaminase interaction: A method for antienzyme purification. *Eur. J. Biochem.*, **11**, 28–36.

Pegg, R. K. and Saunders, M. S. (1994). Heterofunctional crosslinking agent for immobilizing reagents on plastic substrates. *US Patent 5,279,955.*

Persson, L.-O. and Olde, B. (1988). Synthesis of ATP-polyethylene glycol and ATP-dextran and their use in the purification of phosphoglycerate kinase from spinach chloroplasts using affinity partitioning. *J. Chromatogr.*, **457**, 183–193.

Pesce, A. J. *et al.* (1977). Binding of protein to polystyrene in solid-phase immunoassays. *Biochim. Biophys. Acta*, **492**, 399–407.

Peterman, J. H. *et al.* (1988). The immunochemistry of sandwich-ELISAs. IV. The antigen capture capacity of antibody covalently attached to bromoacetyl surface-functionalized polystyrene. *J. Immunol. Methods*, **111**, 271–275.

Peterson, C. and Kwei, T. K. (1961). The kinetics of polymer adsorption onto solid surfaces. *J. Phys. Chem.*, **65**, 1330–1333.

Phillips, T. M. *et al.* (1985). Protein A-coated glass beads. Universal support medium for high-performance immunoaffinity chromatography. *J. Chromatogr.*, **327**, 213–219.

Pillai, V. N. R. *et al.* (1980). New, easily removable poly (ethylene glycol) supports for the liquid-phase method of peptide synthesis. *J. Org. Chem.*, **45**, 5364–5370.

Pirrung, M. C. and Huang, C.-Y. (1996). A general method for the spatially defined immobilization of biomolecules on glass surfaces using 'caged' biotin. *Bioconjugate Chem.*, **7**, 317–321.

Pişkin, A. K. (1988a). Immobilized enzymes in therapy. In *Analytical Uses of Immobilized Biological Compounds for Detection, Medical and Industrial Uses* (Guilbault, G. G. and Mascini, M., eds), pp. 39–46, D. Reidel Publishing Co., Dordrecht, Netherlands.

Pişkin, A. K. (1988b). Immobilization of bioagents by radiation. In *Analytical Uses of Immobilized Biological Compounds for Detection, Medical and Industrial Uses* (Guilbault, G. G. and Mascini, M., eds), pp. 17–23, D. Reidel Publishing Co., Dordrecht, Netherlands.

Plant, A. L. *et al.* (1988). Liposome-enhanced flow injection immunoanalysis. *Bio/Technology*, **6**, 266–269.

Pollak, A. and Whitesides, G. M. (1976). Organic synthesis using enzymes in two-phase aqueous ternary system. *J. Am. Chem. Soc.*, **98**, 289–291.

Pope, M. R. *et al.* (1996). Specific activity of polypyrole nanoparticulate immunoreagents: comparison of surface chemistry and immobilization options. *Bioconjugate Chem.*, **7**, 436–444.

Pope, N. M. *et al.* (1993). New application of silane coupling agents for covalently binding antibodies to glass and cellulose solid supports. *Bioconjugate Chem.*, **4**, 166–171.

Poznansky, M. J. and Juliano, R. L. (1984). Biological approaches to the controlled delivery of drugs: a critical review. *Pharmacol. Rev.*, **36**, 277–336.

Pressman, D. *et al.* (1942). The agglutination of intact azo-erythrocytes by antisera homologous to the attached groups. *J. Immunol.*, **44**, 101–105.

Pundir, C. S. *et al.* (1993). Immobilization of barley oxalate oxidase onto alkylamine glass for determining urinary oxalate. *Clin. Chem.*, **39**, 1750–1751.

Quash, G. A. *et al.* (1989). Diagnostic and therapeutic procedures with haptens and glycoproteins (antigens and antibodies) coupled covalently by specific sites to insoluble supports. In *Covalently Modified Antigens and Antibodies in Diagnosis and Therapy*, 155–186, Marcel Dekker, New York.

Ramirez, C. *et al.* (1993). A bifunctional affinity linker to couple antibodies to cellulose. *Bio/Technology*, **11**, 1570–1573.

Ranucci, E. and Ferruti, P. (1990). A new synthetic method for amino-terminated poly(ethylene glycol) derivatives. *Synth. Commun.*, **20**, 2951–2957.

Ratner, B. D. *et al.* (1981). Analysis of the organisation of protein films on solid surfaces by ESCA. *J. Colloid Interface Sci.*, **83**, 630–643.

Regnier, F. E. and Noel, R. (1976). Glycerolpropylsilane bonded phases in the steric exclusion chromatography of biological macromolecules. *J. Chromatogr. Sci.*, **14**, 316–320.

Reimer, C. B. *et al.* (1978). In *Immunofluorescence and Related Staining Techniques* (Knapp, W. *et al.*, eds), p. 189, Elsevier/North Holland Biomedical Press, Amsterdam.

Rembaum, A. *et al.* (1976). Functional polymeric microspheres based on 2-hydroxyethyl methacrylate for immunochemical studies. *Macromolecules*, **9**, 328–336.

Reisberg, M. A. *et al.* (1970). A method for preparing specific fluorescent-conjugated antibody reagents using bentonite immunoadsorbents. *J. Immunol.*, **105**, 1151–1161.

Riesel, E. and Katchalski, E. (1964). Preparation and properties of water-insoluble derivatives of urease. *J. Biol. Chem.*, **239**, 1521–1524.

Roberts, J. C. *et al.* (1990). Using starburst dendrimers as linker molecules to radiolabel antibodies. *Bioconjugate Chem.*, **1**, 305–308.

Robinson, J. E. *et al.* (1990). A novel enzyme-linked immunosorbent assay (ELISA) for the detection of antibodies to HIV-1 envelope glycoproteins based on immobilization of viral glycoproteins in microtiter wells coated with concanavalin A. *J. Immunol. Methods*, **132**, 63–71.

Robinson, P. J. *et al.* (1971). Porous glass as a solid support for immobilisation or affinity chromatography of enzymes. *Biochim. Biophys. Acta*, **242**, 659–661.

Roederer, J. E. and Baastians, G. J. (1983). Microgravimetric immunoassay with piezoelectric crystals. *Anal. Chem.*, **55**, 2333–2336.

Rosemeyer, H. and Seela, F. (1984). Polymer-linked acycloguanosine. *Makromol. Chem.*, **185**, 687–695.

Rotmans, J. P. and Delwel, H. R. (1983). Cross-linking of *Schistosoma mansoni* antigens and their covalent binding on the surface of polystyrene microtitration trays for use in ELISA. *J. Immunol. Methods*, **57**, 87–98.

Rotmans, J. P. and Scheven, B. A. A. (1984). The effect of antigen cross-linking on the sensitivity of the enzyme-linked immunosorbent assay. *J. Immunol. Methods*, **70**, 53–64.

Rozsnyai, L. F. *et al.* (1992). Photolithographic immobilisation of biopolymers on solid supports. *Angew. Chem., Int. Ed. Engl.*, **31**, 759–761.

Rozzell, D. (1987). Immobilized aminotransferases for amino acid production. *Methods Enzymol.*, **136**, 479–497.

Rubinstein, M. *et al.* (1978). Process for the cross-linking of proteins. *US Patent 4,101,380*, to Research Products Rehovot Ltd.

Saito, T. and Nagai, F. (1983). Immobilization of antibody to a plastic surface by toluene 2,4-diisocyanate and its application to radioimmunoassay. *Clin. Chim. Acta*, **133**, 301–310.

Sakai, H. *et al.* (1997). Surface modification of hemoglobin vesicles with poly (ethylene glycol) and effects on aggregation, viscosity, and blood flow during 90% exchange transfusion in anesthetized rats. *Bioconjugate Chem.*, **8**, 23–30.

Salonen, E.-M. and Vaheri, A. (1979). Immobilization of viral and mycoplasma antigens and of immunoglobulins on polystyrene surface for immunoassays. *J. Immunol. Methods*, **30**, 209–218.

Sänger, M. *et al.* (1992). Light-induced coupling of aqueous-soluble proteins to liposomes formed from carbene-generating phospholipids. *Bioconjugate Chem.*, **3**, 308–314.

Sarver, R. W., Jr and Krueger, W. C. (1993). Infrared investigation of the conformation of proteins deposited on polyethylene films. *Anal. Biochem.*, **212**, 519–525.

Scheffel, U. *et al.* (1972). Albumin microspheres for the study of the reticuloendothelial system. *J. Nucleic Med.*, **13**, 498–503.

Schellenberger, A. and Ulbrich, R. (1989). Protein stabilization by blocking the native unfolding nucleus. *Biomed. Biochim. Acta*, **48**, 63–67.

Schneider, C. *et al.* (1982). A one-step purification of membrane proteins using a high efficiency immunomatrix. *J. Biol. Chem.*, **257**, 10766–10769.

Sehon, A. H. (1992). Suppression of antibody responses by conjugates of antigens and monomethoxypoly(ethylene glycol). In *Poly(Ethylene Glycol) Chemistry: Biotechnical and Biomedical Applications* (Harris, J. M., ed.), pp. 139–151, Plenum Press, New York.

Sevastianov, V. I. *et al.* (1983). *In vitro* assessment of the hemocompatible properties of polymers. *Artif. Organs*, **7**, 126–133.

Shafer, S. and Harris, J. M. (1986). Preparation of a cyanuric chloride-activated poly(ethylene glycol). *J. Polym. Sci., Polym. Chem.*, **24**, 375–378.

Shastri, R. and Roe, R.-J. (1970). Adsorption of bovine serum albumin on polyoxymethylene single crystals. *Org. Coat. Plastic. Chem.*, **40**, 820–823.

Shenoy, N. R. *et al.* (1992). Carboxylic acid-modified polyetheylene: a novel support for the covalent immobilization of polypeptides for C-terminal sequencing. *Protein Sci.*, **1**, 58–67.

Shih, L. B. *et al.* (1990). A fluorouridine–anti-CEA immunoconjugate is therapeutically effective in a human colonic cancer xenograft model. *Int. J. Cancer*, **46**, 1101–1106.

Shirahama, H. and Suzawa, T. (1985). Adsorption of bovine serum albumin onto styrene/acrylic acid copolymer latex. *Colloid Polym. Sci.*, **263**, 141–146.

Singer, J. M. *et al.* (1963). Mechanism of particulate carrier reactions. V. Adsorption of human gamma globulin to 0.2 micron diameter latex particles and their agglutination by rheumatoid factor. *Ann. Rheum. Dis.*, **22**, 424–428.

Singer, S. J. (1990). The structure and insertion of integral proteins in membranes. *Annu. Rev. Cell Biol.*, **6**, 247–296.

Singh, P. *et al.* (1996). Starburst™ dendrimers: a novel matrix for multifunctional reagents in immunoassays. *Clin. Chem.*, **42**, 1567–1569.

Sisson, T. H. and Castor, C. W. (1990). An improved method for immobilizing IgG antibodies on Protein A-agarose. *J. Immunol. Methods*, **127**, 215–220.

Siu Chong, E. D. and Chang, T. M. S. (1974). *In Vivo* effects of intraperitoneally injected L-asparaginase solution and L-asparaginase immobilized within semipermeable nylon microcapsules with emphasis on blood L-asparaginase, 'body' L-asparaginase, and plasma L-asparaginase levels. *Enzyme*, **18**, 218–239.

Smalla, K. *et al.* (1988). Influence of salts on the covalent immobilization of proteins to modified copolymers of 2-hydroxyethyl methacrylate with ethylene dimethacrylate. *Biotechnol. Appl. Chem.*, **10**, 21–31.

Sondergård-Andersen, J. *et al.* (1990). Covalently linked peptides for enzyme-linked immunosorbent assay. *J. Immunol. Methods*, **131**, 99–104.

Southern, E. M. and Maskos, U. (1988). Support-bound polynucleotides. *International Patent PCT GB 89/01114*.

Southern, E. M. *et al.* (1992). Analyzing and comparing nucleic acid sequences by hybridization to arrays of oligonucleotides: evaluation using experimental models. *Genomics*, **13**, 1008–1017.

Spitznagel, T. M. *et al.* (1993). Random and site-specific immobilization of catalytic antibodies. *Enzyme Microb. Technol.*, **15**, 916–921.

Stavitsky, A. B. and Arquilla, E. R. (1958). Studies of proteins and antibodies by specific hemagglutination and hemolysis of protein-conjugated erythrocytes. *Int. Arch. Allergy*, **13**, 1–38.

Steck, T. L. (1972). Cross-linking the major proteins of the isolated erythrocyte membrane. *J. Mol. Biol.*, **66**, 295–305.

Stein, T. and Gerisch, G. (1996). Oriented binding of a lipid-anchored cell adhesion protein onto a biosensor surface using hydrophobic immobilization and photoactive crosslinking. *Anal. Biochem.*, **237**, 252–259.

Sundaram, P. V. (1974). A new method of coupling proteins to insoluble polymers. *Biochem. Biophys. Res. Commun.*, **61**, 717–722.

Sundaram, P. V. (1978). Derivatives of nylon for coupling protein conjugates and other ligands. In *Enzyme Labelled Immunoassay of Hormones and Drugs. Proceedings of the International Symposium on Enzyme Labelled Immunoassay of Hormones and Drugs, Ulm, West Germany, July 10 and*

11, 1978 (Pal, S. B., ed.), pp. 107–127, Walter de Gruyter & Co., Berlin.

Sundaram, P. V. and Hornby, W. E. (1970). Preparation and properties of urease chemically attached to nylon tube. *FEBS Lett.*, **10**, 325–327.

Sundberg, S. A. *et al.* (1995). Spatially-addressable immobilization of macromolecules on solid supports. *J. Am. Chem. Soc.*, **117**, 12050–12057.

Sungur, S. *et al.* (1992). Studies on immobilization of urease in gelatin by cross-linking. *Biomaterials*, **13**, 795–800.

Suter, M. and Butler, J. E. (1986). The immunochemistry of sandwich ELISAs. II. A novel system prevents the denaturation of capture antibodies. *Immunol. Lett.* **13**, 313–316.

Suter, M. *et al.* (1989). The immunochemistry of sandwich ELISAs. III. The stoichiometry and efficacy of the protein–avidin-biotin capture (PABC) system. *Mol. Immunol.*, **26**, 221–230.

Takei, Y. G. *et al.* (1993a). Temperature-responsive bioconjugates. 1. Synthesis of temperature-responsive oligomers with reactive end groups and their coupling to biomolecules. *Bioconjugate Chem.*, **4**, 42–46.

Takei, Y. G. *et al.* (1993b). Temperature-responsive bioconjugates. 2. Molecular design for temperature-modulated bioseparations. *Bioconjugate Chem.*, **4**, 341–346.

Takei, Y. G. *et al.* (1994). Temperature-responsive bioconjugates. 3. Antibody–poly(*N*-isopropylacrylamide) conjugates for temperature-modulated precipitations and affinity bioseparations. *Bioconjugate Chem.*, **5**, 577–582.

Ternynck, T. and Avramas, S. (1976). A new method using *p*-benzoquinone for coupling antigens and antibodies to marker substances. *Ann. Immunol. Inst. Pasteur*, **127C**, 197–208.

Thavarungkul, P. *et al.* (1991). Continuous monitoring of urea in blood during dialysis. *Biosensors Bioelectron.*, **6**, 101–107.

Thomas, J. D. R. (1988). Uni-, bi- and tri-enzyme electrodes for analysis. In *Analytical Uses of Immobilized Biological Compounds for Detection, Medical and Industrial Uses* (Guilbault, G. G. and Mascini, M., eds), pp. 141–152, D. Reidel Publishing Co., Dordrecht, Netherlands.

Thomas, V. *et al.* (1990). A novel covalent enzyme-linked immunoassay (CELIA) for simultaneously measuring free and immune complex bound antibodies with a defined specificity. II. Application to immune complexes containing viral antigens in human sera. *J. Immunol. Methods*, **133**, 13–19.

Tilton, R. D. *et al.* (1990). Lateral diffusion of bovine serum albumin adsorbed at the solid–liquid interface. *J. Colloid Interface Sci.*, **137**, 192–203.

Timkovich, R. (1977). Polymerization side reactions during protein modifications with carbodiimide. *Biochem. Biophys. Res. Commun.*, **74**, 1463–1468.

Tseng, Y.-C. *et al.* (1993). Photografting of albumin onto dimethyldichlorosilane-coated glass. *J. Biomater. Appl.*, **7**, 233–249.

Ueno, H. and Fujino, M. (1987). Chemically modified protein and production thereof. *European Patent 236,987*, to Takeda Chemical Industries Ltd.

Ugi, I. (1962). The α-addition of immonium ions and anions to isonitriles accompanied by secondary reactions. *Angew. Chem., Int. Ed. Engl.*), **1**, 8–21.

Ukeda, H. *et al.* (1993). Dynamic analysis of binding process of bovine serum albumin on glutaraldehyde-activated support. *Anal. Sci.*, **9**, 617–623.

Umeda, M. *et al.* (1986). Liposome immune lysis assay (LILA). Application of sandwich method to determine a serum protein component with antibody-bearing liposomes. *J. Immunol. Methods*, **95**, 15–21.

Uzgiris, E. E. and Fromageot, H. P. M. (1976). Thickness and density of protein films by optical mixing spectroscopy. *Biopolymers*, **15**, 257–263.

Vandenbranden, M. *et al.* (1981). Interactions of γ-immunoglobulins with lipid mono- or bilayers and liposomes. Existence of two conformations of γ-immunoglobulins of different hydrophobicities. *Mol. Immunol.*, **18**, 621–631.

Vandenbranden, M. *et al.* (1982). Immunoglobulin–lipid interaction. A model membrane study. *Biochim. Biophys. Acta*, **685**, 177–181.

van Dulm, P. and Norde, W. (1983). The adsorption of human plasma albumin on solid surfaces, with special attention to the kinetic aspects. *J. Colloid Interface Sci.*, **91**, 248–255.

van Dulm, P. *et al.* (1981). Ion participation in protein adsorption at solid surfaces. *J. Colloid Interface Sci.*, **82**, 77–82.

van Oss, C. J. and Singer, J. M. (1966). The binding of immune globulins and other proteins by polystyrene latex particles. *J. Reticuloendothel. Soc.*, **3**, 29–40.

Varga, J. M. *et al.* (1990). Binding of a mouse monoclonal IgE (anti-DNP) antibody to radio-derivatized polystyrene –DNP complexes. *FASEB J.*, **4**, 2678–2683.

Vegarud, G. and Christensen, T. B. (1975). Glycosylation of proteins: a new method of enzyme stabilization. *Biotechnol. Bioeng.*, **17**, 1391–1397.

Veronese, F. M. *et al.* (1985). Surface modification of proteins: activation of monomethoxy-polyethylene glycols by phenylchloroformates and modification of ribonuclease and superoxide dismutase. *Appl. Biochem. Biotechnol.*, **11**, 141–152.

Veronese, F. M. *et al.* (1992). Preparation and properties of monomethoxypoly(ethylene glycol)-modified enzymes for therapeutic applications. In *Poly(Ethylene Glycol) Chemistry: Biotechnical and Biomedical Applications* (Harris, J. M., ed.), pp. 127–137, Plenum Press, New York.

Vogt, R. F., Jr *et al.* (1987). Quantitative differences among various proteins as blocking agents for ELISA microtiter plates. *J. Immunol. Methods*, **101**, 43–50.

Vroman, L. and Adams, A. L. (1969). Findings with the recording ellipsometer suggesting rapid exchange of specific plasma proteins at liquid/solid interfaces. *Surface Sci.*, **16**, 438–446.

Wahlgren, M. and Arnebrant, T. (1991). Protein adsorption to solid surfaces. *Trends Biotechnol.*, **9**, 201–208.

Walsh, J. and Gosling, J. P. (1986). The influence of radioiodination on the adsorption of IgG and serum albumin to polystyrene. *Anal. Biochem.*, **158**, 413–423.

Walton, A. G. and Maenpa, F. C. (1979). Application of fluorescence spectroscopy to the study of proteins at interfaces. *J. Colloid Interface Sci.*, **72**, 265–278.

Wang, C.-Y. and Huang, L. (1987). pH-sensitive immunoliposomes mediate target-cell-specific delivery and controlled expression of a foreign gene in mouse. *Proc. Natl. Acad. Sci. USA*, **84**, 7851–7855.

Wang, K. and Richards, F. M. (1975). Reaction of dimethyl-3,3'-dithiobispropionimidate with intact human erythrocytes. *J. Biol. Chem.*, **250**, 6622–6626.

Wang, R. *et al.* (1997). Light-induced amphiphilic surfaces. *Nature*, **388**, 431–432.

Weetall, H. W. (1970). Preparation and characterization of antigen and antibody adsorbents covalently coupled to an inorganic carrier. *Biochem. J.*, **117**, 257–261.

Weetall, H. W. (1976). Covalent coupling methods for inorganic support materials. *Methods Enzymol.*, **44**, 134–148.

Weetall, H. W. (1988). Enzymes immobilized on inorganic supports. In *Analytical Uses of Immobilized Biological Compounds for Detection, Medical and Industrial Uses* (Guilbault, G. G. and Mascini, M., eds), pp. 1–16, D. Reidel Publishing Co., Dordrecht, Netherlands.

Widder, K. J. *et al.* (1981). Specific cell binding using staphylococcal Protein A magnetic microspheres. *J. Pharm. Sci.*, **70**, 387–389.

Wieder, K. J. *et al.* (1979). Some properties of polyethylene glycol: phenylalanine ammonia-lyase adducts. *J. Biol. Chem.*, **254**, 12579–12587.

Wilkins Stevens, P. and Kelso, D. M. (1995). Estimation of the protein-binding capacity of microplate wells using sequential ELISAs. *J. Immunol. Methods*, **178**, 59–70.

Wilkins Stevens, P. *et al.* (1995). Assessment of adsorption and adhesion of proteins to polystyrene microwells by sequential enzyme-linked immunosorbent assay analysis. *Anal. Biochem.*, **225**, 197–205.

Woghiren, C. *et al.* (1993). Protected thiol-polyethylene glycol: a new activated polymer for reversible protein modification. *Bioconjugate Chem.*, **4**, 314–318.

Wood, W. G. and Gadow, A. (1983). Immobilisation of antibodies and antigens on macro solid phases: A comparison between adsorptive and covalent binding. *J. Clin. Chem. Clin. Biochem.*, **21**, 789–797.

Yan, M. *et al.* (1993). Photochemical functionalization of polymer surfaces and the production of biomolecule-carrying micrometre-scale structures by deep-UV lithography using 4-substituted perfluorophenyl azides. *J. Am. Chem. Soc.*, **115**, 814–816.

Yokoyama, Y. *et al.* (1988). Enzyme immobilization in an asymmetric charged membrane. *J. Membrane Sci.*, **38**, 223–236.

Yoshida, R. *et al.* (1996). Self-oscillating gel. *J. Am. Chem. Soc.*, **118**, 5134–5135.

Yoshioka, T. *et al.* (1981). Specific delivery of mitomycin C to the liver, spleen and lung: nano- and microspherical carriers of gelatin. *Int. J. Pharm.*, **81**, 131–141.

Young, B. R. *et al.* (1988a). Protein adsorption on polymeric biomaterials. I. Adsorption isotherms. *J. Colloid Interface Sci.*, **124**, 28–43.

Young, B. R. *et al.* (1988b). Protein adsorption on polymeric biomaterials. II. Adsorption kinetics. *J. Colloid Interface Sci.*, **125**, 246–260.

Zalipsky, S. (1993). Synthesis of an end-group functionalized polyethylene glycol–lipid conjugate for preparation of polymer-grafted liposomes. *Bioconjugate Chem.*, **4**, 296–299.

Zalipsky, S. (1995). Functionalized poly(ethylene glycol) for preparation of biologically relevant conjugates. *Bioconjugate Chem.*, **6**, 150–165.

Zalipsky, S. and Barany, G. (1990). Facile synthesis of α-hydroxy-ω-carboxymethylpolyethylene oxide. *J. Bioact. Compatible Polym.*, **5**, 227–231.

Zalipsky, S. and Lee, C. (1992). Use of functionalized poly (ethylene glycol)s for modification of polypeptides. In *Poly(Ethylene Glycol) Chemistry: Biotechnical and Biomedical Applications* (Harris, J. M., ed.), pp. 347–370, Plenum Press, New York.

Zalipsky, S. *et al.* (1983). Attachment of drugs to polyethylene glycols. *Eur. Polym. J.*, **19**, 1177–1183.

Zalipsky, S. *et al.* (1984). Esterification of polyethylene glycols. *J. Macromol. Sci. Chem.*, **A21**, 839–845.

Zalipsky, S. *et al.* (1987). A convenient general method for synthesis of N^{α}- or N^{ω}-dithiasuccinoyl (Dts) amino acids and dipeptides: application of polyethylene glycol as a carrier for functional purification. *Int. J. Peptide Protein Res.*, **30**, 740–783.

Zalipsky, S. *et al.* (1992a). Hydrazine containing conjugates of polypeptides and glycopeptides with polymers. *International Patent WO 92/16555.*

Zalipsky, S. *et al.* (1992b). Evaluation of a new reagent for covalent attachment of polyethylene glycol to proteins. *Biotechnol. Appl. Biochem.*, **15**, 100–114.

Zalipsky, S. *et al.* (1994). Preparation and applications of polyethylene glycol–polystyrene graft resin supports for solid-phase peptide synthesis. *React. Polym.*, **22**, 243–258.

Zammatteo, N. *et al.* (1996). Amination of polystyrene microwells: application to the covalent grafting of DNA probes for hybridization assays. *Anal. Biochem.*, **236**, 85–94.

Zucchini, G. L. *et al.* (1989). Study of some immobilization procedures of a monoclonal antibody on a water-insoluble support (nylon 6). Optimization of physico-chemical parameters. *NucCompact*, **20**, 15–22.

Chapter 9

Recombinant Techniques in the Preparation of Protein Conjugates

Leif Bülow

Department of Pure and Applied Biochemistry, Chemical Centre, University of Lund, POB 124, S-221 00 Lund, Sweden

and Christer Lindbladh

Astra Draco AB, POB 34, S-221 00 Lund, Sweden

9.1 Introduction

As discussed in Chapter 1, many procedures in biochemistry, biology and immunology are dependent on protein conjugates. For instance, enzyme immunoassays (EIA) require the linking of marker enzymes to either antigens or antibodies. Usually this is carried out using bifunctional chemical cross-linking agents which react with functional groups on the enzymes and antibodies/antigens. However, chemical cross-linking is not trivial to perform. Large, bulky and heterogeneous conjugates are frequently formed which may result in impaired properties of the conjugates regarding immunoreactivity as well as enzyme activity. Therefore, it has been desirable to develop and investigate alternative conjugation techniques for EIA such as immunological conjugation (Mason and Sammons, 1978) or conjugation methods based on biotin–avidin (Ternynck and Avrameas, 1990). Recently, methods based on genetic engineering have proved to be an attractive alternative in this area. This technique allows the use of the protein conjugates not only *in vitro* but also *in vivo*.

Protein conjugates or fusion proteins produced by recombinant DNA techniques have been utilised in a number of other fields beside EIA. (1) Production procedures of recombinant proteins (protection from proteolysis). (2) Downstream processing. (3) Development of pharmaceuticals (e.g. production of antiserum, drug targetting) (4) Fundamental studies in cell biology. (5) Design of calibration proteins for chromatography. (6) Preparation of bi- and polyfunctional enzymes.

The first eukaryotic proteins that were cloned, somatostatin and the A and B-chain of insulin, were produced as fusion proteins to β-galactosidase, the bacterial origin of the enzyme helping to prevent proteolytic degradation. This strategy of producing proteins as fusion partners applies particularly well to low-molecular-mass proteins, as further exemplified by the production of a flounder antifreeze protein (Peters *et al.*, 1989). This protein could not be produced in *E. coli* as a native non-fused protein because of presumed inefficient translation or rapid degradation in the cell.

Several fusion proteins have also been prepared with the aim of facilitating purification. Thus, fusion proteins carrying different affinity tails such as β-galactosidase, Protein A, polycysteine, phenylalanine, polyhistidine and polyarginine have frequently been employed for these purposes (for a review, see Sassenfeld, 1990; Sherwood, 1991). The introduction of specific cleavage sites between the affinity tail and the protein of interest has also been employed, removal of the tail generating the native protein. This technique has been used for the large scale production of insulin-like growth factor I, which was purified by the aid of a Protein A affinity tail and an immunoglobulin G adsorbent column (Moks *et al.*, 1987a).

Fusion proteins have also found applications in several medical fields. For example, when a snake toxin was fused in-frame (i.e. at the genetic level) with Protein A, the resulting snake fusion protein was found to greatly enhance the immunogenic response when used in the production of antiserum (Ducancel *et al.*, 1989). Moreover, the toxicity of the toxin decreased compared to the native non-fused toxin. The use of Protein A as an affinity tail in the production of this sero-therapeutically important antigen also allowed a one-step purification procedure of the antigen on an immunoglobulin G column.

Other medically interesting fusion proteins have been designed combining a cell-toxic protein with an affinity tail, which directs the toxin to specific cell receptors, antibodies or other antigenic structures on the cell surface. For instance, a fusion between Protein A and the cell toxin ricin A has been made with high retained biological activity on both moieties (Kim and Weaver, 1989). It is tempting to speculate that mixtures of such fusion proteins with antibodies that display a specific affinity to a certain cell type should have a high potential in the design of various specific immunotoxins. Here, chemical cross-linking is rather unfavourable, since it is non-specific and may damage the antigen binding site of the antibody or lower the toxicity of the toxin. Protein A binds to the Fc part of the antibody and therefore reduces these problems. Other examples of peptide toxins which have been successfully fused in-frame with different targetting proteins are a lymphotoxin (Wada *et al.*, 1990), diphtheria toxin (Perentesis *et al.*, 1988; Frankel *et al.*, 1995) and the *Pseudomonas* exotoxin (Chaudhary *et al.*, 1988).

Fusion proteins have also been shown to be useful in fundamental studies of basic cell and membrane biology. By fusing alkaline phosphatase, a periplasmic enzyme that requires an oxidative milieu to be active, to different portions of structurally unknown membrane proteins in bacteria, it is possible to obtain useful information about the membrane protein topology simply by analysing the enzyme activity (Manoil and Beckwith, 1986). Thus, the enzyme is inactive when fused to a cytoplasmic fraction of a membrane protein domain, but active if fused to a domain which is exposed to the periplasmic space of the bacteria. Also, the processing of signal sequences (Summers and Knowles, 1989),

N-terminal methionine excision (Hirel *et al.*, 1989) and some 'rules' that determine the *in vivo* half life of proteins (Bachmair *et al.*, 1986) have been investigated by the use of different fusion proteins.

Well-defined Protein A fusion proteins with well-known molecular masses have been designed for use as standard marker proteins in chromatography. When applied to Western blotting procedures, the blotted proteins and the sample could be developed simultaneously, with the same reagents, thus resulting in a much simpler visualization protocol (Lindbladh *et al.*, 1991a). It has also been possible to construct artificial bi- and polyfunctional enzymes catalyzing consecutive reactions using gene fusion *in vitro*. The recombinant polypeptide chain then carries two active sites. Such fusion enzymes have proved most useful for various enzymatic analyses as well as for modifying or investigating internal cell metabolism (Bülow and Mosbach, 1991; Carlsson *et al.*, 1992; Lindbladh *et al.*, 1994).

9.1.1 Ideal protein conjugates

It is easy to identify a number of different requirements an ideal linkage between two proteins should fulfil. These include: (1) generation of stable linkages; (2) formation of well-defined and homogeneous conjugates; (3) complete cross-linking; (4) chemical and physical properties of the protein moieties should not be affected by the cross-linking; (5) cross-linking should be easy and economical to perform on a large-scale.

No techniques available today can meet all these demands. Instead, it is common practice to choose a suitable cross-linking procedure by trial and error, even if some methods are more recommendable than others. For instance, among the most commonly used enzymes in immunochemical techniques, glutaraldehyde is widely used for conjugation of alkaline phosphatase, while periodate and reagents carrying maleimide groups are recommended for conjugation of horseradish peroxidase and β-galactosidase, respectively (Tijssen, 1988).

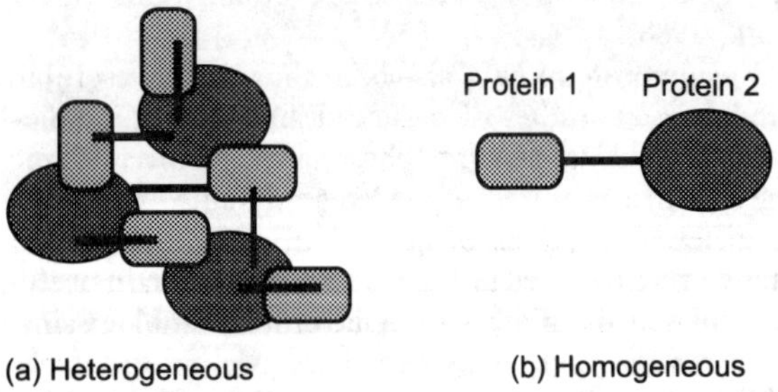

Figure 9.1 Linking two proteins using chemical or genetic techniques result in conjugates with different properties. The genetic method often results in homogeneous 1:1 conjugates (b).

The possibility of structurally linking two or more proteins by means of in-frame fusion between different structural genes has in many respects broken new ground, opening up new possibilities to improve these conjugation techniques, e.g. for coupling enzymes to antigens or antibodies (Figure 9.1). Some basic principles and methods will be discussed in the following paragraphs.

9.2 Gene Cloning: An Introduction

The capability of laboratories to manipulate nucleic acids has been greatly improved in recent years. For instance, several kits for a variety of genetic engineering procedures are commercially available, e.g. for plasmid DNA purification, site-directed mutagenesis and gene expression, implying that even a novice in the field should no longer hesitate to utilise this technology. During the last decade a number of genetic engineering techniques and innovations including automated DNA sequencing systems, PCR and bacterial phage display have also emerged. However, genetic engineering has very much generated a jargon of its own. Learning this field requires a familiarity with this new vocabulary. In this paragraph we give a short introduction to facilitate the reading of the chapter, but for the interested reader we recommend excellent textbooks in the field (Brown, 1990; Sambrook *et al.*, 1989; Watson *et al.*, 1992) as well as various volumes of *Methods in Enzymology* (68, 100, 101, 152, 153, 154, 155, 185, 216, 217, 218).

9.2.1 Plasmids

Plasmids that are able to propagate in *Escherichia coli* have become workhorses in recombinant DNA technology. Plasmids are naturally occurring, circular, extrachromosomal DNA molecules; they often harbour genes specifying novel metabolic activities that are advantageous to the bacterium (Figure 9.2). These genes frequently encode resistance to antibiotics or heavy metals. The plasmids used in the laboratory are often artificial since they are composed of parts having different bacterial origins. Foreign DNA fragments, or passenger DNA, can be inserted into plasmids and propagated as part of the plasmid. Such plasmids, or cloning vectors, then harbour an origin of replication or *ori* sequence allowing an autonomous

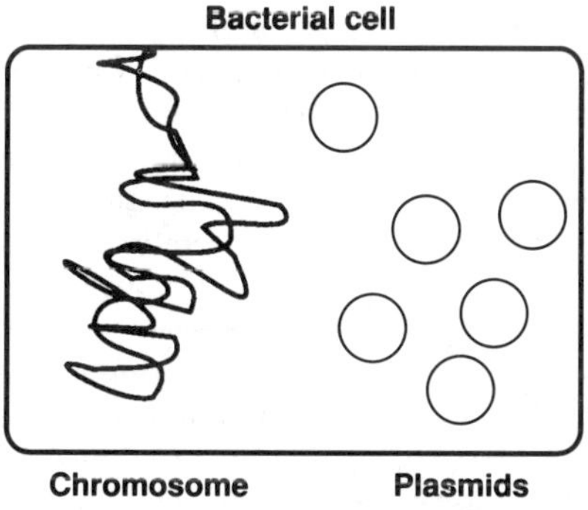

Figure 9.2 *Escherichia coli* can harbour plasmids along with the chromosome as genetic material.

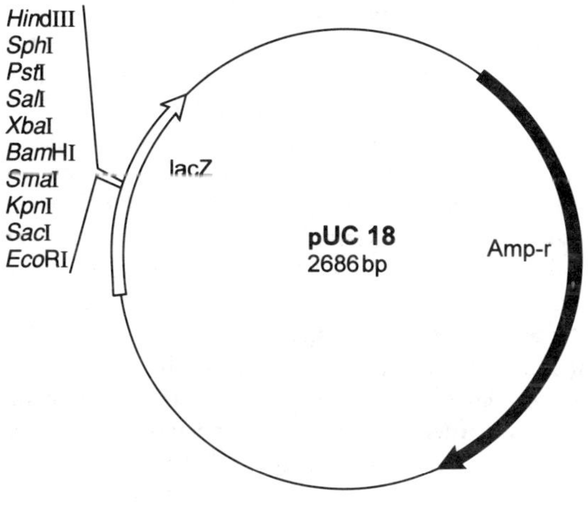

Figure 9.3 Plasmid pUC18 is a frequently used plasmid. It carries an antibiotic resistance gene, Amp-r, a *lac* promoter and a poly-linker region containing several restriction enzyme sites, which can be used for the insertion of foreign DNA fragments.

replication, a selectable marker and suitable restriction enzyme sites where foreign DNA can be inserted. The selectable marker is typically a gene conferring resistance to an antibiotic. Only those cells containing the plasmid will grow in the presence of the antibiotic. The antibiotic therefore selects for cells containing plasmids. Many vectors also contain a poly-linker cloning site, a short DNA fragment carrying numerous restriction sites (Figure 9.3).

9.2.2 Restriction, ligation and transformation

Cleavage with restriction enzymes opens or linearises the circular plasmid which enables a foreign DNA fragment to be inserted. The cleaved plasmid and the foreign DNA fragment are mixed and upon ligation a recombinant plasmid is formed. Recombinant plasmids are also called chimaeric plasmids. The word chimaera is borrowed from mythology and refers to a beast composed of the body of several creatures, usually a lion's head, a goat's body and a serpent's tail. Bacteria often harbour several hundred copies of plasmids per cell (high copy number plasmids) implying that large amounts of a specific DNA fragment can be recovered from a bacterial culture. In principle, any DNA sequence can be cloned and amplified in this way. The only practical limitation is the size of the insert; most plasmids with inserts larger than 10 kb are not efficiently replicated.

The ligation is facilitated if the ends of the plasmid and the insert have complementary, single-stranded overlaps. These ends can base-pair, or anneal. Frequently, the plasmid and the insert are simply cut by the same restriction enzyme. DNA ligase then closes the nicks. However, blunt end ligation is possible but less efficient. Very often it is essential to insert the DNA in a particular orientation, e.g. in order to direct the synthesis of the gene product. To do this, the DNA must be placed downstream from a promoter. When expressed in a host cell, RNA polymerase binds specifically at the promoter region and initiates transcription of the adjacent gene(s).

The plasmids constructed in the laboratory are then used to transform recipient host cells. Most often *E. coli* is used (cf. Section 9.2.4). *E. coli* cells are rendered somewhat permeable to DNA by Ca^{2+} treatment and a brief 42 °C heat shock. Although less than 0.1% of the Ca^{2+}-treated cells become competent for transformation following such treatment, transformed bacteria can easily be selected by their resistance to certain antibiotics (Figure 9.4).

9.2.3 M13

Plasmids are not the only vector molecules available, several attractive alternatives exist. A set of useful vectors are derived from filamentous phages. M13 is very often used and resembles plasmids in having multiple cloning sites and an origin of replication (Yanisch-Peron *et al.*, 1985). Interestingly, foreign DNA gives rise to both double-stranded DNA as well as single-stranded circular copies of one of the original DNA strands. Bacteria transformed with M13 derived recombinant DNA continue to grow slowly, continually secreting phage particles. These phages are excellent for phage display, DNA sequencing and *in vitro* mutagenesis (Figure 9.5).

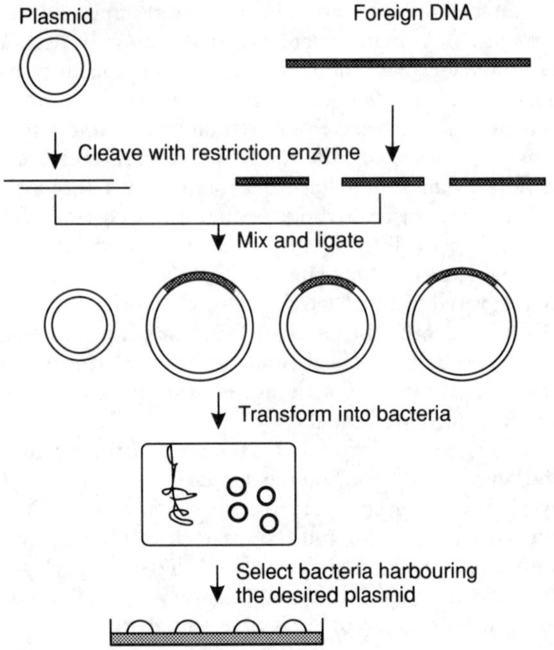

Figure 9.4 Basic principles for the construction and transformation of recombinant molecules.

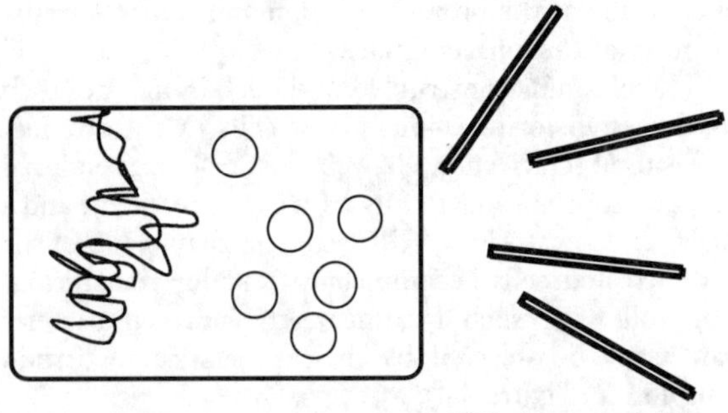

Figure 9.5 M13 is frequently used for the preparation of single-stranded DNA which can easily be isolated from the phages.

9.2.4 Host cells

One of the first questions which is raised when a recombinant protein is going to be made is the choice of host cell. Theoretically, a large number of different cells are available. However, in practice only a few organisms are utilised to produce recombinant proteins. Most frequently, bacterial cells are used since these have the advantage of rapid growth, yielding high cell densities. Large amounts of proteins can therefore be produced in a short period of time. A drawback with bacteria, however, is that proteolysis of the gene product is sometimes observed. Secretion to the growth medium generally simplifies the purification procedure and is therefore preferable to intracellular expression, unless intracellular production gives significantly higher yields. Among the bacterial cells, to date *Escherichia coli* is the most common (Goeddel, 1990). The reason for this is not that *E. coli* is the ultimate organism in terms of its properties, but rather that much is known about its molecular biology. However, other bacterial organisms are currently also used, such as the Gram-positive *Bacillus subtilis* (Henner, 1990) or *Bacillus brevis* (Udaka and Yamagata, 1993), which can secrete their products into the growth medium. When using *E. coli* as host cell, secreted proteins are normally found in the periplasm. However, a few *E. coli* expression systems can induce leakage of periplasmic proteins into the growth medium.

Several eukaryotic proteins are post-translationally modified *in vivo*, as exemplified by glycosylation. Bacterial cells cannot normally glycosylate proteins and are therefore not used for production of glycoproteins. In addition, large eukaryotic proteins are usually obtained in very low yields in bacteria, since the proteins are frequently found in misfolded forms or are not secreted efficiently. Instead, to produce glycoproteins and proteins with complex structures, mammalian cells are used (Levinson, 1990). The drawback with mammalian cells is that they grow slowly, thereby taking a long time to reach high cell densities. Therefore, it is not possible to obtain such high production yields as with bacteria. In addition, mammalian cells require rather complex growth media and are more sensitive to infections than bacteria. The products obtained from mammalian cells will therefore generally be more expensive than the same product derived from bacteria. Beside bacterial and mammalian host cells, it also common to produce recombinant proteins in yeast cells such as *Saccharomyces cerevisiae* (Hitzeman *et al.*, 1990). Yeast cells glycosylate proteins, and the large amounts of protein that can be produced in *E. coli* can frequently also be obtained from yeast. However, yeast glycosylation is different from that observed in mammalian cells.

In this chapter we will focus our discussion on the use of *E. coli* as host cell.

9.2.5 Shuttle vectors

Shuttle vectors are plasmids capable of propagating and transferring (shuttling) genes between two organisms, one of which is most often *E. coli* and the other a eukaryote. Shuttle vectors must have unique origins of replication for each cell type as well as different markers for selection of transformed host cells harbouring the vector. Eukaryotic genes can hence be cloned in a bacterial host and then transferred to a eukaryotic cell for analysis.

9.2.6 cDNA

cDNA consists of copies of mRNA, and is constructed from purified cellular mRNA. Different cell types in eukaryotic organisms express selected subsets of genes: This means that it is possible to enrich a desired mRNA by selecting a certain tissue or certain cells which transcribe this gene more frequently than other cells. Since most eukaryotic mRNAs carry polyA tails in the 3′-terminus, mRNA can be selectively isolated from preparations of total cellular RNA by oligo(dT)-cellulose chromatography. DNA copies can then be prepared by first annealing short oligo(dT) chains to the polyA tails. The oligo(dT) chains serve as primers for reverse transcriptase (Frohman, 1993), which synthesises a DNA strand from the RNA template (Figure 9.6).

9.2.7 Expression vectors

Expression vectors are engineered so that any cloned insert can be transcribed into RNA and translated into protein. Because cDNA lacks introns it is possible to express these eukaryotic copies in prokaryotic host cells which otherwise lack the capacity to remove intervening regions (introns) of nonsense DNA. In order to express a eukaryotic protein in *E. coli*, the eukaryotic cDNA must be cloned in an expression vector that contains regulatory signals for both transcription and translation (Mertens *et al.*, 1995). These vectors thus contain a promoter, where RNA polymerase will initiate transcription, and a ribosome binding site to translate the resulting mRNA. These sites are engineered into the vector just upstream of the inserted foreign DNA. Several strong promoters have been prepared that allow synthesis of foreign protein to levels up to 30% or more of total *E. coli* protein. An example is the p_{tac} promoter which was created by fusing the wild-type *E. coli lac* and *trp* promoters. In bacteria, the p_{tac} promoters are not induced to produce foreign DNA transcripts until the cells are exposed to specific chemical compounds, so called inducers. Analogues of lactose such as isopropyl-β-D-thiogalactoside (IPTG) are most efficient. The expression of the foreign protein is hence easily controlled (Figure 9.7).

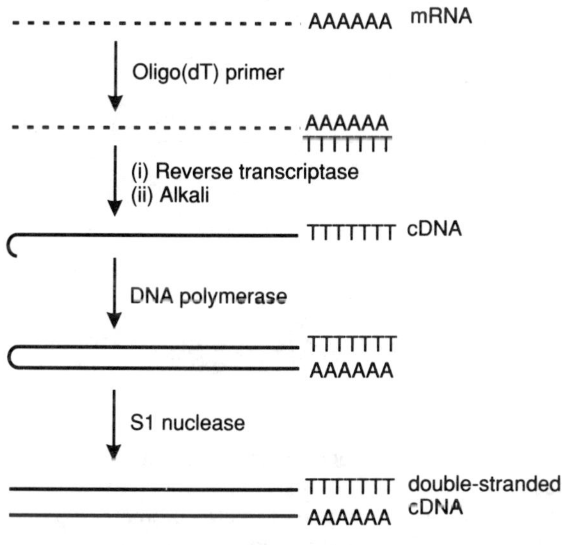

Figure 9.6 One method of preparing cDNA from mRNA using reverse transcriptase.

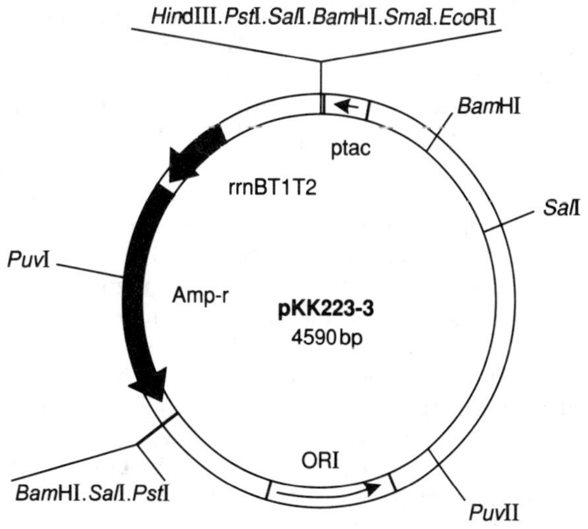

Figure 9.7 Plasmid pKK223-3 is an expression vector using the *tac* promoter. A foreign DNA fragment is inserted in the poly-linker after the promoter.

9.2.8 Polymerase chain reaction (PCR)

The polymerase chain reaction (PCR) is a technique for amplifying a specific DNA fragment (White, 1993; Griffin and Griffin, 1994). A preparation of denatured DNA, for instance chromosomal DNA, containing the

DNA segment of interest serves as a template for DNA polymerase, and two specific primers serve as primers for DNA synthesis. These primers, designed to be complementary to the two 3′-ends of the specific DNA fragments to be amplified, are added in excesses of 1000 times or more. They prime the synthesis of the two complementary strands of the desired DNA fragment, thereby doubling its concentration in solution. The DNA is subsequently heated to denature the DNA duplexes and then cooled so that the primers will anneal to both the newly produced and old strands. Another cycle of DNA ensues (Figure 9.8). The protocol has been automated through the invention of thermal cyclers that alternately heat the reaction mixture to dissociate the DNA, followed by cooling to anneal the primers followed by DNA synthesis. Today several heat-stable DNA polymerases are commercially available which introduce minimal errors during amplification. Theoretically, the amount of target DNA doubles each round, 25 rounds of amplification would increase its concentration more than 30×10^6-fold. In practice, the increase is often about a million times which is more than enough for further gene studies. Thus, starting with a very small amount of chromosomal DNA, a particular DNA fragment can be produced in useful amounts in less than four hours.

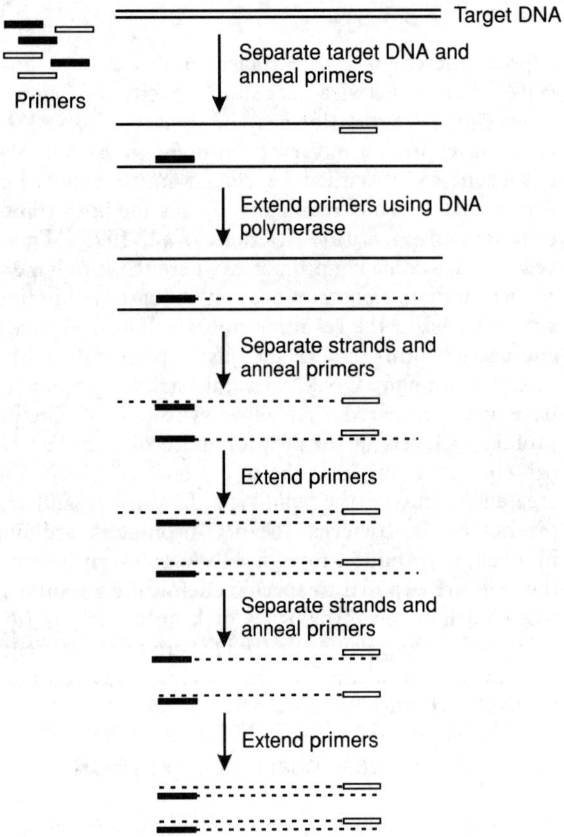

Figure 9.8 Basic principles of the polymerase chain reaction (PCR).

9.3 Recombinant Protein Conjugates

9.3.1 Preparation of recombinant protein conjugates

In order to prepare a protein conjugate the structural genes of the proteins of interest are fused in-frame, generating a recombinant polypeptide chain when the novel gene is expressed in suitable host cells (Figure 9.9). Fusions can be made either to the amino or carboxy-terminal ends of the proteins depending on known restrictions in the tertiary or quaternary structure of the protein. Particularly the use of PCR technology has greatly facilitated the construction of these artificial bifunctional enzymes. Alternatively, suitable restriction enzyme sites necessary for further genetic manipulations can be generated at the 5′- or the 3′-regions of the structural genes by site-directed mutagenesis. Moreover, by using chemically

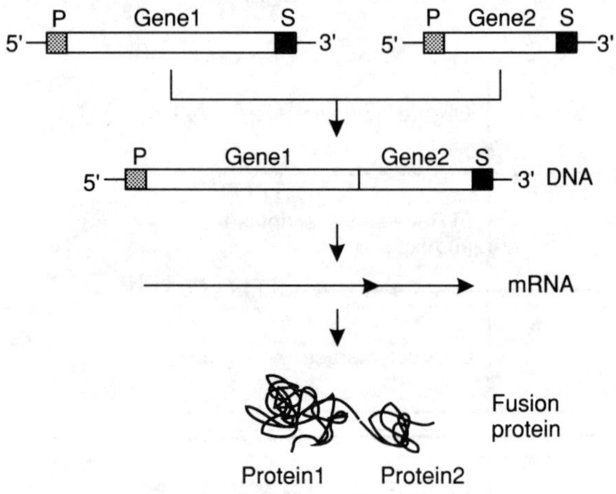

Figure 9.9 Schematic illustration of the gene fusion technique. The translational stop codon at the 3′-terminal end of the first gene and the promoter region at the 5′-terminal end of the second gene are removed followed by an in-frame ligation. Upon expression in a host cell a protein conjugate is produced.

synthesised DNA fragments in the cloning procedure, special properties in the linker region between the enzymes can be introduced by the selection of specific oligonucleotide sequences. The chimaeric gene prepared is then inserted into a proper expression vector and transformed into a suitable host cell. The amount of enzyme conjugate produced when the chimaeric genes are expressed in a bacterial host is dependent on a number of different factors such as choice of expression vector and final location of the product, cytoplasmic or periplasmic. In our experiments we typically obtain between 5 and 15 mg of purified conjugate, produced intracellularly, per litre cell culture and 1–2 mg per litre if produced in the periplasm.

9.3.1.1 Inclusion body formation and proteolysis

One of the problems associated with expression of foreign genes in bacteria is a failure to obtain correct folding of gene products into their native conformation. In fact, some production systems in *E. coli* lead to the formation of aggregates of denatured proteins, often referred to as inclusion bodies, with only a small fraction maturing into the desired native protein. To increase the yield further, this has led to the development of refolding schemes where the insoluble aggregates are dissolved in molar concentrations of strong denaturants, such as urea or guanidine hydrochloride, often with reducing agents. The low solubility of unfolded and reduced proteins requires proteins to be very dilute to avoid re-aggregation and precipitation during the refolding process. If inclusion body formation occurs this, of course, makes genetic engineering approaches complicated and less attractive. However, in many instances these problems can be handled very easily. The temperature of cultivation can be decreased or, alternatively, an expression system with a weaker promoter can be used. Both of these methods will decrease and slow down the expression and will encourage more efficient protein folding. An alternative approach is to make a fusion protein to a hydrophilic and highly soluble protein such as Protein A or thioredoxin (Williams *et al.*, 1995).

Bacteria, as well as eukaryotic cells, have a tendency to degrade rapidly soluble proteins with a highly abnormal structural conformation. The production of altered proteins, such as fusion proteins, is always accompanied by the danger that the proteins could be recognised as foreign and therefore be degraded. It has been demonstrated that the proteolysis of abnormal proteins is largely dependent on an ATP-dependent protease, La, produced from the *lon* gene. Thus, the use of *lon*-negative bacterial strains has been shown to decrease proteolysis of abnormal proteins (Gottesman and Zipser, 1978). Recombinant proteins can also be modified themselves in order to be protected against proteolysis. Interestingly, it has been demonstrated that large bacterial proteins, like β-galactosidase, can protect the other moiety of a fusion protein against proteolysis (Guo *et al.*, 1984).

9.3.1.2 The nature of the linker region

When two proteins are fused there is often a need to introduce a peptide linker, forming a spacer between the two protein moieties. Special properties of the linker can be designed by the selection of a suitable oligonucleotide sequence. The choice of linker region may be very important for the stability, function and production of the fusion protein. We and others have thus observed that the correct folding of fusion proteins can depend on the length and the primary structure of the linker. In this context it is important to study naturally occurring fusion proteins (multifunctional proteins). In these proteins no homology between the sequenced linkers has been observed. The importance of these regions is still unclear, and may reflect the remnants of ancestral genes from the fusion event (Zalkin *et al.*, 1984; Chang and Hammes, 1990). The linkers may function to allow conformational flexibility, permitting the correct folding of the separate domains. For instance, the length of the linker joining the two functional domains of bifunctional yeast tryptophan synthetase has been shown to be critical for function (Crawford *et al.*, 1987). It has furthermore been observed that the introduction of a linker between the two domains of metallothionein does not change the functionality of the enzyme, but longer inter-domain insertions render the protein less stable (Rhee *et al.*, 1990). From examination of inter-domain linkers in native proteins, non-hydrophobic amino acids are considered to be favourable linker constituents (Argos, 1990).

In practice, when artificial bifunctional proteins are prepared, the composition of the linker is often simply determined by base pairs in the restriction enzyme sites necessitated by the gene fusion procedure. A majority of hydrophilic residues should be aimed for, however. Moreover, when designing linkers for engineered proteins it is important not to introduce proteolytic cleavage sites into the linker sequence, for example glycine residues which are particularly prone to proteolytic degradation (Enfors, 1992).

9.3.2 Purification of recombinant protein conjugates

In many instances, fusion protein purification is a rather straightforward process, even if it can occasionally be time-consuming. The purification procedure of choice depends on a number of factors, including the sort of host cell used and the location of the protein, intracellular, periplasmic or extracellular. The desired degree of protein purity is of course also of importance, since in many applications a crude cell homogenate harbouring the fusion protein will be satisfactory.

Depending on these factors, different purification strategies can be chosen (see Chapter 10). These include ion-exchange, size exclusion, or affinity chromatography. These methods take advantage of specific properties of the fusion protein, such as size, charge or various 'handles' suitable for affinity chromatography. For instance, size exclusion chromatography is useful if the size of the conjugate differs significantly from the rest of the cell proteins: β-galactosidase fusion proteins, for example, are frequently very large since the native enzyme has a molecular mass of around 465 kDa. It is also possible to affinity-purify many of the conjugates described in this chapter, e.g. β-galactosidase and luciferase can be purified using affinity chromatography (Ullmann, 1984; Holzman and Baldwin, 1982) (cf. Protocol 9.13). Furthermore, the use of Protein A fusion proteins has been extensive due to its high affinity for IgG (Nilsson and Abrahmsén, 1990)) (cf Protocol 9.14). If fusion protein purification is difficult, the addition of affinity tails to the amino or carboxy terminus can be an attractive solution. For instance, the use of polyhistidine tails, $(His)_{4-6}$, and immobilised metal ion chromatography allows a one-step purification in most instances (Crowe *et al.*, 1994).

9.3.3 Physical and chemical properties of recombinant protein conjugates

Gene fusion results in complete conjugation between the proteins. For instance, if a marker enzyme and an antigen or antibody are linked, it will mean that all the enzyme and immunoreactive molecules exist as fusion partners. This may be important, for example in immunoassay, since even small amounts of free immunoreactive species in the purified conjugate preparation may decrease the analytical response appreciably. Since the stable peptide linkage is located at the N or C-terminal end of the conjugated proteins there is also a higher probability of having an intact active site or antigenic site on the conjugate than if multiple conjugation points exist on the protein. This is especially important if the protein is small.

For many genetically-prepared protein conjugates studied thus far, key properties such as enzymatic and immunological activity have been changed only moderately. Typically, more than 50% of the enzyme activity is retained after correction for the higher molecular mass caused by the fusion. In this respect it is important to consider the fusion site on the enzyme, since it has been observed that only very minor portions of the primary structure of the native enzyme can be removed in order to retain high activity (Fowler and Zabin, 1983; Bülow and Mosbach, 1991). When this general 'rule ' is followed only modest changes have been observed when the K_m, pH optima, thermostability and stability to urea denaturation have been investigated. Moreover, the immunoreactive activity of the antibody/antigen is normally completely retained. Apparently, in most cases it seems that each part of the chimaeric fusion protein is allowed to fold independently into its native active form. The gene fusion technique thus seems to fulfil several of the requirements listed in Section 9.1.1.

9.4 Examples of Recombinant Protein Conjugates

A number of different enzymes have been used in the genetic construction of enzyme conjugates for use in EIA. These include β-galactosidase (*E. coli*), alkaline phosphatase (*E. coli*), both discussed in some detail in Chapter 3, and the bioluminescence enzymes aequorin (*Aequorea victoria*) (Zenno and Inouye, 1990) and luciferase (*Vibrio harveyi*) (Lindbladh *et al.*, 1991b).

9.4.1 β-Galactosidase conjugates

The β-galactosidase structural gene of *E. coli*, *lacZ*, is frequently used for the construction of fusion proteins. The β-galactosidase gene is convenient for making such translational fusions because it is possible to remove its translational initiation region together with up to 27 amino acid codons, without affecting β-galactosidase enzyme activity. The 5′-end of *lacZ* is hence very flexible in accepting new structural genes.

9.4.1.1 LacZ fusion vectors and host cells

Fusions can be made by inserting a DNA fragment (cartridge) containing the β-galactosidase structural

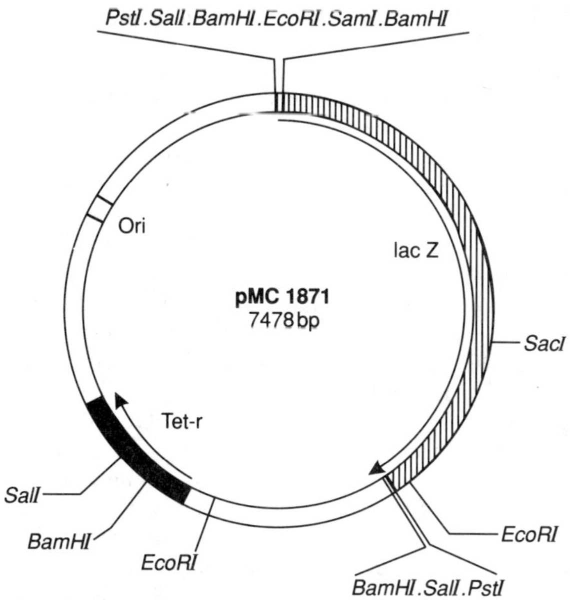

Figure 9.10 Plasmid pMC1871 is a *lacZ* fusion vector which can be used for the preparation of β-galactosidase protein conjugates. The *lacZ* fragment can be isolated using *Bam*HI, *Sal*I or *Pst*I cleavages.

gene into a gene already on a vector (Casadaban *et al.*, 1983; Shapira *et al.*, 1983). For functional fusion, the gene segments must be aligned in their direction of transcription and the amino acid codons must be in phase. Alternatively, a *lacZ* fusion vector containing a promoter and a poly-linker can be used to prepare the conjugate (Casadaban *et al.*, 1983). An example of such a vector is depicted in Figure 9.10.

Several suitable *E. coli* host cells are available. Particularly *E. coli* strain M182 and its derivatives are currently used because their *lac* operon is deleted, they can be efficiently transformed, and they yield good plasmid preparations (Casadaban and Cohen, 1980). β-Galactosidase expression *in vivo* can be detected by different agar media. However, colorimetric substrate media are the most sensitive. X-Gal (5-bromo-4-chloro-3-indolyl-β-D-galactoside) is commonly used since its product from galactosidase hydrolysis dimerises, and forms a blue precipitate that does not diffuse through the agar (see also Chapter 3, Section 3.5.5). X-Gal is dissolved in dimethylformamide solution at 20 mg/ml and can be conveniently stored for months at −20° C. It is used by spreading on the agar surface just before use or by adding to the liquid agar just before pouring (40 μg/ml final concentration).

9.4.1.2 Assay of β-galactosidase conjugates

β-Galactosidase is conveniently assayed by measuring its cleavage of the colorimetric substrate *o*-nitrophenyl-galactoside (ONPG) into galactose and a yellow compound, *o*-nitrophenol (see Section 3.5.5). The reaction can be stopped by addition of 1 M Na_2CO_3 when sufficient product has been generated. β-Galactosidase has several useful features. It is active in a wide range of buffers and can also be assayed in whole cells if they have first been rendered permeable to ONPG by addition of small amounts of toluene. The turnover number of the enzyme is known, so the number of enzyme molecules present in the assay can be calculated. Other enzyme-substrate combinations, e.g. alkaline phosphatase with *p*-nitrophenyl phosphate, employ the same assay principle.

β-Galactosidase synthesis in *E. coli* is induced by growing the cells in the presence of 0.5 mM isopropyl-β-D-thiogalactoside (IPTG) or in a medium with lactose as the sole carbon source. The presence of glucose in the medium inhibits the induction of β-galactosidase synthesis, so glycerol is often utilised as the carbon source when IPTG is used to induce enzyme synthesis. If whole cells are assayed they can be concentrated by centrifugation and resuspended in assay buffer (0.1 M sodium phosphate buffer (pH 7.0) containing 1 mM $MgSO_4$ and 1 mM dithiothreitol). The cells are then made permeable by vortexing for 15 seconds with 1/20 volumes of toluene. A volume of 0.2 ml of ONPG (4 mg/ml in assay buffer) is added to 0.8 ml of cell suspension or protein mixture. The sample is then incubated until a pale yellow colour develops. The reaction is stopped by the addition of 0.5 ml of 1 M Na_2CO_3. Any particulate matter can be removed by centrifugation before reading the absorbance of the yellow product at 420 nm.

Units are often expressed as nanomoles of ONPG cleaved per minute per milligram protein in the extract, where 1 nmol is equivalent to 0.0045 A_{420} unit with a 1 cm light path. Pure enzyme has an activity corresponding to 340,000 U/mg at 30° C in the assay buffer specified.

9.4.1.3 Examples of β-galactosidase conjugates

Fusion enzyme conjugates containing antigens which are difficult and costly to obtain in large quantities have frequently employed β-galactosidase. For instance, a gp 190–β-galactosidase conjugate was successfully employed in the detection of malaria antibodies from infected patients (Früh *et al.*, 1989). This way of producing the malaria major surface protein, gp

190, conjugated to an enzyme is highly advantageous since the production and isolation of this antigen from malaria cultures is hardly feasible. Similarly, detection and quantitation of α_2-interferon has been accomplished by use of a genetically prepared α_2-interferon–β-galactosidase conjugate (Peterhans *et al.*, 1987).

9.4.2 Alkaline phosphatase

The use of *E. coli* alkaline phosphatase instead of β-galactosidase in enzyme conjugates frequently presents advantages. Alkaline phosphatase is a small dimeric enzyme in contrast to β-galactosidase which is a substantially larger tetramer. Furthermore, alkaline phosphatase is a periplasmic enzyme, which greatly facilitates protein purification. Several alkaline phosphatase (*phoA*) fusion vectors in which the enzyme lacks its own amino-terminal signal sequence have been described (Hoffman and Wright, 1985). The screening of bacterial transformants is simple since alkaline phosphatase is active only when secreted. In analogy to β-galactosidase, XP (5-bromo-4-chloro-3-indolyl-β-D-phosphate) can be used for screening of bacterial colonies exhibiting alkaline phosphatase activity (see Section 3.4.7).

9.4.2.1 Examples of alkaline phosphatase conjugates

In the first fusion conjugates successfully employed in enzyme immunoassays, an antigen was fused to alkaline phosphatase as the marker enzyme. For example, a competitive immunoassay for human insulin was set up by use of a genetically prepared proinsulin–alkaline phosphatase conjugate (Lindbladh *et al.*, 1987). In the design of this conjugate the entire structural gene encoding human preproinsulin was fused in-frame to the 5′-terminal end of the gene encoding alkaline phosphatase. Both insulin and alkaline phosphatase require an oxidative milieu in order to form the disulfide bonds necessary for correct structure and biological activity. The fusion protein was therefore designed to be secreted into the periplasm of *E. coli* cells. Employing this conjugate it was also possible to determine proinsulin levels by selecting antibodies with affinity to the C-peptide of proinsulin. The C-peptide is cleaved off from native insulin and is therefore specific for proinsulin. Detection limits for insulin and proinsulin were comparable to or lower than those obtained from other available techniques, based on chemically cross-linked conjugates.

9.4.3 Protein A – enzyme conjugates

Staphylococcal Protein A (SPA) is a protein bound to the cell wall of the pathogenic bacterium *Staphylococcus aureus* (Langone, 1982). It binds to the Fc portion of most mammalian class G immunoglobulins. Therefore, this protein has a unique potential for making 'universal' enzyme conjugates which can be used in various enzyme immunoassays. The nucleotide sequence of the cloned gene shows that the protein consists of three structurally and functionally different regions. Starting at the N-terminus, the first region is the signal sequence consisting of 36 amino acid residues which are cleaved off during translocation to the cell medium. The signal sequence is functional and correctly processed in *E. coli*. The second region consists of the five homologous IgG binding domains, A–E, each with approximately 58 amino acids. The third region is the C-terminal region X, part of which is anchored in the cytoplasmic membrane with the rest located in the cell wall (Guss *et al.*, 1984; Moks *et al.*, 1986) (Figure 9.11). In the gene fusion vectors described here region X has been deleted.

The most important features of SPA are as follows. (1) The tight binding to IgG. The dissociation constant is 2×10^{-8} M for isolated domain B to human IgG. This strong and specific affinity allows the purification of Protein A fusion proteins in a single step by IgG affinity chromatography. (2) SPA is proteolytically stable not only in the homologous host *S. aureus* but also in the cytoplasmic and periplasmic spaces of *E. coli*. (3) The half-life of gene products is often increased *in vivo* when fused to SPA. This protection from proteolysis is a general observation in all host cells and becomes most pronounced for secreted gene products (Moks *et al.*, 1987b).

9.4.3.1 Selection of the fusion point in SPA

The fusion point is frequently selected between domain C of the IgG binding region and region X of SPA. The IgG binding capacity can thereby be maximized by including all five IgG binding domains and the cell wall binding is deleted. When the second gene is placed in the 3′-end of the *spa* gene the expression and signal sequences can be maintained. The fusion point is also placed in an area of the SPA protein which is thought to be flexible.

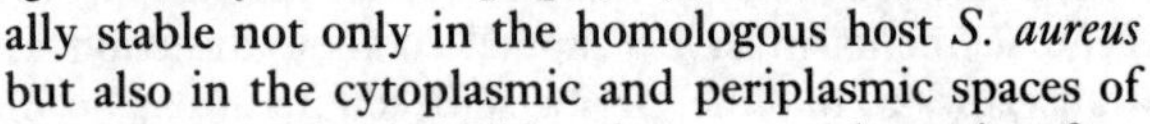

Figure 9.11 Schematic structure of staphylococcal Protein A (SPA).

9.4.3.2 SPA fusion vectors

Two main sets of vectors have been developed, both of which are commercially available. The first set can replicate and be selected for in both *E. coli* and *S. aureus*. These plasmids include the coding sequence of the signal sequence of *spa*. Therefore, the fusion proteins are secreted through the cytoplasmic membrane of either host (Figure 9.12). In the Gram-positive *S. aureus*, the protein will be located extracellularly. In *E. coli*, the expression of SPA fusions will translocate the protein to the periplasmic space (Abrahmsén *et al.*, 1985). These proteins can be recovered after an osmotic shock procedure. In some host cells, the expressed SPA fusion induces a 'leaky phenotype behaviour' in the *E. coli* cell, and the normally periplasmic proteins, including the SPA fusion protein, will leak out to the growth medium. Although the mechanism is not fully understood, it has technical advantages when it occurs, as the fusion protein can be recovered directly from the growth medium.

In the second set of vectors the promoter and the signal sequence of *spa* have been deleted and the transcription is instead initiated from the strong PR promoter of phage λ, which is induced by elevated temperatures (40 °C). SPA fusion proteins will consequently be located intracellularly in *E. coli* when expressed from this type of plasmid. Both sets of vectors have a poly-linker region carrying a number of unique cloning sites (Nilsson and Abrahmsén, 1990).

The choice of vector is dependent on the origin and properties of the gene product to which SPA is to be

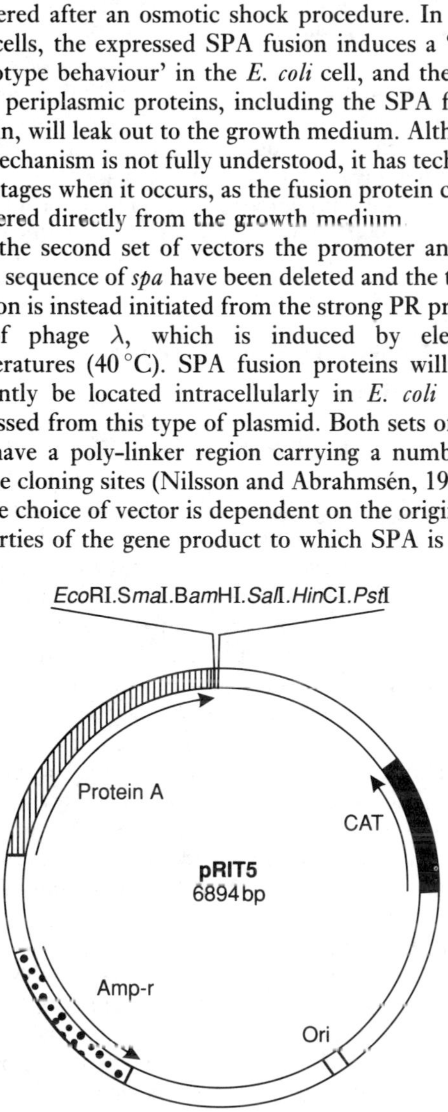

Figure 9.12 Plasmid pRIT5 is an SPA fusion vector.

fused. For the expression of secreted proteins, such as mammalian proteins that are secreted in their natural host, the first type of vector is recommended. On the other hand, proteins which normally are found in the cytoplasm or in a membrane compartment in their natural host are often incompatible with translocation through a bacterial membrane and should therefore be expressed with the second set of vectors.

9.4.3.3 Examples of SPA conjugates

A number of Protein A–enzyme conjugates have been reported. For instance, a Protein A–β-lactamase conjugate was constructed in order to investigate how in-frame gene fusion affects the IgG binding properties and the enzyme activity of the conjugate compared to native proteins (Baneyx and Georgiou, 1989). These investigations demonstrated that the binding of Protein A to IgG and the β-lactamase activity were almost entirely unaffected. The Protein A moiety of the conjugate bound IgG in a ratio of 1:1 and could in fact be used to immobilise β-lactamase onto solid supports with greatly improved enzyme activity, compared to direct immobilisation by conventional chemical methods (Baneyx *et al.*, 1990).

The gene fusion technique also makes it possible to fuse more than two different genes in-frame, hence resulting in enzyme conjugates with several biological activities. For instance, a tripartite fusion protein, Protein A–Protein G–β-galactosidase, was constructed which displayed the biological activities from all three components (Strandberg *et al.*, 1990). This conjugate displayed improved and extended IgG binding properties compared to Protein A conjugates, since Protein G has been shown to exhibit an IgG binding pattern complementary to Protein A.

Construction of enzyme conjugates by conventional chemical cross-linking methods often requires a careful selection of an appropriate cross-linking agent. If not, the cross-linking molecule may also react with reactive amino acids in the active site of the enzymes, potentially resulting in a severe loss of enzyme activity. This can be exemplified by bacterial luciferase, a light-emitting enzyme which has been proposed as an ideal label in immunoassays due to the possibility of monitoring light with extremely high sensitivity. However, the enzyme has many reactive amino acid residues near or around the active site which renders it very sensitive to various chemical cross-linking agents. This is illustrated by the fact that when luciferase is immobilised to solid supports, normally not more than 5–15% of the bioluminescent activity is retained (Wienhausen and DeLuca, 1982). Despite this much effort has been put into the preparation of soluble luciferase conjugates by chemical cross-linking (Jablonsky, 1985). In this particular case in-frame gene fusion represents an interesting alternative to chemical cross-linking since luciferase can easily be fused to other proteins with high retention of bioluminescent activity. Thus, a Protein A–luciferase conjugate retaining more than 50% activity (Lindbladh *et al.*, 1991b) and a galactose dehydrogenase–luciferase conjugate retaining more than 75% activity (Lindbladh *et al.*, 1992) have been prepared. The Protein A–luciferase conjugate had a well-defined molecular structure and exhibited no tendency to form multimeric aggregates. The detection limit for IgG in an immunoassay using this conjugate was estimated to be 90 fmol.

9.4.4 Antibody–enzyme conjugates

Most enzyme conjugates, commercially available, have the enzyme linked to an antibody. Until recently the production of antibody conjugates for immunoassays by gene fusion methods has been limited by difficulties in obtaining and expressing antibody-encoding gene sequences. However, recent developments in recombinant antibody technology have greatly simplified the isolation of DNA fragments encoding the antigen-binding fragment of an antibody, from monoclonal hybridoma cells or from lymphocytes of an immunised (or unimmunised) animal. These recombinant antibody fragments can have the same antigen-binding activities as conventional monoclonal antibodies and have been successfully produced in heterologous systems including bacteria, plants and non-lymphoid animal cells. Normally, only the two antigen-binding domains (V_L and V_H) from the variable heavy and light polypeptide chains need to be isolated, even if cloning, expression and correct folding of larger fragments (V_L-C_L and V_H-C_{H1}) have been demonstrated (Plückthun, 1991, Figure 9.13). The single chain Fv, in which the heavy- and light-chain variable domains are joined by a flexible linker peptide, is a small and extremely versatile recombinant antibody that can be readily conjugated to other moieties by gene fusion techniques (Bird and Walker, 1991). These molecules have, at least in some instances, proven to be more thermostable than conventional Fv fragments and to have the same affinity as the corresponding native immunoglobulin molecule, provided that a flexible linker is used to connect the two polypeptide chains (Alfthan *et al.*, 1995). It is even possible to create bispecific antibodies in which two Fab fragments are covalently linked, each having a unique antigen specificity.

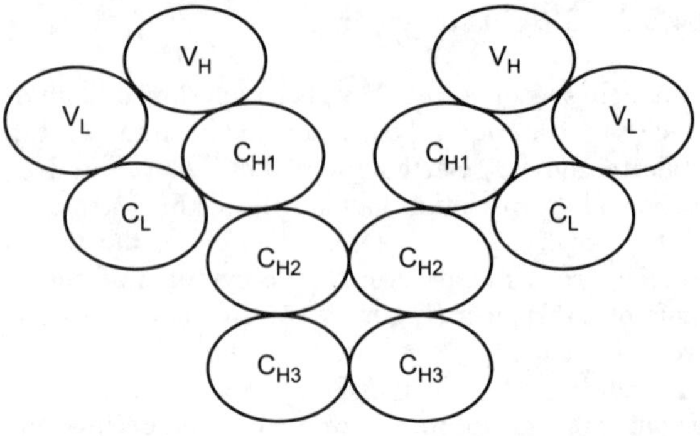

Figure 9.13 Schematic structure of IgG depicting the domain hypothesis.

In early reports it has been demonstrated that the Fab unit of an antibody (V_L, C_L, V_H and C_{H1}) can be fused to enzymes and expressed as a functional enzyme conjugate. Alkaline phosphatase (Ducancel *et al.*, 1993) and a luminescent enzyme, aequorin (Casadei *et al.*, 1990), have been employed for these purposes.

9.5 Biotin-labelled Protein Conjugates

Biotinylation of antigens and antibodies is of great practical importance due to the very tight binding to avidin and streptavidin (see Chapter 5, Section 5.11). Avidin–biotin interactions are widely used in immunotechnology to allow the detection, labelling, purification and immobilisation of biotinylated antibodies. Antibody biotinylation is most often accomplished by chemically cross-linking biotin randomly to the immunoglobulins. This technique leads generally to the preparation of heterogeneous conjugates.

Biotin is a small coenzyme which occurs intracellularly, mainly in a protein-bound state at a specific site. Biotin is hence covalently bound to a protein via an amide linkage to the ϵ-amino group of a unique lysine residue, a reaction catalysed by biotin ligase. In contrast to eukaryotic cells, *E. coli* contains only one *in vivo* biotinylated protein, the biotin carboxy-carrier protein (BCCP) of acetyl-CoA carboxylase.

The gene encoding BCCP has been cloned, and several reports have shown that *in vitro* constructed translational fusion proteins, encoding the biotin-acceptor sequence of BCCP, are efficiently biotinylated in *E. coli*. For instance, β-lactamase, a bacterial periplasmic enzyme, fused to the carboxy-terminal biotin-acceptor domain of BCCP, has been successfully biotinylated *in vivo* and exported in *E. coli* cells (Reed and Cronan, 1991).

In addition to BCCP, the coding region of the 1.3S subunit of *Propionibacterium shermanii* transcarboxylase and that of *Klebsiella pneumoniae* oxaloacetate decarboxylase have also been successfully used to produce biotin-labelled molecules in *E. coli* (Consler *et al.*, 1993). These protein segments required for biotin ligase extend over 100 amino acids. However, using phage display technology, a 13 amino acid residue consensus peptide specifying biotinylation in *E. coli* has been isolated (Schatz, 1993). Moreover, using the same phage technology, a short peptide, the so-called 'strep tag' which contains the core consensus sequence His-Pro-Gln, has been shown to interact well with streptavidin (Kay *et al.*, 1993). However, the strep tag fused to Fab does not behave as well as biotin.

The production of a chimeric biotinylated Fab protein in *E. coli* has been described. This conjugate is secreted into the bacterial culture medium and is efficiently labelled with biotin *in vivo*. This procedure for site-specific labelling permits the production of 0.15 mg/litre of Fab fragments. However, the internal concentration of biotin ligase is normally limiting and therefore this enzyme should be over-expressed at the same time (Cronan, 1990).

9.6 Bacterial Phage Display

M13 phage display systems are being used increasingly (see Section 4.6.3). The system is most often based on fusing the gene of interest to either gene *III*, gene *VI* or gene *VIII*, all of which encode M13 coat proteins (Jespers *et al.*, 1995). Using this technique, the fusion protein will be displayed on the surface of filamentous phage, and the corresponding structural gene will be packed in the phage particle. The system can be used for the selection of phages displaying randomised proteins that interact with a specific ligand (Figure 9.14). For instance, the display of antibodies on M13 has been used to select for immunoreactivity against different antigens. These results are interesting because they clearly demonstrate the possibility of bypassing immunisation in the formation of specific antibodies. Employing this method it is possible to evolve and select specific antibody-carrying phages by affinity purification against the desired antigen (Sanna *et al.*, 1995). All information about the selected antibody fragment is found within the phage genome, facilitating further manipulations. Thus, preparation of enzyme–antibody fusion conjugates with desired specificity will certainly be a reality in a near future.

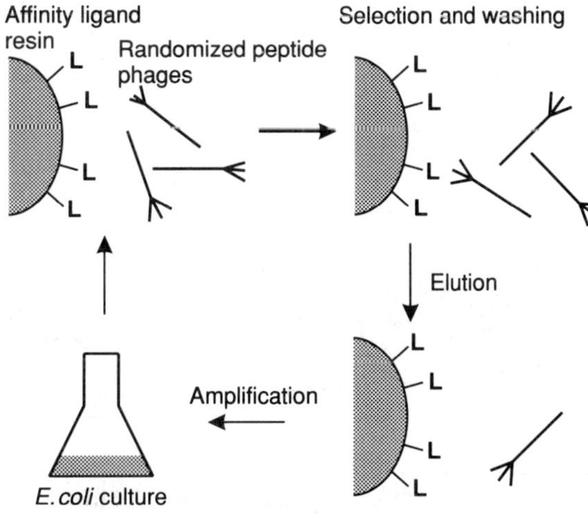

Figure 9.14 Short peptide fragments exhibiting affinity to a desired ligand can be enriched and isolated from a random phage peptide library using a repetitive selection/amplification cycle. The isolated peptides can then be used for the construction of protein conjugates by gene fusion to (for example) a marker enzyme.

Living cells can also be utilised to present the molecule of interest on the cell surface. For instance, antibody fragments have been displayed on the surface of *E. coli* by constructing a peptidoglycan-associated lipoprotein (Fransisco *et al.*, 1993).

9.7 Future Trends

Gene fusion has often been considered to be very labour-consuming compared to conventional chemical cross-linking. However, the recent progress of PCR technology has changed this situation dramatically. The construction of genetic conjugates can often be achieved within a few weeks, even by a novice in the field. Furthermore the purification of these conjugates is often facile, since one of the protein moieties can frequently be used to facilitate a one-step purification of the entire conjugate.

A problem that is often encountered during recombinant production of small proteins is proteolytic degradation within the foreign bacterial host. The fusion of, *e.g.* an antigen to an enzyme, particularly if the latter is of bacterial origin, often stabilises the fusion protein *in vivo*. When expressed in a prokaryotic host the antigen–enzyme conjugate may then constitute more than 10% of the total protein.

In most reported procedures the proteins are linked in a 1:1 ratio, which in many cases may limit the

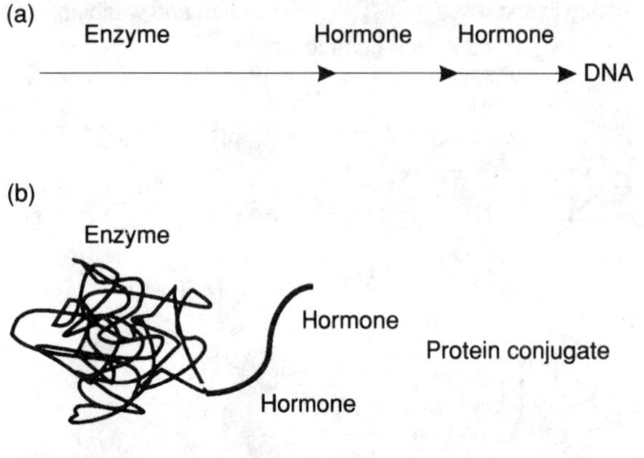

Figure 9.15 The proteins in a genetically produced conjugate need not be fused in a 1:1 ratio. For instance, the fusion *in tandem* of two hormone genes to an enzyme greatly improves the likelihood of antibody binding.

practical applicability of these conjugates. However, by fusing the structural genes *in tandem* the range of activities can be extended (Kuchinke and Müller-Hill, 1985). For instance, the activity of a marker enzyme in a protein conjugate can be increased by linking a further copy of the enzyme gene directly after the first, or by placing the enzyme at both the N and C terminus of the immunoreactive protein (Figure 9.15).

It is relevant to ask how large protein conjugates can be made when employing the fusion technique. In a recent report an artificial tripartite enzyme conjugate, β-galactosidase–galactose dehydrogenase–galactokinase, was constructed and characterised (Ljungcrantz *et al.*, 1990). The hybrid enzyme was active, and formed two major oligomeric aggregation forms comprising four and eight chimaeric subunits respectively, where each chimaeric subunit consisted of approximately 1700 amino acid residues. For comparison, it should be observed that the average molecular mass of a protein in *E. coli* is 50–70 kDa corresponding to 500–600 amino acid residues.

Special precautions should be taken when large fusion conjugates are based on oligomeric proteins. Such fusion proteins may form unfavourable quaternary structures, due to steric hindrances when the fusion subunits associate to form a functional protein. This can result in impaired biological activity, limiting the applicability of the conjugate. Recently, we have investigated the possibility of improving the activity and stability of such bifunctional oligomeric fusion proteins. This was achieved by the co-expression, within the same bacterial cell, of both the fusion protein and the corresponding native subunits of the oligomeric protein contained in the conjugate. Employing this technique, native subunits and fusion subunits assemble in a ratio which releases steric strain, resulting in a fusion conjugate with improved biological activities.

Similarly, large enzyme complexes can be constructed by co-production of different fusion proteins within the same cell. This technique is based on interactions between different fusion subunits of oligomeric hybrid enzymes, which have part of the chimaeric subunit preserved for subunit interaction. Thus, both tripartite hybrid enzyme complexes and bifunctional enzyme complexes with improved physical and chemical properties have been prepared using this method. In the latter case the galactose dehydrogenase activity in a β-galactosidase–galactose dehydrogenase conjugate was improved by approximately 30%, when native galactose dehydrogenase subunits were co-produced with the bifunctional enzyme in the same bacterial cell (Figure 9.16). Both β-galactosidase and

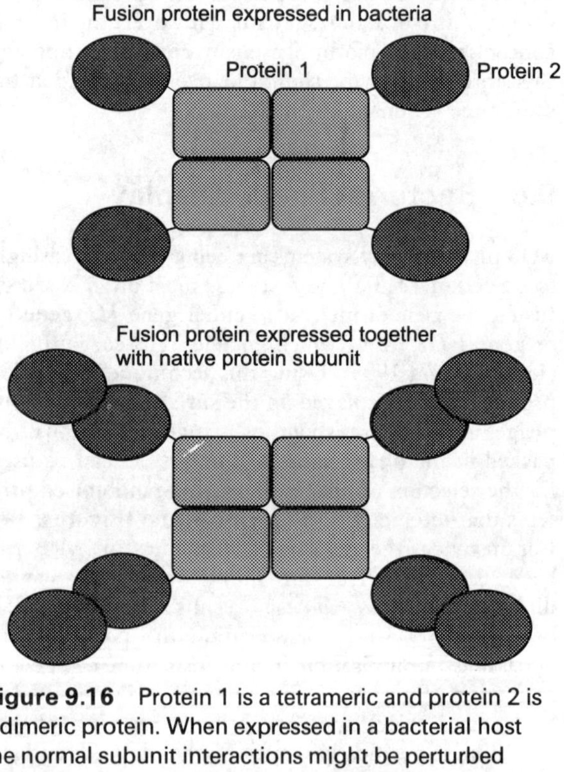

Figure 9.16 Protein 1 is a tetrameric and protein 2 is a dimeric protein. When expressed in a bacterial host the normal subunit interactions might be perturbed resulting in low activity of e.g. protein 2. However, if wild-type protein 2 is co-produced with the protein 1/ protein 2 conjugate within the same bacterial cell, the newly formed complex can display improved activity and topological properties.

galactose dehydrogenase are oligomeric enzymes and the results were explained by an improved subunit interaction in the newly formed hybrid complex. The construction by this approach of fusion proteins which are larger, or have a more optimal quaternary structure, will certainly open up new possibilities. Moreover, well-defined and tailor-made fusion conjugates may be easily produced in large quantities at a low cost.

A number of substances in biological fluids and fermentation broths are quantified using enzymatic methods. Frequently, consecutive enzyme sequences are utilised in analytical procedures to form a monitorable product. For instance, an NAD(P)-dependent dehydrogenase is often used as a last step to allow a simple spectrophotometric determination of a substance. This implies that artificial bifunctional enzymes will be very useful in such applications, since only one enzyme needs to be added to the reaction mixture. The fusions between β-galactosidase and galactose dehydrogenase, and galactose dehydrogenase and luciferase, respectively, have provided enzyme conjugates suitable for lactose and galactose analyses (Carlsson *et al.*, 1994). The fusion enzymes display several advantages compared to the corresponding native enzyme systems, which make them interesting for practical applications. These include proximity effects, changed substrate specificity due to the fusion, and benefits in the purification and production of the enzymes, since only one protein needs to be handled instead of several.

These fusion enzymes can furthermore be seen as ready-made enzyme sequences suitable for immobilisation and use in biosensor applications. The utilisation of fusion enzymes as efficient multi-enzyme analogues is likely to improve detection signals and reduce the time of analysis.

Recently, several interesting and useful proteins have been examined for use in the analysis of gene expression. Some of these proteins will probably also be utilised for the design of protein conjugates. An exciting protein in this respect is *Aequorea victoria* green fluorescent protein (GFP) which emits green light when illuminated with long-wave ultraviolet light. Because no additional co-factors are required, GFP can be used as a very convenient portable marker for gene expression. Vertebrate, invertebrate, yeast and *E. coli* cells producing GFP all glow bright green under the appropriate illumination. In addition, it has been reported that both amino and carboxy-terminal fusions to GFP retain the fluorescence properties, making the protein a very valuable fusion tag for the future (Chalfie *et al.*, 1994).

References

Abrahmsén, L. *et al.* (1985). Analysis of signals for secretion in staphylococcal Protein A gene. *EMBO J.*, **4**, 3901–3906.

Alfthan, K. *et al.* (1995). Properties of a single-chain antibody containing different linker peptides. *Protein Eng.*, **8**, 725–731.

Argos, P. (1990). An investigation of oligopeptides linking domains in protein tertiary structures and possible candidates for general fusion. *J. Mol. Biol.*, **211**, 943–958.

Bachmair, A. *et al.* (1986). *In vivo* half-life of a protein is a function of its amino-terminal residue. *Science*, **234**, 179–186.

Baneyx, F. and Georgiou, G. (1989). Expression, purification, and enzymatic characterization of a Protein A–β-lactamase hybrid protein. *Enzyme Microb. Technol.*, **11**, 559–567.

Baneyx, F. *et al.* (1990). Affinity immobilization of a genetically engineered bifunctional hybrid protein. *Enzyme Microb. Technol.*, **12**, 337–342.

Bird, R. E. and Walker, B. W. (1991). Single chain antibody variable regions. *Trends Biotechnol.*, **9**, 132–137.

Brown, T. A. (1990). *Gene Cloning, an Introduction*, 2nd edn, Chapman and Hall, London.

Bülow, L and Mosbach, K. (1991). Multienzyme systems obtained by gene fusion. *Trends Biotechnol.*, **9**, 226–231.

Carlsson, H. *et al.* (1992). Engineering of lactose metabolism in *E. coli* by introducing β-galactosidase/galactokinase fusion enzymes. *Biotechnol. Lett.*, **14**, 439–444.

Carlsson, H. *et al.* (1994). Use of genetically prepared enzyme conjugates in lactose and galactose analyses. *Anal. Biochem.*, **218**, 278–283.

Casadaban, M. J. and Cohen, S. (1980). Analysis of gene control signals by DNA fusion and cloning in *Escherichia coli*. *J. Mol. Biol.*, **138**, 179–207.

Casadaban, M. J. *et al.* (1983). β-Galactosidase gene fusions for analyzing gene expression in *Escherichia coli* and yeast. *Methods Enzymol.*, **100**, 293–308.

Casadei, J. *et al.* (1990). Expression and secretion of aequorin as chimaeric antibody by means of a mammalian expression vector. *Proc. Natl. Acad. Sci. USA*, **87**, 2047–2051.

Chalfie, M. *et al.* (1994). Green fluorescent protein as a marker for gene expression. *Science*, **262**, 802–805.

Chang, S.-O. and Hammes, G.G. (1990). Structure and mechanism of action of a multifunctional enzyme: fatty acid synthase. *Acc. Chem. Res.*, **23**, 363–369.

Chaudhary, V. K. *et al.* (1988). Selective killing of HIV-infected cells by recombinant human CD4–*Pseudomonas* exotoxin hybrid protein. *Nature*, **335**, 369–372.

Consler, T. G. *et al.* (1993). Properties and purification of an active biotinylated lactose permease from *Escherichia coli*. *Proc. Natl. Acad. Sci. USA*, **90**, 6934–6938.

Crawford, I. P. *et al.* (1987). Crucial role of the connecting region joining the two functional domains of yeast tryptophan synthetase. *J. Biol. Chem.*, **262**, 239–244.

Cronan, J. E., Jr (1990). Biotination of proteins *in vivo*. A post-translational modification to label, purify and study proteins. *J. Biol. Chem.*, **265**, 10327–10333.

Crowe, J. *et al.* (1994). 6xHis-Ni-NTA chromatography as a superior technique in recombinant protein expression/ purification. *Methods Mol. Biol.*, **31**, 371–387.

Ducancel, F. *et al.* (1989). Direct expression in *E. coli* of a functionally active Protein A–snake toxin fusion protein. *Protein Eng.*, **3**, 139–143.

Ducancel, F. *et al.* (1993). Recombinant colorimetric antibodies: construction and characterization of a bifunctional $F(ab)_2$/alkaline phosphatase conjugate produced in *Escherichia coli*. *Bio/Technology*, **11**, 601–605.

Enfors, S.-O. (1992). Control of *in vivo* proteolysis in the production of recombinant proteins. *Trends Biotechnol.*, **10**, 310–315.

Fowler, A. V. and Zabin, I. (1983). Purification, structure and properties of hybrid β-galactosidase proteins. *J. Biol. Chem.*, **258**, 14354–14358.

Frankel, A. E. *et al.* (1995). Clinical trials of targeted toxins. *Semin. Cancer Biol.*, **6**, 307–317.

Fransisco, J. A. *et al.* (1993). Production and fluorescence-activated cell sorting of *Escherichia coli* expressing functional antibody fragment on the external surface. *Proc. Natl. Acad. Sci. USA*, **90**, 10444–10448.

Frohman, M. A. (1993). Rapid amplification of complementary DNA ends for generation of full length cDNAs: thermal RACE. *Methods Enzymol.*, **218**, 340–355.

Früh, K. *et al.* (1989). A new tool for the serodiagnosis of acute *Plasmodium falciparum* malaria in individuals with primary infection. *J. Immunol. Methods*, **122**, 25–32.

Goeddel, D. V. (1990). Systems for heterologous gene expression. *Methods Enzymol.*, **185**, 3–7.

Gottesman, S. and Zipser, D. (1978). Deg phenotype of *Escherichia coli lon* mutants. *J. Bacteriol.*, **133**, 844–851.

Griffin, H. G. and Griffin, A. M. (eds) (1994). *PCR Technology, Current Innovations*, CRC Press, Boca Raton, FL.

Guo, L. H. *et al.* (1984). Synthesis of human insulin. VIII. Construction of expression vectors for fused proinsulin production in *Escherichia coli*. *Gene*, **29**, 251–254.

Guss, B. *et al.* (1984). Region X, the cell-wall-attachment part of staphylococcal Protein A. *Eur. J. Biochem.*, **138**, 413–420.

Henner, D. J. (1990). Inducible expression of regulatory genes in *Bacillus subtilis*. *Methods Enzymol.*, **185**, 223–228.

Hirel, P.-H. *et al.* (1989). Extent of N-terminal methionine excision from *Escherichia coli* is governed by the side-chain length of the penultimate amino acid. *Proc. Natl. Acad. Sci. USA*, **86**, 8247–8251.

Hitzeman, R. A. *et al.* (1990). Use of heterologous and homologous signal sequences for secretion of heterologous proteins in yeast. *Methods Enzymol.*, **185**, 421–440.

Hoffman, C. S. and Wright, A. (1985). Fusions of secreted proteins to alkaline phosphatase: An approach for studying protein secretion. *Proc. Natl. Acad. Sci. USA*, **82**, 5107–5111.

Holzman, T. F. and Baldwin, T.O. (1982). Isolation of bacterial luciferases by affinity chromatography on 2,2-diphenylpropylamine-Sepharose: phosphate-mediated binding to an immobilized substrate analogue. *Biochemistry*, **21**, 6194–6201.

Jablonsky, E. (1985). The preparation of bacterial luciferase conjugates for immunoassay and application to *Rubella* antibody detection. *Anal. Biochem.*, **148**, 199–206.

Jespers, L. S. *et al.* (1995). Surface expression and ligand-based selection of cDNAs fused to filamentous phage gene VI. *Bio/Technology*, **13**, 378–382.

Kay, B. *et al.* (1993). An M13 phage library displaying 38 amino acid peptides as a source of novel sequences with affinity to selected targets. *Gene*, **128**, 59–65.

Kim, J. and Weaver, R. F. (1988). Construction of a recombinant expression plasmid encoding staphylococcal Protein A–ricin a fusion protein. *Gene*, **68**, 315–321.

Kuchinke, W. and Müller-Hill, B. (1985). Fused *lacZ* genes code for di-, tri- and tetra-β-galactosidase in *Escherichia coli*. *EMBO J.*, **4**, 1067–1073.

Langone, J. J. (1982). Protein A of *Staphylococcus aureus* and related immunoglobulin receptors produced by streptococci and pneumococci. *Adv. Immunol.* **32**, 157–241.

Levinson, A. D. (1990). Expression of heterologous genes in mammalian cells. *Methods Enzymol.*, **185**, 485–487.

Lindbladh, C. *et al.* (1987). The design of a simple competitive ELISA using human proinsulin-alkaline phosphatase conjugates prepared by gene fusion. *Biochem. Biophys. Res. Commun.*, **149**, 607–614.

Lindbladh, C. *et al.* (1991a). Standard calibration proteins for western blotting obtained by genetically prepared Protein A conjugates. *Anal. Biochem.*, **197**, 187–190.

Lindbladh, C. *et al.* (1991b). Preparation of a genetically fused Protein A/luciferase conjugate useful for bioluminescent immunoassays. *J. Immunol. Methods*, **137**, 199–207.

Lindbladh, C. *et al.* (1992). Characterization of a recombinant bifunctional enzyme, galactose dehydrogenase/bacterial luciferase, displaying an improved bioluminescence in a three-enzyme system. *Eur. J. Biochem.*, **204**, 241–247.

Lindbladh, C. *et al.* (1994) Metabolic studies on *S. cerevisiae* containing fused citrate synthase/malate dehydrogenase. *Biochemistry*, 33, 11684–11691.

Ljungcrantz, P. *et al.* (1990). Construction and characterization of a recombinant tripartite enzyme, galactose dehydrogenase/β-galactosidase/galactokinase. *FEBS Lett.*, **275**, 91–94.

Manoil, C. and Beckwith, J. (1986). A genetic approach to analyzing membrane protein topology. *Science*, **233**, 1403–1408.

Mason, D. Y. and Sammons, R. (1978). Alkaline phosphatase and peroxidase for double immunoenzymatic labelling of cellular constituents. *J. Clin. Pathol.*, **31**, 454–460.

Mertens, N. *et al.* (1995). Tight transcriptional control mechanism ensures stable high-level expression from T7 promoter-based expression plasmids. *Bio/Technology*, **13**, 175–179.

Moks, T. *et al.* (1986). Staphylococcal Protein A consists of five IgG binding domains. *Eur. J. Biochem.*, **156**, 637–643.

Moks, T. *et al.* (1987a). Large-scale affinity purification of human insulin-like growth factor I from culture medium of *Escherichia coli*. *Bio/Technology*, **5**, 379–382.

Moks, T. *et al.* (1987b). Expression of human insulin-like growth factor I in bacteria: use of optimized gene fusion vectors to facilitate protein purification. *Biochemistry*, **26**, 5239–5244.

Nilsson, B. and Abrahmsén, L. (1990). Fusions to staphylococcal Protein A. *Methods Enzymol.*, **185**, 144–161.

Perentesis, J. P. *et al.* (1988). Expression of diphtheria toxin fragment A and hormone–toxin fusion proteins in toxin-resistant yeast mutants. *Proc. Natl. Acad. Sci. USA*, **85**, 8386–8390.

Peterhans, A. *et al.* (1987). A simple competitive enzyme linked immunosorbent assay using antigen–β-galactosidase fusions. *Anal. Biochem.*, **163**, 470–475.

Peters, I. D. *et al.* (1989). Biosynthesis of winter flounder antifreeze proprotein in *E. coli. Protein Eng.*, **3**, 145–151.

Plückthun, A. (1991). Antibody engineering: advances from the use of *E. coli* expression systems. *Bio/Technology*, **9**, 545–551.

Reed, K. E. and Cronan, J. E., Jr (1991). *Escherichia coli* exports previously folded and biotinylated protein domains. *J. Biol. Chem.*, **266**, 11425–11428.

Rhee, I.-K. *et al.* (1990). Metallothioneins with the inter-domain hinges expanded by insertion mutagenesis. *Protein Eng.*, **3**, 205–213.

Sambrook, J. *et al.* (1989). *Molecular Cloning: A Laboratory Manual*, 2nd edn, Cold Spring Harbor Laboratory Press, Cold Spring Harbor, NY.

Sanna, P. P. *et al.* (1995). Rapid assay of phage-derived recombinant human Fabs as bispecific antibodies. *Bio/Technology*, **13**, 1221–1224.

Sassenfeld, H. M. (1990). Engineering proteins for purification. *Trends Biotechnol.*, **8**, 88–93.

Schatz, P. J. (1993). Use of peptide libraries to map the substrate specificity of a peptide-modifying enzyme: a 13-residue consensus peptide specifies biotinylation in *Escherichia coli. Bio/Technology*, **11**, 1138–1143.

Shapira, S. K. *et al.* (1983). New versatile plasmid vectors for expression of hybrid proteins coded by a cloned gene fused to *lacZ* gene sequences encoding an enzymatically active carboxyl-terminal portion of β-galactosidase. *Gene*, **25**, 71 82.

Sherwood, R. (1991). Protein fusions: bioseparation and application. *Trends Biotechnol.*, **9**, 1–3.

Strandberg *et al.* (1990). Expression and characterization of a tripartite fusion protein consisting of chimeric IgG-binding receptors and β-galactosidase. *J. Biotechnol.*, **13**, 83–96.

Summers, R. G. and Knowles, J. R. (1989). Illicit secretion of a cytoplasmic protein into the periplasm of *Escherichia coli* requires a signal peptide plus a portion of the cognate secreted protein. *J. Biol. Chem.*, **264**, 20074–20081.

Ternynck, T. and Avrameas, S. (1990). Avidin–biotin systems in enzyme immunoassays. *Methods Enzymol.*, **184**, 469–481.

Tijssen, P. (1988). *Practice and Theory of Enzyme Immunoassay, Laboratory Techniques in Biotechnology and Molecular Biology*, Elsevier Science Publishers, Amsterdam.

Udaka, S. and Yamagata, H. (1993). High level secretion of heterologous proteins by *Bacillus brevis. Methods Enzymol.*, **217**, 23–33.

Ullmann, A. (1984). One-step purification of hybrid proteins which have β-galactosidase activity. *Gene*, **29**, 27–31.

Wada, M. *et al.* (1990). Production of a Protein A–lymphotoxin hybrid protein for cancer-targeted therapy. *J. Biotechnol.*, **13**, 325–334.

Watson, J. D. *et al.* (1992). *Recombinant DNA*, 2nd edn, Scientific American Books, New York.

Wienhausen, G. and DeLuca, M. (1982). Bioluminescent assays of picomole levels of various metabolites using immobilized enzymes. *Anal. Biochem.*, **127**, 380–388.

White, B. A. (ed.) (1993). *PCR Protocols: Current Methods and Applications*, Humana Press, Clifton, NJ.

Williams, G. *et al.* (1995). Dissection of the extracellular human interferon receptor alpha-chain into two immunoglobulin-like domains. Production in an *Escherichia coli* thioredoxin gene fusion expression system and recognition by neutralizing antibodies. *Biochemistry*, **34**, 1787–1797.

Yanisch-Peron, C. *et al.* (1985). Improved M13 phage cloning vectors and host strains: nucleotide sequence of the M13mp18 and pUC 19 vectors. *Gene*, **33**, 103–119.

Zalkin, H. *et al.* (1984). Nucleotide sequence of *Saccharomyces cerevisiae* genes TRP2 and TRP3 encoding bifunctional anthranilate synthase: indole-3-glycerol phosphate synthase. *J. Biol. Chem.*, **259**, 3985–3992.

Zenno, S. and Inouye, S. (1990). Bioluminescent immunoassay using a fusion protein of Protein A and the photoprotein aequorin. *Biochem. Biophys. Res. Commun.*, **171**, 169–174.

Chapter 10

The Purification of Protein Conjugates

Alastair H. Dent and Mohammed Aslam

Research and Development Department, Ortho-Clinical Diagnostics, Pollards Wood Laboratories, Nightingales Lane, Chalfont St. Giles HP8 4SP, UK

Occasionally chemical reactions may go fully to completion, and, if stoichiometric quantities of the starting materials have been employed, the only species in the final reaction mixture is the desired product. In practice such examples are extremely rare, however, and particularly so in protein chemistry applications.

It is far more likely that unreacted starting materials and/or byproducts will be present in the mixture; again it is sometimes possible to use such crude products without further processing, but in the vast majority of cases the removal of unwanted species is either essential or desirable. In some of the applications described in the previous chapter (e.g. coupling to a solid phase) this purification step may be facile, but generally the experimenter is faced with separating a protein product from other protein components, or from low-molecular-mass materials such as coupling agents and their byproducts.

Early purification methods relied on simple procedures such as dialysis and precipitation, and even today such techniques are in widespread use. However, the last fifty years or so have seen the development of a range of sophisticated techniques which allow rapid and efficient separation of complex mixtures, and which have led to a variety of new applications for protein conjugates.

The nature of the conjugate, protein–protein, protein–hapten, etc., will dictate the type of purification selected. In some cases a highly specific procedure will be needed, but in practice a considerable overlap in methodologies occurs across nearly all applications.

Modern procedures for protein purification are predominantly branches of chromatography, the components of reaction mixtures being resolved by virtue of their differing behaviour in dynamic liquid–solid systems. The underlying principles of these methods are summarised in Table 10.1. Chromatography has undergone a great deal of development over the last two or three decades, partly in terms of the number of different modes of application which are now available, and partly in the development of the materials to support these methods (Regnier and Gooding, 1980; Regnier, 1983; Dean *et al.*, 1985a). In combination these advances have provided the capability to achieve extremely high resolution in the separation of proteins and their derivatives – although challenges still remain in linking high resolution with large-scale capability.

In this chapter the various techniques available are presented individually, but it is common for multistage purifications to be employed. It is impossible to generalise how these methods should be combined to produce the optimum separation of a given protein reagent, although some useful guidelines have been suggested for protein purification in general (Janson and Hedman, 1987; Berkowitz, 1987; Cartwright, 1987; Sofer and Nyström, 1989a; Asenjo and Patrick, 1990; PerSeptive Biosystems, 1996).

More than any other area described in this book, the approaches available are often defined in practical terms by the offerings of specialist commercial organisations, and it is hard to get far into this topic without mentioning the contributions of Pharmacia, Bio-Rad and other organisations. Although it would be quite

Table 10.1 Summary of the principal methods of protein purification

Purification technique	Basis of separation	Characteristics of method
Precipitation	Solubility	Moderate resolution, high capacity
Dialysis	Molecular size	Moderate resolution, high capacity
Ultrafiltration	Molecular size	Moderate resolution, medium capacity
Partitioning	Solubility	Low resolution, moderate capacity
Size exclusion chromatography	Molecular size	Moderate resolution, low capacity
Ion exchange chromatography	Charge	High resolution, high speed and capacity
Hydrophobic interaction chromatography	Surface hydrophobicity	Good resolution, high speed and capacity
Reversed phase chromatography	Hydrophobic and hydrophilic interactions	Good resolution, intermediate capacity, high speed and risk of denaturation
Affinity chromatography	Biological affinity	Excellent resolution, high speed and capacity
Chromatofocusing	p*I* and charge	

feasible for experimenters to prepare their own size exclusion chromatography (SEC) materials, for example, there is very little point with the current availability of efficient commercial media at a reasonable cost. Therefore this chapter is rather more framed around specific commercial products than earlier ones.

One great advantage of modern techniques is the routine generation of a 'hard copy' of the output from the purification, a chromatogram or electrophoretogram which displays graphically the presence of the desired product, the other components present and their degree of resolution. Because the purification techniques rely on characteristics of the components such as size or charge, conclusions can instantly be made about the nature of the conjugate obtained. However the product peak is detected, its retention properties can be used to estimate its molecular mass or p*I*, for example, or its possession of submolecular structures which cause it to bind to a particular affinity matrix.

With a little attention to detection methods, more information can be deduced. For example where reaction components possess different spectral properties, something of the degree of incorporation in a conjugate can be revealed by monitoring ultraviolet absorption at more than one wavelength.

These techniques can therefore form the basis of the characterisation of the protein conjugate obtained, which forms the topic of the next chapter. The records of purification procedures can only provide so much information about the product, and numerous other techniques have been applied to elucidate conjugate properties. Particularly in medical applications, it is more and more essential to be able to show that the product from a given procedure meets defined criteria of structure and behaviour. The stringent requirements of the pharmaceutical industry have been instrumental in expanding the power of conjugate characterisation, which has otherwise been a rather neglected area.

The principal methods of protein purification are outlined in Table 10.1, and will be described in detail below.

10.1 Precipitation

Fractional precipitation is one of the oldest techniques of chemical purification, and in pre-chromatography days offered one of the few available means for the separation of chemicals from mixtures. Proteins are fairly well suited to precipitation techniques as their solubility in aqueous systems is often fairly marginal. Small changes in pH, temperature, ionic strength or the presence of organic cosolvents can result in the precipitation of proteins from solution, and these effects can be modulated to obtain relatively pure fractions from complex mixtures. Perhaps the most famous application of this approach was Cohn's development of a series of manipulations for the fractionation of serum proteins (Cohn *et al.*, 1946), still in widespread use today.

Although certain broad guidelines can be stated, the identification of a suitable precipitation method for a given application remains highly empirical. There are some very useful general methods for the purification of antibodies by precipitation, which can be relied upon to give material of a fairly high purity; even these, however, may offer improved yield or purity with a little further optimisation.

Where a high-purity product is required, the development of highly individual regimes may be necessary. As an illustration, Erlanger *et al.* (1959) described the following protocol for purification of an estrone–albumin conjugate, already dialysed for 17 hours to remove the bulk of the contaminating steroid: (1) adjust to pH 4.6; (2) refrigerate overnight; (3) collect precipitate by centrifugation; (4) redissolve at pH 7.0; (5) add 300 ml of cold acetone; (6) adjust to pH 4.5; (7) collect precipitate by centrifugation; (8) repeat (4)–(7) twice; (9) redissolve at pH 7.8; (10) remove insoluble material by centrifugation; (11) dialyse against running water for eight hours.

While such processes can undoubtedly yield products of very high purity, a considerable amount of development work may well be necessary to optimise the conditions for the individual case. One can understand how modern chromatographic approaches have come to dominate this type of application!

10.1.1 Key process variables of precipitation

Only the very broadest guidelines can be offered here, and it should be borne in mind that, for a particular protein application, they may often be overridden by other more specific interactions.

Nature of precipitant

The first major choice to be faced is that of the precipitating agent itself. As noted earlier, the stability of soluble protein conformations is marginal and precipitation may be prompted by many different types of change in the environment. Not surprisingly for structures which have (mostly) been evolved to function in aqueous environments at near-neutral pH, the addition of organic cosolvents, acids or bases can readily provoke precipitation. Steps involving acidification, or precipitation with water-miscible solvents such as acetone or ethanol, are indeed frequently encountered in purification protocols over the years (Cohn *et al.*, 1946; Dodgson and Powell, 1959; Nishijima *et al.*, 1977; Heide *et al.*, 1977).

While effective at promoting precipitation these are rather harsh treatments, and can result in considerable damage to protein molecules. The effect of adding inorganic salts is more interesting, and has led to many purification procedures.

Intuitively it might seem likely that protein conformations in solution would be most stable at an ionic strength comparable with that of their natural environment. Any reduction or increase from this level would be likely to promote structural changes, destabilising the native conformation and increasing the likelihood of precipitation. This effect would understandably be variable from one protein to another depending on its *in vivo* environment, and the choice of salt to provide the change in ionic strength might be expected to be a secondary factor.

While adding increasing quantities of a salt to a protein in water does (Figure 10.1; Dixon and Webb, 1961) generally promote an increase in solubility (salting in) followed at higher concentrations by a decrease (salting out), most other aspects of this perception are wrong. Great variations in solubility with ionic strength are seen within proteins from the same milieu; in fact, whether a protein becomes more or less soluble on the addition of salt has been used as a key parameter in the classification of serum proteins. Secondly, the actual identity of the salt is extremely important in determining the outcome of the addition, more so than the ionic strength itself, in fact.

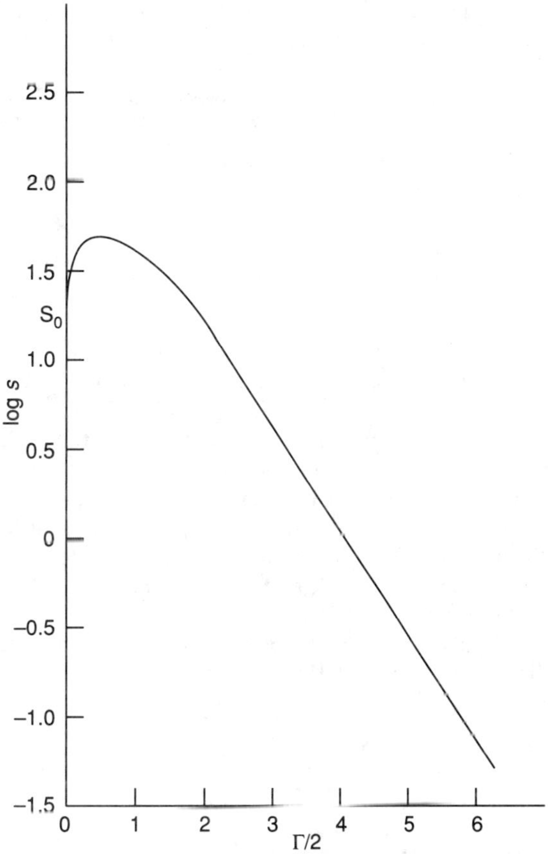

Figure 10.1 Relation between log solubility (*s*) and ionic strength for carboxyhaemoglobin and ammonium sulfate. (Reproduced with permission from: Dixon and Webb (1961). *Adv. Protein Chem.*, **16**, 197–219.)

The fact that different salts have variable effects on protein solubility was recognised by Hofmeister in 1888, and has given rise to the Hofmeister or lyotropic series of salts, an empirical order of effectiveness at salting out. It is the anion which has the greatest effect on the salting-out process, although cations have secondary effects Hofmeister's original work noted that the molarity of salt required to precipitate euglobins decreased in the following order (Von Hippel and Schleich, 1969):

$$NaNo_3 > NaCl > \text{Na acetate} > \text{K acetate} > MgSO_4 > (NH_4)_2SO_4 > Na_2HPO_4 > K_2HPO_4 > Na_2SO_4 > Li_2SO_4 > Na_3\text{citrate}$$

It is difficult to state a definitive lyotropic series for discrete anions, as the effects observed are modulated by the nature and valency of the cation; in addition there are some variations dependent on the type of effect being studied. However, the following series is roughly consistent with many observations (Von Hippel and Schleich, 1969; Melander and Horváth, 1977; Arakawa and Timasheff, 1982; Carrea *et al.*, 1982; Przybycien and Bailey, 1989):

$$SCN^- \approx ClO_4^- > ClO_3^- > I^- > NO_3^- > Br^- > Cl^- > CH_3COO^- > PO_4^{2-} \approx SO_4^{2-} > \text{citrate} > Fe(CN)_6^{4-}$$

The anions at the left hand end of the series such as SCN^- (as well as organic agents such as urea) have salting-in properties, while those at the right hand end (sulfates, acetates) promote salting out. A great deal of work has been carried out over the years to elucidate the rationale for this empirical order (Bruins, 1934; Von Hippel and Wong, 1965; Melander and Horváth, 1977; Arakawa and Timasheff, 1982, 1984a,b).

It is also possible to state a cationic series (Von Hippel and Schleich, 1969):

$$Ba^{2+} > Ca^{2+} > Mg^{2+} > Li^+ > Rb^+ \approx K^+ \approx Na^+ \approx Cs^+ > NH_4^+ > (CH_3)_4N^+$$

There is still no universal explanation for the effects that are observed (Przybycien and Bailey, 1991). In summary, however, salting-out appears to be associated with the ability of the salt to perturb surface tension, quantifiable as a large positive surface tension increment[1]. These salts are preferentially excluded from the immediate vicinity of the protein, which is thus preferentially hydrated. This is an entropically-unfavourable arrangement, precipitation providing a more favourable situation. As there is no direct interaction of these salts with the molecule the structure is generally well-maintained, so they have a stabilising effect as well as promoting precipitation—often, in fact, giving greater stability than is observed with the native protein (Carrea *et al.*, 1982). Not only salts display this behaviour: carbohydrates show salting-out effects too (Franks, 1993).

Salting-in, on the other hand, is associated with low surface tension increments. These salts tend to exhibit binding to the protein molecule, and their presence in the immediate vicinity interferes with hydrogen bonding. Hence their solubilisation effects are often accompanied by structural changes in the protein. Urea and guanidinium salts, of course, are commonly employed as denaturants to unfold protein molecules completely. Changes in secondary structure of proteins have often been observed following treatment with these chaotropic salts: thiocyanate, for example, caused changes in the Raman spectra of a number of different proteins consistent with a conversion of α-helix structure into β-sheet features (Przybycien and Bailey, 1989, 1991).

In terms of reagents for protein precipitation, it is natural that salts should be chosen from the salting-out end of the Hofmeister series: not only are they more effective precipitants but they stabilise the structure during the process. Hence the popularity of salts such as sodium and ammonium sulfate, powerful salting-out agents. On the whole the latter is preferred—although the sodium salt has a higher salting-out ability *per se*, its solubility is much lower than ammonium sulfate. Accordingly, saturated ammonium sulfate solutions are generally more convenient for precipitation as a smaller volume is required. With ammonium sulfate, levels of 40–50% saturation in the final mixture[2] are sufficient to precipitate most immunoglobulins (Harlow and Lane, 1988) and their peroxidase conjugates (Avrameas and Ternynck, 1971). This is a very commonly encountered method for crude purification of antibodies. Sodium sulfate is used in much the same way, Wolters *et al.* (1976) for example precipitated IgG–HRP conjugates with 14% (w/v) sodium sulfate.

[1] Except for divalent cations, where pH-dependent charge interactions have an overriding effect which increases their salting-in tendency (Arakawa and Timasheff, 1984).

[2] Note that a '50% ammonium sulfate cut' requires the addition of an equal volume of saturated ammonium sulfate solution to the sample, not the addition of 50% by volume.

Soluble polymers are another convenient class of precipitants: examples such as polyvinylpyrrolidone (PVP) have been used for this purpose since the 1950s (Burstein, 1957). This is also another of the many useful roles of polyethylene glycol (PEG), which is a very mild precipitating agent and has found widespread application (Harrington *et al.*, 1971; Creighton *et al.*, 1973; Lei, 1974; Ingham, 1984). PEG concentrations of between 2 and 12% have been used for the precipitation of monoclonal antibodies for example (Neoh *et al.*, 1986; Neoh and Zola, 1990). Sequential PEG 6000 and ammonium sulfate precipitation gave very pure mAbs from cell culture media (Brooks *et al.*, 1992).

Caprylic (octanoic) acid (or more accurately its sodium salt) has also been widely used for protein purification since its introduction for this purpose by Reid and Jones (1951). Treatment of crude albumin at 68 °C caused other proteins to become insoluble while the albumin remained in solution. This reagent is also commonly employed cold to harvest immunoglobulins from serum (Steinbuch and Audran, 1969). The addition of a couple of percent of caprylic acid to a mixture of serum and acetate buffer (overall pH 4.8) causes the precipitation of most non-immunoglobulin protein in the sample. The remaining Ig solution can be better than 90% pure. This approach has also been extended successfully to the purification of monoclonal antibodies from ascites and cell culture media (Russo *et al.*, 1983).

Trichloroacetic acid (TCA) is a powerful precipitant, most commonly met in analytical applications where the aim is to precipitate the entire protein fraction. It has also been used preparatively in the purification of protein–small molecule conjugates, where fractionation of protein components is not required. Staros *et al.* (1986), for example, used TCA precipitation to purify glycine–haemocyanin conjugates. Excess acid is washed out with ethanol.

pH

Within a moderate pH range, the solubility of a protein is generally at a minimum at its isoelectric point (p*I*), where its overall charge is zero. Therefore precipitation can be encouraged by adjusting the pH to this value. This effect can be seen in the Erlanger example above: albumin has a p*I* of around five which will have been reduced somewhat by the derivatisation of amine groups, and the precipitation steps in the example were carried out at pH 4.5–4.6.

Useful separations can sometimes be obtained by manipulation of pH alone. For example, the tumoricidal complex SMANCS is prepared from the protein neocarzinostatin and styrene–maleic anhydride (SMA) polymers (Maeda *et al.*, 1984), and can be purified from unreacted starting materials simply by pH shifts. At pH 4, both the SMA and SMANCS are precipitated while the unreacted neocarzinostatin remains in solution; adjusting to pH 6–7 the SMA remains insoluble while the SMANCS goes into solution.

Time

Precipitation is a process which cannot always be hurried. While it is generally quite possible to identify conditions which will cause a rapid, high-yield precipitation of protein from a mixture, this is typically at the expense of purity.

Optimisation of a precipitation process is normally carried out to provide a balance between yield and purity: using a milder precipitant for a longer time can help this balance to be achieved.

Temperature

As a general rule the likelihood of precipitation can be increased by lowering the temperature. Protein precipitations are often carried out under refrigeration, therefore. Note that if organic cosolvents are being used, it may be possible to cool solutions some way below 0 °C without any risk of freezing.

Mode of addition of precipitants

One final variable to mention is the manner in which precipitants such as acetone or saturated ammonium sulfate solution are added to the protein sample. The key point is to keep the mixture as homogeneous as possible, so that portions of the protein solution do not experience anomalous concentrations of precipitant. Thus it is more reproducible to add precipitants dropwise to a vigorously stirred protein solution, than it is to make a single bulk addition.

10.1.2 Affinity precipitation

An interesting variation of this theme is affinity precipitation (Larsson *et al.*, 1984). In common with other affinity techniques which will be discussed later in this chapter, the approach relies on the specific affinity of a target protein for a particular ligand. Here, the fact that the solubility of the protein may be altered by ligand complexation is exploited—it may be possible to precipitate a protein in its complexed state when the native form remains soluble. In particular, where the protein is oligomeric the use of bifunctional ligands can lead to the establishment of a large protein–ligand network with low solubility.

For example, Larsson *et al.* (1984) described the precipitation of oligomeric dehydrogenases in the presence of a reagent containing two NAD groups. The well-known precipitation of an immunoglobulin as a complex with its specific antigen also falls within this category.

10.2 Dialysis and Ultrafiltration

Dialysis offers one of the cheapest and simplest techniques available for the separation of proteins from low-molecular-mass impurities. Despite some serious limitations it has remained popular (Saltonstall, 1992), particularly for very small volumes of protein solution where losses may be impractically large with more elegant techniques.

The principle underlying dialysis is very simple. The protein solution to be purified is contained within a semi-permeable membrane (generally manufactured from cellulose or cellulose acetate), whose pores are small enough to retain the protein but large enough to allow free diffusion of small ions and molecules.

The sample of impure protein solution, securely sealed within the membrane, is placed in a surrounding volume of water or an aqueous solution. The small species present will diffuse both ways through the membrane until an equilibrium situation is reached, in other words, until the concentration of all the small species is identical both inside and outside the membrane.

Clearly, if the volume of 'clean' solution outside the membrane is very large compared to that of the sample, the concentration of impurities within the membrane can be reduced to a very small fraction of its original value. Repeating the process with a fresh solution reduces the remaining impurities by the same factor, so that extremely low levels can be reached within a relatively few changes[1].

The same factors obviously apply to the buffer salts present inside and outside the membrane, so that dialysis also offers a technique for buffer exchange.

10.2.1 Key process variables of dialysis

Sample-to-buffer ratio and number of changes

The choice of buffer volume and the number of changes is obviously critical in defining the maximum possible effiency of the process in removing solutes. As just discussed, very good theoretical purifications are achievable from small numbers of changes, so long as the relative buffer volume is sufficiently high. Dialysis against a running stream of buffer is an interesting variation on multiple changes, but is really only convenient for dialyses against tap water.

Agitation

The speed of dialysis is improved by providing sufficient agitation of both the sample and the external solution, for example by magnetic stirring. This prevents any build-up of concentration gradients in the vicinity of the dialysis membrane which would otherwise work against osmotic exchange.

Time

Dialysis is not a rapid technique. Even with a small sample and adequate agitation, equilibrium is unlikely to be reached in less than a few hours, and much longer may be required in less ideal circumstances. The rate of dialysis is itself affected by a number of parameters. (1) The concentration difference of each component across the membrane; in the absence of other factors the rate of transport is given by:

$$Rate = Ak(C_2 - C_1)$$

where A is the membrane area, k is a constant, and C_1 and C_2 are the concentrations on each side. (2) The size of the solute molecules which must diffuse out of the sample; the larger the molecule, the smaller its diffusion coefficient and hence the slower its progress. This effect is compounded by the greater probability of large molecules being obstructed by the membrane, even if they are smaller than the nominal cut-off. As a consequence of these effects, small components reach equilibrium faster than larger ones.

Temperature

Temperature (T) has a fairly significant effect on dialysis as the diffusion coefficient of a particular solute (D) is controlled (Hamill *et al.*, 1966) by the equation:

$$D = \frac{RT}{Nf}$$

where N is the Avogadro number, R is the gas constant, and f is the frictional coefficient of the solute. Therefore the diffusion coefficient of a solute increases with temperature.

[1] Dialysing a 10 ml sample against 10 litres of buffer reduces the level of impurities to 0.1% of the original value; with only two further changes of buffer, the residual impurity level will, in the absence of any effects such as binding to the protein, be one part per billion of the starting concentration.

However, because process times up to several days can be required, it is common for dialyses to be carried out under refrigeration. This minimises both microbiological growth and any instability of the protein in solution.

Membrane material

Early dialysis membranes were typically derived from animal sources, but synthetic membranes have predominated for many decades. A cellulose nitrate membrane termed Collodion was one of the first to achieve widespread use, but suffers from high protein adsorption.

Regenerated cellulose membranes (e.g. Cupraphan) were introduced and displayed less adsorption problems; cellulose membranes remain in common use to this day. Cellulose acetate membranes can provide better transport properties than cellulose, and are also popular; protein adsorption on this material is very low.

Membrane molecular mass cut-off

It is tempting to think of dialysis membranes as having a sharp cut-off of molecular mass (Figure 10.2(a)), all molecules above a certain mass being retained, and all those below being able to pass through the membrane without hindrance.

In practice, however, the permselectivity of dialysis membranes is nowhere near as simple as this. Sigmoid curves are obtained of the type shown in Figure 10.2(b), and this behaviour should be borne in mind when designing dialysis experiments. To achieve a high degree of separation between retained product and dialysed impurity, it is important that the molecular mass ratio is large, ideally 100:1 or greater.

The cut-offs quoted by manufacturers are based on a defined degree of retention after a given time (e.g. 90% after 17 hours; Saltonstall, 1992), so in the curve shown in Figure 10.2(b) a cut-off of 10,000 Da might be quoted. Note, however, that a significant proportion of components with molecular masses as low as a few hundred are still retained after this time.

Membrane pre-treatment

The cellulose membranes used for dialysis may well be contaminated with impurities from the manufacturing process, and soaking in multiple changes of water (or preferably the buffer to be used in the dialysis) beforehand is certainly recommended. For the determination of protein–ligand binding affinities by equilibrium dialysis, admittedly a more stringent application than many purifications, Klotz (1989) has recommended a series of heating and soaking stages, plus extraction of trace metals by acid treatment or chelators if this is important.

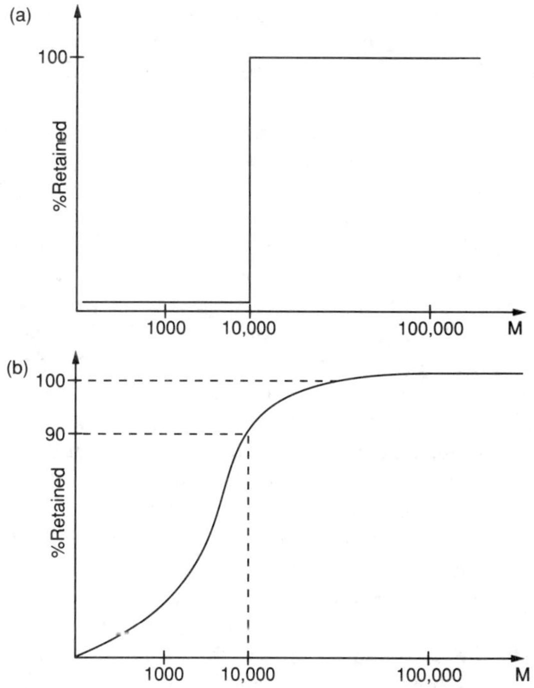

Figure 10.2 Dialysis membrane molecular mass cut-offs in (a) theory and (b) practice.

Despite the availability of low-adsorption materials, adsorption of protein to the membrane is also still perceived as a potential problem with dialysis. This is aggravated by the high surface area-to-volume ratios which may be encountered, and which in fact facilitate rapid diffusion. This is often countered by soaking the membrane in a solution of a protein before use, so that adsorption sites are blocked. Washing of the membrane after this treatment is obviously important too.

Buffer composition

A disadvantage of this technique for protein purification is illustrated by the use, just mentioned, of equilibrium dialysis to measure binding affinities. This technique—still recognised as the 'gold standard' for some diagnostic determinations, is based on the fact that solutes which exhibit binding to the protein will, at equilibrium, remain at a higher concentration within the membrane than would be expected from diffusion arguments alone. The higher the binding affinity, the less effective dialysis will be as a means of removing small solutes. It has frequently been reported that better separations have been achieved by other techniques such as size exclusion chromatography (Burke and Shakespear, 1975; Tesser *et al.*, 1975; Mattox *et al.*, 1979).

A special case of this type of effect arises if there are electrostatic attractions between the protein and the impurities to be dialysed out. This phenomenon (the Donnan effect) can be minimised by working at a reasonably high ionic strength (Klotz, 1989), and also by manipulating the charge of the protein. If the likely charge state of the impurity is known, a pH can be chosen (the 'right' side of the p*I*) which will give the protein the same charge and minimise any electrostatic attraction. If not, or if there are a number of possible effects, it is safest to work close to the p*I* so that at least the protein is not strongly charged in either direction.

The choice of buffer can also have dramatic effects on sample volume. Water molecules are capable of passing through the membrane just as well as solutes, and their distribution too will change towards equalising the concentrations on either side of the membrane. Therefore, if the sample is more concentrated in solutes than the surrounding buffer, this osmotic effect will result in dilution and an increase in sample volume.

The reverse obviously holds true, and concentrated dialysis buffers are quite often employed as a simple way of increasing the concentration of protein solutions. This effect can be exploited most neatly by using solutes which are themselves too large to pass through the membrane, so that there is no movement of solute into the sample. Polyethylene glycol (PEG) polymers are most frequently encountered in this application. Even more extreme is dialysis against a solid such as sucrose. Extremely rapid (if rather messy) volume reductions can be obtained at the expense of some sugar ending up in the sample.

Aside from the considerations just discussed, and the need to avoid highly basic buffers ($pH > 9$) for ester membranes such as cellulose acetate, virtually any buffer compatible with the stability of the protein can be employed for dialysis.

Sample presentation

The last major variable of the dialysis process is the means by which the sample is presented within the semi-permeable membrane, typically a cylindrical length of cellulose tubing. Obviously a good seal is required to ensure that leakage of the sample does not occur. Klotz (1989) recommends making two knots at each end of the dialysis tube, and tying a piece of fibreglass thread between each pair. Plastic clips are also commercially available, which can be used to seal the tubing with rather less physical manipulation.

The 'hands-on' nature of this aspect of dialysis makes the process rather more difficult to validate than, for example, chromatographic alternatives. As a consequence of this, and the convenience of chromatography for repetitive purifications, dialysis has lost favour as a manufacturing process.

The other main choice in sample presentation is how much tubing to use relative to the sample volume, a variable which defines the surface area-to-sample volume ratio among other things. Various factors may influence this decision.

A large surface area will increase the opportunity for diffusion of solutes, at least to the extent that the sample is in contact with the membrane, and speed up the dialysis process. To maximise contact, it is helpful to squeeze almost all the air out before the sample is sealed.

Where the volume of the sample increases during dialysis due to the osmotic effects discussed above, having some spare capacity available is advisable. Bursting of the membrane due to osmotic pressure is rare but not unheard of, but in any case it is not easy to remove the contents of a tautly filled dialysis bag.

These factors both militate in favour of using a large bag volume: as a guideline, double or triple that of the sample itself. But the largest area of tubing also provides the maximum opportunity for loss of protein due to adsorption, and for small or precious samples this may be the overriding concern. In these cases speed of dialysis may be sacrificed to minimise the membrane area. One technique is to place the sample in a small non-adsorbing plastic vial, usually with a conical base to facilitate complete recovery, and to fasten a single sheet of membrane carefully over the aperture of the vial using an elastic band.

Perhaps more securely, commercial devices are available for dialysing small-volume samples in a rapid and convenient fashion, the Microdialyzer™ range from Pierce, for example, which allow multiple dialyses of volumes down to 20 μl to be carried out in less than an hour. Membranes for use with these devices are available in a range of molecular mass cut-offs from 1 to 50 kDa. Pierce also supply disposable cassettes, under the Slide-A-Lyzer™ tradename, for dialysis of 0.5–3 ml samples. The molecular mass cut-off is 10 kDa.

10.2.2 Applications of dialysis

Protein chemistry applications of dialysis are typified by the separation of the entire protein product from low-molecular-mass species. These include some of the most commonly encountered purifications in protein conjugation. (1) separation of protein–small molecule conjugates, both from unreacted ligand and from any low-molecular-mass coupling agents, cosolvents, etc.; (2) separation of derivatised proteins from unreacted derivatising agents, cosolvents, etc.; (3) exchange of any protein reagent from one buffer matrix to another.

Examples of all these categories are legion. However they have all declined to some extent in the face of competition from size exclusion chromatography, particularly in manufacturing applications.

The use of dialysis for concentration of protein solutions is a fourth, slightly separate application, which again has suffered from competition with more modern techniques such as ultrafiltration.

10.2.3 Ultrafiltration

Probably the biggest disadvantage of dialysis is the time required for the process to take place, constrained as it is by the kinetics of molecular diffusion. Ultrafiltration can be viewed as a straightforward adaptation of dialysis, in which the semi-permeable properties of a membrane are still employed to effect the separation between large and small molecules, but in which pressure is employed to force the sample through the membrane. As a result, roughly similar separations can be obtained in a fraction of the time required for dialysis.

This process can be used just to concentrate a protein solution, in which case the sample is simply forced through the membrane until the concentration in the residual fraction is sufficiently high. If the desire is to emulate a dialysis purification, however, it is necessary to top up the sample with a suitable buffer a number of times, so that the original matrix is continuously replaced. Because a reasonable volume of liquid should always be maintained in the sample cell, this top-up operation may need to be carried out several times to achieve a good separation from the components of the original matrix. However, as the duration of each of these steps can be measured in minutes or even seconds, the overall process is still very fast.

An ultrafiltration apparatus consists simply of a pressurisable cell for the sample (generally also allowing some form of stirring), a membrane through which the sample is forced, and a collection vessel. For small-scale applications the pressure is often supplied by connection to a source of compressed gas, such as a nitrogen or compressed air cylinder.

Especially for larger-scale applications, a build-up of back-pressure can be problematical, particularly if the sample is not totally free from particulates, etc. A pre-filtration to remove this type of contamination can go a long way towards avoiding the issue.

10.3 Ultracentrifugation Using Density Radients

Centrifugation is one of the commonest laboratory techniques for separating solids, colloids, and biological materials such as cells from liquid suspensions by exploiting density differences.

In ultracentrifugation, the same principle is extended to the molecular level, where separations between proteins of different molecular mass are achieved by centrifugation for long times and/or at very high speeds. In order to amplify the fractionation effect, the approach was developed of establishing a density gradient in the centrifugation tube, and carrying out the centrifugation of the sample into this matrix. These gradients could be set up, for example, by mixing high and low-concentration sucrose solutions at a ratio which was gradually changed from the bottom to the top of the tube.

Once the centrifugation has been completed, a hole can be made in the bottom of the tube, and the contents eluted out and collected as fractions. Ultracentrifugation of proteins in a sucrose gradient was introduced by Martin and Ames (1961), and the fractionation of lysozyme, alcohol dehydrogenase and

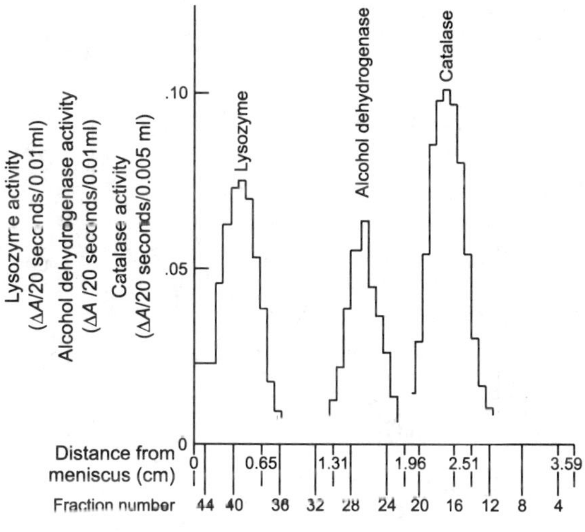

Figure 10.3 Fractionation and analysis (continuous line) of lysozyme, catalase and yeast alcohol dehydrogenase on sucrose gradient. (Reproduced with permission from: Martin and Ames (1961). *J. Biol. Chem.*, **236**, 1372–1379.)

catalase on a 5–20% gradient is demonstrated by the analysis of the fractions collected (Figure 10.3).

Ultracentrifugation on density gradients is predominantly an analytical technique, and is therefore covered in greater detail in Chapter 11, Section 11.2. Before the development of more modern techniques, in particular size exclusion chromatography and gel electrophoresis, ultracentrifugation was the principal method for the estimation of protein molecular masses *via* their sedimentation behaviour. A photograph of the centrifuged sample against suitable lighting allowed component bands to be visualised.

In view of the care required to carry out these gradient ultracentrifugations reproducibly, it is not surprising that the newer chromatographic and electrophoretic techniques have almost totally replaced them for molecular mass determination.

This approach can also be applied preparatively. Before the widespread adoption of chromatographic techniques, it was not uncommon to find reports of ultracentrifugation for the fractionation of protein conjugates (Jolley *et al*., 1967; Sternberger, 1967; Van Weemen and Schuurs, 1971). The elution profile of a conjugation between human chorionic gonadotrophin (HCG) and HRP, using a 20–60% sucrose gradient, is shown in Figure 10.4.

The practicality of carrying out preparative density gradient ultracentrifugations was improved by developments in rotor design, the zonal rotor allowing gradients to be established while the rotor was in motion, and large sample volumes to be handled. Shepherd (1980) reported that human serum lipoproteins could be fractionated from samples as large as 150 ml using this approach.

On the whole there is little to recommend ultracentrifugation as an alternative to preparative chromatographic techniques, and it is no longer encountered routinely for the purification of protein reagents.

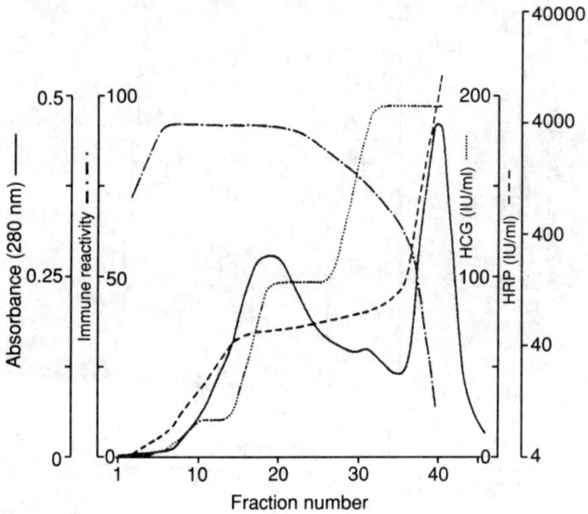

Figure 10.4 Fractionation of HRP–HCG conjugate by density gradient centrifugation. (Reproduced with permission from: Van Weemen and Schuurs (1971). *FEBS Lett.*, **15**, 232–236.)

10.4 Liquid–Liquid Partitioning

Aqueous solutions containing an appropriate mixture of two polymers (e.g. dextran and methylcellulose) will form a two-phase system (Figure 10.5). The separation between the components is not complete, but each polymer will be concentrated to some extent in one of the phases.

The basic principle of liquid–liquid partitioning as a protein purification technique (Albertsson, 1960) is very simple. If a protein sample is added to the system and the molecules spread equally between the two phases, in other words the partition coefficient is unity, then nothing has been gained. However, if a system can be developed in which the protein is concentrated in one phase or the other, there is the basis for a purification method.

Albertsson's original description described many different two-phase polymer systems, the parameters controlling the phase partition, and the behaviour of cells, viruses and proteins. Although in many cases the partition coefficient was close to unity, optimisation of pH and ionic strength could quite often provide a concentration of a target component. Albertsson also described countercurrent approaches where sequential separations provided greater separations.

Liquid–liquid partitioning is very closely related to the much commoner liquid–solid partitioning approaches (chromatography) which will occupy much of this chapter. Albertsson's work was carried out at around the same time as the blossoming of understanding in chromatographic techniques such as size

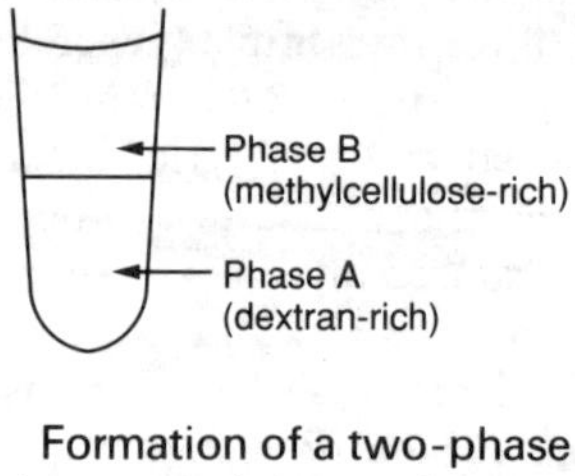

Figure 10.5 Formation of a two-phase system of aqueous solutions containing a mixture of polymers.

exclusion and ion exchange, while electrophoresis was also entering the mainstream of protein purification.

Liquid–liquid partitioning never achieved the same success as these other technologies, a remarkable number of which originated, like this one, from the laboratory of Arne Tiselius at the University of Uppsala in Sweden. Nevertheless it offers some potential for protein purification: it is particularly suited to bulk-scale operations, and can provide economic processes once the effort to optimise the system has been put in.

10.4.1 Simple liquid–liquid partitioning

In its simplest form liquid–liquid partitioning still resembles the basic system shown in Figure 10.5. Simple combinations of polyethylene glycol (PEG) and dextrans can provide useful separations, optimisation of the molecular masses of these species as well as their relative and overall concentrations being required.

Kirchberger and Kopperschläger (1990), for example, described a crude preparation method for alkaline phosphatase using a PEG–4000/dextran M70 system, which concentrated 88% of the enzyme in the top phase.

10.4.2 Affinity partitioning

Greater power can be obtained from partitioning by using an affinity approach, where one of the polymers is derivatised with a species capable of binding to the target protein (Johansson, 1984). Albertsson (1973) observed that the addition of a short-chain ethylene oxide-based detergent to a PEG–dextran system caused a hydrophobic protein to concentrate in the PEG layer. This observation was rationalised on the basis that the hydrophobic portion of the detergent molecule bound to the protein, and the ethylene oxide portion associated predominantly with the structurally similar PEG phase. The overall effect, therefore, was to concentrate the protein in this layer.

Flanagan and Barondes (1975) took this principle further: with an antibody specific for dinitrophenyl (DNP) groups as the target, they prepared a DNP-derivatised PEG which was incorporated in a PEG–dextran system. This caused a large shift in the antibody distribution towards the PEG phase where its ligand was localised.

There have been many other applications of this affinity partitioning approach: for example Kirchberger and Kopperschläger (1990) partitioned the crude alkaline phosphatase described above, using PEG conjugated to a dye with a high affinity for the enzyme. Anderson *et al.* (1991) used a PEG–phosphate system to separate PEG–derivatised topoisomerase–DNA complexes from unbound DNA. The complexes favour the PEG–rich phase, while the DNA has a strong affinity for phosphate. Hence both components are subject to affinity concentration in opposite phases.

10.5 Size Exclusion Chromatography

During electrophoresis studies in the 1950s, various observations were made regarding the behaviour of macromolecules migrating through solid or gelatinous materials. In some systems, e.g. polysaccharides migrating through collodion membranes, larger molecules were retarded relative to smaller ones (Mould and Synge, 1954). However, with other systems such as proteins in starch columns the opposite effect was seen, with smaller molecules being retained longer (Flodin and Porath, 1954).

Flodin and Porath's work simply identified the starch column as a promising matrix for electrophoresis. In 1955, however, Lathe and Ruthven in London carried out chromatographic separations on starch columns (i.e. in the absence of any electrophoretic effect), and observed a clear separation of carbohydrates on the basis of molecular mass, over a range of 180 Da to 160 kDa. Lathe and Ruthven proposed this as a new form of partition system in which the volume available to each solute was determined by its ability to penetrate the starch granules.

These findings were echoed by Porath and Flodin at Uppsala, using cross-linked dextran gels developed for electrophoresis which they named Sephadex™. These proved to exhibit the behaviour observed by Lathe and Ruthven, while having a more defined composition and displaying better stability and flow properties. In 1959 Flodin and Porath published a landmark paper on the use of Sephadex for the new technique they termed gel filtration, whereby large molecules such as proteins could be separated from smaller species due to their different retention properties on a Sephadex column (Porath and Flodin, 1959). They described, for example, the quantitative separation of ammonium sulfate from serum proteins, and it is indicative of the utility of this technique that essentially identical separations are routinely carried out, worldwide, a third of a century after their original description.

This technique has attracted a particularly large number of synonyms since its introduction, and the terms gel filtration, size exclusion chromatography (SEC), molecular sieving, and gel permeation

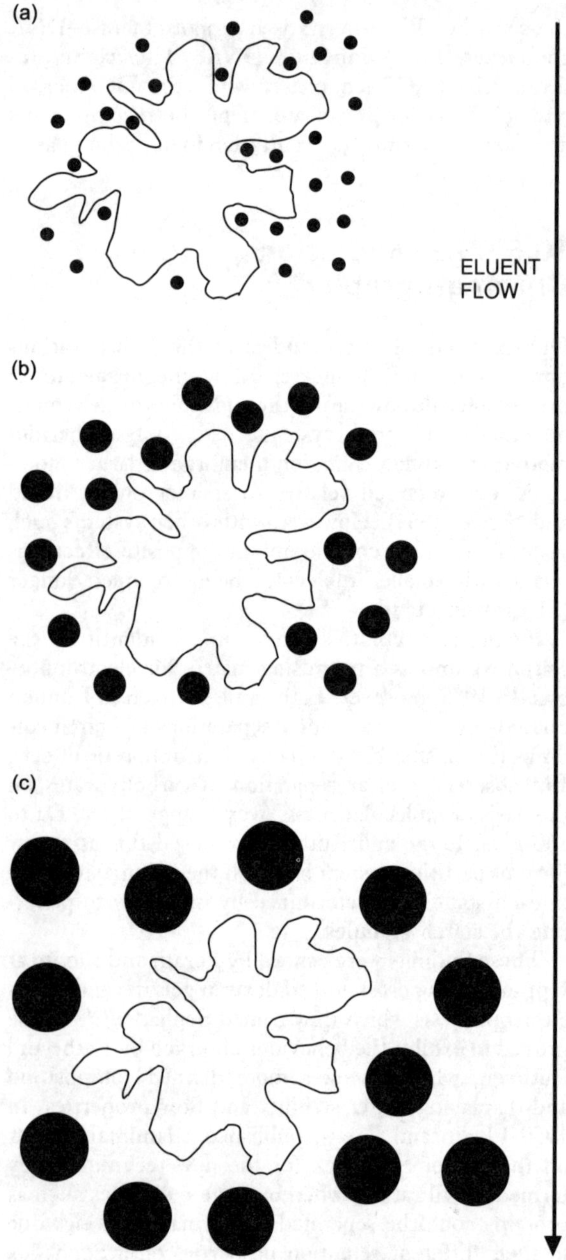

Figure 10.6 The mechanism of SEC: small solutes can diffuse readily in and out of the porous matrix (a); the ability of larger molecules to penetrate the matrix is lower (b), while those above a certain limit are excluded (c). The diffusional path of small molecules through the column is thus longer, and this is reflected in an increased retention time.

chromatography (GPC) will all be encountered. The last of these terms is generally applied to synthetic polymers rather than to proteins, although the principle is identical for both uses.

The term size exclusion is preferred here, as it more accurately reflects the mechanism of separation. It seems counter-intuitive to equate filtration or sieving with a procedure where the large components are allowed through while the small ones are retained. The basic principle of SEC is shown in Figure 10.6.

SEC provided an attractive alternative to dialysis for desalting and buffer-exchange of protein solutions, and was rapidly taken up by the scientific community, Rinderknecht (1962), for example, reported a 'spectacular separation' of fluorophore-labelled proteins from unreacted dye using Sephadex G-25, while Hunter and Ludwig (1962) used the same matrix to desalt proteins derivatised with imidates.

Andrews (1962) suggested that SEC could be used to estimate the molecular mass of globular proteins. Correlations between molecular mass and retention properties were observed by other workers (e.g. Wieland *et al.*, 1963), and Andrews (1965) published a detailed analysis of protein behaviour on Sephadex G-200. Over a molecular mass range of 3.5–820 kDa, most proteins fitted a smooth curve of elution volume *versus* log of molecular mass (Figure 10.7). The main exceptions were certain glycoproteins such as ovomucoid, fibrinogen and γ-globulins, which were retained less efficiently—an effect attributed to the extensive hydration of the carbohydrate moieties, resulting in a more expanded overall structure. The molecular mass of such materials would therefore be overestimated from the 'normal' model. Whitaker (1963) observed other examples of non-ideal behaviour; in the case of lysozyme he suggested that the dextran gel sufficiently resembled the carbohydrate substrate of the enzyme for formation of a weak complex to cause retardation, while haemoglobin was assumed to associate into subunits during the size exclusion process.

These exceptions notwithstanding, SEC offers a very simple means of purifying protein reagents on the basis of their molecular size, and this technique has remained one of the cornerstones of protein purification ever since its introduction.

Low-resolution SEC is a straightforward method for the separation of bulk protein from small contaminants. It therefore offers an attractive alternative for the applications described above for dialysis, which is often very time-consuming by comparison. Perhaps more importantly, there are frequent reported instances where SEC has been able to provide a more complete removal of small molecules than dialysis

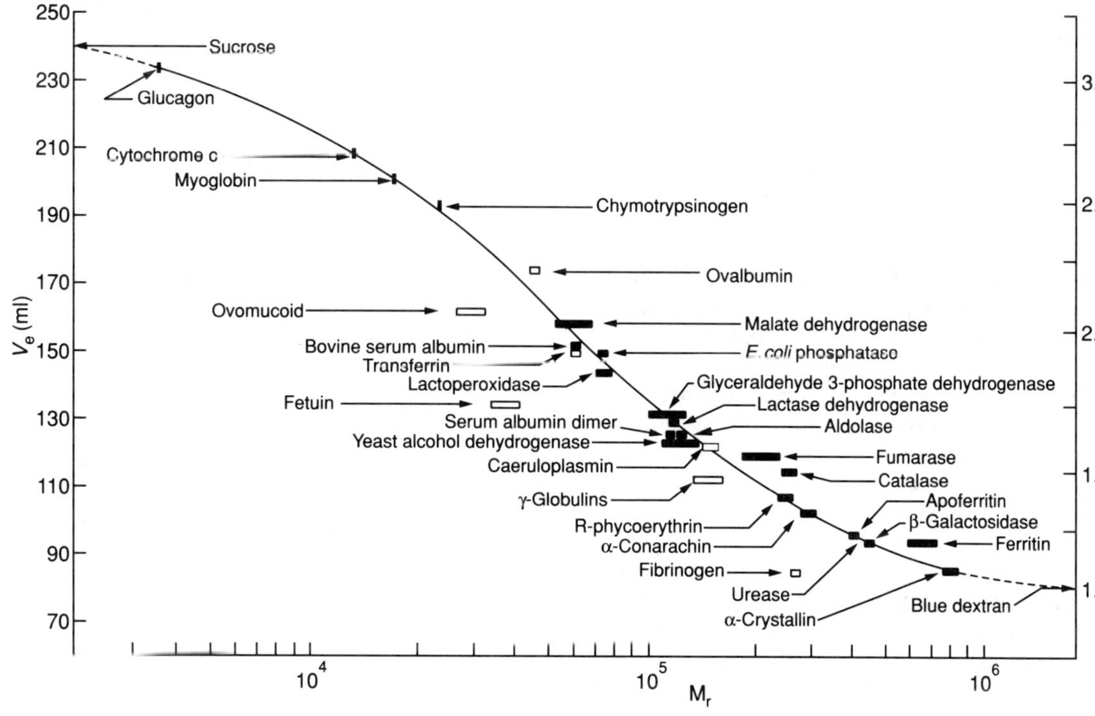

Figure 10.7 Plot of elution volume (V_e) versus log (M_r) for proteins on a Sephadex G-200 column. (Reproduced with permission from: Andrews (1965). *Biochem. J.*, **96**, 595–606.)

(Burke and Shakespear, 1975; Tesser *et al.*, 1975; Mattox *et al.*, 1979).

This effect is particularly marked where there is a degree of non-covalent binding of the small molecule to the protein (e.g. thyronine hormones with albumin; Burke and Shakespear, 1975). Presumably during dialysis, any ligand dissociating from the protein is more likely to reassociate with the binding site than to diffuse away. In SEC the ligand may enter pores in the gel before this reassociation occurs, hence the more complete separation. The capacity of some matrices such as Sephadex to adsorb aromatic and heterocyclic species may also be a factor in separations of this type (Reiland, 1971). Where the association between protein and ligand is particularly strong, material can remain non-covalently bound throughout SEC too, and more rigorous adsorptive techniques may be required to achieve the desired separation.

Because the separation of protein and small molecule begins to occur quite rapidly after the sample is applied to the column, this step is often used to stop reactions in much the same way as a chemical quenching agent (e.g. Brown and Matthews, 1979). For low resolution applications rigid, highly cross-linked gels such as the original Sephadex range and their analogues remain applicable.

High resolution applications of SEC are generally aimed at the fractionation of protein species, a much more demanding challenge. Soft gels such as the higher Sephadex media offer some limited resolution, but the particle size is fairly large and their mechanical properties are unsuitable for high flow-rates. The development of more robust polymeric media has offered improvements in this area, and acceptable separations can be achieved in many applications using these materials.

However the greatest resolution is achieved by the use of very small particles, a few μm only in diameter. To allow flow-rate to be maintained these must be highly robust, ruling out many potential matrices.

Particles of glass and silica offer excellent mechanical properties, but their protein-adsorption tendencies are problematic. Attention focussed therefore on modifying the surface of these materials to reduce their adsorption properties. Coatings of controlled-pore glass with hydrophilic materials such as PEG (Mizutani and Mizutani, 1975) or glycerol derivatives such

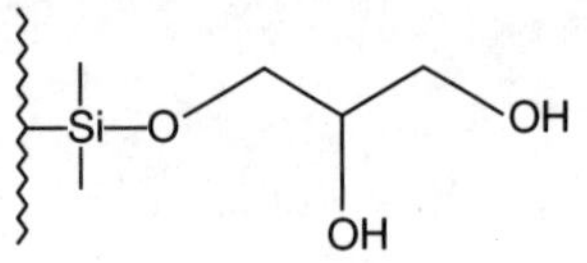

Figure 10.8 Glycerolpropylsilane (GPS)-coated surface.

as glycerolpropylsilane (GPS, Figure 10.8; Crone and Dawson, 1976; Regnier and Noel, 1976) demonstrated good performance. GPS matrices were quickly commercialised by many suppliers, for example as the Glycophase-G range from Corning (Fischer *et al.*, 1978, LiChrosorb Diol from Merck (Roumeliotis and Unger, 1979, 1981), the TSK-Gel SW and SW$_{XL}$ ranges from Toyo Soda (Kato *et al.*, 1980, 1987), and Protein-Pak from Waters (Meyerson and Abraham, 1986).

These high performance SEC media rapidly achieved widespread popularity (Unger, 1984) and this has been maintained to the present day. Optimisation of performance still continues, for example with the introduction of polymeric media such as the TSK Gel PW and Toyopearl series from Toyo Soda. These offer greater chemical robustness than silica-based matrices.

These media are best suited to analytical or small-scale preparative applications. Large columns can be prohibitively expensive, so obtaining high resolution at a large scale remains a serious challenge.

10.5.1 The mechanism of size exclusion chromatography

As the correlation between molecular size and retention on gels became clear, various theories were put forward to explain this behaviour. These were based on decreased diffusion of macromolecules within the gels (Ackers and Steere, 1962), or on exclusion of macromolecules from a fraction of the gel volume (Lathe and Ruthven, 1956; Pedersen, 1962; Porath, 1963).

Later investigations demonstrated that the observed behaviour could be explained entirely on the basis of exclusion (Van Kreveld and Van den Hoed, 1973), and these models have come to predominate. Theories based on this concept have used various models for gel structure such as a random network of polymer fibres (Laurent and Killander, 1964), or a combination of cones, cylinders and crevices (Squire, 1964). Rather more satisfactorily Ackers (1967) employed a simple probability distribution for penetrable volume, which allowed data from a variety of sources to be fitted to the basic model without attempting to describe the gel structure.

These concepts led on to the simple empirical approaches which are commonly encountered, as described for example by Reiland (1971). The total volume of a SEC column (V_t) is considered as being made up of three parts: (1) the volume of the solid polymer matrix (V_p); (2) the volume of liquid within the pores of the matrix (V_i); (3) the volume of liquid between the matrix particles (V_o).

In other words:

$$V_t = V_p + V_i + V_o$$

For a given solute, the elution volume (V_e) is defined by the void volume V_o and the fraction (K_d) of V_i which is available to it, a function of its molecular size and shape as discussed earlier. Thus:

$$V_e = V_o + K_d V_i$$

or:

$$K_d = \frac{V_e - V_o}{V_i}$$

Because the pore volume V_i is not obtained very conveniently, a revised expression has been defined where V_i is replaced by $V_i + V_p$:

$$K_{av} = \frac{V_e - V_o}{V_i + V_p} = \frac{V_e - V_o}{V_t - V_o}$$

The value of K_{av} (a term first used by Laurent and Killander (1964)) can be defined for a given solute using only the volume of the polymer matrix, the elution volume of the solute itself and the void volume (i.e. the elution volume of a solute large enough to be totally excluded from the pores).

While K_d is a genuine partition coefficient describing the distribution of solutes between the liquid within the pores and that outside, K_{av} is distorted by including the solid polymer volume in the 'within-pore' term.

For totally excluded large solutes K_{av} is zero, while for small molecules which can penetrate virtually the entire pore volume, K_{av} is approximately one[1].

[1] Values of K_{av} greater than one indicate that adsorption of the solute to the matrix is taking place, while negative values suggest a physical problem with the column packing.

K_{av} is the parameter most commonly plotted against the log of solute molecular mass in the construction of selectivity curves, as it is independent of column dimensions, flow rate, etc.

10.5.2 Key process variables of size exclusion chromatography

Eluent composition

Among chromatographic techniques, SEC is unusual in that the solutes do not need to be adsorbed onto, or desorbed from, the matrix surface. As a consequence it is probably the least affected by the composition of the eluent.

Many SEC media contain ionic groups, typically ionised carboxyl or silanol functions, which can induce cation-exchange effects. To overcome these, eluents containing at least 0.02–0.2 M of salts are generally employed, so that pure size exclusion behaviour is encouraged. The necessary ionic strength varies from one matrix to another, depending on the quantity of ionic groups present: lower limits of 0.02 and 0.15 M respectively are recommended for Sepharose and Superdex, for example (Pharmacia, 1993d).

These ionic effects have, in fact, occasionally been deliberately exacerbated by pH variation at low ionic strength (Kopaciewicz and Regnier, 1982), or by incorporating zwitterionic compounds (Ovalle, 1995). The result, termed non-ideal size-exclusion chromatography, is a mixed-mode separation based on a combination of molecular size and charge.

On the whole pH has very little effect on SEC separations, and, within the chemical stability constraints of the matrix and solutes, is determined largely for the convenience of the user. Often a pH will be chosen which reflects the subsequent use of the protein, either for storage or for further processing. In fact SEC offers one of the most convenient ways of changing a protein solution from one buffer matrix to another: buffer exchange on a SEC column is frequently employed even when no actual purification of the protein is required.

For the vast majority of SEC applications, roughly neutral buffers containing at least 0.1 M of inorganic salts are wholly satisfactory. The mild nature of such eluents make SEC extremely attractive for retention of protein structure and function, a characteristic probably heightened by the non-adsorptive nature of the technique.

As with other chromatographic techniques, the eluents for SEC may include specific additives to enhance product stability—for example magnesium and zinc salts for alkaline phosphatase conjugates (Nakagami *et al.*, 1991).

Sample volume

Again because adsorption of solutes onto the matrix is not involved, the sample volume is a rather important process variable for SEC, especially when compared to other techniques such as ion exchange or affinity chromatography.

For high-resolution separations, where the difference in K_{av} may be less than 0.1, it may be necessary to keep the sample size as low as 0.5% of the total column volume. Obviously this can necessitate the use of some very large columns if sample sizes are high, and represents a serious drawback of SEC for large scale work.

Many applications, however, are nowhere near as sensitive to sample volume. For buffer exchange, or for desalting where a protein is to be separated simply from low-molecular-mass species, a matrix is generally chosen where ΔK_{av} approaches 1, i.e. where the protein is completely excluded and the other species retarded to the maximum possible extent. In such cases samples as large as 40% of total column volume may give perfectly adequate separations; as a general guideline, using 10–20% generally gives satisfactory results.

Therefore these low-resolution applications may be totally practicable at a large scale, while high-resolution work is best limited to analytical and small-scale preparative uses.

It should be noted that SEC, unlike some other techniques, is bound to result in dilution of the sample. The dilution factor may be anything from around 20% upwards, depending on the column dimensions, etc.

Sample concentration and viscosity

Sample concentration is unimportant except for its influence on viscosity, otherwise being limited only by solubility constraints.

If the viscosity of the sample is very different from that of the eluent system, the solute zones will be hydrodynamically unstable and dramatic losses of resolution may be observed (Figure 10.9). In practice (Pharmacia, 1993d) problems may be encountered if the viscosity of the sample is more than twice that of the eluent. For typical proteins in a dilute aqueous buffer system this corresponds to a concentration of about 70 mg/ml.

Flow rate

K_{av} has generally been demonstrated to be independent of flow rate (Van Kreveld and Van den Hoed, 1973; Sofer and Nyström, 1989a).

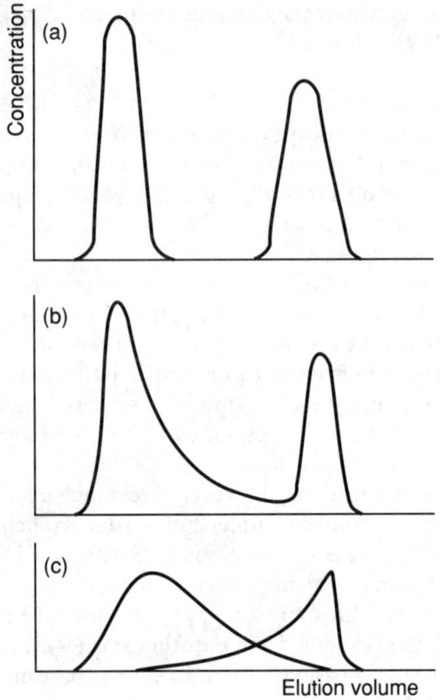

Figure 10.9 Elution diagrams obtained when haemoglobin and NaCl were separated under conditions of increasing viscosity. (Reproduced with permission from Amersham-Pharmacia Biotech)

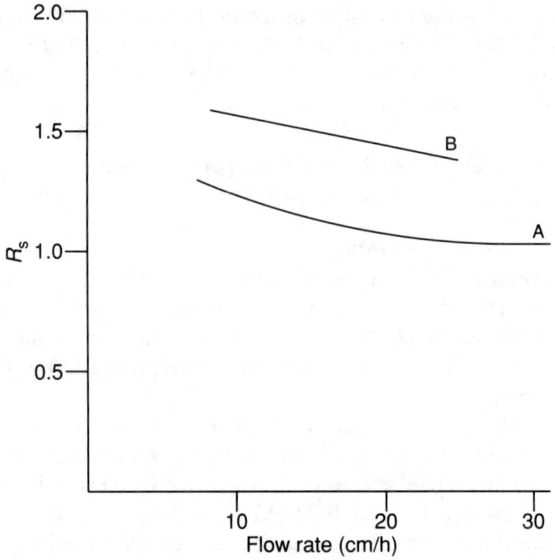

Figure 10.10 Resolution (R_s) as a function of flow rate in columns of 90 cm (A) and 180 cm (B) height. Medium, Sephacryl S200 HR; sample, human albumin and IgG; sample temperature, 4 °C. (Reproduced with permission from: Sofer and Nyström (1989). *Process Chromatography. A Practical Guide*, Academic Press, New York.)

Resolution, however, is affected by flow rate and there is an optimal flow rate range which will depend on both the matrix and the sample (Figure 10.10). For large molecules it is generally accepted that a lower flow rate is better for high resolution (Pharmacia, 1993d).

In general linear flow rates of the order of 5–100 cm/h are appropriate for SEC work, the optimum for resolution being at the lower limit of this range, and higher rates being acceptable where sub-optimal resolution is acceptable.

When scaling up or scaling down SEC separations the linear flow rate, column length and sample concentration should ideally be maintained to obtain comparable performance.

Some gels, particularly the earliest developed, are soft in nature and will tend to pack down in use if high flow rates are employed[1]. Back-pressures also tend to be high with these gels, and may limit the flow rates that can be chosen. The development of more mechanically stable gels has helped to alleviate these limitations. Agarose beads allowed higher flow rates than earlier dextran or polyacrylamide gels, while maintaining good selectivity properties (Locascio *et al.*, 1969). More modern matrices such as Sephacryl™ from Pharmacia are rigid enough to allow very high flow rates without prohibitive back-pressures or deformation problems.

Pore size

Given the mechanism of SEC, it is not surprising that pore size is a critical variable. A diagram of the useful range of pores sizes for solutes of various molecular weights is shown in Figure 10.13. For a 10 kDa protein this corresponds to around 60–100 Å, while at 1000 kDa molecular mass the useful range is roughly 100–3000 Å. In practice, the choice of pore size is normally made indirectly, *via* the fractionation limits quoted by the manufacturers for individual matrices.

Particle size

As noted earlier, the smaller the particle size the larger the surface area within a given volume, and the more efficient the separation – at the expense of lower flow

[1] This effect is minimised in practice by packing the column at a higher pressure than it will experience in routine use.

rates and/or higher back-pressures. Many SEC media are supplied in a range of particle sizes, allowing a trade-off to be made between resolution and throughput. The coarser media (particle size 100 μm or greater) are obviously the most practical for low-resolution applications such as desalting, while at the opposite extreme columns with very high resolving power (e.g. the TSK™ columns) may employ particles of less than 10 μm diameter.

Column dimensions

Broadly, increasing the column length leads to improved resolution: the resolution increases with the square-root of the length, so that doubling the column length may yield around a 40% increase in resolution. The advantages of this improved resolution may however be offset by greater problems in column packing, or back-pressure in operation.

Choosing the column length may also be influenced by achieving the best throughput: a longer column may allow the flow rate to be increased while retaining adequate resolution. On the other hand, where high resolution is *not* required a short, wide column may be the best option, as very high flow rates may be achievable.

With a column length which provides the desired separation, the column diameter can be chosen to keep the sample size as an acceptably small fraction of the overall volume (see above).

Finally, the column dimensions may be affected by the nature of the matrix itself. With soft gels wider columns may be inappropriate, as they will not support such high operating pressures as a narrower bore.

Column packing

One of the most important variables for SEC, and one of the hardest to define in practical terms, is the quality of the column packing. A perfectly-packed column will be completely homogeneous; any imperfections may lead to zone broadening and a loss of resolution. Carefully following manufacturers' instructions on column packing (e.g. Pharmacia, 1993d) will minimise such problems, but some variation in performance can still be expected.

The most secure approach is to measure the efficiency of the column after packing, for example as plates/metre, ensuring that it is sufficient to meet the needs of the separation. Where it is necessary to carry out packings repetitively to obtain consistent results, e.g. as part of a manufacturing process, it may be useful to prepare deliberately badly-packed columns. These can be used to define a baseline efficiency which provides an acceptable separation, and this value can then form an acceptance criterion for individual packed columns.

Buying prepacked columns from a manufacturer may be advantageous, as they may well have been packed under more rigorously controlled conditions. Suppliers will often quote a minimum efficiency specification which can be useful when the column is to be replaced.

10.5.3 Commercially available media for size exclusion chromatography

Since the pioneering studies with Sephadex, many SEC media have been developed by Pharmacia and other manufacturers which offer a variety of performance characteristics. The most important features of a SEC matrix include the following. (1) The molecular mass range over which fractionation can be achieved, depicted most conveniently by the selectivity curve of K_{av} *versus* log of molecular mass. (2) The physicochemical stability of the matrix, which defines the conditions that can be tolerated by the column during separations and other operations such as cleaning, sanitization and storage. For health-related applications, for example, sanitization is often an essential procedure, and the ability to tolerate treatment with sodium hydroxide is a valuable attribute. (3) The linear flow rate range that can be tolerated without deformation or prohibitive back-pressures. (4) Any properties, hydrophobic, ionic or other interactions, which may promote chromatographic mechanisms other than 'pure' size exclusion. For example, certain categories of compound have been observed to show adsorption to Sephadex (Reiland, 1971), including aromatic and heterocyclic species and cyclodextrins. Reiland also observes that borates can form complexes with the matrix, altering the chromatographic properties.

SEC can be carried out on any porous support, and a variety of matrices have been incorporated in commercial products. The original Sephadex matrices consist of linear dextrans cross-linked with epichlorohydrin to form spherical beads. Lower 'G-numbers' denote higher levels of cross-linking, which are reflected by a lower exclusion limit and a greater rigidity. Various particle sizes are available too: the larger facilitate higher flow rates at the expense of some resolution.

Agarose spontaneously forms gels as it cools from a hot solution, the lower the percentage of agarose in the gel, the larger the pore size and the higher the fractionation cut-off. Pore sizes can be very large, hence these gels offer fractionation up to molecular masses of tens of millions. Agarose gels are held together by hydrogen bonds, and are rather less robust than most other media; they melt above 40 °C, and are unstable outside the pH range 4–9 and in the presence of oxidising and chaotropic agents. Stability can be improved by cross-linking. Examples of agarose gels, cross-linked and otherwise, include Pharmacia's Sepharose and Superose series, and the Bio-Gel A range from Bio-Rad. Attractive properties can be achieved by combining dextran and agarose, for example the Superdex range from Pharmacia.

Other SEC media are based on synthetic polymers; these include the polyacrylamide Bio-Gel P range from Bio-Rad, and the Sephacryl series from Pharmacia, an acrylamide–dextran copolymer.

As noted earlier in this section, the development of high peformance SEC methods has required the introduction of microparticulate silica or polymeric media, of which there are many representatives in this list.

One final note of caution: prepacked columns such as the PD-10 from Pharmacia are very convenient for small-scale research and development applications, samples of up to 2.5 ml being desalted with dilution factors as low as 1.4 (Pharmacia, 1993d). Because they are so easily used and there is no chromatogram, however, an element of blind faith tends to take hold, such that baseline separation is assumed to occur in any desalting application. In fact there is very little leeway for error, and if it is important to achieve a complete removal of low-molecular-mass species, a wise precaution is to collect and analyse fractions of the product to ensure that a satisfactory separation is obtained (e.g. Annunziato *et al.*, 1993).

Some of the more commonly encountered size exclusion media available from commercial suppliers are listed in Table 10.2.

10.6 Ion Exchange Chromatography

Ion exchange chromatography (IEC) relies on the charge properties of proteins and protein reagents as the basis for their fractionation. It therefore offers an excellent complementary approach to size-based SEC, and is even more commonly encountered in the protein chemistry field (Bonnerjea *et al.*, 1986).

10.6.1 The mechanism of ion exchange chromatography

Simply speaking, IEC utilises a charged solid phase. Sample solutes are eluted or retained depending on their electrostatic properties, relative to those of the solid matrix: a protein with an abundance of negatively charged groups will tend to be retained strongly on a positively charged (anion exchange) matrix, but poorly on a negatively charged (cation exchange) column. The principles of IEC are shown in Figure 10.11.

The concept of 'exchange' arises because the charged groups on the solid matrix will be associated, *ab initio*, with oppositely charged counter-ions from the surrounding aqueous phase, and these must be displaced by the protein species for retention to occur. Similarly the protein must be replaced by other charged solutes when it is eluted.

In SEC there is no actual adsorption of solutes onto the solid phase, the different species are simply slowed down more or less during their passage through the column because of the varying volumes accessible to them. As a consequence, as noted in the previous section, SEC always results in the dilution of the sample. In IEC, on the other hand, it is quite feasible to choose conditions which result in strong adsorption of solutes, which can then only be removed by a change in the eluent characteristics. Under such conditions, protein can continue being added to, and retained by, the column, until the ion exchange capacity of the matrix is used up. As a result, ion exchange can be used as a concentrating technique, and the purified product can be obtained in a volume much *smaller* than the sample. Particularly for large-scale applications, this is a very great advantage compared to SEC.

The phenomenon of ion exchange by solid matrices was probably first reported by Thompson (1850), who noted the differing capacity of various soil types to adsorb and release inorganic compounds. Most of a century later, synthetic ion exchange resins were developed, such as the Amberlite™ series from Rohm and Haas; again these were employed primarily for inorganic applications. By the 1940s these were being examined for resolution of protein mixtures (McColloch and Kertesz, 1945; Sober *et al.*, 1949) and by 1952 a reasonably-well characterised ion exchange purification of cytochrome *c* had been reported (Margoliash, 1952).

The development of ion exchange matrices for protein chromatography mirrored that of SEC media. Soft gels based on cellulose, dextran, agarose and polyacrylamide were introduced by the mid-1950s, and many of these have remained in use up to the present day. High performance ion exchange media based on small, rigid particles appeared in the 1970s (Chang *et al.*, 1976) as HPLC became a widely accepted vehicle for protein purification.

Table 10.2 Commercially available gel filtration matrices

Supplier	Matrix	Fraction range (kDa)	Bead size (μm)
Pharmacia			
Sephadex G10	Cross-linked dextran (epichlorohydrin)	0–0.7	40–120
Sephadex G15		1–1.5	40–120
Sephadex G25 C		1–5	100–300
Sephadex G25 M		1–5	50–150
Sephadex G25 F		1–5	20–80
Sephadex G25 SF		1–5	10–40
Sephadex G50 C		1.5–30	100–300
Sephadex G50 M		1.5–30	50–150
Sephadex G50 F		1.5–30	20–80
Sephadex G50 SF		1.5–30	10–40
Sephadex G75		3–80	40–120
Sephadex G75 SF		3–70	10–40
Sephadex G100		4–150	40–120
Sephadex G100 SF		4–100	10–40
Sephadex G150		5–300	40–120
Sephadex G150 SF		5–150	10–40
Sephadex G200		5–600	40–120
Sephadex G200 SF		5–250	10–40
Sepharose 6B	Agarose	50–4000	40–165
Sepharose 4B		60–20,000	40–165
Sepharose 2B		70–40,000	60–200
Sepharose CL 6B	Agarose cross-linked by 2,3-dibromopropanol	10–4000	45–165
Sepharose CL 4B		60–20,000	45–165
Sepharose CL 2B		70–40,000	60–200
Superdex 30 Prep	Agarose–dextran (cross-linked)	0–10	24–44
Superdex 75 Prep		3–70	24–44
Superdex 75		3–70	11–15
Superdex 200 Prep		10–600	24–44
Superdex 200		10–600	11–15
Superose 12 Prep	Cross-linked agarose	1–300	20–40
Superose 12		1–300	10–11
Superose 6 Prep		5–5000	20–40
Superose 6		5–5000	13–15
Sephacryl HR S100	Allyl dextran-*N*, *N*-methylene-bisacrylamide	1–100	25–75
Sephacryl HR S200		5–250	25–75
Sephacryl HR S300		10–1500	25–75
Sephacryl HR S400		20–8000	25–75
Sephacryl HR S500		40–20,000	25–75
Sephacryl SF S1000		500–100,000	40–105
BioRad			
Biogel P2 F	Polyacrylamide	0.1–1.8	45–90
Biogel P2 EF		0.1–1.8	<45
Biogel P4 M		0.8–4	90–180
Biogel P4 F		0.8–4	45–90
Biogel P4 EF		0.8–4	<45
Biogel P6 M		1–6	90–180
Biogel P6 F		1–6	45–90
Biogel P6 EF		1–6	<45
Biogel P6 D4		1–6	90–180
Biogel P10 M		1.5–20	90–180
Biogel P10 F		1.5–20	45–90
Biogel P30 M		2.5–40	90–180

Table 10.2 (contd.)

Supplier	Matrix	Fraction range (kDa)	Bead size (μm)
Biogel P30 F	Polyacrylamide	2.5–40	45–90
Biogel P60 M		3–60	90–180
Biogel P60 F		3–60	45–90
Biogel P100 M		5–100	90–180
Biogel P100 M		5–100	45–90
Biogel A-0.5 M C	Agarose	<10–500	150–300
Biogel A-0.5M M			75–150
Biogel A-0.5M F			38–75
Biogel A-1.5M C		<10–500	150–300
Biogel A-1.5M M			75–150
Biogel A-1.5 M F			38–75
Biogel A-5M C		10–5000	150–300
Biogel A-5M M			75–150
Biogel A-5M F			38–75
Biogel A-15 M C		40–15000	150–300
Biogel A-15M M			75–150
Biogel A-15M F			38–75
Biogel A-50M C		100–50000	150–300
Biogel A-50M M			75–150
Biogel A-150M C		1000–150000	150–300
Biogel A-150M M			75–150
Waters			
Protein-Pak 60	Coated silica GPS	1–20	10
Protein-Pak 125		2–80	10
Protein-Pak 300 SW		10–300	10
Protein-Pak 200 SW		0.5–60	10
Shodex 802.5		0.1–50	7
Shodex 803		0.1–150	7
Shodex 804		0.5–600	7
Toso Haas			
TSK Gel G- 2000SWX	GPS-coated silica	5–150	5
TSK Gel G-2000SW		5–100	10
TSK Gel G-2000SWL		5–100	13
3000SWXL.			
TSK Gel G-		10–500	5
TSK Gel G-3000SW		10–500	10
TSK Gel G-3000SWL		10–500	13
TSK Gel G-		20–10000	8
4000SWXL			
TSK Gel G-4000SW		20–7000	13
TSK Gel G-4000SWL		20–7000	17
TSK Gel G1000 PW		<2	10, 17
TSK Gel G2000 PW		<5	10, 17, 20
TSK Gel G2500 PWXL		<8	6
TSK Gel G2500 PW		<8	10, 17, 20
TSK Gel G3000 PWXL		0.5–800	6
TSK Gel G3000 PW		0.5–800	10, 17, 20
TSK Gel G4000 PWXL		10–1500	10
TSK Gel G4000 PW			17, 22
TSK Gel G5000 PWXL		<10,000	10
TSK Gel G5000 PW			17, 20, 22
TSK Gel G6000 PWXL		<200,000	13
TSK Gel G6000 PW			17, 25

Table 10.2 (contd.)

TSK Gel GMPWXL	GPS-coated silica	<200,000	13
TSK Gel GMPW			17
Toyopearl HW 40S	Ethylene glycol/ ethylacrylamide polymer	0.1–10	20–40
Toyopearl HW 40F			30–60
Toyopearl HW 40L			50–100
Toyopearl HW 50S		0.5–80	20–40
Toyopearl HW 50F			30–60
Toyopearl HW 55S		1–700	20–40
Toyopearl HW 55F			30–60
Toyopearl HW 65S		40–5000	20–40
Toyopearl HW 65F			30–60
Toyopearl HW 75S		500–5000	30–60
Toyopearl HW 75F			
Alltech Microspheres			
GPC 60	Diol-coated silica	0.25–28	7
GPC 100		2.5–350	7
GPC 150		4–500	7
GPC 300		7.5–1200	7
GPC 500		20–75,000	7
GPC 1000		30–25,000	7
GPC 4000			7
BioSepra			
Ultragel ACA202	Polyacrylamide/agarose	1000–15,000	60–140
Ultragel ACA54	Polyacrylamide/agarose	5000–70,000	60–140
Ultragel ACA44	Polyacrylamide/agarose	10,000–130,000	60–140
Ultragel ACA34	Polyacrylamide/agarose	20,000–350,000	60–140
Ultragel ACA22	Polyacrylamide/agarose	100,000–1,200,000	60–140
Ultragel A2	Agarose	120,000–25,000,000	60–140
Ultragel A4	Agarose	55,000–9,000,000	60–140
Ultragel A6	Agarose	25,000–2,400,000	60–140
Trisacryl GFO5	Hydrophilic copolymer	200–2500	40–80 M grade 80–160–S grade
Beckman			
UltraSpherogel 2000	Silica bonded with polyether phase	1–250	5
UltraSpherogel 3000		5–700	5
UltraSpherogel 4000		10–2000	5

This still represents the general situation with respect to availability of media, although the properties of the solid phases have continued to improve in terms of physical and chemical robustness. There have been many developments in the methods that are employed with them, too: approaches based on displacement techniques or ion-pairing agents, for example, have opened up new possibilities in the field.

A final introductory point should be made regarding ion exchange chromatography: while it is relatively easy to obtain ideal SEC behaviour in many applications, this is by no means the case for IEC.

On the one hand many of the soft IEC gels are based on the size exclusion matrices described in the previous section, and are therefore almost bound to display some size exclusion properties. And even with new high-performance media designed specifically for ion exchange, porous particles are favoured to increase the effective surface area; although minimised by the use of wide-pore media (typically greater than 200 Å), size exclusion behaviour is again observed.

On the other hand, the groups introduced into the surface of ion exchange media do not consist exclusively of charged species. Generally they are organic compounds possessing lesser or greater degrees of

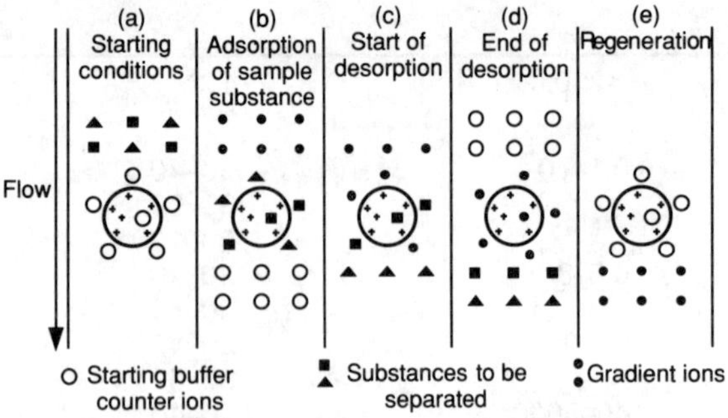

Figure 10.11 The principle of IEC, in this case using an anion exchange matrix. The positive charge of the surface is initially balanced by buffer counter-ions (a). Sample solutes bearing negative charges are adsorbed (b), then an elution buffer is used to desorb them in sequence of their charge (c, d). Finally, the matrix is regenerated using the starting buffer.

hydrophobic character, and these will also be a factor in defining the overall chromatographic performance.

Mixed-mode chromatography is sometimes talked about as if it is a discrete technique in itself, but in fact most ion exchange purifications are mixed-mode to some extent. The difference between approaches to protein chromatography is often determined more by the perception of the user, and by the marketing of the individual media by their suppliers, than by a simple difference in the separation mode.

The multiple mechanisms underlying most ion exchange separations have been deliberately exploited in a number of commercial media such as the Alltech Mixed-Mode™ series and the Dynamax™ 300A-AX from Rainin.

10.6.2 Key process variables of ion exchange chromatography

Charge properties of the solid phase

As mentioned at the start of this section, ion exchange media can possess either acidic or basic groups, giving them cation or anion exchange properties respectively. These groups possess their own pK_a values, which will define the pH range within which the surface is charged (see Section 2.2.1).

A strong cation exchanger possesses strongly acidic groups such as sulfonic acid residues, with a pK_a approaching zero. The surface of such a matrix is almost totally deprotonated (and negatively charged) at any pH higher than around 1–2, and therefore provides a stable surface charge at all pH conditions relevant to protein chemistry. A weak anion exchanger contains groups with a significantly higher pK_a, for example carboxymethyl functions with a pK_a around 4; these will only be predominantly charged at pH 6 and above. The charge properties of this surface can be modulated by adjusting the eluent pH in the mildly acidic region.

Strong and weak anion exchange matrices behave conversely: a strong resin containing groups such as quaternary ammonium salts (pK_a 12 or above) will be fully protonated at all pHs below around 10, while weakly basic groups such as diethylaminoethyl (DEAE) residues (pK_a around 8.5) are fully charged only under fairly acidic conditions.

The arguments above are based on the assumption that all the charged groups of interest possess the same pK_a. If this is not the case, then the pH range over which the charge properties of the surface varies will be wider still, this is true for example of polyethyleneimine (PEI) matrices which possess a mixture of primary, secondary and tertiary amines, and display very broad pH titration curves (Alpert and Regnier, 1979).

These surface charge effects are depicted for the various types of ion exchange matrix in Figure 10.12. It should be noted that the terms 'strong' and 'weak' in the ion exchange field refer to the degree of acidity or basicity of the charged groups; they do not imply that the relative affinities of these matrices are 'generally'

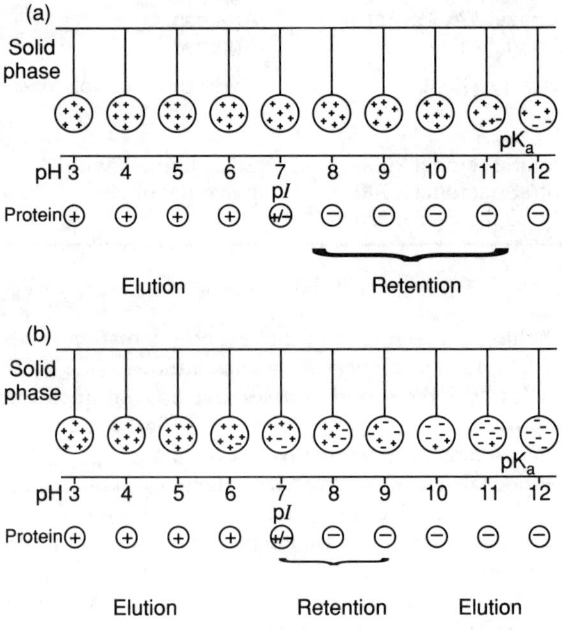

Figure 10.12 Surface charge effects for (a) strong and (b) weak ion exchange matrix.

higher or lower, a concept which can in any case have no meaning without specifying at least the nature of the solute and the pH of the eluent system.

It is hard to find comparisons of weak and strong ion exchange using chemically comparable systems, but Gooding and Schmuck (1985) compared two media based on tertiary (weak) and quaternary (strong) versions of a PEI matrix. They observed, as expected, that the weak system was more pH-dependent in the neutral range. Below around pH 7, for example, ovalbumin was retained more strongly on the weak matrix.

The behaviour of the matrix also depends to some extent on the density of groups on the surface, as well as their individual charge state. Henry (1990) has calculated that the charge density may be as high as 1 charged group per 20 Å^2 in a resin consisting totally of a polymer with ion exchange properties. Henry also points out that the charge density may give an over-simplified view of the surface behaviour, as other effects such as hydrogen bonding and van der Waals attractions may come into play as the protein molecule approaches the surface ligands.

It is generally accepted that protein molecules need to be bound by a number of sites to the ion exchange surface, and that either too low or too high a ligand density may give sub-optimal retention. In a series of experiments with weak anion exchangers of the same type but different surface densities, Kopaciewicz *et al.* (1985) observed that myoglobin was retained at moderate density, but not at the higher or lower levels. Ferritin was readily retained in all cases, but could not be eluted from a high-density matrix.

Other properties of the solid phase

It has already been mentioned that porous media are commonly employed in ion exchange media, and that size exclusion effects therefore need to be taken into account. Figure 10.13 (Henry, 1990) shows the general relationship between pore size and size exclusion behaviour. For fractionation of proteins by SEC, the central segment, where there is differential behaviour dependent on molecular mass, is the useful area. For ion exchange, on the other hand, the aim is to minimise these effects, and ideally the pore size should place the proteins of interest in the upper (total inclusion) segment. From the graph it can be observed that a pore size of 1000 Å is required to obtain this behaviour for proteins up to around 200 kDa. With smaller pores or larger solutes, some degree of size exclusion will be observed.

Non-porous media are available, which do not suffer from size exclusion effects at all, and also provide very rapid separations. Their drawback is their small surface area, and hence their low capacity. Overloading and peak-broadening can thus be a problem unless samples are kept small.

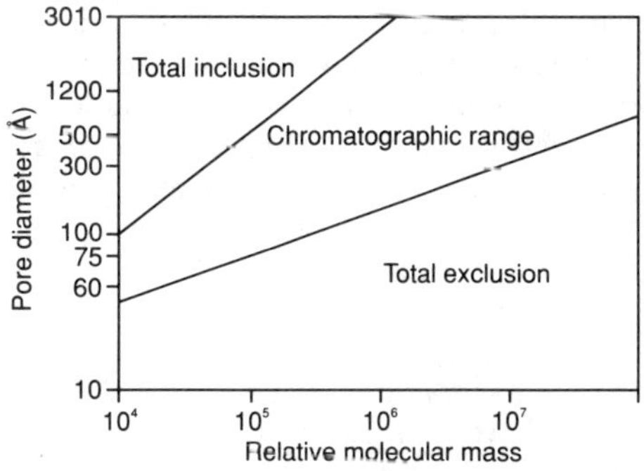

Figure 10.13 Pore diameter limits for total exclusion and total inclusion of polymers as a function of their molecular masses. (Reproduced with permission from: Henry (1990). In *High Performance Liquid Chromatography in Biotechnology* (Hancock, W.S., ed.), pp. 205–261, John Wiley, New York.)

Some of the earlier fibrous cellulose matrices were also non-porous in nature, although the large size of the fibres offered poor resolution properties. Later cross-linked microcrystallline cellulose media lost this non-porous structure.

Hydrophobic effects have also been mentioned earlier, and are not as simple to quantify: they depend far more on the nature and density of the surface moieties, and the nature of the eluent and solutes. Various methods have been described to determine whether a given matrix is displaying hydrophobic properties (Henry, 1990): these are based on looking for reduced or increased retention in the presence of organic cosolvents or high-salt formulations respectively, or on the denaturation of retained proteins as a function of time (see Dwell Time, below).

These hydrophobic effects are often observed (see, for example, Kopaciewicz *et al.*, 1985). Various synthetic approaches have been taken to minimise the hydrophobic nature of ion exchange media, for example the use of media coated with hydrophilic polypeptides (Alpert, 1983), or alkoxysilanes (Ramsden, 1983), but in general it should be borne in mind that these effects may be operating.

The swelling properties of soft gels, which merely necessitate a pre-treatment stage in SEC, are more of a serious issue in ion exchange. Dramatic changes in gel volume can occur either when salt gradients are applied, or when the charge state of the surface alters (Himmelhoch, 1971). These are obviously an unwelcome feature of gels of this type.

The other features of ion exchange matrices which are of interest are common to all chromatographic techniques: chemical robustness to allow a wide range of operating conditions (both for the separations themselves and for cleaning and sanitization), and physical robustness to allow good flow properties. The particle size debate of resolution *vs.* flow rate applies here too.

Properties of the solute proteins

The size and hydrophobicity of the solute molecules help determine the magnitude of the size exclusion and hydrophobic interaction effects just discussed, but it is clearly the charges of the solute protein(s) which most define their ion exchange behaviour on a given matrix. The most convenient measure is the p*I*, and this can be used as a simple basis for a purification strategy.

The protein molecule will be cationic below its p*I* and anionic above it, and is therefore amenable to cation and anion exchange respectively under those conditions. For proteins with a moderate p*I* it may be equally feasible to adopt either approach. Where the p*I* is more extreme only one route may be practicable, e.g. a protein such as lysozyme with a p*I* of 11 will be cationic under all but the most alkaline conditions, and anion exchange is not really a favourable option.

Proteins, therefore, can be purified by IEC roughly on the basis of their p*I*. However this is an over-simplification, and it is worth considering deviations from this model in a little detail.

Regnier and his colleagues carried out an illuminating study on this topic (Kopaciewicz *et al.*, 1983). They described the hypothetical relationship between pH and chromatographic retention that would be expected from the simple 'net charge' model, represented by the retention map shown in Figure 10.14(a). They then examined a number of real proteins, and some of the retention maps observed are shown in Figure 10.14(b)–(d).

A few of the proteins displayed near-ideal behaviour: human IgG for example (Figure 10.14(b)) showed the retention dropping to zero reasonably close to the p*I* for both anion and cation exchange. Many other examples showed greater retention ranges than would be expected, however: β-lactoglobulin was retained on an anion exchanger not only at the p*I* (where net charge = 0) but even up to one pH unit below it, where the overall charge of the protein is the same as that of the matrix. The same effect was noted with cation exchange.

Having eliminated the possibility that these observations were caused by hydrophobic effects, Kopaciewicz *et al.* attributed them to local charge concentrations within the protein molecule. Thus although the overall charge of the protein might be positive, there may be areas within it which are sufficiently negative to form a basis for anion exchange. This phenomenon had in fact been known empirically much earlier (see, for example, Himmelhoch, 1971).

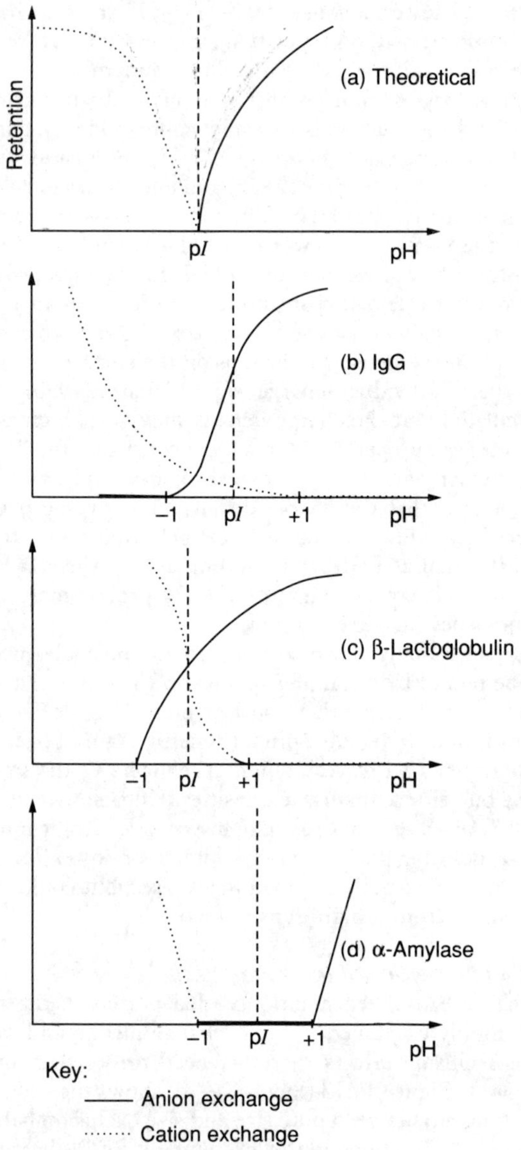

Figure 10.14 Retention maps of various proteins on cation and anion exchangers. (Reproduced with permission from: Kopaciewicz *et al.* (1983), *J. Chromatogr.*, **266**, 3–21.)

This investigation was backed up by estimating the number of charged groups[1] (Z) involved in the association of the protein with the solid phase (see Boardman and Partridge, 1955). For β-lactoglobulin a Z value of 3.6 was observed at the pI, supporting the premise that charged groups were still capable of binding although the overall charge was zero.

This charge distribution parameter would be expected to vary with the exact topography of the particular protein molecule, creating a deviation from the net charge model, and therefore complicating the simple pI-based assessment of binding outlined at the start of this section.

Finally it should be noted that the behaviour of the proteins studied by Kopaciewicz *et al.* was by no means clear-cut, and there were many variations in the types of behaviour observed both in terms of retention (see for example α-amylase, Figure 10.14(d)), and in its relationship with protein charge.

Eluent pH

Unlike size exclusion, where the pH and ionic strength of the eluent can be varied widely without significantly affecting the chromatographic performance, IEC is strongly influenced by both parameters.

Starting with pH, much time has already been spent discussing the charge properties of the protein and those of the matrix surface. However, both of these are totally dependent on the pH of the eluent, which can be modulated to obtain the desired separation.

The most simple case is for a strong exchanger (e.g. a quaternary ammonium-based anion exchanger, with a pK_a of 12, where the surface charge remains positive at all pH values below about 11). Considering the purification of a protein with a pI of 7 (Figure 10.12), it should be possible to retain it on the column over quite a wide pH range, at least 8–11, where the protein is fully and oppositely charged to the matrix. Lowering the pH will reduce and eventually reverse the charge as it passes through the pI of the protein, which can therefore be expected to elute from the column.

To draw a simple general rule from this example, the pH at which the strongest retention will be obtained is between the pK_a of the matrix and the pI of the protein. This applies most straightforwardly to strong exchangers, both cationic and anionic. For weak exchangers the situation is less clear-cut: Figure 10.12 shows the situation when the same purification is attempted on a DEAE matrix, for example. Here there is a much broader range over which the matrix is only partially charged. The best pH for retention of the protein remains between the pI and the pK_a, but the matrix still bears some negative charge under these conditions. This *may* cause weaker binding but at the same time it may be necessary to go to a lower pH to obtain elution. Note that at the pI, both protein and matrix are amphoteric in nature and there may be insufficient repulsion to disturb the binding.

The more broad the range over which the matrix charge varies, and the closer the pK_a of the matrix to the pI of the protein of interest, the harder it is to be definitive about the pH at which retention and elution will be observed.

Finally, the usual type of problem faced by chromatographers involves the separation of one protein from a mixture of others, all with their own pI values which may well be unknown. Therefore it would be naive to imagine that the optimum choice of retention and elution pH can always be arrived at without a considerable element of trial and error.

Eluent ionic strength

Partly because elution pH is not simple to predict, and partly because of practical inconveniences when eluent pH is varied during a run, other strategies have been developed for ion exchange purifications based around ionic strength effects.

During the consideration of SEC (Section 10.5) it was noted that ion exchange effects could be observed due to charged groups in the media, and that these were suppressed by working at a reasonably high ionic strength; buffer salt molarities as high as 0.15 M are required for some media before ideal size exclusion behaviour is observed. This forms the basis for retention and elution of proteins by modulation of ionic strength: retention is achieved by working at low molarities (typically less than 25 mM), the choice of pH being exactly the same as just discussed. But instead of shifting the pH towards or beyond the pI, the strategy now is to raise the ionic strength of the medium so that the competition for the charged groups on the surface is greater. This undermines the electrostatic binding of the protein, offering another route for elution. Buffer molarities up to 2 or 3 M may be employed to achieve this.

While this approach avoids pH changes during the run, and is much more frequently encountered in practice, it does have one major drawback. It has already been pointed out that many ion exchange matrices possess significant hydrophobic properties, and these are exacerbated by working at high ionic strength. Therefore while increasing the buffer

[1] Because two charged groups are involved in each binding event, the actual number of binding sites is $Z/2$.

molarity reduces the electrostatic interactions between the protein and the surface, these may be replaced by equally strong hydrophobic interactions. As a net result, it is possible that the protein can never be eluted at all by this route, and this scenario probably explains the disappearance of many proteins—never to be seen again – on ion exchange columns!

Finally, interesting differences in retention properties can be observed if the ionic species in the elution buffer are altered: as early as 1949 Kunin and McGarvey determined that, for small solutes, there was an order of displacing ability from a strong anion exchanger as follows: citrate > sulfate > oxalate > bromide > chloride > formate > fluoride > acetate.

Kopaciewicz *et al.* (1983) studied this phenomenon in more detail, and observed a roughly similar order for some proteins but not for others. Not only were there significant differences in retention for a single protein (with ovalbumin on an anion exchanger, for example, citrate gave half the retention time of acetate), but selectivity between two proteins could be modified as well: with bromide the relative retention ratio of ovalbumin and soybean trypsin inhibitor (STI) was 1.46, while with citrate it was 2.61.

Even on anion exchange, alteration of the cationic component caused significant effects too: the relative retention of STI with calcium and lithium was 0.47 and 0.82 respectively.

Similar retention time variations were observed with other proteins on a cation exchanger, again from both the cationic and anionic components, although selectivity differences were smaller.

The actual choice of ionic species, then, can also be used to modify the retention properties of the buffer, but Kopaciewicz's *et al.* results suggest that optimisation might be necessary for individual examples.

Eluent profile

For ion exchange separations there are three major categories of eluent profile—isocratic, continuous gradient, and step gradient.

In isocratic systems the entire run is carried out using a single eluent buffer (Figure 10.15(a)). This approach is employed where one set of pH and ionic strength conditions can be identified, which will allow elution of all the components of interest while providing sufficient differential retention to elicit the desired fractionation. Not surprisingly this may be completely impossible for some mixtures, and may require considerable optimisation for others. Also, it does not offer the capacity to concentrate the sample. However, once achieved it provides an extremely simple and reproducible process.

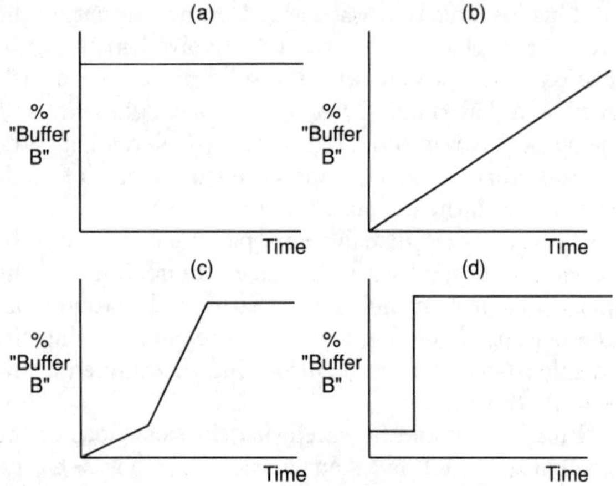

Figure 10.15 Categories of elution profiles: (a) isocratic; (b) continuous gradient; (c) discontinuous gradient; (d) step gradient.

Ekström and Jacobson (1984) analysed the dependence of protein resolution on eluent composition, and demonstrated that it was much sharper than for small molecules. As a consequence it is unlikely to find a single eluent which resolves a multi-protein mixture, whereas such systems are common in work with small peptides.

Yao and Hjertén (1987) investigated the factors which determine whether or not isocratic elution is possible for protein fractionations, and proposed four major requirements for it to be successful: small particles, symmetrical peaks (both maintaining high resolution), negligible non-specific interactions, and low ligand density on the matrix. The latter parameter limits the maximum number of binding sites by which a single protein molecule can be attached ($Z/2$), and therefore reduces the range between low and high-binding solutes.

To achieve adequate resolution of protein mixtures, it is usually necessary to vary the eluent composition, so that either the pH, the ionic strength, or both are altered as the run progresses. In a typical separation based on ionic strength, the sample will be loaded in a low molarity buffer at a pH which allows the protein of interest to be retained, while hopefully allowing some impurities to pass straight through. The eluent composition is then changed, a continuous gradient being established to allow a gradually increasing quantity of a second (high molarity) buffer to be introduced. As the change is gradual, the retained proteins are encouraged to elute on a fractional basis. At the end of the gradient, the ionic strength should be sufficiently high to elute all the adsorbed species. A slowly-run, linear continuous gradient (Figure 10.15(b)) is particularly useful where the behaviour of the solutes is unknown, often providing a useful first shot at fractionating a mixture. Although a shallower gradient is often supposed to yield better fractionation under all circumstances, this is not necessarily the case: Novotný (1971) formulated some basic rules for deciding gradient conditions, and concluded that this was only true at eluted protein concentrations that were millimolar or higher. At lower concentrations, better fractionation could be obtained by a steeper gradient.

Once the basic properties of the mixture have been determined in this way, a more refined gradient may be established which gives a better separation. It may be decided to run the gradient more slowly over regions where there are multiple components evident, and more quickly over segments of the formulation where no components are eluted. The gradient may also be started higher to prevent fast-running impurities from retaining, and stopped earlier if no further products are going to be eluted. After sufficient optimisation the type of elution profile shown in Figure 10.15(c) may be arrived at.

The extreme of this type of approach is to jump straight from one buffer composition to another, yielding a true step gradient (Figure 10.15(d)). If conditions have been optimised to provide the desired separation by this route, it may be the most convenient approach, and certainly offers the greatest potential for concentration of the sample. The use of multiple steps does have one drawback, however, in that artefactual fractionation is often observed. Sometimes, every step change will provoke some elution of components, but many of these supposedly different fractions turn out to contain similar components. The more gradual the profile, the less likely this sort of behaviour, and the greater the confidence that each peak eluted really is different in identity from its neighbours.

Finally, displacement of bound components by an eluent containing strongly binding solutes offers another means of elution (see Section 10.6.4).

Dwell time

Where the sample is concentrated onto the column in a buffer which gives complete retention of the protein, a considerable dwell time may elapse before elution takes place. If purely electrostatic binding is at work, there will be very little effect on the protein during its retained stage. If, however, hydrophobic interactions are involved, there can be some serious consequences. Benedek *et al.* (1984) have demonstrated that some rapid denaturation of the protein takes place as soon as hydrophobic binding occurs, and this supports the many similar observations from the field of protein immobilisation (see Section 8.2). Once this process starts, hydrophobic regions which would normally be buried within the protein structure start to become exposed, and these may bind to the surface (giving stronger binding), or to similar areas in adjacent molecules (giving aggregation).

As a result a number of phenomena may be observed. Because the hydrophobic binding of the protein grows stronger as the dwell time increases, sample may be lost which has been retained on the column for longer, therefore poor yields may be observed as only the most recently-applied sample can be eluted. Secondly the product which is eluted may be heterogeneous, containing fractions which are modified to a greater or lesser degree depending on their individual dwell times (Hearn *et al.*, 1985). Finally, there may be serious run-to-run variation if the loading time is not closely controlled, a factor which might be particularly easy to neglect during a scale-up exercise, for example.

If problems of this type are suspected, they can be minimised by keeping dwell times (and buffer molarities) as low as possible.

Sample characteristics

The importance of sample characteristics depends on the type of elution profile chosen. Sample volume, for example, is relatively unimportant in itself if a concentrating system is being used; the amount of protein which can be adsorbed is only limited by the capacity of the matrix. But in an isocratic set-up, or a gradient without a concentrating stage, large sample volumes will lead to poor performance.

Sample pH and ionic strength, on the other hand, can be critical. Ideally they should be identical to the starting eluent, so that there is no interference with the chosen chromatographic behaviour. Because, particularly at the initial stages, ion exchange tends to be carried out at low buffer molarities, sample properties are not swamped out easily. A sample whose pH or ionic strength is significantly different from the eluent composition can cause a serious local variation, easily enough to prevent retention for example.

The ways to avoid these problems are fairly obvious. Either the sample can be buffer exchanged into the eluent before being introduced to the column, or it must be adjusted or diluted sufficiently that its effects are negligible.

Column dimensions

In isocratic systems, or others where there is no complete retention of the protein of interest, the normal chromatographic considerations apply: the usual choices of column length and increased resolution *versus* low flow rates or high back pressures need to be made. For proteins, however, the dependence of resolution on column length in IEC is small (Ekström and Jacobson, 1984) and there is often little to be gained by using longer columns.

In systems where the protein of interest is fully retained, however, the column dimensions can be based primarily on the required ion exchange capacity. This can be very high for modern media, and column sizes can be extremely small compared to the amount of sample.

The capacity of the matrix can be determined by experiment, but manufacturers' estimates are good enough for many purposes. It should be borne in mind that the capacities of columns can decrease with repeated use. For manufacturing processes where multiple runs are to be carried out, it is therefore wise to work at a reasonably small fraction of the column capacity.

Without capacity information, some very rough rules of thumb can be applied: for soft gels Himmelhoch (1971) recommended trying 10 grams of adsorbent per gram of protein as a starting estimate.

Column packing

For non-concentrating systems, column packing is as important as for any other continuous chromatographic technique. For concentrating systems the effect is less acute, but imperfections can still lead to peak broadening during elution, and to reductions in effective capacity.

Column equilibration

The effectiveness of column (re-)equilibration in the eluent of interest is especially important for gradient systems, commonly encountered in ion exchange methods. Henry (1990) has given a good description of the events which must take place during this process. (1) Fast exchange of buffer outside pores, rapidly gives the correct pH and ionic strength. (2) Slow diffusion of ions into and out of the pores, adjusts buffer pH and ionic strength, and the charge state of the surface. (3) Even slower diffusion as the concentration gradients fall, eventually gives the desired buffer and surface characteristics.

The actual times required for re-equilibration will depend on a number of factors. High flow rates, large pore sizes and strong exchangers are all favourable, for example. In summary it should be borne in mind that this is not an instant process, and this should be reflected in the duration of equilibration stages, and in the rate at which eluent gradients (particularly pH) are altered.

10.6.3 Commercially available media for ion exchange chromatography

Some of the more commonly encountered ion exchange media available from commercial suppliers are listed in Table 10.3; many incorporate the groups shown in Figure 10.16. Their key properties are listed where known.

10.6.4 Displacement ion exchange chromatography

In IEC, like any other form of adsorptive chromatography, there exists the possibility of using a displacement strategy to elute retained solutes, which are constantly in equilibrium between the bound and unbound forms. The reason that a solute is well-retained is that this equilibrium is strongly in favour of the bound form, but desorption is constantly occurring. In the absence of any other binding species, a desorbed molecule has a very strong chance of re-adsorbing to the matrix before it has had a chance to travel far in the eluent flow, hence the strong retention of such species.

However, if a large concentration of a competitive displacer is introduced, particularly one whose affinity is higher as well, then a desorbed solute is highly likely to be replaced by one of these molecules. This forms the basis of displacement chromatography (Peterson and Torres, 1984).

For IEC, displacers tend naturally to be highly-charged species, e.g. carboxymethyl dextrans or sulfate-containing carbohydrates for anion exchange (Peterson and Torres, 1984; Liao *et al.*, 1987); this approach can provide excellent elution characteristics such as concentrated product fractions, and the elimination of peak tailing.

Even more powerful is the use of spacer displacers, which possess affinities intermediate between product components to be resolved; these can form a buffering layer between the components and thus improve resolution. Separations use a range of graded displacers of different affinities (e.g. carboxymethyl dextrans of different acid content, or ampholyte molecules; see Section 10.12.2) to separate the target solutes, and culminate with a powerful final displacer which drives them sequentially off the column—an interesting alternative to standard ion-exchange elution strategies (Torres *et al.*, 1987; Torres and Peterson, 1992).

Table 10.3 Commercially available ion-exchange matrices

Supplier	Matrix/Ligand	Bead size (μm)
Whatman		
DE51	DEAE-Cellulose	60–130
DE52	DEAE-Cellulose	60–130
DE53	DEAE-Cellulose	60–130
QA52	Quat. amine-Cellulose	60–130
DE32	DEAE-Cellulose	60–130
DE23	DEAE-Cellulose	60–130
Express Ion D	DEAE-Cellulose	60–130
Express Ion Q	Quat. amine-Cellulose	60–130
CM52	Carboxymethyl-Cellulose	60–130
SE52	Sulfoxyethyl-Cellulose	60–130
SE53	Sulfoxyethyl-Cellulose	60–130
Express Ion C	Sulfoxyethyl-Cellulose	60–130
Express Ion S	Sulfoxyethyl-Cellulose	60–130
CM32	Carboxymethyl-Cellulose	60–130
CM23	Carboxymethyl-Cellulose	60–130
P11	Phospho -Cellulose	60–130
TosoHaas		
TSKgel DEAE-5PW	DEAE-Hydrophilic resin	10
TSKgel DEAE-NPR	DEAE-Hydrophilic resin	2.5
TSKgel DEAE-2SW	DEAE-Silica	5
TSKgel DEAE-3SW	DEAE-Silica	10
TSKgel CM-5PW	Carboxymethyl-Hydrophilic resin	10
TSKgel CM-2SW	Carboxymethyl-Silica	5
TSKgel CM-3SW	Carboxymethyl-Silica	10
TSKgel SP-5PW	Sulfoxypropyl-Hydrophilic resin	10
TSKgel SP-NPR	Sulfoxypropyl-Hydrophilic resin	2.5
TSKgel SP-2SW	Sulfoxypropyl-Hydrophilic resin	5
Toyopearl DEAE-650S	DEAE-Hydrophilic resin	20–50
Toyopearl DEAE-650M	DEAE-Hydrophilic resin	40–90
Toyopearl DEAE-650C	DEAE-Hydrophilic resin	50–150
Toyopearl super Q-650S	Quat.amine-Hydrophilic resin	20–50
Toyopearl super Q-650M	Quat.amine-Hydrophilic resin	40–90
Toyopearl super Q-650C	Quat.amine-Hydrophilic resin	50–150
Toyopearl QAE-550C	Quat.amine-Hydrophilic resin	50–150
Toyopearl CM-650S	Carboxymethyl-Hydrophilic resin	20–50
Toyopearl CM-650M	Carboxymethyl-Hydrophilic resin	40–90
Toyopearl CM-650C	Carboxymethyl-Hydrophilic resin	50–150
Toyopearl SP-650S	Sulfonypropyl-Hydrophilic resin	20–50
Toyopearl SP-650M	Sulfoxypropyl-Hydrophilic resin	40–90
Toyopearl SP-650C	Sulfoxypropyl-Hydrophilic resin	50–150
Toyopearl SP-550C	Sulfoxypropyl-Hydrophilic resin	50–150
TSK gel AminoPak	Amino PS–DVB	5
TSK gel Sugar AX	Sugar PS–DVB	8 & 10
TSK gel SAX	Sulfo-PS–DVB	5
TSK gel SCX	Sulfo-PS–DVB	5
TSK gel QA pak A/P	Quat.amine-Hydrophilic resin	5 & 10
Boehringer-Mannheim		
Poros HQ H, M, & P-series	PEI-PS–DVB	50
Poros QE, M & P-series	PEI-PS–DVB	50
Poros PI M & P-series	PEI-PS–DVB	50
Poros HS M & P-series	Sulfoxypropyl-PS–DVB	50

Table 10.3 (contd.)

Supplier	Matrix/Ligand	Bead size (μm)
Poros SP H, F, M & P-series	Sulfoxypropyl-PS–DVB	50
Poros S H, F, M & P-series	Sulfoxyethyl-PS–DVB	50
Poros CM M & P-series	Carboxymethyl-PS–DVB	50
Pierce		
CM ion exchange cellulose	Carboxymethyl-Cellulose	75–150
DEAE ion exchange cell.	DEAE-Cellulose	75–150
BioRad		
Mono Prep DEAE	DEAE-Polymer	50
Mono Prep high Q	Quat.amine-Polymer	50
Mono Prep Q	Quat.amine-Polymer	50
Mono Prep high S	Sulfo-Polymer	50
Mono Prep S	Sulfo-Polymer	50
Mono Prep CM	Carboxymethyl-Polymer	50
DEAE Bio-Gel A	DEAE-Bio-Gel A-5m	80–150
CM Bio-Gel A	Carboxymethyl-Bio-Gel A-5m	80–150
BioSepra		
Q Hyper DF	Quat.amine-Rigid composite	35
Q Hyper DM	Quat.amine-Rigid composite	60
S Hyper DF	Sulfo-Rigid composite	35
S Hyper DM	Sulfo-Rigid composite	60
DEAE Hyper DF	DEAE-Rigid composite	35
Q Hyper D 10/20	Quat.amine-Rigid composite	10 or 20
S Hyper D 10/20	Sulfo-Rigid composite	10 or 20
CM Hyper DM	Carboxymethyl-Rigid Composite	35
CM Hyper DF	Carboxymethyl-Rigid Composite	60
DEAE Spherosil M	DEAE-Silica	40–100
QMA Spherosil M	Quat.amine Silica	40–100
DEAE Spherosil LS	DEAE-Silica	100–300
QMA Spherosil LS	Quat.amino-Silica	100–300
DEAE Trisacryl M	DEAE-Polymer	40–80
CM Trisacryl M	Carboxymethyl-Polymer	40–80
SP Trisacryl M	Sulfoxyprophyl-Polymer	
QA Trisacryl M	Quat.amine-Polymer	
DEAE Trisacryl LS	DEAE-Polymer	80–160
CM Trisacryl LS	Carboxymethyl-polymer	80–160
SP Trisacryl LS	Sulfoxypropyl-Polymer	80–160
QA Trisacryl LS	Quat.amine-Polymer	80–160
DEAE Trisacryl Plus M	DEAE Polymer	40–80
SP Trisacryl Plus M	Sulfoxyprophyl-Polymer	40–80
DEAE Trisacryl Plus LS	DEAE-Polymer	80–160
SP Trisacryl Plus LS	Sulfoxyprophyl-Polymer	80–160
DEAE Spherodex M	DEAE-Silica–dextran	40–80
CM Spherodex M	Carboxymethyl-Silica–dextran	40–80
SP Spherodex M	Sulfoxypropyl-Silica–dextran	40–80
DEAE Spherodex LS	Sulfoxypropyl-Silica–dextran	100–300
CM Spherodex LS	Sulfoxypropyl-Silica–dextran	100–300
SP Spherodex LX	Sulfoxypropyl-Silica–dextran	100–300
Waters (Millipore)		
Protein-Pak Q HR	Quat.amine-Polymethacrylate	8, 15, 40
Protein-Pak DEAE HR	DEAE-Polymethacrylate	8, 15, 40

Table 10.3 (contd.)

Protein-Pak CM HR	Carboxymethyl-Polymethacrylate	8, 15, 40
Protein-Pak SP HR	Sulfoxypropyl-Polymethacrylate	8, 15, 40
Accell Plus QMA Beads	Quat.amine-Polymer-coated silica	40
Accell Plus QMA Beads	Quat.amine-Polymer-coated silica	40
Alltech		
Microsphere 300 (WAX-DEAE	DEAE-Silica	7
WCX-COOH	Carboxy-Silica	7
SAX-Quat. amino	Quat.amine-Silica	7
SCX-sulfuric acid	Sulfo-Silica	7
Pharmacia (mono bead)		
Mono Q	Quat.amine-Mono beads	10
Mono S (5, 10, 10)	Sulfo-Mono beads	10
Mono P (5, 5/20)	Phospho-Mono beads	10
Source 15Q	Quat.amine PS–DUB	15
Source 15S	Sulfo-PS-DUB	15
Q Sepharose	Quat.amine agarose	24–44
SP Sepharose	Sulfopropyl agarose	24–44
Q Sepharose fast flow	Quat.amine agarose	45–165 (90 mean)
SP Sepharose fast flow	Sulfopropyl agarose	45–165 (90 mean)
DEAE Sepharose fast flow	DEAE Cross linked agarose	45–165 (90 mean)
CM Sepharose fast flow	Carboxymethyl-Cross-linked agarose	45–165 (90 mean)
DEAE Sepharose CL-6B	DEAE-Cross-linked agarose	45–165 (90 mean)
CM Sepharose CL-6B	Caboxymethyl-crosslinked agarose	45–165 (90 mean)
DEAE Sephacel	DEAE-Cellulose	40–160
DEAE Sephadex A-25	DEAE Cross-linked dextran	40–160
CM Sephadex C-25	Carboxymethyl-Cross-linked dextran	40–160
QAE Sephadex A-25	Quat.Amine-cross-linked dextran	40–160
SP Sephadex A-25	Sulfoprophyl-Cross-linked dextran	40–160
DEAE Sephadex A-25	DEAE-Cross-linked dextran	40–160
CM Sephadex A-25	Carboxymethyl-Cross-linked dextran	40–160
QAE Sephadex A-50	Quat.amine-Cross-linked dextran	40–160
SP Sephadex C-50	Sulfoprophyl-Cross-linked dextran	40–160

Kundu and Cramer (1997), for example, showed good resolution of bovine and equine cytochrome *c* on a cation-exchange system using an arginine derivative as a displacer, despite very similar individual properties.

10.6.5 Dynamic ion exchange chromatography

An interesting variation of ion exchange chromatography is based on generating the charged matrix from components in the eluent. If a detergent is added to the eluent for a reversed phase column (Section 10.8), for example, the molecules will be adsorbed *via* their hydrophobic regions, and the ionic moieties will face outwards to generate an ion exchange surface *in situ* (Figure 10.17).

While the majority of references have dealt with amino acid or oligopeptide solutes (Hearn *et al.*, 1979; Lam-Thanh *et al.*, 1981), a number of protein applications have also been described. Hancock and colleagues (Hancock *et al.*, 1981; Hancock and Sparrow, 1981), for example, demonstrated the separation of many protein mixtures by this approach.

The phenomena occurring during dynamic separations of this kind can be very complex, and the reader is referred to reviews such as that of Sorel and Hulshoff (1983) for detailed treatment. It should be noted that this technique is intimately related to ion-pairing strategies for the separation of charged molecules on hydrophobic surfaces (Figure 10.18), in fact the mechanisms are virtually indistinguishable.

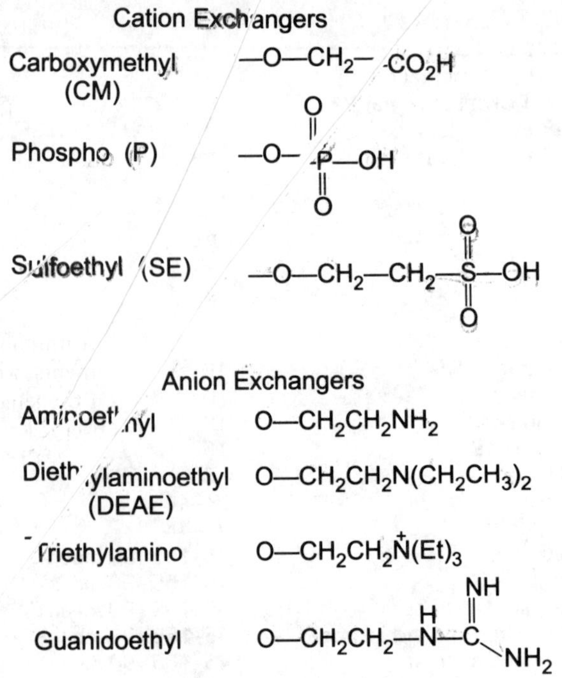

Figure 10.16 Some common cation and anion exchangers.

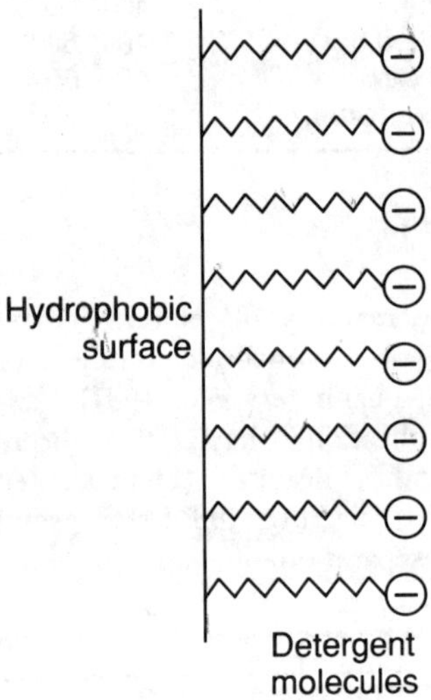

Figure 10.17 Dynamic ion exchange chromatography: if a detergent is added to the eluent in a reversed-phase system, the hydrophobic portions of the detergent molecules will coat the surface, learing the changed regions pointing outwards and providing the capacity for ion exchange.

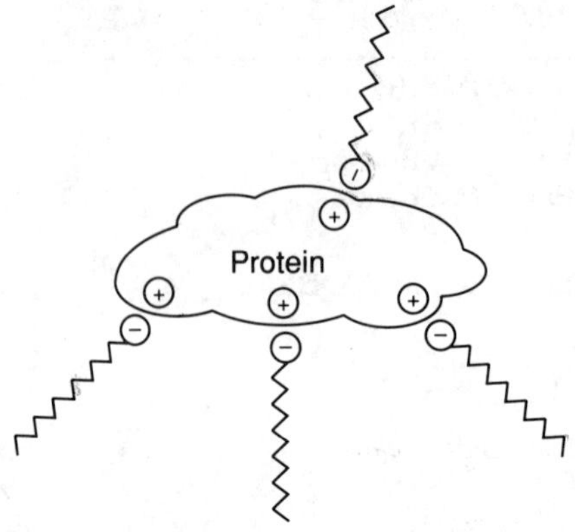

Figure 10.18 Ion pairing of a charged solute renders it hydrophobic, allowing it to be retained on a reversed-phase surface. The mechanism of retention is identical with that of dynamic ion exchange chromatography (Figure 10.17).

10.7 Hydrophobic Interaction Chromatography

The concept of adsorption of proteins to a surface *via* hydrophobic interactions has come up frequently already in this book, both in consideration of protein immobilisation (Chapter 8) and in the discussion of other chromatographic techniques in this chapter. For IEC in particular, the role of hydrophobic effects in defining retention and elution behaviour is quite significant, and it is not surprising that they have come to form the basis of chromatographic techniques in their own right. There are nominally two branches of chromatography based on this approach, but the distinction between them owes more to the history of their development than to any clear difference in mechanism.

Reverse-phase chromatography (RPC) was developed for organic chemistry, particularly to offer separation techniques for polar compounds which were too strongly bound on 'normal-phase' silica: the silica was coated with organic modifiers to provide a hydrophobic surface, and aqueous–organic eluents were employed to elute the bound species. It was only a matter of time before this approach was applied to proteins, and some useful separation properties were discovered.

In parallel, however, protein chromatographers had been developing techniques for their own purposes based on their hydrophobic properties – again using media coated with organic modifiers, but generally using ionic strength gradients to control retention and elution. This technique was termed hydrophobic interaction chromatography (HIC).

The mechanism underlying these techniques is identical, and the only practical differences are ones of degree (Pharmacia, 1993a). RPC columns tend to be more hydrophobic (using higher ligand densities, and modifiers with chain lengths from C_4 to C_{18}), and therefore require the use of organic cosolvents to elute bound molecules; this has made its use most common for small proteins which are stable to these conditions. An important point is that, particularly if the cosolvent is present during the binding stage, it may induce partial denaturation of the protein molecule, exposing hydrophobic regions which would not otherwise be accessible to the matrix. The adsorption properties therefore may reflect the hydrophobic nature of the denatured rather than the native conformation, and this can give rise to major differences between HIC and RPC separations of the same proteins (Fausnaugh *et al.*, 1984).

HIC columns are rather gentler, having a lower surface coverage of generally less hydrophobic modifiers (typically incorporating chain lengths of C_2–C_8, or phenyl groups); this often allows elution to be achieved without the use of organic cosolvents, and the approach is thus applicable to a wider range of proteins. The adsorption properties will generally reflect the native conformation, and in fact HIC can offer a useful means of separating intact from denatured protein.

The separation between the techniques is not complete however: it is not unusual to encounter HIC elution gradients incorporating organic cosolvents too (see for example Neumeier and Maurer, 1982), and these may trigger some degree of solvent-derived protein denaturation.

10.7.1 The mechanism of hydrophobic interaction chromatography

The basic principle of HIC is shown in Figure 10.19. In 1948 Tiselius noted that the adsorption of proteins onto solid matrices could be encouraged by increasing the salt content of their solutions, although the adsorbents studied (silica and filter paper) were hardly typical of the modern HIC approach. The exploitation of this phenomenon was not really followed up until the early 1970s, when a number of independent reports appeared virtually simultaneously.

Yon (1972) coupled hydrophobic groups such as diaminodecane to Sepharose, creating a matrix with a mixed hydrophobic–ionic nature. This was used to purify hydrophobic proteins.

Er-El *et al.* (1972) prepared a number of immobilised glycogen matrices for affinity experiments with glycogen phosphorylase *b*, differing only in the length of the hydrophobic linker chains. They observed that the enzyme was bound only to the matrix with the longer-chain linker, and followed this up with a more detailed investigation (Shaltiel and Er-El, 1973). On this basis they proposed 'hydrophobic chromatography' as a generally applicable technique.

Hofstee (1973) observed that many proteins bound strongly to agarose substituted with 4-phenylbutylamine. A combination of ion exchange and hydrophobic effects appear to have been operating, as elution was facilitated both by high salt, and by the incorporation of ethylene glycol as an organic cosolvent. Hofstee termed this approach 'hydrophobic affinity chromatography' and demonstrated its use to fractionate a protein mixture.

These initial reports were all based on matrices bearing significant quantities of chargeable groups such as amines, or isoureas from cyanogen bromide activation, hence their mixed-mode properties. Porath *et al.* (1973) were the first to describe a more 'pure' hydrophobic solid phase, from which proteins were eluted more readily at low salt molarities, and the development of matrices of this type was subsequently very rapid (Hjertén *et al.*, 1974; Rosengren *et al.*, 1975; Hofstee, 1976; Hofstee and Otillio, 1978).

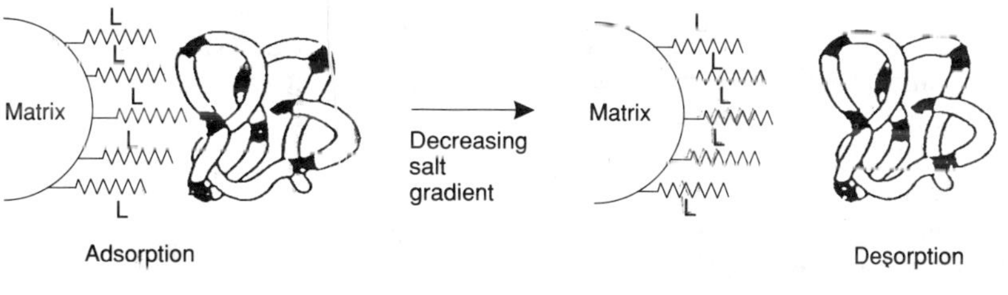

Figure 10.19 Principle of HIC: at high ionic strength the protein is adsorbed *via* hydrophobic regions (shaded) to hydrophobic ligands (L) attached to the matrix. Lowering the ionic strength weakens the interaction and promotes desorption.

Various theories have been proposed to explain the driving forces behind the interaction of hydrophobic species, which echo many of those already encountered for solid phase immobilisation (Chapter 8).

The first category (Porath *et al.*, 1977) invokes an increase in the entropy of water molecules, the disorder of the solvent increasing as the protein molecules are adsorbed onto the solid matrix. However, this effect is counterbalanced by the enthalpic contribution from these bonds breaking, and some studies have indicated (Shinoda, 1977) that these largely cancel each other out. The significance of this water entropy effect remains a topic of discussion (Porath, 1986), but is still often quoted as the driving force for hydrophobic adsorption.

A second approach (Melander and Horvath, 1977) was based on the increase in surface tension of water when salt dissolution occurred. The consideration of individual salts by this approach, which considered both hydrophobic and electrostatic contributions, yielded a convincing explanation for the order of Hofmeister's lyotropic series (Section 10.1.1). This received support from other workers such as Arakawa and Timasheff (1982, 1984a,b).

This type of approach was pursued by Srinivasan and Ruckenstein (1980), in a treatment based on van der Waals forces, and effects on the electrostatic double layer surrounding any charged groups in the matrix or protein. Very briefly, increasing the ionic strength (regardless of the nature of the ionic species used) reduces the effective thickness of the double layers around the protein and/or matrix, facilitates their close approach, and favours retention. This effect has reached saturation by buffer strengths of less than 1 M.

Above this level, the effect of different ionic species was discussed. Ions having a high charge density (i.e. those which are small and/or multiply charged, such as NH_4^+, Na^+, or SO_4^{2-}) create a more ordered solvent system, which reduces the solubility of the protein and/or increases the binding to the adsorbent. Ions having a low charge density (e.g. SCN^-, ClO_4^-), as well as organic cosolvents such as ethylene glycol, break up the water structure, in other words exhibiting chaotropic properties, and promote the solubilisation of the protein from the matrix. These 'structure-making' and 'structure-breaking' effects closely parallel the surface tension considerations discussed by Melander and Horvath, above. Some ions (e.g. $(CH_3CH_2)_4N^+$) also facilitate solubilisation due to dispersive van der Waals effects.

Finally the relevance of pH was considered, particularly in terms of possible conformational effects. These could have a dramatic influence on the surface hydrophobicity of the protein, and therefore on its adsorption behaviour.

Overall, then, there are many effects which control the hydrophobic interactions of proteins (see also Norde, 1986).

10.7.2 Key process variables of hydrophobic interaction chromatography

Properties of the solid phase

The key characteristics of the solid phase for HIC have largely been touched on already. The nature of the hydrophobic ligands is obviously critical in promoting the desired interactions with the protein. Examples of commonly encountered hydrophobic ligands are shown in Figure 10.20. The density of these ligands will also determine how strongly the protein is bound, *via* the number of binding sites involved per molecule. This is important for this approach, based as it is on relatively weak binding. Jennissen (1976) showed cooperative binding of a protein molecule by 3–5 butyl residues on a HIC matrix, while higher numbers of interactions were observed on a methyl matrix.

Any chargeable groups present may give rise to ion exchange effects, with dramatic consequences for the behaviour of adsorbed solutes under salt gradients. Similarly, any size exclusion properties of the matrix must also be taken into account, plus the usual considerations of physical and chemical robustness and particle size.

Even excluding ion exchange and size exclusion properties, the selectivity of HIC matrices is not necessarily simple. An assessment of modern HIC media was carried out by Kårsnäs and Lindblom

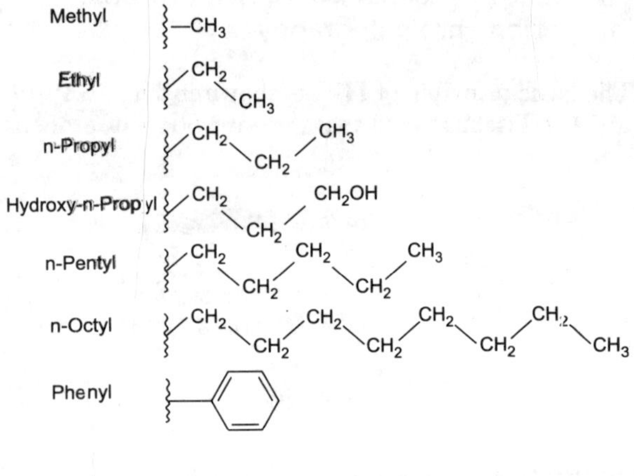

Figure 10.20 Examples of commonly encountered hydrophobic ligands.

(1992): while some matrices (e.g. those bearing octyl functions) showed a correlation between retention and protein hydrophobicity, other ligands (butyl, hexyl, phenyl) showed instead a strong inverse dependence on the number of charges present, or a combination of both modes. In other words, while an Octyl Sepharose™ CL4B column will retain the most hydrophobic protein most strongly, a Butyl Sepharose™ 4B matrix will favour retention of the protein bearing the least charge. Aromatic HIC ligands will also display some binding on account of π-complexation; these effects can in fact be deliberately enhanced by nitration (Porath and Larsson, 1978; Porath, 1986).

More than with many other chromatographic techniques, perhaps, the best choice of HIC column for a given application can only be established by empirical means.

The nature of the eluent

Where a gradient elution is employed, the most straightforward approach is to bind the sample in a buffer containing a high molarity of a good 'structure-forming' salt such as ammonium sulfate or sodium sulfate[1], then to bring about elution by a reduction in the salt concentration. Note that the agents which promote precipitation and binding to HIC media are one and the same – so that the solubility of the protein(s) may be a limiting factor in the choice of binding buffer molarity.

If the protein cannot be eluted by a simple reduction in buffer molarity, it may be necessary to incorporate an organic cosolvent, a detergent or a chaotropic agent. However, the ability of these reagents to break down the solvent structure around the adsorbed solutes may also result in some damage to the protein molecules themselves. An alternative is to employ a salt with weaker structure-forming properties such as sodium chloride as the binding agent, or a less powerful HIC matrix.

While continuous or step gradient approaches are by far the most commonly encountered in HIC separations, it is possible to use isocratic conditions on columns with low ligand density: Hjertén *et al.* (1986) reported good isocratic resolution of protein mixtures by HIC, even with quite high buffer molarities.

The pH of the eluent has little effect on hydrophobic behaviour *per se*, but it can of course affect the properties of the solute proteins so that they behave differently. Not only can a pH change alter retention properties *via* the charge state of the protein molecules, but these ionisations or deionisations may also be accompanied by conformational alterations, such that different hydrophobic areas are exposed.

This can therefore offer another means of modulating retention behaviour; to some extent the effects are unpredictable, but increasing the net charge by moving away from the p*I* will generally reduce the binding strength. Hjertén *et al.* (1986) observed, accordingly, that for many proteins the biggest changes were seen under the more extreme pH conditions, below pH 5 or above pH 8.5. In practice high pH conditions are frequently encountered as a means of encouraging elution of bound species.

Hjertén *et al.* (1986) recommended examining three different pH conditions (in the ranges 2.5–4, 6–8, and 9–11), to ascertain whether useful separation behaviour can be obtained by pH modulation in a given application. Note that ammonium sulfate yields ammonia at high pH, and is therefore not practical under these conditions: Narhi *et al.* (1989) have described the use of alternatives such as sodium sulfate and monosodium glutamate for high pH work.

Temperature

All the proposed mechanisms for HIC, whether based on entropy changes, van der Waals forces, or other hydrophobic interactions, predict stronger binding at a higher temperature, and these effects are observed in practice. Thus a reduction of temperature can be used to encourage elution of adsorbed proteins (Sofer and Nyström, 1989a), although this effect is often too weak to be useful on its own (Roe, 1989).

Like pH, temperature may also affect the conformation of protein solutes in an unpredictable way. This may explain reported instances of temperature dependences counter to the predictions above, i.e. where protein binding is weaker at the higher temperature (Visser and Strating, 1975).

Sample properties

The importance of sample properties in HIC is comparable to that described for IEC. Sample volume is relatively unimportant if a fully retaining system is employed, more critical if this is not the case.

Sample pH may interfere unpredictably with binding if it is different to that of the initial eluent, and low-salt samples can suffer problems with poor binding in the same way as high-salt samples in IEC. Once again the means to correct these problems are obvious.

Dwell time

The potential problems caused by long or variable dwell times for hydrophobically adsorbed proteins

[1] This makes HIC a convenient purification after precipitation with a salt of this type, the sample already being in the ideal matrix for binding.

have been discussed adequately in the section on IEC (Section 10.6.2). Because of the mildly hydrophobic nature of HIC matrices, this may not be a significant problem: Hjertén *et al.* (1986) observed no change to the elution profile of a protein mixture after 16 hours residence on an octyl-agarose column. However the possibility of such effects should be borne in mind, particularly if possible denaturants are included in the binding buffer. Perry and Kirby (1990) noted that HIC generally causes a greater loss of activity in monoclonal antibodies than ion exchange or Protein A chromatography.

10.7.3 Commercially available media for hydrophobic interaction chromatography

Some commercially supplied media for HIC and their key properties are listed in Table 10.4.

10.7.4 Hydrophilic interaction chromatography

Hydrophilic interaction chromatography (HILIC), as the name suggests, applies the opposite principles to HIC: interactions of hydrophilic regions in proteins

Table 10.4 Some commercially available hydrophobic interaction chromatography matrices

Supplier	Functional group	Bead size(μm)	Matrix
Boehringer Mannheim			
Poros HP2, M & P-series	High density phenyl	20	PS–DVB
Poros PE, M & P-series	Phenyl ether	20/50	PS–DVB
Poros ET, M & P-series	Ethyl ether	20	PS–DVB
Toso Haas			
TSKgel Phenyl-5PW	Phenyl	20–40	Hydrophobic resin
TSKgel Ether-5PW	Ether	20–40	Hydrophobic resin
TSKgel Butyl-NPR	Butyl	20–40	Hydrophobic resin
TSKgel Ether-5PW (Prep)	Ether	20–40	Hydrophobic resin
TSKgel Phenyl-NPR	Phenyl	20–40	Hydrophobic resin
Toyopearl Ether-650S	Ether	20–50	Hydrophobic resin
Toyopearl Ether-650M	Ether	40–90	Hydrophobic resin
Toyopearl Phenyl-650S	Phenyl	20–50	Hydrophobic resin
Toyopearl Phenyl-650M	Phenyl	40–90	Hydrophobic resin
Toyopearl Phenyl-650C	Phenyl	50–150	Hydrophobic resin
Toyopearl Butyl-650S	Butyl	20–50	Hydrophobic resin
Toyopearl Butyl-650M	Butyl	40–90	Hydrophobic resin
Toyopearl Butyl-650C	Butyl	50–150	Hydrophobic resin
BioRad			
Macro-Prep Methyl	Methyl	50	Methylacrylate
Macro-Prep Butyl	*t*-Butyl	50	Methylacrylate
Waters (Millipore)			
Protein HIC PH-814	Phenyl	8	Polymethacrylate
Protein-Pak Phenyl 5PW	Phenyl	10	Polymethacrylate
Pharmacia			
Phenyl Superose	Phenyl	24–44 (34 mean)	Cross-linked agarose
Alkyl Superose			
Phenyl Sepharose HP			
Phenyl Sepharose 6 fast flow	Phenyl	45–165 (95 mean)	Cross-linked agarose
Butyl Sepharose 4 fast flow	Butyl	45–165 (90 mean)	Cross-linked agarose
Octyl Sepharose 4 fast flow	Octyl	45–165 (90 mean)	Cross-linked agarose
Butyl Sepharose 4B	Butyl	45–165 (90 mean)	
Octyl Sepharose CL-4B	Octyl	45–165 (90 mean)	
Phenyl Sepharose CL-4B	Phenyl	45–165 (90 mean)	

with a polar column (e.g. poly(hydroxyethylaspartamide or poly(sulfoethylaspartamide)) are encouraged by the use of an organic cosolvent such as acetonitrile or propanol (Alpert, 1990).

This approach has been put to occasional good effect in challenging applications such as the removal of SDS and Coommassie Blue from electrophoresed protein samples (Jenö *et al.*, 1993).

10.8 Reversed Phase Chromatography

Reversed phase chromatography (RPC), despite its origins in synthetic organic chemistry, has proved to offer exceptional resolution for proteins—particularly smaller ones and those of a hydrophobic nature (Blondin, 1979). It is sometimes possible to resolve proteins differing only by a single amino acid residue (Wood, 1990).

10.8.1 The mechanism of reversed phase chromatography

The mechanism of reversed phase chromatography (RPC) has already been discussed in the previous section, differing only from that of HIC in the greater involvement of properties of the partially denatured protein, rather than the native conformation. The practical differences from HIC revolve around the use of more strongly hydrophobic matrices, and consequently of stronger eluents containing organic cosolvents.

10.8.2 Key process variables of reversed phase chromatography

The nature of the solid phase

There are two major types of RPC matrix: those based on silica with hydrophobic ligands attached, and those consisting of hydrophobic polymers which require no further derivatisation.

For derivatised silica columns, the ligands are both very hydrophobic (typically including C_4–C_{18} chains) and densely packed, reflecting the origin of these columns for chromatography of small molecules. Clearly the result when applied to protein separations is to offer the possibility not only of very powerful individual binding events, but also of high numbers of binding events per protein molecule.

As for HIC matrices, the choice of RPC column for a given protein application must be largely empirical.

The nature of the eluent

RPC separations are almost always run using gradient systems, the eluent consisting of a mixture of an aqueous component and a gradually-increasing proportion of organic cosolvent. With silica columns, exposed silanol groups can perturb the hydrophobic nature of the matrix[1], an effect which normally manifests itself in poor peak shape ('tailing'). To overcome this problem, the aqueous component normally contains an acid such as trifluoroacetic acid (TFA), hexafluorobutyric acid (HFBA) or phosphoric acid to lower the pH below 3. Under these conditions the silanol groups are unionised and better peak shapes are obtained, although particularly acid-sensitive proteins may suffer some denaturation.

The most commonly employed organic cosolvents (Sofer and Nyström, 1989b) are acetonitrile, isopropanol and methanol. Isopropanol has strong eluting properties but gives high back-pressures; methanol is a weak eluent giving high back-pressures; acetonitrile, preferred for most applications, is a strong eluent giving low back-pressures and good column efficiencies.

An eluotropic series of decreasing polarity (and hence increasing elution power for RPC) reads as follows: water < methanol < acetonitrile < ethanol < tetrahydrofuran (THF) < *n*-propanol < isopropanol < *n*-butanol. During an RPC run, the proportion of organic cosolvent may be increased to levels as high as 80%, provided the robustness and solubility of the protein(s) concerned are adequate. Such high cosolvent levels are more commonly encountered in small peptide purifications.

Detergent-type ion-pairing agents such as heptanesulfonate or tetraalkyl ammonium salts can be used (Roe, 1989) to increase the hydrophobicity of solutes by binding to oppositely-charged groups (Figure 10.21), effectively 'replacing' them with hydrophobic ligands.

Organic acids such as TFA, used to suppress silanol interactions, can also exhibit ion-pairing properties.

Other key process variables

The other variables discussed for HIC all have some relevance for RPC, but to different degrees. Temperature is much less important due to the greater strength and (particularly) number of binding events between the protein molecules and the adsorbent. Its weak effect on retention is even less pronounced here. Sample characteristics are generally less relevant, as the protein samples are almost always in aqueous buffers

[1] 'Capping' agents such as chlorotrimethylsilane are often used to block these silanol residues, but some may remain exposed.

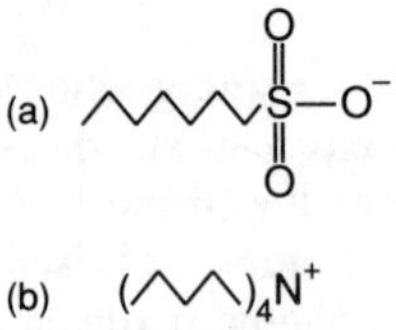

Figure 10.21 Ion pairing agents: (a) anionic—heptane sulfonate; (b) cationic—tetra-n-butylammonium.

which display weak elution properties; pH is only an issue if the ionic strength is sufficiently high to overcome the (generally) strongly acidic eluents, to cause local peak shape problems. Dwell times, on the other hand, may need to be kept as low as possible if denaturation of proteins is to be avoided; the powerful interactions of proteins with RPC matrices have been shown to cause time-dependent conformational changes (Hearn *et al.*, 1985).

10.8.3 Commercially available media for reversed phase chromatography

Some commercially available media for RPC are shown in Table 10.5, along with their key properties.

10.9 Affinity Chromatography

Affinity chromatography is a very broad term, loosely encompassing any chromatographic technique which is based on specific binding properties of proteins, rather than the general properties of size, charge and hydrophobicity which underpin the methods described so far. The basic separation approach is extremely simple (Figure 10.22). (1) A solid phase is obtained with specific binding properties for the protein of interest. (2) Conditions are chosen which facilitate binding of the target protein, while allowing other species to pass through. (3) Conditions are altered to allow elution of the target protein.

In its truest sense, affinity chromatography employs ligands with extremely high selectivity for the target protein, e.g. a monoclonal antibody with specificity for a particular antigen purified on a column bearing the antigen itself as the ligand, or an enzyme purified against an immobilised substrate analogue. It is difficult to imagine more specific purification methods for these target proteins. Products of extremely high purity can be obtained by this type of approach, even if the protein of interest occurs only as a small fraction of a complex mixture. The price to be paid for such high purification efficiency is the work required to prepare

Table 10.5 Some commercially available reverse phase matrices

Supplier	Bead size (*μm*)	Matrix/Ligand
Whatman		
Partisphere C18	5	C_{18}-Monomeric
Partisphere C8	5	C_{18}-Monomeric
Partisil 10 ODS	10	C_{18}-polymeric carbon
Partisil 10 ODS2	5	C_{18}-polymeric carbon
Partisil 5 10 ODS3	5	C_{18}-polymeric carbon
Partisil 5, 10 C8	50	C_{18}-Monomeric carbon
Partisil 5, 10 PAC	50	Polar amine cyano
Boehringer-Mannheim		
Poros R1, H, M & F-series	50	Phenyl-PS–DVB base
Poros R2, H, M & F-series	50	Phenyl ether-PS–DVB base
Toso Haas		
TSKgel ODS-80TM	5, 10, 20	C_{18}-Silica
TSKgel ODS-Prep	5, 10, 15, 30	C_{18}-Silica
TSKgel ODS-120T	5, 10, 20	C_{18}-Silica
TSKgel ODS-120A	5, 10	C_{18}-Silica
TSKgel TMS-250	10	TMS-Silica
TSKgel Oligo DNA RP	5	TMS-Silica
TSKgel Carbon 500	5	Unmodified carbon
TSKgel Phenyl-5PW RP	7	Phenyl-TSK base matrix
TSKgel Octadecyl-4PW	2.5	C_{18}-TSK base matrix
TSKgel Octadecyl-NPR	10	C_{18}-TSK base matrix

Table 10.5 (contd.)

Waters (Millipore)		
Delta-Pak C18	15	C_{18}-Silica
Delta-Pak C4	15	C_4-Silica
Bondapak C18	15–20	C_{18}-Silica
Bondapak C8	15–20	C_8-Silica
Alltech		
Microsphere 300 C4	5, 7	C_4-Silica
Microsphere 300 C8	5, 7	C_8-Silica
Microsphere 300 C18	5, 7	C_{18}-Silica
Beckman		
Ultrasphere C18	5	C_{18}-Silica (end-capped)
Ultrasphere C8	5	C_8-Silica (end-capped)
Ultrapore C3	5	C_3-Silica (end-capped)
Ultrapore C8	5	C_8-Silica (end-capped)
Alltech		
Mixed Mode (RP/IEX)		
C4/Anion	5 and 7	C_4-Spherical silica bonded with IEX
C8/Anion	5 and 7	C_8-Spherical silica bonded with IEX
C18/Anion	5 and 7	C_{18}-Spherical silica bonded with IEX
Phenyl/Anion	5 and 7	Phenyl-Spherical silica bonded with IEX
C4/Cation	5 and 7	C_4-Spherical silica bonded with IEX
C8/Cation	5 and 7	C_8-Spherical silica bonded with IEX
C18/Cation	5 and 7	C_{18}-Spherical silica bonded with IEX
Phenyl/Cation	5 and 7	Phenyl-Spherical silica bonded with IEX
C8/Anion	5 and 7	C_8-Spherical silica bonded with IEX
C18/Anion	5 and 7	C_{18}-Spherical silica bonded with IEX
C8/Cation	5 and 7	C_8-Spherical silica bonded with IEX
C18/Cation	5 and 7	C_{18}-Spherical silica bonded with IEX
Pharmacia		
Source RPC	15	C_{18}-PS–DVB
PepRPC	5 or 15	C_{18}-Silica
ProRPC	5 or 15	C_{18}-Silica
Sephasil C8	5 or 12	C_8-Silica
Sephasil C18	5 or 12	C_{18}-Silica
SuperPac Pep-S	5 or 12	C_{18}-Silica

a suitably-specific affinity matrix: it will almost certainly have to be prepared in-house, and some optimisation of the preparation conditions may be necessary before acceptable performance is obtained. Many manufacturers supply suitably 'activated' matrices, which make the covalent attachment of ligands a relatively straightforward exercise – nevertheless this route is still labour-intensive compared to many other techniques. Finally, it stands to reason that specific binding properties of the target protein must be clearly understood if such an approach is to be practical.

Moving away from this extreme, many applications of affinity chromatography rely on interactions with broader specificities. Thus to purify a monoclonal antibody it may well be sufficient to use a matrix bearing ligands such as Protein A or Protein G, bacterial proteins with high affinities for many classes of antibody from many different species. Such approaches are very useful where the target protein is the only significant member of the binding class in the sample, as is often the case with crude cell culture products for example. The applicability of this approach is sufficiently broad that there is a large market for ready-made matrices of this type: Protein A or Protein G-derivatised media are commercially available in a variety of formats. Therefore, albeit at the cost of a fairly expensive commercial product, there is no requirement for in house matrix preparation.

Proteins which bind strongly to carbohydrate moieties, for example lectins, are another commonly

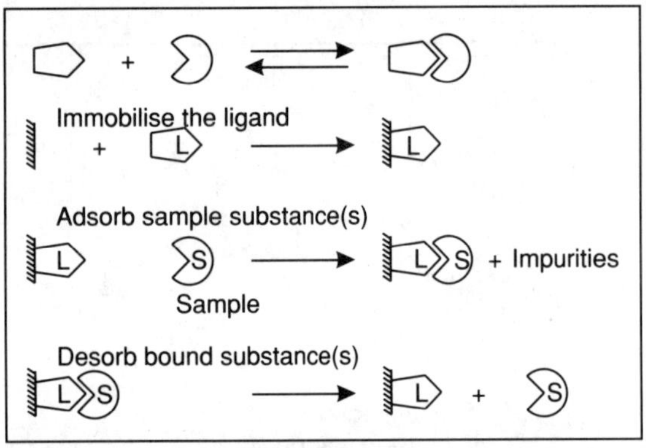

Figure 10.22 Principle of affinity chromatography. The target solute (S) is adsorbed *via* specific interactions with an immobilised ligand (L), allowing impurities to be washed through. Changing the conditions to disrupt the interactions promotes desorption.

encountered class of affinity ligand. They exhibit quite a wide specificity for glycoproteins, and again their application is broad enough to warrant the availability of commercial media.

Other branches of affinity chromatography are still further from the 'pure' model outlined earlier. For example, many dye compounds exhibit high affinities for various protein molecules, and dye-derivatised matrices are sold as affinity media. Their selectivity for different proteins is almost impossible to predict. It is certainly not an exploitation of the natural, predictable protein–ligand affinity of an antibody–antigen or enzyme–substrate pair.[1] No proteins have ever been evolved *in vivo* to bind dyestuffs first synthesised in the last 100 years! The binding powers of these matrices are serendipitous in nature, and their applicability in a given application relies very largely on trial and error. Consequently there is often little advantage in possessing a detailed knowledge of the target protein's properties, it is unlikely to help decide whether an immobilised dye matrix will yield a useful separation.

In these respects their use is really closer to approaches such as HIC, than to pure affinity chromatography. In fact, the structure of dye molecules often includes large aromatic regions of high hydrophobicity, plus highly charged groups such as sulfonic acid residues. Immobilised dye media are probably best viewed as mixed hydrophobic/ion exchange matrices which happen to be highly coloured, so the distinction between affinity chromatography and the techniques discussed in the last few sections is really not so well-defined at all.

Early in the century the principle of affinity purification was applied in the purification of amylase by adsorption onto insoluble starch (Starkenstein, 1910), but it was a long time before the approach was developed further.

The precipitation of antigens by the formation of insoluble complexes with specific antibody was described in some of the classic reports of early immunology (Heidelberger, 1939), but careful control of stoichiometry was required to avoid dissolution of the complexes. Insoluble antigen preparations were subsequently developed for diagnostic applications (Pressman *et al.*, 1942), and a logical advance was to use immobilised antigens for the purification of antibodies. In an important advance, Campbell *et al.* (1951) described the activation of cellulose by the introduction of diazonium groups, followed by derivatisation with albumin. A column of the albumin–cellulose matrix was used to adsorb specific antibody, which was then desorbed by acid treatment. The procedure described by Campbell is affinity chromatography in all but name, the only difference from modern procedures being the removal of the matrix from the column for elution.

This methodology gained widespread acceptance, and over the next two decades there was a considerable amount of work carried out both on applications, and on the procedures for protein immobilisation (Silman and Katchalski, 1966; Robbins *et al.*, 1967; Behrens *et al.*, 1967; Axén *et al.*, 1967; Porath *et al.*, 1967).

In 1968 Cuatrecasas, Wilchek and Anfinsen published the first report referring to affinity chromatography by name: these authors coupled a specific inhibitor of staphylococcal nuclease to Sepharose™, activated with cyanogen bromide, and then used this matrix to adsorb the enzyme from a crude culture mixture (Cuatrecasas *et al.*, 1968). The enzyme was eluted at pH 3, and recoveries of over 90% of very pure product were obtained. Similar results were obtained with other enzyme–inhibitor pairs.

With the rapid emergence of chromatography techniques in general, the subsequent development of affinity methods was very fast (Cuatrecasas, 1970, 1972). As with the other techniques described already, affinity media were developed based on microparticulate silica, allowing high-performance separations to be carried out on HPLC systems (Ohlson *et al.*, 1978; Ernst-Cabrera and Wilchek, 1988; Larsson *et al.*, 1983). By 1982, there were several hundred published

[1] Although some applications of immobilised dyes are based on their structural similarity to nucleotide cofactors, and are thus closer to affinity chromatography in the truest sense.

uses of affinity chromatography per year (Wilchek *et al.*, 1984), and by 1986 this was reported to be the second commonest means of protein chromatography (Bonnerjea *et al.*, 1986).

10.9.1 The mechanism of affinity chromatography

It is difficult to talk of a single mechanism for affinity chromatography in the way that is possible for SEC or ion exchange. The forces involved in forming bonds between the proteins and their affinity ligands are generally a complex combination of hydrophobic interactions, electrostatic effects, hydrogen bonds, etc.—all the same forces which occur in non-covalent complexes of proteins *in vivo* (see Chapter 2, Section 2.2.8).

The mechanism of binding is based simply on exploiting naturally occuring or serendipitous combinations of these forces, which are strong enough to cause adsorption of the protein. The mechanism of elution is to disrupt these bonds so that the protein is released.

10.9.2 Key process variables of affinity chromatography

While some general guidelines can be stated describing the key process variables of the binding stage, it is again difficult to be too definitive about elution variables. These will depend on the combination of forces responsible for the specific protein–ligand interaction of interest: if these are principally hydrophobic in nature the discussion of eluent variables in Section 10.7.2 may apply, but if they are principally ionic Section 10.6.2 may be more relevant. Thus elution may be encouraged by reducing buffer molarity in one instance, and by increasing it in another. Where, as is commonly the case, both modes are operating simultaneously, ionic strength may not have a significant effect either way.

The nature of the support

As with all the other branches of chromatography discussed in Sections 10.6–10.8, a porous matrix is usually favoured to give the greatest possible surface area in a given volume, but the pores should be large enough to allow free access of protein solutes and (in systems with continuous elution) to avoid size exclusion effects. Physical and chemical robustness are also as important in this as in any other chromatographic methodology.

The ideal affinity matrix should show no non-specific adsorption of solutes, and should therefore be devoid of charged or hydrophobic groups; however it must possess sufficient reactivity to facilitate the attachment of affinity ligands. These requirements severely limit the range of suitable matrices, and those possessing hydroxyl functions dominate the field. Except under the most alkaline conditions these are uncharged and hydrophilic, yet reactive enough for derivatisation to be practicable.

Conveniently, the commonest chromatography supports have traditionally been based on either carbohydrates such as dextran or agarose, or on silica, all of which bear hydroxyl groups. These matrices are therefore well suited to affinity work, the soft carbohydrate gels for general purposes, and silica for high-performance applications with small particle sizes. The silica is generally treated with glycerylpropylsilane (GPS) or a similar reagent to suppress the ionic nature of the native silanol groups.

The nature of the affinity ligand

There is an optimum range for ligand affinity in this technique: if the affinity is too low, there will be insufficient binding of the target protein compared to other species; if it is too high, on the other hand, it may be impossible to free the protein from the protein–ligand complex without using conditions so severe that they cause damage in themselves. This is a serious constraint of affinity chromatography, as the types of interaction that create the protein–ligand bonds are also those which maintain the tertiary structure of the protein itself. Obtaining differential disruption of these interactions can be challenging, therefore. There is a wide measure of agreement that the useful upper limit of affinity constant for this technique is around 10^{-8} M^{1} (Pharmacia, 1993b; Larsson *et al.*, 1983; Jack and Wade, 1987; Phillips, 1989; Pharmacia, 1993b).

At the low end, limits in the order of 10^{-4}–10^{-5} M^{-1} have been quoted (Pharmacia, 1993b; Larsson *et al.*, 1983; Philips 1989), although successful adsorption at the 10^{-3} M^{-1} level has been reported (Cuatrecasas, 1972). Where concentrating conditions are used so that the target protein builds up on the matrix, a property of equilibrium systems aids the binding process—a factor which is probably significant in low-affinity situations. Rather grandly termed 'the Concept of Progressively Perpetuating Effectiveness' by Cuatrecasas (1972), this effect is based on the fact that a higher concentration of protein in the buffer surrounding the matrix drives the equilibrium towards greater levels of binding. Therefore as the concentration builds up in the loading zone the retention grows stronger, and this in turn helps to prevent the loading zone from spreading. Subject to limitations discussed below, simply increasing the ligand density also favours binding, and may be an important factor in low-affinity situations.

It should be borne in mind that the affinity constant of the ligand may be altered by the immobilisation process. Depending on the immobilisation chemistry employed, this may cause a serious reduction in affinity from the native value: this will be discussed in the next section.

Where the affinity of the ligand is too high for useful chromatographic performance, an option is to modify it chemically: Murphy *et al.* (1976), for example, used residue-specific modification reagents to modify specific Trp, Tyr and Met residues in immobilised antigen or antibody. This allowed elutions to be carried out using alkaline buffers rather than those based on guanidinium salts.

Without knowing affinity constant values it may be possible to select suitable ligands empirically while avoiding having to immobilise them. Calton (1984), for example, described an immunoassay technique for selecting low-affinity Mabs from a population: these were subsequently used in immunoaffinity adsorbents, from which elution of the antigen could be achieved under extremely mild conditions.

Ligand immobilisation: effects on affinity

Once a suitable ligand has been identified, it must be immobilised to form the chromatography matrix. Ideally this process should have no effect on the affinity of the ligand, but this is not necessarily the case in practice.

For a ligand of low molecular mass (Figure 10.23), the number of possible derivatisation sites is small and (probably) known. Therefore it may be possible to carry out the immobilisation reaction using a part of the molecule which is not critical to its protein-binding properties, and, if there is only a single group of the correct reactivity, with identical orientation of the bound species. In this case the affinity of the immobilised ligand may be relatively uniform, and very similar to that of the native molecule (Figure 10.23(a) and (b)).

However, as the surface of the solid phase is generally porous (or at least highly irregular at the Ångstrom level), there will probably be some reduction in affinity due to steric hindrance. While the most exposed ligands may display behaviour close to that of the mobile molecule, the more deeply buried molecules will display progressively lower affinities – hence

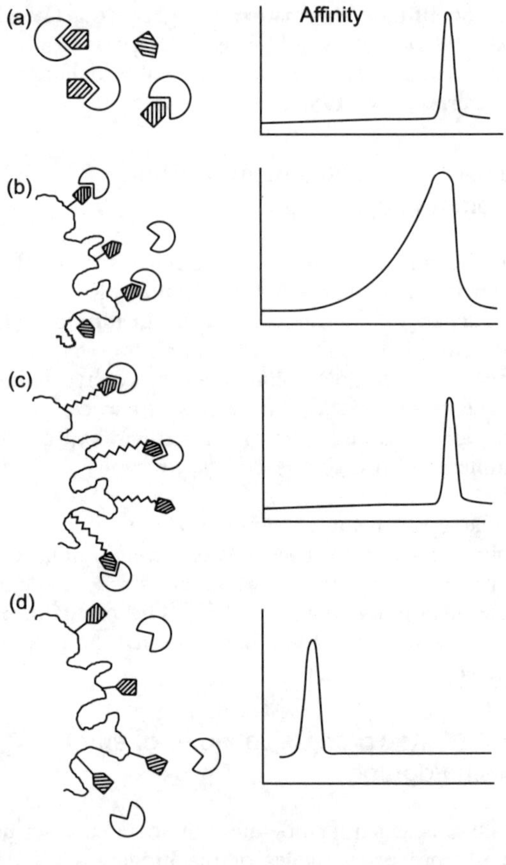

Figure 10.23 Effect on the affinity of a small ligand under various conditions; see text for details.

the tailing of the affinity plot shown in Figure 10.23(b).

To counteract this effect, it is common practice to employ a spacer group (Figure 10.23(c)), which distances the ligand from the matrix surface. This approach yields the best possibility of observing affinity equivalent to the mobile ligand: the relative improvement will depend totally on the binding site topology. Some effective reduction in affinity may still be observed due to ligand density effects (see below).

The nature of the spacer group itself may be very important with regard to properties such as length, flexibility, and hydrophobicity (Larsson *et al.*, 1983; Walters, 1985; Pharmacia, 1993b). Lowe *et al.* (1973) carried out a systematic study with nucleotide-binding enzymes, and observed that there was an optimal spacer length of 8–10 Å. A flexible spacer may, if long enough, allow the ligand to fold back onto the solid matrix, reducing the benefit of the approach. A spacer which is hydrophobic or which bears charged groups, on the other hand, may provide an alternative mechanism for binding: indeed, hydrophobic interaction chromatography (Section 10.7) was founded on observations of such behaviour from affinity matrices (Er-El *et al.*, 1972; Hofstee, 1973).

Needless to say, immobilising a small ligand *via* a site critical to its binding behaviour can have catastrophic effects on affinity (Figure 10.23(d)), regardless of whether a spacer is employed.

With a large protein ligand, the situation is often rather different. It is sometimes possible to identify a single point of derivatisation remote from the binding site (and hence affinity performance similar to that just described for small molecules); very often, however, the immobilisation may be carried out *via* groups such as amines, which are present in large numbers at a variety of distances from the binding site. Multiple binding of ligands may occur too, and this may result in conformational strains which lower affinity.

In this situation (Figure 10.24), the bound molecules will probably display a broad population of affinity values, from very poor binding at one extreme, to performance close to that of the mobile molecule at the other (Figure 10.24(b)). Using spacers may well still improve the affinities generally.

The stylised affinity plots depicted in Figures 10.23 and 10.24 represent fairly extreme cases. In real life the performance obtained may range anywhere between them, depending on size, immobilisation chemistry, binding site topology, etc.

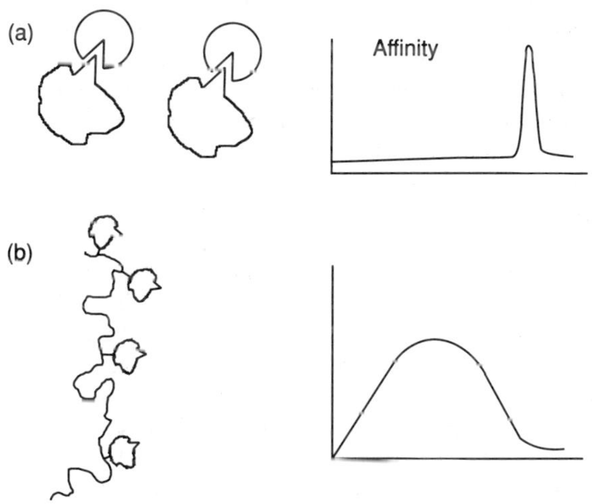

Figure 10.24 Effect on the affinity of a large ligands on immobilisation; see text for details.

Ligand immobilisation: effects of ligand density

For small molecules, a monolayer is equivalent to a surface concentration of a few μmol/m^2, while for a fairly large protein (100 × 100 Å) the concentration is of the order of 0.01 μmol/m^2 (Walters, 1985). Therefore a single protein molecule could theoretically occupy an area of matrix containing a few hundred separate ligands.

The number of sites on the protein which exhibit binding to the ligand may well be very low, therefore there is no need for ligand densities this high. In fact it is far more likely that such an excess of ligands will hinder the desired interactions, and accordingly there is generally an optimal range of ligand density for affinity chromatography.

Where the size of the affinity ligand is of the same order of magnitude as the target protein—generally when the ligand is a protein too—less pronounced effects might be expected, as monolayers of both proteins will have approximately similar surface concentrations. The situation at the matrix surface, however, is probably much more complex than this monolayer model, and many other factors may affect the observed relationship between binding capacity and surface coverage. This is borne out by the rather varied findings that have been reported from studies on this relationship, even within a fairly limited field such as the preparation of Sepharose–IgG matrices for immunoaffinity purification.

Thus Weston and Scorer (1978) observed an optimum binding capacity at 3–4 mg IgG/g Sepharose (approximately 1 mg/ml), with a reduction in binding at higher levels. Pfeiffer *et al.* (1987) reported similar qualitative behaviour: media containing 8–9 mg/ml IgG displayed approximately double the binding capacity of those bearing 10 mg/ml. Eveleigh and Levy (1977), however, saw increased binding up to around 15 mg/ml, and no reduction in capacity at the highest IgG loadings (22 mg/ml) obtained[1].

General guidelines on ligand density have been published (Pharmacia, 1993b): 1–20 μmol of ligand per ml of gel (typically 2 μmol/ml), or (for protein ligands) 5–10 mg/ml gel. As noted earlier, higher ligand densities may be advisable where the affinity for the target protein is low, although non-specific binding may also increase too.

[1] These gels with high ligand density were able to bind just as great a mass of antigen as less derivatised matrices: their molar binding capacity was however reduced, from 60% of the theoretical maximum (the best observed with low-density gels) to around 10%. Thus, although there appeared to be no reduction in binding capacity due to steric effects, the use of antibody was less efficient at these high loadings.

Ligand immobilisation: chemistry

Some of the individual chemistries applied to the preparation of affinity media will be discussed later, but some general factors will be considered briefly here.

The most obvious need is for a means to connect reactive groups in the matrix and in the ligand. Because many of the most popular matrices are based on carbohydrates or silica, the commonest approaches rely on activation of hydroxyl functions with powerful electrophiles, or under strongly alkaline conditions where deprotonation yields the highly nucleophilic RO^- anion. In the ligands, the reactive groups available are exactly the same as those discussed in Chapter 5 (for proteins) or Chapter 6 (for small molecules), and therefore functionalities targetted at amines, carboxylic acids and thiols predominate.

Overall, then, the commonest methods for the formation of bonds between the solid phase and the ligands employ bifunctional reagents capable of forming hydroxyl–amine, hydroxyl–acid or hydroxyl–thiol linkages. It will be noted from earlier chapters that reagents capable of forming such links in an aqueous environment are fairly rare (due primarily to competition from the hydroxyl groups of the water). However, where the hydroxylic component is a solid non-proteinaceous species it is generally possible to derivatise the hydroxyl groups as a separate stage under driving conditions; the reactive functionalities thus introduced can then be used in a milder environment to couple the ligands of interest.

Media whose hydroxyl functions have been activated in this way can often be stored for long periods of time, and many such materials are commercially available, e.g. the cyanogen bromide (CNBr)-activated media which still dominate the field.

The preparation of affinity media from these products is a simple, one-step reaction which can be carried out in an aqueous environment.

Ideally there would be no loss (leakage) of the affinity ligand from the matrix during use, as this can lead to product contamination problems. However it has been stated that 'there is no such thing as leak-proof media' (Sofer and Nyström, 1989b), and this possibility needs to be considered. Thorough washing of the affinity matrix before use is obviously the simplest expedient to minimise such problems. For applications where product purity is of paramount importance, such as the purification of pharmaceutical proteins, sensitive detection techniques are employed to demonstrate the absence of stripped affinity ligand in the product (Sofer and Nyström, 1989b).

Finally, the coupling of ligands to the activated solid phase may leave some reactive groups intact, and it is beneficial to block these by a suitable quenching reaction—generally employing a low-molecular-mass reagent of appropriate reactivity, whose incorporation will not lead to non-specific adsorption problems.

The nature of the eluent

By far the most common way to execute an affinity separation is to use a simple two-stage step gradient: in the first stage the protein of interest is adsorbed from the sample onto the affinity matrix, while all other components ideally pass through unbound; in the second the target protein is eluted from the matrix.

The requirements of the eluent for the adsorption stage are not too demanding, simply that it facilitates the desired protein–ligand interaction while suppressing non-specific effects as far as possible. For many applications a simple buffer of neutral pH, with an ionic strength high enough to suppress any residual ion exchange effects, will suffice. However some ligand–protein interactions may require a particular pH at which the binding is optimal, or the presence of cofactors such as specific metal ions. Such considerations need to be taken into account too, but it is impossible to generalise these requirements.

The eluting buffer, on the other hand, must meet the challenge of disrupting the protein–ligand bonds without causing irreversible damage to the protein itself. While a multitude of different protein–ligand interactions may be encountered in affinity chromatography, the forces underlying them are relatively few, and so are the means of disrupting them. Therefore there is a large degree of commonality in the non-specific approaches to elution.

Where harsh conditions are employed, such as extreme pH or denaturants, it is advisable to keep the exposure time of the target protein down to a minimum. To some extent this can be influenced by the elution profile, but it may also be wise to treat the product immediately after collection to neutralise the harmful properties of the matrix. For pH this may be a simple adjustment back to neutral conditions (although too rapid a pH shift can be harmful in its own right), while with denaturants, chaotropes, etc., the product can either be diluted into a 'safe' buffer, or buffer-exchanged by SEC or some other rapid means.

As an alternative to non-specific procedures, an eluent can be employed which exploits the specific binding properties of the ligand such that the target protein is displaced from the matrix. A solute which shows a stronger affinity for the matrix than the product (or for the product than the matrix) is ideal in this respect, but the nature of binding equilibria means that a weaker binder will also work adequately provided it is present at a reasonably high concentration. Specific elution is a mild option compared to many of the non-specific procedures, and has been described as the best approach (Wilchek *et al.*, 1984): its successful application may also be useful confirmation that 'genuine' biospecific affinity is responsible for adsorption. Specific elution does have some kinetic drawbacks (discussed below), and may also pose a new purification problem in the separation of the product from the competing reagent.

Below are listed some of the different eluent characteristics employed to elute bound proteins from affinity matrices. Note that almost all of these have been encountered in previous sections, reflecting the fact that affinity adsorption is made up principally of a combination of the hydrophobic, hydrogen-bonding and electrostatic effects underlying all the other adsorptive chromatography techniques.

Ionic strength

It is difficult to predict the effect of increasing or lowering ionic strength, as it will depend on whether electrostatic or hydrophobic forces are most important in the protein–ligand interaction. Using an increasing salt gradient is most commonly recommended however (Pharmacia, 1993b), for example up to 1 M sodium chloride.

If successful this approach is a lot milder than many of the other non-specific elution techniques, and may well be worth trying before harsher reagents are employed, particularly if the protein of interest is labile.

Temperature

Another mild means of achieving elution is to modulate the temperature, but again it may have to be empirically determined whether a warmer or cooler environment gives the desired effect. Hydrophobic forces are lowered by a reduction in temperature (see Section 10.7.2), and immunoadsorbents have been described where the antibody–antigen complex has been dissociated by moving from ambient temperature to 4 °C (Dawes, 1986). On the other hand, the increased kinetic energy of the system at a higher temperature may favour dissociation over association, so this approach too may be worth trying (Larsson *et al.*, 1983). Finally, temperature-dependent conformation changes may also facilitate elution, and again the change may be brought about either by raising or lowering the temperature, depending on the protein(s) involved.

pH

Altering the pH of a protein changes the charge state of a number of its amino acids, which in turn can alter the conformation due to the disappearance of existing attractive or repellent forces between neighbouring residues, and the appearance of new ones. Therefore a pH change can alter the protein's structure sufficiently to interfere with ligand binding.

With a protein ligand this effect may be exacerbated by the simultaneous alteration of its own structure, so that both binding partners are disrupted. Even with a small ligand, a shift through the pK_a of a critical group can have similar effects.

Altering pH is one of the most commonly encountered means of eluting bound solutes from affinity matrices. In principle an alteration in either direction, away from the optimum binding pH, is equally likely to succeed. However acidic eluents are much more commonly encountered in practice: for silica-based columns this is partly related to the instability of the support under alkaline conditions, but in general proteins are more susceptible to irreversible damage at high pH. Eluents with a pH between 2 and 4 are often used, but even more acidic conditions are sometimes employed – particularly for high-performance work where run times may be very short. With silica columns Phillips (1985) recommended altering the pH by a steep gradient rather than a step change, in this way degradation of the matrix by acid dissolution was reduced.

Another way of minimising the exposure of the column, and of the bound proteins, is to use a discontinuous pulse approach (Larsson *et al.*, 1983). Only a short time at low pH may be required to effect elution, and this can be conveniently achieved by injecting the acid buffer as if it were a sample, rather than by modifying the pump conditions. Thus a small, acidic pulse moves through the column and elutes the product; the column rapidly reequilibrates to the original conditions and is quickly ready for the next purification sample.

Chaotropes

The ability of chaotropes to disrupt hydrophobic interactions has been discussed in Section 10.7, and these often make very effective eluents for affinity chromatography. Typically high concentrations (3 M) of reagents such as sodium iodide or sodium thiocyanate are employed. This approach has the advantage that standard conditions will disrupt all protein–ligand interactions to some extent. There is generally no need to optimise for a given application, unlike some of the parameters discussed so far.

Sodium chloride is mildly chaotropic, and this property may be a contributing factor in the use of NaCl gradients to achieve elution of bound solutes (see Ionic Strength, above).

Organic cosolvents

Again as discussed in earlier sections, organic cosolvents can be used to disrupt hydrophobic interactions by lowering the polarity of the eluent, and also perhaps by effects on protein conformation. This approach too can be exploited for affinity chromatography; proportions of dioxane up to 10% or ethylene glycol up to 50% have been recommended (Pharmacia, 1993b), based on work by Andersson *et al.* (1979) on immunoadsorbents.

Denaturants

The use of chaotropes or organic cosolvents will generally cause a degree of denaturation of the target protein, contributing to their eluting powers albeit with some risk of irreparable damage to the product. Harsher still is the use of powerful denaturants such as 6 M guanidinium hydrochloride, or 8 M urea. These agents are capable of disrupting the conformation of the protein beyond recognition, and will certainly break down protein–ligand interactions.

This does not necessarily guarantee that the target protein will be eluted smoothly, as (particularly where the ligand is proteinaceous too) there is ample scope for interactions between the newly exposed residues. New hydrophobic interactions in particular may arise, as many of the less polar residues are generally 'buried' in the native protein structure. On the whole though, denaturants are powerful eluents.

Denaturation of the product is a heavy price to pay, however, and this approach may not be practicable for less robust proteins. In all cases it is wise to minimise the exposure as far as possible. Partial denaturation may well be sufficient to promote elution and is much more likely to be reversible.

Refolding of denatured proteins is a topic with a large literature of its own, particularly from the field of recombinant protein production, and cannot be covered here. As a general rule however, it is often better to lower the denaturant concentration gradually, e.g. by dialysis. This gives the molecules more time to form the correct intramolecular bonds as the the environment becomes slowly milder. A rapid removal of denaturant, on the other hand, is likely to result in the formation of intermolecular links between the wide variety of exposed residues, before the correct intramolecular bonds have a chance to take shape. A large proportion of aggregated product can be the result.

Affinity elution: displacement by specific competitors

The last major category of elution technique differs from those just described, in using specific rather than non-specific reagents to disrupt the protein–ligand bond. There are two main approaches, using a reagent which either (Figure 10.25(a)) competes with the target protein for ligand sites, or (Figure 10.25(b)) competes with the immobilised ligand for the protein.

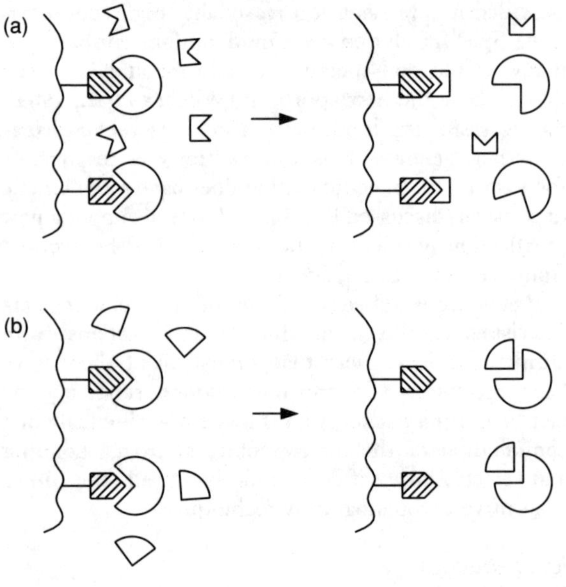

Figure 10.25 Affinity elution: the interaction between ligand and protein is disrupted by competition for the ligand (a) or for the binding site on the protein (b).

In each case the bound protein is released into the eluate, either mixed with the competitive binder or bound to the competitive ligand, depending which approach has been employed. Unless the product can be used in this form a further purification must then be carried out to yield pure protein; in this respect using a competitive binder rather than a competitive ligand may be advantageous, as there is no protein–ligand bond to be broken in the eluted product.

Affinity elution relies on the equilibrium nature of the protein–ligand bond, and the rate of displacement by the competitor is limited by the rate of dissociation of this complex. This is independent of the concentration or affinity of the competitor (Cuatrecasas, 1972), being a first-order process governed only by the affinity constant of the specific interaction. The competitive binder cannot compete for ligand sites until this dissociation has taken place. Its greater concentration will then make it more likely to occupy the newly exposed ligand site than the original protein molecule, whose reassociation is therefore prevented.

The competitive binder will also undergo equilibrium dissociation at a rate determined by the affinity constant of its specific interaction with the ligand. This frees up ligand sites, which can interfere with the elution process by re-binding the newly released protein. This effect can be minimised, however, by using a high concentration of the competitor so that reassociation of the target protein is a very rare event.

If the concentration of the competitor is high enough, it does not matter if its affinity for the ligand is somewhat lower than that of the target protein. Higher-affinity competitors, conversely, are effective at lower concentrations.

Because the rate of the displacement process is governed by the initial dissociation of bound protein, the elution is gradual in nature. The product cannot be wholly eluted until every molecule has dissociated from the ligand, and the time required for this to happen is governed by the dissociative rate constant of the interaction. The values of these constants can vary over an extremely wide range (10^3–10^{-5} s^{-1}) for affinity interactions, giving dissociation half-lives ranging from milliseconds to hours (Walters, 1985). Therefore, even if the concentration and affinity of the competitor are great enough to prevent any reassociation, the time for product elution may be impracticably large for high-affinity situations. Band-broadening is the result if the dissociation half-life is too long.

This is in marked contrast to a non-specific technique such as a pH shift: the conformation of the protein molecules will change within a time limited only by the ability of the buffer to penetrate the matrix fully, and by the speed of the conformational change itself—both rapid processes. All the protein molecules will then display a greatly reduced affinity for the matrix ligands, so dissociation will occur rapidly, reassociation will be rare, and the product will elute in a sharp band.

It follows from the discussion above that fast flow rates will exacerbate the band-broadening problem, and the slowest practicable rates are generally employed for such elutions. Taking this line still further, a useful approach may be to stop the flow altogether with the competitor-containing eluent present, and allow a prolonged period before elution (Cuatrecasas, 1972). In this way desorption of high-affinity complexes with large dissociation half-lives may be feasible.

The dissociative rate constant is heavily dependent on temperature, and working in a warmer environment is another means of increasing the rate of affinity elution. Note that it is unimportant that the associative rate constant may also increase, because the excess of competitor will discourage reassociation as discussed above.

Other elution techniques

A number of other techniques have been used for elution in affinity chromatography, many applicable only in very specific circumstances.

Applying an external electric field to the matrix is a powerful means of encouraging elution, and this electrophoretic approach has been widely employed.

Where the bound protein can form a ternary complex with two ligands, one of these can be immobilised to provide the affinity matrix. When it is desired to elute the bound protein, adding a mixture of the two ligands can be very effective compared to simple competition approaches, the reason being that the protein can form a strong ternary complex on dissociation from the solid phase, losing its affinity for the original immobilised ligand. For example, Larsson *et al.* (1983) adsorbed lactate dehydrogenase (LDH) to an AMP-silica column: while they were unable to elute the enzyme with NAD, a mixture of NAD and oxalate (capable of forming a ternary complex with LDH) was effective.

Proteins whose structure is stabilised by specific metal ions can be dealt with very elegantly (Liebman *et al.*, 1985). Monoclonal antibodies are selected which recognise only the metal-stabilised conformation of the protein, and used as immunoadsorbents. Elution can be achieved by adding a chelating agent such as EDTA, which removes the metal ions from the system; the subsequent conformation change in the protein releases it from the antibody matrix, as the Mab is no longer capable of binding it.

If the ligand is bound to the solid phase by a cleavable linker, the product can be released by carrying out the appropriate cleavage reaction. For example, if a linker incorporating a disulfide link is employed, the product–ligand complex can be eluted by treatment with a suitable thiol such as dithiothreitol. This is an inconvenient approach if the protein product is required in a pure form, as the protein–ligand bond still remains in place; simple techniques such as size exclusion may be successful in separating these complexes, however (Miller *et al.*, 1997).

Along the same lines, and particularly for recombinant fusion proteins where extra residues can be introduced at will, it may be possible to place a cleavable tag at one terminus of the amino acid sequence. This can then be cleaved to free the product.

10.9.3 Ligand immobilisation techniques for affinity chromatography

The approaches to ligand immobilisation for affinity chromatography are, in practical terms, influenced heavily by the convenience of various commercially available reagents for the purpose. It is quite possible to prepare affinity matrices completely *ab initio*, starting with a matrix such as cross-linked agarose for soft-gel work, or microparticulate silica for high performance applications.

For small ligands it may be possible to synthesise an electrophilic derivative which will react directly with the solid phase. Because these matrices are fairly robust, much more driving conditions can be used than those discussed for protein derivatisation in Chapter 6: in particular the ability to work at a raised temperature, in non-aqueous media, facilitates reaction with the relatively passive hydroxyl groups. Rosengren and Glad (1975), for example, have described a number of *O*-acylations and *O*-alkylations of cross-linked agarose under such conditions, using electrophiles such as isocyanates, and alkyl and acyl halides.

Such approaches can be very efficient in terms of ligand consumption, and may therefore be worth examining for materials which are scarce or difficult to prepare. However, organic synthesis in itself can be time-consuming and wasteful, so the applicability of this approach will depend strongly on whether the ligand bears any suitably electrophilic groups in its native form, or the ease with which it can be activated. Finally, because reaction occurs directly with the solid matrix, if an extra spacer arm is required it can only be introduced by incorporating it in the reactive ligand itself, again adding to the complexity of the synthetic stage.

Although the direct approach has some applicability for small ligands, then, it does not enjoy widespread popularity, particularly where an extensive synthetic stage is necessitated. For protein ligands this approach is virtually useless. The simplest alternative is to activate the solid phase in some way, which will render it reactive to the ligand of interest under appropriately mild conditions. Some of the chemistries which can be applied in this way will be discussed briefly in Section 10.9.3.1; the best-known is probably that based on cyanogen bromide (CNBr), first described for solid phase activation by Axén *et al.* in 1967. Cyanogen bromide activation of hydroxylic residues generates a number of species with a usefully wide reactivity. Cyanogen bromide itself, however, is an unpleasant and toxic compound, and most chromatographers are happy to avoid working with it directly.

Commercial suppliers have addressed this desire by marketing ready-activated media (e.g. CNBr-activated Sepharose™ 4B from Pharmacia); many other such products are available. Thus at the extra expense of paying for these materials, the chemistry required to immobilise the ligand can be reduced to a convenient minimum. This is by far the most popular approach to the preparation of affinity matrices.

Finally, some affinity media are so widely applicable that there is a market for them *per se*; these contain ligands with a relatively broad specificity such as Protein A and Protein G for antibody purification, lectins for glycoproteins, or nucleotides for NAD^+ or $NADP^+$-dependent enzymes.

In summary then, affinity matrices can be obtained by a variety of means, and the relative importance of cost *vs.* convenience may determine the route chosen. In general the simplicity with which ready-activated or ready-derivatised media can be used has made them a very popular option. For commercial manufacturing applications, particularly in the more regulated health markets, it may also be much easier to rely on an established proprietary product than to optimise and validate an in-house immobilisation procedure.

10.9.3.1 Chemistries for derivatisation of hydroxylic media

Cyanogen bromide and related procedures

Cyanogen bromide was first described for the activation of agarose gels by Axén *et al.* (1967), while Porath *et al.* (1967) demonstrated that chymotrypsin could be immobilised in this way with reasonable retention of activity. Wilchek *et al.* (1984) published a good review of cyanogen bromide activation chemistry, which is summarised briefly here.

The carbon atom in cyanogen bromide is electrophilic in nature, and is reactive towards nucleophiles such as hydroxyl groups, particularly at high pH where alkoxide ions are present; HBr is eliminated and the product is a cyanate ester (Figure 10.26(a)). This still contains a highly electrophilic carbon, which reacts readily with nucleophiles such as amines. The major product is probably an isourea derivative as shown in Figure 10.26(b) (Svensson, 1973), although other products are possible which create a link to the protein.

However, the cyanate ester is susceptible to a number of side reactions. First, since the cyanogen bromide activation is typically carried out under aqueous conditions, a proportion of the esters hydrolyse to inert carbamates (Figure 10.26(c)). Secondly, the electrophilic cyanate may be able to react with another nearby (intra- or interchain) hydroxyl group, yielding a less reactive imidocarbonate. The likelihood of the latter reaction depends on the particular matrix: Kohn

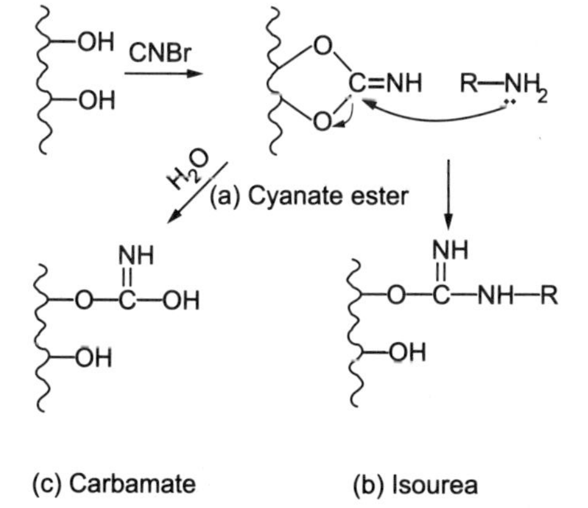

Figure 10.26 Chemistry of CNBr activation; see text for details.

and Wilchek (1982), for example, demonstrated that coupling *via* cyanate esters predominated with Sepharose, while imidocarbonates formed the main route in Sephadex.

While cyanate esters are the most reactive products of CNBr activation, they are only stable below pH 4. For efficient immobilisation of an amine-bearing ligand, the coupling reaction must be carried out at a pH where there is a reasonable population of deprotonated NH_2 groups. For protein amines this necessitates a neutral to basic pH, so there is a mismatch between the two requirements.

As a consequence, the aminolysis reaction is typically carried out under conditions where the proportion of unhydrolysed cyanate esters may be dropping rapidly. Under these conditions reaction with imidocarbonates may predominate, as these are more stable under alkaline conditions. Derivatisation of the protein with a hydrazide or aromatic amine group is a possible answer: due to their low pK_a they can be coupled at acidic pH.

Using cyanogen bromide to couple mAbs to Sepharose, Pfeiffer *et al.* (1987) observed the best results using high CNBr concentrations (100 mg/ml); at this level 90% of the available antibody was coupled within one hour at any pH from 6.4 to 8.7. Using lower CNBr concentrations brought out a greater degree of pH dependence: at 25 mg/ml, overnight reaction was required to obtain 78% incorporation at pH 6.4.

Typically coupling is carried out for a couple of hours at around pH 8; bicarbonate and borate buffers are recommended (Pharmacia, 1993b; Dean *et al.*, 1985b), while those bearing nucleophilic groups (e.g. Tris, glycine, ammonium salts) should definitely be avoided. The stability of the activated matrix is improved by working at 4 °C.

Wilchek *et al.* (1984) described more efficient procedures for the preparation of Sepharose cyanate esters, for example by using triethylamine in place of inorganic base. This yielded a product which contained no imidocarbonates or carbamates, and which contained cyanate levels far in excess of earlier methods. Other approaches used different sources of active cyanogen, such as *p*-nitrophenyl cyanate (a conveniently handled non-volatile solid), *N*-cyanotriethylammonium tetrafluoroborate and 1-cyano-4-dimethyl aminopyridinium tetrafluoroborate.

Regardless of the means of generating the cyanate esters, affinity matrices derived from these groups have certain drawbacks. The N-substituted isoureas typically formed by reaction with an amine-containing ligand are somewhat labile, particularly in the presence of nucleophiles. They also have a high pK_a (around 9.4), meaning that they are positively charged at neutral pH and providing the matrix with unwelcome anion exchange capabilities.

Other techniques for activation of hydroxylic media

There are many other routes by which hydroxylic media can be activated for reaction with proteins and other ligands. Mostly these provide an electrophilic matrix which will react with amines, thiols, etc. Some of these are categorised in Figure 10.27.

One category are those which require that the matrix is derivatised under non-aqueous conditions, for example carbonyldiimidazole (Bethell *et al.*, 1979; Dean *et al.*, 1985b), tresyl and tosyl chloride (Nilsson and Mosbach, 1981; Dean *et al.*, 1985b), chloroformates (Wilchek *et al.*, 1984) and FMP (Ngo, 1986).

The second class are those which can be carried out under aqueous conditions. These include cyanuric chloride and other triazines (Lowe and Pearson, 1984; Dean *et al.*, 1985b), bromoacetyl bromide (Egly and Porath, 1978), divinylsulfone (Scopes and Porath, 1990; Pepper, 1992), benzoquinone (Brandt *et al.*, 1975; Egly and Porath, 1978), bis-oxiranes (Dean *et al.*, 1985b; Cusack and Beynon, 1989; Pepper, 1992), epichlorohydrin (Scopes and Porath, 1990) and periodate oxidation (Dean *et al.*, 1985b; Pepper, 1992).

Further derivatisation of activated hydroxylic media

The chemistries outlined above offer a wide variety of reactive functions for ligand immobilisation, but this can be extended by carrying out further derivatisation on the primary products. As an example (Ernst-Cabrera and Wilchek, 1988; Larsson *et al.*, 1983), the

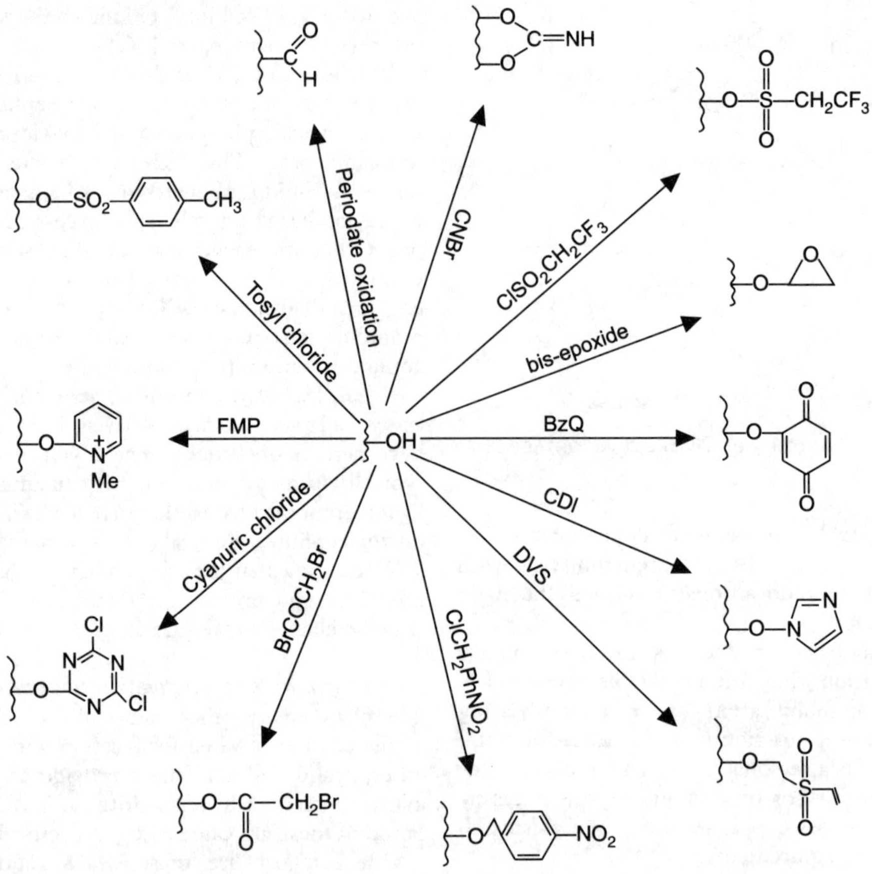

Figure 10.27 Routes for activation of hydroxylic media.

oxirane-derivatised silica prepared by reaction with γ-glycidoxypropyltrimethoxysilane can be treated in a number of different ways to generate new reactive media (Figure 10.28).

Periodate-oxidised Sepharose can be reacted with polymeric hydrazides (Wilchek *et al.*, 1984), yielding a hydrazide solid phase.

CNBr-activated Sepharose can be reacted with 6-aminohexanoic acid to yield a carboxylated matrix, and with 1,6-diaminohexane to introduce amine groups (Figure 10.29). These not only can be used directly, e.g. *via* carbodiimide coupling, but also can form the basis for still further derivatisation as illustrated (Pharmacia, 1993b). Groups such as active esters, thiols, haloacetamides, and diazonium salts thus become available for ligand immobilisation.

Active ester-derivatised solid phases have, of course, attracted considerable attention (Cuatrecasas and Parikh, 1972; Dean *et al.*, 1985c; Wilchek and Miron, 1987; Wilchek *et al.*, 1994).

With this availability of reactive moieties, there is really no limit to the groups which can be incorporated if so desired, provided one is prepared to carry out the necessary chemistry. All the chemistries described in Chapters 5 and 6 are generally applicable here, including complex heterobifunctional approaches: Aithal *et al.* (1988), for example, immobilised a cytosolic protein (Figure 10.30) *via* thiolation of the ligand, which was subsequently coupled to maleimide groups introduced onto an amine-bearing solid phase with the heterobifunctional reagent sulfo-MBS.

Although a large majority of immobilisation techniques for protein ligands conveniently exploit lysine residues, the more uncommon techniques are useful where this approach is precluded for structural reasons. Cytochrome *c*, for example, contains essential lysine residues in the binding site: Azzi *et al.* (1982) therefore immobilised this ligand *via* a free cysteine moiety, using a disulfide-containing matrix.

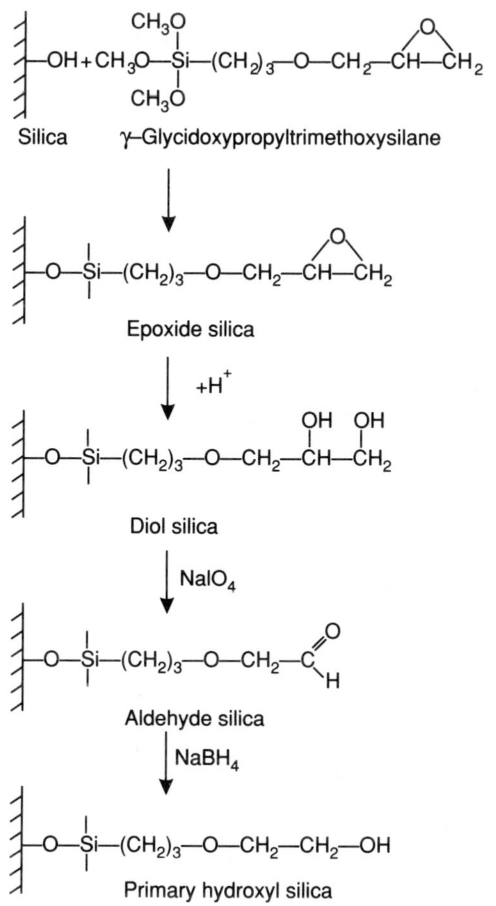

Figure 10.28 Reaction of silanol groups with γ-glycidoxypropyltrimethoxysilane and further derivatisation to generate new reactive media.

10.9.3.2 Commercially available activated media for ligand immobilisation

Some commercially available media for ligand immobilisation are shown in Table 10.6, along with their key properties. As discussed these offer, at a price, the most convenient route to almost all ligand immobilisation requirements. They can either be used in the form provided, or after further derivatisation as just described.

10.9.4 Commonly encountered affinity ligands

With the enormous popularity of affinity chromatography for the purification of protein reagents, an extensive literature has developed describing a multitude of different ligands. Many of these are highly specific in their binding behaviour: avidin columns for biotin-containing species (Bodanszky and Bodanszky, 1970), phosphonates for alkaline phosphatase (Mössner *et al.*, 1980), thiogalactosides for β-galactosidase (Cuatrecasas, 1972). Wilchek *et al.* (1984) summarised a large number of applications of affinity chromatography up to that date; Sofer and Nyström (1989b) presented a list of applications carried out at an industrial scale.

However a certain number of approaches have sufficiently broad applicability to warrant a little more attention.

10.9.4.1 Antibodies

Antibodies are an extremely common class of affinity ligand, as they often offer specificities few paralleled by other techniques. The general approach has a literature dating back several decades, as the use of immunoadsorbents predated their application in chromatographic procedures.

Immunoaffinity purifications of specific antigens require specific antibodies, generally raised deliberately for the purpose; monoclonal antibodies allow greater flexibility, as they can be specifically selected to offer an appropriate level of affinity (Calton, 1984). The application of immunoaffinity purification in general is described widely elsewhere – see for example Hill *et al.* (1989).

Direct immobilisation of antibody molecules can be inefficient due to random ordering. Protein A (see below) has been used to overcome this problem, the Ig molecules being bound predominantly *via* the Fc region. Once the Ig is bound, chemical cross-linking is often employed to prevent leakage (Gersten and Marchalonis, 1978; Phillips *et al.*, 1985). Improvements in antibody activity by up to 90% have been observed by this approach, compared to direct CNBr immobilisation (Gyka *et al.*, 1983).

An interesting variation on the immunoaffinity theme is the use of negative immunopurification: the use of an immunoaffinity matrix to remove a specific contaminant, rather than to adsorb the target antigen. This approach has the advantage that the product does not have to be desorbed from the affinity matrix, and may be worth considering if the number of significant impurities is small and their identity is known.

The number of different antibodies used for immunoaffinity separations is enormous; however one category is particularly useful. These are anti-Ig antibodies, antibodies which themselves recognise the

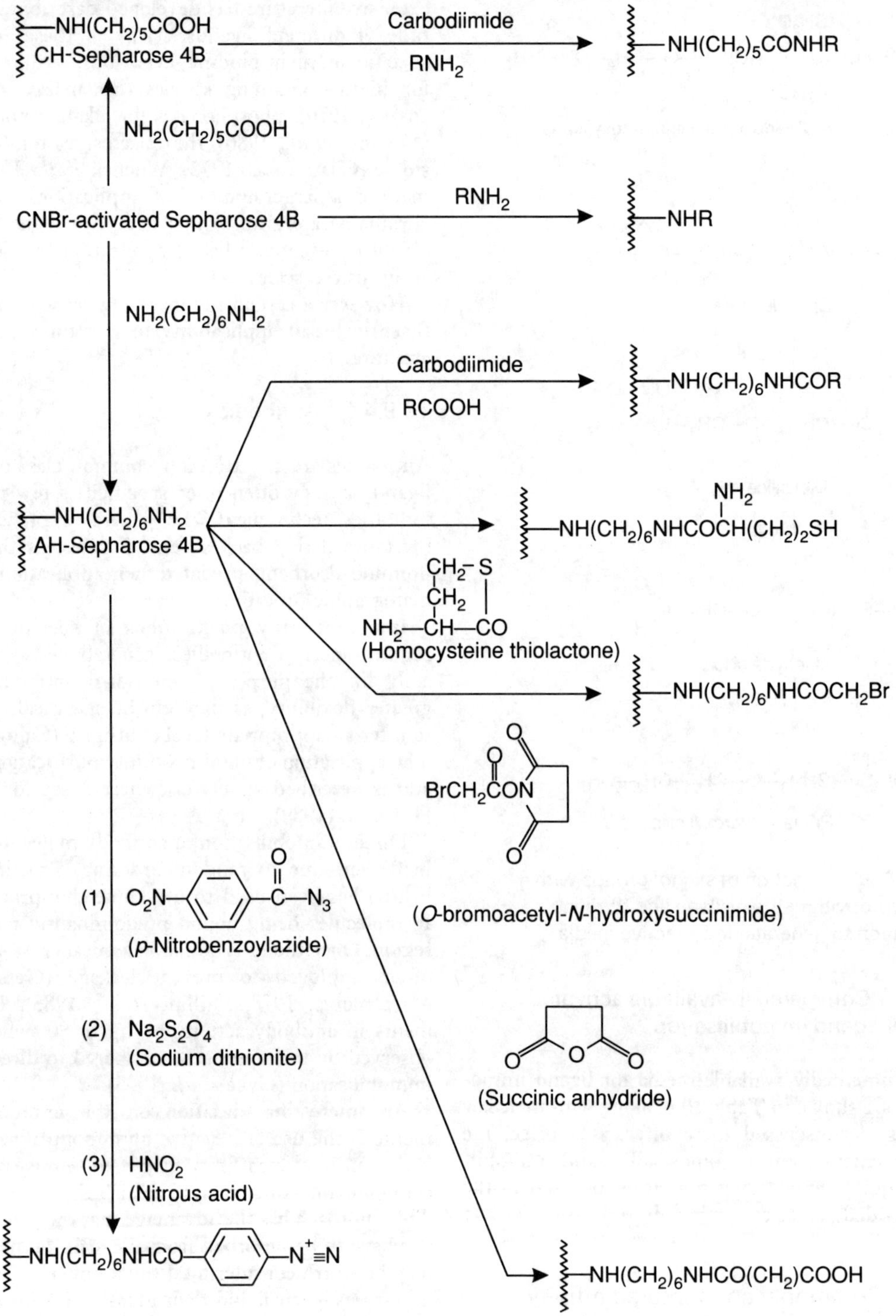

Figure 10.29 Treatment of CNBr-activated Sepharose with 6-aminohexanoic acid and 1,6-diaminohexane and further derivatisation to generate new reactive media. (Reproduced with permission from Amersham Pharmacia Biotech.)

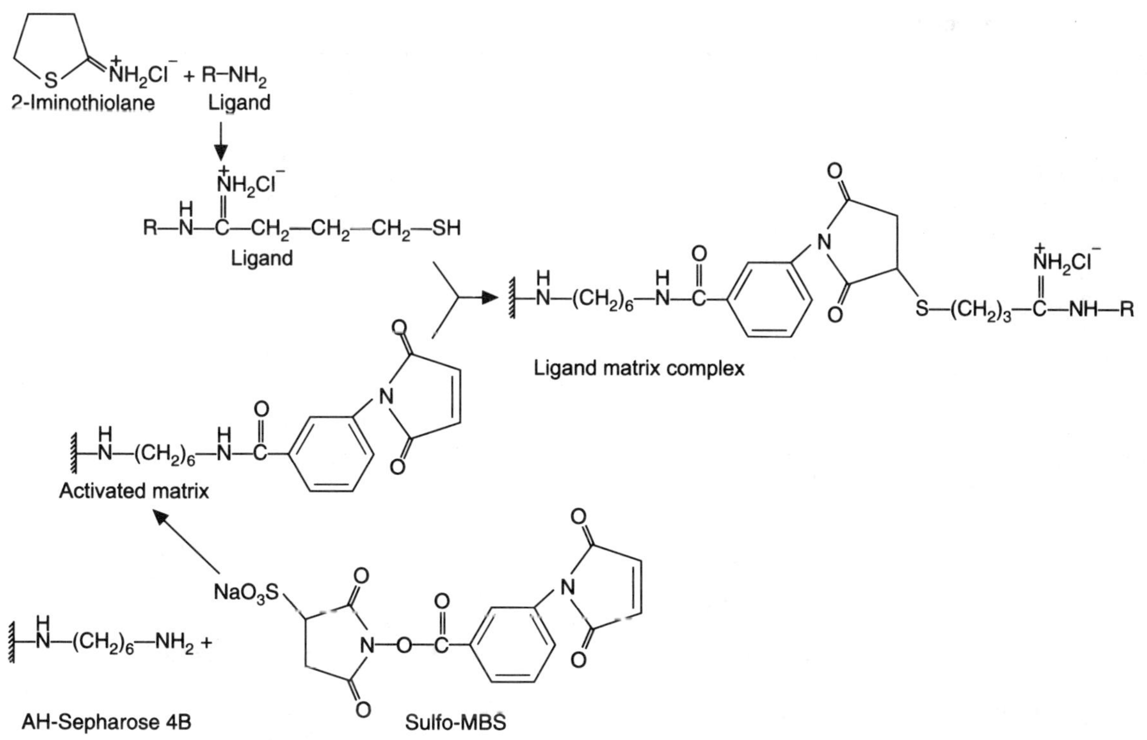

Figure 10.30 Chemical reactions for preparation of the affinity support.

constant regions of other immunoglobulins, and which therefore offer a convenient approach to their purification.

Where the desire is, for example to purify all the IgG in a sample from other components, a polyclonal anti-IgG antibody may be appropriate. This approach may also be suitable for mAb preparations, where the target antibody is present in very large concentrations compared with any other IgG molecules (Goding, 1980; Perry and Kirby, 1990). Polyclonal antisera, however, are likely to contain a fraction of material whose affinity is too high for chromatographic purposes; precautions should therefore be taken to avoid irreversible binding of some of the product. Typically an 'irrelevant' antibody mixture from the same species is used to block such sites, then the elution cycle is performed to free up those of a more useful affinity.

Again the use of monoclonal anti-Ig antibodies for this purpose allows an appropriate, homogeneous level of affinity to be obtained, although care should be taken that the Mab chosen is not too specific for the purpose.

Where a higher level of specificity is required, for example the selective purification of a particular sub-class, polyclonal antisera can still be employed. However pretreatment such as adsorption with irrelevant antigens may well be necessary to increase their specificity (Van der Giessen, 1974; Goosen, 1980). Monoclonal antibodies offer distinct advantages because of their inherent specificity, and there have been many reports of the use of mAbs for selective isolation of IgG subclasses (Bird *et al.*, 1984; Persson, 1987).

An interesting immunoaffinity approach for recombinant products is the incorporation of specific flag peptide sequences in the protein (Hopp *et al.*, 1988). These allow the recombinant protein to be immobilised on an affinity matrix bearing an antibody with appropriate specificity for the oligopeptide. Any sequence of suitable length to yield an epitope could be employed in this model, but the Asp-Tyr-Lys-Asp-Asp-Asp-Asp-Lys flag has the added advantage that it can be cleaved by the protease enterokinase, thus facilitating removal of the tag after purification.

Similar systems have been commercialised for purification of recombinant proteins. The RPAS™ product from Pharmacia, based on the 13-residue 'E-Tag', allows isolation of tagged single-chain antibody fragments.

Table 10.6 Some commercially available affinity matrices

Supplier	Ligand	Bead size (μm)	Matrix
Boehringer-Mannheim			
Poros A, M & P series	Recombinant Protein A	50	PS–DVB-base
Poros G, M & P series	Recombinant Protein G	50	PS–DVB-base
Poros MC, M & P series	Imido diacetate	50	PS–DVB-base
Poros HE, M & P series	Heparin	50	PS–DVB-base
BioSepra			
Protein A HyperD	Protein A	20	Rigid composite
Heparin HyperD	Heparin	20	Rigid composite
Blue HyperD	Dye	20	Rigid composite
Blue Trisacryl M	Dye	40–80	Trisacryl
Blue Trisacryl Plus LS	Dye	40–80	Trisacryl
Gelatin Ultragel	Gelatin	40–80	Agarose
Phospho Ultragel A6R	Phosphate	40–80	Agarose
Pierce			
Immobilised Protein A	Protein A	45–165	Agarose, Trisacryl GF-2000, bisacrylamide azlactone
Immobilised Recomb™ Protein A	Protein A recombinant fragment	45–165	Agarose
Immobilised Protein G	Protein G	45–165	Agarose, bisacrylamide azlactone
Immobilised Protein A/G	Protein A/G fusion protein	45–165	Agarose bisacrylamide azlactone
T-Gel	Thioether/sulfone		T-Gel absorbent
Immobilised jacalin	Jacalin	45–165	Agarose, bisacrylamide azlactone
Immobilised MBP	Mannan-binding Protein	45–165	Agarose, bisacrylamide azlactone
Immobilised avidin	Avidin	45–165	Agarose, bisacrylamide azlactone
Immobilised neutravidin	Modified, carbohydrate-free avidin		Bisacrylamide azlactone
Immobilised streptavidin	Streptavidin	45–165	Agarose, bisacrylamide azlactone
Immobilised iminobiotin	Iminobiotin	45–165	Agarose
Biotin-cellulose	Biotin		Cellulose
Immobilised D-biotin	Biotin	45–165	Agarose
Immobilised P-aminophenylphosphoryl-choline	*P*-Aminophenyl-phosphorylcholine	45–165	Agarose
Immobilised heparin on agarose	Heparin	45–165	Agarose
Immobilised iminodiacetic acid	Iminodiacetic acid	45–165	Agarose TSK HW-65F
Immobilised pepstatin	Pepstatin	45–165	Agarose
Immobilised *N*-acetyl-D-lactosamine	*N*-Acetyl-D-galactosamine	45–165	Agarose
Immobilised boronic acid	Phenylboronic acid		Polyacrylamide
Immobilised D-Galactose	D-Galactose	45–165	Agarose
Immobilised D-Mannose	D-Mannose	45–165	Agarose
TosoHaas			
ABA-5PW	*p*-Aminobenzamidine	10	TSK Gel
Blue-5PW	Cibacron Blue F3 GA	10	TSK Gel
Boronate-5PW	*m*-Aminophenylboronate	10	TSK Gel
Heparin-5PW	Heparin	10	TSK Gel
Chelate-5PW	Iminodiacetic acid	10	TSK Gel

Table 10.6 (contd.)

Toyopearl Activated resins			
AF-Protein A-650M	Protein A	65	TSK Gel
AF-Blue-650ML	Cibacron Blue F3 GA	65	TSK Gel
AF-Blue-650MH	Cibacron Blue F3 GA	65	TSK Gel
AF-Red-650ML	Procion Red HE-3B	65	TSK Gel
AF-Heparin-650M	Heparin	65	TSK Gel
AF-Chelate-650M	Iminodiacetic acid	65	TSK Gel
BioRad			
Ready to use affinity supports			
Affi-Gel Protein A	Protein A 2	75–300	Cross-linked agarose
Affi-Prep Protein A	Protein A 2	50	Pressure stable polymer
Affi-Gel blue	Cibracon Blue F3 GA	150–300	Cross-linked agarose
DEAE Affi-Gel blue	Cibracon Blue F3 GA	75–300	Cross-linked agarose
CM Affi-Gel blue	Cibracon Blue F3 GA	75–300	Cross-linked agarose
Affi-Gel heparin	Heparin	75–300	Cross-linked agarose
Affi-Gel polymyxin	Polymyxin	50	Pressure stable polymer
Affi-Gel 601	Boronate 1.05	75–300	Polyacrylamide gel
Pharmacia			
HiTrap Protein A	Protein A	34	
HiTrap Protein G	Protein G	34	
HiTrap Heparin	Heparin	34	
HiTrap Blue	Blue dye	34	
HiTrap Chelating	Chelator	34	
Protein A Superose		13±2	Base Superose 12
Protein A fast flow	Protein A	45–165	Cross-linked agarose
Protein G Superose	Protein A	13±2	Base Superose 12
mAb Trap G II	Recomb. Protein G	34	Cross-linked agarose
Protein G Sepharose 4 fast flow	Recomb. Protein G	45–165	Cross-linked agarose

10.9.4.2 Protein A

An alternative to the use of anti-Ig antibodies for immunoglobulin purification, and one which has become extremely popular, is to make use of the binding properties of certain bacterial proteins. At least five types of such proteins, isolated from *Streptococci*, *Pneumococci* and *Staphylococci*, have been shown to display an affinity for Ig molecules (Langone, 1982), principally *via* the Fc region.

The Fc-binding properties of Protein A from *Staphylococcus aureus*, sometimes referred to as staphylococcal Protein A (SPA), have been known since the mid-1960s (Forsgren and Sjöquist, 1966), and its use as an adsorbent for antibodies soon became popular (Hjelm *et al.*, 1972; Gersten and Marchalonis, 1978). The power of this approach was demonstrated by Ey *et al.* (1978), who were able for the first time to resolve IgG_1, IgG_{2a} and IgG_{2b} from mouse serum using a pH gradient on Protein A. Subsequently Protein A, which binds antibodies from a wide variety of mammalian species, assumed widespread popularity for affinity purifications.

The major drawback of Protein A in this application is its inability to bind certain important subclasses, notably: mouse, IgG_1 (weak binding only); rat, IgG_{2a}, IgG_{2b}; human, IgG_3. Other Ig classes such as IgM, IgA and IgE may show some binding to Protein A, but this is generally weak compared to that of IgG.

The Protein A molecule bears five similar extracellular domains (E, D, A, B, C) capable of binding Fc groups and a cell-wall binding domain (X; see Figure 9.11). Considerable structural work has been carried out, particularly on the B domain and its binding to Fc: a crystal structures of the B domain Fc complex (Diesenhofer, 1981) and a solution structure of the free B domain (Gouda *et al.*, 1992) have been published.

A gene coding for the B fragment has been described (Saito *et al.*, 1989), while tryptic digestion of Protein A has also been employed to yield active fragments (Solomon *et al.*, 1992). Both these tryptic materials and recombinant fragments have been used as affinity ligands. Site-directed mutagenesis has also been used to generate recombinant variants with different binding properties (Bottomley *et al.*, 1995). Pharmacia have recently started marketing matrices derivatised with a recombinant Protein A, from which some non-essential regions have been deleted, and containing an extra cysteine residue to facilitate orientation on the matrix.

Although Protein A is generally described as binding *via* the Fc region of antibodies, it also exhibits some affinity for Fab regions of various mouse and human immunoglobulin molecules of a variety of classes, in particular those containing domains encoded by V_H^3 genes (Hillson *et al.*, 1993).

The elution of antibodies from Protein A is typically obtained by reducing the pH or the ionic strength of the eluent: Ey *et al.* reported that murine IgG_1 eluted from Protein A–Sepharose at pH 6.0–7.0, IgG_{2a} at pH 4.5–5.0 and IgG_{2b} at pH 3.5–4.0: mouse monoclonals generally obey these general guidelines, but anomalous behaviour may be observed (Perry and Kirby, 1990). Tyrosyl residues are known to be involved in the binding of Protein A to Fc, and affinity elution using reagents such as glycyltyrosine may offer an alternative to the normal non-specific approaches (Pharmacia, 1993b).

10.9.4.3 Protein G

The binding limitations of Protein A led to the evaluation of similar proteins as 'all-purpose' IgG affinity ligands. Protein G is another bacterial cell-wall protein, isolated from human group G streptococci (Åkerstrom and Björck, 1986; Eliasson *et al.*, 1988). This protein binds a much broader range of antibodies, including all the key exceptions (human, mouse, rat) for Protein A listed above. Its affinity for bovine and sheep IgG is also stronger than Protein A, while its binding to non-IgG antibody classes is generally poor.

Not surprisingly then, Protein G has quickly become a popular alternative to Protein A for affinity purification of IgG, and is commercially available in a number of immobilised formats. These include recombinant forms of the protein, often with an albumin-binding region deleted.

Binding conditions are similar to those of Protein A. Elution of bound IgG, however, may require a lower pH than with Protein A; buffers with pH 2.5–3.0 have been recommended in general guidelines (Pharmacia, 1993b).

Low-affinity binding of antibodies from many classes *via* the Fab region (as seen with Protein A) has been reported for Protein G, and it has been used as the basis of purifying Fab and $F(ab')_2$ fragments (see Proudfoot *et al.*, 1992, for example).

10.9.4.4 Protein A/G

Eliasson *et al.* (1988) constructed a fusion protein containing the binding domains of both Protein A and Protein G, which retained both types of binding ability. This 'Protein A/G' hybrid has been commercialised by Pierce, for example.

10.9.4.5 Concanavalin A and other lectins

Lectins are a class of proteins found in organisms widely distributed through the plant and animal kingdoms. They are linked by the ability to bind to carbohydrate moieties, often with a high degree of specificity for particular saccharide structures. Because so many proteins are glycosylated, lectins offer a useful category of broad-affinity ligands for their purification (Sharon and Lis, 1972; Gallagher, 1989).

The first important application of lectins was in blood group testing, where their properties were used to agglutinate red blood cells on the basis of their surface glycosylation. For this reason they are often referred to as haemagglutinins, or phytohaemagglutinins if they are of plant origin.

Probably the best-known of these proteins for affinity chromatography is concanavalin A (Con A), which was first isolated from the jack bean *Canavalia ensiformis* by Sumner (1919). Con A is a metalloprotein, requiring the presence of both Ca^{2+} and a transition metal ion such as Mn^{2+} for the formation of the saccharide-binding site.The Con A subunit has a molecular mass of around 27 kDa (Wang *et al.*, 1971), but the number of subunits in the complete molecule is pH-dependent (McKenzie *et al.*, 1972): below pH 5.6 it is dimeric, between pH 5.6 and 7.0 it is tetrameric, and above pH 7.0 aggregation occurs. Conformational changes are also seen below pH 4.5, possibly connected with loss of metal ions, but do not affect the subunit composition.

Con A will bind to glycoproteins bearing hydroxyl groups on the C-3, C-4 and C-5 carbon atoms (Goldstein *et al.*, 1965), notably α-D-mannopyranosyl and (less strongly) α-D-glucopyranosyl residues; these are very common glycan components so Con A has a broad glycoprotein specificity.

Affinity elution of bound species is often achieved using simple sugars such as methyl α-D-mannoside or α-D-glucoside, sometimes enhanced by an organic cosolvent such as ethylene glycol to reduce hydrophobic interactions (Sutton, 1989). The use of saccharide eluents is interesting, as the affinity constants of lectin–glycoprotein complexes can be very large, and affinity elution would be expected to be extremely slow). On the contrary, the introduction of simple saccharides often causes a rapid disruption of binding, and it appears to be the case that these act by promoting a conformation change in the lectin rather than by prevention of reassociation (Cuatrecasas, 1973). Borate ions may form complexes with carbohydrates, and borate buffers have also been employed as eluents for Con A (Kennedy and Rosevear, 1973). Con A binding is badly affected by detergents, more so than many other lectins (Sutton, 1989), so care needs to be taken if any such reagents are present in the sample.

Even if there are several glycoprotein components in a mixture, they often exhibit sufficiently different binding to Con A to allow resolution. Arends (1978), for example, demonstrated that over 90% of free IgG could be removed from an IgG–HRP conjugate on a Con A column: while the free IgG was eluted in the acetate starting buffer, its conjugate with HRP (a more highly glycosylated protein) was retained until a methylglucoside buffer was introduced.

Many other lectins have been employed as affinity ligands, and some of these are commercially available in immobilised form. These lectins differ in their specificity, both for the identity of the carbohydrate residues, and for the general structure and manner of attachment of the glycan chain (Sutton, 1989). For example, while both peanut agglutinin and some castor bean lectins have an affinity for galactose residues, the former bind more strongly to *O*-linked glycan chains and the latter to *N*-linked chains, particularly of the complex and hybrid-type.

Some of the more commonly-encountered lectins and their key properties are listed in Table 10.7.

Table 10.7 Properties of some commonly encountered lectins

Lectin	M_r (kDa)	No. of subunits	Specificity
Concanavalin A (Con A)	54/108	2 or 4	α-D-Man α-D-Glc α-D-GlcNAc
Lentil lectin	49	2	α-D-Man α-D-Glc
Soybean lectin (SBA)	110–120	4	α-D-GalNAc β-D-GalNAc α-D-Gal
Castor bean lectins			
RCA_{60}	60	2	D-GalNAc
RCA_{120}	120	4	β-D-Gal, D-Gal
Wheat germ agglutinin (WGA)	36	2	$(\beta\text{-D-GlcNAc})_3$ $(\beta\text{-D-GlcNAc})_2$
Peanut agglutinin (PNA)	120	4	β-D-Gal (1–3)-D-GalNAc
Pea lectin	49	4	α-D-Man
Phytohaemagglutinin (PHA)	128	4	D-GalNAc
Bandeirea simplificifolia			
BS-I	114	4	α-D-Gal
BS-II	113	4	α-D-GlcNAc
Dolichos biflorus agglutinin (DBA)	111	4	α-D-GalNAc
Jacalin	40	4	α-D-Gal
Helix pomatia lectin	79	6	α-D-GlcNAc
Limulin	400	18	α-*N*-Acetylneuraminic acid

10.9.4.6 Boronates

Boronates are capable of forming complexes with molecules bearing *cis*-diol moieties, which include many glycan chains, so boronate-derivatised matrices offer an alternative approach for glycoprotein purification. The most commonly-encountered ligands of this type are phenylboronates (Figure 10.31); these have been described for affinity use (Weith *et al.*, 1970; Hageman and Kuehn, 1992), and are commercially available in immobilised form. Binding can be achieved at slightly basic pH in the presence of Mg^{2+} ions, and affinity elution using sorbitol is often employed.

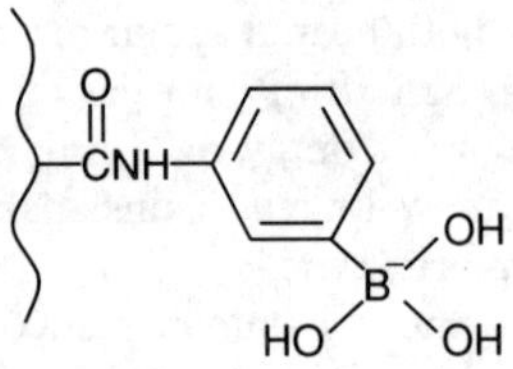

Figure 10.31 Phenylboronates.

10.9.4.7 Dyes

From the 1950s onwards, a number of textile dyes were developed based on mono- or dichlorotriazine structures (Figure 10.32), and marketed under the series names Procion™ (ICI) or Cibacron™ (Ciba-Geigy). During electrophoresis experiments Kopperschläger *et al.* (1968) observed binding of some proteins to Blue Dextran, a dye-labelled marker material based on Cibacron Blue F3G-A, and many reports of similar phenomena followed. This resulted in the commercialisation of Blue Dextran columns by Pharmacia as an affinity matrix (Easterday and Easterday, 1974).

Since then dye affinity chromatography has become a widely employed technique (Dean and Watson, 1979; Clonis and Lowe, 1980; Lowe and Pearson, 1984), with Cibacron Blue F3G-A remaining the best-known ligand. As discussed in the introduction to this section, the binding of proteins by such dyes may be more or less specific in nature. All their interactions with proteins are serendipitous in origin, unlike the *in vivo* basis of binding by lectins, Protein A, etc. However, their polycyclic structures bear some resemblance to nucleotides, and this has made them particularly useful for the purification of enzymes dependent on a nucleotide cofactor such as NAD or ATP—a category including a sizeable proportion of all known enzymes. Cibacron F3G-A has been shown to

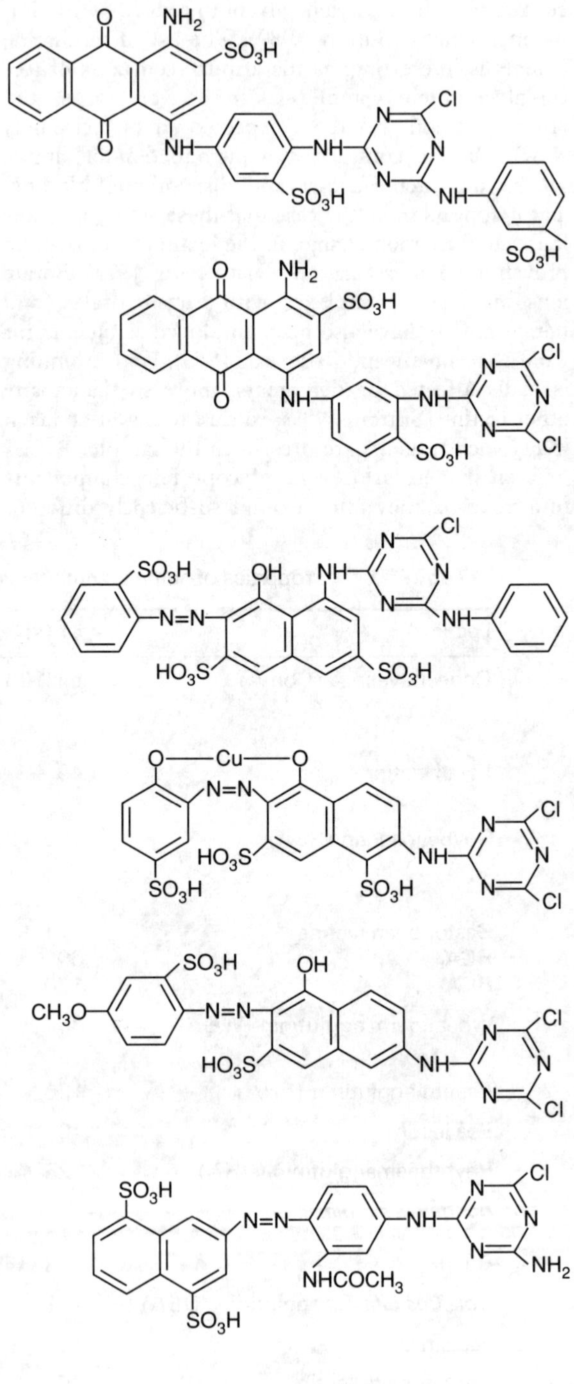

Figure 10.32 Structures of several mono- and dichlorotriazine dyes used in affinity chromatography.

bind to many enzymes, although not all, possessing a 'dinucleotide fold' domain (Thompson *et al.*, 1975), an arrangement of five or six parallel β-chains connected above and below the plane by α-helices. The anionic sulfonate groups of such dyes have also been proposed to mimic phosphates in the binding of alkaline phosphatase, and this hypothesis has been taken further by the deliberate, biomimetic design of dyes bearing phosphonate groups (Lindner *et al.*, 1989). These were successfully used for AP purification.

On the other hand, many of their protein-binding properties are impossible to rationalise on this basis (Qadri, 1985), and appear to be due to a more fortuitous mixture of hydrophobic and electrostatic effects. They have for example been employed for many purifications of all-α and all-β proteins (Lascu *et al.*, 1984), including applications such as immunotoxins (Knowles and Thorpe, 1987; Thorpe *et al.*, 1988). Their combination of large hydrophobic cyclic regions and strongly charged sulfonate moieties certainly offers plenty of scope for mixed-mode binding, albeit in a rather unpredictable fashion.

The choice of dye for a given application will often need to be determined empirically (Clonis *et al.*, 1981), and fast screening approaches have been described, for example using small centrifugal affinity columns (Berg and Scouten, 1990). Commercial kits have also been introduced specifically to facilitate this screening exercise, e.g. from Amicon and Sigma. A little guidance may be obtained from the rule of thumb quoted by Clonis and Lowe (1980), that affinity constants tend to be greater for dyes with low-wavelength absorbance maxima, blue and green dyes broadly giving stronger binding than red, orange or yellow.

The reactivity of chlorotriazines is a convenient feature of these dyes, allowing them to be coupled directly to carbohydrate matrices. Their chromophoric nature is also a useful property in monitoring binding or leakage, and both the dyes themselves and the ether linkages formed with hydroxylic media are chemically robust. A disadvantage of using some of these dyestuffs, particularly if they have not become accepted chromatographic reagents, is that their standard of characterisation and nomenclature is commensurate with the textile industry rather than highly-applied biotechnology applications. The structure of many such dyes remains proprietary, they may contain mixtures of isomers and/or byproducts, and the same Colour Index (C.I.) names and numbers may describe different isomeric compositions.

Adsorption of proteins to immobilised dyes may be assisted by a range of additives such as metal ions, chelators or cosolvents, depending on the specific application (Lowe and Pearson, 1984). Elution has employed all the usual non-specific techniques, but the mixed-mode nature of the binding is reflected in the unpredictability of the behaviour observed by varying pH, ionic strength or temperature. For nucleotide-binding proteins, affinity elution can also be achieved using nucleotides or other dyes (Lowe and Pearson, 1984). The enhancement of elution where a ternary complex can be formed has been mentioned in Section 10.9.2.

10.9.4.8 Immobilised metal ions (IMAC)

Immobilised metal ion affinity chromatography (IMAC) is a useful approach for purification of proteins by virtue of the presence of residues such as histidine, cysteine and tryptophan, which are capable of binding to specific metal ions (Porath *et al.*, 1975; Sulkowski, 1985). His and Cys residues bind most strongly, but the latter are often tied up in disulfide links – so it is the surface histidine content of a protein which primarily defines its metal-binding capabilities (cf. Section 6.19). The binding is generally strongest around neutral pH.

Metal ions are immobilised for IMAC using chelating ligands, which bind the ions to the solid phase while retaining their protein-binding capability: iminodiacetic acid (IDA) and nitrilotriacetic acid (NTA) are commonly employed (Figure 10.33). Many different metal ions, notably from the transition series, display suitable binding behaviour, but Cu^{2+}, Zn^{2+} (often weaker than Cu^{2+}) and Ni^{2+} have probably enjoyed the widest popularity. The use of Fe^{3+}–IDA matrices has been shown to offer some interesting binding behaviour (Porath and Olin, 1983; Porath, 1986).

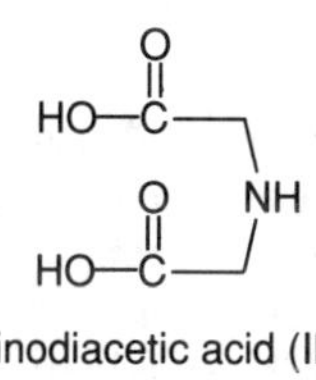

Iminodiacetic acid (IDA)

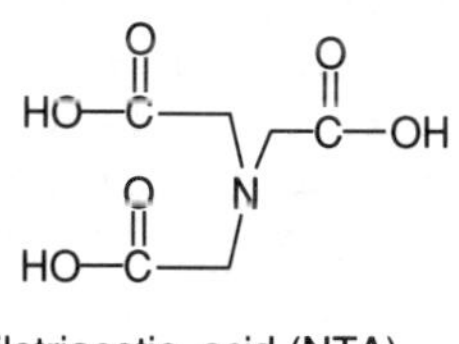

Nitrilotriacetic acid (NTA)

Figure 10.33 Chelating ligands used in IMAC.

Adsorption to IMAC matrices is generally carried out at neutral pH. The ionic strength of buffers for both binding and elution should be kept reasonably high (e.g. by the inclusion of 0.5–1 M sodium chloride) to suppress ion exchange effects, despite the involvement of metal ions, IMAC binding essentially parallels HIC in being enhanced by salting-out buffers (Porath, 1986).

Note that ionic strength changes can affect the binding mechanism in IMAC: Jiang and Hearn (1996) reported that lysozyme binding to a Cu^{2+}–IDA based system showed positive cooperativity at salt concentrations above 0.5 M.

Elution can often be achieved simply by slight acidification: many proteins elute between pH 4.2 and 6.0 (Pharmacia, 1993b). Affinity elution with mild chelators such as imidazole or glycine may also be used; more powerful examples such as EDTA will probably remove all metal-bound material from the column, and are thus useful for difficult elutions or for column regeneration.

IMAC has found a particularly popular niche in the purification of recombinant proteins, where an oligopeptide 'tail' can be introduced into the product specifically to facilitate its purification (cf. flag peptides, Section 10.9.4.1). The introduction of a polyhistidine tail provides a convenient binding capacity for IMAC matrices, and purification of recombinant proteins tagged in this way is now becoming commonplace (Janknecht *et al.*, 1991; Crowe *et al.*, 1994; Dhanabal *et al.*, 1995).

More classical separations such as the removal of free enzyme from enzyme–antibody conjugates have also been demonstrated by IMAC (Sorensen, 1993)

10.9.4.9 Thiophilic and thioaromatic ligands (T-gels and M-gels)

Various ligands have been identified which display poorly understood protein-binding properties, which are nonetheless useful enough to have found some application. Many proteins possess regions of unknown structure (described as thiophilic) which have an affinity for sulfone and thioether moieties which neighbour each other. Therefore media bearing such groups (T-gels) can be used as thiophilic affinity matrices, although the determination of their effectiveness remains almost wholly empirical.

T-gels have been prepared using groups such as that shown in Figure 10.34 (Porath *et al.*, 1985; Hutchens and Porath, 1986, 1987), and have shown some interesting properties: Belew *et al.* (1987), for example, described a one-step purification of mAbs from ascites or cell culture fluid.

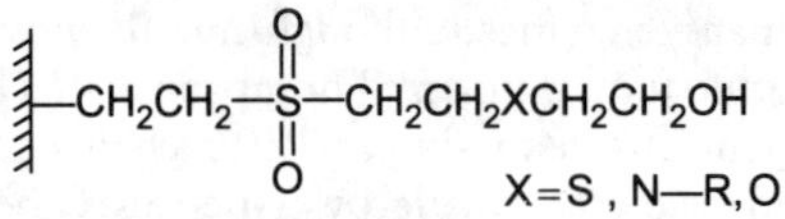

Figure 10.34 Thiophilic ligands.

The binding involved is encouraged by high-salt buffers (Porath, 1986), and elution can be promoted, cf. HIC, by a reduction in ionic strength. For example, Yurov *et al.* (1994) successfully employed thiophilic chromatography to fractionate the components of a Mab fragmentation: a number of proteins were eluted under the binding conditions (15% ammonium sulfate), then $F(ab')_2$ fragments (12.5%), Fc fragments (10%), and finally intact IgG (no ammonium sulfate).

Ligands which contain sulfur atoms linked to an aromatic group bind to regions described as thioaromatic, and these too can be used for affinity matrices termed M-gels. Binding to such media is again promoted by high-salt conditions (Porath, 1986), and the mechanism may be related to charge-transfer effects (Porath, 1978; Porath and Larsson, 1978).

Note that T-gel and M-gel purifications are based on non-covalent interactions with the target proteins; they should not be confused with covalent chromatography (Section 10.9.5), where the chemical reactivity of thiols and related reagents is exploited.

10.9.4.10 Other affinity ligands

Some other ligands are of sufficiently wide applicability to warrant a very brief mention here, if only because they have become commercially available in an immobilised form as a consequence of their usefulness. Heparin binds a wide range of proteins (Kjellén and Lindahl, 1991) either by true bioaffinity (e.g. for growth factors and protease inhibitors), or by a combination of this and the cation exchange properties of its sulfonate moieties; benzamidine is useful for trypsin and related enzymes and plasminogen activator; lysine for plasminogen, plasmin and plasminogen activator; AMP and other nucleotides and oligonucleotides for DNA and RNA-binding proteins and nucleotide-dependent enzymes; phosphorylcholine for C-reactive protein; phosphonates for phosphatases; pepstatin for cathepsin; methotrexate for dihydrofolate reductase.

10.9.4.11 Commercially available matrices for affinity chromatography

Some commercially-available media bearing ready-to-use immobilised ligands or reactive groups are listed in Table 10.6.

10.9.5 Covalent chromatography

This approach is generally listed as a branch of affinity chromatography, although it differs in a very fundamental way from those described so far. As suggested by the name, the technique is based on the formation of covalent bonds between the immobilised ligands and the target proteins, rather than the non-covalent interactions underpinning ion exchange, HIC, RPC, and affinity chromatography in general.

The reasons why such a radically different technique is so rarely encountered are obvious, if the particular requirements of the application are considered. Not only does the chosen ligand need to demonstrate sufficiently rapid reaction with the target protein that it can be stripped from the eluent during the loading stage; the reaction must also be readily reversible, under conditions which regenerate the original structure without damage. Finally, the use of a chemical dissociation step makes it likely that all bound species will be desorbed simultaneously, so a reasonably unusual functionality of the protein needs to be targetted. A method for binding and then simultaneously desorbing all the proteins in a mixture *via* their lysine residues, for example, would be of very little separative value.

Taking these requirements into account, there are very few reactions which lend themselves to covalent chromatography, and the only ones which have attracted serious attention are the highly specific reactions of protein thiols with ligands such as disulfides and organomercurials (Figure 10.35). Free protein thiols are sufficiently uncommon to make separation on this basis worthwhile. The reactions are rapid, and they can be reversed by simple thiol-containing reducing agents, so long as the treatment is not so vigorous as to disrupt any disulfide bridges within the protein structure itself.

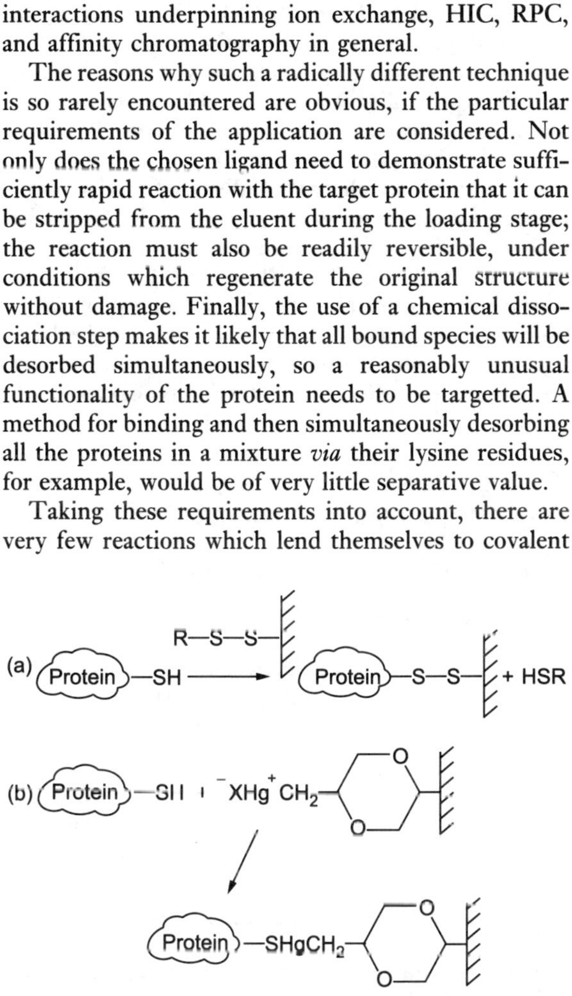

Figure 10.35 Covalent chromatography coupling using thiol chemistry. Thiol-bearing proteins are bound *via* disulphides (a) or mercurials (b)

The first matrices of this type bore mercurial groups (Eldjarn and Jellum, 1963; Shainoff, 1968; Sluyterman and Wijdenes, 1970). Brocklehurst *et al.* (1973) introduced the use of pyridyldisulfide matrices for this application, and these have since come to dominate the field: Carlsson and Svensson (1974), for example, employed this approach for purification of mercaptalbumin. The reaction of disulfides with thiols is extremely specific, and occurs readily (in most cases) at neutral or mildly basic pH where there is a sufficient proportion of deprotonated thiolate (S^-) anions. The pK_a of protein thiol groups varies over quite a wide range, however, and this should be borne in mind when choosing eluent conditions. Extra selectivity of binding can be obtained for thiols with abnormally low pK_a values, such as those in the active site of papain and other proteases, by working under acidic conditions where other thiols will be protonated and unreactive. With pyridyldisulfide matrices, the binding of a thiol-bearing solute is accompanied by the release of a chromophoric thiopyridone, which can be a useful indicator. The binding stage can be carried out with the flow stopped to allow sufficient time for reaction to take place.

The elution of bound thioproteins is effected by the addition of reducing agents such as (in approximately increasing order of strength) L-cysteine, glutathione, 2-mercaptoethanol or dithiothreitol, at concentrations up to around 50 mM at neutral pH (Hillson, 1981). Once again stopping the flow can provide longer reaction times for the reduction to take place. The presence of a chelator such as EDTA in the eluent prevents metal-catalysed atmospheric oxidation of the thiols as they are released, and of the thiols in the buffer itself.

The buffer thiols will react with the remaining pyridyldisulfide groups on the matrix, too, releasing thiopyridones into the product. However, all these low-molecular-mass contaminants are readily removed by SEC.

A related approach for the specific purification of proteins possessing two sequentially close thiol groups has also been described (Kalef *et al.*, 1993), based on the formation of stable dithioarsine derivatives from an immobilised arsine oxide compound.

10.10 Hydroxyapatite Chromatography

Calcium phosphate has an extremely long history in protein purification, being used over a century ago as an adsorbent for pepsin (Brücke, 1861); chromatographic enzyme separations were reported as early as the 1930s (Agner, 1938). Tiselius and his colleagues at the University of Uppsala (Tiselius *et al.*, 1956) concentrated on a particularly stable form of crystalline calcium phosphate, hydroxyapatite (HA), which has the formula $Ca_5(PO_4)_3OH$. This possessed sufficient physical robustness to allow it to be used for chromatographic purposes.

Working in phosphate buffers at around pH 6.8, Tiselius *et al.* reported that many different proteins could be adsorbed onto HA columns at low ionic strength, then eluted as the phosphate concentration was raised. With some proteins, such as γ-globulin, the presence of sodium chloride was required for adsorption.

10.10.1 The mechanism of hydroxyapatite chromatography

HA chromatography has remained in use ever since, although its mechanism has remained poorly understood compared to other techniques. Work to elucidate this mechanism (Bernardi, 1971; Bernardi *et al.*, 1972; Gorbunoff, 1984a,b) has resulted in a fairly complex and somewhat contradictory set of conclusions. (1) Basic proteins (p$I > 8$) can be eluted at pH 6.8 with phosphates, fluorides, chlorides, chaotropes, or low concentrations of Ca^{2+} ion (3 mM). (2) Neutral proteins (pI 7–8) can be eluted by fluorides, phosphates and chlorides, but there is a considerably higher dependence on the nature of the anion than for basic proteins: the strength of elution decreases in the order just listed. No elution is seen with SCN^- or Ca^{2+}. (3) Acidic proteins (p$I < 7$) are eluted roughly equally by phosphate and fluoride, but not by chloride or Ca^{2+}, the latter even at 3 M concentration. (4) The charge content of the protein is its key attribute in determining binding to HA. The presence of high positive charge densities facilitates binding of basic proteins, and the loss of negative charges has the same effect for all protein types.

Overall, the mechanism can be stated to relate to the surface charge of the protein, but, as the oppositely charged calcium and phosphate ions are closely packed in the HA structure, it is the microstructure of the protein molecule which is probably salient. The binding of a given protein is therefore difficult to predict, and suitability must generally be determined empirically.

HA shows good binding to nucleic acids, particularly the double-stranded form, and has been widely used for DNA and RNA fractionations. This affinity is attributed to the high charge density of nucleic acids, and the good surface contact allowed by their planar structure.

The relatively planar structure of immunoglobulins, compared to common contaminants such as albumin and transferrin, has been suggested as a basis for the effective purifications of these molecules on HA (Smith *et al.*, 1984). Very impressive separations of mixed antibody populations have been reported from this approach (Juarez-Salinas *et al.*, 1986).

One of the main drawbacks of HA is the fragility of the crystalline material, which limits its longevity. Bio-Rad have introduced ceramic forms of HA (Macro-Prep™) which alleviate these problems. HA-coated microparticulate media for high-resolution HA chromatography have also been introduced.

Adsorption and elution conditions for HA chromatography have generally varied little from the work of Tiselius, relying on increasing phosphate gradients at near-neutral pH (Roe, 1989). Acidic and neutral proteins are eluted first, and phosphate concentrations as high as 500 mM may be required for very basic species. As might be expected, however, pH can have a useful effect too. Retention times can sometimes be increased under more acidic conditions.

10.10.2 Some commercially available media for hydroxyapatite chromatography

Some of these media are listed in Table 10.8.

Table 10.8 Some commercially available hydroxyapatite matrices

Supplier	Bead size (μm)	Matrix
BioSepra		
HA Ultragel	60–80	Cross-linked agarose and hydroxyapatite
BioRad		
Mono-Prep Ceramic Hydroxyapatite I	20, 40, 80	Hydroxyapatite
Mono-Prep Ceramic Hydroxyapatite II	20, 40, 80	Hydroxyapatite
Bio-Gel HTP	20, 40, 80	Hydroxyapatite
Bio-Gel HTP DNA Grade	20, 40, 80	Hydroxyapatite
Bio-Gel HT	20, 40, 80	Hydroxyapatite

10.11 Chromatofocussing

Chromatofocussing is related to ion exchange, but employs pH gradients to aid band sharpening. It is a mild technique in that (in its usual format) the protein never experiences a pH greater than its p*I*.

10.11.1 The mechanism of chromatofocussing

In ion exchange chromatography, proteins are separated on the basis of their charge in relation to that of the solid phase. Where it is desired to use pH changes to effect a fractionation, this is accomplished by altering the pH of the eluent. IEC columns do not display significant buffering capacity, allowing pH changes in the eluent to be quickly transmitted through the system (Figure 10.36(a)).

However if the buffer has appreciable buffering capacity, a similar change of eluent pH will only be gradually transmitted to the matrix, and, assuming that the charge state of the solid phase can be adjusted over the desired range, it is therefore possible to generate a pH gradient within the column (Figure 10.36(b)). Importantly, and in contrast to IEC, the rate of change of the column pH is thus significantly slower than the elution rate. This principle is the foundation of the technique of chromatofocussing (Sluyterman and Elgersma, 1978; Sluyterman and Wijdenes, 1981a,b).

The behaviour of solutes on such a column is best illustrated by example. With a strong anion exchanger as the solid phase, the column charge remains fixed over the pH range to be considered (7–10); it will be positively charged throughout. The column is first equilibrated to pH 10, then a second buffer at pH 7 is introduced, and a gradient is gradually established ranging from 7 at the top, to 10 at the bottom (Figure 10.37(a)). The behaviour of injected solutes will be governed by their p*I*. (1) Any solutes with a p*I* of less than 7 will be negatively charged when they first enter the column, and will thus be bound at the top of the column (Figure 10.37(b)). (2) Any solutes with a p*I* of greater than 10 will be positively charged when they enter the column at pH 7, and will remain so as their passage continues and the pH gradually rises to 10; they will therefore be eluted without binding (Figure 10.37(c)). (3) A solute with an intermediate p*I* (9, for example) will be positively charged when it enters the column, and will therefore begin to elute as in (b). However, when it reaches the point in the column where the pH is 9, its charge will become negative and it will bind. It is therefore focussed at a point on the column determined by its p*I* (Figure 10.37(d)). Note that, roughly speaking, it will never move beyond the pH equivalent to its p*I*. (4) As more of

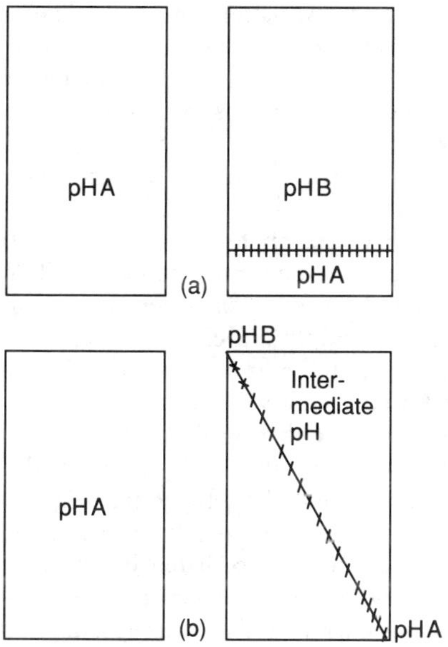

Figure 10.36 Principle of chromatofocussing vs. IEC. (a) pH changes are transmitted rapidly through the system as the eluent is changed. With a higher buffer capacity, changes are transmitted slowly (b) and a pH gradient is established.

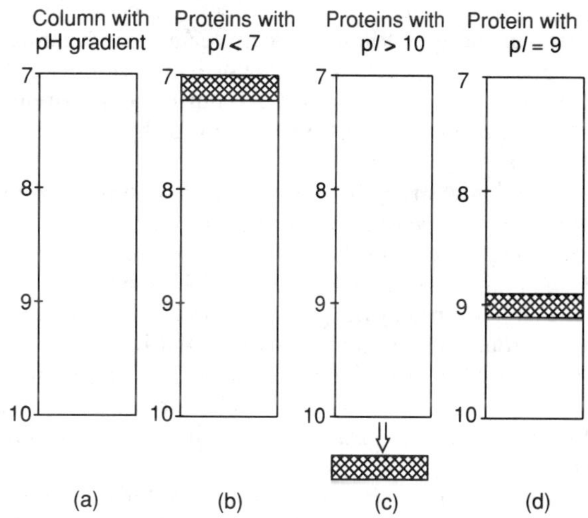

Figure 10.37 Behaviour of solutes of different charge during chromatofocussing on a pH 7–10 gradient. See text for details.

the solute in (c) is injected, it will be able to catch up with the material which has already bound at the point defined by its p*I*. All the injected material with the same p*I* can therefore be focussed at the same point, regardless of when it is injected.

The analysis above ignores the fact that the gradient itself is moving, albeit more slowly than the eluent and the sample proteins within it. The buffering capacity of the column is of course limited, and eventually the gradient itself will be eluted as the whole column reaches pH 7. Therefore the focussed proteins are eluted too, at a pH reflecting their p*I*.

The best focussing performance is attained for proteins with a p*I* close to the pH of the eluting buffer (i.e. slightly greater than 7 in the previous sample), and these can be injected in large volumes while still giving a sharply focussed band. For proteins with a p*I* close to the initial value (just below 10 in the example), only a small volume can be accommodated without overloading the available capacity. The sample volume which can be focussed adequately, for a given solute in a given system, is termed the distribution volume.

10.11.2 Key process variables of chromatofocussing

Sluyterman's original studies on chromatofocussing employed simple ion exchange reagents, DEAE-Sepharose and acetate buffers. However, better results are obtained by employing both matrices and buffers designed to facilitate gradient formation. Pharmacia market a series of matrices (PBE™) and ampholyte buffers (Pharmalyte™, Polybuffer™) specifically for this purpose, although their Mono-P™ anion exchange matrix is recommended for some uses. Recommendations on useful gel/buffer combinations for chromatofocussing have been given by Pharmacia (1993c).

The ionic strength of the eluent should be kept constant throughout the elution profile, so that the buffering capacity of the column is used up in a linear fashion and a linear pH gradient is obtained.

The pH range of the gradient should not be greater than around three units, and the initial buffer should have a pH about 0.4 units greater than the desired starting point. This allows for pH fluctuations caused by the release of protons from the matrix during startup.

A narrow pH range gives the best resolution, as does a slow flow rate. The exact pH range can obviously only be chosen if the p*I* of the target protein is known. The gradient should be centred around this pH, or somewhat higher to give the largest distribution volume. If the p*I* is unknown, trial and error will need to be employed to determine the best conditions; it may be easier to run an IEF gel to determine this value before chromatofocussing is attempted.

Bed volumes of around 20–30 ml are recommended for samples of up to 200 mg protein per pH unit of the gradient.

10.12 Preparative Electrophoresis

Electrophoresis is the separation of different species in an electric field, by virtue of their different charge properties. At its simplest level, negatively charged species in a sample move towards the anode, and positively charged species towards the cathode, at rates related to the amount of charge.

As the overall charge of a protein is determined by the pH of its surroundings (and specifically the relationship of its p*I* to this pH), electrophoresis at a carefully chosen pH can be a powerful fractionation tool. Working at a pH between the p*I* values of two mixed proteins, for example, will cause them to move in opposite directions through the field. This clearly offers a potentially powerful route for protein fractionation on the basis of charge, and has been employed as such for many decades.

Many different variations on the basic electrophoresis theme have been developed, allowing separations to be carried out in a number of complementary ways. However, electrophoresis is overwhelmingly employed as an analytical-scale technique, and as such will be discussed in greater detail in Chapter 11, Sections 11.5–11.11.

The primary reason for this is that electric fields generate energy in the form of Joule heat. The physical layout of an electrophoresis experiment is constrained by the need to supply sufficient cooling. Thin, flat gels of relatively limited area are therefore the commonest vehicles for such separations.

Joule heat can be harmful in two ways: convection currents distort the movement of solutes by altering their mobilities, and the temperature rise can also be sufficient to cause damage to the proteins themselves. This latter effect is exacerbated by the fact that the amount of Joule heat generated is equivalent to i^2/κ, where i is the current and κ is the specific conductance of the medium. Proteins have poor conductance, so κ is low and the Joule heat correspondingly high.

The problems of heat dispersion have been a major hindrance in the development of preparative electrophoresis. However, analytical-scale methods can be used with the isolation of a product in mind—whether to call a method 'preparative' or not is, after all, determined only by the ambitions of the experimenter. Many preparative methods described in the literature provide a few mg of product by simple adaptations of the analytical procedures described in Chapter 11 (Andrews, 1986a). Examples of this type are common, and the formats widespread – see for example Morgan *et al.*, 1978; Roth *et al.*, 1996.

The only extra requirement is to harvest the separated material from the gel[1]. This can be achieved by a number of different means (Shuster, 1971; Hames, 1981; Waterborg and Matthews, 1994): homogenisation of a band scraped from the gel with a suitable buffer, electrophoretic elution of the band from a selected section of gel, or on-line elution of the product into a buffer stream as it is electrophoresed off the end of the intact gel.

However gram-scale preparations have been described occasionally, where sufficient effort has been put into the design of equipment that the effects of Joule heat are minimised as far as possible. To obtain good results at a large scale it is advisable to stick to reasonably simple separations (Chrambach and Rodbard, 1981).

10.12.1 Continuous flow electrophoresis

In continuous zone electrophoresis methods a single buffer is used, and the solutes simply move through the solid phase at a rate related to their charge. Simple techniques such as paper electrophoresis were employed for small-scale preparative separation of protein conjugates (Borek and Silverstein, 1961), though even the authors themselves noted the limitations of this approach.

Materials more amenable to scaling-up were therefore examined, such as silica gel, paper powder, glass beads, asbestos fibres, or starch (Flodin and Porath, 1954). A range of different preparative applications were developed on such media (Bier, 1962), but attention shifted gradually to matrices based on sucrose density gradients, which displayed greater stability towards convection problems. These methods were capable of being used in a number of useful preparative applications (Heimburger and Schmidtberger, 1964).

However it was the introduction of gel matrices based on polymers such as polyacrylamide that allowed the greatest advances: Hjertén *et al.* (1969) were able to demonstrate gram-scale gel electrophoresis of proteins with good resolution.

The mid-1960s also saw the introduction of multiphasic or discontinuous (disc) zone methods, in which the sample solutes are concentrated into a narrow band at the boundary of two buffer systems of different mobility. This allowed much greater resolution of solutes, although scaling-up faced new challenges.

Systems capable of providing high resolution on tens of mg of sample were described (Nees, 1974; Nees *et al.*, 1976). One particular variant of the discontinuous approach attracted particular attention for preparative work, because the solute components remain tightly focussed at zone boundaries throughout the run. This is isotachophoresis (Holloway and Battersby, 1984; Andrews, 1986c; see also Chapter 11, Section 11.5.2). It has formed the basis of many preparative separations at scales of up to a few hundred mg (Ramsden and Louis, 1972; Routs, 1973; Brogren and Svendsen, 1975). Separations of mixtures have also been demonstrated at the two gram scale (Nguyen and Chrambach, 1981).

Development of instrumentation for zone electrophoresis has continued, allowing for example the possibility of recycling to achieve greater resolution (Hymer *et al.*, 1987; Wenger *et al.*, 1988), with production rates reaching 250 mg per hour. Intriguingly, the low-gravity environment of spaceflight has been used to improve the performance of preparative electrophoresis (Hymer *et al.*, 1987), unfortunately not a facility available to most experimenters!

10.12.2 Isoelectric focussing

Isoelectric focussing is closely related in concept to chromatofocussing (Section 10.11), but uses an entirely different approach to generating a pH gradient (Andrews, 1986b). A gel is constructed from a mixture of synthetic components (ampholytes) whose p*I* values span a useful pH range. Under the influence of an electric field these ampholytes order themselves between the electrodes depending on their charge state at the operating pH, which is in turn determined by their p*I*. So the most positively charged ampholytes will migrate furthest towards the cathode, and the most negatively charged the furthest towards the anode (Figure 10.38). This effectively establishes a pH gradient.

As discussed in Chapter 11, an alternative to a gradient made up of mobile ampholytes is to create a *fixed* gradient, by casting the gel from a graded mixture of acrylamide monomers containing carboxyl or tertiary amine groups. These immobilised pH gradients

[1] For some uses this may not even be necessary, e.g. proteins for use as immunogens can be employed quite successfully while still adsorbed to the gel.

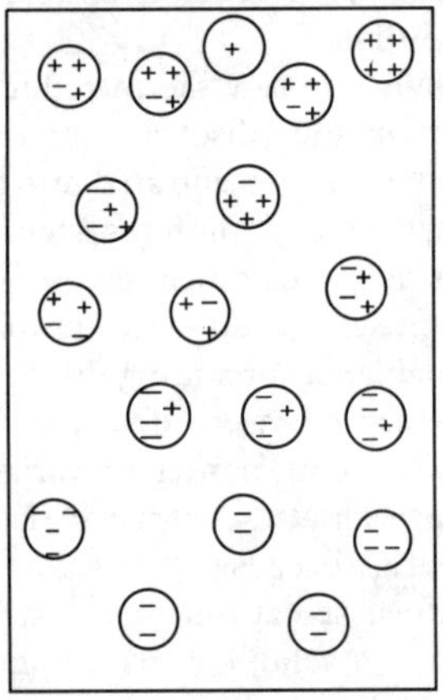

Figure 10.38 Principle of isoelectric focusing (IEF) a pH gradient is established by the arrangement of multi-charged ampholytes between electrodes, and can then be used to focus solutes (cf. chromatofocussing).

(IPGs) have been used for preparative purifications at the milligram scale (Righetti and Bossi, 1997).

A protein solute entering the gel at a pH below its p*I* will migrate towards the cathode, but the pH it experiences will be changing as it passes through the gradient. When it reaches a pH equivalent to its p*I*, it has no net charge, and therefore loses its electrophoretic mobility. So it is focussed at a point in the gel whose pH corresponds to the p*I* of the protein—a result highly analogous to that obtained from chromatofocussing.

In preparative terms the focussing aspect of this approach is very attractive: within the capacity of the matrix large quantities of dilute sample can be introduced, and the fractionated components will concentrate at the points in the gel corresponding to their respective p*I*. IEF has therefore received considerable attention as a preparative technique (Radola, 1984).

Vesterberg (1971) described the use of preparative IEF for quantities of up to 25 mg of protein in vertical columns; Graesslin and Weise (1974) described a flat-bed arrangement with a capacity of 40 mg, while in 1984 Radola noted several reported preparations at the 0.5–1 g scale.

More recently, preparative application of IEF has relied primarily on recycling partially-purified product through equipment carefully designed to dissipate Joule heat (Bier and Egen, 1979; Rodkey and Huff, 1990). This approach has demonstrated very impressive fractionations of proteins whose p*I* values are separated only by small amounts. Binion *et al.* (1982), for example, resolved monoclonal antibody populations from a polyclonal mixture, while Knisley and Rodkey (1990) demonstrated complete separation of ovalbumin (p*I* 4.6–4.8) from β-lactoglobulin (p*I* 5.1–5.4); the resolution obtained was better than that observed with zone electrophoresis methods, as discussed in the previous section. Bondoc *et al.* (1997) resolved recombinant human interleukin-10 from closely related variants using this approach. Serendipitous effects are sometimes observed in IEF preparations, for example the electrodissociation of aggregated material (Kruzel *et al.*, 1989).

Ampholyte molecules may bind to the target protein molecules with varying affinities and be eluted with them, but Rodkey (1988) demonstrated that they could often be dissociated by dialysis or SEC.

Appendix 10.1: Column Efficiency

The aim of chromatography is to resolve two or more components, and in any given situation the resolution is controlled by a combination of selectivity (determined by the choice of solid phase and eluent, and their relative combined effects on the various solutes), the capacity of the column for the solutes of interest (determined predominantly by characteristics of the solid phase and the solutes), and the efficiency of the column itself (determined principally by the nature of the solid phase particles, and the homogeneity of the packing procedure).

Only the latter parameter is substantially independent of the individual solutes being examined, and the efficiency of a given column is a useful characteristic in predicting its resolution capabilities.

Simply speaking, for any solute the efficiency can be calculated by examining the width of a pure peak at a known elution volume. The efficiency N (number of theoretical plates) is expressed by:

$$N = 16(V_e/W)^2$$

where V_e is the elution volume, and W is the base width of the peak (Figure 10.39). Efficiency increases in proportion to column length, and therefore expressions are often encountered which are normalised for this effect. These are the height equivalent to a theoretical plate (HETP; L/N), or the number of theoretical plates per unit length (N/L), where L is the column length.

These parameters can be determined very simply for a given column by injecting a pure, solute, and measuring the width of its eluted peak. The choice of solute will be determined by the nature of the chro-

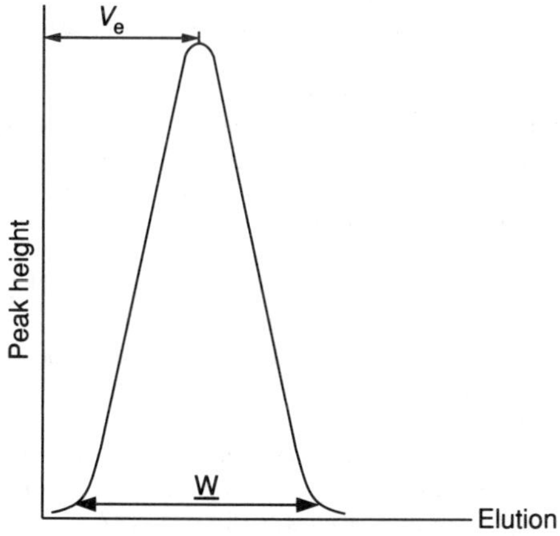

Figure 10.39 Measurement of column plate number using the base width of the peak (W) at elution volume V_e.

matographic system, so that it is retained long enough to give a good peak shape, but is still eluted within a practically short time. Also, the choice may be influenced by the wish to avoid contamination of the column, and by the availability of a suitable detector system. For SEC, for example, a convenient means of measuring column efficiency is to inject a shot of a concentrated salt, detecting the peak by conductivity.

When interpreting plate height data based on small molecules, it should be borne in mind that efficiency will be lower for macromolecules, because their diffusion coefficients are smaller by 2–3 orders of magnitude (Chang *et al.*, 1986). Therefore plate numbers will be lower (or plate heights higher) for proteins than for small molecules. This can only be counteracted by maximising the other factors affecting the resolution, in particular the selectivity of the system. High selectivity allows good separations to be obtained even at relatively low column efficiencies.

References

Ackers, G. K. (1967). A new calibration procedure for gel filtration columns. *J. Biol. Chem.*, **24**, 3237–3238.

Ackers, G. K. and Steere, R. L. (1962). Restricted diffusion of macromolecules through agar-gel membranes. *Biochim. Biophys. Acta*, **59**, 137–149.

Agner, K. (1938). The preparation and properties of a highly active catalase from horse liver. *Biochem. J.*, **32**, 1702–1705.

Aithal, H. N. *et al.* (1988). An alternate method utilizing small quantities of ligand for affinity purification of monospecific antibodies. *J. Immunol. Methods*, **112**, 63–70.

Åkerström, B. and Björck, L. (1986). A physicochemical study of Protein G, a molecule with unique immunoglobulin G-binding Properties. *J. Biol. Chem.*, **261**, 10240–10247.

Albertsson, P.-Å. (1960). *Partition of Cell Particles and Macromolecules*, Almqvist & Wiksell, Stockholm.

Albertsson, P.-Å. (1973). Application of the phase partition method to a hydrophobic membrane protein, phospholipase A1 from *Escherichia coli. Biochemistry*, **12**, 2525–2530.

Alpert, A. J. (1983). Cation-exchange high-performance liquid chromatography of proteins on poly(aspartic acid)-silica. *J. Chromatogr.*, **266**, 23–37.

Alpert, A. J. (1990). Hydrophilic-interaction chromatography for the separation of peptides, nucleic acids and other polar compounds. *J. Chromatogr.*, **499**, 177–196.

Alpert, A. J. and Regnier, F. E. (1979). Preparation of a porous microparticulate anion-exchange chromatography support for proteins. *J. Chromatogr.*, **185**, 375–392.

Anderson, R. J. *et al.* (1991). A method for the purification of DNA/protein complexes applied to DNA topoisomerase II cleavage sites. *Anal. Biochem.*, **193**, 101–111.

Andersson, K. K. *et al.* (1979). The effects of organic solvents and temperature on the desorption of yeast 3-phosphoglycerate kinase from immunoadsorbent. *J. Immunol. Methods*, **25**, 375–381.

Andrews, A. T. (1986a). Preparative gel electrophoresis. In *Electrophoresis. Theory, Techniques, and Biochemical and Clinical Applications*, 2nd edn., pp. 178–204, Clarendon Press, Oxford.

Andrews, A. T. (1986b). Iso-electric focusing. In *Electrophoresis. Theory, Techniques, and Biochemical and Clinical Applications*, 2nd edn., 241–288, Clarendon Press, Oxford.

Andrews, A. T. (1986c). Isotachophoresis (displacement electrophoresis). In *Electrophoresis. Theory, Techniques, and Biochemical and Clinical Applications*, 2nd edn., pp. 289–302, Clarendon Press, Oxford.

Andrews, P. (1962). Estimation of molecular weights of proteins by gel filtration. *Nature*, **196**, 36–39.

Andrews, P. (1965). The gel-filtration behaviour of proteins related to their molecular weights over a wide range. *Biochem. J.*, **96**, 595–606.

Annunziato, M. E. *et al.* (1993). *p*-Maleimidophenyl isocyanate: a novel heterobifunctional linker for hydroxyl to thiol coupling. *Bioconjugate Chem.*, **4**, 212–218.

Arakawa, T. and Timasheff, S. N. (1982). Preferential interactions of proteins with salts in concentrated solutions. *Biochemistry*, **21**, 6545–6552.

Arakawa, T. and Timasheff, S. N. (1984a). Mechanism of protein salting in and salting out by divalent cation salts: balance between hydration and salt binding. *Biochemistry*, **23**, 5912–5923.

Arakawa, T. and Timasheff, S. N. (1984b). The mechanism of action of Na glutamate, lysine HCl, and piperazine-*N*,*N*′-bis(2-ethanesulfonic acid) in the stabilization of tubulin and microtubule formation. *J. Biol. Chem.*, **259**, 4979–4986.

Arends, J. (1978). Purification of peroxidase-conjugated antibody for enzyme immunoassay by affinity chromatography on concanavalin A. *J. Immunol. Methods*, **25**, 171–175.

Asenjo, J. A. and Patrick, I. (1990). Large-scale protein purification. In *Protein Purification Applications: A Practical Approach* (Harris, E. L. V. and Angal, S., eds), pp. 1–28, IRL Press, Oxford.

Avrameas, S. and Ternynck, T. (1971). Peroxidase labelled antibody and Fab conjugates with enhanced intracellular penetration. *Immunochemistry*, **8**, 1175–1179.

Axén, R. *et al.* (1967). Chemical coupling of peptides and proteins to polysaccharides by means of cyanogen halides. *Nature*, **214**, 1302–1304.

Azzi, A. *et al.* (1982). Affinity chromatography purification of cytochrome *c* binding enzymes. *Proc. Natl. Acad. Sci. USA*, **79**, 2447–2450.

Behrens, M. M. *et al.* (1967). Protein–cellulose derivatives for use as immunoadsorbents: preparation employing an active ester intermediate. *Arch. Biochem. Biophys.*, **119**, 411–419.

Belew, M. *et al.* (1987). A one-step purification method for monoclonal antibodies based on salt-promoted adsorption chromatography on a thiophilic adsorbent. *J. Immunol. Methods*, **102**, 173–182.

Benedek, K. *et al.* (1984). Kinetics of unfolding of proteins on hydrophobic surfaces in reversed-phase liquid chromatography. *J. Chromatogr.*, **317**, 227–243.

Berg, A. and Scouten, W. H. (1990). Dye–ligand centrifugal affinity chromatography. *Bioseparations*, **1**, 23–31.

Berkowitz, S. A. (1987). Linear multidimensional liquid chromatography in protein separation. *J. Liq. Chromatogr.*, **10**, 2771–2787.

Bernardi, G. (1971). Chromatography of proteins on hydroxyapatite. *Methods Enzymol.*, **22**, 325–339.

Bernardi, G. *et al.* (1972). Chromatography of polypeptides and proteins on hydroxyapatite columns: Some new developments. *Biochim. Biophys. Acta*, **278**, 409–420.

Bethell, G. S. *et al.* (1979). A novel method of activation of cross-linked agaroses with 1,1′-carbonyldiimidazole which gives a matrix for affinity chromatography devoid of additional charged groups. *J. Biol. Chem.*, **254**, 2572–2574.

Bier, M. (1962). Preparative electrophoresis. *Methods Enzymol.*, **5**, 33–50.

Bier, M. and Egen, N. B. (1979). Large-scale recycling electrofocussing. In *Electrofocus/78* (Haglund *et al.*, eds), pp. 35–48, Elsevier North Holland, Amsterdam.

Binion, S. B. *et al.* (1982). Rapid purification of antibodies to single band purity using recycling isoelectric focusing. *Electrophoresis*, **3**, 284–288.

Bird, P. *et al.* (1984). The separation of human serum IgG into subclass fractions by immunoaffinity chromatography and assessment of specific antibody activity. *J. Immunol. Methods*, **71**, 97–105.

Blondin, G. A. (1979). Resolution of the mitochondrial *N,N′*-dicyclohexylcarbodiimide binding proteolipid fraction into three similar sized proteins. *Biochem. Biophys. Res. Commun.*, **87**, 1087–1094.

Boardman, N. K. and Partridge, S. M. (1955). Separation of neutral proteins on ion-exchange resins. *Biochem. J.*, **59**, 543–552.

Bodanszky, A. and Bodanszky, M. (1970). Sepharose–avidin column for the binding of biotin or biotin-containing peptides. *Experientia*, **26**, 327.

Bondoc, L. L., Jr. *et al.* (1997). Resolution of recombinant human interleukin 10 from variants by recycling free flow focusing. *Anal. Biochem.*, **246**, 234–238.

Bonnerjea, J. *et al.* (1986). Protein purification: the right step at the right time. *Bio/Technology*, **4**, 954–958.

Borek, F. and Silverstein, A. M. (1961). Characterization and purification of ferritin–antibody globulin conjugates. *J. Immunol.*, **87**, 555–561.

Bottomley, S. P. *et al.* (1995). Elution of human IgG from affinity columns containing immobilised variants of protein A. *J. Immunol. Methods*, **182**, 185–192.

Brandt, J. *et al.* (1975). Covalent attachment of proteins to polysaccharide carriers by means of benzoquinone. *Biochim. Biophys. Acta*, **386**, 196–202.

Brocklehurst, K. *et al.* (1973). Covalent chromatography. preparation of fully active papain from dried papaya latex. *Biochem. J.*, **133**, 573–584.

Brogren, C.-H. and Svendsen, P. J. (1975). Preparative isotachophoresis combined with biospecific interaction and neuraminidase treatment in purification of human serum cholinesterase. *Protides Biol. Fluids*, **22**, 685–689.

Brooks, D. A. *et al.* (1992). An improved method for the purification of IgG monoclonal antibodies from culture supernatants. *J. Immunol. Methods*, **155**, 129–132.

Brown, R. D. and Matthews, K. S. (1979). Chemical modification of lactose repressor protein using *N*-substituted maleimides. *J. Biol. Chem.*, **254**, 5128–5134.

Brücke, E. (1861). Beiträge zur Lehre von der Verdauung. *Sitzber. Math.-Naturw. Kl. Bayer Akad. Wiss., München*, **43**, 601–623.

Bruins, E. M. (1934). Quantitative Beziehungen in den Lyotropen-Reihen. II. Die Lyotropen zahlen und Ihre Erklärung. *Rect. Trav. Chim. Pays-Bas*, **53**, 292–307.

Burke, C. W. and Shakespear, R. A. (1975). Rapid purification of tri-iodothyronine and thyroxine protein conjugates for antibody production. *J. Endocrinol.*, **65**, 133–138.

Burstein, M. (1957). Sur la précipitation des β-lipoprotéines et des euglobulines du sérum par le polyvinyl-pyrrolidone. *C. R. Hebd. Acad. Sci.*, **244**, 3189–3191.

Calton, G. J. (1984). Immunosorbent separations. *Methods Enzymol.*, **104**, 381–387.

Campbell, D. H. *et al.* (1951). Immunologic adsorbents. I. Isolation of antibody by means of a cellulose–protein antigen. *Proc. Natl. Acad. Sci. USA*, **37**, 575–578.

Carlsson, J. and Svenson, A. (1974). Preparation of bovine mercaptalbumin by means of covalent chromatography. *FEBS Lett.*, **42**, 183–186.

Carrea, G. *et al.* (1982). The effect of hofmeister anions and protein concentration on the activity and stability of some immobilized NAD-dependent dehydrogenases. *Biotechnol. Bioeng.*, **24**, 1–7.

Cartwright, T. (1987). Isolation and purification of products from animal cells. *Trends Biotechnol.*, **5**, 25–30.

Chang, S. H. *et al.* (1976). High-performance liquid chromatography of proteins. *J. Chromatogr.*, **125**, 103–114.

Chrambach, A. and Rodbard, D. (1981). 'Quantitative' and preparative polyacrylamide gel electrophoresis. In *Gel Electrophoresis of Proteins: A Practical Approach* (Hames, B. D. and Rickwood, D., eds), pp. 93–143, IRL Press, Oxford.

Clonis, Y. D. and Lowe, C. R. (1980). Triazine dyes, a new class of affinity labels for nucleotide-dependent enzymes. *Biochem. J.*, **191**, 247–251.

Clonis, Y. D. *et al.* (1981). The interaction of yeast hexokinase with procion green H-4G. *Biochem. J.*, **197**, 203–211.

Cohn, E. J. *et al.* (1946). Preparation and properties of serum and plasma Proteins. IV. A system for the separation into fractions of the protein and lipoprotein components of biological tissues and fluids. *J. Am. Chem. Soc.*, **68**, 459–475.

Creighton, W. D. *et al.* (1973). Detection of antibodies and soluble antigen–antibody complexes by precipitation with polyethylene glycol. *J. Immunol.*, **111**, 1219–1227.

Crone, H. D. and Dawson, R. M. (1976). Residual anionic properties of a covalently substituted controlled-pore gglass, glyceryl-CPG. *J. Chromatogr.*, **129**, 91–96.

Crowe, J. *et al.* (1994). 6XHis-Ni-NTA chromatography as a superior technique in recombinant protein expression/ purification. *Methods Mol. Biol.*, **31**, 371–387.

Cuatrecasas, P. (1970). Agarose derivatives for purification of protein by affinity chromatography. *Nature*, **228**, 1327–1328.

Cuatrecasas, P. (1972). Affinity chromatography of macromolecules. *Adv. Enzymol.*, **35**, 29–89.

Cuatrecasas, P. (1973). Interaction of wheat germ agglutinin and concanavalin A with isolated fat cells. *Biochemistry*, **12**, 1312–1323.

Cuatrecasas, P. and Parikh, I. (1972). Adsorbents for affinity chromatography. Use of *N*-hydroxysuccinimide esters of agarose. *Biochemistry*, **11**, 2291–2299.

Cuatrecasas, P. *et al.* (1968). Selective enzyme purification by affinity chromatography. *Proc. Natl. Acad. Sci. USA*, **61**, 636–643.

Cusack, M. and Beynon, R. J. (1989). Purification by exploitation of affinity: preparation of adsorbents using oxirane acrylic beads and their use. In *Protein Purification Methods: A Practical Approach* (Harris, E. L. V. and Angal, S., eds), pp. 245–292, IRL Press, Oxford.

Dawes, J. (1986). Immunopurification process. *World Patent WO 86/00910*, to Celltech Ltd.

Dean, P. D. G. and Watson, D. H. (1979). Protein purification using immobilised triazine dyes. *J. Chromatogr.*, **165**, 301–319.

Dean, P. D. G. *et al.* (1985a). Matrix preparations and applications. In *Affinity Chromatography: A Practical Approach* (Dean, P. D. G. *et al.*, eds), pp. 1–30, IRL Press, Oxford.

Dean, P. D. G. *et al.* (1985b). Activation procedures. In *Affinity Chromatography: A Practical Approach* (Dean, P. D. G. *et al.* eds), pp. 31–59, IRL Press, Oxford.

Dean, P. D. G. *et al.* (1985c). Cross-linking agents for coupling matrices to spacers. In *Affinity Chromatography: A Practical Approach* (Dean, P. D. G. *et al.*, eds), pp. 61–73, IRL Press, Oxford.

Dhanabal, M. *et al.* (1995). A novel method to purify immunotoxins from free antibodies using modified recombinant toxins. *J. Immunol. Methods*, **182**, 165–175.

Diesenhofer, J. (1981). Crystallographic refinement and atomic models of human Fc fragment and its complex with fragment B of Protein A from *Staphylococcus aureus* at 2. 9- and 2. 8-Å Resolution. *Biochemistry*, **20**, 2361–2370.

Dixon, M. and Webb, E. C. (1961). Enzyme fractionation by salting-out: a theoretical note. *Adv. Protein Chem.*, **16**, 197–219.

Dodgson, K. S. and Powell, G. M. (1959). Studies on sulphatases. 27. The partial purification and properties of the arylsulphatase of the digestive gland of *Helix pomatia*. *Biochem. J.*, **73**, 672–679.

Easterday, R. L. and Easterday, I. M. (1974). Affinity chromatography of kinases and dehydrogenases on Sephadex® and Sepharose® dye derivatives. *Adv. Exp. Med. Biol.*, **42**, 123–133.

Egly, J. M. and Porath, J. (1978). Gel matrices, coupling methods and charge transfer chromatography. In *Affinity Chromatography* (Hoffmann-Osterhof, O. *et al.*, eds), pp. 5–22, Pergamon Press, New York.

Ekstrom, B. and Jacobson, G. (1984). Differences in retention behaviour between small and large molecules in ion-exchange chromatography and reversed-phase chromatography. *Anal. Biochem.*, **142**, 134–139.

Eldjarn, L. and Jellum, E. (1963). Organomercurial-polysaccharide, a chromatographic material for the separation and isolation of SH-proteins. *Acta Chem. Scand.*, **17**, 2610–2621.

Eliasson, M. *et al.* (1988). Chimeric IgG-binding receptors engineered from staphylococcal Protein A and streptococcal Protein G. *J. Biol. Chem.*, **263**, 4323–4327.

Er-El, Z. *et al.* (1972). Hydrocarbon-coated sepharoses. Use in the purification of glycogen phosphorylase. *Biochem. Biophys. Res. Commun.*, **49**, 383–390.

Erlanger, B. F. *et al.* (1959). Steroid–Protein conjugates. II. Preparation and characterization of conjugates of bovine serum albumin with progesterone, deoxycorticosterone, and estrone. *J. Biol. Chem.*, **234**, 1090–1094.

Ernst-Cabrera, K. and Wilchek, M. (1988). High-performance affinity chromatography. *Trends Anal. Chem.*, **7**, 58–63.

Eveleigh, J. W. and Levy, D. E. (1977). Immunochemical characteristics and preparative application of agarose-based immunosorbents. *J. Solid-Phase Biochem.*, **2**, 45–78.

Ey, P. L. *et al.* (1978). Isolation of Pure IgG_1, IgG_{2a} and IgG_{2b} immunoglobulins from mouse serum using Protein A-Sepharose. *Immunochemistry*, **15**, 429–436.

Fausnaugh, J. L. *et al.* (1984). Comparison of hydrophobic-interaction and reversed-phase chromatography of proteins. *J. Chromatogr.*, **317**, 141–155.

Fischer, L. J. *et al.* (1978). High performance liquid chromatographic separation of insulin, glucagon, and somatostatin. *Anal. Chem.*, **50**, 2143–2144.

Flanagan, S. D. and Barondes, S. H. (1975). Affinity partitioning. A method for purification of proteins using specific polymer-ligands in aqueous polymer two-phase systems. *J. Biol. Chem.*, **250**, 1484–1489.

Flodin, P. and Porath, J. (1954). Zone electrophoresis in starch columns. *Biochim. Biophys. Acta*, **13**, 175–182.

Forsgren, A. and Sjöquist, J. (1966). 'Protein A' from *S. aureus*. I. Pseudo-immune reaction with human γ-globulin. *J. Immunol.*, **97**, 822–827.

Franks, F. (1993). Solution properties of proteins. In *Protein Biotechnology* (Franks, F., ed.), pp. 133–189, Humana Press, Totowa, NJ.

Gallagher, J. T. (1989). Affinity chromatography of complex carbohydrates using lectins. In *HPLC of Macromolecules: A Practical Approach* (Oliver, R. W. A., ed.), pp. 209–227, IRL Press, Oxford.

Gersten, D. M. and Marchalonis, J. J. (1978). A Rapid, novel method for the solid-phase derivatisation of IgG antibodies for immune-affinity chromatography. *J. Immunol. Methods*, **24**, 305–309.

Goding, J. W. (1980). Antibody production by hybridomas. *J. Immunol. Methods*, **39**, 285–308.

Goldstein, I. J. *et al.* (1965). Protein–carbohydrate interaction. II. Inhibition studies on the interaction of concanavalin A with polysaccharides. *Biochemistry*, **4**, 876–883.

Gooding, K. M. and Schmuck, M. N. (1985). Comparison of weak and strong high-performance anion-exchange chromatography. *J. Chromatogr.*, **327**, 139–146.

Goosen, P. C. M. (1980). Isolation of subclasses of human IgG with affinity chromatography. *J. Immunol. Methods*, **37**, 89–93.

Gorbunoff, M. J. (1984a). The interaction of proteins with hydroxyapatite. I. Role of protein charge and structure. *Anal. Biochem.*, **136**, 425–432.

Gorbunoff, M. J. (1984b). The interaction of proteins with hydroxyapatite. II. Role of acidic and basic groups. *Anal. Biochem.*, **136**, 433–439.

Gouda, H. *et al.* (1992). Three-dimensional solution structure of the B domain of staphylococcal Protein A: comparisons of the solution and crystal structures. *Biochemistry*, **31**, 9665–9672.

Graesslin, D. and Weise, H. C. (1974). Preparative gel isoelectric focusing. In *Electrophoresis and Isoelectric Focusing in Polyacrylamide Gel* (Allen, R. C. and Maurer, H. R., eds), pp. 199–205, Walter de Gruyter & Co., Berlin.

Gyka, G. *et al.* (1983). Crosslinkage of antibodies to staphylococcal Protein A matrices. *J. Immunol. Methods*, **57**, 227–233.

Hageman, J. H. and Kuehn, G. D. (1992). Boronic acid matrices for the affinity purification of glycoproteins and enzymes. *Methods Mol. Biol.*, **11**, 45–71.

Hames, B. D. (1981). An introduction to polyacrylamide gel electrophoresis. In *Gel Electrophoresis of Proteins: A Practical Approach* (Hames, B. D. and Rickwood, D., eds), pp. 1–91, IRL Press, Oxford.

Hamill, W. H. *et al.* (1966). Surface chemistry. In *Principles of Physical Chemistry*, 2nd edn., pp. 319–344, Prentice-Hall, Englewood Cliffs, NJ.

Hancock, W. S. and Sparrow, J. T. (1981). Use of mixed-mode, high-performance liquid chromatography for the separation of peptide and protein mixtures. *J. Chromatogr.*, **206**, 71–82.

Hancock, W. S. *et al.* (1981). The use of reversed-phase high-performance liquid chromatography with radial compression for the analysis of peptide and protein mixtures. *J. Chromatogr.*, **206**, 59–70.

Harlow, E. and Lane, D. (1988). Ammonium sulfate precipitation. In *Antibodies: A Laboratory Manual*, 298–299, cold spring harbor publications, cold spring harbor, NY.

Harrington, J. C. *et al.* (1971). Polymer-induced precipitation of antigen–antibody complexes: 'Precipiplex' reactions. *Immunochemistry*, **8**, 413–421.

Hearn, M. T. W. *et al.* (1979). High-performance liquid chromatography of amino acids, peptides and proteins. XX. Investigation of the effect of pH and ion-pair formation on the retention of peptides on chemically-bonded hydrocarbonaceous stationary phases. *J. Chromatogr.*, **185**, 429–444.

Hearn, M. T. W. *et al.* (1985). High-performance liquid chromatography of amino acids, peptides and proteins. LXVI. Investigation of the effects of chromatographic dwell in the reversed-phase high-performance liquid chromatographic separation of proteins. *J. Chromatogr.*, **327**, 47–66.

Heide, K. *et al.* (1977). Plasma protein fractionation. In *The Plasma Proteins. Structure, Function and Genetic Control* (Putnam, F. W., ed.), 2nd edn, vol. 3, pp. 545–597, Academic Press, New York.

Heidelberger, M. (1939). Quantitative absolute methods in the study of antigen–antibody reactions. *Bacteriol. Rev.*, **3**, 49–95.

Heimburger, N. and Schmidtberger, R. (1964). Moderne Verfahren der Chromatographie und Elektrophorese zur Präparation von Plasmaproteinen. *Behringwerk Mitt.*, **43**, 83–128.

Henry, M. P. (1990). High-performance ion-exchange chromatography of proteins: A review of methods and mechanisms. In *High Performance Liquid Chromatography in Biotechnology* (Hancock, W. S., ed.), pp. 205–261, John Wiley, New York.

Hill, C. R. *et al.* (1989). Purification by exploitation of activity: immunopurification. In *Protein Purification Methods – A Practical Approach* (Harris, E. L. V. and Angal, S., eds), pp. 245–292, IRL Press, Oxford.

Hillson, D. A. (1981). Resolution of thiol-containing proteins by sequential-elution covalent chromatography. *J. Biochem. Biophys. Methods*, **4**, 101–111.

Hillson, J. L. *et al.* (1993). The structural basis of germline-encoded V_H3 immunoglobulin binding to staphylococcal Protein A. *J. Exp. Med.*, **178**, 331–336.

Himmelhoch, S. R. (1971). Chromatography of proteins on ion-exchange adsorbents. *Methods Enzymol.*, **22**, 273–286.

Hjelm, H. *et al.* (1972). Protein A from *Staphylococcus aureus*. Its isolation by affinity chromatography and its use as an immunosorbent for isolation of immunoglobulins. *FEBS Lett.*, **28**, 73–76.

Hjertén, S. *et al.* (1969). Apparatus for large-scale preparative polyacrylamide gel electrophoresis. *Anal. Biochem.*, **27**, 108–129.

Hjertén, S. *et al.* (1974). Hydrophobic interaction chromatography. The synthesis and the use of some alkyl and aryl derivatives of agarose. *J. Chromatogr.*, **101**, 281–288.

Hjertén, S. *et al.* (1986). Gradient and isocratic high- performance hydrophobic interaction chromatography of proteins on agarose columns. *J. Chromatogr.*, **359**, 99–109.

Hofmeister, F. (1888). Zur Lehre von der Wirkung der Salze. *Arch. Exp. Pathol. Pharmakol.*, **24**, 247–260.

Hofstee, B. H. J. (1973). Hydrophobic affinity chromatography of proteins. *Anal. Biochem.*, **52**, 430–448.

Hofstee, B. H. J. (1976). Hydrophobic adsorption chromatography of proteins. In *Methods of Protein Separation* (Catsimpoolas, N., ed.), vol. 2, pp. 245–278, Plenum Press, New York.

Hofstee, B. H. J. and Otillio, N. F. (1978). Non-ionic adsorption chromatography of proteins. *J. Chromatogr.*, **159**, 57–69.

Holloway, C. J. and Battersby, R. V. (1984). Preparative isotachophoresis. *Methods Enzymol.*, **104**, 281–301.

Hopp, T. P. *et al.* (1988). A short polypeptide marker sequence useful for recombinant protein identification and purification. *Bio/Technology*, **6**, 1204–1210.

Hunter, M. J. and Ludwig, M. L. (1962). The reaction of imidoesters with proteins and related small molecules. *J. Am. Chem. Soc.*, **84**, 3491–3504.

Hymer, W. C. *et al.* (1987). Continuous flow electrophoretic separation of proteins and cells from mammalian tissues. *Cell Biophys.*, **10**, 61–85.

Ingham, K. C. (1984). Protein precipitation with polyethylene glycol. *Methods Enzymol.*, **104**, 351–369.

Jack, G. W. and Wade, H. E. (1987). Immunoaffinity chromatography of clinical products. *Trends Biotechnol.*, **5**, 91–95.

Janknecht, R. *et al.* (1991). Rapid and efficient purification of native histidine-tagged protein expressed by recombinant vaccinia virus. *Proc. Natl. Acad. Sci. USA*, **88**, 8972–8976.

Janson, J.-C. and Hedman, P. (1987). On the optimization of process chromatography of proteins. *Biotechnol. Prog.*, **3**, 9–13.

Jennissen, H. P. (1976). Evidence for negative cooperativity in the adsorption of phosphorylase *b* on hydrophobic agaroses. *Biochemistry*, **15**, 5683–5692.

Jenö, P. *et al.* (1993). Desalting electroeluted proteins with hydrophilic interaction chromatography. *Anal. Biochem.*, **215**, 292–298.

Jiang, W. and Hearn, M. T. W. (1996). Protein interaction with immobilized metal ion affinity ligands under high ionic strength conditions. *Anal. Biochem.*, **242**, 45–54.

Johansson, G. (1984). Affinity partitioning. *Methods Enzymol.*, **104**, 356–364.

Jolley, W. B. *et al.* (1967). Ultracentrifugation using acrylamide gel. *Anal. Biochem.*, **21**, 454–461.

Juarez-Salinas, H. *et al.* (1986). Separation of IgG idiotypes by high-performance hydroxyapatite chromatography. *Methods Enzymol.*, **121**, 615–622.

Kalef, E. *et al.* (1993). Arsenical-based affinity chromatography of vicinal dithiol-containing proteins: purification of L1210 leukemia cytoplasmic proteins and the recombinant rat c-erb Aβ_1 T_3 receptor. *Anal. Biochem.*, **212**, 325–334.

Kårsnäs, P. and Lindblom, T. (1992). Characterization of hydrophobic interaction and hydrophobic interaction chromatography media by multivariate analysis. *J. Chromatogr.*, **599**, 131–136.

Kato, Y. *et al.* (1980). Separation range and separation efficiency in high-speed gel filtration on TSK-Gel SW columns. *J. Chromatogr.*, **190**, 297–303.

Kato, Y. *et al.* (1987). New high-performance gel filtration columns for protein separation. *J. Chromatogr.*, **404**, 333–339.

Kennedy, J. F. and Rosevear, A. (1973). An assessment of the fractionation of carbohydrate on concanavalin A Sepharose 4B by affinity chromatography. *J. Chem. Soc., Perkin Trans. 1*, pp. 2041–2046.

Kirchberger, J. and Kopperschläger, G. (1990). An improved purification procedure of alkaline phosphatase from calf intestine by applying partition in aqueous two-phase systems and dye–ligand chromatography. *Bioseparation*, **1**, 33–41.

Kjellén, L. and Lindahl, U. (1991). Proteoglycans: structures and interaction. *Annu. Rev. Biochem.*, **60**, 443–475.

Klotz, I. M. (1989). Ligand–protein binding affinities. In *Protein Function: A Practical Approach* (Creighton, T. E., ed.), 25–54, IRL Press, Oxford.

Knisley, K. A. and Rodkey, L. S. (1990). Comparative studies of recycling isoelectric focusing and continuous flow electrophoresis: separation of proteins with minor charge differences. *Electrophoresis*, **11**, 927–931.

Knowles, P. P. and Thorpe, P. E. (1987). Purification of immunotoxins containing ricin A-chain and abrin A-chain using Blue Sepharose CL-6B. *Anal. Biochem.*, **160**, 440–443.

Kohn, J. and Wilchek, M. (1982). Mechanism of activation of Sepharose and Sephadex by cyanogen bromide. *Enzyme Microb. Technol.*, **4**, 161–163.

Kopaciewicz, W. and Regnier, F. E. (1982). Nonideal size-exclusion chromatography of proteins: effects of pH at low ionic strength. *Anal. Biochem.*, **126**, 8–16.

Kopaciewicz, W. *et al.* (1983). Retention model for high-performance ion-exchange chromatography. *J. Chromatogr.*, **266**, 3–21.

Kopaciewicz, W. *et al.* (1985). Stationary phase contributions to retention in high-performance anion-exchange protein chromatography: ligand density and mixed mode effects. *J. Chromatogr.*, **318**, 157–172.

Kopperschläger, G. *et al.* (1968). Some kinetic and molecular properties of yeast phosphofructokinase. *FEBS Lett.*, **1**, 137–141.

Kruzel, M. L. *et al.* (1989). Electrodissociation of high molecular weight complexes of interferon alpha during recycling isoelectric focusing. *J. Biol. Regul. Homeost. Agents*, 3, 79–83.

Kundu, A. and Cramer, S. M. (1997). Low-molecular-weight displacers for high-resolution protein separations. *Anal. Biochem.*, **248**, 111–116.

Kunin, R. and McGarvey, F. X. (1949). Equilibrium and column behavior of exchange resins. Strong base anion exchange resin. *Ind. Eng. Chem.*, **41**, 1265–1268.

Lam-Thanh, H. *et al.* (1981). Reversed-phase, ion-pair separation of L-methionine and L-methionine dipeptides complexed with palladium(II). *J. Liq. Chromatogr.*, **4**, 681–688.

Langone, J. J. (1982). *Staphylococcus aureus* and related immunoglobulin receptors produced by streptococci and pneumococci. *Adv. Immunol.*, **32**, 157–252.

Larsson, P.-O. *et al.* (1983). High-performance liquid affinity chromatography. *Adv. Chromatogr.*, 12, 41–85.

Larsson, P.-O. *et al.* (1984). Affinity precipitation of dehydrogenases. *Methods Enzymol.*, **104**, 364–369.

Lascu, I. *et al.* (1984). Ion-exchange properties of Cibacron Blue 3G-A Sepharose (Blue Sepharose) and the interaction of proteins with Cibacron Blue 3G-A. *J. Chromatogr.*, **283**, 199–210.

Lathe, G. H. and Ruthven, C. R. J. (1956). The separation of substances and estimation of their relative molecular sizes by the use of columns of starch in water. *Biochem. J.*, **62**, 665–674.

Laurent, T. C. and Killander, J. (1964). A theory of gel filtration and its experimental verification. *J. Chromatogr.*, **14**, 317–330.

Lei, J. C. (1974). Application of polyethylene glycol in the preparation of a concentrated, purified foot-and-mouth disease vaccine. Present status of research. *Bull. Office Int. Epizoot.*, **81**, 1169–1199.

Liao, A. W. *et al.* (1987). High performance displacement chromatography of proteins: separation of β-lactoglobulins A and B. *Chromatographia*, **24**, 881–885.

Liebman, H. A. *et al.* (1985). Immunoaffinity purification of factor IX (Christmas Factor) by using conformation-specific antibodies directed against the factor IX–metal complex. *Proc. Natl. Acad. Sci. USA*, **82**, 3879–3883.

Lindner, N. M. *et al.* (1989). Design and application of biomimetic anthraquinone dyes. Purification of calf intestinal alkaline phosphatase with immobilised terminal ring reactive analogues of C. I. Reactive Blue 2. *J. Chromatogr.*, **473**, 227–240.

Locascio, G. A. *et al.* (1969). Estimation of molecular weights of proteins by agarose gel filtration. *J. Chromatogr.*, **40**, 453–457.

Lowe, C. R. and Pearson, J. C. (1984). Affinity chromatography on immobilized dyes. *Methods Enzymol.*, **184**, 97–113.

Lowe, C. R. *et al.* (1973). Some parameters relevant to affinity chromatography on immobilized nucleotides. *Biochem. J.*, **133**, 499–506.

Maeda, H. *et al.* (1984). Tailor-making of protein drugs by polymer conjugation for tumor targeting: A brief review of smancs. *J. Protein Chem.*, **3**, 181–193.

Margoliash, E. (1952). Purification of cytochrome *c*. *Nature*, **170**, 1014–1015.

Martin, R. G. and Ames, B. N. (1961). A method for determining the sedimentation behavior of enzymes: application to protein mixtures. *J. Biol. Chem.*, **236**, 1372–1379.

Mattox, V. R. *et al.* (1979). A comparison of procedures for attaching steroidal glucosiduronic acids to bovine serum albumin. *J. Steroid Biochem.*, **10**, 167–172.

McColloch, R. J. and Kertesz, Z. I. (1945). Pectic enzymes. VI. The use of an ion exchange resin for the complete removal of pectin-methylesterase from commercial pectinases. *J. Biol. Chem.*, **160**, 149–154.

McKenzie, G. H. *et al.* (1972). The molecular weight and stability of concanavalin A. *Biochim. Biophys. Acta*, **263**, 283–293.

Melander, W. and Horváth, C. (1977). Salt effects on hydrophobic interactions in precipitation and chromatography of proteins: An interpretation of the lyotropic series. *Arch. Biochem. Biophys.*, **183**, 200–215.

Meyerson, L. R. and Abraham, K. I. (1986). High performance liquid chromatographic properties of peptides and proteins on a dihydroxyalkyl bonded silica stationary phase. *Peptides*, **7**, 481–489.

Miller, C. E. *et al.* (1997). Recovery of monoclonal antibody from its complex with ligand. *Anal. Biochem.*, **245**, 179–183.

Mizutani, T. and Mizutani, A. (1975). Prevention of adsorption of protein on controlled-pore glass with amino acid buffer. *J. Chromatogr.*, **111**, 214–216.

Morgan, M. R. A. *et al.* (1978). Electrophoretic desorption: preparative elution of steroid specific antibodies from immunoadsorbents. *J. Steroid Biochem.*, **9**, 767–770.

Mössner, E. *et al.* (1980). Purification of human and bovine alkaline phosphatases by affinity chromatography. *Hoppe-Seyler's Z. Physiol. Chem.*, **361**, 543–549.

Mould, D. L. and Synge, R. L. M. (1954). Separations of polysaccharides related to starch by electrokinetic ultrafiltration in collodion membranes. *Biochem. J.*, **58**, 571–585.

Murphy, R. F. *et al.* (1976). Avoidance of strongly chaotropic eluents for immunoaffinity chromatography by chemical modification of immobilized ligand. *Biochim. Biophys. Acta*, **420**, 87–96.

Nakagami, S. *et al.* (1991). Preparation of enzyme-conjugated DNA probe and application to the universal probe system. *Anal. Biochem.*, **198**, 75–79.

Narhi, L. O. *et al.* (1989). Hydrophobic interaction chromatography in alkaline pH. *Anal. Biochem.*, **182**, 266–270.

Nees, S. (1974). Theoretical and practical aspects of high resolution preparative gel electrophoresis using a continuous elution system. In *Electrophoresis and Isoelectric Focusing in Polyacrylamide Gel* (Allen, R. C. and Maurer, H. R., eds), pp. 189–198, Walter de Gruyter & Co., Berlin.

Nees, S. *et al.* (1976). Purification and characterization of *N*-acetylneuraminate lyase from *Clostridium perfringens. Hoppe-Seyler's Z. Physiol. Chem.*, **357**, 839–853.

Neoh, S-H. and Zola, H. (1990). Purification of mouse monoclonal antibodies from ascitic fluid and culture supernatants by precipitation with polyethylene glycol 6000. In *Laboratory Methods in Immunology* (Zola, H., ed.), pp. 73–77, CRC Press, Boca Raton, FL.

Neoh, S.-H. *et al.* (1986). The purification of mouse monoclonal antibodies from ascitic fluid. *J. Immunol. Methods*, **91**, 231–235.

Neumeier, R. and Maurer, H. R. (1982). Isolation of a high molecular mass granulocyte colony-stimulating factor from bovine lung conditioned medium. *Hoppe-Seyler's Z. Physiol. Chem.*, **363**, 1493–1500.

Ngo, T. T. (1986). Facile activation of Sepharose hydroxyl groups by 2-fluoro-1 methylpyridinium toluene-4-sulfonate: Preparation of affinity and covalent chromatographic matrices. *Bio/Technology*, **4**, 134–137.

Nguyen, N. Y. and Chrambach, A. (1981). Gel isotachophoresis. In *Gel Electrophoresis of Proteins: A Practical Approach* (Hames, B. D. and Rickwood, D., eds), pp. 145–155, IRL Press, Oxford.

Nilsson, K. and Mosbach, K. (1981). Immobilization of enzymes and affinity ligands to various hydroxyl group carrying supports using highly reactive sulfonyl chlorides. *Biochem. Biophys. Res. Commun.*, **102**, 449–457.

Nishijima, M. *et al.* (1977). Detergent-resistant phospholipase A of *Escherichia coli* K-12. Purification and properties. *Eur. J. Biochem.*, **73**, 115–124.

Norde, W. (1986). Adsorption of proteins from solution at the solid-liquid interface. *Adv. Colloid Interface Sci.*, **25**, 267–340.

Novotný, J. (1971). Chromatography of proteins and peptides on Sephadex ion-exchangers: Dependence of the resolution on the elution schedule. *FEBS Lett.*, **14**, 7–10.

Ohlson, S. *et al.* (1978). High performance liquid affinity chromatography (HPLAC) and its application to the separation of enzymes and antigens. *FEBS Lett.*, **93**, 5–9.

Ovalle, R. (1995). Nonideal size-exclusion chromatography with zwitterions causes a p*I*-dependent elution of proteins from glycerol propylsilane-modified silica. *Anal. Biochem.*, **229**, 1–7.

Pedersen, K. O. (1962). Exclusion chromatography. *Arch. Biochem. Biophys.*, **Suppl. 1**, 157–168.

Pepper, D. S. (1992). Some alternative coupling chemistries for affinity chromatography. *Methods Mol. Biol.*, **11**, 173–196.

Perry, M. and Kirby, H. (1990). Example purifications: monoclonal antibodies and their fragments. In *Protein Purification Applications: A Practical Approach* (Harris, E. L. V. and Angal, S., eds), pp. 147–166, IRL Press, Oxford.

Perseptive Biosystems, Inc. (1996). *The Busy Researcher's Guide to Biomolecule Chromatography*, Perseptive Biosystems, Inc., Framingham, MA.

Persson, M. A. A. (1987). Preparation of human sera containing one single IgG subclass using affinity chromatography. *J. Immunol. Methods*, **98**, 91–98.

Peterson, E. A. and Torres, A. R. (1984). Displacement chromatography of proteins. *Methods Enzymol.*, **104**, 113–133.

Pfeiffer, N. E. *et al.* (1987). Immunoaffinity chromatography utilizing monoclonal antibodies. *J. Immunol. Methods*, **97**, 1–9.

Pharmacia (1993a). *Hydrophobic Interaction Chromatography. Principles and Methods*, Pharmacia AB, Uppsala, Sweden.

Pharmacia (1993b). *Affinity Chromatography. Principles and Methods*, Pharmacia AB, Uppsala, Sweden.

Pharmacia (1993c). *Chromatofocussing with Polybuffer*™ and PBE™ Pharmacia, Uppsala, Sweden.

Pharmacia (1993d). *Gel Filtration. Principles and Methods.* Pharmacia AB, Uppsala.

Phillips, T. M. (1985). High performance immunoaffinity chromatography. An introduction. *J. Liq. Chromatogr.*, **3**, 962–972.

Phillips, T. M. (1989). High-performance immunoaffinity chromatography. *Adv. Chromatogr.*, **29**, 133–173.

Phillips, T. M. *et al.* (1985). Protein A-coated glass beads. universal support medium for high-performance immunoaffinity chromatography. *J. Chromatogr.*, **327**, 213–219.

Porath, J. (1963). Some recently developed fractionation procedures and their application to peptide and protein hormones. *Pure Appl. Chem.*, **6**, 233–244.

Porath, J. (1978). Explorations into the field of charge-transfer adsorption. *J. Chromatogr.*, **159**, 13–24.

Porath, J. (1986). Salt-promoted adsorption: Recent developments. *J. Chromatogr.*, **376**, 331–341.

Porath, J. and Flodin, P. (1959). Gel Filtration: A method for desalting and group separation. *Nature*, **183**, 1657–1659.

Porath, J. and Larsson, B. (1978). Charge-transfer and water-mediated chromatography. I. Electron-acceptor ligands on cross-linked dextran. *J. Chromatogr.*, **155**, 47–68.

Porath, J. and Olin, B. (1983). Immobilized metal ion affinity adsorption and immobilized metal ion affinity chromatography of biomaterials. Serum protein affinities of gel-immobilized iron and nickel ions. *Biochemistry*, **22**, 1621–1630.

Porath, J. *et al.* (1967). Chemical coupling of proteins to agarose. *Nature*, **215**, 1491–1492.

Porath, J. *et al.* (1973). Salting-out in amphiphilic gels as a new approach to hydrophobic adsorption. *Nature*, **245**, 465–466.

Porath, J. *et al.* (1975). Metal chelate affinity chromatography, a new approach to protein fractionation. *Nature*, **258**, 598–599.

Pressman, D. *et al.* (1942). The agglutination of intact azo-erythrocytes by antisera homologous to the attached groups. *J. Immunol.*, **44**, 101–105.

Proudfoot, K. A. *et al.* (1992). Purification of recombinant chimeric B72. 3 Fab′ and $F(ab')_2$ using streptococcal Protein G. *Protein Express. Purif.*, **3**, 368–373.

Przybycien, T. M. and Bailey, J. E. (1989). Structure–function relationships in the inorganic salt-induced precipitation of α-chymotrypsin. *Biochim. Biophys. Acta*, **995**, 231–245.

Przybycien, T. M. and Bailey, J. E. (1991). Secondary structure perturbations in salt-induced protein precipitates. *Biochim. Biophys. Acta*, **1076**, 103–111.

Qadri, F. (1985). The reactive triazine dyes: their usefulness and limitations in protein purifications. *Trends Biotechnol.*, **3**, 7–12.

Radola, B. J. (1984). High-resolution preparative isoelectric Focusing. *Methods Enzymol.*, **104**, 256–275.

Ramsden, D. B. and Louis, L. (1972). The isolation of human transferrin by isotachophoresis. *Protides Biol. Fluids*, **19**, 521–524.

Ramsden, H. E. (1985). Polyethyleneimine bound chromatographic packing. *US Patent 4,540,486*, to J. T. Baker Chemical Company.

Regnier, F. E. (1983). High-performance liquid chromatography of proteins. *Methods Enzymol.*, **91**, 137–190.

Regnier, F. E. and Gooding, K. M. (1980). High-performance liquid chromatography of proteins. *Anal. Biochem.*, **103**, 1–25.

Regnier, F. E. and Noel, R. (1976). Glycerolpropylsilane bonded phases in the steric exclusion chromatography of biological macromolecules. *J. Chromatogr. Sci.*, **14**, 316–320.

Reid, A. F. and Jones, F. (1951). Fractionation of blood plasma proteins using ion exchange resins. Revised techniques. *Ind. Eng. Chem.*, **43**, 1074–1075.

Reiland, J. (1971). Gel filtration. *Methods Enzymol.*, **22**, 287–321.

Righetti, P. G. and Bossi, A. (1997). Isoelectric focusing in immobilized pH gradients: recent analytical and preparative developments. *Anal. Biochem.*, **247**, 1–10.

Rinderknecht, H. (1962). Ultra-rapid fluorescent labelling of proteins. *Nature*, **193**, 167–168.

Robbins, J. B. *et al.* (1967). Purification of antibodies with immunoadsorbents prepared using bromoacetyl cellulose. *Immunochemistry*, **4**, 11–22.

Rodkey, L. S. (1988). Ionic binding properties of carrier ampholytes. *J. Chromatogr.*, **437**, 147–159.

Rodkey, L. S. and Huff, G. D. (1990). Apparatus and methods for isoelectric focusing. *US Patent 4,963,236*, to Ampholife Technologies.

Roe, S. (1989). Separation based on structure. In *Protein Purification Methods: A Practical Approach* (Harris, E. L. V. and Angal, S., eds), pp. 175–244, IRL Press, Oxford.

Rosengren, J. and Glad, M. (1975). Coupling of ligands to crosslinked Sepharose® in organic solvents. *Protides Biol. Fluids*, **23**, 531–536.

Rosengren, J. *et al.* (1975). Hydrophobic interaction chromatography on non-charged Sepharose derivatives. *Biochim. Biophys. Acta*, **412**, 51–61.

Roth, U. *et al.* (1996). Membrane protein subunit fractionation by means of inverse pore gradient elution polyacrylamide gel electrophoresis. *Anal. Biochem.*, **233**, 67–70.

Roumelotis, P. and Unger, K. K. (1979). Preparative separation of proteins and enzymes in the mean molecular-weight range of 10,000–100,000 on LiChrosorb Diol® dacking by high-performance size-exclusion chromatography. *J. Chromatogr.*, **185**, 445–452.

Roumelotis, P. and Unger, K. K. (1981). Assessment and optimization of system parameters in size exclusion separation of proteins on diol-modified silica columns. *J. Chromatogr.*, **218**, 535–546.

Routs, R. J. (1973). The choice of electrolyte conditions for isotachophoretic separations. *Ann. NY Acad. Sci.*, **209**, 445–456.

Russo, C. *et al.* (1983). Purification of IgG monoclonal antibody by caprylic acid precipitation. *J. Immunol. Methods*, **65**, 269–271.

Saito, A. *et al.* (1989). High level expression of a synthetic gene coding for IgG-binding domain B of staphylococcal Protein A. *Protein Eng.*, **2**, 481–487.

Saltonstall, C. W. (1992). Dialysis. In *Encyclopedia of Immunology* (Roitt, I. M. and Delves, P. J., eds), pp. 463–468, Academic Press, London.

Scopes, R. K. and Porath, J. (1990). Differential salt-promoted chromatography for protein purification. *Bioseparations*, **1**, 3–7.

Shainoff, J. R. (1968). A cellulose mercurial: preparation, properties and use in antibody purification. *J. Immunol.*, **100**, 187–193.

Shaltiel, S. and Er-El, Z. (1973). Hydrophobic chromatography: use for purification of glycogen synthetase. *Proc. Natl. Acad. Sci. USA*, **70**, 778–781.

Sharon, N. and Lis, H. (1972). Lectins: cell-agglutinating and sugar-specific proteins. *Science*, **177**, 949–959.

Shepherd, J. (1980). Separation of plasma lipoproteins in the preparative ultracentrifuge. *Biochem. Soc. Trans.*, **8**, 516–518.

Shuster, L. (1971). Preparative acrylamide gel electrophoresis: continuous and disc techniques. *Methods Enzymol.*, **22**, 412–433.

Silman, I. H. and Katchalski, E. (1966). Water-insoluble derivatives of enzymes, antigens, and antibodies. *Annu. Rev. Biochem.*, **35**, 873–908.

Sluyterman, L. A. A. and Elgersma, O. (1978). Chromatofocussing: isoelectric focusing on ion-exchange columns. I. General principles. *J. Chromatogr.*, **150**, 17–30.

Sluyterman, L. A. A. and Wijdenes, J. (1970). An agarose mercurial column for the separation of mercaptopapain and nonmercaptopapain. *Biochim. Biophys. Acta*, **200**, 593–595.

Sluyterman, L. A. A. and Wijdenes, J. (1981a). Chromatofocussing. III. The properties of a DEAE-agarose anion-exchanger and its suitability for protein separations. *J. Chromatogr.*, **206**, 429–440.

Sluyterman, L. A. A. and Wijdenes, J. (1981b). Chromatofocussing. IV. Properties of an agarose polyethyleneimine ion exchanger and its suitability for protein separations. *J. Chromatogr.*, **206**, 441–447.

Smith, G. J. *et al.* (1984). Lymphoblastoid cell-produced immunoglobulins: preparative purification from culture medium by hydroxyapatite chromatography. *Anal. Biochem.*, **141**, 432–436.

Sober, H. A. *et al.* (1949). Chromatographic analysis of a mixture of proteins from egg white. *Science*, **110**, 564–565.

Sofer, G. K. and Nyström, L.-E. (1989a). *Process Chromatography: A Practical Guide*, pp. 23–54, Academic Press, New York.

Sofer, G. K. and Nyström, L.-E. (1989b). *Process Chromatography: A Practical Guide*, pp. 117–130, Academic Press, New York.

Solomon, B. *et al.* (1992). Affinity purification of antibodies using immobilized FB domain of protein A. *J. Chromatogr.*, **597**, 257–262.

Sorel, R. H. A. and Hulshoff, A. (1983). Dynamic anion-exchange chromatography. *Adv. Chromatogr.*, **12**, 87–129.

Sorensen, K. (1993). Method for isolation and purification of enzyme-antibody conjugates. *US Patent 5, 266, 686*, to Pierce Chemical Company.

Squire, P. G. (1964). A relationship between the molecular weights of macromolecules and their elution volumes based on a model for Sephadex gel filtration. *Arch. Biochem. Biophys.*, **107**, 471–478.

Srinivasan, R. and Ruckenstein, E. (1980). Role of physical forces in hydrophobic interaction chromatography. *Sep. Purif. Methods*, **9**, 267–370.

Starkenstein, E. (1910). Über Fermentwirkung und deren Beeinflussung durch Neutralsalze. *Biochem. Z.*, **24**, 210–218.

Staros, J. V. *et al.* (1986). Enhancement by *N*-hydroxysulfosuccinimide of water-soluble carbodiimide-mediated coupling reactions. *Anal. Biochem.*, **156**, 220–222.

Steinbuch, M. and Audran, R. (1969). The isolation of IgG from mammalian sera with the aid of caprylic acid. *Arch. Biochem. Biophys.*, **134**, 279–284.

Sternberger, L. (1967). Electron microscopic immunocytochemistry; A review. *J. Histochem. Cytochem.*, **15**, 139–159.

Sulkowski, E. (1985). Purification of proteins by IMAC. *Trends. Biotechnol.*, **3**, 1–7.

Sumner, J. B. (1919). The globulins of the jack bean, *Canavalia ensiformis*. *J. Biol. Chem.*, **37**, 137–142.

Sutton, C. (1989). Purification by exploitation of activity: lectin affinity chromatography. In *Protein Purification Methods: A Practical Approach* (Harris, E. L. V. and Angal, S., eds), pp. 245–292, IRL Press, Oxford.

Svensson, B. (1973). Use of isoelectric focusing to characterize the bonds established during coupling of CNBr-activated amylodextrin to subtilisin Type Novo. *FEBS Lett.*, **29**, 167–169.

Tesser, G. I. *et al.* (1975). Some new bifunctional reagents for the conjugation of a protein with an amino compound in water. *Hoppe-Seyler's Z. Physiol. Chem.*, **356**, 1625–1629.

Thompson, H. S. (1850). On the absorbent power of soils. *Agric. Soc. Engl.*, **11**, 68–74.

Thompson, S. T. *et al.* (1975). Blue dextran-Sepharose: An affinity column for the dinucleotide fold in proteins. *Proc. Natl. Acad. Sci. USA*, **72**, 669–672.

Thorpe, P. E. *et al.* (1988). Improved antitumor effects of immunotoxins prepared with deglycosylated ricin A-chain and hindered disulfide linkages. *Cancer Res.*, **48**, 6396–6403.

Tiselius, A. (1948). Adsorption separation by salting out. *Ark. Kemi Min. Geol.*, **26B**, 1–5.

Tiselius, A. *et al.* (1956). Protein chromatography on calcium phosphate columns. *Arch. Biochem. Biophys.*, **65**, 132–155.

Torres, A. R. and Peterson, E. A. (1992). Purification of complex protein mixtures by ion-exchange displacement chromatography using spacer displacers. *J. Chromatogr.*, **604**, 39–46.

Torres, A. R. *et al.* (1987). Preparative high-performance liquid chromatography of proteins on an anion exchanger using unfractionated carboxymethyldextran displacers. *J. Chromatogr.*, **389**, 177–182.

Unger, K. (1984). High-performance size-exclusion chromatography. *Methods Enzymol.*, **104**, 154–169.

Van Der Giessen, M. *et al.* (1974). The production of precipitating antiglobulin reagents specific for the subclasses of human IgG. *Immunology*, **27**, 655–663.

Van Kreveld, M. E. and Van den Hoed, N. (1973). Mechanism of gel permeation chromatography: distribution coefficient. *J. Chromatogr.*, **83**, 111–124.

Van Weemen, B. K. and Schuurs, A. H. W. M. (1971). Immunoassay using antigen–enzyme conjugates. *FEBS Lett.*, 15, 232–236.

Vesterberg, O. (1971). Isoelectric focusing of proteins. *Methods Enzymol.*, **22**, 389–412.

Visser, J. and Strating, M. (1975). Separation of lipoamide dehydrogenase isoenzymes by affinity chromatography. *Biochim. Biophys. Acta*, **384**, 69–80.

Von Hippel, P. H. and Schleich, T. (1969). The effects of neutral salts on the structure and conformational stability of macromolecules in solution. In *Structure and Stability of Biological Molecules* (Timasheff, S. N. and Fasman, G. D., eds), pp. 417–574, Marcel Dekker Inc., New York.

Von Hippel, P. H. and Wong, K-Y. (1965). On the conformational stability of globular proteins. The effects of various electrolytes and nonelectrolytes on the thermal ribonuclease transition. *J. Biol. Chem.*, **240**, 3909–3923.

Walters, R. R. (1985). Affinity chromatography. *Anal. Chem.*, **57**, 1099A–1114A.

Wang, J. L. *et al.* (1971). Unusual fragments in the subunit structure of concanavalin A. *Proc. Natl. Acad. Sci. USA*, **68**, 1130–1134.

Waterborg, J. H. and Matthews, H. R. (1994). The electrophoretic elution of proteins from polyacrylamide gels. *Methods Mol. Biol.*, **32**, 169–175.

Weith, H. L. *et al.* (1970). Synthesis of cellulose derivatives containing the dihydroxyboryl group and a study of their capacity to form specific complexes with sugars and nucleic acid components. *Biochemistry*, **9**, 4396–4401

Wenger, P. *et al.* (1988). Purification of lentil lectins using preparative electrophoresis. *J. Chromatogr.*, **455**, 225–239.

Weston, P. D. and Scorer, R. (1978). Comparison of Sepharose and cellulose as a support for antibody and antigen. In *Affinity Chromatography* (Hoffmann-Osterhof, O. *et al.*, eds), pp. 207–210, Pergamon Press, New York.

Whitaker, J. R. (1963). Determination of molecular weights of proteins by gel filtration on Sephadex. *Anal. Chem.*, **35**, 1950–1953.

Wieland, T. *et al.* (1963). Zum molekulargewicht der Lactat-Dehydrogenasen. Studien des Verhaltens bei der Gelfiltration. *Biochem. Z.*, **337**, 303–311.

Wilchek, M. and Miron, T. (1987). Limitations of *N*-hydroxysuccinimide esters in affinity chromatography and protein immobilization. *Biochemistry*, **26**, 2155–2161.

Wilchek, M. *et al.* (1984). Affinity chromatography. *Methods Enzymol.*, **104**, 3–53.

Wilchek, M. *et al.* (1994). Improved method for preparing *N*-hydroxysuccinimide ester-containing polymers for affinity chromatography. *Bioconjugate Chem.*, **5**, 491–492.

Wolters, G. *et al.* (1976). Solid-phase enzyme-immunoassay for detection of hepatitis B surface antigen. *J. Clin. Pathol.*, **29**, 873–879.

Wood, S. P. (1990). Purification for crystallography. In *Protein Purification Applications–A Practical Approach* (Harris, E. L. V. and Angal, S., eds), pp. 45–58, IRL Press, Oxford.

Yao, K. and Hjertén, S. (1987). Gradient and isocratic high-performance liquid chromatography of proteins on a new aarose-based anion exchanger. *J. Chromatogr.*, **385**, 87–98.

Yon, R. J. (1972). Chromatography of lipophilic proteins on adsorbents containing mixed hydrophobic and ionic groups. *Biochem. J.*, **126**, 765–767.

Yurov, G. K. *et al.* (1994). Thiophilic adsorption: rapid purification of $F(ab)_2$ and Fc fragments of IgG_1 antibodies from murine ascitic fluid. *J. Immunol. Methods*, **177**, 29–33.

Chapter 11

The Characterisation of Protein Conjugates

Alastair H. Dent and Mohammed Aslam

Research and Development Department, Ortho-Clinical Diagnostics, Pollards Wood Laboratories, Nightingales Lane, Chalfont St. Giles HP8 4SP, UK

11.1 Introduction

Protein conjugates are prepared for many different reasons, but underlying all of these is the desire to combine the properties of the component species. Many of their applications are sensitive to the exact nature of these conjugates, both at the macro- and micromolecular level: the first in terms of their stoichiometry, and their heterogeneity if there are multiple components of different stoichiometry; the second in the degree of success that has been achieved in maintaining the integrity of the individual protein molecules, and hence their functional activity.

In order to rationalise the behaviour of protein conjugates, and especially to compare one conjugate to another, it is necessary to understand these factors. It is common in the literature of conjugation chemistry to encounter conclusions of the type: 'Chemistry A is better than chemistry B for this application', but it is far rarer that the conjugates are characterised well enough to support any great generality in the statement. The reaction conditions employed may have resulted in the described differences in behaviour, but only *via* their influence on the properties of the conjugates, their homogeneity and stoichiometry, and the molecular integrity of their components.

It would be more enlightening to say that: 'Conjugates with properties $A_1, B_1, C_1 \cdots$ are better than conjugates with properties $A_2, B_2, C_2 \cdots$ for this application', the specific chemistries employed only being the means of achieving these properties. Such conclusions however require adequate characterisation of the products, and this degree of rigour is not particularly common in the field. It is still much more common, in other words, to compare the recipes rather than their products.

A major factor behind this state of affairs is the high degree of heterogeneity often observed in protein conjugates. Compare the situation with that of an organic compound of low molecular mass. New molecules of this type are synthesised every day in universities and pharmaceutical laboratories around the world, and it is axiomatic that they must be taken to an appropriate level of purity and characterisation before any assessment is made of their properties. Carefully-chosen synthetic strategies, traditional techniques such as solvent extraction and recrystallisation, and the arsenal of modern chromatographic methods are employed to reduce impurities to insignificant levels. Virtually identical isomeric species are rigorously resolved. Once this stage has been reached, analytical techniques of immense power are employed to establish the exact molecular structure of the compound. To virtually all intents and purposes this compound is now fully characterised, and any properties which it possesses in its pure state are independent of its means of preparation.

Even for relatively simple protein molecules, the availability of reasonable quantities of material at a high level of purity is a comparitively recent circumstance. Until the advent of techniques such as the use of hybridomas to generate Mabs, and the recombinant production of proteins in host cells, all proteins had to be purified from their original *in vivo* source. Not only were the available methods of purification much less advanced than those of organic chemistry, but so too were the analytical techniques which demonstrated whether or not homogeneity had been achieved—not surprisingly in view of the much greater complexity of such molecules. Despite the fact that even the most basic principles of protein structure have been established for barely half a century, huge advances have been made in this field. Driven in particular by pharmaceutical applications, the characterisation of pure proteins has been developed to the point where it approaches that of organic compounds. A protein, even one which has never existed before *in vivo*, can be expressed by recombinant means, essentially in a homogeneous state, and in large quantities. This facilitates efficient purification by the techniques described in the preceding chapter, and the product can then be characterised in terms of its amino acid sequence, glycosylation state, content of other components such as metal ions, its secondary, tertiary and quaternary structure, and its functional activity.

Moving finally to protein conjugates, the chemistries employed in their preparation rarely lead to products that even approach homogeneity: unless groups of unique reactivity are employed on each partner, a mixed product is a statistical inevitability. The purification methods described in Chapter 10 struggle to achieve complete fractionation of the resulting mixtures, so that they are almost always employed in their impure form. The pharmaceutical use of protein conjugates such as immunotoxins is inevitably driving greater attention towards purity and characterisation, but for many other applications (including most diagnostic uses) these considerations are viewed as secondary compared to functional performance.

The recombinant approaches to conjugate preparation described in Chapter 9 offer big advantages in these respects: stoichiometry and coupling positions can be built in *ab initio*, a homogeneous major product can be obtained, and synthetic tags can be attached to facilitate purification from other species. Thus it is likely that such approaches will be dominant in the pharmaceutical field: not only are the pressures for purity and characterisation higher in these applications, but the investment required to establish large-scale recombinant production of a fusion protein is more likely to be economical.

However such approaches remain in the minority, and heterogeneous conjugates are therefore still the norm. For this reason the characterisation methods listed in this chapter are of two types, those which incorporate an element of fractionation, and those which do not. Not surprisingly, the first category show a degree of overlap with the purification strategies described in the preceding chapter. They differ from preparative techniques only in terms of scale, and (perhaps more importantly) in that they are permitted to be destructive in nature; in these applications it is information rather than the protein which is the desired product.

Non-fractionating techniques can be applied to provide information on specific aspects of a conjugate; with a homogeneous product this will be 'real' information describing a single species, while with a mixture the information will generally reflect the mean properties of the individual components. Analysis of conjugate incorporations (e.g. by UV–visible spectrophotometry) is a commonly encountered example of this second type; although statements such as 'a conjugate with an incorporation of 2.0 was obtained' are often encountered, they almost always mean that a mixture of conjugates was obtained with a weighted mean incorporation of that level.

As a general point, many of the approaches to the characterisation of protein reagents are less constrained by sample concentration than their preparative counterparts. However it may sometimes still be necessary to concentrate samples before analysis, in which case it is particularly important to achieve quantitative recovery of the sample components to avoid a distortion of the composition. Approaches have been described for the quantitative precipitation of proteins from dilute solution down to μg levels, even in the presence of detergents, lipids, salts, etc. (Wessel and Flügge, 1984).

Finally, there are two areas of analysis which this chapter does not attempt to cover. The first is the determination of protein structure: amino acid sequencing methods for establishing the primary sequence, which although considerable ingenuity may be required to use them to the full (Fothergill-Gilmore, 1993), are nowadays determined mainly by automated instrumentation; the spectroscopic methods for determination of secondary structure which include circular dichroism (CD), optical rotatory dispersion (ORD), and Fourier transform infrared (FT-IR) and Raman spectroscopy (Yang *et al.*, 1986; Susi and Byler, 1986; Williams, 1986), and the methods for elucidating the tertiary structure of crystalline proteins, where the power of 'traditional' X-ray diffraction has been enhanced by the evolution of techniques such as neutron diffraction and electron cryomicroscopy—the latter at last challenging the need for crystalline samples for such analyses (Böttcher *et al.*, 1997; De Rosier, 1997).

This exclusion also applies to the methods which have been employed to detect structural changes in immobilised proteins, and which have been fundamental in understanding protein adsorption behaviour. These techniques include ellipsometry, total internal reflection fluorescence (TIRF), surface enhanced Raman spectroscopy (SERS), surface plasmon resonance, and X-ray photoelectron spectroscopy (XPS or ESCA) (Axelrod *et al.*, 1984; Paynter and Ratner, 1984; Cotton, 1984; Andrade and Hlady, 1986; Davies *et al.*, 1994). The advances in scanning microscopy techniques are also opening up exciting new prospects in this area, for example the application to protein immobilisation of scanning tunnelling microscopy (STM) (Feng *et al.*, 1988, 1989, Davies *et al.*, 1994, 1995) and atomic force microscopy (AFM) (Allen *et al.*, 1977)

The second area omitted is that of the non-covalent interactions of protein, including inter-protein associations and those with other molecules. It has been seen that many of the covalent coupling techniques described in this book can be applied in the elucidation of such relationships (Section 1.5.3), but there are a range of instrumental techniques for probing these interactions. These include methods based on ultracentrifugation (Rivas and Minton, 1993), chromatography (Winzor and Scheraga, 1963; Wood and Cooper, 1970; Koltisko and Walton, 1984; Aubry and McGann, 1994; Laitinen *et al.*, 1996), electrophoresis (Darawshe *et al.*, 1993), surface plasmon resonance (O'Shannessy *et al.*, 1993), and radiation target analysis (Kempner, 1993).

11.2 Ultracentrifugation

This technique, described in its rarer preparative applications in Section 10.3, was one of the mainstays of protein size determination until the advent of more modern approaches. It has now been very largely superseded for this purpose by techniques based on chromatography, electrophoresis, amino acid sequence data, and more recently, mass spectrometry. It remains a viable technique for the investigation of reversible associations in solution (Rivas and Minton, 1993; Basi and Rebois, 1997), where the small sample size required and absence of a solid fractionating matrix offer certain advantages. New instrumentation has in fact been introduced in the last few years to facilitate studies of this type. However for the characterisation of protein conjugates, ultracentrifugation can only offer very limited information, and the technique will only be dealt with here briefly for completeness.

Centrifugation is, of course, a long-standing preparative technique for the separation of solid and colloidal material from suspensions; due to the large density differences in such cases, they can often becarried out very simply in a few minutes using modest equipment. Ultracentrifugation is an extension of the basic approach, very high rotation rates being used to generate large centripetal forces; coupled with long run times this allows the fractionation of soluble species, such as proteins of different molecular masses.

The fundamental relationships controlling the behaviour of solutes under centrifugation were established more than half a century ago (Svedberg and Pedersen, 1940). Solutes are characterised in terms of their sedimentation rate, described by the Svedberg equation:

$$v = \frac{dx}{dt} = \frac{\phi(\rho_p - \rho_m)}{f}\omega^2 x$$

where: v = radial velocity of solute (cm/s), ϕ = volume of solute (cm^3), ρ_p = density of solute (g/cm^3), ρ_m = density of matrix (g/cm^3), f = frictional coefficient (g/s), ω = angular velocity of centrifuge (rad/s) and x = radial distance from axis of rotation (cm)

For a particular solute the sedimentation constant, s can be defined, which represents the velocity of sedimentation per unit of centrifugal field, and is described by the equation:

$$s = \frac{v}{\omega^2 x} = \frac{\phi(\rho_p - \rho_m)}{f}$$

Using the units above the value of s is expressed in seconds, but for convenience these constants are generally quoted in Svedberg units (S), where $1S = 10^{-13}$ seconds. Before the availability of reasonably accurate methods for assessment of molecular mass values, it was common to characterise protein components in terms of these values: a protein of molecular mass around 40 kDa might be described as a 3.5S species, while one of 150 kDa might exhibit a sedimentation constant of 8S[1].

For small solutes, sedimentation behaviour may be complicated by significant diffusion effects; these can, however, be dealt with in the analysis of data (Yphantis and Waugh, 1956).

11.2.1 Density gradient ultracentrifugation

Carrying out ultracentrifugation in simple aqueous solutions is hindered by the instability of the medium. Not only will diffusion effects be large, but convection currents and physical disturbances can very readily disrupt the sample, destroying the order introduced by the ultracentrifugation process.

Density gradient techniques have therefore been described (Anderson, 1955; de Duve *et al.*, 1959), which employ solutes such as sucrose to increase the robustness of the centrifugation matrix. In this stabilising gradient approach (Pickels, 1943), a density gradient is established by increasing the solute content from the top to the bottom of the centrifuge tube. Such gradients are fairly stable under centrifugation conditions.

Solutes will move through the gradient differentially on the basis of their sedimentation properties.

[1] This comparison does not imply that sedimentation behaviour equates directly with molecular mass: like most indirect techniques it is affected by other conformational properties of the protein molecule.

When a relatively low density gradient is employed principally to provide robustness, the centrifugation run will be stopped before equilibrium is reached, the solutes having reached positions determined by their individual sedimentation rates. If the run is continued too long all the solutes will continue to the bottom of the gradient and fractionation will be lost. Problems can also be caused if the sample is more dense than the top of the gradient, it is likely to 'stream' through the matrix and disrupt the structure. Introducing the sample itself *via* a density gradient can minimise such effects (Britten and Roberts, 1960).

An alternative approach is isopycnic gradient centrifugation (Behrens, 1938), where a steeper gradient is employed with the deliberate aim of bracketing the density properties of the sample solutes. Under these conditions a solute will only move through the gradient until it reaches the point at which the matrix density equals its own; it will then remain focused at this position for the remainder of the run. In theory the isopycnic approach yields a separation which is independent of sedimentation rate, the final position of the solute being determined only by its density.

The predominant sucrose-based methods (Martin and Ames, 1961) employ pre-formed gradients; the difference in density between sucrose and water is not sufficiently large to allow a gradient to be set up in a reasonable time under centrifugation conditions. With a denser solute such as caesium chloride, however, such an approach is feasible (Meselson *et al.*, 1957) and has been widely employed for some preparative applications.

The relationships governing sedimentation behaviour under gradient conditions are complicated by the variations in density and viscosity which exist, and by the osmotic pressures which tend to oppose the maintenance of the gradient. Centrifuge geometry is also a complex variable. These factors have been discussed in detail by de Duve *et al.* (1959) and will not be entered into here.

11.2.2 Detection of solutes in ultracentrifugation

Approaches to solute detection fall into two categories, those carried out *in situ*, and those involving fractionation.

The first category are severely constrained, as the detection techniques must be applicable to solutes contained within a centrifuge tube. These are predominantly optical in nature; older approaches were generally based on the visualisation of Schlieren lines which indicate zone boundaries within the matrix, while newer equipment has facilitated the use of UV–visible spectrophotometry. The most recent instrumental advances (Rivas and Minton, 1993) allow UV-visible absorbance to be measured as a function of radial position within the centrifuge rotor, over a broad range of wavelength and sensitivity.

Alternatively the contents of the centrifuge tube can be sequentially eluted after the completion of the run, the base being pierced to provide a small aperture. Once fractions have been collected from this elution procedure they can be analysed by any suitable technique such as radiometry, measurement of enzyme activity, etc. Microfractionation equipment is now available (Rivas and Minton, 1993) which allows minute (sub-microlitre) fractions to be collected from runs carried out in commonly-available preparative ultracentrifuges.

11.2.3 Ultracentrifugation for characterisation of protein conjugates

As noted at the start of this section, analytical ultracentrifugation is an outmoded technique for characterisation of covalent protein conjugates; however a single example will be discussed briefly. Some preparative approaches are also mentioned in Section 10.3.

Mannik and Downey (1973) prepared Fab–HRP conjugates by a glutaraldehyde method, employing ^{125}I-labelled Fab fragments. The product was analysed by ultracentrifugation in a density gradient (5–15% sucrose in a borate buffer); fractions were collected and the radioactivity in each plotted against the elution position, as a percentage of total gradient volume. The results from three conjugations of increasing HRP–Fab ratio are shown in Figure 11.1. Unreacted Fab fragments (3.5S, 41 kDa), Fab–HRP (5.4S, 79 kDa), Fab–HRP_2 (7.1S, 119 kDa) and Fab–HRP_3 (8.8S, 162 kDa) could be observed in the products, the proportion of unreacted Fab decreasing as the ratio was increased.

Such an analysis would now be carried out much more conveniently by high performance SEC.

11.3 Size Exclusion Chromatography

Of the chromatographic techniques described in Chapter 10, SEC lends itself most readily to analytical applications. Provided the ionic strength of the eluent is high enough to suppress any electrostatic binding

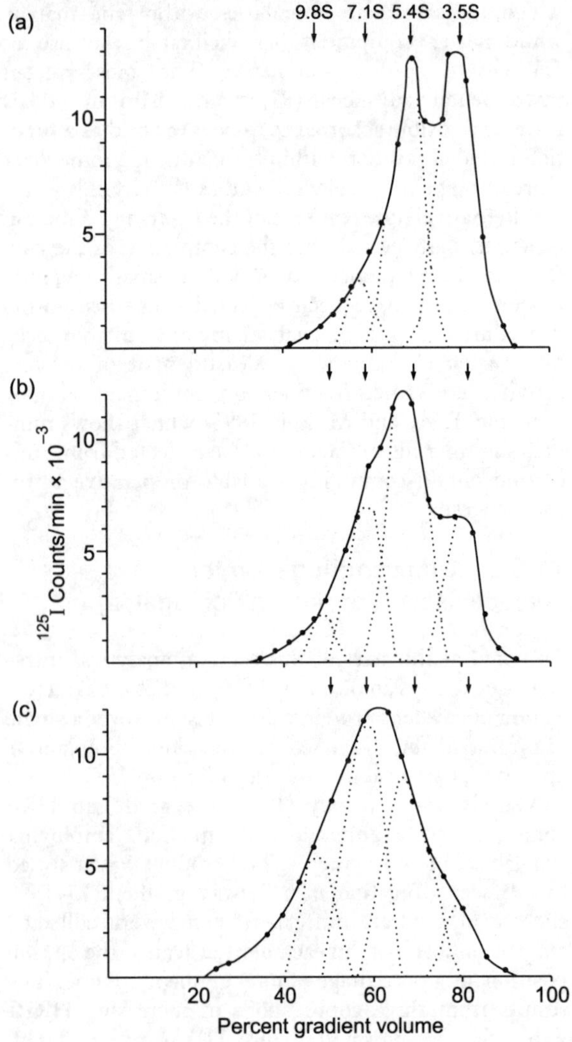

Figure 11.1 Density gradient ultracentrifugation patterns of Fab–HRP conjugates. Sucrose gradients 5–15% were used. The 5.4S, 7.1S and 8.8S peaks show Fab–HRP_1, Fab–HRP_2 and Fab–HRP_3 conjugates. The HRP: Fab molar ratio during conjugation was 8.3 in (a), 16.6 in (b) and 24.9 in (c). (Reproduced with permission from: Mannik and Downey (1973). *J. Immunol. Methods*, **3**, 233–242.)

effects, ideal size exclusion behaviour is fairly easy to obtain: with a few exceptions the retention times of protein molecules on an SEC matrix correlate well with their size (Andrews, 1970).

11.3.1 Fractionation by size exclusion chromatography

Whenever a conjugate is purified by SEC, the accompanying chromatogram provides some information on the molecular mass and heterogeneity of the products. However the information obtained from low-resolution gels is fairly limited in both respects, and high-resolution media are clearly preferable for analytical work. In practice this means the high-performance matrices described in Section 10.5.3 (Table 10.2), typically with bead sizes below 10 μm.

Selectivity curves such as those shown in Figure 10.7 allow the molecular mass of a protein conjugate to be estimated to some extent, but for greater accuracy marker proteins should be employed to calibrate a given combination of column and eluent. Pain and Surolia (1981), for example, used SEC to demonstrate that a Protein A–peroxidase conjugation essentially yielded a homogeneous, 1:1 product (Figure 11.2). This calibration exercise is particularly important for ageing columns, which may be beginning to lose their selectivity.

The mechanism and key process variables of SEC have been discussed at length in Section 10.5, and will not be revisited here. However, this technique provides only the means of fractionation of the product: a detection system is also required to provide a chromatogram, and the choice of this system influences the analytical information which is obtained.

One parameter which is perhaps more critical in analytical applications is the necessity to be certain

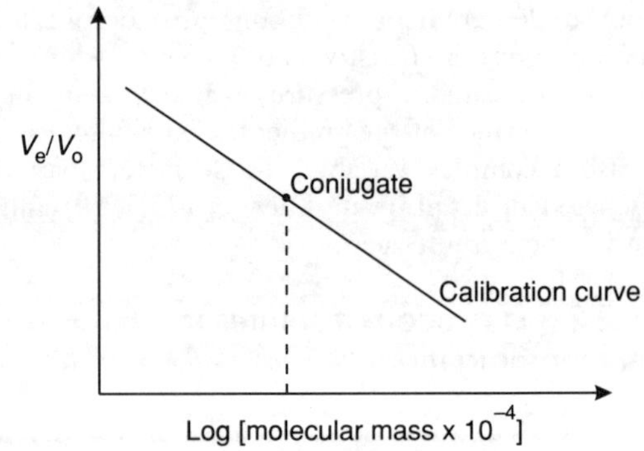

Figure 11.2 Molecular mass of a conjugate determined by size exclusion chromatography: comparison of the conjugate with a calibration series suggested a 1:1 Protein A–HRP product. (Adapted from: Pain and Surolia (1981), *J. Immunol. Methods*, **40**, 219–230.)

that no significant quantity of material is adsorbed irretrievably to the column. In preparative SEC this may result simply in a loss of yield, but for analytical purposes it may lead to an unacceptable distortion ofthe information obtained. On the whole SEC is a good technique in this respect. Serious adsorption ofproteins to the matrix is a relatively uncommon occurrence.

11.3.2 Detection systems for size exclusion chromatography, and other chromatographic approaches

It is beyond the scope of this chapter to attempt a detailed discussion of chromatography detection systems, but the topic deserves brief consideration. UV–visible spectrophotometry is probably the most commonly-encountered technology, and is fairly well suited to protein applications: almost all proteins of a reasonable size can be relied upon to absorb to some extent at 280 nm, a wavelength high enough to be safely clear of aqueous buffer absorbances. Conveniently, some proteins also possess chromophoric groups with specific absorbances well away from 280 nm, and in such cases dual-wavelength detection can offer considerably more information; the absorbance of the haem group in haemoproteins such as HRP (λ_{max} = 401 nm) is a good example.

Single-wavelength UV–visible detection is as good as any other technique for demonstrating that a protein product has eluted at a given retention time. However, it has limitations when it is desired to confirm the homogeneity of a given peak: two or more products eluting at the same point may appear homogeneous from a single-wavelength trace, especially if one of them is present in a dominant concentration. Dual-wavelength detection can offer a greater degree of confidence, as the appearance of symmetrical, proportionate peaks at both wavelengths is better evidence for homogeneity. A logical extension of this argument is to use detection at a broad range of wavelengths simultaneously, and diode array detectors are capable of just this. Coupled with analytical software, systems of this type are extremely powerful in characterising the full composition of a mixed sample, even down to minor impurities.

These approaches can be used to determine with acceptable confidence that a given peak is truly homogeneous. However, it is often desired to expressthe purity of a product in percentage terms. UV–visible spectrophotometry is often used for this purpose, but there are limitations here too (see Section 11.13.2). Although many proteins absorb to a roughly similar extent at 280 nm, this is by no means a safe assumption[1]: each molecule possesses its own extinction coefficient at this wavelength, and accurate purity determinations are impossible without knowing these values for all the components present. This is not normally the case, and it is often assumed in practice that all the species present absorb equally. Where the levels of impurity are small this may well be a justifiable assumption, but it becomes a more significant factor at lower product purities.

Where baseline resolution between sharp peaks is obtained, it is straightforward to assign the number of peaks and their relative areas even by single-wavelength detection. Where the product contains an overlapping series of components, however, peak assignment and integration are much more hazardous operations. While integration software is a very useful facility, and often yields very reliable information, a considerable amount of 'judgment' is often required in dealing with broad peaks, shoulders, baseline fluctuations, etc. To some extent this can be set up by the operator, but it should be borne in mind that quite different purity assessments can sometimes be obtained from the same sample by varying these somewhat subjective parameters.

Finally, it is sometimes assumed that the a sample will display the same peak height when examined on a number of different chromatography units. This is certainly not the case in many instances: the detection systems are often designed to give a reliable indication that a peak is present, not an output which is quantitatively comparable from one unit to another. Integration of peaks from a mixture should in theory still give the sameresults on any given unit, but the sensitivity of the detector may need to be adjusted to a different level to obtain reasonable peak heights on the chromatogram.

Other detection systems for chromatography systems may be based on other techniques such as refractive index or light-scattering (Takagi, 1990; Dollinger *et al.*, 1992), all of which have their own advantages and disadvantages. When used in combination they can provide a powerful source of information, for example allowing on-line determination of molecular mass, and of the carbohydrate content of glycoproteins (Wen *et al.*, 1996).

[1] As a rule of thumb, a 1 mg/ml solution will generally give an A_{280} of around 1–1.5. However, some calcium-binding proteins such as parvalbumins have essentially no absorbance at 280 nm, while a 1 mg/ml solution of lysozyme gives an A_{280} of 2.65 (Dunn, 1989).

11.4 Other Chromatographic Approaches

11.4.1 Column chromatography

None of the other column chromatographic techniques described in Chapter 10 is quite so suited to analytical applications as SEC, because none has a mechanism so closely tied to a single useful parameter. This is not to say that they cannot be used in the same way: Lampson and Tytell (1965), for example, described the use of a standard ion exchange chromatography system to estimate the p*I* values of a number of proteins.

However, unlike SEC where the elution conditions are straightforward, useful analytical systems using other techniques generally need to be optimised for a given application. Of course analytical chromatography of proteins is an immensely common procedure, and all the different branches described in Chapter 10 can be used in this way (Dwyer, 1993). Once such a system is established it may offer a very valuable approach to conjugate characterisation, but it is impossible to generalise on how this may be achieved in a given instance.

There are two particular ways in which other chromatographic approaches may be useful, however. First, in any system it should be possible to compare the elution characteristics of a product with a reference sample of authentic material, giving valuable confirmation that its composition is indeed similar. Secondly, within certain constraints, any of these techniques can be used for quantitative purity assessments in the same way as that described for SEC. Indeed, confirmation of a purity assessment by a second complementary technique is powerful support for a determination by SEC.

Two major constraints need to be considered in such determinations. The first, the possibility of irretrievable loss of product on the column was already mentioned for SEC. But because the other branches ofchromatography are adsorptive in nature, losses of this kind are somewhat more likely and great care should be taken to ensure that protein recovery is sufficiently quantitative. Note that for hydrophobic interactions in particular, dwell time has sometimes been shown to be a significant variable in affecting protein properties. This should therefore be minimised (and consistent from run to run) in analytical methods where this type of binding is implicated in the mechanism.

Secondly, many of the other chromatographic techniques employ non-continuous eluent systems such as step gradients to achieve elution. Such elution profiles are notorious for generating spurious peaks, which in fact reflect partial elution of a single product at more than one eluent composition. This can obviously distort any assessment of conjugate purity very severely, and the most reliable results are only obtained from isocratic systems, or shallow, continuous gradients. Baseline shifts are another possible problem with all gradient systems.

11.4.2 Thin-layer chromatography

Chromatography carried out in thin layers of matrix on a flat surface is much more associated with the analysis of organic compounds, but there is no particular reason why it should not also be applied to protein conjugates. There have been many advances in the basic thin-layer chromatography (TLC) technology since its origins in the 1950s (Kalász and Báthori, 1997).

With standard silica TLC plates, proteins are likely to be adsorbed firmly at the origin, while many organic components will exhibit higher R_f values. This may therefore provide a qualitative means to establish the formation of a protein–small molecule conjugate. Meares *et al.* (1984), for example, analysed conjugates between IgG and polycarboxylic chelators by this means, using a water–methanol–ammonium acetate eluent (Figure 11.3). The presence of the chelator was determined by autoradiography of bound ^{57}Co;

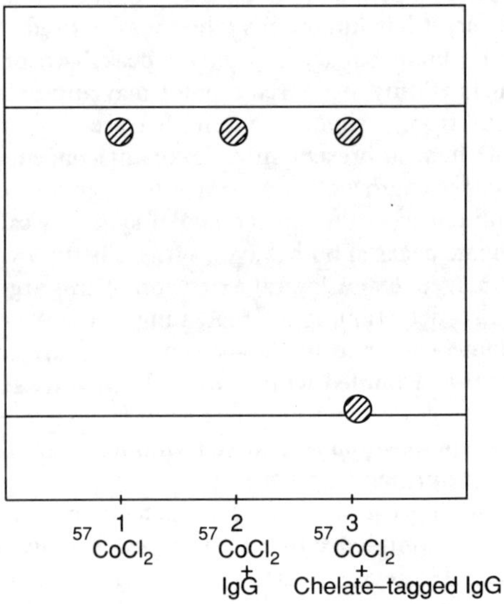

Figure 11.3 Analysis of IgG–chelator conjugates by TLC. See text for details.

the free cobalt reagent had a high R^f (lane 1) which was unaffected by the presence of free IgG (lane 2). However, in the presence of IgG–chelator conjugate some of the ^{57}Co activity remained with the protein band at the origin (lane 3).

Where it is desired to separate protein components on a TLC gel, it is necessary to employ a matrix withsuitable resolving properties. Kennedy *et al.* (1976), for example, prepared plates of Sephadex 2.5pt>TM G-200 and used them to resolve protein conjugates.

More complex applications include the use of thin-layer affinity matrices, prepared by adsorbing suitable ligands onto a membrane (Laitinen *et al.*, 1996); these techniques really overlap with the immunodiffusion methods discussed in Section 11.11.1.

In general, column approaches are much more convenient for analytical chromatography of protein reagents, and, despite potential advantages in terms of low cost, rapidity and the small sample size required, TLC is seldom encountered for this use.

11.5 Introduction to Electrophoretic Techniques

All electrophoretic techniques are based on the separation of solutes under the influence of an electric field, and their resolving properties are therefore linked to the charge of the solutes under the operational conditions. At its isoelectric point a protein molecule is of course uncharged, and under ideal conditions will not display any significant tendency to move towards the anode or the cathode. At a pH below its p*I* it bears an overall positive charge and will move towards the cathode, above the p*I* its negative charge will cause it to move towards the anode. These simple principles underly all electrophoretic methods, but a large number of variations on the basic theme have been developed, which permit a wide variety of information to be obtained.

The relatively rare preparative applications of these methods have been dealt with briefly in Section 10.12, but (particularly due to advances in chromatography) these are now predominantly analytical techniques. Some of the constraints have already been discussed in the earlier chapter, notably the challenges of heat transfer which result from the generation of Joule heat.However some of the most useful analytical techniques are destructive towards proteins, and are wholly precluded from preparative applications on this basis.

11.5.1 Matrices for electrophoresis

Electrophoresis can be carried out on any matrix which will support an electrolyte solution. In its most primitive manifestations this may be as simple as a sheet of filter paper, which is soaked in the electrolyte and placed between the two electrodes. To increase the scale of such approaches, various 'inert' media have been employed in three-dimensional formats such as stacked paper, or beds or columns of paper powder, asbestos fibres or glass beads (Flodin and Porath, 1954).

Over half a century ago (Tiselius, 1937) it was recognised that it was advantageous to maintain a high solution density during electrophoresis, both to minimise the disruptive effects of molecular diffusion, and (more importantly) of convection currents resulting from the Joule heat. Development of this theme led to the incorporation of gelling agents such as starch in electrophoresis matrices (Smithies, 1955); thus the technology of gel electrophoresis was born, and has remained in popular use ever since.

A characteristic of gels is their porosity, and hence their tendency to fractionate passing solutes on the basis of their size. This therefore became a topic of investigation in the development of media for gel electrophoresis, and led to the development of SEC as a chromatographic technique (Porath and Flodin, 1959). In electrophoresis applications, this property of the gels could either be avoided as far as possible to allow ideal electrophoretic behaviour to take place, or exploited to build in some element of size-dependent separation within the gel (Rodbard *et al.*, 1971).

Gels based on carbohydrates such as cellulose and agarose have been employed for electrophoresis, and are still encountered today: agarose, for example, has very wide pores which are amenable to very large analytes. However, polyacrylamide (Raymond and Weintraub, 1959) has found the widest popularity, to the extent that polyacrylamide gel electrophoresis (PAGE) is often referred to as a discrete technology (Chrambach and Rodbard, 1971).

These gels are formed by the copolymerisation of acrylamide monomers with a cross-linking agent such as *N,N′*-methylene bisacrylamide. Various concentrations of monomer and cross-linker can be employed to produce gels of different properties (Dunn, 1989). Pore sizes up to 800 Å can be produced by reducing the amount of monomer at a fixed cross-linker concentration, or up to 2500 Å by stepping up the cross-linker level, which is thought to promote the formation of bead-like structures.

Apart from its pore-forming properties, polyacrylamide has other advantages over some matrices which contain charged groups under operational conditions. With such matrices a phenomenon described as electroosmosis or electroendosmosis is observed: the charged surface will be coated with a layer of oppositely charged counterions, which form a 'skin' around the electrolyte solution. This may cause the solution to take on an electrophoretic mobility of its own, which can override the mobility of the solutes within it. Thus a negatively charged surface will induce a positive charge at the surface of the electrolyte solution, which will therefore tend to migrate towards the cathode. Even negatively charged solutes may be carried along by the electroosmotic flow if they are slow-moving, and thus migrate in the opposite direction to that predicted from their electrophoretic mobility. Polyacrylamide gels do not contain charged groups and are therefore free of electroendosmosis problems.

These gels can be formed under a single set of polymerisation conditions to yield a uniform gel, often characterised by the percentage of acrylamide which they contain: a 3% gel will allow the free passage of solutes with a molecular mass greater than 100 kDa, for example, while a 15% gel has a cutoff at 15 kDa (Dunn, 1989).

However, there are sometimes certain advantages to the use of gradient gels, which are prepared by modifying the polymerisation conditions as the gel is constructed. The usual mode of application is to increase the polyacrylamide concentration (and thus reduce the pore size) in the direction of flow of the solutes: all solutes start off with more-or-less unrestricted mobility, but as the pore size diminishes the largest molecules begin to be hindered by the matrix. Once they reach this pore limit, the molecules move only very slowly, while smaller solutes can continue to move at their original rate. These gels, therefore, allow powerful resolution of solutes over a molecular mass range wider than would be possible from a uniform gel[1]. This technique is sometimes referred to as pore limit electrophoresis.

While the use of gels for electrophoresis reduces theimpact of molecular diffusion and convection currents, these factors can still result in heterogeneous performance. The use of gel matrices can itself lead to other problems, such as adsorptive interactions with solutes, or 'eddy migration' through channels in an imperfect gel.

The most recent major development in electophoresis matrices has been the introduction of capillary electrophoresis (Mikkers *et al.*, 1979; Jorgenson and Lukacs, 1981a,b, 1983). In this approach the dissipation of Joule heat is tackled by reducing the volume of the matrix to a feasible minimum, the inside of a capillary with an internal diameter as low as 50 μm. Heat transfer from such a small volume is very efficient *via* the walls of the capillary, which allows very high voltages to be applied to the system without convection problems. Samples can therefore be resolved rapidly, and molecular diffusion (now the dominant cause of band-broadening) is thus kept to a minimum. Very high resolutions can be obtained by this approach, the fractionated products being eluted through the capillary past a detector of suitably small dimensions and high sensitivity.

With glass capillaries (Jorgenson and Lukacs, 1981a), the surface bears negative charges under most pH conditions, due to ionised silanol groups. The electroosmotic flow generated (see above) is sufficiently high that even molecules bearing multiple negative charges migrate at a reasonable rate towards the cathode, so that both anionic and cationic solutes can be eluted in a single short run. Thus electroosmosis, generally a problem in gel electrophoresis, is exploited to yield rapid separations in capillary systems; the resolution is generally high enough to fractionate components adequately, even though the entire solute mixture is constantly moving towards the cathode under the influence of electroosmotic flow. This effect can be modulated by derivatising the surface of the capillaries to reduce its anionic content (Jorgenson and Lukacs, 1983): greater resolution can therefore be obtained at the expense of longer run times. With protein solutes, derivatisation of this type is also beneficial in minimising adsorption to the capillary surface.

If the solute concentration is too high compared to that of the electrolytes, the electric field gradient within the capillary can be distorted in the vicinity of the sample and cause asymmetric peak shapes (Mikkers *et al.*, 1979), therefore sample concentrations need to be kept small. This factor increases the demands on the detection system still further. However, extremely powerful resolution is obtainable from capillary electrophoresis, and its popularity is growing rapidly as the instrumentation improves.

[1] In a uniform gel, in fact, fractionation on the basis of size would occur in the reverse order (cf. gel filtration), as the smaller solutes are able to access a greater volume of the gel and, in the absence of electrophoretic effects, would therefore move through it more slowly. Without the powerful resolving effect of a gradient gel, however, such effects are generally second-order compared to the relative electrophoretic mobilities of the solutes.

11.5.2 Buffer systems for electrophoresis

Returning to embodiments of electrophoresis using simple matrices such as paper or uniform polyacrylamide gels, the other major variable is the buffer system employed. It is quite possible to use a single buffer (continuous zone electrophoresis, CZE). Obviously the pH of this buffer will dictate the charge states of the protein solutes, and therefore their electrophoretic mobilities. It is generally most convenient if all the solutes in the sample move in the same direction from one of the electrodes. A common approach (Dunn, 1989) is to work at pH 8–9, where most proteins are above their p*I* and negatively charged, and to apply the sample at the cathode; all anionic solutes will therefore move towards the anode. Basic proteins with high p*I* values will fail to enter the gel, however, and where these are the focus of interest, it is better to work under acidic conditions and to apply the sample at the anode.

Alternatively, to resolve two components with fairly close p*I* values, it may be chosen to work at an intervening pH value and to apply the sample at the midpoint between the electrodes. In the absence of electroosmotic effects the solutes will then move in opposite directions.

To minimise the effect of Joule heating, the ionic strength of the buffer employed should be kept low, with molarities ideally below 100 mM (Maizel, 1971).

CZE approaches are simple to carry out, and allow close control on the pH conditions experienced by the solutes. Their major disadvantage is that they do not concentrate the solutes in any way, and are therefore of little value for sample concentrations much below 1 mg/ml, or for large sample volumes. To address these deficiencies, discontinuous (disc) or multiphasic zone electrophoresis (MZE) techniques were developed, which employ more than one buffer system.

The most popular MZE buffer systems were developed by Ornstein (1964) and Davis (1964), and their key characteristic is their stacking ability, which allows the components of a sample to be focused into a narrow, concentrated band before the electrophoretic separation is carried out. The physicochemical basis of MZE approaches has been dealt with in detail by Jovin (1973), and a brief summary only will be given here.

The stacking concept relies fundamentally on the establishment of stable moving boundaries, which separate different but homogeneous regions (phases) of electrolytes (Figure 11.4). A common charged constituent is present in both regions, but the leading and trailing phases contain electrolytes, both of the opposite charge to the common species, with different

Figure 11.4 A moving boundary between leading (L) and trailing (T) electrolyte phases. (Reproduced with permission from: Chrambach *et al.* (1976). *In Methods of Protein Separation* (Catsimpoolas, N., ed.), vol. 2, Plenum Press, New York.)

mobilities. Because the mobility of the leading phase is greater, it will move ahead of the trailing phase under the influence of the electric field.

Phase boundaries of this type will tend to be broken down by diffusion; if this is to be avoided, certain requirements must be met with respect to the relative velocities of the leading and trailing components (Chrambach *et al.*, 1976), particularly in the circumstance where diffusion into the 'wrong' phase has started to occur. Thus the buffer systems used in MZE must be carefully chosen to lend themselves to the establishment of steady-state moving boundaries.

Once such a boundary has been established, a solute which bears the same polarity as the leading and trailing electrolytes will be sandwiched between the two phases, forming a new phase of its own. Gel conditions can be chosen such that sample solutes concentrate at the initial moving boundary wherever they start from, but usually the sample is introduced at the boundary anyway. When a number of solutes are introduced, e.g. a mixture of protein conjugates, they will move to positions determined by their individual relative mobilities. The difference in mobility between protein solutes is generally rather small due to their high molecular mass and moderate charge density (Chrambach *et al.*, 1976). It is thus not difficult to create the initial moving boundary using a suitable choice of leading and trailing electrolytes, bracketing all the protein solutes in the sample, which are all therefore concentrated at the initial moving boundary. It is thisproperty of MZE systems which allows dilute samples to be focussed into a narrow, concentrated band Figure 11.5).

A further convenience is that the different solutes will also order themselves within the stack, on the basis of their relative mobilities. The operating

Figure 11.5 A protein sample (P) focussed between the two electrolyte phases. (Reproduced with permission from: Chrambach *et al.* (1976). *In Methods of Protein Separation* (Catsimpoolas, N., ed.), vol. 2, Plenum Press, New York.)

Figure 11.6 Components (P_1, P_2) within the sample band are stacked in order of their mobility. (Reproduced with permission from: Chrambach *et al.* (1976). *In Methods of Protein Separation* (Catsimpoolas, N., ed.), vol. 2, Plenum Press, New York.)

conditions during the establishment of the stack are chosen such that the overall rate of migration is very low; therefore ordering within the sample band is a very powerful means of improving resolution. The sample can effectively have its components arranged in the order of elution before the 'main' electrophoresis operation even starts (Figure 11.6).

Once this stage has been reached, two main options are possible. On the one hand, conditions can bemaintained such that the steady-state moving boundaries remain in place. The entire stack therefore moves through the gel as a series of discrete phases, and the electrophoresis is simply continued until sufficient separation between solutes has been achieved. This approach is termed isotachophoresis; it has particular attractions for preparative applications (Chrambach *et al.* 1976; Andrews, 1986) on account of the high protein concentrations in the moving steady-state phases[1], which are maintained until the product fractions are eluted (see Section 10.12 for preparative applications of isotachophoresis).

Apart from the sample itself, extra components can be deliberately introduced into the stack for various purposes. A dye can be chosen to run faster than all the sample solutes, and therefore gives a useful indication of the position of the stack front as it moves through the gel. Spacer solutes such as ampholytes can also be introduced, whose mobilities lie between those of sample components of interest. They can therefore improve the resolution of such solutes by separating them in the stack.

The Greek origins of the term isotachophoresis reflect the fact that all the constituents move though the gel at the same rate, as they are constrained to do by the maintenance of the steady-state moving boundaries. Often however, the operational conditions may be changed deliberately once the stack has been established, such that the requirements for steady-state boundaries are lost. Electrolytes can now diffuse though the phase boundaries, and their rates must no longer remain equal: under the new operating conditions the trailing ions may, for example, continuously overtake the sample solutes to establish a gradual voltage gradient across the gel. The pore size of the gel may also be altered at this point, so that faster rates are possible, or that some degree of separation on the basis of size is introduced.

In these approaches the electrophoresis operation is split into discrete sections, which often employ different gels as well as different operating conditions, a stacking gel where the sample solutes are concentrated and ordered, and a running or resolving gel, where the stacked components are separated.

From the discussions in this and the previous section, it should be clear that the essentially simple concept of electrophoresis can be applied in a variety of different ways, depending principally on the design of the matrix, and the composition of the electrolyte solutions used. Within each of these basic techniques there is huge scope for optimisation in terms of choiceof electrolytes, operating pH and voltage, gel pore size, etc., and with sufficient attention it is possible to achieve excellent resolution of very complex systems.

On the other hand, many workers do not wish to invest this effort in each application, and there is a demand for simple, convenient systems which are broadly applicable. Commercial suppliers have helped to answer this need by introducing easy-to-use electrophoresis apparatus, often with preformed gels and buffers, and pre-programmed methods for various applications. This approach is fast and convenient, the preformed reagents are generally very reproducible in performance, and the methods are sufficiently powerful for very many applications. However if there is sufficient motivation to achieve fractionation of a difficult mixture, it should be remembered that a great deal more optimisation is possible outside the constraints of these commercial systems.

Both approaches to electrophoresis will undoubtedly remain popular depending on the application.

Capillary approaches have been able to employ buffers at a higher ionic strength than would be feasible in gel methods (0.5 M), as a result of the superior heat dissipation of capillary systems (Chen *et al.*, 1992). Their use helps to overcome ionic interactions with charged surface groups. Otherwise the choices of buffer for CE methods are much the same as for gels.

11.5.3 Detection systems for electrophoresis

An appropriate combination of matrix and buffer system chosen from the preceding sections will–under a voltage gradient–resolve sample components into

[1] This concentration is controlled by the Kohlrausch regulating function, dependent only on the relative mobilities of the electrolytes in the various leading and trailing phases.

different positions on a gel, or elute them at different positions through a capillary. These techniques therefore provide the fractionation element of a characterisation problem, but it still remains to demonstrate the presence of the solutes, and possibly some further attributes of their behaviour.

Numerous methods are available to achieve this. A good review was published recently by Wirth and Romano (1995).

11.5.3.1 On-gel staining approaches for gel electrophoresis

To detect the fractionated bands on a gel, the commonest approach is to employ a staining agent with a broad affinity for proteins. Generally the protein bands are first fixed with a reagent such as 20% (w/v) trichloroacetic acid, immobilising the solutes onto the gel.

The commonest protein stain for gel electrophoresis is probably Coomassie 2.5pt>TM Brilliant Blue (CBB) R-250 (ICI), a triphenylamine dye, or its less soluble dimethyl analogue, CBB G-250 (Figure 11.7). Dyes of this type were shown to be more sensitive than those such as Amido black (Amidoscharz) and Fast Green which were popular earlier (Fazekas de St. Groth *et al.*, 1963; Maizel, 1966), although the latter are still used quite widely.

Not surprisingly in view of their combination ofhydrophobic and anionic regions, the CBB dyes bind principally *via* basic and aromatic moieties (Compton and Jones, 1985), although later studies have suggested that the aromatic component is insignificant in most cases (Congdon *et al.*, 1993). The specific degree of binding observed with a given protein will therefore depend predominantly on the surface concentration of residues such as arginine and lysine: Congdon *et al.* observed a roughly threefold variation

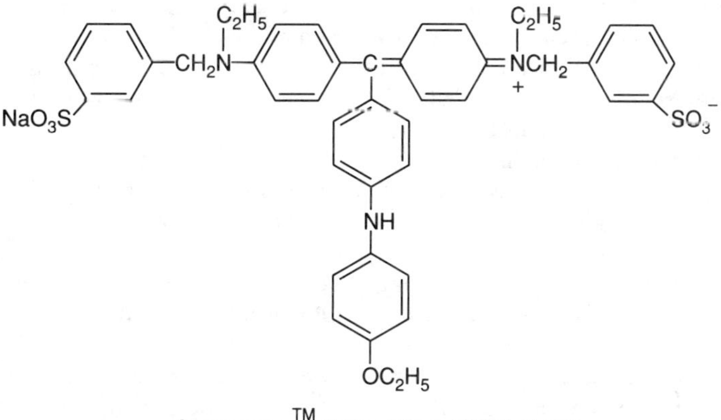

Coomassie™ Brilliant Blue (CBB) R-250

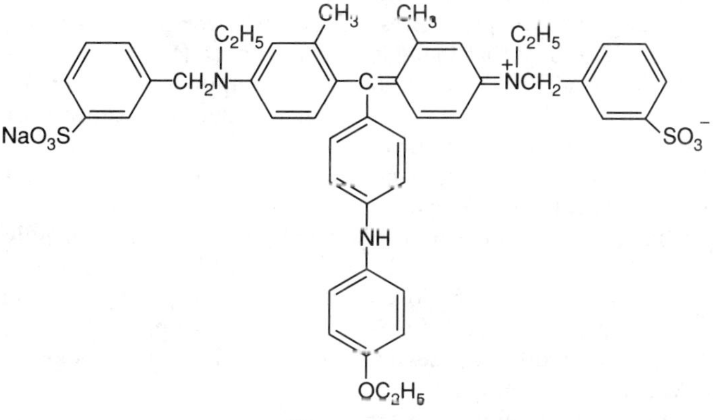

Coomassie™ Brilliant Blue (CBB) G-250

Figure 11.7 Coomassie™ Brilliant Blue (CBB) R-250 and G-250.

in specific binding capacity over a range of ten different proteins.

A typical Coomassie Blue staining solution for gel electrophoresis contains 0.1% (w/v) CBB R-250 in a methanol–water–acetic acid (45:45:10) solvent (Dunn, 1989), and the staining procedure may take a few hours depending on the thickness of the gel and its composition. Destaining, which reduces the background colour, is carried out using the same solvent without the dye; this may take up to 24 hours, but can be speeded up by agitation or by frequent changes of solvent.

Under strongly acidic conditions CBB dyes are converted to their leuco-form, and this colour change is conveniently reversed when the dye adsorbs to the protein surface. Thus Reisner *et al.* (1975) proposed the use of solvent systems containing perchloric acid, which both immobilises protein bands and decolorises unbound dye.

Many other variations of the basic Coomassie Blue procedure, which can detect protein at a surface concentration of approximately 500 ng cm^{-2}, have been described. Many of these have been compared by Neuhoff *et al.* (1985), and an optimised procedure reported by these authors employed 0.1% CBB G-250 in 2% (w/v) phosphoric acid containing 6% (w/v) ammonium sulfate. An increase in sensitivity of nearly tenfold over the basic method was observed, to 70 ng cm^{-2}. Modifications continue to be reported (Choi *et al.*, 1996).

Where greater sensitivity is required silver staining techniques (Switzer *et al.*, 1979; Merril *et al.*, 1984; Merril, 1987; Rabilloud, 1990) have attained widespread popularity, and these are capable of detecting surface protein concentrations as low as 5 ng cm^{-2} or 0.1 ng per band (Dunn, 1989; Dunn and Crisp, 1994).

Fluorescent reagents can also provide similar sensitivity for on-gel protein staining, for example fluorescamine (Udenfriend *et al.*, 1972), *o*-phthalaldehyde (Weidekamm *et al.*, 1973) and the SYPRO dyes from Molecular Probes (Steinberg *et al.*, 1996).

As noted above, CBB dyes bind principally *via* basic, and to some extent perhaps aromatic, moieties in the protein. Not unexpectedly then, some glycoproteins bearing acidic functions are not stained by this approach (Fairbanks *et al.*, 1971), and the periodic acid–Schiff reagent (PAS) technique was employed by these authors to visualise such molecules. Aldehydes are generated from the saccharide moieties by periodate oxidation (cf. Sections 5.2.3 and 6.7.2), and allowed to react with Schiff reagent, an acidic solution of fuchsin reduced with sodium metabisulfite. This PAS staining yields a rose-pink coloration with glycoproteins, and has been a popular method for such applications (Gander, 1984). Other staining methods for glycoproteins, lipoproteins and phosphoproteins are given by Andrews (1986).

Negative staining of gels is another possibility, where the protein forms a colourless, insoluble band against an opaque background: examples include the precipitation of SDS–protein complexes with metals such as potassium (Nelles and Bamburg, 1976) and of native proteins with methyl trichloroacetate (Candiano *et al.*, 1996).

Pre-staining of protein components *before* electrophoresis is another approach (Merril *et al.*, 1986), although great care must be taken to avoid changes in mobility from the derivatisation process.

11.5.3.2 Other on-gel detection approaches for gel electrophoresis

The staining approaches just described can demonstrate the presence of a protein band at a given position in the gel, but do not yield any other useful information about this component: even relative quantitation compared to other bands is an extremely haphazard process due to the different uptake levels observed from one protein to another. More information can be gained by using detection systems which highlight a specific property of the protein(s) of interest.

Autoradiography is a particularly useful approach in circumstances where it is feasible to radiolabel one or more of the sample components (Merril *et al.*, 1986). Obviously this introduces a level of complexity to the preparation, but the specificity of the information obtained in this way is often unsurpassed. The danger of interference from other components is negligible compared to many of the non-specific staining techniques just described. The use of multiple isotopes can also yield complex information from a single set of experiments (see for example Mosser *et al.*, 1975). Possible approaches to obtaining radiolabelled components for such experiments are summarised in Section 11.13.3.

γ-Emitters such as ^{125}I, and more powerful β-emitters such as ^{32}P, can be detected by direct autoradiography from the gel, generally by simple proximity of X-ray film. Laskey and Mills (1977) described the use of an X-ray intensifying screen[1] to increase the sensitivity of this approach. Radiation which has passed through the film (producing an autoradiographic

[1] Such screens are routinely in use for enhancement of medical X-ray images, but the longer exposure times for autoradiography can lead to background problems. These are minimised by pre-sensitising the film, and by carrying out the development at −70 °C.

image) is absorbed by a screen containing a fluorescer such as calcium tungstate, energy transfer generating light. This superimposes a photographic image on the autoradiographic one already produced, increasing sensitivity substantially.

Other isotopes such as ^{14}C and ^{35}S are weaker β-emitters, and, while still powerful enough to be detected directly, are more often quantified by scintillation counting (also referred to as scintillation autography or fluorography), where the β-energy is again used to generate light from a suitable fluorescent substrate (Bonner, 1984). This can be achieved by dissolving the gels of interest in a suitable solvent before transferring to a commercial scintillation reagent, or (Bonner and Laskey, 1974) by soaking the dehydrated gel in 2,5-diphenyloxazole (PPO).

Alternatively, and generally without any prior treatment of the protein components such as radiolabelling, it may be possible to exploit their specific binding properties to facilitate detection by a suitable complementary reagent. This approach of affinity detection requires a reagent which: (1) binds to the target protein, and (2) is detectable. For example if the target protein is an antigen, a specific antibody may be used to detect its presence. In an unfixed gel, even an unlabelled antibody may be sufficient to allow detection of the target antigen, as the complex will precipitate under the right conditions. Once this immunofixation has taken place, other proteins can be washed away and the complex stained with a general reagent. With a fixed gel, it is necessary to use a labelled antibody of some kind, e.g. a ^{125}I-labelled Mab of high specificity, which can then be used to detect the position of the antigen by autoradiography.

If the target proteins are enzymes, it may be possible to detect their presence by exploiting their catalytic properties to produce an insoluble coloured product (Heeb and Gabriel, 1984). The gel can be immersed directly in the substrate solution (plus any cofactors, etc.), and the presence of the enzyme will be indicated by the generation of the precipitated product in the gel itself. Alternatively the substrate mixture can be soaked onto a suitable support such as filter paper, and laid over the gel. The coloured product is therefore 'printed' onto the support. This approach can be used even if the enzyme-catalysed reaction product is soluble, as the support limits diffusion to a large extent.

11.5.3.3 Blotting approaches for gel electrophoresis

An extension of this 'printing' approach is to transfer the sample bands themselves onto another support, rather than just the signal products which they have generated. This concept underlies the highly popular blotting approach, where the fractionated (but unfixed) bands are transferred onto a membrane (e.g. nitrocellulose, nylon or polyvinyldifluoride (PVDF)) for further manipulation (Towbin *et al.*, 1979; Hayes *et al.*, 1989; Wisdom, 1994).

Transferred protein bands are more readily accessible on a membrane than in a gel, so allowing greater sensitivity of detection. Membranes are also much easier to manipulate than the electrophoresis gels themselves, lending themselves more conveniently to specific detection techniques. Procedures such as staining, destaining and drying are considerably faster than in the original gel.

Simple proximity of the blotting membrane to the gel can bring about some transfer of the bands, but it is more efficient to drive this process in some way. One of the most popular techniques is electroblotting (Towbin *et al.*, 1979; Towbin and Gordon, 1984), in which an electric field is applied across the gel and membrane, the electrophoretic mobility of the solute bands driving their transfer in the desired direction. Transfer times will depend on the design of the electroblotting apparatus (in particular the voltage which can be applied without excessive heating problems), the nature of the gel and the size of the solutes – sometimes blotting times of up to 24 hours may be required. Coloured marker proteins can be used to confirm that the transfer process has taken place.

Once on the membrane, general staining procedures can be employed to detect the presence of protein bands, in much the same way as described in Section 11.5.3.1. However, the specific staining method used will be influenced by the binding properties of the particular membrane type. Thus on PVDF membranes Coomassie Blue stains tend to give unacceptably high background levels (Dunn, 1989), and different reagents such as colloidal gold particles (Fowler, 1994) are more useful. The sensitivity of this technique can be improved by silver enhancement (Moeremans *et al.*, 1984). Gold or silver staining gives good results on nitrocellulose (Moeremans *et al.*, 1985), while on nylon, iron colloid approaches have given good results (Moeremans *et al.*, 1986).

In a variant of the direct staining approach, solubilisation of protein–dye complexes with a solvent such as dimethyl sulfoxide allows solution-phase spectrophotometric quantitation (Goldring and Ravaioli, 1996).

Much more information can often be gained from affinity detection, however, by using a specific probe to identify particular proteins. The commonest version of this approach is to use a specific antibody to detect the protein of interest (Towbin *et al.*, 1979; Towbin and Gordon, 1984), a procedure termed immunoblotting or Western blotting (see Chapter 1).

In this type of detection method using a protein reagent, it is necessary to block the remaining sites on the membrane with another protein such as BSA, or (so long as it does not displace bound materials) a detergent such as Tween 20. The specific detection approaches used to visualise the probe antibody on Western blots parallel those described earlier for directapplication on the gel. Antibodies can be labelled, for example, with radioactivity, fluorophores, gold sols, or enzymes capable of producing a highly-coloured product. Luminescent techniques—notably the ECL™ system from Amersham Life Sciences—have gained widespread popularity for sensitive detection (Chevallet *et al.*, 1997; Patel *et al.*, 1997). Luminescent europium complexes also appear to carry some promise as sensitive reagents for direct detection of proteins on a variety of membrane types (Lim *et al.*, 1997). Indirect approaches are very common, such as those based on biotin–avidin (Willison *et al.*, 1986) or a second antibody (Crisp and Dunn, 1994).

Of course, approaches of this type are not limited to antibodies: any other reagent with a suitable affinity can be used for the detection of bound species. Labelled lectins have been used for the detection of glycoproteins, for example (Clegg, 1982) and Protein G–HRP for antibodies (Hu *et al.*, 1997).

Specific chemical reactions can also be applied in the same fashion. Immobilised glycoproteins can be oxidised with periodate, and are then capable of reacting with hydrazide-derivatised probes. Gershoni *et al.* (1985), for example, used hydrazides of AP and HRP with suitable substrates to detect 10 ng of fetuin.

More complicated approaches may involve the derivatisation of the sample before electrophoresis, in such a way that specific components can be detected by affinity labels on the blot. O'Shannessy *et al.* (1987), for example, specifically derivatised sialoglycoproteins in a mixture with a biotin hydrazide, which allowed them to be detected with a labelled streptavidin reagent after electrophoresis and blotting. Similarly, glycoproteins can be derivatised with a hapten such as digoxigenin before electrophoresis, then detected with a labelled, digoxigenin-specific antibody (Haselbeck and Hösel, 1993).

Affinity blotting is a variation on this general approach, in which the membrane itself is derivatised with an affinity ligand, such that it binds only the proteins of interest. In affinity immunoblotting, for example, an antigen-labelled membrane can be used to pick up specific antibodies from a mixture after fractionation by isoelectric focusing (Knisley and Rodkey, 1986; Hamilton *et al.*, 1987).

11.5.3.4 On-line detection systems for capillary electrophoresis

Capillary electrophoresis is very different to gel electrophoresis in its detection requirements; because the sample components are eluted through a column, the detection geometry is more akin to chromatographic equipment, and the very small sample volumes and concentrations dictated by the capillary format call for extremely high sensitivity. Detection systems familiar from chromatography have often been employed, such as UV–visible absorbance and conductivity measurement (Einarsson *et al.*, 1984). Absorbance detectors can detect micromolar concentrations of solutes (Wallingford and Ewing, 1991).

One important application of capillary electrophoresis has been in rapid DNA sequencing, and this has provided a major impetus for the development of more sensitive detection systems. The dominant approach is now that of laser-induced fluorescence (LIF), first described by Gassmann *et al.* (1985). This technique is capable of truly prodigious sensitivity with the right equipment design: Wu and Dovichi (1989) have reported the detection of picomolar solutions of fluorescent amino acids and nucleotides, with samples containing as little as 2 zmol, roughly 1000 molecules.

This level of sensitivity makes capillary electrophoresis capable of the detection of labelled complexes from immunoassays; Chen and Pentoney (1994) reported the use of this approach to quantify phycoerythrin-labelled digoxigenin in a homogeneous immunoassay format.

The coupling of the output of capillary electrophoresis to a mass spectrometer is an extremely powerful combination of techniques (Smith and Udseth, 1988), allowing detailed analysis of difficult components such as oligosaccharides and glycopeptides. This 'hyphenated' approach has been applied for example by Kelly *et al.* (1993), who used CE–MS to investigate protein glycosylation patterns.

11.6 Native Gel Electrophoresis

Some commonly encountered forms of electrophoresis employ procedures or agents which destroy the three-dimensional structure of proteins, either *via* dissociation of subunits or denaturation (e.g. boiling, sodium dodecyl sulfate (SDS), urea), or *via* cleavage of disulfide bonds (e.g. 2-mercaptoethanol). In many applications, however, it is desired to preserve the structure of the protein as far as possible, so that specific affinity detection techniques can be applied, for example[1]. In these cases electrophoresis may have to be carried out under non-denaturing and/or non-dissociating (native) conditions.

As discussed in Section 11.5, a wide variety of different approaches can be taken with regard to gel type and buffer system (Blackshear, 1984). If it is desired to fractionate the mixture predominantly by mass then a gradient gel should be used (e.g. 3–30% polyacrylamide) to maximise resolution on this basis—a non-denaturing alternative to SDS-PAGE (Section 11.7.1). If charge differences are the main point of interest, then a uniform gel is probably more appropriate, with a pore size large enough to prevent size-related effects from being significant. Even with the most porous polyacrylamide gels the upper molecular mass limit is around 500 kDa, and if largermolecules are to be fractionated without size effects, acrylamide–agarose gels (Uriel, 1966) may be better.

With regard to buffer composition, with CZE applications it is simply a matter of choosing a pH which gives the best fractionation and then a buffer system to provide it. Working close to the p*I* values of the target proteins tends to maximise charge differences, and is a useful basis for choice if the range present is not too broad; bands may however be broad if the molecule bears very little charge (Maizel, 1971).

Something of a trial-and-error approach is often necessary to obtain the best performance. A fairly acidic or basic buffer may be used first to determine the p*I* range of the sample components, and then a second pH chosen on this basis to improve the resolution. A summary of convenient buffers for various pH ranges is given in Table 11.1 (Andrews, 1986). As noted earlier, ionic strengths should be kept below 100 mM to minimise Joule heating.

If the sample is too large or too weak for CZE, or if insufficient resolution can be obtained by this means, it may be necessary to examine a discontinuous (MZE) approach. Choice of buffer systems for MZE is more complex, but many useful examples have been described in the literature since the original methods of Ornstein (1964) and Davis (1964). Chrambach and colleagues (Chrombach and Rodbard, 1971; Chrambach and Jovin, 1983) listed selections of formulations generated from computer modelling which should cover a wide variety of needs. Hames (1981) summarised instructions for the execution of non-denaturing MZE systems under basic, neutral and acidic conditions, both with respect to preparation of the gels and the buffers. These are reprinted in Table 11.2. The same systems have been quoted widely elsewhere (Maizel, 1971; Blackshear, 1984).

Table 11.1 Summary of convenient buffers for various pH ranges

Approximate pH range	Primary buffer constituent	pH adjusted to the desired value with
2.4–6.0	0.1 M Citric acid	1 M NaOH
2.8–3.8	0.05 M Formic acid	1 M NaOH
4.0–5.5	0.05 M Acetic acid	1 M NaOH or Tris
5.2–7.0	0.05 M Maleic acid	1 M NaOH or Tris
6.0–8.0	0.05 M KH_2PO_4 or NaH_2PO_4	1 M NaOH
7.0–8.5	0.05 M Na diethyl-barbiturate (veronal)	1 M HCl
7.2–9.0	0.05 Tris	1 M HCl or glycine
8.5–10.0	0.015 M $Na_2B_4O_7$	1 M HCl or NaOH
9.0–10.5	0.05 M Glycine	1 M NaOH
9.0–11.0	0.025 M $NaHCO_3$	1 M NaOH

Ammmonium persulfate is a common initiator for polyacrylamide gel formation, but it is an extremely powerful oxidising agent and residual traces may cause damage to some proteins (Blackshear, 1984). In CZE these traces can be removed by carrying out a preliminary electrophoresis step, driving them out of the gel before the sample is introduced. Otherwise the polymerisation reaction can be initiated by an alternative approach, such as photoactivated riboflavin.

In denaturing applications, powerful detergents such as SDS are often used to facilitate the dissolution of poorly-soluble proteins, and to prevent their aggregation or precipitation during electrophoresis. While these agents should be avoided for native applications, it may be possible to employ milder, non-ionic detergents such as 0.5% (w/v) Tween 80, Triton X-100 or Brij 35 without causing serious denaturation (Dunn, 1989).

To retain protein activity during electrophoresis it may sometimes be necessary to pay more attention to cooling than usual, for example by working in a cold-room or with cooled water circulating through the heat-exchanger of the electrophoresis apparatus (Blackshear, 1984).

[1] Although immunoblot detection techniques have been successfully applied after surprising levels of insult to the target protein, such as 20% SDS treatment (cytochrome P-450: Baskin and Yang, 1982) or even boiling with 4% SDS and 10% 2-mercaptoethanol (viral antigens: Shors and Winston, 1989). Despite such examples, it is to be expected that the immunoreactivity of more labile epitopes, such as those defined by their three-dimensional conformation, may be lost during such procedures.

Table 11.2 Buffers for nondissociating discontinuous systems

Buffer	Preparation
High pH discontinuous: stacks at pH 8.3, separates at pH 9.5	
Stacking gel buffer (Tris–HCl at pH 6.8)	6.0 g of Tris in 40 ml titrated to pH 6.8 with 1 M HCl; bring to 100 ml
Resolving gel buffer (Tris–HCl at pH 8.8)	36.3 g of Tris and 48.0 ml of 1 M HCl brought to 100 ml; titrate to pH 8.8 with HCl, if necessary
Reservoir buffer (Tris–glycine at pH 8.3)	At the correct concentration for use; 3 g of Tris and 14.4 g of glycine in 1 litre
Neutral pH discontinuous: stacks at pH 7.0, separates at pH 8.0	
Stacking gel buffer (Tris–phosphate at pH 5.5)	4.95 g of Tris in 40 ml titrated to pH 5.5 with 1 M phosphoric acid; bring to 100 ml
Resolving gel buffer (Tris–HCl at pH 7.5)	6.85 g of Tris in 40 ml of water titrated to pH 7.5 with 1 M HCl; bring to 100 ml
Reservoir buffer (Tris–diethylbarbiturate at pH 7.0)	5.52 g of diethylbaribituric acid and 1.0 g of Tris to 1 litre
Low pH discontinuous: stacks at pH 5.0, separates at pH 3.8	
Stacking gel buffer (acetic acid–KOH at pH 6.8)	48.0 ml of 1 M KOH and 2.9 ml of glacial acetic acid; bring to 100 ml
Resolving gel buffer (acetic acid–KOH at pH 4.3)	48.0 ml of 1 M KOH and 17.2 ml of glacial acetic acid are mixed and diluted to 100 ml
Reservoir buffer (acetic acid–β-alanine at pH 4.5)	31.2 g of β-alanine and 8.0 ml of glacial acetic acid to 1 litre

Source: Hames, B. D. (1981). In *Gel Electrophoresis of Proteins: A Practical Approach* (Hames, B. D. and Rickwood, D., eds), p. 30, IRL Press, Oxford.

11.6.1 Applications of native gel electrophoresis in conjugate characterisation

The most primitive forms of CZE approaches, such as paper electrophoresis in a single buffer system, are fairly frequently encountered in early reports of protein conjugation, for both preparative and analytical purposes (Singer and Schick, 1961; Borek and Silverstein, 1961). These were able, for example, to provide useful information on the formation of ferritin–IgG conjugates for electron microscopy.

To a large extent the use of such techniques was necessitated by a lack of choice. Developments since this time have offered a number of more convenient alternatives: high-performance chromatographic procedures can offer comparable information, while among electrophoretic techniques, the most popular manifestations are those which can be correlated most directly with a physicochemical attribute such as molecular size (SDS-PAGE, Section 11.7.1) or charge (IEF, Section 11.8).

Analytical applications of native electrophoresis in this field are therefore relatively infrequent. The most common category are probably those employed to fractionate a product before an affinity detection procedure such as Western blotting with a specific antibody. Gels containing a fairly high percentage of polyacrylamide can be used to provide a size-based separation even in the absence of SDS, therefore retaining the integrity of the analyte protein.

Gleizes *et al.* (1994), for example, demonstrated the conjugation of a digoxigenin derivative (molecular mass 613 Da) to a recombinant growth factor (molecular mass 16.5 kDa) by PAGE on an 18% polyacrylamide–tricine gel, noting a slight shift towards higher molecular mass compared to the growth factor itself. Detection of the conjugate down to fmol levels was achieved by blotting onto a PVDF membrane, incubation with an AP-labelled specific antibody, and chemiluminescent detection.

Such an analysis is at the limit of resolution of size-based fractionation techniques. Other approaches have

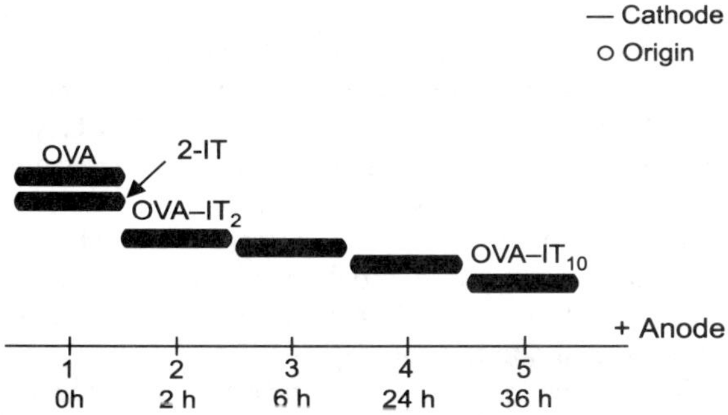

Figure 11.8 Analysis of the sequential addition of 2-IT to ovalbumin and electrophoretic behaviour of the conjugates. (Adapted with permission from: Kamps-Holtzapple *et al.* (1993), *J. Immunol. Methods*, **164**, 245–253.)

instead focussed on detecting charge changes, particularly for conjugation of proteins to small molecules. Kamps-Holtzapple *et al.* (1993), for example, studied a number of protein–hapten conjugations using non-denaturing conditions on high-resolution agarose gels at pH 8.8. In one of these studies, ovalbumin was derivatised with 2-iminothiolane, which replaces amine moieties (largely positively charged at pH 8.8) with thiols (largely neutral or negatively charged at pH 8.8) and thus increases the mobility of the protein towards the anode. The derivatisation of the ovalbumin (OVA) by different numbers of IT molecules was clearly demonstrated in the electrophoretic behaviour of the conjugates (Figure 11.8).

While the popularity of SDS-PAGE and IEF derives from the ability to correlate behaviour directly with simple attributes such as molecular size and p*I*, it should be borne in mind that simple native approaches can sometimes offer very powerful resolution of closely similar components: for example, Stroupe and Foster (1973) used an acidic MZE system to resolve two isomers of bovine mercaptalbumin, with only slight differences in hydrogen ion titration properties.

11.7 Non-native Gel Electrophoresis

If it is not a requirement to retain the functional activity of the analyte protein(s), a number of more vigorous approaches become applicable. Discussed here under the broad heading of non-native techniques, it should be remembered that they cover a continuum of magnitude, from a relatively mild dissociation of subunits, to complete denaturation of secondary and tertiary structure.

These dissociating or denaturing approaches may be chosen for a variety of reasons. First it may be a specific aim of the investigation to disrupt the quaternary structure of an oligomeric protein, such that properties of the individual subunits can be examined; similarly the secondary and tertiary structure of subunits or intact molecules may deliberately be destroyed, to compare the properties with the native species.

On the other hand, structural disruption may be an inevitable consequence of techniques applied for another reason, in particular the solubilisation of 'difficult' samples such as cell membrane proteins, and the dissociation of non-covalently linked components within them. Various reagents such as detergents, or chaotropes such as urea, have been used to achieve solubilisation of such samples. These generally bring with them the reversible or irreversible denaturation of the sample analytes, even if this was not a specific aim of the experiment.

Finally, one corollary of detergent pre-treatment was so useful that it generated a powerful new methodology of its own, SDS-PAGE, dealt with in Section 11.7.1. Here the solubilisation properties of the detergent are extremely useful in allowing all the components of a sample to enter the electrophoresis gel, but denaturation is also required to allow the greatest possible access of detergent to the protein core.

11.7.1 SDS-PAGE

Sodium dodecyl sulfate (SDS; Figure 11.9) is a powerful anionic detergent; it has come to form the basis of one of the most widely-applied branches of analytical electrophoresis, due to its ability to bind avidly to proteins in a manner virtually independent of all parameters except molecular size. A wide range of proteins bind around 140% of SDS by mass (Blackshear, 1984).

In parallel with the evolution of size exclusion chromatography in the late 1950s, it was recognised that the size of protein molecules could be estimated by their rate of diffusion through porous gels (Allison and Humphrey, 1959). Smithies (1962) combined this

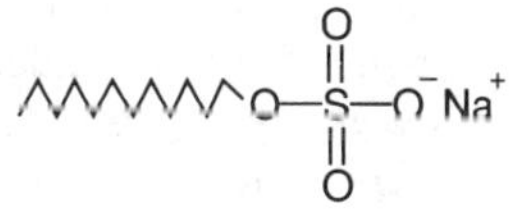

Figure 11.9 Sodium dodecyl sulfate (SDS).

concept with the gel electrophoresis approach, and arrived at a size-based electrophoretic separation. Smithies also recognised that the mobility of the protein molecules would be affected by their conformation, and included urea and a thiol in his formulation to promote unfolding and cleavage of disulfide bonds.

SDS was originally included in electrophoresis formulations to dissociate and solubilise protein complexes (Maizel, 1966; Shapiro *et al.*, 1966). But in 1967, Shapiro *et al.* demonstrated that the complexation of SDS with proteins yielded species whose electrophoretic mobility was sufficiently independent of the original p*I* to offer a good approximation to molecular mass. Even over a p*I* range of 4–11, a linear plot of migration *versus* log M_r could be drawn up, with very few exceptions.

This approach was confirmed using a wider range of proteins by Weber and Osborn (1969), who determined that M_r values in the range of at least 15–100 kDa could be estimated to within 10%. The new technique was quickly adopted, along with homobifunctional cross-linking reagents, as a means of investigating the topology of subunits in oligomeric proteins (Davies and Stark, 1970; Carpenter and Harrington, 1972), and of proteins in cell membranes (Fairbanks *et al.*, 1971; Steck, 1972). Application to smaller polypeptides (molecular mass 1.2–10 kDa) was also demonstrated (Swank and Munkres, 1971), albeit with poorer precision. Occasional examples of proteins which did not display 'ideal' SDS-PAGE behaviour were reported (Dunker and Rueckert, 1969).

Laemmli (1970) described the version of SDS-PAGE most commonly encountered today, which combined Shapiro and Maizel's use of SDS with the improved resolution of discontinuous (MZE) systems (1.5 to 2-fold: Maizel, 1971). This approach has remained in use with only minor modifications to the present day.

The earliest applications of SDS-PAGE employed 2-mercaptoethanol (typically 1%) to reduce disulfide bonds, so that there is the greatest possible access of the detergent to the protein structure. The incorporation of this step has been reported to reduce aggregation-related problems such as streaking and entrapment at the origin (Fairbanks *et al.*, 1971). However, it is quite possible to carry out SDS-PAGE on samples which have not undergone disulfide cleavage, as was necessary in topographical studies using cleavable linkers which possess a disulfide link themselves, such as DSP (Bragg and Hou, 1975). Denaturation by boiling and/or treatment with 8 M urea generally provides enough access for the detergent to give acceptable SDS-PAGE performance, without disruption of disulfides. Maizel (1971) noted that SDS treatment for one minute at 100°C was sufficient to disrupt all non-covalent aggregates. Once denaturation has been achieved with a suitable level (1–2%) of SDS, the electrophoresis run can generally be run at a much lower detergent level, typically 0.1%. This level may rise during the electrophoresis run due to zone effects.

The choice of whether to incorporate a reducing thiol should be driven by the information required from the experiment; in particular for proteins with a linked-subunit structure, whether it is more useful to observe the intact molecule or the discrete subunits. Often the information provided is complementary, and it is beneficial to employ both approaches.

Two other components are generally added to an SDS-PAGE sample: a fast-moving dye such as pyronine Y or bromophenol blue, to track migration during the run, and (for gels incorporating a stacking zone) a high-density material such as glycerol to promote sinking to the bottom of the sample well (Blackshear, 1984).

As noted in Section 11.5, the exclusion limit of polyacrylamide gels is around 500 kDa (Ji, 1979); where larger molecules are of interest, it may be beneficial to use gels containing a proportion of agarose. By this means the exclusion limit can be extended to 3000 kDa or higher.

Commercial suppliers have developed products which offer convenient means for SDS-PAGE analysis, e.g. the Pharmacia Phast™ system. At the expense of rather limited protocols (and rather limited gel dimensions), this allows gels to be generated very rapidly and simply using pre-formed reagents.

11.7.1.1 Applications of SDS-PAGE in conjugate characterisation

Predictably, SDS-PAGE has been used most frequently for the analysis of protein–protein conjugates, where the change in molecular size on conjugation is most distinct from the starting materials. It should be borne in mind that the presence of cross-links may interfere with the unfolding of the protein molecule(s), and hence with mobility (Steele and Nielsen, 1978), but these effects are relatively minor. These applications are typified by the following examples.

Perez *et al.* (1985) used SDS-PAGE under both reducing and non-reducing conditions to examine the product from the conjugation of two separate Mabs by SPDP coupling. Under non-reducing conditions, it was possible to observe dimers, trimers and larger oligomers of the IgG molecules. Under reducing conditions, essentially all the components of the product were indistinguishable from the starting materials: not only were the IgG molecules themselves reduced to their constituent heavy and light chains, but the disulfide bonds linking the conjugated molecules were destroyed also. Worrell *et al.* (1986) similarly used SDS-PAGE to show the fragmentation of disulfide-linked immunoconjugates.

Wood and O'Dorisio (1985) used a variety of cross-linking agents to couple ^{125}I-labelled vasoactive intestinal peptide (VIP: molecular mass ca 3 kDa) to its lymphocyte receptor. SDS-PAGE analysis of the products demonstrated that the receptor protein had a molecular mass of around 47 kDa; this was unchanged on treatment with a reducing thiol, hence it was concluded that the receptor contained no interchain disulfide links.

Neblock *et al.* (1992) coupled two antibody Fab′ fragments *via* a bis-maleimide linker, and demonstrated the presence of the desired 100 kDa $F(ab')_2$ conjugate by SDS-PAGE under non-reducing conditions. The electrophoretic analysis also detected the presence of unreacted Fab′, and products of intermediate molecular mass thought to arise from the coupling of Fab′ to reduced heavy or light chains.

McIntyre *et al.* (1994) used SDS-PAGE under non-reducing conditions to establish the predominant formation of a 1:1 conjugate from interleukin-2 and gelonin. Smaller quantities of 2:1 and 1:2 conjugates were also observed, as well as dimers of both starting materials. Individual fractions from an SEC purification were electrophoresed, improving the resolution obtained.

11.7.1.2 Alternatives to SDS

Although other anionic detergents may display similar properties, SDS is the most effective of its class (Maizel, 1971): shorter chain alkyl sulfates bind less strongly to proteins, while longer homologues may suffer solubility problems. SDS has therefore retained its popularity in this application for thirty years.

However, there are certain instances where SDS binding is incomplete, e.g. with extremely acidic or basic proteins, or those which are heavily glycosylated (Blackshear, 1984). In these cases, cationic detergents have been employed in a similar fashion, obviously with the direction of electrophoresis reversed so that the sample components are fractionated towards the cathode. Detergents used in this way include cetyltrimethylammonium bromide (CTAB) (Schmatchenko and Varshavsky, 1978) and cetylpyridinium chloride (Blackshear, 1984).

11.7.2 Other non-native systems for gel electrophoresis

SDS-PAGE is particularly popular due to its specific fractionation properties, but other systems have been developed principally to allow the solubilisation of sample components – they do not share the same property of size fractionation however.

Urea, included in Smithies' original description of size-based electrophoresis (1962), is a powerful chaotrope which disrupts hydrogen bonds, and therefore can dissociate or denature protein structures[1]. Because the conformations of protein molecules are maintained by very many hydrogen bonds, the urea concentration must be high enough (typically 8 M) to disrupt a large number of these simultaneously. For poorly-soluble analytes such as membrane and viral proteins, urea is still less efficient than SDS: many components may remain insoluble under the pre-treatment conditions (Maizel, 1971).

There are other problems with urea: its solubility is insufficient at low temperatures to achieve denaturing concentrations, and therefore work must be carried out under ambient conditions. Generation of cyanate ions is also an issue. These can react with protein amines to form carbamates, distorting the charge distribution.

If this effect is avoided (for example by incorporating a competing amine in the buffer), the major advantage of urea over SDS is that charge differences between the analytes are maintained, and electrophoretic separations can be carried out which are not wholly size-dependent. It does not, therefore, offer an alternative approach for molecular sizing (Dunker and Rueckert, 1969). Urea-PAGE is really a branch of 'normal' electrophoresis where denaturation may provide advantageous resolution.

Other systems have been described for dissociation of sample proteins (Maizel, 1971; Blackshear, 1984), such as 68% acetic acid, phenol–acetic acid–water (2.1.1), or milder detergents such as Triton X-100 or sodium deoxycholate. Often these systems have been employed in combination with urea, and are also branches of 'denatured PAGE'. The mildest of these systems, however, may promote solubility while maintaining protein structure to allow native electrophoresis.

[1] Guanidinium salts are also powerful chaotropes, often encountered as protein denaturants. However, their ionic nature under all useful pH conditions renders them unsuitable for electrophoretic applications.

11.8 Isoelectric Focusing

Isoelectric focusing (IEF) has already been discussed briefly as a preparative technique in Section 10.12.2. It is based on the establishment of a fixed pH gradient in the electrophoresis cell (Svensson, 1961; Vesterberg and Svensson, 1966; Vesterberg, 1968, 1971). This is achieved by ordering a mixture of ampholytes, synthetic polymers of varying p*I*, in the electric field (see Figure 10.38).

The fixed pH gradient gives IEF some very particular properties, which render it highly attractive for charge-based fractionation. Suppose that a protein analyte with a p*I* of 7 is introduced at the anode of an IEF gel with a pH gradient of 5–8 (Figure 11.10(a)). Entering the gel at pH 5, the protein is below its p*I* and therefore positively charged, and so moves through the pH gradient towards the cathode. However, when it reaches the position in the gradient at which the pH equals its p*I*, the protein molecule bears no net charge. It therefore loses its mobility, and remains at this point in the gel.

Conversely (Figure 11.10(b)), if the same protein is introduced at the cathode, it will enter the gel at pH 8 where it is above its p*I* and positively charged. It will thus migrate towards the anode until it reaches the pH corresponding to its p*I*, and the overall result is the same in both cases. Any tendency for the analyte to diffuse away from its p*I*-determined position will result in it regaining an overall charge, its mobility then promoting its return. So long as endosmosis effects are controlled (e.g. by maintaining the ampholyte level above around 2% when a gel format is used: Awdeh *et al.*, 1968) the protein will remain at this position of the gel for the rest of the run.

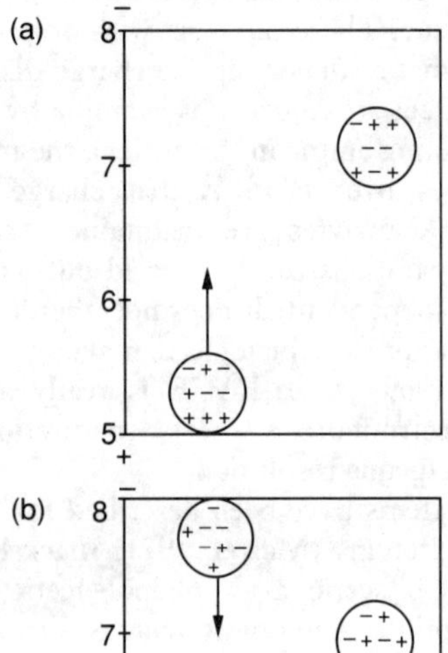

Figure 11.10 Migration of proteins based on p*I* in IEF. See text for details

Proteins in a mixed sample will behave in the same way, essentially independently of each other; this system therefore provides a powerful means of fractionating protein analytes on the basis of their p*I* values.

In the other forms of electrophoresis discussed above, the analytes continuously move towards one or other of the electrodes, and the run must be stopped at a suitable point to prevent them overrunning the gel. IEF, by contrast, is a steady-state method: within reasonable limits the gel can be left running indefinitely without disturbing the focused bands. This allows particularly good resolution to be obtained, as the run can be continued until focusing has reached its full conclusion. The resolution of components of different p*I* is limited chiefly by the quality of the pH gradient; under optimal conditions differences as small as 0.001 can be detected, and up to 120 analyte bands may be resolved in a single gel (Radola, 1984).

As demonstrated in Figure 11.10, the position at which the sample is introduced onto an IEF gel is irrelevant to the eventual focusing points of its components. Therefore there is no problem with choosing the right position for sample introduction, indeed it is quite feasible to use the entire length of the gel for this purpose, giving excellent sample capacities. Due to the focusing properties of the system, sample concentration is relatively unimportant too.

The position at which an analyte component is focused is directly correlated with its p*I*[1], a very fundamental and useful parameter. This p*I* can be determined by measuring the pH at the point of focus, or (much more usually) by incorporating standards of known p*I* onto the gel. In view of the major advantages outlined above, it is not surprising that IEF has

[1] IEF in fact yields a more accurate estimate of the genuine p*I* than other methods such as continuous electrophoresis (Vesterberg, 1968), as there is less interference from ionic associations; p*I* values obtained by continuous methods are typically lower than IEF results. The p*I* determined by IEF should also correspond closely to the isoionic point of the protein (Sorensen, 1926), the pH which is unaltered by addition of further protein to a pure protein solution.

become the predominant charge-based analytical method. It should be remembered that p*I* values often vary with temperature (perhaps being 0.5 pH units higher at 4 °C than at 25 °C, although the effect varies from one protein to another), and various approaches have been proposed to ensure consistency of measurement (Righetti and Caravaggio, 1976).

Because of the property of independent behaviour, non-covalent complexes may well dissociate under IEF: the movement of the complex itself may be defined by its 'mean' p*I*, but each time an equilibrium dissociation event takes place, the components will begin to behave according to their individual properties. This will act to discourage reassociation, and given enough time the complex will dissociate completely, the lower the affinity constant, the shorter the time. Such behaviour was observed with antibody–antigen complexes (Maidment *et al.*, 1980).

As with other electrophoretic techniques, steps must be taken to minimise the convection effects caused by Joule heat. This has been accomplished by three main approaches (Vesterberg, 1971). (1) The use of density gradients in vertical columns, principally based on sucrose up to 50% (w/v). Other reagents such as glycerol, dextrans or Ficoll 2.5pt>TM (Pharmacia) have also been described in this application (Vesterberg, 1971); (2) Zone convection electrofocusing (Valmet, 1968) is where a narrow band of electrolyte and solutes is contained between two cooled, ridged plates; thermal diffusion causes density gradients to be set up in the 'valleys' between each pair of ridges. (3) Gel IEF, using polyacrylamide gels (Wrigley, 1968), has proved as practical for this technique as for other branches of electrophoresis. This approach dominates analytical IEF applications today.

The ampholytes used to create the pH gradient must possess certain properties (Svensson, 1961, 1962): they must possess a degree of buffering capacity around their p*I*, as well as a reasonable conductance. They should be hydrophilic to minimise binding to proteins, and of low molecular mass to facilitate separation if binding does occur, for example, by dialysis or SEC (Rodkey, 1988). Vesterberg (1971) described a series of aliphatic polyamino-polycarboxylic acids, whose general structure is shown in Figure 11.11. These were further developed by LKB to produce the Ampholine 2.5pt>TM reagents, now marketed by Amerham Pharmacia Biotech.

These ampholytes exert a useful solubilising effect on many proteins (cf. the effect of similarly zwitterionic amino acids on protein solubility; Grönwall, 1942). However, stability can sometimes be improved further by incorporating low levels of uncharged thiols in the electrophoresis mixture (Wadström and Hisatsune, 1970).

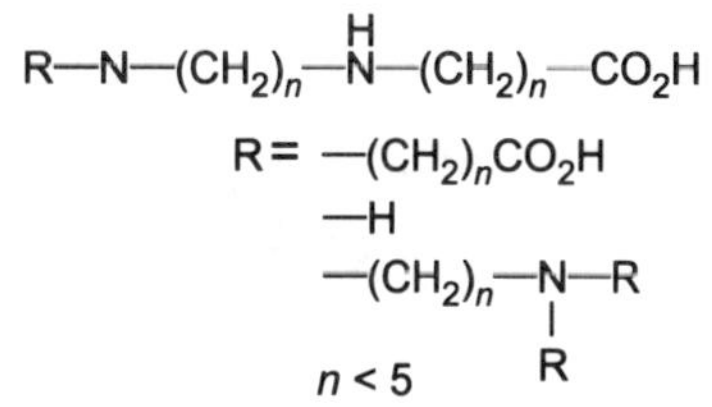

Figure 11.11 Aliphatic polyamino-polycarboxylic acids as ampholytes.

An alternative to a gradient made up of mobile ampholytes is to create a fixed gradient, by casting the gel from a graded mixture of acrylamide monomers containing carboxyl or tertiary amine groups. This can provide a more stable gradient, while the gel retains the ability to transmit the necessary current. These Immobiline 2.5pt>TM gels (Bjellqvist *et al.*, 1982) and other immobilised pH gradients (IPGs) have since achieved popularity in many IEF applications (Altland and Rossmann, 1985; Rimpilainen and Righetti, 1985; Righetti and Bossi, 1997).

Urea is sometimes incorporated in IEF systems, either to improve the solubility of analytes, or for the study of dissociated subunits (Vesterberg, 1971). The presence of urea reduces the activity of hydrogen ions, however, giving apparently higher p*I* values for the ampholytes (Righetti and Caravaggio, 1976); correction factors may therefore need to be applied (Ui, 1971).

11.8.1 Applications of IEF in conjugate characterisation

Because of the extremely high resolution of IEF, it is capable of highlighting the composition of conjugates in great detail. Indeed the amount of fine structure observed within the IEF electrophoretogram of a 'single' protein can be very impressive. In the analysis of antibodies, Awdeh *et al.* (1968) were the first to note the heterogeneity of polyclonal antisera displayed by this technique. Later, the work of Rodkey and his colleagues at the University of Texas School of Medicine has demonstrated exquisite resolution of reactive clonotypes within these populations (Knisley and Rodkey, 1986, 1988). Examples of such analyses, visualised by immunoaffinity blotting, are shown in Figure 11.12(a) and (b). Even monoclonal antibodies routinely display at least 5–10 bands when analysed in this way (Hamilton *et al.*, 1987), as shown in

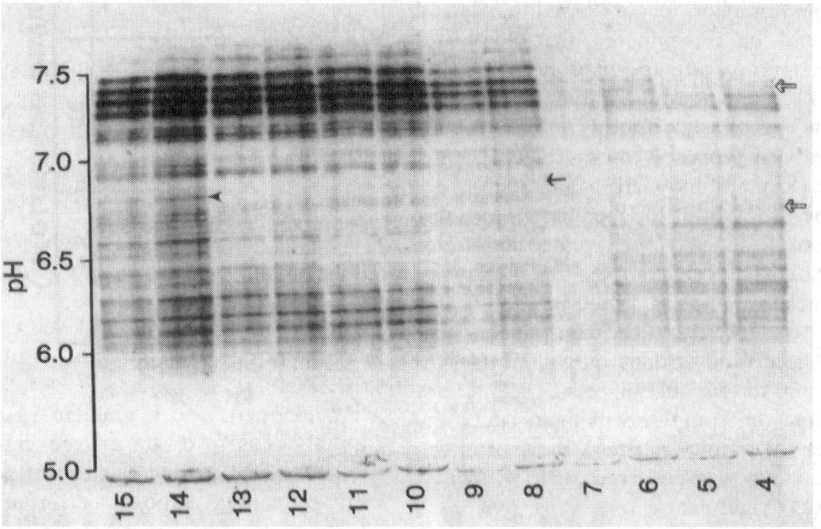

Figure 11.12a Schematic representation of an affinity immunoblot analysis of anti-micrococcal antisera. Antiserum was collected from immunised rabbits beginning on the 4th week of the immunisation programme and weekly intervals thereafter. The bleeding date (weeks) is shown on the abscissa. (Reproduced with permission from: Knisley and Rodkey (1988). *Electrophoresis*, **9**, 183–186.)

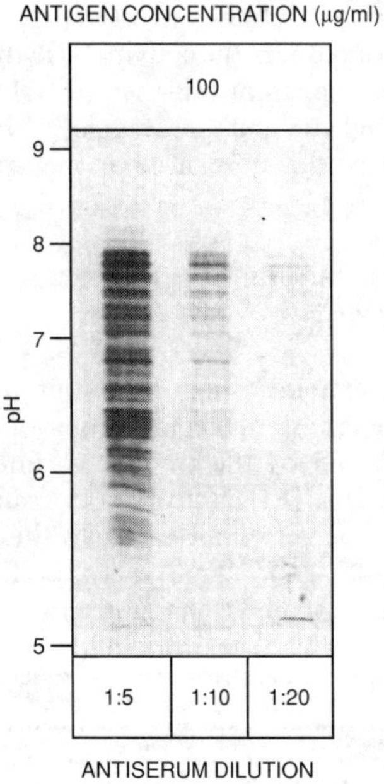

Figure 11.12b Schematic representation of the influence of antigen density on affinity immunoblot analysis. (Reproduced with permission from: Knisley and Rodkey (1986). *J. Immunol. Methods*, **95**, 79–87.)

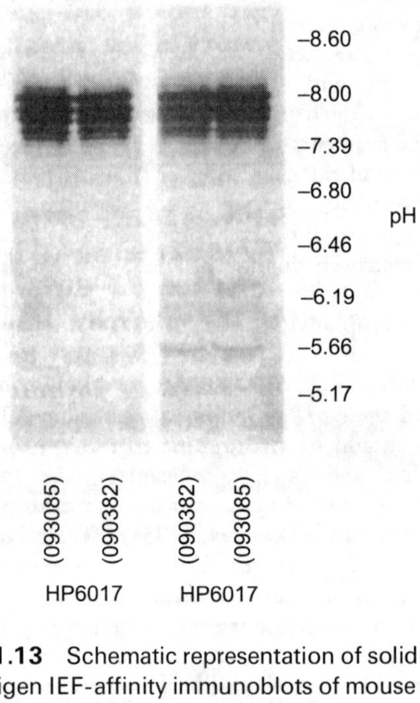

Figure 11.13 Schematic representation of solid-phase antigen IEF-affinity immunoblots of mouse anti-human IgG monoclonal antibody for determination of immunoreactivity and p*I* fingerprints. (Reproduced with permission from: Hamilton *et al.* (1987) *Hybridoma*, **6**, 205–217.)

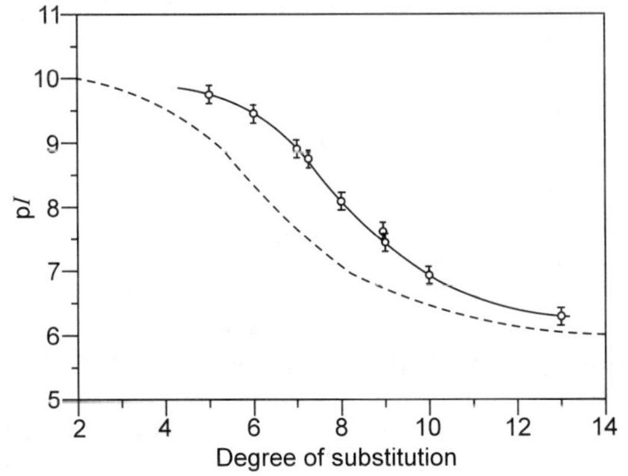

Figure 11.14 Calibration curve that relates the p*I* of avidin–PLP conjugates to the degree of substitution (○). Error bars indicate 1 SD ($n = 3$). The broken line is the calibration curve predicted by the theoretical model. (Reproduced with permission from: Barbarakis and Bachas (1991). *Clin. Chem.*, **37**, 87–90.)

Figure 11.13. With starting materials of this heterogeneity, the prospect of interpreting the IEF gels of mixed conjugation products is a daunting one.

Nevertheless some useful applications have been made in this field. Svensson (1973) used column IEF to determine the nature of the bonds formed when protein amines are coupled to cyanogen bromide activated carbohydrates: no significant change in protein pK_a was observed, a finding consistent with the formation of isoureas rather than the alternative carbamates or imidocarbonates.

Barbarakis and Bachas (1991) investigated IEF for the determination of incorporation levels in a model protein–ligand conjugate, between avidin and pyridoxal 5-phosphate (PLP). The conjugation was carried out *via* the amine groups of the protein, the Schiff base formed with the pyridoxal derivative at pH6 subsequently being reduced with cyanoborohydride.

These authors calculated the predicted effect of sequential substitutions on the p*I* of avidin, from its native value of 10. A sigmoidal curve was obtained (Figure 11.14) showing the p*I* dropping to around 6 as the PLP incorporation level reached 12–14, and the avidin amines were replaced with the less basic moieties of the pyridoxal.

A similar approach was employed by Pham *et al.* (1995) to determine the incorporation of a DTPA derivative into various proteins, in this case replacing their amines with strongly-acidic ligands. Once again, incorporations were confirmed spectrophotometrically, and, using the same type of plot as Barabarakis and Bachas, the results shown in Figure 11.15 were obtained. Although Pham *et al.* (1995) interpreted these results on the basis of a linear correlation, some evidence for a sigmoidal relationship can possibly be detected in the data. Electrophoretograms from this study are depicted in Figure 11.16; the heterogeneity of the Mabs used, as discussed above, is obviously a hindrance to accurate p*I* assignments.

Overall the IEF approach appears to offer some promise for conjugate characterisation, which has yet to be fully exploited.

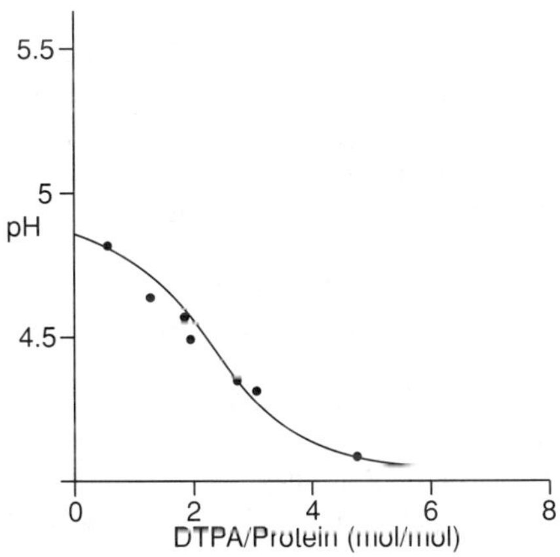

Figure 11.15 Calibration curve of HSA–DTPA derived from the IEF. (Adapted from: Pham *et al.* (1995). *Bioconjugate Chem.*, **6**, 313–315.)

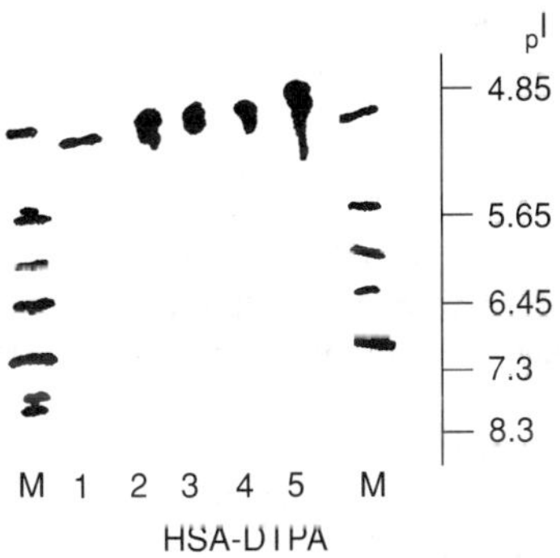

Figure 11.16 IEF analysis of HSA–DTPA conjugate. (Reproduced with permission from: Pham *et al.* (1995). *Bioconjugate Chem.*, **6**, 313–315.)

11.9 Other Applications of Gel Electrophoresis

11.9.1 Affinity electrophoresis

It is interesting to compare the various manifestations of electrophoresis with those of chromatography. Native electrophoresis effects separations primarily on the basis of the charge of the protein molecule, and can, to this extent at least, be equated to ion exchange chromatography. Carrying out the separation in a porous matrix allows separation on the basis of size to be achieved electrophoretically, equating quite closely to size exclusion chromatography (although the direct 'conversion' of mass into charge by SDS treatment is much more widespread). But what of the other major branches of protein chromatography covered in Chapter 10, hydrophobic interaction, reverse phase and affinity chromatography?

In fact it is also possible to carry out electrophoretic equivalents of these techniques too, by incorporating components in the gel which possess an affinity for the protein solute(s) of interest. This constitutes affinity electrophoresis (Hořejší, 1984; Righetti *et al.*, 1986; Andrews, 1986).

If it is desired to investigate hydrophobic interactions, for example, it is possible to incorporate hydrophobic groups into the gel, for example by using suitably-derivatised acrylamide monomers (Nakamura *et al.*, 1979; Chen and Morawetz, 1981). This allows an electrophoretic separation loosely comparable to that of HIC or RPC to be attempted.

However it is equally possible to use any other type of affinity ligand; this technique has been applied fairly widely, for example, using lectins (Righetti *etal.*, 1986), immobilised metal ions (Holmes *et al.*, 1992; Goubran-Botros *et al.*, 1992), and antibodies or antigens.

11.9.2 Two-dimensional gel electrophoresis

While a single electrophoresis operation can offer substantial information about a mixture of proteins or protein reagents, carrying out a subsequent complementary procedure can offer further insight still. This principle underlies the concept of two-dimensional gel electrophoresis (2DGE), in which the resolved product of the first fractionation becomes the sample for the second (Andrews, 1986; Dunn and Burghes, 1986)

There are many possible variations on this theme, but the best-known is that popularised by O'Farrell (1975), which elegantly combined IEF and SDS-PAGE to create a two-dimensional map of a protein population.

A map obtained for normal mouse serum (Manil *et al.*, 1986) is shown in Figure 11.17(a), displaying regions corresponding to IgG heavy and light chains, albumin, transferrin, and α-acidic proteins, along with a number of other minor components; an ascites sample (Figure 11.17(b)) shows large quantities of a specific MAb overlaid on the standard host proteins. Figure 11.17(c) and 11.17(d) display the separation of a (different) MAb from other ascitic components by Protein A chromatography: Figure 11.17(c) shows the unbound fraction at pH 8.5, which contains virtually no IgG, while Figure 11.17(d)) shows highly-purified IgG eluted from the column at pH 5. This approach, with simultaneous assessment of charge and size changes, appears to have a good deal of potential for the characterisation of conjugate mixtures.

Numerous other manifestations of 2DGE have been employed for various purposes, many to resolve complex mixtures such as ribosomal proteins: basic native PAGE followed by acidic native PAGE (Kaltschmidt and Wittmann, 1970); acidic native PAGE followed by SDS-PAGE (Martini and Gould, 1971); SDS-PAGE before and after cleavage of linked subunits (Wang and Richards, 1974; Traut and Kenny, 1977), etc. New applications continue to be reported (Pacher *et al.*, 1996; Wilkins *et al.*, 1996). The possibilities of this approach are extremely broad, and 2DGE seems to offer a largely unexploited potential for the characterisation of protein conjugates.

11.10 Capillary Electrophoresis

As discussed in Section 11.5, capillary electrophoresis (CE) presents a radically different vehicle for the fractionation of protein reagents, and the scale is ideally suited to analytical applications.

In its simplest format CE consists simply of a buffer-filled tube in an electric field, which allows approaches to be applied which correspond to the CZE[1] and MZE forms of gel electrophoresis. CZE approaches have been used to resolve fluorescent-labelled antigens from complexes with their specific antibodies, for example (Schultz and Kennedy, 1993; Chen and Pentoney, 1994). MZE methods were in use as early as 1984 for the characterisation

[1] Unfortunately the abbreviation CZE is also now being applied to capillary zone electrophoresis, as well as its original use for continuous zone electrophoresis (regardless of the format used). Only the older application will be used here.

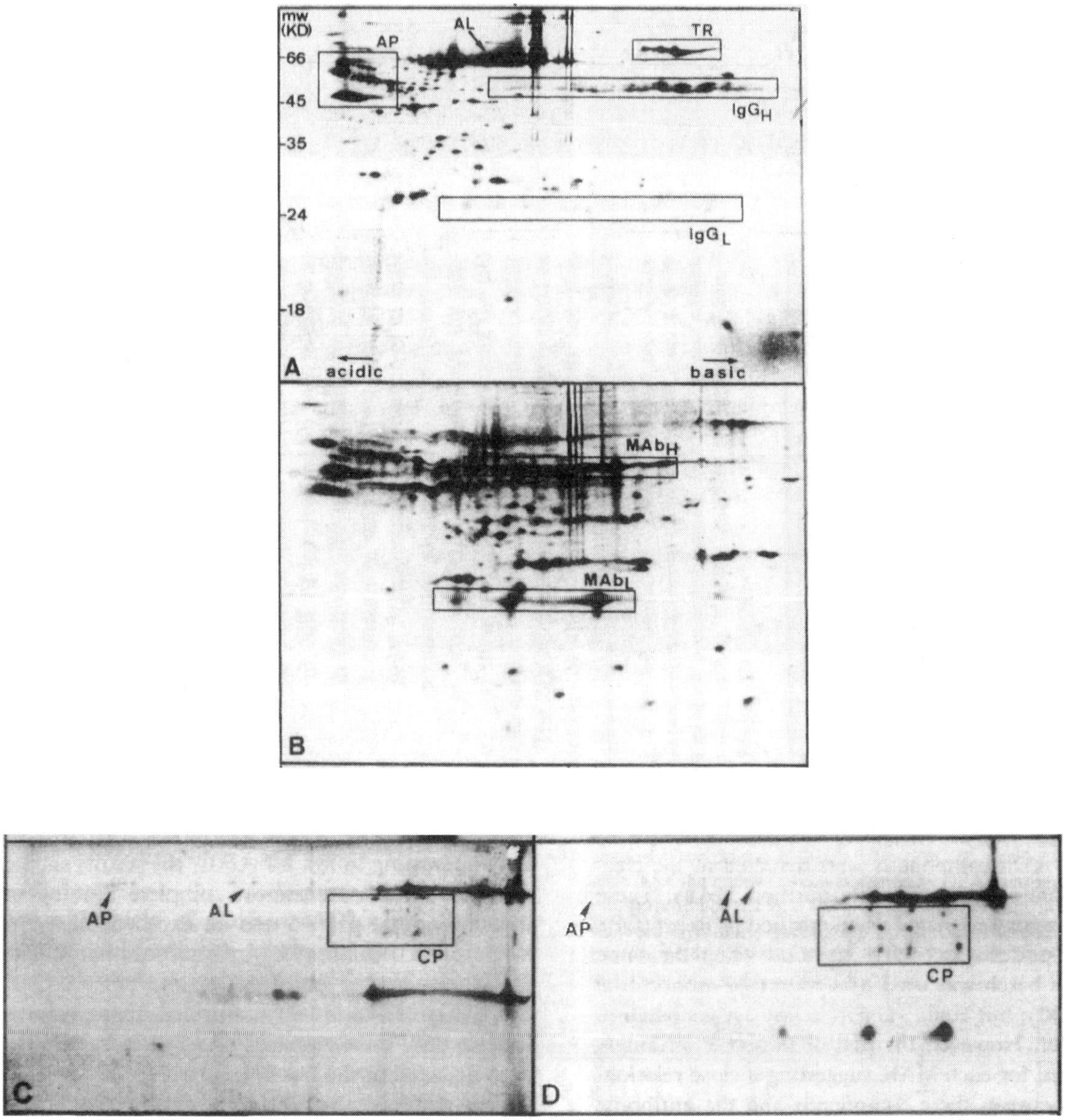

Figure 11.17 2DGE patterns of normal mouse serum (a) and crude ascites (b). Separation of a different Mab from other ascitic components by protein A chromatography is shown in (c) and (d). AL, albumin; TR, transferrin; AP, α-acidic protein; IgG$_H$, IgG heavy chains; IgG$_L$, IgG light chains. (Reproduced with permission from: Manil *et al.* (1986). *J. Immunol. Methods,* **90**, 25–37.)

of protein conjugates (Einarsson *et al.*, 1984), capillary isotachophoresis was used to determine the degree of substitution of albumin with fluorescein, trinitrobenzenesulfonic acid and PEG, higher incorporations leading to an increase in the zone widths observed (Figure 11.18).

However, packing the capillary with a gel opens up the possibility of CE-based equivalents of techniques such as IEF (Yang *et al.*, 1996a) and SDS-PAGE (Cohen *et al.*, 1987), the latter having attracted particular attention. This approach (SDS-CAGE) has been applied to conjugate characterisation: Liu *et al.* (1995)

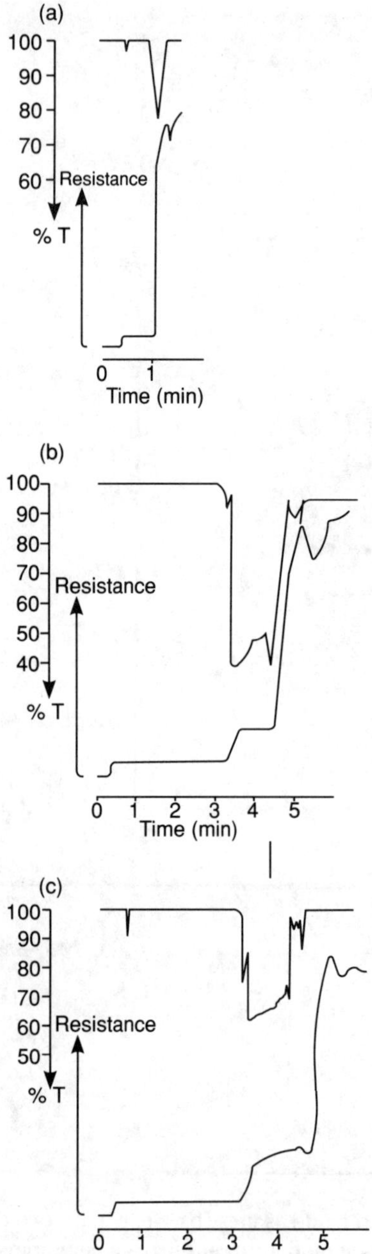

Figure 11.18 Analytical isotachophoretic patterns of mPEG–DA conjugates. (a) Electrolyte system; (b) DA; (c) DA $(mPEG)_{12}$; (d) DA $(mPEG)_{22}$; (e) $DA(mPEG)_{36}$. T_{280}, transmission at 280 nm; R, increasing resistance. (Reproduced with permission from: Einarsson *et al.* (1984). *J. Chromatogr.*, **284**, 143–147.)

resolved Mab-doxorubicin immunoconjugates under denaturing and reducing conditions, observing various combinations of conjugated heavy and light chains. UV spectrophotometry was used to detect all protein species, while fluorimetry picked up only those components bearing the fluorescent doxorubicin moiety.

The incorporation of soluble polymer-bound affinity ligands into the capillary allows separations to be carried out which are analogous to affinity chromatography (Shimura and Kasai, 1997). Haupt *et al.* (1996), for example, used immobilised metal-ion affinity capillary electrophoresis (IMACE) to resolve derivatised chymotrypsin components.

An area of still further potential within the CE format, as yet principally applied to small molecules, is to carry out chromatographic separations within the capillary, using the electroosmotic flow to provide rapid run times. The first descriptions of this type employed a form of liquid–liquid chromatography, termed micellar electrokinetic chromatography, in which solutes are partitioned between the eluent and detergent micelles (Terabe *et al.*, 1985). Under electrophoretic conditions the micelles are slowed down compared to the water phase, and components are fractionated on the basis of their relative residence times inside and outside the micelles.

Electrochromatography techniques have provided separations of prodigiously high efficiency for organic compounds, with plate numbers as high as 40,000,000 plates per metre. Although such approaches are still in their infancy, it seems very likely that these too will be developed to achieve high-resolution analysis of protein reagents. Commercial companies such as Hypersil (Runcorn, UK) are now marketing capillaries with hydrophobic surfaces which should obviate the need for micelles in capillary electrochromatography.

Overall, capillary electrophoresis techniques seem destined to become more and more widespread in the characterisation of protein conjugates. CE offers significant advantages in terms of small sample size, high resolution and rapid run times, and the development of the instrumentation is now, in contrast to earlier manifestations, reaching a point where there is a high level of user-friendliness.

11.11 Immunoelectrophoresis and Other Immunochemical Techniques

One specialised but important category of techniques should be mentioned briefly, which are based on the interactions of antibodies and antigens. These interactions can of course be extremely specific, differentiation of proteins differing by only a single amino acid residue being achievable with a suitable antibody preparation even before the advent of monoclonal methods (Noble *et al.*, 1972).

There are many ways in which the antibody antigen reaction can be employed in an analytical method. They can be applied to the characterisation of coupled protein reagents, for example to determine whether a derivatised protein has retained its antigenicity, or to confirm that a particular antigenic group (such as a hapten) has been introduced.

Some of these are based on the precipitin reaction: within certain ratio ranges, antibodies and their specific antigens will form insoluble immune complexes which, for example, are readily detectable within a gel. Although a number of different immunoprecipitation formats have been described, they all rely on the movement of an antibody into a matrix containing specific antigen, or *vice versa*: in either case the formation of a precipitin deposit at their point of contact is indicative of a specific immune reaction, and to some extent this can be carried out in a quantitative fashion.

Other immunochemical approaches rely on different means to determine the extent of antigen–antibody interactions.

11.11.1 Immunodiffusion techniques

Probably the simplest embodiments of immunochemical analysis rely on the diffusion of the antibody and/or antigen through a suitable matrix such as an agar gel.

In Ouchterlony double diffusion (Kaminski and Ouchterlony, 1951), a central reservoir of antiserum is allowed to diffuse outwards to meet a number of peripheral antigen samples diffusing towards it (Figure 11.19). Specific immune reaction is indicated

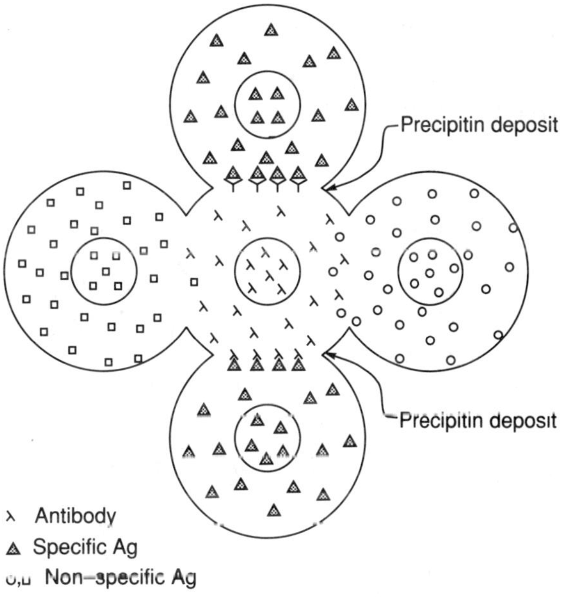

Figure 11.19 Ouchterlony double diffusion principle. A precipitin deposit is only formed where the diffusing antibody meets a specific antigen.

by the formation of a precipitin deposit at the intersections of the diffusion zones. This simple technique has been used since the early days of antibody conjugation to confirm the retention of immunoreactivity. Hsu *et al.* (1963), for example, demonstrated that anti-pneumococcal antibody retained its activity on conjugation to ferritin; precipitin lines were observed when the conjugates were diffused against specific pneumococcal antigens.

Alternatively, this approach has been used to demonstrate the presence of one or more specific proteins in a conjugate. For example, by diffusing ferritin–rabbit IgG[1] conjugates against specific antisera, Vogt and Kopp (1964) were able to demonstrate the success of the incorporation reaction (Figure 11.20). The technique is semi-quantitative: in this example it can be seen that purification significantly reduces the amount of IgG in the conjugate, but has little effect on the ferritin level. This suggests that the purification is removing free IgG from the conjugate; free ferritin is either not present or not resolved.

A more quantitative approach is that of single radial immunodiffusion (RID), first described by Mancini *et al.* (1964, 1965): in this technique the antigen is allowed to diffuse radially from a central source into a matrix containing free antibody. (*cf.* thin-layer affinity chromatography, Section 11.4.2). Complexes start to form, and at a particular ratio become insoluble. Therefore a precipitin ring is formed, the diameter of which is directly related to the amount of antigen in the sample. The mathematics of this relationship have been considered in some detail (Mancini *et al.*, 1965; Vaerman *et al.*, 1969a) and the approach can be

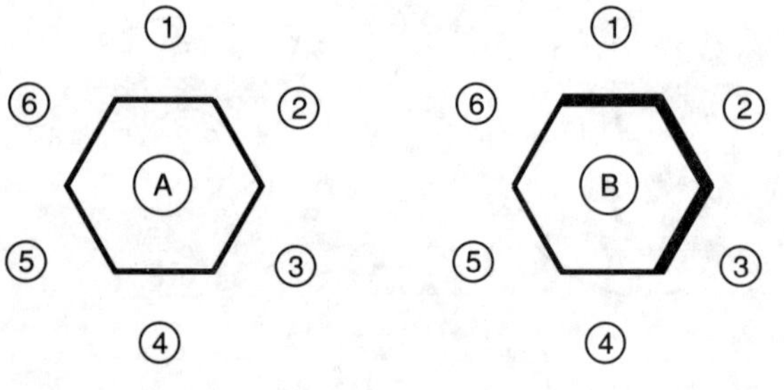

Figure 11.20 A, antiserum to ferritin; B, antiserum to rabbit IgG; 1–3, crude ferritin–rabbit IgG conjugates; 4–6 purified ferritin–rabbit IgG conjugates. Each conjugate shows reactivity to both antiferritin and anti-(rabbit IgG) antisera, confirming that conjugation has occurred.

[1] Note that this approach can be used for any conjugate where both partners are antigenic – not just antibody conjugates.

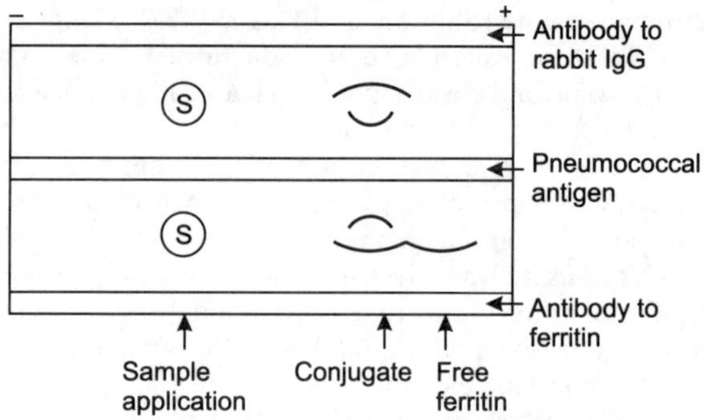

Figure 11.21 Demonstration of the properties of a rabbit IgG–ferritin conjugate by immunodiffusion. See text for details.

employed to quantify either the amount of antigen in the sample, or the titre of the antibody used (Becker, 1969).

Clearly, it is equally possible to reverse the design of RID experiments, such that the antibody is diffused into antigen-containing gels (Stiehm, 1967; Vaerman *et al.*, 1969b; Reimer, 1970).

RID is still very commonly used in the estimation of antibody or antigen concentration, but may also offer information on the immunoreactive concentration of such species within a protein conjugate.

In a two-dimensional approach analogous to those described in Section 11.9.2, immunodiffusion has been carried out on mixtures already fractionated by electrophoresis. First described for the analysis of serum proteins (Grabar and Williams, 1953), this approach has also been employed to characterise protein conjugates. Samples are electrophoresed between long 'troughs' containing the desired antibodies or antigens; once electrophoresis has been completed, immunodiffusion identifies the position of the various immunoreactive species. Hsu *et al.* (1963) used this approach to demonstrate the presence of ferritin and rabbit IgG in a conjugate, as well as the ability of the IgG to detect its specific pneumococcal antigen (Figure 11.21). Note that this analysis also showed the presence of free ferritin, which reacted with neither the specific antigen nor the antiserum to rabbit IgG.

11.11.2 Immunoelectrophoresis techniques

The technique just described has sometimes been referred to as immunoelectrophoresis, but is more properly viewed as the application of immunodiffusion as a detection technique following fractionation of a mixture by electrophoresis. Another class of

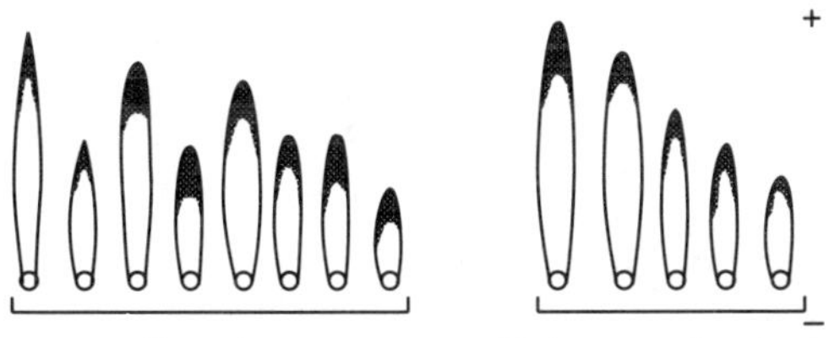

Figure 11.22 Principle of rocket electrophoresis. Antibody or antigen from the sample is driven electrophoretically through a gel containing its specific ligand; precipitin deposits from 'rockets' whose dimensions are related to the quantity of the specific component and can be roughly quantified by comparison with standards.

techniques qualify more convincingly for the title; in these, electrophoresis is carried out in a matrix already containing the relevant antibody or antigen, which therefore actually has an impact on the mobility of immunoreactive components. This is a branch, by far the most widely-encountered, of affinity electrophoresis (Section 11.9.1).

These techniques are akin to radial immunodiffusion, in that they involve the formation of a precipitin rim whose extent is quantitatively related to the amount of antibody (or antigen) in the sample. However the movement of the sample components is driven electrophoretically through the gel, rather than by radial diffusion. Therefore the rim is not circular, but instead appears as a characteristic 'rocket' shape (Figure 11.22). This approach is therefore generally referred to as rocket immunoelectrophoresis; it was first introduced for the quantification of antigens in an antibody-containing gel by Laurell (1966), and for the quantification of antibodies in an antigen-containing gel (reversed rocket immunoelectrophoresis) by Axelsen and Svendsen (1973). In either case the height of the rocket is roughly proportional to the concentration of the component in the sample, offering a fairly quantitative method.

This method is based on the immune component in the gel remaining essentially immobile during the electrophoresis of the sample, or at least moving much more slowly than the antigen. Too much movement of the 'immobile' component may lead to a loss of sharpness in the rocket obtained. This therefore necessitates some care in identifying appropriate conditions of pH and ionic strength, e.g. Axelsen and Svendsen (1973) ran their reversed rocket immunoelectrophoresis of anti-albumin antibodies at pH 4.85, the p*I* of albumin. Under these conditions the majority of the antibody molecules in the antiserum are positively charged and move sharply towards the cathode.

It may be more difficult to apply immunoelectrophoresis techniques where the p*I* of the antigen and antibody are similar. Chemical derivatisation of the immobile component to alter its p*I* is a possibility, but (particularly for antigens) is fraught with the danger of altering the specificity of the immune reaction.

Crossed immunoelectrophoresis (Ressler, 1960; Laurell, 1965) is a variant of 2D-gel electrophoresis (Section 11.9.2) in which a simple electrophoretic separation is carried out in the first dimension, and rocket immunoelectrophoresis in the second; this can be a powerful method for identifying immunoreactive components from a mixture. Crossed immunoaffinoelectrophoresis (Bøg-Hansen, 1973) is a sub-variant in which the first dimension comprises an affinity electrophoresis separation.

Many reviews of immunoelectrophoretic methods have been published (Andrews, 1986; Heegaard and Bøg-Hansen, 1986; Laine, 1992a–d).

11.11.3 Affinity constant determination

Whenever an antibody is incorporated into a conjugate of some kind, the retention of the antibody's binding behaviour is likely to be a parameter of interest, and its evaluation may form part of the characterisation of the conjugate. The affinity constant of a derivatised antibody for its specific antigen can be determined in the same way as for any other ligand–binder interaction. However, unless the conjugate is fully homogeneous with respect to both stoichiometry and position of derivatisation, the results obtained are likely to be complicated by the variations in behaviour of the component species. One should not necessarily expect, therefore, to obtain 'clean' results from analyses of this type, but they may still offer useful information, particularly when comparing one conjugate to another.

The methodologies required here have already been discussed in Sections 4.2.2 and 4.2.3. It is worth recalling the pitfalls of such calculations, and noting the added complexity of analysing a mixed conjugate population. The more a conjugate can be fractionated before such analyses are carried out, the more straightforward it will be to interpret the data obtained.

11.11.4 Immuno-dot blots and slot blots

In a much simpler way, the immunological activity of conjugates can be compared by employing them in the formation of an immobilised antibody–antigen

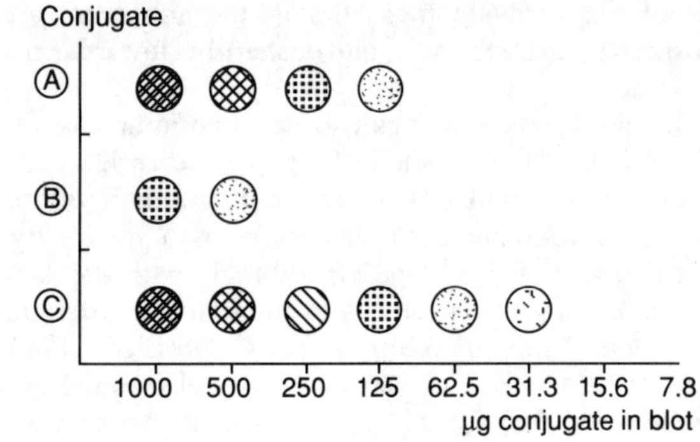

Figure 11.23 Dot blot.

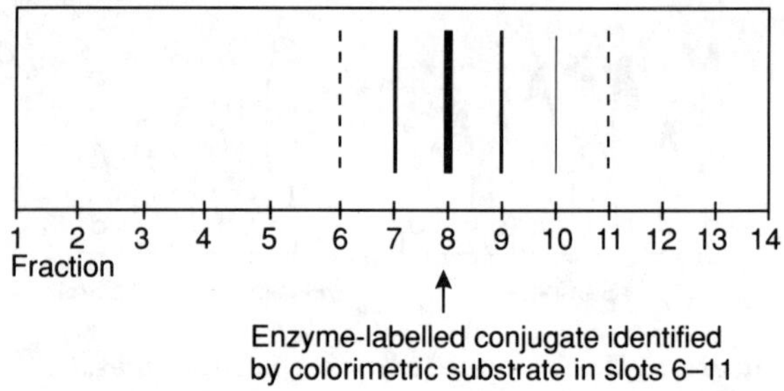

Figure 11.25 Slot blot.

complex, visualised by a suitable means of detection. This concept has already been met for selective visualisation of mixtures fractionated by electrophoresis (Western blotting, etc.); immuno-dot blotting is essentially the same approach without the electrophoresis (Towbin and Gordon, 1984).

In one format, specific antigen is blotted onto a membrane which is then blocked to prevent further non-specific protein binding. Samples of conjugate are then allowed to react with the membranes, which are washed thoroughly to remove unbound conjugate. The presence of any specifically bound conjugate is determined by a suitable visualisation technique. This approach can be rendered semi-quantitative by spotting different quantities of antigen onto the membrane, and establishing a detection limit for the conjugate in question (Figure 11.23). In the example shown, Conjugate C clearly has the best immunoreactivity for the specific antigen.

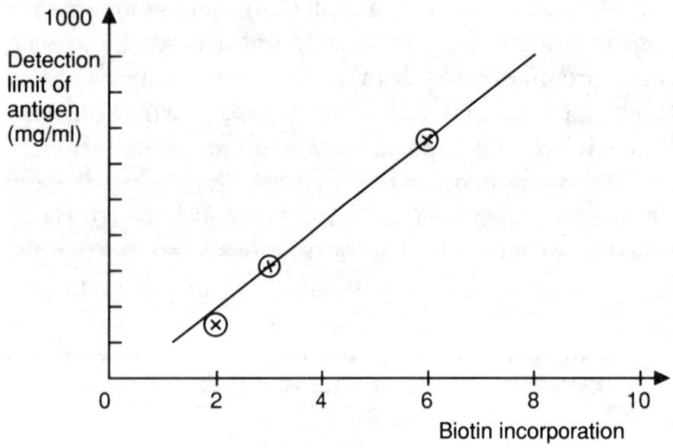

Figure 11.24 Semi-quantitative application of immuno blot to determine conjugate immunoreactivity. (Adapted with permission from: Mirralles *et al.* (1991). *J. Immunol. Methods,* **140**, 191–196.)

Miralles *et al.* (1991) described a semi-quantitative application of this type for assessing the immunoreactivity of biotinylated Mabs: sequential dilutions of specific antigen were spotted onto nitrocellulose, blocked with BSA, incubated with the biotinylated Mab conjugates, washed, and the conjugate visualised with alkaline phosphatase–streptavidin and a BCIP–tetrazolium substrate mixture. The detection limits observed indicated that the immunoreactivity of the conjugate was impaired by increased biotinylation *via* amines (Figure 11.24); similar analysis of Mabs biotinylated on the carbohydrate regions revealed a greater retention of reactivity.

Slot blots are essentially identical to dot blots in everything but shape: the solution is applied to the membrane *via* a device which allows a linear aliquot of sample to be immobilised (Ludeke, 1992). This can be a convenient format for comparison of a number of samples, for example fractions from a chromatographic separation. A typical slot blot is shown in Figure 11.25.

11.11.5 Immunoassays

A final way in which immunochemical techniques can be applied to conjugate characterisation is to use one or other of the conjugate components as the analyte in a specific test such as an immunoassay. These tests are of course in widespread use for the quantification of protein and non-protein analytes in applications such as clinical diagnostics and environmental monitoring.

If a suitable immunoassay is available, it can occasionally be used to confirm the presence of a specific molecule in a conjugate. However, this is strongly dependent on the design of the assay, and results must be interpreted with great caution. Also, such assays are generally designed for the detection of free analyte in a specific medium, and it is not generally possible to use them to quantitate a conjugate. This approach, then, is of very limited value—as a trouble-shooting aid, for example.

More usefully, however, it is quite possible to build protein conjugates into a simple, specific assay[1], and this may provide a useful and highly quantitative means of comparing the performance of examples prepared by different methods. The scope of such approaches is very wide, and is really limited only by how much effort one is prepared to invest in developing an appropriate method.

The design of the assay will determine whether it provides information on the presence or functional activity of one or other component, or a combination of such data.

Paxton *et al.* (1992), for example, used an enzyme immunoassay to demonstrate that the immunoreactivity of a carboranyl peptide–Mab conjugate was unchanged from that of the free antibody (Figure 11.26(a)). Sequential dilutions of both free and conjugated Mab gave identical dose–response curves (Figure 11.26(b)).

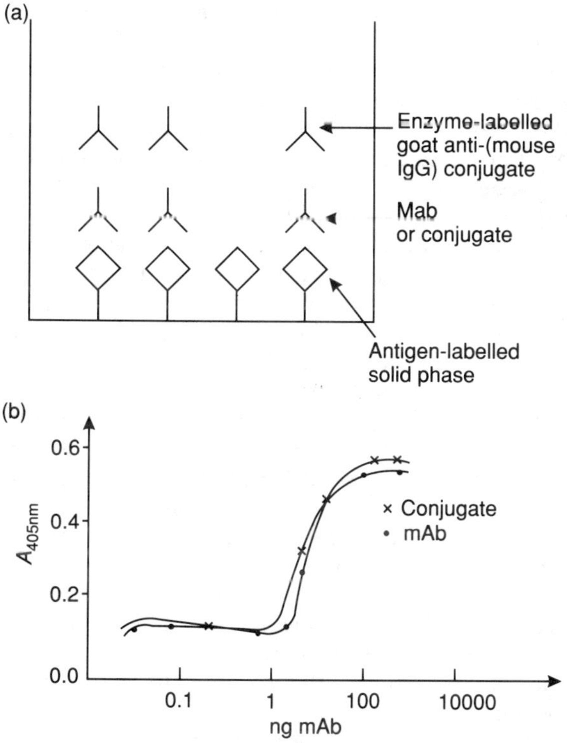

Figure 11.26 Demonstration of conjugate reactivity using an immunometric assay (a). Conjugated and unconjugated Mabs gave identical dose–response curves (b) using a colorimetric substate. (Reproduced with permission from: Paxton (1992). *Bioconjugate Chem.*, **3**, 241–247.)

[1] Of course in diagnostic applications this may be the intended function of the conjugate, rather than a test designed for characterisation purposes.

Table 11.3 Comparison of IgG–enzyme conjugates in terms of detection limit of the enzyme by a colorimetric assay, and detection limit of AFP in an immunoassay using the same signal system (Porstmann *et al.*, 1985)

Enzyme	Enzyme detection limit (ng/l)	AFP detection limit (μg/l)
HRP	28	1.0
AP	70	5.2
β-Gal	1500	5.2

Porstmann *et al.* (1985), on the other hand, carried out a colorimetric immunoassay for α-1-fetoprotein (AFP) using IgG conjugates coupled to HRP, AP and β-Gal, and obtained AFP detection limits of 1.0, 5.2, and 5.2 μg/l respectively (Table 11.3). However the similarity of behaviour of AP and β-Gal in this respect masked large differences in their detection limits in terms of enzyme activity; in other words the final immunoassay was relatively insensitive to variations in the specific activity of the conjugates.

Because immunoassay designs can readily incorporate a labelled solid phase, they can also be employed to characterise the amount and activities of protein immobilised by a coating procedure. Again the nature of the immunoassay will determine the information provided: probing an immobilised mouse Mab with an anti-(mouse IgG) conjugate will quantify total, largely intact antibody. Using a specific antigen conjugate, however, will quantify only immunoreactive antibody, which may well only be a small fraction of the total quantity. Both pieces of information may be of interest, and a combination of immunoassay approaches is often rewarding.

The determination of binding capacity for a solid phase (e.g. a microtitre well) by immunoassay has been subject to fairly thorough mathematical analysis (Underwood and Steele, 1991; Wilkins Stevens and Kelso, 1995; Wilkins Stevens *et al.*, 1995).

11.12 Other Techniques Based on Specific Protein Functions

Antibodies are just one category of proteins, their immunoreactivity just one form of protein function, and the formation of antibody–antigen complexes just one branch of protein–ligand binding. Other proteins also have specific properties which can be exploited in the characterisation of their conjugates, either affinities for specific ligands, or other readily detectable attributes such as enzyme activity.

11.12.1 Techniques based on other ligand–binder interactions

For any ligand–binder interaction it is possible to apply techniques analogous to those described in Section 11.11, with the proviso that not all protein–ligand interactions will conveniently result in the formation of insoluble complexes in the manner of the precipitin reaction. Therefore techniques akin to immunodiffusion and immunoelectrophoresis may be impracticable for many ligand–binder pairs.

However, affinity constant determinations, dot blots, and immunoassay analogues can all be applied to other ligand–binder couples, in exactly the same fashion as for the immunochemical approaches described above.

As usual the biotin–avidin interaction is one of the most commonly encountered couples in this application; biotin is a small enough molecule to be feasible for deliberate introduction as a label, and can subsequently be detected by a suitable avidin reagent. For example Jerry (1993) incorporated biotinylated peptides into BSA or KLH immunogens, then determined the biotin incorporation by an avidin-based immunoassay analogue. The immunogens could also be visualised in a slot-blot using an avidin–HRP conjugate with a colorimetric substrate.

Biotin labels have also been incorporated in protein immobilisation experiments, where they provided a more stable reagent than the radiolabelled equivalent (Underwood and Steele, 1991; Wilkins Stevens and Kelso, 1995). Again enzyme-labelled avidin or streptavidin reagents were used to quantify the bound label.

11.12.2 Determination of enzyme activity

Where an enzyme is incorporated in a conjugation reaction, a very obvious and informative characterisation technique is to quantify the catalytic activity which has been retained in the final product. This can determine the degree to which the enzyme has been damaged by the conditions it has encountered during the conjugation process—generally a more critical parameter than the simple conjugate stoichiometry.

Such determinations can employ any quantitative procedure based on the catalytic properties of the enzyme involved, although those generating a coloured or fluorescent product are most commonly encountered. Substrates for some of the most commonly encountered enzymes are discussed in Chapter 3.

It is worth reiterating here the usual caveats to such analyses: enzyme activities may differ considerably from one substrate system to another, and, even with a common substrate, may be significantly affected by small changes in key variables of the reaction environment (e.g. pH, temperature). Therefore it is virually essential to run appropriate standards alongside the conjugate, typically the native enzyme and perhaps a previously prepared reference conjugate. Similarly, comparisons between the retained activity of a number of conjugates are best made in a single experiment. Comparison of a number of such studies done at different times is most relevant if the retained activity is expressed as a percentage of the initial value.

This raises another issue which can hamper this essentially simple methodology. In order to compare the enzyme activity of the conjugate with native enzyme, it is necessary to know its concentration in the sample with a reasonable degree of accuracy. This must be determined by an appropriate non-functional technique such as spectrophotometric absorbance, with or without chromatography. Even where an enzyme possesses a conveniently unusual chromophore (e.g. the absorbance of HRP at 403 nm), it must be borne in mind that the extinction coefficient may have been affected by the chemistry of derivatisation. This can therefore lead to an error in the determination of retained activity.

Steps can be taken to minimise the effect of such changes, for example by establishing the new extinction coefficient on an empirical basis. One method for estimating such values is to carry out the chemistry on a radiolabelled sample of enzyme, which can therefore be quantified independently of spectrophotometric changes. Once the extinction coefficient has been established in this way, it can be employed to determine enzyme concentrations in conjugates without the need constantly to resort to radiolabelled materials.

Where both partners in a conjugate possess similar spectrophotometric properties, quantification is more difficult. A combination of fractionation and characterisation techniques may be required to provide an accurate non-functional assessment of enzyme concentration. For everyday applications, compromises will often be made to yield a usable value. For example it may be assumed that there is no change to extinction coefficients on conjugation, and the mean stoichiometry of the conjugate may be estimated from a relatively simple technique such as SEC or SDS-PAGE.

Combining these pieces of information allows the enzyme concentration to be estimated from the total absorbance of a conjugate solution at an appropriate wavelength, typically 280 nm.

The usual statistical limitations of analytical techniques are, of course, relevant here too. The figure obtained for retained enzyme activity is subject to imprecision both from the estimates of enzyme concentration, and from the assessment of enzyme activity; both of these should be understood when interpreting comparative data.

In summary, retained enzyme activity is such a useful parameter that it is worth coping with the complications of estimating the enzyme concentration in the conjugate, and the statistical considerations. Even if a simplified approach is applied, valuable comparisons can still be made between one conjugate and another. Ford *et al.* (1978), for example, used a colorimetric substrate to compare the retention of AP activity during glutaraldehyde coupling to IgG, over a range of glutaraldehyde concentrations and reaction times (Figure 11.27). These data are obviously very powerful in guiding the choice of reaction conditions.

While simple comparative assessments under fixed conditions are most frequently encountered, it is quite possible to assess conjugates in more depth using the standard approaches for quantitation of enzyme activity. Kraehenbuhl *et al.* (1974) for example plotted Lineweaver–Burk plots for a microperoxidase conjugate, compared to the microperoxidase itself

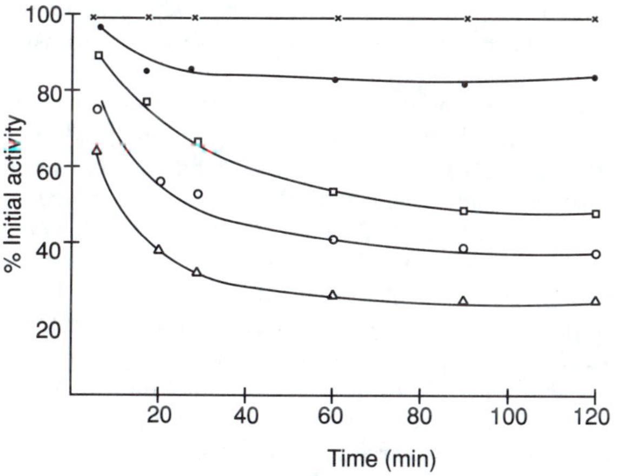

Figure 11.27 Loss of enzymatic activity of alkaline phosphatase–goat anti-human IgG conjugates as a function of time of reaction with glutaraldehyde and its concentration. (Reproduced with permission from: Ford *et al.* (1978). *Immunochemistry*, **15**, 237–243.)

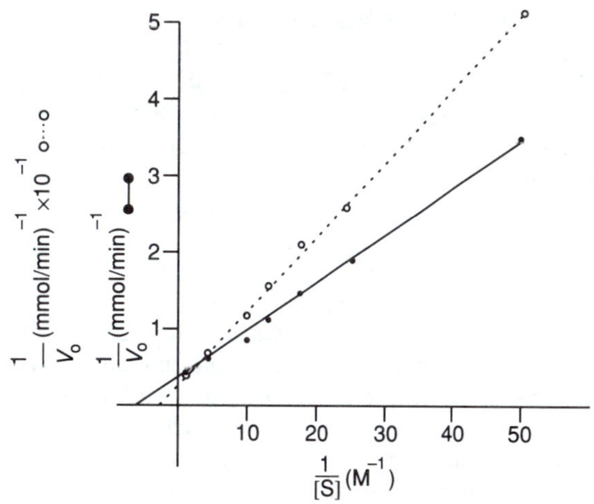

Figure 11.28 Lineweaver–Burk plot for the peroxidatic reaction catalysed by 8-MP (●——●) or by 8-MP-Mab (o · · · o) at pH 7.0 and 25 °C. (Reproduced with permission from: Kraehenbuhl *et al.* (1974). *J. Exp. Med.*, **139**, 208–223.)

(Figure 11.28). These revealed that conjugation caused an increase in K_m from 0.2 to 0.4 M, and a reduction in V_{max} from 4 to 0.4 mmol/min per mg.

11.13 Other Non-fractionating Characterisation Techniques

The bulk of this chapter has dealt with analytical techniques which include an element of fractionation of the sample – most notably those based on chromatography or electrophoresis. The last two sections, however, have included some methodologies (such as immunodiffusion or determination of enzyme activity) which do not incorporate significant physical fractionation of sample components: they rely instead on the specificity of the detection procedure to provide useful information.

There are a number of other frequently encountered techniques which are also non-fractionating, and provide information of a generally less specific nature. Sometimes (e.g. with a total protein assay or a measurement of spectrophotometric absorbance at 280–nm)these may yield a single piece of information describing the mean bulk properties of the sample. Sometimes (for example by consideration of absorbance ratios at different wavelengths) they may offer more than this.

11.13.1 Total protein assays

One of the most obvious aspects of characterisation, and yet one which is not always simple to achieve in practice, is the estimation of total protein content in a sample.

11.13.1.1 Gravimetry

In organic chemistry, a product is typically isolated in essentially pure form by removal of bulk solvents, followed by careful drying to constant weight at a suitable temperature. Yield can be determined very accurately by a simple weighing of the product. It is by no means impossible to apply an analogous approach to a protein solution (see for example Chattoraj and Bull, 1959), although there are certain important provisos. First, it is often inappropriate to remove such products completely from their aqueous milieu, as this may cause serious denaturation and loss of functional activity. Proteins have not generally been evolved to endure dry conditions, and even in organisms which provide exceptions to this rule (see Section 12.4), the proteins are never allowed to dry in a pure state.

For protein conjugates, therefore, it is generally only appropriate to apply such techniques to a sample of the product which can be sacrificed, the bulk product remaining safely in solution. This leads on to the second proviso: to achieve accurate gravimetric results it is really necessary to have at least a few milligrams of protein to work with. Such a loss may be significant in many small-scale experiments.

However, where sufficient sample is available, there is nothing to stop gravimetry being applied to conjugate yield determination. One technique is to dialyse the sample exhaustively into water, then remove the solvent by lyophilisation; a final secondary drying stage can be employed to minimise the final moisture content before weighing. It is necessary to take suitable precautions to avoid losses of the sample components during this process, for example by adequate blocking of dialysis tubing; working at a reasonably high protein concentration (say greater than 1 mg/ml) helps to avoid problems here too. Note that many protein products will precipitate from solution on dialysis into pure water, but this is unimportant so long as the precipitate is transferred quantitatively to the drying stage, along with the remaining liquid sample.

Although not very often encountered, this approach has advantages in providing reasonably unequivocal information. The main alternatives to be discussed below are all to some extent beset by protein-to-protein variations, which can distort a total yield assessment. Gravimetric determinations are relatively free of such uncertainties, and it is perhaps a little surprising that they are so rarely encountered.

11.13.1.2 Nitrogen determination

Again applying an approach derived from analytical chemistry, the nitrogen content of a sample can be determined, and used as the basis for an estimate of protein concentration.

The Kjeldahl method is the most commonly encountered; essentially this involves digestion of the protein with sulfuric acid to generate ammonia (often in the presence of a catalyst such as mercuric oxide or copper sulfate), distillation of the ammonia from the reaction vessel, and titration with acid. A nominal conversion factor of 6.25 is generally applied to convert the quantity of nitrogen determined to a protein yield.

Various embodiments of the Kjeldahl method have been described for protein determination over the years (Cullen and Van Slyke, 1920; Campbell and Hanna, 1937; Ma and Zuazaga, 1942; Miller and Houghton, 1945). It has been used for the determination of conjugate yields (Tawde *et al.*, 1963), but is rarely encountered in this application nowadays.

11.13.1.3 The biuret, Folin–Ciocalteu and Lowry reactions

A number of colorimetric methods were derived for protein determination between the 1920s and 1950s, based on various mechanisms. These are interlinked and the terminology is somewhat confusing.

Cu^{2+} ions in alkaline solution form coloured complexes with the peptide chains of proteins (the biuret reaction), forming the basis of a rather insensitive assay method for proteins (Robinson and Hogden, 1940; Kingsley, 1942; Gornall *et al.*, 1949). This method is sometimes encountered in relatively recent references (Hiratsuka, 1988). Note that biuret itself (Figure 11.30) is not employed in this reaction; it yields the same colour reaction with the alkaline copper solution, and in the past was sometimes used as a standard. The mechanism is not well understood, but probably involves reduction of Cu^{2+} ions to Cu^{+} following chelation by peptide groups in the protein (Chou and Goldstein, 1960; Smith *et al.*, 1985).

The Folin–Ciocalteu reagent, a complex mixture of phosphoric, molybdic and tungstic acid, was originally described by Folin and Denis (1912) as a colour reagent for the determination of phenols and related

$$3H_2O \cdot P_2O_5 \cdot 13WO_3 \cdot 5MoO_3 \cdot 10H_2O$$
$$3H_2O \cdot P_2O_5 \cdot 14WO_3 \cdot 4MoO_3 \cdot 10H_2O$$

Figure 11.29 Species observed in the Folin–Ciocalteu reagent.

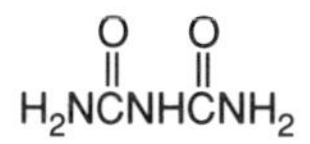

Figure 11.30 Biuret.

compounds. The complex has been shown to contain the species shown in Figure 11.29 (Peterson, 1983):

This is a redox reaction: the reduced form of the acid complex has a λ_{max} of 750 nm. This property was exploited to provide a colorimetric test for protein through the content of tyrosine and tryptophan (Wu, 1920, 1922; Folin and Ciocalteu, 1927), which achieved widespread use (Pressman, 1943; Heidelberger and MacPherson, 1943).

Herriott (1941) observed that the Folin–Ciocalteu reaction was strongly enhanced by the presence of a small amount of a Cu^{2+} salt, the Cu^+ generated by the protein interaction presumably acting as the reducing agent to drive the colour reaction[1]. This approach was developed by Lowry *et al.* (1951) into a highly successful method for protein determination. The Lowry method is basically a combination of two reactions of the Folin–Ciocalteu reagent (Chou and Goldstein, 1960). (1) There is a copper-independent reduction of the acid complex by the side-chains of tyrosine, tryptophan, and, to a lesser extent, cysteine and histidine. (2) There is a reduction of the acid complex mediated by copper complexation to the peptide chain, principally in the region of histidine, arginine or glutamic acid residues. This complexation is closely analogous to the biuret reaction.

The Lowry procedure is subject to a wide range of interferences from substances such as thiols, metal ions, carbohydrates, and buffer salts (Bradford, 1976; Peterson, 1983), and the modifications to avoid these problems (Bennett, 1967) are laborious. The Folin–Ciocalteu reagent is highly unstable in the alkaline matrix in which the reaction must be carried out, andthe response from protein to protein is also somewhat variable. Despite its complexity and limitations, the Lowry method was very widely adopted for protein determination, and has remained in use with a few modifications ever since (Bennett, 1967; Hartree, 1972; Peterson, 1983; Waterborg and Matthews, 1994).

[1] This is consistent with another parallel use of the Folin–Ciocalteu reagent and similar acid complexes, for the detection of reducing sugars (Folin and Wu, 1919; Nelson, 1944). Alkaline copper salts are employed in this reaction, presumably acting as electron transfer agents between the sugar and the acid complex.

11.13.1.4 The BCA method

Bicinchoninic acid (BCA) or 4,4′-dicarboxy-2,2′-biquinoline (Figure 11.31(a)), while not winning any prizes for easy pronunciation, is a sensitive and specific reagent for the determination of Cu^+ ions (Tikhonov and Mustafin, 1965) forming a 2:1 complex with λ_{max} 562 nm (Figure 11.31(b)). BCA has also been used in analytical techniques for reducing sugars (Gindler, 1970).

Smith *et al.* (1985) of the Pierce Chemical Co. applied this principle to the detection of the Cu^+ ions generated by the biuret reaction, thus offering an approach partially analogous to the Lowry method (Walker, 1994).

Yielding sensitivity down to the μg/ml level, BCA also offered a number of advantages over the use of Folin–Ciocalteu reagent in the Lowry method: both the working reagent and the chromophoric product are more stable, allowing quicker formats, and there is less interference from detergents (even SDS and Triton X-100 at the 1% level), and from denaturing agents. Interferences were noted from a number of other

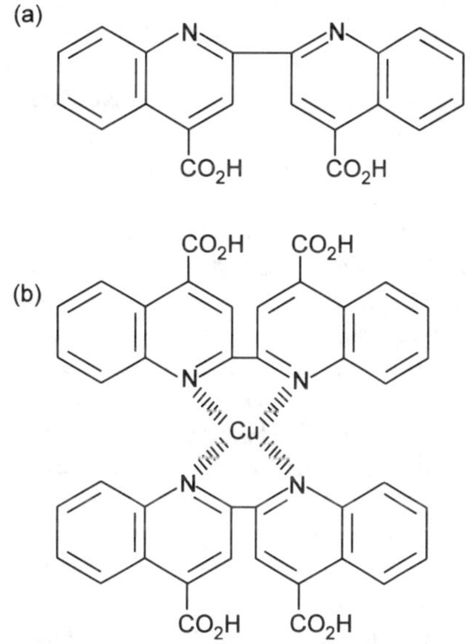

Figure 11.31 BCA and its complex formed with Cu^+ ions.

reagents, however, including EDTA (>10 mM), glucose, glycine, Tris (>0.25 M), ammonium sulfate, acetate (2 M) and phosphate (1 M). The effect of some of these interferents such as glycine, Tris or ammonium sulfate could be ameliorated by working at a strongly basic pH. Not surprisingly, reducing agents such as thiols are also powerful interferents (Pierce, 1995).

The protein-to-protein variability of the BCA method is similar to that of the Lowry approach. As well as providing a protein assay that has gained fairly widespread popularity, this approach has been adapted to on-line detection of proteins in HPLC (see for example Wolfe and Hage, 1994).

11.13.1.5 The Bradford method and other dye-binding approaches

The binding of dyes to proteins has been met in a number of applications in this book, including the detection of proteins on electrophoresis gels or blotting membranes (Section 11.5). The formation and measurement of protein–dye complexes in solution (or after precipitation) offers a route to protein quantification too. A number of dyes have been applied in this role over the years, including Orange G (Kaul *et al.*, 1969; Colenbrander and Martin, 1971), Amido Black (Schaffner and Weissman, 1973; Sherbon, 1974), and mercuric Bromophenol Blue (Swaminathan *et al.*, 1973).

The use of Coomassie Brilliant Blue (CBB) dyes as electrophoresis stains has already been discussed in Section 11.5.3. During work of this type with CBB G250, Reisner *et al.* (1975) noted that the use of an acid solvent converted the dye to its pale orange leuco form (λ_{max} 465 nm) but that the original intense blue form (λ_{max} 565 nm) was regenerated on binding to protein molecules.

Not only did this provide a gel staining method with significantly reduced background problems; it also prompted Bradford (1976) to adapt the technique for use as a protein assay. Using CBB G250 in an aqueous ethanol–phosphoric acid matrix, protein content could be quantified by a single addition step followed by determination of absorbance at 565 nm. This differentiation in signal between bound and unbound dye is a significant advantage over other dye-based approaches. The interaction is principally between the anionic form of the dye and arginine groups, although there is some interaction with histidine, lysine, tryptophan, tyrosine and phenylalanine residues (Compton and Jones, 1985).

The Bradford method is sufficiently sensitive to detect protein at μg/ml levels, giving a fourfold improvement over the Lowry method in this respect. The precision of the method also appeared superior, giving a tighter dose response curve from a range of protein samples.

No significant interference was observed from sodium, potassium or magnesium chloride, ethanol or ammonium sulfate; small effects were seen with Tris, acetic acid, mercaptoethanol, sucrose, glycerol, EDTA and traces of detergents, but these could be overcome by running appropriate controls; only rather high levels of detergent (e.g. 1% SDS or Triton X-100) caused insuperable problems. Overall the resistance of the Bradford method to interference is better than that of the redox-dependent approaches such as Lowry and BCA.

The high sensitivity and precision of the Bradford procedure, and its simplicity, have brought it widespread popularity; it was soon commercialised by Bio-Rad, and is consequently sometimes referred to as the Bio-Rad assay. It has remained one of the most popular techniques for protein quantification over the last twenty years, including countless determinations of conjugation yields (e.g. Gadkari *et al.*, 1985; Butler and O'Kennedy, 1994).

Nevertheless, there are still issues with protein-to-protein variation with this method, and modifications such as working at high temperatures have been described to alleviate this effect (Sharma and Tihon, 1988). These can cause problems with other interferences, however (Kruger, 1994), and the

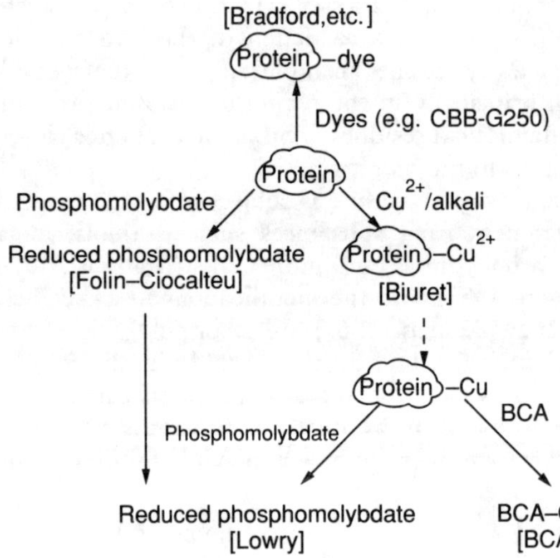

Figure 11.32 Protein determination methods and their inter-relationships.

original method remains popular. Modifications to the Bradford method to improve its performance continue to be reported, such as using wavelength ratios that offer greater linearity (Zor and Selinger, 1996).

The major methods for protein determination are summarised in Figure 11.32.

11.13.2 Ultraviolet–visible spectrophotometry

UV–visible spectrophotometry is a convenient, non-destructive technique which has found extremely widespread application for protein quantification. The UV–visible spectrum of a protein is determined primarily by the individual contributions of amino acid residues, plus any other chromophoric groups present, which may offer valuable specific absorbances at characteristic wavelengths.

Generally applicable techniques must exploit absorbances found in all proteins, however; a typical protein spectrum is shown in Figure 11.33. As can be seen, the strongest of these absorbances are those occurring in the 180–220 nm (far-UV) region, which are attributed to the peptide bonds (Goldfarb *et al.*, 1951). At 205 nm, for example, each peptide bond makes a contribution to the overall molar extinction coefficient[1] of around 3000 (Goldfarb *et al.*, 1951). This is not a particularly strong absorbance *per se*, but as there are hundreds or even thousands of such bonds in a single protein molecule, the overall value of ϵ may be 10^6 or higher.

It should be noted that any particulates contained in a sample solution will give rise to light scattering, which can interfere with spectrophotometry of solutes. Ideally sample filtration should obviate this problem, but scatter-correction methods based on measurement of multiple wavelengths can sometimes be employed (Owen, 1998).

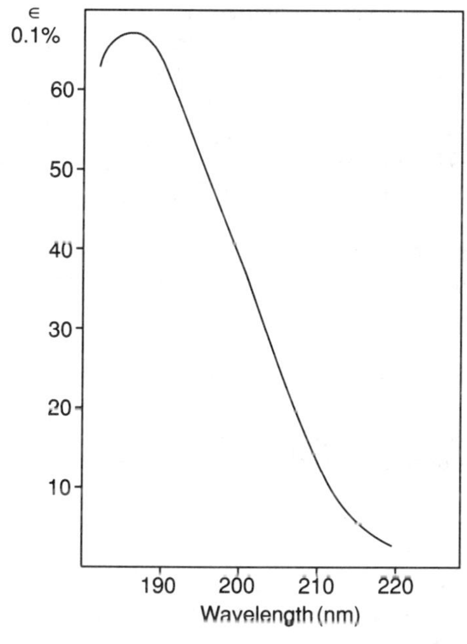

Figure 11.33 A typical UV–Visible spectrum of protein.

11.13.2.1 Protein determination by measurement of absorbance in the far-UV region

Measurement of far-UV absorbances has been used for determining protein concentration in a number of ways, e.g. A_{210} (Tombs *et al.*, 1959), $\Delta(A_{215} - A_{225})$ (Waddell, 1956), A_{191} (Webster, 1970) and A_{187} (Woods and O'Bar, 1970). Webster's study demonstrated that protein concentrations down to the microgram level could be determined from A_{191}, and, perhaps more importantly, that there was relatively little variation in absorbance over a range of seven different proteins. Woods and O'Bar demonstrated that operating at a low temperature (6 °C) provided still greater sensitivity.

Working at these wavelengths, then, might seem to offer the most attractive route to protein determination. However there are certain disadvantages to operating at such a low wavelength, in particular the high risk of interference from other non-protein species, and the relatively poor precision obtained. For these reasons, far-UV protein determination has failed to catch on significantly, despite its inherent sensitivity.

11.13.2.2 Protein determination by measurement of absorbance at 280 nm

Typically, spectrophotometric quantification of proteins is carried out at 280 nm, where the absorbance is due to residues such as tryptophan and tyrosine (Beaven and Holiday, 1952). Tryptophan

[1] The molar extinction coefficient, ϵ, is a concentration-independent measure of UV absorbance for a given substance. It is related to absorbance (A) by the equation $A = \epsilon c L$, where c is the molar concentration and L is the path length in cm (usually standardised as 1). Other types of ϵ may be expressed, such as $\epsilon_{0.1\%}$, the coefficient for a 0.1% solution; such expressions are convenient in some circumstances. All ϵ values given here are molar unless otherwise stated.

has two major absorbance maxima at around 288 nm ($\epsilon = 4600$) and 279 nm ($\epsilon = 5500$), essentially independent of pH. Tyrosine has a maximum at 275 nm ($\epsilon = 1340$) under acid conditions and at 294 nm ($\epsilon = 2330$) in base (Little and Donahue, 1968). While measurements can be made at 280 nm with high precision, and avoiding many non-protein interference issues, the inherent sensitivity is poor due to the relative scarcity of these amino acids. Even a large protein may have no more than 30 of such residues, giving an overall ϵ of 10^5 or lower.

Put another way, a 1 mg/ml solution of BSA yields an A_{280} of 0.63, against an A_{191} of 84 (Webster, 1970). Measurement at 280 nm is really only sensitive enough to determine protein concentrations down to around 100 μg/ml.

Also, there is considerable protein-to-protein variation between the numbers of tryptophan and tyrosine residues, and their relative degrees of shielding. This has a dramatic effect on the extinction coefficient observed: the absorbance of lysozyme at 280 nm is higher than that of BSA by around 300% (cf. 8% at 191 nm; Webster, 1970).

Most typically proteins display an absorbance of 1–1.5 at 280 nm, from a 1 mg/ml solution. Lysozyme is at the top end of the absorbance range, a 1 mg/mlsolution giving an A_{280} of 2.3–2.7 (Little and Donahue, 1968; Webster, 1970); at the other extreme parvalbumins and other calcium-binding proteins havevirtually no absorbance at 280 nm (Dunn, 1989).This obviously makes it impossible to apply a 'generic' extinction coefficient, which would allow direct quantitation of unknown proteins from A_{280} data.

This is true both for straightforward sample analysis, and for UV-based detection methods in chromatography. Where the purity of a protein is assessed by comparison of peak areas, this can cause misleading results. To take a simple but common example, the absorbance of albumin is typically somewhat less thanhalf that obtained from the same concentration of IgG; if this factor is ignored in comparing peak areas detected at 280 nm, an IgG sample of 90% puritymay give an apparent value of greater than 95%.

The overall message with regard to protein determination by A_{280}, therefore, is that such analyses should be interpreted with great care; for the most reliable information the identity of individual components and their extinction coefficients must be known. Calibration with authentic material(s) of known concentration is the safest approach.

11.13.2.3 Protein determination by measurement of absorbance at other wavelengths

Many proteins have no significant absorbance above 280–290 nm, and therefore the choices are limited to those already described. However, there are many examples which contain a chromophoric prosthetic group, which can provide a very useful quantification tool.

Typical of this category are haemoproteins such as horseradish peroxidase, HRP: the haem group in this enzyme (isoenzyme C) has a λ_{max} of 401 nm ($\epsilon = 95,000$), the absorbance at this wavelength being higher than the A_{280} of the protein by a factor of around 3.5 (Shannon *et al.*, 1966). Contamination of the HRP with other (non-haem) proteins lowers this ratio, which has therefore been employed as a measure of the enzyme's purity (purity number or Reinheitszahl, *Rz*).

HRP can be quantified quite simply by measuring the absorbance in the vicinity of 401 nm, in the presence of a variety of other species including non-haem proteins. Even in a case such as this it is advisable to use a calibration plot for concentration estimates, ideally using the same batch of enzyme: lot-to-lot variations can be substantial, especially between different suppliers.

Finally, moving out of the realm of UV–visible spectroscopy into the near-infrared (780–2500 nm), advances in instrumentation and data handling are allowing the complex vibrational spectra observed in this region to be analysed with greater confidence. One application has been as the basis for the quantification of specific proteins in a complex medium (Harthun *et al.*, 1997); this powerful approach seems likely to have increasing importance in the future.

11.13.2.4 Detection of changes in protein structure and environment by absorbance measurements

Both the position (λ_{max}) and strength (ϵ) of UV–visible absorbances are wholly dependent on the distribution of electrons in the protein molecule. This is influenced by a multitude of factors including the number, location and juxtaposition of charged and aromatic groups, the degree of hydrogen bonding, etc.; it is not surprising, therefore, that absorbance characteristics can respond in quite a sensitive way to conformational changes, even without any covalent alteration to the structure.

This can sometimes be usefully exploited. When biotin binds to avidin, for example, a UV absorbance maximum in the avidin spectrum shifts towards longer wavelength, resulting in an increase in ϵ at 233 nm of 25,000 per mole of biotin bound. This provides one method for the quantification of biotin–avidin binding or of avidin itself (Green, 1970; Wei, 1970), particularly if the result is expressed as the ratio of A_{233} to the absorbance at 282 nm, which is substantially unaffected by biotin binding.

However, sensitivity of absorbance characteristics to the protein microenvironment can also add unwelcome complications to 'simple' quantification techniques based on absorbance at a single wavelength. It has already been noted (Section 11.13.2.2) that tyrosine, with a hydroxylic pK_a of around 10, has quite different absorbance properties above and below this pH value, so pH can influence the A_{280} of a protein solution quite significantly. In fact the insensitivity of protein absorbance spectra to pH changes in this region has been employed to demonstrate that derivatisation has taken place *via* the tyrosine hydroxyl groups (Cuatrecasas *et al.*, 1969).

While their effect on A_{280} may be smaller, there are many other important pK_as in the roughly neutral pH range (histidine $N \Rightarrow NH^+$ (6.5–7.0), terminal $NH_2 \Rightarrow NH_3^+$ (7.5–8.0), cysteine $SH \Rightarrow S^-$ (8.5–9.0 but very variable)). Changes in this region can therefore modify the charge distribution, and consequently the absorbance properties of the protein. Ionic strength changes, too, can provoke alterations in conformation, again with possible effects on absorbance. And even when both buffer molarity and pH are kept constant, itis still possible to observe absorbance differences from the buffer system chosen; on moving from Tris to carbonate–bicarbonate (both 0.05 M, pH 9.0), thehaem Soret band of horseradish peroxidase[1] isoenzyme A-1 shifts (λ_{max}) from 405 to 412 nm, accompanied by a significant increase in ϵ (Shannon *et al.*, 1966).

It is important, therefore, that quantification methods based on absorbance are used under well-controlled conditions, and that relevant sensitivities are taken into account.

11.13.2.5 Absorbance measurements for determination of conjugate incorporation ratios

One of the most useful ways in which UV–visible spectrophotometry can be applied to conjugate characterisation is in the estimation of incorporation ratios.

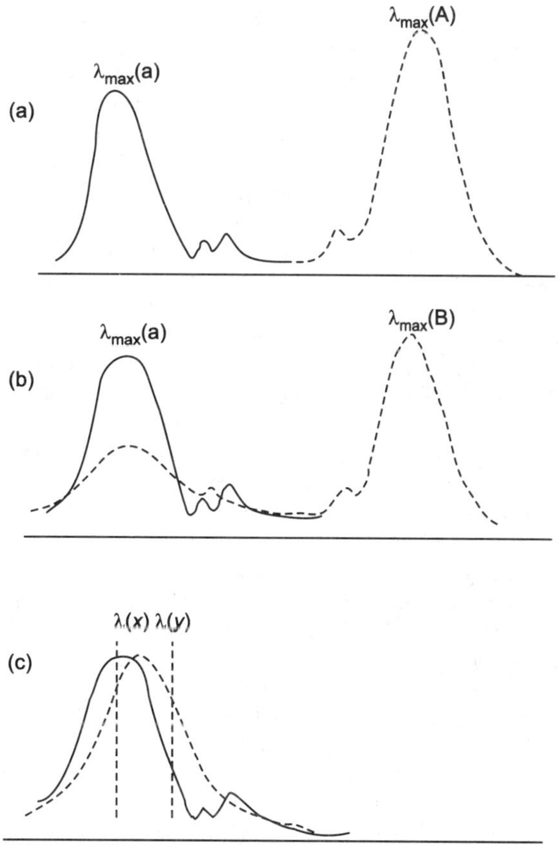

Figure 11.34 Situations where the two components of a conjugate have unique absorbances (a), one unique and one shared absorbance (b) and no absorbances (c).

Taking a model A–B conjugate, the ideal situation would be for A and B each to possess at least one absorbance maximum at which there is no significant absorbance from the other component (Figure 11.34(a)). Leaving aside for a moment possible effects on absorbance from the conjugation chemistry (i.e. assuming that the spectrum of the conjugate reflects additively the spectra of the individual components), the incorporation ratio of such a conjugate could be determined simply by calculating the concentration of each component at the appropriate λ_{max}. Such cases are unfortunately very uncommon: many proteins possess their only useful absorbances in the near-UV range, and it is unusual to find a conjugation partner which does not absorb to some extent in this region too.

[1] The basic isoenzymes of HRP, including the commonly used isoenzyme C, are much less sensitive to changes of buffer pH and composition.

Calculation of incorporation ratios where one component has a unique absorbance

The simplest cases commonly encountered in practice are where one of the conjugation partners has an absorbance which is not overlapped significantly by the spectrum of the other (Figure 11.34(b)). In the example shown the concentration of component B can be determined directly from the absorbance at its λ_{max}, while the absorbance at λ_{max}(a) must be corrected for the contribution from B before it is used to calculate the concentration of A.

Again ignoring any changes due to the conjugation chemistry, such calculations can be carried out using simple equations combining the extinction coefficients at the relevant wavelengths. Such approaches have been applied since the earliest days of protein conjugation, for example to quantify the incorporation of aromatic isocyanates into horse serum albumin (Creech and Jones, 1940, 1941).

Taking a typical case, Fujiwara *et al.* (1981) conjugated daunomycin to mercaptosuccinyl(MS)-BSA to form an immunogen. Daunomycin has an absorbance at 480 nm ($\epsilon = 11,600$), where the MS-BSA has no significant absorbance. However daunomycin also absorbs at 280 nm ($\epsilon = 8600$) where the MS-BSA ($\epsilon = 45,000$) was to be quantified.

The daunomycin concentration could be quantified directly as $A_{480}/11,600$.

The contribution of the daunomycin to A_{280} could be calculated on the basis of the relative extinction coefficients at 280 and 480 nm:

$$A_{280}(\text{daunomycin}) = A_{480}\frac{8600}{11,600}$$

The A_{280} of the MS-BSA could then be calculated by subtracting this value from the total A_{280}, and the MS-BSA concentration could then be calculated as A_{280} (MS-BSA)/45,000.

Overall, then:

$$[\text{daunomycin}] = A_{480}/11,600$$

and $$[\text{MS-BSA}] = \frac{A_{280} - (A_{480}8600/11,600)}{45,000}$$

Combining these equations to obtain an expression of daunomycin incorporation gives:

$$\frac{[\text{daunomycin}]}{[\text{MS-BSA}]} = \frac{A_{480}}{11,600} \times \frac{45,000}{A_{280} - (A_{480}8600/11,600)} = \frac{45,000A_{480}}{11,600A_{280} - 8600A_{480}}$$

There are many different ways that such calculations can be expressed, for example using absorbances at a given component concentration rather than molar extinction coefficients, or including ratios of absorbance at different wavelengths.

Rector *et al.* (1978), for example, prepared dinitrophenyl–IgG conjugates. The dinitrophenyl (DNP) groups were estimated to have an ϵ_{360} of 12,550 and an A_{360}/A_{280} of 3.07. The IgG had no significant absorbance at 360 nm and an ϵ_{280} of 210,000.

The incorporation of the conjugates could therefore be expressed as:

$$\frac{[\text{DNP}]}{[\text{IgG}]} = \frac{A_{360}}{12,550} \times \frac{210,000}{A_{280} - (A_{360}/3.07)} = \frac{16.73A_{360}}{A_{280} - (A_{360}/3.07)}$$

The principle of all such methods, however, remains the same. They can be applied wherever one component of the conjugate has a unique absorbance. Assessing the incorporation of fluorescent sulfoindocyanine dyes is a fairly recent example (Mujumdar *et al.*, 1993)

Even where varying ratios of two separate products may be formed with different overlapping absorbance characteristics, this approach can still be applied *via* the solution of simultaneous equations. Where the tyrosine and histidine residues of a protein are derivatised by a single equivalent each of a diazo-reagent (Tabachnick and Sobotka, 1960), the quantities of monoderivatised histidine (X) and tyrosine (Y) could (under specified conditions) be determined from the simultaneous equations:

$$A_{460} = 16.50X + 9.60Y$$

and $$A_{500} = 2.65X + 10.50Y$$

Calculation of incorporation ratios where neither component has a unique absorbance

Frequently there will be no useful unique absorbance in either conjugation partner (Figure 11.34(c)), for example with the majority of protein–protein conjugates. If the absorbance spectra are too similar, it is impossible to quantify the individual contributions of the components, and other means of characterisation must be sought. However, even rather unpromising situations such as that depicted in Figure 11.34(c) can provide useful incorporation data. While the absorbance maxima of the two components are too close to be used for relative quantification, it is often possible to choose two wavelengths whose relative ratio is very different in each case.

In the example shown, λ_{max}(a) provides one wavelength ($\lambda(x)$). At a second wavelength (shown as $\lambda(y)$), the absorbance of component A is very low by comparison ($A(y)/A(x) = 0.2$). For component B, however, the absorbances at the same two wavelengths are much more similar ($A(y)/A(x) = 0.9$).

Therefore the relative contributions of A and B to a conjugate can be estimated by examining this ratio. For the best results it is advisable to prepare a series of $A + B$ mixtures at varying ratios, measure $A(x)$ and $A(y)$, and calculate the $A(y)/A(x)$ ratios. This provides a calibration curve for converting ratios into incorporations.

Alternatively this approach can be put in equation form by a simple extension of the arguments used in the last section. Taking another example, Wessendorf *et al.* (1990) used this approach to determine the incorporation of the fluorophore AMCA in IgG conjugates. The wavelengths chosen were 350 nm for MCA and 280 nm for IgG. The incorporation is given by:

$$\frac{[\text{AMCA}]}{[\text{IgG}]} = \frac{(\epsilon_{280}(\text{IgG})A_{350}) - (\epsilon_{350}(\text{IgG})A_{280})}{(\epsilon_{350}(\text{AMCA})A_{280}) - (\epsilon_{280}(\text{AMCA})A_{350})}$$

Any two wavelengths can be chosen as a basis of calculations of this type; examination of overlaid spectra of the two components will help to identify wavelengths where useful differentiation may be obtained. However the usefulness of this approach in a given situation will depend on the relative molar extinction coefficients of the two components at the relevant wavelengths, as well as their ratio. Where incorporation ratios are low, it is best to choose $\lambda(x)$ where the molar extinction coefficients of the two components are similar. Where very high incorporations of one component are expected, on the other hand, it is wise to choose a wavelength where ϵ of this component is lower.

Determination of conjugate incorporations by this ratio approach has been used for many protein-small molecule conjugates, such as protein–triazine conjugates characterised by A_{260}/A_{280} (Goodrow *et al.*, 1990), and protein–uracil conjugates characterised by determination of A_{238}/A_{280} (Szurdoki *et al.*, 1992).

Difference spectra

Another approach which can be taken in any of the cases described in this section is to create a difference spectrum, in which the spectrum of one component is subtracted from that of the entire conjugate. This can sometimes provide more information than simple ratio comparisons.

For example, Fox *et al.* (1994) prepared conjugates of cytochrome *c* peroxidase (CCP) with $a_5Ru^{II}H_2O$ complexes, *via* histidine residues in the protein. CCP (as its cyanide adduct at pH 11) has an absorbance at 422 nm, with an extinction coefficient (103,000) which was assumed to remain unchanged on derivatisation, and was therefore used as a basis for quantifying the protein content of the conjugates. The spectrum of an equivalent quantity of CCP was subtracted from each conjugate spectrum. Where these displayed a broad maximum at 370 nm, this was taken to indicate the presence of the desired ruthenium–histidine complex, and the ϵ_{370} of a_5RuHis (3400) was used to calculate the incorporation level.

In this example the unique CCP absorbance allowed the protein concentration of the conjugates to be calculated spectrophotometrically. The incorporation could simply have been calculated by an equation of the type described above, but plotting out the difference spectrum allowed the λ_{max} at 370 nm to be observed, offering support for the hypothesis that the expected Ru–His bonds were being formed.

Difference spectra can also be used in a more approximate fashion where a simple non-spectrophotometric estimate is made as to the protein concentration, e.g. by assuming that no protein has been lost during the conjugation process. A solution of the protein at the starting concentration then provides a blank for analysis of the product. More satisfactorily in some cases, a dummy conjugation can be carried out in which the second component is omitted. The conjugate collected from this preparation can then be used as an analysis blank.

Where the spectrum of the protein component is subtracted in this fashion, the concentration of the derivatising molecule can then be determined by measuring the absorbance at its λ_{max}. This approach is likely to suffer worse precision than those where one component possesses a unique absorbance, particularly where the absorbance of the derivatising molecule is low compared to that of the protein. Moderate agreements were obtained when this approach was compared with other techniques for a series of steroid–BSA conjugates (Lieberman *et al.*, 1959).

Combination of absorbance measurements with other methods

As just discussed, estimating protein content by an independent means can allow incorporations to be determined by spectrophotometric methods. Combination methods are useful, particularly for complex cases such as conjugates of three separate components. For example, Pietersz *et al.* (1988) prepared methotrexate–albumin–IgG conjugates; the albumin concentration was estimated from the incorporation of a radioactive label, the methotrexate from A_{370}, and the IgG from A_{280} after adjustment for the contribution from both other components. Noguchi *et al.* (1992) coupled mitomycin C to IgG *via* a dextran linker; IgG was determined by the Lowry method, dextran by a colorimetric technique, and mitomycin C by A_{363}.

Changes in absorbance characteristics on conjugation

For the sake of convenience, the approaches described above have made the assumption that the molar extinction coefficients of the components are not altered by the conjugation procedure. If this is not the case, it is impossible to apply simple equations which relate absorbance ratios to incorporation: the change in ϵ adds another variable which cannot be quantified simultaneously.

Unfortunately absorbance changes have often been observed, even for direct conjugation methods where no linker group is incorporated. It was noted in Section 11.13.2.4 that even non-covalent binding of proteins can alter their spectrophotometric properties significantly, and this applies to small molecules too: some steroids exhibit a shift to lower wavelength on dissolution in protein solutions (Westphal, 1957), a phenomenon attributed to possible hydrogen bonding between the protein and the α, β-unsaturated system of the steroid (Lieberman *et al.*, 1959). For some other steroids, shifts to higher wavelength have been observed on direct conjugation to BSA (Lieberman *et al.*, 1959), attributed to phenolic electrons interacting with particular amino acid residues.

Establishing whether the assumption of additive absorbance properties is justified generally requires comparison with a less equivocal technique, such as incorporation of a radiolabelled component. For example a series of protein–hapten conjugates can be prepared using a radiolabelled hapten, which allows their mean incorporation to be determined with a high degree of accuracy. The absorbance properties of these conjugates can be compared with those expected from the simple additive model, to establish whether this is realistic or not, and if it is not, may themselves provide a basis for future estimations (Figure 11.35).

More realistic results may be obtained from the simple additive model if extinction coefficients are

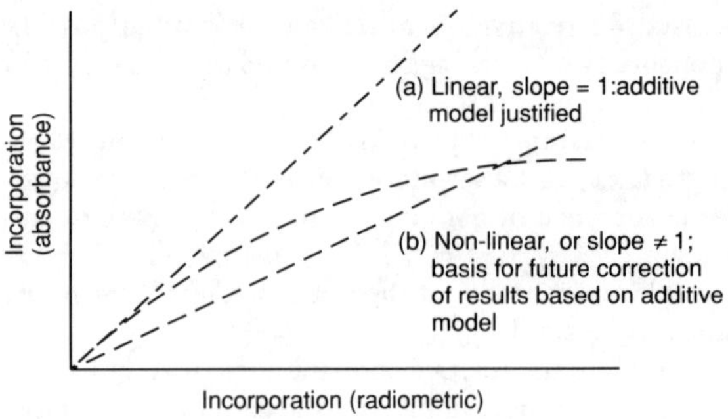

Figure 11.35

obtained for suitable model compounds, rather than using those of the starting materials themselves. This applies particularly to small molecules: if it is known that a compound (e.g. an active ester) is likely to be linked predominantly to lysyl amines in the conjugate, a model derivative can be prepared using lysine, or (even better) a lysine-containing oligopeptide, in either case with a protected α-NH_2 group. Isolation of the resulting model conjugate in pure, solid form, and determination of relevant ϵ values, may give a much better estimate of the absorbance properties in the final conjugate than simply basing the calculation on the properties of the compound itself. The same applies to the spectrophotometric properties of the protein itself, although these are somewhat less sensitive to change.

Where homo- or heterobifunctional linkers are used to generate a conjugate, the situation may be more complicated still, and great care needs to be applied in interpreting UV–visible data. The absorbance properties of the linker groups too need to be taken into account, including any changes to their behaviour on linking to both conjugation partners. While a detailed understanding of such factors can be approached by the preparation of suitable model compounds, the complexity of such situations often justifies simple approximations in interpreting data, provided the limitations of such approaches are justified.

If it is possible to fractionate the conjugate into homogeneous components in reasonable quantities, it is possible to calculate genuine ϵ values by relating their absorbance properties back to a more reliable technique such as incorporation of a radiolabel, or gravimetry. These values can then be used as a basis for future analysis of absorbance data. Such labour-intensive approaches to ensure the rigorousness of UV-based incorporation data are very rarely encountered in practice.

11.13.3 Radiometry and radiography

Radioactive isotopes with suitable decay rates and energies provide an extremely powerful approach to conjugate characterisation, when incorporated into one of the conjugate components. Three major advantages are offered by this methodology. The first is the unlikelihood of signal arising from any other source except the deliberately labelled molecule, so that background (non-specific) signals are generally vanishingly small. The second is the sensitivity of the detection methods used, which can easily reach femtomolar levels. The quantities of component present in protein conjugate are typically vastly greater than this, and it is only necessary to include a small 'spike' of radiolabelled material in the unlabelled ('cold') bulk. The third is that, for high-energy emitters at least, the signal is wholly unaffected by the protein microenvironment, the coupling chemistry, etc.

These combined advantages can provide much more robust data in many cases than can be obtained from UV–visible spectrophotometry, for example—particularly in some of the more complex situations discussed above.

Radiolabelling approaches have disadvantages too, apart from the inherent handling issues: the first is that the radiolabelled molecule must be synthesised unless it can be purchased directly from a supplier, and the second is that the incorporation of a radiolabel may affect the properties of the molecule in a way which distorts the experiment being carried out. For a radiolabelling experiment to be meaningful, unless 'invisible' isotopes such as ^{35}S or ^{14}C are incorporated without changing the molecular structure, it must be established that this is not the case. Coupled with the synthetic work required to produce the radiolabelled material in the first place, this necessitates a reasonably significant investment of effort. However, when properly carried out, radiolabelling techniques can provide unparalleled information on conjugate incorporations, surface concentrations after immobilisation, etc.

Detection is by radiometry where the signal is counted in a suitable instrument, or radiography where the decay particles are detected by their effect on photographic film.

11.13.3.1 The choice of radioisotopes for analytical radiolabelling

There are a large number of radioisotopes available from commercial sources, but the number of these which are useful for radiolabelling is fairly small. The defining properties of a radioisotope include the type and energy of the radiation produced, and the rate of radioactive decay, often characterised by theradioactive half-life of the isotope involved.

Radioisotopes which emit α-particles are not generally favoured because they are so readily absorbed, which can make measurement difficult: much of the radiation is likely to be absorbed by the matrix before reaching the detector. Therefore β- or γ-emitters are more commonly encountered. The transmission of γ-radiation through most materials is sufficiently large that it can be measured directly from the sample. Thin sample vessels of glass or plastic will cause no significant loss of signal. The transmission of β-particles is relatively poor, and while direct detection is possible it is more usual to transfer the energy to a suitable fluorescent molecule, then detect the light produced. This approach is variously referred to as scintillation counting, scintillation autography or fluorography.

Among β- and γ-emitters utility for analytical radiolabelling is determined further by the half-life, values of week or months being the most convenient. While a shorter half-life corresponds to a greater decay rate and the possibility of greater sensitivity, supply and storage become increasingly impractical; on the other hand long half-life isotopes are unlikely to provide enough signal.

The final major considerations are how the radioisotope is to be incorporated into the molecule of interest, and the effect that this is likely to have on its behaviour. The ideal radiolabel should be incorporated 'invisibly', with no extraneous atoms being introduced. This can only be achieved by replacing one or more atoms in the molecule (whose isotopic distribution will reflect that occurring naturally) with a radioactive isotope of the same element. This consideration is mainly responsible for the predominance of isotopes such as ^{3}H, ^{14}C and ^{35}S for most applications, ^{32}P for nucleic acid work in particular, and ^{125}I for rare examples such as thyroid hormones.

For small molecules it may be possible to buy material already incorporating a radiolabel, or it may be necessary to incorporate a radioactive material into a synthetic procedure. Such organic synthesis procedures are beyond the scope of this book, but a general principle is to introduce the radioactive component as late as possible in the synthetic route, so that the bulk of the handling is carried out on 'cold' material. It may also be possible to carry out an exchange reaction on the compound of interest, either simple equilibrium exchange of ^{3}H for labile protons, or two-stage reactions such as bromination of aromatic groups followed by reduction with $^{3}H_2$ over a suitable catalyst.

For proteins and peptides it is more difficult to incorporate an 'invisible' radiolabel directly into the molecule, although in some cases this can be achieved by providing radioactive amino acids as nutrients for biosynthesis (Laskey *et al.*, 1972).

It is much more common to radiolabel proteins by the introduction of an extraneous group, typically ^{125}I. The introduction of an iodine atom, which is highly hydrophobic and of comparable dimensions to a phenyl group, is likely to disrupt the properties of a small organic molecule dramatically. However, its effect on a large protein molecule may be insignificant, particularly if the conditions are designed such that that singly labelled molecules predominate. Low incorporations of radiolabel are also advisable to minimise radiolytic damage to the protein molecules.

Various approaches to the incorporation of ^{125}I atoms into proteins have been discussed in Section 6.2.1, the most common being oxidative iodination of tyrosine and histidine residues, or derivatisation of lysine groups with a radiolabelled active ester such as Bolton–Hunter reagent. An alternative strategy is the reductive alkylation of amines with ^{14}C or ^{3}H-labelled formaldehyde and/or ^{3}H-labelled borohydride (Section 6.2.1).

Another category of radioisotopes which have a specific application are metal ions such as ^{111}In, which are commonly incorporated into therapeutic protein reagents such as radioimmunoconjugates. These metals are typically coupled *via* a chelating group such as a polycarboxylic molecule, and can therefore offer a convenient means of characterising protein–chelator conjugates.

The properties of some of the more commonly encountered isotopes for analytical radiolabelling are summarised in Table 11.4.

11.13.3.2 Radiolabelling experiments for determination of conjugate incorporation ratios

Once one of the components of a conjugation has been obtained in radiolabelled form, incorporation data can be obtained very simply after the coupling reaction. Generally the specific activity of a radiolabelled material is so high that only a small fraction needs to be included in the reaction mixture, the bulk of the reagent being cold material. Because the molar quantity of radiolabelled material is insignificant, it is only necessary to choose a suitable quantity to give a convenient signal level after losses in yield have been accounted for. With ^{125}I-labelling, for example, one may include enough radiolabelled material to yield around 10^4 counts per minute in the final product.

So long as the radiolabelled material has identical chemical reactivity to its 'cold' analogue, it will be incorporated in a random fashion in the conjugate, and will thus accurately reflect the overall incorporation of this component. The activity of the purified conjugate is compared with that of a sample of starting material of known mass, ideally on the same counter and at roughly the same level of counts, to avoid errors from differences in detection efficiency. A simple calculation then provides the amount of this component in the sample, and, when coupled with the concentration of the second component derived by a complementary technique, the incorporation.

Table 11.4 Properties of some of the common isotopes for radiolabelling

Nuclide	Half-life	Type of emission	Pricipal energies (MeV)
Tritium	12.43 years	β	0.0186
Carbon-14	5730 years	β	0.156
Sulfur-35	87.4 days	β	0.167
Iodine-125	60.0 days	γ and	0.035
		X-ray	0.027–0.032
Selenium-75	119.8 days	Electron capture	–
Technetium-99m	6.02 hours	Internal electron	–
Technetium-99	2.1×10^5 years	β	0.204
			0.293
Cobalt-56	77.7 days	β	0.40
			1.5
		Electron capture	–
Chromium-51	27.7 days	Electron capture	–
Indium-111	2.8 days	Electron capture	–
Phosphorus-32	14.3 days	β	1.709

Examples of analytical radiolabelling for conjugate characterisation are very common in the literature, and the following examples are only meant to demonstrate some typical applications.

Abraham *et al.* (1971) incorporated ^{3}H-steroids in an immunogen preparation to allow the incorporation to be determined.

Mannik and Downey (1973) used ^{125}I-labelled Fab fragments in glutaraldehyde conjugations with HRP, and were therefore able to quantify the incorporation in the product, e.g. demonstrating that maximal Fab incorporation was obtained at HRP–Fab ratios of 25:1 or higher, and at reaction times of at least 72 hours.

Lomant and Fairbanks (1976) used a ^{35}S-containing bis-succinimide ester DSP (Figure 11.36) to derivatise haemoglobin, and were able to follow the rapid incorporation of this reagent into the protein. Also, the loss of ^{35}S label could be used to demonstrate the quantitative cleavage of the disulfide bond by reduction.

Eruk *et al.* (1984) employed ^{125}I-thyroxine (T4) in a number of conjugation procedures with alkaline phosphatase, and were able to demonstrate molar incorporation ratios of T4 ranging from 0.1 to 50. The presence of the radioactive label in the conjugates also allowed their binding by anti-T4 antisera to be readily quantified.

Paik *et al.* (1985) used $^{111}InCl_3$ to assess the incorporation of DTPA chelator groups into antibodies, obtaining values of 0.1–11. This approach is slightly less definitive than those discussed so far, in that the radiolabel is not actually incorporated into the conjugation product directly. It assumes quantitative radiolabelling of chelator groups, employing a large excess of (mainly unlabelled) ions to ensure that this is achieved.

Non-ideal behaviour of radiolabelled material

For a homogeneous reagent, radiolabelled without altering the molecular structure except for isotopic changes, it is reasonable to assume without proof that its behaviour will faithfully reflect that of its 'cold' analogue. However, even in such a favourable case it is possible on storage for radiolytic damage to cause changes. Where there is some doubt as to the homogeneity of the material (as is often the case for proteins), or where the radiolabelling procedure has altered the molecular structure in some way, non-ideal behaviour can occur which may cause distorted results.

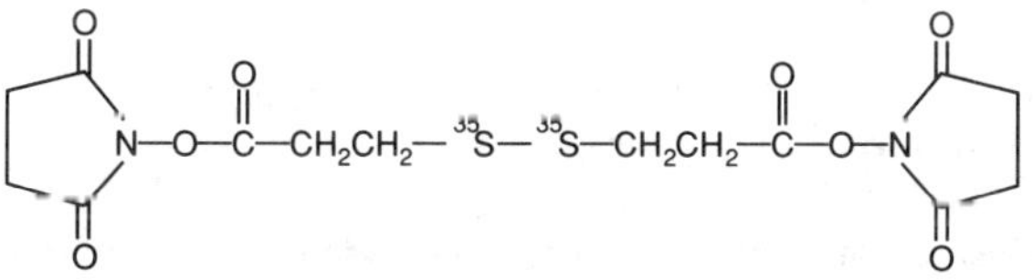

Figure 11.36 ^{35}S-Labelled DSP.

Such behaviour is of two basic types, which may be observed singly or in combination; their possible effects on labelling experiments will now be considered. It will be assumed that a fraction of radiolabelled material has been incorporated in the reagent of interest, which is present in a large overall excess compared to the other conjugation component.

In the first model the radiolabel is homogeneous but differs in reactivity from the cold compound, for example being more reactive. This will be reflected by a greater-than-expected abundance of this material in the final product, and hence an over-estimation of the overall incorporation ratio of the conjugate. If the unlabelled reagent is present in a very large excess over the labelled fraction, most of the individual reaction events will occur totally unaffected by the presence of the hot fraction, most available reaction sites will be taken up by unlabelled molecules despite their lower reactivity, due to the much greater number of such encounters which occur. Thus the extra reactivity of the hot material can only have an impact on a very limited number of sites, and the distortion from the expected answer will be small.

As the proportion of labelled material is increased, each hot molecule has more chance of being involved in an encounter with a reactive site before this is taken up by the excess of cold material, and its enhanced reactivity will therefore have more chance to be reflected in the final composition of the product. So changing the ratio of labelled to unlabelled material in the reagent will result in a different apparent incorporation, higher levels of hot material giving an increasingly distorted answer.

On the other hand, if the radiolabelled material is less reactive than its cold analogue, the same logic means that a more distorted answer will be obtained at lower fractions of labelled material, the already-disadvantaged radioactive molecules being further hindered by an increasingly small number of chances to be involved in a reactive encounter. Thus lowering the fraction of radiolabelled component will result in an increasingly severe under-estimate of overall incorporation.

To establish that effects of this type are not taking place, it is good practice to carry out a sample determination at a number of different ratios of labelled to unlabelled material, covering as wide a range as is practicable. A lack of significant difference between the results is good confirmation that the reactivity of the labelled material accurately reflects that of the unlabelled analogue. If a significant difference is seen, the reagent is not really suitable for the purpose. Note that with both under- and over-reactive radiolabelled reagents, an increase in the hot fraction is likely to result in an increase in the apparent conjugate incorporation, but without knowing which cause is responsible, it is difficult to deduce whether the higher or lower estimates are more likely to be correct.

A second case of non-ideal behaviour is where the radiolabelled material is heterogeneous. This may arise because the bulk material itself is heterogeneous, in which case the hot material is merely reflecting (as it should) the bulk properties. More seriously the radiolabelling procedure itself may lead to a heterogeneous product, for example one containing a fraction of completely unreactive material. Supposing that the reactivity of the intact fraction accurately reflects that of the cold reagent, then the presence of unreactive material will result in an underestimate of the overall incorporation. Also, unlike the case above, this will not depend on the relative populations of labelled and unlabelled material, and so will not be detected by varying this ratio.

To establish whether or not this effect is occurring, it is necessary to determine whether the labelled material can be completely incorporated in the desired product or not, for example by pushing the reaction conditions to a plateau where no further increase in incorporation is observed. It is important to choose the experimental design carefully here, so that complete reaction is truly achievable, e.g. increasing the time of an active ester reaction may not drive it to completion, but this is more likely to reflect the increasing hydrolysis of the ester than any heterogeneity. Changes to more fundamental conditions such as pH or temperature, on the other hand, may distort the reactivity in either direction so that this experiment is no longer indicative of the real situation. In general, increasing reaction ratios is less prone to such artefacts than other parameters.

If 100% of the labelled material is incorporated in such an experiment, then there cannot be an unreactive component. If on the other hand the incorporation plateaus at a lower value, it is reasonable to assume that the remainder of the labelled material is an unreactive fraction. An example of such a situation is shown in Figure 11.37 (Mannik and Downey, 1973): increasing the molar ratio of HRP to ^{125}I-Fab resulted in a plateau where only 90% of the labelled material was incorporated, the remaining 10% apparently being completely resistant to reaction.

If such an effect is observed, it is not straightforward to determine whether it is due to heterogeneity of the bulk material, or just of the radiolabelled fraction; fractionation of the material before and after radiolabelling may well be required to resolve this issue. However, if such an effect is absent, it is good evidence that the reagent will give reliable results.

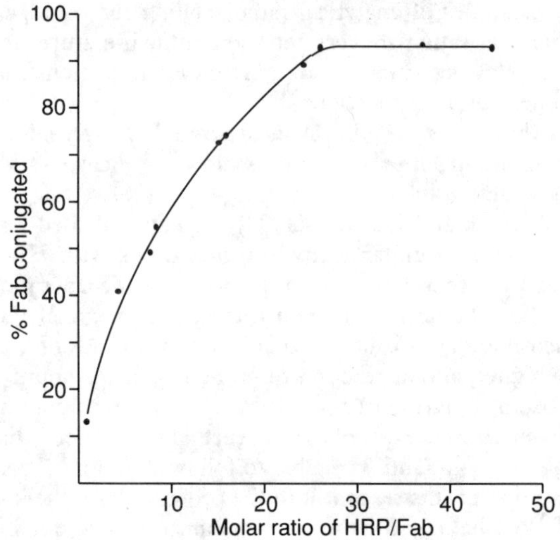

Figure 11.37 Increased conjugation of Fab fragments with increased molar excess of HRP. (Reproduced with permission from: Mannik and Downey (1973). *J. Immunol. Methods*, **3**, 233–242.)

While more complex combinations of these two types of effect could be envisaged, e.g. a mixture of components of varying reactivity, they are all likely (except in the most artificial circumstances) to show up in one of the two checks outlined above. Any analytical technique based on the use of radiolabelled components is much more secure, if a little effort is put into demonstrating that a constant result is obtained over a range of hot:cold ratios, and that essentially all the material is reactive under the conditions of interest. A final trap to avoid is the assumption that radioactive materials will not undergo changes in their properties on storage, Underwood and Steele (1991) have shown dramatic changes in the binding behaviour of labelled proteins, even at -70°C. An understanding of storage effects adds further security to techniques of this kind.

11.13.3.3 Radiolabelling experiments for the determination of immobilised protein

Designs analogous to those discussed above can be applied to immobilisation methods, in fact, the facile separation of unreacted from reacted material in these cases makes such approaches particularly convenient. The basic designs could not be simpler: radiolabelled protein is included in the immobilisation procedure, and the amount of label ending up on the solid phase reflects the immobilised protein concentration.

Both cases of non-ideal behaviour described in the last section can occur here too, causing similar distortions in results. Even low levels of radioiodination have been shown to cause changes in properties of this type (Walsh and Gosling, 1986).

For detecting changes in binding affinity in the radiolabelled material, the same check of examining behaviour at different hot:cold ratios is applicable (Van der Scheer *et al.*, 1978).

Detection of a non-binding fraction is often somewhat more convenient here, in that the protein solution can be retrieved following a coating step and transferred to a second solid phase where the binding behaviour of the remaining protein can be examined. Carrying out a sequence of such transfers will establish whether the entire radiolabelled fraction can be bound or not. Underwood and Steele (1991) observed levels of around 20% of non-binding radioactive material in several different proteins iodinated by the Chloramine-T method. They also reported substantial increases in this percentage on storage. A mouse antibody stored at 4 °C for four weeks contained virtually no bindable ^{125}I, for example.

Underwood and Steele (1991) also described a method for determining protein adsorption which is independent of the presence of bound fraction, this is based on the observation that a fixed percentage of bindable protein is generally immobilised under low-concentration conditions (the 'region of independence', Canterero *et al.*, 1980). Comparison of the amounts immobilised by two sequential incubations allows the overall binding capacity of the solid phase for this material to be calculated, without any interference from non-binding fractions.

11.13.4 Mass spectrometry

Mass spectrometry (MS) has been practised as an analytical technique since the early part of the century Its principle is relatively simple, charged species are generated from an atom or molecule of interest, and separated on the basis of their mass- to-charge (m/z) ratio. The latter process must take place under a high vacuum to minimise interference from atmospheric molecules.

The development of mass spectrometry, summarised recently by Niwa (1995a), has been defined by advances in two key areas. The first is the process of generating charged, gas-phase species from the atom or molecule of interest; the second is the mass range of particles which can be detected.

In line with these these constraints, mass spectrometry started off being applied to small, gaseous species: as early as 1912 J. J. Thomson demonstrated the existence of the hitherto-unknown ^{22}Ne isotope using this technique. Application to small organic molecules was commonplace by the 1950s, but the high molecular masses and low volatilities of proteins rendered them inaccessible to MS analysis until the late 1980s. The methodological advances which have finally facilitated this breakthrough are summarised very briefly below; the reader is referred to reviews such as that by Niwa (1995b) for more detailed explanations.

One final aspect of MS which gives it advantages over many other detection methods is its applicability to carbohydrate residues, which are particularly amenable to analysis by this means (Laine, 1992e). Glycans cleaved from glycoproteins are conveniently analysed by MS methods (Sutton *et al.*, 1994 Küster *et al.*, 1997; Sheeley *et al.*, 1997).

11.13.4.1 Ionisation, separation and detection techniques for mass spectrometry

Ionisation techniques

Small, volatile molecules have traditionally been ionised by electron ionisation (EI), where they are subjected to a stream of high-velocity electrons from a source such as tungsten or rhenium. The impact will typically eject an electron from the target molecule, generating the molecular (M^+) ion whose mass reflects that of the native molecule. However, the energy supplied by the collision may also be sufficient to cause cleavage of covalent bonds, so that fragmented daughter ions are observed in the mass spectrum. Such processes often predominate to the extent that the molecular ion is a very minor component of the spectrum; for pure samples of low molecular mass this isthe ideal outcome, allowing detailed structural information to be obtained.

Chemical ionisation (CI), introduced in the 1960s, is a 'softer' technique where ions are generated from a reagent gas such as methane (in this case typically CH_5^+, $C_2H_5^+$ and $C_3H_5^+$), and then react with the target molecule. This approach causes less fragmentation than EI, giving a correspondingly higher yield of the molecular ion.

Neither EI nor CI, however, can be used with non-volatile samples. Techniques have been developed for dealing with such molecules: ionisation of liquid samples by a beam of neutral high-energy atoms (fast atom bombardment, FAB) or ions (secondary ion mass spectrometry, SIMS) allowed the molecular mass limit for MS to be pushed out beyond 5 kDa. Further advances were obtained by the use of extremely high electric fields (field desorption, FD), fission fragments from a radioactive source (plasma desorption, PD), or, most recently, laser energy (laser desorption ionisation, LDI), often enhanced by the use of an organic matrix (matrix-assisted laser desorption ionisation, MALDI). Other approaches are based on the generation of ionic species from solvent molecules in a liquid matrix, these include thermospray ionisation (TSI), where the sample molecules are incorporated in charged droplets of thermally vaporised liquid, and electrospray ionisation (ESI) where the charged droplets are generated by an electric field. Subsequent evaporation of the solvent results in the formation of charged gas-phase species, which may have already been present in the sample solution, or have been generated by transfer processes.

Separation and detection techniques

A range of separation and detection techniques have also been developed for mass spectrometry, typically culminating with the detection of charged species by an electron multiplier device. The fundamental requirement is to resolve the ionised components on the basis of their m/z ratio, either by a physical separation or by an inherent characteristic of the ratio itself.

The classical separation method for MS is the magnetic sector approach, where the beam of ionised components is swung away from its path by a magnetic field. The degree to which a particular component is affected by the field is determined by its m/z ratio, so that resolution is obtained. In practice it is usual to vary the magnetic field strength across a suitable range, such that components of different m/z are sequentially aimed through a slit at the detector. Only low resolution is obtainable by this approach, and double-focusing methods have been developed where a second electrostatic focusing stage is incorporated before or after the magnetic field. These offer much greater resolution, typically up to molecular mass values of around 10 kDa (Krull *et al.*, 1995), although the instrumentation required is bulky and expensive.

Quadrupole methods are based on the application of a radio frequency across an electric field between four parallel rods. This causes oscillation of ions passing through the quadrupole. Only those of a certain m/z displaying stable oscillations which allow them to emerge; the quadrupole therefore acts as a mass filter. By varying the voltage characteristics of the quadropole, components of different m/z can be resolved. Ion trap methods are based essentially on the reverse principle: ions displaying stable oscillations remain trapped within the electrodes, while those with unstable trajectories are ejected for detection. These methods allow cheap, compact instrumentation and can be operated under relatively low vacuum conditions. Their resolution is not very high, however, and they have typically been applied only up to molecular mass limits of around 4 kDa (Krull *et al.*, 1995), or even lower for ion trap methods.

Recent advances in instrumentation, however, have put them into contention for detecting singly-charged MALDI products up to 30–50 kDa (Jonscher and Yates, 1997).

Time-of-flight (TOF) methods rely on the fact that, under a constant accelerating voltage, ions will travel at a rate determined by the m/z ratio. The detection principle is simple, smaller components being detected before larger ones as the beam travels from the sample. This is a low-resolution technique, but is otherwise particularly suited to protein work: there is no upper molecular-mass limit for TOF methods, and ions can be detected fairly rapidly after generation. This makes TOF compatible with ionisation techniques such as MALDI where the ions generated are short-lived.

Finally, Fourier transform (FT) methods are based on ion cyclotron resonance phenomena; these resonances are set up in a superconductive magnetic field, at frequencies determined by the m/z ratio, and are detected by the image current which they induce in a nearby receiver plate. FT methods have very high resolution, moderate molecular mass cutoffs (typically 8 kDa; Krull *et al.*, 1995), and are rapid because no scanning is required. Their main drawback is the need for liquid nitrogen and helium to maintain the superconducting environment.

11.13.4.2 Electrospray ionisation (ESI) for protein mass spectrometry

Two of the ionisation techniques described above, ESI and MALDI, have proved sufficiently powerful to be used successfully for the characterisation of proteins, with molecular masses in the 10^5 Da range. Their characteristics are somewhat complementary.

ESI was based on studies by Dole *et al.* (1968). It can be applied directly to liquid samples, and can therefore be coupled conveniently to separation techniques such as liquid chromatography or capillary electrophoresis (Smith and Udseth, 1988). It tends to generate multiple charges, perhaps as many as 100 on a

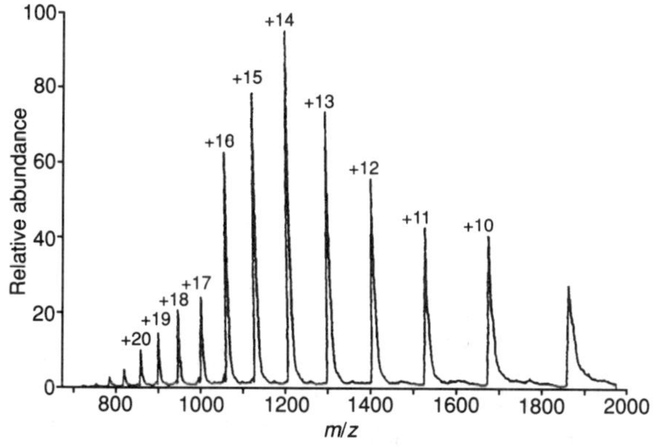

Figure 11.38 Positive-ion electrospray ionization mass spectrum pattern of myoglobin. (Reproduced with permission from: Krull *et al.* (1995). *LC–GC Int.* 8, 14–20.)

large protein, and the spectrum of a single component appears as a roughly Gaussian distribution ('envelope') of peaks corresponding to different m/z ratios (Figure 11.38). Because of this multi-charging effect, these ratios are very low compared to other ionisation techniques. This means that much higher molecular masses can be examined than would be expected from the typical detector cutoff values outlined in the last section. This brings large proteins within the range of simple commercial detectors such as those based on quadrupole design.

Its combination of extremely high resolution (differences of molecular mass of 0.1% being detectable) and mass accuracy (0.01% or better) make it extremely powerful for protein characterisation: van Dorsselaer *et al.* (1990), for example, were able to detect down to 1% of closely related contaminants in a 15 kDa recombinant protein preparation. ESI-MS is capable of defining protein structures more accurately than older techniques. With BSA Hirayama *et al.* (1990) obtained a mass value from ESI-MS differing from that calculated from sequence information by 199 Da; this discrepancy subsequently revealed that, even for one of the most widely used proteins in existence, the sequence data had omitted one residue and misassigned several others.

By 1990 over one hundred peptides and proteins had been analysed by ESI-MS, reaching molecular masses of 133 kDa (Krull *et al.*, 1995). For pure samples such as recombinant protein products, it is relatively straightforward to interpret the envelope spectra. However, the spectra obtained from impure or mixed samples may be intractably complex, which means that thorough fractionation of individual components beforehand may be essential. A recent example of such an approach is the resolution of rat glutathione *S*-transferase subunits by reversed-phase HPLC, followed by mass assignment by ESI-MS (Rouimi *et al.*, 1995). The deconvolution of the envelopes to obtain ausable mass spectrum is demonstrated in Figure 11.39.

ESI is gaining increasing popularity for conjugate analysis: Lewis *et al.* (1994) used this technique to estimate the incorporation of chelator molecules in a cytochrome *c* conjugate (Figure 11.40), and other recent examples include studies on conjugates of antibodies (Bennett *et al.*, 1996), rhodopsins (Hufnagel *et al.*, 1996) and haemoglobins (Adamczyk and Gebler, 1997). Adamczyk *et al.* (1996) reported results from ESI-MS somewhat superior to MALDI-MS in the analysis of protein–hapten conjugates.

11.13.4.3 Matrix-assisted laser desorption ionisation (MALDI) for protein mass spectrometry

Conventional LDI (Honig and Woolston, 1963) used a laser-photon-absorbing solid phase to transmit ionising energy to the analyte molecules, and allowed molecules up to around 1 kDa to be volatilised. This molecular mass range was extended dramatically by incorporating the sample in a photoabsorbing matrix material which evaporates on laser irradiation, taking with it the dispersed protein. The original descriptions of this approach used finely dispersed cobalt metal in an organic solvent (Tanaka *et al.*, 1988), or dry organic acids evaporated from an aqueous solution (Karas and Hillenkamp, 1988); molecular ion peaks up to 100 kDa were observed. Unlike ESI very little multi-charging was observed, and there was virtually no evidence of fragmentation. The spectra were accordingly much simpler (Figure 11.41), the major advantage of the MALDI approach; this is also a rapid technique, spectra being obtainable within 15 minutes.

Subsequent development of the organic acid approach (Karas *et al.*, 1989a,b, 1990) demonstrated that femtomolar quantities of proteins with molecular mass beyond 300 kDa could be detected. The requirements of the matrix have been discussed in some depth (Karas *et al.*, 1990, 1991): these include the reduction of intermolecular forces by isolation of sample molecules, the absorbance of laser energy at a convenient wavelength leading to destruction of the crystalline phase, vacuum stability, water solubility, and a possible photochemical contribution to the analyte ionisation process. Matrices which have proved useful for

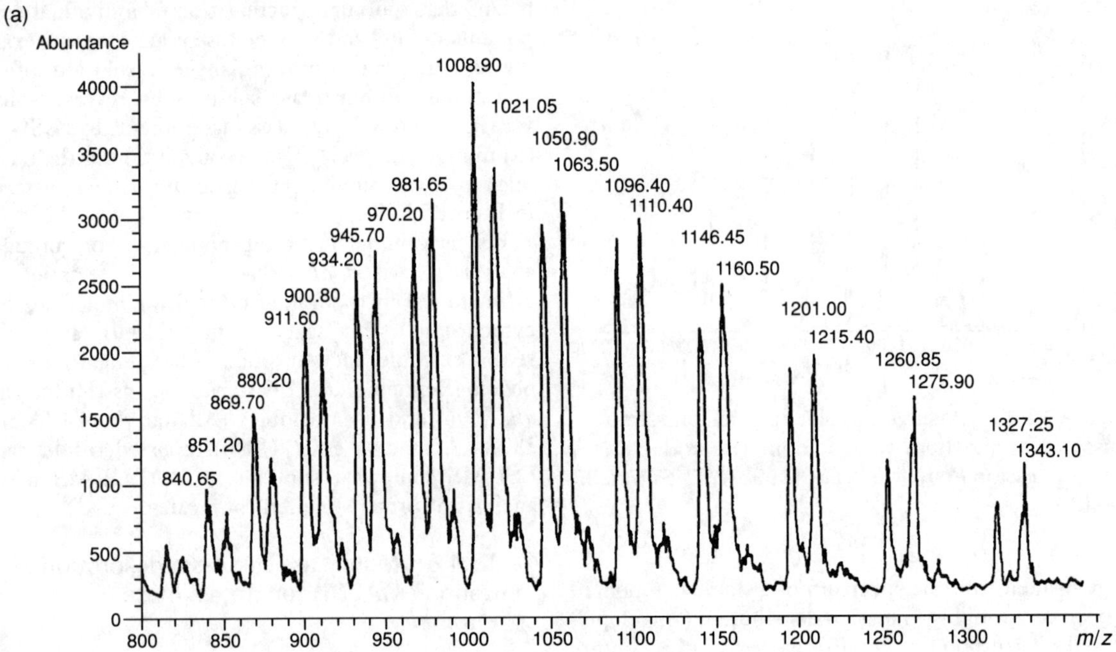

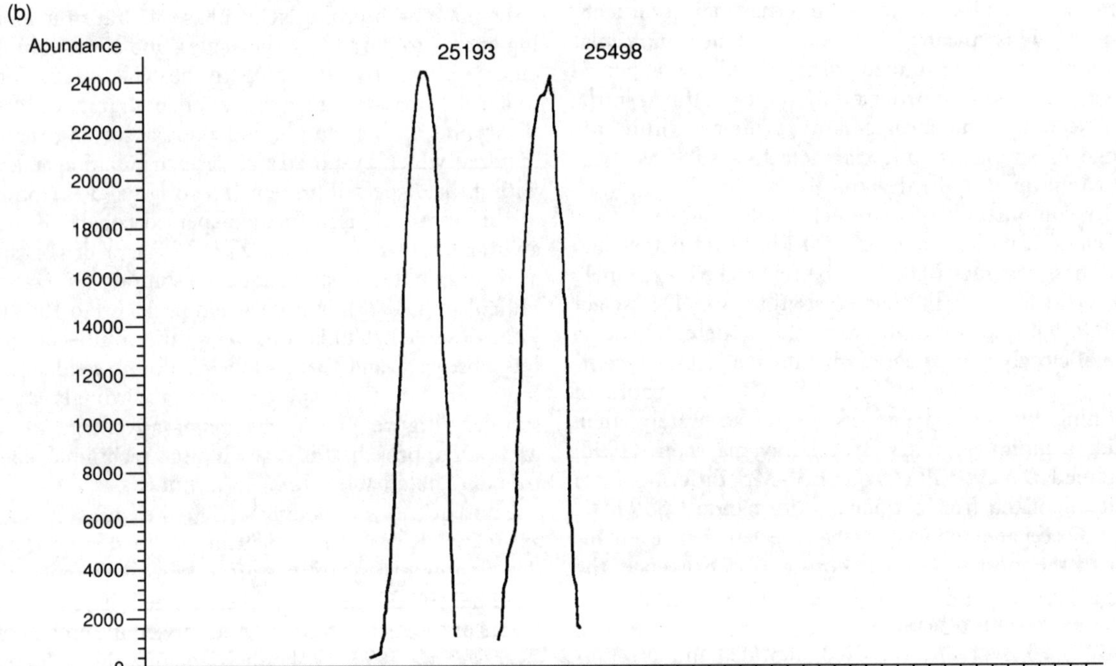

Figure 11.39 Native Sprague–Dawley rat liver GST 1–2. (a) ESI mass spectrum; (b) deconvoluted mass spectrum. (Reproduced with permission from: Rouimi *et al.* (1995). *Anal. Biochem.*, 299, 304–312.)

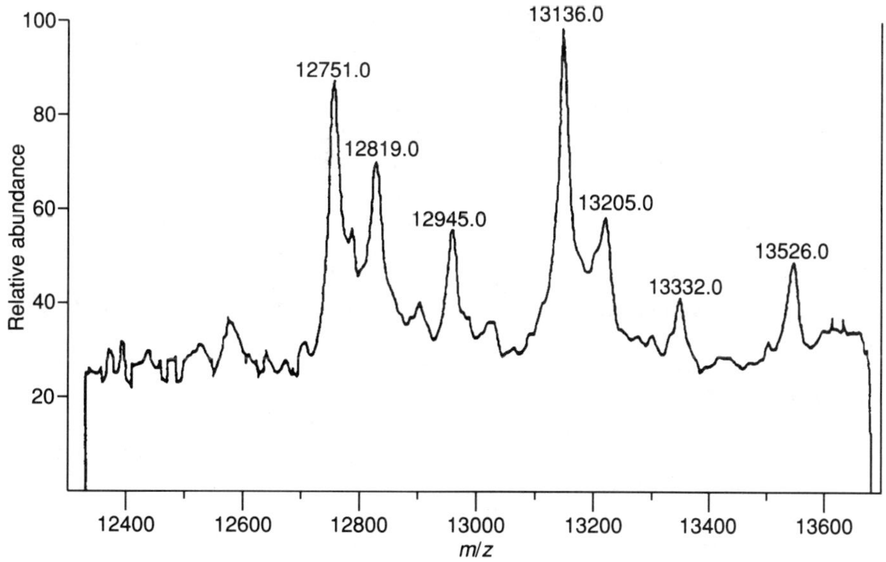

Figure 11.40 Electrospray ionization mass spectrum of cytochrome *c*-DOTA conjugate. (Reproduced with permission from: Lewis *et al.* (1994). *Bioconjugate Chem.*, **5**, 565–576.)

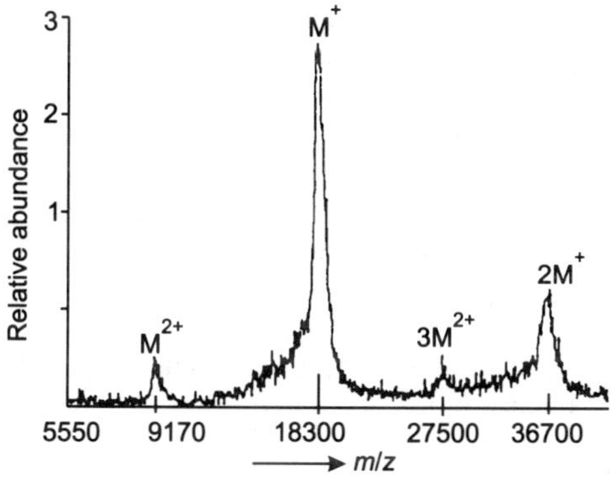

Figure 11.41 MALDI mass spectrum of β-lactoglobulin A. (Reproduced with permission from: Karas and Hillenkamp (1986). *Anal. Chem.*, **60**, 2301–2303.)

MALDI work include nicotinic acid, pyrazinecarboxylic acid and 3-nitrobenzyl alcohol (all absorbing at 266 nm, the emission wavelength of a frequency-quadrupled Nd-YAG laser), and cinnamic acid derivatives such as sinapinic acid, which also absorb at higher wavelengths (e.g. 337 nm for N_2 lasers). Useful matrix compounds are shown in Figure 11.42.

Because of the large m/z ratios generated by MALDI, detection is inefficient using ion-trap

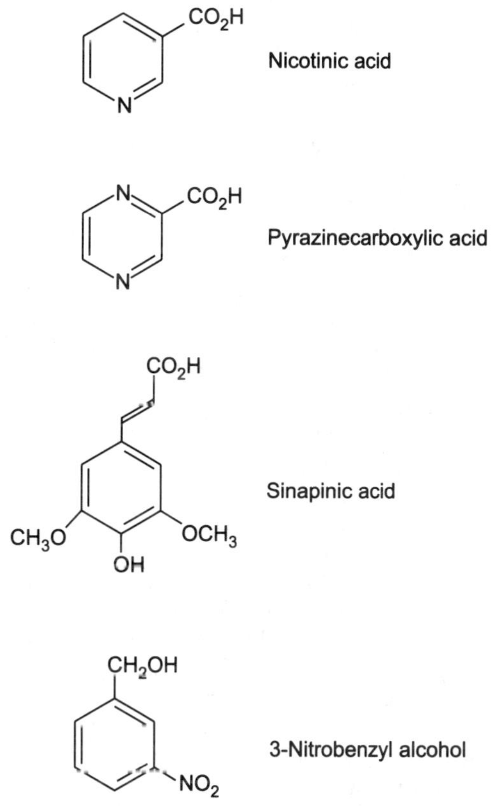

Figure 11.42 MALDI matrix compounds.

methods (Chambers *et al.*, 1993), although improvements in performance are beginning to change this situation (Jonscher and Yates, 1997). Reasonable results have been achieved for small proteins with magnetic sector instruments (Annan *et al.*, 1992), but the best performance is typically achieved by TOF methods.

Using TOF instruments, resolution of about 0.1% by mass can be obtained (e.g. 40 Da on a 40 kDa protein), which is adequate for many purposes. However, higher-resolution FT detection has also begun to be employed, with molecular mass values as high as 87 kDa being accommodated (Köster *et al.*, 1992; Castoro and Wilkins, 1993).

Some interesting variations on the MALDI theme have been described by Hutchens and Yip (1993). In one of these (surface-enhanced neat desorption, SEND) energy-absorbing molecules such as cinnamic acid analogues were bound chemically to the solid phase, remaining bound when the ionisation of analyte took place. This obviously represents a departure from the recognised role of the absorbing molecules outlined earlier, a point supported by the successful use of a compound which was ineffective in the standard MALDI approach. The spectra obtained featured less multiple-charging than MALDI, and suffered no interference from matrix or matrix–analyte complex ions.

In the second variation (surface-enhanced affinity capture, SEAC), the surface was modified to bear an affinity ligand for the protein of interest (in this case single-stranded DNA for lactoferrin), such that this protein was selectively adsorbed from a biological sample. Standard MALDI desorption was then applied. Further developments along this affinity capture route have included the direct analysis of immunoaffinity-bound antibodies (Papac *et al.*, 1994), epitope mapping by identification of antibody-bound peptide fragments (Zhao and Chait, 1994) and an MS-based immunoassay (Nelson *et al.*, 1995).

MALDI has been used analogously to ESI-MS for the assignment of accurate molecular mass values for macromolecules. Beavis *et al.* (1992), for example, determined the molecular mass of an artificial recombinant protein by this technique, not only demonstrating major errors in SDS-PAGE analysis of these materials, but also detecting two alanine-to-valine mutations in the expressed protein, cf. the expected product. However, the simplicity of MALDI spectra has led to a number of applications in the characterisation of protein conjugates, which will be described in the next section.

11.13.4.4 MALDI-MS for determination of conjugate incorporation ratios

Siegel *et al.* (1991, 1993) used MALDI-MS to investigate a number of aspects of the labelling of monoclonal antibodies with chelator and drug molecules. The molecular masses obtained from DNA sequence information were compared with the MS values, the difference being used to estimate carbohydrate content (around 2–3%, essentially all on the heavy chains). The mean incorporations of chelators and drugs were then determined by analysis of the average molecular mass values of the conjugates. The width of the peaks also was used to estimate the spread of incorporation, which appeared to fit a Lorentzian distribution (Figure 11.43). The incorporation values in many cases showed significant differences from those obtained by other techniques: the chelator values were almost always higher than those determined by incorporation of radioactive metals, possibly indicating that radioactive labelling was not quantitative. Some stable drug conjugates gave data which compared reasonably with UV spectrophotometric estimates, sometimes indicating an overestimation by UV, perhaps from non-covalently bound material. MALDI determinations of conjugates with labile compounds such as calicheamicin, on the other hand, were less successful—even this mild ionisation approach appearing to cause fragmentation of the drug molecule. Later work by these workers (Siegel *et al.*, 1993) suggested that the MALDI-MS approach was most applicable for heterogeneous proteins, and for reasonably high incorporations (>3).

Wengatz *et al.* (1992) used the same approach to estimate incorporation of a range of haptens into HRP,

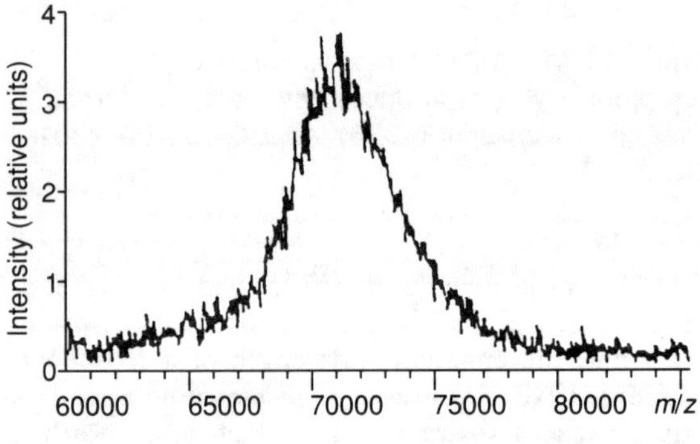

Figure 11.43 Peak shape of HSA–aminopterin conjugate after baseline subtraction. (Reproduced with permission from: Siegel *et al.* (1993). *Biol. Mass Spectrom*, **22**, 369–376.)

Table 11.5 Incorporations of hapten conjugates determined by MALDI-MS and by UV spectrophotometry (Wengatz *et al.*, 1992)

Conjugate	Incorporation	
	MALDI-MS	UV Spec.
HRP–pyrethroid	1.3	4
BSA–atrazine	20.6	33
BSA–methabenzthiazuron[1]	4.9	11
Ovalbumin–pyrethroid	1.4	10

[1] Purified by SEC; others by dialysis.

alkaline phosphatase, BSA and ovalbumin. Again, the values obtained by MS were considerably lower than those from UV, suggesting that the removal of non-covalently bound material from the conjugates was incomplete by dialysis or SEC. Examples of the differences observed for various conjugates are shown in Table 11.5.

Applications of MALDI-MS to conjugate characterisation are slowly becoming more common: Shoyama *et al.* (1993), Adamczyk *et al.* (1994) and Wring *et al.* (1994) used this technique for estimating the incorporation of various BSA immunogens, and Lewis *et al.* (1994) for cytochrome *c*–chelator conjugates. An example of a MALDI/TOF mass spectrum of a BSA immunogen from Wring *et al.* is shown in Figure 11.44.

A variation on the analysis of intact conjugates by MALDI is to carry out an initial digestion, e.g. with trypsin. Identification of the mass of the tryptic fragments can provide information about the site of derivatisation or cross-linking, as well as the level of incorporation (Yang *et al.*, 1996b; Yasuzawa and Tomer, 1997).

Finally, the applicability of MALDI to the detection of oligosaccharides from glycoproteins is worthy of mention (Küster *et al.*, 1997; Sheeley *et al.*, 1997).

As the available instrumentation improves it seems likely that MS techniques will become increasingly accepted for conjugate characterisation work. Many other protein-related applications continue to be described, for example tracing cleavage reactions to locate specific residues within the molecule (Wu *et al.*, 1996), or to monitor enzymatic degradation (Parker *et al.*, 1996).

11.13.5 Reactive group determination

One approach to assessing the incorporation of a protein–small molecule conjugate is to infer this value from a change in the number of reactive groups in the protein molecule. Typically for example the number of amine groups before and after conjugation with an active ester might be examined, and the incorporation taken as the difference between these values (Figure 11.45). Such an approach can employ any suitable method for determining the molar content of the reactive group of interest; most usually these employ specific chromophoric reagents or those bearing a radioactive label. Like all other techniques for analysis of conjugate incorporation, this one is subject to certain constraints.

11.13.5.1 Constraints on determinations of conjugate incorporation ratios by differences in reactive group content

Considerations of reactivity

First of all it is based on important assumptions about the reactivities of the species involved. On the one hand an assumption must be made about the sites in the protein which will react with its conjugation partner. This may be justifiable for a reactive group of high specificity such as a disulfide or an active ester, however it is not really appropriate for groups of broader reactivity such as haloacetamides.

Also, it is assumed that the reactivities of the conjugation partner and the analytical reagent are similar with respect to the reactive groups in the protein. The safest situation is one in which the analytical reagent can be shown beforehand to react with all the available residues in the protein. Any change from this value following derivatisation may then be taken as reliable evidence that some of these groups have been derivatised. If there is a danger that the reactivity of the analytical reagent is lower than that of the conjugation partner, i.e. that some sites are inaccessible to the analytical reagent but accessible for conjugation, then the incorporation values obtained are not secure. It may well be worthwhile to denature the protein samples before analyses of this type are carried out, to obtain the maximum exposure of reactive sites to the analytical reagent[1]. Choice of an analytical reagent resembling the conjugation partner as closely as possible, both in terms of geometry and reactivity, is another means of minimising such effects. Large differences in hydrophobicity should also be avoided if possible, as they may affect the ability of the molecule to enter the protein microenvironment.

[1] Note that it does not matter if sites are exposed which would never be available for reaction under the conjugation conditions; the difference-based incorporation value will be independent of this factor.

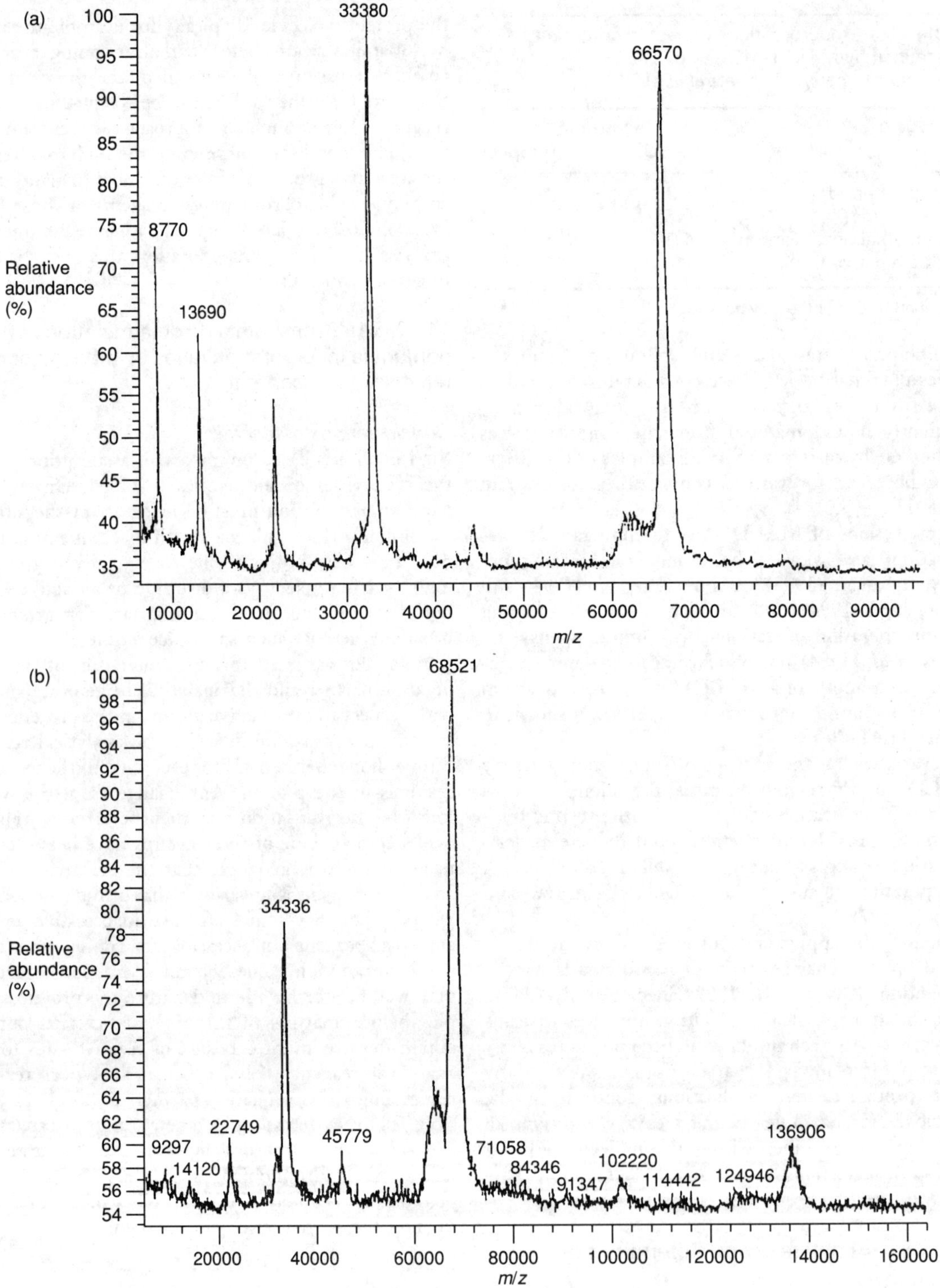

Figure 11.44 MALD/TOF mass spectra for (a) BSA and (b) BSA immunogen. (Reproduced with permission from: Wring *et al.* (1994). *J. Pharm. Biomed. Anal.*, **12**, 1573–1583.)

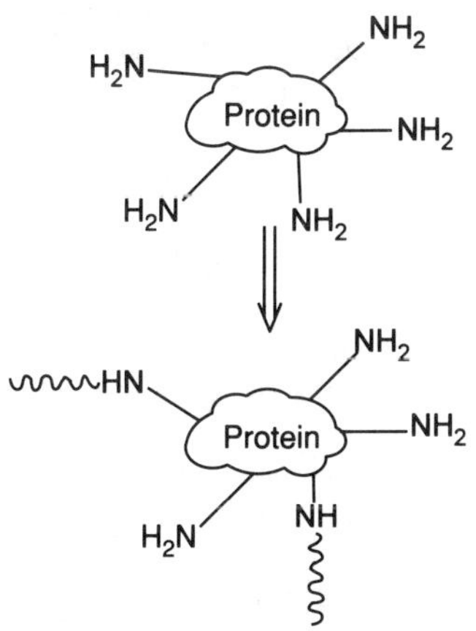

Figure 11.45 Reactive group determination: the loss of two out of five reactive amines during conjugation infers incorporation of two in the product.

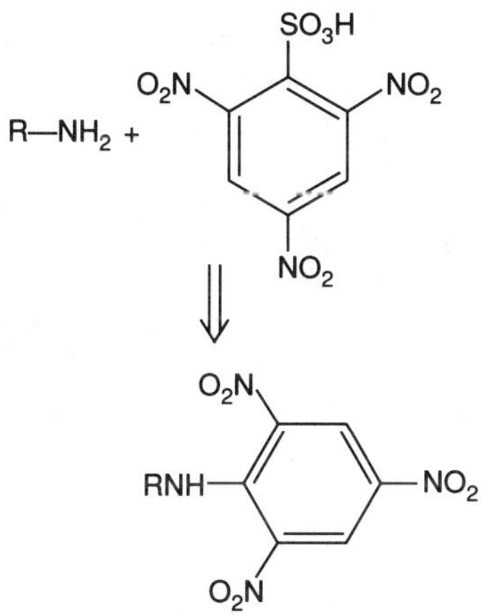

Figure 11.46 Amine determination using TNBS.

While these simple precautions may seem self-evident, it is cautionary to note that one of the commonest reagents for amine determination is trinitrobenzenesulfonic acid (Figure 11.46), which is not only a very hindered molecule but relies on a reaction mechanism quite unlike most of those used for derivation procedures.

Considerations of conjugation type

A second pitfall relates to the category of conjugation reaction being investigated. Direct reactions where no linker groups are involved lend themselves most readily to determinations of this kind, e.g. the derivatisation of protein thiol groups by a disulfide-containing hapten. In such a case the loss of a reactive group in the protein can fairly safely be assumed to correlate with the conjugation reaction of interest.

If a bifunctional reagent is used to link the two conjugation partners the situation is obviously more complex, and the rationale for assessing incorporation needs to be thought through quite carefully. Take, for example, the conjugation of a thiol-containing hapten to a protein using a heterobifunctional amine-thiol reagent. Unless it can be established that all of the heterobifunctional has been derivatised by hapten, e.g. by using a large excess of coupling reagent over thiol in a preliminary stage, there is a danger that a measured loss of amine groups in the conjugate may reflect an incorporation of linker but not of hapten (Figure 11.47). The situation with homobifunctional reagents is even worse, as ensuring that each reactive linker molecule bears a hapten can only be achieved at the expense of a considerable yield of (wasted) cross-linked hapten.

It is not impossible to carry out determinations by this approach for bifunctional methods. In the reaction shown in Figure 11.47, for example, the protein could be derivatised first with an excess of heterobifunctional, purified, and the content of thiol-reactive groups established by a suitable technique. Change in

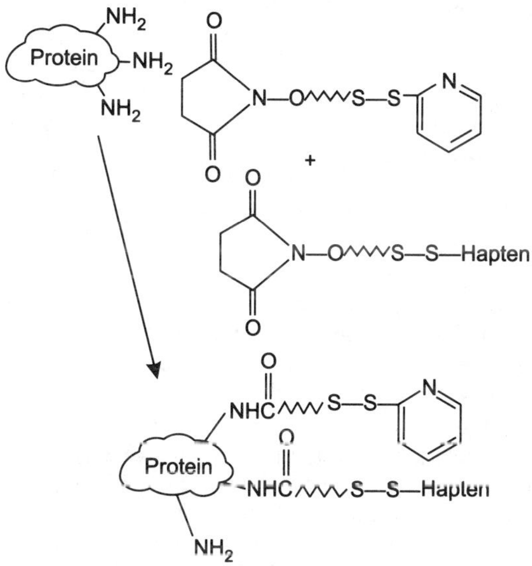

Figure 11.47 Reactive group determination can yield a false result if it is wrongly assumed that each loss of an amine is accompanied by incorporation of the target molecule into the product.

this value on derivatisation with the thiol-containing hapten could yield a reliable estimate of incorporation.

Alternatively using the route shown in the diagram, it would be possible to assess the content of unreacted pyridyldisulfide in the conjugate, by releasing the chromophoric thiopyridone by thiol treatment. Subtraction of this value from that obtained from the loss of amine groups would again offer a more realistic assessment of hapten incorporation.

Considerations of precision and accuracy

All analytical methods are constrained to some extent by limitations of precision and accuracy, but these are more particularly relevant in this application. For a start many of these techniques rely on colorimetric methods whose inherent precision may often be no better than 5–10%, and secondly the use of a difference method means that the imprecision of two determinations must be taken into account.

This consideration can be minimised by carrying out the determinations of derivatised and underivatised protein in the same assay run; the accuracy of the measurement can also be maximised by the incorporation of suitable standards, if the reactive group content of the underivatised protein itself is not well established.

However there still remains the issue of ensuring that equal quantities of derivatised and underivatised protein are compared in the reactive group assay; this will generally rely on a second, independent quantification technique (e.g. UV–visible spectrophotometry at a suitable wavelength), which will also be subject to precision constraints. These too must be considered in the overall precision of the determination of incorporations by this route.

11.13.5.2 Methods for reactive group determination

In the simplest cases, the methods used to measure conjugate incorporation in this way are based on quantifying the loss of the reactive groups in the protein which have been derivatised.

Their principle of operation is simple: the group-specific incorporation of a readily-quantifiable moiety into the protein, or (more rarely) the group-specific release of such a moiety into solution.

Where amines are the target group, the reagents used include 1-fluoro-2, 4-dinitrobenzene (FDNB) (Sanger, 1945; Nakane and Kawaoi, 1974), 2, 4, 6-trinitrobenzenesulfonic acid (TNBS) (Okuyama and Satake, 1960; Fields, 1972), fluorescamine (Udenfriend *et al.*, 1972) and the Bolton–Hunter reagent (Bolton and Hunter, 1973). These offer a choice of colorimetric, fluorometric or radiometric detection. Morçöl *et al.* (1997) recently described a method based on immobilisation of the derivatised protein on to a membrane bearing succinimide esters, which compared favourably with TNBS determination.

Numerous methods exist for thiol determination, based on maleimides (Alexander, 1958; Roberts and Rouser, 1958), mercurials (Boyer, 1954); disulfides (Ellman, 1959; Grassetti *et al.*, 1969; Habeeb, 1972; Sharma *et al.*, 1990) and benzoxadiazoles (Toyo' oka *et al.*, 1989).

Carboxyl functions are a rarer native target for protein derivatisation; the relevant determination methods are typically based on carbodiimide activation (Hoare and Koshland, 1967).

The oxidation of carbohydrates to form aldehydes is a common route for derivatisation, and methods for aldehyde determination include reductive alkylation with a radioactive amine (Wilson and Nakane, 1978) or reaction with a hydrazide-containing dye (Morehead *et al.*, 1991).

Tyrosyl and histidyl residues can be determined by diazonium reagents such as diazonium-1*H*-tetrazole (Takenaka *et al.*, 1969). There are, of course, many reagents for determination of other specific residues in proteins, but they are rarely applicable with respect to monitoring derivatisation reactions.

Many other types of reactive group are introduced artificially into proteins for conjugation purposes, and methods exist for the determination of many of these. Examples include procedures for maleimides (Singh, 1994), isothiocyanates (Zhang *et al.*, 1996) and biotin (Green, 1970; Der-Balian *et al.*, 1990; Shah *et al.*, 1994; Rao *et al.*, 1997).

11.13.6 Release of conjugated groups by hydrolysis, etc.

The last approach to conjugate characterisation, applicable mainly to the protein–small molecule category, is perhaps one of the simplest in concept: releasing the coupled molecule from the conjugate and quantifying it directly.

Many coupling reagents are available which incorporate cleavable groups such as disulfides, and these can be exploited to free the bound molecules, albeit generally in somewhat modified form (Figure 11.48). These can generally be separated from the protein readily enough, by a size-dependent technique such as SEC. Once freed from the conjugate, quantification may be achievable by a simple technique such as spectrophotometry, which would be inconvenient to apply to the intact conjugate.

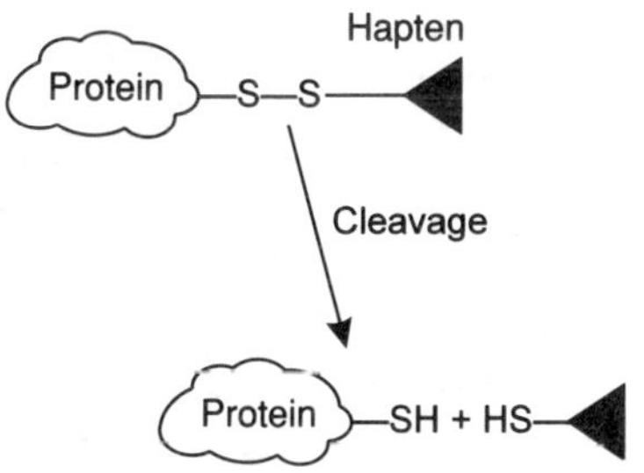

Figure 11.48 Conjugate characterisation by cleavage.

However, cleavable linker groups are often unpopular for the very reason that they are cleavable. For many applications the most robust link possible is required, and stable linkages such as amides and thioethers are favoured. In these cases release of bound species may require more vigorous treatment such as acid or alkaline hydrolysis. The fact that this will destroy the protein structure is unimportant in an analytical sample; it is more critical to establish that the bound molecule of interest can be recovered quantitatively, either in its native form, or in a consistently modified state.

Needless to say, if the recovery of conjugated material is incomplete (either in the cleavage process or in the subsequent assay method), the value obtained will fail to reflect the incorporation of the conjugate. Quantitative recoveries are often difficult to confirm without resorting to a secondary measurement technique, and these approaches are too cumbersome to have been widely adopted for conjugate characterisation. A number of examples will be described briefly, however, to indicate typical applications.

Exley and Abuknesha (1978) used acid hydrolysis to free 6-keto-estradiol-17β from a β-galactosidase conjugate, and then determined the free steroid in a specific radioimmunoassay.

Bernatowicz and Matsueda (1986) coupled synthetic cysteine-containing peptides to bromoacetylated proteins. Acid hydrolysis of the product produced *S*-carboxymethylcysteine at each point of derivatisation, and this material was quantitated by standard amino acid chromatography. Kolodny and Robey (1990) used a parallel approach, coupling bromoacetylated peptide to protein thiols, and carried out analysis in the same way.

Tatake *et al.* (1991) freed the drug sulfamethoxazole from protein conjugates by mild alkaline hydrolysis, quantitating the released material by reversed-phase chromatography.

It is worth mentioning a final area where approaches of this type can be very useful, in identifying the specific residues derivatised by a conjugation reaction. The example from Kolodny and Robey above is an example of this category: the specific identification of modified cysteine is excellent evidence for derivatisation of these residues in the conjugate. Similar approaches have proved powerful in determining the binding sites of affinity ligands: Chesebro and Metzger (1972), for example, coupled a radiolabelled phosphorylcholine analogue to an antibody, hydrolysed the product, and determined that coupling had taken place predominantly at tyrosine residues.

References

Abraham, G. E. *et al.* (1971). Radioimmunoassay of plasma progesterone. *J. Clin. Endocrinol. Metab.*, **32**, 619–624.

Adamczyk, M. and Gebler, J. C. (1997). Electrospray mass spectrometry of α and β chains of selected hemoglobins and their TNBA and TNB conjugates. *Bioconjugate Chem.*, **8**, 400–406.

Adamczyk, M. *et al.* (1994). Characterization of protein–hapten conjugates. 1. Matrix-assisted laser desorption ionization mass spectrometry of immuno BSA–hapten conjugates and comparison with other characterization methods. *Bioconjugate Chem.*, **5**, 631–635.

Adamczyk, M. *et al.* (1996). Characterization of protein-hapten conjugates. 2. Electrospray mass spectrometry of bovine serum albumin–hapten conjugates. *Bioconjugate Chem.*, **7**, 475–481.

Alexander, N. M. (1958). Spectrophotometric assay for sulfhydryl groups using *N*-ethylmaleimide. *Anal. Chem.*, **30**, 1292–1294.

Allen, S. *et al.* (1997). Atomic force microscopy in analytical biotechnology. *Trends Biotechnol.*, **15**, 101–105.

Allison, A. C. and Humphrey, J. H. (1959). Estimation of the size of antigens by gel diffusion methods. *Nature*, **183**, 1590–1592.

Altland, K. and Rossmann, U. (1985). Hybrid isoelectric focusing in rehydrated immobilized pH gradients with added carrier ampholytes: demonstration of human globins. *Electrophoresis*, **6**, 314–325.

Anderson, N. G. (1955). Studies on isolated cell components. VIII. High resolution gradient differential ultracentrifugation. *Exp. Cell. Res.*, **9**, 446–459.

Andrade, J. D. and Hlady, V. (1986). Protein adsorption andmaterials biocompatability: a tutorial review andsuggested hypotheses. *Adv. Polym. Sci.*, **79**, 1–63.

Andrews, A. T. (1986). *Electrophoresis. Theory, Techniques, and Biochemical and Clinical Applications*, Clarendon Press, Oxford.

Andrews, P. (1970). Estimation of molecular size and molecular weights of biological compounds by gel filtration. *Methods Biochem. Anal.*, **18**, 1–53.

Annan, R. S. *et al.* (1992). Matrix-assisted laser desorption using a fast-atom bombardment ion source and a magnetic mass spectrometer. *Rapid Commun. Mass Spectrom.*, **6**, 298–302.

Aubry, A.-F. and McGann, A. (1994). Applications of biochromatography to the determinations of drug–protein binding interactions. *LC–GC Int.*, **7**, 389–394.

Awdeh, Z. L. *et al.* (1968). Isoelectric focusing in polyacrylamide gel and its application to immunoglobulins. *Nature*, **219**, 66–67.

Axelrod, D. *et al.* (1984). Total internal reflection fluorescence. *Annu. Rev. Biophys. Bioeng.*, **13**, 247–268.

Axelsen, N. H. and Svendsen, P. J. (1973). Reversed rocket immunoelectrophoresis. *Scand. J. Immunol.*, **2 (Suppl. 1)**, 55–157.

Barbarakis, M. S. and Bachas, L. G. (1991). Isoelectric focusing electrophoresis of protein–ligand conjugates: effect of the degree of substitution. *Clin. Chem.*, **37**, 87–90.

Basi, N. S. and Rebois, R. V. (1997). Rate zonal sedimentation of proteins in one hour or less. *Anal. Biochem.*, **251**, 103–109.

Baskin, L. S. and Yang, C. S. (1982). Identification of cross-linked cytochrome p-450 in rat liver microsomes by enzyme-immunoassay. *Biochem. Biophys. Res. Commun.*, **108**, 700–707.

Beaven, G. H. and Holiday, E. R. (1952). Ultraviolet absorption spectra of proteins and amino acids. *Adv. Protein. Chem.*, **7**, 319–386.

Beavis, R. C. *et al.* (1992). Analysis of artificial proteins by matrix-assisted laser desorption mass spectrometry. *J. Am. Chem. Soc.*, **114**, 7584–7585.

Becker, W. (1969). Determination of antisera titres using the single radial immunodiffusion method. *Immunochemistry*, **6**, 539–546.

Behrens, M. (1938). Über die Lokalisation der Hefenucleinsäure in pflanzlichen Zellen. *Hoppe-Seyler's Z. Physiol. Chem.*, **253**, 185–192.

Bennett, K. L. *et al.* (1996). Rapid characterization of chemically-modified proteins by electrospray mass spectrometry. *Bioconjugate Chem.*, **7**, 16–22.

Bennett, T. P. (1967). Membrane filtration for determining protein in the presence of interfering substances. *Nature*, **213**, 1131–1132.

Bernatowicz, M. S. and Matsueda, G. R. (1986). Preparation of peptide–protein immunogens using *N*-succinimidyl bromoacetate as a heterobifunctional crosslinking agent. *Anal. Biochem.*, **155**, 95–102.

Bjellqvist, B. *et al.* (1982). Isoelectric focusing in immobilized pH gradients: principle, methodology and some applications. *J. Biochem. Biophys. Methods*, **6**, 317–339.

Blackshear, P. J. (1984). Systems for polyacrylamide gel electrophoresis. *Methods Enzymol.*, **104**, 237–255.

Bøg-Hansen, T. C. (1973). Crossed immunoaffinoelectrophoresis. A method to predict the result of affinity chromatography. *Anal. Biochem.*, **56**, 480–488.

Bolton, A. E. and Hunter, W. M. (1973). The labelling of proteins to high specific radioactivities by conjugation to a I-containing acylating reagent. *Biochem. J.*, **133**, 529–539.

Bonner, W. M. (1984). Fluorography for the detection of radioactivity in gels. *Methods Enzymol.*, **104**, 460–465.

Bonner, W. M. and Laskey, R. A. (1974). A film detection method for tritium-labelled proteins and nucleic acids in polyacrylamide gels. *Eur. J. Biochem.*, **46**, 83–88.

Borek, F. and Silverstein, A. M. (1961). Characterization and purification of ferritin–antibody globulin conjugates. *J. Immunol.*, **87**, 555–561.

Böttcher, B. *et al.* (1997). Determination of the fold of the core protein of hepatitis B virus by electron cryomicroscopy. *Nature*, **386**, 88–91.

Boyer, P. D. (1954). Spectrophotometric study of the reaction of protein sulfhydryl groups with organic mercurials. *J. Am. Chem. Soc.*, **76**, 4331–4337.

Bradford, M. M. (1976). A rapid and sensitive method for the quantitation of microgram quantities of protein utilizing the principle of protein–dye binding. *Anal. Biochem.*, **72**, 248–254.

Bragg, P. D. and Hou, C. (1975). Subunit composition, function and spatial arrangement in the Ca^{2+} and Mg^{2+}-activated adenosine triphosphatases of *Escherichia coli* and *Salmonella typhimurium*. *Arch. Biochem. Biophys.*, **167**, 311–321.

Britten, R. J. and Roberts, R. B. (1960). High-resolution density gradient sedimentation analysis. *Science*, **131**, 32–33.

Butler, M. and O'Kennedy, R. (1994). Use of non-radioactive iodine-labelled antibody for an immunoassay, and comparison of detection methods. *Anal. Lett.*, **27**, 681–692.

Campbell, W. R. and Hanna, M. I. (1937). The determination of nitrogen by modified Kjeldahl methods. *J. Biol. Chem.*, **119**, 1–7.

Candiano, G. *et al.* (1996). Negative staining of proteins in polyacrylamide gels with methyl trichloroacetate. *Anal. Biochem.*, **243**, 245–248.

Carpenter, F. H. and Harrington, K. T. (1972). Intermolecular cross-linking of monomeric proteins and cross-linking of oligomeric proteins as a probe of quaternary structure. *J. Biol. Chem.*, **247**, 5580–5586.

Castoro, J. A. and Wilkins, C. L. (1993). Ultrahigh resolution matrix-assisted laser desorption/ionization of small proteins by Fourier transform mass spectrometry. *Anal. Chem.*, **65**, 2621–2627.

Chambers, D. M. *et al.* (1993). Matrix-assisted laser desorption of biological molecules in the quadrupole ion trap mass spectrometer. *Anal. Chem.*, **65**, 14–20.

Chattoraj, D. K. and Bull, H. B. (1959). Electrophoresis of adsorbed protein. *J. Am. Chem. Soc.*, **81**, 5128–5133.

Chen, F. A. *et al.* (1992). Use of high ionic strength buffersfor the separation of proteins and peptides with capillary electrophoresis. *J. Liq. Chromatogr. A*, **15**, 1143–1161.

Chen, F.-T. A. and Pentoney, S. L., Jr. (1994). Characterization of digoxigenin-labeled B-phycoerythrin by capillary electrophoresis with laser-induced fluorescence. Application to homogeneous digoxin immunoassay. *J. Chromatogr.*, **A680**, 425–430.

Chen, J.-L. and Morawetz, H. (1981). Affinity electrophoresis in gels containing hydrophobic substituents. *J. Biol. Chem.*, **256**, 9221–9223.

Chesebro, B. and Metzger, H. (1972). Affinity labeling of a phosphorylcholine binding mouse myeloma protein. *Biochemistry*, **11**, 766–771.

Chevallet, M. *et al.* (1997). A nonradioactive double detection method for the assignment of spots in two-dimensional blots. *Anal. Biochem.*, **251**, 69–72.

Choi, J.-K. *et al.* (1996). A modified Coomassie Blue staining of proteins in polyacrylamide gels with Bismark Brown R. *Anal. Biochem.*, **236**, 82–84.

Chou, S.-C. and Goldstein, A. (1960). Chromogenic groupings in the Lowry protein determination. *Biochem. J.*, **75**, 109–115.

Chrambach, A. and Jovin, T. M. (1983). Selected buffer systems for moving boundary electrophoresis on gels at various pH values, presented in a simplified manner. *Electrophoresis*, **4**, 190–204.

Chrambach, A. and Rodbard, D. (1971). Polyacrylamide gel electrophoresis. *Science*, **172**, 440–451.

Chrambach, A. *et al.* (1976). Analytical and preparative polyacrylamide gel electrophoresis. An objectively defined fractionation route, apparatus and procedures. In *Methods of Protein Separation* (Catsimpoolas, N., ed.), vol. 2, pp. 27–144, Plenum Press, New York.

Clegg, J. C. S. (1982). Glycoprotein detection in nitrocellulose transfers of electrophoretically separated mixtures using concanavalin A and peroxidase: application to arenavirus and flavivirus proteins. *Anal. Biochem.*, **127**, 389–394.

Cohen, A. S. *et al.* (1987). High-performance capillary electrophoresis using open tubes and gels. *Chromatographia*, **24**, 15–24.

Colenbrander, V. F. and Martin, T. G. (1971). Orange G dye binding for determination of protein of sow's milk. *J. Dairy Sci.*, **54**, 531–533.

Compton, S. J. and Jones, C. G. (1985). Mechanism of dye response and interference in the Bradford protein assay. *Anal. Biochem.*, **151**, 369–374.

Congdon, R. W. *et al.* (1993). The binding interaction of Coomassie Blue with proteins. *Anal. Biochem.*, **213**, 407–413.

Cotton, T. M. (1984). Surface enhanced raman spectroscopy of biological macromolecules. In *Surface and Interfacial Aspects of Biomedical Polymers. Protein Adsorption* (Andrade, J. D., ed.), vol. 2, pp. 161–187, Plenum Press, New York.

Creech, H. J. and Jones, R. N. (1940). The conjugation of horse serum albumin with 1,2-benzanthryl isocyanates. *J. Am. Chem. Soc.*, **62**, 1970–1975.

Creech, H. J. and Jones, R. N. (1941). The conjugation of horse serum albumin with isocyanates of certain polynuclear aromatic hydrocarbons. *J. Am. Chem. Soc.*, **63**, 1661–1669.

Crisp, S. J. and Dunn, M. J. (1994). Detection of proteins on protein blots using chemiluminescent systems. *Methods Mol. Biol.*, **32**, 233–237.

Cuatrecasas, P. *et al.* (1969). Affinity labeling of the active site of staphylococcal nuclease. Reactions with bromoacetylated substrate analogues. *J. Biol. Chem.*, **244**, 4316–4329.

Cullen, G. E. and Van Slyke, D. D. (1920). Determination of the fibrin, globulin, and albumin nitrogen of blood plasma. *J. Biol. Chem.*, **41**, 587–597.

Darawshe, S. *et al.* (1993). Sedimentation equilibrium–quantitative polyacrylamide gel electrophoresis (SE-QPAGE): A new technique for the detection of associations in multicomponent solutions. *Anal. Biochem.*, **215**, 236–242.

Davies, G. E. and Stark, G. R. (1970). Use of dimethyl suberimidate, a cross-linking reagent, in studying the subunit structure of oligomeric proteins. *Proc. Natl. Acad. Sci. USA*, **66**, 651–656.

Davies, J. *et al.* (1994). Use of scanning probe microscopy and surface plasmon resonance as analytical tools in the study of antibody-coated microtiter wells. *Langmuir*, **10**, 2654–2661.

Davies, J. *et al.* (1995). Scanning tunnelling microscopy and dynamic contact angle studies of the effects of partial denaturation on immunoassay solid phase antibody. *J. Immunol. Methods.*, **186**, 111–123.

Davis, B. (1964). Disc electrophoresis: II. Method and application to human serum proteins. *Ann. NY Acad. Sci.*, **121**, 404–427.

de Duve, C. *et al.* (1959). Gradient centrifugation of cell particles. Theory and applications. *Prog. Biophys. Biophys. Chem.*, **9**, 325–369.

Der-Balian, G. P. *et al.* (1990). A fluorometric method for determining the degree of biotinylation of proteins. *J. Immunol. Methods*, **126**, 281–285.

De Rosier, D. J. (1997). Who needs crystals anyway? *Nature*, **386**, 26–27.

Dole, M. *et al.* (1968). Molecular beams of macroions. *J. Chem. Phys.*, **49**, 2240–2249.

Dollinger, G. *et al.* (1992). Practical on-line determination of biopolymer molecular weights by high-performance liquid chromatography with classical light-scattering detection. *J. Chromatogr.*, **592**, 215–228.

Dunker, A. K. and Rueckert, R. R. (1969). Observations on molecular weight determinations on polyacrylamide gel. *J. Biol. Chem.*, **244**, 5074–5080.

Dunn, M. J. (1989). Initial planning: electrophoretic analysis methods. In *Protein Purification Methods: A Practical Approach* (Harris, E. L. V. and Angal, S., eds), pp. 1–66. IRL Press, Oxford.

Dunn, M. J. and Burghes, A. H. M. (1986). High resolution two-dimensional polyacrylamide-gel electrophoresis. In *Gel Electrophoresis of Proteins* (Dunn, M. J., ed.), pp. 203–261, Wright, Bristol.

Dunn, M. J. and Crisp, S. J. (1994). Detection of proteins in polyacrylamide gels using an ultrasensitive silver staining technique. *Methods Mol. Biol.*, **32**, 113–118.

Dwyer, J. L. (1993). Analytical chromatography of amino acids, peptides, and proteins. In *Protein Biotechnology* (Franks, F., ed.), pp. 49–90, Humana Press, Totowa, NJ.

Einarsson, R. *et al.* (1984). Use of capillary isotachophoresis to measure the degree of chemical substitution of dog serum albumin. *J. Chromatogr.*, **284**, 143–147.

Ellman, G. L. (1959). Tissue sulfhydryl groups. *Arch. Biochem. Biophys.*, **82**, 70–77.

Eruk, A. A. *et al.* (1984). The preparation and characterization of alkaline phosphatase–thyroxine conjugates and their use in the enzyme immunoassay of thyroxine. *Ann. Clin. Biochem.*, **21**, 434–443.

Exley, D. and Abuknesha, R. (1978). A highly sensitive enzyme-immunoassay method for eestradiol-17β. *FEBS Lett.*, **91**, 162–165.

Fadden, P. and Haystead, T. A. J. (1995). Quantitative and selective fluorophore labeling of phosphoserine on peptides and proteins: characterization at the attomole level by capillary electrophoresis and laser-induced fluorescence. *Anal. Biochem.*, **225**, 81–88.

Fairbanks, G. *et al.* (1971). Electrophoretic analysis of the major polypeptides of the human erythrocyte membrane. *Biochemistry*, **10**, 2606–2617.

Fazekas de St. Groth, S. *et al.* (1963). Two new staining procedures for quantitative estimation of proteins on electrophoretic strips. *Biochim. Biophys. Acta*, **71**, 377–391.

Feng, L. *et al.* (1988). Scanning tunneling microscopic images of adsorbed serum albumin on highly oriented pyrolytic graphite. *J. Colloid Interface Sci.*, **126**, 650–653.

Feng, L. *et al.* (1989). Scanning tunneling microscopy of proteins on graphite surfaces. *Scanning Microsc.*, **3**, 399–410.

Fields, R. (1972). The rapid determination of amino groups with TNBS. *Methods Enzymol.*, **25**, 464–468.

Flodin, P. and Porath, J. (1954). Zone electrophoresis in starch columns. *Biochim. Biophys. Acta*, **13**, 175–182.

Folin, O. and Ciocalteu, V. (1927). On tyrosine and tryptophane determinations in proteins. *J. Biol. Chem.*, **73**, 627–650.

Folin, O. and Denis, W. (1912). On phosphotungstic–phosphomolybdic compounds as color reagents. *J. Biol. Chem.*, **12**, 239–243.

Folin, O. and Wu, H. (1919). A system of blood analysis. *J. Biol. Chem.*, **38**, 81–110.

Ford, D. J. *et al.* (1978). Characterization of glutaraldehyde coupled alkaline phosphatase–antibody and lactoperoxidase–antibody conjugates. *Immunochemistry*, **15**, 237–243.

Fothergill-Gilmore, L. A. (1993). In *Protein Biotechnology* (Franks, F., ed.), p. 257, Humana Press, Totowa, NJ.

Fowler, S. J. (1994). The detection of proteins on blots using gold or immunogold. *Methods Mol. Biol.*, **32**, 239–255.

Fox. T. *et al.* (1994). Derivatization of yeast cytochrome *c* peroxidase with pentaammineruthenium(II). *Bioconjugate Chem.*, **5**, 14–20.

Fujiwara, K. *et al.* (1981). Novel preparation method of immunogen for hydrophobic hapten, enzyme immunoassay for daunomycin and adriamycin. *J. Immunol. Methods*, **45**, 195–203.

Gadkari, D. A. *et al.* (1985). Enzyme–antibody conjugation by a heterobifunctional reagent and its application in enzyme-linked immunosorbent assay (ELISA) for the detection of hepatitis B surface antigen. *J. Virol. Methods*, **10**, 215–224.

Gander, J. E. (1984). Gel protein stains: glycoproteins. *Methods Enzymol.*, **104**, 447–451.

Gassmann, E. *et al.* (1985). Electrokinetic separation of chiral compounds. *Science*, **230**, 813–814.

Gershoni, J. M. *et al.* (1985). Blot analyses of glycoconjugates: enzyme–hydrazide: a novel reagent for the detection of aldehydes. *Anal. Biochem.*, **146**, 59–63.

Gindler, E. M. (1970). Automated determination of glucose via reductive formation of lavender Cu(I)–2,2′-bicinchonate chelate. *Clin. Chem.*, **16**, 519.

Gleizes, P.-E. *et al.* (1994). Labeling of basic cibroblast growth factor with digoxigenin: A nonradioactive probe for biochemical and cytological applications. *Anal. Biochem.*, **219**, 360–367.

Goldfarb, A. R. *et al.* (1951). The ultraviolet absorption spectra of proteins. *J. Biol. Chem.*, **193**, 397–404.

Goldring, J. P. D. and Ravaioli, L. (1996). Solubilization of protein–dye complexes on nitrocellulose to quantify proteins spectrophotometrically. *Anal. Biochem.*, **242**, 197–201.

Goodrow, M. H. *et al.* (1990). Hapten synthesis, antibody development, and competitive inhibition enzyme immunoassay for *s*-triazine herbicides. *J. Agric. Food Chem.*, **38**, 990–996.

Gornall, A. G. *et al.* (1949). Determination of serum proteins by means of the biuret reaction. *J. Biol. Chem.*, **177**, 751–766.

Goubran-Boutros, H. *et al.* (1992). Immobilized metal ion affinity electrophoresis. A study with several model proteins containing histidine. *J. Chromatogr.*, **597**, 357–364.

Grabar, P. and Williams, C. A. (1953). Méthode permettant l'étude conjuguée des propriétés électrophorétiques et immunochimiques d'un mélange de protéines. Application au sérum sanguin. *Biochim. Biophys. Acta*, **10**, 193–194.

Grassetti, D. R. *et al.* (1969). The interaction of 6, 6'-dithiodinicotinic acid with thiols and with Ehrlich ascites tumor cells. *Biochem. Pharmacol.*, **18**, 603–611.

Green, N. M. (1970). Spectrophotometric determination of avidin and biotin. *Methods Enzymol.*, **18A**, 418–424.

Grönwall, A. (1942). Studies on the solubility of lactoglobulin. I. The solubility of lactoglobulin in dilute solutions of sodium chloride at varying ionic strength and hydrogen ion activity. *C. R. Lab. Carlsberg*, **24**, 185–200.

Habeeb, A. F. S. A. (1972). Reaction of protein sulfhydryl groups with Ellman's reagent. *Methods Enzymol.*, **25**, 457–464.

Hamilton, R. G. *et al.* (1987). Quality control of murine monoclonal antibodies using isoelectric focusing affinity immunoblot analysis. *Hybridoma*, **6**, 205–217.

Harthun, S. *et al.* (1997). Determination of recombinant protein in animal cell culture supernatant by near-infrared spectroscopy. *Anal. Biochem.*, **251**, 73–78.

Hartree, E. F. (1972). Determination of protein: A modification of the Lowry method that gives a linear photometric response. *Anal. Biochem.*, **48**, 422–427.

Haselbeck, A. and Hösel, W. (1993). Immunological detection of glycoproteins on blots based on labeling with digoxigenin. *Methods Mol. Biol.*, **14**, 161–173.

Haupt, K. *et al.* (1996). Immobilized metal ion affinity capillary electrophoresis of proteins: A model for affinity capillary electrophoresis using soluble polymer-supported ligands. *Anal. Biochem.*, **234**, 149–154.

Hayes, P. C. *et al.* (1989). Blotting techniques for the study of DNA, RNA and proteins. *Br. Med. J.*, **299**, 965–968.

Heeb, M. J. and Gabriel, O. (1984). Enzyme localization in gels. *Methods Enzymol.*, **104**, 416–439.

Heegaard, P. M. H. and Bøg-Hansen, T. C. (1986). Immunoelectrophoretic methods. In *Gel Electrophoresis of Proteins* (Dunn, M. J., ed.), pp. 262–322, Wright, Bristol.

Heidelberger, M. and MacPherson, C. F. C. (1943). Quantitative micro-estimation of antibodies in the sera of man and other animals. *Science*, **97**, 405–406.

Herriott, R. M. (1941). Reaction of Folin's reagent with proteins and biuret compounds in presence of cupric ion. *Proc. Soc. Exp. Biol. Med.*, **46**, 642–644.

Hiratsuka, T. (1988). Cross-linking of three heavy-chain domains of myosin adenosinetriphosphatase with a trifunctional alkylating agent. *Biochemistry*, **27**, 4110–4114.

Hirayama, K. *et al.* (1990). Rapid confirmation and revision of the primary structure of bovine serum albumin by ESIMS and frit-FAB LC/MS. *Biochem. Biophys. Res. Commun.*, **173**, 639–646.

Hoare, D. G. and Koshland, D. E., Jr. (1967). A method for the quantitative modification and estimation of carboxylic acid groups in proteins. *J. Biol. Chem.*, **242**, 2447–2453.

Holmes, L. D. *et al.* (1992). Metal-affinity electrophoresis of histidine-containing proteins. *Methods: Companion Methods Enzymol.*, **4**, 103–108.

Honig, R. E. and Woolston, J. R. (1963). Laser-induced emission of electrons, ions and neutral atoms from solid surfaces. *Appl. Phys. Lett.*, **2**, 138–139.

Hořejší, V. (1984). Affinity electrophoresis. *Methods Enzymol.*, **104**, 275–281.

Hsu, K. C. *et al.* (1963). Fluorescent, electron microscopic, and immunoelectrophoretic studies of labeled antibodies. *Science*, **142**, 1471–1473.

Hu, G. R. *et al.* (1997). High resolution autoantibody detection by optimized Protein G–horseradish peroxidase immunostaining in a Western blotting assay. *J. Immunol. Methods*, **202**, 113–121.

Hufnagel, P. *et al.* (1996). Electrospray ionization mass spectrometry of genetically and chemically modified bacteriorhodopsins. *Anal. Biochem.*, **243**, 46–54.

Hutchens, T. W. and Yip, T.-T. (1993). New desorption strategies for the mass spectrometric analysis of macromolecules. *Rapid Commun. Mass Spectrom.*, **7**, 576–580.

Jerry, D. J. (1993). Monitoring coupling of peptides to carrier proteins using biotinylated peptide. *BioTechniques*, **14**, 464–469.

Ji, T. H. (1979). The application of chemical crosslinking for studies on cell membranes and the identification of surface reporters. *Biochim. Biophys. Acta*, **559**, 39–69.

Jonscher, K. R. and Yates, J. R., III. (1997). The quadupole ion trap mass spectrometer—a small solution to a big challenge. *Anal. Biochem.*, **244**, 1–15.

Jorgenson, J. W. and Lukacs, K. DeA. (1981a). Zone electrophoresis in open-tubular glass capillaries. *Anal. Chem.*, **53**, 1298–1302.

Jorgenson, J. W. and Lukacs, K. DeA. (1981b). Free-zone electrophoresis in glass capillaries. *Clin. Chem.*, **27**, 1551–1553.

Jorgenson, J. W. and Lukacs, K. DeA. (1983). Capillary zone electrophoresis. *Science*, **222**, 266–272.

Jovin, T. M. (1973). Multiphasic zone electrophoresis. I. Steady-state moving-boundary systems formed by different electrolyte combinations. *Biochemistry*, **12**, 871–898.

Kalász, H. and Báthori, M. (1997). Present status and future perspectives of thin layer chromatography. *Liq. Chromatogr. Gas Chromatogr. Int.*, **July 1997 issue**, 440–445.

Kaltschmidt, E. and Wittmann, H. G. (1970). Ribosomal proteins. VII. Two-dimensional polyacrylamide gel electrophoresis for fingerprinting of ribosomal proteins. *Anal. Biochem.*, **36**, 401–412.

Kaminski, M. and Ouchterlony, O. (1951). Études immunochimiques, qualitatives et quantitatives des protéides du blanc d'oeuf. *Bull. Soc. Chim. Biol.*, **33**, 758–770.

Kamps-Holtzapple, C. *et al.* (1993). Analysis of hapten-carrier protein conjugates by nondenaturing gel electrophoresis. *J. Immunol. Methods*, **164**, 245–253.

Karas, M. and Hillenkamp, F. (1988). Laser desorption ionization of proteins with molecular masses exceeding 10,000 daltons. *Anal. Chem.*, **60**, 2299–2301.

Karas, M. *et al.* (1989a). Laser desorption/ionization mass spectrometry of proteins of mass 100,000 to 250,000 dalton. *Angew. Chem., Int. Ed. Engl.*, **28**, 760–762.

Karas, M. *et al.* (1989b). Ultraviolet-laser desorption/ionization mass spectrometry of femtomolar amounts of large proteins. *Biomed. Environ. Mass Spectrom.*, **18**, 841–843.

Karas, M. *et al.* (1990). Principles and applications of matrix-assisted UV-laser desorption/ionization mass spectrometry. *Anal. Chim. Acta*, **241**, 175–185.

Karas, M. *et al.* (1991). Matrix-assisted laser desorption ionization mass spectrometry. *Mass Spectrom. Rev.*, **10**, 335–357.

Kaul, A. K. *et al.* (1970). The macro and micro dye-binding techniques of estimating the protein quality in food samples. *J. Food Sci. Technol.*, **7**, 11–16.

Kelly, J. F. *et al.* (1993). Analysis of protein glycoforms by capillary electrophoresis–electrospray mass spectrometry. *Beckman Discovery Newslett.*, **2**, 1–6.

Kempner, E. S. (1993). Novel predictions from radiation target analysis. *Trends Biochem. Sci.*, **18**, 236–239.

Kennedy, J. H. *et al.* (1976). The application of thin-layer gel filtration chromatography in the rapid assessment of protein–protein coupling reactions. *J. Chromatogr.*, **118**, 337–343.

Kingsley, G. R. (1942). The direct biuret method for the determination of serum proteins as applied to photoelectric and visual colorimetry. *J. Lab. Clin. Med.*, **27**, 840–845.

Knisley, K. A. and Rodkey, L. S. (1986). Affinity immunoblotting. High resolution isoelectric focusing analysis of antibody clonotype distribution. *J. Immunol. Methods*, **95**, 79–87.

Knisley, K. A. and Rodkey, L. S. (1988). Isoelectric focusing analysis of antibody clonotype changes occurring during immune responses using immobilized pH gradients. *Electrophoresis*, **9**, 183–186.

Kolodny, N. and Robey, F. A. (1990). Conjugation of synthetic peptides to proteins: quantitation from *S*-carboxymethylcysteine released upon acid hydrolysis. *Anal. Biochem.*, **187**, 136–140.

Koltisko, B. and Walton, A. (1984). Chromatographic analysis of protein adsorption. In *Surface and Interfacial Aspects of Biomedical Polymers Protein Adsorption* (Andrade, J. D., ed.), vol. 2, pp. 217–239, Plenum Press, New York.

Köster, C. *et al.* (1992). High-resolution matrix-assisted laser desorption/ionization of biomolecules by Fourier transform mass spectrometry. *J. Am. Chem. Soc.*, **114**, 7572–7574.

Kraehenbuhl, J. P. *et al.* (1974). Preparation and characterization of an immunoelectron microscope tracer consisting of a heme-octapeptide coupled to Fab. *J. Exp. Med.*, **139**, 208–223.

Kruger, N. J. (1994). The Bradford method for protein quantitation. *Methods Mol. Biol.*, **32**, 9–15.

Krull, I. S. *et al.* (1995). Biopolymer detection, part I: mass spectrometry. *LC–GC. Int.*, **8**, 14–20.

Küster, B. *et al.* (1997). Sequencing of N-linked oligosaccharides directly from protein gels: in-gel deglycosylation followed by matrix-assisted laser desorption/ionization mass spectrometry and normal-phase high-performance liquid chromatography. *Anal. Biochem.*, **250**, 82–101.

Laemmli, U. K. (1970). Cleavage of structural proteins during the assembly of the head of bacteriophage T4. *Nature*, **227**, 680–685.

Laine, A. (1992a). Quantitative and qualitative immunoelectrophoresis. *Methods Mol. Biol.*, **10**, 195–200.

Laine, A. (1992b). Rocket immunoelectrophoresis technique or electroimmunodiffusion. *Methods Mol. Biol.*, **10**, 201–205.

Laine, A. (1992c). Crossed immunoelectrophoresis. *Methods Mol. Biol.*, **10**, 207–213.

Laine, A. (1992d). Crossed immunoaffinoelectrophoresis. *Methods Mol. Biol.*, **10**, 215–219.

Laine, R. A. (1992). Mass spectrometry of carbohydrates. In *Glycoconjugates. Composition, Structure and Function.* (Allen, H. J. and Kisailus, E. C., eds.), pp. 103–119, Marcel Dekker, Inc., New York.

Laitinen, M. P. A. *et al.* (1996). Thin-layer affinity chromatography of protein-ligand affinity. *Anal. Biochem.*, **243**, 279–282.

Lampson, G. P. and Tytell, A. A. (1965). A simple method for estimating isoelectric points. *Anal. Biochem.*, 11, 374–377.

Laskey, R. A. and Mills, A. D. (1977). Enhanced autoradiographic detection of ^{32}P and ^{125}I using intensifying screens and hypersensitized film. *FEBS Lett.*, **82**, 314–316.

Laskey, R. A. *et al.* (1972). Translation of encephalomyocarditis viral RNA in oocytes of *Xenopus laevis*. *Proc. Natl. Acad. Sci. USA*, **69**, 3665–3669.

Laurell, C.-B. (1965). Antigen antibody crossed electrophoresis. *Anal. Biochem.*, **10**, 358–361.

Laurell, C.-B. (1966). Quantitative estimation of proteins by electrophoresis in agarose gel containing antibodies. *Anal. Biochem.*, **15**, 43–52.

Lewis, M. R. *et al.* (1994). A facile, water-soluble method for modification of proteins with DOTA. Use of elevated temperature and optimized pH to achieve high specific activity and high chelate stability in radiolabeled immunoconjugates. *Bioconjugate Chem.*, **5**, 565–576.

Lieberman, S. *et al.* (1959). Steroid–protein conjugates: their chemical, immunochemical and endocrinological properties. *Recent Prog. Hormone Res.*, **15**, 165–200.

Little, J. R. and Donahue, H. (1968). Spectral properties of proteins and small molecules of immunological interest. *Methods Enzymol. Immunochem.*, **2**, 343–364.

Lim, M. J. *et al.* (1997). A luminescent europium complex for the sensitive detection of proteins and nucleic acids immobilized on membrane supports. *Anal. Biochem.*, **245**, 184–195.

Liu, J. *et al.* (1995). Analysis of monoclonal antibody chimeric BR96–doxorubicin immunoconjugate by sodium dodecyl sulfate-capillary gel electrophoresis with ultraviolet and laser-induced fluorescence detection. *Anal. Biochem.*, **229**, 221–228.

Lomant, A. J. and Fairbanks, G. (1976). Chemical probes of extended chemical structures: synthesis and properties of the cleavable protein cross-linking reagent [^{35}S]dithiobis(succinimidyl propionate). *J. Mol. Biol.*, **104**, 243–261.

Lowry, O. H. *et al.* (1951). Protein measurement with the Folin phenol reagent. *J. Biol. Chem.*, **193**, 265–275.

Ludeke, B. I. (1992). An immuno-slot-blot assay for detection and quantitation of alkyldeoxyguanosines in DNA. *Methods Mol. Biol.*, **10**, 307–319.

Ma, T. S. and Zuazaga, G. (1942). Micro-Kjeldahl determination of nitrogen. A new indicator and an improved rapid method. *Ind. Eng. Chem.*, **14**, 280–282.

Macdonald, R. A. *et al.* (1988). The measurement of relative antibody affinity by ELISA using thiocyanate elution. *J. Immunol. Methods*, **106**, 191–194.

Maidment, B. W., Jr. *et al.* (1980). Isoelectric focusing: A new approach to the study of immune complexes. *J. Immunol. Methods*, **35**, 297–306.

Maizel, J. V., Jr (1966). Acrylamide-gel electrophoretograms by mechanical fractionation: radioactive adenovirus proteins. *Science*, **151**, 988–990.

Maizel, J. V., Jr. (1971). Polyacrylamide gel electrophoresis of viral proteins. *Methods Virol.*, **5**, 179–246.

Mancini, G. *et al.* (1964). A single-radial-diffusion method for the immunological quantitation of proteins. *Protides Biol. Fluids*, **11**, 370–373.

Mancini, G. *et al.* (1965). Immunochemical quantitation of antigens by single radial immunodiffusion. *Immunochemistry*, **2**, 235–254.

Manil, L. *et al.* (1986). Evaluation of protocols for purification of mouse monoclonal antibodies. Yield and purity in two-dimensional gel electrophoresis. *J. Immunol. Methods*, **90**, 25–37.

Mannik, M. and Downey, W. (1973). Studies on the conjugation of horseradish peroxidase to Fab fragments. *J. Immunol. Methods*, 3, 233–242.

Martin, R. G. and Ames, B. N. (1961). A method for determining the sedimentation behavior of enzymes: application to protein mixtures. *J. Biol. Chem.*, **236**, 1372–1379.

Martini, O. H. W. and Gould, H. J. (1971). Enumeration of rabbit reticulocyte ribosomal proteins. *J. Mol. Biol.*, **62**, 403–405.

McIntyre, G. D. *et al.* (1994). Preparation and characterization of interleukin-2–gelonin conjugates made using different cross-linking reagents. *Bioconjugate Chem.*, **5**, 88–97.

Meares, C. F. *et al.* (1984). Conjugation of antibodies with bifunctional chelating agents: isothiocyanate and bromoacetamide reagents, methods of analysis, and subsequent addition of metal ions. *Anal. Biochem.*, **142**, 68–78.

Merril, C. R. (1987). Detection of proteins separated by electrophoresis. In *Advances in Electrophoresis* (Chrambach, A. *et al.*, eds), vol. 1, 111–139. VCH, Weinheim, Germany.

Merril, C. R. *et al.* (1984). Gel protein stains: silver stain. *Methods Enzymol.*, **104**, 441–447.

Merril, C. R. *et al.* (1986). Protein staining and detection methods. In *Gel Electrophoresis of Proteins* (Dunn, M. J., ed.), pp. 323–362, Wright, Bristol.

Meselson, M. *et al.* (1957). Equilibrium sedimentation of macromolecules in density gradients. *Proc. Natl. Acad. Sci. USA*, **43**, 581–588.

Mikkers, F. E. P. *et al.* (1979). High-performance zone electrophoresis. *J. Chromatogr.*, **169**, 11–20.

Miller, L. and Houghton, J. A. (1945). The micro-Kjeldahl determination of the nitrogen content of amino acids and proteins. *J. Biol. Chem.*, **159**, 373–383.

Miralles, F. *et al.* (1991). Comparison of carbohydrate and peptide biotinylation on the immunological activity of IgG1 murine monoclonal antibodies. *J. Immunol. Methods*, **140**, 191–196

Moeremans, M. *et al.* (1984). Sensitive visualization of antigen–antibody reactions in dot and blot immune overlay assays with immunogold and immunogold–silver staining. *J. Immunol. Methods*, **74**, 353–360.

Moeremans, M. *et al.* (1985). Sensitive colloidal metal (gold or silver) staining of protein blots on nitrocellulose membranes. *Anal. Biochem.*, **145**, 315–321.

Moeremans, M. *et al.* (1986). FerriDye: colloidal iron binding followed by Perls' reaction for the staining of proteins transferred from sodium dodecyl sulfate gels to nitrocellulose and positively charged nylon membranes. *Anal. Biochem.*, **153**, 18–22.

Morehead, H. W. *et al.* (1991). Optimization of oxidation of glycoproteins: An assay for predicting coupling to hydrazide chromatographic supports. *J. Chromatogr.*, **587**, 171–176.

Morçöl, T. *et al.* (1997). Dot-blot analysis of the degree of covalent modification of proteins and antibodies at amino groups. *J. Immunol. Methods*, **203**, 45–53.

Mosser, A. G. *et al.* (1975). Polypeptide composition of spleen necrosis virus, a reticuloendotheliosis virus. *J. Virol.*, **15**, 1088–1095.

Mujumdar, R. B. *et al.* (1993). Cyanine dye labeling reagents: sulfoindocyanine succinimidyl esters. *Bioconjugate Chem.*, **4**, 105–111.

Nakamura, K. *et al.* (1979). Study of the interaction between phosphorylase and hydrophobic groups by means of affinity electrophoresis. *J. Chromatogr.*, **171**, 89–99.

Nakane, P. K. and Kawaoi, A. (1974). Peroxidase-labeled antibody—a new method of conjugation. *J. Histochem. Cytochem.*, **22**, 1084–1091.

Neblock, D. S. *et al.* (1992). Conjugation and evaluation of 7E3 X P4B6, a chemically cross-linked bispecific $F(ab')_2$ antibody which inhibits platelet aggregation and localizes tissue plasminogen activator to the platelet surface. *Bioconjugate Chem.*, 3, 126–131.

Nelles, L. P. and Bamburg, J. R. (1976). Rapid visualization of protein–dodecyl sulfate complexes in polyacrylamide gels. *Anal. Biochem.*, **73**, 522–531.

Nelson, N. (1944). A photometric adaptation of the Somogyi method for the determination of glucose. *J. Biol. Chem.*, **153**, 375–380.

Nelson, R. W. *et al.* (1995). Mass spectrometric immunoassay. *Anal. Chem.*, **67**, 1153–1158.

Neuhoff, V. *et al.* (1985). Clear background and highly sensitive protein staining with Coomassie Blue dyes in polyacrylamide gels: A systematic analysis. *Electrophoresis*, **6**, 427–448.

Niwa, T. (1995a). Clinical mass spectrometry. Introduction. *Clin. Chim. Acta*, **241–2**, 3–12.

Niwa, T. (1995b). Clinical mass spectrometry. Basic theory of mass spectrometry. *Clin. Chim. Acta*, **241–2**, 15–71.

Noble, R. W. *et al.* (1972). Studies on antibodies directed toward single antigenic sites on globular poteins. *Biochemistry*, **11**, 3326–3332.

Noguchi, A. *et al.* (1992). Preparation and properties of the immunoconjugate composed of anti-human colon cancer monoclonal antibody and mitomycin C–dextran conjugate. *Bioconjugate Chem.*, **3**, 132–137.

O'Farrell, P. H. (1975). High resolution two-dimensional electrophoresis of proteins. *J. Biol. Chem.*, **250**, 4007–4021.

Okuyama, T. and Satake, K. (1960). On the preparation and properties of 2,4,6-trinitrophenyl-amino acids and peptides. *J. Biochem.*, **47**, 454–466.

Ornstein, L. (1964). Disc electrophoresis. I. background and theory. *Ann. NY Acad. Sci.*, **121**, 321–349.

O'Shannessy, D. J. *et al.* (1987). Quantitation of glycoproteins on electroblots using the biotin–streptavidin complex. *Anal. Biochem.*, **163**, 204–209.

O'Shannessy, D. J. *et al.* (1993). Determination of rate and equilibrium binding constants for macromolecular interactions using surface plasmon resonance: use of nonlinear least squares analysis methods. *Anal. Biochem.*, **212**, 457–468.

Owen, A. J. (1998). Quantitative UV–visible analysis in the presence of scattering. *Spectrosc. Eur.*, **10**, 26–28.

Packer, N. H. *et al.* (1996). Characterization of human plasma glycoproteins by two-dimensional gel electrophoresis. *Biotechnology*, **14**, 66–70.

Paik, C. H. *et al.* (1985). Relative reactivity of DTPA, immunoreactive antibody–DTPA conjugates, and nonimmunoreactive antibody–DTPA conjugates toward indium-111. *J. Nucl. Med.*, **26**, 482–487.

Pain, D. and Surolia, A. (1981). Preparation of Protein A–peroxidase monoconjugate using a heterobifunctional reagent, and its use in enzyme immunoassays. *J. Immunol. Methods*, 40, 219–230.

Papac, D. I. *et al.* (1994). Direct analysis of affinity-bound analytes by MALDI/TOF MS. *Anal. Chem.*, **66**, 2609–2613.

Parker, C. E. *et al.* (1996). Monitoring cleavage of fusion proteins by matrix-assisted laser desorption ionization/mass spectrometry: recombinant HIV-1_{IIIB} p26. *Anal. Biochem.*, **239**, 25–34.

Patel, S. *et al.* (1997). A simple SDS-PAGE immunoblotting technique using an enhanced chemiluminescence detection system to identify polyclonal antibody responses to complex cryptosporidial antigen preparations following a monoclonal antibody retest and image overlay technique. *J. Immunol. Methods*, **205**, 157–161.

Paxton, R. J. *et al.* (1992). Carboranyl peptide–antibody conjugates for neutron-capture therapy: preparation, characterization, and *in vivo* evaluation. *Bioconjugate Chem.*, **3**, 241–247.

Paynter, R. W. and Ratner, B. D. (1984). The study of interfacial proteins and biomolecules by X-ray photoelectron spectroscopy. In *Surface and Interfacial Aspects of Biomedical Polymers. Protein Adsorption* (Andrade, J. D., ed.), vol. 2, pp. 189– 216, Plenum Press, New York.

Perez, P. *et al.* (1985). Specific targeting of cytotoxic T cells by anti-T3 linked to anti-target cell antibody. *Nature*, **316**, 354–356.

Peterson, G. L. (1983). Determination of total protein. *Methods Enzymol.*, **91**, 95–119.

Pham, D. T. *et al.* (1995). Electrophoretic method for the quantitative determination of a benzyl-DTPA ligand in DTPA monoclonal antibody conjugates. *Bioconjugate Chem.*, **6**, 313–315.

Pickels, E. G. (1943). Sedimentation in the angle centrifuge. *J. Gen. Physiol.*, **26**, 341–360.

Pierce Chemical Co. (1995). *Life Science, Analytical & Molecular Biology Products Catalog*, pp. L67–L68. Pierce Chemical Co., Rockford, IL.

Pietersz, G. A. *et al.* (1988). Specific *in Vitro* anti-tumour activity of methotrexate-monoclonal antibody conjugates prepared using human serum albumin as an intermediary. *Immunol. Cell Biol.*, **66**, 43–49.

Porath, J. and Flodin, P. (1959). Gel filtration: A method for desalting and group separation. *Nature*, **183**, 1657–1659.

Porstmann, B. *et al.* (1985). Which of the commonly used marker enzymes gives the best results in colorimetric and fluorimetric enzyme immunoassays: horseradish peroxidase, alkaline phosphatase or β-galactosidase? *J. Immunol. Methods*, **79**, 27–37.

Pressman, D. (1943). Determination of microquantities of certain proteins. A colorimetric method. *Ind. Eng., Anal. Ed.*, **15**, 357–359.

Pullen, G. R. *et al.* (1986). Antibody avidity determination by ELISA using thiocyanate elution. *J. Immunol. Methods*, **86**, 83–87.

Rabilloud, T. (1990). Mechanisms of protein silver staining in polyacrylamide gels: A 10-year synthesis. *Electrophoresis*, **11**, 785–794.

Radola, B. J. (1984). High-resolution preparative isoelectric focusing. *Methods Enzymol.*, **104**, 256–275.

Rao, S. V. *et al.* (1997). Determination of the extent of protein biotinylation by fluorescence binding assay. *Bioconjugate Chem.*, **8**, 94–98.

Raymond, S. and Weintraub, L. (1959). Acrylamide gel as a supporting medium for zone electrophoresis. *Science*, **130**, 711.

Rector, E. S. *et al.* (1978). A method for the preparation of protein–protein conjugates of predetermined composition. *J. Immunol. Methods*, **24**, 321–336.

Reimer, C. B. *et al.* (1970). Comparative evaluation of commercial precipitating antisera against human IgM and IgG. *J. Lab. Clin. Med.*, **76**, 949–960.

Reisner, A. H. *et al.* (1975). The use of Coomassie Brilliant Blue G250 perchloric acid solution for staining in electrophoresis and isoelectric focusing on polyacrylamide gels. *Anal. Biochem.*, **64**, 509–516.

Ressler, N. (1960). Two dimensional electrophoresis of protein antigens with an antibody-containing buffer. *Clin. Chim. Acta*, **5**, 795–800.

Righetti, P. G. and Bossi, A. (1997). Isoelectric focusing in immobilized pH gradients: recent analytical and preparative developments. *Anal. Biochem.*, **247**, 1–10.

Righetti, P. G. and Caravaggio, T. (1976). Isoelectric points and molecular weights of proteins. A table. *J. Chromatogr.*, **127**, 1–28.

Righetti, P. G. *et al.* (1986). Conventional isoelectric focusing and immobilised pH gradients. In *Gel Electrophoresis of Proteins* (Dunn, M. J., ed.), 141–202, Wright, Bristol.

Rimpilainen, M. A. and Righetti, P. G. (1985). Membrane protein analysis by isoelectric focusing in immobilized pH gradients. *Electrophoresis*, **6**, 419–422.

Rivas, G. and Minton, A. P. (1993). New developments in the study of biomolecular associations *via* sedimentation equilibrium. *Trends Biochem. Sci.*, **18**, 284–287.

Roberts, E. and Rouser, G. (1958). Spectrophotometric assay for reaction of *N*-ethylmaleimide with sulfhydryl groups. *Anal. Chem.*, **30**, 1291–1292.

Robinson, H. W. and Hogden, C. G. (1940). The biuret reaction in the determination of serum proteins. *J. Biol. Chem.*, **135**, 707–725.

Rodbard, D. *et al.* (1971). Pore gradient electrophoresis. *Anal. Biochem.*, **40**, 135–157.

Rodkey, L. S. (1988). Ionic binding properties of carrier ampholytes. *J. Chromatogr.*, **437**, 147–159.

Rouimi, P. *et al.* (1995). Electrospray ionization-mass spectrometry as a tool for characterization of glutathione *S*-transferase isozymes. *Anal. Biochem.*, **229**, 304–312.

Sanger, F. (1945). The free amino groups of insulin. *Biochem. J.*, **39**, 507–515.

Schaffner, W. and Weissman, C. (1973). A rapid, sensitive, and specific method for the determination of protein in dilute solution. *Anal. Biochem.*, **56**, 502–514.

Schultz, N. M. and Kennedy, R. T. (1993). Rapid immunoassays using capillary electrophoresis with fluorescence detection. *Anal. Chem.*, **65**, 3161–3165.

Shah, D. *et al.* (1994). Determination of biotin in biotin-conjugated protein by measuring fluorescence polarization. *Clin. Chem.*, **40**, 2112–2113.

Shannon, L. M. *et al.* (1966). Peroxidase isozymes from horseradish roots. I. Isolation and physical properties. *J. Biol. Chem.*, **241**, 2166–2172.

Shapiro, A. L. *et al.* (1966). Polyribosomal synthesis and Assembly of the H and L chains of gamma globulin. *Proc. Natl. Acad. Sci. USA*, **56**, 216–221.

Shapiro, A. L. *et al.* (1967). Molecular weight estimation of polypeptide chains by electrophoresis in SDS-polyacrylamide gels. *Biochem. Biophys. Res. Commun.*, **28**, 815–820.

Sharma, H. K. and Tihon, C. (1988). Determination of avidin and streptavidin by a modified Bradford method. *Anal. Biochem.*, **170**, 135–139.

Sharma, P. *et al.* (1990). A spectrophotometric method for the estimation of polymer-supported sulfhydryl groups. *Anal. Biochem.*, **189**, 173–177.

Sheely, D. M. *et al.* (1997). Characterization of monoclonal antibody glycosylation: comparison of expression systems and identification of terminal α-linked galactose. *Anal. Biochem.*, **247**, 102–110.

Sherbon, J. W. (1974). Pro-milk method for the determination of protein in milk by dye binding. *J. Assoc. Off. Anal. Chem.*, **57**, 1338–1341.

Shimura, K. and Kasai, K. (1997). Affinity capillary electrophoresis: a sensitive tool for the study of molecular interactions and its use in microscale analyses. *Anal. Biochem.*, **251**, 1–16.

Shmatchenko, V. V. and Varshavsky, A. J. (1978). A technique of low-pH gel electrophoresis of chromosomal proteins which does not require preliminary removal of DNA. *Anal. Biochem.*, **85**, 42–46.

Shors, S. T. and Winston, V. (1989). Neutralizing antibodies for infectious hematopoietic necrosis virus in eggs of steelhead trout (*Salmo gairdneri*). *Am. J. Vet. Res.*, **50**, 232–234.

Shoyama, Y. *et al.* (1993). Direct determination of forskolin–bovine serum albumin conjugate by matrix-assisted laser desorption ionization mass spectrometry. *Org. Mass Spectrom.*, **28**, 987–988.

Siegel, M. M. *et al.* (1991). Matrix-assisted UV-laser desorption/ionization mass spectrometric analysis of monoclonal antibodies for the determination of carbohydrate, conjugated chelator, and conjugated drug content. *Anal. Chem.*, **63**, 2470–2481.

Siegel, M. M. *et al.* (1993). Determination of the loading values for high levels of drugs and sugars conjugated to proteins by matrix-assisted ultraviolet laser desorption/ionization mass spectrometry. *Biol. Mass Spectrom.*, **22**, 369–376.

Singer, S. J. and Schick, A. F. (1961). The properties of specific stains for electron microscopy prepared by the conjugation of antibody molecules with ferritin. *J. Biochem. Biophys. Cytol.*, **9**, 519–537.

Singh, R. (1994). A sensitive assay for maleimide groups. *Bioconjugate Chem.*, **5**, 348–351.

Smith, P. K. *et al.* (1985). Measurement of protein using bicinchoninic acid. *Anal. Biochem.*, **150**, 76–85.

Smith, R. D. and Udseth, H. R. (1988). Capillary zone electrophoresis–MS. *Nature*, **331**, 639–640.

Smithies, O. (1955). Zone electrophoresis in starch gels: group variations in the serum proteins of normal human adults. *Biochem. J.*, **61**, 629–641.

Smithies, O. (1962). Molecular size and starch–gel electrophoresis. *Arch, Biochem. Biophys.* Suppl. 1, 125–131.

Sørensen, S. P. L. (1926). Studies on proteins. IX. The influence of salt concentration on the acid-binding capacity of egg-albumin. *C. R. Trav. Lab. Carlsberg*, **16**, 1–55.

Steck, T. L. (1972). Cross-linking the major proteins of the isolated erythrocyte membrane. *J. Mol. Biol.*, **66**, 295–305.

Steele, J. C. II., Jr. and Nielsen, T. B. (1978). Evaluation of cross-linked polypeptides in SDS gel electrophoresis. *Anal. Biochem.*, **84**, 218–224.

Steinberg, T. H. *et al.* (1996). SYPRO orange and SYPRO red protein gel stains: One-step fluorescent staining of denaturing gels for detection of nanogram levels of protein. *Anal. Biochem.*, **239**, 223–237.

Stiehm, E. R. (1967). Radial diffusion technique for the quantitative estimation of precipitating antibody. *J. Lab. Clin. Med.*, **70**, 528–534.

Stroupe, S. D. and Foster, J. F. (1973). Further studies on the sulfhydryl-catalyzed isomerization of bovine mercaptalbumin. *Biochemistry*, **12**, 3824–3830.

Susi, H. and Byler, D. M. (1986). Resolution-enhanced Fourier transform infrared spectroscopy of enzymes. *Methods Enzymol.*, **130**, 290–311.

Sutton, C. W. *et al.* (1994). Site-specific characterization of glycoprotein carbohydrates by exoglycosidase digestion and laser desorption mass spectrometry. *Anal. Biochem.*, **218**, 34–46.

Svedberg, T. and Pedersen, K. O. (1940). *The Ultracentrifuge*, Clarendon Press, Oxford.

Svensson, B. (1973). Use of isoelectric focusing to characterize the bonds established during coupling of CNBr-activated amylodextrin to subtilisin type Novo. *FEBS Lett.*, **29**, 167–169.

Svensson, H. (1961). Isoelectric fractionation, analysis, and characterization of ampholytes in natural pH gradients. I. The differential equation of solute concentrations at a steady state and its solution for simple cases. *Acta, Chem. Scand.*, **15**, 325–341.

Svensson, H. (1962). Isoelectric fractionation, analysis, and characterization of ampholytes in natural pH gradients. II. Buffering capacity and conductance of isoionic ampholytes. *Acta Chem. Scand.*, **16**, 456–466.

Swaminathan, K. *et al.* (1973). Rapid photometric method for the determination of "true" protein content of potatoes based on non-selective dye-binding capacity. *Indian J. Exp. Biol.*, **11**, 63–64.

Swank, R. T. and Munkres, K. D. (1971). Molecular weight analysis of oligopeptides by electrophoresis in polyacrylamide gel with sodium dodecyl sulfate. *Anal. Biochem.*, **39**, 462–477.

Switzer, R. C. *et al.* (1979). A highly sensitive stain for detecting proteins and peptides in polyacrylamide gels. *Anal. Biochem.*, **98**, 231–237.

Szurdoki, F. *et al.* (1992). Synthesis of haptens and conjugates for an enzyme immunoassay of the herbicide bromacil. *J. Agric. Food Chem.*, **40**, 1459–1465.

Tabachnick, M. and Sobotka, H. (1960). Azoproteins. II. A spectrophotometric study of the coupling of diazotized arsanilic acid with proteins. *J. Biol. Chem.*, **235**, 1051–1054.

Takagi, T. (1990). Application of low-angle laser light scattering detection in the field of biochemistry. Review of recent progress. *J. Chromatogr.*, **506**, 409–416.

Takenaka, A. *et al.* (1969). States of amino acid residues in proteins. XVIII. A revised way of using diazonium-1*H*-tetrazole for reactivity examination of histidine and tyrosine residues. *Biochim. Biophys. Acta*, **194**, 293–300.

Tanaka, K. *et al.* (1988). Protein and polymer analyses up to m/z 100000 by laser ionization time-of-flight mass spectrometry. *Rapid Commun. Mass Spectrom.*, **2**, 151–153.

Tatake, J. G. *et al.* (1991). Synthesis and characterization of protein and polylysine conjugates of sulfamethoxazole and sulfanilic acid for investigation of sulfonamide drug allergy. *Bioconjugate Chem.*, **2**, 124–132.

Tawde, S. S. *et al.* (1963). Physicochemical and immunochemical studies on the reaction of bovine serum albumin with *p,p′*-difluoro-*m,m′*-dinitrophenylsulfone. *Arch. Biochem. Biophys.*, **100**, 270–278.

Terabe, S. *et al.* (1985). Electrokinetic chromatography with micellar solution and open-tubular capillary. *Anal. Chem.*, **57**, 834–841.

Tikhonov, V. N. and Mustafin, I. S. (1965). Photometric determination of copper in magnesium metal and magnesium alloys by means of bicinchoninic acid. *Zh. Anal. Khim.*, **20**, 390–392.

Tiselius, A. (1937). A new apparatus for electrophoretic analysis of colloidal mixtures. *Trans. Faraday Soc.*, **33**, 524–531.

Tombs, M. P. *et al.* (1959). The spectrophotometric determination of protein at 210 nm. *Biochem. J.*, **73**, 167–171.

Towbin, H. and Gordon, J. (1984). Immunoblotting and dot immunobinding: current status and outlook. *J. Immunol. Methods*, **72**, 313–340.

Towbin, H. *et al.* (1979). Electrophoretic transfer of proteins from polyacrylamide gels to nitrocellulose sheets: procedure and some applications. *Proc. Natl. Acad. Sci. USA*, **76**, 4350–4354.

Toyo'oka, T. *et al.* (1989). Fluorogenic reagent for thiols: 4-(*N*, *N*-dimethylaminosulphonyl)-7-fluoro-2, 1, 3-benzoxadiazole. *Analyst*, **114**, 413–419.

Traut, R. R. and Kenny, J. W. (1977). Crosslinking of ribosomes by cleavable bifunctional mercaptoimidates. *Adv. Exp. Med. Biol.*, **86A**, 215–233.

Udenfriend, S. *et al.* (1972). Fluorescamine: a reagent for assay of amino acids, peptides, proteins, and primary amines in the picomole range. *Science*, **178**, 871–872.

Ui, N. (1971). Isoelectric points and conformation of proteins. I. Effect of urea on the behavior of some proteins in isoelectric focusing. *Biochim. Biophys. Acta*, **229**, 567–581.

Underwood, P. A. and Steele, J. G. (1991). Practical limitations to estimation of protein adsorption to polymer surfaces. *J. Immunol. Methods*, **142**, 83–94.

Uriel, J. (1966). Méthode d'électrophorèse dans des gels d'acrylamide–agarose. *Bull. Soc. Chim. Biol.*, **48**, 969–982.

Vaerman, J.-P. *et al.* (1969a). Further studies on single radial immunodiffusion: I. direct proportionality between area of precipitate and reciprocal of antibody concentration. *Immunochemistry*, **6**, 279–285.

Vaerman, J.-P. *et al.* (1969b). Further studies on single radial immunodiffusion: II. The reversed system: diffusion of antibodies in antigen-containing gels. *Immunochemistry*, **6**, 287–293.

Valmet, E. (1968). Demonstration of the microheterogeneity of human serum gamma-globulin by isoelectric focusing. *Sci. Tools*, **15**, 8–11.

Van Der Scheer, A. *et al.* (1978). The feasibility of radiolabeling for human serum albumin (HSA) adsorption studies. *J. Colloid Interface Sci.*, **66**, 136–145.

Van Dorsselaer, A. *et al.* (1990). Application of electrospray mass spectrometry to the characterization of recombinant proteins up to 44 kDa. *Biomed. Environ. Mass Spectrom.*, **19**, 692–704.

Vesterberg, O. (1968). Isoelectric focusing of proteins. *Svensk Kem. Tidskr.*, **80**, 213–225.

Vesterberg, O. (1971). Isoelectric focusing of proteins. *Methods Enzymol.*, **22**, 389–412.

Vesterberg, O. and Svensson, H. (1966). Isoelectric fractionation, analysis and characterization of ampholytes in natural pH gradients. *Acta Chem. Scand.*, **20**, 820–834.

Vogt, A. and Kopp, R. (1964). Loss of specific agglutinating activity of purified ferritin-conjugated antibodies. *Nature*, **202**, 1350–1351.

Waddell, W. J. (1956). A simple ultraviolet spectrophotometric method for the determination of protein. *J. Lab. Clin. Med.*, **48**, 311–314.

Wadström, T. and Hisatsune, K. (1970). Bacteriolytic enzymes from *Staphylococcus aureus*. Purification of an *Endo-β-N*-acetylglucosaminidase. *Biochem. J.*, **120**, 725–734.

Walker, J. M. (1994). The bicinchinonic acid (BCA) assay for protein quantification. *Methods Mol. Biol.*, **32**, 5–8.

Wallingford, R. A. and Ewing, A. G. (1991). Capillary electrophoresis. *Adv. Chromatogr.*, **29**, 1–76.

Walsh, J. and Gosling, J. P. (1986). The influence of radioiodination on the adsorption of IgG and serum albumin to polystyrene. *Anal. Biochem.*, **158**, 413–423.

Wang, K. and Richards, F. M. (1974). An approach to nearest neighbor analysis of membrane proteins. Application to the human erythrocyte membrane of a method employing cleavable cross-linkages. *J. Biol. Chem.*, **249**, 8005–8018.

Waterborg, J. H. and Matthews, H. R. (1994). The Lowry method for protein quantification. *Methods Mol. Biol.*, **32**, 1–4.

Weber, K. and Osborn, M. (1969). The reliability of molecular weight determinations by dodecyl sulfate-polyacrylamide gel electrophoresis. *J. Biol. Chem.*, **244**, 4406–4412.

Webster, G. C. (1970). Comparison of direct spectrophotometric methods for the measurement of protein concentration. *Biochim. Biophys. Acta*, **207**, 371–373.

Wei, R.-D. (1970). Assay of avidin. *Methods Enzymol.*, **18A**, 424–427.

Weidekamm, E. *et al.* (1973). A new sensitive, rapid fluorescence technique for the determination of proteins in gel electrophoresis and in solution. *Anal. Biochem.*, **54**, 102–114.

Wen, J. *et al.* (1996). Size-exclusion chromatography with on-line light-scattering, absorbance, and refractive index detectors for studying proteins and their interactions. *J. Immunol. Methods*, **240**, 155–166.

Wengatz, I. *et al.* (1992). Determination of the hapten density of immunoconjugates by matrix-assisted UV laser desorption/ionization mass spectrometry. *Anal. Lett.*, **25**, 1983–1997.

Wessel, D. and Flügge, U. I. (1984). A method for the quantitative recovery of protein in dilute solution in the presence of detergents and lipids. *Anal. Biochem.*, **138**, 141–143.

Wessendorf, M. W. *et al.* (1990). A spectrophotometric method for determination of fluorophore-to-protein ratios in conjugates of the blue fluorophore 7-amino-4-methylcoumarin-3-acetic acid (AMCA). *J. Histochem. Cytochem.*, **38**, 87–94.

Westphal, U. (1957). Steroid–protein interactions. III. Spectrophotometric demonstration of interaction between proteins and progesterone, deoxycortisone and cortisol. *Arch. Biochem. Biophys.*, **66**, 71–90.

Wilkins, M. R. *et al.* (1996). From proteins to proteomes: large scale protein identification by two-dimensional electrophoresis and amino acid analysis. *Biotechnology*, **14**, 61–65.

Wilkins Stevens, P. W. and Kelso, D. M. (1995). Estimation of the protein-binding capacity of microplate wells using sequential ELISAs. *J. Immunol. Methods*, **178**, 59–70.

Wilkins Stevens, P. W. *et al.* (1995). Assessment of adsorption and adhesion of proteins to polystyrene microwells by sequential enzyme-linked immunosorbent assay analysis. *Anal. Biochem.*, **225**, 197–205.

Williams, R. W. (1986). Protein secondary structure analysis using raman amide I and amide III spectra. *Methods Enzymol.*, **130**, 311–331.

Willison, H. J. *et al.* (1986). Staining immunoblots with biotinylated primary antibodies. *IRCS Med. Sci.*, **14**, 201–202.

Wilson, M. B. and Nakane, P. K. (1978). Recent developments in the periodate method of conjugating horseradish peroxidase (HRPO) to antibodies. In *Immunofluorescence and Related Staining Techniques* (Knapp, W. *et al.* eds.), pp. 215–224, Elsevier/North Holland Biomedical Press, Amsterdam.

Winzor, D. J. and Scheraga, H. A. (1963). Studies of chemically reacting systems on Sephadex. I. Chromatographic demonstration of the Gilbert theory. *Biochemistry*, **2**, 1263–1267.

Wirth, P. J. and Romano, A. (1995). Staining methods in gel electrophoresis, including the use of multiple detection methods. *J. Chromatogr. A*, **698**, 123–143.

Wisdom, G. B. (1994). Protein blotting. *Methods Mol. Biol.*, **32**, 207–213.

Wolfe, C. A. C. and Hage, D. S. (1994). Automated determination of antibody oxidation using flow injection analysis. *Anal. Biochem.*, **219**, 26–31.

Wood, C. L. and O'Dorisio, M. S. (1985). Covalent crosslinking of vasoactive intestinal polypeptide to its receptors on intact human lymphoblasts. *J. Biol. Chem.*, **260**, 1243–1247.

Wood, G. C. and Cooper, P. F. (1970). The application of gel filtration to the study of protein-binding of small molecules. *Chromatogr. Rev.*, **12**, 88–107.

Woods, A. H. and O'Bar, P. R. (1970). Absorption of proteins and peptides in the far ultraviolet. *Science*, **167**, 179–181.

Worrell, N. R. *et al.* (1986). Effect of linkage variation on pharmacokinetics of ricin A chain–antibody conjugates in normal rats. *Anti-Cancer Drug Des.*, **1**, 179–188.

Wrigley, C. (1968). Gel electrofocusing: A technique for analyzing multiple protein samples by isoelectric focusing. *Sci. Tools*, **15**, 17–23.

Wring, S. A. *et al.* (1994). The production and evaluation of a radioligand and antiserum for the radioimmunoassay of sub-nanogram per millilitre concentrations of lamivudine. *J. Pharm. Biomed. Anal.*, **12**, 1573–1583.

Wu, H. (1920). Contribution to the chemistry of phosphomolybdic acids, phosphotungstic acids, and allied substances. *J. Biol. Chem.*, **43**, 189–220.

Wu, H. (1922). A new colorimetric method for the determination of plasma proteins. *J. Biol. Chem.*, **51**, 33–39.

Wu, J. *et al.* (1996). A strategy to locate cysteine residues in proteins by specific chemical cleavage followed by matrix-assisted laser desorption/ionization time-of-flight mass spectrometry. *Anal. Biochem.*, **235**, 161–174.

Wu, S. and Dovichi, N. J. (1989). High-sensitivity fluorescence detector for fluorescein isothiocyanate derivatives of amino acids separated by capillary zone electrophoresis. *J. Chromatogr.*, **480**, 141–155.

Yang, J. T. *et al.* (1986). Calculation of protein conformation from circular dichroism. *Methods Enzymol.*, **130**, 208–269.

Yang, L. *et al.* (1996a). Capillary isoelectric focusing-electrospray ionization mass spectrometry for transferrin glycoforms analysis. *Anal. Biochem.*, **243**, 140–149.

Yang, T. *et al.* (1996b). Mapping cross-linking sites in modified proteins with mass spectrometry: an application to cross-linked hemoglobins. *Anal. Biochem.*, **242**, 55–63.

Yasuzawa, T. and Tomer, K. B. (1997). Structural determination of the conjugate of human serum albumin with a mitomycin C derivative, KW-2149, by matrix-assisted laser desorption/ionization mass spectrometry. *Bioconjugate Chem.*, **8**, 391–399.

Yphantis, D. A. and Waugh, D. F. (1956). Ultracentrifugal characterization by direct measurement of activity. II. Experimental. *J. Phys. Chem.*, **60**, 630–635.

Zhang, Y. *et al.* (1996). Quantitative determination of isothiocyanates, dithiocarbamates, carbon disulfide, and related thiocarbonyl compounds by cyclocondensation with 1,2-benzenedithiol. *Anal. Biochem.*, **239**, 160–167.

Zhao, Y. and Chait, B. T. (1994). Protein epitope mapping by mass spectrometry. *Anal. Chem.*, **66**, 3723–3726.

Zor, T. and Selinger, Z. (1996). Linearization of the Bradford protein assay increases its sensitivity: theoretical and experimental studies. *Anal. Biochem.*, **236**, 302–308.

Chapter 12

The Stabilisation of Proteins and Other Biological Materials

Ross H.M. Hatley and Camilo A.L.S. Colaco*

Quadrant Healthcare plc and *Quadrant Research Foundation, Maris Lane, Cambridge CB2 2SY, UK

The long-term stabilisation of biological materials is desirable for two main reasons. Firstly it eliminates the need for repeated characterisation of the isolated biological as, if no loss of functional activity is seen on storage, a single characterisation of the isolated biological is sufficient. Secondly, it enables the bulk preparation of the biological which can be stored and used as required, and thus eliminates the time and cost of repeated purification of the biological in question. The retention of functional activity is thus the most important criterion for the evaluation of the efficacy of any stabilisation method.

In the modern biological laboratory, two different types of stabilisation techniques are required, namely the ability to store isolated biological materials and the ability to preserve intact cells and organisms. The latter has been particularly important for the preservation of stock cultures of hybridomas producing monoclonal antibodies. However, the exponential increase in the use of genetically engineered microbial and mammalian cells for the production of recombinant proteins has focussed attention on techniques to preserve these seed culture stocks. The first part of this chapter will focus on the stabilisation of isolated biological materials and the latter part will cover the techniques currently used to preserve intact cells and organisms. Each section will begin with a brief outline of the theoretical basis concerned, followed by a description of the techniques concerned. In the outline of the theory, detailed analysis (both thermodynamic and mathematical) will be minimal as a number of such treatments are available in the literature (Franks, 1990; Franks *et al.*, 1991; Pikal, 1985, 1990a, 1994). Furthermore, at least with respect to freezing and freeze-drying, there remain limitations to these theoretical treatments; none of these analyses have yet been able to predict the parameters controlling the formation of ice nucleation sites. As the latter is the rate controlling step in the formation of ice crystals, the practical value of these theoretical analyses remains unproved. It should be noted that some of the techniques described are patented technologies (e.g. the use of trehalose as a stabilising agent, QT4 Technology, is the patented technology of Quadrant Holdings Cambridge Ltd.) and may thus require a licence for use.

12.1 Protein Stability and Instability

It has been clearly established that the overall stability of a native protein (i.e. the excess free energy in the native state compared to that of the denatured) is of a very low order of magnitude. For a protein of the size of ribonuclease (molecular mass 13.5 kDa) this is typically in the order of 50 kJ mol^{-1} (Franks, 1988). To put this into perspective, it corresponds to the strength of about four hydrogen bonds, around 1% of the total number of hydrogen bonds in the molecule (Finney, 1982).

The low stability margin possessed by a native protein makes it apparent that the protein molecule has not evolved to have maximal structural stability. Indeed, it would appear that the situation is almost the reverse; the protein molecule has evolved to have maximum flexibility, while under optimum conditions retaining sufficient stability for effective operation. This flexibility is required both to ensure efficient interaction with the substrate and to assist in the rapid turnover of the protein itself, a necessary prerequisite to metabolic regulation. This makes proteins particularly difficult to stabilise since the stabilisation process is working against nature.

Overall it appears that no one single effect can be reliably described as the factor that is responsible for the maintenance of protein structure. The stability of the native protein appears to be a balance of the many competing influences. As a consequence of its very low stability margin, a protein is prone to a number of processes that destroy its native conformation. Two general types of instability are observed in proteins: degradational which involves the breaking of peptide bonds, and denaturational which involves an unfolding of the protein's native structure. The latter is sometimes reversible but both processes cause protein instability during storage. Some of the factors which can result in damage to protein molecules are shown in Table 12.1.

Table 12.1 Factors that adversely affect the biological activity of proteins

Chemical	Detergents Guanidinium chloride Heavy metals Organic solvents Oxidising agents Urea
Physical	Elevated temperatures Extremes of pH Freezing and thawing Irradiation Vigorous agitation
Biological	Carbohydrate activity (on glycoproteins) Proteolytic activity

12.1.1 Degradation

The most common degradation process is hydrolysis of the peptides (Cleland *et al.*, 1993). The process is extremely complicated because the side chains of certain amino acids can exert an influence. The side chains of valine and leucine, for example, protect the peptide bond by hindering the hydronium ion. Serine and threonine are more labile and aspartyl residues extremely labile. Deamination is the hydrolysis of the side chain amide of glutamine and asparagine. This deamination can in turn result in the formation of ammonia which will increase the rate of hydrolysis of the peptide linkages (Cleland *et al.*, 1994; McKerrow and Robinson, 1971). Certain amino acids are sensitive to isomerisation. Asparagine, glutamine, aspartic acid and glutamic acid are notable examples. Oxidation of tryptophan, methionine and cysteine often leads to degradation of a protein (Cleland *et al.*, 1993). Disulfide exchange can lead to the formation of oligomers. Last, but certainly not least important when considering storage stability, is light-induced degradation, particularly of tryptophan.

In practice, the major cause of degradation in isolated biologicals is enzymic due to contaminating protease activity. This is a particular problem with highly purified products as they form the sole substrate for contaminating enzyme. A common solution is the use of metal chelators such as EGTA and EDTA in the storage buffers (2–5 mM) to inhibit the activity of metalloproteases. Other useful protease inhibitors include *N*-ethylmaleimide (10–25 mM) and iodoacetamide (0.5–1 mM) for sulfhydryl proteases, diisopropylfluorophosphate (1–2 mM), aprotinin (0.1–0.3 μM) and phenylmethanesulfonyl fluoride (1–5 mM) for serine proteases, and leupeptin (1–3 μM) for both serine and sulfhydryl proteases. Other features that may affect product degradation are pH, and the addition of bulking agents such as albumin.

12.1.2 Denaturation

Denaturation is a thermodynamically controlled process. It is a common occurrence that proteins, because of their marginal stability, unfold under non-optimal conditions. A simple example is boiling an egg. On heating, the proteins in the egg white unfold and aggregate to form a gelatinous mass. However, under certain conditions, if aggregation can be avoided, an unfolded protein can refold effectively on return to a suitable environment—'unboiling an egg' (Perutz, 1980).

In the biosphere there are some 10^{12} proteins (Richardson, 1981). Each is different and is specific for one function. Their varied amino acid sequences generate significant differences in the three dimensional structure. Homologies in the amino acid sequence and the tertiary structure, however, do not necessarily correlate when examined in different proteins. The conformation of any given structure is dependent both on its basic amino acid structure and the environment in which it exists. The importance of the basic amino acid chain can be seen when proteins are isolated from organisms whose natural habitats are at extremes of environmental conditions, e.g. high temperatures, salts or pH (Argos, 1979; Jaenicke, 1981; Brock, 1986; Eisenberg and Wachtel, 1987). These proteins have distinct alterations in the amino acid chains, when compared to proteins of similar function in other non-stressed organisms. However, of equal importance is the environment itself, since this influences the delicate balance of interactions within a protein that is responsible for the maintenance of its active native structure. Consequently, although it is possible to relate changes in amino acid composition to extremes of environment, it is not yet possible to relate these changes to any particular structural differences that give enhanced stability (Jaenicke, 1988).

Denaturation is regarded as an unfolding of a protein. Although the degree of unfolding is unclear, it is confined to the higher degrees of structure, the covalently bonded amino acid chain remains undisturbed. The stability of a protein can be expressed at any given temperature by the difference in free energy (ΔG) between the native and denatured states. ΔG depends on the balance between entropic (ΔS) and enthalpic (ΔH) forces and can be expressed as $\Delta G = \Delta H - T\Delta S$. Because ΔH and ΔS are temperature dependent then the value of ΔG will vary with temperature, and for any protein a $\Delta G(T)$ profile can be constructed. A negative ΔG indicates that the forces stabilising a protein are greater than those destabilising and consequently the protein adopts its native folded state. However, when ΔG is positive destabilising forces predominate and the protein denatures. Many proteins are bonded internally by disulfide bridges. These appear to have little influence on the three-dimensional structure, but act largely to stabilise the protein. However, although covalent, these bridges do not appear to alter significantly the overall structural stability of the protein (Jaenicke, 1988; Wetzel, 1987). The increase in stability due to the disulfide bonds appears to be partially balanced by increases in entropy within the molecule, as indicated by the flattening of the ΔG profile but no significant increase in the ΔG_{max} (Wetzel, 1987). Disulfide bridges are common in thermophilic proteins but it has not been possible to relate directly the effect of the disulfide bonds to greater thermal stability. There is no doubt that they assist in the stabilisation of the molecule under extremes of temperature. However, the studies to date tend to infer that the stability increase in thermophilic proteins is principally due to small, but numerous, localised changes in bonding, in the tertiary level of the protein structure (Zuber, 1981; Skarzynski *et al.*, 1987).

Most attention has been paid to the heat stability of proteins. Experimental observations have employed perturbants, such as pH, alcohols and salts to study the effects of environment on the temperature of denaturation of mesophilic proteins. The aim is to use the information obtained to analyse the solvent–protein interaction and relate this to the factors that maintain protein stability. This approach has been somewhat limited in providing conclusions. Firstly, solvents will themselves interact with the protein and these effects must be fully accounted for when analysing the data. Secondly, and of most importance, such studies can only be performed over limited temperature ranges. Since it is considered that solvent–protein interactions play an important part in protein stabilisation, and these effects are temperature dependent, the resolution of the interactions is hardly possible. A second approach to obtain thermal stability data has been to use thermophilic organisms (Zuber, 1976; Ljungdahl *et al.*, 1981; Amelunxen and Murdock, 1978; Brock, 1985). The enzymes extracted from such sources generally have a greater resistance to heat and denaturants. Researchers hoped that such enzymes would have a common structural determinant that provided the increased stability. This has, however, not yet been found to be the case. Current evidence indicates that the increased stability is brought about by a range of factors including hydrophobic interactions, hydrogen bonding, salt bridges and disulfide bonding. The relative contributions of these factors vary from protein to protein and no clear relationship can yet be deduced between protein sequence and stability (Walker, 1978; Walker *et al.*, 1980).

12.1.3 Cold denaturation

The studies of heat denaturation have been able to do little more than provide a superficial insight into the possible factors that are responsible for stabilisation and folding. Because of the high speed reaction rates, it is rarely possible to follow folding pathways. However, if the temperature range could be extended and protein unfolding/refolding studied at low temperatures, a broader experimental window would be available and the effects of protein–solvent interactions could be quantified. This approach has indeed been taken by some workers to follow the refolding of heat-denatured proteins. The results of these studies are probably also relevant to the stabilisation of proteins by freezing and freeze-drying.

In vitro, low temperature *per se* (chill) exerts its influence mainly through changes in the physical properties of the aqueous solvent medium, e.g. acid/base ionisation, diffusion and reaction rates and hydrogen bond energies, although under *in vivo* conditions, other processes such as lipid phase transitions are probably also implicated.

Cold denaturation can be defined as a major conformational transition that occurs in a protein during a reduction in temperature. It is effectively the mirror image of heat denaturation, where a major conformation transition takes place with an increase in temperature. A distinction can be made between either cold inactivation (or sensitivity) and cold denaturation. The former results from drastic retardation of reaction rates and possibly subunit dissociation, while cold denaturation, as here defined, is caused by major cooperative transitions in at least the tertiary structure. A denaturation brought about by a reduction in temperature at first inspection seems improbable, as it would result in an increase in entropy with a reduction in temperature. The first indication that such an event may occur followed a detailed thermodynamic analysis of the denaturation of chymotrypsinogen by Brandts (1964). This investigation, with the aid of both acid and urea destabilisation, encompassed a temperature range of some 60 °C. For each set of conditions, a ΔG profile was constructed against temperature. The stability profiles $\Delta G(T)$ could be fitted adequately only by a third-order temperature function. The process could be described by a simple two-state equilibrium of the type $N \leftrightarrow D$, where N and D correspond to the native (folded) and denatured (unfolded) state respectively, and where $\Delta G = (G^D - G^N)$. The full description of $\Delta G(\text{pH}, T)$ could only be achieved by a very complex empirical function to correct for Cl^- and urea concentrations, but Brandts found that, following this correction, at any given temperature ΔH was apparently independent of pH. This enabled him to combine the observed denaturation curves at different pH values and to construct the 'master curves' for $\Delta G(T)$. The experimental results indicated that ΔG shows a maximum at 13 °C before decreasing as the temperature was reduced further. Extrapolation of the data to low temperatures indicates that ΔG becomes zero at some low temperature. Since ΔG is the measure of the relative stabilities of the native and denatured conformations, the results implied that in addition to the measured high temperature denaturation, a low temperature transition from native to denatured (a cold denaturation) would also occur. Before this time it had been assumed that ΔH was a linear function of temperature and consequently the protein would become progressively more stable as the temperature was reduced. Subsequently, many detailed high temperature denaturation investigations have confirmed that $\Delta G(T)$ is parabolic, with additional indirect evidence coming from high pressure denaturation experiments (Hawley, 1971); the low temperature denaturation itself, however, received little attention.

It has been well documented that a crystallised protein exhibits the same conformation as a native protein in an aqueous solution (Finney, 1982). For the denatured state Finney assumes a fully unfolded extended peptide chain. From his analysis it is apparent that the overall stability is the result of the balancing of a number of large entropic and enthalpic contributions. Major contributors to stability include intra-peptide hydrogen-bonding and hydrophobic interactions, while configurational entropy and peptide–solvent hydrogen-bonding are destabilising influences. Individually, these factors are orders of magnitude greater than the overall stability margin of a native protein. Consequently, even minor changes in the environment that lead to a change of only a fraction of a percent in one of these influences can destroy the delicate balance and cause the protein to unfold. It is usually considered that it is adequate to analyse stability only in terms of the major interactions, for example H-bonding. Nevertheless, even the minor interactions may be significant because their possible magnitudes are still in excess of the overall stability margin and additionally, these forces are themselves temperature dependent.

12.1.4 Effects of denaturation on protein structure

When a protein is cold-denatured and then renatured, it is found to have a lower absorbance at 240 nm than the original control sample. This would tend to suggest that there are fewer exposed residues in the renatured protein, i.e. the renatured protein in solution is more efficiently folded. This suggests that a cold unfolding/refolding cycle may allow molecules that were incorrectly folded to unfold and refold correctly (Franks and Hatley, 1985).

The full reversibility and recovery of enzymic activity exhibited by cold denatured proteins (Hatley and Franks, 1986, 1989a, 1989b) suggest a difference in protein structure between that produced by cold denaturation and that by heat denaturation. This difference is probably due to additional transitions in the protein taking place rapidly after thermal denaturation. It has been suggested that the N → D transition is not a single step process but involves a number of intermediate states in which the protein is unfolded to various degrees (Zale and Klibanov, 1986; Sanchez-Ruis *et al.*, 1988; Manly *et al.*, 1985; Hu and Sturtevant, 1987; Keleti, 1985). A model for the thermally irreversible transition may be as follows (Figure 12.1):

$$P_n \leftrightarrow P_d \rightarrow P_{nr} \rightarrow P_{ag}$$

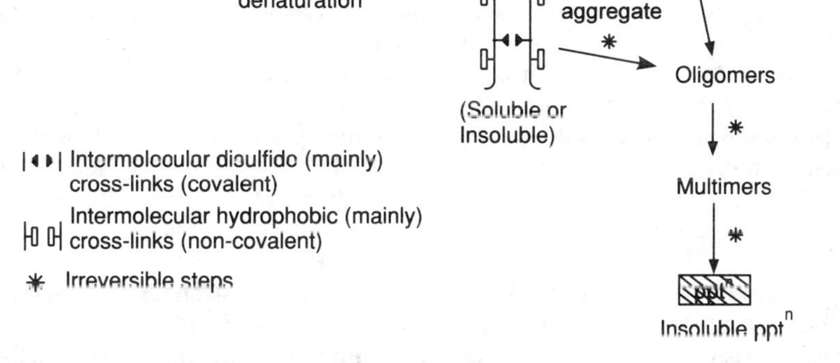

Figure 12.1 Possible routes of protein denaturation: reversible denaturation may be followed by irreversible events, which may lead eventually to an insoluble precipitate.

where: P_n = native protein, P_d = reversibly denatured protein, P_{nr} = irreversibly denatured protein, P_{ag} = aggregated protein.

In this model the initial protein denaturation step is rate limiting. After thermal denaturation the subsequent steps rapidly follow, usually leading to apparently immediate irreversibility. At low temperatures, the environment is not conducive to the $P_d \rightarrow P_{nr}$ transition, so that the protein is trapped in the P_d state. Since reversibility can also be achieved at high temperatures, but only in dilute solution, the $P_d \rightarrow P_{nr}$ step may be dependent on protein–protein interactions. Such interactions could include inter-peptide H-bonding, or repulsion between certain residues resulting in a distortion of the protein conformation from which it can not refold. It has been suggested that *cis–trans* proline tautomerisation may control the refolding of proteins, and it could therefore be that this equilibrium process is inhibited by other protein interactions.

The drop in absorbance found when cold-denatured protein is heated through its original denaturation temperature, without renaturation, is probably a consequence of aggregation (Franks and Hatley, 1985). This observation tends to support the above model. Absorbance data also suggest that the tertiary structure of cold and heat-denatured proteins is unfolded to the same degree: if they were not, then an increase in absorbance (i.e. further residue exposure) would be expected to be seen prior to aggregation.

Based on the analysis of the possible mechanisms of unfolding, outlined above, it is possible that P_d corresponds to a protein form that is largely unfolded in its tertiary structure, but has an intact secondary backbone. This interpretation is in agreement with the current understanding of folding, in which the secondary structure is believed to form first, followed by the organisation of the tertiary structure to create the native protein (Hogberg-Raibaud and Goldberg, 1977; Jaenicke, 1984). The P_d enzyme conformation is able to refold its tertiary structure onto the secondary backbone when conditions favour the native protein conformation. This P_d state is the final form of the cold denatured protein, but exists only momentarily during high temperature denaturation.

12.1.5 Stabilisation by organic co-solvents

The difference in the temperature dependence of the forces involved in cold and heat denaturation provide an insight into the mechanism of stabilisation and solvent–protein interactions at the two temperature extremes. Consider a protein in its native state at its point of maximum stability. Its native conformation is maintained principally by hydrophobic effects providing a greater contribution to stability than the major destabilising force, conformational entropy, provides to instability. Considering the relative changes in the temperature dependence of these influences at the two temperature extremes, it is unlikely that the same interactions are responsible for maintaining the native conformation against both heat and cold denaturation. This, in turn, suggests that the mechanisms of unfolding may be driven in different ways, and differences in the interaction of the unfolded conformations with respect to the solvent may possibly exist at the two temperature extremes.

If we consider the heat denaturation of a protein in the absence of perturbants then, as the temperature increases, hydrophobic interactions strengthen as do salt bridges. This would be expected to stabilise a protein. However, under the same conditions there is a considerable increase in the kinetic energy into the system. This energy is distributed to the bonds within the protein with the consequence that these weaken, destabilising the protein. The change in magnitude of the latter effect is likely to be greater than the former, consequently, the overall effect is a progressive reduction in the stability margin of the protein. At a particular temperature T_H, destabilising influences will outweigh stabilising effects, and the native protein structure will unfold. The solvent then further penetrates the protein molecule and interacts with the core structure, causing this to break apart.

The effects of co-solvents are explained in terms of their modifying effects on this process. It has been claimed that the addition of methanol causes a weakening of the intramolecular hydrophobic interactions (Gerlsma and Stuur, 1974); given this is so it would cause the protein to unfold at a lower temperature. Glycerol conversely is said to stabilise intra-peptide H-bonds and to some extent strengthen the hydrophobic interactions (Gerlsma, 1968; Gerlsma and Stuur, 1974). The result is that configurational entropy does not destabilise the protein until a higher temperature is reached.

Reducing the temperature weakens many of the forces that are responsible for maintaining the protein's stability. Many of the destabilising effects are also weakened but it is noticeable that at least one of the destabilising factors (direct solvent–peptide interactions) becomes stronger. In the absence of solvents it is the reduction in the strength of the intra-protein hydrophobic interactions, principally located in the core structure, and the strengthening of the solvent–peptide interactions that cause residues to react preferentially with the solvent (Finney *et al.*, 1980). This results in subunit dissociation and an unfolding of the tertiary structure. However, the distances, orientations, number and strength of the peptide–peptide hydrogen bonds in the secondary structure are likely to inhibit solvent penetration and prevent further interaction with the solvent. The protein thus maintains its folded backbone. There is evidence that once a secondary structure exists, then the folding to the native configuration can occur, possibly by a specific folding pathway (Kuwajima *et al.*, 1985; Ikeguchi *et al.*, 1986). Hence, upon rewarming, the hydrophobic interactions become stronger and the tertiary structure is able to refold on the secondary structure backbone, thus giving reversibility.

On the assumption that co-solvents act on the protein in the way that has been described in the earlier discussion on heat denaturation, co-solvents will individually produce the same quantitative effects at the low and high temperatures. Methanol will thus weaken further the hydrophobic interactions within the protein and destabilise the protein at a higher temperature. Glycerol, in contrast, will tend to stabilise the maintenance of the hydrophobic interactions and strengthen the peptide–peptide H-bonds, lowering the temperature of denaturation and stabilising the protein. Because of the temperature dependence of the interactions, the magnitude of the effect would be expected to be different at the two temperature extremes. Indeed, the experimental data collected in a study of this type show such a difference (Hatley, 1991).

To summarise, this 'push–pull' model considers the cause of unfolding to be different at the two extremes. At high temperatures it is principally the input of kinetic energy that disrupts the protein molecule. Solvents act on the stabilising/destabilising forces to either raise or lower the temperature at which this occurs. At low temperatures denaturation could be caused principally by a decrease in the strength of the hydrophobic interactions. During cold-induced denaturation the molecule is obviously not disrupted by kinetic energy input. However, the side chains begin to interact with the solvent *via* peptide polar group hydration, resulting in an unfolding of the tertiary structure. The strength of the H-bonds in the protein's secondary structure maintains this intact. Glycerol may have a more pronounced effect at low temperatures, because its influence on the hydrophobic interactions and peptide–peptide H-bonds in the tertiary structure could be far larger than the destabilisation of the hydrophobic interactions brought about by reducing temperature. Hence it would take a large temperature reduction to destabilise the hydrophobic interactions, to a sufficient degree to overcome the additional stabilisation caused by even a low concentration of glycerol. Following the above reasoning, glycerol would, as is found experimentally (Hatley, 1991), be less effective at high temperatures; conformational entropy is large and, consequently, a small rise in temperature would soon overcome any stabilisation brought about by glycerol. Finally there must be a limit to the maximum contribution to the stability that can be supplied by the addition of glycerol to a solvent.

12.2 Stabilisation of Proteins in Solution

Stabilisation of a labile protein in solution reduces the need for the introduction of further stabilisation processes, such as freeze-drying. This in turn reduces the degree of complication and cost of any production process. Nevertheless the selection of a formulation for the stabilisation of a protein in solution is complex (Wang and Hanson, 1988; Hatley *et al.*, 1996). In solution, inactivation is caused either by denaturation or degradation and strategies are needed to stabilise the protein.

Protein denaturation in solution often leads to the usually irreversible aggregation of protein molecules. Denaturation must therefore be avoided. Protein denaturation is promoted by a number of environmental influences including increasing temperature, pH, salt concentration and interactions with surfaces.

Probably the simplest strategy is to reduce the temperature to 4 °C, which minimises the rate of temperature-induced denaturation.

Adsorption to the container surface is a particular problem with protein solutions, since many containers have hydrophobic surfaces. This is a particular problem where the active protein is at low concentration (Sluzky *et al.*, 1991; Lougheed *et al.*, 1983; Thurow and Geisen, 1984), as the available surface area is relatively high under these circumstances. Adsorption of the protein to the container surface can result in both physical loss of the protein by 'binding' to the surface[1], or by denaturation at the container surface that in turn results in aggregation. Surface adsorption can be reduced by using a less hydrophobic container, for example silica in place of glass (Andrade, 1985).

[1] Typical adsorptive losses of this type are of the order of 1 μg cm^{-2} of vial surface. As a consequence, given normal vial geometries, such effects often become significant at protein concentrations below 10–100 μgml^{-1}.

Unfortunately, limited availability and lack of pharmaceutical approval limit the possibilities in this area. A more common approach is to add a second protein to the formulation, such as serum albumin (Suelter and DeLuca, 1983), this protein is then adsorbed onto the container surface in place of much of the active protein. Alternatively surfactants such as Tween™ (Atlas Chemicals) or Triton™ X-100 (Rohm and Haas) may be added to prevent the surface adsorption (Lougheed *et al.*, 1983). Surfactants have the additional advantage that they reduce the hydrophobic nature of gas/liquid interfaces, and so also reduce denaturation at these sites (Henson *et al.*, 1980; Cartwright *et al.*, 1977; Brennan *et al.*, 1985). Further denaturation can occur during storage and transport at the air/solution interface which is also a hydrophobic surface and induces denaturation (Donaldson *et al.*, 1985). This is a particular problem if the product is handled violently and can be minimised by cautious handling of product in the liquid state.

Denaturation occurs in the bulk solution as well as at interfaces. Proteins are extremely sensitive to their environment; the removal of a protein from its *in vivo* environment to *in vitro* promotes unfolding and protein–protein interactions (Timasheff, 1992; Timasheff and Arakawa, 1988; Schellman, 1990). Protein–protein interactions are particularly marked in high salt environments where charge neutralisation on the protein surface occurs. Working at pH values near the protein's isoelectric point can have the same effect. Thus salt concentrations must be kept as low as practically possible whilst allowing for other product criteria such as isotonicity of the final preparation. The pH should also be kept at least 0.5 pH units from the isoelectric point.

Excipients may be added to prevent denaturation. Some salts and amino acids can be used to hydrate preferentially the protein surface and thus increase its stability (Arakawa and Timasheff, 1985). Sugars have the same effect on thermal denaturation but their effect during storage has been found to be unpredictable, in some cases promoting denaturation. Alternatively the excipient can interact with the protein at a molecular level thus directly influencing its stability, as described for polyethylene glycol (PEG) and cyclodextrin (Cleland and Wang, 1993). Cyclodextrin encapsulates the protein molecule in its own microenvironment and is thought to decrease denaturation by interacting with the hydrophobic residues on the protein surface (Stratton, 1991).

In addition to denaturation, proteins suffer deactivation through deamination reactions, particularly of the Asn and Gln residues (Robinson and Rudd, 1974). As for denaturation, pH is important (Pearlman and Nguyen, 1989), and the selection of the appropriate pH depends on the exact mechanism of the deamination process involved. Acid–base deamination can be minimised by using a neutral pH (for example pH 6; Patel and Brochardt, 1990). Storage at low temperatures will also reduce the rates of deamination. Buffer selection is also important: phosphate and borate anions accelerate deamination, but NaCl and $MgSO_4$ have little effect. Surprisingly the K_2HPO_4/KH_2PO_4 buffer system shows a greatly enhanced rate of deamination (Capasso *et al.*, 1991). High buffer strengths have also been shown to enhance the rates of deamination reactions, so the molarity of storage buffers should be kept low if possible (McKerrow and Robinson, 1971).

Oxidation in solution also represents a major formulation challenge (Swallow, 1960; Stadtman, 1990). Again low temperatures can be used to reduce the rate of oxidative reactions. Formulation strategies involve the use of antioxidants (e.g. vitamin E) or reducing agents (e.g. vitamin C), which act as oxygen scavengers and are preferentially oxidised. Any head-space in the container should also be back-filled with an inert, oxygen-free gas to prevent oxygen entering the product during solution (Cleland *et al.*, 1993).

Although no hard and fast rules exist for stabilisation in solution, general recommendations are to reduce ionic strength, optimise pH, minimise adsorption onto surfaces and add an antioxidant if necessary. The addition of a bulking agent such as albumin and a non-ionic detergent (0.1%) is often beneficial for solution storage at 4 °C, as is the addition of 50% glycerol for storage at −20°C or −70°C.

A literature search may well be fruitful in identifying agents which have been demonstrated to improve the stability of a specific protein in solution. Horseradish peroxidase (HRP) reagents, for example, may be stabilised by a diverse range of additives including gentamicin and related compounds (Sun, 1985), calcium salts and polyethylene glycol (Anawis and Lindberg, 1988), haemoproteins (Warren and Boyer, 1991), specific anti-HRP antibodies (Skold *et al.*, 1991), polyvinyl alcohol (Boyd and Yamazaki, 1994; Yamazaki and Boyd, 1996), and aspartate salts in an acidic buffer (Franken *et al.*, 1995). Note that many of these stabilising formulations are patented.

It should be noted that, although storage at low temperatures under appropriate conditions may reduce the rate of deleterious reactions, it cannot eliminate them completely. If the options outlined above fail to achieve acceptable stability, then further processing such as freeze-drying may be the only solution. Once again specific additives (many patented) may enhance the stability obtained from a dry formulation. Returning to HRP reagents, for example, shelflife has been improved by employing a matrix containing polyvalent metal ions such as Fe^{2+} or Mn^{2+} (Dawson *et al.*, 1980), or mixtures of cationic polymers and sugar alcohols (Gibson and Woodward, 1993; Gibson *et al.*, 1993).

With recombinant proteins a final option for improving stability may be to modify the structure by site-directed mutagenesis: Meng *et al.* (1993), for example, replaced aromatic residues in the active site of glucose isomerase from a thermophilic bacterium and observed an improvement in thermal stability. This was attributed to a reduction in the solvent-accessible hydrophobic region of the protein. This route towards stabilisation clearly requires a considerable investment of effort, in this example driven by the industrial importance of the enzyme concerned.

12.3 Stabilisation of Proteins by Freezing and Freeze-drying

The dry state has emerged as the most favoured method of stabilisation since many labile materials show a greater degree of stability than when in solution. Freeze-drying is the most popular method used to achieve this condition. The reasons for the popularity of freeze-drying as the preferred method for achieving the dry state are largely historical: low temperatures are assumed to protect the active material during processing; it can be performed under sterile conditions and the process is a standard practice approved by the pharmaceutical regulatory authorities. Some of the advantages and disadvantages of freeze-drying are summarised in Table 12.2.

In principle, the freeze-drying process is very simple and can be divided into three distinct stages. (1) An aqueous solution of the active material, typically 3–10% (w/w) solids, is frozen (*freezing*), usually to approximately −40 °C. (2) Once frozen, a vacuum is applied and the ice formed during freezing is sublimed (*primary drying*). (3) On completion of sublimation, the final stage of the process is to desorb any residual moisture still in/on the product (*secondary drying*).

The result is a dry product in the form of a plug containing typically 98% (w/w) solids. It should be noted that the above definitions are operational. During production, both within a sample and within a batch, more than one of the above processes can be proceeding at the same time.

Table 12.2 Advantages and disadvantages of the freeze-drying of protein products

Advantages	Disadvantages
Gentle method of protein drying	Expensive equipment
Reduced shipping and distribution costs for dry products	High running costs
Accepted by regulatory authorities as a validated method	Long processing times (typically 3–5 days)
	Irreversible decrease in biological activity of some labile protein products

The equipment requirements for freeze-drying are like the freeze-drying process itself, in theory simple. The primary objective of the equipment is to provide sufficient heat to the product to equal the heat being removed by sublimation of ice. The most basic requirements are a condenser and a vacuum pump to which the sample is attached by an air-tight seal. This formed the basis of the earliest types of freeze-dryers and still today is the basic principle of manifold-type drying systems. For labile products such a simple system is rarely adequate, since temperature control and process monitoring equipment become essential to control the temperature of the product. The dryer requires shelves that can be cooled to freeze the product, and heated to drive sublimation. A vacuum pump capable of lowering the pressure in the system to below the saturated vapour pressure of ice is also a necessity to drive the process. Many products, once dried, are prone to the rapid uptake of moisture from the atmosphere, which in turn causes stability problems. Therefore there must be the facility to seal the vials either under vacuum or with dry gas in the drying chamber itself. For pharmaceuticals a steam sterilisation system and a monitoring system that enables validation of process batches are further requirements.

Although it is possible with some robust materials (mainly organic compounds) that freezing does not result in activity loss, bioactive agents are rarely freeze-dried in the pure form. To minimise damage due to the freezing process, bulking agents such as albumin and carbohydrate excipients are often used. Other chemical components are usually also added to the formulations for specific purposes, such as to adjust pH, enhance solubility or balance osmolarity. When designing a freeze-drying process it is the whole formulation (active agent plus excipients and buffer additives) that determines the parameters of the lyophilisation cycle. Thus, any alteration in the formulation, not just the active agent itself, will require further modification of a cycle. In addition to excipients added for the reasons outlined above, freeze-drying usually requires the addition of further excipients to aid the freeze-drying process itself, or to provide physical strength to the freeze-dried plug during storage. Such excipients are referred to as lyoprotectants or stabilisers (Pikal, 1990b). The informed selection of the appropriate stabiliser is important for the development of effective freeze-drying cycles. A wrong selection can, and does, give rise to unstable products (Roy *et al.*, 1992).

Thus the formulation and process parameters are critical to the stability of a freeze-dried product. Unfortunately the industrial freeze-drying process has only limited degrees of freedom. There are only three variables that can be used to control the quality of the product: temperature, pressure and time, and all of these are limited by engineering or economic constraints. Additionally there are, depending on application, regulatory restraints on the types of excipients that can be used. The results of freeze-drying labile biologicals have been varied, with the formulation and cycle often determined by trial and error. Indeed the view has been expressed that this is the only way to develop formulations: 'Freezing and freeze-drying data can only be determined by trial and error' (Willimer, 1992). This view is now being challenged, and rules to decide formulation and processing conditions are being developed, based on the underlying physicochemistry of the freezing and drying processes (Franks, 1990). Such rules provide the basis for the production of stable, high yielding, easily rehydratable freeze-dried products.

12.3.1 Freezing

12.3.1.1 Nucleation and maturation

The equilibrium freezing point of water is 0 °C. However, water has the capacity to undercool to a high degree, i.e. it can remain in the unfrozen state to temperatures well below the equilibrium freezing point temperature (Hatley *et al.*, 1987). The probability of ice nucleation increases with decreasing temperature; in pure water, the probability of nucleation effectively reaches unity at ca −40 °C (Hobbs, 1974). Water is rarely pure, and generally contains heterogeneous ice nucleating agents which will nucleate ice at higher temperatures, typically −15 °C. Once nucleation has occurred crystal growth proceeds.

In contrast to nucleation, crystal growth and recrystallisation rates decrease with decreasing temperature. Thus as the degree of undercooling increases the probability of ice nucleation increases, and the rate of crystal growth (once nucleation has occurred) decreases. The nucleation temperature is dependent on cooling rate. Thus rapid cooling (for example direct immersion in liquid nitrogen) will give a high degree of undercooling and a low nucleation temperature, resulting in the formation of numerous nuclei. At low temperatures, crystal growth rates will be slow and crystal size will thus be small. At lower rates of cooling, ice nucleation occurs at a higher temperature, crystal growth is more rapid, and larger crystals are produced.

In the freeze-drying process there is, in practice, little that can be done to control the crystal size. The sheer bulk and engineering limitations of freeze-drying plant mean that cooling rates are low and large crystal size is the norm. Even if a product is pre-frozen in liquid nitrogen it is usually transferred to a drier shelf at, say, −40 °C. At these higher temperatures, maturation (recrystallisation) will then commence and large ice crystals will begin to grow from the multitude of smaller crystals formed during the rapid cooling. Thus any advantage of forming the small crystals may be lost. Similarly if a solution is rapidly cooled to yield an aqueous solution at −40 °C and then transferred to −5 °C, the size of the crystals formed will again increase as larger crystals are formed by recrystallisation of the smaller ones. Such size distribution increases can cause significant localised differences within a product during freeze-drying.

Since freezing is a random probability event, the onset of freezing cannot be controlled between vials in a sample batch. Usually there is a spread of 5 °C between sample nucleation temperatures. This can introduce some variability into the freeze-drying process as ice crystal size affects the duration of sublimation (primary drying). A sample containing smaller crystals, i.e. a sample that has undergone a greater degree of undercooling, will dry faster because of the greater ice surface area to volume ratio. Effective process design, and the judicious use of excipients, can nonetheless minimise variation (Pikal, 1985, 1990a,b). It would however be preferable to control nucleation, and examples of such control can be found in nature; the microorganism *Pseudomonas syringae*, for example, has the ability to reduce the degree of undercooling to which its cellular fluid is subject. The proteins of the cell surface of these microorganisms act as ice nucleation sites and, though these have been isolated (Tomchaney *et al.*, 1982), they are not yet available at an economic cost for use in routine freeze-drying processes.

12.3.1.2 Freeze-concentration

A phenomenon associated with freezing is freeze-concentration. When ice forms, the residual phase consisting of unfrozen water and solutes becomes increasingly more concentrated as the temperature is reduced (Hatley and Franks, 1991; Pikal, 1990a). This increase in concentration can have drastic effects on the solutes. Greatly increased rates of reaction can be seen, and in some cases reactions that do not normally proceed can be induced. A reported example of this is a commercially available kit for the measurement of haemoglobin in blood (Hatley *et al.*, 1986a). The assay solution, consisting of potassium ferricyanide and potassium cyanide, is stable in solution at ambient temperatures. On exposure to sub-zero temperatures the ferricyanide is, however, reduced to ferrocyanide. The reaction was studied in controlled conditions at a range of temperatures between −5 °C and −40 °C. It was observed that as the temperature was decreased, and the degree of freeze-concentration concomitantly increased, the rate of the reaction was enhanced. This is a drastic example of the detrimental effects that freezing and freeze-concentration can have on a solution, but other reactions also show the same type of phenomena (Hatley *et al.*, 1986b).

It has been thought that one of the benefits of the freeze-drying process is the protective effects of the low temperatures involved. However, in view of the phenomena just discussed a distinction must be drawn between low temperatures *per se*, and freezing.

This distinction is best shown by considering the enzyme lactate dehydrogenase (LDH) and its stability at sub-zero temperatures in the frozen and unfrozen states. In order to study the stability of LDH at sub-zero temperatures without freezing the process of undercooling was employed (Hatley and Franks, 1986). The process is based on the fact that, as previously mentioned, water can supercool to −40 °C before freezing in the absence of a heterogeneous nucleus (Angell, 1988). An aqueous solution is dispersed, in the form of small droplets, into an inert oil carrier. A small finite number of these droplets will contain heterogeneous nuclei and these will freeze. The vast majority of droplets, however, will remain unfrozen, even at storage temperatures down to −30 °C.

An LDH solution was taken and either dispersed in the oil carrier or frozen as a bulk solution. Periodically enzyme activity was determined. The results are

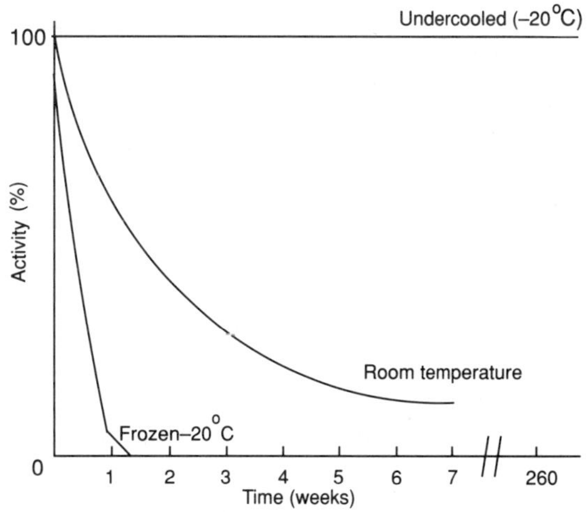

Figure 12.2 The stability of LDH at −20 °C, either frozen or unfrozen (undercooled), compared to room temperature. Note that the frozen sample at −20 °C is less stable than a solution at room temperature. However, when the effects of freezing and freeze-concentration are avoided, stability is enhanced by unfrozen low-temperature storage.

shown in Figure 12.2. In the unfrozen, undercooled, state the protein shows a marked increase in stability compared to storage at room temperature. The stability obtained by low-temperature, unfrozen storage is even more apparent, however, when compared to the severe loss of enzyme activity from the frozen samples.

As well as enhancing reaction rates, freeze-concentration can cause other detrimental effects. Many labile materials are presented for freeze drying in aqueous buffered solutions. As the degree of freeze-concentration increases the buffer salts present can become supersaturated. Should this cause one of the buffer salts to crystallise then drastic alterations in the pH of the solution can occur (Murase and Franks, 1989). The pH shift may cause denaturation of a labile protein. Careful selection of the buffer system depending on the tendency of salts to crystallise is therefore vital.

12.3.1.3 Eutectic crystallisation

In freeze-drying, as the bulk water is removed from the system as ice during freezing, the residual phase becomes more and more concentrated as just discussed. Finally, at its eutectic temperature (T_e), a solute can (and might be expected to) undergo eutectic crystallisation (see Figure 12.3).

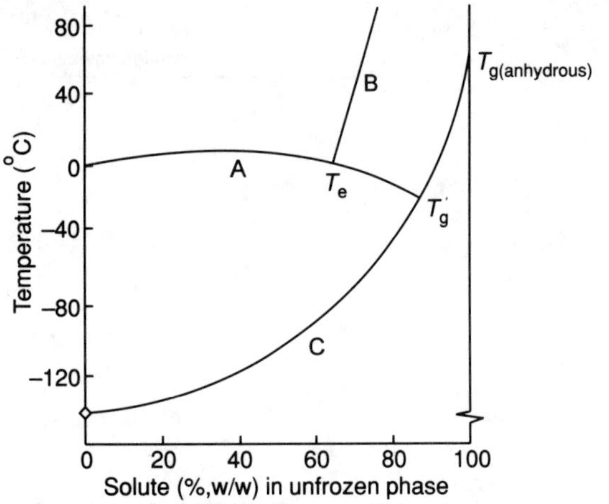

Figure 12.3 The state diagram for a typical glass forming excipient. The three curves shown are the freezing point depression curve (A), the solubility curve (B) and the glass transition curve (C). During freeze-drying a solution (containing typically 5% (w/w) solids) is cooled to the freezing point depression curve. Nucleation begins to occur to give a two-phase system, ice and residual unfrozen water/solute. As the temperature is further reduced, the solution tracks the freezing point depression curve (A) and the residual phase becomes progressively more concentrated. Eventually the concentration of the residual non-ice phase reaches the eutectic composition at T_e, where the freezing point depression curve meets the solubility curve (B). It would be expected that the solute would then crystallise, however in glassy systems the residual phase continues to concentrate (supersaturate) until at T_g' (a unique point on the glass transition curve) the residual phase undergoes a transition to form a stable glass. Once the ice has been sublimed, further moisture can be removed and the glass transition of the product then follows the glass transition curve (C) to higher temperatures.

The ice could then be removed by sublimation under vacuum during primary drying, before raising the temperature to remove the last traces of residual moisture on the crystal surfaces. Unfortunately, the freezing process is rarely so simple: although expected from the state diagram, the eutectic at T_e rarely occurs.

12.3.1.4 Glass formation

In the dry state materials behave as solids, a solid being broadly defined as a material that has the capacity to retain its shape and volume. The solid state is often thought of as corresponding to a crystalline structure, in which the molecules are distributed in a defined arrangement from which all free water is excluded[1]. Solids can however exist in a second form, as an amorphous or glassy structure. Such structures, in contrast to the crystalline state, have no defined order. A glass is defined as a liquid with a viscosity $>10^{14}$ Pa. This viscosity is such that diffusion rates within it are extremely low and can be measured in terms of mm per century. Consequently any active material within a glass is effectively trapped, and deterioration processes are kinetically inhibited.

An example of an amorphous solid is window glass. This displays all the qualitative characteristics associated with a glass, including transparency and brittleness. Other materials which can exist in the amorphous state include many polymers and metal alloys (Wunderlich *et al.*, 1964) and the carbohydrates and carbohydrate polymers (Levine and Slade, 1988).

If liquids or solutions are cooled at a rate that does not permit nucleation to take place, supersaturation occurs, and no crystallisation will be observed at T_e. The solution viscosity will progressively grow greater, until, over a particular narrow temperature range, it suddenly and rapidly increases and the solution becomes a glass. The high viscosity renders the glass stable by kinetically inhibiting crystallisation.

This change is called the glass transition, and the temperature at which it occurs the glass transition point or glass temperature, T_g. The glass transition point of a product is a function of moisture content: increasing the moisture content decreases the glass temperature (Figure 12.3, curve C). The overall glass temperature is also a function of the glass temperatures of the individual solutes, and their respective mass ratios.

T_g' is the point on the glass transition curve at which it intersects the freezing point depression curve (Figure 12.3, curve A); $T_{g(anhydrous)}$ is the glass temperature of the anhydrous product.

Maintaining a solution below T_g' will ensure that the active material remains trapped in the glassy state. For in-house storage where it can be guaranteed that the low product temperatures can be maintained, this provides an effective method of preservation. Nevertheless, the majority of products are not used in-house and are subjected to storage and transport conditions that are not under the control of the producer. A fairly

[1] Although water molecules can be incorporated in stoichiometrically defined quantities in the crystal lattice, as water of crystallisation.

recent article has clearly demonstrated the dangers to product stability that this loss of control can impose (Gray, 1994). A product was shipped on dry ice to maintain an air temperature in a package of 2 to 8 °C. Shipping was supposed to take three days. In reality the delivery took 25 days with the product experiencing temperatures in excess of 30 °C. 'Keep cool' warnings on the carton were ignored.

A further difficulty with transporting frozen solutions below T_g' is the actual temperature required, typically below −40 °C. Conventional packaging and transport practices make it extremely difficult to send products at such temperatures. Thus for effective transport it is necessary to remove the requirement for low temperatures by drying the product, to raise the glass temperature.

12.3.2 Primary drying

Primary drying is the sublimation of the ice phase. The critical factor is the product temperature. In order to protect the active material, a product must be kept in either a crystalline or a glassy state while ice is being sublimed. Thus the ice core temperature must be kept below T_e (crystalline products) or T_g' (glassy products) for the duration of this stage.

Should the product rise above the relevant temperature, the residual glassy or crystalline solute phase will begin to melt back into solution. The result will be loss of physical strength of the 'dry' material as the ice front recedes, and collapse of the product. Degradation of the active material may also be apparent. This collapse phenomenon during primary drying has been known for some time. A critical drying temperature was observed with a number of products (Mackenzie, 1977) although the case of the collapse was not fully understood since the temperature values did not coincide with measured eutectic temperatures. It is interesting however to note that the measured collapse temperatures reported by Mackenzie are the same as the reported glass transitions for the systems.

In order to ensure that the product is kept below T_e or T_g during primary drying, and to maximise the rate of sublimation, it is important to select the correct pressure and time of drying. Additionally it is important to have some understanding of the basic mass and heat transfer processes that take place in the drier.

12.3.2.1 Heat and mass transfer

This subject has been covered in depth by Pikal (1985, 1990a). In the freeze-drying process the objective is to balance the heat transfer from the shelves to the sublimation front with the heat being used for sublimation. The amount of heat that needs to be supplied to the sublimation front is dependent on the rate of ice removal (mass transfer) from the product to the condenser. If the heat transfer to the product exceeds that being used for sublimation, the product temperature can increase and the product collapse. If by contrast too little heat is supplied by the shelves, insufficient heat will reach the sublimation front and sublimation will be slowed. Many of the reported failures of the freeze-drying process can be ascribed to poor heat transfer control. It is common to raise the shelf temperature to ca 20 °C as sublimation is commenced. This can set up a temperature gradient in the product, causing the sample at the base of the vial to heat up and collapse.

Mass transfer of ice from the product to the condenser is influenced by the barriers that the subliming vapour encounters. The first resistance is the product itself. As the ice front recedes a layer of dried product is present above it. The vapour must diffuse through the pores in this layer, which causes a greater and greater resistance as drying proceeds. Thus, in addition to decreasing the surface area-to-volume ratio of a product, a greater product thickness also increases the resistance to sublimation. The practical outcome of this is that doubling a product depth more than doubles the time required for sublimation. The resistance also increases with solute concentration, thus solute concentration should be minimised, but only to the extent that a physically robust product is created. It is rarely possible to reduce the solute concentration below 2% (w/w) in the initial solution.

The next resistance to mass transfer is the stopper. In most processes vials are loaded into the drier with stoppers in the half-open position, to enable automatic sealing *in situ*. The stopper contains pores through which the vapour can pass. A number of stopper designs are available, from a simple cross-cut type to more complex multipore examples. In terms of resistance to vapour flow it is not only the overall pore hole size in a stopper that is important, but also the pore design. Generally one large pore provides less resistance to sublimation than a number of smaller ones. The importance of the stopper in the overall resistance to drying has been analysed (Pikal *et al.*, 1984). Though this study reached the conclusion that as a percentage of total resistance the stopper is not important, this is certainly not the case in practice. The conclusion reached was based on the assumption that the resistances acted in parallel, i.e. increasing the stopper resistance for example by a factor of two, would only result in an increase in stopper resistance from 1 to 2% of the total system resistance. It does however appear that resistances act in series, and it has been observed that a doubling of the stopper resistance significantly increases resistance to sublimation, and prolongs the duration of primary drying.

As has been mentioned, the objective of drying is to set up an efficient heat flow transfer from the shelf to the sublimation front while maintaining the ice core temperature below T_g or T_e. Heat transfer is by three methods: radiation from the shelf above, contact from the shelf below through the tray or vial and through the product, or *via* gas phase conduction. Contrary to the common assumption that contact is the dominant mechanism of heat transfer, gas phase conduction has been shown to be the most significant (Pikal, 1985). This can be optimised by maintaining the pressure in the chamber as high as practically possible. In practice this is determined by the saturation vapour pressure of ice at the maximum allowable drying temperature. Pressure should therefore be set just below the saturation vapour pressure at the drying temperature. Heat transfer by contact should be carefully controlled. As noted above, attempts to increase heat transfer by this method can set up a temperature gradient in the product, hottest at the bottom, and cause collapse. The temperature at the base of the product must be maintained below the collapse temperature and excessively high shelf temperatures should not be used. Nevertheless, there is a loss of heat between the shelf and the vial or tray base and, to compensate for this, the shelf temperature should be raised to 5 to 10 °C above the collapse temperature once the vacuum has been applied. The exact value can only be determined with respect to the individual system use, e.g. vial type. For a detailed analysis of heat and mass transfer and its importance see Pikal (1985, 1990a).

The duration of the primary drying period is important. It is imperative that the shelf temperature is not raised before ice sublimation is complete. Most freeze-driers are equipped with pressure rise testers. In essence the valve between the chamber and the condenser is closed and the pressure rise in the chamber (caused by the subliming ice vapour tending to equilibrium) measured over a given period of time. When the pressure rise is minimal, primary drying can be considered complete. Nevertheless, during formulation development it is useful to be able to calculate the expected primary drying time before launching into full scale trials.

Ice sublimation rate data are available (Mellor, 1978) for both pure ice and ice in solid matrices, for the temperature range 0 to −50 °C. Thus, since the primary drying temperature is known from T_e or T_g, and the amount of ice to be removed from the product can be calculated from the fill volume, the primary drying time can be estimated. Allowance must be made for the solids content of the preparation, the variation of the temperature profile across the drier, and the temperature differential between the chamber and the drier. The latter two factors can only be known once a particular piece of equipment has been fully validated.

12.3.3 Secondary drying

For crystalline systems, secondary drying only involves the removal of any water sorbed on to the crystal surfaces. A rapid increase in shelf temperature is all that is required.

With amorphous systems the situation is considerably more complex. As can be seen in Figure 12.3, once the ice has been removed a residual phase is left. This phase still has a moisture content of 20% (w/w) water and a glass temperature (T'_g), in the case of sucrose, of −32 °C. In order to increase the glass temperature, and enhance the product stability, it is necessary to remove further water. This must be performed with care. If the product temperature rises too rapidly, the temperature rise will exceed the moisture removal rate, and the product temperature will rise above the glass transition curve. The viscosity will rapidly decrease and the product will collapse.

The product temperature must therefore be slowly ramped, balancing moisture removal rate and temperature increase such that the product follows the glass temperature/moisture profile of the glass temperature curve (C). The exact rate of ramping is dependent on the residual moisture content of the product at T'_g. The final moisture content determines the glass temperature of the dry product and thus its temperature of maximum physical stability. This content is principally determined by the final drying temperature and, although it is a commonly recommended practice to lower the chamber pressure during this stage of drying, low pressures seem to have little effect on residual moisture (Pikal, 1985, 1990a).

Sometimes formulations may display hysteresis in their behaviour on cooling and heating. A solution which did not crystallise on freezing, and thus formed a glass, may however crystallise as the temperature moves up through T'_g, and a highly concentrated solution is formed. Such formulations are referred to as doubly unstable (see below).

12.3.4 Determination of glass temperatures and eutectic temperatures

As will be apparent, before a cycle can be devised it is necessary to determine the values of these parameters. There are a number of ways (Evans *et al.*, 1995; Ford and Timmins, 1989) to measure these but all have some drawbacks. For example electrical resistance techniques tend to give a value for T_g some 5 to 10 °C higher than those measured by thermal techniques such as differential scanning calorimetry (DSC), thermogravimetric analysis or thermal mechanical analysis. Since these values are measurements of thermal events, it is recommended that they are measured using a thermal approach such as DSC; the following discussion will therefore concentrate on this technique (Ablett *et al.*, 1992; Hatley, 1992; Her and Nail, 1994).

DSC measures the difference in heat flow to a sample compared to that of an inert reference sample, usually dried sand, over the temperature ranges used in freeze-drying. The output from DSC is presented as a plot of heat flow against temperature. Second order and pseudo-second order transitions such as a glass transition are displayed as a step in the heat flow. First order transitions such as crystallisation or melting are associated with the release/absorption of latent heat and therefore are displayed as peaks or troughs. In this section the example data have been obtained with a Perkin-Elmer DSC, and endothermic events such as melting are shown as peaks on the power–time curve, while exothermic events (crystallisations) appear as troughs. Other manufacturers sometimes use software which displays these the other way round, i.e. endothermic events are reflected by a trough.

The critical values for freeze-drying are the T_g' or the T_e dependent upon system behaviour. If the system is glass-forming then it is also necessary to determine the water content at T_g', expressed either as W_g' (g water/g solid) or C_g' (weight % solids content). All these parameters can be determined by DSC.

A solution is loaded into a calorimeter pan and cooled to at least −40 °C. The solution is then reheated and the power–time curve is recorded. An analysis of the power–time curve, as outlined below, provides the information needed to develop a freeze-drying cycle.

It has already been mentioned that three types of behaviour can be seen when analysing a formulation, glass-forming, eutectic or doubly unstable (Chang and Randell, 1992). Examples of the DSC analysis of each type are shown in Figures 12.4 to 12.6.

Figure 12.4 shows the power–time curve associated with the heating of a prefrozen glass-forming system.

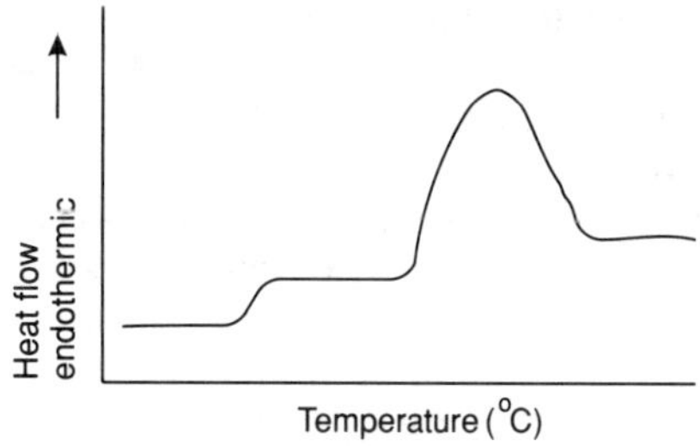

Figure 12.4 A typical power–time curve obtained by DSC when a previously-frozen preparation that forms a glass is reheated. The step in the heat flow curve is the glass transition of the maximally freeze-concentrated product (T_g'). This is the maximum temperature at which primary drying should be performed. The large endothermic peak is the melting of ice.

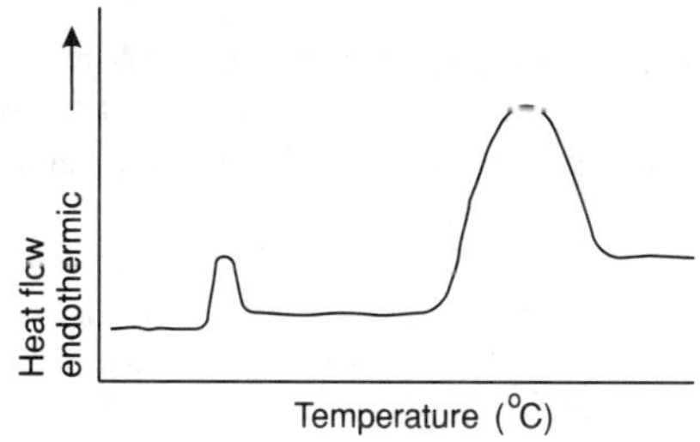

Figure 12.5 The power–time curve for a system that undergoes a eutectic on melting. The lower peak is the eutectic melting at T_e. Ice sublimation should be performed below T_e in such systems.

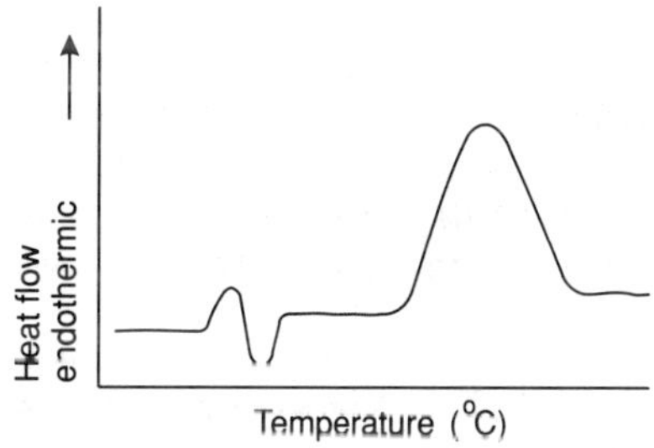

Figure 12.6 The power–time curve obtained when a doubly unstable product is frozen. The solute did not crystallise on freezing. During rewarming the system first undergoes a glass transition. Immediately after the glass transition, as viscous flow commences, the solute crystallises. Doubly unstable products can cause problems in freeze-drying (see the text for details).

The step in the heat flow is the glass transition while the endothermic peak is the melting of ice. The primary drying temperature should be below T'_g, the step in the heat flow.

Figure 12.5 is the power–time curve associated with a binary eutectic system: two endotherms are seen, the ice solute eutectic at ca 258 K, and the ice melting endotherm at 273 K. The primary drying temperature should be below the binary eutectic.

Figure 12.6 represents a doubly unstable system. During freezing the solute did not crystallise but formed a glass. On rewarming however three transitions can be seen: firstly, a step in the heat flow is apparent, corresponding to the glass transition as the non-ice phase transforms from a glass to a viscous solution. This glass transition is, however, interrupted by an exotherm caused by crystallisation of the solute from the viscous solution. Finally there is the ice melting endotherm, on this occasion obscuring the eutectic melting of the solute. It is necessary to ensure that the solute crystallises before drying commences or storage problems can arise (see later). The solution should therefore be cooled to below T'_g, then heated to a temperature just below the T_e of the solute. In the case of Figure 12.6 where the T_e is at ca −2 °C but obscured by the ice melting peak, a temperature of around −10 °C could be used. The sample can then undergo primary drying.

These figures show the three main types of behaviour of solution, but mixtures such as partially glassy, partially crystalline systems can form, which produce power–time curves that contain components of all those shown in Figures 12.4 to 12.6. These are difficult, but not impossible, to freeze-dry because the degree of crystallisation can be variable. If the formulation that produces such behaviour cannot be avoided, then the primary drying temperature should be below the temperature of the lowest thermal event.

12.3.5 Determination of residual water content at the glass temperature

The unfrozen water content of the maximally freeze-concentrated solution, expressed in either of the formats W'_g or C'_g, is the amount of water to be removed during secondary drying, i.e. all the water that did not form ice on freezing. The higher the water content the slower the product temperature should be raised if product collapse is to be avoided. The unfrozen water content can be determined by measuring the area under the melting endotherm, and dividing this by the latent heat of fusion of ice at the ice melting temperature. Since the total amount of water in the initial sample can be determined from its composition data, subtraction of the amount of ice formed from the total water present will give the mass of unfrozen water. This method is prone to some errors, particularly from ignoring the heat of dilution, and from problems in baseline determination for the integration of the endotherm (Hatley *et al.*, 1991; Hatley and Mant, 1993; Willimer, 1992). Nevertheless it provides a sufficiently useful approximation.

12.3.6 Storage stability

The stability of a dried product is determined to a large extent by its glass transition temperature. The most direct way to determine this temperature is by DSC, as discussed in Section 12.3.4, and Figure 12.7 shows the DSC power–time curve for the dry product of a glass-forming preparation. The T_g is 52 °C; if this temperature is exceeded then the product will become deformable, and detrimental diffusion-controlled reactions will commence. In this case the T_g of the product is high, so loss of potency of the active material during storage would not be expected. Many formulations result in much lower glass temperatures; if for example these are ca 25 °C, loss of potency on the laboratory bench or during transport may occur. DSC is however an expensive process and relatively slow for use as a quality control procedure. An alternative is to measure the residual moisture content. The glass temperature for a given formulation is a function of the residual moisture content (see curve C in Figure 12.3). If a moisture/T_g curve is constructed by measuring the glass temperature of the formulation at a number of moisture contents, then moisture content can be used as the quality control criterion. If for example a product has a glass temperature of 30 °C at a moisture content of 3% (w/w) then the quality control specification for a product to be stored at 25 °C could be defined as a moisture content of less than 3% (w/w).

The method of moisture analysis should be carefully considered (Nail and Johnson, 1992). Gravimetric analysis can underestimate the moisture content in materials with a high glass temperature. If the drying temperature is less than the glass temperature then moisture will remain trapped in the glass, even when weight loss appears to have stopped and the material is assumed to be anhydrous. Karl Fischer analysis of moisture content overcomes this difficulty, particularly where the moisture is first extracted using a suitable solvent such as methanol or formamide. Regardless of which method is used it is important that it is specified, since moisture contents determined by different methods are not directly comparable. The guidelines for

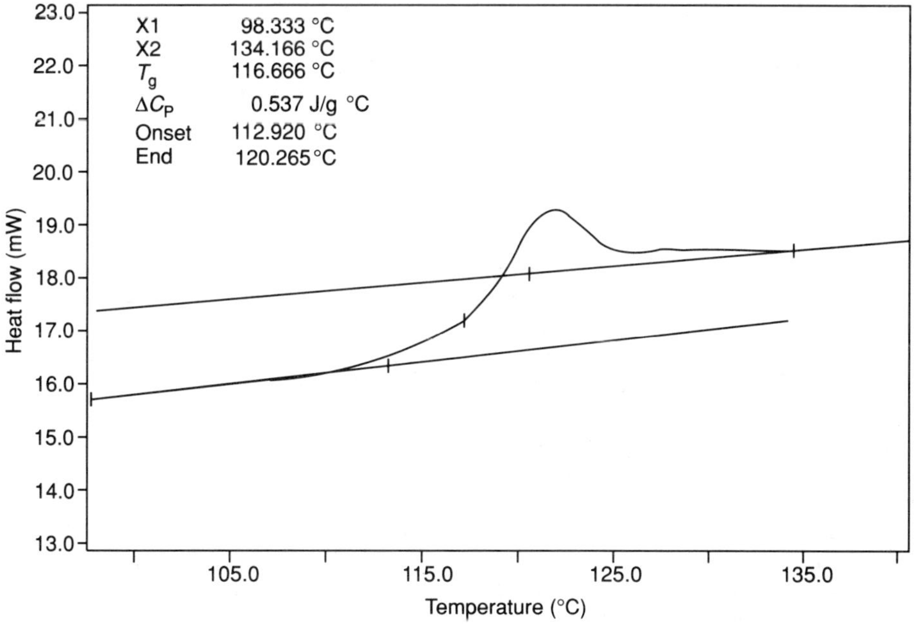

Figure 12.7 Power–time curve for dried glassy product. The transition corresponds to the $T_{g(anhydrous)}$ of trehalose.

moisture determination issued under 21 CFR 10.90 of the FDA can be found in May *et al.* (1992).

A typical dry product will contain ca 2% residual moisture. In a eutectic-forming system the solutes will be crystalline, and stability should not be a problem unless the product is stored above the melting temperature of one of the components. If a doubly unstable excipient has been used without ensuring that it fully crystallised during processing, it is possible it will crystallise during storage (Hatley, 1992). This crystallisation would release additional water into the residual phase (i.e. the active material) lowering its glass temperature and accelerating degradative processes, even at low temperatures that were originally considered safe. This is not a problem if the excipients are made to crystallise completely during processing, for example by introducing a temperature cycling step (annealing) into the freezing process (Mackenzie, 1975).

As stated above, the glass temperature of a dried product represents the temperature above which the rate of deterioration of an active component in a dried preparation rapidly accelerates. It is a temperature that should never be exceeded. However it is becoming apparent that consideration of T_g alone may be inadequate.

12.3.7 T_0: a better predictor of stability than T_g

Recent results show that physical stability of the amorphous formulation and chemical stability of an active component are not the same thing, and the predominant influences may differ (Colaco *et al.*, 1994). Thus though the relationship between the physical stability of the amorphous formulation and T_g is now established, it is becoming apparent that T_g is not the upper limit of chemical stability: reactions can still proceed below this temperature. The amount of degradation varies depending on the glass former, and though the rates of reaction are orders of magnitude lower than those found either above the glass temperature or in solution, they are very significant over the timescales employed during storage (Roy *et al.*, 1992; Kerr *et al.*, 1993; Pikal, 1994). Examples include the observed stability of antibody conjugates and catalase, which cannot be related to the glass temperature of the stabilising excipient (Roy *et al.*, 1992; Pikal, 1994), and the poor stability of human growth hormone in dextran despite its high glass temperature (Pikal, 1996). Thus glass formation, while necessary, is clearly insufficient *per se* for stabilisation of labile components.

Although mechanisms of degradation below T_g have not been fully elucidated, they may result from side-chain flexibility of a labile component in the glass, or diffusion of small molecules such as water or oxygen through the glassy matrix causing hydrolysis or oxidation (Hancock *et al.*, 1995; Pikal, 1996). An additional degradative mechanism which might occur is an interaction between the labile material and the glass-forming excipient, e.g. the glycation of lysine vasopressin in the presence of reducing sugars (Tarelli *et al.*, 1994).

Angell (1995) has recently proposed the concept of 'strong' and 'fragile' glasses. Strong glasses show a broad glass transition by DSC, indicating a gradual loss of rigidity as configurational entropy increases with increasing temperature. Fragile glasses, by contrast, show a sharp glass transition indicating a sudden change from the glassy to a more fluid, rubbery state. Thus the more fragile a glass, the more sensitive it is to changes around the glass transition; above T_g, softening occurs much more rapidly for a given increase in temperature than with a stronger glass. For both fragile and strong systems, as the temperature is reduced below T_g there is a loss of configurational entropy. At some temperature, T_0, the configurational entropy reaches zero; this is the zero mobility temperature, since, in the absence of entropy, molecular motion effectively stops (Angell, 1991, 1995; Moynihan, 1993).

For degradational processes to occur below T_g, there must be mobility or molecular freedom, thus by definition T_0 (as the zero mobility temperature) and not T_g must represent the limiting safe storage temperature for an active material in the amorphous state if all degradation is to be stopped. Below the glass transition fragility is an advantage, as, compared to a stronger glass, the viscosity increases more rapidly for a given drop in temperature. Thus for a fragile glass the difference $T_g - T_0$ is smaller, and T_0 would occur at a higher temperature compared to a stronger example with the same T_g value.

It should be remembered that not all degradative processes can proceed between T_0 and T_g: translational movement of proteins is an example that correlates strongly with T_g, and does not take place below this temperature. Though there are insufficient data available at present to determine where between T_0 and T_g different reaction processes cease, it seems likely that the smaller the reacting species, the closer to T_0 the system needs to be stored to prevent degradation.

For practical purposes it is thus useful to calculate T_0 as well as measuring the T_g value. Glasses are classified as fragile or strong according to their fragility parameter, D, which is derived from the structural relaxation time (Angell, 1991, 1995; Moynihan, 1993). Strong glasses have a larger D value than more fragile examples. In turn, T_0 is related to T_g and D by the equation:

$$T_0 = T_g/(1 + 0.0255D)$$

Consequently a strong glass (large D) exhibits a large $T_g - T_0$, the difference being smaller for a fragile glass (small D). More detailed descriptions of calculations of D and T_0 are available in the literature (Angell, 1991, 1995; Moynihan, 1993; Pikal, 1996). The calculations derived by Angell use Moynihan's observations that the width of the glass transition is related to the effective activation energy for viscous flow, and that this is almost indistinguishable from the activation energy for structural relaxation, ΔH^*.

The accuracy of the calculation of D by this method is highly dependent on the accuracy of measurement of ΔT_g at the glass transition; it relies on equations originally calculated from measurements on high-T_g inorganic glasses, and there may therefore be deviations in the absolute values of T_0 calculated. Nonetheless, Pikal (1996) found this to be less than 10 °C when the value calculated for sucrose was compared with that obtained *via* enthalpy relaxation.

Our comparison of T_0 values calculated for sucrose and trehalose shows inter-sample variance of +0.15 °C, giving a maximum variance in the width of the glass transition of ±0.3 °C. In this study we calculate the T_0 value for sucrose as 3.5 °C. The $T_g - T_0$ difference is 61.5 °C; this is in good agreement with the value of 60 °C found by Hancock *et al.* (1995). The calculations of Hancock *et al.* were based on a more direct approach using enthalpy relaxation, so this agreement helps to validate the more indirect (but simpler) approach based on the width of the glass transition, described above.

It is also interesting to note that while the sample used by Hancock was dry, that employed in this study contained a small amount of residual moisture. Therefore it would appear that the moisture content, at low levels at least, does not affect fragility, since $T_g - T_0$ remains constant. Finally, the calculated T_0 value for trehalose is 44 °C, and this could help to explain the reported rating by Pikal (1994, 1996) of the stabilising properties of a number of carbohydrates as follows: trehalose > sucrose > dextran. Although we have not analysed dextran in detail, it has a broad glass transition indicating a strong glass; Pikal's ranking may thus correspond to fragility: the more fragile the glass, the better it performs as a stabilising agent.

In summary, in order to form an amorphous matrix with good stabilising properties, a material should have a high T_g but also be a fragile glass. High T_g without fragility may be unsatisfactory, as may fragility without a high T_g.

12.3.8 Excipients

An excipient is any material added to a formulation other than the active material itself. Many excipients such as buffers or salts are added for specific purposes not related to freeze-drying, but these must be taken into account when formulating for drying. It is rarely possible to freeze-dry a product without the addition of further excipients. The functions of excipients added for freeze-drying are various, but fall into two main categories. (a) Bulking agents are added to increase the solids content of a dried product to create a stronger plug. These are generally selected because they crystallise on freezing. (b) Stabilisers are added to glass-forming systems to assist in the glass formation. Additionally they help to protect the active material from the detrimental effects of freeze-concentration.

For systems in which the active material crystallises, a crystallising excipient should be added if required, purely to act as a bulking agent. With active materials that do not crystallise, as is common for many proteins and biotechnology products, a glass-forming excipient is required. Leaving aside any regulatory requirements for pharmaceutical products, the glass former should be selected on the basis of (1) its compatibility with the end-use application of the active material, and (2) its glass temperature.

Table 12.3 lists the glass temperatures of a number of salts, proteins, anhydrous carbohydrates and polymers and also the glass temperatures of the freeze-concentrates T'_g. As can be seen there is considerable variation, although a general trend correlating higher molecular mass and higher glass temperature is apparent. For effective freeze-drying an excipient should have a high T'_g, allowing primary drying to be performed at as high temperature as possible, and therefore facilitating faster sublimation. Additionally, a glass former with a high glass temperature when anhydrous should be selected, as this will permit the finished product, once dried, to be stored in the glassy state.

Table 12.3 T_g and T'_g of various salts, sugars and proteins.

Glass former	T_g (°C)	T'_g (°C)
Dextran	83	16
Ficoll	110	−21
Fructose	13	−42
Galactose	38	−42
Gelatin	?	−12
Glucose	30	−43
Glycerol	−93	−65
Lactose	101	−36
Maltose	43	−30
Maltotriose	76	−24
Mannitol	crystallises	−28
PVP	118	−25
Ribose	−10	−57
Sorbitol	−2	−44
Sucrose	77	−32
Trehalose	120	−30
Histidine		−33
Sodium citrate		−41
TAPS		−49
NaH_2PO_4		−45
Citric acid		−54
Tris-maleate		−49
Tris-acetate		−54
Potassium citrate		−62
$CuSO_4$		−43
Sodium acetate		−64
Potassium acetate		−76
$MnCl_2$		−83
$ZnCl_2$		−88
Glutamic acid		−17
Sodium pyrophosphate		−37
NaCl		—
$NaHCO_3$		−52
Magnesium sulfate		−56
KH_2PO_4		−55
Tris base		−51
Hepes		−63
Glycine		
$K_2HPO_4 \cdot 6H_2O$		−65
Tris-HCl		−65
β-Alanine		−65
NaSCN		−80
Guanidine-HCl		−80
$CaCl_2$		−95
$MgCl_2$		—

A substantial literature exists on the use of cryoprotectants as stabilising excipients (Carpenter *et al.*, 1994; Crowe *et al.*, 1990; Levine and Slade, 1988; Pikal, 1990a,b, 1994). Sucrose for example has a T'_g of −32 °C and a $T_{g(\mathrm{anhydrous})}$ of 77 °C, allowing ice sublimation to be performed at a reasonable temperature, but also giving a finished product with a high glass temperature and good storage stability. It should be noted that because of residual moisture in the product (say 2%), its glass temperature will be slightly below the value $T_{g(\mathrm{anhydrous})}$.

The use of sorbitol will never give a product that is stable in anything other than deep-freeze conditions, as its $T_{g(anhydrous)}$ is $-2\,°C$. Again given that the product will contain some residual moisture, its actual glass temperature may be much lower.

In all systems to be freeze-dried the overall glass temperature is a function of the glass temperatures of the individual components and their respective mass ratios, and will lie within the range of the individual T_g' values. Thus in order to optimise a system, the mass of materials such as salts that have a low T_g' should be minimised as far as practically possible, while the proportion of materials such as carbohydrates, with higher glass temperatures, should be maximised.

The residual water content at T_g' determines the required secondary drying temperature ramp rate. The glass temperature of the dry product represents the maximum temperature to which it should be exposed if deterioration is to be prevented.

12.4 Stabilisation of Proteins by Ambient-temperature Drying

As alluded to above, despite its widespread use, freeze-drying can be a complex and variable technique, and many freeze-dried products are still unstable at ambient temperatures (Carpenter *et al.*, 1994; Pikal, 1994). The freezing step (with its consequent freeze-concentration, which cannot be avoided) remains a major variable, and is suggested to be the major cause of protein damage during freeze-drying (Carpenter *et al.*, 1994; Crowe *et al.*, 1990; Pikal, 1994).

Some recent attention has thus focused on the techniques of ambient temperature drying, as these not only eliminate the freezing step but are more rapid and energy-efficient in the removal of water (Colaco *et al.*, 1992; Crowe *et al.*, 1990; Roser, 1991). Two patented approaches (Roser, 1987; Franks and Hatley, 1990) have been developed to achieve this aim and these are described below.

12.4.1 Permazyme technology

Surprisingly it has been found that a product, formulated with a glass forming excipient, can be dried directly to the glassy state by controlled evaporative drying, without the need for freezing (Franks and Hatley, 1993).

The degree of denaturation of a protein during drying is a function of both the time taken to reach the glass state and the temperature of processing. Both freeze-drying and controlled evaporative drying give the same end product, a labile material stored in a glass. Although freeze-drying takes place at lower temperatures, which should be an advantage if the protein can be protected from the detrimental effects of freezing, the process is slow (days). Controlled evaporative drying compensates for the higher temperatures by being more rapid (seconds to hours). Avoiding the freezing step provides additional advantages in the stabilisation of freeze-sensitive labile materials. Indeed, there is some evidence that the drying process itself provides a more chemically stable product than those currently produced by freeze-drying (Skrabrabanja *et al.*, 1994). Restriction enzymes, as marketed by Pharmacia under the Ready-To-Go™ brand name, are examples of freeze-sensitive materials that have been processed by this technique to give a product stable at room temperature.

The Permazyme technique employs the use of carbohydrate and carbohydrate polymer excipients to yield products with high T_g' values (Franks *et al.*, 1991). However the physical instability of many carbohydrates can make formulation difficult and long term stability uncertain (Franks, 1995). This problem does not seem to exist with the disaccharide trehalose as used in QT4 technology (Hatley *et al.*, 1996).

12.4.2 QT4 technology

QT4 technology utilises the principles learnt from nature, where there exist natural organisms that have evolved the remarkable ability to dry out completely at ambient temperatures, remain metabolically dormant for long periods under harsh environmental conditions and yet regain full metabolic activity when rehydrated (Crowe and Clegg, 1973; Coutinho *et al.*, 1988; Gadd *et al.*, 1988; Roser and Colaco, 1993). These cryptobiotic or anhydrobiotic organisms are found in many phyla in both the plant and animal kingdoms and the better known examples are the brine shrimp *Artemia salina*, the soil nematode *Ditylenchus dipsaci*, the tardigrade *Adoribiotus coronifer*, the resurrection plant *Selaginella lepidophylla* and baker's yeast *Saccharomyces cerevisiae* (Leopold, 1986). Studies on cryptobiosis in the baker's yeast, *S. cerevisiae*, have shown that the synthesis of a simple disaccharide, trehalose (α-D-glucopyranosyl-α-D-glucopyranoside, Figure 12.8), is necessary for the protection of all the biomolecules in the organism during drying (Coutinho *et al.*, 1988; Gadd *et al.*, 1988). Recent work has shown that this protection can be replicated *in vitro* by drying biological molecules in the presence of trehalose at ambient

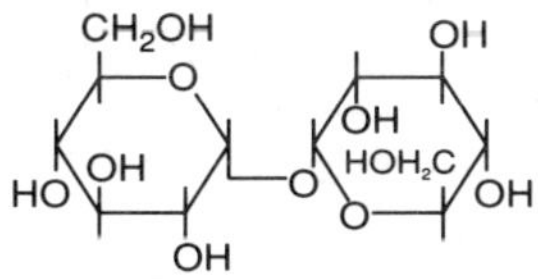

Figure 12.8 Structure of trehalose (α-D-glucopyranosyl-α-D-glucopyranoside).

temperatures (Blakeley *et al.*, 1990; Colaco *et al.*, 1990, 1992, 1994; Crowe *et al.*, 1990; Roser, 1991). Early work has shown that antibodies air-dried in the presence of trehalose are undamaged, and full biological activity is recovered on rehydration, even after several years storage at room temperature or 37 °C (Blakeley *et al.*, 1990; Colaco *et al.*, 1990; Roser, 1991). Similar results were obtained with a variety of enzymes, hormones and blood coagulation factors, suggesting that this process was probably generally applicable to biological molecules (Colaco *et al.*, 1990, 1994; Roser, 1991).

As a stringent test of this technology for the preservation of labile biological molecules, some of the notoriously fragile enzymes used in molecular biology, which are thus usually transported and stored below −20 °C, were studied in detail. It has been shown that both restriction endonucleases and DNA modifying enzymes can be dried from trehalose solutions at ambient temperatures, without loss of activity (Colaco *et al.*, 1992, 1994; Roser, 1991). Furthermore these authors demonstrated that the dried enzymes showed unexpected stability on storage for extended periods, even at elevated temperatures.

Concordant with the fact that the dehydration of cryptobionts occurs at ambient or higher temperatures, this reproducible stabilisation of biomolecules dried at ambient temperatures is in sharp contrast with previously reported work using trehalose as an excipient during freeze-drying (Carpenter and Crowe, 1988a; Carpenter *et al.*, 1994; Colaco *et al.*, 1992, 1994; Crowe *et al.*, 1990; Roser, 1991).

It is interesting to consider the results of studies on the mechanism of action of trehalose, as this is of general relevance to the stabilisation of proteins. Until recently, two main hypotheses have been postulated with respect to this mechanism (Burke, 1985; Clegg, 1985; Colaco *et al.*, 1994). The water replacement theory states that, being a polyol, trehalose can make multiple external hydrogen bonds. It is therefore capable of replacing the essential structural water molecules hydrogen-bonded to the biomolecules, thus maintaining their molecular structure (Clegg, 1985; Colaco *et al.*, 1994). The glassy state theory invokes the normal stabilising effects of glass formation, discussed above (Burke, 1985; Colaco *et al.*, 1994).

The published reports on trehalose stabilisation (Carpenter and Crowe, 1988b; Colaco *et al.*, 1992, 1994; Crowe *et al.*, 1987, 1990; Roser, 1991) are not consistent with either hypothesis being a sufficient sole explanation for the mechanism of action. Most previous work on the stabilisation of dry protein formulations has emphasised the ability of stabilising excipients to replace structural water, and provide an amorphous matrix in the dry state. However, the work on trehalose emphasises that the chemical reactivity of these excipients is an extremely important feature that has hitherto been neglected (Colaco *et al.*, 1994). This is particularly important with respect to extending the shelf-life of dry products, as most of the degradative chemistries we observe occur during the storage of the dry formulation, rather than during the drying process itself.

In summary, it is probably the unique combination of the properties of water replacement, glass transformation and chemical inertness that explains the superiority of trehalose as a stabilising excipient for the production of dry, shelf-stable biomolecule formulations.

12.5 Stabilisation of Cells and Organisms

The requirement for homogeneous stocks of living cells extends from the culture of hybridomas for the production of monoclonal antibodies, to the use of microorganisms and animal cell lines for the production of genetically engineered biomolecules, both key requirements underpinning the current growth of the biotechnology industry. The only alternative to preservation of these living cells is serial culture. This is however not just expensive in time and materials, but can lead to the serious problems of: (1) culture contamination, and (2) the selection of variants resulting from spontaneous genetic mutations. Though some microorganisms can be stored at ambient temperatures for extended periods on agar slopes or slabs, the most widely applied technique for the preservation of living cells is cryopreservation, or the freezing and stabilisation of cells at sub-zero temperatures. It is also, to date, the only method for the successful preservation of mammalian cells, and has also been used to stabilise small multicellular organisms and fertilised embryos.

Although widely used, the mechanisms underlying cryopreservation technology are poorly understood, not least because the freezing processes involved are complex phenomena, and the wide range of cryoprotectants often used to prevent damage during freezing appear to have contradictory properties (Ashwood-Smith and Farrant, 1980; Muldrew and McGann, 1990). The general concept in cryobiology centres around the essential role of water in the cellular metabolism of biological systems. The effective removal of water by freezing results in the cessation of biological activity, until thawing takes place. In addition, the reduction of temperature also results in the lowering of metabolic activity (Ashwood-Smith and Farrant, 1980; Nei, 1969).

The freezing process is thus the most important step in the cryopreservation of living cells. Freezing involves the removal of water from the system by the formation of ice crystals. As with any crystallisation process, the formation of the ice crystals is mediated by the nucleation of seed crystals followed by the growth (or polymerisation) of the crystal by addition of more ice at the crystal front (Muldrew and McGann, 1990). The rate-limiting step of ice crystal formation is thus the nucleation stage, and this makes the process inherently unpredictable. In addition the growth of the ice crystals themselves is governed by two opposing forces: crystal growth is energetically favoured by decreasing temperature, but the increased viscosity of the supercooled medium results in the concomitant slowing of growth of the ice crystal front (Ashwood-Smith and Farrant, 1980; Muldrew and McGann, 1990).

In the freezing of cells and microorganisms in a culture medium, the control of the freezing step is further complicated by the need to balance the effects of freezing on both the internal contents of the cell or organism, and the medium in which it is being frozen. This paradox is illustrated simply by consideration of the effects of the freezing and the rate of cooling. As the freezing point of the extracellular medium is less depressed than that of the cellular cytoplasm, during freezing the water in the extracellular medium will form ice before that in the cell. This will lead to an osmotic imbalance leading to migration of water through the cell membrane and out of the cell (Ashwood-Smith and Farrant, 1980; Nei, 1969; Souzu, 1989).

If the deleterious effects of an extracellular osmotic imbalance are minimised by rapid cooling and freezing, this will leave more water within the cell. However, this maximises the risk of damage from intracellular ice crystal formation and recrystallisation during thawing. And though slow cooling will reduce the risk of damage due to recrystallisation of ice during thawing, it results in the formation of larger intracellular ice crystals again, increasing the risk of cell membrane damage during the freezing step (Ashwood-Smith and Farrant, 1980; Nei, 1969; Souzu, 1989).

In practice, however, the empirical determination of the optimal cooling rate for any cell or organism is less daunting than it appears. A large body of literature exists, covering practically every cell type and most useful organisms, from which information can be extracted to minimise the experimental conditions that need to be examined (Agnelini *et al.*, 1992; Ashwood-Smith and Farrant, 1980; Borrelli *et al.*, 1995; De Fazio *et al.*, 1986; Groth *et al.*, 1991). In addition, a number of useful cryoprotectants have been developed which greatly enhance the recovery of viable cells after freezing (Ashwood-Smith and Farrant, 1980; Balundini *et al.*, 1993; Groth *et al.*, 1991; Pakkanen, 1994). In practice, based on this body of knowledge, a number of critical factors can be described that need to be considered to enable the successful freezing, storage and recovery of viable cells and organisms. Besides the cooling rate mentioned briefly above, these include the warming rate, the concentration of any cryoprotectant excipient, and its cellular permeability and associated osmotic effects. These will be described briefly below to give general practical guidelines.

12.5.1 Effect of cooling rate

Prior to considering the effects of cooling rate, it should be mentioned that cooling itself may damage some cells and organisms due to cold or thermal shock. This is mostly due to the ionic imbalances that result from the disproportionate reduction in activity of cellular ion-transport pumps compared to passive diffusion, but it may sometimes be related to the rapidity of cooling. The use of cryoprotectants can, however, help preserve cellular viability in both these cases. In general, rapid cooling is detrimental to cell survival, probably due to damage during thawing, as evidenced by the variation of survival rates with thawing conditions. Two important exceptions to this generalisation are the freezing of spermatozoa and of human erythrocytes.

However, as mentioned above, a slow cooling rate leads to shrinkage of cells due to osmotic imbalances. This can be damaging to the cell, though some shrinkage is beneficial as it reduces intracellular water and thus reduces formation of ice crystals both during freezing and thawing, and damage due to these. This may be a particularly important factor, as there is little evidence to suggest that extracellular ice crystal formation is damaging to cells, most damage during slow cooling being due to high solute concentrations.

In most cases around 1 °C per minute is the optimal rate of cooling, and is generally the average figure around which cooling rate experiments are performed. This rate was due largely to the work of Mazur on sea urchin eggs (Ashwood-Smith and Farrant, 1980; Nei, 1969; Meryman, 1966). More recently however, the two-step cooling approach developed by Farrant *et al.* has proved to be a more effective general procedure (Ashwood-Smith and Farrant, 1980; Elliot and Whelan, 1988). This involves minimising freezing injury by first 'pre-freezing' the cell to a sub-zero temperature of −20 or −70 °C for a short period, then freezing rapidly to storage temperatures of −80 °C or −160 °C. It appears that the extracellular freezing during the first step increases the external solute concentration enough to induce shrinkage, which is sufficient to prevent or significantly reduce intracellular ice crystal formation during the second step. Overall cell survival is therefore increased.

12.5.2 Effect of warming rate

As with the cooling rate, the general conclusion that rapid thawing often results in the improved recovery of viable cells and organisms is not an unequivocal statement. Two well studied exceptions are mammalian embryos and human red blood cells, though the effect of rapid thawing on the latter is more pronounced when glycerol is used as a cryoprotectant.

The effect of the warming rate is thought to be mainly due to ice recrystallisation effects, which result in the growth of large intracellular crystals during rewarming. These effects are more pronounced during slow thawing. Consistent with this mechanistic postulate, the effect of warming rate is much more critical in samples which have been subjected to rapid cooling rates during freezing, and thus contain greater quantities of intracellular ice.

In addition to the consideration of intracellular ice crystal damage, the osmotic effects of thawing and the subsequent dissolution of media salts and excipients also need to be considered. These osmotic effects can be minimised by the rapid dilution and washing of the thawed samples. In practice, as rapid thawing at 37 °C is much more effective than at 0 or 4 °C, this step is often combined with an immediate dilution by addition of the partially thawed sample to a large volume of pre-warmed culture media. Besides increasing the thawing rate, this also minimises potentially deleterious osmotic effects.

12.5.3 Vitrification

It has been explained above how glass formation can aid in the stabilisation of labile macromolecules during freeze-drying; unfortunately, the stabilisation of cellular systems presents additional problems. While it is possible to modify the suspending medium such that it forms a glass, the difficulty arises because of the cell membrane. The membrane prevents the penetration of the majority of glass-forming solutes into the cell, so that the cell interior is not in the glassy state during drying. Also, penetrating materials (cryoprotectants) such as DMSO or glycerol are used to reduce intracellular ice damage, but prevent glass formation since they have low glass temperatures.

Vitrification attempts to overcome this problem by using solvents such as butane-2,3-diol (Boutron, 1990) and rapid cooling rates (typically 600 °C min^{-1}) to prevent ice nucleation, allowing the cells to be stored in the glassy state at low (usually liquid nitrogen) temperatures. To perform the process the cell suspension is first perfused with a suitable cryoprotectant, and then cooled at a rate, the critical cooling rate, that does not permit ice nucleation before the vitreous state is reached. This rate is dependent on the composition of the solution used to perfuse the cells, and its determination is complex. Emulsion isothermal calorimetry is used to produce time–temperature–transition (TTT) curves. An emulsified solution is quench-cooled in a calorimeter and the time taken for the nucleation of ice, as detected by an exothermic transition, determined. The TTT curve enables the time required for a given volume of solution to crystallise as a function of temperature, and from this the critical cooling rate can be determined. The time taken for nucleation decreases as the sample is quenched to progressively lower temperatures, until it reaches a minimum when it again begins to increase. The data obtained can be plotted as a TTT curve (Figure 12.9). The shape of the curve results from a balance between the greater tendency for water to crystallise at low temperatures, and the reduced molecular mobility also found under colder conditions. Once the maximum rate of crystallisation (the nose of the TTT curve) is known, an appropriate cooling rate can be selected. Although this technique has been successfully applied to get cells into the vitreous state at low temperatures, few reports of successful regeneration of the cells following recovery from storage have been reported (Sutton, 1991). The formation of ice crystals during rewarming, where heating rates cannot be effectively controlled, seems to be the difficulty. For further details of this technique the reader is referred to work by Steponkus (1992).

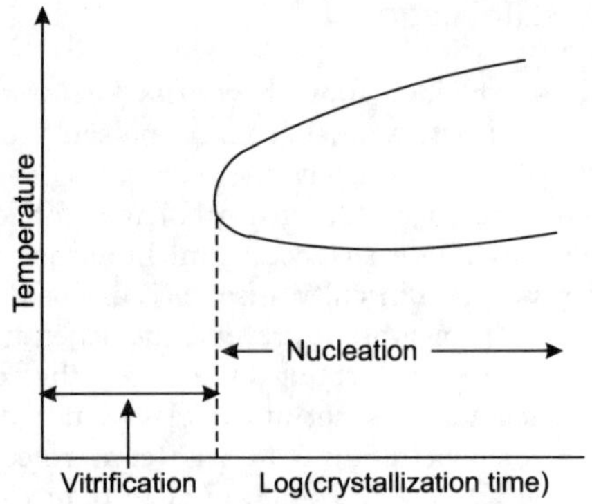

Figure 12.9 A schematic TTT curve. The log of the time taken for crystallisation is plotted against temperature. Cooling rates greater than the crystallisation rate at the 'nose' of the curve enable the sample to be vitrified before ice nucleates. The required cooling rate can be calculated from the TTT curve, as described by Sutton (1991).

12.5.4 Cryoprotectants

The use of glycerol as a cryoprotectant (Polge and Rowson, 1952) opened up a new avenue in the preservation of cells and organisms (Ashwood-Smith and Farrant, 1980; Elliot and Whelan, 1988). A number of reagents including serum, solvents and sugars have also been used as cryoprotectants, either alone or in combination. Glycerol is still often the cryoprotectant of choice, though the use of dimethylsulfoxide (DMSO) is now more widespread (Ashwood-Smith and Farrant, 1980; Elliot and Whelan, 1988). Though the precise mode of action of cryoprotectants is still a matter of much debate, a number of features appear to be common to the most effective cryoprotectants. These are: high solubility in water, lack of toxicity and the ability to penetrate the cell (Ashwood-Smith and Farrant, 1980; Balundini *et al.*, 1993). The choice of cryoprotectant used with any particular cell or organism is often determined on the basis of these parameters. Thus while glycerol may be preferred on account of its low toxicity with most cells, the more penetrating DMSO is often used with larger or more complex cells and organisms (Agnelini *et al.*, 1992; Ashwood-Smith and Farrant, 1980; Balundini *et al.*, 1993; Nei, 1969).

High water solubility is thought to be an important factor in the mechanism by which cryoprotectants protect cells and organisms against freeze damage. As the presence of solutes depresses the freezing point of water in a concentration-dependent manner, a high concentration of the cryoprotectant will lower the temperature at which freezing occurs. This minimises solution effects, and results in dehydration at lower, less harmful temperatures, prior to intracellular freezing. In the case of cryoprotectants that also penetrate the cell, the consequent depression of the intracellular freezing point will aid dehydration during cooling; it will thus minimise intracellular ice crystal formation, as well as directly inhibiting ice recrystallisation during thawing. A number of non-penetrating excipients such as hydroxyethyl starch (HES), serum and polyethylene glycols (PEG) show additional cryoprotective effects when used with DMSO and glycerol (Agnelini *et al.*, 1992; Ashwood-Smith and Farrant, 1980; Groth *et al.*, 1991). It is thus possible that the depression of freezing point is not the sole mechanism of action of all cryoprotectants, the minimising of osmotic effects, especially during thawing, may also be critical to their performance. Cryoprotectants can thus affect both the cooling and thawing rates, emphasising the interactive nature of the conditions for cell preservation at low temperatures.

12.5.5 Practical tips

Although obvious, it is necessary to emphasise that the first step to practical cryopreservation is a good literature search to identify previous work on similar or related cells or microorganisms. These provide the best framework for experimental conditions that merit testing. Given the vast reservoir of work in this area, it is quite likely that only minor changes will need to be made to published protocols to achieve good survival. An illustrative example is the preservation of mouse monoclonal hybridomas, where a two-stage freezing protocol may need to be customised to individual clones, but the use of 10% DMSO and 90% fetal calf serum is the optimal cryoprotectant mixture for most hybridomas. An often unappreciated point is that although the recovery of only 50% viable cells may be disappointing it may be satisfactory, as the random nature of freeze damage seldom results in the preferential selection of genetic variants. Thus as long as the genetic fidelity of the viable cells can be established, even a 50% survival rate may be adequate.

The most important practical step is probably the choice of cryoprotectant, and it is necessary, therefore, first to establish the degree of toxicity of the excipients used with the cell or microorganism to be preserved. It is important to use the highest purity, sterile excipients, as any minor contaminants may themselves be toxic and result in the erroneous titration of cryoprotectant toxicity. In general, the use of glycerol and

DMSO will be satisfactory for most systems, but newborn or fetal calf serum may also be beneficial in many systems. Although HES and PEG may substitute for the cryoprotectant properties of the latter, nutrient needs require some serum to be added to most systems. In all these cases the various excipients will need to be titrated for cellular toxicity, to establish the maximum concentrations at which they can be used, both singly and in combination. It is important to remember that the minimal total time of exposure to the excipients during both cooling and thawing is unlikely to be less than 15 min, and this should be the minimum time-point for the toxicity trial. In general the toxicity at both 4 °C and 37 °C should be examined, in anticipation of the requirement for rapid thawing for maximum cell recovery. In all these experiments, the most rapid and reproducible addition of cryoprotectant is obtained by adding an equal volume of sterile filtered medium (containing twice the required concentration of cryoprotectant) to the test cell suspension.

Once cryoprotectant toxicity levels have been established, the highest acceptable concentrations should be used in experiments to establish the optimal cooling and thawing rates. As mentioned above, an intermediate cooling rate or a two-step freezing rate are generally the least damaging conditions, and should thus be investigated in the first instance. A variety of sophisticated precision cooling and thawing equipment is available and sample temperatures can be measured by thermocouples. However, for most optimisation purposes, the used of fixed volumes of test cell suspension in standard freezing vials, placed at various freezing temperatures, is sufficient—even though the precise numerical cooling rate may be undetermined. A simple system for attaining a variety of low temperatures is the use of various alcohols cooled by solid CO_2. Many workers, however, prefer to freeze vials in lagged containers and, as this will affect the cooling rate, the use of freezers set at the required temperatures is required in these experiments. It is also important to examine the effect of shifting the vials from the experimental freezing temperature to the storage temperature (usually liquid nitrogen), as this is often a neglected variable in experimental design. In effect the latter parameter will also determine if a suitable two-step freezing procedure can be utilised.

As the effects of the freezing rate cannot be assessed without thawing the cells, in practice these experiments will examine the combination of both freezing and thawing rates, and should be designed to optimise both. As mentioned above, a rapid thawing rate is generally less damaging to most cells and the effects of rapid thawing rates should be examined initially. Again, the use of various fixed temperatures in combination with fixed volumes of frozen cultures in standard freezing vials is a generally satisfactory way of optimising thawing rates, without the precise numerical thawing rate being determined. The best control for such thawing rates is achieved by the direct immersion of the frozen vial in a refrigerated, circulating water-bath, with the use of various coolants if subzero temperatures are required.

In protocols 12.1–12.3 are given the general conditions for the preservation of the most commonly used cells and organisms in biotechnology, namely hybridomas, and competent and recombinant *E. coli*.

References

Ablett, S. *et al.* (1992). Calorimetric study of the thermal behaviour of sucrose solutions. *J. Chem. Soc., Faraday Trans.*, **88**, 789–794.

Agnelini, A. *et al.* (1992). Evaluation of four different methods of platelet freezing. *Vox Sang.*, **62**, 142–151.

Amelunxen, R. E. and Murdock, A. L. (1978). Mechanisms of thermophily. *CRC Crit. Rev. Microbiol.*, **6**, 343–393.

Anawis, M. A. and Lindberg, R. E. (1988). Stabilized horseradish peroxidase composition. *US Patent 4,782,023*, to Abbott Laboratories.

Andrade, J. D. (1985). Principle of protein absorption, surface and interfacial. In *Aspects of Biomedical Polymers*, pp. 1–80, Plenum Press, New York.

Angell, C. A. (1988). Supercooled water. In *Water: A Comprehensive Treatise* (Franks, F., ed.), vol. 7, pp. 1–82, Plenum Press, New York.

Angell, C. A. (1991). Relaxation in liquids, polymers and plastic crystals – patterns and problems. *J. Non-Cryst. Solids*, **131**, 13–31.

Angell, C. A. (1995). Formation of glasses from liquids and biopolymers. *Science*, **267**, 1924–1935.

Arakawa, T. and Timasheff, S.N. (1985). Mechanism of polyethylene glycol interactions with proteins. *Biochemistry*, **24**, 6756–6762.

Argos, P. (1979). Thermal stability and protein structure. *Biochemistry*, **18**, 5698–5703.

Ashwood-Smith, M. J. and Farrant, J. (eds) (1980). *Low Temperature Preservation in Medicine and Biology*, Pitman Medical Press, Bath.

Balundini, C. L. *et al.* (1993). Cryopreservation of human platelets using dimethyl sulfoxide and glycerol–glucose: effects on *in vitro* platelet function. *Haematologia*, **78**, 101–104.

Blakeley, D. *et al.* (1990). Dry, instant blood typing plate for bedside use. *Lancet*, **336**, 854–855.

Borrelli, M. J. *et al.* (1995). Protocol for freezing thermotolerant cells and maintaining thermotolerance following thawing. *Int. J. Hyperthermia*, **11**, 389–396.

Boutron, P. (1990). Levo- and dextro-2,3-butanediol and their racemic mixtures: very efficient solutes for vitrification. *Cryobiology*, **27**, 55.

Boyd, S. and Yamazaki, H. (1994). Use of polyvinyl alcohol as a stabilizer of peroxidase–antibody conjugate for enzyme immunoassay. *Biotechnol. Techn.*, **8**, 123–128.

Brandts, J. F. (1964). The thermodynamics of protein denaturation. I. The denaturation of chymotrypsinogen. *J. Am. Chem. Soc.*, **86**, 4291–4301.

Brennan, J. R. *et al.* (1985). Pump-induced insulin aggregation: a problem with the biostator. *Diabetes*, **34**, 353–359.

Brock, T. D. (1985). Life at high temperatures. *Science*, **230**, 132–138.

Brock, T. D. (ed.) (1986). *Thermophiles: General Molecular and Applied Microbiology*, Wiley Interscience, New York.

Burke, M. J. (1985). The glassy state and survival of anhydrous biological systems. In *Membranes, Metabolism and Dry Organisms* (Leopold, A.C., ed.), pp. 358–363, Cornell University Press, Ithaca, NY.

Capasso, S. *et al.* (1991). Deamination *via* cyclic imide of asparaginyl peptides: dependence on salts, buffers and organic solvents. *Peptide Res.*, **4**, 234–238.

Carpenter, J. F. and Crowe, J. H. (1988a). The mechanism of cryopreservation of proteins by solutes. *Cryobiology*, **25**, 244–255.

Carpenter, J. F. and Crowe, J. H. (1988b). Modes of stabilization of a protein by organic solutes during desiccation. *Cryobiology*, **25**, 459–470.

Carpenter, J. F. *et al.* (1994). Interactions of stabilisers with proteins during freezing and drying. *Am. Chem. Soc. Symp.*, **567**, 134–147.

Cartwright, T. *et al.* (1977). The mechanism of inactivation of human fibroblast interferon by mechanical stress. *J. Gen. Virol.*, **36**, 317–321.

Chang, B. S. and Randell, C. (1992). Use of subambient thermal analysis to optimise protein lyophilisation. *Cryobiology*, **29**, 632–656.

Clegg J. S. (1985). The physical properties and metabolic status of *Artemia* cysts at low water contents: the 'water replacement hypothesis'. In *Membranes, Metabolism and Dry Organisms* (Leopold, A. C., ed.), pp. 169–187, Cornell University Press, Ithaca, NY.

Cleland, J. L. *et al.* (1993). The development of stable protein formulations: a close look at protein aggregation, deamination and oxidation. *Crit. Rev. Therapeut. Drug Carrier Syst.*, **10**, 307–377.

Cleland, J. L. and Wang, D. L. C. (1993). Cosolvent effects on refolding and aggregation. *Am. Chem. Soc. Symp.*, **516**, 151–176.

Colaco, C. *et al.* (1992). Extraordinary stability of enzymes dried in trehalose: simplified molecular biology. *Bio/Technology*, **10**, 1007–1011.

Colaco, C. A. L. S. *et al.* (1990). Trehalose stabilisation of biological molecules. In *Biotechnology International 1990*, pp. 345–350, Century Press, London.

Colaco, C. A. L. S. *et al.* (1994). The chemistry of rational protein drug stabilisation. *Am. Chem. Soc. Symp.*, **567**, 222–240.

Coutinho, C. *et al.* (1988). Trehalose as cryoprotectant for preservation of yeast strains. *J. Biotechnol.*, **7**, 23–32.

Crowe, J. *et al.* (1987). Stabilization of dry phospholipid bilayers and proteins by sugars. *Biochem. J.*, **242**, 1–10.

Crowe, J. H. and Clegg, J. S. (1973). *Anhydrobiosis*, Dowden, Hutchinson and Ross, Stroudsburg, PA.

Crowe, J. H. *et al.* (1990) Are freezing and dehydration similar stress vectors? A comparison of the modes of interaction of stabilising solutes with biomolecules. *Cryobiology*, **27**, 219–231.

Dawson, E. C. *et al.* (1980). Stabilization of peroxidase. *US Patent 4,228,240*, to Akzona Inc.

De Fazio, S.R. *et al.* (1986). Mouse skin graft prolongation with donor-strain bone marrow and antilymphocyte serum. Effect of bone marrow storage. *Transplantation*, **41**, 26–28.

Donaldson, T. T. *et al.* (1985). Kinetics of protein denaturation at liquid gas interfaces. *J. Colloid Interface Sci.*, **74**, 441–448.

Eisenberg, H. and Wachtel, E. (1987). Structural studies of halophilic proteins, ribosomes and organelles of bacteria adapted to extreme salt concentration. *Annu. Rev. Biophys. Phys. Biochem.*, **16**, 69–92.

Elliot, K. and Whelan, J. (1988). *The Freezing of Mammalian Embryos*, Elsevier-North Holland, Amsterdam.

Evans, S. A. *et al.* (1995). Dielectric characterisation of thermodynamic first order events in model frozen systems intended for lyophilization. *J. Pharm. Sci. Tech.*, **49**, 2–8.

Finney, J. L. *et al.* (1980). Solvent effects and polar interactions in the structural stability and dynamics of globular proteins. *Biophys. J.*, **32**, 17–33

Finney, J. L. (1982). In *Biophysics of Water* (Franks, F. and Mathias, S.F., eds), John Wiley and Sons, Chichester.

Ford, J. L. and Timmins, P. (1989). *Pharmaceutical Thermal Analysis*, Ellis Horwood, Chichester.

Franken, N. *et al.* (1995). Storable protein solution. *US Patent 5,460,944*, to Boehringer Mannheim.

Franks, F. (ed.) (1988). *Characterization of Proteins*, Humana Press, Clifton, NJ.

Franks, F. (1990). Freeze-drying: from empiricism to predictability. *Cryo-Letters*, **11**, 93–110.

Franks, F. (1995). Using water soluble glasses as a strategy for bioproduct stabilization. *Genet. Eng. News*, August, 20.

Franks, F. and Hatley, R. H. M. (1985). Low temperature unfolding of chymotrypsinogen. *Cryo-Letters*, **6**, 171–180.

Franks, F. and Hatley, R. H. M. (1990). Storage of materials. *European Patent 383,569*, to Pafra Ltd.

Franks, F. and Hatley, R. H. M. (1993). Stable enzymes by water removal. In *Stability and Stabilization of Enzymes* (Van den Tweel, W. J. J. *et al.*, eds), pp. 45–54, Elsevier, Amsterdam.

Franks, F. *et al.* (1991). Materials science and the production of shelf stable biologicals. *Biopharmaceutics* **4**, 38–55.

Gadd, G. M. *et al.* (1988). The role of trehalose in dehydration resistance of *Saccharomyces cerevesiae*. *FEMS Microbiol. Lett.*, **48**, 249–254.

Gerlsma, S. Y. (1968). Reversible denaturation of ribonuclease in aqueous solution as influenced by polyhydric alcohols and some other additives. *J. Biol. Chem.*, **243**, 987–996.

Gerlsma, S. Y. and Stuur, E. R. (1974). The effect of polyhydric alcohols on the heat-induced denaturation of lysozyme and ribonuclease. *Int. J. Peptide Protein Res.*, **6**, 65–71.

Gibson, T. D. and Woodward, J. R. (1993). Enzyme stabilisation. *US Patent 5,240,843*, to Cranfield Biotechnology Ltd.

Gibson, T. D. *et al.* (1993). Preservation of shelf life of enzyme based analytical systems using a combination of sugars, sugar alcohols and cationic polymers or zinc ions. *Anal. Chim. Acta*, **279**, 185–192.

Gray, R. (1994). The bug in the box. *Pharm. Tech. Eur.*, **6**, 20–32.

Groth, J. *et al.* (1991). Freezing of cells: replacement of serum by oxypolygelatine. *J. Immunol. Methods*, **141**, 105–109.

Hancock, B. C. *et al.* (1995). Molecular mobility of smorphous pharmaceutical solids below their glass transition temperatures. *Pharm. Res.*, **12**, 799–806.

Hatley, R. H. M. (1991). The effect of subzero temperatures on the conformation and function of proteins in unfrozen solutions. PhD Thesis, CNAA, London.

Hatley, R. H. M. (1992). The effective use of differential scanning calorimetry in the optimisation of freeze-drying processes and formulations. *Dev. Biol. Stand.*, **74**, 105–122.

Hatley, R. H. M. and Franks, F. (1986). Denaturation of lactate dehydrogenase at subzero temperatures. *Cryo-Letters*, **7**, 226–235.

Hatley, R. H. M. and Franks, F. (1989a). The effect of aqueous methanol cryo-solvents on the heat and cold-induced denaturation of lactate dehydrogenase. *Eur. J. Biochem.*, **184**, 237–240.

Hatley, R. H. M. and Franks, F. (1989b). The cold-induced denaturation of lactate dehydrogenase at sub-zero temperatures in the absence of perturbants. *FEBS Lett.*, **257**, 171–173.

Hatley, R. H. M. and Franks, F. (1991). Applications of DSC in the development of improved freeze-drying processes for labile biologicals. *J. Thermal Anal.*, **37**, 1905–1914.

Hatley, R. H. M. and Mant, A. (1993). Determination of the unfrozen water content of maximally freeze-concentrated carbohydrate solutions: a reappraisal of current methods. *Int. J. Biol. Macromol.*, **15**, pp. 227–232.

Hatley, R. H. M. *et al.* (1986a). Sub-zero preservation of reactive fluids in the undercooled state. I. The reduction of potassium ferricyanide by potassium cyanide. *Biophys. Chem.*, **24**, 41–46.

Hatley, R. H. M. *et al.* (1986b). Sub-zero preservation of reactive fluids in the undercooled state. II. The effect on the oxidation of ascorbic acid of freeze-concentration and undercooling. *Biophys. Chem.*, **24**, 187–192.

Hatley, R. H. M. *et al.* (1987). The stabilisation of labile biochemicals by undercooling. *Process Biochem.*, **22**, 169–172.

Hatley, R. H .M. *et al.* (1991). The unfrozen water content of maximally freeze-concentrated carbohydrate solutions. *Cryo-Letters*, **12**, 113–124.

Hatley, R. H. M. *et al.* (1996). Stabilisation of a pharmaceutical drug substance by freeze-drying: a case study. *Drug Stability*, **1**, 2–14.

Hawley, S. A. (1971). Reversible pressure–temperature denaturation of chymotrypsinogen. *Biochemistry*, **10**, 2436–2442.

Henson, A. F. *et al.* (1980). The surface coagulation of proteins during shaking. *J. Colloid Interface Sci.*, **74**, 441–446.

Her, L. and Nail, S. L. (1994). Measurement of glass transition temperatures of freeze-dried solutes by differential scanning calorimetry. *Pharm. Res.*, **11**, 1–5.

Hobbs, P. V. (1974). *Ice Physics*, Oxford University Press, Oxford.

Hogberg-Raibaud, A. and Goldberg, M. E. (1977). Isolation and characterisation of independently folding regions of the beta chain of *Escherichia coli. biochemistry*, **16**, 4014–4020.

Hu, C. Q. and Sturtevant, J. M. (1987). Thermodynamic study of yeast phosphoglycerate kinase. *Biochemistry*, **26**, 178–182.

Ikeguchi, M. *et al.* (1986). Ca^{2+}-induced alteration in the unfolding behavior of alpha-lactalbumin. *J. Biochem.*, **99**, 1191–1201.

Jaenicke, R. (1981). Enzymes under extreme physical conditions. *Annu. Rev. Biophys. Bioeng.*, **10**, 1–69.

Jaenicke, R. (1984). Protein folding and protein association. *Angew. Chem., Int. Ed. Engl.*, **23**, 395–466.

Jaenicke, R. (1988). Stability and self organisation of proteins. *Naturwissenschaft*, **75**, 606–610.

Keleti, T. (1985). Stability, heat stability and heat sensitivity of proteins: thermodynamic considerations. *Acta Biochem. Biophys. Acad. Sci. Hung.*, **20**, 183–186.

Kerr, W. L. *et al.* (1993). Chemical reaction kinetics in relation to glass temperatures in frozen food polymer solutions. *J. Sci. Food Agric.*, **61**, 51–56.

Kuwajima, K. *et al.* (1985). Comparison of the transient folding intermediate in lysozyme and alpha-lactalbumin. *Biochemistry*, **24**, 874–881.

Leopold, A. C. (ed.) (1986). *Membranes, Metabolism and Dry Organisms*, Cornell University. Press, Ithaca, NY.

Levine, H. and Slade, L. (1988). Non-equilibrium behaviour of small carbohydrate–weater systems. *Pure Appl. Chem.*, **60**, 1841–1864.

Ljungdahl, L. G. *et al.* (1981). Some aspects of thermophilic adaption. *Basic Life Sci.*, **18**, 397–419.

Lougheed, W. D. *et al.* (1983). Physical stability of insulin formulations. *Diabetes*, **32**, 424–432.

Mackenzie, A. P. (1975). Collapse during freeze drying: qualitative and quantitative aspects. In *Freeze-drying and Advanced Food Technology* (Goldblilith, S.A. *et al.*, eds), pp. 277–307, Academic Press, New York.

Mackenzie, A. P. (1977). Physicochemical basis for the freeze-drying process. *Dev. Biol. Stand.*, **36**, 51–67.

Manly, S. P. *et al.* (1985). Thermal denaturation of the core protein of *lac* repressor. *Biochemistry*, **24**, 3842–3846.

May, J. C. *et al.* (1992). Measurement of final container residual moisture in freeze-dried biological products. *Dev. Biol. Stand.*, **74**, 153–164.

McKerrow, J.H. and Robinson, A.B. (1971). Deamination of asparaginyl residues as a hazard in experimental protein and peptide procedures. *Anal. Biochem.*, **42**, pp. 565–568.

Mellor, J. D. (1978). *Fundamentals of Freeze-drying*, chapt. 2, Academic Press, New York,

Meng, M. *et al.* (1993). Thermal stabilization of xylose isomerase from *Thermoanaerobacterium thermosulfurigenes*. *Bio/Technology*, **11**, 1157–1161.

Meryman, H. T. (1966). *Cryobiology*, Academic Press, New York.

Moynihan, C. T. (1993). Correlation between the width of the glass transition region and the temperature dependence of the viscosity of high-T_g glasses. *J. Am. Ceram. Soc.*, **76**, 1081–1087.

Muldrew, K. and McGann, L. E. (1990). Mechanisms of intracellular ice formation. *Biophys. J.*, **57**, 525–532.

Murase, N. and Franks, F. (1989). Salt precipitation during freeze-concentration of phosphate buffer solutions. *Biophys. Chem.*, **34**, 293–300.

Nail, S. L. and Johnson, J. W. (1992). Methodology for in-process determination of residual moisture in freeze dried products. *Dev. Biol. Stand.*, **74**, 137–151.

Nei, T. (ed.) (1969). *Freezing and Drying of Microorganisms*, University of Tokyo Press, Tokyo.

Pakkanen, R. (1994). Bovine colostrum ultrafiltrate supplemented with adult bovine serum and transferrin. An effective FBS substitute for cultivation of Vero and CHO-K1 cells. *In Vitro Cell. Dev. Biol.*, **30A**, 295–299.

Patel, K. and Brochardt, R. T. (1990). Chemical pathways of peptide degradation. II. Kinetics of deamination of an asparaginyl residue in a model hexapeptide. *Pharm. Res.*, 7, 703–711.

Pearlman, R. and Nguyen, T.H.(1989). Formulation strategies for recombinant protein: human growth hormone and tissue plasminogen activator. *Curr. Commun. Mol. Biol.*, **23**, 23–40.

Perutz, M. F. (1980) In *Protein Folding* (Jaenicke, R., ed.), Elsevier-North Holland, Amsterdam.

Pikal, M. J. (1985). Use of laboratory data in freeze drying process design: heat and mass transfer coefficients and the computer simulation of freeze drying. *J. Parenter. Sci. Technol.*, **39**, 115–138.

Pikal, M. J. (1990a). Freeze-drying of proteins. I. Process design. *Biopharmaceuties.*, 3, 18–26.

Pikal, M. J. (1990b). Freeze-drying of proteins. II. Formulation selection. *Biopharmaceuties.*, **3**, 27–30.

Pikal, M. J. (1994). Freeze-drying of proteins: process, formulation and stability. *Am. Chem. Soc. Symp.*, **567**, 120–133.

Pikal, M. J. (1996). Freeze drying of proteins. In *Peptide and Protein Delivery* (Lee, V.H.L., ed.), 2nd edn, Marcel Dekker, New York.

Pikal, M. J. *et al.* (1984). Mass and heat transfer in vial freeze drying of pharmaceuticals: role of the vial. *Parenter. Sci.*, 3, 1224–1237.

Polge, C. and Rowson, L. E. A. (1952). Fertilizing capacity of bull spermatazoa after freezing at −79 °C. *Nature*, **169**, pp. 626–628.

Richardson, J. (1981). The anatomy and taxonomy of protein structure. *Adv. Protein Chem.*, **4**, 167–339.

Robinson, A. B. and Rudd, C. J., (1974). Deamination of glutaminyl and asparaginyl residues in peptides and proteins. *Curr. Top. Cell. Regul.*, **8**, 247–295.

Roser, B. J. (1987). Protection of proteins and the like. *World Patent WO 87/00196*, to Quadrant Bioresources Ltd.

Roser, B. J. (1991). Trehalose drying: a novel replacement for freeze-drying. *Biopharmaceuties.*, **5**, 44–53.

Roser, B. and Colaco, C. (1993). A sweeter way to fresher food. *New Sci.*, **138**, 24–28.

Roy, M. L. *et al.* (1992). The effects of formulation and moisture on the stability of freeze-dried monoclonal antibody–Vinca conjugate: a test of the WLF glass transition theory. *Dev. Biol. Standard.*, **74**, 323–340.

Sanchez-Ruis, J. M. *et al.* (1988). Differential scanning calorimetry of the irreversible thermal denaturation of thermolysin. *Biochemistry*, **27**, 1648–1652.

Schellman, J. A. (1990). A simple model for solvation in mixed solvents. Applications to the stabilisation and destabilisation of macromolecular structures. *Biophys. Chem.*, **37**, 121–140.

Skarzynski, T. *et al.* (1987). Structure of hologlyceraldehyde-3-phosphate dehydrogenase from *Bacillus stearothermophilus* at 1.8 Å resolution. *J. Mol. Biol.*, **193**, 171–187.

Skold, C. N. *et al.* (1991). Method of stabilizing enzyme conjugates. *European Patent 487,301*, to Syntex (USA) Inc.

Skrabrabanja, A. T. P. *et al.* (1994). Lyophilization of biotechnology products. *J. Pharm. Sci. Technol.*, **48**, pp. 311–317.

Sluzky, V. *et al.* (1991). Kinetics of insulin aggregation in aqueous solutions upon agitation in the presence of hydrophobic surfaces. *Proc. Natl. Acad. Sci. USA*, **88**, 9377–9381.

Souzu, H. (1989). Changes in chemical structure and function in *Escherichia coli* cell membranes caused by freeze–thawing. I. Change of lipid state in bilayer vesicles and in the original membrane fragments depending on rate of freezing. *Biochim. Biophys. Acta*, **978**, 105–111.

Stadtman, E. R. (1990). Metal ion catalysed oxidation of proteins: biochemical mechanism and biological consequences. *Free Rad. Biol. Med.*, **9**, 315–325.

Steponkus, P. L. (1992). Cryopreservation of plant tissues by vitrification. *Adv. Low Temp. Biol.*, **1**, 1–62.

Stratton, C. E. (1991). Cyclodextrins and biological macromolecules. *Biopharmaceutics*, **4**, 44–48.

Suelter, C. H. and DeLuca, M. (1983). How to prevent losses of protein by absorption to glass and plastic. *Anal. Biochem.*, **135**, 112–119.

Sun, M. (1985). Stabilized peroxidase compositions. *US Patent 4,504,579*, to Abbott Laboratories.

Sutton, R. L. (1991). Critical cooling rates to avoid ice crystallisation in solutions of cryoprotective agents. *J. Chem. Soc., Faraday Trans.*, **87**, 101–105.

Swallow, A. J. (1960). Effects of ionizing radiation on proteins, RCO groups, peptide bond cleavage, inactivation: SH oxidation. In *Radiation Chemistry of Organic Compounds* (Swallow, A.J., ed.), pp. 211–252, Pergamon Press, New York.

Tarelli, E. *et al.* (1994). Lysine vasopressin undergoes rapid glycation in the presence of reducing sugars. *J. Pharm. Biomed. Anal.*, **12**, 1355–1361.

Thurow, H. and Geisen, K. (1984). Stabilisation of dissolved proteins against denaturation at hydrophobic interfaces. *Diabetologica*, **27**, 212–218.

Timasheff, S. N. and Arakawa, T. (1988). Mechanism of protein precipitation and stabilisation by co-solvents. *J. Cryst. Growth*, **90**, 39–44.

Timasheff, S. N. (1992). Solvent effects on protein stability. *Curr. Opin. Struct. Biol*, **2**, 35–72.

Tomchaney, A. P. *et al.* (1982). Purification, composition and physical properties of thermal hysteresis 'antifreeze' protein from larvae of the beetle, *Tenebrio molitor. Biochemistry*, **21**, 716–721.

Walker, J. E. (1978). In *Proteins: Structure, Function and Industrial Applications* (Hoffmann, E., ed.), p. 211, Pergamon Press, Oxford.

Walker, J. E. *et al.* (1980). Heat stability of tetrameric enzyme D-glyceraldehyde-3-phosphate dehydrogenase. *Eur. J. Biochem.*, **108**, 581–586.

Wang, Y.-C. J. and Hanson, M. A. (1988). Parenteral formulations of proteins and peptides: stability and stabilisers. *J. Parenter. Sci. Technol.*, **42**, S4–S26.

Warren, H. C. and Boyer, B. P. (1991). Use of heme-containing proteins as stabilizers for enzyme-labeled immunoreactants. *European Patent 456,309*, to Eastman Kodak Company.

Wetzel, R. (1987). Harnessing disulfide bonds using protein engineering. *Trends Biochem. Sci.*, **12**, 478–482.

Willimer, H. (1992). Measurements of temperatures, ice evaporation rates and residual moisture contents in freeze-drying. *Dev. Biol. Standard.*, **74**, 123–136.

Wunderlich, B. *et al.* (1964). Theory and measurements of the glass transformation interval of polystyrene. *J. Appl. Phys.*, **95**, 35–45.

Yamazaki, H. and Boyd, C. S. (1996). Stabilization of peroxidase conjugates in solution. *US Patent 5,512,448.*

Zale, S. E. and Klibanov, A. M. (1986). Why does ribonuclease irreversibly inactivate at high temperature? *Biochemistry*, **25**, 5432–5444.

Zuber, H. (1976). *Enzymes and Proteins from Thermophilic Microorganisms*, Birkhauser Verlag, Basel.

Zuber, H. (1981). Thermophily. *Basic Life Sci.*, **18**, 499–512.

Chapter 13

The Manufacturability of Coupled–protein Reagents

Alastair H. Dent and Mohammed Aslam

Research and Development Department, Ortho-Clinical Diagnostics, Pollards Wood Laboratories, Nightingales Lane, Chalfont St. Giles HP8 4SP, UK

13.1 Introduction

The earlier chapters of this book have described the many approaches that are available for forming links between protein molecules and a variety of coupling partners (other proteins, small organic molecules, solid matrices, etc.), as well as techniques for purifying, characterising and stabilising the products.

At one extreme, it may only be desired to carry out such a sequence of operations as a single, small-scale research experiment. Of course it is important that such experiments are carried out under suitably controlled conditions, so that the results may be interpreted with confidence. However, there are few limitations on such an experiment other than the cost and availability of materials, equipment, and so on. Within reason, anything can be tried.

On the other hand, it may be desired to produce the same protein reagent repeatedly, at a large enough scale to supply many hundreds or thousands of users, and with a high degree of run-to-run consistency with respect to product characteristics. Under such circumstances, the preparation method is properly viewed as a manufacturing process, and a number of factors need to be considered which are important in making it successful in this role.

13.1.1 Manufacturing cost

If the protein reagent forms part of a commercial product, the economics of the preparation process may be critical to the overall profitability. Raw material issues (inventory level, cost *versus* purity decisions) and equipment requirements (capital investment and depreciation, running and maintenance costs) may well be very significant in this respect. However, it is often the case that labour expenses form the largest contribution to the manufacturing cost. These are in turn determined by the nature of the process itself. The obvious ideal is a short process requiring little manual intervention, and especially one which does not require highly skilled or qualified operators. The choice of batch size can also be an important factor in determining overall cost; this decision is not always straightforward given the expected changes in sales volume as a product moves through its lifecycle, and the uncertainty of projected sales figures.

Beyond these extremely broad generalisations, it is not intended here to attempt guidelines on how profitability can be built into a production process. A single point might be borne in mind, however: if the potential profitability of a process is not considered early enough in the development programme, it will become more and more difficult and expensive to correct this situation later on.

The relationship between process development and profitability has seldom been explored in depth, but data suggest that the impact can be very large (Pisano, 1997).

At a more global level, a great deal of debate has taken place (particularly over the last 50 years) on the philosophy of manufacturing, and a number of new approaches have been introduced aimed at maximising the efficiency (and profitability) of production operations. The contributions of 'quality' gurus such as Deming, Juran, Taguchi, Goldratt *et al.* have been fundamental in bringing about some quite radical changes throughout the manufacturing industry, often with dramatic effects on profitability. It is not appropriate within a volume of this kind to attempt a review of these manufacturing philosophies.

13.1.2 Regulatory factors

The production of reagents for use in medical applications is subject to the scrutiny of many regulatory organisations around the world, in particular pharmaceuticals and diagnostic products. One role of such agencies is to enforce the application of accepted Good Manufacturing Practice (GMP) in the production of such reagents, and they have been a major driving force in establishing standards worldwide. It is understood that the accepted view of 'good practice' does not remain fixed over time, and the term CGMP (to denote current GMP) is sometimes employed to indicate this fact. The interpretation of GMP does vary somewhat from agency to agency as well, although major international initiatives have been under way for some years to harmonise the requirements of individual countries.

Some specific requirements are aimed particularly at controlling the quality of products which can have a significant effect on the health of an individual. Generally, however, GMP is intended to ensure that manufacturing practices consistently yield product with the expected performance characteristics, and this aim is valid in all manufacturing applications, regardless of the end-use.

Much effort has also gone into the interpretation of regulatory requirements from various countries on the standards for medical products, not least in terms of what is considered adequate to demonstrate the robustness of a manufacturing process. Once again it is not the intention here to revisit these assessments, and the reader is referred to specific volumes on this subject: see, for example, DeSain (1993).

13.1.3 Robustness of manufacturing processes

At the molecular level, predicting the likelihood of an individual binding event (covalent or non-covalent) is a rather haphazard affair. The chances of a potentially reactive molecule colliding with a suitable coupling partner are controlled principally by the relative concentrations of the species concerned, and their rate of movement through the medium. When such a collision occurs, the formation of a bond will again depend on probabilistic arguments: the likelihood of each molecule being correctly oriented and (for proteins) in the correct conformation, the distribution of energy within the potentially reactive bonds, etc. And these energetic variations will lead to some probability of dissociation too, so that the effect of the coupling reaction may be reversed.

For a small number of molecules, therefore, it would be difficult to predict the exact composition of the product after a given reaction time. But at the scales of interest to preparative biochemists the numbers of molecules involved are enormous, far greater than the level at which probabilistic variations would be expected to have an observable effect. So, not to labour the point any further, it is fully justifiable to expect that a protein-coupling reaction carried out using the same reagents under the same conditions should yield exactly the same product.

The existence of troubleshooting functions attached to manufacturing operations the world over suggests that the products obtained from such reactions are not always exactly the same, with the obvious corollary that (however much this was the intention) they cannot have been carried out using the same reagents under the same conditions.

The aim of this chapter is very simple: to summarise some of the considerations which may affect the choice or development of processes for the production of coupled-protein reagents, such that they meet the basic requirement of robustness in a manufacturing environment. The success of this approach can only really be observed in the behaviour of the process in routine manufacturing. A good goal is to provide a process which can be carried out repeatedly over a number of years, consistently yielding a product with the desired performance characteristics. An indication of even greater success is if this goal is achieved with a method which is readily understood and executed by the manufacturing operators, requires minimal rework or troubleshooting effort, and, lastly, does not call into question the sanity of the individual or group responsible for its development.

Finally, it will be observed that there are very few references to literature reports in this chapter. The concerns summarised below are instead the result of many years of experience at the interface between R&D and manufacturing functions, influenced of course by the philosophical and regulatory factors noted earlier. It is hoped that their consideration will help to avoid some commonly-encountered pitfalls in the transition from a bench-scale preparation to a fully fledged manufacturing operation.

13.2 Process Variation

For the purposes of this discussion, any of the multitude of possible methods for creating coupled-protein reagents can be viewed as a simple process, which converts a number of starting materials to a number of products *via* a number of stages, sometimes termed unit process or sub-processes (Figure 13.1); it is immaterial whether the process is one for creating a radioimmunoconjugate for imaging, a protein-coated biosensor for environmental monitoring, or a DNA probe for *in vitro* diagnostics. The decision on what constitutes a 'stage' is always rather arbitrary, as each individual sub-process can generally be broken down further into a number of 'sub-sub-processes', and so on.

13.2.1 Key process variables

Taking any individual sub-process, it is possible to identify a number of variables which may (or may not) influence the product obtained. Take for example any sub-process in which a protein in aqueous solution is derivatised by a suitably-reactive compound such as a succinimide ester. A list could very easily be drawn up of parameters which might be expected to have a significant effect on the outcome of the reaction: protein concentration; ester concentration; reaction time; temperature; pH. Obviously it is a matter of judgment as to which process variables to include on this list; other 'possibles' (e.g. ionic strength) have been omitted on the basis that they are less likely to have an effect. Assume for now that the list is more or less correct; these form a family which will be termed key process variables (KPVs).

In order to carry out a comprehensive assessment of potential KPVs, it is good practice to flow-chart

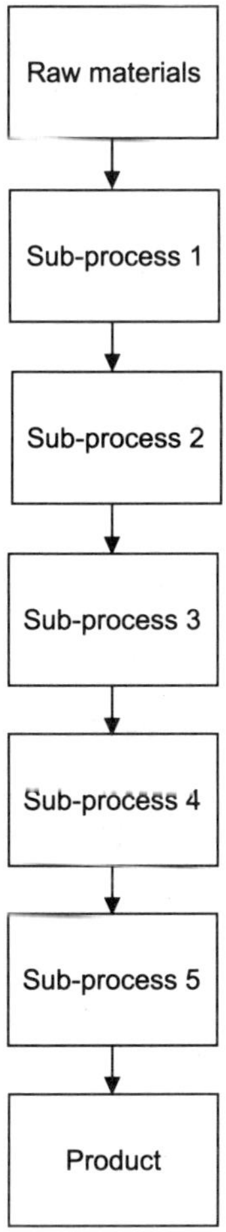

Figure 13.1 The conversion of raw materials into product: the overall process can be broken down into a number of unit processes or sub-processes.

the entire process at a much greater level of detail than that shown in Figure 13.1. If this is carried out rigorously then it is easier to ensure that all the stages of the process are considered with regard to their likely criticality. Many potential KPVs are clearly identifiable and are unlikely to be neglected, temperatures, reaction ratios, chromatography conditions, etc. But there are others which are perhaps less obvious, and it is worth considering some of these briefly.

Some times in the process fall in the 'obvious' category, e.g. the duration of a reaction stage. However times *between* stages can often be critical too, and are more easily neglected. Common examples include the time between preparation of a reagent solution and its use: in the derivatisation described above, for example, the length of time between preparation and use of the active ester solution should undoubtedly be added to the list of KPVs. The age of the protein solution is generally less important, and may well be added to the 'non-critical' list. However this is a good example of where judgment is necessary: a protein with poor solution stability or a tendency towards aggregation or precipitation may deserve greater attention.

Another related category of variables which is sometimes neglected consists of times between stages, which also incorporate a change in temperature. These may consist simply of a period allowing a solution stored at one temperature to equilibrate to another, or they may incorporate a phase change as well, typically the thawing of a frozen solution.

The conditions under which these stages are carried out, with or without temperature change, can cause a number of possible effects. The most obvious are physicochemical changes: the hydrolysis of an active ester solution, the precipitation of an antibody, etc. However there is another more insidious class of effects that can be crucially important. Microorganisms such as bacteria, yeasts, fungi, etc., will flourish under any conditions which allow them to do so, and their effects on a protein reagent can be dramatic and unpredictable. The presence of microbiological contamination in pharmaceutical reagents (either live organisms or their toxic products) obviously has a special significance of its own, and the manufacturers (and regulatory authorities) rightly go to great lengths to ensure that this is not allowed to happen. However, unwanted effects of microbiological contamination are not limited to pharmaceuticals, and this area should be considered carefully for all classes of protein reagent. If nothing else, the potential of such organisms to release proteases is sufficient cause to worry that their presence might adversely affect a reaction stage.

When the flow-charted process is considered, therefore, the likelihood of an individual stage to invite microbiological problems should be considered in determining whether there may be KPVs related to it. The dissolution of a small organic molecule in sterile water, and the use of the resulting solution within ten minutes, can be safely passed over in terms of microbiological concerns; on the other hand, blocking of antibody-coated tubes by overnight incubation at 37 °C with a 5% albumin solution should set a few alarm bells ringing. The potential for such problems should obviously be minimised in the design of the process, e.g. by carrying out long incubation stages at reduced temperatures, by including appropriate filtration procedures, and/or by incorporating biocidal agents in the formulation.

13.2.2 Fluctuation of key process variables

When any process is carried out, it is impossible to control any of the KPVs completely; there will always be a certain level of imprecision around each parameter. Returning to the derivatisation of a protein with an active ester, the accuracy[1] with which the ester solution is prepared on a single occasion, for example, can be no better than that obtainable from the instruments used for measuring out the masses and/or volumes concerned (Figure 13.3(a)). On a second occasion, the combination of these variations may result in a different final concentration, i.e. a different inaccuracy (Figure 13.3(b)). Over many occasions the spread of inaccuracies obtained from this preparation can be plotted out, to obtain an estimate of the overall precision of the solution preparation process (Figure 13.3c).

In the example shown the process is accurate (there is no bias of the mean from the target value), and the population approximates to a normal distribution. The standard deviation of the population can be calculated from the data, and used to predict the likelihood of the actual concentration falling within a given range. A typical statement might be that 'on 99% of occasions it is expected that the concentration will fall within ± 2.8% of the target value'. Alternatively, if the accu-

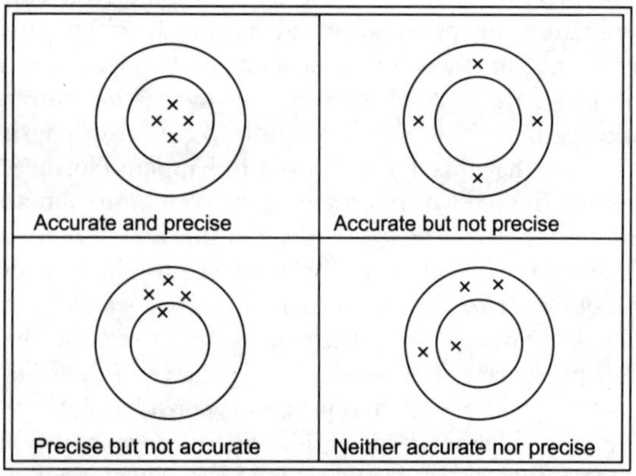

Figure 13.2 Accuracy and precision.

[1] The terms accuracy and precision sometimes cause confusion. If a process is carried out a large number of times and the mean value obtained is very close to the target, the process is accurate; however, the same mean value can be obtained from a tight population of individual results (accurate and precise) or from a wide spread (accurate but imprecise). An inaccurate process, similarly, can be either precise or imprecise. The widely-used analogy of shots aimed at a target (Figure 13.2) probably remains the clearest visualisation of the difference between the two terms.

The result of one individual occasion can only be accurate or inaccurate; the likelihood of a particular accuracy being achieved is controlled by the overall precision of the process.

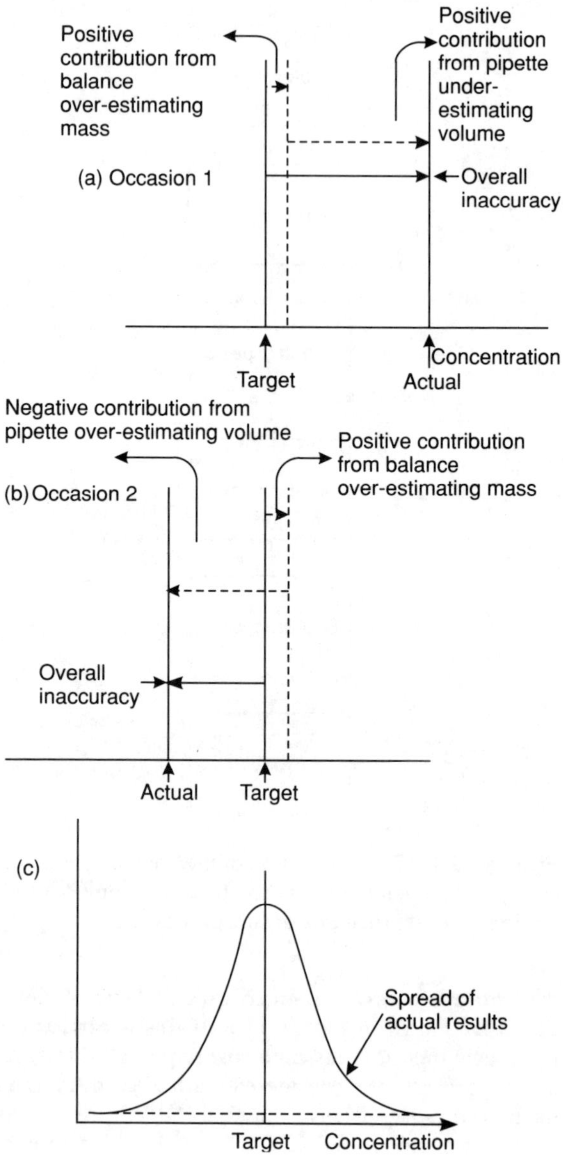

Figure 13.3 The contribution of two individual factors (mass of active ester, volume of buffer) on the actual concentration obtained; on two occasions (a, b) the combination of inaccuracy in these factors yield an overall inaccuracy in the ester concentration. Plotting a large number of such individual observations (c) gives an indication of the overall precision of the process, in terms of ester concentration obtained.

racy and precision of the various measurements of mass and volume are known, these can be combined 'on paper' to arrive at an estimate of the overall characteristics of the process.

Taking another example, if the buffer used to control the reaction pH is prepared from a well-developed gravimetric recipe, similar predictions can be made about the expected range of concentrations of the buffer salts within the final formulation (Figure 13.4(a)). However this will probably not correlate with the pH obtained, which may well be relatively unaffected by small changes of this kind (Figure 13.4(b)). Hence it is necessary to understand the variation of pH with formulation to make predictions about the outcome of this process from the known

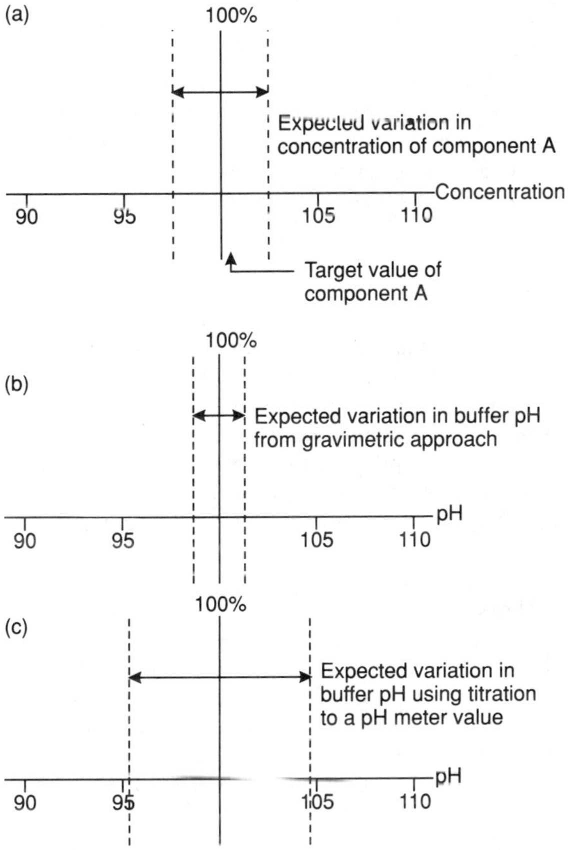

Figure 13.4 The expected variation of a component concentration in a buffer (a) may only yield a small variation in pH (b). In a gravimetric approach this may limit the pH precision obtained. However, where the buffer is titrated to a target pH value, the limit is instead defined by the performance of the pH meter (c).

precision of mass and volume measurements. On the other hand, a common practice is to titrate such formulations to a given pH using a pH meter. With this approach the precision of the mass and volume measurements becomes irrelevant, and instead the precision of the pH determination itself governs the outcome. If the pH meter (or more likely the electrode) is not well maintained, this may actually introduce a wider spread in the expected pH range than would be obtained by a gravimetric approach (Figure 13.4(c)). To make a prediction of this range which will be useful over a long-term manufacturing lifetime, it is necessary to obtain an estimate of the likely variation in the performance of pH meters and electrodes under realistic conditions, which may involve studying many different meters and electrodes over many occasions.

The simple examples just discussed illustrate the fluctuations in KPVs which may be expected when a process is carried out many times, and some of the approaches which may be taken to estimate the extent of these fluctuations. In order to determine whether these fluctuations are likely to be problematical in terms of product performance, it obviously makes sense to investigate the process by deliberately altering KPVs and observing the effect—latitude or guard band studies. However, before this phase begins it is necessary to decide how the effect should be observed.

13.2.3 Choice of a measurable characteristic

Coupled-protein reagents are manufactured with a particular end-use in mind, and it may or may not be practicable to measure the performance of such a reagent using the final application itself as a test. For example, an enzyme–IgG conjugate may be prepared for testing the capacity of antigen-coated microtitre wells, by measuring the UV absorbance obtained from a suitable colorimetric substrate. It might be chosen to investigate the performance of conjugates prepared under different conditions simply by measuring, under well-controlled conditions, the signal obtained from a reference batch of wells and of substrate solution. So long as this signal fell within an appropriate range, any variations brought about by varying KPVs might be viewed as acceptable. In such a simple application, an approach of this type may be justified.

Suppose, however, that a radioimmunoconjugate is being prepared for *in vivo* experiments.[1] Determining

[1] As noted above, for human applications in particular there are well-defined procedures for demonstrating the safety and effectiveness of such reagents, promoted and policed by regulatory agencies. The approach outlined below for demonstrating process robustness should be viewed as a contribution to this overall process.

directly whether the conjugate is successful in the end-use is a complex and time-consuming operation, and the effort required to test a large number of conjugates in this way is likely to be prohibitive. The most likely approach in such circumstances is to choose one or more surrogate characteristics, which are readily measured, and which bear a useful relationship to the performance in the final application. With an immunoconjugate it may well be decided initially to measure the molar incorporation ratio (MIR) by a combination of UV spectrophotometry and radiometry. A series of conjugates of differing MIR may be prepared, and used to assess the acceptable range in the application of interest, this may lead to the conclusion, for example, that a MIR in the range 1.0–2.5 gives acceptable performance[1].

Of course there is a danger in this approach: what if the measurable characteristic chosen does not adequately reflect the performance in the end-use? In these circumstances the result of subsequent experiments may be misleading. The MIR of the immunoconjugate, for example, would not detect whether the affinity of the antibody for its specific antigen had been reduced by a particular variation to the conjugation process, and a false sense of security might be obtained from the MIR study.

A more secure approach is to use a number of measurable characteristics chosen to cover all the key attributes of the conjugate. The disadvantage here, of course, is that the amount of work required to define acceptable ranges in the final application escalates. Carefully chosen conjugation conditions and experimental design in the assessment can limit the amount of effort required, but this can still be considerable[2].

The choice between measuring end-use performance, or surrogate characteristics such as incorporation or functional activity, will always require a degree of judgment. This choice will be affected by a number of factors. The more complex the establishment of effective performance in the end-use, the more attractive it will be to use a surrogate measure for challenge work. On the other hand, the more critical the end-use, the more one may be inclined to carry out comprehensive studies using end-use performance as the acceptance criterion.

There is another benefit, however, which is always gained from the investigation of surrogate parameters: an increased understanding of the effects of KPV changes. While it may be possible to establish that the process for preparation of an immunoconjugate is or is not acceptably robust simply by *in vivo* studies, it is difficult to be definitive about the mechanism of any loss in performance seen. A great deal more could be learnt by breaking down the observed changes into effects on MIR, effects on immunoreactivity, effects on stability, etc. The data generated from investigations of this kind can be extremely valuable in influencing the course of development, and the study of surrogate characteristics which help to define overall performance is highly recommended for this reason if no other.

13.2.4 Process latitude studies

Having established a list of KPVs, an estimate of how much each KPV is likely to fluctuate in a manufacturing environment, a suitable measure (or measures) of product performance, and an acceptable range for this measure, it is possible to carry out a study to determine whether the process is sufficiently robust to give consistent performance. Such latitude studies are simple in principle: each KPV is varied, and examined to assess whether their performance is acceptable.

These studies can be used in two main ways. The first is the optimisation mode, where the process design is still being defined. Examination of the robustness of a candidate process to KPV variations is a very worthwhile activity, rather than optimising on end-use performance of a few specimen preparations alone. The second is the challenge mode, where a process has already been chosen and it is desired to show definitively that it is robust to stressing of known KPVs. In the first mode products with unacceptable performance are fully expected as the understanding of KPVs is built up; in the second it is anticipated that all the products will meet the acceptance criteria.

There is in fact a considerable degree of overlap between the two activities—a well-structured optimi-

[1] Note that the precision of the measurement method used to determine this range must be taken into account, otherwise it may be meaningless in practical terms.

[2] A compromise route is to apply a two-tier strategy, which treats parameters differently depending on whether they are *expected* to change as the KPVs are modified within the small range of interest. Thus with the radioimmunoconjugate, it might be decided that the MIR is likely to change as the conjugation conditions are altered, but that the immunoreactivity of the antibody is not. An approach could then be taken where an acceptable MIR range was established by *in vivo* experiments, the robustness of the process to KPV changes demonstrated using this range, then the immunoreactivity of the challenge conjugates assessed to confirm that there were no unexpected changes in this parameter. In this way the amount of *in vivo* work is minimised. The strategy is also failsafe: if the assessment of the challenge conjugates does reveal an unexpected variation in immunoreactivity, then one can return to the approach of establishing *in vivo* an acceptable range for this parameter.

sation programme will answer many of the questions which might be posed in the challenge phase, and will influence the design of such a study.

While the experimental designs may vary somewhat, the basic principles are the same, and this section is relevant to both modes.

As noted earlier the overall process is made up of a series of rather arbitrarily defined sub-processes, as shown in Figure 13.1, and a number of different approaches can be applied to the latitude studies depending on how much one wishes to break down the whole. At one extreme a large study can be applied using a single set of acceptance criteria, which examines all the sub-processes at once (Figure 13.5(a)). This has some advantages in terms of the consistency of assessment, but is likely to generate a very complex experimental design if a reasonable number of KPVs are to be included.

Alternatively the process can be addressed in a modular approach, separate sets of experiments investigating the robustness of a mixing stage, a chemical derivatisation, a purification step, etc. (Figure 13.5(b)). The separate modules may employ quite different acceptance criteria: the robustness of a mixing step may be investigated by the achievement of a homogeneous solution *via* UV absorbance measurements, while the purification of a derivatised protein may be assessed in terms of product purity and yield, etc.

Apart from yielding more manageable experimental designs, the modular approach may offer some economy of effort. If, for example, the activation of an enzyme with a heterobifunctional reagent is used in a number of different preparations, the demonstration of robustness may be carried out just once using a suitable surrogate characteristic such as the MIR. Providing the requirements are comparable, the robustness of this sub-process in a number of different settings can then be assumed without further work. The same might apply to a mixing process used within certain constraints of time, volume, equipment, etc., or to a desalting procedure using size exclusion chromatography. One can imagine, therefore, a process built up of a number of 'off-the-shelf' sub-processes whose robustness has already been demonstrated individually.

In practice, the approach taken may well fall between these two models, with any widely used sub-processes being dealt with *via* generic studies, and any remaining product-specific sub-processes being addressed in a single study.

In any cases where a sizeable number of KPVs is to be examined, it is worth considering the experimental design carefully. First of all, it may well be more economical to carry out the work at a small scale relative to that intended for production. In fact, in many cases it may be totally impractical to carry out work of this kind at full scale. But of course the reduced scale must reasonably reflect the behaviour of the final procedure, otherwise the result will not be representative. Secondly, the way the KPVs are combined for study deserves a little thought. A full factorial design of each possible combination of variables generates an enormous amount of information, particularly concerning the interactions between multiple variables. However the full factorial approach becomes unwieldy at a very small total number of KPVs, even if each is only studied at two levels. In any case, full factorial designs generally produce far more information than is necessary.

One alternative approach is to take each KPV of interest in sequence, fix all other variables at the target value, and investigate the effect of varying the chosen KPV above and below the target figure. This generates information which is very simple to interpret, but is reasonably intensive (two experiments per variable) and provides no information about the effect of varying more than one KPV simultaneously.

Another useful technique is to employ 'reduced factorial' designs which provide a balanced array of conditions within a fairly small number of experiments. Approaches such as Taguchi designs, Plackett–Burman plots (Phadke, 1989), etc. allow investigations to be carried out on a large number of KPVs simultaneously. The results can be used simply to show that all combinations of conditions provide conjugates meeting the required acceptance criteria; however, they can also be interpreted to identify which KPVs have the largest effects on the process and which can be safely ignored.

Software is now available for the convenient application of these experimental design techniques, such as the JMP® package from the SAS Institute (Cary, NC) and RS/Discover® and Cornerstone® from Domain Solutions Corporation (Cambridge, MA).

Much of the potential to observe the effect of interactions between multiple variables is lost with the reduction of the full factorial design. However it is still possible to obtain some limited information of this type, and to some extent KPVs of particular interest can be placed in the design such that they are strongly represented among the 'visible' interactions.

A problem of reduced factorial designs is that they may 'over-stress' the process under examination. One particular set of experimental conditions in a conjugate challenge may, for example, have all the KPVs stressed

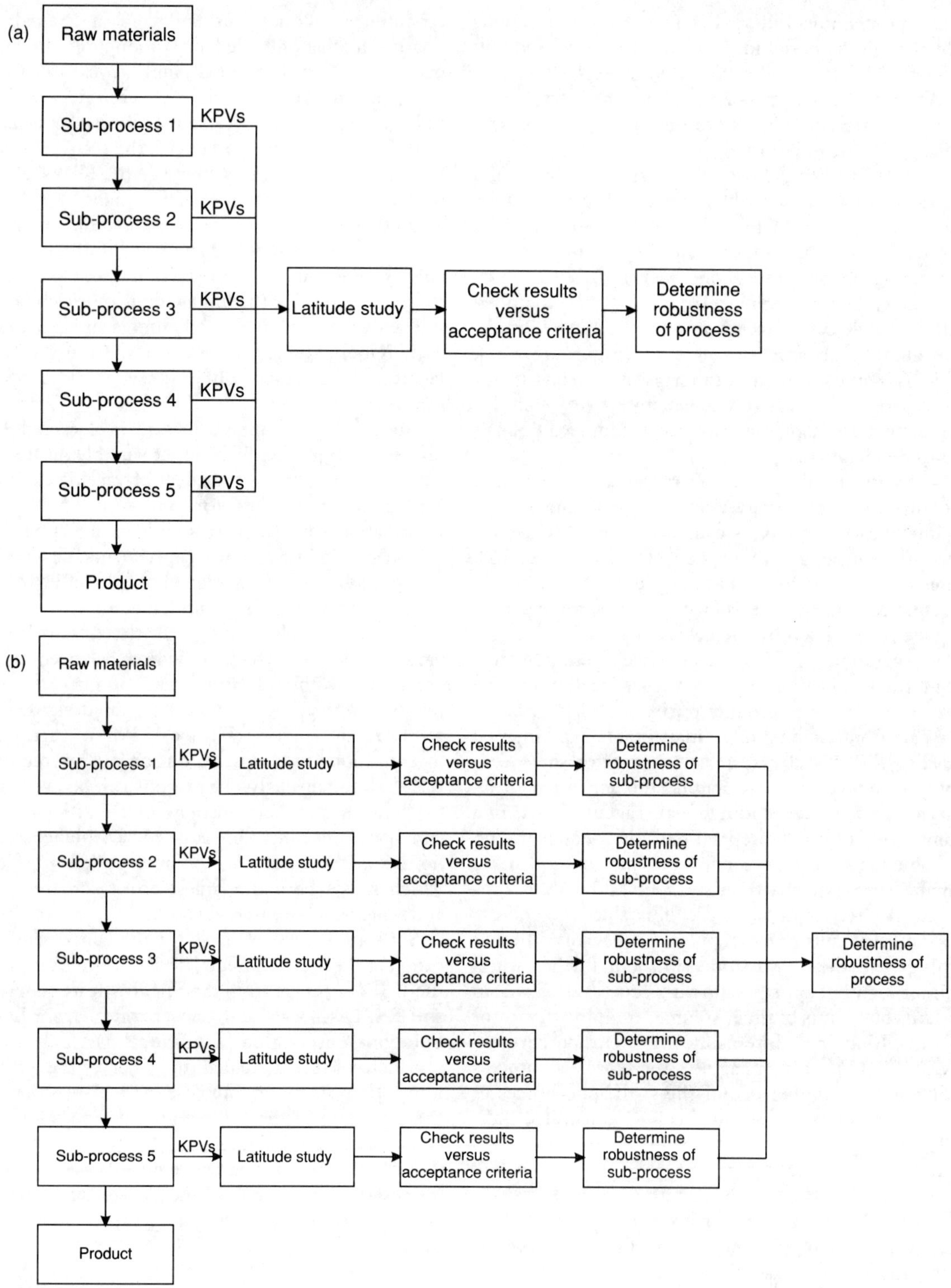

Figure 13.5 Two approaches to investigating the robustness of a process: variables in each of the sub-processes can be stressed in a single large study (a) or in a series of smaller individual studies (b).

in the direction of over-incorporation: high overall concentration, high concentration of activating reagent, long time, high temperature, high pH, etc. Not surprisingly such combinations of conditions can sometimes cause the characteristics of the product to fall outside the acceptable range, although the simultaneous shift of all these variables in the same direction is a very unlikely event statistically. This, however, is the price to be paid for labour-saving designs such as reduced factorial arrays; again such effects can be minimised to some extent by a particular choice of array, or placement of variables.

13.2.5 Handover and process performance monitoring

Once the process latitude studies have been completed successfully, the method may be considered close to readiness for handover into the manufacturing environment. It is wise not to attempt this handover until a number of (if possible) full-scale preparations have been carried out in R&D using the final procedure, and their performance assessed—ideally both in the end-use application, and in terms of surrogate characteristics such as incorporation (for conjugates), surface coverage (for protein-coated solid phases), etc.

These results will begin to confirm that the latitude studies have done their job, and that the process is genuinely and consistently capable of yielding a product with the expected performance characteristics, time after time. This stage can be particularly important if the latitude studies have incorporated major assumptions, e.g. if they have been carried out at a very small scale compared to the final process, or if the process consists largely of sub-process modules whose robustness has been challenged individually using surrogate characteristics. Obtaining the required end-use performance repeatedly from the final full-scale process is another important milestone in demonstrating that the development has been carried out in a satisfactory manner.

The way in which the procedure is handed over into the manufacturing environment can itself be important in terms of understanding and buy-in. Generally different groups of operators are involved in the R&D and manufacturing roles, and they may well be geographically separated too. Although the method should be described well enough that further explanation is unnecessary, the presence of the staff who have actually developed the process for the first few manufacturing runs is generally a very worthwhile investment. Also, the more communication there has been with manufacturing personnel during the development of the process, the less likelihood there will be that the final method handed over will come as a nasty surprise—so involvement of this kind should really start very early in the development programme.

As noted earlier, elements of judgment are required in the demonstration of robustness of a coupled-protein reagent, both in terms of which process variables are likely to be significant and thus constitute KPVs, and in which measurable characteristics are likely to provide a good reflection of end-use performance. While a group with appropriate expertise can be expected to make a good assessment of these questions, it is not impossible that a KPV can be 'missed': a variable which was not considered likely to be important in fact turns out to have a significant effect.

If such a KPV first manifests itself in the manufacturing environment the result may be a failed run, which constitutes a demonstrable failure of the procedure to build in robustness.

It is very wise to carry out continuous monitoring of performance as the reagents are repeatedly prepared in the manufacturing environment—again, both in terms

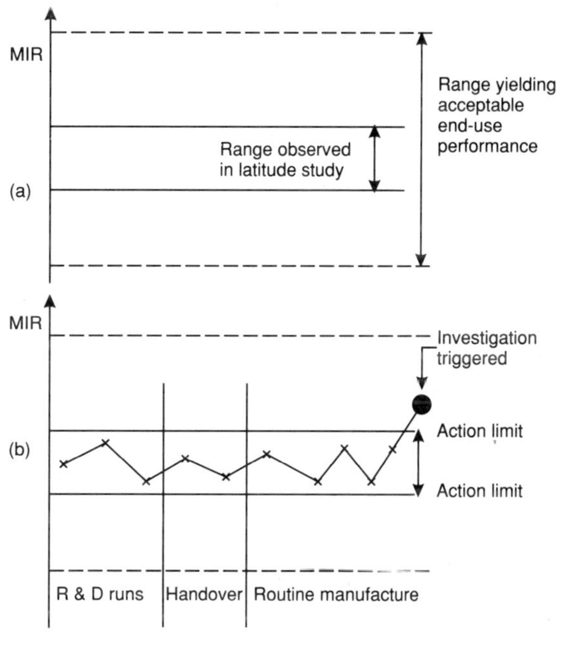

Figure 13.6 The range of molar incorporation ratio (MIR) observed in the challenge study for a conjugation should fall within the range known to give acceptable end-use performance (a). This range also provides a useful basis for action limits, which trigger an investigation if breached (b).

of end-use performance and useful surrogate characteristics. Using the information which has been generated during the R&D phase, a range of performance can be defined which has been observed from the latitude studies, and is known to fall within the wider band of acceptable end-use performance (Figure 13.6(a)). Hopefully, the observed performance of the manufactured reagents will fall repeatedly well within the band observed in the latitude studies (Figure 13.6(b)). This will serve as continuing confirmation that the development of a robust process was successful, and that no KPVs have been neglected. It is good practice to define action limits, however, so that if the performance starts to approach unacceptable levels an investigation is triggered. As a rule of thumb the extremes observed in the latitude study form a good initial guideline for these action limits, but they may be narrowed if the performance is consistently well within this band. By this approach a missed KPV may perhaps be picked up before it has a chance to cause a failed run, and the new information gained can be fed back into future developments, to refine the judgment process.

This approach, of course, is an example of statistical process control (SPC), which is commonly employed in a multitude of manufacturing situations (Ishikawa, 1986).

13.3 Raw Material Variation

When an action limit is breached and an investigation is carried out as to the cause, it is quite often the case that the process itself turns out not to be at fault. Section 13.2 has quite deliberately not mentioned the possibility of variability in the raw materials, because this is not strictly a failing of the process. But if the raw materials show a batch-to-batch or time-dependent variation with a significant effect on product performance, all the efforts put in to develop a robust process may be wasted.

Proteins are of course biological reagents derived from variable living sources, and this applies to a greater or lesser extent even to the more reproducible examples such as monoclonal antibodies obtained from ascites or cell culture, or recombinant proteins expressed by yeast, *E. coli*, etc. Therefore a certain amount of variation can be expected in raw materials of this type, and this needs to be considered during the development of a manufacturing process.

There are really only a few approaches that can be taken to minimise such problems.

13.3.1 Sourcing of raw materials

Although recombinant or cell-culture derived proteins may still exhibit some variation, they are likely to be considerably more reproducible batch-to-batch than an equivalent material extracted from tissue sources, for example. Therefore materials of this type should be considered wherever possible. Using this approach to avoid the use of materials from human or animal origin can also head off various regulatory or ethical problems.

Secondly, as much care as possible can be put into choosing an appropriate supplier for raw materials, ensuring for example that they meet established quality standards, that they use appropriate equipment and well-defined methods, that their operators are well-trained, and that their own supply of raw materials is adequately controlled too. As well as carrying out periodic audits to establish that these criteria are being maintained, it is often beneficial to establish a working relationship with major suppliers so that a good understanding is maintained, and potential problems can be headed off before they are realised. Of course, it is possible to bring the manufacture of raw materials in-house, and this may be appropriate in some instances. Often however it is simply more convenient or cost-effective for such operations to be contracted out to an external supplier.

13.3.2 Acceptance tests for raw materials

Suppliers will often carry out a series of analyses on their products, and this can provide a useful basis for assessing suitability. The usual problem is to correlate the data provided by the supplier with the requirements of the process. Quite frequently the connection between the two is not obvious, but this can be addressed from both ends. On the one hand, the development of the process can incorporate the establishment of key raw material variables, in much the same way as KPVs are handled; a conjugation process may be demonstrated to be robust to a certain purity and functional activity of raw material, as quoted by the supplier. And if the significance of additional raw material variables becomes apparent during the development process, the supplier may be persuaded to carry out extra testing to demonstrate, for each batch supplied, that the material possesses the appropriate characteristics. Again the benefit of a good working relationship with the supplier is obvious here.

If sufficient testing cannot be carried out by the supplier, the alternative is to demonstrate the acceptability of the material after its receipt. This is somewhat less ideal, if only that the financial responsibilities in the event of failure can be more of a problem. However, this approach can work quite satisfactorily—so long as material of appropriate quality is readily obtainable. There is little point having a stringent set of acceptance criteria if it is never possible to find materials which pass them!

Once again the acceptance criteria should be linked as closely as possible to the development of the process(es) in which the material is used. The ideal test will fail all batches of material which will give unacceptable performance, while passing all batches which will behave satisfactorily. As in Section 13.2.3, there is often a choice between testing the material in the end-use, which is often time-consuming and expensive, or using a surrogate test, whose relationship with end-use performance should be understood as clearly as possible.

However much effort is put into deriving acceptance criteria, it is always a good idea to examine a reasonable number of batches of material during the development programme. In fact a good rule of thumb is that the more difficult it is to characterise the key attributes of a raw material, the more batches should be examined during the development process. Even if it is not obvious how measurable variables of a raw material relate to the end-use performance, some confidence can be taken from the observation that a large number of batches from several different suppliers have all shown comparable and acceptable performance in the end-use. Entering routine manufacture when only one or two batches of a poorly understood material have been examined is a very risky enterprise.

13.3.3 Stock approaches for raw materials

Sometimes it is genuinely impossible to obtain multiple batches of a raw material, e.g. if a unique polyclonal antiserum is identified from a particular stage of an immunisation programme. Such situations obviously put unusual pressures on the security of raw material supply, and there may be no option but to lay down a stock of this material which is believed to be sufficient to last the expected lifetime of the product. The stability of the material obviously becomes of paramount importance under such circumstances, and it is worth putting a reasonable amount of thought into the best conditions for storage, including splitting of stocks to avoid catastrophic loss in the event of equipment breakdown, fire, etc.

This bulk stock philosophy can of course be applied in less demanding circumstances, and if the stability of the material is adequate it may be considered to provide stock security for important items. However there are a number of disadvantages with this approach. In terms of the cost of inventory maintained, the storage space requirements, the need for material stability, and the ability to avoid wastage if demand drops, it is preferable to apply the opposite approach—a small stock level and frequent replacement. The bulk stock route is really best limited to cases where the replacement of the stock is expected to be particularly troublesome, or where lead times are very long.

Once again, good supplier relationships can sometimes open up the possibility of achieving the best of both worlds. A supplier may for example hold a large batch of material for the customer, and allow it to be purchased in small quantities as the need arises.

13.3.4 Other raw material considerations

The factors discussed above are most commonly considered for protein raw materials, as these might be expected to show the highest susceptibility to variations. However, the quality of other raw materials such as coupling agents should also be given serious consideration. It is not unusual to find significant fluctuations in the purity of such reagents, which can obviously translate directly into a significant and unwelcome variation in the outcome of a reaction. As well as controlling the purity of incoming raw materials as much as possible, and attempting to influence this factor *via* the supplier, it may be advantageous to carry out a determination of active component content for each batch received, and to include this factor in the calculation of how much material is to be used in the derivatisation reaction.

With regard to the storage of raw materials before use, it goes without saying that adequate control of environmental conditions such as temperature and humidity must be maintained, and that shelflives are set to match material stability under the specified conditions. Additionally, it is accepted good practice to ensure that the materials are physically secure, for example by arranging that they are accessible only to appropriate staff. A level of segregation is also generally expected between materials which have been approved for use, those which are expired or have failed acceptance criteria, and those which are quarantined pending approval.

Water is a particularly important raw material, both because it is used in so many key processes, and because it represents a possible route for the introduction of microbiological contamination. Plenty of advice on the procedures necessary for control of water systems can be obtained from GMP references or from contractors who supply such systems.

Finally, if the process monitoring procedure described in Section 13.2.5 is drawn up in such a way that key raw material changes are clearly highlighted, the action limit approach may head off problems originating from this source too.

13.4 Equipment Variation

Almost all the processes of interest here rely on equipment of some sort for their execution. This may range, for example, from simple magnetic or overhead stirrers, through chromatography setups, to complex, purpose-built plant for a multi-stage protein coating process. Measurement equipment such as pH meters, chart recorders, etc., may be equally important in allowing decisions to be made about the product, or during the process.

13.4.1 Qualification and upkeep of equipment

Unless the performance of these pieces of equipment is both known and consistent, a perfectly-controlled process with perfectly controlled raw materials may still yield irreproducible results. For this reason a certain amount of effort is necessary to ensure that the key attributes of manufacturing equipment are routinely checked. Often a comprehensive programme of equipment qualification[1] will be carried out initially (and generally repeated on a relatively long, fixed-term basis) to ensure that appropriate acceptance criteria are met, then a system of regular checking or calibration of key criteria will be implemented so that any drift in performance is detected.

For all but the simplest pieces of equipment, a choice must be made as to which parameters are to be included in the qualification programme only, and which are to be monitored routinely thereafter; this choice will be determined by the criticality of a particular parameter and the likelihood of variation. With a pH meter, for example, it is commonplace to check the calibration of the particular meter/electrode combination against reference buffers very frequently—perhaps every time a determination is made. But other features of pH meter performance may only be checked infrequently, e.g. the ability of automatic correction mechanisms to deal adequately with solution temperature fluctuations. With a chromatography system, it may be decided to run a check on the efficiency of the particular column being used on every run, but a much wider range of parameters may be checked on a longer-term basis: pump flow rates, linearity of gradient controllers, chart recorder behaviour, etc.

Really the activities carried out as part of the qualification exercise, and those executed as part of a regular checking programme, should be viewed as a continuum, ranging from frequent simple tests of critical and/or variable behaviour, to more occasional tests of more robust parameters, or those which are expected to have a less significant effect.

One particular aspect of equipment control is worthy of further mention; it has become more and more commonplace for equipment to be computerised and operated *via* software; this may range in complexity from a simple programme for controlling a fraction collector or integrator in a chromatography setup, to an integrated system for process control of an entire full-scale manufacturing plant. Data handling is almost universally carried out *via* software packages too, e.g. the analysis of data from process checks or product test procedures. The ability of these software programmes to function correctly, in particular to carry out the right operation or calculation from a given input is of course just as important as the integrity of the physical attributes of the system. Development of software must be carried out to in such a fashion that its reliable operation is assured, and as this development is normally carried out by commercial suppliers or their contractors, it is advisable for such assurance to be sought as part of the supply contract.

Another aspect of equipment upkeep is that of maintenance: in order for continuous reliable performance to be achieved it is often necessary to specify a regular programme of engineering adjustments, replacement of wearable parts, etc. This may include both in-house activities and scheduled service visits from contractors. It is also important that any changes to the equipment arising from maintenance, overhaul after breakdown, etc., are considered in terms of whether they may affect any qualification work carried out, requalification may be necessary if the changes are sufficiently major. Similarly, it is important to keep a

[1] The nomenclature of equipment qualification activities has been the cause of some confusion, and conflicting definitions have been employed by various authorities. In a relatively common model, equipment qualification is viewed as being made up of three constituent activities: installation qualification (IQ), which demonstrates that the equipment has been assembled and installed correctly; operational qualification (OQ), which demonstrates that the key performance attributes of the equipment are achieved (e.g. pump flow rates, mixer rotation speeds), and performance qualification (PQ), which demonstrates that the equipment is actually capable of performing processes representative of those for which it is to be used.

record of any changes to equipment layout, etc.; this is relatively trivial for small, self-contained items, but for larger pieces of plant a comprehensive and accurate set of drawings should be maintained for this purpose.

Another area which requires careful thought is that of cleaning and sanitization of equipment before or after use: the procedures and reagents used to carry out these operations can be highly critical in their effects on production processes. On the one hand it is important to ensure that the cleaning process is effective, so that the equipment is returned to its original state before it is next used; also that microbiological contamination is minimised by the sanitization method. In particular the possibility of cross-contamination must be carefully dealt with when the same equipment is used for many different processes.

On the other hand, agents used in these operations might typically include strong alkali, detergents and organic solvents, and these are more than capable of wrecking the preparation of a protein reagent in themselves. Therefore the procedures used to flush away the cleaning agents before the next production run are at least as important as the cleaning methods *per se*.

13.4.2 Relationship of equipment variation to process latitudes

In Section 13.3 it was noted that key variables in raw materials such as purity and functional activity should be considered as inputs into the assessment of process robustness. Exactly the same applies to the variation of equipment performance characteristics, and the likely magnitude of such variations is an important factor in deciding the design of process latitude studies. If it is known that a manufacturing dispenser can only (with a reasonable degree of confidence) be expected to deliver 100 ($\pm$ 5)% of the target volume in a tube coating operation, it is necessary to challenge the robustness of product performance to variations of at least this magnitude at the R&D stage.

Similarly if it has been established that, under normal operating conditions, a pH meter/electrode combination is reliably capable of determining pH to $\pm$ 0.1 units, then the latitude study for a process where such a determination is critical should be designed to challenge at least this range.

For the examples quoted it is fairly simple to define which parameters of the equipment are likely to be critical, and to mimic them using simple, bench-scale equipment. This is not always the case: the efficiency of mixing operations, for example, depends on a wide range of factors. For an overhead stirrer, these are likely to include paddle design, vessel geometry, and exact paddle position within the vessel, as well as the more obvious variables such as rotation rate, mixing time, solution volume, etc. As may be imagined, it is not always straightforward to mimic such operations on the small scale. Sometimes there is no alternative to carrying out the process studies on the full-scale equipment itself, or pilot plant which has been demonstrated to behave in the same manner. Operations involving aspiration of solutions from filled vessels are another difficult area, their performance can be heavily affected the exact design and geometry of the equipment, pressure supply, interdependence of multiple aspirators, etc. Once again it may be necessary to use the final equipment itself to assess process robustness during such stages, unless appropriate pilot plant is available.

The availability or non-availability of suitable pilot plant for such experiments is a key factor in assuring the robustness of manufacturing processes.

Another point worth noting is that it is becoming less acceptable to take a 'black box' approach to the behaviour of manufacturing or pilot equipment. An understanding of the KPVs which underlie the performance of the equipment itself gives a greater degree of security than simply basing predictions on its observed output performance to date.

One final point on this topic: while it is commonplace for manufacturing plant to receive detailed attention as to its performance characteristics, it is less frequent for R&D equipment to be treated in the same way, the rationale presumably being that it is somehow 'less important'. However, unrecognised differences in performance between R&D and manufacturing equipment are one of the commonest causes of unforeseen problems, when a process is eventually handed over into the manufacturing environment. Out-of-control equipment may invalidate an otherwise perfect development programme in terms of manufacturability issues. It is thoroughly recommended that R&D equipment is given an equally high priority in terms of the reliability of its performance.

13.5 Environmental Variation

13.5.1 Temperature

It has already been noted that temperature can commonly form a KPV, and at the small scale this variable is often controlled using a suitable piece of equipment such as an adjustable bench-scale incubator. This, of course, is subject to the usual requirements of qualification, checking and calibration, and maintenance. A particular feature of the qualification of such equipment is the need to check temperature at a number of different positions within the incubator; this temperature mapping ensures that there are no 'hotspots' or 'coldspots' which fall outside the desired range.

At a larger scale it is common to provide rooms controlled within a defined temperature range: fixed 2–8 °C coldrooms, adjustable-temperature facilities, etc. In some ways these can simply be viewed as larger incubators, and they can be treated in much the same way. For some operations the temperature may simply be defined by the ambient working conditions of the laboratory or factory. This may be controlled to the same degree as the purpose-built facilities just described, but in many cases the control systems in place are designed primarily to provide a reasonably comfortable working environment for the operators. In such cases there may well be significant seasonal fluctuations in temperature within the facility, and there may also be effects from other factors such as the number of operators present, the operation of pieces of equipment which generate heat[1], the amount of air transport to and from adjacent facilities, etc. If there is no other means of control of the temperature at which an operation is carried out, the potential range of such variations must be understood and taken into account in process latitude studies.

13.5.2 Humidity

Humidity is another parameter which is often defined principally by the working environment; it may not be a key variable for processes carried out in the solution phase, but it can certainly be critical for operations such as drying, storage of dried components, etc. It too can be affected strongly by atmospheric conditions, by the operation of equipment in the laboratory which generate water vapour, and by the *modus operandi* of the air handling system itself. Once again the potential range in humidity that may be observed in the manufacturing environment should be taken into account as a possible KPV, when the challenge of processes such as drying is addressed.

13.5.3 Light

A final parameter which falls in the environmental category is light: UV–visible radiation can cause changes to protein-based reagents, and should therefore also be considered as a potential KPV. This is obviously particularly important where in-process or finished components are exposed to daylight or artificial light for long periods, especially where there may be differences in exposure within or between batches. If light-sensitivity is known to be a possibility, steps should be taken to minimise any potential problems in manufacturing by building in appropriate shielding, etc. and ensuring that exposure times are both minimal and consistent within and between batches. Some care has to be put into studying light effects, as there can obviously be a wide variation in the type of light involved, wavelength distribution, intensity, etc. Intensity falls off rapidly with distance from source, so it is useful in the first instance to carry out challenge experiments using a reasonably powerful source quite close to the 'target' (but beware of heating effects). If there is no observed effect then it is a reasonable assumption that the process is robust to light; if a loss of performance is detected more detailed studies will be necessary to mimic the manufacturing environment more closely.

As with equipment, it is important to pay attention to the environmental conditions under which R&D is carried out. Again, problems can be introduced by performing R&D work under conditions of temperature, humidity or light exposure which do not reflect the manufacturing environment.

13.5.4 Microbiology

It has already been mentioned that microbiological contamination can be a cause of process problems. It is mentioned again here briefly, simply to observe that individual environments have their own particular distribution of microorganisms. This difference from one environment to another may be a source of process variation, if the procedures which have been successfully put in place to control contamination on one site are undermined by the presence of particularly resistant or damaging organisms at another. It is sound practice to build up and maintain a good knowledge of the specific microbiological profile of the manufacturing environment(s), such that the organisms typically encountered can be considered when control measures are decided.

Such problems can obviously be minimised by the use of environmentally controlled areas where bioburden is extremely low, and by good practice in operating procedures, sanitization regimes, etc.

[1] These factors may also be relevant for some of the more closely controlled facilities mentioned above, if they are sufficient to stress the air supply and control systems present.

13.6 Method Definition and Operator Training

Sections 13.2 to 13.5 have listed a large number of factors that can affect the robustness of a manufacturing process in the routine working environment. If sufficient care has been put into controlling the behaviour of the process, the raw materials, the equipment and the environment, it might be considered that consistent performance is assured, and this might possibly be the case for a fully automated process.

However, most processes for the manufacture of protein-based reagents require a considerable level of operator intervention, and consequently a set of instructions must be prepared which instruct the operator what to do, and when. A quite significant proportion of manufacturing problems stem from the way in which these instructions are written, and this is an area where a little more care can be a very worthwhile investment.

Note that this does not imply anything wrong with the operators' powers of comprehension. Rather it is a comment on (a) the imperfection of everyday language as a means of describing technical procedures and (b) the difference between writing a procedure and having to follow it.

13.6.1 Definition of process steps

The most common sources of error are where the description of a particular operation is not clear enough for it to be followed unambiguously. If an instruction states 'filter the solution to 0.22 μm' there is obviously considerable latitude for interpretation on the part of the operator. What design of filter should be used? What type of membrane material? Should the filter be pre-flushed before use, and if so how? If the filter becomes blocked is it permissible to use a second filter? It is apparent that a little more care in the writing of these instructions can add considerably to the chances of the process being carried out reproducibly, even by a number of different operators.

Note that the development work itself must have been carried out in a suitably thoughtful way to allow 'contingency' instructions to be made. Although filtration methods should be introduced which bear little risk of blocking, it is a useful challenge experiment to check that the product is not adversely affected by the use of multiple filters: a clear, justifiable statement can then be made to instruct the operator on what to do in the event of such an occurrence.

The more difficult it is to define a process step unambiguously, the more care needs to be put into framing the instruction carefully. Therefore problems commonly arise from stages where fairly detailed information is required to ensure consistency, such as mixing, filtration, or the choice of fractions from a chromatography run. Obviously writing clear instructions is facilitated if the process step itself is simple, and again it is worthwhile putting a bit of extra effort into designing process stages which are easy to carry out in themselves, and are consequently easy to document accurately.

Similarly it is helpful to rationalise procedures so that (as far as possible) the same approach is applied wherever a particular operation is required. Communication within R&D departments is often somewhat imperfect, and a frequent cause of complaint from manufacturing groups is the number of different ways which have been dreamt up to carry out essentially identical processes. Even in a step as simple as formulating a buffer of given pH and molarity, it is not uncommon to find more than one recipe and formulation procedure. This can only make it more difficult for manufacturing operators to build up a familiarity with the processes which they need to use.

In terms of demonstrating that a process is straightforward to carry out *per se*, and that the instructions for its execution have been written in a clear and unambiguous fashion, there is nothing to beat carrying out the procedure a number of times before it is transferred to the manufacturing environment. Unless there are issues with equipment scale this can be done fairly simply within the R&D environment. It is particularly useful if a number of different operators are involved in these trial runs. The person who has written the process is generally the least effective at spotting possible ambiguities in the instructions!

13.6.2 Generic procedures and operator training

It is never possible to include every single level of detail in the operator instructions, at least if the size of such documents is to be kept to a manageable level. There must also be a level of generic expertise which can be captured by documented standard procedures, and realised by comprehensive operator training programmes. The specific instructions for a process should provide enough information for the operator, given the generic procedures which are in place, and which the operator has been trained in. For example, it is clearly unnecessary to write out the full instructions for a pH determination every time such a step is incorporated in the method; an instruction such as 'determine the pH of the solution as described in Standard Procedure XXX' is good enough, provided that Standard Procedure XXX itself is clear enough in its instructions, and the operator is well-trained in its requirements. Note that this economy of wording in the instruction document is quite different from the 'filter to 0.22 μm' type, which leaves the operator to decide the details without sufficient help.

Quite rightly, the level of training of manufacturing operators (in all procedures, generic and specific) is a key interest of regulatory authorities, simply because it can have such a direct effect on product quality.

13.7 Scaleability

As noted earlier, it is not always straightforward to estimate the optimal scale for the manufacture of a coupled-protein reagent, given the uncertainty in demand. Therefore it makes sense to carry out the initial development on the basis that the scale may have to be altered (ideally upwards) at a later date. Scaleability issues are also relevant where, for example, it is desired to carry out process latitude studies at a reduced scale to economise on time and materials.

Although it is impossible to generalise too much over scaleability issues, it is perhaps worth making a few points that may highlight potential problem areas.

Bulk solution steps at a fixed temperature normally scale up or down very well. So long as concentrations are maintained, it is generally possible to obtain the same results from a larger or smaller total volume. This is particularly true if the vessel material and geometry are kept as constant as possible.

Steps involving a temperature change, however, are notorious for causing scaleability problems. If a reaction mixture is transferred from ambient to 37 °C for a 60-minute incubation, the actual temperature experienced by the solution may vary greatly depending on the scale involved. A small, thin glass vial of solution may equilibrate quite rapidly to the desired temperature, and thus spend most of the 60 minutes close to 37 °C. However, with a large volume of solution contained in a thick bottle made of a more strongly insulating plastic, the contents may never even reach the desired temperature by the end of the incubation period. Clearly this can cause scale-related differences in the products obtained from the reaction.

Temperature equilibrations involving a phase change, such as freezing and thawing are also subject to these scale problems. They may be particularly significant for freezing operations, where the rate of heat transfer may affect the freezing rate, and consequently the size of ice crystals formed, and the amount of damage caused to protein molecules (see Chapter 12, Section 12.3.1).

The importance of mixing during temperature equilibrations can also vary with scale. Mixing the solution may affect the equilibration time significantly where this is achieved slowly, but may have little effect on the outcome if equilibration occurs rapidly. In general, mixing can be problematic to scale up, due to the complexity of interactions between vessel and stirrer geometry, etc.

Filtration processes require a little thought. The pore size itself is not an issue; the biggest potential problems are generally loss of material onto the filter, or leaching of extractables into the solution , or blocking of filters during a run. Scale changes should be carried out with this in mind, e.g. a small-scale latitude study might be executed using a filter area-to-volume ratio higher than that which will be experienced at the full scale. This will represent an appropriate challenge of possible material loss or leaching of extractables—but may well fail to detect incipient blockage problems. Challenging a *lower* filter area-to-volume ratio will address this concern, so it may well be advisable to include both extremes in the study.

Chromatography is another area that can cause difficulties. For non-adsorbing methods such as size exclusion chromatography the rules of thumb are to maintain the linear flow rate of the system (flow rate/cross-sectional area) and the injection volume (as a fraction of column volume). As a secondary consideration it is wise to keep the column geometry as similar as possible; this may have to be somewhat approximate due to the availability of appropriate components, etc.

For many other systems the component of interest is concentrated onto the column in one step, then removed from it in a second. This is commonly the approach for ion exchange, affinity chromatography, etc. In these circumstances the process is generally much less scale-dependent: very large volumes of crude material can be passed down a small column so long as the binding capacity of the component of interest is not exceeded. For this reason, techniques of this type are much better suited to large-scale work than size exclusion, and should be strongly considered for use: if large scale-ups are required in future it will be much easier using a method of this type. Also the volumes of elution buffers required for such methods are much smaller than for size exclusion, etc., and the time savings in buffer preparation generally outweigh any extra complexity from using multi-step elution profiles.

13.8 Overview on Manufacturability of Coupled-protein Reagents

It would be possible to devote a large volume simply to manufacturability issues, and the treatment in this chapter has been necessarily somewhat superficial. However, it is believed that the vast majority of manufacturing problems can be avoided if sufficient care is paid to investigating the robustness of the process to changes in variables (including those arising from raw materials, equipment and the environment), if the instruction methods are written in an unambiguous manner, and if the operators are well trained.

13.8.1 Validation

The concept of validation is frequently encountered, particularly with regard to regulatory requirements; a great deal of debate has taken place on the exact nature of what constitutes validation, and will no doubt continue to do so (DeSain and Sutton, 1994; Levin, 1996; Donowa, 1997). Not least this is because the terminology used by regulatory agencies has not been well defined, and has changed over the years. The term has been deliberately avoided so far in this chapter, in the belief that it is easier to explain once all its constituent parts have been dealt with. The concept of qualification has already been raised with regard to equipment, where it is readily understood in terms of showing that the equipment (including any associated software) has the necessary attributes to carry out the required function. The notion of qualification can be extended over all the other areas discussed: one can consider the facility itself to be qualified in terms of its environmental control, the processes to be qualified in terms of their robustness, the raw materials to be qualified in terms of their properties, and the manufacturing operators to be qualified in terms of their training.

Only once all these qualified elements have been put together, can one consider that the overall manufacturing process has been validated, providing a high degree of assurance that a product with the desired properties will consistently result from the defined procedures.[1]

1 Note that this does not imply that the overall product offering presented to the customer will necessarily do the job expected of it. This is the job of design validation, which rightly now receives an equal degree of attention, but falls outside the scope of this discussion.

Finally, adequate change control mechanisms are necessary to ensure that any future modifications are carried out to the same standards.

With all these elements in place, it is a reasonable hope that the process will meet its expectations in terms of delivery. It is hoped that processes developed in this way will provide a trouble-free manufacturing lifetime of consistent product performance.

Reference

DeSain, C. (1993). *Drug, Device and Diagnostic Manufacturing. The Ultimate Resource Handbook*, 2nd edn, Interpharm Press, Buffalo Grove, IL.

DeSain, C. and Sutton, C. V. (1994). *Validation for Medical Device and Diagnostic Manufacturers*, Interpharm Press, Buffalo Grove, IL.

Donowa, M. E. (1997). Book review: *Validation for Medical Device and Diagnostic Manufacturers* (DeSain, C. and Sutton, C.V.). *Med. Device Technol.*, **November 1997 issue**, 33–35.

Ishikawa, K. (1986). *Guide to Quality Control*, 2nd revised edn. edited for clarity, Asian Productivity Organisation, Tokyo and Quality Resources, New York.

Levin, M. (1996). Validation. In *How to Fail an FDA Quality Audit*, pp. 69–74, Mort Levin Inc., Natick, MA.

Phadke, M. S. (1989). *Quality Engineering Using Robust Design*, Prentice Hall, Engelwood Cliffs, NJ.

Pisano, G. P. (1997). *The Development Factory*, Harvard Business School Press, Boston, MA

Appendix: Protocols

Protocols 3.1–3.4, 4.1–4.3, 5.1–5.4, 6.1–6.3, 8.1

Alastair H. Dent and Mohammed Aslam

Research and Development Department, Ortho-Clinical Diagnostics, Pollards Wood Laboratories, Nightingales Lane, Chalfont St. Giles HP8 4SP, UK

Protocols 7.1–7.4

Adrian Simmonds and Martin Cunningham

Nycomed Amersham plc, White Lion Road, Amersham HP7 9LL,UK
Thanks are due to Ian Durrant for *in situ* hybridisation and biotin phosphoramidite protocols and to Martin Harris, Frank Craig and Bronwen Harvey for contributions to other protocols.

Protocols 9.1–9.14

Leif Bülow

Department of Pure and Applied Biochemistry, Chemical Centre, University of Lund, POB 124, S-221 00 Lund, Sweden

Christer Lindbladh

Astra Draco AB,
POB 34, S-221 00 Lund, Sweden

Protocols 12.1–12.3

Ross H.M. Hatley and Camilo A.L.S. Colaco*

Quadrant Healthcare plc and *Quadrant Research Foundation, Maris Lane, Cambridge CB2 2SY, UK

3.1 Colorimetric Determination of Horseradish Peroxidase Activity (Adapted from Boehringer Mannheim, 1990)

This procedure uses ABTS™ (2,2′-azino-di[3–ethyl-benzthiazoline sulfonate (6)] to calculate horseradish peroxidase (HRP) activity in free enzyme or in conjugates. It is best used as a comparative procedure, for example comparing one batch of HRP with another, or a conjugate with its 'parent' HRP. Often A_{403} can be used in parallel (Protocol 3.2) to estimate the *concentration* of HRP present, and this can be used to normalise the activity readings to give a comparable specific activity, or the percentage of activity retained in a conjugate.

Solutions

ABTS solution: ABTS™ (Boehringer Mannheim product No. 122661) dissolved at 5.0 mg/ml in 100 mM potassium phosphate buffer, pH 5.0; peroxide solution: hydrogen peroxide solution (30%) diluted 1:1000 in redistilled water; buffer A: 40 mM potassium phosphate buffer, pH 6.8 + 0.25% (w/v) BSA + 0.5% (w/v), ice cold.

Protocol

(1) Sample: Dissolve or dilute HRP-containing material to 0.1 mg/ml (HRP) in buffer A. Dilute 1:2000 in buffer A. Note—if the concentration is initially lower than 0.1 mg/ml, use a correspondingly lower dilution to arrive at the same concentration.

(2) Pipette into cuvettes:
  (a) ABTS solution (2.90 ml) + buffer A (0.05 ml);
  (b) ABTS solution (2.90 ml) + sample (0.05 ml).

(3) Mix and incubate at 25 °C until there is no sign of further pre-reaction.

(4) Add peroxide solution (0.10 ml) to each cuvette and mix. Place each cuvette in one cell of the spectrophotometer.

(5) Record the increase in A_{405} (sample − blank) over several minutes. Plot and calculate the rate of change as ΔA_{405}/min from the linear portion of the curve. A range of around 0.025–0.035 is expected for HRP itself.

Notes

(1) A twin-beam spectrophotometer is assumed for this protocol, but it can be adapted simply for use with a single-beam instrument by running sample and blank sequentially.

(2) If the absorbance is too low under these conditions, repeat using a lower dilution, and include the dilution difference in the calculation of activity relative to the sample.

(3) ABTS solutions are very vulnerable to contamination: new or thoroughly cleaned glassware is recommended.

3.2 Determination of Horseradish Peroxidase Concentration by A_{403} and Calculation of Rz

HRP has a strong absorbance with λ_{max} at around 403 nm, depending on the exact conditions. As this is well removed from many common absorbances, it can often be used as a convenient means of estimating HRP concentration.

The protocol is facile: simply measure A_{403} of the sample using a suitable spectrophotometer, diluting if necessary to fall within the working range of the instrument.

A 1 mg/ml solution of HRP typically gives an A_{403} of around 1.6–1.9; the figure varies somewhat from batch to batch. For detailed work it is therefore useful to carry out a calibration exercise for the specific batch being used, ideally in the same buffer as the samples of interest.

Note that the position and intensity of this absorbance can be affected by derivatisation or chemical treatment.

Rz (A_{403}/A_{280}) gives an indication of HRP purity (or level of incorporation in an HRP conjugate with another protein not absorbing in this region). Pure HRP has an Rz approaching 3.5.

3.3 Colorimetric Determination of Alkaline Phosphatase Activity (adapted from Boehringer Mannheim, 1990)

This protocol uses *p*-nitrophenyl phosphate (PNPP) as an indicator of alkaline phosphatase (AP) activity. Again (cf. Protocol 3.1) it is best used as a comparative procedure.

Solutions

PNPP solution: *p*-nitrophenyl phosphate disodium salt (Boehringer Mannheim product No. 004847), 250 mg/ml in redistilled water; buffer A: 1 M diethanolamine, pH 9.8 + 0.5 mM $MgCl_2$; incubate at 37 °C; buffer B: 100 mM TEA, pH 7.6 + 1 mM $MgCl_2$ + 0.1 mM $ZnCl_2$ (N.B. the zinc chloride solution should be prepared freshly before use).

Protocol

(1) Sample: dissolve or dilute AP-containing material to 0.1 mg/ml (AP) in buffer A. Dilute 1:5000 in buffer B. Note—if the concentration is initially lower than 0.1 mg/ml, use a correspondingly lower dilution to arrive at the same concentration. Allow the sample to stand for 15–20 min at room temperature before use.

(2) Pipette into each of two cuvettes PNPP solution (0.05 ml) + buffer A (3.00 ml). Mix and incubate at 37 °C.

(3) Add sample (0.05 ml) to one cuvette and buffer B (0.05 ml) to the other. Mix. Place each cuvette in one cell of the spectrophotometer.

(4) Record the increase in A_{405} (sample − blank) over several minutes. Plot and calculate the rate of change as ΔA_{405}/min from the linear portion of the curve.

Notes

(1) A twin-beam spectrophotometer is assumed for this protocol, but it can be adapted simply for use with a single-beam instrument by running sample and blank sequentially.

(2) If the absorbance is too low under these conditions, repeat using a lower dilution, and include the dilution difference in the calculation of activity relative to the sample.

3.4 Colorimetric Determination of β-Galactosidase Activity (adapted from Boehringer Mannheim, 1990)

This is a procedure analogous to Protocol 3.3, using *o*-nitrophenyl-β-D-galactopyranoside (ONPG) as a substrate for β-galactosidase (β-Gal).

Solutions

ONPG solution: *o*-nitrophenyl-β-D-galactopyranoside (Boehringer Mannheim product No. 810690), 6 mg/ml in 50 mM potassium phosphate buffer, pH 7.8; This solution should only be slightly yellow coloured; 10 M mercaptoethanol: 2-mercaptoethanol (69.8 ml) made up to 100 ml with redistilled water; $MgCl_2$ solution: 10 mM $MgCl_2$ in redistilled water; buffer A: 50 mM potassium phosphate buffer, pH 7.8 (keep part ice cold, incubate remainder at 37 °C).

Protocol

(1) Sample: dissolve or dilute β-Gal-containing material to 0.1 mg/ml (β-Gal) in ice-cold buffer A. Dilute 1:50 in ice-cold buffer A, immediately prior to measurement. Note—if the concentration is initially lower than 0.1 mg/ml, use a correspondingly lower dilution to arrive at the same concentration. Sample buffer effects may become significant at dilutions much below this level, however.

(2) Pipette into each of two cuvettes buffer A (2.20 ml), $MgCl_2$ solution (0.30 ml), ONPG solution (0.40 ml) and 10 M mercaptoethanol (0.03 ml). Mix and incubate at 37 °C.

(3) Add sample (0.05 ml) to one cuvette and buffer A (0.05 ml) to the other. Mix. Place each cuvette in one cell of the spectrophotometer.

(4) Record the increase in A_{405} (sample − blank) over several minutes. Plot and calculate the rate of change as ΔA_{405}/min from the linear portion of the curve.

Notes

(1) A twin-beam spectrophotometer is assumed for this protocol, but it can be adapted simply for use with a single-beam instrument by running sample and blank sequentially.

(2) If the absorbance is too low under these conditions, repeat using a lower dilution, and include the dilution difference in the calculation of activity relative to the sample.

4.1 Fragmentation of IgG to Yield Fab Fragments

As discussed in Chapter 4, Section 4.5.1.2, the resistance of antibodies to papain digestion varies widely with subclass and species origin. For mouse IgG, the order of resistance to digestion is $IgG_{2b} < IgG_3 < IgG_{2a} < IgG_1$, and for human IgG the order is IgG_1 and $IgG_3 < IgG_4 <$ IgG2.

The following is a protocol which will provide good results within the time window studied for many examples of human and mouse IgG. Some options for optimising the procedure in the event of failure are given in the notes.

Fab fragments contain free thiols which can form a useful site for derivatisation. Alternatively, alkylation (see Protocol 4.3) can be employed to yield unreactive fragments which are stable to recombination.

Solutions

Papain solution: papain (e.g. Calbiochem product No. 5125), 1 mg/ml in 100 mM phosphate buffer + 20 mM cysteine hydrochloride + 10 mM EDTA, pH 6.2, incubated at 37 °C. To obtain the correct additive molarities, add cysteine hydrochloride hydrate (M_r 157.6) at 3.2 mg/ml and EDTA disodium salt dihydrate (M_r 372.2) at 3.7 mg/ml.

Protocol—test fragmentation

(1) Prepare a 2–3 mg sample of IgG in 100 mM phosphate buffer + 20 mM cysteine hydrochloride + 10 mM EDTA, pH 6.2, ideally at 10–15 mg/ml. Presentation in similar buffers with the correct concentrations of cysteine and EDTA additives is also acceptable. The pH specificity of papain is broad, and successful fragmentations have been carried out over the range 5–8. Lower IgG concentration is also acceptable, but will give longer reaction times.

(2) To the IgG sample add sufficient papain solution to give an IgG:papain ratio of 100:1 (w/w); e.g. for a 2 mg IgG sample add 20 μl of papain solution to yield 0.02 mg of papain.

(3) After 1 h, pipette out a 50 μl sample and analyse immediately (see step 5).

(4) Remove further samples after 2, 4, 8, 24 and 48 h, in each case.

(5). Analyse the samples to track the progress of the fragmentation process. Ideally this can be carried out using SDS-PAGE under non-reducing conditions. The PhastTM system (Amersham Pharmacia Biotech UK) provides a convenient means of carrying out the analysis. Sample preparation for SDS-PAGE involves taking a 20 ml sample of the quenched fragmentation sample, adding 20 ml of a solution of sodium dodecyl sulfate (25 mg/ml) and incubating for 10 min.

If using the Phast system, run the digestion samples on a Gradient 10–15 PhastGel, along with a sample of the whole IgG prepared in the same way, and molecular mass markers covering at least 15–200 kDa. Stain with Coomassie Blue.

Fragmentation of the IgG is indicated by the loss of the 150–160 kDa band of intact IgG, and the appearance of Fab by a new band at 100–120 kDa. The Fc product is typically broken down into small peptide fragments and often will not be observed as a discrete band on the gel.

(6) Choose the earliest set of conditions where the fragmentation has gone to completion. Some degree of compromise may be necessary between under-digestion (intact IgG remaining) and over-digestion (further fragmentation of $F(ab')_2$ product).

Protocol—bulk fragmentation

(7) Repeat the fragmentation at the required bulk scale, using the incubation time identified in the test protocol. Take particular care to ensure that the fragmentation temperature reflects that of the test protocol by thorough equilibration of the citrate buffer at 37 °C.

(8) Purification of the fragmented IgG can be achieved by size exclusion chromatography, the 100–120 kDa $F(ab')_2$ fragment being readily separable from peptide fragments (pFc) and pepsin (M_r 35 kDa). Use of a buffer at pH 7.0–7.5 is particularly convenient if reduction to Fab′ is to follow.

Notes

(1) In the event of over-digestion under the conditions specified, try repeating the test protocol at a higher pH (e.g. 4.0 or 4.5), or reducing the pepsin concentration. Note that the activity of pepsin drops rapidly with increasing pH.

(2) In the event of under-digestion, increase the pepsin concentration or employ a longer incubation time.

4.2 Fragmentation of IgG to Yield $F(ab')_2$ Fragments, and Reduction to *Fab′*

As discussed in Chapter 4, Section 4.5.1.2, the resistance of antibodies to peptic digestion varies widely with subclass and species origin. For mouse IgG, the order of resistance to digestion is IgG2b < IgG_3 < IgG_{2a} < IgG_1. With IgG_{2b} it is not generally possible to prepare $F(ab')_2$ fragments; instead, a Fab/c fragment combining intact heavy chain with one Fab fragment may be obtained (Parham, 1986).

The following is a protocol which will provide good results within the time window studied for many examples of human IgG and mouse IgG_1 and IgG_{2a}. For mouse IgG_3 there is little point examining time courses greater than 1–2 h and the protocol should be amended accordingly if it is known that this subclass is the target for fragmentation. Some options for optimising the procedure in the event of failure are given in the notes.

A procedure for reduction to Fab′ is also provided—this yields free thiols which can form a useful site for derivatisation. Alternatively alkylation (see Protocol 4.3) can be employed to yield unreactive fragments which are stable to recombination.

Solutions

Pepsin solution: pepsin (e.g. Calbiochem product No. 516360), 25 μg/ml in 100 mM citrate buffer, pH 3.5, incubated at 37 °C; Tris buffer: 2 M Tris buffer, pH 8.6.

Protocol—test fragmentation

(1) Prepare a 2–3 mg sample of IgG in 100 mM citrate, pH 3.5, ideally at 10–15 mg/ml. Presentation in low ionic strength buffer at a different pH (e.g. 10 mM phosphate, pH 7.0) is also acceptable. Lower IgG

concentration is also acceptable, but will give longer reaction times.

(2) Add 423 μl of pepsin solution to 77 μl of IgG solution and incubate at 37 °C.

(3) After 15 min, pipette out a 50 μl sample and add to a test tube containing 50 μl of Tris buffer. Mix immediately by vortexing.

(4) Remove further samples after 30 min, and 1, 2, 4, 8, 24 and 48 h, in each case quenching the reaction by adding to a tube containing 50 ml of Tris buffer and mixing.

(5) Analyse the samples to track the progress of the fragmentation process. Ideally this can be carried out using SDS-PAGE under non-reducing conditions. The PhastTM system (Amersham Pharmacia Biotech UK) provides a convenient means of carrying out the analysis. Sample preparation for SDS-PAGE involves taking a 20 ml sample of the quenched fragmentation sample, adding 20 ml of a solution of sodium dodecyl sulfate (25 mg/ml) and incubating for 10 min.

If using the Phast system, run the digestion samples on a Gradient 10–15 PhastGel, along with a sample of the whole IgG prepared in the same way, and molecular mass markers covering at least 15–200 kDa. Stain with Coomassie Blue.

Fragmentation of the IgG is indicated by the loss of the 150–160 kDa band of intact IgG, and $F(ab')_2$ by the appearance of a new band at 100–120 kDa. The Fc product is typically broken down into small peptide fragments and often will not be observed as a discrete band on the gel.

(6) Choose the earliest set of conditions where the fragmentation has gone to completion. Some degree of compromise may be necessary between under-digestion (intact IgG remaining) and over-digestion (further fragmentation of $F(ab')_2$ product).

Protocol—bulk fragmentation

(7) Repeat the fragmentation at the required bulk scale, using the incubation time identified in the test protocol. Take particular care to ensure that the fragmentation temperature reflects that of the test protocol by thorough equilibration of the citrate buffer at 37 °C.

(8) Purification of the fragmented IgG can be achieved by size exclusion chromatography, the 100–120 kDa $F(ab')_2$ fragment being readily separable from peptide fragments (pFc) and pepsin (M_r 35 kDa). Use of a buffer at pH 7.0–7.5 is particularly convenient if reduction to Fab′ is to follow.

Notes

(1) In the event of over-digestion under the conditions specified, try repeating the test protocol at a higher pH (e.g. 4.0 or 4.5), or reducing the pepsin concentration. Note that the activity of pepsin drops rapidly with increasing pH.

(2) In the event of under-digestion, increase the pepsin concentration or employ a longer incubation time.

Reduction to Fab′

Mild reduction of the $F(ab')_2$ product can cleave the disulfide bonds linking the heavy chain fragments, while leaving the intrachain and heavy-light links intact.

(9) For reduction the $F(ab')_2$ fragments should be presented in a high ionic strength buffer (100 mM) at around pH 7.0–7.5. If the product is not already in such a buffer, buffer-exchange or adjust the pH accordingly. Add EDTA disodium salt dihydrate (M_r 372.2) at 3.7 mg/ml (10 mM) and stir to dissolve. Incubate the solution at 37 °C.

(10) Prepare an 11.4 mg/ml (100 mM) solution of 2 mercaptoethylamine hydrochloride (M_r 113.6) in a high ionic strength (100 mM) buffer at around pH 7.0–7.5, typically the buffer already employed for fragment presentation. Check the pH and readjust if necessary. Incubate the solution at 37 °C.

(11) Add mercaptoethylamine solution at 10% (v/v) to the $F(ab')_2$ solution, giving a final concentration of 10 mM.

(12) Incubate at 37 °C for 2 h. If desired the progress of the reduction can be tracked using SDS-PAGE as described above; the $F(ab')_2$ band at 100–120 kDa will be replaced by a Fab′ band at 50–60 kDa.

(13) Purify as necessary.

4.3 Fragmentation of IgG to Yield Half-molecules and Alkylation with Iodoacetamide

The method just described for reduction of $F(ab')_2$ fragments can also be used for intact IgG; because the disulfide bonds linking the heavy and light chains are generally the most sensitive to reduction, mild treatment can generate 'half-molecules' (IgG/2) combining one intact heavy chain and one intact light chain. This protocol also includes an alkylation step, providing fragments which are stable to recombination—this may be useful if it is not desired to use the fragment thiols for chemical derivatisation. The method can also be applied to Fab or Fab′ fragments as generated from Protocols 4.1 and 4.2.

Protocol

Reduction

(1) For reduction the IgG should be presented in a high ionic strength buffer (100 mM) at around pH 7.0–7.5. If the product is not already in such a buffer, buffer-exchange or adjust the pH accordingly. Add EDTA disodium salt dihydrate (M_r 372.2) at 3.7 mg/ml (10 mM) and stir to dissolve. Incubate the solution at 37 °C.

(2) Prepare an 11.4 mg/ml (100 mM) solution of 2-mercaptoethylamine hydrochloride (M_r 113.6), for example Aldrich product No. 12,292–0, in a high ionic strength (100 mM) buffer at around pH 7.0–7.5. Typically use the buffer already employed for IgG presentation, or 100 mM phosphate buffer, pH 7.5. Check the pH and readjust if necessary. Incubate the solution at 37 °C.

(3) Add mercaptoethylamine solution at 10% (v/v) to the IgG solution, giving a final concentration of 10 mM.

(4) Incubate at 37 °C for 2 h. If desired, the progress of the reduction can be tracked using SDS-PAGE as described above; the IgG band at 150–160 kDa will be replaced by an IgG/2 band at 75–80 kDa.

(5) Purify as necessary, or continue to the alkylation step. Simple size exclusion methods are adequate for separation of IgG/2 fragments from mercaptoethylamine, and may also resolve the product from traces of unreacted IgG.

Alkylation

(6) Add iodoacetamide (M_r 185.0), for example Aldrich product No. I-670-9, to the IgG solution at 11.1 mg/ml, giving a final concentration of 60 mM. Stir to dissolve. Check the pH and readjust if necessary.

(7) Incubate at 37 °C for 2 h. If desired, the progress of the alkylation can be tracked using a suitable thiol assay—for example a dipyridyl disulfide method using Aldrithiol-4™ from Aldrich (product No. 14,305–7).

(8) Purify as necessary. Simple size exclusion methods are adequate for the separation of alkylated fragments from iodoacetamide and low molecular weight reaction products, and may also resolve the product from traces of unreacted IgG.

5.1 General Method for Coupling Two Proteins by Thiol–Maleimide Coupling, Incorporating a General Procedure for Protein Thiolation

Obviously it is difficult to write a generic protocol to cover the wide range of protein–protein conjugations that might be attempted. However, if both proteins can be presented at a reasonable concentration in a neutral buffer, this procedure should at least give a product of some sort, and can act as a basis for further optimisation. Some suggestions are given on how to approach reoptimisation.

No comment can be made on the purification of the conjugate as this will depend totally on the proteins concerned.

Solutions

Buffer A: 100 mM phosphate buffer, pH 7.5; buffer B: 100 mM phosphate buffer, pH 6.0 + 5 mM EDTA; SATA solution: SATA (e.g. Calbiochem product No. 573100, Pierce product No. 26102), 40 mg/ml in DMF; deprotection reagent: 2.5 M hydroxylamine + 50 mM EDTA, adjusted to pH 7.0; SMCC solution: SMCC (e.g. Calbiochem product No. 573114, Pierce product No. 22360), 5 mg/ml in DMF.

Protocol

(A) Thiolation of Protein X

(1) Dissolve 5 mg of Protein X in 1 ml of buffer A. Other similar presentations at around pH 7–8 are acceptable, and concentrations down to around 1 mg/ml can be used without affecting the reaction rate significantly.

(2) If the molecular mass of Protein X is known, calculate the quantity of SATA (M_r 231.2) required to provide a 50:1 excess over the 5 mg sample.

(3) Calculate the volume of SATA solution required to provide this quantity, and add to the Protein X solution. If the molecular mass is not known, add 10 μl of SATA solution.

(4) Incubate for 1 h at 25 °C.

(5) Add 100 μl of deprotection reagent and incubate for 15 min at 25 °C.

(6) Separate the thiolated Protein X from low molecular mass byproducts, ideally by a small desalting (size exclusion) column (e.g. PD-10 from Amersham Pharmacia Biotech UK). Equilibrate and elute with buffer B.

(B) Maleimide derivatisation of Protein Y

(1) Dissolve 5 mg of Protein Y in 1 ml of buffer A. Other similar presentations at around pH 7–8 are acceptable, and concentrations down to around 1 mg/ml can be used without affecting the reaction rate significantly.

(2) If the molecular mass of Protein Y is known, calculate the quantity of SMCC (M_r 334.3) required to provide a 50:1 excess over the 5 mg sample.

(3) Calculate the volume of SMCC solution required to provide this quantity, and add to the Protein Y solution. If the molecular mass is not known, add 10 μl of SATA solution.

(4) Incubate for 1 h at 25 °C

(5) Separate the maleimide-derivatised Protein Y from low molecular mass byproducts, as in (A).

(C) Conjugation

(1) Mix the thiolated Protein X solution and the maleimide-derivatised Protein Y solution.

(2) Incubate for 120 minutes at 25 °C.

(3) Purify the conjugate as required.

Notes

(1) If the concentration of either protein is below 1 mg/ml, it may be beneficial to increase the quantity of SATA or SMCC added.

(2) In the event of under-conjugation, increase the quantities of SATA and/or SMCC and repeat.

(3) In the event of over-conjugation, reduce the quantities of SATA and/or SMCC.

(4) Variation in the Protein X:Protein Y ratio is a useful means of modulating conjugate properties: retain the conditions in (A) and (B) but try conjugations with varying combinations of subaliquots of activated protein solutions, e.g. covering a range of Protein X:Protein Y ratios from 10:1 to 1:10. Thiolating Protein Y and maleimide-derivatising Protein X is another alternative.

(5) Other thiol-reactive succinimide esters from Chapter 5, Section 5.7.1, can be substituted for SMCC, including other maleimide-containing examples, or the disulfide-based SPDP.

5.2 General Method for Coupling Two Proteins by Two-step Glutaraldehyde Coupling

As noted in Chapter 5, Section 5.3.1, glutaraldehyde is a poorly characterised coupling agent and its chemistry is complex and often unpredictable. This protocol is included, however, owing to the enduring popularity of these methods.

It describes a mild two-step conjugation procedure, with fairly low pH and short reaction times. Owing to the complexity of the reaction, no attempt is made to relate the quantities used here to molar excesses.

As for Protocol 5.1, this procedure should at least give a product of some sort, and can act as a basis for further optimisation. Some suggestions are given on how to approach reoptimisation.

No comment can be made on the purification of the conjugate as this will depend totally on the proteins concerned.

Solutions

Buffer A: 100 mM phosphate buffer, pH 7.2; glutaraldehyde solution: 25% (w/v) solution in water. These solutions are generally stored frozen, and should be thawed at 2–8 °C before use. Agitate after thawing to ensure homogeneity.

Protocol

(A) Glutaraldehyde activation of Protein X

(1) Dissolve 5 mg of Protein X in 1 ml of buffer A. Other similar presentations at around pH 7–8 are acceptable.

(2) Add 10 μl of glutaraldehyde solution. This gives a final glutaraldehyde concentration of around 0.25%

(3) Incubate for 1 h at 25 °C.

(4) Separate the glutaraldehyde-activated Protein X from low molecular mass byproducts, ideally by a small desalting (size exclusion) column (e.g. PD-10 from Amersham Pharmacia Biotech UK). Equilibrate and elute with buffer A.

(B) Conjugation

(5) If Protein Y is available in solid form, simply add the eluate from (A) directly to 5 mg of Protein Y. Alternatively, presentations of Protein Y in buffers at pH 7–10 are acceptable, e.g. dialysis or buffer exchange into buffer A.

(6) Incubate for 4 h at 25 °C.

(7) Quench by adding 10 μl of ethanolamine.

(8) Purify as necessary.

Notes

(1) In the event of under-conjugation, increase the incubation times for the activation and/or conjugation stages, or increase the concentration of glutaraldehyde added in stage (A).

(2) In the event of over-conjugation, reduce the concentration of glutaraldehyde added in stage (A), or reduce the incubation times for the activation and/or conjugation stages.

(3) Variation in the Protein X:Protein Y ratio is a useful means of modulating conjugate properties: retain the

conditions in (A), but vary the ratio of activated Protein X added to Protein Y, e.g. covering a range of Protein X:Protein Y ratios from 10:1 to 1:10. Activating Protein Y instead of Protein X is another possibility.

5.3 General Method for Coupling Two Proteins (at Least One a Glycoprotein) by Periodate Coupling

Periodate coupling has disadvantages in the level of control achievable, often evidenced by a heterogeneous conjugate population. However, it allows glycoproteins to be coupled without derivatisation of amines.

As for Protocol 5.1, this generic procedure should at least give a product of some sort, and can act as a basis for further optimisation. This protocol describes a mild periodate method using a low concentration of the oxidising agent; this avoids unnecessary damage to the protein structure. Some suggestions are given on how to approach reoptimisation.

No comment can be made on the purification of the conjugate as this will depend totally on the proteins concerned.

Solutions

Buffer A: 100 mM acetate buffer, pH 5.5, incubated at 2–8 °C; buffer B: 100 mM carbonate/bicarbonate buffer, pH 9.5; periodate solution: sodium periodate, 21.4 mg/ml in water (= 100 mM); borohydride solution: sodium borohydride, 3.8 mg/ml in water (= 100 mM).

Protocol

(A) Periodate activation of Glycoprotein X

(1) Dissolve, dilute or buffer exchange 5 mg of Glycoprotein X into 1 ml of buffer A.

(2) Add 20 μl of periodate solution, to give a final concentration of 2 mM periodate.

(3) Incubate for 20 min in the dark.

(4) Separate the periodate-oxidised Glycoprotein X from low molecular mass byproducts, ideally by a small desalting (size exclusion) column (e.g. PD-10 from Amersham Pharmacia Biotech UK). Equilibrate and elute with buffer B.

(B) Conjugation

(5) If Protein Y is available in solid form, simply add the eluate from (A) directly to 5 mg of Protein Y. Alternatively, presentations of Protein Y in buffers at pH 9–10 are acceptable, e.g. dialysis or buffer exchange into buffer B.

(6) Incubate for 180 min at 2–8 °C.

(7) Add 20 μl of borohydride solution.

(8) Incubate for 15 min at 2–8 °C.

(9) Add sufficient 1 M sodium dihydrogen orthophosphate solution to bring the pH below 7.0; this will quench excess borohydride.

(10) Purify as required.

Notes

(1) In the event of under-conjugation, increase the quantity of periodate added, or the incubation time, in stage (A).

(2) In the event of over-conjugation, reduce the quantity of periodate added, or the incubation time, in Stage (A).

(3) Variation in the Glycoprotein X:Protein Y ratio is a useful means of modulating conjugate properties: retain the conditions in (A), but vary the ratio of activated Protein X added to Protein Y, e.g. covering a range of Protein X:Protein Y ratios from 10:1 to 1:10. If both proteins are glycoproteins, activating Protein Y instead of Protein X is another possibility.

5.4 General Method for Coupling Two Proteins by Benzoquinone Coupling

Benzoquinone coupling may offer a useful alternative to some of the better-known conjugation procedures—products with quite different properties are sometimes obtained.

As for Protocol 5.1, this procedure should at least give a product of some sort, and can act as a basis for further optimisation. Some suggestions are given on how to approach reoptimisation.

No comment can be made on the purification of the conjugate as this will depend totally on the proteins concerned.

Solutions

Buffer A: 100 mM phosphate buffer, pH 6.0; buffer B: 100 mM carbonate–bicarbonate buffer, pH 9.5; benzoquinone (BQ) solution: *p*-benzoquinone, 30 mg/ml in DMF. Prepare immediately before use and protect from light.

Protocol

(A) Benzoquinone activation of Protein X

(1) Dissolve 5 mg of Protein X in 1 ml of buffer A. Other similar presentations at around pH 6–7 are acceptable, and concentrations down to around 1 mg/ml can be used without affecting the reaction rate significantly.

(2) If the molecular mass of Protein X is known, calculate the quantity of BQ (M_r 108.1) required to provide a 300:1 excess over the 5 mg sample.

(3) Calculate the volume of BQ solution required to provide this quantity, and add to the Protein X solution. If the molecular mass is not known, add 35 μl of BQ solution.

(4) Incubate for 1 h at 25 °C in the dark.

(5) Separate the BQ-activated Protein X from low molecular mass byproducts, ideally by a small desalting (size exclusion) column (e.g. PD-10 from Amersham Pharmacia Biotech UK). Equilibrate and elute with buffer B. Note that the column will become discoloured owing to reaction of the media carbohydrate residues with BQ.

(B) Conjugation

(6) If Protein Y is available in solid form, simply add the eluate from (A) directly to 5 mg of Protein Y. Alternatively, presentations of Protein Y in buffers at pH 9–10 are acceptable, e.g. dialysis or buffer exchange into buffer B.

(7) Incubate for 16–24 h at 37 °C.

(8) Purify as required.

Notes

(1) In the event of under-conjugation, increase the quantity of BQ added, or the reaction time in (A). Increasing the pH in (A) is a last resort, but may lead to extensive cross-linking of Protein X.

(2) In the event of over-conjugation, reduce the quantity of BQ added, or the reaction time, in (A).

(3) Variation in the Protein X:Protein Y ratio is a useful means of modulating conjugate properties: retain the conditions in (A), but vary the ratio of activated Protein X added to Protein Y, e.g. covering a range of Protein X:Protein Y ratios from 10:1 to 1:10. Activating Protein Y instead of Protein X is another possibility.

(4) Because BQ will react with carbohydrate groups under aqueous conditions, glycoproteins with a high carbohydrate content may exhibit a higher than typical degree of reaction.

6.1 General Method for Coupling a Carboxylic Small Molecule to a Protein by DCC–NHS Coupling

This is an excellent method for coupling carboxylic molecules (referred to here as 'hapten') to proteins. Under the anhydrous conditions of stage (A), quantitative conversion to the succinimide ester can be expected, generally leading to good yields in the subsequent reaction with protein.

This protocol employs a lower conjugation pH than many in the literature. These neutral conditions are preferred as they allow perfectly adequate reaction of amines with the active ester function (cf. Protocol 5.1), while minimising stress to the protein molecule.

Solutions

NHS solution: NHS, 5 mg/ml in dry DMF; DCC solution: DCC, 10 mg/ml in dry DMF; buffer A: 100 mM phosphate, pH 7.0.

Protocol

(A) Activation of hapten

(1) Calculate the mass of hapten representing a 50:1 excess over 5 mg of the protein to be labelled. Dissolve this quantity of hapten in a minimal quantity of dry DMF.

(2) Calculate the mass of NHS (M_r 115.1) representing a 1.1–fold molar excess over the amount of hapten.

(3) Calculate the volume of NHS solution required to yield this mass, and add this volume to the hapten solution.

(4) Calculate the mass of DCC (M_r 206.3) representing a 1.1–fold molar excess over the amount of hapten.

(5) Calculate the volume of DCC solution required to yield this mass, and add this volume to the hapten–NHS solution.

(6) Mix and incubate the mixture for 3 h at 25 °C. Ensure that the reaction mixture is tightly sealed to prevent the ingress of moisture.

(7) Some dicyclohexylurea (DCU) may have precipitated by this time; this may be filtered off, but this is not essential.

(8) It is important that the active ester solution is used quickly once exposed to the atmosphere.

(B) Conjugation

(9) Dissolve 5 mg of the protein in 1 ml of buffer A. Other similar presentations at around pH 7–9 are acceptable, and concentrations down to around 1 mg/ml can be used without affecting the reaction rate significantly.

(10) Add the entire activated hapten solution and incubate the mixture for 1 h at 25 °C. Some cloudiness is often observed on mixing.

(11) Purify as required. An initial filtration is advisable to remove precipitated DCU. Size exclusion methods are often perfectly adequate for separating the labelled protein from unreacted hapten and other low molecular mass byproducts.

Notes

(1) In the event of under-derivatisation, increase the hapten:protein ratio, or (failing this) the pH in stage (B).

(2) In the event of over-derivatisation, reduce the hapten:protein ratio.

(3) Good-quality DCC and NHS are key to successful results with this method—they should both be stored desiccated, and used within a reasonably short period after purchase. Resublimation of DCC can restore quality of deteriorated material.

(4) Similarly, DMF containing moisture is disastrous—again, this material should be stored carefully.

6.2 General Method for Coupling a Carboxylic Small Molecule to a Protein by Mixed Anhydride Coupling with Isobutyl Chloroformate

The mixed anhydride approach using isobutyl chloroformate (IBCF) is another reliable procedure for coupling carboxylic molecules (here referred to as 'hapten') to proteins.

Solution

Buffer A: 100 mM phosphate, pH 7.0.

Protocol

(A) Activation of hapten

(1) Calculate the mass of hapten representing a 50:1 excess over 5 mg of the protein to be labelled. Dissolve this quantity of hapten in a minimal quantity of dry, ice-cold DMF.

(2) Calculate the mass of tri-*n*-butylamine (TBA) (M_r 185.4) representing a 1.1-fold molar excess over the amount of hapten.

(3) Calculate the volume of TBA (ρ = 0.778 g/ml) required to yield this mass, and add this volume to the hapten solution.

(4) Calculate the mass of IBCF (M_r 136.6) representing a 1.1-fold molar excess over the amount of hapten.

(5) Calculate the volume of IBCF (ρ = 1.053 g/ml) required to yield this mass, and add this volume to the hapten–NHS solution.

(6) Mix and incubate the mixture for 20 min, ideally at −5 °C, although ice-bath temperatures or 2–8 °C refrigeration are often adequate. Ensure that the reaction mixture is tightly sealed to prevent the ingress of moisture.

(7) It is important that the mixed anhydride solution is used quickly once exposed to the atmosphere.

(B) Conjugation

(8) Dissolve 5 mg of the protein in 1 ml of buffer A. Other similar presentations at around pH 7–9 are acceptable, and concentrations down to around 1 mg/ml can be used without affecting the reaction rate significantly.

(9) Add the entire activated hapten solution and incubate the mixture for 1 h at 2–8 °C.

(10) Purify as required. Size exclusion methods are often perfectly adequate for separating the labelled protein from unreacted hapten and other low molecular mass byproducts.

Notes

(1) In the event of under-derivatisation, increase the hapten:protein ratio, or (failing this) the pH in stage (B).

(2) In the event of over-derivatisation, reduce the hapten:protein ratio.

(3) Good-quality IBCF and TBA are key to successful results with this method—they should both be stored in a moisture-free environment such as a vacuum desiccator, and used within a reasonably short period after purchase.

(4) Similarly, DMF containing moisture is disastrous—again, this material should be stored carefully.

6.3 General Method for Coupling a Small Molecule Containing an Amine Group to a Protein Using a Bis-succinimide Ester

This protocol provides a relatively simple method for coupling small amines to proteins, although it is somewhat inefficient in terms of hapten usage. The homobifunctional reagent used in this case is disuccinimidyl suberate (DSS), although many others could be substituted.

Solution
Buffer A: 100 mM phosphate buffer, pH 7.0.

Protocol

(A) Activation of hapten
(1) Calculate the mass of hapten representing a 50:1 excess over 5 mg of the protein to be labelled. Dissolve this quantity of hapten in a minimal quantity of dry DMF.

(2) Calculate the mass of DSS ($M_r = 368.4$) equimolar to the quantity of hapten, and add this amount of DSS (e.g. Pierce product No. 21555) to the solution.

(3) Mix and incubate for 40 min at room temperature.

(B) Conjugation
(4) Dissolve 5 mg of the protein in 1 ml of buffer A. Other similar presentations at around pH 7–9 are acceptable, and concentrations down to around 1 mg/ml can be used without affecting the reaction rate significantly.

(5) Incubate for 1 h at 25 °C.

(6) Purify as required. Size exclusion methods are often perfectly adequate for separating the labelled protein from unreacted hapten and other low molecular mass byproducts.

Notes
(1) In the event of under-derivatisation, increase the hapten:protein ratio, increase the DSS:hapten ratio or alter the incubation time in (A). Lengthening or shortening this time may be beneficial, depending on whether under-reaction or dimerisation of the hapten is the cause of poor incorporation in the conjugate.

(2) In the event of over-derivatisation, reduce the hapten:protein ratio, or reduce the incubation time in (A).

(3) DMF containing moisture will reduce yields from the activation—this material should be stored carefully.

(4) DSS can be substituted by many of the bis-succinimide esters described in Chapter 5, Section 5.3.5; these can provide linkers of varying length and polarity (see Figure 5.59), with consequent effects on conjugate behaviour.

6.4 General Method for Coupling a Small Molecule to a Protein by the Mannich Reaction

The Mannich reaction provides a means for coupling many 'difficult' molecules to proteins: see Chapter 6, Section 6.20 for details of candidate structures. Some literature protocols employ extremely long reaction times, but the reaction is often complete within a couple of hours—hence the design here.

Solutions
Buffer A: 650 mM sodium acetate buffer, pH 6.0; formaldehyde solution: 37% (w/v).

Protocol
(1) Dissolve 5 mg of the protein in 5 ml of buffer A.

(2) Calculate the quantity of hapten representing a 50:1 molar excess over the protein, and dissolve this quantity in the protein solution.

(3) Calculate the quantity of formaldehyde (M_r30.0) representing a 2000–fold excess over the protein, and the volume of 37% solution required to yield this mass. Given the large excess, it is unnecessary to take into account the density of this solution in the calculation: assume $\rho = 1$.

(4) Add this volume of formaldehyde solution to the protein–hapten mixture.

(5) Incubate for 2 h at room temperature.

(6) Purify as required. Size exclusion methods are often perfectly adequate for separating the labelled protein from unreacted hapten and other low molecular mass byproducts.

Notes
(1) In the event of under-derivatisation, increase the hapten:protein ratio, or the quantity of formaldehyde added.

(2) In the event of over-derivatisation, reduce the hapten:protein ratio.

(3) Formaldehyde fumes are highly toxic and this reagent should be handled accordingly.

7.1 Direct Labelling of DNA with Horseradish Peroxidase and Enhanced Chemiluminescence Detection in Blotting Applications

Solutions
Components for labelling, hybridisation and enhanced chemiluminescent detection are available in kit format as the ECL™ direct nucleic acid labelling and detection kit (Amersham Pharmacia Biotech UK Ltd). Detailed protocols are given with each kit. The following components (plus controls) are included in the system. Labelling reagent; glutaraldehyde solution (1.5% v/v in water); 'Gold' hybridisation buffer;

blocking agent; ECL detection reagent 1; ECL detection reagent 2.

Additional solutions and reagents

Sodium chloride (analytical grade); 20×SSC (0.3 M trisodium citrate, 3.0 M sodium chloride, pH 7.0); primary wash buffer with urea (360 g urea, 4 g sodium dodecyl sulfate (SDS), 25 ml 20×SSC; make up to one litre); primary wash buffer without urea (4 g sodium dodecyl sulfate, 25 ml 20×SSC; make up to one litre); secondary wash buffer (2×SSC); Blue-sensitive autoradiography film (e.g. Hyperfilm-ECL); SaranWrap™ (Dow Chemical Company).

Protocol

Probe labelling

(1) Dilute the nucleic acid to be labelled to a concentration of 10 ng/μl in a 1.5 ml microcentrifuge tube. Close the tube and incubate for five minutes in a boiling water bath to denature the nucleic acid.

(2) Immediately cool the sample on ice for five minutes, then centrifuge the tube briefly (ten seconds) to collect the contents at the bottom.

(3) Add a volume of labelling reagent equal to that of the sample and mix briefly by taking up the solution in the tip of a micropipette and gently expelling.

(4) Add a volume of glutaraldehyde solution equal to that of the labelling reagent. Again mix gently by pipette.

(5) Incubate at 37 °C for ten minutes. The probe is now labelled with HRP and can be stored on ice for up to 15 minutes before use.

Hybridisation

(6) Prepare the hybridisation buffer as follows. Take the required volume of 'Gold' hybridisation buffer and add solid sodium chloride to a final concentration which is most suitable for effective probe hybridisation. This may vary for different probes although 0.5 M NaCl has been found to give satisfactory results in most cases. Add blocking agent to a final concentration of 5% (w/v). Rapidly mix the blocking agent into the buffer to avoid the formation of clumps. Continue to mix thoroughly using a magnetic stirrer at 42 °C for 30–60 minutes. A few undissolved particles of blocking agent will not affect the hybridisation.

(7) Place the blot containing the target nucleic acid sequence onto the surface of the fully prepared hybridisation buffer prewarmed to 42 °C. A volume of buffer equivalent to 0.25 ml/cm^2 of membrane is satisfactory for most applications. Allow the blot to saturate fully before submerging in the buffer. Prehybridise at 42 °C for 15–60 minutes in a shaking water bath with gentle agitation.

(8) Add the labelled probe from step 5 above to the prehybridisation buffer at a concentration of either 10 ng/ml or 2–5 ng/ml depending on application (see above). Avoid placing the probe directly on the membrane and mix gently. Continue incubation with gentle agitation for between two hours and overnight at 42 °C.

Stringency washes

(9) Preheat enough primary wash buffer to 42 °C to allow two washes of 2 ml buffer/cm^2 membrane.

(10) Carefully transfer the blot from the hybridisation container to half of the prewarmed primary wash buffer in a clean container. Incubate at 42 °C in a shaking water bath for 20 minutes. Discard the wash buffer, replace with the remaining buffer and incubate with agitation for a further 20 minutes at 42 °C.

(11) Discard the wash buffer, place the blot in a fresh container, and add an excess of secondary wash buffer (at least 2 ml buffer/cm^2 membrane). Incubate for five minutes at room temperature with gentle agitation. Discard the wash buffer and repeat this step with fresh secondary wash buffer. The blot may be temporarily stored at this stage if required.

Detection

(12) Combine equal volumes of detection reagents 1 and 2 so that there is sufficient for 0.125 ml combined reagents/cm^2 membrane.

(13) Using forceps, carefully remove the blot from the final secondary wash buffer. Drain off the excess wash buffer by touching the bottom edge of the blot onto a paper towel. Place the blot, sample side uppermost, onto a sheet of SaranWrap on the bench.

(14) Cover the blot evenly with the freshly mixed detection reagents and leave for one minute. From this stage it is advisable to work as quickly as possible to minimise the delay between addition of substrates and exposure to film.

(15) Using forceps to hold the blot, drain off the excess detection reagents by touching the bottom edge of the blot onto a paper towel. Place the blot nucleic acid side down onto a sheet of SaranWrap in a film cassette.

(16) Fold over the SaranWrap to wrap the blots completely, but ensure that only a single layer covers the sample side. Turn the blots so that the sample side is uppermost.

(17) In a darkroom with a red safelight, place a piece of blue-light sensitive film on top of the blot, shut the cassette and expose at room temperature for the required length of time (please see notes). Develop film under normal conditions.

Notes

Numbering refers to the stages of the Protocol.

(1a) Between 10 μl (100 ng) and 150 μl (1.5 μg) probe may be labelled in each tube. Labelling efficiency will be reduced if the volume of nucleic acid exceeds 150 μl per tube.

(1b) Complete denaturation of double stranded probes is essential for maximal labelling. Extending the length of this step should be considered when using heating blocks as these can be less efficient at denaturation than boiling waterbaths. The denaturation step is not strictly necessary for single stranded probes although, in general, denaturation of RNA is advisable to avoid secondary structure.

(1c) Since labelling is achieved in part through an ionic interaction between the negatively charged nucleic acid and the positively charged labelling reagent, the concentration of salt in the sample should not exceed 10 mM.

(3) It is important to use equivalent volumes of nucleic acid and labelling reagent for maximum labelling efficiency. Accurate assessment of probe concentration (10 ng/μl) is essential to achieve maximum sensitivity.

(5) Probes once labelled are single-stranded and may be used in any membrane hybridisation application. If they are to be stored for more than 15 minutes, sterile glycerol should be added to a final concentration of 50% (v/v) and the probe stored at −20 °C until required. Probes stored for up to twelve months have been used in hybridisation with no appreciable loss of sensitivity.

(7a) Blots greater than 100 cm^2 should be pre-wet in 5×SSC before addition to the hybridisation buffer. The highest sensitivity detection is obtained with charged nylon hybridisation membranes such as Hybond-N+ (Amersham Pharmacia Biotech UK Ltd). However, for many applications, satisfactory results can be obtained with uncharged nylon (e.g. Hybond-N) or nitrocellulose (e.g. Hybond-ECL).

(7b) For hybridisation carried out in boxes a volume of buffer equivalent to 0.25 ml/cm^2 membrane is recommended. If the box is significantly larger than the blot, then the volume used should correspond to the area of the bottom of the box. Several blots may be hybridised together if adequate circulation of the buffer is maintained so that the blots can move freely. Increased backgrounds will otherwise result.

(7c) Hybridisations may also be carried out in bags or tubes. In these cases a buffer volume of 0.125 ml/cm^2 may be sufficient. When using tubes, it is important to exclude any air bubbles trapped between the hybridisation tube and the membrane as this can lead to patches of high background following detection. Nylon mesh may be used if there is significant overlap of the blot, to ensure good access of probe.

(8) Final probe concentration (10 ng/ml) and overnight hybridisation will give maximum sensitivity and is necessary for high sensitivity work such as the detection of single copy genes on genomic Southerns. However, for low sensitivity applications such as many plaque or colony screens or for detection of PCR products, it may be possible to use a lower probe concentration (2–5 ng/ml) in conjunction with a two-hour hybridisation.

(10) Stringency is controlled in the primary washes by the concentration of SSC. The primary wash recipe given below contains 5×SSC, but higher stringency and hence more specific hybridisation signal can be obtained with lower concentrations of SSC such as 0.1×SSC. Generally this washwill contain 6 M urea and is performed at 42 °C. However, it is possible to achieve similar results using primary wash without urea with incubation at 55 °C. In this case, the total wash time should not exceed 20 minutes in order to conserve peroxidase activity.

(11a) The secondary washes are included to ensure removal of SDS which may interfere with the enhanced chemiluminescence reaction.

(11b) Blots may be stored for up to 24 hours at 4 °C, wetted in secondary wash buffer and wrapped in SaranWrap. After storage, the blots should be rinsed briefly in fresh secondary wash buffer just before detection.

(12) The detection reagents should ideally be kept separate until immediately before detection to avoid gradual chemical decomposition of the substrates. However, if not used immediately it is possible to store the prepared solution on ice for up to 30 minutes.

(17) For initial experiments an exposure of 30–60 minutes is recommended for low target applications. An alternative approach is to carry out a short 1–2 minute initial exposure and use this to judge the length of the subsequent exposure. For plaque and colony screens in contrast, it is important to use the minimum film exposure time which enables positive colonies/plaques

to be unambiguously determined in order to avoid false positives.

7.2 Random Prime Labelling with Fluorescein-11-dUTP and Alkaline Phosphatase-mediated Dioxetane Chemiluminescence Detection in Blotting Applications

Solutions and reagents

Components for labelling and dioxetane chemiluminescent detection are available in kit format as the Fluorescein Gene Images labelling and detection system (Amersham Pharmacia Biotech UK Ltd). The following components (plus controls) are included in the system. Nucleotide mix (5× stock solution of fluorescein-11-dUTP (Fl-dUTP), dATP, dCTP, dGTP and dTTP in Tris-HCl, pH 7.8, 2-mercaptoethanol and $MgCl_2$); primers (random nonamers); enzyme solution (5 units/μl exonuclease-free Klenow); water; liquid block (for use in the recommended hybridisation buffer and for membrane blocking prior to addition of antibody conjugate); anti-fluorescein alkaline phosphatase conjugate (5000× stock); dioxetane detection reagent; detection bags for use during application of the detection reagent and subsequent exposure to film.

Additional solutions and reagents

20×SSC (3 M NaCl, 0.3 M sodium citrate, pH 7.0); hybridisation buffer (5×SSC, 1 in 20 dilution of liquid block reagent, 0.1% (w/v) sodium dodecyl sulfate (SDS), 5% (w/v) dextran sulfate) ±100 μg/ml sheared, denatured heterologous DNA (*e.g.* herring sperm DNA); TE buffer (10 mM Tris-HCl, pH 8.0, 1 mM EDTA); 0.2 M or 0.5 M EDTA, pH 8.0; 10% or 20% (w/v) SDS; buffer A (100 mM Tris-HCl, pH 7.5, 300 mM NaCl). Autoclave in 500 ml aliquots in one litre bottles for 15 min at 105 kPa (15 psi) (once opened after autoclaving, do not use for more than one day); dextran sulfate (molecular mass 500 kDa, for example Sigma Product no. D-6001); bovine serum albumin (BSA) Fraction V (for example, Sigma Product no. A-2153, BDH Product no. 44155); Tween™ 20 (polyoxyethylene sorbitan monolaurate; for example, Sigma Product no. P-1379, BDH Product no. 66368); autoradiography film such as Hyperfilm-MP™ (Amersham Pharmacia Biotech UK Ltd).

Protocol

Preparation of labelled probe

This protocol allows the labelling of 50 ng of template DNA. The standard reaction can be used to label 25 ng–2 μg of template DNA. Synthesis of labelled probe is most efficient at 50 ng of template, although net synthesis tends to increase with the amount of template.

(1) Dilute the DNA to be labelled to a concentration of 2–25 ng/μl in either distilled water or TE buffer (10 mM Tris-HCl, 1 mM EDTA, pH 8.0).

(2) Place the nucleotide mix, primers and water in an ice bath to thaw. Leave the enzyme at −20 °C until required, and return it to the freezer immediately after use.

(3) Denature the DNA sample by heating for five minutes in a boiling water bath, then chill on ice. It is advisable to denature in a volume of at least 20 μl.

(4) To a 1.5 ml microcentrifuge tube, placed in an ice bath, add the appropriate volume of each reagent in the following order: water, to a final volume of 50 μl; nucleotide mix, 10 μl; primers, 5 μl; denatured DNA, 50 ng; enzyme solution, 1 μl; (exo-free Klenow, 5 units/μl).

(5) Mix gently by pipetting up and down and cap the tube. Spin briefly in a microcentrifuge to collect the contents at the bottom of the tube.

(6) Incubate the reaction mix at 37 °C for one hour.

(7) If any of the probe is to be stored, rather than being used in a hybridisation immediately, then the reaction should be terminated by the addition of EDTA to a final concentration of 20 mM. Probes can then be stored in a freezer at −20 °C in the dark for at least six months. Do not use a frost-free freezer.

(8) The yield of probe after labelling 25 or 50 ng of template will typically be between 6 and 8 ng/μl. With higher template levels, while the amount of labelled template produced will increase, the overall reaction will be less efficient. At this stage, the rapid labelling assay (see protocol below) can be used as a check to ensure that the probe has been successfully labelled.

Hybridisation and stringency washes

The hybridisation and wash conditions given in the following protocol are appropriate for a majority of probes, allowing detection of single copy mammalian genes without significant cross-hybridisation to non-homologous sequences. However, if these conditions are found to be insufficiently stringent for particular probes, then hybridised filters can be washed in 0.2× or 0.1×SSC at 60 °C. Alternatively, stringency can be increased by raising the hybridisation or wash temperature to 65 °C. Such alterations may lead to some decrease in specific signal, although an overall improvement in signal-to-noise should result.

(9) Prepare the hybridisation buffer as follows: 5×SSC, 01% (w/v) SDS, 5% (w/v) dextran sulfate, 20-fold dilution of liquid block (supplied), 100 μg/ml denatured sheared heterologous DNA (optional). Combine all the components and make up to the required volume. Gentle heating with continuous stirring will be required to completely dissolve the dextran sulfate. The liquid block has been shown to contain low enough ribonuclease activity to be used in the hybridisation buffer for Northern blots. However, if required, liquid block can be autoclaved, on a liquid cycle, but must be sub-aliquoted into a suitable autoclavable container prior to sterilisation.

(10) Preheat the required volume of hybridisation buffer (0.3 ml/cm^2 for small blots and 0.125 ml/cm^2 or less for large blots) to 60 °C, then place the blots in the buffer and prehybridise for at least 30 minutes at 60 °C with constant, gentle agitation.

(11) Remove the required amount of probe to a clean microcentrifuge tube. If the volume is less than 20 μl, make up to this volume with water or TE buffer. Denature the probe by boiling for five minutes and snap-cool on ice.

(12) Centrifuge the denatured probe briefly, then add to the prehybridisation buffer (avoiding placing it directly on the membrane), and mix gently.

(13) Hybridise overnight at 60 °C with gentle agitation.

(14) Prepare stringency wash solutions of 1×SSC, 0.1% (w/v) SDS and 0.5×SSC, 0.1% (w/v) SDS and preheat both solutions to 60 °C. Carefully transfer the blots to the 1 SSC, 0.1% (w/v) SDS and wash for 15 minutes at 60 °C with gentle agitation. Carry out a further wash in preheated 0.5×SSC, 0.1% (w/v) SDS at 60 °C for 15 minutes. These wash solutions should be used in excess volumes of approximately 2–5 ml/cm^2 of membrane.

Blocking, antibody incubation and washes

The following steps are performed at room temperature and all the incubations require constant agitation of the blots.

(15) Following the hybridisation washes, briefly rinse the blots in an excess (2 ml/cm^2) of buffer A at room temperature.

(16) With gentle agitation, incubate blots for one hour at room temperature in approximately 0.75–1.0 ml/cm^2 of a 1 in 10 dilution of liquid blocking agent in buffer A.

(17) Briefly rinse blots in buffer A as in 15.

(18) Dilute the anti-fluorescein–AP conjugate 5000-fold in freshly-prepared 0.5% (w/v) bovine serum albumin in buffer A. Incubate the blots in diluted conjugate (0.3 ml/cm^2 of membrane) with gentle agitation at room temperature for 1 hour.

(19) Remove unbound conjugate by washing for 3 × 10 minutes in 0.3% (v/v) Tween-20 in diluent buffer at room temperature with agitation. An excess volume is again used (2–5 ml/cm^2).

(20) Rinse blots in buffer A as in 15.

Signal generation and detection

Please read through this whole section before proceeding. If possible, wear powder-free gloves.

(21) After completion of the antibody incubation stage and subsequent washes, drain off any excess wash buffer from the blots (by touching the corner of the blot against the box used for washing the blots or other convenient clean surface) and place them on a sheet of clear plastic.

(22) Pipette detection reagent onto the blots (30 to 40 μl per cm^2) and leave for 2–5 minutes.

(23) Drain off excess detection reagent, and transfer the blots to one of the bags as follows. (a) Cut a section from the bag large enough to cover the blot(s) with a small border of at least 1 cm. (b) Cut along the sides of the bag, leaving one side uncut. Open out the bag and after draining off the blots, arrange them sample side up on one half of the opened bag. Do not allow the blots to dry out. Fold the bag over the blots.

(24) Ensure the outside of the bag is dry before exposing to film. Place the bag containing the blots (sample side up) in a film cassette and take to a darkroom.

(25) Switch off the lights and place a sheet of autoradiography film on top of the blots. Close the cassette and expose for 1–2 hours for high target/low sensitivity applications. Longer exposures (4–16 hours) should be used for higher sensitivity applications. Remove film and develop.

Notes

Numbering refers to the stage of the Protocol

(1) If the DNA solution is too dilute to be used directly, it should be concentrated by ethanol precipitation followed by redissolution in an appropriate volume of water or TE buffer. If an intact plasmid probe is to be used it may be necessary to correct for the proportion of plasmid sequence present during the hybridisation.

(3) Closed circular double-stranded DNA can be linearised to avoid more rapid denaturation following this step, although this is not usually necessary.

(4) A volume of DNA solution up to 25 μl may be used for fragments in low melting point agarose.

(6) The temperature of incubation and the reaction time can be chosen for convenience as the reaction reaches a plateau and does not decline significantly overnight at room temperature. However, the rate of reaction does depend to some extent on DNA purity so, for samples of lower purity such as miniprep DNA or DNA in agarose, a longer incubation period may be required. With purified DNA samples, an increase in probe yield can be obtained with longer reaction times (up to four hours at 37 °C).

(7) Labelled probe can be used directly in a hybridisation, following denaturation, without removal of unincorporated nucleotide. For long term storage, it is advisable to keep probes in the dark as the fluorescein molecule is light-sensitive.

(9) The hybridisation buffer can be stored in suitable aliquots at −20 °C for at least 12 months. The addition of sheared, denatured heterologous DNA to the hybridisation buffer can reduce non-specific hybridisation although some reduction in sensitivity may also be observed.

(10) Hybridisation can be carried out in either boxes, bags or tubes provided there is sufficient buffer to allow adequate access of probe to the blot. With larger blots (greater than 50 cm^2), or if using minimal hybridisation volumes in bags or tubes, the blots should be pre-wetted in 5× SSC. It is also possible to hybridise several blots in the same solution, again providing there is adequate volume and circulation of buffer. A maximum of two blots should be hybridised together in a single bag and, when using more than one blot in a tube, it is important that the blots do not overlap one another. High backgrounds can result if there is insufficient hybridisation buffer or if a blot is not totally immersed. Agitation improves access of buffer to the blot, particularly if several blots are to be hybridised in the same solution.

(11) Recommended probe concentrations are 10–20 ng/ml for low target applications (for a 50 ng template reaction, this is approximately 1–2 μl of reaction product per ml of hybridisation buffer) and 1–5 ng/ml for high target applications.

(12) Alternatively, a small aliquot of the hybridisation buffer may be removed and mixed with the probe before returning the mixture to the bulk of the buffer. For high target applications it is possible to use shorter hybridisation times. Some loss of sensitivity will result, but if necessary such loss can be offset by the use of higher probe concentrations.

(14) During the second wash the SSC concentration can be varied in the range 0.1–1× SSC to achieve the desired stringency. The temperature of these washes may also be increased to achieve the desired stringency. Several blots can be washed in the same solution provided that they can move freely.

(15) Buffer A is an autoclaved solution of 100 mM Tris-HCl, 300 mM NaCl (pH 7.5). Once opened after autoclaving, unused buffer A should be discarded at the end of that working day to avoid contamination with exogenous alkaline phosphatases.

(16) Diluted block can be stored frozen (in aliquots) for several weeks and is easily thawed when required. However, it is recommended that no aliquot is subjected to repeated thawing and re-freezing. Also, it may be convenient to separate any remaining undiluted liquid block into appropriate aliquots for re-freezing.

(18) Diluted conjugate should be used immediately. Loss of sensitivity occurs with diluted conjugate that has been stored or subjected to a freeze–thaw cycle. Several blots can be incubated together but it is important that there is free access of solution to the blot. With larger blots, it is possible to carry out this stage in hybridisation bags or tubes to help minimise volume. For a 20 cm × 20 cm blot, a volume of 50 ml can be used. It may be found that the dilution of conjugate can be optimised further. Greater dilutions will give lower signal but may reduce background, while more concentrated conjugate will give the opposite effect. Bovine serum albumin Fraction V (for example, Sigma Product no. A-2153) should be used in the buffer.

(21) A sheet of SaranWrap™ can be used in place of the detection bags, although we have found bags to give the most satisfactory results.

(24) If using SaranWrap, ensure no liquid can escape.

(25) If required, expose a second film for an appropriate length of time. Depending on the application, the high sensitivity of this system may mean that overnight exposures of the blot to film result in too dark an image. By the second day, however, the light output will have stabilised at a significantly higher level than that emitted during the first few hours after application of the detection reagent, allowing significantly shorter optimum exposure times. Once

the light output has stabilised, it is easier to judge the optimum exposure time.

7.3 Use of Fluorescein-labelled RNA with Alkaline Phosphatase-based Colorimetric Detection for *in situ* Hybridisation Applications

The labelling process set out below is for the generation of fluorescein-labelled RNA probes. All the rest of the procedures regarding the treatment of samples, controls, hybridisation and detection are applicable to DNA, RNA or oligonucleotide probes. Reagents for labelling each type of probe with a fluorescein-labelled nucleotide, for carrying out the *in situ* hybridisation process and subsequent detection are available, as complete systems, from Amersham Pharmacia Biotech UK Ltd.

Solutions and reagents

Probe labelling
Transcription buffer: 200 mM Tris-HCl (pH 7.5), 30 mM $MgCl_2$, 10 mM spermidine, 0.05% (w/v) BSA; human placental ribonuclease inhibitor (HPRI): 20 U/μl; 0.2 M DTT: freshly prepared solution, sterilised by filtration; linearised DNA template: typically 0.5–1 μg/μl; SP6, T7 or T3 RNA polymerase, supplied, for example, by Amersham Pharmacia Biotech UK Ltd; nucleotide solution: 1.25 mM ATP, 1.25 mM GTP, 1.25 mM CTP, 0.312 mM UTP and 0.312 mM fluorescein-11-UTP (available from Amersham Pharmacia Biotech UK Ltd) in sterile water (see Note 1).

Post-labelling purification and processing
RNase-free DNase I: (Amersham Pharmacia Biotech UK Ltd E2210Y), freshly diluted to 10 units/80 μl in sterile water; 0.4 M $NaHCO_3$: sterilised by autoclaving; 0.6 M Na_2CO_3: sterilised by autoclaving; glacial acetic acid; 3 M sodium acetate (pH 5.2): sterilised by autoclaving; 10 mg/ml yeast tRNA: for example, Sigma Product No. R8759.

Hybridisation and stringency washes
Hybridisation buffer: standard hybridisation buffers (see Note 2) are usable but a preformed, optimised buffer, available from Amersham Pharmacia Biotech UK Ltd, is recommended which consists of 4×SSC, 600 μg/ml herring testes DNA, 2× Denhardt's solution and a proprietary rate enhancement compound (this buffer should be diluted 1:1 with deionised formamide before use; once made, the working buffer can be stored at −20 °C); 20×SSC stock: 3 M sodium chloride, 0.3 M trisodium citrate (pH 7.0); 10% (w/v) SDS.

Blocking, antibody incubations and washes
TBS: 100 mM Tris-HCl (pH 7.5), 400 mM NaCl; block buffer: 0.5% (w/v) blocking agent (RPN3023, Amersham Pharmacia Biotech UK Ltd) in TBS; anti-fluorescein–alkaline phosphatase conjugate: available from Amersham Pharmacia Biotech UK Ltd.

Detection
Detection buffer: 100 mM Tris-HCl (pH 9.5), 100 mM NaCl, 50 mM $MgCl_2$; NBT solution: 50 mg/ml nitroblue tetrazolium in dimethylformamide; BCIP solution: 75 mg/ml bromochloroindolyl phosphate in 70% (v/v) dimethylformamide.

Protocols

Probe labelling
(1) To a 1.5 ml microcentrifuge tube at room temperature (see Note 3), add the following reaction components in the order given: water, to a final volume of 20 μl; transcription buffer, 4 μl; 0.2 M DTT, 1 μl; HPRI, 20 units; nucleotide solution, 8 μl; linearised DNA template, 1 μg; RNA polymerase, 25 units.

(2) Mix gently by pipetting up and down.

(3) Incubate for a minimum of two hours; for SP6 incubate at 40 °C and for T7 and T3 incubate at 37 °C.

(4) The resulting labelled RNA probes can be stored at −20 °C (see Note 4).

Post-labelling purification and processing
(5) Add 10 units of RNase-free DNase I to the RNA probe mixture (see Note 5) and incubate at 37 °C for ten minutes.

(6) Add 20 μl of 0.4 M $NaHCO_3$, 20 μl of 0.6 M Na_2CO_3 and 60 μl sterile water. Mix gently and incubate at 60 °C to allow alkaline hydrolysis to occur for a time based on the transcript length and the probe size required (see Note 6).

(7) Add 1.3 μl of glacial acetic acid, 20 μl of 3 M sodium acetate (pH 5.2), 2 μl of 10 mg/ml yeast tRNA and 500 μl ethanol. Allow the RNA to precipitate at −20 °C for at least two hours.

(8) Pellet the RNA at maximum speed in a microcentrifuge for 15 min. Remove the supernatant.

(9) Rinse the pellet in 70% ethanol, kept at −20 °C. Centrifuge for five minutes. Remove and discard the supernatant.

(10) Redissolve the RNA probe in sterile water at a concentration of 10–20× that required in the hybridisation experiment. Store the RNA probe at −20 °C.

Pretreatment of cells and tissues

This stage is vital to the successful outcome of any ISH experiment (see Note 7) and each individual system will require its own optimised pre-treatment protocol. The pretreatments are designed to balance the requirements for probe and antibody conjugate access to their target molecules, with retention of the morphology and the target molecule within the tissue. In addition, the pre-treatment schedule may also be used to establish some of the controls that are required for ISH experiments (see Note 8).

Hybridisation and stringency washes

The control of stringency may be achieved in the hybridisation step by manipulation of the temperature, salt concentration and formamide concentration. However, it is more conveniently controlled by use of these parameters in the post-hybridisation stringency washes.

(11) Briefly warm the supplied hybridisation buffer at 37 °C for five minutes. Dilute 1:1 with deionised formamide and mix thoroughly.

(12) Add the required amount of probe to the hybridisation buffer, pre-warmed at 55 °C. A probe concentration of 300–600 ng/ml will be suitable for most applications (see Note 9).

(13) Apply a suitable volume of hybridisation buffer/probe mix over the section (see Note 10).

(14) Place slides on a hot plate set at 95 °C for 3–6 minutes dependent on the target (see Note 11).

(15) Hybridise in an humidified chamber at the desired temperature (typically 55 °C) for the required length of time (see Note 12).

(16) Wash the slides at the desired stringency. Typical washes would consist of 1× SSC, 0.1% (v/v) SDS for 2 × 5 minutes at room temperature followed by 0.2× SSC, 0.1% (v/v) SDS at 55 °C for 2 × 10 minutes.

(17) RNA probes can be further treated to reduce background by incubation in RNase A. This removes single stranded, un-hybridised material only. Rinse slides in 2 × SSC for two minutes. Incubate slides in pre-warmed 10 μg/ml RNase A in 2×SSC at 37 °C for 20–30 minutes. Rinse slides in 2×SSC before proceeding to the next stage.

Blocking, antibody incubations and washes

The following series of incubations should be performed with gentle shaking and at room temperature unless otherwise stated. Note, the TBS used in these steps has a higher salt concentration than standard TBS.

(18) Wash slides in TBS for five minutes.

(19) Incubate the slides in block solution (0.5% (w/v) blocking agent in TBS) for one hour.

(20) Rinse slides in TBS for one minute. Drain the buffer off the slides and, if necessary, dry the surface of each slide around the section (see Note 13).

(21) Dilute the antibody 1:1000 in 0.5% (w/v) BSA fraction V in TBS (see Note 14). Cover the section with the antibody solution, typically using 100 μl for each section. Incubate for one hour without shaking.

(22) Wash the slides in TBS for 3 × 5 minutes.

Detection

(23) Wash the slide in detection buffer for five minutes. Drain the buffer off each slide and, if necessary, dry the surface of the slide around the section (see Note 15).

(24) Add 45 μl NBT stock and 35 μl BCIP stock to 10 ml detection buffer. Apply 500 μl of detection reagent to each section and leave to develop in the dark for 4–24 hours (see Note 16).

(25) Rinse the slides for 2 × 2 minutes in distilled water.

(26) Counterstain, if required, and apply a mountant solution and a cover slip (see Note 17).

(27) View in a suitable light microscope (see Note 18).

Notes

Numbering refers to the stage of the Protocol

(1) Alternative hapten labelled nucleotides can be used, including biotin-UTP (Sigma Product No. B8280) and digoxigenin-UTP (Boehringer Product No. 1209 256). Final concentrations of CTP, GTP and ATP in the reaction mix should be 0.5–1 mM. The combined final concentration of UTP and labelled UTP should be 0.25–1 mM. Depending on the label used, the yield of RNA probe may benefit from optimisation of the nucleotide ratios within the reaction mixture.

(2) The hybridisation buffer supplied by Amersham Pharmacia Biotech UK Ltd is a 2× stock and requires 1:1 dilution with deionised formamide. Working solutions of standard buffer formulations would include an optimised combination of the following components; 2–4×SSC, 1–5×SSC Denhardt's solution, non-specific DNA, non-specific RNA, 25–50% deionised formamide and 5–10% dextran sulfate.

(3) The labelling reaction is set up at room temperature to avoid precipitation of the template DNA by the spermidine in the reaction buffer.

(4) Although it is generally difficult to monitor probe yields when using non-radioactive labels, it is possible, with fluorescein as the label to design a simple and rapid assay making use of its fluorescent properties. This assay is able to indicate that probe synthesis has occurred and gives an estimate of probe yield. The test should be performed on an aliquot withdrawn from the *in vitro* transcription reaction mixture. Further processing of the transcript, notably alkaline digestion, may reduce the fluorescence, but this does not appear to affect the antigenicity of the label which is important since the detection system is based upon recognition by anti-fluorescein antibody.

(5) When using fluorescein-labelled RNA probes there is no absolute requirement to purify the probes further, provided that the transcript is already of a suitable size for *in situ* hybridisation. The presence of unincorporated nucleotides does not appear to have any significant effect on the background observed with this system. However, it may still be advantageous to remove the template DNA by enzymic digestion. If alkaline hydrolysis is not required then the probe can be ethanol precipitated after this process.

(6) The presence of the fluorescein label does not interfere with alkaline hydrolysis when this is required. As a consequence of this process the unincorporated nucleotides are also largely removed, but that is not the prime reason for performing this procedure. The time required for the limited alkaline hydrolysis of the primary transcript is derived from the following equation:

$$t\,(\text{min}) = \frac{L_o - L_f}{kL_oL_f}$$

where L_o = primary transcript length (kbases), L_f = average probe length required (kbases), and the rate constant $k = 0.11$ cuts/kb/min. For *in situ* hybridisation the optimal probe size is 200–400 bases.

(7) Pre-treatments cover all the processes required before hybridisation can occur including slide coating, sample preparation and fixation. A general scheme would include the following steps: (a) Rapid fixation or rapid freezing of the biological sample. (b) Sectioning onto coated slides. (c) Sample pretreatments that include a combination of incubations in acids, detergents and proteases. The most important step is a Proteinase K treatment; the concentration and incubation time of Proteinase K have to be optimised empirically and different tissues may require different conditions. Cells may not require any treatment at all. (d) Treatment to remove endogenous alkaline phosphatase activity.

(8) It is essential to perform a number of negative controls alongside the main reactions used for *in situ* hybridisation. These can take various forms, for example: (a) Incubation in hybridisation buffer alone (no probe); allows an assessment to be made of the level of signal due to the hybridisation buffer components and non-specific binding of the antibody. (b) DNase or RNase pre-treatment of the slide prior to hybridisation; destroys the target DNA or RNA respectively and should show that the probe binds to the predicted target species. (c) Hybridisation with a totally non-specific probe; indicates that the stringency conditions are sufficient to remove non-specific probes. (d) With RNA probes labelled transcripts can be made that are either complementary or identical to an mRNA target. Hybridisation with the identical strand probe should show that the stringency conditions are sufficient to remove a probe with the same GC content and same length as the complementary, specific, probe. (e) If in doubt probes should be validated on Northern blots.

(9) The hybridisation time and probe concentration are linked. Short hybridisations are possible with high probe concentrations (600–1000 ng/ml). Overnight incubations may be possible with a lower probe concentration (200–400 ng/ml).

(10) The hybridisation buffer volume is dependent on the size of the section and on whether cover slips are used. Small sections can be covered easily with 25–50 μl without using a cover slip; 10 μl may be sufficient with a cover slip.

(11) If the target is DNA, then the slide must be denatured by placing onto a hot plate at 95 °C for 5–6 minutes. If the target is RNA this procedure, whilst not strictly required, is recommended. An incubation on the hot plate, for 3–4 minutes, is thought to remove the secondary structure of the target mRNA thus increasing the efficiency of hybridisation.

(12) The chamber should be sealed and kept at a high humidity throughout the hybridisation to avoid hybridisation buffer evaporation. In this way it may be possible to avoid the need to seal cover slips onto the slide or even to avoid the use of cover slips entirely. At present most experiments are performed in sealed boxes placed in incubators, but recently intelligent heating block technology has been used to design machines that can be used for this purpose (e.g.

OmniSlide™, Hybaid). Typical experiments require an overnight hybridisation. However, it is possible to perform 2–4 hour hybridisations· using high probe concentrations with high levels of target.

(13) It is important to avoid the presence of excess buffer around the section as this may lead to over-dilution of the antibody conjugate.

(14) A dilution of 1:1000 of antibody conjugate should be suitable for most applications. The volume of diluted antibody added to each section may also need to be altered to account for the size of the sections and for whether cover slips are used. These alterations will need to be determined empirically.

(15) It is important to avoid the presence of excess buffer around the section as this may lead to over-dilution of the substrates.

(16) The volume of detection reagent applied to each section must be enough to flood the section. The length of time required to obtain a result will depend on the target level but, typically, an overnight incubation is required for maximum sensitivity. Other substrates for alkaline phosphatase can also be used, for example Fast Red, although in our laboratory the NBT/BCIP system gives the best signal:noise.

(17) A histochemical stain can be used before mounting in order to provide some contrast across the tissue section. However, care must be taken that the stain does not interfere with the coloured reaction product already deposited on the slide. With NBT/BCIP substrates it is possible to counterstain the section utilizing a light green counterstain, such as 0.5% (w/v) Methyl Green, or 0.5% (w/v) Fast Green, in water for one minute. A number of standard mountant solutions are suitable for use with this system. Aqueous-based mountants include PBS/glycerol (CitiFluor) and glycerol/gelatin (Sigma). However, these cannot be used with counterstains as the stain does fade, and the NBT/BCIP colour may fade, with time. Non-aqueous mountants include DPX (BDH/Merck) and Histotech (Serotec). It is important to ensure that the sections are washed in graded ethanols, ending in 100% ethanol, and are dried before applying the non-aqueous mountant as the presence of water in the section will lead to loss of colour with time. The Histotech mountant does not require a cover slip as it is baked onto the section.

(18) A good quality microscope is recommended with a transmitted white light source and ×10 and ×40 objective lenses. In our laboratory the microscope is fitted with a 35 mm camera and Kodak Ektar™ 25 colour print film is used to record the results.

7.4 Preparation of Biotin-labelled Oligonucleotides and Their Use as PCR Primers

The following protocol describes the synthesis of biotinylated oligonucleotides using biotin phosphoramidite (Misiura *et al.*, 1990). Following purification these oligonucleotides can be substituted for normal primers in PCR reactions (Innis *et al.*, 1990). PCR products prepared with biotinylated primers can then be selectively isolated, for example using streptavidin-linked separation columns, or can be detected using streptavidin–HRP or streptavidin–alkaline phosphatase with appropriate colorimetric, fluorescent or chemiluminescent detection. It is also possible to use such products as low sensitivity hybridisation probes with detection again using streptavidin-HRP or streptavidin–alkaline phosphatase.

Using the recommended biotin phosphoramidite below, it is possible to introduce a biotin residue anywhere in the growing nucleotide chain, although 5′-end labelled oligonucleotides are most frequently used. Multiple residues can also be introduced in order to maximise detection sensitivity. As biotin phosphoramidite does not introduce additional nucleotides, the effect on stringency of multiple labelling should be minimal.

Solutions, reagents and equipment

Biotin phosphoramidite (for example, Amersham Pharmacia Biotech UK Ltd RPN2012); standard reagents for oligonucleotide synthesis as recommended by supplier of synthesiser; acetonitrile (anhydrous); 35% (v/v) ammonia; 80% (v/v) acetic acid; 0.1 M ammonium acetate; HPLC; SpeedVac concentrator.

Protocols

Preparation of the automated DNA synthesiser

The oligonucleotide synthesiser should be set up according to the operating instructions, and following the safety guidelines, of the machine manufacturer.

(1) Dissolve biotin phosphoramidite (0.12 g) in 1.25 ml anhydrous acetonitrile (see Note 1) shaking the bottle gently until material is fully dissolved.

(2) Place bottle on a spare port on the machine following the machine operating procedure.

Oligonucleotide synthesis

(3) Key in the desired sequence, specifying the biotin phosphoramidite at the required position(s) (see Note 2).

(4) Perform synthesis in the 'Trityl-on' mode (see Note 3).

(5) The oligonucleotide is removed from the column in 35% (v/v) ammonia solution following the manufacturer's end procedure, or manually.

Deprotection

(6) Incubate oligonucleotide in 35% (v/v) ammonia solution at 55 °C for a minimum of five hours.

(7) Evaporate sample to dryness. This is best achieved by evaporation under vacuum in a centrifuge such as a Savant SpeedVac concentrator or similar machine.

(8) Redissolve in 300 μl of 80% (v/v) acetic acid and incubate at room temperature for 30 minutes. This process removes the final trityl group.

(9) Evaporate to dryness as in step 7 above.

(10) Redissolve in 500 μl sterile water.

(11) Filter the suspension through a 0.45 μm filter.

Purification

It is recommended that biotin-labelled oligonucleotides are purified by HPLC as detailed below. It is possible to utilise oligonucleotide purification cartridges but under their standard operating instructions, very low yields of purified material may be obtained. For some applications it may be possible to use the biotin labelled oligonucleotides without further purification.

Time	Buffer A (%)	Buffer B (%)
0	100	0
5	100	0
35	70	30
40	50	50
45	50	50
50	100	0

(12) HPLC-purify samples by applying to a C-18 cross-linked reverse phase column under a gradient buffer system. Buffer A is 0.1 M ammonium acetate solution and buffer B consists of 20% buffer A/80% acetonitrile. The gradient programme is set out in the table above.

A flow rate of 1 ml/min is recommended for columns such as a 5 μm, 4.6 mm × 15 cm Ultrasphere-ODS (Beckman). Other columns may require alternative conditions.

(13) Collect the peak corresponding to the biotin-labelled oligonucleotide (see Note 4).

(14) Evaporate to dryness as described in step 7.

(15) Redissolve in sterile water and determine the concentration of oligonucleotide assuming that an absorbance at 260 nm of 1.0 is equivalent to a concentration of 33 μg/ml (see Note 5).

Notes

(1) Care should be taken to avoid introducing moisture into bottles containing phosphoramidites. Follow machine manufacturer's operating instructions for dissolving nucleotide phosphoramidites.

(2) The reaction conditions for biotin phosphoramidite are similar to the conditions used for standard nucleotide phosphoramidites. If required, longer coupling times, e.g. 300 seconds, may be used to increase coupling efficiency further.

(3) The final trityl group must be present on the oligonucleotide during the deprotection procedure.

No. of biotins	Retention time (min)	Buffer B (%)
0	18–20	14.5
1	21–22	16.5
2	23–24	18.5
4	25–26	20.5
8	30–31	25.5
10	31–32	26.0

(4) Using the HPLC programme and column described, non-biotinylated sequences have a retention time of approximately 19 minutes. Biotinylated oligonucleotides are eluted progressively later depending on the number of biotins attached (Misiura *et al.*, 1990), as shown in the table above.

(5) If an accurate value for the oligonucleotide concentration is required then the following formula is used to calculate the molar extinction coefficient at 260 nm:

$$\epsilon_M = (A \times 15\,200) + (G \times 12\,010) + (C \times 7050) + (T \times 8400)$$

where A, G, C and T represent the number of times each base occurs in the oligonucleotide sequence. The concentration of the oligonucleotide solution is determined by the measurement of the absorbance at 260 nm.

8.1 General Method for Physisorption of Proteins to Polystyrene Microtitre Wells

As noted in Chapter 8, Section 8.2, physisorption of proteins to hydrophobic surfaces such as polystyrene is extremely facile. Optimal conditions for both coating and retention of protein functionality will differ from one protein to another, but these basic conditions will provide reasonably efficient immobilisation in many cases.

Solutions

Buffer A: 100 mM Tris buffer, pH 8.5; buffer B: 100 mM Tris buffer + 0.1% Tween-20.

Protocol

(1) Dissolve or dilute the protein to 25 μg/ml in buffer A.

(2) Pipette or dispense an appropriate volume of the protein solution into each well.

(3) Incubate for 1 h at 25 °C.

(4) Aspirate or decant the coating solution.

(5) Pipette or dispense an appropriate volume of buffer B into each well, then aspirate or decant.

(6) Repeat step 5 twice.

(7) Repeat step 5, but using buffer A.

(8) Dry the wells, for example using a controlled-humidity incubator, or by simple air-drying with a fan to promote air movement.

(9) Store the wells with a desiccant or in a low-moisture environment.

Notes

(1) In the event of under-coating, investigate altering the pH of the coating reaction. It is possible to increase the coating concentration too, but this often provides little extra benefit. Note that apparent under-coating is often in fact attributable to denaturation of the protein on the surface.

(2) In the event of over-coating reduce the coating concentration, or consider diluting the coating solution with a second, inert protein.

(3) This protocol is readily adaptable to other hydrophobic solid phases such as polystyrene tubes and latex particles.

(4) Bear in mind the possibility of desorption of protein from immobilisation procedures of this type, and the consequences which this may have under the conditions of use.

9.1 Quantitation of DNA in Solution

Both DNA and RNA absorb ultraviolet light so efficiently that optical absorbance can be used as an accurate and rapid measure of their concentration, even at levels as low as 2.5 μg/ml. A solution containing 50 μg/ml of double-stranded DNA or RNA has an absorbance of 1 at 260 nm. Similarly, a solution containing 40 μg/ml of single-stranded DNA or RNA has an absorbance of 1 at 260 nm.

Spectrophotometry is also most useful to estimate the purity of nucleic acid preparations. The ratio $A_{260} : A_{280}$ of a pure double-stranded DNA preparation should be 1.65–1.85. Higher ratios are often due to RNA contamination and lower values to protein contamination.

9.2 Agarose Gel Electrophoresis

Agarose gels are frequently used to separate and analyse a DNA preparation (see Section 11.6). Agarose gels are prepared by melting the agarose (DNA grade) in the presence of the desired buffer until a transparent solution is obtained. The melted agarose solution is subsequently poured into a mould and allowed to harden. The agarose then forms a matrix. When an electric field is applied across the gel DNA will migrate towards the anode, due to its negative charge at neutral pH.

Linear DNA fragments migrate at different rates depending on their size. Larger molecules migrate more slowly. The rates are inversely proportional to the logarithm of the number of base pairs. It is also important to realise that the DNA conformation can influence the migration. Covalently closed circular, open circular and linear DNAs of the same molecular masses will hence move through agarose gels at different rates. After separation of DNA fragments for identification purposes, it is also possible to cut out specific fragments from the gel and use them for further manipulations such as DNA ligation.

(1) Prepare a 0.8% agarose gel by adding 0.2 g agarose to a 100-ml Erlenmeyer flask containing 20 ml H_2O and boil and swirl the solution until the agarose is completely dissolved. Add 2.5 ml 10 × TAE electrophoresis buffer (0.4 M Tris-acetate, 0.2 M sodium acetate and 10 mM EDTA, pH 8.1) and adjust the final volume to 25 ml with H_2O.

(2) Prepare a mini-gel by adding the hot (approximately 60 °C) agarose to a glass plate (8 × 10 cm). Surface tension will hold the agarose solution on the plate. Place a well-former at the end of the glass plate

and let the agarose harden for 30 minutes at room temperature. Carefully remove the well-former.

(3) Fill the reservoirs with 1× TAE electrophoresis buffer.

(4) Place each DNA sample into a single well in the gel. It is convenient to use lambda DNA digested with *Hin*dIII and *Eco*RI as size markers. These fragments have the following sizes: 21,226, 5148, 4973, 4268, 2027, 1904, 1584, 1375, 947, 831, 564, 125 base-pairs (bp).

(5) Electrophorese at 100–150 V/gel for 90–120 minutes. The bromophenol blue marker of samples should have migrated about 6 cm.

(6) Turn off the voltage. Stain the gel in an ethidium bromide solution (1 μg/ml) for 5–10 minutes. **NB: Ethidium bromide is a potential carcinogenic substance.**

(7) Expose the stained gel to ultraviolet light and photograph the fluorescence of the stained gel. For maximum sensitivity, it is beneficial to use a red–orange filter between the gel and the camera. Polaroid™ film is by far the most practical for gel photography. Film types 107 and 667 have adequate sensitivity and resolution. When using ultraviolet illumination of the gels be careful to wear eye and skin protection.

9.3 Mini-preparations of Plasmid DNA

The isolation of plasmid DNA is essential for many molecular cloning experiments. Today several alternative kits are commercially available for plasmid DNA preparations. Most of these are convenient but they are all fairly expensive. The following procedure is a simple and reliable alternative.

(1) Transfer a single bacterial colony to 5 ml LB medium (10 g tryptone, 5 g yeast extract and 10 g NaCl per litre H_2O) containing the appropriate antibiotic(s) in a sterile test tube. Incubate overnight at 37 °C with vigorous shaking.

(2) Pour 1.0 ml of the culture into a microfuge tube. Centrifuge at 12,000 g for 60 seconds.

(3) Remove as much as possible of the supernatant.

(4) Resuspend the pellet in 100 μl ice-cold GTE buffer (4 mg/ml lysozyme, 50 mM glucose, 25 mM Tris-HCl, 10 mM EDTA, pH 8.0). Vortex.

(5) To the cell suspension add 200 μl of a freshly prepared solution composed of 0.2 M NaOH and 1% SDS. Mix by inverting the tube. Store on ice for five minutes.

(6) Add 150 μl of ice-cold potassium acetate solution (60 ml 5 M potassium acetate, 11.5 ml glacial acetic acid, 28.5 ml H_2O). Vortex vigorously. Store on ice for five minutes.

(7) Centrifuge at 12,000*g* for five minutes. Transfer the supernatant to a fresh microfuge tube.

(8) Add an equal volume of phenol chloroform (1:1). Mix by vortexing. After centrifugation for two minutes at 12 000 *g* transfer the supernatant to a fresh microfuge tube.

(9) Precipitate the plasmid DNA with two volumes of ethanol at room temperature. Incubate for two minutes at room temperature.

(10) Centrifuge at 12 000 *g* for five minutes at 4 °C. Remove the supernatant. Stand the tube upside-down on a paper towel to allow the ethanol to drain away.

(11) Rinse the pellet with 1 ml of 70% ethanol. Remove the ethanol as in the previous step.

(12) Dissolve the pellet in 50 μl TE buffer (10 mM Tris-HCl, 1 mM EDTA, pH 8.0) containing 20 μg/ml DNase-free pancreatic RNase.

9.4 Restriction Enzyme Digests

Restriction enzymes are extremely useful tools both to analyse a DNA sample and to construct recombinant DNA molecules. Restriction enzymes are classified into three groups; only type II enzymes are widely used for DNA manipulations, however, and several commercial suppliers are available. The majority of type II restriction enzymes recognise specific sequences that are four, five or six nucleotides in length and display two-fold symmetry.

(1) Transfer a DNA solution containing 0.2–1 μg of DNA to a sterile microfuge tube. Add sterile H_2O to a final volume of 18 ml.

(2) Add 2 μl of the appropriate 10× restriction enzyme buffer (normally supplied with the restriction enzyme). Mix.

(3) Add 1–5 units of restriction enzyme. Mix.

(4) Incubate the tube for one hour at the appropriate temperature.

(5) If the DNA is to be analysed by electrophoresis, add 4 μl gel loading buffer (40% (w/v) sucrose in water containing 0.25% bromophenol blue). Mix gently.

(6) Load the sample into the agarose gel.

9.5 DNA Ligation

(1) After digestion of the plasmid and the insert with the proper restriction enzyme(s), isolate the desired

DNA fragments by gel electrophoresis and redissolve the DNAs in TE buffer (10 mM Tris-HCl, 1 mM EDTA, pH 7.6) at a concentration of 100 μg/ml.

(2) Set up a ligation mixture composed of approximately 0.1 μg of vector DNA and an equimolar amount of foreign DNA in a sterile microfuge tube. Add H_2O to 8.5 μl and warm the solution at 45 °C for 5–10 minutes, to prevent any annealing of cohesive termini in the vectors or inserts themselves. Chill the tube on ice. It is advisable to prepare two controls containing the vector and the foreign DNA alone.

(3) Add 1 μl 10× ligation buffer (200 mM Tris-HCl, pH 7.6, 50 mM $MgCl_2$, 50 mM dithiothreitol, 5 mM ATP; this buffer is frequently supplied with the ligase by many manufacturers). Add 0.1 U T4 DNA ligase.

(4) Incubate for 2–4 hours at 15 °C.

(5) Use 5 μl of the ligation mixture to transform competent *E. coli* cells.

9.6 Preparation of Competent *E. coli* Cells

(1) Pick a single colony of *E. coli* cells from a fresh Petri dish. Transfer it to 40 ml of LB broth in a 500 ml shaker flask. Incubate the culture for approximately 2 hours at 37 °C with vigorous shaking. Monitor A_{550} which is a simple means of determining the number of viable cells. When A_{550} reaches 0.4, transfer the culture to a sterile SS-34 centrifuge tube.

(2) Centrifuge at 6000 g for five minutes.

(3) Discard the supernatant and add 20 ml ice-cold 0.1 M $CaCl_2$.

(4) Store on ice for 20 minutes.

(5) Centrifuge again at 6000 *g* for five minutes. Discard the supernatant. Add 4 ml ice-cold 0.1 M $CaCl_2$.

9.7 Transformation of *E. coli* Cells

(1) Transfer 200 μl of competent cells from protocol Section 9.8.6 to a sterile microfuge tube. Add DNA (50–100 ng in a volume of approximately 10 μl). Mix. Store on ice for 30 minutes.

(2) Transfer the tube to a water bath at 42 °C. Incubate for 120 seconds.

(3) Transfer the tube to an ice bath again. Add the cell suspension to culture tubes containing 1 ml of LB medium. Incubate at 37 °C for 45 minutes.

(4) Spread 100 μl of the cell suspension over the surface of Petri dish with LB medium containing the appropriate antibiotics.

(5) Incubate overnight at 37 °C.

9.8 Polymerase Chain Reaction (PCR)

PCR is used to amplify a DNA segment placed between two regions of known DNA. Two oligonucleotides are used as primers for a series of synthetic reactions that are catalysed by DNA polymerase. These primers should be at least 16 (preferably 20–24) nucleotides in length to avoid annealing of the primers at wrong locations on the target DNA. However, the oligonucleotides should be short enough to prevent formation of stable hybrids between the primers at the temperature used for polymerisation (approximately 72 °C).

A limitation of the method is its relatively high rate of incorporation of the wrong nucleotide. The frequently used *Taq* DNA polymerase lacks editing functions and incorporates an incorrect nucleotide at a rate of 2×10^{-4} nucleotides per cycle in PCR reactions. Several alternative thermostable DNA polymerases have recently become commercially available.

It should be noted that the commercial application of PCR technology is strictly controlled by a number of patents and licensing agreements.

(1) In a sterile 0.4 ml microfuge tube, mix in the following order: 15 μl sterile H_2O; 5 μl 10× PCR buffer (500 mM KCl, 100 mM Tris-HCl (pH 8.3), 15 mM $MgCl_2$, 0.1% gelatin); 8 μl nucleotide mix of all four dNTPs (each 1.25 mM); 100 pmol primer 1; 100 pmol primer 2; template DNA (up to 1 μg); H_2O to a final volume of 50 μl.

(2) Heat the tube at 94 °C for five minutes to denature the DNA.

(3) Add 2 U of *Taq* DNA polymerase while the tube is still at 94 °C.

(4) Overlay the reaction mixture with 50 μl of light mineral oil to prevent evaporation.

(5) Perform the amplification reactions. Temperatures and times may vary between different amplifications (see table). Up to 30 subsequent cycles may be used.

(6) Analyse the reaction mixture by agarose gel electrophoresis.

Cycle	Denaturation	Annealing	Polymerization
First	5 min/94 °C	2 min/50 °C	3 min/72 °C
Others	1 min/94 °C	2 min/50 °C	3 min/72 °C

9.9 Cell Cultivation and Fusion Protein Expression

In *E. coli* the expression of fusion proteins can be performed either by using overnight cultures, or by growing the cultures for a shorter period of time. Overnight cultures typically express higher levels of fusion proteins as the production continues into the stationary growth phase. The expression levels of fusion protein from many types of vectors vary in the range 2 to 100 mg/litre in a shake flask. In a fermenter, where the growth medium is strictly controlled, expression levels exceeding 1 g/litre have been obtained in many instances.

(1) Seed cultures of 5 ml are grown overnight at 37 °C in LB (10 g/litre tryptone, 5 g/litre yeast extract, 10 g/litre NaCl) supplemented with a suitable selection antibiotic (e.g. 100 mg/l ampicillin).

(2) 0.5 ml is used as inoculum to 100 ml of fresh LB medium containing the same antibiotic(s). Incubate at 37 °C under vigorous shaking. Follow cell growth by monitoring the absorbance at 550 nm. Grow for approximately four hours, until mid or late log phase is reached. This often corresponds to an A_{550} of approximately 2 in a shake flask.

(3) Induce the culture. When expression plasmids based on the p_{lac} or p_{tac} promoters are used add 0.1 mM IPTG and continue the cultivation for another two hours. This induction must often be optimised in order to achieve a high the level of gene product.

(4) Harvest the cells.

9.10 Cell Lysis and Preparation of a Crude Cell Homogenate

When protein is produced intracellularly in *E. coli*, a crude cell homogenate can be prepared by the following method.

(1) Harvest the cells by centrifugation for five minutes at 6000 *g* and 4 °C.

(2) Wash the cells in 0.1 M sodium phosphate buffer (pH 7.3) containing 1 mM EDTA. Centrifuge as above.

(3) Dissolve the cells in the same phosphate buffer. The cells are frequently at least ten-fold concentrated compared to the cell medium. If proteolysis is a problem a protease inhibitor can be added at this step.

(4) Add lysozyme to a final concentration of 0.5 mg/ml and put the tube on a rocking table for 30 minutes at ambient temperature. This will break down the cell walls of the bacteria. Usually this cell breaking step needs to be optimised depending on the bacterial production system used. Thus, the addition of a sonication step after the lysozyme treatment can frequently improve the yield substantially.

(5) The cell extract is clarified by centrifugation at 12 000 *g* for five minutes.

(6) The protein in the supernatant can be concentrated and enriched by ammonium sulfate precipitation. Fusion proteins are frequently precipitated at low ammonium sulfate concentrations (20%) implying that a substantial degree of purification can be obtained in this step. The fusion protein is then often dialysed before further purification steps.

9.11 Isolation of Periplasmic Fusion Proteins

An osmotic shock procedure is often used to release periplasmic proteins from the *E. coli* cells. This procedure (Randall and Hardy, 1986) prevents contamination from cytoplasmic proteins.

(1) Harvest the cells by centrifugation.

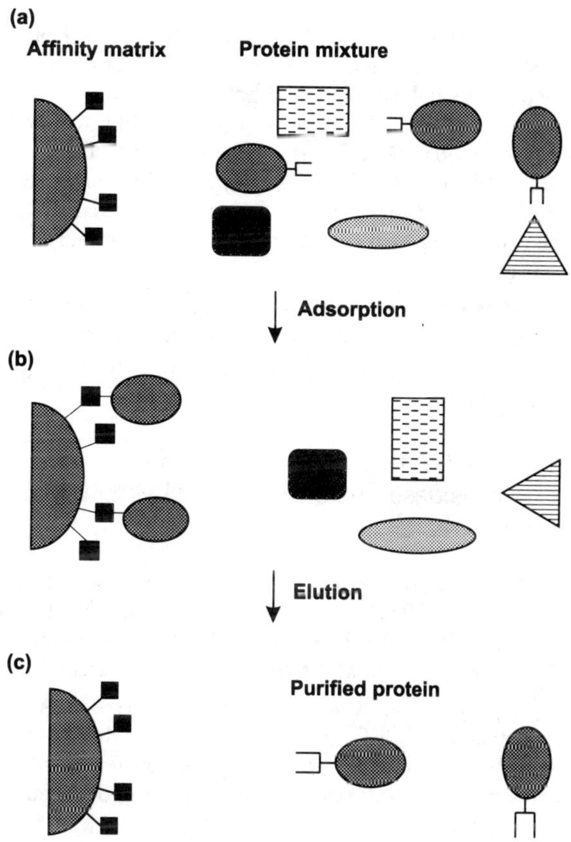

Figure 9.17 SPA displays tight binding to IgG. Therefore, it IgG is coupled to a carrier material, protein A conjugates can easily be adsorbed from the rest of the proteins in a crude cell homogenate. After washing, the conjugate can eluted by lowering the pH.

(2) Resuspend the cell pellet in 5 ml ice-cold sucrose buffer (0.5 M sucrose, 0.1 M Tris-HCl, pH 8.2, 1 mM EDTA). Incubate on an ice bath for ten minutes.

(3) Add 80 μl of lysozyme solution (10 mg/ml in H_2O) and immediately thereafter add 5 ml of ice-cold H_2O. Mix by vortexing.

(4) Incubate in an ice/water bath for five minutes.

(5) Add 180 μl of 1 M $MgSO_4$ to stabilise the spheroplasts.

(6) Centrifuge at 10 000 *g* for 20 minutes.

(7) Load the clear supernatant onto a suitable affinity column.

9.12 Purification of Protein A Fusion Proteins (See Figure 9.17)

(1) Load (by gravity) the crude cell extract from a 100 ml cell culture onto a small column (2 ml) of IgG Sepharose 6 Fast Flow™ (Pharmacia, Sweden). The gel should previously have been equilibrated with 20 ml of 50 mM Tris-HCl (pH 7.4), 150 mM NaCl and 0.05% (v/v) Tween 20. The binding of Protein A fusion proteins to IgG columns is usually very fast meaning that the flow rates are normally not critical.

(2) Wash the column with 40 ml of the same Tris buffer followed by 10 ml of 1 mM ammonium acetate.

(3) Elute the fusion protein with 2 ml of 0.5 M acetic acid (pH 3.3).

9.13 One-step Purification of β-Galactosidase Fusion Proteins (Ullmann, 1984)

(1) Mix bacterial extracts (15–25 mg of protein per ml, prepared in 20 mM Tris-HCl, 10 mM $MgCl_2$, 1 mM dithiothreitol (pH 7.4)) with 5 M NaCl to obtain a final concentration of 1.6 M NaCl in the extract.

(2) Apply the solution to a *p*-aminophenyl-β-D-thiogalactoside (TPEG)–agarose (Sigma) column previously equilibrated with the same buffer (1 ml of affinity matrix will retain more than 2 mg of β-galactosidase).

(3) Wash the column extensively with the equilibration buffer until no more material appears in the flow-through.

(4) Elute the fusion protein from the column with 100 mM sodium borate, 1 mM dithiothreitol (pH 10).

9.14 Adsorption of Biotinylated Fusion Proteins on Streptavidin–Agarose

(1) Mix 4 ml of concentrated bacterial supernatant (from a 500 ml culture; in 20 mM Tris-HCl (pH 8.0), 50 mM NaCl, 1 mM EDTA, 0.2 mM PMSF) with 50% streptavidin–agarose (Pierce; 0.25 ml, in 10 ml of 20 mM Tris-HCl (pH 8.0), 50 mM NaCl, 1% BSA). Incubate for one hour at room temperature on a rocking table.

(2) Wash the beads twice with 20 mM Tris-HCl (pH 8.0), 0.5 M NaCl, 0.1% NP-40.

(3) Wash the beads with 20 mM Tris-HCl (pH 8.0), 50 mM NaCl.

12.1 Storage Conditions for Hybridoma Cells

These are optimally preserved by a two-step freezing process in 10% DMSO and 90% heat-inactivated fetal calf serum. The freezing mixture is made by adding the required amount of DMSO to heat-inactivated fetal calf serum to give a final concentration of 10%; this freezing mixture is sterilised by filtration through a 0.22 μm filter and cooled on ice for 10 min before adding to the cells.

The cells are pelleted in 50 ml polypropylene conical bottom tubes, excess medium tipped off and the cell pellet flicked to resuspend partially the cells in the residual medium. The required amount of freezing mixture is then added to give a final cell concentration of 10^6–10^7 ml^{-1}. (If more convenient, the cells may be resuspended in medium to a final concentration of 10^9 ml^{-1}, and a hundred volumes of cold freezing mixture of 11% DMSO and 90% heat-inactivated fetal calf serum added.) The cells are then gently but rapidly resuspended using a 2 ml pipette, and transferred immediately to 2 ml polypropylene freezing vials. The vials are then placed in a lagged polystyrene container, transferred to a −70 °C cabinet and frozen for 12–16 hours, after which the vials are transferred to the liquid nitrogen storage vessel. Rapid thawing and immediate dilution of the frozen cells are required for optimal recovery of viable cells. This is achieved by transferring the cells to a 37 °C water bath for a few minutes, partially thawing the surface of the frozen culture in contact with the side of the vial, which then enables the rest of the frozen contents to be transferred directly into 50 ml of medium plus 10% heat-inactivated fetal calf serum prewarmed to 37 °C. (The medium used for the thawing step is generally the growth medium of choice.) The contents are then gently mixed to ensure complete thawing. The cells are immediately pelleted, and resuspended to the desired density in complete growth medium, with the necessary additives for normal culture.

12.2 Storage Conditions for Competent *E. coli* Cells

These are optimally preserved by rapid freezing as a cell pellet, and thus do not require the addition of any cryoprotectants. Bacteria made competent by the $CaCl_2$ method are pelleted at 4 °C in 1.5 ml polypropylene microfuge tubes, the excess supernatant removed with a micropipette, and the tube containing the cell pellet transferred immediately to a dewar containing liquid nitrogen. The frozen cell pellet is then stored in a freezer at −70 °C. The cells are thawed by addition of 1 ml of ice-cold Luria-Broth (LB), and the DNA for transfection is added immediately. This snap freezing also works for competent cells prepared by the modified Hanahan method (Ausubel *et al.*, 1989) and the cells can be frozen directly in the polypropylene Falcon 2059 tubes used to prepare the competent cells.

12.3 Storage Conditions for Recombinant *E. coli* Cells

These are optimally preserved by slow freezing in 20% glycerol and 80% LB medium. A freezing mixture of 40% glycerol in LB is sterile filtered through a 0.22 μm filter and cooled on ice before use. Saturated cultures are pelleted, resuspended in 1/5 volume of ice-cold LB, and transferred to 1.5 ml polypropylene microfuge tubes. An equal volume of cold freezing mixture is added and the cultures transferred to a −70 °C freezer cabinet for freezing and storage. Optimal thawing is achieved by placing the frozen tubes in a water bath at 37 °C to thaw, and immediately transferring the thawed contents to 40 volumes of LB at 37 °C. The cells are then pelleted and resuspended in the desired volume of the medium of choice, containing the appropriate antibiotic for selection of recombinant bacterial cells. In general these frozen stock cultures are rarely thawed: more often, a scraping of the frozen culture surface is taken and plated on an appropriate agar plate for growth and selection of individual recombinant bacterial colonies.

References

Ausubel, F.M. *et al.* (eds) (1989). *Current Protocols in Molecular Biology*, Wiley Interscience, New York.

Boehringer Mannheim (1990). *Immunologicals for the Diagnostic Industry 1990/91*, Boehringer Mannheim Biochemica, Mannheim, Germany.

Innis, M.A. *et al.* (1990). *PCR Protocols: A Guide to Methods and Applications*, Academic Press, San Diego, CA.

Misiura, K. *et al.* (1990). Biotinyl and phosphotyrosinyl phosphoramidite derivatives useful in the incorporation of multiple reporter groups on synthetic oligonucleotides. *Nucleic Acids Res.*, **18**, 4345–4354.

Parham, P. (1986). Preparation and purification of active fragments from mouse monoclonal antibodies. In *Handbook of Experimental Immunology, 4th edn, Volume 1: Immunochemistry* (Weir, D.M., ed.), pp. 14.1–14.23, Blackwell Scientific Publications, Oxford.

Randall, L.L. and Hardy, S.J.S. (1986). Correlation of competence for export with lack of tertiary structure of the mature species: a study *in vivo* of maltose-binding protein in *E. coli*. *Cell*, **46**, 921.

Ullmann, A. (1984). One-step purification of hybrid proteins which have β-galactosidase activity. *Gene*, 29, 27–31.

Index